CLASSIFICATION		DESCRIBED SPECIES
Class Branchiopoda	Fairy shrimps, brine shrimps, tadpole shrimps, clam shrimps, cladocerans	
Class Malacostraca	Crabs, shrimps, pillbugs, beach hoppers, etc.	
Class Maxillopoda	Barnacles, ostracods, etc.	
Subphylum Cheliceriformes		
Class Chelicerata	Horseshoe crabs, spiders, scorpions, mites, ticks, etc.	
Class Pycnogonida	Sea spiders	
Subphylum Myriapoda		
Class Diplopoda	Millipedes	
Class Chilopoda	Centipedes	
Class Pauropoda	Pauropodans	
Class Symphyla	Symphylans	
Subphylum Hexapoda		
Class Entognatha	Springtails, proturans, diplurans	
Class Insecta	Insects	
Phylum Mollusca	Molluscs	**93,195**
Class Aplacophora	Caudofoveatans and Solenogasters	
Class Monoplacophora	Monoplacophorans	
Class Polyplacophora	Chitons	
Class Gastropoda	Snails and slugs	
Class Bivalvia (= Pelecypoda)	Clams	
Class Scaphopoda	Tusk shells	
Class Cephalopoda	Octopuses, squids, *Nautilus*	
Phylum Phoronida	Phoronids	**20**
Phylum Ectoprocta	Ectoprocts	**4,500**
Class Phylactolaemata		
Class Stenolaemata		
Class Gymnolaemata		
Phylum Brachiopoda	Brachiopods, lamp shells	**335**
Class Inarticulata		
Class Articulata		
Phylum Echinodermata	Echinoderms	**7,000**
Class Crinoidea	Sea lilies and feather stars	
Class Asteroidea	Sea stars and sea daisies	
Class Ophiuroidea	Brittle stars and basket stars	
Class Echinoidea	Sea urchins and sand dollars	
Class Holothuroidea	Sea cucumbers	
Phylum Chaetognatha	Chaetognaths, arrow worms	**100**
Phylum Hemichordata	Hemichordates	**85**
Class Enteropneusta	Acorn, or tongue worms	
Class Pterobranchia	Pterobranchs	
Class Planctosphaeroidea	*Planctosphaera pelagica*	
Phylum Chordata	Chordates	**49,693**
Subphylum Urochordata (= Tunicata)	Tunicates	
Class Ascidiacea	Ascidians, sea squirts	
Class Thaliacea	Pelagic tunicates, salps	
Class Appendicularia (= Larvacea)	Larvaceans	
Class Sorberacea	Sorberaceans	
Subphylum Cephalochordata (= Acrania)	Lancelets, amphioxus	
Subphylum Vertebrata		
Class Myxini	Hagfishes	
Class Cephalaspidomorphi	Lampreys	
Class Chondrichthyes	Sharks, skates, rays	
Class Osteichthyes	Bony fishes	
Class Amphibia	Frogs, salamanders, etc.	
Class Reptilomorpha (= Sauropsida)	Reptiles and birds	
Class Mammalia	Mammals	

Invertebrates

SECOND EDITION

Invertebrates

SECOND EDITION

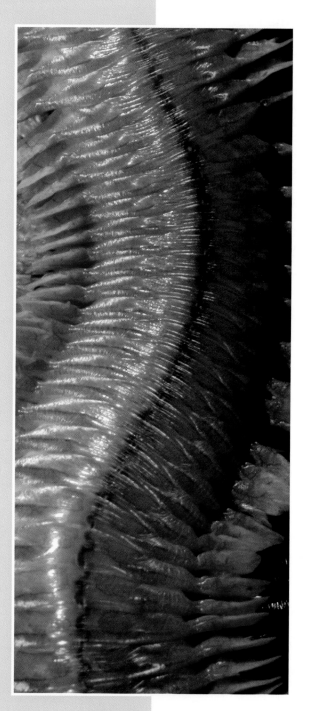

Richard C. Brusca

Director of Conservation and Science,
Arizona-Sonora Desert Museum

Gary J. Brusca

Late, Professor of Zoology,
Humboldt State University

with illustrations by Nancy Haver

 Sinauer Associates, Inc., Publishers
Sunderland, Massachusetts 01375

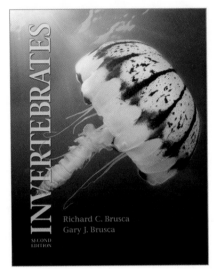

About the cover:
Pelagia, a large semaeostoman medusa.
©D. J. Wrobel/Biological Photo Service

Title Page Photo:
Close-up of a polychaete worm, *Nereis* sp.
Courtesy of Larry Jon Friesen.

Invertebrates, Second Edition

Sinauer Associates, Inc.
23 Plumtree Road/PO Box 407
Sunderland, MA 01375 U.S.A.
FAX: 413-549-1118
Email: publish@sinauer.com
www.sinauer.com

Library of Congress Catologing-in-Publication Data
Brusca, Richard C.
        Invertebrates / Richard C. Brusca, Gary J. Brusca.-- 2nd ed.
            p. cm.
        Includes bibliographical references.
ISBN 0-87893-097-3 (hardcover)
1. Invertebrates. I. Brusca, Gary J. II. Title.
QL362 .B924 2002
592--dc21   2002154089

Printed in U.S.A.

10 9 8 7 6 5 4 3

To the Memory of Gary

Who had a passion for teaching and for invertebrates.
And who loved tide pools on fog-shrouded mornings, a good fishing trip, and a good poker game.

We had some great times.

Brief Contents

Contents

4 Animal Development, Life Histories, and Origins 93

5 The Protists 121

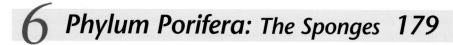

6 *Phylum Porifera: The Sponges* 179

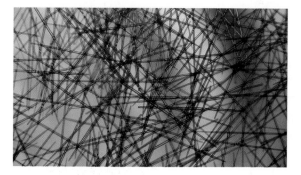

7 *Four Phyla of Uncertain Affinity* 209

8 Phylum Cnidaria 219

9 Phylum Ctenophora: The Comb Jellies 269

10 Phylum Platyhelminthes 285

11 Phylum Nemertea: The Ribbon Worms 319

12 Blastocoelomates and Other Phyla 337

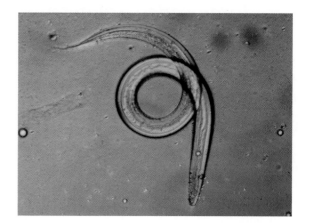

13 *Phylum Annelida: The Segmented Worms* **387**

14 *Sipuncula and Echiura* **445**

15 *The Emergence of the Arthropods: Onychophorans, Tardigrades, Trilobites, and the Arthropod Bauplan* **461**

16 Phylum Arthropoda: The Crustacea 511

17 Phylum Arthropoda: The Hexapoda (Insects and Their Kin) 589

18 Phylum Arthropoda: The Myriapods (Centipedes, Millipedes, and Their Kin) 637

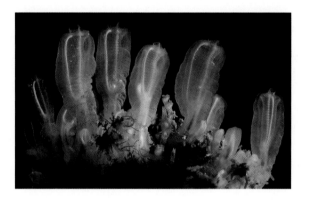

Preface

It is not now nor will it ever be given to one man to observe all the things recounted in the following pages.

Waldo L. Schmitt
Crustaceans, 1965

During the revision of *Invertebrates* my brother Gary passed away. For a while the project was stalled. But, buoyed by the support of family, friends, and colleagues, I eventually returned to the task, which, at times, seemed overwhelming. The field of invertebrate biology is so vast, and cuts across so many disciplinary lines, that even in a book of this size it is necessary to generalize about some topics and to slight others. As university instructors, my brother and I realized early on that the teaching of invertebrate zoology should not be compartmentalized. Thus, in planning this book we were concerned about two potential dangers. First, the text might become an encyclopedic list of "facts" about one group after another, the sort of "flash-card" approach that we wanted to avoid. Second, the book might be a rambling series of stories or vignettes about randomly selected animals (or "model organisms") and their ways of life. The first book would be dull, would encourage rote memorization instead of understanding, and might give the misconception that there is little left to discover. The second book might be full of interesting "gee whiz" stuff but would seem disorganized and without continuity or purpose to the serious student. Either approach could fail to present the most important aspects of invertebrates—their phenomenal diversity, their natural history, and their evolutionary relationships. We also held to the belief that what we *know* about these animals is not as important as how we *think* about them. You should be prepared to assimilate much new material, but you should also be prepared for a great deal of uncertainty and mystery, as much remains to be discovered.

To avoid the pitfalls noted above, and to establish threads of continuity in our discussions about invertebrates, we developed our book around two fundamental themes: **unity** and **diversity**. The first theme we approach by way of functional body architecture, or what we call the bauplan concept. The second theme we approach through the principles of phylogenetic biology. Our hope was that weaving the book tightly to these themes would provide a meaningful flow as readers move from one phylum to the next. The first four chapters provide background for these themes and thus provide an important foundation upon which the rest of the book rests. Please read these chapters carefully and refer back to them throughout your study.

The bulk of this book (Chapters 5–23) is devoted to a phylum-by-phylum discussion of invertebrates. Fairly detailed classifications or taxonomic synopses for each phylum are included in separate sections of each chapter to serve as references. A consistent organization is maintained throughout each chapter, although we did yield to the important and sometimes different lessons to be learned by investigating the special attributes of each group of animals. In addition, because of their size and diversity, some taxa receive more attention than others—although this does not mean that such groups are more "important" biologically than smaller or more homogeneous ones. (Five chapters are devoted to the arthropods and their kin.) In certain chapters more than one phylum is covered. In some cases the phyla covered are thought to be closely related to one another; in other cases the phyla merely represent a particular grade of complexity and their inclusion in a single chapter facilitates our comparative approach.

Certain aspects of this book have, of course, been influenced by our own biases; this is especially true of the discussions on phylogeny. We use a combination of phylogenetic trees (cladograms) and narrative discussions to talk about animal evolution. Cladograms are used when appropriate, because they provide the least ambiguous statements that can be made about animal relationships. We always knew that some of you, professors and students both, would disagree with our methods and ideas to various degrees—at least we *hoped* that you would. Never placidly accept what you see in a textbook, or anyplace else for that matter, but try to be critical in your reading.

The book's final chapter is a phylogenetic summary of the animal kingdom. It reinforces the point that much remains to be explored and learned about the evolutionary relationships of invertebrates. Like all scientific knowledge, we are dealing here with provisional, transient "truths" that always remain open to challenge and revision. And, of course, scientists disagree. It is this disagreement and the constant challenging of hypotheses that enliven the field and push the frontiers of knowledge forward.

There are a few other things you should know about this book. A brief historical review of the classification of each major group is provided. We felt this material was not only interesting but also served to imbue students with a sense of the dynamic nature of taxonomy and the development of our understanding of each group. Unless otherwise indicated, the Classification section in each chapter deals only with extant taxa. Descriptions of taxa in these annotated classifications are written in somewhat telegraphic style to save space; we never expected these sections to be "read"—they are for reference. Important new words, when first defined, are set in boldface type. These boldfaced terms are also indicated by boldfaced page references in the index; thus the index can also be used as a glossary. We tried hard to be consistent in our usage of zoological terminology, but the existence of similar terms for entirely different structures in certain groups is notoriously troublesome—these are noted in the text.

For this second edition of *Invertebrates*, of course, we have tried to be as current as possible with the research literature, but even as this book goes into production important new publications appear daily. It has been estimated that the volume of scientific information is doubling about every 10 years (or faster). A half-million nonclinical biology papers are published annually. As Professor George Bartholomew noted, "If one equates ignorance with the ratio between what one knows and what is available to be known...each biological investigator becomes more ignorant with every passing day." My goal has been to provide sufficient reference material to lead the interested student quickly into the heart of the relevant literature. Most of the references cited in the text will be found at the end of the corresponding chapter. However, to conserve space and eliminate redundancy, in a number of cases (especially in figure citations) references of a general nature may be listed only once, usually in the introductory chapters. You will also notice citations of a fair number of references that are quite old, some from the nineteenth century. These are included not out of whimsy, but because many of these

are benchmark research papers or they stand out as some of the best available descriptions for the subject at hand. (It is surprising how many of the illustrations in modern biology texts can be traced back to origins in nineteenth-century publications.) It is distressing to see how commonplace it has become for researchers to ignore the excellent (and important) work of past decades. For example, many phylogenetic research papers completely ignore 150 years of careful embryological research that was published, largely in the German and American literature, in the nineteenth and twentieth centuries. For some scientists, biological research seems to be little more than "sound bites" from the past decade. Sadly, today, this "sound bite research culture" is often imbued in graduate students—a shocking and dangerous trend that encourages dilettantes. To understand animals requires a thorough understanding of their overall biology, and the dedication of a career, not just dabbling.

Since the first edition of this book, there has been an explosion of research in the field of molecular biology. Much of this has been in molecular phylogenetics, but huge strides are also being made in the area of molecular developmental biology. Papers in these fields now appear at such a pace that it is difficult to write about them in a textbook, for fear the ideas will be obsolete in six months. There have been many new phylogenetic hypotheses proposed on the basis of DNA sequence analyses since the fist edition of this book. Many of the molecular phylogenetic trees that were published before 2000 were quirky and troublesome, due to the simple fact that the field is still new and emerging. Because most of these trees are relatively new and still await rigorous testing with independent data, we do not discuss them all. However, we do discuss molecular-based hypotheses that have a growing body of support or have received widespread attention. But, in general we have taken a conservative approach in this regard—we are only just beginning to discover which genes are appropriate for different levels of phylogenetic analysis, and how best to analyze them.

These things being said, I hope you are now ready to forge ahead in your study of invertebrates. The task may at first seem daunting, and rightly so. I hope that this book will make this seemingly overwhelming task a bit more manageable. If I succeed in enhancing your enjoyment and appreciation of invertebrates, then my efforts will have been worthwhile.

R.C.B.

Tucson, Arizona
December 2002

Acknowledgments

I have benefited immeasurably from the careful and professional work of many conscientious reviewers for this second edition (listed below), most of whom went far beyond simply correcting factual errors. Many sent extended commentaries on difficult points, reprints of their recent work, and even photographs for the book. I extend to these reviewers my most sincere gratitude; this book is far better for their efforts. Where weaknesses and errors remain, they are of course solely my responsibility. Some people have taken a special interest in this book, providing thoughtful discussion and insight over the years, and their contributions are especially appreciated, including Steve Gould, Todd Haney, Robert Higgins, Jens Høeg, Reinhardt Kristensen, Jody Martin, Jim Morin, Diana Lipscomb, and Wendy Moore.

A very special note of appreciation goes to Diana Lipscomb, who, with the help of her student Kristen Kivimaki, essentially rewrote the protist chapter. Diana and Kristen devoted many weeks in that effort, and I am deeply grateful to them.

I cannot list here all of the colleagues who sent photographs or allowed me to rummage through their files of slides and prints and never complained about how long I kept their pictures. I thank them all, but I am particularly indebted to Gita Bodner, Judith Connor, Peter Fankboner, Rainer Foelix, Larry Friesen, Todd Haney, Jens Høeg, Alex Kerstitch, Wayne Maddison, Gary McDonald, and Jim Morin. In addition, I am grateful to the authors and publishers who allowed me to reproduce material from their books, journals, and other publications. A special note of appreciation is due to Katja Schultz—thanks so much for sharing the results of your literature searches for phylogenetic papers for the Tree of Life project. Special thanks also to Ken Rinehart, for all those unexpected phone calls from sundry airports that resulted in rendezvous in remote regions of the world to collect invertebrates. Some authors have strongly influenced my thinking, including Paul Dayton, Niles Eldredge, Steve Gould, Ed Ricketts, Warren Stauls, John Steinbeck, and Jerome Tichnor.

Most of the original artwork in this text was done from our own sketches or from other sources by Nancy Haver, supported by our publisher, Sinauer Associates. Ms.

Haver's extraordinary talent is exceeded only by her patience with our nit-picking; I am proud to have her outstanding work in this book. The production staff at Sinauer truly are magicians with photographs and layouts, and they never lost patience with my last-minute changes and requests—thanks especially to Chris Small and his expert production team, including Janice Holabird, who was responsible for turning the manuscript into beautiful page spreads, to Jefferson Johnson for his extraordinary design work, and to David McIntyre for his exceptional ability to track down hard-to-find images. All of the editors at Sinauer Associates have been critical and passionate about their work (seemingly a company policy), and I am indebted to them all, including Carol Wigg, Kerry Falvey, Nan Sinauer, Sydney Carroll, Chelsea Holabird, and Kathaleen Emerson. I especially want to thank Kathaleen, who stewarded this second edition through to closure with extraordinary style and good humor. Andy Sinauer originally became a part of this project in the mid-1980s, at a time when my brother and I wondered if the book would ever see the light of day. He has been an unwavering friend since that day. Were it not for Andy's gentle manner, wise council, good business sense, and clever prodding, this book would probably not exist.

I want to offer my apologies to those whose work I might have missed in my attempts to keep up with the ever-expanding literature on invertebrates. And I want to thank those who have been thoughtful enough to send me their publications over the years. Keeping track of published research on invertebrates is an all-consuming task.

Finally, I thank my colleagues, students, friends, and family for encouragement and patience, especially Bob Edison (Director of the Arizona-Sonora Desert Museum) who graciously allowed me the freedom from other obligations to devote time to this project. Most important of all, of course, has been Wendy Moore, who helped me in more ways than I can count through the years of revising this book. Wendy's phylogenetic expertise and knowledge of arthropods is exceeded only by her good humor, unwavering patience, and steadfast friendship.

Scientific Reviewers

In addition to again thanking the reviewers for the first edition of *Invertebrates*, I wish to thank the following people who reviewed parts of this second edition.

GERALD J. BAKUS
University of Southern California, Los Angeles, CA

BILL BIRKY
University of Arizona, Tucson, AZ

NICOLE BOURY-ESNAULT
Centre d'Océanologie de Marseille, Marseille, France

STEVE CAIRNS
Smithsonian Institution, Washington, D.C.

JOSE LUIS CARBALLO
*National University of Mexico
Estación Mazatlán, Mazatlán, Mexico*

ANNE COHEN
Bodega Bay, CA

JOHN O. CORLISS
Albuquerque, NM

RICHARD DEMAREE
California State University, Chico, CA

NILES ELDREDGE
American Museum of Natural History, New York, NY

DAPHNE FAUTIN
University of Kansas, Lawrence, KS

PETER GLYNN
University of Miami, Miami, FL

MICHAEL J. GREENBERG
University of Florida, St. Augustine, FL

RICHARD GROSBERG
University of California, Davis, CA

TODD HANEY
Natural History Museum, Los Angeles, CA

GERHARD HASZPRUNAR
Zoologische Staatssammlung Muenchen, Munich, Germany

DAVID J. HORNE
University of Greenwich, Kent, England

M. A. HOUCK
Texas Tech University, Lubbock, TX

KRISTEN KIVIMAKI
The George Washington University, Washington, D.C.

DIANA LIPSCOMB
The George Washington University, Washington, D.C.

JOEL MARTIN
Natural History Museum, Los Angeles, CA

WENDY MOORE
University of Arizona, Tucson, AZ

JAMES MORIN
Shoals Marine Laboratory, Cornell University, Ithaca, NY

KATJA SCHULTZ
University of Arizona, Tucson, AZ

WILLIAM SHEAR
Hampden-Sydney College, Hampden-Sydney, VA

ROWLAND SHELLEY
North Carolina State Museum of Natural Sciences, Raleigh, NC

TIM STEBBINS
Ocean Monitoring Program, San Diego, CA

JEAN VACELET
Centre Oceanographie, Marseille, France

JOHANN-WOLFGANG WÄGELE
Ruhr-Universitaet Bochum, Germany

GREG WRAY
Duke University, Durham, NC

JILL YAGER
Antioch College, Yellow Springs, OH

1

Introduction

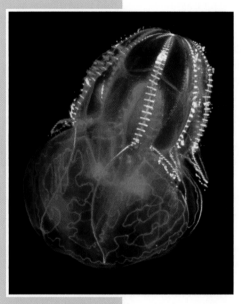

For a gentleman should know something of invertebrate zoology, call it culture or what you will, just as he ought to know something about painting and music and the weeds in his garden.

Martin Wells, *Lower Animals*, 1968

One of the first evolutionary trees of life conceived from a Darwinian (genealogical) perspective was published by Ernst Haeckel in 1866 (Figure 1.1). Haeckel's famous tree of life began a tradition of depicting phylogenetic hypotheses as branching diagrams, or trees, a tradition that has persisted since that time. We discuss various ways in which these trees are developed in Chapter 2. Since Haeckel's day, many names have been coined for the larger branches that sprout from these trees. We will not burden you with all of these names, but a few of them need to be defined here, before we launch into our study of the invertebrates. Some of these names refer to groups of organisms that are probably natural phylogenetic groups (i.e., groups that include an ancestor and all of its descendants), such as **Metazoa** (the animal kingdom). Other names refer to unnatural, or composite, groupings of organisms, such as "**microbes**" (i.e., any organism that is microscopic in size, such as bacteria, most protists, and unicellular fungi) and "**protozoa**" (a loose assemblage of primarily unicellular heterotrophic eukaryotes).

The discovery that organisms with a cell nucleus constitute a natural group divided the living world neatly into two categories, the **prokaryotes** (those organisms lacking membrane-enclosed organelles and a nucleus, and without linear chromosomes), and the **eukaryotes** (those organisms that do possess membrane-bound organelles and a nucleus, and linear chromosomes). Investigations by Carl Woese and others, beginning in the 1970s, led to the discovery that the prokaryotes actually comprise two distinct groups, called **Eubacteria** and **Archaea** (= **Archaebacteria**), both quite distinct from eukaryotes (Box 1A). Eubacteria corre-

---

BOX 1A   *The Six Kingdoms of Life*

THE PROKARYOTES (the "domains" Eubacteria and Archaea)[a]

Kingdom Eubacteria (Bacteria)

The "true" bacteria, including Cyanobacteria (or blue–green algae) and spirochetes. Never with membrane-enclosed organelles or nuclei, or a cytoskeleton; none are methanogens; some use chlorophyll-based photosynthesis; with peptidoglycan in cell wall; with a single known RNA polymerase.

Kingdom Archaea (Archaebacteria)

Anaerobic or aerobic, largely methane-producing microorganisms. Never with membrane-enclosed organelles or nuclei, or a cytoskeleton; none use chlorophyll-based photosynthesis; without peptidoglycan in cell wall; with several RNA polymerases.

THE EUKARYOTES (the "domain" Eukaryota, or Eukarya)

Cells with a variety of membrane-enclosed organelles (e.g., mitochondria, lysosomes, peroxisomes) and with a membrane-enclosed nucleus. Cells gain structural support from an internal network of fibrous proteins called a cytoskeleton.

Kingdom Fungi

The fungi. Probably a monophyletic group that includes molds, mushrooms, yeasts, and others. Saprobic, heterotrophic, multicellular organisms. The earliest fossil records of fungi are from the Middle Ordovician, about 460 mya. The 72,000 described species are thought to represent only 5–10 percent of the actual diversity.

Kingdom Plantae (= Metaphyta)

The multicellular plants. Photosynthetic, autotrophic, multicellular organisms that develop through embryonic tissue layering. Includes some groups of algae, the bryophytes and their kin, and the vascular plants (about 240,000 of which are flowering plants). The described species are thought to represent about half of Earth's actual plant diversity.

Kingdom Protista

Eukaryotic single-celled microorganisms and certain algae. A polyphyletic grouping of perhaps 18 phyla, including euglenids, green algae, diatoms and some other brown algae, ciliates, dinoflagellates, foraminiferans, amoebae, and others. Many workers feel that this group should be split into several separate kingdoms to better reflect the phylogenetic lineages of its members. The 80,000 described species probably represent about 10 percent of the actual protist diversity on Earth today.

Kingdom Animalia (= Metazoa)

The multicellular animals. A monophyletic taxon, containing 34 phyla of ingestive, heterotrophic, multicellular organisms. About 1.3 million living species have been described; estimates of the number of undescribed species range from lows of 10–30 million to highs of 100–200 million.

---

[a]Portions of the old "Kingdom Monera" are now included in the Eubacteria and the Archaea. Viruses (about 5,000 described "species") and subviral organisms (viroids and prions) are not included in this classification.

---

spond more or less to our traditional understanding of bacteria. Archaea strongly resemble Eubacteria, but they have genetic and metabolic characteristics that make them unique. For example, Archaea differ from both Eubacteria and Eukaryota in the composition of their ribosomes, in the construction of their cell walls, and in the kinds of lipids in their cell membranes. Some Eubacteria conduct chlorophyll-based photosynthesis, a trait that is never present in Archaea. Not surprisingly, due to their great age,* the genetic differences among

prokaryotes are much greater than those seen among eukaryotes, even though these differences do not typically reveal themselves in gross anatomy. Current thinking favors the view that prokaryotes ruled Earth for at least 2 billion years before the modern eukaryotic cell appeared in the fossil record. In fact, it seems likely that a significant portion of Earth's biodiversity, at the level of both genes and species, resides in the "invisible" prokaryotic world. About 4,000 species of prokaryotes have been described, but there are an estimated 1 to 3 million undescribed species living on Earth today.

Evolutionary change in the prokaryotes gave rise to metabolic diversity and the evolutionary capacity to explore and colonize every conceivable environment on Earth. Many Archaea live in extreme environments, and this pattern is often interpreted as a refugial lifestyle—in other words, these creatures tend to live in places where they have been able to survive without confronting

---

*The date of the first appearance of life on Earth remains debatable. The oldest evidence consists of 3.8-billion-year-old trace fossils from Australia, but these fossils have recently been challenged, and opinion is now split on whether they are traces of early bacteria or simply mineral deposits. Uncontestable fossils occur in rocks 2 billion years old, but these fossils already include multicellular algae, suggesting that life must have evolved well before then.

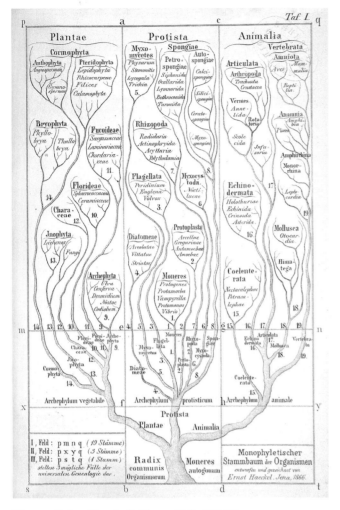

Figure 1.1   Haeckel's Tree of Life (1866).

sure).* Molecular phylogenetic studies now suggest that some of these extremophiles, particularly the thermophiles, lie close to the "universal ancestor" of all life on Earth.

It has recently been suggested that the three main divisions of life (Eubacteria, Archaea, Eukaryota) should be recognized at a new taxonomic level, called **domains**. However, fundamental questions remain about these three "domains," including how many natural groups (kingdoms) exist in each domain, whether the domains themselves represent natural (= monophyletic) groups, and what the phylogenetic relationships are among these domains and the kingdoms they contain. Current evidence suggests that eukaryotes are a natural group, defined by the unique trait of a nucleus and linear chromosomes, whereas Eubacteria and Archaea may not be natural groups.

Courses and texts on invertebrates often include discussions of two eukaryotic kingdoms, the Animalia (= Metazoa) and certain "animal-like" (i.e., heterotrophic) protist phyla loosely referred to as "protozoa." Following this tradition, we treat 34 phyla of Metazoa and 18 phyla of protists (many of which have traditionally been viewed as "protozoa") in this text. The vast majority of kinds (species) of living organisms that have been described are animals. The kingdom **Animalia,** or **Metazoa,** is usually defined as the multicellular, ingestive, heterotrophic[†] eukaryotes. However, its members possess other unique attributes as well, such as an acetylcholine/cholinesterase-based nervous system, special types of cell–cell junctions, and a unique family of connective tissue proteins called collagens. Over a million species of living animals have been described, but estimates of how many living species remain to be discovered and described range from lows of 10–30 million to highs of 100–200 million.[‡] Among the Metazoa are some species that possess a backbone (or vertebral column), but most do not. Those that possess a backbone constitute the subphylum Vertebrata of the phylum Chordata, and account for less than 5 percent (about 46,670 species) of all described animals. Those

competition with more highly derived life forms. Many of these "extremophiles" are anaerobic chemoautotrophs, and they have been found in a variety of habitats, such as deep-sea hydrothermal vents, benthic marine cold seeps, hot springs, saline lakes, sewage treatment ponds, certain sediments of natural waters, and the guts of humans and other animals. One of the most astonishing discoveries of the 1980s was that extremophile Archaea (and some fungi) are widespread in the deep rocks of Earth's crust. Since then, a community of hydrogen-eating Archaea has been found living in a geothermal hot spring in Idaho, 600 feet beneath Earth's surface, relying on neither sunshine nor organic carbon. Other Archaea have been found at depths as great as 2.8 km, living in igneous rocks with temperatures as high as 75°C. Extremophiles include **halophiles** (which grow in the presence of high salt concentrations), **thermophiles** and **psychrophiles** (which live at very high or very low temperatures), **acidiphiles** and **alkaliphiles** (which are optimally adapted to acidic or basic pH values), and **barophiles** (which grow best under pres-

*One of the most striking examples of a thermophile is *Pyrolobus fumarii*, a chemolithotrophic archaean that lives in oceanic hydrothermal vents at temperatures of 90°–113°C. (Chemolithotrophs are organisms that use inorganic compounds as energy sources.) On the other hand, *Polaromonas vacuolata* grows optimally at 4°C. *Picrophilus oshimae* is an acidiphile whose growth optimum is pH 0.7 (*P. oshimae* is also a thermophile, preferring temperatures of 60°C). The alkaliphile *Natronobacterium gregoryi* lives in soda lakes where the pH can rise as high as 12. Halophilic microorganisms abound in hypersaline lakes such as the Dead Sea, Great Salt Lake, and solar salt evaporation ponds. Such lakes are often colored red by dense microbial communities (e.g., *Halobacterium*). *Halobacterium salinarum* lives in the salt pans of San Francisco Bay and colors them red. Barophiles have been found living at all depths in the sea, and one unnamed species from the Mariana Trench has been shown to require at least 500 atmospheres of pressure in order to grow.

[†]Heterotrophic organisms are those that consume other organisms or organic materials as food.

that do not possess a backbone (the remainder of the phylum Chordata, plus 33 additional animal phyla) constitute the **invertebrates**. Thus we can see that the division of animals into invertebrates and vertebrates is based more on tradition and convenience, reflecting a dichotomy of zoologists' interests, than it is on the recognition of natural biological groupings. About 10,000 to 13,000 new species are named and described by biologists each year, most of them invertebrates.

Where Did Invertebrates Come From?

The incredible array of extant (= living) invertebrates is the outcome of billions of years of evolution on Earth. Indirect evidence of prokaryotic organisms has been found in some of the oldest sediments on the planet, suggesting that life first appeared in Earth's seas almost as soon as the planet cooled enough for it to exist.[§] A remarkable level of metabolic sophistication had been achieved by the end of the Archean eon, about 2.5 billion years ago. Hydrocarbon biomarkers suggest that the first eukaryotic cells might have appeared 2.7 billion years ago. However, we know very few details about the origin or early evolution of the eukaryotes. Even though the eukaryotic condition appeared early in Earth's history, it probably took a few hundred million more years for evolution to invent multicellular organisms. Molecular clock data (tenuous as they are) suggest that the last common ancestor of plants and animals existed about 1.6 billion years ago—long after the initial appearance of eukaryotes and long before a de-

finitive fossil record of metazoans, but in line with trace fossil evidence. The fossil record tells us that metazoan life had its origin in the Proterozoic eon, at least 600 million years ago, although trace fossils suggest that the earliest animals might have originated more than 1.2 billion years ago.

The ancestors of both plants and animals were almost certainly protists, suggesting that the phenomenon of multicellularity arose independently in the Metazoa and Metaphyta. Indeed, genetic and developmental data suggest that the basic mechanisms of pattern formation and cell–cell communication during development were independently derived in animals and in plants. In animals, segmental identity is established by the spatially specific transcriptional activation of an overlapping series of master regulatory genes, the homeobox (**Hox**) genes. The master regulatory genes of plants are not members of the homeobox gene family, but belong to the MADS box family of transcription factor genes. There is no evidence that the animal homeobox and MADS box transcription factor genes are homologous.

Although the fossil record is rich with the history of many early animal lineages, many others have left very few fossils. Many were very small, some were soft-bodied and did not fossilize well, and others lived where conditions were not suitable for the formation of fossils. Therefore, we can only speculate about the abundance of members of most animal groups in times past. However, groups such as the echinoderms (sea stars, urchins), molluscs (clams, snails), arthropods (crustaceans, insects), corals, ectoprocts, brachiopods, and vertebrates have left rich fossil records. In fact, for some groups (e.g., echinoderms, brachiopods, ectoprocts, molluscs), the number of extinct species known from fossils exceeds the number of known living forms. Representatives of nearly all of the extant animal phyla were present early in the Paleozoic era, more than 500 million years ago (mya). Life on land, however, did not appear until fairly recently, by geological standards, and terrestrial radiations began only about 470 mya. Apparently it was more challenging for life to invade land than to first evolve on Earth! The following account briefly summarizes the early history of life and the rise of the invertebrates.

The Dawn of Life

It used to be thought that the Proterozoic was a time of only a few simple kinds of life; hence the name. However, discoveries over the past 20 years have shown that life on Earth began early and had a very long history throughout the Proterozoic. It is estimated that Earth is about 4.6 billion years old, although the oldest rocks found are only about 3.8 billion years old. The oldest evidence of possible life on Earth consists of 3.8-billion-year-old, debated, biogenic traces suspected to represent anaerobic sulfate-reducing prokaryotes and

---

[‡]Our great uncertainty about how many species of living organisms exist on Earth is unsettling and speaks to the issue of priorities and funding in biology. We know approximately how many genes are in organisms from yeast (about 6,000 genes) to humans (about 10,000 genes), but taxonomic research has lagged behind other disciplines. At our current rate of species descriptions, it would take us 2,000–8,000 years to describe the rest of Earth's life forms. Not all of these new species are invertebrates—in fact, just between 1990 and 2002, 38 new primate species were discovered and named. If prokaryotes are thrown into this mix, the numbers become even larger (one recent estimate suggested that a ton of soil could contain as many as 4 million species, or "different taxa," of prokaryotes). However, at our current rate of anthropogenic-driven extinction, an estimated 90 percent of all species could go extinct before they are ever described. In the United States alone, at least 5,000 species are threatened with extinction, and an estimated 500 species have already gone extinct since people first arrived in North America. Globally, the United Nations Environment Programme estimates that by 2030 nearly 25 percent of the world's mammals could go extinct.

[§]There are three popular theories on how life first evolved on Earth. The classic "primeval soup" theory, dating from Stanley Miller's work in the 1950s, proposes that self-replicating organic molecules first appeared in Earth's early atmosphere and were deposited by rainfall in the ocean, where they reacted further to make nucleic acids, proteins, and other molecules of life. More recently, the idea of the first synthesis of biological molecules by chemical and thermal activity at deep-sea hydrothermal vents has been suggested. The third proposal is that organic molecules first arrived on Earth from another planet, or from deep space, on comets or meteorites.

perhaps cyanobacterial stromatolites. The first certain traces of prokaryotic life (secondary chemical evidence of Cyanobacteria) occur in rocks dated at 2.5 billion years, although fossil molecular residues of Cyanobacteria have been found in rocks 2.7 billion years old. The first actual fossil traces of eukaryotic life (benthic algae) are 1.7 to 2 billion years old, whereas the first certain eukaryotic fossils (phytoplankton) are 1.4 to 1.7 billion years old. Together, these bacteria and protists appear to have formed diverse communities in shallow marine habitats during the Proterozoic eon. Living stromatolites (compact layered colonies of Cyanobacteria and mud) are still with us, and can be found incertain high evaporation/high-salinity environments in such places as Shark Bay (Western Australia), Scammon's Lagoon (Baja California), the Persian Gulf, the Paracas coast of Peru, the Bahamas, and Antarctica.

The Ediacaran Epoch and the Origin of Animals

One of the most perplexing unsolved mysteries in biology is the origin and early radiation of the Metazoa. We now know that by 600 mya, at the beginning of a period in the late Proterozoic known as the Ediacaran epoch, a worldwide marine invertebrate fauna had already made its appearance. If any animals existed before this time, they left no known unambiguous fossil record. The **Ediacaran fauna** (600–570 mya) contains the first evidence of many modern phyla, although the precise evolutionary relationships of many of these fossils are still being debated.* The modern phyla thought to be represented among the Ediacaran fauna include Porifera, Cnidaria, Echiura, Mollusca, Onychophora, Echinodermata, a variety of annelid-like forms (including possible pogonophorans), and quite probably arthropods (soft-bodied trilobite-like organisms, anomalocarids and their kin, etc.). However, many Ediacaran animals cannot be unambiguously assigned to any living taxa, and these animals may represent phyla or other high-level taxa that went extinct at the Proterozoic–Cambrian transition.[†]

Ediacaran fossils were first reported from sites in Newfoundland and Namibia, but the name is derived from the superb assemblages of these fossils discovered at Ediacara in the Flinders Ranges of South Australia. Most of the Ediacaran organisms were preserved as shallow-water impressions on sandstone beds, but some of the 30 or more worldwide sites represent deepwater and continental slope communities. The Ediacaran fauna was almost entirely soft-bodied, and there have been no heavily shelled creatures reported from these deposits. Even the molluscs and arthropod-like creatures from this fauna are thought to have had relatively soft (unmineralized, or lightly calcified) skeletons. A few chitinous structures developed during this time, such as the jaws of some annelid-like creatures (and the chitinous sabellid-like tubes of others) and the radulae of early molluscs.[‡] In addition, siliceous spicules of hexactinellid sponges have been reported from Australian and Chinese Ediacaran deposits. Many of these Proterozoic animals appear to have lacked complex internal organ structures. Most were small and possessed radial symmetry. However, at least by late Ediacaran times, large animals with bilateral symmetry had appeared, and some almost certainly had internal organs (e.g., the segmented, sheetlike *Dickinsonia*, which reached a meter in length; Figure 1.2). The Ediacaran epoch was followed by the Cambrian period and the great "explosion" of skeletonized metazoan life associated with that time (see below). Why skeletonized animals appeared at that particular time, and in such great profusion, remains a mystery.

Geological evidence tells us that Earth's earliest atmosphere lacked free oxygen, and clearly the radiation of the animal kingdom could not have begun under those conditions. Free oxygen probably accumulated over many millions of years as a by-product of photosynthetic activity in the oceans, particularly by the Cyanobacterial (blue-green algae) stromatolites. However, the evidence on free oxygen levels in the Proterozoic is still a little murky. Significant atmospheric O_2 levels may have been achieved fairly early in the Proterozoic, 1.5 to 2.8 billion years ago, or perhaps even earlier. Proterozoic seas might have been oxic near the surface, but anoxic in deep waters and on the bottom. Some workers suggest that the absence of metazoan life in the early fossil record is due to the simple fact that the first animals were small, lacked skeletons, and did not fossilize well, not to the absence of oxygen. The discovery of highly diverse communities of metazoan meiofauna[§] in the Proterozoic strata of south China and in deposits from the Middle and Upper Cambrian (e.g.,

---

*During the 1980s, some workers believed that most of the Ediacaran biota was unrelated to modern phyla—that it was a "failed experiment" in the evolution of life on Earth. There was even the suggestion that Ediacaran organisms be referred to a new phylum or even a new kingdom, the "Vendozoa," which was said to contain "quilted" organisms that lacked mouths and guts and presumably received energy by absorbing dissolved organic molecules or by harboring photosynthetic or chemosynthetic symbionts. Today we know that this biota (now sometimes called the "Vendobionta") actually represents only a portion of the Ediacaran biota, and the entire fauna includes many species now viewed as primitive members of extant phyla.

[†]The largest mass extinctions occurred at the ends of the Proterozoic era (Ediacaran epoch) and the Ordovician, Devonian, and Permian periods, and in the Early Triassic, Late Triassic, and end-Cretaceous. Most of these extinction events were experienced by both marine and terrestrial organisms. An excellent review of Ediacaran/Cambrian animal life can be found in Lipps and Signor (1992).

[‡]Chitin is a cellulose-like family of compounds that is widely distributed in nature, especially in invertebrates, fungi, and yeasts, but it is apparently uncommon in deuterostome animals and higher plants, perhaps due to the absence of the chitin synthase enzyme.

[§]Meiofauna is usually defined as the interstitial animals that pass through a 1 mm mesh sieve, but are retained by a 0.1 mm mesh sieve.

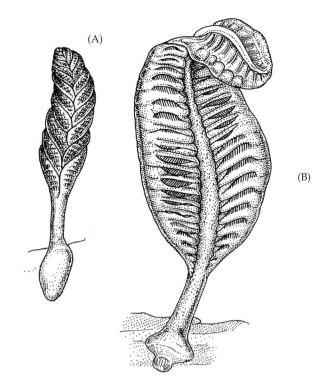

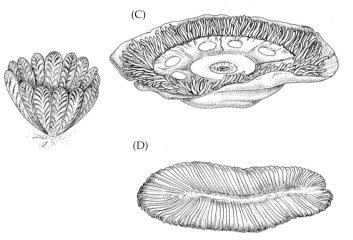

Figure 1.2   Some Ediacaran (late Proterozoic) animals. (A) *Charnia* and *Charniodiscus*, two Cnidaria resembling modern sea pens (Anthozoa, Pennatulacea). (B) A bushlike fossil of uncertain affinity (suggestive of a cnidarian). (C) *Ediacara*, a cnidarian medusa. (D) *Dickinsonia*, probably a polychaete annelid. (E) One of the numerous soft-bodied trilobites known from the Ediacaran period (some of which also occurred in the Early Cambrian).

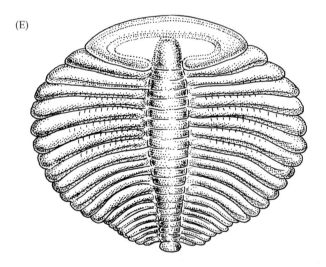

Cambrian heralded the beginning of the Phanerozoic eon. The Ediacaran fauna seems to have included primarily passive suspension and detritus feeders; very few of these animals appear to have been active carnivores or herbivores. Only a few Ediacaran species are known to have spanned the transition to the Cambrian period. Early Cambrian animal communities, on the other hand, included most of the trophic roles found in modern marine communities, including giant predatory arthropods.

The Paleozoic Era (570–250 mya)

The Phanerozoic eon was ushered in with the almost simultaneous appearance in the Lower Cambrian of well-developed calcareous body skeletons in numerous groups, including archaeocyathans, molluscs, ectoprocts, brachiopods, crustaceans, and trilobites. The appearance of mineralized animal skeletons thus defines the beginning of the Cambrian, and it was an event of fundamental importance in the history of life. The newly skeletonized animals radiated quickly and filled a multitude of roles in all shallow-water marine environments. The other major event at the Proterozoic–Cambrian transition was the explosion of bilaterally symmetrical animals. Most of our modern metazoan phyla and classes were established as distinct lineages at this time.

Much of what we know about early Cambrian life comes from the Lower Cambrian Chengjiang fossil deposits of the Yunnan Province of southern China and similarly aged (although less well preserved) deposits spread across China and the Siberian Platform. The

the Swedish Orsten fauna) lends support to the idea that many of the first animals were microscopic. In thinking about the earliest metazoans, it is not difficult to imagine a microscopic primordial animal resembling a colony of choanoflagellate protists whose cell–cell connections were enhanced by metazoan cell junctions and whose inner and outer cells became separated and specialized. However, large animals are not uncommon among the Ediacaran and early Cambrian faunas.

It has also been proposed that the advent of predatory lifestyles was the key that favored the first appearance of animal skeletons (as defensive structures), leading to the "Cambrian explosion." The rapid appearance and spread of diverse metazoan skeletons in the early

Chengjiang deposits are the oldest Cambrian occurrences of well-preserved soft-bodied and hard-bodied animals, and they include a rich assemblage of exquisitely preserved arthropods, onychophorans, medusae (Cnidaria), and brachiopods, many of which appear closely related to Proterozoic Ediacaran species.

In the Middle Cambrian (e.g., the Burgess Shale fauna of western Canada and similar deposits elsewhere; Figure 1.3) polychaetes and tardigrades made their first positive appearances in the fossil record, and the first complete echinoderm skeletons appeared. In the Upper Cambrian (e.g., the Orsten deposits of southern Sweden and similar strata), the first pentastomid Crustacea and the first agnathan fishes made their appearances. By the end of the Cambrian, nearly all of the major animal phyla had appeared.

The early Paleozoic also saw the first xiphosurans, eurypterids, trees, and teleost fishes (in the Ordovician). The first land animals (arachnids, centipedes, myriapods) appeared in the Upper Silurian. By the middle Paleozoic (the Devonian), life on land had begun to proliferate. Forest ecosystems became established and began reducing atmospheric CO_2 levels (eventually terminating an earlier Paleozoic greenhouse environment). The first insects also appeared in the middle Paleozoic fossil record. Insects developed flight in the Lower Carboniferous, and they began their long history of co-evolution with plants shortly thereafter (at least by the mid-Carboniferous, when tree fern galls first appeared in the fossil record). During the Carboniferous period, global climates were generally warm and humid, and extensive coal-producing swamps existed.

The late Paleozoic experienced the formation of the world supercontinent Pangaea in the Permian period (about 270 mya). The end of the Permian (250 mya) was brought about by the largest mass extinction known, in which 85 percent of Earth's marine species (and 70 percent of the terrestrial vertebrate genera) were lost over a brief span of a few million years. The Paleozoic reef corals (Rugosa and Tabulata) went extinct, as did the once dominant trilobites, never to be seen again. The driving force of the Permian extinction is thought to have been a huge asteroid impact, probably coupled with massive Earth volcanism, and perhaps degassing of stagnant ocean basins. The volcanism may have been the same event that created the massive flood basalts known as the Siberian Traps in Asia. This event may have led to atmospheric "pollution" in the form of dust and sulfur particles that cooled Earth's surface or massive gas emissions that led to a prolonged greenhouse warming.

The Mesozoic Era (250–65 mya)

The Mesozoic era is divided into three broad periods: the Triassic, Jurassic, and Cretaceous. The Triassic began with the continents joined together as Pangaea. The land was high, and few shallow seas existed. Global climates were warm, and deserts were extensive. The terrestrial flora was dominated by gymnosperms, with angiosperms first appearing in the latest part of the period. The oldest evidence of a flowering plant is from 130 mya. Vertebrate diversity exploded in the Triassic. On land, the first mammals appeared, as well as the turtles, pterosaurs, plesiosaurs, and dinosaurs. In Triassic seas, the diversity of predatory invertebrates and fishes increased dramatically, although the paleogeological data suggest that deeper marine waters might have been too low in oxygen to harbor much (or any) multicellular life. The end of the Triassic witnessed a sizable global extinction event, perhaps driven by the combination of asteroid impact and widespread volcanism that created the Central Atlantic Magmatic Province of northeastern South America 200 mya.

The Jurassic saw a continuation of warm, stable climates, with little latitudinal or seasonal variation and probably little mixing between shallow and deep oceanic waters. Pangaea split into two large land masses, the northern Laurasia and southern Gondwana, separated by a circumglobal tropical seaway known as the Tethys Sea. Many tropical marine families and genera today are thought to be direct descendants of inhabitants of the pantropical Tethys Sea. On land, modern genera of many gymnosperms and advanced angiosperms appeared, and birds began their dramatic radiation. Leaf-mining insects (lepidopterans) appeared by the late Jurassic (150 mya), and other leaf-mining orders appeared through the Cretaceous, coincident with the radiation of the vascular plants.

In the Cretaceous, large-scale fragmentation of Gondwana and Laurasia took place, resulting in the formation of the Atlantic and Southern Oceans. During this period, land masses subsided and sea levels were high; the oceans sent their waters far inland, and great epicontinental seas and coastal swamps developed. As land masses fragmented and new oceans formed, global climates began to cool and oceanic mixing began to move oxygenated waters to greater depths in the sea. The end of the Cretaceous was marked by the Cretaceous–Tertiary mass extinction, in which an estimated 50 percent of Earth's species were lost, including the dinosaurs and all of the sea's rich Mesozoic ammonite diversity. There is strong evidence that this extinction event was driven by a combination of two factors: a major asteroid impact (probably in the Yucatan region of modern Mexico) and massive Earth volcanism associated with the great flood basalts of India known as the Deccan Traps.

The Cenozoic Era (65 mya–present)

The Cenozoic era dawned with a continuing worldwide cooling trend. As South America decoupled from Antarctica, the Drake Passage opened to initiate the circum-Antarctic current, which eventually drove the formation of the Antarctic ice cap, which in turn led to our modern cold ocean bottom conditions (in the Miocene).

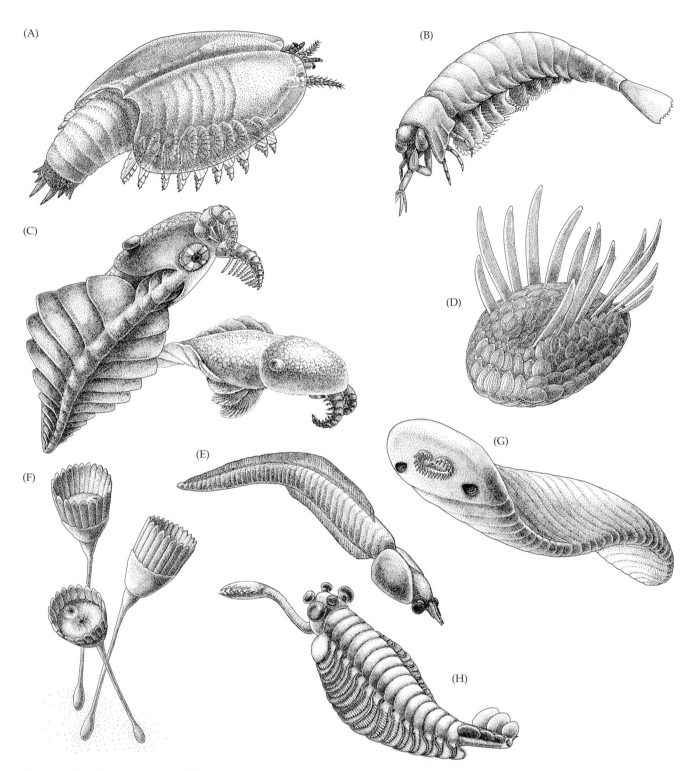

Figure 1.3   Some Cambrian life forms from the Burgess Shale deposits of Canada. (A) *Canadaspis*, an early malacostracan crustacean. (B) *Yohoia*, an arthropod of uncertain classification. (C) Two species of *Anomalocaris*, *A. nathorsti* (above) and *A. canadensis*. Anomalocarids were once thought to represent an extinct phylum of segmented animals, but are now regarded by many workers as primitive crustaceans dating back to the Ediacaran. (D) *Wiwaxia*, a Burgess Shale animal with no clear affinity to any known metazoan phylum (although some workers regard it as a polychaete annelid). (E) *Nectocaris*, another creature that has yet to be classified into any known phy-lum (despite its strong chordate-like appearance). (F) *Dinomischus*, a stalked creature with a U-shaped gut and with the mouth and anus both placed on a radially symmetrical calyx. Although superficially resembling several extant phyla, *Dinomischus* is now thought to belong to an unnamed extinct phylum of sessile Cambrian animals. (G) The elusive *Odontogriphus*, an appendageless flattened vermiform creature of unknown affinity. (H) One of the more enigmatic of the Burgess Shale animals, *Opabinia*; this segmented creature was probably an ancestral arthropod. Notice the presence of five eyes, a long prehensile "nozzle," and gills positioned dorsal to lateral flaps.

India moved north from Antarctica and collided with southern Asia (in the early Oligocene). Africa collided with western Asia (late Oligocene/early Miocene), separating the Mediterranean Sea from the Indian Ocean and breaking up the circumtropical Tethys Sea. Relatively recently (in the Pliocene), the Arctic ice cap formed, and the Panama isthmus rose, separating the Caribbean Sea from the Pacific and breaking up the last remnant of the ancient Tethys Sea about 3.5 mya. Modern coral reefs (scleractinian-based reefs) appeared early in the Cenozoic, reestablishing the niche once held by the rugose and tabulate corals of the Paleozoic.

This textbook focuses primarily on invertebrate life at the very end of the Cenozoic, in the Recent (Holocene) epoch. However, evaluation of the present-day "success" of animal groups also involves consideration of the history of modern lineages, the diversity of life over time (numbers of species and higher taxa), and the abundance of life (numbers of individuals). The predominance of certain kinds of invertebrates today is unquestionable. For example, of the 1,335,188 or so described species of animals (1,288,518 of which are invertebrates), 85 percent are arthropods. Most arthropods today are insects, probably the most successful group of animals on Earth. But the fossil record tells us that arthropods have always been key players in the biosphere, even before the appearance of the insects. Box 1B conveys a general idea of the levels of diversity among the animal phyla today.

Where Do Invertebrates Live?

Marine Habitats

Earth is a marine planet—salt water covers 71 percent of its surface. The vast three-dimensional world of the seas contains 99 percent of Earth's inhabited space. Life almost certainly evolved in the sea, and the major events described above leading to the diversification of invertebrates occurred in late Proterozoic and early Cambrian shallow seas. Many aspects of the marine world minimize physical and chemical stresses on organisms. The barriers to evolving gas exchange and osmotic regulatory structures that can function in freshwater and terrestrial environments are formidable, and relatively few lineages have escaped their marine origins to do so. Thus, is not surprising to find that the marine environment continues to harbor the greatest diversity of higher taxa and major body plans. Some phyla (e.g., echinoderms, sipunculans, chaetognaths, cycliophorans, placozoans, echiurans, ctenophores) have remained exclusively marine. Productivity in the world's oceans is very high, and this also probably contributes to the high diversity of animal life in the sea (the total primary productivity of the seas is about 48.7×10^9 metric tons of carbon per year). Perhaps the most significant factor, however, is the special nature of seawater itself.

Water is a very efficient thermal buffer. Because of its high heat capacity, it is slow to heat up or cool down.

BOX 1B   *Approximate Numbers of Known Extant Species in Various Groups*

Specialists in certain groups estimate that the known kinds of organisms probably represent only a small fraction of actual existing species. Note that we have broken down the phyla Arthropoda and Chordata into their respective subphyla, and that we have lumped the protist phyla together (see Chapter 5 for a complete classification of the protists). Of the 1,335,188 estimated described species of Animalia (excluding protists), 1,288,518 (96 percent) are invertebrates.

Kingdom Protista (80,000)	Nematomorpha (230)	Hexapoda (948,000: estimates range from 870,000 to 1,500,000)
Porifera (5,500)	Priapula (16)	
Cnidaria (10,000)	Acanthocephala (700)	
Ctenophora (100)	Cycliophora (1)	Myriapoda (11,460)
Placozoa (1)	Entoprocta (150)	Mollusca (93,195)
Monoblastozoa (1)	Loricifera (10)	Brachiopoda (335)
Rhombozoa (70)	Annelida (16,500)	Ectoprocta (4,500)
Orthonectida (20)	Echiura (135)	Phoronida (20)
Platyhelminthes (20,000)	Sipuncula (320)	Chaetognatha (100)
Nemertea (900)	Tardigrada (600)	Echinodermata (7,000)
Gnathostomulida (80)	Onychophora (110)	Hemichordata (85)
Rotifera (1,800)	Arthropoda (1,097,631)	Chordata (49,693)
Gastrotricha (450)	Cheliceriformes (70,000)	Urochordata (3,000)
Kinorhyncha (150)	Crustacea (68,171)	Cephalochordata (23)
Nematoda (25,000)		Vertebrata (46,670)

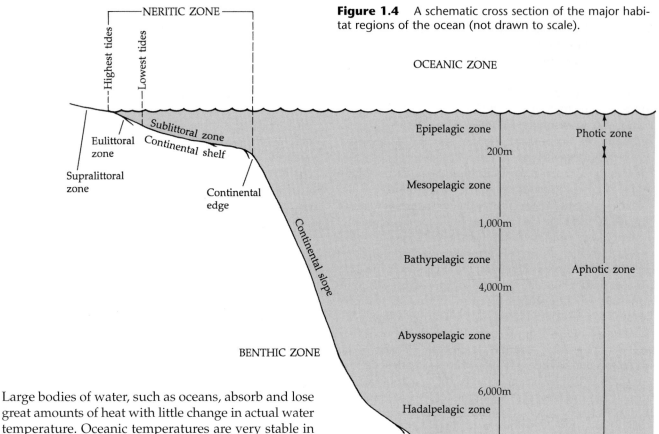

Figure 1.4 A schematic cross section of the major habitat regions of the ocean (not drawn to scale).

Large bodies of water, such as oceans, absorb and lose great amounts of heat with little change in actual water temperature. Oceanic temperatures are very stable in comparison with those of freshwater and terrestrial environments. Short-term temperature extremes occur only in intertidal and estuarine habitats; invertebrates living in such areas must possess behavioral and physiological adaptations that allow them to survive these temperature changes, which are often combined with aerial exposure during low tide periods.

The saltiness, or salinity, of seawater averages about 3.5 percent (usually expressed as parts per thousand, 35‰). This property, too, is quite stable, especially in areas away from shore and the influence of freshwater runoff. The salinity of seawater gives it a high density, which enhances buoyancy, thereby minimizing energy expenditures for flotation. Furthermore, the various ions that contribute to the total salinity occur in fairly constant proportions. These qualities result in a total ionic concentration in seawater that is similar to that in the body fluids of most animals, minimizing the problems of osmotic and ionic regulation (see Chapter 3). The pH of seawater is also quite stable throughout most of the ocean. Naturally occurring carbonate compounds participate in a series of chemical reactions that buffer seawater at about pH 7.5–8.5. However, today's anthropogenic changes in atmospheric CO_2 threaten to alter the carbonate buffering capacity of the world's seas.

In shallow and nearshore waters, carbon dioxide, various nutrients, and sunlight are generally available in quantities sufficient to allow high levels of photosynthesis, either seasonally or continuously (depending on latitude and other factors). Dissolved oxygen levels rarely drop below those required for normal respiration, except in stagnant waters such as might occur in certain estuarine or ocean basin habitats, or where anthropogenic activities have created eutrophic conditions.

Because the marine realm is home to most of the animals discussed in this book, some terms that describe the subdivisions of that environment and the categories of animals that inhabit them will be useful. Figure 1.4 illustrates a generalized cross section through an ocean. The shoreline marks the **littoral region**, where sea, air, and land meet and interact (Figure 1.5A). Obviously, this region is affected by the rise and fall of the tides, and we can subdivide it into zones or shore elevations relative to the tides. The supralittoral zone, or splash zone, is rarely covered by water, even at high tide, but it is subjected to storm surges and spray from waves. The eulittoral zone, or true intertidal zone, lies between the levels of the highest and lowest tides. It can be subdivided by its flora and fauna, and by mean monthly hours of aerial exposure, into high, mid-, and low intertidal zones. The sublittoral zone, or subtidal zone, is never uncovered, even at very low tides, but it is influenced by tidal action (e.g., by changes in turbulence, turbidity, and light penetration).

Organisms that inhabit the world's littoral regions are subjected to dynamic and often demanding conditions, and yet these areas commonly are home to exceptionally high numbers of species. As noted above, most animals and plants are more or less restricted to particular elevations along the shore, a condition resulting in the phenomenon of **zonation**. Such zones are visible as distinct bands or communities of organisms along the shoreline. The upper elevational limit of an intertidal organism is commonly established by its ability to tolerate

Figure 1.5 A few of Earth's major habitat types. (A) Exposed rocks and algae in the intertidal zone, northern California. (B) A tidal flat and bordering salt marsh in northern California. (C) A mangrove swamp at low tide, in Mexico. (D) A freshwater stream in a tropical wet forest ("rain forest"), Costa Rica. (E) Flowering trees in a tropical dry forest, Costa Rica. (F) The Sonoran desert.

(A)

(B)

(C)

(D)

(E)

(F)

conditions of exposure to air (e.g., desiccation, temperature fluctuations), whereas its lower elevational limit is often determined by biological factors (competition with or predation by other species). There are, of course, many exceptions to these generalizations.

Extending seaward from the shoreline is the **continental shelf**, a feature of most large land masses. The continental shelf may be only a few kilometers wide, or it may extend up to 1,000 km from shore (50 to 100 km is average for most areas). It usually reaches a depth of 150–200 m. These nearshore shelf areas are among the most productive environments of the ocean, being rich in nutrients and shallow enough to permit photosynthesis over much of the area.

The outer limit of the continental shelf—called the **continental edge**—is indicated by a relatively sudden increase in the steepness of the bottom contour. The "steep" parts of the ocean floor, the **continental slopes**, actually have slopes of only 4–6 percent (although the slope is much steeper around volcanic islands). The continental slope continues from the continental edge to the deep ocean floor, which forms the expansive, relatively flat **abyssal plain**. The abyssal plain lies an average of about 4 km below the sea's surface, but it is interrupted by a variety of ridges, sea mounts, abyssal mountain ranges, trenches, and other formations. The bottoms of deep-sea trenches can exceed 10 km in depth.

Organisms that inhabit the water column are known as **pelagic organisms**, whereas those living on the sea bottom anywhere along the entire contour shown in Figure 1.4 are referred to as **benthic organisms**. Both the variety and the abundance of life tend to decrease with increasing depth, from the rich littoral and continental shelf environments to the deep abyssal plain. However, an overgeneralization of this relationship can be misleading. For example, although pelagic biomass declines exponentially with depth, both diversity and biomass increase again near the bottom, in a thick layer of resuspended sediments called the **benthic boundary layer**. Also, shelf and slope habitats in temperate regions are often characterized by low animal density but high species diversity. In many areas, benthic diversity increases abruptly below the continental edge (100–300 m depth), peaks at 1,000 to 2,000 m depth, and then decreases gradually. Species diversity in the benthic abyssal region itself may be surprisingly high. The first impression of early marine scientists—that the deep sea bed was an environment able to sustain only a few species in impoverished simple communities—was simply wrong.

Benthic animals may live on the surface of the substratum (**epifauna**, or **epibenthic** forms, such as most sea anemones, sponges, many snails, and barnacles) or burrow within soft substrata (**infauna**). Infaunal forms include many relatively large invertebrates, such as clams and various worms, as well as some specialized, very tiny forms that inhabit the spaces between sand grains, termed **interstitial** organisms (the smallest of which are **meiofauna**, animals smaller than 0.5 mm). Benthic animals may also be categorized by their locomotor capabilities. Animals that are generally quite motile and active are described as being **errant** (e.g., crabs, many worms), whereas those that are firmly attached to the substratum are **sessile** (e.g., sponges, corals, barnacles). Others are unattached or weakly attached, but generally do not move around much (e.g., crinoids, solitary anemones, most clams); these animals are said to be **sedentary**.

The region of water extending from the surface to near the bottom of the sea is called the **pelagic zone**. The pelagic region over the continental shelf is called the **neritic zone**, and that over the continental slope and beyond is called the **oceanic zone**. The pelagic region can also be subdivided into increments on the basis of water depth (Figure 1.4) or the depth to which light penetrates. The latter factor is, of course, of paramount biological importance. Only within the **photic zone** does enough sunlight penetrate that photosynthesis can occur, and (except in a few special circumstances) all life in the deeper, **aphotic zone** depends ultimately upon organic input from the overlying sunlit layers of the sea. Notable exceptions are the restricted deep-sea hydrothermal vent and benthic cold seep communities, in which sulfur-fixing microorganisms serve as the basis of the food chain.* The photic zone can be up to 200 m deep in the clear waters of the open ocean, decreasing to about 40 m over continental shelves and to as little as 15 m in some coastal waters. Note that some oceanographers restrict the term "aphotic zone" to depths below 1,000 m, where absolutely no sunlight penetrates; the region between this depth and the photic zone is then called the **disphotic zone**.

Organisms that inhabit the pelagic zone are often described in terms of their relative powers of locomotion. Pelagic animals that are strong swimmers, such as fishes and squids, constitute the **nekton**. Those pelagic forms that simply float and drift, or generally are at the mercy of water movements, are collectively called the **plankton**. Many planktonic animals (e.g., small crustaceans) actually swim very well, but they are so small that they are swept along by prevailing currents in spite of their swimming movements, even though those movements may serve to assist them in feeding or escaping predators. Both plants (**phytoplankton**) and animals (**zooplankton**) are included among the plankton,

---

*In addition to deep-sea hydrothermal vents, nonphotosynthetic chemoautotroph-based communities have recently been discovered in a cave (Movile Cave in Romania). The base of the food chain in this unique ecosystem consists of autotrophic microorganisms (bacteria and fungi) thriving in thin mats in and near geothermal waters that contain high levels of hydrogen sulfide. These communities, which sustain dozens of microbial and invertebrate species, create pockets of oxygen-poor, CO_2- and methane-rich air. It is thought that the hydrogen sulfide originates from a deep magmatic source, similar to that seen in deep-sea vents.

the latter being represented by invertebrates such as jellyfishes, comb jellies, arrow worms, many small crustaceans, and the pelagic larvae of many benthic adults. Planktonic animals that spend their entire lives in the pelagic realm are called **holoplanktonic** animals; those whose adult stage is benthic are called **meroplanktonic** animals.

Estuaries and Coastal Marshlands

Estuaries usually occur along low-lying coasts and are created by the interaction of fresh and marine waters, typically where rivers enter the sea. Here one finds an unstable blending of freshwater and saltwater conditions, moving water, tidal influences, and drastic seasonal changes. Estuaries receive high concentrations of nutrients from terrestrial runoff in their freshwater sources and are typically highly productive environments. Temperature and salinity vary greatly with tidal activity and with season. Depending on tides and turbulence, the waters of estuaries may be relatively well mixed and more or less homogeneously brackish, or they may be distinctly stratified, with fresh water floating on the denser salt water below.

The amount of dissolved oxygen in an estuary may also change markedly throughout a 24-hour cycle as a function of temperature and the metabolism of autotrophs. In many cases, hypoxic conditions may occur on a daily basis, especially in the early morning hours. Animals inhabiting these areas must be capable of migrating to regions of higher oxygen levels, be able to store oxygen bound to certain body fluid pigments, or be able to switch temporarily to metabolic processes that do not require oxygen-based respiration. Furthermore, vast amounts of silt borne by freshwater runoff are carried into the waters of estuaries; most of this silt settles out and creates extensive tidal flats (Figure 1.5B). In addition to the natural stresses common to estuarine existence, the inhabitants of estuaries are also subject to stresses resulting from human activity—pollution, thermal additions from power plants, dredging and filling, excessive siltation resulting from coastal and upland deforestation and development, and storm drain discharges are some examples.

Most coastal swamps and marshlands, such as salt marshes and mangrove swamps, are characterized by stands of **halophytes** (flowering plants that flourish in saline conditions; Figure 1.5B,C). Salt marshes and mangrove swamps are alternately flooded and uncovered by tidal action within the estuary, and are thus subjected to the fluctuating conditions described above. The dense halophyte stands and the mixing of waters of different salinities create an efficient nutrient trap. Instead of being swept out to sea, most dissolved nutrients entering an estuary (or generated within it) are utilized there, yielding some of the most productive regions in the world. This great productivity does eventually enter the sea in two principal ways: as plant detritus (mainly from halophyte debris), and via the nektonic animals that migrate in and out of the estuary. The contribution of estuaries to general coastal productivity can hardly be exaggerated. The organic matter produced by plants of the Florida Everglades, for example, forms the base of a major detritus food web that culminates in the rich fisheries of Florida Bay. Furthermore, an estimated 60–80 percent of the world's commercial marine fishes rely on estuaries directly, either as homes for migrating adults or as protective nurseries for the young. Estuaries and other coastal wetlands are also of prime importance to both resident and migratory populations of water birds.

A large number of invertebrates have adapted to life in these dynamic environments. In general, animals have but two alternatives when encountering stressful conditions: either they migrate to more favorable environments, or they remain and tolerate (accommodate to) the changing conditions. Many animals migrate into estuaries to spend only a portion of their life cycle, whereas others move in and out on a daily basis with the tides. Other species remain in estuaries throughout their lives, and these species show a remarkable range of physiological adaptations to the environmental conditions with which they must cope (Chapter 3).

Freshwater Habitats

Because bodies of fresh water are so much smaller than the oceans, they are much more readily and drastically influenced by extrinsic environmental factors, and thus are relatively unstable environments (Figure 1.5D). Changes in temperature and other conditions in ponds, streams, and lakes may occur quickly and be of a magnitude never experienced in most marine environments. Seasonal changes are even more extreme, and may include complete freezing during the winter and complete drying in the summer. Ponds that hold water for only a few weeks during and after rainy seasons are called ephemeral pools (or vernal pools). They typically contain a unique and highly specialized invertebrate fauna capable of producing resting, or **diapause**, stages (usually eggs or embryos) that can survive for months or even years without water. As stressful as this sounds, ephemeral pools contain rich communities of plant and animal life, especially endemic species of crustaceans (e.g., King et al. 1996). Diapause is a form of dormancy in which invertebrates in any stage of development before the adult, including the egg stage, cease their growth and development. Diapause is genetically determined. Some species are programmed to enter diapause when certain environmental conditions provide the proper cues (often a combination of temperature and length of daylight). **Hibernation** and **aestivation** are two other types of dormancy, but they are not genetically programmed and may occur irregularly, or not at all, during any stage of an animal's development. Hibernation is a temporary response to cold, while aestivation is a temporary response to heat.

The very low salinity of fresh water (rarely more than 1‰) and the lack of constant relative ion concentrations subject freshwater inhabitants to severe ionic and osmotic stresses. These conditions, along with other factors such as reduced buoyancy, less stable pH, and rapid nutrient input and depletion produce environments that support far less biological diversity than the ocean does. Nonetheless, many different invertebrates do live in fresh water and have solved the problems associated with this environment. Special adaptations to life in fresh water are summarized in Chapter 3 and discussed in relation to the groups of invertebrates that have such adapations in later chapters.*

Terrestrial Habitats

Life on land is in many ways even more rigorous than life in fresh water. Temperature extremes are usually encountered on a daily basis, water balance is a critical problem, and just physically supporting the body requires major expenditures of energy. Water provides a medium for support, for dispersing gametes, larvae, and adults, and for diluting waste products, and is a source of dissolved materials needed by animals. Animals living in terrestrial environments do not enjoy these benefits of water, and must pay the price. Adaptations to terrestrial living are discussed in Chapter 3.

Relatively few higher taxa have successfully invaded the terrestrial world. Invertebrate success on land is exemplified by the arthropods, notably the terrestrial isopods, insects, spiders, mites, scorpions, and other arachnids. These arthropod groups include truly terrestrial species that have invaded even the most arid environments (Figure 1.5F). Except for some snails and nematodes, all other land-dwelling invertebrates, including such familiar animals as earthworms, are largely restricted to relatively moist areas. In a very real sense, many smaller terrestrial invertebrates survive only through the permanent or periodic presence of water.

A Special Type of Environment: Symbiosis

Many invertebrates live in intimate association with other animals or plants. This kind of association is termed a symbiotic relationship, or simply **symbiosis**. Symbiosis was first defined in 1879 by German mycologist H. A. DeBary as "unlike organisms living together." In most symbiotic relationships, a larger organism (called the **host**) provides an environment (its body, burrow, nest, etc.) on or within which a smaller organism (the **symbiont**) lives. Some symbiotic relationships are rather transient—for example, the relationship between

ticks or lice and their vertebrate host—whereas others are more or less permanent. Some symbionts are opportunistic (facultative), whereas others cannot survive without their host (obligatory).

Symbiotic relationships can be subdivided into several categories based on the nature of the interaction between the symbiont and its host. Perhaps the most familiar type of symbiotic relationship is **parasitism**, in which the symbiont (a parasite) receives benefits at the host's expense. Parasites may be external (**ectoparasites**), such as lice, ticks, and leeches; or internal (**endoparasites**), such as liver flukes, some roundworms, and tapeworms. Other parasites may be neither strictly internal nor strictly external; rather, they may live in a body cavity or area of the host that communicates with the environment, such as the gill chamber of a fish or the mouth or anus of a host animal (**mesoparasites**). Some parasites live their entire adult lives in association with their hosts and are **permanent parasites**, whereas **temporary**, or **intermittent parasites**, such as bedbugs, only feed on the host and then leave it. Parasites that parasitize other parasites are **hyperparasitic**. Temporary parasites, such as mosquitoes and aegiid isopods, are often referred to as **micropredators**, in recognition of the fact that they usually "prey" on several different host individuals. **Parasitoids** are insects, usually flies or wasps, whose immature stages feed on their hosts' bodies, usually other insects, and ultimately kill the host. A **definitive host** is one in which the parasite reaches reproductive maturity. An **intermediate host** is one that is required for parasite's development, but in which the parasite does not reach reproductive maturity.

A few groups of invertebrates are predominantly or exclusively parasitic, and almost all invertebrate phyla have at least some species that have adopted parasitic lifestyles. Many texts and courses on parasitology pay particular attention to the effects of these animals on humans, crops, livestock, and economic conditions. Here we also try to focus on parasitism from "the parasite's point of view," that is, as a particular lifestyle suited to a specific environment, requiring certain adaptations and conferring certain advantages. It has been estimated that 50 to 70 percent of the world's species are parasitic, making parasitism the most common way of life. Since insects are the most diverse group of organisms on Earth, and since all insects harbor numerous parasites, it is fair to say that the most common mode of life on Earth is that of an insect parasite.

Mutualism is another form of symbiosis that is generally defined as an association in which both host and symbiont benefit. Such relationships may be extremely intimate and important for the survival of both parties; for example, the bacteria in our own large intestine are important in the production of certain vitamins and in processing material in the gut. In fact, beneficial associations with specific bacterial symbionts characterize many, if not all, animal species, although most of these relationships

---

*Freshwater habitats are some of the most threatened environments on Earth. Throughout the United States, people destroy 100,000 acres of wetlands annually. Rare aquatic habitats such as ephemeral pools and subterranean rivers are disappearing faster than they can be studied. Underground, or **hypogean**, habitats are often aquatic, and these habitats are quickly being destroyed by pollution and groundwater overdraft.

have not been studied. Another example is the relationship between termites and certain protists that inhabit their digestive tracts and are responsible for the breakdown of cellulose into compounds that can be assimilated by their insect hosts. Other mutualistic relationships may be less binding on the organisms involved. Cleaner shrimps, for example, inhabit coral reef environments, where they establish "cleaning stations" that are visited regularly by reef-dwelling fishes that present themselves to the shrimps for the removal of parasites. Obviously, even this rather loose association results in benefits for the shrimps (a meal) as well as for the fishes (removal of parasites). The mutualistic relationships between plants and their pollinators are essential to the survival of most flowering plants and their insect partners (and, in some cases, their bird or nectar-feeding bat partners).

A third type of symbiosis is called **commensalism**. This category is something of a catch-all for associations in which significant harm or mutual benefit is not obvious. Commensalism is usually described as an association that is advantageous to one party (the symbiont) but leaves the other (the host) unaffected. For instance, among invertebrates there are numerous examples of one species inhabiting the tube or burrow of another (**inquilism**); the former obtains protection, food, or both with little or no apparent effect on the latter. A special type of commensalism is **phoresis**, wherein the two symbionts "travel together," but there is no physiological or biochemical dependency on the part of either participant. Usually one **phoront** is smaller than the other and is mechanically carried about by its larger companion.

There is a good deal of overlap among the categories of symbiosis described above, and many animal relationships have elements of two or even of all the categories, depending on life history stage or environmental conditions. Taken in its broad sense, the concept of symbiosis has profound implications for understanding Earth's biodiversity. It has been said that at least half the planet's species are symbionts and that all species have symbiotic partnerships—concepts suggesting that every individual is an ecosystem.

Some Comments On Evolution

Fitness By Any Other Name
Would Be As Loose

> A group inept
> Might better opt
> To be adept
> And so adopt
> Ways more apt
> To wit, adapt.

> John Burns
> *Biograffiti*, 1975

This book takes evolution as its central theme. However, the paradigms that have guided evolutionary biology for the past 60 years are presently in the midst of a major reevaluation. This reevaluation has been precipitated by three phenomena. First is the revolution in molecular biology, which has produced dramatic discoveries since the late 1970s and will no doubt continue to do so for many decades to come. Second is the continuing development of an explicit method of inferring phylogenies, called phylogenetic systematics or cladistics (see Chapter 2). Third is the development of some very new and different ideas regarding the operation of evolution itself. Most students are familiar with Darwin's theory of natural selection, and with the fundamental genetic mechanisms that underlie adaptation, but fewer students are familiar with more recent ideas that have been proposed outside the framework of natural selection and adaptation. We briefly review some of these interesting new thoughts, all of which have implications for the processes represented by the phylogenetic trees appearing in the following chapters.

There are three fundamental patterns we see when we examine evolutionary history: anagenesis, speciation, and extinction. **Anagenesis** seems to be driven by those neo-Darwinian processes often referred to as **microevolution**—the within-species, generation-by-generation evolution of populations and groups of populations over the "lifetime" of a species. Natural selection and adaptation are powerful driving forces at this level. **Speciation** is the "birth" of a species, and **extinction** is the "death" (termination) of a species. Speciation and extinction engage processes outside the natural selection–adaptation paradigm—processes often referred to as **macroevolution**. The mechanisms that initiate and sculpt each of these patterns differ. Most college courses today focus primarily on microevolution, or anagenesis, and most students reading this book already know a great deal about population genetics and natural selection. However, the view that all of evolution can be understood solely on the basis of microevolutionary phenomena is being reexamined in light of new ideas regarding evolutionary change. Consequently, we would like to introduce readers to some ideas with which they might be less familiar. We will do so by first discussing within-species processes (presented here under the term "microevolution"), and then speciation and extinction (grouped under the heading "macroevolution").

Microevolution

The neo-Darwinian evolutionary model, or so-called modern synthesis, that resulted from the integration of Mendelian genetics into Darwinian natural selection theory dominated evolutionary biology through the twentieth century. Basically, the neo-Darwinian view holds that all evolutionary changes result from the action of natural selection on variation within populations (see John Burns's poem above). This view has been called the "adaptationist paradigm." The theory focuses on adaptation and deals primarily with genes and changes in allelic frequencies within populations. These genetic variations come about primarily by recombination and mutation, although the random phenomena of genetic drift and the founder effect are also part of the neo-Darwinian synthesis.

Evolution by natural selection is viewed as a deterministic process, even though certain elements of chance are accepted within the theory (e.g., mutation, random mating, the founder effect). The theory of natural selection implies that, given a complete understanding of the environment and genetics, evolutionary outcomes should be largely predictable. The theory of natural selection further implies that virtually all of the characteristics animals possess are products of adaptations leading to increased fitness (ultimately, to increased reproductive success). An adaptationist view might lead one to assume that every aspect of an animal's phenotype is the product of natural selection working to increase the fitness of a species in a particular environment.

Microevolution is thus seen to be a deterministic, within-species phenomenon that affects population genetics on a generation-to-generation basis to produce changes and patterns in gene frequencies within and among populations. The modern synthesis deals almost exclusively with evolution at this level.

Macroevolution

Macroevolution is the focus of some of the most interesting debates among evolutionists today. Macroevolutionary phenomena include such things as the origin of species and radiations of species lineages (**cladogenesis**), "explosive" radiations that appear to be linked to the opening up of new ecological arenas or niches, transgenic events, major shifts in developmental processes that might result in new body plans, various karyotypic alterations (e.g., polyploidy and polyteny), geological events that profoundly alter the distributions of species, and mass extinctions (and the subsequent new biotic proliferations).

Mass extinction events in Earth's history have played major roles in reshaping the directions of animal evolution in unpredictable ways. The largest of these extinction events wiped out a majority of life forms on Earth. In the Permian–Triassic event described above, for example, an estimated 85 percent of all marine species went extinct (although no *phylum* is known to have gone extinct since the start of the Cambrian). Mass extinctions thus are profound macroevolutionary events that can abruptly (in geological time) terminate millions of species and lineages.

In contrast to microevolution, macroevolution is evolutionary change, often rapid, that produces phylogenetic patterns formation *above* the species level (e.g., the patterns depicted on the phylogenetic trees in this book). The fossil record suggests that speciation events (one species giving rise to one or more new species)

tend to be rapid, or geologically instantaneous. Analysis of the fossil record also shows that the number of species has increased, perhaps exponentially, since the end of the Proterozoic, with this diversification periodically interrupted by mass extinctions. And mass extinctions have always been followed by periods of rapid speciation and radiation at higher taxonomic levels (i.e., macroevolution).

Newer views suggest that speciation might not be initiated by natural selection, but rather by processes outside the natural selection paradigm—most frequently by purely stochastic processes. Microevolution can be thought of as a within-species process that *maintains* genomic continuity and continually "fine-tunes" populations and species to their changing environment. A reasonable analogy might be the basic metabolic activities that keep your own body "fine-tuned" to the environment—a background process that is always at work maintaining a level of homeostasis (within your body, or within a species' gene pool). A macroevolutionary event, on the other hand, is typically a processes that *disrupts* genomic, or reproductive, continuity in a species and may thus initiate speciation events. Following the above analogy, macroevolutionary events disrupt the homeostasis of species' gene pools.

One of the most fundamental new approaches to evolutionary biology is the consideration of **stochastic** processes or events—those that occur at random or by chance. Some examples of stochastic events are described below.

The geneticist Goldschmidt, the paleontologist Schindewolf, and the zoologists Jeannel, Cuénot, and Cannon all maintained until the 1950s that neither evolution within species nor simple allopatric speciation could fully explain macroevolution. They advanced an idea called **saltation theory**—the sudden origin of wholly new types of organisms—the "hopeful monsters" of Goldschmidt—in great leaps of change. It has been proposed that one way such rapid changes might occur is through **transgenic** events, involving the "lateral transfer" of genetic material from one species to another. Two mechanisms of lateral genetic transfer have been implicated as possible agents of saltation: transposable genetic elements and symbiogenesis.

Transposable elements (TEs) are specialized DNA segments that move (transpose) from one location to another, either within a cell's DNA, between individuals in a species, or even between species. They were discovered in maize (*Zea mays*) by the Nobel laureate Barbara McClintock in the 1950s, but little was known about them until recently. With the growth of molecular genetics, hundreds of TEs now have been identified—over 40 different ones are known from the laboratory fruit fly *Drosophila melanogaster* alone. The mechanisms of TE transfer between organisms are not yet well understood. However, the transfer of genetic elements from one species to another is suspected to be by way of viruses, bacteria, arthropod parasites, or other vectors. There is strong evidence that parasitic mites have been responsible for the lateral transfer of genetic elements among *Drosophila* species.

The movement of a TE within a genome is mediated by a TE-encoded protein called a **transposase**, probably interacting in complex ways with certain cellular factors. A transposase recognizes the ends of the TE, breaks the DNA at these ends to release the TE from its original position, and joins the ends to a new target sequence. The transposition of some TEs from bacteria to bacteria, and from bacteria to plant cells, is partially understood, and we know that the introduction of these DNA segments can contribute powerful mutagenic qualities to the new host's genome. Recent work suggests that a great deal of such "gene swapping" took place during the early evolution of the prokaryotes. Although TEs have been best studied in prokaryotes, they have been found in most organisms that have been examined, including insects, mammals, flowering plants, sponges, and flatworms. Although we lack specific evidence, there is reason to suspect that TEs could have been responsible for some of the major genetic innovations that have taken place in the history of life.

Another way in which evolutionary novelties can arise is through symbiosis. The Russian biologist Konstantin Mereschkovsky (1855–1921) developed the "two-plasm" (cell within a cell) theory, claiming that chloroplasts originated from blue-green algae (Cyanobacteria). For this process, he invented the term **symbiogenesis**. In Chapter 5, we describe the symbiogenic origin of the eukaryotic cell, which probably arose by way of incorporation of once free-living prokaryotes that came to be what we recognize today as mitochondria, chloroplasts, cilia, flagella, and other organelles. Although symbiogenesis is an old idea, it was Lynn Margulis who most vigorously championed it in the twentieth century.

Beyond the origin of the eukaryotic cell, symbiogenesis may be at work in many other systems, but we have little knowledge of how genetic material might be shared by or influenced among animals in such relationships. In extremely intimate symbiotic partnerships, however, the two symbionts could have profound effects on each other's genetic evolution. Many invertebrates are invovled in such relationships, including the corals and other cnidarians that serve as hosts for symbiotic dinoflagellates (called zooxanthellae) that live within their tissues. Various animals that harbor (and exploit) tetrodotoxin-secreting bacteria (many chaetognaths, the blue-ringed octopus, a sea star, and a horseshoe crab, and certain tetraodontid fishes), and squids with luminous bacteria, and lichens (an intimate association between fungi and Cyanobacteria or green algae) are other examples. That symbionts can affect the evolution of their hosts in unexpected ways can be seen in parasites that enhance their own chances for survival by altering aspects of their host's lives—for example, parasites that

increase the likelihood that their intermediate host will fall prey to their definitive host by changing the intermediate host's size, color, biochemistry, or behavior in ways that make it more vulnerable to predation.

Another revelation in our thinking about macroevolution has come from the discovery of **homeobox (Hox) genes**. These master regulatory genes modulate other sets of developmental genes and, in doing so, "select" the developmental pathways that are followed by dividing cells. Hox genes have two functions in the early development of embryos: (1) they encode short regulatory proteins that bind to a particular sequences of bases in DNA and either enhance or repress gene expression, and (2) they encode proteins that are expressed in complex patterns that determine the basic geometry of the organism. The term **Hox genes** refers specifically to those genes that are clustered in an array on the chromosome and function primarily in establishing regional or segmental identities. In all Metazoa that have been examined, regional or segmental specialization is controlled by the spatially localized expression of these genes, which play crucial roles in determining body patterns. They underlie such fundamental attributes as anterior–posterior differentiation (in both invertebrates and vertebrates) and the positioning of body wall outgrowths (e.g., limbs).

Hox genes have been conserved to a remarkable degree throughout the animal kingdom, and they are now known from all animal phyla that have been examined. There is a striking correlation between the order of these genes on their chromosomes and the position of their expression in the developing animal along the main body axis. Hox proteins regulate the genes that control the cellular processes involved in morphogenesis. In doing so, they demarcate relative positions in animals—they do not specify the precise nature of particular structures. For example, in arthropods, Hox genes regulate where body appendages form, and they can either suppress limb development or modify it (in concert with other regulatory genes) to create unique appendage morphologies. Mutations in Hox genes, and other developmental genes, can create gross mutations (homeotic mutations or homeosis).

There is a growing body of evidence suggesting that Hox genes have played major roles in the evolution of new body plans among the Metazoa. The evolutionary potential of Hox genes lies in their hierarchical and combinatorial nature. We now know that a single Hox gene can modulate the expression of dozens of interacting downstream genes, the products of which determine developmental outcomes. Variation in the output of these multigene networks can arise at many levels simply through changes in the relative timing of developmental gene expression (i.e., by heterochrony; see Chapter 4), or through interactions between genes in the regulatory network. To understand the profound potential of Hox genes to drive evolutionary change, consider that, within the genome of *Drosophila*, 85–170 different genes are regulated by the product of the Hox gene *Ultrabithorax* (*Ubx*) alone (Carroll 1995). Changes in the *Ubx* protein could potentially alter the regulation of all these genes! In some families of sea spiders (Pycnogonida), Hox gene mutations appear to have produced spurious segment/leg duplications, creating polymerous lineages (see Chapter 19). Another example of the potential of Hox genes is seen in the abdominal limbs of insects. Abdominal limbs ("prolegs") occur on larvae of various insects in several orders, and they are ubiquitous in the Lepidoptera (e.g., caterpillars). These limbs were probably present in insect ancestors, hence prolegs may have reappeared through the de-repression of an ancestral limb developmental program (i.e., they are a Hox gene mediated atavism). Proleg formation appears to involve a change in the regulation and expression of a single gene (*abd-A*) during embryogenesis.

In summary, the processes of microevolution (e.g., natural selection) act on individuals and populations, maintain genomic continuity, and create anastomosing patterns of relationship over time (Figure 1.6). Macroevolutionary processes (e.g., speciation and extinction), on the other hand, act on species and lineages, disrupt genomic continuity, and create ascending, bifurcating patterns of relationship over time (Figure 1.6). In a cladogram of species, the line segments represent the places where anagenesis (microevolution) is taking place within a given species. The nodes in the cladogram represent macroevolutionary events, speciation and extinction. Although Darwin titled his book *On the Origin of Species*, he dealt primarily with the maintenance of adaptations. In fact, the nature of the relationship between anagenesis and cladogenesis is still not well understood. Evidence for the disengagement of natural selection and speciation comes from the fossil record, which suggests that most species do not change significantly throughout their existence; rather, they remain phenotypically stable for millions of years, then undergo a rapid change in which they essentially "replace themselves" with one or more new and different species. These new species, in turn, remain phenotypically static for millions more years. The fossil record suggests that most species of marine invertebrates persist more or less unchanged for 5–10 million years, whereas the time required for significant anatomical change seems to be only a few thousand years or less. This idea of speciation in rapid bursts, sandwiched be-

Figure 1.6 Microevolution and macroevolution depicted graphically. The highlighted portion of the cladogram (on the right) is shown in detail in the drawing to the left.

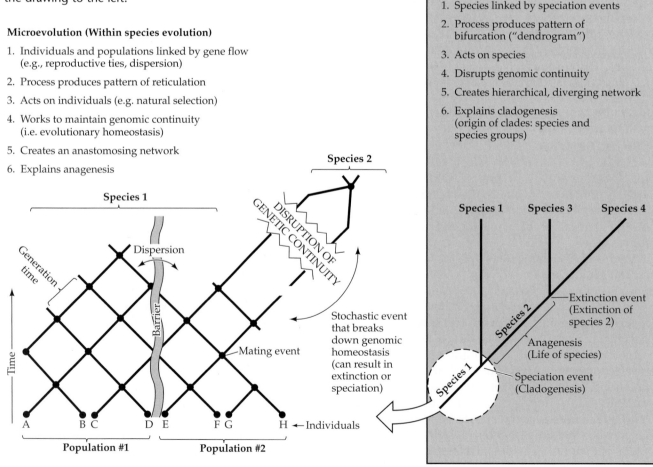

Microevolution (Within species evolution)

1. Individuals and populations linked by gene flow (e.g., reproductive ties, dispersion)
2. Process produces pattern of reticulation
3. Acts on individuals (e.g. natural selection)
4. Works to maintain genomic continuity (i.e. evolutionary homeostasis)
5. Creates an anastomosing network
6. Explains anagenesis

Macroevolution (Species/clade evolution)

1. Species linked by speciation events
2. Process produces pattern of bifurcation ("dendrogram")
3. Acts on species
4. Disrupts genomic continuity
5. Creates hierarchical, diverging network
6. Explains cladogenesis (origin of clades: species and species groups)

tween long periods of species stasis, was explained in the **punctuated equilibrium model** of Eldredge and Gould (1972).

Biologists are still a long way from understanding all the causes and mechanisms of the evolutionary process. That evolution has occurred and is occurring is well documented and consistent with all of the available data. We are developing excellent methods for analyzing the patterns or the history of evolution (e.g., phylogenetics). The current debates concern the process—the nature of the evolutionary mechanisms themselves. It seems probable that different processes, working at different levels, have created the patterns we see in the world today. Despite the many evolutionary questions currently being discussed, and despite whatever evolutionary processes are now at work (probably all of these and other processes are), biologists are quite able to continue their efforts at reconstructing the evolutionary history of life on Earth, because the processes of evolution (whatever they entail) result in new organisms that are distinct by virtue of various unique new characters or attributes that they have acquired. Their descendants retain these attributes and in time acquire still others, which are retained by their descendants. In this fashion, the living world provides us with an analyzable hierarchical pattern consisting of nested sets of features recognizable both in fossils and in living organisms. Those features, in turn, are the data (i.e., the "characters") with which we can reconstruct a history of the descent of life. We will have much more to say regarding this reconstruction process in the following chapter, because understanding what characters are and how they are evaluated is fundamental to comparative biology and to an appreciation of the invertebrate world.

A Final Introductory Message to the Reader

If you have not already done so, please read the Preface to this text, which explains this book's limitations, describes what it is about, and outlines what sort of information we intend to convey. Because of our comparative approach, it is critical that you become familiar with the initial chapters (Chapters 1–4) before attempting to study and comprehend the sections dealing with individual animal groups. These first four chapters are designed to accomplish several goals: (1) to define some basic terminology, (2) to introduce a number of important concepts, and (3) to describe in detail the themes that we use throughout the rest of the book.

The fundamental theme of this book is evolution, and we approach invertebrate evolution primarily through the field of comparative biology . In Chapter 2 we provide an explanation of how biologists derive evolutionary schemes and classifications, how theories about the phylogeny of animal groups grow and change, and how the information presented in this text has been used to construct theories on how life evolved on Earth. In Chapters 3 and 4 we lay out the fundamental anatomical and morphological designs and developmental strategies of invertebrates. Like the features of organisms, these designs and strategies are not random, but form patterns. Recognition and analysis of these patterns constitute the basic building blocks of this book. We then proceed in the "animal chapters" to explore the evolution of the invertebrates in light of various combinations of these basic functional body plans and lifestyles. With this background, you should be able to follow the evolutionary changes and branchings among the invertebrate phyla, their body systems, and their various pathways to success on Earth.

Through our approach, we hope to add continuity to the massive subject of invertebrate zoology, which is often covered (in texts and lectures) by a sort of "flash-card" method, in which the primary goal is to have the student memorize animal names and characteristics and keep them properly associated, at least until after the examination. Thus, we urge you to look back frequently at these first few chapters as you read ahead and explore how invertebrates are put together, how they live, and how they evolved.

Selected References

General References

Adams, E. 1987. Invertebrate collagens. Science 202: 591–598.

Barnes, R. D. and E. Ruppert. 1994. *Invertebrate Zoology*, 6th Ed. Saunders, Philadelphia.

Beklemishev, W. N. 1969. *Principles of Comparative Anatomy of Invertebrates*. 2 vols. University of Chicago Press, Chicago. [Translated from Russian; a different view of the subject, quite unlike Western texts.]

Bengtson, S. and Y. Zhao. 1997. Fossilized metazoan embryos from the earliest Cambrian. Science 277: 1645–1648.

Boardman, R. S., A. H. Cheetham and A. J. Rowell. 1987. *Fossil Invertebrates*. Blackwell, London. [A good distillation of fossil invertebrate zoology.]

Briggs, D. E. G., D. E. Erwin and F. J. Collier. 1994. *The Fossils of the Burgess Shale*. Smithsonian Institution Press, Washington, D.C.

Brusca, R. 2000. Unraveling the history of arthropod biodiversification. Ann. Missouri Bot. Garden 87(1): 13–25.

Buchsbaum, R., M. Buchsbaum, J. Pearse and V. Pearse. 1987. *Animals without Backbones*, 3rd Ed. University of Chicago Press, Chicago. [The third edition of this classic book provides delightful reading and many excellent photographs.]

Carefoot, T. 1977. *Pacific Seashores: A Guide to Intertidal Ecology*. University of Washington Press, Seattle. [A very clear and readable account of ecological concepts as they apply to temperate seashores; the emphasis is on invertebrates.]

Coleman, D. C. and P. F. Hendrix (eds.). 2000. *Invertebrates as Webmasters in Ecosystems*. Oxford University Press.

Combes, C. 2001. *Parasitism: The Ecology and Evolution of Intimate Interactions*. University Chicago Press, Chicago. [Brings a unifying approach to a subject usually presented in a fragmented fashion.]

Crawford, C. S. 1981. *Biology of Desert Invertebrates*. Springer-Verlag, New York.

Curtis, T. P., W. T. Sloan and J. W. Scannell. 2002. Estimating prokary- otic diversity and its limits. Proc. Natl. Acad. Sci. U.S.A. 99(16): 10494–10499. [Also see comment by B. Ward, same issue, 10234–10236.]

Donoghue, M. J. and W. S. Alverson. 2000. A new age of discovery. Ann. Missouri Bot. Garden 87(1): 110–126.

Ekman, S. 1953. *Zoogeography of the Sea*. Sedgwick and Jackson, London. [Excellent review of marine invertebrate distributions; dated, but a benchmark work.]

Erwin, T. L. 1991. How many species are there? Revisited. Conserv. Biol. 5: 1–4. [See also F. Ødegaard. 2000. How many species of arthropods? Erwin's estimate revised. Biol. J. Linn. Soc. 71: 583–597.]

Fredrickson, J. K. and T. C. Onstott. 1996. Microbes deep inside the Earth. Sci. Am. 275: 68–73.

Freeman, W. H. and B. Bracegirdle. 1971. *An Atlas of Invertebrate Structure*. Heinemann Educational Books, London. [A good labo- ratory aid; treats gross morphology, anatomy, and histology, with illustrations and photographs.]

Giese, A. and J. S. Pearse (eds.). 1974–1987. *Reproduction of Marine Invertebrates*. Vols. 1–5, 9. Academic Press, New York. [With more on the way; excellent review articles.]

Gilbert, S. F. and A. M. Raunio (eds.). 1997. *Embryology: Constructing the Organism*. Sinauer Associates, Sunderland, MA. [The best con- temporary work on the subject.]

Grassé, P. (ed.). 1948. *Traité de Zoologie*. Masson et Cie, Paris. [Work continues on this multivolume enterprise covering the animal kingdom; one of the best single reference sources on invertebrates.]

Hardy, A. C. 1956. *The Open Sea*. Houghton Mifflin, Boston. [Still the best introduction to the world of plankton.]

Harrison, F. W. (ed.). 1991–1997. *Microscopic Anatomy of Invertebrates*. Wiley-Liss, New York. [A 20-book series providing up-to-date, de- tailed treatments of anatomy, histology, and ultrastructure.]

Hawkesworth, D. L. 1995. *Biodiversity: Measurement and Estimation*. Chapman & Hall, London.

Haywood, V. E. 1996. *Global Biodiversity Assessment*. Cambridge University Press, Cambridge. [A whole-earth encyclopedia of bio- diversity.]

Hedgpeth, J. W. (ed.). 1957. *Treatise on Marine Ecology and Paleoecology*. Geological Society of America Memoir 67. [Still frequently consult- ed and cited; excellent reviews of major aspects of marine biology.]

Hoppert, M. and F. Mayer. 1999. Prokaryotes. Amer. Sci. 87: 518–525.

Hyman, L. H. 1940–1967. *The Invertebrates*. 6 vols. McGraw-Hill, New York. [This series has probably ended. Naturally, some of the ma- terial in early volumes is out of date, but they still remain among the best references available.]

Kaestner, A. 1967–1970. *Invertebrate Zoology*. Vols. 1–3. Wiley- Interscience, New York. [Translated from German.]

King, J. L., M. A. Simovich and R. C. Brusca. 1996. Species richness, en- demism and ecology of crustacean assemblages in northern California vernal pools. Hydrobiologia 328: 85–116. [A detailed look at a rare and threatened environment.]

Lankester, R. (ed.). 1900–1909. *A Treatise on Zoology*. Adam and Charles Black, London. [A classic multivolume work on invertebrates. Despite its age, this remains a valuable resource.]

Larwood, G. and B. Rosen (eds.). 1979. *Biology and Systematics of Colonial Organisms*. Academic Press, New York.

Lincoln, R. J., G. A. Boxshall, and P. F. Clark. 1982. *A Dictionary of Ecology, Evolution and Systematics*. Cambridge University Press, New York. [Excellent.]

Madigan, M. 2000. Extremophilic bacteria and microbial diversity. Ann. Missouri Bot. Garden 87(1): 3–12.

Madigan, M. and B. Marrs. 1997. Extremophiles. Sci. Am. 276: 82–87.

Marshall, N. B. 1980. *Deep Sea Biology: Development and Perspective*. Garland STPM Press, New York.

Moore, J. 1984. Parasites that change the behavior of their host. Sci. Am. (May): 108–115.

Moore, R. C. (ed.). 1952–present. *Treatise on Invertebrate Paleontology*. Geological Society of America and University of Kansas Press, Lawrence. [Detailed coverage of fossil forms; many volumes still pending.]

Norse, E. (ed.). 1993. *Global Marine Biological Diversity*. Island Press, Washington, D.C.

Panganiban, G. et al. 1997. The origin and evolution of animal ap- pendages. Proc. Natl. Acad. Sci. U.S.A. 94: 5162–5166.

Parker, S. P. (ed.). 1982. *Synopsis and Classification of Living Organisms*. 2 vols. McGraw-Hill, New York. [Encyclopedic; dry.]

Poore, G. C. B. and G. D. F. Wilson. 1993. Marine species richness. Nature 361: 597–598.

Price, P. W. 1980. *Evolutionary Biology of Parasites*. Princeton University Press, Princeton, NJ.

Prosser, C. L. (ed.). 1991. *Environmental and Metabolic Animal Physiology*. Wiley-Liss, New York. [One of the best accounts of comparative physiology.]

Roberts, L. S. and J. Janovy, Jr. 1996. *Foundations of Parasitology*, 5th Ed. Wm. C. Brown, Dubuque, IA. [One of the better of a generally dis- appointing field of textbooks on parasitology.]

Sarbu, S. M. and T. C. Kane. 1995. A subterranean chemoautotrophi- cally based ecosystem. NSS Bull. 57: 91–98.

Seilacher, A., P. K. Bose and F. Pflüger. 1998. Triploblastic animals more than one billion years ago: Trace fossil evidence from India. Science 282: 80–83.

Stachowitsch, M. 1992. *The Invertebrates: An Illustrated Glossary*. Wiley- Liss, New York.

Stephensen, T. A. and A. Stephensen. 1972. *Life Between Tide Marks on Rocky Shores*. W. H. Freeman, San Francisco. [A summary of the au- thors' life work on the subject; primarily deals with algae and in- vertebrates; global in coverage.]

Wilson, E. O. 1992. *The Diversity of Life*. Belknap Press, Harvard University Press, Cambridge, MA. [Outstanding writing by one of the greatest living American naturalists.]

Wilson, E. O., and F. M. Peter (eds.). 1988. *Biodiversity*. National Academy Press, Washington, D.C.

Wray, G. A., J. S. Levinton and L. H. Shapiro. 1996. Molecular evidence for deep Precambrian divergences among metazoan phyla. Science 274: 568–573.

Zhuravlev, A. and R. Riding (eds.). 2001. *The Ecology of the Cambrian Radiation*. Columbia University Press, New York.

Manuals and Field Guides for Identification of Invertebrates

We have included here only a few of the scores of identification guides, booklets, and the like. Guides to particular taxa are listed in appropriate chapters.

Allen, G. R. and R. Steene. 1994. *Indo-Pacific Coral Reef Field Guide*. Tropical Reef Research, Singapore.

Allen, R. 1969. *Common Intertidal Invertebrates of Southern California*. Peek Publications, Palo Alto, CA. [The only compilation of keys to southern California invertebrates ever published; out of print and hard to find.]

Bright, T. J. and L. H. Pequegnat (eds.). 1974. *Biota of the West Flower Garden Bank*. Gulf Publishing Co., Houston, TX.

Brusca, G. J. and R. C. Brusca. 1978. *A Naturalist's Seashore Guide: Common Marine Life of the Northern California Coast and Adjacent Shores*. Mad River Press, Eureka, CA. [Currently being revised.]

Brusca, R. C. 1980. *Common Intertidal Invertebrates of the Gulf of California*, 2nd Ed. University of Arizona Press, Tucson. [A fairly exhaustive treatment of the subject, including keys, descriptions, and figures for over 1,300 species; currently being revised.]

Colin, P. I. 1978. *Caribbean Reef Invertebrates and Plants*. T. F. H. Publications, Neptune City, NJ.

Edmondson, W. T., H. B. Ward and G. C. Whipple (eds.). 1959. *Freshwater Biology*. Wiley, New York. [Good keys to freshwater in- vertebrates.]

Fielding, A. 1982. *Hawaiian Reefs and Tidepools*. Oriental, Honolulu.

Fish, J. D. and S. Fish. 1996. *A Student's Guide to the Seashore*, 2nd Ed. Cambridge University Press, Cambridge. [An excellent survey of the invertebrates of British shores.]

Gosliner, T. M., D. W. Behrens and G. C. Williams. 1996. *Coral Reef Animals of the Indo-Pacific*. Sea Challengers, Monterey, CA.

Gosner, K. L. 1971. *Guide to the Identification of Marine Estuarine Invertebrates*. Wiley-Interscience, New York. [For use on the north- eastern coast of the United States.]

Hayward, P. and J. S. Ryland (eds.). 1995. *Handbook of the Marine Fauna of North-West Europe*. Oxford University Press, Oxford.

Hobson, E. and E. H. Chave. 1990. *Hawaiian Reef Animals*. University of Hawaii Press, Honolulu.

Humann, P. 1992. *Reef Creature Identification: Florida, Caribbean, Bahamas*. New World Publications, Jacksonville, FL.

Humann, P. 1993. *Reef Coral Identification: Florida, Caribbean, Bahamas*. New World Publications, Jacksonville, FL.

Kaplan, E. 1982. *A Field Guide to Coral Reefs of the Caribbean and Florida*. Houghton Mifflin, Boston. [Excellent; one of the Peterson Field Guides.]

Kozloff, E. 1974. *Keys to the Marine Invertebrates of Puget Sound, the San Juan Archipelago, and Adjacent Regions*. University of Washington Press, Seattle.

Kozloff, E. 1987. *Marine Invertebrates of the Pacific Northwest*. University of Washington Press, Seattle.

Laboute, P. and Y. Magnier. 1979. *Underwater Guide to New Caledonia*. Les Editions Pacifique, Papeete, Tahiti.

Luther, W. and K. Fiedler. 1976. *A Field Guide to the Mediterranean Sea Shore*. Collins, London.

McConnaughey, B. H. and E. McConnaughey. 1985. *Pacific Coast: The Audubon Society Nature Guides*. Chanticleer Press, New York.

Morris, R. H., D. P. Abbott and E. C. Haderlie. 1980. *Intertidal Invertebrates of California*. Stanford University Press, Stanford, CA.

Newell, G. and R. Newell. 1973. *Marine Plankton: A Practical Guide*. Hutchinson, London.

Pennak, R. W. 1989. *Fresh-Water Invertebrates of the United States: Protozoa to Mollusca, 3rd Ed.* John Wiley & Sons, New York.

Ricketts, E. F., J. Calvin, J. W. Hedgpeth and D. W. Phillips. 1985. *Between Pacific Tides*, 4th Ed. Stanford University Press, Stanford, CA. [A standard reference for natural history of Pacific coast intertidal invertebrates.]

Riedl, R. (ed.). 1983. *Fauna und Flora des Mittelmeeres*. Verlag Paul Parey, Hamburg. [Perhaps the best field guide for the Mediterranean.]

Ruppert, E. E. and R. S. Fox. 1988. *Seashore Animals of the Southeast: A Guide to Common Shallow-Water Invertebrates of the Southeastern Atlantic Coast*. University of South Carolina Press, Columbia.

Sefton, N. and S. K. Webster. 1986. *A Field Guide to Caribbean Reef Invertebrates*. Sea Challengers, Monterey, CA.

Smith, R. I., and J. Carlton (eds.). 1975. *Light's Manual: Intertidal Invertebrates of the Central California Coast*, 3rd Ed. University of California Press, Berkeley. [A product of the editors' devotion to the task; includes keys; well referenced.]

Sterrer, W. (ed.). 1986. *Marine Fauna and Flora of Bermuda*. Wiley, New York. [A comprehensive guide.]

Thorp, J. H. and A. P. Covich (eds.). 1991. *Ecology and Classification of North American Freshwater Invertebrates*. Academic Press, San Diego, CA.

Todd, C. D., M. S. Laverack and G. A. Boxshall. 1996. *Coastal Marine Zooplankton*, 2nd Ed. Cambridge University Press, Cambridge.

Wirtz, P. 1995. *Unterwasserführer Madeira Kanaren/Azoren*. Verlag Stephanie Naglschmid, Stuttgart. [An underwater guide to the Canary Islands, Azores, and Madeira.]

Some Recommended References on Evolution

Avers, C. J. 1989. *Process and Pattern in Evolution*. Oxford University Press, New York.

Ayala, F. J. (ed.). 1976. *Molecular Evolution*. Sinauer Associates, Sunderland, MA.

Ayala, F. J. and J. W. Valentine. 1979. *The Theory and Processes of Organic Evolution*. Benjamin/Cummings, Menlo Park, CA.

Benton, M. J. 1995. Diversification and extinction in the history of life. Science 268: 52–58.

Berg, D. E. and M. M. Howe (eds.). 1989. *Mobile DNA*. American Society of Microbiology, Washington, D.C.

Bermudes, D. and L. Margulis. 1987. Symbiont acquisition as neoseme: Origin of species and higher taxa. Symbiosis 4: 185–198.

Brock, J. J., G. A. Logan, R. Buick and R. E. Summons. 1999. Archean molecular fossils and the early rise of eukaryotes. Science 285: 1033–1036.

Brooks, D. R. and E. O. Wiley. 1988. *Evolution as Entropy*, 2nd Ed. University of Chicago Press, Chicago. [A view of evolution that challenges traditional thinking on the subject.]

Bush, G. L. 1975. Modes of animal speciation. Annu. Rev. Ecol. Syst. 6: 339–364. [One of the classic review studies on "traditional" speciation models.]

Dyer, B. D. and R. Obar (eds.). 1985. *The Origin of Eukaryotic Cells: Benchmark Papers in Systematic and Evolutionary Biology*. Van Nostrand Reinhold, New York.

Eldredge, N. 1982. Phenomenological levels and evolutionary rates. Syst. Zool. 31: 338–347.

Eldredge, N. 1985a. *Time Frames*. Simon & Schuster, New York. [With this book, and a series of subsequent books on macroevolutionary topics, Eldredge has synthesized many of our changing views on evolution.]

Eldredge, N. 1985b. *Unfinished Synthesis: Biological Hierarchies and Modern Evolutionary Thought*. Oxford University Press, New York. [An important, thought-provoking look at the "modern synthesis" of evolution, its shortcomings, and some alternative ideas on evolutionary theory.]

Eldredge, N. 1989. *Macroevolutionary Dynamics: Species, Niches, and Adaptive Peaks*. McGraw-Hill, New York. [Excellent reading.]

Eldredge, N. 1991. *The Miner's Canary: Unraveling the Mysteries of Extinction*. Princeton University Press, Princeton, NJ.

Eldredge, N. (ed.). 1992. *Systematics, Ecology, and the Biodiversity Crisis*. Columbia University Press, New York.

Eldredge, N. and S. J. Gould. 1972. Punctuated equilibria: An alternative to phyletic gradualism. *In* T. J. M. Schopf (ed.), *Models in Paleobiology*. Freeman, Cooper, San Francisco, pp. 82–115.

Eldredge, N. and S. N. Salthe. 1984. Hierarchy and evolution. *In* R. Dawkins and M. Ridley (eds.), *Oxford Surveys in Evolutionary Biology*, vol. 1. Oxford University Press, Oxford, pp. 182–206.

Eldredge, N. and S. M. Stanley (eds.). 1984. *Living Fossils*. Springer Verlag, New York.

Fisher, A. 1989. Endocytobiology: The wheels within wheels in the superkingdom Eucaryota. Mosaic 20: 2–13.

Futuyma, D. J. 1986. *Evolutionary Biology*, 2nd Ed. Sinauer Associates, Sunderland, MA. [An enjoyable approach to the subject; de-emphasizes phylogeny and macroevolution but excellent on the subjects of anagenesis and adaptation.]

Futuyma, D. J. and G. C. Mayer. 1980. Non-allopatric speciation in animals. Syst. Zool. 29: 254–271.

Gillespie, J. H. 1991. *The Causes of Molecular Evolution*. Oxford University Press, New York. [Good background reading for those with a serious interest in molecular evolution.]

Gould, S. J. 1977. *Ontogeny and Phylogeny*. Harvard University Press, Cambridge, MA. [A scholarly treatment of the myriad relationships postulated, over the past 100 years, to exist between ontogeny and phylogeny, including recapitulation, paedomorphosis, and neoteny; highly recommended. Also see G. J. Nelson's 1978 article, "Ontogeny, phylogeny, paleontology and the biogenetic law" (Syst. Zool. 27: 324–345).]

Gould, S. J. 1989. *Wonderful Life*. W.W. Norton, New York.

Gould, S. J. and N. Eldredge. 1977. Punctuated equilibria: The tempo and mode of evolution reconsidered. Paleobiology 3: 115–151.

Gould, S. J. and R. C. Lewontin. 1979. The spandrels of San Marco and the Panglossian paradigm: A critique of the adaptationist programme. Proc. R. Soc. Lond. Ser. B 205: 581–598. [A highly recommended read.]

Gray, J. and W. Shear. 1992. Early life on land. Am. Sci. 80: 444–456.

Haeckel, E. 1866. *Generelle Morphologie der Organismen*. Georg Reimer, Berlin.

Hall, B. K. 1996. Baupläne, phylotypic stages, and constraint: Why are there so few types of animals? Evol. Biol. 29: 215–257.

Hallam, A. 1978. How rare is phyletic gradualism and what is its evolutionary significance? Evidence from Jurassic Bivalvia. Paleobiology 4: 16–25.

Hallam, A. and P. B. Wignall. 1997. *Mass Extinctions and Their Aftermath*. Oxford University Press, Oxford.

Hsü, K. J. et al. 1982. Mass mortality and its environmental and evolutionary consequences. Science 216: 249–256.

Jablonski, D. 1986. Larval ecology and macroevolution in marine invertebrates. Bull. Mar. Sci. 39: 565–587.

Kidwell, M. G. 1993. Lateral transfer in natural populations of eukaryotes. Ann. Rev. Genet. 27: 235–256.

Kimura, M. 1983. *The Neutral Theory of Molecular Evolution*. Cambridge University Press, New York.

Lenski, R. E. and J. E. Mittler. 1993. The directed mutation controversy and neo-Darwinism. Science 259: 188–194.

Levin, D. A. 2000. *The Origin, Expansion, and Demise of Plant Species*. Oxford Univesity Press, New York.

Lewontin, R. C. 1974. *The Genetic Basis of Evolutionary Change*. Columbia University Press, New York. [A solid treatment of evolutionary genetics.]

Li, W.-H. and D. Graur. 1991. *Fundamentals of Molecular Evolution*. Sinauer Associates, Sunderland, MA.

Lim, J. K. and M. J. Simmons. 1994. Gross chromosome rearrangements mediated by transposable elements in *Drosophila melanogaster*. BioEssays 16(4): 269–275.

Lipps, J. H. and P. W. Signor (eds.). 1992. *Origin and Early Evolution of the Metazoa*. Plenum Press, New York. [Excellent contributed chapters dealing with the early evolution of Metazoa.]

Lipscomb, D. 1985. The eukaryote kingdoms. Cladistics 1: 127–140.

Margulis, L. 1989. *Symbiosis in Cell Evolution*. W. H. Freeman, San Francisco.

Margulis, L. and R. Fester (eds.). 1991. *Symbiosis as a Source of Evolutionary Innovation: Speciation and Morphogenesis*. MIT Press, Cambridge, MA.

Margulis, L. and L. Olendzenski (eds.). 1992. *Environmental Evolution: Effects of the Origin and Evolution of Life on Planet Earth*. MIT Press, Cambridge, MA.

Margulis, L. and D. Sagan. 1986. *Origins of Sex: Three Billion Years of Genetic Recombination*. Yale University Press, New Haven, CT.

Otte, D. and J. A. Endler (eds.). 1989. *Speciation and Its Consequences*. Sinauer Associates, Sunderland, MA.

Patterson, C. 1999. *Evolution. 2nd ed*. Cornell University Press, Ithaca, New York. [One of the best, most concise descriptions of evolution and classification.]

Patterson, C. (ed.). 1987. *Molecules and Morphology in Evolution: Conflict or Compromise?* Cambridge University Press, Cambridge.

Patterson, C., D. M. Williams and C. J. Humphries. 1993. Congruence between molecular and morphological phylogenies. Ann. Rev. Ecol. Syst. 24: 153–188.

Price, P. W. 1980. *Evolutionary Biology of Parasites*. Princeton University Press, Princeton, NJ.

Radicella, J. P., P. U. Park, and M. S. Fox. 1995. Adaptive mutation in *Escherichia coli*: A role for conjugation. Science 268:418–420.

[Support for recent, controversial claims that not all mutations are random—i.e., the evolutionary watchmaker isn't always blind!]

Raff, R. A. 1996. *The Shape of Life: Genes, Development, and the Evolution of Animal Form*. University of Chicago Press, Chicago.

Raff, R. A. and T. C. Kaufman. 1983. *Embryos, Genes, and Evolution*. Macmillan, New York. [Excellent reading.]

Raup, D. M. 1983. On the early origins of major biologic groups. Paleobiol. 9: 107–115.

Raup, D.M., S. J. Gould, T. M. Schopf and D. S. Simberloff. 1973. Stochastic models of phylogeny and the evolution of diversity. J. Geol. 81: 525–542.

Raup, D. M. and J. J. Sepkoski. 1982. Mass extinctions in the marine fossil record. Science 215: 1501–1503.

Rensch, B. 1959. *Evolution Above the Species Level*. Columbia University Press, New York. [The English translation of Rensch's classic treatment of a conceptually challenging subject. Though now showing its age, the original text (1947; 1954) had considerable influence on the development of modern evolutionary theory.]

Ridley, M. 1996. *Evolution*. Blackwell Science, Cambridge, MA. [One of the best general evolution texts available.]

Schopf, J. W. (ed.). 1983. *Earth's Earliest Biosphere: Its Origin and Evolution*. Princeton University Press, Princeton, NJ. [Good review of Precambrian biology.]

Schopf, T. M. J. (ed.). 1972. *Models in Paleobiology*. Freeman, Cooper, San Francisco.

Shixing, Z. and C. Huineng. 1995. Megascopic multicellular organisms from the 1700-million-year-old Tuanshanzi Formation in the Jixian area, north China. Science 270: 620–622.

Stanley, S. M. 1979. *Macroevolution: Pattern and Process*. W. H. Freeman, San Francisco.

Stanley, S. M. 1982. Macroevolution and the fossil record. Evolution 36: 460–473.

Swofford, D. L. 1991. When are phylogeny estimates from molecular and morphological data incongruent? *In* M. M. Miyamoto and J. Cracraft (eds.), *Phylogenetic Analysis of DNA Sequences*. Oxford University Press, New York, pp. 295–333.

Woese, C. R. 1981. Archaebacteria. Sci. Am. 244(6): 98–122.

Woese, C. R., O. Kandler and M. L. Wheelis. 1990. Towards a natural system of organisms: Proposal for the domains Archaea, Bacteria, and Eucarya. Proc. Natl. Acad. Sci. U.S.A. 87: 4576–4579.

Wray, G. A., J. S. Levinton and L. H. Shapiro. 1996. Molecular evidence for deep Precambrian divergences among metazoan phyla. Science 274: 568–573.

[Also see evolution references in Chapter 2.]

2 Classification, Systematics, and Phylogeny

Our classifications will come to be, as far as they can be so made, genealogies.
Charles Darwin, *The Origin of Species*, 1859

And you see that every time I made a further division, up came more boxes based on these divisions until I had a huge pyramid of boxes. Finally you see that while I was splitting the cycle up into finer and finer pieces, I was also building a structure. This structure of concepts is formally called a hierarchy and since ancient times has been a basic structure for all Western knowledge.
Robert M. Pirsig, *Zen and the Art of Motorcycle Maintenance*, 1974

This book deals with the field of **comparative biology**, or what may be called the science of the diversity of life. To understand invertebrate zoology, one must understand comparative biology, the tasks of which are to describe the characteristics and patterns of living systems and to explain those patterns by the scientific method. When those patterns have resulted from evolutionary processes, they illuminate the history of life on Earth. Biologists have been undertaking comparative studies of anatomy, morphology, embryology, physiology, and behavior for over 150 years. Many biologists, particularly systematists, do so with the specific intent of recovering the history of life. Because we cannot directly observe that history, we must rely on the strength of the scientific method to reconstruct it, or infer it. This chapter provides an overview of this process. Comparative biology, then, in its attempt to understand diversity in the living world, deals with three distinguishable elements: (1) descriptions of organisms, particularly in terms of similarities and differences in their characteristics; (2) the phylogenetic history of organisms through time; and (3) the distributional history of organisms in space.

The field of biological systematics has experienced a revolution in its theory and application in the past 30 years, especially with regard to phylogenetic reconstruction. Some philosophical aspects and operating principles of this exciting field are described in this chapter. It is essential that biology students have a basic grasp of how classifications are developed and phylogenetic relationships inferred, and we urge you to reflect carefully on the ideas presented below.

Biological Classification

The term **biological classification** has two meanings. First, it means the *process* of classifying, which consists of delimiting, ordering, and ranking organisms in groups. Second, it means the *product* of this process, or the classificatory scheme itself. The living world has an objective structure that can be empirically documented and described. One goal of biology is to discover and describe this structure, and classification is one way of doing this. Carrying out the process of biological classification constitutes one of the principal tasks of the systematist (or taxonomist).

The construction of a classification may at first appear straightforward; basically, the process consists of analyzing patterns in the distribution of characters among organisms. On the basis of such analyses, specimens are grouped into **species** (the word "species" is both singular and plural); related species are grouped into **genera** (singular, **genus**); related genera are grouped to form **families**; and so forth. The grouping process creates a system of subordinated, or nested, **taxa** (singular, **taxon**) arranged in a **hierarchical** fashion following basic set theory. If the taxa are properly grouped according to their degree of shared similarity, the hierarchy will reflect patterns of evolutionary descent—the "descent with modification" of Darwin.

The concept of similarity is fundamental to taxonomy, the classificatory process, and comparative biology as a whole. **Similarity**, evaluated on the basis of characteristics shared among organisms, is generally accepted by biologists to be a measure of biological (evolutionary) relatedness among taxa. The concept of **relatedness**, or genealogical kinship, is also fundamental to systematics and evolutionary biology. Patterns of relatedness are usually displayed by biologists in branching diagrams called **trees** (e.g., phylogenetic, genealogical, or evolutionary trees). Once constructed, such trees can then be converted into classification schemes, which are a dynamic way of representing our understanding of the history of life on Earth. Thus, trees and classifications are actually hypotheses of the evolution of life and the natural order it has created.

Classifications are necessary for several reasons, not the least of which is to efficiently catalog the enormous number of species of organisms on Earth. Over 1.7 million different species of prokaryotes and eukaryotes have been named and described. The insects alone comprise nearly a million named species, and over 350,000 of these are beetles! Classifications provide a detailed system for storage and retrieval of names. Second, and most important to evolutionary biologists, classifications serve a descriptive function. This function is served not only by the descriptions that define each taxon, but also, as noted above, by the detailed hypotheses of evolutionary relationships among the organisms that inhabit Earth. In other words, classifications are (or should be) constructed from evolutionary relationships; that is, from the patterns of ancestry and descent depicted in phylogenetic trees.

So, we see that a biological classification scheme is really a set of hypotheses defined and summarized by a phylogenetic tree. Thus, classifications, like other hypotheses and theories in science, have a third function, that of prediction. The more precise and less ambiguous the classification, the greater its predictive value. Predictability is another way of saying testability, and it is testability that places an endeavor in the realm of science rather than in the realm of art, faith, or rhetoric. Like other theories, classifications are always subject to refutation, refinement, and growth as new data become available. These new data may be in the form of newly discovered species or characteristics of organisms, new tools for the analysis of characters, or new ideas regarding how characteristics are evaluated. Changes in classifications reflect changes in our view and understanding of the natural world.

Nomenclature

The names employed within classifications are governed by rules and recommendations that are analogous to the rules of grammar that govern the use of the English language. The primary goals of **biological nomenclature** are the creation of classifications in which (1) any single kind of organism has one and only one correct name, and (2) no two kinds of organisms bear the same name. Nomenclature is an important tool of biologists that facilitates communication and stability*

Prior to the mid-1700s, animal and plant names consisted of one to several words or often simply a descriptive phrase. In 1758 the great Swedish naturalist Carl von Linné (Carolus Linnaeus, in the Latinized form he preferred) established a system of naming organisms now referred to as **binomial nomenclature**. Linnaeus's system required that every organism have a two-part scientific name—a **binomen**. The two parts of a binomen are the generic, or genus, name and the **specific epithet** (= **trivial name**). For example, the scientific name for one of the common Pacific Coast sea stars is *Pisaster giganteus*. These two names together constitute the binomen; *Pisaster* is the animal's generic (genus) name, and *giganteus* is its specific epithet. The specific epithet is never used alone, but must be preceded by the generic name, and the animal's "species name" is thus the complete binomen. Use of the first letter of a genus name preceding the specific epithet is also acceptable once the name has appeared spelled out on the page or in a short article (e.g., *P. giganteus*).

---

*We generally avoid using common, or vernacular, names in this book, simply because they are frequently misleading. Most invertebrates have no specific common name, and those that do typically have more than one name. For example, several dozen different species of sea slugs are known as "Spanish dancers." All manner of creatures are called "bugs," most of which are not true bugs (Hemiptera) at all (e.g., "ladybugs," "sowbugs," "potato bugs").

The 1758 version of Linnaeus's system is actually the tenth edition of his famous *Systema Naturae*, in which he listed all animals known to him at that time and included critical guidelines for classifying organisms. Linnaeus distinguished and named over 4,400 species of animals, including *Homo sapiens*. Linnaeus's *Species Plantarum* (in which he named over 8,000 species) had done the same for the plants in 1753. Linnaeus was one of the first naturalists to emphasize the use of *similarities* among species or other taxa in constructing a classification, rather than using *differences* among them. In doing so, he unknowingly began classifying organisms by virtue of their genetic, and hence evolutionary, relatedness. Linnaeus produced his *Systema Naturae* 100 years prior to the appearance of Darwin and Wallace's theory of evolution by natural selection (1859), and thus his use of similarities in classification foreshadowed the subsequent emphasis by biologists on evolutionary relationships among taxa.

Binomens are Latin (or Latinized) because of the custom followed in Europe prior to the eighteenth century of publishing scientific papers in Latin, the universal language of educated people of the time. For several decades after Linnaeus, names for animals and plants proliferated, and there were often several names for any given species (different names for the same organism are called **synonyms**). The name in common use was usually the most descriptive one, or often it was simply the one used by the most eminent authority of the time. In addition, some generic names and specific epithets were composed of more than one word each. This lack of nomenclatural uniformity led, in 1842, to the adoption of a code of rules formulated under the auspices of the British Association for the Advancement of Science, called the Strickland code. In 1901 the newly formed International Commission on Zoological Nomenclature adopted a revised version of the Strickland code, called the **International Code of Zoological Nomenclature (I.C.Z.N.)**. Botanists had adopted a similar code for plants in 1813, the Théorie Elémentaire de la Botanique, which became in 1930 the International Code of Botanical Nomenclature.

The I.C.Z.N. established January 1, 1758 (the year the tenth edition of Linnaeus's *Systema Naturae* appeared) as the starting date for modern zoological nomenclature. Any names published the same year, or in subsequent years, are regarded as having appeared after the *Systema*. The I.C.Z.N. also slightly changed the description of Linnaeus's naming system, from binomial nomenclature (names of two parts) to **binominal nomenclature** (names of two names). However, one still sees the former designation in common use. This subtle change implies that the system must be truly binary; that is, both generic and trivial names can be only one word each. Although the system is binary, it also accepts the use of **subspecies** names, creating a **trinomen** (three names) within which is contained the mandatory binomen. For example, the sea star *Pisaster giganteus* is known to have a distinct form occurring in the southern part of its range, which is designated as a subspecies, *Pisaster giganteus capitatus*.

All codes of biological nomenclature share the following six basic principles:

1. Botanical and zoological codes are independent of each other. It is therefore permissible, although not recommended, for a plant genus and an animal genus to bear the same name (e.g., the name *Cannabis* is used for both a plant genus and a bird genus).
2. A taxon can bear one and only one correct name.
3. No two genera within a given code can bear the same name (i.e., generic names are unique); and no two species within one genus can bear the same name (i.e., binomens are unique).
4. Scientific names are treated as Latin, regardless of their linguistic origin, and hence are subject to Latin rules of grammar.
5. The correct or valid name of a taxon is based on priority of publication (first usage).
6. For the categories of superfamily in animals and order in plants, and for all categories below these, taxon names must be based on type specimens, type species, or type genera.*

When strict application of a code results in confusion or ambiguity, problems are referred to the appropriate commission for a "legal" decision. Rulings of the International Commission on Zoological Nomenclature are published regularly in its journal, the *Bulletin of Zoological Nomenclature*. Note that the international commissions rule only on nomenclature or "legal" matters, not on questions of scientific or biological interpretation; these latter problems are the business of systematists.

The hierarchical categories recognized by the I.C.Z.N. are as follows:

Kingdom
 Phylum
  Superclass
   Class
    Subclass
     Cohort
      Superorder
       Order
        Suborder
         Superfamily
          Family
           Subfamily
            Tribe
             Genus
              Subgenus
               Species
                Subspecies

---

*When a biologist first names and describes a new species, he or she takes a typical specimen, declares it a type specimen, and deposits it in a safe repository such as a large natural history museum. If later workers are ever uncertain about whether they are working with the same species described by the original author, they can compare their material to the type specimen. Although of substantially less value, the designation of a "typical" or type species for a genus, or a type genus for a family, serves a somewhat similar purpose in establishing, a "typical" species or genus upon which a genus or family is based.

The above names represent **categories**; the actual animal group that is placed at any particular categorical level forms a **taxon**. Thus, the taxon Echinodermata is placed at the hierarchical level corresponding to the category phylum—Echinodermata is the taxon; phylum is the category. All categories (and taxa) above the species level are referred to as the **higher categories** (and higher taxa), as distinguished from the **species group categories** (species and subspecies).

The common Pacific sea star *Pisaster giganteus* is classified as follows:

Category	Taxon
Phylum	Echinodermata
Class	Asteroidea
Order	Forcipulatida
Family	Asteriidae
Genus	*Pisaster*
Species	*Pisaster giganteus* (Stimpson, 1857)

Notice that a person's name follows the species name in this classification. This is the name of the **author** of that species—the person who first described the organism (*Pisaster giganteus*) and gave it its name. In this particular case the author's name is in parentheses, which indicates that this species is now placed in a different genus than originally assigned by Professor Stimpson. Authors' names usually follow the first usage of a species name in the **primary literature** (i.e., articles published in professional scientific journals). In the **secondary literature**, such as textbooks and popular science magazines, authors' names are rarely used.

The names given to animals and plants are usually descriptive in some way, or perhaps indicative of the geographic area in which the species occurs. Others are named in honor of persons for one reason or another. Occasionally one runs across purely whimsical names, or even names that seem to have been formulated for seemingly diabolical reasons.*

The **biological species definition** (or **genetical species concept**), as codified by Ernst Mayr, defines species as groups of interbreeding (or potentially interbreeding) natural populations that are reproductively isolated from other such groups. Obviously, this definition fails to accommodate nonsexual species. Hence, G. G. Simpson and E. O. Wiley developed the **evolutionary species concept**, which states that a species is a single lineage of ancestor–descendant populations that maintains its identity separate from other such lineages and that has its own evolutionary tendencies and historical fate. In reality, of course, biologists rely heavily on anatomical and morphological aspects of organisms as surrogates in gauging these conceptual views of species. That is, we conceive of species as genetic or evolutionary entities, but we recognize them primarily by their phenotypic characters. Hence, an understanding of these characters is of great importance (see below).

Higher taxa (categories and taxa above the species level) are natural groups of species (or lineages) chosen by biologists for naming in order to reflect our state of knowledge regarding their evolutionary relationships. Higher taxa, if correctly constructed, represent ancestor-descendant lineages that, like species, have an origin, a common ancestry and descent, and eventually a death (extinction of the lineage); thus they too are evolutionary units with definable boundaries. There are no rules for how many species should make up a genus—only that it be a natural group. Nor are there rules about how many genera constitute a family, or whether any group of genera should be recognized as a family, or a subfamily, or an order, or any other categorical rank. What matters is simply that the named group (the taxon) be a natural group. Hence, it is incorrect to assume that families of insects are in some way evolutionarily comparable to families of molluscs, or orders of worms comparable to orders of crabs. Nor are there any rules about categori-

---

*Among the many clever names given to animals are *Agra vation* (a tropical beetle that was extremely difficult for Dr. Terry Erwin to collect) and *Lightiella serendipida* (a small crustacean; the generic name honors the famous Pacific naturalist S. F. Light, 1886–1947, while the trivial name is taken from "serendipity," a word coined by Walpole in allusion to the tale of "The Three Princes of Serendip," who in their travels were always discovering, by chance or sagacity, things they did not seek—the term is said to aptly describe the circumstances of the initial discovery of this species). The nineteenth-century British naturalist W. E. Leach erected numerous genera of isopod crustaceans whose spellings were anagrams of the name Caroline. Exactly who Caroline was (and the nature of her relationship with Professor Leach) is still being debated, but the prevailing theory implicates Caroline of Brunswick, who was in the public eye at this time in history. It is said that Caroline was badly treated by her husband (the Prince Regent, later George IV), and that she was herself a lady of questionable fidelity. Leach, from Devon, may have taken the side of support for Caroline by honoring her with a long series of generic names, including *Cirolana, Lanocira, Rocinela, Nerocila, Anilocra, Conilera, Olincera,* and others. A light-hearted attitude toward naming organisms has not always been without Freudian overtones, as there also exist *Thetys vagina* (a large, hollow, tubular pelagic salp), *Succinea vaginacontorta* (a hermaphroditic snail whose vagina twists in corkscrew fashion), *Phallus impudicus* (a

slime-covered mushroom), and *Amanita phalloides* and *Amanita vaginata* (two species of highly toxic mushrooms around which numerous aboriginal ceremonies and legends exist). The hoopoe (a bird), *Upupa epops,* is euphoniously named for its call. The fish *Zappa confluentus* was named by a fan of Frank Zappa's, and the Grateful Dead have a fly named in their honor (*Dicrotendipes thanatogratus*). There is a bivalve named *Abra cadabra,* a bloodsucking spider *Draculoides bramstokeri,* and a wasp *Aha ha.* Even Linnaeus created a curious name for a common ameba, *Chaos chaos.* And, in a stroke of whimsy, the entomologist G. W. Kirkaldy created the bug genera *Polychisme* ("Polly kiss me"), *Peggichisme, Marichisme, Dolychisme,* and *Florichisme.* There are fish genera named *Zeus, Satan, Zen, Batman,* and *Sayonara.* There are insect genera named *Cinderella, Aloha, Oops,* and *Euphoria.* Some other clever binomens include *Leonardo davincii* (a moth), *Phthiria relativitae* (a fly), and *Ba humbugi* (a snail). A few biologists have gone overboard in erecting names for new animals, and many binomens exceed 30 letters in length, including those of the chaetognath *Sagitta pseudoserratadentatoides* (31 letters) and the common North Pacific sea urchin *Strongylocentrotus drobachiensis* (31 letters). Amphipod crustaceans probably win the grand prize in the longest-name category, with *Siemienkiewicziechinogammarus siemienkiewitschii* (47 letters) and *Cancelloidokytodermogammarus* (*Loveninsuskytodermogammarus*) *loveni* (61 letters, including the subgeneric name).

cal rank and geological or evolutionary age. These aspects of higher taxa are often misunderstood. Interestingly, this being said, family-level taxa often tend to be the most stable taxonomic groupings, usually recognizable even to laypersons—think, for example, of cats (Felidae), dogs (Canidae), abalone (Haliotidae), ladybird beetles (Coccinellidae), mosquitoes (Culicidae), octopuses (Octopodidae), or shore crabs (Grapsidae). This stability seems to be an artifact of the history of taxonomy, but it nonetheless makes families convenient higher taxa to study and discuss. However, biologists err when they compare equally ranked higher taxa between phyla in ways that presuppose them to be somehow equivalent.

Systematics

The science of **systematics** (or **taxonomy**) is the oldest and most encompassing of all fields of biology. The eminent biologist George Gaylord Simpson referred to systematics as "the study of the kinds and diversity of life on Earth, and of any and all relationships between them." The modern systematist is a natural historian of the first order. His or her training is broad, cutting across the fields of zoology and botany, genetics, paleontology, biogeography, geology, historical biology, ecology, and even ethology, chemistry, philosophy, and cellular and molecular biology. Ernst Mayr said that the field of systematics can be thought of as a continuum, from the routine naming and describing of species through the compilation of large faunal compendia and monographs to more synthetic studies, such as the fitting of these species into classifications that depict evolutionary relationships, biogeographic analyses, studies of population biology and genetics, and evolutionary and speciation studies. Mayr designated three stages of study within this continuum, which he called alpha, beta, and gamma, corresponding to the three general levels of complexity he perceived in systematics. When a group of organisms is first discovered or is in a poorly known state, work on that group is necessarily at the alpha level (e.g., the describing of new species). It is only when most, or at least many, species in a taxon become known that the systematist is able to work at the beta or gamma levels within that group (e.g., to perform evolutionary studies). Some biologists choose to refer to those people working at the alpha level as taxonomists, reserving the term "systematist" for those engaging in studies at the beta or gamma level. Although this may be an instructive way to scrutinize the spectrum of endeavors systematists engage in, it is actually a gross oversimplification.* These stages in systematic study overlap and cycle back on themselves

in a highly iterative fashion. In sum, the role of systematics is to document and understand Earth's biological diversity, to reconstruct the history of this biodiversity, and to develop natural (evolutionary) classifications of living organisms.

Systematists use a great variety of tools to study the relationships among taxa. These tools include not only the traditional and highly informative techniques of comparative and functional anatomy, but also the methods of embryology, serology, physiology, immunology, biochemistry, population and molecular genetics, and molecular gene sequencing. A sound classification lies at the root of any study of evolutionary significance, as does a thorough appreciation for the enormous diversity of life. Without systematics, the science of biology would grind to a halt, or worse yet, would drift off into pockets of isolated reductionist or deterministic schools with no conceptual framework or continuity.

The field of systematics is currently experiencing a welcome revival in popularity. Within the worldwide literature, there are now about 200 scientific journals publishing specifically in the fields of systematics and evolution, and another 1,500 or so cover the general field of natural history. As of 1991, about one new phylogeny per day was being published; today the number is probably twice that. There are at least three causes for this revived interest in systematic biology. First is the growing awareness that too few systematists have been trained over the past 30 years. As the previous generation's cadre of systematists retires, few systematists are left to continue work on important taxa and evolutionary problems. For many groups of organisms today, there are simply no working specialists anywhere! Second is the recent discovery of a great many naturally occurring anticancer, antibiotic, and other pharmacologically important compounds in animals and plants. About 90 percent of the prescriptions written in North America contain active compounds first discovered in living organisms. Many of the most "active" plants and animals that chemists are discovering come from the most poorly known regions of the world, such as rain forests and coral reefs, where most species have yet to be named and described. Third is the rapidly deteriorating state of affairs in the tropics, which are thought to harbor about 80 percent of the total animal and plant species on Earth. These regions are being destroyed by humans at the rate of 50 million acres per year (an area larger than the state of Kansas). Estimates of anthropogenic extinctions in the tropics of terrestrial species alone range as high as 50 percent of the total world fauna and flora by the year 2050, if present trends of human exploitation continue. The extirpation of millions of animal and plant species is not only an outrageous insult to the natural environment, but also represents an enormous loss of potential food, drug, timber, and other product sources, and it is damaging the global biosphere to the point of reducing the quality of life for all creatures, including humans.

---

*Europeans tend to use the terms "systematics" and "biosystematics" for the field as a whole, whereas North Americans tend to use "taxonomy" more frequently. In this text, we use the terms "taxonomy" and "systematics" interchangeably.

Important Concepts and Terms

One of the concepts most crucial to our understanding of biological systematics and evolutionary theory in general is monophyly. A **monophyletic group** is a group of species that includes an ancestral species and all of its descendants—that is, a *natural group* (Figure 2.1). In other words, a monophyletic taxon is a group of species whose members are related to one another through a unique history of descent (with modification) from a common ancestor—a single evolutionary lineage.

A group whose member species are all descendants of a common ancestor, but that does not contain *all* the species descended from that ancestor, is called a **paraphyletic group**. Paraphyly implies that for some reason (e.g., lack of knowledge, purposeful manipulation of the classification) some members of a natural group have been placed in a different group. As we will see below, many paraphyletic taxa exist within animal classifications today, to the consternation of those who prefer to recognize only monophyletic taxa.

A third possible kind of taxon is a **polyphyletic group**—a group comprising species that arose from two or more different immediate ancestors. Such composite taxa have been established primarily because of insufficient knowledge concerning the species in question. One of the principal goals of systematists is to discover such polyphyletic or "artificial" taxa and, through careful study, reclassify their members into appropriate monophyletic taxa. These three kinds of taxa or species groups are illustrated diagrammatically in Figure 2.1.

There are many examples of known or suspected polyphyletic taxa in the zoological literature. For example, the old phylum Gephyrea contained what we now recognize as three distinct phyla—Sipuncula, Echiura, and Priapula. Another example is the old group Radiata, which included all animals possessing radial symmetry (e.g., cnidarians, ctenophores, and echinoderms). Still another example is the former "phylum Protozoa," whose members are now distributed among many phyla (see Chapter 5). Protozoa comprise no more than

a loose assemblage of heterotrophic, single-celled eukaryotes. Polyphyletic taxa usually are established because the features or characters used to recognize and diagnose them are the result of evolutionary convergence in different lineages, as discussed below. Convergence can be discovered only by careful comparative embryological or anatomical studies, sometimes requiring the efforts of several generations of specialists.

Characters are the attributes, or features, of organisms or groups of organisms (taxa) that biologists rely on to indicate their relatedness to other similar organisms (or other taxa) and to distinguish them from other groups. Characters are the observable products of the genotype, and they can be anything from the actual amino acid sequences of the genes themselves to the phenotypic expressions of the genotype. A character can be any genetically based feature that taxonomists can examine and measure; it can be a morphological, anatomical, developmental, or molecular feature of an organism, its chromosomal makeup (karyotype) or biochemical "fingerprint," or even an ecological, physiological, or ethological (behavioral) attribute. Several biochemical and molecular techniques for measuring similarity among organisms have been developed over the past 30–40 years; these include DNA hybridization,

Figure 2.1 Two dendrograms, illustrating three kinds of taxa. Taxon W, comprising three species, is monophyletic because it contains all the descendants (species C and D) of an immediate common ancestor (species B), plus that ancestor. Taxon X is paraphyletic because it includes an ancestor (species A), but only *some* of its descendants (species E through I, leaving out species B, C, and D). Taxon Y is polyphyletic because it contains taxa that are *not* derived from an immediate common ancestor; species M and P may look very much alike as a result of evolutionary convergence or parallelism, and therefore may have been mistakenly placed together in a single taxon. Taxon Z is paraphyletic. In this case, further work on species J through P should eventually reveal the correct relationships among these taxa, resulting in species M being classified with species K and L, and species P with species N and O.

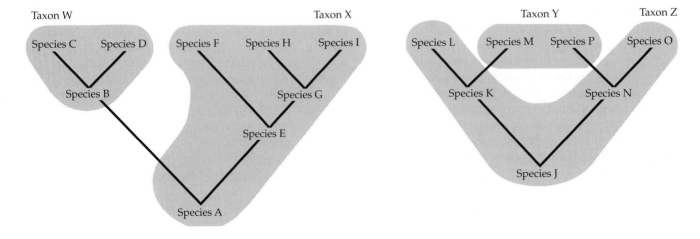

starch gel electrophoresis of proteins and amino acids, immunological similarity indices, and most recently, nucleotide sequencing of genes. All of these kinds of "characters" have been used to create phylogenetic trees. Thus, a variety of kinds of data are available to provide systematists with characters with which to define and compare species and higher taxa.

The fundamental basis for comparative biology is the concept of **homology**. Characters that share descent from a common ancestor are called **homologues**. In other words, homologues are characters that are present in two or more taxa, but are traceable phylogenetically and ontogenetically to (i.e., share genetic and developmental bases with) the *same* character in the common ancestor of those taxa. In order to compare characters among different organisms or groups of organisms, it must be established that the characters being compared are homologous. Our ability to recognize anatomical homologues usually depends on developmental or embryological evidence and on the relative position of the anatomical structure in adults (see Chapter 4).

Homology is an absolute relationship: characters either are, or are not, homologous. Homology doesn't come in degrees. Homology is also completely independent of function. The functions of homologous structures may be similar or different, but this has no bearing on the underlying homology of the structures involved. Genes, like anatomical structures, may be homologous characters if they are derived from a common ancestral gene either by duplication (which generates **paralogous genes**) or as simple copies passed on via speciation events (**orthologous genes**). The process of evolutionary descent with modification has produced a hierarchical pattern of homologies that can be traced through lineages of living organisms. It is this pattern that we use to reconstruct the history of life.

Homology is a concept that is applicable to anatomical structures, to genes, and to developmental processes. However, homology at one of these levels does not necessarily indicate homology at another. Biologists should always be clear regarding the level at which they are inferring homology: genes, their expression patterns, their developmental roles, or the structures to which they give rise. Recently, some investigators have interpreted similar patterns of regulatory gene expression as evidence of homology among structures. This is a mistake because it ignores the evolutionary histories of the genes and of the structures in which they are expressed. The fact is, the functions of homologous genes (orthologues or paralogues), just like those of homologous structures, can diverge from one another through evolutionary time. Similarly, the functions of non-homologous genes can converge over time. Therefore, similarly of function is not a valid criterion for the determination of homology of either genes or structures. For example, the phenomenon of gene recruitment (co-option) can lead to situations in which truly orthologous genes are expressed in nonhomologous structures during development. Most regulatory genes play several distinct roles during development, and homologous genes can be independently recruited to superficially similar roles. A classic example is the regulatory gene *Distal-less*, which is expressed in the distal portion of appendages of many animals during their embryogeny (e.g., arthropods, echinoderms, chordates). Although the domains of *Distal-less* gene expression might reflect a homologous role in specifying proximodistal axes of appendages, the appendages themselves are clearly not homologous.

Attempts to relate two taxa by comparing nonhomologous characters will result in errors. For example, the hands of chimpanzees and humans are homologous characters (i.e., homologues) because they have the same evolutionary and developmental origin; the wings of bats and butterflies, although similar in some ways, are not homologous characters because they have completely different origins. The concept of homology has nothing to do, in the strict sense, with similarity or degree of resemblance. Some homologous features look very different in different taxa (e.g., the pectoral fins of whales and the arms of humans; the forewings of beetles and of flies). Again, the concept of homology is related to the level of analysis being considered. The wings of bats and birds are homologous as tetrapod forelimbs, but they are not homologous as "wings," because wings evolved independently in these two groups (i.e., the wings of bats and birds do not share a common ancestral wing). Homology is a powerful concept, but we must always remember that homologies are really hypotheses, open to testing and possible refutation.

Through the phenomenon of **convergent evolution**, similar-appearing structures may evolve in entirely unrelated groups of organisms in quite different ways. For example, early biologists were misled by the superficial similarities between the vertebrate eye and the cephalopod eye, the bivalve shells of molluscs and of brachiopods, and the sucking mouthparts of true bugs (Hemiptera) and of mosquitoes (Diptera). Structures such as these, which appear superficially similar but that have arisen independently and have separate genetic and phylogenetic origins, are called **convergent characters**. Failure to recognize convergences among different groups of organisms has led to the creation of many "unnatural," or polyphyletic, taxa in the past.

Convergence is often confused with **parallelism**. Parallel characters are similar features that have arisen more than once in different species within a lineage, but that share a common genetic and developmental basis.* Parallel evolution is the result of "distant" or underlying homology; for parallel evolution to occur, the genetic

---

*Parallelism in this context is not to be confused with the evolution of species (or characters within species) "in parallel," that is, when two species (or characters) change more or less together over time. Host–parasite coevolution is an example of "evolution in parallel."

potential for certain features must persist within a group, thus allowing the feature to appear and reappear in various taxa. Parallelism is commonly encountered in characters of morphological "reduction," such as reduction in the number of segments, spines, fin rays, and so on in many different kinds of animals. It is also common among the segmented animals, annelids, and arthropods. The phenomena of convergence and parallelism might be thought of as a kind of "evolutionary redundancy." A third phenomenon in this general category is **evolutionary reversal**, wherein a feature reverts back to a previous, ancestral condition. Together, these three evolutionary processes (convergence, parallelism, reversal) constitute the phenomenon known as **homoplasy**—the recurrence of similarity in evolution (Figure 2.2). As you might guess, for systematists, homoplasy is both fascinating and irritating!

When comparing homologues among species, one quickly sees that variation in the expression of a character is the rule, rather than the exception. The various conditions of a homologous character are often referred to as its **character states**.* A character may have only two contrasting states, or it may have several different states within a taxon. **Polymorphic species** are those that show a range of phenotypic or genetic variation as a result of the presence of numerous character states for the features being examined. A simple example is hair color in humans; black, brown, red, and blond are all states of the character "hair color." Not only can characters vary within a species, but they also typically have several states among groups of species within higher taxa, such as patterns of body hair among various primates or the spine patterns on the legs of crustaceans.

It is important to understand that a character is really a hypothesis—that two attributes that appear different in different organisms are simply alternative states of the same feature (i.e., they are homologues). Note that convergences are not homologies, whereas parallelisms and reversals do represent an underlying genetic homology. In other words, some kinds of homoplastic characters are homologues, and others are not. The recognition and selection of proper characters is clearly of primary importance in biological systematics, and a great deal has been written on this subject. Systematics is, to a great extent, a search for the homologues that define natural evolutionary lineages.

Another important concept in systematics and comparative biology is the dendrogram. A **dendrogram** is a branching diagram, or tree, depicting the relationships among groups of organisms. It is a graphical means of expressing relationships among species or other taxa. Most dendrograms are intended to depict evolutionary

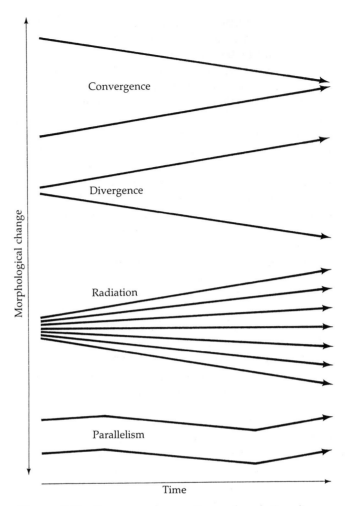

Figure 2.2   Some common patterns of evolution displayed by independent lineages. Convergence occurs when two or more lineages (or characters) evolve independently toward a similar state. Convergence generally refers to unrelated (or very distantly related) taxa and to characters sharing no common genetic (phylogenetic or ontogenetic) basis. Divergence occurs when two or more lineages (or characters) evolve independently to become less similar. Radiations are multiple divergences from a common ancestor that result in more than two descendant lineages. Parallel evolution occurs when two or more species (or lineages) change similarly so that, despite evolutionary activity, they remain similar in some ways, or become more similar over time. Parallelism generally refers to closely related taxa, usually species, within which the characters or structures in question share a common genetic basis.

---

*In practical usage the terms "character" and "character state" are often used interchangeably when comparing species. This practice can be a bit confusing. When the term "character" is used in a discussion of two or more homologues, it is typically being used in the same sense as "character state."

relationships, with the base representing the oldest (earliest) ancestors and the higher branches indicating successively more recent divisions of evolutionary lineages. But dendrograms can be constructed with different goals in mind. The traditional dendrograms drawn by biologists were called **evolutionary trees**, and they were meant to depict a variety of ideas concerning the evolution of the organisms in question. Such trees often had (at least implied) a time component as the vertical axis and genetic or morphological divergence as the horizontal axis. Three examples of evolutionary trees are given in Figure 2.3. Recall from our earlier discussion that classification schemes are ultimately derived from trees of some sort. Various kinds of dendrograms are discussed in further detail in succeeding pages, and they also appear throughout this book to provide the reader with current theories on the evolution of various invertebrate taxa.

When examining dendrograms and classifications derived from them, it is important to understand the concept of grades and clades. As depicted in Figure 2.4, a **clade** is a monophyletic group or branch of a tree, which may undergo very little or a great deal of diversification. A clade, in other words, is a group of species related by direct descent. A **grade**, on the other hand, is a group of species (or higher taxa) defined by somewhat more abstract measures. In fact, it is a group defined by a particular level of functional or morphological complexity. Thus, a grade can be polyphyletic, paraphyletic, or monophyletic (in the latter case, it is also a clade). A good example of a grade is the large group of gastropod taxa that have achieved shellessness. These "slugs," however, do not constitute a clade, because shell loss has occurred independently in several different lineages; thus "slugs" are a polyphyletic group. An example of a monophyletic grade is the subphylum Vertebrata (animals with backbones).

One last concept important to our understanding of systematics is that of primitive versus advanced character states. **Primitive character states** are attributes of species that are relatively "old" and have been retained from some remote ancestor; in other words, they have been around for a long time, geologically or genealogically speaking. Character states of this kind are often referred to as **ancestral**. **Advanced character states**, on the other hand, are attributes of species that are of relatively recent origin—often called **derived** character states. *Within* the phylum Chordata, for example, the possession of hair, milk glands, and three middle ear bones are derived character states whose evolutionary appearance marked the origin of the mammals (thus distinguishing them from all other chordates). Within a subset of the Mammalia, however, such as the primates, these same features represent retained ancestral features, whereas possession of an opposable thumb is a defining, derived trait.

It should be apparent from the preceding paragraph that the designations "primitive" and "advanced" are relative, and that any given character state or attribute can be viewed as *either* ancestral or derived, depending on the level of the phylogenetic tree or classification being examined. Opposable thumbs may be a derived trait defining primates within the mammal lineage, but it is not a derived character state *within* the primate line itself (all primates have opposable thumbs). Thus, in the primate genus *Homo*, "opposable thumbs" is a primitive

(ancestral) feature, and certain features of the nervous system that distinguish humans from the "lower apes" would be considered derived (such as Broca's center in the human brain). Thus it behooves us to more precisely define the concepts of primitive and advanced. The most unambiguous way to describe and use these important concepts is to define the exact place in the history of a group of organisms at which a character actually undergoes an evolutionary transformation from one state to another. At the specific point on a phylogenetic tree where such a transformation takes place, the new (derived) character state is called an **apomorphy** and the former (ancestral) state a **plesiomorphy**. Use of these terms thus implies a precise phylogenetic placement of the character in question, and this placement constitutes a testable phylogenetic hypothesis in and of itself.

Constructing Phylogenies and Classifications

From what you have read so far in this chapter, it should be evident that comparative biologists, particularly systematists, spend a great deal of their time seeking to identify and unambiguously define two natural entities, homologues and monophyletic groups. Biologists may present their ideas on such matters of relationship in the form of trees, classifications, or narrative discussions (evolutionary scenarios). In all three contexts, these presentations represent sets of evolutionary hypotheses—hypotheses of common ancestry (or ancestor–descendant relationships).

The least ambiguous (most testable) way to present evolutionary hypotheses is in the form of a dendrogram, or branching tree. Although classification schemes are ultimately derived from such dendrograms, they do not always reflect precisely the arrangement of natural groups in the tree. Discrepancies between phylogenetic trees and classifications derived from them most commonly occur when biologists purposely choose to establish or recognize paraphyletic taxa. Whereas most systematists advocate that only monophyletic taxa be recognized in a formal classification, some paraphyletic taxa seem to persist if for no other reason than tradition. For example, the long-recognized group Reptilia is certainly paraphyletic because it excludes one of that group's most distinct lineages, the birds. As we will see in Chapter 13, the classes Polychaeta and Oligochaeta are probably also paraphyletic groups. The issue of how to deal with such long-standing, well-known paraphyletic taxa in classification schemes is still being debated. One way of doing this might be to indicate their paraphyletic status by a code in the classification scheme (e.g., some type of notation beside the name). This code would inform readers that to view the precise phylogenetic relationships of such taxa, they must look

(A)

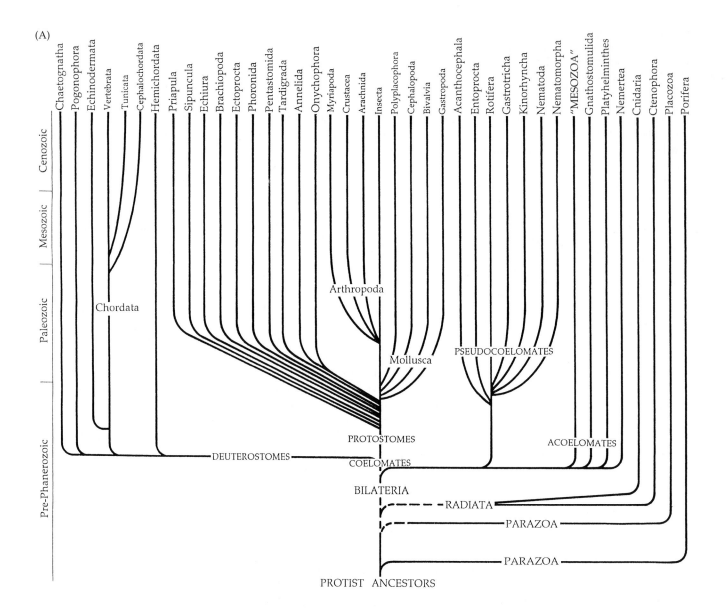

to the phylogenetic tree. Of course, the other way to deal with such taxa is to eliminate them altogether, which in some cases (e.g., the Polychaeta) could require major classificatory revisions.

Most workers today use a method known as **phylogenetic systematics**, or **cladistics**, when construing biological dendrograms and their resultant classifications. Phylogenetic systematics had its origin in 1950 in a textbook by the German biologist Willi Hennig; the English translation (with revisions) appeared in 1966. Its popularity has grown steadily since that time. Through the years, cladistics has evolved well beyond the framework Hennig originally proposed. Its detailed methodology has been formalized and expanded and will probably continue to be elaborated for some time to come. (For good discussions of cladistic systematics see Nelson and Platnick 1981, Eldredge and Cracraft 1980, and Wiley 1981.) The goal of phylogenetic systematics is to produce explicit and testable hypotheses of genealogical relationships among monophyletic groups of organisms. As a system-

atic methodology, cladistics is based entirely on *recency of common descent* (i.e., genealogy). The dendrograms used by phylogenetic systematists are called **cladograms**, and they are constructed to depict only genealogy, or ancestor–descendant relationships. The term **cladogenesis** refers to splitting; in the case of biology, this means the splitting of one species (or lineage) into two or more species (or lineages). It is this splitting process that produces genealogical (ancestor–descendant) relationships.

Phylogenetic systematists rely heavily on the concept of ancestral versus derived character states discussed earlier. They identify these homologies in the strict sense, as plesiomorphies and apomorphies. An apomorphy restricted to a single species is referred to as an **autapomorphy**, whereas an apomorphic character state that is shared between two or more species (or other taxa) is called a **synapomorphy**. Identifying synapomorphies (also known as shared derived characters, or evolutionary novelties) is the phylogenetic systematist's most powerful means of recognizing close evolutionary

Figure 2.3 Three types of traditional evolutionary trees (*not* cladograms) that depict phylogeny among the Metazoa.

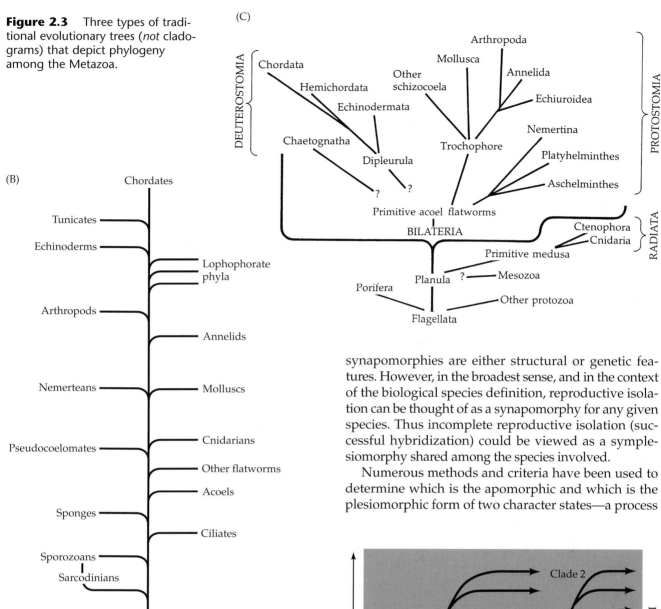

synapomorphies are either structural or genetic features. However, in the broadest sense, and in the context of the biological species definition, reproductive isolation can be thought of as a synapomorphy for any given species. Thus incomplete reproductive isolation (successful hybridization) could be viewed as a symplesiomorphy shared among the species involved.

Numerous methods and criteria have been used to determine which is the apomorphic and which is the plesiomorphic form of two character states—a process

(genealogical) relationships. Because synapomorphies are shared homologues inherited from an immediate common ancestor, all homologues may be considered synapomorphies at one (but only one) level of phylogenetic relationship, and they therefore constitute **symplesiomorphies** at all lower levels. As noted earlier, hair, milk glands, and so forth are synapomorphies uniquely defining the appearance of the mammals within the vertebrates, but these are symplesiomorphies *within* the group Mammalia. Jointed legs are a synapomorphy of the Arthropoda, but *within* the arthropods jointed legs are a symplesiomorphy. The keystone of phylogenetic systematics is the recognition that all homologues define monophyletic groups *at some level*. The challenge is, of course, recognizing the level at which each character state is a unique synapomorphy. Generally speaking,

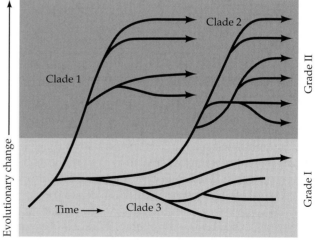

Figure 2.4 Clades and grades. Clades are monophyletic branches that may undergo various degrees of diversification. Grades are groups of animals classified together on the basis of levels of functional or morphological complexity. Grades may be monophyletic, paraphyletic, or polyphyletic. In this figure, grade I is monophyletic, encompassing only a single clade (clade 3); grade II is polyphyletic, because the associated level of complexity has been achieved independently by two separate lineages, clades 1 and 2.

referred to as **character state polarity analysis**. No method is foolproof, but some may be better than others under specific circumstances. Only three methods appear to have a strong evolutionary basis *and* provide a reasonably powerful means for recognizing the relative place of origin of a synapomorphy on a tree: out-group analysis (seeking clues to ancestral character states in groups thought to be more primitive than the study group), developmental studies (ontogenetic analysis, or seeking clues to ancestral character states in the embryogeny of the study group), and study of the fossil record. Out-group analysis identifies the states of the characters in question in taxa that are closely related to the study group, but are not part of it. Ontogenetic analysis identifies character changes that occur during the development of a species (see the discussion of ontogeny and phylogeny in Chapter 4). And the use of fossils and associated dating and stratigraphic techniques provides direct historical information. However, the fossil record is very incomplete, and such fragmentary data can be misleading. These techniques of polarity analysis are not discussed in detail here; we refer those with a serious interest in systematics, evolution, and comparative biology to the readings listed at the end of this chapter.

A cladistic analysis often comprises four steps: (1) identifying homologous characters among the organisms being studied, (2) assessing the direction of character change or character evolution (character state polarity analysis), (3) constructing a cladogram of the taxa possessing the characters analyzed, and (4) testing the cladogram with new data (new taxa, new characters, new character interpretations, etc.). Cladograms depict only one kind of event: the origin or sequence of appearances of a unique derived character state (synapomorphy). Hence, cladograms may be thought of in the most fundamental sense as nested synapomorphy patterns. However, biologists define and categorize taxa by the character states they possess. Thus, in a larger sense, the sequential branching of nested sets of evolutionary novelties (synapomorphies) in a cladogram creates a "family tree"—*an evolutionary pattern of hypothesized monophyletic lineages.*

Phylogenetic systematists have adopted the **principle of logical parsimony*** and thus generally prefer the tree containing the smallest number of evolutionary transformations (character state changes). Typically this will also be the tree with the least evolutionary redundancy (= homoplasy). Although parsimony is the only inference method currently used for analyses of nonmolecular data, the use of gene sequence data has spawned a new family of model-based methods that incorporate hypotheses of nucleotide evolution. In these methods (i.e., maximum likelihood and distance methods), DNA nucleotide sequences from organisms in the study group are analyzed within a framework of assumptions based on how we believe nucleotides operate and change over time.

Construction of a cladogram can be a time-consuming process. The number of mathematically possible cladograms for more than a few species is enormous—for three taxa there are only four possible cladograms, but for ten taxa there are about 280 million possible cladograms, 34 million of which are fully dichotomous. Needless to say, a thorough analysis of a family of several dozen species and determination of the most parsimonious tree is not possible without the aid of a computer. Algorithms for computer-assisted cladogram construction began appearing in the late 1970s. These programs generate cladograms by clustering taxa on the basis of nested sets of synapomorphies. There are several good programs available for phylogenetic analyses. The cladograms in this text were generated with the program PAUP (Swofford 2001).

By identifying the precise points at which synapomorphies occur, cladograms unambiguously define monophyletic lineages. Hence, cladograms are called **explicit phylogenetic hypotheses**. Being explicit, they can be tested (and potentially falsified) by anyone. The synapomorphies are markers that identify specific places in the tree where new monophyletic taxa arise. For phylogenetic systematists, a phylogeny consists of a genealogical branching pattern expressed as a cladogram. Each split or dichotomy produces a pair of newly derived taxa called **sister taxa**, or **sister groups** (for example, sister species). Sister groups always share an immediate common ancestor. In Figure 2.5, set W is the sis-

---

*Parsimony is a method of logic in which economy in reasoning is sought. The principle of parsimony, also known as Ockham's razor, has strong support in science. William of Ockham (Occam), the fourteenth-century English philosopher, stated the principle as, "Plurality must not be posited without necessity." Modern renderings would read, "An explanation of the facts should be no more complicated than necessary," or, "Among competing hypotheses, favor the simplest one." Scientists in all disciplines follow this rule daily, and it can be viewed as a consequence of deeper principles that are supported by statistical inferences. Thus, parsimonious solutions or hypotheses are those that explain the data in the simplest way. Evolutionary biologists rely on the principle of logical parsimony for the same reason other scientific disciplines rely on it: doing so presumes the fewest ad hoc assumptions and produces the most testable (i.e., the most easily falsified) hypotheses. If evidential support favored only one hypothesis, we would have little need for parsimony as a method. The reason we must rely on parsimony in science is that there is virtually always more than one hypothesis that can explain our data. Parsimony considerations come into play most strongly when a choice must be made among equally supported hypotheses.

In phylogenetic reconstruction, any given data set can be explained by a great number of possible trees. A three-taxon data set has 3 possible dichotomous (all lines divide into just two branches) trees that explain it. A four-taxon data set has 15 possible bifurcating trees, a five-taxon data set has 105 possible trees, and so on. Thus, the evidence alone does not sufficiently narrow the class of admissible hypotheses, and some extraevidential criterion (parsimony) is required. The virtue of choosing the shortest (i.e., most parsimonious) tree among a universe of possible trees lies in its simplicity, or testability. William of Ockham, by the way, also denied the existence of universals except in the minds of humans and in language. This notion resulted in a charge of heresy from the Church, after which he fled to Rome and, alas, died of the Black Plague.

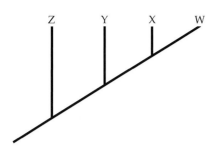

Figure 2.5 A cladogram of four taxa, illustrating the concept of sister groups. Taxon W is the sister group of taxon X; taxon W + X is the sister group of taxon Y; taxon W + X + Y is the sister group of taxon Z.

ter group of set X; set W + X is the sister group of set Y; and set W + X + Y is the sister group of set Z. This nested-set pattern of hierarchical relationships results from the fact that cladogenesis is a historical process.

Like all scientific hypotheses, cladistic analyses and their resulting cladograms are tested by the discovery of new data. As new characters or new species are identified and their character states elucidated, new data matrices are developed, and new analyses are undertaken. Cladograms are also tested when characters are reassessed, which can lead to changes in character state (= homologue) interpretation. Cladograms can also be tested with different kinds of data (e.g., today molecular phylogenies are being used to test earlier generations of phylogenies based on anatomical, morphological, and embryological data). Hypotheses (branches of the tree) that consistently resist refutation are said to be highly corroborated. For example, the clade called Arthropoda has been examined in scores of cladistic analyses using a great variety of data, and it has consistently been shown to constitute a monophyletic group (i.e., it is a highly corroborated phylogenetic hypothesis).

The final step in a cladistic analysis may be the conversion of the cladogram into a classification scheme. Strict phylogenetic systematists strive to convert their cladograms *directly* into classifications strictly on the basis of the branching sequence depicted. They use only as much information for the construction of the classification as is contained in the cladogram. Thus, phylogenetic systematists erect classifications based solely on genealogy. Phylogenetic systematists give no taxonomic consideration to the *degree of difference* between taxa (i.e., the number and kinds of characters used to separate taxa), to differential rates of change in various groups, or to evolutionary events other than those involving the origin of new apomorphies.

Figure 2.6 shows a cladogram that is believed by both phylogenetic systematists and traditional taxonomists to represent the phylogeny of the vertebrates. However, these two groups of systematists have derived three different classification schemes from this cladogram, incorporating different hierarchical arrange-

ments of the taxa within it. The difference is due entirely to the fact that the the phylogeneticist (or "cladist") view considers only the branching sequence, whereas the traditional view considers the overall degree of difference between taxa. In doing so, traditional taxonomists are willing to accept paraphyletic taxa.

As depicted on a cladogram, the product of cladogenesis (or the splitting of a taxon) is two (or more) new lineages that constitute sister groups. Another way of stating this is to say that the two subsets of any set defined by a synapomorphy constitute sister groups. A good example of the sister-group concept can be seen in a series of four families of marine isopod crustaceans (Figure 2.7). These four families show an evolutionary trend from free-living (the Cirolanidae) to parasitic lifestyles (the Cymothoidae). The Cymothoidae (a family of isopods that are obligatory parasites on fishes) is the sister group of Aegidae (a family of temporary fish parasites); together they constitute a sister group of the Corallanidae (micropredators on fishes); and all three constitute a sister group of the Cirolanidae (carnivorous predators and scavengers). Each of these nested sister-group pairs shares one or more unique synapomorphies that defines them. In Figure 2.7, the synapomorphies that define the sister group Cirolanidae + Corallanidae + Aegidae + Cymothoidae become symplesiomorphies higher in the cladogram (i.e., for each of the separate families). Sister groups are monophyletic by definition.

As illustrated in Figure 2.6 (classification scheme B), some phylogenetic systematists early on suggested that every lineage depicted in a tree should be designated by a formal name and categorical rank, and that each member of a sister-group pair must be of the *same* categorical rank. A moment's thought reveals that giving names to every branching point in a cladogram would result in an enormous and unacceptable proliferation of names and ranks. Other phylogenetic systematists have proposed a method of avoiding such name proliferation, called the **phylogenetic sequencing convention**. When this convention is used, *linear* sequences of taxa can all be given equal categorical designations (e.g., they can all be classified as genera, or all as families, and so on), so long as they are listed in the classification scheme in the precise sequence in which the branches appear on the cladogram (classification scheme C). Thus, either method of creating a classification scheme allows one to convert the classification scheme directly back into a cladogram—that is, to visualize the phylogenetic branching pattern it depicts.

One of the most illustrative examples of the difference of opinion between phylogenetic systematists and traditional taxonomists regarding the categorical ranking of sister groups is the case of the crocodilians and the birds, which may be more recently descended from a common ancestor than either is from any other group. Because of this relationship, the crocodilians and birds form a sister group to most other reptiles (see the clado-

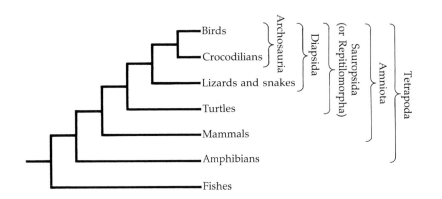

Classification scheme A

Phylum Chordata
 Subphylum Vertebrata
  Class Pisces (fishes)
  Class Amphibia (amphibians)
  Class Reptilia (turtles, crocodilians,
    snakes, lizards)
  Class Aves (birds)
  Class Mammalia (mammals)

Classification scheme B

Phylum Chordata
 Vertebrata
  Tetrapoda
   Lissamphibia (amphibians)
   Amniota
    Mammalia (mammals)
    Sauropsida (Reptilomorpha)
     Anapsida (turtles)
     Diapsida
      Lepidosaura (snakes, lizards)
      Archosauria
       Crocodilia (crocodilians)
       Aves (birds)

Classification scheme C

Phylum Chordata
 Subphylum Pisces
  Class Pisces
  Class Amphibia
  Class Mammalia
  Class Anapsida (turtles)
  Class Lepidosaura (snakes, lizards)
  Class Crocodilia (crocodilians)
  Class Aves (birds)

Figure 2.6 Sometimes genealogy and overall morphological similarity/dissimilarity can lead to conflicting conclusions about classification. The conflict between phylogenetic systematists (for whom genealogy has priority) and traditional systematists (who emphasize overall similarity/dissimilarity) is exemplified in the case of the birds and reptiles. The cladogram in this figure depicts the generally accepted view of the relationships among the major groups of living vertebrates. Classification scheme A depicts a traditional classification of the vertebrates, in which crocodilians are classified with lizards, snakes, and turtles in the taxon Reptilia, while birds are retained as a separate taxon, Aves. Traditional systematists, in their desire to express both branching patterns and degree of overall similarity/dissimilarity in classifications, are willing to accept paraphyletic taxa (e.g., Reptilia) in order to formally distinguish what they view as "similar" groups of vertebrates (actually grades, not clades). Schemes B and C are phylogenetic systematic classifications. Scheme B strictly reflects the branching pattern of the cladogram; thus, the reptiles are broken into separate taxa in recognition of their genealogical relationships, and the birds and crocodilians are classified together as a separate sister group to the reptiles (called Archosauria in this scheme). Scheme C also strictly mirrors the tree, but uses the phylogenetic sequencing convention. In schemes B and C, all taxa are monophyletic. Notice that scheme C requires four fewer taxonomic names than scheme B.

gram in Figure 2.6). In other words, birds originated from the branch of reptiles that also gave rise to the crocodilians. By cladistic methodology (on genealogical grounds), birds and crocodilians should therefore be ranked together, separate from the "other reptiles" (phylogenetic systematists recognize such a group, calling it the Archosauria), or birds should be classified with "reptiles" and the definition of that group expanded to include birds (phylogenetic systematists also recognize this grouping, often referring to it as the Sauropsida, or Reptilomorpha). Traditional systematists argue that even though birds and crocodilians may be "most closely related" on a genealogical basis (sister groups in a cladogram), birds are very different from reptiles, and hence the two groups should be placed in entirely different taxa (classification scheme A in Figure 2.6). Furthermore, traditional systematists argue that, taking all attributes into consideration, the crocodilians are clearly members of the reptilian grade (and should be retained within the Reptilia), whereas the birds have evolved many new attributes and belong to a separate avian grade. In other words, the crocodilians have retained more primitive reptilian features (symplesiomorphies) than the birds have, and for this reason the crocodilians should be classified with the other reptiles, not with the birds. The phylogenetic sequencing convention (Figures 2.6 and 2.7 scheme C) is one solution to this dilemma.

One criticism of phylogenetic systematics occasionally heard is that it always depicts the speciation process as the splitting of an ancestral species into two sister species, despite the probability that numerous other speciation modes exist (Figure 2.8). In a cladogram, once a new species appears, a "split" must be placed on the tree, and the two branches represent sister groups,

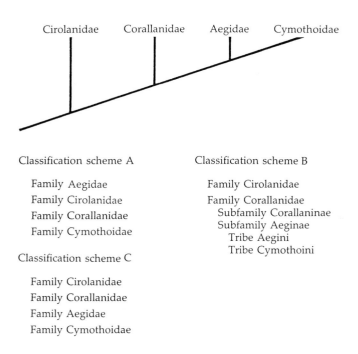

Cirolanidae   Corallanidae   Aegidae   Cymothoidae

Classification scheme A

   Family Aegidae
   Family Cirolanidae
   Family Corallanidae
   Family Cymothoidae

Classification scheme C

   Family Cirolanidae
   Family Corallanidae
   Family Aegidae
   Family Cymothoidae

Classification scheme B

   Family Cirolanidae
   Family Corallanidae
      Subfamily Corallaninae
      Subfamily Aeginae
        Tribe Aegini
        Tribe Cymothoini

Figure 2.7 A dendrogram of four closely related groups of isopod crustaceans (marine "pillbugs"; see Chapter 16). The dendrogram can be viewed as either a cladogram or a traditional tree. In this particular example, the four taxa listed constitute an interesting "evolutionary series," from the free-living carnivorous Cirolanidae through micropredators and temporary fish parasites (Corallanidae and Aegidae) to obligatory fish parasites (Cymothoidae). Classification scheme A depicts the classification developed by traditional systematists and currently in use. The taxa can be listed in any order (here, alphabetically) in the classification, so the order of listing does not necessarily reflect their arrangement in the tree. Scheme B views the tree as a cladogram, arranging the taxa in a *subordinated (hierarchical) classification* and depicting precisely the arrangement of the cladogram. Scheme C also views the dendrogram as a cladogram and utilizes the phylogenetic sequencing convention to arrange the taxa in the exact sequential order in which they appear on the tree. There is no way to convert the traditional classification of scheme A directly into the tree from which it was derived; hence phylogenetic relationships cannot be ascertained from the classification. Schemes B and C can be directly converted back to the tree from which they were derived, because they precisely reflect the genealogical (phylogenetic) relationships of the taxa.

whether or not the original species has in fact changed at all. Some biologists have claimed that this practice is misleading. This criticism, however, is unfounded, and it derives from simple lack of understanding. First of all, cladograms are not always completely dichotomous; they can have branching points that are trichotomous or even polytomous (Figure 2.8D). Second, a terminal taxon on a cladogram may lack any defining synapomorphies, thus indicating that it is not only the sister group of its adjacent lineage, but also the actual ancestor of that lineage. The cladogram of annelids (see Figure 13.40 in Chapter 13) is an example. The oligochaetes (earthworms and their kin) lack any unique defining synapomorphies; hence they are depicted as the hypothetical direct ancestors of the hirudinidans (leeches and

their kin)—that is, leeches probably evolved from an oligochaetous ancestor. Thus, "Oligochaeta" constitutes a paraphyletic taxon. A cladogram can express any kind of speciation event; it simply does so in a restricted way—by way of branches depicting a pattern of nested synapomorphies.

The methods of phylogenetic systematics force the systematist to be explicit about groups *and* characters. The method is also largely independent of the biases of the discipline in which it is applied. In its fundamental principles, it is not restricted to biology, but is applicable to a variety of fields in which the relations that characterize groups are comparable to the homology concept and possess a hierarchical nature. Thus, cladistic analyses have been applied to other historical systems, such

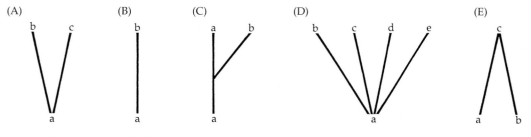

Figure 2.8 Common models of speciation. (A) One species splits into two new species. (B) One species is transformed into another. This type of speciation may be viewed as either gradual or rapid. (C) One species remains unchanged, while an isolated peripheral population evolves into a distinct new species. This model probably represents for most evolutionists the most common mode of speciation. (D) "Explosive radiation," in which one species suddenly splits into many new species. Speciation events represented

by this model are predicted to occur when a species is suddenly confronted with a vast new array of habitats or "unfilled niches" to exploit, resulting in rapid specialization and reproductive isolation as the new niches are filled. Explosive radiation might also occur when the range of a widespread species is fragmented into numerous smaller, isolated populations. (E) A new species is "created" by hybridization of two other species; this type of speciation appears to be rare and may occur primarily in plants and protists.

as linguistics and textual criticism (in which the "homologues" are shared tongues or texts), and even to the classification of musical instruments. They are also used in biogeographic analyses, wherein taxa are replaced by their appropriate areas of endemism, and the "homologues" are thus sister groups shared by geographic areas. Although the information stored in a cladogram is restricted to genealogy, such trees are often used to test other kinds of hypotheses, such as modes of speciation, historical relationships among geographic areas, and coevolution in host–parasite lineages.

As stressed earlier, the concept of similarity plays a central role in phylogenetic systematics. There are really only three kinds of evolutionary similarity expressed among organisms: (1) shared evolutionary novelties inherited from an immediate common ancestor; (2) similarity inherited from some more remote ancestor (any number of descendant taxa may retain such similarity); and (3) similarity due to evolutionary convergence. Phylogenetic systematists accept only the first kind of similarity (synapomorphies) as valid evidence of close affinity (common ancestry) between two taxa. Traditional systematists also rely heavily on synapomorphies, but consider the second kind of similarity (symplesiomorphies) in their analyses as well. They also use "degree of difference" (i.e., the numbers and types of similarities distinguishing a lineage) to classify organisms. The third kind of similarity (convergence) holds no value at all in phylogenetic analyses, and its use serves only to create chaos.

It is worth noting that the concept of shared derived characters has been around for many decades, and a careful review of the work produced by the most critical systematists through time will reveal that most were striving to delimit monophyletic taxa and construct phylogenetic trees based, as cladistics prescribes, on nested sets of synapomorphies. However, many existing older classifications are still based in part on symplesiomorphies rather than solely on synapomorphies, and these classifications are destined to be revised as more cladistic studies are accomplished.

There has also been a trend over the past 30 years toward redefining taxa so that they are based strictly upon "positive characters," or the possession of distinct recognizable features. Formerly recognized taxa based on "negative characters" (the absence of features) have largely been redefined and reorganized, or are simply no longer considered valid. The most obvious example of a group based on negative characters is, of course, the "Invertebrata." The invertebrates are a group of convenience, useful for didactic purposes but no more. They are *not* evolutionarily related by their *lack* of a backbone—whereas vertebrates *are* related by their possession of a backbone (a vertebrate synapomorphy). "Invertebrata" is a paraphyletic group.

There have been very few phylogenetic methodological tests of *known* evolutionary histories, although a few strains of laboratory animals, plant cultivars, and microorganisms have been examined in this way. Methods of phylogenetic reconstruction can be tested with such known phylogenies (or with computer models of simulated phylogenies). So far, such tests have shown that cladistic methods (i.e., reconstructions based on genealogical histories and parsimony) come close to recapturing actual evolutionary histories.

There is no doubt that the future of biological systematics will be an exciting one. Biological systematics is now beginning to play key roles in such diverse fields as ecology, conservation biology, biological pest control, and natural products chemistry. As our present methodologies and philosophies are refined, and as new tools are discovered, they will interact with our view of evolution and stimulate continued growth and improvement in our understanding of biological diversity and the history of life.

Selected References

Ayers, D. M. 1972. *Bioscientific Terminology: Words from Latin and Greek Stems*. University of Arizona Press, Tucson. [A wonderful little workbook to learn the basics of biological terminology.]

Blackwelder, R. E. 1967. *Taxonomy: A Text and Reference Book*. Wiley, New York. [A pleasant little text on practical taxonomy, identification of specimens, curatorial practices, the use of names, the use of taxonomic literature, and publication; a considerable portion of the text is devoted to the intricacies of nomenclature and the rules for name publication, changes, etc.]

Bremer, K. 1994. Branch support and tree stability. Cladistics 10: 295–304.

Carpenter, J. M. 1988. Choosing among equally parsimonious cladograms. Cladistics 4: 291–296.

Cracraft, J. and N. Eldredge (eds.). 1979. *Phylogenetic Analysis and Paleontology*. Columbia University Press, New York. [Somewhat dated, but still with excellent discussions of phylogenetic reconstruction; an eclectic overview of systematics and evolution; good reading.]

Croizat, L. 1958. *Panbiogeography*. Published by the author. Caracas, Venezuela.

Croizat, L. 1964. *Space, Time, Form: The Biological Synthesis*. Published by the author. Caracas, Venezuela. [Croizat's two books initiated a paradigm shift in the field of biogeography.]

Croizat, L., G. Nelson, and D. E. Rosen. 1974. Centers of origins and related concepts. Syst. Zool. 23: 265–287.

Cummings, M. P., S. P. Otto and J. Wakeley. 1995. Sampling properties of DNA sequence data in phylogenetic analysis. Mol. Biol. Evol. 12(5): 814–822.

DeJong, R. 1980. Some tools for evolutionary and phylogenetic studies. J. Syst. Zool. Evol. Res. 18: 1–23.

Duncan, T. and T. Stuessy (eds.). 1984. *Perspective on the Reconstruction of Evolutionary History*. Columbia University Press, New York.

Eldredge, N. and J. Cracraft. 1980. *Phylogenetic Pattern and the Evolutionary Process: Method and Theory in Comparative Biology*. Columbia University Press, New York. [Still one of the best texts on the theory of cladistic analysis and classification.]

Felsenstein, J. 1983. Parsimony in systematics: Biological and statistical issues. Annu. Rev. Ecol. Syst. 14: 313–333.

Felsenstein, J. 1985. Phylogenies and the comparative method. Am. Nat. 126: 1–25.

Felsenstein, J. 2002. *Inferring Phylogenies*. Sinauer Associates, Sunderland, MA.

Frizzell, D. L. 1933. Terminology of types. Am. Midland Nat. 14(6): 637–668. [A listing of every kind of "type" designation Frizzell could locate, the vast majority of which have no particular nomenclatural validity.]

Gould, S. J. and R. C. Lewontin. 1979. The spandrels of San Marco and the Panglossian paradigm: A critique of the adaptationist programme. Proc. R. Soc. Lond. Ser. B 205: 581–598.

Hall, B. K. 1994. *Homology. the Hierarchical Basis of Comparativve Biology.* Academic Press, San Diego, CA.

Hall, B. G. 2001. Phylogenetic Trees Made Easy: A How-To Manual for Molecular Biologists. Sinauer Associates, Sunderland, MA.

Harvey, P. H. and M. D. Pagel. 1991. *The Comparative Method in Evolutionary Biology.* Oxford University Press, New York. [A detailed elucidation of the method.]

Hennig, W. 1979. *Phylogenetic Systematics.* University of Illinois Press, Urbana. [The "third edition" of Hennig's original 1950 text on cladistic classification theory; be aware that the philosophy and methodology of cladistics has changed/grown a great deal since Hennig's original ideas.]

Hillis, D. M., J. P. Huelsenbeck and C. W. Cunningham. 1994. Application and accuracy of molecular phylogenies. Science 264: 671–677.

Hillis, D., C. Mortiz and B. Mable. 1996. *Molecular Systematics*, 2nd Ed. Sinauer Associates, Sunderland, MA.

Huelsenbeck, J. P. and J. J. Bull. 1996. A likelihood ratio test to detect conflicting phylogenic signal. Syst. Biol. 45: 92–98.

Huelsenbeck, J. P. and B. Rannala. 1997. Phylogenetic methods come of age: Testing hypotheses in an evolutionary context. Science 276: 227–232.

International Commission on Zoological Nomenclature. 2000. *International Code of Zoological Nomenclature*, 4th Ed. The International Trust for Zoological Nomenclature, London. [The nomenclatural rule book.]

Jablonski, D. and D. Bottjer. 1991. Environmental patterns in the origins of higher taxa: The post-Paleozoic fossil record. Science 252: 1831–1833.

Jefferys, W. H. and J. O. Berger. 1992. Ockham's razor and Bayesian analysis. Am. Sci. 80: 64–72.

Jeffrey, C. 1977. *Biological Nomenclature*, 2nd Ed. Crane, Russak & Co., New York.

Kluge, A. G. and A. J. Wolf. 1993. Cladistics: What's in a word? Cladistics 9: 183–200.

Maddison, D. R. 1991. The discovery and importance of multiple islands of most-parsimonious trees. Syst. Zool. 40: 315–328.

Maddison, D. R. and W. P. Maddison. 2000. *MacClade 4.0.* Sinauer Associates, Sunderland, MA.

Maddison, W. P. 1996. Molecular approaches and the growth of phylogenetic biology. Pp. 47–63 in J. D. Ferran's and S. R. Palumbi (eds.), *Molecular Zoology. Advances, Strategies, and Protocols.* Wiley-Liss, New York.

Maddison, W. P. 1997. Gene trees in species trees. Syst. Biol. 46: 523–536.

Maddison, W. P., M. J. Donoghue and D. R. Maddison. 1984. Outgroup analysis and parsimony. Syst. Zool. 33(1): 83–103.

Mayr, E. and P. D. Ashlock. 1991. *Principles of Systematic Zoology*, 2nd Ed. McGraw-Hill, New York. [The "bible" of non-cladistic systematics.]

Mickevich, M. E. and S. J. Weller. 1990. Evolutionary character analysis: Tracing character change on a cladogram. Cladistics 6: 137–170.

Mindell, D. P. and C. E. Thacker. 1996. Rates of molecular evolution: Phylogenetic issues and applications. Annu. Rev. Ecol. Syst. 27: 279–303.

Nelson, G. and N. I. Platnick. 1981. Systematics and biogeography: Cladistics and vicariance. Cladistics 5: 167–182.

Nelson, G. and N. I. Platnick. 1981. *Systematics and Biogeography: Cladistics and Vicariance.* Columbia University Press, New York. [Excellent review of the history and development of systematics and biogeography, as well as a thorough, although theoretical, treatment of cladistics and vicariance biogeography.]

Page, R. D. M. and E. C. Holmes. 1998. *Molecular Evolution: A Phylogenetic Approach.* Blackwell Science Ltd., Oxford.

Patterson, C. 1982. Morphological characters and homology. *In* K. Joysey and A. Friday, *Problems of Phylogenetic Reconstruction.* Systematics Association Special Volume no. 21. Academic Press, New York, pp. 21–74.

Patterson, C. 1990. Reassessing relationships. Nature 344: 199–200.

Philippe, H., A. Chenuil and A. Adoutte. 1994. Can the Cambrian explosion be inferred through molecular phylogeny? Development (Suppl.), 15–25.

Raff, R. A. 1996. *The Shape of Life.* University of Chicago Press, Chicago.

Rose, M. R. and G. V. Lauder. 1996. *Adaptation.* Academic Press, San Diego, CA.

Rosen, D. E. 1985. Geological hierarchies and biogeographic congruence. Ann. Missouri Bot. Garden 72: 636–659.

Sanderson, M. J. 1990. Flexible phylogeny reconstruction: A review of phylogenetic inference packages using parsimony. Syst. Zool. 39: 414–420.

Sanderson, M. J. 1995. Objections to bootstrapping phylogenies: A critique. Syst. Biol. 44: 299–320.

Sanderson, M. J. and L. Hufford. *Homoplasy. The Recurrence of Similarity in Evolution.* Academic Press, San Diego, CA.

Savory, T. 1962. *Naming the Living World.* The English Universities Press, Ltd., London.

Schuh, R. T. 2000. *Biological Systematics: Principles and Applications.* Comstock Publishing Associates, Ithaca, NY.

Sepkoski. J. J., R. K. Bambach, D. M. Raup and J. W. Valentine. 1981. Phanerozoic marine diversity and the fossil record. Nature 293: 435–437.

Simpson, G. G. 1961. *Principles of Animal Taxonomy.* Columbia University Press, New York. [Aging fast but still provides a sound look at non-cladistic thinking on evolution, the species concept, and traditional classification methods.]

Stanley, S. M. 1982. Macroevolution and the fossil record. Evolution 36: 460–473.

Stevens, P. F. 1980. Evolutionary polarity of character states. Annu. Rev. Ecol. Syst. 11: 333–358.

Swofford, D. 2001. *PAUP: Phylogenetic Analysis Using Parsimony, Ver. 4.0+.* Sinauer Associates, Sunderland, MA.

Watrous, L. E. and Q. D. Wheeler. 1981. The out-group comparison method of character analysis. Syst. Zool. 30(1): 1–11.

Wiley, E. O. 1981. *Phylogenetics: The Theory and Practice of Phylogenetic Systematics.* Wiley, New York. [An excellent review of cladistic methodology; less theoretical and more operational in its approach than the Nelson/Platnick and Eldredge/Cracraft volumes on the same subject; also includes a good discussion of species and speciation concepts.]

Wiley, E. O. 1988. Vicariance biogeography. Annu. Rev. Ecol. Syst. 19: 513–542.

Wiley, E. O., D. Siegel-Causey, D. R. Brooks, and V. Funk. 1991. *The Compleat Cladist: A Primer of Phylogenetic Procedures.* Special Publication no. 19. Museum of Natural History, University of Kansas, Lawrence.

Wilkins, A. S. 2002. *The Evolution of Developmental Pathways.* Sinauer Associates, Sunderland, MA.

Winston, J. E. 1999. Describing Species: Practical Taxonomic Procedure for Biologists. Columbia University Press, New York.

3 Animal Architecture and the Bauplan Concept

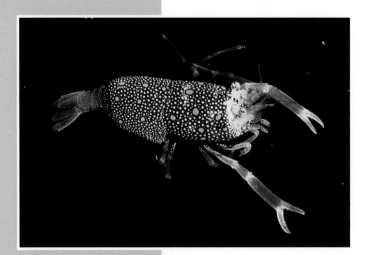

The business of animals is to stay alive until they reproduce themselves, and . . . the business of zoologists is to try to understand how they do it.

E. J. W. Barrington,
Invertebrate Structure and Function, 1967

The German language includes a wonderful word that expresses the essence of animal architecture: ***Bauplan*** (pl. *Baupläne*); we will use the Anglicized spelling, bauplan (pl. bauplans). The word means, literally, "a structural plan or design," but a direct translation is not entirely adequate. An animal's bauplan is, in part, its "body plan," but it is more than that. The concept of a bauplan captures in a single word the essence of structural range and architectural limits, as well as the functional aspects of a design (Box 3A). If an organism is to "work," *all* of its body components must be both structurally *and* functionally compatible. The entire organism encompasses a definable bauplan, and the specific organ systems themselves also encompass describable bauplans; in both cases the structural and functional components of the particular plan establish both capabilities and limits. Thus, the bauplan determines the major constraints that operate at both the organismic and the organ system levels.

The diversity of form in the biological world is dazzling, yet there are real limits to what may be successfully molded by evolutionary processes. All animals must accomplish certain basic tasks in order to survive and reproduce. They must acquire, digest, and metabolize food and distribute its usable products throughout their bodies. They must obtain oxygen for cellular respiration, while at the same time ridding themselves of metabolic wastes and undigested materials. The strategies employed by animals to maintain life are extremely varied, but they rest upon relatively few biological, physical, and chemical principles. Within the constraints imposed by particular bauplans, animals have a limited number of options available to accomplish life's tasks. For this reason a few recurring fundamental themes become apparent. This chapter is a general review of these themes: the structural/functional aspects of invertebrate bauplans and the basic survival

BOX 3A *The Bauplan and Related Concepts*

Stability of organismal morphology is a deep-seated notion that dates at least from the eighteenth century; the idea of a limited number of plans, types, or archetypes of animal forms is thus an old one. Richard Owen introduced the term **archetype** in 1848 to represent a model organism, or the sum of the features shared by a group of related organisms. The concept of the embryological archetype, and the fact that adults are nothing more than the accumulation of features added during their development, was formalized by Karl von Baer and Ernst Haeckel in the second half of the nineteenth century. Today this concept has grown into the notion of conserved body plans, or *Baupläne* (bauplans, in the Anglicized form used in this book). The concept speaks to a stability in form that maintains itself through evolutionary time and phylogenetic divergence.

The term *Bauplan* (German for "ground plan" or "blueprint") was introduced as a technical term in zoology in 1945 by the embryologist-turned-philosopher Joseph Henry Woodger. More recently, Niles Eldredge (1989) discussed the bauplan as the common structural plan within a monophyletic taxon; Valentine (1986) distinguished bauplans as assemblages of homologous architectural and structural features distinguishing phyla and classes; and Gould (1977, 1980, 1992) spoke of structural constraints leading to fundamental ground plans of anatomy. And, in his review of the reunion of developmental and evolutionary biology, Atkinson (1992) claimed that "the single most critical concept of the reunion is that of the bauplan."

The concept of the bauplan expresses both a notion of morphological stability and the fact that some aspects of embryonic and/or adult morphology are more free to vary than are others. That is, some stages of development are more constrained than others. The most striking evidence of developmental constraints is the simple fact that, despite the great variety of animals, there are relatively few basic types of animal body plans.

Developmental canalization, sometimes called *developmental buffering* or *genetic homeostasis*, is a form of constraint that channels ontogeny into restricted sets of pathways that lead to a standard phenotype in spite of genetic or environmental disturbances. The concept can be viewed at the genomic or organismal level, or even at a character-by-character level. The more highly canalized a character the less it will vary among individuals, and characters that define bauplans are highly canalized. The preservation of *Hox* gene function across phyla is a good example of developmental canalization. In fact, we are beginning to realize that many of the basic body patterns that evolved during the Precambrian/Cambrian origins of animal phyla represent the outcomes of conserved genes and developmental plans.

The characteristics of an organism's bauplan are not the same thing as its phylogenetically unique features, or synapomorphies. Instead, bauplans must be viewed as nested sets of conserved body plans, as would be predicted within an ancestor–descendant hierarchical system such as animal phylogenesis. For example, snakes possess a bauplan that differs from the bauplans of lizards, turtles, or crocodiles—yet each shares the reptilian bauplan. Reptiles, birds, and mammals each have individual bauplan but share the vertebrate bauplan. Thus, bauplans consist of a mix of ancestral and derived characters. To understand their origin requires knowledge of adult, larval, and embryonic phases of the life cycle.

Woodger explicitly argued, as had von Baer, that the most basic structures defining the bauplan develop early in embryonic life. Consequently, deviations early in development would have much more drastic consequences for morphology than deviations later in development. Mechanisms that establish bauplans buffer development against environmental and genetic perturbations. They constrain development. Ernst Mayr repeatedly drew attention to the importance of such constraints, specifically in relation to bauplans, conserved morphological features, and the taxonomic features used in classification. Heterochrony (see Chapter 4) may be one powerful force that can alter or overcome the inertia of bauplans.

The field of molecular evolutionary developmental biology is just emerging, but already its discoveries are shedding new light on these old ideas. For example, we now know that much of an animal's initial embryogeny is under maternal cytoplasmic control rather genomic control. However, at some point early in embryogenesis, the zygote's parental (nuclear) genome takes primary control of development. Recent work suggests that this may be one of several pivotal points in the control of animal ontogeny—occasions that demarcate the fixation of bauplans. The point at which the zygotic genome takes over control of embryogenesis has been referred to by several names, but perhaps the most fitting term in the literature is the **zootypic stage.** It may be here that the *Hox* genes establish the most basic, or primary, animal body patterning (e.g., the anterior–posterior axis and dorsal–ventral surfaces).

At a later stage of embryogenesis another critical point is reached, which has been called the phylotypic (phyletic) stage. The *phylotype* theoretically represents the stage when the genes responsible for secondary patterning of a body plan are first fully expressed and the adult morphogenetic fields are positioned. This juncture is not well understood. Anderson (1973) identified the blastula as the phylotypic stage for the annelids and arthropods, whereas Sander (1976, 1983) identified the germ band stage (a 20-segmented larval stage with head, thorax, and abdomen already delineated and segmented) as the phylotypic stage of insects. Cohen (1977, 1979), on the other hand, distinguished *phylotypic larvae* (the trochophore of annelids, for instance) from *adaptive larvae*. Phylotypic stage larvae have a simple morphology determined more by developmental (genetic) programs than by physiological requirements.

The phylotypic stage is usually thought of as the stage at which embryos within a phylum show the greatest level of morphological similarity. Beyond this stage, the zygotic genome begins moving embryos down the individual tracks of the various lineages. In other words, early developmental stages of closely related taxa converge on a phylotype in the course of their ontogeny, only to diverge again as the adult form unfolds.

Thus it seems likely that there are several fundamental levels of body patterning during ontogeny, and closely related taxa share critical junctures of this process in ways that hearken back to Haeckel's "law of recapitulation." The next several decades will see the elaboration of more explicit descriptions of these hierarchical developmental patterns. However, it is already becoming clear that such developmental stages (e.g., the zootypic stage and the phylotypic stage) canalize ontogenetic events to produce the adult bauplan. This canalization comes about, in part, due to developmental constraints* that work to maintain overall body plans. Such constraints are also only just beginning to be understood, but several have been proposed, including:

1. Structural constraints (e.g., constraints imposed by the limitations of patterns in early developmental stages)
2. Genetic constraints (e.g., rates of mutation and recombination of individual alleles)
3. Direct developmental constraints

(e.g., obligatory tissue interactions)
4. Cellular constraints (e.g., limits to the rate and number of cell divisions, secretions of cell products, cell migration)
5. Metabolic constraints (e.g., dependence on particular metabolic pathways
6. Functional constraints (e.g., the interconnectedness of parts of different organ systems involved in critical functions)

Constraint is perhaps not the best term, for to constrain is not to restrain evolution. Constraints set *limits* to evolution, especially morphological evolution, but groups with constrained characters are among the most adaptively successful and speciose animal taxa. For example, the number of segments in insects is highly constrained, but insects are both "advanced" and highly successful. So constraints work well with selection and adaptive radiation, and in fact are presumably themselves a consequence of past selection.

strategies employed within each. It is a description of how invertebrates are put together and how they manage to survive and reproduce. Each subject discussed here reflects fundamental principles of animal mechanics, physiology, and adaptation.

Keep in mind that even though this chapter is organized on the basis of what might be called the "components" of animal structure, whole animals are integrated functional combinations of these components. Furthermore, there is a strong element of predictability in the concepts discussed here. For example, given a particular type of symmetry, one can make reasonable guesses about other aspects of an animal's structure that should be compatible with that symmetry—some combinations work, others do not. Herein are explained many of the concepts and terms used throughout this book, and we encourage you to become familiar with this material now as a basis for understanding the remainder of the text.

Body Symmetry

A fundamental aspect of an animal's bauplan is its overall shape or geometry. In order to discuss invertebrate architecture and function, we must first acquaint ourselves with a basic aspect of body form: **symmetry**. Symmetry refers to the regular arrangement of body structures relative to the axis of the body. Animals that can be bisected or split along at least one plane, so that the resulting halves are similar to one another, are said to be **symmetrical**. For example, a shrimp can be bisected vertically through its midline, head to tail, to produce right and left halves that are mirror images of one another. A few animals have no body axis and no plane of symmetry, and are said to be **asymmetrical**. Many sponges, for example,

have an irregular growth form and lack any clear plane of symmetry. Similarly, many protists, particularly the ameboid forms, are asymmetrical (Figure 3.1).

One form of symmetry is **spherical symmetry**. It is seen in creatures whose bodies lack an axis and have the form of a sphere, with the body parts arranged concentrically around, or radiating from, a central point (Figure 3.2). A sphere has an infinite number of planes of symmetry that pass through its center to divide it into like halves. Spherical symmetry is rare in nature; in the strictest sense, it is found only in certain protists. Organisms with spherical symmetry share an important functional attribute with asymmetrical organisms, in that both groups lack **polarity**. That is, there exists no clear differentiation along an axis. In all other forms of symmetry, some level of polarity has been achieved; and with polarity comes specialization of body regions and structures.

A body displaying **radial symmetry** has the general form of a cylinder, with one main axis around which the various body parts are arranged (Figure 3.3). In a body displaying perfect radial symmetry, the body parts are arranged equally around the axis, and any plane of sectioning that passes along that axis results in similar halves (rather like a cake being divided and subdivided into equal halves and quarters). Nearly perfect radial symmetry occurs in some sponges and in many cnidarian polyps (Figure 3.3A,B). Perfect radial symmetry is relatively rare, however, and most radially symmetrical animals have evolved modifications on this theme. **Biradial symmetry**, for example, occurs where portions of the body are specialized and only two planes of sectioning can divide the animal into perfectly similar halves. Common examples of biradial organisms are ctenophores and many sea anemones (Figure 3.3C).

(A)

(B)

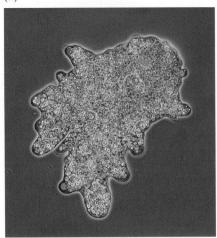

Figure 3.1 Examples of asymmetrical invertebrates. (A) An assortment of sponges. (B) An ameba.

Further specializations of the basic radial body plan can produce nearly any combination of multiradiality. For example, many jellyfishes possess **quadriradial symmetry** (Figure 3.3D). Most echinoderms are said to display **pentaradial symmetry** (Figure 3.3E,F), although many multiarmed sea stars are also known In fact, the presence in sea stars of certain organs (e.g., the madreporite) allows for only one plane by which perfectly matching halves exist, and thus sea stars actually possess a form of pentaradial bilaterality. But this is splitting hairs. The adaptive significance of body symmetry operates at a much grosser level than organ position, and in this regard most echinoderms, including sea stars, are functionally radially symmetrical.

A radially symmetrical animal has no front or back end; rather it is organized about an axis that passes through the center of its body, like an axle through a wheel. When a gut is present, this axis passes through the mouth-bearing (**oral**) surface to the opposite (**aboral**) surface. Radial symmetry is most common in sessile and sedentary animals (e.g., sponges, sea stars, and sea anemones) and drifting pelagic species (e.g., jellyfishes and ctenophores). Given these lifestyles, it is clearly advantageous to be able to confront the environment equally from a variety of directions. In such creatures the feeding structures (tentacles) and sensory receptors are distributed at equal intervals around the periphery of the organisms, so that they contact the environment more or less equally in all directions. Furthermore, many bilaterally symmetrical animals have become functionally radial in certain ways associated with sessile lifestyles. For example, their feeding structures may be in the form of a whorl of radially arranged tentacles, an arrangement allowing more efficient contact with their surroundings.

The body parts of **bilaterally symmetrical** animals are oriented about an axis that passes from the front (**anterior**) to the rear (**posterior**) end. A single plane of symmetry—the **midsagittal plane** (or median sagittal plane)—passes along the axis of the body to separate right and left sides. Any longitudinal plane passing perpendicular to the midsagittal plane and separating the upper (**dorsal**) from the underside (**ventral**) is called a

(A)

(B)

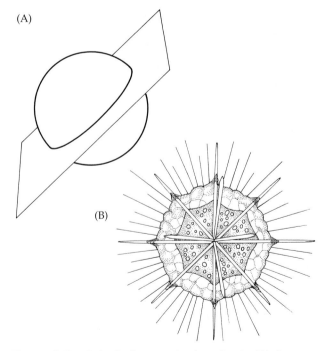

Figure 3.2 Spherical symmetry in animals. (A) An example of spherical symmetry; any plane passing through the center divides the organism into like halves. (B) A radiolarian (protist).

Figure 3.3 Radial symmetry in invertebrates. The body parts are arranged radially around a central oral–aboral axis. (A) Representation of perfect radial symmetry. (B) The sponge *Xetospongia*. (C) The sea anemone *Epiactus,* whose mouth alignment and internal organization produce biradial symmetry. (D) The hydromedusa *Scrippsia*, with quadriradial symmetry. (E) The sea star *Patiria*, with pentaradial symmetry. (F) The sea bisquit, *Clypeaster*, with pentaradial symmetry.

frontal plane. Any plane that cuts across the body perpendicular to the main body axis and the midsagittal plane is called a **transverse plane** (or, simply, a cross section) (Figure 3.4). In bilaterally symmetrical animals the term **lateral** refers to the sides of the body, or to structures away from (to the right and left of) the midsagittal plane. The term **medial** refers to the midline of the body, or to structures on, near, or toward the midsagittal plane.

Whereas spherical and radial symmetry are typically associated with sessile or drifting animals, bilaterality is generally found in animals with controlled mobility. In these animals, the anterior end of the body confronts the environment first. Associated with bilateral symmetry and unidirectional movement is a concentration of feeding and sensory structures at the anterior end of the body. The evolution of a specialized "head," containing those structures and the nervous tissues that innervate them, is called **cephalization**. Furthermore, the surfaces of the animal differentiate as dorsal and ventral regions, the latter becoming locomotory and the former being specialized for protection. A variety of secondary asymmetrical modifications of bilateral (and radial) symmetry have occurred, for example, the spiral coiling of snails and hermit crabs.

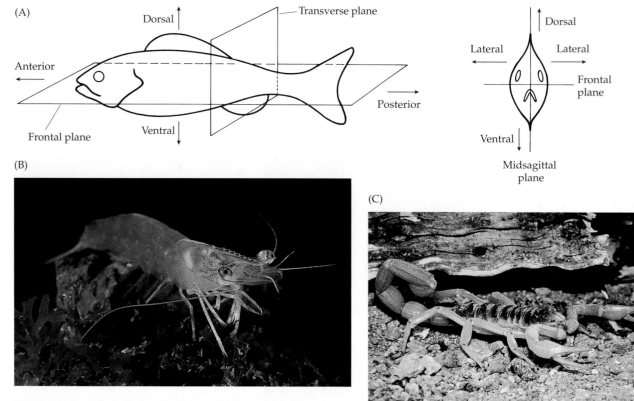

(A)

Dorsal

Transverse plane

Anterior

Frontal plane

Ventral

Posterior

Dorsal

Lateral   Lateral

Frontal
plane

Ventral

Midsagittal
plane

(B)

(C)

Figure 3.4 Bilateral symmetry in animals; a single plane—the midsagittal plane—divides the body into equal halves. (A) Diagrammatic illustration of bilateral symmetry, with terms of orientation and planes of sectioning. A Pacific shrimp (B) and Sonoran Desert scorpion (C) show obvious bilateral symmetry.

Cellularity, Body Size, Germ Layers, and Body Cavities

One of the main characteristics used to define grades of animal complexity is the presence or absence of true **tissues**. Tissues are aggregations of morphologically and physiologically similar cells that perform a specific function. The Protista (Chapter 5) do not possess tissues, but occur only as single cells or as simple colonies of cells. In a sense, they are all at a **unicellular grade** of construction. Beyond the protista is the vast array of multicelled animals, the **Metazoa**. The Metazoa can be divided into three major levels, or grades: mesozoa, parazoa, and eumetazoa. These names do not represent formal taxa, but may be used to group the Metazoa by their level of overall structural complexity.* The mesozoa and parazoa are not generally considered to possess true tissues, and for this reason they are separated from the rest of

---

*The term *metazoans* is used in this text to indicate a formal taxonomic entity—those organisms belonging to the kingdom Metazoa, or Animalia.

the Metazoa. The eumetazoa pass through distinct embryonic stages during which tissue layers form (Chapter 4). Box 3B provides an outline of these general grades of body architecture.

Each of these grades of body complexity is associated with inherent constraints and capabilities, and within each grade there are obvious limits to size. As the British biologist D'Arcy Thompson wrote, "Everything has its proper size . . . men and trees, birds and fishes, stars and star-systems, have . . . more or less narrow ranges of absolute magnitudes." As a cell (or an organism) increases in size, its volume increases at a rate faster than the rate of increase of its surface area (surface area increases as the square of linear dimensions; volume increases as the cube of linear dimensions). Because a cell ultimately relies on transport of material across its plasma membrane for survival, this disparity quickly reaches a point at which the cytoplasm can no longer be adequately serviced by simple cellular diffusion. Some unicellular forms develop complexly folded surfaces or are flattened or threadlike in shape. Such creatures can be quite large, but eventually a limit is reached; thus we have no meter-long protists.

To increase in size, ultimately the only way around the surface-to-volume dilemma is to increase the number of cells constituting a single organism; hence the Metazoa. But size increase in the Metazoa is also limit-

BOX 3B *Organization of the Protista and Animal Phyla on the Basis of their Body Construction*[a]

I. Unicellular organisms (Protista): Phyla **Euglenoida, Kinetoplastida, Ciliophora, Apicomplexa, Dinoflagellata, Stramenopiles, Rhizopoda, Actinopoda, Granuloreticulosa, Diplomonadida, Parabasilida, Chlorophyta, Cryptomonada, Microspora, Ascetospora, Myxozoa, Opalinida, Choanoflagellata.**

II. Multicellular organisms: the **Metazoa**, or Animalia

    A. Without true tissues

        1. The mesozoa
            Phyla **Orthonectida, Rhombozoa, Placozoa,** and **Monoblastozoa**[b]

        2. The parazoa
            Phylum **Porifera** (sponges)

    B. With true tissues: the eumetazoa

        1. The diploblastic eumetazoa (lacking true mesoderm): the "radiata"
            Phyla **Cnidaria, Ctenophora**

        2. The triploblastic eumetazoa (with true mesoderm): the "bilateria"

            a. Acoelomates: without a body space other than the digestive tract; mesenchyme and muscle fill region between gut and epidermis.
               Phyla **Platyhelminthes, Gastrotricha, Entoprocta, Gnathostomulida**

            b. Blastocoelomates: with a persistent blastocoel between gut and body wall[c]
               Phyla **Acanthocephala, Kinorhyncha, Loricifera, Nematoda, Nematomorpha, Rotifera**

            c. Coelomates (or eucoelomates). With a true coelom (= mesodermal cavity).
               Phyla **Nemertea, Phoronida, Ectoprocta, Brachiopoda, Sipuncula, Echiura, Mollusca, Priapula, Onychophora, Tardigrada, Annelida, Arthropoda, Echinodermata, Chaetognatha, Hemichordata, Chordata**[d]

---

[a]Only those phyla set in boldface type are recognized taxa; other names are simply designations used to group various taxa by the level of body complexity they have achieved.

[b]Monoblastozoa (*Salinella*) is a phylum of questionable validity (see Chapter 6).

[c]Our view of the blastocoelomate (= "pseudocoelomate") condition has changed markedly since the 1980s, and several phyla formerly viewed as pseudocoelomates are now viewed as acoelomates (e.g., Gastrotricha, Entoprocta, Gnathostomulida). See Chapter 12 for discussions of these groups.

[d]Some of these groups (e.g., Arthropoda, Mollusca) have greatly reduced coelomic spaces; often the main body cavity is a bloodfilled space called a hemocoel, and is associated with an open circulatory system.

---

ed. Those Metazoa lacking complex specializations of tissues and organs must rely on diffusion into and out of the body, and this is inadequate to sustain life unless a majority of the body's cells are near or in contact with the external environment. In fact, diffusion is an effective method of oxygenation only when the diffusion path is less than about 1.0 mm. So here, too, there are limits. An animal simply cannot increase indefinitely in volume when most of its cells must lie close to the body surface. Primitive animals solve this problem to some degree by arranging their cellular material so that diffusion distances from cell to environment are comfortably short. One method of accomplishing this is to pack the internal bulk of the body with nonliving material, such as the jelly-like mesoglea of medusae and ctenophores. Another is to assume a body geometry that maximizes the surface area. Increase in one dimension leads to a vermiform body plan, like that of ribbon worms (Nemertea). Increase in two dimensions results in a flat, sheetlike body like that of the flatworms (Platy-

helminthes). In both cases the diffusion distances are kept short. Sponges effectively increase their surface area by a process of complex branching and folding of the body, both internally and externally. This folding keeps most of the body cells close to the environment.

If these were the only solutions to the surface-to-volume dilemma, the natural world would be filled with tiny, thin, flat animals and convoluted, spongelike creatures. However, many organisms increase in size by one to several orders of magnitude during their ontogeny, and life forms on earth span about 19 orders of magnitude in mass. Thus, another solution arose during the course of animal evolution that allowed for increases in body size. This solution was to bring the "environment" functionally closer to each cell in the body by the use of internal transport and exchange systems with large surface areas. A significant three-dimensional increase in body size thus necessitated the development of sophisticated internal transport mechanisms (e.g., circulatory systems) for nutrients, oxygen, waste products, and so on. These evolving transport structures became the organs and organ systems of higher animals. For example, the body volume of humans is so large that we require a highly branched network of gas exchange surfaces (our lungs) to provide an adequate surface area for gas diffusion. This network has about 1,000 square feet of surface—as much area as half a tennis court! The same constraints apply to food absorption surfaces; hence the evolution of very long, highly folded, or branched guts.

The embryonic tissue layers of eumetazoa are called **germ layers** (from the Latin *germen*, "a sprout, bud, or embryonic primordium"), and it is from these germ layers that all adult structures develop. Chapter 4 presents the details of germ layer formation and other aspects of metazoan developmental patterns. Here we need only point out that the germ layers initially form as outer and inner sheets or masses of embryonic tissue, termed **ectoderm** and **entoderm** (or endoderm), respectively. In the embryogeny of the radiate phyla Cnidaria and Ctenophora, only these two germ layers develop (or if a middle layer does develop, it is produced by the ectoderm, is largely noncellular, and is not considered a true germ layer). These animals are regarded as **diploblastic** (Greek *diplo*, "two"; *blast*, "bud" or "sprout"). In the embryogeny of most animals, however, a third cellular germ layer, the **mesoderm**, arises between the ectoderm and the entoderm; these metazoan groups are said to be **triploblastic**.

The evolution of a mesoderm greatly expanded the evolutionary potential for animal complexity. As we shall see, the triploblastic phyla have achieved many more highly sophisticated bauplans than are possible within the confines of a diploblastic body plan. Simply put, a developing triploblastic embryo has more building material than does a diploblastic embryo.

One of the major trends in the evolution of the triploblastic Metazoa has been the development of a fluid-filled cavity between the outer body wall and the digestive tube; that is, between the derivatives of the ectoderm and the entoderm. The evolution of this space created a radically new architecture, a tube-within-a-tube design in which the inner tube (the gut and its associated organs) was freed from the constraint of being attached to the outer tube (the body wall), except at the very ends. The fluid-filled cavity not only served as a mechanical buffer between these two largely independent tubes, but also allowed for the development and expansion of new structures within the body, served as a storage chamber for various body products (e.g., gametes), provided a medium for circulation, and was in itself an incipient hydrostatic skeleton. The nature of this cavity (or the absence of it) is associated with the formation and subsequent development of the mesoderm, as discussed in detail in Chapter 4.

Three major grades of construction are recognizable among the triploblastic Metazoa: **acoelomate**, **blasto-**

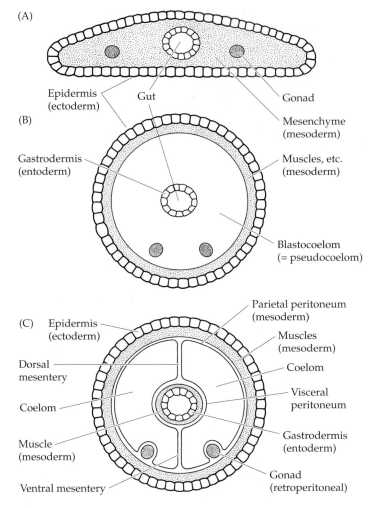

Figure 3.5   Principal body plans of triploblastic Metazoa (diagrammatic cross sections). (A) The acoelomate body plan. (B) The blastocoelomate body plan. (C) The eucoelomate body plan.

coelomate (formerly called **pseudocoelomate**), and **eucoelomate**. The acoelomate grade (Greek *a*, "without"; *coel*, "hollow, cavity") occurs in several triploblastic phyla: Platyhelminthes, Entoprocta, Gnathostomulida, and Gastrotricha. In these animals, the mesoderm forms a more or less solid mass of tissue, sometimes with small open spaces (lacunae), between the gut and body wall (Figure 3.5A). In nearly all other triploblastic animals, an actual space develops as a fluid-filled cavity between the body wall and the gut. In many phyla (e.g., annelids and echinoderms), this cavity arises within the mesoderm itself and is completely enclosed within a thin lining called the **peritoneum**, which is derived from the mesoderm. Such a cavity is called a true **coelom** (eucoelom). Notice that the organs of the body are not actually free within the coelomic space itself, but are separated from it by the peritoneum (Figure 3.5C). Peritoneum is usually a squamous epithelial layer, at least that portion of it covering the gut and internal organs.

Several groups of triploblastic Metazoa (e.g. rotifers, roundworms, and others) possess small or large body cavities that are neither formed from the mesoderm nor fully lined by peritoneum or any other form of mesodermally derived tissue. Such a cavity used to be called a **pseudocoelom** (Greek *pseudo*, "false"; *coel*, "hollow, cavity") (Figure 3.5B). The organs of these animals actually lie free within the body cavity and are bathed directly in its fluid. In most cases the space represents persistent remnants of the embryonic blastocoel, and since there is nothing "false" about it, we use the more descriptive term **blastocoelom** in this text.

Within the constraints inherent in each of the basic body organizations discussed above, animals have evolved a multitude of variations on these themes. Each additional level of complexity that evolved opened new avenues for potential variation and adaptation. Throughout the remainder of this chapter we describe the fundamental organizational plans of major body systems as they have evolved within these basic bauplans. In subsequent chapters, we describe how members of the various phyla have modified these basic plans through their own particular evolutionary program or direction.

Locomotion and Support

As life progressed from the single-celled stage to multicellularity, body size increased dramatically. And this increase in body size, coupled with directed movement, was accompanied by the evolution of a variety of support structures and locomotor mechanisms. Because these two body systems evolved mutually and usually work in a complementary fashion, they are conveniently discussed together.

There are four fundamental locomotor patterns in protists and Metazoa: ameboid movement, ciliary and flagellar movement, hydrostatic propulsion, and locomotor limb movement. There are three basic kinds of support systems: structural endoskeletons, structural exoskeletons, and hydrostatic skeletons. In this section we briefly describe the basic architecture and mechanics of the various combinations of these systems.

Most invertebrates live in water, and aquatic environments present obstacles and advantages to support and locomotion that are quite different from those of terrestrial environments. Just staying in one place in the face of swiftly moving water, without being damaged or dislodged, requires both suport and flexibility. Animals moving through water (or moving water over their bodies—the effect is the same) face problems of fluid dynamics created by the interaction between a solid body and a surrounding liquid. What happens during this interaction is tied to the concept of **Reynolds number**, a unitless value based on the experiments of Osborne Reynolds (1842–1912). Reynolds number represents a

ratio of inertial force to viscous force. At higher Reynolds numbers, inertial force predominates and determines the behavior of water flow around an object. At lower Reynolds numbers, viscous force predominates and determines the behavior of the water flow. The importance of this concept is being increasingly recognized and applied to biological systems. Although there is still a great deal to be done in this area, some interesting generalizations can be made about locomotion of aquatic animals and, as we discuss later, aquatic suspension feeding. Reynolds number is expressed by the following equation:

$$R_e = \frac{plU}{v}$$

where *p* equals the density of fluid, *l* is some measurement of the size of the solid body, *U* equals the relative velocity of the fluid over the body surface, and *v* is the viscosity of the fluid. The formula was derived by Reynolds to describe the behavior of cylinders in water. Of course, since animals' bodies are not perfect cylinders, the size variable (*l*) is difficult to standardize. Nonetheless, meaningful relative values can be derived and applied to living creatures in water.

Without belaboring this issue beyond its importance here, it turns out that the problems of a large animal swimming through water are very different from those of a small animal. Large animals such as fishes, whales, or even humans, by virtue of their size or high velocity or both, move in a world of high Reynolds numbers. With increased body size, fluid viscosity becomes less and less significant as far as the animal's energy output during locomotion is concerned. At the same time, however, inertia becomes more and more important. A large animal must expend more energy than a small animal does to put its body in motion. But, by the same token, inertia works in favor of the moving large animal by carrying it forward when the animal stops swimming. When large animals move at high Reynolds numbers, the effect of inertia also imparts motion to the water around the animal's body. Thus, as the Reynolds number increases, a point is reached at which the flow of water changes from laminar to turbulent, decreasing swimming efficiency.

Small organisms generally move in a world of low Reynolds numbers. For example, a larva 1 mm in diameter, moving at a speed of 1 mm/sec, has a Reynolds number of about 1.0. Inertia and turbulence are virtually nonexistent, but viscosity becomes important—increasingly so as body size and velocity decrease (i.e., as the Reynolds number decreases). Small organisms swimming through water have been likened to a human swimming through liquid tar or thick molasses. The effect of this situation is that tiny creatures, such as ciliate and flagellate protists and many small Metazoa, start and stop instantaneously, and the motion of the water set up by their swimming also ceases immediate-

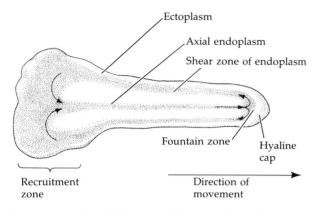

Figure 3.6 Ameboid locomotion: pseudopod formation in an ameba.

ly if the animal stops moving. Thus, small creatures neither pay the price nor reap the benefits of the effects of inertia. The organism only moves forward when it is expending energy to swim; as soon as it stops moving its cilia, or flagella, or appendages, it stops—and so does the fluid surrounding it. Tiny organisms swimming at low Reynolds numbers (i.e., <1.5) must expend an incredible amount of energy to propel themselves through their "viscous" surroundings.

Ameboid Locomotion

Ameboid movement is used principally by certain protists and by numerous kinds of ameboid cells that occur internally, *within* the bodies of most Metazoa. Ameboid cells possess a gel-like **ectoplasm**, which surrounds a more fluid **endoplasm** (Figure 3.6). Movement is facilitated by changes in the states of these regions of the cell. At one (or several) points on the cell surface, **pseudopodia** develop; and as endoplasm flows into a growing pseudopodium, the cell creeps in that direction. This seemingly simple process actually involves complex changes in cell fine structure, chemistry, and behavior. The innermost endoplasm moves "forward" while the outermost endoplasm takes on a granular appearance and remains fairly stable. The advancing portion of endoplasm pushes forward and then becomes semirigid ectoplasm at the tip of the advancing pseudopodium. Concurrently, endoplasm is recruited from the trailing end of the cell, from whence it streams forward to join in the "growing" pseudopodium.*

Although biologists have been studying ameboid locomotion for over 100 years, the precise mechanism is not yet fully understood. The molecular basis of ameboid movement may be essentially the same as that of vertebrate muscle contraction, involving actin, myosin, and ATP. Two principal theories exist to explain the process. Perhaps the more popular of the two ideas has the actin molecules floating freely in the endoplasm,

polymerizing into their filamentous form at the point of active pseudopodium growth, where they interact with myosin molecules. The resultant contraction literally pulls the streaming endoplasm forward, while at the same time converting it to the ectoplasm that rings the forward-streaming pseudopodium. The second theory suggests that the actin–myosin interaction takes place at the rear of the cell, where it produces a contraction of the ectoplasm. The contraction squeezes the cell like a tube of toothpaste, causing the endoplasm to stream forward and create a pseudopodium directly opposite the point of ectoplasmic contraction.

These and several other theories have been proposed to explain pseudopodial movement, but the definitive answer to the question of how a simple single-celled ameba moves remains elusive. Some modifications of pseudopodial movement are discussed in Chapter 5.

Cilia and Flagella

Cilia or **flagella** or both occur in virtually every animal phylum (with the qualified exception of the Arthropoda). Structurally, cilia and flagella are nearly identical (and clearly homologous), but the former are shorter and tend to occur in relatively larger numbers (in patches or tracts), whereas the latter are long and generally occur singly or in pairs. During cell development, each new flagellum or cilium arises from an organelle called the **basal body** (sometimes called a kinetosome or a blepharoplast), to which it remains anchored.

The movement of cilia and flagella creates a propulsive force that either moves the organism through a liquid medium or, if the animal (or cell) is anchored, creates a movement of fluid over it. Such action always occurs at very low Reynolds numbers. When the animal is large, the viscosity is increased by secretion of mucus, which lowers the Reynolds number. The general structure of a flagellum or cilium consists of a long, flexible rod, the outer covering of which is an extension of the plasma membrane of the cell (Figure 3.7A). Inside is a circle of nine paired **microtubules** (often called doublets) that runs the length of the flagellum or cilium. One microtubule of each doublet bears two rows of projections, the **dynein arms**, directed toward the adjacent doublet.[†] Flagella and cilia move as the microtubules slide up or down against one another, bending the flagellum or cilium in one direction or another. Microtubule sliding is driven by the dynein arms, particularly by protein complexes called **radial spokes** that arise from the arms. The radial spokes attach to each doublet microtubule immediately adjacent to the inner row of dynein arms and project centrally. Down the center of the doublet circle is an additional pair of microtubules. This familiar 9+2 pattern is characteristic of nearly all flagella and cilia (Figure 3.7B).

---

*Singular, pseudopodium; plural, pseudopodia. The diminutive is often used: singular, pseudopod; plural, pseudopods.

[†]Dyneins are a family of adenosine triphosphatases that cause microtubule sliding in ciliary and flagellar axonemes.

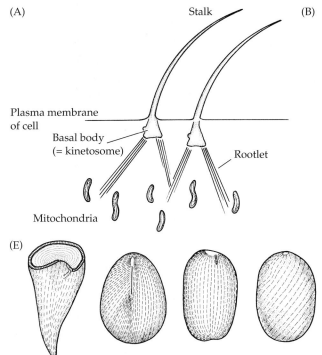

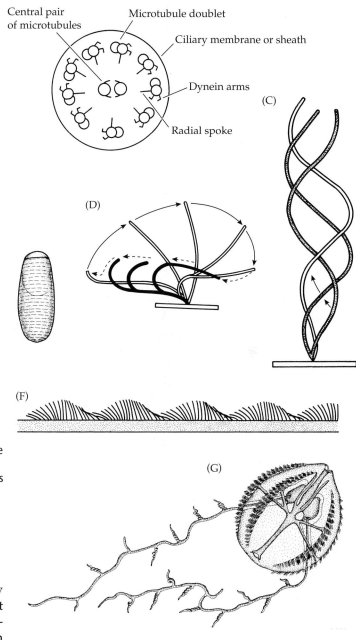

Figure 3.7 Cilia and flagella. (A) Structures of two adjacent cilia. (B) Cross section of a cilium. (C) Three successive stages in the undulatory movement of a flagellum. (D) Successive stages in the oarlike action of a cilium. The power stroke is shown in white, the recovery stroke in black. (E) Examples of ciliary tract patterns in various ciliate protists (tracts indicated by dashed lines). (F) Appearance of metachronal waves of a line of cilia. (G) The comb jellies (ctenophores) are the largest animals known to rely primarily on cilia for locomotion. Shown here is the rather small ctenophore, *Pleurobrachia bachei* (about two cm in diameter).

Flagellar/ciliary microtubules are modified hollow tubules similar to those present in the matrix of most cells. The principal function of these cellular tubules appears to be support. Just as the ectoplasm helps retain the shape and integrity of a protozoan cell (acting as a type of rudimentary "exoskeleton"), so the cytoplasmic microtubules act as a sort of simple "endoskeleton" that help protists (and other cells) retain their shape. Microtubules are also components of the spindle and so distribute the chromosomes during cell division. The movement of microtubules in dividing cells is being studied intensely and may help develop future models of both flagellar and cytoplasmic movement.

In addition to the locomotor function seen in some protists and small Metazoa, cilia and flagella have an enormous variety of functions in many other animals. For example, they create feeding and gas exchange currents; they line digestive tracts and facilitate food movement; and they propel sex cells and larvae. They also form sensory structures of many kinds. Here we focus on their use as locomotor structures.

Analysis by high-speed photography reveals that the movement of these structures is complex and differs among taxa, and even at different locations on the same organism. Some flagella beat back and forth, while others beat in a helical rotary pattern that drives flagellate protist cells something like the propeller of an outboard motor (Figure 3.7C). Depending on whether the undulation moves from base to tip or from tip to base, the effect will be, respectively, to push or pull the cell along.

Some flagella possess tiny, hairlike side branches called **mastigonemes** that increase the surface area and thus improve the propulsive capability. The beat of a cilium is generally simpler, consisting of a **power stroke** and a relaxed **recovery stroke** (Figure 3.7D). When many cilia are present on a cell, they often occur in distinct tracts, and their action is integrated, with beats usually moving in **metachronal waves** over the cell surface (Figure 3.7E,F). Since at any one time some cilia are always performing a power stroke, metachronal coordination ensures a uniform and continuous propulsive force.

It was once suggested that ciliary tracts on individual cells were coordinated by a primitive sort of cellular "nervous system," but this hypothesis was never confirmed. Current thinking suggests that the coordinated beating of cilia is probably due to hydrodynamic constraints imposed on them by the interference effects of the surrounding water layers and by the simple mechanical stimulation of moving, adjacent cilia. Nevertheless, some ciliary responses in animals are clearly under neural control, for example, reversal of power stroke direction.

Ciliated protists are the swiftest of the single-celled organisms. Flagellated protists are the next most rapid, and amebas are the slowest. Most amebas move at rates around 5 μm/sec (about 2 cm/hr), or about 100 times slower than most ciliates. Cilia are also used for locomotion by members of several metazoan groups (including the mesozoa, ctenophores, platyhelminths, rotifers, and some gastropods), and by the larval stages of many taxa.

Muscles and Skeletons

Almost all animals have some sort of a skeleton, the major functions of which are to maintain body shape, provide support, serve as attachments for muscles, transmit the forces of muscle contraction to perform work, and extend relaxed muscles. These functions may be attained either by hard tissues or secretions, or even by the turgidity of body fluids or tissues under pressure. Muscles, skeletons, and body form are closely integrated, both developmentally and functionally. When rigid skeletal elements are present, they can serve as fixed points for muscle attachment. For example, the rigid and jointed exoskeleton of arthropods allows for a complex system of levers that results in very precise and restricted limb movements. Many invertebrates lack hard skeletons and can change their body shape by alternate contraction and relaxation of various muscle groups attached to tough connective tissues or to the inside of the body wall. These "soft-bodied" invertebrates usually have a **hydrostatic skeleton**.

The hydrostatic skeleton. The performance of a hydrostatic skeleton is based on two fundamental properties of liquids: their incompressibility and their ability to assume any shape. Because of these features, body fluids transmit pressure changes rapidly and equally in all directions. It is important to realize a basic physical limitation concerning the action of muscles—they can only perform work by getting shorter (contracting).* To extend or protrude a body region,

the contractile force of a muscle is usually imparted to a fluid-filled body compartment, creating a hydrostatic pressure that displaces the wall of the compartment. Such indirect muscle actions can be compared to squeezing a rubber glove filled with water, thereby extending and stiffening the fingers. The enclosure of a fluid-filled chamber (e.g., a coelom) within sets of opposing muscle layers establishes a system in which muscles in one part of the body can contract, forcing body fluids into another region of the body, where the muscles relax; the body is thus extended or otherwise changed in shape.

In the most common plan, two muscle layers surround a fluid-filled body cavity, and the fibers of the layers run in different directions (i.e., a circular muscle layer and a longitudinal muscle layer). A soft-bodied invertebrate can move forward by using its hydrostatic skeleton in the following way. The circular muscles at the posterior end of the animal contract, so the hydrostatic pressure generated there pushes anteriorly to extend the relaxed longitudinal muscles of the front of the body. Then contraction of the posterior longitudinal muscles pulls the rear end of the body forward. This sequence of muscle contractions results in a directed and controlled movement forward. Such movement requires that the posterior end be anchored when the anterior end is extended, and that the anterior end be anchored when the posterior end is pulled forward. This system is commonly used for locomotion by many worms that generate posterior-to-anterior metachronal waves of muscle action, resulting in what is called **peristalsis**. A similar hydrostatic system can be used to temporarily or intermittently extend selected parts of the body, such as the feeding proboscis of most worms, tube feet of echinoderms, and siphons of clams.

The contraction of the circular muscles at one end of a vermiform animal may actually have four possible effects: the contracting end may elongate; the opposite end may elongate, or it may thicken; or both ends may elongate. The event that transpires depends not on the contraction of the circular muscles of the contracting end, but on the state of contraction of the longitudinal and circular muscles in other parts of the body (Figure 3.8). Such combinations of muscle contraction and relaxation create a versatile movement system based on relatively simple principles.

Reliance on only circular and longitudinal muscles could result in twists and kinks when a hydrostatic system engages itself against the resistance of the substratum. Hence, most animals that rely on hydrostatic movement also have helically wound, diagonal muscle fibers—a left- and right-handed set, intersecting at an angle between 0 and 180 degrees. The diagonal muscles allow extension and contraction, even at a constant volume, without stretching and while preventing kinking and twisting. A good analogy is the children's toy—the helically woven straw cylinder into which one young-

---

*Bear in mind that muscles can contract isometrically and do no work.

ster convinces another to insert his two index fingers; pushing your fingers together increases the diameter of the cylinder (and decreases the length), pulling them apart decreases the diameter (and increases the length). All of this is accomplished without an appreciable stretching or compression of the straw fibers (or the diagonal muscles of an invertebrate). When a cylinder is extended, the fiber angle decreases; when it is compressed, the angle increases.

The volume of the working fluid in a hydrostatic skeleton should remain constant; thus any leakage should not be greater than the rate at which the fluid can be replaced. Body fluids must be retained despite "holes" in the body wall, such as the excretory pores of many coelomate animals or the mouth openings of cnidarians. Such openings are often encircled by **sphincter muscles** that can close and control the loss of body fluids.

One way in which movement by a hydrostatic skeleton can be made more precise is to divide an animal up into a series of separate compartments. For example, in annelid worms, partitioning of the coelom and body muscles into segments with separate neural control enables body expansions and contractions to be confined to a few segments at a time. By manipulating particular sets of segmental muscles, most annelids not only can move forward and backward but can turn and twist in complex maneuvers.

The rigid skeleton. In "hard-bodied" invertebrates, a fixed or rigid skeletal system prevents the gross changes in body form seen in soft-bodied invertebrates. This trade-off in flexibility gives hard-bodied animals several advantages: the capacity to grow larger (an advantage that is especially useful in terrestrial habitats, which lack the buoyancy provided by aquatic environments), more precise or controlled body movements, better defense against predators, and often greater speed of movement.

Hard skeletons can be broadly classed as either **endoskeletons** or **exoskeletons**. Endoskeletons are generally derived from mesoderm, whereas exoskeletons are derived from ectoderm; both usually have organic and inorganic components. It has been hypothesized that rigid skeletons may have originated by chance, as by-products of certain metabolic pathways. By sheer accident (preadaptation, or exadaptation), for example, the accumulation of nitrogenous wastes and their incorporation into complex organic molecules might have resulted in the evolution of the chitinous exoskeleton so common among invertebrates. Similar speculation suggests that a metabolic system that originally functioned to eliminate excess calcium from the body might have produced the first calcareous shell of molluscs. In any event, marine invertebrates are capable of forming, through their various biological activities, a vast array of minerals, some of which cannot be formed inorganical-

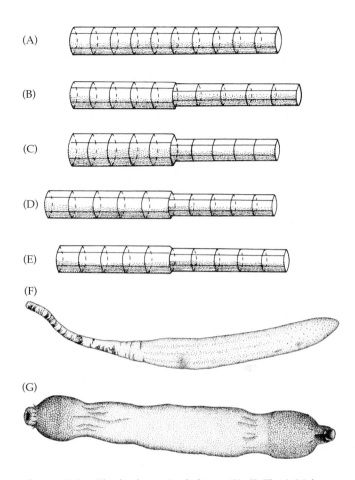

Figure 3.8 The hydrostatic skeleton. (A–E) The initial state and the four possible results of contraction of the circular muscles at one end of a cylindrical animal with a hydrostatic skeleton. (A) The muscles are all relaxed. (B) The circular muscles of the right-hand end have contracted and this end has elongated; the left-hand end has remained unaltered. (C) The length of the right-hand end has remained the same but the diameter of the left-hand end has increased. (D) the length of the right-hand end and the diameter of the left-hand end have remained the same, but the length of the left-hand end has increased. (E) The lengths of both ends have increased, but their respective diameters have remained the same. (F,G) Two animals that rely on hydrostatic skeletons for support and locomotion. (F) The sipunculan worm *Phascolosoma*. (G) The echiuran worm *Urechis caupo*.

ly in the biosphere. Indeed, the ever-increasing amounts of these **biominerals** have radically altered the character of the biosphere since the origin of hard skeletons in the earliest Cambrian. Most common among these biominerals are various carbonates, phosphates, halides, sulfates, and iron oxides.

Invertebrate skeletons may be of the articulating type (e.g., the exoskeletons of arthropods, clams, and brachiopods and the endoskeleton of some echinoderms), or they may be of the nonarticulating type, as seen in the simple one-piece exoskeletons of snails and the rigid

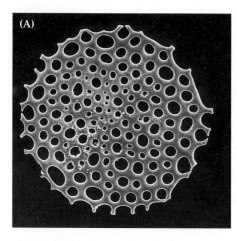

Figure 3.9 Some invertebrate endo-skeletons. (A) An ossicle (skeletal element) from a sea cucumber. (B) Isolated sponge spicules. (C) A deepwater glass sponge from the eastern Pacific; the long, siliceous spicules can be seen protruding from the body. (D) The rigid test of a sea urchin.

endoskeletons composed of interlocking fused plates of sea urchins and sand dollars. Animal endoskeletons may be as simple as the microscopic calcareous or siliceous spicules embedded in the body of a sponge, cnidarian, or sea cucumber, or they may be as complex as the bony skeleton of vertebrates (Figure 3.9). Hard skeletons of calcium carbonate have evolved in many animal (and some algal) phyla. Vertebrate skeletal tissues include a calcium phosphate-collagen matrix. In invertebrates, collagen often forms a substratum upon which calcareous spicules or other skeletal structures form, but with a single exception (certain gorgonians) collagen is never incorporated directly into the calcareous skeletal material.

In the broadest sense, virtually every group of invertebrates has developed an exoskeleton of sorts (Figure 3.10). Even cells of protists possess a semirigid ectoplasm, and some have surrounded themselves with a **test** comprising bits of sand or other foreign matter glued together. Other protists build a test made from chemicals that they either extract from sea water or produce themselves.

From their epidermis, many Metazoa secrete a nonliving external layer called the **cuticle**, which serves as an exoskeleton. The cuticle varies in thickness and complexity, but it often has several layers of differing structure and composition. In the arthropods, for example, the cuticle is a a complex combination of the polysaccharide **chitin*** and various proteins. This skeleton may be strengthened by the formation of internal cross-linkages (a process called tanning) and by the addition of calcium. In most insects, the outermost layer is impregnated with wax, which decreases its permeability to water. The cuticle is often ornamented with spines, tubercles, scales, or striations; frequently it is divided into rings or segments, a feature lending flexibility to the body. Other examples of exoskeletons are the calcareous shells of many molluscs and the casings of corals (Figure 3.10).

Most skeletons act as body elements against which muscles operate and by which muscle action is converted to body movement. Because muscles cannot elongate by themselves, they must be stretched by antagonistic forces—usually other muscles, hydrostatic forces, or elastic structures. In animals possessing rigid but articulated skeletons, antagonistic muscles often appear in pairs, for example, **flexors** and **extensors**. These muscles extend across a joint and are used to move a limb or

---

*The term **chitin** refers to a family of closely related chemical compounds, which, in various forms, are produced by and incorporated into the cuticles of many invertebrates. Certain types of chitin are also produced by some fungi and diatoms. Chitins are high-molecular-weight, nitrogenous polysaccharide polymers that are tough yet flexible (Figure 3.10F). In addition to its supportive and protective functions in the formation of exoskeletons, chitin is also a major component of the teeth, jaws, and grasping and grinding structures of a wide variety of invertebrates. That chitin is one of the most abundant macromolecules on Earth is evidenced by the estimated 10^{11} tons produced annually in the biosphere—most of it in the ocean.

(A)
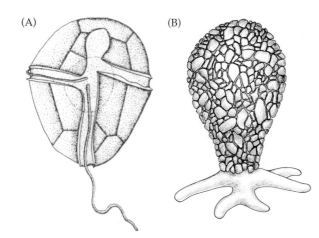

(B)

Figure 3.10 Some invertebrate exoskeletons. (A) The dinoflagellate protist *Gonyaulax*, encased in cellulose plates. (B) The ameba *Difflugia*, with a test of minute sand grains. (C) The foraminiferan *Cyclorbiculina*, with a calcareous, multichambered shell. (D) An assassin bug, with a jointed, chitinous exoskeleton. (E) The giant clam, *Tridacna*, among corals. These two very different animals both have calcareous exoskeletons. (F) Chemical structure of the polysaccharide chitin.

(C)

(D)
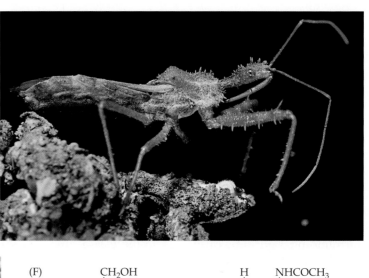

(E)

(F)
$$CH_2OH \qquad\qquad H \quad NHCOCH_3$$

and the forearm. Movement of a limb toward the body is brought about by flexor muscles, of which the biceps is an example (Figure 3.11A,B). The muscle antagonistic to the biceps is the triceps, an extensor muscle whose contraction extends the forearm away from the body. Other common sets of antagonistic muscles and actions are **protractors** and **retractors**, which respectively cause anterior and posterior movement of entire limbs at their place of juncture with the body; and **adductors** and **abductors**, which move a body part toward or away from a particular point of reference. Although vertebrates have endoskeletons and arthropods have exoskeletons, most muscles of arthropods are arranged in antagonistic sets similar to those seen in vertebrates (Figure 3.11C). The muscles of arthropods attach to the inside of the skeletal parts, whereas those of vertebrates attach to the outside, but they both operate systems of levers.

other body part (Figure 3.11). Most muscles have a discrete **origin**, where the muscle is anchored, and an **insertion**, which is the point of major body or limb movement. A classic vertebrate example of this system is the biceps muscle of the human arm, in which the origin is on the scapula and the insertion is on the radius bone of the forearm; contraction of the biceps causes flexion of the arm by decreasing the angle between the upper arm

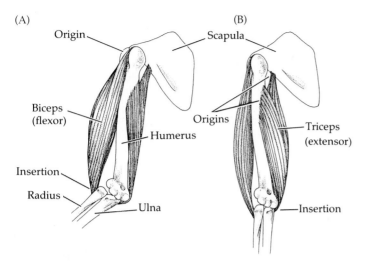

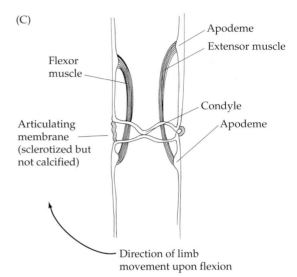

Figure 3.11 How antagonistic muscles work. (A) The biceps is contracted and the triceps relaxed; this combination flexes the forearm. (B) The biceps is relaxed and the triceps is contracted; this combination extends the forearm. (C) A diagrammatic representation of an arthropod joint, illustrating a similar relationship between flexor and extensor. In this animal, however, the muscles attach to the inside of the skeleton.

Not all muscles attach to rigid endo- or exoskeletons. Some form masses of interlacing muscle fibers, like those in the body wall of a worm, the foot of a snail, or the muscle layers in the walls of "hollow" organs (like those surrounding a gut tube or uterus). In these cases, the muscles have no definite origin and insertion but act on each other and the surrounding tissues and body fluids to affect changes in the shape of the body or body parts.

The basic physiology and biochemistry of muscle contraction is the same in vertebrates and invertebrates, although a variety of specialized variations on the basic model have evolved. For example, the adductor muscle of a clam (the muscle that holds the shell closed) is divided into two parts. One part is heavily striated and used for rapid shell closure (the phasic, or "quick" muscle); the other is smooth, or tonic, and is used to hold the shell closed for hours or even days at a time (the "catch" muscle). Brachiopods have a similar adductor muscle specialization—a good example of convergent evolution. Other specializations are found in crustacean muscle innervation, which differs from that typically seen in other invertebrates, and in certain insect flight muscles that are capable of contracting at frequencies far higher than can be induced by nerve impulses alone.

Feeding Mechanisms

Intracellular and Extracellular Digestion

Virtually all Metazoa and heterotrophic Protista must locate, select, capture, ingest, and finally digest and assimilate food. Although the physiology of digestion is similar at the biochemical level, considerable variation exists in the mechanisms of capture and digestion as a result of constraints placed on organisms by their overall bauplans.

Digestion is the process of breaking down food by hydrolysis into units suitable to the nutrition of cells. When this breakdown occurs outside the body altogether, it is called **extracorporeal digestion**; when it occurs in a gut chamber of some sort, it is referred to as **extracellular digestion**; and when the process occurs *within* a cell, it is called **intracellular digestion**. Regardless of the site of digestion, all organisms are ultimately faced with the fundamental challenge of cellular capture of nutritional products (food, digested or not). This cellular challenge is met by the process of **phagocytosis** (literally, "eating by cells") and **pinocytosis** ("drinking by cells"). These processes, collectively called **endocytosis**, are mechanically simple and involve the engulfment of food "particles" at the cell surface.

In 1892 the great comparative anatomist Elie Metchnikoff made a discovery that led to his receiving the Nobel Prize 16 years later. Metchnikoff discovered the process by which certain ameboid cells in the coelomic fluid of sea stars engulf and destroy foreign matter such as bacteria. He called this process **phagocytosis**. In phagocytosis, extensions of a cell's plasma membrane encircle the particle to be captured (whether it be food or a foreign microbe), form an inpocketing on the cell surface, and then pinch off the pocket inside the cell (Figure 3.12A). The resultant intracellular membrane-bounded structure is called a **food vacuole**. Because the food particle is inside a chamber formed and bounded by a piece of the original plasma membrane of the cell, some biologists consider that it is not actually "inside" the cell. This point is irrelevant. The plasma membrane surrounding the food vacuole is, of course, no longer part of the cell's outer membrane and in this sense it and whatever is in the vacuole are now "inside" the cell, and the subsequent digestive process-

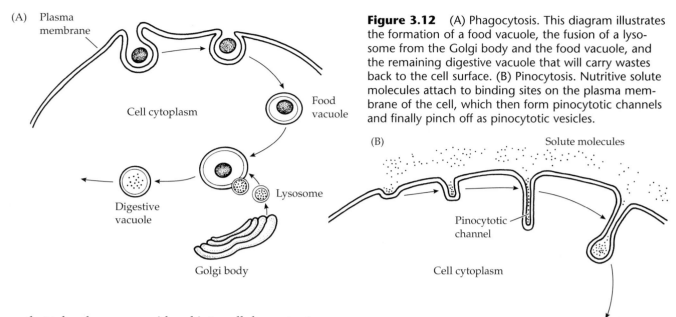

Figure 3.12 (A) Phagocytosis. This diagram illustrates the formation of a food vacuole, the fusion of a lysosome from the Golgi body and the food vacuole, and the remaining digestive vacuole that will carry wastes back to the cell surface. (B) Pinocytosis. Nutritive solute molecules attach to binding sites on the plasma membrane of the cell, which then form pinocytotic channels and finally pinch off as pinocytotic vesicles.

es that take place are considered intracellular, not extracellular. However, food inside the food vacuole is not actually incorporated into the cell's cytoplasm until it is digested and the resultant molecules are released.

Protists and sponges rely on phagocytosis as a feeding mechanism, and the digestive cells of metazoan guts take up food particles in the same fashion. Once a cell has phagocytosed a food particle and intracellular digestion has been completed, any remaining waste particles may be carried back to the cell surface by what remains of the old food vacuole, which fuses with the plasma membrane to discharge its wastes in a sort of reverse phagocytosis, called **exocytosis**.

Pinocytosis can be thought of as a highly specialized form of phagocytosis, in which molecule-sized particles are taken up by the cell. Such molecules are always dissolved in some fluid (e.g., a body fluid, or sea water). During pinocytosis, minute invaginations (**pinocytotic channels**) form on the cell surface, fill with liquid from the surrounding medium (which includes the dissolved nutritional molecules), and then pinch off to enter the cytoplasm as **pinocytotic vesicles** (Figure 3.12B). Pinocytosis generally occurs in cells lining some body cavity (e.g., the gut) in which considerable extracellular digestion has already taken place and nutritive molecules have been released from the original food source. In some cases, however, nutritional molecules may be taken up directly from sea water, and there is growing evidence that many invertebrates rely substantially on the direct uptake of dissolved organic matter (DOM) from their environment.

Metazoa generally possess some sort of an internal digestive tract into which food passes. In some (e.g., cnidarians and flatworms) there is only one opening through which food is ingested and undigested materials eliminated. These animals are said to have an **incomplete** or **blind gut**. Most other Metazoa have both mouth and anus (a complete gut), an arrangement that

allows the one-way flow of food and the specialization of different gut regions for functions such as grinding, secretion, storage, digestion, and absorption. As the noted biologist Libbie Hyman so aptly put it, "The advantages of an anus are obvious."

The overall anatomy and physiology of an animal's gut are closely tied to the type and quality of food consumed. In general, the guts of herbivores are long and often have specialized chambers for storage, grinding, and so on because vegetable matter is difficult to digest and requires long residence times in the digestive system. Carnivores tend to have shorter, simpler guts; the animal foods they consume are higher quality and easier to digest.

Feeding Strategies

Just as body architecture influences and limits the digestive modes of invertebrates, it is also intimately associated with the processes of food location, selection, and ingestion. Animals and animal-like protists are generally defined as **heterotrophic** organisms (as opposed to autotrophs and saprophytes); they ingest organic material in the form of other organisms, or parts thereof. However, in several groups of protists (e.g., many euglenoids and chlorophytans), both photosynthesis and heterotrophy can occur as nutritional strategies. In addition, many nonphotosynthetic invertebrate groups have developed intimate symbiotic relationships with single-celled algae, especially with certain species of dinoflagellates. These invertebrates use photosynthetic by-products as an accessory (or occasionally as the primary) food source. Notable in this regard are reef-building corals, giant clams (tridacnids), and certain flatworms, sea slugs, hydroids, ascidians, sea anemones, freshwater sponges, and even some species of *Paramecium*. However, the overwhelming majority of invertebrates lead strictly heterotrophic lives.

Biologists classify heterotrophic feeding strategies in a number of ways. For example, organisms can be considered **herbivores, carnivores,** or **omnivores**; or they can be classed as **grazers, predators,** or **scavengers**. Organisms can also be classified as **microphages** or **macrophages** by the comparative size of their food or prey, or they can be classified by the environmental source of their food as **suspension feeders, deposit feeders,** or **detritivores**. In the remainder of this section we define some important feeding-strategy terms and explain some common themes of feeding.

Few animals are strictly herbivores or carnivores, even though most show a clear preference for either a vegetable or a meat diet. For example, the Atlantic purple sea urchin *Arbacia punctulata* usually feeds on micro- and macroalgae. However, in certain portions of its range, where algae may become seasonally scarce, epifaunal animals constitute the bulk of this urchin's diet. Omnivores, of course, must have the anatomical and physiological capability to capture, handle, and digest both plant and animal material. Among invertebrates, there are two large categories of feeding strategies in which omnivory prevails: suspension feeding and deposit feeding.

Suspension feeding. **Suspension feeding** is the removal of suspended food particles from the surrounding medium by some sort of capture, trapping, or filtration mechanism. It has three basic steps: transport of water past the feeding structures, removal of particles from the water, and transport of the captured particles to the mouth. It is a major mode of feeding in sponges, ascidians, appendicularians, brachiopods, ectoprocts, entoprocts, phoronids, most bivalves, and many crustaceans, polychaetes, and gastropods. The main food selection criterion is particle size, and the size limits of food are determined by the nature of the particle-capturing device. In some cases potential food particles may also be "sorted" on the basis of their specific gravity, or even their perceived nutritional quality.

Suspension-feeding invertebrates generally consume bacteria, phytoplankton, zooplankton, and some detritus. All suspension feeders probably have optimal ranges of particle size; but some are capable, experimentally, of preferentially selecting "enriched" artificial food capsules over "nonenriched" (nonfood) capsules, an observation suggesting that chemosensory selectivity may occur *in situ* as well. To capture food particles from their environment, suspension feeders either must move part or all of their body through the water, or water must be moved over their feeding structures. As with locomotion in water, the relative motion between a solid and liquid during suspension feeding creates a system that behaves according to the concept of Reynolds numbers. Virtually all suspension-feeding invertebrates capture particles from the water at low Reynolds numbers. The flow rates in such systems are very low and the feeding structures are small (e.g., cilia, flagella, setae).

Recall that at low Reynolds numbers viscous forces dominate, and water flow over small feeding structures is laminar and nonturbulent and ceases instantaneously when energy input stops. Thus, in the absence of inertial influence, suspension feeders that generate their own feeding currents expend a great deal of energy. Some suspension feeders conserve energy by depending to various degrees on prevailing ambient water movements to continually replenish their food supplies (e.g., barnacles on wave-swept shores and mole crabs in the wash zone on sandy beaches). For most organisms, however, the effort expended for feeding is a major part of their energy budget.

Only a relatively few suspension feeders are true filterers. Because of the principles outlined above, it is energetically extremely costly to drive water through a fine-meshed filtering device. For small animals, this is somewhat analogous to moving a fine-mesh filter through thick syrup. Such actual sieving does occur, most notably in many bivalve molluscs, many tunicates, some larger crustaceans, and some worms that produce mucous nets. However, most suspension feeders employ a less expensive method of capturing particles from the water, one that does not involve continuous filtration. Many invertebrates expose a sticky surface, such as a coating of mucus, to flowing water. Suspended particles contact and adhere to the surface and then are moved to the mouth by ciliary tracts (as in crinoids), setal brushes (as in certain crustaceans), or by some other means of transport. Other "contact" suspension feeders living in still water may simply expose a sticky surface to the rain of particulate material settling downward from the water above, thus letting gravity do much of the work of food-getting. Some oysters are suspected of this feeding strategy, at least on a part-time basis. Several other "contact" methods of suspension feeding may occur, but all eliminate the costly activity of actual sieving in the highly viscous world of low Reynolds numbers.

Another nonfiltering suspension feeding method is called "scan-and-trap" (LaBarbera 1984). The general strategy here is to move water over part or all of the body, detect suspended food particles, isolate the particles in a small parcel of water, and process only that parcel by some method of particle extraction. The animal thus avoids the energetic expense of continuously driving water over the feeding surface at low Reynolds numbers. The precise methods of particle detection, isolation, and capture vary among different invertebrates that use the scan-and-trap technique; but this basic strategy is probably employed by certain crustaceans (e.g., planktonic copepods), many ectoprocts, and a variety of larval forms.

(A)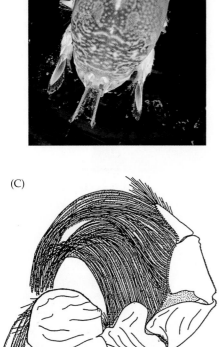

(B)

(C)

(D)

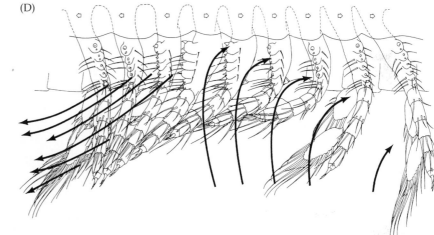

Figure 3.13 Setal-net suspension-feeding invertebrates. (A) The sand crab *Emerita*. (B) A goose barnacle, *Pollicipes,* with feeding appendages extended. (C) The third maxilliped of the porcelain crab *Petrolisthes elegans*. Note the long, dense setae used in feeding. (D) A portion of the trunk (sagittal view) of a cephalocarid crustacean during the metachronal cycle of the feeding limbs. The arrows indicate the direction of water currents; the arrow above each trunk limb indicates the limb's direction of movement.

The trick of removing small food particles from the surrounding environment is achieved through four fundamentally different mechanisms. Because there are a limited number of ways in which animals can suspension feed, it is not surprising that a great deal of evolutionary convergence has appeared among their feeding mechanisms.

Among some crustaceans, certain limbs are equipped with rows of feather-like setae adapted for removing particles from the water (Figure 3.13). The size of the particles captured is often directly proportional to the "mesh" size of the interlaced setae on the food-capture structure. In sessile crustaceans such as barnacles, the feeding appendages are swept through the water or held taut against moving water. In either case, sessile animals are dependent upon local currents to continually replenish their food supply. Motile **setal-net feeders**, like many larger planktonic crustaceans and certain benthic crustaceans (e.g., porcelain crabs), may have modified appendages that generate a current across the feeding appendages that bear the capture setae. Sometimes these same appendages serve simultaneously for locomotion. In cephalocarid and many branchiopod crustaceans, for example, complex coordinated movements of the highly setose thoracic legs propel the animal forward and also produce a constant current of water (Figure 3.13D). These appendages simultaneously capture food particles from the water and collect them in a median ventral food groove at the leg bases, where they are passed forward to the mouth region.

A second suspension-feeding device is the **mucous net**, or **mucous trap**, wherein patches or a sheet of mucus are used to capture suspended food particles. Most mucous-net feeders consume their net along with the food and recycle the chemicals used to produce it. Again, sessile and sedentary species often rely largely on local currents to keep a fresh supply of food coming

(A)

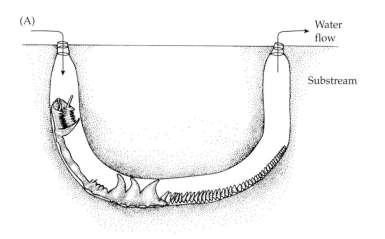

Water flow

Substream

Figure 3.14   Some mucous-net and ciliary-mucus suspension feeders. (A) The annelid worm *Chaetopterus* in its burrow. Note the direction of water flow through its mucous net. (B) The solitary ascidian *Styela* has incurrent and excurrent siphons through which water enters and leaves the body. Inside, the water passes through a sheet of mucus covering holes in the wall of the pharynx. (C) A maldanid polychaete, *Praxillura maculata*. This animal constructs a membranous tube that bears 6–12 stiff radial spokes. A mucous web hangs from these spokes and passively traps passing food particles. The worm's head is seen sweeping around the radial spokes to retrieve the mucous web and its trapped food particles.

(B)

(C)

their way. Some, however, especially benthic burrowers, actively pump water through their burrow or tube, where it passes across or through the mucous sheet. A classic example of mucous-net feeding is seen in the annelid worm *Chaetopterus* (Figure 3.14A). This animal lives in a U-shaped tube in the sediment and pumps water through the tube and through a mucous net. As the net fills with trapped food particles, it is periodically manipulated and rolled into a ball, which is then passed to the mouth and swallowed. An example of mucous-trap feeding is seen the tube-building gastropods (family Vermetidae). These wormlike snails construct colonies of meandering calcareous tubes in the intertidal zone. Each animal secretes a mucous trap that is deployed just outside the opening of the tube, until nearly the entire colony surface is covered with mucus. Suspended particulate matter settles and becomes trapped in the mucus. At periodic intervals, each animal withdraws its mucous sheet and swallows it, whereupon a new sheet is immediately constructed.

Another type of suspension feeding is the **ciliary-mucous mechanism**, in which rows of cilia carry a mucous sheet across some structure while water is passed through or across it. Ascidians (sea squirts; Figure 3.14B) move a more or less continuous mucous sheet across their sievelike pharynx, while at the same time pumping water through it. Fresh mucus is secreted at one side of the pharynx while the food-laden mucus at the other

side is moved into the gut for digestion. Several polychaete groups also make use of the ciliary-mucous feeding technique (Figure 3.14C). For example, some species of tube-dwelling fan worms feed with a crown of tentacles that are covered with cilia and mucus and bear ciliated grooves that slowly move captured food particles to the mouth. Many sand dollars capture suspended particles, especially diatoms, on their mucus-covered spines; food and mucus are transported by the tube feet and ciliary currents to food tracts, and then to the mouth.

Still another kind of suspension feeding is **tentacle** or **tube feet suspension feeding**. In this strategy, some sort of tentacle-like structure captures larger food particles, with or without the aid of mucus. Food particles captured by this mechanism are generally larger than those captured by setal or mucous traps or sieves. Examples of tentacle or tube feet suspension feeding are most commonly encountered in the echinoderms (e.g., many brittle stars and crinoids) and cnidarians (e.g., certain sea anemones and corals) (Figure 3.15).

Figure 3.15 Tube feet suspension feeding. Food-particle capture in the brittle star *Ophiothrix fragilis.* The photographs show two views of a captured food particle being transported by the arm tentacles to the mouth.

Much research has been done on suspension feeding in the past 20 years, and we now know what size range of particles many animals feed on and what kinds of capture rates they have. In general, feeding rates increase with food particle concentration to a plateau, above which the rate levels off. At still higher particle concentrations, entrapment mechanisms may become overtaxed or clogged and feeding is inhibited or simply ceases. In sessile and sedentary suspension feeders, for example, pumping rates decrease quickly as the amount of suspended inorganic sediment (mud, silt, and sand) increases beyond a given concentration. For this reason, the amount of sediment in coastal waters limits the distribution and abundance of certain invertebrates such as clams, corals, sponges, and ascidians. Many tropical coral reefs are dying as a result of increased coastal sediment loads generated by run-off from land areas subjected to deforestation or urban development.

Deposit feeding. The deposit feeders make up another major group of omnivores. These animals obtain nutrients from the sediments of soft-bottom habitats (muds and sands) or terrestrial soils, but their techniques for feeding are diverse. **Direct deposit feeders** simply swallow large quantities of sediment—mud, sand, soil, organic matter, everything. They may consume up to 500 times their body weight daily. The usable organics are digested and the unusable materials passed out the anus. The resultant fecal material is essentially "cleaned dirt." This kind of deposit feeding is seen in many polychaete annelids (Figure 3.16A), some snails, some sea urchins, and most earthworms.

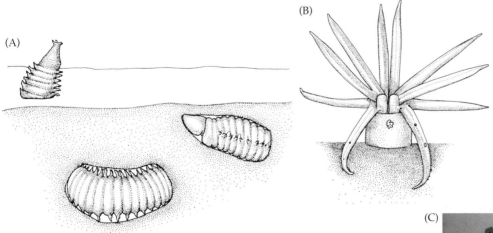

Figure 3.16 Some deposit feeding invertebrates. (A) A lumbrinerid polychaete burrowing in the sediment. This worm is a subsurface deposit feeder. (B) The sabellid polychaete *Manayunkia aestuarina* in its feeding posture. A pair of branchial filaments are being used to feed. The large particle falling in front of the tube has just been expelled from the branchial crown by a rejection current. (C) A surface deposit-feeding holothurian (*Euapta*).

(A)

(B)

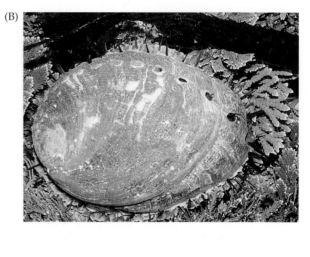

(C)

(D)

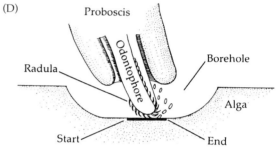

Figure 3.17 Representative herbivorous invertebrates. (A) The common land snail *Helix*, munching on some foliage. (B) The red abalone *Haliotis rufescens*. (C) The radula, or rasping organ, of *H. rufescens*. (D) The action of a radula (sagittal section). (E) The tropical Pacific sea urchin *Toxopneustes roseus*.

(E)

Some deposit feeders utilize tentacle-like structures to consume sediment, such as some sea cucumbers, most sipunculans, certain clams, and several types of polychaetes (Figure 3.16B,C). Tentacle-utilizing deposit feeders preferentially remove only the uppermost deposits from the sediment surface and thus consume a far greater percentage of living (especially bacteria, diatoms, and protozoa) and detrital organic material that accumulates there than do the burrowing deposit feeders. These animals are generally called **selective deposit feeders**. Aquatic deposit feeders may also rely to a significant extent on fecal material that accumulates on the bottom, and many will actively consume

their own fecal pellets (**coprophagy**), which may contain some undigested or incompletely digested organic material as well as microorganisms. Studies have shown that only about half of the bacteria ingested by marine deposit feeders is digested during passage through the gut. In all cases, deposit feeders are microphagous.

The ecological role of deposit feeding in sediment turnover is a critical one. When burrowing deposit feeders are removed from an area, organic debris accumulates, subsurface oxygen is depleted by bacterial decomposition, and anaerobic sulfur bacteria eventually bloom. On land, earthworms and other burrowers are important in maintaining the health of agricultural and garden soils.

Herbivory. The following discussion deals with **macroherbivory**, or the consumption of macroscopic plants. Herbivory is common throughout the animal kingdom. It is most dramatically illustrated when certain invertebrate herbivores undergo a temporary population explosion. Famous examples are outbreaks of locust, which can destroy virtually all plant material in their path of migration. In a similar fashion, herbivory by extremely high numbers of the Pacific sea urchin *Strongylocentrotus* results in the wholesale destruction of kelp beds. Unlike suspension- and deposit-feeding herbivory, in which mostly single-celled and microscopic plant matter is consumed, macroherbivory requires the ability to "bite and chew" large pieces of vegetable matter. Although the evolution of biting and chewing mechanisms has taken place within the architectural framework of a number of different invertebrate lineages, it is always characterized by the development of hard (usually calcified or chitinous) "teeth," which are manipulated by powerful muscles. Members of a number of major invertebrate taxa have evolved macroherbivorous lifestyles, including molluscs, polychaetes, arthropods, and sea urchins.

Most molluscs have a unique structure called a **radula**, which is a muscularized, beltlike rasp armed with chitinous teeth. Herbivorous molluscs use the radula to scrape algae off rocks or to tear pieces of algal fronds or the leaves of terrestrial plants. The radula acts like a curved file that is drawn across the feeding surface (Figure 3.17C,D). Some polychaetes such as nereids (family Nereidae) have sets of large chitinous teeth on an eversible pharynx or proboscis. The proboscis is protracted by hydrostatic pressure, exposing the teeth, which by muscular action tear or scrape off pieces of algae that are swallowed when the proboscis is retracted. As might be expected, the toothed pharynx of polychaetes is also suited for carnivory, and some primarily herbivorous polychaetes can switch to meat-eating when algae are scarce.

Macroherbivory in arthropods is best illustrated by certain insects and crustaceans. Both of these large groups have powerful mandibles capable of biting off pieces of plant material and subsequently grinding or chewing them before ingestion. Some macroherbivorous arthropods are able to temporarily switch to carnivory when necessary. This switching is rarely seen in the terrestrial herbivores because it is almost never necessary; terrestrial plant matter can almost always be found. In marine environments, however, algal supplies may at times be very limited. Some herbivorous invertebrates cause serious damage to wooden man-made structures (like homes, pier pilings, and boats) by burrowing through and consuming the wood (Figure 3.18).

Carnivory and scavenging. The most sophisticated methods of feeding are those that require the active capture of live animals, or **predation**.* Most carnivorous predators will, however, consume dead or dying animal matter when live food is scarce. Only a few generalizations about the many kinds of predation are

---

*Although in the broad sense even herbivory is a form of "predation," for clarity of discussion we restrict the use of these terms to vegetable eating and animal eating, respectively.

Figure 3.18 Wood from the submerged part of an old dock piling, split open to show the work of the wood-boring bivalve shipworm *Teredo navalis*. The shell valves are so reduced that they can no longer enclose the animal; instead they are used as "auger blades" in boring. The walls of the burrow are lined with a smooth, calcareous, shell-like material.

presented here; detailed discussions of various taxa are presented in their appropriate chapters.

Active predation often involves five recognizable steps: prey location (predator orientation), pursuit (usually), capture, handling, and finally, ingestion. Prey location usually requires a certain level of nervous system sophistication in which specialized sense organs are present (discussed later in this chapter). Many carnivorous invertebrates rely primarily on chemosensory location of prey, although many also use visual orientation, touch, and vibration detection. Chemoreceptors tend to be equally distributed around the bodies of radially symmetrical carnivores (e.g., jellyfish) but, coincidentally with cephalization, most invertebrates have their gustatory and olfactory receptors ("tasters" and "smellers") concentrated in the head region.

Predators may be classified by how they capture their prey—as **motile stalkers, lurking predators** (ambushers), **sessile opportunists**, or **grazers** (Figure 3.19). Stalkers actively pursue their prey; they include members of such disparate groups as ciliate protists, polyclads, nemerteans, polychaete worms, gastropods, octopuses and

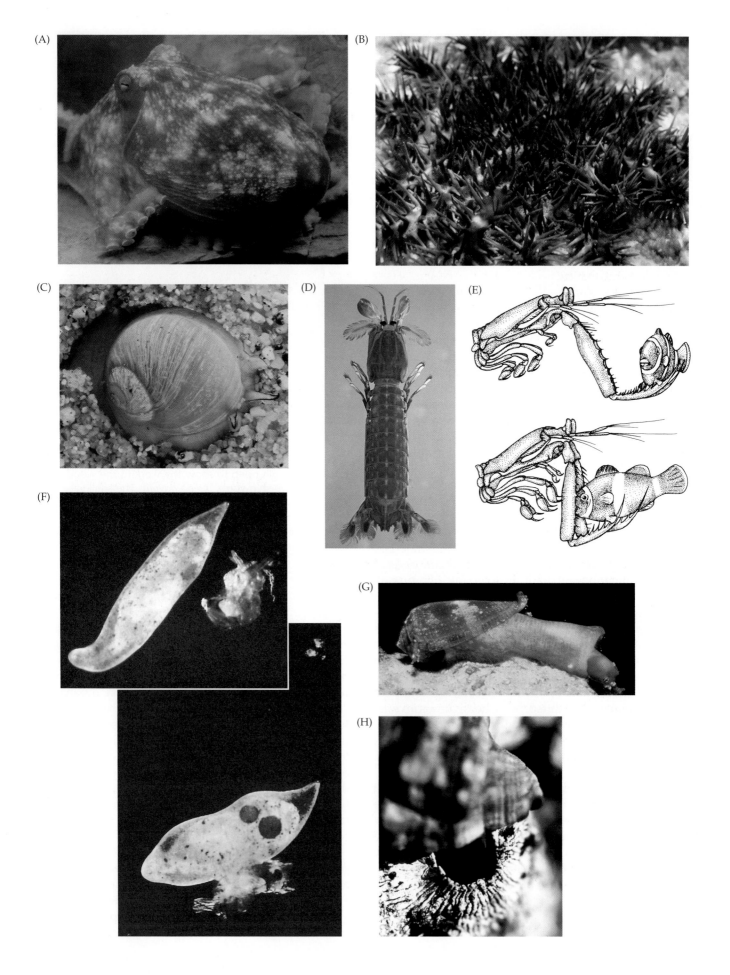

◀ Figure 3.19 Some predatory invertebrates. (A) Most octopuses are active hunting predators; this one is a member of the genus *Eledone*. (B) The crown-of-thorns sea star, *Acanthaster*, feeds on corals. (C) The moon snail, *Polinices*, drills holes in the shells of bivalve molluscs to feed on the soft parts. (D) A mantis shrimp (stomatopod); the two drawings (E) depict its raptorial strike to capture a passing fish. (F) The predatory flatworm *Mesostoma* attacking a mosquito larva. (G) A cone snail (*Conus*) eating a fish. (H) *Acanthina*, a predatory gastropod feeding on small barnacles.

squids, crabs, and sea stars. In all these groups, chemosensation is highly important in locating potential prey, although some cephalopods are known to be the most highly visual of all the invertebrate predators.

Lurking predators are those that sit and wait for their prey to come within capture distance, whereupon they quickly seize the victim. Many lurking predators, such as certain species of mantis shrimps (stomatopods), crabs, snapping shrimp (Alpheidae), spiders, and polychaetes, live in burrows or crevices from which they emerge to capture passing prey. There are even ambushing planarian flatworms, which produce mucous patches that form sticky traps for their prey. The cost of building traps is significant. Ant lions, for example, may increase their energy consumption as much as eightfold when building their sand capture pits, and energy lost in mucus secretion by planarians may account for 20 percent of the worm's energy. Predatory invertebrates, especially lurking predators, tend to be more or less territorial.

Sessile opportunists operate in much the same fashion as lurking predators do, but they lack the mobility of the latter. The same may be said for drifting opportunists, such as jellyfishes. Many sessile predators, such as some protists, barnacles, and cnidarians, are actually suspension feeders with a strong preference for live prey.

Grazing carnivores move about the substratum picking at the epifauna. Grazers may be indiscriminate, consuming whatever happens to be present, or they may be fairly choosy about what they eat. In either case, their diet consists largely of sessile and slow-moving animals, such as sponges, ectoprocts, tunicates, snails, small crustaceans, and worms. Most grazers are omnivorous to some degree, consuming plant material along with their animal prey. Many crabs and shrimps are excellent grazers, continuously moving across the bottom and picking through the epifauna for tasty morsels. Sea spiders (pycnogonids) and some carnivorous sea slugs can also be classed as grazers on hydroids, ectoprocts, sponges, tunicates, and other sessile epifauna. Ovulid snails (family Ovulidae) inhabit, and usually mimic, the gorgonians and corals upon which they slowly crawl about, nipping off polyps as they go.

One special category of carnivory is **cannibalism**, or intraspecific predation. Gary Polis (1981) examined over 900 published reports describing cannibalism in about 1,300 different species of animals. In general, he found that species of large animals (and also larger individuals in any given species) are the most likely to be cannibals. By far, the majority of the victims are juveniles. However, in a number of invertebrate groups the tables turn and cannibalism occurs when smaller individuals band together to attack and consume a larger individual. Furthermore, females tend generally to be more cannibalistic than males, and males tend to be eaten far more often than females. In many species, filial cannibalism is common, in which a parent eats its dying, deformed, weak, or sick offspring. Polis concluded that cannibalism is a major factor in the biology of many species and may influence population structure, life history, behavior, and competition for mates and resources. He goes so far as to point out that *Homo sapiens* may be "the only species capable of worrying whether its food is intra- or extraspecific."

Dissolved organic matter. The total living biomass of the world's oceans is estimated to be about 2×10^9 tons of organic carbon (roughly 500 times the amount of organic carbon in the terrestrial environment). Furthermore, an additional 20×10^9 tons of particulate organic matter is estimated to occur in the seas, and another 200×10^9 tons of organic carbon (C) may occur in the seas as **dissolved organic matter** (**DOM**). Thus, at any moment in time, only a small fraction of the organic carbon in the world's seas actually exists in living organisms. Amino acids and carbohydrates are the most common dissolved organics. Typical oceanic values of DOM range from 0.4 to 1.0 mg C/liter, but may reach 8.0 mg C/liter near shore. Pelagic and benthic algae release copious amounts of DOM into the environment, as do certain invertebrates. Coral mucus, for example, is an important fraction of suspended and dissolved organic material over reefs, and it contains significant amounts of energy-rich and nitrogen-rich compounds, including mono- and polysaccharides and amino acids. Other sources of DOM include decomposing tissue, detritus, fecal material, and metabolic by-products discharged into the environment.

The idea that DOM may contribute significantly to the nutrition of marine invertebrates dates at least from the turn of the century. However, after nearly 100 years of research the issue is still not fully resolved. Marine microorganisms are well known to use DOM, but the relative role of dissolved organic matter in the nutrition of aquatic Metazoa is problematic. Available data strongly suggest that members of all marine taxa (except perhaps arthropods and vertebrates) are capable of absorbing DOM to some extent, and in the case of ciliary-mucous suspension feeders, marine larvae, many echinoderms, and mussels, the ability to rapidly take up dissolved free amino acids from a dilute external medium is well established. But because of the complex chemical nature of dissolved organics, and the difficulty of measuring their rates of influx and loss, we still lack strong evidence of the actual use, or relative nutritional importance, of DOM to invertebrates.

Evidence from numerous studies indicates that absorption of DOM occurs directly across the body wall of invertebrates, as well as via the gills. Also, inorganic particles of colloidal dimensions provide a surface on which small organic molecules are concentrated by adsorption, thereon to be captured and utilized by suspension-feeding invertebrates. Interestingly, most freshwater organisms seem incapable of removing small organic molecules from solution at anything like the rates characteristic of marine invertebrates. In fresh water, the uptake of DOM is probably retarded by the processes of osmoregulation. Also, with the exception of the aberrant hagfish, marine vertebrates seem not to utilize DOM to any significant extent.

Chemoautotrophy. A special form of autotrophy that occurs in certain bacteria relies not on sunlight and photosynthesis as a source of energy to make organic molecules from inorganic raw materials (**photoautotrophy**), but rather on the oxidization of certain inorganic substances. This special case is called **chemoautotrophy**. Chemoautotrophs use CO_2 as their carbon source, obtaining energy by oxidizing hydrogen sulfide (H_2S), ammonia (NH_3), methane (CH_4), ferrous ions (Fe^{2+}), or some other chemical, depending on the species. These prokaryotes are not uncommon in aerated soils, and certain species live as symbionts in the tissues of a few marine invertebrates.

Some of the most interesting of these chemoautotrophic organisms derive their energy from the oxidation of hydrogen sulfide released at hot water vents on the deep-sea floor—where, in fact, they are the sole primary producers in the ecosystem. In this environment, chemoautotrophic bacteria inhabit the tissues of certain mussels, clams, and vestimentiferan tube worms, where they produce organic compounds that are utilized by their hosts. Similar invertebrate–bacteria relationships have recently been discovered in shallow cold-water petroleum and salt (brine) seeps, where the chemoautotrophic microorganisms live off the methane- and hydrogen sulfide-rich waters associated with such sea floor phenomena. In all these cases, the bacteria actually live within the cells of their hosts. In bivalves, the bacteria inhabit the gill cells and extract methane or other chemicals from the water that flows by those structures In the case of the tube worms, the host must transport the H_2S to their bacterial partners, which live in tissues deep within the animals' body. The worms have a unique type of hemoglobin that transports not only oxygen (for the worm's metabolism) but sulfide as well.

Excretion and Osmoregulation

Excretion is the elimination from the body of metabolic waste products, including carbon dioxide and water (produced primarily by cellular respiration) and excess nitrogen (produced as ammonia from deamination of amino acids). The excretion of respiratory CO_2 is generally accomplished by structures that are separate from those associated with other waste products and is discussed in the section that follows.

The excretion of nitrogenous wastes is usually intimately associated with **osmoregulation**—the regulation of water and ion balance within the body fluids—so these processes are considered together here. Excretion, osmoregulation, and ion regulation serve not only to rid the body of potentially toxic wastes, but also to maintain concentrations of the various components of body fluids at levels appropriate for metabolic activities. As we shall see, these processes are structurally and functionally tied to the overall level of body complexity and construction, the nature of other physiological systems, and the environment in which an animal lives. We again emphasize the necessity of looking at whole animals, the integration of all aspects of their biology and ecology, and the possible evolutionary histories that could have produced compatible and successful combinations of functional systems.

Nitrogenous Wastes and Water Conservation

The source of most of the nitrogen in an animal's system is amino acids produced from the digestion of proteins. Once absorbed, these amino acids may be used to build new proteins, or they may be deaminated and the residues used to form other compounds (Figure 3.20). The excess nitrogen released during deamination is typically liberated from the amino acid in the form of **ammonia** (NH_3), a highly soluble but quite toxic substance that either must be diluted and eliminated quickly or converted to a less toxic form. The excretory products of vertebrates have been studied much more extensively than those of invertebrates, but the available data on the latter allow some generalizations. Typically, one nitrogenous waste form tends to predominate in a given species, and the nature of that chemical is generally related to the availability of environmental water.

The major excretory product in most marine and freshwater invertebrates is ammonia, since their environment provides an abundance of water as a medium for rapid dilution of this toxic substance. Such animals are said to be **ammonotelic**. Being highly soluble, ammonia diffuses easily through fluids and tissues, and much of it is lost straight across the body walls of some ammonotelic animals. Animals that do not possess definite excretory organs (e.g., sponges, cnidarians, and echinoderms) are more or less limited to the production of ammonia and thus are restricted to aquatic habitats.

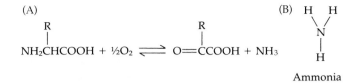

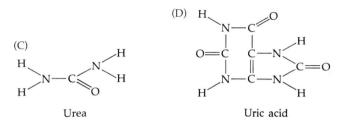

Ammonia

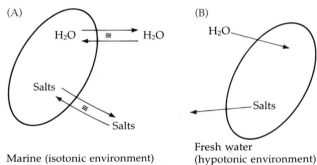

Urea   Uric acid

Figure 3.20 Nitrogenous waste products. (A) The general reaction for deamination of an amino acid producing a keto acid and ammonia. (B–D) the structures of three common excretory compounds. (B) Ammonia. (C) Urea. (D) Uric acid.

Terrestrial invertebrates (indeed, all land animals) have water conservation problems. They simply cannot afford to lose much body water in the process of diluting their wastes. These animals convert their nitrogenous wastes to more complex but far less toxic substances. These compounds are energetically expensive to produce, but they often require relatively little or no dilution by water, and they can be stored within the body prior to excretion.

There are two major metabolic pathways for the detoxification of ammonia: the urea pathway and the uric acid pathway. The products of these pathways, **urea** and **uric acid**, are illustrated in Figure 3.20, along with ammonia for comparison. **Ureotelic** animals include amphibians, mammals, and cartilaginous fishes (sharks and rays); urea is a relatively rare and insignificant excretory compound among invertebrates. On the other hand, the ability to produce uric acid is critically associated with the success of certain invertebrates on land. **Uricotelic** animals have capitalized on the relative insolubility (and very low toxicity) of uric acid, which is generally precipitated and excreted in a solid or semisolid form with little water loss. Most land-dwelling arthropods and snails have evolved structural and physiological mechanisms for the incorporation of excess nitrogen into molecules of uric acid. We emphasize that various combinations of these and other forms of nitrogen excretion are found in most animals. In some cases, individual animals can actually vary the proportion of these compounds they produce, depending on short-term environmental changes affecting water loss.

Osmoregulation and Habitat

In addition to its relationship to excretion, osmoregulation is directly associated with environmental condi-

tions. As mentioned in Chapter 1, the composition of sea water and that of the body fluids of most invertebrates is very similar, both in terms of total concentration and the concentrations of many ions. Thus, the body fluids of many marine invertebrates and their habitats are close to being **isotonic**. We hasten to add, however, that probably no animal has body fluids that are exactly isotonic with sea water, and therefore all are faced with the need for some degree of ionic and osmoregulation. Nonetheless, marine invertebrates certainly do not face the extreme osmoregulatory problems encountered by land and freshwater forms.

As shown in Figure 3.21, the body fluids of freshwater animals are strongly **hypertonic** with respect to their environment, and thus they face serious problems of water influx as well as the potential loss of precious body salts. Terrestrial animals are exposed to air and thus to problems of water loss. The evolutionary invasion of land and fresh water was accompanied by the development of mechanisms that solved these problems, and only a relatively small number of invertebrate groups have managed to do this. Animals inhabiting

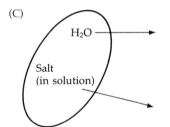

Marine (isotonic environment)   Fresh water (hypotonic environment)

Terrestrial ("dry" environment)

Figure 3.21 Relative osmotic and ionic conditions existing between marine, freshwater, and terrestrial invertebrates and their environments. The arrows indicate the directions in which water and salts move passively in response to concentration gradients. Remember that in each of these cases movement occurs in both directions, but it is the potential *net* movement along the gradient that is important and against which freshwater and terrestrial animals must constantly battle. For marine invertebrates, the body fluids and the environment are nearly isotonic to one another and there is little net movement in either direction. (A) The organism is isotonic to its environment. (B) The organism is hypertonic. (C) The organism is hypotonic.

freshwater and terrestrial habitats generally have excretory structures that are responsible for eliminating or retaining water as needed, and they often possess modifications of the body wall to reduce overall permeability. The most successful invertebrate bauplans on land, and in some ways of all environments, are those of the arthropods and gastropods. Their effective excretory structures and thickened exoskeletons provide them with physiological osmoregulatory capabilities plus a barrier against desiccation.

Osmoregulatory problems of aquatic animals are, of course, determined by the salinity of the environmental water relative to the body fluids (Figure 3.21). Organisms respond physiologically to changes in environmental salinities in one of two basic ways. Some, such as most freshwater forms (certain crustaceans, protists, and oligochaetes), maintain their internal body fluid concentrations regardless of external conditions and are thus called **osmoregulators**. Others, including a number of intertidal and estuarine forms (mussels and some other bivalves, and a variety of soft-bodied animals), allow their body fluids to vary with changes in environmental salinities; they are appropriately called **osmoconformers**. Again, even the body fluids of marine, so-called osmoconformers are not exactly isotonic with respect to their surroundings; thus these animals must osmoregulate slightly. Neither of these strategies is without limits, and tolerance to various environmental salinities varies among different species. Those that are restricted to a very narrow range of salinities are said to be **stenohaline**, while those that tolerate relatively extensive variations, such as many estuarine animals, are **euryhaline**.

Although the preceding discussion may seem clearcut, it is an oversimplification. Experimental data from whole animals tell only part of the story of osmoregulation. When a whole marine animal is placed in a hypotonic medium, it tends to swell (if it is an osmoconformer) or to maintain its normal body volume (if it is an osmoregulator). Even at this gross level, most invertebrates usually show evidence of both conforming and regulating. For instance, an osmoconformer generally swells for a period of time in a lowered salinity environment and then begins to regulate. Its swollen volume will decrease, although probably not to its original size. The same is true of most osmoregulators when faced with a decrease in environmental salinity, but the degree of original swelling is much reduced. In both cases, the swelling of the body is a result of an influx of environmental water into the extracellular body fluids (blood, coelomic fluids, and intercellular fluids). Within limits, this excess water is handled by excretory organs and various surface epithelia of the gut and body wall. However, the second part of the osmoregulatory phenomenon takes place at the cellular level.

As the tonicity of the body fluids drops with the entrance of water, the cells in contact with those fluids are placed in conditions of stress—they are now in hypotonic environments. These stressed cells swell to some degree because of the diffusion of water into their cytoplasm, but not to the degree one might expect given the magnitude of the osmotic gradient to which they are subjected. Cellular-level osmoregulation is accomplished by a loss of dissolved materials from the cell into the surrounding intercellular fluids. The solutes released from these cells include both inorganic ions and free amino acids. The actual mechanisms involved remain somewhat elusive. The point here is that osmoconformers are not passive animals that inactively tolerate extremes of salinities. Nor are marine invertebrates free from osmotic problems just because we read statements that they are "98 percent water" or other such comments.

Excretory and Osmoregulatory Structures

Water expulsion vesicles.   The form and function of organs or systems associated with excretion and osmoregulation are related not only to environmental conditions, but also to body size (especially the surface-to-volume ratio) and other basic features of an organism's bauplan. In very small creatures, notably the protists, most metabolic wastes diffuse easily across the body covering because these organisms have sufficient body surface (environmental contact) relative to their volume. However, this high surface area-to-volume ratio presents a distinct osmoregulatory problem, particularly for freshwater forms. Freshwater protists (and even some marine species) typically possess specialized organelles called **contractile vacuoles**, or **water expulsion vesicles** (**WEVs**), which actively excrete excess water (Figure 3.22). These structures accumulate cytoplasmic water and expel it from the cell. Both of these activities apparently require energy, as suggested in part by the large numbers of mitochondria typically associated with WEVs. The idea that WEVs are primarily osmoregulatory in function is supported by a good deal of evidence. Most convincing is the fact that their rates of filling and emptying change dramatically when the cell is exposed to different salinities. For example, the marine flagellate *Chlamydomonas pulsatilla* lives in supralittoral tidal pools and is exposed to low salinities during rainy periods, at which times it regulates its cell volume and internal osmotic pressure via the action of WEVs (which increase in activity as the salinity of their rock pool drops). Interestingly, WEVs also occur in freshwater sponges, where they probably perform similar osmoregulatory functions.

Nephridia.   Although certain metazoan invertebrates possess no known excretory structures, most have some sort of ectodermally derived **nephridia** that serve for excretion or osmoregulation, or both. The evolution of various types of invertebrate nephridia and their relationships to other structures were discussed by E. S. Goodrich in 1945 in a classic

(A)

(B)

(C)

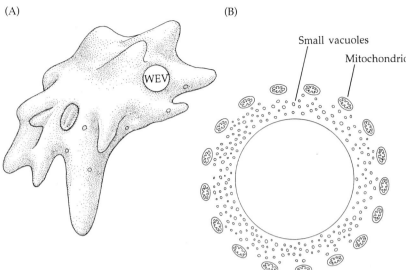

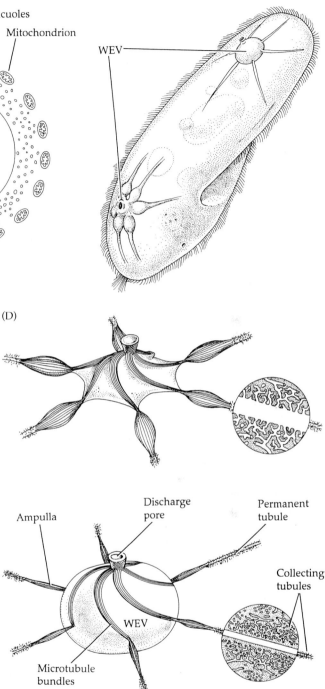

Small vacuoles

Mitochondrion

WEV

(D)

Figure 3.22 Water expulsion vesicles (WEV), or "contractile vacuoles." (A) An ameba with a single WEV. Here the vesicle is transitory and may form anywhere within the cell. (B) The WEV of an ameba, and its association with mitochondria. The numerous small vacuoles accumulate water and then contribute their contents to the main WEV. (C) *Paramecium*. Note the positions of two fixed WEV surrounded by arrangements of collecting canals that pass water to the vesicle. (D) The WEV of *Paramecium*, in filled (bottom) and emptied (top) conditions. Enlarged areas show details of a collecting canal surrounded by cytoplasmic tubules that accumulate cell water. The water is passed into the main vesicle, which is collapsed by the action of contractile fibrils, thereby expelling the water through a discharge channel to the outside.

Discharge pore

Permanent tubule

Ampulla

Collecting tubules

Microtubule bundles

WEV

paper, "The Study of Nephridia and Genital Ducts since 1895."

Probably the earliest type of nephridium to appear in the evolution of animals was the **protonephridium** (Figure 3.23A). Protonephridial systems are characterized by a tubular arrangement opening to the outside of the body via one or more **nephridiopores** and terminating internally in closed unicellular units. These units are the **cap cells** (or **terminal cells**) and may occur singly or in clusters. Each cell is folded into a cup shape, creating a concavity leading to an excretory duct (**nephridioduct**) and eventually to the nephridiopore. Two generally recognized types of protonephridia are **flame bulbs**, bearing a tuft of numerous cilia within the cavity, and **solenocytes**, usually with only one or two flagella. There is some evidence that several different types of flame bulb protonephridia have been independently derived from solenocyte precursors, but the details of nephridial evolution are highly controversial.

The cilia or flagella drive fluids down the nephridioduct, thereby creating a lowered pressure within the tubule lumen. This lowered pressure draws body fluids, carrying wastes, across the thin cell membranes and into the duct. Selectivity is based primarily on molecular size. Protonephridia are common in adult acoelomates (flatworms), many blastocoelomates (rotifers), and some annelids, but are rare among adult eucoelomates (although they occur frequently in various larval types). Protonephridia are probably more important in osmoregulation than in excretion. In most of these animals, nitrogenous wastes are expelled primarily by diffusion across the general body surface.

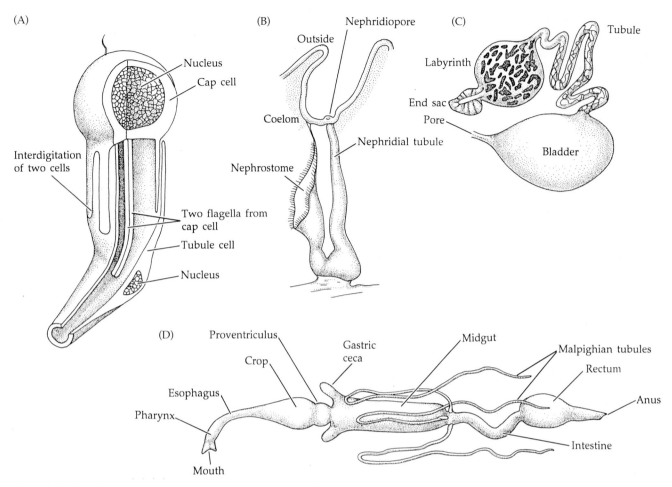

Figure 3.23 Some invertebrate excretory structures. (A) A single protonephridium, with the cap cell and tubule cell (cutaway view). (B) A simple metanephridium from a polychaete worm. The nephrostome opens to the coelom, and the pore opens to the exterior. (C) The internally closed nephridium (antennal gland) of a crustacean. (D) An insect's digestive tract. Excretory Malpighian tubules extract wastes from the hemocoel and empty them into the gut.

A second and probably more advanced type of excretory structure among invertebrates is the **metanephridium** (Figure 3.23B). There is a critical structural difference between protonephridia and metanephridia: both open to the outside, but metanephridia are open internally to the body fluids as well. Metanephridia are also multicellular. The inner end typically bears a ciliated funnel (**nephrostome**), and the duct is often elongated and convoluted and may include a bladder-like storage region. Metanephridia function by taking in large amounts of body fluid through the open nephrostome and then selectively absorbing most of the nonwaste components back into the body fluids through the walls of the bladder or the excretory duct.

In very general terms, we can relate the structural and functional differences between proto- and metanephridia to the bauplans with which they are commonly associated. Whereas protonephridia can adequately serve animals that have solid bodies (acoelomates), body cavities of small volume (blastocoelomates), or very small bodies (e.g., larvae), metanephridia cannot. Open funnels would be ineffective in acoelomates, and would quickly drain small blastocoelomates of their body fluids. Conversely, protonephridia are generally not capable of handling the relatively large body and fluid volumes typical of eucoelomate invertebrates. Thus, in many large coelomate animals (e.g., annelids, molluscs, sipunculans, and echiurans) one or more pairs of metanephridia are found.

We have very broadly interpreted the terms *protonephridia* and *metanephridia* in the above discussion, and we use them as explained above throughout this text unless specified otherwise. However, there are more complications than our simple usage suggests. For example, there is a frequent association of nephridia, especially metanephridia, with structures called coelomoducts. **Coelomoducts** are tubular connections arising from the coelomic lining and extending to the outside via special pores in the body wall. Their inner ends are frequently funnel-like and ciliated, resembling the nephrostomes of metanephridia. Coelomoducts may have arisen evolutionarily as a means of allowing the escape of gametes to the outside; they are, in fact, considered homologous to the reproductive ducts of many invertebrates. Primitively, the coelomoducts and nephridia were separate units; however, through evolution they have in many cases fused in various fashions to become what are called **nephromixia**.

Generally speaking, there are three types of nephromixia. When a coelomoduct is joined with a protonephridium and they share a common duct, the structure is called a **protonephromixium**. When a coelomoduct is united with a metanephridium, the result is either a **metanephromixium** or **mixonephridium**, depending on the structural nature of the union. Whereas coelomoducts originate from the coelomic lining, the nephridial components arise from the outer body wall, so nephromixia are a combination of mesodermally and ectodermally derived parts. Obviously there is some confusion at times about which term applies to a particular "nephridial" type if the precise developmental origin is not clear. We do not wish to belabor this point, so we leave it here to be resurrected periodically in later chapters.

Other organs of excretion. Not all Metazoa possess excretory organs that are clearly proto- or metanephridia. In some taxa (e.g., sponges, echinoderms, chaetognaths, and cnidarians), no definite excretory structures are known. In such cases wastes are eliminated across the surface of the skin or gut lining, perhaps with the aid of ameboid phagocytic cells that collect and transport these products. Other groups possess excretory organs that may represent highly modified nephridia or secondarily derived ("new") structures. For example, the **antennal** and **maxillary glands** of crustaceans appear to be derived from metanephridia, whereas the **Malpighian tubules** of insects and spiders arose independently (Figure 3.23C,D). The details of these structures are discussed in appropriate later chapters.

Circulation and Gas Exchange

Internal Transport

The transport of materials from one place to another within an organism's body depends on the movement and diffusion of substances in body fluids. Nutrients, gases, and metabolic waste products are generally carried in solution or bound to other soluble compounds within the body fluid itself or sometimes in loose cells (such as blood cells) suspended in fluid. Any system of moving fluids that reduces the functional diffusion distance that these products must traverse may be referred to as a **circulatory system**, regardless of its embryological origin or its ultimate design. The nature of the circulatory system is directly related to the size, complexity, and lifestyle of the organism in question. Usually the circulatory fluid is an internal, extracellular, aqueous

medium produced by the animal. There are, however, a few instances in which circulatory functions are accomplished at least partly by other means. For instance, in most protists the protoplasm itself serves as the medium through which materials diffuse to various parts of the cell body, or between the organism and the environment. Sponges and most cnidarians utilize water from the environment as a circulatory fluid, sponges by passing the water through a series of channels in their bodies, and cnidarians by circulating water through the gut (Figure 3.24A,B).

In all Metazoa, the intercellular tissue fluids play a critical role as a transport medium. Even where complicated circulatory plumbing exists, tissue fluids are still necessary to bring dissolved materials in contact with cells, a vital process for life support. In some animals (e.g., flatworms), there are no special chambers or vessels for body fluids other than the gut and intercellular spaces through which materials diffuse on a cell-to-cell level. This condition limits these animals to relatively small sizes or to shapes that maintain low diffusion distances. Most animals, however, have some specialized structure to facilitate the transport of various body fluids and their contents. This structure may include the body cavities themselves or actual circulatory systems of vessels, chambers, sinuses, and pumping organs. Actually, many animals employ both their body cavity and a circulatory system for internal transport.

Blastocoelomate invertebrates use the fluids of the body cavity for circulation (Figure 3.24C). Most of these animals (e.g., rotifers and roundworms) are quite small, or are long and thin, and adequate circulation is accomplished by the movements of the body against the body fluids, which are in direct contact with internal tissues and organs. Several types of cells are generally present in the body fluids of blastocoelomates. These cells may serve in activities such as transport and waste accumulation, but their functions have not been well studied. A few eucoelomate invertebrates (e.g., sipunculans and most echinoderms) also depend largely on the body cavity as a circulatory chamber.

Circulatory Systems

Beyond the relatively rudimentary circulatory mechanisms discussed above, there are two principal designs or structural plans for accomplishing internal transport (exceptions and variations are discussed under specific taxa). These two organizational plans are **closed** and **open circulatory systems**, both of which contain a circulatory fluid, or **blood**. In closed systems the blood stays in distinct vessels and perhaps in lined chambers; exchange of circulated material with parts of the body occurs in special areas of the system such as capillary beds (Figure 3.24D). Since the blood itself is physically separated from the intercellular fluids, the exchange sites must offer minimal resistance to diffusion; thus one finds capillaries typically have membranous walls that

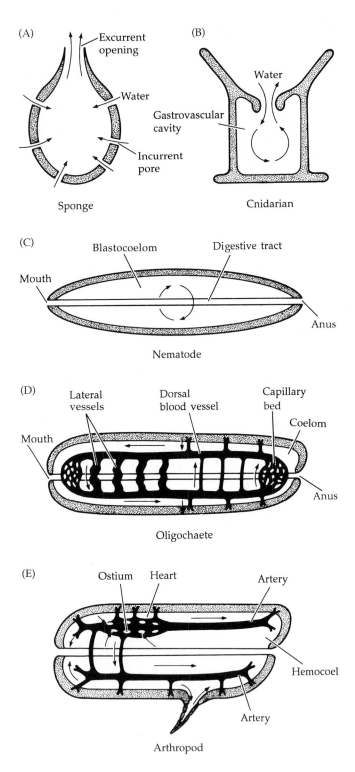

Figure 3.24 Invertebrate circulatory systems. Sponges (A) and cnidarians (B) utilize environmental water as their circulatory fluid. (C) Blastocoelomates (e.g., rotifers and nematodes) use their body cavity fluid for internal transport. (D) The closed circulatory system of an earthworm contains blood that is kept separate from the coelomic fluid. (E) Arthropods are characterized by an open circulatory system, in which the blood and body cavity (hemocoelic) fluid are one and the same.

function. For example, blood may transport nutrients and gases, while coelomic fluid may accumulate metabolic wastes for removal by nephridia and also serves as a hydrostatic skeleton.

It takes power to keep a fluid moving through a plumbing system. Many invertebrates with closed systems rely on body movements and the exertion of coelomic pressure on vessels (often containing one-way valves) to move their blood. These activities are frequently supplemented by muscles of the blood vessel walls that contract in peristaltic waves. In addition, there may be special heavily muscled pumping areas along certain vessels. These regions are sometimes referred to as hearts, but most are more appropriately called **contractile vessels**.

Open circulatory systems are associated with a reduction of the adult coelom, including a secondary loss of most of the peritoneal lining around the organs and inner surface of the body wall. The circulatory system itself usually includes a distinct heart as the primary pumping organ and various vessels, chambers, or ill-defined sinuses (Figure 3.24E). The degree of elaboration of such systems depends primarily on the size, complexity, and to some extent the activity level of the animal. This kind of system, however, is "open" in that the blood, often called the **hemolymph**, empties from vessels into the body cavity and directly bathes the organs. The body cavity is called a **hemocoel**. Open circulatory systems are typical of arthropods and noncephalopod molluscs, and such animals are sometimes referred to as being **hemocoelomate**.

Just because the open circulatory system seems a bit sloppy in its organization, it should not be viewed as poorly "designed" or inefficient. In fact, in many groups this type of system has assumed a variety of functions beyond circulation. For example, in bivalves and gastropods, the hemocoel functions as a hydrostatic skeleton for locomotion and certain types of burrowing activities. In aquatic arthropods, it also serves a hydrostatic function when the animal molts and temporarily loses its exoskeletal support. In large terrestrial insects, the transport of respiratory gases has been largely assumed by the tracheal system, and one of the primary responsibilities taken on by the open circulatory system appears to be thermal regulation. In most spiders, the limbs are extended by forcing hemolymph into the appendages.

are only a single cell-layer thick. Closed circulatory systems are common in animals with well developed or spacious coelomic compartments (e.g., annelids, echiurans, phoronids, and vertebrates). Such arrangements facilitate the transport from one body area to another of materials that might otherwise be isolated by the mesenteries or peritoneum of the body cavity. In such situations the blood and coelomic fluid may be quite different from one another, both in composition and in

Hearts and Other Pumping Mechanisms

Circulatory systems, open or closed, generally have structural mechanisms for pumping the blood and maintaining adequate blood pressures. Beyond the influence of general body movements, most of these structures fall into the following categories: **contractile vessels** (as in annelids); **ostiate hearts** (as in arthropods); and **chambered hearts** (as in molluscs and vertebrates). The method of initiating contraction of these different pumps (the pacemaker mechanisms) may be intrinsic (originating within the musculature of the structure itself) or extrinsic (originating from motor nerves arising outside the structure). The first case describes the **myogenic hearts** of molluscs and vertebrates; the second describes the **neurogenic hearts** of most arthropods and, at least in part, the contractile vessels of annelids.

Blood pressure and flow velocities are intimately associated not only with the activity of the pumping mechanism but also with vessel diameters. Energetically, it costs a good deal more to maintain flow through a narrow pipe than through a wide pipe. This cost is minimized in animals with closed circulatory systems by keeping the narrow vessels short and using them only at sites of exchange (i.e., capillary beds), and by using the larger vessels for long-distance transport from one exchange site to another. In the human circulatory system, for example, arteries have an average radius of 2.0 mm, veins 2.5 mm, and capillaries 0.006 mm. But reducing the diameter of a single vessel increases flow velocity, which poses problems at an exchange site. This problem is solved by the presence of large *numbers* of small vessels, the total cross-sectional area of which exceeds that of the larger vessel from which they arise. The result is that blood pressure and total flow velocity actually decrease at capillary exchange sites. A drop in blood pressure and a relative rise in blood osmotic pressure along the capillary bed facilitate exchanges between the blood and surrounding tissue fluids. In open systems, both pressure and velocity drop once the blood leaves the heart and vessels and enters the spacious hemocoel.

Gas Exchange and Transport

One of the principal functions of most circulatory fluids is to carry oxygen and carbon dioxide through the body and exchange these gases with the environment. With few exceptions, oxygen is necessary for cellular respiration. Although a number of invertebrates can survive periods of environmental oxygen depletion—either by dramatically reducing their metabolic rate or by switching to anaerobic respiration—most cannot; they depend upon a relatively constant oxygen supply.

All animals can take in oxygen from their surroundings while at the same time releasing carbon dioxide, a metabolic waste product of respiration. We define the uptake of oxygen and the loss of carbon dioxide at the surface of the organism as **gas exchange**, reserving the term *respiration* for the energy-producing metabolic activities within cells. Some authors distinguish these two processes with the terms external respiration and cellular (internal) respiration.

Gas exchange in nearly all animals operates according to certain common principles regardless of any structural modifications that serve to enhance the process under different conditions. The basic strategy is to bring the environmental medium (water or air) close to the appropriate body fluid (blood or body cavity fluid) so that the two are separated only by a wet membrane across which the gases can diffuse. The system must be moist because the gases must be in solution in order to diffuse across the membrane. The diffusion process depends on the concentration gradients of the gases at the exchange site; these gradients are maintained by the circulation of internal fluids to and away from these areas (Figure 3.25).

Gas exchange structures. Protists and a number of invertebrates lack special gas exchange structures. In

Figure 3.25 Gas exchange in animals. Oxygen is obtained from the environment at a gas exchange surface (A), and is transported by a circulatory body fluid (B) to the body's cells and tissues (C), where cellular respiration occurs (D). Carbon dioxide follows the reverse path. See text for details.

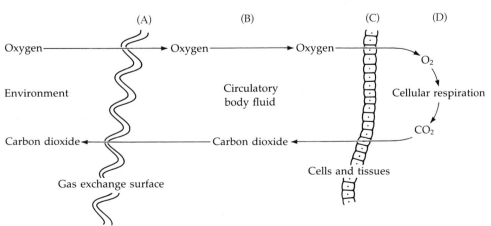

such animals gas exchange is said to be **integumentary** or **cutaneous**, and occurs over much of the body surface. Such is the case in many tiny animals with very high surface-to-volume ratios and in some larger soft-bodied forms (e.g., cnidarians and flatworms). Most animals with integumentary gas exchange are restricted to aquatic or damp terrestrial environments where the body surface is kept moist. Integumentary gas exchange also supplements other methods in many animals, even certain vertebrates (e.g., amphibians).

Most marine and many freshwater invertebrates possess **gills** (Figure 3.26A,B,C,G), which are external organs or restricted areas of the body surface specialized for gas exchange. Basically, gills are thin-walled processes, well supplied with blood or other body fluids, that promote diffusion between this fluid and the environment. Gills are frequently highly folded or digitate, increasing the diffusive surface area. A great number of nonhomologous structures have evolved as gills in different taxa, and they often serve other functions in addition to gas exchange (e.g., sensory input and feeding). By their very nature, gills are permeable surfaces that must be protected during times of osmotic stress, such as occur in estuaries and intertidal environments. In these instances, the gills may be housed within chambers or be retractable.

A few marine invertebrates employ the lining of the gut as the gas exchange surface. Water is pumped in and out of the hindgut, or a special evagination thereof, in a process called **hindgut irrigation**. Many sea cucumbers and echiuran worms use this method of gas exchange (Figure 3.26F).

As we have defined them, protruding gills will not work on dry land. Here the gas exchange surfaces must be internalized to keep them moist and protected and to prevent body water loss through the wet surfaces. The lungs of terrestrial vertebrates are the most familiar example of such an arrangement. Among the invertebrates, the arthropods have managed to solve the problems of "air-breathing" in two basic ways. Spiders and their kin possess **book lungs**, and most insects, centipedes, and millipedes possess **tracheae** (Figure 3.26D,E). Book lungs are blind inpocketings with highly folded inner linings across which gases diffuse between the hemolymph and the air. Tracheae, however, are branched, usually anastomosed invaginations of the outer body wall and are open both internally and externally.

The tracheae of most insects allow diffusion of oxygen from air directly to the tissues of the body; the blood plays little or no role in gas transport. Rather, intercellular fluids extend part way into the tracheal tubes as a solvent for gases. Atmospheric pressure tends to prevent these fluids from being drawn too close to the external body surface where evaporation is a potential problem. In addition, the outside openings (**spiracles**) of the tracheae are often equipped with some mechanism of closure. In many insects, especially large ones, special muscles ventilate the tracheae by actively pumping air in and out. Terrestrial isopod crustaceans (e.g., sowbugs and pillbugs) have invaginated gas exchange structures on some of their abdominal appendages. These inpocketings are called **pseudotrachea**, but are probably not homologous to the trachea or the book lungs of insects and spiders.

The only other major group of terrestrial invertebrates whose members have evolved distinct air-breathing structures is the molluscan subclass Pulmonata—the land snails and slugs (Figure 3.26H). The gas exchange structure here is a **lung** that opens to the outside via a pore called the **pneumostome**. This lung is derived from a feature common to molluscs in general, the **mantle cavity**, which in other molluscs houses the gills and other organs.

Gas transport. As illustrated in Figure 3.25, oxygen must be transported from the sites of environmental gas exchange to the cells of the body, and carbon dioxide must get from the cells where it is produced to the gas exchange surface for release. Generally, groups displaying marked cephalization circulate freshly oxygenated blood through the "head" region first, and secondarily to the rest of the body.

Invertebrates vary considerably in their oxygen requirements. In general, active animals consume more oxygen than sedentary ones. In slow-moving and sedentary invertebrates, oxygen consumption and utilization are quite low. For example, no more than 20 percent oxygen withdrawal from the gas exchange water current has ever been demonstrated in sessile sponges, bivalves, or tunicates. The amount of oxygen available to an organism varies greatly in different environments. The concentration of oxygen in dry air at sea level is uniformly about 210 ml/liter, whereas in water it ranges from near zero to about 10 ml/liter. This variation in aquatic environments is due to such factors as depth, surface turbulence, photosynthetic activity, temperature, and salinity (oxygen concentrations drop as temperature and salinity increase). With the exception of certain areas prone to oxygen depletion (e.g., muds rich in organic detritus), most habitats provide adequate sources of oxygen to sustain animal life. Also, the relatively low capacity of body fluids to carry oxygen in solution is greatly increased by binding oxygen with complex organic compounds called **respiratory pigments**.

Respiratory pigments differ in molecular architecture and in their affinities for oxygen, but all have a metal ion (usually iron, sometimes copper) with which the oxygen combines. In most invertebrates, these pigments occur in solution within the blood or other body fluid, but in some invertebrates (and virtually all vertebrates), they may be in specific blood cells. In general, the pigments respond to high oxygen concentrations by "loading" (combining with oxygen) and to low oxygen con-

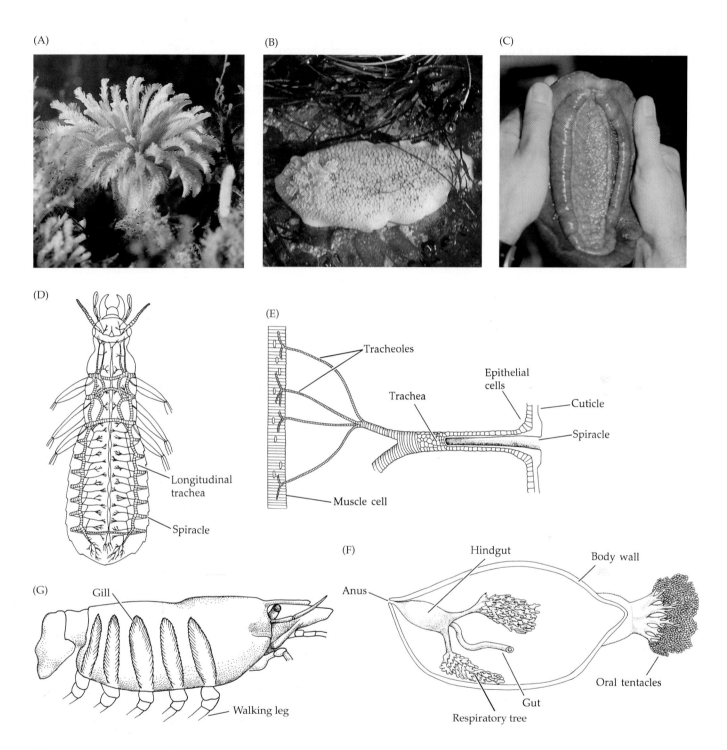

Figure 3.26 Some gas exchange structures in invertebrates. (A) The tube-dwelling polychaete worm *Eudistylia,* with its feeding–gas exchange tentacles extended. (B) A sea slug (nudibranch) displaying its branchial plume. (C) The gills of the giant gumboot chiton (*Cryptochiton stelleri*) are visible along the right side of its foot. (D) A general plan of the tracheal system of an insect. (E) A single insect trachea and its branches (tracheoles), which lead directly to a muscle cell. (F) A sea cucumber dissected to expose the paired respiratory trees, which are flushed with water by hindgut irrigation. (G) The placement of gills beneath the flaps (carapace) of the thorax in a crustacean (lateral view). (H) A terrestrial banana slug has a pneumostome that opens to the air sac, or "lung."

TABLE 3.1 *Properties of oxygen-carrying respiratory pigments*

Pigment	Molecular weight	Metal	Ratio of metal to O_2	Metal associate
Hemoglobin	65,000	Fe	1:1	Porphyrin
Hemerythrin	40,000–108,000	Fe	2:1	Protein chains
Hemocyanin	40,000–9,000,000	Cu	2:1	Protein chains
Chlorocruorin	3,000,000	Fe	1:1	Porphyrin

centrations by "unloading" or dissociating from oxygen (releasing oxygen). The loading and unloading qualities are different for various pigments in terms of their relative saturations at different levels of oxygen in their immediate surroundings, and are generally expressed in the form of dissociation curves. Respiratory pigments load at the site of gas exchange, where environmental oxygen levels are high relative to the body fluid, and unload at the cells and tissues, where surrounding oxygen levels are low relative to the body fluid. In addition to simply carrying oxygen from the loading to the unloading sites, some pigments may carry reserves of oxygen that are released only when tissue levels are unusually low. Other factors, such as temperature and carbon dioxide concentration, also influence the oxygen-carrying capacities of respiratory pigments.

Hemoglobin is among the most common respiratory pigments in animals. There are actually several different hemoglobins. Some function primarily for transport, while others store oxygen and then release it during times of low environmental oxygen availability. Hemoglobins are reddish pigments containing iron as the oxygen-binding metal. They are found in a variety of invertebrates and, with the exception of a few fishes, in all vertebrates. Among the major groups of invertebrates, hemoglobin occurs in many annelids, some crustaceans, some insects, and a few molluscs and echinoderms. Interestingly, hemoglobin is not restricted to the Metazoa; it is also produced by some protists, certain fungi, and in the root nodules of leguminous plants. Among animals, hemoglobin may be carried within red blood cells (**erythrocytes**), in coelomic cells called **hemocytes** (in a few echinoderms), or it may simply be dissolved in the blood or coelomic fluid.

Hemocyanins are the most commonly occurring respiratory pigments in molluscs and arthropods, and they occur only in members of these phyla. Among arthropods, hemocyanin occurs in chelicerates, a few myriapods, and the "higher Crustacea." There is indirect evidence that it also occurred in trilobites. Hemocyanin has been found in most classes of molluscs. Although hemocyanins, like hemoglobins, are proteins, they display significant structural differences, contain copper rather than iron, and tend to have a bluish color when oxygenated. The oxygen binding site on a hemocyanin molecule is a pair of copper atoms linked to amino acid side chains. Unlike most hemoglobins, hemocyanins tend to release oxygen easily and provide a ready source of oxygen to the tissues as long as there is a relatively high concentration of available environmental oxygen. Hemocyanins are always found in solution, never in cells, a characteristic probably related to the necessity for rapid oxygen unloading. Hemocyanins often give a bluish tint to the hemolymph of arthropods, although the presence of carotenoid pigments (beta-carotene and related molecules) commonly impart a brown or orange coloration.

Two other types of respiratory pigments occur incidentally in certain invertebrates; these are **hemerythrins** and **chlorocruorins**, both of which contain iron. The former is violet to pink when oxygenated; the latter is green in dilute concentrations but red in high concentrations. Chlorocruorins generally function as efficient oxygen carriers when environmental levels are relatively high; hemerythrins function more in oxygen storage. Chlorocruorin is structurally similar to hemoglobin and may have been derived from it. Chlorocruorin occurs in several families of polychaete worms; hemerythrin is known from sipunculans, at least one genus of polychaetes, and some priapulans and brachiopods.

Table 3.1 gives some of the basic properties of oxygen-carrying pigments. There seems to be no obvious phylogenetic rhyme or reason to the occurrence of these pigments among the various taxa. Their sporadic and inconsistent distribution suggests that some of them may have evolved more than once, through parallel or convergent evolution. Respiratory pigments are rare among insects and are known only from the occurrence of hemoglobin in chironomid midges, some notonectids, and certain parasitic flies of the genus *Gastrophilus*. The absence of respiratory pigments among the insects reflects the fact that most of them do not use the blood as a medium for gas transport, but employ extensive tracheal systems to carry gases directly to the tissues. In those insects without well developed tracheae, oxygen is simply carried in solution in the hemolymph.

Respiratory pigments raise the oxygen-carrying capacity of body fluids far above what would be achieved by transport in simple solution. Similarly, carbon dioxide levels in body fluids (and in sea water) are much higher than would be expected strictly on the basis of its solubility. The enzyme **carbonic anhydrase** greatly accelerates the reaction between carbon dioxide and water, forming carbonic acid:

$$CO_2 + H_2O \rightleftharpoons H_2CO_3$$

Furthermore, carbonic acid ionizes to hydrogen and bicarbonate ions, so a series of reversible reactions takes place:

$$CO_2 + H_2O \rightleftharpoons H_2CO_3 \rightleftharpoons H^+ + HCO_3^-$$

By "tying up" CO_2 in other forms, the concentration of CO_2 in solution is lowered, thus raising the overall CO_2-carrying capacity of the blood. This set of reactions responds to changes in pH, and in the presence of appropriate cations (e.g., Ca^{2+} and Na^+) it shifts back and forth, serving as a buffering mechanism by regulating hydrogen ion concentration.

Nervous Systems and Sense Organs

Invertebrate Sense Organs

All living cells respond to some stimuli and conduct some sort of "information," at least for short distances. Thus, even when no real nervous system is present—the condition found in protists and sponges—coordination and reaction to external stimulation do occur. The regular metachronal beating of cilia in ciliate protists and the responses of certain flagellates to varying light intensities are examples. In addition, most protists are known to respond to gradients of various environmental factors by moving to or away from areas of high concentration. For example, when subjected to conditions of low oxygen concentration (hypoxia), paramecia move to regions of lower water temperature, thus lowering their metabolic rate and presumably their oxygen need. But the integration and coordination of bodily activities in Metazoa are in large part due to the processing of information by a true nervous system. The functional units of nervous systems are **neurons**: cells that are specialized for high-velocity impulse conduction.

The generation of an impulse within a true nervous system usually results from a stimulus imposed on the nervous elements. The source of stimulation may be external or internal. A typical pathway of events occurring in a nervous system is shown in Figure 3.27. A stimulus received by some **receptor** (e.g., a sense organ) generates

an impulse that is conducted along a **sensory nerve (afferent nerve)** via a series of adjacent neurons to some coordinating center or region of the system. The information is processed and an appropriate response is "selected." A **motor nerve (efferent nerve)** then conducts an impulse from the central processing center to an **effector** (e.g., a muscle), where the response occurs. Once an impulse is initiated within the system, the mechanism of conduction is essentially the same in all neurons, regardless of the stimulus. The wave of depolarization along the length of each neuron and the chemical neurotransmitters crossing the synaptic gaps between neurons are common to virtually all nervous conduction. How then is the information interpreted within the system for response selection? The answer to this question involves three basic considerations.

First is the occurrence of a point called a **threshold**, which corresponds to the minimum intensity of stimulation necessary to generate an impulse. Receptor sites consist of specialized neurons whose thresholds for various kinds of stimuli are drastically different from one another because of structural or physiological qualities. For example, a sense organ whose threshold for light stimulation is very low (compared with other potential stimuli) functions as a light sensor, or **photoreceptor**. In any such specialized sensory receptor, the condition of differential thresholds essentially screens incoming stimuli so that an impulse normally is generated by only one kind of information (e.g., light, sound, heat, or pressure). Second is the nature of the receptor itself. Receptor units (e.g., sense organs) are generally constructed in ways that permit only certain stimuli to reach the impulse-generating cells. For example, the light-sensitive cells of the human eye are located beneath the eye surface, where stimuli other than light would not normally reach them.

And third, the overall "wiring" or circuitry of the entire nervous system is such that impulses received by the integrative (response-selecting) areas of the system from any particular nerve will be interpreted according to the kind of stimulus for which that sensory pathway is specialized. For example, all impulses coming from a photoreceptor are understood as being light-induced. Threshold and circuitry can be demonstrated by introducing false information into the system by stimulating a specialized sense organ in an inappropriate manner: if

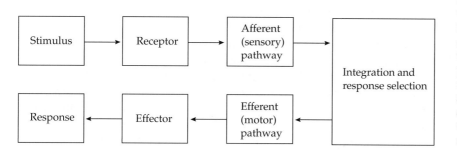

Figure 3.27 A generalized pathway within the nervous system. A stimulus initiates an impulse within some sensory structure (the receptor); the impulse is then transferred to some integrative portion of the nervous system via sensory nerves. Following response selection, an impulse is generated and transferred along motor nerves to an effector (e.g., muscle), where the appropriate response is elicited.

(A)

Tactile bristle

(B)

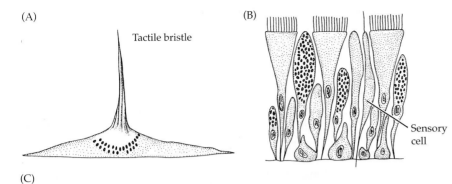

Sensory cell

Figure 3.28   Some invertebrate tactile receptors. (A) Tactile organ of *Sagitta bipunctata* (an arrow worm, phylum Chaetognatha). (B) A sensory epithelial cell of a nemertean worm. (C) Long, touch-sensitive setae (and stout grasping setae) on the leg of the isopod, *Politolana* (SEM).

(C)

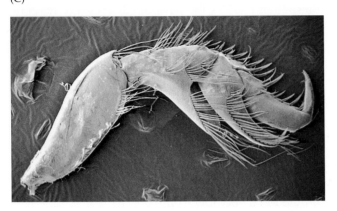

The activities of receptor units represent the initial step in the usual functioning of the nervous system; they are a critical link between the organism and its surroundings. Consequently, the kinds of sense organs present and their placement on the body are intimately related to the overall complexity, mode of life, and general bauplan of any animal. The following general review provides some concepts and terminology that serve as a basis for more detailed coverage in later chapters. The first five categories of sense organs may all be viewed as mechanoreceptors, in that they respond to mechanical stimuli (e.g., touch, vibrations, and pressure). The last three are sensitive to nonmechanical input (e.g., chemicals, light, and temperature).

Tactile receptors.   Touch or **tactile receptors** are generally derived from modified epithelial cells associated with sensory neurons. The nature of the epithelial modifications depends a great deal on the structure of the body wall. For instance, the form of a touch receptor in an arthropod with a rigid exoskeleton must be different from that in a soft-bodied cnidarian. Most such receptors, however, involve projections from the body surface, such as bristles, spines, setae, tubercles, and assorted bumps and pimples (Figure 3.28). Objects in the environment with which the animal makes contact move these receptors, thereby creating mechanical deformations that are imposed upon the underlying sensory neurons to initiate an impulse.

Virtually all animals are touch-sensitive, but their responses are varied and often integrated with other sorts of sensory input. For example, the gregarious nature of many animals may involve a positive response to touch (positive thigmotaxis) combined with the chemical recognition of members of the same species. Some touch receptors are highly sensitive to mechanically induced vibrations propagated in water, loose sediments, through solid substrata, or other materials. Such **vibration sensors** are common in certain tube-dwelling polychaetes that retract quickly into their tubes in response to movements in their surroundings. Some crustacean ambush-predators are able to detect the vibrations induced by nearby potential prey animals, and web-building spiders sense struggling prey in their webs through vibrations of the threads.

photoreceptors in the eye are stimulated by electricity or pressure, the nervous system will interpret this input as light. Remember that an impulse can be generated in any receptor by nearly any form of stimulation if the stimulus is intense enough to exceed the relevant threshold. A blow to the eye often results in "seeing stars," or flashes of light, even when the eye is closed. In such a situation, the photoreceptor's threshold to mechanical stimulation has been reached. By the same token, the application of extreme cold to a heat receptor may feel hot.

Nervous systems in general operate on the principles outlined above. However, this description applies largely to nervous systems that have structural centralized regions. Following a discussion below of the basic types of sense organs (receptor units), we discuss centralized and noncentralized nervous systems and their relationships to general body architecture.

Sense Organs

Invertebrates possess an impressive array of receptor structures through which they receive information about their internal and external environments. An animal's behavior is in large part a function of its responses to that information. These responses often take the form of some sort of movement relative to the source of a particular stimulus. A response of this nature is called a **taxis** and may be positive or negative depending on the reaction of the animal to the stimulus. For example, many animals tend to move away from bright light and are thus said to be negatively phototactic.

Georeceptors. **Georeceptors** respond to the pull of gravity, giving animals information about their orientation relative to "up and down." Most georeceptors are structures called **statocysts** (Figure 3.29). Statocysts usually consist of a fluid-filled chamber containing a solid granule or pellet called a **statolith**. The inner lining of the chamber includes a touch-sensitive epithelium from which project bristles or "hairs" associated with underlying sensory neurons. In aquatic invertebrates, some statocysts are open to the environment and thus are filled with water. In some of these the statolith is a sand grain obtained from the animal's surroundings. Most statoliths, however, are secreted within closed capsules by the organisms themselves.

Because of the resting inertia of the statolith within the fluid, any movement of the animal results in a change in the pattern or intensity of stimulation of the sensory epithelium by the statolith. Additionally, when the animal is stationary, the position of the statolith within the chamber provides information about the organism's orientation to gravity. The fluid within statocysts of at least some invertebrates (especially certain crustaceans) also acts something like the fluid of the semicircular canals in vertebrates. When the animal moves, the fluid tends to remain stationary—the relative "flow" of the fluid over the sensory epithelium provides the animal with information about its linear and rotational acceleration relative to its environment.

Whether stationary or in motion, animals utilize the input from georeceptors in different ways, depending on their habitat and lifestyle. The information from these statocysts is especially important under conditions where other sensory reception is inadequate. For example, burrowing invertebrates cannot rely on photoreceptors for orientation when moving through the substratum, and some employ statocysts for that purpose. Similarly, planktonic animals face orientation problems in their three-dimensional aqueous environment, especially in deep water and at night; many such creatures possess statocysts.

There are a few exceptions to the standard statocyst arrangements described above. For example, a number of aquatic insects detect gravity by using air bubbles trapped in certain passageways (e.g., tracheal tubes). The bubbles move according to their orientation to the vertical, much like the air bubble in a carpenter's level, and stimulate sensory bristles lining the tube in which they are located.

Proprioceptors. Internal sensory organs that respond to mechanically induced changes caused by stretching, compression, bending, and tension are called **proprioceptors**, or simply **stretch receptors**. These receptors give the animal information about the movement of its body parts and their positions relative to one another. Proprioceptors have been most

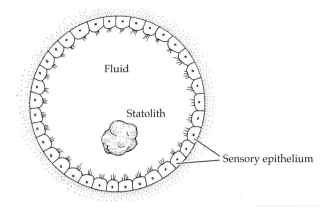

Figure 3.29 A generalized statocyst, or georeceptor (section).

thoroughly studied in vertebrates and arthropods, where they are associated with appendage joints and certain body extensor muscles. The sensory neurons involved in proprioception are associated with and attached to some part of the body that is stretched or otherwise mechanically affected by movement or muscle tension. These parts may be specialized muscle cells, elastic connective tissue fibers, or membranes that span joints. As these structures are stretched, relaxed, and compressed, the sensory endings of the attached neurons are distorted accordingly and thus stimulated. Some of these receptor arrangements can detect not only changes in position but also in static tension.

Phonoreceptors. General sensitivity to sound—**phonoreception**—has been demonstrated in a number of invertebrates (certain annelid worms and a variety of crustaceans), but true auditory receptors are known only in a few groups of insects and perhaps some arachnids and centipedes. Crickets, grasshoppers, and cicadas possess phonoreceptors called **tympanic organs** (Figure 3.30). A rather tough but flexible **tympanum** covers an internal air sac that allows the tympanum to vibrate when struck by sound waves. Sensory neurons attached to the tympanum are stimulated directly by the vibrations. Most arachnids possess structures called **slit sense organs,** which, although poorly studied, are suspected to perform auditory functions; at least they appear to be capable of sensing sound-induced vibrations. Certain centipedes bear so-called **organs of Tömösvary**, which some workers believe may be sensitive to sound.

Baroreceptors. The sensitivity of invertebrates to pressure changes—**baroception**—is not well understood, and no structures for this purpose have been positively identified. However, behavioral responses to pressure changes have been demonstrated in several pelagic invertebrates including medusae,

Figure 3.30   An arthropod phonoreceptor, or auditory organ, of the fork-tailed katydid, *Scudderia furcata*. Note the position of the right-side tympanum on the tibia of the first walking leg.

ctenophores, squids, and copepod crustaceans, as well as in some planktonic larvae. Aquatic insects also sense changes in pressure, and may use a variety of methods to do so. Some intertidal crustaceans coordinate daily migratory activities with tidal movements, perhaps partly in response to pressure as water depth changes.

Chemoreceptors.   Many animals have a **general chemical sensitivity**, which is not a function of any definable sensory structure but is due to the general irritability of protoplasm itself. When they occur in sufficiently high concentrations, noxious or irritating chemicals can induce responses via this general chemical sensitivity. In addition, most animals have specific **chemoreceptors**.

Chemoreception is a rather direct sense in that the molecules stimulate sensory neurons by contact, usually after diffusing in solution across a thin epithelial covering. The chemoreceptors of many aquatic invertebrates are located in pits or depressions, through which water may be circulated by ciliary action. In arthropods, the chemoreceptors are usually in the form of hollow "hairs" or other projections, within which are chemosensory neurons. While chemosensitivity is a universal phenomenon among invertebrates, a wide range of specificities and capabilities exists.

The types of chemicals to which particular animals respond are closely associated with their lifestyles. Chemoreceptors may be specialized for tasks such as general water analysis, humidity detection, sensitivity to pH, prey tracking, mate location, substratum analysis, and food recognition. Probably all aquatic organisms leak small amounts of amino acids into their environment through the skin and gills as well as in their urine and feces. These released amino acids form an organism's "body odor," which can create a chemical pic-

ture of the animal that others detect to identify such characteristics as species, sex, stress level, distance and direction, and perhaps size and individuality. Amino acids are widely distributed in the aquatic environment, where they provide general indicators of biological activity. Many aquatic animals can detect amino acids with much greater sensitivity than our most sophisticated laboratory equipment.

Photoreceptors.   Nearly all animals are sensitive to light, and most have some kind of identifiable photoreceptors. Although members of only a few of the metazoan phyla appear to have evolved eyes capable of image formation (Cnidaria, Mollusca, Annelida, Arthropoda, and Chordata), virtually all animal photoreceptors share structurally similar light receptor molecules that probably predate the origin of discrete structural eyes. Thus, the structural photoreceptors of animals share the common quality of possessing **light-sensitive pigments**. These pigment molecules are capable of absorbing light energy in the form of photons, a process necessary for the initiation of any light-induced, or **photic**, reaction. The energy thus absorbed is ultimately responsible for stimulating the sensory neurons of the photoreceptor unit.

Beyond this basic commonality, however, there is an incredible range of variation in complexity and capability of light-sensitive structures. Arthropods, molluscs, and some polychaete annelids possess eyes with extreme sensitivity, good spatial resolution, and, in some cases, multiple spectral channels. Most classifications of photoreceptors are based upon grades of complexity, and the same categorical term may be applied to a variety of nonhomologous structures, from simple pigment spots (found in protists) to extremely complicated lensed eyes (found in squids and octopuses). Functionally, the capabilities of these receptors range from simply perceiving light intensity and direction to forming images with a high degree of visual discrimination and resolution.

Certain protists, particularly flagellates, possess subcellular organelles called **stigmata**, which are associated with simple spots of light-sensitive pigment (Figure 3.31A). The simplest metazoan photoreceptors are unicellular structures scattered over the epidermis or concentrated in some area of the body. These are usually called **eyespots**. Multicellular photoreceptors may be classified into three general types, with some subdivisions. These types include **ocelli** (sometimes called simple eyes or eyespots), **compound eyes** (found in many arthropods), and **complex eyes** (the "camera" eyes of cephalopod molluscs and vertebrates). In multicellular ocelli, the light-sensitive (**retinular**) cells may face outward; these ocelli are then said to be **direct**. Or the light-sensitive cells may be **inverted**. The inverted type is common among flatworms and nemerteans and is made up of a cup of reflective pigment and retinular

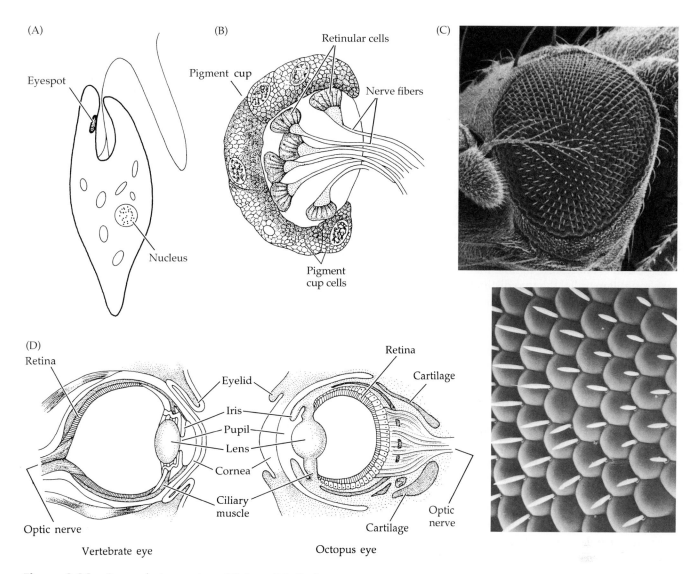

Figure 3.31 Some photoreceptors. (A) A protist, *Euglena*. Note the position of the stigma. (B) An inverted pigment-cup ocellus of a flatworm (section). (C) An insect's compound eye. A single unit is called an ommatidium. (D) A vertebrate eye (left) and a cephalopod eye (right) (vertical sections).

cells (Figure 3.31B). The light-sensitive ends of these neurons face into the cup. Light entering the opening of the pigment cup is reflected back onto the retinular cells. Because light can enter only through the cup opening, this sort of ocellus gives the animal a good deal of information about light direction as well as variations in intensity.

Compound eyes are composed of a few to many distinct units called **ommatidia** (Figure 3.31C). Although eyes of multiple units occur in certain annelid worms and some bivalve molluscs, they are best developed and best understood among the arthropods. Each ommatidium is supplied with its own nerve tract leading to a large optic nerve, and apparently each has its own dis-

crete field of vision. The visual fields of neighboring ommatidia overlap to some degree, with the result that a shift in position of an object within the total visual field causes changes in the impulses reaching several ommatidial units; based in part on this phenomenon, compound eyes are especially suitable for detecting movement. Compound eyes are described in more detail in Chapter 15.

The complex eyes of squids and octopuses (Figure 3.31D) are probably the best image-forming eyes among the invertebrates. Cephalopod eyes are frequently compared with those of vertebrates, but they differ in many respects. The eye is covered by a transparent protective **cornea**. The amount of light that enters the eye is controlled by the **iris**, which regulates the size of the slitlike **pupil**. The lens is held by a ring of **ciliary muscles** and focuses light on the **retina**, a layer of densely packed photosensitive cells from which the neurons arise. The receptor sites of the retinal layer face in the direction of the light entering the eye. This **direct eye** arrangement is quite different from the **indirect eye** condition in vertebrates, where the retinal layer is inverted. Another difference is that in many vertebrates, focusing is accomplished by the action of muscles that change the shape of the lens, whereas in cephalopods it is achieved by moving the lens back and forth with the ciliary muscles and by compressing the eyeball.

A good deal of work suggests that metazoan photoreceptors evolved along two lines (see Eakin 1963). On one hand are photoreceptor units derived from or closely associated with cilia (e.g., in cnidarians, echinoderms, and chordates). These types of eyes are called **ciliary eyes**. On the other hand are photoreceptors derived from microvilli or microtubules and referred to as **rhabdomeric eyes** (e.g., in flatworms, annelids, arthropods, and molluscs). It is not yet known if these different categories represent actual lineages of homologous structures, but it is interesting to note that the two groups of taxa roughly parallel the two distinct lines of evolutionarily related taxa known as deuterostomes and protostomes (see Chapters 4 and 24).

Thermoreceptors. The influence of temperature changes on all levels of biological activity is well documented. Every student of general biology has learned about the basic relationships between temperature and rates of metabolic reactions. Furthermore, even the casual observer has noticed that many organisms' activity levels range from lethargy at low temperatures to hyperactivity at elevated temperatures, and that thermal extremes can result in death. The problem is determining whether the organism is simply responding to the effects of temperature at a general physiological level, or whether discrete thermoreceptor organs are also involved.

There is considerable circumstantial evidence that at least some invertebrates are capable of directly sensing differences in environmental temperatures, but actual receptor units are for the most part unidentified. A number of insects, some crustaceans, and the horseshoe crab (*Limulus*) apparently can sense thermal variation. The only nonarthropod invertebrates that have received much attention in this regard are certain leeches, which apparently are drawn to warm-blooded hosts by some heat-sensing mechanism. Other ectoparasites (e.g., ticks) of warm-blooded vertebrates may also be able to sense the "warmth of a nearby meal," but little work has been done on this subject.

Independent Effectors

Independent effectors are specialized sensory response structures that not only receive information from the environment but also elicit a response to the stimulus directly, without the intervention of the nervous system per se. In this sense, independent effectors are like closed circuits. As discussed in later chapters, the stinging capsules (nematocysts) of cnidarians and the adhesive cells of ctenophores are, at least under most circumstances, independent effectors.

Bioluminescence

Bioluminescence, the production of light by living creatures, occurs in a variety of organisms. Some bacteria and fungi bioluminesce, as do as certain dinoflagellates, cnidarians, annelids, molluscs, arthropods, echinoderms, tunicates, and fishes. Most bioluminescence seen in the sea is produced by dinoflagellates emitting rapid (one-tenth of a second) flashes. But the patient nighttime observer will also discover that flashes of light are produced by some species of medusae, ctenophores, copepods, benthic ostracods, brittle stars, sea pansies and sea pens, chaetopterid and syllid polychaetes, limpets, clams, tunicates, and others.

Luminescence is the emission of light without heat. It involves a special type of chemical reaction in which the energy, instead of being released as heat as occurs in most chemical reactions, is used to excite a product molecule that releases energy as a photon. In all cases, the reaction involves the oxidation of a substrate called **luciferin**, catalyzed by an enzyme called **luciferase**. The structures of these chemicals differ among taxa, but the reaction is similar. The color of light varies from deep blue (shrimp and dinoflagellates) to blue-green or green (ostracods and tunicates), to yellow and even red (fireflies). Bioluminescence serves several functions, including offense, defense, prey attraction, and intraspecific communication. In some cases, the luminescent organs of metazoans (particularly fishes) are not intrinsic but are symbiotic colonies of microorganisms.

Nervous Systems and Body Plans

The nervous system is always receiving information via its associated receptors, processing this information, and eliciting appropriate responses. We limit our discussion at this point to those conditions in which distinct systems of identifiable neurons exist, leaving the special situations in protists and sponges for later chapters.

The structure of the nervous system of any animal is related to its bauplan and mode of life. Consider first a radially symmetrical animal with limited powers of locomotion, such as a planktonic jellyfish or a sessile sea anemone. In such animals the major receptor organs are more or less regularly (and radially) distributed around the body; the nervous system itself is a noncentralized, diffuse meshwork generally called a **nerve net** (Figure 3.32A). Radially symmetrical animals tend to be able to respond equally well to stimuli coming from any direction—a useful ability for creatures with either sessile or free-floating lifestyles. Interestingly, at least in cnidarians, there are both polarized and nonpolarized synapses within the nerve net. Impulses can travel in either direction across the nonpolarized synapses because the neuronal processes on both sides are capable of releasing synaptic transmitter chemicals. This capability, coupled with the gridlike form of the nerve net, enables impulses to travel in all directions from a point of stimulation. From this brief description, it might be as-

(A)

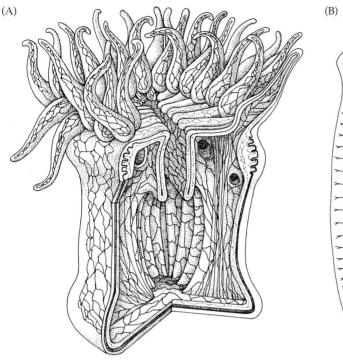

(B)

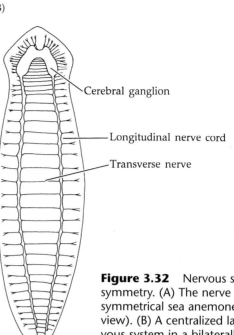

Cerebral ganglion

Longitudinal nerve cord

Transverse nerve

Figure 3.32 Nervous systems and symmetry. (A) The nerve net in a radially symmetrical sea anemone (cutaway view). (B) A centralized ladder-like nervous system in a bilaterally symmetrical flatworm.

sumed that such a simple and "unorganized" nervous system would not provide enough integrated information to allow complex behaviors and coordination. In the absence of a structurally recognizable integrating center, the nerve net does not fit well with our earlier description of the sequence of events from stimulus to response. But many cnidarians are in fact capable of fairly intricate behavior, and the system works, often in ways that are as yet unexplained. In any case, symmetry, sense organ distribution, nervous system organization, and lifestyles are clearly correlated to one another.

The tremendous evolutionary success of bilateral symmetry and unidirectional locomotion must have depended in large part on associated changes in the organization of the nervous system and the distribution of sense organs. The trend has been to centralize and concentrate the major coordinating elements of the nervous system. This **central nervous system** is usually made up of an anteriorly located neuronal mass (ganglion) from which arise one or more longitudinal nerve cords that often bear additional ganglia (Figure 3.32B). The anterior ganglion is referred to by a variety of names. We largely abandon the term *brain* for such an organ because of the multifaceted implications of that word and adopt the more neutral term **cerebral ganglion** (or ganglia) for the general case. In many instances, a term of its relative position to some other organ is applied. For example, the cerebral ganglion commonly lies dorsal to the anterior portion of the gut and is thus a **supraenteric** (or supraesophageal, or suprapharyngeal) ganglion.

In addition to the cerebral ganglion, most bilaterally symmetrical animals have many of the major sense or-gans placed anteriorly. The concentration of these organs at the front end of an animal is called **cephalization**—the formation of a head region. Even though cephalization may seem an obvious and predictable outcome of bilaterality and mobility, it is nonetheless extremely important. It simply would not do to have information about the environment gathered by the trailing end of a motile animal, lest it enter adverse and potentially dangerous conditions unawares. Hunting, tracking, and other forms of food location are greatly facilitated by having the appropriate receptors placed anteriorly—toward the direction of movement.

Longitudinal nerve cords receive information through peripheral sensory nerves from whatever sense organs are placed along the body, and they carry impulses from the cerebral ganglion to peripheral motor nerves to effector sites. Additionally, nerve cords and peripheral nerves often serve animals in reflex actions and in some highly coordinated activities that do not depend on the cerebral ganglion. The most primitive centralized nervous system may have been similar to that seen today in some free-living flatworms, with pairs of longitudinal cords attached to one another by a series of transverse connectives (Figure 3.32B). This arrangement is commonly referred to as a **ladder-like nervous system**. Among those Metazoa that have developed active lifestyles, (e.g., errant polychaetes, most arthropods, cephalopod molluscs, and vertebrates), the nervous system has become increasingly centralized through a reduction in the number of longitudinal nerve cords. However, a number of invertebrates (e.g., ectoprocts, tunicates, and echinoderms) have secondarily taken up sedentary or sessile modes of existence. Within these groups there has been a corresponding decentralization of the nervous system and a general reduction in and dispersal of sense organs.

Hormones and Pheromones

We have stressed the significance of the integrated nature of the parts and processes of living organisms and have discussed the general role of the nervous system in this regard. Organisms also produce and distribute within their bodies a variety of chemicals that regulate and coordinate biological activities. This very broad description of what may be called **chemical coordinators** obviously includes almost any substance that has some effect on bodily functions. One special category of chemical coordinators is the **hormones**. This term refers to any chemicals that are produced and secreted by some organ or tissue, and are then carried by the blood or other body fluid to exert their influence elsewhere in the body. In vertebrates, we associate this type of phenomenon with the **endocrine system**, which includes well known glands as production sites. For our purposes we may subdivide hormones into two types. First are **endocrine hormones**, which are produced by more or less isolated glands and released into the circulatory fluid. Second are **neurohormones**, which are produced by special neurons called **neurosecretory cells**.

Much remains unknown concerning hormones in invertebrates. Most of our information comes from studies on insects and crustaceans, although hormonal activity has been demonstrated in a few other taxa and is suspected in many others. Among the arthropods, hormones are involved in the control of growth, molting, reproduction, eye pigment migration, and probably other phenomena; in at least some other taxa (e.g., annelid worms), hormones influence growth, regeneration, and sexual maturation.

Hormones do not belong to any particular class of chemical compounds, nor do they all produce the same effects at their sites of action: some are excitatory, some are inhibitory. Because endocrine hormones are carried in the circulatory fluid, they reach all parts of an animal's body. The site of action, or **target site**, must be able to recognize the appropriate hormone(s) among the myriad other chemicals in its surroundings. This recognition usually involves an interaction between the hormone and the cell surface at the target site. Thus, under normal circumstances, even though a particular hormone is contacting many parts of the body, it will elicit activity only from the appropriate target organ or tissue that recognizes it.

In a general sense, **pheromones** are substances that act as "interorganismal hormones." These chemicals are produced by organisms and released into the environment, where they have an effect on other organisms.

Most pheromone research has been on intraspecific actions, especially in insects, where activities such as mate attraction are frequently related to these airborne chemicals. We may view intraspecific pheromones as coordinating the activities of populations, just as hormones help coordinate the activities of individual organisms. There is also a great deal of evidence for the existence of interspecific pheromones. For example, some predatory species (e.g., some sea stars) release chemicals into the water that elicit extraordinary behavioral responses on the part of potential prey species, generally in the form of escape behavior. We discuss examples of various pheromone phenomena for specific animal groups throughout the book.

Reproduction

As noted in the passage from Barrington that introduces this chapter, the biological success of any species depends upon its members staying alive long enough to reproduce themselves. The following account includes a discussion of the basic methods of reproduction among invertebrates and leads to the account of embryology and developmental strategies provided in Chapter 4.

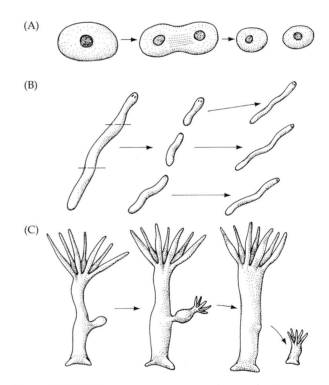

Figure 3.33   Some common asexual reproductive processes. (A) Simple mitotic binary fission; this process occurs in most protists. (B) Fragmentation, followed by regeneration of lost parts. This process occurs in a number of vermiform invertebrates. (C) Budding may produce separate solitary individuals, as it does in *Hydra* (shown here); or it may produce colonies (see Figure 3.34).

Asexual Reproduction

Asexual reproductive processes do not involve the production and subsequent fusion of haploid cells, but rely solely on vegetative growth through mitosis. Cell division itself is a common form of asexual reproduction among the protists, and many invertebrates engage in various types of body fission, budding, or fragmentation, followed by growth to new individuals (Figure 3.33). These asexual processes depend largely on the organism's "reproductive exploitation" of its ability to **regenerate** (regrow lost parts). Even wound healing is a form of regeneration, but many animals have far more dramatic capabilities. The replacement of a lost appendage in familiar animals such as sea stars and crabs is a common example of regeneration. However, these regenerative abilities are not "reproduction" because no new individuals result, and their presence does not imply that an animal capable of replacing a lost leg can necessarily reproduce asexually. Examples of organisms that possess regenerative abilities of a magnitude permitting asexual reproduction include protists, sponges, many cnidarians (corals, anemones, and hydroids), certain types of worms, and sea squirts.

In many cases asexual reproduction is a relatively incidental process and is rather insignificant to a species' overall survival strategy. In others, however, it is an integral and even necessary step in the life cycle. There are important evolutionary and adaptive aspects to asexual reproduction. Organisms capable of rapid asexual reproduction can quickly take advantage of favorable environmental conditions by exploiting temporarily abundant food supplies, newly available living space, or other resources. This competitive edge is frequently evidenced by extremely high numbers of asexually produced individuals in disturbed or unique habitats, or in other unusual conditions. In addition, asexual processes are often employed in the production of resistant cysts or overwintering bodies, which are capable of surviving through periods of harsh environmental conditions. When favorable conditions return, these structures grow to new individuals.

A word about colonies. A frequent result of asexual reproduction, particularly some forms of budding, is the formation of **colonies**. This phenomenon is especially common in certain taxa (e.g., cnidarians, ascidians, ectoprocts) (Figure 3.34). The term *colony* is not easy to define. It may initially bring to mind ant or bee colonies, or even groups of humans; but these examples are more appropriately viewed as social units rather than as colonies, at least in the context of our discussions. We accept Barrington's (1967) definition that "True colonies can be defined as ... associations in which the constituent individuals are not completely

(A)

(B)

(C)

(D)

Figure 3.34 Representative invertebrate colonies. (A) *Botryllus*, a colonial ascidian. (B) *Lophogorgia*, a colonial gorgonian. (C) Three species of coral. (D) *Aglaophenia*, a colonial hydroid.

separated from each other, but are organically connected together, either by living extensions of their bodies, or by material that they have secreted." This definition will suffice for now; we describe the nature of particular examples of colonial life in later chapters.

The formation of colonies not only may enhance the benefits of asexual reproduction in general, but also produces overall functional units that are much greater in size than mere individuals; thus this growth habit may be viewed as a partial solution to the surface-to-volume dilemma. Increased functional size through colonialism can result in a number of advantages for animals; it can increase feeding efficiency, facilitate the handling of larger food items, reduce chances of predation, increase the competitive edge for food, space, and other resources, and allow groups of individuals within the colony to specialize for different functions.

Sexual Reproduction

Although reproduction is critical to a species' survival, it is the one major physiological activity that is not essential to an individual organism's survival. In fact, when animals are stressed, reproduction is usually the first activity that ceases. Sexual reproduction is especially energy costly, yet it is the characteristic mode of reproduction among multicellular organisms.*

Given the advantages of asexual reproduction, one might wonder why all animals do not employ it and abandon sexual activities entirely. The most frequently given explanation for the popularity of sexual reproduction (aside from anthropomorphic views, of course) focuses on the long-term benefits of genetic variation. Recombination allows for the maintenance of high genetic heterozygosity in individuals and high polymorphism in populations. Through regular meiosis and recombination, a level of genetic variation is maintained generation after generation, within and among populations; thus species are thought to be more "genetically prepared" for environmental changes, including both shifts in the physical environment and the changing milieu of competitors, predators, prey, and parasites.

Although this advantage must surely be real, does it satisfactorily explain the role of sex in short-term selection (i.e., generation by generation)? Presumably even in the short term an advantage lies in the maintenance of genetic variability. That is, genetic variability in both individuals and populations may increase their chances of adapting to environmental fluctuations, predators, parasites, and disease. Leigh Van Valen (1973) proposed the idea that in order just to "keep up" with changing environments, populations must continually access new

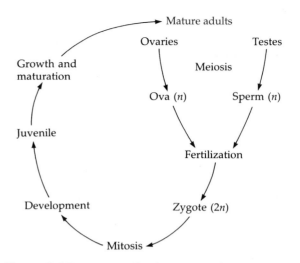

Figure 3.35 A generalized metazoan life cycle.

and different gene combinations through the process of natural selection—a notion called the "red queen hypothesis" after the Red Queen in *Alice in Wonderland*, who commanded her courtiers to run continuously just to stay in the same place.

Sexual reproduction involves the formation of haploid cells through meiosis and the subsequent fusion of pairs of those cells to produce a diploid **zygote** (Figure 3.35). The haploid cells are **gametes**—sperm and eggs—and their fusion is the process of **fertilization**, or **syngamy**. (Exceptions to these general terms and processes are common among protists as discussed in Chapter 5.) The production of gametes is accomplished by the **gonads**—**ovaries** in females and **testes** in males—or their functional equivalents. The gonads are frequently associated with reproductive systems that may include various arrangements of ducts and tubes, accessory organs such as yolk glands or shell glands, and structures for copulation. The different levels of complexity of these systems are related to the developmental strategies used by the organisms in question, as discussed in Chapter 4 and described in the coverage of each phylum. The variation in such matters is immense, but at this point we introduce some basic terminology of structure and function.

Many invertebrates simply release their gametes into the water in which they live (**broadcast spawning**), where external fertilization occurs. In such animals the gonads are usually simple, often transiently occurring structures associated with some means of getting the eggs and sperm out of the body. This release is accomplished through a discrete plumbing arrangement (coelomoducts, metanephridia, or gonoducts—sperm ducts and oviducts), or by temporary pores in or rupture of the body wall. In such animals, synchronous spawning is critical, and marine species rely largely on this synchrony and the water currents to achieve fertilization. Water temperature, light, phytoplankton abun-

---

*In thinking of animals, we typically view "sex" and "reproduction" as one in the same. However, at the cellular level these two processes are opposites: reproduction is the division of one cell to form two, whereas the sexual process includes two cells fusing to form one.

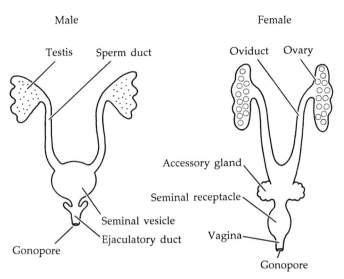

Figure 3.36 Schematic and generalized male and female reproductive systems. See text for explanation.

dance, lunar cycle, and the presence of conspecifics have all been implicated in synchronized spawning events of invertebrates.

On the other hand, invertebrates that pass sperm directly from the male to the female, where fertilization occurs internally, must have structural features to facilitate such activities. Figure 3.36 illustrates stylized male and female reproductive systems. A general scenario leading to internal fertilization in such systems is as follows. Sperm are produced in the testes and transported via the **sperm duct** to a precopulatory storage area called the **seminal vesicle**. Prior to mating, many invertebrates incorporate groups of sperm cells into sperm packets, or **spermatophores**. Spermatophores provide a protective casing for the sperm and facilitate transfer with minimal sperm loss. In addition, many spermatophores are themselves motile, acting as independent sperm carriers. Some sort of male copulatory or intromittent organ (e.g., **penis, cirrus, gonopod**) is inserted through the female's **gonopore** and into the **vagina**. Sperm are passed through the male's **ejaculatory duct** directly, or by way of a copulatory organ, into the female system, where they are received and often stored by a **seminal receptacle**.

In the female, eggs are produced in the ovaries and transported into the region of the **oviducts**. Sperm eventually travel into the female's reproductive tract, where they encounter the eggs; fertilization often takes place in the oviducts. Among invertebrates, the sperm may move by flagellar or ameboid action, or by locomotor structures on the spermatophore packet; they may be aided by ciliary action of the lining of the female reproductive tract. Various **accessory glands** may be present both in males (such as those that produce spermatophores or seminal fluids) and in females (such as those that produce yolk, egg capsules, or shells). This simple

sequence is typical (although with many elaborations) of most invertebrates that rely on internal fertilization.

Animals in which the sexes are separate, each individual being either male or female, are termed **gonochoristic**, or **dioecious**. However, many invertebrates are **hermaphroditic**,* or **monoecious**: each animal contains both ovaries and testes and thus is capable of producing both eggs and sperm (though not necessarily at the same time). Although self-fertilization may seem to be a natural advantage in this condition, such is not the case. In fact, with some exceptions, self-fertilization in hermaphrodites is usually prevented. Fertilizing one's self would be the ultimate form of inbreeding and would presumably result in a dramatic decrease in potential genetic variation and heterozygosity. The rule for many hermaphroditic invertebrates is **mutual cross fertilization**, wherein two individuals function alternately or simultaneously as males and exchange sperm, and then use the mate's sperm to fertilize their own eggs. The real advantage of hermaphroditism now becomes clear: a single sexual encounter results in the impregnation of two individuals, rather than only one as in the gonochoristic condition.

A common phenomenon among hermaphroditic invertebrates is that of **protandric hermaphroditism**, or simply **protandry** (Greek *proto*, "first"; *andro*, "male"), where an individual is first a functional male, but later in life changes sex to become a functional female. The reverse situation, female first and then male, is called **protogynic hermaphroditism**, or simply **protogyny** (Greek *gynos*, "female"). At least some invertebrates alternate regularly between being functional males and females, as explained by Jerome Tichenor (*Poems in Contempt of Progress*, 1974):

> Consider the case of the oyster,
> Which passes its time in the moisture;
> Of sex alternate,
> It chases no mate,
> But lives in self-contained cloister.

In addition to the clever oysters immortalized by Professor Tichenor, some other taxa in which the hermaphroditic condition is common include barnacles, arrow worms (Chaetognatha), flatworms, oligochaetous annelids, leeches, tunicates, and advanced gastropods.

The sexual conditions of colonial animals incude myriad variations on the themes described above. Colonies may include only one sex, both sexes, or the individuals may be hermaphroditic. A major contribution toward clarifying and classifying all the possible conditions has recently been made by Wasson and Newberry (1997) and the interested reader should consult their work.

---

*Hermaphroditus, the beautiful son of Hermes and Aphrodite, was united with a water nymph at the Carian fountain. Thus his body became both male and female.

Parthenogenesis

Parthenogenesis (Greek *partheno*, "virgin"; *genesis*, "birth") is a special reproductive strategy in which unfertilized eggs develop into viable adult individuals. Parthenogenetic species are known in many invertebrate (and vertebrate and plant) groups, including gastrotrichs, rotifers, tardigrades, nematodes, gastropods, certain insects, and various crustaceans. The taxonomic distribution of parthenogenesis is very spotty; it is rare to find a whole genus, let alone any higher taxon, that is wholly parthenogenetic. Some higher taxa that are largely parthenogenetic (e.g., aphids, cladocerans) are cyclical parthenogens, and they punctuate their life histories with sex.* The bdelloid rotifers are an exception, comprising a whole order for which males are totally unknown. There are also a number of protist higher taxa for which sex has yet to be described. Among the invertebrates, parthenogenesis usually occurs in small-bodied species that are parasites, or are free-living but inhabit extreme or highly variable habitats such as temporary freshwater ponds. There is a general trend for parthenogenesis to become more prevalent as one moves toward higher latitudes or into harsher environments. Overall, it appears that parthenogenetic taxa arise from time to time and succeed in the short run due to certain immediate advantages, but they are probably in the long run condemned to extinction through competition with their sexual relatives.

In most species that have been studied, parthenogenetic periods alternate with periods of sexual reproduction. In temperate freshwater habitats, parthenogenesis often occurs during summer months, with the population switching to sexual reproduction as winter approaches. In some species, parthenogenesis takes place for many generations, or several years, eventually to be punctuated by a brief period of sexual reproduc-

tion. In some rotifers, parthenogenesis predominates until the population attains a certain critical size, at which time males appear and a period of sexual reproduction ensues. Cladocerans switch from parthenogenesis to sexual reproduction under a number of conditions, such as overcrowding, adverse temperature, food scarcity, or even when the nature of the food changes. Many parasitic species alternate between a free-living sexual stage and a parasitic parthenogenetic one; this arrangement is seen in some nematodes, thrips (Thysanoptera), gall wasps, aphids, and certain other homopterans.

One of the most interesting examples of parthenogenesis occurs in honeybees; in these animals the queen is fertilized by one or more males (drones) at only one period of her lifetime, in her "nuptial flight." The sperm are stored in her seminal receptacles. If sperm are released when the queen lays eggs, fertilization occurs and the eggs develop into females (queens or workers). If the eggs are not fertilized, they develop parthenogenetically into males (drones).

The question of the existence or prevalence of purely parthenogenetic species has been debated for decades. Many species once thought to be entirely parthenogenetic have proved, upon closer inspection, to alternate between parthenogenesis and brief periods of sexual reproduction. In some species purely parthenogenetic populations apparently exist only in some localities. In other species, parthenogenetic lineages have been traced to sexual ancestral populations occupying relictual habitats. Nevertheless, for some parthenogenetic animals, males have yet to be found in any population, and these may indeed be purely clonal species. One cannot help but wonder how long such species can exist in the face of natural selection without the benefits of any genetic exchange. One would predict that, as with any form of asexual reproduction, obligatory parthenogenesis would eventually lead to genetic stagnation and extinction. There may, however, be some as yet unexplained genetic mechanisms to avoid this, because some parthenogenetic animals (e.g., some earthworms, insects, and lizards) are capable of inhabiting a wide range of habitats. Presumably they either have a significant level of genetic adaptability or possess "general purpose genotypes."

---

*A number of fishes and amphibians are parthenogenetic, but none has overcome the need for their egg to be penetrated by a sperm in order to initiate development. The parthenogenetic females mate with a male of another species, providing sperm that trigger development (a behavior called **pseudogamy**). A few lizards apparently have no need for a sperm to trigger parthenogenetic development. No wild birds or mammals are parthenogenetic.

Selected References

General References on Form and Function

Atkinson, J. 1992. Conceptual issues in the reunion of development and evolution. Synthese 91: 93–110.

Barrington, E. J. W. 1967. *Invertebrate Structure and Function*. Halstead Press, New York.

Beklemishev, V. N. 1969. *Principles of Comparative Anatomy of Invertebrates* (2 vols.). Trans. J. M. MacLennan; ed. Z. Kabata. University of Chicago Press, Chicago. [One of the best coverages of form and function, although written from a very non-Western perspective.]

Calow, P. 1981. *Invertebrate Biology: A Functional Approach*. Halstead Press, New York.

Clark, R. B. 1964. *Dynamics in Metazoan Evolution*. Clarendon Press, Oxford. [A functional approach to phylogeny of the Metazoa; dated, but still good reading.]

Clark, R. B. and J. B. Cowey. 1958. Factors controlling the change in shape in certain nemertean and turbellarian worms. J. Exp. Biol. 35: 731–748.

Cohen, J. 1963. *Living Embryos*. Pergamon Press, Oxford.

Cohen, J. 1977. *Reproduction*. Butterworth, London.

Cohen, J. 1979. Maternal constraints in development. *In* D. R. Newth and M. Balls (eds.), *Maternal Effects in Development*. Cambridge University Press, Cambridge, pp. 1–28.

Denny, M. 1995. Survival in the surf zone. Am. Sci. 83: 166–173.

Denny, M. 1996. *Air and Water: The Biology and Physics of Life's Media*. Princeton University Press, Ewing, NJ. [A great read.]

Denny, M. W., T. L. Daniel and M. A. R. Koehl. 1985. Mechanical limits to size in wave-swept organisms. Ecol. Monogr. 55 (1): 69–102.

Fretter, V. and A. Graham. 1976. *A Functional Anatomy of Invertebrates*. Academic Press, New York. [Somewhat dated, but still an excellent reference for the serious student.]

Gould, S. J. 1977. *Ontogeny and Phylogeny*. The Belknap Press of Harvard University Press, Cambridge, MA.

Gould, S. J. 1980. The evolutionary biology of constraint. Daedalus 109: 39–52.

Gould, S. J. 1992. Constraint and the square snail: Life at the limits of a covariance set. The normal teratology of *Cerion disforme*. Biol. J. Linnean Soc. 47: 4077–437.

Hall, B. K. 1996. Baupläne, phylotypic stages, and constraint. Why are there so few types of animals? *In* M. K. Hecht et al. (eds.), *Evolutionary Biology*, Vol. 29. Plenum, New York, pp. 215–261.

Hickman, C. S. 1988. Analysis of form and function in fossils. Am. Zool. 28: 775–793.

Hyman, L. H. 1940, 1951. *The Invertebrates*. Vol. 1, Protozoa through Ctenophora; Vol. 2, Platyhelminthes and Rhynchocoela: The Acoelomate Bilateria. McGraw-Hill, New York. [These two volumes of Hyman's series are especially useful in their discussion of body architecture in lower Metazoa.]

Jackson, J. B. C., L. W. Buss and R. E. Cook (eds.). 1985. *Population Biology and Evolution of Clonal Organisms*. Yale University Press, New Haven, CT.

Jacobs, M. H. 1967. *Diffusion Processes*. Springer-Verlag, New York.

Keegan, B. F., P. O. Ceidigh and P. J. S. Boaden (eds.). 1971. *Biology of Benthic Organisms*. Pergamon Press, New York.

Nicol, J. A. C. 1960. *The Biology of Marine Animals*. Putnam, New York. [One of the best summaries of ecological physiology in print; a shame it was never revised.]

Prosser, C. L. (ed.). 1993. *Comparative Animal Physiology*, 4th. Ed. John Wiley and Sons, NY. [An excellent and classic reference.]

Rockstein, M. (ed.). 1965–1974. *Physiology of Insecta*. Academic Press, New York.

Sander, K. 1976. Specification of the basic body pattern in insect embryogenesis. Advances in Insect Physiology 12: 125–238.

Sander, K. 1983. The evolution of patterning mechanisms: Gleanings from insect embryogenesis and spermatogenesis. *In* B. C. Goodwin, N. Holder and C. C. Wylie (eds.), *Development and Evolution*. Cambridge University Press, Cambridge pp. 137–160.

Schmidt-Nielsen, K. 1996. *Animal Physiology*, 4th Ed. Cambridge University Press, New York.

Schmidt-Nielsen, K., L. Bolis, and S. H. P. Maddrell (eds.). 1978. *Comparative Physiology: Water, Ions and Fluid Mechanics*. Cambridge University Press, New York.

Seidel, F. 1960. Körpergrundgestalt und Keimstruktur. Eine Erörterung über die Grundlagen der vergleichenden und experimentallen Embryologie und deren Gültigkeit bei phylogenetischen Überlegungen. Zool. Anz. 164: 245–305.

Thompson, D'Arcy. 1942. *On Growth and Form*, Rev. Ed. Macmillan, New York.

Tombes, A. S. 1970. *An Introduction to Insect Endocrinology*, 2nd Ed. Academic Press, New York.

Valentine, J. W. 1986. Fossil record of the origin of Baupläne and its implications. *In* J. H. Lipps and P. W. Signor (eds.), *Origin and Early Evolution of the Metazoa*. Plenum, New York, pp. 209–222.

Vogel, S. 1981. *Life in Moving Fluids: The Physical Biology of Flow*, 2nd Ed. Princeton University Press, Ewing, NJ.

Vogel, S. 1988. How organisms use flow-induced pressures. Am. Sci. 76: 28–34.

Vogel, S. 1988. *Life's Devices. The Physical World of Animals and Plants*. Princeton University Press, Ewing, NJ.

Waterman, T. H. (ed.). 1960. *Physiology of Crustacea*. Academic Press, New York.

Wigglesworth, V. B. 1974. *Insect Physiology*. Chapman and Hall, London.

Wilbur, K. M. and C. M. Yonge (eds.). 1964, 1967. *Physiology of Mollusca*. Academic Press, New York.

Woodger, J. H. 1945. On biological transformations. *In* W. E. Le Gros Clark and P. B. Medawar (eds.), *Essays on Growth and Form*

Presented to D'Arcy Wentworth Thompson. Cambridge University Press, Cambridge, pp. 95–120.

Yeates, D. K. 1995. Groundplans and exemplars: Paths to the tree of life. Cladistics 11: 343–357.

Locomotion and Support

Alexander, R. M. and G. Goldspink (eds.). 1977. *Mechanics and Energetics of Animal Locomotion*. Chapman and Hall, London.

Bereiter-Hahn, J., A. G. Matoltsy and K. S. Richards (eds.). 1984. *Biology of the Integument*. Vol. 1, Invertebrates. Springer-Verlag, New York.

Blake, J. R. and M. A. Sleigh. 1974. Mechanics of ciliary locomotion. Biol. Rev. 49: 85–125.

Bray, D. and J. G. White. 1988. Cortical flow in animal cells. Science 239:883–888.

Chapman, G. 1958. The hydrostatic skeleton in the invertebrates. Biol. Rev. 33: 338–371.

Herreid, C. T. II and C. R. Rourtner (eds.). 1981. *Locomotion and Energetics in Arthropods*. Plenum, New York.

Huxley, T. 1965. The mechanism of muscle contraction. Sci. Am. 213: 18–27.

Jahn, T. L. and E. C. Bovee. 1967. Motile behavior of protozoa. *In* T. Chen (ed.), *Research in Protozoology*. Pergamon Press, New York, pp. 41–200.

Jahn, T. L. and E. C. Bovee. 1969. Protoplasmic movements within cells. Phys. Rev. 49(4): 830–862.

Jeffrey, D. J. and J. D. Sherwood., 1980. Streamline patterns and eddies in low Reynolds number flow. J. Fluid Mech. 96: 315–334.

Jones, A. R. 1974. *The Ciliates*. St. Martin's, New York.

Koehl, M. A. R. 1984. How do benthic organisms withstand moving water? Am. Zool. 24: 57–70.

Lowenstam, H.A. 1981. Minerals formed by organisms. Science 211: 1126–1131.

Muzzarelli, R. 1977. *Chitin*. Pergamon Press, New York.

Satir, P. 1974. How cilia move. Sci. Am. 231: 44–54.

Sellers, J. R. and B. Kachar. 1990. Polarity and velocity of sliding filaments: Control of direction by actin and of speed by myosin. Science 249: 406–408.

Sleigh, M. A. (ed.). 1974. *Cilia and Flagella*. Academic Press, New York.

Stossel, T. P. 1990. How cells crawl. Am. Sci. 78: 408–423.

Tennekes, H. 1996. *The Simple Science of Flight: From Insects to Jumbo Jets*. M.I.T Press, Cambridge, MA.

Trueman, E. R. 1975. *The Locomotion of Soft-Bodied Animals*. American Elsevier, New York.

Warner, F. D. and P. Satir. 1974. The structural basis of ciliary bend formation. Radial spoke positional changes accompanying microtubule sliding. J. Cell Biol. 63: 35–63.

Yates, G. T. 1986. How microorganisms move through water. Am. Sci. 74: 358–365.

Feeding

Anderson, J. M. and A. MacFayden (eds.). 1976. *The Role of Terrestrial and Aquatic Organisms in Decomposition Processes*. Blackwell Scientific, Oxford.

American Zoologist 22(3). 1982. "The Role of Uptake of Organic Solutes in Nutrition of Marine Organisms." Pp. 611–733. [A collection of ten papers.]

Blake, J. R., N. Liron and G. K. Aldis. 1982. Flow patterns around ciliated microorganisms and in ciliated ducts. J. Theor. Biol. 98: 127–141.

Case, T. J. and R. K. Washino. 1979. Flatworm control of mosquito larvae in rice fields. Science 206: 1412–1414.

Deibel, D., M.-L. Dickson and C. V. L. Powell. 1985. Ultrastructure of the mucous feeding filter of the house of the appendicularian *Oikopleura vanhoeffeni*. Mar. Ecol. Prog. Ser. 27: 79–86.

Emlet, R. B. and R. R. Strathmann. 1985. Gravity, drag, and feeding currents of small zooplankton. Science 228: 1016–1017.

Fauchald, K. and P. A. Jumars. 1979. The diet of worms: A study of polychaete feeding guilds. Oceanogr. Mar. Biol. Annu. Rev. 17: 193–284.

Ford, M. J. 1977. Energy costs of the predatory strategy of the web-spinning spider *Lepthyphantes zimmermanni* Bertkau (Linyphiidae). Oecologia 28: 341–349.

Gerritsen, J. and K. G. Porter. 1982. The role of surface chemistry in filter feeding by zooplankton. Science 216: 1225–1227.

Hopkins, T.L. 1985. Food web of an Antarctic midwater ecosystem. Mar. Biol. 89:197–212.

Hughes, R. N. (ed.). 1993. *Diet Selection: An Interdisciplinary Approach to Foraging Behaviour*. Blackwell Science, Boston.

Jennings, J. B. 1973. *Feeding, Digestion and Assimilation in Animals*, 2nd Ed. Pergamon Press, New York.

Jorgensen, C. B. 1966. *Biology of Suspension Feeding*. Pergamon Press, New York.

Jorgensen, C. B. 1976. August Putter, August Krogh, and modern ideas on the use of dissolved organic matter in aquatic environments. Biol. Rev. 51: 291–329. [For an introduction to the literature on dissolved organic matter and its role in invertebrate nutrition, see *American Zoologist* 22 (3) (1982), "The Role of Uptake of Organic Solutes in Nutrition of Marine Organisms."]

Jorgensen, C. B. 1982. Uptake of dissolved amino acids from natural sea water in the mussel *Mytilus edulis* L. Ophelia 21: 215–221.

Jorgensen, C. B., T. Kiorboe, J. Mohlenberg and H. U. Riisgard. 1984. Ciliary and mucus-net filter feeding, with special reference to fluid mechanical characteristics. Mar. Ecol. Prog. Ser. 15: 283–292.

Koehl, M. A. R. and J. R. Strickler. 1981. Copepod feeding currents: Food capture at low Reynolds number. Limnol. Oceanogr. 26: 1062–1093.

Krogh, A. 1931. Dissolved substances as food of aquatic organisms. Biol. Rev. 6: 412–442.

LaBarbera, M. 1984. Feeding and particle capture mechanisms in suspension feeding animals. Am. Zool. 24: 71–84.

Lehman, J. T. 1976. The filter-feeder as an optimal forager and the predicted shapes of feeding curves. Limnol. Oceanogr. 21: 501–516.

Lindstedt, K. J. 1971. Chemical control of feeding behavior. Comp. Biochem. Physiol. 39A: 553–581.

Madin, L. P. 1988. Feeding behavior of tentaculate predators: In situ observations and a conceptual model. Bull. Mar. Sci. 43(3): 413–429.

Manahan, D. T., S. H. Wright, G. C. Stephens and M. A. Rice. 1982. Transport of dissolved amino acids by the mussel, *Mytilus edulis*: Demonstration of net uptake from natural sea water. Science 215: 1253 1255.

Mechiorri-Santolini, U. and J. W. Hopton (eds.). 1972. Detritus and its role in aquatic ecosystems. Proceedings of an IBP-UNESCO Symposium Mem. Dell'Instituto Italiano di Idrobiologia. Vol. 29 (Suppl.), 1972.

Menge, B. 1972. Foraging strategies in starfish in relation to actual prey availability and environmental predictability. Ecol. Monogr. 42: 25–50.

Meyers, D. G. and J. R. Stickler (eds.). 1984. *Trophic Interactions within Aquatic Ecosystems*. AAAS Selected Symp., Vol 85. Westview Press, Boulder, CO.

Meyhofer, E. 1985. Comparative pumping rates in suspension-feeding bivalves. Mar. Biol. 85: 137–142.

Nicol, E. A. 1930. The feeding mechanism, formation of the tube and physiology of digestion in *Sabella pavonina*. Trans. Royal Soc. Edinburgh 56: 537–596.

O'Brien, W. J., D. Keetle and H. Riessen. 1979. Helmets and invisible armour: Structures reducing predation from tactile and visual planktivores. Ecology 60: 287–294.

Polis, G. A. 1981. The evolution and dynamics of intraspecific predation. Annu. Rev. Ecol. Syst. 12: 225–251.

Poulet, S.A. and P. Marsot. 1975. Chemosensory grazing by marine calanoid copepods (Arthropoda: Crustacea). Science 200: 1403–1405.

Reid, R. G. B. and F. R. Bernard. 1980. Gutless bivalves. Science 208: 609–610.

Roth, L. E. 1960. Electron microscopy of pinocytosis and food vacuoles in *Pelomyxa*. J. Protozool. 7: 176–185.

Schoener, T. W. 1971. Theory of feeding strategies. Annu. Rev. Ecol. Syst. 2: 369–404.

Sleeper, H. L., V. J. Paul and W. Fenical. 1980. Alarm pheromones from the marine opisthobranch *Navanax inermis*. J. Chem. Ecol. 6: 57–70.

Smith, D., L. Muscatine, and O. Lewis. 1969. Carbohydrate movement from autotrophs to heterotrophs in parasitic and mutualistic symbioses. Biol. Rev. 44: 17–90.

Smith, D. C. and Y. Tiffon (eds.). 1980. *Nutrition in the Lower Metazoa*. Pergamon Press, Elmsford, NY.

Spielman, L. A. 1977. Particle capture from low-speed laminar flows. Annu. Rev. Fluid Mech. 9: 297–319.

Strathmann, R. R., T. L. Jahn and J. R. C. Fonesca. 1972. Suspension feeding by marine invertebrate larvae: Clearance of particles by ciliated bands of a rotifer pluteus, and trochophore. Biol. Bull. 142: 505–519.

Strathmann, R. R. 1978. The evolution and loss of feeding larval stages of marine invertebrates. Evolution 32: 894–906.

Stephens, G. C. 1981. The trophic role of dissolved organic material. *In* A. L. Longhurst (ed.), *Analysis of Marine Ecosystems*. Academic Press, New York.

Yonge, C. M. 1928. Feeding mechanisms in the invertebrates. Biol. Rev. 3: 21–76.

Zaret, T. M. 1980. *Predation in Freshwater Communities*. Yale University Press, New Haven.

Excretion and Osmoregulation

Bartolomaeus, T. and P. Ax. 1992. Protonephridia and metanephridia—their relation within the Bilateria. Z. Zool. syst. Evolut.-forsch. 30: 21–45.

Fisher, R. S., B. E. Persson and K. R. Spring. 1981. Epithelial cell volume regulation: Bicarbonate dependence. Science 214: 1357–1359.

Giles, R. and A. Pequeux. 1981. Cell volume regulation in crustaceans: Relationship between mechanisms for controlling extracellular and intracellular fluids. J. Exp. Zool. 215: 351–362.

Goodrich, E. S. 1945. The study of nephridia and genital ducts since 1895. Q. J. Micros. Sci. 86: 113–392. [Goodrich's classic paper on the evolution of coelomoducts, gonoducts, and nephridia.]

Hellebust, J. A., T. Mérida and I. Ahmad. 1989. Operation of contractile vacuoles in the euryhaline green flagellate *Chlamydomonas pulsatilla* (Chlorophyceae) as a function of salinity. Mar. Biol. 100: 373–379.

Kormanik, G. A. and J. N. Cameron. 1981. Ammonia excretion in animals that breath water: A review. Mar. Biol. Letters 2:11–23.

Oglesby, L. C. 1981. Volume regulation in aquatic invertebrates. J. Exp. Zool. 215: 289–301. [Excellent summary paper.]

Pierce, S. K. 1982. Invertebrate cell volume control mechanisms: A coordinated use of intracellular amino acids and inorganic ions as osmotic solute. Biol. Bull. 163: 405–419. [Osmoregulation at the cellular level; an important paper.]

Potts, W. T. W. and G. Parry. 1964. *Osmotic and Ionic Regulation in Animals*. Pergamon Press, New York.

Ramsay, J. A. 1954. Movements of water and electrolytes in invertebrates. Symp. Soc. Exp. Biol. 8: 1–15.

Ramsay, J. A. 1961. The comparative physiology of renal function in invertebrates. *In* J. A. Ramsay and V. B. Wigglesworth (eds.), *The Cell and the Organism*, Cambridge University Press, London. pp. 158–174.

Ruppert, E. E. and P. R. Smith. 1988. The fundamental organization of filtration nephridia. Biol. Rev. 6: 231–258.

Walsh, P. and P. Wright (eds.). 1995. *Nitrogen Metabolism and Excretion*. CRC Press.

Wilson, R. A. and L. A. Webster. 1974. Protonephridia. Biol. Rev. 49: 127–160.

Wright, S. H. 1982. A nutritional role for amino acid transport in filter-feeding marine invertebrates. Am. Zool. 22:621–634.

Yancey, P. H., M. E. Clark, S. C. Hand, R. D. Bowlus and G. N. Somero. 1982. Living with water stress: Evolution of osmolyte systems. Science 217: 1214–1222.

Circulation and Gas Exchange

Bird, R. B., W. E. Stewart and N. Lightfoot. 1960. *Transport Phenomena*. Wiley, New York.

Florkin, M. and B. Sheer (eds.). 1967–1978. *Chemical Zoology*. Academic Press, New York. [Ten volumes, which include articles by numerous authors on animal physiology and biochemistry.]

LaBarbera, M. 1990. Principles of design of fluid transport systems in zoology. Science 249: 992–1000.

LaBarbera, M. and S. Vogel. 1982. The design of fluid transport systems in organisms. Am. Sci. 70: 54–60.

Malvin, G. M. and S. C. Wood. 1992. Behavioral hypothermia and survival of hypoxic protozoans *Paramecium caudatum*. Science 2556: 1423–1426.

McMahon, B. R. and L. E. Burnett. 1990. The crustacean open circulatory system. A reexamination. Physiol. Zool. 63: 35–71.

Mill, P. J. 1972. *Respiration in the Invertebrates*. St. Martin's Press, London.

Ratcliffe, N. A. and A. F. Rowley (eds.). 1981. *Invertebrate Blood Cells*. Vols. 1–2. Academic Press, New York. [For a good introduction to the literature on circulation and gas exchange in invertebrates, see *American Zoologist* 19 (1) (1979), "Comparative Physiology of Invertebrate Hearts;" for a good introduction to the literature on respiratory pigments, see *American Zoologist* 20 (1) (1980), "Respiratory Pigments."]

Nervous Systems, Endocrines, and Behavior

Ali, M. A. (ed.). 1982. *Photoreception and Vision in Invertebrates*. Plenum, New York.

Atwood, H. L. and D. C. Sandeman (eds.). 1982. *The Biology of Crustacea*. Vol. 3, Neurobiology: Structure and Function. Academic Press, New York.

Autrum, H., R. Jung, W. R. Loewenstein, D. M. Mackay and H. L. Teuber (eds.). 1972–1981. *Handbook of Sensory Physiology*. Vols. 1–7. Springer-Verlag, New York. [This multivolume work includes the efforts of over 400 authors, and contains excellent coverage.]

Barrington, E. J. W. 1963. *An Introduction to General and Comparative Endocrinology*. Clarendon Press, Oxford.

Bullock, T. and G. Horridge. 1969. *Structure and Function of the Nervous System of Invertebrates*. W. H. Freeman, San Francisco. [This two-volume work is still one of the best summaries available on the subject.]

Bullock, T., R. Ork and A. Grinnel. 1977. *Introduction to Nervous Systems*. W. H. Freeman, San Francisco.

Carthy, J. and G. Newell (eds.). 1968. *Invertebrate Receptors*. Academic Press, New York. [Excellent.]

Corning, W. C., J. A. Dyal and A. O. D. Willows (eds.). 1973. *Invertebrate Learning* (2 vols.). Plenum, New York.

Cronin, T. W. 1986. Photoreception in marine invertebrates. Am. Zool. 26: 403–415.

Dumont, J. P. C. and R. M. Robertson. 1986. Neuronal circuits: An evolutionary perspective. Science 233: 849–853.

Eakin, R. M. 1963. Lines of evolution of photoreceptors. *In* D. Mazia and A. Tyler (eds.), *General Physiology of Cell Specialization*. McGraw-Hill, New York, pp. 393–425.

Fein, A. and E. Z. Szuts. 1982. *Photoreceptors: Their Role in Vision*. Cambridge University Press, New York.

Hanstrom, B. 1928. *Vergleichende Anatomie des Nervensystems der Wirbellosen Tiere*. Springer-Verlag, Berlin. Reprinted in 1968 by A. Asher and Co., Amsterdam.

Highnam, K. C. and L. Hill. 1977. *The Comparative Endocrinology of the Invertebrates*, 2nd Ed. University Park Press, Baltimore.

Jennings, H. S. 1976. *Behavior of the Lower Organisms*. Indiana University Press, Bloomington.

Laverack, M. S. 1968. On the receptors of marine invertebrates. Oceanogr. Mar. Biol. Annu. Rev. 6: 249–324.

Mill, P. J. 1976. *Structure and Function of Proprioceptors in the Invertebrates*. Halsted Press, New York. [The author assumes a very broad definition of proprioceptors, hence this large book also includes fine descriptions of a number of mechanoreceptors.]

Oceanus 23 (3). 1980. [This issue contains a number of introductory review articles on senses of marine animals.]

Salanki, J. (ed.). 1973. *Neurobiology of Invertebrates*. Academiai Riado, Budapest.

Sandeman, D. C. and H. L. Atwood (eds.). 1982. *The Biology of Crustacea*. Vol. 4, Neural Integration and Behavior. Academic Press, New York.

Wolken, J. J. 1971. *Invertebrate Photoreceptors*. Academic Press, New York.

Reproduction

Adiyodi, K. G. and R. G. Adiyodi (eds.). 1983. *Reproductive Biology of Invertebrates*. Vol. 1. Oogenesis, Oviposition, and Oosorption. Wiley, New York.

American Zoologist 19 (3). 1979. "Ecology of Asexual Reproduction." [This collection provides an introduction to the literature on the evolutionary implications of asexual reproduction.]

Boardman, R. S., A. M. Cheetham and W. A. Oliver (eds.). 1973. *Animal Colonies: Development and Function Through Time*. Dowden, Hutchinson and Ross, Stroudsburg, PA.

Braverman, M. H. and R. G. Schrandt. 1966. Colony formation of a polymorphic hydroid as a problem in pattern formation. *In* W. J. Rees (ed.), *The Cnidaria and their Evolution*. Symp. Zool. Soc. Lond. 16: 169–198.

Giese, A. C. and J. A. Pearse (eds.). 1974–1987. *Reproduction of Marine Invertebrates*. Vols. 1–5, 9. Academic Press, New York. [Volumes 6–8 are not yet published as we write this; Volume 9 has outstanding reviews of most aspects of invertebrate gametogenesis and development.]

Lockwood, G. and B. R. Rosen (eds.). 1979. *Biology and Systematics of Colonial Animals*. Academic Press, New York. Published for The Systematics Association, Special Vol. No. 11.

Maynard Smith, J. 1978. *The Evolution of Sex*. Cambridge University Press, Cambridge.

Morse, A. N. C. 1991. How do planktonic larvae know where to settle? Am. Sci. 79: 154–167.

Policansky, D. 1982. Sex change in plants and animals. Annu. Rev. Ecol. Syst. 13: 471–496.

Rose, S. M. 1970. *Regeneration*. Appleton-Century-Crofts, New York.

Van Valen, L. 1973. A new evolutionary law. Evol. Theory 1: 1–30.

Wasson, K. and A. T. Newberry. 1997. Modular metazoans: Gonochoric, hermaphroditic, or both at once? Invert. Reprod. Devel. 31: 159–175.

[Also see Chapter 4 references.]

4 *Animal Development, Life Histories, and Origins*

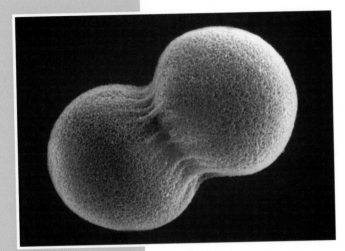

He who sees things grow from their beginning will have the finest view of them.

Aristotle

nimals, or Metazoa, are multicellular, as opposed to protists, (i.e., members of the kingdom Protista), which are usually viewed as being unicellular (see Chapter 5). This distinction, however, is sometimes blurred, for there are a number of protists that form complex colonies with some division of labor among different cell types. Thus, the Metazoa possess certain qualities that must be considered in concert with the basic idea of multicellularity. The cells of animals are organized into functional units, generally as tissues and organs, with specific roles that support the life of the whole animal. These cell types are interdependent and their activities are coordinated into predictable patterns and relationships. Structurally, the cells of animals are organized as layers that develop through a series of events early in an organism's embryogeny. These embryonic tissues, or **germ layers**, form the framework upon which metazoan body plans are constructed (see Chapter 3). Thus, the cells of animals (i.e., the Metazoa) are *specialized, interdependent, coordinated in function, and develop through layering during embryogeny*. This combination of features is absent from the protists.

Eggs and Embryos

The attributes that distinguish the Metazoa are the result of their embryonic development. To put it another way, adult phenotypes result from specific sequences of developmental stages, and evolutionary patterns reveal themselves in large part through ontogenies. Therefore, both animal unity and diversity are as evident in patterns of development as they are in the architecture of adults. The

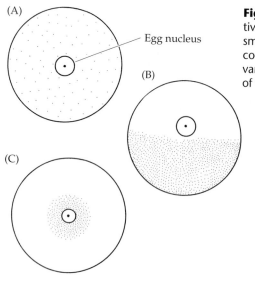

(A)

Egg nucleus

(B)

(C)

Figure 4.1   Types of ova. The stippling denotes the distribution and relative concentration of yolk within the cytoplasm. (A) An isolecithal ovum has a small amount of yolk distributed evenly. (B) The yolk in a telolecithal ovum is concentrated toward the vegetal pole. The amount of yolk in such eggs varies greatly. (C) A centrolecithal ovum has yolk concentrated at the center of the cell.

patterns of development discussed below reflect this unity and diversity, and serve as a basis for understanding the sections on embryology in later chapters.

Eggs

Biological processes in general are cyclical. The production of one generation after another through reproduction exemplifies this generality, as the term *life cycle* implies. At what point one begins describing such things is more or less a matter of convenience. For our purposes in this chapter we choose to begin with the **egg**, or **ovum**, a single remarkable cell capable of developing into a new individual. Once the egg is fertilized, all of the different cell types of an adult animal are derived during embryogenesis from this single totipotent cell. A fertilized egg contains not only the information necessary to direct development but also some quantity of nutrient material called **yolk**, which sustains the early stages of life.

Eggs are polarized along what is called the **animal–vegetal axis**. This polarity may be apparent in the egg itself, or it may be recognizable only as development proceeds. The vegetal pole is commonly associated with the formation of nutritive organs (e.g., the digestive system), whereas the animal pole tends to produce other regions of the embryo. These and many other manifestations of the egg's polarity are more completely explored throughout this chapter.

Animal ova are categorized primarily by the amount and location of yolk within the cell (Figure 4.1), two factors that greatly influence certain aspects of development. **Isolecithal eggs** contain a relatively small amount of yolk that is more or less evenly distributed throughout the cell. Ova in which the yolk is concentrated at one end (toward the vegetal pole) are termed **telolecithal eggs**; those in which the yolk is concentrated in the center are called **centrolecithal eggs**. The actual amount of yolk in telolecithal and centrolecithal eggs is highly variable. Yolk production (**vitellogenesis**) is typically the longest

phase of egg production, although its duration varies by orders of magnitude among species. Rates of yolk production depend on the specific vitellogenic mechanism used. In general, so-called *r*-selected (opportunistic) species have evolved vitellogenic pathways for the rapid conversion of food into egg production, while *K*-selected (specialist) species utilize slower pathways.

Cleavage

The stimulus that initiates development in an ovum is usually provided by the penetration of a sperm cell and the subsequent fusion of the male and female nuclei to produce a fertilized egg, or **zygote**. The initial cell divisions of a zygote are called **cleavage**, and the resulting cells are called **blastomeres**. Certain aspects of the patterns of early cleavage are determined by the amount and placement of yolk, while other features are inherent in the genetic programming of the particular organism. Isolecithal and weakly to moderately telolecithal ova generally undergo **holoblastic cleavage**. That is, the cleavage planes pass completely through the cell, producing blastomeres that are separated from one another by thin cell membranes (Figure 4.2A). Whenever very large amounts of yolk are present (as in strongly telolecithal eggs), the cleavage planes do not pass readily through the dense yolk, so the blastomeres are not fully separated from one another by cell membranes. This pattern of early cell division is called **meroblastic cleavage** (Figure 4.2B). The pattern of cleavage in centrolecithal eggs is dependent on the amount of yolk and varies from holoblastic to various modifications of meroblastic (for some examples, see the descriptions of arthropod development in Chapter 15).

Orientation of cleavage planes.   A number of terms are used to describe the relationship of the planes of cleavage to the animal–vegetal axis of the egg and the relationships of the resulting blastomeres to each other. Figure 4.3 illustrates the patterns described below. Cell divisions during cleavage are often referred to as either **equal** or **unequal**, the terms indicating the comparative sizes of groups of blastomeres. The term **subequal** is used when blastomeres are only slightly different in size. When cleavage is distinctly unequal, the larger cells are called **macromeres** and usually lie at the vegetal pole. The smaller cells are called **micromeres** and are usually located at the animal pole.

Cleavage planes that pass through or parallel to the animal–vegetal axis produce **longitudinal** (= **meridional**) divisions; those that pass at right angles to the axis

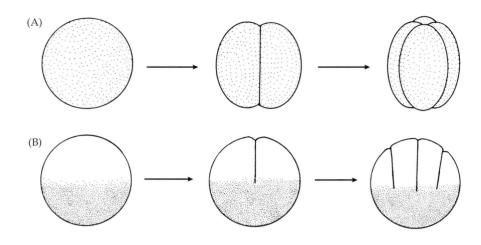

Figure 4.2 Types of early cleavage in developing zygotes. (A) Holoblastic cleavage. The cleavage planes pass completely through the cytoplasm. (B) Meroblastic cleavage. The cleavage planes do not pass completely through the yolky cytoplasm.

Radial and spiral cleavage. Most invertebrates display one of two cleavage patterns defined on the basis of the orientation of the blastomeres about the animal–vegetal axis. These patterns are called **radial cleavage** and **spiral cleavage** and are illustrated in Figure 4.4. Radial cleavage involves strictly longitudinal and transverse divisions. Thus, the blastomeres are arranged in rows either parallel or perpendicular to the animal–vegetal axis. The placement of the blastomeres shows a radially symmetrical pattern in polar view.

Spiral cleavage is quite another matter. Although not inherently complex, it can be difficult to describe. The first two divisions are longitudinal, generally equal or subequal. Subsequent divisions, however, result in the displacement of blastomeres in such a way that they lie in the furrows between one another. This condition is a result of the formation of the mitotic spindles at acute angles rather than parallel to the axis of the embryo; hence the cleavage planes are neither perfectly longitudinal nor perfectly transverse. The division from four to eight cells involves a displacement of the cells near the animal pole in a clockwise (**dextrotropic**) direction (viewed from the animal pole). The next division, from eight to sixteen cells, occurs with a displacement in a counterclockwise (**levotropic**) direction; the next is clockwise, and so on—alternating back and forth until approximately the 64-cell stage. We hasten to add that divisions are frequently not synchronous; not all of the cells divide at the same rate. Thus, a particular embryo may not proceed from four cells to eight, to sixteen, and so on, as neatly as in our generalized example.

produce **transverse** divisions. Transverse divisions may be **equatorial**, when the embryo is separated equally into animal and vegetal halves, or simply **latitudinal**, when the division plane does not pass through the equator of the embryo.

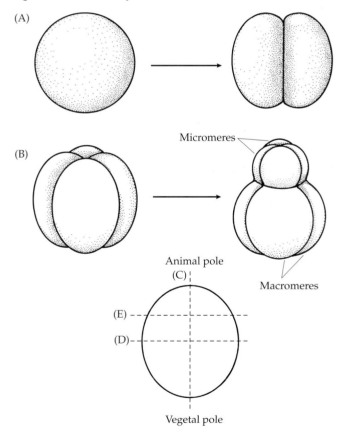

Figure 4.3 Planes of holoblastic cleavage. (A) Equal cleavage. (B) Unequal cleavage produces micromeres and macromeres. (C–E) Planes of cleavage relative to the animal–vegetal axis of the egg or zygote. (C) Longitudinal (= meridional) cleavage parallel to the animal–vegetal axis. (D) Equatorial cleavage perpendicular to the animal–vegetal axis and bisecting the zygote into equal animal and vegetal halves. (E) Latitudinal cleavage perpendicular to the animal–vegetal axis but not passing along the equatorial plane.

An elaborate coding system for spiral cleavage was developed by E. B. Wilson (1892) during his extensive studies on the polychaete worm *Neanthes succinea* conducted at the Marine Biological Laboratory at Woods Hole. Wilson's system is usually applied to spiral cleavage in order to trace cell fates and compare development among species. The following account of spiral cleavage is a general one, but it will provide a point of reference for later consideration of the patterns in different groups of animals. Wilson's code is a simple and elegant means of following the developmental lineage of each and every cell in an embryo.

At the 4-cell stage, following the initial longitudinal divisions, the cells are given the codes of A, B, C, and D,

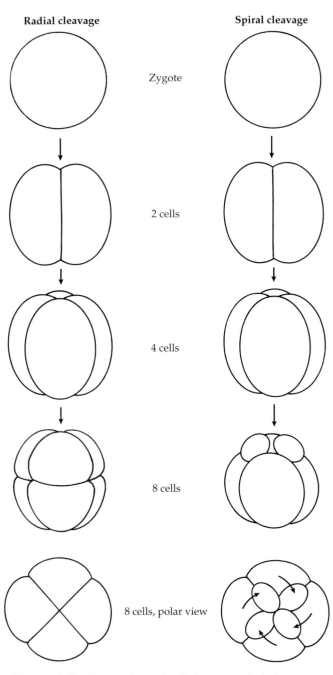

Radial cleavage   **Spiral cleavage**

Zygote

2 cells

4 cells

8 cells

8 cells, polar view

Figure 4.4   Comparison of radial versus spiral cleavage through the 8-cell stage. During radial cleavage, the cleavage planes all pass either perpendicular or parallel to the animal–vegetal axis of the embryo. Spiral cleavage involves a tilting of the mitotic spindles, commencing with the division from 4 to 8 cells. The resulting cleavage planes are neither perpendicular nor parallel to the axis. The polar views of the resulting 8-cell stages illustrate the differences in blastomere orientation.

and are labeled clockwise in that order when viewed from the animal pole (Figure 4.5A). These four cells are referred to as a quartet of macromeres, and they may be collectively coded as simply Q. The next division is more or less unequal, with the four cells nearest the animal pole being displaced in a dextrotropic fashion, as

explained above. These four smaller cells are called the first quartet of micromeres (collectively the lq cells) and are given the individual codes of 1a, 1b, 1c, and 1d. The numeral "1" indicates that they are members of the first micromere quartet to be produced; the letters correspond to their respective macromere origins. The capital letters designating the macromeres are now preceded with the numeral "1" to indicate that they have divided once and produced a first micromere set (Figure 4.5B). We may view this 8-celled embryo as four pairs of daughter cells that have been produced by the divisions of the four original macromeres as follows:

$$A \begin{cases} \nearrow 1a \\ \searrow 1A \end{cases} \quad B \begin{cases} \nearrow 1b \\ \searrow 1B \end{cases} \quad C \begin{cases} \nearrow 1c \\ \searrow 1C \end{cases} \quad D \begin{cases} \nearrow 1d \\ \searrow 1D \end{cases}$$

It should be mentioned that even though the macromeres and micromeres are sometimes similar in size, these terms are nonetheless always used in describing spiral cleavage. Much of the size discrepancy depends upon the amount of yolk present at the vegetal pole in the original egg; this yolk tends to be retained primarily in the larger macromeres.

The division from 8 to 16 cells occurs levotropically and involves cleavage of each macromere and micromere. Note that the only code numbers that are changed through subsequent divisions are the prefix numbers of the macromeres. These are changed to indicate the number of times these individual macromeres have divided, and to correspond to the number of micromere quartets thus produced. So, at the 8-cell stage, we can designate the existing blastomeres as the 1Q (= 1A, 1B, 1C, 1D) and the lq (= 1a, 1b, 1c, 1d). The macromeres (1Q) divide to produce a second quartet of micromeres (2q = 2a, 2b, 2c, 2d), and the prefix numeral of the daughter macromeres is changed to "2." The first micromere quartet also divides and now comprises eight cells, each of which is identifiable not only by the letter corresponding to its parent macromere but now by the addition of superscript numerals. For example, the 1a micromere (of the 8-cell embryo) divides to produce two daughter cells coded the $1a^1$ and the $1a^2$ cells. The cell that is physically nearer the animal pole of the embryo receives the superscript "1," the other cell the superscript "2." Thus, the 16-cell stage (Figure 4.5C) includes the following cells:

Derivatives of the 1q $\begin{cases} 1a^1 \quad 1b^1 \quad 1c^1 \quad 1d^1 \\ 1a^2 \quad 1b^2 \quad 1c^2 \quad 1d^2 \end{cases}$

Derivatives of the 1Q $\begin{cases} 2q = 2a \quad 2b \quad 2c \quad 2d \\ 2Q = 2A \quad 2B \quad 2C \quad 2D \end{cases}$

The next division (from 16 to 32 cells) involves dextrotropic displacement. The third micromere quartet (3q) is formed, and the daughter macromeres are now given the prefix "3" (3Q), and all of the 12 existing mi-

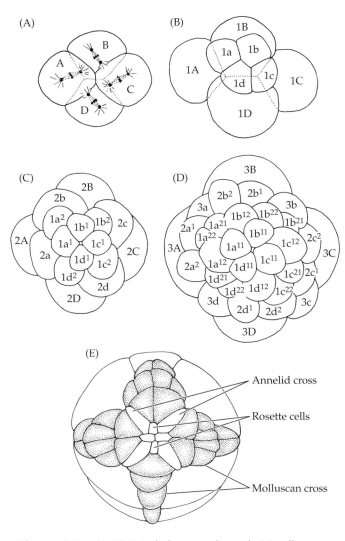

embryo relative to its sister cells. The 32-cell state (Figure 4.5D) is composed of the following:

$$
\text{Derivatives of the 1q}
\begin{cases}
1a^{11} & 1b^{11} & 1c^{11} & 1d^{11} \\
1a^{12} & 1b^{12} & 1c^{12} & 1d^{12} \\
1a^{21} & 1b^{21} & 1c^{21} & 1d^{21} \\
1a^{22} & 1b^{22} & 1c^{22} & 1d^{22}
\end{cases}
$$

$$
\text{Derivatives of the 2q}
\begin{cases}
2a^{1} & 2b^{1} & 2c^{1} & 2d^{1} \\
2a^{2} & 2b^{2} & 2c^{2} & 2d^{2}
\end{cases}
$$

$$
\text{Derivatives of the 2Q}
\begin{cases}
3q = & 3a & 3b & 3c & 3d \\
3Q = & 3A & 3B & 3C & 3D
\end{cases}
$$

The division to 64 cells follows the same pattern, with appropriate coding changes and additions of superscripts. The displacement is levotropic and results in the following cells:

$$
\text{Derivatives of the 1q}
\begin{cases}
1a^{111} & 1b^{111} & 1c^{111} & 1d^{111} \\
1a^{112} & 1b^{112} & 1c^{112} & 1d^{112} \\
1a^{121} & 1b^{121} & 1c^{121} & 1d^{121} \\
1a^{122} & 1b^{122} & 1c^{122} & 1d^{122} \\
1a^{211} & 1b^{211} & 1c^{211} & 1d^{211} \\
1a^{212} & 1b^{212} & 1c^{212} & 1d^{212} \\
1a^{221} & 1b^{221} & 1c^{221} & 1d^{221} \\
1a^{222} & 1b^{222} & 1c^{222} & 1d^{222}
\end{cases}
$$

$$
\text{Derivatives of the 2q}
\begin{cases}
2a^{11} & 2b^{11} & 2c^{11} & 2d^{11} \\
2a^{12} & 2b^{12} & 2c^{12} & 2d^{12} \\
2a^{21} & 2b^{21} & 2c^{21} & 2d^{21} \\
2a^{22} & 2b^{22} & 2c^{22} & 2d^{22}
\end{cases}
$$

$$
\text{Derivatives of the 3q}
\begin{cases}
3a^{1} & 3b^{1} & 3c^{1} & 3d^{1} \\
3a^{2} & 3b^{2} & 3c^{2} & 3d^{2}
\end{cases}
$$

$$
\text{Derivatives of the 3Q}
\begin{cases}
4q = & 4a & 4b & 4c & 4d \\
4Q = & 4A & 4B & 4C & 4D
\end{cases}
$$

Notice that no two cells share the same code, so exact identification of individual blastomeres and their lineages is always possible.

Late in the spiral cleavage of certain animals, distinctive cell patterns appear, formed by the orientation of some of the apical first-quartet micromeres (Figure 4.5E). The topmost cells ($1q^{111}$ micromeres) lie at the embryo's apex and form the **rosette**. In some groups (e.g., annelids and echiuran worms), other micromeres ($1q^{112}$ micromeres) produce an **annelid cross** roughly at right angles to the rosette cells. In molluscs and sipunculan worms, the annelid cross persists (often called peripheral rosette cells in these groups), but an additional **molluscan cross** forms from the $1q^{12}$ cells and their deriva-

Figure 4.5 (A–D) Spiral cleavage through 32 cells (assumed synchronous) labeled with E. B. Wilson's coding system (all diagrams are surface views from the animal pole). (E) Schematic diagram of a composite embryo at approximately 64 cells showing the positions of the rosette, annelid cross, and molluscan cross.

cromeres divide. Superscripts are added to the derivatives of the first and second micromere quartets according to the rule of position as stated above. Thus, the $1b^{1}$ cell divides to yield the $1b^{11}$ and $1b^{12}$ cells; the $1a^{2}$ cell yields the $1a^{21}$ and $1a^{22}$ cells; the $2c$ yields the $2c^{1}$ and $2c^{2}$, and so on. Do not think of these superscripts as double-digit numbers (i.e., "twenty-one" and "twenty-two"), but rather as two-digit sequences reflecting the precise lineage of each cell ("two-one" and "two-two").

The elegance of Wilson's system is that each code tells the history as well as the position of the cell in the embryo. For instance, the code $1b^{11}$ indicates that the cell is a member (derivative) of the first quartet of micromeres, that its parent macromere is the B cell, that the original 1b micromere has divided twice since its formation, and that this particular cell rests uppermost in the tives. The arms of the molluscan cross lie between the cells of the annelid cross (Figure 4.5E). Recently some phylogenetic significance has been given to the appearance of these crosses, as we discuss in later chapters.

The Problem of Cell Fates

Tracing the fates of cells through development has been a popular and productive endeavor of embryologists for over a century. Such studies have played a major role in enabling researchers not only to describe development but also to establish homologies among attributes in different animals. Although the cells of embryos eventually become established as functional parts of tissues or organs, there is a great deal of variation in the timing of the establishment of cell fates and in how firmly fixed the fates eventually become. Even in the adult stages of some animals (e.g., sponges), cells retain the ability to change their structure and function, although under normal conditions they are relatively specialized. Furthermore, many groups of animals have remarkable power to regenerate lost parts, wherein cells may dedifferentiate and then generate new tissues and organs. In other cases cell fates are relatively fixed and cells are able only to produce more of their own kind.

By carefully watching the development of any animal, it becomes clear that certain cells predictably form certain structures. The emerging field of molecular developmental biology has shown that many molecular components of development are also widely conserved throughout the animal kingdom. For example, transcription factors and cell signaling systems from widely divergent phyla are clearly homologous and probably operate in much the same way. On the other hand, these highly conserved molecular components can also be used in diverse ways by embryos. Even such basic developmental features as adult body axis formation and cleavage geometry differ among the metazoan phyla. Such fundamental developmental variations have presumably been essential in fabricating the highest levels of animal bauplans.

In some cases, cell fates are determined very early during cleavage—as early as the 2- or 4-cell stage. If one experimentally removes a blastomere from the early embryo of such an animal, then that embryo will fail to develop normally; the fates of the cells have already become fixed, and the missing cell cannot be replaced. Animals whose cell fates are established very early are said to have **determinate cleavage**. On the other hand, the blastomeres of some animals can be separated at the 2-cell, 4-cell, or even later stages, and each separate cell will develop normally; in these cases the fates of the cells are not fixed until relatively late in development. Such animals are said to have **indeterminate cleavage**. Eggs that undergo determinate cleavage are often called **mosaic ova**, because the fates of regions of undivided cells can be mapped. Eggs that undergo indeterminate cleavage are called **regulative ova**, in that they can "reg-

ulate" to accommodate lost blastomeres and thus cannot easily be predictably mapped prior to division.

In any case, formation of the basic body plan is generally complete by the time the embryo comprises about 10^4 cells (usually after one or two days). By this time, all available embryonic material has been apportioned into specific cell groups, or "founder regions." These regions are relatively few, each forming a territory within which still more intricate developmental patterns unfold. As these zones of undifferentiated tissue are established, the unfolding genetic code drives them to develop into their "preassigned" body tissues, organs, or other structures. Graphic representations of these regions are called **fate maps**.

In the past, mosaic eggs and determinate cleavage have been equated with spirally cleaving embryos, and regulative ova and indeterminate cleavage with radially cleaving embryos. However, surprisingly few actual tests for determinacy have been performed, and what evidence is available suggests that there are many exceptions to this generalization. That is, some embryos with spiral cleavage appear indeterminate, and some with radial cleavage appear determinate. Much more work remains to be done on these matters, and for the present the relationships among these features of early development are questionable. (For additional information see Costello and Henley 1976; Siewing 1980; and Ivanova-Kazas 1982.)

In spite of the variations and exceptions, there is a remarkable underlying consistency in the fates of blastomeres among embryos that develop by typical spiral cleavage. Many examples of these similarities are discussed in later chapters, but we illustrate the point by noting that the germ layers of spirally cleaving embryos tend to arise from the same groups of cells. The first three quartets of micromeres and their derivatives give rise to ectoderm (the outer germ layer), the 4a, 4b, 4c, and 4Q cells to entoderm (the inner germ layer), and the 4d cell to mesoderm (the middle germ layer). Many students of embryology view this uniformity of cell fates as strong evidence that taxa sharing this pattern are related to one another in some fundamental way and that they share a common evolutionary heritage. We will have much more to say about this idea throughout this book.

Blastula Types

The product of early cleavage is called the **blastula**, which may be defined developmentally as the embryonic stage preceding the formation of embryonic germ layers. Several types of blastulae are recognized among invertebrates. Holoblastic cleavage generally results in either a hollow or a solid ball of cells. A **coeloblastula** (Figure 4.6A) is a hollow ball of cells, the wall of which is usually one cell-layer thick. The space within the sphere of cells is the **blastocoel**, or **primary body cavity**. A **stereoblastula** (Figure 4.6B) is a solid ball of blastomeres; obviously there is no blastocoel at this stage.

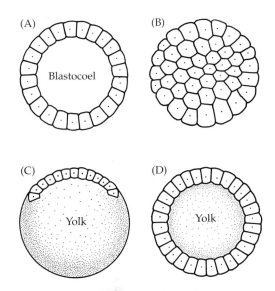

Figure 4.6 Types of blastulae. These diagrams represent sections along the animal–vegetal axis. (A) Coeloblastula. The blastomeres form a hollow sphere with a wall one cell layer thick. (B) Stereoblastula. Cleavage results in a solid ball of blastomeres. (C) Discoblastula. Cleavage has produced a cap of blastomeres that lies at the animal pole, above a solid mass of yolk. (D) Periblastula. Blastomeres form a single cell layer enclosing an inner yolky mass.

Meroblastic cleavage sometimes results in a cap or disc of cells at the animal pole over an uncleaved mass of yolk. This arrangement is appropriately termed a **discoblastula** (Figure 4.6C). Some centrolecithal ova undergo odd cleavage patterns to form a **periblastula**, similar in some respects to a coeloblastula that is centrally filled with noncellular yolk (Figure 4.6D).

Gastrulation and Germ Layer Formation

Through one or more of several methods the blastula develops toward a multilayered form, a process called **gastrulation** (Figure 4.7). The structure of the blastula dictates to some degree the nature of the process and the form of the resulting embryo, the **gastrula**. Gastrulation is the formation of the embryonic germ layers, the tissues on which all subsequent development eventually depends. In fact, we may view gastrulation as the embryonic analogue of the transition from protozoan to metazoan grades of complexity. It achieves separation of those cells that must interact directly with the environment (i.e., locomotor, sensory, and protective functions) from those that process materials ingested from the environment (i.e., nutritive functions).

The initial inner and outer sheets of cells are the entoderm and ectoderm, respectively; in most animals a third germ layer, the mesoderm, is produced between the ectoderm and the entoderm. One striking example of the unity among the Metazoa is the consistency of the fates of these germ layers. For example, ectoderm always forms the nervous system and the outer skin and its derivatives;

entoderm the main portion of the gut and associated structures; and mesoderm the coelomic lining, the circulatory system, most of the internal support structures, and the musculature. The process of gastrulation, then, is a critical one in establishing the basic materials and their locations for the building of the whole organism.

Coeloblastulae often gastrulate by **invagination**, a process commonly used to illustrate gastrulation in general zoology classes. The cells in one area of the surface of the blastula (frequently at or near the vegetal pole) pouch inward as a sac within the blastocoel (Figure 4.7A). These invaginated cells are now called the entoderm, and the sac thus formed is the embryonic gut, or **archenteron**; the opening to the outside is the **blastopore**. The outer cells are now called ectoderm, and a double-layered hollow **coelogastrula** has been formed. Note that the diagrams in Figure 4.7 represent 3-dimensional embryos. Thus, the coelogastrula (Figure 4.7A) actually resembles a balloon with a finger poking into it.

The coeloblastulae of many cnidarians undergo gastrulation processes that result in solid gastrulae (**stereogastrulae**). Usually the cells of the blastula divide such that the cleavage planes are perpendicular to the surface of the embryo. Some of the cells detach from the wall and migrate into the blastocoel, eventually filling it with a solid mass of entoderm. This process is called **ingression** (Figure 4.7B) and may occur only at the vegetal pole (unipolar ingression) or more or less over the whole blastula (multipolar ingression). In a few instances (e.g., certain hydroids), the cells of the blastula divide with cleavage planes that are parallel to the surface, a process called **delamination** (Figure 4.7C). This process produces a layer or a solid mass of entoderm surrounded by a layer of ectoderm.

Stereoblastulae that result from holoblastic cleavage generally undergo gastrulation by **epiboly**. Because there is no blastocoel into which the presumptive entoderm can migrate by any of the above methods, gastrulation involves a rapid growth of presumptive ectoderm around the presumptive entoderm (Figure 4.7D). Cells of the animal pole proliferate rapidly, growing down and over the vegetal cells to enclose them as entoderm. The archenteron typically forms secondarily as a space within the developed entoderm.

Figure 4.7E illustrates gastrulation by **involution**, a process that usually follows the formation of a discoblastula. The cells around the edge of the disc divide rapidly and grow beneath the disc, thus forming a double-layered gastrula with ectoderm on the surface and entoderm below. There are several other types of gastrulation, mostly variations or combinations of the above processes. These gastrulation methods are discussed in later chapters.

During the gastrulation process, subtle shifts in timing of the expression of regulatory gene products, the timing of cell fate specification, or the movement of cells can generate distinct developmental pathways. Such

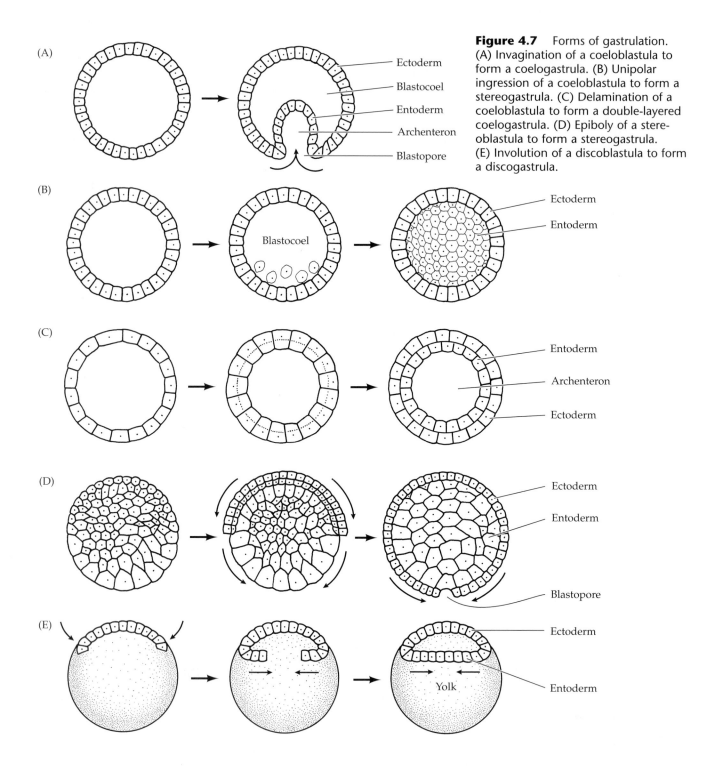

Figure 4.7   Forms of gastrulation. (A) Invagination of a coeloblastula to form a coelogastrula. (B) Unipolar ingression of a coeloblastula to form a stereogastrula. (C) Delamination of a coeloblastula to form a double-layered coelogastrula. (D) Epiboly of a stereoblastula to form a stereogastrula. (E) Involution of a discoblastula to form a discogastrula.

(A)

Ectoderm
Blastocoel
Entoderm
Archenteron
Blastopore

(B)

Blastocoel

Ectoderm
Entoderm

(C)

Entoderm
Archenteron
Ectoderm

(D)

Ectoderm
Entoderm
Blastopore

(E)

Ectoderm
Yolk
Entoderm

developmental divergences may dramatically shift larval or even adult formation in a lineage. For example, it has been hypothesized that sea urchin larvae have switched from planktotrophy to lecithotrophy at least 20 times in the history of this echinoderm clade. In the non-feeding larvae, egg size is greater, cleavage is significantly altered, and the larval life span is shorter.

Mesoderm and Body Cavities

Some time during or following gastrulation, a middle layer forms between the ectoderm and the entoderm.

This middle layer may be derived from ectoderm, as it is in members of the diploblastic phylum Cnidaria, or from entoderm, as it is in members of the triploblastic phyla. In the first case the middle layer is said to be **ectomesoderm**, and in the latter case **entomesoderm** (or "true mesoderm"). Thus, the triploblastic condition, by definition, includes entomesoderm. In this text, and most others, the term *mesoderm* in a general sense refers to entomesoderm rather than ectomesoderm.

In diploblastic and certain triploblastic phyla (the acoelomates), the middle layer does not form thin sheets

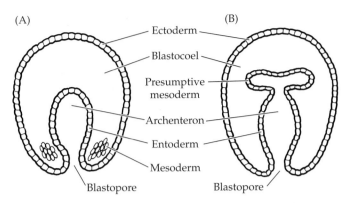

Figure 4.8 Methods of mesoderm formation in late gastrulae (frontal sections). (A) Mesoderm formed from derivatives of a mesentoblast. (B) Mesoderm formed by archenteric pouching.

of cells; rather it produces a more-or-less solid but loosely organized **mesenchyme** consisting of a gel matrix (the **mesoglea**) containing various cellular and fibrous inclusions. In a few cases (e.g., the hydrozoans) a virtually noncellular mesoglea lies between the ectoderm and entoderm (see Chapter 8).

In most animals, the area between the inner and outer body layers includes a fluid-filled space. As discussed in Chapter 3, this space may be either a blastocoelom, which is a cavity not completely lined by mesoderm, or a true coelom, which is a cavity fully enclosed within thin sheets of mesodermally derived tissue. Mesoderm generally originates in one of two basic ways, as described below (Figure 4.8); modifications of these processes are discussed in later chapters. In most phyla that undergo spiral cleavage (e.g., flatworms, annelids, and molluscs), a single micromere—the 4d cell, called the **mesentoblast**—proliferates as mesoderm between the developing archenteron (entoderm) and the body wall (ectoderm) (Figure 4.8A). The other cells of the 4q (the 4a, 4b, and 4c cells) and the 4Q cells generally contribute to entoderm. In some other taxa (e.g., echinoderms and chordates) the mesoderm arises from the wall of the archenteron itself (that is, from preformed entoderm), either as a solid sheet or as pouches (Figure 4.8B).

In addition to giving rise to other structures (such as the muscles of the gut and body wall), in coelomate animals mesoderm is intimately associated with the formation of the body cavity. In those instances where mesoderm is produced as solid masses derived from a mesentoblast, the body cavity arises through a process called **schizocoely**. Normally in such cases, bilaterally paired packets of mesoderm gradually enlarge and hollow, eventually becoming thin-walled coelomic spaces (Figure 4.9A,B). The number of such paired coeloms varies among different animals and is frequently associated with segmentation, as it is in annelid worms (Figure 4.9C).

The other general method of coelom formation is called **enterocoely**; it accompanies the process of mesoderm formation from the archenteron. In the most direct sort of enterocoely, mesoderm production and coelom formation are one and the same process. Figure 4.10A illustrates this process, which is called **archenteric pouching**. A pouch or pouches form in the gut wall. Each pouch eventually pinches off from the gut as a complete coelomic compartment. The walls of these pouches are defined as mesoderm. In some cases the mesoderm arises from the wall of the archenteron as a solid sheet or plate that later becomes bilayered and hollow (Figure 4.10B). Some authors consider this process to be a form of schizocoely (because of the "splitting" of the mesodermal plate), but it is in fact a modified form of enterocoely. Enterocoely frequently results in a tripartite arrangement of the body cavities, which are designated **protocoel**, **mesocoel**, and **metacoel** (Figure 4.10C).

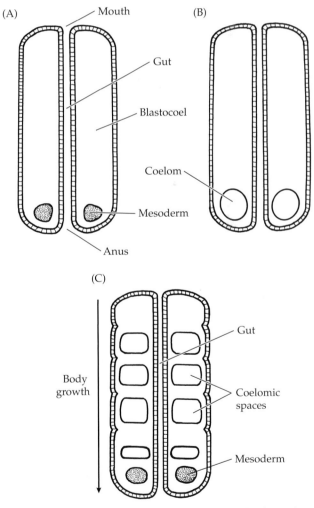

Figure 4.9 Coelom formation by schizocoely (frontal sections). (A) Precoelomic conditions with paired packets of mesoderm. (B) Hollowing of the mesodermal packets to produce a pair of coelomic spaces. (C) Progressive proliferation of serially arranged pairs of coelomic spaces. This process occurs in metameric annelids.

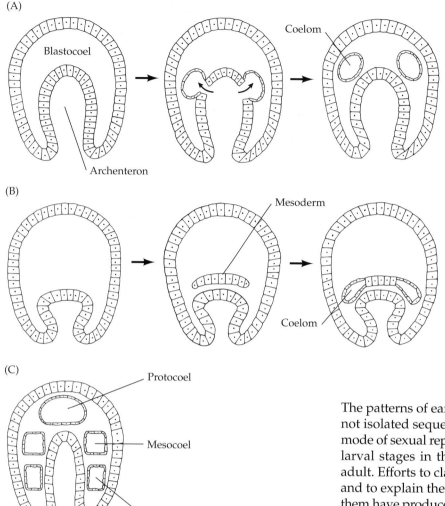

Figure 4.10 Coelom formation by enterocoely (frontal sections). (A) Archenteric pouching. (B) Proliferation and subsequent hollowing of a plate of mesoderm from the archenteron. (C) The typical tripartite arrangement of coeloms in a deuterostome embryo.

factors, and special substances called morphogens. The second method involves actual contact between the surfaces of adjacent cells. Cells selectively recognize other cells, adhering to some and migrating over others. Of all the stages of ontogeny, we know least about morphogenesis.

Life Cycles: Sequences and Strategies

The patterns of early development described above are not isolated sequences of events, but are related to the mode of sexual reproduction, the presence or absence of larval stages in the life cycle, and the ecology of the adult. Efforts to classify various invertebrate life cycles and to explain the evolutionary forces that gave rise to them have produced a large number of publications and a lot of controversy. Most of these studies concern marine invertebrates, on which we center our attention first. We then present some comments on the special adaptations of terrestrial and freshwater forms.

Classification of Life Cycles

Our discussion of life cycles focuses on sexually reproducing animals. Sexual reproduction with some degree of gamete dimorphism is nearly universal among eukaryotes. Male and female gametes may be produced by the same individual (hermaphroditism, monoecy, or cosexuality) or by separate individuals (dioecy or gonochory). Most terrestrial animals are dioecious, but hermaphroditism is widespread among marine invertebrates (and among land plants). Mechanisms of sex determination are diverse, but the most common involve structurally distinct sex chromosomes, usually male heterogametes, in which males carry X and Y sex chromosomes and females are XX. With female heterogamy, females are ZW and males ZZ. There is typically little or no recombinational exchange between X and Y chromosomes (or between Z and W) because there is almost no genetic homology between the sex chromosomes. Most of the Y (or W) chromosome is devoid of functional gene loci, other than a few RNA

Following germ tissue establishment, cells begin to specialize and sort themselves out to form the organs and tissues of the body—a poorly understood process known as **morphogenesis**. Cell movements are an essential part of morphogenesis. In addition, in order to sculpt the organs and systems of the body, cells need to know when to stop growing and even die. For example, in nematode worms the vas deferens first develops with a closed end; the cell that blocks the end of this tube helps the vas deferens link up to the cloaca. But once the connection has been made, this terminal cell dies, its death creating the opening to the cloaca.

Recent research suggests that the same families of molecules that guide the earliest stages of embryogenesis—setting up the elements of body patterning—also play vital roles during morphogenesis. Communication among adjacent cells is also critical to morphogenesis, and there are two ways cells "talk" to one another during this process. The first is via diffusible signaling molecules released from one cell and detected by the adjacent cell. These substances include hormones, growth

genes and some genes required for male (or female) fertility and sex determination. In any event, the fusion of male and female gametes initiates the process of ontogeny and a new cycle in the life history of the organism.

A number of classification schemes for life cycles have been proposed over the past four decades (see papers by Thorson, Mileikovsky, Chia, Strathmann, Jablonsky, Lutz, and McEdward). We have generalized from the works of various authors and suggest that most animals display some form of one of the three following basic patterns (Figure 4.11).

1. **Indirect development**. The life cycle includes free spawning of gametes followed by the development of a free larval stage (usually a swimming form), which is distinctly different from the adult and must undergo a more or less drastic metamorphosis to reach the juvenile or young adult stage. In aquatic groups, two basic larval types can be recognized.

   a. Indirect development with **planktotrophic larvae**. The larva survives primarily by feeding, usually on plankton. (The feeding larvae of some deep-sea species are demersal and feed on detrital matter, never swimming very far off the bottom.)

   b. Indirect development with **lecithotrophic larvae**. The larva survives primarily on yolk supplied to the egg by the mother.

2. **Direct development**. The life cycle does not include a free larva. In these cases the embryos are cared for by the parents in one way or another (generally by brooding or encapsulation) until they emerge as juveniles.

3. **Mixed development**. The life cycle involves brooding or encapsulation of the embryos at early stages of development and subsequent release of free planktotrophic or lecithotrophic larvae. The initial source of nutrition and protection is the adult.

Not every species can be conveniently categorized into just one of the above developmental patterns. For example, some species have free larvae that depend on yolk for a time, but begin to feed once they develop the capability to do so. Some species actually display different developmental strategies under different environmental conditions—convincing evidence that embryogenies are adaptable, evolutionarily plastic, and subject to selection pressures (as are adults).

These life cycle patterns provoke three basic questions. First, how do these different sequences relate to other aspects of reproduction and development such as egg types and mating or spawning activities? Second, how do the overall developmental sequences relate to the survival strategies of the adults? Third, what evolutionary mechanisms are responsible for the patterns seen in any given species? Given the large number of interacting factors to be considered, these are very complex questions, and our understanding is still incomplete. However, by first examining cases of direct and indirect development, we can illustrate some of the principles that underlie their relationships to different ecological situations. Then we briefly address some ideas about mixed development.

Indirect Development

Consider first a life cycle with planktotrophic larvae (Figure 4.11A). The metabolic expense incurred on the part of the adults involves only the production and release of gametes. Animals with fully indirect development generally do not mate; instead, they shed their eggs and sperm into the water, thus divorcing the adults from any further responsibility of parental care. Such animals typically undergo synchronous (**epidemic**) broadcast spawning of very large numbers of gametes, thereby ensuring some

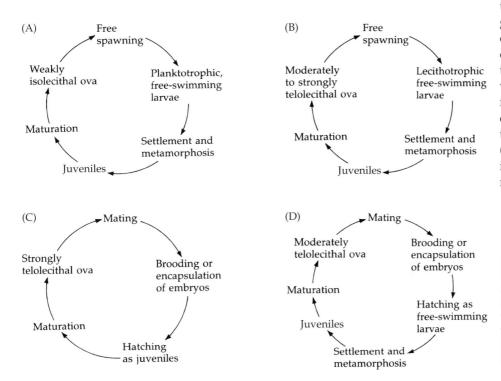

Figure 4.11 Some generalized invertebrate life cycle strategies. (A) Indirect development with planktotrophic larvae. (B) Indirect development with lecithotrophic larvae. (C) Direct development. (D) Mixed life cycle.

level of successful fertilization. This pattern of development is relatively common in *r*-selected, opportunistic marine species that are capable of rapid production of high numbers of gametes.

The eggs are usually isolecithal and individually inexpensive to produce. The cost—and it is a significant one—to the parent is in the production of very high numbers of eggs. Being supplied with little yolk, the embryos must develop quickly into feeding larvae to survive. Mortalities among the embryos and larvae are extremely high and can result from a variety of factors, including lack of food, predation, or adverse environmental conditions. Each successful larva must accumulate enough nutrients from feeding to provide the energy necessary for immediate survival and for the processes of settling and metamorphosis from larva to juvenile or subadult. That is, they must feed to excess as they prepare for a new lifestyle as a juvenile. Survival rates from zygote to settled juvenile are often less than one percent; such high mortalities are compensated for by the initial high production of gametes. By the same token, the mortalities *compensate for* this high production of gametes—if all of these zygotes survived, Earth would quickly be covered by the offspring of animals with indirect development.

What are the advantages and limitations of such a life history, and under what circumstances might it be successful? This sort of planktotrophic development is most common and largely predominates among benthic marine invertebrates in relatively shallow water and the intertidal zones of tropical and warm temperate seas. Here the planktonic food sources are more consistently available (although often in low concentration) than they are in colder or deeper waters, thus reducing the danger of starvation of the larvae. Such meroplanktonic life cycles allow animals to take advantage of two distinct resources (plankton in the upper water column as larvae; benthos and bottom plankton as adults). This arrangement reduces or eliminates competition between larvae and adults. Indirect development also provides a mechanism for dispersal, a particularly significant benefit to species that are sessile or sedentary as adults. There is good evidence to suggest that animals with free-swimming larvae are likely to recover more quickly from damage to the adult population than those engaging in direct development. A successful set of larvae is a ready-made new population to replace lost adults.

The disadvantages of planktotrophic development result from the unpredictability of larval success. Excessive larval deaths can result in poor recruitment and the possibility of invasion of suitable habitats by competitors. Conversely, unusually high survival rates of larvae can lead to overcrowding and intraspecific competition upon settling.

Animals that produce fully lecithotrophic larvae (Figure 4.11B) must produce yolky and thus more metabolically expensive eggs. This built-in nutrient supply releases the larvae from dependence on environmental food supplies and generally results in reduced mortalities. It is not surprising that these animals produce somewhat fewer ova than those with planktotrophic larvae. The eggs are either spawned directly into the water or are fertilized internally and released as zygotes. Again, the adults' parental responsibility ends with the release of gametes or zygotes into the environment. Although survival rates of lecithotrophic larvae are generally higher than those of planktotrophic types, they are low compared with those of embryos that undergo direct development.

There is a tendency for marine invertebrates that live in relatively deep benthic environments to produce lecithotrophic larvae. Here some of the advantages of indirect development are realized, but without depending on environmental food supplies and without subjecting the larvae to the intense predation commonly encountered in surface water. The trade-off is clear: in deeper water the trend is to produce fewer (more expensive) zygotes that have a higher survival rate than planktotrophic larvae and an ability to survive where planktotrophic larvae cannot.

Settling and Metamorphosis

Of particular importance to the successful completion of life cycles with free larval stages, and thus to the perpetuation of such species, are the processes of settlement and metamorphosis. Settling and metamorphosis are crucial and dangerous times in an animal's life cycle, when the organism is changing habitats and lifestyles. Typically, the free-swimming larva metamorphoses into a benthic juvenile and must survive this transformation in form and function as it adopts a new mode of life. Throughout its free-swimming life the larva has been "preparing" for these events, until it reaches a condition in which it is physiologically capable of metamorphosis; such a larva is then termed **competent**. The duration of the free-swimming period varies greatly among invertebrate larvae and depends on factors such as original egg size, yolk content, the availability of food for planktotrophic forms, various environmental factors (e.g., water temperature), and locating a suitable substratum for settlement.

Once a larva becomes competent, it generally begins to respond to certain environmental cues that induce settling behavior. Metamorphosis is often preceded by settling, although some species metamorphose prior to settling and still others engage in both processes simultaneously. In any case, larvae typically become negatively phototactic and/or positively geotactic and swim toward the bottom. Once contact with a substratum is made, a larva tests it to determine its suitability as a habitat. This act of substratum selection may involve processing physical, chemical, and biological information about the immediate environment. A number of studies show that important factors include substratum texture, composition, and particle size; presence of conspecific adults (or dominant competitors); presence of key chemical cues; presence of appropriate food sources; and the nature of bottom currents or turbulence. Many invertebrate larvae touch down on the bottom for a few minutes, then launch themselves back up into the current again and again until a suitable substratum is found. Assuming an appropriate situation is encountered, metamorphosis is induced and proceeds to completion. Interestingly, some feeding larvae are able to resume planktonic life and postpone metamorphosis if they initially encounter an unsuitable substratum. In such cases, however, the larvae become gradually less selective; eventually, metamorphosis ensues regardless of the availability of a proper substratum. The ability to prolong the larval period until conditions are favorable for settlement has obvious survival advantages, and invertebrates differ greatly in this capability. Those that can postpone settlement may do so by several hours, days, or even months (based on laboratory experiments).

Direct Development

Direct development avoids some of the disadvantages but misses some of the advantages of indirect development. A typical scenario involves the production of relatively few, very yolky eggs, followed by some sort of mating activity and internal fertilization (Figure 4.11C). The embryos receive prolonged parental care, either directly (by brooding in or on the parent's body) or indirectly (by encapsulation in egg cases provided by the parent). Animals that simply deposit their fertilized eggs, either freely or in capsules, are said to be **oviparous**. A great number of invertebrates as well as some vertebrates (amphibians, many fishes, reptiles, and birds) display oviparity. Animals that brood their embryos internally and nourish them directly, such as placental mammals, are described as **viviparous**. **Ovoviviparous** animals brood their embryos internally but rely on the yolk within the eggs to nourish their developing young. Most internally brooding invertebrates are ovoviviparous.

The large, yolky eggs of most invertebrates with direct development are metabolically expensive to produce. But even though only a few can be afforded, the investment is protected and survival rates are relatively high. The dangers of planktonic larval life and metamorphosis are avoided and the embryos eventually hatch as juveniles.

What sorts of environments and lifestyles might result in selection for such a developmental sequence? At the risk of overgeneralizing, we can say that there is a tendency for *K*-selected specialist species to display direct development. Another situation in which direct development occurs is when the adults have no dispersal problems. We find, for example, that holoplanktonic species with pelagic adults (e.g., arrow worms, phylum Chaetognatha) often undergo direct development, either by brooding or by producing floating egg cases. A second situation is one in which critical environmental factors (e.g., food, temperature, water currents) are highly variable or unpredictable. There is a trend among benthic invertebrates to switch from planktotrophic indirect development to direct development at increasingly higher latitudes. The relatively harsh conditions and strongly seasonal occurrence of planktonic food sources in polar and subpolar areas partially explain this tendency.

In addition to avoiding some of the danger of larval life, direct development has another distinct advantage. The juveniles hatch in suitable habitats where the adults brooded them or deposited the eggs in capsules. Thus, there is a reasonable assurance of appropriate food sources and other environmental factors for the young.

Mixed Development

As defined earlier, mixed life histories involve some period of brooding prior to release of a free larval stage. Costly, yolky zygotes are protected for some time and then are released as larvae, exploiting the advantages of dispersal. This developmental pattern is often ignored when classifying life histories, but in fact it is widespread and extremely popular among invertebrates (e.g., in many gastropods, insects, crustaceans, sponges, cnidarians, and a host of other groups). Some workers view mixed development as either the "best" or the "worst" of both worlds (i.e., fully indirect or direct). Others suggest that such sequences are evolutionarily unstable, and that local environmental pressures are driving them toward direct or indirect development (see Caswell 1981 for a review). There are, however, other possible explanations. It may very well be that under some environmental situations a brooding period followed by a larval phase *is* adaptive and stable.

Furthermore, there is evidence that at least some species show a sort of built-in variability in the relative lengths of time the embryo exists in a brooded versus a free larval phase. If this variability responds to local environmental pressures, then clearly such a species might adapt quickly to changing conditions, or even exploit this ability by extending its geographic range to live under a variety of settings. In this regard, mixed life histories are an area deserving of further investigation.

Our short description of life history strategies certainly does not explain all observable patterns in nature. The historical and evolutionary forces acting on invertebrates (and their larvae) are highly complex. For example, larvae are subject to all manner of oceanographic variables (e.g., diffusion, lateral and vertical transport, sea floor topography, storms) as well as their self-directed vertical movements, seasonality, and biotic factors (predators, prey, competition, nutrient availability). Life history predictions based strictly on environmental conditions do not always hold true. Invertebrates living in the deep sea and at the poles do not always brood (as once thought). We now know that all life history strategies occur in these regions, and many deep-sea and polar species release free-swimming larvae, and even planktotrophic larvae. Even some invertebrates of deep-sea hydrothermal vent communities produce free-swimming larvae. In many cases, this may be due to evolutionary constraints: vent gastropods, for example, belong to lineages that are almost strictly lecithotrophic, regardless of latitude or habitat. Thus, vent gastropods are apparently constrained by their phylogenetic histories. Other vent species that release free larvae, however, are not so constrained: mytilid bivalves, for example, possess a wide range of reproductive modes, and tend to release planktotrophic larvae in deep-sea and vent environments. Furthermore, reproductive cycles in many abyssal invertebrates appear to be seasonal, perhaps cued by annual variations in surface water productivity. There is still much to be learned.

Adaptations to Land and Fresh Water

The foregoing account of life cycle strategies applies largely to marine invertebrates. Many invertebrates, however, have invaded land or fresh water, and their success in these habitats requires not only adaptation of the adults to special problems, but also adaptation of the developmental forms. As discussed in Chapter 1, terrestrial and freshwater environments are more rigorous and unstable than the sea, and they are generally unsuitable for reproductive strategies that involve free spawning of gametes or the production of delicate larval forms. Most groups of terrestrial and freshwater invertebrates have adopted internal fertilization followed by direct development, while their marine counterparts often produce free-swimming larvae. A notable exception is the insects, many of which have evolved elaborate mixed development life histories. In these cases, the larvae are highly adapted to their freshwater or aerial environment and almost certainly evolved as secondary features rather than from any marine larval ancestor.

Parasite Life Cycles

There is no doubting the success of parasitism as a lifestyle. Most parasites have rather complicated life cycles, and specific examples are given in later chapters. At this point, however, we can view the situation in a general way and examine the strategies of parasitism in terms of parasite life cycles, at the same time introducing some basic terminology.

As outlined in Chapter 1, parasites may be classed as **ectoparasites** (living upon the host), **endoparasites** (living internally, within the host), or **mesoparasites** (living in some cavity of the host that opens directly to the outside, such as the oral, nasal, anal, or gill cavities). While associated with a host, an adult parasite engages in sexual reproduction, but the eggs or embryos are usually released to the outside via some avenue through the

host's body. The problems at this point are very similar to those encountered during indirect development: some mechanism(s) must be provided to ensure adequate survival through the developmental stages, and the sequence of events must bring the parasite back to an appropriate host (the proper "substratum") for maturation and reproduction. Many parasites are also parthenogenetic.

Parasites exploit at least two different habitats in their life cycles. This practice is essential because their hosts eventually die. Thus, the developmental period from zygote to adult parasite involves either the invasion of another host species or a free-living period. When more than one host species is utilized for the completion of the life cycle, the organism harboring the adult parasite is called the **primary** or **definitive host**, and those hosts in which any developmental or larval forms reside are called **intermediate hosts**. The completion of such a complex life cycle often requires elaborate methods of transfer from one host to the other, and of surviving the changes from one habitat to another. Losses are high, and it is common to find life cycle stages that compensate by engaging in periods of rapid asexual reproduction in addition to the sexual activities of the adult.

Thus, we find that many parasites enjoy some of the benefits of indirect development (e.g., dispersal and exploitation of multiple resources) while being subjected to accompanying high mortalities and the dangers of very specialized lifestyles.

We emphasize again that the above discussions of life cycles are generalities to which there are many exceptions. But given these basic patterns, you should recognize and appreciate the adaptive significance of life history patterns of the different invertebrate groups discussed later. You might also be able to predict the sorts of sequences that would be likely to occur under different conditions. For example, given a situation in which a particular species is known to produce very high numbers of free-spawned, isolecithal ova, what might you predict about cleavage pattern, blastula and gastrula type, presence or absence of a larval stage, type of larva, adult lifestyle, and ecological settings in which such a sequence would be advantageous? We hope you will develop the habit of asking these kinds of questions and thinking in this way about all aspects of your study of invertebrates.

The Relationships between Ontogeny and Phylogeny

Of the many fields of study from which we draw information used in phylogenetic investigations, embryology has been one of the most important. The construction of phylogenies may be accomplished and subsequently tested by several different methods (see Chapter 2). But regardless of method, one of the principal problems of phylogeny reconstruction—in fact, central to the process—is separating true homologies from similar character traits that are the result of evolutionary convergence. Even when these problems involve comparative adult morphology, one must often seek answers in studies of the development of the organisms and structures in questions. The search is for developmental processes or structures that are homologues and thus demonstrate relationships between ancestors and descendants. Changes that take place in developmental stages are not trivial evolutionary events. It has been effectively argued that developmental phenomena may themselves provide the evolutionary mechanisms by which entire new lineages originated (see Chapter 1). As Stephen Jay Gould (1977) has noted,

> Evolution is strongly constrained by the conservative nature of embryological programs. Nothing in biology is more complex than the production of an adult . . . from a single fertilized ovum. Nothing much can be changed very radically without discombobulating the embryo.

Indeed, the persistence of distinctive body plans throughout the history of life is testimony to the resistance to change of complex developmental programs. (See Hall 1996 for an excellent analysis of these issues.)

Although few workers would argue against a significant relationship between ontogeny and phylogeny, the exact nature and extent of the relationship have historically been subjects of considerable controversy, a good deal of which continues today. (Gould 1977 presents a fine analysis of these debates.) Central to much of the controversy is the concept of recapitulation.

The Concept of Recapitulation

In 1866 Ernst Haeckel, a physician who found a higher calling in zoology and never practiced medicine, introduced his **law of recapitulation** (or **the biogenetic law**), most commonly stated as "ontogeny recapitulates phylogeny." Haeckel suggested that a species' embryonic development (ontogeny) reflects the adult forms of that species' evolutionary history (phylogeny). According to Haeckel, this was no accident, but a result of the mechanistic relationship between the two processes: phylogenesis is the actual *cause* of embryogeny. Restated, animals have an embryogeny *because of* their evolutionary history. Evolutionary change over time has resulted in a continual adding on of morphological stages to the developmental process of organisms. The implications of Haeckel's proposal are immense. Among other things, it means that to trace the phylogeny of an animal, one need only examine its development to find therein a sequential or "chronological" parade of the animal's adult ancestors.

Ideas and disagreement concerning the relationship between ontogeny and phylogeny were by no means new even at Haeckel's time. Over 2,000 years ago Aristotle described a sequence of "souls" or "essences" of increasing quality and complexity through which animals pass in their development. He related these conditions to the adult "souls" of various lower and higher organisms, a notion suggestive of a type of recapitulation.

Descriptive embryology flourished in the nineteenth century, stimulating vigorous controversy regarding the relationship between development and evolution. Many of the leading developmental biologists of the time were in the thick of things, each proposing his own explanation (Meckel 1811; Serres 1824; von Baer 1828; and others). It was Haeckel, however, who really stirred the pot with his discourse on the "law" of recapitulation. He offered a focal point around which biologists argued pro or con for 50 years; sporadic skirmishes still erupt periodically. Walter Garstang critically examined the biogenetic law and gave us a different line of thinking. His ideas, presented in 1922, are reflected in many of his poems (published posthumously in 1951). Garstang made clear what a number of other biologists had suggested: that evolution must be viewed not as a succession of ancestral *adult* forms, but as a succession of *ontogenies*. Each animal is a result of its own developmental processes, and any change in an adult must represent a change in its ontogeny. So what we see in the embryogeny of a particular species are not tiny replicas of its adult ancestors, but rather an evolved pattern of development in which clues or traces of ancestral ontogenies, and thus phylogenetic relationships to other organisms, may be found.

Arguments over these matters did not end with Garstang, and they continue today in many quarters. In general, we tend to agree with the approach (if not all of the details) of Gosta Jägersten in *Evolution of the Metazoan Life Cycle* (1972). Recapitulation per se should not categorically be accepted or dismissed as an "always" or "never" phenomenon. The term must be clearly defined in each case investigated, not locked in to Haeckel's original definition and implications. For instance, similar, distinctive, homologous larval types within a group of animals reflect some degree of shared ancestry (e.g., crustacean nauplii or molluscan veligers). And we may speculate on such matters at various taxonomic levels, even when the adults are quite different from one another (e.g., the similar trochophore larvae of polychaetes, molluscs, and sipunculans). These phenomena may be viewed as developmental evidence of relatedness through shared ancestry, and thus they are examples of "recapitulation" in a broad sense.

Jägersten's example of vertebrate gill slits is particularly appropriate because, to him, it provides a case in which Haeckel's strict concept of recapitulation is manifest. In writing of this feature Jägersten (1972) stated,

> The fact remains . . . that character which once existed in the adults of the ancestors but was lost in the adults of the descendants is retained in an easily recognizable shape in the embryogenesis of the latter. This is my interpretation of recapitulation (the biogenetic 'law').

Hyman (1940) perhaps put it most reasonably when she wrote,

> Recapitulation in its narrow Haeckelian sense, as repetition of adult ancestors, is not generally applicable; but ancestral resemblance during ontogeny is a general biological principle. There is no need to quibble over the word recapitulation; either the usage of the word should be altered to include any type of ancestral reminiscence during ontogeny, or some new term should be invented.

Other authors, however, are not comfortable with such flexibility and have made great efforts to categorize and define the various possible relationships between ontogeny and phylogeny, of which strict recapitulation is considered only one (see especially Chapter 7 of Gould 1977). Although much of this material is beyond the scope of this book, we discuss a few commonly used terms here because they bear on topics in later chapters. We have drawn on a number of sources cited in this chapter to mix freely with our own ideas in explaining these concepts.

Heterochrony and Paedomorphosis

When comparing two ontogenies, one often finds that some feature or set of features appears either earlier or later in one sequence than in the other. Such temporal displacement is called **heterochrony**. When comparing suspected ancestral and descendant embryogenies, for example, one may find the very rapid (accelerated) development of a particular feature and thus its relatively early appearance in the descendant species or lineage. Conversely, the development of some trait may be slower (retarded) in the descendant than in the ancestor and thus appear later in the descendant's ontogeny. This retardation may be so pronounced that a structure may never develop to more than a rudiment of its ancestral condition. For excellent reviews of heterochrony and its impact on phylogeny see Gould (1977) and McKinney and McNamara (1991).

Particular types of heterochrony result in a condition known as **paedomorphosis**, wherein sexually mature adults possess features characteristically found in early developmental stages of related forms (i.e., juvenile or larval features). Paedomorphosis results when the reproductive structures develop before completion of the development of all the nonreproductive (somatic) structures. Thus, we find a reproductively functional animal retaining what in the ancestor were certain embryonic, larval, or juvenile characteristics. This condition can result from two different heterochronic processes. These are **neoteny**, in which somatic development is retarded, and **paedogenesis**, in which reproductive development is accelerated. These two terms are frequently used interchangeably; certainly it is not always possible to know which process has given rise to a particular paedomorphic condition. Recognition of paedomorphosis may play a significant role in examining phylogenetic hypotheses concerning the origins of certain taxa. For

example, the evolution of precocious sexual maturation of a planktonic larval stage (that would "normally" continue developing to a benthic adult) might result in a new diverging lineage in which the descendants pursue a fully pelagic existence. Such a scenario, for example, may have been responsible for the origin of the crustacean subclass Maxillopoda (see Chapter 18). In a different lineage, paedomorphosis plays a major role in certain theories about the origin of the vertebrates.

Myriad questions about the role of embryogenesis in evolution and the usefulness of embryology in constructing and testing phylogenies persist. As the following accounts show, different authors continue to hold a variety of opinions about these matters.

Origins of Major Groups of Metazoa

One theme we develop throughout this book is the evolutionary relationships within and among the invertebrate taxa. Life has probably existed on this planet for nearly 4 billion years; humans have been observing it scientifically for only a few hundred years, and evolutionarily for only about 150 years. Thus, the thread of evolutionary continuity we actually see around us today may look much like frazzled ends, representing the legions of successful animals that survive today. It is only through conjecture, study, inference, and the testing of hypotheses that we are able to trace phylogenetic strands back in time, joining them at various points to produce hypothetical pathways of evolution. We do not operate blindly in this process, but use rigorous scientific methodology to draw upon information from many disciplines in attempts to make our evolutionary hypotheses meaningful and (we hope) increasingly closer to the truth—to the actual history of life on Earth.

In Chapter 1 we briefly reviewed the history of life on Earth, as inferred from the fossil record. In Chapter 24 we present a phylogenetic analysis of the animal phyla based on living animal taxa. However, many workers have not been satisfied to develop phylogenetic analyses based solely upon known (extant and extinct) animal phyla, but have felt compelled to speculate on hypothetical ancestors that might have occurred along the evolutionary road to modern life. A variety of evolutionary stories have been proposed to describe these sequences of hypothetical metazoan ancestors. We discuss some of these below, and some key works are cited in the references at the end of this chapter and Chapter 24. Our goal in this section is to introduce the reader only to the major ideas concerning the origins of metazoan grades. We reserve our discussion of the origins of specific taxa for later chapters.

Origin of the Metazoa

The origin of the metazoan condition has received considerable attention for more than a century. One of the most spectacular phenomenon in the fossil record is the abrupt appearance and diversification of nearly all of the metazoan phyla living today in a brief span of 30 million years, at the Precambrian–Cambrian transition (approximately 570–600 million years ago). There is little doubt that animals—the Metazoa—arose as a monophyletic group from a protist ancestor, perhaps 700 million years ago or earlier (see Chapter 1). The debates concern which protist group was ancestral to the first Metazoa, what these first animals were like, what environments they inhabited, and how the changes from unicellularity to multicellularity took place. Historically, the two hypotheses that have enjoyed most support are usually referred to as the **syncytial theory** and the **colonial theory**.

The syncytial theory was consolidated in the 1950s and 1960s by J. Hadzi and E. D. Hanson who suggested that the metazoan ancestor was a multinucleate, bilaterally symmetrical, ciliated protist that assumed a benthic lifestyle, crawling about on the bottom with its oral groove directed toward the substratum. In a major evolutionary step, the surface nuclei became partitioned off from one another by the formation of cell membranes, producing a cellular epidermis surrounding an inner syncytial mass. The result of this and other changes was an acoel flatworm-like creature (phylum Platyhelminthes). Figure 4.12 illustrates the steps in this ciliate-to-acoel hypothesis. It is important to note that the cellular nature of the acoel inner mass is a relatively recent discovery. It had been previously believed to be syncytial, and we can forgive Hadzi and company for relying in part on that widespread misconception.

The principal arguments in support of the syncytial theory rest upon certain similarities between modern ciliates and acoel flatworms: size (large ciliates can actually exceed small acoels in size), shape, symmetry, mouth location, and surface ciliation. Most of the objections to this hypothesis concern developmental matters and differences in general levels of adult complexity. Like all flatworms, acoels undergo a complex embryonic development; nothing of this sort occurs in ciliates. And, of course, their innards are cellular, not syncytial. In addition, recent work suggests that the acoels are probably not the most primitive flatworms (see Chapter 10). The syncytial theory suggests that the flatworms (acoelomate triploblastic bilateria) were the first and thus the ancestral Metazoa, leaving us to somehow derive from them the seemingly more primitive cnidarians and ctenophores, as well as more advanced groups. The syncytial theory enjoys little support today.

The colonial theory is based upon ideas first expressed by Ernst Haeckel (1874), who suggested that a colonial flagellated protist gave rise to a planuloid metazoan ancestor (the planula is the basic larval type of cnidarians; see Chapter 8). The ancestral protist in this theory was a hollow sphere of flagellated cells that developed some degree of anterior–posterior locomotor orientation, and also evolved some level of specialization of cells into separate somatic and reproductive

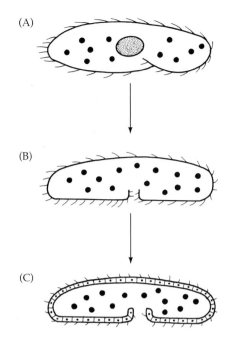

Figure 4.12 The syncytial theory. Schematic representation of the hypothetical transition from a multinucleate, ciliated protist to an acoel-like flatworm. (A) A ciliated protist. (B) The hypothetical metazoan precursor (sagittal section) as it assumed a benthic, crawling lifestyle and developed a ventral mouth and simple pharynx. (C) The hypothetical metazoan precursor (sagittal section) after it achieved the acoel grade via cellularization of the epidermis, which surrounded a syncytial entodermis.

functions. As we explain in Chapter 5, such conditions are common in living colonial protists. Haeckel called this hypothetical protometazoan ancestor the **blastea** and supported its validity by noting the widespread occurrence of coeloblastulae among modern animals. In this scenario, the first Metazoa arose by invagination of the blastea; the resulting animals had a double-layered, gastrula-like body (Haeckel's **gastrea**) with a blastopore-like opening to the outside (Figure 4.13B) similar to the gastrulae of many modern animals. Haeckel believed that these ancestral creatures (the blastea and gastrea) were recapitulated in the ontogeny of modern animals, and the gastrea was viewed as the metazoan precursor to the cnidarians. In addition, the colonial theory has been supported by the argument that the body walls of many lower animals (e.g., members of the phyla Porifera and Cnidaria) bear monoflagellated or monociliated cells.

Haeckel's original ideas have been modified over the years by various authors investigating a colonial protist ancestry to the metazoan condition (Metschnikoff 1883; Hyman 1940). Some have argued that the transition to a layered construction occurred by ingression rather than by invagination, and that the original Metazoa were solid, not hollow (Figure 4.13C). This idea is based in large part on the view that ingression is the primitive form of gastrulation among cnidarians.

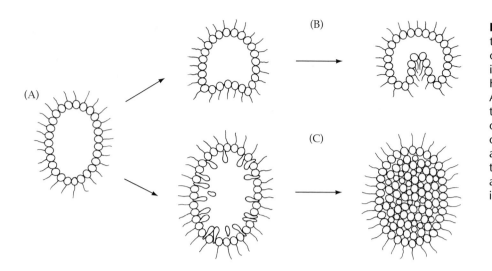

Figure 4.13 Two versions of the colonial theory of the origin of the Metazoa. (A) The hypothetical colonial flagellate ancestor, Haeckel's "blastea" (section). (B) According to Haeckel, the transition to a multicellular condition occurred by invagination, a developmental process that resulted in a hollow "gastrea." (C) According to Metschnikoff, the formation of a solid "gastrea" occurred by ingression.

For some reason, people tend to search among living creatures for possible ancestral types and "missing links." When attempting to support the colonial theory (particularly Haeckel's blastea) in this way, investigators long considered the volvocines (freshwater, colonial, photosynthetic flagellates such as *Volvox*; see Figure 4.15A) to be the most likely candidates. Consequently, most of the arguments against the colonial theory have dealt with the implied plantlike nature and freshwater habitat of the presumed ancestor.

An interesting offshoot of the colonial theory was presented by Otto Bütschli in 1883. He suggested that the primitive metazoan was a bilaterally symmetrical flattened creature of two layers of cells; he called this hypothetical animal a **plakula**. According to Bütschli, the plakula crawled about ingesting food through its "ventral" cell layer. Eventually the animal became somewhat hollow by the separation of its dorsal and ventral cell layers; this development allowed an invagination of the nutritive cells (Figure 4.14). This formation of a "gut" chamber increased the digestive surface area and at the same time produced inner and outer cell layers, an arrangement approaching the metazoan grade of complexity. As we will soon see, this old idea has been revitalized and may have some merit. In any case, the evidence is strong that the Metazoa had their origin in a flagella-bearing protist.

Most evidence today points to the protist phylum Choanoflagellata as the likely ancestral group from which the Metazoa arose. Choanoflagellates possess collar cells essentially identical to those found in sponges. Genera such as *Proterospongia*, *Sphaeroeca*, and others are animal-like colonial choanoflagellates (Figure 4.15C,D) and are commonly cited as typifying a potential metazoan precursor.

Among all of the ideas concerning the evolutionary origin of the metazoan condition, there exists a common problem: the search for intermediates between the Protista and Metazoa. Some authors have chosen to design logical but hypothetical forms of life for this purpose, while others rummage among extant types, arguing the advantages of using "real" organisms. Even though it is very probable that the actual precursor of the Metazoa long ago joined the ranks of the extinct, the presence of modern-day forms that somewhat combine protist and metazoan traits keeps debate alive. These organisms include not only various multinucleate ciliates and the colonial *Volvox*, but also several other protists and some enigmatic little multicellular animals of uncertain position. Figure 4.15 illustrates some of these creatures for comparative purposes.

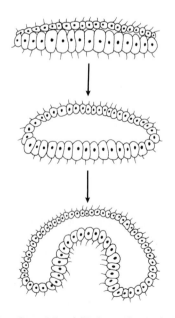

Figure 4.14 Otto Bütschli's hypothetical plakula and its transformation to a metazoan "gastrea" by invagination of a digestive chamber.

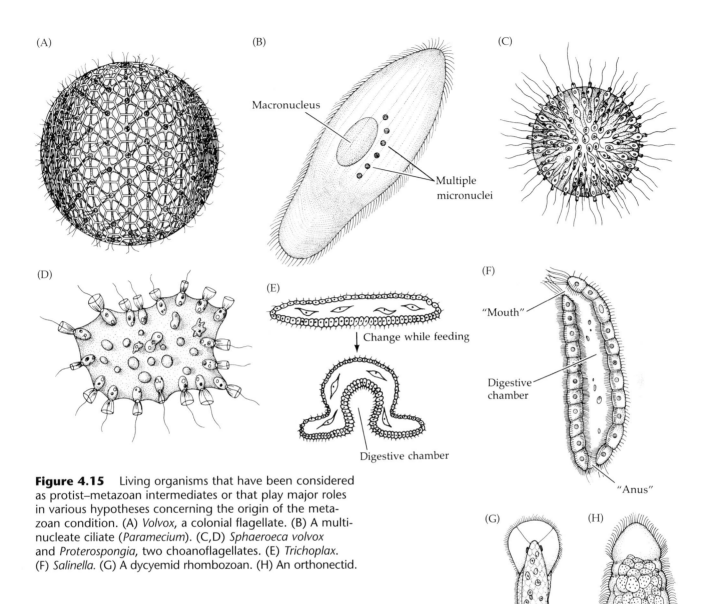

Figure 4.15 Living organisms that have been considered as protist–metazoan intermediates or that play major roles in various hypotheses concerning the origin of the metazoan condition. (A) *Volvox*, a colonial flagellate. (B) A multinucleate ciliate (*Paramecium*). (C,D) *Sphaeroeca volvox* and *Proterospongia*, two choanoflagellates. (E) *Trichoplax*. (F) *Salinella*. (G) A dycyemid rhombozoan. (H) An orthonectid.

About a century ago a tiny, flagellated, but multicellular creature was discovered in a marine aquarium. This animal, named *Trichoplax adhaerens*, was placed in its own phylum, the Placozoa (see Chapter 7 and papers by Grell). For many years, *Trichoplax* was thought to be a larval stage of some invertebrate, but workers are now convinced that it is an adult of uncertain affinity at the mesozoan grade of construction. *Trichoplax* has an outer, partly flagellated epithelium surrounding an inner mesenchymal cell mass. Its body margins are irregular, and it changes shape like an ameba. There is definitely some division of labor among various cells and areas of the body. When feeding, *Trichoplax* "hunches up" to form a temporary digestive chamber on its underside (Figure 4.15E)—producing a form strikingly similar to Bütschli's hypothetical plakula. Through discoveries of such real animals, hypothetical creatures gain credence. *Trichoplax* may represent the most primitive of all living Metazoa and perhaps even be a conservative descendant line of the ancestral metazoan type.

In 1892, J. Frenzel described a tiny organism reportedly collected from salt beds in Argentina. He named this mesozoan animal *Salinella* (Figure 4.15F). Although *Salinella* does not possess the layered construction of the Metazoa, it appears to display a higher level of functional organization than colonial protists. A single layer of cells forms the entire body wall and separates the digestive cavity from the outside. The digestive cavity was described as open at both ends as a mouth and anus,

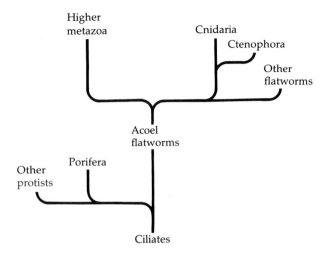

Figure 4.16 Scheme of possible metazoan relationships based on a ciliate ancestor and the syncitial theory.

and the animal fed on organic detritus. The phylum name Monoblastozoa has been proposed for this odd animal. Sadly, *Salinella* has not been seen since Frenzel's original report, and many zoologists now suspect that Frenzel seriously misinterpreted whatever creature he saw.

Finally, we mention briefly two other so-called mesozoan phyla, Rhomboza and Orthonectida (Figure 4.15 G,H). These animals are structurally rather simple but display complicated life cycles; they are all endoparasites of invertebrates. Some workers think they are derived from another group (or groups) of Metazoa, perhaps from parasitic trematodes (flukes), which undergo similarly complex life cycles. Other authors suggest that they are primitively simple and thus may have been derived from early metazoan or premetazoan stock.

Origin of the Bilateral Condition

We discussed the functional significance of bilaterality briefly in Chapter 3. The evolution of an anterior–posterior body axis, unidirectional movement, and cephalization almost certainly coevolved to some degree, and probably coincided with the invasion of benthic environments and the development of creeping locomotion. Furthermore, it is likely that the origin of the triploblastic condition took place soon after the appearance of the first bilateral forms. At least, bilaterality and triploblasty generally occur together in modern-day invertebrates.

There are several ideas concerning the origin of bilateral symmetry within the Metazoa. If one accepts Hadzi's syncytial theory or Bütschli's plakula, then the problem is already solved, since bilaterality would characterize the first Metazoa. On the other hand, supporters of the colonial theory generally assume that the first bilateral animals arose from a gastrula-like ancestor, presumably one with spherical or radial symmetry, or its planuloid descendant. Jägersten suggested that the

blastea took up a benthic lifestyle, invaginated, and then assumed bilaterality. In this scenario, the first metazoon was a bilaterally symmetrical gastrea (Jägersten's **bilaterogastrea**) from which all other major groups arose. This idea, and many others, can be found in Jägersten's works (1955, 1959, 1972).

Figures 4.16 and 4.17 illustrate two evolutionary schemes that show the implications of some of the ideas discussed above. Most proposed animal phylogenies are based on the widely held view that two major phyletic lines arose during the evolution of bilateral symmetry and the triploblastic condition, particularly among the coelomate animals (Figure 4.18). All coelomate animals (including those that have secondarily lost the coelom) can be placed along one or the other of these lineages, although some fit much more conveniently than others. Animals on these two evolutionary lines are called the protostomes and the deuterostomes and are distinguished from one another on the basis of several relatively consistent differences in development (Box 4A).

Origin of the Coelomic Condition

No less controversial than the origin of the metazoan condition is the evolutionary appearance of the coelom. The various hypotheses concerning this matter are summarized in R. B. Clark's fine book *Dynamics in Metazoan Evolution* (1964). Clark's personal approach is a functional one that emphasizes the adaptive significance of the coelom as the central criterion for evaluating ideas concerning its origin.

When early soft-bodied, bilaterally symmetrical animals larger than a few millimeters or so assumed a ben-

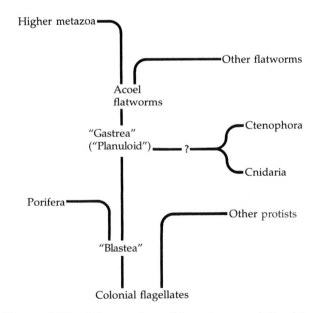

Figure 4.17 Scheme of possible metazoan relationships based on a colonial flagellate ancestor and the colonial theory.

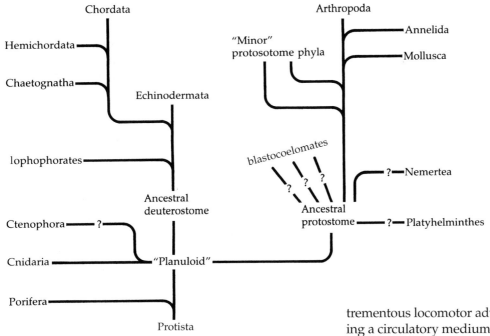

Figure 4.18 A commonly depicted evolutionary scheme of the Metazoa, showing some of the lineages from which major phyla and groups of phyla may have evolved.

thic, crawling, or burrowing lifestyle, a fluid (hydrostatic) skeleton was essential for certain types of movement. The evolution of a body cavity filled with fluid against which muscles could operate would have offered a trementous locomotor advantage in addition to providing a circulatory medium and space for organ development. How might such spaces have originated?

Most of the ideas concerning the evolutionary origin of the coelom were developed from the mid-nineteenth to early twentieth century, during the heyday of comparative embryology. Most of these hypotheses shared the premise of monophyly—that the coelomic condition arose only once. The inherent problem with a mono-

---

BOX 4A *Protostomes and Deuterostomes*

Developmental differences between protostomes and deuterostomes, and some representative phyla.

PROTOSTOMES	*DEUTEROSTOMES*
Spiral cleavage	Radial cleavage
Blastopore becomes mouth	Blastopore does not become mouth (often becomes the anus)
Nerve cords of central nervous system ventral	Nerve cords of central nervous system not ventral
Mesoderm derived from mesentoblast (usually the 4d cell)	Mesoderm arises from wall of archenteron
Subepidermal musculature derived, at least in part, from 4d mesoderm	Sheets of subepidermal muscles derived, at least in part, from archenteric mesoderm
Schizocoelous coelom formation	Enterocoelous coelom formation
Embryogeny results in adult coeloms as a single pair or metamerically arranged pairs, or adult coelom reduced	Embryogenesis results in tripartite arrangement of body cavities (protocoel, mesocoel, and metacoel), except in phylum Chordata
Examples: Nemertea, Sipuncula, Echiura, Annelida, Onychophora, Arthropoda, Mollusca	Examples: Echinodermata, Hemichordata, Chordata

Note: Some authors regard the acoelomate flatworms (Platyhelminthes) as protostomes because they show all the developmental traits of the latter, except they do not possess a coelom, do not have a circumenteric nervous system, and lack an anus.

phyletic approach is the difficulty of relating existing coelomate animals to a single common coelomate ancestor. Considering the advantages of possessing a coelom, the very different methods of embryonic development (schizocoely and various forms of enterocoely), and the variety of adult coelomic bauplans, it is more biologically reasonable to suggest that the coelomic condition arose twice. There are several currently debated ideas about how this might have happened, and a number of others have mostly been discarded as being incompatible with existing evidence or with our standard definition of the coelom.

The coelom may have originated by the pinching off and isolation of embryonic gut diverticula as occurs in the development of many extant enterocoelous animals (Figure 4.19). This so-called **enterocoel theory** (in several versions) has enjoyed relatively strong support by many authors since it was originally proposed by Lankester in 1877 (e.g., Lang 1881; Sedgwick 1884; Masterman 1897; Hubrecht 1904; Jägersten 1955). An obvious point in favor of this general idea is that enterocoely does occur in many living animals, thus retaining the hypothetical ancestral process. In addition, various authors cite examples of noncoelomate animals (anthozoans and turbellarian flatworms) in which gut diverticula exist in arrangements that resemble possible ancestral patterns.

Another popular idea concerning coelom origin is the **gonocoel theory** (Bergh 1885; Hatschek 1877, 1878; Meyer 1890, 1901; Goodrich 1946). This hypothesis suggests that the first coelomic spaces arose by way of mesodermally derived gonadal cavities that persisted subsequent to the release of gametes (Figure 4.20). The serial arrangement of gonads, as seen in animals such as flatworms and nemerteans, could have resulted in serially arranged coelomic spaces and linings such as occurs in annelids, where (at least primitively) they still produce and store gametes. A major argument against this hypotheses is the fact that in no modern-day coelomate animals do gonads develop *before* coelomic spaces. As we have seen, however, heterochrony can account for such turnabouts.

Another idea on coelom origin is called the **nephrocoel theory** (Lankester 1874; Ziegler 1898, 1912; Faussek 1899, 1911; Snodgrass 1938). The association between the coelom and excretion has prompted different versions of this hypothesis through about 75 years of moderate support. One idea is that the protonephridia of flatworms expanded to coelomic cavities, arguing that the coelom first arose from ectodermally derived structures. Another view is that coelomic spaces arose as cavities within the mesoderm and served as storage areas for waste products. Certainly the coelomic cavities of many animals are related to excretory functions, but there is no convincing evidence that this relationship was the primary selective force in the origin of the coelomate condition.

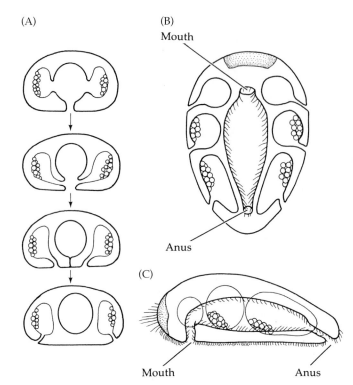

Figure 4.19 Jägersten's bilaterogastrea theory, according to which the coelomic compartments arise by enterocoelic pouching. (A) The formation of paired coeloms from the wall of the archenteron. The slitlike blastopore of the bilaterogastrea closes midventrally, leaving mouth and anus at opposite ends (B). (B,C) The tripartite coelomic condition in Jägersten's hypothetical early coelomate animal (ventral and lateral views).

Clark (1964) speculated that schizocoely as we know it today could have evolved by the formation of spaces within the solid mesoderm of acoelomate animals and then have been retained in response to the positive selection for the resulting hydrostatic skeleton. This is a very straightforward and parsimonious view, in part because, like the enterocoel theory, it accommodates a real developmental process.

As we mentioned earlier, these hypotheses share the fundamental constraint of arguing a monophyletic origin to all coelomate animals. The basic developmental differences between the two clades of coelomate animals (the protostomes and the deuterostomes) suggest that the coelom may have arisen separately in these two lineages; we explore this further in Chapter 24. Given the strong similarities between the coelomate protostomes and acoelomate flatworms and nemerteans, it is easy to envision the protostome clade arising from a triploblastic acoelomate ancestor. Hollowing of the mesoderm in such a precursor to produce fluid-filled hydrostatic spaces can be easily explained both developmentally (modern-day schizocoely) and functionally (peristaltic burrowing, increased size, and so on). On the other hand, schizocoely and the origin of mesoderm in

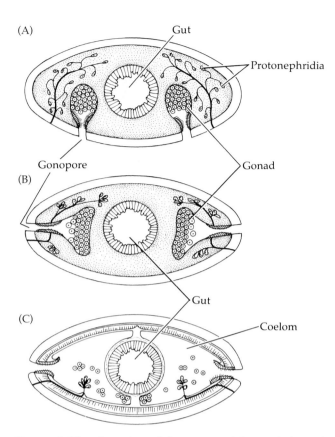

Figure 4.20 A version of the gonocoel theory (schematic cross sections). (A) The condition in flatworms, which have mesodermally derived gonads leading to ventral gonopores. (B) The condition in nemerteans, which have serially arranged gonadal masses leading to laterally placed gonopores. (C) The condition in polychaetes, in which the linings of the gonads have expanded to produce coelomic spaces with coelomoducts to the outside.

the ancestral protostome may have arisen simultaneously, and the acoelomate condition may be secondarily derived from coelomate precursors (see Rieger 1985, 1986).

To derive deuterostome and protostome clades from a common immediate coelomate ancestor creates a complicated scenario. The most parsimonious hypothesis might be to view the deuterostome ancestor as a diploblastic animal, perhaps a planuloid form, in which enterocoely occurred. Deriving the deuterostome lineage *separately* from the evolution of spiral cleavage and the other features of protostomes avoids many of the complications inherent in a monophyletic view of coelom origin. Imagine a hollow, invaginated, gastrula-like metazoon swimming with its blastopore trailing, as do the planula larvae of some cnidarians. Enterocoely may have accompanied a tendency toward benthic life, giving the animal a peristaltic burrowing ability. The archenteron may have then opened anteriorly as a mouth, and the new coelomate creature adopted a deposit-feeding lifestyle. If such a story began at the level

of diploblastic Metazoa (e.g., cnidarians), then the radial cleavage seen today in deuterostomes was also present in the ancestor to this group. A diphyletic origin of the coelomic condition from larval-like ancestors has been presented by Nielsen and Nørrevang (1985). In their hypothesis, a pelagic gastrea gave rise to the cnidarians and to a second lineage—another larval-like creature they call a **trochaea**. This ancestor was the precursor to the protostomes and the deuterostomes, but the coelom arose separately in each group.

Prominent among contemporary workers who have speculated on origins of major metazoan groups—and introduced yet more hypothetical ancestors (and names)—is Claus Nielsen. Nielsen envisions the two major metazoan clades radiating from an ancient common ancestor that conforms to Haeckel's radially symmetrical gastrea. From this planktonic ancestor there evolved two separate lines. One line led to the protostomes via a series of at least two hypothetical ancestors, called by Nielsen the **gastroneuron** and the **trochaea**. The other line, to the deuterostomes, was by way of a hypothetical **notoneuron** ancestor. (The names *gastroneuron* and *notoneuron* refer to the ventral versus dorsal positions of the major nerve cords in most protostomes and deuterostomes, respectively.) Nielsen claims that the notoneuron ancestor (and its descendant deuterostomes) retained the monociliated cell condition of the gastraea ancestor, whereas the gastroneuron ancestor (and its descendant protostomes) evolved a more advanced condition of multiciliated cells. In addition, the gastroneuron line came to rely on "downstream ciliary feeding," in which the larvae capture food particles from the water on the downstream side of the ciliary feeding bands, whereas the notoneuron line developed "upstream feeding," in which the larvae capture food particles from the water on the upstream side of the ciliary feeding bands.

As you can see, evolutionary analysis at the level of phyla, when it attempts to describe hypothetical ancestors, can be convoluted and problematical, and many different viewpoints exist. We trust, however, that you have gained some insights not only into the particular hypotheses discussed here, but also into evolutionary speculation. A fundamental caveat should be kept in mind: any number of evolutionary pathways can be proposed and made to appear convincing on paper by imagining appropriate hypothetical ancestors or intermediates, but one must always ask whether these hypothetical creatures would have worked. Do they possess realistic bauplans? Clark (1964) spends a good deal of time on this point and emphasizes it in his conclusion with the following passage (p. 258):

> The most important and least considered of these [principles] is that hypothetical constructs which represent ancestral, generalized forms of modern groups, or stem forms from which several modern phyla diverge, must be possible animals. In other words, they

must be conceived as living organisms, obeying the same principles that we have discovered in existing animals.

It is in such terms that evolutionary hypotheses can be evaluated. From a cladistic point of view, it is best to avoid initial speculation on what a hypothetical ancestor might have looked like, and instead rely on phylogenetic trees (cladograms) of known taxa to establish genealogical relationships or branching patterns. Once a cladogram has been constructed, the pattern of features associated with the taxa on the tree will themselves predict the nature (character combination) of the ancestor for each branch. This method attempts to avoid the potential problem of circular reasoning, in which a hypothetical ancestor is established first and hence constrains and predicts the nature of the taxa descended from it. In either case, for the hypotheses to be truly scientific, they must be testable with new data gathered outside the framework of that used to construct the initial hypotheses.

These and other speculations on animal phylogeny have been complemented (and complicated) in recent years by the emergence of phylogenetic hypotheses based on molecular research, particularly DNA sequence data. The unsettled nature of molecular phylogenetic research on phyletic affinities is due to several facts: the field of molecular phylogenetics is still young; few species have actually been studied, most research has focused on a single gene (the 18S ribosomal gene), which has yielded conflicting results at the phylum level; we lack reliable information on rates of molecular change and degree of true "homology" of seemingly similar DNA sequences; and, finally, genes and species need not evolve in perfect sequence. In later chapters we include some of the phylogenetic ideas being generated by "unraveling the double helix," and in Chapter 24 we review this rapidly changing field as it is being applied to the reconstruction of phylum-level lineages among the Metazoa.

Selected References

General Invertebrate Embryology

Adiyodi, K. G. and R. G. Adiyodi (eds.). 1983–1998. *Reproductive Biology of Invertebrates*. Vols. 1–8. Wiley, New York.

Brusca, G. J. 1975. *General Patterns of Invertebrate Development*. Mad River Press, Eureka, CA.

Conn, D. B. 1991. *Atlas of Invertebrate Reproduction*. Wiley-Liss, New York. [Includes photographs of developmental stages of most major groups.]

Cooke, J. 1988. The early embryo and the formation of body pattern. Am. Sci. 76: 35–41.

Costello, D. P. and C. Henley. 1976. Spiralian development: A perspective. Am. Zool. 16: 277–291. [The introduction to an entire issue of *American Zoologist* (Spring 1976) devoted to an examination of those creatures showing spiral cleavage.]

Dawydoff, C. 1928. *Traité d'Embryologie Comparée des Invertébrés*. Masson et Cie, Libraires de l'Academie de Medicine, Paris.

Eckelbarger, K. J. 1994. Diversity of metazoan ovaries and vitellogenic mechanisms: Implications for life history theory. Proc. Biol. Soc. Wash. 107: 193–218.

Giese, A. C. and J. S. Pearse (and V. B. Pearse, Vol. 9) (eds.). 1974–1987. *Reproduction of Marine Invertebrates*. Vols. 1–5, 9. Blackwell Scientific, Palo Alto, CA. Vol. 6. Echinoderms and Lophophorates, Boxwood Press, Pacific Grove, CA. [An outstanding series of volumes containing reviews of every invertebrate phylum.]

Gilbert, S. F. 2000. *Developmental Biology*. 6th Ed. Sinauer Associates, Sunderland, MA.

Gilbert, S. F. and A. M. Raunio (eds.). 1997. *Embryology: Constructing the Organism*. Sinauer Associates, Sunderland, MA. [Includes up-to-date chapters on the development of most major invertebrate phyla.]

Hall, B. K. 1992. *Evolutionary Developmental Biology*. Chapman & Hall, New York.

Harrison, F. W. and R. R. Cowden (eds.). 1982. *Developmental Biology of Freshwater Invertebrates*. Alan R. Liss, New York.

Korschelt, E. and K. Heider. 1900. *Lehrbuch der vergleichenden Entwicklungsgeschichte der wirbellosen Tiere*. Fischer, Jena.

Kume, M. and K. Dan. 1957. *Invertebrate Embryology*. (English translation published by NOLIT Publishing House, Belgrade, Yugoslavia.)

Malacinski, G. M. (ed.). 1984. *Pattern Formation: A Primer in Developmental Biology*. Macmillan, New York.

Marthy, H. J. (ed.). 1990. *Experimental Embryology in Aquatic Plants and Animals*. Plenum, New York.

Masterman, A. 1897. On the Diplochordata. 1. The structure of *Actinotrocha*. 2. The structure of *Cephalodiscus*. Q. J. Microsc. Sci. 40: 281–366.

Pflugfelder, O. 1962. *Lehrbuch der Entwicklungsgeschichte und Entwicklungsphysiologie der Tiere*. Fischer, Jena.

Raff, R. A. 1996. *The Shape of Life: Genes, Development, and the Evolution of Animal Form*. University of Chicago Press, Chicago.

Reverberi, G. 1971. *Experimental Embryology of Marine and Fresh-Water Invertebrates*. American Elsevier Publisher, New York.

Sawyer, R. H. and R. M. Showman (eds.). 1985. *The Cellular and Molecular Biology of Invertebrate Development*. University of South Carolina Press, Columbia.

Strathmann, M. F. 1987. *Reproduction and Development of Marine Invertebrates of the Northern Pacific Coast*. University of Washington Press, Seattle.

Wilson, E. B. 1892. The cell lineage of *Nereis*. J. Morphol. 6: 361–480. [Wilson's classic work establishing the coding system for spiral cleavage.]

Wilson, W. H., S. Stricker, and G.L. Shinn (eds.). 1994. *Reproduction and Development of Marine Invertebrates*. Johns Hopkins University Press, Baltimore, MD. [A collection of recent work on a host of developmental topics.]

Life Histories

Ayal, Y. and U. Safriel. 1982. *r*-Curves and the cost of the planktonic stage. Am. Nat. 119: 391–401.

Cameron, R. A. (ed.). 1986. Proceedings of the Invertebrate Larval Biology Workshop held at Friday Harbor Laboratories, University of Washington, 26–30 Mar. 1985. Bull. Mar. Sci. 39: 145–622. [Thirty-seven papers on larval biology.]

Caswell, H. 1978. Optimal life histories and the age-specific cost of reproduction. Bull. Ecol. Soc. Am. 59: 99. [This paper and the two below include interesting discussions of the adaptive qualities of various life cycle patterns, especially those that do not fit the typical direct or indirect definitions.]

Caswell, H. 1980. On the equivalence of maximizing fitness and maximizing reproductive value. Ecology 61: 19–24.

Caswell, H. 1981. The evolution of "mixed" life histories in marine invertebrates and elsewhere. Am. Nat. 117(4): 529–536.

Charlesworth, B. 1991. The evolution of sex chromosomes. Science 251: 1030–1033.

Charnov, E. L., J. M. Smith, and J. J. Bull. 1976. Why be an hermaphrodite? Nature 263: 125–126.

Chia, F. S. 1974. Classification and adaptive significance of developmental patterns in marine invertebrates. Thalassia Jugosl. 10: 121–130.

Chia, F. S. and M. Rice (eds.). 1978. *Settlement and Metamorphosis of Marine Invertebrate Larvae.* Elsevier/North-Holland, New York. [Proceedings of the Symposium on Settlement and Metamorphosis of Marine Invertebrate Larvae, Am. Zool. Soc. Meeting, Toronto, Ontario, Canada, Dec. 27–28, 1977.]

Christiansen, F. and T. Fenchel. 1979. Evolution of marine invertebrate reproductive patterns. Theor. Pop. Biol. 16: 267–282.

Crisp, D. 1974. Energy relations of marine invertebrate larvae. Thalassia Jugosl. 10: 103–120.

Crisp, D. 1974. Factors influencing the settlement of marine intertidal larvae. *In* P. T. Grant and A. M. Macie (eds.), *Chemoreception in Marine Organisms.* Academic Press, New York, pp. 177–265.

Dawydoff, C. 1928. *Traité d'Embryologie Comparée des Invertébrés.* Masson et Cie, Libraires de l'Academie de Médicine, Paris. [In many ways out of date, yet still a useful benchmark work.]

Eckelbarger, K. J. 1994. Diversity of metazoan ovaries and vitellogenic mechanisms: Implications for life history theory. Proc. Biol. Soc. Wash. 107: 193–218.

Eckelbarger, K. J. and L. Watling. 1995. Role of phylogenetic constraints in determining reproductive patterns in deep-sea invertebrates. Invert. Biol. 114(3): 256–269.

Ecology of Asexual Reproduction in Animals. 1979. Am. Zool. 19(3): 667–797. [Eleven papers from the symposium by the same name.]

Emlet, R. B. and E. E. Ruppert (eds.). 1994. Symposium: Evolutionary morphology of marine invertebrate larvae and juveniles. Am. Zool. 34: 479–585.

Gilbert, L. I. and E. Frieden (eds.). 1981. *Metamorphosis: A Problem in Developmental Biology,* 2nd Ed. Plenum, New York. [A largely biochemical approach.]

Grosberg, R. K. 1981. Competitive ability influences habitat choice in marine invertebrates. Nature 290: 700–702.

Hadfield, M. G. 1978. Metamorphosis in marine molluscan larvae: An analysis of stimulus and response. *In* F. S. Chia and M. E. Rice (eds.), *Marine Natural Products Chemistry.* Plenum, New York, pp. 165–175.

Hadfield, M. G. 1984. Settlement requirements of molluscan larvae: New data on chemical and genetic roles. Aquaculture 39: 283–298.

Jablonsky, D. and R. A. Lutz. 1983. Larval ecology of marine benthic invertebrates: Paleobiological implications. Biol. Rev. 58: 21–89. [An excellent review of larval ecology from an evolutionary perspective.]

Jeffrey, W. R. and R. A. Raff (eds.). 1982. *Time, Space, and Pattern in Embryonic Development.* Alan R. Liss, New York.

Kohn, A. J. and F. E. Perron. 1994. *Life History and Biogeography: Patterns in* Conus. Clarenton Press, Oxford.

Lutz, R. A., D. Jablonski and R. D. Turner. 1984. Larval development and dispersal at deep-sea hydrothermal vents. Science 226: 1451–1454.

McEdward, L. R. (ed.). 1995. *Ecology of Marine Invertebrate Larvae.* CRC Press, Boca Raton, FL.

Mileikovsky, S. 1971. Types of larval development in marine bottom invertebrates, their distribution and ecological significance: A re-evaluation. Mar. Biol. 10: 193–213. [An excellent treatment of the subject.]

Pechenick, J. A. 1979. Role of encapsulation in invertebrate life histories. Am. Nat. 114: 859–870. [Ideas concerning mixed life cycles.]

Pechenick, J. A. 1990. Delayed metamorphosis by larvae of benthic marine invertebrates: Does it occur? Is there a price to pay? Ophelia 32: 63–94.

Perron, F. and R. Carrier. 1981. Egg size distribution among closely related marine invertebrate species: Are they bimodal or unimodal? Am. Nat. 118: 749–755. [Explains, in part, how egg size varies with other aspects of the developmental pattern.]

Sammarco, P. W. and M. L. Heron (eds.). 1994. *The Bio-Physics of Marine Larval Dispersal.* Am. Geophysical Union, Wash. DC. [An excellent blending of physical oceanography and biology.]

Starr, M., J. H. Himmelman and J. C. Therriault. 1990. Direct coupling of marine invertebrate spawning with phytoplankton blooms. Science 247: 1071–1074.

Steidinger, K. A. and L. M. Walker (eds.). 1984. *Marine Plankton Life Cycle Strategies.* C.R.C. Press, Boca Raton, FL, pp. 93–120.

Strathmann, R. 1977. Egg size, larval development and juvenile size in benthic marine invertebrates. Am. Nat. 111: 373–376.

Strathmann, R. 1978. The evolution and loss of feeding larval stages of marine invertebrates. Evolution 32(4): 894–906.

Strathmann, R. 1985. Feeding and nonfeeding larval development and life history evolution in marine invertebrates. Ann. Rev. Ecology and Systematics 16: 339–361.

Strathmann, R. and M. Strathmann. 1982. The relationship between adult size and brooding in marine invertebrates. Am. Nat. 119: 91–101.

Thorson, G. 1946. Reproduction and larval development of Danish marine bottom invertebrates with special reference to the planktonic larvae in the South (Oresund). Medd. Danm. Fisk., Havunders., Ser. Plankton, 4.

Thorson, G. 1950. Reproduction and larval ecology of marine bottom invertebrates. Biol. Rev. 25: 1–45. [These two works by G. Thorson laid the foundation for modern studies concerning the classification of invertebrate life cycles and their significance.]

Todd, C. D. and R. W. Doyle. 1981. Reproductive strategies of marine benthic invertebrates: A settlement-timing hypothesis. Mar. Ecol. Prog. Ser. 4: 75–83.

Wray, G. A. and R. A. Raff. 1991. The evolution of developmental strategy in marine invertebrates. Trends Ecol. Evol. 6: 45–50.

Young, C. M. 1990. Larval ecology of marine invertebrates: A sesquicentennial history. Ophelia 32: 1–48.

Young, C. M. and K. J. Eckelbarger (eds.). 1994. *Reproduction, Larval Biology, and Recruitment of the Deep-Sea Benthos.* Columbia University Press, New York.

On the Origins of Major Invertebrate Lines

Alberch, P., S. J. Gould, G. F. Osta, and D. B. Wake. 1979. Size and shape in ontogeny and phylogeny. Paleobiology 5(3): 296–317.

Bergh, R. S. 1885. Die Exkretionsorgane der Würmer. Kosmos, Lwow 17: 97–122.

Bergstrom, J. 1989. The origin of animal phyla amd the new phylum Procoelomata. Lethalia 22: 259–269.

Bütschli, O. 1883. Bemerkungen zur Gastrea Theorie. Morph. Jahrb. 9.

Carter, G. S. 1954. On Hadzi's interpretations of animal phylogeny. Syst. Zool. 3: 163–167. [An analysis, sometimes quite pointed, of Hadzi's views.]

Clark, R. B. 1964. *Dynamics in Metazoan Evolution.* Oxford University Press, New York. [A fine functional approach to metazoan evolution, especially concerning the origin of the coelom and metamerism.]

Dougherty, E. C. (ed.). 1963. *The Lower Metazoa: Comparative Biology and Phylogeny.* University of California Press, Berkeley.

Eaton, T. H. 1953. Paedomorphosis: An approach to the chordate–echinoderm problem. Syst. Zool. 2: 1–6.

Faussek, V. 1899. Über die physiologische Bedeutung des Cöloms. Trav. Soc. Nat. St. Petersberg 30: 40–57.

Faussek, V. 1911. Vergleichend–embryologische Studien. (Zur Frage über die Bedeutung der Cölom-hölen). Z. Wiss. Zool. 98: 529–625. [Faussek's works include his views on the nephrocoel theory.]

Frenzel, J. 1892. *Salinella.* Arch. Naturgesch. 58, Pt. 1.

Garstang, W. 1922. The theory of recapitulation. J. Linn. Soc. Lond. Zool. 35: 81–101. [Garstang's revolutionary ideas on Haeckel's recapitulation concept.]

Garstang, W. 1985. *Larval Forms and Other Zoological Verses.* University of Chicago Press, Chicago. [A wonderful collection of prose and poetry by Garstang, published after his death. The biographical sketch by Sir Alister Hardy and the Foreword by Michael LaBarbera chronicle many of Garstang's contributions to our understanding of the relationships between ontogeny and phylogeny and serve as a delightful introduction to the 26 poems in this little volume. This new edition of the original (1951) version also includes Garstang's famous address on "The Origin and Evolution of Larval Forms."]

Goodrich, E. S. 1946. The study of nephridia and genital ducts since 1895. Q. J. Microsc. Sci. 86: 113–392. [One of the great classics concerning the origin of the coelom and related evolutionary matters.]

Gould, S. J. 1977. *Ontogeny and Phylogeny*. Harvard University Press, Cambridge, MA. [A scholarly coverage of ideas concerning recapitulation and other interactions between development and evolution.]

Grell, K. G. 1971. *Trichoplax adhaerens* F. E. Schulze, und die Entstehung der Metazoen. Naturwiss. Rundsch. 24(4): 160–161.

Grell, K. G. 1971. Embryonalentwicklung bei *Trichoplax adhaerens* F. E. Schulze. Naturwiss. 58: 570.

Grell, K. G. 1972. Formation of eggs and cleavage in *Trichoplax adhaerens*. Z. Morphol. Tiere 73(4): 297–314.

Grell, K. G. 1973. *Trichoplax adhaerens* and the origin of the Metazoa. Actualite's Protozooligiques. IVe. Cong. Int. Protozoologie. Paul Couty, Clermont–Ferrand.

Grell, K. G. and G. Benwitz. 1971. Die Ultrastruktur von *Trichoplax adhaerens* F. E. Schulze. Cytobiologie 4(2): 216–240.

Gutman, W. F. 1981. Relationships between invertebrate phyla based on functional–mechanical analysis of the hydrostatic skeleton. Am. Zool. 21: 63–81.

Hadzi, J. 1953. An attempts to reconstruct the system of animal classification. Syst. Zool. 2: 145–154. [Odd ramblings about lumping all animals into a few phyla, the author's views on the origins of the metazoan condition, and other things.]

Hadzi, J. 1963. *The Evolution of the Metazoa*. Macmillan, New York. [Overkill. But then, any book that begins with the sentence, "It was in 1903, 58 years ago, that I, then a young man who had just left the classical grammar school at Zagreb, went to Vienna to study natural sciences and above all my beloved Zoology at Vienna University," can't be all bad!]

Haeckel, E. 1866. *Generelle Morphologie der Organismen: Allgemeine Grundzüuge der organischen Formen-Wissenschaft mechansch begrüundet durch die von Charles Darwin reformierte Descendenz-Theorie*. Vols. 1–2. George Reimer, Berlin.

Haeckel, E. 1874. The gastrea-theory, the phylogenetic classification of the animal kingdom and the homology of the germ-lamellae. Q. J. Microscop. Sci. 14: 142–165; 223–247. [Haeckel's concepts of recapitulation and blastea-gastrea idea of metazoan origin. A translation of the original German paper that introduced the colonial theory of metazoan origin (Jena. Z. Naturwiss. 8: 1–55).]

Hall, B. K. 1996. Baupläne, phylotypic stages, and constraints. Why are there so few types of animals? *In* M. K. Hecht, et al. (eds.), *Evolutionary Biology*, Vol. 29. Plenum, New York, pp. 215–261.

Hanson, E. D. 1958. On the origin of the eumetazoa. Syst. Zool. 7: 16–47. [Support for Hadzi's views.]

Hanson, E. D. 1977. *The Origin and Early Evolution of Animals*. Wesleyan University Press, Middletown, CT.

Hatschek, B. 1877. Embryonalentwicklung und Knospung der *Pedicellina echinata*. Z. Wiss. Zool. 29: 502–549. [Some early thoughts on the gonocoel theory.]

Hatschek, B. 1878. Studien üuber Entwicklungsgeschichte der Anneliden. Ein Beitrag zur Morphologie der Bilaterien. Arb. Zool. Inst. Wien 1: 277–404.

House, M. R. (ed.). 1979. *The Origin of Major Invertebrate Groups*. Academic Press, New York. The Systematics Association Special Vol. No. 12.

Hubrecht, A. 1904. Die Abstammung der Anneliden und Chordaten und die Stellung der Ctenophoren und Platyhelminthen im System. Jena. Z. Naturwiss. 39: 152–176. [Odd ideas about the cnidarian ancestry of the annelids and chordates relating to the enterocoel theory.]

Hyman, L. H. 1940–1967. *The Invertebrates*. Vols. 1–6. McGraw-Hill, New York. [All volumes include especially fine discussions on embryology of the included taxa. Volumes 1 and 2 include the author's views on the origin of the Metazoa, bilaterality, and coelom, and other related matters.]

Inglis, W. G. 1985. Evolutionary waves: Patterns in the origins of animal phyla. Aust. J. Zool. 33: 153–178.

Ivanova-Kazas, O. M. 1982. Phylogenetic significance of spiral cleavage. Soviet J. Mar. Biol. 7(5): 275–283.

Jablonski, D. and D. J. Bottjer. 1991. Environmental patterns in the origins of higher taxa: The post-Paleozoic fossil record. Science 252: 1831–1833.

Jägersten, G. 1955. On the early phylogeny of the Metazoa. The bilaterogastrea theory. Zool. Bidr. Uppsala 30: 321–354.

Jägersten, G. 1959. Further remarks on the early phylogeny of the Metazoa. Zool. Bidr, Uppsala 33: 79–108.

Jägersten, G. 1972. *Evolution of the Metazoan Life Cycle*. Academic Press, London. [The phylogeny of the Metazoa according to Jägersten, based in part on his bilaterogastrea hypothesis. Included are some of the author's thoughts on recapitulation.]

Jefferies, R. P. S. 1986. *The Ancestry of the Vertebrates*. British Museum (Natural History), London.

Lang, A. 1881. Der Bau von *Gunda segmentata* und die Verwandtschaft der Platyhelminthen mit Coelenteraten und Hirundineen. Mitt. Zool. Sta. Neapel. 3: 187–251.

Lang, A. 1903. Beiträuge zu einer Trophocoltheorie. Jena. Z. Naturw. 38: 1–373. [Lang's 1881 paper was in support of the enterocoel theory, suggesting that the coelom arose from pinched-off gut diverticula in flatworms; this opinion was based upon his study of the turbellarian *Gunda* (now *Procerodes*). However, Lang eventually switched his allegiance to the gonocoel theory (1903).]

Lankester, E. R. 1874. Observations on the development of the pond snail (*Lymnaea stagnalis*), and in the early stages of other Mollusca. Q. J. Microsc. Sci. 14: 365–391. [Some of the author's ideas about coelom origin.]

Lankester, E. R. 1877. Notes on the embryology and classification of the animal kingdom; comprising a revision of speculations relative to the origin and significance of the germ layers. Q. J. Microsc. Sci. 17: 399–454. [In addition to the ambitious title, this work includes thoughts about the gonocoel theory.]

Løvtrup, S. 1975. Validity of the Protostomia–Deuterostomia theory. Syst. Zool 24: 96–108. [Arguments against the concept of protostomes and deuterostomes.]

Marcus, E. 1958. On the evolution of the animal phyla. Q. Rev. Biol. 33: 24–58.

Margulis, L. 1981. *Symbiosis in Cell Evolution: Life and Its Environment on the Early Earth*. W. H. Freeman, San Francisco.

Masterman, A. 1897. On the theory of archimeric segmentation and its bearing upon the phyletic classification of the Coelomata. Proc. R. Soc. Edinburgh 22: 270–310. [Masterman was generally a proponent of the enterocoel theory.]

McKinney, M. L. and K. J. McNamara. 1991. *Heterochrony: The Evolution of Ontogeny*. Plenum Press, NY.

Meckel, J. 1811. Entwurf einer Darstellung der zwischen dem Embryozustande der höheren Tiere und dem Permanenten der niedere stattfindenen Parallele: Beitrüage zur vergleichenden Anatomie, Vol. 2. Carl Heinrich Reclam., Leipzig, pp. 1–60.

Meckel, J. 1811. Über den Charakter der allmüahligen Vervollkommung der Organisation, oder den Unterschied zwischen den höheren und niederen Bildungen: Beytrüage zur vergleichenden Anatomie, Vol. 2. Carl Heinrich Reclam., Leipzig, pp. 61–123. [Works by Meckel contain interesting pre-Haeckelian concepts of relationships between development and evolution as it was understood before Darwin.]

Metschnikoff, E. 1883. Untersuchungen über die intracellulare Verdauung bei wirbellosen Thieren. Arb. Zool. Inst. Wien. 5: 141–168. [Translated into English and published as, "Researches on the intracellular digestion of invertebrates," Q. J. Microsc. Sci. (1884) 24: 89–111. This paper includes some of the studies that led Metschnikoff and eventually others to conclude that ingression was the animal form of gastrulation.]

Meyer, E. 1890. Die Abstimmung der Anneliden. Der Ursprung der Metamerie und die Bedeutung des Mesoderms. Biol. Cbl. 10: 296–308. [An English translation appeared in Am. Natur. 24: 1143–1165.]

Meyer, E. 1901. Studien üuber den Körperbau der Anneliden. V. Das Mesoderm der Ringelwüurmer. Mitt. Zool. Sta. Neapel. 14:

247–585. [The two papers by Meyer include coverage of the gonocoel theory.]

Morris, S. C. J. D. George, R. Gibson and H. M. Platt (eds.). 1985. *The Origins and Relationships of Lower Invertebrates*. Clarenton Press, Oxford. Published for the Systematics Association, Special Vol. 28.

Nielsen, C. 1979. Larval ciliary bands and metazoan phylogeny. Forschr. Zool. Syst. Evolutionsforsch. 1: 178–184.

Nielsen, C. 1985. Animal phylogeny in light of the trochaea theory. Biol. J. Linn. Soc. London 25: 243–299.

Nielsen, C. 1987. Structure and function of metazoan ciliary bands and their phylogenetic significance. Acta Zool. 68: 205–262.

Nielsen, C. 1994. Larval and adult characters in animal phylogeny. Am. Zool. 34: 492–501.

Nielsen, C. 1995. *Animal Evolution: Interrelationships of the Living Phyla*. Oxford University Press, Oxford.

Nielsen, C. and A. Nørrevang. 1985. The trochea theory: An example of life cycle phylogeny. *In* Morris et al., *The Origins and Relationships of Lower Invertebrates*, Clarenton Press, Oxford, pp. 28–41.

Patterson, C. 1990. Reassessing relationships. Nature 344: 199–200.

Popkov. D. V. 1993. Polytrochal hypothesis of origin and evolution of trochophora type larvae. Zool. Zh. 72: 1–17.

Raff, R. A. and T. C. Kaufman. 1983. *Embryos, Genes, and Evolution*. Macmillan, New York.

Remane, A. 1963. The evolution of the Metazoa from colonial flagellates *vs*. plasmodial ciliates. *In* E. C. Dougherty et al. (eds.), *The Lower Metazoa: Comparative Biology and Phylogeny*. University of California Press, Berkeley, pp. 78–90.

Rieger, R. M. 1985. The phylogenetic status of the acoelomate organization within the Bilateria: A histological perspective. *In* Morris et al. (eds.), *The Origins and Relationships of Lower Invertebrates*. Clarenton Press, Oxford, pp. 101–122.

Rieger, R. M. 1986. Uber den Ursprung der Bilateria: die Bedeutung der Ultrastrukturforschung fur ein neues Verestehen der Metazoenevolution. Verh. Dtsch. Zool. Ges. 79: 31–50.

Rieger, R. M. 1994. The biphasic life cycle—A central theme of metazoan evolution. Am. Zool. 484–491.

Salvini-Plawen, L. 1980. Was ist eine Trochophora? Eine Analyse der Larventypen mariner Protostomier. Zool. Jb., Anat. 103: 389–423.

Salvini-Plawen, L. V. 1982. A paedomorphic origin of the oligomerous animals? Zool. Scr. 11: 77–81. [Dubious and difficult reading.]

Sarvaas, A. E. du Marchie. 1933. La theorie du coelome. Thesis, University of Utrecht. [Some ideas on the schizocoel theory that never quite took hold.]

Schleip, W. 1929. *Die Determination der Primitiventwicklung*. Akad. Verlags, Leipzig. [The origin of the concept of the "Spiralia."]

Sedgwick, A. 1884. On the nature of metameric segmentation and some other morphological questions. Q. J. Microsc. Sci. 24: 43–82.

[This work provided the main driving force behind the idea that the coelom arose (via enterocoely) from cnidarian gut pouches rather than by a pinching off of the diverticula in flatworm digestive tracts.]

Serres, E. R. A. 1824. Explication de système nerveux des animaux invertébrés. Ann. Sci. Nat. 3: 377–380.

Serres, E. R. A. 1830. Anatomie transcendante—Quatrieme mémoire: Loi de symétrie et de conjugaison du systéme sanguin. Ann. Sci. Nat. 21: 5–49.

Siewing, R. 1980. Das Archichelomatenkonzept. Zool. Jahrb. Abt. Anat. Ontog. Tiere 8,103: 439–482.

Simonetta, A. M. and S. Conway Morris (eds.). 1989. *The Early Evolution of Metazoa and the Significance of Problematic Taxa*. Cambridge Univ. Press, Cambridge.

Smith, J. III and S. Tyler. 1985. The acoel turbellarians: Kingpins of metazoan evolution or a specialized offshoot? *In* Morris et al. (eds.), *The Origins and Relationships of Lower Invertebrates*. Clarenton Press, Oxford, pp. 123–142.

Snodgrass, R. E. 1938. Evolution of the Annelida, Onychophora and Arthropoda. Smithson. Misc. Collections 97(6)1–159. [A proponent of the nephrocoel theory explains his views.]

Thiele, J. 1902. Zur Colomfrage. Zool. Anz. 25: 82–84.

Thiele, J. 1919. Über die Auffassung der Leibeshöhle von Mollusken und Anneliden. Zool. Anz. 35: 682–695.

Valentine, J. 1973. Coelomate superphyla. Syst. Zool. 22: 97–102.

Valentine, J. 1975. Adaptive strategy and the origin of grades and ground-plans. Am. Zool. 15: 391–404.

Valentine, J., S. M. Awramik, P. S. Signor and P. M. Sadler. 1991. The biological explosion at the Precambrian–Cambrian boundary. *In* M. K. Hecht, B. Wallace and R. J. Macintyre (eds.), *Evolutionary Biology*, Vol. 25. Plenum, New York, pp. 279–356.

Vecchia, G. L., R. Valvassori and M. D. C. Carnevali (eds.). 1995. [Body cavities: Function and phylogeny. Proceedings of the International Symposium on Body Cavities, Varese.] Collana U.Z.I. Selected Symposia and Monographs No. 8. Mucchi Editore, Modena, Italy.

von Baer, K. E. 1828. *Entwicklungsgeschichte der Thiere: Beobachtung und Reflexion*. Borntrager, Konigsberg.

Wilson, E. B. 1898. Considerations in cell-lineage and ancestral reminiscence. Ann. N. Y. Acad. Sci. 11: 1–27.

Ziegler, H. E. 1898. Über den derzeitigen Stand der Colomfrage. Verh. Dtsch. Zool. Ges. 8: 14–78.

Ziegler, H. E. 1912. Leibeshöhle. Handwörterbuch Naturwiss. 6: 148–165.

[For additional references on metazoan phylogeny, see Chapter 24.]

5

The Protists

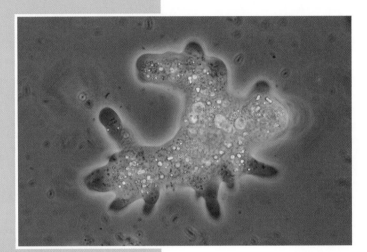

My excrement being so thin, I was at divers times persuaded to examine it; and each time I kept in mind what food I had eaten, and what drink I had drunk, and what I found afterwards. I have sometimes seen animalcules a-moving prettily...

A. van Leeuwenhoek,
November 4, 1681

Although the term "protozoa" has been around for nearly 200 years, and was used in the context of a phylum for about 100 years, it is now apparent that it defines no more than a loose assemblage, or grade, of primarily single-celled, heterotrophic, eukaryotic organisms—it does not define a monophyletic assemblage meriting single-phylum status. The kingdom Protista (or "Protoctista") includes those organisms traditionally called protozoa, as well as some autotrophic groups. But the kingdom Protista itself is not united by any unique distinguishing features or synapomorphies, and the boundary between autotrophy and heterotrophy is blurred in these organisms. *Protists are definable only as a confederation of eukaryotes lacking the tissue level of organization seen in plants, animals, and fungi.* Invertebrate biology texts have traditionally treated the "protozoa," and for many biology students this coverage will be their only detailed exposure to this important group of organisms. In this chapter we include those groups traditionally lumped under the name "protozoa" (i.e., the heterotrophic protists), and for completeness, we also briefly cover the commonly recognized autotrophic protist phyla (Box 5A).

The bauplans of protists demonstrate a remarkable diversity of nonmetazoan form, function, and survival strategies. Most, but not all, are unicellular. They carry out all of life's functions using *only the organelles* found in the "typical" eukaryotic cells of animals. Many of the fundamentally unicellular protist phyla also contain species that form colonies. Others are multicellular, though lacking the cell-tissue specialization seen in the Metazoa or Plantae. No protists undergo the embryonic tissue layering process that occurs in metazoans and plants. Protists include an awesome array of shapes and functional types, and there are

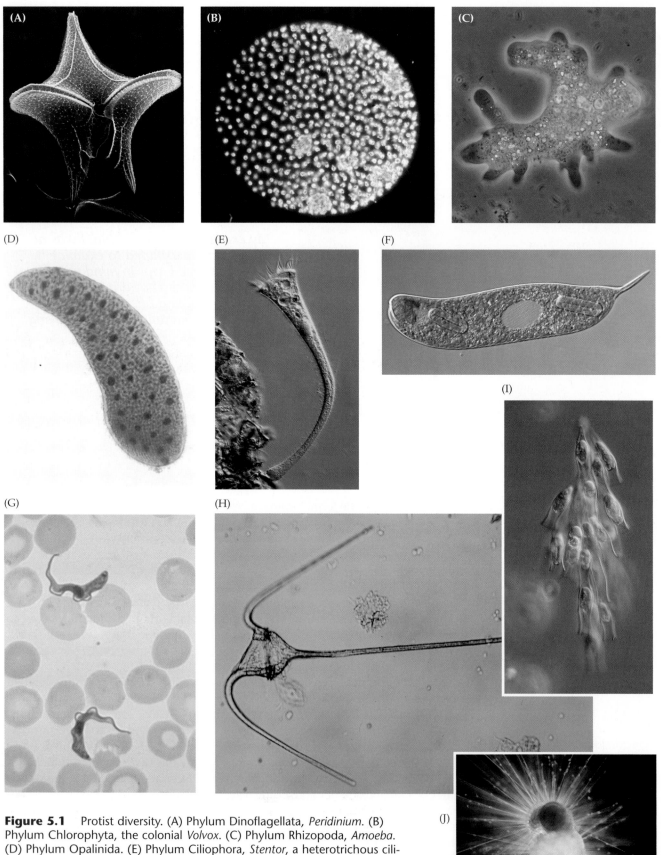

Figure 5.1 Protist diversity. (A) Phylum Dinoflagellata, *Peridinium*. (B) Phylum Chlorophyta, the colonial *Volvox*. (C) Phylum Rhizopoda, *Amoeba*. (D) Phylum Opalinida. (E) Phylum Ciliophora, *Stentor*, a heterotrichous ciliate. (F) Phylum Euglenida, *Euglena*. (G) Phylum Kinetoplastida, *Trypanosoma* in blood smear. (H) Phylum Dinoflagellata, *Ceratium*. (I) Phylum Stramenopiles, *Dinobrion*, a colonial golden alga. (J) Phylum Granuloreticulosa, *Globigerinella*, a foram (note the calcareous spines that radiate out from the body).

many tens of thousands of species yet to be discovered. Figure 5.1 illustrates some of this variety. Most unicellular protists are microscopic, ranging in size from about 2 to 200 μm. A few, such as the foraminiferans, are much larger and are commonly visible to the naked eye; many protists are actually larger than the smallest of the Metazoa (e.g., some gastrotrichs, kinorhynchs, nematodes, loriciferans, and others). Protists include marine, freshwater, terrestrial, and symbiotic species, the last category including many serious pathogens. Humans are hosts to over 30 species of protistan symbionts, many of which are pathogenic.

Taxonomic History and Classification

Antony van Leeuwenhoek is generally credited with being the first person to report seeing protists, in about 1675. In fact, Leeuwenhoek was the first to describe a number of microscopic aquatic life forms (e.g., rotifers), referring to them as animalcules (little animals). For nearly 200 years, protists were classified along with a great variety of other microscopic life forms under various names (e.g., Infusoria). The name protozoon (Greek, *proto*, "first"; *zoon*, "animal") was coined by Goldfuss in 1818 as a subgrouping of a huge assemblage of animals known at that time as the Zoophyta (protists, sponges, cnidarians, rotifers, and others). Following the discovery of cells in 1839, the distinctive nature of protists became apparent. On the basis of this distinction, von Siebold, in 1845, restricted the name Protozoa to apply to all unicellular forms of animal life. It was the great naturalist Ernst Haeckel who united the algae and protozoa into a single group, the Protista, which is today usually regarded as one of the kingdoms of life.

Throughout most of the twentieth century, a relatively standard classification scheme developed for the heterotrophic protists, or "protozoa" (see Brusca and Brusca 1990). This scheme was based on the idea that the different groups could be classified primarily by their modes of nutrition and locomotion. Thus, they were divided into the Mastigophora (locomotion with flagella), Ciliophora (locomotion with cilia), Sarcodina (locomotion with pseudopodia), and Sporozoa (parasites with no obvious locomotory structures). Flagellated protists were further divided into the zooflagellates (heterotrophs) and phytoflagellates (photosynthetic autotrophs). While these divisions might accurately describe protists' roles in ecosystems, they do not accurately reflect evolutionary relatedness. Pseudopodia and flagella are present in many different kinds of cells (including plant and animal cells) and their presence does not indicate unique relatedness (i.e., they are shared primitive features, or symplesiomorphies). Photosynthetic protists contain many different types of chlorophylls and have differently constructed chloroplasts, indicating that they are not closely related. (For discussions of changes in protist classification see recent papers by Patterson, Lipscomb, and Corliss.) Although there continues to be much debate over how these enigmatic organisms are related to each other, most experts agree that there are several well-defined groups of protists (herein treated as phyla), which are summarized below. Readers are cautioned, however, that the field of protist systematics is dynamic, and major changes can be expected for some time to come. One of the most exciting recent discoveries, for example, is that the former protist phylum Myxozoa actually comprises a group of highly modified cnidarians, parasitic in certain invertebrates and vertebrates (Chapter 8). This revelation was made possible by the discovery of certain metazoan and cnidarian features (e.g., collagen, nematocysts) in these animals, as well as ribosomal DNA data. Some of the most enigmatic groups of protists are those phyla containing photosynthetic species, such as the Euglenida, Dinoflagellata, Stramenopiles, and Chlorophyta, all of which have been claimed by botanists, in one form or another, in the past.

Readers will notice an absence of classification above and below the level of phylum for the protists. Ideas about protist phylogeny and classification are in such a state of flux that any scheme used at the time of this writing would be out-of-date by the time this text reached the bookshelves. Thus, we chose to restrict our treatment to the phylum level for this edition of *Invertebrates*.

CLASSIFICATION OF THE PROTISTA

PHYLUM EUGLENIDA (1,000 species): The euglenids and their kin; previously classified in the old phylum "Sarcomastigophora" (e.g., *Ascoglena, Colacium, Entosiphon, Euglena, Leocinclis, Menodium, Peranema, Phacus, Rhobdomonas, Stromonas, Trachelomonas*).

PHYLUM KINETOPLASTIDA (600 species): The trypanosomes and their kin; previously classified in the old phylum "Sarcomastigophora" (e.g.,. *Bodo, Cryptobia, Dimastigella, Leishmania, Leptomonas, Procryptobia, Rhynchomonas, Trypanosoma*).

PHYLUM CILIOPHORA (12,000 species): The ciliates (e.g., *Balantidium, Colpidium, Didinium, Euplotes, Laboea, Paramecium, Stentor, Tetrahymena, Vorticella*).

PHYLUM APICOMPLEXA (5,000 species): Gregarines, coccidians, heamosporidians, and piroplasms (e.g., *Cryptosporidia, Diaplauxis, Didymophyes, Eimeria, Gregarina, Haemoproteus, Lecudina, Leucocytozoon, Plasmodium, Strombidium, Stylocephalus, Toxoplasma*).

PHYLUM DINOFLAGELLATA (4,000 species): Dinoflagellates; previously classified in the old phylum "Sarcomastigophora" (e.g., *Amphidinium, Ceratium, Kofoidinium, Gonyaulax, Nematodinium, Nematopsides, Noctiluca, Peridinium, Polykrikos, Protoperidinium, Zooxanthella*).

PHYLUM STRAMENOPILA (9,000 species): Diatoms, brown algae, golden algae (Chrysophytes), silicoflagellates, labyrinthulids (slime nets), oomycetes, and hyphochytridiomycetes.

PHYLUM RHIZOPODA (200 species): The rhizopodans, or amebas; previously classified in the old phylum "Sarcomastigophora" (e.g., *Acanthamoeba, Amoeba, Arcella, Centropyxis, Chaos, Difflugia, Endolimax, Entamoeba, Euhyperamoeba, Flabellula, Hartmanella, Iodamoeba, Mayorella, Nuclearia, Pamphagus, Pelomyxa, Pompholyxophrys, Thecamoeba*).

PHYLUM ACTINOPODA (4,240 species): The polycystines (= radiolarians), phaeodarians, heliozoans, and acantharians; previously classified in the old phylum "Sarcomastigophora" (e.g., *Acanthocystis, Actinophrys, Actinosphaerium, Heterophrys, Lithocolla, Sticholonche*).

PHYLUM GRANULORETICULOSA (40,000 species): Foraminiferans; previously classified in the old phylum "Sarcomastigophora" (e.g., *Allogramia, Astrorhiza, Biomyxa, Elphidium, Glabratella, Globigerina, Gromia, Iridia, Microgromia, Nummulites, Rhizoplasma, Rotaliella, Technitella, Tretomphalus*).

PHYLUM DIPLOMONADIDA (100 species): Diplomonads; previously classified in the old phylum "Sarcomastigophora"

(e.g., *Enteromonas, Giardia, Hexamida, Octonitis, Spironucleus, Trimitus*).

PHYLUM PARABASILIDA (300 species): Hypermastigotes and trichomonads; previously classified in the old phylum "Sarcomastigophora" (e.g., *Dientamoeba, Histomonas, Monocercomonas, Pentatrichomonas, Trichomonas, Trichonympha, Tritrichomonas*).

PHYLUM CRYPTOMONADA: Cryptomonads; previously classified in the old phylum "Sarcomastigophora" (e.g., *Chilomonas*).

PHYLUM MICROSPORA (800 species): Microsporans (e.g., *Encephalitozoon, Metchnikorella, Nosema*).

PHYLUM ASCETOSPORA: Ascetosporans (e.g., *Haplosporidium, Marteilia, Paramyxa*).

PHYLUM CHOANOFLAGELLATA: Choanoflagellates; previously classified in the old phylum "Sarcomastigophora" (e.g., *Codosiga, Monosiga, Proterospongia*).

PHYLUM CHLOROPHYTA: The green algae (e.g., *Chlamydomonas, Eudorina, Polytoma, Polytonella, Volvox*).

PHYLUM OPALINIDA: Opalinids; previously classified in the old phylum "Sarcomastigophora" (e.g., *Cepedea, Opalina, Protopalina*).

GENUS *STEPHANOPOGON*: Incertae sedis.

The Protist Bauplan

While realizing that the protists do not represent a monophyletic clade, it is still advantageous to examine them together from the standpoint of the strategies and constraints of a unicellular bauplan. Remember that within the limitations imposed by unicellularity, these creatures still must accomplish all of the basic life functions common to the Metazoa.

Body Structure, Excretion, and Gas Exchange

As we discussed in Chapter 3, most life processes are dependent on activities associated with surfaces, notably with cell membranes. Even in the largest multicellular organisms, the regulation of exchanges across cell membranes and the metabolic reactions along the surfaces of various cell organelles are the phenomena on which all life ultimately depends. Consequently, the total area of these important surfaces must be great enough relative to the volume of the organism to provide adequate exchange and reaction sites. Nowhere is the "lesson" of the surface area-to-volume ratio more clearly demonstrated than among the protists, where it reveals the impossibility of massive, 100 kg amebas (1950s horror movies notwithstanding). Lacking both an efficient mechanism for circulation within the body and the presence of membrane partitions (multicellularity) to enhance and regulate exchanges of materials, protists must remain relatively small. The diffusion distances between protists' cell membranes (the "body surface")

BOX 5A **Protist Groups Treated in This Chapter**

Phylum Euglenida: Euglenids

Phylum Kinetoplastida: Trypanosomes and their kin

The Alveolata Phyla

  Phylum Ciliophora: Ciliates

  Phylum Apicomplexa: Gregarines, coccidians, haemosporidians, and piroplasms

  Phylum Dinoflagellata: Dinoflagellates

Phylum Stramenopiles: Diatoms, brown algae, golden algae (Chrysophytes), silicoflagellates, labyrinthulids (slime nets), oomycetes, and hyphochytridiomycetes

Phylum Rhizopoda: Amebas

Phylum Actinopoda: Polycystines (radiolarians), phaeodarians, heliozoans, and acantharians

Phylum Granuloreticulosa: Foraminiferans

Phylum Diplomonadida: Diplomonads

Phylum Parabasilida: Hypermastigotes and trichomonads

Phylum Cryptomonada: Cryptomonads

Phylum Microspora: Microsporans

Phylum Ascetospora: Ascetosporans

Phylum Choanoflagellata: Choanoflagellates

Phylum Chlorophyta: Green algae

Phylum Opalinida: Opalinids

Genus *Stephanopogon*

and the innermost parts of their bodies can never be so great that it prevents adequate movement of materials from one place to another within the cell. Certainly there are structural elements (e.g., microtubules, endoplasmic reticula) and various processes (e.g., protoplasmic streaming, active transport) that supplement passive phenomena. But the fact is, unicellularity mandates that a high surface area-to-volume ratio be maintained by restricting shape and size. This is the principle behind the fact that the largest protists (other than certain colonies) assume shapes that are elongate, thin, or flattened—shapes that maintain small diffusion distances.

The formation of membrane-bounded pockets, or vesicles, is common in protists, and these structures help maintain a high surface area for internal reactions and exchanges. The elimination of metabolic wastes and excess water, especially in freshwater forms living in hypotonic environments, is facilitated by **water expulsion vesicles** (Chapter 3, Figure 3.22). As explained in Chapter 3, these vesicles (frequently called **contractile vacuoles**) release their contents to the outside in a more or less controlled fashion, often counteracting the normal diffusion gradients between the cell and the environment.

Support and Locomotion

The cell surface is critical not only in providing a means of exchange of materials with the environment but also in providing protection and structural integrity to the cell. The plasma membrane itself serves as a mechanical and chemical boundary to the protist "body," and when present alone (as in the asymmetrical naked amebas), it allows great flexibility and plasticity of shape. However, many protists maintain a more or less constant shape (spherical, radial, or bilaterally symmetrical) by thickening the cell membrane to form a **pellicle**, by secreting scales or a shell-like covering called a **test**, by accumulating particles from the environment, or by other skeletal arrangements. Furthermore, locomotor capabilities are also ultimately provided by interactions between the cell surface and the surrounding medium. Pseudopodia, cilia, and flagella provide the means by which protists push or pull themselves along.

Nutrition

Various types of nutrition occur among protists—they may be either autotrophic or heterotrophic, and some may be both. Photosynthetic protists have chloroplasts and are capable of photosynthesis, although not all use the same chloroplast pigments, and they often differ in chloroplast structure (Figure 5.2).* All heterotrophic protists acquire food through some interaction between the cell surface and the environment. Heterotrophic forms may be **saprobic**, taking in dissolved organics by diffusion, active transport, or pinocytosis. Or they may be **holozoic**, taking in solid foods—such as organic detritus or whole prey—by phagocytosis. Many heterotrophic protists are symbiotic on or within other organisms. Those protists that engage in pinocytosis or phagocytosis rely on the formation of membrane-bounded vesicles called **food vacuoles** (Figure 5.3). These structures may form at nearly any site on the cell surface, as they do in the amebas, or at particular sites associated with some sort of "cell mouth," or **cytostome**, as they do in most protists with more or less fixed shapes. The cytostome may be associated with further elaborations of the cell surface that form permanent invaginations or feeding structures (discussed in more detail below, under specific taxa).

Once a food vacuole has formed and moved into the cytoplasm, it begins to swell as various enzymes and other chemicals are secreted into it. The vacuole first becomes acidic, and the vacuolar membrane develops numerous inwardly directed microvilli (Figure 5.3). As digestion proceeds, the vacuolar fluid becomes increasingly alkaline. The cytoplasm just inside the vacuolar membrane takes on a distinctive appearance from the products of digestion. Then the vacuolar membrane forms tiny vesicles that pinch off and carry these products into the cytoplasm. Much of this activity resembles surface pinocytosis. The result is numerous, tiny, nutrient-carrying vesicles offering a greatly increased surface area for absorption of the digested products into the cell's cytoplasm. During this period of activity, the original vacuole gradually shrinks and undigested materials eventually are expelled from the cell. In some protists (e.g., many amebas), the spent vacuole may discharge anywhere on the cell surface. But in ciliates and others in which a relatively impermeable covering exists around the cell, the covering bears a permanent pore (**cytoproct**) through which the vacuole releases material to the outside.

In protists, as in other eukaryotic organisms, the organelles responsible for most ATP production are the mitochondria. The mitochondria of protists, like all mitochondria, have two membranes, but the inner membranes, or cristae, have different forms—tubular, discoidal, and lamellar (Figure 5.4).

Activity and Sensitivity

Many protists display remarkable degrees of sensitivity to environmental stimuli and are capable of some fairly complex behaviors. But, unlike that of animals, protists' entire stimulus–response circuit lies within the confines of the single cell. Response behavior may be a function of the general sensitivity and conductivity of proto-

---

*Like mitochondria, chloroplasts have an intricate internal structure of folded membranes. But the chloroplast's membranes are not continuous with the inner membrane of the chloroplast envelope. Instead, the internal membranes lie in flattened disclike sacs called **thylakoids**. Each thylakoid consists of an outer thylakoid membrane surrounding an inner thylakoid space. Thylakoids are piled up like plates. Each stack is called a **granum**, and a chloroplast may contain many grana.

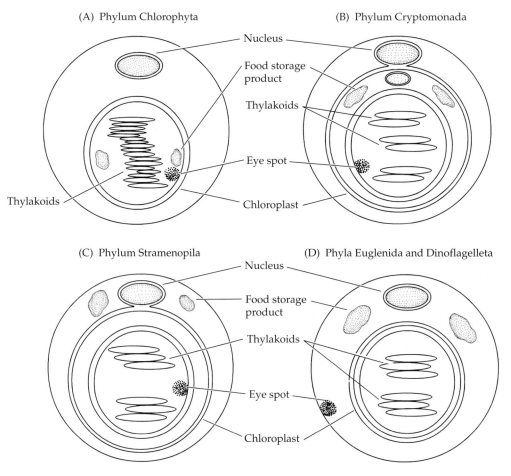

Figure 5.2 Variations in protist chloroplast morphology. (A) Phylum Chlorophyta. As in higher plants, the chloroplast in Chlorophyta is surrounded by two membranes and the thylakoids are arranged in irregular stacks, or grana. Also as in higher plants, the primary photosynthetic pigments in chlorophytes are chlorophylls a and b, and food reserves are stored as starch inside the chloroplast. (B) Phylum Cryptomonada. In cryptomonads, the chloroplast is surrounded by four membranes and the thylakoids occur in stacks of two. The inner two membranes enclose the thylakoids and eyespot; the outer two membranes enclose the nucleus, food storage products, and nucleomorph. The nucleomorph is thought to be the nucleus of an ancient endosymbiont that eventually became the chloroplast. Food reserves are stored as starch and oils, and the primary photosynthetic pigments are chlorophylls a and c_2; accessory pigments include phycobilins and alloxanthin. (C) Phylum Stramenopila. In stramenopiles, the chloroplast is surrounded by four membranes and the thylakoids occur in stacks of three. The inner two membranes enclose the thylakoids and eyespot (if present); the outer two membranes enclose the nucleus. Food reserves are stored as liquid polysaccharide (usually laminarin) and oils, which are located in the cytoplasm. The primary photosynthetic pigments are chlorophylls a, c_1, and c_2. (D) Phyla Euglenida and Dinoflagellata. In both of these phyla, the chloroplasts are surrounded by three membranes and the thylakoids are arranged in stacks of three. Also in both, the food storage products (starch and oils) and the eye spots are located outside of the chloroplast. The primary photosynthetic pigments in euglenids are chlorophylls a and b. Food reserves are stored as paramylon (unique to euglenids). In dinoflagellates, the photosynthetic pigments include chlorophylls a and c_2; accessory pigments include the xanthophyll peridinin, which is unique to dinoflagellates. Note that in some dinoflagellates, the eye spot is located inside the chloroplast rather than in the cytoplasm The food storage products are starch and oils.

plasm, or it may involve special organelles. Sensitivity to touch often involves distinctive locomotor reactions in motile protists and avoidance responses in many sessile forms. Cilia and flagella are touch-sensitive organelles; when mechanically stimulated, they typically stop beating or beat in a pattern that moves the organism away from the point of stimulus. These responses are most dramatically expressed by sessile stalked ciliates, which display very rapid reactions when the cilia of the cell body are touched. Contractile elements within the stalk shorten, pulling the animal's body away from the source of the stimulus.

Many protists have **extrusomes**, membrane-bound (exocytotic) organelles containing various chemicals. Extrusomes have a variety of functions (e.g., protection, food capture, secretion), but they have one feature in common: they readily, and sometimes explosively, discharge their chemical contents when subjected to stim-

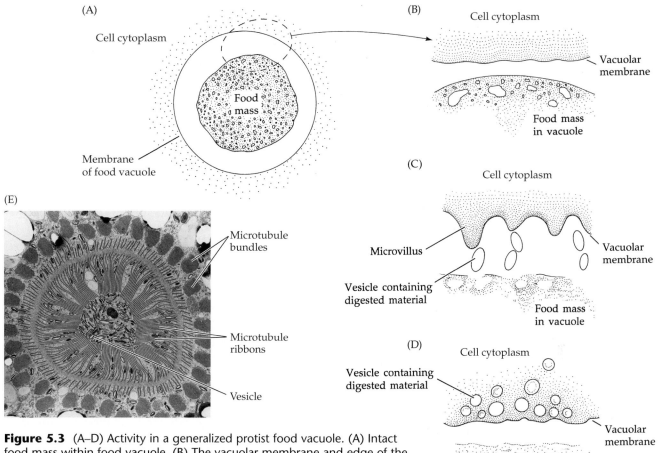

Figure 5.3 (A–D) Activity in a generalized protist food vacuole. (A) Intact food mass within food vacuole. (B) The vacuolar membrane and edge of the food mass (magnified view). (C) Formation of microvilli and vesicles of vacuolar membrane. (D) Uptake of vesicles containing products of digestion into the cytoplasm. (E) Cross section through the cytostome of the ciliate *Helicoprorodon*, showing the area of food vacuole formation at center. Microtubules provide support to the mouth.

uli. The best known extrusome is the trichocyst of ciliates such as *Paramecium*, but about ten different types are known from a variety of protists. Some workers have suggested they are related to the rhabdites of flatworms.

Thermoreception is known to occur in many protists but is not well understood. Under experimental conditions, most motile protists will seek optimal temperatures when given a choice of environments. This behavior probably is a function of the general sensitivity of the organism and not of special receptors. Evidence suggests that thermoreception in protists may be under electrophysiological control. Chemotaxic responses are probably similarly induced. Most protists react positively or negatively to various chemicals or concentrations of chemicals. For example, amebas are able to distinguish food from nonfood items and quickly egest the latter from their vacuoles. Many ciliates, especially predators, have specialized patches of sensory cilia that aid in finding prey, and even filter feeders use cilia located around the cytostome to "taste" and then accept or reject items as food.

Photosynthetic protists typically show a positive taxis to low or moderate light intensities, an obviously advantageous response for these creatures. They usually become negatively phototactic in very strong light. Specialized light-sensitive organelles are known among many flagellates, especially the photosynthetic ones. These **eye spots**, or **stigmata** (sing. stigma) are frequently located at or near the anterior end. Some, however, are found associated with the chloroplasts. Eye spots vary in complexity, ranging from very simple pigment spots to complex, lens-like structures.

Reproduction

A major aspect of protist success is their surprising range of reproductive strategies. Most protists have been able to capitalize on the advantages of both asexual and sexual reproduction, although some apparently reproduce only asexually. Many of the complex cycles seen in certain protists (especially parasitic forms) involve alternation between sexual and asexual processes, with a series of asexual divisions between brief sexual phases.

(A)

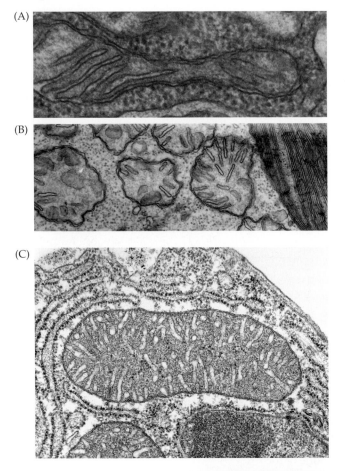

(B)

(C)

Figure 5.4 Protist mitochondria, showing variation in the inner membrane (i.e., cristae). (A) Lamellar cristae from the mitochondrion of the choanoflagellate *Stephanocea* (× 80,000). (B) Discoidal cristae from the mitochondrion of the euglenid *Euglena spirogyra* (× 40,000). (C) Tubular cristae from the mitochondrion of the chlorophyte *Pteromonas lacertae* (× 27,000). (D) Dilated tubular cristae from the mitochondrion of *Apusomonas proboscidea*, an enigmatic flagellate of uncertain affinity (× 97,000).

(D)

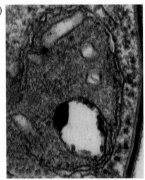

Protists undergo a variety of strictly asexual reproductive processes including **binary fission, multiple fission**, and **budding**. Binary fission involves a single mitotic division, resulting in two daughter cells. During multiple fission, the nucleus undergoes several multiple divisions prior to **cytokinesis** (partitioning of the cytoplasm), resulting in many daughter cells. This is common in parasitic protists such as the Apicomplexa. Some engage in a process called **plasmotomy**, considered by some to be a form of budding, in which a multinucleate adult simply divides into two multinucleate daughter

cells. Other members of the Apicomplexa undergo a type of internal budding called **endopolyogeny**, during which daughter cells actually form within the cytoplasm of the mother cell.

The advantage of sexual reproduction is thought to be the generation and maintenance of genetic variation within populations and species. Protists have evolved a variety of methods that achieve this end, not all of which result in the immediate production of additional individuals. If we expand our traditional definition of meiosis to include any nuclear process that results in a haploid condition, then meiosis can be considered a protist, as well as a metazoan, phenomenon. This disclaimer is necessary because protist "meiosis" is more variable than that seen in animals, and it is certainly less well understood. Nonetheless, reduction division does occur, and haploid cells or nuclei of one kind or another are produced and then fuse to restore the diploid condition. This production and subsequent fusion of gametes in protists is called **syngamy**. Protist cells responsible for the production of gametes are usually called **gamonts**. Syngamy may involve gametes that are all similar in size and shape (**isogamy**), or the more familiar condition of gametes of two distinct types (**anisogamy**). Thus, as in the Metazoa, both haploid and diploid phases are produced in the life histories of sexual protists. The meiotic process may immediately precede the formation and union of gametes (prezygotic reduction division), or it may occur immediately after fertilization (postzygotic reduction division), as it does in many lower plants. Other sexual processes that result in genetic mixing by the exchange of nuclear material between mates (**conjugation**) or by the re-formation of a genetically "new" nucleus within a single individual (**autogamy**) are best known among the ciliates and are discussed below for that phylum.

There is also considerable variability in mitosis among protists (Box 5B). Different mitotic patterns are primarily distinguished on the basis of persistence of the nuclear membrane (= envelope), and the location and symmetry of the spindle (Figure 5.5). The terms *open, semi-open,* and *closed* refer to the persistence of the nuclear envelope. If mitosis is **open**, the nuclear membrane breaks down completely; if **semi-open**, the nuclear envelope remains intact except for small holes (**fenestrae**) where the spindle microtubules penetrate the nuclear envelope; if **closed**, the nuclear envelope remains completely intact throughout mitosis. The terms *orthomitosis* and *pleuromitosis* refer to the symmetry of the spindle. During **orthomitosis**, the spindle is bipolar and symmetrical, and an equatorial plate usually forms. During **pleuromitosis**, the spindle is asymmetrical, and an equatorial plate does not form. The terms *intranuclear* and *extranuclear* refer to the location of the spindle. During **intranuclear mitosis**, the spindle forms inside of the nucleus; during **extranuclear mitosis**, the spindle forms outside of the nucleus (see Raikov 1994 for more detailed descriptions of protist mitosis).

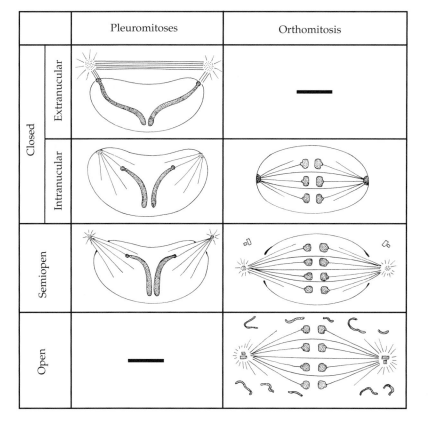

		Pleuromitoses	Orthomitosis
Closed	Extranucular		—
Closed	Intranucular		
	Semiopen		
	Open	—	

Figure 5.5 Mitosis in protists. In open mitosis the nuclear membrane breaks down completely. In semi-open mitosis the nuclear envelope remains intact except for small holes where the spindle microtubules penetrate the nuclear envelope. In closed mitosis the nuclear envelope remains completely intact throughout mitosis, the spindle forming either inside the nucleus (intranuclear mitosis) or outside the nucleus (extranuclear mitosis). Orthomitosis refers to the spindle being bipolar and symmetrical, usually with the formation of an equatorial plate. Pleuromitosis occurs when the spindle is asymmetrical and an equatorial plate does not form.

Protist nuclei also show remarkable diversity. The most common type is the **vesicular nucleus**, and it is characterized as being between 1–10 μm in diameter, round (usually), with a prominent nucleolus, and uncondensed chromatin. **Ovular nuclei** are characterized as being large (up to 100 μm in diameter), with many peripheral nucleoli, and uncondensed chromatin. **Chromosomal nuclei** are characterized by the tendency of the chromosomes to remain condensed during interphase, and for there to be one nucleolus that is associated with a chromosome. Ciliates have two unique kinds of nuclei: small **micronuclei** (with no nucleoli and dispersed chromatin) and large **macronuclei** (with many prominent nucleoli and compact chromatin). In summary, protist diversity and success is reflected by the tremendous variation within the unicellular bauplan. The following accounts of protist phyla explore this variation in some detail.

Phylum Euglenida

Most euglenids occur in fresh water, but a few marine and brackish-water species are known. The majority are noncolonial, but some colonial forms exist (e.g., *Colacium*). Euglenids come in a wide variety of shapes—elongate, spherical, ovoid, or leaf-shaped (Figures 5.1F and 5.6). This group includes organisms such as the familiar genus *Euglena*, which has been used extensively in research laboratories and is commonly studied in introductory biology and invertebrate zoology courses.

The phyla Euglenida and Kinetoplastida appear to be closely related, even though many euglenids are photosynthetic and most kinetoplastids (see below) are parasitic heterotrophs. The morphological features they share include linked microtubules underlying the cell membrane, discoidal cristae in a single large mitochondrion, flagella containing a lattice-like supportive rod, and a similar pattern of mitosis (Box 5C). Molecular studies, using sequences from the gene for ribosomal DNA, also indicate that these are closely related groups (e.g., Schlagel 1994).

BOX 5B *Six Categories of Mitosis in Protists*

Open mitosis
1. *Open orthomitosis.* The nuclear envelope breaks down completely; the spindle is symmetrical and bipolar; an equatorial plate occurs.

Semi-open mitosis
2. *Semi-open orthomitosis.* The nuclear envelope persists except for small fenestrae through which the spindle microtubules enter the nucleus; the spindle is symmetrical and bipolar; an equatorial plate occurs.

3. *Semi-open pleuromitosis.* The nuclear envelope persists except for small fenestrae through which the spindle microtubules enter the nucleus; the spindle is asymmetrical; an equatorial plate does not form.

Closed mitosis with an intranuclear spindle
4. *Intranuclear orthomitosis.* Nuclear envelope persists throughout mitosis; spindle is symmetrical, bipolar, and forms inside the nucleus; an equatorial plate usually forms.

5. *Intranuclear pleuromitosis.* Nuclear envelope persists throughout mitosis; spindle is asymmetrical and forms inside the nucleus; an equatorial plate does not form.

Closed mitosis with an extranuclear spindle
6. *Extranuclear pleuromitosis.* Nuclear envelope persists throughout mitosis; spindle is asymmetrical and forms outside of the nucleus; an equatorial plate does not form.

(A)

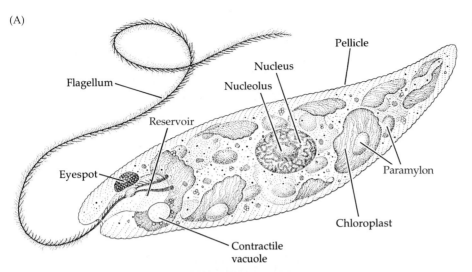

Flagellum

Pellicle

Nucleus

Nucleolus

Reservoir

Paramylon

Eyespot

Chloroplast

Contractile
vacuole

(B)

(C)

Figure 5.6 Phylum Euglenida. (A) Diagram of *Euglena's* anatomy. (B) *Peranema* exhibiting metaboly. (C) *Entosiphon*.

Euglenids are commonly found in bodies of water rich in decaying organic matter. As such, some of them are useful indicator organisms of water quality (e.g., *Leocinclis, Phacus, Trachelomonas*). Some species of *Euglena* have been used in experiments for waste water treatment and have been reported to extract heavy metals such as magnesium, iron, and zinc from sludge. Other euglenids, however, are environmental pests, and some have been shown to produce toxic substances which have been associated with diseases in trout fry. Others are responsible for toxic blooms, which have caused massive destruction of fishes and molluscs in Japan.

Support and Locomotion

The shape of euglenids is maintained by a **pellicle** consisting of interlocking strips of protein lying beneath the cell membrane. The stripes that can sometimes be seen on a euglenid are the seams between the long protein strips winding around the cell. Often the pellicle is supported by regularly arranged microtubules lying just underneath it. The rigidity of the pellicle is variable. Some (e.g., *Menodium, Rhobdomonas*) have protein strips that are fused together into a rigid pellicle, while others (e.g., *Euglena*) have protein strips that are articulated to produce a flexible pellicle. Those euglenids with a flexible pellicle undergo **euglenoid movement,** or **metaboly,** in which the cell undulates as it rapidly extends and contracts (Figures 5.6 and 5.7). Although this type of movement is not fully understood, it is thought to be accomplished by the sliding of microtubules against the protein strips (Figure 5.8).

BOX 5C   *Characteristics of the Phylum Euglenida*

1. Shape of cell maintained by pellicle, formed by interlocking strips of protein beneath plasma membrane; pellicle generally associated with microtubules arranged in regular pattern

2. Most with two flagella of unequal lengths for locomotion, supported by paraxonemal rods; each flagellum with single row of hairs. Region between axoneme and basal body of flagellum appears structureless and hollow

3. Single mitochondrion has discoidal cristae

4. With single chromosomal nucleus. Nuclear membrane is not continuous with endoplasmic reticulum (unusual among protists)

5. Nuclear division occurs by closed intranuclear pleuromitosis without centrioles; organizing center for mitotic spindle is not obvious

6. Asexual reproduction by longitudinal binary fission (symmetrogenic)

7. May be strictly asexual—neither meiosis nor sexual reproduction has been confirmed

8. Photosynthetic forms have chlorophylls a and b; usually appear grass-green in color. Thylakoid membranes arranged in stacks of three, three membranes surround the chloroplast; outermost membrane is not continuous with nuclear membrane

9. Food reserves stored in cytoplasm as the unique starch paramylon

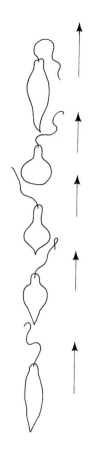

Figure 5.7 Phylum Euglenida. Euglenoid movement in *Euglena*.

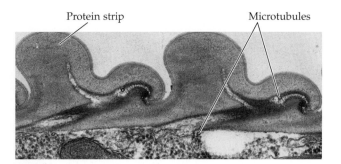

Figure 5.8 Phylum Euglenida. Cross section (Transmission electron micrograph) through the pellicle of *Euglena* showing the protein strips and microtubules.

A few euglenids (e.g., *Ascoglena, Colacium, Stromonas, Trachelomonas*) secrete a **lorica**, or envelope, exterior to the cell membrane. The lorica is formed by the mucous secretions of small organelles called **mucocysts,** which are located under the cell membrane along the seams between the protein strips of the pellicle. Secretions of mucocysts are also used to form protective coverings when environmental conditions become unfavorable.

Locomotion in euglenids is primarily by flagella. They have two flagella, but one may be very short or represented by just a kinetosome.* The flagella originate in an invagination at the anterior end of the cell called a reservoir (= flagellar pocket). The longer, anteriorly directed flagellum propels the cell through the water or across surfaces (Figure 5.6). The shorter flagellum either trails behind or does not emerge from the reservoir at all. Both flagella have a single row of hairs on their surface and a lattice-like supporting rod, called the paraxonemal rod, lying adjacent to the microtubules within the shaft.

Nutrition

Euglenids are quite variable in their nutrition. Approximately one-third have chloroplasts and are photoautotrophic. These are positively phototaxic and have a swelling near the base of the anterior flagellum that acts as a photoreceptor. The chloroplast is surrounded by three membranes and has thylakoids that are arranged in stacks of three (see Figure 5.2). The photosynthetic pigments include: chlorophylls a and b, phycobilins, β-carotene, and the xanthophylls neoxanthin and diadinoanthin. Apparently, not all nutritional requirements are satisfied by photosynthesis, and all euglenids which have been studied, even those with chloroplasts, require at least vitamins B1 and B12, which must be obtained from the environment.

Approximately two-thirds of the described species of euglenids lack chloroplasts and are thus obligate heterotrophs, and even phototrophic forms can lose their chloroplasts and switch to heterotrophy. A few parasitic species have been reported in invertebrates and amphibian tadpoles. Most euglenids take in dissolved organic nutrients by saprotrophy. This is generally restricted to parts of the cell not covered by the pellicle, such as the reservoir. Some euglenids also ingest particulate food items by phagocytosis of relatively large (sometimes comparatively huge) food materials. These have a cytostome located near the base of a flagellum where food vacuoles form (e.g., *Peranema*; Figure 5.9A). The cytostome usually leads to a tube called the cytopharynx that extends from the cytostome deep into the cytoplasm. The walls of the cytopharynx are often reinforced by highly organized bundles of microtubules (e.g., *Entosiphon, Peranema*) (Figure 5.9B). Extrusomes are often found near the cytostome and presumably aid in prey capture.

Reproduction

Asexual reproduction in euglenids is by longitudinal cell division. Division occurs along the longitudinal plane (Figure 5.10). Nuclear division occurs by closed intranuclear pleuromitosis. During mitosis, the nucleolus remains distinct and no obvious microtubular organizing center is evident. Sexual reproduction has been reported in one species, but this has not been confirmed.

Phylum Kinetoplastida (The Trypanosomes and Their Relatives)

There are about 600 described species of kinetoplastids. This group includes two major subgroups: the bodonids and the trypanosomes (Figures 5.1G and 5.11A). The bodonids are primarily free living in marine and freshwater environments rich in organic material. Because bodonids have a strict oxygen requirement they usually aggregate at a particular distance from the water surface. The trypanosomes are exclusively parasitic, and they occur in the digestive tracts of invertebrates,

*The kinetosomes of protists have the characteristic nine peripheral elements and are virtually indistinguishable from the centrioles of other eukaryotic cells. In protists, they often lie at the bottom of flagellar pockets or reservoirs.

(A)

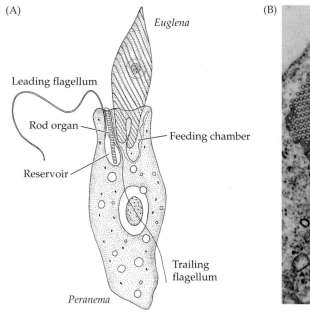

(B)

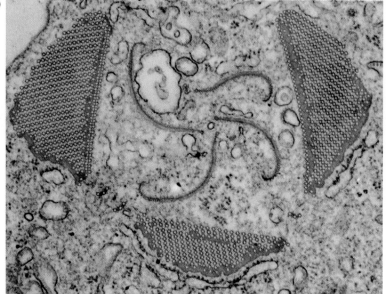

Figure 5.9 Phylum Euglenida. (A) *Paranema* feeding on *Euglena*. *Paranema* possesses an expandable feeding pocket separate from the reservoir in which the flagella arise. The rod organ can be extended to pierce prey and either pull it into the feeding apparatus or hold it while the contents are sucked out. (B) TEM through the cytopharyngeal rods of the euglenid *Entosiphon*.

phloem vessels of certain plant species, and the blood of vertebrates. *Leptomonas* exhibits the simplest cycle, in which an insect is the sole host and transmission occurs by way of an ingested cyst. In humans, *Leishmania* and *Trypanosoma* cause several debilitating and often fatal diseases.

Kinetoplastids are best known as agents of disease in humans and domestic animals. Species of *Leishmania* cause a variety of ailments collectively called **leishmaniasis**, and these include kala-azar (a visceral infection that particularly affects the spleen), oriental sore (characterized by skin boils), and several other skin and mucous membrane infections. Leishmaniasis strikes over a

million humans annually but due to effective treatment, it only kills about 1,000 people each year. Leishmaniasis is transmitted almost exclusively by the bite of sandflies (Diptera: Phlebotominae).*

More serious diseases are caused by members of the genus *Trypanosoma*, all of which are parasites of all classes of vertebrates. *T. brucei* is a nonlethal parasite that lives in the bloodstream of African hoofed animals, in which it causes a disease called **nagana**. Unfortunately, it also attacks domestic livestock, including horses, sheep, and cattle; in the latter case it is often fatal, making it impossible to raise livestock on more than 4.5 million square miles of the African continent (an area larger than the United States). Two other African species (often considered subspecies of *T. brucei*) are *T. gambiense* and *T. rhodesiense*, both of which cause sleeping sickness in humans. These parasites are introduced into the blood of humans from the salivary glands of the blood-sucking tsetse fly (*Glossina*). From the blood, trypanosomes can enter the lymphatic system and ultimately the cere-

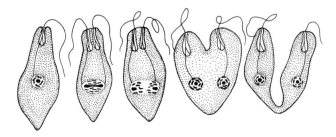

Figure 5.10 Phylum Euglenida, asexual reproduction. Longitudinal fission in *Euglena*, in which the flagella and reservoir duplicate prior to cell division.

*Species of *Leishmania* are difficult to differentiate morphologically, and their taxonomy is unsettled. Most widespread are *Leishmania tropica* and *L. major*, which occur in Africa and southern Asia and are transmitted by species of the genus *Phlebotomus*. These species produce the cutaneous ulcers variously known as oriental sore, cutaneous leishmaniasis, Jericho boil, Aleppo boil, and Delhi boil. *Leishmania donovani* is endemic to southern Asia but also occurs in low levels in Latin America and the Mediterranean region; it is the etiological agent of Dum-Dum fever, or kala-azar. Kala-azar can result in extreme and even grotesque skin deformations. *Leishmania braziliensis* is endemic to Brazil, where it causes espundia, or uta, which often leads to such severe destruction of the skin and associated tissues that complete erosion of the lips and gums ensues. *Leishmania mexicana* occurs in northern Central America, Mexico, Texas, and probably some Caribbean islands, where it mostly affects agricultural or forest laborers. Infections of *L. mexicana* cause a cutaneous disease called chiclero ulcer, because it is so common in "chicleros," men who harvest the gum of chicle trees.

brospinal fluid. Sleeping sickness is often fatal and kills about 65,000 people annually.

Chagas' disease (common in Central and South America, and Mexico) is caused by *T. cruzi* and is transmitted to humans by cone-nosed hemipteran bugs (also known as assassin or kissing bugs; family Reduviidae, subfamily Triatominae). These bugs feed on blood and often bite sleeping humans. They commonly bite around the mouth, hence the vernacular name. After feeding they leave behind feces that contain the infective stage, which invades through mucous membranes or the wound caused by the insect's bite. Occasionally the bugs bite around the eyes of the sleeping vicims, and subsequent rubbing leads to conjunctivitis and swelling of a particular lymph node, a symptom known as **Romaña's Sign**. The parasites migrate to the bloodstream, where they circulate and invade other tissues. In chronic human infections, *T. cruzi* can cause severe tissue destruction including the enlargment and thinning of walls of the heart. In Central and South America the incidence of Chagas' disease is high, and an estimated 15 to 20 million persons are infected at any given time. A study in Brazil attributed a 30 percent mortality rate to Chagas' disease. In the United States, at least 14 species of mammals may serve as reservoirs (including dogs, cats, opossums, armadillos, and wood rats), although the U.S. strain of *T. cruzi* is considerably less pathogenic than the Central and South American strains and the incidence of disease is very low.

The kinetoplastids have a single, elongate mitochondrion with a uniquely conspicuous dark-staining concentration of mitochondrial DNA (mDNA) called the **kinetoplast**—hence the phylum name. Generally the kinetoplast is found in the part of the mitochondrion lying close to the kinetosomes, although there is no known relationship between these two structures. The size, shape, and position of the kinetoplast is important in the taxonomy of trypanosomes and bodonids, and it is used in distinguishing between different stages in the life cycle (Box 5D). Kinetoplast DNA (kDNA) is organized into a network of linked circles, quite unlike the DNA in mitochondria of other organisms.

Support and Locomotion

The shape of a cell is maintained by a pellicle consisting of the cell membrane and a supporting layer of microtubules. In bodonids, the pellicular microtubules consist of three microtubular bands, whereas in trypanosomes the pellicular microtubules are evenly spaced and form a corset which envelops the entire body. In trypanosomes, a layer of glycoprotein (12 to 15 μm thick) coats the outside of the cell and acts as a protective barrier against the host's immune system. The composition of the glycoprotein coat is changed cyclically. A consequence of this is that the trypanosome is able to avoid the host's immune system. This has been well studied in pathogenic trypanosomes such as *Trypanosoma brucei*

BOX 5D *Characteristics of the Phylum Kinetoplastida*

1. Shape of cell maintained by pellicle consisting of corset of microtubules beneath plasma membrane

2. With two flagella for locomotion. In bodonids, both flagella bear axonemes; in trypanosomes, only one flagellum bears an axoneme. Flagella associated with paraxonemal rod, with characteristic transition zone

3. Single, elongate mitochondrion has discoidal cristae and conspicuous discoidal concentration of mDNA (the kinetoplast). Shape of cristae can change as organism progresses through life cycle, but predominantly discoidal

4. With single vesicular nucleus; prominent nucleolus typically evident

5. Nuclear division occurs by closed intranuclear pleuromitosis without centrioles. Plaques on inside of nuclear envelope may act as organizing centers for mitotic spindle

6. Asexual reproduction by longitudinal binary fission (symmetrogenic)

7. Neither meiosis nor sexual reproduction has been confirmed

8. Without plastids and storage carbohydrates

9. Mitochondrial DNA forms aggregates, collectively known as the kinetoplast, readily seen with the light microscope

---

(the causative agent of African sleeping sickness). When the trypanosome enters the host's body, the immune system recognizes the glycoprotein as foreign (an antigen) and specific antibodies are made against it. Although most of the trypanosome population is destroyed, a few are able to evade the immune system by changing their glycoprotein coat so that the new coat is unrecognizable to the host's antibodies. Once a new antibody is produced by the host, another new glycoprotein is produced by the trypanosome, and so on. About 1,000 genes code for the surface glycoproteins, although only one gene is expressed at a time. The ability of trypanosomes to change their glycoprotein coat makes treatment of trypanosome infections difficult.

Both bodonids and trypanosomes move using flagella which, like those of euglenids, usually emerge from an inpocketing and contain a paraxonemal rod. Trypanosomes have two kinetosomes, but only one has a flagellum. In many forms this flagellum lies against the side of the cell and its outer membrane is attached to the cell body's membrane. When the flagellum beats, the membrane of the cell is pulled up into a fold and looks like a waving or undulating membrane (Figure

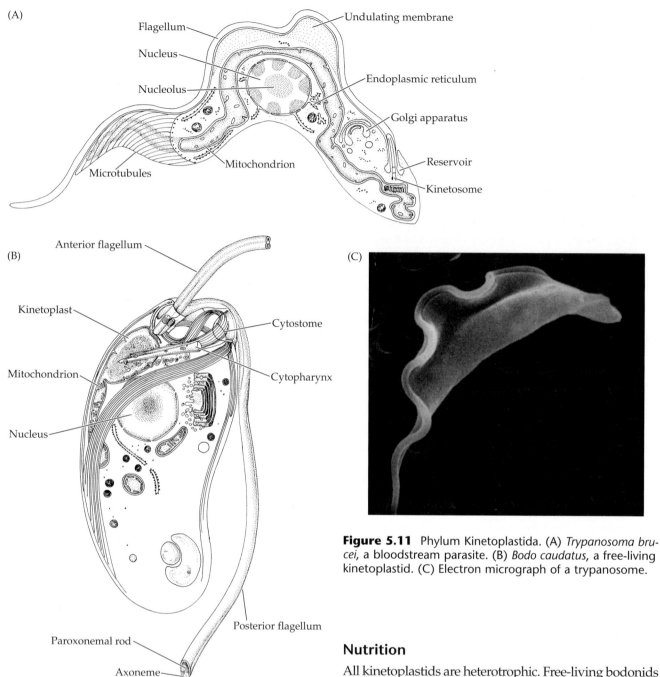

(A)

Flagellum

Nucleus

Nucleolus

Microtubules

Mitochondrion

Undulating membrane

Endoplasmic reticulum

Golgi apparatus

Reservoir

Kinetosome

(B)

Anterior flagellum

Kinetoplast

Mitochondrion

Nucleus

Cytostome

Cytopharynx

Paroxonemal rod

Axoneme

Posterior flagellum

(C)

Figure 5.11 Phylum Kinetoplastida. (A) *Trypanosoma brucei*, a bloodstream parasite. (B) *Bodo caudatus*, a free-living kinetoplastid. (C) Electron micrograph of a trypanosome.

5.11). This arrangement appears to be relatively efficient in moving the cell through viscous media (such as blood). Although trypanosomes can change the direction of flagellar beat in response to chemical or physical stimuli, usually the beat begins at the tip of the flagellum and proceeds toward the kinetosome. This is the reverse of the way flagella usually beat (from base to tip). Bodonids have two flagella, but only the one that is extended anteriorly is used in locomotion; the other trails behind and may be partially attached to the body in some species (e.g., *Dimastigella*, *Procryptobia*).

Nutrition

All kinetoplastids are heterotrophic. Free-living bodonids capture particulate food, primarily bacteria, with the aid of their anterior flagellum and ingest through a permanent cytostome. The cytostome leads to a cytopharynx which is supported by microtubules. At the base of the cytopharynx, food is enclosed in food vacuoles by endocytosis.

Unfortunately, little is known about feeding mechanisms in trypanosomes, all of which are parasitic. Some trypanosomes have a cytostome–cytopharyngeal complex through which proteins are ingested. The proteins are taken into food vacuoles by pinocytosis at the base of the cytopharynx. It has also been reported that some trypanosomes can take in proteins by pinocytosis from the membrane lining the flagellar pocket or by some sort of cell membrane-mediated mechanism.

Reproduction and Life Cycles

Although sexual reproduction has never been observed in kinetoplastids, there is indirect genetic evidence that it occurs. Asexual reproduction occurs by longitudinal binary fission or budding, as in *Euglena*. Nuclear division is by closed intranuclear pleuromitosis. During pleuromitotic mitosis, the nucleolus remains distinct and plaques on the inside of the nuclear envelope appear to organize the spindle (centrioles are absent). An unusual feature of kinetoplastid mitosis is that condensed chromosomes cannot be identified when the nucleus is dividing, even though they are typically conspicuous during interphase.

The life cycles of trypanosomes are complex and involve at least one host, but usually more. Trypanosomes which have only one host are said to be **monoxenous** while those which occupy more than one host are said to be **heteroxenous**. Monoxenous trypanosomes usually are found infecting the digestive tracts of arthropods and annelids. Most heteroxenous forms live for part of their life cycle in the blood of vertebrates and the remaining part of their life cycle in the digestive tracts of blood-sucking invertebrates, usually insects. As a trypanosome progresses through its life cycle, the shape of the cell undergoes different body form changes, depending on the phase of the cycle and the host it is parasitizing. Not all of these forms (Figure 5.12) occur in all genera. For example, in *Trypanosoma cruzi*, only the epimastigote, amastigote, and trypomastigote forms occur. These body forms differ in shape, position of the kinetosome and kinetoplast, and in the development of the flagellum.

The alveolata phyla. Three protist phyla (ciliates, apicomplexans, and dinoflagellates) have an **alveolar membrane system**, which comprises flattened membrane-bound sacs (alveoli) lying beneath the outer cell membrane. The presence of this system, together with evidence from molecular sequence comparisons, indicate that these three protist phyla are closely related evolutionarily.

Phylum Ciliophora (The Ciliates)

There are about 12,000 described species of ciliates. Ciliates are very common in benthic and planktonic communities in marine, brackish, and freshwater habitats, as well as in damp soils. Both sessile and errant types are known, and many are ecto- or endosymbionts, including a number of parasitic species. Most occur as single cells, but branching and linear colonies are known in several species. The ciliates shown in Figures 5.1E and 5.13 illustrate a variety of body forms within this large and complex group of protists.

Ciliates are important mutualistic endosymbionts of ruminants such as goats, sheep, and cattle. They are found by the millions in the digestive tracts, feeding on plant matter ingested by the host and converting it into a form that can be metabolized by the ruminant. In addition, some ciliates are found parasitizing fish and at least one (*Balantidium coli*) is known to be an occasional endoparasite of the human digestive tract. Ciliates such as *Tetrahymena* and *Colpidium* have been used as model organisms in experiments to evaluate the effects of chemicals on protists. Others are widely used as indicators of water quality and have been used to clarify water in sewage treatment plants (Box 5E).

Support and Locomotion

The fixed cell shape of ciliates is maintained by the alveolar membrane system and an underlying fibrous layer called the **epiplasm**, or **cortex** (Figure 5.14). A few types (e.g., tintinnids) secrete external skeletons, or loricae, which have been documented in the fossil record as

Figure 5.12 Phylum Kinetoplastida. Body plans of various trypanosomes. (A) *Leishmania* (amastigote form). (B) *Crithidia* (choanomastigote form). (C) *Leptomonas* (promastigote form). (D) *Herpetomonas* (opisthomastigote form). (E) *Trypanosoma* (trypomastigote form*)*.

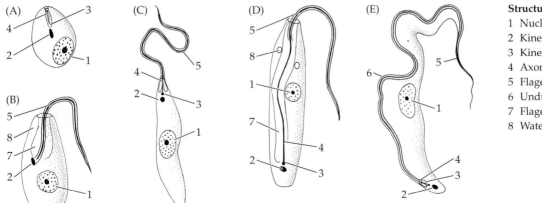

Structures

1 Nucleus
2 Kinetoplast
3 Kinetosome
4 Axoneme
5 Flagellum
6 Undulating membrane
7 Flagellar pocket
8 Water expulsion vesicle

(A)

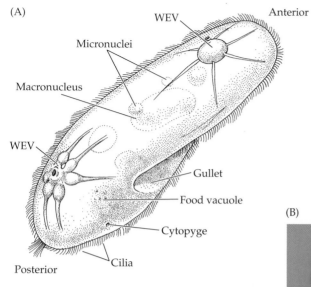

(B)

(C)

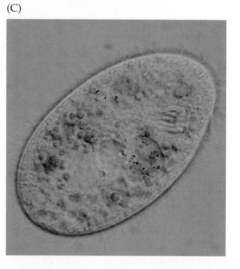

(D)

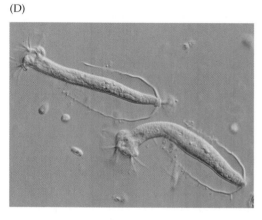

(E)

Cirri

AZM

AZM

(F)

AZM

(G)

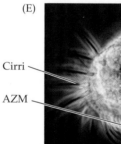

Peristomial
area

Buccal cavity

Cytostome

Micronuclei

Macronucleus

WEV

Attachment
base

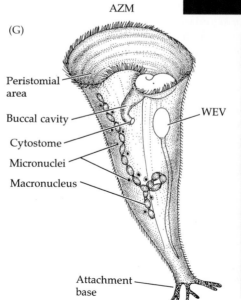

Figure 5.13 Phylum Ciliophora. Representative ciliates. (A) *Paramecium* (a peniculine). (B) *Loxophyllum* (a haptorid). (C) *Nassula* (a nassulid). (D) *Vaginacola* (a loricated peritrich). (E) *Euplotes* (a hypotrich); note the prominent AZM. (F) *Euplotes*; note the AZM. (G) *Stentor* (a heterotrich).

BOX 5E *Characteristics of the Phylum Ciliophora*

1. Shape of cell maintained by pellicle consisting of alveolar vesicles and fibrous layer of epiplasm beneath plasma membrane

2. With cilia for locomotion. Associated with the basal bodies (kinetosomes) of cilia are two microtubular roots and one fibrous root; these roots plus basal bodies are collectively known as infraciliature

3. Mitochondria with tubular cristae

4. With two distinct types of nuclei, a hyperpolyploid macronucleus and a diploid micronucleus

5. Micronucleus divides by closed intranuclear orthomitosis (in most) without centrioles. Electron-dense bodies inside nucleus act as organizing centers for mitotic spindle. Macronucleus divides amitotically by simple constriction

6. Asexual reproduction by transverse binary fission (homothetogenic)

7. Sexual reproduction by conjugation: pair of ciliates fuse and exchange micronuclei through a cytoplasmic connection at point of joining

8. Without plastids

9. Carbohydrates sorted as glycogen

gle, or **simple cilia**, or the kinetosomes may be grouped close together to form **compound ciliature** (e.g., cirri, membranelles). Ciliatologists have developed a detailed and complicated terminology for talking about their favorite creatures. This special language reaches almost overwhelming proportions in matters of ciliature, and we present here only a necessary minimum of new words in order to adequately describe these organisms. Beyond this, we offer the reference list at the end of this chapter, especially the extensive and illustrated glossary in J. O. Corliss's *The Ciliated Protozoa* (1979).

The cilia are also, of course, the locomotory organelles of ciliate protists. Their structural similarities to flagella are well known, and many workers treat cilia simply as short specialized flagella; but ciliates do not move like protists with flagella. The differences are due in large part to the facts that cilia are much more numerous and densely distributed than flagella, and the patterns of ciliation on the body are extremely varied and thus allow a range of diverse locomotor behaviors not possible with just one or a few flagella.

As discussed in Chapter 3, each individual cilium undergoes an effective (power) stroke as it beats. The cilium does not move on a single plane, but describes a distorted cone as it beats (Figure 5.15A,B). The counterclockwise movement (when viewed from the outside of the cell) is an intrinsic feature of cilia and occurs even when they are isolated. The beating of a ciliary field occurs in metachronal waves that pass over the body surface (Figure 5.15C). The coordination of these waves is apparently due largely to hydrodynamic effects generated as each cilium moves. Microdisturbances created in the water by the action of one cilium stimulate movement in the neighboring cilium, and so on over the cell surface.

The effective stroke of the cilium, coupled with the direction of the metachronal waves, results in three main patterns of metachronal coordination (Figure 5.15D–F). Fields of cilia that are spaced closely together tend to show symplectic metachrony, in which the metachronal

early as the Ordovician (500 million years ago). Another common group (*Coleps* and its relatives) has calcium carbonate plates in their alveoli.

The cilia are in rows called kineties, and the different patterns these rows create are used as taxonomic characters for identification and classification. Associated with the kinetosomes are three fibrillar structures—two microtubular roots, the postciliary microtubules and the transverse microtubules, and one fibrous root, the kinetodesmal fiber. These roots, together called the infraciliature, anchor the cilium and provide additional support for the cell surface (Figure 5.14).

The cilia can be grouped into two functional and two structural categories. Cilia associated with the cytostome and the surrounding feeding area are the **oral ciliature**, whereas cilia of the general body surface are the **somatic ciliature**. The cilia in both of these categories may be sin-

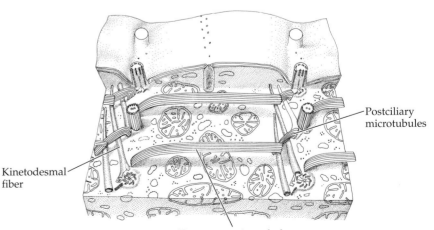

Figure 5.14 Phylum Ciliophora. Fine structure of cortex of *Tetrahymena pyriformis*.

Kinetodesmal fiber

Postciliary microtubules

Transverse microtubules

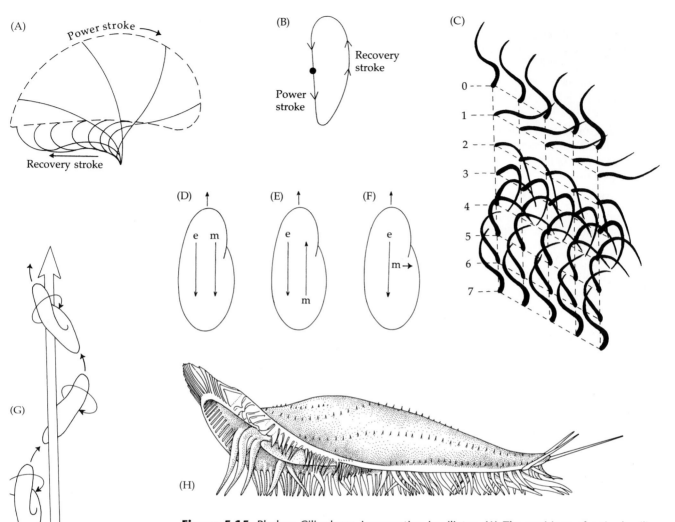

Figure 5.15 Phylum Ciliophora. Locomotion in ciliates. (A) The positions of a single cilium during the effective (power) and recovery strokes. (B) Flattened oval described by the tip of a beating cilium. (C) A ciliary field fixed during metachronal beating. Rows 0–2 are engaged in the power stroke, whereas rows 3–7 are at various stages of the recovery stroke. (D) Symplectic metachrony. (e, direction of effective stroke; m, direction of metachronal wave). (E) Antiplectic metachrony. (F) Diaplectic metachrony. (G) Helical pattern of forward movement of *Paramecium*. (H) *Stylonychia* uses cirri for "walking."

wave passes in the same direction as the effective stroke: anterior to posterior. In some metazoans that have widely spaced ciliary fields, the cilia show antiplectic metachrony, in which the metachronal wave and the effective stroke pass in opposite directions. The most common metachronal pattern is diaplectic metachrony. In this case, the effective stroke is perpendicular to the metachronal wave. There are two subtypes of diaplectic metachrony. If the tip of the cilium follows a clockwise path during the recovery stroke, the pattern is laeoplectic; this occurs in molluscs. If the tip of the cilium follows a counterclockwise path during the recovery stroke, the pattern is dexioplectic; this is the most common pattern found in ciliates and results in a spiraling motion as the organism swims (Figure 5.15G). Many ciliates (e.g., *Didinium*, *Paramecium*) can vary the direction of ciliary beating and metachronal waves. In such forms, complete reversal of the body's direction of movement is possible by simply reversing the ciliary beat and wave directions.

Perhaps more than any other protist group, the ciliates have been studied for their complex locomotory behavior. *Paramecium*, a popular laboratory animal, has received most of the attention of protozoological behaviorists. When a swimming *Paramecium* encounters a mechanical or chemical environmental stimulus of sufficient intensity, it begins a series of rather intricate response activities. The animal first initiates a reversal of movement, effectively backing away from the source of the stimulus. Then, while the posterior end of the body remains more or less stationary, the anterior end swings around in a circle. This action is appropriately called the cone-swinging phase. The *Paramecium* then proceeds forward again, usually along a new pathway. Much of the literature refers to this behavior in terms of simple "trial and error," but the situation is not so easily explained. The response pattern is not constant, because the cone-swinging phase may not always occur; sometimes the animal simply changes direction in one movement and swims forward again. Furthermore, the cone-swinging phase occurs even in the absence of recognizable stimuli and thus may be regarded as a phenomenon of "normal" locomotion. Recent studies suggest that *Paramecium* swimming behavior is governed by the cell's membrane potential. When the membrane is "at rest," the cilia beat posteriorly and the cell swims forward. When the membrane becomes depolarized, the cilia beat in a reverse direction (ciliary reversal) and the cell backs up. In *Paramecium* genetic mutations are known to result in abnormal behavior characterized by prolonged periods of continuous ciliary reversal (the so-called "paranoiac *Paramecium*").

An interesting form of locomotion in ciliates is exhibited by the hypotrich ciliates (e.g., *Euplotes*). In this group, the somatic cilia are arranged in bundles called cirri, which they use to "crawl" or "walk" over surfaces (Figures 5.13E and 5.15H). Sessile ciliates are also capable of movement in response to stimuli. The attachment stalk of many peritrichs (e.g., *Vorticella*) contains contractile myonemes that serve to pull the cell body against the substratum. Similar myonemes are found in the cell walls of other ciliates (e.g., *Stentor*), and they are capable of contracting and extending the entire cell. Other ciliates (e.g., *Lacrymaria*) use sliding microtubules to contract.

Nutrition

The ciliates include many different feeding types. Some are filter feeders, others capture and ingest other protists or small invertebrates, many eat algal filaments or diatoms, some graze on attached bacteria, and a few are saprophytic parasites. In almost all ciliates, feeding is restricted to a specialized oral area containing the cytostome, or "cell mouth." Food vacuoles are formed at the cytostome and then are circulated through the cytoplasm as digestion occurs (Figure 5.16). Because of the different ciliate feeding types, however, there are a variety of structures associated with, and modifications of, the cytostome.

Holozoic ciliates that ingest relatively large food items usually possess a nonciliated tube, called the cytopharynx, which extends from the cytostome deep into the cytoplasm. The walls of the cytopharynx are often

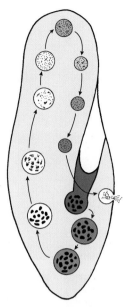

Figure 5.16 Phylum Ciliophora. Formation of and digestion within a food vacuole in *Paramecium caudatum*. The sequence of digestive events may be followed by staining yeast cells with Congo red dye and allowing the stained cells to be ingested by the protist. The changes in color from red to red-orange to blue-green reflects the change to an acid condition within the food vacuole and thus the initial stage of the digestive process. The change back to red-orange occurs as the vacuole subsequently becomes more alkaline. The pattern of movement of the food vacuole (arrows) is typical of this animal and is often termed cyclosis.

reinforced with rods of microtubules (nematodesmata). In a few forms, most notably *Didinium*, the cytopharynx is normally everted to form a projection that sticks to prey and then inverts back into the cell, thus pulling the prey into a food vacuole. In this way, *Didinium* can engulf its relatively gigantic prey, *Paramecium* (Figure 5.17A). Other ciliates, such as the hypostomes, have complex nematodesmal baskets in which microtubules work together to draw filaments of algae into the cytostome, reminiscent of the way a human sucks up a piece of spaghetti (Figure 5.17B). In most of these ciliates, the cilia around the mouth are relatively simple.

Other ciliates, including many of the more familiar forms (e.g., *Stentor*) are suspension feeders (Figure 5.18). These often lack or have reduced cytopharynxes. Instead, they have elaborate specialized oral cilia for creating water currents, and filtering structures or scraping devices. Their cytostomes often sit in a depression on the cell surface. The size of the food eaten by such ciliates depends on the nature of the feeding current and, when present, the size of the depression. The oral ciliature often consists of compound ciliary organelles, called the **adoral zone of membranelles,** or simply the **AZM** (Figure 5.13E,F), on one side of the cytostome, and a row of closely situated paired cilia which is frequently

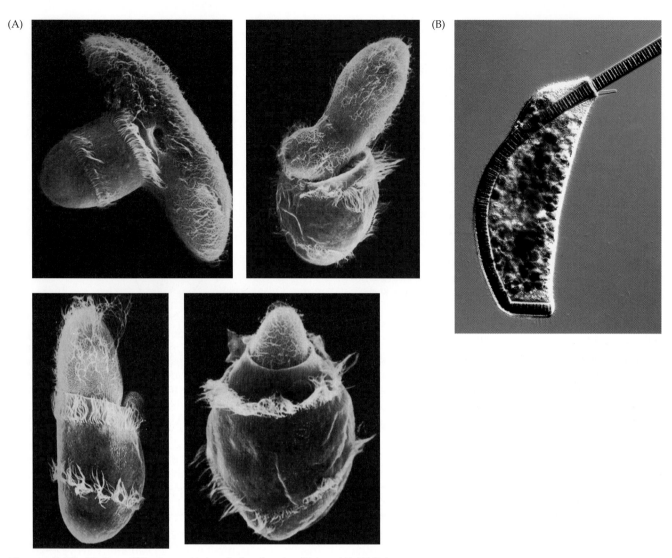

(A)

(B)

Figure 5.17 Phylum Ciliophora. Holozoic feeding in ciliates. (A) *Didinium* consuming *Paramecium*. (B) *Nassulopsis* ingesting blue-green algae.

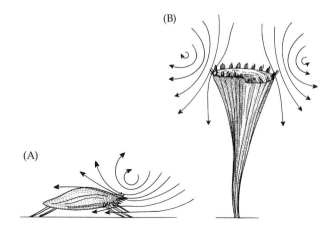

Figure 5.18 Phylum Ciliophora. Feeding currents produced by two ciliates. (A) *Euplotes*. (B) *Stentor*. The ciliary currents bring suspended food to the cell, where it can be ingested.

called the **paroral membrane** on the other side. Ciliates that feed like this include such common genera as *Euplotes*, *Stentor*, and *Vorticella*. Many hypotrichs (e.g., *Euplotes*) that move about the substratum with their oral region oriented ventrally use their specialized oral ciliature to swirl settled material into suspension and then into the buccal cavity for ingestion.

Among the most specialized ciliate feeding methods are those used by the suctorians, which lack cilia as adults and instead have knobbed feeding tentacles (Figure 5.19). A few suctorians have two types of tentacles, one form for food capture and another for ingestion. The swellings at the tips of the tentacles contain extrusomes called **haptocysts**, which are discharged upon contact with a potential prey. Portions of the haptocyst penetrate the victim and hold it to the tentacle. Sometimes prey are actually paralyzed after contact with haptocysts, presumably by enzymes released dur-

Figure 5.19 Phylum Ciliophora. Feeding in the suctorian ciliate *Acineta*. (A) *Acineta* has capitate feeding tentacles; note the absence of cilia. (B–D) Schematic drawings of enlarged feeding tentacles, showing the sequence of events in prey capture and ingestion. (B) Contact with prey and firing of haptocysts into prey. (C) Shortening of tentacle and formation of a temporary feeding duct within a ring of microtubules. (D) Drawing of contents of prey into duct and formation of food vacuole.

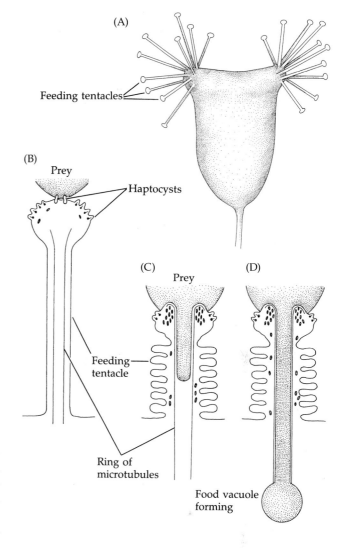

ing discharge. Following attachment to the prey, a temporary tube forms within the tentacle, and the contents of the prey are sucked into the tentacle and incorporated into food vacuoles (Figure 5.19B–D).

In addition to haptocysts, several other types of extrusomes are present in ciliates. Some predatory ciliates have tubular extrusomes, called toxicysts, in the oral region of the cell (Figure 5.20A). During feeding, the toxicysts are extruded and release their contents, which apparently include both paralytic and digestive enzymes. Active prey are first immobilized and then partially digested by the discharged chemicals; this partially digested food is later taken into food vacuoles. Some ciliates have organelles called mucocysts located just beneath the pellicle (Figure 5.20B). Mucocysts discharge mucus onto the surface of the cell as a protective coating; they may also play a role in cyst formation. Others have trichocysts, which contain nail-shaped structures that can be discharged through the pellicle. Most specialists suggest that these structures are not used in prey capture, but serve a defensive function.

A number of ciliates are ecto- or endosymbionts associated with a variety of vertebrate and invertebrate hosts. In some cases these symbionts depend entirely upon their hosts for food. Some suctorians, for example, are true parasites, occasionally living within the cytoplasm of other ciliates. A number of hypostome ciliates are ectoparasites on freshwater fishes and may cause significant damage to their hosts' gills. *Balantidium coli*, a large vestibuliferan ciliate, is common in pigs and occasional-

Figure 5.20 Phylum Ciliophora. Extrusomes in ciliates. (A) Toxicyst (longitudinal section) from *Helicoprorodon*. (B) Mucocyst (longitudinal section) from *Colpidium*. (C) The pellicle of *Nassulopsis elegans*, showing mucocysts (raised dots) just below the surface.

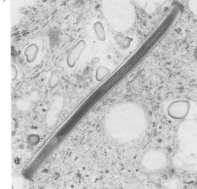
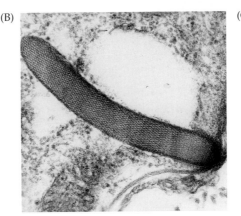
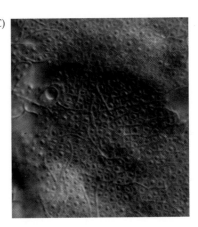

ly is acquired by humans, where it can cause intestinal lesions. The rumen of ungulates contains whole communities of ciliates, including species that break down the grasses eaten by the host, bactivorous species, and even predators preying on the other ciliates. Members of the order Chonotrichida are mostly ectosymbiotic on crustaceans (and occasionally on whales). Chonotrichs are sessile, attaching to their hosts by a stalk produced from a special adhesive organelle. Other ciliates are symbiotic on a variety of hosts, including bivalve and cephalopod molluscs, polychaete worms, and perhaps mites.

A few ciliates (e.g., *Laboea*, *Strombidium*) sequester photosynthetically functional chloroplasts derived from ingested algae. The chloroplasts lie free in the cytoplasm, beneath the pellicle, where they actively contribute to the ciliate's carbon budget. This unusual practice has also been documented in foraminiferans (see

Trench 1980). The cellular mechanisms by which the prey chloroplasts are removed, sequestered, and cultured are not known. During the summer, in some areas, "photosynthetic ciliates" can comprise a majority of the freshwater planktonic ciliate fauna.

Reproduction

Ciliates have two types of nuclei in each cell. The larger type—the **macronucleus**—controls the general operation of the cell. The macronucleus is usually hyperpolyploid (containing many sets of chromosomes) and may be compact, ribbon-like, beaded, or branched. The smaller type—the **micronucleus**—has a reproductive function, synthesizing the DNA associated with reproduction. It is usually diploid.

Asexual reproduction in ciliates is usually by binary fission, although multiple fission and budding are also known (Figure 5.21). Binary fission in ciliates is usually transverse. The micronucleus is the reservoir of genetic material in ciliates. As such, each micronucleus within the cell (even when there are many) forms an internal mitotic spindle during fission, thus distributing daughter micronuclei equally to the progeny of division. Macronuclear division is highly variable, although the nuclear envelope never breaks down. The large, sometimes multiple, macronuclei usually condense into a single macronucleus which divides by constriction. Some macronuclei have internal microtubules that appear to push daughter nuclei apart, but there is never a clear, well organized spindle. Since many ciliates are anatomically complex and frequently bear structures that are not centrally or symmetrically placed on the body (especially structures associated with the cytostome), a significant amount of reconstruction must occur following fission. Such re-formation of parts or special ciliary fields does not take place haphazardly; it apparently is controlled, at least in part, by the macronucleus.

Binary fission is typical of the colonial and solitary peritrichs. In colonial species, the division is equal, with both daughter cells remaining attached to the growing colony, but in solitary species divisions may be unequal and may involve a swimming phase. A type of unequal division frequently referred to as budding occurs in a variety of sessile ciliates, including chonotrichs and suctorians. In these cases the ciliated bud is released as a so-called **swarmer** that swims about before adopting the adult morphology and lifestyle. In some cases, several buds are formed and released simultaneously.

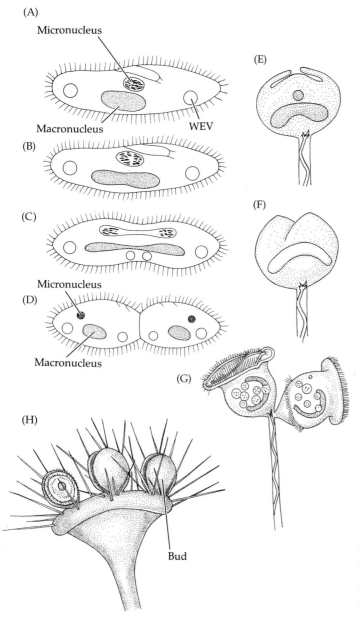

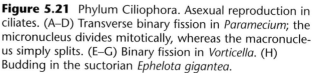

Figure 5.21 Phylum Ciliophora. Asexual reproduction in ciliates. (A–D) Transverse binary fission in *Paramecium*; the micronucleus divides mitotically, whereas the macronucleus simply splits. (E–G) Binary fission in *Vorticella*. (H) Budding in the suctorian *Ephelota gigantea*.

True multiple fission is known in a few groups of ciliates and typically follows the production of a cyst by the prospective parent. Repeated divisions within the cyst produce numerous offspring, which are eventually released with breakdown of the cyst coating.

Sexual reproduction (or more precisely, genetic recombination) by ciliates is usually by conjugation, less commonly by autogamy. Conjugation is perhaps most easily understood by first describing it in *Paramecium*. As with any sexual process, the biological "goal" of the activity is genetic mixing or recombination, and it is accomplished during conjugation by an exchange of micronuclear material. The following account (Figure 5.22) is of *Paramecium caudatum*—details vary in other species of the genus.

As paramecia move about and encounter one another, they recognize compatible "mates" (i.e., members of another clone). After making contact at their anterior ends, the "mates"—called **conjugants**—orient themselves side by side and attach to each other at their oral areas. In each conjugant, the micronucleus undergoes two divisions that are equivalent to meiosis and reduce the chromosome number to the haploid condition.

Three of the daughter micronuclei in each conjugant disintegrate and are incorporated into the cytoplasm; the remaining haploid micronucleus in each cell divides once more by mitosis. The products of this postmeiotic micronuclear division are called **gametic nuclei**. One gametic nucleus in each conjugant remains in its "parent" conjugant while the other is transferred to the other conjugant via a cytoplasmic connection formed at the point of joining. Thus, each conjugant sends a haploid micronucleus to the other, thereby accomplishing the exchange of genetic material. Each migratory gametic nucleus then fuses with the stationary micronucleus of the recipient, producing a diploid nucleus, or **synkaryon**, in each conjugant. This process is analogous to mutual cross-fertilization in metazoan invertebrates.

Following nuclear exchange, the cells separate from each other and are now called **exconjugants**. The process is far from complete, however, for the new genetic combination must be incorporated into the macronucleus if it is to influence the organism's phenotype. This is accomplished as follows. The macronucleus of each exconjugant has disintegrated during the meiotic and exchange processes. The newly formed

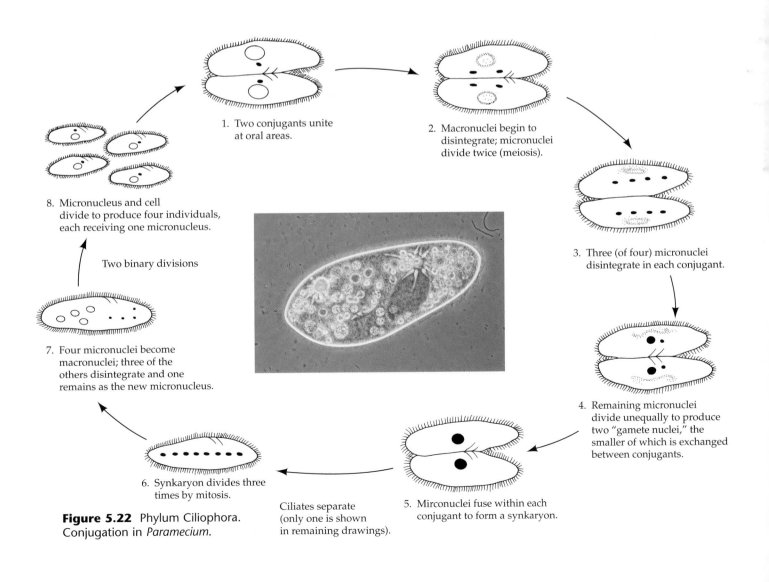

1. Two conjugants unite at oral areas.

2. Macronuclei begin to disintegrate; micronuclei divide twice (meiosis).

3. Three (of four) micronuclei disintegrate in each conjugant.

4. Remaining micronuclei divide unequally to produce two "gamete nuclei," the smaller of which is exchanged between conjugants.

5. Micronuclei fuse within each conjugant to form a synkaryon.

6. Synkaryon divides three times by mitosis.

7. Four micronuclei become macronuclei; three of the others disintegrate and one remains as the new micronucleus.

8. Micronucleus and cell divide to produce four individuals, each receiving one micronucleus.

Two binary divisions

Ciliates separate (only one is shown in remaining drawings).

Figure 5.22 Phylum Ciliophora. Conjugation in *Paramecium*.

Figure 5.23 Phylum Ciliophora. Sexual processes in cili-ates. (A,B) *Ephelota gemmipara* (a suctorian); two mating partners of unequal sizes are attached to each other, appar-ently following chemical recognition. Both have undergone nuclear meiosis. The smaller mate detaches from its stalk and is absorbed by the larger one, then the gametic nuclei fuse. Subsequent nuclear divisions produce the multimi-cronuclear and macronuclear components of the normal individual. (C,D) Unequal divisions of *Vorticella campanula* results in macro- and microconjugants; conjugation follows. (E) Schematic diagrams of sexual activities in certain per-itrichs. Unequal divisions result in macro- and microga-monts; the latter detach from their stalks and become free-swimming organisms; eventually the free-swimming microgamont attaches itself to a sessile macrogamont (1–2). The macronuclei begin to disintegrate (2) and ulti-mately disappear (9). The micronucleus of the macroga-mont divides twice (2–3) and the micronucleus of the microgamont divides three times (2–3). All but one of the micronuclei in each gamont disintegrate, and the remain-ing micronucleus of the microgamont moves to fuse with the micronucleus of the macrogamont (4–5). As the zygot-ic nucleus (synkaryon) begins to divide, the microgamont is absorbed into the cytoplasm of the macrogamont. The synkaryon divides three times (6–8); one of the daughter nuclei becomes the micronucleus and the others eventually form the new macronucleus (9). It should be noted that the sequence of nuclear activities and numbers of divisions vary among different peritrichs.

diploid synkaryon divides mitotically three times, pro-ducing eight small nuclei (all, remember, containing the combined genetic information from the two original conjugants). Four of the eight nuclei then enlarge to be-come macronuclei. Three of the remaining four small nuclei break down and are absorbed into the cytoplasm. Then the single remaining micronucleus divides twice mitotically as the entire organism undergoes two binary fissions to produce four daughter cells, each of which receives one of the four macronuclei and one micronu-cleus. Thus, the ultimate product of conjugation and the subsequent fissions is four new diploid daughter organ-isms from each original conjugant.

Variations on the sequence of events described above for *Paramecium* include differences in the number of di-visions, which seem to be determined in part by the nor-mal number of micronuclei present in the cell. Even when two or more micronuclei are present, they typical-ly all undergo meiotic divisions. All but one disinte-grate, however, and the remaining micronucleus di-vides again to produce the stationary and migratory gametic nuclei.

In most ciliates the members of the conjugating pair are indistinguishable from each other in terms of size, shape, and other morphological details. However, some species, especially in the Peritrichida, display distinct

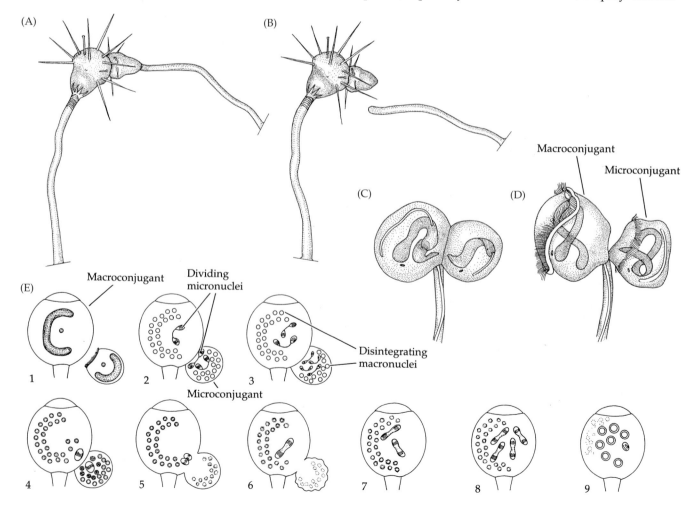

and predictable differences between the two conjugants, particularly in size. In such cases, we refer to the members of the mating pair as the **microconjugant** and the **macroconjugant** (Figure 5.23). The formation of the microconjugant generally involves one or a series of unequal divisions, which may occur in a variety of ways. The critical difference between conjugation of similar mates and that of dissimilar mates is that in the latter case, there is often a one-way transfer of genetic material. The microconjugant alone contributes a haploid micronucleus to the macroconjugant; thus only the larger individual is "fertilized." Following this activity, the entire microconjugant usually is absorbed into the cytoplasm of the macroconjugant (Figure 5.23E). A similar process occurs in most chonotrichs. One conjugant appears to be swallowed into the cytostome of another, after which nuclear fusion and reorganization take place. There are several other modifications on this complex sexual process in ciliates, but all have the same fundamental result of introducing genetic variation into the populations.

One other aspect of conjugation that deserves mention is that of **mating types**. Individuals of the same genetic mating type (e.g., members of a clone produced by binary fission) cannot successfully conjugate with one another. In other words, conjugation is not a random event but can occur only between members of different mating types, or clones. This restriction presumably ensures good genetic mixing among individuals.

The second basic sexual process in ciliates is autogamy. Among ciliates in which it occurs (e.g., certain species of *Euplotes* and *Paramecium*), the nuclear phenomena are similar if not identical to those occurring in conjugation. However, only a single individual is involved. When the point is reached at which the cell contains two haploid micronuclei, these two nuclei fuse with one another, rather than one being transferred to a mate. Autogamy is known in relatively few ciliates, although it may actually be much more common than demonstrated thus far. Its significance in terms of genetic variation is not clear.

Phylum Apicomplexa (The Gregarines, Coccidians, Haemosporidians, and Piroplasms)

The phylum Apicomplexa includes about 5,000 species that are all parasitic and characterized by the presence of a unique combination of organelles at the anterior end of the cell called the **apical complex** (Figure 5.24A; Box 5F). The apical complex apparently attaches the parasite to a host cell and releases a substance that causes the host cell membrane to invaginate and draw the parasite into its cytoplasm in a vacuole. The anterior end is frequently equipped with hooks or suckers for attachment to the host's epithelia (Figure 5.24B). Gregarines occupy the guts and body cavities of several kinds of invertebrates, including annelids, sipunculans, tunicates, and arthropods. The genus *Gregarina* itself has nearly a thousand described species, mostly from insects. The coccidians are parasites of several groups of animals, mostly vertebrates. They typically reside within the epithelial cells of their host's gut, at least during some stages, and many are pathogenic. Some coccidians pass their entire life cycle within a single host; many others require an intermediate host that serves as a vector. Coccidians are responsible for a variety of diseases, including coccidiosis in rabbits, cats, and birds, and toxoplasmosis and malaria in people. Piroplasms and haemosporidians are both parasites of vertebrates. The piroplasms are transmitted by ticks and are responsible for some serious diseases of domestic animals, including red-water fever in cattle.

(B)

(C)

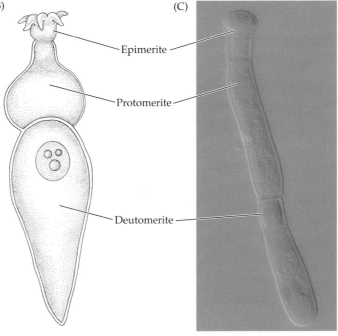

(A)

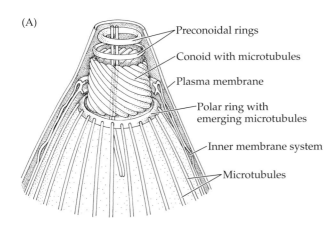

Figure 5.24 Phylum Apicomplexa. (A) Microstructure of the apical complex. (B,C) The body of a gregarine is commonly divided into three recognizable regions.

BOX 5F Characteristics of the Phylum Apicomplexa

1. Shape of cell maintained by pellicle consisting of alveolar vesicles beneath plasma membrane

2. Locomotion characterized as gliding, but precise mechanism of movement not well understood. Cilia absent, but some species produce flagellated or ameboid gametes

3. Mitochondria with vesicular cristae

4. With single vesicular nucleus

5. Nuclear division occurs by semi-open pleuromitosis in all except gregarines, which show much diversity in their mitoses: mitotic spindle organizers are either disks, plugs, or electron-dense crescents on nuclear envelope in the location of fenestrae. In coccidians, centrioles are associated with spindle organizers; in haemosporidians and piroplasms, centrioles absent

6. Asexual reproduction by binary fission, multiple fission, or endopolygeny

7. Sexual reproduction is gametic; gametes either isogametous or anisogametous. Meiosis involves single division after formation of zygote

8. Without plastids

9. Carbohydrates stored as paraglycogen (= amylopectin)

10. With unique system of organelles, the apical complex, in anterior region of cell

Haemosporidians are blood parasites of vertebrates that are transmitted by flies and include the organisms which cause malaria and malaria-like ailments in humans and birds.

Many species of apicomplexan protists are responsible for serious diseases in domestic animals and humans. Malaria, which affects people in 102 countries, is caused by several species of the genus *Plasmodium* (Box 5G). Although malaria was greatly reduced worldwide during the 1960s, it is making an alarming comeback, and it is one of the most prevalent and severe health problems in the developing world. Nearly 500 million people are stricken annually with malaria, 90 percent of them in Africa. Malaria kills one to three million people annually, half of them children. The development of a vaccine continues to elude researchers. One of the principal causes of this resurgence is the dramatic rise in the numbers of pesticide-resistant (particularly DDT-resistant) strains of *Anopheles* mosquitoes, the insect vector for *Plasmodium*. (DDT was banned in the United States in 1972, but it is still used in many countries.) By 1968, 38 strains or species of *Anopheles* in India alone had been identified as largely pesticide-resistant; and between 1965 to 1975 the incidence of malaria in Central America tripled. In many parts of the world, *Plasmodium falciparum*—the most deadly species—is now resistant to chloroquine, once a mainstay of malarial drug therapy. Researchers have recently discovered that *Plasmodium* species, like *Trypanosoma*, have the ability to avoid detection by the human immune system by switching among as many as 150 genes that code for different versions of the protein that coats the surface of their cells (it is this protein coat that the human immune system relies on as a recognition factor). Furthermore, because the parasite sequesters itself inside red blood cells, it is largely protected from most drugs. Recent evidence also suggests that the parasite may affect its mosquito host by inducing it to bite more frequently than uninfected mosquitoes. Other malaria-like genera (e.g., *Haemoproteus*, *Leucocytozoon*) are parasites of birds and reptiles but have life cycles similar to *Plasmodium* and also utilize mosquitoes as vectors.

Several other apicomplexan genera are worth mentioning. The coccidian genus *Eimeria* causes a disease known as cecal coccidiosis in chickens. In 1999, this disease was estimated by the USDA to cost American poultry farmers $600 million dollars a year in lost animals, medication, and additional labor. *Toxoplasma gondii*, another coccidian, is found worldwide and is known to parasitize many animals, including pigs, rodents, primates, birds, cats, and humans. In humans, *Toxoplasma* usually produces no symptoms, or only mild symptoms, but it and another genus, *Cryptosporidia*, has been increasingly problematic in AIDS patients and other immunosuppressed people. Transmission of *Toxoplasma* parasites is thought to be via raw or undercooked meat (beef, pork, lamb), through fecal contamination from a pet cat, or by way of flies and cockroaches (which can carry the *T. gondii* cysts from a cat's litter box to the table). In 1982 Martina Navratilova lost the U.S. Open Tennis Championship (and a half-million dollars) when she had toxoplasmosis.

Support and Locomotion

The fixed shape of apicomplexans is maintained by a pellicle composed of chambers, or alveoli, which lie just beneath the plasma membrane. Microtubules originate at the apical complex and run beneath the alveoli, providing additional support. Apicomplexa do not have cilia, flagella or pseudopodia. Nevertheless, they can be observed gliding and flexing. What causes these movements is not well understood, but microtubules and microfilaments underneath the alveoli may play a role.

Nutrition

The alveoli are interrupted at both the anterior and posterior ends, and at tiny invaginations of the cell membrane called microspores, which have been implicated in feeding. Nutrient ingestion is thought to occur primarily by pinocytosis or phagocytosis at the microspores. In the haemosporidians, ingestion of the host's cytoplasm through the microspores has been observed. Absorption of nutrients has also been reported in some gregarines at the point where the parasite attaches to the host's cell.

Reproduction and Life Cycles

Asexual reproduction in Apicomplexa is by binary fission, multiple fission, or endopolygeny. Mitosis is a semi-open pleuromitosis in all apicomplexans except for some gregarines. Gregarines undergo a variety of mitoses, depending on the species. For example, *Diaplauxis hatti* and *Lecudina tuzetae* undergo semi-open orthomitosis, *Monocystis* sp. and *Stylocephalus* sp. undergo open orthomitosis, and *Didymophyes gigantea* undergoes closed intranuclear orthomitosis. Sexual reproduction occurs by a union of haploid gametes, which can be the same size (isogametous) or different sizes (anisogametous), and may be flagellated or form pseudopodia.

The life cycle varies somewhat between the different groups of apicomplexan protists, but can be divided into three general stages: (1) gamontogony (the sexual phase), (2) sporogony (the spore-forming stage), and (3) the growth phase. A great deal has been written about the life cycles of these protists and a full description is beyond the scope of this text. The life cycles of the gregarine *Stylocephalus* and the haemosporidan *Plasmodium* are given as examples to illustrate the basic themes and variations in apicomplexan reproduction.

The life cycle of gregarines is usually monoxenous—it involves only one host. Some of the best-studied gregarines are those found in coleopterans (beetles), and the life cycles of these forms are well understood (the life cycle of *Stylocephalus longicollis* is diagrammed in Figure 5.25). Sexual reproduction in gregarines usually involves the enclosure of a mating pair of gamonts, the mature trophozoites destined to produce gametes, within a cyst or capsule. This encapsulation is known as **syzygy**. Each gamont undergoes multiple fission to produce many gametes. Both isogamy and anisogamy are known among different gregarines. Each zygote that is formed by the fusion of two gametes becomes a spore, which divides to produce as many as eight **sporozoites**. Each sporozoite enters a period of growth, to mature as a **trophozoite**, which eventually becomes a sexual gamont, completing the cycle.

The life cycle of haemosporidians is heteroxenous, involving two hosts, usually a vertebrate and an invertebrate. In *Plasmodium*, the vertebrate host is commonly humans, and the invertebrate host is the *Anopheles* mosquito (Figure 5.26). When a human is bitten by a female *Anopheles* mosquito, sporozoites, which reside in the salivary glands of the mosquito, are released into the bloodstream where they migrate to the liver and enter the hepatocytes. Once in the liver cells, the sporozoites undergo multiple fission (schizogony) until the liver cell ruptures, releasing forms called **merozoites** which enter red blood cells (erythrocytes) and transform into trophozoites. In erythrocytes, the trophozoites either transform into gamonts or undergo schizogony, forming more merozoites. If the latter occurs, the cell eventually bursts, releasing merozoites that subsequently infect other red blood cells. The rapid destruction of red blood cells and the release of metabolic by-products are responsible for the characteristic chills, fever, and anemia that are common symptoms of malaria. If the trophozoite becomes a gamont, it will be morphologically distinguishable as either a macrogamont or a microgamont. The life cycle of gamonts does not proceed

---

BOX 5G   *Malaria*

Human malaria has been known since antiquity, and descriptions of the disease are found in Egyptian papyruses and temple hieroglyphs (outbreaks followed the annual flooding of the Nile). The relationship between the disease and "swamp land" led to the belief that it could be contracted by breathing "bad air" (*mal aria*). It is likely that malaria was brought to the New World by the conquering Spaniards and their African slaves. By the time the Panama Canal was built, malaria (and yellow fever) were well established in the Neotropics. The presence of these diseases was the principal reason the French failed in their attempt to construct the canal. William Gorgas, the medical officer in charge during the U.S. phase of the canal's construction, became a hero when his mosquito-control efforts allowed for its successful completion by U.S. engineers. The president made Gorgas Surgeon General, Oxford University gave him an honorary doctorate, and the King of England knighted him.

Among the many symptoms of malaria are cyclical paroxysms, wherein the patient feels intensely cold as the hypothalamus (the body's thermostat) is activated; body temperature then rises rapidly to 104–106°F. The victim suffers intense shivering. Nausea and vomiting are usual. Copious perspiration signals the end of the hot stage, and the temperature drops back to normal within 2 or 3 hours; the entire paroxysm is over within 8 to 12 hours. It is thought that these episodes are stimulated by the appearance of the waste products of parasites feeding on erythrocytes, which are released when the blood cells lyse. Secondary symptoms include anemia due to the red blood cell destruction. In extreme cases of falciparum malaria, massive lysis of erythrocytes results in high levels of free hemoglobin and various break-down products that circulate in the blood and urine, resulting in a darkening of these fluids, hence the condition called blackwater fever.

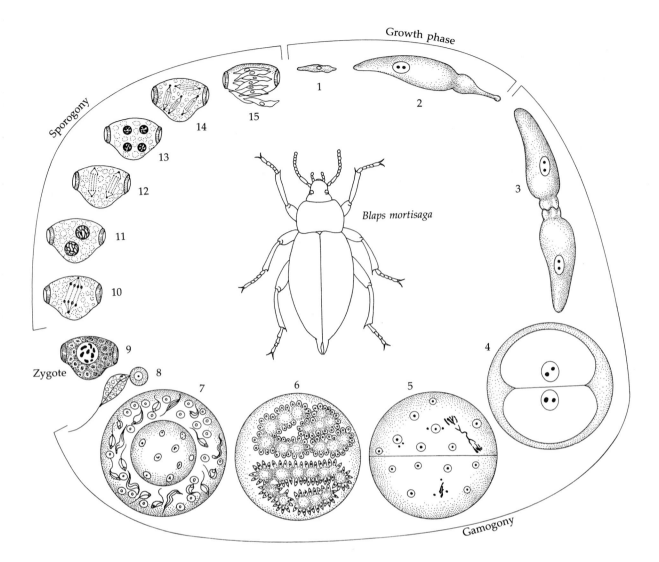

Figure 5.25 Phylum Apicomplexa. Life cycle of the gregarine *Stylocephalus longicollis*, a gut parasite of the coleopteran *Blaps mortisaga*. Stages 1–4 take place within the host, 5–15 outside the host. The spores (15) are ingested by the beetle and release sporozoites within the gut lumen. Each sporozoite grows into a gamont (2); the gamonts subsequently mate (3–4), becoming enclosed within a mating cyst, which leaves the host with the feces. Repeated mitotic divisions within the cyst produce anisogametes (5–7); these ultimately fuse (8) to produce a zygote (9), which eventually becomes a spore. The first divisions of the spore cell are meiotic (10), so all subsequent stages leading back to gamete fusion are haploid.

any further unless they are taken in by a mosquito during its blood meal. Once in the gut of the mosquito, the macrogamont becomes a spherical **macrogamete**, while the microgamont undergoes three nuclear divisions and develops eight projections (**microgametes**), which each receive a nucleus. The microgametes break away and each fertilizes a single macrogamete to form a diploid zygote called an **ookinete**. The ookinete then actively burrows through the mosquito's stomach and secretes a covering around itself on the outside of the stomach, forming an oocyst. Inside the **oocyst**, the zygote undergoes a meiotic reduction division followed by schizogony to form sporozoites that are released from the oocyst into the gut, to migrate to the salivary gland where they remain until the next time the insect feeds.

Phylum Dinoflagellata

There are approximately 4,000 described species of dinoflagellates, many of which are known only as fossils. Although unquestionable fossil dinoflagellates date back to the Triassic (240 million years ago), evidence from organic remains in Early Cambrian rocks suggest that they were abundant as early as 540 million years ago. These protists are common in all aquatic environments, but about 90 percent of the described species are planktonic in the world's seas. Approximately half of the living species of dinoflagellates are photosynthetic, and these are important primary producers in many aquatic environments. They can be quite beautiful and many are capable of bioluminescence (e.g., *Gonyaulax*) using a luciferin–luciferase system. Although most are

Figure 5.26 Phylum Apicomplexa. Life cycle of *Plasmodium*, the causative agent of malaria in humans. When a female anopheline mosquito takes a blood meal, she releases sporozoites into the victim's bloodstream (1). These sporozoites enter the host's liver cells and undergo multiple fission, producing many merozoites (2); each sporozoite may produce as many as 20,000 merozoites in a single liver cell. The infected liver cells rupture, releasing the merozoites into the blood where they invade red blood cells (3). Through continued multiple fission, more merozoites are produced. The red blood cells eventually burst, releasing merozoites, which enter other red blood cells. Some merozoites differentiate to become gametocytes (4), which are picked up by mosquitoes. The female gametocyte forms a single macrogamete; the male gametocyte typically undergoes multiple fission to produce several motile, flagellated microgametes within the gut of the mosquito (5). After fertilization occurs, the zygote migrates to the mosquito's salivary glands and divides to form numerous sporozoites, thereby completing the life cycle.

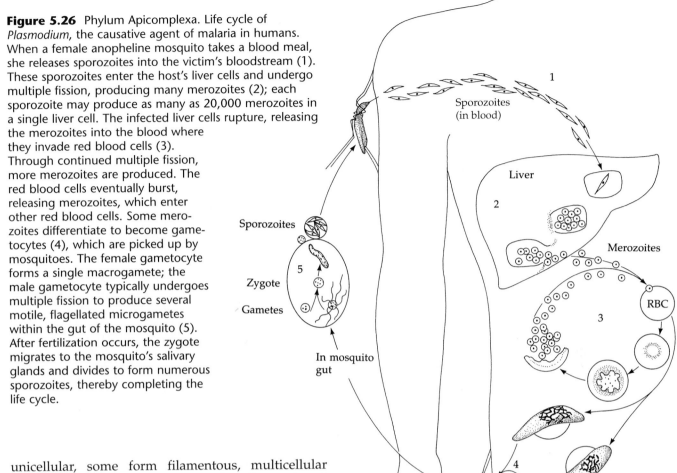

unicellular, some form filamentous, multicellular colonies. Dinoflagellates have two flagella, postioned such that they whirl or spin as they swim (Figures 5.1A, H and 5.27), the attribute for which they are named (Greek *dinos*, "whirling, turning") (Box 5H).

Endosymbiotic marine dinoflagellates that occur as coccoid cells when inside their invertebrate or protist hosts, but produce motile cells periodically, are called

---

BOX 5H *Characteristics of the Phylum Dinoflagellata*

1. Shape of cell maintained by pellicle consisting of alveolar vesicles beneath plasma membrane; alveoli may be filled with cellulose

2. With two flagella for locomotion: One is transverse and has single row of hairs, the other is longitudinal and has two rows of hairs; both are supported by a paraxonemal rod. Flagella oriented in longitudinal groove and equatorial groove (a diagnostic feature of this group)

3. Mitochondria with tubular cristae

4. Nuclei contain permanently condensed chromosomes; no histone proteins associated with DNA

5. Nuclear division occurs by closed extranuclear pleuromitosis without centrioles. No obvious organizing center for mitotic spindle

6. Asexual reproduction by binary fission along the longitudinal plane (symmetrogenic)

7. Sexual reproduction occurs in some. Meiosis involves two divisions, one just after nuclei from pair of gametes fuse and one after cell undergoes period of dormancy

8. Photosynthetic forms have chlorophylls a and c_2, thylakoids in stacks of three, and three surrounding membranes. Outer membrane not continuous with nuclear membrane. Accessory pigments often give cells brownish color

9. Food reserves stored as starch and oils

10. With unique system of pusules for osmoregulation, excretion, or buoyancy regulation

(A)

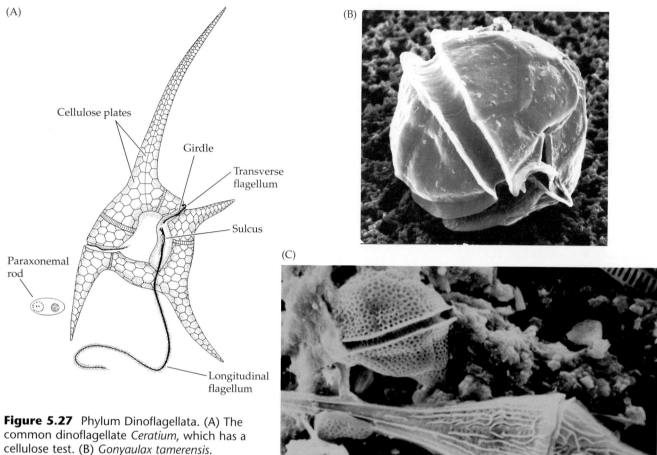

Cellulose plates

Girdle

Transverse flagellum

Sulcus

Paraxonemal rod

Longitudinal flagellum

Figure 5.27 Phylum Dinoflagellata. (A) The common dinoflagellate *Ceratium*, which has a cellulose test. (B) *Gonyaulax tamerensis*. (C) *Peridineum* (upper left) and *Ceratium*.

(B)

(C)

zooxanthellae. They belong to the poorly understood but vastly important genera *Zoochlorella* (symbionts of various freshwater organisms), *Zooxanthella* (symbionts of radiolarians), and *Symbiodinium* (symbionts of cnidarians and some other metazoans). Species of *Symbiodinium* are best known as critically important mutualistic symbionts of hermatypic corals (Chapter 8). There are several species of *Symbiodinium* in corals. All are photosynthetic and provide nutrients to the corals and help create the internal chemical environment necessary for the coral to secrete its calcium carbonate skeleton. Zooxanthellae also occur in many cnidarians other than scleractinian corals, such as milleporinids, chondrophorans, sea anemones, and various medusae (Chapter 8).

Some planktonic dinoflagellates occasionally undergo periodic bursts of population growth and are responsible for a phenomenon known as **red tide**.* Red tides have nothing to do with actual tides, and they are only rarely red. A red tide is simply a streak or patch of ocean water discolored (generally a pinkish orange) by the

presence of billions of dinoflagellates. During a red tide, densities of these dinoflagellates may be as high as 10 to 100 million cells per liter of sea water. Exactly why the population explosions of these specific organisms occur is not entirely clear, but organic pollutants from terrestrial runoff may be a key culprit.

Many red tide organisms manufacture highly toxic substances. One group of toxins produced by dinoflagellate species such as *Alexandrium* spp., *Gymnodinium catenatum*, and *Pyrodinium bahamense* is called **saxitoxins**. Saxitoxins block the sodium–potassium pump of nerve cells and prevent normal impulse transmission. When suspension feeders such as mussels and clams eat these protists, they store the toxins in their bodies. Extremely high concentrations of toxic dinoflagellates will even kill suspension feeders and occasionally also fish caught in the thick of the bloom. The shellfish feeding on the dinoflagellates become toxic to animals that eat them. In humans, the result is a disease known as paralytic shellfish poisoning (PSP). Extreme cases of PSP result in muscular paralysis and respiratory failure. Over 300 human deaths worldwide have been docu-

---

*Some red tides are also caused by diatoms (Stramenopila).

mented from PSP, and this number is growing as red tides become increasingly frequent around the world (presumably linked to anthropogenic disturbances of the coastal environment).

Gymnodinium breve releases a family of toxins called **brevetoxins** that result in neurotoxic shellfish poisoning (NSP). Humans consuming animals that have this toxin accumulated in their tissues experience uncomfortable gastrointestinal side effects such as diarrhea, vomiting, and abdominal pain, and also neurological problems, including dizziness and reversal of temperature sensation. Though temporarily incapacitating, no human deaths have been reported from NSP. Ocean spray containing *G. breve* toxins can blow ashore and cause temporary health problems for seaside residents and visitors (skin, eye, and throat problems). *G. breve* is responsible for producing devastating red tides that have produced massive fish kills in Florida and along the Gulf of Mexico.

In recent years, a newly discovered species of dinoflagellate called *Pfiesteria piscicida* has been creating havoc in the coastal areas of the eastern United Sates, from Delaware to North Carolina. It was particularly problematic during the summer of 1997, when it caused massive fish kills on the eastern shore of Maryland and prompted the closure of several waterways for weeks at a time. *Pfiesteria piscicida* is known to have a complex life cycle that includes at least 24 forms, only a few of which are toxic. Normally it exists in a benign state and can even photosynthesize if it eats another organism with chloroplasts. However, with the proper stimulation, which is believed to be fish oils or excrement in the water, *P. piscicida* becomes a voracious predator. It first produces a toxin that causes the fish to become lethargic and then releases other toxins that cause sores to open on the fish's body, exposing the tissues on which it feeds. The toxins of *P. piscicida* have been reported to affect humans but have caused no known deaths. Dinoflagellate toxins are among the strongest known poisons. In crystalline form, an aspirin-sized tablet of saxitoxin from *G. catenella* would be strong enough to kill 35 persons. The toxin of *Pfiesteria* is 1,000 times more powerful than cyanide.*

Support and Locomotion

The shape of dinoflagellates is maintained by alveoli beneath the cell's surface, and a layer of supporting microtubules. In some, the alveoli are filled with polysaccharides, typically cellulose, and these dinoflagellates are said to be **thecate,** or **armored** (e.g., *Protoperidinium*). Dinoflagellates that have empty alveoli are said to be **athecate,** or **naked** (e.g., *Noctiluca*). The part of the theca above the girdle is called the **epitheca** in armored species and **epicone** in naked species; the part below the girdle is the **hypotheca** in armored species and the **hypocone** in naked species.

Dinoflagellates possess two flagella that enable their locomotion. A transverse flagellum with a row of slender hairs wraps around the cell in a groove, or **girdle** (Figure 5.27). When it beats, this flagellum spins the cell around, effectively pushing it through the water like a screw. The second, longitudinal flagellum has two rows of hairs and also lies in a grove on the cell's surface, called the **sulcus.** It extends posteriorly behind the cell and its beat adds to the forward propulsion of the cell. Both flagella are supported by a **paraxonemal rod** similar to that found in the kinetoplastids and euglenids.

Osmoregulation

Most freshwater and some marine dinoflagellates have a unique system of double-membrane–bound tubules called **pusules,** which open to the outside via a canal. The two membranes of the pusules distinguish them from water expulsion vesicles, but apparently these membranes have a similar function—osmoregulation.

Nutrition

Dinoflagellates exhibit wide variation in feeding habits; many are both autotrophic and heterotrophic. Approximately half of the living species are photosynthetic, but even most of these are heterotrophic to some extent, and some dinoflagellates with functional chloroplasts can switch entirely to heterotrophy in the absence of sufficient light.

The chloroplasts are surrounded by three membranes, and thylakoids are arranged in stacks of three (Figure 5.2). Some contain eye spots (stigmata) that can be very simple pigment spots or more complex organelles with lens-like structures that apparently focus the light. Photosynthetic pigments include chlorophylls a and c_2, phycobilins, carotenoids (e.g., β-carotene), and also the xanthophylls (e.g., peridinin, found only in dinoflagellates), neoperidinin, dinoxanthin, and neodinoxanthin. These xanthophylls mask the chlorophyll pigments and account for the golden or brown color that is commonly seen in dinoflagellates.

Some dinoflagellates always lack chloroplasts and are obligate heterotrophs. Most of these are free living, but some parasitic species are known. The feeding mechanisms of heterotrophic dinoflagellates are quite diverse. Both free-living and endoparasitic dinoflagellates that live in environments rich in dissolved organic compounds take in dissolved organic nutrients by saprotrophy. Other dinoflagellates ingest food particles by phagocytosis. Many, in fact, are voracious predators that ingest other protists and microinvertebrates or use

---

*Outbreaks of *Pfiesteria* are thought to be linked to large-scale hog farming in North Carolina, which is second only to Iowa in hog production. The industry dumps hundreds of millions of gallons of untreated hog feces and urine into earthen lagoons along the coast that often leak or collapse. In 1995, 25 million gallons of liquid swine manure (more than twice the size of the Exxon Valdez oil spill) flowed into the New River when a lagoon was breached.

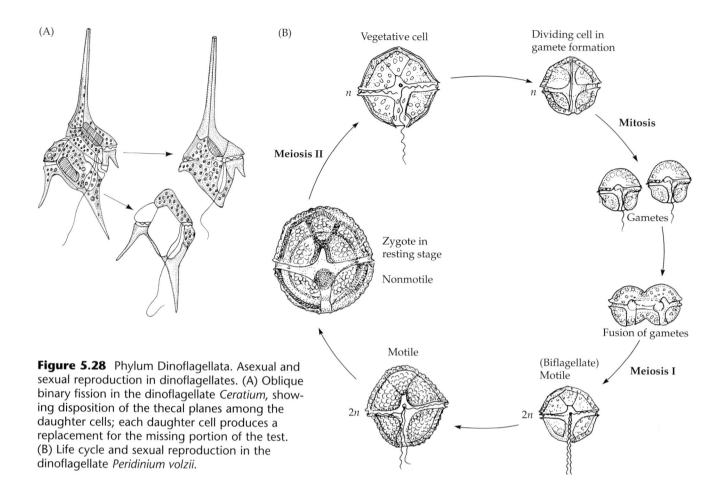

Figure 5.28 Phylum Dinoflagellata. Asexual and sexual reproduction in dinoflagellates. (A) Oblique binary fission in the dinoflagellate *Ceratium*, showing disposition of the thecal planes among the daughter cells; each daughter cell produces a replacement for the missing portion of the test. (B) Life cycle and sexual reproduction in the dinoflagellate *Peridinium volzii*.

specialized appendages to pierce prey and suck out their cytoplasmic contents.

A few dinoflagellates (e.g., *Kofoidinium* and *Noctiluca*) have a permanent cell mouth or cytostome supported by sheets of microtubules. The cytostome is often surrounded by extrusomes. There are three types: **trichocysts**, **mucocysts**, and **nematocysts**. The most common are trichocysts (similar to those found in ciliates), believed to be fired in defense or to capture and bind prey. Sac-like mucocysts secrete sticky mucous material onto the surface of the cell. This may aid in attachment to substrata (e.g., *Amphidinium*) or may help to capture prey (e.g., *Noctiluca*). Other dinoflagellates (e.g., *Nematodinium*, *Nematopsides*, *Polykrikos*) have nematocysts that resemble, but are not homologous to, the stinging organelles of cnidarians of the same name.

Reproduction

The nuclei of dinoflagellates have three unusual features: (1) they contain five to ten times the amount of DNA that is found in most eukaryotic cells; (2) the five histone proteins that are typically associated with the DNA of other eukaryotic cells are absent; and (3) the chromosomes of dinoflagellates remain condensed and the nucleolus remains intact during interphase and mitosis. Most dinoflagellates (except *Noctiluca*) spend

much of their lives as haploid cells (called vegetative cells to distinguish them from haploid gametes).

Nuclear division is by closed extranuclear pleuromitosis. No centrioles are present, and the organizing center for the mitotic spindle is not obvious. Asexual reproduction occurs by oblique, longitudinal fission, beginning at the posterior end of the cell. The thecate forms may divide the thecal plates between the two daughter cells (e.g., *Ceratium*), or they may shed the thecal plates prior to cell division (Figure 5.28A). In the former case, each daughter cell synthesizes the missing plates; in the latter case, each daughter cell synthesizes all of the thecal plates.

Sexual reproduction begins when the haploid vegetative cells divide by mitosis to produce two flagellated daughter cells, which act as gametes. When a pair of gametes fuses to form a zygote, a fertilization tube develops beneath the basal bodies of its flagella. The nucleus from each gamete enters the tube where they fuse. The first meiotic division follows shortly after nuclear fusion. Over the next few weeks, the zygote grows in size and then enters a resting stage, or cyst. The cyst develops a resistant outer wall and remains dormant for an indefinite period of time. Eventually the second meiotic division occurs, all but one of the nuclei disintegrate, and a haploid vegetative cell emerges from the cyst (Figure 5.28B).

(A)

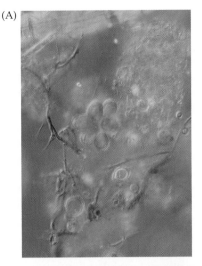

(C)

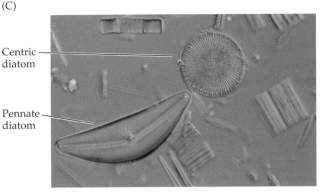

Centric diatom

Pennate diatom

Figure 5.29 Phylum Stramenopila. A diversity of stramenopiles. (A) *Synura*, a colonial golden alga. (B) A kelp bed in California waters (*Macrocystis*). (C) Centric and pennate diatoms.

(B)

Stramenopiles (Latin *stamen*, "straw"; *pilus*, "hair") refers to the appearance of these hairs (Figure 5.30).

Stramenopiles are found in a variety of habitats. Freshwater and marine plankton are rich in diatoms and chrysophytes, and they can also occur in moist soils, sea ice, snow, and glaciers. They have even been found living in clouds in the atmosphere! Heterotrophic free-living stramenopiles are also found in marine, estuarine, and freshwater habitats. A few are symbiotic on algae in marine or estuarine environments. Many produce calcite or silicon scales, shells, cysts, or tests, which are preserved in the fossil record. The oldest of these fossils are from the Cambrian/Precambrian boundary, about 550 million years ago. Diatoms are key compo-

Phylum Stramenopila

The stramenopiles consist of some 9,000 species including diatoms, brown and golden algae (the Chrysophytes), some heterotrophic flagellates, labyrinthulids (slime nets), and Oomycetes and Hyphochytridiomycetes (formerly classified as fungi) (Figures 5.1I and 5.29) (Box 5I). The photosynthetic members are usually called "chromophytes" by botanists, who rarely study the heterotrophic members of the group. A few stramenopiles, such as some of the brown seaweeds and kelps, form complex, rigid colonies and may reach extremely large sizes. The diversity of form in this group is staggering, and at first glance it may be difficult to imagine diatoms and kelp being closely related. However, their relatedness seems certain on the basis of several synapomorphies, including the fact that almost all have unique, complex, three-part tubular hairs on the flagella at some stage in the life cycle. The name

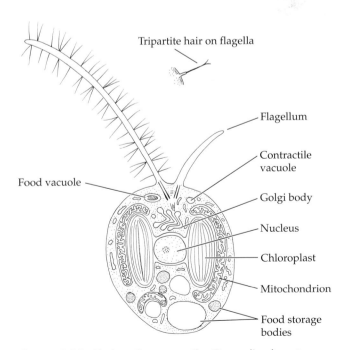

Tripartite hair on flagella

Flagellum

Contractile vacuole

Food vacuole

Golgi body

Nucleus

Chloroplast

Mitochondrion

Food storage bodies

Figure 5.30 Phylum Stramenopila. Generalized anatomy (*Ochromonas*). Note the tripartite hairs on the flagellum.

BOX 5I Characteristics of the Phylum Stramenopila

1. Cell surrounded by plasma membrane, which may be supported by silicon, calcium carbonate, or protein shells, scales, or tests

2. Two flagella with three-part hairs present at some stage in the life cycle. Sometimes only the reproductive cells are flagellated and the trophic cells lack any obvious mode of locomotion.

3. Mitochondria with short tubular cristae

4. With single vesicular nucleus

5. Nuclear division occurs by open orthomitosis without centrioles

6. Asexual reproduction by binary fission

7. Sexual reproduction gametic; gametes usually isogametous

8. Photosynthetic forms have chlorophylls a, c_1, and c_2; thylakoids in stacks of three; four membranes surrounding the chloroplast with the outermost membrane continuing around nucleus. Yellow and brown xanthophylls give them a brownish-green color that has earned them the common name "golden algae"

9. Food reserves stored as liquid polysaccharide (usually laminarin) or oils

nents of marine ecosystems and extremely important for the biogeochemical cycling of silica and as contributors to global fixed carbon.

Photosynthetic stramenopiles are the essential base of many food webs, responsible for about 50 percent of the primary production in the oceans and a large percentage of total global primary production. Brown seaweeds form an integral base to many coastal food webs, especially on temperate coasts. Algin, extracted from certain brown algae (kelp), is used as an emulsifier in everything from paint to baby food to cosmetics. Benthic deposits of the siliceous shells of dead marine diatoms can, over geologic time, result in massive uplifted land formations that are mined as diatomaceous earth. This material has many industrial uses (e.g., in paint as a spreader; as filtration material in food production and water purification). Silicon deposits produced by diatoms and other stramenopiles are also used in geology and limnology as markers of different stratigraphic layers of the Earth. None of the stramenopiles are serious parasites or agents of disease. Because some secrete fishy-smelling aldehydes, they may become a nuisance when they occur in great quantities, but stramenopiles only rarely cause fish kills or foul drinking water.

Support and Locomotion

Stramenopile support structures are highly varied. Their cells are covered with a cell membrane, and because they may also possess shells, tests, and other support structures, they have many different shapes and appearances. Like all protists, they lack chitin in the cell wall (further distinguishing them from fungi). Some chrysophytes produce small discs of calcite, protein, or even silicon in their cells. These are then packaged in endoplasmic reticulum vesicles and secreted onto the surface of the cell to form a layer of distinctive scales. Some of these can be quite elaborate and beautiful (Figure 5.31C,D). The calcite scales, called **coccoliths**, may accumulate in marine sediments in large numbers and eventually form huge chalk beds, such as can be seen in England's famous White Cliffs of Dover. Another group, called silicoflagellates, has a distinctive internal skeleton of tubular pieces of silicon associated with a central nucleus, and a complex lobed body that contains many chloroplasts.

Diatoms also secrete silicon in the form of an internal test, or **frustule,** which consists of two parts, called **valves.** Beneath the test is the cell membrane, enclosing the nucleus, chloroplasts, and the rest of the cytoplasm. There are two different forms: **centric diatoms** have radially symmetrical frustules, and since one valve is slightly larger than the other, they resemble a petri dish (Figure 5.31A); **pennate diatoms** are bilaterally symmetrical and often have longitudinal grooves on the valves (Figure 5.31B).

Stramenopiles exhibit **heterokont flagellation.** That is, they possess two flagella, one directed anteriorly, the other usually extended posteriorly. The anteriorly directed flagellum has a bilateral array of tripartite, tubular hairs, while the posterior is either smooth or has a row of fine, filamentous hairs (Figure 5.30). The tripartite, tubular hairs are stiff and reverse the direction of the thrust of the flagellum so that, even though the flagellum is beating in front of the cell, the cell is still drawn forward.

Labyrinthulids are commonly called "slime nets," and because of their unique lifestyle and locomotion, in the past they have been classified as a separate phylum. When it was discovered that they can produce cells with two heterokont flagella, they were reclassified as Stramenopiles. The nonflagellated stage of the labyrinthulid's life cycle forms complex colonies of spindle-shaped cells that glide rapidly along a membrane-bound ectoplasmic network. This network contains a calcium-dependent contractile system of actin-like proteins that is responsible for shuttling the cells through the net.

Nutrition

As you have no doubt already guessed, stramenopiles exhibit a wide variety of feeding habits. Some are photosynthetic, others are ingesting heterotrophs, and still

(A)

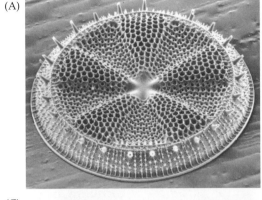

(B)

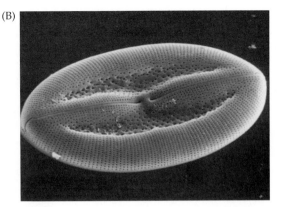

(C)

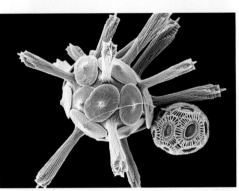

Figure 5.31 Phylum Stramenopila. Stramenopile skeletons. (A) Exterior valve of the centric diatom *Actinoptychus*. (B) The pennate diatom *Navicula*. (C) Two coccolithophores (color added): *Rhabdosphaera clavigera* (green) and *Emiliania huxleyi* (purple).

others are saprophytic-like fungi. Those forms that are photosynthetic have chlorophylls a, c_1, and c_2; thylakoids in stacks of three; and four membranes surrounding the chloroplast (the outermost membrane continues around the nucleus) (see Figure 5.2). Yellow and brown accessory pigments (primarily xanthophylls like fucoxanthin, but carotenoids are present as well) give them a brownish-green color that has earned them the common name "golden algae." There is usually an eyespot associated with the region of the chloroplast near the basal bodies. Interestingly, a similar eyespot is seen in the flagellated stage of the nonphotosynthetic labyrinthulids.

Many of the heterotrophic stramenopiles use the anteriorly directed flagellum with tripartite hairs to capture food particles, which are engulfed by small pseudopodia near the base of the flagellum. Other heterotrophic forms feed saprophytically by excreting enzymes that digest food items outside the cell and then absorbing nutrients through small pores on the cell surface. This mode of nutrition is similar to that of true fungi and is the reason that the labyrinthulids, Oomycetes ("water molds"), and Hyphochytridiomycetes were once mistakenly classified as fungi. The presence of a heterokont-flagella stage in these organisms makes it clear that they are stramenopiles.

Reproduction

Mitosis in most stramenopiles is characterized as open pleuromitosis without centrioles. During division, the basal bodies of the two flagella separate and a spindle forms adjacent to them or adjacent to the striated root at the base of each. In those forms with scales, the scaly armor appears to be added to the surface of the daughter cells as division proceeds. In diatoms, each daughter cell gets one of the silicon valves and makes a new second valve to complete the frustule.

Sexual reproduction is poorly studied in most forms but appears to almost always occur by the production of haploid gametes, which then fuse to form a zygote. In many, the gametes are undifferentiated, but in a few (such as diatoms), one of the gametes is flagellated and motile, and the other is stationary.

Phylum Rhizopoda (Amebas)

The small phylum Rhizopoda consists of approximately 200 species. Most are free living, but some endosymbiotic groups are known, including some pathogenic forms. The most obvious characteristic of rhizopodans is that they form temporary extensions of the cytoplasm, called pseudopodia ("false feet"), that are used in feeding and locomotion (Figures 5.1C, 5.32, 5.33) (Box 5J). In fact, the name of the phylum, Rhizopoda, is based on this feature and means "root-like foot." Rhizopods are ubiquitous creatures that can be found in nearly any moist or aquatic habitat: in soil or sand, on aquatic vegetation, on wet rocks, in lakes, streams, glacial meltwater, tidepools, bays, estuaries, on the ocean floor, and afloat in the open ocean. Many are ectocommensals on aquatic organisms and some are parasites of diatoms, fishes, molluscs, arthropods, and mammals. Some rhizopodans have intracellular symbionts such as algae, bacteria, and viruses, though the nature of these relationships is not well understood. Rhizopodans are often used in laboratories as experimental organisms for studies of cell locomotion (*Amoeba proteus*), nonmuscle contractile systems

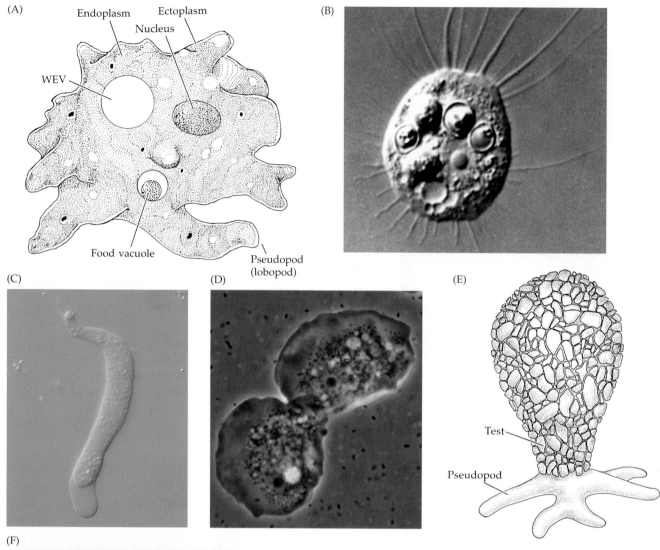

(A)
Endoplasm
Ectoplasm
Nucleus
WEV
Food vacuole
Pseudopod (lobopod)

(E)
Test
Pseudopod

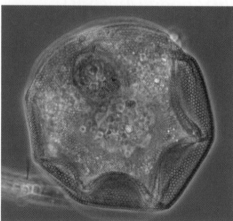

Figure 5.32 Phylum Rhizopoda. Rhizopodan diversity. (A–D) Naked amebas; (E,F) Testate amebas. (A) Anatomy of an ameba; note the multiple lobopods; (B) *Nuclearia*, with filose pseudopods. (C) *Hartmanella*, with single finger-like lobopod. (D) *Vannella*, with fan-shaped pseudopod. (E) *Difflugia* (with a test of microscopic mineral grains). (F) *Arcella*; the granular texture of the manufactured test can be seen surrounding the cytoplasm of the cell.

(*Acanthamoeba*), and the effects of removing and transplanting nuclei.

Although most rhizopodans are harmless, free-living creatures, some are endosymbiotic, and many of these are considered to be parasites. They occur most commonly in arthropods, annelids, and vertebrates (including humans). Three cosmopolitan species are commensals in the large intestines of humans: *Endolimax nana*, *Entamoeba coli*, and *Iodamoeba buetschlii*. All three feed on other microorganisms in the gut. *E. coli* infection levels reach 100 percent in some areas of the world. *E. coli* often coexists with *E. histolytica*; transmission (via cysts) is by the same methods, and the trophozoites of the two species are difficult to differentiate. *Iodamoeba buetschlii* infects humans, other primates, and pigs. *Entamoeba gingivalis* was the first ameba of humans to be described. Like *E. coli*, it is a harmless commensal, residing only on the teeth and gums, in gingival pockets near the base of the teeth, and occasionally in the crypts of the tonsils. It also occurs in dogs, cats, and other primates. No cyst is

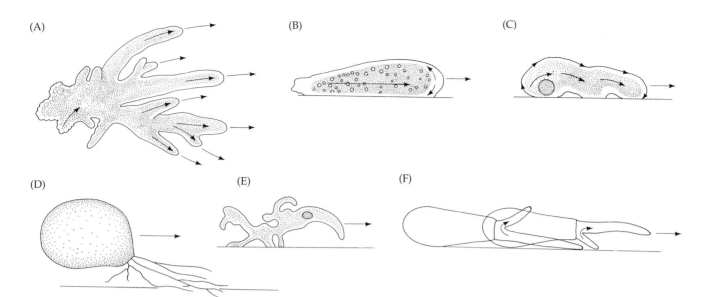

Figure 5.33 Phylum Rhizopoda. Locomotion in amebas. (A) Typical ameboid movement by lobopodia in *Amoeba proteus*. (B) Creeping, "limax" form of movement. (C) Rolling, treadlike movement. (D) Filopodial creeping in *Chlamydophorus*. (E) "Walking" locomotion in certain naked amebas. (F) "Bipedal-stepping" in *Difflugia*.

formed and transmission is direct, from one person to another. An estimated 50 percent of the human population with healthy mouths harbors this ameba.

Under severe stress, symbiotic gut rhizopodans (e.g., *Entamoeba coli*) that are normally harmless can increase to abnormally high numbers and cause temporary mild gastrointestinal distress in people. *Entamoeba histolytica* is a serious pathogen in humans (Box 5K). This species causes amebic dysentery, an intestinal disorder resulting in destruction of cells lining the gut. The parasite is usually ingested in its cyst stage, and is acquired by way of fecal contamination. Emergence of individuals in the active (motile) stage (trophozoites) takes place quickly once in the host's gut, and it is these individuals that release the histolytic enzymes that break down the epithelium of the large intestine and rectum. *Naegleria fowleri* (= *N. aerobia*) is the major agent of a disease called primary amebic meningoencephalitis (PAM), or simply "amebic meningitis." PAM is an acute, fulminant, rapidly fatal illness usually affecting young people who have been exposed to water harboring the free-living trophozoites, most commonly in lakes and swimming pools (but this ameba has even been isolated from bottled mineral water in Mexico). It is thought that the amebas are forced into the nasal passages when the victim dives into the water. Once in the nasal passages, they migrate along the olfactory nerves, through the cribiform plate, and into the cranium. Death from brain destruction is rapid. They do not form cysts in the host.

Support and Locomotion

Rhizopodans may be surrounded only by a plasma membrane. These are the so-called **naked amebas** (Figure 5.32A–D). Others, known as **testate amebas**, have the plasma membrane covered by some sort of test (Figure 5.32E,F). The tests of rhizopodans may be composed of particulate material either gathered from the environment (e.g., *Difflugia*) or secreted by the cell itself (e.g., *Arcella*). Some of the naked amebas (e.g., the genus *Amoeba*) may secrete a mucopolysaccharide layer, called the **glycocalyx**, on the outside of the plasma membrane. Sometimes, there may be flexible, sticky structures protruding from the glycocalyx, which are

---

BOX 5J **Characteristics of the Phylum Rhizopoda**

1. Cell surrounded by plasma membrane, which may be coated with sticky layer of glycoprotein; some also form external test

2. With temporary extensions of cytoplasm (pseudopodia) for locomotion; pseudopodia can be blunt (lobopodia) or slender (filopodia)

3. Mitochondria with tubular cristae

4. Most with single vesicular nucleus

5. Mitotic patterns extremely variable

6. Asexual reproduction by binary fission and multiple fission

7. Sexual reproduction reported but not confirmed

8. Without plastids

9. Some store glycogen (e.g., Pelomyxa), but most do not appear to store carbohydrates

BOX 5K Amebic Dysentery

Entamoeba histolytica is the etiological agent of amebic dysentery, a disease that has plagued humans throughout all of recorded history. *E. histolytica* is the third most common cause of parasitic death in the world. About 500 million people in the world are infected at any one time, with up to 100,000 deaths annually. Interestingly, *E. histolytica* comes in two sizes. The smaller race (trophozoites 12 to 15 μm in diameter, cysts 5 to 9 μm wide) is nonpathogenic, and some workers consider it a separate species (*E. hartmanni*). The larger form (trophozoites 20 to 30 μm in diameter, cysts 10 to 20 μm wide) is sometimes pathogenic and other times not. Another species of *Entamoeba, E. moshkovskii*, is identical in morphology to *E. histolytica*, but is not a symbiont; it lives in sewage and is often mistaken for *E. histolytica*.

When swallowed, the cysts of *E. histolytica* pass through the stomach unharmed. When they reach the alkaline medium of the small intestine, they break open and release trophozoites that are swept to the large intestine. They can survive both anaerobically and in the presence of oxygen. Trophozoites of *E. histolytica* may live and multiply indefinitely within the crypts of the mucosa of the large intestine, apparently feeding on starches and mucous secretions. In order to absorb food in this setting, they may require the presence of certain naturally-occurring gut bacteria. However, they can also invade tissues by hydrolyzing mucosal cells of the large intestine, and in this mode they need no help from their bacterial partners to feed. *E. histolytica* produces several hydrolytic enzymes, including phosphatases, glycosidases, proteinases, and an RNAse. They erode ulcers into the intestinal wall, eventually entering the bloodstream to infect other organs such as the liver, lungs, or skin. Cysts form only in the large intestine and pass with the host's feces. Cysts can remain viable and infective for many days, or even weeks, but are killed by desiccation and temperatures below 5°C and above 40°C. The cysts are resistant to levels of chlorine normally used for water purification.

Symptoms of amebiasis vary greatly, due to the strain of *E. histolytica* and the host's resistance and physical condition. Commonly, the disease develops slowly, with intermittent diarrhea, cramps, vomiting, and general malaise. Some infections may mimic appendicitis. Broad abdominal pain, fulminating diarrhea, dehydration, and loss of blood are typical of bad cases. Acute infections can result in death from peritonitis, the result of gut perforation, or from cardiac failure and exhaustion. Hepatic amebiasis results when trophozoites enter the mesenteric veins and travel to the liver through the hepato-portal system; they digest their way through the portal capillaries and form abscesses in the liver. Pulmonary amebiasis usually develops when liver abscesses rupture through the diaphragm. Other sites occasionally infected are the brain, skin, and penis (possibly acquired through sexual contact).

Although amebiasis is most common in tropical regions, where up to 40 percent of the population may be infected, the parasite is firmly established from Alaska to Patagonia. Transmission is via fecal contamination, and the best prevention is a sanitary lifestyle. Filth flies, particularly the common housefly (*Musca domestica*), and cockroaches are important mechanical vectors of cysts, and houseflies' habit of vomiting and defecating while feeding is a key means of transmission. Human carriers (cyst passers) handling food are also major sources of transmission. The use of human feces as fertilizer in Asia, Europe, and South America contributes heavily to transmission in those regions. Although humans are the primary reservoir of *E. histolytica*, dogs, pigs, and monkeys have also been implicated.

---

thought to aid in the capture and ingestion of bacteria during feeding.

Rhizopodans use pseudopodia for locomotion. Pseudopodia can vary in shape, especially in the smaller rhizopodans, and pseudopod form is an important taxonomic feature. There are two primary types of pseudopodia—lobopodia and filopodia (sometimes called "rhizopodia"). Lobopodia are blunt and rounded at the tip (Figure 5.33A,E). Filopodia are thin and tapering (Figure 5.33D). Lobopodia are the most common type of pseudopodia found within the Rhizopoda, perhaps best known in the genus *Amoeba*, while filopodia are found in relatively few taxa (e.g., *Nuclearia* and *Pompholyxophrys*).

Some rhizopodans that form lobopodia produce several pseudopods that extend in different directions at the same time. Probably the most familiar organism that produces multiple lobopodia is *Amoeba proteus* (Figure 5.33A). Similar pseudopodia are formed by some testate rhizopodans such as *Arcella, Centropyxis,* and *Difflugia*.

Some rhizopodans that produce multiple lobopodia also produce subpseudopodia on the surfaces of their lobopodia. This situation is found in the genus *Mayorella*, which forms finger-like subpseudopodia, and *Acanthamoeba*, which forms thin subpseudopodia called **acanthopodia**.

Some rhizopodans produce only a single lobopod. One such group is the so-called **limax rhizopodans**. These species form a large, single, finger-like "anterior" lobopod (giving the organism a sluglike, or *Limax*, appearance) (Figure 5.33B). Limicine locomotion is commonly found in rhizopods that dwell in soil (e.g., *Chaos, Euhyperamoeba, Hartmanella, Pelomyxa*). Other rhizopodans that produce a single lobopod include the genera *Thecamoeba* and *Vannella*. In *Vannella*, the lobopod is shaped such that it gives the body a fan-like appearance (Figure 5.32D), while in *Thecamoeba*, the lobopod has a somewhat indefinite shape and creates the impression that the cell rolls like the tread of a tractor or tank, the leading surface adhering temporarily to the substratum as the organism progresses.

As described in Chapter 3, the physical processes involved in pseudopodial movement are not fully understood. It is likely that more than one method of pseudopod formation occurs among the different rhizopodans; certainly the gross mechanics involved in the actual use of pseudopodia vary greatly (Figure 5.33). Recall the typical differentiation of the cytoplasm into ectoplasm (plasmagel) and endoplasm (plasmasol), the latter being much more fluid than the former. The formation of broad lobopodia results from the streaming of the inner plasmasol into areas where the constraints of the plasmagel have been temporarily relieved. In contrast, the formation of filopodia and reticulopodia typically does not involve sol–gel interactions, but simply rapid streaming of the cytoplasm to form single or branching pseudopodial extensions.

While many rhizopods move by "flowing" into their pseudopodia or by "creeping" with numerous filopodia, some engage in more bizarre methods of getting from one place to another. Some hold their bodies off the substratum by extending pseudopodia downward; leading pseudopodia are then produced and extended sequentially, pulling the organism along in a sort of "multilegged" walking fashion. Some of the shelled rhizopods (e.g., *Difflugia*) that possess a single pylome extend two pseudopodia through the aperture (Figure 5.32E and 5.33F). By alternately extending and retracting these pseudopodia, the organism "steps" foreward. During locomotion, one pseudopodium is extended and used to "pull" the organism along, trailing the other pseudopodium behind the cell.

Nutrition

While there is little doubt that rhizopodans take up dissolved organics directly across the cell membrane, the most common mechanisms of ingestion are pinocytosis and phagocytosis (Figure 5.34). The size of the food vacuoles varies greatly, depending primarily on the size of the food material ingested. Generally, ingestion can occur anywhere on the surface of the body, there being no distinct cytostome. Most rhizopodans are carnivores and are frequently predaceous. Some, such as *Pelomyxa*, inhabit soils or muds and are predominantly herbivorous, but they are known to ingest nearly any sort of organic matter in their environment. As explained earlier, a food vacuole forms from an invagination in the cell surface—sometimes called a food cup—that pinches off and drops inward. This process, sometimes called **endocytosis**, occurs in response to some stimulus at the interface between the cell membrane and the environment. Vacuole formation in rhizopodans may be induced by either mechanical or chemical stimuli; even nonfood items may be incorporated into food vacuoles, but they are soon egested.

Not only the size of a food item, but also the amount of water taken in during feeding determine the size of the food vacuole. Frequently the pseudopodia that form the food cup do not actually contact the food item; thus, a packet of the environmental medium is taken in with the food. In other cases, the walls forming the vacuole press closely against the food material; thus, little water is included in the vacuole. Food vacuoles move about the cytoplasm and sometimes coalesce. If live prey have

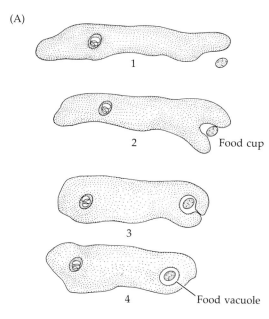

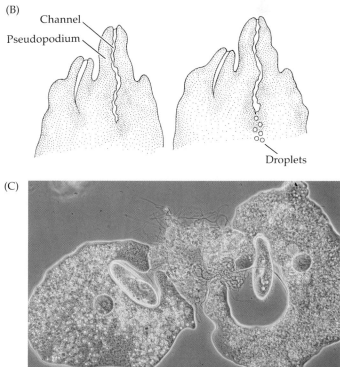

Figure 5.34 Phylum Rhizopoda. Feeding in amebas. (A) Sequence of events during which a lobopodium engulfs a food particle. (B) Uptake of dissolved nutrients through a pinocytotic channel in *Amoeba*. (C) Two soil amebas, *Vahlkampfia*, ingesting ciliates by phagocytosis.

(A) 1 2 Food cup 3 4 Food vacuole

(B) Channel Pseudopodium Droplets

been ingested, they generally die within a few minutes from the paralytic and proteolytic enzymes present. Undigested material that remains within the vacuole eventually is expelled from the cell when the vacuole wall reincorporates into the cell membrane. In most rhizopodans this process of cell defecation may occur anywhere on the body, but in some active forms it tends to take place at or near the trailing end of the moving cell.

Feeding in rhizopodans with skeletal elements varies with the form of the test and the type of pseudopodia. Those with a relatively large single aperture or opening, such as *Arcella* and *Difflugia,* feed much as described above. By extending lobopodia through the aperture, they engulf food in typical vacuoles.

Reproduction

Simple binary fission is the most common form of asexual reproduction, differing only in minor details among the different groups (Figure 5.35). In the naked rhizopodans, nuclear division occurs first and then cytoplasmic division follows. During cytoplasmic division, the two potential daughter cells form locomotor pseudopodia and pull away from each other. In those species with an external test, the shell itself may divide more or less equally in conjunction with the formation of daughter cells (e.g., *Pamphagus*); or, as occurs more frequently, the shell may be retained by one daughter cell, the other producing a new shell (e.g., *Arcella*). The relative density and rigidity of the test determine which process occurs. Multiple fission is also known among rhizopodans. Certain endosymbiotic naked species, including *Entamoeba histolytica*, produce cysts in which multiple fission takes place.

Cyst formation during unfavorable environmental conditions is well developed in some rhizopodans, including all testate amebas, most soil amebas, and the parasitic amebas. In the parasitic amebas (e.g., *Entamoeba*), cysts protect the organism as it passes through the digestive tract of the host.

Mitotic patterns in rhizopodans vary and have been used as a criterion for classification within the Rhizopoda. In most species, mitosis is characterized as open orthomitosis without centrioles; in some, the break-

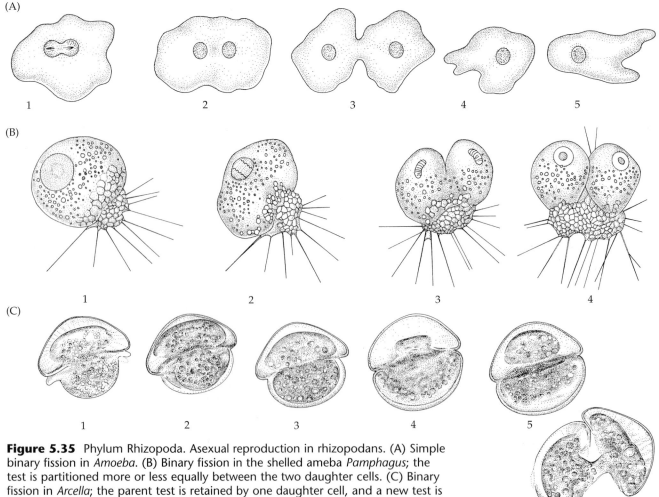

Figure 5.35 Phylum Rhizopoda. Asexual reproduction in rhizopodans. (A) Simple binary fission in *Amoeba*. (B) Binary fission in the shelled ameba *Pamphagus*; the test is partitioned more or less equally between the two daughter cells. (C) Binary fission in *Arcella*; the parent test is retained by one daughter cell, and a new test is produced by the other daughter cell.

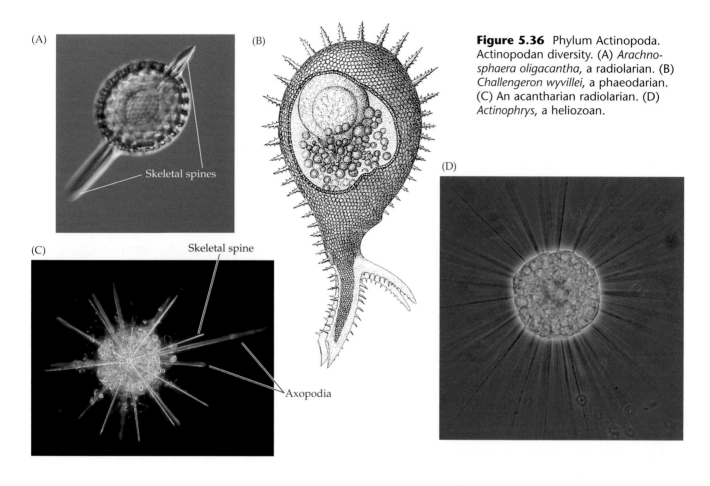

Figure 5.36 Phylum Actinopoda. Actinopodan diversity. (A) *Arachnosphaera oligacantha*, a radiolarian. (B) *Challengeron wyvillei*, a phaeodarian. (C) An acantharian radiolarian. (D) *Actinophrys*, a heliozoan.

down of the nucleus and nucleolus is delayed. Closed intranuclear orthomitosis with a persistent nucleolus occurs in some rhizopodans, such as *Entamoeba*.

Phylum Actinopoda

Approximately 4,240 species of Actinopoda have been described. The phylum includes four major groups: the Polycystina (= Radiolaria), Phaeodaria, Heliozoa, and Acantharia (Figure 5.36) (Box 5L). Most have internal siliceous skeletons that preserve well in the fossil record. Polycystines, phaeodarians, and acantharians are all planktonic; they are found exclusively in marine habitats and are most abundant in warm waters (26° to 37°C). Heliozoa predominantly occur in fresh water and are often found attached to the benthos or to submerged objects via a proteinaceous peduncle or a cytoplasmic base.

The name Actinopoda means "ray feet" and refers to the **axopodia**, which radiate from the bodies of these beautiful protists. Axopodia are slender pseudopodia supported by an inner core of microtubules that extends from a central region called the axoplast (Figure 5.37). The pattern of the microtubule arrangement within the **axopodia** varies and is an important taxonomic feature. Axopodia function primarily in feeding and locomotion. The cytoplasm exhibits a characteristic bidirection-

BOX 5L **Characteristics of the Phylum Actinopoda**

1. Cell surrounded by plasma membrane, which may be supported by skeleton secreted by the cell and usually internal; variable composition of skeleton

2. Locomotion mostly passive; some movement can be accomplished with special organelles called axopodia

3. Mitochondria with tubular cristae (in most)

4. Most have single vesicular nucleus; some have single ovular nucleus; some have multiple nuclei

5. Except heliozoans, nuclear division occurs by closed intranuclear pleuromitosis. Electron-dense plaques act as organizers for mitotic spindle. Pair of centrioles are located outside nucleus and situated near plaques. Amorphous structures called polar caps located in cytoplasm act as mitotic spindle organizers. In heliozoans, nuclear division occurs by semi-open orthomitosis

6. Asexual reproduction by binary fission, multiple fission, or budding

7. Sexual reproduction only known in some heliozoans and occurs by autogamy. Meiosis involves two divisions prior to formation of gametes

8. Without plastids

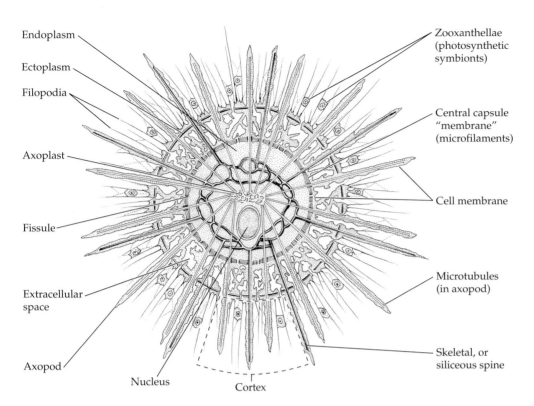

Endoplasm
Ectoplasm
Filopodia
Axoplast
Fissule
Extracellular space
Axopod
Nucleus
Cortex

Zooxanthellae (photosynthetic symbionts)
Central capsule "membrane" (microfilaments)
Cell membrane
Microtubules (in axopod)
Skeletal, or siliceous spine

Figure 5.37 Phylum Actinopoda. General anatomy of a pelagic (polycystine) actinopodan, showing the axopodia (radiating spines that aid in flotation) and other structures.

al movement (like the Granuloreticulosa), circulating substances in the cytoplasm between the pseudopodia and the main body of the cell (Figure 5.38).

Actinopoda, with an exception of some heliozoans, have little use in laboratory experiments since they cannot be maintained in culture for more than a few weeks. Heliozoans can be maintained in culture and have been used in cell biology studies to examine locomotion, feeding, and certain biochemical features. Probably the most useful aspect of actinopodans for humans is related to the nature of their skeletons. For example, the strontium sulfate skeletons of acantharians have been used by scientists to measure amounts of natural or anthropogenic radioactivity in marine environments. The siliceous skeletons of polycystines and phaeodarians do not dissolve under great pressure and therefore accumulate—along with diatom tests—as deposits called siliceous ooze on the floors of deep ocean basins (between 3,500 and 10,000 m deep). These skeletons date back to the Cambrian and have been used as paleoenvironmental indicators.

Support and Locomotion

In *Acantharia*, *Phaeodaria*, and the Polycystina, the cytoplasm is divided into two regions, the endoplasm and the ectoplasm, which are separated by a wall that is composed (usually) of mucoprotein. The central endo-

plasm is granular and dense, and contains most of the organelles: nucleus, mitochondria, Golgi apparatus, pigmented granules, digestive vacuoles, crystals, and the axoplast. Axopodia emerge from the axoplast in the endoplasm through pores in the capsule wall. The pore pattern is variable. In polycystines, for example, there are many pores in the capsule wall, all of which are associated with collar-like structures called fusules. In phaeodarians, there are only three pores in the capsule wall. The largest pore, the **astropyle**, is associated with fusules. Axopodia emerge from the two smaller pores, the **parapyles**, which are not associated with fusules.

Most polycystines, phaeodarians, and acantharians have skeletons for support. In these organisms, the skeleton is formed and housed within the endoplasm and is therefore internal. In Polycystina and Phaeodaria, the skeleton is composed primarily of siliceous elements that are solid in polycystines and hollow in phaeodarians. In acantharians, the skeleton is composed of strontium sulfate embedded in a proteinaceous matrix. These skeletons vary greatly in construction and ornamentation, and frequently bear radiating spines that aid in flotation. In the Acantharia, there is a strict arrangement of 20 radial spicules, which is a diagnostic feature of this group.

The ectoplasm, often called the **calymma**, lies outside the capsule wall and contains mitochondria, large digestive vacuoles, extrusomes, and (in some) algal symbionts. The calymma has a rather foamy appearance due to the presence of a large number of vacuoles (Figure 5.38C). The vacuoles, some of which house oil droplets and other low-density fluids, aid in flotation in free-living species. When surface water conditions be-

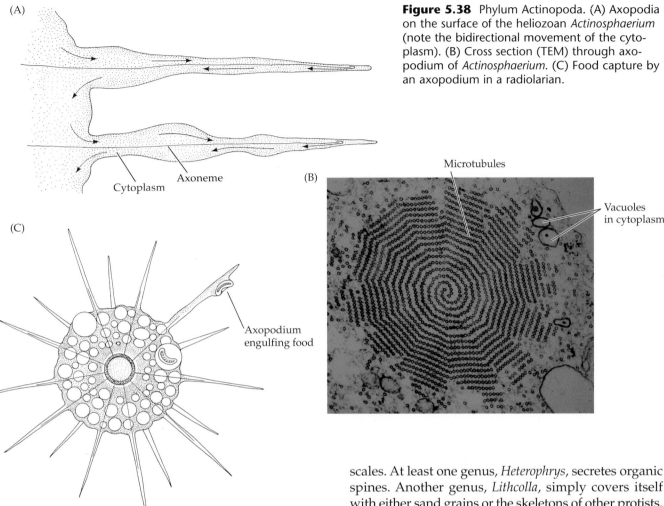

Figure 5.38 Phylum Actinopoda. (A) Axopodia on the surface of the heliozoan *Actinosphaerium* (note the bidirectional movement of the cytoplasm). (B) Cross section (TEM) through axopodium of *Actinosphaerium*. (C) Food capture by an axopodium in a radiolarian.

come rough and potentially dangerous to these delicate protists, the calymma expels some of its contents and the creature sinks to calmer depths. Eventually the cell replaces the oils and other fluids, and the organism rises toward the surface again. Unique to the ectoplasm of phaeodarians are balls of waste products called **phaeodium**, after which this group was named. The ectoplasm of acantharians is covered with a net-like cortex that is anchored to the apex of the spicules by contractile myonemes. Both the cortex and the presence of myonemes in association with the skeletal spicules are distinctive features of the acantharians.

Although heliozoans lack a capsular wall dividing the cytoplasm into a distinct endoplasm and ectoplasm, there is usually a division between a granular central region and a vacuolated region near the surface. These beautiful protists bear numerous axopodia radiating from their spherical bodies, inspiring the common name "sun animals." The surface of heliozoans is covered by a cell coat that is between 0.05 and 0.5 μm thick. None has an internal skeleton, and most lack a skeleton altogether, but some (e.g., *Acanthocystis*) have a skeleton embedded in the cell coat, consisting of secreted siliceous spines or scales. At least one genus, *Heterophrys*, secretes organic spines. Another genus, *Lithcolla*, simply covers itself with either sand grains or the skeletons of other protists.

Locomotion in the Actinopoda is limited. Most drift passively in the water column using the axopodia, skeletal spines (if present), and ectoplasmic vacuoles as flotation devices. In some cases, however, the axopodia and spines play a more active role in locomotion. For example, the axopodia may also help these organisms maintain their position in the water column by the expansion and contraction of vacuoles between the axopodia. This has been suggested because it has been observed in polycystines that when the ectoplasm and axopodia are lost during cell division, the organisms sink. In the Heliozoa, it has been observed that the axopodia are used to roll among algae and in at least one genus, *Sticholonche*, they appear to be used as tiny oars. In acantharians, it is thought that contraction of the myonemes that are attached to the spicules may somehow regulate buoyancy.

Nutrition

All actinopodans are heterotrophic, obtaining food by phagocytosis, and many are voracious predators (Figure 5.38C). Prey items include bacteria, other protists (e.g., ciliates, diatoms, flagellates), and even small invertebrates (e.g., copepods). Actinopoda use their axopodia as traps for prey. The axopodia are usually

(A)

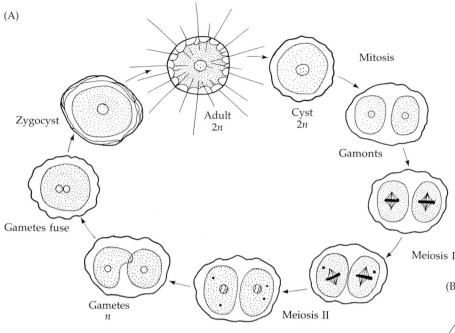

Zygocyst

Adult
2n

Cyst
2n

Mitosis

Gamonts

Gametes fuse

Gametes
n

Meiosis II

Meiosis I

Figure 5.39 Phylum Actinopoda. Reproduction in actinopodans. (A) Autogamy in the heliozoan *Actinophrys*. The adult enters a cyst stage and undergoes mitosis to produce a pair of gamonts. The nucleus of each gamont undergoes meiosis, but only one haploid nucleus survives in each cell. The gametes (and their haploid nuclei) fuse to produce an encysted zygote, which eventually grows into a new individual. (B) Mass of swarmers produced by multiple fission within the central capsule of the radiolarian *Thalassophysa*.

(B)

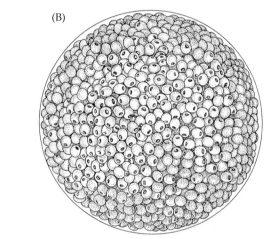

equipped with extrusomes such as mucus-producing **mucocysts,** and **kinetocysts** (found only in this group), which eject barbed, thread-like structures. Prey items adhere to mucus (discharged by mucocysts) that covers the extended axopodia, or they become attached to the axopodia by the discharged kinetocysts.

The size and motility of the prey determines the particular feeding mechanism used. Small prey are engulfed in food vacuoles directly, whereas large prey may be partially digested extracellularly by the action of secretory lysosomes in the mucous coat or broken into pieces by the action of large pseudopodia. The extracellular food is drawn toward the cell body by cytoplasmic streaming, eventually enclosed within food vacuoles, and completely digested in the central portion of the cell. In polycystines and heliozoans, it has been observed that when large, fast-moving prey (especially those with skeletons, such as diatoms) contact the axopodia, the axopodia actually collapse, drawing the prey into the cell body where it is engulfed by thin filopodia and then enclosed in a food vacuole (Figure 5.38C). The collapse of the axopodia is thought to involve microtubule disassembly.

An interesting feeding arrangement is found in the phaeodarians. As mentioned earlier, they have only three openings in the capsule wall: two parapylae and the single astropyle. Prey become trapped on the axopodia. Then a large pseudopod, formed from the astropyle, engulfs the prey item into a food vacuole where it is digested in the ectoplasm. Because of this behavior, some workers have referred to the astropyle as a cytostome.

Many polycystines, heliozoans, and acantharians live near the water's surface. These protists often have algal symbionts, including chlorophytes and dinoflagellates, which presumably provide them with additional nutri-

ents. Phaeodarians do not have algal symbionts, which is not surprising since they tend to be found in water depths unsuitable for photosynthesis.

Reproduction

Asexual reproduction occurs by binary fission, multiple fission, or budding. In the heliozoans, binary fission occurs along any plane through the body; in the Polycystina and various shelled forms, however, division occurs along planes predetermined by body symmetry and skeletal arrangement. The same basic mode of multiple fission is seen in all groups. A polyploid nucleus results from numerous mitotic divisions. The nucleus fragments, producing many biflagellate individuals called **swarmers,** which eventually lose their flagella and develop into adults (Figure 5.39B). In polycystines, swarmers have a crystal of strontium sulfate in their cytoplasm. In most polycystine and acantharian species, multiple fission is the only mode of asexual reproduction.

Sexual reproduction is apparently rare in actinopodans. A few genera of heliozoans (e.g., *Actinophrys* and *Actinosphaerium*) undergo autogamy, or self-fertilization (Figure 5.39A). Autogamy is usually triggered by either

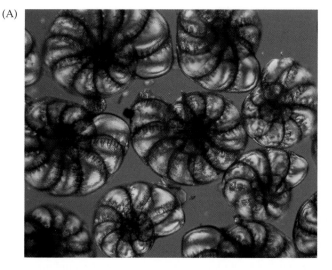

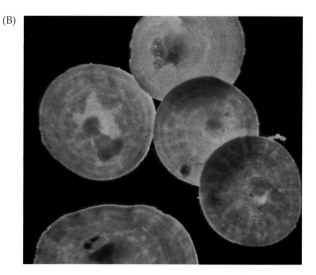

Figure 5.40 Phylum Granuloreticulosa. Foraminiferan skeletons. (A) *Elphideum* sp. (B) Soritid foram.

a lack of food or, conversely, it follows heavy feeding. First the cell encysts and undergoes a mitotic division to produce two gamonts. Each gamont nucleus divides by meiosis without cytokinesis. All of the nuclei but two disintegrate. The two surviving haploid nuclei fuse while still inside the cyst, forming a diploid zygote that later emerges from the cyst when environmental conditions become more favorable.

Nuclear division, in all actinopodans except the Heliozoa, occurs by closed intranuclear pleuromitosis. Electron-dense plaques located on the inner surface of the nuclear envelope act as organizers for the mitotic spindle. A pair of centrioles, located outside of the nucleus are found near the plaques. In the Heliozoa, nuclear division occurs by semi-open orthomitosis. Amorphous structures called polar plaques, which are located in the cytoplasm, act as mitotic spindle organizers.

Phylum Granuloreticulosa (Foraminifera and Their Kin)

The phylum Granuloreticulosa contains about 40,000 described species, many of which are fossils. Members of

this phylum are ubiquitous in all aquatic habitats from the poles to the equator. Some are planktonic, living near the water's surface, but most are benthic. The phylum consists of two major groups: the Athalamida and Foraminiferida (including the monothalamids). The Foraminiferida (e.g., *Globigerina*), also called Foraminifera or forams, are the most common and well-known members of the Granuloreticulosa (Figures 5.1J and 5.40) (Box 5M). They are most frequently found in marine and brackish water and are characterized by the presence of a test or skeleton with one to two chambers and a complex life cycle that involves an alternation of generations. In addition, the pseudopodia emerge from the test at one or two fixed openings. Most forams are benthic and have flattened tests. A small number, however, are pelagic and have calcareous spines that aid in prey capture (e.g., *Globigerinella*, Figure 5.1J). Athalamids are found in fresh water, soil, and marine environments and are distinguished from the forams in that they lack a test and the pseudopodia can emerge any place on the body.

The tests of forams leave an excellent fossil record dating back to the Lower Cambrian. The tests of plank-

---

BOX 5M *Characteristics of the Phylum Granuloreticulosa*

1. Cell surrounded by plasma membrane, which may by supported by organic, agglutinated, or calcareous test; test always formed outside plasma membrane

2. Locomotion involves cytoplasmic extensions called reticulopodia

3. Mitochondria with tubular cristae

4. Nuclei either ovular or vesicular; many are multinucleate; some exhibit nuclear dualism

5. Nuclear division occurs by closed intranuclear pleuromitosis

6. Asexual reproduction by budding and/or multiple fission

7. Sexual reproduction known in most. The life cycle is usually complex, involving alternation of an asexual form (agamont) and a sexual form (gamont).

8. Without plastids

tonic forams are used by geologists as paleoecological and biostratigraphic indicators, and deposits of benthic foram tests are often used by petroleum geologists to search for oil. Foraminiferan tests are not only very abundant in recent and fossil deposits but also extremely durable. On the island of Bali, the tests of one species are mined and used as gravel in walks and roads. Much of the world's chalk, limestone, and marble is composed largely of foraminiferan tests or the residual calcareous material derived from the tests. Most of the stones used to build the great pyramids of Egypt are foraminiferan in origin. Before they get buried on the sea floor, the tests of foraminiferans function as homes and egg-laying sites for many minute metazoan species, such as small sipunculans, polychaetes, nematodes, copepods, isopods, and others.

Support and Locomotion

Although little is known about locomotion in the Granuloreticulosa, the reticulopodia are believed to be involved (Figure 5.41). Most Granuloreticulosa have tests covering their plasma membrane. The Althalamida lack a test and instead are covered by a thin, fibrous envelope. The Foraminifera have tests that are usually constructed as a series of chambers of increasing size, with a main opening, or **aperture**, in the largest chamber from which the reticulopodia emerge.

There are three types of tests in the Foraminifera—(1) **organic**, (2) **agglutinated**, and (3) **calcareous**. The nature of the test is a taxonomic feature used to classify forams. Organic tests are composed of complexes of proteins and mucopolysaccharides. These tests are flexible and allow the organisms that secrete them (e.g., *Allogramia*) to change shape rapidly. Cytoplasm emerging from the aperture(s) forms the reticulopodial network and often forms a layer covering the outside of the test.

Agglutinated tests are composed of materials gathered from the environment (e.g., sand grains, sponge spicules, diatoms, etc.) that are embedded in a layer of mucopolysaccharide secreted by the cell. The test may be made rigid by calcareous and iron salts. Some forams that have agglutinated tests are highly selective about the building materials used to build their tests (e.g., *Technitella*), while others are not (e.g., *Astrorhiza*).

Calcareous tests are composed of an organic layer reinforced with calcite ($CaCO_3$). The arrangement of calcite crystals give the tests a characteristic appearance, and three major categories of calcareous tests are recognized: (1) **porcelaneous**, (2) **hyaline**, and (3) **microgranular**. Porcelaneous tests appear shiny and white, like fired porcelain in reflected light, and are probably the most familiar to the introductory student (Figure 5.40). These tests generally lack perforations and the reticulopodia emerge from a single aperture. Hyaline tests have a glass-like appearance in reflected light and often are perforated with tiny holes. Microgranular tests have a sugary (granular) appearance in reflected light.

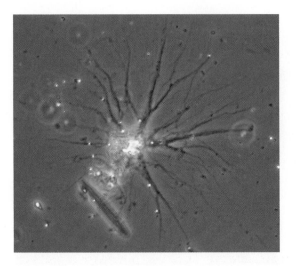

Figure 5.41 Phylum Granuloreticulosa. An unidentified athalamid with reticulopodia.

Planktonic foraminiferans can occur in such high numbers that the calcareous tests of dead individuals constitute a major portion of the sediments of ocean basins. In some parts of the world, these sediments—called foraminiferan ooze—are hundreds of meters thick. Such sediments are restricted to depths shallower than about 3,000 to 4,000 m, however, because $CaCO_3$ dissolves under high pressure.

Nutrition

All Granuloreticulosa are heterotrophic and feed by phagocytosis. The prey can vary, depending on the species. Some are herbivores, other are carnivores, and still others are omnivorous. Planktonic herbivores can be tuned to the bloom of certain algae, such as diatoms or chlorophytes, and feed heavily at those times. All use their reticulopodia to trap their prey. Vesicles at the tip of the reticulopodia secrete a sticky substance that the prey adhere to upon contact. The prey are engulfed in food vacuoles, into which digestive enzymes are secreted. The food vacuoles are then carried to the main part of the body, where digestion is completed. Benthic species trap prey by spreading their reticulopodia out on the lake or ocean bottom.

Both shallow-water benthic forams and planktonic forams that live near the water's surface often harbor endosymbiotic algae such as diatoms, dinoflagellates, and red and green algae, which can migrate out of the reticulopodia to expose themselves to more sunlight. These forams are particularly abundant in warm tropical seas. Studies suggest that nutrient and mineral recycling may occur between the forams and their algal symbionts. Furthermore, it has been shown that the symbionts may enhance the test-building capacities of forams and that their presence often allows their hosts to grow to very large sizes (e.g., the Eocene foram *Nummulites gizehensis* reached 12 cm in diameter), even in nutrient-poor waters. The "giant" forams, such as *Nummulites*, are much more common in fossil deposits than they are today.

Reproduction and Some Life Cycles

The life cycles of granuloreticulosans are frequently complex, and many are incompletely understood. These cycles often involve an alternation of sexual and asexual phases (Figure 5.42). However, some smaller species apparently only reproduce asexually, by budding and/or multiple fission. Nuclear division in both sexually and asexually reproducing species occurs by intranuclear pleuromitosis. In those that reproduce sexually, it is not uncommon to find individuals of the same foraminiferan species differing greatly in size and shape at different phases of the life cycle. The size difference is generally determined by the size of the initial shell chamber (the **proloculum**), produced following a particular life-cycle event. Often the proloculum that is formed following asexual processes is significantly larger than one formed after syngamy. Individuals with large prolocula are called the **macro-** or **megaspheric generation**; individuals with small prolocula are the microspheric generation.

During the sexual phase of the life cycle, the haploid individuals (gamonts) undergo repeated divisions to produce and release bi- or triflagellated isogametes, which pair and fuse to form the asexual individuals. Asexual, diploid individuals (called agamonts) undergo meiosis and produce haploid gamonts—the sexual individuals. The means of return to the diploid condition varies. In many foraminiferans (*Elphidium*, *Iridia*, *Tretomphalus*, and others), flagellated gametes are produced and released; fertilization occurs free in the sea water to produce a young agamont. In others, such as *Glabratella*, two or more gamonts come together and temporarily attach to one another. The gametes, which may be flagellate or ameboid, fuse within the chambers of the paired tests. The shells eventually separate, releasing the newly formed agamonts. True autogamy occurs in *Rotaliella*: each gamont produces gametes that pair and fuse within a single test, and the zygote is then released as an agamont.

Phylum Diplomonadida

The diplomonads were one of the first protist groups ever to be observed and recorded. Antony van Leeuwenhoek described a diplomonad protist, now known as *Giardia intestinalis*, from his own diarrheic stool as early as 1681 (see opening quote; Figure 5.43A, Box 5N) About 100 species of diplomonads are known today (Box 5O). This is a group of predominantly symbiotic flagellates, but a few free-living genera are known. Those that are free living tend to be found in organically polluted waters. Most diplomonads live as harmless commensals within the digestive tracts of animals, but a few are pathogenic.

The diplomonads have their name because the first species described from this group had a twofold symmetry defined by a pair of karyomastigont systems (Figure 5.43B). It was later discovered that some genera (the enteromonads) have only one. The **karyomastigont system** is composed of fibers that originate at the basal bodies of the flagella and form an intimate association with the nucleus. A pair of anteriorly located nuclei, along with their nucleoli, makes the organism look like it has eyes that peer up as it is being observed through the microscope (these are the eyes van Leeuwenhoek saw looking at him in 1681).

As noted above, some diplomonads are pathogenic. *Hexamida salmonis,* a parasite of fish, causes many deaths in salmon and trout hatcheries. Infestations of *H.*

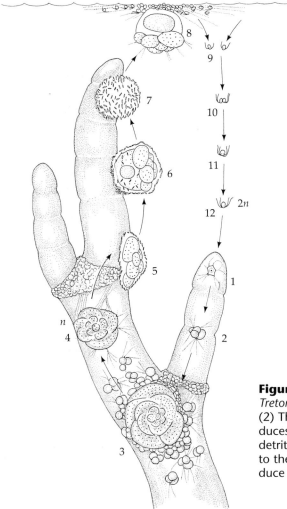

Figure 5.42 Phylum Granuloreticulosa. Life cycle of the foraminiferan *Tretomphalus bulloides*. (1) The settled zygote is shell-less and ameboid. (2) The cell grows and matures as an agamont, which (3) asexually produces young gamonts. Each mature gamont (4) accumulates particles of detritus (5–6) to produce a flotation chamber (7). (8) The gamont floats to the surface and produces and releases gametes (9), which fuse to produce a swimming zygote (10–12).

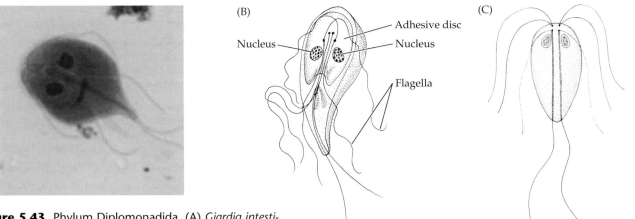

Figure 5.43 Phylum Diplomonadida. (A) *Giardia intestinalis* from human stool (length 12–15 μm). (B) Schematic drawings of *Giardia* and *Hexamita* (C), illustrating the paired nuclei and numerous flagella.

meleagridis in turkey farms annually costs the turkey industry millions of dollars in the United States (Figure 5.43C). *Giardia intestinalis* is a very common intestinal parasite in humans that causes diarrhea, dehydration, and intestinal pain. Although it is not fatal if treated promptly, giardiasis is one of the top ten most common parasite diseases.

Support and Locomotion

The cell is surrounded by a plasma membrane, but some rigidity is provided by three microtubular roots that are associated with the basal bodies. These roots include a **supranuclear fiber** that passes over or in front of the nucleus, an **infranuclear fiber** that extends beneath or behind the nuclei, and a band of microtubules that parallels the recurrent flagellum. The relative development of these fibers varies, depending on the genus.

BOX 5N *Giardia*

The genus *Giardia* is notable in lacking mitochondria, smooth endoplasmic reticulum, Golgi bodies, and lysosomes. For many years, this has been interpreted as a primitive trait that placed the genus near the point of divergence between prokaryotes and eukaryotes (hence the genus has been referred to as a "missing link"). However, recent phylogenetic analyses suggest that this genus is actually highly derived and that the absence of mitochondria represents a secondary loss. There are probably five valid species in the genus: *G. lamblia* (= *G. intestinalis*, = *G. duodenalis*) and *G. muris* from mammals, *G. ardeae* and *G. psittaci* from birds, and *G. agilis* from amphibians. The closely related genus *Hexamita* has no human parasites, but *H. meleagridis* is a common parasite of the guts of young galliform birds (e.g., turkey, quail, pheasant), and it causes millions of dollars in loss to the U.S. turkey industry annually.

Giardia lamblia is a cosmopolitan species that occurs most commonly in warm climates. However, in recent years it has been introduced by hikers and campers throughout the warm temperate zone of the United States. It is the most common flagellated protist of the human digestive tract. Over 30,000 cases of giardiasis are reported annually in the United States, where animal reservoirs of *G. lamblia* include beavers, dogs, cats, and sheep. Treatment with quinacrine or metronidazole ("Flagyl") usually effects complete cure within a few days.

The teardrop-shaped organism is dorsoventrally flattened, the ventral surface bearing a concave bilobed adhesive disc with which the cell adheres to the host tissue. Five flagella arise from kinetosomes located between the anterior portions of the two nuclei. The flagella facilitate rapid swimming. Members of this genus also possess a unique pair of large, curved, dark-staining median bodies lying posterior to the adhesive discs; their function is unknown. In severe infections, the free surface of nearly every cell in the infected portion of the gut is covered by a parasite. A single diarrheic stool can contain up to 14 billion parasites, facilitating the rapid spread of this very common protist. Some infections show no evidence of disease, whereas others cause severe gastritis and associated symptoms, no doubt due to differences in host susceptibility and strains of the parasite. The dense coating of these protists on the intestinal epithelium interferes with absorption of fats and other nutrients. Stools are fatty, but never contain blood. The parasite does not lyse host cells, but appears to feed on mucous secretions. Some protective immunity can apparently be acquired.

Lacking mitochondria, the tricarboxylic acid cycle and cytochrome system are absent in *Giardia*, but the organisms avidly consume oxygen when it is present. Glucose is apparently the primary substratum for respiration, and the parasites store glycogen. However, they also multiply when glucose is absent. Trophozoites divide by binary fission. As with trypanosomes, *G. lamblia* exhibits antigenic variation, with up to 180 different antigens being expressed over 6 to 12 generations.

Some genera have additional fibrous structures that are associated with the basal bodies. For example, the genus *Giardia* attaches to the host's intestinal epithelium with an adhesive disc that is constructed in part from the microtubular bands of the cytoskeleton. The disc is delimited by a ridge or lateral crest that is composed of actin and is used to bite into the host's tissue. The contractile proteins myosin, actinin, and tropomyosin have all been reported around the periphery of the disc and may be involved with attaching to the host. The supranuclear fiber in *Giardia* is composed of a single ribbon of microtubules that connects to the plasma membrane of the disc. Each microtubule of the supranuclear fiber is associated with a ribbon of protein that extends into the cytoplasm.

Each karyomastigont system typically has four kinetosomes—two anterior and two posterior. One of the posterior flagella is recurrent and trails passively behind. Locomotion is accomplished by the coordinated beat of the eight flagella. It has been suggested that the flagella may also be involved with creating a suction force beneath the adhesive disc in *Giardia*, enabling it to attach to its host.

Nutrition

Most diplomonads are phagotrophic and feed on bacteria. These forms have a cytostome through which endocytosis of the bacteria occurs. In *Spironucleus* and *Hexamida*, for example, the two intracellular channels in which the recurrent flagella lie function as cytostomes.

Other genera such as *Giardia* and *Octonitis* lack cytostomes and are saprozoic, feeding by pinocytosis on mucous secretions of the host's intestinal tissue.

Reproduction

Asexual reproduction is the only mode of reproduction known in diplomonads. Division occurs along the longitudinal plane. Nuclear division involves semi-open orthomitosis and is synchronous between the two nuclei (if there are two). Replicated basal bodies act as organizing centers for the mitotic spindle. Most symbiotic diplomonads form cysts at some point during their life cycle. Those that form cysts alternate between a motile trophozoite form and a dormant encysted form. *Giardia intestinalis*, for example, will form a thick protective covering that resists desiccation as it passes from the host's small intestine into the large intestine, where it is prone to dehydration. Once it leaves the digestive system through the anus, it must be swallowed by another host, where it will travel through the digestive system until it reaches the duodenum of the small intestine where it will excyst. Giardiasis is very contagious and prevention depends on maintaining high levels of hygiene. Cyst formation is not known to occur in free-living diplomonads.

Phylum Parabasilida (The Trichomonads and Hypermastigotes)

There are approximately 300 described species of parabasilids. This group includes two major subgroups: the trichomonads and the hypermastigotes (Figure 5.44) (Box 5P). All parabasilid groups that have been studied are endosymbionts of animals. The hypermastigotes (e.g., *Trichonympha*) are obligate mutualists in the digestive tracts of wood-eating insects such as termites and wood roaches. Trichomonads are symbionts in the digestive, reproductive, and respiratory tracts of vertebrates, including humans. The parabasilids get their name from a fiber, called the **parabasal fiber**, which extends from the basal bodies to the Golgi apparatus. Several other fibers are associated with the basal bodies (an atractophore, an axostyle, and a pelta) and their presence, along with the parabasal fiber, is a diagnostic feature of this group.

The most interesting aspect of parabasilid biology may be their symbiosis with other organisms. As noted below, hypermastigotes are symbionts within the digestive system of wood-eating insects such as wood roaches, cockroaches, and termites. Although these insects eat wood, they lack the enzymes necessary to break it down. The hypermastigote protists produce the enzyme cellulase, which breaks down cellulose in wood into a form that the insect can metabolize—a genuinely mutualistic relationship.

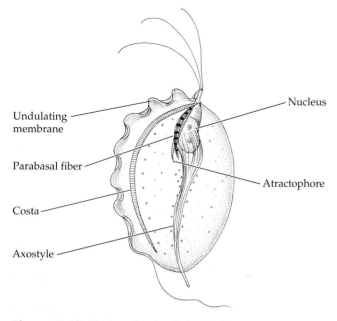

Undulating membrane

Nucleus

Parabasal fiber

Atractophore

Costa

Axostyle

Figure 5.44 Phylum Parabasilida. *Trichomonas murius*, a trichomonad inhabiting the large intestine of mice.

Trichomonads have been the focus of much research because there are four species found as symbionts in humans (*Dientamoeba fragilis*, *Pentatrichomonas homonis*, *Trichomonas tenax*, and *Trichomonas vaginalis*) and several species that parasitize domestic stock (*Tritrichomonas foetus*). In humans, only *Trichomonas vaginalis* is known to be a serious pathogen. *Trichomonas vaginalis* is a cosmopolitan species found in the vagina and urethra of women, and in the prostate, seminal vesicles, and urethra of men. It is transmitted primarily by sexual intercourse, although it has been found in newborn infants. Its occasional presence in very young children suggests that the infection can also be contracted from shared washcloths, towels, or clothing. Most strains of *T. vaginalis* are of such low pathogenicity that the victim is virtually asymptomatic. However, other strains cause intense inflammation with itching and a copious greenish-white discharge (leukorrhea) that is swarming with the parasite. *Trichomonas vaginalis* feeds on bacteria, leukocytes, and cell exudates.

O. F. Müller discovered *T. tenax* in 1773, when he examined a culture of tartar from his own teeth. *Dientamoeba fragilis* is a fairly common parasite of human intestinal tracts, where it lives in the large intestine and feeds mainly on debris. Although traditionally considered a harmless commensal, recent studies suggest that infections by this protist routinely result in abdominal stress (e.g., diarrhea, abdominal pain). *Tritrichomonas foetus* is a parasite in cattle and other large mammals and is one of the leading causes of abortion in these animals; it is common in the United States and Europe. *Histomonas meleagridis* is a cosmopolitan parasite of gallinaceous fowl. Histomoniasis in chickens and turkeys causes about a million dollars in losses annually.

Support and Locomotion

In parabasilids, the cell body is surrounded only by plasma membrane, but some rigidity is provided by a system of supporting fibers and microtubules that are associated with the kinetosomes. There are two striated fibrous roots, a **parabasal fiber** and an **atractophore**, and two microtubular roots, an **axostyle** and a **pelta** (Figure 5.44). The number of parabasal fibers is variable. In small trichomonads, there are only a few, whereas in hypermastigotes, such as *Trichonympha*, there can be over a dozen. The atractophore extends toward the nucleus from the basal bodies. The axostyle is a rod-like bundle of microtubules that originates near the basal bodies and curves around the nucleus as it extends to the posterior region of the cell. The pelta is a sheet of microtubules that encloses the flagellar bases. In trichomonads, an additional striated fiber called the **costa** is present. This fiber originates at the bases of the flagella and extends posteriorly beneath the undulating membrane. These fibers, along with the flagella and the nucleus, comprise the karyomastigont system (similar to that found in diplomonads).

Locomotion is accomplished by the beats of flagella. Trichomonads tend to have only a few flagella, typically four or five, located in the anterior region of the body. In *Trichomonas vaginalis*, for example, four free flagella form

---

BOX 5P   *Characteristics of the Phylum Parabasilida*

1. Body surrounded only by plasma membrane; some rigidity provided by four roots associated with the flagella

2. With flagella for locomotion. Number of flagella can vary between four to thousands; two fibrous roots (parabasal fiber, atractophore) and two microtubular roots (axostyle, pelta) associated with basal bodies of the flagella

3. Without mitochondria; anaerobic activity occurs in organelles called hydrogenosomes

4. Hypermastigotes possess either a single chromosomal or vesicular nucleus with a prominent nucleolus. Trichomonads possess a single vesicular nucleus with a minute nucleolus

5. Nuclear division occurs by closed extranuclear pleuromitosis without centrioles

6. Asexual cell division by longitudinal binary fission (symmetrogenic)

7. Sexual reproduction occurs in some hypermastigotes, but is unknown in trichomonads. In hypermastigotes, sexual reproduction varies, occurs by either gametogamy, gamontogamy, or autogamy

8. Without plastids

a tuft in the anterior region of the cell. The fifth flagellum is attached to the cell body at regular attachment sites so that when it beats, the cell membrane in that region of the body is pulled up into a fold, forming an undulating membrane. As in the kinetoplastids, the flagellum–undulating membrane complex seems to be efficient in moving the organism through viscous media. Hypermastigotes usually have dozens or even hundreds of flagella occurring all over the body. In these protists, the basal bodies of the flagella are arranged in parallel rows and are connected by microfibrils. The beat of the flagella is synchronized, forming metachronal waves.

Some trichomonads (e.g., *Dientamoeba fragilis*, *Histomonas meleagridis*, and *Trichomonas vaginalis*) form pseudopodia. These pseudopodia function primarily in phagocytosing food particles, but they are also used in locomotion.

Nutrition

All parabasilids are heterotrophic and lack a distinct cytostome. In some trichomonads (e.g., *Tritrichomonas*), fluid is taken up by pinocytosis in depressions on the cell surface. Most parabasilids, however, take in particulate matter by phagocytosis. In hypermastigotes, pseudopodia formed in a sensitive region at the posterior end of the cell engulf wood particles. Trichomonads also form pseudopodia that engulf bacteria, cellular debris, and leukocytes.

Reproduction

Asexual reproduction is by longitudinal binary fission (Figure 5.45A). Nuclear division occurs by closed extranuclear pleuromitosis with an external spindle. The atractophores are thought to act as microtubular organizing centers.

Sexual reproduction is unknown for the trichomonads, but has been observed in some hypermastigotes. In these organisms, sexual reproduction is well understood, thanks to the work of L. R. Cleveland in the 1950s. A variety of sexual processes are exhibited by hypermastigotes, including gametogamy, gamontogamy, and autogamy. In addition, meiosis can occur in one or two divisions, depending on the species. Hypermastigotes spend most of their lives as haploids in the digestive tract of wood-eating insects, dividing asexually by mitosis. Sexual reproduction is stimulated when the host insect molts and produces the molting hormone, ecdysone.

An example of a life cycle involving gametogamy is seen in *Trichonympha* (Figure 5.45B). In this group, the gametes are anisogametous, the male gamete being smaller than the female gamete. In some other species that undergo gametogamy, the gametes are isogametous. The haploid individual encysts and transforms into a gamont. While still encysted, the gamont divides by mitosis to produce a pair of flagellated gametes, one male and one female, which escape from the cyst. The posterior end of the female gamete is modified to form a fertilization cone through which the male gamete enters the cell. Once the male enters, its body is absorbed by the female gamete. Nuclear fusion produces a diploid zygote. Within a few hours, the zygote undergoes a two-division meiosis, resulting in four haploid cells. Because *Trichonympha* are obligate anaerobes in the guts of insects, encysting prior to host molting may allow the insects to maintain their protist symbionts.

Figure 5.45 Phylum Parabasilida. Reproduction in parabasilids. (A) Longitudinal binary (asexual) fission in the trichomonad *Devescovina*. (B) Sexual reproduction in *Trichonympha* (a hypermastigote) (C) Mating activity (fertilization) in *Eucomonympha* (a hypermastigote), in which individuals act as gametes.

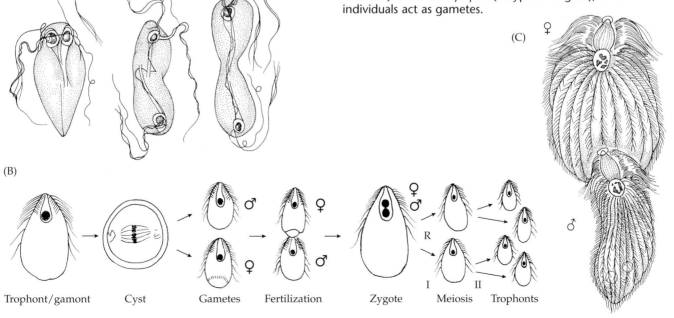

(A)

(B)

Trophont/gamont   Cyst   Gametes   Fertilization   Zygote   Meiosis   Trophonts

(C)

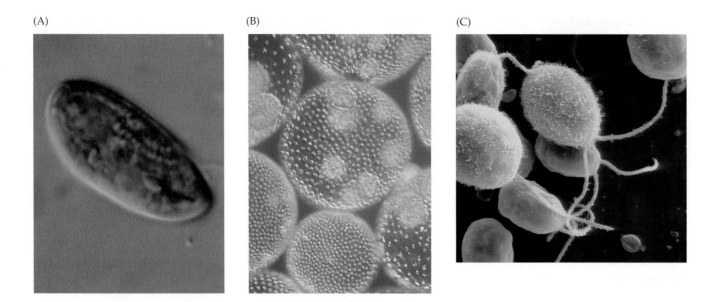

Figure 5.46 Phyla Cryptomonada and Chlorophyta. (A) *Cryptomonas*, a cryptomonad. (B) *Volvox*, a colonial chlorophyte. (C) *Chlamydomonas reinhardtii*, an autotrophic, unicellular species.

Phylum Cryptomonada

Cryptomonads occur in marine and freshwater habitats, and one genus, *Chilomonas*, is a commonly used research tool in biological laboratories. Cryptomonads are biflagellated cells with a large flagellar pocket, a semirigid cell surface supported by proteinaceous plates (called the periplast), and a single large mitochondrion with cristae that appear to be flattened tubes. They contain what is thought to be a reduced photosynthetic eukaryotic endosymbiont, expressed as a double membrane–bound region containing ribosomes, starch grains, a chloroplast, and a unique double membrane–bounded structure containing DNA called the **nucleomorph**, which is supposedly the remnant of the endosymbiont's nucleus (Figure 5.46).

Phylum Microspora

The 800 or so species of the phylum Microspora (Figure 5.47) are intracellular parasites occurring in nearly every phylum of animals; some are even found in other protists, including gregarines and ciliates. Much remains to be learned about the various stages in the life histories of microsporans; there is even disagreement about whether any sexual phases are present. The common visible form that occurs in host tissues is the spore, and it is on the basis of the details of spore structure that the phylum is defined and the species characterized. The microsporans occur as unicellular, uni- or binucleate spores within a typically multilayered cyst. The spore bears a polar cap and a single polar tube (the polar tube has also been referred to as the polar filament). When a spore is ingested by a potential host, the polar tube extends, carrying with it the spore cytoplasm, and attach-

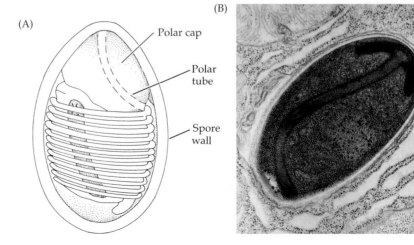

Figure 5.47 Phylum Microspora. (A) Spore of *Thelohania californica* with its coiled polar tube. (B) Spore of microsporan from the body of the ciliate, *Spathidiopsis* (TEM).

es to and penetrates cells of the host's gut wall. These organisms cause damage to host tissues, and species that infect commercially important insects (e.g., silkworms, honeybees) are of considerable economic importance. Recent molecular evidence suggests they are related to the fungi.

Phylum Ascetospora

The ascetosporans are another exclusively parasitic group of protists. We have little knowledge of their life histories and they, like microsporans, are characterized on the basis of spore structure. The spores of most species are multicellular or distinctly bicellular, although some species are unicellular. The spores lack polar filaments or tubes.

Phylum Choanoflagellata

The choanoflagellates are stalked, sessile cells existing singly or in colonies (Figure 5.48). They are distinctive in that they are seemingly identical to choanocytes, the flagellated feeding cells of sponges. Like choanocytes, they have a single flagellum that is encircled by a basket-like transparent collar. The collar acts as a food-catching net; feeding is accomplished when food particles are swept into the collar by the beating of the flagellum, pressed down against the cell surface, and engulfed by small pseudopodia. The choanoflagellates have long been viewed as a transitional link between the flagellated protists and the sponges, or more specifically, as the actual ancestors of the Metazoa, and DNA molecular sequence data support this hypothesis. On the other hand, some protozoologists have suggested that, because they are not obviously related to any other protist group, they might actually be highly reduced sponges! Common genera include *Codosiga*, *Monosiga*, and *Proterospongia*.

Phylum Chlorophyta

Arguing whether to classify the chlorophytes as protists or plants is a favorite pastime of biologists. Because some are unicellular and have the protist bauplan, they are often studied along with the protists. However, chlorophytes are clearly a paraphyletic group. The Chlorophyta are commonly called the "green algae" because of their grassy-green chloroplasts. These chloroplasts are very similar to those seen in the multicellular plants (the kingdom Metaphyta, or Plantae). They have chlorophylls a and b, and store food as starch (amylose/amylopectin) in their chloroplasts. The chloroplasts are bounded by two membranes, and the thylakoids are in many-layered stacks (Figure 5.2). Most unicellular

Figure 5.48 Phylum Choanoflagellata. The choanoflagellate *Salpingoeca*.

chlorophytes have a cell wall or scales, but not always made of cellulose. *Chlamydomonas*, *Eudorina*, and *Volvox* are common genera (Figure 5.46).

Some unicellular chlorophytes form colonies, the best known of which is *Volvox*. The individual cells of a *Volvox* colony are embedded in the gelatinous surface of a hollow sphere that may reach a diameter of 0.5 to 1 mm. Each cell has a nucleus, a pair of flagella, a single large chloroplast, and an eye spot. Adjacent cells are connected with each other by cytoplasmic strands. But only a few of the cells are responsible for reproduction. These colonial forms are often used as an example of the beginnings of a division of labor and an experiment toward true multicellularity.

Several lineages of chlorophytes have lost their photosynthetic ability (e.g., *Polytoma*, *Polytonella*, and several groups of *Chlamydomonas*). In these cases, chlorophyll may no longer be present (although the carotenoid pigments usually remain) and the organisms have switched to heterotrophy, often relying on decaying organic matter for their nutrition.*

Phylum Opalinida

Once classified as protociliates, then as zooflagellates, their placement here as a separate phylum is meant to draw attention to the enigmatic nature of the opalinids. Their numerous oblique rows of cilia differ from the rows in ciliates in that they lack the kinetidal system. During asexual reproduction, the fission plane parallels

---

*Photosynthetic plastids (with chlorophyll) are called chloroplasts, and contain their own DNA (cpDNA). Plastids without chlorophyll are usually called leucoplasts (and generally contain lpDNA). Both cpDNA and lpDNA are transcribed and translated just like nuclear DNA, except these organelle genes are effectively haploid.

Figure 5.49 Phylum Opalinida. *Opalina.*

these rows; thus it is longitudinal (as it is in flagellates) rather than transverse (as it is in ciliates). Some opalinids are binucleate, others multinucleate, but all are **homokaryotic** (i.e., the nuclei are all identical).

There are about 150 species of opalinids, in several genera, almost all being endosymbiotic in the hindgut of anurans (frogs and toads), where they ingest food anywhere on their body surface. Sexual reproduction is by syngamy and asexual reproduction is by binary fission and plasmotomy, the latter involving cytoplasmic divisions that produce multinucleate offspring. *Opalina* and *Protopalina* are two genera most commonly encountered (Figure 5.49). Opalinids are often found in routine dissections of frogs in the classroom; their large size and graceful movements through the frog's rectum make them a pleasant discovery for students.

Genus **Stephanopogon**

We must briefly mention the enigmatic genus *Stephanopogon* (Figure 5.50). These organisms have played an important role in phylogenetic speculations regarding protist evolution. Several theories have implicated *Stephanopogon* not only in the origin of the ciliates from a flagellate ancestor and in the origin of the ciliate binuclear condition, but also in the origin of the Metazoa from a ciliate protist line. Until recently, *Stephanopogon* was classified in the phylum Ciliophora because they have a conspicuous cytostome and rows of cilia. In 1982, however, D. L. Lipscomb and J. O. Corliss provided evidence based on ultrastructural studies that these protists have little in common with ciliates and are probably more closely related to euglenids. Lipscomb and Corliss found that the two (or up to 16) nuclei of *Stephanopogon* are identical, rather than differentiated into macro- and micronuclei as they are in the ciliates (this long-ignored fact was actually first noticed in the 1920s). Nuclear division is very much like that seen in euglenids and kinetoplastids. While *Stephanopogon* cells have an unusually

short kinetosome at the base of each flagellum, these are not associated with the typical ciliate kinetidal system. Lipscomb and Corliss pointed out that *Stephanopogon* appears to be far from the main trunk on any phylogenetic tree that depicts the origin of ciliates, and they also demonstrated that the use of *Stephanopogon* to derive the Metazoa from a ciliate ancestry is no longer plausible.

Protist Phylogeny

We can do no more than touch upon the myriad questions and interesting points of view concerning the origin and evolution of the protists. Beyond the problems of relationships among the various protist groups themselves, we are faced on one hand with questions about the very origin of eukaryotic life on Earth, and on the other with interpreting the ancestral forms of the rest of the living world. The origin of eukaryotic cells probably took place 2 to 2.5 billion years ago, and this event marked the origin of the protist grade of life. Although there are over 30,000 known fossil species of protists, they are of little use in establishing the origin or subsequent evolution of the various protist groups. Only those with hard parts have left us much of a fossil record, and only the foraminiferans and radiolarians have well established records in Precambrian rocks (and there is some debate even about this). The origin of the eukaryotic condition was, of course, a momentous event in the biological history of the Earth, for it enabled life to escape from the limitations of the prokaryotic bauplan by providing the various subcellular units that have formed the basis of specialization among the Protista and the Metazoa.

Of the number of hypotheses explaining how eukaryotes might have evolved, the most popular is the

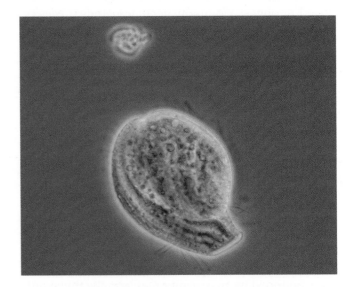

Figure 5.50 The enigmatic protist, *Stephanopogon.*

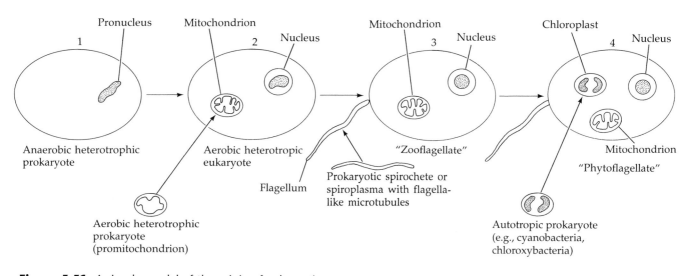

Figure 5.51 A simple model of the origin of eukaryotic cells by symbiosis (the serial endosymbiotic theory). The three major events depicted are: acquisition of an aerobic heterotrophic prokaryote (origin of mitochondrion); acquisition of a spirochate or spiroplasma-like prokaryote (origin of flagellum); and acquisition of an autotrophic prokaryote (origin of chloroplast).

serial endosymbiotic theory (SET). Figure 5.51 outlines the basic steps in the transition from a prokaryotic to a eukaryotic condition according to this theory. The premise in this model is that the eukaryotes arose through intimate symbiotic relationships between various prokaryotic cells, a series of macroevolutionary events called **symbiogenesis**. Unlike prokaryotic cells, all eukaryotic cells contain several kinds of membrane-bounded organelles that harbor distinct genetic systems. Three classes of organelles—mitochondria, cilia/flagella, and plastids — are hypothesized to have once been free-living prokaryotes that were acquired symbiotically and in a certain sequence by another (host) prokaryote. For example, molecular genomic evidence suggests that eukaryotic mitochondria evolved by way of a symbiotic relationship with a class of free-living prokaryotes called α-proteobacteria. Hence, the functions now performed by these various eukaryotic organelles are postulated to have evolved long before the eukaryotic cell itself evolved. SET suggests that a prokaryotic heterotroph ingested other, mitochondrion-like prokaryotes and, roughly at the same time, began forming an organized nucleus. Subsequently, this non-motile cell established a symbiotic relationship with yet another prokaryote in the form of a spirochete or spiroplasma bacterium attached to the outside of the cell. Such bacteria contain protein microtubules and are capable of something like flagellar activity; thus a protoflagellate evolved. Eventually a photosynthetic prokaryote was engulfed by this now-eukaryotic organism, the "prey" representing a chloroplast precursor—thus, the origin of the "phytoflagellates" (autotrophic

flagellates). Lynn Margulis believes this photosynthetic prokaryote might have been a bacterium that evolved in an anaerobic environment very early in the history of life. The type of photosynthesis that produces oxygen would have evolved later, and oxygen-respiring organisms still later, after photosynthetically produced oxygen began accumulating in the environment.

There is now a great deal of evidence in support of SET. For example, there are prokaryotic organisms that are very similar to those SET views as symbionts in this story. Comparisons of rRNA sequences provide some support for the endosymbiotic origin of mitochondria from aerobic bacteria. From this symbiotic basis the modern protist groups may be derived in a variety of ways (e.g., Figure 5.51). Although the eukaryotes probably arose only once, most workers believe that the protists are a paraphyletic group (i.e., many different groups that evolved out of the Protista are not classified as protists). The Protista contains a heterogeneous mix of organisms at intermediate levels of organization and with equivocal boundaries with multicellular taxa. There is no agreed-upon view of how protists are interrelated or how they should be classified, and our taxonomic arrangement is just one of many competing views.

The origins of the various multicelled groups probably lie among several different protist ancestors, and some protist groups are probably no more than unicellular members of lineages that spawned multicellular taxa. For example, other than the embryonic formation of tissues, there seems to be no clear, single boundary between the unicellular chlorophytes and the green plants, or the choanoflagellates and sponges, or the chrysophytes and multicellular brown algae. Of course, not all unicellular forms are related to one of the multicellular kingdoms. There is evidence that some protists (e.g., rhizopodans, euglenids, ciliates) are independent lineages and are not related to any multicellular organisms. It is clear that the commonly used six-kingdom classification scheme and traditional categories (such as zooflagellate, phytoflagellate, and sarcodine) are oversimplified and inadequate for describing the true history and nature of diversity at this level. Ultimately, multiple new kingdoms—based on very different criteria than have traditionally been used—may have to be erected. Determining what and how many of these categories there might be is not an easy task. Perhaps the most empirical approach to these problems has been Diana Lipscomb's attempts to develop cladograms of the eukaryotic "kingdoms." Lipscomb's analyses suggest that the eukaryotes comprise 13 distinct lineages, each perhaps warranting "kingdom" status.

We realize that we have slighted the protists, and especially their evolution, in our necessarily limited treatment here. However, we hope that you have gained enough information about their complexity and diversity to appreciate their success, that you have an appreciation of the unicellular bauplan, and that you are now aware of the many unsolved problems that await further work in the realm of protist biology.

Selected References

General References

Allen, R. D. 1967. Fine structure, reconstruction, and possible functions of the cortex of *Tetrahymena pyriformis*. J. Protozool. 14: 553–565.

Allen, R. D. 1970. Comparative aspects of amoeboid movement. Acta Protozool. 7: 291–299.

Allen, R. D., D. Francis and R. Zek. 1971. Direct test of the positive pressure gradient theory of pseudopod extension and retraction in amoebae. Science 174: 1237–1240.

Anderson, O. R. 1983. *Radiolaria*. Springer-Verlag, New York.

Anderson, O. R. 1988. *Comparative Protozoology: Ecology, Physiology, Life History*. Springer-Verlag.

Baker, J. R. (ed.). 1991. *Parasitic Protozoa,* 2nd Ed. Academic Press, London.

Bamforth, S. S. 1980. Terrestrial protists. J. Protozool. 27: 33–36.

Bardele, C. F. 1972. A microtubule model for ingestion and transport in the suctorian tentacle. Z. Zellforsch. Mikrosk. Anat. 130: 219–242. [One view on how suctorians may eat.]

Be, A. 1982. Biology of planktonic Foraminifera. University of Tennessee Studies in Geology 6: 51–92.

Blank, R. J. and R. K. Trench. 1985. Speciation and symbiotic dinoflagellates. Science 229: 656–658.

Borror, A. C. 1973. Marine flora and fauna of the northeastern United States. Protozoa: Ciliophora. NOAA Tech. Rpt. NMFS Circular No. 378, U.S. Department of Commerce.

Bovee, E. C. and T. L. Jahn. 1966. Mechanisms of movement in taxonomy of Sarcodina. III. Orders, suborders, families, and subfamilies in the superorder Lobida. Syst. Zool. 25: 229–240.

Bovee, E. C. and T. K. Sawyer. 1979. Marine flora and fauna of the northeastern United States. Protozoa: Sarcodina: Amoebae. NOAA Tech. Rpt. NMFS Circular No. 419, U.S. Department of Commerce.

Brehm, P. and R. Eckert. 1978. An electrophysiological study of the regulation of ciliary beating frequency in *Paramecium*. J. Physiol. 282: 557–568.

Bricheux, G. and G. Brugerolle. 1987. The pellicular complex of Euglenoids. II. Protoplasma 140: 43–54.

Brugerolle, G., J. Lom, E. Noh'ynková and L. Joylon. 1979. Comparison et evolution des structures cellulaires chez plusieurs especes de Bodonides et Cryptobiides appartenant aux genes *Bodo, Cryptobia* et *Trypanoplasm* (Kinetoplastida, Mastigophora). Protistologica 15: 197–221.

Byrne, B. J. and B. C. Byrne. 1978. An ultrastructural correlate of the membrane mutant "Paranoiac" in *Paramecium*. Science 199: 1091–1093.

Buetow, D. E. (ed.) 1968, 1982. *The Biology of Euglena*. Vols. 1–2 and 3. Academic Press, New York.

Capriulo, G. M. (ed.). 1990. *Ecology of Marine Protozoa*. Oxford Univ. Press, New York.

Cavalier-Smith, T. 1981. Eukaryote kingdoms: Seven or nine? BioSystems 14: 461–481.

Chapman-Andresen, C. 1973. Endocytic processes. *In* K. Jeon (ed.), *The Biology of Amoeba*, Academic Press, New York, pp. 319–348.

Chen, T. 1950. Investigations of the biology of *Peranema trichophorum*. J. Microsc. Sci. 9: 279–308.

Chen, T. (ed.). 1967–1972. *Research in Protozoology*. Pergamon Press, Oxford.

Corliss, J. O. 1979. *The Ciliated Protozoa: Characterization, Classification and Guide to the Literature*, 2nd Ed. Pergamon Press, New York. [A landmark effort by a distinguished protozoologist. The extensive illustrated glossary is especially useful to those of us not so well versed in ciliate biology.]

Corliss, J. O. 1982. Protozoa. *In* S. Parker (ed.), *Synopsis and Classification of Living Organisms*, Vol. 1. McGraw-Hill, New York, pp. 491–637. [Includes coverage of all protist taxa by several specialists.]

Corliss, J. O. 1989. Protistan diversity and origins of multicellular/multitissued organisms. Boll. Zool. 56: 227–234.

Corliss, J. O. 1994. An interim utilitarian ("user-friendly") hierarchical classification and characterization of the protists. Acta Protozoologica 33: 1–52.

Corliss. J. O. and D. L. Lipscomb. 1982. Establishment of a new order in kingdom Protista for *Stephanopogon*, long-known "ciliate" revealed now as a flagellate. J. Protozool. 29: 294.

Cox, E. R. (ed.). 1980. *Phytoflagellates*. Elsevier/North-Holland, New York. [An excellent review of the major groups, by appropriate specialists; includes a number of articles addressing the origin and phylogeny of various phytoflagellate taxa.]

Cullen, K. J. and R. D. Allen. 1980. A laser microbeam study of amoeboid movement. Exp. Cell Res. 128: 353–362.

Curds, C. R. 1992. Protozoa in the water industry. Cambridge Univ. Press, Cambridge.

Dodge, J. D. and R. M. Crawford. 1970. A survey of thecal fine structure in the Dinophyceae. Bot. J. Linn. Soc. 63: 53–67.

Dolan, J. R. 1991. Microphagous ciliates in mesohaline Chesapeake Bay waters: estimates of growth rates and consumption by copepods. Mar. Biol. 111: 303–309.

Dodge, J. D. 1973. The fine structure of algal cells. Academic Press, London.

Donelson, J. E. and M. J. Turner. 1985. How the trypanosome changes its coat. Sci. Am. 252(2): 44–51.

Eckert, R. 1972. Bioelectric control of ciliary activity. Science 176: 473–481.

Elliott, A. M. (ed.). 1973. *The Biology of Tetrahymena*. Dowden, Hutchinson, and Ross, Stroudsburg, Pennsylvania.

Farmer, J. N. 1980. *The Protozoa: An Introduction to Protozoology*. Mosby, St. Louis.

Fenchel, T. 1980. Suspension feeding in ciliated protozoa: Structure and function of feeding organelles. Arch. Protistenkd. 123: 239–260. [A fine review of the subject.]

Fenchel, T. 1987. *Ecology of Protozoa: The Biology of Free-Living Phagotrophic Protists*. Springer-Verlag, New York.

Fenchel, T. and B. J. Finlay. 1995. Communities and evolution in anoxic worlds. Oxford Univ. Press, Oxford.

Foissner, W. 1987. Soil protozoa: fundamental problems, ecological significance, adaptations in ciliates and testaceans, bioindicators, and guide to the literature. Prog. Protistol. 2: 69–212.

Gall, J. G. (ed.). 1986. *The Molecular Biology of Ciliated Protozoa*. Academic Press, Orlando, Florida.

Gates, M. A. 1978. An essay on the principles of ciliate systematics. Trans. Am. Microsc. Soc. 97: 221–235.

Gibbons, I. R. and A. V. Grimstone. 1960. On flagellar structure in certain flagellates. J. Biophys. Biochem. Cytol. 7: 697–716.

Gojdics, M. 1953. *The Genus Euglena*. University of Wisconsin Press, Madison. [Includes keys, descriptions, and figures of all species known at the time.]

Gooday, A. 1984. Records of deep-sea rhizopod tests inhabited by metazoans in the north-east Atlantic. Sarsia 69: 45–53.

Grain, J. 1986. The cytoskeleton of protists. Int. Rev. Cytol. 104: 153–249.

Grassé, P. (ed.). 1952. *Traite de Zoologie*. Vol. 1, pts. 1–2. Masson et Cie, Paris.

Grell, K. B. 1973. *Protozoology*. Springer-Verlag, New York. [This translation from the original German text is a classic and among the very best of protozoology texts. Although beginning to fall out of date on certain subjects, it remains a most valuable resource.]

Grimes, G. W. 1982. Pattern determination in hypotrich ciliates. Am. Zool. 22: 35–46.

Groto, K., D. L. Laval-Martin and L. N. Edmunds Jr. 1985. Biochemical modeling of an autonomously oscillatory circadian clock in *Euglena*. Science 228: 1284–1288.

Hanson, E. D. 1967. Protist development. *In* M. Florkin and B. J. Sheer (eds.), *Chemical Zoology*, Vol. 1. Academic Press, New York.

Hanson, E. D. 1977. *The Origin and Early Evolution of Animals*. Wesleyan University Press, Middletown, Connecticut, and Pitman Publishing Ltd., London. [Contains nearly 300 pages on the protists.]

Harrison, F. W. and J. O. Corliss. 1991. Protozoa. *In*, F.W. Harrison (ed.), *Microscopic Anatomy of Invertebrates*, Vol. 1. Wiley-Liss, New York.

Hausmann, K. and R. Peck. 1979. The mode of function of the cytopharyngeal basket of the ciliate *Pseudomicrothorax dubius*. Differentiation 14: 147–158.

Hausmann, K. and P. C. Bradbury. 1996. *Ciliates: Cells as Organisms*. Gustav Fischer, Stuttgart.

Hausmann, K. and N. Hülsmann. 1996. *Protozoology*, 2nd Ed. Georg Thieme Medical Publishers, Inc., New York.

Hedley, R. H. and C. G. Adams (eds.). 1974. *Foraminifera*. Academic Press, New York.

Hoek, C. van den, D. G. Mann and H. M. Jahns. 1995. *Algae*. Cambridge Univ. Press, Cambridge.

Hyman, L. H. 1940. *The Invertebrates*. Vol. 1, Protozoa through Ctenophora. McGraw-Hill, New York.

Issi, I. V. and S. S. Shulman. 1968. The systematic position of Microsporidia. Acta Protozool. 6: 121–135.

Jahn, T. L. 1961. The mechanism of ciliary movement. I. Ciliary reversal and activation by electric current; the Ludloff phenomenon in terms of core and volume conductors. J. Protozool. 8: 369–380.

Jahn, T. L. 1962. The mechanism of ciliary movement. II. Ion antagonism and ciliary reversal. J. Cell. Comp. Physiol. 60: 217–228.

Jahn. T. L. 1967. The mechanism of ciliary movement. III. Theory of suppression of reversal by electric potential of cilia reversed by barium ions. J. Cell. Physiol. 70: 79–90.

Jahn, T. L. and E. C. Bovee. 1964. Protoplasmic movements and locomotion of Protozoa. *In* S. H. Hutner (ed.), *Biochemistry and Physiology of Protozoa*, Vol. 3. Academic Press, New York.

Jahn, T. L. and E. C. Bovee. 1967. Motile behavior of Protozoa. *In* T.-T. Chen (ed.), *Research in Protozoology*, Vol. 1. Pergamon Press, Oxford.

Jahn, T. L., E. C. Bovee and F. F. Jahn. 1979. *How to Know the Protozoa*, 2nd Ed. W. C. Brown, Dubuque, IA.

Jahn, T. L., W. M. Harmon and M. Landman. 1963. Mechanisms of locomotion in flagellates. I. *Ceratium*. J. Protozool. 10: 358–363.

Jahn, T. L., M. Landman and J. R. Fonesca. 1964. The mechanism of locomotion of flagellates. II. Function of the mastigonemes of *Ochromonas*. J. Protozool. 11: 291–296.

Jeon, K. W. (ed.). 1973. *The Biology of Amoeba*. Academic Press, New York. [21 of the world's ameba specialists contributed to this fine book dealing with all aspects of biology of free-living amebas.]

Jepps, M. W. 1956. *The Protozoa: Sarcodina*. Oliver and Boyd, Edinburgh.

Jones, A. R. 1974. *The Ciliates*. St. Martin's Press, New York.

Jurand, A. and G. G. Selman. 1969. *The Anatomy of Paramecium aurelia*. Macmillan, London, and St. Martin's Press, New York. [All you ever wanted to know about the anatomy of *Paramecium*. Many fine illustrations and micrographs.]

Kreier, J. P. 1991–1994. *Parasitic Protozoa* [eight volumes], 2nd Ed. Academic Press, New York.

Kudo, R. R. 1966. *Protozoology*, 5th Ed. Charles C. Thomas, Springfield, Illinois. [Dated, but still a good treatment.]

Laybourn-Parry, J. 1985. *A Functional Biology of Free-Living Protozoa*. University of California Press, Berkeley.

Lee, J. J., S. H. Hunter and E. C. Bovee (eds.). 1985. *An Illustrated Guide to the Protozoa*. Allen Press, Lawrence, Kansas. Published for the Society of Protozoologists. [Includes coverage of systematics and phylogenetics, as well as keys to most groups.]

Lee, J. J. and O. R. Anderson. 1991. *Biology of Foraminifera*. Academic Press, London.

Leedale, G. F. 1967. *Euglenoid Flagellates*. Prentice-Hall, Englewood Cliffs, NJ.

Levandowsky, M. and S. H. Hutner (eds.). 1979–1981. *Biochemistry and Physiology of Protozoa*. Vols. 1–4. Academic Press, New York. [A scholarly series of state-of-the-art reviews on various aspects of protist physiology and chemistry.]

Levine, N. D. 1985. *Veterinary Protozoology*. Iowa State Univ. Press, Ames, IA.

Levine, N. D. 1988. *The Protozoan Phylum Apicomplexa*. Vols. 1–2. CRC Press, Boca Raton, Florida. [Provides diagnoses of all genera and species.]

Levine, N. D. and 15 others. 1980. A newly revised classification of the Protozoa. J. Protozool. 276(1): 37–58. [A glimpse at the world of "protist" systematics in 1980; remains a valuable source of information.]

Lumsden, W. H. R. and D. A. Evans (eds.). 1976, 1979. *Biology of the Kinetoplastida*. Vols. 1 and 2. Academic Press, New York.

Lynn, D. H. and P. Didier. 1978. Caractéristiques ultrastructurales du cortex somatique et buccal du cilié *Colpidium campylum* quant à la position systématique de Turaniella. Can. J. Zool. 56: 2336–2343.

Lynn, D. H. and E. B. Small. 1988. An update on the systematics of the phylum Ciliophora: the implications of kinetid diversity. Bio Systems 21: 317–322.

Mackinnon, D. L. and R. S. Hawes. 1961. *An Introduction to the Study of Protozoa*. Clarendon Press, Oxford.

Margulis, L. and K. V. Schwartz. 1988. *Five Kingdoms*. 2nd Ed. W. H. Freeman, New York.

Margulis, L., J. O. Corliss, M. Melkonian and D. J. Chapman (eds.). 1989. *Handbook of Protoctista*. Jones and Bartlett, Boston.

Mazier, D., R. and 11 other authors. 1985. Complete development of hepatic stages of *Plasmodium falciparum* in vitro. Science 227: 440–442.

Møestrup, Ø. 1982. Flagellar structure in algae: a review with new observations particularly on the Chrysophycea, Phaeophyceae, Euglenophyceae and *Reckertia*. Phycol. 21: 427–528.

Mohr, J. L., H. Matsudo and Y.-M. Leung. 1970. The ciliate taxon Chonotricha. Oceanogr. Mar. Biol. Annu. Rev. 8: 415–456.

Moldowan, J. M. and N. M. Talyzina. 1998. Biogeochemical evidence for dinoflagellate ancestors in the early Cambrian. Science 281: 1168–1170.

Murray, J. W. 1971. *An Atlas of British Recent Foramaniferids*. Heinemann, London.

Murray, J. W. 1973. *Distribution and Ecology of Living Benthic Foraminiferida*. Crane, Russak, New York.

Nigrini, C. and T. C. Moore. 1979. *A Guide to Modern Radiolaria*. Special Publ. No. 16, Cushman Foundation for Foraminiferal Research, Washington, DC.

Nisbet, B. 1983. *Nutrition and Feeding Strategies in Protozoa*. Croom Helm Publishers, London.

Ogden, C. G. and R. H. Hedley. 1980. *An Atlas of Freshwater Testate Amoebae*. Oxford University Press, Oxford. [A book of magnificent SEM photographs of ameba shells accompanied by descriptions of the species.]

Olive, L. S. 1975. *The Mycetozoans*. Academic Press, New York.

Page, F. C. 1984. *Gruberella flavescens* (Gruber, 1889), a multinucleate lobose marine amoeba (Gymnamoebia). J. Mar. Biol. Assoc. U.K. 64: 303–316.

Patterson, D. J. 1980. Contractile vacuoles and associated structures: Their organization and function. Biol. Rev. 55: 1–46.

Patterson, D. J. 1996. *Free-Living Freshwater Protozoa. A Colour Guide.* John Wiley & Sons, NY.

Patterson, D. J. and J. Larsen. 1991. *The Biology of Free-Living Heterotrophic Flagellates.* Clarendon Press, Oxford.

Patterson, D. J. and S. Hedley. 1992. *Free-Living Freshwater Protozoa.* CRC Press, Boca Raton, FL.

de Puytorac, P. and J. Grain. 1976. Ultrastructure du cortex buccal et évolution chez les Ciliés. Protistologica 12: 49–67.

Raikov, I. B. 1982. *The Protozoan Nucleus: Morphology and Evolution.* Cell Biol. Monogr. 9. Springer-Verlag, New York.

Raikov, I. B. 1994. The diversity of forms of mitosis in protozoa: a comparative review. European J. Protistology 30: 253–259.

Roberts, L. S. and J. Janovy, Jr. 1996. *Foundations of Parasitology*, 5th Ed. Wm. C. Brown. Chicago, IL.

Roth, L. E. 1960. Electron microscopy of pinocytosis and food vacuoles in *Pelomyxa*. J. Protozool. 7: 176–185.

Sandon, H. 1963. *Essays on Protozoology.* Hutchinson Educational Ltd., London. [We like the way this little book is written; it is lively and entertaining reading and provides a nice introduction to the protists.]

Schmidt, G. D. and L. S. Roberts. 1989. *Foundations of Parasitology,* 4th Ed. Times Mirror/Mosby College Publishing, NY.

Seliger, H. H. (ed.). 1979. *Toxic Dinoflagellate Blooms.* Elsevier/North Holland, NY.

Seravin, L. N. 1971. Mechanisms and coordination of cellular locomotion. Comp. Physiol. Biochem. 4: 37–111.

Siddall, M. E., D. S. Martin, D. Bridge, S. S. Desser and D. K. Cone. 1995. The demise of a phylum of protists: phylogeny of Myxozoa and other parasitic Cnidaria. J. Parasitol. 81: 961–967.

Sleigh, M. A. (ed.). 1973. *Cilia and Flagella.* Academic Press, London.

Sleigh, M. A. 1989. *Protozoa and Other Protists.* 2nd Ed. Edward Arnold, London. [This is the second edition of Sleigh's *Biology of Protozoa*.]

Smothers, J. F., C. D. von Dohlen, L. H. Smith, Jr. and R. D. Spall. 1994. Molecular evidence that the myxozoan protists are metazoans. Science 265: 1719–1721.

Spector, D. (ed.). 1984. *Dinoflagellates.* Academic Press, NY.

Spoon, D. M., G. B. Chapman, R. S. Cheng and S. F. Zane. 1976. Observations on the behavior and feeding mechanisms of the suctorian *Heliophyra erhardi* (Reider) Matthes preying on *Paramecium*. Trans. Am. Microsc. Soc. 95: 443–462.

Sprague, V. 1977. Classification and phylogeny of the Microsporidia. *In* L. A. Bulla and T. Cheng (eds.), *Comparative Pathobiology,* Vol. 2, Systematics of the Microsporida. Plenum, New York.

Steidinger, K. A. and E. R. Cox. 1980. Free-living dinoflagellates. *In* E. R. Cox (ed.), *Phytoflagellates.* Elsevier/North-Holland, New York.

Steidinger, K. A. and K. Haddad. 1981. Biologic and hydrographic aspects of red tides. BioScience 31(11): 814–819.

Stoecker, D. K., M. W. Silver, A. E. Michaels and L. H. Davis. 1988. Obligate mixotrophy in *Laboea strobila*, a ciliate which retains chloroplasts. Mar. Biol. 99: 415–423.

Tartar, V. 1961. *The Biology of Stentor.* Pergamon Press, New York.

Taylor, F. J. R. 1987. *The Biology of Dinoflagellates.* Blackwell Scientific Publications, Palo Alto, CA.

Taylor, G. T. 1982. The role of pelagic heterotrophic protozoa in nutrient cycling: A review. Ann. Inst. Oceanogr., Paris 58: 227–241.

Trench, R. K. 1980. Uptake, retention and function of chloroplasts in animal cells. *In* W. Schwemmler and H. Schenk (eds.), *Endocytobiology.* Vol. I. Walter de Gruyter, Berlin, pp. 703–730.

van Wagtendonk, W. J. (ed.). 1974. *Paramecium: A Current Survey.* Elsevier, New York.

Vickerman, K. and F. E. G. Cox. 1967. *The Protozoa.* Houghton Mifflin, Boston.

Vickerman, K. and L. Tetley. 1977. Recent ultrastructural studies on trypanosomes. Ann. Soc. Blege Med. Trop. 57: 444–455.

Wefer, G. and W. H. Berger. 1980. Stable isotopes in benthic Foraminifera: Seasonal variation in large tropical species. Science 209: 803–805.

Wichterman, R. 1986. *The Biology of Paramecium*, 2nd Ed. Plenum, New York. [A major reference.]

Protist Phylogeny

Cavalier-Smith, T. 1975. The origin of nuclei and of eukaryotic cells. Nature (London) 256: 463–468.

Corliss, J. O. 1972. The ciliate Protozoa and other organisms: Some unresolved questions of major phylogenetic significance. Am. Zool. 12: 739–753.

Corliss, J. O. 1974. Time for evolutionary biologists to take more interest in protist phylogenetics? Taxon 23: 497–522.

Corliss, J. O. 1975. Nuclear characteristics and phylogeny in the protistan phylum Ciliophora. BioSystems 7: 338–349.

Corliss, J. O. 1981. What are the taxonomic and evolutionary relationships of the protozoa to the Protista? BioSystems 14: 445–459.

Gray, M. W. 1992. The endosymbiont hypothesis revisited. Internat. Rev. Cytol. 141: 223–357.

Gray, M. W. and W. F. Doolittle. 1982. Has the endosymbiont hypothesis been proven? Microbiol. Rev. 46: 1–42.

Jeon, K. W. (ed.). 1983. *Intracellular Symbiosis.* Int. Rev. Cytol. (Suppl.) 14: 1–379.

John, P. and F. W. Whatley. 1975. *Paracoccus dentrificans*: A present-day bacterium resembling the hypothetical free-living ancestor of the mitochondrion. Symp. Soc. Exp. Biol. 29: 39–40.

Kabnick, K. S. and D. A. Peattie. 1991. *Giardia*: a missing link between prokaryotes and eukaryotes. Amer. Scientist 79: 34–43.

Lipscomb, D. L. 1985. The eukaryote kingdoms. Cladistics 1: 127–140.

Lipscomb, D. L. 1989. Relationships among the eukaryotes. *In* B. Fernholm, F. Bremer and H. Jornvall (eds.), *Hierarchy of Life.* Excerpta Medica, Amsterdam, pp. 161–178.

Lipscomb, D. L. 1991. Broad classification: the kingdoms and the protozoa. *In* J. R. Baker (ed.), *Parasitic Protozoa,* 2nd Ed. Academic Press, London, pp. 81–136.

Lipscomb, D. L. and J. O. Corliss. 1982. *Stephanopogon*, a phylogenetically important "ciliate," shown by ultrastructural studies to be a flagellate. Science 215: 303–304.

Lipscomb, D. L., J. S. Farris, M. Källersjö and A. Tehler. 1998. Support, ribosomal sequences and the phylogeny of the eukaryotes. Cladistics 14(4): 303–338.

Mahler, H. A. and R. A. Raff. 1975. The evolutionary origin of the mitochondrion: A non-symbiotic model. Int. Rev. Cytol. 43: 1–124. [An argument against the serial symbiosis theory.]

Margulis, L. 1970. *Origin of Eukaryotic Cells.* Yale University Press, New Haven, CT.

Margulis, L. 1976. The genetic and evolutionary consequences of symbiosis. Exp. Parsitol. Rev. 39: 277–349.

Margulis, L. 1978. Microtubules in prokaryotes. Science 200: 1118–1124.

Margulis, L. 1980. Flagella, cilia, and undulipodia. BioSystems 12: 105–108.

Margulis, L. 1981. *Symbiosis in Cell Evolution.* W. H. Freeman, San Francisco. [An assessment of the serial endosymbiotic theory and a review of the evolution of life on Earth.]

Pohley, H. J., R. Dornhaus and B. Thomas. 1978. The amoebo-flagellate transformation: A system-theoretical approach. BioSystems 10: 349–360.

Raff, R. A. and H. A. Mahler. 1975. The symbiont that never was: An inquiry into the evolutionary origin of the mitochondrion. Symp. Soc. Exp. Biol. 29: 41–92.

Ragan, M. A. and D. J. Chapman. 1978. *A Biochemical Phylogeny of the Protists.* Academic Press, New York. [An exhaustive treatment of the biochemical data available on protists and their evolutionary implications.]

Raikov, I. B. 1976. Evolution of macronuclear organization. Annu. Rev. Genet. 10: 413–440.

Schwartz, R. M. and M. Dayhoff. 1978. Origins of prokaryotes, eukaryotes, mitochondria, and chloroplasts. Science 199: 395–403.

Spoon, D. M., C. J. Hogan, and G. B. Chapman. 1995. Ultrastructure of a primitive, multinucleate, marine cyanobacteriophagous amoeba (*Euhyperamoeba biospherica* n. sp.) and its possible significance in the evolution of the eukaryotes. Invertebrate Biol. 114 (3): 189–201.

Taylor, F. J. R. 1976. Flagellate phylogeny: A study in conflicts. J. Protozool. 23:2 8–40.

Williams, A. G. and G. S. Coleman. 1992. *The Rumen Protozoa.* Springer Verlag, Berlin.

6 *Phylum Porifera:*
The Sponges

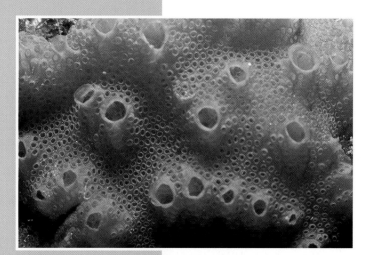

Sponges have made no progress in the formation of an anterior end or a head.

Libbie Hyman,
The Invertebrates, Vol. 1, 1940

The phylum Porifera (Latin *porus*, "pore"; *ferre*, "to bear") contains those animals commonly called sponges. Figures 6.1 and 6.2 illustrate a variety of sponge body forms and some sponge anatomy. Box 6A lists the major characteristics of sponges. Poriferans are sessile, suspension-feeding, multicellular animals that utilize flagellated cells called choanocytes to circulate water through a unique system of water canals. Porifera is the only phylum at the **parazoan** grade of body construction (i.e., Metazoa lacking true embryological germ layering). Not only are true tissues absent, but most of the body cells are totipotent—they are capable of changing form and function. Despite the fact that sponges are large-bodied multicellular animals, they function largely like organisms at the unicellular grade of complexity. As you will discover in this chapter, their nutrition, cellular organization, gas exchange, and response to environmental stimuli are all very protist-like.

About 5,500 living species of sponges have been described, nearly all of which are restricted to benthic marine environments. They occur at all depths, but unpolluted littoral and tropical reef habitats harbor especially rich sponge faunas. Most littoral sponges grow as thick or thin layers on hard surfaces. Benthic sponges that live on soft substrata are often upright and tall, thus avoiding burial by the shifting sediments of their environment. Some sponges reach considerable size (up to 2 m in height on Caribbean reefs, and even larger in the Antarctic) and may constitute a significant portion of the benthic biomass. In Antarctica, sponges make up almost 75 percent of the total benthic biomass at a depth of 100–200 m. Subtidal and deeper water species that do not confront strong tidal currents or surge are usually large and exhibit a stable, even symmetrical, external form. The deeper water hexactinellid sponges often assume unusual shapes, many being delicate glasslike structures, others round and massive, and still others ropelike. A few species in the class Demispongiae inhabit fresh waters. (Figure 6.1).

Sponges display nearly every color imaginable, including bright lavenders, blues, yellows, crimsons, and white. Many species harbor symbiotic bacteria or unicellular algae that may color the sponge's body.

Taxonomic History and Classification

The sessile nature of sponges and their generally amorphous (asymmetrical) growth form convinced early naturalists that they were plants. It was not until 1765, when the nature of their internal water currents was described, that sponges were recognized as animals. The great naturalists of the late eighteenth and early nineteenth centuries (Lamarck, Linnaeus, and Cuvier) classified the sponges under Zoophytes or Polypes, regarding them as allied to anthozoan cnidarians. Throughout much of the nineteenth century they were placed with cnidarians under the name Coelenterata or Radiata. The morphology and physiology of sponges were first adequately understood by R. E. Grant. Grant created for them the name Porifera, although other names were frequently used (e.g., Spongida, Spongiae, Spongiaria). Huxley (1875) and Sollas (1884) first proposed the separation of sponges from "higher" Metazoa.

Historically, the classes of Porifera have been defined by the nature of their internal skeletons. Until recently, four classes were recognized: Calcarea, Hexactinellida, Demospongiae, and Sclerospongiae. The class Sclerospongiae included those species that produce a solid, calcareous, rocklike matrix on which the living animal grows. These poriferans are also known as coralline sponges; about 15 living species have been described. This class, however, was abandoned over a decade ago, and its members relegated to the Calcarea and Demospongiae (Vacelet 1985).* Demospongiae is the largest sponge class, comprising about 95 percent of the living species. Because of its size and variability, the Demospongiae presents the most problems to taxonomists. In a series of papers published between 1953 and 1957, Lévi proposed an important reappraisal of the Demospongiae, incorporating reproductive characteristics for the first time.

Although the mainstay of sponge taxonomy has traditionally been the anatomy of the spicules, these skeletal structures have proven inadequate for developing stable phylogenetic hypotheses and classifications. Indeed, some sponge species lack spicules altogether. Hence specialists are now using embryological, bio-

> ### BOX 6A *Characteristics of the Phylum Porifera*
>
> 1. Metazoa at the cellular grade of construction, without true tissues; adults asymmetrical or superficially radially symmetrical
> 2. Cells totipotent
> 3. With unique flagellated cells—choanocytes—that drive water through canals and chambers constituting the aquiferous system
> 4. Adults are sessile suspension feeders; larval stages are motile and usually lecithotrophic
> 5. Outer and inner cell layers lack a basement membrane (except perhaps in the subclass Homoscleromorpha)
> 6. Middle layer—the mesohyl—variable, but always includes motile cells and usually some skeletal material
> 7. Skeletal elements, when present, composed of calcium carbonate or silicon dioxide (typically in the form of spicules), and/or collagen fibers

chemical, histological, and cytological methods to diagnose sponge taxa. The great variability in sponge morphology and the difficulty in precisely setting the limits of a sponge species have probably driven many potential poriferologists to frustration (and to other taxa) early in their careers. Even the great sponge taxonomist Arthur Dendy was known to frequently end a species diagnosis with a question mark. This state of affairs was summarized by one student regarding California sponge studies (Ristau 1978):

> The study of California sponges has not generated a fervor of activity over the years, nor has the literature been saturated with information about this little-studied . . . phylum. Probably the greatest interest generated by sponges occurred recently, when several news agencies reported that giant, and presumably mutant, sponges were found growing on undersea nuclear waste storage containers (*San Francisco Chronicle*, September 14, 1976). It has been rumored that the Japanese are now planning a motion picture in which a sleeze[†] of giant sponges rises from the depths of the Farallon Islands and phagocytizes the North Beach area of San Francisco. Undoubtedly, when this epic materializes, research and interest in California sponges will increase. Until that time, however, those interested in the sponge fauna of this area must be content with the paucity of scientific literature on this subject.

Recently a host of important bioactive compounds has been discovered in sponges, many having potential pharmacological significance (e.g., antimicrobial, anti-

---

*It has also been suggested that the Hexactinellida should be removed from the Porifera and assigned to a separate phylum, Symplasma, but this proposal has not received much support.

†"Sleeze" is a term coined by Ristau for an aggregation of sponges; the usage is comparable to other such collective nouns that define animal groups (e.g., flock, herd, gaggle).

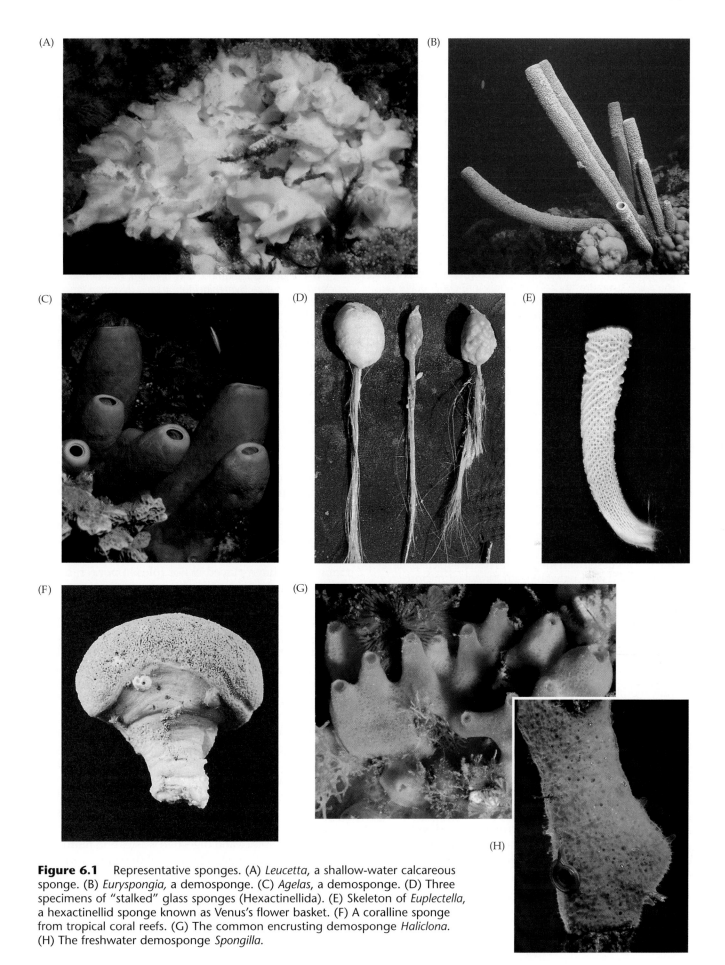

Figure 6.1 Representative sponges. (A) *Leucetta*, a shallow-water calcareous sponge. (B) *Euryspongia*, a demosponge. (C) *Agelas*, a demosponge. (D) Three specimens of "stalked" glass sponges (Hexactinellida). (E) Skeleton of *Euplectella*, a hexactinellid sponge known as Venus's flower basket. (F) A coralline sponge from tropical coral reefs. (G) The common encrusting demosponge *Haliclona*. (H) The freshwater demosponge *Spongilla*.

inflammatory, antitumor, cytotoxic, and anti-fouling compounds). The discovery of these natural products in sponges has led to a renewed interest in this group and a call for the training of more sponge taxonomists, whose numbers have dwindled over the past few decades.

PHYLUM PORIFERA

CLASS CALCAREA: Calcareous sponges (Figure 6.1A). Spicules of mineral skeleton composed entirely of calcium carbonate laid down as calcite; skeletal elements often not differentiated into megascleres and microscleres; spicules usually 1, 3, or 4-rayed; body with asconoid, synconoid, or leuconoid construction; all marine.

SUBCLASS CALCINEA: Free-living larvae are hollow coeloblastulae, flagellated, and can become solid parenchymula-like structures by cellular ingression; choanocyte nuclei located basally; flagellum arises independent of nucleus; with regular triradiate spicules; spicules free, but some species (e.g. *Murrayona*) with massive calcite skeleton. (e.g., *Clathrina, Dendya, Leucascus, Leucetta, Soleniscus*)

SUBCLASS CALCARONEA: Free-living larvae are partly flagellated amphiblastulae; choanocyte nuclei apical; flagellum arises directly from nucleus; spicules free or fused. (e.g., *Amphoriscus, Grantia, Leucilla, Leucosolenia, Petrobiona, Scypha* [= *Sycon*])

CLASS HEXACTINELLIDA: Glass sponges (Figure 6.1D,E). Spicules siliceous and basically 6-rayed (hexactinal); both megascleres and microscleres always present; body wall cavernous, with trabecular network; external pinacoderm absent and replaced by a noncellular dermal membrane; choanocyte layer may be syncytial; exclusively marine; primarily deep-water.

SUBCLASS AMPHIDISCOPHORA: Body never attached to a hard substratum but anchored in soft sediments by a basal tuft or tufts of spicules; megascleres discrete spicules, never fused into a rigid network; with birotulate microscleres, never hexasters; mostly deep-water. (e.g., *Hyalonema, Monorhaphis, Pheronema*)

SUBCLASS HEXASTEROPHORA: Usually attached to hard substrata, but sometimes attached to sediments by a basal spicule tuft or mat; microscleres are hexasters; megascleres sometimes free, but usually fused into a rigid skeletal framework, in which case sponge may assume large and elaborate morphology. (e.g., *Aphrocallistes, Caulophacus, Euplectella, Hexactinella, Leptophragmella, Lophocalyx, Rosella, Sympagella*)

CLASS DEMOSPONGIAE: Demosponges (Figure 6.1B,C,F–H). With siliceous spicules; spicules not 6-rayed; spicule skeleton may be supplemented or replaced by an organic collagenous network ("spongin"); marine, brackish, or freshwater sponges, occurring at all depths.

SUBCLASS HOMOSCLEROMORPHA: Embryos incubated, larvae amphiblastulae-like ("cinctoblastula"); differentiation of spicules into mega- and microscleres not evident; all spicules very small (usually less than 100 mm) and distributed in large numbers throughout body, with little regional organization; with a "pseudobasal membrane" underlying the pinacoderm. Usually littoral, but some occuring at shelf and slope depths. (e.g., *Corticium, Oscarella, Plakina, Plakortis, Pseudocorticium*)

SUBCLASS TETRACTINOMORPHA: Reproduction typically oviparous, but incubation with direct development occurs in one order; larvae, when present, typically parenchymulae, with distinct megascleres and microscleres; megascleres organized into distinct patterns, either axial or radial; numerous orders and families. (e.g., *Asteropus, Chondrilla, Chondrosia, Cliona, Cryptotethya, Geodia, Polymastia, Rhabderemia, Stelletta, Suberites, Tethya, Tetilla*). This subclass now contains some sponges from the recently abandoned Sclerospongiae: the merliids (*Merlia*) and at least some tabulates (*Acanthochaetetes*).

SUBCLASS CERACTINOMORPHA: Mostly viviparous, with incubation of parenchymulla larvae; distinct microscleres and megascleres present; spongin present in all but one family (Halisarcidae); includes the freshwater families Spongillidae and Potamelepidae. (e.g., *Adocia, Agelas, Aplysilla, Aplysina* [= *Verongia*], *Asbestopluma, Axinella, Axociella, Callyspongia, Clathria, Coelosphaera, Halichondria, Haliclona, Halisarca, Hymeniacidon, Ircinia, Lissodendoryx, Microciona, Mycale, Myxilla, Spongia, Spongilla, Tedania*). The Ceractinomorpha now contain some sponges previously assigned to the Sclerospongiae, including the stromatoporids (e.g., *Astrosclera, Calcifibrospongia*), the ceratoporellids (e.g., *Ceratoporella, Stromatospongia, Hispidopetra, Goreauiella*), and the enigmatic *Vaceletia crypta*.

The Poriferan Bauplan

In Chapter 3 we discussed some of the limitations of the parazoan grade of construction, in which true tissues and organs are absent. Now we discuss the various ways in which sponges have overcome the handicaps imposed by their primitive level of organization. You will notice a striking resemblance to the protists in many regards. Two unique organizational attributes define sponges and have played major roles in poriferan success: the water current channels, or **aquiferous system** (and its choanocytes), and the highly totipotent nature of sponge cells. The tremendous diversity among sponges in size and shape has occurred both evolutionarily and individually and is largely derived from these two unique characteristics. Increases in size and surface area are accomplished by folding of the body wall into a variety of patterns. Furthermore, variation in the overall shapes of sponges results from different growth patterns in various environments. This general plasticity in size, shape, and construction, plus the fact that most individual sponge cells are capable of radically altering their form and function as needed, compensate in part for the absence of tissues and organs. The aquiferous system brings water through the sponge and close to the cells responsible for food gathering and gas exchange. At the same time, excretory and digestive wastes and reproductive products are expelled by way of the water currents. The volume of water moving through a sponge's aquiferous system is remarkable. A 1×10-cm individual of the complex sponge *Leuconia* pumps about 22.5 l of water through its body daily. Researchers have recorded sponge pumping rates that range from 0.002 to 0.84 ml of water per second per cubic centimeter of sponge body. A large sponge filters its own volume of water every 10 to 20 seconds.

Early workers treated sponges as essentially colonial animals. The other view, and the one that we prefer, holds that the entirety of a sponge—that is, any and all sponge material bounded by a continuous outer covering—constitutes a single individual. The fact is, a whole sponge grows as a whole body, dictated largely by environmental factors (e.g., water flow dynamics, substratum contours). Changes in body form can arise anywhere in or on the organism in response to these environmental pressures. Sponges grow by continually adding new cells that differentiate as needed; this is not usually viewed as colonial asexual reproduction. The existence of some coordinated behavior in sponges (e.g., cessation of choanocyte pumping, synchronous oscular contractions) further supports the view that each sponge is, in its entirety, an "individual."

Body Structure and the Aquiferous System

The outer surface cells of a sponge make up the **pinacoderm** and are called **pinacocytes**. Most of the inner surfaces comprise the **choanoderm** and are composed of flagellated cells called **choanocytes**. Both of these layers are a single cell thick. Between these two thin cellular sheets is the **mesohyl**, which may be very thin in some simple sponges, or massive and thick in larger species (Figure 6.2). The pinacoderm is perforated by small holes called **dermal pores** or **ostia** (singular, ostium), depending on whether the opening is surrounded by several cells or one cell, respectively (Figure 6.3). Water is pulled through these openings and is driven across the choanoderm by the beating of the choanocyte flagella. The choanocytes pump large volumes of water through the sponge body at very low pressures, establishing the water current (aquiferous) system.

A cuticle, or layer of coherent collagen, may cover (or even replace) the pinacoderm in some species. The pinacoderm itself can be a simple external sheet, but typically it also lines some of the internal cavities of the aquiferous system where choanocytes do not occur. Pinacoderm cells that line internal canals are called **endopinacocytes**. The choanoderm also can be simple and continuous, or folded and subdivided in various ways. The mesohyl varies in thickness and plays vital roles in digestion, gamete production, secretion of the skeleton, and transport of nutrients and waste products by special ameboid cells. The mesohyl includes a noncellular colloidal mesoglea in which are embedded collagen fibers, spicules, and various cells; as such, it is really a type of mesenchyme. A great number of cell types may be found in the mesohyl. Most of these cells are able to change from one type to another as required; but some differentiate irreversibly, such as those that commit themselves to reproduction or to skeleton formation.

The mobility of all cells, including pinacocytes and choanocytes, has been demonstrated by dramatic time-lapse cinematography. The cells of the pinacoderm and choanoderm are more stable than those of the mesohyl, but in general, the whole structure may be thought of as a continuously mobile system. In fact, recent observations by Bond (1997, 1998) confirm a report of nearly half a century ago that some sponges actually do move from one place to another. Ameboid cells along the base of the sponge "crawl" as others bring spicules as support for the leading edge of the sponge. Bond reported that some ameobocytes actually broke free from the sponge and moved about on their own for a time, eventually returning to the parent sponge body. This locomotion in sponges is not sufficient to provide them with a quick escape mechanism from predators, however; in Bond's words, "The champion speedster regularly moved more than four millimeters a day."

During growth, the pinacoderm and choanoderm are each only one cell thick. By increasing their folding as mesohyl volume increases, these layers maintain a surface area-to-volume ratio sufficient to sustain adequate nutrient and waste exchange throughout the whole individual. The one-cell thick choanoderm may remain simple and continuous (the **asconoid** condition), or it may become folded (the **syconoid** condition), or it may become greatly subdivided into separate flagellated chambers (the **leuconoid** condition) (Figure 6.3).

The asconoid condition is found in some adult, radially symmetrical calcareous sponges (e.g., *Clathrina*, *Leucosolenia*) and in the early growth stage (**olynthus**) of newly settled calcareous sponges (Figure 6.4A). Asconoid sponges rarely exceed 10 cm in height and remain as simple, vase-shaped, tubular units. The thin walls enclose a central cavity called the **atrium** (= **spongocoel**), which opens to the outside via a single **osculum**. The pinacoderm of asconoid and very simple syconoid sponges has specialized cells called **porocytes**. During embryogeny, each porocyte elongates and rolls to form a cylindrical tube. The porocyte extends all the way through the pinacoderm, the thin mesohyl, and the choanoderm into the atrium, emerging between adjacent choanocytes (Figures 6.4B and 6.7C). The external opening of the porocyte canal is called an **ostium** or **incurrent pore**. The choanoderm is a simple, unfolded layer of choanocytes lining the entire atrium. Water moving through an asconoid sponge flows through the following structures: ostium → spongocoel (over the choanoderm) → osculum.

Simple folding of the pinacoderm and choanoderm produces the syconoid condition, within which several levels of complexity are possible (Figure 6.3B,C). As complexity increases, the mesohyl may thicken and appear to have two layers. The outer "cortical region," or **cortex**, often contains skeletal elements that are different from those found in the interior portion of the mesohyl.

Figure 6.2 Sponge body forms. (A) The unusual demosponge *Coelosphaera hatchi* (height in life 27 mm). (B) The coralline sponge *Merlia normani* (vertical section) has a basal calcareous matrix within which individual compartments are filled by secondary deposition. The superficial soft tissue contains the choanocyte chambers and is supported by tracts of siliceous spicules. (C) The demosponge *Haliclona permollis*, a sponge with a tubular type of architecture; three successive levels of magnification are shown, from left to right. (D) *Microciona prolifera*, a demosponge with a more solid type of architecture; three successive levels of magnification are shown, from left to right.

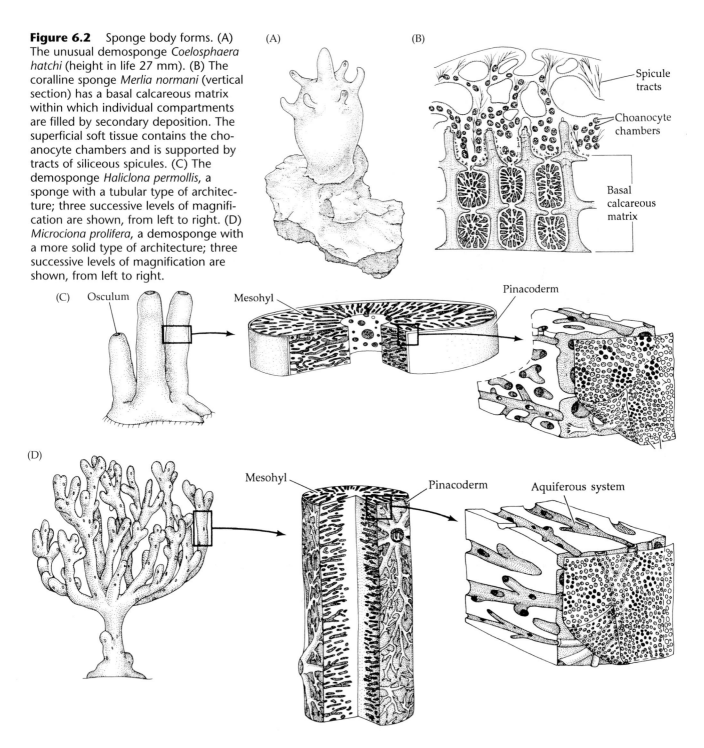

In those sponges with a cortex, the incurrent openings are lined by several cells (not formed by a single porocyte) and are referred to as **dermal pores**. In the syconoid condition, choanocytes are restricted to specific chambers or diverticula of the atrium called **choanocyte chambers** (or **flagellated chambers**, or **radial canals**). Each choanocyte chamber opens to the atrium by a wide aperture called an **apopyle**. Syconoid sponges with a thick cortex possess a system of channels or **incurrent canals** that lead from the dermal pores through the mesohyl to the choanocyte chambers. The openings from these channels to the choanocyte chambers are called **prosopyles**. In such a complex syconoid sponge, water moving from the surface into the body flows along the following route: incurrent (dermal) pore → incurrent canal → prosopyle → choanocyte chamber → apopyle → atrium → osculum. Syconoid construction is found in many calcareous sponges (e.g., *Scypha*, also known as *Sycon*). Some syconoid sponges appear radially symmetrical, but their complex internal organization is largely asymmetrical.

The leuconoid condition is produced by additional folding of the choanoderm and further thickening of the mesohyl by cortical growth. These modifications are ac-

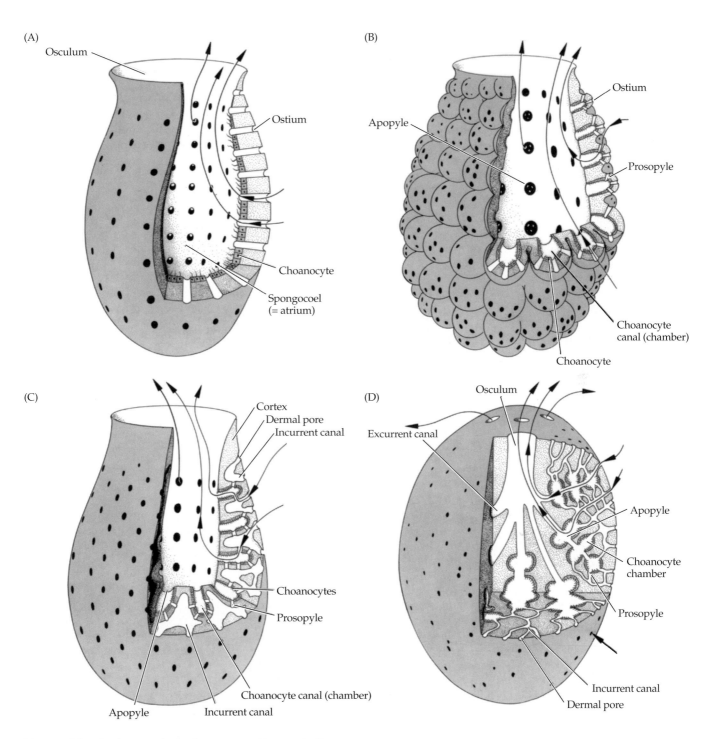

Figure 6.3 Body complexity in sponges. (Arrows indicate flow of water.) (A) The asconoid condition. (B) A simple syconoid condition. (C) A complex syconoid condition with cortical growth. (D) A leuconoid condition.

companied by subdivision of the flagellated surfaces into discrete oval choanocyte chambers (Figure 6.3D). In the leuconoid condition, one finds an increase in number and a decrease in size of the choanocyte chambers, which typically cluster in groups in the thickened mesohyl. The atrium is reduced to a series of **excurrent canals** (or **exhalent canals**) that carry water from the cho-

anocyte chambers to the oscula (Figure 6.5). The flow of water through a leuconoid sponge is: dermal pore → incurrent canal → prosopyle → choanocyte chamber → apopyle → excurrent canal → osculum. Leuconoid organization is typical of most calcareous sponges and all members of the Demospongiae.

It is important to realize that the flow rate is not uniform through the various parts of the aquiferous system. Functionally, it is critical that water be moved very slowly over the choanoderm, allowing time for ex-

(A)

(B)

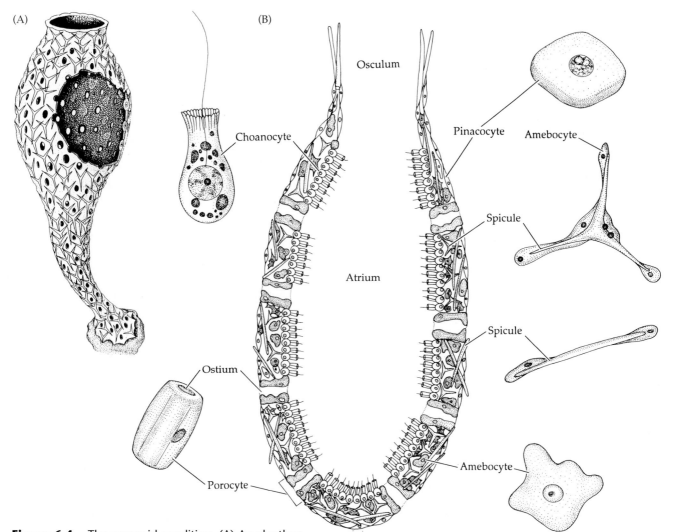

Osculum

Choanocyte

Pinacocyte

Amebocyte

Spicule

Atrium

Spicule

Ostium

Amebocyte

Porocyte

Figure 6.4 The asconoid condition. (A) An olynthus, the asconoid form that follows larval settlement in calcareous sponges. (B) Major cell types in an asconoid sponge. (C) The simple calcareous sponge *Leucosolenia* shows the asconoid body form and skeleton of $CaCO_3$ spicules.

(C)

changes of nutrients, gases, and wastes between the water and the choanocytes. The changes in water flow velocity through this plumbing are a function of the effective accumulated cross-sectional diameters of the channels through which the water moves (see Chapter 3, or your old physics notes). Water flow velocity decreases as the cross-sectional diameter increases; thus, in a sponge, velocities are lowest over the choanoderm. Furthermore, water leaving the oscula must be carried far enough away to prevent it being recycled by the sponge. In environments of relatively high turbulence, currents, or wave action, this potential recycling of wastes is not a problem. However, sponges that reside in relatively calm water rely on the maintenance of high velocities of water flow through the oscula (or on modified body shapes) to push the excurrent water far enough away from the sponge to avoid the incoming

currents. In an irregularly shaped leuconoid sponge living in quiet water, the combined cross-sectional diameter of all the incurrent pores is far less than that of all the choanocyte chambers. But the total oscular diameter is even less than that of the incurrent pores. Simply put,

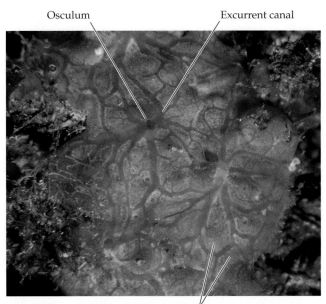

Osculum   Excurrent canal

Dermal (incurrent) pores

Figure 6.5 The surface of a living demosponge (*Clathria*). The complex system of ostia opens into underlying incurrent canals, and large oscula receive several excurrent canals.

the water enters at some velocity *x*, slows to a small fraction of *x* as it passes over the choanoderm, then exits the sponge at a velocity much greater than *x*. In complex sponges, the differences in velocity are dramatic. Flow rate regulation is also facilitated in some sponges, in part, by the activity of ameboid cells (called **central cells**) that reside near the apopyles of the choanocyte chambers. These cells can slow or speed the exit of water from the chambers by changing shape and position across the apopyle (see Figure 6.7I).

The recognition of the various levels of organization and complexity among poriferans allows one to quickly and simply describe a sponge's basic anatomical plan. There is very little evidence, however, that the asconoid plan is necessarily the most primitive, or that all sponge lineages have moved through these three levels of complexity during their evolution. Nor do all sponges pass through three such developmental stages. In addition, gradations of and intermediates between the three basic plans are common. Nonetheless, among adult sponges, the simplest organizations (asconoid and syconoid) occur only in the class Calcarea, which is thought to be the most primitive class of living poriferans. Furthermore, calcareous sponges of the leuconoid condition do pass through asconoid and syconoid stages as they grow, and it is only in this class that all three organizational body plans occur.

The hexactinellid sponges. The hexactinellids differ considerably from calcareous sponges and demosponges (Figure 6.6). The bodies of hexactinellid sponges

display a greater degree of radial, or superficial radial, symmetry than any other group. There is no pinacoderm or its equivalent in hexactinellids. A **dermal membrane** is present, but it is extremely thin; no discrete or continuous cellular structure supports it. Incurrent pores are simple holes in this dermal membrane. Cellular material is sparsely distributed and forms a **trabecular network** stretching across interconnecting internal cavities called **subdermal lacunae** (Figure 6.6A). The thimble-shaped flagellated chambers are arranged in a single layer and are supported within the trabecular network. Both the trabecular network and the walls of the flagellated chambers appear to be syncytial (i.e., discrete choanocytes do not exist). Water enters the incurrent pores, passes into the subdermal lacunae, and from there enters the choanocyte chambers via the prosopyles.

The unique structure of hexactinellids is so striking that some workers (e.g., Bergquist 1985) have even suggested the hexactinellids might be regarded as a separate phylum (the Symplasma). However, as explained in

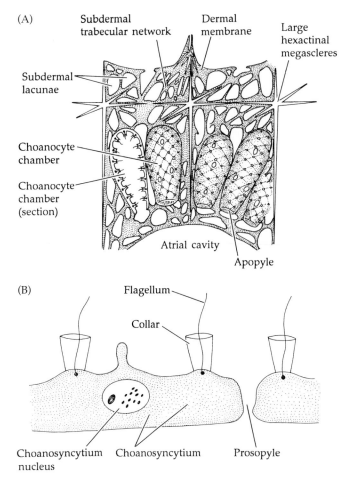

(A)   Subdermal trabecular network   Dermal membrane   Large hexactinal megascleres

Subdermal lacunae

Choanocyte chamber

Choanocyte chamber (section)

Atrial cavity

Apopyle

(B)   Flagellum

Collar

Choanosyncytium nucleus   Choanosyncytium   Prosopyle

Figure 6.6 Internal anatomy of Hexactinellida. (A) The body wall of *Euplectella* (transverse section). A dermal layer covers the trabecular network. (B) The choanosyncytium of *Aphrocallistes vastus* (vertical section).

Chapter 2, phylogenetic relationships are best sought in similarities among groups, not in differences, and by this reasoning we treat the hexactinellids as poriferans. Also, the syncytial nature of hexactinellids has recently been questioned.

Cell Types

Because of the nontissue nature of sponges and because cellular totipotency plays a major role in poriferan biology, considerable effort has gone into describing and classifying sponge cell types. Prior to the 1970s, texts generally recognized only a few basic kinds of poriferan cells. However, subsequent detailed histochemical and ultrastructural studies have revealed a host of cell types. These discoveries, combined with the dynamic and totipotent nature of sponge cells, makes succinct classification of their cells difficult. We present below an abbreviated version of Bergquist's (1978) cell classification.

Cells that line surfaces. Pinacoderm forms a continuous layer on the external surface of sponges and also lines all incurrent and excurrent canals. The **pinacocytes** that make up this layer are usually flattened and often overlapping (Figure 6.7A,B). Internal, canal-lining pinacocytes (**endopinacocytes**) are usually more fusiform in shape and have less overlap than outer exopinacocytes. Furthermore, ciliated endopinacoderm occurs in the large excurrent canals of some leuconoid sponges. Although the endopinacoderm is "epithelial" in function, and probably phagocytic as well, the apparent absence of a basal membrane distinguishes sponge pinacoderm from the true tissue epithelia of the higher Metazoa.* External cells of the basal or attaching region of a sponge surface are called **basopinacocytes**. These flattened, T-shaped cells are responsible for secreting a fibrillar collagen–polysaccharide complex called the basal lamina, which is the actual attachment structure. In freshwater sponges, the basopinacocytes are active in feeding and extend ameba-like "filopodia" to engulf bacteria. Freshwater sponge basopinacocytes also play an active role in osmoregulation and contain large numbers of water expulsion vesicles, or contractile vacuoles.

Porocytes are cylindrical, tubelike cells of the pinacoderm that form the ostia (Figure 6.7C,D). They are contractile and can open and close the pore and regulate the ostial diameter; however, no microfilaments have been observed in them and their precise method of contraction and expansion is unknown. Some can produce across the ostial opening a diaphragm-like cytoplasmic membrane that also regulates pore size.

Choanocytes are the flagellated cells that make up the choanoderm and create the currents that drive water through the aquiferous system (Figure 6.7F–H). Choanocytes are not coordinated in their beating, not

---

*Recent work suggests that a basal membrane may be present in the Homoscleromorpha.

Figure 6.7 Cells that line sponge surfaces. (A) A pinacocyte from the surface of the demosponge *Halisarca* (drawn from an electron micrograph). The outer surface is covered with a polysaccharide-rich coat. The cell is fusiform and overlaps adjacent pinacocytes. (B) Pinacoderm from a calcareous sponge (section). T-shaped pinacocytes alternate with fusiform pinacocytes. (C,D) A porocyte from the calcareous asconoid sponge *Leucosolenia*. (C) Cross section. (D) Side view. (E) Myocytes surrounding a prosopyle. (F) A section of choanoderm, showing three choanocytes; arrows indicate direction of water current. (G) A choanocyte. (H) Ultrastructure of a choanocyte (longitudinal section, drawn from an electron micrograph). (I) A choanocyte chamber opening into an excurrent canal in a demosponge.

even within a given chamber. However, they are aligned such that the flagella are directed toward the apopyle and beat from base to tip. Water is thus drawn into the chamber through the prosopyles, driven across the choanoderm, and then out the apopyle into the atrium or an excurrent canal. The long flagellum is always surrounded by a so-called **collar**, which is made up of 20 to 55 cytoplasmic microvilli (= **villi**). The villi have microfilament cores and are connected to one another by anastomosing mucous strands (a mucous reticulum). Choanocytes rest on the mesohyl, held in place by interdigitation of adjacent basal surfaces. In keeping with their central role in phagocytosis and pinocytosis, choanocytes are highly vacuolated.

Cells that secrete the skeleton. There are several types of ameboid cells in the mesohyl, some of which secrete the various elements of sponge skeletons. In almost all sponges, the entire supportive matrix is built on a framework of fibrillar collagen. The cells that secrete this material are called **collencytes**, **lophocytes**, and **spongocytes**. Collencytes are morphologically nearly indistinguishable from pinacocytes, whereas lophocytes are large, highly motile cells that can be recognized by a collagen tail they typically trail behind them (Figure 6.8C). The primary function of both cell types is to secrete the dispersed fibrillar collagen found intercellularly in virtually all sponges. Spongocytes produce the fibrous supportive collagen referred to as spongin (Figure 6.10A). Spongocytes operate in groups and are always found wrapped around a spicule or spongin fiber (Figure 6.8D).

Sclerocytes are responsible for the production of calcareous and siliceous sponge spicules (Figure 6.8A,B). They are active cells that possess abundant mitochondria, cytoplasmic microfilaments, and small vacuoles. Numerous types of sclerocytes have been described; these cells always disintegrate after spicule secretion is complete.

Contractile cells. Contractile cells in sponges, called **myocytes**, are found in the mesohyl (Figure 6.7E). They are

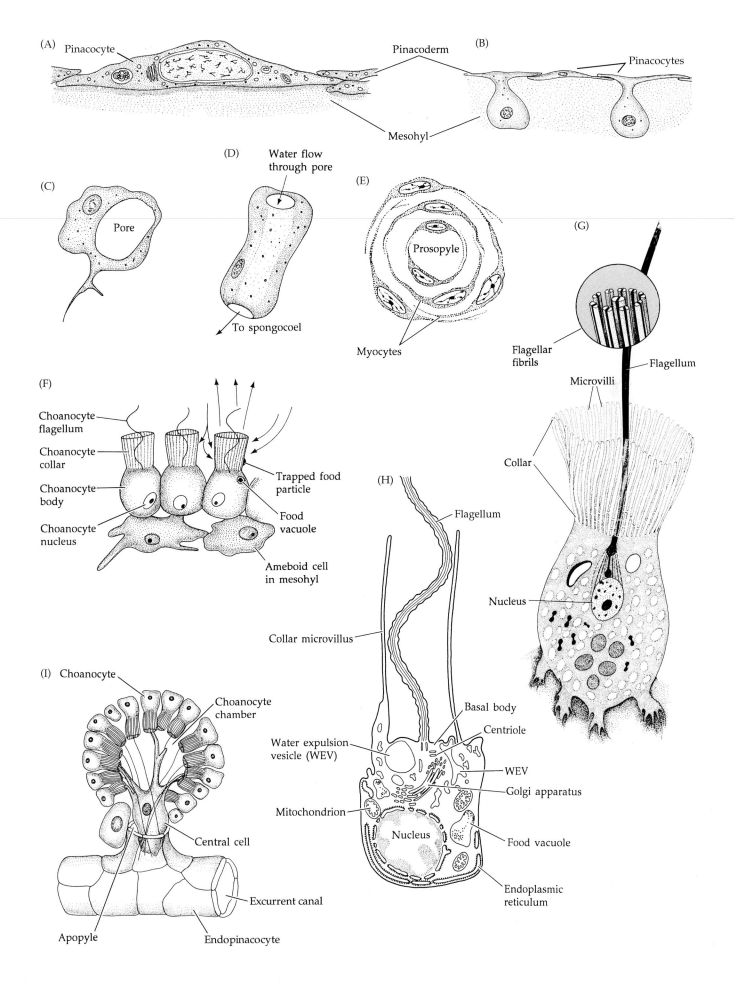

(A) Pinacocyte

(B) Pinacoderm / Pinacocytes

Mesohyl

(C) Pore

(D) Water flow through pore

To spongocoel

(E) Prosopyle

Myocytes

(G) Flagellar fibrils / Flagellum

Microvilli

Collar

Nucleus

(F)
Choanocyte flagellum
Choanocyte collar
Choanocyte body
Choanocyte nucleus

Trapped food particle

Food vacuole

Ameboid cell in mesohyl

(H) Flagellum

Collar microvillus

Water expulsion vesicle (WEV)

Basal body

Centriole

WEV

Golgi apparatus

Mitochondrion

Nucleus

Food vacuole

Endoplasmic reticulum

(I) Choanocyte

Choanocyte chamber

Central cell

Excurrent canal

Apopyle

Endopinacocyte

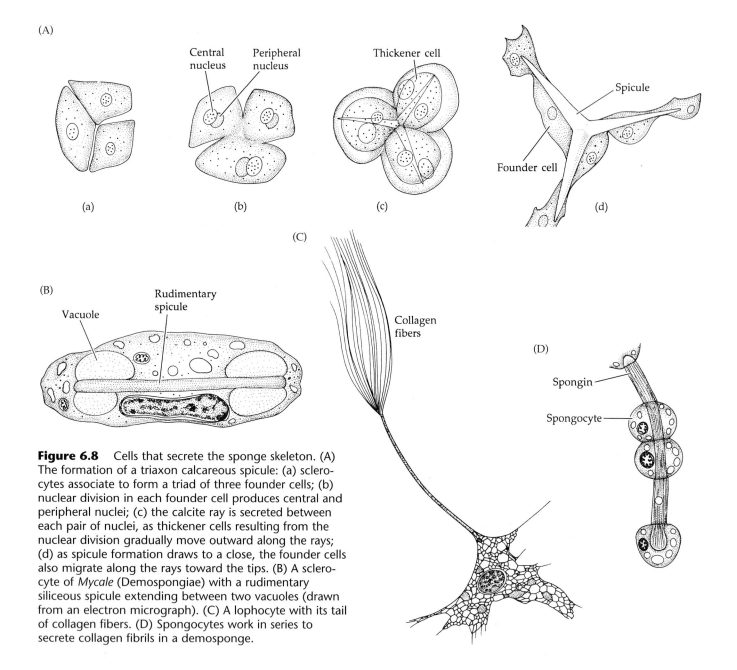

Figure 6.8 Cells that secrete the sponge skeleton. (A) The formation of a triaxon calcareous spicule: (a) sclerocytes associate to form a triad of three founder cells; (b) nuclear division in each founder cell produces central and peripheral nuclei; (c) the calcite ray is secreted between each pair of nuclei, as thickener cells resulting from the nuclear division gradually move outward along the rays; (d) as spicule formation draws to a close, the founder cells also migrate along the rays toward the tips. (B) A sclerocyte of *Mycale* (Demospongiae) with a rudimentary siliceous spicule extending between two vacuoles (drawn from an electron micrograph). (C) A lophocyte with its tail of collagen fibers. (D) Spongocytes work in series to secrete collagen fibrils in a demosponge.

usually fusiform and grouped concentrically around oscula and major canals. Myocytes are distinguished by the great numbers of microtubules and microfilaments contained in their cytoplasm. Because of the nature of their filament arrangement, it has been suggested that myocytes are homologous with the smooth muscle cells of higher invertebrates. Myocytes are independent effectors with a slow response time, and, unlike neurons and true muscle fibers, they are insensitive to electrical stimuli.

Some other cell types. **Archaeocytes** are ameboid cells that are capable of differentiating and giving rise to virtually any other cell type. Archaeocytes are large, highly motile cells that play a major role in digestion and food transport (Figure 6.9). These cells possess a variety of digestive enzymes (e.g., acid phosphatase, protease, amylase, lipase) and can accept phagocytized material from the choanocytes. They also phagocytize material directly through the pinacoderm of water canals. As the principal macrophage of a sponge, archaeocytes carry out much of the digestive, transport, and excretory activities. As cells of maximum totipotency, archaeocytes are essential to the developmental program of sponges and to various asexual processes (e.g., gemmule formation).

Spherulous cells are large mesohyl cells containing various chemical inclusions. These cells often contain the secondary metabolites which are so abundant in sponges.

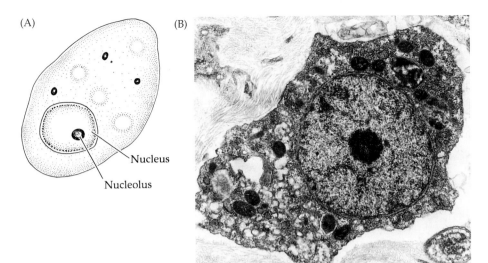

Nucleus

Nucleolus

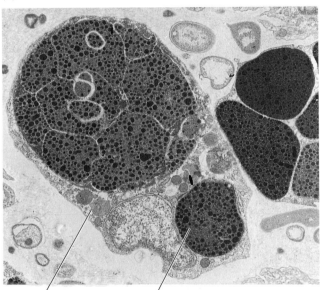

Archaeocyte   Nucleus of archaeocyte

Figure 6.9 Archaeocytes. (A) A typical archaeocyte with a large nucleus and a prominent nucleolus. (B) Photo of a typical archaeocyte. (C) An archaeocyte engages in phagocytosis.

Atlantic "red beard" sponge (*Microciona prolifera*) are pressed through fine cloth, the separated cells immediately begin to reorganize themselves by active cell migration. Within 2 to 3 weeks, a functional sponge re-forms and the original cells return to their respective functions. Furthermore, if cell suspensions of two different sponge species are mixed, the cells sort themselves out and reconstitute individuals of each separate species—evidence of the ability of self recognition.

Support

The skeletal elements of sponges are of two types, organic and inorganic. The former is always collagenous and the latter either siliceous (hydrated silicon dioxide) or calcareous (calcium carbonate in the form of calcite or aragonite). Sponges are the only animals that use hydrated silica as a skeletal material.

Collagen is the major structural protein in invertebrates; it is found in virtually all metazoan connective tissues. In sponges, it is either dispersed as thin **fibrils** in the intercellular matrix or organized as a fibrous framework called **spongin** in the mesohyl. True spongin is found only in members of the class Demospongiae; dispersed collagen fibrils are found in all sponges. The amount of this fibrillar collagen varies greatly from species to species. In hexactinellids it is quite sparse, whereas in demosponges it is abundant and may form dense bands in the cortex.

Traditionally, the sponge organic skeleton has been termed spongin. This term, however, should be restricted to the form of collagen that constitutes a distinct organized network in the mesohyl of demosponges (Figure 6.10A). The network often contains very thick fibers, and may incorporate siliceous spicules into its structure. Spongin often cements siliceous spicules together at their points of intersection. The encysting coat of the asexual gemmules of freshwater (and some marine) sponges is also composed largely of spongin.

Mineral skeletons of silica or calcium are found in almost all sponges, except certain members of the class Demospongiae. Several demosponge genera lack both spongin and a spicule skeleton (e.g., *Chondrosia, Euspongia, Halisarca, Oscarella*). Sponges lacking mineral

Several other cell types have been identified in sponges, but most of these have been characterized only morphologically and their functions remain unknown.

Cell Aggregation

Around the turn of the twentieth century, H. V. Wilson first demonstrated the remarkable ability of sponge cells to reaggregate after being mechanically dissociated. Although this discovery was interesting in itself, lending insight into the plasticity and cellular organization of sponges, it also foreshadowed more far-reaching cytological research. Recent studies on sponges have shed light on the basic questions of how cells adhere, segregate, and specialize. Many sponges that are dissociated and maintained under proper conditions will form aggregates, and some will eventually reconstitute their aquiferous system. For example, when pieces of the skeletons possess only fibrous collagen networks. They are still used as bath sponges, despite the prevalence nowadays of synthetic "sponges."

Sponges have been harvested for millennia; Homer and other ancient Greek writers mention an active Mediterranean sponge trade. Prior to the 1950s, an active natural sponge fishery thrived in south Florida, the Bahamas, and the Mediterranean. The industry peaked in 1938, when the world's annual sponge catch (including cultivated sponges) exceeded 2.6 million pounds, 700,000 pounds of which came from the United States and the Bahamas. Almost all commercial sponges belong to the genera *Hippospongia* and *Spongia,* but these sponges have been largely "fished out" in the traditional sponge hunting grounds of the Mediterranean and Florida.

Sponge **spicules** (Figure 6.10) are produced by special mesohyl cells called sclerocytes, which are capable of accumulating calcium or silicate and depositing it in an organized way. In some cases, one sclerocyte produces one spicule; in others, several sclerocytes work together to produce a single spicule, often two cells per spicule ray (Figure 6.8A–D). The construction of a siliceous spicule begins with the secretion of an organic axial filament within an elongated vacuole in a sclerocyte. As the axial filament elongates at both ends, hydrated silica is secreted into the vacuole and deposited around the filament. Unlike siliceous spicules, calcium carbonate spicules do not have an organic axial structure. Calcareous spicules are produced extracellularly, in intercellular spaces bounded by a number of sclerocytes. Each spicule is essentially a single crystal of calcite or aragonite.

Considerable taxonomic weight has been given to spicule morphology, and an elaborate nomenclature exists to classify these skeletal structures. According to their morphology, spicules are termed either **microscleres** or **megascleres**. The former are small to minute reinforcing (or packing) spicules; the latter are large structural spicules. The demosponges and hexactinellids have both types; calcareous sponges often have only megascleres. Descriptive terms that designate the number of axes in a spicule end in the suffix *-axon* (e.g., monaxon, triaxon). Terms that designate the number of rays end in the suffixes *-actine* or *-actinal* (e.g., monactinal, hexactinal, tetractinal). In addition, there is a detailed nomenclature specifying shape and ornamentation of various spicules (Figure 6.10).

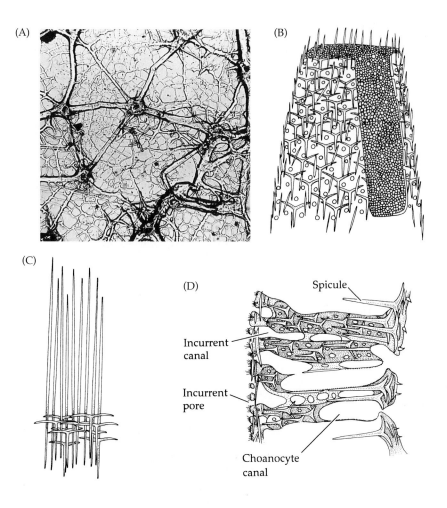

A spicular skeleton may be viewed as a supplemental supporting structure. If the amount of inorganic material is increased in relation to organic material, the sponge becomes increasingly solid until the texture approaches that of a rock, as it does in members of the demosponge orders Choristida and Lithistida. In contrast to discrete spicules, the massive calcareous skeletons of some species (the coralline sponges and "sclerosponges") have a polycrystalline microstructure; they are composed of needles ("fibers") of either calcite or aragonite embedded in an organic fibrillar matrix. The advantage of incorporating organic matter into the calcareous framework has been compared to lathe-and-plaster, or reinforced concrete. The mix of organic and inorganic materials probably yields fibrous calcites and aragonites that are less prone to fracture while also producing substances that are more easily molded by the organism.

Nutrition, Excretion, and Gas Exchange

Although sponges lack the complex organs and organ systems seen in the higher Metazoa, they are nevertheless a highly successful group of animals. Their success appears to be due largely to their cell totipotency, the aquiferous system, and the general plasticity of their body form.

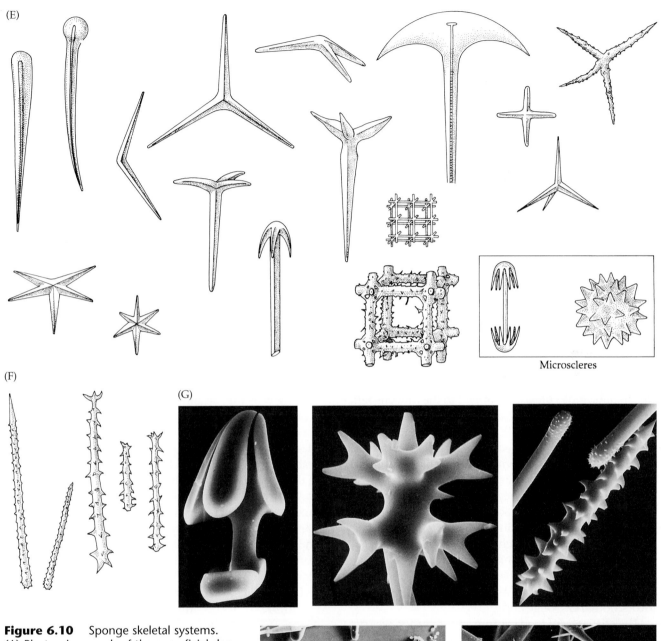

(E)

Microscleres

(F)

(G)

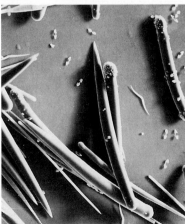

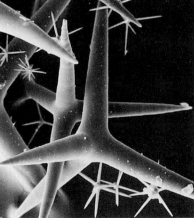

Figure 6.10 Sponge skeletal systems. (A) Photomicrograph of the superficial dermal spongin–fiber skeleton typical of the demosponge family Callyspongiidae. (B) Arrangement of calcareous triaxon spicules near the oscular opening in *Leucosolenia*. (C) Arrangement of monaxon and triaxon calcareous spicules near the oscular opening of *Scypha*. (D) Cross-section of a simple syconoid calcareous sponge (atrium on right) illustrating placement of triaxon spicules. (E) Some common types of siliceous spicules from demosponges. (F) Some siliceous spicules from sclerosponges. (G) Various spicule types (SEMs).

Unlike most Metazoa, nearly all sponges rely on intracellular digestion, and thus on phagocytosis and pinocytosis as means of food capture. The aquiferous system has already been described; sponges more or less continuously circulate water through their bodies, bringing with it the microscopic food particles upon which they feed. They are size-selective particle feeders, and the arrangement of the aquiferous system creates a series of "sieves" of decreasing mesh size (e.g., inhalant ostia or dermal pores → canals → prosopyles → choanocyte villi → intertentacular mucous reticulum). The upper limit of the diameter of incurrent openings is usually around 50 μm, so larger particles do not enter the aquiferous system. A few species have larger incurrent pores, reaching diameters of 150 to 175 μm, but in most species the incurrent openings range from 5 to 50 μm in diameter. Internal particle capture in the 2 to 5 μm range (e.g., bacteria, small protists, unicellular algae, organic detritus) is by phagocytic motile archaeocytes that move to the lining of the incurrent canals. Then, as water passes over the choanoderm, eddies are formed around the choanocyte collars. This water passes between the villi, into the collar, and is driven out the collar opening. Particles in the 0.1 to 1.5 μm range (e.g., bacteria, large free organic molecules) are trapped in the mucous reticulum between the collar villi. The distance between adjacent villi is consistently 0.1 to 0.2 μm . Undulations of the collar move the trapped food particles down to the choanocyte cell body, where they are ingested by phagocytosis or pinocytosis.

In the case of archaeocyte phagocytosis, digestion takes place in the food vacuole formed at the time of capture. In the case of choanocyte capture, food particles are partly digested in the choanocytes and then quickly passed on to a mesohyl archaeocyte (or other wandering amebocyte) for final digestion. In both cases, the mobility of the mesohyl cells assures transport of nutrients throughout the sponge body.

The efficiency of food capture and digestion was dramatically shown in a study by Schmidt (1970) using fluorescence-tagged bacteria fed to the freshwater sponge *Ephydatia fluviatilis*. By monitoring the movement of the fluorescent material, Schmidt determined that 30 minutes elapsed from the onset of feeding until the bacteria had been captured by choanocytes and moved to the base of the cells. Transfer of the fluorescent material to the mesohyl commenced 30 minutes later. Twenty-four hours later, fluorescent wastes began to be discharged into the water, and no fluorescent material remained in the sponges after 48 hours. Additional studies on this same species led to an estimate of 7,600 choanocyte chambers per cubic millimeter of sponge body, each chamber pumping approximately 1,200 times its own volume of water daily. More complex leuconoid sponges have as many as 18,000 choanocyte chambers per cubic millimeter. In some thin-walled asconoid and simple syconoid sponges, a distinctive mesohyl is hard-

ly present. In these sponges the choanocytes assume both capture and digestive/assimilative functions.

Sponges also take up significant amounts of dissolved organic matter (DOM) by pinocytosis from the water within the aquiferous system. Studies by Reiswig in the 1970s on Jamaican sponges showed that 80 percent of the organic matter taken in by these sponges was of a size below that resolvable by light microscopy. The other 20 percent comprised primarily bacteria and dinoflagellates.

Recent studies show that at least some sponges form simple fecal pellets. Experiments on the cosmopolitan species *Halichondria panicea* have revealed that undigested material is expelled as discrete capsules coated with a thin layer of mucus.

Although the phylum Porifera is characterized by filter feeding, members of the demosponge family Cladorhizidae display an entirely different and unique mode of feeding. Species in this group have lost the characteristic choanocyte-lined aquiferous system and instead feed as macrophagous carnivores! They do so by trapping small prey on hook-shaped spicules that protrude from the surfaces of tentacle-like structures (Figure 6.11). Trapped prey are gradually enveloped by migrating feeding cells that accomplish digestion and absorption. Although most cladorhizids live at great depths, one species of *Asbestopluma* lives in shallow caves in the Mediterranean, where it has been the subject of considerable study (Vacelet and Boury-Esnault 1995). Another of the remarkable cladorhizid sponges, an undescribed species of *Cladorhiza*, has been discovered to harbor methanotrophic bacterial symbionts in its cells, such as seen in certain animals inhabiting hydrothermal vents and cold seeps. The sponge thus feeds both by predation and by direct consumption of its microbial symbionts (Vacelet et al. 1998).

Excretion (primarily ammonia) and gas exchange are by simple diffusion, much of which occurs across the choanoderm. We have already seen how folding of the body, combined with the presence of an aquiferous system, overcomes the surface-to-volume dilemma posed by an increase in size. The efficiency of the poriferan bauplan is such that diffusion distances never exceed about 1.0 mm, the distance at which gas exchange by diffusion becomes inefficient. In addition, water expulsion vesicles (contractile vacuoles) occur in freshwater sponges and presumably aid in osmoregulation.

Activity and Sensitivity

There is no conclusive evidence that sponges possess neurons or discrete sense organs. Furthermore, action potentials have never been recorded in sponges, and nothing resembling the synaptic connections of higher Metazoa are known in these animals. However, they are capable of responding to a variety of environmental stimuli by closure of the ostia or oscula, canal constriction, backflow, and reconstruction of flagellated cham-

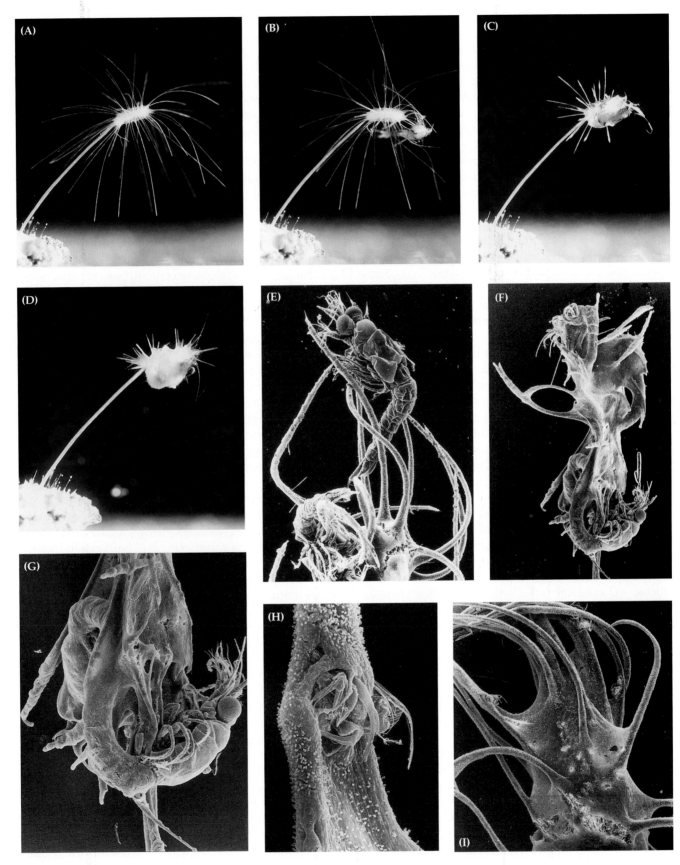

Figure 6.11 These remarkable SEMs and color photographs show predation in the carnivorous sponge *Asbestopluma*. (A–D) *Asbestopluma* in ambush posture (A), followed by capture of a mysid. (E) Fifteen minutes after capture of a mysid on its tentacle-like feeding filaments. (F–H) The mysid prey has been partly engulfed by the sponge. (I) The prey is entirely engulfed.

bers. The usual effect of most of these actions is to reduce or stop the flow of water through the aquiferous system. For example, when suspended particulates become too large or too concentrated, sponges typically respond by closing the incurrent openings and immobilizing the choanocyte flagella. Direct physical stimulation will also elicit this reaction, which is easily observed by simply running one's finger across a sponge surface and observing the dermal pore or oscular contractions with a hand lens or low-power microscope.

Activity also varies with certain endogenous factors. For example, during a major growth phase, such as canal or chamber reorganization, activity levels typically fall and pumping rates drop. Periods of reproductive activity also cause a substantial decrease in water pumping, because many choanocytes are expended in the reproduction process (see the next section). Even under normal conditions, variations in pumping rates occur. Important studies by Reiswig (see references) on Caribbean sponges documented a number of endogenous activity patterns. Some sponges cease pumping activity periodically, for a few minutes or for hours at a time; others cease activity for several days at a time.

The switch from full pumping activity to complete cessation requires at least several minutes; considering the organism, however, this is a fairly short response time. The spread of stimulation and response in sponges appears to be by simple mechanical stimulation from one cell to the adjacent cells, and perhaps also by diffusion of certain chemical messengers associated with the irritability of cytoplasm in general. The contractile myocytes of sponges act as independent effectors; they are organized into a network formed by contacts between filopodial extensions of adjacent myocytes and pinacocytes. Response time of myocytes is relatively slow. Latency periods average 0.01 to 0.04 seconds, and conduction velocities are typically less than 0.04 cm/sec (except in the hexactinellids, where velocities of 0.30 cm/sec have been recorded). Conduction is always unpolarized and diffuse. Considerable research once focused on the myocytes in attempts to shed light on the possible presence of a sponge nervous system analogous or homologous to that in higher Metazoa. But, in spite of these efforts, there has been no verification of such a system.

One study on sponge activity hypothesized a diffuse conduction system in the hexactinellid sponge *Rhabdocalyptus* (Lawn et al. 1981). Both mechanical and electrical stimulation elicited a diffuse all-or-none response wherein pumping activity ceased within 20 to 50 seconds. Conduction velocities of 0.17 to 0.30 cm/sec were estimated. Although the authors agreed that this is too slow for a true neuronal system, they felt it was too fast for conduction by simple chemical diffusion.

Reproduction and Development

All sponges appear to be capable of sexual reproduction, and several types of asexual processes are also common. Many of the details of these processes are unknown, however, largely because sponges lack distinct or localized gonads (gametes and embryos occur throughout the mesohyl). Furthermore, within any species and population, there is a marked asynchrony among individuals in terms of reproductive activity; at any given moment, reproductive activity may be taking place in only a small number of individuals in any area.

Asexual reproduction. Probably all sponges are capable of regenerating viable adults from fragments. Some branching species "pinch off" branch ends by a process of cellular reorganization. The dislocated pieces fall off and regenerate into new individuals. This regenerative ability used to be used by Florida commercial sponge farmers, who propagated their sponges by attaching "cuttings" to submerged cement blocks. Additional asexual processes of poriferans include formation of gemmules and reduction bodies, budding, and possibly formation of asexual larvae.

In freshwater sponges of the family Spongillidae, small spherical structures called **gemmules** are produced at the onset of winter (Figure 6.12). These dormant overwintering bodies are invested with a thick collagenous coat in which supportive siliceous microscleres are embedded. Gemmules are highly resistant to both freezing and drying. The gemmules of some species can withstand exposure to –70°C for up to an hour, while others experience mass mortality at –10°C (Ungemach et al. 1997).

The formation and eventual growth of gemmules are remarkable examples of poriferan cell totipotency. As winter approaches, archaeocytes aggregate in the mesohyl and undergo rapid mitosis. "Nurse cells" called **trophocytes** stream to the archaeocyte mass and are engulfed by phagocytosis. The result is a mass of archaeocytes containing food reserves stored in elaborate **vitelline platelets**. This entire mass eventually becomes surrounded by a three-layered spongin covering. Developing amphidisc spicules are transported by their parent cells to the growing gemmule and incorporated into the spongin envelope. The final bit of the gemmule to be enclosed by the spongin case is covered only by a single layer of spongin that is devoid of spicules; this single-layered patch is the **micropyle**. Thus formed, hibernation of the gemmule commences, while the parent sponge usually dies and disintegrates.

When environmental conditions are again favorable, the micropyle opens and the first archaeocytes begin to flow out (Figure 6.12C). They immediately flow over the gemmule and onto the substratum, whereupon they begin to construct a framework of new pinacoderm and choanoderm. The second wave of archaeocytes to leave the gemmule colonizes this framework. In the course of gemmule "hatching," archaeocytes give rise to every cell type of the adult sponge. Gemmule dormancy appears to be of two types, a quiescence and a true diapause.

(A)

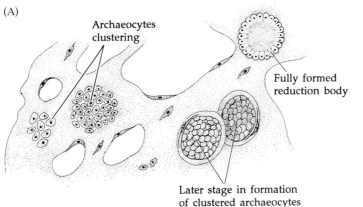

Archaeocytes
clustering

Fully formed
reduction body

Later stage in formation
of clustered archaeocytes

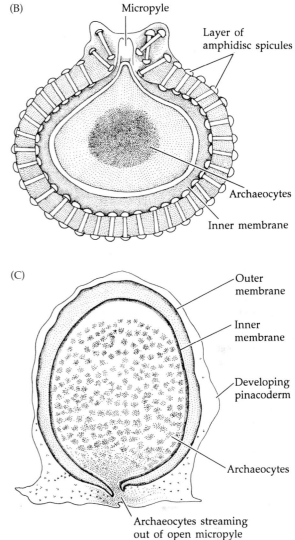

(B)

Micropyle

Layer of
amphidisc spicules

Archaeocytes

Inner membrane

(C)

Outer
membrane

Inner
membrane

Developing
pinacoderm

Archaeocytes

Archaeocytes streaming
out of open micropyle

Figure 6.12 (A) Reduction bodies forming in a marine sponge. (B) A gemmule (in section) of a freshwater sponge (Spongillidae). (C) A gemmule (in section) of the freshwater sponge *Spongilla* in the process of hatching.

Quiescence is imposed by generally unfavorable conditions, including low temperatures, and ends when suitable conditions return. Diapause, on the other hand, is imposed by a combination of endogenous mechanisms and adverse environmental conditions. The breaking of a diapause state typically requires exposure to very low temperatures for a prescribed number of days.

No other sponge group produces gemmules as complex as those of the Spongillidae. However, many marine species produce asexual reproductive bodies (called **reduction bodies**) that are roughly similar to freshwater gemmules but incorporate a variety of amebocytes and have a less complex wall structure.

Many marine sponges produce buds of various types. They appear as squat or elongate club-shaped protrusions arising on the sponge surface. The buds fall from the parent sponge surface and may be carried about by water currents for a brief period before they adhere to the substratum to form a new individual. Some members of the family Clionidae produce unique armored buds that are rich in stored foods and can drift in the plankton for extended periods of time.

Some sponges are reported to be capable of producing larvae by asexual means. This little-studied and controversial process has been suggested as a means of assuring production of a free dispersal stage even when fertilization has failed.

Sexual processes. Most sponges are hermaphroditic, but they produce eggs and sperm at different times. This sequential hermaphroditism may take the form of protogyny or protandry, and the sex change may occur only once, or an individual may repeatedly alternate between male and female. In some species individuals appear to be permanently male or female. In still other species, some individuals are permanently gonochoristic, whereas some in the same population are hermaph-

roditic. In all cases, cross-fertilization is probably the rule.

Sperm appear to arise primarily from choanocytes; eggs arise from choanocytes or archaeocytes. Spermatogenesis usually occurs in distinct **spermatic cysts** (= **sperm follicles**), which form either when all the cells of a choanocyte chamber are transformed into spermatogonia or when transformed choanocytes migrate into the mesohyl and aggregate there (Figure 6.13A). Little is known about oogenesis, although available information suggests that solitary oocytes develop within cysts surrounded by a layer of follicle cells and nurse cells (trophocytes). Meiosis commences after an oogonium has accumulated a sufficient quantity of food reserves, presumably supplied by feeding on the trophocytes (Figure 6.13B).

There is only one, rather brief, account of the embryogeny of a hexactinellid (Okada 1928). Therefore, our discussion is restricted to generalities about the Demospongiae and Calcarea. Mature sperm and oocytes are released into the environment through the aquifer-

(A)

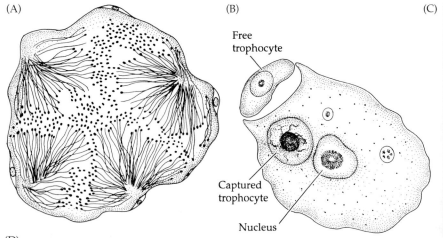

(B)

Free
trophocyte

Captured
trophocyte

Nucleus

(C)

(D)

Figure 6.13   Sexual reproduction in sponges. (A) Sperm follicle (in section) containing mature spermatozoa. (B) An oocyte (in section) of *Ephydatia fluviatilis* (Demospongiae) is phagocytizing a trophocyte. Inside the oocyte is a trophocyte that was recently ingested. (C) Sperm release from a tubular West Indian sponge, *Aplysina archeri* (Demospongiae). The sponge is about 1.5 m tall. (D) Oocyte release in the sponge *Agelas* (Demospongiae). The individual in the foreground is covered by cords of yellow mucus that surround the oocytes during their early development; two specimens in the center show no sign of oocyte release.

ous system. The rapid release of sperm from sponge oscula is dramatic, and such individuals are often referred to as "smoking sponges" (Figure 6.13C). Sperm release may be synchronized in a local population or restricted to certain individuals. Fertilization usually takes place in the water (ovipary) with subsequent planktonic larvae. However, some sponges practice vivipary, and in these species sperm are taken into the aquiferous system of neighboring oocyte-containing individuals. They must then cross the cellular barrier of the choanoderm, enter the mesohyl, locate the oocytes, penetrate the follicular barrier, and finally fertilize the egg. In at least some species, this impressive feat involves sperm capture by choanocytes and enclosure in an intracellular vesicle (somewhat like the formation of a food vacuole during feeding). The choanocyte then loses its collar and

flagellum and migrates through the mesohyl as an ameboid cell, transporting the sperm to the oocyte (Figure 6.14). The migratory choanocyte is called a **carrier cell**, or **transfer choanocyte**. Choanocytes no doubt regularly consume and digest the unlucky sperm of different species of sponges and other benthic invertebrates but, by some as yet undiscovered recognition mechanism, they respond with a remarkably different behavior to sperm of their own kind.

In viviparous species, embryos are typically released as mature swimming larvae. Release of the larva is through either the excurrent plumbing of the aquiferous system or a rupture in the parent's body wall. Larvae may settle directly, they may swim about for several hours or a few days before settling, or they may simply crawl about the substratum until ready to attach. In all known cases, the larvae are lecithotrophic. In general, littoral sponges tend to produce planktonic larvae, whereas subtidal species' larvae tend to settle directly or move about on the ocean floor for a few days before beginning growth into a new adult individual.

(A)

Oocyte

Mesohyl

Choanocyte

Sperm

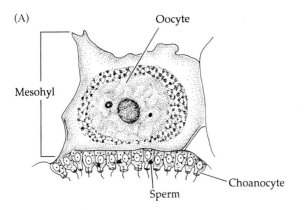

(B)

Mesohyl

Oocyte

Choanocyte

Sperm

Transfer
choanocyte

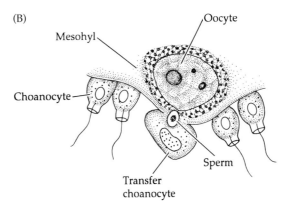

Figure 6.14 Fertilization in the calcareous sponge *Grantia*. (A) Sperm are trapped by choanocytes; an egg is lying in the mesohyl adjacent to the choanoderm. (B) A transfer choanocyte gives up its sperm to the egg; note that the egg lies next to the choanoderm and that the choanocyte has lost its flagellum.

Three basic larval types have been described in sponges: "**coeloblastula**" larvae (= "**blastula**" larvae), **parenchymula larvae** (= **parenchymella larvae**), and **amphiblastula larvae**. Most demosponges incubate embryos until a late stage, producing a solid parenchymula larva with an outer surface of monoflagellated cells and an inner mesohyl-like core of matrix and cells (Figure 6.15). Parenchymula larvae have a short planktonic life, usually just a few days. During this swimming phase the larvae of at least some species can change shape rapidly from elongate to ovoid to flat.

Following settling, the external flagellated cells disappear and flagellated choanocytes appear internally, as choanoderm. This process has long been attributed to a unique embryological inversion process wherein external cells drop their flagella, migrate to the inner cell layer, then re-form the flagella. However, recent work has challenged the existence of this inversion process and suggests that the external flagellated cells are simply shed or phagocytized during larval metamorphosis, the internal choanocytes subsequently forming anew from archaeocytes. In any case, the result of this postsettlement metamorphosis is a tiny leuconoid form called a **rhagon**.

Calcareous sponges (and a few demosponges) often release their embryos early, as free-swimming "coeloblastula" larvae (Figure 6.16A). These larvae may undergo one of two developmental processes. In the simplest case, transformation of the larva involves an inward migration of surface cells that have lost their flagella; these same cells subsequently regain their flagella as they metamorphose into choanocytes.

A more complex embryonic development produces two distinct cell types

(A)

(B)

(C)

(D)

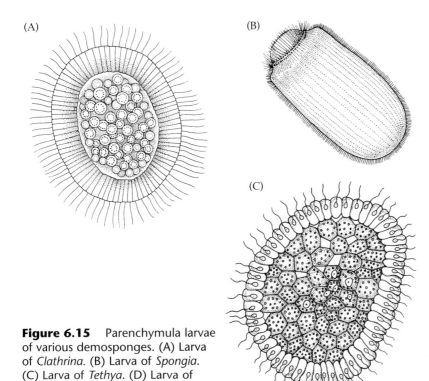

Figure 6.15 Parenchymula larvae of various demosponges. (A) Larva of *Clathrina*. (B) Larva of *Spongia*. (C) Larva of *Tethya*. (D) Larva of *Lissodendoryx isodictyalis*.

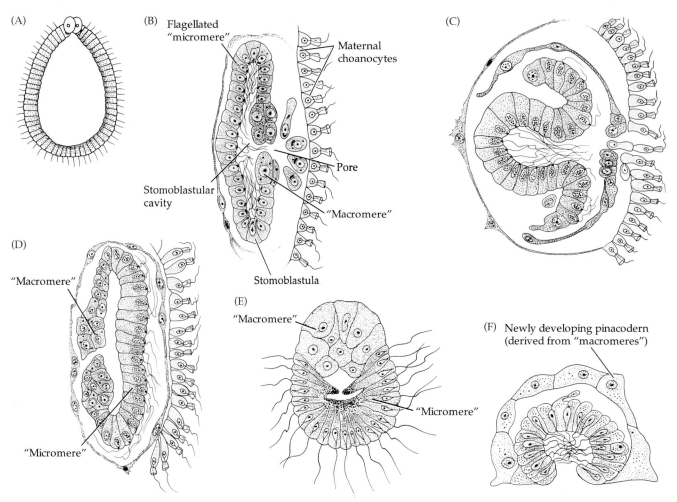

(A)

(B) Flagellated "micromere"

Maternal choanocytes

Pore

"Macromere"

Stomoblastular cavity

Stomoblastula

(C)

(D)

"Macromere"

"Micromere"

(E)

"Macromere"

"Micromere"

(F) Newly developing pinacoderm (derived from "macromeres")

Figure 6.16 "Coeloblastula" and amphiblastula larvae (in section). (A) Typical "coeloblastula" larva with its posterior "macromeres." (B–D) During the remarkable process of inversion in *Scypha*, the stomoblastula turns itself inside out to form an amphiblastula larva with externally directed flagella. (E) A typical amphiblastula larva (*Scypha*). (F) Settled young sponge (*Scypha*) after invagination of flagellated cells.

resembling the macromeres and micromeres of some true Metazoa. At the 16-cell stage, eight large round cells ("macromeres") rest at one pole, and eight smaller cells ("micromeres") form most of the hollow embryo. The larger cells are destined to be future pinacoderm and mesohyl, and the smaller cells become the choanoderm. The "micromeres" divide rapidly and develop flagella that extend into the embryo's cavity. The "macromeres" remain undivided for some time and never develop flagella; in the center of the "macromere" cell cluster is a pore to the outside. This stage is called the **stomoblastula**. While still within the mesohyl of the adult sponge, the stomoblastula ingests nutrient-rich amebocytes. As development proceeds, a remarkable process of inversion takes place, in which the stomoblastula turns inside out through the pore, moving the flagella from the inside to the outside and producing a hollow, flagellated, amphiblastula larva (Figure 6.16B–D). This larva subsequently is released from the parent sponge. There is no known counterpart to this process in any other sponge group, or in the higher Metazoa.

The initiation of settlement and metamorphosis in sponges is poorly understood, especially since they apparently lack formal sensory receptors and neurons at any life stage. Recent work by Woollacott and Hadfield

(1996) shows that certain chemicals (KCl and CsCl) induce metamorphosis in the larvae of the demosponge *Aplysilla*, but the mechanism of this phenomenon is still a mystery.

After a free-swimming period, the amphiblastula larva settles on its flagellated end. Metamorphosis involves a rapid proliferation of the "macromeres" to form pinacoderm that overgrows the flagellated hemisphere. The flagellated cells pocket inward to form a chamber lined with cells destined to become choanocytes (Figure 6.16F). An osculum breaks through, and the tiny asconoid-like sponge becomes capable of circulating water and feeding. This initial functional stage is called an **olynthus** (Figure 6.4A). After further growth, it will become an asconoid, syconoid, or leuconoid adult.

The preceding account of development and larval types is drastically simplified. In fact, sponges show more variation in embryological development than many other animal groups. We recommend Bergquist (1978) as a good starting place if you wish to learn more details.

Some Additional Aspects of Sponge Biology

Some basic sponge ecology has been presented in the previous sections of this chapter. However, because sponges play such important roles in so many marine habitats, we add here some special aspects of their natural history.

Distribution and Ecology

Certain distributional patterns are evident among the three classes of sponges. Calcareous sponges (and coralline demosponges) are far more abundant in shallow waters (less than 200 m), although they are not uncommon at slope depths, and a few species (particularly *Scypha*) have even been reported from depths to 3,800 m. Hexactinellids, which were common in shallow seas of past eras, are now largely restricted to depths below 200 m, except in extremely cold environments (such as Antarctica), where they occur in shallow waters. The demosponges live at all depths. Calcareous sponges are probably restricted largely to shallow waters because they require a firm substratum for attachment. On the other hand, many demosponges and hexactinellids grow on soft sediments, attaching by means of rootlike spicule tufts or mats. The coralline sponges, once a predominant group on shallow tropical reefs, are now largely restricted to shaded crevices and caves, or to subreef depths, where their potential competitors (the hermatypic corals) cannot grow. They are thought to be relics of major reef-constructing groups of Mesozoic and Paleozoic seas.

Although sponges are very sensitive to suspended sediment in their environment, they seem to be quite resistant to hydrocarbon and heavy metal contamination. Many species can actually accumulate these contaminants without apparent harm. The capacity of certain species to accumulate metals at far higher levels than that of the environment has been suggested as a possible defense mechanism (antipredation, antifouling). Detergents also do not appear to affect many sponges, and in fact may even serve as a source of nutrition for these amazingly adaptable animals.

Sponges are the dominant animals in a great many benthic marine habitats. Most rocky littoral regions harbor enormous numbers of sponges, and recent work indicates that they even occur in large numbers (and larger size) around Antarctica. Although many animals prey on sponges, the amount of serious damage they do is usually slight. Some tropical fishes and turtles crop certain kinds of sponges, and small predators (mainly opisthobranchs) consume limited amounts of sponge "tissue" in both warm and temperate seas. Overall, however, sponges appear to be very stable and long-lived animals, probably in part due to their spicules and toxic and/or distasteful compounds that discourage potential predators.

Biochemical Agents

Even a casual seashore explorer or SCUBA diver will quickly notice that sponges are just about everywhere. Most grow on open rock or occasionally sand/mud surfaces, where they are obviously exposed to potential predation. Clearly, some mechanism(s) must be working to prevent these animals from being cropped excessively by predators. The primary defense mechanisms in sponges are mechanical (skeletal structures) and biochemical. Studies over the past two decades show that sponges manufacture a surprisingly broad spectrum of biotoxins, some of which are quite potent. A few, such as *Tedania* and *Neofibularia,* can cause painful skin rashes in humans.

Research in sponge biochemistry has also revealed the widespread occurrence of antimicrobial agents in sponges. Sponges appear to use "chemical warfare" not only to reduce predation and prevent infection by microbes, but also to compete for space with other sessile invertebrates such as ectoprocts, ascidians, and even other sponges. Different species have evolved chemicals (**allelochemicals**) that may be species-specific deterrents or actually lethal weapons for use against competing sessile and encrusting organisms. For example, the coral-inhabiting sponge *Siphonodictyon* releases a toxic chemical into the mucus exuded from its oscula, thus preventing potential crowding by maintaining a zone of dead coral polyps around each osculum (Figure 6.17).

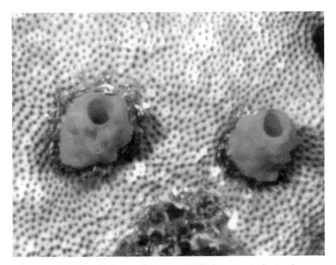

Figure 6.17 *Siphonodictyon coralliphagum* infests the hermatypic coral *Siderastrea siderea* on a Caribbean reef. Note the "dead zone" between the oscular chimneys of the sponge and the coral polyps.

Many of the chemicals produced by sponges and other marine invertebrates are being closely studied by natural products chemists and biologists interested in their potential as pharmaceutical agents. Compounds with respiratory, cardiovascular, gastrointestinal, anti-inflammatory, antitumor, and antibiotic activities have been identified from many marine sponges. One New Zealand sponge (*Halichondria moorei*) has long been used by native Maoris to promote wound healing and was recently discovered to contain remarkably high concentrations (10 percent of the sponge dry weight) of the potent anti-inflammatory agent potassium fluorosilicate. Sponge antimicrobial compounds are also of potential use to humans. For example, a compound that is active against the herpes virus (belonging to a class of chemicals called **arabinosides**) has been found in the tropical sponge *Cryptotethya crypta*. Some sponges, including the west Pacific species *Luffariella variabilis*, produce a remarkable terpenoid compound called **manoalide** that is not only an extremely powerful antibacterial compound but also acts as both an analgesic and anti-inflammatory agent. One study (Bergquist and Bedford 1978) found that 87 percent of the temperate sponges, and 58 percent of the tropical species examined in New Zealand produced extracts with antibacterial activity. Sponges of the genera *Halichondria* and *Pandaros* are known to produce potent antitumor compounds belonging to a group of chemicals called **halichondrins**. The coming decades will undoubtedly witness the emergence of many new pharmacological compounds of poriferan origin.

Growth Rates

Little is known regarding growth rates in sponges, but available data suggest that rates vary widely among species. Some species are annuals (especially small-bodied calcareous sponges of colder waters); hence they grow from larvae or gemmules to reproductive adulthood in a matter of months. Others are perennials and grow so slowly that almost no change can be seen from one year to the next; this growth pattern is especially true of tropical and polar demosponges. Age estimates of perennial species range from 20 to 100 years.

Some sponges are capable of very rapid growth, and they regularly overgrow neighboring flora and fauna. For example, the tropical encrusting sponge *Terpios* grows over both living and nonliving substrata. In Guam this sponge grows at rates averaging 23 mm per month over almost every live coral in the area as well as over hydrocorals, molluscs, and many algae. Experiments have shown that *Terpios* is toxic to living corals, and presumably to many other animals. Still another physiological trick of some sponges is the ability to rapidly produce copious amounts of mucus when disturbed. On the west coast of North America, the beautiful red-orange *Plocamia karykina* covers itself with a thick layer of mucus when injured or disturbed. Yet the little red sea slug *Rostanga pulchra* has evolved the ability to live and feed inconspicuously on this and other sponges, and even lays its camouflaged red egg masses on the sponge's exposed surface without eliciting the mucous reaction.

Symbioses

Commensalism is common among sponges of all kinds. It would be difficult to find a sponge that is not utilized by at least some smaller invertebrates and often by fishes (e.g., gobies and blennies) as refuge. The porous nature of sponges makes them ideally suited for habitation by opportunistic crustaceans, ophiuroids, and various worms. A single specimen of *Spheciospongia vesparia* from Florida was found to have over 16,000 alphaeid shrimps living in it, and a study from the Gulf of California found nearly 100 different species of plants and animals in a 15 × 15-cm piece of *Geodia mesotriaena*.

Most symbionts of sponges use their hosts only for space and protection, but some rely on the sponge's water current for a supply of suspended food particles. A classic example of this phenomenon is the male–female pair of shrimp (*Spongicola*) that inhabit hexactinellid sponges known as Venus's flower basket (*Euplectella*; Figure 6.1E). The shrimp enter the sponge when they are young, only to become trapped in their host's glasslike case as they grow too large to escape. Here they spend their lives as "prisoners of love." Appropriately, this sponge (with its guests) is a traditional wedding gift in Japan—a symbol of the lifetime bond between two partners.

Other even more intimate symbiotic relationships with sponges are common. Some snails and clams characteristically have specific sponges encrusting their shells, and many species of crabs (hermits and brachyurans) collect certain sponges and cultivate them on their shell or carapace. Demosponges, such as *Suberites*, are commonly involved in these commensalistic relationships. The sponge serves primarily as protective camouflage for its host, and it perhaps benefits by being carried about to new areas. And the sponge no doubt feeds off small bits of animal matter dislodged during the feeding activities of its host.

Another spectacular example of poriferan symbiosis are certain sponge–bacteria and sponge–algae associations that appear to be mutualistic. For example, a typical member of the demosponge order Verongida contains a mesohyl bacterial population accounting for some 38 percent of its body's volume, far exceeding the actual sponge-cell volume of only 21 percent. Presumably, the sponge matrix provides a rich medium for bacterial growth, and the host benefits by being able to conveniently phagocytize the bacteria for food. Similar relationships are common between poriferans and various cyanobacteria. Recent evidence suggests that some products of normal cyanobacterial metabolism (e.g., glycerol and certain organic phosphates) are translocated directly to the sponge for nutrition. In many sponges, both regular bacteria and cyanobacteria occur, the former in deeper cellular regions, the latter closer to the surface where light is available. In a remarkable study, C. R. Wilkinson (1983) showed that 6 of the 10 most common sponge species on the forereef slope of Davies Reef (Great Barrier Reef) are actually net primary producers, with three times more oxygen produced by photosynthesis (by their symbionts) than consumed by respiration. In some areas of the Caribbean and Great Barrier Reef, sponges are second only to corals in overall biomass, and they appear to owe their rapid growth to the presence of large populations of symbiotic cyanobacteria. Most freshwater spongillids maintain similar relationships with zoochlorellae (symbiotic green algae; Chlorophyta). These sponges grow larger and more rapidly than specimens of the same species that are kept in dark conditions. Some marine sponges (e.g., the boring sponges *Cliona* and *Spheciospongia*) harbor commensal zooxanthellae similar to those of corals. Commensalistic relationships have also been reported between sponges and red algae, filamentous green algae, and diatoms.

Not all sponge symbioses are commensal or mutualistic. In fact, some are clearly harmful—for example, the boring demosponges that excavate complex galleries in calcareous material such as corals and mollusc shells (Figure 6.18). The phenomenon of boring, known as **bioerosion**, causes significant damage to commercial oysters as well as to natural coral, clam, and scallop populations. The active boring process involves a chemical and mechanical removal of fragments or chips of the calcareous material by specialized archaeocytes

called **etching cells**. The use of carbonic anhydrase has been implicated in this process. The chips are expelled in the excurrent canal system and can contribute significantly to local sediments.

Sponge bioerosion has a significant impact on coral reefs. Perhaps even more important than actual erosion is the weakening of attachment regions of large corals. This action may result in much coral loss during heavy tropical storms. Boring sponges do not appear to gain any direct nourishment from their host coral; rather they use it as a protective casing in which they reside. If you carefully examine the shells of dead bivalves along any beach, you will discover that many of them are perforated with minute holes and galleries of boring sponges. These poriferans are responsible for a major portion of the initial breakdown of such calcareous structures, and thus they set the stage for their eventual decomposition and recycling through Earth's geochemical cycle.

Poriferan Phylogeny

The Origin of Sponges

Sponges are an ancient group, and the important events in their origin and early evolution lie hidden in Precambrian time. The unique nature of the poriferan bauplan is clear, however, and it is strikingly revealed by its aquiferous system, cellular totipotency and reproductive flexibility, and by the lack of true tissues, reproductive organs, body polarity, and basement membrane. These features, in combination with the prevalence and importance of flagellated (monociliated) cells in sponges, strongly suggest a direct protistan ancestry. It would seem that the poriferans share as many similari-

(A)

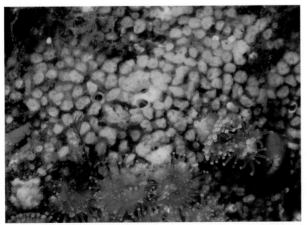

(B)

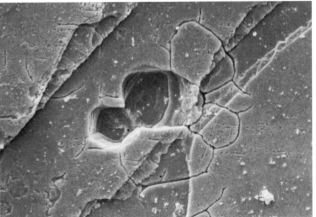

Figure 6.18 Boring sponges. (A) Surface of a coral (stellate openings) infected by the sponge *Cliona* (circular oscula); (B) A close view (SEM) of the surface of a clam shell, showing six eroded "chips," two of which have been entirely removed and four that are only partly etched by *Cliona*.

ties with protists than they do with the higher Metazoa (e.g., cellular totipotency; excretory, respiratory, and osmoregulatory strategies; reliance on flagellated cells in a variety of ways, including feeding; strictly intracellular digestion). At the same time, sponges appear to stand apart from all other Metazoa in their possession of the unique aquiferous system, which represents a key synapomorphy defining this phylum. Sponges thus constitute a biological grade that contributes to our understanding of the transition from unicellular to multicellular life.

Current opinion views the Porifera as originating from flagellated protist ancestors—either a simple hollow, free-swimming colonial form or a colonial choanoflagellate. The choanoflagellates possess certain features that seem to ally them strongly to sponges. For example, the collar cells of sponges are strikingly similar to the collar cells of choanoflagellates. Curiously, similar collar cells have also been found in some widely divergent Metazoa (certain echinoderm larvae, in the oviducts of some sea cucumbers, and in certain corals). However, these discoveries do little to diminish the force of the argument for a choanoflagellate ancestry to the sponges.

The mesohyl is generally viewed as originating (evolutionarily) by simple ingression of surface cells, as seen in the embryogeny of many living sponges. Adoption of a benthic lifestyle by the earliest sponges could have fostered increased body size. Increase in size led to surface-to-volume problems that were overcome by the evolution of the syconoid and leuconoid bauplans, increasing the surface area of the choanoderm-lined areas and maintaining small diffusion distances as the aquiferous system became more complex.

The solutions that poriferans evolved to problems of survival created a group of animals unlike any other. Sponges achieved multicellularity and large body size without such typically metazoan traits as embryological tissue layering, neuronal coordination, extracellular digestion, excretory structures, or fixed reproductive structures. Taken together, these and other poriferan attributes suggest that the sponges arose very early in metazoan cladogenesis.

Evolution within the Porifera

Sponges are such an ancient and enigmatic phylum that their phylogeny has largely eluded biologists. There is no generally agreed-upon phylogenetic hypothesis of relationships among the classes. We do know that sponges probably evolved during the Precambrian (Figure 6.19). Their hard skeletal components have left good fossil records for all three extant classes, beginning in the Cambrian and extending to the present. Well over 1,000 fossil genera have been described, about 20 percent of which are still extant. The early Paleozoic witnessed the growth of massive tropical reefs composed largely of four spongelike groups: the archaeocyathans, stromatoporoids, sphinctozoans, and chaetetids. The oldest group, Archaeocyatha (Figure 6.20), had a short life within the Cambrian (550–500 mya). Sphinctozoans also appeared in the Cambrian (about 540 mya), while chaetetids and stromatoporoids first appeared in the Ordovician (about 480 mya). The affinities of these four groups have been debated for the past 100 years, and various alliances with cyanobacteria, red algae, ectoprocts, cnidarians, and foraminiferans have been proposed. The discovery of living coralline sponges led most workers to conclude that the majority of species in these four groups were primitive, but true, sponges.

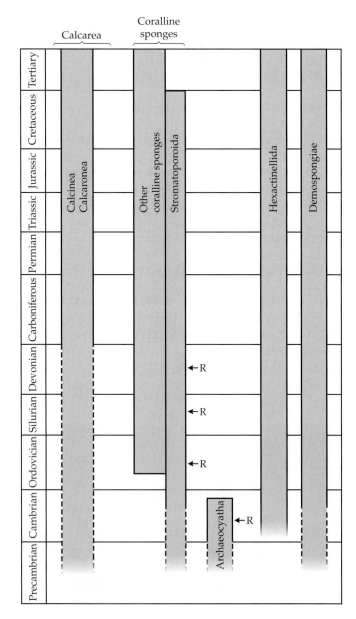

Figure 6.19 Fossil record of the three sponge classes, the coralline sponges, and Archaeocyatha. Dashed lines indicate suggested occurrence, even though fossils have not yet been found. "R" indicates the times when the group in question is known to have been an important marine reef builder.

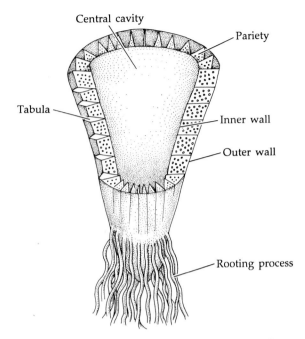

Figure 6.20 A typical archaeocyathan. A vertical section has been partly cut away to show the structure between the inner and outer walls (i.e., vertical parieties and horizontal tabula).

Stromatoporoids have a long geological history, extending from late Cambrian times to the present. The nature of these reef-building invertebrates has been hotly debated. That the fossil stromatoporoids were true sponges is most strongly suggested by the apparent homology of structures called **astrorhizae**, found in their calcareous skeletons, to similar stellate impressions in the skeletons of recent coralline sponges ("sclerosponges"). In living sponges these stellate marks are the traces of the converging exhalent canal systems (be-

neath the oscula). The absence of siliceous spicules in the calcareous skeletons of fossil stromatoporoids has been cited as evidence against a true sponge relationship. However, not all recent stromatoporoids have siliceous spicules; and in some that do, these spicules are never incorporated into the calcareous basal skeleton that would be fossilized anyway.

The coralline sponges have also had a long history, and many distinctive early lines have died out since their origins in Cambrian and Precambrian times. Interestingly, some of these early calcareous forms were important reef builders during the Permian, but like stromatoporoids, the few surviving genera of those lines have all retreated to marine cave habitats in modern tropical seas.

Unlike the coralline sponges, which have decreased in abundance and diversity since the Mesozoic, the remaining calcareous sponges and the demosponges appear to have increased in diversity throughout their history. Hexactinellids, on the other hand, were most diverse and abundant during the Cretaceous. The oldest hexactinellid fossils, of the early Cambrian, were all thin-walled, saclike sponges with a dispersed surface spicule layer that probably could not support a thick body wall. During the Paleozoic, hexactinellids were common in shallow-water environments. Since then, however, they have become restricted largely to the deep oceans. The demosponges were well established by the mid-Cambrian, from which the earliest fossils are known, and all known orders of modern demosponges are found in Cretaceous rocks. But only recently have the complex relationships among the dozen or so orders of Demospongiae been critically examined. The challenge of unraveling the relationships in this ponderous group of sponges is still in an early phase.

Selected References

General References

Bakus, G. J. 1964. The effects of fish-grazing on invertebrate evolution in shallow tropical waters. Occ. Pap. Allan Hancock Fndtn. 27: 1–29.

Bakus, G. J. 1969. Energetics and feeding in shallow marine waters. Int. Rev. Gen. Exp. Zool. 4: 275–369.

Bakus, G. J., N. M. Targett and B. Schulte. 1986. Chemical ecology of marine organisms: An overview. J. Chem. Ecol. 12: 951–987.

Bayer, F. M. and H. B. Owre. 1968. *The Free-Living Lower Invertebrates.* Macmillan, New York.

Bergquist, P. R. 1978. *Sponges.* University of California Press, Berkeley.

Bergquist, P. 1985. Poriferan relationships. *In* S. C. Morris et al. (eds.), *The Origin and Relationships of Lower Invertebrates.* Syst. Assoc. Spec. Vol. 28, Oxford, pp. 14–27.

Bergquist, P. R. and J. J. Bedford. 1978. The incidence of antibacterial activity in marine Demospongiae; systematic and geographic considerations. Mar. Biol. 46: 215–221.

Bergquist, P. R. and R. J. Wells. 1983. Chemotaxonomy of the Porifera: The development and current status of the field. *In Marine Natural Products, III.* Academic Press, New York, pp. 1–50.

Bond, C. 1997/1998. Keeping up with the sponges. Nat. Hist. 106(11): 22–25.

Borojevic, R. 1970. Différentiation cellulaire dans l'embryogenèse et la morphogenèse chez les spongiaires. Symp. Zool. Soc. Lond. 25: 467–490.

Borojevic, R., W. G. Fry, W. C. Jones, C. Lévi, R. Rasmont, M. Sarà and J. Vacelet. 1967. Mise au point actuelle de la terminologie des Éponges. (A reassessment of the terminology for sponges.) Bull. Mus. Hist. Nat. Paris (2) 39: 1224–1235.

Boury-Esnault, N. 1973. L'exopinacoderme des Spongiaires. Bull. Mus. Nat. D'Hist. Natur. (Zool.) 117: 1193–1206.

Boury-Esnault, N. 1977. A cell type in sponges involved in the metabolism of glycogen. The gray cells. Cell Tiss. Res. 175: 523–539.

Boury-Esnault, N., L. deVos, C. Donadey and J. Vacelet. 1985. Ultrastructure of choanosome and sponge classification. 3rd Internat. Sponge Conf., 1985, pp. 237–244.

Bowerbank, J. S. 1861, 1862. On the anatomy and physiology of the Spongiidae. Part 1: On the spicula. Philos. Trans. R. Soc. Lond. 148: 279–332. Part 2: Proc. R. Soc. Lond. 11: 372–375.

Bowerbank, J. S. 1861. On the anatomy and physiology of the Spongiidae. Philos. Trans. R. Soc. Lond. 152: 747–829, 1087–1138.

Bowerbank, J. S. 1864, 1866, 1874. A monograph of the British Spongiidae. Vols. 1, 2, and 3. Ray Society, London.

Brien, P. 1968. The sponges, or Porifera. *In* M. Florkin and B. T. Scheer (eds.), *Chemical Zoology,* Vol. 2. Academic Press, New York, pp. 1–30.

Brien, P., C. Lévi, M. Sara, O. Tuzet and J. Vacelet. 1973. Spongiaires. *In* P. Grassé (ed.), *Traité de Zoologie* 3(1): 716. Masson et Cie, Paris.

Brill, B. 1973. Untersuchungen zur Ultastruktur der Choanocyte von *Ephydatia fluviatilis,* L. Z. Zellforsch. 144: 231–245.

Burkholder, P. R. 1973. The ecology of marine antibiotics and coral reefs. *In* O. A. Jones and R. Endean (eds.), *Biology and Geology of Coral Reefs,* Vol. 1. Biology, 1. Academic Press, New York, pp. 117–182.

Burkholder, P. R. and K. Ruetzler. 1969. Antimicrobial activity of some marine sponges. Nature Lond. 222: 983–984.

Burton, M. 1932. Sponges. Discovery Rep. 6: 327–392.

Carballo, J. L., J. E. Sanchez-Moyano and J. C. Garcia-Gomez. 1994. Taxonomic and ecological remarks on boring sponges (Clionidae) from the Strait of Gibraltar: Tentative bioindicators? Zool. J. Linn. Soc. 112: 407–424.

Cox, G. and A. W. D. Larkum. 1983. A diatom apparently living in symbiosis with a sponge. Bull. Mar. Sci. 33(4): 943–945.

Curtis, A. 1979. Individuality and graft rejection in sponges: A cellular basis for individuality in sponges. *In* G. Larwood and B. Rosen (eds.), *Biology and Systematics of Colonial Organisms.* Academic Press, New York, pp. 39–48.

Dayton, P. K. 1979. Observations of growth, dispersal and population dynamics of some sponges in McMurdo Sound, Antarctica. *In* C. Lévi and N. Boury-Esnault (eds.), *Biologie des Spongiaires,* Centre Nat. Recherche Scient., Paris, pp. 271–282.

de Laubenfels, M. W. 1955. Porifera. *In* R. C. Moore (ed.), *Treatise on Invertebrate Paleontology. Archaeocyatha and Porifera,* E21–E112. Geological Society of America and the University of Kansas Press, Lawrence.

De Vos, L., K. Rützler, N. Boury-Esnault, C. Donadey and J. Vacelet. 1991 *Atlas of Sponge Morphology.* Smithsonian Institution Press, Washington, D.C.

Faulkner, D. J. 1973. Variabilin, an antibiotic from the sponge *Ircinia variabilis.* Tetrahedron Let. 29: 3821–3822.

Faulkner, D. J. 1977. Interesting aspects of marine natural products chemistry. Tetrahedron Let. 33: 1421–1443.

Fautin, D. G. (ed.). 1988. *Biomedical Importance of Marine Organisms.* Mem. Calif. Acad. Sci. No. 13.

Fell, P. E. 1974. Porifera. *In* A. C. Giese and J. S. Pearse (eds.), *Reproduction of Marine Invertebrates,* Vol. 1. Academic Press, New York, pp. 51–132.

Fell, P. E. 1976. Analysis of reproduction in sponge populations: An overview with specific information on the reproduction of *Haliclona loosanoffi. In* F. W. Harrison and R. R. Cowden (eds.), *Aspects of Sponge Biology.* Academic Press, New York, pp. 51–67.

Fell, P. E. 1995. Deep diapause and the influence of low temperature on the hatching of the gemmules of *Spongilla lacustris* (L.) and *Eunapius fragilis* (Leidy). Invert. Biol. 114(1): 3–8.

Fell, P. E. 1997. Porifera: The sponges. *In* S. F. Gilbert and A. M. Raunio (eds.), *Embryology: Constructing the Organism.* Sinauer Associates, Sunderland, MA, pp. 39–54.

Finks, R. M. 1970. The evolution and ecologic history of sponges during Palaeozoic times. Symp. Zool. Soc. Lond. 25: 3–22.

Frost, T. M. 1976. Sponge feeding: A review with a discussion of continuing research. *In* F. W. Harrison and R. R. Cowden (eds.), *Aspects of Sponge Biology.* Academic Press, New York, pp. 283–298.

Fry, W. G. (ed.). 1970. *The Biology of the Porifera.* Academic Press, New York.

Garrone, R. 1978. *Phylogenesis of Connective Tissue.* S. Karger, Basel.

Garrone, R. and J. Pottu. 1973. Collagen biosynthesis in sponges: Elaboration of spongin by spongocytes. J. Submicrosc. Cytol. 5: 199–218.

Goodwin, T. W. 1968. Pigments of Porifera. *In* M. Florkin and B. T. Scheer (eds), *Chemical Zoology,* Vol. 2. Academic Press, New York, pp. 53–64.

Harrison, F. W. and R. R. Cowden (eds.). 1976. *Aspects of Sponge Biology.* Academic Press, New York.

Harrison, F. W. and L. De Vos. 1990. Porifera. *In* F. W. Harrison and J. A. Westfall (eds.), *Microscopic Anatomy of Invertebrates,* Vol. 2. Alan R. Liss, New York.

Hartman, W. D. 1958. Natural history of the marine sponges of southern New England. Bull. Peabody Mus. Nat. Hist. 12: 1–155.

Hartman, W. D. 1982. Porifera. *In* S P. Parker (ed.), *Synopsis and Classification of Living Organisms,* Vol. 1. McGraw-Hill, New York, pp. 641–666.

Hartman, W. D. and T. F. Goreau. 1975. A Pacific tabulate sponge, living representative of a new order of sclerosponges. Postilla 167: 1–21.

Hartman, W. D. and H. M. Reiswig. 1973. The individuality of sponges. *In* R. S. Boardman, A. H. Cheetham and W. A. Oliver (eds.), *Animal Colonies.* Dowden, Hutchinson & Ross, Inc., Stroudsburg, PA, pp. 567–584.

Hartman, W. D., J. W. Wendt and F. Wiedenmayer. 1980. *Living and Fossil Sponges.* (Compiled by R. N. Ginsburg and P. Reid.) Sedimenta VIII (Comparative Sedimentology Lab, Div. Mar. Geol. and Geophysics, University of Miami).

Hildemann, W. H., I. S. Johnson and P. L. Jobiel. 1979. Immunocompetence in the lowest metazoan phylum: Transplantation immunity in sponges. Science 204: 420–422.

Hill, D. 1972. Archaeocyatha. *In* R. C. Moore (ed.), *Treatise on Invertebrate Paleontology* 1: 158. Geological Society of America and the University of Kansas Press, Lawrence.

Hill, D. and E. C. Strumm. 1956. Tabulata. *In* R. C. Moore (ed.), *Treatise on Invertebrate Paleontology,* F: 444–477. Geological Society of America and the University of Kansas Press, Lawrence.

Hooper, J. N. A., R. J. Capon, C. P. Keenan, D. L. Parry, and N. Smit. 1992. Chemotaxonomy of marine sponges. Families Microcionidae, Raspailiidae and Axinellidae, and their relationships with other families in the orders Poecilosclerida and Axinellida (Porifera: Demospongiae). Invertebr. Taxon. 6: 261–01.

Humphreys, T. 1963. Chemical dissolution and *in vitro* reconstruction of sponge cell adhesions. I. Isolation and functional demonstration of components involved. Dev. Biol. 8: 27–47.

Humphreys, T. 1970. Species-specific aggregation of dissociated sponge cells. Nature 228: 685–686.

Hyman, L. H. 1940. *The Invertebrates,* Vol. 1, Protozoa through Ctenophora. McGraw-Hill, New York, pp. 284–364.

Jackson, J. B. C. 1977. Competition on marine hard substances: The adaptive significance of solitary and colonial strategies. Am. Nat. 111: 743–767.

Jackson, J. B. C. and L. Buss. 1975. Allelopathy and spatial competition among coral reef invertebrates. Proc. Natl. Acad. Sci. USA 72: 5160–5163.

Jefford, C. W., K. L. Rinehart, and L. S. Shield (eds). 1987. *Pharmaceuticals and the Sea.* Technomic Publ. Co., Basel, Switzerland.

Jones, W. C. 1962. Is there a nervous system in sponges? Biol. Rev. 37: 1–50.

Kaye, H. and T. Ortiz. 1981. Strain specificity in a tropical marine sponge. Mar. Biol. 63: 165–173.

Kázmierczak, J. 1984. Favositid tabulates: Evidence for poriferan affinity. Science 225: 835–837.

Kázmierczak, J. and S. Kempe. 1990. Modern cyanobacterial analogs of Paleozoic stromatoporoids. Science 250: 1244–1248.

Koltun, V. M. 1968. Spicules of sponges as an element of the bottom sediments of the Antarctic. *In Symposium on Antarctic Oceanography,* Scott Polar Res. Inst., Cambridge, pp. 121–123.

Kuhns, W., G. Weinbaum, R. Turner and M. Burger. 1974. Sponge cell aggregation: A model for studies on cell–cell interactions. Ann. N.Y. Acad. Sci. 234: 58–74.

LaBarbera, M. and G. E. Boyajian. 1991. The function of astrorhizae in stromatoporoids: Quantitative tests. Paleobiology 17(2): 121–132.

Lang, J. C., W. D. Hartman and L. S. Land. 1975. Sclerosponges: Primary framework constructors on the Jamaican deep forereef. J. Mar. Res. 33: 223–231.

Lawn, I. D., G. O. Mackie, and G. Silver. 1981. Conduction system in a sponge. Science 211: 1169–1171.

Lecompte, M. 1956. Stromatoporoidea. *In* R. C. Moore (ed.), *Treatise on Invertebrate Paleontology*, F: F107–F114. Geological Society of America and the University of Kansas Press, Lawrence.

Lévi, C. 1957. Ontogeny and systematics in sponges. Syst. Zool. 6: 174–183.

Lévi, C. and N. Boury-Esnault (eds.). 1979. *Sponge Biology*. Colloques Internationaux du Centre National de la Recherche Scientifique. Ed. Cen. Nat. Resch. Sci. No. 291.

Li, C.-W., J.-Y. Chen and T.-E. Hua. 1998. Precambrian sponges with cellular structures. Science 279: 879–882.

Minchin, E. A. 1900. Sponges. *In* E. R. Lankester (ed.), *A Treatise in Zoology*, Pt. 2. Adam and Charles Black, London, pp. 1–178.

Neigel, J. E. and G. P. Schmahl. 1984. Phenotypic variation within histocompatibility-defined clones of marine sponges. Science 224: 413–415.

Paine, R. T. 1964. Ash and calorie determinations of sponge and opisthobranch tissue. Ecology 45: 384–387.

Palumbi, S. R. 1984. Tactics of acclimation: Morphological changes of sponges in an unpredictable environment. Science 225: 1478–1480.

Randall, J. E. and W. D. Hartman. 1968. Sponge-feeding fishes of the West Indies. Mar. Biol. 1: 216–225.

Reitner, J. and H. Deupp (eds.). 1991. *Fossil and Recent Sponges*. Springer-Verlag, Berlin.

Reiswig, H. 1975. Bacteria as food for temperate-water marine sponges. Can. J. Zool. 53: 582–589.

Rezvoi, P. D., I. T. Zhuravleva and V. M. Koltun. 1971. Phylum Porifera. *In* Y. A. Orlov and B. S. Sokolov (eds.), *Fundamentals of Paleontology*, Vol. 1, Pt. II. Porifera, Archaeocyatha, Coelenterata, Vermes. Israel Program for Scientific Translations, Jerusalem, pp. 5–97.

Rinehart, K. L., Jr., and 25 others. 1981. Marine natural products as sources of antiviral, antimicrobial, and antineoplastic agents. Pure Appl. Chem. 53: 795–817.

Rützler, K. 1970. Spatial competition among Porifera: Solution by epizoism. Oecologia 5: 85–95.

Rützler, K. (ed.) 1990. *New Perspectives in Sponge Biology*. Smithsonian Institution Press, Washington, D.C.

Sara, M. 1970. Competition and cooperation in sponge populations. Symp. Zool. Soc. Lond. 25: 273–285.

Sara, M. 1974. Sexuality in the Porifera. Bull. Zool. 41: 327–348.

Sara, M. and J. Vacelet. 1973. Ecologie des Demosponges. *In* P. Grassé (ed.), *Traite de Zoologie* 3(1): 462–576. Masson et Cie, Paris.

Schwab, D. W. and R. E. Shore. 1971. Fine structure and composition of a siliceous sponge spicule. Biol. Bull. 140: 125–136.

Sharma, G. H. and B. Vig. 1972. Studies on the antimicrobial substances of sponges. VI. Structure of two antibacterial substances isolated from the marine sponge *Dysidea herbacea*. Tetrahedron Lett. 28: 1715–1718.

Shore, R. E. 1972. Axial filament of siliceous sponge spicules, its organic components and synthesis. Biol. Bull. 143: 689–698.

Simpson, T. L. 1984. *The Cell Biology of Sponges*. Springer-Verlag, New York.

Stearn, C. W. 1975. The stromatoporoid animal. Lethaia 8: 89–100.

Stearn, C. W. 1977. Studies of stromatoporoids by scanning electron microscopy. Mem. Bur. Rech. Geol. Min. 89: 33–40.

Sullivan, B., D. J. Faulkner and L. Webb. 1983. Siphonodictidine, a metabolite of the burrowing sponge *Siphonodictyon* sp. that inhibits coral growth. Science 221: 1175–1176.

Tuzet, O. 1963. The phylogeny of sponges according to embryological, histological and serological data, and their affinities with the Protozoa and Cnidaria. *In* E. C. Dougherty, Z. N. Brown, E. D. Hanson and W. D. Hartman (eds.), *The Lower Metazoa: Comparative Biology and Phylogeny*. University of California Press, Berkeley, pp. 129–148.

Vacelet, J. 1985. Coralline sponges and the evolution of the Porifera. *In* S. C. Morris et al. (eds.), *The Origins and Relationships of Lower Invertebrates*. Syst. Assoc. Spec. Vol. No. 28, Oxford, pp. 1–13.

Vacelet, J. 1988. Indications de profondeur données par les Spongiaires dans les milieux benthiques actuels. Géolog. Médierranéene, 15(1): 13–26.

Vacelet, J. and N. Boury-Esnault. 1995. Carnivorous sponges. Nature 373: 333–335.

Vacelet, J., N. Boury-Esnault, A. Flala-Medioni and C. R. Fisher. 1998. A methanotrophic carnivorous sponge. Nature 377: 296.

Van de Vyver, G. 1975. Phenomena of cellular recognition in sponges. *In* A. Moscona and A. Monroy (eds.), *Current Topics in Developmental Biology*, Vol. 4. Academic Press, New York, pp. 123–140.

Van Soest, R. W. M., T. M. G. van Kempen and J. C. Braekman (eds.). 1994. *Sponges in Time and Space: Biology, Chemistry, Paleontology*. Balkema, Rotterdam.

Warburton, F. E. 1966. The behaviour of sponge larvae. Ecology 47: 672–674.

Weissenfels, N. 1989. *Biologie und mikroskopische Anatomie der Süsswasserschwämme (Spongillidae)*. Gustav Fischer, Stuttgart.

Westinga, E. and P. C. Hoetjes. 1981. The intrasponge fauna of *Speciospongia vesparia* at Curaçao and Bonaire. Mar. Biol. 26: 139.

Wilkinson, C. R. 1983. Net primary productivity in coral reef sponges. Science 219: 410–412.

Wilkinson, C. R. and J. Vacelet. 1979. Transplantation of marine sponges to different conditions of light and current. J. Exp. Mar. Biol. Ecol. 37: 91–104.

Wilson, H. V. 1891. Notes on the development of some sponges. J. Morphol. 5: 511–519.

Wolfrath, B. and D. Barthel. 1989. Production of fecal pellets by the marine sponge *Halichondria panicea* Pallas, 1766. J. Exp. Mar. Biol. Ecol. 129: 81–94.

Wood, R. 1990. Reef-building sponges. Am. Sci. 78: 224–235.

Calcarea

Borojevic, R. 1979. Evolution des spongiaires Calcarea. *In* C. Lévi and N. Boury-Esnault (eds.), *Sponge Biology*, Colloques Internat. C.N.R.S. 291: 527–530.

Borojevic, R., B.-E. Esnault and J. Vacelet. 1990. A revision of the supraspecific classification of the subclass Calcinea (Porifera, Class Calcarea). Bull. Mus. Natn. Hist. Nat. 12: 243–276.

Burton, M. 1963. A revision of the classification of the calcareous sponges. Brit. Mus. Nat. Hist., London.

Hartman, W. D. 1958. A re-examination of Bidder's classification of the Calcarea. Syst. Zool. 7: 97–110.

Jones, W. C. 1965. The structure of the porocytes in the calcareous sponge *Leucosolenia complicata* (Montagu). J. R. Microsc. Soc. 85: 53–62.

Jones, W. C. 1970. The composition, development, form and orientation of calcareous sponge spicules. Symp. Zool. Soc. Lond. 25: 91–123.

Ledger, P. W. and W. C. Jones. 1978. Spicule formation in the calcareous sponge *Sycon ciliatum*. Cell Tiss. Res. 181: 553–567.

Tuzet, O. 1973. Éponges Calcaires. *In* P. Grassé (ed.), *Traité de Zoologie* 3(1): 27–132. Masson et Cie, Paris.

Ziegler, B. and S. Rietschel. 1970. Phylogenetic relationships of fossil calcisponges. Symp. Zool. Soc. Lond. 25: 23–40.

Hexactinellida

Okada, Y. 1928. On the development of a hexactinellid sponge *Farrea sollasii*. Tokyo University Fac. Sci. J., Sect. IV, 2: 1–27.

Reiswig, H. M. 1979. Histology of Hexactinellida (Porifera). *In* C. Lévi and N. Boury-Esnault (eds.), *Sponge Biology*, Colloques Internat. C.N.R.S. 291: 173–180.

Demospongiae

Ayling, A. L. 1980. Patterns of sexuality, asexual reproduction and recruitment in some subtidal marine Demospongiae. Biol. Bull. 158: 271–282.

Ayling, A. L. 1983. Growth and regeneration rates in thinly encrusting Demospongiae from temperate waters. Biol. Bull. 165: 343–352.

Bergquist, P. R. and J. H. Bedford. 1978. The incidence of antibacterial activity in marine Demospongiae: Systematic and geographic considerations. Mar. Biol. 46: 215–221.

Bergquist, P. R. and W. D. Hartman. 1969. Free amino acid patterns and the classification of the Demospongiae. Mar. Biol. 3: 247–268.

Bergquist, P. R. and J. J. Hogg. 1969. Free amino acid patterns in Demospongiae: A biochemical approach to sponge classification. Cah. Biol. Mar. 10: 205–220.

Bergquist, P. R., M. E. Sinclair and J. J. Hogg. 1970. Adaptation to intertidal existence: Reproductive cycles and larval behaviour in Demospongiae. Symp. Zool. Soc. Lond. 25: 247–271.

Bryan, P. G. 1973. Growth rate, toxicity and distribution of the encrusting sponge *Terpios* sp. (Hadromerida: Suberitidae) in Guam, Mariana Islands. Micronesica 9: 237–242.

Cimino, G., S. D'Stefano, L. Minale and G. Sodano. 1975. Metabolism in Porifera. III. Chemical patterns and the classification of the Demospongiae. Comp. Biochem. Physiol. 50B: 279–285.

Connes, R., J.-P. Diaz and J. Paris. 1971. Choanocytes et cellule centrale chez la Démosponge *Suberites massa* Nardo. C. R. Hebd. Séanc. Acad. Sci. 273: 1590–1593.

de Laubenfels, M. W. 1948. The order Keratosa of the phylum Porifera: A monographic study. Occ. Pap. Allan Hancock Fndtn. 3: 1–217.

Elvin, D. W. 1976. Seasonal growth and reproduction of an intertidal sponge, *Haliclona permollis* (Bowerbank). Biol. Bull. 151: 108–125.

Fell, P. E. 1969. The involvement of nurse cells in oogenesis and embryonic development in the marine sponge *Haliclona ecobasis*. J. Morphol. 127: 133–149.

Fell, P. E. 1976. The reproduction of *Haliclona loosanoffi* and its apparent relationship to water temperature. Biol. Bull. 150: 200–210.

Fell, P. E. and K. B. Lewandrowski. 1981. Population dynamics of the estuarine sponge *Halichondria* sp., within a New England eelgrass community. J. Exp. Mar. Biol. Ecol. 55: 49–63.

Finks, R. M. 1967. The structure of *Saccospongia laxata* Bassler (Ordovician) and the phylogeny of the Demospongiae. J. Paleontol. 41: 1137–1149.

Frost, T. M. and C. E. Williamson. 1980. *In situ* determination of the effect of symbiotic algae on the growth of the freshwater sponge *Spongia lacustris*. Ecology 61: 1361–1370.

Gerrodette, T. and A. O. Fleschig. 1979. Sediment-induced reduction in the pumping rate of the tropical sponge *Verongia lacunosa*. Mar. Biol. 55: 103–110.

Guida, V. G. 1976. Sponge predation in the oyster reef community as demonstrated with *Cliona celata* Grant. J. Exp. Mar. Biol. Ecol. 25: 109–122.

Hatch, W. I. 1980. The implication of carbonic anhydrase in the physiological mechanism of penetration of carbonate substrata by the marine burrowing sponge *Cliona celata* (Demospongiae). Biol. Bull. 159: 135–147.

Lévi, C. 1956. Étude des *Halisarca* de Roscoff. Embryologie et systématiques des Démosponges. Arch. Zool. Exp. Gen. 93: 1–181.

Lévi, C. 1973. Systématique de la classe des Demospongiaria (Démosponges). *In* P. Grassé (ed.), *Traité de Zoologie* 3(1): 577–631. Masson et Cie, Paris.

Penney, J. T. 1960. Distribution and bibliography (1892–1957) of the fresh water sponges. University of South Carolina Publ. Biol. 3: 1–97.

Penney, J. T. and A. A. Racek. 1968. Comprehensive revision of a worldwide collection of freshwater sponges (Porifera: Spongillidae). Bull. U.S. Nat. Mus. No. 272: 1–184.

Pond, D. 1992. Protective-commensal mutualism between the queen scallop *Chlamys opercularis* (Linnaeus) and the encrusting sponge *Suberites*. J. Moll. Studies 58: 127-134.

Rasmont, R. 1962. The physiology of gemmulation in fresh water sponges. *In* D. Rudnick (ed.), *Regeneration, 20th Growth Symposium*. Ronald Press, New York, pp. 1–25.

Reiswig, H. M. 1970. Porifera: Sudden sperm release by tropical Demospongiae. Science 170: 538–539.

Reiswig, H. M. 1971a. *In situ* pumping activities of tropical Demospongiae. Mar. Biol. 9: 38–50.

Reiswig, H. M. 1971b. Particle feeding in natural populations of three marine Demosponges. Biol. Bull. 141: 568–591.

Reiswig, H. M. 1973. Population dynamics of three Jamaican Demospongiae. Bull. Mar. Sci. 23: 191–226.

Reiswig, H. M. 1974. Water transport, respiration, and energetics of three tropical sponges. J. Exp. Mar. Biol. Ecol. 14: 231–249.

Reiswig, H. M. 1975. The aquiferous systems of three marine Demospongiae. J. Morphol. 145(4): 493–502.

Ristau, D. A. 1978. Six new species of shallow-water marine demosponges from California. Proc. Biol. Soc. Wash. 91: 203–216.

Rützler, K. 1975. The role of burrowing sponges in bioerosion. Oecologia 19: 203–216.

Rützler, K. and G. Reiger. 1973. Sponge burrowing: Fine structure of *Cliona lampa* penetrating calcareous substrata. Mar. Biol. 21: 144–162.

Schmidt, I. 1970. Phagocytose et pinocytose chez les Spongillidae. Z. Vgl. Physiol. 66: 398–420.

Simpson, T. L. and J. J. Gilbert. 1973. Gemmulation, gemmule hatching and sexual reproduction in freshwater sponges. I. The life cycle of *Spongilla lacustris* and *Tubella pennsylvanica*. Trans. Am. Microsc. Soc. 92: 422–433.

Sollas, W. J. 1884. On the origin of freshwater faunas: A study in evolution. Trans. R. Soc. Dublin 2(3): 87–118.

Woollacott, R. M. 1990. Structure and swimming behavior of the larva of *Halichondria melanadocia* (Porifera: Demospongiae). J. Morph. 205: 135–145.

Woollacott, R. M. and M. G. Hadfield. 1996. Induction of metamorphosis in larvae of a sponge. Invert. Biol. 115(4): 257–262.

Ungemach, L. F., K. Souza, P. E. Fell and S. H. Loomis. 1997. Possession and loss of cold tolerance by sponge gemmules: A comparative study. Invert. Biol. 116(1): 1–5.

7

Four Phyla
of Uncertain Affinity

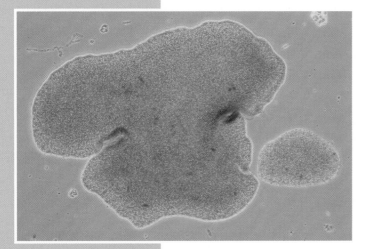

The only solid piece of scientific truth about which I feel totally confident is that we are profoundly ignorant about nature.

Lewis Thomas,
The Medusa and the Snail, 1979

As we discussed in Chapter 4, there are a number of extant Metazoa that exist at the mesozoan grade of body construction, and whose relationships seem to defy understanding. Currently four phyla contain animals in this category: Placozoa (*Trichoplax adhaerens*), Monoblastozoa (*Salinella*), Rhombozoa (e.g., *Dicyema, Pseudicyema*), and Orthonectida (e.g., *Rhopalura*). Together, these creatures number only 100 or so species. *Trichoplax* is marine (first discovered in marine aquaria); *Salinella* lives in salt beds; rhombozoans are symbionts in the nephridia of cephalopod molluscs; and orthonectids are parasitic in a variety of invertebrates, including echinoderms, molluscs, nemerteans, free-living flatworms, and polychaete worms. All of these animals have been the subjects of a great deal of taxonomic and phylogenetic controversy, much of which will no doubt remain unresolved for some time to come.

Taxonomic History

The first group of these enigmatic animals to be discovered was the Rhombozoa, described and named by A. Krohn (1839) in Germany. But it was not until 1876 that a careful study of these creatures was published by the Belgian zoologist Edouard van Beneden. He was convinced that these odd parasites represented a true link between the protists and the Metazoa and coined the name Mesozoa (= middle animals) to emphasize his point of view. By the end of the nineteenth century, *Trichoplax, Salinella*, and the orthonectids had also been discovered, and the phylum Mesozoa had become a dumping ground for a great variety of multicellular but presumed nonmetazoan (or at least, ancient metazoan) organisms. Over time, all of these animals except the rhombozoans and orthonectids were removed to other taxa as further studies revealed them to be protists, larval stages, or simply unrelated to one another or to other established groups. In the past, *Salinella* has been treated by some workers as a protist and by others as a larval stage, and *Trichoplax* was once thought to be a hydrozoan planula larva. These suggestions, however, have been rejected by most workers. Based on our current understanding of these organisms, it seems most appropriate to leave them as sole members of two separate (monotypic) phyla until and unless more information dictates otherwise.

The rhombozoans and orthonectids were for many years treated as closely related groups composing a monophyletic taxon of phylum rank: the Mesozoa. These organisms were usually assigned to two orders, sometimes in a single class called the Moruloidea to characterize their "ball-of-cells" grade of construction. Eventually, some authors assigned them to two classes rather than to two orders. However, other workers became convinced that Rhombozoa and Orthonectida are not at all closely related and should not even be classified together in a single phylum. We agree with this point of view and abandon the name Mesozoa as a formal taxon. We use the term *mesozoa* to represent organisms at a particular grade of complexity without implying that they represent a monophyletic clade or are even necessarily closely related. Thus, Rhombozoa, Orthonectida, Placozoa, and Monoblastozoa are treated here as separate phyla, each containing animals belonging to the mesozoan grade of complexity.

Mesozoan Bauplans

You will recall that multicellularity is only one of the criteria by which the metazoan grade of complexity is established. While mesozoa certainly satisfy this requirement, they do not have the typical layered construction of higher animals, and they do not obviously pass through any developmental stage that may be equated unequivocally with gastrulation. Also, they lack true tissues and organs. Their survival strategies are varied and are reflected partly in the body plans of the adults (Figure 7.1). In each case, certain advantages are realized by division of labor among their component cells, but there has been no development of true tissues or complex organ systems.

Phylum Placozoa

Trichoplax adhaerens was discovered in 1883 in a seawater aquarium at the Graz Zoological Institute in Austria. Specimens have subsequently been found in marine situations around the world. In recent years this organism has been studied extensively by Grell and Ruthmann (see references). The body of *Trichoplax* is only 2 to 3 mm in diameter, although it consists of several thousand cells arranged as a simple double-layered plate (Figure 7.1A–C). It lacks anterior–posterior polarity and symmetry. The cells of the upper and lower layers differ in shape, and there is a consistent dorsal–ventral orientation of the body relative to the substratum. The dorsal cells are flattened, monociliate, and contain lipid droplets. Most of the ventral cells are also monociliate, but they are all more columnar and lack distinct oil droplets. Furthermore, the ventral epithelium can be temporarily invaginated, presumably for feeding (Figure 4.16E). This observation supports the notion that there are functional as well as structural differences between the two cell layers. Between these two epithelial sheets is a mesenchymal layer of stellate ameboid cells embedded in a supportive gel matrix. Grell (1982) considers *Trichoplax* to be a true diploblastic metazoon and suggests that the upper and lower epithelia are homologous to ectoderm and entoderm, respectively. However, a basement membrane has not yet been identified beneath the epithelia, which suggests that *Trichoplax* may be closer to the Porifera in organization than it is to the diploblastic eumetazoa.

Trichoplax moves by ciliary gliding along a solid surface, aided by irregular, ameba-like shape changes along the body edges. Very small, presumably young individuals can swim, while larger individuals crawl. Most evidence suggests that *Trichoplax* feeds by phagocytosis of organic detritus. Phagocytosis occurs only in the invaginated cells of the ventral epithelium. Although there is no evidence for extracellular digestion, *Trichoplax* may secrete digestive enzymes onto its food within the ventral digestive pocket.

Trichoplax reproduces asexually by fission of the entire body into two new individuals and by a budding process that yields numerous multicellular flagellated "swarmers," each of which forms a new individual. Sexual reproduction is also known, followed by a developmental period of holoblastic cell division and growth. Eggs have been observed within the mesenchyme, but their origin is unknown. *Trichoplax* has very little DNA (about as much as a bacterium or protist), and its chromosomes are very small.

Phylum Monoblastozoa

Salinella has apparently not been studied since its reported discovery in 1892 by Frenzel, who found it in cultures of Argentine salt-bed material. There is serious question about the accuracy of the original description, and *Salinella* may have existed more in Frenzel's imagination than in Argentina's salt beds. According to Frenzel, the body wall of *Salinella* consists of but a single layer of cells. The inner cell borders line a cavity, which is open at both ends (Figure 7.1D). The openings function as an anterior "mouth" and posterior "anus," both of which are ringed by bristles. The rest of the body, inside and out, is densely ciliated.

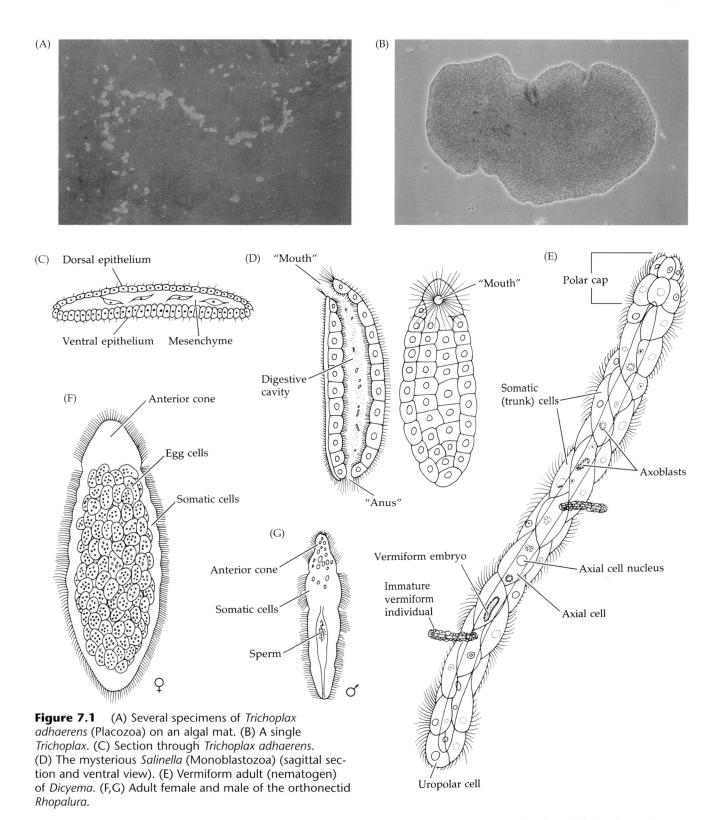

Figure 7.1 (A) Several specimens of *Trichoplax adhaerens* (Placozoa) on an algal mat. (B) A single *Trichoplax*. (C) Section through *Trichoplax adhaerens*. (D) The mysterious *Salinella* (Monoblastozoa) (sagittal section and ventral view). (E) Vermiform adult (nematogen) of *Dicyema*. (F,G) Adult female and male of the orthonectid *Rhopalura*.

The animal was said to move by ciliary gliding, much like ciliated protists and small flatworms. *Salinella* was thought to feed by ingesting organic detritus through the "mouth" and digesting it in the internal cavity. Undigested material would be carried to the "anus" by ciliary action. Asexual reproduction was said to take place by transverse fission of the body, and sexual reproduction was suspected to occur as well. The true nature of this animal, including its very existence, remains elusive.

Phylum Rhombozoa

Stunkard (1982) considered the taxon Rhombozoa as a class comprising the orders Dicyemida and Hetero-

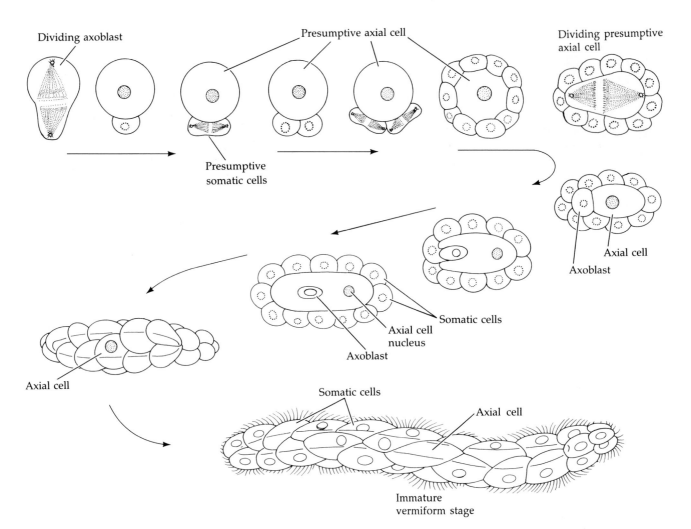

Figure 7.2 A young vermiform embryo develops from an axoblast within the axial cell of the vermiform adult *Dicyema* (Rhombozoa).

cyemida. While we treat the Rhombozoa as a phylum, we retain the ordinal level assignment of the two subtaxa. The rhombozoan bauplan includes a solid body construction. An outer layer of somatic/nutritive cells surrounds an inner core of reproductive cells (or often a single reproductive cell).

Dicyemids are more common and better understood than are the heterocyemids. Members of both groups are obligate symbionts in the nephridia of cephalopod molluscs. Various terminologies have been applied to the rhombozoans, especially to the different stages in their life cycles. We have drawn from several sources in an attempt to use the most descriptive terms. Some of the frequently used alternative terms are also noted.

The Dicyemida. Adult dicyemids, called **vermiform adults** or **nematogens**, are only 0.5 to 2.5 mm long. The body of a nematogen consists of an outer sheath of ciliated **somatic cells**, the number of which has been constant for most, but not all, species that have been examined.* Within the covering of somatic cells

*A constancy in the number of cells (in a given organ, or in the entire body of an animal) is called **eutely** and is a common feature of many microscopic and near-microscopic organisms.

lies a single long **axial cell** (Figure 7.1E). Eight or nine somatic cells at the anterior end form a distinctive **polar cap**. Immediately behind the polar cap are two **parapolar cells**. The rest of the 10 to 15 somatic cells are sometimes called **trunk cells**; the two most posterior cells are the **uropolar cells**.

Young dicyemids are motile and swim about in the host's urine by ciliary action. The adults, however, attach to the inner lining of the nephridia by their polar caps. There is no conclusive evidence that these animals cause damage to their hosts, but when present in very high numbers they may interfere with the normal flow of fluids through the nephridia. Nematogens consume particulate and molecular nutrients from the host's urine by phagocytotic and pinocytotic action of their somatic cells. Once the adult has attached to the host, the somatic cilia probably serve to keep fluids moving over the body, bringing nutrients in contact with the surface cells. Although in nature dicyemids appear to be obligatorily associated with cephalopods, they have been successfully maintained in experimental nutrient media (Lapan and Morowitz 1972).

What we know so far about dicyemid life history is rather bizarre. The stages of the dicyemid life cycle that occur outside the host are still incompletely known. However, the host-dwelling portion of the life cycle includes both asexual and sexual processes, but without a regular alternation between them. In a curious cellular arrangement, the cytoplasm of the axial cell of the vermiform adult contains numerous tiny cells called **axoblasts**. Immature vermiform organisms are produced asexually by a sort of embryogeny of individual axoblasts within the parent axial cell (Figure 7.2). The first division of an axoblast is unequal and produces a large presumptive axial cell and a small presumptive somatic cell. The presumptive somatic cell divides repeatedly, and its daughter cells move by an epiboly-like process to enclose the presumptive axial cell, which has not yet divided. When this inner cell finally does divide, it does so unequally, and the smaller daughter cell is then engulfed by the larger one! The larger cell becomes the progeny's axial cell proper with its single nucleus, and the smaller engulfed cell becomes the progenitor of all future axoblasts within that axial cell. The "embryo," which now consists of its own central axial cell surrounded by somatic cells, elongates and the somatic cells develop cilia. The resulting structure is a miniature vermiform organism. The immature vermiform organism leaves the parent vermiform adult and swims about in the nephridial fluids. Eventually it attaches to the host and enters the adult stage of the life cycle.

The initiation of sexual reproduction in dicyemids may be a density-dependent phenomenon associated with high numbers of vermiform individuals within the host's nephridia. Lapan and Morowitz (1972) suggested that the switch from asexual to sexual processes might be a response to some chemical factor that accumulates in the urine of the host. Other workers suggest that sexual reproduction in dicyemids is brought on by the sexual maturation of the host (e.g., Hyman 1940; Stunkard 1982; Hochberg 1983). In any event, as the vermiform adults become sexually "motivated," their somatic cells usually enlarge as they become filled with yolky material; the name **rhombogen** is often applied to the individuals in this stage (Figure 7.3). However, because the reported differences between rhombogens and nematogens are not consistently found, it is probably best to simply call them sexual and asexual vermiform adults, respectively.

The axoblasts of sexual vermiform adults develop into multicellular structures called **infusorigens**, consisting of an outer layer of ova and an inner mass of sperm (Figure 7.3B). The infusorigens are retained within the parent's axial cell. They have been likened to separate hermaphroditic individuals or to transient double-sexed gonads. The centrally located sperm fertilize the peripherally arranged ova, and each zygote develops

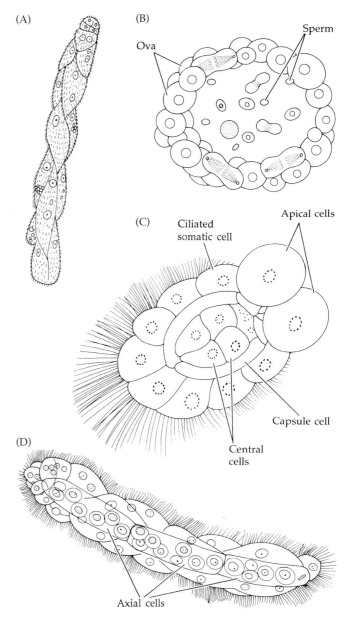

Figure 7.3 Sexual reproduction in dicyemid rhombozoans. (A) Sexual (rhombogen) form of vermiform adult. (B) Infusorigen of sperm and ova formed within the axial cell of the vermiform adult. (C) Infusoriform larva produced by fertilization. (D) Stem nematogen with three axial cells.

into a ciliated **infusoriform larva** (Figure 7.3C). This larva has a fixed number of cells; the two anteriormost cells—called **apical cells**—contain high-density substances within their cytoplasm. The rest of the surface cells are ciliated and form a sheath around a ring of **capsule cells**, which in turn enclose four central cells. The infusoriform larvae escape from the parent vermiform adult and pass out of the host's body with the urine.

The events of the dicyemid life cycle that occur outside the cephalopod host remain a mystery. Some workers have held to the view that the infusoriform larva en-

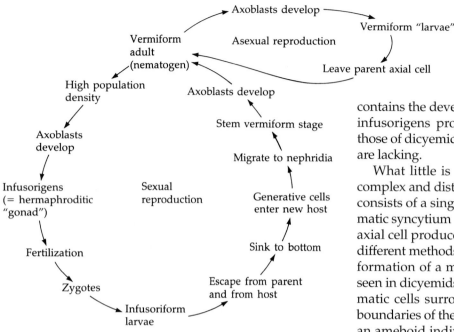

Figure 7.4 Life cycle of a dicyemid.

ters an intermediate host (presumably some benthic invertebrate), but most of the evidence to date suggests that this is not the case. While much remains to be learned, the following scenario seems most plausible. After leaving the host, the larva sinks to the bottom—the dense contents of the apical cells serving as ballast. The larva, or some persisting part of the larva (perhaps the innermost four cells), enters another cephalopod host. This infectious individual travels through the host, probably via the circulatory system, and enters the nephridia, where it becomes a so-called **stem nematogen** (Figure 7.3D). The stem nematogen is similar to the vermiform adult except that the former has three axial cells rather than one. Axoblasts within the axial cells of the stem nematogen give rise to more vermiform adults, just as the axoblasts within the adults described earlier did. The vermiform adults produce more individuals like themselves until the onset of sexual reproduction is triggered again, presumably by the high population density. This putative life cycle is schematically represented in Figure 7.4.

The Heterocyemida. Only two species are included in this group of rhombozoans. *Conocyema polymorpha* lives in the nephridia of octopuses, and *Microcyema gracile* in cuttlefishes of the genus *Sepia*. These two heterocyemids differ from each other in certain respects.

The vermiform adult of *Conocyema* bears a polar cap of four enlarged cells and has a trunk of somatic cells around an inner axial cell; all the cells of the body lack cilia (Figure 7.5A). The axial cell contains axoblasts, which give rise to ciliated "larvae" that escape from the parent, lose their cilia, and grow into more vermiform

adults within the host. The individuals that produce the infusorigens lack a polar cap. They have only a very thin layer of somatic cells surrounding the axial cell, which contains the developing infusorigens (Figure 7.5B). The infusorigens produce infusoriform larvae similar to those of dicyemids. Details of the life cycle of *Conocyema* are lacking.

What little is known about *Microcyema* suggests a complex and distinctive life cycle. The vermiform adult consists of a single inner axial cell surrounded by a somatic syncytium (Figure 7.6A). The axoblasts within the axial cell produce more vermiform adults by two very different methods. One sequence of events involves the formation of a multicellular "embryo" similar to that seen in dicyemids (Figure 7.6B). As the presumptive somatic cells surround the axial cell precursor, the cell boundaries of the somatic cells break down, resulting in an ameboid individual in which a syncytial mass surrounds the growing axial cell (Figure 7.6C,D). This individual apparently develops into a new vermiform adult. Another asexual process involves the formation of ciliated **Wagener's larvae** from the axoblasts (Figure 7.6E). These larvae leave the parent, swim about in the host's nephridial fluids, and eventually attach and metamorphose into more vermiform adults. *Microcyema* adults also produce infusorigens and infusoriform larvae much like those of the dicyemids. The infusoriform larvae apparently leave the host via the urine, but nothing is known about the stages of the life cycle outside the

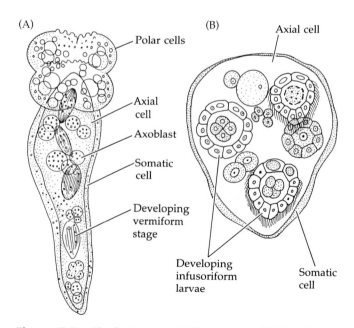

Figure 7.5 The heterocyemid *Conocyema*. (A) Vermiform adult. (B) During the reproductive phase, infusoriform larvae are formed within the adult's axial cell (cross section).

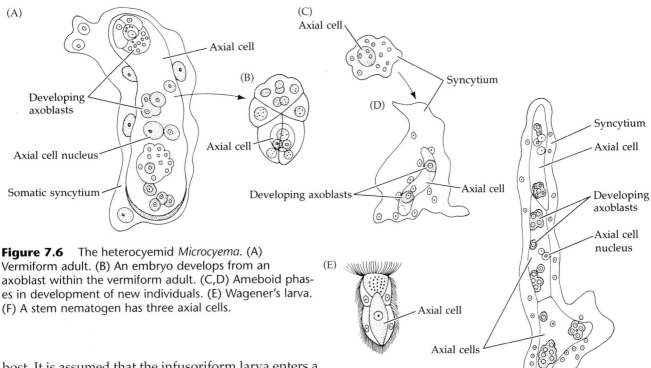

Figure 7.6 The heterocyemid *Microcyema*. (A) Vermiform adult. (B) An embryo develops from an axoblast within the vermiform adult. (C,D) Ameboid phases in development of new individuals. (E) Wagener's larva. (F) A stem nematogen has three axial cells.

host. It is assumed that the infusoriform larva enters a host and matures into a ciliated nematogen, which has three axial cells. Such stem nematogens have been observed in host animals. Eventually, the cilia and the cell boundaries between adjacent somatic cells are lost (Figure 7.6F), and the animal develops into another vermiform adult.

Phylum Orthonectida

The life cycles of some orthonectids are well known and in certain respects differ markedly from those of the rhombozoans. Asexual individuals dominate the life cycle; they are ameboid syncytial forms, commonly called **plasmodial stages** (Figure 7.7A). Some plasmodia grow and spread to such an extent that they cause severe damage to the host. For example, *Rhopalura ophiocomae* (in the brittle star *Amphipholis squamata*) and *R. granosa* (in the bivalve mollusc *Heteranomia squamula*) destroy the gonads of their hosts.

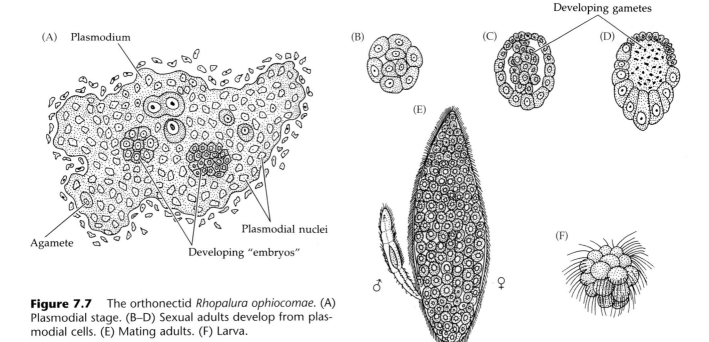

Figure 7.7 The orthonectid *Rhopalura ophiocomae*. (A) Plasmodial stage. (B–D) Sexual adults develop from plasmodial cells. (E) Mating adults. (F) Larva.

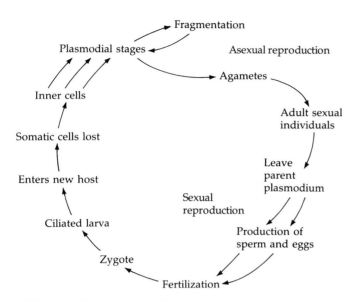

Figure 7.8   Generalized life cycle of an orthonectid.

The plasmodium produces more syncytial masses by fragmentation, and it also gives rise to the sexual individuals. Certain nuclei within the plasmodium (frequently called **agametes**) accumulate and partition off a small amount of cytoplasm (Figure 7.7A). These cellular units within the plasmodium undergo cleavage (Figure 7.7B–D) and eventually form the sexual individuals. In most species, a single plasmodium produces only males or only females. In a few species, however, both sexes are produced from one plasmodium. The sexual organism consists of an outer layer of ciliated somatic cells and an inner mass of gametes (Figures 7.1F,G and 7.7D,E). Between the gametes and the outer somatic layer are what appear to be contractile cells. Upon maturation, the sexual forms leave the parent plasmodium; then they leave the host and swim about in the environment. In the gonochoristic species, the male attaches to the female and deposits sperm through a small genital pore located near the posterior end of her body (Figure 7.7E). The sperm fuse with ova to form zygotes. Each zygote develops into a ciliated larva (Figure 7.7F), which escapes from the female's body and eventually enters another host animal. Once inside the host, the larva loses its somatic cells and releases the mass of inner cells, each of which apparently develops into a new plasmodium. The life cycle is shown in Figure 7.8.

Mesozoan Phylogeny

Some of the controversies about the origins and evolution of mesozoan creatures have already been mentioned. We cannot, of course, solve these problems here, but we can offer some hypotheses that provide a basis for discussion, speculation, and further testing.

If *Salinella* is real, it is difficult to relate it to any known mesozoan or metazoan taxon. Structurally and

functionally, it seems barely removed from the protistan grade. Perhaps *Salinella* arose from some colonial flagellate ancestor that took up benthic life and assumed a somewhat bilateral form. By this hypothesis, the most striking apomorphic feature of *Salinella* is its complete digestive cavity with inwardly directed cilia (which Frenzel may have misinterpreted). The animal's nutritive activities are assumed by the inside surfaces of the single layer of body cells, while its locomotor functions are performed by the outer cell surfaces. Thus, *Salinella* seems to be little more than a colony of totipotent, multifunctional cells.

Its short chromosomes and small DNA content suggest that *Trichoplax* may indeed be a very primitive animal; it is tempting to view this organism as a surviving descendant of some protometazoan ancestor. The most acceptable hypotheses concerning the origin of the metazoan condition depend upon the evolution of a layered construction through some form of gastrulation (Chapter 4). Perhaps *Trichoplax* is a descendant of an early protometazoan "experiment" in body layering. Perhaps several such evolutionary experiments took place, one of which eventually led to the origin of the modern eumetazoa. One of the events in this experimentation could have been the formation of a temporary digestive pouch by invagination or simple inpocketing, thus increasing the surface area for feeding. *Trichoplax* may be an extant remnant of that ancient event. We know that certain colonial flagellates tend to have groups of cells that are somewhat specialized for various functions; it would seem to be a relatively small evolutionary step to turn the nutritive cells inward during feeding. Such an event would have been particularly advantageous to benthic animals.

The rhombozoans and orthonectids present phylogenetic problems that have been argued enthusiastically for decades. Some authors suggest that the relatively simple construction of these animals is primitive, and that the rhombozoans and orthonectids arose as a side branch from an early protometazoan lineage, perhaps even from some ancient planuloid form. Other specialists have championed the idea that these animals are descended from established metazoan stock, and that their simple construction represents an anatomical "degeneration" associated with their parasitic habits. Lameere (1922) suggested that the orthonectids may have arisen from echiuran worms, apparently because some echiurans (e.g., *Bonellia*) show extreme sexual dimorphism, with tiny, reduced males. Lameere likened this feature to the dimorphic nature of some orthonectids, but his hypothesis never gained much favor. The most likely candidates for such an ancestral group are probably found among the parasitic trematodes (Platyhelminthes), an idea most strongly supported by Stunkard (1954, 1972). This contention is based largely on general morphological features and on the complex life cycles of these animals. Space does not permit a complete examination of Stunkard's views, but arguments against them have been presented by several authors (Dodson 1956; Kozloff 1969; Lapan and Morowitz 1972; Hochberg 1983). In our opinion, one can view the complex life cycles of rhombozoans, orthonectids, and trematodes as examples of convergence associated with their parasitic life styles. Such life cycles are common among internal parasites of all grades of complexity.

Our treatment of the orthonectids and rhombozoans as unrelated taxa is based upon several considerations. Their similarities appear to be superficial results of con-vergence related to their similar life styles and general level of complexity. In terms of body construction, rhombozoans and orthonectids are similar only in their basic strategy of relegating reproductive cells to the inside while maintaining an outer covering of somatic cells that function in nutrition, protection, and locomotion. Many workers have compared the sexual adult stages of orthonectids to the asexual vermiform adults of dicyemids. However, the sexual stages are drastically different from one another, and this difference is a major criterion for separating the two groups.

Selected References

Atkins, D. 1933. *Rhopalura granosa* sp. nov. an orthonectid parasite of a lamellibranch *Heteranomia squamula* L. with a note on its swimming behavior. J. Mar. Biol. Assn. U.K. 19: 233–252.

Austin, C.R. 1964. Gametogenesis and fertilization in the Mesozoan *Dicyema aegira*. Parasitol. 54: 597–600.

Caullery, M. 1961. Classe des Orthonectides. *In* P. Grassé (ed.), *Traité de Zoologie*. Masson et Cie, Paris, pp. 695–706.

Caullery, M. and A. Lavellée. 1908. La fécondation et le développement de l'oeuf des Orthonectides. I. *Rhopalura ophiocomae*. Arch. Zool. Exp. Gén., Series 4(8): 421–469.

Caullery, M. and F. Mesnil. 1901. Recherches sur les Orthonectides. Arch. Anat. Microsc. 4: 381–470.

Dodson, E. O. 1956. A note on the systematic position of the Mesozoa. Syst. Zool. 5(1): 37–40.

Frenzel, J. 1892. Untersuchungen über die mikroskopische Fauna Argentiniens. Arch. Naturgesch. 58: 66–96.

Grassé, P. 1961. Classe des Dicyémides. *In* P. Grassé (ed.), *Traité de Zoologie*. Masson et Cie, Paris, pp. 707–729.

Grell, K. 1971a. Embryonalentwicklung bei *Trichoplax adhaerens* F.E. Schulze. Naturwiss. 58: 570.

Grell, K. 1971b. *Trichoplax adhaerens* F.E. Schulze und die Entstehung der Metazoen. Naturwiss. Rundsch. 24: 160–161.

Grell, K. 1972. Eibildung und Furchung von *Trichoplax adhaerens* F.E. Schulze (Placozoa). Z. Morphol. Tiere 73: 297–314.

Grell, K. 1973. *Trichoplax adhaerens* and the origin of the Metazoa. Actualites Protozoologiques. I^{ve} Cong. Int. Protozoologie. Paul Couty, Clermont-Ferrand.

Grell, K. 1982. Placozoa. *In* S. Parker (ed.), *Synopsis and Classification of Living Organisms*. McGraw-Hill, New York, p. 639.

Grell, K. and G. Benwitz. 1971. Die Ultrastruktur von *Trichoplax adhaerens* F.E. Schulze. Cytobiologie 4: 216–270.

Grell, K. and G. Benwitz. 1974. Elektronenmikroskopische Beobachtungen über das Wachstum der Eizelle und die Bildung der "Befruchtungsmembran" von *Trichoplax adhaerens* F. E. Schulze (Placozoa). Z. Morphol. Tiere 79: 295–310.

Harrison, F. and J. Wesfall (eds.). 1991. *Microscopic Anatomy of Invertebrates*, Vol. 2: *Placozoa, Porifera, Cnidaria, and Ctenophora*. Wiley-Liss, New York.

Hatschek, B. 1888. *Lehrbuch der Zoologie*. Jena.

Hochberg, F. G. 1983. The parasites of cephalopods: A review. Mem. Natl. Mus. Victoria 44: 109–145.

Horvath, P. 1997. Dicyemid mesozoans. *In* S. Gilbert and A. Raunio (eds.), *Embryology: Constructing the Organism*. Sinauer Associates, Sunderland, MA, pp 31–38.

Hyman, L. 1940. *The Invertebrates*. Vol. 1, Protozoa through Ctenophora. McGraw-Hill, New York.

Kozloff, E. 1965. *Ciliocincta sabellariae* gen. and sp. n., an orthonectid mesozoan from the polychaete *Sabellaria cementarium* Moore. J. Parasitol. 51(1): 37–44.

Kozloff, E. 1969. Morphology of the orthonectid *Rhopalura ophiocomae*. J. Parasitol. 55(1): 171–195.

Kozloff, E. 1971. Morphology of the orthonectid *Ciliocincta sabellariae*. J. Parasitol. 57(3): 585–597.

Lameere, A. 1916. Contributions á la connaissance des Dicyémides. Première partie. Bull. Sci. Fr. Belg. 59: 1–35.

Lameere, A. 1918. Contributions á la connaissance des Dicyémides. Deuxieme partie. Bull. Biol. 51: 347–390.

Lameere, A. 1922. L'histoire naturelle des Dicyémides. Brussels Acad. R. Belg. Bul. Cl. Sci. Ser. 5(8): 779–792.

Lapan, E. and H. Morowitz. 1972. The Mesozoa. Sci. Am. 227(6): 94–101.

McConnaughey, B. H. 1963. The Mesozoa. *In* E. C. Dougherty (ed.), *The Lower Metazoa*. University of California Press, Berkeley, pp. 151–168.

Nouvel, H. 1948. Les Dicyémides. 2° partie: Infusoriforme, teratologie, spécificité du parasitisme, affinitiés. Arch. Biol. Paris 59: 147–223.

Ruthmann, A. 1977. Cell differentiation, DNA content, and chromosomes of *Trichoplax adhaerens* F. E. Schulze. Cytobiologie 15: 58–64.

Ruthmann, A. and H. Wenderoth. 1975. Der DNA-Gehalt der Zellen bei dem primitiven Metazoon *Trichoplax adhaerens* F. E. Schulze. Cytobiologie 10: 421–431.

Stunkard, H. 1954. The life-history and systematic relations of the Mesozoa. Q. Rev. Biol. 29: 220–244.

Stunkard, H. 1972. Clarification of taxonomy in Mesozoa. Syst. Zool. 21(2): 210–214.

Stunkard, H. 1982. Mesozoa. *In* S. Parker (ed.), *Synopsis and Classification of Living Organisms*. McGraw-Hill, New York, pp. 853–855.

van Beneden, É. 1876. Recherches sur les Dicyémides. Bull. Acad. Belg. Cl. Sci. Series 2(41): 1160–1205; 2(42): 35–97.

van Beneden, É. 1882. Contribution a l'histoire des Dicyémides. Arch. Biol. Paris 3: 195–228.

Whitman, C. O. 1882. A contribution to the embryology, life history, and classification of the dicyemides. Mitt. Zool. Sta. Neapel. 4: 1–89.

8 Phylum Cnidaria

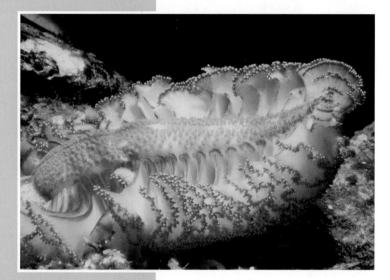

"Cyanea!" I cried.
"Cyanea! Behold the Lion's Mane!"
Sherlock Holmes,
The Adventure of the Lion's Mane

The phylum Cnidaria is a highly diverse assemblage that includes jellyfish, sea anemones, corals, and the common laboratory *Hydra*, as well as many less familiar forms such as hydroids, sea fans, siphonophores, zoanthids, and myxozoans (Figure 8.1). There are about 11,000 extant species of cnidarians. Much of the striking diversity seen in this phylum results from two fundamental aspects of their lifestyle. First is the tendency to form colonies by asexual reproduction; the colony can achieve dimensions and forms unattainable by single individuals. Second, many species of cnidarians exhibit a **dimorphic life cycle** that includes two entirely different adult morphologies: a **polypoid form** and a **medusoid form**. The dimorphic life cycle has major evolutionary implications touching on nearly every aspect of cnidarian biology.*

Cnidarians are diploblastic Metazoa at a tissue grade of construction. They possess primary radial symmetry, tentacles, stinging or adhesive structures called **cnidae**, an entodermally derived incomplete gastrovascular cavity as their only "body cavity," and a middle layer (called mesenchyme, or mesoglea[†]) derived primarily from ectoderm. They lack cephalization, a centralized nervous system, and discrete respiratory, circulatory, and excretory organs (Box 8A). This basic bauplan is retained in both the polypoid and medusoid forms (Figure 8.2). The primitive nature of the cnidarian bauplan is exemplified by the fact that they have fewer cell types than any other animals except the sponges and mesozoans. In fact, cnidarians contain fewer cell types than does a single *organ* in many other Metazoa.

---

*When both phases are present in a species' life cycle, it is said to undergo an alternation of generations, or as it is sometimes called, "metagenesis."

(A)

(B)

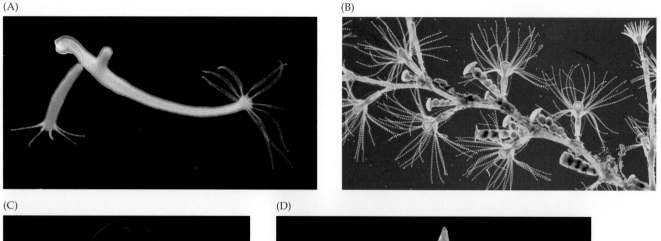

(C)

(D)

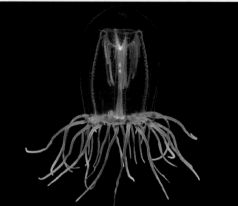

Figure 8.1   Some cnidarians. (A–D) Hydrozoa. (A) *Hydra*, an aberrant freshwater anthomedusan (shown here budding). (B) A colony of the leptomedusan *Gonothyrea*. (C) The medusa of *Polyorchis*, an anthomedusan. (D) A chondrophoran, *Velella* ("by-the-wind sailor"). (E–F) Scyphozoa. (E) *Pelagia*, a large semaeostoman medusa. (F) *Haliclystis*, the strange, sessile stauromedusan. (G–K) Anthozoa. (G) An actinarian, the giant sea anemone *Metridium*. (H) The Caribbean elkhorn coral, *Acropora palmata*. (I) The sea pen, *Ptilosarcus* (Pennatulacea). (J) A large sea fan, or gorgonacean. (K) *Renilla*, the sea pansy (Pennatulacea).

Cnidarians are mostly marine, but a few groups have successfully invaded fresh waters. Most are sessile (polyps) or planktonic (medusae) carnivores, although some employ suspension feeding and many species harbor symbiotic intracellular algae from which they may derive energy. Cnidarians range in size from nearly microscopic polyps and medusae to individual jellyfish

†There exists a suite of terms in zoological literature that is frequently confused, misused, and generally messy. These terms include mesenchyme, mesoglea, collenchyme, parenchyme, and coenenchyme. In this book, these terms are used in the following ways. **Mesenchyme** (Greek, literally "middle juices") refers to a primitive connective tissue derived wholly or in part from ectoderm and located between the epidermis and the gastrodermis (entodermis). Mesenchyme generally consists of two components: a noncellular, jelly-like matrix called **mesoglea**, and various cells and cell products (e.g., fibers). When no cellular material is present, this layer is properly called mesoglea. Mesenchyme is the typical middle layer of sponges (where it is called the mesohyl) and of members of the phyla Cnidaria and Ctenophora. In these diploblastic groups, where no true (ento-) mesoderm exists, the mesenchyme is fully ectodermally derived. When cellular material is sparse or densely packed, mesenchyme may sometimes be designated as **collenchyme** or **parenchyme**, respectively. The term parenchyme is sometimes used for the mesenchymal layer of triploblastic acoelomate animals (such as flatworms), in which the dense layer includes tissues derived from both ecto- and entomesoderm.

In some colonial cnidarians, particularly anthozoan polyps, the individuals are embedded in and arise from a mass of mesenchyme perforated with gastrovascular channels that are contin-

uous among the members of the colony. The term **coenenchyme** refers to this entire matrix of common basal material, which is itself covered by a layer of epidermis.

Adding to the potential confusion, the term mesenchyme is used in a second, very different way by some biologists. Vertebrate embryologists use the term to refer to that part of true (ento-) mesoderm from which all connective tissues, blood vessels, blood cells, the lymphatic system, and the heart are derived. Thus, to an embryologist, the term "mesenchymal cell" often denotes any undifferentiated cell found in the embryonic mesoderm that is capable of differentiating into such tissues. Because of this confusion, some authors prefer to use the term mesoglea in lieu of mesenchyme when referring to the middle layers of sponges and diploblastic Metazoa. However, we strictly adhere to the former definition of mesenchyme and hope that this note will lessen rather than add to the muddle.

A word of caution regarding spelling: the meanings of some of these terms can be altered by changing the terminal "e" to an "a." The termination "-chyme" is preferred for animals, "-chyma" for plants. **Mesenchyma** refers to tissue lying between the xylem and phloem in plant roots; **collenchyma** refers to certain primordial leaf tissues. **Parenchyma** is a very general botanical term used in reference to various supportive tissues. Unfortunately, the same spelling is occasionally (improperly) used by zoologists.

(E)

(F)

(G)

(H)

(I)

(J)

(K)

BOX 8A *Characteristics of the Phylum Cnidaria*

1. Diploblastic Metazoa with ectoderm and entoderm separated by a (primarily) ectodermally derived acellular mesoglea or partly cellular mesenchyme

2. Possess primary radial symmetry, often modified as biradial, quadriradial, or other form; the primary body axis is oral–aboral

3. Possess unique stinging or adhesive structures called cnidae; each cnida resides in and is produced by one cell, a cnidocyte. The most common cnidae are called nematocysts

4. Musculature formed largely of myoepithelial cells (= epitheliomuscular cells), derived from ectoderm and entoderm (adult epidermis and gastrodermis)

5. Exhibit alternation of asexual polypoid and sexual medusoid generations; but there are many variations on this basic theme

6. The entodermally derived gastrovascular cavity (coelenteron) is the only "body cavity." The coelenteron is saclike, partitioned, or branched, but has only a single opening, which serves as both mouth and anus

7. Without a head, centralized nervous system, or discrete gas exchange, excretory, or circulatory structures

8. Nervous system is a simple nerve net(s), composed of naked and largely nonpolar neurons

9. Typically have planula larvae (ciliated, motile, gastrula larvae)

2 m wide and with tentacles 25 m long. Colonies, such as corals, may be many meters across. The phylum dates from the Precambrian, and its members have played important roles in various ecological settings throughout their long history, just as modern coral reefs are important today.

Taxonomic History and Classification

As is the case with sponges, the nature of cnidarians was long debated. In reference to their stinging tentacles, Aristotle called the medusae Acalephae (*akalephe*) and the polyps Cnidae (*knide*), both names derived from terms meaning "nettle." Renaissance scholars considered them plants, and it was not until the eighteenth century that the animal nature of the cnidarians was widely recognized. Nineteenth-century naturalists classified them along with the sponges and a few other groups under Linnaeus's Zoophytes, a category for organisms deemed somewhere between plants and animals. Lamarck instituted the group Radiata (or "Radiaires") for medusoid cnidarians, ctenophores, and echinoderms. In the early nineteenth century the great naturalist Michael Sars demonstrated that medusae and polyps were merely different forms of the same group of organisms. Sars also demonstrated that the genera *Scyphistoma*, *Strobila*, and *Ephyra* actually represented stages in the life history of certain jellyfish (scyphozoans). The names have been retained and are now used to identify these stages in the life cycle. Leuckart eventually recognized the fundamental differences between the two great "radiate" groups, the Porifera/Cnidaria/Ctenophora and the Echinodermata, and in 1847 created the name Coelenterata (Greek *koilos*, "cavity"; *enteron*, "intestine") for the former group in his recognition of the "intestine" as the sole body cavity. In 1888 Hatschek split Leuckart's Coelenterata into the three phyla recognized today: Porifera, Cnidaria, and Ctenophora. Although some workers have been inclined to retain the cnidarians and ctenophores together in the Coelenterata (or even the Radiata), these two groups are almost universally recognized as distinct phyla, a view upheld by recent molecular analyses. The older term

Figure 8.2 Tissue layer homologies in cnidarians. (A) A hydrozoan polyp. (B) An anthozoan polyp. (C) A hydrozoan medusa, shown upside down for similar orientation. The outer tissue layer is ectodermal (= epidermis); the inner tissue layer is entodermal (= gastrodermis); and the middle layer is the mesenchyme/mesoglea.

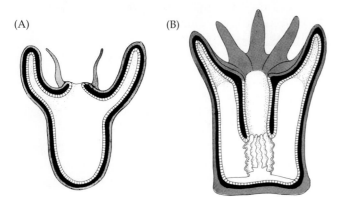

(A)   (B)

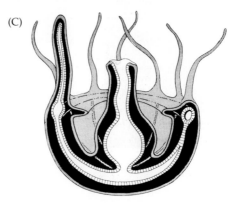

(C)

"Coelenterata" is still preferred by some specialists, who regard it as a synonym of Cnidaria. The most recent major advance in cnidarian systematics is the realization that the myxozoans, formerly classified as protists, are highly derived parasitic cnidarians.

PHYLUM CNIDARIA

CLASS HYDROZOA: Hydroids and hydromedusae* (Figure 8.1A–D). Alternation of generations occurs in most genera (typically asexual benthic polyps alternate with sexual planktonic medusae), although one or the other generation may be suppressed or lacking; medusoids often retained on the polyp; polyps usually colonial, with interconnected coelenterons; often polymorphic, individual polyps modified for various functions (e.g., gastrozooids feed, gonozooids are reproductive, dactylozooids are for defense and prey capture); exoskeleton usually of chitin or occasionally calcium carbonate (hydrocorals); coelenteron of polyps and medusae lacks a pharynx and mesenteries; mesoglea acellular; tentacles solid or hollow; cnidae occur only in epidermis; gametes arise from epidermal cells; medusae mostly small and transparent, nearly always craspedote (with a velum) and with a ring canal; mouth typically borne on pendant manubrium; medusae lack rhopalia. About 3,200 species in 5 extant orders; includes some freshwater groups.

ORDER HYDROIDA: Hydroids and their medusae. Polypoid generation often predominant; polyps may have a chitinous exoskeleton; oral tentacles filiform or capitate, rarely branched or absent; colonies often polymorphic; many do not release free medusae but release gametes from sporosacs or sessile attached medusoids (= medusoid buds, or gonophores) on colony; colonies dioecious. A large group, with over 75 described families. Hydroids occur at all depths; the polypoid forms are very common in the littoral zone.

SUBORDER ANTHOMEDUSAE (= GYMNOBLASTEA OR ATHECATA): Polyps solitary or colonial; hydranths and gonozooids lack exoskeleton; gonozooids produce free or sessile medusae; some groups produce gametes in transient sporosacs; free medusae tall and bell-shaped, without statocysts, with or without ocelli; medusae form gametes on subumbrella or manubrium. (e.g., *Bougainvilla, Calycopsis, Eleutheria, Eudendrium, Hydra, Hydractinia, Hydrocoryne, Janaria, Lar, Pennaria, Polyorchis, Sarsia, Staurocladia, Stylactis, Tubularia*)

Two anthomedusan families were formerly considered separate orders of hydrocorals, the Milleporidae (one extant genus, *Millepora*) and the Stylasteridae. The milleporids are called "fire corals" because of their potent stinging nematocysts. Milleporids are distinguished by forming massive or encrusting calcareous coral-like skeletons, with the calcareous matrix covered by thin epidermal layer; gastrozooids with short capitate tentacles; each gastrozooid is surrounded by 4–8 discrete dactylozooid-like tentacles, each tentacle in a separate skeletal cup; gonophores housed in pits (ampullae) in skeleton; small, free medusae lack mouth, tentacles, and velum. Like the true (stony) corals, milleporids rely on a commensal relationship with zooxanthellae and are thus restricted to the photic zone. Stylasterids (e.g., *Allopora, Stylaster*) also form erect or encrusting calcareous colonies, and are often brightly colored (purple, red, yellow). Skeleton secreted within the epidermis and covered by a thick epidermal layer; a calcareous style often rises from base of polyp cup, hence the name "stylasterine"; polyps may have tentacles; free medusae not produced, but sessile medusoid gonophores retained in shallow chambers (ampullae) of colony; several dactylozooids surround each gastrozooid, although polyp pits are joined.

SUBORDER LEPTOMEDUSAE (= CALYPTOBLASTEA OR THECATA): Polyps always colonial; hydranths and gonozooids encased in exoskeleton; free medusae usually absent, but when present flattened and with statocysts; medusae form gametes on subumbrella beneath radial canals; gonozooids (= gonangia) with blastostyle that produces medusae buds. (e.g., *Abietinaria, Aequorea, Aglaophenia, Bonneviella, Campanularia, Cuvieria, Gonionemus, Gonothyrea, Lovenella, Obelia, Plumularia, Sertularia*)

ORDER TRACHYLINA: Trachyline medusae. Polypoid generation greatly reduced or absent; medusae produce planula larvae that usually develop directly into actinula larvae, which metamorphose into adult medusae; medusae craspedote, with tentacles often arising from exumbrellar surface, well above bell margin; medusae mostly dioecious. This order is now thought to include microscopic parasitic forms (e.g., *Henneguya, Myxidium, Myxobdus, Sphaerospora*) previously assigned to the protist phylum Myxozoa (Siddall et al. 1995). The trachylines are probably a polyphyletic group, and currently include three suborders: Laingiomedusae, Narcomedusae and Trachymedusae. (e.g., *Aegina, Botrynema, Craspedacusta, Cunina, Gonionemus, Hydroctena, Liriope, Polypodium, Rhopalonema, Solmissus*)

ORDER SIPHONOPHORA: Siphonophorans. Polymorphic swimming or floating colonies, with a number of distinct types of polyps and attached modified medusae; most have a gas-filled flotation zooid. (e.g., *Agalma, Apolemia, Eudoxoides, Nectocarmen, Physalia, Rhizophysa, Sphaeronectes*)

ORDER CHONDROPHORA: Chondrophorans. Enigmatic group viewed either as colonies comprising gastrozooids, gonozooids, and dactylozooids, or as a solitary but highly specialized polypoid individual; "zooids" are attached to a chitinous, multichambered, disclike float, that may or may not have an oblique sail; "gonozooids" bear medusiform gonophores that are released and shed gametes; most are richly supplied with zooxanthellae. Once considered a highly modified group of siphonophorans, their position within the Hydrozoa is still debated. (e.g., *Porpita, Velella*)

ORDER ACTINULIDA: Actinulidans. Free-living, solitary, minute (to 1.5 mm), motile, polypoid hydrozoans; no medusa stage; interstitial, using cilia to swim and crawl among sand grains; no sexual reproduction has been recorded. (e.g., *Halammohydra, Otohydra*)

CLASS ANTHOZOA: Anemones,[†] corals, sea pens (Figure 8.1G–K). Exclusively marine; solitary or colonial; without a medusoid stage; cnidae epidermal and gastrodermal; coelenteron divided by longitudinal (oral–aboral) mesenteries, the free edges of which form thick, cordlike mesenterial filaments;

---

*Several recent papers have proposed major revisions of the Hydrozoa, and readers are cautioned that this is an area of current debate.

[†]We use the term "anemone" in a general sense for all anthozoan polyps, and restrict the term "sea anemone" to the true anemones of the order Actiniaria.

mesenchyme thick; tentacles usually number 8 or occur in multiples of 6 and contain extensions of the coelenteron; stomodeal pharynx (= actinopharynx) extends from the mouth into the coelenteron and bears one or more ciliated grooves (siphonoglyphs); polyps may reproduce both sexually and asexually; gametes arise from gastrodermis. About 6,225 species divided into three subclasses.

SUBCLASS OCTOCORALLIA (= ALCYONARIA): Octocorals. Polyps with 8 hollow, marginal, pinnate tentacles, and 8 complete (perfect) mesenteries, each with retractor muscle on sulcal side, facing the single siphonoglyph; with free or fused calcareous sclerites embedded in mesenchyme; stolons or coenenchyme connect the polyps; new polyps are usually budded from stolons. All but one of the 8 orders is colonial.

ORDER ALCYONACEA: Soft corals. Colonies encrusting or erect, often massive; usually fleshy and flexible, although the coenenchyme is sclerite-filled; fleshy distal portions of polyps retractable into more compact basal portion. (e.g., *Alcyonium, Anthomastus, Ceratocaulon, Gersemia, Parerythropodium*)

ORDER GASTRAXONACEA: Monotypic order containing only the family Pseudogorgiidae and the species *Pseudogorgia godeffroyi*, known from shallow sand bottoms along the southeastern Australian coast. The unusual, bladelike colony has a single axial polyp extending along its entire length; lateral polyps occur only along the upper portion of the colony.

ORDER GORGONACEA: Sea fans and sea whips. Colonies typically brightly-colored and arborescent and may be several meters across; firm internal axial skeleton composed of horny proteinaceous material (gorgonin); occasionally the skeleton is calcareous, as in "precious coral," after which the color coral was named; colonies always covered with a thin layer of sclerite-filled mesoglea; one family (*Isidae*) has calcareous "segments" that alternate with thin, horny intercalary plates, giving flexibility to the otherwise rigid colony; polyps interconnected by gastrodermal solenia. A large and diverse group, with 18 recognized families. (e.g., *Acanthogorgia, Briareum, Corallium, Eugorgia, Eunicella, Gorgonia, Isis, Leptogorgia, Lophogorgia, Muricea, Parisis, Psammogorgia, Swiftia*)

ORDER HELIOPORACEA: Helioporaceans. Colonies produce rigid calcareous skeletons of aragonite crystals (not fused sclerites) similar to those of milleporids and stony corals; polyps monomorphic. Two genera: *Epiphaxum* and *Heliopora* ("blue coral").

ORDER PENNATULACEA: Sea pens and sea pansies. Colonies complex and polymorphic; adapted for life on soft benthic substrata; often luminescent; elongate primary axial polyp extends length of the colony (to 1 m) and consists of a basal bulb or peduncle for anchorage and a distal stalk, the latter giving rise to dimorphic secondary polyps; coelenteron of axial polyp with skeletal axes of calcified horny material in canals. (e.g., *Anthoptilum, Balticina, Cavernularia, Funiculina, Pennatula, Ptilosarcus, Renilla, Stylatula, Umbellula, Virgularia*)

ORDER PROTOALCYONARIA: Protoalcyonarians are solitary deep-water octocorals that reproduce exclusively by sexual means. Five genera in two families. (e.g., *Haimea, Hartea, Monoxenia, Psuchastes, Taiaroa*)

ORDER STOLONIFERA: Stoloniferans. Simple polyps arise separately from ribbon-like stolon that forms an encrusting sheet or network; oral disc and tentacles retractable into

anthostele (stiff proximal portion of polyp); mesenchyme with sclerites; horny external skeleton covers polyps and stolons; including the organ-pipe "corals" (*Tubipora*), in which sclerites fuse to form a calcareous skeleton. Three families. (e.g., *Clavularia, Cornularia, Sarcodictyon, Tubipora*)

ORDER TELESTACEA: Telestaceans. Colonies usually branched; polyps simple, cylindrical, very tall, and typically bud off lateral polyps; polyps connected at base and grow from a creeping stolon; axis never solid, although axial spicules may be somewhat fused, providing rigidity. (e.g., *Coelogorgia, Paratelesto, Telesto, Telestula*)

SUBCLASS HEXACORALLIA (= ZOANTHARIA): Anemones and true corals. Solitary or colonial; naked, or with calcareous skeleton or chitinous cuticle, but never with isolated sclerites; mesenteries usually paired and in multiples of six; mesenteries bear longitudinal retractor muscles arranged so that those of each pair either face toward each other or away from each other; mesenterial filaments typically trilobed, with two ciliated bands flanking a central one bearing cnidocytes and gland cells; one to several circles of hollow tentacles arise from endocoels (the spaces between the members of each mesentery pair) and exocoels (the spaces between adjacent mesentery pairs); pharynx may have 0, 1, 2, or many siphonoglyphs; cnidae very diverse; entodermal zooxanthellae may be profuse.

ORDER ACTINIARIA: The true sea anemones. Solitary or clonal, but never colonial; calcareous skeleton lacking, although some species secrete a chitinous cuticle; some harbor zooxanthellae; column often with specialized structures, such as warts or verrucae, acrorhagi, pseudotentacles, or vesicles; oral tentacles conical, digitiform, or branched, usually hexamerously arrayed in one or more circles; typically with two siphonoglyphs. About 800 species in 41 families, the largest being the Actiniidae. (e.g., *Actinia, Adamsia, Aiptasia, Alicia, Anthopleura, Anthothoe, Bartholomea, Bunodactis, Calliactis, Condylanthus, Diadumene, Edwardsia, Epiactis, Halcampa, Haliplanella, Heteractis (= Radianthus), Liponema, Metridium, Peachia, Phyllodiscus, Ptychodactis, Stichodactyla, Stomphia, Triactis*)

ORDER SCLERACTINIA (= MADREPORARIA): True or stony corals. Mostly colonial; polyp morphology almost identical to that of Actiniaria, except corals lack siphonoglyphs and ciliated lobes on the mesenterial filaments; zooxanthellae present in about half the known species; colony forms delicate to massive calcareous (aragonite) exoskeleton, with platelike skeletal extensions (septa). Over 1,300 extant species, in 24 extant families.* (e.g., *Acropora, Agaricia, Astrangia, Balanophyllia, Dendrogyra, Flabellum, Fungia, Goniopora, Letepsammia, Meandrina, Montipora, Oculina, Pachyseris, Porites, Psammocora, Siderastraea, Stylophora*)

ORDER ZOANTHIDEA: Zoanthids. Polyps arise from a basal mat or stolon containing gastrodermal solenia or canals; new polyps bud from gastrodermal solenia of stolons; pharynx flattened, with one siphonoglyph; mesenteries numerous, but with weak musculature; tentacles never pinnate; without intrinsic skeleton, but many species incorporate sand, sponge spicules, or other debris into the thick body wall; most with a thick cuticle; zooxanthellae abundant in some species; many species epizootic. (e.g., *Epizoanthus, Isaurus, Isozoanthus, Palythoa, Parazoanthus, Thoracactus, Zoanthus*)

---

*The taxonomy of the Scleractinia is in a state of confusion; many higher taxa appear to be nonmonophyletic, and intraspecific polymorphism seems to be common.

ORDER CORALLIMORPHARIA: Solitary or colonial polyps, without a skeleton; lack siphonoglyphs and ciliated bands on mesenterial filaments. (e.g., *Amplexidiscus, Corynactis, Rhodactis, Ricordea*)

SUBCLASS CERIANTIPATHARIA: Ceriantipatharians. Mesenteries complete, but with feeble musculature; with six primary mesenteries, but others added immediately opposite the single siphonoglyph.

ORDER ANTIPATHARIA: Black or thorny corals. Gorgonian-like colonies up to 6 m tall; hard axial skeleton, usually brown or black and covered by a thin coenosarc bearing small polyps, usually with 6 (but up to 24) nonretractable tentacles; with feeble mesenteries; skeleton produces thorns on its surface. (e.g., *Antipathes*)

ORDER CERIANTHARIA: Cerianthids or tube anemones. Large, solitary, elongate polyps living in vertical tubes in soft sediments; tube constructed of interwoven specialized cnidae (ptychocysts) and mucus; aboral end lacks a pedal disc and possesses a terminal pore; long thin tentacles arise from margin of oral disc, fewer shorter labial tentacles encircle mouth; mesenteries complete; gonads occur only on alternate mesenteries; protandric hermaphrodites. (e.g., *Arachnanthus, Botruanthus, Ceriantheomorphe, Cerantheopsis, Cerianthus, Pachycerianthus*)

CLASS CUBOZOA: Sea wasps and box jellyfish (Figure 8.15). Medusae 15–25 cm tall, largely colorless; polyps each produce a single medusa by complete metamorphosis (strobilation does not occur), medusa bell nearly square in cross section; hollow interradial tentacle(s) hang from bladelike pedalia, one at each corner of umbrella; unfrilled bell margin drawn inward to form a velum-like structure (the velarium) into which diverticula of the gut extend. Their sting is very toxic, in some cases fatal to humans, hence the name "sea wasps." The single order Cubomedusae (= Carybdeida), with about three dozen species, was formerly placed in the class Scyphozoa. Cubozoans occur in all tropical seas but are especially abundant in the Indo-West Pacific region. Almost all are tropical to subtropical in their range, but a large temperate species (*Carybdea alata*) occurs on the Skeleton Coast of southwestern Africa. (e.g., *Carybdea, Chironex, Tamoya, Tripedalia*)

CLASS SCYPHOZOA: Jellyfish (Figures 8.1E,F). Medusoid stage predominates; polypoid individuals (scyphistomae) are small and inconspicuous but often long-lived; polyps lacking in some groups; polyps produce medusae by asexual budding (strobilation); coelenteron divided by four longitudinal (oral–aboral) mesenteries; medusae acraspedote (without a velum), typically with a thick mesogleal (or collenchymal) layer, distinct pigmentation, filiform or capitate tentacles, and marginal notches producing lappets; sense organs occur in notches and alternate with tentacles; gametes arise from gastrodermis; cnidae present in epidermis and gastrodermis; mouth may or may not be on a manubrium; usually without a ring canal. Scyphozoans are exclusively marine; planktonic, demersal, or attached. About 200 species are divided into four orders.

ORDER STAUROMEDUSAE: Small, sessile individuals that develop directly from benthic planula larvae; with stalked adhesive disc by which individuals attach to substratum; with eight tentacle-bearing "arms"; sexual reproduction only. Occur in shallow water at high latitudes. Long considered to be sessile medusae, current opinion views them as polyps. (e.g., *Haliclystis, Lucernaria*)

ORDER CORONATAE: High bell divided into upper and lower regions by a coronal groove encircling exumbrella; margin of bell deeply scalloped by gelatinous thickenings termed pedalia, which give rise to tentacles, rhopalia, and marginal lappets; gonads present on the four gastrovascular septa. Small to moderate in size; primarily bathylpelagic; some contain zooxanthellae. (e.g., *Atolla, Linuche, Nausithoe, Periphylla, Stephanoscyphus, Tetraplatia*)

ORDER SEMAEOSTOMAE: Corners of mouth drawn out into four broad, gelatinous, frilly lobes; stomach with gastric filaments; hollow marginal tentacles contain extensions of radial canals; without coronal furrow or pedalia; gonads on folds of gastrodermis. This order contains most of the typical jellyfish of temperate and tropical seas; moderate to very large forms. (e.g., *Aurelia, Chrysaora, Cyanea, Pelagia, Sanderia, Stygiomedusa*)

ORDER RHIZOSTOMAE: Lack a central mouth; frilled edges of the four oral lobes are fused over the mouth so that many suctorial "mouths" (ostioles) open from a complicated canal system on eight branching armlike appendages; bell without marginal tentacles or pedalia; stomach without gastric filaments; gonads on folds of gastrodermis. Small to large jellyfish that swim vigorously using a well developed subumbrellar musculature; primarily occur in low latitudes. (e.g., *Cassiopea, Cephea, Eupilema, Mastigias, Rhizo-stoma, Stomolophus*)

The Cnidarian Bauplan

As true Metazoa, the cnidarians show marked advances over the groups covered thus far. However, they possess only two embryonic germ layers—the ectoderm and the entoderm—which become the adult epidermis and gastrodermis, respectively. In fact, the terms "ectoderm" and "entoderm" were originally coined as names for the outer and inner tissues of cnidarians, and many specialists still use them in that way. The middle mesoglea or mesenchyme in adults is derived largely from ectoderm and never produces the complex organs seen in triploblastic Metazoa.*

The essence of the cnidarian bauplan is radial symmetry (Figure 8.3). As discussed in Chapter 3, radial symmetry is associated with various architectural and strategic constraints. Cnidarians are either sessile, sedentary, or pelagic, and do not engage in the active unidirectional movement seen in bilateral, cephalized creatures. Radial symmetry demands certain anatomical arrangements, particularly of those parts that interact directly with the environment, such as feeding structures and sensory receptors. Thus, we typically find a ring of tentacles that can collect food from any direction, and a diffuse, noncentralized nerve net with radially

---

*Whether or not a true basement membrane (= basal lamina) exists in cnidarians is debatable. We define a basement membrane as a thin sheet of extracellular matrix upon which an epithelial layer may rest; it contains collagen and other proteins. By this definition, sponges and mesozoa lack a basement membrane, cnidarians and ctenophores possess one by way of the mesenchyme, and the bilateria possess a well developed, highly proteinaceous basement membrane.

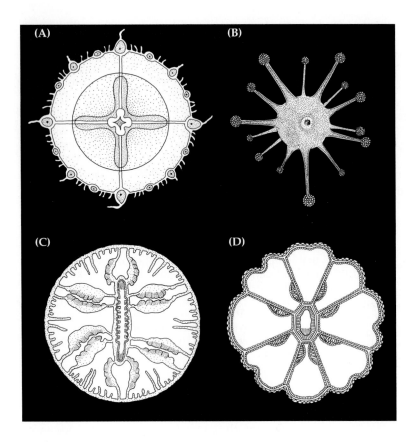

Figure 8.3 Cnidarian radial symmetries. (A) Quadriradial symmetry of a hydromedusa. (B) Radial symmetry of a hydrozoan polyp. (C) Biradial symmetry of an actiniarian polyp (a sea anemone). (D) Biradial symmetry of an octocoral polyp (Anthozoa).

boring cells interconnect, often forming longitudinal and circular sheets capable of contracting like true muscle layers. Other sorts of myoepithelial cells do occur in some other animals; recall the contractile myocytes of sponges. Similar cells are known even among mammals, where they are found in association with certain secretory tissues.

Some cnidarians also possess subepidermal mesenchymal muscles, apparently derived from the contractile elements of the myoepithelial cells. In anemones, for example, cordlike sphincters are sunk below the epithelium and reside as distinct muscles wholly within the mesenchyme.

In addition to epitheliomuscular cells, the epidermis contains sensory cells, cnida-bearing cells called **cnidocytes**, gland cells, and **interstitial cells**. The last are undifferentiated and capable of developing into other types of cells. The gatrodermis is histologically somewhat similar to the epidermis (Figure 8.4). Along with the nutritive-muscular cells it also contains cnidocytes (except in the Hydrozoa) and gland cells.

In hydrozoans the middle layer is a rather simple, gel-like, largely acellular mesoglea. Scyphomedusae and cubomedusae have very thick mesogleal layers with scattered cells. In anthozoans the middle layer is often a thick and richly cellular mesenchyme.

The polypoid form. Polyps are much more diverse than medusae are, largely as a result of their capacities for asexual reproduction and colony formation (see Figures 8.5 through 8.12). The polypoid stage occurs in all four classes of cnidarians, although it is greatly reduced in the Scyphozoa and Cubozoa. Polyps are tubular structures with an outer epidermis, an inner gut sac (coelenteron) lined with gastrodermis, and a layer of jelly-like mesoglea or mesenchyme in between. Most polyps are small, but some species of sea anemones get quite large; the largest is the tropical Indo-Pacific *Stichodactyla mertensii*, which can exceed a meter in diameter, and the northeast Pacific *Metridium giganteum*, which can extend its column to a meter in height.

The basic polypoid symmetry is radial, although as a result of subtle modifications most species possess a biradial or quadriradial symmetry. The main body axis runs longitudinally through the mouth (oral end) to the

distributed sense organs. These and other implications of radial symmetry are explored further throughout this chapter.

In spite of the limitations of a diploblastic, radially symmetrical bauplan, cnidarians are a very successful and diverse group. Much of their success has resulted from the apparent evolutionary plasticity of their dimorphic life histories, the alternation between polypoid and medusoid phases. Although polyps and medusae are very different in appearance, they are really variations on the basic cnidarian bauplan. But, the two stages are vastly different ecologically, and their presence in a single life history allows an individual species to exploit different environments and resources, leading a "double life." This particular kind of dimorphic life cycle is unique to the Cnidaria.

The Body Wall

Cnidarian epithelia—the outer epidermis and inner gastrodermis—include **myoepithelial cells** (Figures 8.4 and 8.19), viewed by many workers as the most primitive muscle cells in the Metazoa. These columnar cells bear flattened, contractile, basal extensions called **myonemes**. In the epidermis, these cells are referred to as **epitheliomuscular cells**, and in the gastrodermis are called **nutritive-muscular cells**. The myonemes rest against the middle mesoglea or mesenchyme, and the opposite ends of the cells form the outer body and gut surfaces. The myonemes run parallel to free surfaces and contain contractile myofibrils. Myonemes of neigh-

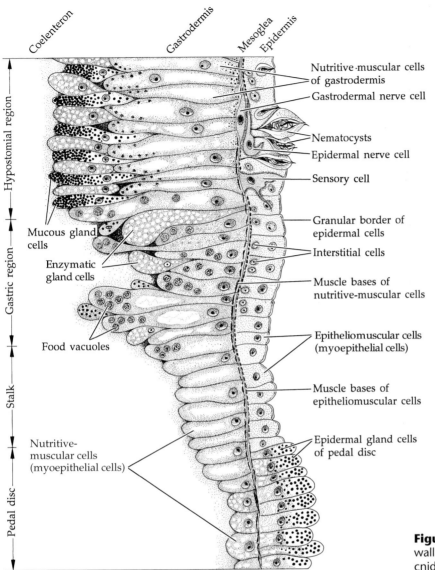

Coelenteron Gastrodermis Mesoglea Epidermis

Nutritive·muscular cells of gastrodermis

Gastrodermal nerve cell

Nematocysts

Epidermal nerve cell

Sensory cell

Granular border of epidermal cells

Interstitial cells

Muscle bases of nutritive-muscular cells

Epitheliomuscular cells (myoepithelial cells)

Muscle bases of epitheliomuscular cells

Epidermal gland cells of pedal disc

Hypostomial region

Gastric region

Stalk

Pedal disc

Mucous gland cells

Enzymatic gland cells

Food vacuoles

Nutritive-muscular cells (myoepithelial cells)

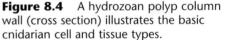

Figure 8.4 A hydrozoan polyp column wall (cross section) illustrates the basic cnidarian cell and tissue types.

base (aboral end) of the polyp. The aboral end may be a **pedal disc** for attaching to hard substrata (as in most common sea anemones); it may be a rounded structure—called a **physa**—adapted for digging and anchoring in soft substrata (as in burrowing anemones); or it may arise from a common mat, stalk, or stolon in colonial forms.

The mouth may be set on an elevated **hypostome** or **manubrium** as in hydrozoans, or it may be on a flat **oral disc** as in anthozoans (Figures 8.5 and 8.6). In anthozoans the mouth is usually slitlike and leads to a muscular, ectodermally derived pharynx that extends into the coelenteron. The pharynx usually bears from one to several ciliated grooves called **siphonoglyphs**, which drive water into the gut cavity (Figure 8.6). It is in part the presence of siphonoglyphs that gives these polyps a secondary biradial symmetry. The side of an anthozoan polyp that bears a single siphonoglyph is called the **sulcal** side, and the opposite side is called the **asulcal** side.

The coelenteron, or **gastrovascular cavity**, serves for circulation as well as digestion and distribution of food. In hydrozoan polyps, the coelenteron is a single, uncompartmentalized tube. In scyphozoan polyps (**scyphistomae**), it is partially subdivided by four longitudinal, ridgelike **mesenteries**; and in anthozoan polyps, it is extensively compartmentalized by mesenteries. Anthozoan mesenteries are projections of the inner body wall and thus are lined with gastrodermis and filled with mesenchyme. They extend from the inner body wall toward the pharynx, some or all of them fusing with it as **complete mesenteries**. Those that do not connect to the pharynx are called **incomplete mesenteries**. In anthozoan polyps, the free inner edge of each mesentery below the pharynx has a thickened, cordlike margin armed with cnidae, cilia, and gland cells and is called the **mesenterial filament** (Figures 8.6 and 8.20). In some sea anemones these filaments give rise to long threads, called **acontia**, that hang free in the gastrovas-

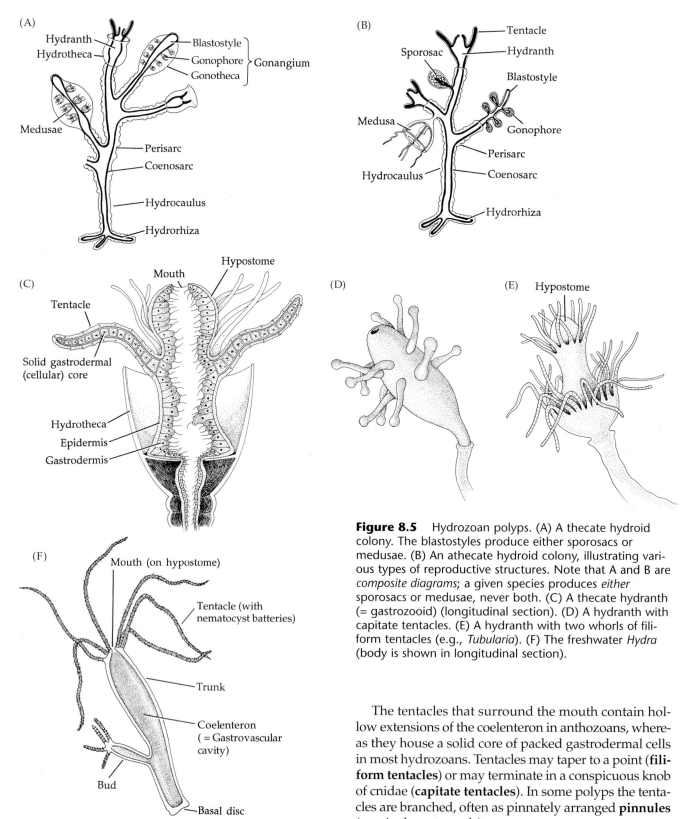

Figure 8.5 Hydrozoan polyps. (A) A thecate hydroid colony. The blastostyles produce either sporosacs or medusae. (B) An athecate hydroid colony, illustrating various types of reproductive structures. Note that A and B are *composite diagrams*; a given species produces *either* sporosacs or medusae, never both. (C) A thecate hydranth (= gastrozooid) (longitudinal section). (D) A hydranth with capitate tentacles. (E) A hydranth with two whorls of filiform tentacles (e.g., *Tubularia*). (F) The freshwater *Hydra* (body is shown in longitudinal section).

The tentacles that surround the mouth contain hollow extensions of the coelenteron in anthozoans, whereas they house a solid core of packed gastrodermal cells in most hydrozoans. Tentacles may taper to a point (**filiform tentacles**) or may terminate in a conspicuous knob of cnidae (**capitate tentacles**). In some polyps the tentacles are branched, often as pinnately arranged **pinnules** (e.g., in the octocorals).

Branched hydrozoan colonies grow in two patterns (Figure 8.7). In **monopodial growth**, the first polyp elongates continuously from a growth zone at the distal end of the hydrocaulus. This primary (axial) polyp may even lose its hydranth and persist merely as a stalk. The primary hydrocaulus gives rise to secondary polyps by lat-

cular cavity. They function in defense and feeding, as discussed later. In most colonial anthozoans the cellular mesenchyme unites individual zooids (see Figure 8.12). In some, such as the soft corals, gastrovascular cavities are connected to one another by canals called **solenia**.

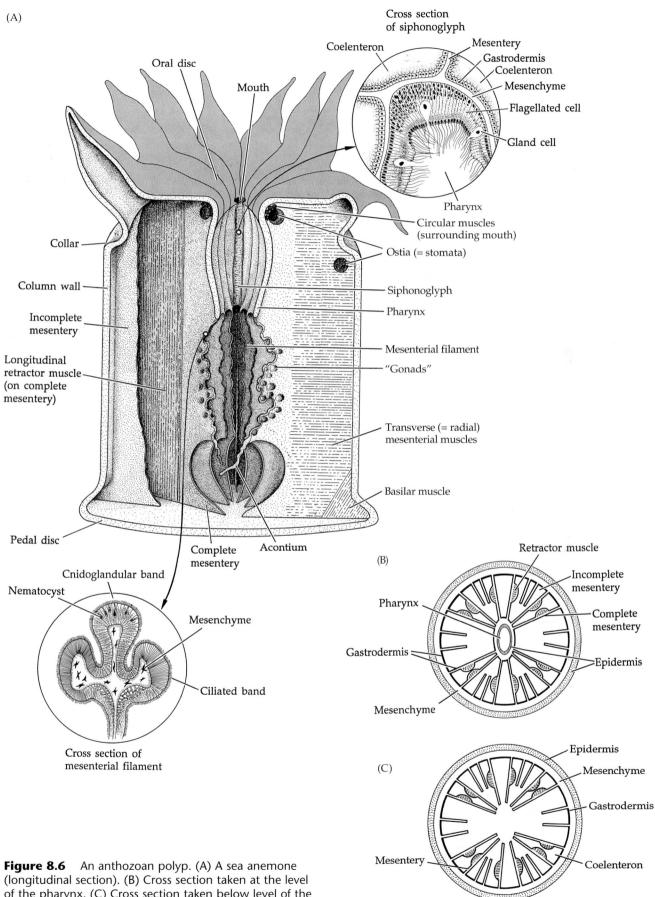

Figure 8.6 An anthozoan polyp. (A) A sea anemone (longitudinal section). (B) Cross section taken at the level of the pharynx. (C) Cross section taken below level of the pharynx.

(A)

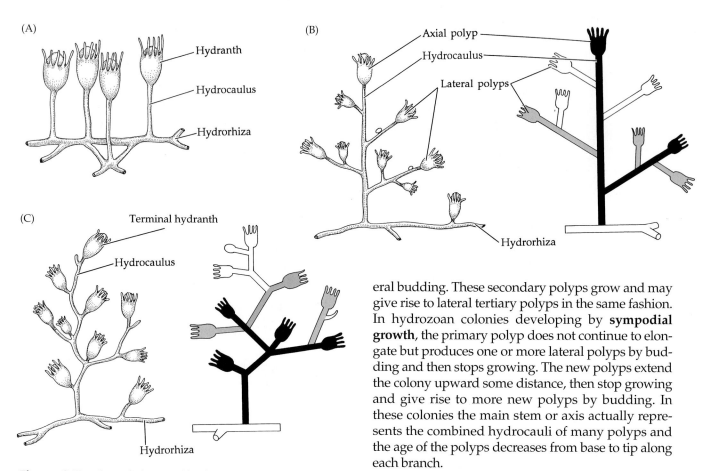

(B)

- Axial polyp
- Hydrocaulus
- Lateral polyps
- Hydrorhiza

(A) labels:
- Hydranth
- Hydrocaulus
- Hydrorhiza

(C)
- Terminal hydranth
- Hydrocaulus
- Hydrorhiza

Figure 8.7 Growth forms of hydrozoan colonies. (A) Hydrorhizal colony. (B) Colony displaying monopodial growth. (C) A colony displaying sympodial growth. The models next to B and C illustrate age of polyps in the colony; oldest polyps are shown in black, youngest in white.

eral budding. These secondary polyps grow and may give rise to lateral tertiary polyps in the same fashion. In hydrozoan colonies developing by **sympodial growth**, the primary polyp does not continue to elongate but produces one or more lateral polyps by budding and then stops growing. The new polyps extend the colony upward some distance, then stop growing and give rise to more new polyps by budding. In these colonies the main stem or axis actually represents the combined hydrocauli of many polyps and the age of the polyps decreases from base to tip along each branch.

Most marine hydroids are surrounded, at least in part, by a nonliving protein–chitin exoskeleton secreted by the epidermis and called the **perisarc** (Figure 8.5). However, this outer covering is absent in freshwater hydroids. The living tissue inside the perisarc is

Figure 8.8 Diversity of form among the colonial Hydrozoa. (A) *Proboscidactyla*, a two-tentacled hydroid that lives around the open end of polychaete worm tubes. (B) *Monobrachium*, a one-tentacled hydroid that lives on clam shells. (C) *Hydractinia*, a colonial hydroid commensal on shells inhabited by hermit crabs. (D) The chondrophoran *Porpita* (aboral view). (E) A colony of the calcareous mille-porinid hydrocoral *Millepora*. (F) A siphonophore, *Physalia* ("man-of-war"). (G) A colony of the calcareous stylasterine hydrocoral *Allopora*. (H) *Nectocarmen antonioi*, a colonial calycophoran siphonophore from California. (I) Another siphonophore.

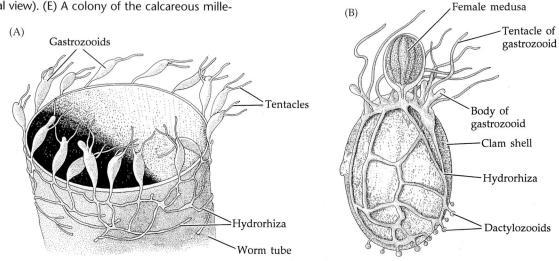

(A)
- Gastrozooids
- Tentacles
- Hydrorhiza
- Worm tube

(B)
- Female medusa
- Tentacle of gastrozooid
- Body of gastrozooid
- Clam shell
- Hydrorhiza
- Dactylozooids

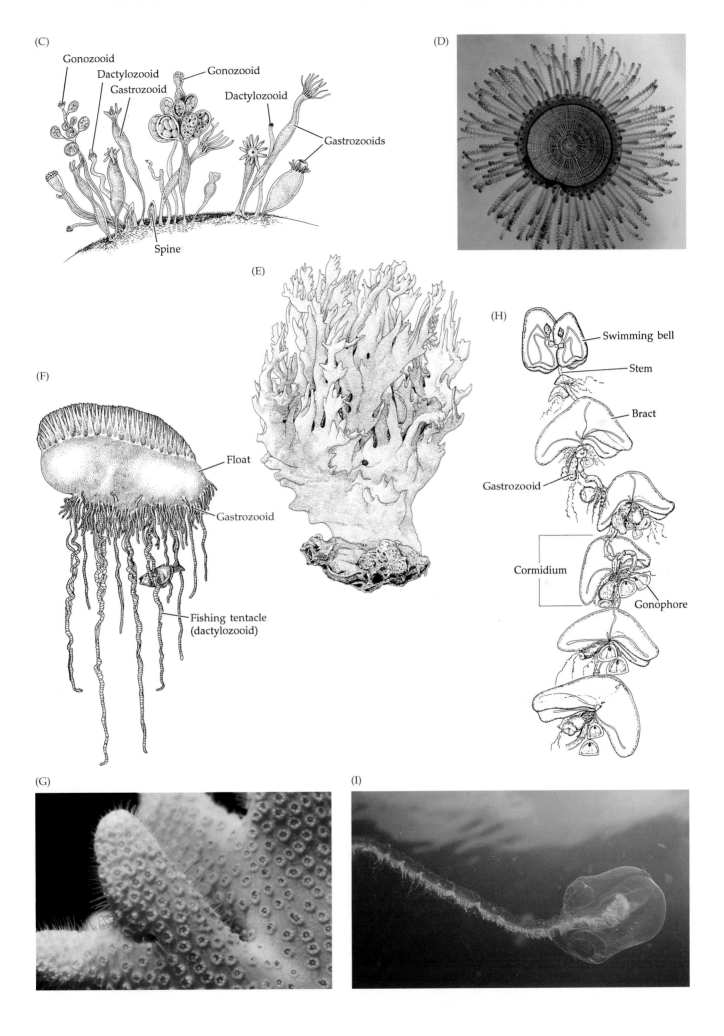

(C)

Gonozooid

Dactylozooid

Gastrozooid

Gonozooid

Dactylozooid

Gastrozooids

Spine

(D)

(E)

(H)

Swimming bell

Stem

Bract

Gastrozooid

Cormidium

Gonophore

(F)

Float

Gastrozooid

Fishing tentacle
(dactylozooid)

(G)

(I)

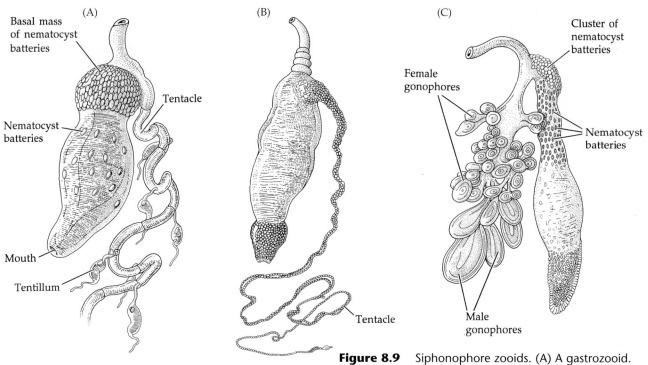

(A)

Basal mass
of nematocyst
batteries

Tentacle

Nematocyst
batteries

Mouth

Tentillum

(B)

Tentacle

(C)

Cluster of
nematocyst
batteries

Female
gonophores

Nematocyst
batteries

Male
gonophores

Figure 8.9   Siphonophore zooids. (A) A gastrozooid.
(B) A dactylozooid. (C) A gonozooid.

termed the **coenosarc**. The perisarc may extend around
each hydranth and gonozooid as a **hydrotheca** and
gonotheca, respectively. When this occurs, the hydroids
are said to be **thecate**; hydroids whose perisarcs do not
extend around the zooids are **athecate**.

A complex terminology has been developed to describe hydrozoan polyps, or **hydroids** (Figures 8.5 and
8.7). One reason for this special nomenclature is that hydroid colonies are usually polymorphic, containing
more than one kind of polyp, or zooid. The term **hydranth** or **gastrozooid** refers to feeding zooids, which
typically bear tentacles and a mouth. Other commonly
occurring polyp types include defense polyps (**dactylozooids**) and reproductive polyps (**gonozooids** or **gonangia**). Each zooid typically arises from a stalk, called
a **hydrocaulus**. In most colonial hydrozoans, the individual polyps are anchored in a rootlike stolon called a
hydrorhiza, which grows over the substratum. From
the hydrorhiza arise hydrocauli, bearing polyps singly
or in clusters.

Gastrozooids capture and ingest prey and provide
energy and nutrients to the rest of the colony, including
all the nonfeeding polyps. Dactylozooids, which occur
in a variety of sizes and shapes, are heavily armed with
cnidae. Often several dactylozooids surround each gastrozooid and serve for both defense and food capture.
Gonozooids produce medusa buds called **gonophores**
that are either released or retained on the colony.
Whether released as free medusae or retained as gonophores, they produce gametes for the sexual phase of
the hydrozoan life cycle. The living tissue (coenosarc) of
the gonozooid is called the **blastostyle**; the gonophores
arise from this tissue. When a gonotheca surrounds the
blastostyle, the zooid is called a **gonangium**.

The most dramatic examples of polymorphism among
polyps are seen in the hydrozoan order Siphonophora
and the anthozoan order Pennatulacea. Siphonophorans
(Figures 8.8F,H,I and 8.9) are hydrozoan colonies composed of both polypoid and medusoid individuals, with
as many as a thousand zooids in a single colony. This
large order includes a great variety of unusual and poorly understood forms, including the famous man-of-war,
Physalia (Figure 8.8F). The gastrozooids of siphonophorans are highly modified polyps with a large mouth and
one long, hollow tentacle that bears many cnidae (Figure
8.9). This long feeding tentacle reaches lengths of 13 m in
the Atlantic species *Physalia physalis*. The nonfeeding
dactylozooids also bear one long (unbranched) tentacle.
The gonozooids are usually branched; they produce sessile gonophores that are never released as free medusae.

Siphonophorans use a swimming bell (**nectophore**)
or a gas-filled float (**pneumatophore**), or both, to help
maintain their position in the water. The nectophore is a
true medusoid individual with many of the structures
common to free-swimming medusae, although it has
lost its mouth, tentacles, and sense organs. The pneumatophore, once also thought to be a modified medusa,
is now known to be derived directly from the larval
stage and probably represents a highly modified polyp.
Pneumatophores are double-walled chambers lined
with chitin. Each float houses a gas gland, which consists of a mitochondria-laden glandular epithelium lining a pit or chamber. The gland secretes a gas usually
similar to air in composition, although in *Physalia* it apparently includes a surprisingly high proportion of carbon monoxide. Many siphonophorans have mechanisms by which they regulate gas in their floats to keep

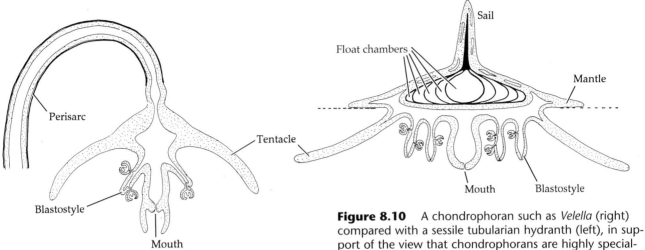

Figure 8.10 A chondrophoran such as *Velella* (right) compared with a sessile tubularian hydranth (left), in support of the view that chondrophorans are highly specialized, solitary tubularian zooids.

the colony at a particular depth, much like the swim bladders of fishes.

Siphonophorans are grouped into three suborders on the basis of colony structure: the suborder Calycophorae includes colonies with swimming bells but no float; members of the suborder Physonectae have a small float and a long train of swimming bells; and those of the suborder Cystonectae have a large float and no bell. Calycophorans have a long tubular **stem** extending from the swimming bell, from which various types of zooids bud in groups called **cormidia** (Figure 8.8H). Each cormidium acts as a colony-within-a-colony, and is usually composed of a shieldlike **bract**, a gastrozooid, and one or more gonophores that may function as swimming bells. The cormidia commonly break loose from the parent colony to live an independent existence, at which time they are termed **eudoxids**. The physonectans have an apical float with a long stem bearing a series of nectophores followed by a long train of cormidia. The cystonectans, including *Physalia*, usually have a large pneumatophore with a prominent budding zone at its base, which produces the various polyps and medusoids (Figure 8.8F).

The hydrozoan order Chondrophora is composed of colorful oceanic organisms that drift about on the sea surface in enormous flotillas, occasionally washing ashore to coat the beach with their bluish-purple bodies (Figures 8.1D and 8.10). Current opinion holds these animals to be large, solitary, athecate hydranth polyps floating upside down at the surface instead of sitting on a stalk attached to the bottom. The aberrant medusae of chondrophorans are short-lived and do not possess a functional mouth or gut, probably relying instead on their symbiotic zooxanthellae for nutrition. The aboral sail in *Velella* (the "by-the-wind-sailor") has no counterpart in sessile hydroids. In its ability to sail at an angle to the wind, *Velella* resembles the siphonophore *Physalia*, a similarity attributed to convergent evolution. Figure 8.10 compares a chondrophoran and a sessile hydroid, such as *Tubularia* or *Corymorpha*.

The pennatulaceans are the most complex and polymorphic members of the class Anthozoa (Figure 8.1I,K and 8.11D,F). The colony is built around a main supportive stem, which is actually the primary polyp and buds lateral polyps in a regular fashion. The base of the primary polyp is anchored in sediment, but the upper, exposed portion (the **rachis**) produces polyps in whorls or rows, or sometimes united in crescent-shaped "leaves." Often these polyps are of two distinct types. **Autozooids** bear tentacles and function in feeding; **siphonozooids** are small, have reduced tentacles, and serve to create water currents through the colony. In sea pens the rachis is elongated and cylindrical; in sea pansies it is flattened and shaped like a large leaf (Figure 8.1K). In the odd deep-sea genus *Umbellula*, the secondary polyps radiate outward to give the colony the appearance of a pinwheel set on the end of a tall narrow stalk. The first deep benthic photos of *Umbellula* had biologists scratching their heads for years, wondering to which phylum this preposterous creature might belong.

Gorgonians are also colonial anthozoans (Figures 8.1J and 8.12). Some grow in bushy shapes, whereas others are planar; size and shape of the colony are often mediated by the hydrodynamics of the local surges and currents. Where prevailing currents are more or less in one plane (although they may move in two directions back and forth), the branches of the colony tend to grow largely in one plane also—perpendicular to the flow. In regions of mixed currents, the same species tends to grow in two planes.

The medusoid form. Free medusae occur in all cnidarian classes except the Anthozoa. Although variation in form exists, medusae are far less diverse than polyps and it is much easier to generalize about their anatomy. The relative uniformity of medusae is a result in part of their usually similar lifestyles in open water, and of their inability to form colonies by asexu-

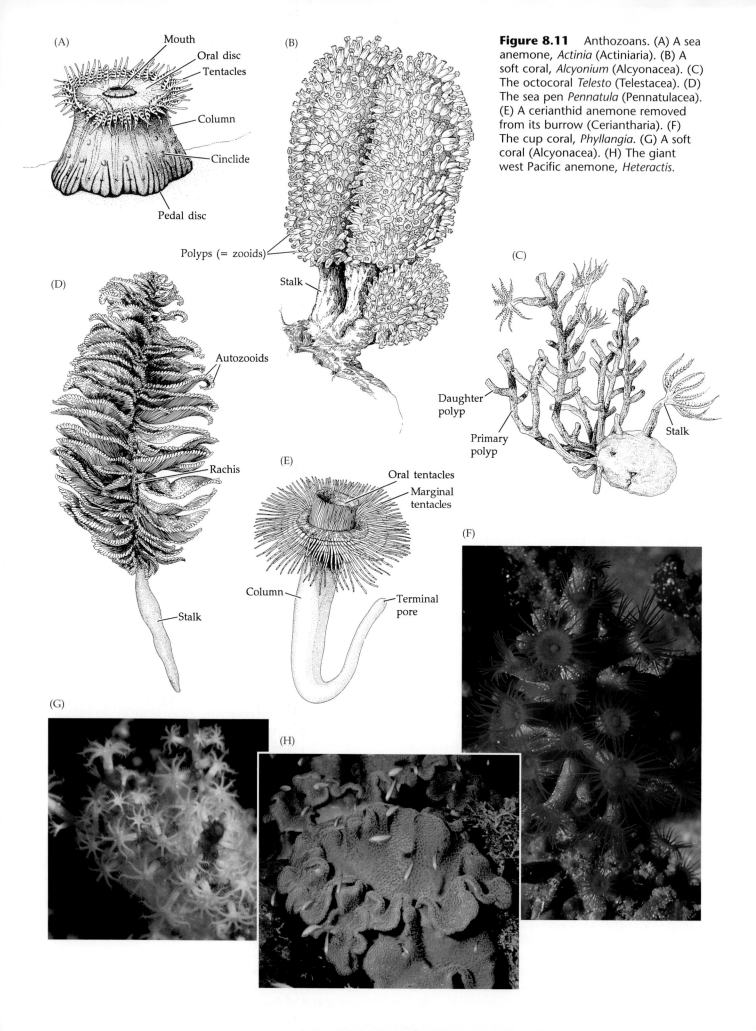

(A)
Mouth
Oral disc
Tentacles
Column
Cinclide
Pedal disc

(B)
Polyps (= zooids)
Stalk

Figure 8.11 Anthozoans. (A) A sea anemone, *Actinia* (Actiniaria). (B) A soft coral, *Alcyonium* (Alcyonacea). (C) The octocoral *Telesto* (Telestacea). (D) The sea pen *Pennatula* (Pennatulacea). (E) A cerianthid anemone removed from its burrow (Ceriantharia). (F) The cup coral, *Phyllangia*. (G) A soft coral (Alcyonacea). (H) The giant west Pacific anemone, *Heteractis*.

(C)
Daughter polyp
Primary polyp
Stalk

(D)
Autozooids
Rachis
Stalk

(E)
Oral tentacles
Marginal tentacles
Column
Terminal pore

(F)

(G)

(H)

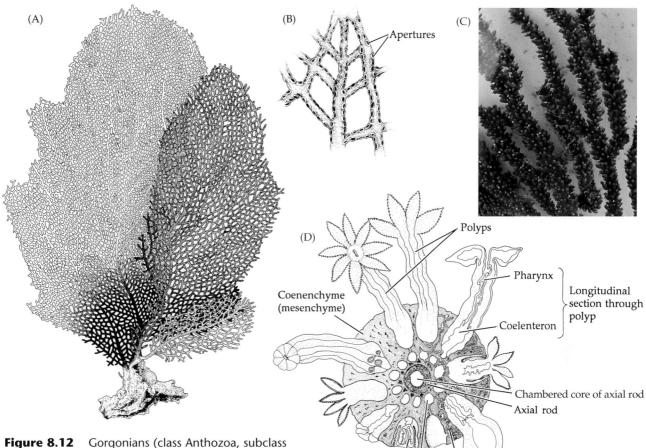

(A)

(B) Apertures

(C)

(D)

Polyps

Pharynx

Longitudinal section through polyp

Coenenchyme (mesenchyme)

Coelenteron

Chambered core of axial rod

Axial rod

Cross section of retracted polyp

Longitudinal canal (solenium)

Fine gastodermal canals connect polyps with each other and with major (longitudinal) canals (solenia)

Figure 8.12 Gorgonians (class Anthozoa, subclass Alcyonaria, order Gorgonacea). (A) The sea fan *Gorgonia* has lacelike branches. (B) Apertures of retracted polyps are visible on the branches of *Gorgonia*. (C) The Pacific gorgonian, *Muricea californica,* releasing round white eggs from reproductive polyps. (D) A branch of the sea fan *Pseudoplexaura* (cross section). (E) Polyps of *Psammogorgea*.

(E)

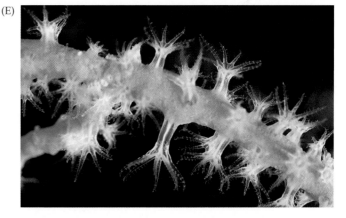

al reproduction. They are participants in colonial life only insofar as some remain attached to hydrozoan colonies as sessile gonophores. Sessile benthic forms are rare (but see Figure 8.1F).

Even though the body walls of medusae and polyps are similar and both adhere to the general cnidarian bauplan outlined earlier, their gross morphologies are adapted to their very different lifestyles. Medusae are bell-, dish-, or umbrella-shaped, and usually embued with a thick, jelly-like mesogleal layer (hence the name jellyfish). The convex upper (aboral) surface is called the **exumbrella**; the concave lower (oral) surface is the **subumbrella**. The mouth is located in the center of the subumbrella, often suspended on a pendant, tubular extension called the **manubrium**, which is almost always present on hydromedusae (Figure 8.13), but usually reduced or absent in scyphomedusae (Figure 8.14).

The coelenteron or gastrovascular cavity occupies the central region of the umbrella and extends radially via **radial canals**. In most hydromedusae, a marginal **ring canal** within the rim of the bell connects the ends of the

radial canals. The presence of four radial canals and of tentacles in multiples of four (in hydromedusae) and the division of the stomach by mesenteries into four gastric pouches (in scyphomedusae) give most jellyfish a **quadriradial** (= **tetramerous**) symmetry (Figure 8.3A). Most hydromedusae have a thin circular flap of tissue, the **velum**, within the margin of the bell (Figure 8.13). Such medusae are termed **craspedote**. Those lacking a velum, such as scyphomedusae, are said to be **acraspe-**

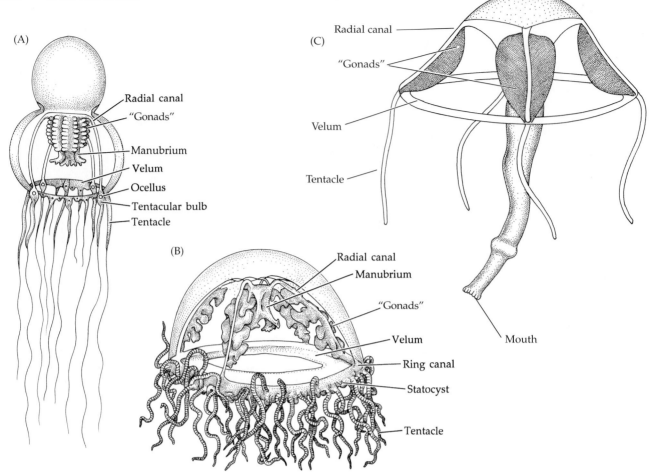

Figure 8.13 Hydrozoan medusae. (A) A typical anthomedusa. (B) A typical leptomedusa. (C) A trachyline medusa (*Liriope*)

dote (Figures 8.14 and 8.15). Like that of polyps, the external surfaces of medusae are covered with epidermis, and the internal surfaces (coelenteron and canals) are lined with gastrodermis. The bulky, gelatinous middle layer is either a largely acellular mesoglea or a partly cellular mesenchyme.

Support

Cnidarians employ a wide range of support mechanisms. Polypoid forms rely substantially on the hydrostatic qualities of the water-filled coelenteron constrained by circular and longitudinal muscles of the body wall. In addition, the mesenchyme may be stiffened with fibers, particularly in the anthozoans. Colonial anthozoans may incorporate bits of sediment and shell fragments onto the column wall for further support. Many colonial hydrozoans produce a flexible, horny perisarc, composed largely of chitin secreted by the epidermis. In medusae, the principal support mechanism is the middle layer, which ranges from a fairly thin and flexible mesoglea to an extremely thick and stiffened fibrous mesenchyme, which may be almost cartilaginous in consistency.

In addition to these soft or flexible support structures, there is an impressive array of hard skeletal structures of three fundamental types: horny or woodlike axial skeletal structures, calcareous sclerites, and massive calcareous frameworks. Horny **axial skeletons** occur in several groups of colonial anthozoans such as gorgonians, sea pens and antipatharian corals (Figures 8.11 and 8.12). Amebocytes in the coenenchyme secrete a flexible or stiff internal axial rod as a supportive base embedded in the coenenchymal mass. Axial rods are protein–mucopolysaccharide complexes (called gorgonin in the order Gorgonacea), but little is known of their chemistry. In the antipatharians (black coral), the axial skeleton is so hard and dense that it is ground and polished to make jewelry.

In most octocorals, mesenchymal cells called **scleroblasts** secrete calcareous **sclerites** of various shapes and colors (Figure 8.16). It is usually these sclerites that give gorgonians their characteristic color and texture. In many species, the sclerites become quite dense and may even fuse to form a more-or-less solid calcareous framework. The precious red coral *Corallium* is actually a gorgonian with fused red coenenchymal sclerites. In the stoloniferan organ-pipe corals (*Tubipora*), the sclerites of the body walls of the individual polyps are fused into rigid tubes. Invertebrate calcium carbonate skeletons do not usually have collagen incorporated into their framework, as occurs in vertebrates. However, in at least some gorgonians (e.g., *Leptogorgia*) the calcareous spicules do include a collagen component.

Figure 8.14 A typical scyphozoan medusa. (A) Cutaway side view. (B) Oral view.

Figure 8.15 Comparison of cubomedusae and scyphomedusae. (A) A cubomedusa. (B) A coronate scyphomedusa (order Coronatae). (C) A semaeostome scyphomedusa (order Semaeostomae). (D) A rhizostome scyphomedusa (order Rhizostomae). (E) A sessile stauromedusa (order Stauromedusae).

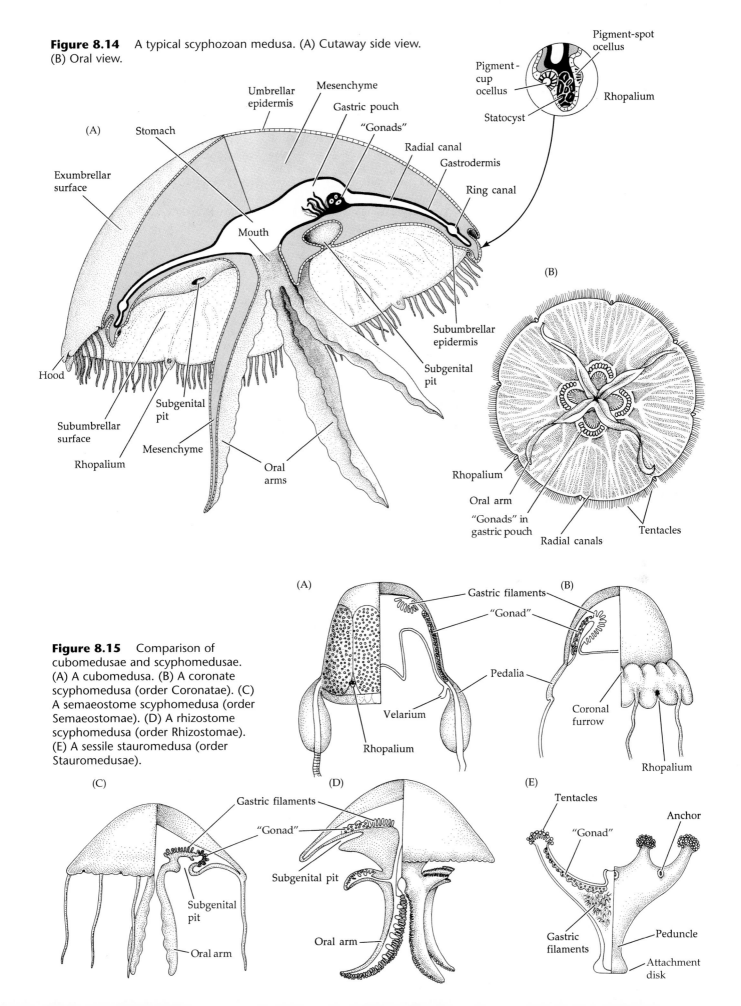

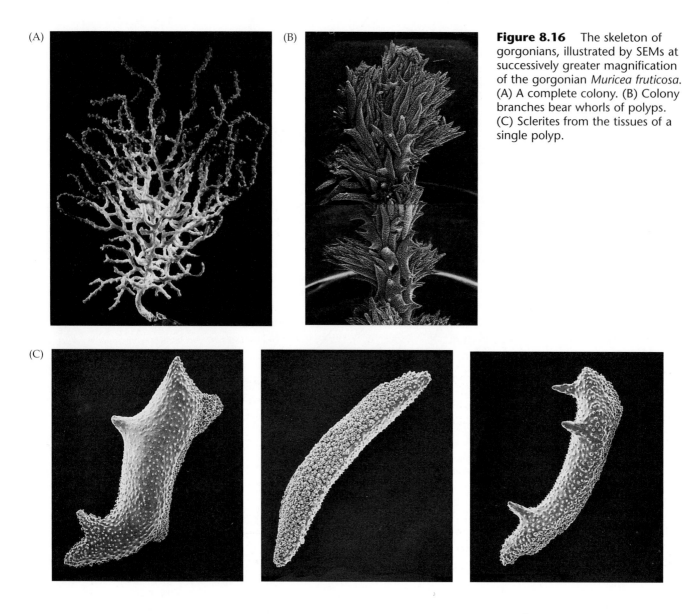

(A)

(B)

(C)

Figure 8.16 The skeleton of gorgonians, illustrated by SEMs at successively greater magnification of the gorgonian *Muricea fruticosa*. (A) A complete colony. (B) Colony branches bear whorls of polyps. (C) Sclerites from the tissues of a single polyp.

Massive calcareous skeletons are found in only certain groups of Anthozoa and Hydrozoa. The best known are the true, or stony, anthozoan corals (order Scleractinia), in which epidermal cells on the lower half of the column secrete a calcium carbonate skeleton (Figure 8.17). The skeleton is covered by the thin layer of living epidermis that secretes it, and thus it might technically be considered to be an internal skeleton. However, because the coral colony generally sits atop a large nonliving calcareous framework, most authors speak of the skeleton as being external.

The entire skeleton of a scleractinian coral is termed the **corallum**, regardless of whether the animal is solitary or colonial; the skeleton of a single polyp, however, is called a **corallite**. The outer wall of the corallite is the **theca**; the floor is the **basal plate** (Figure 8.17). Rising from the center of the basal plate is often a supportive skeletal process called the **columella**. The basal plate and inner thecal walls give rise to numerous radially arranged calcareous partitions, the **septa**, which project inward and support the mesenteries of the polyp.

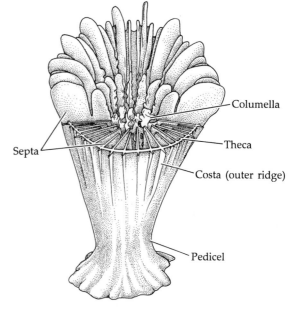

Figure 8.17 The corallite of a solitary scleractinian coral (diagrammatic), illustrating morphological features.

Polyps occupy only the uppermost surface of the corallum. Skeletal thickness increases as polyps grow, and the bottoms of the corallites are sealed off by transverse calcareous partitions called **tabulae**, each of which becomes the new basal support of a polyp. The corallum can assume a great variety of shapes and sizes, from simple cup-shaped structures in solitary corals to large branching forms in colonial species. The corallum of colonial forms may be upright, low and massive, or even encrust on other hard substrata.

Members of the hydrozoan families Milleporidae and Stylasteridae also produce calcareous exoskeletons, and they are often referred to as the **hydrocorals**. Like true corals, milleporid colonies may assume a variety of shapes, from erect branching forms to encrustations. The milleporid exoskeleton, termed a **coenosteum**, is perforated by pores of two sizes that accommodate two kinds of polyps (Figure 8.18). The gastrozooids live in large holes, or **gastropores**, and are surrounded by a circle of smaller **dactylopores**, which house the dactylozooids. Canals lead downward from the pores into the coenosteum and are closed off below by transverse calcareous tabulae. As growth proceeds and the colony thickens, new tabulae are formed, keeping the polyp pores at a more or less fixed depth. Hydrocoral colonies thus differ

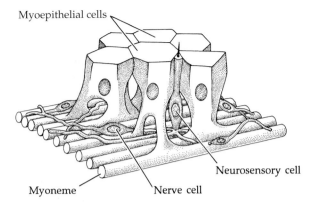

Myoepithelial cells

Neurosensory cell

Myoneme   Nerve cell

Figure 8.19   Myoepithelial cells and the nerve net of cnidarian epithelium.

from scleractinian colonies in having the skeleton penetrated by living tissue. The stylasterine skeleton is similar to the milleporid skeleton, but the margins of the gastropores often bear notches that serve as dactylopores, and the gastrozooids and dactylozooids are supported by calcareous, spine-like **gastrostyles** and low ridges called **dactylostyles**, respectively. Stylasterine gonophores arise in chambers called **ampullae**, which connect to the feeding zooids through the coenosteum.

Movement

The contractile elements of cnidarians are derived from their myoepithelial cells (Figure 8.19). In spite of the epithelial origin of these elements, for convenience we use the terms "muscles" and "musculature" for the sets of longitudinal and circular fibrils.

In polyps, these two muscle systems work in conjunction with the gastrovascular cavity as an efficient hydrostatic skeleton, as well as providing a means of movement. However, unlike the fixed-volume hydrostatic skeletons of many animals (e.g., many worms), water can enter and leave the coelenteron of cnidarians, adding to its versatility as a support device. Polyp body musculature is most highly specialized and well developed in the anthozoans, particularly the sea anemones, and many muscles lie in the mesenchyme. In anemones, the muscles of the column wall are largely gastrodermal, although epitheliomuscular cells occur in the tentacles and oral disc. Bundles of longitudinal fibers lie along the sides of the mesenteries and act as **retractor muscles** for shortening the column (Figure 8.20). **Circular muscles** derived from the gastrodermis of the column wall are also well developed. In most anemones, the circular muscles form a distinct sphincter at the junction of the column and the oral disc. Circular fibers also occur in the tentacles and the oral disc, and circular muscles surrounding the mouth can close it completely. When an anemone contracts, the upper rim of the column is pulled over to cover the oral disc. In many sea anemones, a circular fold—the **collar**, or **parapet**—occurs near the sphincter to further cover and protect the delicate oral surface upon contraction.

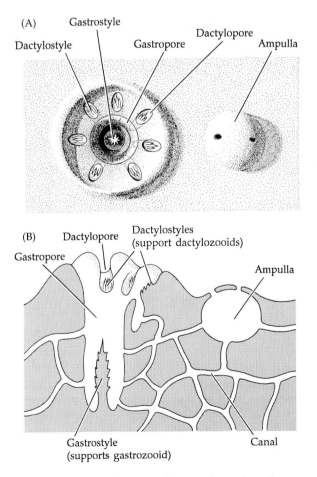

(A)   Gastrostyle

Dactylostyle   Gastropore   Dactylopore   Ampulla

(B)   Dactylopore   Dactylostyles (support dactylozooids)

Gastropore

Ampulla

Gastrostyle (supports gastrozooid)   Canal

Figure 8.18   Hydrozoan skeletons. The stylasterine hydrocoral *Allopora* has a calcareous skeleton. Plane view, from above (A) and cross section through the skeleton (B).

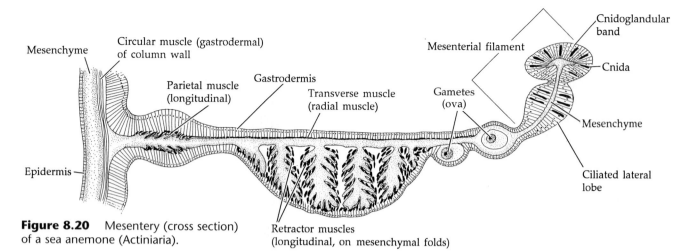

Figure 8.20   Mesentery (cross section) of a sea anemone (Actiniaria).

Most polyps are sedentary or sessile. Their movements consist mainly of food-capturing actions and the withdrawal of the upper portion of the polyp during body contractions. These activities are accomplished primarily by the epidermal muscles of the tentacles and oral disc, and by the strong gastrodermal muscles of the column. Circular muscles work in conjunction with the hydrostatic skeleton to distend the tentacles and body.

A variety of locomotor methods have evolved among polyps (Figure 8.21). Most can creep about slowly by using their pedal disc musculature. In some solitary hydrozoan polyps (e.g., *Hydra*), the column can bend far enough to allow the tentacles to contact and temporarily adhere to the substratum, whereupon the pedal disc releases its hold and the animal somersaults or moves like an inchworm. A few sea anemones can detach from the substratum and actually swim away by "rapid" flexing or bending of the column (e.g., *Actinostola, Stomphia*); others swim by thrashing the tentacles (e.g., *Boloceroides*). These swimming activities are temporary behaviors, generally elicited by the approach or contact of a predator. In a few species of sea anemones, the basal disc may detach and secrete a gas bubble, permitting the polyp to float away to a new location.

Many species of small anthozoans can float hanging upside-down on the sea surface by using water surface-tension forces (e.g., *Epiactis, Diadumene*). Sea anemones of one family (Minyadidae) are wholly pelagic and float upside down in the sea by means of a gas bubble enclosed within the folded pedal disc. *Hydra* also is known to float upside down by means of a mucus-coated gas bubble on the bottom of its pedal disc. One of the oddest forms of polyp locomotion is that of the sea anemone *Liponema brevicornis* of the Bering Sea, which is capable of drawing itself into a tight ball that can be rolled around the sea floor by the bottom currents (Figure 8.21D). Even colonial sea pansies (Pennatulacea) are motile, using their muscular peduncle to move to different depths.

Ceriantharians are burrowing, tube-building organisms (Figure 8.11E). They differ from the true sea anemones (Actiniaria) in several important ways. They have no sphincter muscle, and their weak longitudinal gastrodermal muscles do not form distinct retractors in the mesenteries. As a result, cerianthids cannot retract the oral disc and tentacles as they withdraw into their tubes. In contrast to other anemones, however, they possess a complete layer of longitudinal epidermal muscles in the column, which allows a very rapid withdrawal response. The mere shadow of a passing hand will cause a cerianthid to rapidly pull itself deep into its long, buried tube.

In medusae, epidermal and subepidermal musculatures predominate, and the gastrodermal muscles that are so important in polyps are reduced or lacking. The epidermal musculature is best developed around the bell margin and over the subumbrellar surface. Here the muscle fibers usually form circular sheets called **coronal muscles** that are partly embedded in the mesenchyme or mesoglea. Contractions of the coronal muscles produce rhythmic pulsations of the bell, driving water out from beneath the subumbrella and moving the animal by jet propulsion. The stiffened cellular collenchyme of scyphomedusae and cubomedusae includes elastic fibers that provide the antagonistic force to restore the bell shape between contractions. Many medusae also possess radial muscles that aid in opening the bell between pulses. In craspedote forms, the velum serves to reduce the size of the subumbrellar aperture, thus increasing the force of the water jet (Figure 8.13). The velarium of the fast-swimming cubomedusae has the same effect (Figure 8.15A), and the evolutionary forces that produced these two convergent features were probably similar.

Most medusae spend their time swimming upward in the water column, then sinking slowly down to capture prey by chance encounter, thereafter to pulsate upward once again. Some medusae have the ability to change direction as they swim, however, and many are strongly attracted to light (especially those harboring symbiotic zooxanthellae). At least some medusae house their zooxanthellae in small pockets, that expand during

(A)

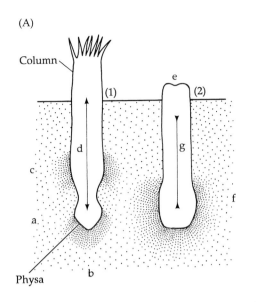

Column

Physa

(1) (2)

e

d g

c

a.

b

f

(B)

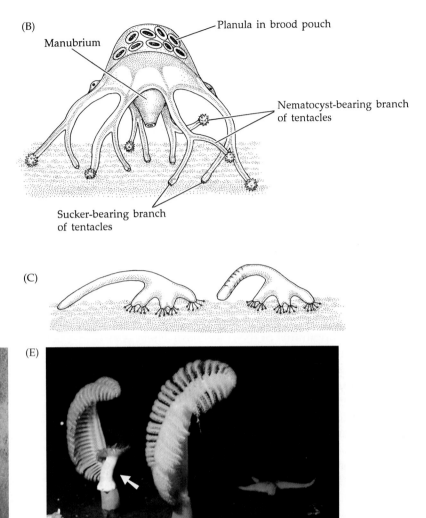

Manubrium

Planula in brood pouch

Nematocyst-bearing branch
of tentacles

Sucker-bearing branch
of tentacles

(C)

(D)

(E)

Figure 8.21 Benthic locomotion in some cnidarians.
(A) A sea anemone burrowing: (1) eversion of the physa
with displacement of sand (a) and further penetration (b)
into substratum; the anemone is held by a column anchor
(c) as extension (d) follows retraction in (2); with the tentacles folded inward (e), the physa is swollen to form an
anchor (f), which allows retractor muscles (g) to pull the
anemone into the sand. (B) The hydromedusan *Eleutheria*,
which creeps about on its tentacles. (C) The scyphomedu-
san *Lucernaria*, which also creeps about on its tentacles.
(D) *Liponema brevicornis*, a sea anemone that folds itself
into a "ball" and rolls about on the sea floor with the bottom currents. (E) The sea anemone *Stomphia* (note arrow)
swimming off the substratum by undulatory back-and-
forth contractions of the column—an escape response to
the predatory sea star *Gephyreaster swifti*, visible in this
photo (Puget Sound, Washington).

the day, exposing the algae to light. The pockets contract
at night.

Medusae can be abundant in certain localities. Some,
such as the moon jelly *Aurelia* (Figure 8.22), are known
to aggregate at temperature or salinity discontinuity
layers in the sea, where they feed on small zooplankters,

which also concentrate at these boundaries. A few unusual groups of medusae are benthic. Some hydromedusae (e.g., *Eleutheria*, *Gonionemus*) crawl about on algae
or sea grasses by adhesive discs on their tentacles
(Figure 8.21B). Members of the scyphozoan order

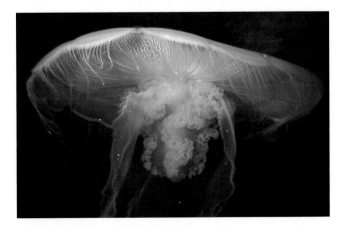

Figure 8.22 The semaeostoman medusa *Aurelia* (moon
jellies) often forms large swarms.

Stauromedusae (e.g., *Haliclystis*) develop directly from the polypoid stage and affix to algae and other substrata by an aboral adhesive disc (Figure 8.1F).

Cnidae

Before considering feeding and other aspects of cnidarian biology, it is necessary to present some information on the structure and function of cnidae. **Cnidae**, often referred to collectively as nematocysts in older works, are unique to cnidarians. They have a variety of functions, including prey capture, defense, locomotion, and attachment. They are produced inside cells called **cnidoblasts**, which develop from interstitial cells in the epidermis and, in many groups, in the gastrodermis. Once the cnida is fully formed, the cell is properly called a **cnidocyte**. During formation of a cnida, the cnidoblast produces a large internal vacuole in which a complex but poorly understood intracellular reorganization takes place. Cnidae may be complex secretory products of the Golgi apparatus of the cnidoblast. There is also some evidence that cnidae might have originated symbiogenetically from some ancient protist(s), and cnida-like structures have been reported from such diverse groups as dinoflagellates, "sporozoans," and microsporans (see Shostak and Kolluri, 1995).

Cnidae are among the largest and most complex intracellular structures known. When fully formed, they are cigar- or flask-shaped capsules, 5–100 μm or more long, with thin walls composed of a collagen-like protein. One end of the capsule is turned inward as a long, hollow, coiled, eversible tubule (Figure 8.23). The outer capsule wall consists of globular proteins of unknown function. The inner wall is composed of bundles of collagen-like fibrils having a spacing of 50–100 nm, with cross-striations every 32 nm (in the nematocysts of *Hydra*). The distinct pattern of mini-collagen fibers provides the tensile strength necessary to withstand the high pressure in the capsule. The entire structure is anchored to adjacent epithelial cells (**supporting cells**) or to the underlying mesenchyme.

When sufficiently stimulated, the tube everts from the cell. In members of the classes Hydrozoa, Scyphozoa, and perhaps Cubozoa, the capsule is covered by a hinged lid, or **operculum**, which is thrown open when the cnida discharges. In members of these three classes, each cnida bears a long cilium-like bristle called a **cnidocil**, a mechanoreceptor that elicits discharge when stimulated. The cnidocil responds to specific water-borne vibration frequencies. Chemoreceptors on the adjacent supporting cells may actually "tune" the cnidocil to the proper reception frequency for available prey (see Watson and Hessinger 1989). Anthozoan cnidae lack a cnidocil and have a tripartite apical flap instead of an operculum. Cnidocytes are most abundant in the epidermis of the oral region and the tentacles, where they often occur in clusters of wartlike structures called "nematocyst batteries."

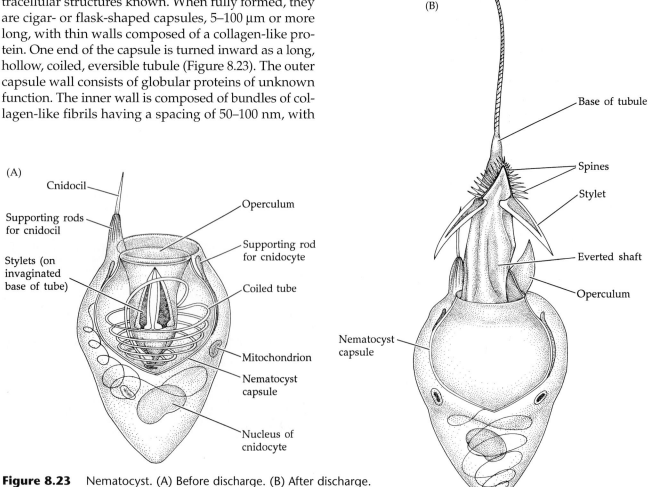

Figure 8.23 Nematocyst. (A) Before discharge. (B) After discharge.

About 30 kinds of cnidae have been described (Figures 8.24 and 8.25), but they can be assigned to three basic types. True **nematocysts** have double-walled capsules containing a toxic mixture of phenols and proteins. The tubule of most types is armed with spines or barbs that aid in penetration of and anchorage in the victim's flesh. The toxin is injected into the victim through a terminal pore in the thread or is carried into the wound on the tubule surface (see Lotan et al. 1995). **Spirocysts** have single-walled capsules containing mucoprotein or glycoprotein. Their adhesive tubules wrap around and stick to the victim rather than penetrating it. The capsule tubules of spirocysts never have an apical pore. Nematocysts occur in members of all four cnidarian classes; spirocysts occur only in the zoantharian Anthozoa. The third kind of cnidae, the **ptychocyst**, differs morphologically and functionally from both nematocysts and spirocysts. The capsule tubule of a ptychocyst lacks spines and an apical pore and is strictly adhesive in nature. In addition, the tubule is folded into pleats rather than coiled within the capsule. Ptychocysts occur only in the ceriantharians and function in forming the unique tube in which these animals reside.

Cnidae have usually been viewed as independent effectors, and, indeed, they often discharge upon direct stimulation. However, experimental evidence suggests that the animal does have at least some control of the action of its cnidae. For example, starved anemones seem to have a lower firing threshold than satiated animals. It has also been demonstrated that stimulating discharge

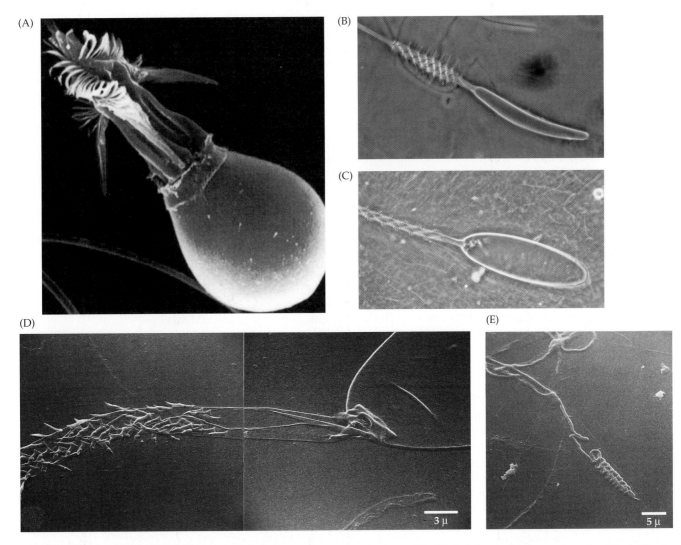

Figure 8.24 Discharged nematocysts. (A) The base of a discharged nematocyst from the hydrozoan *Hydra* (SEM). (B) A nematocyst of the anthozoan *Corynactis californica* (Corallimorpharia). The nematocyst has been "stopped" when partially everted; the everting tubule can be seen passing up through the already external region (light micrograph). (C) A fully everted nematocyst of *C. californica* (light micrograph). (D) A fully everted nematocyst of *C. californica* (SEM of the base of the everted thread and the tip of the capsule). (E) Everting nematocyst of the anthozoan coral *Balanophyllia elegans*.

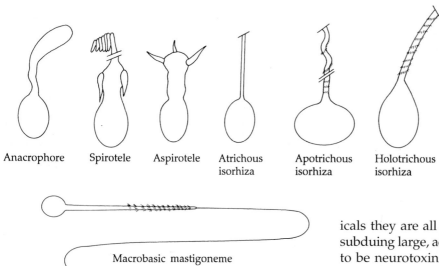

Figure 8.25 Some types of cnidae and their specialized nomenclature.

of cnidae in one area of the body results in discharge in surrounding areas. Still, either chemical and mechanical stimuli, initially perceived by the cnidocil or a similar structure, cause most cnidae to fire. Cnidarians are known to discharge their cnidae in the presence of various sugars, low-molecular-weight amino compounds, and occasionally glutathione—the latter being a chemical that is liberated when animals are injured or when tissue breaks down. Reduced glutathione also causes feeding tentacles and gastrozooids to become active, writhe about, and prepare for feeding.

The ejection of the tubule from a cnida is called **exocytosis**, and an individual cnida can be fired only once. Three hypotheses have been proposed to explain the mechanism of firing: (1) the discharge is the result of increased hydrostatic pressure caused by a rapid influx of water (the **osmotic hypothesis**); (2) intrinsic tension forces generated during cnidogenesis are released at discharge (the **tension hypothesis**); and (3) contractile units enveloping the cnida cause the discharge by "squeezing" the capsule (the **contractile hypothesis**). Because of the small size of cnidae and the extreme speed of the exocytosis process, these hypotheses have been difficult to test. Recent work using ultra high-speed microcinematography suggests that both the osmotic and tension models may be at work, and that capsules have very high internal pressures. The coiled capsular tubule is forcibly everted and thrown out of the bursting cell to penetrate or wrap around a portion of the unwary victim. It takes only a few milliseconds for the cnida to fire, and the everting tubule may reach a velocity of 2 m/sec—an acceleration of about 40,000 *g*—making it one of the fastest cellular processes in nature.

Most nematocysts contain several different toxins that vary in activity and strength, but as a class of chemicals they are all potent biological poisons capable of subduing large, active prey, including fish. Most appear to be neurotoxins. The toxins of some cnidarians are powerful enough to affect humans (e.g., those of some jellyfish; certain colonial hydroids, such as *Lytocarpus*; many hydrocorals, such as *Millepora*; many siphonophores, such as *Physalia*). The toxin of most cubomedusans (box jellies) is more potent than cobra venom. In tropical Australia, twice as many people die annually from box jellies as from sharks. Stings by *Chironex* (the sea wasp) and *Chiropsalmus* usually result in severe pain at the least, and fatal respiratory or cardiac failure at worst. In northern Australia, twice as many people have been killed by sea wasps as by sharks. Both acidic and alkaline environments suppress nematocyst firing. Thus, if you come out of the surf with a jellyfish tentacle on you, douse it with urine (acidic) or baking soda (alkaline) to reduce the impact. Meat tenderizer also works, presumably by denaturing the toxins, but the amount needed could damage your skin. If you want to swim in an area known to be frequented by dangerous jellies, you can always do what lifeguards in northern Australia do—don a pair of pantyhose, which seem to offer some protection. Lotions are also being developed that inhibit nematocyst discharge.

Feeding and Digestion

Most cnidarians are carnivores. Typically, nematocyst-laden feeding tentacles capture animal prey and carry it to the mouth region where it is ingested whole (Figure 8.26). Digestion is initially extracellular in the coelenteron. The gastrodermis is abundantly supplied with enzyme-producing cells that facilitate digestion (Figure 8.4). In many groups gastrodermal cilia (or flagella) aid in mixing of the gut contents. In the absence of a true circulatory system, the gastrovascular cavity distributes the partially digested material. The larger the cnidarian, the more extensively branched or partitioned is its coelenteron. The product of this preliminary breakdown is a soupy broth, from which polypeptides, fats, and carbohydrates are taken into the nutritive-muscular cells by phagocytosis and pinocytosis. Digestion is completed intracellularly within food vacuoles. Undigested wastes in the coelenteron are expelled through the mouth.

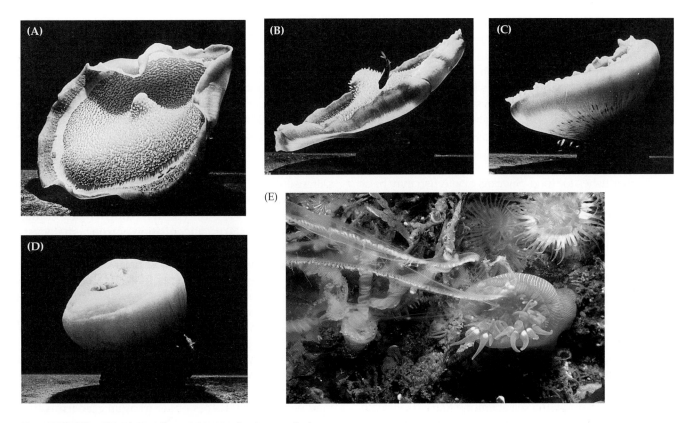

Figure 8.26 (A–D) Feeding sequence in the tropical sea anemone *Amplexidiscus senestrafer*. (A) An expanded oral disc has a tentacle-free area near the periphery, and an oral cone. (B) An expanded disc (side view). (C) Closure one-third complete, 1 second after stimulation of the oral disc. (D) Complete closure, 3 seconds after stimulation. (E) The temperate sea anemone *Epiactis prolifora* capturing a jellyfish (*Aequorea*?).

In anthozoans, the free edges of most of the gastrovascular mesenteries are thickened to form three-lobed mesenterial filaments (Figures 8.6 and 8.20). The lateral lobes are ciliated and aid in circulating the digestive juices in the coelenteron. The middle lobe, called the **cnidoglandular band**, bears cnidae and gland cells. In some sea anemones (e.g., *Aiptasia, Anthothoe, Calliactis, Diadumene, Metridium, Sagartia*), the cnidoglandular band continues beyond the base of the septum as a free thread called an **acontium**, which hangs freely in the coelenteron. The cnida-bearing acontia not only subdues live prey within the coelenteron, but may be shot out through the mouth or pores (**cinclides**) in the body wall when the animal contracts violently; when this occurs, the acontia presumably play a defensive role.

Most scyphomedusae capture food using nematocysts on the tentacles, the oral arms, or both. *Pelagia noctiluca*, an open-sea diurnal migrator, follows the other migrating macrozooplankton upon which it feeds. *Pelagia* uses its marginal tentacles to paralyze and capture moving prey, then transports it to the oral arms dangling from the center of the subumbrella. The oral arms transport the prey to the mouth. Motionless prey may also be captured by the oral arms directly, through chance contact.

Several groups of cnidarians have adopted feeding methods other than the direct use of nematocyst-laden tentacles. One group of large tropical anemones in the order Corallimorpharia (e.g., *Amplexidiscus*) lacks nematocysts on the external surfaces of most tentacles. These remarkable anemones capture prey directly with the oral disc, which can envelop crustaceans and small fishes, rather like a fisherman's cast-net (Figures 8.26A–D).

In addition to tentacular feeding on small plankters, many corals are capable of mucous-net suspension feeding, which is accomplished by spreading thin mucous strands or sheets over the colony surface and collecting fine particulate matter that rains down from the water. The food-laden mucus is driven by cilia to the mouth. In a few corals (e.g., members of the family Agariciidae), the tentacles are greatly reduced or absent, and all direct feeding is by the mucous-net suspension method. The amount of mucus produced by corals is so great that it is an important food source for certain fishes and other reef organisms, which feed directly off the coral or recover mucus sloughed into the surrounding sea water. Coral mucus released into the sea contains a variable mixture of macromolecular components (glycoproteins, lipids, and mucopolysaccharides) or a mucous lipoglycoprotein of specific character for a given species. These loose mucous webs, or flocs, are usually enriched by bacterial colonies and entrapped detrital materials, further enhancing their nutritional value.

The role of cnidarians as potentially significant members of food webs depends largely on location and circumstance. Corals obviously hold critical trophic positions in tropical reef environments, as do zoanthids and gorgonians in many tropical and subtropical habitats. In many warm and temperate areas sea pens and sea pansies dominate benthic sandy habitats. Large scyphomedusae (e.g., *Aurelia, Cyanea, Pelagia*) often occur in great swarms and may consume high numbers of larvae of commercially important fishes, as well as competing with other fishes for food. Swarms of jellyfish may be so dense that they clog and damage fishing nets and power plant intake systems.

Hydromedusae are also major components of temperate pelagic food webs. Members of several hydrozoan genera occur in huge congregations in tropical seas, where they are important carnivores in the neuston food web. Best known among these are the chondrophorans *Porpita* (which actively feeds on motile crustaceans, such as copepods) and *Velella* (which feeds on relatively passive prey, such as fish eggs and crustacean larvae), and the siphonophoran *Physalia* (which actively catches and consumes fishes).

Cnidarians play numerous roles in folklore around the world. In Samoa, the sea anemone *Rhodactis howesii*, known as mata malu, is served boiled as a festive holiday dish. Eaten raw, however, mata malu causes death and is a traditional device in Samoan suicide. Hawaiians refer to the zoanthid *Palythoa toxica* as "limu-make-o-Hana" ("the sacred deadly seaweed of Hana"). Hawaiians used to smear their spear tips with this cnidarian, the toxin from which is called **palytoxin**. Interestingly, palytoxin may be produced by an unidentified symbiotic bacterium, not by the cnidarian itself. It is one of the most powerful toxins known, being more deadly than that of poison arrow frogs (batrachotoxin) and paralytic shellfish toxin (saxitoxin).

Defense, Interactions, and Symbiosis

There are so many interesting aspects of cnidarian biology that do not fall neatly into our usual coverage of each group that we present this special section. The following discussion also points out the surprising level of sophistication possible at the relatively primitive diploblastic, radiate grade of complexity.

In most cnidarians defense and feeding are intimately related. The tentacles of most anemones and jellyfish usually serve both purposes, and the defense polyps (dactylozooids) of hydroid colonies often aid in feeding. In some cases, however, the two functions are performed by distinctly separate structures (as in many siphonophorans).

Some species of acontiate sea anemones (e.g., *Metridium*) bear two types of tentacles: feeding tentacles and defense tentacles. Whereas the former usually move in concert to capture and handle prey, the defense tentacles move singly, in a so-called searching behavior, in which they extend to three or four times their resting length, gently touch the substratum, retract, and extend once more. Defense tentacles are used in aggressive interactions with other sea anemones, either those of a different species or nonclonemates of the same species. The aggressive behavior consists of an initial contact with the opponent followed by autonomous separation of the defense tentacle tip, leaving the tip behind attached to the other sea anemone. Severe necrosis develops at the site of the attached tentacle tip, occasionally leading to death. Defense tentacles develop from feeding tentacles and tend to increase under crowded conditions. The development involves loss of typical feeding tentacle cnidae (largely spirocysts) and acquisition of true nematocysts and gland cells, which dominate in defense tentacles.

The **acrorhagi** (= **marginal spherules**) that ring the collar of some sea anemones (e.g., *Anthopleura*) also have a defensive function. These normally inconspicuous vesicles at the base of the tentacles bear nematocysts and usually spirocysts. In *A. elegantissima,* contact of an acrorhagi-bearing sea anemone with non-clonemates or other species causes the acrorhagi in the area of contact to swell and elongate. The expanded acrorhagi are placed on the victim and withdrawn; the application may be repeated. Pieces of acrorhagial epidermis that remain on the victim result in localized necrosis. Interclonal strips of bare rock are maintained by this aggressive behavior, and may help prevent overcrowding (Figure 8.27A). In addition to this behavior, the acrorhagi are exposed as a ring of nematocyst batteries around the top of the constricted column whenever an acrorhagi-bearing sea anemone contracts in response to violent stimulation. Other competitive interactions are known among stony corals (Figure 8.27B).

There are many of examples of associations between cnidarians and other organisms, some of which are truly symbiotic, others of which are less intimate. Few groups of cnidarians are truly parasitic, although several species of hydroids infest marine fishes. The polyps of some of these hydroids lack feeding tentacles and occasionally even nematocysts. The basal portion of the polyp erodes the fish's epidermis and underlying tissues, and nutrients are absorbed directly from the host. One species invades the ovaries of Russian sturgeons (a caviar feeder!).

A recent addition to the phylum Cnidaria is a group of about 1,200 species of tiny parasites, previously classified among the protists as the phylum Myxozoa. Morphological data, 18S ribosomal gene sequences, and the presence of Hox genes, all provide evidence that these strange creatures are indeed cnidarians, probably related to the hydrozoan order Trachylina. The coiled "polar filaments" housed within "polar capsules" are now viewed as typical nematocysts (Figure 8.28).

Myxozoan cnidarians infect annelids and various poikilothermic vertebrates, especially fishes. The nematocysts are presumably used for attachment to the host.

(A)

(B)

Figure 8.27 (A) Defensive acrorhagi (white-tipped tentacles) on two sea anemones (*Anthopleura elegantissima*), engaging in territorial chemical combat. (B) Competition between true corals (Scleractinia) in the Virgin Islands. The coral *Isophyllia sinuosa* is seen extruding its mesenterial filaments and externally digesting the edge of a colony of *Porites astereoides*.

Once ingested by a suitable host, the parasite emerges from its spore-like casing, penetrates the gut wall, and migrates to a final site of infection. Some species include both a vertebrate and an invertebrate host in their life cycle. *Myxobolus cerebralis*, a parasite of freshwater fishes (especially trout), devours the host's cartilage, leaving the fish deformed. Inflammation resulting from the infection puts pressure on nerves and disrupts balance, causing the fish to swim in circles—a condition known as **whirling disease**. When an infected fish dies, *M. cerebralis* "spores" are released from the decaying carcass and may survive for up to 30 years in mud. Eventually,

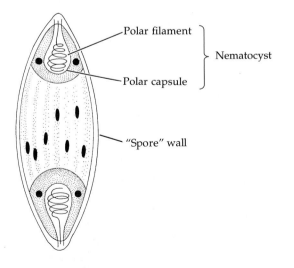

Figure 8.28 Previously considered to be protists, myxozoans are now viewed as highly specialized, parasitic cnidarians.

the spores are consumed by *Tubifex* worms (oligochaete annelids). They reside in this intermediate host until eaten by a new host fish.

Mutualism is common among cnidarians. Many species of hydroids live on the shells of various gastropods, hermit crabs, and other crustaceans. The hydroid gets a free ride and the host perhaps gains some protection and camouflage. Hydroids of the genus *Zanclea* are epizootic on ectoprocts, where they sting small predators and adjacent competitors, helping the ectoproct to survive and overgrow competing species. The ectoproct lends protection to the hydroid with its coarse skeleton, and the mutualism allows both taxa to cover a larger area than either could individually. The aberrant hydroid *Proboscidactyla* lives on the rim of polychaete worm tubes (Figure 8.8A) and dines on food particles dislodged by the host's activities.

Some sea anemones attach to snail shells inhabited by hermit crabs. These partnerships are mutualistic; the sea anemone gains motility and food scraps while protecting the hermit crab from predators. The most extreme case of this mutualism is that of the cloak anemones (e.g., *Adamsia*, *Stylobates*), which wrap themselves around the hermit crab's gastropod shell and grow as the crab does (Figure 8.29). Initially, the anemone's pedal disc secretes a chitinous cuticle over the small gastropod shell occupied by the hermit. Such fortunate crabs need not seek new, larger shells as they grow, for the cloak anemone simply grows and provides the hermit with a living protective cnidarian "shell," often dissolving the original gastropod shell over time. As if it were itself a gastropod, the sea anemone grows to produce a flexible coiled house called a **carcinoecium**. These odd anemone "shells" were initially described and classified as flexible gastropod shells. A similar relationship exists between *Parapagurus* and certain species of *Epizoanthus*. The hydroid *Janaria mirabilis* secretes a shell-like casing that is inhabited by hermit crabs and, in an odd case of evolutionary convergence, so does the ectoproct *Hippoporida calcarea* (Figure 8.30).

(A)

(B)

Figure 8.29   The golden "cloak anemone" (Anthozoa, Actiniaria) *Stylobates aenus*. (A,B) The anemone is forming a "shell," or carcinoecium, around the hermit crab *Parapagurus dofleini*. (C) The empty carcinoecium of *S. aenus*.

(C)

Several groups of animals utilize the cnidae of cnidarians for their own defense. Several aeolid sea slugs consume cnidarian prey, ingesting their unfired nematocysts and storing them in processes on their dorsal surfaces. Once the nematocysts are in place, the sea slugs use them for their own defense. The ctenophore *Haeckelia rubra* feeds on certain hydromedusae and incorporates their nematocysts into its tentacles. The freshwater turbellarian flatworm *Microstoma caudatum* feeds on *Hydra*, risking being eaten itself, and then uses the stored nematocysts to capture other prey. Several species of hermit crabs and true crabs carry sea anemones (e.g., *Calliactis*, *Sagartiomorphe*) on their shells or claws and use them as living weapons to deter would-be predators. The hermit crabs transfer their anemone partners to new shells, or the anemones do so on their own, when the hermits take new shells. Some hermit crabs of the genus *Pagurus* often have their shell covered by a mat of symbiotic colonial hydroids (e.g., *Hydractinia*, *Podocoryne*). The presence of the hydroid coat deters more aggressive hermits (*Clibinarius*) from commandeering the paguroid's shell.

Several cases of fish–cnidarian symbiosis have been documented. The well known association of anemone fishes (clownfishes) and their host sea anemones serves an obvious protective function for the fishes. About a dozen species of sea anemones participate in this interesting relationship. The fish's ability to live among the sea anemone's tentacles is still not fully understood. However, the

Figure 8.30   A case of remarkable evolutionary convergence. (A,B) The hydrozoan colony *Janaria mirabilis* (Athecata) forms a shell-like corallum inhabited by hermit crabs. (C) The ectoproct *Hippoporida calcarea*, which forms a similar structure, is also inhabited by hermit crabs.

(A)

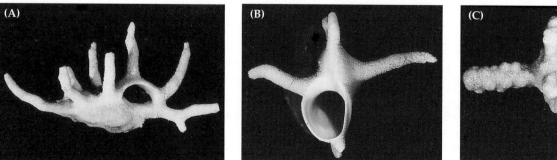

(B)

(C)

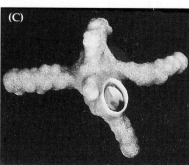

sea anemone does not voluntarily fail to spend its nema-tocysts on its fish partner; rather, the fish alters the chemi-cal nature of its own mucous coating, perhaps by accumu-lating mucus from the sea anemone, thereby masking the normal chemical stimulus to which the anemone's cnidae would respond. *Neomus* is a small fish that lives symbioti-cally among the tentacles of *Physalia* and appears to sur-vive by simply avoiding direct contact with the beast. When stung accidentally, however, it shows a much high-er survival rate than do other fishes of the same size. *Neomus* feeds on prey captured by its host.

A number of associations are known between cnidar-ians and crustaceans. Nearly all amphipods of the sub-order Hyperiidea are symbionts on gelatinous zoo-plankters, including medusae. The nature of many of these associations is unclear, but various species of the amphipods are known to use their hosts as a nursery for the young and perhaps for dispersal. Some actually live among and eat the nematocyst-bearing parts of the host, such as the tentacles or oral arms. Many are commonly found inside the medusa's coelenteron, where they seem unaffected by host's digestive enzymes. In a rela-tionship similar to that of anemone fish, a few cases of

anemone shrimp are known, at least one that is obligate for the shrimp (*Pericimenes brevicarpalis*).

One of the most noteworthy evolutionary achieve-ments of cnidarians is their close relationship with unicel-lular photosynthetic partners. The relationship is wide-spread and occurs in many shallow-water cnidarians. The symbionts of freshwater hydrozoans (e.g., *Chlor-hydra*) are single-celled species of green algae (Chloro-phyta) called "zoochlorellae." In marine cnidarians, the protists are unicellular cryptomonads and dinoflagellates called "zooxanthellae" (probably many genera including *Zooxanthella* [= *Symbiodinium*]) (Figure 8.31). These algae may be capable of living free from their hosts, and per-haps do so normally, but very little is known about their natural history. The algae typically reside in the host's gastrodermis or epidermis, although some cnidarians harbor extracellular zooxanthellae in the mesoglea. It is usually the algal symbionts that give cnidarians their green, blue-green, or brownish color. Corals that are reef-builders (called **hermatypic corals**) typically harbor zooxanthellae (i.e., they are **zooxanthellate corals**). Resident populations of zooxanthellae in these corals may reach a density of 30,000 algal cells per cubic mil-

Figure 8.31 (A) An octocoral with zooxanthellae distributed throughout the gastrodermis (schematic section). (B) Cells of zooxanthellae in tissue of the giant green sea anemone *Antho-pleura xanthogrammica*. (C) *Mas-tigias*, a rhizostoman medusa, harbors zooxanthellae in its cells.

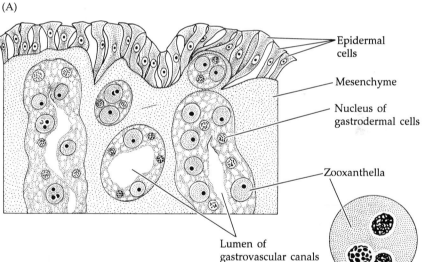
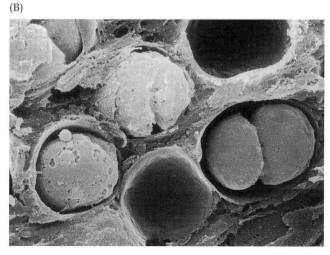

limeter of host tissue (or from 1 to 2×10^6 cells per square centimeter of coral surface!). Zooxanthellae also occur in many tropical gorgonians, anemones, and zoanthids. They also occur in some scyphozoans, and in some cases they are thought to produce much of the energy required by the host (e.g., *Cassiopea, Linuche, Mastigias*).

It has only been in the last 30 years or so that the physiological and adaptive nature of this relationship has been described to any extent. Some of this information comes from studies on scyphomedusa *Mastigias* (Figure 8.31C), which lives in marine lakes on the islands of Palau, where it may occur in densities exceeding 1,000 per m^3. In these lakes, *Mastigias* makes daily vertical migrations between the oxygenated, nutrient-poor upper layers and the anoxic, nutrient-rich lower layers. This behavior appears to be related to the light and nutrient requirements of its symbiotic zooxanthellae. Unlike the zooxanthellae in benthic cnidarians, which tend to reproduce more-or-less evenly over a 24-hour period, the zooxanthellae of *Mastigias* show a distinct reproductive peak during the hours when their host occupies a position in the deeper nitrogen-rich layers of the lakes. This reproductive peak may be a result of the alga's use of free ammonia as a nutrient source.

Many cnidarians seem to derive only modest nutritional benefit from their algal symbionts, but in many others a significant amount of the hosts' nutritional needs appears to be provided by the algae. In such cases, a large portion of the organic compounds produced by photosynthesis of the symbiont may be passed on to the cnidarian host, probably as glycerol but also as glucose and the amino acid alanine. In return, metabolic wastes produced by the cnidarian provide the alga with nitrogen and phosphorus. In corals, the symbiosis is thought to be important for rapid growth and for efficient deposition of the calcareous skeleton, and many corals can only form reefs when they maintain a viable dinoflagellate population in their tissues. Although the precise physiological-nutritional link between corals and their zooxanthellae has been elusive, the algae may increase the rate of calcium carbonate production (precipitation) by utilizing CO_2 produced by the host. Corals and other cnidarians can be deprived of their algal symbionts by experimentally placing the hosts in dark environments. In such cases the algae may simply die, they may be expelled from the host, or they may (to a limited extent) actually be consumed directly by the host. Because they are dependent on light, zooxanthellate corals can live to depths of only 90 m or so. For unknown reasons, they also require warm waters, and thus occur almost exclusively in shallow tropical seas. Under stress, such as unusually high temperatures, corals may lose their zooxanthellae—a process known as **coral bleaching**. Long assumed detrimental, a recent theory suggests this might be an adaptive mechanism providing opportunity for acquiring new types of zooxanthellae better adapted to the changing environment.

Circulation, Gas Exchange, Excretion, and Osmoregulation

There is no independent circulatory system in cnidarians. The coelenteron serves in this role to a limited extent by circulating partly digested nutrients through the interior of the body, absorbing metabolic wastes from the gastrodermis, and eventually expelling waste products of all types through the mouth. But large anemones and large medusae confront a serious surface area:volume dilemma. In such cases, the efficiency of the gastrovascular system as a transport device is enhanced by the presence of mesenteries in the anemones and the radially arranged canal system in the medusae. Cnidarians also lack special organs for gas exchange or excretion. The body wall of most polyps is either fairly thin or has a large internal surface area, and the thickness of many medusae is due largely to the gel-like mesoglea or mesenchyme. Thus, diffusion distances are kept to a minimum. Gas exchange occurs across the internal and external body surfaces. Facultative anaerobic respiration occurs in some species, such as anemones that are routinely buried in soft sediments. Nitrogenous wastes are in the form of ammonia, which diffuses through the general body surface to the exterior or into the coelenteron. In freshwater species there is a continual influx of water into the body. Osmotic stress in such cases is relieved by periodic expulsion of fluids from the gastrovascular cavity, which is kept hypoosmotic to the tissue fluids.

Nervous System and Sense Organs

Consistent with their radially symmetrical bauplan, cnidarians have a diffuse, noncentralized nervous system. The neurosensory cells of the system are the most primitive in the animal kingdom, being naked and largely nonpolar. Usually the neurons are arranged in two reticular arrays, called **nerve nets**, one between the epidermis and the mesenchyme and another between the gastrodermis and the mesenchyme (Figure 8.32). The subgastrodermal net is generally less well developed than the subepidermal net, and is absent altogether in some species. Some hydrozoans possess one or two additional nerve nets.

A few nerve cells and synapses are polarized (bipolar) and allow for transmission in only one direction, but most cnidarian neurons and synapses are nonpolar—that is, impulses can travel in either direction along the cell or across the synapse. Thus, sufficient stimulus sends an impulse spreading in every direction. In some cnidarians where both nerve nets are well developed, one net serves as a diffuse slow-conducting system of nonpolar neurons, and the other as a rapid through-conducting system of bipolar neurons.

Polyps generally have very few sensory structures. The general body surface has various minute hairlike structures developed from individual cells. These serve as mechanoreceptors, and perhaps as chemoreceptors, and are most abundant on the tentacles and other re-

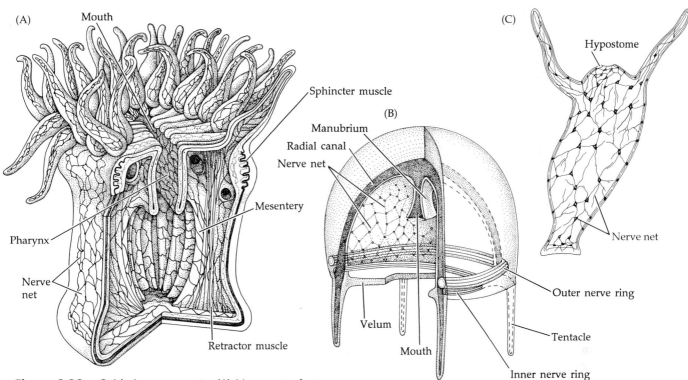

Figure 8.32 Cnidarian nerve nets. (A) Nerve net of a typical sea anemone (Anthozoa). (B) Nerve net in a hydromedusa (Hydrozoa). (C) Nerve net of *Hydra* (Hydrozoa).

gions where cnidae are concentrated. They are involved in behavior such as tentacle movement toward a prey or predator and in general body movements. Some appear to be associated specifically with discharged cnidae, such as the **ciliary cone apparatus** of anthozoan polyps, which is believed to function like the cnidocil in hydrozoan and scyphozoan nematocysts (Figure 8.33). Oddly, these structures do not appear to be connected directly to the nerve nets. In addition, most polyps show a general sensitivity to light, not mediated by any known receptor but presumably associated with neurons concentrated in or just beneath the translucent surface of epidermal cells.

As might be expected, motile medusae have more sophisticated nervous systems and sense organs than do the sessile polyps (Figure 8.34). In many groups, especially the hydromedusae, the epidermal nerve net of the bell is condensed into two **nerve rings** near the bell margin. These nerve rings connect with fibers enervating the tentacles, muscles, and sense organs. The lower ring stimulates rhythmic pulsations of the bell. This ring is also connected to statocysts, when present, on the bell margin, which is supplied with **general sensory cells** and with radially distributed ocelli and (probably) chemoreceptors. The general sensory cells are neurons whose receptor processes are exposed at the epidermal surface. The ocelli are usually simple patches of pigment and photoreceptor cells organized as a disc or a pit. Statocysts may be in the form of pits or closed vesicles, the latter housing a calcareous statolith adjacent to

a sensory cilium. When one side of the bell tips upward, the statocysts on that side are stimulated. Statocyst stimulation inhibits adjacent muscular contraction and the medusa contracts muscles on the opposite side. Many medusae maintain themselves in a particular photo-

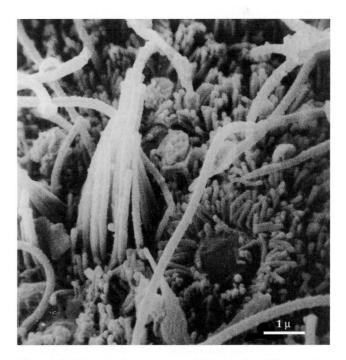

Figure 8.33 A ciliary cone on the tentacle of the corallimorpharian anemone *Corynactis californica* lies adjacent to cnidocyte (the circle of microvilli).

regime by directed swimming behaviors. This action is seen especially in those medusae harboring large populations of zooxanthellae, such as the medusa *Cassiopea*, which lies upside down on the shallow sea floor, exposing to light the dense zooxanthellae population residing in tissues of its tentacles and oral arms.

In the physonectid siphonophores, a linear condensation of the nerve net produces longitudinal "giant

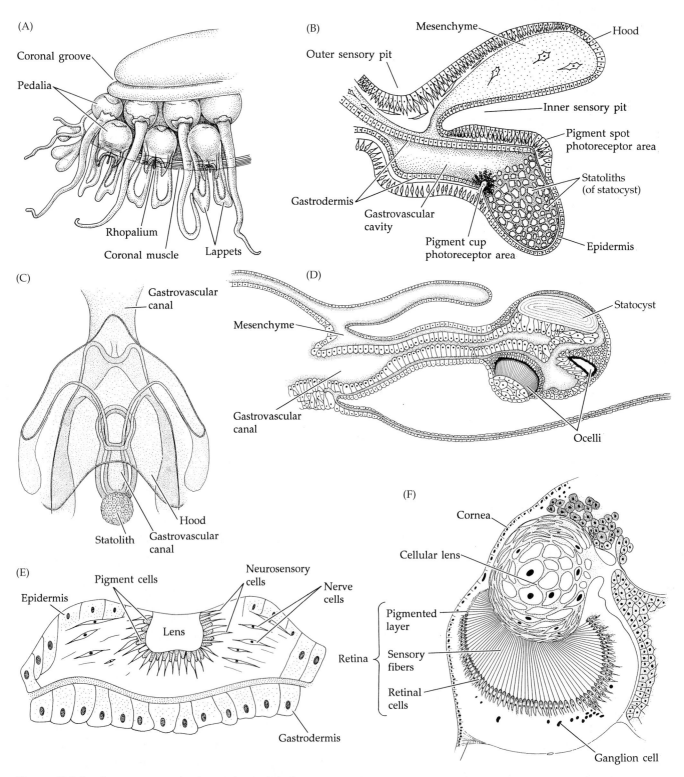

Figure 8.34   Sensory structures in medusae. (A) The rhopalia of the scyphomedusa *Atolla* are situated between the marginal lappets. (B) A rhopalium (section) has various sensory regions. (C) A rhopalium of *Aurelia* (diagrammatic). A portion of the gastrovascular canal has been cut away. (D) A cubozoan rhopalium (section). (E) A pigment-cup ocellus (cross section) of a hydrozoan medusa. (F) The eye of a cubozoan (*Carybdea*) (cross section).

axons" in the stem and the nerve tracts in the tentacles. This longitudinal "giant axon" is actually a neuronal syncytium that originates by fusion of neurons from the nerve net of the stem. The high-speed impulses in these large diameter nerve tracts enable physonectids to contract rapidly and initiate a fast escape reaction.

Cubomedusae possess as many as 24 well-developed eyes located near the bell margin. The most complex of these have a true epidermal cornea, a spherical cellular lens, and a retina (Figure 8.34F). The retina is multilayered, containing a sensory layer, a pigmented layer, a nuclear layer, and a region of nerve fibers. There are roughly 11,000 sensory cells in each of these remarkable eyes. Cubomedusae only 3 cm tall swim at speeds up to 6 m/min and can orient accurately to the light of a match as far away as 1.5m. This combination of speed and sensitivity to light may enable them to locate and feed on luminescent prey at night.

Cubomedusae and scyphomedusae generally lack well-developed nerve rings (although they are present in members of the order Coronatae). The bell margins of cubomedusae and scyphomedusae usually bear club-shaped structures—called **rhopalia**—that are situated between a pair of flaps, or **lappets** (Figure 8.34). The rhopalia are sensory centers, each containing a concentration of epidermal neurons, a pair of chemosensory pits, a statocyst, and often an ocellus. One pit is located on the exumbrellar side of the hood of the rhopalium, the other on the subumbrellar side.

In addition to the neuronal system just described, cnidarians are said to possess a "cytoplasmic conducting system" similar in nature to that of sponges. Epidermal cells and muscle elements appear to be the principal components of the system. Impulse conduction, which travels very slowly and in a highly diffuse fashion, is not well understood but probably relies primarily on physical contact and stimulation of adjacent cells.

Bioluminescence is common in cnidarians and has been documented in all classes except the Cubozoa. In some forms (e.g., many hydromedusae), luminescence consists of single flashes in response to a local stimulus. In others, bursts of flashes propagate as waves across the body or colony surface (e.g., sea pens and sea pansies). The most complicated luminescent behaviors occur in hydropolyps, where a series of multiple flashes is propagated. Propagated luminescence is probably controlled by the nervous system, although this phenomenon is not well understood. In at least one hydromedusa (*Aequorea*), luminescence appears not to be the result of the usual luciferin-luciferase reaction. Rather, a high-energy protein, named aequorin, emits light in the presence of calcium (Shimomura 1995).

Reproduction and Development

Reproductive processes in cnidarians are intimately tied to the alternation of generations that characterizes this phylum. As you have already learned, cnidarian life cy-

cles often involve an asexually reproducing polyp stage, alternating with a sexual medusoid stage that produces a characteristic **planula larva.** Thus, we generally find a complex indirect or mixed life history that includes phases of asexual reproduction. There are, however, many variations on this life cycle. The four classes are discussed separately below.

Hydrozoan reproduction. Hydrozoan polyps reproduce asexually by budding. This is a rather simple process wherein the body wall evaginates as a bud, incorporating an extension of the gastrovascular cavity with it. A mouth and tentacles arise at the distal end, and eventually the bud either detaches from the parent and becomes an independent polyp, or, in the case of colonial forms, remains attached. Medusa buds, or gonophores, are also produced by polyps in a similar fashion, although the process is sometimes quite complex. A rather special kind of budding occurs in the siphonophores, in which the floating colonies produce chains of individuals called **cormidia**, which may break free to begin a new colony.

Certain hydromedusae also undergo asexual reproduction, either by the direct budding of young medusae (Figure 8.35), or by longitudinal fission. The latter process often involves the formation of multiple gastric pouches (**polygastry**), followed by longitudinal splitting, which produces two daughter medusae. In some species (e.g., *Aequorea macrodactyla*), **direct fission** may take place. Polygastry does not occur during this process; instead, the entire bell folds in half, severing the stomach, ring canal, and velum (Figure 8.36). Eventually the entire medusa splits in half and each part regenerates the missing portions.

Cnidarians in general have a great capacity for regeneration, as exemplified by experiments on *Hydra*. The eighteenth-century naturalist Abraham Trembley had the clever idea of turning a *Hydra* inside out, which he did. To his delight, the animal survived quite well, with the gastrodermal cells functioning as the "new epidermis" and vice versa. Cells removed from the body of a *Hydra* also have a modest degree of reaggregative ability, like that seen so dramatically in sponges. In some cases, entire animals can be reconstituted from cells taken only from the gastrodermis or only from the epidermis. Although *Hydra* is an unusual and atypical cnidarian, this great capacity for cellular reorganization is a reflection of the primitive state of tissue development in the animals belonging to this phylum.

A typical *Hydra* consists of only about 100,000 cells of roughly a dozen different types. Although distinct epidermis and gastrodermis exist, these tissues are very similar to one another, comprising mainly epitheliomuscular cells. The nervous system is, of course, also very simple. It takes only a few weeks for all the cells in a *Hydra* to reproduce themselves, or "turn over." These attributes make *Hydra* an ideal creature for studies of de-

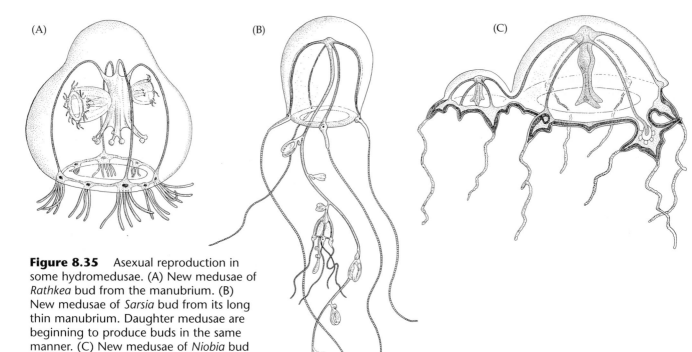

Figure 8.35 Asexual reproduction in some hydromedusae. (A) New medusae of *Rathkea* bud from the manubrium. (B) New medusae of *Sarsia* bud from its long thin manubrium. Daughter medusae are beginning to produce buds in the same manner. (C) New medusae of *Niobia* bud from the tentacular bulbs.

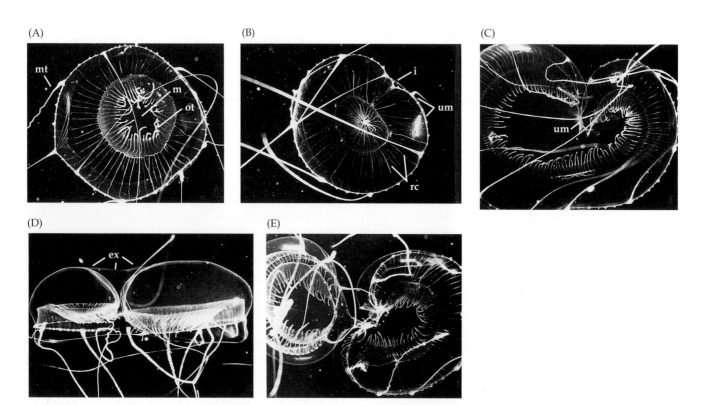

Figure 8.36 Asexual reproduction in the hydromedusa *Aequorea*. The sequence of photographs shows the direct fission of *A. macrodactyla*. (A) This oral view shows a nondividing medusa with its marginal fishing tentacles (mt) deployed. (B) Initiation of invagination (i). (C–E) A progression of the direct fission process. The oral (C) and marginal (D) views illustrate the severing of the umbrellar margin (um) and the separation of exumbrellar halves; (E) shows the exumbrellar surface (ex) beginning to pull apart, producing free-swimming daughter medusae; healing is nearly complete in the smaller daughter medusa on the left. ot, oral tentacles; m, mouth; rc, radial canals.

velopmental biology, histogenesis, and morphogenesis. This work has shed light on many fundamental biological phenomena, as summarized in Lenhoff and Loomis (1961), Gierer (1974), and Lenhoff (1983).

All hydrozoan cnidarians have a sexual phase in their life cycle (Figure 8.37). In solitary species (e.g., *Hydra*) and some colonial forms, the medusoid phase is suppressed or absent. The polyps develop simple, transient, epidermal gamete-producing structures called **sporosacs** (Figures 8.5A,B). Most colonial hydroids, however, produce medusa buds (gonophores) either from the walls of the hydranths or from separate gonozooids. The gonophores may grow to medusae that are released as free-living, sexually reproducing individuals, or they may remain attached to the polyps as incipient medusae that produce gametes.

In free-living hydromedusae, germinal cells arise from interstitial epidermal cells that migrate to specific sites on the bell surface, where they consolidate into a temporary gonadal mass. Subsequently, gametogenic tissue appears on the surface of the manubrium, beneath the radial canals, or on the general subumbrellar surface. Hydromedusae are usually dioecious, with either sperm or eggs usually being released directly into the water, where fertilization occurs. In some, only sperm are released and fertilization occurs on or in the female medusa's body.

Although several cleavage patterns occur in the Hydrozoa, it is generally radial and holoblastic. A coeloblastula forms, which gastrulates by uni- or multipolar ingression to a stereogastrula. The interior cell mass is entoderm; the exterior cell layer is ectoderm (Figure 8.38). The stereogastrula elongates to form a unique, elongate, solid or hollow, nonfeeding, free-swimming planula larva (Figure 8.39). The planula larva is radially symmetrical, but it swims with a distinct "anterior–posterior" orientation. The ectodermal cells are monociliated and destined to become the adult epidermis; the entoderm is destined to become the adult gastrodermis. The trailing end of the larva (of all cnidarians) becomes the oral end of the adult, and even in the larval stage a mouth sometimes develops at this end. Hydrozoan planulae swim about for a few hours, a few days, or a few weeks before settling by attaching at the leading end. If the larva is still solid, then the entoderm hollows to form the coelenteron. The mouth opens at the unattached oral end and tentacles develop as the larva metamorphoses into a young solitary polyp.

This overview of the hydrozoan reproductive cycle includes some minor variations on the basic theme. However, far more variety actually exists than we have space to discuss in detail (Figure 8.37). For example, in some trachylines, the polypoid stage is apparently lost altogether. The medusae produce planula larvae that develop into **actinula larvae**, which metamorphose into adult medusae, bypassing any sessile polypoid phase.

Some trachylines and some siphonophorans undergo direct development, bypassing the larval stage altogether. The order Actinulida includes minute interstitial polyps that lack a medusoid stage and have suppressed the larval phase. The adult polyp is ciliated and resembles an actinula larva (hence the name).

Scyphozoan reproduction. The asexual form of scyphozoan cnidarians is a small polyp called the scyphistoma (= scyphopolyp; Figure 8.40A). It may produce new scyphistomae by budding from the column wall or from stolons. At certain times of the year, generally in the spring, medusae are produced by repeated transverse fission of the scyphistoma, a process called **strobilation** (Figure 8.40B). During this process the polyp is known as a **strobila**. Medusae may be produced one at a time (**monodisc strobilation**), or numerous immature medusae may stack up like soup bowls and then be released singly as they mature (**polydisc strobilation**). Immature and newly released medusae are called **ephyrae**. An individual scyphistoma may survive only one strobilation event, or it may persist for several years, asexually giving rise to more scyphistomae and releasing ephyrae annually.

Ephyrae are very small animals with characteristically incised bell margins (Figure 8.40C). The ephyral arms, or primary tentacles, mark the position of the adult lappets and rhopalia. In some groups (e.g., *Aurelia*) the number of ephyral arms is quite variable (Figure 8.40D). Maturation involves growth between these arms to complete the bell. Development into sexually mature adult scyphomedusae takes a few months to a few years, depending on the species.

The gamete-forming tissue in adult scyphomedusae is always derived from the gastrodermis, usually on the floor of the gastric pouches, and gametes are generally released through the mouth. Most species are dioecious. Fertilization takes place in the open sea or in the gastric pouches of the female. Cleavage and blastula formation are similar to the processes in hydrozoans. Gastrulation is by ingression or invagination, and results in a mouthless, double-layered planula larva; when invagination occurs, the blastopore closes. The planula larva eventually settles and grows into a new scyphistoma.

The medusa phase clearly dominates the life cycles of most scyphozoans. The polyp stage is often significantly suppressed or absent altogether. For example, many pelagic scyphomedusae have eliminated the scyphistoma, and the planula larva transforms directly into a young medusa (e.g., *Atolla, Pelagia, Periphylla*). In others, the larvae are brooded, developing in cysts on the parent medusa's body (e.g., *Chrysaora, Cyanea*). A few genera have branching colonial scyphistomae with a supportive skeletal tube and an abbreviated medusoid stage (e.g., *Nausithoe, Stephanoscyphus*). In none, howev-

Figure 8.37 Some hydrozoan life cycles. (A) Life cycle of *Hydra*. Sperm produced by the male polyp (a) fertilizes the eggs of the female polyp (b). During cleavage, the eggs secrete a chitinous theca about themselves. After hatching, the embryos (c) grow into polyps that reproduce asexually by budding (d), until environmental conditions again trigger sexual reproduction. (B) Life cycle of *Obelia*, a thecate hydroid with free medusae. (C) Life cycle of *Tubularia*, an athecate hydroid that does not release free medusae. The polyp (a) bears many gonophores, whose eggs develop *in situ* into planulae (b) and then into actinula larvae (c) before release (d); the liberated actinula larvae (d) settle and transform directly into new polyps (e), which each proliferate to form a new colony (f). (D) Life cycle of a trachyline hydrozoan medusa without a polypoid stage (*Aglaura*). After fertilization, a dioecious adult (a) releases a planula larva (b), which adds a mouth and tentacles (c) to become an actinula larva (d). Subsequently the actinula larva becomes a young medusa (e). (E) Life cycle of a trachyline hydrozoan with a polypoid stage, the freshwater *Limnocnida*. Dioecious medusae (a) release fertilized eggs (b) that grow into planula larvae (c). Planula larvae settle to form small hydroid colonies (d), which bud off new medusae (e).

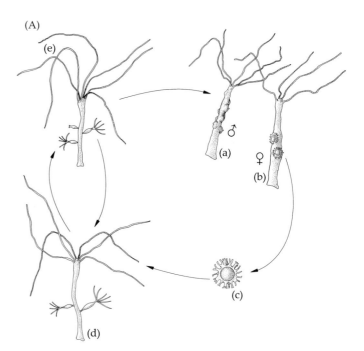

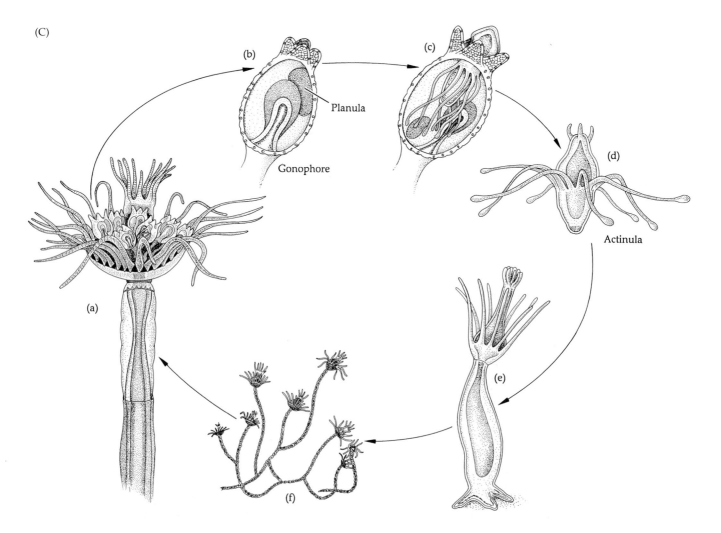

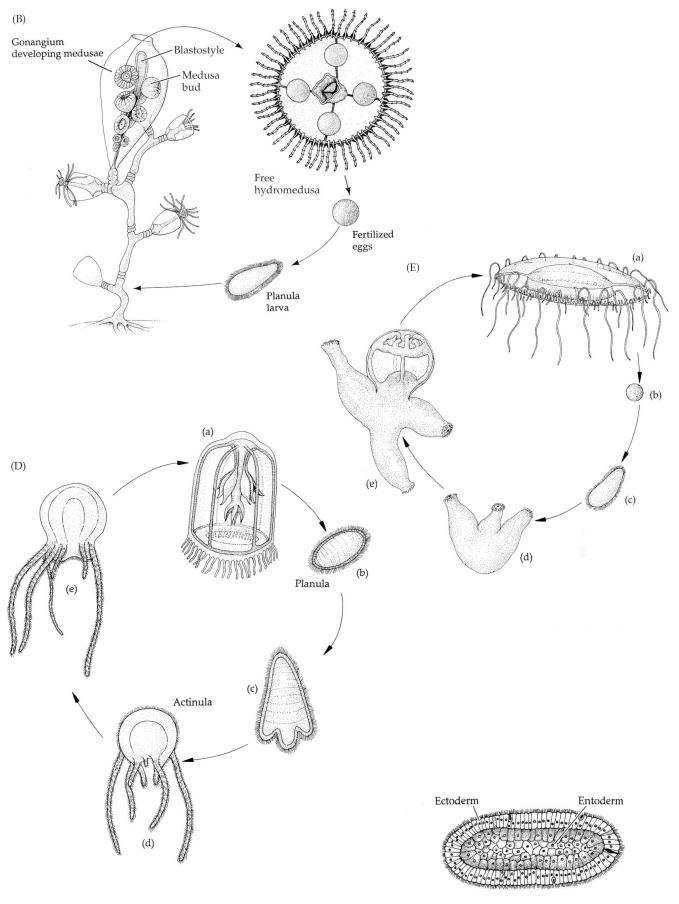

(B)

Gonangium developing medusae

Blastostyle

Medusa bud

Free hydromedusa

Fertilized eggs

Planula larva

(E)

(a)

(b)

(c)

(d)

(e)

(D)

(a)

(b)

Planula

Actinula

(c)

(d)

(e)

Ectoderm

Entoderm

Figure 8.38 A typical solid hydrozoan planula larva resulting from ingression.

Figure 8.39 The hollow planula larva of the hydroid *Gonothyraea* (longitudinal section).

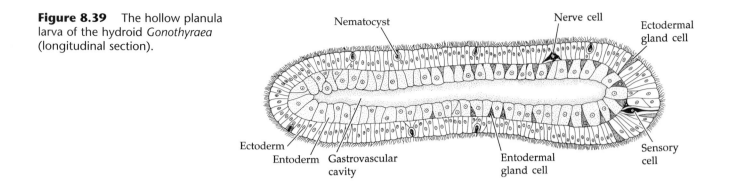

er, has the medusoid stage been lost altogether. Some scyphozoan life cycles are shown in Figure 8.41.

Cubozoan reproduction. The biology of cubozoans is not yet well known and the polyps of only a few species have been described. Apparently, each polyp metamorphoses directly into a single medusa, rather than undergoing strobilation as the scyphozoan polyps do. Some cubozoan medusae engage in a form of copulation, in which sperm are transferred directly from the male to an adjacent female in the water column.

Anthozoan reproduction. Members of this class are exclusively polypoid. Asexual reproduction is common in anthozoan polyps. **Longitudinal fission** can result in large groups, or clones, of genetically identical individuals (e.g., seen in some species of *Anthopleura*, *Diadumene*, and *Metridium*), as can the less common process of **pedal laceration** (e.g., seen in some acontiate sea anemones: *Diadumene*, *Haliplanella*, *Metridium*). In the latter phenomenon, the pedal disc spreads, and the anemone simply moves away, leaving behind small fragments from the disc, each of which develops into a young sea anemone. In addition to these two common modes of asexual reproduction, a few species of sea anemones are known to undergo **transverse fission**, and one family of sea anemones produces new individuals from tentacle buds (e.g., *Boloceroides*). Additionally, certain anemones and one scleractinian coral, *Pocillopora damicornis*, are known to produce planula larvae parthenogenetically and brood them until release. Most

(A)

(B)

(C)   (D)

Figure 8.40 Scyphozoan (*Aurelia*) scyphistoma (and one strobila) (A), and strobila (B). (C) A "typical" 8-armed ephyra. (D) A 12-armed ephyra.
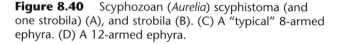

Figure 8.41 Scyphozoan life cycles. (A) Life cycle of *Aurelia*. The fertilized egg (b) is released to develop into a planula larva (c), which settles to grow into a polyp, the scyphistoma (d). The scyphistoma either buds off new polyps (e) or produces ephyrae by strobilation (f); ephyra (g) grows into an adult medusae. (B) Life cycle of *Pelagia*, a scyphomedusa lacking the polyploid stage. (C) Life cycle of the "cannonball jellyfish," *Stomolophus meleagris*.

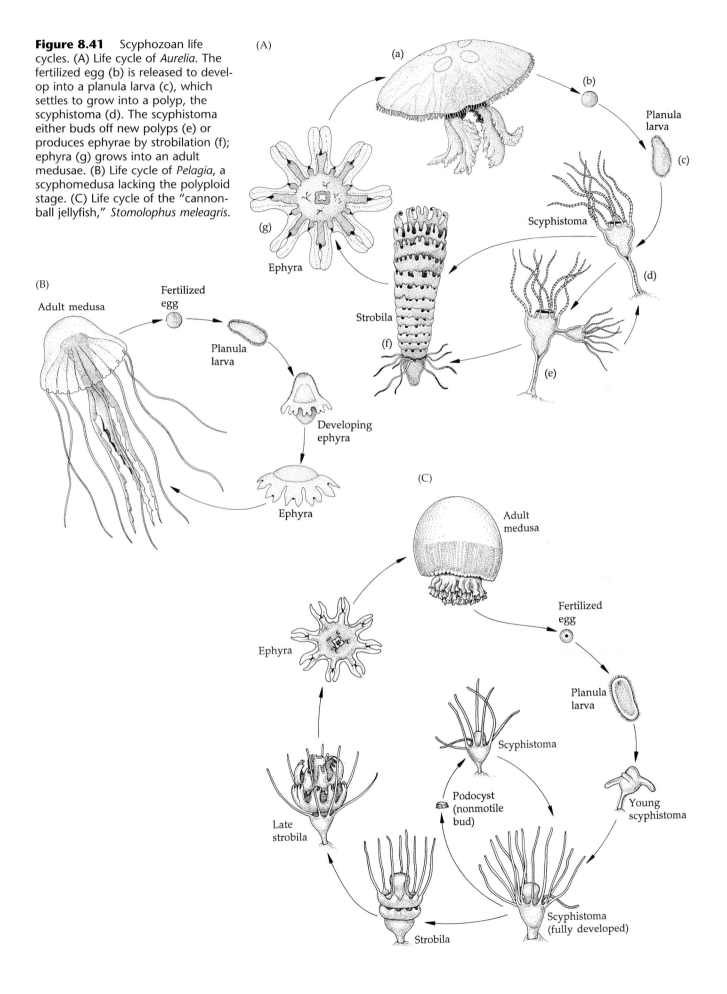

surprising is the recent discovery that some sea anemones internally brood young that are asexually produced, by a mechanism not yet understood.

Little is known about the reproductive biology of most alcyonarians and ceriantipatharians. Zoantharians (anemones and corals) may be dioecious or hermaphroditic. In some species, the colony can contain males, females, and hermaphrodites. The gametes arise from patches of tissue on the gastrodermis of all or only some mesenteries. Eggs are fertilized either in the coelenteron, followed by early development in the gut chambers, or more commonly outside the body, in the sea. A number of anemones brood the developing embryos internally or on the external body surface. The northeast Pacific sea anemone *Aulactinia incubans* releases its brooded young through a pore at the tip of each tentacle! Some corals undergo mixed development involving internal fertilization, brooding, and then release of planula larvae. Some alcyonaceans and gorgonaceans (e.g., *Briareum, Parerythropodium*) brood their embryos in a mucous coat on the body surface; then the planula larvae escape. Others shed their gametes and rely on external fertilization and indirect development. Some coral planula larvae are long-lived, spending several weeks or months in the plankton, an obvious means of dispersal. Other corals release benthic planulae that crawl away from the parent and settle nearby. Recently, corals have been shown to undergo synchronous spawning over large areas on reefs. In some cases this synchrony is restricted to colonies of a single species, but in other cases the synchronous spawning of as many as 105 *different* coral species has been reported (on the Great Barrier Reef; Babcock et al. 1986). Such synchronous spawning events may lead to high levels of hybridization among scleractinian corals, and this may be one explanation for the great range of polymorphism seen in many species. There are even some verified cases of hybridization between members of different genera!

Sagartia troglodytes is the only sea anemone known to copulate. The coupling starts when a female glides up to a receptive male, whereupon their pedal discs are pressed together in such a way as to create a chamber into which the gametes are shed and fertilization occurs. The copulatory position is maintained for several days, presumably until planula larvae have developed. This behavior may be an adaptation to areas of great water movement that might otherwise scatter gametes and reduce the probability of successful fertilization.

Cleavage in anthozoan embryos is radial and usually holoblastic, resulting in a coeloblastula that undergoes gastrulation by ingression or, more frequently, by invagination to form a ciliated planula larva. When invagination occurs, the blastopore remains open and sinks inward, drawing with it a tube of ectoderm that becomes the adult pharynx. Many anthozoan planulae are planktotrophic, although very yolky ones do not feed. The ability of some larvae to feed allows them a potentially longer larval life and enhances dispersal. As development proceeds, eight complete mesenteries develop in the planula, producing the so-called **edwardsia stage**, named after the octamesenterial genus *Edwardsia*. The larva eventually settles on its aboral end and tentacles grow around the upwardly directed mouth and oral disc. A typical anthozoan life cycle is shown in Figure 8.42.

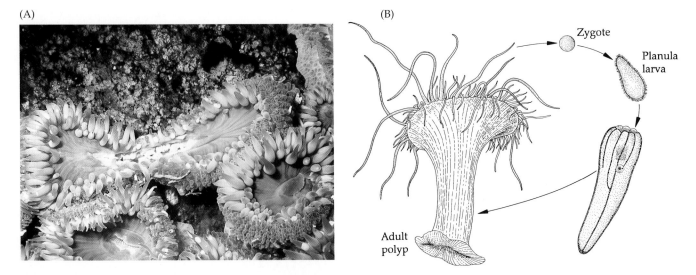

Figure 8.42 Reproduction in Anthozoa. (A) Asexual reproduction by longitudinal fission in the aggregating anemone *Anthopleura elegantissima*. (B) A typical anthozoan sexual life cycle: the adult polyp releases gametes which fuse externally, or fertilized eggs are released, and zygotes develop into a planula larvae; the larvae settle and transform directly into young polyps.

Cnidarian Phylogeny

The cnidarians have one of the longest fossil histories among the Metazoa. The first documented cnidarian fossils are from the famous Ediacara Hills of South Australia, which contain several kinds of medusae and sea pens that lived nearly 600 million years ago. The only suggestions of metazoan life preceding the appearance of cnidarians are trace fossils and estimates based on molecular clock calculations. Indeed, the origin of the cnidarians is intimately tied to the origin of the Metazoa themselves. (Theories on the origin of the Metazoa are discussed in Chapter 4.) The colonial theory depicts a colonial flagellated protist giving rise to a hollow metazoan ancestor, termed a blastea, which in turn gave rise to a diploblastic planuloid animal called a gastrea. On the other hand, the syncytial theory implies that the ancestors of the cnidarians were triploblastic, acoelomate organisms, perhaps something like rhabdocoel turbellarians, that underwent "degenerative evolution" to produce what we recognize today as the cnidarians. This view, sometimes called the **turbellarian theory**, holds the Anthozoa to be the most primitive cnidarian class and cites the "remnants" of bilateral symmetry in that class as evidence of a bilateral ancestry.

The turbellarian theory is not generally promoted by most contemporary zoologists, and we find it weak on several counts. "Degenerative evolution" (a poor choice of words) is a phenomenon primarily associated with the evolution of parasites or the exploitation of smallness (e.g., interstitial forms), wherein it may result in the reduction of certain systems and the specialized adaptive development of others. General loss of fundamental body architecture in other kinds of free-living animals seems to be an unlikely event. We view the idea of a free-living, triploblastic, bilateral, motile flatworm taking up a sessile existence and transforming into a radially symmetrical, diploblastic, anthozoan polyp to be such an unlikely evolutionary scenario. The adoption of radiality (or at least "functional" radiality) of bilaterally symmetrical animals is well-documented in some taxa (e.g., Echinodermata), but does not involve the kinds of "degeneration" required by the turbellarian theory. To us the transformation suggested by the turbellarian theory simply involves the loss or drastic simplification of too many complex systems (especially the reproductive system), and major changes in fundamental body design. Both larvae and adults of extant cnidarians maintain a basic radial symmetry. The so-called remnant bilaterality of anthozoan polyps is not true bilaterality at all, but biradiality about an oral–aboral axis, which develops late in the ontogeny of these animals. The turbellarian theory is also weak on embryological grounds, such as differences in cleavage patterns and germ layer formation.

In Chapter 4 we reviewed the important embryological differences between the coelomate metazoan clades known as the protostomes and deuterostomes. To the extent that these traits occur in noncoelomate Metazoa, it is of phylogenetic importance to note them. For example, radial cleavage is characteristic of the deuterostomes, but probably arose very early in metazoan evolution; it occurs in cnidarians and, in a slightly different form, in sponges. Clearly it is the pleisiomorphic type of cleavage among animals, whereas spiral cleavage appears as a synapomorphy for a clade that includes the flatworms and protostomes.

Less clear is the fate of the blastopore in cnidarians. We traditionally use the term "mouth" for the single opening to the gastrovascular cavity. When that opening arises early in development (e.g., in anthozoan gastrulation), it arises from the blastopore. If we accept that this opening is indeed homologous to the mouth of some higher Metazoa, then this feature is shared with the flatworm–protostome clade. If however, we ignore the implications of conventional terminology, then it is also not unreasonable to interpret the opening of the cnidarian gastrovascular cavity as being homologous to the anus of deuterostomes.

There are two major competing theories concerning the nature of the ancestral cnidarian; these ideas focus on whether the first cnidarian was polypoid or medusoid in form. One view is the **medusa theory**. If the Cnidaria did arise from a swimming or creeping, flagellated or ciliated, planuloid ancestor, then the development of tentacles could have produced an animal resembling an actinula larva. The transition from planula to actinula in the modern medusa form can be seen today in the life cycle of certain hydrozoans. Asexual reproduction, such as budding, by a benthic actinula larva could have led to the establishment of a distinct polypoid stage. If so, the polyp can be viewed as an extended larval form specialized for asexual reproduction and benthic existence. A likely scenario is that once the polypoid form became established, some cnidarians began to suppress the medusoid phase of their life cycles, various degrees of which can be seen among the hydrozoans. The epitome of this trend is seen in the class Anthozoa, whose members have no medusa stage.

Other zoologists hold the polyp to be the original cnidarian body form; they view the medusa as a derived dispersal stage that could have evolved independently among the hydrozoans and the scyphozoans. This **polyp theory** is supported in part by certain fundamental differences between scyphomedusae and hydromedusae

If the medusa theory is valid, then the hydrozoan order Trachylina appears to be the most primitive group. The life cycle is dominated by a relatively simple medusoid form and the polyp stage is absent. Gastrulation in this group is by ingression, yielding a solid stereogastrula. The precnidarian may have been a solid-bodied, ciliated, planuloid form. Figure 8.43A is a cladogram depicting a phylogeny based on the medusa theory.

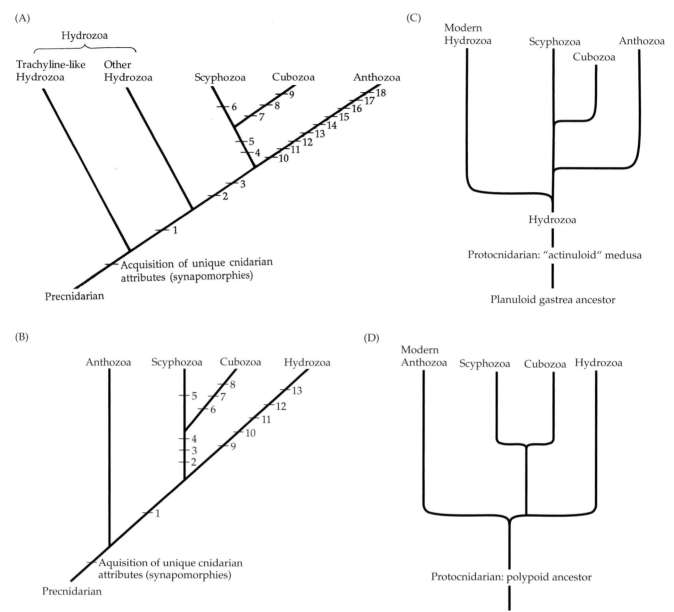

Figure 8.43 Evolution of the Cnidaria. (A) Cladogram depicting the phylogeny of cnidaria according to the medusa theory. The unique cnidarian attributes (synapomorphies) indicated at the base of the cladogram include evolution of the radial, medusoid body form; cnidae; planula larvae; and perhaps the cnidarian coelenteron with mouth surrounded by tentacles. Structures such as the cnidarian nerve net system and simple gut may be cnidarian synapomorphies, or may be primitive features retained from an earlier ancestor (i.e., symplesiomorphies). Numbered synapomorphies on the cladogram are as follows: (1) evolution of the polypoid body form and "alternation of generations;" (2) gonads relocated in gastrodermis; (3) partitions (mesenteries) appear, subdividing the coelenteron; (4) polyp stage secondarily reduced or lost; (5) evolution of the rhopalium; (6) evolution of strobilation; (7) acquisition of a boxlike medusa body; (8) evolution of complex lensed rhopalial eyes; (9) invention of velarium; (10) complete suppression of the medusoid stage; (11) development of hexaradial and octaradial symmetry; (12) evolution of the cnidarian actinopharynx; (13) evolution of the siphonoglyph; (14) coelenteron acquires mesenterial filaments; (15) loss of the cnidal operculum; (16) loss of the cnidocil; (17) evolution of tripartite series of flaps on cnidae; (18) evolution of special ciliary cones associated with cnidae. (B) Cladogram based on the polyp theory, assuming the ancestral position of the anthozoans. This scenario assumes that the first cnidarians were exclusively polypoid, and that the medusa phase arose later. Numbered synapomorphies on this tree are: (1) evolution of the cnidocil (plus the loss of uniquely anthozoan traits possessed by the ancestor); (2) reduction or loss of the polyp phase; (3) evolution of the acraspedote medusa form; (4) evolution of rhopalia; (5) strobilation; (6) boxlike medusa form; (7) lensed rhopalial eyes; (8) appearance of the velarium; (9) relocation of gamete-forming tissue to the epidermis; (10) loss of gut mesenteries; (11) simplification of the middle layer to an acellular mesoglea; (12) evolution of the craspedote medusa form; (13) loss of gastrodermal nematocysts. (C and D are conventional evolutionary trees based on cladograms A and B, respectively.)

The lack of unique synapomorphies for the hydrozoan line also implies that the ancestral cnidarian was in fact what we would today classify as a member of the class Hydrozoa. By this line of reasoning, subsequent to the appearance of the first medusoid cnidarian, a polyp phase evolved, and the ancestral form began to experience an alternation between the two life history stages. From this ancestral line evolved a group of cnidarians with increased specializations of the mesenchyme (e.g., cellularity), gastrodermis and gastrovascular system, and nervous system. Among these events were the movement of the gonadal tissue to the gastrodermis, formation of mesenteries subdividing the coelenteron, and the origin of internal nematocyst-bearing filaments. In addition to allowing greater specialization of the digestive cavity, these changes set the stage for an increased size in individuals rather than simply for colonies of small zooids. These and other evolutionary processes resulted in an animal that we envision as ancestral to the two major lineages we recognize today as the Anthozoa and the Scyphozoa/Cubozoa. The resemblance of the scyphistoma to anthozoan polyps also supports the idea of a common ancestry for the scyphozoans and anthozoans.

As the two lines diverged, one emphasized the medusoid form and the polyp stage was greatly reduced or lost. Emphasis on the medusoid form favored the development of complex sensory units, the rhopalia. The main clade of this medusa-dominated lineage resulted in the Scyphozoa, the members of which tend towards large body size and a fully pelagic life style. The retention of sessile polyps (scyphistomae) by many species was eventually accompanied by the evolution of a unique form of budding, strobilation, as a means of asexually producing new medusae. A side branch from this medusoid line led to the Cubozoa, distinguished by the evolution of a suite of distinguishing features, including the boxlike shape, complex lensed rhopalial eyes, and the velarium. The velarium is probably a convergence to the velum of hydromedusae, and the structures differ in certain ways. For example, the cubomedusan velarium contains extensions of the gastrovascular canal system, whereas the hydromedusan velum does not.

The other main line diverging from the early hydrozoans led to the anthozoans through reduction and eventual loss of the medusoid form. The anthozoans are highly specialized and demarcated in our cladogram (Figure 8.43A) by several unique synapomorphies: complete loss of the medusoid stage; development of hexaradial or octaradial symmetry (transformed into biradial symmetry in most); evolution of an actinopharynx, siphonoglyphs, and unique mesenterial filaments in the coelenteron; loss of the cnidal operculum and cnidocil; evolution of the tripartite flaps on the cnidae; and evolution of special ciliary cones associated with the cnidocytes. Other noteworthy trends from the ancestral hydrozoan condition to the anthozoans are progressively more complex gastrovascular and nervous systems, and a marked increase in the degree of cellularity of the mesenchyme.

The alternative notion, that the exclusively polypoid anthozoans are closest to the ancestral cnidarian, has recently received increasing support. A cladogram depicting the **polyp theory** is shown in Figure 8.43B. There are also some convincing molecular data that support this point of view. For example, Bridge et al. (1992, 1995) show that among cnidarians, only the anthozoans possess circular mitochondrial DNA, a trait they share with other Metazoa. Members of the Cubozoa, Scyphozoa, and Hydrozoa all have linear mtDNA, viewed as a derived condition *within* the cnidarians. These authors present other genetic and molecular evidence that supports the polyp theory.

In the polyp theory, the ancestral cnidarian possessed some or all of the traits that define modern-day Anthozoa. An absence of synapomorphies on the anthozoan line suggests that the Anthozoa is a paraphyletic taxon. Alternatively, some of the features used to define the class Anthozoa may have arisen after the origin of the first cnidarian (see synapomorphies 10–18 in cladogram A). In either case, the clade comprising the the Cubozoa, Scyphozoa, and Hydrozoa is defined by the evolution of the cnidocil (and the loss of whatever uniquely anthozoan traits were present in the ancestor). The monophyly of this lineage is reinforced by the presence of linear mtDNA, as mentioned earlier. Some workers also place the origin of the medusa here, suggesting that all cnidarian medusae are homologous. Others are convinced that hydromedusae and scyphomedusae (and cubomedusae) arose independently, as depicted on cladogram B.

The Cubozoa and Scyphozoa are defined by the same synapomorphies as in cladogram A, plus the evolution of acraspedote medusae. However, the class Hydrozoa is now defined by craspedote medusae plus a suite of synapomorphies viewed as ancestral in cladogram A. These traits include: loss of gut mesenteries; loss of gastrodermal nematocysts; loss of cells from the mesoglea; and movement of the gamete-forming tissue to the epidermis.

Certainly other characters exist that can be used to describe the evolution of the cnidarian classes. But even the obvious features used here provide two competing phylogenies in this phylum. Figure 8.43 C and D are generalized evolutionary trees derived from the two cladograms.

Phylogeny within each of the cnidarian classes is equally interesting but largely beyond the scope of this text. However, a few generalizations can be made about some important events. Coloniality has been a common and important evolutionary theme among the Cnidaria. Coloniality in the hydrozoans probably arose by retention of young polyps during asexual reproduction, and this development ultimately led to the highly specialized colonial groups such as the Siphonophora, Chondrophora, Milleporidae, Stylasteridae, and, of course, the hydroids. In the class Scyphozoa, evolution has clearly favored increasing specialization of the pelagic medusoid form and diminishing importance of the polypoid stage in their life cycle. Scyphomedusae have evolved large size, special musculature, a cellular or fibrous mesenchyme, a complex gastrovascular system, and a fairly sophisticated sensory system.

Among members of the class Anthozoa, evolution has produced a grand series of experiments in colonial polypoid living, resulting in such "super-organisms" as corals, gorgonians, pennatulaceans, and zoanthidians. True corals (scleractinians) first appear in the fossil record in the Triassic period, about 237 million years ago. The first scleractinians were not reef builders.

Scleractinian phylogeny is still not well understood. Although the orthodox opinion views corals evolving out of the Actiniaria, or the Corallimorpharia, an oposing hypothesis suggests just the opposite—that both of these groups evolved from scleractinian ancestors by way of skeletal loss. An increase in polyp size within the Anthozoa has occurred, with the evolution of complex structural components of the mesenchyme and with efficient musculature. These cnidarians have also exploited the commensal relationship with zooxanthellae to a greater degree than members of the other classes. Convergent evolution has occurred frequently throughout the Cnidaria, as witnessed by such features as colonies, calcareous skeletons, the velum-velarium structures, and various means of suppressing the medusoid or polypoid stage in the life cycle.

Selected References

General References

Anderson, C. L., E. U. Canning and B. Okamura. 1998. A triploblast origin for Myxozoa? Nature 392: 346.

Bayer, F. M. and H. B. Owre. 1968. *The Free-Living Lower Invertebrates.* Macmillan, New York.

Benson, A. A. and R. F. Lee. 1975. The role of wax in oceanic food chains. Sci. Am. 232(3): 76–86.

Blank, R. J. and R. K. Trench. 1985. Speciation and symbiotic dinoflagellates. Science 229: 656–658.

Boardman, R. S., A. H. Cheetham and W. A. Oliver (eds.). 1973. *Animal Colonies.* Dowden, Hutchinson and Ross, Stroudsburg, Pennsylvania.

Bridge, D., C. W. Cunningham, B. Schierwater, R. DeSalle and L.W. Buss. 1992. Class-level relationships in the phylum Cnidaria: Evidence from mitochondrial genome structure. Proc. Natl. Acad. Sci. 89: 8750–8753.

Bridge, D., C. W. Cunningham, R. DeSalle and L.W. Buss. 1995. Class-level relationships in the phylum Cnidaria: Molecular and morphological evidence. Mol. Biol. Evol. 12(4): 679–689.

Bullock, T. H. 1965. Coelenterata and Ctenophora. *In* T. H. Bullock and G. A. Horridge (eds.), *Structure and Function in the Nervous System of Invertebrates*, Vol. 1. W. H. Freeman, San Francisco, pp. 459–534.

Cairns, S. D. and 11 other authors. 1991. *Common and scientific names of aquatic invertebrates form the United States and Canada: Cnidaria and Ctenophora.* Spec. Publ. No. 22, Amer. Fisheries Soc., MD.

Cairns, S. D. and I. G. Macintyre. 1992. Phylogenetic implications of calcium carbonate minerology in the Stylasteridae (Cnidaria: Hydrozoa). Palaios 7: 96–107.

Campbell, R. D. 1974. Cnidaria. *In* A. C. Giese and J. S. Pearse (eds.), *Reproduction of Marine Invertebrates*, Vol. 1. Academic Press, New York, pp. 133–200.

Carre, D. 1980. Hypothesis on the mechanism of cnidocyst discharge. J. Cell Biol. 20: 265–271.

Cheng, L. 1975. Marine pleuston — animals at the sea–air interface. Oceanogr. Mar. Biol. Annu. Rev. 13: 181–212.

Chia, F. S. and L. R. Bickell. 1978. Mechanisms of larval attachment and the induction of settlement and metamorphosis in coelenterates: A review. *In* F. S. Chia and M. Rice (eds.), *Settlement and Metamorphosis of Marine Invertebrate Larvae.* Elsevier/North-Holland Biomedical Press.

Conklin, E. J. and R. N. Mariscal. 1977. Feeding behavior, ceras structure, and nematocyst storage in the aeolid nudibranch, *Spurilla neapolitana* (Mollusca). Bull. Mar. Sci. 27(4): 658–667.

Connor, J. L. and N. L. Deans. 2002. *Jellies. Living Art.* Monterey Bay Aquarium, Monterey, CA.

Crowell, S. (ed.). 1965. Behavioral physiology of coelenterates. Am. Zool. 5: 335–589.

Doumene, D. (ed.). 1993. Cnidaria — Cténaires. *In* P. Grassé (ed.). *Traité de Zoologie*, Vol. III. Masson, Paris. (A multiauthored contribution and wonderful source of information.)

Dunn, D. F. 1982. Cnidaria. *In* S. P. Parker (ed.), *Synopsis and Classification of Living Organisms*, Vol. 1. McGraw-Hill, New York, pp. 669–706.

Elder, H. Y. 1973. Distribution and functions of elastic fibers in the invertebrates. Biol. Bull. 144: 43–63.

Florkin, M. and B. T. Scheer (eds.). 1968. *Chemical Zoology*, Vol. 2, Porifera, Coelenterata, and Platyhelminthes. Academic Press, New York, pp. 81–284.

Gladfelter, W. B. 1973. A comparative analysis of locomotor system of medusoid Cnidaria. Helgol. Wiss. Meeresunters. 25: 228–272.

Hand, C. 1959. On the origin and phylogeny of the coelenterates. Syst. Zool. 8: 191–202.

Hanson, E. D. 1958. On the origin of the Eumetazoa. Syst. Zool. 7: 16–47.

Harrison, F. W. and J. Westfall (eds.). 1991. *Microscopic Anatomy of Invertebrates*, Vol. 2. Placozoa, Porifera, Cnidaria, and Ctenophora. Alan Liss, New York.

Hinsch, G. W. 1974. Comparative ultrastructure of cnidarian sperm. Am. Zool. 14: 457–465.

Holstein, T. and P. Tardent. 1983. An ultrahigh-speed analysis of exocytosis: Nematocyst discharge. Science 223: 830–833.

Holstein, T. W., M. Benoit, G. v. Herder, G. Wanner, C. N. David and H. E. Gaub. 1994. Fibrous mini-collagens in *Hydra* nematocysts. Science 265: 402–404.

Hyman, L. H. 1940. *The Invertebrates*, Vol. 1, Protozoa through Ctenophora. McGraw-Hill, New York.

Hyman, L. H. 1959. Coelenterata. *In* W. T. Edmondson, H. B. Ward and G. C. Whipple (eds.), *Freshwater Biology*, 2nd Ed. Wiley, New York, pp. 313–344.

Jagersten, G. 1955. On the early phylogeny of the Metazoa: The bilateogastraea theory. Zool. Bidr. Uppsala 33: 79–108.

Kramp, P. L. 1957. On development through alternating generations, especially in Coelenterata. Viddensk. meddr. dansk naturh. Foren. 107: 13–32.

Kramp, P. L. 1961. Synopsis of the medusae of the world. J. Mar. Biol. Assoc. U.K. 40: 1–469. [No keys or figures, but a landmark taxonomic listing; excellent bibliography.]

Lenhoff, H. M. and L. Muscatine (eds.). 1971. *Experimental Coelenterate Biology.* University of Hawaii Press, Honolulu.

Lotan, A., L. Fishman, Y. Loya and E. Zlotkin. 1995. Delivery of a nematocyst toxin. Nature 375: 456.

Mackie, G. O. (ed.). 1976. *Coelenterate Ecology and Behavior.* Plenum Press, New York. [Although over 25 years old, still one of the best treatments of the subject.]

Mackie, G. O. and L. M. Passano. 1968. Epithelial conduction in hydormedusae. J. Gen. Physiol. 52; 600–608.

Madin, L. P. 1988. Feeding behavior of tentaculate predators: In situ observations and a conceptual model. Bull. Mar. Sci. 43: 413–429.

Mariscal, R. N. 1974. Nematocysts. *In* L. Muscatine and H. M. Lenhoff (eds.), *Coelenterate Biology. Reviews and New Perspectives.* Academic Press, N.Y. pp. 129–178.

Mariscal, R. N. 1984. Cnidaria: Cnidae. *In* J. Bereiter-Hahn, A. G. Batoltsy and K. Sylvia Richards (eds.), *Biology of the Integument: I. Invertebrates.* Springer Verlag, Berlin, pp. 57–68 .

Mariscal, R. N. E. J. Conklin and C. H. Bigger. 1977. The ptychocyst, a major new category of cnida used in tube construction by a cerianthid anemone. Biol. Bull. 152: 392-405.

Mariscal, R. N., R. B. McLean and C. Hand. 1977. The form and function of cnidarian spirocysts. 3. Ultrastructure of the thread and function of spirocysts. Cell Tissue Res. 178: 427–433.

Marshall, A. T. 1996. Calcification in hermatypic and ahermatypic corals. Science 271: 637–639.

Miller, R. L. and C. R. Wyttenbach (eds.). 1974. The developmental biology of the Cnidaria. Am. Zool. 14: 440–866.

Moore, R. C. (ed.). 1956. *Treatise on Invertebrate Paleontology. Coelenterata,* Vol. F. Geol. Soc. Am. and University of Kansas Press, Lawrence.

Moore, R. E. and P. J. Scheuer. 1971. Palytoxin: A new marine toxin from a coelenterate. Science 172: 495–498.

Muscatine, L. and H. M. Lenhoff (eds.). 1974. *Coelenterate Biology. Reviews and New Perspectives.* Academic Press, New York. [A highly useful volume with excellent reviews of histology, skeletal systems, cnidae, development, symbiosis, and bioluminescence.]

Muscatine, L., R. R. Pool and R. K. Trench. 1975. Symbiosis of algae and invertebrates: Aspects of the symbiont surface and the host–symbiont interface. Trans. Am. Microsc. Soc. 94(4): 450–469.

Nielsen, C. 1987. *Haeckelia* (= *Euchlora*) and *Hydroctena* and the phylogenetic interrelationships of Cnidaria and Ctenophora. Z. Zool. Syst. Evolutionsforsch. 25: 9–12.

Rees, W. J. (ed.). 1966. *The Cnidaria and Their Evolution.* Zool. Soc. Lond. Symp. No. 16, Academic Press, London. [A benchmark publication with major papers by most specialists on the subject of cnidarian evolution.]

Russell, F. S. 1954, 1970. *Medusae of the British Isles,* Vols. 1 and 2. Cambridge University Press, London.

Shimomura, O. and F. H. Johnson. 1975. Chemical nature of bioluminescence systems in coelenterates. Proc. Natl. Acad. Sci. USA 72: 1546–1549.

Shostak, S. and V. Kolluri. 1995. Symbiogenic origins of cnidarian cnidocysts. Symbiosis 19: 1–29.

Siddall, M.E., D. S. Martin, D. Bridge, S. S. Desser and D. K. Cone. 1995. The demise of a phylum of protists: Phylogeny of Myxozoa and other parasitic Cnidaria. J. Parasitol. 81(6): 961–967.

Tardent, P. and R. Tardent (eds.). 1980. *Developmental and Cellular Biology of Coelenterates.* Elsvier/North-Holland Biomedical Press.

Taylor, D. L. 1973. The cellular interactions of algal–invertebrate symbioses. Adv. Mar. Biol. 11: 1–56.

Trench, R. K. 1979. The cell biology of plant–animal symbioses. Ann. Rev. Plant Physiol. 30: 485–531.

Tursch, B., J. C. Braeckman, D. Daloze and M. Kaisin. 1978. Terpenoids from coelenterates. *In* P. Scheuer (ed.), *Marine Natural Products,* Vol. 2. Academic Press, New York, pp. 347–396.

Weill, R. 1934. Contribution a l'étude des Cnidaires et de leurs nématocystes. I. Recherches sur les nématocystes (morphologie-physiologie-développement). II. Valeur taxonomique du cnidome. Travaux Station Zoologique de Wimereux, Volumes X-XI. [In this paper, Weill created a classification scheme for cnidocytes that remains in use today.]

Westfall, J. A. 1973. Ultrastructural evidence for neuromuscular systems in coelenterates. Am. Zool. 13: 237–246.

Williams, R. B., P. F. S. Cornelius, R. G. Hughes and E. A. Robson (eds.). 1991. *Coelenterate Biology: Recent Research on Cnidaria and Ctenophora.* Kluever Acad. Publ.

Wyttenbach, C. R. (ed.). 1974. The developmental biology of the Cnidaria. Am. Zool. 14(2): 540–866. [Papers presented at a 1972 symposium.]

Hydrozoa

Alvariño, A. 1978. *Nectocarmen antonioi,* a new Prayinae, Calycophorae, Siphonophorae, from California. Proc. Biol. Soc. Washington 96: 339–348.

Bellamy, N. and M. J. Risk. 1982. Coral gas: Oxygen production in *Millepora* on the Great Barrier Reef. Science 215: 1618–1619.

Bieri, R. 1970. The food of *Porpita* and niche separation in three neuston coelenterates. Publ. Seto Mar. Biol. Lab. 27: 305–307.

Biggs, D. C. 1977. Field studies of fishing, feeding and digestion in siphonophores. Mar. Behav. Physiol. 4: 261–274.

Bouillon, J. 1985. Essai de classification des Hydropolypes-Hydroméduses (Hydrozoa-Cnidaria). Indo-Malayan Zool. 1: 29-243.

Burnett, A. L. (ed.). 1973. *Biology of Hydra.* Academic Press, New York.

Cairns, S. D. and J. L. Barnard. 1984. Redescription of *Janaria mirabilis,* a calcified hydroid from the eastern Pacific. Bull. South. Calif. Acad. Sci. 83: 1–11.

Eakin, R. M. and J. A. Westfall. 1962. Fine structure of photoreceptors in the hydromedusan, *Polyorchis penicillatus.* Proc. Natl. Acad. Sci. USA 48: 826–833.

Edwards, C. 1966. *Velella velella* (L.): The distribution of its dimorphic forms in the Atlantic Ocean and the Mediterranean, with comments on its nature and affinities. *In* H. Barnes (ed.), *Some Contemporary Studies in Marine Science.* Allen and Unwin, London, pp. 283–296.

Fields, W. G. and G. O. Mackie. 1971. Evolution of the Chondrophora: Evidence from behavioural studies on *Velella.* J. Fish. Res. Bd. Can. 28: 1595–1602.

Francis, L. 1985. Design of a small cantilevered sheet: The sail of *Velella velella.* Pac. Sci. 39(1): 1–15.

Freeman, G. 1983. Experimental studies on embryogenesis in hydrozoans (Trachylina and Siphonophora) with direct development. Biol. Bull. 165: 591–618.

Gierer, Z. 1974. *Hydra* as a model for the development of biological form. Sci. Am. 231(6): 44–54.

Lane, C. E. 1960. The Portuguese man-of-war. Sci. Am. 202: 158–168.

Lenhoff, H. M. 1983. *Hydra: Research Methods.* Plenum, New York.

Lenhoff, H. M. and W. F. Loomis (eds.). 1961. *The Biology of Hydra and of Some Other Coelenterates: 1961.* University of Miami Press, Coral Gables, Florida.

Lentz, T. L. 1966. *The Cell Biology of Hydra.* Wiley, New York.

Mackie, G. O. 1959. The evolution of the Chondrophora (Siphonophora: Disconanthae): New evidence from behavioral studies. Trans. R. Soc. Can. 53: 7–20.

Mackie, G. O. 1960. The structure of the nervous system in *Velella.* Q. J. Microsc. Sci. 101: 119–133.

Martin, W. E. 1975. *Hydrichthys pietschi,* new species (Coelenterata) parasitic on the fish, *Ceratias holboelli.* Bull. South. Calif. Acad. Sci. 74: 1–6. [A parasitic hydroid from California.]

Naumov, D. V. 1960. Hydroids and hydromedusae of marine, brackish, and fresh water basins of the U.S.S.R. Opred. Faune SSSR, No. 70, 585 pp.

Pardy, R. L. and B. N. White. 1977. Metabolic relationships between green hydra and its symbiotic algae. Biol. Bull. 153: 228–236.

Petersen, K. W. 1990. Evolution and taxonomy in capitate hydroids and medusae (Cnidaria: Hydrozoa). Zool. J. Linnean Soc. 100: 101–231.

Purcell, J. E. 1980. Influence of siphonophore behavior upon their natural diets: Evidence for aggressive mimicry. Science 209: 1045–1047.

Purcell, J. E. 1984. The functions of nematocysts in prey capture by epipelagic siphonophores (Coelenterata, Hydrozoa). Biol. Bull. 166: 310–327.

Shimomura, O. 1995. A short story of aequorin. Biol. Bull. 189: 1–5.

Singla, C. L. 1975. Statocysts of hydromedusae. Cell Tissue Res. 158: 391–407.

Stretch, J. J. and J. M. King. 1980. Direct fission: An undescribed reproductive method in hydromedusae. Bull. Mar. Sci. 30: 522–526.

Totton, A. K. 1960. Studies on *Physalia physalis* (L.). Part 1, natural history and morphology. Discovery Rpt. 30: 301–367.

Wahle, C. M. 1980. Detection, pursuit, and overgrowth of tropical gorgonians by milleporid hydrocorals: Perseus and Medusa revisited. Science 209: 689–691.

West, D. A. 1978. The epithelio-muscular cell of hydra: Its fine structure, three-dimensional architecture and relationship to morphogenesis. Tissue Cell 10: 629–646.

Scyphozoa

Alexander, R. M. 1964. Visco-elastic properties of the mesoglea of jellyfish. J. Exp. Biol. 41: 363–369.

Anderson, P. A. V. and G. O. Mackie. 1977. Electrically coupled photosensitive neurons control swimming in jellyfish. Science 197: 186–188.

Arai, M. N. 1997. A Functional Biology of Scyphozoa. Chapman and Hall, London.

Berrill, M. 1963. Comparative functional morphology of the Stauromedusae. Can. J. Zool. 41: 741–752.

Calder, D. R. 1971. Nematocysts of Aurelia, Chrysaora and Cyanea and their utility in identification. Trans. Am. Microscop. Soc. 90: 269–274.

Calder, D. R. 1982. Life history of the cannonball jellyfish, Stomolophus meleagris L. Agassiz, 1860 (Scyphozoa, Rhizostomida). Biol. Bull. 162: 149–162.

Fancett, M. S. 1988. Diet and prey selectivity of scyphomedusae from Port Phillip Bay, Australia. Mar. Biol. 98: 503–509.

Fancett, M. S. and G. P. Jenkins. 1988. Predatory impact of scyphomedusae on ichthyoplankton and other zooplankton in Port Phillip Bay. J. Exp. Mar. Biol. Ecol. 116: 63–77.

Gladfelter, W. B. 1972. Structure and function of the locomotory system of the scyphomedusa Cyanea capillata. Mar. Biol. 14: 150–160.

Horridge, A. 1954. Observations on the nerve fibers of Aurelia aurita. Q. J. Microsc. Sci. 95: 85–92.

Horridge, A. 1956. The nervous system of the ephyra larva of Aurelia aurita. Q. J. Microsc. Sci. 97: 59–73.

Larson, R. J. 1987. Trophic ecology of planktonic gelatinous predators in Saanich Inlet, British Columbia: diets and prey selection. J. Plankt. Res. 9: 811–820.

Mayer, A. G. 1910. The Medusae of the World. I,II, Hydromedusae. III, Scyphomedusae. Carnegie Inst. Washington Publ. 109. [Reprinted in 1977 by A. Asher, Amsterdam.]

Möller, H. 1984. Reduction of a larval herring population by jellyfish predator. Science 224: 621–622.

Purcell, J. E. 1985. Predation on fish eggs and larvae by pelagic cnidarians and ctenophores. Bull. Mar. Sci. 37: 739–755.

Rottini Sandrini, L. and M. Avian. 1989. Feeding mechanism of Pelagia noctiluca (Scyphozoa: Semaeostomeae); laboratory and open sea observations. Mar. Biol. 102: 49–55.

Russell, F. S. 1970. The Medusae of the British Isle. II. Pelagic Scyphozoa with a Supplement to the First Volume on Hydromedusae. Cambridge Univ. Press, Cambridge.

Sandrini, L. R. and M. Avian. 1989. Feeding mechanism of Pelagia noctiluca, laboratory and open sea observations. Mar. Biol. 102: 49–55.

Shushkina, E. A. and E. I. Musayeva. 1983. The role of jellyfish in the energy system of Black Sea plankton communities. Oceanology 23: 92–96.

Cubozoa

Conant, F. S. 1900. The Cubomedusae. Mem. Biol. Lab. Johns Hopkins University 4(1): 1–52.

Pearse, J. S. and V. B. Pearse. 1978. Vision in cubomedusan jellyfishes. Science 199: 458.

Anthozoa

Adey, W. H. 1978. Coral reef morphogenesis: A multidimensional model. Science 202: 831–837.

Anderson, P. A. V. and J. F. Case. 1975. Electrical activity associated with luminescence and other colonial behavior in the pennatulid Renilla kollikeri. Biol. Bull. 149: 80–95.

Arai, M. N. 1965. The ceriantharian nervous system. Am. Zool. 5: 424–429.

Babcock, R., G. Bull, P. Harrison, A. Heyward, J. Oliver, C. Wallace and B. Willis. 1986. Synchronous spawnings of 105 scleractinian coral species on the Great Barrier Reef. Mar. Biol. 90: 379–394.

Barham, E. G. and I. E. Davies. 1968. Gorgonians and water motion studies in Gulf of California. Underwater Nat. [Bull. Am. Littoral Soc.] 5(3): 24–28, 42.

Batham, E. J. 1960. The fine structure of epithelium and mesoglea in a sea anemone. Quart. J. Microscop. Sci. 101: 481–485.

Batham, E. J., C. F. A. Pantin and E. A. Robson. 1960. The nerve-net of the sea-anemone Metridium senile: The mesenteries and the column. Q. J. Microsc. Sci. 101: 487–510.

Bayer, F. M., M. Grasshoff and J. Verseveldt (eds.). 1983. Illustrated Trilingual Glossary of Morphological and Anatomical Terms Applied to Octocorallia. E. J. Brill, Leiden.

Benayahu, Y. and Y. Loya. 1983. Surface brooding in the Red Sea soft coral Parerythropodium fulvum fulvum (Forskål, 1775). Biol. Bull. 165: 353–369.

Bigger, C. H. 1980. Interspecific and intraspecific acrorhagial aggressive behavior among sea anemones: A recognition of self and not-self. Biol. Bull. 159: 117–134.

Bigger, C. H. 1982. The cellular basis of the agressive acrorhagial response of sea anemones. J. Morph. 173 (3): 259–278.

Birkeland, C. 1974. Interactions between a sea pen and seven of its predators. Ecol. Monogr. 44(2): 211–232.

Burnett, J. W. and J. S. Sutton. 1969. The fine structural organization of the sea nettle fishing tentacles. J. Exp. Zool. 172: 335–348.

Coffroth, M. A. 1984. Ingestion and incorporation of coral mucus aggregates by a gorgonian soft coral. Mar. Ecol. Prog. Ser. 17: 193–199.

Cook, C. B., C. F. D'Elia, and G. Muller-Parker. 1988. Host feeding and nutrient sufficiency for zooxanthellae in the sea anemone Aptasia pallida. Mar. Biol. 98: 253–262.

Dana, T. F. 1975. Development of contemporary eastern Pacific coral reefs. Mar. Biol. 33: 355–374.

Darwin, C. 1842. The Structure and Distribution of Coral Reefs. Reprinted in 1984 by the University of Arizona, Tucson.

Dubinski, Z. (ed.). 1990. Coral Reefs. Ecosystems of the World 25. Elseview, Amsterdam.

Ducklow, H. W. and R. Mitchell. 1979. Composition of mucus released by coral reef coelenterates. Limnol. Oceanogr. 24(4): 706–714.

Dunn, D. F. 1977. Dynamics of external brooding in the sea anemone Epiactis prolifera. Mar. Biol. 39: 41–49.

Dunn, D. F., D. M. Devaney, and B. Roth. 1980. Stylobates: A shell-forming sea anemone (Coelenterata, Anthozoa, Actiniidae). Pac. Sci. 34: 379–388.

Fautin, D. G. 1991. The anemone fish symbiosis: What is known and what is not. Symbiosis. 10: 23–46.

Fautin, D. G., C.-C. Guo and J.-S. Hwang. 1995. Costs and benefits of the symbiosis between the anemone shrimp Periclimenes brevicarpalis and its host Entacmaea quadricolor. Mar. Ecol. Prog. Ser. 129: 77–84.

Fautin, D. G. and J. Lowenstein. 1992. Phylogenetic relationships among scleractinians, actinians, and corallimorpharians. Proc. 7th Internat. Coral Reef Symp. 2: 665–670.

Fenical, W., R. K. Okuda, M. M. Bandurraga, P. Culver and R. S. Jacobs. 1981. Lophotoxin: A novel neuromuscular toxin from Pacific sea whips of the genus Lophogorgia. Science 212: 1512–1514.

Francis, L. 1973a. Clone specific segregation in the sea anemone Anthopleura elegantissima. Biol. Bull. 144: 64–72.

Francis, L. 1973b. Intraspecific aggression and its effect on the distribution of Anthopleura elegantissima and some related sea anemones. Biol. Bull. 144: 73–92.

Francis, L. 1976. Social organization within clones of the sea anemone Anthopleura elegantissima. Biol. Bull. 150: 361–376.

Francis, L. 1979. Contrast between solitary and colonial lifestyles in the sea anemone Anthopleura elegantissima. Am. Zool. 19: 669–681.

Fricke, H. and L. Hottinger. 1983. Coral biotherms below the euphotic zone in the Red Sea. Mar. Ecol. Prog. Ser. 11: 113–117.

Gladfelter, E. H. 1983. Circulation of fluids in the gastrovascular system of the reef coral Acropora cervicornia. Biol. Bull. 165: 619–636.

Glynn, P. W. 1976. Some physical and biological determinants of coral community structure in the eastern Pacific. Ecol. Monogr. 46: 431–456.

Glynn, P. W. 1980. Defense by symbiotic Crustacea of host corals elicited by chemical cues from predators. Oecologia 47: 287–290.

Glynn, P. W. 1982. Coral communities and their modifications relative to past and prospective Central American seaways. Adv. Mar. Biol. 19: 91–132.

Glynn, P.W. 1993. Coral reef bleaching: Ecological perspectives. Coral Reefs 12: 1–17.

Glynn, P. W., G. M. Wellington and C. Birkeland. 1978. Coral reef growth in the Galapagos: Limitation by sea urchins. Science 203: 47–49.

Godknechy, A. and P. Tardent. 1988. Discharge and mode of action of the tentacular nematocysts of *Anemonia sulcata*. Mar. Biol. 100: 83–92.

Goreau, T. F. 1959. The physiology of skeleton formation in corals. Biol. Bull. 116: 59–75.

Goreau, T. F. 1963. Calcium carbonate deposition by coralline algae and corals in relation to their roles as reef builders. Ann. N.Y. Acad. Sci. 109: 127–167.

Grigg, R. W. 1965. Ecological studies of black coral in Hawaii. Pac. Sci. 19(2): 244–260.

Grigg, R. W. 1972. Orientation and growth form of the sea fans. Limnol. Oceanogr. 17: 185–192.

Grigg, R. W. 1977. Population dynamics of two gorgonian corals. Ecology 58: 279–290.

Grimstone, A. V., R. W. Horne, C. F. A. Pantin and E. A. Robson. 1958. The fine structure of the mesenteries of the sea-anemone *Metridium senile*. Q. J. Microscop. Sci. 99: 523–540.

Hamner, W. M. and D. F. Dunn. 1980. Tropical Corallimorpharia (Coelenterata: Anthozoa): Feeding by envelopment. Micronesica 16: 37–41.

Hand, C. and K. R. Uhlinger. 1995. Asexual reproduction by transverse fission and some anomalies in the sea anemone *Nematostella vectensis*. Invert. Biol. 114: 9–18.

Harrison, P. L., R. C. Babcock, G. D. Bull, J. K. Oliver, C. C. Wallace and B. L. Willis. 1984. Mass spawning in tropical reef corals. Science 232: 1186–1189.

Howe, N. R. and Y. M. Sheikh. 1975. Anthopleurine: A sea anemone alarm pheromone. Science 189: 386–388.

Jennison, B. L. 1979. Gametogenesis and reproductive cycles in the sea anemone *Anthopleura elegantissima*. Can. J. Zool. 57: 403–411.

Jones, O. A. and R. Endean (eds.). 1973–1976. *Biology and Geology of Coral Reefs*. Vol. I–III. Academic Press, New York.

Josephson, R. K. 1974. The strategies of behavioral control in a coelenterate. Am. Zool. 14: 905–915.

Josephson, R. K. and S. C. March. 1966. The swimming performance of the sea-anemone *Boloceroides*. J. Exp. Biol. 44: 493–506.

Kastendiek, J. 1976. Behavior of the sea pansy *Renilla kollikeri* Pfeffer and its influence on the distribution and biological interactions of the species. Biol. Bull. 151: 518–537.

Kingsley, R. J., M. Tsuzaki, N. Watabe and G. L. Mechanic. 1990. Collagen in the spicule organic matrix of the gorgonian *Leptogorgia virgulata*. Biol. bull. 179: 207–213.

Lang, J. 1973. Interspecific aggression by scleractinian corals: Why the race is not only to the swift. Bull. Mar. Sci. 23: 269–279.

Lewis, D. H. and D. C. Smithe. 1971. The autotrophic nutrition of symbiotic marine coelenterates with special reference to hermatypic corals. Proc. R. Soc. Lond. Ser. B 178: 11–129.

Lewis, J. B. 1977. Processes of organic production on coral reefs. Biol. Rev. 52: 305–347.

Lewis, J. B. and W. S. Price. 1975. Feeding mechanisms and feeding strategies of Atlantic reef corals. J. Zool. 176: 527–544.

Lewis, J. B. and W. S. Price. 1976. Patterns of ciliary currents in Atlantic reef corals and their functional significance. J. Zool. 178: 77–89.

Mariscal, R. N. 1970. Nature of symbiosis between Indo–Pacific anemone fishes and sea anemones. Mar. Biol. 6: 58.

Mariscal, R. N. 1974. Scanning electron microscopy of the sensory epithelia and nematocysts of corals and a corallimorpharian sea anemone. Proc. Second Int. Coral Reef Symp. 1: 519–532.

Marshall, A. T. 1996. Calcification in hermatypic and ahermatypic corals. Science 271: 637–639.

Mauzey, K. P., C. Birkeland, and P. Dayton. 1968. Feeding behavior of asteroids and escape responses of their prey in the Puget Sound region. Ecology 49: 603–619.

Meyer, J. L., E. T. Schultz, and G. S. Helfman. 1983. Fish schools: An asset to corals. Science 220: 1047–1049.

Muscatine, L. and C. Hand. 1958. Direct evidence for the transfer of materials from symbiotic algae to the tissues of a coelenterate. Proc. Natl. Acad. Sci. USA 44: 1259–1263.

Muscatine, L. and J. W. Porter. 1977. Reef corals: Mutualistic symbioses adapted to nutrient-poor environments. Biol. Sci. 27: 454–459.

Newell, N. D. 1972. The evolution of reefs. Sci. Am. 226(6): 54–65.

Patton, J. S., S. Abraham and A. A. Benson. 1977. Lipogenesis in the intact coral *Pocillopora capitata* and its isolated zooxanthellae: Evidence for a light-driven carbon cycle between symbiont and host. Mar. Biol. 44: 235–247.

Pearson, R. G. 1981. Recovery and recolonization of coral reefs. Mar. Ecol. Prog. Ser. 4: 105–122.

Proceedings of the Symposium on Coral Reefs. 1974. Published by The Great Barrier Reef Committee, Brisbane, Australia. [In 3 vols.]

Proceedings of the Third International Coral Reef Symposium. 1977. Published by the Rosenstiel School of Marine and Atmospheric Science University of Miami, Coral Gables, Florida.

Proceedings of the Fourth International Coral Reef Symposium, Vol. 1. 1980. Marine Sciences Center, University of the Philippines. [This volume, and the two symposium sets cited above, are major sources of information on corals and coral reefs.]

Purcell, J. E. 1977. Aggressive function and induced development of catch tentacles in the sea anemone *Metridium senile* (Coelenterata: Actiniaria). Biol. Bull. 153: 355–368.

Purcell, J. E. and C. L. Kitting. 1982. Intraspecific aggression and population distributions of the sea anemone *Metridium senile*. Biol. Bull. 162: 345–359.

Reese, E. S. 1981. Predation on corals by fishes of the family Chaetodontidae: Implications for conservation and management of coral reef ecosystems. Bull. Mar. Sci. 31(3): 594–604.

Reimer, A. A. 1971. Feeding behavior in the Hawaiian zoanthids *Palythoa* and *Zoanthus*. Pac. Sci. 25(4): 257–260.

Richmond, R. H. 1985. Reversible metamorphosis in coral planula larvae. Mar. Ecol. Prog. Ser. 22: 181–185.

Rinkevich, B. and Y. Lola. 1983. Short-term fate of photosynthetic products in a hermatypic coral. J. Exp. Mar. Biol. Ecol. 73: 175–184.

Romano, S. L. and S. R. Palumbi. 1996. Evolution of scleractinian corals inferred from molecular systematics. Science 271: 640-642.

Ross, D. M. 1970. Behavioral and ecological relationships between sea anemones and other invertebrates. Annu. Rev. Mar. Biol. Oceanogr. 5: 291–316.

Sandberg, D. M., P. Kanciruk and R. N. Mariscal. 1971. Inhibition of nematocyst discharge correlated with feeding in a sea anemone, *Calliactis tricolor* (Leseur). Nature 232: 263–264.

Schmidt, H. 1972. Die Nesselkapseln der Anthozoa und ihre Bedeutung für die Phylogenetische systematik. Helgol. Wiss. Meeresunters. 23: 422–458.

Schmidt, H. 1974. On the evolution of the Anthozoa. Proc. Second Internat. Coral Reef Symp. 1: 533–560.

Schuhmacher, H. and H. Zibrowius. 1985. What is hermatypic? A redefinition of ecological groups in corals and other organisms. Coral Reefs 4: 1–9.

Sebens, K. P. 1981a. Recruitment in a sea anemone population: Juvenile substrate becomes adult prey. Science 213: 785–787.

Sebens, K. P. 1981b. Reproductive ecology of the intertidal sea anemones *Anthopleura xanthogrammica* (Brandt) and *A. elegantissima* (Brandt): Body size, habitat and sexual reproduction. J. Exp. Mar. Biol. Ecol. 54: 225–250.

Sebens, K. P. 1984. Agonistic behavior in the intertidal sea anemone *Anthopleura xanthogrammica*. Biol. Bull. 166: 457–472.

Sebens, K. P. and K. DeRiemer. 1977. Diel cycles of expansion and contraction in coral reef anthozoans. Mar. Biol. 43: 247–256.

Shick, J. M. 1981. Heat production and oxygen uptake in intertidal sea anemones from different shore heights during exposure to air. Mar. Biol. Lett. 2: 225–236.

Schick, J. M. 1991. *A Functional Biology of Sea Anemones*. Chapman and Hall, New York.

Shick, J. M. and J. A. Dykens. 1984. Photobiology of the symbiotic sea anemone *Anthopleura elegantissima*: Photosynthesis, respiration, and behavior under intertidal conditions. Biol. Bull. 166: 608–619.

Spaulding, J. G. 1974. Embryonic and larval development in sea anemones (Anthozoa: Actiniaria). Am. Zool. 14: 511–520.

Stoddart, D. R. 1969. Ecology and morphology of recent coral reefs. Biol. Rev. 44: 433–498.

Stoddart, J. A. 1983. Asexual production of planulae in the coral *Pocillopora damicornis*. Mar. Biol. 76: 279–284.

Veron, J. 2000. *Corals of the World* (3 vols.). Aust. Inst. Mar. Sci., Townsville, Australia.

Walsh, G. E. 1967. An annotated bibliography of the families Zoanthidae, Epizoanthidae, and Parazoanthidae (Colenterata, Zoantharia). University of Hawaii, Hawaii Inst. Mar. Biol. Tech. Rpt. 13: 1–77.

Ware, J. R., D. G. Fautin and R. W. Buddemeier. 1996. Patterns of coral bleaching: modeling the adaptive bleaching hypothesis. Ecol. Modeling 84: 199–214.

Watson, G. M. and D. A. Hessinger. 1989. Cnidocyte mechanoreceptors are tuned to the movements of swimming prey by chemoreceptors. Science 243: 1589–1591.

Watson, G. M. and R. N. Mariscal. 1983a. Comparative ultrastructure of catch tentacles and feeding tentacles in the sea anemone *Haliplanella*. Tissue Cell, 15(9): 939–953.

Watson, G. M. and R. N. Mariscal. 1983b. The development of a sea anemone tentacle specialized for aggression: Morphogenesis and regression of the catch tentacle of *Haliplanella luciae* (Cnidaria, Anthozoa). Biol. Bull. 164: 506–517.

Westfall, J. A. 1965. Nematocysts of the sea anemone *Metridium*. Am. Zool. 5: 377–393.

Williams, R. B. 1975. Catch-tentacles in sea anemones: Occurrence in *Haliplanella luciae* (Verrill) and a review of current knowledge. J. Nat. Hist. 9: 241–248.

Yonge, C. M. 1963. The biology of coral reefs. Adv. Mar. Biol. 1: 209–260. [Although somewhat dated, still one of the most readable reviews of coral reefs available.]

Yonge, C. M. 1973. The nature of reef-building (hermatypic) corals. Bull. Mar. Sci. 23(1): 1–15.

Yonge, C. M. 1974. Coral reefs and molluscs. Trans. R. Soc. Edinburgh 69(7): 147–166.

9 *Phylum Ctenophora:*
The Comb Jellies

Their power of destruction is not surprising once we see their method of obtaining food.

Sir Alister Hardy,
(speaking of ctenophores, 1965)

Ctenophores (Greek *cten*, "comb"; *phero*, "to bear")—commonly called comb jellies, sea gooseberries, or sea walnuts—are transparent, gelatinous animals. Most of them are planktonic, living from surface waters to depths of at least 3,000 meters; a few species are epibenthic. Their transparency and fragile nature make them difficult to capture or observe by traditional sampling methods such as towing or trawling with nets, and until the recent advent of manned submersibles and "blue water" SCUBA techniques they were thought to be only modestly abundant. However, they are now known to form a major portion of the planktonic biomass in many areas of the world, and they may periodically be the predominant zooplankters in some areas. About 100 species have been described, but there are probably many deep-sea forms yet to be discovered.

The ctenophores are radially (biradially) symmetrical, diploblastic (or perhaps triploblastic) animals, resembling cnidarians in several respects. This similarity is immediately obvious, for example, in features such as symmetry, a gelatinous mesenchyme or collenchyme (formed from ectomesoderm), the absence of a body cavity between the gut and the body wall, and a relatively simple, netlike nervous system. Some zoologists, however, view these similarities as convergent features resulting from adaptations to pelagic lifestyles. Ctenophores are significantly different from cnidarians in their more extensively organized digestive system, their wholly mesenchymal (perhaps mesodermal) musculature, and certain other features (Box 9A).

Ctenophores also differ fundamentally from cnidarians in that they are monomorphic throughout their life histories, are never colonial, and lack any

(A)

(B)

(C)

(D)

(E)

(F)

(G)

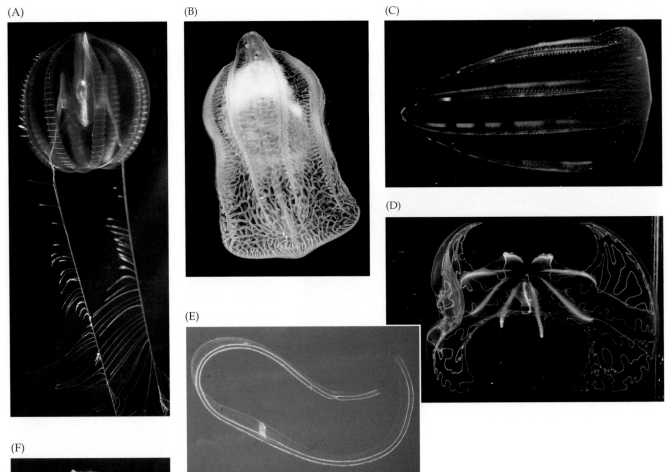

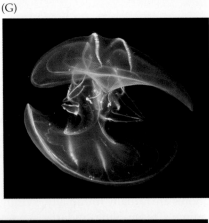

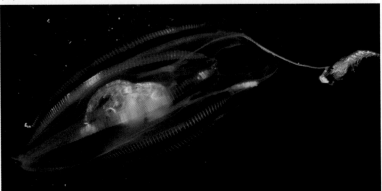

Figure 9.1 Representative ctenophores. (A) *Pleurobrachia* (order Cydippida). (B) *Beröe forskali* (order Beroida). (C) *Beröe*, bioluminescing. (D) An unidentified lobate ctenophore from 780m depth in the Caribbean Sea. (E) *Cestum* (order Cestida). (F) *Leucothea* (order Lobata). (G) *Mnemiopsis* (order Lobata). (H) An Antarctic lobate ctenophore with two krill in its gut, and a third one being captured.

(H)

trace of an attached sessile stage. Ctenophores lack a hard skeleton, an excretory system, and a respiratory system. Most are simultaneous hermaphrodites capable of self-fertilization—a relatively unusual quality among the Metazoa. A distinctive larval stage, the cydippid larva, is usually produced. Ctenophores are exclusively marine. They display a wonderful variety of shapes, and they range in size from less than 1 cm in height to ribbon-shaped forms 2 m long (Figure 9.1). Some have evolved rather bizarre body forms, and a few have taken up a benthic creeping existence. They occur in all the world's seas and at all latitudes. Desiccated specimens of the genus *Pleurobrachia* are often found washed ashore after storms. However, in their planktonic environment, ctenophores are some of the most elegant and graceful creatures in the sea, and observing them in their natural habitat is a memorable experience.

Taxonomic History and Classification

Perhaps because many well known ctenophores are brilliantly luminescent and are commonly seen from ships, the group has been known since ancient times. The first recognizable figures of ctenophores were drawn by a ship's doctor and naturalist in 1671. Linnaeus placed them in his group Zoophyta, along with various other "primitive" invertebrates. Cuvier classified them with medusae and anemones in Zoophytes. In the early nineteenth century, Eschscholtz designed the first rational classification of pelagic medusae and ctenophores by creating the orders Ctenophorae (for comb jellies), Discophorae (for all the solitary cnidarian medusae), and Siphonophorae (for the colonial siphonophorans and chondrophorans). Eschscholtz viewed these orders as subdivisions of the class Acalepha, regarding them as intermediate between Zoophytes and Echinodermata (on the basis of the common presence of radial symmetry). Recall that it was Leuckart who, in 1847, first separated the Coelenterata from the echinoderms, although his Coelenterata also included sponges and ctenophores. Vosmaer (1877) was responsible for removing the sponges and Hatschek (1889) for removing the ctenophores as separate groups. Until recently, two classes were recognized: Nuda, for species lacking tentacles (the single order Beroida), and Tentaculata, for those species with tentacles (all other orders). However, the monophyly of these two groups has been questioned, and we follow Harbison and Madin (1983) in simply dividing the ctenophores into 7 orders (containing 19 families). Figure 9.2 illustrates the general anatomy of the major groups.

PHYLUM CTENOPHORA

ORDER BEROIDA: (Figures 9.1B, 9.2D,E) Pelagic; body cylindrical or thimble-shaped and strongly flattened in tentacular plane; tentacles and sheaths absent; aboral end rounded (*Beröe*) or with two prominent keels (*Neis*); stomodeum greatly enlarged; aboral sense organ well developed; comb rows present; meridional canals with numerous side branches; paragastric canals simple or with side branches. Two genera: *Beröe* and *Neis*.

ORDER CESTIDA: (Figures 9.1E, 9.2I) Pelagic; body extremely compressed in tentacular plane, and greatly elongated in stomodeal plane, producing a ribbon-like form up to 1 m long in some species; substomodeal comb rows elongated, ex-

---

BOX 9A Characteristics of the Phylum Ctenophora

1. Diploblastic (or possibly triploblastic) Metazoa, with ectoderm and entoderm separated by a cellular mesenchyme

2. Biradial symmetry; the body axis is oral–aboral

3. With adhesive exocytotic structures called colloblasts

4. Gastrovascular cavity (gut) is the only "body cavity"; gut with stomodeum and canals that branch complexly throughout body; gut ends in two small anal pores

5. Without discrete respiratory, excretory, or circulatory systems (other than the gut)

6. Nervous system in the form of a nerve net or plexus, but more specialized than that of cnidarians

7. Musculature always formed of true mesenchymal cells

8. Monomorphic, without alternation of generations and without any kind of an attached sessile life stage

9. With eight rows of ciliary plates (combs or ctenes) at some stage in their life history; comb rows controlled by unique apical sense organ

10. Some adults and most juveniles with a pair of long tentacles, often retractable into sheaths

11. Most are hermaphroditic; typically with a characteristic cydippid larval stage

---

tending along entire aboral edge; subtentacular meridional canals arise under subtentacular comb rows (*Cestum*) or equatorially from interradial canals (*Velamen*); paragastric canals extend along oral edge and fuse with meridional canals; tentacles and tentacle sheaths present. Two genera: *Cestum* and *Velamen*.

ORDER CYDIPPIDA: (Figures 9.1A; 9.2A,B) Pelagic; with well developed comb rows; tentacles long and retractable into sheaths; body globular or ovoid, occasionally flattened in the stomodeal plane; meridional canals end blindly, paragastric canals (when present) end blindly at mouth. (e.g., *Aulococtena, Bathyctena, Callianira, Dryodora, Euplokamis, Hormiphora, Lampea, Mertensia, Pleurobrachia, Tinerfe*)

ORDER GANESHIDA: (Figure 9.2C) Pelagic; body form somewhat intermediate between Cydippida and Lobata, compressed in tentacular plane; tentacles branched and with sheaths; interradial canals arise from infundibulum and divide into adradial canals, which join the aboral ends of the meridional canals; meridional canals and paragastric canals join and form a circumoral canal (as in Beroida); mouth large and expanded in tentacular plane; without auricles or oral lobes. One genus, *Ganesha*, with two known species.

ORDER LOBATA: (Figure 9.1D,F–H,9.2J–L) Pelagic; body compressed in tentacular plane; with a pair of characteristic oral

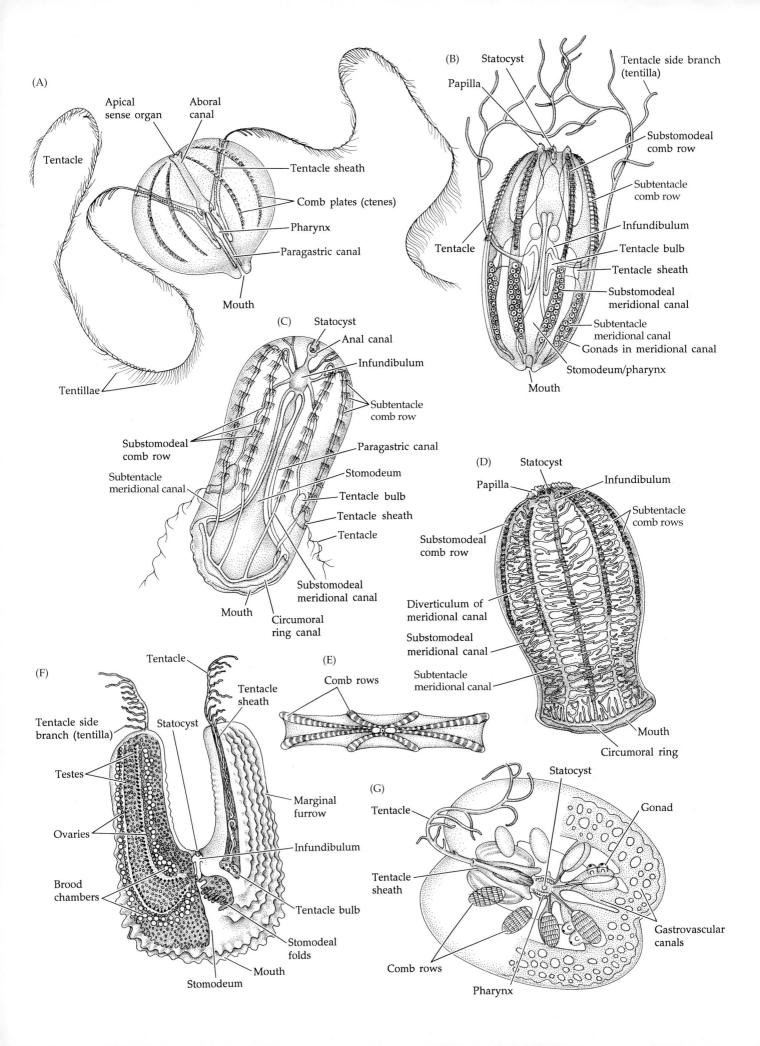

(A)

Tentacle

Apical
sense organ

Aboral
canal

Tentacle sheath

Comb plates (ctenes)

Pharynx

Paragastric canal

Mouth

Tentillae

(B) Statocyst

Papilla

Tentacle side branch
(tentilla)

Substomodeal
comb row

Subtentacle
comb row

Infundibulum

Tentacle bulb

Tentacle sheath

Substomodeal
meridional canal

Subtentacle
meridional canal

Gonads in meridional canal

Stomodeum/pharynx

Tentacle

Mouth

(C) Statocyst

Anal canal

Infundibulum

Subtentacle
comb row

Paragastric canal

Stomodeum

Tentacle bulb

Tentacle sheath

Tentacle

Substomodeal
comb row

Subtentacle
meridional canal

Mouth

Substomodeal
meridional canal

Circumoral
ring canal

(D) Statocyst

Papilla

Infundibulum

Subtentacle
comb rows

Substomodeal
comb row

Diverticulum of
meridional canal

Substomodeal
meridional canal

Subtentacle
meridional canal

Mouth

Circumoral ring

(F)

Tentacle

Tentacle side
branch (tentilla)

Tentacle
sheath

Statocyst

Testes

Ovaries

Marginal
furrow

Brood
chambers

Infundibulum

Tentacle bulb

Stomodeal
folds

Mouth

Stomodeum

(E) Comb rows

(G) Statocyst

Tentacle

Gonad

Tentacle
sheath

Gastrovascular
canals

Comb rows

Pharynx

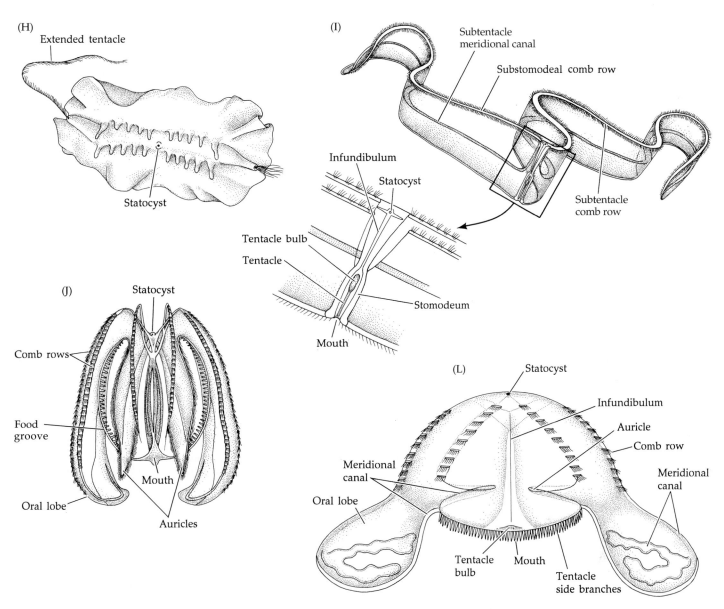

(H) Extended tentacle

Statocyst

(I)

Subtentacle meridional canal

Substomodeal comb row

Infundibulum

Statocyst

Tentacle bulb

Tentacle

Subtentacle comb row

Stomodeum

Mouth

(J)

Statocyst

Comb rows

Food groove

Oral lobe

Mouth

Auricles

(L)

Statocyst

Infundibulum

Auricle

Comb row

Meridional canal

Meridional canal

Oral lobe

Tentacle bulb

Mouth

Tentacle side branches

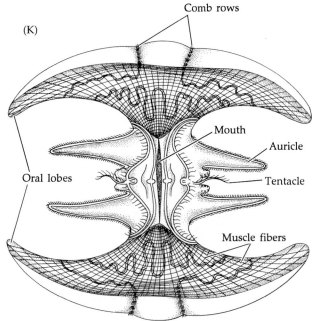

Comb rows

(K)

Mouth

Auricle

Tentacle

Oral lobes

Muscle fibers

Figure 9.2 General anatomy of some major groups of ctenophores; see also Figure 9.3. (A) Order Cydippida, *Pleurobrachia*. The extensive gastrovascular canal system is not shown here completely. (B) Order Cydippida, *Tinerfe*, with gametes developing in the meridional canals. (C) Order Ganeshida, *Ganesha*. Note the circumoral ring canal that connects the meridional and paragastric (pharyngeal) canals. (D,E) Order Beroida, *Beröe*. (D) Side view. The aboral surface has sensory papillae and the meridional canals are branched. (E) Aboral view. Members of this order are extremely compressed on the tentacular plane. (F) Order Platyctenida, the odd-shaped *Lyrocteis*, shown here in layered cutaway view exposing various internal structures. (G) Order Platyctenida, *Ctenoplana* (aboral view). Only one tentacle is shown. (H) Order Platyctenida, *Coeloplana*. This ctenophore is a benthic form. (I) Order Cestida, *Cestum*. This ctenophore exhibits an extreme modification of body form. (J,K) Order Lobata, *Mnemiopsis* (J) Side view. *Mnemiopsis* has oral lobes and auricles. (K) Oral view. Note the greatly expanded oral lobes with their distinctive pattern of muscle fibers. (L) Order Lobata, *Deiopea*.

lobes and four flaplike auricles; a ciliated auricular groove extends to base of auricles from each side of each tentacle base; paragastric and subtentacular meridional canals unite orally. (e.g., *Bolinopsis, Deiopea, Leucothea, Mnemiopsis, Ocyropsis*)

ORDER PLATYCTENIDA: (Figure 9.2F–H) Planktonic or benthic; most species greatly flattened, with part of stomodeum everted as a creeping sole; often with tentacle sheaths; tentacle canals bifid; gastrovascular system complexly anastomosing; most species possess anal pores; many are ectocommensals on other organisms. Unlike most ctenophores, fertilization is often internal, and many platyctenids brood their embryos to the larval stage; asexual reproduction is common. (e.g., *Coeloplana, Ctenoplana, Lyrocteis, Savangia, Tjalfiella*)

ORDER THALASSOCALYCIDA: Pelagic; body extremely fragile, expanded orally into medusa-like bell, to 15 cm along tentacular axis; body slightly compressed in stomodeal plane; tentacle sheaths absent; tentacles arise near mouth and bear lateral filaments; comb rows short; mouth and pharynx borne on central conical peduncle; meridional canals long, describing complex patterns in bell; all meridional canals end blindly aborally. Monotypic: *Thalassocalyce inconstans.*

The Ctenophoran Bauplan

Although ctenophores are among the most primitive living animals, they do possess true tissues. Between the epidermis and the gastrodermis is a well developed middle layer, which is always a cellular mesenchyme. Within this mesenchyme true muscle cells develop, a condition that also characterizes the triploblastic Metazoa.

As we noted in the preceding chapter, a critical essence of the cnidarian and ctenophoran bauplans is radiality; we have explained some of the structural constraints and advantages that derive from this symmetry. Thus, predictably, the nervous system of ctenophores is in the form of a simple, noncentralized nerve net, the locomotor structures are arranged radially about the body, and so on. Other features that characterize the Ctenophora include: retractile tentacles and often tentacle sheaths; anal pores; adhesive prey-capturing structures called colloblasts; locomotor structures called ctenes or comb plates, arranged in comb rows; and an apical sense organ containing a statolith that regulates the activity of the comb rows. The sheathed tentacles, colloblasts, comb plates, and nature of the apical sense organ are unique features of ctenophores.

Most ctenophores are spherical or ovoid in shape, although some species have evolved flattened shapes through compression and elongation in one of the two planes of body symmetry (Figures 9.1 and 9.2). The general body plan can best be understood by first examining a generalized cydippid ctenophore (Figure 9.3A). Most specialists consider the cydippids to be primitive within the phylum. As in cnidarians, the principal axis is oral–aboral. The mouth is at the oral pole; the aboral pole bears the apical sense organ. On the surface of the body are eight equally spaced meridional rows of **comb plates**. Each comb plate, or **ctene**, is composed of a transverse band of long, fused (= compound) cilia. On

each side of the body of many species is a deep, ciliated epidermal pouch, the **tentacle sheath**, from whose inner wall a tentacle arises. The tentacles are typically very long and contractile, and bear lateral branches called filaments, or **tentillae**. The epidermis of both the tentacle and the lateral tentillae is richly armed with colloblasts. Most species can retract the tentacles into the sheaths by muscles. It is the tentacles and certain aspects of the internal anatomy that give ctenophores a biradial symmetry. The elongate stomodeum lies on the oral–aboral axis of the body. It is distinctly flattened in one plane of body symmetry, the **stomodeal plane** (Figure 9.3B). Bisecting the animal along the stomodeal plane separates the two tentacular halves of the body. The second plane of body symmetry, called the **tentacular plane**, is defined by the position of the tentacle sheaths.

Some variations of the basic ctenophoran body plan are illustrated in Figures 9.1 and 9.2. In members of the unique order Lobata (Figure 9.2J–L and opening photo on page 269), the body is compressed in the tentacular plane and the oral end is expanded on each side into rounded, contractile **oral lobes**. The mouth sits on an elongate **manubrium**, the base of which bears four long flaps called **auricles**. The tentacles are reduced and lack sheaths. From either side of each tentacle base, a ciliated **auricular groove** arises and extends to the auricles. Members of the order Cestida are also compressed in the tentacular plane and extremely elongated in the stomodeal plane, giving these ctenophores a striking snake- or ribbon-like appearance. The sheathed tentacles are reduced and shifted alongside the mouth. Beroids are thimble-shaped and also flattened in the tentacular plane. They lack tentacles and sheaths. In the single species of Thalassocalycida (*Thalassocalyce inconstans*), the body is expanded around the mouth to form a medusa-like bell.

The oddest ctenophores are members of the order Platyctenida. Platyctenids are benthic and small, often less than 1 cm in length, and, in contrast to most pelagic ctenophores, they are pigmented rather than transparent. The body is oval and markedly flattened. Despite these unusual features, early naturalists recognized them as ctenophores by the presence of an apical sense organ, comb rows, and a pair of tentacles. Detailed studies have shown that the flattened oral surface is actually an everted portion of the pharynx! The platyctenid pharynx was, in a sense, preadapted to conversion to a creeping foot or sole by its intrinsic musculature. Most of these animals crawl about on the sea bottom, but some are ectocommensals on alcyonarian cnidarians, echinoderms, or pelagic salps.

Support and Locomotion

Ctenophores rely primarily on their elastic mesenchyme for structural support. The watery gelatinous mesenchyme makes up most of the body mass; ctenophore

(A)

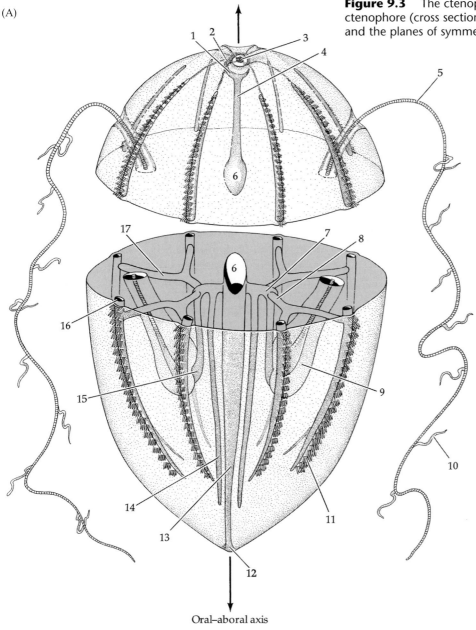

Oral–aboral axis

Figure 9.3 The ctenophoran bauplan. (A) A cydippid ctenophore (cross section). (B) Ctenophoran biradiality and the planes of symmetry (oral view).

Key
1 Anal canal
2 Anal pore
3 Apical sense organ
4 Aboral canal
5 Tentacle
6 Infundibulum
7 Transverse canal
8 Interradial canal
9 Tentacle sheath
10 Tentilla
11 Ctenes of comb row
12 Mouth
13 Pharynx
14 Pharyngeal canal
15 Tentacle canal
16 Meridional canal
17 Adradial canal

(B)

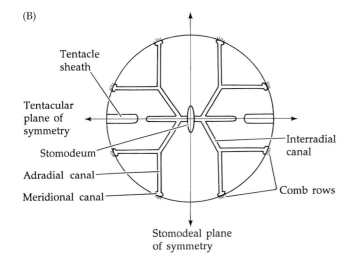

dry weights are only about 4 percent of their live wet weights. The mesenchyme contains both elastic supportive cells and muscle cells, the general tonus of the latter being primarily responsible for maintaining body shape. Figure 9.4 shows a highly stylized cutaway section of a cydippid ctenophore and illustrates the arrangement of the supportive mesenchymal muscle fibers. Tension in the looped muscles tends to maintain the spherical geometry. Action of the radial muscles diminishes the radius and hence the circumference, and also serves to open the pharynx. These two muscle sets work antagonistically to one another.

Most ctenophores are pelagic. The gelatinous body and low specific gravity maintain a relatively neutral buoyancy, allowing these creatures to float about with the ocean currents. Neutral buoyancy appears to be maintained by passive osmotic accommodation. Be-

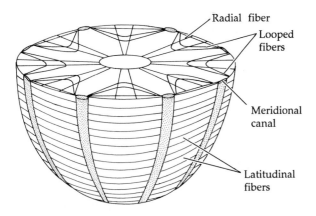

Figure 9.4 Stereogram of the arrangement of muscle fibers in *Pleurobrachia*, a cydippid ctenophore. The diagram depicts a transverse section through the region of the pharynx; the gastrovascular system and tentacle sheaths have been omitted for clarity.

cause buoyancy adjustments take time, ctenophores may temporarily accumulate at discontinuity layers in the sea, where a water mass of one density overlies a water mass of a slightly different density.

The beating of the ctenes provides most of the modest locomotor power that allows ctenophores to move up and down in the water column and to locate richer feeding grounds or preferred environmental conditions. Each comb row comprises many ctenes. Each ctene consists of a transverse band of hundreds of very long, partly fused cilia (to 3.5 mm in length) that beat together as a unit. Ctenophores are the largest animals known to use

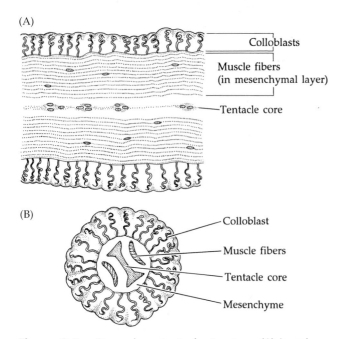

Figure 9.5 Ctenophore tentacle structure. (A) Longitudinal section of tentacle. (B) Cross section of a lateral filament (tentilla) of a tentacle.

cilia for locomotion. Each cilium has a typical 9+2 microtubule structure, but each also possesses a unique set of lamellae at the 3 and 8 doublets; these lamellae protrude to link together the adjacent cilia. Because of their size, ctenophore ctene plates move at low Reynolds numbers (where the flow is laminar), while the entire animal moves at high Reynolds numbers (where turbulent flow is dominant).

The mesenchymal musculature is used to maintain body shape and assist in feeding; it is involved in behaviors such as prey swallowing, pharyngeal contractions, and tentacle movements. Usually both longitudinal and circular muscles are present just beneath the epidermis. In the benthic and epifaunal platyctenids, stomodeal musculature facilitates a creeping locomotion. In the snakelike cestids, body muscles may generate graceful swimming undulations. The lobate ctenophores swim by muscular flapping of their two oral lobes, and perhaps also by use of the four paddle-like auricles. The lobate species *Leucothea* can swim either by typical slow ctene propulsion or by rapid ctene propulsion; the latter is accomplished by an increased ciliary beat that produces a vortex wake, resulting in jet propulsion. Giant smooth muscle fibers—the first to be discovered in ctenophores—have been found in *Beröe*.

Feeding and Digestion

Comb jellies are, so far as is known, entirely predatory in their habits. The long tentacles of cydippids (and of the larvae of most other forms) have a muscular core with a colloblast-laden epidermal covering (Figure 9.5). The tentacles trail passively or are "fished" by various swirling movements of the body. Upon contact with zooplankton prey, the colloblasts (sometimes called **lasso cells**) burst and discharge a strong adhesive material. Each colloblast develops from a single cell and consists of a hemispherical mass of secretory granules attached to the muscular core of the tentacle by a spiral filament coiled around a straight filament (Figure 9.6). The straight filament is actually the highly modified nucleus of the colloblast cell. The spiral filament, which uncoils upon discharge, adheres to the prey by the sticky material produced in the secretory granules. As the tentacles accumulate prey, they are periodically wiped across the mouth by muscular contractions, occasionally combined with a coordinated somersaulting action of the animal that brings the mouth to the trailing tentacle. In members of the orders Lobata and Cestida, which bear very short tentacles, small zooplankton are trapped in mucus on the body surface and then carried to the mouth by ciliary currents (along the ciliated **auricular grooves** in lobate forms and ciliated **oral grooves** in cestids). Most of the benthic platyctenids also feed by capturing zooplankton in a somewhat similar fashion. In some areas of the world's seas, ctenophores may be the dominant macrozooplankters and planktonic predators (e.g., *Mertensia ovum* in the Arctic region).

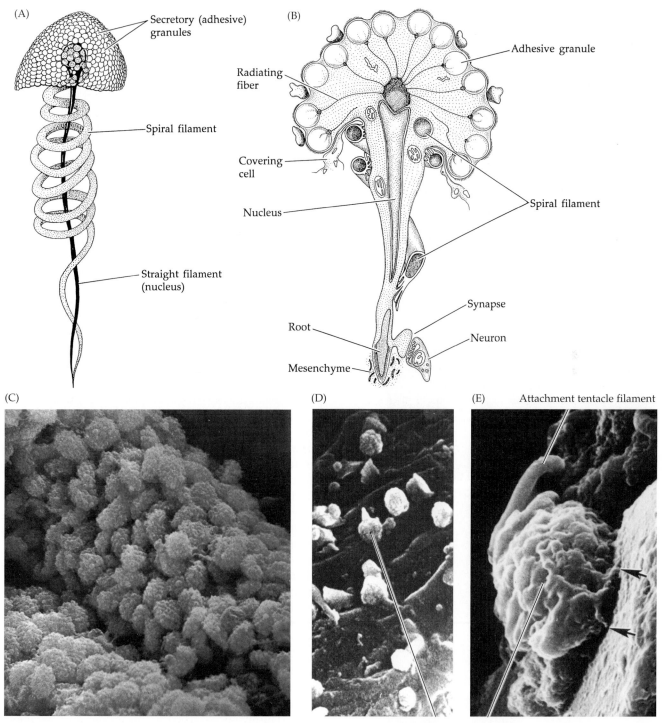

Figure 9.6 Colloblasts. (A) The functional parts of a colloblast. (B) Longitudinal section. (C) Colloblasts on the lateral tentacle filaments (tentillae) of *Pleurobrachia* (SEM). (D) Fired colloblasts of *Pleurobrachia*, showing adhesive granules attached to fragments of a copepod (small crustacean). (E) Fired colloblasts are still attached to the tentacle filament. The adhesive ends of the coiled filaments are stuck (arrows) to a bit of copepod.

Some ctenophores prey upon larger animals, especially gelatinous forms. The cydippid *Lampea* (formerly *Gastrodes*), for example, lives embedded in the body of pelagic tunicates of the genus *Salpa*, upon which it feeds. Figure 9.7 is a series of remarkable photographs showing the cydippid ctenophore *Haeckelia* eating the tentacles of the trachyline hydromedusa *Aegina*. After consuming the tentacles one by one, *Haeckelia* retains the prey's unfired nematocysts, incorporates them into its epidermis, and uses them for its own defense. This phenomenon,

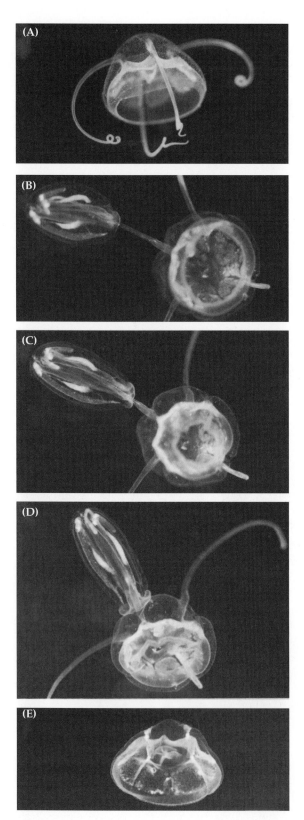

Figure 9.7 The cydippid ctenophore *Haeckelia rubra* (= *Euchlora rubra*) feeding on the trachyline hydromedusa *Aegina citrea*. (A) Intact specimen of *Aegina*, with all four tentacles present. (B) *Haeckelia* begins to consume one of *Aegina's* tentacles. (C) Most of the first tentacle of the medusa has been ingested. (D) Same animals, 2 minutes after feeding began. (E) *Aegina* has lost all four of its tentacles to a hungry *Haeckelia*.

known as kleptocnidae (see Chapter 8), occurs in several unrelated groups who prey on cnidarians.

Ctenophores were center stage in an ecological drama that recently played out in the Black Sea. In the 1980s, the predatory northwest Atlantic ctenophore *Mnemiopsis leidyi* was accidentally introduced into the Black Sea by way of ship ballast water. It quickly underwent an explosive population growth, reaching biomass levels in excess of one kilogram per cubic meter by 1989, devastating the food web of the entire Black Sea Basin and causing a collapse of the anchovy fishery (one of the favored prey of *M. leidyi*). Then, in 1997, another ctenophore, *Beröe ovata*, was accidentally introduced to the Black Sea, probably also from ballast water. *Beröe ovata* feeds almost exclusively on *Mnemiopsis*, and its introduction resulted in a precipitous decline (perhaps extirpation) of *M. leidyi* in the lake, followed by the disappearance of *Beröe* itself.

The ctenophoran mouth opens into an elongate, highly folded, flattened, muscular, stomodeal pharynx. The epithelium of the pharynx is richly endowed with gland cells that produce the digestive juices. Large food items are tumbled within the pharynx by ciliary action. Digestion takes place extracellularly, mostly in the pharynx. The largely digested food passes via a small chamber (the **infundibulum**, **funnel**, or **stomach**) from the pharynx into a complex system of radiating gastrovascular canals (Figures 9.2 and 9.3). The details of the arrangement of the canals vary among different groups; the following description applies to the arrangement in a cydippid.

Two **paragastric** or **pharyngeal canals** recurve and lie parallel to the pharynx. Two **transverse canals** depart at right angles to the stomodeal plane and divide into three more branches. The middle branch of each triplet, the **tentacle canal**, leads to the base of the tentacle sheath. Each of the other two branches (the **interradial canals**) bifurcates to form a total of four **adradial canals** on each side of the animal. These in turn connect to the eight **meridional canals**, one beneath each comb row. Finally, an **aboral canal** passes from the infundibulum to the aboral pole, where it divides beneath the apical sense organ into four short canals, two ending blindly and two (the **anal canals**) opening to the outside via small **anal pores**. The anal pores serve as a primitive anus, assisting the mouth in the voiding of indigestible wastes. They may also serve as an exit for metabolic wastes.

Within this very complicated gastrovascular canal system, digestion is completed, nutrients are distributed through the body, and absorption takes place. Minute pores lead from the various canals into the mesenchyme (Figure 9.8). Surrounding these pores are circlets of ciliated gastrodermal cells called **cell rosettes**, which appear to regulate the flow of the digestive soup and perhaps also play a role in excretion. Except for the stomodeal pharynx, the gastrovascular system is lined by a simple epithelium of entodermal origin.

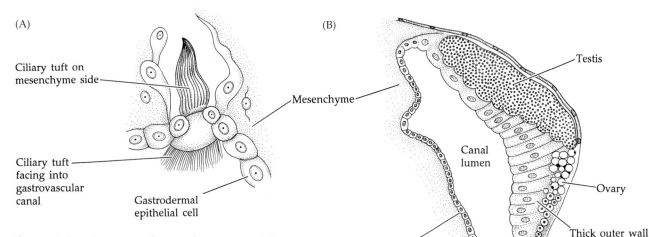

Figure 9.8 Gastrovascular canal structures. (A) A rosette cell from the gastrovascular canal of *Coeloplana*. (B) A meridional canal (in section) of a ctenophore. The gonads are strips of cells in the outer wall of the meridional canal.

Circulation, Excretion, Gas Exchange, and Osmoregulation

There is no independent circulatory system in ctenophores; as in cnidarians, the gastrovascular canal system serves in this role by distributing nutrients to most parts of the body. The gastrovascular system probably also picks up metabolic wastes from the mesenchyme for eventual expulsion out of the mouth or anal pores. The cell rosettes may also transport wastes to the gut. Gas exchange occurs across the general body surface and across the walls of the gastrovascular system. All of these activities are augmented by diffusion through the gelatinous mesenchyme. Movement of water over the body surface is enhanced by the beating of the comb plates. Thus, the extensive canal system and the ciliary bands help to overcome the problem of long diffusion distances.

Nervous System and Sense Organs

Although the nervous systems of both ctenophores and cnidarians are noncentralized nerve nets, there are certain important differences. In a ctenophore, nonpolar neurons form a diffuse subepidermal plexus. Beneath the comb rows, the neurons form elongate plexes or meshes such that they produce nervelike strands. The bases of the ctenes are thus in contact with a rich array of nerve cells. A similar concentrated plexus surrounds the mouth. However, as in cnidarians, no true ganglia occur, a condition that contrasts markedly with the presence of a centralized nervous system in bilateral Metazoa.

The apical sense organ is a statolith that functions in balance and orientation. The calcareous statolith is supported by four long tufts of cilia called **balancers** (Figure 9.9). The whole structure is enclosed in a transparent dome that is apparently derived from cilia. From each balancer arises a pair of **ciliated furrows** (= **ciliated grooves**), each of which connects with one comb row. Thus, each balancer innervates the two comb rows of its particular quadrant. Tilting the animal causes the statolith to press more heavily on the downside balancers, and the resulting stimulus elicits a vigorous beating of the corresponding comb rows to right the body.

The two comb rows in each quadrant innervated by a single ciliated furrow beat synchronously. If a ciliated furrow is cut, the beating of the two corresponding comb rows becomes asynchronous. The normal direction of ciliary power strokes is toward the aboral pole, so that the animal is driven forward oral end first. The beat in each row, however, begins at the aboral end of the comb row and proceeds in metachronal waves toward the oral end (i.e., antiplectic metachrony). Stimulation of the oral end reverses the direction of both the wave and the power stroke. Removal of the apical sense organ or statolith results in an overall lack of coordination of the comb rows, and the injured ctenophore loses its ability to maintain a vertical position. The comb rows are very sensitive to contact; when a comb row is touched, many species retract it into a groove formed in the jelly-like body.

In cydippids and beroids, the stimulation for any given ctene to beat is triggered mechanically, by hydrodynamic forces arising from the movements of the preceding plate. However, in the lobate ctenophores, the ctenes are not coordinated in this mechanical fashion. In these animals a narrow tract of shorter cilia—the **interplate ciliated groove**—runs between successive ctenes and is responsible for coordinating their activity. It is not known how the cilia of the groove are coordinated or how the grooves stimulate the appropriate comb row, so these actions may also be mechanical. The interplate ciliated grooves develop only as the lobate ctenophores mature to adulthood; the free-swimming larvae resemble cydippids and lack the grooves.

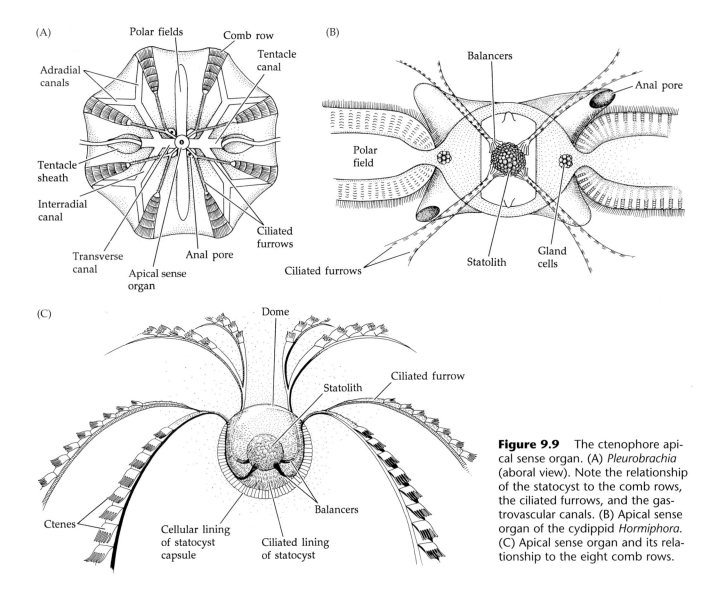

Figure 9.9 The ctenophore apical sense organ. (A) *Pleurobrachia* (aboral view). Note the relationship of the statocyst to the comb rows, the ciliated furrows, and the gastrovascular canals. (B) Apical sense organ of the cydippid *Hormiphora*. (C) Apical sense organ and its relationship to the eight comb rows.

In some ctenophores, two oval tracts of cilia called **polar fields** lie on the stomodeal plane of the aboral surface (Figure 9.9B). These structures are presumed to be sensory in function.

Reproduction and Development

Asexual reproduction and regeneration. Ctenophores can regenerate virtually any lost part, including the apical sense organ. Entire quadrants and even whole halves will regenerate. Speculation that ctenophores may reproduce by fission or budding is still under investigation. Platyctenids reproduce asexually by a process that resembles pedal laceration in sea anemones; small fragments break free as the animal crawls, and each piece can regenerate into a complete adult.

Sexual reproduction and development. Most ctenophores are hermaphroditic, but a few gonochoristic species are known (e.g., members of the genus *Ocyropsis*). The gonads arise on the walls of the meridional canals (Figure 9.8). Pelagic ctenophores generally shed their gametes via the mouth into the surrounding sea water, where either self-fertilization or cross-fertilization takes place. Special sperm ducts occur in at least some platyctenid species. The eggs are centrolecithal and formed in association with nurse cell complexes. Polyspermy is common. Those that free-spawn typically produce embryos that grow quickly to planktotrophic **cydippid larvae**, although species in the order Beroida lack this larval phase (Figure 9.10). Development is thus indirect, although growth to the adult is gradual rather than metamorphic. In the benthic *Coeloplana* and *Tjalfiella*, fertilization is internal and embryos are brooded until a cydippid larva is formed and released. This mixed life history provides a means of dispersal for these benthic, sedentary animals.

Ctenophoran cleavage cannot easily be classified as either spiral or radial. During early cleavage, the first four blastomeres arise by the usual two meridional cleavages, which mark the adult planes of symmetry. The third division is also nearly vertical and results in a

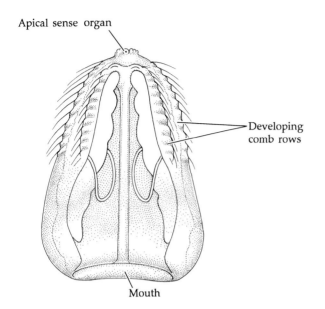

Figure 9.10   A typical young cydippid larva.

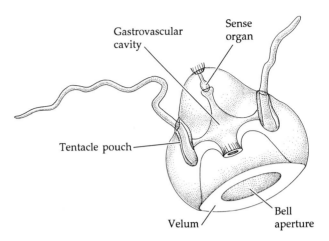

Figure 9.11   The aberrant cnidarian trachyline medusa *Hydroctena*, which superficially resembles a ctenophore in its possession of an apical sensory structure and tentacle pouches.

curved plate of eight cells (macromeres). The next division is latitudinal and unequal, giving rise to micromeres on the concave side of the macromere plate. The micromeres continue to divide and spread by epiboly over the aboral pole and eventually over the macromeres. The latter also invaginate into the interior, so the gastrula arises through a combination of epiboly and invagination. Thus, the micromeres become ectoderm and the macromeres become entoderm. Just prior to gastrulation, the macromeres divide and produce additional micromeres on the oral side of the embryo. Whereas the aboral micromeres become ectoderm, these oral micromeres are incorporated into the entoderm and, in at least some species, give rise to photoreceptor cells. There is some question about the fate of all of these oral micromeres. Metschnikoff (1885) suggested that these cells may contribute to the mesenchyme and may thus be viewed as true entomesoderm. More recently, Harbison (1985) also made a case for a triploblastic condition in ctenophores.

As the micromeres cover the embryo to form the epidermis, four interradial bands of small, rapidly dividing cells become apparent. Eventually, each of these thickened ectodermal bands differentiates into two of the comb rows. The aboral ectoderm differentiates into the apical sense organ and its related parts; the oral ectoderm invaginates to form the stomodeum. The gastrovascular system develops from entodermal outgrowths and the tentacle sheaths arise as ectodermal invaginations from the points where the tentacles sprout. The embryo eventually develops into a free-swimming cydippid larva (Figure 9.10) that closely resembles adult ctenophores of the order Cydippida. Some authors have taken this as evidence that Cydippida is the most primitive of the extant ctenophore orders.

The development of ctenophores differs markedly from that of cnidarians. In the latter group, early cleavage results in an irregular mass of cells whose fates are not clearly predictable until later development, and the mesenchyme is strictly ectodermal in origin. In the ctenophores, on the other hand, development is determinate and a very precise cleavage pattern unfolds, in which the ultimate morphology is definitely mapped. In fact, if the two blastomeres of a 2-cell embryo are experimentally separated, the "half-embryos" develop into adults with exactly half the normal set of adult structures. Furthermore, ctenophores lack the planula larva that characterizes cnidarians; instead, they produce a cydippid larval type having no obvious counterpart among the cnidarians.*

Ctenophoran Phylogeny

Although the ctenophores and cnidarians are widely regarded as belonging to the same general grade of construction, it is difficult to derive ctenophores from any existing cnidarian group. Some zoologists suggest that ctenophores arose from the hydrozoans, by way of an intermediate medusa possessing an aboral statocyst and two tentacle sheaths, such as is seen today in the aberrant trachyline medusa *Hydroctena* (Figure 9.11). In this medusa, the number of tentacles has been reduced to two, and these are set high on the bell, like the tentacles of trachylines in general. The tentacles also arise from deep epidermal pockets that resemble the tentacle

*Komai (1922, 1963) reported the presence of a brief "planula" stage in the development of the parasitic cydippid *Gastrodes parasiticum*, which was said to burrow into the test of host salps, where it then developed into a free-swimming cydippid. This has not been confirmed by any subsequent workers and the nature of Komai's ctenophoran "planula" remains unsettled.

sheaths of ctenophores. Furthermore, *Hydroctena* has a single apical sense organ, although its construction differs from that of ctenophores. Several other trachyline medusae also have solitary aboral sense organs. Although these similarities may suggest a relationship between ctenophores and trachyline cnidarians, ctenophores also show certain similarities to the scyphozoans and anthozoans, such as the stomodeum and the highly cellular mesenchyme, and the four-lobed gastrovascular cavity of the cydippid larva. As we have seen, however, ctenophores are really quite different from cnidarians in many fundamental ways. These differences are evident both in adult morphology and in patterns of development. Many of the similarities between ctenophores and medusae may well be convergences reflecting adaptations to their similar lifestyles and, in fact, many gelatinous zooplankters show superficial similarities in body form and construction.

The presence of mesenchymal muscle cells and gonoducts in some species, along with certain features of early cleavage, have led some zoologists to suggest a relationship between the ctenophores and the flatworms (Platyhelminthes; Chapter 10). Some workers view the ctenophores as ancestral to the flatworms; but a reverse scenario has also been suggested. The presence of ben-thic, crawling ctenophores (e.g., *Ctenoplana* and *Coeloplana*) is used as evidence that ecological and anatomical intermediates between the two groups are plausible. Harbison (1985) reviews these matters and concludes that there is no more evidence to link the ctenophores to the flatworms than to the cnidarians, and this is also the conclusion indicated by our phylogenetic analysis in Chapter 24.

There is also disagreement about evolution within the Ctenophora, centering largely on whether the tentaculate or atentaculate condition is primitive, or whether the atentaculate lineage (the Beroida) arose from somewhere among the tentaculate groups. Without a clearer picture of the origin of the phylum, it is difficult to resolve such questions. The known fossil record offers virtually no help in these matters, as it consists of only two questionable records, one from the Devonian and the other from the mid-Cambrian. On this sparse and controversial information, we can only hypothesize that the ctenophores are a monophyletic group that probably arose early in the evolution of the metazoan condition. The present classification of ctenophores is not based on phylogenetic analysis, and more information is needed to understand the relationships among the ctenophoran orders.

Selected References

Abbott, J. F. 1907. Morphology of *Coeloplana*. Zool. Jahrb. Abt. Anat. Ontog. Tiere 24.

Arai, M. N. 1976. Behavior of planktonic coelenterates in temperature and salinity discontinuity layers. *In* G. O. Mackie (ed.), *Coelenterate Ecology and Behavior*. Plenum, New York, pp. 211–218.

Bigelow, H. B. 1912. Reports on the scientific results of the expedition to the eastern tropical Pacific, in charge of Alexander Agassiz, by the U.S. Fish Commission Steamer *Albatross*, from October, 1904, to March 1905, Lieutenant Commander L. M. Garrett, U.S.N., commanding. XXVI. The Ctenophores. Bull. Mus. Comp. Zool. Harvard College 54(12): 369–404.

Carré, C. and D. Carré. 1980. Les cnidocysts du ctenophore *Euchlora rubra* (Kolliker 1853). Cah. Biol. Mar. 21: 221–226.

Carré, D., C. Rouvière and C. Sardet. 1991. *In vitro* fertilization in ctenophores: Sperm entry, mitosis, and the establishment of bilateral symmetry in *Beroe ovata*. Dev. Biol. 147: 381–391.

Coonfield, B. R. 1936. Regeneration in *Mnemiopsis*. Biol. Bull. 71.

Dawydoff, C. 1963. Morphologie et biologie des *Ctenoplana*. Arch. Zool. Exp. Gen. 75.

Farfaglio, G. 1963. Experiments on the formation of the ciliated plates in ctenophores. Acta Embryol. Morphol. Exp. 6: 191–203.

Franc, J.-M. 1978. Organization and function of ctenophore colloblasts: An ultrastructural study. Biol. Bull. 155: 527–541.

Freeman, G. 1976. The effects of altering the position of cleavage planes on the process of localization of developmental potential in ctenophores. Dev. Biol. 51: 332–337.

Freeman, G. 1977. The establishment of the oral–aboral axis in the ctenophore embryo. J. Embryol. Exp. Morphol 42: 237–260.

Harbison, G. R. 1985. On the classification and evolution of the Ctenophora. *In* Morris et al. (eds.), *The Origins and Relationships of Lower Invertebrates*. Syst. Assoc. Spec. Vol. No. 28, pp. 78–100.

Harbison, G. R. and L. P. Madin. 1979. A new view of plankton biology. Oceanus 22(2): 18–27.

Harbison, G. R. and L. P. Madin. 1983. Ctenophora. *In* S. P. Parker (ed.), *Synopsis and Classification of Living Organisms*, Vol. 1. McGraw-Hill, New York, pp. 707–715.

Harbison, G. R., L. P. Madin and N. R. Swanberg. 1984. On the natural history and distribution of oceanic ctenophores. Deep-Sea Res. 25: 233–256.

Harbison, G. R. and R. L. Miller. 1986. Not all ctenophores are hermaphrodites. Studies on the systematics, distribution, sexuality and development of two species of *Ocyropsis*. Mar. Biol. 90: 413–424.

Hernandez, M.-L. 1991. Ctenophora. *In* F. Harrison and J. Westfall (eds.), *Microscopic Anatomy of Invertebrates*, Vol. 2, Placozoa, Porifera, Cnidaria and Ctenophora. Wiley-Liss, New York, pp. 359–418.

Horridge, G. A. 1965a. Macrocilia with numerous shafts from the lips of the ctenophore *Beroe*. Proc. Roy. Soc. London B 162: 351–364.

Horridge, G. A. 1965b. Relations between nerves and cilia in ctenophores. Am. Zool. 5: 357–375.

Horridge, G. A. 1974. Recent studies on the Ctenophora. *In* L. Muscatine and H. M. Lenhoff (eds.), *Coelenterate Biology*. Academic Press, New York, pp. 439–468.

Hyman, L. H. 1940. *The Invertebrates*, Vol. 1, Protozoa through Ctenophora. McGraw-Hill, New York, pp. 662–696.

Komai, T. 1922. Studies on two aberrant ctenophores—*Coeloplana* and *Gastrodes*. Kyoto. [Published by the author.]

Komai, T. 1934. On the structure of *Ctenoplana*. Kyoto Univ. Col. Sci. Mem. (Ser. B) 9.

Komai, T. 1936. Nervous system, *Coeloplana*. Kyoto Univ. Col. Sci. Mem. (Ser. B) 11.

Komai, T. 1963. A note on the phylogeny of the Ctenophora. *In* E. C. Dougherty (ed.), *The Lower Metazoa: Comparative Biology and Phylogeny*. University of California Press, Berkeley, pp. 181–188.

Komai, T. and T. Tokioka. 1940. *Kiyohimea aurita* n. gen., n. sp., type of a new family of lobate Ctenophora. Annot. Zool. Japan. 19: 43–46.

Komai, T. and T. Tokioka. 1942. Three remarkable ctenophores from the Japanese seas. Annot. Zool. Japan. 21: 144–151.

Kremer, P. 1977. Respiration and excretion by the ctenophore *Mnemiopsis leidyi*. Mar. Biol. 44: 43–50.

Kremer, P., M. F. Canino and R. W. Gilmer. 1986. Metabolism of epipelagic tropical ctenophores. Mar. Biol. 90: 403–412.

Kremer, P., M. R. Reeve and M. A. Syms. 1986. The nutritional ecology of the ctenophore *Bolinopsis vitrea*: Comparisons with *Mnemiopsis mccradyi* from the same region. J. Plankton Res. 8: 1197–1208.

Krumbach, T. 1925. Ctenophora. *In* W. Kukenthal and T. Krumbach (eds.), *Handbuch des Zoologie* 1: 905–995.

Mackie, G. O., C. E. Mills and C. L. Singla. 1988. Structure and function of the prehensile tentilla of *Euplokamis* (Ctenophora, Cydippida). Zoomorphology 107: 319–337.

Madin, L. P. and G. R. Harbison. 1978a. *Bathocyroe fosteri* gen. et sp. nov., a mesopelagic ctenophore observed and collected from a submersible. J. Mar. Biol. Assoc. U. K. 58: 559–564.

Madin, L. P. and G. R. Harbison. 1978b. *Thalassocalyce inconstans*, new genus and species, an enigmatic ctenophore representing a new family and order. Bull. Mar. Sci. 28(4): 680–687.

Main, R. J. 1928. Observations on the feeding mechanism of a ctenophore, *Mnemiopsis leidyi*. Biol. Bull. 55: 69–78.

Martindale, M. Q. 1987. Larval reproduction in the ctenophore *Mnemiopsis mccradyi* (order Lobata). Mar. Biol. 94: 409–414.

Martindale, M. Q. and J. Q. Henry. 1995. Diagonal development: Establishment of the anal axis in the ctenophore *Mnemiopsis leidyi*. Biol. Bull. 189: 190–192.

Martindale, M. Q. and J. Q. Henry. 1997. Ctenophorans, the comb jellies. *In* S. F. Gilbert and A. M. Raunio (eds.), *Embryology: Constructing the Organism*. Sinauer Associates, Sunderland, MA. pp. 87–111.

Martindale, M. Q. and J. Q. Henry. 1996. Development and regeneration of comb plates in the ctenophore *Mnemiopsis leidyi*. Biol. Bull. 191: 290–292.

Matsumoto, G. I. 1991. Functional morphology and locomotion of the Arctic ctenophore *Mertensia ovum* (Fabricius) (Tentaculata: Cydippida). Sarsia 76: 177–185.

Matsumoto, G. I. and W. M. Hamner. 1988. Modes of water manipulation by the lobate ctenophore *Leucothea* sp. Mar. Biol. 97: 551–558.

Mayer, A. G. 1912. *Ctenophores of the Atlantic Coast of North America*. Carnegie Inst. Washington Publ. No. 162.

Metschnikoff, E. 1885. Gastrulation und mesodermbildung der Ctenophoren. Z. Wiss. Zool. 42.

Mills, C. E. 1984. Density is altered in hydromedusae and ctenophores in response to changes in salinity. Biol. Bull. 166: 206–215.

Mills, C. E. 1987. Revised classification of the genus *Euplokamis* Chun, 1880 (Ctenophora: Cydippida: Euplokamidae n. fam.) with a description of the new species *Euplokamis dunlapae*. Can. J. Zool. 65: 2661–2668.

Mills, C. E. and R. L. Miller. 1984. Ingestion of a medusa (*Aegina citrea*) by the nematocyst-containing ctenophore *Haeckelia rubra* (formerly *Euchlora rubra*): Phylogenetic implication. Mar. Biol. 78: 215–221.

Mills, C. E. and R. G. Vogt. 1984. Evidence that ion regulation in hydromedusae and ctenophores does not facilitate vertical migration. Biol. Bull. 166: 216–227.

Mortensen, T. 1912. Ctenophora. Danish Ingolf-Expedition Vol. 5, Pt. 2: 1–95. Zoological Museum, University of Copenhagen.

Mortensen, T. 1913. Regeneration in ctenophores. Vidensk. Medd. Dan. Naturhist. Foren. Khobenhavn 66.

Nielsen, C. 1987. *Haeckelia* (= *Euchlora*) and *Hydroctena* and the phylogenetic interrelationships of the Cnidaria and Ctenophora. Z. Zool. Syst. Evolutionsforsch. 25: 9–12.

Ortolani, G. 1989. The ctenophores: A review. Acta. Embryol. Morphol. Exp. 10: 13–31.

Pianka, H. D. 1974. Ctenophora. *In* A. C. Giese and J. S. Pearse (eds.), *Reproduction of Marine Invertebrates*, Vol. 1. Academic Press, New York, pp. 201–265.

Picard, J. 1955. Les nematocystes du ctenaire *Euchlora rubra* (Kolliker, 1953). Recl. Trav. Stn. Mar. Endoume-Marseille Fasc. Hors. Ser. Suppl. 15: 99–103.

Rankin, J. J. 1956. The structure and biology of *Vallicula multiformis* gen. et sp. nov. a platyctenid ctenophore. Zool. J. Linn. Soc. 43: 55–71.

Reeve, M. R. and L. D. Baker. 1974. Production of two planktonic carnivores (chaetognath and ctenophore) in South Florida inshore waters. Fish. Bull. U.S. 73: 238–248.

Reeve, M. R. and M. A. Walter. 1976. A large-scale experiment on the growth and predation of ctenophore populations. *In* G. Mackie (ed.), *Coelenterate Ecology and Behavior*. Plenum, New York, pp. 187–199.

Reeve, M. R. and M. A. Walter. 1978. Nutritional ecology of ctenophores: A review of recent research. *In* F. S. Russell and M. Yonge (eds.), *Advances in Marine Ecology*, Vol. 15. Academic Press, New York, pp. 249–289.

Reeve, M. R., M. A. Walter and T. Ikeda. 1978. Laboratory studies of ingestion and food utilization in lobate and tentaculate ctenophores. Limnol. Oceanogr. 23: 740–751.

Robilliard, G. A. and P. K. Dayton. 1972. A new species of platyctenean ctenophore, *Lyrocteis flavopallidus* sp. nov., from McMurdo Sound, Antarctica. Can. J. Zool. 50: 47–52.

Siewing, R. 1977. Mesoderm bei Ctenophoren. Z. Zool. Syst. Evolutionsforsch. 15 (1): 1–8.

Stanlaw, K. A., M. R. Reeve and M. A. Walter. 1981. Growth rates, growth variability, daily rations, food size selection and vulnerability to damage by copepods of the early life history stages of the ctenophore *Mnemiopsis mccradyi*. Limnol. Oceanogr. 26: 224–234.

Stanley, G. D., Jr. and A. Sturmer. 1983. The first fossil ctenophore from the lower Devonian of West Germany. Nature 303: 518–520.

Stretch, J. J. 1982. Observations on the abundance and feeding behavior of the cestid ctenophore, *Velamen parallelum*. Bull. Mar. Sci. 32(3): 796 799.

Sullivan, B. K. and M. R. Reeve. 1982. Comparison of estimates of the predatory impact of ctenophores by two independent techniques. Mar. Biol. 68: 61–65.

Tamm, S. L. 1973. Mechanisms of ciliary coordination in ctenophores. J. Exp. Biol. 59: 231–245.

Tamm, S. L. 1980. Ctenophores. *In* G. A. B. Shelton (ed.), *Electrical Conduction and Behavior in Invertebrates*. Oxford University Press, New York.

Tamm, S. L. and S. Tamm. 1985. Visualization of changes in ciliary tip configuration caused by sliding displacement of microtubules in macrocilia of the ctenophore *Beröe*. J. Cell Sci. 79: 161–179.

Totton, A. 1954. Egg-laying in Ctenophora. Nature 174: 360.

Wiley, A. 1896. *Ctenoplana*. Q. J. Microsc. Sci. 39.

10 *Phylum Platyhelminthes*

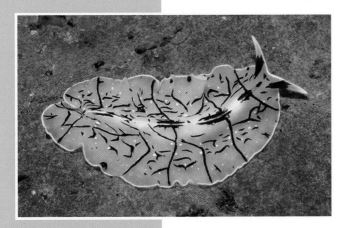

Identification of these animals requires microscopic examination of the reproductive system from thin slices (sections) prepared according to procedures somewhat reminiscent of alchemy.

J. W. Hedgpeth,
Introduction to Seashore Life, 1962

*T*he phylum Platyhelminthes (Greek *platy*, "flat"; *helminth*, "worm") includes about 20,000 extant species of free-living and parasitic worms. These animals are at a grade of complexity that may be called the triploblastic acoelomate bilateria. Platyhelminths display a variety of body forms (Figure 10.1) and are successful inhabitants of a wide range of environments. The majority of flatworms are parasitic members of the classes Trematoda and Monogenea (the flukes) and Cestoda (the tapeworms). The class Turbellaria includes primarily free-living forms in marine and freshwater benthic habitats; a few are terrestrial and some are symbiotic in or on other invertebrates. Marine turbellarians are often the most colorful and graceful creatures found in tidepools. As their name suggests, most flatworms are strikingly flattened dorsoventrally, although the body shape varies from broadly oval to elongate and ribbonlike; a few bear short tentacles at the anterior end or have other elaborations of the body surface. The free-living forms range from less than 1 mm to about 30 cm long, although most are 1–3 cm. long. The largest of all flatworms are certain tapeworms that attain lengths of several meters.

The combined features of the platyhelminths represent a suite of attributes marking major advancements in the evolution of the Metazoa (Box 10A), although some recent work suggests that these animals might have had a coelomate ancestry (see later section on phylogeny). Coupled with a third germ layer (mesoderm), bilateral symmetry, and cephalization are some sophisticated organs and organ systems and a trend toward centralization of the nervous system. The solid (acoelomate) bauplan usually includes a relatively dense mesenchyme (parenchyme) between the gut and the body wall. The mesenchyme is not homogeneous, but comprises a multitude of differentiated cell types and small lacunae.

Within the mesenchyme of most flatworms are discrete excretory/osmoregulatory structures. These structures, **protonephridia**, are found in a number of invertebrate taxa, especially among protostomes. Most flatworms possess complex reproductive systems and an incomplete yet complex gut with a single opening serving for both ingestion and egestion. The mouth leads to a pharynx of varying complexity and thence to a blind intestine. The digestive area contains no permanent cavity in the turbellarian order Acoela, and the gut is entirely lacking in tapeworms.

Taxonomic History and Classification

In his first edition of *Systema Naturae* (1735), Linnaeus established two phyla to encompass all of the known invertebrates. To one he assigned the insects and to the other the rest of the invertebrates. Linnaeus called this latter taxon Vermes (Greek, "worms"). By the thirteenth edition of *Systema Naturae* (1788), the various groups of flatworms were placed together in the order Intestina. During the early 1800s, several biologists, including Lamarck and Cuvier, questioned and rejected the concept of the phylum Vermes, although the taxon continued to surface from time to time and actually persisted into the twentieth century as a dumping ground for almost any creatures with wormlike bodies (and many that were not so wormlike).

During the nineteenth century, the flatworms were eventually separated from most other groups of worms and wormlike creatures. In 1851, Vogt isolated the flatworms and the nemerteans as a single taxon, which he called the Platyelmia, a name changed to Platyelminthes by Gegenbaur in 1859. (Unfortunately, Gegenbaur also resurrected the phylum Vermes.) Gegenbaur's Platyelminthes (now Platyhelminthes) was eventually raised to the rank of phylum, comprising four classes: Turbellaria, Nemertea, Trematoda, and Cestoda. In 1876 Minot dropped the nemerteans from this assemblage, although many workers did not accept this change for several decades.

It is now generally agreed that the flukes represent two classes, the Trematoda (digenetic flukes) and the Monogenea (monogenetic flukes). However, the classification of flatworms is still a matter of considerable controversy and is subjected to frequent revisions. The interested student is directed to the references at the end of this chapter, especially papers by Ulrich Ehlers. Ehlers has produced several phylogenetic classifications for flatworm groups based on a variety of data, including ultrastructure. No single scheme is likely to be acceptable to all workers at the present time. One change in the higher classification of flatworms, however, seems imminent. The flukes and tapeworms share a unique synapomorphy—the neodermis, discussed later—that clearly separates them from the turbellarians. A number of specialists suggest that the Trematoda, Monogenea, and Cestoda should be united under a single taxon, the Neodermata, based on this feature. While we retain a more conservative and traditional classification, a reorganization may well be forthcoming. There are about 4,500 species of turbellarians, 9,000 species of flukes, and 5,000 species of tapeworms.

PHYLUM PLATYHELMINTHES

CLASS TURBELLARIA: Free-living flatworms (Figure 10.1A–E,H,I). Predominately free-living and aquatic; not strobilated; mouth leads to a stomodeal pharynx, (the structure of which differs among orders) and thence to a primarily closed gut region; epidermis cellular and usually ciliated. The orders of turbellarians were previously grouped into two superorders on the basis of whether yolk is deposited within the cytoplasm of the ova (entolecithal ova) or separately, outside the ova (ectolecithal ova). Those with entolecithal ova were placed in the superorder Archoöphora and those with ectolecithal ova in the Neoöphora. Even though these names have been largely abandoned as formal taxa, the placement of yolk still provides an additional character for describing the orders and has important implications in the early development of these animals. There are currently 12 recognized orders.

ORDER ACOELA: Acoels. Pharynx, when present, is simple; no permanent gut cavity; mouth or pharynx leads to a solid syncytial or cellular entodermal mass; entolecithal ova. These small (1–5 mm), common flatworms inhabit marine and brackish water sediments; a few are planktonic or symbiotic. (e.g., *Amphiscolops*, *Convoluta*, *Haplogonaria*, *Polychoerus*)

ORDER CATENULIDA: Catenulids. Simple pharynx; simple, saclike gut; mesenchyme sometimes reduced to a fluid matrix; with entolecithal ova. Catelunids are elongate freshwater and marine forms. (e.g., *Catenula*, *Paracatenula*, *Stenostomum*)

ORDER HAPLOPHARYNGIDA: Minute turbellarians (up to 6 mm long) with a simple proboscis and pharynx; proboscis separate from pharynx and beneath the anterior tip of the body (reminiscent of nemerteans); anal pore weakly developed, but permanent; brain encapsulated by a unique membrane. One genus (*Haplopharynx*) and two species; sometimes placed in Macrostomida.

ORDER LECITHOEPITHELIATA: Pharynx variable; gut simple. About 30 species united on the basis of an intermediate condition between entolecithal and ectolecithal ova. (e.g., *Gnosonesima*, *Prorhynchus*)

Figure 10.1 Representative flatworms. (A–E,H,I) Members of the class Turbellaria. (A) The terrestrial triclad *Bipalium*. (B) SEM of *Cheliplana*, an interstitial rhabdocoel. (C) Unidentified, intertidal polyclad flatworm from the Sea of Cortez, Mexico. (D) The strikingly colored polyclad *Pseudoceros ferrugineus*. (E) The familiar freshwater triclad *Dugesia*. (F) The liver fluke *Fasciola hepatica* (class Trematoda, subclass Digenea). (G) Anterior end of the tapeworm *Taenia* (class Cestoda, subclass Eucestoda). (H) The polyclad flatworm *Thysanozoon*, a predator on small invertebrates, including barnacles. (I) The marine polyclad, *Eurylepta californica*.

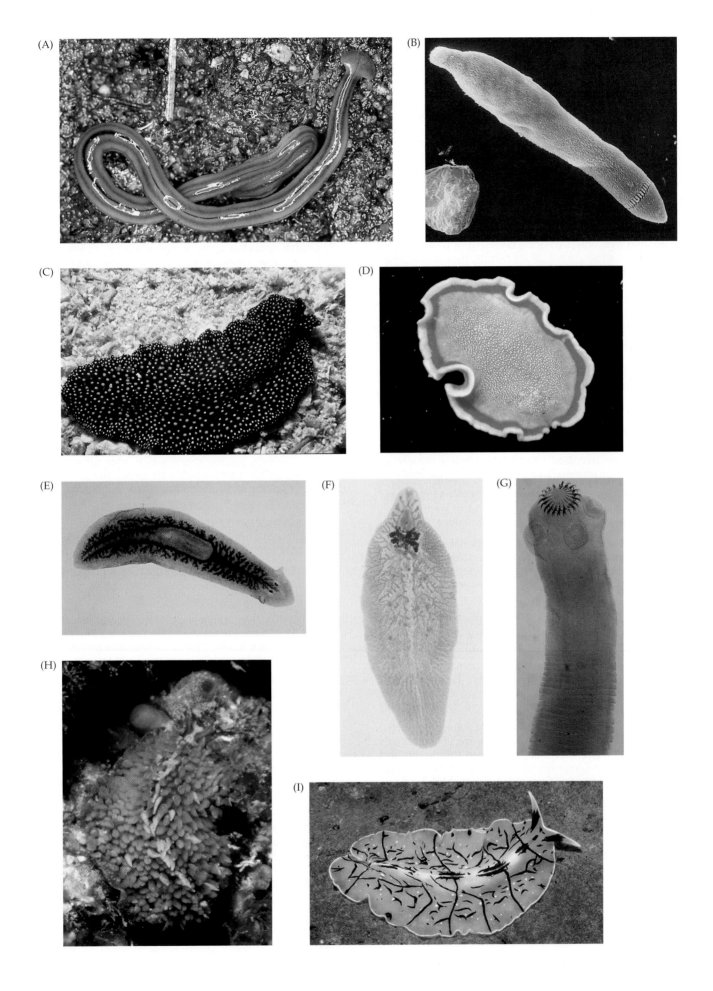

ORDER MACROSTOMIDA: Macrostomids. Simple pharynx; simple, saclike gut; with entolecithal ova. These turbellarians are small and predominately interstitial; marine and freshwater species. (e.g., *Macrostomum, Microstomum*)

ORDER NEMERTODERMATIDA: Nemertodermatids. Mouth and pharynx present or absent; pharynx simple when present; gut cavity with interdigitating processes from intestinal lining; uniflagellate sperm (all other flatworms possess sperm with 0 or 2 flagella); with entolecithal ova. These small turbellarians inhabit subtidal marine muds and sands. One genus (*Meara*) is parasitic in sea cucumbers. (e.g., *Flagellophora, Meara, Nemertoderma*)

ORDER POLYCLADIDA: Polyclads. Most with ruffled plicate pharynx; gut multibranched with diverticula; entolecithal ova. Polyclads are a diverse group of relatively large turbellarians. Nearly all are marine; they are common in littoral zones throughout the world, especially in the tropics; predominately benthic and free-living. Some are so large and colorful as to be easily mistaken for sea slugs (opisthobranchs). Many can swim by graceful undulations of the body margins. A few are pelagic or symbiotic. (e.g., *Eurylepta, Hoploplana, Leptoplana, Notoplana, Planocera, Prostheceraeus, Pseudoceros, Stylochus, Thysanozoon*)

ORDER PROLECITHOPHORA: Prolecithophorans. Pharynx plicate or bulbous; gut simple; with ectolecithal ova. Small, free-living, marine and fresh water. (e.g., *Plagiostomum, Urostoma*)

ORDER PROPLICASTOMATA: Based upon only a few known specimens of a single species, *P. jenseni*, from 180 m in fortune Bay, Greenland. Resembles the Acoela, but with an elongate plicate pharynx; without statocysts; entolecithal ova.

ORDER PROSERIATA: Proseriatans. Cylindrical plicate pharynx; simple gut; ectolecithal ova. Most are free-living marine species. (e.g., *Nemertoplana, Octoplana, Taboata*)

ORDER RHABDOCOELA: Rhabdocoels. Bulbous pharynx; simple saclike gut without diverticula; ectolecithal ova produced by ovaries that are usually fully separate from the yolk glands. This extremely large and diverse group is divided into 4 suborders.

SUBORDER DALYELLIOIDA: Dalyellioids. Anterior mouth; free-living or ecto- or entosymbionts of marine and freshwater invertebrates. (e.g., *Callastoma, Graffilla, Pterastricola*)

SUBORDER TYPHLOPLANOIDA: Typhloplanoids. Mouth not anterior; free-living marine and freshwater species. (e.g., *Kytorhynchus, Mesostoma, Typhlorhynchus*)

SUBORDER KALYPTORHYNCHIA: Kalyptorhynchs. Mouth not anterior; with a complex eversible proboscis at anterior end that is separate from the mouth and pharynx; free-living marine and freshwater species. (e.g., *Cheliplana, Cystiplex, Gnathorhynchus, Gyratrix*)

SUBORDER TEMNOCEPHALIDA: Temnocephalids. Small symbionts on freshwater decapod crustaceans (a few live on other invertebrates or on turtles); with posterior sucker and anterior tentacles used for attachment and inchworm-like locomotion. (e.g., *Temnocephala*)

ORDER TRICLADIDA: Triclads. Cylindrical plicate pharynx; gut three-branched with numerous diverticula; ectolecithal ova. Marine, freshwater, and some terrestrial species. Most are free-living, including the familiar planarians. (e.g., *Bdelloura, Bipalium, Crenobia, Dugesia* [formerly *Planaria*], *Geoplana, Polycelis, Procotyla*)

BOX 10A *Characteristics of the Phylum Platyhelminthes*

1. Parasitic or free-living, unsegmented worms (The subclass Eucestoda, in the class Cestoda, is strobilated)

2. Triploblastic, acoelomate, bilaterally symmetrical; flattened dorsoventrally

3. Spiral cleavage and 4d mesoderm

4. Complex, though incomplete, gut usually present; gut absent in some parasitic forms (Cestoda)

5. Cephalized, with a central nervous system comprising an anterior cerebral ganglion and (usually) longitudinal nerve cords connected by transverse commissures (ladder-like nervous system)

6. With protonephridia as excretory/osmoregulatory structures

7. Hermaphroditic, with complex reproductive system

CLASS MONOGENEA: Monogenetic flukes (Figure 10.3C). Body covered by a tegument; oral sucker reduced or absent; acetabulum absent; with anterior prohaptor and posterior hooked opisthaptor; life cycle involves only one host. Most are ectoparasitic, usually on fishes (some occur on turtles, frogs, hippos, copepods, or squids); a few are entoparasitic in ectothermic vertebrates.

SUBCLASS MONOPISTHOCOTYLEA: Opisthaptor simple and single, but sometimes divided by septa; oral sucker reduced or absent. (e.g., *Gyrodactylus, Polystoma*)

SUBCLASS POLYOPISTHOCOTYLEA: Opisthaptor complex, with multiple suckers; oral sucker absent. (e.g., *Diplozoon*)

CLASS TREMATODA: Digenetic and aspidogastrean flukes (Figures 10.1F, 10.3A,B,D,E). Body covered by a tegument; with one or more suckers; lacking prohaptor and opisthaptor. Most have 2 or 3 hosts during the life cycle; most are entoparasitic.

SUBCLASS DIGENEA: With 2 to 3 hosts during life cycle; first intermediate host a mollusc, final host a vertebrate; with oral and usually a ventral (acetabulum) sucker. (e.g., *Echinostoma, Fasciola, Microphallus, Opisthorchis* [= *Clonorchis*], *Sanguinicola, Schistosoma*)

SUBCLASS ASPIDOGASTREA: Most with a single host (a mollusc) in life cycle; second host, when present, is a fish or turtle; oral sucker absent, ventral sucker large, divided by septa as a row of suckers. (e.g., *Aspidogaster, Cotylaspis, Multicotyl*)

CLASS CESTODA: Tapeworms (Figures 10.1G, 10.4). Exclusively entoparasitic; body covered by a tegument; in most, the body consists of an anterior scolex, followed by a short neck, and then a strobila composed of a series of "segments" or proglottids; digestive tract absent. There are two subclasses.

SUBCLASS CESTODARIA: Cestodarians. Small group of uncommon, flattened tapeworms lacking scolex and proglottids (not strobilated); some with suckers; first larval stage (lycophore larva) bears ten hooks. Entoparasites in the guts or coelomic cavities of cartilaginous and certain primitive bony fishes, and less commonly in turtles. Their life cycles are poorly understood. (e.g., *Amphilina, Gyrocotyl, Gyrometra*)

SUBCLASS EUCESTODA: Eucestodes. Often very large (some over 10 m long), strobilated tapeworms; almost all with well developed scolex, neck, and strobila; first larval stage (onchosphere, or hexacanth larva) with six hooks. Entoparasitic in the guts of various vertebrates; most require one or more intermediate hosts during the life cycle. (e.g., *Diphyllobothrium, Dipylidium, Hymenolepis, Moniezia, Taenia*)

The Platyhelminth Bauplan

Compared with taxa discussed in preceding chapters, flatworms display some of the most important advances found in the animal kingdom. They represent the acoelomate bilateria and, according to most hypotheses, the basic bauplan from which many other triploblastic animals (the protostomes) were ultimately derived.

The evolution of the triploblastic condition and bilateral symmetry almost certainly occurred in concert with the evolution of sophisticated internal "plumbing" (organs and organ systems) and the tendency to cephalize, to centralize the nervous system, and to develop specialized units within the nervous system for sensory, integrative, and motor activities. With these features came unidirectional movement and a more active lifestyle than that of radially symmetrical animals. The primary evolutionary advantages of these coincidental changes derived chiefly from the ability of these "new" creatures to move around more or less freely and thus exploit survival strategies theretofore impossible.

These strategies can be appreciated by examining the rather complex structural features displayed by the free-living turbellarians (Figure 10.2). The presence of mesoderm allows the formation of a fibrous and muscular mesenchyme that provides structural support and allows patterns of locomotion not possible in diploblastic radiates. Elaborate reproductive systems evolved in the platyhelminths, providing for internal fertilization and enhancing the production of yolky and encapsulated eggs. Most flatworms have abandoned indirect development for mixed and direct life histories. Osmoregulatory structures in the form of protonephridia were instrumental in the invasion of fresh water.

This bauplan is not without constraints, however. Higher energy demands accompany an active lifestyle. The major limiting factor for flatworms, functionally, is the absence of an efficient circulatory mechanism to move materials throughout the body. This problem is compounded by the lack of any special structures for gas exchange. These problems relate, of course, to the surface-to-volume dilemma discussed in Chapter 3. In the absence of circulatory and gas exchange structures, flatworms (particularly the free-living ones) are constrained in terms of size and shape. They have remained relatively small and flat, with shapes that maintain short diffusion distances. The largest free-living flatworms have highly branched guts that assume much of the responsibility of internal transport.

Having a high surface-to-volume ratio and using the entire body surface for gas exchange create potential problems of ionic balance and osmoregulation in freshwater and terrestrial species, and of desiccation in intertidal and terrestrial habitats. The permeable body surface must be kept moist; thus, flatworms have invaded land rarely and only in very damp areas. They have, however, exploited a variety of marine and freshwater habitats, and are particularly successful as parasites and commensals, enjoying the benefits of living on or in their hosts.

It is generally assumed that the ancestral flatworm was a free-living form from which the present-day turbellarians evolved and diversified. The flukes and tapeworms were undoubtedly derived from within this varied turbellarian assemblage, as discussed in more detail later. Thus, in each of the following sections we first examine the basic features of the turbellarians and set the stage for understanding not only the diversity within that class but the derivation of the specialized parasitic taxa as well.

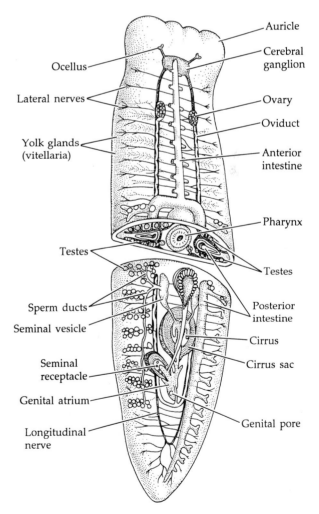

Figure 10.2 A generalized freshwater turbellarian (order Tricladida).

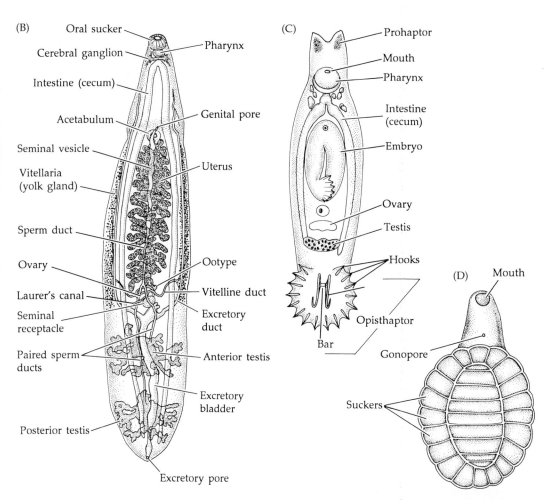

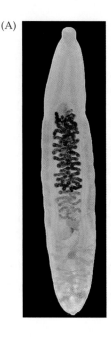

Figure 10.3 Representative flukes. (A,B) *Opisthorchis* (=*Clonorchis*) *sinensis*, a digenetic fluke that inhabits human livers. (C) *Gyrodactylus* (class Monogenea), an ectoparasite on fishes. (D) The trematode *Cotylaspis* (subclass Aspidogastrea). (E) The human blood fluke, *Schistosoma mansoni* (a copulating male and female).

The anatomy of turbellarians, flukes, and tapeworms is shown in Figures 10.2, 10.3, and 10.4. Turbellarians vary in shape from broadly oval to ribbon-like, and are typically flattened dorsoventrally, although very small ones may be nearly cylindrical. The head is usually ill-defined, except for the presence of sense organs. The mouth is located ventrally, either near the middle of the body or more anteriorly. Most flukes (Figure 10.3) are oval or leaf-shaped and bear external attachment organs such as hooks and suckers. As their common name suggests, the tapeworms are typically elongate and ribbon-like (Figure 10.4). Their anterior end is a tiny **scolex**, modified for attachment within the host; the rest of the body is essentially a reproductive machine.

Tapeworms live in the guts of vertebrates. Most species belong to the subclass Eucestoda and possess three distinguishable regions of the body. The scolex serves for attachment and is usually armed with hooks and suckers. Immediately behind the scolex is a short region called the **neck**, followed by an elongated, segmented trunk, or **strobila**, consisting of individual **proglottids**. The proglottids bud (**strobilate**) from a germinal zone in the neck (or at the base of the scolex when a neck is absent). As new proglottids arise, older ones move posteriorly and mature, become inseminated, and fill with embryos. Strobilation in tapeworms is thus not

by way of teloblastic growth (Chapter 13), and it is clearly not homologous to the true segmentation seen in annelids and arthropods.

Tapeworms of the subclass Cestodaria are somewhat flukelike in appearance. They lack a scolex, and the body is not divided into proglottids. They are placed within the Cestoda because of the absence of a digestive tract and because of certain features of the life cycle.

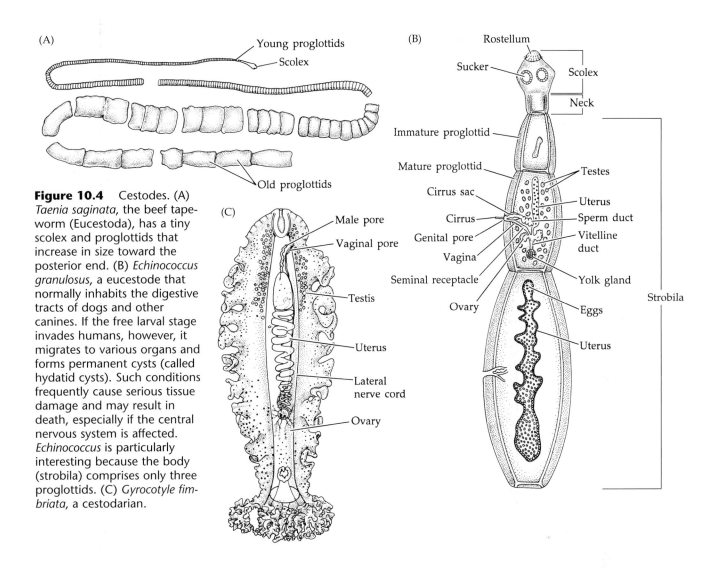

Figure 10.4 Cestodes. (A) *Taenia saginata*, the beef tapeworm (Eucestoda), has a tiny scolex and proglottids that increase in size toward the posterior end. (B) *Echinococcus granulosus*, a eucestode that normally inhabits the digestive tracts of dogs and other canines. If the free larval stage invades humans, however, it migrates to various organs and forms permanent cysts (called hydatid cysts). Such conditions frequently cause serious tissue damage and may result in death, especially if the central nervous system is affected. *Echinococcus* is particularly interesting because the body (strobila) comprises only three proglottids. (C) *Gyrocotyle fimbriata*, a cestodarian.

They may represent the primitive, prestrobilation body plan of the Cestoda.

Body Wall

Turbellarians. The body wall of turbellarians is multilayered and complex (Figure 10.5). The epidermis is composed of a wholly or partially ciliated, syncytial or cellular epithelium, with gland cells and sensory nerve endings distributed in various patterns. Beneath the epidermis is a basement membrane, which is often thick enough to lend some structural support to the body. In the orders Acoela, Catenulida, and Macrostomida, the basement membrane is apparently absent, but this condition is viewed as secondarily derived. Internal to the basement membrane are smooth muscle cells, frequently arranged in rather loosely organized outer circular, middle diagonal, and inner longitudinal layers. The area between the body wall and the internal organs is usually filled with a mesenchyme (often called a parenchyme) that includes a variety of loose and fixed cells, muscle fibers, and connective tissue. Most acoels, and perhaps the macrostomids, lack a cellular mesenchyme.

The gland cells of the body wall are generally derived from ectoderm. When mature, many of these cells lie in the mesenchyme with a "neck" extending between epidermal cells to the body surface. These cells produce mucous secretions that serve a number of functions. In semiterrestrial and intertidal turbellarians, the mucus forms a moist covering that provides protection from desiccation and aids in gas exchange. Most benthic flatworms possess a ventral concentration of mucous gland cells that secrete a slime that aids in locomotion. Mucous secretion around the mouth aids in prey capture and swallowing. Other gland cells or complexes of cells provide adhesives for temporary attachment. In some ectocommensal forms (e.g., *Bdelloura* and various temnocephalids; Figure 10.6) these adhesive glands are associated with special plates or suckers for attachment to the host.

Most turbellarians possess epidermal structures called **rhabdoids** (Figure 10.5B). These unique rod-shaped inclusions generally are produced by epithelial cells and then stored in packets within the epidermis. Upon release, rhabdoids produce copious amounts of mucus that may help protect the animal from desicca-

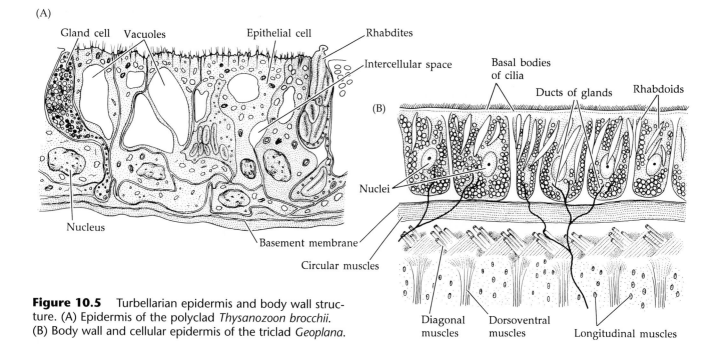

Figure 10.5 Turbellarian epidermis and body wall structure. (A) Epidermis of the polyclad *Thysanozoon brocchii*. (B) Body wall and cellular epidermis of the triclad *Geoplana*.

tion and from possible predators. Rhabdoids that are produced by gland cells in the mesenchyme are called **rhabdites**. These structures can reach the body surface through intercellular spaces in the epidermis (Figure 10.5A) and also contribute to mucus production. They

may be responsible for the release of noxious defense chemicals by some turbellarians. Some turbellarians have prominent tubercles covering the dorsal surface; these structures probably have a defensive role. In some species, unfired nematocysts from hydroid prey are transported to

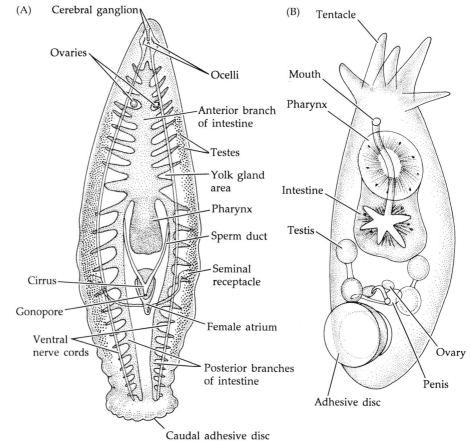

Figure 10.6 Two symbiotic turbellarians with adhesive attachment organs. (A) *Bdelloura candida,* a triclad ectocommensal on horseshoe crabs (*Limulus*). (B) *Temnocephala caeca,* a rhabdocoel ectocommensal on *Phreatoicopis terricola* (a freshwater isopod).

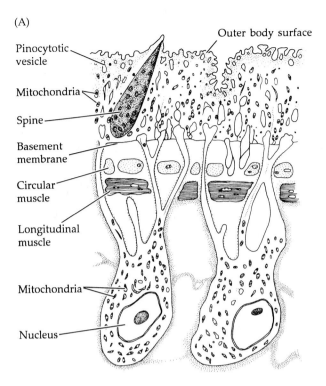

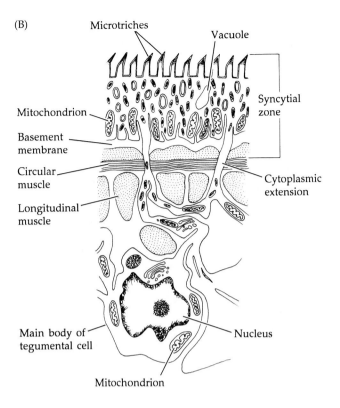

Figure 10.7 (A) The tegument and underlying body wall of a digenetic fluke (*Fasciola hepatica*; longitudinal section). (B) The tegument and body wall of a cestode (cross section).

the tubercles. In others, such as *Thysanozoon*, the tubercles appear to release a powerful acid that may deter would-be predators.

Flukes and tapeworms. Modifications of the outer body covering are common among parasites, and platyhelminths are no exception. Unlike the turbellarians, flukes and tapeworms possess an external covering called a **tegument**, formed of nonciliated cytoplasmic extensions of large cells whose cell bodies actually lie in the mesenchyme (Figure 10.7). The tegument not only provides some protection but is an important site of exchange between the body and the environment. Gases and nitrogenous wastes move across this surface by diffusion, and some nutrients, especially amino acids, are taken in by pinocytosis. In tapeworms, the uptake of nutrients occurs solely across the body wall, and the surface area of the tegument is greatly increased by many tiny folds called **microtriches** (Figure 10.7B). As one of nature's more remarkable adaptations, these folds may interdigitate with the intestinal microvilli of the host and aid in the absorption of nutrients.

The nature of the tegument in flukes and tapeworms is viewed by some zoologists as unique and of major phylogenetic importance (e.g., see papers by Ehlers). The larvae of these parasitic worms have a "normal" cil-

iated epidermis over at least part of their bodies. However, this epidermis is shed, and postlarval stages develop a new, syncytial body covering—the **neodermis**, or tegument. According to Ehlers (1985) and others, this phenomenon occurs in no other animals and should be viewed as a key synapomorphy uniting the Trematoda, Monogenea, and Cestoda as a monophyletic taxon that Ehlers calls the Neodermata (in reference to the "new skin" of these animals).

The tegument is underlain by a basement membrane, beneath which is the mesenchyme. Most flukes and tapeworms have circular and longitudinal muscles within the mesenchyme, and sometimes diagonal, transverse, and dorsoventral muscles as well. The mesenchyme varies from masses of densely packed cells to syncytial and fibrous networks with fluid-filled spaces. In some digenetic flukes, spaces form vessels through the mesenchyme called **lymphatic channels**, which contain free cells that have been likened to lymphocytes. The mesenchyme also contains gland cells with connections to the surface of the body through the tegument. These gland cells are few in number compared with those of turbellarians, and they are primarily adhesive in nature and associated with certain organs of attachment.

One of the least explored yet most interesting attributes of tapeworms, and indeed of all intestinal parasites, is their ability to thrive in an environment of hydrolytic enzymes without being digested. One popular hypothesis is that gut parasites produce enzyme inhibitors (sometimes called "antienzymes"). One study showed that *Hymenolepis diminuta* (a common tapeworm in rats and mice) releases proteins that appear to inhibit trypsin activity. This tapeworm can also regulate the pH of its immediate environment to about 5.0 by excreting organic acids; this acidic output also may inhibit the activity of trypsin (Uglem and Just 1983).

Support, Locomotion, and Attachment

Only a very few flatworms possess any sort of special skeletal elements. In a few turbellarians, tiny calcareous plates or spicules are embedded in the body wall. Body support in all other flatworms is provided by the hydrostatic qualities of the mesenchyme, the elasticity of the body wall, and the general body musculature.

Turbellarians. Most benthic turbellarians move on their ventral surface by cilia-powered gliding. Mucus provides lubrication as the animal moves and serves as a viscous medium against which the cilia act. Some of the larger or more elongate forms also use muscular contractions. The ventral surface of the body is thrown into a series of alternating transverse furrows and ridges that move as waves along the animal, propelling it forward. Muscular undulations of the lateral body margins allow some large polyclads to swim for brief periods of time. Muscular action allows the body to twist and turn, providing steerage. Some interstitial forms are highly elongate and use the body wall muscles to slither between sand grains. Many of these types of flatworms possess adhesive glands, the secretions of which provide temporary stickiness and enable the animals to gain purchase and leverage as they move. Very small turbellarians (e.g., acoels) swim or glide by the action of cilia that cover the entire body surface.

Flukes. Adult flukes lack external cilia, and their movement depends on the flukes' own body wall muscles or on the body fluids of their host. Some move about slowly on or within their host by muscle action, and a few (e.g., blood flukes) are carried in the host's circulatory system. However, certain larval stages are highly motile and do swim using ciliary action.

Once established within or on a host, it is advantageous for a fluke to stay more or less in one place. In that regard, nearly all of them are equipped with external organs for temporary or permanent attachment (Figures 10.3C and 10.8). Monogenetic flukes typically have an anterior and a posterior adhesive organ called the **prohaptor** and the **opisthaptor**, respectively. The prohaptor consists of a pair of adhesive structures, one on each side of the mouth, bearing suckers or simple adhesive pads. The opisthaptor is usually the major organ of attachment, and includes one or more well developed suckers with hooks or claws.

The digenetic flukes possess two hookless suckers. One, the **oral sucker**, surrounds the mouth, and the other, the **acetabulum**, is located on the ventral surface (Figure 10.3B). These suckers are usually supplied with adhesive gland cells, although the well developed ones operate mainly on suction produced by muscle action. The aspidogastrean flukes lack an oral sucker but have a large, subdivided ventral sucker (Figure 10.3D).

Tapeworms. Adult tapeworms do not move around much, but they are capable of muscular undulations of the body. They remain fixed to the host's intestinal wall by the scolex (or, in the case of members of the subclass Cestodaria, by an anterior adhesive organ) and by the microtriches.

The details of scolex anatomy (Figure 10.9) are extremely variable and of critical importance in the taxonomy of the Eucestoda. The tip of the scolex in many cestodes (e.g., *Taenia*) is equipped with a movable hook-bearing **rostellum**, which is sometimes retractable into the scolex. In others (e.g., *Cephalobothrium*) the anterior end bears a protrusible sucker, or adhesive pad, called a **myzorhynchus**. The rest of the scolex bears various suckers or sucker-like structures and sometimes hooks

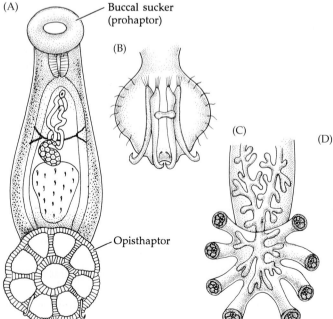

(A) Buccal sucker (prohaptor)

(B)

(C)

Opisthaptor

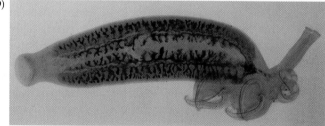

(D)

Figure 10.8 Some attachment organs of monogenetic flukes. (A) *Anoplocotyloides papillata.* (B,C) Opisthaptors from monogenetic flukes. (D) An unidentified fluke with suckered prohaptor and elaborate opisthaptor.

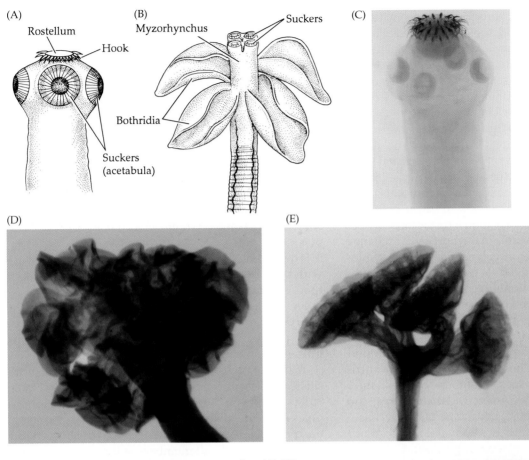

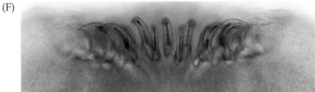

Figure 10.9   Scoleces of various eucestodes. (A) "Typical" scolex with rostellum, hooks, and suckers (*Taenia solium*). (B) Complex scolex with suckered myzorhynchus and leaflike bothridia (*Myzophyllobothrium*). (C–F) Photos of four different scoleces.

or spines. There are three categories of adhesive suckers upon which ordinal and subordinal classification of cestodes is partially based. **Bothria** are elongate, longitudinal grooves on the scolex. They possess weak muscles but are capable of some sucking action. Bothria occur as a single pair and are typical of the order Pseudophyllidea (e.g., *Diphyllobothrium*). Members of the order Tetraphyllidea (e.g., *Acanthobothrium, Phyllobothrium*) bear four symmetrically placed **bothridia** around the scolex. These foliose structures are often equipped with suckers at their anterior ends. The third and most familiar type of attachment structures on the scolex are true suckers, or **acetabula**. They are identical in structure and are probably homologous to the acetabula of digenetic trematodes. There are usually four acetabula, placed symmetrically around the circumference of the scolex. They are characteristic of many members of the order Cyclophyllidea (e.g., *Dipylidium, Taenia*).

Feeding and Digestion

Turbellarians.   Most turbellarians are carnivorous predators or scavengers, feeding on nearly any available animal matter. A few are herbivorous on microalgae, and some species switch from herbivory to carnivory as they mature. Their prey includes almost any invertebrate small enough to be captured and ingested (e.g., protists, small crustaceans, worms, tiny gastropods). Some species graze on sponges, ectoprocts, and tunicates, while others consume the flesh of barnacles, leaving behind the empty shell. Most turbellarians locate food by chemoreception. Land planarians capture and consume earthworms (e.g., *Bipalium*), land snails (e.g. *Platydesmus, Endeavouria*), and insects (e.g., *Rhynchodemus, Microplana*).

More than 100 species of turbellarians are known to be symbiotic with other invertebrates. Some of these are simply commensals that derive some protection from their associations, showing only physical modifications for temporary attachment. Others, however, feed upon their hosts, causing various degrees of damage and displaying true physiological dependency on the relationship. While we can devote space to mentioning only a few examples of symbiotic turbellarians, recognition of these situations is of considerable importance. First, it emphasizes the evolutionary adaptability of the turbellarian bauplan; and second, it provides some essential foundation for our later discussion of the origins of the flukes and tapeworms. (For an excellent survey of the symbiotic turbellarians, see Jennings 1980.)

Most of the symbiotic turbellarians belong to the order Rhabdocoela (suborders Dalyellioida and Temnocephalida). One notable exception is the triclad *Bdelloura* (Figure 10.6A), an ectocommensal on the gills of *Limulus*, the horseshoe crab. The temnocephalids (Figure 10.6B) are ectocommensals within the branchial chambers of freshwater decapod crustaceans, where they feed on microorganisms in the host's gas exchange currents. Several families of dalyellioids include symbiotic members. The umagillids (e.g., *Syndesmis*) live within the gut and coelomic fluid of echinoids (Figure 10.10). These tiny flatworms feed on protists and bacteria, and some may devour cells of their hosts. Graffillid dalyellioids (*Graffilla* and *Paravortex*) include several species of parasites in the digestive tracts of gastropod and bivalve molluscs. These worms derive their nutrients from the host tissues. Members of the family Fecampiidae (*Fecampia, Kronborgia, Glanduloderma*) are parasites in marine crustaceans and certain polychaete worms. They reside in the host's body fluids and absorb soluble organic nutrients. One dalyellioid, *Oekiocolax* (family Provorticidae), is a parasite of other marine turbellarians. An undescribed species of *Prosthiostomum* (Polycladida) is a parasite on the Hawaiian coral *Montipora* (Jokiel and Townsley 1974).

The general plan of the turbellarian digestive system includes a mouth and a pharynx, which lead to an intestine, or enteron. Like that of cnidarians, the turbellarian gut is incomplete, bearing a single opening, and thus may be called a gastrovascular cavity. The mouth varies in position from midventral to anterior. The pharynx is derived from embryonic ectoderm, so it is a stomodeum lined with epidermis. Epithelial **pharyngeal glands** are associated with the lumen of the pharynx; they produce mucus that aids in feeding and swallowing, and (in some species) proteolytic enzymes that initiate digestion outside the body.

The feeding methods of free-living turbellarians vary with the size of the animal and the complexity of their food-getting apparatus, especially the pharynx. As noted in the classification scheme, the nature of the pharynx varies greatly among taxa. There are three basic pharynx types among the turbellarians: **simple**, **bulbous**, and **plicate** (Figure 10.11).

A simple pharynx (or **pharynx simplex**) is a short, ciliated tube connecting the mouth and intestine (Figure 10.12). This type of pharynx is considered plesiomorphic within the phylum Platyhelminthes and is found in the orders Nemertodermatida, Acoela, Macrostomida, and Catenulida. In all members of these orders except the Acoela, the pharynx leads to a simple saclike or elongate intestine generally lacking extensive diverticula. Members of the order Acoela lack any permanent digestive cavity; instead the pharynx leads to a solid syncytial or cellular mass of internal digestive tissue (Figure 10.13A). It was long thought that the syncytial nature of the digestive mass of some acoel turbellarians was a sign of their primitiveness, but newer evidence suggests that this is a secondarily derived feature. This discovery bears significantly on certain phylogenetic hypotheses of flatworm origin and corresponding ideas regarding the origin of the Metazoa (see Chapter 4).

Turbellarians with a simple tubular pharynx are generally quite small, with the mouth located more or less midventrally. They usually feed by sweeping small organic particles and tiny prey into the pharynx by ciliary action. Those with an eversible pharynx usually fold their body around the prey or other food source and cover it with mucus from the epidermal glands. Then the pharynx is everted over or into the food item.

Some turbellarians, especially triclads, secrete digestive enzymes externally via special glands that empty through the pharyngeal lumen or from the tip of the pharynx. The food is partially digested and reduced to a soupy consistency prior to swallowing. Many other turbellarians swallow their food whole by the action of powerful pharyngeal muscles.

Rhabdocoels typically possess a slightly protractile, muscular, bulbous pharynx and a simple saclike gut

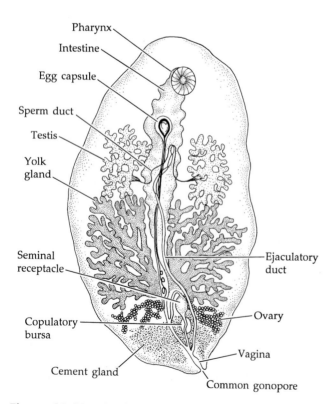

Pharynx
Intestine
Egg capsule
Sperm duct
Testis
Yolk gland
Seminal receptacle
Copulatory bursa
Cement gland
Ejaculatory duct
Ovary
Vagina
Common gonopore

Figure 10.10 *Syndesmis*, a rhabdocoel from the gut of a sea urchin.

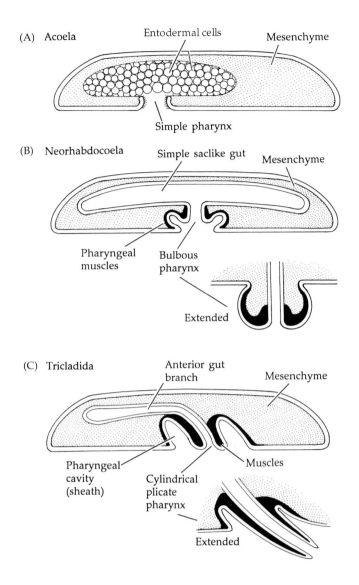

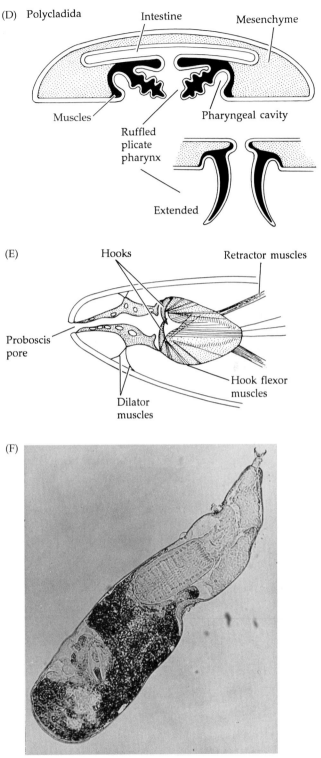

Figure 10.11 (A–D) The pharynges of four turbellarians (sagittal sections). (E) The proboscis apparatus (sagittal section) of *Gnathorhynchus* (order Kalyptorhynchia). (F) *Cheliplana*, another kalyptorhynch, with jawed proboscis extended (top right).

(Figure 10.13B). Members of the orders Proseriata, Tricladida, and Polycladida have eversible plicate pharynges. The eversible portion of a plicate pharynx lies within a space called the **pharyngeal cavity**, which is produced by a muscular fold of the body wall (Figures 10.11 and 10.13).

Proseriates and triclads possess cylindrical plicate pharynges oriented along the body axis. Most polyclads have a ruffled, skirtlike plicate pharynx attached dorsally within the pharyngeal cavity. During feeding, a plicate pharynx is protruded by a squeezing action of extrinsic pharyngeal muscles. Once extended, the pharynx can be moved about by intrinsic muscles of its wall. Retractor muscles pull the pharynx back inside the cavity.

Active prey can be subdued in several ways. Some turbellarians produce mucus, which, in addition to en-

tangling the prey, may contain poisonous or narcotic chemicals. A few flatworms use the sharp stylet of the copulatory organ to stab prey; one cannot help but concede the remarkable adaptive capacity of the flatworms. Members of the suborder Kalyptorhynchia (order Rhabdocoela) are unique among turbellarians in their possession of a muscular proboscis that is situated at the anterior end of the body and is separate from the mouth

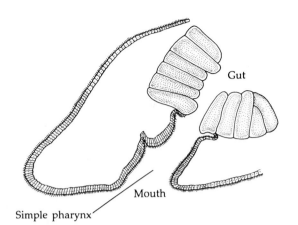

Figure 10.12 Saggital section through anterior end of *Macrostomum* (class Turbellaria, order Macrostomida), which has a simple tubular pharynx.

(Figure 10.11E,F); the proboscis, which in some species is armed with hooks, can be everted to grab prey.

Most of the turbellarians that possess a plicate pharynx are relatively large, especially the triclads and the polyclads. Associated with this large body size is an elaboration of the intestine. The triclad intestine comprises three main branches, one anterior and two posterior, each with numerous diverticula; the intestine of polyclads is multibranched with diverticula (Figure 10.13C,D). These ramifications of the intestine provide not only an increased surface area for digestion and absorption, but also are a means of distributing the products of digestion in the absence of a circulatory system. The lining of the intestine is a single cell layer of phagocytic nutritive cells and enzymatic gland cells (Figure 10.14). In some groups, the gastrodermis is ciliated.

In most turbellarians, digestion begins extracellularly with the action of endopeptidases secreted by the pha-

ryngeal glands or by the enzymatic gland cells of the intestine. The partially digested material is distributed throughout the gut, then phagocytized by the intestinal cells, wherein final digestion occurs. There are, however, some notable exceptions to this sequence. In some acoels, temporary spaces form within the gastrodermal mass. Primary digestion occurs within these spaces, and the products are phagocytized by the surrounding cells. Bowen (1980) described an interesting phagocytic process in the small freshwater triclad *Polycelis tenuis*. Following the ingestion of tiny food particles or the preliminary extracellular digestion of larger food, the intestinal phagocytic cells extend processes into the gut lumen, nearly occluding the digestive cavity. These processes interdigitate to form a complex web, forcing food material into the phagocytes, where digestion is completed. Certain polyclads in the suborder Cotylea apparently digest their food entirely extracellularly, and phagocytosis of particulate matter is unknown in this group.

Since the flatworm gut is generally incomplete, any undigested material must be expelled through the mouth. As discussed in Chapter 3, the major limitation of single-opening guts is the restriction on regional specialization. However, an incipient anus occurs in several flatworms, suggesting that evolutionary "experimentation" with a complete gut began in this group. One macrostomid, *Haplopharynx rostratus*, possesses a minute anal pore, and some polyclads have pores at the ends of gut branches; some proseriates (e.g., *Taboata*) may form a temporary anus.

Flukes and tapeworms. Adult flukes feed on host tissues and fluids or, in some cases, material within the host's gut. Most of the food is taken in through the mouth by a pumping action of the muscular pharynx, but some organic molecules are picked up across the

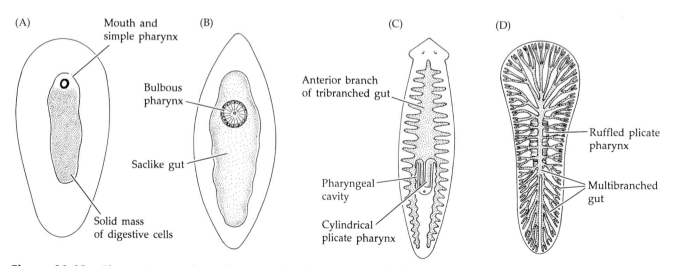

Figure 10.13 Pharynx type and gut shape combinations among turbellarians. (A) Acoela. (B) Rhabdocoela. (C) Tricladida. (D) Polycladida.

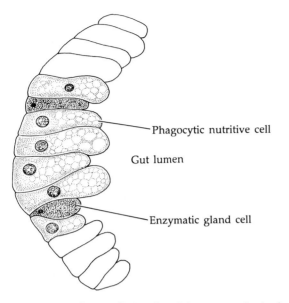

Figure 10.14 The gut lining (partial cross section) of a freshwater triclad contains enzymatic gland cells and phagocytic nutritive cells.

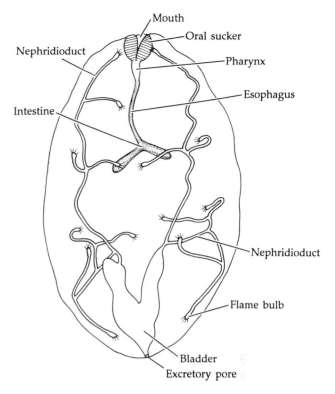

Figure 10.15 Gut and protonephridial system of *Microphallus* (subclass Digenea; see also Figure 10.3.) In most monogenetic flukes, the protonephridial ducts are separate and terminate anteriorly in separate pores.

tegument by pinocytosis. The anterior part of the digestive system includes a mouth, a muscular pharynx, and a short esophagus. The esophagus leads to a pair of intestinal ceca (occasionally, a single cecum), which extend(s) posteriorly in the body (Figures 10.3 and 10.15). The lining of the ceca includes absorptive nutritive cells and enzymatic gland cells. Digestion is at least partly extracellular. Some flukes secrete enzymes from the gut out the mouth, or from the suckers, to partially digest host tissue prior to ingestion.

Cestodes lack any vestige of a mouth or digestive tract. All nutrients must be taken into the body across the tegument. Uptake probably occurs by pinocytosis and by diffusion across the increased surface area of the microtriches. Some work suggests that tapeworms are unable to take in large molecules and thus rely to a considerable extent on the digestive processes of their hosts and the secretion of enzymes outside their bodies to chemically reduce the size of potential nutrient material. It has also been proposed that the surface of the scolex may absorb host tissue fluids through the site of attachment to the gut wall.

Circulation and Gas Exchange

As mentioned earlier, except for the lymphatic channels in some flukes, flatworms lack special circulatory or gas exchange structures. This condition imposes restrictions on size and shape. The key to survival with such limitations and a generally solid mesenchyme is the maintenance of small diffusion distances. Thus, the flatness of their bodies facilitates gas exchange across the body wall, between the tissues and the environment; nutrients are distributed internally by the digestive system and by diffusion, which is aided by general body movements.

The entoparasitic flatworms are capable of surviving in areas of their host where oxygen is absent. In such cases, they rely on anaerobic metabolism, producing a variety of reduced end products (e.g., lactate, succinate, alanine, and long-chain fatty acids). These adaptable animals also possess the appropriate enzymes for and are capable of aerobic respiration in the presence of oxygen.

Excretion and Osmoregulation

One of the major advances of flatworms over diploblastic animals is the development of protonephridia. These structures occur in all turbellarians except members of the orders Nemertodermatida and Acoela, and some marine catenulids. Turbellarian protonephridia are flame bulbs and may occur singly (as they do in some catenulids) or in pairs (from one to many pairs in different taxa). The protonephridia are connected to networks of collecting tubules that lead to one or more nephridiopores (Figure 10.16). Turbellarian protonephridia function primarily as osmoregulatory structures. Freshwater turbellarians tend to have more protonephridia and more complex tubule systems than do their marine counterparts. Although a small amount of ammonia is released via the protonephridia, most metabolic wastes are lost by diffusion across the body wall.

Flukes also possess variable numbers of flame bulb protonephridia. Two nephridioducts drain the nephridia

(A)

Protonephridial
network

Nephridiopores

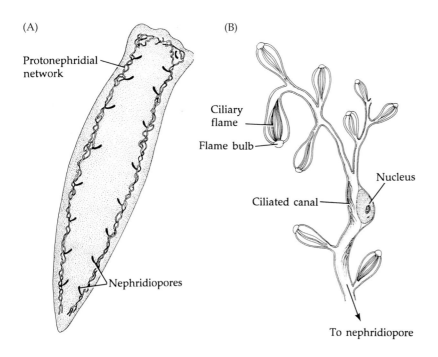

(B)

Ciliary
flame

Flame bulb

Ciliated canal

Nucleus

To nephridiopore

Figure 10.16 (A) The protonephridial system in a freshwater triclad. (B) The nephridial arrangement in a turbellarian that has anucleate flame bulbs attached to collecting tubules.

and lead to a storage area, or **bladder**, which in turn connects with a single posterior nephridiopore in the digenetic flukes or a pair of anterior pores in the monogenetic types (Figure 10.15). Nitrogenous waste in the form of ammonia is excreted largely across the tegument, and the protonephridia are primarily osmoregulatory.

Tapeworms possess numerous flame bulb protonephridia throughout the body. The flame bulbs drain to pairs of dorsolateral and ventrolateral nephridioducts that run the length of the body (Figure 10.17). Although some variation in plumbing occurs, the ventral ducts are typically connected to one another by transverse tubules near the posterior end of each proglottid. In relatively young worms that have not lost any proglottids (see the section on reproduction), the excretory ducts lead to a collecting bladder in the most posterior proglottid. Once this terminal proglottid is lost, the nephridioducts open separately to the outside on the posterior margin of the remaining hindmost proglottid.

There is still much to be learned about the protonephridia of cestodes. They probably function both in excretion and osmoregulation. They may also serve to eliminate certain organic acid products of anaerobic cellular metabolism. Some experimental work indicates that tapeworms are capable of precipitating and storing some wastes within their proglottids.

Nervous System and Sense Organs

Turbellarians. The nervous system of turbellarians varies from a simple netlike nerve plexus with only a minor concentration of neurons in the head (e.g., acoels) to a distinctly bilateral arrangement with a well developed cerebral ganglion and longitudinal nerve cords connected by transverse commissures (Figure 10.18). The more advanced condition is referred to as a

ladder-like nervous system. Even many of those turbellarians that possess distinctly centralized nervous systems have a plexus formed by the repeated branching of nerve endings (e.g., polyclads). In general, larger flatworms show an increasing concentration of the peripheral nerves into fewer and fewer longitudinal cords and an accumulation of neurons in the head as an associative center or cerebral ganglion. Furthermore, they show a tendency to separate the elements of the nervous system into distinct sensory and motor pathways and to develop a circuitry that operates primarily on unidirectional impulse transmission.

The turbellarian nervous system and sense organs evolved in association with bilateral symmetry and unidirectional movement. The result is a general concentration of sense organs at the anterior end of the body and an elaboration of those receptor types that are compatible with the turbellarian lifestyle. Tactile receptors are abundant over much of the body surface as sensory bristles projecting from the epidermis. These receptors tend to be concentrated at the anterior end and around

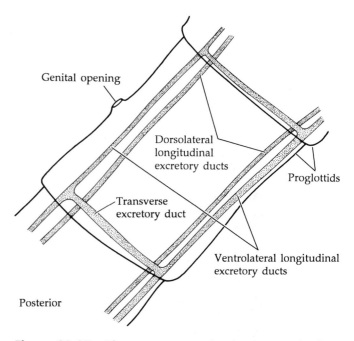

Genital opening

Dorsolateral
longitudinal
excretory ducts

Proglottids

Transverse
excretory duct

Ventrolateral longitudinal
excretory ducts

Posterior

Figure 10.17 The arrangement of major protonephridial ducts in a eucestode proglottid (ventral view).

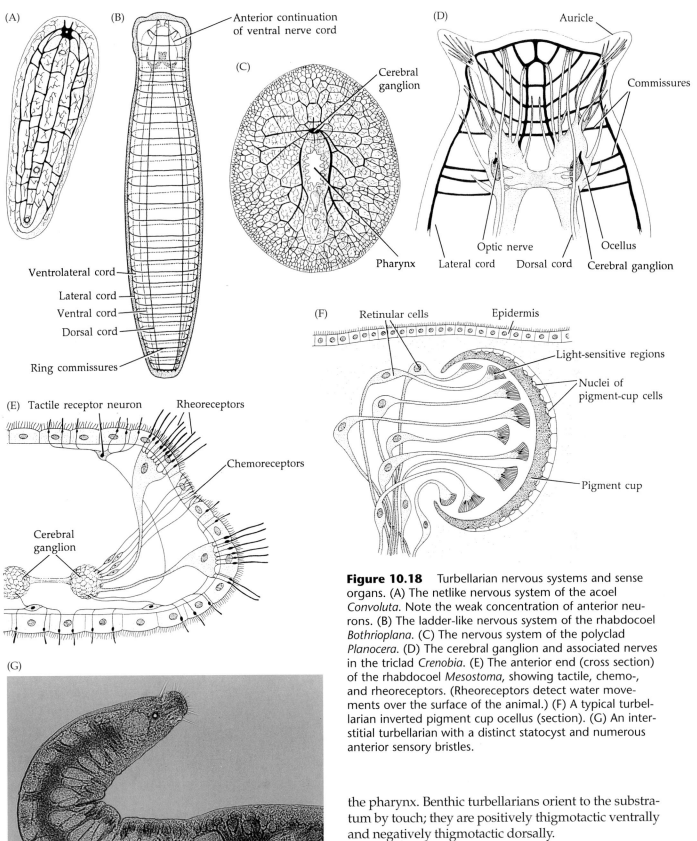

Figure 10.18 Turbellarian nervous systems and sense organs. (A) The netlike nervous system of the acoel *Convoluta*. Note the weak concentration of anterior neurons. (B) The ladder-like nervous system of the rhabdocoel *Bothrioplana*. (C) The nervous system of the polyclad *Planocera*. (D) The cerebral ganglion and associated nerves in the triclad *Crenobia*. (E) The anterior end (cross section) of the rhabdocoel *Mesostoma*, showing tactile, chemo-, and rheoreceptors. (Rheoreceptors detect water movements over the surface of the animal.) (F) A typical turbellarian inverted pigment cup ocellus (section). (G) An interstitial turbellarian with a distinct statocyst and numerous anterior sensory bristles.

the pharynx. Benthic turbellarians orient to the substratum by touch; they are positively thigmotactic ventrally and negatively thigmotactic dorsally.

Most turbellarians are equipped with chemoreceptors that aid in food location. Although sensitive over most of the body, turbellarians have distinct concentrations of chemoreceptors anteriorly, particularly on the sides of the head. Some forms, such as the familiar freshwater planarians, have the chemoreceptors located in flaplike processes called **auricles** on the head (Figures 10.2 and 10.18D), whereas others have these sense organs in ciliated pits, on tentacles, or distributed over much of the anterior end of the body. The epithelium bearing the chemoreceptors is often ciliated and frequently forms depressions or grooves. The cilia are the receptor organelles, but also circulate water, thus facilitating sensory input from the environment.

The utilization of chemoreception in locating food has been demonstrated in many turbellarians. Some are known to home in on concentrations of dissolved chemicals associated with potential food. Others, such as *Dugesia*, "hunt" by waving the head back and forth as they crawl forward, exposing the auricles to any chemical stimulus in their path. When exposed to diffuse chemical attractants, some turbellarians begin a trial-and-error behavior pattern. If unable to determine the direction of the attractant, the worm begins moving in a straight line. If the stimulus weakens, the animal makes apparently random turns until it encounters sufficient stimulus, then moves toward it in a straight line. This behavior can eventually bring the animal near enough to the food source to home in on it directly. Some turbellarians orient to water movements by rheoreceptors located on the sides of the head (Figure 10.18E).

Statocysts are common in certain turbellarians, notably in members of the Nemertodermatida, Acoela, Catenulida, and Proseriata. These orders include mostly swimming and interstitial forms in which orientation to gravity could not be accomplished by touch. When present, the statocyst is usually located on or near the cerebral ganglion. Ehlers (1991) presents details on the ultrastructure of some flatworm statocysts.

Most turbellarians possess photoreceptors in the form of inverted pigment-cup ocelli (Figure 10.18F). A few types of acoels and macrostomids possess simple pigment-spot ocelli, which are presumed to be primitive within the flatworms. Many turbellarians bear a single pair of ocelli on the head; but some, such as certain polyclads and terrestrial triclads, may have many pairs of eyes. In a few terrestrial forms (e.g., *Geoplana mexicana*) and many of the large tropical polyclads, numerous eyes extend along the edges of the body. Most free-living turbellarians are negatively phototactic. The dorsal placement of the eyes and the orientation of the pigment cups facilitate the detection of light direction as well as intensity.

Larvae of the flatworm *Pseudoceros canadensis* possess two dissimilar kinds of eyes. The right eye appears to be microvillar (i.e., rhabdomeric), but the left one has components of both microvillar and ciliary origin (Eakin and Brandenberger 1980). The phylogenetic lineages of these two eye types were noted in Chapter 3. The discovery of both types of eyes in a flatworm larva suggests to some researchers the possibility that this animal stands at a major point of evolutionary divergence.

Some other aspects of photoreceptor ultrastructure in turbellarians are discussed by Sopott-Ehlers (1991).

Neurosecretory cells have been known in turbellarians for more than three decades, and work continues on exploring their functions. These special cells are generally located in the cerebral ganglion, but they also occur along major nerve cords in at least some species. Neurosecretions probably play important roles in regeneration, asexual reproduction, and gonad maturation.

Flukes and tapeworms.   The nervous system of flukes is distinctly ladder-like and very similar to that in many turbellarians (Figure 10.19). The cerebral ganglion comprises two well defined lobes connected by a dorsal transverse commissure. Nerves from the cerebral ganglion extend anteriorly to supply the area of the mouth, adhesive organs, and any cephalic sense organs. Extending posteriorly from the cerebral ganglion are up to three pairs of longitudinal nerve cords with transverse connectives. A pair of ventral cords is usually most well developed, and dorsal cords are present in the digenetic flukes. Most flukes also have a pair of lateral nerve cords.

The suckers of flukes bear tactile receptors in the form of bristles and small spines. There is also some evidence of reduced chemoreceptors. Nearly all monogenetic flukes possess a pair of rudimentary pigment-cup ocelli near the cerebral ganglion.

The cerebral ganglion of cestodes is usually a complex **nerve ring** located in the scolex (Figure 10.20). The

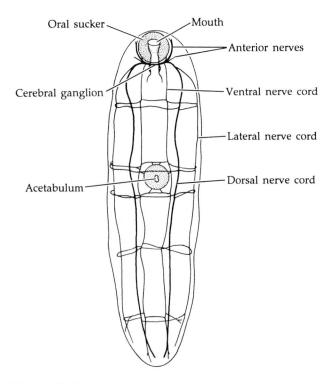

Figure 10.19   A generalized ladder-like nervous system of a fluke (ventral view).

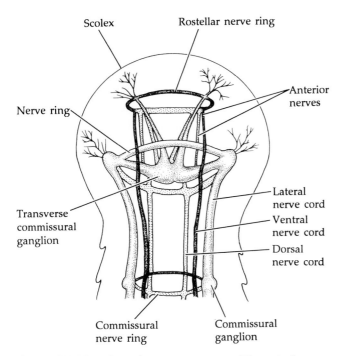

Figure 10.20 Cestode nervous system. The anterior end of *Moniezia*. The longitudinal cords extend the length of the animal.

ring bears ganglionic swellings and gives rise to a number of nerves. Anterior nerves, in the form of a ring or plexus, serve the rostellum (when present) and other attachment organs. Lateral cerebral ganglionic swellings give rise to a pair of major lateral longitudinal nerves, which extend the length of the animal. In each proglottid, these nerves bear additional ganglia from which transverse commissures arise and connect the two longitudinal cords. Additional longitudinal nerves are often present; the most typical pattern includes two pairs of accessory lateral cords—a pair of dorsal cords and a pair of ventral cords. As might be expected, sense organs are greatly reduced in cestodes and are limited to abundant tactile receptors in the scolex.

Reproduction and Development

Asexual processes. Asexual reproduction is common among freshwater and terrestrial turbellarians, and it generally occurs by transverse fission. In the catenulids and macrostomids, an odd sort of multiple transverse fission occurs wherein the individuals thus produced remain attached to one another in a chain until they mature enough to survive alone (Figure 10.21A). Some freshwater triclads (e.g., *Dugesia*) split in half behind the pharynx, and each half goes its own way, eventually regenerating the lost parts. A few (e.g., *Phagocata*) reproduce by fragmentation, each part encysting until the new worm forms.

The remarkable regenerative abilities of turbellarians have been studied intensely for many years. Much of the experimental work has been conducted on the com-

mon triclad *Dugesia*, a familiar animal to beginning zoology students. Underlying all of the bizarre results of various surgeries performed on these animals (Figure 10.21B,C) is the fact that the cells of organisms like *Dugesia* are not totipotent; an anterior–posterior body polarity exists in terms of the regenerative capabilities of the cells. However, the cells in the midbody region are less fixed in their potential to produce other parts of the body than are those toward the anterior or posterior ends. Thus, if the flatworm is cut through the middle of the body (as it is in normal transverse fission), each half will regenerate the corresponding lost part. However, if the animal is cut transversely near one end—say, separating a small piece of the tail from the rest of the body—the larger piece will grow a new posterior end, but the piece of tail lacks the capability to produce an entire new anterior end. This gradient of cell potency has been of particular interest to cell biologists and medical researchers because of its relevance to healing and regeneration potential in higher animals.

Asexual reproduction is an important feature of the life cycle of flukes, where the ability to reproduce asexually helps ensure survival, particularly when potential mates may not be nearby.

Sexual reproduction: Turbellarians. Turbellarians are hermaphroditic and possess complex and highly diverse reproductive systems (Figure 10.22). The male system includes single (e.g., macrostomids), paired

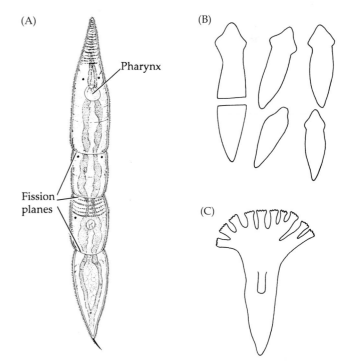

Figure 10.21 (A) Asexual reproduction by transverse fission in the catenulid rhabdocoel *Alaurina*. (B,C) Regeneration after experimental injuries in planarians.

Figure 10.22 Turbellarian reproductive systems. (A) Generalized acoel condition, without separate yolk glands (archoöphoran condition). (B) Generalized triclad condition with separate ovaries and yolk glands (neoöphoran condition). (C) The copulatory structures of a triclad (sagittal section).

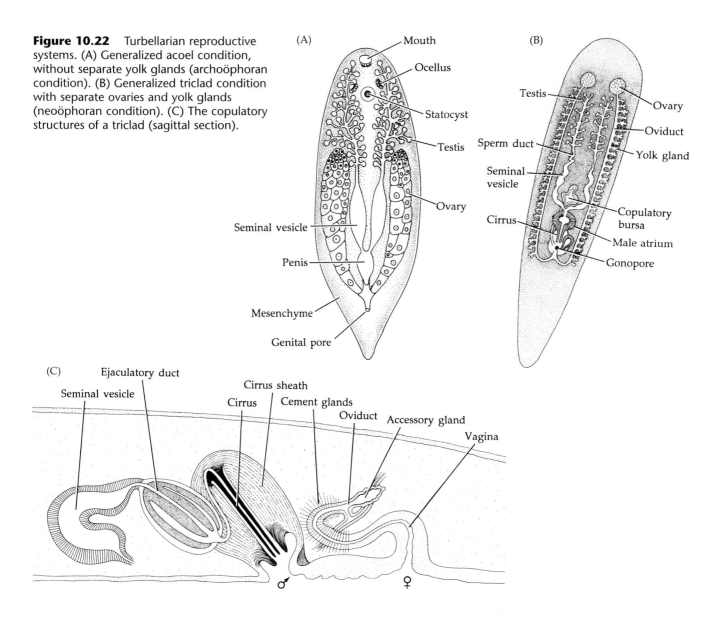

(e.g., many rhabdocoels), or multiple (e.g., polyclads) testes. The testes are generally drained by collecting tubules that unite to form one or two sperm ducts, which often lead to a precopulatory storage area or seminal vesicle. **Prostatic glands**, which supply seminal fluid to the sperm, are often associated with and empty into the seminal vesicle. The seminal vesicle is typically part of a muscular chamber called the **male atrium**, which houses the copulatory organ. The actual organ of sperm transfer may be a papilla-like penis or an eversible **cirrus**, through which sperm are forced by muscular action of the atrium.

The female reproductive system is more variable than that of the male. Much of the variation is related to whether the flatworm in question produces entolecithal or ectolecithal ova—that is, whether the worm is described as archoöphoran or neoöphoran. The archoöphorans (members of the orders Nemertodermatida, Acoela, Macrostomida, and Polycladida) typically possess an organ that produces both eggs and yolk. The

final product is entolecithal ova. Such an organ is called a **germovitellarium**, and may occur either singly or paired. In the neoöphorans (members of the orders Rhabdocoela, Prolecithophora, and Tricladida), the ovary (**germarium**) is separate from the yolk gland (**vitellarium**). Yolk-free eggs are produced by the ovary and then yolk is transported through a vitelline duct and deposited alongside the ova inside the eggshell, a process resulting in ectolecithal ova.

In both cases, the eggs are typically moved via an oviduct toward the **female atrium**, which often bears special chambers for receipt and storage of sperm (i.e., **copulatory bursa** and **seminal receptacle**). Associated with this arrangement may be a variety of accessory glands, such as cement glands, for the production of shells and egg cases.

The male and female gonopores are often separate, the female opening usually located posterior to the male pore. In some species, however, the two systems share a common genital opening, and in a few the male atrium

(A)

Egg capsule

Hatching

(B)

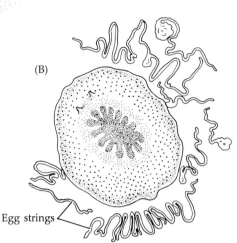

Egg strings

Figure 10.23 Mating and egg laying in turbellarians. (A) Mating, egg cluster, and hatching in a freshwater planarian. (B) Egg laying by the polyclad *Stylochus*.

opens just inside the mouth. In the latter case, the mouth is referred to as an **orogenital pore**.

Mating is usually by mutual cross-fertilization. The two mates align themselves so that the male gonopore of each is pressed against the female gonopore of the mate (Figure 10.23A). The male copulatory organ (penis or cirrus) is everted by hydrostatic pressure caused by the muscles surrounding the atrium and then is inserted into the mate's female atrium, where sperm are deposited. The mates then separate, each going its own way and carrying foreign sperm. Fertilization usually occurs as the eggs pass into the female atrium or within the oviduct itself. The zygotes are frequently stored for a period of time in special parts of the female system or in enlarged oviducts; any such storage area is called a **uterus**. A few turbellarians exhibit **hypodermic impregnation**, whereby the male copulatory organ is thrust through the body wall of the mate; the sperm are forcibly injected into the mesenchyme. By some method not yet understood, the sperm find their way to the female system and fertilize the eggs.

Once fertilization is accomplished, the zygotes are either retained by the parent within the uteri of the female reproductive tract or laid in various sorts of gelatinous or encapsulated egg masses (Figure 10.23B). Thus, most maternal turbellarians are obliged to contribute substantially toward the care of their embryos; they may be described as oviparous or ovoviviparous. Some freshwater triclads produce special overwintering zygotes, which are encapsulated and retained within the female reproductive tract until spring.

The general strategy of the vast majority of turbellarians is to produce relatively few zygotes, which are protected by brooding or encapsulation and undergo direct development. A few polyclads produce **Müller's larva**, which swims about for a few days prior to settling and metamorphosing (Figures 10.24 and 10.25). This larva is equipped with eight ventrally directed ciliated lobes, by means of which it swims. A few species of parasitic polyclads of the genus *Stylochus* produce a **Götte's larva**, which bears four rather than eight lobes; and members of the freshwater catenulid genus *Rhynchoscolex* pass through a vermiform stage in their development that has been referred to as a larval form (Ruppert 1978).

Early embryogeny differs greatly between the archoöphoran and neoophöran turbellarians. The entolecithal ova of the archoöphorans (e.g., acoels, polyclads, macrostomids) undergo some form of spiral cleavage, the details of which are described in Chapter 4. The pattern and cell fates in many of these archoö-

Figure 10.24 "Face-on" view of Müller's larva of a polyclad (*Planocera*).

Figure 10.25 Polyclad development. (A) A Müller's larva (sagittal section). (B) A later larval stage, showing formation of the pharyngeal apparatus (sagittal section).

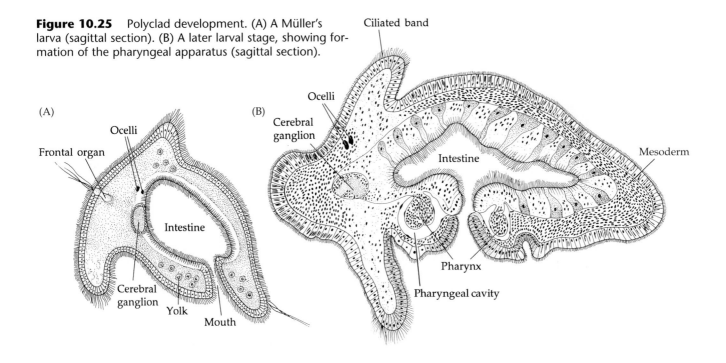

phoran embryos are distinctly protostome-like, and appear to represent an evolutionary step or precursor to the protostome line. (We examine this relationship more fully at the end of this chapter.)

The typical spiral cleavage in polyclads has been well studied. Quartets of cells are produced, the fates of which may be described using Wilson's coding system (see Chapter 4). By the end of spiral cleavage, the embryo is considered a stereoblastula, oriented with the derivatives of the first micromere quartet at the animal pole and the macromeres at the vegetal pole. The lq cells become the anterior ectoderm, cerebral ganglion, and most of the rest of the nervous system. The 2q derivatives contribute to ectoderm and ectomesoderm, particularly that of the pharyngeal apparatus and its associated musculature. The remainder of the ectoderm and probably some ectomesoderm are formed from the derivatives of the third micromere quartet. The 4d cell, normally associated solely with endomesoderm in typical protostomes, divides to produce a $4d_1$ and $4d_2$ cell in polyclads. The $4d_1$ gives rise to entoderm and thus to the intestine; the $4d_2$ produces the entomesoderm from which the body wall and mesenchymal muscles, much of the mesenchymal mass, and most of the reproductive system are derived. The remaining cells (4a, 4b, 4c, and the 4Q) include most of the yolk and are incorporated into the developing archenteron as embryonic food.

Gastrulation is by epiboly of the presumptive ectoderm derived from some of the cells of the first three micromere quartets. The ectoderm grows from the animal pole to the vegetal pole, surrounding the 4q and 4Q cells. At the vegetal pole the ectoderm turns inward as a stomodeal invagination, which later elaborates as the pharynx and connects with the developing intestine

(Figure 10.25). As development proceeds, the embryo flattens, with the mouth directed ventrally, and hatches as a tiny polyclad. If development is mixed, the larva emerges about the time the intestine is hollowing.

Members of the order Acoela are unique in that spiral cleavage occurs by the production of duets rather than quartets of cells (Figure 10.26). Thus, spiral displacement begins during the division from two to four cells, and a 32-cell stereoblastula is formed. The 4A, 4B, 4a, and 4b cells move inward from the vegetal pole and contribute to the majority of the inner cellular mass, particularly the digestive cells. The rest of the interior of the animal apparently derives from ectomesoderm arising from the second and third micromere duets, while the first duet divides to form an outer ectoderm and the nervous system. The inner cellular mass is never readily divisible into entoderm and mesoderm; and, because no 4d cell is produced, no "true" entomesoderm forms.

Because of the deposition of yolk on the surface of ectolecithal ova, the development of neoöphoran turbellarians is highly modified from the plan described

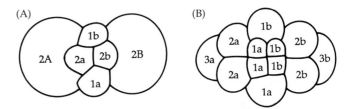

Figure 10.26 Spiral cleavage in the acoel *Polychoerus*. Animal pole views of 6-cell (A) and 16-cell (B) stages, illustrating the formation of duets rather than quartets of micromeres.

above. Certain species of rhabdocoels and triclads have been most extensively studied, and the two groups differ—especially in the early stages. In both cases, cleavage is so distorted that cell fates and germ layer formation cannot easily be compared with the typical spiralian pattern. In the rhabdocoels, early cleavage leads to the formation of three masses of cells positioned along the presumptive ventral surface of the embryo beneath the mass of yolk (Figure 10.27A). The cell masses then produce a layer of cells that extends around to enclose the yolk. This covering thickens to several cell layers, the innermost eventually becoming the intestinal lining (and enclosing the yolk), the outermost becoming the epidermis. The anterior cell mass produces the nervous system, the middle cell mass the pharynx and associated muscles. The posterior cell mass forms the rear portion of the worm and the reproductive system.

Early development in triclads differs from that of other neoöphorans. During early cleavage, the blastomeres are loose within a surrounding mass of fluid yolk. A few of the blastomeres migrate away from the others and flatten to produce a thin membrane enclosing a packet of the yolk including the remaining blastomeres (Figure 10.27B,C). Additional yolk cells are produced as a syncytial mass around a group of developing embryos and encapsulated, as many as 40 per capsule. Through migration and differentiation of various blastomeres, each embryo forms a temporary intestine, pharynx, and mouth, through which it ingests the yolky syncytium. The embryonic mouth eventually closes and the wall of the embryo thickens to form anterior, middle, and posterior cell masses, whose fates are similar to those in rhabdocoels.

Sexual reproduction: Flukes. Like the turbellarians, flukes are hermaphroditic and typically engage in mutual cross-fertilization. Self fertilization occurs only in rare cases. There is a great deal of variation in the details of the reproductive systems among flukes, but most are built around a common plan similar to that in certain turbellarians (Figure 10.28). The male system includes a variable number of testes (usually many in the monogenetic flukes and two in the digenetic flukes), all of which drain to a common sperm duct that leads to a **copulatory apparatus**, usually an eversible cirrus. The lumen of the cirrus is continuous with that of the sperm duct, and their junction is frequently enlarged as a seminal vesicle. Prostatic glands are typically present, opening into the cirrus lumen near the seminal vesicle. All of these terminal structures are housed within a muscular **cirrus sac**, the contraction of which causes eversion of the cirrus as an intromittent organ (Figure 10.28C). The common genital pore opens ventrally near the anterior end of the animal and leads to a shallow atrium, usually shared by both the male and female systems. Many monogenetic flukes have simpler male systems than that just described, often lacking much elaboration of the terminal structures and possessing a simple penis papilla rather than an eversible cirrus.

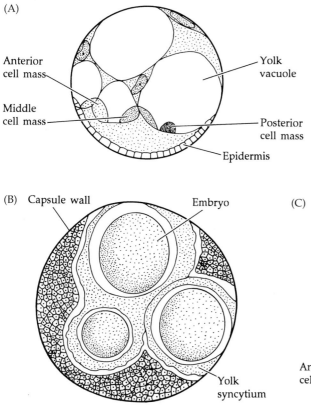

(A)

Anterior cell mass

Middle cell mass

Yolk vacuole

Posterior cell mass

Epidermis

Figure 10.27 Neoöphoran development. (A) The embryo of a typical rhabdocoel has three cell masses with large, vacuolated external yolk cells. (B) A triclad egg capsule containing three embryos surrounded by yolk syncytium. (C) This single triclad embryo has ingested the yolk through the temporary embryonic pharynx, and shows the three cell masses.

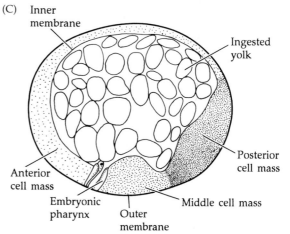

(B) Capsule wall

Embryo

Yolk syncytium

(C) Inner membrane

Ingested yolk

Posterior cell mass

Middle cell mass

Outer membrane

Embryonic pharynx

Anterior cell mass

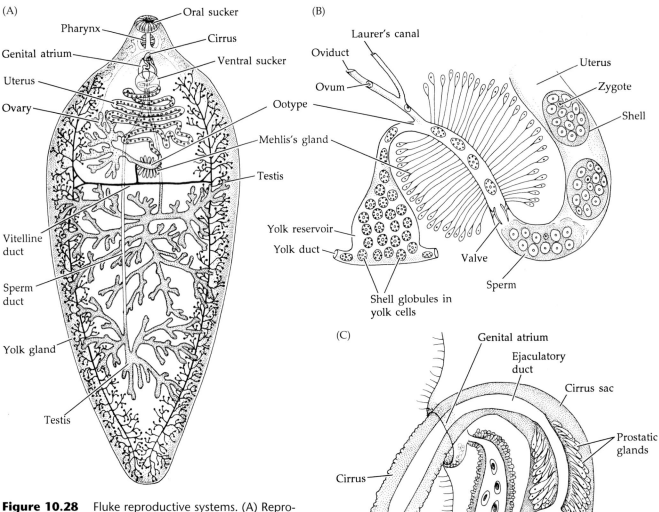

Figure 10.28 Fluke reproductive systems. (A) Reproductive structures of *Fasciola hepatica*. (B) The region of the oötype in *F. hepatica*. (C) Male copulatory apparatus with cirrus extended.

The female reproductive system (Figure 10.28B) usually bears a single ovary connected by a short oviduct to a region known as the **oötype**. The oviduct is joined by a **yolk duct** (= vitelline duct) formed by the union of paired ducts, which carry yolk from the multiple laterally placed yolk glands. A seminal receptacle is usually present as a blind pouch off the oviduct. Extending anteriorly to the genital atrium is a single uterus, which is sometimes modified as a **vagina** near the female gonopore.

Sperm are produced in the testes and stored prior to copulation in the seminal vesicle (Figure 10.28C). During mating, two flukes align themselves such that the cirrus of each can be inserted into the female orifice of the other. Sperm, along with semen from the prostatic glands, are ejaculated into the female system by muscular contractions. The sperm move to, and are stored within, the seminal receptacle, and the mates separate. As eggs pass through the oviduct to the oötype, they are fertilized by sperm released from the seminal receptacle into the oviduct.

Flukes produce ectolecithal ova. The yolk glands produce yolk, which is deposited outside the eggs along with secretions that form a tough shell around the zygote. Thus encapsulated, the zygotes move from the oötype into the uterus, probably aided by secretions from clusters of unicellular **Mehlis's glands**. The zygotes may be stored within the uterus for various lengths of time prior to release through the female gonopore.

Some flukes possess an additional canal that arises from the oviduct and serves as a special copulatory duct. This duct, called **Laurer's canal**, opens on the dorsal body surface and receives the male cirrus during mating. A few polyclad and triclad turbellarians also possess a Laurer's canal.

High fecundity is a general rule among parasites, and the flukes are no exception. The dangers of complex life cycles and host location result in extremely high mortality rates that must be offset by increased zygote production or asexual processes. Flukes may produce as many as 100,000 times as many eggs as free-living turbellarians.

The early stages of development in flukes are usually highly modified because of the ectolecithal nature of the ova. In species where little yolk is present, cleavage is holoblastic, and cell fates and germ layer formation have been traced accurately. Development is virtually always mixed, involving one or more independent larval stages.

The life cycles of monogenetic flukes are relatively simple and involve only a single host. Most of the adults are ectoparasites on fishes, although some attach to turtles, various amphibians, and even some invertebrates. A few members of the Monogenea have taken up mesoparasitic life and reside in host body chambers that open to the environment (e.g., gill chambers, mouth, bladder, cloacal cavity). When the embryos are released from the uterus they often attach to the host tissue by means of special adhesive threads on the shell. Upon hatching, a larval stage called an **oncomiracidium** is released to the environment (Figure 10.29). The oncomiracidium is densely ciliated and swims about until it encounters another appropriate host. The prohaptor and especially the opisthaptor develop during the larval stage and facilitate attachment to the new host, whereupon the larva metamorphoses to a juvenile trematode. It is about this time that the ciliated larval skin is shed and the tegument (= **neodermis**) forms. There are many variations on this basic life cycle among members of the class Monogenea; we present two in outline form in Figure 10.30.

The subclass Digenea includes some of the most successful parasites known. A good deal of variation exists not only in adult morphology but also in life cycles (Figure 10.31). In general, eggs are produced by adult worms in their definitive host. After fertilization, the zygotes are eventually discharged via the host's feces, urine, or sputum. Upon reaching the water they either are eaten by an intermediate host or hatch as free-swimming ciliated larvae called **miracidia**, which actively penetrate an intermediate host. Several asexual generations of larval forms occur in the intermediate host, eventually producing free-swimming forms called **cercaria**. The cercaria usually encyst within a second intermediate host, becoming **metacercaria**. Infection of the definitive host occurs when the metacercaria are eaten or, if there is no second intermediate host, when the cercaria penetrate directly. The larval skin is lost in the definitive host and the syncytial tegument develops.

In their adult stages, nearly all of the digenetic flukes are entoparasites of vertebrates. They are known to inhabit nearly every organ of the body, and

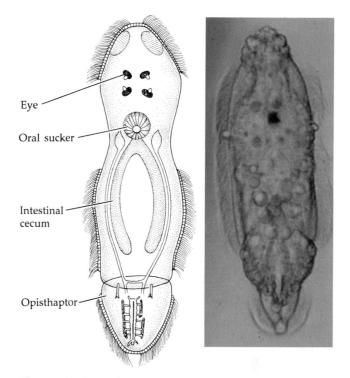

Figure 10.29 Photograph and drawing of an oncomiracidium larva of a monogenetic fluke.

many are serious pathogens of humans and livestock. The intermediate hosts of most digenetic flukes are gastropods, although some are known to use other invertebrates or even certain vertebrates. Most species of snails host one or more species of digenetic flukes; the common California tidal flat gastropod *Cerithidea californica* serves as intermediate host to nearly two dozen species of digenetic flukes, most of which ultimately infect shore birds. Space does not permit an account of more than a few of the life cycles of these worms. We begin below with a general case, using the Chinese liver fluke *Opisthorchis sinensis* as an example. This trematode is widespread in the Far East. It displays all the common stages found in the life cycles of most digenetic flukes.

The adult liver fluke usually lives within the branches of the bile duct in humans. This animal may reach several centimeters in length and in high numbers causes serious problems. While still in the uterus of the female reproductive tract, the zygotes develop to miracidia, each housed within its original egg case. Once released from the female system and passed out of the host with the feces, the miracidia are eaten by the first intermediate host, a snail of the genus *Parafossarulus*. The ciliated, swimming miracidium hatches from its egg case in the gut of the snail and migrates into the digestive gland. Here each miracidium becomes an asexually active form called a **sporocyst**, within which germinal cells become yet another larval form called a **redia**. Subsequently, germinal cells within the redia produce

Figure 10.30 Life cycles of two monogenetic trematodes. (A) The life cycle of *Dactylogyrus vastator*, a parasite of freshwater cyprinodont fishes. (B) The life cycle of *Polystoma integerrimum*, a parasite in the urinary bladders of frogs. This fascinating life history demonstrates the rather dramatic influences exerted by the developmental stage of the host on the development of the parasite. Under normal conditions, the adult fluke resides in the bladder of adult frogs. The fluke releases fertilized eggs into the water, where they hatch as oncomiracidia. These in turn become so-called gyrodactylid larvae, which attack the tadpole larval stages of the host. If the tadpole is very young, the fluke larvae attach to the external gills of the host and undergo precocious sexual maturation to produce more zygotes; these flukes die upon metamorphosis of the host. However, if the fluke larvae encounter more advanced tadpoles, they enter the branchial chambers and attach to the host's internal gills, where they reside until the host undergoes metamorphosis. At that time, the flukes leave the branchial chamber, migrate to the cloacal pore, and enter the host's bladder. Here the flukes live and grow, but they do not become sexually active until they are influenced by the host's sex hormones. Thus, sexual reproduction of the host and its parasites are synchronized—a pattern that guarantees availability of larval hosts for larval parasites!

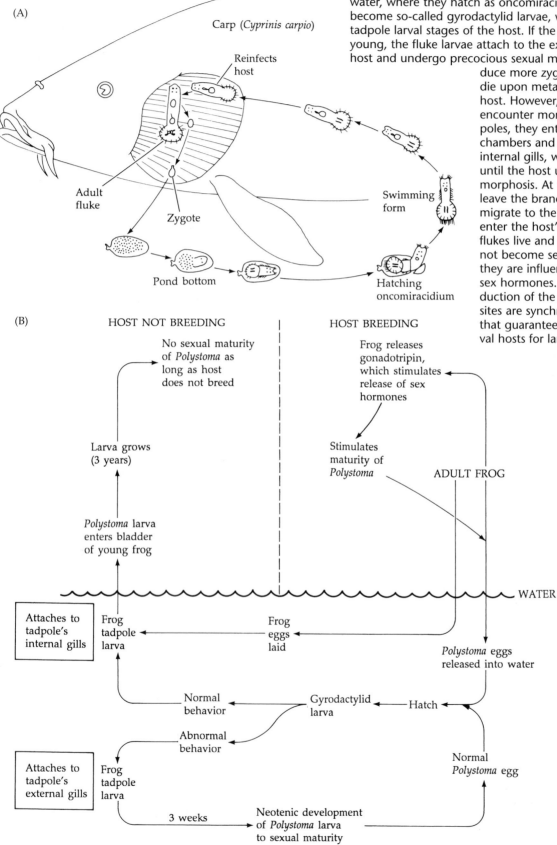

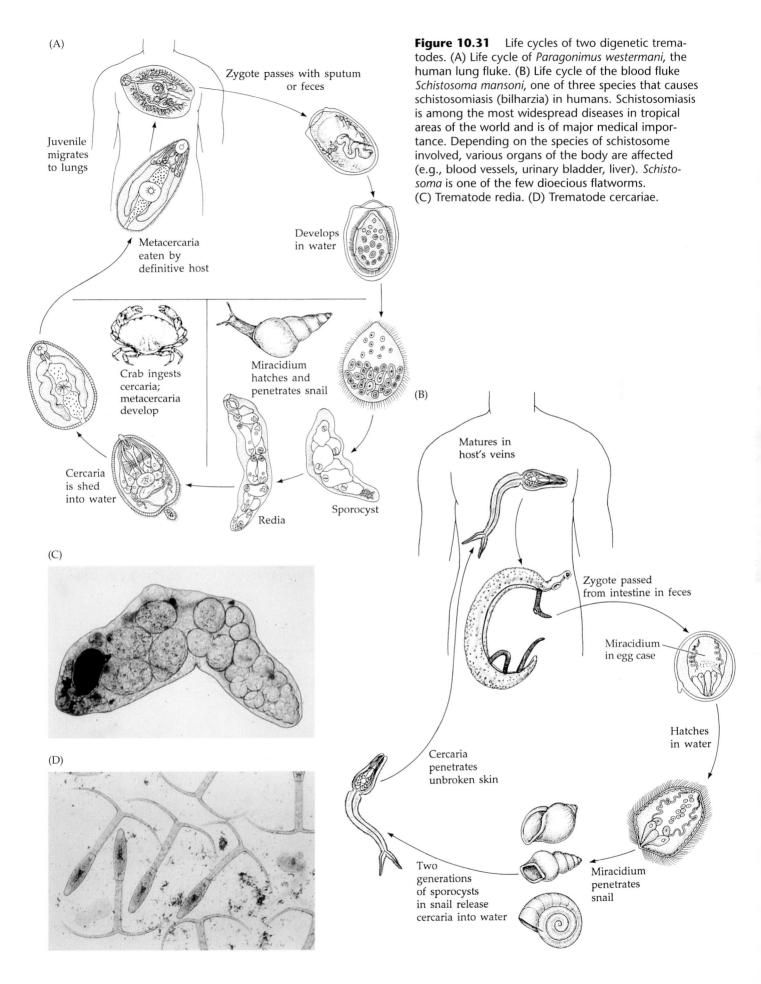

(A)

Zygote passes with sputum or feces

Juvenile migrates to lungs

Metacercaria eaten by definitive host

Develops in water

Crab ingests cercaria; metacercaria develop

Miracidium hatches and penetrates snail

Cercaria is shed into water

Redia

Sporocyst

Figure 10.31 Life cycles of two digenetic trematodes. (A) Life cycle of *Paragonimus westermani*, the human lung fluke. (B) Life cycle of the blood fluke *Schistosoma mansoni*, one of three species that causes schistosomiasis (bilharzia) in humans. Schistosomiasis is among the most widespread diseases in tropical areas of the world and is of major medical importance. Depending on the species of schistosome involved, various organs of the body are affected (e.g., blood vessels, urinary bladder, liver). *Schistosoma* is one of the few dioecious flatworms. (C) Trematode redia. (D) Trematode cercariae.

(B)

Matures in host's veins

Zygote passed from intestine in feces

Miracidium in egg case

Hatches in water

Cercaria penetrates unbroken skin

Two generations of sporocysts in snail release cercaria into water

Miracidium penetrates snail

(C)

(D)

the cercaria larvae. This double sequence of rapid asexual reproduction results in perhaps 250,000 cercariae from each original miracidium!

The cercariae leave the snail, swim about, and enter the second intermediate host, the Chinese golden carp (*Macropodus opercularis*). The cercariae of *Opisthorchis* burrow through the skin of the fish and encyst in the muscle tissue as metacercariae. If the fish is insufficiently cooked and then eaten by a human, the metacercariae survive and are released from their cysts by the action of the host's digestive enzymes. Once freed, they migrate into the bile duct, metamorphose into juvenile worms, mature, and complete the cycle. With this general life cycle in mind, we refer you to Figure 10.30 for a brief overview of two additional examples.

The critical point here is the strategy for survival displayed by these parasites. The advantages of such high specialization are efficiency and reduced competition with other species (once established in a proper host). It is, however, a costly strategy. There is, of course, no assurance of finding the proper hosts at the proper times, and mortalities are incredibly high. As we have emphasized earlier, the compensation for these mortalities is high fecundity coupled with asexual reproduction—and therein lies the expense.

Sexual reproduction: Tapeworms. That the "business of animals is to reproduce themselves" is a lesson the cestodes demonstrate well. Most of their time, energy, and body mass are devoted to the production of more tapeworms. Like other flatworms, cestodes are hermaphroditic and practice mutual cross-fertilization when mates are available. However, many eucestodes are known to self-fertilize. The cestodarians possess a single male and a single female reproductive system, whereas the eucestodes contain complete systems repeated in each proglottid. There is a good deal of variation in the details of these systems; the following is a generalized description of the male and female systems as they occur in a single proglottid of a eucestode (Figure 10.32).

The testes are numerous. Some are scattered throughout the mesenchyme, but most are concentrated along the lateral margins. Collecting tubules lead from the testes to a single coiled sperm duct, which extends laterally (as a seminal vesicle) to a cirrus housed within a muscular **cirrus sac**. The male system empties into a common genital atrium.

The female system usually includes two ovaries from which an oviduct extends to an ootype surrounded by shell glands. The uterus is a branched blind sac extending from the ootype. A duct extends from the female gonopore in the genital atrium to the oviduct; its junction with the oviduct, near the tube, is swollen as a seminal receptacle. The portion of the duct near the genital atrium is called the vagina. A diffuse yolk gland empties via a vitelline duct into the oviduct.

During mutual cross-fertilization the cirrus of each mate is inserted into the vagina of the other. Many tapeworms double back on themselves so that two proglottids of the same worm cross-fertilize; in some species, self-fertilization is known to occur within a single proglottid. Sperm are injected into the vaginal duct and are stored in the seminal receptacle. Eggs are fertilized as they move through the oviduct from the ovaries to the ootype. Capsule material and yolk cells are deposited around each zygote, and the zygotes are moved into the uterus for temporary storage. The reproductive systems

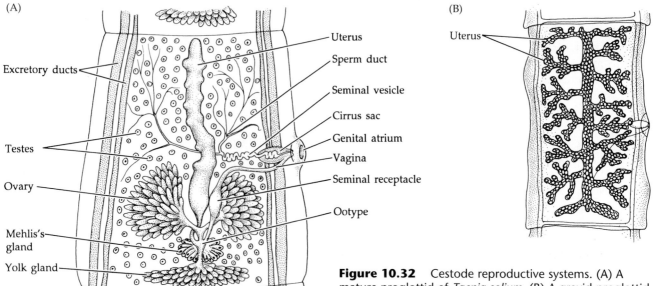

(A)

Excretory ducts

Testes

Ovary

Mehlis's gland

Yolk gland

Uterus
Sperm duct
Seminal vesicle
Cirrus sac
Genital atrium
Vagina
Seminal receptacle
Ootype

(B)

Uterus

Figure 10.32 Cestode reproductive systems. (A) A mature proglottid of *Taenia solium*. (B) A gravid proglottid with expanded uterus. See also Figure 10.4.

mature and become functional with age as they are moved more posteriorly along the body by the production of new proglottids. If mating has occurred, the proglottids toward the posterior end become filled with an expanded uterus engorged with developing embryos (Figure 10.32B). These proglottids eventually break free from the body and are lost from the host with the feces, although in some cases the proglottids release embryos inside the host and the embryos pass out with the feces. Early development of tapeworm embryos is drastically modified from the turbellarian pattern and varies somewhat among different groups. The ectolecithal ova have lost most vestiges of spiral cleavage, and even germ layer formation is often difficult or impossible to trace.

Most adult tapeworms live in the digestive tracts of vertebrates and usually require one or more intermediate hosts to complete their life cycles. A few can complete their life cycle in a single host. Depending on the number of hosts and other factors, tapeworm life cycles are quite variable, and we describe only two examples.

Taenia saginata (order Cyclophyllidea) is commonly known as the beef tapeworm, since cattle are the intermediate host. The adults, which may exceed 1 m in length, reside in human small intestines (Figure 10.33A). As proglottids mature, they are released in the host's feces. The fertilized eggs break free into the environment as the proglottids disintegrate. By this time each zygote has developed to a stage called an **oncosphere** surrounded by a resistant coat called an **embryophore**, which allows the embryo to remain in the environment for two or three months. Usually six tiny hooks are evident in the embryo; thus the oncosphere is sometimes called a **hexacanth**.

If grazing cattle ingest the oncosphere, it is released from its covering and is carried by the circulatory system to the cow's skeletal muscle. Here the oncosphere develops into a stage called the **cysticercus**, or **bladder worm**, which encysts in the connective tissue within the muscle of the intermediate host. Each cysticercus contains an invaginated developing scolex. If raw or poorly cooked infected beef is eaten by a human, then the scolex evaginates and attaches to the lining of the new host's small intestine, where the adult worm grows and matures. (Another tapeworm, *Taenia solium*, utilizes pigs as its intermediate host and follows a similar life cycle to that just described.)

The life cycles of some cestodes involve two or more host, such as that of *Diphyllobothrium latum*, the so-called broad fish tapeworm (order Pseudophyllidea) (Figure 10.33B). Nearly any fish-eating mammal, including humans, can serve as the definitive host for this tapeworm. Encapsulated zygotes are released from mature proglottids and shed in the host's feces. After one or two weeks in water, the embryos develop to the oncosphere (hexacanth) stage. At this time, each oncosphere is encased in a ciliated embryophore and it hatches as a free-swimming larva called a **coracidium**.

To successfully continue the life cycle, the coracidium must be eaten by the first intermediate host, a copepod (Crustacea). The cilia are shed and the released oncosphere bores through the gut wall into the host's body cavity, where it develops into a **procercoid** stage. Certain species of freshwater fish can serve as the second intermediate host. The fish eats the copepod, the procercoid bores through the gut and into the fish's muscle tissue, and there it grows into a segmented **plerocercoid** stage, complete with a tiny scolex. When a human consumes raw or undercooked infected fish, the plerocercoid attaches to the intestinal wall and matures.

Platyhelminth Phylogeny

Ideas about the origin of flatworms, their relationship to other taxa, and evolution within the group have been hotly debated for decades. We hinted at some of this controversy earlier in this chapter, and we discussed some of its implications in Chapter 4. In this section we first explore some hypotheses about flatworm origin and then examine some views on the relationships of taxa within the phylum. As you will see, opinions on these matters differ greatly and reflect some extremely diverse views about the position of flatworms in animal evolution.

There have been several popular hypotheses concerning the origin of flatworms. The ciliate-to-acoel hypothesis (discussed in Chapter 4 as part of the syncytial theory of Hadzi and others) has been abandoned by most modern zoologists. It is no longer tenable in its original form, due in part to the discovery that the syncytial nature of the entoderm of many acoels is probably secondarily derived from an ancestor with a cellular gut.

Another hypothesis has been called the ctenophore–polyclad theory. Some workers have suggested that the ctenophores (Chapter 9) gave rise to polyclad turbellarians. This scenario envisions a flattened ctenophore that assumed a benthic, crawling lifestyle, with the mouth directed against the substratum. By reducing the tentacles and moving them forward along with the apical sense organ, a bilateral condition was achieved. Couple these events with increased gut branching and the formation of a plicate pharynx, and a polyclad bauplan is approximated—at least on paper. This hypothesis, too, no longer has much popular support.

Having dispensed with the above proposals, at least temporarily, we can examine the major persisting ideas on the origin of the flatworm body plan. We can safely assume that the original flatworm was turbellarian-like, although not necessarily assignable to any extant order. Peter Ax, Tor Karling, and Ulrich Ehlers have presented various versions of the turbellarian archetype. These hypothetical ancestral forms are envisioned as having had a simple pharynx and a saclike gut without diverticula (Figure 10.34A,B). Another popular version of the arche-

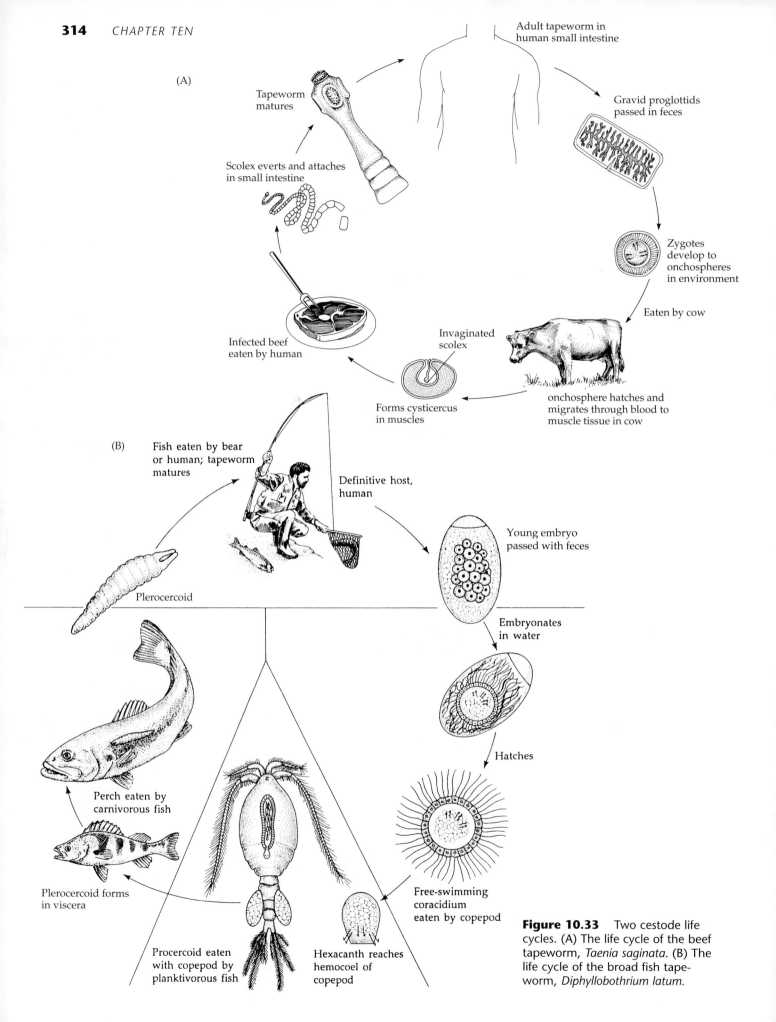

(A)

Tapeworm matures

Adult tapeworm in human small intestine

Gravid proglottids passed in feces

Scolex everts and attaches in small intestine

Zygotes develop to onchospheres in environment

Eaten by cow

Infected beef eaten by human

Invaginated scolex

Forms cysticercus in muscles

onchosphere hatches and migrates through blood to muscle tissue in cow

(B)

Fish eaten by bear or human; tapeworm matures

Definitive host, human

Young embryo passed with feces

Plerocercoid

Embryonates in water

Hatches

Perch eaten by carnivorous fish

Plerocercoid forms in viscera

Free-swimming coracidium eaten by copepod

Procercoid eaten with copepod by planktivorous fish

Hexacanth reaches hemocoel of copepod

Figure 10.33 Two cestode life cycles. (A) The life cycle of the beef tapeworm, *Taenia saginata*. (B) The life cycle of the broad fish tapeworm, *Diphyllobothrium latum*.

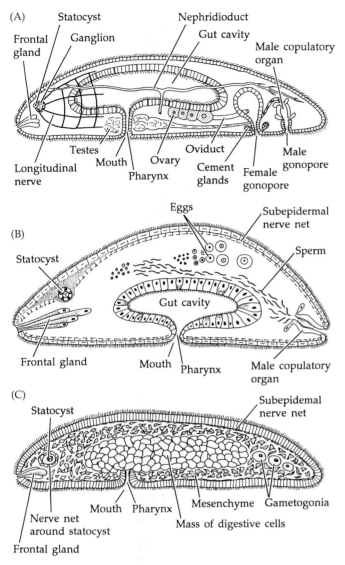

Figure 10.34 Three examples of hypothetical turbellarian archetypes. (A) Macrostomid-like archetype suggested by Ax (1963). (B) Archetype proposed by Karling (1974). (C) Acoel-like archetype proposed by several authors.

within the lineage leading to the protostomes. Taking this idea further, the flatworms represent the "first" spiralian taxon, and perhaps the stem group from which the protostome clade arose. An evolutionary tree depicting these events is shown in Figure 10.35A.

Second, in 1963 Peter Ax suggested that flatworms represent a series of reductions from a vermiform coelomate ancestor. Since that time, others have presented mounting evidence that the platyhelminth acoelomate condition is secondarily derived from a coelomate ancestor. This implies that the origin of the protostome mesoderm probably occurred simultaneously with the origin of the coelom, and that acoelomate animals (and probably blastocoelomate animals as well) branched from that coelomate clade (Figure 10.35B). Some workers suggest that these lineages may have arisen through neoteny from developmental stages of protostomes prior to the embryonic appearance of the coelomic cavi-

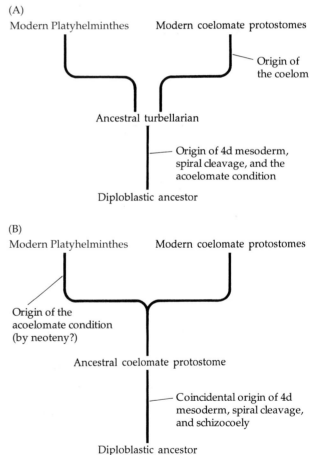

Figure 10.35 Two views on the origin of flatworms and the acoelomate condition. (A) An evolutionary tree based on the assumption that the acoelomate condition is primitive within the triploblastic spiralians, making Platyhelminthes the "first" descendant group on this lineage. (B) An evolutionary tree based on the coincidental origin of the coelom, spiral cleavage, and mesoderm; here the flatworms are viewed as arising (by neoteny) from a coelomate ancestor.

type is an acoel form, lacking any gut cavity (Figure 10.34C), although Ax, Karling, and Ehlers all hold that the solid acoel condition was secondarily derived. There is also general agreement that the primitive turbellarian was an archoöphoran with spiral cleavage and possessed a single-layered, completely ciliated epidermis.

Considering the above, we are left today with at least two very different ideas on the origin of the phylum Platyhelminthes. First, as discussed in Chapter 4, is the hypothesis that turbellarian flatworms arose from some diploblastic, radially symmetrical, planuloid ancestor. Various versions of this scenario enjoy some support, but all share a common theme: that the very first triploblastic, bilaterally symmetrical organisms were like flatworms. This view implies that the acoelomate condition is primitive within triploblastic phyla, at least ties—that is, from larval or other stages that had solid bodies or still contained the blastocoel. Controversies such as this (especially when reduction, reversion, or neoteny are invoked) illustrate why it is difficult to derive formal cladograms at the phylum level.

Recent DNA data (18S rDNA sequences) have suggested that the acoels (order Acoela) may be rather distinct from other Platyhelminthes, perhaps even warranting their own separate phylum. The unique version of spiral cleavage in acoels lends support to this idea, and preliminary phylogenetic analyses suggest their placement between the diploblastic phyla (Cnidaria, Ctenophora) and the triploblasts.

Deserving mention at this point is a strange flatworm-like creature named *Xenoturbella bocki*. These small, ciliated, marine worms were first discovered in mud sediments off the coast of Sweden, and have been the subject of controversy ever since. Variously considered to be a flatworm, a cnidarian, or a hemichordate, *Xenoturbella* is currently viewed as the sole species of an independent taxon not assignable to any known phylum. This proposal is accompanied by the hypothesis that *Xenoturbella* arose from a common ancestor to all the bilaterally symmetrical animals, and that it may re-tain some of the most primitive features of the Metazoa (for details see Ehlers and Sopott-Ehlers 1997).

Being uncertain about the origin of the flatworms obviously creates difficulties in analyzing phylogeny within the phylum. These problems are compounded by the absence of clear synapomorphies defining the Platyhelminthes. Ehlers (1986, 1995) suggests that the Platyhelminthes are defined by the absence of mitosis in somatic cells in all postembryonic stages, and the possession of multiciliated epidermal cells lacking centrioles. The first trait is based on evidence suggesting that somatic growth—as during regeneration—is provided by totipotent, undifferentiated "stem" cells rather than by proliferation of previously differentiated ones. Whether this condition is unique to flatworms, or is a retained symplesiomorphy, remains to be seen. The second character is based on the assumption that the flatworms share a common ancestor with the Gnathostomulida (see Chapter 12), which possess monociliated epidermal cells, rendering the multiciliated condition derived.

These and other ideas are still in the early stages of analysis and all of these hypotheses should be taken as tentative at the present time. We have opted for a conservative approach. We continue to view the flatworms as lacking solid synapomorphies, thus supporting the notion that an ancient flatworm was not only the precursor of modern-day forms, but the ancestor of the protostomes as well. Clearly, this view suggests that the phylum Platyhelminthes, as herein defined, is a paraphyletic group.

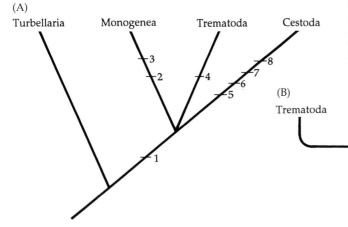

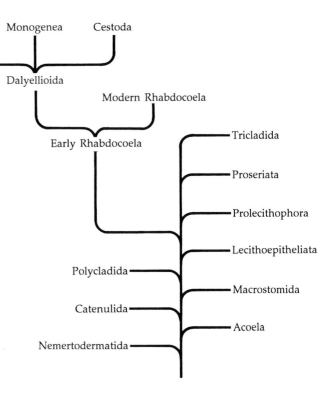

Figure 10.36 (A) A cladogram to the classes of flatworms. The absence of synapomorphies for the turbellarians suggests that the ancestral platyhelminth was itself a turbellarian, and that some member of that class (perhaps a dalyellioid) was ancestral to the three living parasitic groups. The classes Monogenea, Trematoda, and Cestoda share the unique synapomorphy of a tegument that arises during development through replacement of the larval epidermis by extensions of mesenchymal cells (1). This arrangement is strengthened by the exclusively parasitic nature of members of these three classes. We cannot unambiguously resolve the trichotomy leading to these three taxa, but each line is defined by synapomorphies as follows: prohaptor (2); opisthaptor (3) (Monogenea); ventral sucker or acetabulum in most (4) (Trematoda); tegumental microtriches (5); scolex (6); loss of digestive tract (7); and strobilation of the body (8) (Cestoda). (B) A traditional evolutionary tree depicting general ideas about the relationships among the orders of Turbellaria and the parasitic classes.

Figure 10.36A is a cladogram that emphasizes the uniqueness of the tegument as a synapomorphy uniting the Monogenea, Trematoda, and Cestoda, as discussed earlier in the chapter. The absence of synapomorphies for the Turbellaria suggest that it is a paraphyletic group. Our cladogram is admittedly a simple one, based on very few characters (some of which are questionable), and several publications listed in the references present alternative hypotheses or refinements of the tree shown here.

Figure 10.36B is a traditional (orthodox) evolutionary tree illustrating one set of hypotheses about overall platyhelminth relationships, including the proposition that the parasitic taxa arose from a dalyellioid-like turbellarian ancestor. Again, much of this is highly speculative. The situation is in such a state of flux that we anticipate substantial changes in the next few years, perhaps leading to a fundamental restructuring of flatworm classification.

Selected References

We have included only a few major references on the trematodes and cestodes, and refer you to texts on general parasitology and helminthology for additional literature lists.

General References

Auladell, C., J. Garcia-Valero and J. Baguña. 1993. Ultrastructural localization of RNA in the chromatoid bodies of undifferentiated cells (neoblasts) in planarians by the RNase-gold complex technique. J. Morph. 216: 319–326.

Ax, P. 1984. *Das phylogenetische System. Systematisierung der lebenden Natur aufgrund ihrer Phylogenese.* Gustav Fischer Verlag, Stuttgart.

Ax, P. 1996. *Multicellular Animals: A New Approach to the Phylogenetic Order in Nature.* Springer-Verlag, Berlin.

Boeger, W. A. and D. C. Kritsky. 1993. Phylogeny and revised classification of the Monogena Bychowsky, 1937 (Platyhelminthes). Syst. Parasitol. 26: 1–32.

Brooks, D. R. 1982. Higher classification of parasitic Platyhelminthes and fundamentals of cestode classification. *In* D. F. Mettrick and S. S. Dresser (eds.), *Parasites: Their World and Ours.* Elsevier Biomedical Press, Amsterdam.

de Beauchamp, P., M. Caullery, L. Euzet, P. Grasse and C. Joyeux. 1961. Plathelminthes et Mesozoaires. *In* P. Grasse (ed.), *Traité de Zoologie*, Vol. 4, Pt. 1. Masson et Cie, Paris, pp. 1–729.

Burt, D. R. R. 1970. *Platyhelminthes and Parasitism.* American Elsevier, New York.

Dougherty, E. C. 1963. *The Lower Metazoa: Comparative Biology and Phylogeny.* University of California Press, Berkeley and Los Angeles. [See in particular Sections I, II, and III.]

Ehlers, U. 1984. *Das phylogenetische System der Platyhelminthes.* Akademie der Wissenschaften und der Literatur, Mainz, and Gustav Fischer Verlag, Stuttgart.

Ehlers, U. 1986. Comments on a phylogenetic system of the Platyhelminthes. Hydrobiologia 132: 1–12.

Ehlers, U. 1995. The basic organization of the Plathelminthes. Hydrobiologica 305: 21–26.

Ehlers, U. and B. Sopott-Ehlers. 1997. Ultrastructure of the subepidermal musculature of *Xenoturbella bocki*, the adelphotaxon of the Bilateria. Zoomorphology 117: 71–79.

Ellis, C.H. and A. Fausto-Sterling. 1997. Platyhelminthes: The flatworms. *In* S. F. Gilbert and A. M. Raunio (eds.), *Embryology: Constructing the Organism.* Sinauer Associates, Sunderland, MA, pp. 115–130.

Florkin, M. and B. T. Scheer. 1968. *Chemical Zoology*, Vol. 2. Academic Press, New York.

Hyman, L. H. 1951. *The Invertebrates*, Vol. 2, Platyhelminthes and Rhynchocoela: The Acoelomate Bilateria. McGraw-Hill, New York.

Llewellyn, J. 1965. The evolution of parasitic platyhelminths. *In* A. E. R. Taylor (ed.), *Evolution of Parasites*, 3rd Ed. Symp. Soc. Parasitol. Blackwell, Oxford, pp. 47–78.

Llewellyn, J. 1986. Phylogenetic inference from platyhelminth life-cycle stages. Int. J. Parasitol. 17: 281–289.

Malakhov, V. V. and N. V. Trubitsina. 1998. Embryonic development of the polyclad turbellarian *Pseudoceros japonicus* from the Sea of Japan. Russian J. Mar. Biol. 24 (2): 106–113.

Martin, G. G. 1978. Ciliary gliding in lower invertebrates. Zoomorphologie 91: 249–262.

Morris, S. C., J. D. George, R. Gibson and H. M. Platt (eds.). 1985. *The Origins and Relationships of Lower Invertebrates.* The Systematics Association, Spec. Vol. No. 28. Oxford. [Includes several papers addressing recent views on flatworm phylogeny; see especially papers by Ax, Ehlers, and Smith and Tyler.]

Rieger, R. M. 1986. Über den Ursprung der Bilateria: Die Bedeutung der Ultrastrukturforschung für ein neues Verstehen der Metazoenevolution. Verh. Dtsch. Zool. Ges. 79: 31–50.

Ruppert, E. E. and P. R. Smith. 1988. The functional organization of filtration nephridia. Biol. Rev. 63:231–258.

Turbellaria

Ax, P. and G. Apelt. 1965. Die "Zooxanthellen" von *Convoluta convoluta* (Turbellaria, Acoela) entstehen aus Diatomen. Naturwissenschaften 52: 444–446.

Baguña, J. and C. Boyer. 1990. Experimental embryology of the Turbellaria: Present knowledge, open questions, and future trends. *In* H. J. Marthy (ed.), *Experimental Embryology in Aquatic Plants and Animals.* Plenum, New York.

Baguña, J., E. Salo, R. Romero, J. García-Fernández, D. Bueno, A. M. Muñoz-Marmol, J. R. Byascas-Ramírez and A. Casali. 1994. Regeneration and pattern formation in planarians: Cells, molecules, and genes. Zool. Sci. 11: 781–795.

Bowen, I. D. 1980. Phagocytosis in *Polycelis tenuis*. *In* D. C. Smith and Y. Tiffon (eds.), *Nutrition in the Lower Metazoa.* Pergamon Press, Oxford, pp. 1–14.

Boyer, B. C. 1971. Regulative development in a spiralian embryo as shown by cell deletion experiments on the acoel *Childia*. J. Exp. Zool. 176: 97–105.

Boyer, B. C. 1989. The role of the first quartet micromeres in the development of the polyclad *Hoploplana inquilina*. Biol. Bull. 177: 338–343.

Boyer, B.C., J.Q. Henry and M.Q. Martindale. 1996. Dual origins of mesoderm in a basal spiralian: Cell lineage analysis in the polyclad turbellarian *Hoploplana inquilina*. Dev. Biol. 179: 329–338.

Crezee, M. 1982. Turbellaria. *In* S. P. Parker (ed.), *Synopsis and Classification of Living Organisms*, Vol. 1. McGraw Hill, New York, pp. 718–740.

Eakin, R. M. and J. L. Brandenberger. 1980. Unique eye of probable evolutionary significance. Science 211: 1189–1190.

Ehlers, U. 1991. Comparative morphology of statocysts in the Platyhelminthes and Xenoturbellida. Hydrobiologia 227: 263–271.

Ehlers, U. 1992a. Frontal glandular and sensory structures in *Nemertoderma* (Nemertodermatida) and *Paratomella* (Acoela): Ultrastructure and phylogenetic implications for the monophyly of the Euplathelminthes (Plathelminthes). Zoomorph. 112: 227–236.

Ehlers, U. 1992b. On the fine structure of *Paratomella rubra* Rieger & Ott (Acoela) and the postition of the taxon *Paratomella* Dorjes in a phylogenetic system of the Acoelomorpha (Plathelminthes). Microfauna Marina 7: 265–293.

Heitkamp, C. 1977. The reproductive biology of *Mesostoma ehrenbergii*. Hydrobiologia 55: 21–32.

Henley, C. 1974. Platyhelminthes (Turbellaria). *In* A. C. Giese and J. S. Pearse (eds.), *Reproduction of Marine Invertebrates*, Vol. 1, Acoelomate and Pseudocoelomate Metazoans. Academic Press, New York.

Hurley, A. C. 1976. The polyclad flatworm *Stylochus tripartitus* Hyman as a barnacle predator. Crustaceana 3(1): 110–111.

Hooge, M. D. and S. Tyler. 1999. Musculature of the faculative parasite *Urastoma cyprimae* (Platyhelminthes). J. Morphol. 241: 207–216.

Jennings, J. B. 1980. Nutrition in symbiotic Turbellaria. *In* D. C. Smith and Y. Tiffon (eds.), *Nutrition in the Lower Metazoa.* Pergamon Press, Oxford, pp. 45–56.

Jokiel, P. L. and S. J. Townsley. 1974. Biology of the polyclad *Prosthiostomum* sp., a new coral parasite from Hawaii. Pacific Sci. 28(4): 361–375.

Karling, T. G. 1966. On nematocysts and similar structures in turbellarians. Acta Zool. Fenn. 116: 1–21.

Karling, T. G. and M. Meinander (eds.). 1978. The Alex Luther Centennial Symposium on Turbellaria. Acta Zool. Fenn. 154: 193–207.

Kato, K. 1940. On the development of some Japanese polyclads. Japan. J. Zool. 8: 537–573.

Lauer, D. M. and B. Fried. 1977. Observations on nutrition of *Bdelloura candida*, an ectocommensal of *Limulus polyphemus*. Am. Midl. Nat. 97(1): 240–247.

McKanna, J. A. 1968. Fine structure of the protonephridial system in planaria. I. Female cells. Z. Zellforsch. Mikrosk. Anat. 92: 509–523.

Moraczewski, J. 1977. Asexual reproduction and regeneration of *Catenula*. Zoomorphologie 88: 65–80.

Moraczewski, J., A. Czubaj and J. Bakowska. 1977. Organization and ultrastructure of the nervous system in Catenulida. Zoomorph. 87: 87–95.

Mueller, J. F. 1965. Helminth life cycles. Am. Zool. 5: 131–139.

Muscatine, L., J. E. Boyle and D. C. Smith. 1974. Symbiosis of the acoel flatworm *Convoluta roscoffensis* with the alga *Platymonas convolutae.* Proc. R. Soc. Lond. 187(1087): 221–234.

Nentwig, M. R. 1978. Comparative morphological studies after decapitation and after fission in the planarian *Dugesia dorotocephala*. Trans. Am. Microsc. Soc. 97: 297–310.

Ogren, R. E. 1995. Predation behavior of land planarians. Hydrobiologia 305: 105–111.

Prudhoe, S. 1985. *A Monograph on the Polyclad Turbellaria*. Oxford University Press, New York.

Riser, N. W. and M. P. Morse (eds.). 1974. *Biology of the Turbellaria*. Libbie H. Hyman Memorial Volume. McGraw-Hill, New York.

Ruppert, E. E. 1978. A review of metamorphosis of turbellarian larvae. *In* F. S. Chia and M. Rise (eds.), *Settlement and Metamorphosis of Marine Invertebrate Larvae*. Elsevier/North-Holland Biomedical Press, Amsterdam, pp. 65–81.

Ruiz-Trillo, I. et al. 1999. Acoel flatworms: Earliest extant bilaterian metazoans, not members of Platyhelminthes. Science 283: 1919–1923.

Salo, E., A. M. Muñoz-Marmol, J. R. Byascas-Ramírez, J. García-Fernández, A. Mirales, Z. Casali, M. Cormominas and J. Baguña. 1995. The freshwater planarian *Dugesia* (G.) *tigrina* contains a great diversity of homeobox genes. Hydrobiologia 305: 269–275.

Smith, J., S. Tyler, M. B. Thomas and R. M. Rieger. 1982. The morphology of turbellarian rhabdites: Phylogenetic implications. Trans. Am. Microsc. Soc. 101: 209–228.

Smith, J. III and S. Tyler. 1985. The acoel turbellarians: Kingpins of metazoan evolution or a specialized offshoot. *In* S. C. Morris, J. D. George, R. Gibson and H. M. Platt (eds.), *The Origins and Relationships of Lower Invertebrates*. Syst. Assoc. Spec. Vol. No. 28, Oxford, pp. 123–142.

Sopott-Ehlers, B. 1991. Comparative morphology of photoreceptors in free-living platyhelminths: A survey. Hydrobiologia 227: 231–239.

Steinbock, O. 1966. Die Hofsteniiden (Turbellaria Acoela): Grundsüatzliches zur Evolution der Turbellarien. Z. Zool. Syst. Evolutionsforsch. 4: 58–195.

Tyler, S. and R. M. Rieger. 1975. Uniflagellate spermatozoa in *Nemertoderma* and their phylogenetic significance. Science 188: 730–732.

Trematoda and Monogenea

Brooks, D. R., S. M. Bandoni, C. A. Macdonald and R. T. O'Grady. 1989. Aspects of the phylogeny of the Trematoda Rudolphi, 1808 (Platyhelminthes: Cercomeria). Can. J. Zool. 67: 2609–2624.

Bychowsky, B. E. 1957. *Monogenetic Trematodes: Their Systematics and Phylogeny*. American Institute of Biological Sciences, Washington, DC. [Translated from the Russian.]

Combes, C. et al. 1980. The world atlas of cercariae. Mem. Mus. Natl. Hist. Nat. Ser. A Zool. 115: 1–235.

Dawes, D. 1956. *The Trematoda*. Cambridge University Press, New York.

Erasmus, D. A. 1972. *The Biology of Trematodes*. Crane, Russak, New York.

Martin, W. E. 1972. An annotated key to the cercariae that develop in the snail *Cerithidea californica*. Bull. South. Calif. Acad. Sci. 71(1): 39–43.

Pearson, J. C. 1972. A phylogeny of life-cycle patterns of the Digenea. Adv. Parasitol. 10: 153–189.

Schell, S. C. 1982. Trematoda. *In* S. P. Parker (ed.), *Synopsis and Classification of Living Organisms*. Vol. 1. McGraw-Hill, New York, pp. 740–807.

Smyth, J. D. and D. W. Halton. 1985. *The Physiology of Trematodes*, 2nd Ed. W. H. Freeman, San Francisco, CA.

Sproston, N. G. 1946. A synopsis of the monogenetic trematodes. Trans. Zool. Soc. Lond. 25: 185–600.

Yamaguti, S. 1963. *Systema Helminthum*, Vol. 4, Monogenea and Aspidocatylea. Interscience, New York.

Yamaguti, S. 1971. *Synopsis of Digenetic Trematodes of Vertebrates*, Vols. 1, 2. Keigaku, Tokyo.

Yamaguti, S. 1975. *A Synoptical Review of Life Histories of Digenetic Trematodes of Vertebrates*. Keigaku, Tokyo.

Cestoda

Aral, H. P. (ed.). 1980. *Biology of the Tapeworm Hymenolepsis diminuta*. Academic Press, New York.

Biserova, N. M., K. S. Margaretha, M. R. Gustafsson and N. B. Terenina. 1996. The nervous system of the pike-tapeworm *Triaenophorus nodulosus* (Cestoda: Pseudophyllidea): Ultrastructure and immunocytochemical mapping of aminergic and peptidergic elements. Invert. Biol. 115: 273–285.

Brooks, D. R., E. P. Hoberg and P. J. Weekes. 1991. Preliminary phylogenetic systematic analysis of the major lineages of the Eucestoda (Platyhelminthes: Cercomeria). Proc. Biol. Soc. Wash. 104(4): 651–668.

Schmidt, G. D. 1982. Cestoda. *In* S. P. Parker (ed.), *Synopsis and Classification of Living Organisms*, Vol. 1. McGraw-Hill, New York, pp. 807–822.

Smyth, J. D. 1969. *The Physiology of Cestodes*. W. H. Freeman, San Francisco.

Uglem, G. L. and J. J. Just. 1983. Trypsin inhibition by tapeworms: Antienzyme secretion or pH adjustment. Science 220: 79–81.

Wardle, R., J. McLeod and S. Radinovsky. 1974. *Advances in the Zoology of Tapeworms*. University of Minnesota Press, Minneapolis.

Yamaguti, S. 1959. *Systema Helminthum*, Vol. 2, The Cestodes of Vertebrates. Interscience, New York.

Xylander, W. E. R. 1987. Das Protonephridialsystem der Cestoda: Evolutive Veränderungen und ihre mögliche funktionelle Bedeutung. Verh. Dtsch. Zool. Ges. 80: 257–258.

General Parasitology

Baer, J. G. 1952. *Ecology of Animal Parasites*. University of Illinois Press, Urbana.

Bogitsh, B. J. and T. C. Cheng. 1990. *Human Parasitology*. Saunders College Publishing, New York.

Marquardt, W. C., R. S. Demaree and R. B. Grieve. 2000. *Parasitology and Vector Biology*. 2nd Ed. Academic Press, San Diego, CA.

Noble, E., G. Noble, G. A. Schad and A. J. MacInnes. 1989. *Parasitology*, 6th Ed. Lea and Febiger, Philadelphia.

Olsen, O. W. 1974. *Animal Parasites: Their Life Cycles and Ecology*, 3rd Ed. University Park Press, Baltimore.

Rohde, K. 1982. *Ecology of Marine Parasites*. University of Queensland Press, St. Lucia.

Roberts, L. S. and J. Janovy, Jr. 1999. *Foundations of Parasitology*, 6th Ed. McGraw-Hill, New York.

Smyth, J. D. 1995. *Introduction to Animal Parasitology*, 3rd Ed. Cambridge Univ. Press, New York.

11

Phylum Nemertea:
The Ribbon Worms

Nemerteans are rewarding and satisfying animals to work with, and there is still so very much that we do not know about them.

Ray Gibson,
Nemerteans, 1972

Members of the phylum Nemertea (Greek, "a sea nymph") or Rhynchocoela (Greek *rhynchos*, "snout"; *coel*, "cavity") are commonly called ribbon worms. Figure 11.1 illustrates a variety of body forms within this taxon and the major features of their anatomy. These unsegmented vermiform animals are usually flattened dorsoventrally and are moderately cephalized; they possess highly extensible bodies. Many ribbon worm species are rather drab in appearance, but many others are brightly colored and distinctively marked (e.g. the tropical eastern Pacific species *Baseodiscus mexicanus*, above).

About 900 species of nemerteans have been described. They range in length from less than 1 cm to several meters. Many can stretch easily to several times their contracted lengths (one specimen of *Lineus longissimus* reportedly measured 60 m in length). They are predominately benthic marine animals. A few, however, are planktonic, and some are symbiotic in molluscs or other marine invertebrates. A few freshwater and terrestrial species are known, the latter being found in greenhouses.

Many features of the nemertean body plan (Box 11A) are similar to the conditions seen in flatworms, and the two taxa are often grouped together as the triploblastic acoelomate bilateria. There are similarities in the overall architecture of the nervous systems, the types of sense organs, and the protonephridial excretory structures. However, in other respects ribbon worms differ greatly from flatworms: nemerteans possess an anus (a complete, one-way digestive tract), a closed circulatory system, and an eversible **proboscis** surrounded by a hydrostatic cavity called the **rhynchocoel**. They probably also possess remnants of a schizocoelous coelom. The structure of the proboscis apparatus is unique to nemerteans and represents a novel synapomorphy that distinguishes the Nemertea from all other invertebrate taxa.

(A)

(B)

(C)

(D)

(E)

Cephalic slit

(G)

Cephalic blood vessel

Cerebral ganglion

Mouth
Esophagus
Network of blood lacunae

Lateral blood vessel

Intestinal diverticulum

Genital pore

Dorsal blood vessel

Gonad

(F)

Proboscis

Eggs

Gut

Caudal cirrus

Anus

Sucker

Anus

Connecting vessel

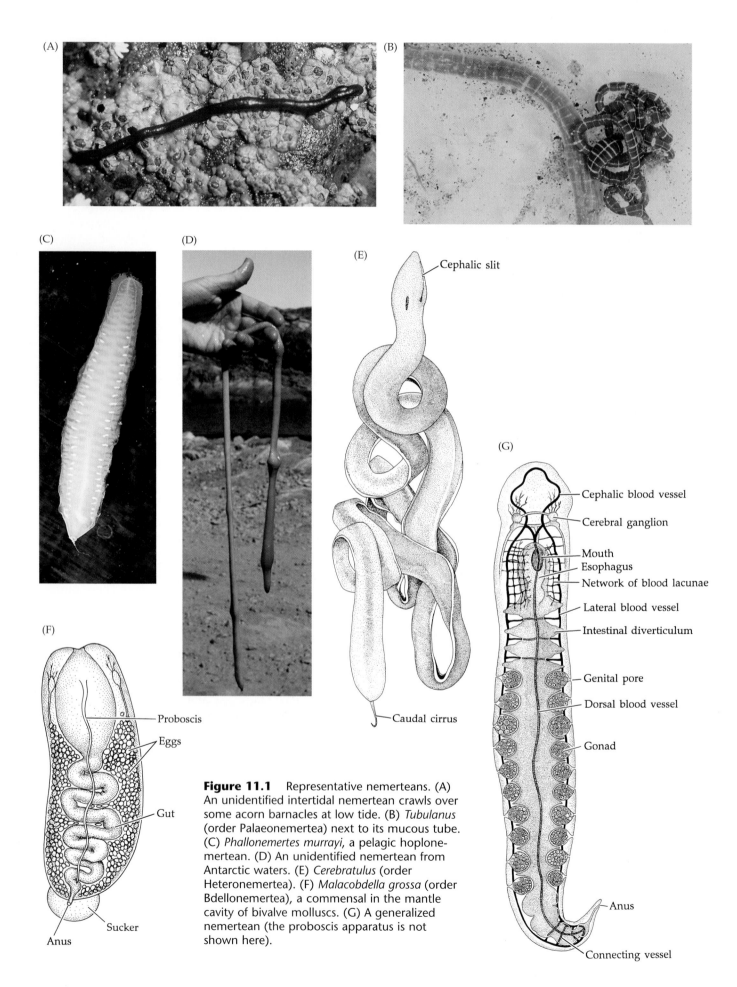

Figure 11.1 Representative nemerteans. (A) An unidentified intertidal nemertean crawls over some acorn barnacles at low tide. (B) *Tubulanus* (order Palaeonemertea) next to its mucous tube. (C) *Phallonemertes murrayi*, a pelagic hoplonemertean. (D) An unidentified nemertean from Antarctic waters. (E) *Cerebratulus* (order Heteronemertea). (F) *Malacobdella grossa* (order Bdellonemertea), a commensal in the mantle cavity of bivalve molluscs. (G) A generalized nemertean (the proboscis apparatus is not shown here).

BOX 11A Characteristics of the Phylum Nemertea

1. Marine, freshwater, or terrestrial
2. Triploblastic, [probably] coelomate, bilaterally symmetrical unsegmented worms
3. Digestive tract complete, with an anus
4. With protonephridia
5. With bilobed cerebral ganglion that surrounds proboscis apparatus (not the gut), and two or more longitudinal nerve cords connected by transverse commissures
6. With two or three layers of body wall muscles arranged in various ways
7. With a unique proboscis apparatus lying dorsal to the gut and surrounded by a coelom-like hydrostatic chamber called the rhynchocoel
8. With a closed circulatory system; some species with hemoglobin
9. Most are dioecious; cleavage holoblastic; early development typically spiralian, either direct or indirect
10. Asexual reproduction by fragmentation is common

Taxonomic History and Classification

The earliest report of a nemertean was that of Borlase (1758), who described his specimen as "the sea long-worm" and categorized it "among the less perfect kind of sea animals." For nearly a century, most authors placed the ribbon worms with the turbellarian flatworms, although other writers suggested that they were allied to the annelids, the sipunculans, the roundworms, and even the molluscs and the insects. It was during this period that Cuvier (1817) described a particular ribbon worm and called it *Nemertes*, from which the phylum name was eventually derived. However, it was not until 1851 that substantial evidence for the distinctive nature of the ribbon worms was published by Max Schultze, who described the functional morphology of the proboscis, established the presence of nephridia and an anus, and discussed many other features of these animals. Schultze even proposed the basis for classifying these worms which is still employed today by most authorities. Interestingly, he persisted in considering them to be turbellarians, but he coined the names Nemertina and Rhynchocoela. Minot separated the nemerteans from the flatworms in 1876, but it was not until the mid-twentieth century (see Coe 1943 and Hyman 1951) that the unique combination of characters displayed by the ribbon worms was fully accepted. Since that time they have been treated as a valid phylum.

Classification

Since the publication of Schultze's classic accounts, the primary effort of nemertean taxonomists has been to refine the details of his scheme, with surprisingly few controversies. The classification scheme used here is a traditional one established by Coe (1943) and followed by most specialists. In it the phylum is subdivided into two classes, each with two clearly differentiated orders.

The principal features used to distinguish between the classes and orders of nemerteans include proboscis armature, mouth location relative to the position of the cerebral ganglion, gut shape, layering of the body wall muscles, and position of the longitudinal nerve cords.*

PHYLUM NEMERTEA (= RHYNCHOCOELA)

CLASS ANOPLA: Unarmed nemerteans (Figure 1.1B,E). Proboscis not armed with stylets and not morphologically specialized into three regions. Mouth separate from proboscis pore and located directly below or somewhat posterior to cerebral ganglion; longitudinal nerve cords located within epidermis, dermis, or muscle layers of body wall (not within the mesenchymal mass internal to the body wall muscles).

ORDER PALAEONEMERTEA: Two or three layers of body wall muscles, from external to internal either circular-longitudinal or circular-longitudinal-circular; dermis thin and gelatinous, or absent; longitudinal nerve cords epidermal, dermal, or intramuscular within the longitudinal layer; cerebral organs and ocelli frequently lacking. Palaeonemerteans are marine, primarily littoral forms. (e.g., *Carinoma, Cephalothrix, Hubrechtella, Tubulanus*)

ORDER HETERONEMERTEA: Three layers of body wall muscles, from external to internal longitudinal-circular-longitudinal; dermis usually thick, partly fibrous; longitudinal nerve cords intramuscular, between outer longitudinal and middle circular layers; cerebral organs and ocelli usually present; development indirect. These nemerteans are primarily marine littoral forms. (e.g., *Baseodiscus, Cerebratulus, Lineus, Micrura, Paralineus*)

CLASS ENOPLA: Typically armed nemerteans (Figure 11.1C,F). Proboscis usually armed with distinct stylets and

---

*Iwata (1960) proposed an alternative classification scheme based on certain embryological and morphological features, but his proposal has not been widely accepted. Notice that, according to our cladogram (Figure 11.16C), the class Anopla is not a monophyletic group, but a paraphyletic one.

morphologically specialized into three regions (except in Bdellonemertea); mouth and proboscis pore usually united into a common aperture; mouth located anterior to cerebral ganglion; longitudinal nerve cords within mesenchyme, internal to body wall muscles.

ORDER HOPLONEMERTEA: Proboscis armed; trunk lacks a posterior sucker; gut more or less straight but nearly always with numerous lateral diverticula; cerebral organs and ocelli usually present.

SUBORDER MONOSTILIFERA: Stylet apparatus consists of a single main stylet and two or more sacs housing accessory (replacement) stylets. Most species are marine and benthic, but freshwater, terrestrial, and ectoparasitic forms are known. (e.g., *Amphiporus, Annulonemertes, Carcinonemertes, Emplectonema, Geonemertes, Paranemertes*)

SUBORDER POLYSTILIFERA: Stylet apparatus consists of many small stylets borne on a basal shield. All species are marine, either benthic or pelagic. (e.g., *Hubrechtonemertes, Nectonemertes, Pelagonemertes*)

ORDER BDELLONEMERTEA: Proboscis unarmed; trunk with large posterior sucker; proboscis apparatus opens into foregut; gut convoluted and lacks lateral diverticula. Bdellonemerteans are commensal in the mantle cavities of marine bivalves and, in one species, a freshwater gastropod. Monogeneric: *Malacobdella*.

The Nemertean Bauplan

In Chapter 3 we discussed some of the limitations of the acoelomate body plan, and we have seen the results of these constraints in our examination of the flatworms. It might be said that the ribbon worms have made the best of a rather difficult situation. Even though it is probable that nemerteans have true coelomic cavities (the rhynchocoel and certain blood vessels), these worms have relatively solid bodies. Thus they are at least "functionally" acoelomate. Recall that many of the problems inherent in the acoelomate architecture are related to restricted internal transport capabilities. The evolution of a circulatory system in nemerteans has largely eased this problem, and the functional anatomy of many other systems is related directly or indirectly to the presence of this circulatory mechanism. For example, the nemertean protonephridia are usually intimately associated with the blood, from which wastes are drawn, rather than with the mesenchymal tissues as are flatworm protonephridia.

The increased capabilities for internal circulation and transport have allowed a number of developments that would otherwise be impossible. First, the circulatory system provides a solution to the surface-to-volume dilemma, and as a result nemerteans tend to be much larger and more robust than flatworms, having been largely relieved of the constraints of relying on diffusion for internal transport and exchange. Second, the digestive tract is complete and somewhat regionally specialized. With a one-way movement of food materials

through the gut, and a circulatory system to absorb and distribute digested products, the anterior region of the gut has been freed for feeding and ingestion. Third, since the animal does not have to rely on diffusion for transport through a loosely organized mesenchyme, that general body area is freed for the development of other structures, notably the well developed layers of muscles. In summary, the development of a circulatory system in concert with these other changes has resulted in relatively large, robust, active animals, capable of more complex feeding and digestive activities than seen in platyhelminths.

This general bauplan is enhanced by the presence of the unique proboscis apparatus (which usually functions in prey capture), the distinctly anterior location of the mouth, and well developed cephalic sensory organs for prey location. Thus, while variation exists, the "typical" nemertean may be viewed as an active benthic hunter/tracker that moves among nooks and crannies preying on other invertebrates.

Body Wall

The body wall of nemerteans comprises an epidermis, a dermis, relatively thick muscle layers surrounding the gut and other internal organs, and a mesenchyme of varying thickness (Figure 11.2). The epidermis is a ciliated columnar epithelium (Figure 11.2C). Mixed among the columnar cells are sensory cells (probably tactile), mucous gland cells, and basal replacement cells that may extend beneath the epidermis. Below the epidermis is the dermis, which varies greatly in thickness and composition. In some ribbon worms (e.g., the palaeonemerteans) the dermis is extremely thin or composed of only a homogeneous gel-like layer; in others (e.g., the heteronemerteans), it is typically quite thick and densely fibrous and usually includes a variety of gland cells. Beneath the dermis are well developed layers of circular and longitudinal muscles. As indicated in the classification above, the organization of these muscles varies among taxa and may occur in either a two- or three-layered plan (Figure 11.3). The layering arrangement may also vary to some degree along the body length of individual animals. Internal to the muscle layers is a dense, more or less solid mesenchyme, although in some nemerteans the muscle layers are so thick that they nearly obliterate this inner mass. The mesenchyme includes a gel matrix and often a variety of loose cells, fibers, and dorsoventrally oriented muscles. Figure 11.3 depicts cross-sectional views of the four orders, showing mesenchyme thickness, muscles, placement of longitudinal nerve cords, major longitudinal blood vessels, and other features.

Support and Locomotion

In the absence of any rigid skeletal elements, the support system of nemerteans is provided by the muscles

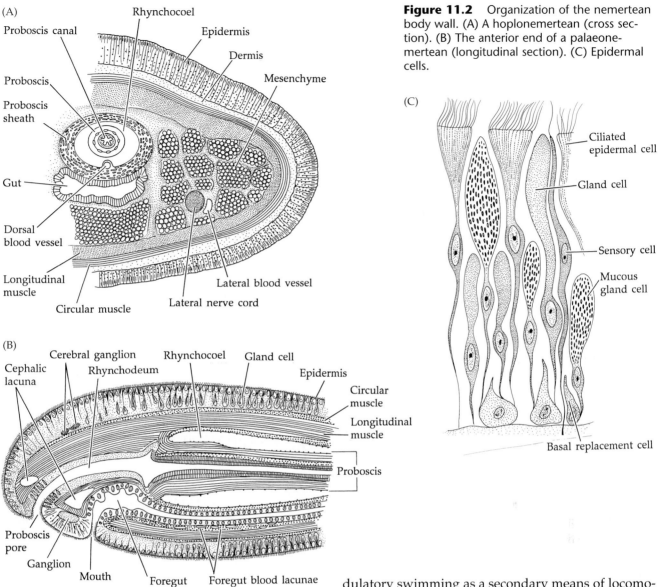

(A)

Rhynchocoel
Proboscis canal
Epidermis
Dermis
Mesenchyme
Proboscis
Proboscis sheath
Gut
Dorsal blood vessel
Longitudinal muscle
Lateral blood vessel
Lateral nerve cord
Circular muscle

(B)

Cerebral ganglion
Rhynchocoel
Gland cell
Cephalic lacuna
Rhynchodeum
Epidermis
Circular muscle
Longitudinal muscle
Proboscis
Proboscis pore
Ganglion
Mouth
Foregut
Foregut blood lacunae

(C)

Ciliated epidermal cell
Gland cell
Sensory cell
Mucous gland cell
Basal replacement cell

Figure 11.2 Organization of the nemertean body wall. (A) A hoplonemertean (cross section). (B) The anterior end of a palaeonemertean (longitudinal section). (C) Epidermal cells.

and other tissues of the body wall and by the hydrostatic qualities of the mesenchyme. These features permit dramatic changes in both length and cross-sectional shape and diameter, characteristics that are closely associated with locomotion and accommodation to cramped quarters. Most very small benthic ribbon worms are propelled by the action of their epidermal cilia. A slime trail is produced by the body wall mucous glands and provides a lubricated surface over which the worm slowly glides. Small nemerteans commonly live among the interstices of filamentous algae or in the spaces of other irregular surfaces such as those found in mussel beds and sand, mud, or pebble bottoms. Larger epibenthic ribbon worms and most of the burrowing forms employ peristaltic waves of the body wall muscles to propel them over moist surfaces or through soft substrata. Some of the larger forms (e.g., *Cerebratulus*) use un-

dulatory swimming as a secondary means of locomotion, and perhaps as an escape reaction to benthic predators. Fully pelagic nemerteans (certain hoplonemerteans) generally drift or swim slowly. Some of the terrestrial forms produce a slime sheath through which they glide by ciliary action, and some use their proboscis for rapid escape responses.

Feeding and Digestion

Feeding behavior. Most ribbon worms are active predators on small invertebrates, but some are scavengers and others feed on plant material (at least under laboratory conditions). There is evidence to suggest that species of the commensal genus *Malacobdella*, which inhabit the mantle cavity of bivalve molluscs, feed largely on phytoplankton captured from their host's feeding and gas exchange currents. Field observations indicate that the diets of predatory forms may be either extremely varied or

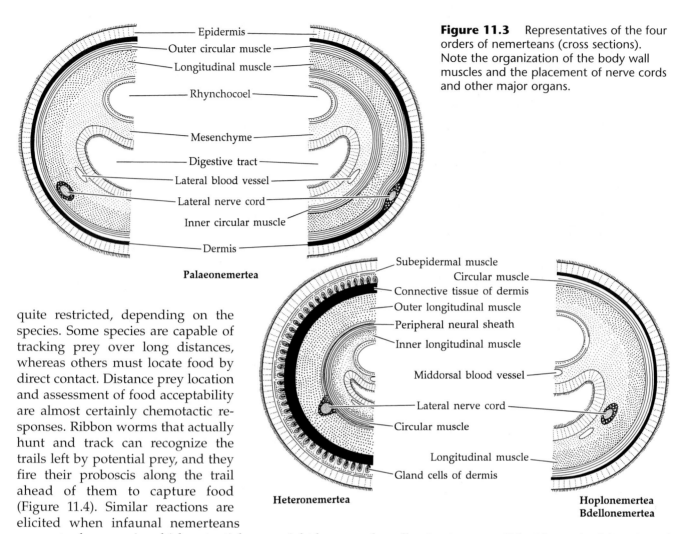

Figure 11.3   Representatives of the four orders of nemerteans (cross sections). Note the organization of the body wall muscles and the placement of nerve cords and other major organs.

Labels (Palaeonemertea): Epidermis, Outer circular muscle, Longitudinal muscle, Rhynchocoel, Mesenchyme, Digestive tract, Lateral blood vessel, Lateral nerve cord, Inner circular muscle, Dermis

Palaeonemertea

Labels (Heteronemertea / Hoplonemertea): Subepidermal muscle, Circular muscle, Connective tissue of dermis, Outer longitudinal muscle, Peripheral neural sheath, Inner longitudinal muscle, Middorsal blood vessel, Lateral nerve cord, Circular muscle, Longitudinal muscle, Gland cells of dermis

Heteronemertea

Hoplonemertea Bdellonemertea

quite restricted, depending on the species. Some species are capable of tracking prey over long distances, whereas others must locate food by direct contact. Distance prey location and assessment of food acceptability are almost certainly chemotactic responses. Ribbon worms that actually hunt and track can recognize the trails left by potential prey, and they fire their proboscis along the trail ahead of them to capture food (Figure 11.4). Similar reactions are elicited when infaunal nemerteans encounter burrows in which potential prey might be located. Surface hunters that live in intertidal areas generally forage during high tides or at night, and thus avoid the threats of desiccation and visual predators. However, members of some genera (e.g., *Tubulanus, Paranemertes, Amphiporus*) may frequently be seen during low tides on foggy mornings, gliding over the substratum in search of prey.

The behavior involved in the capture and ingestion of live prey is significantly different from that associated with scavenging on dead material. In predation, the proboscis is employed both in capturing prey and in moving it to the mouth for ingestion. The proboscis is everted and wrapped around the victim (Figure 11.4). The prey is not only physically "held down" by the proboscis but may be subdued or killed by its toxic secretions. In the Pacific species *Paranemertes peregrina*, which feeds primarily on nereid polychaetes, the glandular epithelium of the everted proboscis secretes a potent neurotoxin. Nemerteans with an armed proboscis (Hoplonemertea) actually use the stylets to pierce the prey's body (often numerous times) to introduce the toxin. Once captured, the prey is drawn to the mouth by retraction and manipulation of the proboscis; it is usually swallowed whole. The mouth is expanded and pressed against the food,

and swallowing is accomplished by peristaltic action of the body wall muscles aided by ciliary currents in the anterior region of the gut. Scavenging, in contrast, usually does not involve the proboscis. The worm simply ingests the food directly by muscular action of body wall and foregut. In some predatory hoplonemerteans (those in which the lumen of the proboscis is connected with the anterior gut lumen), the foregut itself may be everted for feeding on animals too large to be swallowed whole. In such cases, fluids and soft tissues are generally sucked out of the prey's body. Polychaete worms and amphipod crustaceans seem to be favorite food items for many predatory nemerteans.

Species of the hoplonemertean genus *Carcinonemertes* are ectoparasites (egg predators) on brachyuran crabs. Different species inhabit different regions of the host's body, but all migrate to the egg masses on gravid female crabs and feed on the yolky eggs. In high numbers, these egg predators can kill all of the embryos in the host's clutch. Recent studies (Wickham 1979, 1980) report 99 percent infestation rates of *Carcinonemertes errans* on the commercially important Pacific Dungeness crab (*Cancer magister*), with up to 100,000 worms per host. This parasite has been implicated in the general collapse of the central California Dungeness crab fishery.

Figure 11.4 *Paranemertes peregrina* (order Hoplonemertea) capturing a nereid polychaete. The proboscis is coiled around the polychaete.

The proboscis apparatus. The proboscis apparatus is a complex arrangement of tubes, muscles, and hydraulic systems (Figure 11.5). The proboscis itself is an elongate, eversible, blind tube, and either is associated with the foregut or opens through a separate **proboscis pore**. The proboscis may be regionally specialized and bear **stylets** in various arrangements (Figures 5E–H). Nemertean stylets are nail-shaped structures that typically reach lengths of 50–200 μm. Each stylet is composed of a central organic matrix surrounded by an inorganic cortex that contains crystalline calcium and phosphorus. The stylets are formed within large epithelial cells called **styletocytes**. Because growing ribbon worms must replace their stylets with new larger ones, and because they often lose the stylet during prey capture, new stylets are continuously produced in **reserve stylet sacs** and stored until needed, whereupon they are transported and affixed in their proper position.

The basic structure and action of the proboscis are most easily described where the apparatus is entirely separate from the gut. As shown in Figures 11.5A and B,

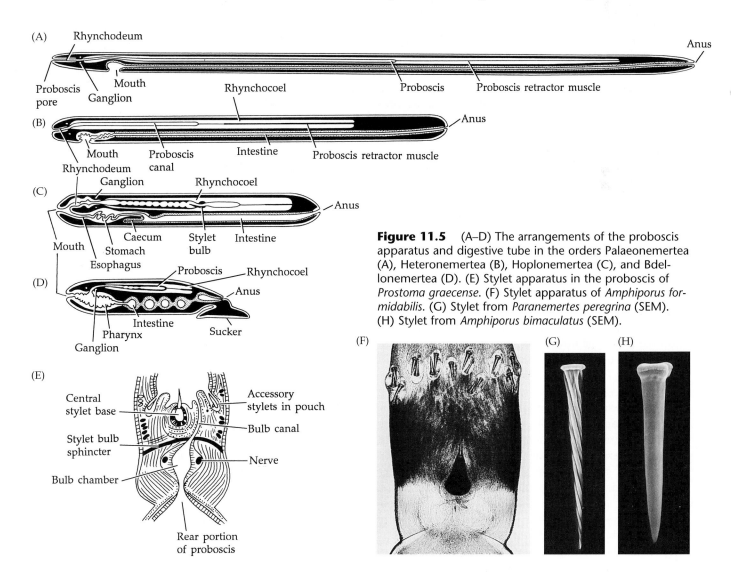

Figure 11.5 (A–D) The arrangements of the proboscis apparatus and digestive tube in the orders Palaeonemertea (A), Heteronemertea (B), Hoplonemertea (C), and Bdellonemertea (D). (E) Stylet apparatus in the proboscis of *Prostoma graecense*. (F) Stylet apparatus of *Amphiporus formidabilis*. (G) Stylet from *Paranemertes peregrina* (SEM). (H) Stylet from *Amphiporus bimaculatus* (SEM).

(A)

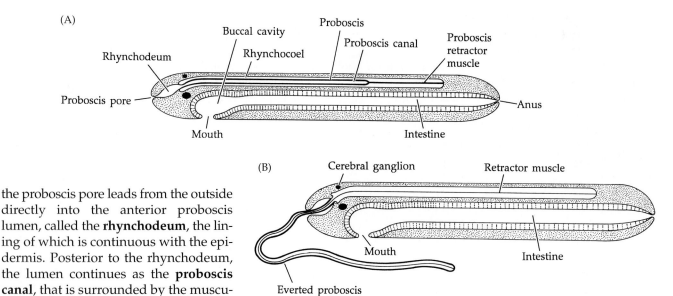

(B)

Figure 11.6 A retracted (A) and an extended (B) proboscis of a hoplonemertean.

the proboscis pore leads from the outside directly into the anterior proboscis lumen, called the **rhynchodeum**, the lining of which is continuous with the epidermis. Posterior to the rhynchodeum, the lumen continues as the **proboscis canal**, that is surrounded by the muscular wall of the proboscis itself; these muscles are derived from the muscles of the body wall. The proboscis is surrounded by a closed, fluid-filled, perhaps coelomic space called the **rhynchocoel**, which in turn is surrounded by additional muscle layers. The inner blind end of the proboscis is connected to the posterior wall of the rhynchocoel by a **proboscis retractor muscle**. In a few taxa (e.g., *Gorgonorhynchus*), there is no retractor muscle, and eversion *and* retraction are accomplished hydrostatically.

Eversion of the proboscis (Figure 11.6) is accomplished by contraction of the muscles around the rhynchocoel; this increases the hydrostatic pressure within the rhynchocoel itself, squeezing on the proboscis and causing its eversion. The everted proboscis moves with the muscles in its wall; the proboscis is retracted back inside the body by the coincidental relaxation of the muscles around the rhynchocoel and contraction of the proboscis retractor muscle. The retracted proboscis may extend nearly to the posterior end of the worm, and usually only a portion of it is extended during eversion.

Digestive system. In contrast to the flatworms, nemerteans possess an anus (Figure 11.7). Associated with the one-way movement of food from mouth to anus we find various degrees of regional specialization (both structural and functional) in the guts of ribbon worms. The mouth leads inward to an ectodermally derived foregut (stomodeum) consisting of a bulbous **buccal cavity**, sometimes a short **esophagus**, and a **stomach**. The stomach leads to an elongate **intestine** or midgut, which is more or less straight but usually bears numerous lateral diverticula. In *Malacobdella*, the intestine is loosely coiled and lacks diverticula; diverticula are also lacking in the strange, "segmented" *Annulonemertes* (Berg 1985). At the posterior end of the intestine is a short ectodermally derived hindgut (proctodeum) or **rectum**, which terminates in the anus. Elaborations on this basic plan

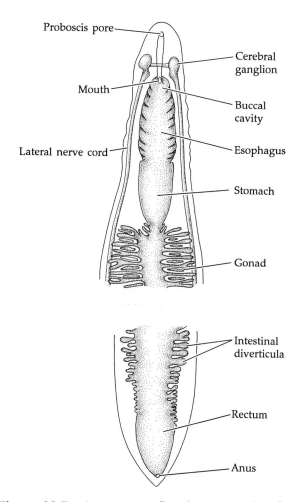

Figure 11.7 A nemertean digestive system. Anterior and posterior regions of the gut of *Carinoma* (ventral view).

are common in certain taxa and may include various ceca arising from the stomach or from the intestine at its junction with the foregut.

The entire digestive tube is ciliated, the foregut more densely than the midgut. The gut epithelium is basically columnar, mixed with gland cells. The foregut contains a variety of mucus-producing cells, sometimes multicellular mucous glands, and occasionally enzymatic gland cells in the stomach region. The midgut is lined with vacuolated ciliated columnar cells; these are phagocytic and bear microvilli, greatly increasing their surface area. Enzymatic gland cells are abundantly mixed with the ciliated cells of the midgut. The hindgut typically lacks gland cells. Food is moved through the digestive tract by cilia and by the action of the body wall muscles; there are usually no muscles in the gut wall itself, except in the foregut of some heteronemerteans.

The process of digestion in carnivorous nemerteans is a two-phase sequence of protein breakdown. The first step involves the action of endopeptidases released from gland cells into the gut lumen. This extracellular digestion is quite rapid and is followed by phagocytosis (and probably pinocytosis) of the partially digested material by the ciliated columnar cells of the midgut. Protein digestion is completed intracellularly by exopeptidases within the food vacuoles of the midgut epithelium. Lipases have been discovered in at least one species (*Lineus ruber*), and carbohydrases are known in the omnivorous commensal *Malacobdella*. Food is stored primarily in the form of fats, and to a much lesser extent as glycogen, in the wall of the midgut. Transportation of digested materials throughout the body is accomplished by the circulatory system, which absorbs these products from the cells lining the intestine. Undigestible materials are moved through the gut and out the anus.

Circulation and Gas Exchange

We have mentioned briefly the evolutionary and adaptive significance of the circulatory system in nemerteans and its general relationship to other systems and functions. This closed system consists of vessels and thin-walled spaces called **lacunae** (Figure 11.2B). At least some of these vessels are thought to be homologous to coelomic cavities. There is a good deal of variation in the architecture of nemertean circulatory systems (Figure 11.8). The simplest arrangement occurs in certain palaeonemerteans in which a single pair of longitudinal vessels extends the length of the body, connecting anteriorly by a **cephalic lacuna** and posteriorly by an **anal lacuna**. Elaboration on this basic scheme may include transverse vessels between the longitudinal vessels, enlargement and compartmentalization of the lacunar spaces, and the addition of a middorsal vessel. The walls of the blood vessels are only slightly contractile, and general body movements generate most of the blood flow. There is no consistent pattern to the movement of blood through the system; it may flow either

anteriorly or posteriorly in the longitudinal vessels, and currents often reverse directions.

The blood consists of a colorless fluid in which various cells are suspended. These cells can include pigmented corpuscles (yellow, orange, green, red), at least some of which contain hemoglobin, and a variety of so-called lymphocytes and leukocytes of uncertain function. The anatomical association of the circulatory system with other structures, as well as the composition of the blood, suggest several circulatory functions. Although conclusive evidence is lacking, the circulatory system appears to be involved with the transport of nutrients, gases, neurosecretions, and excretory products. Some intermediary metabolism probably occurs in the

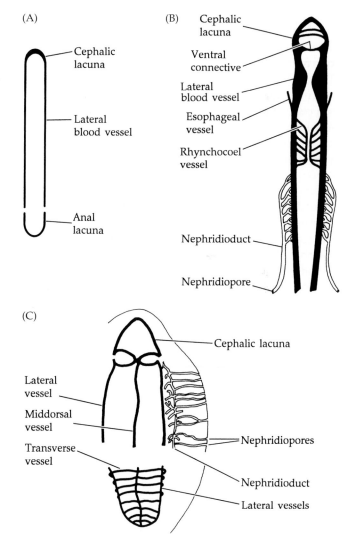

Figure 11.8 Nemertean circulatory systems. (A) The simple circulatory loop of *Cephalothrix* consists of a pair of lateral blood vessels connected by cephalic and anal lacunae. (B) The complex circulatory system of *Tubulanus*. Note the intimate association of the nephridial system with the lateral blood vessels. (C) The circulatory system of *Amphiporus* includes a middorsal vessel and numerous transverse vessels.

blood, as several appropriate enzymes have been identified in solution. The blood may also serve as an aid to body support through changes in hydrostatic pressure within the vessels and lacunar spaces. There is some evidence to support the idea that the blood may also function in osmoregulation.

Gas exchange in nemerteans is epidermal and does not involve any special structures. Oxygen and carbon dioxide diffuse readily across the moist body surface, which is usually covered with mucous secretions. Some robust forms (e.g., *Cerebratulus*) augment this passive exchange of gases across the skin with regular irrigation of the foregut, where there is an extensive system of blood vessels. In those species in which hemoglobin occurs, this pigment probably aids in oxygen transport or storage within the blood.

Excretion and Osmoregulation

The excretory system of most nemerteans consists of two to thousands of flame bulb protonephridia (Figures 11.8 and 11.9) similar to those found in turbellarian flatworms. However, apparently none occurs in the deepsea pelagic hoplonemerteans. The flame bulbs are usually intimately associated with the lateral blood vessels or less commonly with other parts of the circulatory system. The nephridial units are often pressed into the blood vessel walls, and in some instances the walls are actually broken down so that the nephridia are bathed directly in blood. In the simplest case, a single pair of flame bulbs leads to two nephridioducts, each with its own laterally placed nephridiopore. More complex conditions include rows of single flame bulbs or clusters of flame bulbs with multiple ducts. In some species the walls of the nephridioducts are syncytial and lead to

hundreds or even thousands of pores on the epidermis. The most elaborate conditions occur in certain terrestrial nemerteans where approximately 70,000 clusters of flame bulbs (six to eight in each cluster) lead to as many surface pores. In some heteronemerteans (e.g., *Baseodiscus*), the excretory system discharges into the foregut.

The functioning of nemertean protonephridia in the excretion of metabolic wastes has not been well studied. The close association of the flame bulbs with the circulatory system suggests that nitrogenous wastes (probably ammonia), excess salts, and other metabolic products are removed from the blood as well as from the surrounding mesenchyme by the nephridia. If such is the case, it explains again the significance of the circulatory system in overcoming surface-to-volume problems and the constraints of simple diffusion on body size. Relatively active animals produce large amounts of metabolic wastes. Dependence on diffusion alone would seriously limit any increase in body bulk, but the transport of these wastes from the tissues to the protonephridial system by circulatory vessels greatly eases this limitation. One of the most remarkable evolutionary achievements of the nemerteans has been their ability to grow to great size, particularly in length, without segmentation or the development of a large body cavity.

There is some morphological and experimental evidence that the protonephridia also play an important role in osmoregulation, especially in freshwater and terrestrial ribbon worms. It is in some of these forms, which are subjected to extreme osmotic stress, that the most elaborate excretory systems are found; these systems are probably associated with water balance. Furthermore, it appears that there may be a very complex interaction between the nervous system (neurose-

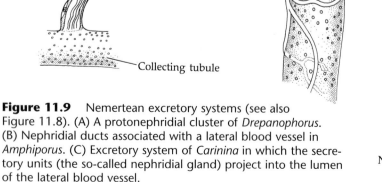

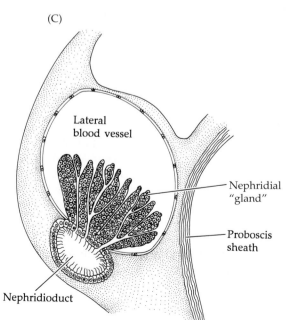

Figure 11.9 Nemertean excretory systems (see also Figure 11.8). (A) A protonephridial cluster of *Drepanophorus*. (B) Nephridial ducts associated with a lateral blood vessel in *Amphiporus*. (C) Excretory system of *Carinina* in which the secretory units (the so-called nephridial gland) project into the lumen of the lateral blood vessel.

TABLE 11.1 *Location of the lateral longitudinal nerve cords in nemertean orders*

Taxon	Epidermal	Dermal	Within body wall muscles	Internal to body wall muscles (mesenchymal)
Palaeonemertea	X	X	X	
Heteronemertea			X	
Hoplonemertea				X
Bdellonemertea				X

cretions), the circulatory system, and the nephridia to facilitate osmoregulatory mechanisms, but the details remain to be studied. Some members of all nemertean orders except the Palaeonemertea have invaded fresh water and must combat water influx from their strongly hypotonic surroundings. Members of some genera (e.g., *Geonemertes*) are terrestrial, although restricted to moist shady habitats where they avoid serious problems of desiccation. In addition, they tend to cover their bodies with a mucous coat that reduces water loss. Those forms that inhabit marine subtidal or deep-water environments, or are endosymbiotic (one genus of Heteronemertea, several genera of Hoplonemertea, and the Bdellonemertea), face little or no osmotic stress. But the many species found intertidally do face periods of exposure to air and to lowered salinities. Their soft bodies are largely unprotected, and they are relatively intolerant of fluctuations in environmental conditions. Intertidal nemerteans rely strongly on behavioral attributes to survive periods of potential osmotic stress and remain in moist areas during low tide periods. Burrowing in soft, water-soaked substrata, or living among algae or mussel beds, in cracks and crevices, or other areas that retain sea water at low tide are lifestyles illustrating how habitat preference and behavior prevent exposure to stress. In addition, most intertidal nemerteans are somewhat negatively phototactic, and many restrict their activities to night hours or to foggy or overcast mornings and evenings.

Nervous System and Sense Organs

The basic organization of the nemertean nervous system reflects a relatively active lifestyle. Nemerteans are somewhat more cephalized than flatworms are, especially in the anterior placement of the mouth and feeding structures, and we find related concentrations of sensory and other nervous elements in the head. The central nervous system of ribbon worms consists of a complex cerebral ganglion from which arises a pair of ganglionated, longitudinal (lateral) nerve cords (Figure 11.10A). The cerebral ganglion is formed of four attached lobes that encircle the proboscis apparatus (not the gut, as in many other invertebrates). Each side of the cerebral ganglion includes a dorsal and a ventral lobe; the two sides are attached to one another by dorsal and ventral connectives. Several pairs of sensory nerves pro-

vide input directly to the cerebral ganglion from various cephalic sense organs. The main longitudinal nerve cords arise from the ventral lobes of the cerebral ganglion and pass posteriorly; they attach to each other at various points by branched transverse connectives and terminally by an **anal commissure**. The longitudinal nerves also give rise to peripheral sensory and motor nerves along the length of the body. Elaboration on this basic plan includes additional longitudinal nerve cords, frequently a middorsal one arising from the dorsal commissure of the cerebral ganglion, and a variety of connectives, nerve tracts, and plexi.

As noted in the classification scheme, the positions of the major longitudinal nerve cords vary among the four orders (Figure 11.3). These differences are summarized in Table 11.1. These changes in the position of the nerve cords from epidermal to mesenchymal correspond to general increases in body complexity and tendencies toward specialization. Most workers agree that these differences reflect a plesiomorphic (epidermal) to apomorphic (subepidermal) trend among these taxa.

Ribbon worms possess a variety of sensory receptors, many of which are concentrated at the anterior end and associated with an active, typically hunting lifestyle and with other aspects of their natural history. Nemerteans are very sensitive to touch. This tactile sensitivity plays a role in food handling, avoidance responses, locomotion over irregular surfaces, and mating behavior. Several types of modified ciliated epidermal cells are scattered over the body surface (especially abundant at the anterior and posterior ends) and are presumed have a tactile function. The cells occur either singly or in clusters; some of the latter types are located in small depressions and can be thrust out from the body surface.

The eyes of ribbon worms are located anteriorly and number from two to several hundred; they can be arranged in various patterns (Figure 11.10B). Most of these ocelli are of the inverted pigment-cup type, similar to those seen in flatworms, although a few species possess lensed eyes. As discussed in Chapter 3, these types of eyes typically are sensitive to light intensity and light direction. They help the nemerteans avoid bright light and potential exposure to predators or environmental stresses.

Much of the sensory input important to nemerteans is chemosensory. These worms are very sensitive to dis-

(A)

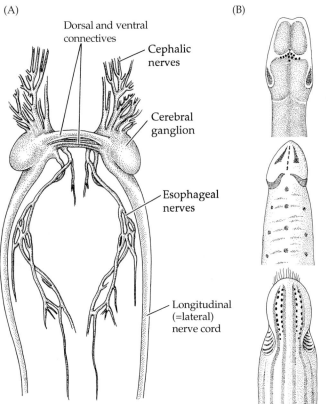

Dorsal and ventral connectives

Cephalic nerves

Cerebral ganglion

Esophageal nerves

Longitudinal (=lateral) nerve cord

(B)

(C)

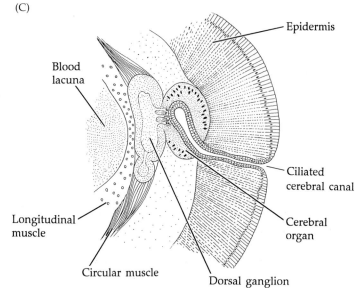

Blood lacuna

Longitudinal muscle

Circular muscle

Epidermis

Ciliated cerebral canal

Cerebral organ

Dorsal ganglion

Figure 11.10 Nervous system and sense organs of nemerteans. (A) Anterior portion of the nervous system of *Tubulanus*; see text for explanation of variations. (B) The cephalic slits and grooves and eye spots are visible on the heads of three nemerteans. (C) The cerebral organ of *Tubulanus* (cross section). Note the association of the organ with the cerebral canal, the nervous system, and the blood system. (D) Clusters of frontal glands occur in the anterior end of a hoplonemertean (longitudinal section).

(D)

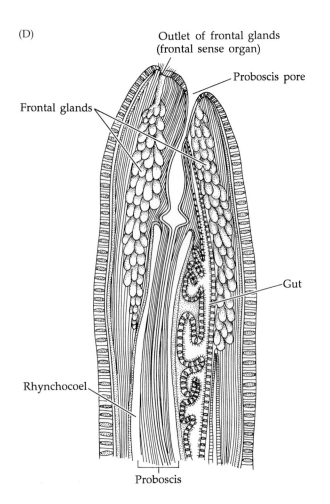

Outlet of frontal glands (frontal sense organ)

Proboscis pore

Frontal glands

Gut

Rhynchocoel

Proboscis

solved chemicals in their environment and employ this sensitivity in food location, probably mate location, substratum testing, and general water analysis. Probably all nemerteans respond to contact with chemical stimuli, and many are capable of distance chemoreception of materials in solution. At least three different nemertean structures have been implicated (some through speculation) in the initiation of chemotactic responses; these are **cephalic slits** or **grooves**, **cerebral organs**, and **frontal glands** (= **cephalic glands**) (Figures 11.10B,C,D). Cephalic slits are deep or shallow furrows that occur laterally on the heads of many ribbon worms (see also Figure 11.1E). These furrows are lined with a ciliated sensory epithelium supplied with nerves from the cerebral ganglion. Water is circulated through the cephalic slits and over this presumably chemosensory epithelial lining.

Most nemerteans possess a pair of the remarkably complex cerebral organs (Figure 11.10C). The core of each cerebral organ is a ciliated epidermal invagination (the **cerebral canal**), which is expanded at its inner end. These canals lead laterally to pores within the cephalic slits (when present) or else directly to the outside via separate pores on the head. The inner ends of the canals are surrounded by nervous tissue of the cerebral ganglion, and by glandular tissue, and they are often intimately associated with lacunar blood spaces. Cilia in the cerebral canal circulate water through the open portion of the organ; this activity intensifies in the presence of food. Nemerteans presumably use this mechanism when hunting and tracking prey or in other chemotactic responses. The association of the cerebral canals with glandular, nervous, and circulatory structures has led some workers to suggest an endocrine/neurosecretory function for the cerebral organs. Other suggestions have included auditory, gas exchange, excretory, and tactile activities. Cerebral organs are absent in several genera, including the symbiotic *Carcinonemertes* and *Malacobdella* and the pelagic hoplonemerteans.

In the region anterior to the cerebral ganglion, large frontal glands open to the outside through a pitlike **frontal sense organ** (Figure 11.10D). These structures receive nerves from the cerebral ganglion and appear to be chemosensory, but solid evidence for this suggestion is lacking. Finally, statocysts have also been found in some nemerteans, including pelagic forms where geotaxis is an obvious advantage.

Reproduction and Development

Asexual processes. Many nemerteans show remarkable powers of regeneration, and nearly all species can regenerate at least posterior portions of the body. Those with the greatest regenerative abilities are certain species of *Lineus*, which engage in asexual reproduction on a regular basis by undergoing multiple transverse fission into numerous fragments (Figure 11.11). The fragments are often extremely small and the process is sometimes referred to simply as fragmentation. The small pieces often form mucous cysts within which the new worm regenerates; larger pieces grow into new animals without the protection of a cyst. In some nemerteans only anterior fragments can regenerate into new worms.

Sexual reproduction. Nemerteans constitute a relatively small phylum, but they show remarkable varia-

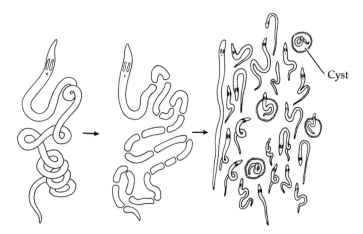

Figure 11.11 Fragmentation and regeneration in a nemertean, *Lineus vegetus*. Each fragment regenerates into a complete worm. Small fragments may form cysts.

tion in reproductive and developmental strategies. Most ribbon worms are dioecious, although protandric and even simultaneous hermaphrodites are known. Unlike the complex reproductive systems of flatworms, the reproductive system of nemerteans has gonads that are simply specialized patches of mesenchymal tissue arranged serially along each side of the intestine and alternating with the midgut diverticula (Figure 11.12). In *Malacobdella* and a few others, the gonads are more-or-less packed within the mesenchyme (Figure 11.1F). In most nemerteans the development of gonads occurs along nearly the entire length of the body, but in a few species they are restricted to certain regions, usually toward the anterior end. The gonads begin to enlarge and hollow just prior to the onset of breeding activities. Specialized cells in the walls of the rudimentary ovaries and testes prolif-

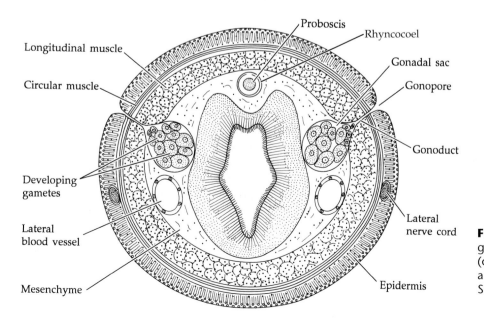

Figure 11.12 Arrangement of gonads in the nemertean *Carinina* (cross section). Note the position of a pair of gonads in the mesenchyme. See also Figures 11.1F,G and 11.7.

erate eggs and sperm into the lumina of the enlarging gonadal sacs. In females additional special cells are responsible for yolk production. There is evidence that maturation is under neurosecretory hormonal control, at least in some species. The secretions are probably from the cerebral organ complex (see publications by Bierne).

With the proliferation of gametes, the gonadal sacs expand to almost fill the area between the gut and the body wall. When the animals are nearly ready to spawn, mating behavior is initiated and the worms become increasingly active. As mentioned earlier, mate location probably depends on chemotactic responses. The same is apparently true of spawning itself, at least for some species, because the presence of a ripe conspecific stimulates the release of gametes from other mature individuals. Experimental evidence indicates that physical contact is not necessary for such a spawning response; thus, some sort of pheromone is probably involved. In nature, however, spawning usually occurs in concert with actual physical contact; tactile responses evidently follow chemotactic mate location. During such mating activities, veritable knots of scores of worms may writhe in a mucus-covered mating mass. The coordinated release of ripe gametes under such conditions ensures successful fertilization. The gametes are extruded through temporary pores or through ruptures in the body wall. Rupture occurs by contraction of the body wall muscles or of special mesenchymal muscles surrounding the gonads.

Fertilization is often external, either free in the sea water or in a gelatinous mass of mucus produced by the mating worms. In the latter situation, actual egg cases are frequently formed, and part or all of the embryonic development occurs within them (Figure 11.13). Internal fertilization occurs in certain nemerteans. In some cases the sperm are released into the mucus surrounding the mating worms and then move into the ovaries of the female; once fertilized, the eggs are usually deposited in egg capsules, where they develop. Some terrestrial species are ovoviviparous; the embryos are retained within the body of the female and development is fully direct—an obvious advantage for surviving on land. Ovoviviparity is also known in a few other nemerteans, including deep-sea pelagic forms. Since the population densities of these pelagic worms are extremely low, they must presumably capitalize on the relatively infrequent encounters of males and females and ensure successful fertilization. In a few cases the males are equipped with suckers, which are used to clasp the female, or, rarely, with a protrusible penis, which is used to transfer sperm.

Regardless of the method of fertilization, development through the gastrula is similar among most of the nemerteans studied to date. Cleavage is holoblastic and spiral, producing either three (*Tubulanus*) or, more typically, four quartets of micromeres. A coeloblastula forms, and this often shows the rudiments of an apical

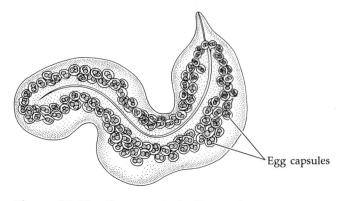

Figure 11.13   Egg capsules in *Lineus ruber*.

ciliary tuft associated with a slight thickening of the blastula wall at or near the animal pole. Gastrulation is usually by invagination of the macromeres and the fourth micromere quartet to produce a coelogastrula. In at least one genus (*Prostoma*, a hermaphroditic freshwater form), gastrulation is by unipolar ingression of the vegetal macromeres; this movement produces a stereogastrula, which later hollows. Mesoderm may originate in several ways, and in some cases the processes are poorly understood. In a few instances it appears that at least some of the middle layer arises from micromere ectodermal precursors, hence producing ectomesoderm. In *Tubulanus*, which produces only three quartets of micromeres, the mesoderm apparently derives from the 3D macromere; in others it is probably from the usual 4d mesentoblast. Unraveling the embryogeny of nemerteans has led to the recent discovery that the rhynchocoel is formed by schizocoely, and it therefore represents a true coelomic cavity.

Developmental strategies are also varied among the nemerteans. Members of the orders Palaeonemertea, Hoplonemertea, and Bdellonemertea undergo direct development within egg cases. The embryos in these three orders pass through stages much like the larval stages of certain flatworms, but they develop gradually to juvenile worms without any abrupt metamorphosis. These embryos are nourished by yolk until they hatch, whereupon they commence feeding. The hatching forms of some of these nemerteans resemble macrostomids (Figure 11.14). This is especially true of the palaeonemerteans, in which the proboscis apparatus is not fully formed at hatching (the proboscis of heteronemerteans and bdellonemerteans is functional at the time of hatching).*

The heteronemerteans undergo a bizarre and fascinating pattern of indirect development. Most species of this order produce a free-swimming, planktotrophic larva called the **pilidium** (Figure 11.15). At this stage the gut is incomplete, consisting of a mouth located between a pair of flaplike ciliated lobes, a stomodeal foregut, and a blind intestine; the anus forms later as a proctodeal in-

---

*Hubrechtella dubia is classified as a palaeonemertean but possesses a typical larval stage. It is not known whether this is a developmental oddity or an indication that a taxonomic reassignment is warranted for this species.

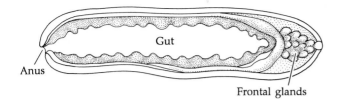

Figure 11.14 Hatching form produced by direct development in *Prosorhochmus*, a hoplonemertean.

vagination. Interestingly, the intestinal diverticula of nemerteans do not form as evaginations of the gut wall but are produced by medial encroachments of mesenchyme, which press in the gut wall, thus creating the diverticula. As the pilidium swims and feeds, a series of invaginations in the larval ectoderm (Figure 11.15B) eventually pinch off internally to produce the presumptive adult ectoderm. Thus, the metamorphosed juvenile develops *within* a larval skin while the animal is still planktonic (Figure 11.15C). In this way the animal prepares for benthic life before it faces the rigors of settlement. When development is completed, the larval skin is shed and the juvenile assumes its life on the bottom.

Another free-swimming larval form, called the **Iwata larva**, is known among certain heteronemerteans. This larval stage was named after Fumio Iwata of Hokkaido University, who has contributed much to our knowledge of nemertean development. The Iwata larva derives from a yolky egg and undergoes lecithotrophic development. It passes through a stage of ectodermal invaginations and eventually sheds its larval skin, much like the processes described for the pilidium. In many species the emerging juveniles eat their larval ectoderm.

Some heteronemerteans undergo direct development via the production of a larval form (the **Desor larva**) that passes its entire developmental life within a protective egg case. The Desor larva was named after E. Desor, who saw and described this pattern of development in *Lineus* in 1848. Prior to hatching from its egg capsule, the Desor larva undergoes ectodermal inpocketing like

that in other heteronemerteans. For this reason, many authorities categorize Desor development as indirect (since a metamorphosis does occur), exemplifying one of the semantic problems of defining life cycle patterns.

Nemertean Phylogeny

The fossil record is of no use in establishing the origin of nemerteans in geological time. They obviously diverged sometime after the origin of the spiralian bilateral condition. Their origin is puzzling, and one must take into account controversies about the position of the flatworms, the origin of the acoelomate condition (Chapter 10), and the probable coelomic nature of the nemertean rhynchocoel and certain blood vessels. An older view is that nemerteans arose from early archoöphoran turbellarian stock, perhaps sharing common ancestry with the macrostomid flatworms. The nemerteans and turbellarians display a number of similarities including protonephridia, types of ocelli, certain histological characteristics (especially of the epidermis), and the general organization of the nervous system. Furthermore, various ciliated slits and depressions among turbellarians may be homologous to the cephalic slits and similar structures of the nemerteans. Some flatworms possess frontal (cephalic) glands thought to be homologous to those of ribbon worms. Some of the strongest evidence for a close flatworm-nemertean relationship comes from examina-tion of their embryogenies. The spiral cleavage of archoöphoran turbellarians (especially of macrostomids and some polyclads) is certainly paralleled in the early development of nemerteans. This commonality is enhanced by a consideration of the similarities among various later developmental and hatching stages.

But, if nemerteans are descendants of flatworms, or if they are sister groups, there is no doubt that they have "done far more" (evolutionarily) with their acoelomate bauplan than have their platyhelminth cousins. Furthermore, the similarities shared by platyhelminths and nemerteans are not unique, because they occur in other taxa as well; in other words, their similarities are symplesiomorphies. Their union to a common ancestor (or

Figure 11.15 Development of a pilidium larva. (A) Pilidium larva (section). (B) A pilidium larva (transverse section) during invagination of larval ectoderm to form adult skin. (C) Late pilidium larva with juvenile formed within.

(A)

(B)

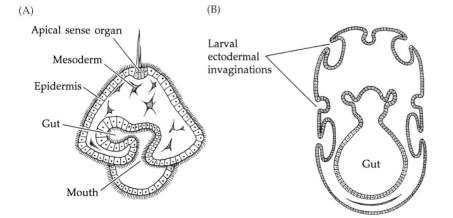

(C)

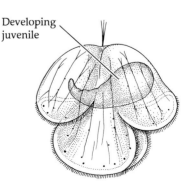

Figure 11.16 Some ideas about nemertean phylogeny. (A) A cladogram based on the hypothesis that the platyhelminths and nemerteans constitute a true monophyletic clade, and that both share the acoelomate condition. However, we may interpret this cladogram in two very different ways. First, we may view the events represented by (1) as the origin of the triploblastic, spiralian condition to produce an acoelomate ancestor (a). This condition, viewed here as plesiomorphic within the spiralian line, was retained in ancestor (b) and in the Platyhelminthes, whereas the nemerteans acquired a set of synapomorphies (3), including the proboscis apparatus, a circulatory system, and an anus. In this scenario, the coelom arose (schizocoely) along the line (4) leading to the modern coelomate protostomes. Alternatively, we may view the assumed acoelomate condition of the flatworms and nemerteans as having arisen secondarily from a coelomate, protostome ancestor. In this case, we add schizocoely to the list of synapomorphies defining (a) and occurring at (1). The common ancestor (b) to the platyhelminths and nemerteans, then, becomes defined by a loss (or reduction) of the coelom (2) to achieve the acoelomate condition.

(B) A cladogram based on the hypothesis that the platyhelminths and nemerteans do not constitute a monophyletic clade. Again, this arrangement can be interpreted in more than one way. Assuming (1) to be the origin of the triploblastic, spiralian condition leading to ancestor (a), then the flatworms retain this ancestral condition and are not distinguished by any unique synapomorphies. The remaining spiralian phyla are distinguished by the origin of an anus and a circulatory system (3), and ancestor (b) is an acoelomate creature. The characteristics of ancestor (b) are retained by the nemerteans that acquire the nemertean synapomorphies (4), and the coelom arises at (5). The other interpretation would view the coelom arising at either (1) or (3). In the first case, the flatworms

must have lost the coelom (2) to become secondarily acoelomate. In the second case, the flatworms are viewed as primitively acoelomate. In either case, we may view the nemerteans as either losing the coelom (if they are acoelomate) or retaining it in the form of the rhynchocoel (and some blood vessels) (4) from their coelomate ancestor (b), which also gave rise to the modern coelomate protostomes.

(C) This cladogram to the orders of nemerteans is applicable regardless of which hypothesis of nemertean origin is accepted. The phylum can be defined by at least the presence of the unique proboscis apparatus (1), and by other things, depending on how one views the taxon's origin (see above). The ancestral nemertean was probably much like a modern-day member of the order Palaeonemertea, and that order has retained many of the ancestral traits. In fact, the lack of any unique synapomorphies distinguishing the Palaeonemertea suggests that the ancestral nemertean was itself what we would recognize as a palaeonemertean. The heteronemerteans are distinguished anatomically by their unique arrangement of body wall muscles in outer longitudinal, middle circular, and inner longitudinal bands (2), and by the appearance of indirect development and free-living larvae (3). The origin of the Enopla is defined by movement of the mouth anterior to the cerebral ganglion (4), movement of the longitudinal nerve cords to a mesenchymal position (5), and fusion of the proboscis pore with the buccal region (6). The hoplonemerteans are defined by the appearance of proboscis armature (7) and the division of the proboscis into distinct regions (8). The bdellonemerteans are characterized by a suite of synapomorphies associated, in large part, with their endocommensalistic lifestyles. These include a reduction in sensory organs (9), the appearance of a posterior sucker (10), the elongation and coiling of the gut (11), and the loss of intestinal diverticula (12).

deriving one from the other) is usually supported by treating them both as acoelomates. However, we have seen in this chapter that the nemerteans probably have coeloms, and in Chapter 10 we explored the possibility that the acoelomate condition may even be secondarily derived from some spiralian, schizocoelomate ancestor.

We are left with various ways to interpret the evidence about nemertean and flatworm relationships. The alternative cladograms shown in Figures 16A and B illustrate these possibilities. First, the nemerteans and flatworms may (Figure 11.16A) or may not (Figure 11.16B) share an immediate common ancestor. Second, the acoelomate condition (in one or both groups) may be primitive within triploblastic, spiralian taxa, or it may be secondarily derived from some protostomous, schizocoelomate ancestor. These views encompass ideas about when the coelom arose relative to the origins of these groups and are explained more fully in the legend for Figure 11.16. And, we should point out that a preliminary 18s rRNA sequence analysis has suggested that the nemerteans are derived well up the spiralian tree, somewhere near the molluscs (Turbeville et al. 1992). See Chapter 24 for a discussion of these phyla in the broad context of the Metazoa.

The phylogenetic relationships among the various taxa of the Nemertea are also difficult to assess. We propose one hypothesis, illustrated and explained in Figure 11.16C. One of the principal structural trends among the nemerteans is the internalization of the major longitudinal nerve cords as depicted in Table 1. In our hypothesis, we assume that the earliest ribbon worms possessed epidermal nerve cords, as do some modern palaeonemerteans. The palaeonemerteans and heteronemerteans retain the plesiomorphic feature of the placement of the mouth posterior to the cerebral ganglion, a feature shared with most turbellarians. The relatively simple and unarmed proboscis and the placement of the nerve cords external to the mesenchyme suggest further that these orders are primitive among the ribbon worms. The heteronemerteans diverged from this clade with their adoption of indirect development, the unique formation of the double larval and adult ectoderm during metamorphosis, and the evolution of their unique arrangement of body wall muscles. The encapsulation, and thus functionally direct development, of those heteronemerteans with a Desor larva is almost certainly a secondary abandonment of free larval life. The divergence of the heteronemerteans apparently occurred

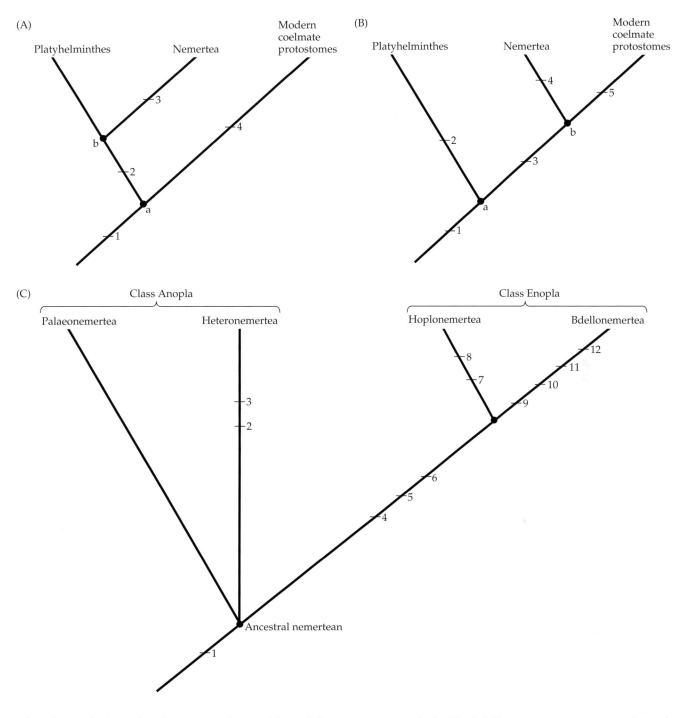

after the evolution of an intramuscular position of the nerve cords among the palaeonemertean ancestors.

The hoplonemerteans show some distinct changes from the members of orders mentioned above. Most notable are the regional specialization and armature of the proboscis, the movement of the nerve cords to a mesenchymal position, and the movement of the mouth more anteriorly. The bdellonemerteans are a specialized offshoot that displays significant modification for an endosymbiotic lifestyle, including simplification of the proboscis, coiling and increased relative length of the gut (probably associated with their herbivorous habits), a posterior body sucker, and decreased body length.

Selected References

Berg, G. 1985. *Annulonemertes* gen. nov., a new segmented hoplonemertean. *In* C. Morris et al. (eds.), *The Origins and Relationships of Lower Invertebrates*. Syst. Assoc. Spec. Vol. No. 28, Oxford, pp. 200–209.

Bianchi, S. 1969a. On the neurosecretory system of *Cerebratulus marginatus* (Heteronemertini). Gen. Comp. Endocr. 12: 541–548.

Bianchi, S. 1969b. The histochemistry of the neurosecretory system in *Cerebratulus marginatus* (Heteronemertini). Gen. Comp. Endocr. 13: 206–210.

Bierne, J. 1966. Localisation dans les ganglions cérébroïdes du centre regulateur de la maturation sexuelle chez la femelle de *Lineus ruber* Muller (Hétéronémertes). C. r. Hebd. Séanc. Acad. Sci. Paris 262: 1572–1575.

Borlase, W. 1758. *The Natural History of Cornwall*. Jackson, Oxford. [Includes the first "description" of a ribbon worm.]

Cantell, C.-E. 1969. Morphology, development, and biology of the pilidium larvae from the Swedish west coast. Zool. Bidrag. Fran Uppsala 38: 61–111.

Coe, W. R. 1931. A new species of nemertean (*Lineus vegetus*) with asexual reproduction. Zool. Anz. 94: 54–60.

Coe, W. R. 1934. Regeneration in nemerteans. IV. Cellular changes involved in restitution and reorganisation. J. Exp. Zool. 67: 283–314.

Coe, W. R. 1940. Revision of the nemertean fauna of the Pacific coasts of North, Central, and northern South America. Allan Hancock Pac. Exped. 2(13): 247–323.

Coe, W. R. 1943. Biology of the nemerteans of the Atlantic coast of North America. Trans. Conn. Acad. Sci. 35: 129–328. [This is one of Coe's classic works; see also his dozens of other publications from 1895 through the late 1950s.]

Desor, E. 1848. On the embryology of *Nemertes*, with an appendix on the embryonic development of *Polynoe*, and remarks upon the embryology of marine worms in general. Boston J. Nat. Hist. 6: 1–18.

Eakin, R. M. and J. A. Westfall. 1968. Fine structure of nemertean ocelli. Am. Zool. 8: 803 (abstract).

Fleming, L. C. and R. Gibson. 1981. A new genus and species of monostiliferous hoplonemerteans, ectohabitant on lobsters. J. Exp. Mar. Biol. Ecol. 52: 79–93.

Gibson, R. 1968. Studies on the biology of the entocommensal rhynchocoelan *Malacobdella grossa*. J. Mar. Biol. Assoc. U.K. 48: 637–656.

Gibson, R. 1972. *Nemerteans*. Hutchinson University Library, London. [An excellent summary of what was known and what was not known in the early 1970s.]

Gibson, R. 1982. *A Synopsis of British Nemerteans*. Synopses of the British Fauna No. 24. Published for the Linnean Society of London, Cambridge Univ. Press, London.

Gibson, R. 1982. Nemertea. *In* S. Parker (ed.), *Synopsis and Classification of Living Organisms*, Vol. 1. McGraw-Hill, New York, pp. 823–846.

Gibson, R. 1985. The need for a standard approach to taxonomic descriptions of nemerteans. Am. Zool. 25: 5–14.

Gibson, R. and J. Jennings. 1969. Observations on the diet, feeding mechanism, digestion and food reserves of the ectocommensal rhynchocoelan *Malacobdella grossa*. J. Mar. Biol. Assoc. U.K. 49: 17–32.

Gibson, R. and J. Moore. 1976. Freshwater nemertines. Zool. J. Linn. Soc. 58: 177–218.

Gontcharoff, M. 1961. Nemertiens. *In* P. Grassé (ed.), *Traité de Zoologie*, Vol. 4, Pt. 1. Masson et Cie, Paris, pp. 783–886.

Harrison, F. W. and B. J. Bogitsh (eds.). 1991. *Microscopic Anatomy of Invertebrates. Vol. 3. Platyhelminthes and Nemtinea*. Wiley-Liss, New York.

Henry, J. Q. and M. Q. Martindale. 1996a. The origins of mesoderm in the equally-cleaving nemertean worm, *Cerebratulus lacteus*. Biol. Bull. 191: 286–288.

Henry, J. Q. and M. Q. Martindale. 1996b. The establishment of embryonic axial properties in the nemertean worm, *Cerebratulus lacteus*. Dev. Biol. 180: 713–721.

Hylbom, R. 1957. Studies on palaeonemerteans of the Gullmar Fiord area (west coast of Sweden). Ark. Zool. 10: 539–582.

Hyman, L. H. 1951. *The Invertebrates*, Vol. 2, Playthelminthes and Rhynchocoela: The Acoelomate Bilateria. McGraw-Hill, New York.

Iwata, F. 1958. On the development of the nemertean *Micrura akkenshiensis*. Embryologia 4: 103–131.

Iwata, F. 1960. Studies on the comparative embryology of nemerteans with special reference to their interrelationships. Publ. Akkeshi Mar. Biol. Stn. 10: 1–51.

Iwata, F. 1985. Foregut formation of the nemerteans and its role in nemertean systematics. Am. Zool. 25: 23–36.

Jennings, J. B. 1960. Observations in the nutrition of the rhynchocoelan *Lineus ruber*. Biol. Bull. 119(2): 189–196.

Jennings, J. B. and R. Gibson. 1969. Observations on the nutrition of seven species of rhynchocoelan worms. Biol. Bull. Mar. Biol. Lab. Woods Hole 136: 405–433.

Jensen, D. D. 1960. Hoplonemertines, myxinoids and deuterostome origins. Nature 188: 649–650.

Jensen, D. D. 1963. Hoplonemertines, myxinoids, and vertebrate origins. *In* E. C. Dougherty (ed.), *The Lower Metazoa: Comparative Biology and Phylogeny*. Univ. Calif. Press, Berkeley, pp. 113–126.

Jespersen, A. and J. Lützen. 1988a. The fine structure of the protonephridial system in the land nemertean *Pantinonemertes californiensis*. Zoomorphol. 108: 69–75.

Jespersen, A. and J. Lützen. 1988b. Ultrastructure and morphological interpretation of the circulatory system of nemerteans. Vidensk. Meddr. Dansk. Naturh. Foren. 147: 47–66.

Karling, T. G. 1965. *Haplopharynx rostratus* Meixner mit den Nemertinen vergleichen. Z. Zool. Syst. Evol. 3: 1–18.

McDermott, J. J. 1976a. Predation of the razor clam *Ensis directus* by the nemertean worm *Cerebratulus lacteus*. Chesapeake Sci. 17(4): 299–301.

McDermott, J. J. 1976b. Observations on the food and feeding behavior of estuarine nemertean worms belonging to the order Hoplonemertea. Biol. Bull. 150: 57–68.

McDermott, J. J. and P. Roe. 1985. Food, feeding behavior, and feeding ecology of nemerteans. Am. Zool. 25: 113–125.

Moore, J. and R. Gibson. 1981. The *Geonemertes* problem (Nemertea). J. Zool. Lond. 194: 175–201. [A revision of the world's terrestrial nemerteans.]

Riser, H. W. 1974. Nemertinea. *In* A. Giese, A. and J. Pearse (eds.), *Reproduction of Marine Invertebrates*, Vol. 1. Academic Press, New York, pp. 359–389.

Riser, H. W. 1985. Epilogue: Nemertinea, a successful phylum. Am. Zool. 25: 145–151.

Riser, N. W. 1989. Speciation and time-relationships of the nemertines to the acoelomate metazoan Bilateria. Bull. Mar. Sci. 45: 531–538.

Roe, P. 1970. The nutrition of *Paranemertes peregrina*. I. Studies on food and feeding behavior. Biol. Bull. 139: 80–91.

Roe, P. 1976. Life history and predator-prey interactions of the nemertean *Paranemertes peregrina* Coe. Biol. Bull. 150: 80–106.

Schultze, M. S. 1851. *Beiträage zur Naturgeschichte der Turbellarien*. C. A. Koch, Greifswald.

Stricker, S. A. 1982. The morphology of *Paranemertes sanjuanensis* sp. n. (Nemertea, Monostylifera) from Washington, U.S.A. Zool. Scr. 11(2): 107–115.

Stricker, S. A. 1983. S.E.M. and polarization microscopy of nemertean stylets. J. Morph. 175: 153–169.

Stricker, S. A. 1985. The stylet apparatus of monostyliferous hoplonemerteans. Am. Zool. 25: 87–97.

Stricker, S. A. and R. Cloney. 1982. Stylet formation in nemerteans. Biol. Bull. 162: 387–403.

Stricker, S. A. and R. A. Cloney. 1983. The ultrastructure of venom-producing cells in *Paranemertes peregrina* (Nemertea, Hoplonemertea). J. Morphol. 177: 89–107.

Turbeville, J. M. 1986. An ultrastructural analysis of coelomogenesis in the hoplonemertine *Prosorchochmus americanus* and the polychaete Magelona sp. J. Morphol. 187: 51–60.

Turbeville, J. M., K. C. Field and R. A. Raff. 1992. Phylogenetic position of phylum Nemertini, inferred from 18S rRNA sequences: Molecular data as a test of morphological character homology. Mol. Biol. Evol. 9: 235–249.

Turbeville, J. M. and E. E. Ruppert. 1985. Comparative ultrastructure and the evolution of nemertines. Am. Zool. 25: 53–71.

Wickham, D. E. 1979. Predation by the nemertean *Carcinonemertes errans* on eggs of the Dungeness crab, *Cancer magister*. Mar. Biol. 55: 45–53.

Wickham, D. E. 1980. Aspects of the life history of *Carcinonemertes errans* (Nemertea: Carcinonemertidae), an egg predator of the crab *Cancer magister*. Biol. Bull. 159: 247–257.

Wourms, J. P. 1976. Structure, composition, and unicellular origin of nemertean stylets. Am. Zool. 16: 213 (abstract).

12 *Blastocoelomates and Other Phyla*

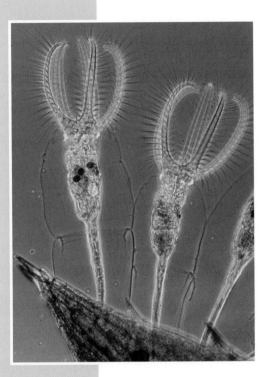

In the face of this bewildering array of conflicting opinions about the interrelationships of the aschelminth phyla, it is impossible to form a coherent picture of the evolution of the animals.

R. B. Clark,
Dynamics in Metazoan Evolution, 1964

I n this chapter we examine eleven phyla of Metazoa, placing them together both for convenience and because they provide some interesting examples of comparative biology, the basic subject of this book. The phyla discussed here do not constitute a monophyletic clade, but are an odd assemblage of animals, most of which have been referred to in the past as "pseudocoelomates" or "aschelminths." These phyla are: Rotifera, Gastrotricha, Kinorhyncha, Nemata, Nematomorpha, Acanthocephala, Entoprocta, Priapula, Gnathostomulida, Loricifera, and the recently discovered Cycliophora. Space does not permit us to cover these groups in great detail, but you should be aware that the literature on some of them (e.g., rotifers and nematodes) is vast. We will explore several of the concepts explained in the introductory chapters, including systematic problems, convergent evolution, and the exploitation of limiting bauplans. While the members of each phylum represent a success story in their own right, some are far more abundant and diverse than others. Some of these phyla comprise only a few known species (in one case, a single species), whereas others (such as Nemata) include thousands of described species.

Taxonomic History

Three of the phyla dealt with in this chapter were established quite recently and thus do not figure in much of the complex history of the others. The group Gnathostomulida was originally described as an order of Turbellaria by Peter Ax (1956) and subsequently raised to phylum status by R. J. M. Riedl (1969). The phylum Loricifera was established in 1983 by Reinhardt Kristensen, and the phylum Cycliophora was described in 1995 by Peter Funch and R. Kristensen.

Many of the other creatures discussed here were discovered centuries ago. Aristotle discussed the intestinal parasite *Ascaris lumbricoides* (phylum Nemata) around 350 B.C., as did the Egyptians before him, in 1550 B.C. For our purposes we may consider their taxonomic history beginning in the late nineteenth century. By that time the infrastructure of the classification of most of these groups had been established, and their evolutionary affinities were being examined. Much of this work was brought into focus by Grobben (1908), who inferred a blastocoelomate (= pseudocoelomate) nature for the rotifers, kinorhynchs, gastrotrichs, nematodes, nematomorphans, and acanthocephalans; he used this condition to justify placing these groups as classes within the phylum Aschelminthes (previously Nemathelminthes). Grobben proposed a dozen or so characteristics of these taxa that he considered *unique-in-combination*. The general concept of a phylum Aschelminthes was retained by Hyman (1951), but she dropped the Acanthocephala and added the Priapula, a group about which there is still considerable debate.

Although a few authors continue to recognize the phylum Aschelminthes, most do not. It is now generally accepted that the blastocoelomate condition probably arose several times in the Metazoa. That is to say, the aschelminths represent a grade (or grades), not a single clade. Recent morphological and molecular data support the idea that this assemblage is indeed polyphyletic.

The discovery of entoprocts, or kamptozoans, dates back to the late eighteenth century. By about 1840 their superficial resemblance to the Ectoprocta (Chapter 21) was recognized, and the two groups were placed together in the phylum Bryozoa. Hatschek (1888) eventually undertook a detailed embryological study of certain entoprocts, noting fundamental organizational differences between them and the ectoprocts. On the basis of his observations, he raised the entoprocts to the rank of a separate phylum. In 1921, A. H. Clark also proposed separate phylum status for the entoprocts (under the name Calyssozoa). Claus Nielsen (1971, 1977) argues that the ectoprocts and entoprocts are closely related to one another, and interested readers should consult his publications for an explanation of that controversial proposal.

The Blastocoelomate Condition

Before discussing each phylum separately, it is essential to note some aspects of the blastocoelomate condition itself. In the blastocoelomate phyla, the body cavity is a blastocoel that persists into adulthood. It is, therefore, a retained embryonic feature and may be justly considered a paedomorphic characteristic. Of course, the appearance of an embryonic blastocoel (hence a transient or incipient blastocoelom) is not unique to the groups under consideration here. That is, most Metazoa have a blastocoel at some stage in their ontogeny; it is a more or less universal trait of animals at the metazoan grade of complexity. For example, the larval body cavities of some animals (e.g., polychaete annelids) are blastocoels, even though the adults are truly coelomate. Many workers today view the adult blastocoelom as a paedomorphic condition secondarily derived from coelomate ancestors rather than as a primitive, precoelomic stage in metazoan evolution (e.g., Ruppert 1991).

The advantages of body cavities were discussed in Chapter 3. The benefits related to internal transport, space for organ development, hydrostatic support systems, and pressure-operated protrusion of parts should come to mind in this regard. However, in the absence of an effective circulatory system, these cavity-conferred benefits are generally only realized when body size is kept relatively small or when body shape maintains small diffusion distances. Exceptions to these principles involve either structural modifications of the body cavity or the establishment of a lifestyle (such as endoparasitism) that reduces the problems of the low surface-to-volume ratio.

As we shall see, not all of these animals retain a spacious fluid-filled body cavity. In some cases (e.g., gastrotrichs, entoprocts), the space has been to various degrees invaded by mesenchyme, which sometimes obliterates it completely to produce a solid body construction, functionally similar to that in acoelomates. The trade-off in such instances may be the acquisition of support and storage facilities at the expense of transport and hydrostatic qualities.

Phylum Rotifera: The Rotifers

The phylum Rotifera (Latin *rota*, "wheel"; *fera*, "to bear") includes more than 1,800 described species. These tiny Metazoa, first seen by early microscopists such as Antony van Leeuwenhoek in the late seventeenth century, were at first lumped with the protists as *animalcules*. A few species reach lengths of 2–3 mm, but most are less than a millimeter long. In fact, some ciliated protists have been incorrectly described as rotifers, and the error has only recently been exposed (Turner 1995).

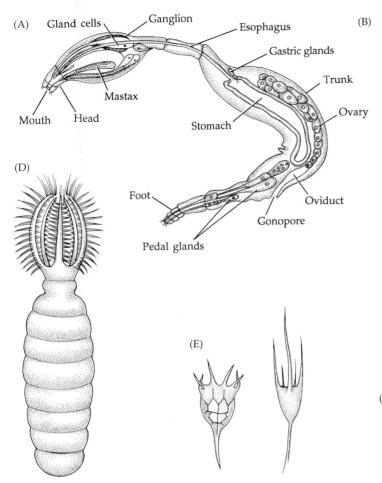

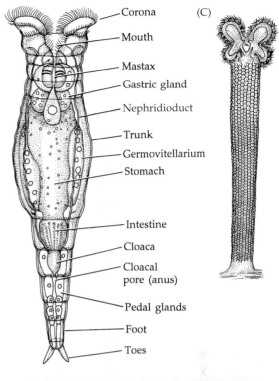

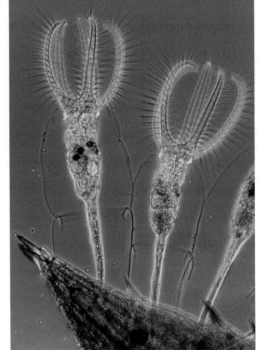

Figure 12.1 Representative rotifers. (A) *Seison annulatus* (class Seisonidea), a marine rotifer from the gills of *Nebalia*. (B) *Philodina roseola* (class Bdelloidea). (C–F) Members of the class Monogononta. (C) *Floscularia*, a sessile rotifer. (D) *Stephanoceros*, one of the strange collothecacean rotifers with the corona modified as a trap. (E) The cuticular loricae from two loricate rotifers. (F) Live specimens of *Stephanoceros*.

Despite their small size, rotifers are quite complex and display a variety of body forms (Figure 12.1). Most are solitary, but some sessile forms are colonial, a few of which secrete gelatinous casings into which the individuals can retract. They are most common in fresh water, but many marine species are also known, and others live in damp soil or in the water film on mosses. They often comprise an important component of the plankton of fresh and brackish waters.

The body comprises three general regions, the head, trunk, and foot. The anterior end bears a ciliary organ called the **corona**. When active, the coronal cilia often give the impression of a pair of rotating wheels, hence the derivation of the phylum name; the rotifers were historically called "wheel animalcules." The members

of this phylum are further characterized by being blastocoelomate, having a complete gut (usually), protonephridia, showing a tendency to eutely, and often having syncytial tissues or organs (Box 12A). The pharynx is modified as a **mastax** comprising sets of internal jaws, or **trophi**.

*PHYLUM ROTIFERA**

CLASS DIGONATA: Varied habitats; with paired germovitellaria.

> **ORDER SEISONIDEA:** (Figure 12.1A) Epizoic on marine leptostracan crustaceans (e.g., *Nebalia*); corona reduced to bristles; trophi fulcrate (piercing); males fully developed; sexual females produce only mictic ova. One genus: *Seison*.

> **ORDER BDELLOIDEA:** (Figure 12.1B) Freshwater, moist soils and foliage, marine, and terrestrial; corona typically well developed; trophi ramate (grinding); males unknown; reproduction strictly by parthenogenesis. (e.g., *Adineta, Ceratotrocha, Embata, Habrotrocha, Philodina, Rotaria*)

CLASS MONOGONONTA: (Figure 12.1C–F) Predominately freshwater, some are marine; swimmers, creepers, or sessile; corona and trophi variable; males typically short-lived and reduced in size and complexity; sexual reproduction probably occurs at some point in the life history of all species; mictic and amictic ova produced in many species; single germovitellarium. (e.g., *Asplanchna, Collotheca, Euchlanis, Floscularia, Monostyla, Stephanoceros, Testudinella*)

General External Anatomy and Details of the Corona

The body surface of many rotifers is annulated, allowing flexibility. The surface often bears spines, tubercles, or other sculpturing, and may be developed as a thickened casing, or **lorica** (Figure 12.1E). Many rotifers bear single dorsal and paired lateral sensory antennae arising from various regions of the body. In most species the foot is elongate, with cuticular annuli that permit a telescoping action. The distal portion of the foot often bears spines, or a pair of "toes" through which the ducts from **pedal glands** pass. The secretion from the pedal glands enables the rotifer to attach temporarily to the substratum. The foot is absent from some swimming forms (e.g., *Asplanchna*) and is modified for permanent attachment in sessile types (e.g., *Floscularia*).

The corona is the most characteristic external feature of rotifers. Its morphology varies greatly and has been used extensively in taxonomic and phylogenetic investigations. The presumed primitive condition is shown in Figure 12.2A. A well developed patch of cilia surrounds the anteroventral mouth. This patch is the **buccal field**, or **circumoral field**, and it extends dorsally around the head as a ciliary ring called the **circumapical field**. The extreme anterior part of the head bordered by this ciliary ring is the **apical field**.

The corona has evolved to a variety of modified forms in different rotifer taxa. The most familiar coronal form is that seen in the bdelloids. Here the buccal field is quite reduced, and the circumapical field is separated into two ciliary rings, one slightly anterior to the other (Figure 12.2B). The anteriormost ring is called the **trochus**, the other the **cingulum**. In many rotifers the

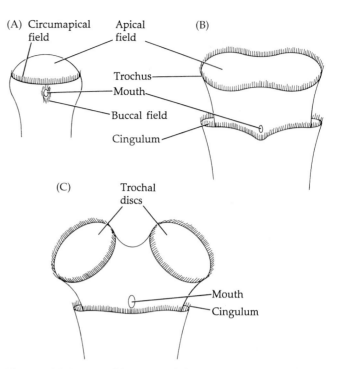

Figure 12.2 Modifications of the corona among selected rotiferan types. (A) The presumed plesiomorphic condition has buccal and circumapical fields. (B) The circumapical field is separated into trochus and cingulum. The trochus is lobed, like that of *Floscularia*. (C) The trochus is separated into two trochal discs.

BOX 12A *Characteristics of the Phylum Rotifera*

1. Triploblastic, bilateral, unsegmented blastocoelomates

2. Gut complete and regionally specialized

3. Pharynx modified as a mastax, containing jawlike elements called trophi

4. Anterior end bears variable ciliated fields as a corona

5. Posterior end often bears toes and adhesive glands

6. Epidermis syncytial, with fixed number of nuclei; secretes extracellular cuticle and intracellular skeletal lamina (the latter forming a lorica in some species)

7. With protonephridia, but no special circulatory or gas exchange structures

8. With unique retrocerebral organ

9. Males generally reduced or absent; parthenogenesis common

10. With modified spiral cleavage

11. Inhabit marine, freshwater, or semiterrestrial environments; sessile or free-swimming

trochus is a pair of well defined anterolateral rings of cilia called **trochal discs** (Figure 12.2C), which may be retracted or extended for locomotion and feeding. The metachronal ciliary waves along these trochal discs impart the impression of rotating wheels.

Many organs and tissues of rotifers display **eutely**: cell or nuclear number constancy. This condition is established during development, and there are no mitotic cell divisions in the body following ontogeny.

Body Wall, Body Cavity, Support, and Locomotion

Most rotifers possess a gelatinous cuticle outside the syncytial epidermis. Within the epidermis is a dense laminar layer, the **skeletal lamina**, which produces the lorica (when present) and other surface structures. The epidermal nuclear number is consistent for each species (usually from about 900 to 1,000 nuclei). Beneath the epidermis are various circular and longitudinal muscle bands (Figure 12.3); there are no sheets or layers of body wall muscles. The internal organs lie within a typically spacious, fluid-filled blastocoelom.

In the absence of a thick, muscular body wall, body support and shape are maintained by the intro-epidermal skeletal lamina and the hydrostatic skeleton provided by the body cavity. In most cases the cuticle is only flexible enough to allow slight changes in shape, so increases in hydrostatic pressure within the body cavity can be used to protrude body parts (e.g., foot, trochal discs). These parts are protracted and retracted by various muscles (Figure 12.3), each consisting of only one or two cells.

Although a few rotifers are sessile, most are motile and quite active, moving about by swimming or inchworm-like creeping. Some are exclusively either swimmers or crawlers, but many are capable of both methods of locomotion. Swimming is accomplished by beating the coronal cilia, forcing water posteriorly along the body and driving the animal forward, sometimes in a spiral path. When creeping, a rotifer attaches its foot with secretions of the pedal glands, then elongates its body and extends forward. It attaches the extended anterior end to the substratum, releases its foot, and draws its body forward by muscular contraction.

Feeding and Digestion

Rotifers display a variety of feeding methods, depending upon the structure of the corona (Figure 12.2) and the mastax trophi (Figure 12.4). Ciliary suspension feeders have well developed coronal ciliature and a grinding mastax. These forms include the bdelloids, which have trochal discs and a **ramate mastax**, and a number of monogonontan rotifers, which have separate trochus and cingulum and a **malleate mastax**. These forms typically feed on organic detritus or minute organisms. The feeding current is produced by the action of the cilia of the trochus (or trochal discs), which beat in a direction

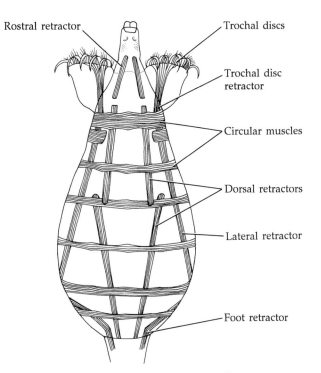

Figure 12.3 Major muscle bands of the bdelloid *Rotaria* (dorsal view).

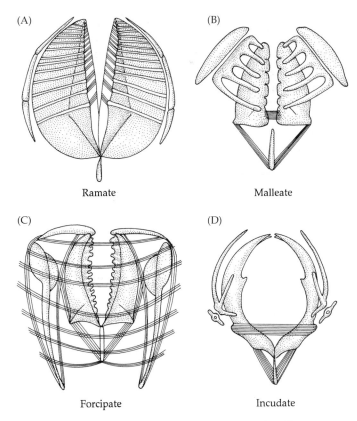

Ramate

Malleate

Forcipate

Incudate

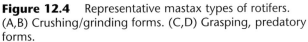

Figure 12.4 Representative mastax types of rotifers. (A,B) Crushing/grinding forms. (C,D) Grasping, predatory forms.

opposite to that of the cilia of the cingulum. Particles are drawn into a ciliated food groove that lies between these opposing ciliary bands and are carried to the buccal field and mouth.

Raptorial feeding is common in many species of Monogononta. Coronal ciliation in these rotifers is often reduced or used exclusively for locomotion. Raptorial feeders obtain food by grasping it with protrusible, pincer-like mastax jaws; most possess either a **forcipate mastax** (non-rotating) or an **incudate mastax** (rotating 90–180 degrees during protrusion) (Figure 12.4C,D). Raptorial rotifers feed mainly on small animals but are known to ingest plant material as well. They may ingest their prey whole and subsequently grind it to smaller particles within the mastax, or they may pierce the body of the plant or animal with the tips of the mastax jaws and suck fluid from the prey.

Some monogonontan rotifers have adopted a trapping method of predation. In such cases the corona usually bears spines or setae arranged as a funnel-shaped trap (Figure 12.1D,F). The mouth in these trappers is located more or less in the middle of the ring of spines (rather than in the more typical anteroventral position); thus, captured prey are drawn to it by contraction of the trap elements. The mastax in trapping forms is often reduced.

A few rotifers have adopted symbiotic lifestyles. As noted in the classification scheme, seisonids live on marine leptostracan crustaceans of the genus *Nebalia*. These rotifers (*Seison*) crawl around the carapace and limbs of their host, feeding on detritus and on the host's brooded eggs. Some bdelloids (e.g., *Embata*) also live on the gills of crustaceans, particularly amphipods and decapods. There are isolated examples of entoparasitic rotifers inhabiting hosts such as *Volvox* (a colonial protist), freshwater algae, snail egg cases, and the body cavities of certain annelids and terrestrial slugs. Little is known about nutrition in most of these forms.

The digestive tract of most rotifers is complete and more or less straight (Figure 12.5A). (The anus has been secondarily lost in a few species, and some have a moderately coiled gut.) The mouth leads inward to the pharynx (mastax) either directly or via a short, ciliated **buccal tube**. Depending on the feeding method and food sources, swallowing is accomplished by various means, including ciliary action of the buccal field and buccal tube, or a piston-like pumping action of certain elements of the mastax apparatus. The mastax is lined with cuticle and is ectodermal in origin. Opening into the gut lumen just posterior to the mastax are ducts of the **salivary glands**. There are usually two to seven such glands; they are presumed to secrete digestive enzymes and perhaps lubricants aiding the movement of the mastax trophi.

A short esophagus connects the mastax and stomach. A pair of **gastric glands** opens into the posterior end of the esophagus; these glands apparently secrete digestive enzymes. The walls of the esophagus and gastric glands are often syncytial. The stomach is generally thick-walled and may be cellular or syncytial, usually comprising a specific number of cells or nuclei in each species (Figure 12.5B). The intestine is short and leads to the anus, which is located dorsally near the posterior end of the trunk. Except for *Asplanchna*, which lacks a hindgut, an expanded **cloaca** connects the intestine and anus. The oviduct and usually the nephridioducts also empty into this cloaca.

Digestion probably begins in the lumen of the mastax and is completed extracellularly in the stomach, where absorption occurs. In one large and enigmatic group of bdellids the stomach lacks a lumen. Although much remains to be learned about the digestive physiology of rotifers, some experimental work indicates that diet has multiple and important effects on various aspects of their biology, including the size and shape of individuals as well as some life cycle activities (Gilbert 1980).

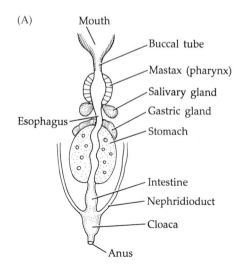

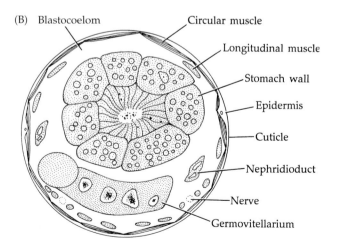

Figure 12.5 (A) Digestive system of a rotifer. (B) Cross section through the trunk.

Circulation, Gas Exchange, Excretion, and Osmoregulation

There are no special organs for internal transport or for the exchange of gases between tissues and the environment. The blastocoelomic fluid provides a medium for circulation within the body, which is aided by general movement and muscular activities. Small body size reduces diffusion distances and facilitates the transport and exchange of gases, nutrients, and wastes. These activities are further enhanced by the absence of linings and partitions within the body cavity, so the exchanges occur directly between the organ tissues and the body fluid. Gas exchange probably occurs over the general body surface wherever the cuticle is sufficiently thin.

Most rotifers possess one pair of flame bulb protonephridia, located far forward in the body. A nephridioduct leads from each flame bulb to a collecting bladder, which in turn empties into the cloaca via a ventral pore. In some forms, especially the bdelloids, the ducts open directly into the cloaca, which is enlarged to act as a bladder (Figure 12.5A). The protonephridial system of rotifers is primarily osmoregulatory in function, and is most active in freshwater forms. Excess water from the body cavity and probably from digestion is also pumped out via the anus by muscular contractions of the bladder. This "urine" is significantly hypotonic relative to the body fluids. It is likely that the protonephridia also remove nitrogenous excretory products from the body. This form of waste removal is probably supplemented by simple diffusion of wastes across permeable body wall surfaces.

Some rotifers (especially the freshwater and semiterrestrial bdelloids) are able to withstand extreme environmental stresses by entering a state of metabolic dormancy. They have been experimentally desiccated and kept in a dormant condition for as long as four years—reviving upon the addition of water. Some have survived freezing in liquid helium at –272°C and other severe stresses dreamed up by biologists.

Nervous System and Sense Organs

The cerebral ganglion of rotifers is located dorsal to the mastax, in the neck region of the body. Several nerve tracts arise from the cerebral ganglion, some of which bear additional small ganglionic swellings (Figure 12.6A). There are usually two major longitudinal nerves positioned either both ventrolaterally or one dorsally and one ventrally.

The coronal area generally bears a variety of touch-sensitive bristles or spines and often a pair of ciliated pits thought to be chemoreceptors (Figure 12.6B). The dorsal and lateral antennae are probably tactile. Some rotifers bear sensory **flosculi**, which are arranged as a cluster of micropapillae encircling a pore. These flosculi may be tactile or chemosensory. Most of the errant rotifers possess at least one simple ocellus embedded in the cerebral ganglion. In some, this cerebral ocellus is accompanied by one or two pairs of lateral ocelli on the coronal surface, and sometimes by a pair of apical ocelli in the apical field. The lateral and apical ocelli are multicellular epidermal patches of photosensitive cells. Clément (1977) described possible baro- or chemoreceptors in the body cavity that may help regulate internal pressure or fluid composition.

Associated with the cerebral ganglion is the so-called **retrocerebral organ**. This curious glandular structure gives rise to ducts that lead to the body surface in the apical field (Figure 12.6B). Once thought to be sensory in function, more recent work suggests that it may secrete mucus to aid in crawling.

Reproduction and Development

Parthenogenesis is probably the most common method of reproduction among rotifers, but other forms of asex-

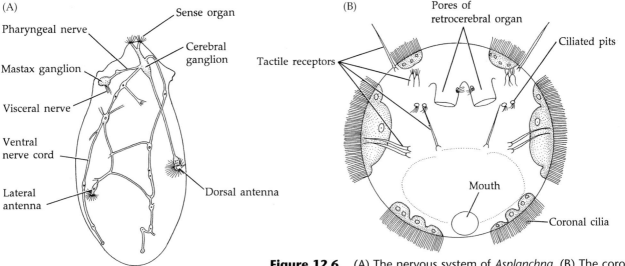

Figure 12.6 (A) The nervous system of *Asplanchna*. (B) The coronal area of *Euchlanis* (apical view). Note the various sense organs.

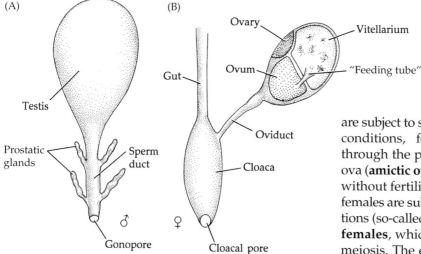

Figure 12.7 Male and female reproductive systems from a generalized monogonontan rotifer.

ual reproduction are unknown, and most groups show only very weak powers of regeneration. Most rotifers are dioecious; but, except for the genus *Seison*, males are either reduced in abundance, size, and complexity (Monogononta), or are completely unknown (Bdelloidea). If you find a rotifer, the chances are good that it is a female.

The male reproductive system (Figure 12.7A) includes a single testis (paired in *Seison*), a sperm duct, and a posterior gonopore whose wall is usually folded to produce a copulatory organ. Prostatic glands are sometimes present in the wall of the sperm duct. The males are short-lived and possess a reduced gut unconnected to the reproductive tract.

The female system includes paired (Bdelloidea) or single (Monogononta) syncytial **germovitellaria** (Figure 12.7B). Eggs are produced in the ovary and receive yolk directly from the vitellarium before passing along the oviduct to the cloaca; in those forms that have lost the intestinal portion of the gut (e.g., *Asplanchna*), the oviduct passes directly to the outside via a gonopore. In the Seisonidea, there are no yolk glands.

In rotifers with a male form, copulation occurs either by insertion of the male copulatory organ into the cloacal area of the female or by hypodermic impregnation. In the latter case, males attach to females at various points on the body and apparently inject sperm directly into the blastocoelom (through the body wall). The sperm somehow find their way to the female reproductive tract, where fertilization takes place. The number of eggs produced by an individual female is determined by the original, fixed number of ovarian nuclei—usually 20 or fewer, depending on the species. Once fertilized, the ova produce a series of encapsulating membranes and are then either attached to the substratum or carried externally or internally by the brooding female.

Parthenogenesis is clearly the rule among the bdelloids, but it is also a common and usually seasonal occurrence in the monogonontans, where it tends to alternate with sexual reproduction. This cycle (Figure 12.8A) is an adaptation to freshwater habitats that are subject to severe seasonal changes. During favorable conditions, females reproduce parthenogenetically through the production of mitotically derived diploid ova (**amictic ova**). These eggs develop into more females without fertilization. However, when ova from amictic females are subjected to particular environmental conditions (so-called **mixis stimuli**), they develop into **mictic females**, which then produce **mictic** (haploid) **ova** by meiosis. The exact stimulus apparently varies among different species and may include such factors as changes in day length, temperature, food resources, or increases in population density. Although these cycles are commonly termed "summer" and "autumn cycles," this is a bit misleading because mixes can also occur during warm weather and many populations have several periods of mixis each year. Mictic ova may develop by parthenogenesis to haploid males, which produce sperm by mitosis. These sperm fertilize other mictic ova, producing diploid, thick-walled, resting zygotes. The resting zygotic form is extremely resistant to low temperatures, desiccation, and other adverse environmental conditions. When favorable conditions return, the zygotes develop and hatch as amictic females (Figure 12.8B), completing the cycle.

Only a few studies have been conducted on the embryogeny of rotifers (see especially Pray 1965). In spite of the paucity of data, and some conflicting interpretations in the literature, it is generally accepted that rotifers have modified spiral cleavage. The isolecithal ova undergo unequal holoblastic early cleavage to produce a stereoblastula. Gastrulation is by epiboly of the presumptive ectoderm and involution of the entoderm and mesoderm; the gastrula gradually hollows to produce the blastocoel, which persists as the adult body cavity. The mouth forms in the area of the blastopore. Definitive nuclear numbers are reached early in development for those organs and tissues displaying eutely.

Errant rotifers undergo direct development, hatching as mature or nearly mature individuals. Sessile forms pass through a short dispersal phase, sometimes called a "larva," which resembles a typical swimming rotifer. The "larva" eventually settles and attaches to the substratum. In all cases, there is a total absence of cell division during postembryonic life (i.e. they are eutelic).

Many rotifers exhibit **developmental polymorphism**, a phenomenon also seen in some protists, insects, and primitive crustaceans. It is the expression of alternative morphotypes under different ecological conditions, by organisms of a given genetic constitution (the differentiation of certain castes in social insects is one of the most

(A)

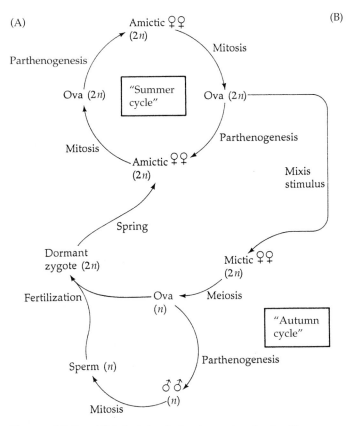

(B)

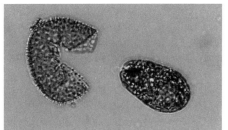

Figure 12.8 (A) Mictic/amictic alternation in the life cycle of a monogonontan rotifer. (B) Micrograph of an amictic female hatching from an overwintering phase.

remarkable examples of developmental polymorphism). In all such animals studied to date, the alternative adult morphotypes appear to be products of flexible developmental pathways, triggered by environmental cues and often mediated by internal mechanisms such as hormonal activities. In one well studied genus of rotifers (*Asplanchna*), the environmental stimulus regulating which of several adult morphologies is produced is the presence of a specific molecular form of vitamin E—alpha tocopherol. *Asplanchna* obtains tocopherol from its diet of algae or other plant material, or when it preys on other herbivores (animals do not synthesize tocopherol). The chemical acts directly on the rotifer's developing tissues, where it stimulates differential growth of the syncytial hypodermis after cell division has ceased. Predator-induced morphologies also occur among rotifers. *Keratella slacki* eggs, in the presence of the predator *Asplanchna* (both are rotifers), are stimulated to develop into larger-bodied adults with an extra long anterior spine, thus rendering them more difficult to eat.

Phylum Gastrotricha: The Gastrotrichs

The phylum Gastrotricha (Greek *gasteros*, "stomach"; *trichos*, "hair") comprises about 450 species of small ma-

rine, brackish, and freshwater Metazoa. Most species are less than 1 mm long, although a few reach 3 mm in length. Many gastrotrichs bear a superficial resemblance to rotifers or even large ciliate protists (for which they are often mistaken). Figure 12.9 illustrates some of the body forms within this phylum and some features of their external anatomy, and Box 12B lists the distinguishing features of this group. Many gastrotrichs are meiofaunal and live in the interstitial spaces of loose sediments. Others are found in surface detritus or among the filaments of aquatic plants; a few are planktonic.

The gastrotrich body is typically divisible into a head and a trunk. A few possess an elongate "tail" (e.g., *Urodasys*; Figure 12.9C). Externally, these animals bear vari-

---

BOX 12B **Characteristics of the Phylum Gastrotricha**

1. Triploblastic, bilateral, unsegmented

2. Area between gut and body wall filled with loose organs and mesenchyme, effectively creating an acoelomate condition

3. With one to many pairs of adhesive tubes

4. Cuticle well developed, often forming plates and spines; outer cuticle composed of several layers of unit-membrane-like structures

5. Epidermis partly cellular, partly syncitial

6. External ciliation restricted to ventral surface; with monociliate epidermal cells; external cilia covered by outer layers of cuticle

7. Gut complete

8. With protonephridia, but without special circulatory or gas exchange structures

9. Hermaphroditic, or, if dioecious, only females are known

10. Cleavage seemingly radial, but not well studied

11. Inhabit marine and freshwater environments; errant

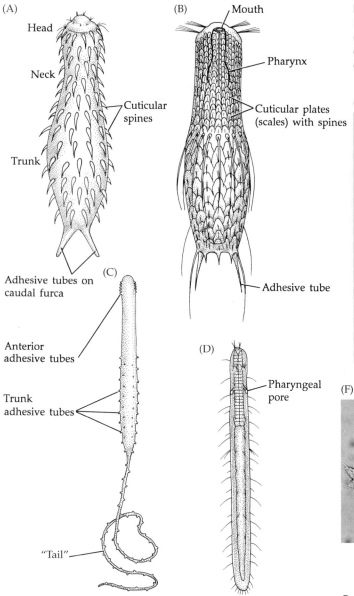

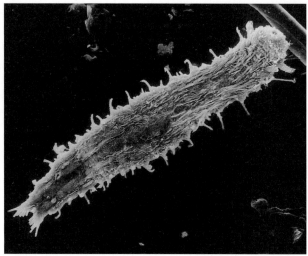

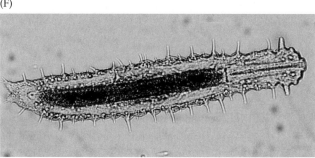

Figure 12.9 Representative gastrotrichs. (A,B) Members of the class Chaetonotidea. (A) *Chaetonotus*. (B) *Aspidophorus*. (C–F) Members of the class Macrodasyida. (C) *Urodasys*. (D) *Pleurodasys*. (E) SEM and (F) light micrograph of *Turbanella*.

ous arrangements of spines, bristles, and scales or plates derived from the cuticle. The body bears two or many **adhesive tubes** equipped with glands that secrete attachment and releaser substances used for temporary adherence to objects in the environment. The few gastrotrichs that lack adhesive tubes are planktonic swimmers. Gastrotrichs have a complete digestive tract and protonephridia. There is, however, no body cavity surrounding the internal organs. The "space" once thought to be present is apparently an artifact caused by certain common methods of fixation. Adult gastrotrichs are more appropriately viewed as functionally acoelomate.

PHYLUM GASTROTRICHA

ORDER MACRODASYIDA: (Figure 12.9C–F). Marine and estuarine gastrotrichs, usually bearing numerous adhesive tubes along the head and trunk; pharyngeal pores present; usually with several pairs of protonephridia; hermaphroditic. (e.g.,

Dactylopodola, Macrodasys, Platydasys, Pleurodasys, Turbanella, Urodasys)

ORDER CHAETONOTIDA: (Figure 12.9A,B). Primarily fresh water, but also some marine, estuarine, and semiterrestrial species; with a variable number of adhesive tubes; pharyngeal pores absent; one pair of protonephridia usually present; most are parthenogenetic (males are unknown).

SUBORDER MULTITUBULATINA: Marine; with posterior, anterior, and numerous lateral adhesive tubes; hermaphroditic. Monogeneric: *Neodasys*, with two known species.

SUBORDER PAUCITUBULATINA: Mostly freshwater species; usually with two adhesive tubes at the ends of posterior caudal furca, but adhesive tubes are lacking in some; hermaphroditic or (usually) parthenogenetic females only. (e.g., *Aspidophorus, Chaetonotus, Dasydytes, Lepidodermella*)

Body Wall, Support, and Locomotion

The body is covered by a cuticle of varying thickness and complexity. The outer part of the cuticle comprises from one to many layers; the inner part is fibrous and produces the spines, scales, plates, and other covering structures (Figure 12.9). The gastrotrich cuticle is unique in that its outer portion is made up of several layers of membrane-like structures. The epidermis is partly cellular and partly syncytial and is ciliated ventrally (thus the derivation of the phylum name: "hairy belly"). Some of the epidermal cells are characteristically monociliated, a feature shared with the gnathostomulids. Oddly, the outer cuticle extends over these external cilia. Internal to the epidermis are bands of circular and longitudinal muscles. There is no body cavity in gastrotrichs, and the internal organs are rather tightly packed together.

Support is provided by the cuticle and the compact body construction. Most move by ciliary gliding, although some of the more flexible forms "inch" along by alternately attaching the posterior and anterior adhesive tubes (Figure 12.9E,F). The ventral surface is very densely ciliated.

Feeding and Digestion

Most gastrotrichs feed by pumping small food items into the gut by action of the muscular pharynx, or by ciliary currents in the foregut. In the macrodasyids, **pharyngeal tubes** connect the pharynx lumen to the outside (Figure 12.9D) and allow the release of excess water taken in with the food. Gastrotrichs feed on nearly any organic material, alive or dead, of appropriately small size (protists, unicellular algae, bacteria, detritus).

The mouth is terminal and often surrounded by a ring of oral spines that aid in food capture. A short buccal cavity connects the mouth to an elongate muscular pharynx lined with cuticle (a stomodeum) (Figure 12.10). The entodermally derived portion of the gut is a straight tube, which may be differentiated into a stomach and an intestine. The rectum is a proctodeum and leads to the anus on the dorsal surface. In a few species of *Urodasys*, the anus is absent. Digestion and absorption are quite rapid and occur in the midgut, although details of these processes are wanting.

Circulation, Gas Exchange, Excretion, and Osmoregulation

There are no special circulatory or gas exchange structures in gastrotrichs. These functions are accomplished by simple diffusion, as they are in so many other tiny Metazoa. Hemoglobin has been detected in one species of *Neodasys*, and at least some gastrotrichs are capable of anaerobic respiration.

One pair of protonephridia are present in the chaetonotid gastrotrichs; macrodasyids generally possess several pairs. The protonephridia lie on each side of the gut as single cells or clusters of cells that lead to long, coiled nephridioducts (Figure 12.10). These ducts lead to a pair of nephridiopores located near the ventral midline of the body. Each protonephridial cell is a **solenocyte** (sometimes called a **cyrtocyte**), which is characterized in part by bearing only one or two flagella rather than the many flagella found in flame bulbs.

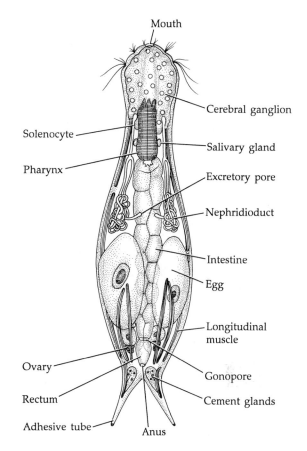

Figure 12.10 Internal anatomy of a chaetonotid gastrotrich (*Chaetonotus*).

Nervous System and Sense Organs

The cerebral ganglion is a relatively large, bilobed mass connected dorsally over the anterior portion of the pharynx (Figure 12.10). A ganglionated, lateral longitudinal nerve cord arises from each lobe of the cerebral ganglion and extends to the posterior end of the body.

Sensory receptors are predominantly tactile. They occur as spines and bristles over much of the body but are concentrated on the head. Some species bear ciliated depressions on the sides of the head, which may be chemosensory pits. A few species contain pigmented ocelli in the cerebral ganglion.

Reproduction and Development

Although asexual reproduction is unknown in gastrotrichs, recent studies by O. G. Manylov (1995) demonstrate regenerative capability in *Turbanella*. Manylov's experimental work is the first conclusive evidence of regeneration in gastrotrichs.

Most species of gastrotrichs are hermaphroditic or are known only as parthenogenetic females. The hermaphroditic reproductive structures are often complex and have been best studied in members of the order Macrodasyida (Figure 12.11). The male system comprises one or two testes with associated sperm ducts that lead to a single gonopore on the ventral midline. In a few instances paired male pores are present; rarely, the

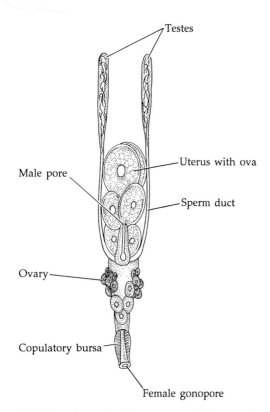

Testes

Uterus with ova

Male pore

Sperm duct

Ovary

Copulatory bursa

Female gonopore

Figure 12.11 Reproductive system of a hermaphroditic gastrotrich (*Macrodasys*).

male system connects with the hindgut. In a very few forms, (e.g., *Macrodasys*), a copulatory structure called a **caudal organ** is present near the male gonopore.

The female system includes one or two ovaries, lying just posterior to the testes in the hermaphroditic forms. The ovaries are rather diffuse organs, not bounded by the typical capsule or tissue layer. As eggs are produced they are released into an ill-defined space called the uterus, which is adjacent to the ovaries and associated with a yolk-producing tissue or vitellarium. The eggs are eventually carried to an oviduct or, more commonly, to a saclike receiving area called the **X organ**, which connects with the female gonopore.

Following mutual cross-fertilization, the zygotes are released singly or a few at a time by rupture of the body wall. Gastrotrichs produce both thick-shelled, dormant eggs and thin-shelled, rapidly developing eggs. Parthenogenesis predominates in the freshwater forms, where the appearance of hermaphroditic individuals (and the production of sperm) seems to take place only infrequently. The appearance of hermaphrodites is probably associated with certain environmental conditions, or perhaps with an internal genetic clock.

Few embryological studies have been conducted on gastrotrichs. Cleavage is holoblastic and seemingly radial; the early blastomeres arise such as to produce a bilaterally symmetrical embryo. A coeloblastula forms, and gastrulation occurs by the movement of two cells into

the blastocoel from the presumptive ventral surface. These "ingressed" cells presumably contribute to the formation of entoderm and, ultimately, of the midgut. Stomodeal and proctodeal invaginations form and connect with the developing gut. Two additional cells drop from the surface to the interior as the presumptive germ cells from which gametes eventually arise.

Development in gastrotrichs is direct, and juveniles hatch from the egg capsules. Although developmental time to hatching varies with the type of egg produced, maturation is rapid and the animals usually are sexually mature within a few days after hatching.

Phylum Kinorhyncha: The Kinorhynchs

Among the most intriguing of the "little animalcules" are the members of the phylum Kinorhyncha (Greek *kineo*, "movable"; *rhynchos*, "snout"), previously called the Echinodera (Greek *echinos*, "spiny"; *dere*, "neck") or Echinoderida. Since their discovery in 1841 on the northern coast of France, about 150 species of kinorhynchs have been described from around the world, nearly all of which are less than 1 mm in length. Most live in marine sands or muds, from the intertidal zone to a depth of 8,000 meters. Some are known from algal mats or holdfasts, sandy beaches, and brackish estuaries; others live on hydroids, ectoprocts, or sponges.

Externally, most kinorhynchs appear similar (Figure 12.12), and most of the specific differences are in the details of spination and the arrangement and structure of their thick cuticular plates. The body comprises a distinct proboscis or "head," which is retractable into the anterior portion of the neck, and a trunk. The body is divided into 13 clearly defined **zonites**. Many specialists view these zonites as true segments, although the exact nature of their formation is uncertain (i.e., are they produced by teloblastic development?).

The head is formed by segment 1 and bears a retractable **oral cone**. The mouth is borne on the oral cone and is surrounded by a ring of anteriorly directed spines called **oral stylets**. Behind the oral cone, the head bears up to seven rings of posteriorly directed spines called **scalids**. Up to 90 scalids occur in some species, each one articulating at a basal joint but none with intrinsic musculature. Segment 2 forms the "neck," which consists of a series of plates called **placids** that fold over the head when it is retracted. The remaining eleven segments form the trunk. Most trunk segments are made up of a dorsal (**tergal**) plate and a pair of ventral (**sternal**) plates (Figure 12.12C). The anus is located on the last segment and is usually flanked by strong **lateral end spines**.

The kinorhynchs are arguably either acoelomate or blastocoelomate (Box 12C). As for the gastrotrichs, previous reports of large blastocoelomic spaces ("pseudo-

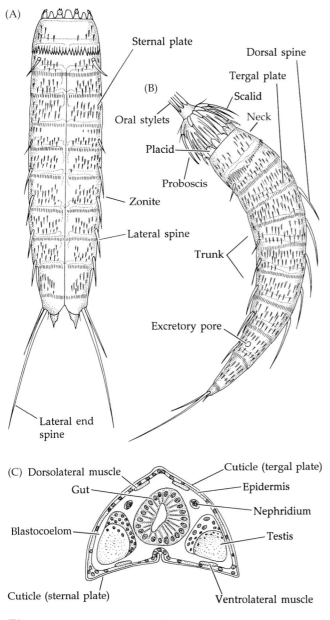

Figure 12.12 (A,B) External anatomy of the kinorhynch *Echinoderes*. (A) Ventral view with head retracted. (B) Lateral view with head extended. (C) A trunk zonite (cross section) of a kinorhynch. Note the arrangement of the body wall structures and the organs within the blastocoelom. (D) A meiofaunal kinorhynch with oral stylets extended.

BOX 12C **Characteristics of the Phylum Kinorhyncha**

1. Triploblastic, bilateral, segmented

2. Body probably blastocoelomate

3. Body divided into 13 segments by articulating cuticular plates (dorsal tergites and ventral sternites). Segmentation reflected internally by arrangement of epidermal glands, muscles and ganglia on nerve cord

4. Introvert with hooked spines

5. Epidermis cellular; without external ciliation

6. Gut complete

7. With one pair protonephridia, but without special circulatory or gas exchange structures

8. Early growth accompanied by periodic shedding of the cuticle

9. Dioecious

10. Embryonic cleavage patterns not well understood

11. Inhabit marine, interstitial environments

---

coeloms") in kinorhynchs probably resulted from artifacts of preservation. They appear to possess only small spaces, or lacunae rather than a spacious body cavity. Their possible metameric condition has led to many controversies about their affinities with other taxa. Other similarities with the arthropods are striking and include a cellular nonciliated epidermis, intersegmental muscle bands, "segmentally" arranged ganglia, and the nature of the cuticle. Kinorhynchs also periodically shed their cuticle, but whether or not it is mediated by the hormone ecdysone (as it is in arthropods) is not yet known.

PHYLUM KINORHYNCHA

ORDER CYCLORHAGIDA: With 14–16 placids as a closing apparatus, although in one group a clamshell-like first trunk segment may assist; trunk oval, round, or slightly triangular in cross section; numerous cuticular trunk spines; adhesive tubes present; trunk segments usually with dense covering of cuticular hairs or denticles. Habitats vary. (e.g., *Cateria, Centroderes, Echinoderes*)

ORDER HOMALORHAGIDA: With 4–8 placids, but ventral plate(s) of first trunk segment also assists in closing; trunk triangular in cross section; trunk spines and cuticular hairs few in number or rudimentary. Most live in subtidal muds. (e.g., *Kinorhynchus, Neocentrophyes, Pycnophyes*)

Body Wall

Beneath the thick, well developed, chitinous cuticle is a nonciliated epidermis (Figure 12.12C), which contains

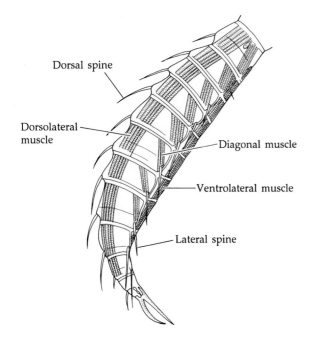

Dorsal spine

Dorsolateral muscle

Diagonal muscle

Ventrolateral muscle

Lateral spine

Figure 12.13 The trunk muscles of a kinorhynch (lateral view).

elements of the nervous system. Largely internal to the epidermis but still attached to the cuticle are bands of dorsolateral and ventrolateral intersegmental muscles (Figure 12.13). The muscles are cross-striated and arranged intersegmentally. Some of the anterior longitudinal muscle bands serve as head retractors. A series of metamerically arranged dorsoventral muscles creates the increased hydrostatic pressure that protracts the head and oral cone when the retractor muscles relax. The internal organs lie within small blastocoelomic spaces, which are greatly reduced posteriorly due to the presence of organs and abundant amebocytes. A more spacious body cavity occurs anteriorly, facilitating retraction of the oral cone and head region.

Support and Locomotion

Body shape in kinorhynchs is more or less fixed by the rigid plates of the supportive cuticular exoskeleton, but the animals are able to flex and even twist at the points of articulation between adjacent segments. In the absence of external cilia, burrowing is accomplished by extending the head into the substratum, anchoring the anterior spines, and then pulling the rest of the body forward.

Feeding and Digestion

Kinorhynchs are probably direct deposit feeders, ingesting the substratum and digesting the organic material or eating unicellular algae contained therein; they have been found with their guts full of benthic diatoms. However, the details of feeding and the exact nature of their food are not known.

The mouth leads into a buccal cavity located within the oral cone and thence to a muscular pharynx (Figure 12.14A). The buccal cavity, pharynx, and esophagus are lined with cuticle and represent a stomodeum. Various paired glands are often associated with the esophagus, but their functions are uncertain. The esophagus connects with an elongate, straight midgut, which leads to a short, cuticle-lined proctodeal hindgut (rectum) and the anus on the last segment. So-called digestive glands often arise from the midgut. Nothing is known about the digestive physiology in kinorhynchs, but ultrastructural details are discussed by Neuhaus (1994).

Circulation, Gas Exchange, Excretion, and Osmoregulation

Circulation within the body is by diffusion through the body cavity. Gas exchange is by diffusion across the body wall. Body movements aid diffusion in accomplishing internal transport. Kinorhynchs possess one pair of solenocyte protonephridia in the tenth segment, each with a short nephridioduct that opens on the eleventh segment (Figure 12.14B). Excretory and osmoregulatory physiology remains largely unstudied, although some species can tolerate low salinities for short periods, and some live in salinities as low as 6 parts per thousand in the Gulf of Finland.

Nervous System and Sense Organs

The central nervous system of kinorhynchs is relatively simple and is intimately associated with the epidermis. A series of ten connected ganglia is arranged in a ring around the pharynx. Each ganglion probably primitively gave rise to one longitudinal nerve cord, eight of which are retained by most species. The two midventral nerve cords are most prominent, and they bear ganglia in each segment.

Sensory receptors include tactile bristles, spines, and flosculi on the body. Microvillar eyespots are present on the pharyngeal nerve ring in at least some species.

Reproduction and Development

Kinorhynchs are dioecious and possess relatively simple reproductive systems (Figure 12.14C). Externally, males and females are usually indistinguishable from one another. In both sexes, paired saclike gonads lead to short gonoducts that open separately on the thirteenth segment. In the males, the gonopore is associated with two or three cuticular, hollow, penile spines (**spicules**) that presumably aid in copulation. The female gonads comprise both germ cells and nutritive (yolk-producing) cells. Each oviduct bears a diverticulum that forms as a seminal receptacle prior to ending at a gonopore between zonites 12 and 13.

Mating has never been observed, and egg laying and early development (e.g., cleavage patterns) have not been adequately studied. Fertilized ova are deposited in the environment in egg cases. The embryos develop di-

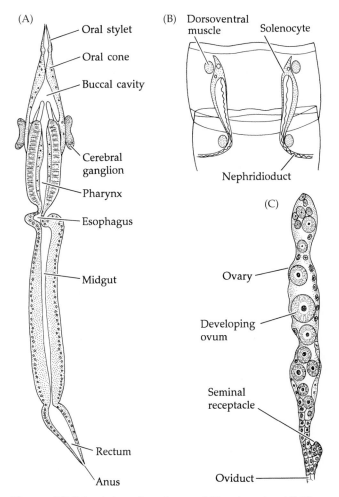

Figure 12.14 Internal anatomy of kinorhynchs. (A) The digestive tract of *Pycnophyes*. (B) The nephridial arrangement in *Pycnophyes*. (C) The simple female reproductive system of a kinorhynch.

rectly to juveniles, which emerge from the egg cases with 11 of the 13 body segments already formed (Kozloff 1972). They do not hatch as unsegmented "larvae" that add segments by a sequence of molts (as reported in earlier studies). Juveniles do molt periodically, passing through six juvenile stages to adulthood. Segments 12 and 13 are added during this molt period. Adults apparently do not molt.

Phylum Nemata: The Nematodes

An enormous amount of literature exists on the nematodes—roundworms and threadworms—much of it dealing with the parasitic species of economic or medical importance. Many of the large parasitic forms have been known since ancient times, but the small, free-living types were not discovered until after the invention of the microscope. Most authorities on the group now prefer the phylum name Nemata (Greek *nema*, "thread"), although Nematoda is still commonly used. With nearly

25,000 described species (and probably several times that many undescribed species), they are one of the most abundant groups of Metazoa; one study revealed some 90,000 nematodes in a single rotting apple, and another turned up 1,074 individuals in 6.7 cc of coastal mud. Good farmland soil in the United States typically harbors from 3 to 9 *billion* nematodes per acre.

Whereas most of the free-living species are microscopic, many of the parasitic forms are much larger; and members of one species attain lengths of 8 m. Nematodes are known from virtually every habitat in the seas, fresh water, and on land. Some are generalists, but many have very specific habitats. One species is known only from felt coasters under beer mugs in a few towns in eastern Europe. Marine nematodes are among the most common and widespread groups of animals, occurring from the shore to the abyss. Where they are found, they are often the most numerous Metazoa, in both number of species and number of individuals. Some environments yield as many as 3 million threadworms per square meter. Despite their abundance, most free-living nematodes are poorly known, and their importance in marine benthic systems is little appreciated.

An exception to this relative obscurity is the soil nematode *Caenorhabditis elegans*, which has been targeted by molecular and developmental biologists as a "model organism." Scientists around the world focus their research on *C. elegans* with the goal of fully understanding every aspect of its biology and the developmental fate of every embryonic cell. *Caenorhabditis elegans* lends itself to such research because it is free-living, easy to culture, matures in just 3 days, and has a fixed number of cells in most of its organs. In 1998, scientists announced that they had successfully mapped the complete genome of *C. elegans*—97 million bases! (*Science*, 11 Dec. 1998.)

Nematodes parasitize nearly all groups of animals and plants. Some cause serious damage to crops and livestock, and some are pathogenic in humans. Most pet owners eventually encounter parasitic nematodes, as they are commonly seen in the feces and vomit of dogs and cats. One species, *Onchocerca volvulus*, causes an eye disease in humans called "river blindness" and is thought to infect nearly 20 million people in Latin America and Africa.

The Nemata are vermiform blastocoelomates with thin, unsegmented bodies that are usually distinctly round in cross section (Box 12D). To the untrained and unaided eye, most nematodes look very much alike, but there are variations in external body form (Figure 12.15).

PHYLUM NEMATA (= NEMATODA)

It is beyond the scope of this text to present much of the exhaustive classification scheme of the nematodes. The two classes, Adenophorea and Secernentea, are usually subdivided into two and three subclasses, respectively. Each contains several orders and many families. For a more detailed classification see Maggenti 1982. Argument exists about the monophyly of the

(A)

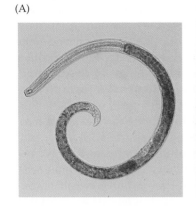

(B)

(C)

(D)

(E)

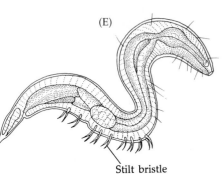

Stilt bristle

(F)

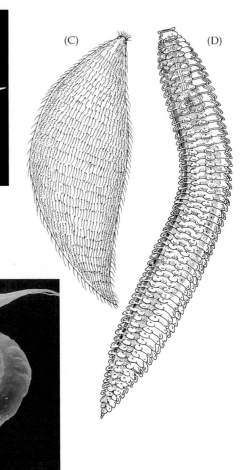

Figure 12.15 Representative examples of the phylum Nemata. (A) *Mononchus* (subclass Spiruria), a predatory soil dweller. (B) Male and female *Ascaris suum* (subclass Rhabditia), parasites in the gut of pigs. (C) *Greeffiella* (subclass Chromadoria), a free-living marine roundworm. (D) *Criconema* (subclass Diplogasteria), a plant parasite with ornate cuticular plates. (E) A member of the subclass Chromadoria (class Adenophorea), has locomotor stilt-bristles that contain adhesive glands. (F) SEM of a free-living interstitial nematode from marine sand. (G) *Desmoscloex*, a meiofaunal nematode.

(G)

nematodes (Adamson 1987). However, recent molecular work by Kampfer et al. (1998) supports the conventional view that these two classes represent a valid clade.

CLASS ADENOPHOREA (= APHASMIDA): (Figures 12.15C, E,G). With cephalic chemoreceptors called amphids, but lacking caudal phasmids; excretory system comparatively simple, not cuticularized, and without collecting tubules; most are free-living (marine, fresh water, terrestrial), although parasitic species are known. Includes the subclasses Enoplia and Chromadoria. (e.g., *Dioctophyme, Strongyloides, Trichinella, Trichuris*)

CLASS SECERNENTEA (= PHASMIDA): (Figure 12.15A,B,D). With cephalic amphids and caudal phasmids; excretory system comparatively complex in some, with cuticularized duct and well developed collecting tubules; most are parasitic; those that are free-living are predominantly terrestrial. Includes the subclasses Rhabditia, Spiruria, and Diplogasteria. (e.g., *Anchylostoma, Ascaris, Necator, Wuchereria*).

Body Wall, Support, and Locomotion

The nematode body is covered by a well developed and complexly layered cuticle secreted by the epidermis (Figure 12.16D). The cuticle is responsible in part for allowing the invasion of hostile environments, such as dry terrestrial soils and the digestive tracts of hosts, for it drastically reduces the permeability of the body wall. Predominantly terrestrial or parasitic nematodes (class

BOX 12D *Characteristics of the Phylum Nemata*

1. Triploblastic, bilateral, vermiform, unsegmented, blastocoelomate

2. Body round in cross section and covered by a layered cuticle; growth in juveniles usually accompanied by cuticular shedding

3. With unique cephalic sense organs called amphids; some have caudal sense organs called phasmids

4. Gut complete; various mouth structures arranged in radially symmetrical pattern

5. Most with unique excretory system, comprised of one or two renette cells or a set of collecting tubules

6. Without special circulatory or gas exchange structures

7. Body wall has only longitudinal muscles (no circular muscles)

8. Epidermis cellular or syncitial, forming longitudinal cords housing nerve cords

9. Dioecious; males commonly with "hooked" posterior end

10. With unique cleavage pattern; not unambiguously radial or spiral

11. Inhabit marine, freshwater, and terrestrial environments; some are free-living, some parasitic

Secernentea) usually have a dense, fibrous inner layer of the cuticle, whereas most of the free-living marine and freshwater forms (class Adenophorea) lack this inner layer. The texture of the cuticle is highly variable among nematodes. It may be relatively smooth, or covered with sensory setae and wartlike bumps. The cuticle in many roundworms is ringed (Figure 12.15G) or marked with longitudinal ridges and grooves. In many marine forms, the cuticle contains radially arranged rods or other inclusions of various shapes. As a nematode grows, it sheds its cuticle and grows a new one through a series of four molts during its lifetime.

The epidermis varies among the different taxa from cellular to syncytial and is often thickened as dorsal, ventral, and lateral longitudinal cords (Figure 12.16A,C). The dorsal and ventral thickenings house longitudinal nerve cords; the lateral thickenings contain excretory canals (when present, as they are in some secernenteans) and neurons. Internal to the epidermis is a relatively thick layer of obliquely striated longitudinal muscle arranged in four quadrants. The muscles are connected to the dorsal and ventral nerve cords by unique extensions called **muscle arms** (Figure 12.16

A,B). This arrangement is different from the usual neuromuscular junctions in most other animals; in nematodes the connections are made by extensions of the muscle cells rather than of the neurons. Oddly, a similar condition apparently exists in the cephalochordate *Branchiostoma* (Chapter 23), presumably a case of convergent evolution. Also, in nematomorphs (horsehair worms) and gastrotrichs, the longitudinal muscles bear extensions suggested to be a possible homologue of the nematode muscle arms. There is no circular muscle layer in nematodes (or in nematomorphs), a condition viewed as homologous by some workers.

The fluid-filled blastocoelom is not spacious. The apparently large body cavity seen in many laboratory specimens is an artifact caused by shrinkage of tissues in alcohol. Modern microscopy techniques reveal that the organs of most nematodes occupy nearly all of the internal space. The cuticle provides most of the body support in nematodes. In the absence of circular body wall muscles, some types of locomotion, such as peristaltic burrowing, are impossible. The typical pattern of nematode locomotion involves contractions of longitudinal muscles, producing a whiplike undulatory motion (Figure 12.16E). Among the free-living nematodes this movement pattern relies on contact with environmental substrata, against which the body pushes. The muscles act against the hydrostatic skeleton and the cuticle, which serve as antagonistic forces to the muscle contractions. The crossed collagenous fibers of the cuticle are nonelastic, but their arrangement allows shape changes as the body undulates. When placed in a fluid environment and deprived of contact with solid objects, benthic nematodes thrash about rather inefficiently. Some actually do swim (but not very well), and some are able to crawl along using various cuticular spines, grooves, ridges, and glands to gain purchase on the substratum (Figure 12.15E).

Feeding and Digestion

Nematodes are extremely diverse in habits and habitats, and have evolved a variety of feeding strategies that are often reflected in anatomical features of the mouth area. Labial flaps, spines, teeth, jaws, and other armature are arranged in radially symmetrical patterns (Figure 12.17). Many infaunal nematodes are direct deposit feeders. Others are detritivores or microscavengers, living in or on dead organisms or fecal material. Many of these species apparently do not feed directly on the carcasses they inhabit, but on fungi and bacteria growing in the decomposing organic matter. Many free-living nematodes are predatory carnivores, feeding on a variety of other small animals. Others feed on diatoms, algae, and bacteria. Plant parasitic nematodes use an oral stylet to pierce individual root cells and suck out the contents.

Bacterial symbioses occur in some nematodes. For example, species of *Artomonema* and *Parastomonema* that

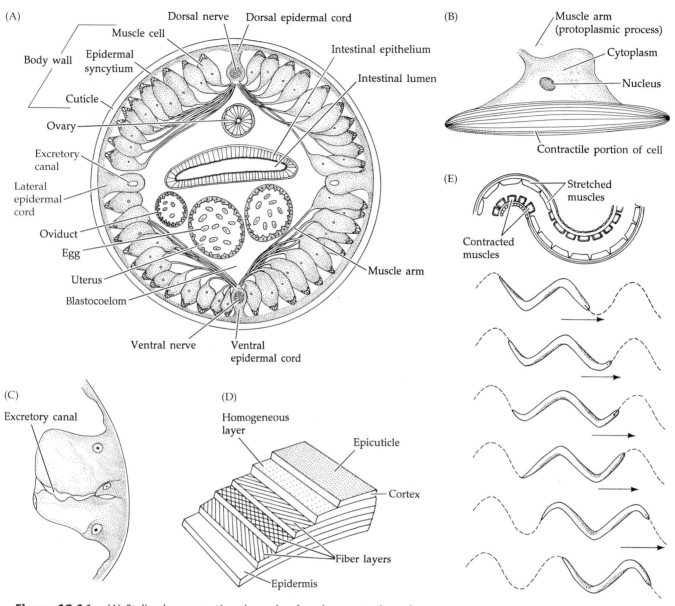

Figure 12.16 (A) Stylized cross section through a female nematode such as *Ascaris* (subclass Rhabditia). (B) A single longitudinal muscle cell, illustrating the origin of the muscle arm. (C) The lateral epidermal cord of *Cucullanus* (Rhabditia). (D) The layers of the cuticle. (E) Undulatory locomotion in a free-living nematode results from the action of the longitudinal muscle fibers. The concave areas along the body represent positions of muscle contraction; the convex areas are regions of muscle stretching. Leverage is gained against surrounding objects or the substratum in the environment.

inhabit sulfur-rich sediments harbor chemoautotrophic bacteria in their highly reduced guts. These worms meet their nutritional requirements by absorbing some of the metabolic products of the bacteria. Some other nematodes (subfamily Stilbonematinae) inhabit sulfide-rich, oxygen-poor sediments, where symbiotic bacteria coat their cuticle. These nematodes are able to twist themselves in such a way that they can feed on the bacteria "farmed" on their body surface.

The myriad parasitic nematodes are known from nearly every group of plants and animals. In inverte-brates and vertebrates (including humans), nematodes parasitize a variety of body fluids and organs, where they may cause extreme tissue damage.

Nematode digestive tracts vary greatly in complexity and regional specialization (Figure 12.18). The anteriorly located mouth leads to a short buccal cavity connected to a stomodeal, cuticle-lined esophagus (often called a pharynx). The esophagus is elongate and may be subdivided into distinct muscular and glandular regions, the details of which are of considerable taxonomic importance (Figure 12.18A–G). The muscles of the esopha-

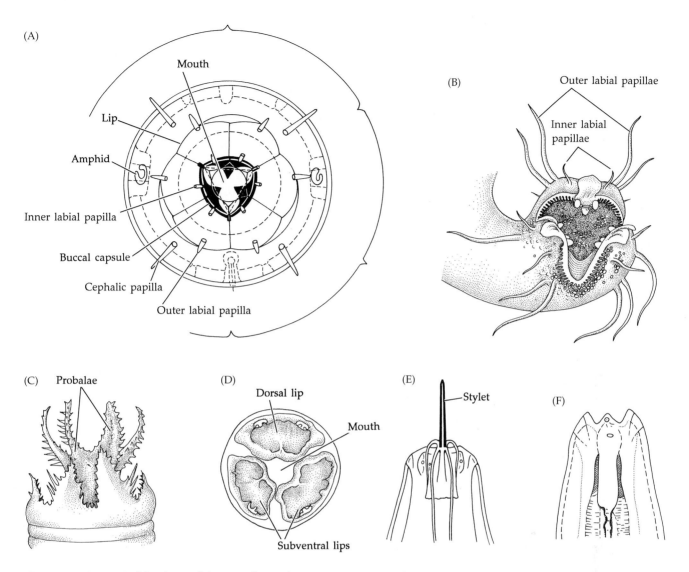

Figure 12.17 Modifications of the anterior end among selected nematodes. (A) The front end of a generalized nematode (anterior view). Note the basic radial symmetry of the parts. (B) The anterior end of a free-living marine nematode. (C) The anterior end of *Acrobeles* (subclass Rhabditia), a soil nematode bearing modified labia (probolae) apparently used in burrowing or food sorting. (D) The tripartite lip of *Ascaris* (subclass Rhabditia) is used to attach the parasite to the host's intestinal wall. (E) The anterior end of *Nygolaimus* (subclass Enoplia) has a protruded stylet, used to puncture prey. The stylet is then retracted and the prey's body fluids sucked out. (F) The simple anterior end of the free-living nematode *Panagrolaimus* (subclass Rhabditia).

gus pump food material from the buccal cavity into the intestine, which leads to a short proctodeal rectum and a subterminal anus on the ventral surface of the body. In males the rectum is a cloaca, thus also receiving products of the reproductive system. The esophageal glands and perhaps the midgut lining secrete digestive enzymes into the gut lumen. Initial digestion is extracellular; final intracellular digestion occurs in the midgut,

following absorption across the surfaces of the microvilli of the midgut cells (Figure 12.18I).

Circulation, Gas Exchange, Excretion, and Osmoregulation

There are no special circulatory or gas exchange structures in nematodes. As in many of the other groups considered in this chapter, these functions are accomplished by diffusion and movement of body cavity fluids. Some parasitic nematodes possess a form of hemoglobin in these fluids that presumably transports and stores oxygen. Both aerobic and anaerobic metabolic pathways are found among the nematode groups, and many of these worms are able to shift from one mechanism to the other according to environmental oxygen concentrations. Facultative anaerobiosis is surely significant in parasitic nematodes and those that live in other anoxic environments.

Nematode excretory structures are unique and apparently not homologous to any of the protonephridial types found in other Metazoa. There exists a rather clear

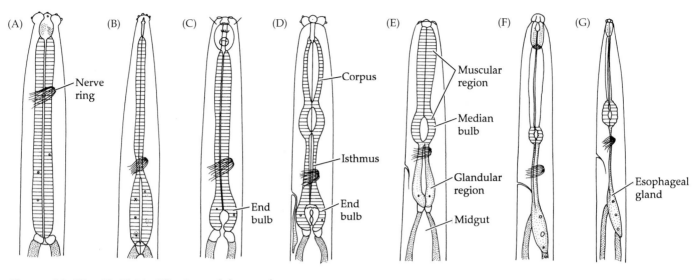

Figure 12.18 (A–G) Modifications of the esophagus among different nematodes. Note the different degrees of regional specialization. (A) Cylindrical esophagus (*Mononchus*: subclass Enoplia). (B) Dorylaimoid esophagus (*Dorylaimus*: Enoplia). (C) Bulboid esophagus (*Ethmolaimus*: subclass Chromadoria). (D) Rhabditoid esophagus (*Rhabditis*: subclass Rhabditia). (E) Diplogasteroid esophagus (*Diplogaster*: subclass Diplogasteria). (F) Tylenchoid esophagus (*Helicotylenchus*: Diplogasteria). (G) Aphelencoid esophagus (*Aphelenchus*: Diplogasteria). (H) Digestive tract and reproductive system of a female *Rhabditis*. (I) intestinal epithelium of *Ascaris* (Rhabditia).

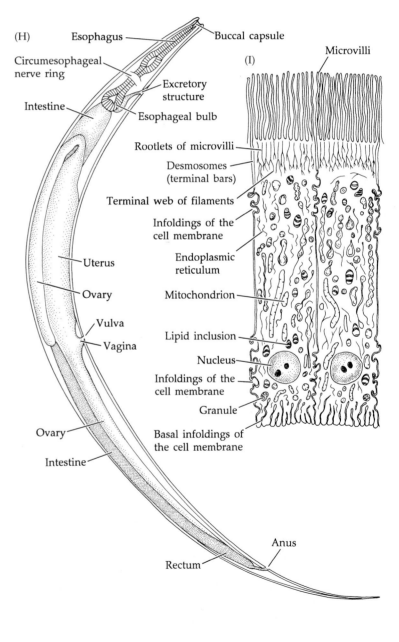

evolutionary sequence of different excretory structures among nematodes (Figure 12.19). The presumed plesiomorphic condition occurs in certain free-living taxa and has been modified among other groups, especially within specialized parasitic forms. In many free-living threadworms, the system comprises one or two glandular **renette cells** that connect directly to a midventral excretory pore, and sometimes a third cell forming an **ampulla** at the opening (Figure 12.19A). Modifications to this system often include various arrangements of intracellular collecting ducts within the cytoplasm of extensions of the renette cells (Figure 12.19B). In many advanced parasitic species the renette cell bodies are lost completely, leaving only the system of tubules in an H or inverted Y pattern (Figure 12.19C,D). Many members of the subclass Enoplia lack renette cells altogether. Instead they have numerous unicellular units distributed along the entire length of the body. Each cell opens to the outside via a duct and a pore. If these cells are excretory in function, they may represent nonciliated protonephridia.

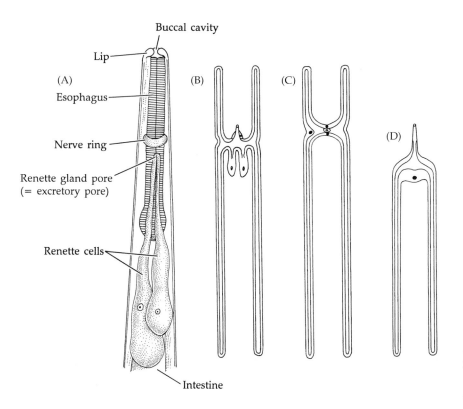

Figure 12.19 Nematode excretory systems. (A) A pair of renette cells (= renette gland) leading to the excretory pore (*Rhabditis*). (B) Schematic of the excretory system of *Oesophagostomum* (subclass Rhabditia), wherein the renette cells are associated with lateral excretory canals. (C) The so-called H-system of collecting canals remaining after loss of the glandular renette cell bodies (*Camallanus*: subclass Spiraria). (D) Modification of the H-system with anterior excretory pore and lateral canals (in many ascarids).

Most nematodes are ammonotelic, although some excrete increased amounts of urea when in a hypertonic environment. Apparently much of the loss of nitrogenous wastes is across the wall of the midgut, and the renette cells are primarily osmoregulatory. Water balance is also aided by the activities of other tissues, organs, and structures. The cuticle of at least some threadworms is differentially permeable to water in that it allows water to enter but not to leave the body. This condition is advantageous under conditions of potential desiccation, but it presents problems in hypotonic environments where excess water must be eliminated. Such elimination is apparently accomplished by the renette cells (when present), by the gut lining, and by the epidermis. Marine species do not osmoregulate well and desiccate rapidly when exposed to air.

Nervous System and Sense Organs

With some variation, the structure of the central nervous system of nematodes is similar throughout the phylum (Figure 12.20A,B). The cerebral ganglion is made up of a circumesophageal nerve ring and various associated ganglia that contain the majority of the nerve cell bodies. A wreath of sensory and motor nerves extends anteriorly from the nerve ring and serves the cephalic sense organs and mouth structures. Via a series of associated ganglia, longitudinal nerves extend posteriorly through the epidermal cords (Figure 12.16A). The major nerve trunk is ventral and includes both motor and sensory fibers. It is formed from the union of paired nerve tracts that arise ventrally on the nerve ring and fuse posteriorly, where the main trunk bears ganglia. The dorsal nerve

cord is motor, and the less well developed lateral nerve tracts are predominantly sensory. Lateral commissures connecting some or all of the longitudinal nerves occur in many nematodes.

The most abundant sense organs of nematodes are the papillae and setae that serve as tactile receptors in these worms' highly touch-oriented world (e.g., interstitial, parasitic, and soil habitats). **Amphids** are paired organs located laterally on the head. They consist of an external pore leading inward to a short duct and **amphidial pouch**. The pouch is associated with a unicellular gland and an **amphidial nerve** from the cerebral nerve ring (Figure 12.20A,C), although there is some variation in structural details among species. The receptor sites of amphids are derived from modified cilia, but motile cilia do not occur in nematodes. Specialists think that the amphids are chemosensory in function. Most members of the class Secernentea (the parasitic forms) possess a posteriorly located pair of glandular structures called **phasmids** (Figure 12.20D). These structures also are considered to be chemoreceptors. Some freshwater and marine free-living nematodes possess a pair of anterior pigment-cup ocelli, and at least some nematodes contain proprioceptor cells in the lateral epidermal cords. These sensory cells contain a cilium and apparently monitor bending of the body during locomotion (Hope and Gardiner 1982).

Reproduction, Development, and Life Cycles

Most nematodes are dioecious and show some degree of sexual dimorphism (Figures 12.15B and 12.21). The female reproductive system (Figure 12.21A) usually consists of paired elongate ovaries that gradually hollow as oviducts, then enlarge as uteri. The uteri converge to form a short vagina connected to the single gonopore. The female gonopore is completely separate from the anus, opening on the ventral surface near the middle of the body.

(A)

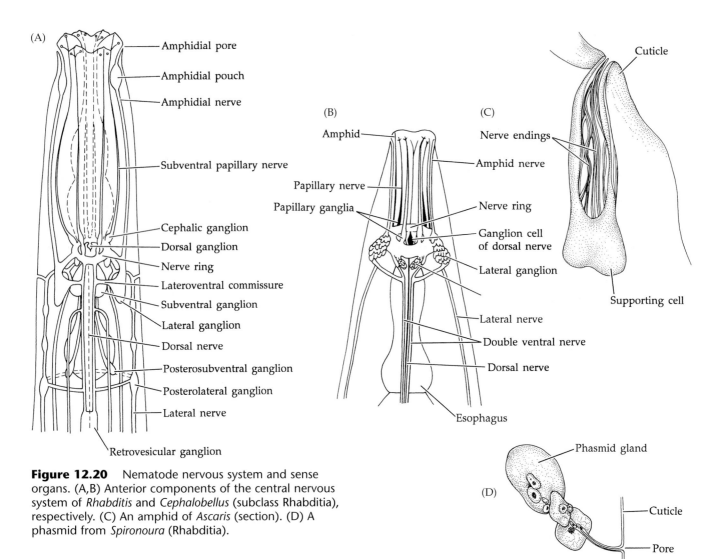

— Amphidial pore

— Amphidial pouch

— Amphidial nerve

— Subventral papillary nerve

— Cephalic ganglion
— Dorsal ganglion
— Nerve ring
— Lateroventral commissure
— Subventral ganglion
— Lateral ganglion
— Dorsal nerve
— Posterosubventral ganglion
— Posterolateral ganglion
— Lateral nerve

— Retrovesicular ganglion

(B)

Amphid —

Papillary nerve —
Papillary ganglia —

(C)

Cuticle

Nerve endings —

— Amphid nerve

— Nerve ring

— Ganglion cell
of dorsal nerve

— Lateral ganglion

Supporting cell

— Lateral nerve
— Double ventral nerve

— Dorsal nerve

Esophagus

(D)

Phasmid gland

Cuticle

Pore

Figure 12.20 Nematode nervous system and sense organs. (A,B) Anterior components of the central nervous system of *Rhabditis* and *Cephalobellus* (subclass Rhabditia), respectively. (C) An amphid of *Ascaris* (section). (D) A phasmid from *Spironoura* (Rhabditia).

Males tend to be smaller than females and are often sharply curved posteriorly. The male reproductive system (Figure 12.21B,C) typically includes one or two threadlike tubular testes, each of which is regionally differentiated into a distal **germinal zone**, a middle **growth zone**, and a proximal **maturation zone** near the junction with the sperm duct. The sperm duct extends posteriorly, where it enlarges as a seminal vesicle leading to a muscular ejaculatory duct that joins the hindgut near the anus. Some species have prostatic glands that secrete seminal fluid into the ejaculatory duct. Most male nematodes possess a copulatory apparatus, including one or two cuticular spines or spicules that can be inserted into the female gonopore to guide the transmission of sperm.

Prior to copulation the males produce sperm and store them in the seminal vesicle, while the females produce eggs that are moved into the hollow uteri. Potential mates make contact (females of some species produce male-attracting pheromones), and the male usually wraps his curved posterior end around the body of the female near her gonopore (Figures 12.15B and 12.21D). Thus positioned, the copulatory spines are inserted into the vagina; sperm are transferred by contractions of the ejaculatory duct. Fertilization usually occurs within the uteri. A relatively thick double-layered shell forms around each zygote; the inner layer is derived from the fertilization membrane and the outer layer is produced by the uterine wall. The zygotes are usually deposited in the environment, where development takes place.

In addition to the general description given above, two relatively uncommon reproductive processes occur in nematodes. In the few known hermaphroditic nematodes, sperm and egg production take place within the same gonad (an **ovitestis**). Sperm formation precedes egg production, so the animals are technically protandric; but they do not engage in cross-fertilization as occurs in most sequential hermaphrodites. Rather, the sperm are stored until ova are produced, and self-fertilization occurs. Parthenogenesis also occurs in a few

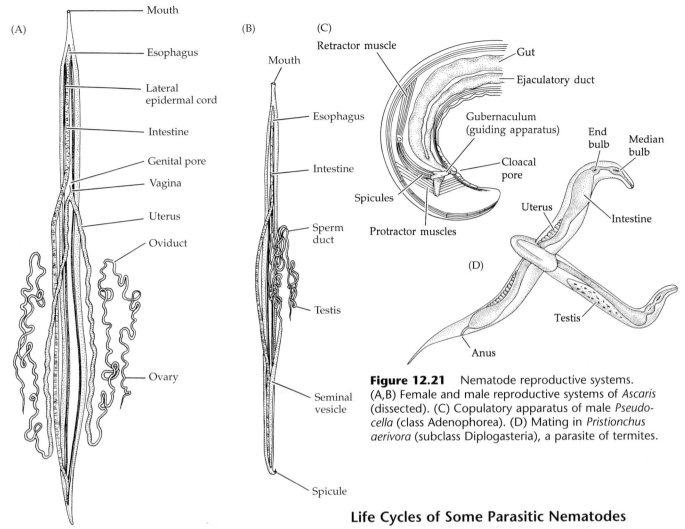

Figure 12.21 Nematode reproductive systems. (A,B) Female and male reproductive systems of *Ascaris* (dissected). (C) Copulatory apparatus of male *Pseudocella* (class Adenophorea). (D) Mating in *Pristionchus aerivora* (subclass Diplogasteria), a parasite of termites.

species of nematodes. Sperm and eggs are produced by separate males and females that then engage in typical copulation. However, the sperm do not fuse with the egg nuclei, but apparently serve only to stimulate cleavage.

Development among free-living nematodes typically is direct, although the term "larva" is often used for juvenile stages. Cleavage is holoblastic and subequal, but the pattern appears to be unique among the Metazoa. The orientation of blastomeres during early cleavage is fairly consistent among those nematodes that have been studied, but it cannot be readily assigned to a clearly radial or spiral pattern. Figure 12.22 illustrates this cleavage pattern and some details of cell fates. A stereoblastula or slightly hollow coeloblastula forms and undergoes gastrulation by epiboly of the presumptive ectoderm combined with an inward movement of presumptive entoderm and mesoderm. After a specific point in development, few nuclear divisions occur, and most subsequent growth, even after hatching, is via the enlargement of existing cells. Four sequential cuticular molts during juvenile life usually accompany growth.

Life Cycles of Some Parasitic Nematodes

The study of parasitic nematode life cycles is a field unto itself, and we present only a few of these interesting life histories here (Figures 12.23 and 12.24). An example of a simple parasitic life cycle is that of the whipworm *Trichuris trichiura* (class Adenophorea) (Figure 12.23A). These relatively large (3–5 cm long) nematodes reside and mate in the human gut, and the fertilized eggs are passed with the host's feces. Reinfection occurs when another host ingests the embryos. A more complicated life cycle is that of another adenophoran nematode, the trichina worm *Trichinella spiralis* (Figure 12.23B). A mammalian host acquires *Trichinella* by ingesting raw or poorly cooked meat containing encysted larvae.

Most parasitic nematodes belong to the class Secernentea, which includes such notables as hookworms, pinworms, and ascarid worms; hookworm and pinworm life cycles are shown in Figure 12.24.

One of the most dramatic nematode infections in humans is **filariasis**, caused by any of a number of secernentean nematodes called the **filarids**. These parasites require an intermediate host, typically a blood-sucking insect (e.g., fleas, biting flies, mosquitoes). One such filarid is *Wuchereria bancrofti*, whose vector is a mosquito. When

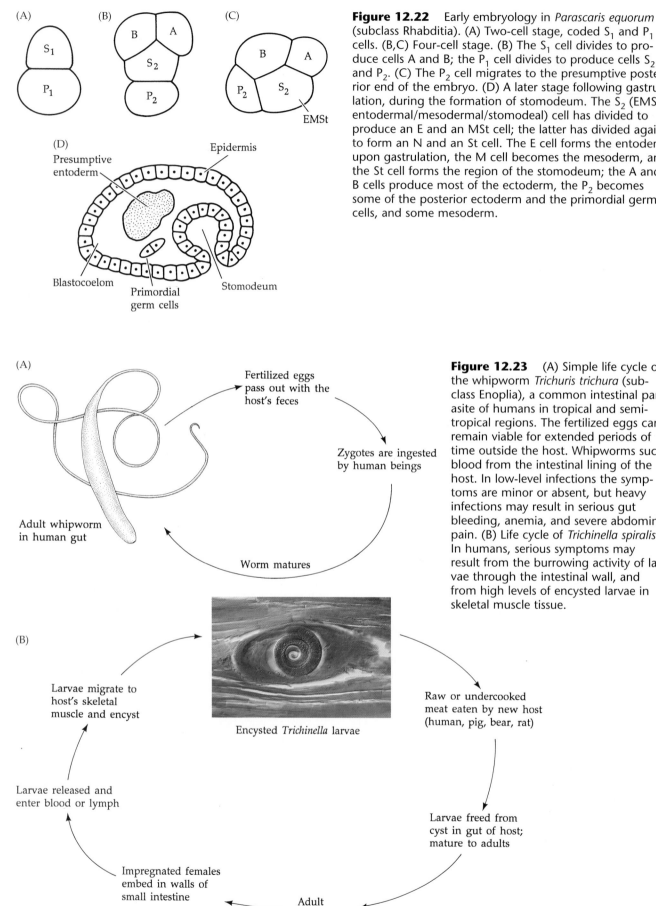

Figure 12.22 Early embryology in *Parascaris equorum* (subclass Rhabditia). (A) Two-cell stage, coded S_1 and P_1 cells. (B,C) Four-cell stage. (B) The S_1 cell divides to produce cells A and B; the P_1 cell divides to produce cells S_2 and P_2. (C) The P_2 cell migrates to the presumptive posterior end of the embryo. (D) A later stage following gastrulation, during the formation of stomodeum. The S_2 (EMSt, entodermal/mesodermal/stomodeal) cell has divided to produce an E and an MSt cell; the latter has divided again to form an N and an St cell. The E cell forms the entoderm upon gastrulation, the M cell becomes the mesoderm, and the St cell forms the region of the stomodeum; the A and B cells produce most of the ectoderm, the P_2 becomes some of the posterior ectoderm and the primordial germ cells, and some mesoderm.

Figure 12.23 (A) Simple life cycle of the whipworm *Trichuris trichura* (subclass Enoplia), a common intestinal parasite of humans in tropical and semi-tropical regions. The fertilized eggs can remain viable for extended periods of time outside the host. Whipworms suck blood from the intestinal lining of the host. In low-level infections the symptoms are minor or absent, but heavy infections may result in serious gut bleeding, anemia, and severe abdominal pain. (B) Life cycle of *Trichinella spiralis*. In humans, serious symptoms may result from the burrowing activity of larvae through the intestinal wall, and from high levels of encysted larvae in skeletal muscle tissue.

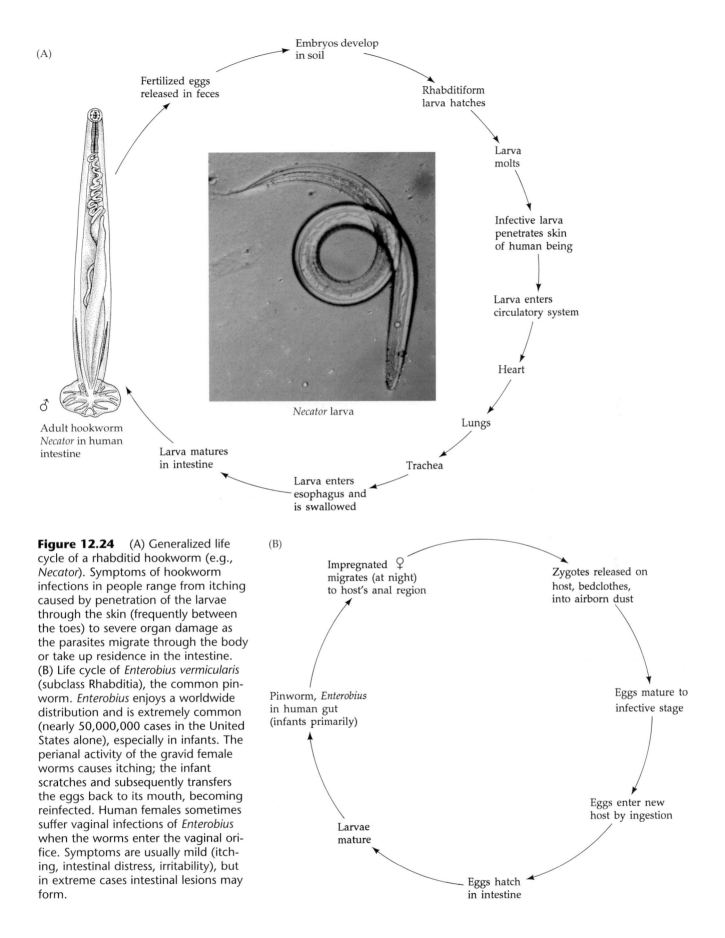

(A)

Embryos develop
in soil

Fertilized eggs
released in feces

Rhabditiform
larva hatches

Larva
molts

Infective larva
penetrates skin
of human being

Larva enters
circulatory system

♂

Heart

Necator larva

Adult hookworm
Necator in human
intestine

Lungs

Larva matures
in intestine

Trachea

Larva enters
esophagus and
is swallowed

Figure 12.24 (A) Generalized life cycle of a rhabditid hookworm (e.g., *Necator*). Symptoms of hookworm infections in people range from itching caused by penetration of the larvae through the skin (frequently between the toes) to severe organ damage as the parasites migrate through the body or take up residence in the intestine. (B) Life cycle of *Enterobius vermicularis* (subclass Rhabditia), the common pinworm. *Enterobius* enjoys a worldwide distribution and is extremely common (nearly 50,000,000 cases in the United States alone), especially in infants. The perianal activity of the gravid female worms causes itching; the infant scratches and subsequently transfers the eggs back to its mouth, becoming reinfected. Human females sometimes suffer vaginal infections of *Enterobius* when the worms enter the vaginal orifice. Symptoms are usually mild (itching, intestinal distress, irritability), but in extreme cases intestinal lesions may form.

(B)

Impregnated ♀
migrates (at night)
to host's anal region

Zygotes released on
host, bedclothes,
into airborn dust

Pinworm, *Enterobius*
in human gut
(infants primarily)

Eggs mature to
infective stage

Eggs enter new
host by ingestion

Larvae
mature

Eggs hatch
in intestine

present in high numbers, masses of adult *Wuchereria* block human lymphatic vessels and cause fluid accumulation (edema) and severe swelling, a condition resulting in grotesque enlargement of body parts and known as **elephantiasis**. Such infections often affect the legs and arms, or the scrotum in males and breasts in females.

In coastal areas, humans occasionally acquire infections of nematodes that normally parasitize only marine animals. Such infections result from consuming raw or undercooked fish, such as sashimi or ceviche, that serve as the intermediate host of the parasite.

Phylum Nematomorpha: Hair Worms and Their Kin

There are about 320 described species in the phylum Nematomorpha (Greek, *nema*, "thread"; *morph*, "shape"), commonly called the hair, horsehair, or gordian worms. The phylum name, and the common name of hair and horsehair worms, derive from the threadlike or hairlike shape of these animals (Figure 12.25A), and from the belief held for some time after their discovery in the fourteenth century that they actually arose from the hairs of horses' tails. They are generally from 1 to 3 mm in diameter and up to 1 m in length. Many of the very elongate

forms tend to twist and turn upon themselves in such a way as to give the appearance of complicated knots, and thus the name gordian worms.* Characteristics of this phylum are listed in Box 12E.

The nematomorphan blastocoelom may be somewhat spacious and fluid-filled or, more commonly, nearly obliterated by the invasion of mesenchyme. The digestive tract is largely nonfunctional, and nematomorphans appear to lack any structural excretory mechanisms. There are no functional cilia in these animals.

The larvae of nematomorphans are parasitic in arthropods. Most adults live in fresh water, among litter and algal mats near the edges of ponds and streams. A few species are semiterrestrial in damp soil, such as in moist gardens and greenhouses. Members of the enigmatic genus *Nectonema* are pelagic in coastal marine environments.

PHYLUM NEMATOMORPHA

ORDER NECTONEMATOIDEA: (Figure 12.25F). Marine, planktonic; with a double row of natatory setae along each side of body; with dorsal and ventral longitudinal epidermal cords; blastocoelom spacious and fluid-filled; gonads single; larvae parasitize decapod crustaceans. Monogeneric: *Nectonema* (4 known species).

ORDER GORDIOIDEA: (Figure 12.25A). Fresh water and semiterrestrial; lack lateral rows of setae; with a single, ventral epidermal cord; blastocoelom filled with mesenchyme in young animals but becomes spacious in older individuals; gonads paired; larvae parasitize aquatic and terrestrial insects, such as grasshoppers and crickets. (e.g., *Chordotes*, *Gordius*, *Paragordius*)

---

*King Gordius of Phrygia tied a formidable knot (the Gordian knot) and declared that whoever might undo it would be the ruler of all Asia. No one could until Alexander the Great cut it through with his broadsword, settling the issue in a style consistent with the rest of his adventures.

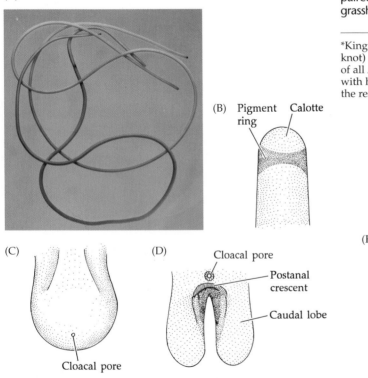

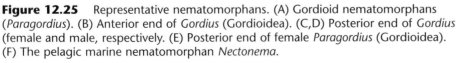

Figure 12.25 Representative nematomorphans. (A) Gordioid nematomorphans (*Paragordius*). (B) Anterior end of *Gordius* (Gordioidea). (C,D) Posterior end of *Gordius* (female and male, respectively). (E) Posterior end of female *Paragordius* (Gordioidea). (F) The pelagic marine nematomorphan *Nectonema*.

BOX 12E *Characteristics of the Phylum Nematomorpha*

1. Triploblastic, bilateral, unsegmented, vermiform; body long and thin

2. Blastocoelom spacious or reduced by mesenchyme

3. Cuticle well developed, periodically shed

4. Body without functional cilia or flagella

5. Body wall has only longitudinal muscles (no circular muscles)

6. Gut reduced to various degrees

7. With unique larva that is parasitic in arthropods

8. Epidermis forms cords housing longitudinal nerves

9. Without special excretory, circulatory, or gas exchange structures

10. Dioecious

11. With unique cleavage pattern; not unambiguously radial or spiral

12. Inhabit freshwater, terrestrial, or planktonic marine environments; larvae parasitic in arthropods

Body Wall, Support, and Locomotion

The general organization of the body wall of nematomorphans is similar in many aspects to that of the nematodes (Figure 12.26A–C). The cuticle, secreted by the epidermis, is very thick (especially in the gordioids) and comprises an outer homogeneous layer and an inner, lamellate, fibrous layer. The homogeneous layer often forms bumps, warts, or papillae (collectively called **areoles**; Figure 12.26D), some of which bear apical spines or pores. The function of the areoles is unknown, but spined ones may be touch-sensitive and pore-bearing ones may produce a lubricant. Some species have two or three **caudal lobes** at the posterior end.

The unciliated epidermis covers the entire body and rests upon a thin basal lamina. The epidermis is produced into a ventral (Gordioidea) or dorsal and ventral (Nectonematoidea) epidermal cords containing longitudinal nerve tracts. Beneath the epidermis is a single layer of longitudinal muscle cells; as in nematodes, there is no layer of circular muscle in the body wall. The longitudinal muscles give rise to hollow tubular extensions, called the **rete system**, that may be homologous to the muscle arms of nematodes.

The blastocoelom of nematomorphans varies from spacious in *Nectonema* to largely mesenchyme-filled in gordioids (see Figure 12.6A,B) and surrounds the inter-nal organs. In most species small spaces, presumably blastocoelic remnants, lie within the mesenchyme near the remnants of the gut. The hydrostatic or structural qualities of the body cavity, along with the well developed cuticle, provide body support.

Locomotion in the planktonic *Nectonema* is by undulatory swimming using the body wall muscles and the natatory bristles (Figure 12.25F) or by passive flotation in nearshore currents, aided by the bristles, which provide resistance to sinking. Freshwater and semiterrestrial nematomorphans (order Gordioidea) use their longitudinal muscles to move by undulations or by coiling and uncoiling movements.

Feeding and Digestion

Until recently, the general impression was that nematomorphans "feed" only during the parasitic larval stage, when they absorb nutrients across their body wall from the host's tissue and fluids, and that adults rely entirely on nutrients stored during their larval and juvenile parasitic life. We now know that adults do take in nutrients by absorbing small organic molecules across the body wall (as do the larvae) and by way of the reduced gut.

The reduced but functional digestive tract of adult nematomorphans is a simple elongate tube running the length of the body (Figure 12.26E). The gut is actively involved with uptake and storage of nutrients, transported to it across the body wall. However, there is usually no mouth opening, and the pharyngeal region is typically solid. The intestine or midgut is a thin-walled tube and may serve an excretory function as well as a digestive one. The hindgut is proctodeal, functions as a cloaca, and receives the reproductive ducts. In *Nectonema*, a tiny mouth and pharynx lead to a midgut that deteriorates posteriorly and is not connected to the cloaca.

Circulation, Gas Exchange, Excretion, and Osmoregulation

Very little is known about the physiology of nematomorphans. Internal transport is undoubtedly by diffusion through the blastocoelom and mesenchyme, and is probably aided by body movement. The free-living gordian worms are presumably obligate aerobes as adults and are restricted to moist environments with ample available oxygen. The threadlike body results in short diffusion distances between the environment and the body organs and tissues.

Excretory and osmoregulatory functions probably operate on a strictly cellular level, there being no protonephridia or other known special structures for these functions. Some workers, however, have speculated that the cells of the midgut may function in the excretion of metabolic wastes, and that they may have a structure similar to the Malpighian tubules of insects (Chapter 17).

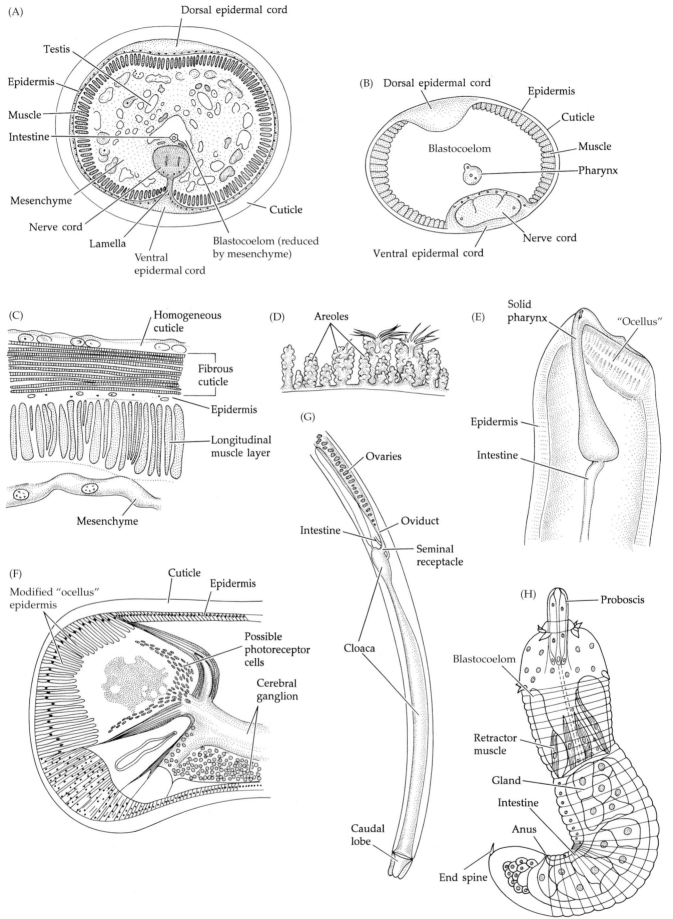

(A)

Dorsal epidermal cord

Testis

Epidermis

Muscle

Intestine

Mesenchyme

Nerve cord

Lamella

Ventral epidermal cord

Blastocoelom (reduced by mesenchyme)

Cuticle

(B)

Dorsal epidermal cord

Epidermis

Cuticle

Muscle

Blastocoelom

Pharynx

Nerve cord

Ventral epidermal cord

(C)

Homogeneous cuticle

Fibrous cuticle

Epidermis

Longitudinal muscle layer

Mesenchyme

(D) Areoles

(E) Solid pharynx

"Ocellus"

Epidermis

Intestine

(G)

Ovaries

Intestine

Oviduct

Seminal receptacle

Cloaca

Caudal lobe

(F)

Cuticle

Epidermis

Modified "ocellus" epidermis

Possible photoreceptor cells

Cerebral ganglion

(H)

Proboscis

Blastocoelom

Retractor muscle

Gland

Intestine

Anus

End spine

Nervous System and Sense Organs

Like the nervous systems of nematodes and some other small Metazoa, the nervous system of nematomorphans is closely associated with the epidermis. The cerebral ganglion is a circumpharyngeal mass of nervous tissue located in a region of the head called the **calotte** (Figures 12.25B and 12.26F). In gordioids a single midventral nerve cord arises from the cerebral ganglion and extends the length of the body. It is attached to the epidermis by a tissue connection called the **epidermal lamella** (Figure 12.26A,B). *Nectonema* possesses an additional dorsal, intraepidermal nerve cord.

All nematomorphans are touch-sensitive, and some are apparently chemosensitive. Adult males are able to detect and track mature females from a distance. However, the structures associated with these sensory functions are a matter of speculation. Presumably, some of the cuticular areoles are tactile, and perhaps others are chemoreceptive. Members of the genus *Paragordius* possess modified epidermal cells in the calotte that contain pigment and may be photoreceptors (Figure 12.26F), although this function has not been confirmed. Other possible sensory structures are four "giant cells" near the cerebral ganglion in *Nectonema* (Schmidt-Rhaesa 1996c).

Reproduction and Development

Nematomorphans are dioecious and display some sexual dimorphism (Figure 12.25). The male reproductive system includes one (*Nectonema*) or two (gordioids) testes (Figure 12.26A). Each testis opens to the cloaca via a short sperm duct, which is sometimes swollen as a seminal vesicle. Female gordioids possess a pair of elongate ovaries that open to the cloaca through a seminal receptacle (Figure 12.26G). *Nectonema* contains no discrete ovary; instead, the germinal cells occur as scattered oocytes in the body cavity.

Mating has been studied in some gordioids. The females remain relatively inactive, but males become highly motile during the breeding season and respond to the presence of potential mates in their environment. Once a male locates a receptive female, he wraps his body around her and deposits a drop of sperm near her cloacal pore. The sperm find their way into the seminal receptacle and are stored while the ova mature. The eggs are fertilized internally and laid in gelatinous strings.

Early development has been studied in only a few species of gordioid nematomorphans. Cleavage is holoblastic, but not clearly spiral or radial. A coeloblastula forms, then gastrulates by invagination of the presumptive entoderm. Mesodermal cells proliferate into the blastocoel from the area around the blastopore. The anus and cloacal chamber also form from the area of the blastopore. A **nematomorphan larva** develops (Figure 12.26H) and emerges from the egg case. The larva will develop normally only in an appropriate arthropod host, which it probably enters by being eaten. Within the host's hemocoel, the larva grows into a juvenile nematomorphan, which in turn leaves the host to mature. A single cuticular molt has been reported in some species, taking place shortly before the young worm emerges. Juveniles are nearly full size when they leave their host and do not grow much as adults.

Phylum Priapula: The Priapulans

Sixteen extant species (and 14 fossil species) are assigned to the phylum Priapula (from *Priapos*, the Greek god of reproduction, symbolized by his enormous penis). These odd creatures were recorded in Linnaeus's *Systema Naturae*, where he mentioned the species *Priapus humanus* (literally, "human penis"). Since that time the priapulans have been allied with several different groups of invertebrates. About the time the aschelminth concept was losing favor, William Shapeero (1961) published findings suggesting that the priapulans were truly coelomate. However, the nature of the body cavity has been questioned again, and recent work indicates that it is not lined by a nucleated epithelium. In the absence of complete embryological data, these questions remain unresolved and the evolutionary relationships of priapulans remain uncertain.

The major characteristics of these animals are given in Box 12F. Priapulans are cylindrical vermiform creatures, from 0.55 mm to 20 cm in length. The body comprises an **introvert** (= proboscis or **presoma**), necklike **collar**, **trunk** (= abdomen), and sometimes a "tail" or **caudal appendage** (Figure 12.27). The introvert has scalid-like spines resembling those of loriciferans and kinorhynchs, and is entirely retractable. When present, the caudal appendage varies in form and function among different species.

Large priapulans are active infaunal burrowers in relatively fine marine sediments and occur primarily in boreal and cold temperate seas. A few species construct tubes. Small meiofaunal forms burrow or live interstitially among sediment particles. One species, *Halicryptus spinulosus*, lives in anoxic, sulfide-rich sediments in the Baltic Sea and shows remarkable abilities to both tolerate and detoxify high sulfide levels in its body.

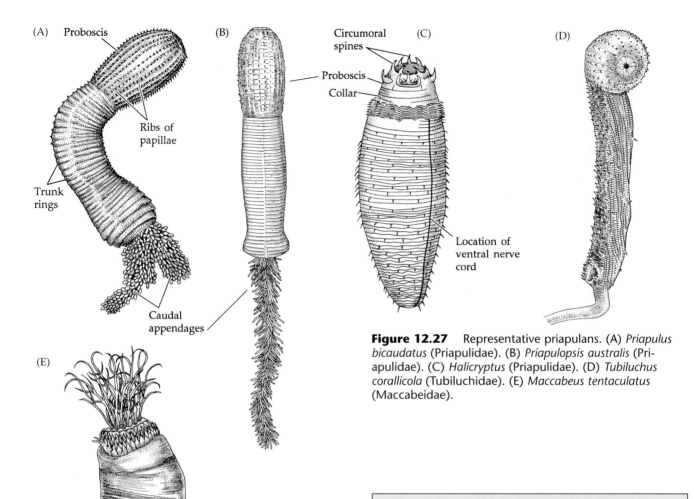

Figure 12.27 Representative priapulans. (A) *Priapulus bicaudatus* (Priapulidae). (B) *Priapulopsis australis* (Priapulidae). (C) *Halicryptus* (Priapulidae). (D) *Tubiluchus corallicola* (Tubiluchidae). (E) *Maccabeus tentaculatus* (Maccabeidae).

Priapulans are rather common in the fossil record. They may have been one of the major predators in Cambrian seas, and they were almost certainly more abundant in Paleozoic times than they are today. The most abundant fossil species, *Ottoia prolifica*, resemble the extant *Halicryptus spinulosus*.

PHYLUM PRIAPULA

Classes and orders are usually not recognized within the Priapula; the phylum is divided into the three extant families (listed below) and five extinct families.

FAMILY PRIAPULIDAE: (Figure 12.27A–C). Relatively large (4–20 cm); abdomen with superficial annulations; caudal appendage (absent from *Halicryptus spinulosus*) either a grapelike cluster of fluid-filled sacs (called vesiculae) or a muscular extension with cuticular hooks. (e.g., *Acanthopriapulus, Halicryptus, Priapulopsis,* and *Priapulus*)

> **BOX 12F Characteristics of the Phylum Priapula**
>
> 1. Triploblastic, bilateral, unsegmented (may be superficially annulated), and vermiform
>
> 2. Body cavity lining probably not peritoneal; presumably blastocoelomate
>
> 3. Introvert with hooked spines
>
> 4. Nervous system radially arranged and largely intraepidermal
>
> 5. Gut complete
>
> 6. With (often multicellular) protonephridia associated with the gonads as a urogenital system
>
> 7. Many with unique caudal appendage (in adults) that may serve for gas exchange
>
> 8. No circulatory system
>
> 9. Thin cuticle is periodically shed
>
> 10. With unique loricate larva, with cuticular lorica that is shed at metamorphosis to adult stage
>
> 11. Dioecious
>
> 12. Cleavage radial-like
>
> 13. Marine and benthic; most are burrowers

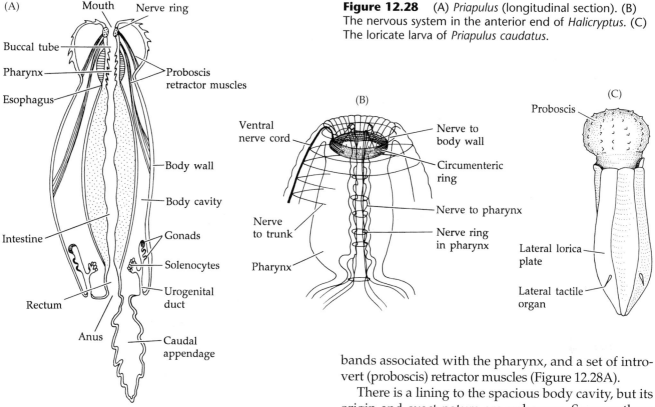

Figure 12.28 (A) *Priapulus* (longitudinal section). (B) The nervous system in the anterior end of *Halicryptus*. (C) The loricate larva of *Priapulus caudatus*.

FAMILY TUBILUCHIDAE: (Figure 12.27D). Small (less than 2 mm long); abdomen not annulated; caudal appendage vermiform and muscular. Tubiluchids live in sediments of shallow tropical waters; four species, two genera (*Tubiluchus* and *Meiopriapulus*).

FAMILY MACCABEIDAE (= CHAETOSTEPHANIDAE): (Figure 12.27E). Small (less than 3 mm long); meiofaunal; abdomen with rings of tubercles and posterior longitudinal ridges with hooks; no caudal appendage; posterior end of abdomen extensible and mobile, used for burrowing (posterior end first). Maccabeids are found in the Mediterranean Sea and Indian Ocean; monogeneric, with only two described species (*Maccabeus tentaculatus* and *M. cirratus*).

Body Wall, Support, and Locomotion

The priapulan body is covered by a thin, flexible cuticle that forms a variety of spines, warts, and tubercles (Figure 12.27). Large hooked spines are often present around the mouth and on the introvert. These spines may be homologous to those of kinorhynchs and loriciferans, but this is not yet certain. In all three groups, the spines function as sensory structures and assist in locomotion. Priapulans move through sediments by means of the introvert and peristaltic body muscle action. The cuticle may contain some chitin and is periodically shed as the animal grows. Beneath the cuticle lies an unciliated epidermis of thin, elongate cells with large fluid-filled intercellular spaces. Beneath the epidermis are well developed layers of circular and longitudinal muscles. There are also complex muscle layers and

bands associated with the pharynx, and a set of introvert (proboscis) retractor muscles (Figure 12.28A).

There is a lining to the spacious body cavity, but its origin and exact nature are unknown. Some authors (e.g., Shapeero 1961) hold that this lining is a cellular peritoneum and that the body cavity is a true coelom, although more recent work suggests that the lining is a simple noncellular membrane secreted by surface cells on the retractor muscles, and that the body cavity is a blastocoelom. In any case, this lining covers not only the inner surface of the body wall but the internal organs as well, and forms mesentery-like extensions. Storch et al. (1989) report that, in contrast to other priapulans, *Meiopriapulus fijiensis* does possess a ring of small compartments in the introvert that are lined by a distinct epithelium (i.e., these spaces may be truly coelomic in nature). The fluid of the body cavity contains motile phagocytic amebocytes and free erythrocytes with hemerythrin.

Maintenance of body form and support are provided by the hydrostatic skeleton of the body cavity. The contraction of circular muscles around this cavity also facilitates protrusion of the introvert by increasing the internal pressure. Priapulans that move through the substratum do so largely by peristaltic burrowing, probably using the various hooks and other cuticular extensions to hold one part of the body in place while the rest is pushed or pulled along. *Maccabeus* is thought to use its ring of posterior cuticular spines for anchorage within its burrow (Figure 12.27E).

Feeding and Digestion

The majority of priapulans (i.e., members of the family Priapulidae) live in soft sediments and prey on soft-bodied invertebrates such as polychaete worms. During feeding, a portion of the toothed, cuticle-lined pharynx is everted through the mouth at the end of the extended introvert. As the prey is grasped, the pharynx is inverted; the introvert then retracts and the prey is drawn into the gut.

Tubiluchus corallicola (Figure 12.27D) lives in coral sediments and feeds on organic detritus. The pharynx is lined with pectinate teeth, which the animal uses to sort food material from the coarse sediment particles. *Maccabeus tentaculatus*, a tube dweller, is a trapping carnivore. Surrounding the mouth are eight short tentacles, presumed to be touch-sensitive. These are in turn ringed by 25 highly branched spines (Figure 12.27E). It is suspected that when a potential prey touches the sensory tentacles, the outer spines quickly close as a trap. Meiofaunal species are probably detritivores or micropredators.

The digestive system is complete and either straight or slightly coiled (Figure 12.28A). The portion of the gut that lies roughly within the bounds of the introvert comprises the buccal tube, pharynx, and esophagus, all of which are lined with cuticle and together constitute a stomodeum. In members of the genus *Tubiluchus*, the stomodeum also includes a region behind the esophagus called a **polythridium**, which bears a circlet of two rows of plates and may be used to grind food. The midgut or intestine is the only entodermally derived section of the digestive tract and is followed by a short proctodeal rectum. The anus is located at the posterior end of the abdomen, either centrally or slightly to one side. Nothing is known about the digestive physiology of priapulans, although it is likely that digestion and nutrient absorption occur in the midgut.

Circulation, Gas Exchange, Excretion, and Osmoregulation

Internal transport takes place through diffusion and movement of the fluid in the body cavity. The presence of the respiratory pigment hemerythrin in the body fluid cells suggests an oxygen transport or storage function, and many priapulans are known to live in marginally anoxic muds. In those species with a vesiculate caudal appendage, the lumen of that structure is continuous with the main body cavity. Such caudal appendages may function as gas exchange surfaces.

Clusters of solenocyte protonephridia lie in the posterior portion of the body cavity and are associated with the gonads as a urogenital system or complex (Figure 12.28A). Priapulan protonephridia are possibly unique in being composed of two or more terminal cells. A pair of urogenital pores opens near the anus. Priapulans inhabit both hyper- and hyposaline environments, so the protonephridia may function in both osmoregulation and excretion.

Nervous System and Sense Organs

The priapulan nervous system is intraepidermal and is constructed for the most part on a radial plan within the cylindrical body (Figure 12.28B). Although a typical cerebral ganglion is absent, there is a circumenteric nerve ring within the buccal tube epithelium. The main ventral nerve cord arises from this ring and gives off a series of ring nerves and peripheral nerves along the body. In addition, longitudinal nerves extend from the main nerve ring along the inner pharyngeal lining and are connected by the ring commissures.

Little is known of sense organs in priapulans. The caudal appendage contains tactile receptors, and so may many of the bumps and spines on the body surface. Members of the family Tubiluchidae bear flosculi in the form of pores encircled by micropapillae, somewhat like those in rotifers.

Reproduction and Development

Priapulans are dioecious, although males are unknown in *Maccabeus tentaculatus*. The reproductive organs are similarly placed and connected in both sexes. The paired gonads are drained by genital ducts, which are joined by collecting tubules from the protonephridia to form a pair of urogenital ducts exiting posteriorly through urogenital pores (Figure 12.28A).

Priapulans free-spawn (first the males and then the females), and fertilization is external. Cleavage is holoblastic, appears to be radial, and results in a coeloblastula (some accounts differ) that undergoes invagination. The origin of the body cavity and many other aspects of morphogenesis remain unknown. Direct development occurs in *Meiopriapulus fijiensis*, in which the females brood the embryos. In all other species a larval form (sometimes called a **loricate larva**) eventually develops (Figure 12.28C), which in some ways resembles loriciferans. The trunk of the larva is encased within a thick cuticular lorica into which the introvert can be withdrawn. The larvae live in benthic muds and are probably detritivores. The lorica is periodically shed as the larva grows; it is finally lost as the animal metamorphoses to a juvenile priapulan. At that time the caudal appendage forms in those species that possess this structure. The larvae of different genera vary somewhat in cuticular shape and ornamentation. Larval development is not well studied but is suspected to be protracted, taking perhaps one to two years.

Phylum Acanthocephala: The Acanthocephalans

As adults, the 1,100 described species of acanthocephalans are obligate intestinal parasites in vertebrates, particularly in freshwater teleost fish. Larval development takes place in intermediate arthropod hosts. The name Acanthocephala (Greek *acanthias*, "prickly"; *cephalo*, "head") derives from the presence of recurved hooks located on an eversible proboscis at the anterior end (Figure 12.29B,D). The rest of the body forms a cylindrical or flattened trunk, often bearing rings of small

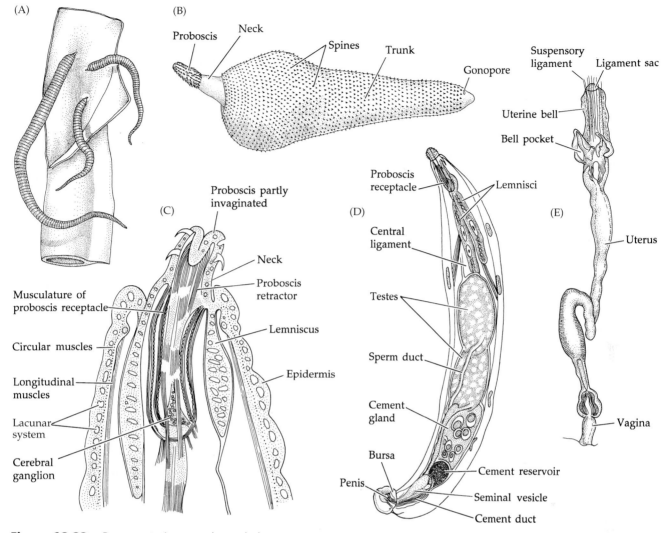

Figure 12.29 Representative acanthocephalans. (A) *Macracanthorhynchus hirudinaceus*, an archiacanthocephalan, attached to the intestinal wall of a pig. (B) *Corynosoma*, a palaeacanthocephalan found in aquatic birds and seals. (C) Longitudinal section through the anterior end of *Acanthocephalus* (class Palaeacanthocephala). (D) An adult male eoacanthocephalan (*Pallisentis fractus*). (E) The isolated female reproductive system of *Bolbosoma*.

spines. Most acanthocephalans are less than 20 cm long, although a few species exceed 60 cm in length; females are generally larger than males. The digestive tract has been completely lost. And, except for the reproductive organs, there is significant structural and functional reduction of most other systems, related to the parasitic lifestyles of these worms (Box 12G). The persisting organs lie within an open blastocoelom, partially partitioned by mesentery-like **ligaments**.

Phylum Acanthocephala

The phylum Acanthocephala is divided into three classes on the basis of the arrangement of proboscis hooks,

the nature of the epidermal nuclei, spination patterns on the trunk, and nature of the reproductive organs. Representative members of these classes—Palaeacanthocephala, Archiacanthocephala, and Eoacanthocephala—are shown in Figure 12.29.

Body Wall, Support, Attachment, and Nutrition

Adult acanthocephalans attach to their host's intestinal wall by their proboscis hooks, which are retractable into pockets, like the claws of a cat (Figure 12.29A). The chemical nature of the hooks is not known. In nearly all species the proboscis itself is retractable into a deep **proboscis receptacle**, enabling the body to be pulled close to the host's intestinal mucosa. Nutrients are absorbed through the body wall. The nature of the body wall has been reinterpreted over the past few years in light of studies on its ultrastructure (for details see Dunagan and Miller 1991). The outer body wall is a multilayered, synctial, living **tegument**, which overlies sheets of circular and longitudinal muscles. The tegument includes layers of dense fibers as well as what appear to be sheets of plasma membrane, and an inner connective tissue-like layer resembling the skeletal lamina of rotifers. The tegument is perforated by numerous canals that connect to a complex set of circulatory channels called the **lacunar system** (Figure 12.29C). The tegumental channels near the body surface may facilitate pinocytosis of nutrients from the host. The body wall organization is such that each species has a distinct external appearance; some even appear to be segmented, but they are not.

At the junction of the proboscis and the trunk, the epidermis extends inward as a pair of hydraulic sacs (**lemnisci**) that facilitate extension of the proboscis; the proboscis is withdrawn by retractor muscles. The lemnisci are continuous with each other and with a ring-shaped canal near the anterior end of the body, whereas their distal ends float free in the blastocoel. This arrangement may help to circulate nutrients and oxygen from the body to the proboscis, although the actual function of the lemnisci is not known.

One or two large sacs lined with connective tissue arise from the rear wall of the proboscis receptacle and extend posteriorly in the body. These structures support the reproductive organs and divide the body into dorsal and ventral **ligament sacs** in the archiacanthocephalans and eoacanthocephalans, or produce a single ligament sac down the center of the body cavity in the palaeacanthocephalans (Figure 12.29D,E). Within the walls of these sacs are strands of fibrous tissue—the ligaments—that may represent remnants of the gut. The space between these internal organs is presumably a blastocoelom.

The body is supported by the fibrous tegument and the hydrostatic qualities of the blastocoelom and lacunar system. The muscles and ligament sacs add some structural integrity to this support system. Most of the muscles are penetrated by canals of the lacunar system.

Circulation, Gas Exchange, and Excretion

Exchanges of nutrients, gases, and waste products occur by diffusion across the body wall (some Archiacanthocephala possess a pair of protonephridia and a small bladder). Internal transport is by diffusion within the body cavity and by the lacunar system, the latter functioning as a unique sort of circulatory system, which invades most body tissues. The lacunar fluid is moved about by action of the body wall muscles.

Nervous System

As in many obligate endoparasites, the nervous system and the sense organs of acanthocephalans are greatly reduced. A cerebral ganglion lies within the proboscis receptacle (Figure 12.29C) and gives rise to nerves to the body wall muscles, the proboscis, and the genital regions. Males possess a pair of genital ganglia. The proboscis bears several structures that are presumed to be tactile receptors, and small sensory pores occur at the tip and base of the proboscis. Males have what appear to be sense organs in the genital area, especially on the penis.

Reproduction and Development

Acanthocephalans are dioecious and females are generally somewhat larger than males. In both sexes, the reproductive systems are associated with the ligament sacs (Figure 12.29D). In males, paired testes (usually arranged in tandem) lie within a ligament sac and are drained by sperm ducts to a common seminal vesicle. Entering the seminal vesicle or the sperm ducts are six or eight cement glands, whose secretions serve to plug the female genital pore following copulation. When nephridia are present, they also drain into this system. The seminal vesicle leads to an eversible penis, which lies within a genital bursa connected to the gonopore. This gonopore is often called a cloacal pore, because the bursa appears to be a remnant of the hindgut.

In females, a single mass of ovarian tissue forms within a ligament sac. Clumps of immature ova are released from this transient ovary and enter the body cavity, where they mature and are eventually fertilized. The female reproductive system comprises a gonopore, a vagina, and an elongate uterus that terminates internally in a complex open funnel called the **uterine bell** (Figure 12.29E). During mating the male everts the copulatory bursa and attaches it to the female gonopore. The penis is inserted into the vagina, sperm are transferred, and the vagina neatly capped with cement. Sperm then travel up the female system, enter the body cavity through the uterine bell, and fertilize the eggs.

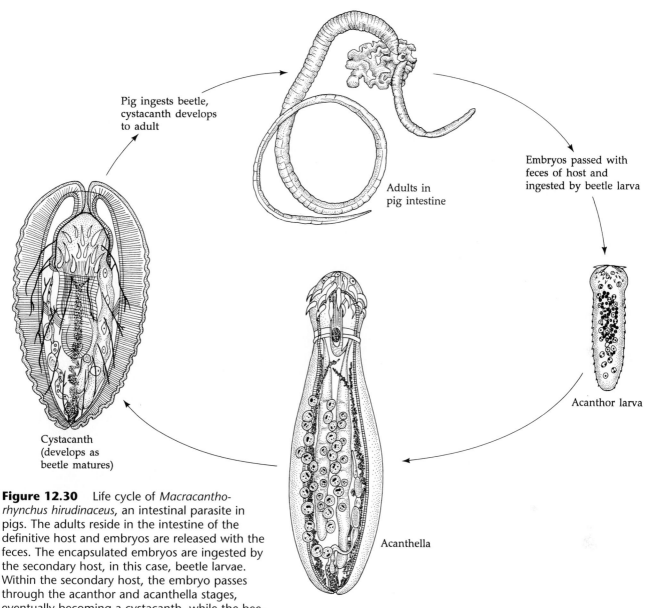

Pig ingests beetle,
cystacanth develops
to adult

Adults in
pig intestine

Embryos passed with
feces of host and
ingested by beetle larva

Acanthor larva

Cystacanth
(develops as
beetle matures)

Acanthella

Figure 12.30 Life cycle of *Macracantho-rhynchus hirudinaceus*, an intestinal parasite in pigs. The adults reside in the intestine of the definitive host and embryos are released with the feces. The encapsulated embryos are ingested by the secondary host, in this case, beetle larvae. Within the secondary host, the embryo passes through the acanthor and acanthella stages, eventually becoming a cystacanth, while the beetle grows. When the beetle is ingested by a pig, the juvenile matures into an adult, thereby completing the cycle.

Much of the early development of acanthocephalans takes place within the body cavity of the female. Cleavage is holoblastic, unequal, and likened to a highly modified spiral pattern. A stereoblastula is produced, at which time the cell membranes break down to yield a syncytial condition. Eventually, a shelled **acanthor larva** is formed (Figure 12.30). At this or an earlier stage the embryo leaves the mother's body. Remarkably, the uterine bell "sorts" through the developing embryos by manipulating them with its muscular funnel; it accepts only the appropriate embryos into the uterus. Embryos in earlier stages are rejected and pushed back into the body cavity, where they continue development. The selected embryos pass through the uterus and out the genital pore and are eventually released with the host's feces.

Once outside the definitive host, the developing acanthocephalan must be ingested by an arthropod intermediate host—usually an insect or a crustacean—to continue its life cycle. The acanthor larva penetrates the gut wall of the intermediate host and enters the body cavity, where it develops into an **acanthella** and then into an encapsulated form called a **cystacanth** (Figure 12.30). When the intermediate host is eaten by an appropriate definitive host, the cystacanth attaches to the intestinal wall of the host and matures into an adult.

Phylum Entoprocta: The Entoprocts

The phylum Entoprocta (Greek *entos*, "inside"; *proktos*, "anus"), or Kamptozoa (Greek *kamptos*, "bent"), includes about 150 species of small, sessile, solitary or colonial

Figure 12.31 Entoprocts. (A) A portion of a *Pedicellina* colony. (B) A colony of *Barentsia* (left) and an unidentified entoproct (right). (C) *Loxosomella* growing on the surface of a sponge. (D) *Loxosoma*; a solitary individual with a bud. (E) A zooid of *Pedicellina* (sagittal section).

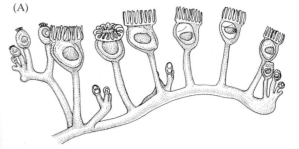

(A)

(B)

(C)

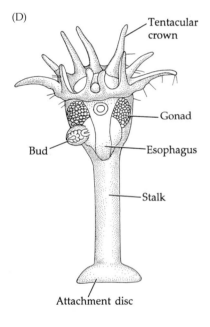

(D)

- Tentacular crown
- Gonad
- Bud
- Esophagus
- Stalk
- Attachment disc

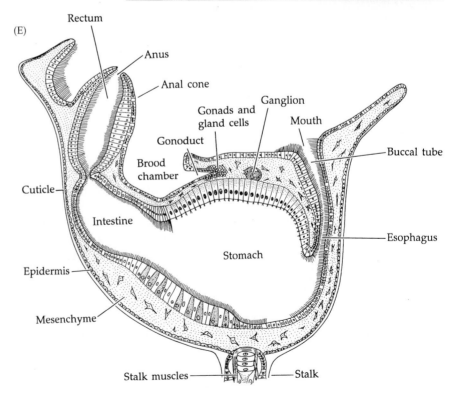

(E)

- Rectum
- Anus
- Anal cone
- Gonads and gland cells
- Ganglion
- Mouth
- Gonoduct
- Brood chamber
- Buccal tube
- Cuticle
- Intestine
- Esophagus
- Epidermis
- Stomach
- Mesenchyme
- Stalk muscles
- Stalk

creatures that superficially resemble cnidarian hydroids and ectoprocts (Figure 12.31). All but a single genus of entoprocts are marine. Colonial forms live attached to various substrata, including algae, shells, and rock surfaces. Solitary species are commensal on a variety of hosts, especially sponges (Figure 12.31C), ectoprocts, polychaetes, sipunculans, and ascidians, and are typically associated with just one or a few host species. Entoprocts are not uncommon intertidally, and some species are known from depths as great as 500 meters. The single genus of freshwater entoprocts, *Urnatella*, forms colonies with beaded stolons fastened to a small attachment disk.

Although technically bilateral, the individual zooids of entoprocts have in many respects assumed a functionally radial form. The body consists of a cuplike **calyx** from which arises a whorl of ciliated tentacles (Figure 12.31). Both the mouth and anus are located on the surface or **vestibule** of the calyx and are surrounded by the **tentacular crown**. The anus is elevated on a distinct papilla, called the **anal cone**. Each calyx is supported by a stalk, which in solitary forms attaches directly to the substratum and in colonial forms attaches to larger branches or to horizontal stolons. Box 12H lists the distinguishing features of entoprocts.

The phylogenetic relationships of entoprocts with other taxa are controversial. During most of the eighteenth and nineteenth centuries, entoprocts were included with ectoprocts in the phylum Bryozoa. Since the discovery of their noncoelomate nature (Clark 1921), a separate phylum status for the Entoprocta has generally been accepted. For many years the entoprocts were considered to be blastocoelomate, albeit with the body cavity fully invested by mesenchyme. However, Claus Nielsen (1971, 1977) again raised the possibility that the entoprocts and ectoprocts (Chapter 21) may be closely related—in fact, that the former may represent the ancestral condition of the latter. To complicate the issue even more, it has been suggested that the entoprocts' solid body structure may be primitively acoelomate rather than representing a secondarily filled blastocoelom. Be that as it may, we present the entoprocts here, along with these other enigmatic groups, until more information becomes available.

The phylum is commonly divided into four families (Figure 12.31). Some specialists recognize orders on the basis of the presence or absence of a septum between the stalk and calyx, or on solitary versus colonial habits.

PHYLUM ENTOPROCTA

FAMILY LOXOSOMATIDAE: Without a septum between stalk and calyx; solitary; often commensal on other invertebrates; some are capable of limited movement on a suckered base; muscles continuous from stalk to calyx. (e.g., *Loxosoma*, *Loxosomella*)

FAMILY LOXOKALYPODIDAE: Without a septum between stalk and calyx; colonial; muscles continuous from stalk to

---

BOX 12H Characteristics of the Phylum Entoprocta

1. Triploblastic, bilateral, unsegmented, functionally acoelomate

2. Sessile and solitary or colonial with zooids borne on stalks

3. Visceral mass housed within a cup-shaped calyx, the ventral surface of which is directed away from the substratum

4. Zooids bear a ring of tentacles that enclose both the mouth and the anus

5. Gut complete and U-shaped

6. With one pair protonephridia

7. Hermaphroditic or dioecious

8. Cleavage spiral

9. All are shallow-water (0–500 m) marine forms, except *Urnatella* (which is freshwater)

---

calyx; ectocommensal on the polychaete *Glycera nana* in the northeastern Pacific. Monotypic: *Loxokalypus socialis*.

FAMILY PEDICELLINIDAE: With incomplete stalk–calyx septum; muscles extend length of stalk, but are not continuous with those of calyx; stalk undifferentiated; colonial. (e.g., *Myosoma*, *Pedicellina*)

FAMILY BARENTSIIDAE: With incomplete stalk–calyx septum; stalk muscles short and discontinuous; stalk differentiated into wide, muscular nodes and narrow, nonmuscular rods; colonial; some live in fresh water. (e.g., *Barentsia*, *Urnatella*)

Body Wall, Support, and Movement

The calyx and stalk are covered by a thin cuticle that does not extend over the ciliated portion of the tentacles or the vestibule. The epidermis is cellular, and the epidermal cells are cuboidal to somewhat flattened. Various subepidermal muscle bands serve to retract the tentacles, compress the body to extend the tentacles, and contract the calyx. Other muscles are located within the stalk and provide the ability to bend. As mentioned above, there is no persistent body cavity, and the area between the gut and body wall is filled with mesenchyme (Figure 12.31E). The mesenchyme and cuticle provide body support.

Feeding and Digestion

Entoprocts are suspension feeders. They extract food particles, mostly phytoplankton, from currents produced by the lateral cilia on the tentacles (Figure 12.32B). These animals are oriented with their ventral side away from the substratum; the dorsal surface is attached to the stalk. Water currents pass from dorsal to ventral, flow-

(A)

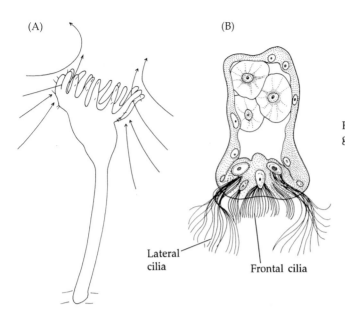

(B)

Lateral cilia

Frontal cilia

(C)

Action of frontal cilia

Action of lateral cilia

Vestibule

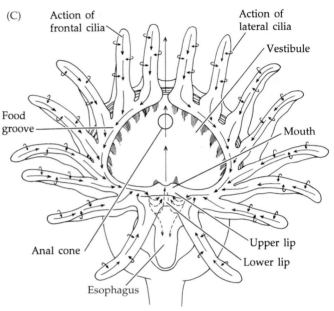

Food groove

Mouth

Anal cone

Upper lip

Lower lip

Esophagus

Figure 12.32   Entoproct feeding. (A) General ciliary currents in *Loxosoma*. (B) A tentacle of *Pedicellina* (cross section). Note the lateral and frontal cilia. (C) *Loxosoma* (top view), illustrating details of feeding currents.

ing between the tentacles (Figure 12.32A). Food is trapped by the lateral cilia and moved in a sheet of mucus to the frontal cilia, which move the mucus and food to ciliated vestibular food grooves at the base of the tentacular ring. Here additional ciliary tracts carry the food to the mouth (Figure 12.32C).

Food and mucus are moved into the gut by cilia lining the buccal tube and by muscular contractions of the esophagus (Figure 12.31D). The esophagus leads to a spacious stomach, from which a short intestine extends to the rectum located within the anal cone. Food is moved through the gut by cilia. The stomach lining secretes digestive enzymes and mucus, which are mixed with the food by a tumbling action caused by the ciliary currents. Digestion and absorption probably occur within the stomach and intestine, where food is held for a time by an intestinal–rectal sphincter muscle.

Circulation, Gas Exchange, and Excretion

The gut also apparently serves as an excretory passage. Cells in the ventral stomach wall accumulate precipitations of uric acid and guanine and release them into the stomach lumen, from whence they are discharged through the anus. Adult entoprocts also possess a pair of flame bulb protonephridia located between the stomach and the vestibule epithelium. The protonephridia drain to a short common nephridioduct that leads to a pore on the surface of the vestibule. The duct is lined by ameboid cells called **athrocytes**; these cells are thought to phagocytize wastes from the mesenchyme and release them to the excretory duct. In most species, protonephridia appear to be present in larvae as well as adults.

Internal transport is largely through the expansive gut; diffusion distances through the mesenchyme are small between its lumen and the body wall. Colonial entoprocts have a so-called **star-cell organ** located near the stalk–calyx junction. This structure functions as a heart by pulsating and pumping fluid from the calyx to the stalk. Gas exchange probably occurs over much of the body surface, particularly at the cuticle-free tentacles and vestibule.

Nervous System

As is often the case in small sessile invertebrates, the nervous system is greatly reduced. A single ganglionic mass lies between the stomach and vestibular surface, and is called the subenteric ganglion (Figure 12.31E). The subenteric ganglion gives rise to several pairs of nerves to the tentacles, calyx wall, and stalk. Unicellular tactile receptors are concentrated on the tentacles and scattered over much of the body surface. Ciliated papillae form lateral sense organs in some loxosomatids.

Reproduction and Development

Colony growth occurs by budding at the base of zooids or on various branches of the stalk (Figure 12.31A). Solitary forms produce buds on the calyx that separate from the parent (Figure 12.31D).

Most, perhaps all, loxostomatids are hermaphroditic and many are protandric. Those that are thought to have separate sexes may also be protandric, but with a long temporal separation of the male and female phases. Colonial forms may have hermaphroditic or dioecious zooids, and colonies may contain one or both sexes. One or two pairs of gonads lie just beneath the surface of the vestibule. Short gonoducts lead from the gonads to a common pore opening to a brood chamber (Figure 12.31E).

(A)

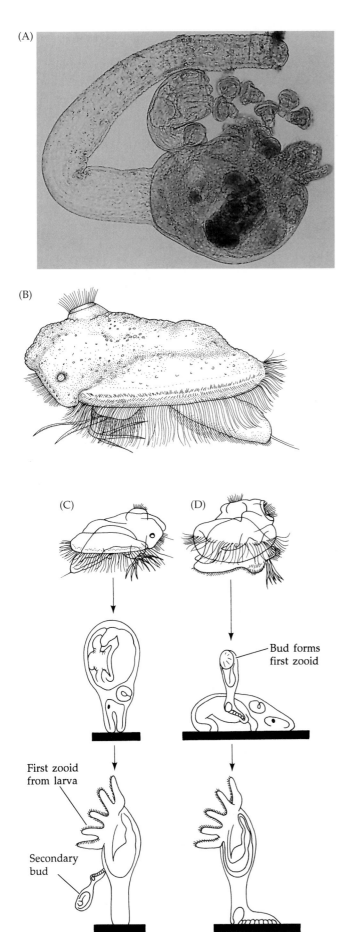

(B)

(C)　　　(D)

Bud forms
first zooid

First zooid
from larva

Secondary
bud

Figure 12.33 Development in entoprocts. (A) A maternal *Loxosomella* bearing a bud and releasing larvae. (B) A larva of *Loxosoma harmeri*. (C) Metamorphosis of *L. harmeri*; the larval body forms the first zooid. (D) Metamorphosis of *Loxosoma leptoclini*; the larva attaches and produces a bud that becomes the first zooid.

Sperm apparently are released into the water and then enter the female reproductive tract, and fertilization occurs in the ovaries or oviducts. As the zygotes move along the oviducts, cement glands secrete stalks by which the embryos are attached to the wall of the brood chamber. A kind of viviparity occurs in some species where the embryos are retained within the ovary. In these and in a few external brooders, special cells of the adult provide nutrition to the developing embryos.

Cleavage in entoprocts is holoblastic and spiral. Nonsynchronous divisions produce five "quartets" of micromeres at about the 56-cell stage. Cell fates are similar to those in typical protostome development, including the derivation of mesoderm from the 4d mesentoblast (see Chapters 4 and 13 for a discussion about protostome development). A coeloblastula forms and gastrulates by invagination. A larva develops (Figure 12.33B) that, according to some authors, is similar to a trochophore, the basic larval type among protostomes (see Chapter 13 for a discussion of the trochophore larva). Most entoproct larvae are free-swimming and planktotrophic, but a few species produce lecithotrophic or benthic crawling larvae. Upon settling, larvae of most species attach by the coronal ciliary band and undergo a remarkable unequal growth of the body mass to direct the ventral, vestibular surface away from the substratum (Figure 12.33C). In some, however, the larva adheres to the substratum and, without rotating, produces an appropriately oriented asexual bud that becomes the adult (Figure 12.33D). Such buds may also be formed earlier, while the larva is still in the brood chamber. These buds may be released by rupture of the larva prior to settling.

Phylum Gnathostomulida: The Gnathostomulids

The phylum Gnathostomulida (Greek, *gnathos*, "jaw"; *stoma*, "mouth") includes 80 or so species of minute vermiform animals (Figure 12.34). These meiofaunal creatures were first described by Peter Ax (1956) as turbellarians, but were later given phylum status by Riedl (1969). Gnathostomulids are found worldwide, interstitially in marine sands, often occurring in high densities in anoxic, sulfide-rich conditions, from the intertidal zone to depths of hundreds of meters. The elongate body (less than 2 mm long) is usually divisible into head, trunk, and narrow tail regions. Distinguishing fea-

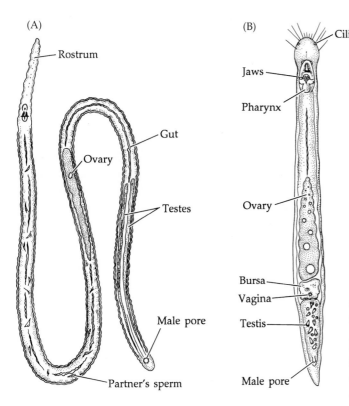

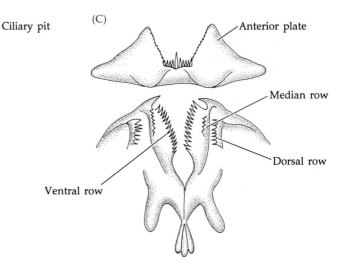

Figure 12.34 Representative gnathostomulids. (A) *Haplognathia simplex*. (B) *Austrognatharia kirsteueri*. (C) The pharyngeal jaw apparatus of *Gnathostomula mediterranea*.

tures of this phylum include a unique jawed pharyngeal apparatus and monociliated epidermal cells (Box 12I). The gnathostomulids are acoelomate, but their relationship to other groups is highly controversial (Sterrer et al. 1985). The 55 species and 25 genera are divided between two orders.

PHYLUM GNATHOSTOMULIDA

ORDER FILOSPERMOIDEA: Body usually very elongate, with slender rostrum; males without penis; sperm filiform; females without vagina. (e.g., *Haplognathia*, *Pterognathia*)

ORDER BURSOVAGINOIDEA: Body usually not extremely elongate relative to width; without slender rostrum; males with penis; sperm not filiform; females with bursa and usually a vagina. (e.g., *Austrognatharia*, *Gnathostomula*)

Body Wall, Support, and Locomotion

Each outer epithelial cell bears a single cilium by which the animal moves in a gliding motion. Movement is aided by body contortions produced by the contraction of thin strands of subepidermal (cross-striated) muscle fibers. These actions, plus reversible ciliary beating, facilitate twisting, turning, and crawling among sand grains, and allow limited swimming in some species. Mucous gland cells occur in the epidermis of at least some species. The body is supported by its more or less

solid construction, with a loose mesenchyme filling the area between the internal organs.

Nutrition, Circulation, Excretion, and Gas Exchange

The mouth is located on the ventral surface at the "head–trunk" junction and leads inward to a complex muscular pharynx armed with an anterior, often comb-like plate and a pair of movable jaws (Figure 12.34C), although the jaw apparatus is lacking in at least one species. Gnathostomulids ingest bacteria and fungi by scraping them into the mouth with the comb plate of the pharynx and swallowing them by the action of the jaws. The pharynx connects with a simple, elongate, saclike gut. In some gnathostomulids there is a tissue connection between the posterior end of the gut and the overlying epidermis. This enigmatic feature has been variously interpreted as a temporary anal connection to the exterior,

---

BOX 12I ***Characteristics of the Phylum Gnathostomulida***

1. Triploblastic, bilateral, unsegmented, vermiform acoelomates

2. Epidermis monolayered; all epithelial cells are monociliate

3. Gut incomplete (anus rudimentary or vestigial)

4. Pharynx houses unique jaw apparatus

5. Without special excretory, circulatory, or gas exchange structures

6. Hermaphroditic

7. Cleavage spiral

8. Inhabit marine interstitial environments

as the remnant of an anus that has been evolutionarily lost, and as an incipient anus that has yet to fully develop.

These animals probably depend largely on diffusion for circulation, excretion, and gas exchange. Isolated solenocytes have been reported in some species, but they resemble epidermal cells and join to "epidermal canal cells" that eventually open on the body surface.

Nervous System

The nervous system is intimately associated with the epidermis and as yet is incompletely described. A host of sensory ciliary pits and sensory cilia occur on the epidermis, especially on the head. Gnathostomulid specialists have attached a formidable array of names to these structures, which are of major taxonomic significance.

Reproduction and Development

Gnathostomulids are protandric or simultaneous hermaphrodites. The male reproductive system includes one or two testes generally located in the posterior part of the trunk and tail; the female system consists of a single large ovary (Figure 12.34). Members of the order Bursovaginoidea possess a vaginal orifice and a sperm-storage bursa, both associated with the female gonopore, and a penis in the male system; members of the order Filospermoidea lack these structures.

Mating has been only superficially studied in gnathostomulids. The penis is glandular and adheres to the partner's body. Although the method of sperm transfer is not certain, suggestions include sperm boring through the body wall and mutual hypodermic impregnation. In any case, these animals appear to be gregarious, to rely on internal fertilization, and to deposit zygotes singly in their habitat. Cleavage is spiral and development is direct, but the details of later development are lacking.

Phylum Loricifera: The Loriciferans

It may be apparent to you by now that interstitial habitats (the meiobenthic realm) are home for a host of bizarre and specialized creatures. Recent studies by two German zoologists, D. Walossek and K. J. Müller, have revealed that a rich meiofaunal ecosystem was already in place as early as the upper Cambrian Era (Müller et al. 1995). Studies on modern meiofauna continue to reveal new animals, previously undescribed taxa, and myriad examples of convergent evolution associated with success in this environment. Among these recently described groups is the Loricifera, first named and described by Reinhardt Kristensen (1983), a Danish zoologist who has studied meiofauna for many years. The name Loricifera (Latin *lorica*, "corset"; *ferre*, "to bear") refers to the well developed cuticular **lorica** encasing most of the body (Figure 12.35; Box 12J). The description of the phylum was initially based upon a single widespread species, *Nanaloricus mysticus*, but several other species have since been described. Most loriciferans have been found at depths of about 300–450 m in coarse marine sediments. One species, *Piciloricus hadalis*, was collected in the western Pacific at a depth of over 8,000 m. Others, as yet undescribed, have been recorded from additional deep-sea muddy bottom locations. To date, all loriciferans have been placed in two families (Nanaloricidae and Pliciloricidae) in a single order, Nanaloricida. All are free-living.

The loriciferan body is minute (115–383 μm long) but complex, containing over 10,000 cells. It is divided into a head (introvert), neck, thorax, and loricate abdomen. The head, neck, and most of the thorax can telescope into the lorica. The mouth is located at the end of an introvert, called the **oral cone**, that projects from the head and contains protrusible **oral stylets** in some species. Nine rings of spinelike **scalids** (200–400 of them) of various shapes protrude from the spherical head, most apparently with intrinsic muscles. In the pliciloricids, some of these projections bear joints near their bases, reminiscent of articulated limbs. The first ring consists of anteriorly directed **clavoscalids**; the remaining eight rings of **spinoscalids** are directed posteriorly. In at least some species, the number of clavoscalids differs between males and females.

The lorica of nanaloricids comprises six plates, which bear anteriorly directed spines around the base of the neck. Beneath the cuticle and the body wall are several muscle bands, including those responsible for retraction of the anterior parts. The body cavity is presumably a blastocoelom and varies from rather spacious in pliciloricids to virtually absent in *Nanaloricus*.

BOX 12J *Characteristics of the Phylum Loricifera*

1. Bilateral, unsegmented, and probably blastocoelomate

2. Body divided into a head, neck, and thorax, all retractable into an abdomen

3. Abdomen housed in cuticular lorica; cellular epidermis underlain by basal lamina

4. Mouth on oral cone beset with spines (stylets); head (introvert) and neck with 7–9 rows of scalids

5. Gut complete

6. No apparent circulatory or gas exchange systems

7. One pair of protonephridia, situated in gonads

8. Dioecious; development includes unique Higgins larva

9. All inhabit marine interstitial environments

(A)

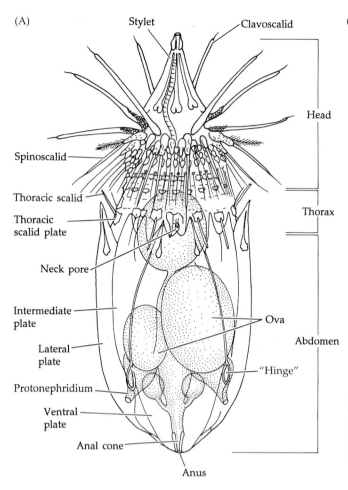

(B)

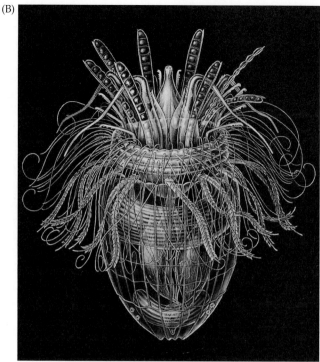

Figure 12.35 Representative loriciferans. (A) An adult female loriciferan, *Nanaloricus mysticus* (ventral view). (B) *P. enigmaticus*. (C) *Pliciloricus gracilis*. (D) The Higgins larva of *P. gracilis*. (E) The Higgins larva of *Armorloricus davidi*. (F) The anterior end of *N. mysticus*.

The digestive tract is complete. A long, tubular, buccal canal leads from the mouth to a muscular pharynx bulb and esophagus (Figure 12.35F). The pharynx has both circular and longitudinal muscles. The lumen of these anterior gut structures is lined with cuticle. Behind the esophagus is a long midgut leading to a short, cuticle-lined rectum and an anus located on an anal cone. One pair of **salivary glands** is associated with the buccal tube. Little is known about feeding in loriciferans, but some apparently eat bacteria.

The central nervous system includes a large, circumpharyngeal ganglion and a number of smaller ganglia associated with the various body regions and parts. A large ventral nerve cord also bears ganglia. At least some of the scalids are probably sensory in function.

Loriciferans are dioecious; males and females differ externally in the form and number of certain scalids. The male reproductive system comprises two dorsal testes in the abdominal body cavity, probably a blastocoelom. The female system includes a pair of ovaries and probably a seminal receptacle. Fertilization is suspected to be internal. Loriciferans have one pair of monoflagellate protonephridia that are actually located within the gonads.

Nothing is yet known about early development in loriciferans. In most species, a feeding **Higgins larva** de-velops. This larva (Figure 12.35D,E) is built along the same general body plan as the adult, but it possesses a pair of "toes" at the posterior end that are used for locomotion. These toes are thought to have adhesive glands at their bases. Early stages apparently contain yolk reserves in the cells of the body cavity. The cuticle is periodically shed as the individuals grow in size. In some species the larva metamorphoses to a so-called postlarva (juvenile), which resembles an adult female but lacks ovaries. In others, larval metamorphosis involves a "molt" directly to an adult. Certain deep-sea species hasten their life cycle by producing neotenous larvae, which produce from two to four additional larvae by parthenogenesis. The larvae of some loriciferans encyst for a period of time before metamorphosis.

Phylum Cycliophora: The Cycliophorans

Students sometimes lament that there is little left to be discovered. Clearly, nothing could be further from the truth, as witnessed by the recent description of a new phylum of tiny metazoans called the Cycliophora (Funch and Kristensen 1995). The only described species, *Symbion pandora*, was discovered living on the mouth-

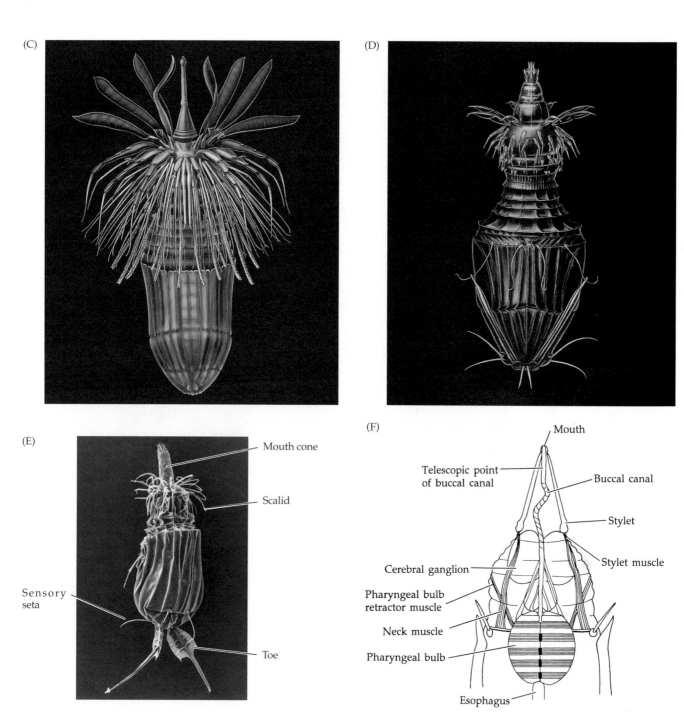

(C)

(E)

Mouth cone

Scalid

Sensory
seta

Toe

(D)

(F)

Mouth

Telescopic point
of buccal canal

Buccal canal

Stylet

Stylet muscle

Cerebral ganglion

Pharyngeal bulb
retractor muscle

Neck muscle

Pharyngeal bulb

Esophagus

parts of Norway lobsters found off the coast of Denmark, Sweden, and other locations in the north Atlantic (Figure 12.36A). It has since been found on other decapod species.

Cycliophorans are apparently acoelomate; the area between the gut and body wall is packed with large mesenchymal cells. The body is divided into an anterior **buccal funnel**, an oval trunk, and a posterior adhesive disc by which the animal attaches to its host's setae. A layered cuticle covers the trunk and adhesive disc, the latter apparently composed entirely of cuticular material.

Feeding adults are about 350 μm in length. They suspension feed by creating water currents with dense cilia that are situated on a ring of modified epidermal cells encircling the open end of the buccal funnel. These ciliated epidermal cells alternate with contractile cells that form a pair of sphincters, permitting closure of the oral area. The U-shaped gut is ciliated along its entire length. The buccal funnel leads to a curved esophagus and thence to a stomach consisting of large gland cells penetrated by a narrow lumen. An intestine extends anteriorly to a short rectum and anus, located dorsally near the base of the buccal funnel (Figure 12.36B). Two

(A)

Figure 12.36 *Symbion pandora*, the type species of the newly described phylum Cycliophora. (A) Some 40 individuals of *S. pandora* can be seen attached to the mouthparts of the Norwegian lobster (*Nephrops norvegicus*). (B) A feeding adult *Symbion* with an attached dwarf male.

(B)

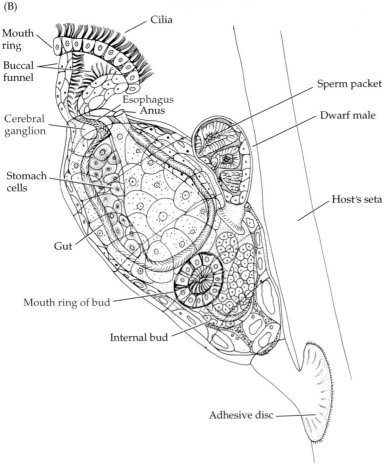

muscle fibers extend from the dorsal base of the buccal funnel to the ventral side of the trunk, and probably serve to move the funnel during feeding. The area between the gut and other organs is packed with a cellular mesenchyme.

The dwarf males (about 85 μm long) lack a gut and buccal funnel; the body is largely filled with mesenchyme and packets of developing sperm. They live attached to the body of mature feeding adults (Figure 12.36B).

Circulation and gas exchange are presumably accomplished by simple diffusion in these tiny animals. A pair of flame bulb protonephridia is present in one of the larval stages (the chordoid larva described below), but has not been identified in adults. A cerebral ganglion lies dorsal to the esophagus, but no other elements of the nervous system have yet been identified.

The life cycle of cycliophorans is still incompletely understood, but Funch and Kristensen (1995) proposed a possible set of events involving a complex and somewhat bizarre series of stages alternating between sexual and asexual phases. Their "hypothetical life cycle" involves females asexually producing **pandora larvae**, which escape from the parent, settle on the same host, and grow into new feeding adults.

Sexual reproduction appears to be timed to coincide with the end of the host's molt cycle. Some cycliophorans form a bud that develops into a dwarf male, which escapes and settles on a nearby feeding adult that contains a presumptive female bud. The female bud, in turn, houses a developing oocyte. Somehow, the male impregnates this female bud, which then escapes carrying the fertilized egg. The freed zygote-bearing female "bud" settles (on the same host), and her embryo develops into the **chordoid larva**. This larva leaves the female and is probably the avenue by which dispersal to new hosts is achieved. The anatomy of the chordoid larva was described in detail by Funch (1996), who suggested that it is a modified trochophore larva, homologous to those of many protostomes (e.g., annelids, molluscs, sipunculans, and echiurans).

Clearly, there are myriad details and unanswered questions yet to be addressed about members of this most recently discovered phylum.

Some Phylogenetic Considerations

We have alluded to a few of the possible phylogenetic relationships of the eleven phyla covered in this chapter to one another and to members of other phyla. Although there has been a renewed interest in exploring these topics, the relationships of these groups remain some of the most challenging mysteries in biology.

As we stressed in the introductory chapters of this book, one of the major difficulties in sorting out phylogenetic hypotheses is evaluating the nature of shared characteristics among groups. Particularly difficult is the separation of homologous traits from convergently similar ones. The exploitation of similar habitats by animals with different phylogenetic histories is frequently accompanied by the appearance of convergent similarities. We have seen this principle in the case of parasitism, and it is no less evident in other special environments, such as those encountered by interstitial animals, or simply those associated with small body size.

It is probable that the phyla discussed in this chapter do in fact represent fewer than eleven independent monophyletic clades (i.e., some are sister groups to one another). But it is also almost certainly true that members of at least some of these phyla must be regarded as having achieved a grade of body organization that reflects considerable convergent evolution. Along these lines, we offer the following sampling of ideas.

The presence of a blastocoelom (= "pseudocoelom") in different kinds of animals is insufficient grounds by itself for hypothesizing that they constitute a monophyletic group, or clade. The retention of a persistent blastocoel

may be nothing more than a paedomorphic characteristic derived convergently in numerous groups. The same may be said of certain other features that tend to accompany the blastocoelomate bauplan, such as small size, eutely, syncytial tissues or organs, reduced nervous systems, and external ciliation. Some of the phyla discussed here contain species with a spacious body cavity, as well as other species in which the cavity is invaded to various degrees by mesenchyme, suggesting plasticity in this developmental feature. We consider the adult blastocoelom *sensu lato* not to be a particularly conservative or useful phylogenetic feature, but rather the repeated product of flexible developmental programs.

Most of the taxa discussed in this chapter appear to show some basic affinity to the protostome line, especially in the tendency toward spiral-like cleavage. This feature, coupled with small size, blastocoel retention, external ciliation patterns, and the presence of protonephridia in adults, have led many workers to speculate that the "aschelminth" groups may have arisen from coelomate protostome ancestors, perhaps by neoteny.

A number of authors have supported a sister-group relationship between the nematodes and nematomorphans. A revolutionary proposal by Aguinaldo et al. (1997) proposes a clade including arthropods and their kin (tardigrades and onychophorans), plus a number of other groups that shed their cuticle—kinorhynchs, priapulans, nematodes, and nematomorphans. This proposed group—called the "Ecdysozoa"—is based on 18S ribosomal DNA sequence analysis and is dramatically different from more conventional views. The homology of the molting process in this proposed clade is yet to be tested (Chapter 24).

The gnathostomulids have been viewed as being related to annelids, rotifers, turbellarians, or gastrotrichs. The latter two groups have a more or less mesenchyme-filled body, external ciliation, and other common features. Possible ties of the gnathostomulids to annelids and rotifers have been based on similarities in the pharynx and jaw apparatus, but these are probably convergent features rather than true homologues. A more traditional hypothesis has allied the gnathostomulids with the turbellarians, but these groups differ in some very basic ways. The gnathostomulids contain rather poorly developed mesenchyme, have monociliated epidermal cells, and have a unique pharynx apparatus, all of which argue against a relationship with the flatworms. Ax (1985), however, made a case for treating the Gnathostomulida and Platyhelminthes as sister taxa. Ahlrichs (1997) proposed the taxon "Gnathifera," a group based in large part on ultrastructural evidence and including the gnathostomulids, acanthocephalans, and rotifers.

Another recent proposal, put forth by Adrianov and Malakhov (1995), suggested uniting the Priapula, Loricifera, Kinorhyncha, and Nematomorpha into a single taxon, to be called the "Cephalorhyncha." On the other hand, Ehlers et al. (1996) and Neuhaus et al. (1996) presented phylogenetic analyses in support of a taxon they called "Nemathelminthes," to include the gastrotrichs, nematodes, nematomorphans, kinorhynchs, priapulans, and loriciferans. According to these authors, the group is defined by synapomorphies such as mouth position, nature of the cuticle, and structure of the pharynx, and is a sister group to the rotifers and acanthocephalans.

The evolutionary relationships of the Acanthocephala are particularly enigmatic. Since no free-living species are known, it is difficult to trace their phylogeny to any free-living ancestors. The phylum may have had its origin in a group that had already established itself in a parasitic lifestyle. Cestodes have been suggested as a possible ancestral group to the acanthocephalans. However, these two taxa differ in fundamental aspects of their anatomy and ontogeny, so there is little to corroborate such a hypothesis. Lorenzen (1985) suggested that acanthocephalans may be highly derived rotifers, and recent morphological and molecular studies (18S rDNA) have supported this view. Rieger and Tyler (1995) supported a sister-group relationship between gnathostomulids and a rotifer–acanthocephalan clade. However, the absence in acanthocephalans of any trace of the rotiferan mastax argues against a kinship with the rotifers. The simplest known acanthocephalans live in fishes, and the most specialized forms in birds and mammals. From this fact it has been suggested that the

acanthocephalans were parasites of the earliest fishes and evolved along with their vertebrate hosts. If so, these parasites could be very old (Silurian or older). The absence of an acanthocephalan fossil record and the reduction of their body parts make phylogenetic studies on this phylum difficult.

The entoproct/ectoproct problem remains controversial. The spiral cleavage, 4d mesoderm, and larval features in entoprocts suggests a protostome affinity, whereas ectoprocts are clearly deuterostome-like in their early development. We do not think that the two groups are closely related to each other, and we explain this point of view in Chapter 21. This issue is now confounded by speculation that the cycliophorans may be related to one or both of these groups.

These and other speculations are attempts to bring order out of the phylogenetic chaos of these groups. Some hope lies in future studies on ultrastructure and development of those animals where information is still incomplete, and with the examination of new (nuclear) genes for molecular phylogenetic analysis. With the limited data available, and so many questions still unanswered, we feel it is appropriate to consider these eleven groups as separate phyla for the time being, although we anxiously await new developments. Additional discussion on these taxa can be found in Chapter 24.

Selected References

General References

Adrianov, A. V. and V. V. Malakhov. 1995. Synopsis of the system of the phylum Cephalorhyncha. Russ. J. Mar. Biol. 21(2): 89–96.

Aguinaldo, A. M. A., J. M. Turbeville, L. S. Linford, M. C. Rivera, J. R. Garey, R. A. Raff and J. A. Lake. 1997. Evidence for a clade of nematodes, arthropods and other moulting animals. Nature 387: 489–493.

Ahlrichs, W. H. 1997. Epidermal ultrastructure of *Seison nebaliae* and *Seison annulatus*, and a comparison of epidermal structures within the Gnathifera. Zoomorph. 117: 41–48.

Boaden, P. J. S. 1989. Meiofauna and the origins of the metazoa. Zool. J. Linn. Soc. 96: 217–227.

Dougherty, E. C. (ed.) 1963. *The Lower Metazoa: Comparative Biology and Phylogeny.* University of California Press, Berkeley. [See Sections IG and IIIA (in part) for papers by various authors dealing with blastocoelomate phylogeny.]

Ehlers, U., W. Ahlrichs, C. Lemburg and A. Schmidt-Rhaesa. 1996. Phylogenetic systematization of the Nemathelminthes (Aschelminthes). Verh. Dtsch. Zool. Ges. 89 (1): 8.

Florkin, M. and B. T. Scheer. 1969. *Chemical Zoology* Vol. 3, Sect. 2, Nematoda and Acanthocephala. Academic Press, New York.

Garey, J. R., T. J. Near, M. R. Nonnemacher, and S. A. Nadler. 1996. Molecular evidence for Acanthocephala as a subtaxon of Rotifera. J. Mol. Evol. 43: 287–292.

Giese, A. C. and J. S. Pearse. 1974. *Reproduction of Marine Invertebrates,* Vol. 1, Acoelomate and Pseudocoelomate Metazoans. Academic Press, New York.

Grassé, P. (ed.) 1965. Némathelminthes, Rotifères, Gastrotriches, et Kinorhynques. *In* P. Grassé, *Traité de Zoologie,* Vol. 4, Pts. 2–3. Masson et Cie, Paris.

Grobben, K. 1908. Die systematische Einteilung des Tierreiches. Verh. K. Zool. Bot. Ges. Wien. 58.

Higgins, R. P. and H. Thiel (eds.) 1988. *Introduction to the Study of Meiofauna.* Smiths. Inst. Press, Washington, DC.

Hulings, N. C. and J. S. Gray (eds.) 1971. A manual for the study of meiofauna. Smithson. Contrib. Zool. 78: 1–83.

Hyman, L. H. 1951. *The Invertebrates,* Vol. 3, Acanthocephala, Aschelminthes, and Entoprocta. McGraw-Hill, New York.

Kristensen, R. M. 1995. Are Aschelminthes pseudocoelomate or acoelomate? *In* G. Lanzavecchia, R. Valvassori and M. D. Candia Carnevali (eds.), *Body Cavities: Function and Phylogeny.* Selected Symposia and Monographs U. Z. I. pp 41–43.

Lorenzen, S. 1985. Phylogenetic aspects of pseudocoelomate evolution. *In* S. C. Morris et al. (eds.), *The Origins and Relationships of Lower Invertebrates.* Syst. Assoc. Spec. Vol. No. 28, Oxford, pp. 210–223.

Malakhov, V. V. 1980. Cephalorhyncha, a new type of animal kingdom uniting Priapulida, Kinorhyncha, Gordiacea, and a system of aschelminthes worms. Zool. Zh. 54(4): 481–499. [In Russian.]

Marquardt, W. C. and R. S. Demaree Jr. 1985. *Parasitology.* Macmillan, New York.

Morris, S. C., J. D. George, R. Gibson and H. M. Platt (eds.). 1985. *The Origins and Relationships of Lower Invertebrates.* Syst. Assoc. Spec. Vol. No. 28, Oxford.

Müller, K. J., D. Walossek and A. Zakharov. 1995. "Orsten" type phosphatized soft-integument preservation and a new record from the Middle Cambrian Kuonamka Formation in Siberia. Neus Jahrb. Geol. Paläontol. Abh. 197: 101–118.

Neuhaus, B., R. M. Kristensen and C. Lemberg. 1996. Ultrastructure of the cuticle of the Nemathelminthes and electron microscopical localization of chitin. Vehr. Dtsch. Zool. Ges. 89 (1): 221.

Nielsen, C. 1998. Sequences lead to a tree of worms. Nature 392: 25–26.

Rieger, R. M. and S. Tyler. 1995. Sister-group relationship of Gnathostomulida and Rotifera–Acanthocephala. Invert. Biol. 114: 186–188.

Roberts, L. S. and J. Janovy, Jr. 1996. *Foundations of Parasitology*, 5th Ed. Wm. C. Brown, Chicago.

Ruppert, E. E. 1991. Introdution to the aschelminth phyla: A consideration of mesoderm, body cavities, and cuticle. *In* F. W. Harrison and E. E. Ruppert (eds.), *Microscopic Anatomy of Invertebrates*, Vol. 4, Aschelminthes. Wiley-Liss, New York, pp.1–17.

Wägele, J. W., T. Erickson, P. Lockkhort, and B. Misof. 1999. The Ecdysozoa. Artifact or Monophylum. J. Zool. Syst. Evol. Res. 37: 211–223.

Wilson, R. A. and L. A. Webster. 1974. Protonephridia. Biol. Rev. 49: 127–160.

Winnepenninckx, B., T. Backeljau, L. Y. Mackey, J. M. Brooks, R. deWachter, S. Kumar and J. R. Garey. 1995. 18SrRNA data indicate that Aschelminthes are polyphyletic in origin and consist of at least three distinct clades. Mol. Biol. Evol. 12: 1132–1137.

Rotifera

Aloia, R. and R. Moretti. 1973. Mating behavior and ultrastructural aspects of copulation in the rotifer *Asplanchna brightwelli*. Trans. Am. Microsc. Soc. 92: 371–380.

Birky, W. 1971. Parthenogenesis in rotifers: The control of sexual and asexual reproduction. Am. Zool. 11: 245–266.

Birky, W. and B. Field. 1966. Nuclear number in the rotifer *Asplancha*: intraclonal variation and environmental control. Science 151: 585–587.

Clément, P. 1977. Ultrastructure research on rotifers. Arch. Hydrobiol. Beih. Ergebn. Limnol. 8: 270–297.

Clément, P. 1985. The relationships of rotifers. *In* S. C. Morris et al. (eds.), *The Origins and Relationships of Lower Invertebrates*. Syst. Assoc. Spec. Vol. No. 28, Oxford, pp. 224–247.

Clément, P. 1987. Movements in rotifers: Correlations of ultrastructure and behavior. Hydrobiologia 147: 339–359.

Clément, P. and E. Wurdak. 1991. Rotifera. *In* F. W. Harrison and E. E. Ruppert (eds.), *Microscopic Anatomy of Invertebrates*, Vol. 4, Aschelminthes. Wiley-Liss, New York, pp. 219–297.

Donner, J. 1966. *Rotifers*. Frederick Warne & Co., New York.

Felix, A., M. E. Stevens and R. L. Wallace. 1995. Unpalatability of a colonial rotifer, *Sinantherina socialis*, to small zooplanktiverous fishes. Invert. Biol. 114: 139–144.

Gilbert, J. J. 1980. Developmental polymorphism in the rotifer *Asplanchna sieboldi*. Am. Sci. 68: 636–646.

Gilbert, J. J. 1983. Rotifera. pp. 181–209 in K. G. and R. G. Adiyodi, *Reproductive Biology of Invertebrates, Vol. 1*. John Wiley, London.

King, C. E. 1977. Genetics of reproduction, variation and adaptation in rotifers. Arch. Hydrobiol. Beih. 8: 187–201.

King, C. E. and M. R. Miracle. 1980. A perspective on aging in rotifers. Hydrobiologia 73: 13–19.

Koste, W. 1978. Rotatoria. Die Rädertiere Mitteleuropas Begründer von Max Voight., Vols. I,II Borntraeger, Berlin. [The best compilation of rotifer species available.]

Nogrady, T. 1982. Rotifera. *In* S. Parker (ed.), *Synopsis and Classification of Living Organisms*, Vol. 1. McGraw-Hill, New York, pp. 866–872.

Nogrady, T. (ed.). 1993. *Rotifera*. SPB Academic Publishing bv, The Hague.

Pray, F. A. 1965. Studies on the early development of the rotifer *Monostyla cornuta* Muller. Trans. Am. Microsc. Soc. 84: 210–216.

Ricci, C., G. Melone and C. Sotgia. 1993. Old and new data on Seisonidea (Rotifera). Hydrobiologia 255/256: 495–511.

Segers, H. 2002. The nomenclature of the Rotifera: Annotated checklist of valid family- and genus-group names. J. Nat. Hist. 36(6): 631–640.

Turner, P.N. 1995. Rotifer look-alikes: Two species of *Colurella* are ciliated protozoans. Invert. Biol. 114: 202–204.

Wallace, R. L. and R. A. Colburn. 1989. Phylogenetic relationships within phylum Rotifera: Orders and genus *Notholca*. Hydrobiologia 186/187: 311–318.

Gastrotricha

Boaden, P. J. S. 1985. Why is a gastrotrich? *In* S. C. Morris et al. (eds.), *The Origins and Relationships of Lower Invertebrates*. Syst. Assoc. Spec. Vol. No. 28, Oxford, pp. 248–260.

Colacino, J. M. and D. W. Kraus. 1984. Hemoglobin-containing cells in *Neodasys* (Gastrotricha: Chaetonotida). II. Respiratory significance. Comp. Biochem. Physiol. 79A: 363–369.

D'Hondt, J. L. 1971. Gastrotricha. Oceanogr. Mar. Biol. Annu. Rev. 9: 141–192.

Hummon, W. 1982. Gastrotricha. *In* S. Parker (ed.), *Synopsis and Classification of Living Organisms*, Vol. 1. McGraw-Hill, New York, pp. 857–863.

Hummon, M. R. 1986. Reproduction and development in a freshwater gastrotrich. 4. Life history traits and the possibility of sexual reproduction. Trans. Am. Microsc. Soc. 105: 97–109. [See also earlier papers by Hummon on gastrotrich reproduction.]

Manylov, O. G. 1995. Regeneration in Gastrotricha. I. Light microscopical observations on the regeneration on *Turbanella* sp. Acta Zool. 76(1): 1–6.

Rieger, G. E. and R. M. Rieger. 1977. Comparative fine structure study of the gastrotrich cuticle and aspects of cuticle evolution within the Aschelminthes. Z. Zool. Syst. Evolutionsforsch. 15: 81–124.

Rieger, R. M. 1976. Monociliated epidermal cells in Gastrotricha: Significance for concepts of early metazoan evolution. Z. Zool. System. Evolutionsforsch. 14(3): 198–226.

Ruppert, E. E. 1978. The reproductive system of gastrotrichs. II. Insemination in *Macrodasys*: A unique mode of sperm transfer in metazoa. Zoomorphologie 89: 207–228.

Ruppert, E. E. 1978. The reproductive system of gastrotrichs. III. Genital organs of Thaumastodermatinae subfam. n. and Diplodasyniae subfam. n. with discussion of reproduction in Macrodasyida. Zool. Scr. 7: 93–114.

Ruppert, E. E. 1982. Comparative ultrastructure of the gastrotrich pharynx and the evolution of myoepithelial foreguts in aschelminthes. Zoomorphologie 99: 181–220.

Ruppert, E. E. Gastrotricha. *In* F. W. Harrison and E. E. Ruppert (eds.), *Microscopic Anatomy of Invertebrates*, Vol. 4, Aschelminthes. Wiley-Liss, New York, pp. 41–109.

Teuchert, G. 1973. Die Feinstruktur des Protonephridial systems von *Turbanella cornuta* Remane, einem marinen Gastrotrich der Ordung Macrodasyoidea. Z. Zellforsch. 136: 277–289.

Teuchert, G. 1977. The ultrastructure of the marine gastrotrich *Turbanella cornuta* Remane (Macrodasyoidea). Zoomorphologie 88: 189–246.

Weiss, M. J. and D. P. Levy. 1980. Hermaphroditism and sperm diversity among freshwater Gastrotricha. Am. Zool. 20: 749.

Kinorhyncha

Adrianov, A. V. and V. V. Malakhov. 1994. *Kinorhyncha: Structure, Development, Phylogeny and Taxonomy*. Nauka Publishing. [In Russian.]

Brown, R. 1988. Morphology and ultrastructure of the sensory appendages of a kinorhynch introvert. Zool. Scr. 18: 471–482.

Higgins, R. P. 1974. Kinorhynchs. *In* A. G. Giese and J. S. Pearse (eds.), *Reproduction of Marine Invertebrates*. Academic Press, New York, pp. 507–518.

Higgins, R. P. 1982. Kinorhyncha. *In* S. Parker (ed.), *Synopsis and Classification of Living Organisms*, Vol. 1. McGraw-Hill, New York, pp. 874–877.

Higgins, R. P. 1990. Zelinkaderidae, a new family of cyclorhagid Kinorhyncha. Smiths. Contr. Zool. 500: 1–26.

Higgins, R. P. and Y. Shirayama. 1992. Dracoderidae, a new family of cyclorhagid Kinorhyncha from the Inland Sea of Japan. Zool. Scr. 7: 939–946.

Kozloff, E. 1972. Some aspects of development in *Echinoderes* (Kinorhyncha). Trans. Am. Microsc. Soc. 91: 119–130.

Kristensen, R. M. and R. P. Higgins. 1991. Kinorhyncha. *In* F. W. Harrison and E. E. Ruppert (eds.), *Microscopic Anatomy of Invertebrates*, Vol. 4, Aschelminthes. Wiley-Liss, New York, pp. 378–404.

Neuhaus, B. 1994. Ultrastructure of alimentary canal and body cavity, ground pattern, and phylogenetic relationships of the Kinorhyncha. Microfauna Marina 9: 61–156.

Neuhaus, B. 1995. Postembryonic development of *Paracentrophyes praedictus* (Homalorhagida): Neoteny questionable among the Kinorhyncha. Zool. Scr. 24: 179–192.

Zelinka, C. 1928. *Monographie der Echinodera*. Wilhelm Engleman, Leipzig.

Nemata

Adamson, M. L. 1987. Phylogenetic analysis of the higher classification of the Nematoda. Can. J. Zool. 65: 1478–1482.

Allgen, C. A. 1947. West American marine nematodes. Vid. Medd. Dansk. Naturhist. Foren. (Copenhagen) 110: 65–219.

Bauer-Nebelsick, M. M. Blumer, W. Urbanick and J. A. Ott. 1995. The glandular sensory organ of Desmodoridae (Nematoda): Ultrastructure and phylogenetic implications. Invert. Biol. 114: 211–219.

Bird, A. F. and J. Bird. 1971. *The Structure of Nematodes*, 2nd Ed. Academic Press, New York.

Bird, A. F. 1984. Growth and moulting in nematodes: Moulting and development of *Rotylenchulus reniformis*. Parasitol. 89: 107–119.

Blaxter, M. L. and eleven others. 1998. A molecular evolutionary framework for the phylum Nematoda. Nature 392: 71–75.

Boveri, T. 1899. Die Entwicklung von *Ascaris* mit besonderer Rucksicht auf die Kernverhaltnisse. Festschr. 70.

Chitwood, B. G. and M. B. Chitwood. 1974. *Introduction to Nematology*. University Park Press, Baltimore.

Crofton, H. D. 1966. *Nematodes*. Hutchinson University Library, London.

Croll, N. A. and B. E. Matthews. 1977. *Biology of Nematodes*. Wiley, New York.

Deutsch, A. 1978. Gut ultrastructure and digestive physiology of two marine nematodes, *Chromadorina germanica* (Butschli, 1874) and *Diplolaimella* sp. Biol. Bull. 155: 317–355.

Ehlers, U. 1994. Absense of a pseudocoel or pseudocoelom in *Anoplostoma vivipara* (Nematoda). Microfauna Marina 9: 345–350.

Harris, J. E. and H. D. Crofton. 1957. Structure and function of nematodes. J. Exp. Biol. 34: 116-155.

Heip, C., M. Vinex and G. Vranken. 1985. The ecology of marine nematodes. Oceanogr. Mar. Biol. Annu. Rev. 23: 399–489.

Hope, W. D. 1967. Free living marine nematodes…from the west coast of North America. Trans. Amer. Micros. Soc. 86: 307–334.

Hope, W. D. and S. L. Gardiner. 1982. Fine structure of a proprioceptor in the body wall of the marine nematode *Donostoma californicum* Steiner and Albin, 1933 (Enoplida = Leptostomatidae). Cell Tissue Res. 225: 1–10.

Hope, W. D. and D. G. Murphy. 1972. A taxonomic hierarchy and checklist of the genera and higher taxa of marine nematodes. Smithson. Contrib. Zool. 137. [Includes an outstanding bibliography of papers dealing with marine nematodes.]

Kampfer, S., C. Sturmbauer and J. Ott. 1998. Phylogenetic analysis of rDNA sequences from adenophorean nematodes and implications for the Adenophorea-Secernentea controversy. Invert. Biol. 117: 29–36.

Lee, D. L. and H. J. Atkinson. 1977. *Physiology of Nematodes*, 2nd Ed. Columbia University Press, New York.

Maggenti, A. R. 1982. Nemata. *In* S. Parker (ed.), *Synopsis and Classification of Living Organisms*. McGraw-Hill, New York, pp. 880–929.

Nicholas, W. L. 1975. *The Biology of Free-Living Nematodes*. Clarendon Press, Oxford.

Poinar, G. O., Jr. 1983. *The Natural History of Nematodes*. Prentice-Hall, Englewood Cliffs, New Jersey.

Roggen, D. R., D. J. Raski and N. O. Jones. 1966. Cilia in nematode sensory organs. Science 152: 515–516.

Schaefer, C. 1971. Nematode radiation. Syst. Zool. 20: 77–78.

Somers, J. A., H. H. Shorey and L. K. Gastor. 1977. Sex pheromone communication in the nematode, *Rhabditis pellio*. J. Chem. Ecol. 3: 467–474.

Welch, D. M. and M. Meselson. 2000. Evidence for the evolution of bdelloid rotifers without sexual reproduction or genetic exchange. Science 288: 1211–1214.

Wright, K. A. 1991. Nematoda. *In* F. W. Harrison and E. E. Ruppert (eds.), *Microscopic Anatomy of Invertebrates*, Vol. 4, Aschelminthes, Wiley-Liss, New York, pp. 111–195.

Yeats, G. W. 1971. Feeding types and feeding groups in plant and soil nematodes. Pedobiologia 11: 173–179.

Zuckerman, B. M., W. F. Mai and R. A. Rhode (eds.) 1971. *Plant Parasitic Nematodes*. Vol. 1, Morphology, Anatomy, Taxonomy, and Ecology. Vol. 2, Cytogenetics, Host–Parasite Interactions, and Physiology. Academic Press, New York.

Nematomorpha

Bresciani, J. 1991. Nematomorpha. *In* F. W. Harrison and E. E. Ruppert (eds.), *Microscopic Anatomy of Invertebrates*, Vol. 4, Aschelminthes, Wiley-Liss, New York, pp. 197–218.

Carvalho, J. C. M. 1942. Studies on some Gordiacea of North and South America. J. Parasitol. 28: 213–222.

Cham, S. A., M. K. Seymour and D. J. Hooper. 1983. Observations of a British hairworm, *Parachordodes wolterstorffii* (Nematomorpha: Gordiidae). J. Zool. Lond. 199: 275–285.

Leslie, H. A., A. Campbell, and G. R. Daborn. 1981. *Nectonema* (Nematomorpha: Nectonematoidea), a parasite of decopod Crustacea in the Bay of Fundy. Can. J. Zool. 59: 1193–1196.

Malakhov, V. V. and S. E. Spiridonov. 19484. The embryogenesis of *Gordius* sp. from Turmenia, with special reference to the position of the Nematomorpha in the animal kingdom. Zool. Z. 63: 1285–1296.

Schmidt-Rhaesa, A. II. 1996a. Monophyly and systematic relationships of the Nematomorpha. Vehr. Dtsch. Zool. Ges. 89: 23.

Schmidt-Rhaesa, A. 1996b. Ultrastructure of the anterior end in three ontogenetic stages of *Nectonema munidae* (Nematomorpha). Acta Zool. 77: 267–278.

Schmidt-Rhaesa, A. 1996c. The nervous system of *Nectonema munidae* and *Gordius aquaticus*, with implications for the ground pattern of the Nematomorpha. Zoomorph. 116: 133–142.

Schmidt-Rhaesa, A. 1997a. Ultrastructural observations of the male reproductive system and spermatozoa of *Gordius aquaticus* L., 1758. Invert. Reprod. and Dev. 32: 31–40.

Schmidt-Rhaesa, A. 1997b. Ultrastructural features of the female reproductive system and female gametes of *Nectonema munidae* Brinkman 1930. Parasit. Res. 83: 77–81.

Schmidt-Rhaesa, A. 1997c. *Nematomorpha*. Gustav Fischer Verlag, Stuttgart.

Schmidt-Rhaesa, A. 1998. Muscular ultrastructure in *Nectonema munidae* and *Gordius aquaticus* (Nematomorpha). Invert. Biol. 117: 37–44.

Skaling, B. and B. M. MacKinnon. 1988. The absorptive surfaces of *Nectonema* sp. (Nematomorpha: Nectonematoidea) from *Pandalus montagui*: Histology, ultrastructure, and absorptive capabilities of the body wall and intestine. Can. J. Zool. 66: 289–295.

Swanson, A. R. 1982. Nematomorpha. *In* S. Parker (ed.), *Synopsis and Classification of Living Organisms*, Vol. 1. McGraw-Hill, New York, pp. 931–932.

Priapula

Alberti, G. and V. Storch. 1988. Internal fertilization in a meiobenthic priapulid worm: *Tubiluchus philippinensis* (Tubiluchidae, Priapulida). Protoplasma 143: 193–196.

Calloway, C. B. 1975. Morphology of the introvert and associated structures of the priapulid *Tubiluchus corallicola* from Bermuda. Mar. Biol. 31: 161–174.

Calloway, C. B. 1982. Priapulida. *In* S. Parker (ed.), *Synopsis and Classification of Living Organisms*, Vol. 1. McGraw-Hill, New York, pp. 941–944.

Calloway, C. B. 1988. Priapulida. *In* P. Higgins and H. Thiel (eds.), *Introduction to the Study of Meiofauna*, Smithsonian Institution Press, Washington, DC. pp. 322–327.

Dawydoff, C. 1959. Classes der Echiuriens et Priapuliens. *In* P. Grassé (ed.), *Traité de Zoologie*, Vol. 5, Pt. 1. Masson et Cie, Paris, pp. 855–926.

Hammon, R. A. 1970. The burrowing of *Priapulus caudatus*. J. Zool. 162: 469–480.

Higgins, R. P. and V. Storch. 1989. Ultrastructural observations of the larva of *Tubiluchus corallicola* (Priapulida). Helgoländer Meeresuntersuchungen 43(1): 1–11.

Higgins, R. P. and V. Storch. 1991. Evidence for direct development in *Meiopriapulus fijiensis*. Trans. Am. Microsc. Soc. 110: 37–46.

Joffe, B. I. and E. A. Kotikova. 1988. Nervous system of *Priapulus caudatus* and *Halicryptus spinulosus* (Priapulida). Proc. Zool. Inst. USSR Acad. Sci. 183: 52–77.

Lang, K. 1948. On the morphology of the larva of *Priapulus*. Arkiv. Zool. 41A, art. nos. 5 and 9.

Lang, K. 1953. Die Entwicklung des Eies von *Priapulus caudatus* Lam. und die systematische Stellung der Priapuliden. Arkiv. Zool. Ser. 2 5(5): 321–348.

Morris, S. C. 1977. Fossil priapulid worms. Palaeontol. Assoc. Lond. Spec. Pap. Palaeontol. 20: 1–95.

Morris, S. C. and R. A. Robison. 1985. Middle Cambrian priapulids and other soft-bodied fossils from Utah and Spain. Univ. Kansas Paleontol. Contr. 117: 1–22.

Oeschger, R. and K. B. Storey. 1990. Regulation of glycolytic enzymes in the marine invertebrate *Halicryptus spinulosus* (Priapulida) during environmental anoxia and exposure to hydrogen sulfide. Mar. Biol. 106(2): 261–266.

Oeschger, R. and R. D. Vetter. 1992. Sulfide detoxification and tolerance in *Halicryptus spinulosus* (Priapulida): A multiple strategy. Mar. Ecol. Prog. Ser. 86: 167–179.

Por, F. D. 1983. Class Seticoronaria and phylogeny of the phylum Priapulida. Zool. Scr. 12: 267–272.

Por, F. D. and H. J. Bromley. 1974. Morphology and anatomy of *Maccabeus tentaculatus* (Priapulida: Seticoronaria). J. Zool. 173: 173–197.

Schreiber, A. and V. Storch. 1992. Free cells and blood proteins of *Priapulus caudatus* Lamarck (Priapulida). Sarsia 76: 261–266.

Shapeero, W. 1961. Phylogeny of Priapulida. Science 133(3455): 879–880.

Shapeero, W. L. 1962. The epidermis and cuticle of *Priapulus caudatus* Lamarck. Trans. Am. Microsc. Soc. 81(4): 352–355.

Shirley, T. C. 1990. Ecology of *Priapulus caudatus* Lamarck, 1816 (Priapulida) in an Alaskan subarctic ecosystem. Bull Mar. Sci. 47(1): 149–158.

Storch, V. 1991. Priapulida. In F. W. Harrison and E. E. Ruppert (eds.), *Microscopic Anatomy of Invertebrates*, Vol. 4, Aschelminthes, Wiley-Liss, New York, pp. 333–350.

Storch, V., R. P. Higgins, and M. P. Morse. 1989. Internal anatomy of *Meiopriapulus fijienses* (Priapulida). Trans. Am. Microsc. Soc. 108: 245–261.

Storch, V., R. P. Higgins and H. Rumohr. 1990. Ultrastructure of the introvert and pharynx of *Halicryptus spinulosus* (Priapulida). J. Morphol. 206: 163–171.

Storch, V., R. P. Higgins, P. Anderson and J. Svavarsson. 1995. Scanning and transmission electron microscopic analysis of the introvert of *Priapulopsis australis* and *Priapulopsis bicaudatus* (Pripulida). Invert. Biol. 114: 64-72.

van der Land, J. 1970. Systematics, zoogeography, and ecology of the Priapulida. Zool. Verh. Rijksmus. Nat. Hist. Leiden 112: 1–118.

van der Land, J. and A. Nørrevang. 1985. Affinities and intraphyletic relationships of the Priapulida. In S. C. Morris et al. (eds.), *The Origins and Relationships of Lower Invertebrates*. Syst. Assoc. Spec. Vol. No. 28, Oxford, pp. 261–273.

Wolter, K. 1947. Submikroskopische Strukturen von Priapulida der Mediterranen Meiofauna. PhD dissertation, University of Wien.

Acanthocephala

Abele, L. G. and S. Gilchrist. 1977. Homosexual rape and sexual selection in acanthocephalan worms. Science 197: 81–83.

Amin, O. M. 1982. Acanthocephala. In S. Parker (ed.), *Synopsis and Classification of Living Organisms*. McGraw-Hill, New York, pp. 933–940.

Baer, J. C. 1961. Acanthocéphales. In P. Grassé (ed.), *Traité de Zoologie*, Vol. 4, Pt. 1. Masson et Cie, Paris, pp. 733–782.

Crompton, D. W. T. and B. B. Nickol. 1985. *Biology of the Acanthocephala*. Cambridge University Press, New York.

Dunagan, T. T. and D. M. Miller. 1991. Acanthocephala. In F. W. Harrison and E. E. Ruppert (eds.), *Microscopic Anatomy of Invertebrates*, Vol. 4, Aschelminthes, Wiley-Liss, New York, pp. 299–332.

Miller, D. M. and T. T. Dunagan. 1985. New aspects of acanthocephalan lacunar system as revealed in anatomical modeling by corrosion cast method. Proc. Helm. Soc. Wash. 52: 221–226.

Nicholas, W. L. 1973. The biology of Acanthocephala. Adv. Parasitol. 11: 671–706.

Whitfield, P. J. 1971. Phylogenetic affinities of Acanthocephala: An assessment of ultrastructural evidence. Parasitology 63: 49–58.

Yamaguti, S. 1963. *Systema Helminthum*, Vol. 5, Acanthocephala. Interscience, New York.

Entoprocta

Emschermann, P. 1969. Ein Kreislauforgan bei Kamtozoen. Zeit. der Zellforsch. 97: 576–607.

Emschermann, P. 1982. Les kamtozoaires. État actuel de nos connaissances sur leur anatomie, leur dévelopement, leur biologie et leur position phylogénétique. Bull. Soc. Zool. France 107(2): 317–344.

Franke, M. 1993. Ultrastructure of the protonephridia of *Loxosomella fauveli, Barentsia matsushimana* and *Pedicellina cernua*. Implications for the protonephridia in the ground pattern of the Entoprocta (Kamptozoa). Microfauna Marina 8: 7–38.

Mariscal, R. N. 1965. The adult and larval morphology and life history of the entoproct *Barentsia gracilis* (M. Sars, 1835). J. Morphol. 116(3): 331–338.

Nielsen, C. 1971. Entoproct life cycles and the entoproct/ectoproct relationship. Ophelia 9(2): 209–341.

Nielsen, C. 1977. The relationships of Entoprocta, Ectoprocta, and Phoronida. Am. Zool. 17: 149–150.

Nielsen, C. 1982. Entoprocta. In S. Parker (ed.), *Synopsis and Classification of Living Organisms*, Vol. 1. McGraw-Hill, New York, pp. 771–772.

Todd, J. A. and P. D. Taylor. 1992. The first fossil entoproct. Naturwiss. 79: 311–314.

Wasson, K. 1997a. Sexual modes in the colonial kamftozoan genus *Barentsia*. Biol. Bull. 193: 163–170.

Wasson, K. 1997b. Systematic revision of colonial kamptozoans (entoprocts) of the Pacific coast of North America. Zool. J. Linn. Soc. 121: 1–63.

Gnathostomulida

Ax, P. 1956. Die Gnathostomulida, eine rätselhafte Wurmgruppe aus dem Meeressand. Abh. Akad. Wiss. Lit. Mainz Math. Naturwiss. Kl. 8: 1–32.

Ax, P. 1965. Zur Morphologie und Systematik der Gnathostomulida. Untersuchungen an *Gnathostomula paradoxa* Ax. Z. Zool. Syst. Evolutionsforsch. 3: 259–296.

Ax, P. 1985. The position of the Gnathostomulida and Platyhelminthes in the phylogenetic system of the Bilateria. In S. C. Morris et al. (eds.), *The Origins and Relationships of Lower Invertebrates*. Syst. Assoc. Spec. Vol. No. 28, Oxford, pp. 168–180.

Durden, C., J. Rodgers, E. Yochelson and R. Riedl. 1969. Gnathostomulida: Is there a fossil record? Science 164: 855–856.

Herlyn, H. and U. Ehlers. 1997. Ultrastructure and function of the pharynx of *Gnathostomula paradoxa* (Gnathostomulida). Zoomorph. 117: 135–145.

Kirsteuer, E. 1969. Gnathostomulida: A new component of the benthic invertebrate fauna of lagoons. Mem. Symp. Int. Lagunas Costeras UNAM/UNESCO, pp. 537–544.

Knauss, E. B. 1979. Indication of an anal pore in Gnathostomulida. Zool. Scr. 8: 181–186.

Knauss, E. B. 1979. Fine structure of the male reproductive system in two species of *Haplognathia* Sterrer (Gnathostomulida, Filospermoidea). Zoomorphologie 94: 33–48.

Kristensen, R. M. and A. Nørrevang. 1977. On the fine structure of *Rastrognathia macrostoma* gen. et sp. n. placed in Rastrognathiidae fam. n. (Gnathostomulida). Zool. Scr. 6: 27–41.

Kristensen, R. M. and A. Nørrevang. 1978. On the fine structure of *Valvognathia pogonostoma* gen. et sp. n. (Gnathostomulida, Onychognathiidae) with special reference to the jaw apparatus. Zool. Scr. 7: 179–186.

Lammert, V. 1984. The fine structure of the spiral ciliary receptors in Gnathostomulida. Zoomorphol. 104: 360–364.

Lammert, V. 1989. Fine structure of the epidermis in the Gnathostomulida. Zoomorphol. 107: 14–28.

Lammert, V. 1991. Gnathostomulida. *In* F. W. Harrison and E. E. Ruppert (eds.), *Microscopic Anatomy of Invertebrates*, Vol. 4, Aschelminthes, Wiley-Liss, New York, pp. 20–39.

Mainitz, M. 1979. The fine structure of gnathostomulid reproductive organs I. New characters in the male copulatory organ of *Scleroperalia*. Zoomorphologie 92: 241–272.

Riedl, R. J. 1969. Gnathostomulida from America. Science 163: 445–452.

Riedl, R. J. 1971. On the genus *Gnathostomula* (Gnathostomulida). Int. Rev. Ges. Hydrobiol. 56: 385–496.

Riedl, R. J. and R. Rieger. 1972. New characters observed on isolated jaws and basal plates of the family Gnathostomulidae (Gnathostomulida). Zool. Morphol. Tiere 72: 131–172.

Rieger, R. M. and M. Mainitz. 1977. Comparative fine structure of the body wall in Gnathostomulida and their phylogenetic position between Platyhelminthes and Aschelminthes. Z. Zool. Syst. Evolutionsforsch. 15: 9–35.

Sterrer, W. 1971a. *Agnathiella beckeri* nov. gen. nov. spec. from southern Florida. The first gnathostomulid without jaws. Hydrobiologia 56: 215–225.

Sterrer, W. 1971b. On the biology of Gnathostomulida. Vie Milieu Suppl. 22: 493–508.

Sterrer, W. 1972. Systematics and evolution within the Gnathostomulida. Syst. Zool. 21: 151–173.

Sterrer, W. 1982. Gnathostomulida. *In* S. Parker (ed.), *Synopsis and Classification of Living Organisms*. McGraw-Hill, New York, pp. 847–851.

Sterrer, W., M. Mainitz and R. M. Rieger. 1985. Gnathostomulida: Enigmatic as ever. *In* S. C. Morris et al. (eds.), *The Origins and Relationships of Lower Invertebrates*. Syst. Assoc. Spec. Vol. No. 28, Oxford, pp. 181–199.

Sterrer, W. 1997. Gnathostomulida from the Canary Islands. Proc. Biol. Soc. Wash. 110: 186–197.

Loricifera

Franzen, Å. and R. M. Kristensen. 1984. Loricifera, en nyupptäckt stam inom djurriket. Fauna Flora 79: 56–60.

Higgins, R. P. and R. M. Kristensen. 1986. New Loricifera from southeastern United States coastal waters. Smithson. Contrib. Zool. 438: 1–70.

Higgins, R. P. and R. M. Kristensen. 1988. Loricifera. *In* R. P. Higgins and H. Theil (eds.), *Introduction to the Study of Meiofauna*. Smiths. Inst. Press, Washington, DC. pp. 319–321.

Kristensen, R. M. 1983. Loricifera, a new phylum with Aschelminthes characters from the meiobenthos. Z. Zool. Syst. Evolutionsforsch. 21: 163–180.

Kristensen, R. M. 1991a. Loricifera. *In* F. W. Harrison and E. E. Ruppert (eds.), *Microscopic Anatomy of Invertebrates*, Vol. 4, Aschelminthes, Wiley-Liss, New York, pp. 351–375.

Kristensen, R. M. 1991b. Loricifera: A general biological and phylogenetic overview. Vehr. Dtsch. Zool. Ges. 84: 231–246.

Kristensen, R. M. and Y. Shirayama. 1988. *Pliciloricus hadalis* (Pliloricidae) a new loriciferan collected from the Izu-Ogasawara Trench, Western Pacific. Zool. Sci. (Japan) 5: 875–881.

Lewin, R. 1983. New phylum discovered, named. Science 222: 149.

Cycliophora

Funch, P. 1996. The chordoid larva of *Symbion pandora* (Cycliophora) is a modified trochophore. J. Morphol. 230: 231–263.

Funch, P. and R. M. Kristensen. 1995. Cycliophora is a new phylum with affinities to Entoprocta and Ectoprocta. Nature 378: 661–662.

Funch, P. and R. M. Kristensen. 1997. Cycliophora. *In* F. W. Harrison and E. E. Ruppert (eds.), *Microscopic Anatomy of Invertebrates*, Vol. 13, Lophophorates, Entoprocta, and Cyclifophora, Wiley-Liss, New York, pp. 409–474.

Morris, S. C. 1995. A new phylum from the lobster's lips. Nature 378: 661–662.

13

Phylum Annelida:
The Segmented Worms

The study of polychaetes used to be a leisurely occupation, practiced calmly and slowly.

Kristian Fauchald,
The Polychaete Worms, 1977

This chapter treats the segmented worms, or phylum Annelida (Greek, *annulatus*, "annulated," or "ringed"), which comprise about 16,500 species. Annelids include such familiar animals as earthworms and leeches as well as various marine "sand worms," "tube worms," and an array of other forms. Some are tiny animals of the meiofauna; others, such as certain southern hemisphere earthworms and some marine species, exceed 3 m in length. The Annelida are archetypical protostomes, and they are often used as a model for understanding this great lineage of the Metazoa.

The annelids have successfully invaded virtually all habitats where sufficient water is available. They are particularly abundant in the sea but also abound in fresh water, and many live in damp terrestrial environments. There are also parasitic, mutualistic, and commensal species. Their success is no doubt due in part to the evolutionary plasticity of the segmented bauplan and to their exploitation of a variety of life history strategies. The basic annelid condition is characterized by a segmented body in which most internal and external parts are repeated with each segment, a situation referred to as **serial homology**. Serial homology refers to body structures with the same genetic and developmental origins that arise repeatedly during the ontogeny of an organism. In annelids (as in arthropods) this repetition of homologous body structures results in **metamerism**, or body segmentation that arises by way of teloblastic development (the proliferation of paired, segmental mesodermal bands from teloblast cells at a posterior growth zone in the embryo; see Chapter 15 for more details). These triploblastic coelomate worms possess a complete gut, a closed circulatory system, a well developed nervous system, and excretory structures in the form of protonephridia or, more commonly, metanephridia (Box 13A). Many marine forms produce a char-

(A)

(C)

Figure 13.1 Representative annelids. (A) *Nereis virens*, an errant polychaete. (B) A large oligochaete from southeastern Australia (the giant Gippsland earthworm, *Megascolides australis*; reported to reach lengths of 3.3 to 3.6 m). (C) A pond leech (subclass Hirudinoidea).

(B)

acteristic **trochophore larva**, a feature shared with several other protostome taxa (e.g., sipunculans, echiurans, many molluscs). The story of annelid diversity and success is one of variation on this basic theme.

Taxonomic History and Classification

As mentioned in earlier chapters, the roots of modern animal classification can be traced to Linnaeus (1758), who placed all invertebrates except insects in the taxon Vermes. In 1809, Lamarck established the taxon Annelida; he had a reasonably good idea of their unity and of their differences from other groups of worms. He and many other workers recognized especially the affinity between polychaetes and oligochaetes, but the hirudinoideans (leeches) were often allied with the trematode platyhelminths. Cuvier (1816) united the annelids and the arthropods under the taxon Articulata, a scheme that remains popular even today. The relationship of the leeches to the other annelids was not solidly established until 1851, by Vogt.

Annelida has traditionally been divided into three or four classes. The class Polychaeta is the largest and most diverse group, with over 10,000 described species. Members of the Myzostomida should probably be considered as aberrant symbiotic polychaetes, although some authorities now view them as meriting separate class rank. Recent phylogenetic research indicates that the Oligochaeta and Hirudinoidea should be united as the class Clitellata, a concept that we herein adopt.

Also included in this chapter is a group of strange worms traditionally known as the beard worms (pogonophorans and vestimentiferans). These worms have long been treated as a separate phylum (or two phyla), but recent work suggests that they are highly modified polychaetes. Work continues on their exact placement within the Polychaeta; we treat them in a separate section preceding our discussion of annelid phylogeny.

PHYLUM ANNELIDA

CLASS POLYCHAETA: Sand worms, tube worms, clam worms, and others (Figures 13.2 and 13.3). With numerous chaetae (often called setae) on the trunk segments; most with well developed parapodia; prostomium and peristomium often bear sensory organs (palps, tentacles, cirri) or extensive feeding and gas exchange tentacular structures; foregut often modified as eversible stomodeal pharynx (proboscis), sometimes armed with chitinous jaws; reproductive structures simple, often transient; without a clitellum; most are dioecious; development often indirect, with a free-swimming trochophore larva; mostly marine; errant, burrowing, tube-dwelling, interstitial, or planktonic; some live in brackish water, a few inhabit fresh water or are parasitic. The class is divided into 25 orders and 87 families,* a few of which are listed below to illustrate the diversity within

---

*Older classification schemes recognized a division into two subclasses, the motile Errantia and the nonmotile Sedentaria. It is now agreed that the superficial similarities used to unite the members of these "subclasses" are the results of convergence; hence the terms no longer have taxonomic validity. A group of minute, interstitial polychaetes, loosely called the "archiannelids," was once considered a separate taxon, but its members have now been reassigned to several orders (e.g., Nerillida, Dinophilida, Polygordiida, Protodrilida); these odd worms comprise a polyphyletic assemblage of paedomorphic polychaetes that retain various larval features, including retention of external ciliary bands in some species. (e.g., *Dinophilus, Nerilla, Polygordius, Protodrilus, Saccocirrus*).

BOX 13A *Characteristics of the Phylum Annelida*

1. Schizocoelous, bilaterally symmetrical, segmented worms

2. Development typically protostomous; segments arise by teloblastic growth

3. Digestive tract complete, usually with regional specialization

4. With a closed circulatory system; respiratory pigments include hemoglobin, chlorocruorin and hemerythrin

5. Nervous system well developed, with a dorsal cerebral ganglion, circumenteric connectives, and ventral ganglionated nerve cord(s)

6. Most possess metanephridia or, less commonly, protonephridia

7. With lateral, segmentally arranged epidermal chaetae

8. Head composed of presegmental prostomium and peristomium

9. Dioecious or hermaphroditic; many have a characteristic trochophore larva (secondarily lost in some groups)

10. Marine, terrestrial, and freshwater species

the class. The annotations are not diagnostic, but merely descriptive synopses. See Fauchald 1977 and Pettibone 1982 for complete listings and diagnoses of all orders and families.

ORDER CAPITELLIDA

FAMILY ARENICOLIDAE: The so-called lugworms have a rather thick, fleshy, heteronomous body divided into two or three distinguishable regions; the pharynx is unarmed but eversible and aids burrowing and feeding. Arenicolids live in J-shaped burrows in intertidal and subtidal sands and muds, where they are direct deposit feeders. (e.g., *Abarenicola, Arenicola*)

FAMILY CAPITELLIDAE: Body long and thin, weakly heteronomous; with unornamented prostomium. Extremely common and abundant burrowing deposit feeders. Most are marine, but a few occur in fresh water. (e.g., *Capitella, Notomastus*)

FAMILY MALDANIDAE: Body long and homonomous except that some mid-trunk segments are elongate, hence their common name, "bamboo worms"; burrow head down and secrete a mucous sheath to which sand particles adhere, thereby forming a tube; proboscis unarmed but eversible, and used in burrowing and selective deposit feeding. (e.g., *Clymenella, Maldane, Praxillella*)

ORDER CHAETOPTERIDA

FAMILY CHAETOPTERIDAE: Body fleshy, relatively large, and distinctly heteronomous, divided into two or three functional regions with highly modified parapodia. Chaetopterids live in more or less permanent U-shaped burrows lined with secretions from the worm; most are mucous-net filter feeders, eating plankton and detritus passed through the tube by water currents. (e.g., *Chaetopterus, Mesochaetopterus, Phylochaetopterus*)

ORDER CIRRATULIDA

FAMILY CIRRATULIDAE: Elongate, relatively homonomous, with up to 350 segments, each with a pair of threadlike branchial filaments; pharynx unarmed and noneversible. Cirratulids are mostly shallow-water burrowers lying just beneath the surface of the sediment, from where they extend their branchiae into the overlying water; most are selective deposit feeders, extracting organic detritus from the surface sediments. (e.g., *Cirratulus, Cirriformia, Dodecaceria*)

ORDER EUNICIDA

FAMILY ARABELLIDAE: Thin, elongate, with reduced parapodia; without head appendages; pharynx with complex jaw apparatus; most burrow in soft substrata aided by secretion of copious amounts of mucus. Predatory carnivores; some are endoparasitic in various worms, including other polychaetes. (e.g., *Arabella, Drilonereis*)

FAMILY EUNICIDAE: Elongate, homonomous, generally large polychaetes, some exceeding 3 m in length; pharynx with complex set of jaw plates. Some are sedentary in mucous or parchment-like tubes; many are gregarious in cracks and crevices in hard substrata; some leave their tube areas to feed; most are predatory carnivores, although many omnivorous species are also known. (e.g., *Eunice, Marphysa, Palola*)

FAMILY ICHTHYOTOMIDAE: Highly modified fish parasites, with scissors-like jaws for attaching to host. (e.g., *Ichthyotomus*)

FAMILY LUMBRINERIDAE: Thin, elongate polychaetes without head appendages and with reduced parapodia; pharynx with complex jaw apparatus of several elements. Most crawl about in algal mats, holdfasts, and small cracks in hard substrata; some burrow in sand or mud; carnivores, scavengers, detritivores, and deposit feeders. (e.g., *Lumbrinerides, Ninoe*)

FAMILY ONUPHIDAE: Body homonomous, but most live in tubes. Some are sessile, others carry their tubes with them; most are scavengers or hunters. (e.g., *Diopatra, Onuphis, Paradiopatra*)

ORDER MYZOSTOMIDA: Includes several groups of flattened, oval, aberrant polychaetes. The body is drastically modified for symbiotic life; with suckers and hooks; ecto- and endosymbionts of echinoderms, mainly crinoids. The monophyly of this group is uncertain. (e.g., *Asteriomyzostomum, Cystimyzostomum, Myzostoma*) (Figure 13.3)

ORDER OPHELIIDA

FAMILY OPHELIIDAE: Homonomous polychaetes with up to 60 segments; general body shape varies from rather short and thick to elongate and somewhat tapered. Most opheliids burrow in soft substrata, but many swim by undulatory body movements; pharynx unarmed; most are direct deposit feeders. (e.g., *Armandia, Euzonus, Ophelia, Polyophthalmus*)

(A)

(B)

(C)

(D)

(E)

(H)

(F)

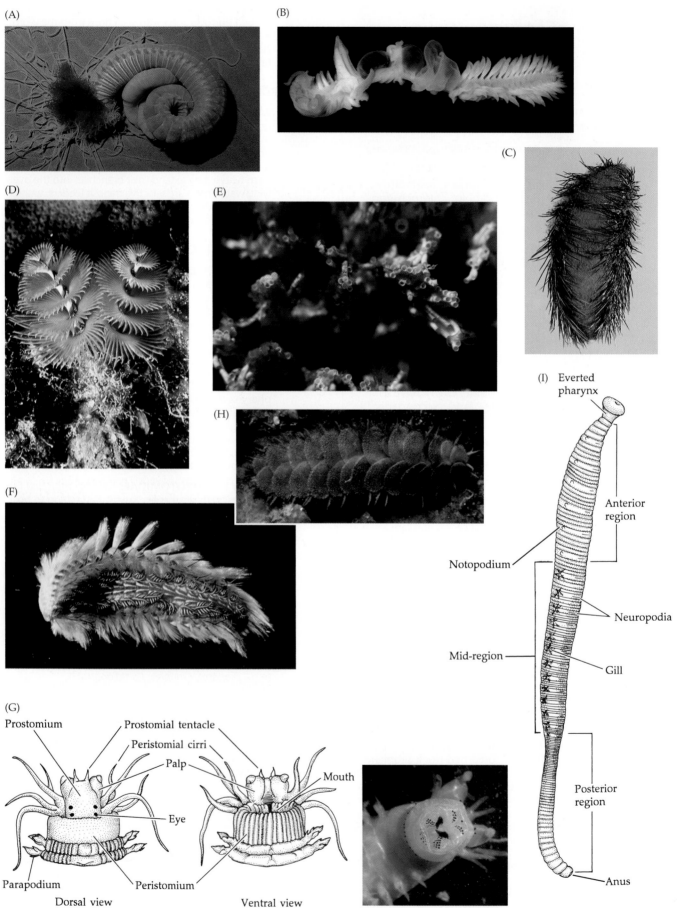

(G)

Prostomium — Prostomial tentacle

Peristomial cirri

Palp

Mouth

Eye

Parapodium — Peristomium

Dorsal view   Ventral view

(I) Everted pharynx

Anterior region

Notopodium

Neuropodia

Mid-region

Gill

Posterior region

Anus

Figure 13.2 Some polychaete worms. (A) *Thelepus*, a deposit-feeding polychaete, removed from its burrow (family Terebellidae). (B) *Chaetopterus*, a filter-feeding polychaete, removed from its burrow (family Chaetopteridae). (C) *Aphrodita*, the "sea mouse" (family Aphroditidae). (D) *Spirobranchus grandis* (family Serpulidae). Note the tentacular feeding and gas exchange crowns. (E) A colony of *Filograna implexa* (family Serpulidae). (F) *Notopygos ornata*, a colorful Pacific polychaete (family Amphinomidae). (G) The head of a nereid polychaete (dorsal and ventral views); photograph shows jaws. (H) *Halosydna*, a scale worm (family Polynoidae). (I) *Arenicola* (family Arenicolidae), a burrower in soft sediments. (J,K) The rare pelagic polychaete *Tomopteris* (family Tomopteridae). (L) *Protodrilus*, an interstitial archiannelid.

ORDER SPIONIDA

FAMILY SCALIBREGMIDAE: Body length and thickness vary, although the anterior end is often enlarged; somewhat heteronomous. Burrowers in soft sediments. (e.g., *Polyphysia*)

FAMILY SPIONIDAE: Body thin, elongate, homonomous; peristomial palps long and coiled; pharynx unarmed. Most burrow, or form delicate sand or mud tubes; a few bore into calcareous substrata, including rocks and mollusc shells; most use the grooved peristomial palps to selectively extract food from the sediment surface. (e.g., *Polydora, Scolelepis, Spio, Spiophanes*)

(J)

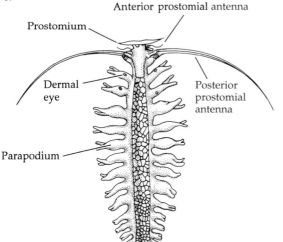

Anterior prostomial antenna

Prostomium

Dermal eye

Posterior prostomial antenna

Parapodium

(K)

(L)

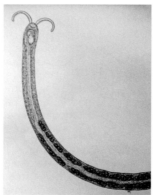

ORDER ORBINIIDA

FAMILY ORBINIIDAE: Body relatively homonomous; prostomium without appendages. Burrowers in marine sediments. (e.g., *Orbinia, Scoloplos*)

ORDER OWENIIDA

FAMILY OWENIIDAE: Small, tube-dwelling polychaetes with prostomium often lobed or folded. (e.g., *Owenia*)

ORDER PHYLLODOCIDA

FAMILY ALCIOPIDAE: Body homonomous, but form varies from short and broad to long and slender; body transparent except for pigment spot in some genera; with pair of huge complex eyes on prostomium; planktonic predators. (e.g., *Alciopa, Alciopina, Torrea, Vanadis*)

FAMILY APHRODITIDAE: Body broad, oval or oblong, with less than 60 segments; with flattened, solelike ventral surface, and rounded dorsum covered with scales (elytra) overlaid by a thick felt- or hairlike layer, giving some the common name of "sea mouse"; slow-moving; epibenthic or burrowers; most are omnivorous. (e.g., *Aphrodita, Pontogeneia*)

FAMILY GLYCERIDAE: Long, cylindrical, tapered, homonomous body; enormous pharynx armed with four hooked jaws used in prey capture; large pharyngeal proboscis also used in burrowing; most are infaunal burrowers in soft substrata. (e.g., *Glycera, Glycerella, Hemipodus*)

FAMILY NEPHTYIDAE: Often large, or long and slender, with well developed parapodia; burrowing in marine sands and muds; reversible, jawed pharynx used in prey capture and burrowing. (e.g., *Aglaophamus, Micronephtyes, Nephtys*)

FAMILY NEREIDAE: Moderate to large polychaetes tending to homonomy; mostly errant predators with well developed parapodia; one pair of large, curved pharyngeal jaws. Epibenthic in protected habitats: found among mussel communities, in holdfasts of algae, in crevices, under rocks, etc. (e.g., *Cheilonereis, Dendronereis, Neanthes, Nereis, Platynereis*)

FAMILY PHYLLODOCIDAE: With thin, elongate bodies of up to 700 homonomous segments; most common as active epibenthic predators on solid substrata; a few burrow in mud. (e.g., *Eteone, Eulalia, Notophyllum, Phyllodoce*)

FAMILY POLYNOIDAE: Most are relatively short and somewhat flattened dorsoventrally; one Antarctic species, *Eulagisca gigantea*, reaches a length of nearly 20 cm and a width of about 10 cm; polynoids tend to have relatively few segments of a more or less fixed number; the dorsum is covered by scales (elytra), hence the common name "scale worms"; pharynx with one pair of jaws; well developed parapodia. Errant but usually cryptic (under stones, etc.) predators; numerous species are commensal on the bodies or in the dwellings of other animals. (e.g., *Arctonoe, Gorgoniapolynoe, Halosydna, Harmothoe, Hesperonoe, Polynoa*)

FAMILY SYLLIDAE: Mostly small, homonomous worms found on various substrata; active predators

(A)

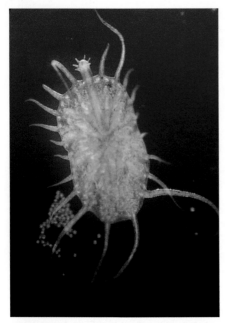

(B)

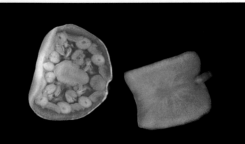

Figure 13.3 Myzostomids. (A) A myzostomid from a crinoid. Note the egg-filled body and the everted proboscis. (B) Dorsal and ventral views of unidentified myzostomid from a tropcial Pacific crinoid. (C) *Myzostoma*. The parapodia are reduced to tiny suckers, often armed with chaetal hooks for attachment to a host. Their reproductive systems are much more complex than are those of most other polychaetes, a feature common to many parasitic animals.

(C)

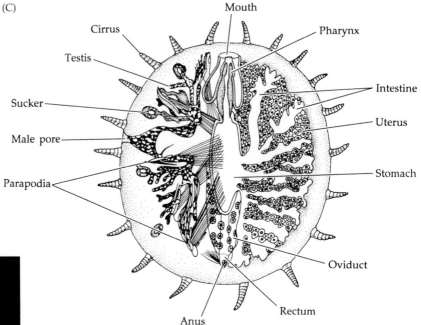

on small invertebrates (some eat diatoms); pharynx armed with a single tooth or a ring of small teeth for grasping prey; a few are interstitial. (e.g., *Autolytus, Brania, Odontosyllis, Syllis, Trypanosyllis*)

FAMILY TOMOPTERIDAE: Body flattened, with finlike parapodia; transparent; planktonic, swimming predators. Monogeneric: *Tomopteris*.

ORDER SABELLIDA

FAMILY SABELLIDAE: Tube-dwelling polychaetes commonly called "fan worms" or "feather-duster worms"; body heteronomous, divided into two regions similar to those of terebellids; pharynx unarmed and noneversible; peristomium bears a crown of branched, feathery tentacles ("cirri") that projects from the tube and functions in gas exchange and ciliary suspension feeding. (e.g., *Bispira, Eudistylia, Fabricia, Myxicola, Sabella, Schizobranchia*)

FAMILY SERPULIDAE: Heteronomous body divided into two regions; calcareous tube dwellers; anterior end bears a tentacular crown as in sabellids, plus a funnel-shaped operculum that can be pulled into the end of the tube when the worm withdraws; ciliary suspension feeders. (e.g., *Filograna, Hydroides, Serpula, Spirobranchus*)

FAMILY SPIRORBIDAE: Small, heteronomous polychaetes living in coiled calcareous tubes attached to hard substrata; anterior end with tentacular crown and operculum similar to those of serpulids. (e.g., *Circeis, Paralaeospira, Spirorbis*)

ORDER TEREBELLIDA

FAMILY PECTINARIIDAE: Body short and conical, with only about 20 segments; live in conical sandy tubes open at both ends (the "ice-cream-cone worms"); feed on detritus extracted from sediment. (e.g., *Amphictene, Pectinaria, Petta*)

FAMILY SABELLARIIDAE: Heteronomous tube dwellers; anterior chaetae modified as operculum; tubes of some may form extensive shelves or "reefs." (e.g., *Phragmatopoma, Sabellaria*)

FAMILY TEREBELLIDAE: Moderate-sized, tube-dwelling polychaetes with fragile, fleshy heteronomous body of two distinct regions; most lack an eversible pharynx and live in various types of permanent tubes (e.g., mud, sand, shell fragments); head bears numerous elongate feeding tentacles; most with 1–3 pairs of well developed branchiae on anterior trunk segments; feed on surface detritus. (e.g., *Amphitrite, Pista, Polycirrus, Terebella, Thelepus*)

CLASS CLITELLATA: Earthworms, leeches, and related forms. No parapodia; chaetae ususally greatly reduced or absent; hermaphroditic, often with complex reproductive systems; with clitellum that functions in cocoon formation; direct

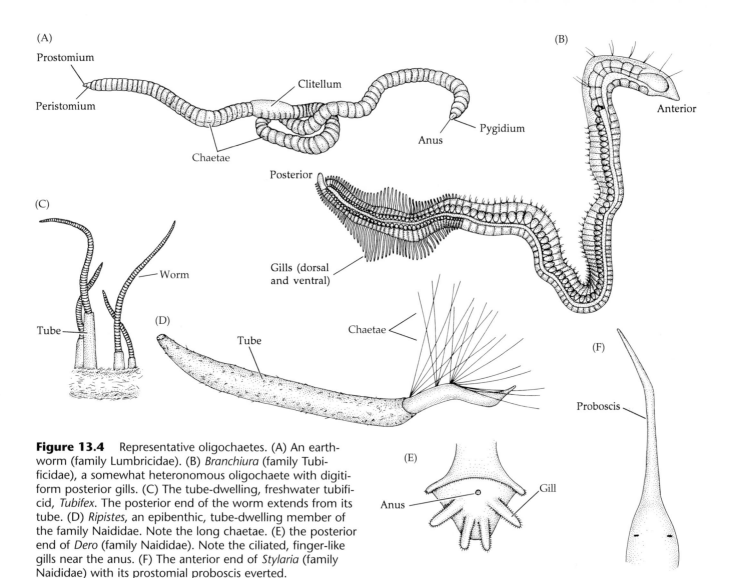

Figure 13.4 Representative oligochaetes. (A) An earthworm (family Lumbricidae). (B) *Branchiura* (family Tubificidae), a somewhat heteronomous oligochaete with digitiform posterior gills. (C) The tube-dwelling, freshwater tubificid, *Tubifex*. The posterior end of the worm extends from its tube. (D) *Ripistes*, an epibenthic, tube-dwelling member of the family Naididae. Note the long chaetae. (E) the posterior end of *Dero* (family Naididae). Note the ciliated, finger-like gills near the anus. (F) The anterior end of *Stylaria* (family Naididae) with its prostomial proboscis everted.

development; inhabit freshwater, marine, and damp terrestrial environments.

SUBCLASS OLIGOCHAETA: Over 6,000 species (in about 25 families) of earthworms, freshwater annelids, and a few marine species (Figure 13.4). With few chaetae; cephalic sensory structures reduced; body externally homonomous except for clitellum. This subclass comprises three orders based in part on details of the male reproductive system—the first two contain a single family each. There is some controversy about the taxonomic arrangement of oligochaetes.

ORDER LUMBRICULIDA (FAMILY LUMBRICULIDAE): Moderate-size, freshwater oligochaetes, many of which are known only from Lake Baikal in the Soviet Union. (e.g., *Lamprodilus*, *Rhynchelmis*, *Stylodrilus*, *Styloscolex*, *Trichodrilus*)

ORDER MONILIGASTRIDA (FAMILY MONILIGASTRIDAE): Presumed primitive terrestrial oligochaetes (some workers consider these as a suborder of the order Haplotaxida; see Jamieson 1988); most known from damp soil in Asia; a few are quite large, exceeding 1 m in length. (e.g., *Desmogaster*, *Moniligaster*)

ORDER HAPLOTAXIDA: Over 25 families; includes the vast majority of oligochaete species; occur in nearly all habitats; diverse body forms.

FAMILY ALMIDAE: Includes about 40 species of freshwater and mud-dwelling species. With a dorsal groove that functions in gas exchange. Most are tropical (e.g., *Alma*, *Callidrilus*)

FAMILY MEGASCOLECIDAE: Represented nearly worldwide and includes some of the largest earthworms. (e.g., *Megascolides*, *Megascolex*, *Pheretima*)

FAMILY TUBIFICIDAE: So-called sludge worms or tubifex worms; up to 2 cm long; freshwater and marine; some very common in areas of high pollution. Some small, gutless species are known from tropical regions (e.g., *Inanidrilus*, *Olavius*). Other genera include *Branchiura*, *Clitellio*, *Limnodrilus*, and *Tubifex*.

FAMILY NAIDIDAE: Many freshwater species; some live in marine or brackish water; some parasitic forms; some build tubes; several species bear an elongate prostomial proboscis; a few possess gills; almost all reproduce asexually, but most possess gonads at some stage of development; these fully aquatic oligochaetes occur worldwide. (e.g., *Branchiodrilus*, *Dero*, *Ripistes*, *Slavina*, *Stylaria*)

FAMILY LUMBRICIDAE: Includes the common terrestrial earthworms; often relatively large; with well developed and complex reproductive systems; most are direct deposit feeders. (e.g., *Allolobophora*, *Diporodrilus*, *Eisenia*, *Lumbricus*)

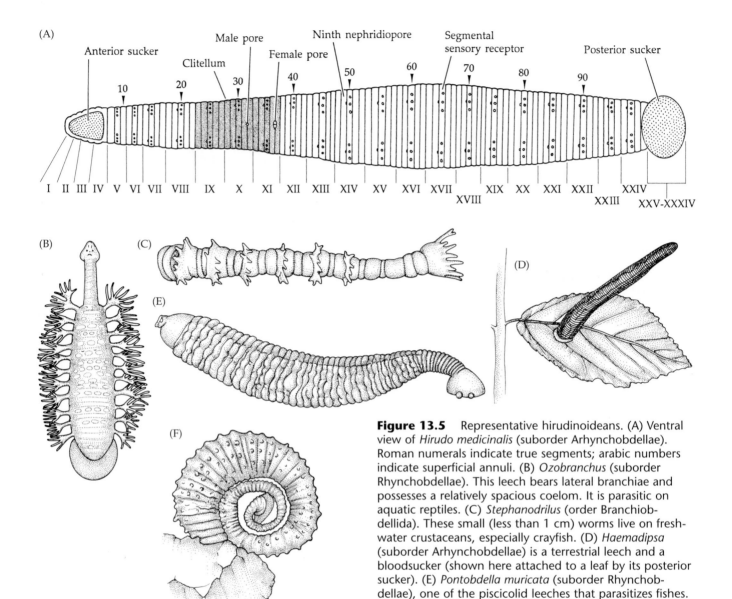

Figure 13.5 Representative hirudinoideans. (A) Ventral view of *Hirudo medicinalis* (suborder Arhynchobdellae). Roman numerals indicate true segments; arabic numbers indicate superficial annuli. (B) *Ozobranchus* (suborder Rhynchobdellae). This leech bears lateral branchiae and possesses a relatively spacious coelom. It is parasitic on aquatic reptiles. (C) *Stephanodrilus* (order Branchiobdellida). These small (less than 1 cm) worms live on freshwater crustaceans, especially crayfish. (D) *Haemadipsa* (suborder Arhynchobdellae) is a terrestrial leech and a bloodsucker (shown here attached to a leaf by its posterior sucker). (E) *Pontobdella muricata* (suborder Rhynchobdellae), one of the piscicolid leeches that parasitizes fishes. (F) *Acanthobdella peledina* (order Acanthobdellida), a part-time parasite of freshwater fishes.

SUBCLASS HIRUDINOIDEA: Leeches (Figure 13.5). Body with fixed number of segments, each with superficial annuli; chaetae few or absent; heteronomous, with clitellum and a posterior and usually an anterior sucker; most live in freshwater or marine habitats, a few are semiterrestrial; ectoparasitic, predaceous, or scavenging. The subclass has traditionally been recognized as comprising three orders (described briefly below), but is currently being reevaluated.

ORDER ACANTHOBDELLIDA: To 3 cm long; found in cold, freshwater lakes; part of the animal's life is spent as an ectoparasite on freshwater fishes (notably trout, char, and grayling), and presumably the rest of the time is spent in vegetation; body with 30 segments; with posterior sucker only; chaetae on anterior segments; coelom partially reduced, but obvious and with intersegmental septa; often considered to represent something of a pre-leech condition. With a single family and species (Acanthobdellidae, *Acanthobdella peledina*).

ORDER BRANCHIOBDELLIDA: Usually less than 1 cm long; ectocommensal or ectoparasitic on freshwater crayfishes; body with 15 segments; with anterior and posterior suckers; chaetae absent; coelom partially reduced, but spacious throughout most of the body; recent work suggests that these worms are probably more closely related to oligochaetes than to leeches, and a revision of their status is underway. With a single family (Branchiobdellidae). (e.g., *Branchiobdella, Cambarincola, Stephanodrilus*)

ORDER HIRUDINIDA: The "true" leeches; most are marine or freshwater forms, a few are semiterrestrial or amphibious; many are ectoparasitic bloodsuckers, others are free-living predators or scavengers; some parasitic forms serve as vectors for pathogenic protozoa, nematodes, and cestodes; body of 34 segments; with anterior and posterior suckers; no chaetae; coelom reduced to a complex series of channels (lacunae); with about 12 families; the two principal suborders are the Rhynchobdellae and the Arhynchobdellae. (e.g., *Erpobdella, Glossiphonia, Haemadipsa, Hirudo, Oxobranchus, Piscicola, Placobdella, Pontobdella*)

The Annelid Bauplan

The annelid body plan is the classic example of the metameric, triploblastic, coelomate bilateria, and provides a good model for comparison with other protostomes. The elongate body is usually cylindrical, but it has become markedly flattened in some groups, notably the leeches. The head is composed of a **prostomium** and a **peristomium**, the latter bearing the mouth. These two regions are primitively presegmental (i.e., already present in the larval stage, anterior of the teloblastic growth zone) and without chaetae, but some groups have secondarily incorporated true segments into the head (in which cases lateral chaetae may be present). A terminal, postsegmental part—the **pygidium**—bears the anus (Figures 13.2 and 13.4). The gut is separated from the body wall by the coelom, except in those species in which the body cavity has been secondarily obliterated.

The trunk segmentation is visible externally as rings, or **annuli**, and is reflected internally by the serial arrangement of coelomic compartments separated from one another by intersegmental septa (Figure 13.7). This basic arrangement has been modified to various degrees among the annelids, particularly by reduction in the size of the coelom or by loss of septa; the latter modification leads to fewer but larger internal compartments. The bodies of some annelids are **homonomous**, bearing segments that are very much alike. Many others, however, have groups of segments specialized for different functions and are thus **heteronomous** (Figures 13.2 and 13.4). Heteronomy is facilitated by the presence of serially repeated segments and has contributed greatly to the morphological diversification among annelids. The hydraulic properties of different coelomic arrangements has allowed corresponding modifications in patterns of locomotion, which are responsible in part for the success of the annelids in a variety of habitats.

Segmentation is reflected internally by the metameric arrangement of organs and system components (i.e., serial homology). In the primitive condition, each segment contains a portion of the circulatory, nervous, and excretory systems, in addition to the coelomic compartments. This arrangement in the annelids reflects the principle of system compatibility, which we have stressed as being critical to the success of any bauplan. That is to say, given the relative isolation of the body segments from one another by the septa, each segment must contain system components to adequately serve its structural and physiological needs. Thus, the origin of the segmented coelomic condition must have involved the coevolution of this serially homologous arrangement of other parts as well. In general, we can view much of annelid success in terms of their evolutionary escape from the limitations of the nonsegmented coelomate bauplan.

The combination of characteristics listed in Box 13A—especially the coelom, segmentation, closed circulatory system, regionally specialized gut, and nature of the excretory structures—act in concert to alleviate many of the constraints imposed by some other body plans. For example, the annelids are not bound to the small size or flattened shapes dictated by the necessity for small diffusion distances. Actions of the body wall musculature do not interfere directly with the internal organs as they do in "solid" body constructions, and active lifestyles are served by the metameric body plan and efficient physiological systems.

Body Forms

Polychaeta. The polychaetes presumably include the most primitive members of the Annelida. Nearly all of these animals are marine, living in habitats ranging from the intertidal zone to extreme depths. But quite a few inhabit brackish or fresh water, at least two species live on land, and there are a number of symbiotic forms. They range in length from less than 1 mm for some interstitial species to over 3 m for some giant errant species. Associated with their success is a great diversity of structural types, all evolved from a basic metameric, homonomous coelomate bauplan.

The body form of polychaetes typically reflects their habits and habitat. Active (errant) hunting predators (e.g., Phyllodocidae, Glyceridae, Syllidae, Nereidae) and some burrowing deposit feeders (e.g., Lumbrineridae) are characterized by a more or less homonomous body construction. Less active (sedentary) polychaetes include various suspension-feeding tube dwellers (e.g., Sabellidae, Serpulidae, Spirorbidae), those that inhabit permanent burrows (e.g., Chaetopteridae), and certain groups of direct and indirect deposit feeders and detritus feeders (e.g., Arenicolidae, Terebellidae). These sedentary polychaetes usually show some degree of heteronomy, with different body regions specialized for particular functions.

The myriad variations in body form among polychaetes can best be described relative to the basic annelid plan of head, segmented trunk, and pygidium. The head comprises the prostomium and peristomium. It often bears appendages in the form of prostomial **palps** and **antennae** (tentacles) and fleshy peristomial **cirri**, or it may be naked, like that of some infaunal burrowers. The nature of these head appendages varies greatly and often reveals clues as to the worms' habits. The trunk may be homonomous or variably heteronomous as noted above, and each segment typically bears a pair of unjointed appendages, called **parapodia**, and bundles of **chaetae** (Figure 13.6).

Chaetae are tiny, spinelike structures derived from single epidermal cells. Recent evidence suggests strongly that annelid chaetae are homologous to those found in echiuran worms (Chapter 14) and in the pogonophorans and vestimentiferans (covered later in this chapter). They are called setae in much of the literature, but the name chaetae is now preferred to emphasize their homology within these groups and to distinguish them from setae such as those found in arthropods.

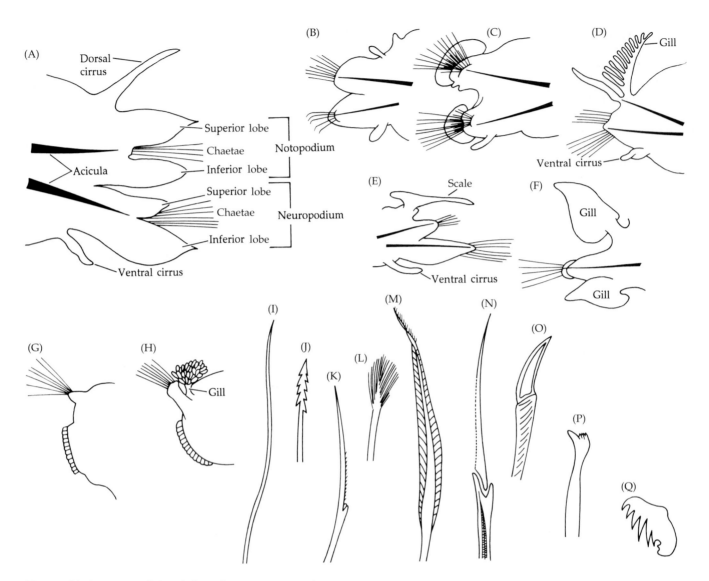

Figure 13.6 Parapodial and chaetal types among poly-chaetes. (A) A stylized parapodium. (B) The parapodium of a glycerid, with reduced lobes. (C) The parapodium of a nephtyid. (D) The parapodium of a eunicid with its modified notopodium; note the dorsal filamentous gill. (E) The parapodium of a polynoid has the dorsal cirrus modified as a scale, or elytron. (F) The parapodium of a phyllodocid; the noto- and neuropodia are modified as gill blades. (G) The very reduced parapodium of a tube-dwelling sabellid. (H) The parapodium of an arenicolid. (I–Q) Chaetae from various polychaetes. The classification of chaetae types rivals that of sponge spicules in complexity and terminology. A few general types are distinguished here as simple chaetae (I–M), compound chaetae (N–O), hooks (P), and uncini (Q).

The parapodia are primitively biramous, with a dorsal **notopodium** and a ventral **neuropodium**, each lobe with its own cluster of chaetae. However, these highly adaptable structures have evolved a diversity of forms and serve a variety of functions (locomotion, gas exchange, protection, anchorage, creation of water currents). In heteronomous polychaetes, the morphology of the parapodia may vary greatly in different body regions. Often, for example, parapodia in one region are modified as gills, in another region as locomotor structures, and elsewhere to assist in food gathering.

Polychaetes are dioecious but, oddly, have simple reproductive structures, generally proliferating gametes from the peritoneum. Many produce free-swimming larvae, but there is a great variety of life history strategies in this class.

Members of the aberrant order Myzostomida are adapted for symbiotic life (Figure 13.3). Most live on the arms of crinoids (sea lilies) or on other echinoderms and ingest suspended material picked up by their hosts. Some are truly parasitic and reside within the gut or body cavity of their hosts, apparently feeding on body fluids or gut contents. They are drastically modified from the basic polychaete plan, with small flattened bodies and chaetal hooks on their parapodia for attachment to their hosts. Their segmentation is reduced and the coelom is greatly reduced (essentially obliterated in most species). Some specialists feel they are so distinct from polychaetes that they warrant their own class.

Oligochaeta. The subclass Oligochaeta comprises well over 6,000 species of clitellate worms, the majority of which live in freshwater and terrestrial environments. Some 200 species of oligochaetes have successfully invaded brackish and marine waters, including both littoral and deep-sea benthic habitats. However, many new species are being described and the number of kinds of marine oligochaetes is probably much greater than currently known. Most oligochaetes are burrowers, although some inhabit layers of epibenthic detritus or live among algal filaments; a few are tube dwellers, and some are parasitic. They range in length from less than 1 mm for some aquatic forms to over 3 m for certain Australian earthworms (Figure 13.1B).

Oligocheates differ from polychaetes in a number of respects, many of which are related to their exploitation of land and freshwater environments and their burrowing habits. They lack parapodia, and generally bear fewer or less elaborate head appendages and fewer chaetae. Furthermore, they are hermaphroditic, and most possess relatively complicated reproductive systems. Oligochaetes typically copulate and produce brooded or encapsulated embryos that develop directly to juveniles. They are far less diverse than the polychaetes in terms of variation on their basic body plan.

Externally, the body may be homonomous or may include some regional specialization, such as localized gills or variation in chaetal length (Figure 13.4). A few segments are always modified as the **clitellum**, a secretory region that functions in reproduction. Chaetae are relatively few, occurring in segmentally paired bundles ranging from one to as many as 25 chaetae per bundle. The chaetae of different species vary in length, shape, and thickness, some being short and quite stout, others long and thin. Most oligochaete chaetae are movable, a capability these animals often employ in burrowing.

In spite of the apparent differences between oligochaetes and polychaetes, both are clearly derived from a common stock and show numerous homologies of major body parts arranged according to the metameric plan. Albeit small, the oligochaete head is formed from a prostomium and peristomium, and it often incorporates some anterior body segments. The prostomium is usually very small, but in a few species it is elongated as a tentacle or proboscis (Figure 13.4F). Posterior to the metameric trunk is the pygidium, which bears the anus.

Hirudinoidea. Most of the 500 or so species in this subclass are hirudinidans (true leeches), and it is to them that we devote most of our attention. The enigmatic branchiobdellids and the acanthobdellids contain only a single family and a single species, respectively, and their special features are mentioned briefly where appropriate.

Like the oligochaetes, the hirudinoideans are clitellate annelids. They possess a fixed number of segments, which are traditionally numbered with Roman numerals. Counting the minute prostomium, the bodies of branchiobdellids are composed of 15 segments, acanthobdellids 30 segments, and hirudinidans 34 segments. These segments are generally obscured by superficial annulations, giving the impression of many more segments (Figure 13.5). Externally, the leeches are characterized by anterior and posterior suckers and a clitellum; they lack parapodia, and chaetae are lacking in all but the Acanthobdellida. Internally, the coelom is typically reduced to a series of interconnected channels and spaces, without serially arranged septa.

We usually think of leeches as the large bloodsuckers popularized in adventure stories and films. Many, however, are free-living predators, and some are scavengers. Most hirudinoideans range from less than 0.5 to about 2 cm in length, although one species from the Amazon basin, *Haementeria ghilianii*, may reach lengths of 45 cm. Leeches occur in both fresh and salt water, and a few even live in moist terrestrial environments. Those that are full- or part-time parasites feed on the body fluids of a variety of vertebrate and invertebrate hosts. Some of these leeches serve as the intermediate hosts and vectors for certain protozoan, nematode, and cestode parasites.

Most leeches are flattened dorsoventrally. The body is usually divisible into five regions, although the points of division are somewhat arbitrary (Figure 13.5A). The anteriormost head region is composed of the much-reduced prostomium and the anterior body segments. A peristomium is not apparent (it may be absent), and the prostomium may be fused with some of the anterior body segments. The **anterior region** usually bears a number of eyes and a ventral mouth surrounded by the oral or anterior sucker. Segments V–VIII form the **preclitellar region**, followed by the **clitellar region** (segments IX–XI). The clitellum is only apparent during periods of reproductive activity. The **postclitellar** or midbody region comprises segments XII–XXVII. The **posterior region** of the body includes the ventrally directed posterior sucker formed from seven fused segments (XXVIII–XXXIV).

Other than the gonopores and nephridiopores, leeches have few distinctive external features. In a few forms the body surface bears tubercles, and members of the family Ozobranchidae possess fingerlike or filamentous gills (Figure 13.5B). Chaetae are present on the first few segments of the acanthobdellid *Acanthobdella peledina*, but they do not occur in any other member of the subclass.

Reproductively, hirudinoideans are similar to oligochaetes; they are hermaphroditic and possess complex reproductive systems. They generally copulate, and their embryos undergo direct development.

Body Wall and Coelomic Arrangement

Polychaeta. The polychaete body is covered by a thin cuticle of scleroprotein and mucopolysaccharide fibers deposited by epidermal microvilli. The epidermis is a columnar epithelium that is often ciliated on certain

(A)

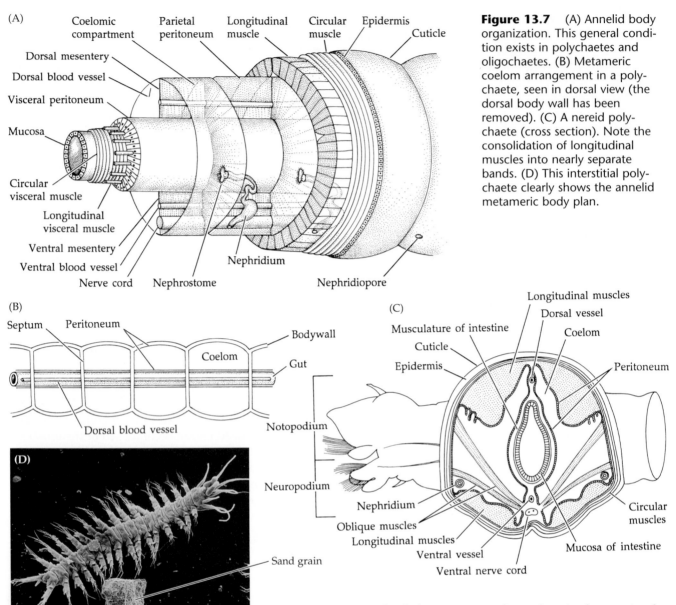

Figure 13.7 (A) Annelid body organization. This general condition exists in polychaetes and oligochaetes. (B) Metameric coelom arrangement in a polychaete, seen in dorsal view (the dorsal body wall has been removed). (C) A nereid polychaete (cross section). Note the consolidation of longitudinal muscles into nearly separate bands. (D) This interstitial polychaete clearly shows the annelid metameric body plan.

parts of the body. Beneath the epidermis lies a layer of connective tissue, circular muscles, and thick longitudinal muscles, the latter often arranged as four bands (Figures 13.7A,C). The circular muscles do not form a continuous sheath, but are interrupted at least at the positions of the parapodia. The inner lining of the body wall is the peritoneum, which surrounds the coelomic spaces and lines the surfaces of internal organs.

The polychaete coelom is primitively arranged as laterally paired (i.e., right and left) spaces, serially (segmentally) arranged within the trunk. Dorsal and ventral mesenteries separate the members of each pair of coeloms, and muscular intersegmental septa isolate each pair from the next along the length of the body. In some polychaetes, the intersegmental septa have been secondarily lost or are perforated, so in these animals the coelomic fluid is continuous among segments. In many small polychaetes, the coelomic lining is entirely lost. Such conditions radically alter the hydraulic qualities of the body; the significance of some of these differences is discussed in the next section.

In addition to the main body wall and septal muscles, other muscles function to retract protrusible and eversible body parts (e.g., branchiae, pharynx), and to operate the parapodia (Figure 13.7C). Each parapodium is an evagination of the body wall and contains a variety of muscles. Movable parapodia are operated primarily by sets of diagonal (oblique) muscles, which have their origin near the ventral body midline. These muscles branch and insert at various points inside the parapodium. Large parapodia typically contain a pair of chitinous and scleroproteinaceous supporting rods called **acicula** (Figure 13.6A), on which some muscles insert and operate. The chaetae are also served by muscles and can usually be retracted and extended (quite unlike the setae of arthropods).

Oligochaeta. The body wall of oligochaetes (Figure 13.8) is constructed on the same plan as that described for polychaetes. A thin cuticle covers the epidermis, which is usually a columnar epithelium containing various mucus-secreting gland cells. Cilia sometimes occur on certain parts of the epidermis—around the prostomium and gills of some small freshwater forms, for example. Beneath the epidermis are the usual circular and longitudinal muscle layers, the latter being bounded internally by the peritoneum. Both muscle layers are usually quite thick, especially in the larger terrestrial oligochaetes. The intersegmental septa are generally well developed and muscular. They are for the most part functionally complete, except at the anterior and posterior ends of the worm, and perforations in the septa are regulated by sphincter muscles. This effective isolation of coelomic compartments from one another plays a major role in oligochaete locomotion.

Many terrestrial oligochaetes bear pores connecting individual coelomic spaces to the outside. These pores are guarded by sphincter muscles and regulate the escape of coelomic fluids onto the body surface. This controlled loss of body fluids is presumed to help maintain a moist film over the body, facilitating gas exchange and preventing desiccation.

Parapodia are absent in oligochaetes, but chaetae do occur in nearly all species. The chaetae are movable by various muscles and play a role in locomotion.

Hirudinoidea. A cross-sectional view of the body wall of a leech is dramatically different from that of other annelids, in large part because of the presence of a thick dermal connective tissue layer beneath the epidermis and the reduction of the coelom (Figure 13.9). A thin cuticle covers a single layer of simple epidermal cells. The epidermis contains mucous gland cells, some of which are quite large and extend well below the surface. The usual circular and longitudinal muscles are present, but they are more loosely organized than in polychaetes and oligochaetes. Distinct bands of dorsoventral muscles are also present, as well as diagonal (oblique) muscles between the circular and longitudinal layers. The dense dermis fills the areas between the muscle bands.

Reduction of the coelom is associated with the proliferation of connective tissue deep beneath the body surface. Septate coelomic compartments are present only in *Acanthobdella* (in the first five segments) and in the midbody region of branchiobdellids. In all other members of this subclass, the coelomic spaces are represented by

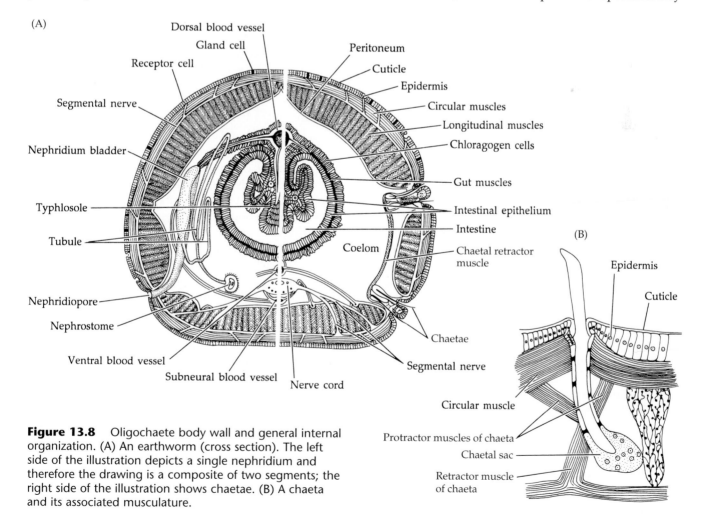

Figure 13.8 Oligochaete body wall and general internal organization. (A) An earthworm (cross section). The left side of the illustration depicts a single nephridium and therefore the drawing is a composite of two segments; the right side of the illustration shows chaetae. (B) A chaeta and its associated musculature.

(A)

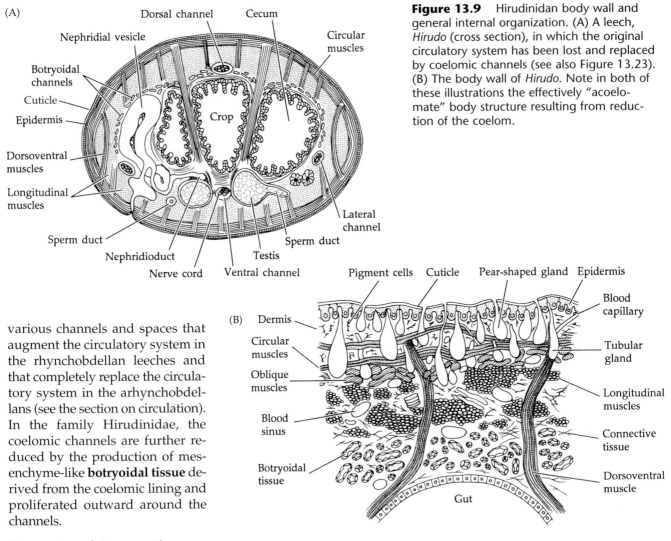

Figure 13.9 Hirudinidan body wall and general internal organization. (A) A leech, *Hirudo* (cross section), in which the original circulatory system has been lost and replaced by coelomic channels (see also Figure 13.23). (B) The body wall of *Hirudo*. Note in both of these illustrations the effectively "acoelomate" body structure resulting from reduction of the coelom.

various channels and spaces that augment the circulatory system in the rhynchobdellan leeches and that completely replace the circulatory system in the arhynchobdellans (see the section on circulation). In the family Hirudinidae, the coelomic channels are further reduced by the production of mesenchyme-like **botryoidal tissue** derived from the coelomic lining and proliferated outward around the channels.

Support and Locomotion

Polychaeta. Polychaetes provide a classic example of the employment of coelomic spaces as a hydrostatic skeleton for body support. Coupled with the well developed musculature, the metameric body plan, and the parapodia, this hydrostatic quality provides the basis for understanding locomotion in these worms. We begin a survey of locomotor patterns by examining *Nereis*, an errant, homonomous polychaete (Figure 13.1A). Keep in mind that in such polychaetes the intersegmental septa are functionally complete, and thus the coelomic spaces in each segment can be effectively isolated hydraulically from each other. Modifications on this fundamental arrangement are discussed later.

In addition to burrowing, *Nereis* can engage in three basic epibenthic locomotor patterns: slow crawling, rapid crawling, and rather inefficient swimming (Figure 13.10). All of these methods of movement depend primarily on the bands of longitudinal muscles, especially the larger dorsolateral bands, and on the parapodial muscles. The circular muscles are relatively thin and serve primarily to maintain adequate hydrostatic pres-

sure within the coelomic compartments. Each method of locomotion in *Nereis* (and similar forms) involves the antagonistic action of the longitudinal muscles on opposite sides of the body in each segment.

During movement, the longitudinal muscles on one side of any given segment alternately contract and relax (and are stretched) in opposing synchrony with the action of the muscles on the other side of the segment (Figure 13.10A). Thus, the body is thrown into undulations that move in metachronal waves from posterior to anterior. Variations in the length and amplitude of these waves combine with parapodial movements to produce the different patterns of locomotion. The parapodia and their chaetae are extended maximally in a power stroke as they pass along the crest of each metachronal wave. Conversely, the parapodia and chaetae retract in the wave troughs during their recovery stroke. Thus, the parapodia on opposite sides of any given segment are exactly out of phase with one another.

When *Nereis* is crawling slowly, the body is thrown into a high number of metachronal undulations of short wavelength and low amplitude (Figure 13.10B). The ex-

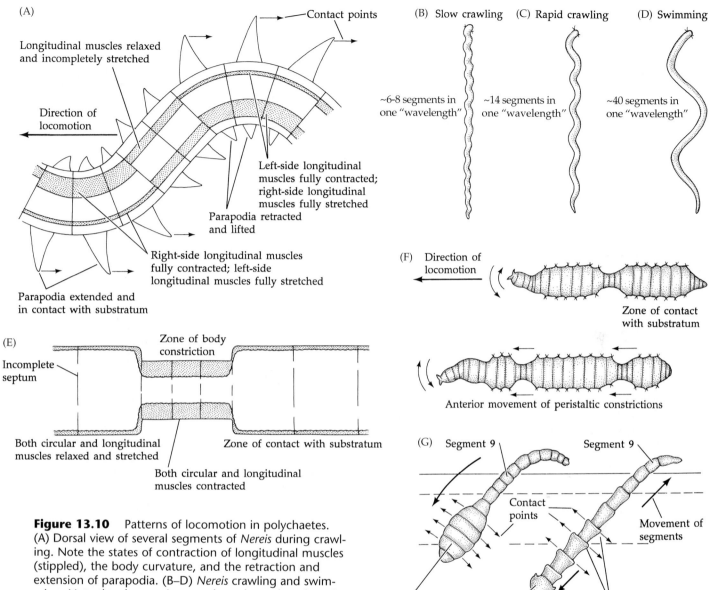

Figure 13.10 Patterns of locomotion in polychaetes. (A) Dorsal view of several segments of *Nereis* during crawling. Note the states of contraction of longitudinal muscles (stippled), the body curvature, and the retraction and extension of parapodia. (B–D) *Nereis* crawling and swimming. Note the changes in metachronal wavelength and amplitude. (E) Midsaggital section through a polychaete. The perforated intersegmental septa allow peristaltic body contractions to cause volumetric changes in segments. (F,G) Burrowing movements in *Polyphysia* (F) and *Arenicola* (G).

Labels in figure:

(A) Contact points
Longitudinal muscles relaxed and incompletely stretched
Direction of locomotion
Left-side longitudinal muscles fully contracted; right-side longitudinal muscles fully stretched
Parapodia retracted and lifted
Right-side longitudinal muscles fully contracted; left-side longitudinal muscles fully stretched
Parapodia extended and in contact with substratum

(B) Slow crawling — ~6-8 segments in one "wavelength"
(C) Rapid crawling — ~14 segments in one "wavelength"
(D) Swimming — ~40 segments in one "wavelength"

(E) Zone of body constriction
Incomplete septum
Both circular and longitudinal muscles relaxed and stretched
Both circular and longitudinal muscles contracted
Zone of contact with substratum

(F) Direction of locomotion
Zone of contact with substratum
Anterior movement of peristaltic constrictions

(G) Segment 9
Contact points
Movement of segments
Body inflated as anchor
Everted proboscis
Annular flanges anchor body

tended parapodial chaetae on the wave crests are pushed against the substratum and serve as pivot points as the parapodium engages in its power stroke. As the parapodium moves past the crest, it is retracted and lifted from the substratum as it is brought forward during its recovery stroke. The main pushing force in this sort of movement is provided by the parapodial muscles.

During rapid crawling, much of the driving force is provided by the longitudinal body wall muscles in association with the longer wavelength and greater amplitude of the body undulations (Figure 13.10C), which accentuate the power strokes of the parapodia.

Nereis can leave the substratum to engage in a rather inefficient swimming behavior (Figure 13.10D). In swimming, the metachronal wavelength and amplitude are even greater than they are in rapid crawling. When watching a nereid swim, however, one gets the impression that the "harder it tries" the less progress it makes, and there is some truth to this. The problem is that, even though the parapodia act as paddles pushing the animal forward on their power strokes, the large metachronal waves continue to move from posterior to anterior and actually create a water current in that same direction; this current tends to push the animal into reverse. The result is that *Nereis* is able to lift itself off the substratum, but then largely thrashes about in the water. This behavior is used primarily as a short-term mechanism to escape benthic predators rather than as a means to get from one place to another.

With these basic patterns and mechanisms in mind, we consider a few other methods of locomotion in polychaetes. *Nephtys* superficially resembles *Nereis*, but its methods of movement are significantly different. Although *Nephtys* is less efficient than *Nereis* at slow walking, it is a much better swimmer; it is also capable of effective burrowing in soft substrata. The large, fleshy parapodia serve as paddles, and, when swimming, *Nephtys* does not produce long, deep metachronal waves. Rather, the faster it swims, the shorter and shallower the waves become, thus eliminating much of the counterproductive force described for *Nereis*. When initiating burrowing, *Nephtys* swims head-first into the substratum, anchors the body by extending the chaetae laterally from the buried segments, and then extends the proboscis deeper into the sand. A swimming motion is then employed to burrow deeper into the substratum.

In contrast to the above descriptions, scale worms (family Polynoidae; Figure 13.2H) have capitalized on the use of their muscular parapodia as efficient walking devices. The body undulates little if at all, and there is a corresponding reduction in the size of the longitudinal muscle bands and their importance in locomotion. In fact, these worms depend almost entirely on the action of the parapodia for walking; polynoids cannot swim.

Many of the highly efficient burrowers have secondarily lost most of the intersegmental septa, or have septa that are perforated (e.g., *Arenicola*, *Polyphysia*). The loss of complete septa means that segments are not of constant volume; in other words, a loss of coelomic fluid from one body region causes a corresponding gain in another (Figure 13.10E). These polychaetes have reduced parapodia. The chaetae, or simply the surface of the expanded portions of the body, serve as anchor points, while the burrow wall provides an antagonistic force resisting the hydraulic pressure. In *Polyphysia*, peristaltic waves move constricted body regions forward while the anchored parts provide leverage (Figure 13.10F). The constricted areas are reduced both in diameter and in length by simultaneous contraction of both the circular and the longitudinal muscles.

Arenicola burrows by first embedding and anchoring the anterior body region in the substratum. The anchoring is accomplished by contracting the circular muscles of the posterior portion of the body, thus forcing coelomic fluid anteriorly and causing the first few segments to swell (Figure 13.10G). Then the posterior longitudinal muscles contract, thereby pulling the back of the worm forward. To continue the burrowing, a second phase of activity is undertaken. As the anterior circular muscles contract and the longitudinal bands relax, the posterior edges of each involved segment are protruded as anchor points to prevent backward movement; the proboscis is thrust forward, deepening the burrow. Then the proboscis is retracted, the front end of the body is engorged with fluid, and the entire process is repeated.

Different burrowing mechanisms are known among other polychaetes. For example, *Glycera*, a long, sleek worm, burrows rapidly using its large, muscular proboscis almost exclusively (Figure 13.14B). The proboscis is thrust into the substratum and swelled; then the body is drawn in by contraction of the proboscis muscles.

Most tube-dwelling polychaetes (Figure 13.11) are heteronomous and have rather soft bodies and relatively weak muscles. The parapodia are reduced, so the chaetae are used to position and anchor the animal in its tube. Movement within the tube is usually accomplished by slow peristaltic action of the body or by chaetae movements. When the anterior end is extended for feeding, it may be quickly withdrawn by special retractor muscles while the unexposed portion of the body is anchored in the tube.

Polychaete tubes provide protection as well as support for these soft-bodied worms, and also keep the animal oriented properly in relation to the substratum. Some polychaetes build tubes composed entirely of their own secretions. Most notable among these tube builders are the serpulids and spirorbids, which construct their tubes of calcium carbonate secreted by a pair of large glands near a fold of the peristomium called the **collar**. The crystals of calcium carbonate are added to an organic matrix; the mixture is molded to the top of the tube by the collar fold and held in place until it hardens.

Some sabellids produce parchment-like or membranous tubes of organic secretions molded by the collar. Others, such as *Sabella*, mix mucous secretions with size-selected particles extracted from feeding currents, then lay down the tube with this material (Figure 13.11B). Numerous other polychaete groups form similar tubes of sediment particles collected in various ways and cemented together with mucus.

A few polychaetes are able to excavate burrows by boring into calcareous substrata, such as rocks, coral skeletons or mollusc shells (e.g., certain members of the families Eunicidae, Spionidae, Sabellidae). In extreme situations, the activity of the polychaetes may have deleterious effects on the "host." For example, species of *Polydora* (Spionidae) can cause serious damage to commercially raised oysters.

Many sedentary polychaetes use modifications of the basic locomotor actions described above to provide means of moving water through their tubes or burrows. Some of these modifications are discussed in the section on feeding.

Oligochaeta. Oligochaetes rely heavily on their well developed hydrostatic skeleton for both support and locomotion. The action of the body wall muscles on the coelomic fluids provides the hydraulic changes associated with the typical pattern of oligochaete locomotion. In the absence of parapodial "paddles," oligochaetes depend on peristalsis and chaetal manipulation for burrowing, for moving through bottom debris, or for crawling over surfaces.

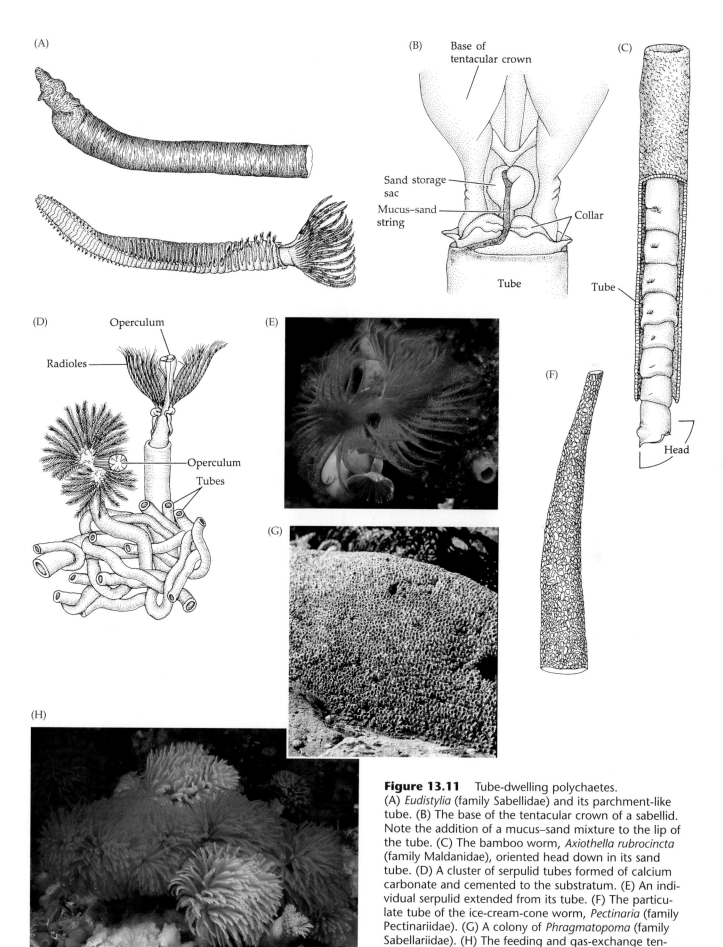

Figure 13.11 Tube-dwelling polychaetes.
(A) *Eudistylia* (family Sabellidae) and its parchment-like tube. (B) The base of the tentacular crown of a sabellid. Note the addition of a mucus–sand mixture to the lip of the tube. (C) The bamboo worm, *Axiothella rubrocincta* (family Maldanidae), oriented head down in its sand tube. (D) A cluster of serpulid tubes formed of calcium carbonate and cemented to the substratum. (E) An individual serpulid extended from its tube. (F) The particulate tube of the ice-cream-cone worm, *Pectinaria* (family Pectinariidae). (G) A colony of *Phragmatopoma* (family Sabellariidae). (H) The feeding and gas-exchange tentacular crowns of *Eudistylia* (family Sabellidae).

There are notable differences between the mechanics of oligochaete and polychaete locomotion, even when comparing relatively similar movement patterns such as burrowing. The most important reason for these differences is that while most burrowing polychaetes have lost intersegmental septa (or at least have evolved perforations in them), most oligochaetes retain functionally complete septa. Each segment of an oligochaete functions more or less independently of the others, and constricting one area of the body does not result in the flow of coelomic fluid to another area. Thus each segment is of fixed volume, so a decrease in the diameter of a particular segment must be accompanied by an increase in its length, and vice versa.

Oligochaete burrowing involves the alternately contracting circular and longitudinal muscles within each segment. The shape of a segment changes from long and thin to short and thick with the respective muscle actions. These shape changes move anteriorly along the body in a peristaltic wave generated by a sequence of impulses from the ventral nerve cord and associated motor neurons. So, at any moment during locomotion, the body of the worm appears as alternating thick and thin regions (Figure 13.12). Without some method of anchoring the body surface, this action would not produce any forward motion. The chaetae provide this anchorage as they protrude like so many barbs from the thick portions of the body. When the longitudinal muscles relax and the circular muscles contract, the body diameter decreases and the chaetae are turned to point posteriorly and lie close to the body. As shown in Figure 13.12, as the anterior end of the body is extended by circular-muscle contraction, the chaetae prevent backsliding, the head is pressed into the substratum, and the worm advances. The anterior end then swells by contraction of the longitudinal muscles, and the rest of the body is pulled along.

There are some variations on this general scheme. For instance, when moving across relatively smooth surfaces, earthworms may employ the mouth as a sort of sucker. The mouth is pressed against the substratum and provides a temporary attachment point against which the muscles can operate in place of the usual chaetal anchorage. Also, giant neurons in many oligochaetes may be stimulated and cause the rapid contraction of longitudinal muscles in many segments, thereby eliciting rapid escape or withdrawal responses.

Hirudinoidea. Body support in leeches is provided by the more or less solid body construction, the fibrous connective tissue and included muscle bands, and the hydrostatic qualities of the coelomic channels. The absence of isolated, spacious, and segmentally arranged coelomic compartments precludes certain kinds of locomotion seen in many polychaetes and oligochaetes. The coelomic spaces in leeches are reduced and continuous, so these animals cannot move

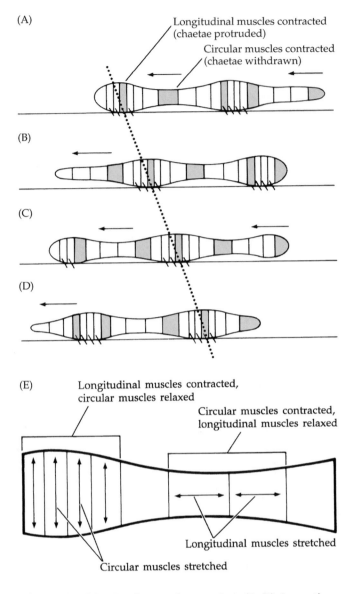

Figure 13.12 Earthworm locomotion. (A–D) An earthworm moving to the left. Every fourth segment is darkened for reference. The dotted line passes through a posteriorly moving point of contact with the substratum. (E) Several segments of an earthworm (sagittal section). Since each segment is a functionally isolated compartment, shortening and elongation accompany the contraction of longitudinal and circular muscles, respectively, while each segment essentially maintains a constant volume.

like a truly segmented worm. We may view the circular and longitudinal muscles as acting antagonistically against a functionally single internal space whose volume remains constant (as does the volume of the whole body).

Leeches do not burrow, but are mostly surface dwellers; thus they move over substrata rather than through them. Without chaetae or parapodia, the suckers serve as the points of contact with the substratum against which the muscle action can operate. Beginning with the posterior sucker attached, the circular muscles are contracted. Given the mechanics of the creature, the only possible result of this action is for the entire body to elongate as its diameter is reduced. Thus, the body is extended forward, and the anterior sucker is attached. Now the posterior sucker is released and the longitudinal muscles contract, shortening the body (and increasing its diameter) and drawing the posterior end forward. The whole business is an "inchworm-like" movement, as depicted in Figure 13.13. Some leeches are also capable of swimming by dorsoventral body undulations; this behavior is an important mechanism for locating and contacting nonbenthic hosts.

Feeding

Polychaeta. The great diversity of form and function among polychaetes has allowed them to exploit nearly all marine food resources in one way or another. For convenience we have categorized polychaetes as raptorial, deposit, and suspension feeders (see Chapter 3). However, there are several feeding methods and dietary preferences within each of these basic designations. Following a discussion of selected examples of these feeding types, we mention a few of the symbiotic polychaetes.

The most familiar raptorial polychaetes are hunting predators (e.g., many phyllodocids, syllids, and nereids). These animals tend toward homonomy and are capable of rapid movement across the substratum. For the most part they feed on small invertebrates. When prey is located by chemical or mechanical means, the worm everts its pharynx by quick contractions of the body wall muscles in the anterior segments, increasing the hydrostatic pressure in the coelomic spaces and causing the eversion. As a result of the design of the pharynx, the jaws gape at the anteriormost end when the pharynx is evert-

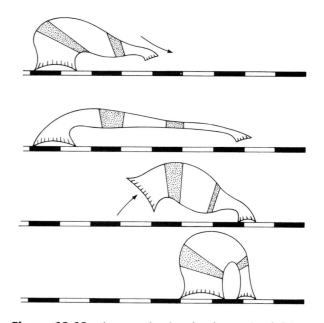

Figure 13.13 Locomotion in a leech, moving left to right, using the anterior and posterior suckers to progress in "inch worm" fashion.

ed (Figure 13.14). Once the prey is positioned within the jaws, the coelomic pressure is released, the jaws collapse on the prey, and the proboscis and captured victim are pulled into the body by large retractor muscles. Many of these raptorial feeders can also ingest plant material and detritus. Some scavenge, feeding on almost any dead organic material they encounter.

Some predatory polychaetes do not actively hunt. Many scale worms (family Polynoidae) sit and wait for passing prey, then ambush it by sucking it into their mouth or grasping it with the pharyngeal jaws. In addition, not all raptorial polychaetes are surface dwellers.

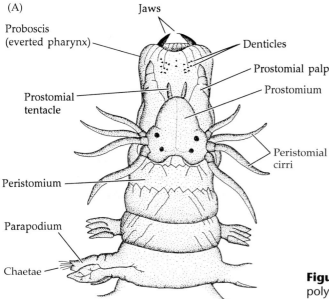

(A)

Proboscis (everted pharynx)
Jaws
Denticles
Prostomial palp
Prostomium
Prostomial tentacle
Peristomial cirri
Peristomium
Parapodium
Chaetae

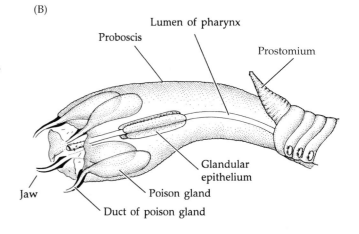

(B)

Lumen of pharynx
Proboscis
Prostomium
Glandular epithelium
Jaw
Poison gland
Duct of poison gland

Figure 13.14 The eversible pharyngeal jaws of two raptorial polychaetes. (A) *Nereis*. (B) *Glycera*.

Some live in tubes (*Diopatra*) or in complex branched burrows (*Glycera*). Such polychaetes detect the presence of potential prey outside their tubes or burrows by chemosensory or vibration-sensory means and extend their everted proboscis to capture the prey. Some leave their residence to hunt for short periods of time. Certain forms (e.g., *Glycera*) have poison glands associated with the jaws (Figure 13.14B).

A number of polychaetes are direct deposit feeders, actually ingesting the substratum and digesting the organic matter contained therein (e.g., members of the families Arenicolidae, Opheliidae, and Maldanidae). The lugworms, such as *Arenicola*, excavate an L-shaped burrow, which they irrigate with water drawn into the open end by peristaltic movements of the worm's body (Figure 13.15A). The water percolates upward through the overlying sediment and tends to "liquefy" the sand at the blind end of the L, near the worm's mouth. This sand is ingested by the muscular action of a bulbous proboscis. The water brought into the burrow also adds suspended organic material to the sand at the feeding site. The worm periodically moves to the open end of its tunnel and defecates the ingested sand outside the burrow in characteristic surface castings.

Some maldanids live in straight vertical burrows, head down, and ingest the sand at the bottom. They periodically move upward (backward) to defecate on the surface. Some maldanids, and perhaps many other polychaetes, take in dissolved organic compounds, especially amino acids, as a significant part of their nutrient supply.

A number of other direct deposit feeders (some opheliids, for example) do not live in constructed burrows

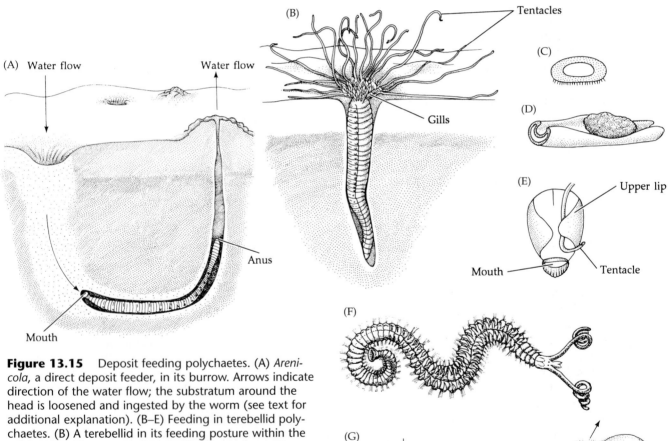

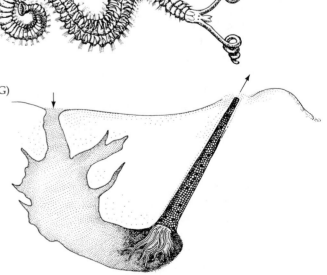

Figure 13.15 Deposit feeding polychaetes. (A) *Arenicola*, a direct deposit feeder, in its burrow. Arrows indicate direction of the water flow; the substratum around the head is loosened and ingested by the worm (see text for additional explanation). (B–E) Feeding in terebellid polychaetes. (B) A terebellid in its feeding posture within the substratum. The prostomial tentacles "creep" over the surface of the substratum and accumulate food, which is then passed to the mouth. (C) A terebellid tentacle (cross section) has cilia on the underside. (D) A section of the tentacle rolls to form a temporary food groove. (E) A tentacle is wiped across the oral area, where food is passed to the mouth and ingested. Such terebellids are indirect (selective) deposit feeders. (F) The spionid *Polydora*, another selective deposit feeder, uses its tentacle-like prostomial palps to obtain food. (G) The ice-cream-cone worm, *Pectinaria*, in feeding position. A water current is created (arrows), liquefying the sand around the tentacled head; organic matter is removed and ingested.

but simply move through the substratum ingesting sediments as they go. In high concentrations, some populations of these polychaetes pass thousands of tons of sediments through their guts each year—which has a significant impact on the nature of the deposits in which they live.

Selective deposit feeders are defined by their ability to effectively sort the organic material from the sediment prior to ingestion (e.g., many members of the families Terebellidae, Spionidae, and Pectinariidae, among others). However, the methods used in these families to sort food differ significantly. Most terebellids (e.g., *Amphitrite, Pista, Terebella*) establish themselves vertically in the sediment, posterior end down, either in shallow burrows or permanent tubes (Figure 13.15B). The feeding tentacles are modified prostomial appendages that are extended over the substratum. These hollow tentacles are extended by ciliary crawling, and can be retracted by muscles. Once extended, the tentacular epithelium secretes a mucous coat to which organic material, sorted from the sediment, adheres. The tentacle edges curl up to form a longitudinal groove along which food and mucus are carried by cilia to the mouth (Figure 13.15C–E). Tube-dwelling spionids engage in a similar method of feeding. In these animals, the feeding structures are more muscular and are derived from the prostomial palps (Figure 13.15F). They are swept through the water or brushed through the surface sediments, extracting food and moving it to the mouth.

Pectinaria, the "ice-cream-cone-worm," lives in a tube constructed of sand grains and shell fragments. The tube is open at both ends. The animal orients itself head down, with the posterior end of the tube projecting to the sediment surface (Figure 13.15G). Head appendages partially sort the sediment, and a relatively high percentage of organic matter is ingested. A number of other polychaetes employ these and other methods of selective deposit feeding.

Various forms of suspension feeding are accomplished by many tube-dwelling polychaetes (e.g., members of the families Serpulidae and Sabellidae), and by some that live in relatively permanent burrows (e.g., Chaetopteridae). The feeding structures of *Sabella* and many related types are a crown of bipinnately branched peristomial tentacles called **radioles**. Some of these worms generate their own feeding currents, whereas others "fish" their tentacles in moving water. As food-laden water passes over the tentacles, the water is driven by cilia upward between the **pinnules** (branches) of the radioles (Figure 13.16A,B). Eddies form on the medial side (inside) of the tentacular crown and between the pinnules, slowing the flow of water, decreasing its carrying capacity, and thus facilitating extraction of suspended particles. The particles are carried, with mucus, along a series of small ciliary tracts on the pinnules to a groove along the main axis of each radiole. This groove is widest at its opening and decreases in width in a step-wise fashion to a narrow slot deep in the groove. By this means, particles are mechanically sorted into three size categories as they are carried into the groove. Typically, the smallest particles are carried to the mouth and ingested, the largest particles are rejected, and the medium-sized ones are stored for use in tube building.

Members of the family Chaetopteridae are among the most heteronomous of all polychaetes; the body is distinctly regionally specialized (Figure 13.16C). Chaetopterids truly filter water for food. These animals reside in U-shaped burrows through which they move water, extracting suspended materials. Each body region plays a particular role in this feeding process. For example, in *Chaetopterus*, segments 14–16 bear greatly enlarged notopodial fans that serve as paddles to create the water current through the burrow. These and a few other segments also bear suckers modified from the neuropodia, which help anchor the worm in position within the burrow. A mucous bag, produced by secretions from segment 12, is held as shown in Figure 13.16C, so that water flows into the open end of the bag and through its mucous wall. Particles as small as 1 μm in diameter are captured by this structure, and there is some evidence that even protein molecules are held in the mucous net (probably by ionic charge attraction rather than mechanical filtering). During active feeding, the bag is rolled into a ball, passed to the mouth by a ciliary tract, and ingested every 15–30 minutes or so; then a new bag is produced.

Symbiotic relationships with other animals occur among several groups of polychaetes. There are some interesting cases that reflect, again, the adaptive diversity of these worms. Many symbiotic polychaetes are hardly modified from their free-living counterparts and do not show the drastic adaptive characteristics often associated with this sort of life. For many, the relationship with their host is a loose one, the polychaete often using the host merely as a protective refuge. We have already mentioned polychaetes that burrow into the shells of other invertebrates and are quite similar to their nonsymbiotic relatives. Among the most common commensalistic polychaetes are certain polynoid scale worms, especially members of the genera *Halosydna* and *Arctonoe* that live on the bodies of various molluscs, echinoderms, and cnidarians (Figure 13.17A). A polynoid has even been discovered living as a commensal in the mantle cavity of giant deep-sea mussels residing near thermal vents on the East Pacific Rise. One scale worm, *Hesperonoe adventor*, inhabits the burrows of the Pacific innkeeper worm, *Urechis caupo* (Echiura; see Chapter 14). Recently several species of scale worms have been found in commensalistic associations with gorgonacean and stylasterine cnidarians (Pettibone 1991). There are many examples of these rather informal associations: certain syllids that live and feed on hydroids, a nereid (*Nereis fucata*) that resides in the shells of hermit crabs, and so on. Most of these animals do not

Figure 13.16 Two strategies of suspension feeding in polychaetes. (A,B) Suspension feeding by a sabellid. (A) Tentacular crown extended from tube and water currents (arrows) passing between tentacles. (B) A portion of a tentacle (radiole) in section. Various ciliary tracts remove particulate matter and direct it to the longitudinal groove on the radiole axis. Here, sorting by size occurs. Most of the largest particles are rejected, the smallest ones are ingested, and the medium-size particles are used in tube building. (C) *Chaetopterus* in its U-shaped burrow. The ventral view shows details of the worm's anterior end. A water current (arrows) is produced through the borrow by fan-shaped parapodia. Food is removed as the water passes through a secreted mucous bag. The bag is eventually passed to the mouth and ingested, food and all. See text for additional details.

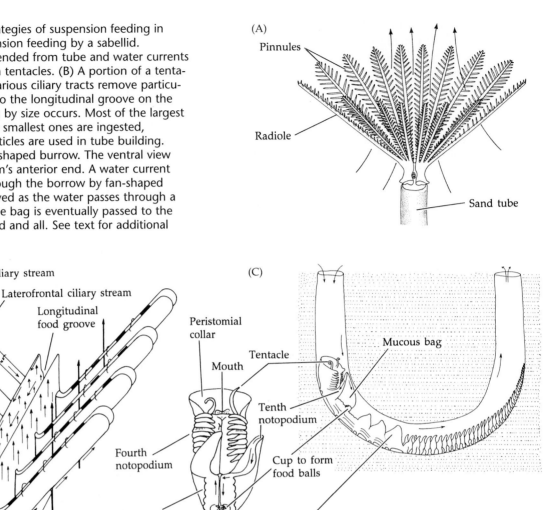

feed upon their hosts, but prey upon tiny organisms that happen into their immediate environment. Others consume detritus or scraps from their host's meals.

A number of other odd associations are known among the polychaetes. Several species of arabellids live in the bodies of echiurans and of other polychaetes. Again, these endosymbionts show little structural modification associated with their lifestyles, other than a tendency for small body size and reduction in the pharyngeal jaws. A clear example of a parasitic polychaete is *Ichthyotomus sanguinarius*. These small (1 cm long) worms attach to eels by a pair of stylets or jaws. The stylets are arranged so that when their associated muscles contract, the stylets fit together like the closed blades of a scissors. The stylets are thrust into the host, and when the muscles relax they open and anchor the parasite to the fish (Figure 13.17B). Species of *Polydora* often excavate galleries in various calcareous substrata (e.g., shells) and have been responsible for killing oys-

ters in commercially harvested areas of Europe, Australia, and North America. The Pacific hydrocoral *Allopora californica* typically harbors colonies of *Polydora alloporis*, whose paired burrow openings are often mistaken for the hydrozoan's polyp cups.

A most unusual symbiotic relationship exists between the strange "Pompeii worm" (*Alvinella pompejana*; named after the deep-sea submersible "Alvin") and a variety of marine chemoautotrophic sulfur bacteria. This polychaete is a member of the deep hydrothermal vent communities of the East Pacific Rise. It lives closer to the hot water extrusions than any other animal in the vent community, building honeycomb-like structures called "snowballs" around the thermal plumes. Temperatures inside the snowballs reach an astonishing 250°C. The bodies of Pompeii worms are covered with unique vent bacteria. Evidence suggests that the worms continually transport these symbiotic bacteria to the mouth, by ciliary mucous tracts, for consumption.

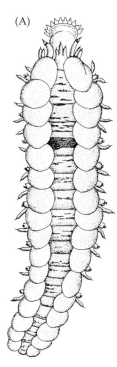

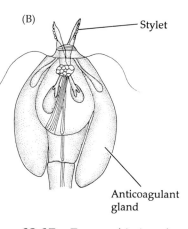

Figure 13.17 Two symbiotic polychaetes. (A) *Arctonoe*, a polynoid that lives in the ambulacral grooves of seastars and the mantle chambers of certain molluscs (with proboscis extended). (B) The anterior end of *Ichthyotomous sanguinarius*, a syllid parasitic on fishes. The stylets anchor the worm to its host and the large glands secrete an anticoagulant.

Oligochaeta. Feeding strategies among the oligochaetes are less diverse than among the polychaetes. This difference is not surprising given the general absence of body elaborations among oligochaetes. The lack of head appendages precludes tentacular feeding, and the absence of parapodia eliminates the ability to sit in one place and generate water currents from which food can be extracted. Nevertheless, oligochaetes do show specialized feeding modes that evolved in association with their particular habitats, and they do exploit a variety of food resources. Most can be classified as predators, detritivores, or direct deposit feeders. Predation occurs in some freshwater oligochaetes, which capture prey by a sucking action of their muscular pharynx. They generally feed on small invertebrates such as other worms and tiny crustaceans. Many are able to evert the dorsal portion of the pharynx (as a proboscis), on which are located mucus-secreting glands; prey are stuck to the everted structure by the mucus and withdrawn into the gut with retraction of the pharynx.

Detritivorous oligochaetes employ a variety of methods. Many live in the surface layer of organic debris on the bottoms of ponds and streams, where they draw small particles of food into the gut by muscular or ciliary action of the foregut. Most such "detritivores" also ingest live microorganisms along with detrital material. A most unusual feeding method is seen in the tube-dwelling members of the freshwater naidid genus *Ripistes* (Figure 13.4D). Long chaetae located on the anterior segments are waved about in the water and small detrital particles adhere to them; food material is then ingested by wiping the chaetae across the mouth.

Most terrestrial and many aquatic oligochaetes are at least in part direct deposit feeders. Earthworms burrow through the soil, ingesting the substratum as they move. As the soil is passed along the digestive tract, the organic material is digested and absorbed from the gut. The inorganic, indigestible material passes out the anus. Earthworms are said to "work" the soil in this manner, loosening and aerating it. Many of these terrestrial burrowers, including the common earthworm *Lumbricus*, also retrieve organic material from the surface. These worms can burrow to the surface of the soil and there use their sucker-like mouth to obtain relatively large pieces of food (e.g., partially decomposed leaves), which they carry back underground for ingestion.

Several species of gutless marine oligochaetes have been described in shallow coral-sand habitats and in anaerobic, sulfide-rich subsurface sediments. These worms typically harbor subcuticular symbiotic bacteria, whose precise role in the host's nutritional regimen is not yet fully understood. The endosymbiotic bacteria may be very important to the worms; they are passed to the fertilized eggs during oviposition from storage areas next to the female's gonopore.

Hirudinoidea. Well over half the known species of the Hirudinoidea are ectoparasites that feed by sucking the blood or other body fluids from their hosts. Most of the remaining members of this subclass are predators on small invertebrates, and there are a few scavengers that feed on dead animal matter. Some families contain members adapted to a particular feeding mode, but more often feeding methods cut across taxonomic lines. Food-getting involves the structures of the foregut, which generally include either a protrusible pharyngeal proboscis or cutting structures in the form of slicing jaws or stylets. Unfortunately, little work has been done on the details of feeding in most of these animals.

The branchiobdellids are tiny worms that live on freshwater crustaceans, especially crayfish. The anterior end of the pharynx bears a pair of toothed jaws. These animals eat other epizoites living on the host, but they also feed on the host's eggs and body fluids. *Acanthobdella peledina*, the only known species of Acanthobdellida, lives on the skin of freshwater fishes in cold, high-elevation lakes, particularly in northern Europe and Alaska. It apparently spends only about four months each year attached to its host; the rest of the time it is presumably free-living.

The two suborders of Hirudinida are distinguished from one another in part on the basis of the structure of the feeding apparatus. Members of the Rhynchobdellae possess a pharyngeal proboscis but lack jaws, whereas members of the Arhynchobdellae lack a proboscis and all but a few possess jaws (Figure 13.20B). Still, predators and parasites are known among both groups. The predatory forms either grasp their prey with the jaws or pierce them with the stiff proboscis. In either case, the prey is typically swallowed whole, although a few rhynchobdellans attack relatively large prey by piercing them with the proboscis and then sucking out the fluids and soft tissues by a pumping action of the pharynx.

Because of their medical importance, much work has been done on the parasitic leeches, especially those that affect livestock, game animals, or humans. Blood-sucking leeches are not especially host-specific, and most do not remain attached to a single host for long periods of time; many of these leeches may feed by other means when not attached to a suitable host.

A few species of leeches feed exclusively on invertebrate hosts, including annelids (even other leeches), gastropods, and crustaceans, but the majority of them parasitize vertebrates. Some leeches are parasitic on members of particular groups of vertebrates. For example, most of the Piscicolidae (Rhynchobdellae) feed on the blood of fishes (including some deep-sea and hydrothermal vent fishes), whereas the Ozobranchidae (another family of Rhynchobdellae) seem to prefer aquatic reptiles such as turtles and crocodilians.

Of all the leeches, none has been more intensively studied than *Hirudo medicinalis*—the medicinal leech. The common name is derived from the practice of using these leeches to draw blood from humans afflicted with particular maladies for which such "bleeding" was once thought to be an effective treatment. Today, leeches are used to reduce hematomas in areas of the body that are difficult to treat surgically, and to avoid leaving scars. The leech produces a powerful anticoagulant called **hirudin** that allows blood to continue to drain even after the leech as been removed. In addition to the anticoagulant, some leeches produce a number of other chemicals, including anesthetics and vasodilators, that are being studied for possible use in human medicine (Conniff 1987).

Hirudo and many other members of the family Hirudinidae are relatively common in tropical and temperate freshwater habitats. Most of them favor warm-blooded hosts and take their preferred blood meals from wading mammals, including humans when available (late-night TV buffs should take note next time *The African Queen* is aired). These leeches will, however, feed on other vertebrates. When feeding, the leech anchors to the host by the suckers and presses the mouth against the surface of the host's body. Most of these leeches possess three bladelike jaws, each shaped like a half circle (Figure 13.20B). The jaws are set at roughly 120-degree angles to one another so that the cutting edges form a Y-shaped incision. Muscles rock the jaws to and fro, making slices in the host's skin. The leech releases an anesthetic as it makes its incisions, then secretes hirudin into the wound, and blood is sucked from the host by the muscular pharynx. The anesthetic desensitizes the victim's skin; those of you who have encountered leeches in the wild will have noticed that the worms can go un-

noticed while they take their blood meal. While the predatory leeches eat frequently, the bloodsuckers probably feed at widely spaced, very irregular intervals, depending on the availability of hosts. These long periods of fasting apparently present no problems to these animals; some can survive well over a year without feeding. When they do feed, they gorge themselves with several times their own weight in blood. The digestive process is very slow.

Digestive System

Polychaeta. The gut of polychaetes is constructed on a basic annelid plan of foregut, midgut, and hindgut; some examples are shown in Figure 13.18. The foregut is a stomodeum and includes the buccal capsule or tube, the pharynx, and at least the anterior portion of the esophagus. It is lined with cuticle, and the teeth or jaws, when present, are derived from scleroprotein produced along this lining. The jaws are often hardened with calcium carbonate or metal compounds. When present, the eversible portion of this foregut (the proboscis) is derived from the buccal tube or the pharynx. Various glands are often associated with the foregut, including poison glands (glycerids), esophageal glands (nereids and others), and mucus-producing glands in several groups

The entodermally derived midgut generally includes the posterior portion of the esophagus and a long, straight intestine, the anterior end of which may be modified as a storage area, or stomach. The midgut may be relatively smooth, or its surface area may be increased by folds, coils, or many large evaginations (or ceca). The midgut is often histologically differentiated along its length. Typically, the anterior midgut (stomach or anterior intestine) contains secretory cells that produce digestive enzymes. The secretory midgut grades to a more posterior absorptive region. Toward the posterior end of the gut, there may be additional secretory cells that produce mucus, which is added to the undigested material during the formation of fecal pellets. Food is moved along the midgut by cilia and by peristaltic action of gut muscles, usually comprising both circular and longitudinal layers. A short proctodeal rectum connects the midgut to the anus, located on the pygidium.

There has been surprisingly little work done on the digestive physiology of polychaetes, but a variety of digestive enzymes are known from different species. Predators tend to produce proteases, herbivores largely carbohydrases. Some omnivorous forms (e.g., *Nereis virens*) produce a mixture of proteases, carbohydrases, lipases, and even cellulase. Digestion is predominantly extracellular in the midgut lumen, although intracellular digestion is known in some groups (e.g., *Arenicola*). Some polychaetes harbor symbiotic bacteria in their guts that aid in the breakdown of cellulose and perhaps other compounds (Plante et al. 1990).

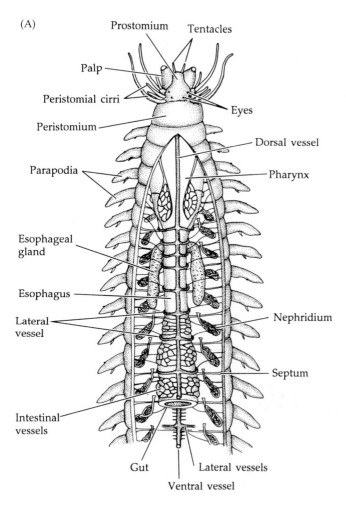

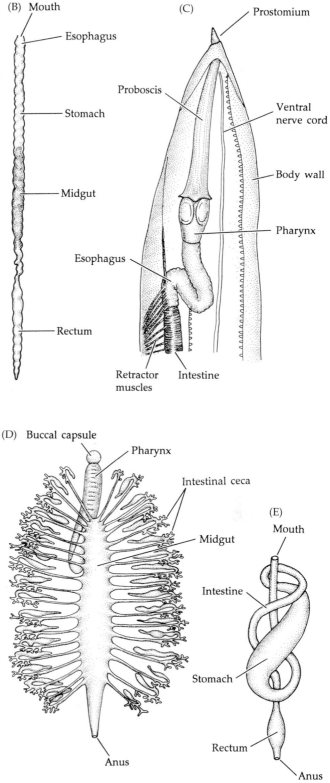

Figure 13.18 Polychaete digestive systems. (A) A dissected nereid (dorsal view). Note the regional specialization of the anterior gut. (B) The simple tubular gut of *Owenia*. (C) A dissected *Glycera* (dorsal view). (D) The multicecate gut of *Aphrodita*. (E) The coiled digestive tract of *Petta*.

Oligochaeta. The oligochaete digestive system is basically a straight tube with various degrees of regional specialization, particularly toward the anterior end (Figure 13.19). In an earthworm, for example, the mouth leads inward to a stomodeal foregut composed of a short buccal tube (or buccal cavity), muscular pharynx, and esophagus. The posterior esophagus often bears enlarged regions forming a **crop**, where food is stored, and one or more muscular **gizzards** lined with cuticle and used to mechanically grind ingested material. The esophagus of many oligochaetes also bears thickened portions of the wall in which are located lamellar evaginations lined with glandular tissue (Figure 13.19D). These **calciferous glands** remove calcium from ingested material. The excess calcium is precipitated by the glands as calcite and then released back into the gut lumen. Calcite is not absorbed by the intestinal wall and so passes out

of the body via the anus. In addition, the calciferous glands apparently regulate the level of calcium ions and carbonate ions in the blood and coelomic fluids, thereby buffering the pH of those fluids.

The primary functions of the foregut are ingestion, transport, storage, and mechanical digestion of food.

Figure 13.19   Oligochaete digestive systems. (A,B) The digestive tract of *Eisenia foetida* (family Lumbricidae). (A) The digestive tract in dorsal view; note the marked regional specialization (hindgut not shown). (B) A portion of the esophagus and calciferous glands, crop, gizzard, and anterior region of the intestine (partial frontal section). (C,D) The foregut of *Lumbricus* (see also Figure 13.8 for a cross-sectional view of the intestine and typhlosole). (C) Note the positional relationship of the anterior gut regions to other organs. (D) The esophagus (cross section) showing lamellar calciferous glands.

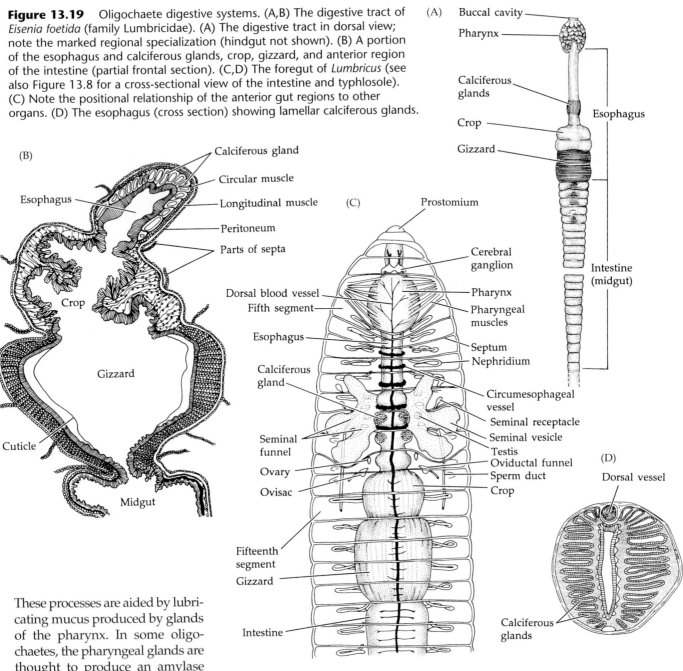

These processes are aided by lubricating mucus produced by glands of the pharynx. In some oligochaetes, the pharyngeal glands are thought to produce an amylase and a protease that initiate chemical breakdown.

The remainder of the digestive tube is dominated by a straight, entodermally derived midgut or intestine leading to a short proctodeal hindgut and anus located on the pygidium. The anterior midgut is predominantly secretory and produces a variety of digestive enzymes that are released into the gut lumen. Various authors have reported that carbohydrases, proteases, cellulase, and chitinase are produced from the midgut epithelium. Digestion is mostly extracellular. Much of the absorption of digested food occurs across the posterior half of the intestinal wall into the blood. Undigested materials pass from the anus as characteristic castings or fecal pellets.

In terrestrial species, the surface area of the intestine is enlarged by a middorsal groove called the **typhlosole** (Figures 13.8A and 13.22B); in addition, the intestines of some oligochaetes bear segmentally arranged lateral diverticula. Food is moved through the digestive tube by peristaltic action of the muscular gut wall and by general body movements associated with locomotion.

Associated with the midgut of many oligochaetes, and some other annelids as well, are masses of pigmented cells called **chloragogen cells**. These modified peritoneal cells contain greenish, yellowish, or brownish globules that impart the characteristic coloration to this **chloragogenous tissue**. This tissue lies within the coelom, but is pressed tightly against the visceral peritoneum of the intestinal wall and typhlosole. This tissue serves as a site of intermediary metabolism (e.g., synthesis and storage of glycogen and lipids, deamination of proteins). It also plays a major role in excretion, as discussed below.

Hirudinoidea. Like that of other annelids, the leech digestive tract includes a stomodeal foregut, entodermal midgut, and short proctodeal hindgut (Figure 13.20). The foregut, as mentioned earlier, includes a mouth, jaws, buccal cavity, proboscis, pharynx, and esophagus. This region is lined with cuticle that provides stiffness to the proboscis and forms the jaws. The stomodeum also contains masses of unicellular **salivary glands** that secrete hirudin in the jawed bloodsuckers and may produce enzymes to aid penetration of the proboscis in those parasitic forms that lack jaws.

Posterior to the esophagus is the enlarged midgut, usually called the stomach or crop. This region bears large ceca in most leeches, providing a large storage capacity as well as a high surface area (Figure 13.20C). In some kinds of leeches, the posterior midgut is structurally differentiated from the anterior portion. A short proctodeal rectum connects the midgut to the anus, located dorsally near the junction of the body and the posterior sucker.

Little is known about digestion in hirudinoideans, except for some fragmentary information on bloodsucking leeches. Midgut enzymes apparently are limited to exopeptidases, which probably accounts for the extremely slow rate of digestion in these animals (a medicinal leech may take several months to digest the contents of a full blood meal). Most leeches, including predatory and parasitic species, harbor a rich bacterial gut flora. These bacteria probably aid in the digestive events and may also provide metabolic products, such as vitamins, that are useful to their host.

Circulation and Gas Exchange

Polychaeta. You will recall the relationship between the presence of a complete, regionally specialized digestive tract combined with a circulatory system as discussed in our coverage of the nemerteans (Chapter 11). The same principle applies to polychaetes, and to annelids in general, but takes on additional significance when viewed in association with the coelomate, segmented bauplan. Given the relatively large size of many polychaetes, the compartmentalization of their coelomic chambers, and the fact that only certain portions of their gut absorb digested food products, it is essential that a circulatory mechanism be present for internal transport and distribution of nutrients. Furthermore, many polychaetes have their gas exchange structures limited to particular body regions; thus they depend on the circulatory system for internal transport of gases.

It is easiest to understand the circulatory system of polychaetes by considering it in concert with their gas exchange structures. In many polychaetes that lack appendages, the entire body surface functions in gas exchange (e.g., lumbrinerids, arabellids). Some of the active epibenthic forms utilize highly vascularized portions of the parapodia as gills. Special gas exchange structures, or **branchiae**, are found in the form of trunk filaments (cirratulids), anterior gills (terebellids), and

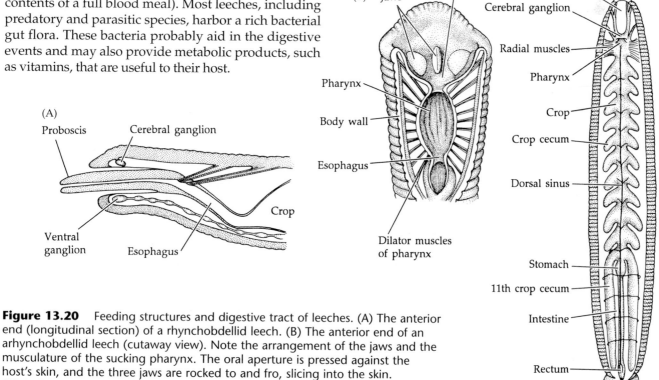

Figure 13.20 Feeding structures and digestive tract of leeches. (A) The anterior end (longitudinal section) of a rhynchobdellid leech. (B) The anterior end of an arhynchobdellid leech (cutaway view). Note the arrangement of the jaws and the musculature of the sucking pharynx. The oral aperture is pressed against the host's skin, and the three jaws are rocked to and fro, slicing into the skin. (C) Basic gut structure of *Hirudo*.

tentacular, or branchial, crowns on the head (sabellids, serpulids, and spirorbids). Since the blood generally carries respiratory pigments, the anatomy of the circulatory system has evolved along with the structure and location of these gas exchange structures.

We again begin our examination with a homonomous polychaete, such as *Nereis*, in which the parapodia are more or less similar to one another and the notopodia function as gills. The major blood vessels include a middorsal longitudinal vessel, which carries blood anteriorly, and a midventral vessel, which carries blood posteriorly. Exchange of blood between these vessels occurs through posterior and anterior vascular networks and serially arranged segmental vessels (Figure 13.21A). Anterior vessel networks are especially well developed around the muscular pharynx and the region of the cerebral ganglion.

The movement of blood in *Nereis* depends on the action of the body wall muscles and on intrinsic muscles in the walls of the blood vessels, especially the large dorsal vessel. There are no special "hearts" or pumping organs. The blood passing through the various segmental vessels supplies the body wall muscles, gut, nephridia, and parapodia, as illustrated in Figure 13.21A. Note that the oxygenated blood is being returned to the dorsal vessel, thus maintaining a primary supply of oxygen to the anterior end of the animal, including the feeding apparatus and cerebral ganglion.

There are many variations on this basic circulatory scheme, and we mention only a few to illustrate the diversity within the polychaetes. Drastic differences are present even among polychaetes of generally similar body forms. Among the homonomous forms, for example, the circulatory system may be reduced or lost. In some cases (e.g., members of the family Syllidae), this reduction is probably associated with small size. This hypothesis, however, cannot be applied to the glycerids, many of which are large and quite active. In these worms and some others, the circulatory system is greatly reduced and has become fused with remnants of the coelom. Glycerids contain red blood cells (with hemoglobin) in the coelomic fluid. Since glycerids have incomplete septa, the coelomic fluid can pass among segments, moved by body activities and ciliary tracts on the peritoneum. In their burrowing lifestyle, enlarged parapodial gills or delicate anterior gills would be disadvantageous; thus the general body surface has probably taken over the function of gas exchange and the coelom the function of circulation. Reduction or loss of the circulatory system has also occurred in a few sedentary polychaetes and in a few nonsedentary types as well (e.g., the terebellid *Polycirrus* and the archianellids).

Compared with *Nereis*, many polychaetes display additional blood vessels, modification of vessels, differences in blood flow patterns, and the formation of large sinuses. As might be predicted, some striking differences are seen among certain heteronomous poly-

chaetes with reduced parapodia and anteriorly located branchiae (e.g., terebellids, sabellids, and serpulids). In many of these worms, in the region of the stomach and anterior intestine, the dorsal vessel is replaced by a voluminous blood space called the **gut sinus** (Figure 13.21C,D). Usually, the dorsal vessel continues anteriorly from this sinus and often forms a ring connecting with the main ventral vessel. In the sabellids and serpulids, a single, blind-ended vessel extends into each branchial tentacle. Blood flows in and out of these branchial vessels, which in some forms (e.g., serpulids) are equipped with valves that prevent backflow into the dorsal vessel. This two-way flow of blood within single vessels is quite different from the capillary exchange system in most closed vascular systems.

Specialized pumping structures have evolved in a number of polychaetes. They are especially well developed in certain tube-dwelling forms and compensate for the reduced effect of general body movements on circulation. These structures, sometimes called "hearts," are often little more than an enlarged and muscularized portion of one of the usual vessels; the dorsal muscular vessel of chaetopterids is such a structure. Terebellids possess a "pumping station" at the base of the gills that functions to maintain blood pressure and flow within the branchial vessels (Figure 13.21B). A variety of similar structures are known.

Most polychaetes contain some respiratory pigment within their circulatory fluid. Those without any such pigment include some very small forms and various syllids, phyllodocids, polynoids, aphroditids, *Chaetopterus*, and a few others. When a pigment is present, it is usually some type of hemoglobin, although chlorocruorin is common in some families (e.g., certain sabellids and serpulids), and hemerythrin occurs in magelonids. Some polychaetes have more than one type of pigment; for example, the blood of some serpulids contains both hemoglobin and chlorocruorin.

Polychaete respiratory pigments may occur in the blood itself, the coelomic fluid, or both. With a few exceptions, blood pigments occur in solution and coelomic pigments are contained within corpuscles. The latter situation is generally associated with a degeneration or loss of the circulatory system (as in glycerids). The incorporation of coelomic pigments, usually hemoglobin, into cells is probably a mechanism to prevent the serious osmotic effects that would result from large numbers of free dissolved molecules in the body fluid. Corpuscular coelomic hemoglobins tend to be of much smaller molecular sizes than those dissolved in the blood plasma. The significance of this difference is not clear, but some interesting ideas on this and related matters were discussed by Mangum (1976) in relation to the oxygen problems of arenicolids.

The types of respiratory pigments and their disposition within the body are related at least in part to the lifestyles of polychaetes. As discussed in Chapter 3, dif-

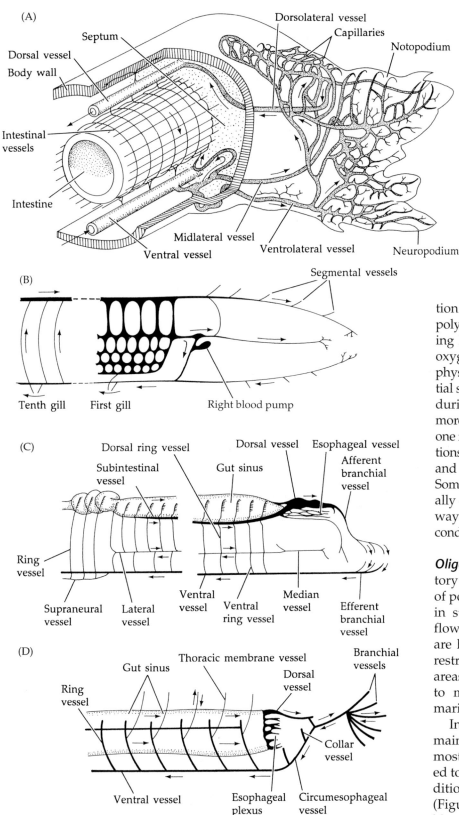

Figure 13.21 Polychaete circulatory and gas exchange systems. (A) A segment and parapodium (cutaway view) of a nereid. Note the major blood vessels and blood flow pattern (arrows). Blood flows anteriorly in the dorsal vessel and posteriorly in the ventral vessel (see also Figure 13.18A). In such polychaetes the flattened parapodia serve as gills. (B–D) Circulatory patterns in an arenicolid (B), a terebellid (C), and a serpulid (D). Major modifications from the basic plan include additional vessels, sinuses associated with the foregut, and branchial vessels serving anterior gills.

tion. A number of intertidal burrowing polychaetes take up and store oxygen during high tides and dissociate the stored oxygen during low tides. This sort of physiological cycle ameliorates the potential stress of oxygen depletion in the body during periods of low tide. Some have more than one pigment type; for example, one form of hemoglobin for normal conditions and another form that stores oxygen and releases it during periods of stress. Some polychaetes (e.g., *Euzonus*) can actually convert to anaerobic metabolic pathways during extended periods of anoxic conditions.

Oligochaeta. The oligochaete circulatory system is generally similar to that of polychaetes, with some modifications in structure and the pattern of blood flow. The differences described below are largely adaptations to living in terrestrial and freshwater environments—areas that generally subject inhabitants to more physiological stress than do marine habitats.

In *Lumbricus* and many others, three main longitudinal blood vessels extend most of the body length and are connected to one another in each segment by additional segmentally arranged vessels (Figure 13.22). The largest longitudinal blood vessel is the **dorsal vessel**; the wall of this vessel is quite thick and muscular, and provides much of the pumping force for blood movement. Suspended in the mesentery beneath the gut is the longitudinal **ventral vessel**. The third longitudinal vessel lies ventral to the nerve cord and is called the **subneural vessel**.

ferent pigments—even different forms of the same pigment—have different oxygen loading and unloading characteristics. The nature of the pigments in a particular worm reflects its ability to store oxygen and then release it during periods of environmental oxygen deple-

Figure 13.22 Circulatory system of *Lumbricus*. (A) Anterior blood vessels (lateral view). (B) The circulatory pattern in one segment (cross section).

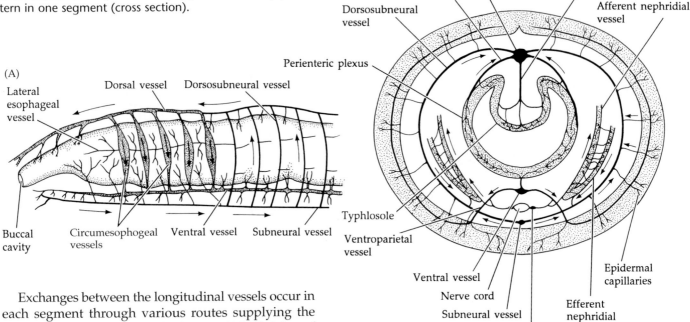

Exchanges between the longitudinal vessels occur in each segment through various routes supplying the body wall, gut, and nephridia (Figure 13.22B). Most of the exchanges between the blood and the tissues take place through capillary beds supplied by afferent and efferent vessels. Blood flows posteriorly in ventral and subneural vessels and anteriorly in the dorsal vessel. Generally, exchange between the dorsal and ventral vessels occurs in each segment, as shown in Figure 13.22. Most oligochaetes also possess from two to five pairs of large, muscular, circumesophageal vessels that carry blood from the dorsal to the ventral vessel region. These vessels aid in propelling the blood and maintaining blood pressure and, along with the dorsal vessel, are often equipped with flap valves to ensure a one-way blood flow.

Most oligochaetes have hemoglobin dissolved in the plasma; members of some families (e.g., Naididae) lack blood pigments. Various phagocytic amebocytes are also present in the circulatory fluids of most of these worms.

A few oligochaetes possess extensions of the body wall that increase the surface area and function as simple gills (e.g., *Branchiura*, *Dero*; Figure 13.4), but most exchange gases across the general body surface. The body surface is kept moist either by the environment, by mucus, or by coelomic fluid released through pores, as described earlier. Most oligochaetes, especially the relatively large ones, have an extensive intraepidermal capillary network derived from the blood vessels within the body wall (Figure 13.22B). These capillaries provide a constant blood supply from the ventral vessel to the body wall and a high surface area for exchange of gases between the blood and the environment.

Many terrestrial oligochaetes are capable of sufficient gas exchange only when exposed to air; they will drown if submerged. (Remember, air contains far more oxygen than does water.) We have all seen earthworms crawl-

ing about the surface following a heavy rain. One particular species of earthworm (*Alma emini*) has evolved a remarkable adaptation that allows it to survive the rainy season in its East African habitat. When rains cause its burrow to flood, the worm moves to the surface of the soil and forms a temporary opening. The worm then projects its posterior end out through the opening and rolls the sides of the body wall into a pair of folds, forming an open chamber that serves as a kind of "lung." The highly vascularized posterior epithelium enhances the exchange of gases. A number of aquatic oligochaetes can tolerate periods of low available oxygen and even anoxic conditions for short periods of time.

Hirudinoidea. Certain features of the hirudinoidean bauplan demand some sort of circulatory system. Many are relatively large and quite active. The drastic reduction of the coelom and invasion by tissue results in a more or less solid body construction. And, the gut is regionally specialized. As we have seen before, these sorts of characteristics typically evolve in concert with some mechanism of internal transport or adaptive modification in shape as solutions to the surface-to-volume dilemma.

Evolutionarily, the leeches have approached this problem in several ways, including flattening of the body and the formation of extensive gut ceca or diverticula, both of which reduce internal diffusion distances. However, the most important adaptations for internal transport are structural circulatory vessels and channels. In most of the rhynchobdellans, this system is a combination of the ancestral annelid circulatory system and the reduced coelomic spaces; in the arhynchobdellans, the original circulatory system is completely replaced by one derived entirely from the reduced coelom (Figure 13.23). In both of these arrangements the circulatory fluid is moved through the system by the action of contractile vessels and by general body movements.

Gas exchange is accomplished by diffusion across the body wall; gills are present only in the ozobranchids (Figure 13.5B). Some leeches possess hemoglobin in solution in the circulatory fluid, thought to account for about 50 percent of their oxygen-carrying capacity.

Excretion and Osmoregulation

In Chapter 3 we discussed the structural types of nephridial organs in invertebrates. Thus far, we have seen various types of protonephridia, especially among the acoelomate and certain blastocoelomate Metazoa. Most annelids possess some sort of nephromixia, often serially arranged as one pair per segment, with the pore in the segment posterior to the nephrostome. However, variations on this theme are many, and we may view the different conditions as having been derived from a basic primitive plan that arose with the evolution of the coelomate metameric bauplan.

We remind you again that the success of an animal with segmentally arranged coelomic compartments depends on the physical and physiological maintenance of those separate segments. The removal of metabolic wastes (predominantly ammonia) and the regulation of osmotic and ionic balance must occur in each functionally isolated coelomic chamber. (See also the discussion and figures pertaining to excretion and osmoregulation in Chapter 3.)

Polychaeta. The presumed primitive condition in polychaetes is a serially homonomous body with a pair of complete coelomic spaces in each segment. Primitively, each coelom is served by two pairs of ducts that lead to the exterior; each pair includes a coelomoduct and a nephridioduct. The inner end of each coelomoduct bears an open, ciliated funnel through which gametes escape, and the inner end of each nephridioduct bears protonephridia that function in excretion and osmoregulation. This primitive condition has been lost in all but a very few extant polychaetes, but it persists in *Vanadis* (family Alciopidae). In the other few hundred species of polychaetes that possess protonephridia, the coelomoduct and nephridioduct are united to form a protonephromixium (e.g., various phyllodocids; Figure 13.24A).

The vast majority of polychaetes, however, possess metanephridia that open to the coelom by a ciliated nephrostome. In some (e.g., the capitellids), these metanephridia are entirely separate from the coelomoducts, but in most, either the coelomic and nephridial ducts are united to form a metanephromixium, or there is a single interior opening that leads to a single duct as a

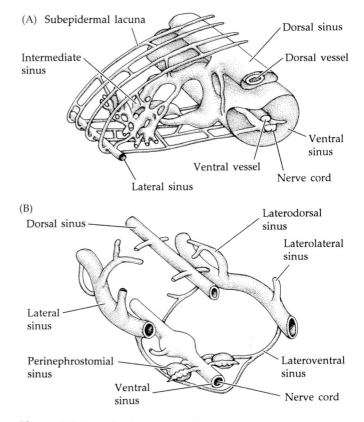

Figure 13.23 Coelomic and circulatory systems of two leeches. (A) A portion of the circulatory and coelomic systems of *Placobdella*, a rhynchobdellan leech in which the circulatory system persists and is associated with the coelomic channels. (B) A portion of the coelomic system in *Hirudo*, an arhynchobdellid leech. Here the circulatory system has been completely replaced by coelomic channels.

mixonephrium (Figure 13.24B). This last case, generally called simply a metanephridium, may represent the complete incorporation of coelomoduct and nephridium into a single organ. However, much of the phylogenetic sequence implied here has not been clearly retained in the ontogeny of living polychaetes. In all of these cases, the functional significance of different arrangements remains much the same in terms of serving a metameric body. That is, each coelomic compartment is equipped with a mechanism for elimination of wastes, for osmoregulation, and for the discharge of gametes.

In certain polychaete groups that possess incomplete septa, or that have lost septa between the coelomic spaces, the number of nephridia is reduced. In some sedentary polychaetes (e.g., serpulids and sabellids) without complete intersegmental septa, there is typically only a single pair of nephridia. These are located at the anterior end of the worm and lead to a single nephridioduct and common pore on the head (Figure 13.24C).

As described in Chapter 3, the open nephrostomes of metanephridia nonselectively pick up coelomic fluids. This action is followed by resorption of materials from

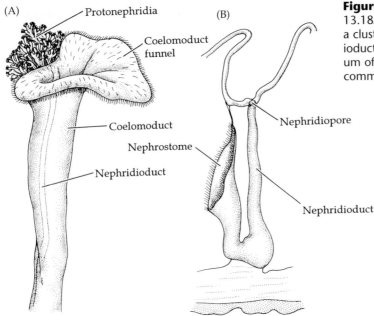

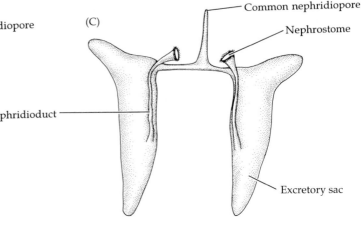

Figure 13.24 Polychaete nephridia (see also Figure 13.18A). (A) A protonephromixium of a phyllodocid. Here a cluster of solenocytic protonephridia sits atop a nephridioduct that joins with the coelomoduct. (B) A mixonephrium of a spionid. (C) A single pair of nephridia joined to a common duct in a serpulid.

the nephridium back into the body, either directly into the surrounding coelomic fluid, or into the blood in cases where extensive nephridial blood vessels are present (e.g., in some nereids and in aphroditids). In either case the composition of the urine is quite different from that of the body fluids; the difference indicates a significant amount of physiological selectivity along the length of the nephridium.

Osmoregulation presents little problem for subtidal polychaetes living in relatively constant osmotic conditions. Intertidal and estuarine forms, however, must be able to withstand periods of stress associated with fluctuations in environmental salinities. There are also a few polychaetes that inhabit fresh water, and a few tropical forms that burrow in damp soil and leaf litter. These animals deal with their osmotic problems by tolerance or regulation, or both. Many species are osmoconformers (e.g., *Arenicola*), allowing the tonicity of their body fluids to fluctuate with changes in the environmental salinity. Most polychaete osmoconformers have relatively simple metanephridia, with comparatively short nephridioducts and correspondingly weaker resorptive and regulatory capacities. Some also have relatively thin body wall musculature, and the body swells when in a hypotonic medium. It is likely that burrowers and tube-dwellers face less osmotic stress than epibenthic forms, because the water in their tubes may be less subject to ionic variation than the overlying water.

Osmoregulators, such as a number of estuarine nereids, often have thicker body walls that tend to resist changes in shape and volume. When water enters the body from a hypotonic surrounding, the increased hydrostatic pressure generated within the coelom works against that osmotic gradient. In addition, regulators are

able to maintain (within limits) a more or less constant internal fluid tonicity because of the greater selective capabilities of their more complex nephridia.

Oligochaeta. Oligochaetes possess paired, segmentally arranged metanephridia, usually in all but the extreme anterior and posterior segments. These nephridia are similar to the mixonephria of polychaetes, but many show various secondary modifications or elaborations.

A typical oligochaete nephridium is composed of a preseptal nephrostome (either open to the coelom or secondarily closed as a bulb), a short canal that penetrates the septum, and a postsegmental nephridioduct that is variably coiled and sometimes dilated as a bladder (Figure 13.25). The nephridiopores are usually located ventrolaterally on each segment.

Aquatic oligochaetes are ammonotelic, but most terrestrial forms are at least partially ureotelic. These wastes are transported to the nephridia via the circulatory system and by diffusion through the coelomic fluid. Uptake of materials into the nephridial lumen is partly nonselective (in those worms with open nephrostomes) from the coelom, and partly selective across the walls of the nephridioduct from the afferent nephridial blood vessels. A significant amount of selective resorption occurs into the efferent blood flow along the distal portion of the nephridioduct, facilitating efficient excretion as well as ionic and osmoregulation (Figure 13.25).

The precise role of the chloragogen cells in excretion is not fully understood. While it is known that protein deamination and nitrogenous waste formation occur within these cells, the method of elimination of this waste is unknown. Individual chloragogen cells break free into the coelom and are probably engulfed by

Figure 13.25 *Lumbricus* nephridia. (A) A single nephridium and its relationship to a septum. (B) Details of the nephrostome. Evidence suggests that earthworm nephridia are highly selective excretory and osmoregulatory units. The nephridioduct is regionally specialized along its length. The narrow tube receives body fluids and various solutes, first from the coelom through the nephrostome and then from the blood via capillaries that lie adjacent to the tube. In addition to various forms of nitrogenous wastes (ammonia, urea, uric acid), certain coelomic proteins, water, and ions (Na⁺, K⁺, Cl⁻) are also picked up. Apparently, the wide tube serves as a site of selective reabsorption (probably into the blood) of proteins, ions, and water, leaving the urine rich in nitrogenous wastes.

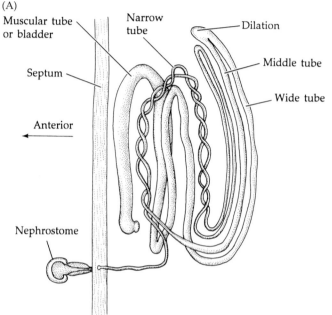

(A)

Muscular tube or bladder

Narrow tube

Dilation

Middle tube

Septum

Wide tube

Anterior

Nephrostome

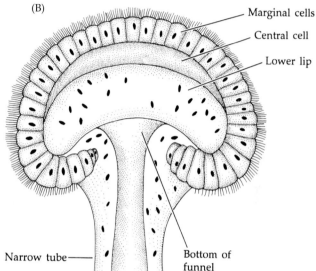

(B)

Marginal cells

Central cell

Lower lip

Narrow tube

Bottom of funnel

phagocytic amebocytes that apparently accumulate wastes in a precipitated form. How, or if, these waste-filled cells are ever actually eliminated from the body remains an unanswered question.

Ionic and osmoregulation are of utmost importance to freshwater and terrestrial soft-bodied invertebrates such as oligochaetes. The moist, permeable surface necessary for gas exchange and the severe osmotic gradients across the body wall present potentially serious problems of water loss to terrestrial forms, and of water gain to freshwater forms; both face the loss of precious diffusible salts. Passive diffusion of water and salts also occurs across the gut wall.

The major organs of water and salt balance in freshwater oligochaetes are, of course, the nephridia. Excess water is excreted and salts are retained by selective and active resorption along the nephridioduct. The problem in terrestrial forms is more serious. Surprisingly, earthworms are not absolute osmoregulators, rather they lose and gain water according to the amount of water in their environment. Various species can tolerate a loss of 20 to 75 percent of their body water and still recover. Under normal conditions, water conservation by earthworms is probably accomplished in several ways. The production of urea allows the excretion of a relatively hypertonic urine compared with that of a strictly ammonotelic animal. There may also be active uptake of water and salts from food across the gut wall. Certainly there are behavioral adaptations for remaining in relatively moist environments in addition to the physiological adaptations that allow these animals to tolerate temporary partial dehydration of their bodies.

Hirudinoidea. The excretory structures of hirudinoideans are structurally different from those of oligochaetes and polychaetes, but they are presumably derived from metanephridia. Leech nephridia are paired and segmentally arranged but are usually absent from several anterior and posterior segments. The nephrostomes are ciliated funnels associated with coelomic circulatory vessels, an arrangement that probably evolved with the reduction in the main body coelom and the loss of septa. Some hirudineans possess clusters of nephrostomes called **ciliated organs**. Each nephridium leads ultimately to a ventrolateral nephridiopore. It is, however, the microscopic structure of each unit between the nephrostome and external pore that is so remarkably different from other metanephridia (Figure 13.26).

The nephrostome leads not into an open duct but into a blind chamber called the **nephridial capsule**. The capsule is connected to a "nephridioduct," uniquely composed of a single row of cells through which an intracellular canal runs. This canal appears to be somewhat transitory, especially near the capsule, in that it forms from the coalescence of tiny intracellular tubules and vacuole-like chambers. Exactly how this arrangement works is unclear, but its structure suggests a good

(A)

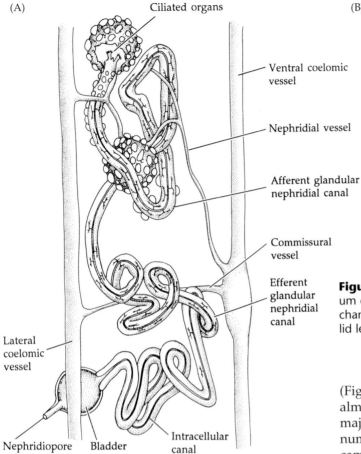

Ciliated organs

Ventral coelomic vessel

Nephridial vessel

Afferent glandular nephridial canal

Commissural vessel

Efferent glandular nephridial canal

Lateral coelomic vessel

Nephridiopore   Bladder   Intracellular canal

(B)

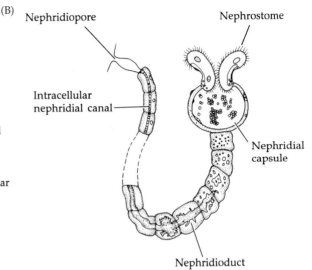

Nephridiopore

Nephrostome

Intracellular nephridial canal

Nephridial capsule

Nephridioduct

Figure 13.26   Leech nephridia. (A) A complex nephridium of *Erpobdella* and its association with the coelomic channels. (B) Details of a nephridium of an arhynchobdellid leech.

deal of selectivity during urine formation. Selective filtration and resorption would be expected in an animal whose excretory units directly drain the circulatory fluid. The intracellular nephridial canal connects with a short chamber derived from an ectodermal invagination at the nephridiopore. In some true leeches a relatively large bladder is formed near the pore (Figure 13.26).

Ammonia is the main nitrogenous waste product eliminated via the nephridia. Apparently, particulate waste materials are engulfed by phagocytes, both in the coelomic fluid and in the "mesenchyme," but the eventual disposition of this material is not known.

The nephridia of freshwater leeches also serve as osmoregulatory organs. The urine is very dilute, a fact suggesting the excretion of excess water and the retention of various salts. In certain terrestrial leeches the urine from the anterior and posterior nephridiopores is released onto the surfaces of the suckers, thereby providing a moist surface for effective suction.

Nervous System and Sense Organs

The fundamental plan of the central nervous system in annelids (as in protostomes in general) includes a dorsal cerebral ganglion, paired circumenteric connectives, and one or more ventral longitudinal nerve cords

(Figure 13.27). The central nervous system of annelids almost certainly arose from a ladder-like system. The major trends in annelids have been a reduction in the number of longitudinal cords as the nervous system became more centralized and concentrated, and the development of segmentally arranged ganglia along the longitudinal cord(s) associated with the metameric bauplan.

Polychaeta.   The cerebral ganglion of polychaetes is usually bilobed and lies within the prostomium. One or two pairs of circumenteric connectives extend from the cerebral ganglion around the foregut and unite ventrally in the subenteric ganglion. Primitively, a pair of longitudinal nerve cords arises from the subenteric ganglion and extends the length of the body (Figure 13.27C). Ganglia are arranged along these nerve cords, one pair in each segment, and are connected by transverse commissures. Lateral nerves extend from each ganglion to the body wall and each bears a so-called **pedal ganglion**. This double nerve cord arrangement is common in certain groups of polychaetes, including sabellids and serpulids. Interestingly, in the amphinomids there are four longitudinal nerve cords, a medial pair and a lateral pair, the latter connecting the pedal ganglia. Some workers consider the amphinomid condition primitive, while others contend that the lateral longitudinal cords have been secondarily derived within this group. Similar but perhaps nonhomologous lateral longitudinal cords appear in some other polychaete taxa that are considered to be relatively advanced. It may be that the genetic potential for additional lateral longi-

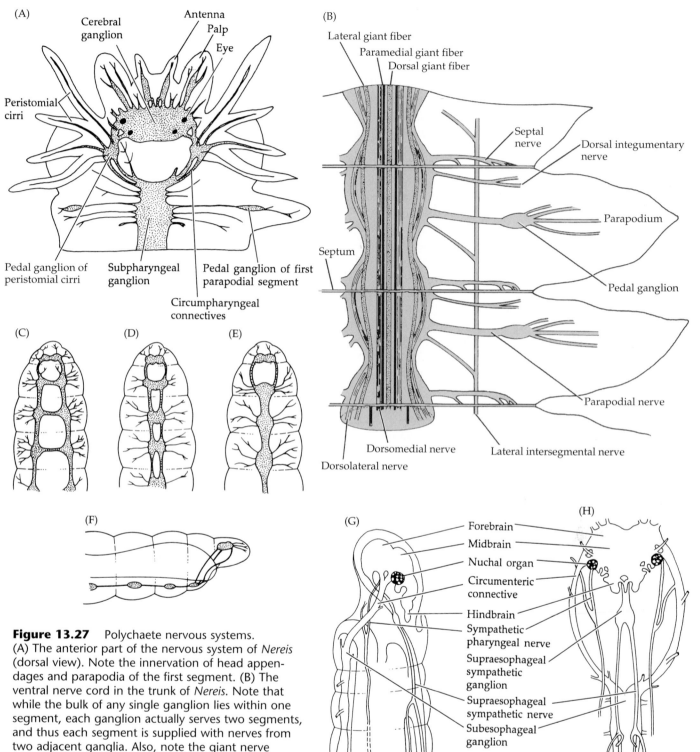

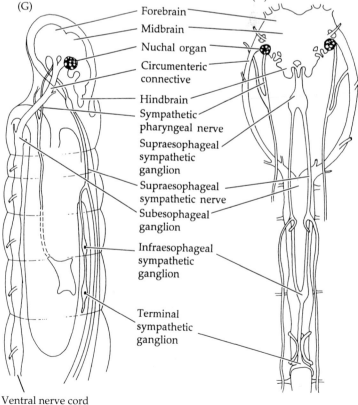

Figure 13.27 Polychaete nervous systems. (A) The anterior part of the nervous system of *Nereis* (dorsal view). Note the innervation of head appendages and parapodia of the first segment. (B) The ventral nerve cord in the trunk of *Nereis*. Note that while the bulk of any single ganglion lies within one segment, each ganglion actually serves two segments, and thus each segment is supplied with nerves from two adjacent ganglia. Also, note the giant nerve fibers. (C–E) Suspected evolutionary sequence of progressive fusion of the ventral nerve cords in various polychaetes. (C) The presumed primitive ladder-like condition. (D) Ganglionic fusion with separate cords. (E) The single nerve cord condition. (F) Lateral view of a generalized polychaete nervous system. Note that the cerebral ganglion is located within the prostomium, unlike the condition in clitellate annelids (see Figure 13.30). (G,H) Some details of the anterior nervous systems of a eunicid polychaete, *Eunice* (lateral and dorsal views). The cerebral ganglion is specialized into fore-, mid-, and hindbrain.

tudinal nerve cords can be activated by some mechanism such as a homeobox "switch."

The evolutionary trend among most polychaetes has been the fusion of the medial nerve cords to form a single midventral longitudinal cord (Figure 13.27C–E). The degree of fusion varies among taxa; some retain separate nerve tracts within the single cord. In addition, the position of the ventral nerve cord varies. Primitively the cord is subepidermal, but in more advanced forms it lies internal to the body wall muscle layers.

The cerebral ganglion is often specialized into three regions, typically called the **forebrain**, **midbrain**, and **hindbrain**. Generally, the forebrain innervates the prostomial palps, the midbrain the eyes and prostomial antennae or tentacles, and the hindbrain the chemosensory **nuchal organs** (Figure 13.27A,G, and H). The circumenteric connectives arise from the fore- and midbrains. The midbrain also gives rise to a complex of motor **stomatogastric nerves** associated with the foregut, especially with the operation of the proboscis or pharynx. The circumenteric connectives often bear ganglia from which nerves extend to the peristomial cirri, or else these appendages are innervated by nerves from the subenteric ganglion. The subenteric ganglion appears to exhibit excitatory control over the ventral nerve cord(s) and segmental ganglia.

The nerves that arise from the segmental ganglia innervate the body wall musculature and parapodia (via the pedal ganglia), and the digestive tract. The ventral nerve cord and sometimes the lateral nerves of most annelids contain some extremely long neurons, or **giant fibers**, of large diameter; these neurons facilitate rapid, "straight-through" impulse conduction, bypassing the ganglia (Figure 13.27B). Giant fibers are apparently lacking in some polychaetes (e.g., syllids), but are well developed in tube-dwellers, such as sabellids and serpulids, permitting rapid contraction of the body and retraction into the tube.

Polychaetes as a group possess an impressive array of sensory receptors. As would be expected, the kinds of sense organs present and the degree of their development vary greatly among polychaetes with different lifestyles. Certainly, the requirements for particular sorts of sensory information are not the same for a tube-dwelling sabellid as they are for an errant predatory nereid or a burrowing arenicolid.

In general, polychaetes are highly touch-sensitive. Crawlers, tube dwellers, and burrowers depend on tactile reception for interaction with their immediate surroundings (locomotion, anchorage within their tube, and so on). Touch receptors are distributed over much of the body surface but are concentrated in such areas as the head appendages and parts of the parapodia. The chaetae are also typically associated with sensory neurons and serve as touch receptors. Some burrowers and tube dwellers have such a strong positive response to contact with the walls of their burrow or tube that the response dominates all other receptor input. Some of these polychaetes will remain in their burrow or tube regardless of other stimuli that would normally produce a negative response.

Most polychaetes possess photoreceptors, although these structures are lacking in many burrowers. The best developed polychaete eyes occur in pairs on the dorsal surface of the prostomium. In some there is a single pair of eyes (e.g., most phyllodocids); in many there are two or more pairs (e.g., nereids, polynoids, hesionids, many syllids). These prostomial eyes are direct pigment cups. They may be simple depressions in the body surface lined with retinular cells, or they may be quite complex, with a distinct refractive body or lens (Figure 13.29A–D). In nearly all cases, the eye units are covered by a modified section of the cuticle that functions as a cornea. The eyes of most polychaetes are capable of transmitting information on light direction and intensity, but in certain pelagic forms (e.g., alciopids) the eyes are huge and possess true lenses capable of accommodation and perhaps image perception (Figure 13.29C,D).

In addition to, or instead of, the prostomial eyes, some polychaetes bear photoreceptors on other parts of the body. Peristomial eyes occur in the dorvilleid *Ophryotrocha*. A few species bear simple eyespots along the length of the body (e.g., the opheliid *Polyophthalmus*). Pygidial eyespots occur in newly settled sabellariids and some adult sabellids (small ones such as *Fabricia*). Interestingly, in these cases the animals crawl backward. Many sabellids and serpulids possess complex eyes or simple ocelli on the branchial crown tentacles and react to sudden decreases in light intensity by retracting into their tubes. This "shadow response" helps these sedentary worms avoid predators and can easily be demonstrated by passing one's hand to cast a shadow over a live worm.

Nearly all polychaetes are sensitive to dissolved chemicals in their environment. Most of the chemoreceptors are specialized cells that bear a receptor process extending through the cuticle (Figure 13.28). Sensory nerve fibers extend from the base of each receptor cell. Such simple chemoreceptors are often scattered over much of the worm's body, but they tend to be concentrated on the head and its appendages. Some polychaetes also possess ciliated pits or slits called **nuchal organs**, which are presumed to be chemosensory (Figure 13.29E,F). These structures are typically paired and lie posteriorly on the dorsal surface of the prostomium. In some forms (e.g., certain nereids) the nuchal organs are simple depressions, whereas in others (e.g., opheliids, most archiannelids) they are rather complex eversible structures equipped with special retractor muscles. In members of the family Amphinomidae, the nuchal organs are elaborate outgrowths of an extension of the prostomium called the **caruncle**.

Statocysts are common in some burrowing and tube-dwelling polychaetes (e.g., certain terebellids, arenicol-

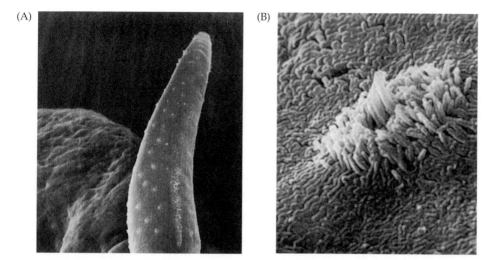

Figure 13.28 Epithelial sense organs (probably chemoreceptors) on the polychaete *Nereis*. (A) A dorsal cirrus of a parapodium showing distribution of sense organs (SEM). (B) A single sense organ (SEM).

ids, and sabellids). A few forms possess several pairs of statocysts, but most have just a single pair, located near the head. These statocysts may be closed or open to the exterior, and the statolith may be a secreted structure or formed of extrinsic material, such as sand grains. It has been demonstrated experimentally that the statocysts of some polychaetes do serve as georeceptors and help maintain proper orientation when the bearer is burrowing or tube building.

A number of other structures of presumed sensory function occur in some polychaetes. These structures are often in the form of ciliated ridges or grooves occurring on various parts of the body and associated with sensory

Figure 13.29 Polychaete photoreceptors and nuchal organs. (A) Simple pitlike eye of a chaetopterid. (B) Lensed pigment cup eye of a nereid. (C) A complex eye (section) of an alciopid (*Vanadis*). (D) The head of *Vanadis* (ventral view). Note the large eye lobes. (E) Nuchal organ of *Arenicola*. (F) Nuchal organs of *Notomastus*.

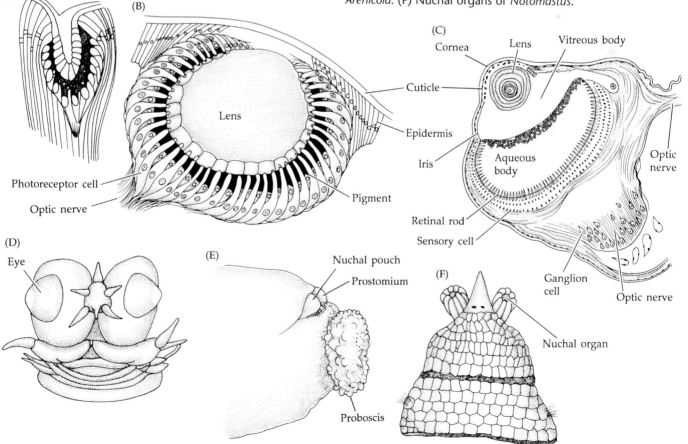

neurons. A variety of names have been applied to these structures, but in most cases their function(s) remains unclear.

Polychaetes also possess organs or tissues of neurosecretory or endocrine functions. Most of the secretions appear to be associated with the regulation of reproductive activities, as discussed in the following section.

Oligochaeta. The central nervous system of oligochaetes consists of the usual annelid components: a supraenteric cerebral ganglion joined to a ganglionated ventral nerve cord by circumenteric connectives and a subenteric ganglion (Figure 13.30). With the reduction in head size, especially of the prostomium, the cerebral ganglion occupies a more posterior position than in the polychaetes, often lying as far back as the third body segment. The paired ventral nerve cords are almost always fused as a single tract in oligochaetes, and it usually contains some giant fibers, similar to those of many polychaetes.

The cerebral ganglion gives rise to several anteriorly directed prostomial nerves, most of which are sensory. The circumenteric connectives and segmental ganglia give rise to sensory and motor nerves to the body wall and various organs in each segment. As in the polychaetes, it is the subenteric ganglion that appears to be the center for motor control of body movements; the cerebral ganglion mediates these activities by inhibitory influences.

The independent but coordinated action of each segment during locomotion depends on a series of stimulus–response reactions involving the segmental ganglia, but these reactions are initiated by the subenteric ganglion. If that ganglion is removed, all movement ceases; but if the cerebral ganglion is removed, normal movement continues but responses to external stimuli are ab-

sent. Oligochaetes also possess special cells within the body wall muscles that serve as stretch receptors. These sensory units supply feedback to the ventral nerve cord about the state of the muscles in each segment, and thus constitute a sophisticated system to coordinate contraction and relaxation of segmental muscles during locomotion and burrowing.

The sense organs of oligochaetes are clearly associated with their habits. The general name of **epithelial sense organs** has been given to a variety of receptor units distributed over most of the body. These receptors can be free nerve endings within the epidermis or clusters of special receptor cells associated with various bumps and tubercles. Many of these structures are undoubtedly tactile in function, providing an important source of information during burrowing and crawling. Others are suspected to be chemoreceptors that supply important information about the relatively unstable freshwater or terrestrial environment of oligochaetes. Many oligochaetes are highly sensitive to changes in pH and to the secretions of other worms. Chemoreception may also play a role in the location and selection of food items.

Some freshwater oligochaetes possess paired pigment cup ocelli at the anterior end; nearly all others bear simple photoreceptors distributed over the entire body surface. These worms are generally negatively phototactic to bright light.

Hirudinoidea. The nervous system of leeches has received a great deal of attention from zoologists. The leech nervous system—even that of large leeches—is composed of very few neurons, and these individual nerve cells are sufficiently large that their circuitry has been traced in great detail. (e.g., Nicholls and Van Essen 1974.)

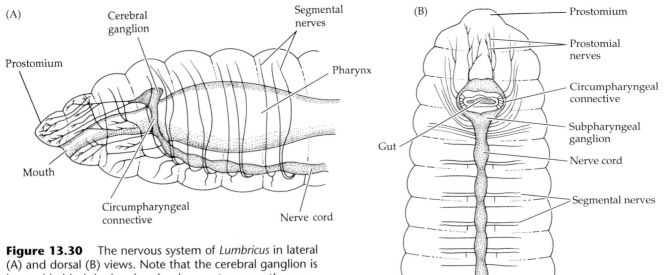

Figure 13.30 The nervous system of *Lumbricus* in lateral (A) and dorsal (B) views. Note that the cerebral ganglion is located behind the head, a development apparently associated with a reduction in the size of the prostomium.

The hirudinoidean nervous system includes a cerebral ganglion that is usually set back from the anterior end of the body at about the level of the pharynx (Figure 13.31). The cerebral ganglion, circumenteric connectives, and subenteric ganglion together form a rather thick nerve ring around the foregut. Two longitudinal nerve cords arise from the ventral portion of this ring and extend posteriorly through the body. The nerve cords are separate in some areas, but the segmental ganglia are fused. Peripheral nerves include abundant sensory neurons from the cerebral ganglion and segmentally arranged motor and sensory neurons from the ventral nerve cord ganglia.

As in other annelids, hirudinoideans employ neurosecretions to control certain activities, including rapid color changes in some species. Reproductive activities may also be under neurosecretory control.

Some hirudinoideans, especially true leeches, are extremely sensitive to certain types of environmental stimuli, although their sensory receptors are relatively simple in structure. These animals possess an array of epidermal sense organs similar to those found in oligochaetes. In addition, leeches have from two to ten dorsal eyes of varying complexity, and special sensory papillae that bear bristles extending from the body surface. The presence of the bristles suggests touch sensitivity, but the exact function of the papillae is unknown.

In fact, except for the eyes, the functions of various leech sense organs are not well understood at all, and most of the information is based on behavioral responses to different stimuli.

Leeches tend to be negatively phototactic. However, some of the blood-sucking species react positively to light when preparing to feed. This behavioral change is presumably adaptive, causing the leech to move into areas where a host encounter is more likely. Most leeches can also detect movement in their surroundings, as evidenced by their responses to shadows passing over them. This reaction has been noted particularly in leeches that attack fishes. Again, the adaptive significance of this behavior may be in facilitating encounters with hosts by responding with increased movement when a fish passes overhead.

Leeches also respond to mechanical stimulation in the forms of direct touch and vibrations in their environments. They are also chemosensitive and attracted to the secretions of potential hosts. Some aquatic and even terrestrial leeches that prefer warm-blooded mammalian hosts are apparently attracted to points of relatively high temperatures in their surroundings, thus aiding in food location. Standing in a leech-inhabited pond is a great way to observe first hand this rapid response, as these animals will detect your presence and begin swimming or crawling toward you within seconds.

Regeneration and Asexual Reproduction

Polychaeta. Polychaetes show various degrees of regenerative capabilities. Nearly all of them are capable of regenerating lost appendages such as palps, tentacles, cirri, and parapodia. Most of them can also regenerate posterior body segments if the trunk is severed.

There are numerous exceptional cases of the regenerative powers of polychaetes. While regeneration of the posterior end is common, most cannot regenerate lost heads. However, sabellids, syllids, and some others can regrow the anterior end. The most dramatic regenerative powers among the polychaetes occur, oddly, in a few forms with highly specialized and heteronomous bodies. In *Chaetopterus*, for example, the anterior end will regenerate a normal posterior end as long as the regenerating part (the anterior end) includes not more than fourteen segments; if the animal is cut behind the fourteenth segment, regeneration does not occur. Furthermore, any single segment from among the first fourteen can regenerate anteriorly and posteriorly to produce a complete worm (Figure 13.32A). An even more dramatic example of regenerative power is known among certain species of *Dodecaceria* (Cirratulidae), which are capable of fragmenting their bodies into individual segments, each of which can regenerate a complete individual!

Regeneration appears to be controlled by neuroendocrine secretions released by the central nervous system at sites of regrowth. It is initiated by severing the elements of the nervous system. Initiation has been

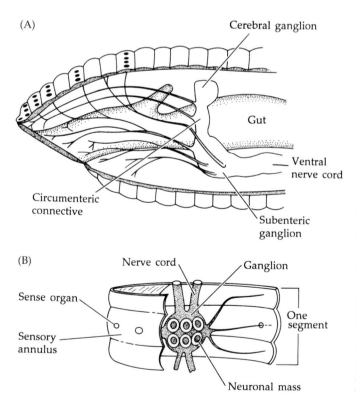

Figure 13.31 Leech nervous system. (A) The anterior nervous system (lateral view). (B) A generalized leech segment comprising three annuli, cut away to show segmental ganglion and innervation of epithelial sense organs.

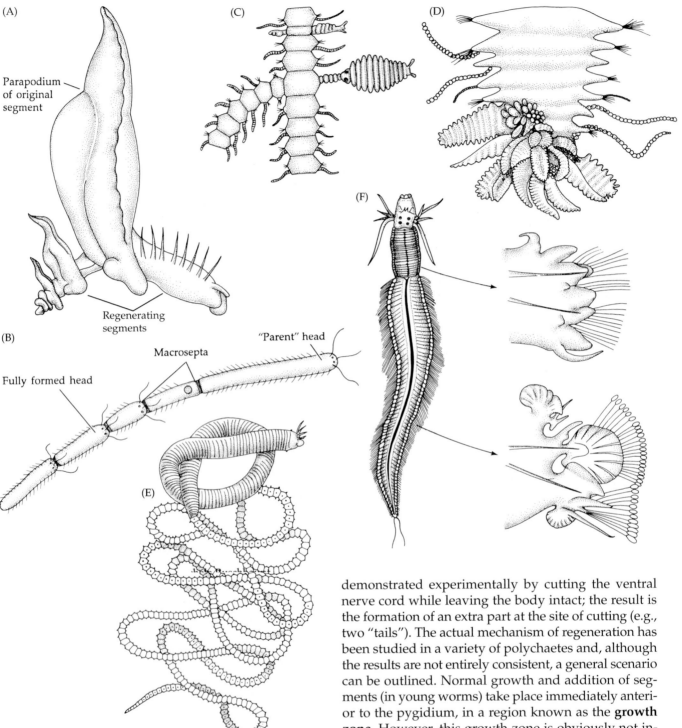

Figure 13.32 Regeneration, asexual reproduction, and epitokey in polychaetes. (A) The remarkable regeneration of a chaetopterid from a single excised segment (in this case, a fan parapodia segment). (B) Asexual reproduction by transverse partitioning in a syllid (see text). (C) A portion of another syllid in which reproductive individuals are budded from the parent's parapodia (*Syllis*). (D) The posterior end of *Typhlosyllis*, bearing a cluster of asexually produced epitokes. (E) The epitokous palolo worm, *Palola viridis*. (F) An epitokous nereid, *Nereis irrorata*. Note the dimorphic condition of the anterior and posterior parapodia.

demonstrated experimentally by cutting the ventral nerve cord while leaving the body intact; the result is the formation of an extra part at the site of cutting (e.g., two "tails"). The actual mechanism of regeneration has been studied in a variety of polychaetes and, although the results are not entirely consistent, a general scenario can be outlined. Normal growth and addition of segments (in young worms) take place immediately anterior to the pygidium, in a region known as the **growth zone**. However, this growth zone is obviously not involved in regeneration. Rather, when the trunk is severed, the cut region heals over and then a patch of generative tissue, or **blastema**, forms. The blastema comprises an inner mass of cells originating from nearby tissues that were derived originally from mesoderm, and an outer covering of cells from ectodermally derived tissues such as the epidermis. These two cell masses act somewhat as a growth zone analogue, proliferating new body parts according to their tissue origins. This process is coupled with the growth of the gut, which contributes parts of entodermal origin.

In addition, workers have shown that relatively undifferentiated cells from mesenchyme-like layers of the body migrate to injured areas and contribute to various (and uncertain) degrees to the regenerative process. These so-called **neoblast cells** are ectomesodermal in origin, because they arise embryonically from presumptive ectoderm. During regeneration, they apparently contribute to tissues and structures normally associated with true mesoderm, and perhaps other germ layers as well. The implication here is that the germ layer of the precursor of a regenerated part may not correspond to the normal origin of that part. For example, regenerated coelomic spaces may be lined with tissue derived originally from ectoderm rather than from mesoderm.

A number of polychaetes use their regenerative powers for asexual reproduction. A few reproduce asexually by multiple fragmentation. We mentioned above the ability of *Dodecaceria* to regenerate complete individuals from isolated segments; this phenomenon occurs spontaneously and naturally in these animals as an efficient reproductive strategy. Spontaneous transverse fragmentation of the body into two or several groups of segments also occurs in certain syllids, chaetopterids, cirratulids, and sabellids (Figure 13.32B). The point (or points) at which the body fragments is typically species-specific, and can be anticipated by an ingrowth of the epidermis that produces a partition across the body called a **macroseptum**. Asexual reproduction results in a variety of regeneration patterns, including chains of individuals, budlike outgrowths, or direct growth to new individuals from isolated fragments. Asexual reproduction in polychaetes may be under the same sort of neurosecretory control as that postulated for nonreproductive regeneration.

Oligochaeta. The ability of oligochaetes to regenerate parts of the body varies greatly among species. Many can regenerate almost any excised body part and can regrow both front and back ends, whereas others, such as *Lumbricus terrestris*, have very weak regenerative powers. A polarity exists from anterior to posterior, with rear segments generally being more easily regrown than front ones. The mechanisms of regeneration may be similar to those of polychaetes.

In contrast to polychaetes, oligochaete species tend to have a more fixed number of body segments, a condition that has implications in the regenerative process. Data from regeneration experiments on certain earthworms indicate that most oligochaetes never regenerate more segments than were possessed by the original worm. In fact, the regenerated worms tend to possess the same number of segments as the original. The mechanism controlling this regeneration of a predictable number of segments is unknown, but Moment (1953) offered an interesting theory. At least some oligochaetes (e.g., earthworms) have a measurable voltage difference along the length of their bodies, each segment having a slightly different electrical potential from the next. Moment suggested that regeneration ceases when the original, overall voltage potential is regained by the regrowth of the proper number of segments. Presumably the normal electrical gradient thus produced imparts an inhibitory effect on the regeneration process.

Most freshwater oligochaetes are capable of asexual reproduction. In some members of the family Naididae, sexual reproduction is very rare, and in a few it may not occur at all. Usually, however, asexual reproduction is a seasonal event, alternating with sexual activity. Rapid asexual reproduction usually occurs in early to mid summer. Thus the worms take advantage of mild conditions and abundant food supplies. The offspring mature and reproduce sexually in the late summer and early fall and produce overwintering stages that hatch in the spring.

Oligochaetes reproduce asexually by one or more forms of transverse fission. In some it is by fragmentation at one or several points along the body, followed by regeneration of each fragment to a new worm. In others, buds are produced on the parent's body as "offspring precursors." Once the new individuals are partially formed, fission occurs, and the offspring break free.

Hirudinoidea. Asexual reproduction is unknown among the hirudinoideans.

Sexual Reproduction and Development

Polychaeta. The great majority of polychaetes are dioecious; hermaphroditism is known in serpulids, certain freshwater nereids, and isolated cases in other families (Berglund 1986; Franke 1986). Some syllids are protogynic, and some eunicids are protandric. The general de-emphasizing of hermaphroditic strategies among polychaetes seems odd, given that they lack permanent gonads or other complex reproductive organs. Rather, the gametes arise by proliferation of cells from the peritoneum, these being released into the coelom as gametogonia or primary gametocytes. Formation of gametes may occur throughout the body or only in particular regions of the trunk. Within a reproductive segment, the production of gametes may occur all over the coelomic lining or only on specific areas.

The gametes generally mature within the coelom and are released to the outside by mechanisms such as gonoducts, coelomoducts, nephridia, or a simple rupture of the parent body wall. Many species release eggs and sperm into the water, where external fertilization is followed by fully indirect development with a planktotrophic larval stage. Others display mixed life history patterns. In these forms, fertilization is internal, followed by brooding or by the production of floating or attached egg capsules. In most instances the embryos are released as free-swimming larvae. Some species brood their embryos on the body surface.

Many of the free-spawning polychaetes have evolved methods that ensure relatively high rates of fertilization. One of these methods is the fascinating phenomenon of **epitoky**, characteristic of many benthic syllids, nereids, and eunicids. This phenomenon involves the production of a sexually reproductive worm called an **epitokous individual**. Epitokous forms may arise from nonreproductive (**atokous**) animals by a transformation of an individual worm, as in most nereids and eunicids, or by the asexual production of new epitokous individuals, as in most syllids. In some forms, the whole body may transform into a sexual individual called a **heteronereid** or **epitoke**. In others, only the posterior body segments (again called the epitoke) become swollen and filled with gametes, and their associated parapodia become enlarged and natatory (Figure 13.32F). In cases where the epitokous worm is asexually produced, the reproductive individual is often without a head and lacks the atokous anterior end; such epitokes are formed as single or clusters of outgrowths from particular body regions (Figure 13.32D).

In any event, the epitokes are gamete-carrying bodies capable of swimming from the bottom upward into the water column, where the gametes are released. Epitoky is controlled by neurosecretory activity, and the upward migration of the epitokes is precisely timed to synchronize spawning within a population. The reproductive swarming of epitokes is linked with lunar periodicity. This activity not only ensures successful fertilization but establishes the developing embryos in a planktonic habitat suitable for the larvae. Perhaps the most famous of the epitokous worms are the palola worms of the South Seas, first described from a series of posterior ends (epitokes) collected by a nineteenth-century British expedition to the Samoas. Native Polynesian islanders have long been known to predict the swarming (typically to the day and hour) and collect the ripe epitokes under a full moon and feast on them.

Early polychaete development exemplifies a classic protostomous, spiralian pattern. The eggs are telolecithal with small to moderate quantities of yolk. Those with a period of encapsulation or brooding prior to larval release generally contain more yolk than those that free spawn and develop quickly to planktotrophic larvae. In any case, cleavage is holoblastic and clearly spiral. A coeloblastula or, in the cases of more yolky eggs, a stereoblastula (Figure 13.33) develops and undergoes gastrulation by invagination, epiboly, or a combination of these two events. Gastrulation results in the internalization of the presumptive entoderm (the 4A, 4B, 4C, 4D and the 4a, 4b, 4c cells) and presumptive mesoderm (the 4d mesentoblast). The derivatives of the first three micromere quartets give rise to ectoderm and ectomesoderm, the latter producing various larval muscles between the body wall and the developing gut. As the entoderm hollows as the archenteron, a stomodeal invagination forms at the site of the blastopore and a proctodeal invagination produces the hindgut.

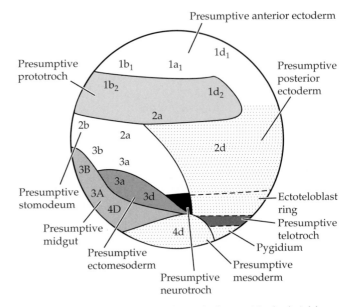

Figure 13.33 Fate map of a polychaete (*Scoloplos*) blastula (viewed from the left side).

In many polychaetes, these early ontogenetic events result in a **trochophore larva**, a form characteristic of many protostomous groups (e.g., molluscs, sipunculans, and echiurans). The early, presegmental trochophore (Figure 13.34A) is characterized by a locomotor ciliary band just anterior to the region of the mouth. This ciliary band, the **prototroch**, arises from special cells, called **trochoblasts**, of the first and second quartets of micromeres. Most trochophore larvae also bear an **apical ciliary tuft** associated with an **apical sense organ** derived from a plate of thickened ectoderm at the anterior end. In addition, there is often a perianal ciliary band called the **telotroch** and a midventral band called the **neurotroch**. By this stage the mesentoblast has divided to form a pair of cells called **teloblasts**, which in turn proliferate a pair of mesodermal bands, one on each side of the archenteron in the region of the hindgut, an area known as the **growth zone** (Figure 13.35B). Many trochophores bear larval sense organs such as ocelli, as well as a pair of larval protonephridia. Many trochophores bear bundles of mobile chaetae, which are known to serve as a defense against predators (Pennington and Chia 1984) and to help retard sinking. Several polychaete larvae are shown in Figure 13.34.

The larva grows and elongates by proliferation of tissue in the growth zone (Figure 13.35), while segments are produced by the anterior proliferation of mesoderm from the teloblast derivatives on either side of the gut. These packets of mesoderm hollow (schizocoely) and expand as paired coelomic spaces, which eventually obliterate the blastocoel. Thus the production of serially arranged coelomic compartments and the formation of segments are one and the same; the anterior and posterior walls of adjacent coelomic compartments form the

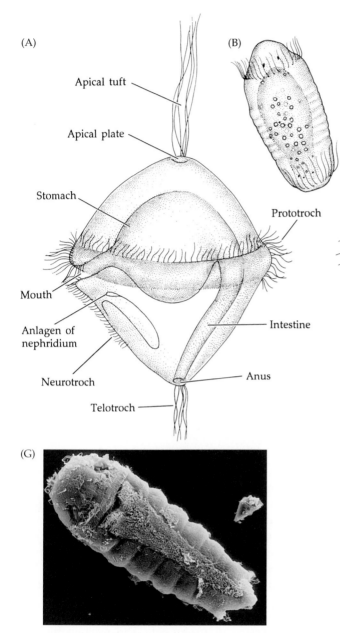

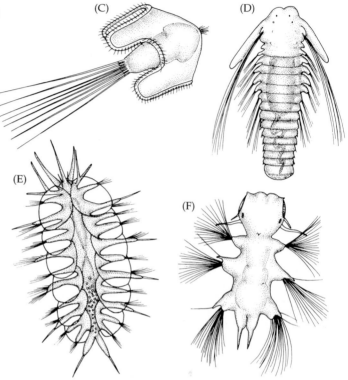

Figure 13.34 Some polychaete larvae—trochophores and beyond. (A) Anatomy of a generalized early trochophore larva. (B) Young trochophore of a capitellid. (C) Trochophore of *Owenia*. (D) Late spionid larva. (E) Late larval stage of a polynoid (scale worm). (F) Three-segmented nectochaeta "larva" of a nereid. (G) SEM of a polychaete polytroch larva, ventral view.

intersegmental septa. Proliferation of segments by this process is called **teloblastic growth**. Externally, additional ciliary bands are added at each segment. These **metatrochal bands** aid in locomotion as the animal increases in size. Such segmented larvae are sometimes called **polytroch larvae**.

The fates of the various larval regions are now apparent (Figure 13.35). The region anterior to the prototrochal ring becomes the prostomium, while the prototrochal area forms the peristomium. Note that these two parts are not involved in the proliferation of segments and are thus presegmental. However, in some polychaetes, one or more of the anterior trunk segments may be incorporated into the peristomium during growth. The segmental, metatrochal portion of the larva forms the trunk, and the growth zone and postsegmental pygidium remain as the corresponding adult body parts. The apical sense organ becomes the cerebral gan-

glion, which is eventually joined with the developing ventral nerve cord by the formation of circumenteric connectives. The body continues to elongate as more segments form, and the juvenile worm finally drops from the plankton and assumes the lifestyle of a young polychaete. This whole affair was beautifully described in verse by the late Walter Garstang (1951), where he explains the development of *Phyllodoce* in the first part of his classic poem, *The Trochophores*:

The trochophores are larval tops the Polychaetes
    set spinning
With just a ciliated ring—at least in the beginning—
They feed, and feel an urgent need to grow
    more like their mothers,
So sprout some segments on behind, first one,
    and then the others.
And since more weight demands more power,
    each segment has to bring
Its contribution in an extra locomotive ring:
With these the larva swims with ease, and,
    adding segments more,
Becomes a *Polytrochula* instead of *Trochophore*.
Then setose bundles sprout and grow, and the
    sequel can't be hid:
The larva fails to pull its weight, and sinks—
    an Annelid.

(A)

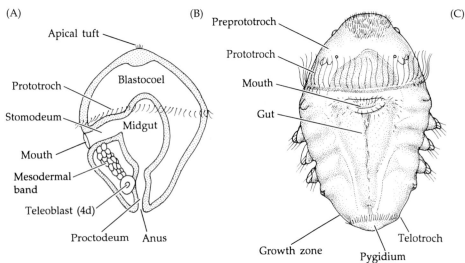

Apical tuft

Prototroch

Blastocoel

Stomodeum

Midgut

Mouth

Mesodermal band

Teleoblast (4d)

Proctodeum   Anus

(B)

Preprototroch

Prototroch

Mouth

Gut

Growth zone

Telotroch

Pygidium

(C)

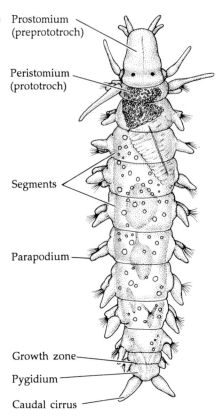

Prostomium (preprototroch)

Peristomium (prototroch)

Segments

Parapodium

Growth zone

Pygidium

Caudal cirrus

Figure 13.35   Growth of a trochophore larva. (A) Trochophore larva of *Arenicola*. Note the teloblastic (4d) mesodermal bands destined to form the metameric coelomic spaces. (B,C) Two stages in the development of *Eteone*. (B) Early-stage segmentation. (C) Juvenile, showing the fates of the larval regions.

Oligochaeta.   Nowhere is the divergence of the clitellate annelids from the polychaetes more apparent than in the major differences in reproductive and life history strategies. The evolution of new reproductive styles unquestionably contributed greatly to the successful invasion of fresh water and land by the oligochaetes and hirudinoideans.

Oligochaetes are hermaphroditic and usually possess distinct and complex reproductive systems, including permanent gonads. Furthermore, various parts of the reproductive apparatus are restricted to particular segments, usually in the anterior portion of the worm (Figure 13.36). The arrangement of the reproductive systems facilitates mutual cross-fertilization followed by encapsulation and deposition of the zygotes.

The male system includes one or two pairs of testes located in one or two specific body segments. Sperm are released from the testes into the coelomic spaces, where they mature or are picked up by storage sacs (seminal vesicles) derived from pouches of the septal peritoneum (Figure 13.36B). There may be a single seminal vesicle or as many as three pairs in some earthworms. When mature, the sperm are released from the seminal vesicles, picked up by ciliated **seminal (sperm) funnels**, and carried by sperm ducts to paired gonopores.

The female reproductive system consists of a single pair of ovaries located posterior to the male system (Figures 13.19C and 13.36B). Again, the ova are released into the adjacent coelomic space and sometimes stored until mature in shallow pouches in the septal wall called the **ovisacs**. Next to each ovisac is a ciliated funnel that carries the mature ova to an oviduct and eventually to the female gonopore. Most oligochaetes also possess one or two or more pairs of blind sacs called **spermath-**

ecae (seminal receptacles) that open to the outside via separate pores (Figure 13.36).

Of major importance to the overall reproductive strategy of oligochaetes is a unique region of glandular tissue called the **clitellum** (Latin for "saddle") (Figure 13.36A). This structure is a principal anatomical feature unifying the Oligochaeta and the Hirudinoidea as the clitellate annelids. The clitellum has the appearance of a thick sleeve that partially or completely encircles the worm's body. It is formed of secretory cells within the epidermis of particular segments. The exact position of the clitellum and the number of segments involved are consistent within any particular species. In freshwater forms the clitellum is located around the position of the

Figure 13.36   The reproductive system of *Lumbricus* ▶ and mating in earthworms (see also Figure 13.19C). (A) External structures associated with reproduction of *Lumbricus* (ventral view). (B) Segments 9–15 of *Lumbricus* (composite lateral view). (C) The clitellum epithelium (section) showing the three types of secretory cells. (D,E) Copulating earthworms. (D) *Pheretima* transfers sperm directly from the male pore, through a penis, into the mate's spermatheca. (E) *Eisenia* uses indirect sperm transfer. As in *Lumbricus*, the sperm leave the male pores and travel along paired seminal grooves to the spermathecal openings of the mate. (F–H) An earthworm forming and releasing a cocoon. As the cocoon slides over the worm, it receives ova and sperm. (I) Engaged copulatory apparatus of *Rhynchelmis*, an oligochaete with direct sperm transfer. (J) Copulating earthworms (*Lumbricus*).

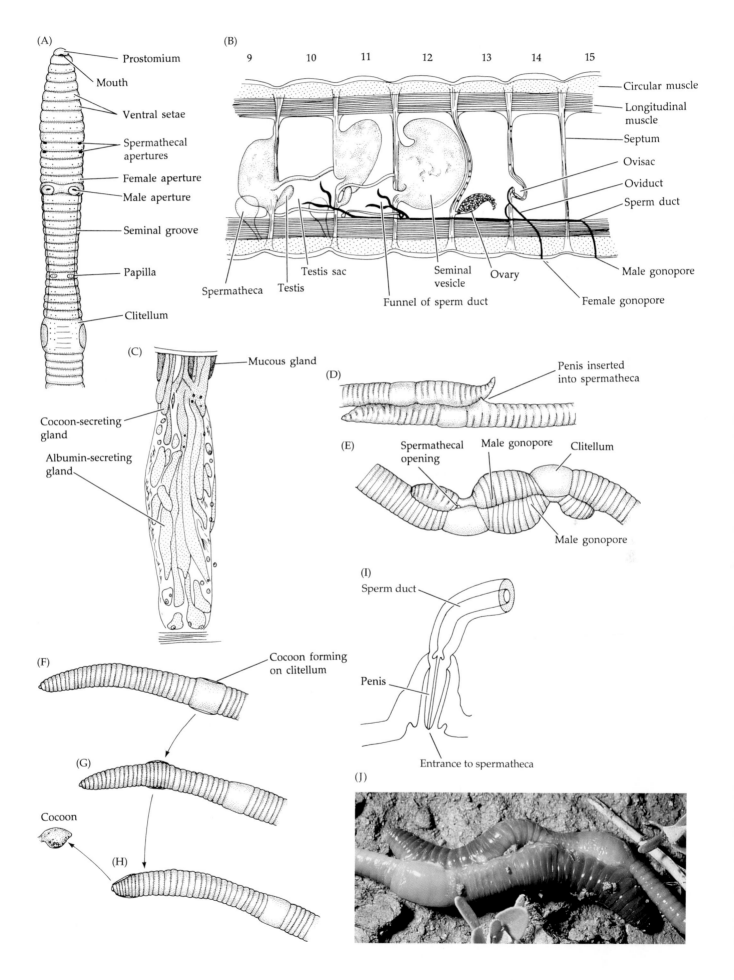

(A)
- Prostomium
- Mouth
- Ventral setae
- Spermathecal apertures
- Female aperture
- Male aperture
- Seminal groove
- Papilla
- Clitellum

(B) 9 10 11 12 13 14 15
- Circular muscle
- Longitudinal muscle
- Septum
- Ovisac
- Oviduct
- Sperm duct
- Male gonopore
- Female gonopore
- Funnel of sperm duct
- Seminal vesicle
- Ovary
- Testis sac
- Testis
- Spermatheca

(C)
- Mucous gland
- Cocoon-secreting gland
- Albumin-secreting gland

(D)
- Penis inserted into spermatheca

(E)
- Spermathecal opening
- Male gonopore
- Clitellum
- Male gonopore

(I)
- Sperm duct
- Penis
- Entrance to spermatheca

(F)
- Cocoon forming on clitellum

(G)

(H)

Cocoon

(J)

gonopores, but in most earthworms it is posterior to the gonopores.

There are three types of gland cells within the clitellum, each secreting a different substance important to reproduction: mucus that aids in copulation; the material forming the outer casing of the egg capsule (or **cocoon**); and albumin deposited with the zygotes inside the cocoon. In most aquatic oligochaetes these cell types all lie in a single layer, but in the terrestrial forms distinct cell layers are present (Figure 13.36C).

Free spawning and indirect larval development simply would not do in the environments of most oligochaetes, and the success of these animals has depended in large part on contact mating, exchange of sperm, and direct development. The high survival rate of zygotes produced by such methods balances the relatively high parental investment. And, as we have seen, hermaphroditism is one way for slow and sluggish animals, who might encounter sexual partners only infrequently, to increase their reproductive success.

During copulation, the mating worms align themselves facing in opposite directions (Figure 13.36D,E) and mucous secretions from the clitellum hold them in this copulatory posture. Many oligochaetes position themselves so that the male gonopores of one are aligned with the spermathecal openings of the other. In such cases, special copulatory chaetae near the male pores or eversible penis-like structures aid in anchoring the mates together (Figure 13.36I). The lumbricid earthworms are not so accurate with their mating, and their copulatory position does not bring the male pores against the spermathecal openings. Instead, lumbricids develop external sperm grooves along which the male gametes must travel prior to entering the spermathecal pores (Figure 13.36A,E). These grooves are actually formed temporarily by muscle contraction, and are covered by a sheet of mucus. By this remarkable anatomical/behavioral adaptation, underlying muscles cause the grooves to undulate, and the sperm are transported along the body to their destination. Following the mutual exchange of sperm to the seminal receptacles of each mate, the worms separate, each functioning as an inseminated female.

From several hours to a few days following copulation, a sheet of mucus is produced around the clitellum and all the anterior segments. Then the clitellum produces the cocoon itself in the form of a leathery, proteinaceous sleeve. The cocoons of terrestrial species are especially tough and resistant to adverse conditions. Albumin is secreted between the cocoon and the clitellar surface. The amount of albumin deposited with the cocoon is much greater in terrestrial species than in aquatic forms.

Thus formed, the cocoon and underlying albumin sheath are moved toward the anterior end of the worm by muscular waves and backward motion of the body. As it moves along the body, the cocoon first receives eggs from the female gonopores, and then sperm previously received from the mate and stored in the seminal receptacles. Fertilization occurs within the albumin matrix inside the cocoon. The open ends of the cocoon contract and seal as they pass off the anterior end of the body (Figure 13.36F–H). The closed cocoons are deposited in benthic debris by aquatic oligochaetes. Terrestrial forms deposit their cocoons in the soil at various depths, the particular depth depending on the moisture content of the substratum. The shape and size of the cocoon are often species-specific.

Oligochaetes produce telolecithal ova, but the amount of yolk varies greatly and inversely with the amount of albumin secreted into the cocoon. The eggs of freshwater oligochaetes contain relatively large amounts of yolk but are encased with only a small quantity of albumin. Conversely, the eggs of terrestrial species tend to be weakly yolked but are supplied with large quantities of albumin on which the developing embryos depend for a source of nutrition. In any case, cleavage is holoblastic and unequal. And, although highly modified, evidence of the ancestral spiralian pattern is still apparent in cell placement and fates (e.g., an identifiable 4d mesentoblast homologue gives rise to the presumptive mesoderm). Development is direct, with no trace of a larval stage. However, the teloblastic production of coelomic spaces and segments is an obvious retained characteristic of the basic annelid developmental program.

Development time varies from about one week to several months, depending on the species and environmental conditions. In climates where relatively severe conditions follow cocoon deposition, development time is usually long enough to ensure that the juveniles hatch in the spring. Under more stable conditions, development time is shorter and reproduction is less seasonal.

The number of zygotes within each cocoon varies from 1 to about 20, again depending on the species. However, when several zygotes are included, only one or a few actually reach the hatching stage.

Hirudinoidea.   Hirudinoideans, like oligochaetes, are clitellate, hermaphroditic annelids and have complex reproductive organs (Figure 13.37A,B). They also undergo direct development, an ontogeny well suited to their environments and lifestyles.

The male reproductive system includes a variable number of paired testes, usually from 5 to 10 pairs in leeches, arranged serially beginning in segment XI or XII (Figure 13.37B). The testes are drained by a pair of sperm ducts that lead to a copulatory apparatus and a single gonopore located midventrally on segment X.

The copulatory apparatus of leeches is often complex and varies in structure among species. Each sperm duct is coiled distally and enlarges as an ejaculatory duct. The two ducts join at a common glandular, muscular atrium. In arhynchobdellids, the atrium is modified as an eversible penis. The rhychobdellids lack a penis and the atrium functions as a chamber in which spermatophores are produced.

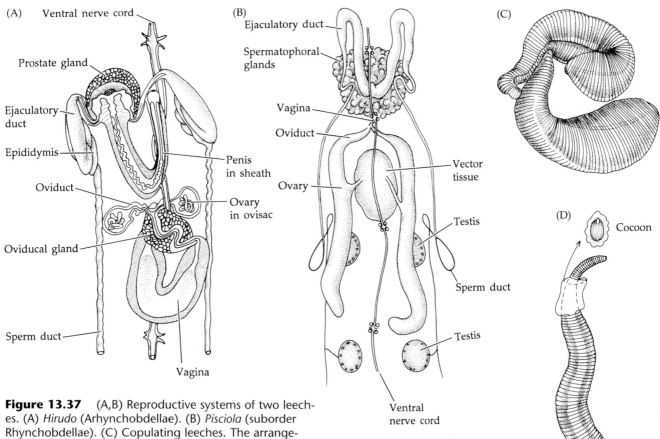

Figure 13.37 (A,B) Reproductive systems of two leeches. (A) *Hirudo* (Arhynchobdellae). (B) *Pisciola* (suborder Rhynchobdellae). (C) Copulating leeches. The arrangement is much like that in oligochaetes. (D) *Erpobdella* with cocoon.

There is a single pair of ovaries in leeches, which may extend through several segments (Figure 13.37A,B). Oviducts extend anteriorly from the ovaries and unite as a common vagina, which leads to the female gonopore on the midventral surface of segment XI, just behind the male pore. In some leeches, an **oviducal gland** surrounds a portion of the oviduct and vagina and apparently functions in egg-laying activity.

Copulation and sperm transfer differ markedly between the rhynchobdellans and arhynchobdellans, in part due to the differences in the male copulatory structures. In those with a penis (Arhynchobdellae), mating worms align themselves so that the male pore of each rests over the female pore of the other. The penis is everted and inserted into the mate's vagina, where sperm are deposited. Fertilization takes place within the female reproductive system.

The rhynchobdellan leeches engage in an unusual form of hypodermic impregnation. These animals grasp one another by their anterior suckers and bring their male pores in alignment with a particular region of the mate's body (Figure 13.37C). In most rhynchobdellans, the spermatophores are released onto the clitellar region of the mate. After the spermatophores are placed on the body of the mate, they penetrate the body surface of the recipient and the sperm emerge beneath the epidermis. The sperm migrate to the ovaries by way of the coelomic channels and sinuses. In some of the piscicolid leeches there is a special "target" area, beneath which is a mass of tissue called **vector tissue** that connects with the ovaries via short ducts.

Cocoon formation in leeches is similar to that in oligochaetes, with the clitellum producing a cocoon wall and albumin (Figure 13.37D). However, as the cocoon slides anteriorly past the female gonopore, it receives the zygotes or young embryos rather than separate eggs and sperm. The cocoons are deposited in damp soil by terrestrial species and even by a few aquatic forms that migrate to land for this process (e.g., *Hirudo*). Most aquatic forms deposit their cocoons by attaching them to the bottom or to algae; a few attach them to their hosts (e.g., some piscicolids). A few freshwater leeches display some degree of parental care for their cocoons. Some of these bury their cocoons and remain over them, generating ventilatory currents. Others attach the cocoons to their own bodies and brood them externally.

The embryogeny of leeches is similar to that described for oligochaetes. Except for a few species, the amount of yolk is relatively small and development time quite short.

Siboglinidae: The Beard Worms

Here we discuss an enigmatic group of tube-dwelling marine worms known as pogonophorans (Greek *pogon*, "beard"; *phor*, "to bear") and vestimentiferans (Latin *vestimentum*, "garment"; *ferre*, "to bear"). Together these worms comprise about 100 species of strange vermiform creatures that have been the subject of taxonomic and phylogenetic debate since their discovery some one hundred years ago. The higher classification of this group is still being debated. We recognize the family name Siboglinidae, but other names (e.g., Lamellibrachiidae) are also seen.

Beard worms live in thin tubes buried in sediments at ocean depths from 100 to 10,000 m. Members of one genus, *Sclerolinum*, reside in wood fragments, pieces of waterlogged rope, and other accumulated benthic debris. Most siboglinids are less than 1 mm in diameter but 10 to 75 cm in length. The tubes may be three or four times as long as the worm's body.

The body of a beard worm is divided into four regions (Figure 13.38A). Each region usually contains a coelomic cavity. The anterior part of the body, called the **cephalic lobe**, bears from 1 to over 200 thin branchial tentacles, which bear tiny side branches called **pinnules**. The cephalic lobe is followed by a short **forepart** and then an elongate **trunk**, the latter bearing various annuli, papillae, and ciliary tracts. Behind the trunk is a short, metamerically segmented **opisthosoma** that contains serially arranged coelomic spaces separated by septa and bears external paired chaetae. The nature of the opisthosoma and the determination of homology between these chaetae and those of annelids is in part responsible for the recent inclusion of the beard worms in the phylum Annelida.

Vestimentiferans (e.g. *Riftia*, *Lamellibranchia*), like other beard worms, are marine, benthic tube dwellers. One thick-bodied species, *Riftia pachyptila*, lives in hydrothermal vent communities and reaches the incredible length of 3 m (Figure 13.38G). The vestimentiferans differ from other beard worms in possession of an anterior first body part, the **obturaculum**. The obturaculum is followed by a forepart that bears the branchial tentacles, a trunk, and a multisegmented opisthosoma. The main trunk of the body bears flaplike or winglike extensions, the **vestimentum**, from which the group gets its name. Other comparisons between the vestimentiferans and the other beard worms are noted in Box 13B. The differences between these animals prompted many workers to treat them as separate phyla for a decade or so (Vestimentifera and Pogonophora). The two groups are now considered to share so many characters as to be inseparable, at least at any higher categorical level, and the current view is to combine them as a family within the polychaetous annelids (Siboglinidae).

Internally, beard worms possess a complex closed circulatory system and well developed nervous system. There are apparently no true nephridia (although coelo-moducts are present), and no digestive tract in the adults. We must also mention that there has been some confusion about the overall body orientation of these animals. Until about a decade ago, it was unclear whether the main longitudinal nerve cord is dorsal or ventral. Jones and Gardiner (1988) conducted developmental studies that demonstrated a transitory gut in at least one species. By examining the orientation of that system, it now seems clear that the nerve cord is ventral, as it is in annelids and other protostomes.

Taxonomic History

Beard worms have a rich and interesting history. They were first studied in the early 1900s, following the expedition of the Dutch research vessel *Siboga* in Indonesia. The specimens (without their opisthosomas) were given to the eminent French zoologist Maurice Caullery, who studied them for nearly 50 years and published several papers describing these strange worms. Eventually Caullery named the originally collected creatures *Siboglinum weberi*. (Although Caullery created the family name Siboglinidae for these worms, he was unable to assign them to any known phylum.) During the middle years of the twentieth century, other zoologists, including Dawydoff, suggested that *Siboglinum* was related to the enteropneusts, a group of vermiform hemichordates with a tripartite body architecture (Chapter 23). Other workers studied additional species and continued to suggest a deuterostome relationship. Some species, however, were likened to sabellid polychaetes, and the name Pogonophora was first coined (as Pogonofora) to be included as a class of Annelida.

Much of the pioneer work on beard worms was done by the Russian specialist A. V. Ivanov, who for years continued to interpret the incomplete specimens as deuterostomes. However, in 1964, the recovery of whole specimens, with opisthosomas intact, led to a new line of thinking about their relationships to other phyla. The discovery of the gigantic vestimentiferans associated with hydrothermal vents further complicated the controversies surrounding these animals.

While the arguments are far from being settled to everyone's satisfaction, current opinion—to which we subscribe—is that these worms should all be placed within the Annelida as a highly specialized polychaete family—Siboglinidae (tentatively assigned to the order Sabellida). As described below, this view is supported by a good deal of anatomical and developmental evidence. McHugh (1997) also adds the strength of molecular data from DNA sequencing that places the beard worms in the Annelida.

The Tube, Body Wall, and Body Cavity

The elongate tubes of beard worms are composed of chitin and scleroproteins secreted by the epidermis. The tubes are often fringed, flared, or otherwise distinctively shaped and are frequently banded with yellow or

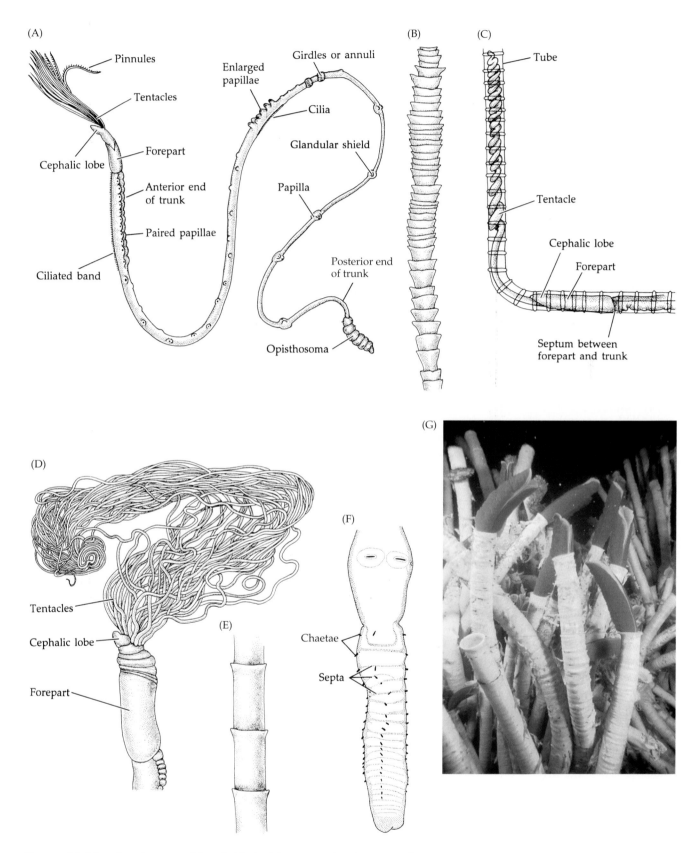

Figure 13.38 Beard worms (Siboglinidae). (A) A generalized pogonophoran. (B) The tube of *Lamellisabella*. (C) Anterior end of the monotentaculate *Siboglinum* in its tube. (D,E) Anterior end and tube of *Polybrachia*. (F) Enlarged view of the opisthosoma of *Polybrachia*. (G) Living specimens of the deep-sea, hydrothermal vent vestimentiferan *Riftia pachyptila*.

BOX 13B *Some Characteristics of the Polychaete Family Siboglinidae (the Beard Worms)*

Vestimentiferans (Lamellibranchia, Ridgeia, Riftia, etc.)

1. Anterior body cavities well developed

2. Vestimental "wings" present

3. Opisthosoma with single, nonganglionated nerve cord

4. Opisthosomal coeloms with medial mesenteries

5. With transitory gut in juveniles

6. Anterior obturaculum present

Other Beard Worms (Birstenia, Polybrachia, Siboglinum, etc.)

1. Anterior body cavities usually not well developed

2. Without vestimental wings

3. Opisthosoma with three nerve cords, all apparently bearing ganglia

4. Opisthosomal coeloms lack medial mesenteries

5. Apparently without gut at any time

6. Obturaculum absent

---

brown pigment rings (Figure 13.38B,C,E). The upper end of the tube projects above the substratum so the tentaculate part of the worm can extend into the water.

The body surface is covered by a flexible cuticle that is thickened in various patterns, including a collar-like ridge that apparently rests on the rim of the tube when the animal is extended. The epidermis is mostly a cuboidal to columnar epithelium and includes various gland cells, papillae, and ciliary tracts. Microvilli extend into the cuticle and produce part of its outer layer. Large glands extend inward from the epidermis, and some of the trunk epithelium is thought to contain absorptive cells. Beneath the epidermis is a thin layer of circular muscle and a thick layer of longitudinal muscle, the latter developed as bands or bundles in some parts of the body.

The body cavities are fairly spacious except in the anterior end, where large muscle bands occur. In at least some species, the cavity of the cephalic lobe includes a small sacciform or U-shaped space with a pair of coelomoducts to the outside. The cavities of the forepart and trunk are paired, with sagittally arranged mesenteries bisecting the body. (The cavities of these first three body regions were likened to the protocoel, mesocoel, and metacoel of deuterostomes until the body orientation was resolved and the opisthosoma was discovered.) The opisthosoma comprises several segments, each with cavities in a typical annelid-like metameric arrangement.

Nutrition

In the absence of a functional digestive tract, the method of nutrition in many of these relatively large worms is indeed puzzling. Experimental work suggests that most beard worms are able to absorb dissolved organic matter (DOM) from the sea water flowing across the tentacles and from the muddy sediments in which the animals are buried. Apparently no digestive enzymes are involved in this process; rather the worms take in glu-

cose, amino acids, and fatty acids directly from the environment. Some species may also be capable of epidermal pinocytosis and phagocytosis.

The absorption of organic nutrients probably also occurs in the giant worms associated with the chemoautotrophic based ecosystems of hydrothermal vent environments. However, work on *Siboglinum fiordicum* indicates that uptake rates of dissolved amino acids are insufficient to account for the animal's metabolic requirements, suggesting augmentation by some other nutritional mechanism. In some forms, including *Riftia*, which inhabit the sulfur-rich waters near hydrothermal vents, symbiotic chemoautotrophic bacteria inhabit certain body tissues (the **trophosome organ**). Apparently, these symbionts generate ATP by carrying out sulfide oxidation and by reducing CO_2 to organic compounds, which may in turn be a source of nutrition for the host worms. In at least some species, the bacteria are taken in by a transitory digestive tract during early juvenile stages of a worm's growth. These bacteria are apparently stored in the worm's midgut, which persists as the trophosome organ after the rest of the gut is lost.

Circulation, Gas Exchange, Excretion, and Osmoregulation

As we saw earlier in this chapter, a body plan that includes internal partitions demands an internal transport mechanism. So it is with the beard worms, which possess a well developed closed circulatory system (Figure 13.39A). The major blood vessels are dorsal and ventral longitudinal vessels that extend nearly the entire length of the body. The dorsal vessel is swollen as a muscular pump in the forepart of the body. As in other annelids, blood flows anteriorly in the dorsal vessel and posteriorly in the ventral vessel. Anteriorly, the vessels branch extensively. Some of these vessels supply parts of the head region and lead to afferent and efferent vessels in the tentacles (Figure 13.39B). Exchange also occurs in

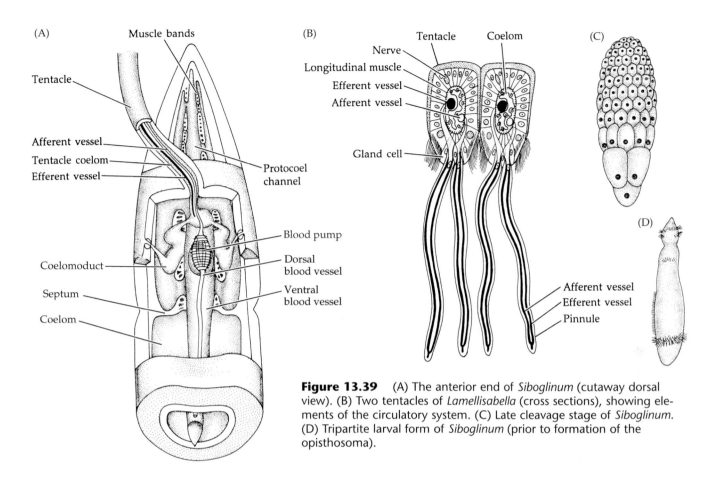

(A) Muscle bands

Tentacle

Afferent vessel

Tentacle coelom

Efferent vessel

Protocoel channel

Coelomoduct

Blood pump

Septum

Dorsal blood vessel

Coelom

Ventral blood vessel

(B) Tentacle Coelom

Nerve

Longitudinal muscle

Efferent vessel

Afferent vessel

Gland cell

(C)

(D)

Afferent vessel

Efferent vessel

Pinnule

Figure 13.39 (A) The anterior end of *Siboglinum* (cutaway dorsal view). (B) Two tentacles of *Lamellisabella* (cross sections), showing elements of the circulatory system. (C) Late cleavage stage of *Siboglinum*. (D) Tripartite larval form of *Siboglinum* (prior to formation of the opisthosoma).

the posterior part of the body through a series of connecting blood rings.

The blood contains hemoglobin in solution and a variety of cells and cell-like inclusions. Gas exchange probably takes place across the thin walls of the tentacles. In the forms that live near hydrothermal vents, problems of oxygen supply and sulfur toxicity are especially critical. These worms, such as *Riftia pachyptila*, live where hot, anoxic, sulfurous vent water mixes with the surrounding cold, oxygenated water. The animals are thus exposed to potentially dramatic fluctuations in oxygen and sulfide availability and in ambient temperature. These worms possess very high concentrations of hemoglobin in the fluid of their body cavities as well as in their blood. This hemoglobin appears to retain its high affinity for oxygen across a wide temperature range. Furthermore, a unique sulfide-binding protein occurs in their blood, which serves to concentrate the sulfur and avoid sulfide toxicity, and also to transport the sulfur to the chemoautotrophic bacteria. Interestingly, similar adaptations are known in the burrowing polychaete *Arenicola*, which typically lives in anoxic muds in shallow marine environments.

Little is known about excretion and osmoregulation in beard worms. Some workers think that the coelomoducts near the front of the body function in these capacities, especially given their close association with the circulatory system.

Nervous System and Sense Organs

The nervous system of beard worms is largely intraepidermal. A well developed nerve ring in the cephalic lobe bears a large ventral ganglion (probably homologous to the subenteric ganglion of other annelids). A single ventral nerve cord arises from this ganglion and extends through all body regions. In many cases there are ganglia (or at least nerve "bulges") at the junctions of the various body regions, and there is an enlargement of the nerve cord in the anterior part of the trunk. In most species, the opisthosoma contains three distinct nerve cords that apparently bear segmental ganglia. The vestimentiferans, however, have a single opisthosomal nerve cord without ganglia.

Sense organs are poorly developed. The tentacles probably contain tactile receptors, and a ventral ciliary tract on the body surface may be chemosensory.

Reproduction and Development

Nothing is known about asexual reproduction or regeneration in these animals. The sexes are separate and each possesses a pair of gonads in the trunk. In males, paired sperm ducts extend from the testes to gonopores located near the anterior end of the trunk. As sperm move along these ducts, they are packaged as spermatophores of various shapes. The female system includes a pair of ovaries from which arise oviducts leading to gonopores on the sides of the trunk. The eggs are elongate and moderately telolecithal.

Fertilization has not been observed. In most species, males apparently release their spermatophores, which drift to the open tubes of nearby females. Fertilization must take place in the female's body or tube, because developing embryos have been found within the tube itself. In some cases the zygotes are deposited in the upper end of the tube, where they are oriented with the animal pole directed upwards. In at least some vestimentiferans the eggs are apparently released, not brooded, but are lecithotrophic (Young et al. 1996).

From what little work has been done on development, it appears that cleavage is holoblastic, unequal, and somewhat irregular. Young et al. (1996) describe cleavage as spiral in at least some vestimentiferans. The blastomeres at the vegetal pole are larger than those at the animal pole, the latter destined to form the worm's anterior end. Cleavage results in a bilaterally symmetrical stereoblastula (Figure 13.39C). It appears that the inner cells give rise to entoderm. A gut apparently never forms in nonvestimentiferans, whereas a transitory digestive tract occurs during development and in juveniles of at least some vestimentiferans. Some of these inner cells also give rise to a single coelomic cavity, from which the body cavities of the cephalic lobe, forepart, and trunk seemingly arise. Inner cells near the presumptive posterior end of the embryo proliferate spaces within the opisthosomal segments as they develop, a process strikingly similar to the teloblastic growth and schizocoely seen in typical polychaetes.

Beard worms produce motile ciliated larvae (Figure 13.39D). Gardiner and Jones (1994) describe the larva of the vestimentiferan *Ridgeia* as bearing a prototroch, metatroch, and neurotroch, as seen in polychaete trochophores. Young et al. (1996) describe the larvae of other species as lacking a telotroch. These larvae swim for a short time before settling and secreting their tubes. Development is thus mixed, with a long brooding phase and a short dispersal phase. The production of a few yolky eggs and lecithotrophic larvae is clearly adaptive in an environment where long-lived planktotrophic larvae have little selective value. The short larval phase allows limited dispersal while reducing the risks associated with feeding in the deep sea.

Annelid Phylogeny

A discussion of annelid phylogeny involves consideration of the origin of the coelomic condition and segmentation. We cannot cover all the myriad details and arguments on these topics; instead we present a set of general ideas about the origin of and radiation within the Annelida and establish a foundation for placing the rest of the protostome phyla in a reasonable and understandable perspective. The scenario described below is expressed in the cladogram and evolutionary tree in Figure 13.40. As indicated in Figure 13.40B, we continue to treat the annelids and arthropods as closely related taxa stemming from a common, segmented ancestor. The presence of teloblastic metameres in these two phyla (and in Onychophora and Tardigrada) is generally viewed as a powerful synapomorphy linking these groups as a monophyletic clade. However, we note that while some molecular phylogenetic work also supports this view, some other studies do not (Chapter 24).

The protostome coelom may have evolved as a response to peristaltic burrowing (Chapter 4). Within the protostomes, this first schizocoelomate creature may have arisen from a triploblastic, acoelomate ancestor or from a diploblastic precursor. In the latter case, the coelom arose coincidentally with 4d mesoderm and other spiralian features. Thus, we may view the protoannelids as homonomous metameric burrowers derived from a segmented, coelomate, ancestral vermiform creature. Mettam (1985) proposed the possibility that the first annelids were tiny interstitial creatures, but Westheide (1985) suggests that they were moderate to large forms.

Paleontological evidence indicates that coelomate burrowers existed well over 1 billion years ago, in Precambrian times. Some of the Precambrian and early Cambrian strata include trace fossils and burrows thought to be of annelid origin. Fossils that are clearly polychaetes appear by the middle Cambrian, and these are the earliest certain records of the Annelida. Thus marine polychaetes are typically considered to be the earliest derived group within the phylum as we know it today. The evolution of parapodia in association with the coelomate metameric bauplan presumably allowed early annelids to "emerge from the muds" and begin to exploit surface locomotion by crawling, as seen in many modern-day polychaetes. Tube-dwelling lifestyles, heteronomy, the loss of complete septa, and most of the other variations within members of the Polychaeta are thus viewed as secondarily derived.

The cladogram in Figure 13.40A depicts the annelids as comprising two great lineages, the Polychaeta and the Clitellata. Although the precise nature of the ancestral group (the "protoannelids") from which these two lines arose is lost in time, we can reasonably hypothesize that it was a relatively simple, homonomous, metameric burrower with a compartmentalized segmented coelom, paired epidermal chaetae, and a head composed of a presegmental prostomium and peristomium. Parapodia and head appendages evolved in the line that became modern polychaetes. The early polychaetes were probably dioecious broadcast spawners with indirect development. Their ancestors had already established the basic protostome plan of development, complete with a trochophore larval stage.

The other line, the clitellate annelids, arose by way of evolution of a complex hermaphroditic life history, permanent gonads, direct development, the clitellum, and

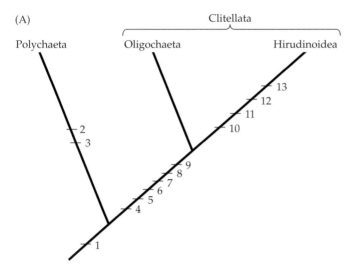

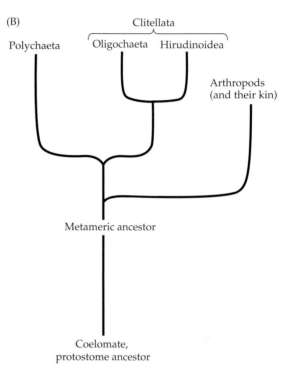

Figure 13.40 Proposed cladogram (A) and evolutionary tree (B) depicting the evolution of annelids. The ancestor to the annelid–arthropod line is viewed as a segmented coelomate protostome. The origin of the annelids was marked by the appearance of (1) the unique annelid head, with its presegmental prostomium and peristomium. The resulting protoannelid was the ancestor from which all modern taxa were ultimately derived. Early radiation from this ancestral annelid led to the polychaete lineage, defined by the evolution of parapodia (2), and the elaboration of the complex head (3).

The clade leading to the clitellate annelids is defined by the appearance of: (4) the obligatory hermaphroditic condition; (5) development of the clitellum and its complex epidermal structures; (6) loss of the free larval stage and reliance on direct development; (7) posterior displacement of the cerebral ganglion into the anterior trunk segments; (8) evolution of distinct, fixed, and complex reproductive organs; (9) the tendency toward a fixed number of body segments. The oligochaetes represent the primitive clitellate condition.

The hirudinoidean line is defined by the appearance of the following synapomorphies: (10) a tendency toward

reduction of the intersegmental septal walls and fusion of the coelomic compartments; (11) appearance of a posterior body sucker; (12) subdivision of the body segments by superficial annuli; (13) a major reduction in body chaetae, initially lost on all but the anterior segments. This is the level at which the acanthobdellidans have remained, and this group can thus be viewed as a primitive "relict" taxon representative of the transition from an oligochaete (or oligochaete-like ancestor) to hirudinidans. Within the hirudinoidean lineage was the eventual evolution of "mesenchymal" tissue and reduction of the coelom. It should be noted that some of the characters we have used are quite complex and may be viewed as multicharacter suites, themselves subject to more detailed analyses.

other features. Within this clade are two major groups: the oligochaetes and the hirudinoideans. However, note that the line leading to the oligochaetes on the cladogram is not defined by any synapomorphies. We are not aware of any features unique to the oligochaetes as they are currently understood. The oligochaetes are defined entirely by their retention of primitive features (symplesiomorphies) and by the *lack* of the features distinguishing the hirudinoidean lineage. Thus, the ancestor of the entire clitellate lineage may be viewed as being an oligochaete. As we noted in Chapter 2, groups such as the Oligochaeta that are defined solely by retention of plesiomorphic features (rather than by synapomorphies) are typically paraphyletic. This is almost certainly the case with the Oligochaeta. Recognition of this fact adds credence to the proposal that the oligochaetes and hirudinoideans be combined into a single class, the

Clitellata. Oligochaetes are simply primitive clitellates, and hirudinoideans are more derived clitellates, evolved from some isolated oligochaete ancestor.

Some workers suggest that the oligochaetes arose quite late in geological time, in association with the evolution of flowering land plants in the Jurassic and Cretaceous periods. However, fossils of oligochaete-like creatures (e.g., *Palaeoscolex*) are known from Cambrian deposits, suggesting that clitellates arose closer to the origin of polychaetes, in line with the sister-group relationship depicted in the cladogram. Actually, it is probable that the clitellate line arose from a preexisting polychaete stock through such evolutionary steps as loss of the parapodia, reduction of the complex head region, and so on. Were this shown to be true, characters (2) and (3) on the cladogram (Figure 13.40A) would drop down to the ancestral line, next to character (1). However, there is little in the way of morphological data to address this idea, although molecular evidence may eventually come to bear on the issue. There is no evidence of transient or rudimentary parapodia in any clitellate annelids.

The evolution of complex hermaphroditic reproductive systems, yolky eggs, and the clitellum that facilitated encapsulation of the developing embryos freed early clitellate annelids from the marine environment and dependency on a larval phase. These events were certainly of major importance in their successful exploitation of freshwater and land habitats.

The acanthobdellids and branchiobdellids show a mixture of characteristics between the oligochaetes and the "true leeches," although at least the branchiobdellids are now viewed as modified oligochaetes and not an evolutionary link to the hirudinidans. The evolution of suckers in many of these groups may have initially provided a means of clinging to objects as they assumed an epibenthic lifestyle. This ability would have been especially useful in streams and rivers. The use of suckers for temporary attachment to substrata probably preadapted some of these creatures for later exploitation of ectoparasitism. Some workers have suggested that the odd Acanthobdellida may be no more than degenerate leeches that have lost the anterior sucker. However, the presence of anterior chaetae in this enigmatic group argues for its position as a primitive hirudinoidean. The evolutionary tree in Figure 13.40B presents the relationships outlined above in a more traditional fashion.

Selected References

General References

Aguinaldo, A. M. A., J. M. Turbeville, L. S. Linford, M. C. Rivera, J. R. Garey, R. A. Raff and J. A. Lake. 1997. Evidence for a clade of nematodes, arthropods and other moulting animals. Nature 387: 489–493.

Anderson, D. T. 1973. *Embryology and Phylogeny in Annelids and Arthropods.* Pergamon Press, Oxford.

Berrill, N. J. 1952. Regeneration and budding in worms. Biol. Rev. 27: 401.

Brasier, M. D. 1979. The Cambrian radiation event. *In* M. R. House (ed.), *The Origin of Major Invertebrate Groups.* Academic Press, London, pp. 103–160.

Clark, R. B. 1969. Systematics and phylogeny: Annelida, Echiura, Sipuncula. *In* M. Florkin and B. T. Scheer (eds.), *Chemical Zoology*, Vol. 4. Academic Press, New York, pp. 1–68.

Clark, R. B. 1979. Radiation of the metazoa. *In* M. R. House (ed.), *The Origin of Major Invertebrate Groups.* Academic Press, London, pp. 55–102.

Dales, R. P. 1967. *Annelids*, 2nd Ed. Hutchinson University Library, London.

Eeckhaut, I., and six others. 2000. Myzostomida: A link between trochozoans and flatworms? Proc. Royal Soc. London B267: 1383–1392.

Eernisse, D. J. 1999. Anthropod and annelid relationships re-examined. *In* R. A. Fortey and R. H. Thomas (eds.), *Arthropod Relationships.* Chapman and Hall, London, pp. 43–56.

Eernisse, D. J., J. S. Albert and F. E. Anderson. 1992. Annelida and Arthropoda are not sister taxa: A phylogenetic analysis of spiralian metazoa morphology. Syst. Biol. 41: 305–330.

Fauvel, P., M. Avel, H. Harant, P. Grassé and C. Dawydoff. 1959. Embranchement des Annélides. *In* P. Grassé (ed.), *Traité de Zoologie*, Vol. 5, Pt. 1. Masson et Cie, Paris, pp. 3–686.

Goodrich, E. S. 1946. The study of nephridia and genital ducts since 1895. Q. J. Microsc. Sci. 86: 113–392.

Halanych, K. M., J. D. Bacheller, A. M. A. Aguinaldo, S. M. Liva, D. M. Hillis, J. A. Lake. 1995. Evidence from 18S ribosomal DNA that the lophophorates are protostome animals. Science 267: 1641–1643.

Herlant-Meewis, H. 1965. Regeneration in annelids. *In* M. Abercrombie and J. H. Brachet (eds.), *Advances in Morphogenesis*, Vol. 4. Academic Press, New York, pp. 155–215.

Lopez, G. I. and J. S. Levinton. 1987. Ecology of deposit-feeding animals in marine sediments. Q. Rev. Biol. 62: 235-260.

Løvtrup, S. 1975. Validity of the protostomia–deuterostomia theory. Syst. Zool. 24: 96–108.

Mangum, C. P. 1976. Primitive respiratory adaptations. *In* R. C. Newell (ed.), *Adaptation to Environment.* Butterworth Group Publishing, Boston, pp. 191–278.

McHugh, D. 1997. Molecular evidence that echiurans and pogonophorans are derived annelids. Proc. Natl. Acad. Sci. USA 94: 8006–8009.

McHugh, D. 1999. Phylogeny of the Annelida: Siddall, et al. (1998) rebutted. Cladistics 15: 85–89.

McHugh, D. 2000. Molecular phylogeny of the Annelida. Canadian J. Zool. 78: 1873–1884.

Mettam, C. 1985. Functional constraints in the evolution of the Annelida. *In* S. C. Morris et al. (eds.), *The Origins and Relationships of Lower Invertebrates.* Syst. Assoc. Spec. Vol. No. 28, Oxford, pp. 297–309.

Mill, P. J. (ed.). 1978. *Physiology of Annelids.* Academic Press, London.

Moment, G. B. 1953. A theory of growth limitation. Am. Nat. 88(834): 139–153.

Pettibone, M. H. 1982. Annelida. *In* S. P. Parker, *Synopsis and Classification of Living Organisms*, Vol. 2. McGraw-Hill, New York, pp. 1–43. [Pettibone's contribution includes an introductory section on the phylum and coverage of the Polychaeta.]

Rouse, G. W. and K. Fauchald. 1995. The articulation of annelids. Zool. Scripta 24: 269–301.

Ruppert, E. E. and P. R. Smith. 1988. The functional organization of filtration nephridia. Biol. Rev. 63: 231–258. [A benchmark paper.]

Siddall, M. E., K. Fitzhugh and K. A. Coates. 1998. Problems determining the phylogenetic position of echiursans and pogonophorans with limited data. Cladistics 14: 401–410.

Siddall, M. E., and 7 others. 2001. Validating Livanow: Molecular data agree that leeches, branchiobdellidans, and *Acanthobdella peledina* form a monophyletic group of ogliochates. Mol. Phylog. and Evol. 21: 364–351.

Wald, G. and S. Rayport. 1977. Vision in annelid worms. Science 196: 1434–1439.

Westheide, W., D. McHugh, G. Purschke and G. Rouse. 1999. Systematization of the Annelida: Different approaches. Hydrobiol. 402: 291–307.

Wheeler, W. C., P. Cartwright and C. Y. Hayashi. 1993. Arthropod phylogeny: A combined approach. Cladistics 9(1): 1–39.

Zrzavý, J., J. V. Hypša and D. F. Tietz. 2001. Myzostomida are not annelids: Molecular and morphological support for a clade of annelids with anterior sperm flagella. Cladistics 17: 170–198.

Polychaeta (including myzostomids)

Anderson, D. T. 1966. The comparative embryology of the Polychaeta. Acta. Zool. Stockholm 47: 1–41.

Baskin, D. G. 1976. Neurosecretion and the endocrinology of nereid polychaetes. Am. Zool. 16: 107–124.

Berglund, A. 1986. Sex change by a polychaete: Effects of social and reproductive costs. Ecology 67: 837–845.

Blake, J. A. 1975. The larval development of Polychaeta from the northern California coast. III. Eighteen species of Errantia. Ophelia 14: 23–84.

Brown, S. C. 1975. Biomechanics of water-pumping by *Chaetopterus variapedatus* Renier: Skeletomusculature and kinematics. Biol. Bull. 149: 136–150.

Brown, S. C. 1977. Biomechanics of water-pumping by *Chaetopterus variapedatus* Renier: Kinetics and hydrodynamics. Biol. Bull. 153: 121 132.

Caspers, H. 1984. Spawning periodicity and habit of the palolo worm *Eunice viridis* in the Samoan Islands. Mar. Biol. 79: 229–236.

Clark, R. B. 1956a. The blood vascular system of *Nephtys* (Annelida, Polychaeta). Q. J. Microsc. Sci. 97: 235–249.

Clark, R. B. 1956b. *Capitella capitata* as a commensal, with a bibliography of parasitism and commensalism in polychaetes. Ann. Mag. Nat. Hist. 9(102): 433–448.

Clark, R. B. and P. J. W. Olive. 1973. Recent advances in polychaete endocrinology and reproductive biology. Ocean. Mar. Bio. Ann. Rev. 11: 175–222.

Clark, R. B. and D. J. Tritton. 1970. Swimming mechanisms of nereidiform polychaetes. J. Zool. London 161: 257–271.

Dales, R. P. 1962. The polychaete stomodeum and the interrelationships of the families of Polychaeta. Proc. Zool. Soc. London 139: 389–428.

Dales, R. P. and G. Peter. 1972. A synopsis of the pelagic Polychaeta. J. Nat. Hist. 6: 55–92.

Desbruyeres, D., F. Gaill, L. Laubier, D. Prieur and G. Rau. 1983. Unusual nutrition of the "pompeii worm" *Alvinella pompejana* (polychaetous annelid) from a hydrothermal vent environment: SEM, TEM, ^{13}C, and ^{15}N evidence. Mar. Biol. 75: 201–205.

Eakin, R. M., G. G. Martin and C. T. Reed. 1977. Evolutionary significance of fine structure and archiannelid eyes. Zoomorphologie 88: 1–18.

Eisig, H. 1906. *Ichthyotomus sanguinarius*, eine auf Aalen schmarotzend Annelide. Fauna Flora Nepal 28.

Evans, S. M. 1973. A study of fighting reactions in some nereid polychaetes. Animal Behavior. 21: 138–146.

Fauchald, K. 1975. Polychaete phylogeny: A problem in protostome evolution. Syst. Zool. 23: 493–506.

Fauchald, K. 1977. The polychaete worms: Definitions and keys to the orders, families, and genera. Nat. Hist. Mus. Los Angeles Co. Ser. 28: 1–190.

Fauchald, K. and P. A. Jumars. 1979. The diet of worms: A study of polychaete feeding guilds. Oceanogr. Mar. Biol. Annu. Rev. 17: 193–284. [This and the above paper (Fauchald 1977) are key works in modern polychaetology.]

Fischer, A. and U. Fischer. 1995. On the life-style and life-cycle of the luminescent polychaete *Odontosyllis enopla* (Annelida: Polychaeta). Invert. Biol. 114: 236–247.

Fischer, A. and H.-D. Pfannenstiel (eds.). 1984. *Polychaete Reproduction. Progress in Comparative Reproductive Biology*. Gustav Fischer Verlag, Stuttgart and New York. [A symposium volume.]

Fitzharris, T. P. 1976. Regeneration in sabellid annelids. Am. Zool. 16: 593–616.

Franke, H. D. 1986. Sex ratio and sex change in wild and laboratory populations of *Typosyllis prolifera* (Polychaeta). Mar. Biol. 90: 197 208.

Glaessner, M. F. 1976. Early Phanerozoic annelid worms and their geological and biological significance. J. Geol. Sci. 132: 259–275.

Gray, J. 1939. Studies in animal locomotion. VIII. The kinetics of locomotion of *Nereis diversicolor*. J. Exp. Biol. 16: 9–17.

Grossmann, S. and W. Reichardt. 1991. Impact of *Arenicola marina* on bacteria in intertidal sediments. Mar. Ecol. Prog. Ser. 77: 85–93.

Hartman, O. 1959 and 1965. Catalogue of the polychaetous annelids of the world. Occ. Papers Allan Hancock Found. Vol. 23. Supplement and Index (1965).

Hartman, O. 1966. Polychaeta, Myzostomidae, and Sedentaria of Antarctica. Contrib. #288, Allan Hancock Foundation, Vol. 7, Antarctic Res. Ser., pp. 1 158.

Hermans, C. O. 1969. The systematic position of the Archiannelida. Syst. Zool. 18: 85–102.

Hermans, C. and R. M. Eakin. 1974. Fine structure of the eyes of an alciopid polychaete *Vanadis tagensis*. Z. Morphol. Tiere 79: 245–267.

Hill, S. D. 1970. Origin of the regeneration blastema in polychaete annelids. Am. Zool. 10: 101–112.

Jacobsen, V. H. 1967. The feeding of the lugworm, *Arenicola marina*. Ophelia 4: 91–109.

Jumars, P. A., R. F. L. Self and A. R. M. Nowell. 1982. Mechanics of particle selection by tentaculate deposit-feeders. J. Exper. Mar. Biol. Ecol. 64: 47–70.

Kristensen, R. M. and D. Eibye-Jacobsen. 1995. Ultrastructure of spermiogenesis and spermatozoa in *Diurodrilus subterraneus* (Polychaeta, Diurodrilidae). Zoomorphologie 115: 117–132.

Kudenov, J. D. 1977. Brooding behavior and protandry in *Hipponoe gaudichaudi* (Polychaeta: Amphinomidae). Bull. So. Calif. Acad. Sci. 76(2): 85–90.

Lewbart, G. A. and N. W. Riser. 1996. Nuchal organs of the polychaete *Parapionosyllis manca* (Syllidae). Invert. Biol. 115: 286–298.

MacGinitie, G. E. 1939. The method of feeding of *Chaetopterus*. Biol. Bull. 77: 115–118.

Mangum, C. 1976. The oxygenation of hemoglobin in lugworms. Physiol. Zool. 49(1): 85–99.

Mettam, C. 1967. Segmental musculature and parapodial movement of *Nereis diversicolor* and *Nephtys hombergi* (Annelida: Polychaeta). J. Zool. Lond. 153: 245–275.

Nicol, E. A. T. 1931. The feeding mechanism, formation of the tube, and physiology of digestion in *Sabella pavonia*. Trans. R. Soc. Edinburgh 56(3): 537–598.

Oglesby, L. C. 1965. Water and chloride fluxes in estuarine nereid polychaetes. Comp. Biochem. Physiol. 16: 437–455.

Orrhage, L. 1980. Structure and homologies of the anterior end of the polychaete families, Sabellidae and Serpulidae. Zoomorphologie 96: 113–168.

Pawlick, J., C. Butman and V. Starczak. 1991. Hydrodynamic facilitation of gregarious settlement of a reef-building tube worm. Science 251: 421–424.

Pennington, T. J. and F. S. Chia. 1984. Morphological and behavioral defenses of trochophore larvae of *Sabellaria cementarium* (Polychaeta) against four planktonic predators. Biol. Bull. 167: 168–175.

Penry, D. L. and P. A. Jumars. 1990. Gut architecture, digestive constraints and feeding ecology of deposit-feeding and carnivorous polychaetes. Oecologia 82: 1-11.

Pettibone, M. H. 1991. Polynoids commensal with gorgonian and stylasterid corals, with a new genus, new combinations, and new species (Polychaeta: Polynoidae: Polynoinae). Proc. Biol. Soc. Wash. 104(4): 699–713.

Plante, C. J., P. A. Jumars and J. A. Baross. 1990. Digestive associations between marine detritivores and bacteria. Annu. Rev. Ecol. Syst. 21: 93–127.

Reish, D. J. 1970. The effects of varying concentrations of nutrients, chlorinity, and dissolved oxygen on polychaetous annelids. Water Res. 4: 721–735.

Reish, D. J. and K. Fauchald (eds.). 1977. *Essays on Polychaetous Annelids in Memory of Dr. Olga Hartman*. Allan Hancock Foundation, Los Angeles.

Rhode, B. 1990. Eye structure of *Ophryotrocha puerilis* (Polychaeta: Dorvilleidae). J. Morph. 205: 147–154.

Rouse, G. W. and K. Fouchald. 1997. Cladistics and Polychaetes. Zool. Scripta 26: 139–204.

Rouse, G. W. and F. Pleijel. 2001. *Polychaetes*. Oxford Univ. Press, Oxford.

Ruby, E. G. and D. L. Fox. 1976. Anerobic respiration in the polychaete *Euzonus* (Thoracophelin) *mucronata*. Mar. Biol. 35: 149–153.

Schroeder, P. C. and C. O. Hermans. 1975. Annelida: Polychaeta. *In* A. C. Giese and J. S. Pearse (eds.), *Reproduction of Marine Invertebrates*, Vol. 3. Academic Press, New York, pp. 1–205.

Smetzer, B. 1969. Night of the palolo. Nat. Hist. 78: 64–71.

Smith, P. R. and E. E. Ruppert. 1988. Nephridia. *In* W. Westheide and C. O. Hermans (eds.), *Microfauna Marina* 4. Gustave Fischer Verlag, Stuttgart and New York, pp. 231–262.

Smith, P. R., E. E. Ruppert and S. L. Gardiner. 1987. A deuterostome-like nephridium in the mitraria larva of *Owenia fusiformis* (Polychaeta, Annelida). Biol. Bull. 172: 315–323.

Smith, R. I. 1963. A comparison of salt loss rate in three species of brackish-water nereid polychaetes. Biol. Bull. 125: 332–343.

Wells, G. P. 1950. Spontaneous activity cycles in polychaete worms. Symp. Soc. Exp. Biol. 4: 127–142.

Westheide, W. 1985. The systematic position of the Dinophilidae and the archiannelid problem. *In* S. C. Morris et al. (eds.), *The Origins and Relationships of Lower Invertebrates*. Syst. Assoc. Spec. Vol. No. 28, Oxford, pp. 310–326.

Whitlatch, R. B. 1974. Food resource partitioning in the deposit feeding polychaete *Pectinaria gouldii*. Biol. Bull. 147: 227–235.

Wilson, E. B. 1892. The cell-lineage of *Nereis*. J. Morphol. 6: 361–480.

Clitellata

Anderson, D. T. 1966. The comparative early embryology of the Oligochaeta, Hirudinea and Onychophora. Proc. Linn. Soc. N.S.W. 91: 10–43.

Brinkhurst, R. O. 1982. Oligochaeta. *In* S. P. Parker, *Synopsis and Classification of Living Organisms*, Vol. 2. McGraw-Hill, New York, pp. 50–61. [Includes coverage of some taxa by R. W. Sims.]

Brinkhurst, R. O. and D. G. Cook (eds.). 1980. *Aquatic Oligochaete Biology*. Plenum, New York.

Brinkhurst, R. O. and S. R. Gelder. 1989. Do the lumbriculids provide the ancestors of the branchiobdellidans, acanthobdellidans and leeches? Hydrobiologia 180: 7–15.

Brinkhurst, R. O. and B. G. Jamieson. 1972. *Aquatic Oligochaeta of the World*. Toronto University Press, Toronto.

Conniff, R. 1987. The little suckers have made a comeback. Discover, August: 85–94.

Cook, D. G. and R. O. Brinkhurst. 1973. *Marine Flora and Fauna of the Northeastern United States. Annelida: Oligochaeta*. NOAA Tech. Rpt. NMFS CIRC-374.

Edwards, C. A. and P. Bohlen. 1996. *Biology and Ecology of Earthworms*, 3rd ed. Chapman and Hall, London.

Edwards, C. A. and J. R. Lofty. 1972. *Biology of Earthworms*. Chapman and Hall, London.

Gelder, S. R. 1996. A review of the taxonomic nomenclature and a checklist of the species of Branchiobdellae (Annelida: Clitellata). Proc. Biol. Soc. Wash. 109: 653–663.

Gelder, S. R. and M. E. Siddall. 2001. Phylogenetic assessment of the Branchiobdellidae using 18S rDNA mitochondrial cytochrome *c* oxidase subunit I, and morphological characters. Zool. Seripta 30: 219–222.

Giere, O. and C. Langheld. 1987. Structural organization, transfer and biological fate of endosymbiotic bacteria in gutless oligochaetes. Mar. Biol. 93: 641–650.

Gray, J. and H. W. Lissmann. 1938. Studies in animal locomotion, VII. Locomotory reflexes in the earthworm. J. Exp. Biol. 15: 506–517.

Gray, J., H. W. Lissmann and R. J. Pumphrey. 1938. The mechanism of locomotion in the leech. J. Exp. Biol. 15: 408–430.

Holt, T. C. 1965. The systematic position of the Branchiobdellida. Syst. Zool. 14: 25–32.

Holt, T. C. 1968. The Branchiobdellida: Epizootic annelids. Biologist 1: 79–94.

Jamieson, B. G. 1981. *The Ultrastructure of the Oligochaeta*. Academic Press, London.

Jamieson, B. G. 1988. On the phylogeny and higher classification of the Oligochaeta. Cladistics 4: 367–410.

Klemm, D. J. 1982. *Leeches (Annelida: Hirudinea) of North America*. U.S. Environ. Protection Agency, EPA–600/3–82–025.

Lasserre, P. 1975. Clitellata. *In* A. C. Giese and J. S. Pearse (eds.), *Reproduction of Marine Invertebrates*, Vol. 3. Academic Press, New York, pp. 215–275.

Laverack, M. S. 1963. *The Physiology of Earthworms*. Macmillan, New York.

Mann, K. H. 1962. *Leeches (Hirudinea), Their Structure, Ecology, and Embryology*. Pergamon Press, New York.

Nicholls, J. G. and D. Van Essen. 1974. The nervous system of the leech. Sci. Am. 230(1): 38–48.

Rota, E., M. de Eguileor and A. Grimaldi. 1999. Ultrastructure of the head organ: a putative compound georeceptor in *Grania* (Annelida, Clitellata, Enchytraeidae). Ital. J. Zool. 66: 11–21.

Satchell, J. E. (ed.). 1983. *Earthworm Ecology: From Darwin to Vermiculture*. Methuen, New York. [A symposium volume.]

Sawyer, R. 1990. In search of the giant Amazon leech. Nat. Hist. 12: 66–67.

Seymour, M. K. 1969. Locomotion and coelomic pressure in *Lumbricus*. J. Exp. Biol. 51: 47.

Stephensen, J. 1930. *The Oligochaeta*. Oxford University Press, New York.

Stuart, J. 1982. Hirudinoidea. *In* S. P. Parker (ed.), *Synopsis and Classification of Living Organisms*, Vol. 2. McGraw-Hill, New York, pp. 43–50.

Tembe, V. B. and P. J. Dubash. 1961. The earthworms: A review. J. Bombay Nat. Hist. Soc. 51(1): 171–201.

Van Gansen, P. 1963. Structures et fonctions du tube digestif du lombricien *Eisenia foetida* Savigny. Ann. Soc. R. Zool. Belgique 93: 1–121.

Van Praagh, B. D., A. L. Yen and P. K. Lillywhite. 1989. Further information on the giant Gippsland earthworm *Megascolides australis* (McCoy 1878). Victorian Nat. 106(5): 197–201.

Wallwork, J. A. 1983. *Earthworm Biology*. Edward Arnold and University Park Press, Baltimore.

Pogonophoridae

Arp, A. J. and J. J. Childress. 1981. Blood function in the hydrothermal vent vestimentiferan tube worm. Science 213: 342–344.

Arp, A. J. and J. J. Childress. 1983. Sulfide bonding by the blood of the hydrothermal vent tube worm *Riftia pachyptila*. Science 219: 295–297.

Bakke, T. 1977. Development of *Siboglinum fiordicum* Webb (Pogonophora) after metamorphosis. Sarsia 63: 65–73.

Bakke, T. 1980. Embryonic and post-embryonic development in the Pogonophora. Zool. Jb. Anat. 103: 276–284.

Caullery, M. 1944. *Siboglinum*. Siboga Exped. Monogr. 25 Bis, Livr. 138.

Cavanaugh, C. M., S. L. Gardiner, M. L. Jones, H. W. Janasch and J. B. Waterburg. 1981. Prokaryotic cells in the hydrothermal vent tube worm *Riftia pachyptila* Jones: Possible chemoautotrophic symbionts. Science 213: 340–342.

Cutler, E. B. 1974. The phylogeny and systematic position of the Pogonophora. Syst. Zool. 24: 512–513.

Cutler, E. B. 1982. Pogonophora. *In* S. P. Parker (ed.), *Synopsis and Classification of Living Organisms*, Vol. 2. McGraw-Hill, New York, pp. 63–64.

Felbeck, H. 1981. Chemoautotrophic potential of the hydrothermal vent tube worm, *Riftia pachyptila* Jones (Vestimentifera). Science 213: 366–338.

Fisher, C. R., Jr. and J. J. Childress. 1984. Substrate oxidation by trophosome tissue from *Riftia pachyptila* Jones (phylum Pogonophora). Mar. Biol. Lett. 5: 171–183.

Fry, B., H. Gest and J. M. Hayes. 1983. Sulphur isotopic compositions of deep-sea hydrothermal vent animals. Nature 306: 51–52.

Gardiner, S. L. and M. L. Jones. 1994. On the significance of larval and juvenile morphology for suggesting phylogenetic relationships of the Vestimentifera. Amer. Zool. 34: 513–522.

George, J. D. and E. C. Southward. 1973. A comparative study of the setae of Pogonophora and polychaetous Annelida. J. Mar. Biol. Assoc. UK 53(2): 403–424.

Grassel, J. F. 1985. Hydrothermal vent animals: Distribution and biology. Science 229: 713–717.

Haymon, R. M., R. A. Koski and C. Sinclair. 1984. Fossils of hydrothermal vent worms from Cretaceous sulfide ores of the Sawail Ophiolite, Oman. Science 223: 1407–1409.

Ivanov, A. V. 1962. *Pogonophora*. Consultants Bureau, New York.

Ivanov, A. V. 1963. *Pogonophora*. Academic Press, London. [English translation by D. B. Carlilse.]

Jägersten, G. 1957. On the larva of *Siboglinum*. Zool. Bidrag. 32.

Jones, M. L. 1980. *Riftia pachyptila*, new genus, new species, the vestimentiferan worm from the Galapagos Rift geothermal vents (Pogonophora). Proc. Biol. Soc. Wash. 93: 1295–1313.

Jones, M. L. 1981. *Riftia pachyptila* Jones: Observations on the vestimentiferan worm from the Galapagos Rift. Science 213: 333–336.

Jones, M. L. 1984. The giant tube worms. *In* P. R. Ryan (ed.), Deep-sea hot springs and cold seeps. Oceanus 27(3): 47–52. [With 13 other papers on various aspects of deep-sea vents and seeps.]

Jones, M. L. 1985. On the Vestimentifera, new phylum: Six new species and other taxa, from, hydrothermal vents and elsewhere. Bull. Biol. Assoc. Wash. 6: 117–158.

Jones, M. L. 1985. Vestimentiferan pogonophorans: Their biology and affinities. *In* S. C. Morris et al. (eds.), *The Origins and Relationships of Lower Invertebrates*. Syst. Assoc. Spec. Vol. No. 28, Oxford, pp. 327–342.

Jones, M. L. (ed.). 1985. Hydrothermal vents of the eastern Pacific: An overview. Bull. Biol. Assoc. Wash. 6: 1–57.

Jones, M. L. 1987. On the status of the phylum name, and other names, of the vestimentiferan tube worms. Proc. Biol. Soc. Wash. 10: 1049–1050.

Jones, M. L. and S. L. Gardiner. 1988. Evidence for a transitory digestive tract in Vestimentifera. Proc. Biol. Soc. Wash. 11: 423–433.

Land, J. van der and A. Nørrevang. 1977. Structure and relationships of *Lamellibrachia* (Annelida, Vestimentifera). Kongel. Dans. Vidensk. Selsk. Biol. Skr. 21: 1–102.

Malakhov, V. V., I. S. Popelyaev and S. V. Galkin. 1996. Microscopic anatomy of *Ridgeia phaeophiale* Jones, 1985. (Pogonophora, Vestimentifera). Parts I–V. Russian J. Mar. Biol. 22: 63–74, 125–136, 189–198, 249–260, 307–313.

Nørrevang, A. 1970. The position of Pogonophora in the phylogenetic system. Z. Zool. Syst. Evolutionforsch. 8: 161–172.

Nørrevang, A. 1970. On the embryology of *Siboglinum* and its implications for the systematic position of the Pogonophora. Sarsia 42: 7016.

Nørrevang, A. (ed.). 1975. The phylogeny and systematic position of Pogonophora. Z. Zool. Syst. Evolutionforsch. [Includes 10 papers resulting from a symposium on the pogonophorans held at the University of Copenhagen in 1973.]

Powell, M. A. and G. N. Somero. 1983. Blood components prevent sulfide poisoning of respiration of the hydrothermal vent tube worm *Riftia pachyptila*. Science 219: 297–299.

Rouse, G. W. 2001. A cladistic analysis of Siboglinidae Caullery, 1914 (Polychaeta, Annelida): Formerly the phyla Pogonophora and Vestimentifera. Zool. J. Linnean Soc. 132: 55–80.

Southward, A. J., E. C. Southward, P. R. Dando, R. L. Barrett and R. Ling. 1986. Chemoautotrophic function of bacterial symbionts in small Pogonophora. J. Mar. Biol. Assoc. U. K. 66: 415–437.

Southward, E. C. 1975. Pogonophora. *In* A. C. Giese and J. S. Pearse (eds.), *Reproduction of Marine Invertebrates*, Vol. 2, Entoprocts and Lesser Coelomates. Academic Press, New York, pp. 129–156.

Southward, E. C. 1978. Description of a new species of *Oligobrachia* (Pogonophora) from the North Atlantic, with a survey of the Oligobrachiidae. J. Mar. Biol. Assoc. U. K. 58: 357–365.

Southward, E. C. 1979. Horizontal and vertical distribution of Pogonophora in the Atlantic Ocean. Sarsia 64: 51–56.

Southward, E. C. 1980. Regionation and metamerism in Pogonophora. *In* R. Siewing (ed.), Structuranalyse und Evolutionforschung: Das Merkmal. Zool. Jahrb. Abt. Anat. Ontog. Tiere 103: 2264–2275.

Southward, E. C. 1982. Bacterial symbionts in Pogonophora. J. Mar. Biol. Assoc. U. K. 62: 889–906.

Southward, E. C. 1984. Pogonophora. *In* J. Bereiter-Hahn, A. G. Matolsky and K. S. Richard (eds.), *Biology of the Integument*, Vol. 1. Springer-Verlag, Berlin, pp. 376–388.

Southward, E. C. and J. K. Cutler. 1986. Discovery of Pogonophora in warm shallow waters of the Florida shelf. Mar. Ecol. Prog. Ser. 28: 287–289.

Webb, M. 1980. The Pogonophora: A critical assessment. Ann. Univ. Stellenbosch Ser. A2 Zool. 2: 29–33.

Young, C. M., E. Vázquez, A. Metaxas and P. A. Tyler. 1996. Embryology of vestimentiferan tube worms from deep-sea methane/sulphide seeps. Nature 381: 514–516.

14 *Sipuncula and Echiura*

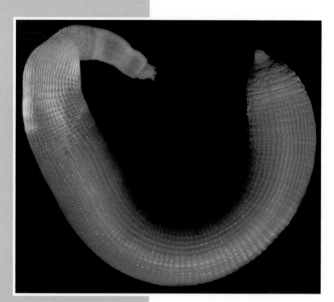

The reader will find the echiurids and sipunculids between the annelids and the arthropods, simply because three-dimensional printing is not practical.

Joel W. Hedgpeth,
Introduction to Seashore Life, 1962

The coelomate worm phyla Sipuncula and Echiura are often dismissed in short fashion as "minor" or "lesser" groups. These two phyla comprise fewer than 400 species that are never as abundant or important ecologically as some other worms, especially the polychaetes and nematodes. Nonetheless, they display body plans that are different from any discussed thus far and provide important lessons in functional morphology. We include them in a single chapter for comparison with each other and with other coelomate groups. The sipunculans and echiurans are clearly protostomes, although we have only recently begun to understand their phylogenetic positions within that assemblage. Sipunculans and echiurans resemble one another in several respects and they are often found in similar habitats. For these reasons, biologists who find interest in one of these groups often study both.

The Sipunculans

The phylum Sipuncula (Greek *siphunculus*, "little tube") includes about 250 species in 17 genera, most of which are commonly called "peanut worms." In many respects sipunculans are built along an annelidan plan, but they show no evidence of segmentation (Box 14A). The body is sausage-shaped and divisible into a retractable **introvert** and a thicker **trunk** (Figure 14.1). It is when the introvert is retracted and the body is turgid that some species resemble a peanut—hence the vernacular name. The anterior end of the introvert bears the mouth and feeding tentacles. The tentacles are derived from the regions around the mouth (**peripheral tentacles**) and around the nuchal organ (**nuchal tentacles**); differ-

BOX 14A Characteristics of the Phylum Sipuncula

1. Bilateral, unsegmented, schizocoelomate worms

2. Mesoderm derived from 4d cell (protostomous)

3. With a complete, U-shaped gut and antero-dorsally positioned anus

4. Mouth area with tentacles derived from region around mouth (peripheral tentacles) or nuchal organ (nuchal tentacles)

5. One pair of metanephridia, or a single metanephridium

6. Nervous system constructed on an annelid-like plan, but simple and with no evidence of segmentation

7. With a unique retractable anterior body region (introvert) supported by compensation system

8. Coelomic fluid with specialized multicellular structures (urns) for waste collection

9. Development is usually indirect, with a trochophore larva and sometimes a second larval stage, the pelagosphera. Early embryogenesis displays the molluscan cross

10. Entirely marine, benthic

Vietnam), large sand-burrowing species are occasionally consumed as human food. Some representative species are illustrated in Figure 14.1.

The sipunculan bauplan is founded on the qualities of the spacious body coelom. Uninterrupted by transverse septa, the coelomic fluid provides an ample circulatory medium for these sedentary worms. The coelom and associated musculature function as a hydrostatic skeleton and as a hydraulic system for locomotion, circulation of coelomic fluid, and introvert extension.

(A)

(B)

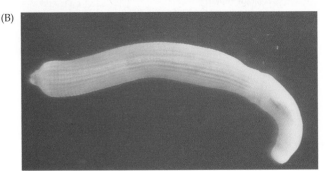

ences in the arrangement are of taxonomic importance. The gut is characteristically U-shaped and highly coiled, and the anus is located dorsally on the body near the introvert–trunk junction. The body surface is usually beset with minute bumps, warts, tubercles, or spines. Sipunculans range in length from less than 1 cm to about 50 cm, but most are 3–10 cm long.

The coelom is well developed and unsegmented, forming a spacious body cavity. Metanephridia are present, with nephridiopores on the ventral body surface. There is no circulatory system, but the coelomic fluid includes cells containing a respiratory pigment. Most sipunculans are dioecious and reproduce by epidemic spawning. Development is usually indirect, is typically protostomous, and includes a free-swimming larva.

Sipunculans are benthic and exclusively marine. They are usually reclusive, either burrowing into sediments or living beneath stones or in algal holdfasts. In tropical waters sipunculans are common inhabitants of coral and littoral communities, where they often burrow into hard, calcareous substrata. Some inhabit abandoned gastropod shells, polychaete tubes, and other such structures. They are found from the intertidal zone to depths of over 5,000 meters. In the Far East (e.g.,

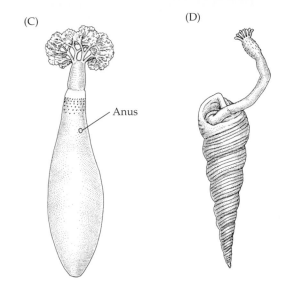

Figure 14.1 Representative sipunculans. (A) *Phascolosoma*, with the tip of the introvert turned inward. (B) *Sipunculus nudus*. (C) *Themiste pyroides* has a short introvert and stalked tentacles. (D) *Phascolion* in a gastropod shell.

Taxonomic History and Classification

The first published illustrations of sipunculans were produced from woodcuts made in the mid-sixteenth century. Linnaeus included these animals in the twelfth edition of his *Systema Naturae* (1767) and placed them in the Vermes, along with so many other odds and ends. In the nineteenth century, Lamarck and Cuvier considered the sipunculans to be relatives of the holothurian echinoderms (sea cucumbers). No separate taxon was established for these worms until 1828, when DeBlainville introduced the name Sipunculida and allied the group with certain parasitic helminths.

In 1847 Quatrefages invented the group Gephyrea to include sipunculans, echiurans, and priapulans. The Greek root *gephyra* means "bridge," as Quatrefages regarded these animals as intermediate between annelids and echinoderms. The gephyrean concept was founded on superficial characteristics, but it persisted well into the twentieth century even though many authors attempted to raise the constituent groups to individual phylum status. Finally, Hyman (1959), recognizing the polyphyletic nature of the Gephyrea, elevated the sipunculans to separate phylum rank; her view was quickly accepted. At that time, however, no classes, orders, or families were recognized, and the phylum was divided into only genera and species. The Herculean effort by Stephen and Edmonds (1972) and subsequent modifications by other workers (e.g., Rice 1982) led to a classification comprising 4 families and 16 genera. More recently, Cutler and Gibbs (1985) applied modern phylogenetic methods to the sipunculans and produced a classification scheme of 2 classes, 4 orders, 6 families, and 17 genera. The classification below is based largely upon that work.

PHYLUM SIPUNCULA

CLASS PHASCOLOSOMIDA: (Figure 14.1A). Without peripheral tentacles around mouth; nuchal tentacles present in an arc around nuchal organ; introvert hooks usually organized into distinct rings; with two metanephridia.

> **ORDER ASPIDOSIPHONIFORMES:** Trunk bears anteriorly (and occasionally posteriorly) located cuticular or calcareous shield; one family (Aspidosiphonidae) and three genera (*Aspidosiphon, Cloeosiphon, Lithacrosiphon*).

> **ORDER PHASCOLOSOMIFORMES:** Without trunk shields; one family (Phascolosomatidae) and three genera (*Antillesoma, Apionsoma, Phascolosoma*).

CLASS SIPUNCULIDA: (Figures 14.1B–D). Peripheral oral tentacles present (sometimes reduced to a flaplike veil) partially or wholly surrounding the mouth; nuchal tentacles present or absent; introvert hooks not organized into distinct rings.

> **ORDER GOLFINGIAFORMES:** Longitudinal muscles of body wall not in distinct bands.

>> **FAMILY THEMISTIDAE:** Peripheral tentacles borne in clusters on stalks produced from oral disc; dorsal retractor muscles absent; with two metanephridia; monogeneric (*Themiste*).

>> **FAMILY PHASCOLIONIDAE:** Peripheral tentacles not borne in stalked clusters; dorsal retractor muscles present; with a single metanephridium; spindle muscle absent; most species asymmetrically coiled; two genera (*Onchnesoma, Phascolion*).

>> **FAMILY GOLFINGIIDAE:** Peripheral tentacles not borne in stalked clusters; dorsal retractor muscles present; with two metanephridia; spindle muscle present; three genera (*Golfingia, Nephasoma, Thysanocardia*).

> **ORDER SIPUNCULIFORMES:** Longitudinal muscles of the body wall in distinct bands; dermis with system of coelomic channels; with two metanephridia; one family (Sipunculidae) and five genera (*Phascolopsis, Siphonomecus, Siphonosoma, Sipunculus, Xenosiphon*).

Body Wall, Coelom, Circulation, and Gas Exchange

The sipunculan body surface is everywhere covered by a well developed cuticle that varies from thin on the tentacles to quite thick and layered over much of the trunk (Figure 14.2). The cuticle often bears papillae, warts, or spines of various shapes. Members of the family Aspidosiphonidae are characterized by anterior and sometimes posterior shields derived from thickened cuticle or calcareous deposits.

Beneath the cuticle lies the epidermis, the cells of which are cuboidal over most of the body but grade to columnar and ciliated on the tentacles. The epidermis contains a variety of unicellular and multicellular glands, some of which project into the cuticle and produce some of the surface papillae. Some of these glands are associated with sensory nerve endings; others are responsible for producing the cuticle or for mucus secretion.

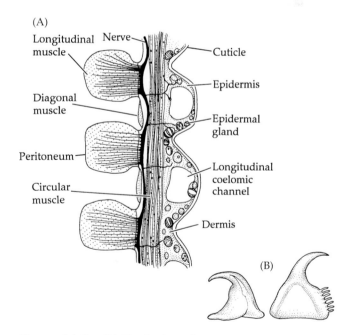

Figure 14.2 (A) The body wall of *Sipunculus nudus* (cross section). (B) Two types of cuticular spines from sipunculans.

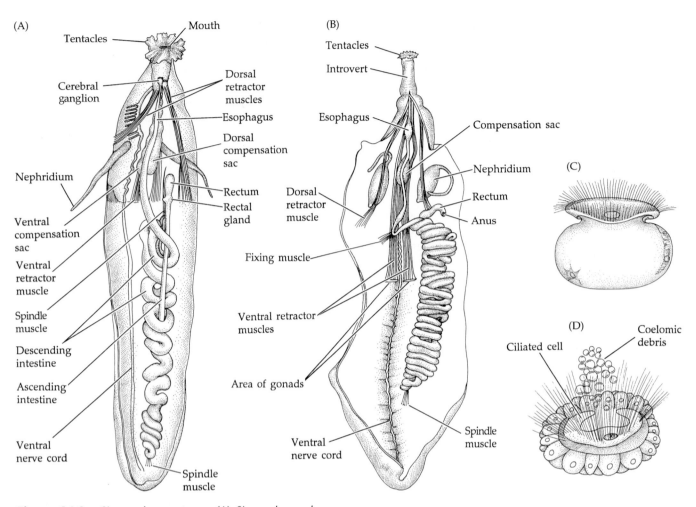

Figure 14.3 Sipunculan anatomy. (A) *Sipunculus nudus.* (B) *Golfingia vulgaris.* (C,D) Free coelomic urns from *Sipunculus* (C) and *Phascolosoma* (D). (E) The gut of *Golfingia* (partial section). Note the ciliated intestinal groove. (F) A nephridium of *Phascolosoma* (section).

Beneath the epidermis, especially where it is raised, is a connective tissue dermis of fibers and loose cells. In the Sipunculidae, which are primarily large, warm-water species, the dermis also houses a system of coelomic extensions or channels (Figure 14.2). These **coelomic channels** may extend entirely through the muscle layers and into the epidermal layer, and have a gas transport function in some species. They are connected to the trunk coelom by pores. The muscles of the body wall include outer circular and inner longitudinal layers and sometimes a thin middle layer of diagonal fibers. The longitudinal muscles form a continuous sheet in many sipunculans, but in some genera these muscles form distinct bundles or bands (e.g., *Phascolosoma* and all members of the family Sipunculidae). This arrangement gives the internal surface of the body wall a ribbed appearance that is often visible through the animal's cuticle. From one to four large **introvert retractor muscles** extend from the body wall into the in-

trovert where they insert on the gut just behind the mouth (Figure 14.3A,B).

A peritoneum lines the body wall and the internal organs. The coelom is a continuous space, but peritoneal mesenteries form incomplete partitions supporting the organs. In addition to the main body coelom and the coelomic channels in the body wall, a separate fluid-filled "coelom" called the **compensation system** is associated with the tentacles. The hollow tentacles contain lined spaces that are continuous with one or two sacs (the **compensation sacs**) that lie next to the esophagus (Figure 14.3A). Upon eversion of the introvert, circular body muscles apply pressure on these sacs and force the contained fluid into the tentacles, causing their erection.

The fluid of the body cavity contains a variety of cells and other inclusions. There are both granular and agranular amebocytes of uncertain function, and red blood cells containing hemerythrin. Also contained in the coelomic fluid are unique and fascinating multicellular structures called **urns**, some of which are fixed to the peritoneum and some of which swim free in the fluid (Figure 14.3C,D). The urns accumulate waste materials and dead cells by trapping them with cilia and mucus.

Gas exchange apparently varies among species. Ruppert and Rice (1995) suggest that rock borers and

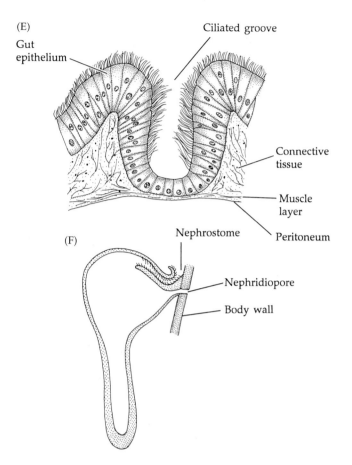

(E)

Gut epithelium

Ciliated groove

Connective tissue

Muscle layer

Peritoneum

(F)

Nephrostome

Nephridiopore

Body wall

those that burrow in sediments of low oxygen content (e.g., *Themiste*) exchange gases largely across the tentacles, which extend into the overlying water. Other burrowers with long introverts may use the body surface of both the tentacles and the introvert. Still others, such as the large-bodied species of the genus *Sipunculus*, may use the entire body surface. The coelomic fluid in the body cavity and in body wall channels provides a circulatory medium, aided by diffusion and body movements. Hemerythrin, carried in red blood cells, stores and transports oxygen.

Support and Locomotion

Sipunculans are sedentary creatures. The general body shape is maintained by the muscles of the body wall and the hydrostatic skeleton established by the large coelom. The body is essentially a fluid-filled bag of constant volume, so any constriction at one point is accompanied by an expansion at another. Burrowing in soft substrata is accomplished by peristalsis, driven by the circular and longitudinal muscles of the body wall and by the action of the introvert. Movement through algal holdfasts and bottom rubble occurs in a similar manner. Some species with an anterior cuticular shield burrow into hard substrata and use the shield as an operculum to close the burrow entrance. Burrowing into hard substrata is probably accomplished by both mechanical and

chemical means, the former using cuticular structures (such as spines and the posterior shield) as rasps, the latter facilitated by the secretions of epidermal glands.

The introvert is extended when the coelomic pressure is increased by contracting the circular muscles of the body wall. Withdrawal is accomplished by the retractor muscles, which pull from the mouth end, turning the introvert inward on itself as the body wall muscles relax. When the introvert is fully extended, the tentacles are erected by increasing the pressure on the compensation sacs.

Sipunculans are highly tactile and strongly thigmotactic, requiring contact with their surroundings. Placed alone in a glass dish, they are rather inactive except for rolling the introvert in and out. However, if several sipunculans are placed together or with small stones or shell fragments, they soon respond by making contact with each other or surrounding objects.

Feeding and Digestion

There is surprisingly little information on the details of sipunculan feeding mechanisms. Indirect evidence from anatomy, gut contents, and general behavior suggests that these animals use different feeding methods in different habitats. Most of the sipunculans that can place their tentacles at a substratum–water interface are selective or nonselective detritivores (e.g., shallow burrowers, algal holdfast dwellers); they use the mucus and cilia on the tentacles to obtain food. Deeper burrowers in sand are direct deposit feeders. Some appear to be ciliary-mucus suspension feeders, using the tentacles to extract organic material from the water. Sipunculans that burrow in calcareous substrata use spines or hooks on their introverts to retrieve organic detritus within reach and ingest the material by retracting the introvert. Limited data suggest that at least some sipunculans take up dissolved organic compounds directly across the body wall. Some workers speculate that up to 10 percent of these animals' nutritional requirements may be met in this fashion.

Because the anus is located anteriorly on the dorsal side of the body, the digestive tract is basically U-shaped, although highly coiled (Figure 14.3A,B). The mouth is at the end of the introvert and is wholly or partially surrounded by the peripheral tentacles (Sipunculida) or lies near the nuchal tentacles (Phascolosomida). The mouth leads inward to a short, muscular stomodeal pharynx, which is followed by an esophagus that extends through the introvert and into the trunk. The midgut consists of a long intestine composed of descending and ascending portions coiled together. It is usually supported by a threadlike **spindle muscle** that extends from the body wall near the anus through the coils to the end of the trunk and by several **fixing muscles** connecting the gut to the body wall. The ascending intestine leads to a short proctodeal rectum, terminating in the anus.

The intestine is ciliated and bears a distinct groove along its length (Figure 14.3E). This ciliated groove leads ultimately to a small pouch or diverticulum (or rectal gland) off the rectum. The function of this groove and diverticulum is unknown. The lumen epithelium of the descending intestine contains a variety of gland cells that are presumably the sources of digestive enzymes.

Excretion and Osmoregulation

Most sipunculans possess one pair of metanephridia (nephromixia; Figure 14.3F). Two genera (*Onchnesoma* and *Phascolion*) have but a single nephridium. Species in these genera tend to be asymmetrically coiled. The nephridiopores are located ventrally on the anterior region of the trunk. The nephrostome lies close to the body wall, near the pore, and leads to a large nephridial sac that extends posteriorly in the trunk.

Sipunculans are ammonotelic. Nitrogenous wastes accumulate in the coelomic fluid and are excreted via the nephridia. The urns also play a major excretory role by picking up particulate waste material in the coelom. The fate of wastes accumulated by urns is unknown, but at least some is probably transported to the nephridia. Urns originate as fixed epithelial cell complexes in the peritoneum, where they trap and remove particulate debris. They are also known to secrete mucus in response to pathogens in the coelomic fluid. Fixed urns regularly detach and become free-swimming in the coelomic fluid. Not only do urns effectively cleanse the coelomic fluid, but they also participate in a clotting process when a sipunculan is injured. Free urns can be seen by preparing a wet slide of fresh coelomic fluid. They are usually obvious, moving about like little bumper cars, trailing strands of mucus and bits of particulate matter.

Sipunculans are basically osmoconformers and they are unknown from fresh and brackish-water habitats. Under normal conditions, the coelomic fluid is nearly isotonic to the surrounding sea water. However, when placed in hypotonic or hypertonic environments, the body volume increases or decreases, respectively. Interestingly, the rates of volume change differ when the animal is exposed to these opposing environments, suggesting that sipunculans are better at preventing water loss than at preventing water gain. This situation may be due to a differential permeability of the cuticle, or perhaps to some active mechanism of the nephridia. In any case, sipunculans rarely face severe osmotic problems in their usual environments, and even in laboratory experiments they are able to recover nicely from most conditions of osmotic stress.

Nervous System and Sense Organs

The general structure of the sipunculan nervous system is similar in many respects to that in annelids. A bilobed cerebral ganglion lies dorsally in the introvert, just behind the mouth. Circumenteric connectives extend from the cerebral ganglion to a ventral nerve cord running along the body wall through the introvert and trunk (Figures 14.3A,B and 14.4A). The ventral nerve cord is single, and there is no evidence of segmental ganglia. In *Phascolosoma agassizii*, a double ventral nerve cord forms initially, but later fuses into a single cord. This ontogenetic example suggests that the ancestral sipunculan condition may have been a double ventral nerve cord like that found in some primitive annelids. Lateral nerves arise from the nerve cord and extend to the body wall muscles and sensory receptors in the epidermis.

Sensory receptors are widespread in sipunculans, but many are poorly understood. Tactile receptor cells are scattered over the body within the epidermis, as would be expected, and are especially abundant on and around the tentacles. Chemosensory nuchal organs are located on the dorsal side of the introvert in many forms (Figure 14.4B). Many species possess a pair of pigment-cup ocelli on the dorsal surface of the cerebral ganglion, and some possess a so-called **cerebral organ**, consisting of a ciliated pit projecting inward to the cerebral ganglion. The cerebral organ may be involved in chemoreception or perhaps neurosecretion, as is the similar structure in nemerteans.

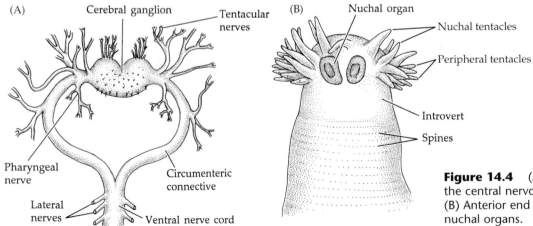

Figure 14.4 (A) Anterior portion of the central nervous system of *Golfingia*. (B) Anterior end of *Golfingia*. Note the nuchal organs.

Reproduction and Development

Sipunculans possess reasonable powers of regeneration. Most species are able to regrow lost parts of the tentacles and even the introvert, and some can regenerate portions of the trunk and the digestive tract.

It was long believed that sipunculans could not reproduce asexually. However, in the 1970s it was discovered that at least some species do possess this capacity. The process takes place by transverse fission of the body, whereby the worm divides into a small posterior fragment and a larger anterior portion. Both portions then regrow the missing parts. Regeneration from the small posterior part is quite remarkable, since most of the trunk, anterior gut, retractor muscles, nephridia, introvert, and so on must be regrown.

Except for *Golfingia minuta*, sipunculans are dioecious. (Facultative parthenogenesis has been reported in one species, *Themiste lageniformis*.) The gametes arise from the coelomic lining, often near the origins of the retractor muscles. Gametes are released into the coelom, where they mature. Ripe eggs and sperm are picked up selectively by the nephridia and stored in the sacs until released. The eggs are encased in a layered, porous covering. Males spawn first, probably in response to some environmental cue; the presence of sperm in the water stimulates females to spawn.

Following external fertilization, the zygotes pass through typical protostomous development. Cleavage is spiral and holoblastic, but the relative sizes of micromeres and macromeres differ among species, depending on the amount of yolk in the egg. Sipunculan embryos display the molluscan cross, a feature that suggests a close relationship with the Mollusca.

The cell fates are the same as those in most other protostomes. The first three quartets of micromeres become ectoderm and ectomesoderm; the 4d cell produces entomesoderm; and 4a, 4b, 4c, and 4Q form the entoderm. The mouth opens at the site of the blastopore, and the surrounding ectodermal cells grow inward as a stomodeum. The anus breaks through secondarily on the dorsal surface (Figure 14.5). The 4d mesoderm proliferates as two bands, as it does in annelids, but yields the major trunk coelom without segmentation.

Four different developmental sequences have been recognized among the sipunculans (see papers by M. Rice). A few species are known to undergo direct development (e.g., *Golfingia minuta*, *Phascolion crypta*, and *Themiste pyroides*). The eggs of these sipunculans are covered by an adhesive jelly and attach to the substratum after fertilization. The embryo develops directly to a vermiform individual that hatches as a minute juvenile sipunculan.

The other three developmental patterns are indirect, involving various combinations of larval stages. In some species (e.g., *Phascolion strombi*), a free-living lecithotrophic trochophore larva develops and metamorphoses into a juvenile worm. The other two developmental patterns involve a second larval stage, the **pelagosphera larva**, that forms after a metamorphosis of the trochophore (Figure 14.5C). In some species, both the trochophore and the pelagosphera forms are lecithotrophic and relatively short-lived (e.g., some species of *Golfingia* and *Themiste*), while in others the pelagosphera larva is planktotrophic and may live for extended periods of time in the plankton (e.g., *Aspidosiphon parvulus*, *Sipunculus nudus*, and members of the genus *Phascolosoma*).

The transformation of the trochophore to the pelagosphera larva involves a reduction or loss of the prototrochal ciliary band and the formation of a single metatrochal band for locomotion. The pelagosphera eventually elongates, settles, and becomes a juvenile sipunculan (Figure 14.5D).

The Echiurans

Members of the phylum Echiura (Greek *echis*, "serpent-like") are also unsegmented, coelomate worms. There are about 135 known species. Like the sipunculans, the echiurans share certain features with the annelid bauplan but lack any indications of metamerism (Box 14B). The vermiform body is divided into an anterior, preoral **proboscis** and an enlarged **trunk** (Figure 14.6). The mouth is located at the anterior end of the trunk at the base of a **proboscis groove**, or **gutter**. The body surface may be smooth or somewhat warty, and sometimes bears chaetae (e.g., *Urechis*).

Many echiurans are quite large. The trunk may be from a few to as many as 40 cm long, but the proboscis may reach lengths of 1–2 meters (e.g., in *Bonellia* and *Ikeda*). The Pacific *Listriolobus pelodes* may be sexually mature when only 7 mm long, whereas the Japanese *Ikeda taenioides* may reach lengths in excess of 2 meters. Some forms, such as the beautiful emerald green *Bonellia*, show drastic sexual dimorphism, wherein "dwarf" males are less than 1 cm long. *Bonellia* is also notable in its production of the green toxin **bonellin**, which probably has an antipredatory role.

The spacious body cavity contains numerous mesenteries but is not completely partitioned by septa, and there is no convincing evidence of segmentation. From one to many metanephridia lie in the anterior trunk region, each usually leading to its own ventral pore. Most echiurans possess a simple closed circulatory system. These worms are dioecious, and their development reflects a clear protostome affinity, including a trochophore larval stage.

Like sipunculans, with which they commonly co-occur, echiurans are benthic and live exclusively in marine or brackish-water habitats. Most burrow in sand or mud, or live in surface detritus or rubble. Some species typically inhabit rock galleries excavated by boring clams or other invertebrates. They are known from intertidal

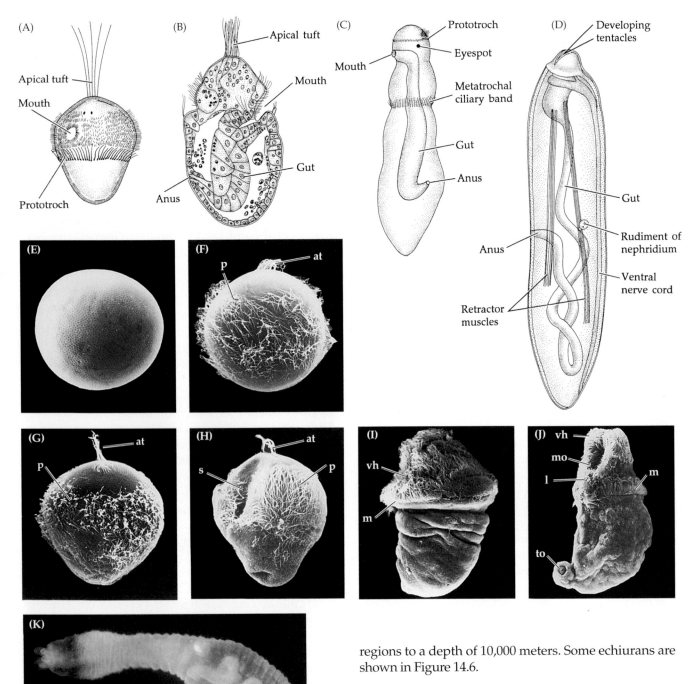

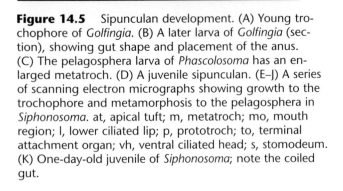

Figure 14.5 Sipunculan development. (A) Young trochophore of *Golfingia*. (B) A later larva of *Golfingia* (section), showing gut shape and placement of the anus. (C) The pelagosphera larva of *Phascolosoma* has an enlarged metatroch. (D) A juvenile sipunculan. (E–J) A series of scanning electron micrographs showing growth to the trochophore and metamorphosis to the pelagosphera in *Siphonosoma*. at, apical tuft; m, metatroch; mo, mouth region; l, lower ciliated lip; p, prototroch; to, terminal attachment organ; vh, ventral ciliated head; s, stomodeum. (K) One-day-old juvenile of *Siphonosoma*; note the coiled gut.

regions to a depth of 10,000 meters. Some echiurans are shown in Figure 14.6.

Taxonomic History and Classification

Echiurans were first reported in the literature during the eighteenth century. The first few species were described (by Pallas) as annelids. Eventually they were placed in the now-abandoned taxon Gephyrea, along with the sipunculans and priapulans. In 1896 Sedgwick suggested raising the sipunculans and priapulans to separate phylum status, but he considered the echiurans to be a class of annelid worms. It was not until 1940—when Newby, and subsequently others, conducted studies on echiuran development—that these worms were established as a separate phylum. The name of the phylum has varied over the years (e.g., Echiurida, Echiuroidea), but Echiura is now the preferred spelling. The phylum contains three orders.

BOX 14B Characteristics of the Phylum Echiura

1. Bilateral, unsegmented, schizocoelomate worms

2. Mesoderm derived from 4d cell (protostomous)

3. With a complete gut and posterior terminal anus

4. With metanephridia

5. Nervous system constructed on an annelid-like plan, but simple and with no evidence of segmentation

6. Most possess a simple closed circulatory system; blood is unpigmented, but red blood cells with hemoglobin occur in the coelomic fluid

7. With a muscular but nonretractable preoral proboscis at the anterior end of the trunk

8. Many with paired epidermal chaetae, similar in chemical composition to the chaetae of annelids

9. Development is indirect, with a trochophore larva

10. Entirely marine, benthic

PHYLUM ECHIURA

ORDER ECHIUROINEA: Body wall muscle layers are (from outer to inner) circular, longitudinal, and oblique; with one to several pairs of metanephridia (some species with a single nephridium); circulatory system present; hindgut not modified as a gas exchange organ; about 130 species in two families, Bonelliidae (e.g., *Achaetobonellia, Bonellia, Bruunella*) and Echiuridae (e.g., *Echiurus, Listriolobus, Thalassema*).

ORDER XENOPNEUSTA: Body wall muscle layers as in Echiuroinea; 2–3 pairs of nephridia; circulatory system absent; hindgut elongate and thin-walled, modified for gas exchange; proboscis very short; one family (Urechidae), monogeneric (*Urechis*), with four species.

ORDER HETEROMYOTA: Body wall muscle layers are (from outer to inner) longitudinal, circular, oblique; 200–400 nephridia; circulatory system present and bearing heartlike enlargement; hindgut not enlarged; one family (Ikedaidae), monotypic (*Ikeda taenioides*).

Body Wall and Coelom

The body wall of echiurans is roughly similar to that of sipunculans, except for variations in muscle arrangement, as outlined in the classification scheme. A thin cuticle covers the epidermis, which is composed of a cuboidal epithelium and contains a variety of gland cells. Paired epidermal chaetae occur in some species. A fibrous dermis lies beneath the epidermis. Layers of circular, longitudinal, and oblique muscles form the bulk of the body wall, which is lined internally by the peritoneum. The epidermis is ciliated along the proboscis groove, or **gutter**.

The coelomic cavity is spacious and occupies most of the trunk. It is interrupted only by partial mesenteries between the gut and the body wall. The coelomic fluid contains red blood cells, with hemoglobin in some species, and various types of amebocytes. Cells likened to chlorogogen cells have been reported in a few species.

Support and Locomotion

The large trunk coelom provides a hydrostatic skeleton against which the body wall muscles operate. The non-septate coelom allows peristaltic movements as the animal burrows or moves through shell rubble or gravel.

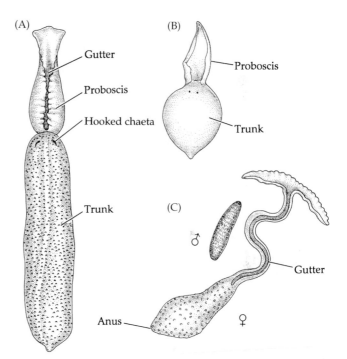

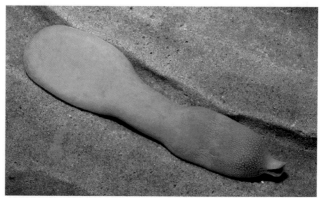

Figure 14.6 Representative echiurans. (A) *Echiurus*. (B) *Listriolobus*, with the proboscis contracted. (C) *Bonellia viridis*. Note the extreme sexual dimorphism between the large female and the tiny male. (D) *Urechis caupo*, the "fat innkeeper."

(A)

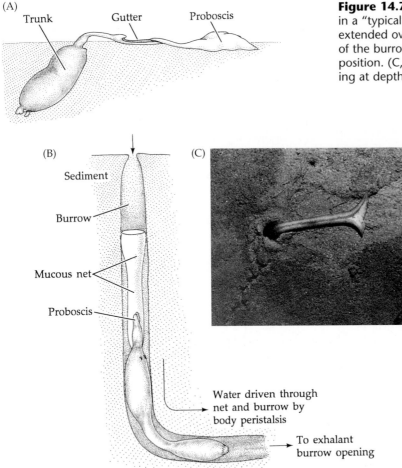

Trunk   Gutter   Proboscis

Sediment

Burrow

Mucous net

Proboscis

Water driven through
net and burrow by
body peristalsis

To exhalant
burrow opening

(B)

(C)

(D)

Figure 14.7   Feeding in echiurans. (A) *Tatjanellia grandis* in a "typical" echiuran feeding posture, with the proboscis extended over the surface of the substratum. (B) A portion of the burrow of *Urechis caupo*. The worm is in its feeding position. (C,D) The proboscides of deep-sea echiurans living at depths of 2635 m and 7570 m, respectively.

The proboscis is capable of shortening and lengthening, but it does not roll in and out as does the introvert of sipunculans. This characteristic, and the more conventional posterior position of the anus, allow one to distinguish quickly between members of these two similar phyla.

Feeding and Digestion

Most echiurans feed on epibenthic detritus. Typically, the animal lies with the trunk more or less buried in the substratum, with the anterior end directed upward and the proboscis extended over the sediment (Figure 14.7A,C,D). Densely packed gland cells of the proboscis epithelium secrete mucus, to which organic detrital particles adhere. The mucous coating and the food are moved along the ventral proboscis gutter by ciliary action into the mouth.

An interesting exception to the above feeding method occurs in *Urechis*. Members of this genus excavate and reside within U-shaped burrows in soft substrata (Figure 14.7B). *Urechis* possesses a very short proboscis, unlike those of more typical echiurans, and engages in mucous-net filter feeding. A ring of glands located near the proboscis–trunk junction produces a funnel-shaped mucous net, which is attached to the burrow wall by the proboscis. Water is drawn through the burrow and the sheet

of mucus by peristaltic movements of the body, and suspended food particles as small as 1 µm are caught in the fine-meshed net. Periodically the animal grasps the food-laden net with its proboscis and ingests it. The whole process is remarkably similar to the feeding behavior of polychaete annelids of the genus *Chaetopterus*.

The digestive tract is generally very long and coiled, leading from the mouth at the base of the proboscis to the posterior anus (Figure 14.8). The foregut is a stomodeum and may be regionally specialized as pharynx, esophagus, gizzard, and stomach, or it may be more or less uniform along its length. The midgut usually bears a longitudinal ciliated groove, or **siphon**, which probably aids the movement of materials through the gut. It may also shunt excess water from the main midgut lumen, thereby concentrating food and facilitating digestion. The hindgut, or cloaca, is a proctodeum and varies in structure among different species. In most echiurans, the cloaca bears a pair of large excretory diverticula called anal vesicles (see below). In some species of *Urechis* the cloaca is enlarged and very thin walled (Figure 14.8C). In such cases water is pumped in and out of the hindgut for gas exchange.

Not much is known about digestive physiology in echiurans. The epithelium of the midgut is rich in gland cells that presumably produce and secrete digestive enzymes. Digestion and absorption occur mainly in the midgut.

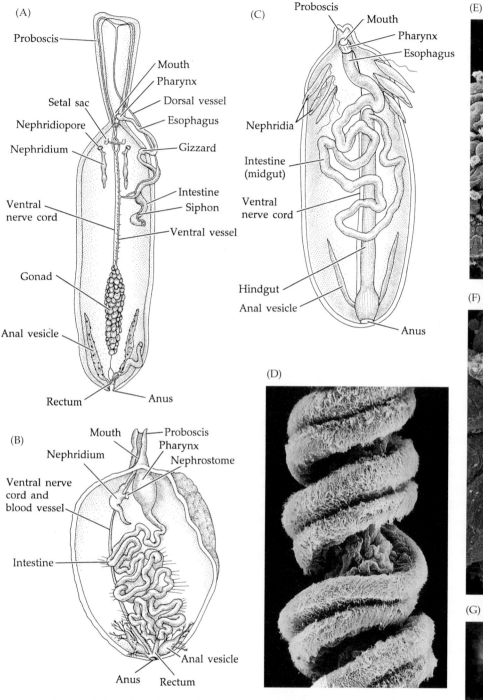

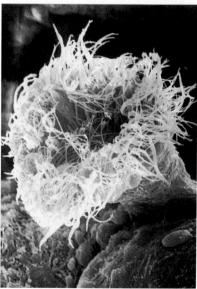

Figure 14.8 Internal anatomy of three echiurans. (A) General internal anatomy, as seen in *Echiurus*. (B) A dissected *Bonellia viridis* female (proboscis cut off). (C) Internal organs of *Urechis*. Note especially the enlarged midgut. (D) Part of the coiled nephridial funnels of *Urechis*, which serve to sort gametes from the coelom (SEM). (E–G) SEMs of anal vesicles and excretory funnels of *Urechis* at 60× (E); 138× (F); and 840× (G).

Circulation and Gas Exchange

Most echiurans possess a simple closed circulatory system, although it is entirely absent from some forms (e.g., *Urechis*). The circulatory system generally includes dorsal and ventral longitudinal vessels in the trunk and median and lateral vessels in the proboscis (Figure 14.8A,B). There is no major pumping organ (except in *Ikeda*); the blood is transported by pressures generated from body movements and by the weak musculature of the vessel walls. The blood is usually colorless. Apparently the main function of the circulatory system is the transport of nutrients and perhaps other metabolites through the body.

The main site of gas exchange is probably the hindgut, which is usually provided with oxygenated water by **cloacal irrigation**—water being pumped in and out of the anus by muscle action. It is likely that some gas exchange also occurs across the general body surface, particularly on the proboscis. The coelomic fluid contains red blood cells with hemoglobin.

Excretion and Osmoregulation

The excretory structures of echiurans include paired metanephridia and anal vesicles. The number of nephridia varies: one pair in *Bonellia*; two pairs in *Echiurus*; three pairs in *Urechis*; and hundreds of pairs in *Ikeda*. When only one or a few pairs are present, the nephridia are located in the anterior region of the trunk and lead to nephridiopores on either side of the ventral midline (Figure 14.8). The degree to which these nephridia function in excretion is debatable. In some, such as *Urechis*, the nephrostomes are long and coiled (Figure 14.8D); they seem to function primarily in picking up gametes from the coelom, having relinquished the major excretory responsibility to the anal vesicles.

The **anal vesicles** are hollow sacs arising as evaginations of the cloaca near the anus (Figure 14.8A–C, E–G). Each vesicle bears from about a dozen to as many as 300 ciliated funnels that open to the coelom. Few studies have been conducted on the function of these structures, but they apparently pick up wastes from the coelomic fluid and remove the material to the hindgut and anus. The anal vesicles may also function in gas exchange, and perhaps in osmoregulation and the control of intracoelomic pressure. Echiurans are relatively poor osmoregulators, but it is unlikely that they encounter severe osmotic stress in their usual marine habitats.

Nervous System and Sense Organs

The nervous system of echiurans is simple, although constructed in a fashion generally similar to the annelidan plan. An anteriorly located nerve ring extends around the gut and dorsally forward into the proboscis. Ventrally, in the trunk, the nerve ring connects with a single ventral nerve cord extending the length of the body. There are no ganglia in this system and no evidence of segmentation. Lateral nerves arise irregularly from the ventral nerve cord and extend to the body wall muscles (Figure 14.8A–C). In a few species, the ventral nerve cord forms as a double cord but fuses during development.

Associated with the simple nervous system and infaunal sedentary lifestyle of echiurans is the absence of major sensory receptors. These animals are mildly touch-sensitive, especially on the proboscis and the chaetae, and they also may possess chemoreceptors.

Reproduction and Development

Asexual reproduction is unknown in echiurans, and little work has been done on the powers of regeneration. At least some display remarkable healing capabilities. For example, *Urechis caupo*—the "fat innkeeper"—is often found in bay muds that are subjected to heavy pressure from clam diggers. In some of these tidal flats, nearly every *Urechis* specimen bears scars, some nearly completely across the body—signs that the animal has survived the onslaught of the clammers' shovels.

The echiuran sexes are separate. The gametes are produced in special "gonadal" regions of the peritoneum, often near the base of the ventral blood vessel, and released into the coelom to mature. When ripe, the gametes accumulate in the nephridia until spawning occurs. The nephridia often swell enormously when packed with eggs or sperm. In most cases, epidemic spawning takes place and is followed by external fertilization.

Echiuran development (Figure 14.9) is basically similar to that of other protostomes described thus far (e.g., annelids and sipunculans). Cleavage is holoblastic and spiral. The cell fates follow the typical protostome pattern. A trochophore larva (Figure 14.9B,C) develops and may drift in the plankton for up to three months as it gradually elongates to produce a young worm (Figure 14.9D,E). The prototroch and the preprototroch regions develop into the proboscis, and the postprototroch forms the enlarged trunk. The 4d mesentoblast proliferates the main trunk coelom.

A number of earlier works, including many general texts, make reference to transitory segmentation of the larval coelom and cite such evidence as supporting an annelid ancestry to the echiurans. It is now generally agreed that no such segmentation occurs. The superficial annulations on the larvae are simply tiers of epithelial cells (Figure 14.9C); thus there is no evidence of segmentation in any stage of echiuran life history.

Mention must be made of a strange case of sexual dimorphism and sex determination in the family Bonelliidae. Female bonelliids are quite large, reaching lengths of up to 2 m, including the proboscis. The males, however, are only a few millimeters long, very reduced in complexity, and often retain remnants of larval ciliation. They live on the female's body or in her nephridia. Evidence suggests that the determination of sex in bonelliids is largely controlled environmentally at the time of larval settlement. If a larva settles on or near the proboscis area of an adult female, it will mature rapidly as a "dwarf" male. However, if the larva settles away from a female, it will burrow and eventually grow and mature as a female. The induction of maleness is apparently caused by a masculinizing hormone produced by the proboscis of the female worm. However, sometimes two larvae clump together prior to settlement, whereupon one becomes a female and

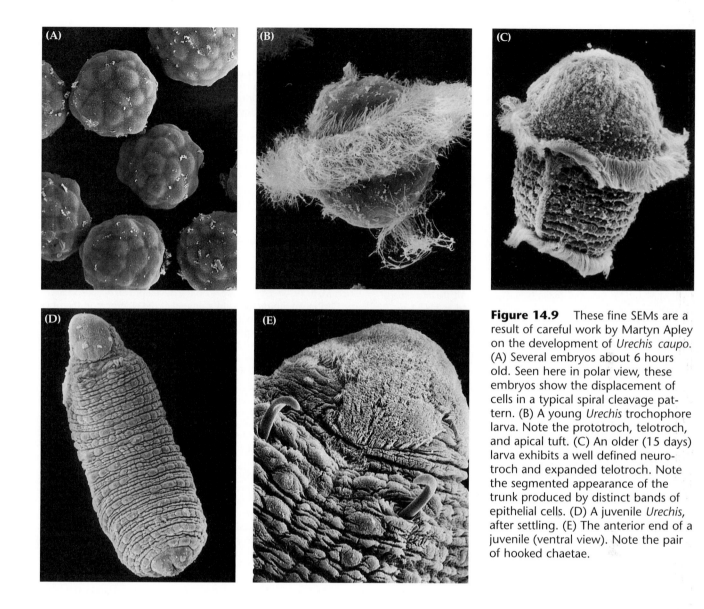

Figure 14.9 These fine SEMs are a result of careful work by Martyn Apley on the development of *Urechis caupo*. (A) Several embryos about 6 hours old. Seen here in polar view, these embryos show the displacement of cells in a typical spiral cleavage pattern. (B) A young *Urechis* trochophore larva. Note the prototroch, telotroch, and apical tuft. (C) An older (15 days) larva exhibits a well defined neurotroch and expanded telotroch. Note the segmented appearance of the trunk produced by distinct bands of epithelial cells. (D) A juvenile *Urechis*, after settling. (E) The anterior end of a juvenile (ventral view). Note the pair of hooked chaetae.

the other a male. Development into a male can also be induced by pH changes that cause a slightly acidic condition of the sea water. Interestingly, a larva that settles on the trunk of a female rather than on the proboscis commonly develops an intermediate sexual condition.

Some Comments on Phylogeny

The embryonic and adult characteristics of sipunculans and echiurans place them solidly within the protostomes. In the past, some workers have treated the sipunculans and echiurans as offshoots of an early polychaete line. In this scenario, one might view the absence of segmentation as a secondary loss of a partitioned coelom associated with the exploitation of a sedentary, burrowing lifestyle. The reduction in sensory receptors and simplification of the nervous system in general are explainable on this same basis. Certainly there are striking similarities among these three phyla, especially between echiurans and annelids. However, metamerism seems not to be easily lost, and many species of annelids have evolved burrowing lifestyles while retaining their basic segmentation. It thus seems more likely that the absence of segmentation is a retained primitive state, or symplesiomorphy. Sipunculan/echiuran attributes such as the trochophore larva, paired ventral nerve cords in some juveniles (fused into a single cord in adults), nephromixia, and serial arrangement of nephridia in certain species all seem to speak of an ancient, preannelid, protostome relationship.

Exactly where these groups arose along the protostome lineage is gradually becoming clearer. As mentioned in Chapter 13, there is strong recent evidence that echiuran epidermal chaetae are homologous with those of annelids. Thus it appears that the echiurans arose higher on the protostome line than did the sipunculans (and molluscs) and are a possible sister-group to the annelids. We discuss the phylogenetic positions of these two phyla in more detail in Chapter 24.

Selected References

General References

Clark, R. B. 1969. Systematics and phylogeny: Annelida, Echiura, Sipuncula. *In* M. Florkin and B. T. Scheer (eds.), *Chemical Zoology*, Vol. 4. Academic Press, New York, pp. 1–68.

Harrison, F. W. and M. E. Rice (eds.). 1993. *Onychophora, Chilopoda, and Lesser Protostomes. Vol. 12. Microscopic Anatomy of Invertebrates*. Wiley–Liss, New York.

Hyman, L. 1959. *The Invertebrates*, Vol. 5, Smaller Coelomate Groups. McGraw-Hill, New York.

Kohn, A. J. and M. E. Rice. 1971. Biology of Sipuncula and Echiura. BioScience 21: 583–584.

MacGinitie, G. E. and N. MacGinitie. 1968. *Natural History of Marine Animals*, 2nd Ed. McGraw-Hill, New York.

McHugh, D. 1997. Molecular evidence that echiurans and pogonophorans are derived annelids. Proc. Natl. Acad. Sci. 94: 8006–8009.

Pilger, J. F. 1997. Sipunculans and Echiurans. *In* S. F. Gilbert and A. M. Raunio, *Embryology. Constructing the Organism*. Sinauer Assoc., Sunderland, MA, pp. 167–168.

Rice, M. E. and M. Todororic (eds.) 1970. *Proceedings of the International Symposium on the Biology of the Sipuncula and Echiura*, Vols. 1 and 2. Nauchno Delo Press, Belgrade; Institute for Biological Research, Yugoslavia; and the Smithsonian Institution, Washington, DC.

Siddall, M. E., K. Fitzhugh and K. A. Coates. 1998. Problems determining the phylogenetic position of echiurans and pogonophorans with limited data. Cladistics 14: 401–410.

Stephen, A. C. and S. J. Edmonds. 1972. *The Phyla Sipuncula and Echiura*. British Museum (Natural History), London.

Storch, V. 1984. Echiura and Sipuncula. *In* J. Bereiter-Hahn, A. G. Matolsky and K. S. Richards (eds.), *Biology of the Integument*, Springer-Verlag, Berlin, pp. 368–375.

Sipuncula

Abercrombie, R. K. and R. M. Bagby. 1984. An explanation for the folding fibers in proboscis retractor muscles of *Phascolopsis* (= *Golfingia*) *gouldi*. II. Structural evidence from SEM of teased, glycerinated muscles and light microscopy of KOH-isolated fibers. Comp. Biochem. Physiol. 77A(1): 31–38.

Bang, B. G. and F. G. Bang. 1980. The urn cell complex of *Sipunculus nudus*: A model for study of mucus-stimulating substances. Biol. Bull. 159: 571–581.

Cutler, E. B. 1973. Sipuncula of the western North Atlantic. Bull. Am. Mus. Nat. Hist. 152: 105–204.

Cutler, E. B. 1977. *Sipuncula: Marine Flora and Fauna of the Northeastern U.S.* NOAA Technical Report NMFS Circular 403. U.S. Government Printing Office, Washington, DC.

Cutler, E. B. 1995. *The Sipuncula: Their Systematics, Biology, and Evolution*. Cornell University Press, New York.

Cutler, E. B. and N. J. Cutler. 1988. A revision of the genus *Themiste* (Sipuncula). Proc. Biol. Soc. Wash. 101: 741–766. [This paper also lists the many other generic revisions published by the Cutlers.]

Cutler, E. B. and P. E. Gibbs. 1985. A phylogenetic analysis of higher taxa in the phylum Sipuncula. Syst. Zool. 34: 162–173.

Dybas, L. 1981. Cellular defense reactions of *Phascolosoma agassizii*, a sipunculan worm: Phagocytosis by granulocytes. Biol. Bull. 161: 104–114.

Fischer, W. K. 1952. The sipunculid worms of California and Baja California. Proc. U.S. Nat. Mus. 102: 371–450.

Gibbs, P. E. 1977. *British Sipunculans*. Synopsis of the British Fauna No. 122. Academic Press, New York.

Hansen, M. D. 1978. Food and feeding behavior of sediment feeders as exemplified by sipunculids and holothurians. Helgol. Wiss. Meeresunters. 31: 191–221.

Hermans, C. O. and R. M. Eakin. 1969. Fine structure of the cerebral ocelli of a sipunculid, *Phascolosoma agassizii*. Z. Zellforsch. Mikrosk. Anat. 100: 325–399.

Mangum, C. P. and M. Kondon. 1975. The role of coelomic hemerythrin in the sipunculid worm *Phascolopsis gouldi*. Comp. Biochem. Physiol. 50A: 777–785.

Nicosia, S. V. 1979. Lectin-induced release in the urn cell complex of the marine invertebrate *Sipunculus nudus* (Linnaeus). Science 206: 698–700.

Pilger, J. F. 1982. Ultrastructure of the tentacles of *Themiste lageniformis* (Sipuncula). Zoomorphologie 100: 143–156.

Pilger, J. F. 1987. Reproductive biology and development of *Themiste lageniformis*, a parthenogenetic sipunculan. Bull. Mar. Sci. 41: 59–67.

Pörter, H.-O., U. Kreutzer, B. Siegmund, N. Heisler and M. K. Grieshaber. 1984. Metabolic adaptation of the intertidal worm *Sipunculus nudus* to functional and environmental hypoxia. Mar. Biol. 79: 237–247.

Rice, M. E. 1967. A comparative study of the development of *Phascolosoma agassizii*, *Golfingia pugettensis*, and *Themiste pyroides* with a discussion of developmental patterns in the Sipuncula. Ophelia 4: 143–171.

Rice, M. E. 1969. Possible boring structures of sipunculids. Am. Zool. 9: 803–812.

Rice, M. E. 1970a. Asexual reproduction in a sipunculan worm. Science 167: 1618–1620.

Rice, M. E. 1970b. Observations on the development of six species of Caribbean Sipuncula with a review of development in the phylum. Proc. Int. Symp. Biol. Sipuncula and Echiura. I. Kotor: 18–25.

Rice, M. E. 1976. Sipunculans associated with coral communities. Micronesica 12: 119–132.

Rice, M. E. 1978. Morphological and behavioral changes at metamorphosis in the Sipuncula. *In* F. S. Chia and M. E. Rice (eds.), *Settlement and Metamorphosis of Marine Invertebrate Larvae*. Elsevier, New York, pp. 83–102.

Rice, M. E. 1981. Larvae adrift: Patterns and problems in life histories of sipunculans. Am. Zool. 22: 605–619.

Rice, M. E. 1982. Sipuncula. *In* S. P. Parker (ed.), *Synopsis and Classification of Living Organisms*, Vol. 2. McGraw-Hill, New York, pp. 67–69.

Rice, M. E. 1985. Sipuncula: Developmental evidence for phylogenetic inference. *In* S. C. Morris et al. (eds.), *The Origins and Relationships of Lower Invertebrates*. Syst. Assoc. Spec. Vol. No. 28, Oxford, pp. 274–296.

Rice, M. E. 1988. Observations on development and metamorphosis of *Siphonosoma cumanense* with comparative remarks on *Sipunculus nudus* (Sipuncula, Sipunculidae). Bull. Mar. Sci. 42: 1–15.

Romero-Wetzel, M. B. 1987. Sipunculans as inhabitants of very deep, narrow burrows in deep-sea sediments. Mar. Biol. 96: 87–91.

Ruppert, E. E. and M. E. Rice. 1995. Functional organization of dermal coelomic canals in *Sipunculus nudus* (Sipuncula) with a discussion of respiratory designs in sipunculans. Invert. Biol. 114: 51–63.

Scheltema, R. S. and M. E. Rice. 1990. Occurrence of teleplanic pelagosphera larvae of sipunculans in tropical regions of the Pacific and Indian Oceans. Bull. Mar. Sci. 47: 159–181.

Stehle, G. 1953. Anatomie und Histologie von *Phascolosoma elongatum*. Ann. Univ. Saraviensis, Naturwiss. 2, No. 3.

Stephen, A. C. 1964. A revision of the classification of the phylum Sipuncula. Ann. Mag. Nat. Hist. Ser. 13, 7: 457–462.

Storch, V. and U. Welsh. 1979. Zur ultrastruktur der metanephridien des landlebenden sipunculiden *Phascolosoma* (*Physcosoma*) *lurco*. Kiel. Meeresforsch. 28(2): 227–231.

Thompson, B. E. 1980. A new bathyl sipunculan from southern California, with ecological notes. Deep-Sea Res. 27A: 951–957.

Uexkull, J. V. 1903. Die biologische Bauplan von *Sipunculus*. Z. Biol. 44.

Valembois, P. and D. Bioledieu. 1980. Fine structure and functions of haemerythrocytes and leucocytes of *Sipunculus nudus*. J. Morphol. 163: 69–77.

Walter, M. D. 1973. Feeding and studies on the gut content in sipunculids. Helgol. Wiss. Meeresunters. 25(4): 486–494.

Williams, J. A. and S. V. Margolis. 1974. Sipunculid burrows in coral reefs: Evidence for chemical and mechanical excavation. Pac. Sci. 28(4): 357–359.

Zuckerkandl, E. 1950. Coelomic pressures in *Sipunculus nudus*. Biol. Bull. 98: 161.

Echiura

Bosch, C. 1981. La musculature, le squelette conjonctif et les mouvements de la trompe de la Bonellie (*Bonellia viridis* Rol., Echiurida). Ann. Sci. Nat. Zool. Paris 3: 203–229.

Dawydoff, C. 1959. Classes des Echiuriens et Priapuliens. *In* P. Grassé (ed.), *Traité de Zoologie*, Vol. 5, Pt. 1. Masson et Cie, Paris.

Edmonds, S. J. 1982. Echiura. *In* S. P. Parker (ed.), *Synopsis and Classification of Living Organisms*, Vol. 2. McGraw-Hill, New York, pp. 65–66.

Fischer, W. K. 1946. Echiuroid worms of the north Pacific Ocean. Proc. U.S. Natl. Mus. 96: 215–292.

Fischer, W. K. 1949. Additions to the echiuroid fauna of the north Pacific Ocean. Proc. U.S. Natl. Mus. 99(3248): 479–497.

Gislen, T. 1940. Investigations on the ecology of *Echiurus*. Lunds Univ. Arsskr. Avd. 36(10): 1–39.

Gould-Somero, M. C. 1975. Echiura. *In* A. C. Giese and J. S. Pearse (eds.), *Reproduction of Marine Invertebrates*, Vol. 3. Academic Press, New York, pp. 277–311.

Gould-Somero, M. C. and L. Holland. 1975. Oocyte differentiation in *Urechis caupo* (Echiura): A fine structural study. J. Morphol. 147: 475–506.

Jaccarini, V., L. Agius, P. J. Schembri and M. Rizzo. 1983. Sex determination and larval sexual interaction in *Bonellia viridis* (Echiura, Bonelliidae). J. Exp. Mar. Biol. Ecol. 66: 25–40.

Jose, K. V. 1964. The morphology of *Acanthobonellia pirotaensis*, n. sp., a bonellid from the Gulf of Kutch, India. J. Morphol. 115: 53.

Newby, W. W. 1940. The embryology of the echiuroid worm *Urechis caupo*. Mem. Am. Philos. Soc. 16: 1–213.

Ohta, S. 1984. Star-shaped feeding traces produced by echiuran worms on the deep-sea floor of the Bay of Bengal. Deep-Sea Res. 31: 1415–1432.

Pilger, J. F. 1978. Settlement and metamorphosis in the Echiura: A review. *In* F. S. Chia and M. E. Rice (eds.), *Settlement and Metamorphosis of Marine Invertebrate Larvae*. Elsevier, New York, pp. 103–111.

Pilger, J. F. 1980. The annual cycle of oogenesis, spawning, and larval settlement of the echiuran *Listriolobus pelodes* off southern California. Pac. Sci. 34: 129–142.

Pilger, J. F. 1993. Echiura. *In* F. W. Harrison and M. E. Rice (eds.), *Microscopic Anatomy of Invertebrates* Vol. 12: *Onychophora, Chilopoda, and Lesser Protostomata*, Wiley-Liss, New York, pp. 185–236.

Pritchard, A. and F. N. White. 1981. Metabolism and oxygen transport in the innkeeper worm *Urechis caupo*. Physiol. Zool. 54: 44–54.

Redfield, A. C. and M. Florkin. 1931. The respiratory function of the blood of *Urechis caupo*. Biol. Bull. 11: 85–210.

Suer, A. L. 1984. Growth and spawning of *Urechis caupo* (Echiura) in Bodega Harbor, California. Mar. Biol. 78: 275–284.

Suer, A. L. and D. W. Phillips. 1983. Rapid, gregarious settlement of the larvae of the marine echiuran *Urechis caupo* Fisher and MacGinitie, 1928. J. Exp. Mar. Biol. Ecol. 67: 243–259.

15

The Emergence of the Arthropods:
Onychophorans, Tardigrades, Trilobites, and the Arthropod Bauplan

Here came great swarms of flies into the house of Pharaoh and into his servants' houses, and in all the land of Egypt the land was ruined by reason of the flies.
Exodus 9:24

If we live out our span of life on earth without ever knowing a crab intimately we have missed having a jolly friendship. Life is a little incomplete if we can look back and recall these small people only as supplying the course after soup and with the Chablis.
William Beebe, *Nonsuch: Land of Water*, 1932

This chapter will introduce you to the vast world of arthropods. It also covers two close arthropod allies, the onychophorans (*Peripatus* and their kin) and the tardigrades (water bears). The close relationship among these three phyla has never been seriously questioned. Virtually all of the morphological and molecular analyses of the past 25 years support a sister-group relationship between the Tardigrada and Arthropoda, and between these groups the Onychophora. Some authors refer to this clade of three phyla as the **Panarthropoda**.

The first arthropods probably arose in ancient Precambrian seas over 600 million years ago, and by the early Cambrian true crustaceans were already well established. The arthropods have undergone a tremendous evolutionary radiation since then, and today they occur in virtually all environments on Earth, exploiting every imaginable lifestyle (Figure 15.1). Modern forms range in size from tiny mites and crustaceans less than 1 mm long to great Japanese spider crabs with leg spans exceeding 3 m. There are an estimated 1,097,289 described living arthropods, although the exact number isn't known (Table 15.1). The arthropods constitute 85 percent of all described animal species. Our inadequate knowledge of Earth's biodiversity is apparent when we review the range of estimates of the number of undescribed species of arthropods, which spans three orders of magnitude, from 3 million to over 100 million species. Most of this undiscovered diversity resides among the insects and mites on land and the crustaceans in the sea. Whether one leans toward the conservative or the liberal estimates, one is struck by the reality that no other group of organisms approaches the magnitude of species richness seen in arthropods. The modern world truly belongs to these creatures. Yet, despite their overwhelming diversity, the arthropods share a suite

Figure 15.1 Arthropods and their close allies, onychophorans and tardigrades. (A) An onychophoran (from the Osa Peninsula, Costa Rica). (B) A water bear (Tardigrada); note the developing embryos inside the body. (C) An orb-weaver spider (Arthropoda: Chelicerata). (D) A sphinx moth (Arthropoda: Hexapoda). (E,F) Two very different crustaceans (Arthropoda: Crustacea): an acorn barnacle (E) and an isopod (F). (G) Fossils of two Silurian trilobites (Arthropoda: Trilobitomorpha).

of fundamental similarities, a distinct unifying bauplan.

Arthropods are so abundant, so diverse, and play such vital roles in all Earth's environments that we devote five chapters to them. The present chapter is divided into five parts, first treating the phyla Onychophora and Tardigrada, sometimes called the "proto-arthropods." Next we introduce the Arthropods themselves, exploring the bauplan and basic unifying features of the phylum and how this combination of features has led to its preeminent success. This is followed by a discussion of the extinct arthropod subphylum Trilobitomorpha. Finally, we provide an overview of arthropod evolution. Detailed treatments of the living arthropod subphyla (Crustacea, Hexapoda, Myriapoda, Cheliceriformes) are provided in chapters 16–19.

Phylum Onychophora

The first living onychophoran (Greek *onycho*, "talon"; *phora*, "bearer") was described by the Reverend Lansdown Guilding in 1826 as a leg-bearing "slug" (a mollusc). Since that initial discovery, 110 or so species of onychophorans have been described, and probably at least that many more remain to be discovered. All the living species are terrestrial. However, we now know that onychophorans were part of the explosive marine diversification in the Early Cambrian (Chapter 1). Their fossils have been found in Middle Cambrian marine faunas at several localities (e.g., *Aysheaia pedunculata* from the famous Middle Cambrian Burgess Shale deposits of British Columbia, Canada, and *Aysheaia prolata* from a similar deposit in Utah), in the remarkable Chengjiang Lower Cambrian (520–530 mya) deposits of China, and in the equally stunning Swedish Upper Cambrian Orsten fauna. Perhaps the most famous onychophoran fossil is the amazing Cambrian *Hallucigenia*, long a mystery because it was originally interpreted in an upside-down orientation, but recently turned right side up and discovered to be an onychophoran with long dorsal spines (Figure 15.2A; Ramsköld and Hou 1991).

Thus, the Onychophora is an old group that has changed very little over the past 530 million years, but at some point in its long history successfully invaded the terrestrial environment (evidence suggests that the terrestrial invasion took place in the Ordovician period). Like annelids and arthropods, onychophorans are segmented animals, and they have features that are somewhat intermediate between those of annelids and arthropods. For these reasons most workers regard onychophorans as "living fossils," or "missing links" between these two phylas. We list some features shared by the annelids, onychophorans, tardigrades, and arthropods in Table 15.1.

Living onychophorans (Figure 15.3) are confined to humid habitats. During dry periods they retire to protective burrows or other retreats and become inactive. During wet periods they can be found by sifting through leaf litter in the regions where they live. Onychophorans probably live for several years, during which time periodic molting takes place, as often as every two weeks in some species. The phylum comprises two families, Peripatidae and Peripatopsidae. The former is circumtropical in distribution, whereas the latter is circumaustral (confined to the temperate Southern Hemisphere).

The Onychophoran Bauplan

Modern onychophorans resemble caterpillars (Figure 15.3), ranging from 5 mm to 15 cm in length. Within a given species, males are always smaller than females and have fewer legs. Little cephalization is present, and the body is strongly homonomous. Three paired appendages adorn the head: one pair of fleshy annulated **antennae**, a single pair of **jaws**, and a pair of fleshy **oral**

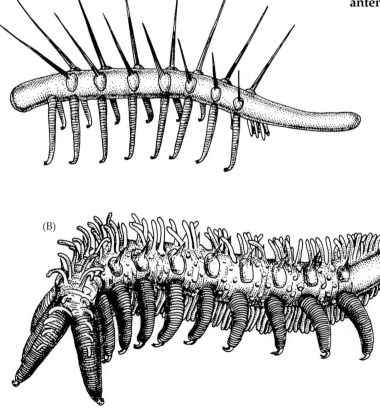

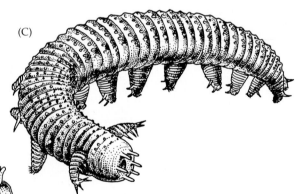

Figure 15.2 Cambrian marine onychophorans. (A) The enigmatic *Hallucigenia sparsa*, with two rows of long dorsal spines. (B) An unnamed onychophoran from the Lower Cambrian Chengjiang deposits of China, with dorsal spines and papillae. (C) *Aysheaia*, from Middle Cambrian shale deposits.

TABLE 15.1 A Comparison of Annelids, Onychophorans, Tardigrades, and Arthropods

	Estimated numbers of described living species	With teloblastic segmentation	*Engrailed* gene expression defines segmental boundaries	With ecdysone-mediated molting (ecdysis)	Coelom
Annelida	16,500	Yes	Yes	No	Well developed
Onychophora	110	Yes	Yes	Yes	Reduced as hemocoel
Tardigrada	800	Yes	Yes	Yes	Reduced as hemocoel
Arthropoda, Crustacea	67,829	Yes	Yes	Yes	Reduced as hemocoel
Arthropoda, Hexapoda	948,000	Yes	Yes	Yes	Reduced as hemocoel
Arthropoda, Myriapoda	11,460	Yes	Yes	Yes	Reduced as hemocoel
Arthropoda, Cheliceriformes	70,000	Yes	Yes	Yes	Reduced as hemocoel
Arthropoda, Trilobitomorpha	4,000 (all extinct)	Yes	Yes	Yes	Reduced as hemocoel

(A)

(C)

(B)

Figure 15.3 Modern onychophorans. (A) *Opisthopatus roseus*, a rare species from Natal, South Africa. (B) *Peripatopsis moseleyi*, from Natal. (C) An undescribed, pure white species of *Opisthopatus*, from Natal.

Malpighian tubules	Muscles isolated as bands	Jointed legs	Compound eyes
Absent	No	No	No
Absent	No	No	No
Present; probably ectodermally derived	To some degree	No	No
Absent	Strongly	Yes	Yes
Present; ectodermally derived	Strongly	Yes	Yes
Present; ectodermally derived	Strongly	Yes	Yes
Present; entodermally derived	Strongly	Yes	Yes
Absent	Strongly	Yes	Yes

The homology of onychophoran head structures with those of annelids and arthropods has long been a matter of debate, and the issue is nowhere near being resolved. The unjointed, fleshy nature of the head appendages, the structure of the jaws, and the lobopod legs appear more like those of polychaetes than those of arthropods. As the serially arranged, clawed lobopodal appendages of onychophorans (including certain enigmatic fossil forms) have no clear counterpart in the animal kingdom.

The body is covered by a thin chitinous cuticle that is molted, as it is in the arthropods. However, like that of the annelids, the cuticle of onychophorans is noncalcified, thin, flexible, very permeable, and not divided into articulating plates or sclerites. Beneath the cuticle is a thin epidermis, which overlies a connective tissue dermis and layers of circular, diagonal, and longitudinal muscles (Figure 15.5). The body surface of onychophorans is covered with wartlike tubercles, usually arranged in rings or bands around the trunk and appendages. The tubercles are covered with minute scales. Most onychophorans are distinctly colored blue, green, orange, or black, and the papillae and scales give the body surface a velvety sheen—hence the common name "velvet worms."

The true coelom, like that of arthropods, is restricted almost entirely to the gonadal cavities. The hemocoel is also arthropod-like, being partitioned into sinuses, including a dorsal **pericardial sinus**.

Locomotion. The paired walking legs of onychophorans are conical, unjointed, ventrolateral lobes with a multispined **terminal claw** (sometimes called hooks). When the animal is standing or walking, each leg rests on three to six distal transverse **pads** (Figure 15.5). The lobopodal legs are filled with hemocoelomic fluid and contain only extrinsic muscle insertions.

papillae ("**slime papillae**"), which lie adjacent to the jaws (Figure 15.4). Circular lips surround the jaws. Beady eyes are located at the bases of the antennae. These anterior appendages are followed by 13–43 pairs of simple saclike (**lobopodal**) walking legs. Although the head appendages and lobopods are superficially annulated, they are not jointed or segmented, nor do they possess intrinsic (segmental) musculature.

Figure 15.4 The "business end" of onychophorans. (A) *Peripatopsis sedgwicki* feeding on a piece of meat. The tips of the jaws are visible within the distended lips. (B) Ventral view of oral region of *Peripatus*.

(A)

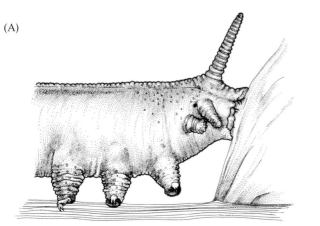

(B)

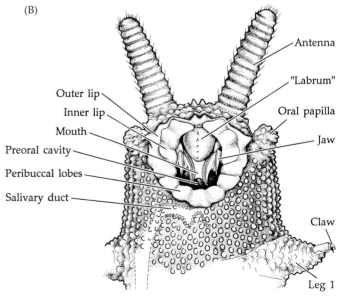

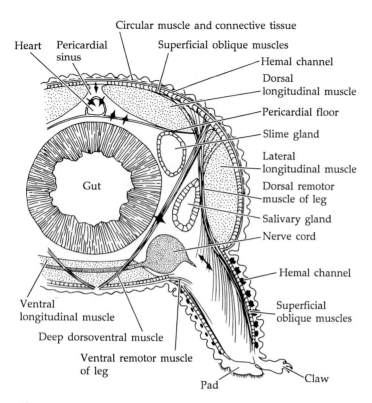

Figure 15.5 Body segment and leg of *Peripatopsis* (transverse section). The arrows indicate the direction of blood flow.

Walking is accomplished by leg mechanics combined with extension and contraction of the body by hydrostatic forces exerted via the hemocoel. Waves of contraction pass from anterior to posterior. When a segment is elongated, the legs are lifted from the ground and moved forward. When a segment contracts, a pulling force is exerted and the more anterior legs are held against the substratum. The overall effect is reminiscent of some types of polychaete locomotion, wherein the parapodia are used mainly for purchase rather than as legs or paddles.

The body muscles are a combination of smooth and obliquely striated fibers and are arranged similarly to those of annelids. The thin cuticle, soft body, and hydrostatic bauplan allow onychophorans to crawl and force their way through narrow passages in their environment. As we saw in the annelids, the efficiency of a hydrostatic skeleton is enhanced by internal longitudinal communication of body fluids. The ancestors of the onychophorans apparently eliminated the intersegmental septa of their putative annelidan predecessors and expanded the blood vascular system at the expense of the coelom, thus converting from a true coelomic hydrostatic skeleton to a hemocoelic hydrostatic skeleton. As we noted in Chapter 13, a somewhat similar trend toward reduction in the size of the coelom and loss of internal septa has occurred in some annelids.

Feeding and digestion. Onychophorans occupy a niche similar to that of centipedes. Almost all are carnivores that prey on small invertebrates such as snails, worms, termites, and other insects, which they pursue into cracks and crevices. Special **slime glands**, thought to be modified nephridia, open at the ends of the oral papillae (Figure 15.6); through these openings an adhesive is discharged in two powerful streams, sometimes to a distance of 30 cm. The adhesive hardens quickly, entangling prey (or would-be predators) for subsequent leisurely dining.

The jaws are used to grasp and cut up prey. Paired salivary glands, also thought to be modified nephridia, open into a median dorsal groove on the jaws (Figure 15.6). Salivary secretions pass into the body of the prey and partly digest it; the semiliquid tissues are then sucked into the mouth. The mouth opens into a chitin-lined foregut, composed of a pharynx and esophagus. A large, straight intestine is the principal site of digestion and absorption. The hindgut (rectum) usually loops forward over the intestine before passing posteriorly to the anus, which is located ventrally or terminally on the last body segment.

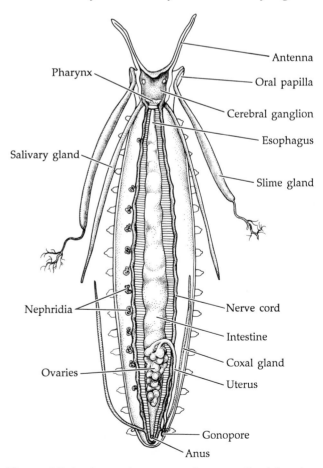

Figure 15.6 Internal anatomy of a generalized female onychophoran.

Circulation and gas exchange. The circulatory system of onychophorans is arthropod-like and linked to the hemocoelic bauplan. A tubular heart is open at each end and bears a pair of lateral ostia in each segment. The heart lies within a pericardial sinus. Blood leaves the heart anteriorly and then flows posteriorly within the large hemocoel via body sinuses, eventually reentering the heart by way of the ostia. The blood is colorless, containing no oxygen-binding pigments. Onychophorans possess a unique system of subcutaneous vascular channels, called **hemal channels** (Figure 15.7C). These channels are situated beneath the transverse rings, or ridges, of the cuticle. A bulge in the layer of circular muscle forms the outer wall of each channel, and the oblique muscle layer forms the inner wall. The hemal channels may be important in the functioning of the hydrostatic skeleton. Thus the superficial annulations of the onychophoran body are external manifestations of the subcutaneous hemal channels.

Gas exchange is by **tracheae** that open to the outside through the many small **spiracles** located between the bands of body tubercles. Each tracheal unit is small and supplies only the immediate tissue near its spiracle (Figure 15.7B). Anatomical data suggest that the tracheal system is not homologous to those of insects, arachnids, or terrestrial isopods, but has been independently derived in the Onychophora.

Excretion and osmoregulation. A pair of nephridia lie in each leg-bearing body segment except the one possessing the genital opening (Figure 15.6, 15.7A). The nephridiopores are situated next to the base of each leg, except in the fourth and fifth leg, whose nephridia open through distal nephridiopores on the transverse pads. Internally, the nephridia are connected to a coelomic end sac (**sacculus**). Each set of sacculus + nephridioduct together is called a **segmental gland**. The nephridioduct, or tubule, enlarges to form a contractile bladder just before opening to the outside via the nephridiopore. The nature of the excretory wastes is not known. The anterior nephridia are thought to be represented by the salivary glands and slime glands, and the posterior ones by gonoducts in females. Recent work suggests that the nephridia are modified metanephridia.

The legs of some onychophorans, such as *Peripatus*, bear thin-walled eversible sacs or vesicles that open to the exterior near the nephridiopores by way of minute pores or slits. These vesicles may function in taking up moisture, as do the coxal glands of many myriapods, insects, and arachnids. They are everted by hemocoelic pressure and pulled back into the body by retractor muscles.

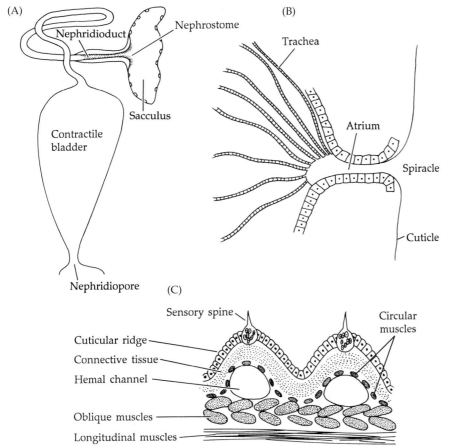

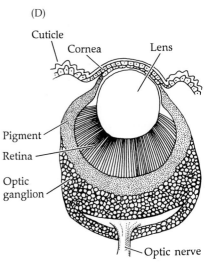

Figure 15.7 Some onychophoran anatomy. (A) Nephridium from *Peripatus capensis*. (B) A tracheal unit of *P. capensis* (cross section). (C) The body wall of *Peripatopsis moseleyi* (section). Note the hemal channels internal to each annular ridge. The ridges bear papillae surmounted by sensory spines. (D) The eye of *Peripatus* (longitudinal section).

Nervous system and sense organs.   The nervous system of onychophorans is ladder-like in structure, and it too is intermediate between the annelid and arthropod plans. A large bilobed cerebral ganglion ("brain") lies dorsal to the pharynx. A pair of ventral nerve cords is connected by transverse segmental commissures. The cerebral ganglion supplies nerves to the antennae, eyes, and oral region. A ganglionic swelling occurs in each segment and gives rise to paired nerves to the appendages and the body wall. The general body surface, especially the larger tubercles, is supplied with sensory hairs, or sensilla, that are probably homologous to those of tardigrades and arthropods. There is a small dorsolateral eye at the base of each antenna. The eyes are of the direct type, with a large chitinous lens and a relatively well-developed retinal layer (Figure 15.7D). Onychophorans are nocturnal and photonegative.

Reproduction and development.   With the exception of one known parthenogenetic species, all onychophorans are dioecious. Females have a pair of largely fused ovaries in the posterior region of the body (Figure 15.6). Each ovary connects to a gonoduct (oviduct), and the gonoducts fuse as a uterus. The end of the uterus opens through a posteroventral gonopore. Males are smaller than females and have a pair of elongate, separate testes. Paired sperm ducts join to form a single tube in which sperm are packaged into spermatophores up to 1 mm in length. The male gonopore is also located posteroventrally.

Copulation has been observed in only a few onychophorans. In the southern African *Peripatopsis* the male deposits spermatophores seemingly at random on the general body surface of the female. The presence of the spermatophores stimulates special amebocytes in her blood to bring about a localized breakdown of the integument beneath the spermatophore. Sperm then pass from her body surface into her hemocoelic fluid, through which they eventually reach the ovaries, where fertilization takes place. In some onychophorans a portion of the uterus is expanded as a seminal receptacle, but sperm transfer in these species is not well understood.

The limited embryological work done on onychophorans suggests that, despite their fundamental similarities to both annelids and arthropods, they also possess some unusual features. For example, onychophorans may be oviparous, viviparous, or ovoviviparous. Females of oviparous species (e.g., *Oöperipatus*) have an ovipositor and produce large, oval, yolky eggs with chitinous shells. Evidence suggests that this is the primitive onychophoran condition, even though living oviparous species are rare. The eggs of oviparous onychophorans contain so much yolk that early, superficial, intralecithal cleavage takes place, with the eventual formation of a germinal disc similar to that seen in many terrestrial arthropods. Most living onychophorans, however, are vi-

viparous and have evolved a highly specialized mode of development associated with small, spherical, nonyolky eggs. Interestingly, most Old World viviparous species, although developing at the expense of maternal nutrients, lack a placenta, whereas all New World viviparous species have a placental attachment to the oviducal wall (Figure 15.8A). Placental development is viewed as the most advanced condition in the onychophorans.

The yolky eggs of lecithotrophic species have a typical centrolecithal organization. Cleavage is by intralecithal nuclear divisions, similar to that seen in many groups of arthropods. Some of the nuclei migrate to the surface and form a small disc of blastomeres that eventually spreads to cover the embryo as a **blastoderm**, thus producing a periblastula. Simultaneously, the yolk mass divides into a number of anucleate "yolk spheres" (Figure 15.8B–D).

Nonyolky and yolk-poor eggs are initially spherical, but once within the oviduct, they swell to become ovate. As cleavage ensues, the cytoplasm breaks up into a number of spheres. The nucleate spheres are the blastomeres, and the anucleate ones are called **pseudoblastomeres** (Figure 15.8E,F). The blastomeres divide and form a saddle of cells on one side of the embryo (Figure 15.8G). The pseudoblastomeres disintegrate and are absorbed by the dividing blastomeres. The saddle expands to cover the embryo with a one-cell-thick blastoderm around a fluid-filled center.

Placental oviparous species have even smaller eggs than do nonplacental species, and the eggs do not swell after release from the ovary. Further, these eggs are not enclosed in membranes. Cleavage is total and equal, yielding a coeloblastula. The embryo then attaches to the oviducal wall and proliferates as a flat **placental plate**. As development proceeds, the embryo moves progressively down the oviduct and eventually attaches in the uterus. Gestation may be quite long, up to 15 months, and the oviduct/uterus often contains a series of developing embryos of different ages.

Development after the formation of the blastula is remarkably similar among the few species of onychophorans that have been studied. Unlike gastrulation in annelids, which involves the movement of presumptive areas of the blastula into their organ-forming positions, gastrulation in onychophorans involves very little actual cell migration. Cells of the presumptive areas undergo immediate organogenesis by direct proliferation. This process involves the proliferation of small cells into the interior of the embryo through and around the yolk mass or fluid-filled center and the production by surface cells of the germinal centers of limb buds and other external structures. All onychophorans have direct development, and there is no strong evidence suggesting that they had larval stages in their evolutionary past. In all species that have been studied, the full complement of segments and adult organ systems is attained before they hatch or are born as juveniles.

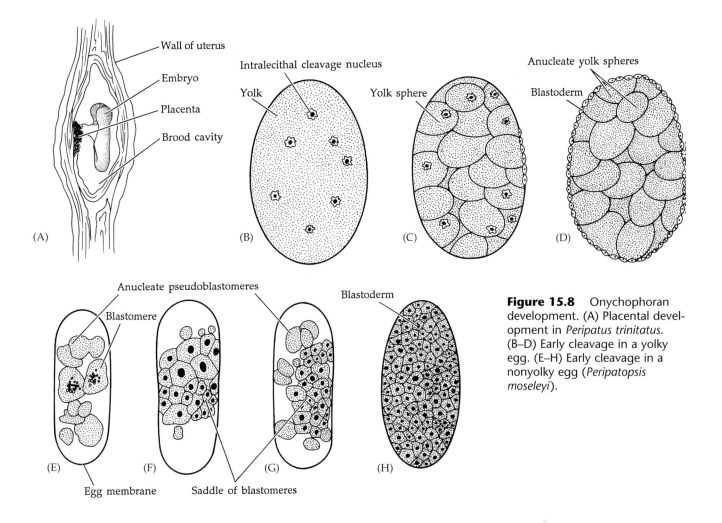

Figure 15.8 Onychophoran development. (A) Placental development in *Peripatus trinitatus*. (B–D) Early cleavage in a yolky egg. (E–H) Early cleavage in a nonyolky egg (*Peripatopsis moseleyi*).

Onychophorans are also unusual in that neither a presegmental acron nor a postsegmental pygidium or telson can be clearly differentiated. Recall that in annelids and arthropods a segmental growth zone occurs in front of the anus and the ectoderm around the anus eventually forms a terminal, appendageless telson or pygidium. In onychophorans, even though growth is teloblastic, the growth zone from which the trunk segments arise appears to be postanal. When the last mesoderm has been formed, the growth-zone ectoderm apparently develops directly into the anal somite with no postsegmental ectoderm remaining. Nevertheless, the last body segment of onychophorans, like that of arthropods, lacks appendages, a fact that leaves the matter a bit unsettled.

Phylum Tardigrada

The first tardigrade was discovered in 1773. Since then, about 800 species have been described. Fossil tardigrades are almost unknown, and until some very recent discoveries in the Lower Cambrian Chengjiang deposits of China and the Middle Cambrian Orsten deposits of Sweden and Siberian limestones, the only specimens were from Cretaceous amber of North America.

Most living tardigrade species are found in semi-aquatic habitats such as the water films that exist on mosses, lichens, liverworts, and certain angiosperms, or in soil and forest litter. Others live in various freshwater and marine benthic habitats, both deep and shallow, often interstitially or among shore algae. A few have been reported from hot springs. Some marine species are commensals on the pleopods of isopods or the gills of mussels; others are parasites on the epidermis of holothurians or barnacles. Tardigrades occasionally occur at high densities, up to 300,000 per square meter in soil and more than 2,000,000 per square meter in moss. All are small, usually on the order of 0.1–0.5 mm in length, although some 1.7 mm giants have been reported.

Under the microscope, tardigrades resemble miniature eight-legged bears, and even move with a lumbering, ursine gait—hence the name Tardigrada (Latin *tardus*, "slow"; *gradus*, "step"). Their locomotion, paunchy body, and clawed legs have earned them the nickname "water bears" (Figure 15.9).

Most tardigrade species are widespread, and many are cosmopolitan. A major factor in their wide distribution may be the fact that their eggs, cysts, and tuns (see below) are light enough and resistant enough to be carried great distances either by winds or on sand and mud clinging to

(A)

(B)

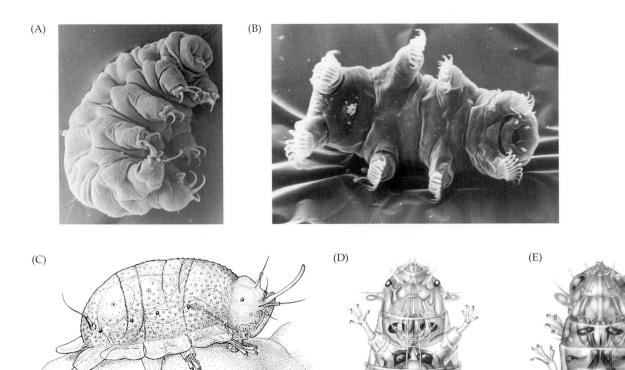

(C)

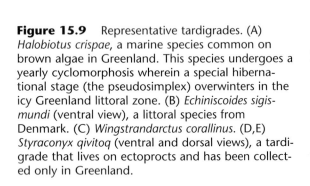

50 m

Figure 15.9 Representative tardigrades. (A) *Halobiotus crispae*, a marine species common on brown algae in Greenland. This species undergoes a yearly cyclomorphosis wherein a special hibernational stage (the pseudosimplex) overwinters in the icy Greenland littoral zone. (B) *Echiniscoides sigismundi* (ventral view), a littoral species from Denmark. (C) *Wingstrandarctus corallinus*. (D,E) *Styraconyx qivitoq* (ventral and dorsal views), a tardigrade that lives on ectoprocts and has been collected only in Greenland.

(D)

(E)

♀

♂

the feet of insects, birds, and other animals. The minute sizes and precarious habitats of water bears have resulted in their acquisition of numerous traits also seen in some blastocoelomate groups that live in similar habitats.

Tardigrades are well known for their remarkable powers of **anabiosis** (a state of dormancy that involves greatly reduced metabolic activity during unfavorable environmental conditions) and **cryptobiosis** (an extreme state of anabiosis in which all external signs of metabolic activity are absent). During dry periods, when the vegetation inhabited by terrestrial tardigrades becomes desiccated, these little creatures can encyst themselves by pulling in their legs, losing body water, and secreting a double-walled cuticular envelope around the shriveled body. Such **cysts** maintain a very low basal metabolism. Further reorganization (or "deorganization") of the body can result in a single-walled **tun** stage, in which body metabolism is undetectable (a cryptobiotic state).

The resistant qualities of the tardigrade tun have been demonstrated by experiments in which individuals have recovered after immersion in extremely toxic compounds such as brine, ether, absolute alcohol, and even liquid helium. They have survived temperatures ranging from +149°C to –272°C, on the brink of absolute zero. They have also survived high vacuums, intense ionizing radiation, and long periods with no environmental oxygen whatsoever.

Following desiccation, when water is again available, the animals swell and become active within a few hours. Many rotifers, nematodes, mites, and a few insects are also known for their anabiotic powers, and these groups often occur together in the surface water of plants such as mosses and lichens. One marine tardigrade (*Echiniscoides*) survives quite well with a life cycle that regularly alternates between active and tun stages, and can even survive an experimentally induced cycle forcing it to undergo cryptobiosis every six hours! Evidence indicates that the tardigrade aging process largely ceases during cryptobiosis, and that by alternating active and cryptobiotic periods, tardigrades may extend the life span to several decades. One rather sensational report described a dried museum specimen of moss that yielded living tardigrades when moistened after 120 years on the shelf!

In certain areas of extreme environmental conditions, marine tardigrades may undergo an annual cycle of **cy-** clomorphosis (rather than the cryptobiosis typical of terrestrial and freshwater forms). During cyclomorphosis, two distinct morphologies alternate. For example, *Halobiotus crispae*, a littoral species from Greenland, (Figure 15.9A), has a summer morph and a winter morph. The latter is a special hibernational stage called the **pseudosimplex** that is resistant to freezing temperatures and perhaps low salinities. In contrast to cryptobiotic tuns, the pseudosimplex is active and motile. Cyclomorphosis is coupled with gonadal development, and in *H. crispae* only the summer morph is sexually mature.

The phylum Tardigrada comprises ten families in three orders: Heterotardigrada, Mesotardigrada, and Eutardigrada. The orders are defined largely on the basis of the details of the head appendages, the nature of the leg claws, and the presence or absence of "Malpighian tubules."

The Tardigrade Bauplan

The body of a tardigrade bears four pairs of ventrolateral legs (Figure 15.10). The legs are short, hollow extensions of the body wall, essentially lobopodal in design, although showing an advance over the lobopodal legs of onychophorans in having intrinsic musculature. Each leg terminates in one or as many as a dozen or so "toes," which end in adhesive pads or discs or in claws resembling those of onychophorans. In some tardigrades, the legs are partially telescopic.

As in onychophorans, the body is covered by a thin, uncalcified cuticle that is periodically molted. It is often

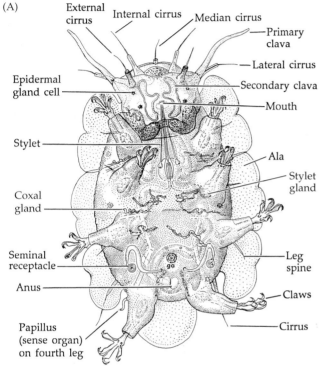

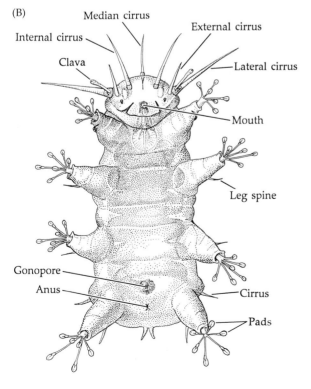

Figure 15.10 Tardigrade anatomy. (A) *Wingstrandarctus corallinus* (ventral view), an inhabitant of shallow, sandy marine habitats in Australia and Florida. (B) *Batillipes noerrevangi* (ventral view). (C) Generalized *Echiniscus* (dorsal view).

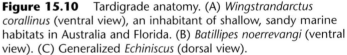

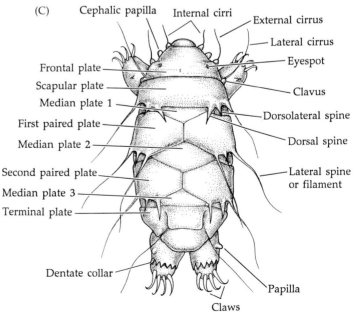

ornamented and occasionally divided into symmetrically arranged dorsal and lateral (rarely ventral) plates (Figure 15.10C). These plates may be homologous with the sclerites of arthropods, but this is not certain. The cuticle shares some features with those of both annelids and arthropods, but it is also unique in certain ways. It comprises up to seven distinguishable layers, contains various sclerotized ("tanned") proteins and chitin, occasionally has a wax layer, and lines the foregut and rectum. The cuticle is secreted by an underlying epidermis composed of a constant cell number in many (but not all) species. Such eutely is common in minute metazoans, and we saw several other examples among the blastocoelomate phyla, notably rotifers. Growth in tardigrades proceeds by molts, as in onychophorans and arthropods, with sexual maturity being attained after three to six instars.

Although the body is quite short, it is nevertheless homonomous and rather weakly cephalized. Nonmarine tardigrades are often colorful animals, exhibiting shades of pink, purple, green, red, yellow, gray, and black. Color is determined by cuticle pigments, the color of the food in the gut, or the presence of granular bodies suspended in the hemocoel.

Like the coelom of arthropods and onychophorans, the coelom of tardigrades is greatly reduced, and in adults it is confined largely to the gonadal cavities. The main body cavity is thus a hemocoel, and the colorless body fluid directly bathes the internal organs and body musculature. The musculature of tardigrades is very different from the annelid-like arrangement seen in onychophorans, in which the body wall muscles are in sheetlike layers; in tardigrades there is no circular muscle layer in the body wall, and the muscles occur in separate bands extending between subcuticular attachment points, as they do in arthropods (Figure 15.11).

It was long thought that tardigrades possessed only smooth muscle, in contrast to the striated muscles of arthropods, and in the past this feature was used as an argument against a close relationship between these two phyla. However, recent work by the Danish zoolo-gist R. M. Kristensen has shown that both smooth and striated muscles occur in tardigrades, the latter predominantly in the most primitive species. The striated muscles are of the arthropod type, being cross-striated, rather than obliquely striated like those of the onychophorans. Numerous fine structural details of the muscle attachment regions are also shared between tardigrades and arthropods. Kristensen has suggested that a partial shift from arthropod-like striated muscle to smooth muscle in some tardigrades might have accompanied a transition from the marine to the terrestrial environment, and might be functionally tied to the phenomenon of cryptobiosis. Furthermore, both slow and fast nerve fibers occur in tardigrades, the former predominating in the somatic musculature and the latter in the leg musculature. However, the leg musculature appears to be entirely extrinsic, like that of onychophorans, with one attachment near the tip of the leg and the other within the body proper. Most of the muscle bands in tardigrades consist of only a single muscle cell or a few large muscle cells each.

Locomotion. The concentration of muscles as discrete units and the thickening of the cuticle in tardigrades resulted in a major shift in locomotor strategy away from the primarily hydrostatic system used in annelids and onychophorans. Instead, tardigrades use a step-by-step gait controlled by independent antagonistic sets of muscles or by flexor muscles that work against hemocoelic pressure. The claws, pads, or discs at the ends of the legs are used for purchase and for clinging to objects, such as strands of vegetation or sediment particles (Figure 15.12). As if to prove the "rule of exceptions," at least one marine species is capable of limited jellyfish-like swimming by use of a bell-shaped expansion of the cuticle margin to keep it suspended just above the sediment.

Feeding and digestion. Water bears usually feed on the fluids inside plant or animal cells by piercing the cell walls with a pair of **oral stylets**. Soil-dwelling species feed on bacteria, algae, and decaying plant matter or are predators on small invertebrates. Carni-

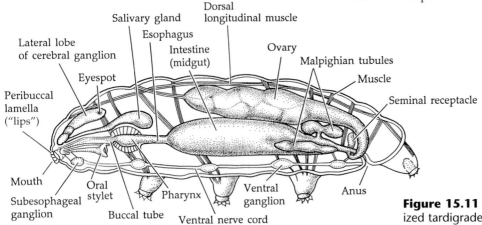

Figure 15.11   Internal anatomy of a generalized tardigrade (lateral cutaway view).

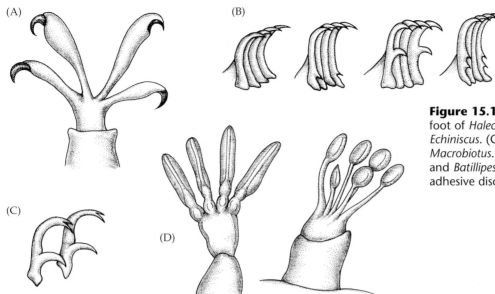

Figure 15.12 Tardigrade feet. (A) The foot of *Halechiniscus.* (B) Claw types from *Echiniscus.* (C) Typical claws from *Macrobiotus.* (D) The feet of *Orzeliscus* (left) and *Batillipes* (right). These genera have adhesive discs or pads on the claws.

vorous and omnivorous tardigrades have a terminal mouth; herbivorous and detritivorous ones have a ventral mouth.

The mouth opens into a short stomodeal buccal tube, which leads to a bulbous, muscular pharynx (Figure 15.11). A large pair of salivary glands flank the esophagus and produce digestive secretions that empty into the mouth cavity; these glands also are responsible for the production of a new pair of oral stylets with each molt (hence they are often referred to as "**stylet glands**"). The muscular pharynx produces suction that attaches the mouth tightly to a prey item during feeding and pumps the cell fluids out of the prey and into the gut. In many species there is a characteristic arrangement of chitinous **rods**, or **placoids**, within an expanded region of the pharynx. These rods provide "skeletal" support for the musculature of that region and may contribute to masticating action. The pharynx empties into an esophagus, which in turn opens into a large intestine (midgut), where digestion and absorption take place. The short hindgut (the cloaca or rectum) leads to a terminal anus. In some species defecation accompanies molting, with the feces and cuticle being abandoned together.

At the intestine–hindgut junction in freshwater species are three large glandular structures that are called Malpighian tubules, each consisting of only about three to nine cells. The precise nature of these organs is not well understood, but they are probably not homologous to the Malpighian tubules of arthropods. In at least one tardigrade genus (*Halobiotus*), the Malpighian tubules are greatly enlarged and have an osmoregulatory function. It is probable that some excretory products are absorbed through the gut wall and eliminated with the feces; other waste products may be deposited in the old cuticle prior to molting.

Circulation and gas exchange. Perhaps because of their small size and moist habitats, tardigrades have lost all traces of discrete blood vessels, gas exchange structures, and metanephridia; consequently, they rely on diffusion through the body wall and the extensive body cavity. The body fluid contains numerous cells credited with a storage function.

Nervous system and sense organs. The nervous system of tardigrades is built on the annelid–arthropod plan and is distinctly metamerous. A large, lobed, dorsal cerebral ganglion is connected to a subesophageal ganglion by a pair of commissures surrounding the buccal tube (Figure 15.11). From the subesophageal ganglion, a pair of ventral nerve cords extends posteriorly, connecting a chain of four pairs of ganglia that serve the four pairs of legs. Sensory bristles or spines occur on the body, particularly in the anterior and ventral region and on the legs (Figure 15.10). The structure of these bristles is essentially homologous to that of arthropod setae (Figure 15.13). A pair of sensory eyespots is often present. Each eyespot consists of five cells, one of which is a pigmented light-sensitive cell. The anterior end of many tardigrades bears long sensory cirri, and most species also have a pair of hollow anterior cirri called **clava** that are probably chemosensory in nature. The clava appear structurally similar to the olfactory setae of many arthropods.

Reproduction and development. Tardigrades are often dioecious, with both sexes possessing a single saclike gonad lying above the gut. In males, the gonad terminates as two sperm ducts, suggesting that the single gonad is derived from an ancestral paired condition. The ducts extend to a single gonopore, which opens

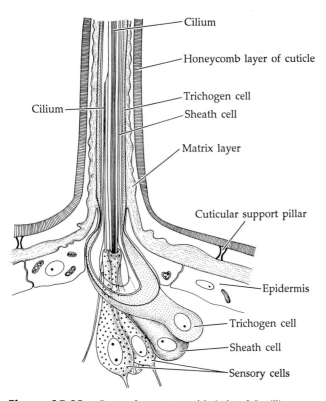

Figure 15.13   Base of an external bristle of *Batillipes noerrevangi* (longitudinal section), showing the relationship between the various cells and the cuticle.

just in front of the anus or into the rectum. In females a single oviduct (right or left) opens either through a gonopore dorsal to the anus or into the rectum (which in this case is called a **cloaca**) (Figure 15.11). There are either two complex seminal receptacles that open separately or a single, small seminal receptacle that opens into the rectum near the cloaca.

Males are unknown in some genera, but most tardigrades that have been studied copulate and lay eggs. Dwarf males have been recently discovered in several genera. In some tardigrades the male deposits sperm directly into the female's seminal receptacle (or cloaca) or into the body cavity by cuticular penetration. In the latter case fertilization takes place in the ovary. In other tardigrades, a wonderfully curious form of indirect fertilization takes place: the male deposits sperm beneath the cuticle of the female prior to her molt, and fertilization occurs when she later deposits eggs in the shed cuticular cast. Several studies have shown tardigrade sperm to be flagellated. In at least a few species, a very primitive courtship behavior exists, wherein the male strokes the female with his cirri. Thus stimulated, the female deposits her eggs on a sand grain, upon which the male then spreads his sperm.

Females lay from 1 to 30 eggs at a time, depending on the species. In strictly aquatic species the fertilized eggs are either left in the shed cuticle or glued to a submerged object. The eggs of terrestrial species bear thick, sculptured shells that resist drying (Figure 15.14). Some species alternate between thin-walled and thick-walled eggs, depending on environmental conditions. Parthenogenesis may be common in some species, notably those in which males are unknown. Hermaphroditism has also been reported in a few genera.

The only reasonably complete studies of tardigrade embryology were published by E. Marcus in the 1920s. A modern look at tardigrade embryogenesis is greatly needed. According to Marcus, whose work is now questioned by some specialists, development is direct and rapid. Cleavage is described as holoblastic. A blastula develops, with a small blastocoel; eventually it proliferates an inner mass of entoderm that later hollows to form the archenteron. Stomodeal and proctodeal invaginations develop, completing the digestive tube. Subsequent to gut formation, five pairs of archenteric coelomic pouches are said to appear off the gut, reminiscent of the enterocoelous development of many deuterostomes. The first pair arises from the stomodeum (ectoderm) and the last four pairs from the midgut (entoderm). The two posterior pouches fuse to form the gonad; the others disappear as their cells disperse to form the body musculature. Development is typically completed in 14 days or less, whereupon the young use their stylets to break out of the shell.

Juveniles lack adult coloration, have fewer lateral and dorsal spines and cirri, and may have reduced numbers of claws. At birth, the number of cells in the body is relatively fixed, and growth is primarily by in-

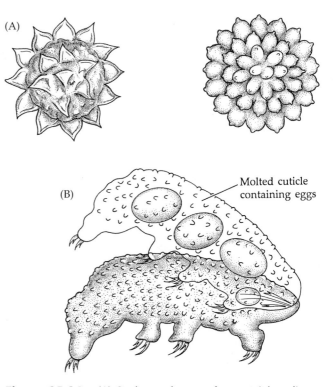

Figure 15.14   (A) Sculptured eggs of terrestrial tardigrades. (B) A female *Hypsibius annulatus* in the process of molting an egg-containing cuticle.

creases in cell size rather than in cell number. In nature, these remarkable animals may live only a few months or may survive for a great many years.

AN INTRODUCTION TO THE ARTHROPODS

The arthropods and annelids, along with the onychophorans and tardigrades, have long been thought to share close evolutionary ties, and most evidence suggests that these four phyla constitute a clade derived from a segmented common ancestor of Precambrian origin. The similarities between annelids and arthropods are reflected most notably in their body metamerism, their embryonic development, and the general architecture of their nervous systems. The major differences between these two phyla derive largely from the invention of a rigid exoskeleton in arthropods, which apparently led to the evolution of an open hemocoel and jointed appendages.

Examples of the principal arthropod groups are shown in Figure 15.1, and the basic features of the arthropod bauplan are listed in Box 15A. Some of these features are unique to the phylum Arthropoda and thus represent defining synapomorphies; others also occur in closely related taxa, such as the onychophorans, tardigrades, and even the annelids, and hence are symplesiomorphies within the arthropod bauplan.

There are five clearly distinguished groups of arthropods, which we recognize as subphyla: Trilobitomorpha (trilobites and their kin, extinct since the end of the Paleozoic), Crustacea (crabs, shrimps, etc.), Hexapoda (insects and their kin), Myriapoda (centipedes, millipedes, and their kin), and Cheliceriformes (horseshoe crabs, eurypterids, arachnids, pycnogonids, etc). It is likely that the Myriapoda and Crustacea are not monophyletic groups, although in both cases the debate is unsettled. The phylogenetic relationships among these five groups are, surprisingly, still unresolved.

Taxonomic History and Classification

As we noted earlier, Linnaeus recognized six major groups of animals (Vermes, Insecta, Pisces, Amphibia, Aves, and Mammalia), placing all of the invertebrates except the insects in a single group, the Vermes. In the early 1800s such famous zoologists as Lamarck and Cuvier presented substantial reorganizations of Linnaeus's earlier scheme, and it was during this period that the various arthropod taxa began to emerge. Lamarck recognized four basic arthropod groups: Cirripedia (barnacles), Crustacea, Arachnida, and Insecta. He placed the ostracods with the brachiopods and, of course, he did not realize the crustacean nature of the barnacles. The

---

BOX 15A *Characteristics of the Phylum Arthropoda*

1. Bilateral, triploblastic protostomes
2. Body segmented, both internally and externally; segments arise by teloblastic growth (showing *en* gene expression)
3. Minimally, body divided into head (cephalon) and trunk regions; commonly with further regional body specialization or tagmosis; typically with a head shield or carapace
4. Head with labrum (or clypeolabrum) (showing *Dll* gene expression) and with nonsegmental acron; *engrailed* (*en*) gene expression suggests that acron and first true head segment develop as single morphological unit
5. Cuticle forms well developed exoskeleton, generally with thick sclerotized plates (sclerites) consisting of dorsal tergites, lateral pleurites, and ventral sternites; cuticle of exoskeleton consists of chitin and protein (including resilin), with varying degrees of calcification; without collagen
6. Each true body segment primitively with a pair of segmented (jointed), ventrally attached appendages, showing a great range of specialization among the various taxa; appendages composed of a proximal protopod and a distal telopod (both multiarticulate); protopodal articles may bear medal endites or lateral exites
7. Cephalon with a pair of lateral faceted (compound) eyes and one to several simple median ocelli; the compound eyes, ocelli, or both have been lost in several groups
8. Coelom reduced to portions of the reproductive and excretory systems; main body cavity is an open hemocoel (= mixocoel)
9. Circulatory system is open; dorsal heart is a muscular pump with lateral ostia for blood return
10. Gut complete, complex, and highly regionalized, with well developed stomodeum and proctodeum; digestive material (and often also the feces) encapsulated in a chitinous peritrophic membrane
11. Nervous system annelid-like, with dorsal (supraenteric) ganglia (= cerebral ganglia), circumenteric (circumesophageal) connectives, and paired, ganglionated ventral nerve cords, the latter often fused to some extent; protocerebrum forms ocular center; deutocerebrum forms antennal center
12. Functional cilia suppressed, except in sperm of a few groups
13. Growth by ecdysone-mediated molting (ecdysis); with cephalic ecdysial glands
14. Muscles metamerically arranged, striated, and grouped in isolated, intersegmental bands; dorsal and ventral longitudinal muscles present; intersegmental tendon system present; without circular somatic musculature
15. Most are dioecious, with direct, indirect, or mixed development; some species are parthenogenetic

close relationship between arthropods and annelids was recognized by Cuvier, who included both in his Articulata (referring to the segmented nature of these animals), and by Lankester, who classified them together with rotifers in his Appendiculata. The great zoologists Hatschek, Haeckel, Beklemishev, Snodgrass, Tiegs, Sharov, and Remane all recognized the Articulata as a discrete phylum, including in it at various times the groups Echiura, Sipuncula, Onychophora, Tardigrada, and Pentastomida. It was Leuckart, in 1848, who separated out the arthropods as a distinct phylum; Von Siebold coined the name Arthropoda in the same year, noting as the group's principal distinguishing attribute its members' jointed legs (Greek *arthro*, "jointed"; *pod*, "foot"). Haeckel published the first evolutionary tree of the arthropods in 1866. Brief diagnoses of the five arthropod subphyla recognized today are provided below, and detailed treatments are presented in the following chapters.

It is important to offer a word of caution about the use of terminology among the various groups of arthropods. Because the Arthropoda is such a vast and diverse assemblage, specialists usually concentrate on only one or a few groups. Thus, over time, slightly different terminologies have evolved for the different groups. Students sometimes feel overwhelmed by arthropod terminology; for example, the hindmost region of the body may be called an abdomen or pleon (as in insects and crustaceans), an opisthosoma (in chelicerates), or a pygidium (in trilobites). But there is a more subtle danger to this mixed terminology. Different terms for similar parts or regions in different taxa do not necessarily imply nonhomology; conversely, the same term applied to similar parts of different arthropods does not always imply homology. To deal with these problems in this text, we have made an effort to achieve consistency in terminology as much as possible, to simplify word use and spelling, and to indicate homologies (and nonhomologies) where known.

Synopses of the Five Arthropod Subphyla

SUBPHYLUM TRILOBITOMORPHA: Trilobites and their kin (Figure 15.1G). About 4,000 described species. Wholly extinct. Body divided into three tagmata: cephalon, thorax, and pygidium (abdomen); segments of cephalon fused, as are those of pygidium; those of thorax free; body demarcated by two longitudinal grooves into a median and two lateral lobes ("trilobite"); cephalon with one pair preoral antennae; all other appendages postoral and more or less similar to one another, with a robust locomotory telopod to which is attached at the base a long filamentous branch (thought to be a protopodal exite); most with compound eyes, the fine structure of which is not well understood.

SUBPHYLUM CRUSTACEA: Crabs, lobsters, shrimps, beach hoppers, pillbugs, etc. About 67,829 described living species. Body usually divided into three tagmata: head (cephalon), thorax, and abdomen (the notable exception being the class Remipedia, which has only head + trunk); appendages uniramous or biramous; 5 pairs of cephalic appendages—the preo-

ral first antennae (antennules) and 4 pairs of postoral appendages: second antennae (which migrate to a "preoral position" in adults), mandibles, first maxillae (maxillules), and second maxillae; cerebral ganglia tripartite (with deutocerebrum); with compound eyes usually having tetrapartite crystalline cone; gonopores located posteriorly on thorax or anteriorly on abdomen. (Chapter 16)

SUBPHYLUM HEXAPODA: Insects and their kin. An estimated 948,000 described living species. Body divided into three tagmata: head (cephalon), thorax, abdomen; with 5 pairs cephalic appendages: first antennae, clypeolabrum, mandibles, maxillae, and labium (fused second maxillae); 3-segmented thorax with uniramous legs; cerebral ganglia tripartite (with deutocerebrum); with compound eyes having tetrapartite crystalline cone; gas exchange by spiracles and tracheae; with ectodermally derived (proctodeal) Malpighian tubules; gonopores open on abdominal segment 7, 8, or 9. (Chapter 17)

SUBPHYLUM MYRIAPODA: Millipedes, centipedes, etc. About 11,460 described living species. Body divided into two tagmata, cephalon and long, homonomous, many-segmented trunk; with 4 pairs cephalic appendages (antennae, mandibles, first maxillae, second maxillae); first maxillae free or coalesced; second maxillae absent or partly (or wholly) fused; all appendages uniramous; cerebral ganglia tripartite (with deutocerebrum); living species may all lack compound eyes (the status of some taxa is still unclear in this regard); with ectodermally derived (proctodeal) Malpighian tubules; gonopores on third or last trunk somite. (Chapter 18)

SUBPHYLUM CHELICERIFORMES: Horseshoe crabs, scorpions, spiders, mites, "sea spiders," etc. About 70,000 described species. Body divided into two tagmata, anterior prosoma (cephalothorax) and posterior opisthosoma (abdomen); opisthosoma with up to 12 segments (plus telson); prosoma of 6 somites, each with a pair of uniramous appendages (chelicerae, pedipalps, 4 pairs of legs); gas exchange by gill books, book lungs, or tracheae; excretion by coxal glands and/or entodermally derived (midgut) Malpighian tubules; with simple medial eyes and lateral compound eyes; cerebral ganglia bipartite (without deutocerebrum). (Chapter 19)

The Arthropod Bauplan and Arthropodization

Remember that the bauplan concept includes not only the themes of body form and function, but also the idea that all components of a system must be compatible to produce a functional animal. In addition, one must consider the constraints imposed on form and lifestyle by various combinations of features. If we are to somehow comprehend the "essence of arthropod," then we must first understand the effect of one of the major synapomorphies of this phylum—the hard, jointed **exoskeleton**—that distinguishes these animals from their soft-bodied relatives, in particular the annelids. Recall that annelids are characterized by segmented bodies, serially arranged coelomic spaces that emerge in front of the pygidium during teloblastic embryogenesis, and well-developed circular and longitudinal muscles in the body wall. Now imagine encasing an annelid-like creature in a rigid exoskeleton. What kinds of structural and functional problems would *have* to be solved in order for such an animal to survive? Although there are a number of possible solutions, the arthropods evolved one particular suite of highly successful adaptations, known as **arthropodization**. Arthropodization has its roots in the Onychophora and Tardigrada, but came to full fruition in the arthropods themselves.

Being encased in an exoskeleton resulted in some obvious constraints on growth and locomotion, including the loss of all mobile body cilia. The fundamental problem of locomotion was solved by the evolution of body and appendage joints and highly regionalized muscles. Flexibility was provided by thin intersegmental areas (joints) in the otherwise rigid exoskeleton, imbued with a unique and highly elastic protein called **resilin**. As the muscles became concentrated into intersegmental bands associated with the individual body segments and appendage joints, the circular muscles were lost almost entirely.

With the loss of peristaltic capabilities resulting from body rigidity and the loss of circular muscles, the coelom became nearly useless as a hydrostatic skeleton. The ancestral body coelom was therefore lost, and an open circulatory system evolved—the body cavity became a **hemocoel**, or blood chamber, in which the internal organs could be bathed directly in body fluids.* The large bodies of these animals still required some way of moving the blood around through the hemocoel; hence the annelid-like dorsal vessel was retained. However, in the absence of body wall muscles to move blood around (as in annelids), the dorsal vessel became a highly muscularized pumping structure—a heart. In contrast to the open metanephridia typical of polychaetes (Chapter 13), the excretory organs became closed internally, thereby preventing the blood from being drained from the body. Surface sense organs (the "arthropod setae") differentiated, becoming numerous and specialized and acquiring various devices for transmitting sensory impulses to the nervous system in spite of the hard exoskeleton. Gas exchange structures evolved in various ways that overcame the barrier of the exoskeleton.

For these animals, now encased in a rigid outer covering, growth was no longer a simple process of gradual increase in body size. Thus the complex process of **ecdysis**, a specific hormone-mediated form of molting, evolved, first appearing in the Onychophora. Through the process of ecdysis, the exoskeleton is periodically shed to allow for an increase in real body size. It is probable that the kind of hormone-mediated ecdysis seen in

arthropods, tardigrades, and onychophorans, regulated by dozens of genes, is a synapomorphy unique to these three phyla and is not homologous to the cuticular shedding known to occur in several other metazoan phyla (e.g., Nemata, Nematomorpha, Kinorhyncha, Priapula, and Loricifera); however, the jury is still out on this question. (See Chapter 24 for more discussion.)

If we add to this complicated suite of events the notion of arthropods invading terrestrial and freshwater environments, the evolutionary challenges are compounded by osmotic and ionic stresses, the necessity for aerial gas exchange, and the need for structural support and effective reproductive strategies.

While the origin of the exoskeleton demanded a host of coincidental changes to overcome the constraints it placed on arthropods, it clearly endowed these animals with great selective advantages, as evinced by their enormous success. One of the key advantages is the protection it provides. Arthropods are armored not only against predation and physical injury, but also against physiological stress. In many cases the cuticle provides an effective barrier against osmotic and ionic gradients, and as such is a major means of homeostatic control. It also provides the strength needed for segmental muscle attachment and for predation on other shelled invertebrates.

The undisputed evolutionary success of arthropods is dramatically reflected in their diversity and abundance. If we start with a generalized, rather homonomous arthropod prototype with a fairly high number of segments, and with paired appendages on each of those segments, we can set the stage for just such a dramatic diversification. Indeed, the diversity seen today has resulted largely from the differential specialization of various segments, regions, and appendages. We saw a hint of this process in our examination of the polychaetes, but nothing of the magnitude evident among the arthropods, in which segment and appendage diversification has reached its zenith. The spectacular radiation within the Arthropoda is not unlike that seen in another highly successful animal group that also exploited appendage modification—the vertebrates.

The arthropod body has itself undergone various forms of regional specialization, or **tagmosis**, to produce segment groups specialized for different functions. These specialized body regions (e.g., the head, thorax, and abdomen) are called **tagmata**. Tagmosis is an extreme form of heteronomy, mediated by Hox genes and the other developmental genes they influence. Our emerging understanding of Hox genes tells us that the most fundamental aspects of animal design arise from spatially restricted expression of these "master developmental genes." However, tagmosis varies among the arthropod groups (see the classification above). The genetic and evolutionary plasticity of regional specialization, like limb variation, has been of paramount importance in establishing the diversity of

---

*This cavity is not a true coelom, either evolutionarily or ontogenetically, but it may be viewed as a persistent blastocoelic remnant. Thus, one might at first reason that the arthropods technically are blastocoelomates. However, the absence of a large body coelom in arthropods is a secondary condition resulting from a loss of the ancestral coelomic body cavity during the evolution of the arthropod bauplan, not a primary condition like that seen in the true blastocoelomates (Chapter 12), at least some of which may never have had a true coelom in their ancestry. A similar secondary loss of the coelom has occurred, in a different way, in the molluscs (Chapter 20).

the arthropods and their dominant position in the animal world.

One of the best examples of arthropod tagmosis is revealed by the expression pattern of the segment polarity gene *engrailed* (*en*) in the head, or cephalon, of Crustacea, Hexapoda, and Myriapoda. In each of these subphyla, the same six head regions emerge during embryogenesis. The most anterior region is the presegmental **acron**, also known as the ocular, or protocerebral, region. Following the acron are the first and second antennal segments, the mandibular segment, and the first and second maxillary segments. (The labrum region shows no *en* expression.) This five-segmented (plus the acron) pattern was long ago used in support of a grouping known as the Mandibulata.

As we discuss the various aspects of the arthropod bauplan below, and in subsequent chapters, do not lose sight of the "whole animal" and the "essence of arthropod" described in this section.

The Body Wall

A cross section through a body segment of an arthropod reveals a good deal about its overall architecture (Figure 15.15). As noted above, the body cavity is an open hemocoel, and the organs are bathed directly in the hemocoelic fluid, or blood. The body wall is composed of a complex, layered **cuticle** secreted by an underlying **epidermis** (Figure 15.16). The epidermis, often referred to in arthropods as the **hypodermis**, is typically a simple cuboidal epithelium. In general, each body segment (or **somite**) is "boxed" by skeletal plates called **sclerites**. Each somite typically has a large dorsal and ventral sclerite, the **tergite** and **sternite** respectively.* The side regions, or **pleura**, are flexible unsclerotized areas in which are embedded various minute, "floating" sclerites, the origins of which are hotly debated. The legs (and wings) of arthropods articulate in this pleural region. Numerous secondary deviations from this plan

exist, such as fusion or loss of adjacent sclerites. Muscle bands are attached at points where the inner surfaces of sclerites project inward as ridges or tubercles, called **apodemes**.

Figure 15.16 illustrates the cuticles of an insect and a marine crustacean. The outermost layer is the **epicuticle**, which is itself multilayered (Figure 15.16D). The external surface of the epicuticle is a protective **lipoprotein layer**—sometimes called the **cement layer**. Beneath this is a **waxy layer** that is especially well developed in arachnids and insects. The waxes in this layer, which are long-chain hydrocarbons and the esters of fatty acids and alcohols, provide an effective barrier to water loss and, coupled with the outer lipoprotein layer, protection against bacterial invasion. These outermost two layers of the epicuticle largely isolate the arthropod's internal environment from the external environment. No doubt the development of the epicuticle was critical to the invasion of land and fresh water by various arthropod lineages. The innermost layer of the epicuticle is a **cuticulin layer**, which consists primarily of proteins and is particularly well developed in insects. The cuticulin layer usually has two components: a thin but dense outer layer and a thicker, somewhat less dense inner layer. The cuticulin layer is involved in the hardening of the exoskeleton, as discussed below, and contains canals through which waxes reach the waxy layer.

Beneath the epicuticle is the relatively thick **procuticle**, which may be subdivided into an outer **exocuticle** and an inner **endocuticle** (Figure 15.16A,B)[†] The procuticle consists primarily of layers of protein and chitin (but no collagen). It is intrinsically tough, but flexible. In fact, certain arthropods possess rather soft and pliable exoskeletons (e.g., many insect larvae, parts of spiders,

---

[†]*Caution*: Some authors use the term *endocuticle* to refer to the entire procuticle of crustaceans, and use the two subdivision terms only when referring to insects.

---

*The terms tergum, sternum, and pleuron are often used interchangeably with tergite, sternite, and pleurite. Technically, however, the term tergum refers more precisely to the dorsal, or tergal region, and sternum to the ventral, or sternal region. Thus we restrict the use of the terms tergite (pl., tergites) and sternite (pl., sternites) to the specific skeletal plates, or sclerites.

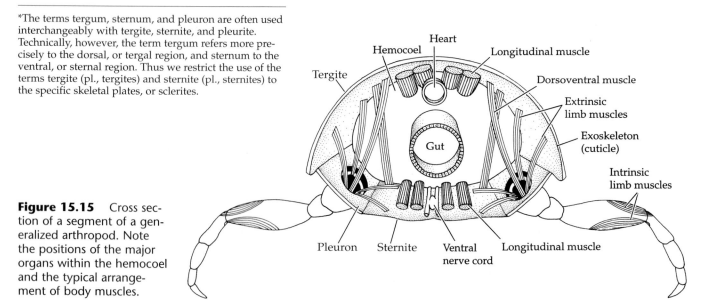

Figure 15.15 Cross section of a segment of a generalized arthropod. Note the positions of the major organs within the hemocoel and the typical arrangement of body muscles.

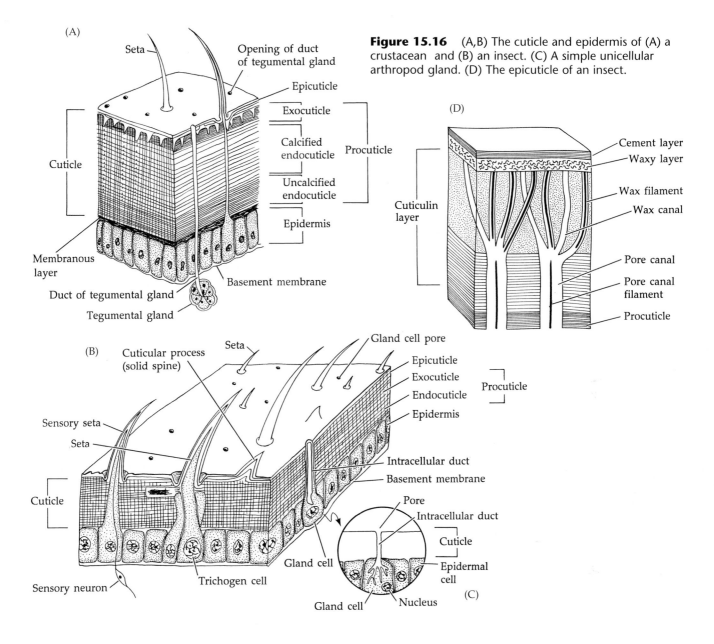

Figure 15.16 (A,B) The cuticle and epidermis of (A) a crustacean and (B) an insect. (C) A simple unicellular arthropod gland. (D) The epicuticle of an insect.

some small crustaceans). However, in most arthropods, the cuticle is hard and inflexible except at the joints, a condition brought about by one or both of two processes: sclerotization and mineralization.

Cuticular hardening by **sclerotization** ("tanning") occurs to various degrees in all arthropods. The layered arrangement of untanned proteins yields a flexible structure. To produce a rigid sclerotized structure, the protein molecules are cross-bonded to one another by orthoquinone linkages. The bonding agent is typically produced from polyphenols and catalyzed by polyphenol oxidases present in the protein layers of the cuticle. Sclerotization generally begins in the cuticulin layer of the epicuticle and progresses into the procuticle to various degrees, where it is associated with a distinct darkening in color. The relationship between cuticular hardening, joints, and molting is discussed in the section dealing with support, locomotion, and growth. **Mineralization** of the skeleton is largely a phenomenon of

crustaceans, and is accomplished by the deposition of calcium carbonate in the outer region of the procuticle.

The epidermis is responsible for the secretion of the cuticle, and as such contains various unicellular glands (Figure 15.16C), some of which bear ducts to the surface of the cuticle. Because the cuticle is secreted by the cells of epidermis, it often bears their impressions in the form of microscopic geometric patterns. The epidermis is underlain by a distinct **basement membrane** that forms the outer boundary of the body cavity or hemocoel.

Arthropod Appendages

Appendage anatomy. In an evolutionary sense, one might be tempted to say that "arthropods are all legs." Certainly, much of arthropod evolution has been about the appendages, modified in myriad ways over the 600-million-year history of this group. The unique combination of body segmentation and serially homologous appendages, in combination with the evolutionary potential of developmental genes, has allowed arthropods to develop modes of locomotion, feeding, and body region/appendage specialization that have been unavailable to the other metazoan phyla. The enormous variety of limb designs in arthropods has, unfortunately, also driven zoologists to create a plethora of terms to describe them. Read on, and we will try to walk you through this terminological jungle in the clearest fashion we can.

Primitively, every true body somite bore a pair of appendages, or limbs. Arthropod appendages are articulated outgrowths of the body wall, equipped with sets of **extrinsic muscles** (connecting the limb to the body) and **intrinsic muscles** (wholly within the limb). These muscles move the various limb segments or pieces, which are called **articles** or **podites**.* The limb articles are organized into two groups, the basalmost group constituting

the **protopod** (= **sympod**) and the distalmost group constituting the **telopod** (Figure 15.17). Whether the protopod is composed of one or more articles, the basalmost article is always called the **coxa** (in living species). The telopod arises from the distalmost **protopodite**, or protopodal article. Sometimes the exoskeleton of the telopodites becomes annulated, forming a **flagellum**, as in the antennae of many arthropods, but these annuli should not be confused with true articles.

A great variety of additional structures can arise from the protopodites, either laterally (collectively called **exites**) or medially (collectively called **endites**) (Figure 15.17A). Evolutionary creativity among the protopodal exites has been exceptional within the arthropods. In crustaceans and trilobites they form a diversity of structures such as gills, gill cleaners, and swimming paddles.

---

*Although some authors refer to the articles of the appendages as "segments," we attempt to restrict the use of the latter term to the true body segments, or somites.

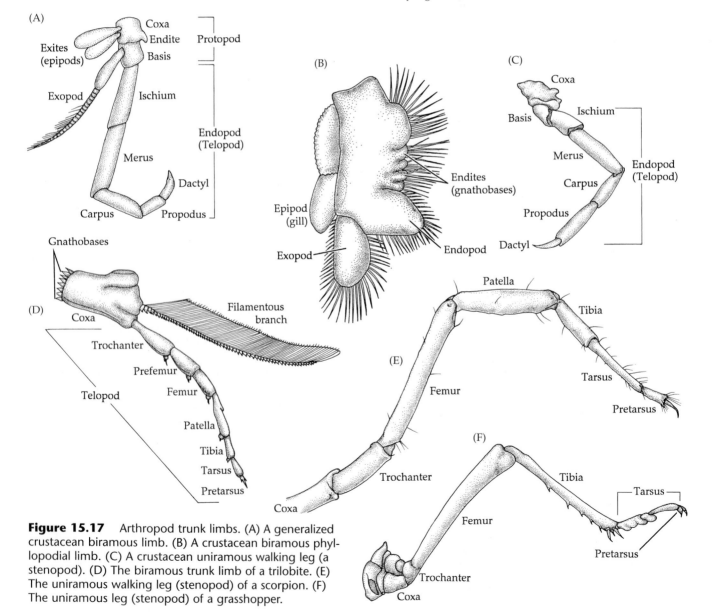

Figure 15.17 Arthropod trunk limbs. (A) A generalized crustacean biramous limb. (B) A crustacean biramous phyllopodial limb. (C) A crustacean uniramous walking leg (a stenopod). (D) The biramous trunk limb of a trilobite. (E) The uniramous walking leg (stenopod) of a scorpion. (F) The uniramous leg (stenopod) of a grasshopper.

Broad or elongate exites that function as gills or gill cleaners are often called **epipods**. Exites may become annulated, like the flagella of antennae. Protopodal exites probably gave rise to the wings of insects. Protopodal endites, on the other hand, often form grinding surfaces, or "jaws," usually termed **gnathobases**. Figure 15.17 illustrates some arthropod appendage types and the terms applied to their parts.*

Appendages with large exites, such as gills, gill cleaners, or swimming paddles (the latter often developed in combination with a paddlelike telopod) are often called **biramous limbs** (or, sometimes, triramous or polyramous limbs). Biramous limbs occur only in crustaceans and trilobites, although their ancestral occurrence in chelicerates is suggested by the gills and other structures that may be derivatives of early limb exites. In crustaceans, the exite on the last protopodite can be as large as the telopod itself, and in these cases it is termed an **exopod**, the telopod then being called the **endopod** (Figure 15.17A). Biramous limbs are commonly associated with swimming arthropods, and in crustaceans in which they are greatly expanded and flattened (e.g., Cephalocarida, Branchiopoda, Phyllocarida), they may also be called **foliacious limbs**, or **phyllopodia** (Greek *phyllo*, "leaf-shaped"; *podia*, "feet") (Figure 15.17B).

Appendages without large exites are called **uniramous limbs** (or **stenopods**; Greek *steno*, "narrow"; *podia*, "feet") (Figure 15.17C). Uniramous limbs are characteristic of the cheliceriforms, hexapods, myriapods, and some crustaceans, although these appendages were probably secondarily derived from biramous limbs on more than one occasion. Uniramous legs are typically ambulatory (walking).

---

*Much of our modern understanding of arthropod limb evolution and homology comes from 30 years of detailed comparative morphology by Jarmila Kukalová-Peck, who studies both fossil and living arthropod limbs. In the Kukalová-Peck model, the ancestral arthropod appendage comprised a series of 11 articles (4 protopodites, 7 telopodites), each of which could theoretically bear an articulated endite or exite. The *number* of articles in her arthropod limb ground plan is not so important as her concept of a single series of articles, with endites and exites that specialized to become the diversity of structures seen in modern taxa. Over evolutionary time, so her theory goes, the basalmost protopodites fused with the pleural region of the body to form pleural sclerites in various taxa. On the thoracic segments of hexapods, the exite of the first protopodite (the epicoxa) migrated dorsally and gave rise to insect wings (see Chapter 17). Many of the earliest known arthropod fossils (including trilobites) have protopods with a single article, as do many living arthropods, suggesting to some workers that multiarticulate protopods might be derived conditions. Kukalová-Peck's hypothesis, however, holds that such uniarticulate protopods represent cases in which protopodal articles have fused together (e.g., in trilobites) or migrated onto the pleural region of the body somites. The number of articles in the telopods of living arthropods varies greatly, reflecting, in Kukalová-Peck's view, various kinds of loss or fusion of articles. The elegance of Kukalová-Peck's theory is that it simply explains the origin of all arthropod limb structures. Viewing the arthropod limb ground plan as a series of articles from which endites and exites were modified in a variety of ways eliminates 100 years of confusion over the nature of uniramous, biramous, and polyramous limbs (these terms now having little phylogenetic significance).

The combination of protopodal and telopodal articles, and their evolutionarily "plastic" endites and exites, has created in arthropods a veritable "Swiss army knife" of appendages. This diversity has no equal in the animal kingdom, and it has played a pivotal role in the evolutionary success of the phylum. As you peruse the following chapters, be sure to notice the phenomenal array of limb morphologies and adaptations among the arthropods.

Appendage evolution. The amazing diversity seen in arthropod limbs has come about through the unique potential of homeobox (Hox) genes and other developmental genes, and the downstream genes they regulate, which are conserved and yet flexible in their expression. We are just beginning to understand how these genes work, and new information in this field is appearing so fast that we hesitate to go into great detail—our understanding of arthropod developmental biology is literally changing from one week to the next! We now know that the fates of arthropod appendages are largely under the ultimate control of Hox genes, which dictate where body appendages form and the general types of appendages that form. Hox genes can either suppress limb development or modify it to create alternative appendage morphologies. These unique genes have played major roles in the evolution of new body plans among arthropods.

A good example of the evolutionary potential of Hox genes is seen in the abdominal limbs of insects. Abdominal limbs (**prolegs**) occur on the larvae (but not the adults) of various insects in several orders, and they are ubiquitous in the order Lepidoptera (i.e., caterpillars). Abdominal limbs were almost certainly present in the ancestry of adult insects. Hence, prolegs may have reappeared in groups such as the Lepidoptera through something as simple as the de-repression of an ancestral limb development program (i.e., they are a Hox gene–mediated atavism). We now know that proleg formation is initiated during embryogenesis by a change in the regulation and expression of the bithorax gene complex (which includes the Hox genes *Ubx*, *abdA*, and *AbdB*).

Molecular developmental biology has also begun to unravel the origins of arthropod appendages themselves. We now know that appendage development is orchestrated by a complex of developmental genes, in particular the genes *Distal-less* (*Dll*) and *Extradenticle* (*Exd*). Evidence suggests that *Exd* is necessary for the development of the proximal region of arthropod limbs (the protopod), whereas *Dll* is expressed in the distal region of developing appendages (the telopod). Thus the protopod and the telopod of arthropod appendages are somewhat distinct, each under its own genetic control and each, presumably, free to respond to the whims and processes of evolution. So, whether an arthropod mandible is a "telomeric," or "whole-limb," appendage (i.e., built of all, or most, of the full complement of articles) or a "gnathobasic" appendage (i.e., built of only the basalmost, or protopodal, articles) depends on whether or not the gene *Dll* is expressed.

Dll is expressed throughout the development of the multiarticulate, telomeric chelicerae and pedipalps of cheliceriforms, but only transiently in myriapod mandibles and in crustacean mandibles lacking palps. It is expressed throughout embryogeny in crustacean mandibles (in the mandibular palp), but not at all in the mandibles of hexapods. This expression pattern suggests that only the cheliceral and pedipalps of cheliceriforms are fully telomeric appendages, although the palp of the crustacean mandible represents the telopod of that limb. It also suggests that the development of a telopod is an evolutionarily flexible feature that can easily show homoplasy (i.e., parallelism). *Dll* is also expressed in the endites of arthropod limbs (e.g., in the phyllopodous limbs of Branchiopoda). In fact, *Dll* is an ancient gene that occurs in many animal phyla, where it is expressed at the tips of ectodermal body outgrowths in such different structures as the limbs of vertebrates, the parapodia and antennae of polychaete worms, the tube feet of echinoderms, and the siphons of tunicates.

So, we see that despite their considerable diversity, all arthropod limbs have a common ground plan and similar genetic mechanisms in their development. Evidence now suggests that the uniramous legs of hexapods (and perhaps myriapods) arose from the biramous (or uniramous) appendages of crustacean ancestors. Although morphologists have struggled to establish homologies among the specific articles, or podites, of arthropod legs, considerable debate still exists on this matter. We may not yet have the developmental genetic tools needed to resolve this issue across the arthropod subphyla.

Support and Locomotion

The arthropods, having largely abandoned the hydrostatic skeleton of their vermiform coelomate ancestors, lack discrete coelomic spaces and the associated muscle sheets that act on them. Instead, they rely on the exoskeleton for support and maintenance of body shape. Muscle sheets simply would not work in the presence of the exoskeleton. Hence, the muscles are arranged as short bands that extend from one body segment to the next, or across the joints of appendages and other regions of articulation. An understanding of the nature of these articulation points—the areas where the cuticle is notably thin and flexible—is crucial to an understanding of the action of the muscles and hence of locomotion.

In contrast to most of the exoskeleton, the articulations, or joints, between body and limb segments are bridged by areas of very thin, flexible cuticle in which the procuticle is much reduced and unhardened (Figure 15.18). These thin areas are called **arthrodial** or **articular membranes**. Generally, each articulation is bridged by one or more pairs of antagonistic muscles. One set of

muscles, the flexors, acts to bend the body or appendage at the articulation point; the opposing set of muscles, the extensors, serves to straighten the body or appendage.

Joints that operate as described above generally articulate in only a single plane (much like your own knee or elbow joints). Such movement is limited not only by the placement of the antagonistic muscle sets, but also by the structure of the hard parts of the cuticle that border the articular membrane. In such cases the articular membrane may not form a complete ring of flexible material, but is interrupted by points of contact between hard cuticle on either side of the joint. These contact points, or bearing surfaces, are called **condyles** and serve as the fulcrum for the lever system formed by the joint. Thus, a **dicondylic joint** allows movement in one plane, but not at angles to that plane. The motion at a joint is also usually limited by hard cuticular processes called **locks** or **stops**, which prevent overextension and overflexion (Figure 15.18C,D).

Some joints are constructed to allow movement in more than one plane, much like a ball-and-socket joint. For example, in most arthropods the joints between walking legs and the body (the coxal–pleural joints) lack large condyles, and the articular membranes form complete bands around the joints. In other cases, two adjacent dicondylic limb joints articulate at 90 degrees to one another, forming a gimbal-like arrangement that facilitates movement in two opposing planes.

Arthropods have evolved a plethora of locomotor devices for movement in water, on land, and in the air (see the collection of papers in Herreid and Fourtner 1981). Only the vertebrates can boast a similar range of abilities, albeit utilizing a far less diverse set of mechanisms. Like so many other aspects of arthropod biology, their methods of movement reflect the extreme evolutionary plasticity and adaptive qualities associated with the segmented body and appendages.

Movement through water involves various patterns of swimming that include smooth paddling by shrimps, jerky stroking by certain insects and small crustaceans, and startling backward propulsion by tail flexion in lobsters and crayfish. Aerial locomotion has been mastered by the pterygote (winged) insects, but is also practiced by certain spiders that drift on threads of silk. Many arthropods burrow or bore into various substrata (e.g., ants, bees, termites, burrowing crustaceans). Some terrestrial arthropods that are normally associated with the ground engage in short-term aerial movements that serve as escape responses. Some, like fleas, simply jump, whereas others jump and glide, giving us possible clues to the evolutionary origin of flight. Some crustaceans jump as well, such as the familiar beach hoppers (amphipods) that bound away over the sand when disturbed. Arthropods that move in contact with the surface of the substratum, under water or on land, by various forms of walking, creeping, crawling, or running are referred to as **pedestrian** or **reptant**.

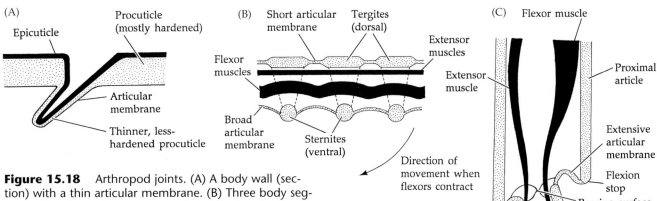

Figure 15.18 Arthropod joints. (A) A body wall (section) with a thin articular membrane. (B) Three body segments like those of a crustacean abdomen (longitudinal section). Note the arrangement of the intersegmental muscles and the articular membranes. In this situation, the segments are capable of ventral flexion only. (C) A generalized limb joint (longitudinal section), showing the arrangement of antagonistic muscles, one condyle, and stops. (D) The extended condition at a simple joint (cutaway view).

All of the common forms of arthropod locomotion except flight depend on the use of typical appendages and thus are based on the principles of joint articulation described above, coupled with specialized architecture of the appendages. Below we discuss some aspects of two fundamental types of appendage-dependent locomotion in arthropods, swimming and pedestrian locomotion, exploring variations on these methods and others in subsequent chapters.

Many examples of swimming arthropods are found among the crustaceans. Most swimming crustaceans (e.g., anostracans and shrimps) and even those that swim only infrequently (e.g., isopods and amphipods) employ ven-

tral, flaplike setose appendages as paddles (Figure 15.17B). The appendages used for swimming may be restricted to particular body regions (e.g., the abdominal swimmerets of shrimps, stomatopods, and isopods; the metasomal limbs of swimming copepods) or may occur along much of the trunk (e.g., the appendages of anostracans, remipedes, and cephalocarids) (Figure 15.19). These appendages engage in a backward power (propulsive) stroke and a forward recovery stroke. In all cases, the appendages are constructed in such a way that on the recovery stroke, they are flexed and the flaps and marginal

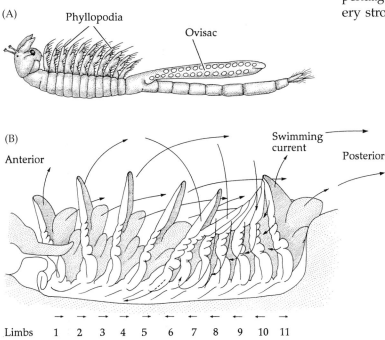

Figure 15.19 Swimming motions in a primitive crustacean. (A) A fairy shrimp (Anostraca) on its back in its normal swimming posture. (B) The appendages "in motion," producing a posteriorly directed flow of water that propels the animal forward. Arrows near the bases of the appendages indicate feeding currents (see Chapter 18). The small arrows below the drawing indicate the direction of movement of each numbered appendage at this moment in the anterior progression of the metachronal wave. Water is drawn into the interlimb spaces as adjacent appendages move away from one another, and water is pressed out of the spaces as adjacent limbs move together. The lateral articles of these phyllopodial appendages are hinged in such a way that they extend on the power stroke to present a large surface area and collapse on the recovery stroke, thereby producing less drag.

setae passively "collapse" to reduce the coefficient of friction (drag). On the power stroke the limbs are held erect, with their largest surface are facing the direction of limb movement, thus increasing thrust efficiency (by increasing the coefficient of friction and the distance through which the limb travels).These swimming appendages typically articulate with the body only on a plane parallel to the body axis. Less sophisticated swimming is accomplished in other arthropods by use of various other appendages, including the antennae of many minute crustaceans and larvae and the thoracic stenopods of many aquatic insects.

Pedestrian locomotion in arthropods is highly variable, both among different groups and even in individual animals. With the exception of a few strongly homonomous "vermiform" types (e.g., centipedes and millipedes), most arthropods are incapable of lateral body undulations. Thus, they cannot amplify the stride length of their appendages by body waves as many polychaetes do. Walking arthropods depend almost entirely on the mobility of specialized groups of appendages. The structure of these ambulatory legs is quite different from that of paddle-like swimming appendages, and their action is much more complex and variable (Figure 15.20).

Consider the general movement of an ambulatory leg as it passes through its power and recovery strokes (Figure 15.20). At the completion of the power stroke, the appendage is extended posteriorly and its tip is in contact with the substratum. The recovery stroke involves lifting the limb, swinging it forward, and placing it back down on the substratum; by then the limb is extended anterolaterally. The power stroke is accomplished by first flexing and then extending the leg while the tip is held in place against the substratum. Thus the body is first pulled and then pushed forward by each limb.

These complicated movements obviously would not be possible if all of the limb joints and limb–body joints were dicondylic articulations in the same plane, parallel to the body axis. The leg must be able to move up and down as well as forward and backward, and the action at each joint must be coordinated with the actions of all the others. In general, the distal limb joints are dicondylic, with articulation (and movement) planes parallel to the limb axis. They allow the appendage to flex and extend, that is, to move the tip closer to (adduction) or farther from (abduction) the point of limb origin. The actions of these joints typically involve the usual sets of antagonistic flexor and extensor muscles described earlier. In some arachnids and a few crustaceans, however, certain limb joints lack extensor muscles, and the limbs are extended by an increase in blood pressure. Raising and lowering of the limb are also accomplished by extensor and flexor muscles, which thus serve as levators

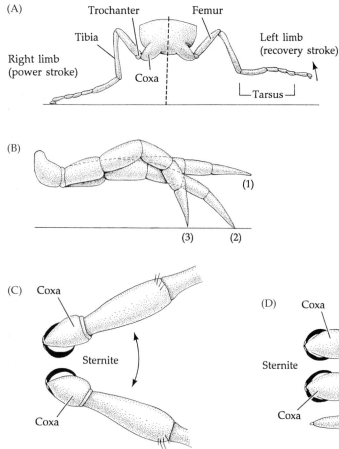

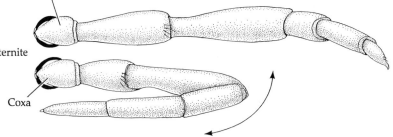

Figure 15.20 Aspects of leg movement in arthropods. (A) Ground-level view of one pair of walking legs on an approaching insect. The leg in contact with the substratum is in its power stroke position, whereas the opposite leg is off the ground in its recovery stroke. (B) Anterior view of a walking limb in various positions during recovery and power strokes: (1) shows the limb extended and raised during the forward-swing recovery stroke; (2) shows the limb extended and lowered against the substratum, as positioned at the beginning or the end of the power stroke; (3) shows the limb flexed and lowered against the substratum in the middle of the power stroke. Notice the change in body to limb tip distance during the power stroke. (C,D) Ventral views of a walking limb, illustrating the range of anterior–posterior (promotor–remotor) and adductor–abductor movements. (C) Rotational movement at coxa–body junction to swing limb forward and backward. (D) Extension and flexion of a walking limb with resultant abduction and adduction of the limb tip relative to the body.

and depressors, respectively; the muscles in the proximal leg joints usually serve these purposes.

Anterior–posterior limb movements are accomplished in two basic ways. First, the ball-and-socket type of joint at the point of limb–body articulation typically carries out these actions in most crustaceans, insects, and myriapods. Promotor and remotor muscles that are associated with these joints rotate the limb forward and backward, respectively. Second, many arachnids accomplish multidirectional limb movements by using only uniplanar dicondylic joints. In these arthropods, one or more of the proximal joints articulate perpendicular to the limb axis, and thus to the rest of the limb joints, providing forward and backward movement.

Understanding how a single limb moves does not, of course, describe the locomotion of the whole animal. The various patterns of pedestrian locomotion in arthropods, called **gaits**, are the result of many factors (e.g., leg number, leg movement sequences, stride lengths, speed). The number of patterns is great, but it is limited by certain biological and physical constraints. Speed is limited by rates of muscle contraction and the necessity for coordinating leg movements to avoid tangling. Furthermore, the animal must maintain an appropriate distribution of legs at all times in various phases of power and recovery strokes so that its weight is fully supported.

The gaits of insects have been more extensively studied than those of other arthropods. Studies on insects and myriapods led to an attempt to establish principles under which all pedestrian arthropod locomotion could be unified. The most frequently used descriptions of arthropod walking, crawling, and running are based on the "metachronal model." The basic idea of this model is that the legs on each side of the body move in metachronal (repeated) waves from back to front and that the waves overlap to various degrees, depending on the speed of movement. This model does work for some arthropods, some of the time, but things are not so simple, and attempts to over generalize have been misleading. A good deal of the work on crustaceans and arachnids (and even insects) indicates that leg movement sequences, stepping patterns, stride lengths, and other characteristics are extremely variable, even within individuals, and depend on a host of factors other than speed. The actions of the joints are coordinated by information supplied to the central nervous system by proprioceptors in the joints themselves. Detailed analyses are beyond the scope of this text, and we refer you again to Herreid and Fourtner (1981) for more information. Additional information also is presented in Chapters 16–19.

Growth

The imposition of a rigid exoskeleton on the arthropods (and their cousins, the onychophorans and tardigrades) precludes growth by means of a gradual increase in external body size. Rather, an overall increase in body size takes place in staggered increments associated with the

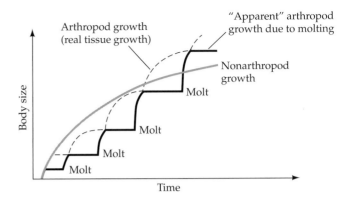

Figure 15.21 Arthropod versus non-arthropod growth. The heavy solid line indicates the incremental ("stair-step") growth pattern of an arthropod as measured by changes in external body size associated with molts. The dotted line depicts real tissue growth in the same arthropod. The gray line depicts the typical growth of a non-arthropod.

periodic loss of the old exoskeleton and the deposition of a new, larger one (Figure 15.21). The process of shedding the exoskeleton is called **molting**, and it is a phenomenon characteristic of arthropods and a few other invertebrates with thick cuticles (onychophorans, tardigrades, kinorhynchs, nematodes, nematomorphans, loriciferans, and priapulans). It is unlikely, however, that the complex physiological basis of molting in all these groups is homologous, although this hypothesis has been suggested (Aguinaldo et al. 1997). The molting process varies in detail even among the arthropods. It has been best studied in certain insects and crustaceans, and the description below is based primarily on those two groups. We first outline the basic steps in the arthropod molt cycle and then briefly discuss the hormonal control of those events. In all arthropods (and onychophorans and tardigrades) molting is regulated by a hormone called **ecdysone**; thus, molting in these groups is referred to as **ecdysis**.

The intermolt stages between molts are called **instars**. It is during these intermolt stages that real tissue growth occurs, although with no increase in external size. When such tissue growth reaches the point at which the body "fills" its exoskeletal case, the animal usually enters a physiological state known as **premolt** or **proecdysis**. During this stage there is active preparation for the molt, including accelerated growth of any regenerating parts. Certain epidermal glands secrete enzymes that begin digesting the old endocuticle, thus separating the exoskeleton from the epidermis. In many crustaceans, some of the calcium is removed from the cuticle during this period and stored within the body for later redeposition. As the old cuticle is loosened and thinned, the epidermis begins secreting a soft new cuticle. Figure 15.22 depicts some of these events.

Once the old cuticle has been substantially loosened and a new cuticle formed, actual molting occurs. The

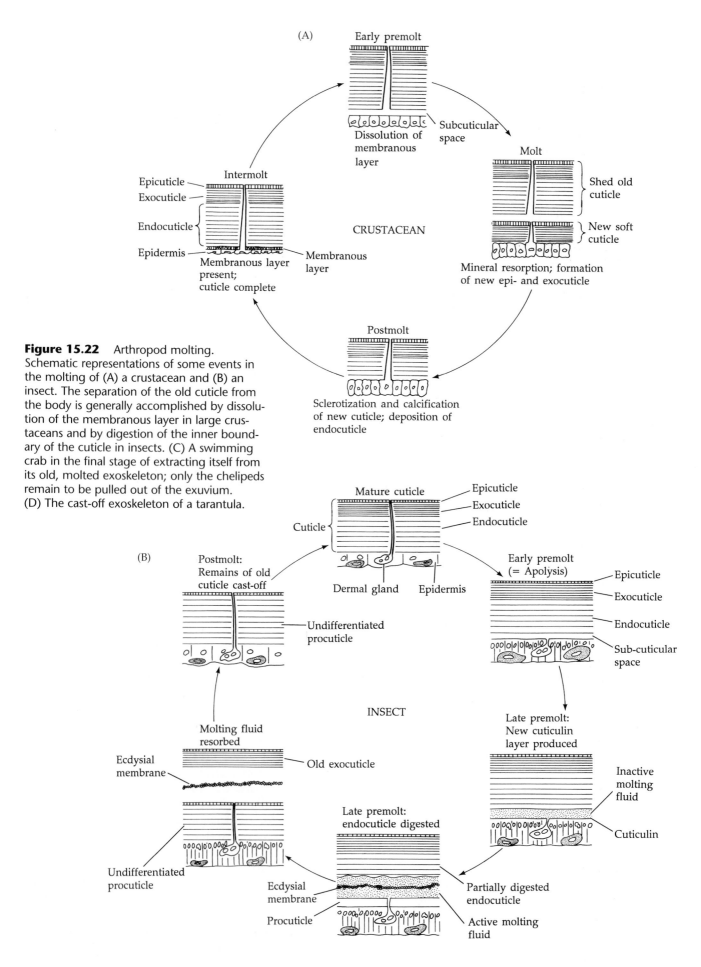

Figure 15.22 Arthropod molting. Schematic representations of some events in the molting of (A) a crustacean and (B) an insect. The separation of the old cuticle from the body is generally accomplished by dissolution of the membranous layer in large crustaceans and by digestion of the inner boundary of the cuticle in insects. (C) A swimming crab in the final stage of extracting itself from its old, molted exoskeleton; only the chelipeds remain to be pulled out of the exuvium. (D) The cast-off exoskeleton of a tarantula.

(C)

(D)

old cuticle splits in such a way that the animal can wriggle free and pull itself out. The lines along which the cuticle splits vary among the arthropods but are consistent within particular groups. It is important to remember that all cuticular linings are lost during ecdysis, including the linings of the foregut and hindgut, the eye surfaces, and the cuticle that lines every pit, groove, spine, and seta on the body surface. When you see a cast-off intact exoskeleton, or **exuvium**, of an arthropod, you are bound to be impressed with its wonderfully perfect detail. It is at first difficult to imagine how the animal could extricate itself from each and every tiny part of the old cuticle (Figure 15.22C,D). The ability to do so depends, of course, on the great flexibility of the body within its new and unhardened exoskeleton.

As soon as the arthropod emerges from its old cuticle, and while the new cuticle is still soft and pliable, its body swells rapidly by taking up air or water. Once the new cuticle is thus enlarged, the animal enters a **post-molt period** (**postecdysis**) during which the cuticle is hardened by sclerotization or the redeposition of calcium salts. The excess water (or air) is then actively pumped from the body, and real tissue growth occurs during the subsequent intermolt period. During the sclerotization process, the cuticle becomes drier, stiffer, and resistant to chemical and physical degradation through the molecular cross-linking process in the protein–chitin matrix described earlier.

We have stressed the adaptive significance of the arthropod exoskeleton in terms of its protective and supportive qualities. However, during the postmolt period, before the new exoskeleton is hardened, the ani-

mal is quite vulnerable to injury, predation, and osmotic stress. Many arthropods become reclusive at this time, hiding in protective nooks and crannies and not even feeding when in this "soft-shell" condition. The time required for hardening of the new exoskeleton varies greatly among arthropods, generally being longer in larger animals. The well known and delectable "soft-shell crabs" of the eastern United States are simply blue crabs (*Callinectes*) caught during their postmolt period.

The events of the molt cycle outlined above are controlled by many genes and a complex hormonal system (Figure 15.23). Several models have been proposed to explain the hormonal pathways involved in molting in insects and crustaceans, and the picture is still incomplete. The hormonal activities of the crustacean ecdysial cycle have been most extensively studied in decapods. In some (e.g., lobsters and crayfish), molting occurs periodically throughout the animal's life, but in many others (e.g., copepods and some crabs), molting, and therefore growth, ceases at some point, and a maximum size is attained. Animals that have engaged in their final molt are said to have entered a state of **anecdysis**, or permanent intermolt—they are in their final instar. Among insects, molting is largely associated with metamorphosis from one developmental stage to the next (e.g., pupa to adult), and, except for the most primitive hexapods, adults do not molt (i.e., they are in anecdysis).

In both crustaceans and hexapods (and probably all arthropods), the initiation of molting, beginning with the events of proecdysis, is brought about by the action of a molting hormone called **ecdysone**. Apparently, however, the pathways controlling the secretion of ecdysone are different in insects and crustaceans, as diagrammed in Figure 15.23. In crustaceans ecdysone is secreted by an endocrine gland called the **Y-organ** located at the base of the antennae or near the mouthparts. The action of the Y-organ is controlled by a complex neurosecretory apparatus located near the eyes or in the eyestalks. During the intermolt period, a **molt-inhibiting hormone (MIH)** is produced by neurosecretory cells of the **X-organ**, located in a region of the eyestalk nerve (or ganglion) called the **medulla terminalis** (Figure 15.23A). MIH is carried by axonal transport to a storage area called the **sinus gland**, which appears to control MIH release into the blood. As long as sufficient levels of MIH are present in the blood, the production of ecdysone by the Y-organ is inhibited.

The active premolt and subsequent molt phases are initiated by sensory input to the central nervous system. The stimulus is external for some crustaceans (e.g., day length or photoperiod for certain crayfishes) and internal for others (e.g., growth of soft tissues in certain crabs). External stimuli are transmitted via the central nervous system to the medulla terminalis and X-organ (Figure 15.23B). Appropriate stimuli inhibit the secretion of MIH, ultimately resulting in the production of ecdysone and the initiation of a new molt cycle.

The sequence of events in insects is somewhat different from that in crustaceans in that a molt inhibitor is apparently not involved. When an appropriate stimulus is introduced to the central nervous system, certain neurosecretory cells in the cerebral ganglia are activated. These cells, which are located in the **pars intercerebralis**, secrete **ecdysiotropin**. This hormone is carried by axonal transport to the **corpora cardiaca**, paired neural masses associated with the cerebral ganglia. Here, **thoracotropic hormone** is produced and carried to the **prothoracic glands**, stimulating them to produce and release ecdysone (Figure 15.23C).

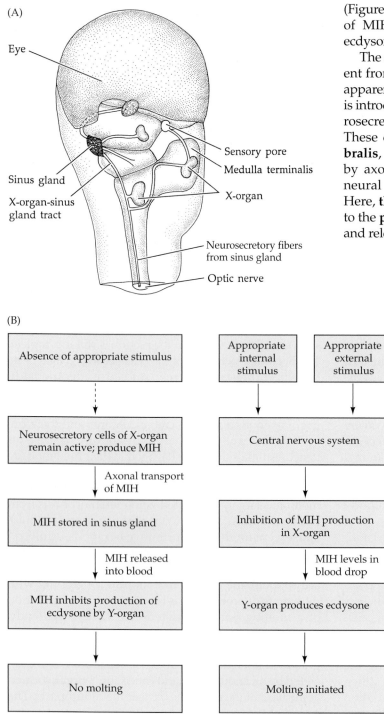

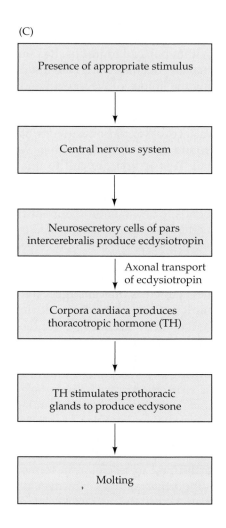

Figure 15.23   (A) The neurosecretory apparatus in a crustacean eyestalk. (B) Flow diagram of events inhibiting and initiating molting in crustaceans. (C) Flow diagram of events initiating molting in an insect.

The Digestive System

It will come as no surprise that the great diversity among arthropods is reflected in their display of nearly every feeding method imaginable. As a group, the only real constraint on arthropods in this regard is the absence of external, functional cilia. Evolutionarily, many arthropods have overcome even this limitation and suspension feed by other means. So varied are arthropod feeding strategies that we postpone discussion of them to the sections and chapters on particular taxa, and here attempt only to generalize about the basic structure and function of arthropod digestive systems.

The digestive tract of arthropods is complete and generally straight, extending from a ventral mouth on the head to a posterior anus. Various appendages (the **mouthparts**) may be associated with processing food and moving it to the mouth. Regional specialization of the gut occurs in most taxa. In almost all cases there is a well-developed, cuticle-lined, stomodeal **foregut** and proctodeal **hindgut**, connected by an entodermally derived **midgut** (Figure 15.24). In general, the foregut serves for ingestion, transport, storage, and mechanical digestion of food; the midgut for enzyme production, chemical digestion, and absorption; and the hindgut for water absorption and the preparation of fecal material. The midgut typically bears one or more evaginations in the form of **digestive ceca** (often referred to as the "**digestive gland**," "**liver**," or "**hepatopancreas**"). The number of ceca and the arrangement of the other gut regions vary among the different taxa. A characteristic feature of arthropods (and tardigrades) is the enclosure of the material being digested in the hindgut within a permeable **peritrophic membrane**, which allows digestive fluids to flow in and water and nutrients to flow out. Arthropod feces are typically "packaged" in the remains of the peritrophic membrane.

The various terrestrial arthropods have convergently evolved many similar features as adaptations to life on land. Many of these convergent structures are associated with (although not necessarily derived from) the gut. For example, excretory structures called Malpighian tubules (see below) that develop from the midguts or hindguts of insects, arachnids, myriapods, and tardigrades appear to be convergences (i.e., nonhomologous structures). The excretory structures of onychophorans also used to be called Malpighian tubules, but they have recently been shown to be complex metanephridia with secondarily derived, closed end sacs. Many unrelated terrestrial taxa have special **repugnatorial glands**, which may or may not be associated with the gut and which produce noxious substances used to deter predators. Many different groups of terrestrial arthropods also have evolved the ability to produce silks or silk-like substances for use outside their bodies. These silk-like fibers are produced by nonhomologous structures among different arthropods. Although they vary greatly in chemical composition, all share a common molecular feature that gives them strength and elasticity: they are composed of regular assemblies of long-chain macromolecules (most being fibrous proteins); many also incorporate collagens. Modified salivary glands are common silk-producing organs, but silks are also secreted by the digestive tract, Malpighian tubules, accessory reproductive glands, and assorted dermal glands. Silk production occurs in chelicerates (false scorpions, spiders, and mites), many insect orders (such as the larvae of the commercial silkworm moths, *Bombyx* and *Anaphe*), and myriapods (chilopods, diplopods, and symphylans). Arthropod silks are used in the production of cocoons, egg cases, webs, larval "houses," flotation rafts, prey entrapment threads, draglines, spermatophore receptacles, intraspecific recognition devices, and other sundry items. The truly spectacular array of silk uses by spiders is discussed in Chapter 19. Silk production and use provide one of the more spectacular examples of evolutionary convergence seen in the arthropods.

Circulation and Gas Exchange

A major aspect of the arthropod bauplan is reflected in the nature of the circulatory system. The open hemocoelic system is in part a result of the imposition of the rigid exoskeleton and the loss of an internally segmented and fluid-filled coelom. We have seen that isolated coelomic spaces (like those in annelids) require a closed circulatory system to service them, but this requirement is eliminated in the arthropods. Furthermore, without a muscular, flexible body wall to augment blood movement, a pumping mechanism becomes necessary, resulting in the elaboration of a muscular heart. The result is a system wherein the blood is driven from the heart chamber through short vessels and into the hemocoel, where it bathes the internal organs. The blood returns to the heart via a noncoelomic **pericardial sinus** and perforations in the heart wall called **ostia** (Figure 15.25). The blood flows back to the heart along a decreasing pressure gradient resulting from lowered pressure within the pericardial sinus as the heart contracts. The complexity of the circu-

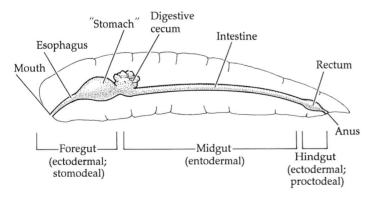

Figure 15.24 The major gut regions of arthropods. The myriad variations on this theme are discussed in subsequent chapters on particular taxa.

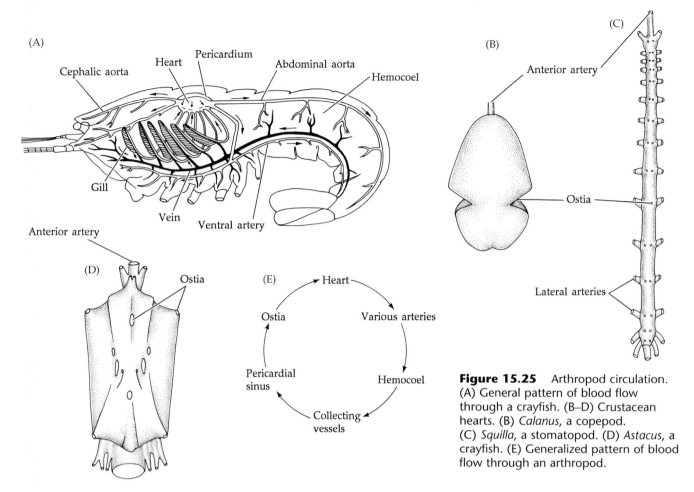

Figure 15.25 Arthropod circulation. (A) General pattern of blood flow through a crayfish. (B–D) Crustacean hearts. (B) *Calanus*, a copepod. (C) *Squilla*, a stomatopod. (D) *Astacus*, a crayfish. (E) Generalized pattern of blood flow through an arthropod.

latory system varies greatly among arthropods, the differences being dependent in large part on body size and shape. These differences include variations in the size and shape of the heart (Figure 15.25B–D), the number of ostia, the length and number of vessels, the arrangement of hemocoelic sinuses, and the circulatory structures associated with gas exchange.

Arthropod blood, or **hemolymph**, serves to transport nutrients, wastes, and usually gases. It includes a variety of types of amebocytes and, in some groups, clotting agents. The blood of many kinds of small arthropods is colorless, simply carrying gases in solution. Most of the larger forms, however, contain hemocyanin, and a few contain hemoglobin. Both pigments are always dissolved in the hemolymph rather than contained within cells. In most groups of arthropods, the circulatory route takes at least some of the blood past the gas exchange surfaces (e.g., gills) before returning to the heart.

One of the major evolutionary problems arising from the acquisition of a relatively impermeable exoskeleton involves gas exchange, particularly for terrestrial arthropods. On land, any increase in cuticular permeability to facilitate gas exchange also increases the threat of water loss. Remember that gas exchange surfaces not only must be permeable but also must be kept moist (Chapter 3). Evolutionarily, the challenge for the arthropods becomes one of disrupting the integrity of the exoskeleton in such a way as to allow gas exchange without seri-

ously jeopardizing the survival of the animal by abandoning the principal benefits of the exoskeleton.

The design of arthropod gas exchange structures has taken one form in aquatic groups and quite another in terrestrial taxa (Figure 15.26). The former is best exemplified by the crustaceans and the latter by the insects and terrestrial chelicerates. Some very tiny crustaceans (e.g., copepods) with a low surface area-to-volume ratio exchange gases cutaneously across the general body surface or at thin cuticular areas such as articulating membranes. However, most of the larger crustaceans have evolved various types of **gills** in the form of thin-walled, hemolymph-filled cuticular evaginations. Gills are commonly branched or folded, providing large surface areas (Figure 15.26A). The gills of some crustaceans (e.g., euphausids) are exposed, unprotected, to the surrounding medium, whereas in others (e.g., crabs and lobsters) the gills are carried beneath protective extensions of the exoskeleton.

The most successful terrestrial arthropods—the insects and arachnids—have evolved gas exchange structures in the form of invaginations of the cuticle, rather than the evaginations seen in aquatic crustaceans. Obviously, external gills would be unacceptable in dry conditions, but placed internally, these gas exchange structures remain moist and act as humidity chambers, allowing oxygen to enter solution for uptake. Many arachnids possess invaginations called **book lungs**,

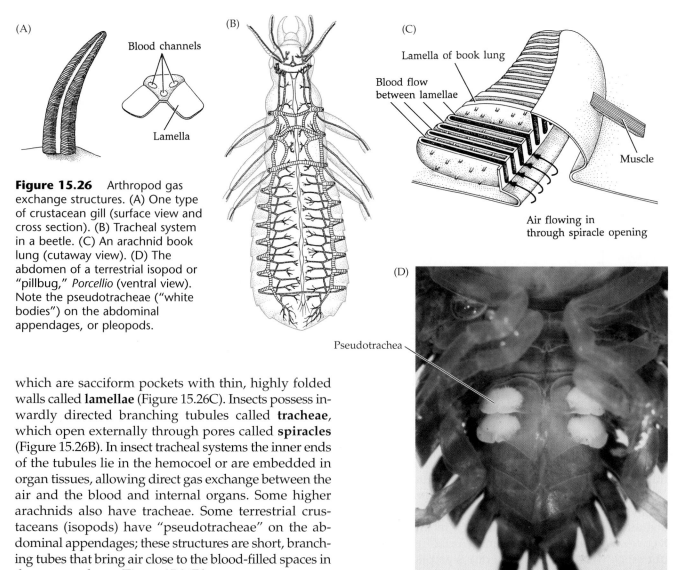

Figure 15.26 Arthropod gas exchange structures. (A) One type of crustacean gill (surface view and cross section). (B) Tracheal system in a beetle. (C) An arachnid book lung (cutaway view). (D) The abdomen of a terrestrial isopod or "pillbug," *Porcellio* (ventral view). Note the pseudotracheae ("white bodies") on the abdominal appendages, or pleopods.

which are sacciform pockets with thin, highly folded walls called **lamellae** (Figure 15.26C). Insects possess inwardly directed branching tubules called **tracheae**, which open externally through pores called **spiracles** (Figure 15.26B). In insect tracheal systems the inner ends of the tubules lie in the hemocoel or are embedded in organ tissues, allowing direct gas exchange between the air and the blood and internal organs. Some higher arachnids also have tracheae. Some terrestrial crustaceans (isopods) have "pseudotracheae" on the abdominal appendages; these structures are short, branching tubes that bring air close to the blood-filled spaces in these appendages (Figure 15.26D).

Excretion and Osmoregulation

With the evolution of a hemocoelic circulatory system in arthropods, nephridia with open nephrostomes became functionally untenable. It simply would not do to drain the blood directly from an open hemocoel to the outside. Arthropods have evolved a variety of highly efficient excretory structures that share a common adaptive feature in that they are internally closed. In addition to this major difference between the nephridia of arthropods and those of segmented and other coelomate protostomes (e.g., annelids, sipunculans, echiurans), there has been a reduction in the overall number of excretory units.

In many arthropods, portions of the excretory units are coelomic remnants and are formed in various segments during development. In most adult crustaceans only a single pair of nephridia (nephromixia) persists, usually associated with particular segments of the head (i.e., as **antennal glands** or **maxillary glands**) (Figure 15.27A). In arachnids there may be as many as four pairs of nephridia (and in onychophorans, many more) opening at the bases of the walking legs (i.e., **coxal glands**).

A second type of excretory structure occurs in four terrestrial arthropod taxa: arachnids, myriapods, insects, and tardigrades. These structures, known as **Malpighian tubules**, arise as blind tubules extending into the hemocoel from the gut wall (Figure 15.27C). However, anatomical and developmental evidence suggests that these Malpighian tubules evolved independently in each of these groups, representing yet another exemplary case of convergent evolution among the Arthropoda.

The excretory physiology of arthropods is a complex and extensively studied topic, and we present only a very general summary here. The various nephridial types of coelomic origin are functionally much more complex and efficient than open metanephridia. The uptake of materials from the hemocoel by the inner ends of these nephridia apparently involves passive movement in response to filtration pressure, as well as active transport. The fluid entering the nephridium is generally similar in composition to the hemolymph it-

(A)

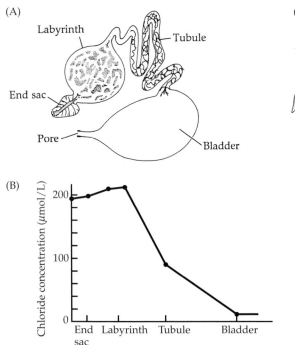

(C)

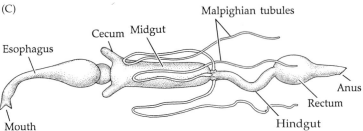

Figure 15.27 Arthropod excretory structures. (A) The antennal gland of a decapod crustacean (section). (B) Changes in chloride content of the excretory fluid in different regions of a decapod antennal gland. Note the active resorptive capabilities of the structure. (C) An insect gut. Note the attachments of Malpighian tubules.

Nervous System and Sense Organs

The general plan of the arthropod nervous system is nearly identical to that of annelids, and many obvious homologies exist (Figure 15.28A). The arthropod "brain" (cerebral ganglia) actually comprises two or three regions, each of which is a separate set of coalesced segmental ganglia. The anterior ganglia form the dorsal (supraesophageal) **protocerebrum** and, when present, the **deutocerebrum**. The posteriormost ganglion, the **tritocerebrum**, usually forms circumenteric connectives around the esophagus to a ventral subesophageal (subenteric) ganglion. The latter is formed by the coalescence of several other head ganglia, usually those associated with the mandibles and maxillae. A double or single, ganglionated ventral nerve cord extends through some or all of the body segments. Crustaceans, hexapods, and myriapods all possess a tripartite brain composed of the two anterior ganglia (protocerebrum and deutocerebrum) and the posterior ganglion (tritocerebrum). In cheliceriforms the deutocerebrum is absent. Each of these regions gives rise to a major pair of nerves to particular head appendages (Figure 15.28B,C).*

The segmental ganglia of the ventral nerve cord show various degrees of linear fusion with one another in different groups of arthropods. Hence, just as tagmosis is reflected in the joining of body segments externally, it is also apparent in the union of groups of ganglia along the ventral nerve cord. These modifications of the central nervous system are examined more closely in the following chapters on the arthropod subphyla.

Although the presence of an exoskeleton has had little evolutionary effect on the structure of the cerebral

self, but as it passes along the plumbing system of the nephridium, a good deal of selective reabsorption occurs, particularly of salts and nutrients such as glucose. Thus the urine exiting the nephridial pores is markedly different from the hemolymph and represents a concentration of nitrogenous waste products (Figure 15.27B).

Malpighian tubules accomplish the same process, but they must rely on assistance from the gut. Malpighian tubule uptake from the hemocoel is relatively nonselective, and the resulting "primary urine" (containing nutrients, water, salts, and so on) is emptied directly into the gut. Very little reabsorption of non-waste material occurs along the length of the tubule itself. The hindgut is mostly responsible for concentrating the urine by reabsorbing the non-waste fractions. The ability of the gut to reabsorb water plays a critical role in osmoregulation in terrestrial and freshwater arthropods. Like most aquatic invertebrates, marine crustaceans excrete most (about 70–90 percent) of their nitrogenous wastes as ammonia; the remainder is excreted in the forms of urea, uric acid, amino acids, and some other compounds. Terrestrial arachnids, myriapods, and insects excrete predominantly uric acid (via the hindgut and anus). In Chapter 3 we reviewed some of the relationships between excretory products and osmoregulation in terms of adaptation to terrestrial habitats. The ability to produce large quantities of uric acid, and thus conserve water, has doubtless contributed significantly to the success of arachnids and insects on land. The crustaceans, on the other hand, have not been able to make a major shift from ammonotelism to uricotelism. Only the terrestrial crustaceans (i.e., isopods—woodlice and pillbugs) show a slight increase in uric acid excretion over that of their marine counterparts.

---

*The concentration of nervous tissue in the arthropod head has been called a brain, cerebrum, cerebral ganglion, and cerebral ganglionic mass. These terms may be somewhat misleading because they may seem anthropomorphic, or they suggest the presence of a single head ganglion. In fact, the cerebral ganglia are composed of clusters of associated ganglia (concentrations of nervous tissue composed primarily of neuronal cell bodies)—hence the term cerebral ganglia is probably the most accurate.

(A)

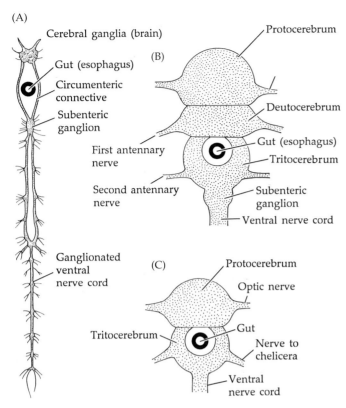

Cerebral ganglia (brain)

Gut (esophagus)

Circumenteric connective

Subenteric ganglion

(B)

Protocerebrum

Deutocerebrum

Gut (esophagus)

Tritocerebrum

First antennary nerve

Second antennary nerve

Subenteric ganglion

Ventral nerve cord

Ganglionated ventral nerve cord

(C)

Protocerebrum

Optic nerve

Tritocerebrum

Gut

Nerve to chelicera

Ventral nerve cord

(D)

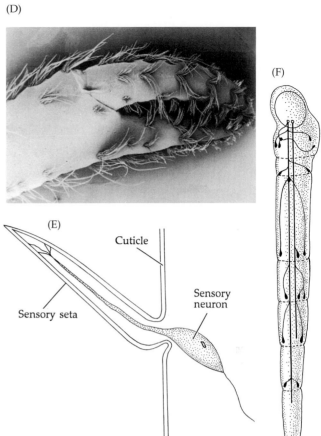

(F)

(E)

Cuticle

Sensory neuron

Sensory seta

Figure 15.28   The arthropod nervous system and some sense organs. (A) The central nervous system of a crayfish, showing the basic annelid-like plan (dorsal view). (B) The brain of a crustacean. (C) The brain of a chelicerate. (D) Sensory setae on the walking leg of a lobster (*Homarus*). (E) A typical arthropod tactile seta. (F) Distribution of proprioceptors in a spider leg.

ganglia and nerve cord, it has had a major effect on the nature of sensory receptors. Unmodified, the exoskeleton would impose an effective barrier between the environment and the epidermal sensory nerve endings. Hence, most of the external mechanoreceptors and chemoreceptors are actually cuticular processes (setae, hairs, bristles), pores, or slits, collectively called **sensilla**. It seems probable that ancestral (evolutionary) remnants of cilia form the nonmotile nerve basis for many of these sensory structures.

Most arthropod tactile receptors (mechanoreceptors) are cuticular projections in the form of movable bristles or setae, the inner ends of which are associated with sensory neurons (Figure 15.28D,E). When the cuticular projections are touched, that movement is translated into a deformation of the nerve ending, thereby initiating a nerve impulse. Sensitivity to environmental vibrations is similar to tactile reception. Sensilla in the form of fine "hairs" or setae are mechanically moved by external vibrations and impart that movement to underlying sensory neurons. Some terrestrial arthropods bear thin, membranous cuticular windows overlying chambers lined with sensory nerves. When struck by airborne vibrations (e.g.,

sound), these windows vibrate in turn and impart the stimulus to the chamber and thence to the nerves below.

We have seen in soft-bodied invertebrates that chemoreception is usually associated with ciliated epithelial structures (e.g., nuchal organs, ciliated pits), across which dissolved chemicals diffuse to nerve endings. In arthropods, in the presence of a relatively impermeable cuticle and in the absence of free cilia, such arrangements are obviously not possible. Thus, many arthropods possess special thin or hollow setae, often associated with the head appendages, with permeable cuticular coverings or minute pores that bring the environment into contact with chemoreceptor neurons.

Proprioception is of particular importance to animals with jointed appendages, such as arthropods and vertebrates. The way in which these stretch receptors span the joints enables them to convey information to the central nervous system about the relative positions of appendage articles or body segments (Figure 15.28F). Through this system, an arthropod (or vertebrate) knows where its appendages are, even without seeing them. Arthropod versions of these "strain gauges" are called **campaniform sensilla** in hexapods, **slit sensilla** in arachnids, and **force-sensitive organs** in most crustaceans. Despite subtle differences in their anatomy, all are linked to the central nervous system in similar ways, and all record exoskeletal strain by means of neuronal stretching or deformation.

(A)

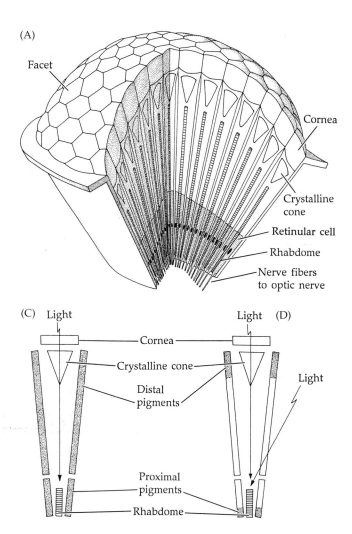

(B)

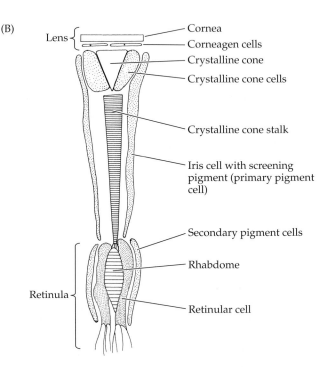

Figure 15.29 Arthropod compound eyes. (A) A compound eye (cutaway view). (B) A single ommatidium. (C,D) Major ommatidial elements in (C) an appositional, or light-adapted, eye and (D) a superpositional, or dark-adapted, eye.

Arthropods possess three basic kinds of photoreceptors, including **simple ocelli**, **complex lensed ocelli**, and faceted or **compound eyes**. Ocelli were described in Chapter 3; although they vary in anatomical detail, their basic structure and operation is consistent in all invertebrates. Compound eyes, though found in all four arthropod subphyla, have been lost or modified in various groups throughout the phylum. Because of their unique structure and function, compound eyes are described here in some detail.

As their name indicates, compound eyes comprise from a few to many distinct photoreceptive units, called **ommatidia**. Each ommatidium is supplied with its own nerve tracts leading to the major optic nerve, and each has its own field of vision through square or hexagonal cuticular **facets** on the eye surface. The visual fields of neighboring ommatidia overlap to some extent, such that a shift in the position of an object within the visual field generates impulses from several ommatidia; hence compound eyes are especially suitable for detecting movement. However, visual acuity is affected by the degree of overlap among the fields of vision of neighboring ommatidia—the greater the overlap, the poorer the visual acuity. In general, compound eyes with many

small facets probably produce higher-resolution images than eyes with fewer, larger facets.

The following discussion describes the structure and function of compound eyes using the crustacean–hexapod model (Figure 15.29). Each ommatidium is covered by a modified portion of the cuticle called the **cornea** (= **corneal lens**); the special epidermal cells that produce the corneal elements are called **corneagen cells**. The corneagen cells may later withdraw to the sides of the ommatidium to form (usually two) **primary pigment cells**. When viewed externally, the facets on the surface of each cornea produce the characteristic mosaic pattern so frequently photographed by microscopists. The core of each ommatidium comprises a group of **crystalline cone cells** and the **crystalline cone** that they produce, sometimes a **crystalline cone stalk**, and a basal **retinula** (= **retinular element**). There are typically four (rarely three or five) crystalline cone cells; an ommatidium with a four-part crystalline cone is highly diagnostic of crustacean–hexapod eyes and is termed a **tetrapartite ommatidium**.* The crystalline cone is a hard, clear structure bordered laterally by the primary pigment cells, or **Iris cells**. The retinula is a complex structure formed from several **retinular cells**, which are the actual photosensitive units that give rise to the sensory nerve tracts. These retinular cells, usually numbering 8 but ranging from 5

*The crystalline cone cells are often called **Semper cells** (especially by entomologists).

to 13 in various derived conditions, are arranged in a cylinder along the long axis of the ommatidium. The retinular cells are surrounded by **secondary pigment cells**, which isolate each ommatidium from its neighbors. The core of the cylinder is the **rhabdome**, which is made up of rhodopsin-containing microtubular folds (microvilli) of the cell membranes of the retinular cells. Each retinular cell's contribution to the rhabdome is called a **rhabdomere**. The microvilli of the rhabdomeres extend toward the central axis of the ommatidium at right angles to the long axis of the retinular cell.

The initiation of an impulse depends upon light striking the rhabdome portion of the retinular element. Light that enters through the facet of a particular ommatidium is directed to its rhabdome by the lenslike qualities of the cornea and the crystalline cone. The lens has a fixed focal length, so accommodation to objects at different distances is not possible. Light is shared among all the rhabdomeres of a given rhabdome, although not necessarily equally.*

In contrast to those of insects and crustaceans, the lateral eyes of most myriapods are probably not true compound eyes, but comprise clusters of simple ocelli. There is some evidence, however, that the eyes of scutigeromorph centipedes (and of some fossil myriapods) may be true compound eyes. The only chelicerates with typical compound eyes, the xiphosurans, have ommatidia that differ in most details from the insect–crustacean design.† There is evidence that the lateral eye groups of terrestrial chelicerates may be derived from reduced and fused compound eye ommatidia, and Silurian scorpions had huge bulging compound eyes.

The versatility of compound eyes is to a large degree a result of the distal and proximal **screening pigments** located in cells that wholly or partially surround the core of the ommatidium (Figure 15.29B–D). **Distal screening pigments** are located in the iris cells, and **proximal pigments** are often located in the retinular cells and secondary pigment cells. In many cases, these screening pigments are capable of migrating in response to varying light conditions and thus changing their positions somewhat along the length of the ommatidium. In bright light, the screening pigments may disperse so that nearly all of the light that strikes a particular rhabdomere must have entered through the facet of its ommatidium. In other words, the screening pigment prevents light that strikes the facets at an angle from passing through one ommatidium and into another. Many crustacean eyes are fixed in this condition. Such **appositional eyes** (= light-adapted eyes) are thought to maximize resolution, in that the image from the visual field of each ommatidium is maintained as a discrete unit. Conversely, under conditions of dim light, screening pigments may concentrate, usually distally, thereby allowing light to pass through more than one ommatidium before striking rhabdomeres. The result is that the image formed by each ommatidium is superimposed on the images formed by neighboring ommatidia. This design has the advantage of producing enhanced irradiances on the retinula, but at the cost of reduced resolution. Many crustacean eyes are fixed in this condition also. Such **superpositional eyes** (= dark-adapted eyes) function as efficient light-gathering structures while sacrificing visual acuity and image formation capabilities.

Some arthropod groups possess compound eyes that are always either appositional or superpositional; thus they lack the ability to switch back and forth with varying light conditions. For example, maxillopodans and branchiopods apparently all possess appositional eyes. However, within the two principal malacostracan clades, Eucarida and Peracarida, both types of eyes occur (e.g., isopods and amphipods have appositional eyes, but mysids have superpositional eyes). Furthermore, crustacean larvae that possess compound eyes almost always have the appositional type, which metamorphose into superpositional eyes in those groups that possess them in adulthood.

Among the arthropods, compound eyes elevated on stalks occur only in certain crustaceans (and perhaps a few Paleozoic trilobites). Biologists have long argued over the derivation of such **stalked eyes**, and the matter is still far from settled—are they the primitive condition in arthropods or in crustaceans, or have they been derived multiple times from sessile-eyed ancestors (the latter seems far more likely). The eyestalk is much more than a device to support and move the eye. Eyestalk movements are produced by up to a dozen or more muscles with complex motor innervation. In most malacostracans the eyestalks contain several optic ganglia separated by chiasmata, as well as important endocrine organs, usually including the sinus gland and X-organ. Thus, neither loss of eyestalks nor convergent recreation of these structures would have been a simple evolutionary feat. The loss of *functional* eyes is a common evolutionary pathway among the Crustacea (and other arthropods), especially in species that inhabit subterranean, deep-sea, or interstitial habitats. But among those clades with stalked eyes, the eyestalk remains even when the eye itself degenerates—testimony to importance of this complex bit of anatomy.

Reproduction and Development

The great diversity of adult form and habit among arthropods is also reflected in their reproductive and de-

---

*The compound eyes of one large group of crustaceans, the class Maxillopoda, differ considerably from those of other crustaceans; see Chapter 16.

†The eyes of chelicerates are relatively larger than those of insects or crustaceans, and their ommatidia have an indeterminate number of cells. The pigment cells are arranged in a cuplike manner, and the bottom of the "cup" is occupied by a sheet of cells secreting a protuberance of the cuticle, which is a functional but not a morphological equivalent of a crystalline cone. A special "eccentric cell"—a large specialized photoreceptor—found in chelicerate ommatidia has no equivalent in the insect or crustacean retina.

velopmental strategies. The extreme evolutionary and ontogenetic plasticity of arthropods has led to a great deal of convergence and parallelism as different groups have developed similar structures under similar selective conditions or pressures.

Nearly all arthropods are dioecious, and most engage in some sort of formal mating. Fertilization is usually, although not always, internal and is often followed by brooding or some other form of parental care, at least during early development. Development is frequently mixed, with brooding and encapsulation followed by larval stages, although direct development occurs in many groups.*

Arthropod eggs are centrolecithal, but the amount of yolk varies greatly and results in different patterns of early cleavage. Cleavage is holoblastic in the relatively weakly-yolked eggs of xiphosurans, some scorpions, and various crustaceans (e.g., copepods and barnacles), and is meroblastic in the strongly-yolked eggs of most insects and many other crustaceans. A number of arthropods exhibit a unique form of meroblastic cleavage that begins with nuclear divisions within the yolky mass (Figure 15.30). These intralecithal nuclear divisions are followed by a migration of the daughter nuclei to the periphery of the cell and subsequent partitioning of the nuclei by cell membranes. These processes typically result in a periblastula that consists of a single layer of cells around an inner yolky mass.

Holoblastic cleavage may appear more or less radial or may show traces of a spiral pattern (e.g., in barnacles). However, the latter pattern appears to differ somewhat from the typical spiralian pattern seen in sipunculans, echiurans, polychaetes, and most molluscs.

One of the most striking features of arthropods is their segmentation and the complex way in which it is embryologically derived. This developmental process is shared with annelids (and with the onychophorans and tardigrades) and provides powerful evidence of shared annelid–arthropod ancestry. The process is called **teloblastic segmental growth** (or simply **teloblasty**) (Greek *telos*, "end"; *blasto*, "bud"). It is characterized by a progressive, anterior-to-posterior addition of segments

*Direct development in arthropods is sometimes called **amorphic development**, and indirect development is often called **anamorphic development**.

from a distinct **posterior growth zone** situated near the anus (i.e., in front of the terminal, or anal, segment—often called the **telson** or **pygidium**). So programmed is the development of the segments that among most annelids and crustaceans the growth zone is composed of a fixed number of uniquely identifiable stem cells (**teloblasts**) whose progeny undergo a predictable and stereotyped sequence of segmentally iterated cell divisions. Teloblasty is further characterized by the formation of secondary body cavities (coelomic cavities) at the growth zone as segments are elaborated. In arthropods, the adult body cavity is derived from the fusion of the blood vascular system with these transitory embryonic coelomic cavities. Although the coelomic cavities (which remain paired and segmentally distinct in the annelids) are secondarily lost during arthropod embryogenesis, they form by way of schizocoely, originating from a pair of caudally situated mesodermal cell bands ultimately derived from the 4d (mesentoblast) cell and situated on either side of the archenteron, just as in annelids.

Not only do the segments form in a similar fashion in annelids and arthropods, but the two groups also appear to use a similar genetic mechanism (especially the products of the *engrailed* gene) to define segmental boundaries and polarity. The *engrailed* (*en*) gene belongs to a class of genes known as segment polarity genes. It plays a key role in determining and maintaining the posterior cell fates of segmental structures in all arthropods (and in annelids and onychophorans). Because this gene occurs in iterated transverse stripes in the posterior portion of each developing segment in the germ band of all annelids and arthropods, it can be used to clarify ambiguities in segment position and number.

In adult annelids, each segment contains a pair of body cavities that retain the simple myoepithelial lining developed during embryogenesis. In arthropods, the transitory embryonic coelomic cavities are also lined

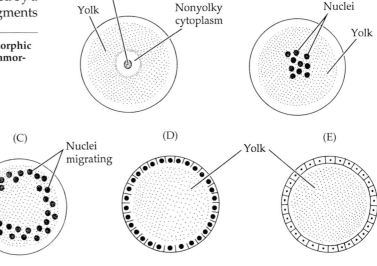

Figure 15.30 Superficial cleavage of a centrolecithal egg and the formation of a periblastula in arthropods. (A) Centrolecithal egg. (B) Intralecithal nuclear divisions following fertilization. (C) Migration of nuclei to the periphery of the cell. (D,E) The periblastula is produced by a partitioning of nuclei as cell membranes form.

with simple epithelium, but this gradually disappears as the embryo develops, and the large adult body cavity comes to be lined by an extracellular matrix. Hence, the body cavity of adult arthropods is derived by the fusion of a primary body cavity (the blood vascular system) and a secondary body cavity (the coelom) and is therefore termed a **hemocoel** (or mixocoel). Segmentally arranged pairs of transitory coelomic cavities, and their embryonic fusion, have been reported in the early developmental stages of virtually every onychophoran and arthropod that has been studied. The transitory coelomic cavities of arthropods are thus regarded as homologous to the coelomic cavities of annelids—recapitulations of the paired and segmental coelomic cavities of the common ancestor of annelids and arthropods.

The Trilobites (Subphylum Trilobitomorpha)

Of all fossil invertebrates, trilobites are perhaps the most symbolic (in the minds of many) of ancient and exotic faunas.* The subphylum Trilobitomorpha (Latin *trilobito*, "three-lobed"; Greek *morph*, "form") includes nearly 4,000 species of arthropods known only from the fossil record (Figure 15.31). They were restricted to, and are characteristic of, Paleozoic seas. Trilobites dominate the fossil record of the Cambrian and Ordovician periods (440 to 550 mya) and continued to be important components of marine communities until the Permo-Triassic mass extinction that marked the end of the Paleozoic era. Because of their hard exoskeletons, great abundances, and broad distributions, the trilobites left a rich fossil record, and more is known about them than about most other extinct taxa. Most of the present world's land areas were submerged during various parts of the Paleozoic, so trilobites are found in marine sedimentary rocks worldwide.

Although trilobites were exclusively marine, they exploited a variety of habitats and lifestyles. Most were benthic, either crawling about over the bottom or plowing through the top layer of sediment. Most benthic species were a few centimeters long, although some giants reached lengths of 60–70 centimeters. A few trilobites appear to have been planktonic; they were mostly small forms, less than 1 cm long and equipped with spines that presumably aided in flotation. Most of the benthic trilobites were probably scavengers or direct deposit feeders, although some species may have been predators that laid partially burrowed in soft sediments and grabbed passing prey. Some workers speculate that trilobites may have suspension fed by using the filamentous parts of their appendages. Some recent evidence suggests that one group of Olenidae may have

had symbiotic relationships with sulfur bacteria. These late Cambrian–early Ordovician trilobites had vestigial mouthparts and large "gill filaments" (epipods) that might have been sites for bacterial cultivation.

General Body Form

The trilobite body was broadly oval and somewhat flattened dorsoventrally. Some trilobites were capable of rolling into a ball—a behavior called **enrollment** or **conglobation** (Figure 15.31F). Two longitudinal furrows on the dorsum divided the body into a median and two lateral lobes—thus the name "trilobite." The lateral lobes (sometimes called "pleural lobes") were produced by outgrowths of the dorsal body wall that extended over the appendages. The exoskeleton of the dorsal surface was much thicker than that of the ventral surface. From front to back, trilobites were divided into three tagmata, **cephalon**, **thorax**, and **pygidium**, with each body region bearing a number of appendages. The pygidium comprised a number of fused or partially budded body segments plus a postsegmental part called the **telson**.

The cephalon was commonly composed of five or six fused segments covered dorsally by a solid cephalic shield (or carapace), which in most species bore well-developed compound eyes on the lateral lobes. The longitudinal furrows divided the cephalon into a median **glabella** and lateral **cheeks**. The glabella was usually further subdivided into three to five lobes by a series of transverse furrows (Figure 15.31A). The cephalic shield was distinctly rolled under along its anterior and lateral edges, thereby producing a concavity on the ventral surface of the cephalon. Ventrally the cephalon commonly bore four or five pairs of appendages plus a median, unpaired flap called the **labrum** projecting posteriorly over the mouth. Only the first pair of appendages, the multiarticulate antennae, was preoral in adults. However, evidence indicates that even the antennae arose postorally during embryogenesis, migrating to a preoral position later in development. Behind the antennae were typically three or four pairs of biramous postoral appendages, similar in structure to the limbs on the rest of the body (Figures 15.31A and 15.32). In some groups the number of cephalic somites (and hence appendages) varied from this typical arrangement. For example, in the highly unusual genus *Marrella* there seems to have been but one postoral somite, whereas in *Rhenops* and *Emeraldella* there were five. With the exception of the antennae and the tagmosis of the body, the trilobites were relatively homonomous.

The thorax comprised a variable number of segments, each supported ventrally by a transverse **tendinous bar** and bearing a pair of biramous appendages. The pygidium included from a few to as many as two dozen appendage-bearing segments. The terminal piece of the trilobite pygidium, the telson, is considered homologous to the postsegmental pygidium of annelids and to the telson of other arthropods.

---

*Although a variety of different taxa are grouped within the phylum Trilobitomorpha, we use the diminutive "trilobites" to collectively refer to them all.

(A)

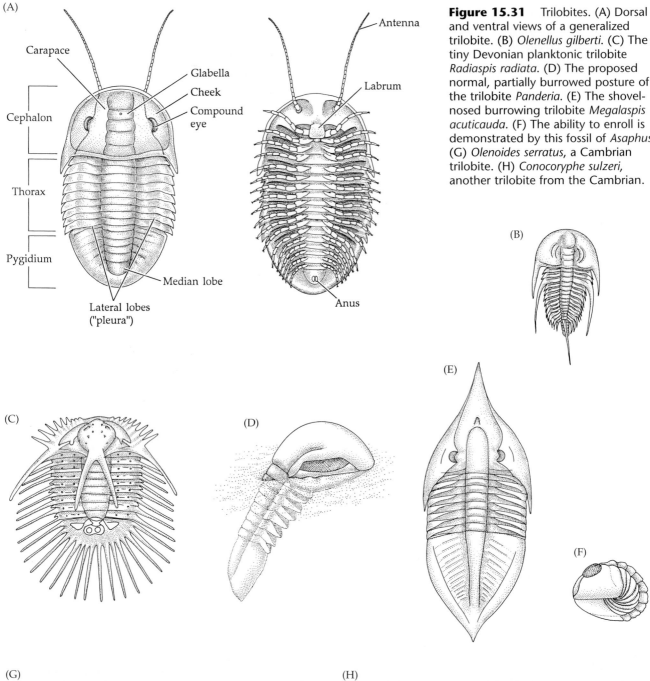

Carapace

Cephalon

Thorax

Pygidium

Glabella
Cheek
Compound eye

Antenna

Labrum

Median lobe

Lateral lobes
("pleura")

Anus

(B)

(C)

(D)

(E)

(F)

Figure 15.31 Trilobites. (A) Dorsal and ventral views of a generalized trilobite. (B) *Olenellus gilberti*. (C) The tiny Devonian planktonic trilobite *Radiaspis radiata*. (D) The proposed normal, partially burrowed posture of the trilobite *Panderia*. (E) The shovel-nosed burrowing trilobite *Megalaspis acuticauda*. (F) The ability to enroll is demonstrated by this fossil of *Asaphus*. (G) *Olenoides serratus*, a Cambrian trilobite. (H) *Conocoryphe sulzeri*, another trilobite from the Cambrian.

(G)

(H)

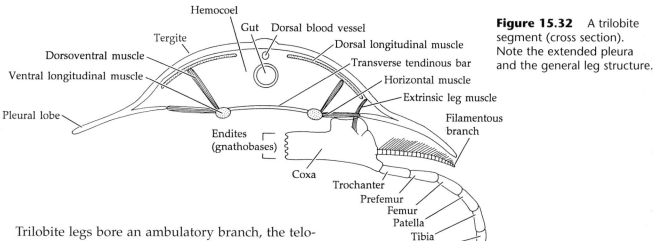

Figure 15.32 A trilobite segment (cross section). Note the extended pleura and the general leg structure.

Trilobite legs bore an ambulatory branch, the telopod, and a laterally directed epipod, presumably gill-bearing, called the **filamentous branch** (formed from a protopodal exite). Trilobite limbs were thus biramous. Both the telopod and the filamentous branch arose from the single protopodal article (called a coxa). The coxae also bore well-developed endites that served as masticatory or grasping gnathobases (Figure 15.32). The homology of the trilobite coxa is unclear, but it may represent several fused protopodites.

Internal Anatomy

X-ray techniques have provided information about the structure of the digestive tract and the arrangement of the muscles in trilobites. As in other arthropods, the muscles occurred in bands associated with body and limb articulations (Figure 15.33A). The body was incapable of lateral bending, but the animals could easily flex ventrally. The digestive tract was relatively simple and straight (Figure 15.33B), extending from the mouth on the ventral side of the cephalon to the anus on the pygidium. From the mouth, the esophagus extended forward in the cephalon and then dorsally to meet an enlarged area variably called the stomach or crop. A digestive cecum apparently arose from the anterior part of the stomach. The intestine arose from the posterior end of the stomach and extended straight to the anus.

Development

Trilobites had a mixed life history pattern. After they hatched, they passed through at least three larval stages of several instars each (Figure 15.33C–F). The larval stages are well represented in the fossil record and appear to be consistent throughout the subphylum. The hatching form, or **protaspis larva**, was less than 1 mm long and consisted of most or all the cephalic segments fused to a "protopygidium" and covered by a dorsal shield. In the next stage (**meraspis larva**) segments were added through several molts as the cephalon, trunk, and pygidium became distinct. Through subsequent molts the animal eventually took on the form of a miniature trilobite (**holaspis larva**). A further series of molts brought the holaspis larva to the juvenile form by the addition of segments and an increase in size.

The Evolution of Arthropods

The Origin of the Arthropoda

A cladogram depicting the phylogeny of the Panarthropoda is shown in Figure 15.34. These relationships are largely noncontroversial. However, the origin of the Panarthropoda itself is highly controversial. There are numerous synapomorphies shared between the Annelida and Arthropoda (and Onychophora and Tardigrada). These synapomorphies include: a characteristic "brain," with anterior mushroom bodies and segmental ganglia; an elongate dorsal tubular heart derived from a longitudinal blood vessel; four to five bands of longitudinal muscles; and dual segmentally iterated *engrailed* gene expression during early segmentation and neurogenesis. However, the strongest argument for the origin of arthropods and annelids from a common ancestor is the complicated way in which the body cavities and associated structures unfold during ontogeny in these two groups. As described in the section on arthropod development above, even though there is a difference between annelids and arthropods in adult body cavity anatomy, the unique metameric teloblastic development and resulting body segmentation in these two groups are essentially identical and presumably homologous, and thus represent a powerful synapomorphy. The transitory coelomic cavities of arthropods, onychophorans, and tardigrades are viewed as recapitulations of the paired and segmental coelomic cavities of the common ancestor of annelids and arthropods. For nearly two centuries this complex embryological pattern has united the annelids and arthropods. In 1817 Cuvier coined the name "Articulata" (*Les articulés*) for these taxa, and today the theory of their common ancestry is known as the **Articulata theory**.

In 1997, based on analyses of 18S rDNA sequence data, Aguinaldo et al. proposed an alternative hypothesis of arthropod relationships: that there is a clade of an-

Figure 15.33 (A) The major body muscles of a trilobite (dorsal cutaway view). (B) A "dissected" trilobite (lateral view), showing the position of the digestive tract (digestive cecum not shown). (C–E) Stages in the development of trilobites. (C,D) Early and late protaspis larvae of *Sao hirsuata*. (E) Meraspis larva of *Shumardia pusilla*. (F) Holaspis larva of *S. pusilla*.

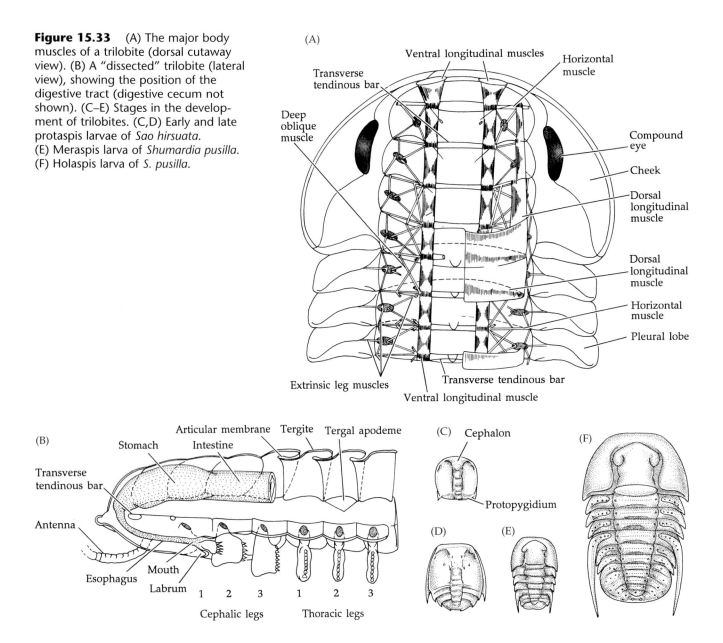

imals that includes arthropods, tardigrades, onychophorans, nematodes, nematomorphans, kinorhynchs, and priapulans, but NOT annelids. They suggested that this clade contains all the known "molting" Metazoa and thus named it **Ecdysozoa**. Subsequent studies based on the 18S rDNA gene, and on alternative genes have yielded mixed results. One of the great weaknesses of the Ecdysozoa hypothesis has been the failure of its proponents to address the morphological, anatomical, and developmental data that stand in opposition to it. None of the annelid–arthropod synapomorphies noted above occur in the broader "ecdysozoan clade." It is difficult to imagine the precise, multi-gene mediated complexity of metameric teloblastic development evolving twice in the animal kingdom. Dozens of studies have been published since 1997 in support or opposition to the Ecdysozoa hypothesis, and this work is briefly reviewed in Chapter 24.

Evolution within the Arthropoda

Perhaps no area of animal research has been more active in recent years than that of arthropod evolutionary biology. Virtually every imaginable phylogenetic tree has been proposed for the arthropods at one time or another; the four most popular are shown in Figure 15.35. Since the mid-1980s, a virtual explosion of new information on this phylum has appeared, much of it concerning phylogeny, paleontology, gene expression, and developmental biology. If we examine the evolution of arthropods in light of these recent discoveries, a very different kind of phylogenetic tree begins to emerge than seen in previous textbooks (including the first edition of this one).

We do not yet have enough details to fill in the evolutionary history of this new tree or generate a well-corroborated cladogram, but Figure 15.36 provides a simple (hypothetical) model of the current thinking. The

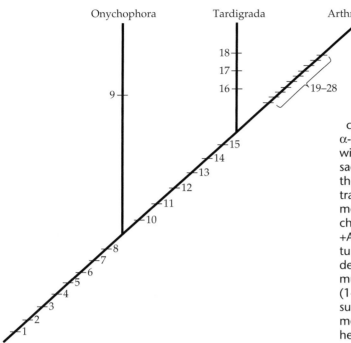

Figure 15.34 Phylogeny of the arthropods and their allies (the Panarthropoda). For synapomorphies uniting this lineage with the Annelida, see Chapter 24. Synapomorphies of the Onychophora + Tardigrada + Arthropoda: (1) arthropod ecdysial molting; (2) reduction of the coelom and creation of a hemocoel; (3) lobopodal legs; (4) leg tips with pads and claws; (5) various chemical similarities of the cuticle (e.g., α-chitin and protein, but lacking collagen); (6) dorsal heart with segmental ostia; (7) modified metanephridia with sacculus; (8) panarthropodan sensilla. Synapomorphies of the Onychophora: (9) hemal channel system; spiracles, tracheae, and oral papillae with slime glands are synapomorphies that define the modern terrestrial lineage of onychophorans. Synapomorphies of the Tardigrada +Arthropoda: (10) articulated limbs with intrinsic musculature; (11) cerebral ganglia differentiated into proto-, deuto-, and tritocerebrum; (12) loss of all sheetlike body musculature; (13) loss of circular muscle layer in body wall; (14) muscles occur in separate bands extending between subcuticular attachment points; (15) gut forms peritrophic membrane. Synapomorphies of the Tardigrada: (16) loss of heart; (17) loss of nephridia; (18) claws of anterior pair of legs modified into stylets and stylet supports. Synapomorphies of Arthropoda: (19) hardened, articulated exoskeleton (body and appendages); (20) head shield/carapace; (21) localized and fully segmental sclerotized cuticular plates (= sclerites); (22) appendages constructed of two distinct regions, a protopod and a distal multiarticulate telopod; (23) coxal endites on anterior appendages function in feeding; (24) resilin; (25) cephalon with antennules and 3–4 pairs of appendages; (26) compound eyes; (27) complete loss of motile cilia (even in metanephridia); (28) unique arthropod hemocyanin polymer.

new tree has a panorama of crustacean-like Precambrian and early Cambrian arthropods at its base—a diverse array of creatures that had typical crustacean (= crustaceamorph) bodies, eyes, development, and naupliar larvae, though perhaps with a smaller number of fused head somites than seen in modern Crustacea. Early in the Cambrian, the trilobites emerged from this crustaceamorph stem line, radiating rapidly to become the most abundant arthropods of Paleozoic seas, but then abruptly disappearing in the Permian–Triassic extinction. Next to appear were probably the Cheliceriformes, in the form of giant marine water scorpions (eurypterids) and their kin, which had appeared at least by the Ordovician; by the Silurian, eurypterids had probably become keystone predators in the marine realm. Also by the Silurian, the chelicerates had invaded land and begun to leave a

fossil record of terrestrial arachnids. By the late Ordovician or early Silurian the first myriapods had evolved, perhaps marine creatures, and about 15 million years later terrestrial millipedes appear in the fossil record. The last major arthropod group to appear was probably the hexapods, making their appearance in the Devonian, or perhaps the Silurian, and radiating rapidly to dominate the terrestrial world, ultimately qualifying the Cenozoic to be called "the age of insects."*

This model of a Paleozoic crustacean stem line spinning off the other arthropod subphyla one after another differs considerably from previous views of arthropod evolution (see Figure 15.35), and many details are still lacking. Nonetheless, a great deal of information now supports this new view—one in which the Crustacea is seen to be an ancient paraphyletic assemblage that was the "mother of all modern Arthropoda." Some details of this new view are provided below.

Arthropods are arguably the most successful animal phylum on Earth. They encompass an unparalleled

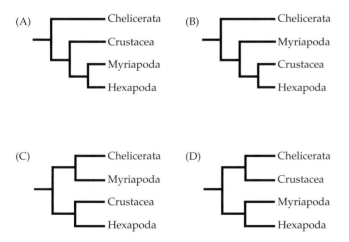

Figure 15.35 The four principal hypotheses of relationships among the four living subphyla of Arthropoda.

---

*Terrestrial arthropods rarely reached the large sizes achieved by their marine ancestors, perhaps because their bodies could not support themselves out of water during the molting process.

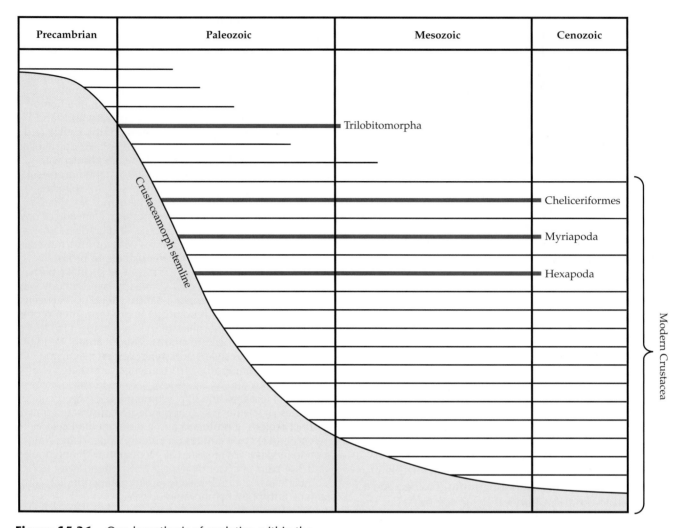

Precambrian	Paleozoic	Mesozoic	Cenozoic

Trilobitomorpha

Crustaceamorph stemline

Cheliceriformes

Myriapoda

Hexapoda

Modern Crustacea

Figure 15.36 One hypothesis of evolution within the Arthropoda, in which all arthropod lineages arose from a Paleozoic crustaceamorph stem line. In this model, the subphylum Crustacea is paraphyletic. (The unlabeled lines shown are not meant to represent any specific crustacean taxa, but merely illustrate the many origins of modern crustacean lineages.)

range of structural and taxonomic diversity, have a rich fossil record, and have become the favored study animals of evolutionary developmental biology. Some arthropod "model systems" (e.g., *Drosophila*) have been studied intensively for many decades. Arthropods were among the earliest animals to evolve; even the Ediacaran fauna of the late Proterozoic included animals regarded by some as Crustacea (anomalocarids and others, some of which had stalked eyes as in many modern Crustacea; see Chapter 1). For all these reasons, we know a lot about arthropods. However, we still do not know, with any great precision, how the arthropods are related to one another, and the tree presented in Figure 15.36 is thus hypothetical.

Arthropods are the first land animals for which we have a paleontological record. Probable marine myri-

apods have been reported from the late Ordovician/ early Silurian, and molecular clock calculations suggest that they might have been present as early as the Cambrian. The first land arthropods also appeared in the late Ordovician or early Silurian (arachnids, millipedes, centipedes), and these fossils represent the first terrestrial invertebrates for which we have direct evidence. Indeed, animal life on land might not have been possible before the late Ordovician, when terrestrial plants first made their appearance. The first insects in the fossil record are 390-million-year-old Devonian springtails (Collembola) and bristletails (Archaeognatha). By the mid-Paleozoic all five arthropod subphyla were in existence and had already undergone substantial radiation. By the Middle Devonian, centipedes, millipedes, mites, amblypygids, opilionids, scorpions, pseudoscorpions, spiders, and hexapods were all well established. Hence, terrestrial arthropods seem to have undergone major radiations in the Silurian. The presence of a wide variety of predatory terrestrial arthropods during the early Paleozoic suggests that complex terrestrial ecosystems were in place at least as early as the late Silurian.

Ever since Darwin, biologists have asked the question, "How has the incredibly successful diversification of the arthropods come about?" Why are there so many arthropods? Is there something "special" about these animals? And what is the phylogenetic history of the Arthropoda? These fundamental questions remain fertile areas of evolutionary research. There have been some significant challenges to biologists working on these questions. Until the 1990s, there were no hypotheses of arthropod evolution based on principles of explicit phylogenetic inference; we still have an incomplete understanding of arthropod development; and there have been very few phylogenetic studies based on fossils from the earliest ages of arthropod evolution (late Proterozoic and early Paleozoic). It is apparent that high levels of homoplasy exist among the arthropods, and we are only beginning to understand the molecular genetic basis of this homoplasy. Despite these hurdles, major discoveries since the mid-1980s have begun to address each of these challenges.

Work by the great comparative biologist Robert Snodgrass in the 1930s established a benchmark in arthropod biodiversity research. Box 15B shows his classification of the arthropods at that time, and it is this classification that one still finds in most biology textbooks. The Snodgrass classification embraces three important hypotheses: (1) arthropods constitute a monophyletic taxon; (2) myriapods and hexapods are sister groups, together forming a taxon called Atelocerata (= Tracheata, or Uniramia according to some authors); and (3) Crustacea and Atelocerata are sister groups, together forming a taxon called the Mandibulata. The Atelocerata were united by several attributes: a tracheal respiratory system, uniramous legs, Malpighian tubules for excretion, and loss of the second antennae (as the name Atelocerata implies). The Mandibulata were united on the basis of the

mandibles (which appear to be homologous in these taxa) and a similar head and head appendage structure.*

It was not until the late 1980s that Snodgrass's longstanding view of arthropod relationships began to be seriously questioned. The reasons for this reevaluation included (1) the discovery of new, exquisitely preserved Precambrian and Cambrian fossils, (2) the appearance of explicit morphological and molecular phylogenetic analysis techniques, (3) new detailed analyses of the arthropod nervous system, (4) the emergence of molecular evolutionary developmental biology, and (5) the discovery of the amazing potential of "master" developmental genes in arthropod development and evolution. Some of these new discoveries are described below.

Emerging Views of Arthropod Relationships

Phylogenetic studies of the arthropods have a long history of controversy. Today there are five principal competing hypotheses of arthropod phylogeny (Figures 15.35 and 15.36). Virtually all modern analyses agree that the arthropods are a monophyletic taxon. Some recent morphological analyses continue to obtain the

---

*For a brief period of time in the mid-twentieth century (1955–1975) the concept of a polyphyletic Arthropoda, championed mainly by S. Manton and D. Anderson, enjoyed some popularity. The Mantonian view recognized four separate phyla with independent origins in annelid ancestors: (1) Crustacea, (2) Trilobita, (3) Chelicerata, and (4) Uniramia (myriapods, hexapods, and onychophorans). This somewhat bizarre idea, based on flawed phylogenetic argumentation and interpretation of data, did not long survive the rigors of scientific testing and modern methods of phylogenetic inference. Today, few people seriously question the monophyly of the Arthropoda, which share many unique features (synapomorphies), as discussed in this chapter and noted in Figure 15.34.

BOX 15B *Classification Schemes for the Arthropods*

The Three-Subphylum Classification of Arthropods (*sensu* Snodgrass 1938)

Phylum Arthropoda

Subphylum Trilobita

Subphylum Chelicerata

  Class Merostomata

    Subclass Xiphosura

    Subclass Eurypterida

  Class Arachnida

  Class Pycnogonida

Subphylum Mandibulata

  Class Crustacea

  Class Atelocerata (= Tracheata)

    Subclass Hexapoda

    Subclass Myriapoda

The Five-Subphylum Classification of Arthropods Followed in This Book

Phylum Arthropoda

Subphylum Trilobitomorpha

Subphylum Cheliceriformes

  Class Chelicerata

    Subclass Merostomata

      Order Eurypterida

      Order Xiphosura

    Subclass Arachnida

  Class Pycnogonida

Subphylum Crustacea

Subphylum Hexapoda

  Class Entognatha

  Class Insecta

Subphylum Myriapoda

Snodgrass view of relationships, retaining the traditional groupings of Atelocerata and Mandibulata, and often combining the trilobites and chelicerates in a clade. An analysis of sequences from the histone H3 and U2 genes and an analysis of combined morphological and molecular data recently suggested support for retention of the traditional views of Mandibulata and Atelocerata (Edgecombe et al. 2000). When fossils are included in morphological analyses, however, different results often emerge, and the Crustacea tend to arise at the very base of the arthropod tree as a paraphyletic sequence of taxa from which the other subphyla emerge—much as shown in Figure 15.36. Molecular phylogenetic studies from at least five nuclear genes, as well as mitochondrial gene arrangements, support this view, and are in agreement that hexapods are *not* the sister group to the myriapods, but are most closely related to crustaceans.

Research in the field of developmental biology and gene expression has revealed that the Arthropoda are rich with homoplasy, and it is now clear that much of the difficulty in reconciling morphological trees and molecular trees is a result of high levels of parallel evolution within the Arthropoda. The rigid, compartmentalized bodies of arthropods have allowed for modes of body region specialization unavailable to other metazoan phyla. We now know that the fates of segmental units and their appendages are under the ultimate orchestration of Hox and other developmental genes. These genes select the critical developmental pathways to be followed by groups of cells during morphogenesis. Hox genes determine such basic body architecture as the dorsoventral and anterior–posterior body axes and where body appendages form. Hox genes can either suppress limb development or modify it to create alternative appendage morphologies. A growing body of evidence suggests that these unique genes have played major roles in the evolution of new body plans among arthropods and the Metazoa in general. The degree to which such developmental genes have been conserved is remarkable, and most of them probably date back at least to the Cambrian. For example, homologues of the developmental gene *Pax-6* seem to dictate where eyes will develop in all animal phyla. *Pax-6* is so similar in protostomes (e.g., insects) and deuterostomes (e.g., mammals) that their genes can be experimentally interchanged and still function more or less correctly (Chapter 1).

The molecular, developmental, and microscopic anatomy of the nervous system also suggest that the Hexapoda are more closely related to the Crustacea than to the Myriapoda, and provide strong evidence that the hexapods arose from within the Crustacea. The implications of this idea are profound. First, it demands that some characters shared between hexapods and myriapods, which have been long assumed to be synapomorphies, be reinterpreted as convergences (or parallelisms), including such things as the tracheal gas exchange system, uniramous legs, Malpighian tubules, organs of Tömösvary, loss of the second antennae, and loss of mandibular palps. Second, it suggests that the Crustacea are paraphyletic, and that the hexapods (and perhaps other arthropod subphyla) are derived lineages emerging out of a crustaceamorph stem line, much as the birds arose from deep within the reptile line (i.e., insects are flying crustaceans, in the same sense that birds are flying reptiles). Not all biologists agree with this view, and the evidence, compelling as it is, remains preliminary. However, evidence for a close Crustacea–Hexapoda relationship is converging from such a broad variety of disciplines that it cannot be ignored.* Below we briefly discuss some of these recent findings.

Neurological features. It has been known since the late nineteenth century that the compound eyes of hexapods and crustaceans possess many complex homologous features, and that they differ markedly from the eyes of myriapods and chelicerates.[†] Recall that in hexapods and crustaceans, each ommatidium consists of a cuticular corneal lens that is, at least partly, secreted by two cells—termed primary pigment cells in Hexapoda and corneagen cells in Crustacea. The crystalline cone, produced by four Semper cells, is fundamentally tetrapartite. A retinula is present, typically composed of eight retinular cells. This common anatomical plan is unique to the Hexapoda + Crustacea and constitutes strong evidence of a close relationship between the two groups. Dohle (2001) has even proposed a name on this basis for the clade, the "Tetraconata."[‡]

Recent research on the anatomy and development of the arthropod central nervous system has identified many other neurological features that appear to be unique to the Hexapoda + Crustacea. In fact, Strausfeld (1998) developed a phylogeny of the Arthropoda based solely on 100 anatomical features of the cerebral ganglia, concluding that the hexapods and crustaceans are sister groups. In particular, numerous anatomical similarities

---

*Although the idea of a Crustacea–Hexapoda sister-group relationship seems new to most of us, the idea was actually proposed at the turn of the twentieth century, when Calman presented detailed arguments from comparative anatomy that argued for an alliance between the Crustacea (particularly the Malacostraca) and the Hexapoda. The "rediscovery" of this idea and a failure to cite the older work is symptomatic of a short-sighted trend in biology to ignore published research more than a few years old.

†In the scutigeromorph myriapods, peculiar "ommatidia" have been described. Beneath the lens is a large conelike vitreous body consisting of numerous confluent parts without visible nuclei. The rhabdome is two-layered, and the distal retinular cells form a ring around the proximal cone tip. The number of cells in this ring is highly variable (7–23). The proximal part consists of 4 retinular cells, 3 of which form a triangular "rhabdome," whereas the fourth is displaced eccentrically. There are some published claims that the complex lateral eyes of some millipedes are fusion products of many ommatidia, but these claims are disputed. And, even if true, this does not necessarily imply that in the pre-fusion condition the ommatidia would have been tetrapartite.

exist in the elements of the optic lobes and "midbrain," and similarities are especially strong between malacostracan crustaceans and hexapods.

Developmental similarities suggest a fundamental distinction between embryonic development in the hexapod + crustacean central nervous system and in that of myriapods and chelicerates. In hexapods and crustaceans, development of the CNS begins with the delamination of enlarged cells, called **neuroblasts**, from the ectoderm (neuroectoderm) of each segment. The neuroblasts aggregate to form the segmental ganglia. The formation of neuroblasts is highly stereotyped and predictable in both Crustacea and Hexapoda. These large neuroblasts are of a special type, regarded as stem cells, and they divide unequally to generate a specific number of neurons. So far, 29–31 neuroblasts have been identified in each segment of hexapods and 25–30 in crustacean segments, many of which appear to be homologous between the two subphyla. Nothing resembling these stem cell neuroblasts has been seen in myriapods. In both hexapods and crustaceans, longitudinal connectives of the CNS originate in the segmental neurons, whereas in myriapods they derive from neurons in the cerebral ganglia that send their axons posteriorly to set up long parallel connectives. Thus, myriapod segmental ganglia receive contributions from more widely distributed neurons.

Molecular phylogenetics. Beginning with the pioneering work of Kathryn Field and her colleagues in 1988 using DNA nucleotide sequence data, and in a stream of subsequent studies, the Crustacea and Hexapoda have consistently shown a sister-group relationship in molecular phylogenetic trees. Most of these analyses have relied on sequences from 18S ribosomal DNA (18S rDNA), a problematic gene for many phylogenetic reconstructions that frequently produces bizarre results. However, recent studies using other genes, such as 12S rDNA, 28S rDNA, elongation factor–1α (EF-1α), RNA polymerase II (POLII), and ubiquitin, have all corroborated the 18S rDNA results for within-arthropod relationships. Innovative studies examining the linear arrangement of mitochondrial genes by J. L. Boore and his colleagues also support a Hexapoda + Crustacea sister-group relationship through the finding of unique gene arrangements. In addition, analyses of the sequences and functions of Hox genes support a Crusatcea–Hexapoda sister-group relationship.

The paleontological data. Unfortunately, most people who study arthropod evolution pay scant attention to the many exquisite early arthropod fossils that are available, which, of course, cannot be included in molecular DNA analyses. The most important ancient arthropod fossils are those in which even the soft parts of the animal were preserved—the so-called ancient *Lagerstätten*, such as the faunas of the Upper Cambrian Orsten deposits of Sweden, the Middle Cambrian Burgess Shale of Canada (and elsewhere), and the Lower Cambrian Chengjiang (530 mya) deposits of China. Recent discoveries from these well-preserved deposits have shown that the fossil record of Crustacea dates at least to the early Cambrian, and probably to the late Precambrian. These extraordinary faunas are now informing us that crustaceans probably predate the appearance of trilobites in the fossil record, and some workers consider the first arthropods to be crustaceans (or "protocrustaceans"). The Chengjiang fauna includes at least a hundred species of animals, many without hard skeletons, including the first known members of many modern groups. However, it is the arthropods that dominate this fauna, including trilobites and bradoriid crustaceans (and also tardigrades and onychophorans). The largest of the Chengjiang animals is *Anomalocaris*, an arthropod (perhaps a crustacean) that is also known from Ediacaran and Middle Cambrian deposits (see Figure 1.3C). The Chengjiang fauna is very similar to that of the Burgess Shale, and it demonstrates that the arthropods were already far advanced by this early date. We now know that arthropods have probably been the dominant animals in terms of species diversity since the Cambrian. In fact, arthropods constitute over one-third of all species described from Lower Cambrian strata.

The spectacular discoveries and research by Klaus Müller and Dieter Walossek since the mid-1980s on microscopic arthropods from the Upper Cambrian Orsten deposits of Sweden have brought to light a rich fauna of crustaceans, many of which closely resemble modern groups such as cephalocarids, mystacocarids, and branchiopods. The exquisite Orsten arthropod fossils show little or no signs of decomposition, and they preserve details less than a micrometer in size (e.g., cuticular pores and setae). The recovery of these three-dimensionally preserved animals and the developmental series that have been found (with successive larval, juvenile, and adult instars) have provided us with information on the detailed anatomy of the body seg-

---

‡Although tetrapartite ommatidia appear to be restricted to the Hexapoda and Crustacea, they may not be a synapomorphy for this clade. We don't yet know when the tetrapartite condition first appeared within the Crustacea or the Arthropoda, but it probably was well before the insects emerged (perhaps in the early Cambrian crustacean stem line). Crustacean fossils from Cambrian *Lagerstätten* deposits have eyes that strongly resemble modern crustacean eyes, at least superficially. In any case, tetrapartite ommatidia are probably a symplesiomorphy retained in both Hexapoda and crown Crustacea (their transformation in myriapods perhaps being a synapomorphy defining that clade). The compound eyes of xiphosurans are very different from those of other Arthropoda, suggesting they might have evolved apart from the tetrapartite condition seen in Crustacea and Hexapoda (or that they are somehow derived from the crustacean tetrapartite condition).

ments and appendages of many ancient stem arthropods. The Orsten fauna shows that Cambrian Crustacea had all the attributes of modern crustaceans, such as compound eyes, a head shield, naupliar larvae (with locomotory first antennae), and biramous appendages on the second and third head somites (the second antennae and mandibles).

Where Are We Now?

Let us now return to our two fundamental questions about arthropod evolution: Why are there so many arthropods, and what is their phylogenetic history? Addressing the first question, it seems there are six overarching explanations, each complex in its own right:

1. The numerical superiority of arthropods is not a recent event. Recent fossil discoveries and molecular clock data tell us that arthropod diversification began very early in the history of the Metazoa, probably in the Precambrian, and by the Cambrian the arthropods were perhaps already the most speciose metazoan phylum on Earth. Arthropods have probably been on a powerful phylogenetic trajectory for at least 600 million years.

2. Their great size range, especially at the smaller end of the scale, adapts arthropods for a great variety of ecological niches. The Cambrian Orsten deposits inform us that a whole fauna of interstitial/meiofaunal arthropods already existed as early as the mid-Cambrian, and this habitat has continued to be rich in adaptive radiation and specialized species ever since. Similar small-body-size niches are filled by arthropods in a great many specialized environments today. We find high diversities of minute arthropods in habitats such as marine sediments, coral reefs, among the fronds of algae, on mosses and other primitive plants, and on the bodies of every kind of animal imaginable. Small insects and mites have exploited virtually every terrestrial microhabitat available.

3. Their close relationships and coevolution with plants (on land) and algae (in aquatic environments) have been powerful forces in the radiation of the arthropods. It is not just the insects that have been on a coevolutionary trajectory with plants; many crustaceans utilize algae as both a living substratum and a food source and show strong evidence of coevolution.

4. The arthropods (insects) were the first flying animals, and the ability to fly led them into niches that other invertebrates simply could not penetrate.

5. Metamerism (the serial repetition of body segments and appendages) in arthropods provides an enormous amount of easily manipulated body plan material upon which evolutionary processes can act. Given the great age and the sheer diversity of arthropods, and our emerging knowledge of developmental and regulatory genes in these animals, a high level of homoplasy is no longer surprising.

6. The potential for major changes in body plans due to variations in Hox and other regulatory genes and the downstream genes they affect is just beginning to be realized, but this potential is clearly enormous. There seems little doubt that changes in Hox, and other developmental genes over time have profoundly affected arthropod evolution. Considering the number and position of limbs in arthropods and the flexibility of homeobox and regulatory switches, it is little wonder that arthropod anatomical diversity seems so endless.

As for the second question—concerning the phylogenetic history of the arthropods—traditional morphological classifications are in conflict with emerging data from a variety of different fields. All the evidence indicates that the arthropods are monophyletic. However, fossil data, comparative neuroanatomy, developmental data, and molecular phylogenetic analyses all suggest that the arthropods comprise a paraphyletic crustacean stem line out of which the other subphyla emerged (Figure 15.36). New evidence suggests, in particular, that the insects arose from within a crustacean stem line. This view of a paraphyletic Crustacea spinning off a series of other major arthropod lineages may explain why morphologists have been unable to come to agreement on the sister-group relationships of the major arthropod lineages. Better clarity of these ideas should emerge within the next decade with a better understanding of the genetic regulation of developmental processes, examination of new nuclear and mitochondrial genes, and more cladistic analyses that include fossil species.

Selected References

Here we provide only a small sampling of the vast literature on arthropods. We have selected works that are broad in coverage as well as those dealing with the onychophorans, tardigrades, and trilobites. Several of these references are collections of papers, and in this and the following arthropod chapters we have generally not cited the papers they contain separately. Lists of readings on the separate subphyla are provided in Chapters 16–19. For references on the Ecdysozoa hypothesis, see Chapter 24.

General References

Alexander, R. D. 1964. The evolution of mating behaviour in arthropods. Symp. R. Entomol. Soc. Lond. 2: 78–94.

Anderson, S. O. and T. Weis-Fogh. 1964. Resilin: A rubberlike protein in arthropod cuticle. Adv. Insect Physiol. 2: 1–66.

Bartolomaeus T. and H. Ruhberg. 1999. Ultrastructure of the body cavity lining in embryos of *Epiperipatus biolleyi* (Onychophora, Peripatidae): A comparison with annelid larvae. Invert. Biol. 118(2): 165–174.

Bennett, D. and S. M. Manton. 1963. Arthropod segmental organs and Malpighian tubules with particular reference to their function in Chilopoda. Ann. Mag. Nat. Hist. (13)5: 545–556.

Bergström, J. 1987. The Cambrian *Opabinia* and *Anomalocaris*. Lethaia 20: 187–188.

Bernhard, G. C. (ed.). 1966. *The Functional Organization of the Compound Eye*. Pergamon Press, New York.

Bowerman, R. F. 1977. The control of arthropod walking. Comp. Biochem. Physiol. 56A: 231–247.

Briggs, D. E. G. 1994. Giant predators from the Cambrian of China. Science 264: 1283–1284.

Briggs, D. E. G., D. H. Erwin and F. J. Collier. 1994. *The fossils of the Burgess Shale*. Smithsonian Institution Press, Washington, D.C.

Cameron, J. N. 1985. Molting in the blue crab. Sci. Am. 252(5): 102–109.

Cellular mechanisms of ion regulation in arthropods. 1984. Am. Zool. 24, no. 1. [Papers presented at a 1982 symposium.]

Chang, E. S. 1985. Hormonal control of molting in decapod Crustacea. Am. Zool. 25: 179–185.

Clarke, K. U. 1973. *The Biology of Arthropoda*. American Elsevier, New York.

Cloudsley-Thompson, J. L. 1958. *Spiders, Scorpions, Centipedes and Mites: The Ecology and Natural History of Woodlice, Myriapods and Arachnids*. Pergamon Press, New York.

Cronin, T. W. 1986. Optical design and evolutionary adaptation in crustacean compound eyes. J. Crust. Biol. 6(1) 1–23.

Edney, E. B. 1957. *The Water Relations of Terrestrial Arthropods*. Cambridge University Press, London.

Eguchi, E. and Y. Tominaga. 1999. *Atlas of Arthropod Sensory Receptors*. Springer-Verlag, New York.

Eguchi, E. and T. H. Waterman. 1966. Fine structure patterns in crustacean rhabdomes. *In* Proceedings of the International Symposium on the Functional Organization of the Compound Eye, Stockholm, Oct. 25–27, 1965. Pergamon Press, New York, pp. 105–124.

Ewing, A. W. 1991. *Arthropod Bioacoustics*. Cornell University Press, Ithaca, NY.

Florkin, M. and B. T. Scheer (eds.). *Chemical Zoology*. Vols. 5 and 6. Academic Press, New York.

French, V., P. Ingham, J. Cooke and J. Smith (eds.). 1988. *Mechanisms of Segmentation*. The Company of Biologists, Ltd., Cambridge.

Gilbert, S. F. and A. M. Raunio. 1997. *Embryology: Constructing the Organism*. Sinauer Associates, Sunderland, MA. [The only modern summary text on the subject.]

Gould, S. J. 1989. *Wonderful Life: The Burgess Shale and the Nature of History*. W. W. Norton, New York. [Classic Gould.]

Gould, S. J. 1995. Of tongue worms, velvet worms, and water bears. Nat. Hist., January, 6–15.

Gupta, A. P. (ed.) 1983. *Neurohemal Organs of Arthropods: Their Development, Evolution, Structures, and Functions*. Charles C. Thomas, Springfield, IL.

Hallberg, E. and R. Elofsson. 1989. Construction of the pigment shield of the crustacean compound eye: A review. J. Crust. Biol. 9(3): 359–372.

Harrison, F. W. and M. E. Rice (eds.), *Microscopic Anatomy of Invertebrates*. Vol. 12. Onychophora, Chilopoda, and Lesser Protostoma. Wiley-Liss, New York.

Hassell, M. P. 1978. *The Dynamics of Arthropod Predator-Prey Systems*. Princeton University Press, Princeton, NJ.

Herreid, C. F. and C. R. Fourtner (eds.) 1981. *Locomotion and Energetics in Arthropods*. Plenum Press, New York.

Herrnkind, W. F. 1972. Orientation in shore-living arthropods, especially the sand fiddler crab. *In* H. E. Winn and B. L. Olla (eds.), *Behavior of Marine Animals*, Vol. 1, Invertebrates. Plenum Press, New York.

Hou, X. and J. Bergström. 1997. *Arthropods of the Lower Cambrian Chengjiang fauna of southwest China*. Fossils and Strata, no. 45 Universitetsforlaget, Oslo.

Kaestner, A. 1968, 1970. *Invertebrate Zoology*. Vols. 2 and 3. Wiley, New York. [These volumes do an excellent job of describing the arthropods; adroitly translated from the German by H. W. and L. R. Levi.]

Kenchington, W. 1984. Biological and chemical aspects of silks and silk-like materials produced by arthropods. S. Pac. J. Nat. Sci. 5: 10–45.

Kunze, P. 1979. Apposition and superposition eyes. *In* H. Autrum (ed.), *Comparative physiology and evolution of vision in invertebrates*. Part A, Invertebrate photoreceptors. *Handbook of Sensory Physiology*, Vol. 7, part 6. Springer-Verlag, Berlin, pp. 441–502.

Linzen, B. et al.. 1985. The structure of arthropod hemocyanins. Science 229: 519–524. [A review of these large copper-based, oxygen-carrying proteins in arthropods.]

Lockwood, A. P. M. 1968. *Aspects of the Physiology of Crustacea*. Oliver and Boyd, London. [Somewhat dated, but still highly regarded.]

Manton, S. M. 1950–1973. The evolution of arthropod locomotory mechanisms. Parts 1–11. All published in Zool. J. Linn. Soc.

Manton, S. M. 1977. *The Arthropoda: Habits, Functional Morphology, and Evolution*. Clarendon Press, Oxford. [Although Manton's polyphyletic hypothesis of arthropod evolution is no longer in favor, her work on comparative/functional anatomy remains useful; see review by Platnick 1978.]

McIver, S. B. 1975. Structure of cuticular mechanoreceptors of arthropods. Annu. Rev. Entomol. 20: 381–397.

Moore, R. C. (ed.). 1969. *Treatise on Invertebrate Paleontology*. Part R, Arthropoda 4. Geological Society of America and University of Kansas Press, Lawrence.

Müller, K. J. and D. Walossek. 1985. A remarkable arthropod fauna from the Upper Cambrian "Orsten" of Sweden. Trans. R. Soc. Edinburgh: Earth Sciences 76: 161–172.

Müller, K. J. and D. Walossek. 1986. Arthropod larvae from the Upper Cambrian of Sweden. Trans. R. Soc. Edinburgh: Earth Sciences 77: 157–179.

Parker, S. P. (ed.) 1982. *Synopsis and Classification of Living Organisms*. Vol. 2. McGraw-Hill, New York. [This volume is devoted to the Arthropoda and its allies.]

Patel, N. H., T. B. Kornberg and C. S. Goodman. 1989. Expression of *engrailed* during segmentation in grasshopper and crayfish. Development 107: 201–212.

Richards, A. G. 1951. *The Integument of Arthropods*. University of Minnesota Press, Minneapolis.

Rivindranth, M. H. 1980. Haemocytes in haemolymph coagulation of arthropods. Biol. Rev. 55: 139–170.

Snodgrass, R. E. 1951. *Comparative Studies on the Head of Mandibulate Arthropods*. Comstock Publishing Associates, Ithaca, NY.

Snodgrass, R. E. 1952. *A Textbook of Arthropod Anatomy*. Cornell University Press, Ithaca, NY. [Obviously dated, but still an excellent introduction to arthropod anatomy.]

Strausfeld, N. J. 1996. Oculomotor control in flies: From muscles to elementary motion detectors. *In* P. S. G. Stein and D. Stuart (eds.), *Neurons, Networks, and Motor Behavior*. Oxford University Press, Oxford, pp. 277–284.

Tickle, C. 1981. Limb regeneration. Am. Sci. 69: 639–646.

Tu, A. T. (ed.). 1984. *Handbook of Natural Toxins*. Vol. 2, Insect Poisons, Allergens, and Other Invertebrate Venoms. Marcel Dekker, New York. [Summarizes information on insects, chilopods, and arachnids.]

Wedeem, C. J. and D. A. Weisblat. 1991. Segmental expression of an *engrailed*-class gene during early development and neurogenesis in an annelid. Development 113: 805–814.

Wehner, R. 1981. Spatial vision in arthropods. *In* H. Autrum (ed.), *Comparative physiology and evolution of vision in invertebrates*. Part C. *Handbook of Sensory Physiology*, Vol. 7, part 6. Springer-Verlag, Berlin, pp. 287–616.

Whitington, P. M., T. Meier and P. King. 1991. Segmentation, neurogenesis, and formation of early axonal pathways in the centipede *Ethmostigmus rubripes* (Brandt). Roux's Arch. Dev. Biol. 199: 349–363.

Zill, S. N. and E. Seyfarth. 1996. Exoskeletal sensors for walking. Sci. Am. 275(1): 86–90.

Onychophorans

Dzik, J. and G. Krumbiegel. 1989. The oldest "onychophoran" *Xenusion*: A link connecting phyla? Lethaia 22: 169–182.

Hoyle, G. and M. Williams. 1980. The musculature of *Peripatus dominicae* and its innervation. Phil. Trans. R. Soc. Lond. Ser. B 288: 481–510.

Manton, S. M. and N. Heatley. 1937. The feeding, digestion, excretion and food storage of *Peripatopsis*. Phil. Trans. R. Soc. Lond. Ser. B 227: 411–464.

Monje-Najera, J. 1995. Phylogeny, biogeography and reproductive trends in the Onychophora. Zool. J. Linn. Soc. 114: 21–60.

Nylund, A., H. Ruhberg, A. Tjoenneland and B. A. Meidell. 1988. Heart ultrastructure in four species of Onychophora, and phylogenetic implications. Zool. Beitr. n.f. 32: 17–30.

Peck, S. B. 1975. A review of the New World Onychophora, with the description of a new cavernicolous genus and species from Jamaica. Psyche 82: 341–358.

Ramsköld, L. 1992a. Homologies in Cambrian Onychophora. Lethaia 25: 443–460.

Ramsköld, L. 1992b. The second leg row of *Hallucigenia* discovered. Lethaia 25: 321–324.

Ramsköld, L. and X. Hou. 1991. New early Cambrian animal and onychophoran affinities for enigmatic metazoans. Nature 351: 225–228.

Ruhberg, H. and W. B. Nutting. 1980. Onychophora: Feeding, structure, function, behaviour and maintenance. Berh. Naturwiss. Ver. Hamburg (NF) 24: 79–87.

Walker, M. H. 1995. Relatively recent evolution of an unusual pattern of early embryonic development (long germ band?) in a South African onychophoran, *Opisthopatus cinctipes* Purcell (Onychophora: Peripatopsidae). Zool. J. Linn. Soc. 114: 61–75.

Tardigrades

Bussers, J. C. and C. Jeuniaux. 1973. Structure et composition de la cuticle de *Macrobiotus* et de *Milnesium tardigradum*. Ann. Soc. R. Zool. Belg. 103: 271–279.

Kinchin, I. M. 1994. *The Biology of Tardigrades*. Portland Press, London.

Kristensen, R. M. l978. On the structure of *Batillipes naserrevangi* Kristensen, 1978. 2. The muscle attachments and the true cross-striated muscles. Zool. Anz. Jena 200 (3/4): 173–184.

Kristensen, R. M. 1981. Sense organs of two marine arthrotardigrades (Heterotardigrada, Tardigrada). Acta Zool. 62: 27–41.

Kristensen, R. M. 1982. The first record of cyclomorphosis in Tardigrada, based on a new genus and species from Arctic meiobenthos. Z. Zool. Syst. Evolutionsforsch. 20: 249–270.

Kristensen, R. M. 1984. On the biology of *Wingstrandarctus corallinus* nov. gen. et spec., with notes on the symbiontic bacteria in the subfamily Florarctinae (Arthrotardigrada). Vidensk. Medd. Dan. Naturhist. Foren. Khobenhavn 145: 201–218.

Kristensen, R. M. and R. P. Higgins. 1984a. A new family of Arthrotardigrada (Tardigrada: Heterotardigrada) from the Atlantic Coast of Florida, U.S.A. Trans. Am. Microsc. Soc. 103: 295–311.

Kristensen, R. M. and R. P. Higgins. 1984b. Revision of *Styraconyx* (Tardigrada: Halechiniscidae), with descriptions of two new species from Disko Bay, west Greenland. Smithsonian Contributions to Zoology 391. Smithsonian Institution Press, Washington, D.C.

Marcus, E. 1929. Tardigrada. In H. G. Bronn (ed.), *Klassen und Ordnungen des Tierreichs*, Bd. 5, Abt. 4. Akad. Verlagsgesellschaft, Frankfurt.

Marcus, E. 1957. Tardigrada. In W. T. Edmondson, H. B. Ward and G. C. Whipple (eds.), *Freshwater Biology*, 2nd Ed. Wiley, New York, pp. 508–521.

Morgan, C. I. and P. E. King. 1976. *British Tardigrades: Keys and Notes for the Identification of the Species*. Synopsis of British Fauna No. 9. Academic Press, New York.

Nelson, D. R. (ed.) 1982. *Proceedings of the Third International Symposium on Tardigrada*. East Tennessee State University Press, Johnson City.

Nelson, D. R. 1991. Tardigrada. In J. H. Thorp and A. P. Covich (eds.), *Ecology and Classification of North American Freshwater Invertebrates*. Academic Press, New York, pp. 501–521. 1991.

Pollock, L. W. 1976. *Marine flora and fauna of the northeastern United States: Tardigrada*. NOAA Technical Report/NMFSCircular 394. National Marine Fisheries Service, Seattle, WA.

Trilobites

Bergstrom, J. 1972. Appendage morphology of the trilobite *Cryptolithus* and its implications. Lethaia 5: 85–94.

Bergstrom, J. B. 1973. *Organization, life, and systematics of trilobites*. Fossils and Strata, no. 2. Universitetsforlaget, Oslo.

Cisne, J. L. 1973. Life history of an Ordovician trilobite, *Triarthrus eatoni*. Ecology 54: 135–142.

Cisne, J. L. 1975. Anatomy of *Triarthrus* and the relationships of the Trilobita. In A. Martinsson (ed.), *Evolution and Morphology of the Trilobita, Trilobitoida and Merostomata*. Fossils and Strata no. 4. Universitentsforlaget, Oslo, pp. 45–63.

Cisne, J. L. 1979. The visual system of trilobites. Palaeontology 22: 1–22.

Eldredge, N. 1977. Trilobites and evolutionary patterns. In A. Hallam (ed.), *Patterns of Evolution*. Elsevier, Amsterdam, pp. 30–332.

Fortey, R. A. 1975. Early Ordovician trilobite communities. In A. Martinsson (ed.), *Evolution and Morphology of the Trilobita, Trilobitoida and Merostomata*. Fossils and Strata no. 4. Universitentsforlaget, Oslo, pp. 339–360.

Fortey, R. A. 2001. Trilobite systematics: The last 75 years. J. Paleontol. 75: 1141–1151.

Levi-Setti, R. 1975. *Trilobites: A Photographic Atlas*. University of Chicago Press, Chicago. [Coffee-table paleontology at its best.]

Martinsson, A. (ed.) 1975. *Evolution and Morphology of the Trilobita, Trilobitoida and Merostomata*. Fossils and Strata, no. 4. Universitentsforlaget, Oslo.

McNamara, K. J. 1986. The role of heterochrony in the evolution of Cambrian trilobites. Biol. Rev. 61: 121–156.

Størmer, L. 1939, 1941, 1951. Studies on trilobite morphology, I, II, and III. Nor. Geol. Tidsskr. 19: 143–273, 21: 49–163, and 29: 108–157.

Stubblefield, C. J. 1926. Notes on the development of a trilobite, *Shumardia pusilla* (Sars). Zool. J. Linn. Soc. 36: 345–472.

Sturmer, W. and J. Bergstrom. 1973. New discoveries on trilobites by X-rays. Paleontology Z. 47 1/2: 104–141.

Towe, K. M. 1973. Trilobite eyes: Calcified lenses in vivo. Science 179: 1007–1009.

Whittington, H. B. 1957. The ontogeny of trilobites. Biol. Rev. 32: 421–469.

Whittington, H. B. 1971. Redescription of *Marrella splendens* (Trilobitoidea) from the Burgess Shale, Middle Cambrian, British Columbia. Bull. Geol. Surv. Can. 209: 1–24.

Whittington, H. B. 1975. Trilobites with appendages from the Burgess Shale, Middle Cambrian, British Columbia. In A. Martinsson (ed.), *Evolution and Morphology of the Trilobita, Trilobitoida and Merostomata*. Fossils and Strata, no. 4. Universitetsforlaget, Oslo, pp. 97–136.

Arthropod Evolution

Aguinaldo, A. M. A., J. M. Turbeville, L. S. Linford, M. C. Rivera, J. R. Garey, R. A. Raff and J. A. Lake. 1997. Evidence for a clade of nematodes, arthropods and other moulting animals. Nature 387: 489–493. [See Chapter 24 for further references on this subject.]

Akam, M. 2000. Arthropods: Developmental diversity within a (super) phylum. Proc. Natl. Acad. Sci. U.S.A. 97: 4438–4441.

Averof, M. and M. Akam. 1995a. Hox genes and the diversification of insect and crustacean body plans. Nature 376: 420–423.

Averof, M. and M. Akam. 1995b. Insect–crustacean relationships: Insights from comparative developmental and molecular studies. Phil. Trans. R. Soc. Lond. Ser. B 347: 293–303.

Averof, M. and N. H. Patel. 1997. Crustacean appendage evolution associated with changes in Hox gene expression. Nature 388: 682–686.

Bergström, J. 1992. The oldest arthropods and the origin of the Crustacea. Acta Zool. 73(5): 287–292.

Boore, J. L., T. M. Collins, D. Stanton, L. L. Daehler and W. M. Brown. 1995. Deducing the pattern of arthropod phylogeny from mitochondrial DNA rearrangements. Nature 376: 163–165.

Boore, J. L., D. V. Lavrov and W. M. Brown. 1998. Gene translocation links insects and crustaceans. Nature 392: 667–668.

Bourdreaux, H. B. 1979. *Arthropod Phylogeny with Special Reference to Insects*. Wiley, New York.

Bowman, T. E. 1984. Stalking the wild crustacean: The significance of sessile and stalked eyes in phylogeny. J. Crust. Biol. 4(1): 7–11.

Briggs, D. E. G. and R. A. Fortey. 1992. The Early Cambrian radiation of arthropods. In J. H. Lipps and P. W. Signor (eds.), *Origin and Early Evolution of the Metazoa*. Plenum Press, New York, pp. 335–373.

Briggs, D. E. G. and H. B. Whittington. 1987. The affinities of the Cambrian animals *Anomalocaris* and *Opabinia*. Lethaia 20: 185–186.

Briggs, D. E. G., R. A. Fortey and M. A. Wills. 1992. Morphological disparity in the Cambrian. Science 256: 1670–1673.

Brusca, R. C. 2000. Unraveling the history of arthropod biodiversification. Ann. Missouri Bot. Garden 87: 13–25.

Brusca, R. C. 2001. Review of *Origin of the Hexapoda*, T. Deuve (ed.). Ann. Soc. Entomol. France 37 (1/2). J. Crustacean Biol. 21(4): 1084–1086.

Budd, G. E. 1996. The morphology of *Opabinia regalis* and the reconstruction of the arthropod stem-group. Lethaia 29: 1–14.

Calman, W. T. 1909. Crustacea. Part VII, Fasicle 3, *In* R. Lankester, *A Treatise on Zoology*. Adam and Charles Black (Londen. Reprinted by A. Asher and Co., Amsterdam, 1964).

Chen, J.-Y., L. Ramsköld and G.-Q. Zhou. 1994. Evidence for monophyly and arthropod affinity of Cambrian giant predators. Science 264: 1304–1308.

Colgan, D. J., A. McLauchlan, G. D. F. Wilson, S. Livingston, G. D. Edgecombe, J. Macaranas, G. Cassis and M. R. Gray. 1998. Histone H3 and U2 snRNA sequences and arthropod molecular evolution. Aust. J. Zool. 46: 419–437.

Cook, C. E., M. L. Smith, M. J. Telford, A. Bastianello and M. Akam. 2001. Hox genes and the phylogeny of the arthropods. Curr. Biol. 11: 759–763.

Delle Cave, L. and A. M. Simonetta. 1991. Early Paleozoic arthropods and problems of arthropod phylogeny, with some notes on taxa of doubtful affinities. *In* A. M. Simonetta and S. Conway Morris (eds.), *The Early Evolution of Metazoa and the Significance of Problematic Taxa*. Cambridge University Press, London, pp. 189–244.

Deuve, T. (ed.) 2001. *Origin of the Hexapoda*. Ann. Soc. Entomol. France 37 (1/2): 1–304. [Papers presented at conference held in Paris, January 1999; see review by Brusca 2001.]

Dohle, W. 1997. Are the insects more closely related to the crustaceans than to the myriapods? Entomol. Scand. (Suppl.) 51: 7–16.

Dohle, W. 2001. Are the insects terrestrial crustaceans? Ann. Soc. Entomol. France 37:85–103.

Edgecombe, G. D. (ed.). 1998. *Arthropod Fossils and Phylogeny*. Columbia University Press, New York.

Edgecombe, G. D. and L. Ramsköld. 1999. Relationships of Cambrian Arachnata and the systematic position of Trilobita. J. Paleontol. 73: 263–287.

Edgecombe, G. D., G. D. F. Wilson, D. J. Colgan, M. R. Gray and G. Cassis. 2000. Arthropod cladistics: Combined analysis of histone H3 and U2 snRNA sequences and morphology. Cladistics 16: 155–203.

Edwards, J. S. and M. R. Meyer. 1990. Conservation of antigen 3G6: A crystalline cone constituent in the compound eye of arthropods. J. Neurobiol. 21: 441–452.

Eldredge, N. and S. M. Stanley (eds.). 1984. *Living Fossils*. Springer-Verlag, New York.

Eriksson, B. J. and G. E. Budd. 2000. Onychophoran cephalic nerves and their bearing on our understanding of head segmentation and stem-group evolution of Arthropoda. Arthropod Structure Dev. 29: 197–209.

Fortey, R. A. and R. H. Thomas (eds.). 1997. *Arthropod Relationships*. Chapman and Hall, London.

Friedrich, M. and D. Tautz. 2001. Arthropod rDNA phylogeny revisited: A consistency analysis using Monte Carlo simulation. Ann. Soc. Entomol. France 37: 21–40.

García-Machado, E., M. Pempera, N. Dennebouy, M. Oliver-Suarez, J. C. Mounolou and M. Monnerot. 1999. Mitochondrial genes collectively suggest the paraphyly of Crustacea with respect to Insecta. J. Mol. Evol. 49: 142–149.

Garey, J. R., M. Krotec, D. R. Nelson and J. Brooks. 1996. Molecular analysis supports a tardigrade-arthropod association. Invert. Biol. 115: 79–88.

Giribet, G. and C. Ribera. 2000. A review of arthropod phylogeny: New data based on ribosomal DNA sequences and direct character optimization. Cladistics 16: 204–231.

Giribet, G., G. D. Edgecombe and W. C. Wheeler. 2001. Arthropod phylogeny based on eight molecular loci and morphology. Nature 413: 157–161.

Harzsch, S. and D. Walossek. 2001. Neurogenesis in the developing visual system of the branchiopod crustacean *Triops longicaudatus* (LeConte, 1846): Corresponding patterns of compound-eye formation in Crustacea and Insecta? Dev. Genes Evol. 211: 37–43.

Hessler, R. R. and W. A. Newman. 1975. A trilobitomorph origin for the Crustacea. *In* A. Martinsson (ed.), *Evolution and Morphology of the Trilobita, Trilobitoida and Merostomata*. Fossils and Strata no. 4. Universitetsforlaget, Oslo, pp. 437–459.

Hou, X. and J. Bergström. 1995. Cambrian lobopodians—ancestors of extant onychophorans? Zool. J. Linn. Soc. 114: 3–19.

Hou, X. and C. Junyuan. 1989. Early Cambrian arthropod-annelid intermediate sea animal, gen. nov. from Chengjiang, Yunnan. Acta Palaeontol. Sinica 28(2): 207–213.

Hou, X. H. and S. Weiguo. 1988. Discovery of Chengjiang fauna at Meichucun, Jinning, Yunnan. Acta Palaeontol. Sinica 27(1): 1–12.

Hou, X., L. Ramsköld and J. Bergström. 1991. Composition and preservation of the Chengjiang fauna—A lower Cambrian soft-bodied biota. Zool.Scripta 20: 395–411.

Hwang, U. W., M. Friedrich, D. Tautz, C. J. Park and W. Kim. 2001. Mitochondrial protein phylogeny joins myriapods with chelicerates. Nature 413: 154–157.

Jeram, A. J., P. A. Selden and D. Edwards. 1990. Land animals in the Silurian: Arachnids and myriapods from Shropshire, England. Science 250: 658–661.

Maas, A. and D. Waloszek. 2001. Cambrian derivatives of the early arthropod stem lineage, pentastomids, tardigrades and lobopodians—an "Orsten" perspective. Zool. Anz. 240: 451–559.

Mangum, C. P. et al. 1985. Centipedal hemocyanin: Its structure and its implications for arthropod phylogeny. Proc. Natl. Acad. Sci. U.S.A. 82: 3721–3725.

McMenamin, M. A. S. 1986. The Garden of Ediacara. Palaios 1: 178–182.

Melzer, R. R., R. Diersch, D. Nicastro and U. Smola. 1997. Compound eye evolution: Highly conserved retinula and cone cell patterns indicate a common origin of the insect and crustacean ommatidium. Naturwissenschaften 84: 542–544.

Melzer, R. R., C. Michalke and U. Smola. 2000. Walking on insect paths? Early ommatidial development in the compound eye of the ancestral Crustacea, *Triops cancriformis*. Naturwissenschaften 87: 308–311.

Mikulic, D. G., D. E. G. Briggs and J. Kluessendorf. 1985a. A new exceptionally preserved biota from the lower Silurian of Wisconsin, U.S.A. Phil. Trans. R. Soc. Lond. Ser. B 311: 75–85.

Mikulic, D. G., D. E. G. Briggs and J. Kluessendorf. 1985b. A Silurian soft-bodied biota. Science 228: 715–717.

Norstad, K. 1987. Cycads and the origin of insect pollination. Am. Sci. 75: 270–278.

Osorio, D., M. Averof and J. P. Bacon. 1995. Arthropod evolution: Great brains, beautiful bodies. Trends Ecol. Evol. 10(11): 449–454.

Paulus, H. F. 2000. Phylogeny of the Myriapoda-Crustacea-Insecta: A new attempt using photoreceptor structure. J. Zool. Syst. Evol. Res. 38: 189–208.

Popadic, A., D. Rusch, M. Peterson, B. T. Rogers and T. C. Kaufman. 1996. Origin of the arthropod mandible. Nature 380: 395.

Platnick, N. I. 1978. Review of *The Arthropoda: Habits, Functional Morphology, and Evolution* by S. M. Manton, 1977, Oxford University Press. Syst. Zool. 27(2): 252–255. [A critical review of the phylogenetic reasoning behind Manton's polyphyletic hypothesis.]

Ramsköld, L. 1991. New early Cambrian animal and onychophoran affinities of enigmatic metazoans. Nature 351: 225–228.

Raven, J. A. 1985. Comparative physiology of plant and arthropod land adaptation. Phil. Trans. R. Soc. Lond. Ser. B 309: 273–288.

Regier, J. C. and J. W. Schultz. 1997. Molecular phylogeny of the major arthropod groups indicates polyphyly of crustaceans and a new hypothesis of the origin of hexapods. Mol. Biol. Evol. 14: 902–913.

Regier, J. C. and J. W. Schultz. 1998. Molecular phylogeny of arthropods and the significance of the Cambrian "explosion" for molecular systematics. Am. Zool. 38: 918–928.

Regier, J. C. and J. W. Schultz. 2001. Elongation factor-2: a useful gene for arthropod phylogenetics. Mol. Phylog. and Evol. 20(1): 136–148.

Retallack, G. J. and C. R. Feakes. 1987. Trace fossil evidence for Late Ordovician animals on land. Science 235: 61–63.

Robison, R. A. 1985. Affinities of *Aysheaia* (Onychophora), with description of a new Cambrian species. J. Paleontol. 59: 226–235.

Rolfe, W. D. I. 1985. Early terrestrial arthropods: A fragmentary record. Phil. Trans. R. Soc. Lond. Ser. B 309: 207–218.

Scholtz, G. 1995. Head segmentation in Crustacea: An immunocytochemical study. Zoology 98: 104–114.

Scholtz, G. 2001. Evolution of developmental patterns in arthropods—the analysis of gene expression and its bearing on morphology and phylogenetics. Zoology 103: 99–111.

Schram, F. R. 1991. Cladistic analysis of metazoan phyla and the placement of fossil problematica. *In* A. M. Simonetta and S. Conway Morris (eds.), *The Early Evolution of the Metazoa and the Significance of Problematic Taxa*. Cambridge University Press, Cambridge, pp. 35–46.

Schultz, J. W. and J. C. Regier. 2000. Phylogenetic analysis of arthropods using two nuclear protein-coding genes supports a crustacean + hexapod clade. Proc. R. Soc. Lond. Ser. B 267: 1011–1019.

Sharov, A. G. 1966. *Basic Arthropodan Stock*. Pergamon Press, New York.

Shear, W. A. and J. Kukalová-Peck. 1990. The ecology of Paleozoic terrestrial arthropods: The fossil evidence. Can. J. Zool. 68: 1807–1834.

Shubin, N. C. Tabin and S. Carroll. 1997. Fossils, genes and the evolution of animal limbs. Nature 388: 639–648.

Simpson, P. 2001. A review of early development of the nervous system in some arthropods: Comparison between insects, crustaceans and myriapods. Ann. Soc. Entomol. France 37(1/2): 71–84.

Snodgrass, R. E. 1938. Evolution of the Annelida, Onychophora and Arthropoda. Smithsonian Misc. Coll. 97: 1–159. [This paper codified the concepts of Mandibulata and Atelocerata.]

Størmer, L. 1977. Arthropod invasion of land during late Silurian and Devonian times. Science 197: 1362–1364.

Strausfeld, N. J. 1998. Crustacean-insect relationships: The use of brain characters to derive phylogeny amongst segmented invertebrates. Brain Behav. Evol. 52: 186–206.

Strausfeld, N. J., E. K. Bushbeck and R. S. Gomez. 1995. The arthropod mushroom body: Its roles, evolutionary enigmas and mistaken identities. *In* O. Breidbach and W. Kutsch (eds.), *The Nervous Systems of Invertebrates: An Evolutionary and Comparative Approach*. Birkhäuser Verlag, Basel, pp. 349–381.

Swain, T. 1978. Plant-animal co-evolution: A synoptic view of the Paleozoic and Mesozoic. *In* J. B. Harborne, *Biochemical Aspects of Plant and Animal Co-evolution*. Academic Press, New York, pp. 3–19.

Wägele, J.-W. and G. Stanjek. 1995. Arthropod phylogeny inferred from partial 12S rRNA revisited: Monophyly of the Tracheata depends on sequence alignment. J. Zool. Syst. Evol. Res. 33: 75–80.

Waggoner, B. M. 1996. Phylogenetic hypotheses of the relationships of arthropods to Precambrian and Cambrian problematic fossil taxa. Syst. Biol. 45: 190–222.

Walossek, D. 1995. The Upper Cambrian *Rehbachiella*, its larval development, morphology and significance for the phylogeny of Branchiopoda and Crustacea. Hydrobiologia 298: 32: 1–13.

Walossek, D. and K. J. Müller. 1990. Upper Cambrian stem-lineage crustaceans and their bearing upon the monophyletic origin of Crustacea and the position of *Agnostus*. Lethaia 23: 409–427.

Walossek, D. and K. J. Müller. 1992. The "Alum Shale Window"–contribution of "Orsten" arthropods to the phylogeny of Crustacea. Acta Zool. 73(5): 305–312.

Walossek, D. and K. J. Müller. 1997. Cambrian "Orsten"-type arthropods and the phylogeny of Crustacea. *In* R. A. Fortey and R. H. Thomas (eds.), *Arthropod Relationships*. Chapman and Hall, New York, pp. 139–153.

Walossek, D., I. Hinz-Schallreuter, J. H. Shergold and K. J. Müller. 1993. Three-dimensional preservation of arthropod integument from the Middle Cambrian of Australia. Lethaia 26: 7–15.

Waterson, C. D. (ed.). 1985. Fossil arthropods as living animals. Trans. R. Soc. Edinburgh Earth Sci. 76: 103–399. [A synopsis of opinions resulting from an international conference held in 1984.]

Whittington, H. B. and D. E. G. Briggs. 1985. The largest Cambrian animal, *Anomalocaris*, Burgess Shale, British Columbia. Phil. Trans. R. Soc. Lond. Ser. B 309: 569–609.

Whittington, P. M., D. Leach and R. Sandeman. 1993. Evolutionary change in neural development within the arthropods: Axonogensis in the embryos of two crustaceans. Development 118: 449–461.

Wills, M. A., D. E. Briggs and R. A. Fortey. 1994. Disparity as an evolutionary index: A comparison of Cambrian and Recent arthropods. Paleobiology 20: 93–130.

Zrzavy, J. and P. Stys. 1997. The basic body plan of arthropods: Insights from evolutionary morphology and developmental biology. J. Evol. Biol. 10: 353–367.

16

Phylum Arthropoda:
The Crustacea

No group of plants or animals on the planet exhibit the range of morphological diversity seen among the extant Crustacea.

Joel W. Martin and George E. Davis, 2001

C rustaceans are one of the most popular invertebrate groups, even among nonbiologists, for they include some of the world's most delectable gourmet fare, such as lobsters, crabs, and shrimps (Figure 16.1). There are more than 67,000 described living species of Crustacea, and probably five or ten times that number waiting to be discovered and named. They exhibit an incredible diversity of form, habit, and size. The smallest known crustaceans are less than 100 μm in length and live on the antennules of copepods. The largest are Japanese spider crabs (*Macrocheira kaempferi*), with leg spans of 4 m and giant Tasmanian crabs (*Pseudocarcinus gigas*) with carapace widths of 46 cm. The heaviest crustaceans are probably American lobsters (*Homarus americanus*), which, before the present era of overfishing, attained weights in excess of 20 kilograms.

Crustaceans are found at all depths in every marine, brackish, and freshwater environment on Earth. A few have become successful on land, the most notable being sowbugs and pillbugs (the terrestrial isopods). Crustaceans are commonly the dominant organisms in aquatic subterranean ecosystems, and new species of these **stygobionts** continue to be discovered as new caves are explored. They also dominate ephemeral pool habitats, where many undescribed species are known to occur.* And, of course, crustaceans are the most widespread, diverse, and abundant animals inhabiting the world's oceans. The biomass of one species, the Antarctic krill (*Euphausia superba*), has been estimated at 500 million tons at any given time, probably surpassing the biomass of any other group of marine

---

*One large study of 58 ephemeral pools in northern California discovered 30 probable undescribed/unnamed crustacean species (King et al. 1996).

(P)

(R)

(T)

(U)

(Q)

(S)

Figure 16.1 Diversity among the Crustacea. (A) The fiddler crab, *Uca princeps* (Brachyura). (B) A porcelain crab, *Petrolisthes armatus* (Anomura). (C) The giant hermit crab, *Petrochirus californiensis* (Anomura). (D) The unusual hermit crab *Pylopagurus varians*, living in a staghorn hydrocoral (Cnidaria, Hydrozoa). (E) The cleaner shrimp, *Lysmata californica* (Caridea). (F) The Hawaiian regal lobster, *Euoplometopus* (Palinura). (G) The unusual rock-boring thalassinid, *Axius vivesi* (Thalassinidea). (H) The pelagic lobsterette, *Pleuroncodes planipes* (Anomura). (I) Acorn barnacles, *Chthamalus anisopoma* (Cirripedia). (J) A remipede, *Speleonectes ondinae* (Remipedia). Note the remarkably homonomous body of this swimming crustacean. (K) A tadpole shrimp (Branchiopoda; Notostraca) from an ephemeral pool. (L) The calanoid copepod *Diaptomus*, a common planktonic genus. (M) *Lepas anatifera* (Cirripedia) hanging from a floating timber. (N) A tantulocarid, *Deoterthron* (Maxillopoda, Tantulocarida), parasitic on other crustaceans. (O) The coconut crab, *Birgus latro* (Anomura), climbing a tree. (P) *Ligia pacifica* (Isopoda), the rock "louse," an inhabitant of the high spray zone on rocky shores. (Q) A clam shrimp (Diplostraca), carrying eggs. (R,S) Two rather strange amphipods: (R) *Cystisoma*, a huge (some exceed 10 cm), transparent, pelagic hyperiid amphipod. (S) *Cyamus scammoni*, a parasitic caprellid amphipod that lives on whales. (T) A cephalocarid. (U) Two female cumaceans; note the eggs in the marsupium of the lower specimen.

BOX 16A *Characteristics of the Subphylum Crustacea*

1. Body composed of a 5-segmented head, or cephalon (plus the acron), and a long postcephalic trunk; trunk divided into two more or less distinct tagmata (e.g., thorax and abdomen) in all but the remipedes and ostracods (Figure 16.2)

2. Cephalon composed of (anterior to posterior): presegmental (indistinguishable) acron, antennular somite, antennal somite, mandibular somite, maxillulary somite, and maxillary somite; one or more anterior thoracomeres may fuse with the head in members of the classes Remipedia, Maxillopoda, and Malacostraca, their appendages forming maxillipeds

3. Cephalic shield or carapace present (highly reduced in anostracans, amphipods, and isopods)

4. Appendages multiarticulate, uniramous or biramous

5. Mandibles usually multiarticulate limbs that function as biting, piercing, or chewing/grinding jaws

6. Gas exchange by aqueous diffusion across specialized branchial surfaces, either gill-like structures or specialized regions of the body surface

7. Excretion by true nephridial structures (e.g., antennal glands, maxillary glands)

8. Both simple ocelli and compound eyes occur in most taxa (not Remipedia), at least at some stage of the life cycle; compound eyes often elevated on stalks

9. Gut with digestive ceca

10. With nauplius larva; development mixed or direct

Metazoa and rivaling that of the world's ants. The range of morphological diversity among crustaceans far exceeds that of even the insects. In fact, because of their taxonomic diversity and numerical abundance, it is often said that crustaceans are the "insects of the sea." We prefer to think of insects as "crustaceans of the land."

Despite the enormous morphological diversity seen among crustaceans (Figures 16.1 to 16.20), they display a suite of fundamental unifying features (Box 16A). In an effort to introduce both the diversity and the unity of this large group of arthropods, we first present a classification and short synopses of the major taxa. We then discuss the biology of the group as a whole, drawing examples from its various members. As you read this chapter, we ask that you keep in mind the general account of the arthropods presented in Chapter 15.

Classification of the Crustacea

Crustaceans have been known to humans since ancient times and have served them as sources of both food and legend. It is somewhat comforting to carcinologists (those who study crustaceans) to note that *Cancer*, one of the two invertebrates represented in the zodiac, is a crab (the other, of course, is *Scorpio*—another arthropod). Our modern view of Crustacea as a taxon can be traced to Lamarck's scheme in the early nineteenth century. He recognized most crustaceans as such, but placed the barnacles and a few others in separate

groups. For many years barnacles were classified with molluscs because of their thick, calcareous outer shell. Crustacean classification as we know it today was more or less established during the second half of the nineteenth century, although internal revisions continue. Martin and Davis (2001) present an excellent overview of crustacean classification, and readers are referred to that publication for a window into the labyrinthine history of this taxon. Our classification, which recognizes five classes, differs in only minor ways from theirs.

SUBPHYLUM CRUSTACEA

CLASS REMIPEDIA: Remipedes. One living order, Nectiopoda (e.g., *Cryptocorynectes, Godzillius, Lasionectes, Pleomothra, Speleonectes*)

CLASS CEPHALOCARIDA: Cephalocarids (e.g., *Chiltoniella, Hampsonellus, Hutchinsoniella, Lightiella, Sandersiella*)

CLASS BRANCHIOPODA: Branchiopods

> **ORDER ANOSTRACA:** Fairy shrimps (e.g., *Artemia, Branchinecta, Branchinella, Streptocephalus*)

> **ORDER NOTOSTRACA:** Tadpole shrimps (*Lepidurus, Triops*)

> **ORDER DIPLOSTRACA:** The "bivalved" branchiopods

>> **SUBORDER LAEVICAUDATA:** Clam shrimps (e.g., *Lynceus*)

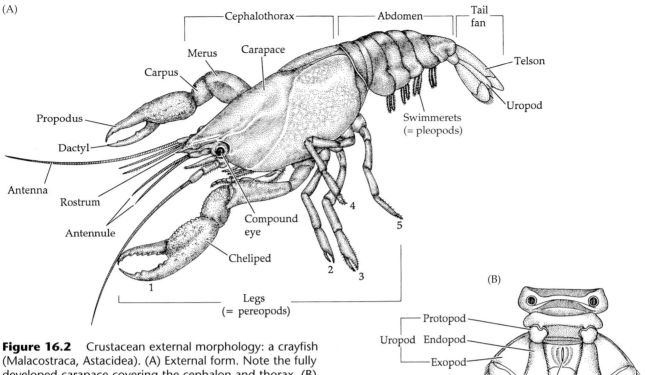

Figure 16.2 Crustacean external morphology: a crayfish (Malacostraca, Astacidea). (A) External form. Note the fully developed carapace covering the cephalon and thorax. (B) The malacostracan tail fan (ventral view). Note the position of the anus on the telson.

SUBORDER SPINICAUDATA: Clam shrimps (e.g., *Cyzicus, Eulimnadia, Imnadia, Metalimnadia*)

SUBORDER CYCLESTHERIDA: Monotypic: *Cyclestheria hislopi*

SUBORDER CLADOCERA: Water fleas (e.g., *Anchistropus, Daphnia, Leptodora, Moina, Polyphemus*)

CLASS MALACOSTRACA

SUBCLASS PHYLLOCARIDA

ORDER LEPTOSTRACA: Leptostracans or nebaliaceans (e.g., *Dahlella, Levinebalia, Nebalia, Nebaliella, Nebaliopsis, Paranebalia*)

SUBCLASS EUMALACOSTRACA

SUPERORDER HOPLOCARIDA

ORDER STOMATOPODA: Mantis shrimps (e.g., *Echinosquilla, Gonodactylus, Hemisquilla, Lysiosquilla, Squilla*)

SUPERORDER SYNCARIDA: Syncarids

ORDER BATHYNELLACEA (e.g., *Bathynella*)

ORDER ANASPIDACEA (e.g., *Anaspides, Psammaspides*)

SUPERORDER EUCARIDA

ORDER EUPHAUSIACEA: Euphausids, or krill (e.g., *Bentheuphausia, Euphausia, Meganyctiphanes, Nyctiphanes*)

ORDER AMPHIONIDACEA: Amphionids. Monotypic: *Amphionides reynaudii*

ORDER DECAPODA: Crabs, shrimps, lobsters, etc.

SUBORDER DENDROBRANCHIATA: Penaeid and sergestid shrimps (e.g., *Lucifer, Penaeus, Sergestes, Sicyonia*)

SUBORDER PLEOCYEMATA

INFRAORDER CARIDEA: Caridean and procaridean shrimps (e.g., *Alpheus, Crangon, Hippolyte, Lysmata, Macrobrachium, Palaemon, Pandalus, Pasiphaea, Procaris*)

INFRAORDER STENOPODIDEA: Stenopodidean shrimps (e.g., *Spongicola, Stenopus*)

INFRAORDER BRACHYURA: "True" crabs (e.g., *Callinectes, Cancer, Cardisoma, Grapsus, Hemigrapsus, Maja, Ocypode, Pachygrapsus, Pinnotheres, Portunus, Uca*)

INFRAORDER ANOMURA: Hermit crabs, galatheid crabs, sand crabs, porcelain crabs, etc. (e.g., *Birgus, Coenobita, Emerita, Galathea, Hippa, Lithodes, Lomis, Paguristes, Pagurus, Petrochirus, Petrolisthes, Pleuroncodes, Pylopagurus*)

INFRAORDER ASTACIDEA: Crayfishes and clawed (chelate) lobsters (e.g., *Astacus, Cambarus, Homarus, Nephrops*)

INFRAORDER PALINURA: Palinurid, spiny, and Spanish (slipper) lobsters (e.g., *Enoplometopus, Evibacus, Ibacus, Jassa, Jasus, Palinurus, Panulirus, Scyllarus, Stereomastis*)

INFRAORDER THALASSINIDEA: Mud and ghost shrimps (e.g., *Axius, Callianassa, Gebiacantha, Thalassina, Upogebia*)

SUPERORDER PERACARIDA

ORDER MYSIDA: Mysids or opossum shrimps (e.g., *Acanthomysis, Hemimysis, Mysis, Neomysis*)

ORDER LOPHOGASTRIDA: Lophogastrids (e.g., *Gnathophausia, Lophogaster*)

ORDER CUMACEA: Cumaceans (e.g., *Campylaspis, Cumopsis, Diastylis, Diastylopsis*)

ORDER TANAIDACEA: Tanaids (e.g., *Apseudes, Heterotanais, Paratanais, Tanais*)

ORDER MICTACEA: Mictaceans (e.g., *Hirsutia, Mictocaris*)

ORDER SPELAEOGRIPHACEA: Spelaeogriphaceans. Three described living species (*Potiicoara brazilienses, Spelaeogriphus lepidops, Mangkurtu mityula*) and two known fossil species (the Carboniferous *Acadiocaris novascotica* and the Upper Jurassic *Liaoningogriphus*)

ORDER THERMOSBAENACEA: Thermosbaenaceans (e.g., *Halosbaena, Limnosbaena, Monodella, Theosbaena, Thermosbaena, Tulumella*)

ORDER ISOPODA: Isopods (sea slaters, rock lice, pillbugs, sowbugs, roly-polies)

SUBORDER ANTHURIDEA (e.g., *Anthura, Colanthura, Cyathura, Mesanthura*)

SUBORDER ASELLOTA (e.g., *Asellus, Eurycope, Jaera, Janira, Microcerberus, Munna*)

SUBORDER CALABOZOIDEA (*Calabozoa*)

SUBORDER EPICARIDEA (e.g., *Bopyrus, Dajus, Hemiarthrus, Ione, Pseudione*)

SUBORDER FLABELLIFERA (e.g., *Aega, Bathynomus, Cirolana, Limnoria, Sphaeroma*)

SUBORDER GNATHIIDEA (e.g., *Gnathia, Paragnathia*)

SUBORDER ONISCIDEA (e.g., *Armadillidium, Ligia, Oniscus, Porcellio, Trichoniscus, Tylos, Venezillo*)

SUBORDER PHREATOICIDEA (e.g., *Mesamphisopus, Phreatoicopis, Phreatoicus*)

SUBORDER VALVIFERA (e.g., *Arcturus, Idotea, Saduria*)

ORDER AMPHIPODA: Amphipods—beach hoppers, sand fleas, scuds, skeleton shrimps, whale lice, etc.

SUBORDER GAMMARIDEA (e.g., *Ampithoe, Anisogammarus, Corophium, Eurythenes, Gammarus, Niphargus, Orchestia, Phoxocephalus, Talitrus*)

SUBORDER HYPERIIDEA (e.g., *Cystisoma, Hyperia, Phronima, Primno, Rhabdosoma, Scina, Streetsia, Vibilia*)

SUBORDER CAPRELLIDEA (e.g., *Caprella, Cyamus, Metacaprella, Phtisica, Syncyamus*)

SUBORDER INGOLFIELLIDEA (e.g., *Ingolfiella, Metaingolfiella*)

CLASS MAXILLOPODA

SUBCLASS THECOSTRACA: Barnacles and their kin

INFRACLASS FACETOTECTA: Monogeneric (*Hansenocaris*): the mysterious "y-larvae," a group of marine nauplii and cyprids for which adults are unknown

INFRACLASS ASCOTHORACIDA: Parasitic thecostracans (e.g., *Ascothorax, Dendrogaster, Laura, Synagoga, Zoanthoecus*)

INFRACLASS CIRRIPEDIA: Cirripedes, the barnacles and their kin

SUPERORDER ACROTHORACICA: Boring "barnacles" (e.g., *Cyptophialus, Trypetesa*)

SUPERORDER RHIZOCEPHALA: Parasitic "barnacles." Two orders, Kentrogonida and Akentrogonida (e.g., *Heterosaccus, Lernaeodiscus, Mycetomorpha, Peltogaster, Sacculina, Sylon*)

SUPERORDER THORACICA: True barnacles. Two orders, Pedunculata (pedunculate or goose barnacles) and Sessilia (sessile or acorn barnacles) (e.g., *Balanus, Chthamalus, Conchoderma, Coronula, Lepas, Pollicipes, Tetraclita, Verruca*)

SUBCLASS TANTULOCARIDA: Deep water, marine parasites (e.g., *Basipodella, Deoterthron, Microdajus*)

SUBCLASS BRANCHIURA: Fish lice, or argulids. A single family (Argulidae) (e.g., *Argulus, Chonopeltis, Dipteropeltis, Dolops*)

SUBCLASS PENTASTOMIDA: Tongueworms. Two orders, numerous families (e.g., *Cephalobaena, Linguatula, Pentastoma, Waddycephalus*)

SUBCLASS MYSTACOCARIDA: Mystacocarids, with a single family (*Derocheilocarididae*), and about a dozen species (e.g., *Ctenocheilocaris, Derocheilocaris*)

SUBCLASS COPEPODA

INFRACLASS PROGYMNOPLEA

ORDER PLATYCOPIOIDA: Platycopioids (e.g., *Antrisocopia, Platycopia*)

INFRACLASS NEOCOPEPODA

ORDER CALANOIDA: Calanoids (e.g., *Bathycalanus, Calanus, Diaptomus, Eucalanus, Euchaeta*)

ORDER CYCLOPOIDA: Cyclopoids (e.g., *Cyclopina, Cyclops, Eucyclops, Lernaea, Mesocyclops, Notodelphys*)

ORDER GELYELLOIDA: Gelyelloids (e.g., *Gelyella*)

ORDER HARPACTICOIDA: Harpacticoids (e.g., *Harpacticus, Longipedia, Peltidium, Porcellidium, Psammus, Sunaristes, Tisbe*)

ORDER MISOPHRIOIDA: Misophriods (e.g., *Boxshallia, Misophria*)

ORDER MONSTRILLOIDA: Monstrilloids (e.g., *Monstrilla, Stilloma*)

ORDER MORMONILLOIDA: Mormonilloids. Monogeneric: *Mormonilla*

ORDER POECILOSTOMATOIDA: Poecilostomatoids (e.g., *Chondracanthus, Erebonaster, Ergasilus, Pseudanthessius*)

ORDER SIPHONOSTOMATOIDA: Siphonostomatoids (e.g., *Clavella, Nemesis, Penella, Pontoeciella, Trebius*)

SUBCLASS OSTRACODA: Ostracods

SUPERORDER MYODOCOPA

ORDER MYODOCOPIDA (e.g., *Cypridina, Euphilomedes, Eusarsiella, Gigantocypris, Skogsbergia, Vargula*)

ORDER HALOCYPRIDA (e.g., *Conchoecia, Polycope*)

SUPERORDER PODOCOPA

ORDER PODOCOPIDA (e.g., *Cypris, Candona, Celtia, Darwinula, Limnocythre*)

ORDER PLATYCOPIDA (e.g., *Cytherella, Sclerocypris*)

ORDER PALAEOCOPIDA (e.g., *Manawa*)

Synopses of Crustacean Taxa

The following descriptions of major crustacean taxa will give you an idea of the range of diversity within the group and the variety of ways in which these successful animals have exploited the basic crustacean bauplan. A diagnosis of each taxon is followed by some general comments.

Subphylum Crustacea

Body composed of a 5-segmented cephalon, or head, and multisegmented postcephalic trunk; trunk divided into thorax and abdomen (except in remipedes and ostracods); segments of cephalon bear first antennae (antennules), second antennae, mandibles, maxillules, and maxillae; one or more anterior thoracomeres may fuse with the head (in some Remipedia, Maxillopoda, and Malacostraca), their appendages forming maxillipeds; cephalic shield or carapace present (secondarily lost in some groups); with antennal glands or maxillary glands (excretory nephridia); both simple ocelli and compound

eyes in most groups, at least at some stage of the life cycle; compound eyes stalked in many groups; with nauplius larval stage (suppressed in some groups), and often a series of additional larval stages.* An estimated 67,829 (living) species.

Class Remipedia

Body of two regions, a cephalon and an elongate homonomous trunk of up to 32 segments, each with a pair of flattened limbs. Cephalon with a pair of sensory preantennular frontal processes; first antennae biramous; trunk limbs laterally directed, biramous, paddle-like, but without large epipods; rami of trunk limbs (exopod and endopod) each of three or more articles; without a carapace, but with cephalic shield covering head; midgut with serially arranged digestive ceca; first trunk segment fused with head and bearing one pair of prehensile maxillipeds; labrum very large, forming a chamber (atrium oris) in which reside the "internalized" mandibles; maxillules function as hypodermic fangs; last trunk segment partly fused dorsally with telson; telson with caudal rami; segmental double ventral nerve cord; eyes absent in living species; male gonopore on trunk limb 14, female on 7; up to 30 mm in length. The above diagnosis is for the 12 known living remipedes (order Nectiopoda); the fossil record is currently based on a single poorly preserved specimen (order Enantiopoda) (Figures 16.1J, 16.3D–F, 16.21D, 16.22F, 16.31E,F).

The discovery of living remipedes, strange vermiform crustaceans first collected from a cavern in the Bahamas, gave the carcinological world a turn (Yager 1981). The combination of features distinguishing these creatures is puzzling, for they possess characteristics that are certainly very primitive (e.g., long, homonomous trunk; double ventral nerve cord; segmental digestive ceca; cephalic shield) as well as some attributes traditionally recognized as advanced (e.g., maxillipeds; nonphyllopodous (though flattened), biramous limbs). They swim about on their backs as a result of metachronal beating of the trunk appendages, a style of locomotion similar to that of anostracans. The remipedes are thus reminiscent of two other primitive classes, the branchiopods and cephalocarids. However, the laterally directed limbs are unlike those of any other crustacean, and the "internalized" mandibles and the poison-injecting hypodermic maxillules are unique. The presence of the preantennular processes is also puzzling, although similar structures are known to occur in a few other crustaceans. Phylogenetic analyses based on morphological data suggest that remipedes may be the most primitive living crustaceans, whereas molecular data remain ambiguous on the subject.

The 12 species of living remipedes discovered thus far are found in caves (usually with connections to the sea) in the Caribbean Basin, Indian Ocean, Canary Islands, and Australia. The water in these caves is often distinctly stratified, with a layer of fresh water overlying

---

*Segments of the **thorax** are called **thoracomeres** (regardless of whether or not any of these segments are fused to the head), whereas appendages of the thorax are called **thoracopods**. The term **pereon** refers to that portion of the thorax *not* fused to the head (when such fusion occurs), and the terms **pereonites** (= **pereomeres**) and **pereopods** are used for the segments and appendages, respectively, of the pereon. Hence, on a crustacean with the first thoracic segment (thoracomere 1) fused to the head, thoracomere 2 is typically called pereonite 1, the first pair of pereopods represents the second pair of thoracopods, and so on. Be assured that we are trying to simplify, not confuse, this issue. Also, we caution you that the homology of the thorax and abdomen among the major crustacean lineages is probably more reverie than reality; the segmental homologies of the thorax and abdomen have not yet been unraveled among the crustacean classes.

(A)

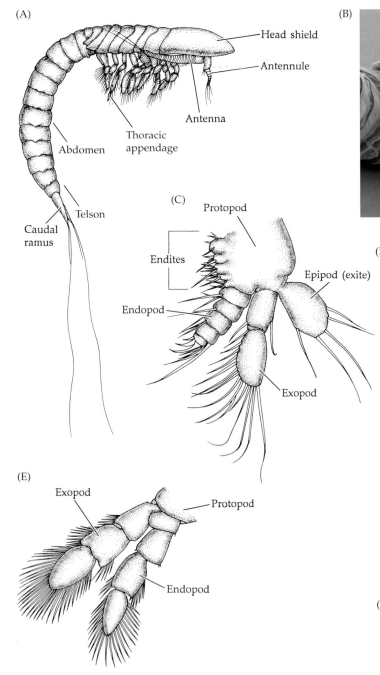

Head shield

Antennule

Antenna

Thoracic appendage

Abdomen

Telson

Caudal ramus

(C)

Protopod

Endites

Epipod (exite)

Endopod

Exopod

(B)

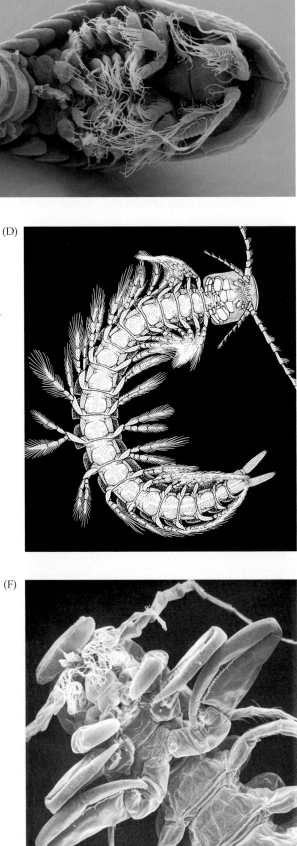

(D)

(E)

Exopod

Protopod

Endopod

(F)

Figure 16.3 Anatomy in the classes Cephalocarida and Remipedia. (A) The cephalocarid *Hutchinsoniella* (lateral view). (B) SEM of head and thorax of a cephalocarid. (C) First trunk limb of the cephalocarid *Lightiella*. (D) The remipede *Speleonectes* (ventral view). (E) Tenth trunk limb of the remipede *Lasionectes*. (F) The anterior end of a remipede (ventral view). In both the cephalocarids and the remipedes, the trunk is a long, homonomous series of somites with biramous swimming appendages. In cephalocarids the first trunk appendages are like all the others, which bear large swimming epipods (exites). In remipedes the first trunk somite is fused to the head, and its appendages are maxillipeds.

the denser salt water in which the remipedes swim. Remipede larvae have not been found, nor are they expected to be (most cave crustaceans have direct development). Juveniles apparently have fewer trunk segments than adults. A key to the living species was given by Yager and Humphries (1996).

Class Cephalocarida

Head followed by an 8-segmented thorax, 11-segmented abdomen, and telson with caudal rami; common gonopore on protopods of sixth thoracopods; carapace absent but head covered by cephalic shield; thoracopods 1–7 biramous and phyllopodous, with large flattened exopods and epipods (exites) and stenopodous endopods; thoracopods 8 reduced or absent; maxillae resemble thoracopods; no maxillipeds; eyes absent; nauplii with antennal glands, adults with maxillary glands and (vestigial) antennal glands (Figures 16.1T, 16.3A–C, 16.21A).

Cephalocarids are tiny, elongate crustaceans ranging in length from 2 to 4 mm. There are 10 species, in 5 genera. All are benthic marine detritus feeders. Most are associated with sediments covered by a layer of flocculent organic detritus, although some have been found in clean sands. They occur from the intertidal zone to depths of over 1,500 m. Researchers agree that cephalocarids are very primitive crustaceans, largely because of their relatively homonomous body form, undifferentiated maxillae, and generalized appendage structure. Some even place them at the base of the (living) crustacean tree.

Class Branchiopoda

Number of segments and appendages on thorax and abdomen vary, the latter usually lacking appendages; carapace present or absent; telson usually with caudal rami; body appendages generally phyllopodous; maxillules and maxillae reduced or absent; no maxillipeds (Figures 16.1K,Q, 16.4, 16.21B, 16.31C, 16.35B).

The branchiopods are difficult to describe in a general way. Most are small freshwater forms with minimal body tagmosis and leaflike legs. Most are short-lived, and those inhabiting ephemeral waters complete their life cycle in just a few weeks. Because of their short life cycle and prediliction for ephemeral waters, many groups produce drought-resistant eggs or zygotes, called cysts, that can survive years or decades, until the next adequate rains appear. As diverse as the branchiopods might appear, both morphological and molecular analyses indicate that they comprise a monophyletic group. Several taxonomic names have been proposed to cluster alleged sister-groups within the Branchiopoda (e.g., Sarsostraca, Calmanostraca, Phyllopoda), but these have been erected more on the basis of intuition than analysis, and we choose not to embed them in the textbook literature until the matter of branchiopod phylogeny is better resolved. Interested readers are referred to Martin and Davis (2001) for a succinct review of these proposed ideas. 900 species have been described.

Order Anostraca. Postcephalic trunk divisible into appendage-bearing thorax of 11 segments (17 or 19 in members of the family Polyartemiidae) and abdomen of 8 segments plus telson with caudal rami; gonopores on genital region of abdomen; trunk limbs biramous and phyllopodous; small cephalic shield present; paired, large, stalked compound eyes and a single median simple (naupliar) eye.

The anostracans are commonly called fairy shrimps (including *Artemia*, the brine shrimp). They differ from other branchiopods in lacking a carapace. There are 270 living species in the order, most of which are less than 1 cm in length, although a few giants attain lengths of 10 cm. These cosmopolitan animals inhabit ephemeral ponds, hypersaline lakes, and marine lagoons. In many areas they are an important food resource for water birds. Anostracans are sometimes united with the extinct order Lipostraca in the subclass Sarsostraca. Their fossil record dates back to the Silurian.

Anostracans swim ventral side up by metachronal beating of the trunk appendages. Many use these limb movements for suspension feeding. Some other species scrape organic material from submerged surfaces, and at least one species (*Branchinecta gigas*) is specialized as a predator on other fairy shrimps.

Order Notostraca. Thorax of 11 segments, each with a pair of phyllopodous appendages; abdomen of "rings," each formed of more than one true segment; each anterior ring with several pairs of appendages, posterior rings lack appendages; telson with long caudal rami; gonopores on last thoracomere; broad, shieldlike carapace fused only with head, but extending to loosely cover thorax and part of abdomen; paired, sessile compound eyes and a single simple eye near anterior midline on carapace.

Notostracans are often called tadpole shrimps. There are about 12 living species placed in a single family, Triopsidae, most of which are 2–10 cm long. The common name derives from the general body shape: The broad carapace and narrow "trunk" give the animals a superficially tadpole-like appearance.

Notostracans inhabit inland waters of all salinities, but none occur in the ocean. Of the two known genera, *Triops* lives only in temporary waters, and its eggs are capable of surviving extended dry periods. Most species of *Lepidurus* live in temporary ponds, but at least one species (*L. arcticus*) inhabits permanent ponds and lakes. However, all species are short-lived, and most complete their lifecycle in just 30 to 40 days. *Triops* is of some economic importance in that large populations often occur in rice paddies and destroy the crop by burrowing into the mud and dislodging young plants. Tadpole shrimps mostly crawl, but they are also capable of swimming for short periods by beating the thoracic limbs. They feed on organic material stirred up from the sediments, although some scavenge or prey on other animals, includ-

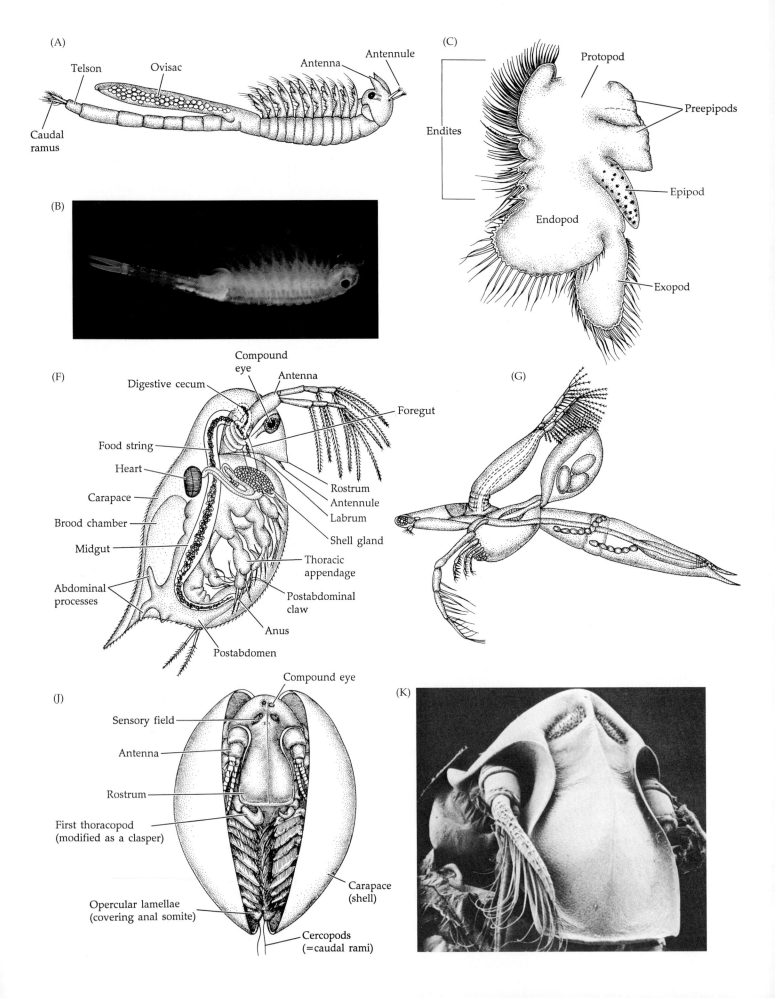

(A)

Telson
Ovisac
Antenna
Antennule
Caudal ramus

(B)

(C)

Protopod
Preepipods
Endites
Epipod
Endopod
Exopod

(F)

Compound eye
Digestive cecum
Antenna
Foregut
Food string
Heart
Rostrum
Carapace
Antennule
Brood chamber
Labrum
Midgut
Shell gland
Thoracic appendage
Abdominal processes
Postabdominal claw
Anus
Postabdomen

(G)

(J)

Compound eye
Sensory field
Antenna
Rostrum
First thoracopod (modified as a clasper)
Opercular lamellae (covering anal somite)
Carapace (shell)
Cercopods (=caudal rami)

(K)

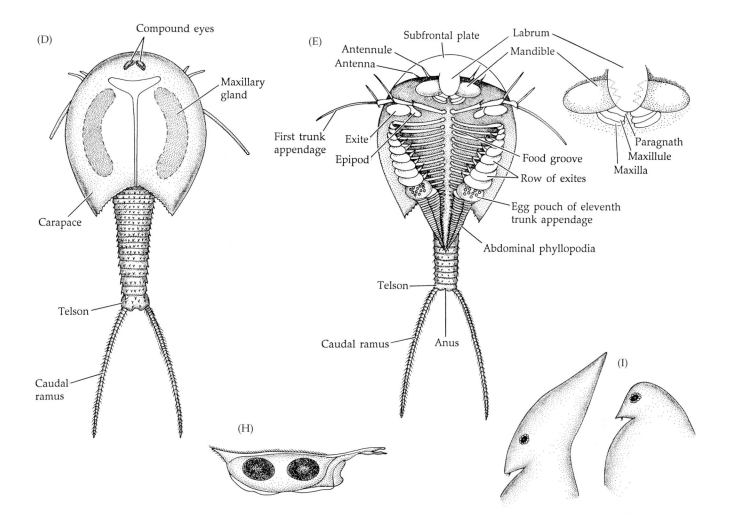

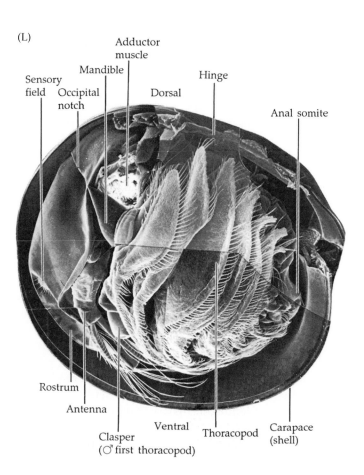

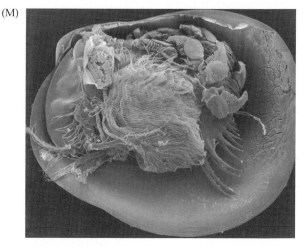

Figure 16.4 Anatomy and diversity in the class
Branchiopoda. (A) An anostracan (*Branchinecta*) in swimming posture. (B) The anostracan *Branchipus schaefferi*
swimming. (C) Trunk limb of an anostracan (*Linderiella*).
(D,E) The notostracan *Triops*: (D) dorsal and (E) ventral
views. (F,G) Two cladocerans: (F) *Daphnia* and (G)
Leptodora. (H) The shed carapace, or ephippium, of
Daphnia, with the embryos enclosed. (I) Two extreme
stages in the seasonal change in head form of *Daphnia*
(cyclomorphosis). (J–L) The clam shrimp (Diplostraca)
Lynceus: (J) The valves are partially open (ventral view).
(K) The head (ventral view). (L) Whole animal. One valve
has been removed. (M) The clam shrimp (Cyclestherida)
Cyclestheria, SEM photo, one valve removed (the specimen is slightly distorted; in life the shell is round).

ing molluscs, other crustaceans, frog eggs, and even frog tadpoles and small fishes. Some species of notostracans are exclusively dioecious, but others may include hermaphroditic populations (often those populations living at high latitudes). Some European populations appear to reproduce solely by parthenogenesis.

Order Diplostraca. Clam shrimps and cladocerans comprise four suborders of closely related branchiopods known as the Diplostraca. They share the feature of a uniquely developed, large, "bivalved" carapace that covers all or most of the body.

Clam shrimps. In the clam shrimps (suborders Laevicaudata, Spinicaudata, Cyclestherida—formerly lumped together as the Conchostraca) the body is divided into cephalon and trunk, the latter with 10–32 segments, all with appendages, and with no regionalization into thorax and abdomen; trunk limbs phyllopodous, decreasing in size posteriorly; males with trunk limbs 1, or 1–2, modified for grasping females during mating; trunk typically terminates in spinous anal somite or telson, usually with robust caudal rami (cercopods); gonopores on eleventh trunk segment; bivalved carapace completely encloses body; valves folded (Spinicaudata, Cyclestherida) or hinged (Laevicaudata) dorsally; usually with a pair of sessile compound eyes and a single, median, simple eye.

The common name derives from the clamlike appearance of the valves, which usually bear concentric growth lines reminiscent of bivalved molluscs. The 221 species of clam shrimps live primarily in ephemeral freshwater habitats worldwide, except in Antarctica. *Cyclestheria hislopi*, the only member of the Cyclestherida, also inhabits permanent freshwater habitats throughout the world's tropics, and is one of the most widespread nondomestic animals on Earth. *Cyclestheria* is also the only clam shrimp with direct development, the larval and juvenile stages being passed within the brood chamber. Recent evidence suggests *Cyclestheria* may actually be most closely related to the Cladocera. Most diplostracans are benthic, but many swim during reproductive periods. Some are direct suspension feeders, whereas others stir up detritus from the substratum and feed on suspended particles, and others scrape pieces of food from the sediment.

Cladocerans. In cladocerans (suborder Cladocera) the carapace is never hinged (only folded dorsally, like a taco) and never covers the entire body, and appendages do not occur on all the trunk somites. The body segmentation is generally reduced. The thorax and abdomen are fused as a "trunk" bearing 4–6 pairs of appendages anteriorly and terminating in a flexed "postabdomen" with clawlike caudal rami. Trunk appendages are usually phyllopodous. The carapace usually encloses the entire trunk, but not the cephalon, serving as a brood chamber (and greatly reduced to this function) in some species; a single median compound eye is always present.

The cladocerans, or "water fleas," include about 400 species of predominantly freshwater crustaceans, although several American marine genera and species are known (e.g., *Evadne*, *Podon*). Most cladocerans are 0.5–3 mm long, but *Leptodora kindtii* reaches 18 mm in length. Except for the cephalon and large natatory antennae, the body is enclosed by a folded carapace, which is fused with at least some of the trunk region. The carapace is greatly reduced in members of the families Polyphemidae and Leptodoridae, in which it forms a brood chamber.

Cladocerans are distributed worldwide in nearly all inland waters. Most are benthic crawlers or burrowers; others are planktonic and swim by means of their large antennae. One genus (*Scapholeberis*) is typically found in the surface film of ponds, and another (*Anchistropus*) is ectoparasitic on *Hydra*. Most of the benthic forms feed by scraping organic material from sediment particles or other objects; the planktonic species are suspension feeders. Some (e.g., *Leptodora*, *Bythotrephes*) are predators on other cladocerans.

In sexual reproduction, fertilization generally occurs in a brood chamber between the dorsal surface of the trunk and the inside of the carapace. Most species have direct development. In the family Daphnidae the developing embryos are retained by a portion of the shed carapace, which functions as an egg case called an **ephippium** (Figure 16.4H), whereas in the Chydoridae the ephippium remains attached to the entire shed carapace. *Leptodora* have free-living larvae (metanauplii hatch from the shed resting eggs).

Cladoceran life histories are often compared with those of animals such as rotifers and aphids. Dwarf males occur in many species in all three groups, and parthenogenesis is common. Members of two cladoceran families that undergo parthenogenesis (Moinidae and Polyphemidae) produce eggs with very little yolk. In these groups the floor of the brood chamber is lined with glandular tissue that secretes a fluid rich in nutrients, which are absorbed by the developing embryos. Periods of overcrowding, adverse temperatures, or food scarcity can induce parthenogenetic females to produce male offspring. Occasional periods of sexual reproduction have been shown to occur in most parthenogenetic species. Many planktonic cladocerans undergo seasonal changes in body form through succeeding generations of parthenogenetically produced individuals, a phenomenon known as **cyclomorphism** (Figure 16.4I).

Class Malacostraca

Body of 19–20 segments, including 5-segmented cephalon, 8-segmented thorax, and 6-segmented pleon (7-segmented in leptostracans), plus telson; with or without caudal rami; carapace covering part or all of thorax, or reduced, or absent; 0–3 pairs of maxillipeds; thora-

Figure 16.5 Anatomy in leptostracans (class Malacostraca, subclass Phyllocarida). (A) General anatomy of *Nebalia*. (B) Phyllopodous swimming limb of *Nebalia*. (C) SEM of *Nebalia*. (D) Anterior end of an ovigerous *Nebalia*.

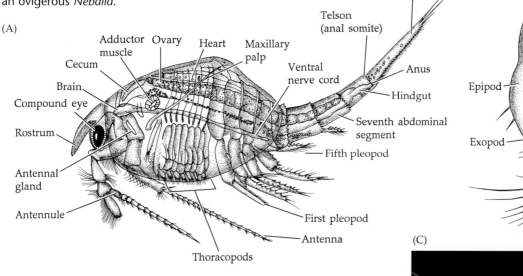

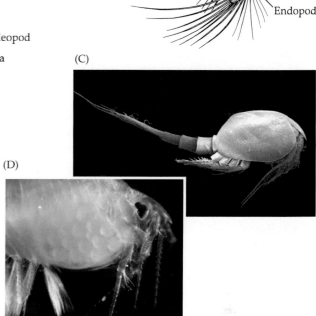

copods primitively biramous, uniramous in some groups, phyllopodous only in members of the subclass Phyllocarida; antennules and antennae usually biramous; abdomen (pleon) usually with 5 pairs of biramous pleopods and 1 pair of biramous uropods; eyes usually present, compound, stalked or sessile; mostly dioecious; female gonopores on sixth, and male pores on eighth thoracomeres. When uropods are present, they are often broad and flat, lying alongside the broad telson to form a **tail fan**.

Most classification schemes divide the more than 40,200 species of malacostracans into two subclasses, Phyllocarida and Eumalacostraca. The phyllocarids are typically viewed as representing the primitive malacostracan condition (5-8-7 body segments plus telson; Figure 16.5). The basic eumalacostracan bauplan, characterized by the 5-8-6 (plus telson) arrangement of body segments, was recognized in the early 1900s by W. T. Calman, who termed the defining features of the Eumalacostraca "caridoid facies" (Figure 16.6). Much work has been done since Calman's day, but the basic elements of his caridoid facies are still present in all members of the subclass Eumalacostraca.*

Subclass Phyllocarida

Order Leptostraca. With the typical malacostracan characteristics, except notable for presence of seven free pleomeres (plus telson) rather than six, generally taken to represent the primitive condition for the class. Also, with phyllopodous thoracopods (all similar to one another); no maxillipeds; large carapace covering thorax and compressed laterally so as to from an unhinged bivalved "shell," with an adductor muscle; cephalon with a movable, articulated rostrum; pleopods 1–4 similar and biramous, 5–6 uniramous; no uropods; paired stalked compound eyes; antennules biramous; antennae uniramous; adults with both antennal and maxillary glands (Figures 16.5, 16.21C).

The subclass Phyllocarida includes about 36 species in 10 genera and three families. Most are 5–15 mm long, but *Nebalioposis typica* is a giant at nearly 5 cm in length. The leptostracan body form is distinctive, with its loose bivalved carapace covering the thorax, a protruding rostrum, and an elongate abdomen. All leptostracans are marine, and most are epibenthic from the intertidal zone to a depth of 400 m; *Nebaliopsis typica* is bathypelagic. Most species seem to occur in low-oxygen environments. One species, *Dahlella caldariensis*, is associated

---

*Because hoplocarids possess several striking apomorphies, some workers recommend removing them from the Eumalacostraca and elevating them to a third subclass within the Malacostraca. However, based on their shared similarities, we retain them in the Eumalacostraca.

Figure 16.6 The basic eumalacostran bauplan and the "caridoid facies." Note the thick (muscled) abdomen and the tail fan, which work in combination to produce a powerful tail flip escape reaction.

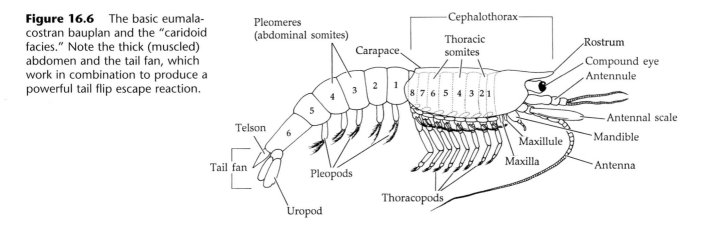

with the hydrothermal vents of the Galapagos and the East Pacific Rise. *Speonebalia cannoni* is known only from marine caves.

Most leptostracans suspension feed by stirring up bottom sediments. They are also capable of grasping relatively large bits of food directly with the mandibles. Some are carnivorous scavengers, and some are known to aggregate in areas on the sea floor where large amounts of detritus accumulate. In many species the antennae or antennules of males are modified to hold females during copulation.

Subclass Eumalacostraca

Head, thorax, and abdomen of 5-8-6 somites respectively (plus telson); with 0, 1, 2, or 3 thoracomeres fused with head, their respective appendages usually modified as maxillipeds; antennules and antennae primitively biramous; antennae often with scalelike exopod; most with well developed carapace, secondarily reduced in syncarids and some peracarids; gills primitively as thoracic epipods; tail fan composed of telson plus paired uropods; abdomen long and muscular. Four superorders: Hoplocarida, Syncarida, Eucarida, and Peracarida.

Superorder Hoplocarida, Order Stomatopoda. Carapace covering portion of head and fused with thoracomeres 1–4; head with movable, articulated rostrum; thoracopods 1–5 uniramous and subchelate, second pair massive and raptorial (all five are sometimes called "maxillipeds" or gnathopods because they are involved in feeding); thoracopods 6–8 biramous, ambulatory; pleopods biramous, with dendrobranchiate-like gills on exopods; antennules triramous; antennae biramous, with large, paired, stalked compound eyes (Figures 16.7A–C, 16.27D, 16.33J).

All 350 living hoplocarids are placed in the order Stomatopoda, known as "mantis shrimps." They are relatively large crustaceans, ranging in length from 2 to 30 cm. Compared with that of most malacostracans, the muscle-filled abdomen is notably thick and robust.

Most stomatopods are found in shallow tropical or subtropical marine environments. Nearly all of them live in burrows excavated in soft sediments or in cracks and crevices, among rubble, or in other protected spots. They are raptorial carnivores, preying on fishes, molluscs, cnidarians, and other crustaceans. The large, distinctive subchelae of the second thoracopods act either as crushers or as spears (Figure 16.7C).

Stomatopods crawl about using the posterior thoracopods and the flaplike pleopods. They also can swim by metachronal beating of the pleopods (the swimmerets). For these relatively large animals, living in narrow burrows requires a high degree of maneuverability. The short carapace and the flexible, muscular abdomen allow these animals to twist double and turn around within their tunnels or in other cramped quarters. This ability facilitates an escape reaction whereby a mantis shrimp darts into its burrow rapidly head first, then turns around to face the entrance.

Stomatopods are one of only two groups of malacostracans that possess pleopodal gills. Only the isopods share this trait, but the pleopods are quite different in the two groups. The tubular, thin, highly branched gills of stomatopods provide a large surface area for gas exchange in these active animals.

Superorder Syncarida. Without maxillipeds (Bathynellacea) or with one pair of maxillipeds (Anaspidacea); no carapace; pleon bears telson with or without furcal lobes; at least some thoracopods biramous, eighth often reduced; pleopods variable; compound eyes present (stalked or sessile) or absent (Figures 16.7D,E).

There are about 200 described species of syncarids in two orders, Anaspidacea and Bathynellacea.* To many workers, the syncarids represent a key group in eumalacostracan evolution, and they may represent an ancient relictual taxon that now inhabits refugial habitats. Through studies of the fossil record and extant members of the order Anaspidacea (e.g., *Anaspides*), it

---

*Until recently a third syncarid order was recognized, the Stygocaridacea, endemic to the Southern Hemisphere. Most workers now agree that the stygocarids should be reduced to the rank of family within the order Anaspidacea.

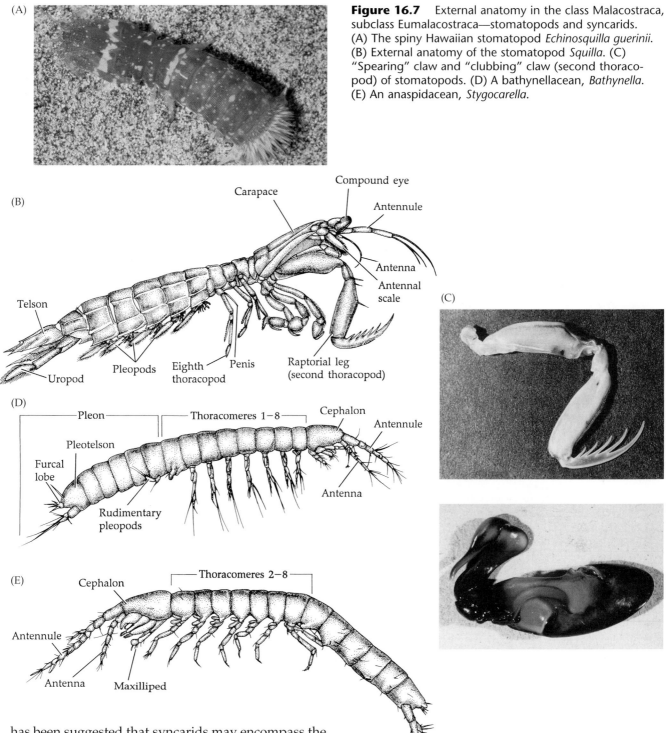

Figure 16.7 External anatomy in the class Malacostraca, subclass Eumalacostraca—stomatopods and syncarids. (A) The spiny Hawaiian stomatopod *Echinosquilla guerinii.* (B) External anatomy of the stomatopod *Squilla.* (C) "Spearing" claw and "clubbing" claw (second thoracopod) of stomatopods. (D) A bathynellacean, *Bathynella.* (E) An anaspidacean, *Stygocarella.*

has been suggested that syncarids may encompass the most primitive living eumalacostracan bauplan. Among the living syncarids, bathynellaceans occur worldwide in interstitial or groundwater habitats, whereas the anaspididaceans are strictly Gondwanan in distribution. Many Anaspidacea are endemic to Tasmania, where they inhabit freshwater environments, such as open lake surfaces, streams, ponds, and crayfish burrows. No syncarids are marine. These reclusive eumalacostracans show various degrees of what some have regarded as paedomorphism, including small size (Anaspididae includes members to 5 cm,

whereas most others are less than 1 cm long), eyelessness, and reduction or loss of pleopods and some posterior pereopods. Bathynellaceans are small (1–3 mm long), possess 6 or 7 pairs of long, thin swimming legs, and have a pleotelson formed by the fusion of the telson to the last pleonite.

Syncarids either crawl or swim. Little is known about the biology of most species, although some are considered omnivorous. Unlike most other crustaceans, which carry the eggs and developing early embryos, syncarids lay their eggs or shed them into the water following copulation.

Superorder Eucarida.

Telson without caudal rami; 0, 1, or 3 pairs of maxillipeds; carapace present, covering and fused dorsally with head and entire thorax; usually with stalked compound eyes; gills thoracic. Although members of this group are highly diverse, they are united by the presence of a complete carapace that is fused with all thoracic segments, forming a characteristic cephalothorax. Most species (several thousand) belong to the order Decapoda. The other two orders are the Euphausiacea (krill), and the monotypic Amphionidacea.

Order Euphausiacea.

Euphausids are distinguished among the eucarids by the absence of maxillipeds, the exposure of the thoracic gills external to the carapace, and the possession of biramous pereopods (the last 1 or 2 pairs sometimes being reduced). They are shrimplike in appearance. Adults have antennal glands.

Most of them have photophores on the eyestalks, the bases of the second and seventh thoracopods, and between the first 4 pairs of abdominal limbs.

The 90 or so species of euphausids are all pelagic and range in length from 4 to 15 cm. The pleopods function as swimmerets. Euphausids are known from all oceanic environments to depths of 5,000 m. Most species are distinctly gregarious, and where they occur in huge schools (krill) they provide a major source of food for larger nektonic animals (baleen whales, squids, fishes) and even some marine birds. Krill densities, particularly for *Euphausia superba*, often exceed 1,000 animals/m³ (614 g wet weight/m³).* Generally, euphausids are suspension feeders, although predation and detritivory also occur (Figures 16.8A,B, 16.21E).

Order Amphionidacea.

The single known species of the order Amphionidacea, *Amphionides reynaudii*, possesses an enlarged cephalothorax covered by a thin, almost membranous carapace that extends to enclose the thoracopods. The thoracopods are biramous with

---

*Where krill densities exceed about 100 grams per cubic meter, they now are often fished commercially.

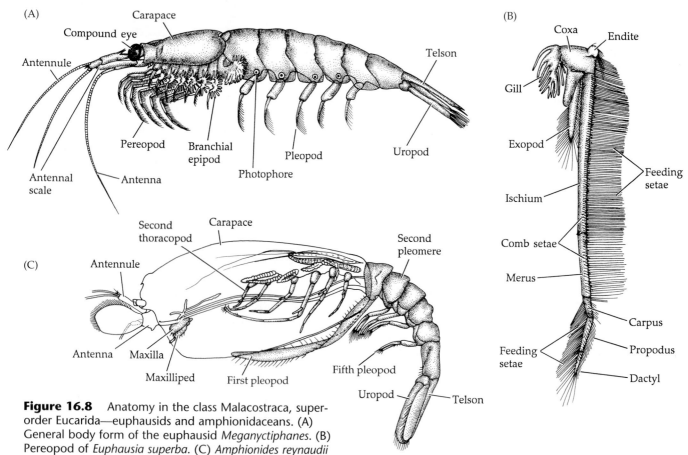

Figure 16.8 Anatomy in the class Malacostraca, superorder Eucarida—euphausids and amphionidaceans. (A) General body form of the euphausid *Meganyctiphanes*. (B) Pereopod of *Euphausia superba*. (C) *Amphionides reynaudii* (female), the only living species of the Amphionidacea.

short exopods. The first pair is modified as maxillipeds, and the last pair is absent in females. Some of the mouthparts are highly reduced in females. The pleopods are biramous and natatory, except that the first pair in females is uniramous and greatly enlarged, perhaps functioning to form a brood pouch extending under the thorax. Females have a reduced gut and apparently do not feed. *Amphionides* is a worldwide member of marine oceanic plankton and occurs to a depth of 1,700 m (Figure 16.8C).

Order Decapoda. The decapods are among the most familiar eumalacostracans. They possess a well developed carapace enclosing a branchial chamber, but they differ from other eucarid orders in always possessing 3 pairs of maxillipeds, leaving 5 pairs of functional uniramous or weakly biramous pereopods (hence the name, Decapoda); one (or more) pairs of anterior pereopods are usually clawed (chelate). Adults have antennal glands. Rearrangement of the subtaxa within this order is a popular carcinological pastime (see Martin and Davis 2001 for an entry into the vast literature on decapod classification). In vernacular terms, nearly every decapod may be recognized as some sort of shrimp, crab, lobster, or crayfish.

We do not want to belabor the issue of decapod gill nomenclature. However, the gills play a prominent role in the taxonomy of this group; thus, we provide brief descriptions of the basic types. All decapod gills arise as thoracic coxal exites (epipods), but their final placement varies. Those that remain attached to the coxae are **podobranchs** (= "foot gills"), but others eventually become associated with the articular membrane between the coxae and body and are thus called **arthrobranchs** (= "joint gills"). Some actually end up on the lateral body wall, or surface of the thoracic pleura, as **pleurobranchs** (= "side gills"). The sequence by which some of these gills arise ontogenetically varies. For example, in the Dendrobranchiata and the Stenopodidea, arthrobranchs appear before pleurobranchs, whereas in members of the Caridea the reverse is true. In most of the other decapods the arthrobranchs and pleurobranchs tend to appear simultaneously. These developmental differences may be minor heterochronic dissimilarities and of less phylogenetic importance than actual gill anatomy.

Among the decapods, the gills can also be one of three basic structural types, described as **dendrobranchiate**, **trichobranchiate**, and **phyllobranchiate** (Figure 16.28B–D). All three of these gill types include a main axis carrying afferent and efferent blood vessels, but they differ markedly in the nature of the side filaments or branches. Dendrobranchiate gills bear two principal branches off the main axis, each of which is divided into multiple secondary branches. Trichobranchiate gills bear a series of radiating unbranched tubular filaments. Phyllobranchiate gills are characterized by a double series of platelike or leaflike branches

from the axis. The occurrences of these gill types among various taxa are presented below. Close inspection of the proximal parts of the pereopods usually reveals another decapod feature: In most forms, the basis and ischium are fused (as a basi-ischium), with the point of fusion often indicated by a suture line.

The 14,000 or so living species of decapods comprise a highly diverse group. They occur in all aquatic environments at all depths, and a few spend most of their lives on land. Many are pelagic, but others have adopted benthic sedentary, errant, or burrowing lifestyles. Their feeding strategies include suspension feeding, predation, herbivory, scavenging, and more. Most workers recognize two suborders: Dendrobranchiata and Pleocyemata.

Suborder Dendrobranchiata. This group includes about 450 species of decapods, most of which are penaeid and sergestid shrimps. As the name indicates, these decapods possess dendrobranchiate gills (Figure 16.28B), a unique synapomorphy of the taxon. One genus, *Lucifer*, has secondarily lost the gills completely. The dendrobranchiate shrimps are further characterized by chelae on the first three pereopods, copulatory organs modified from the first pair of pleopods in males, and ventral expansions of the abdominal tergites (**pleural lobes**). Generally, none of the chelipeds is greatly enlarged. In addition, females of this group do not brood their eggs. Fertilization is external, and the embryos hatch as nauplius larvae (see the section on development below). Many of these animals are quite large, over 30 cm long. The sergestids are pelagic and all marine, whereas the penaeids are pelagic or benthic, and some occur in brackish water. Some dendrobranchiates (e.g., *Penaeus, Sergestes, Acetes*) are of major commercial importance in the world's shrimp fisheries, most of which are now being exploited beyond sustainable levels (Figures 16.9A, 16.33F).

Suborder Pleocyemata. All of the remaining decapods belong to the suborder Pleocyemata. Members of this taxon never possess dendrobranchiate gills. The embryos are brooded on the female's pleopods and hatch at some stage later than the nauplius larva. Included in this suborder are several kinds of shrimps and the crabs, crayfish, lobsters, and a host of less familiar forms. Most current workers recognize seven infraorders within the Pleocyemata, as we have done below, but a number of other schemes have been proposed and persist in the literature. One older approach divided decapods into two large groups, called the Natantia and Reptantia—the swimming and walking decapods, respectively. Although these terms have largely been abandoned as formal taxa, they still serve a useful descriptive purpose (much as the adjectives *errant* and *sedentary* do for polychaete

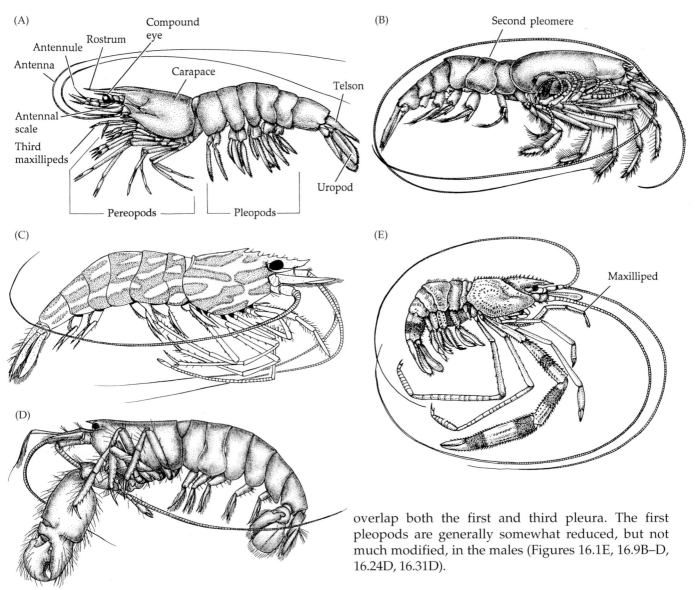

(A) Antenna, Antennule, Rostrum, Compound eye, Antennal scale, Third maxillipeds, Carapace, Telson, Uropod, Pereopods, Pleopods

(B) Second pleomere

(C)

(D)

(E) Maxilliped

Figure 16.9 External anatomy and diversity in shrimps (Decapoda). (A) A penaeid shrimp (Dendrobranchiata), *Penaeus setiferus*. (B) A procarid shrimp (Pleocyemata, Caridea), *Procaris ascensionis*. (C) A hippolytid shrimp (Caridea, Hippolytidae), *Lysmata californica*. (D) An alpheid, or snapping shrimp (Caridea, Alpheidae), *Alpheus*. (E) A stenopodid shrimp (Stenopodidea), *Stenopus*.

worms), and one continues to see references to natant decapods and reptant decapods.

Infraorder Caridea. The nearly 2,500 living species in this infraorder are generally referred to as the caridean shrimps. These swimming decapods have phyllobranchiate gills. The first 1 or 2 pairs of pereopods are chelate and variably enlarged (except in the unique genus *Procaris*, which lacks chelation of any limbs). The second abdominal pleura are distinctly enlarged to overlap both the first and third pleura. The first pleopods are generally somewhat reduced, but not much modified, in the males (Figures 16.1E, 16.9B–D, 16.24D, 16.31D).

Infraorder Stenopodidea. The two dozen or so species in this infraorder belong to two families, Stenopodidae and Spongicolidae. The first 3 pairs of pereopods are chelate, and the third pair is significantly larger than the others. The gills are trichobranchiate. The first pleopods are uniramous in males and females, but are not strikingly modified. The second abdominal pleura are not expanded as they are in carideans (Figure 16.9E, 16.31B).

These colorful shrimps are usually only a few centimeters long (2–7 cm). Most species are tropical and associated with benthic environments, especially with coral reefs. Many are commensal, and the group includes the cleaner shrimps (e.g., *Stenopus*) of tropical reefs, which are known to remove parasites from local fishes. Stenopodids often occur as male–female couples. Perhaps the most noted example of this bonding is associated with the glass sponge shrimp, *Spongicola venusta*: A young male and female shrimp enter the atrium of a host sponge, eventually growing too large to escape and thus spending the rest of their days together.

Infraorder Brachyura. These are the so-called "true crabs." The abdomen is symmetrical but highly reduced and flexed beneath the thorax, and uropods are usually absent. The body, hidden beneath a well developed carapace, is distinctly flattened dorsoventrally and often expanded laterally. The gills are phyllobranchiate. The first pereopods are chelate and usually enlarged. Pereopods 2 to 5 are typically simple, stenopodous walking legs. The eyes are positioned lateral to the antennae. Males lack pleopods 3 to 5. The larval carapace is spherical and bears a ventrally directed rostral spine (or no spine) (Figures 16.1A, 16.10, 16.27I, 16.28F,G, 16.29C, 16.32, 16.33G,H).

Brachyuran crabs are mostly marine, but freshwater, semi-terrestrial, and moist terrestrial species occur in the tropics. The land crabs (certain species in the families Gecarcinidae, Ocypodidae, Grapsidae, etc.) are still dependent on the ocean for breeding and larval development. Freshwater crabs (classified into about a dozen families) have direct development, incubate their embryos, and are independent of sea water. Some freshwater crabs are intermediate hosts of *Paragonimus*, a cosmotropical parasitic human lung fluke, and others are obligate phoretic hosts of larval black flies (*Simulium*), the vector for *Onchocerca volvulus* (the causative agent of river blindness). There are 10,500 described species.

Infraorder Anomura. This group includes hermit crabs, galatheid crabs, king crabs, porcelain crabs, mole crabs, and sand crabs. The abdomen may be soft and asymmetrically twisted (as in hermit crabs) or symmetrical, short, and flexed beneath the thorax (as in porcelain crabs and others). Those with twisted abdomens typically inhabit gastropod shells or other empty "houses" not of their own making. Carapace shape and gill structure vary. The first pereopods are chelate; the third pereopods are never chelate. The second, fourth, and fifth pairs are usually simple, but occasionally they are chelate or subchelate. The fifth pereopods (and sometimes the fourth) are generally much reduced and do not function as walking limbs; the fifth pereopods function as gill cleaners and often are not visible externally. The pleopods are reduced or absent. The eyes are positioned medial to the antennae. The nauplius larva is longer than broad, with the rostral spine directed anteriorly. Most anomurans are marine, but a few freshwater and semi-terrestrial species are known (Figures 16.1B, C,D,H,O, 16.11C–H, 16.24A–C, 16.31A, 16.33I).

Infraorder Astacidea. The crayfish and clawed lobsters are among the most familiar of all decapods (Figure 16.2). As in most other decapods, the dorsoventrally flattened abdomen terminates in a strong tail fan. The gills are trichobranchiate. The first 3 pairs of pereopods are always chelate, and the first pair is greatly enlarged. Most crayfish live in fresh water, but a few species live in damp soil, where they may excavate extensive and complex burrow systems. *Homarus americanus*, the "American" or Maine lobster, is strictly marine and is the largest living crustacean by weight (the record weight being over 20 kilograms) (Figures 16.27E,H, 16.29B).

Infraorder Palinura. This group includes the spiny lobsters and slipper lobsters. The flattened abdomen bears a tail fan; the carapace may be cylindrical or flattened dorsoventrally; the gills are trichobranchiate. The chelation of the pereopods varies: The first 4 pairs, only the fifth pair, or no pereopods may be chelate. All species are marine, and they are found in a variety of habitats throughout the tropics. Most produce sounds by rubbing a process (the **plectrum**) at the base of the antennae against a "file" on the head (Figures 16.1F, 16.11B, 16.30A,C, 16.33K).

Infraorder Thalassinidea. The mud and ghost shrimps are particularly difficult to place within the decapods. Sometimes they are included with the crayfish and chelate lobsters (Astacidea), and sometimes they are grouped with the hermit crabs and their relatives (Anomura). We retain them in a separate infraorder. These decapods have a symmetrical abdomen that is flattened dorsoventrally and extends posteriorly as a well developed tail fan. The carapace is somewhat compressed laterally, and the gills are trichobranchiate. The first 2 pairs of pereopods are chelate, and the first pair is generally much enlarged. Most of these animals are marine burrowers or live in coral rubble. They generally have a rather thin, lightly sclerotized cuticle, but some (e.g., members of the family Axiidae) have thicker skeletons and are more lobster-like in appearance. Thalassinids often occur in huge colonies on tidal flats, where their burrow holes form characteristic patterns on the sediment surface (Figures 16.1G, 16.11A).

Superorder Peracarida. Telson without caudal rami; 1 (rarely 2–3) pair of maxillipeds; maxilliped basis typically produced into an anteriorly directed, bladelike endite; mandibles with articulated accessory processes in adults, between molar and incisor processes, called the **lacinia mobilis**; carapace, when present, not fused with posterior pereonites and usually reduced in size; gills thoracic or abdominal; with unique, thinly flattened thoracic coxal endites, called **oostegites**, that form a ventral brood pouch or marsupium in all species except members of the order Thermosbaenacea (the latter using the carapace to brood embryos); young hatch as **mancas**, a prejuvenile stage lacking the last pair of thoracopods (no free-living larvae occur in this group) (Figures 16.12–16.15).

The roughly 21,558 species of peracarids are divided among nine orders. The orders Mysida and Lopho-

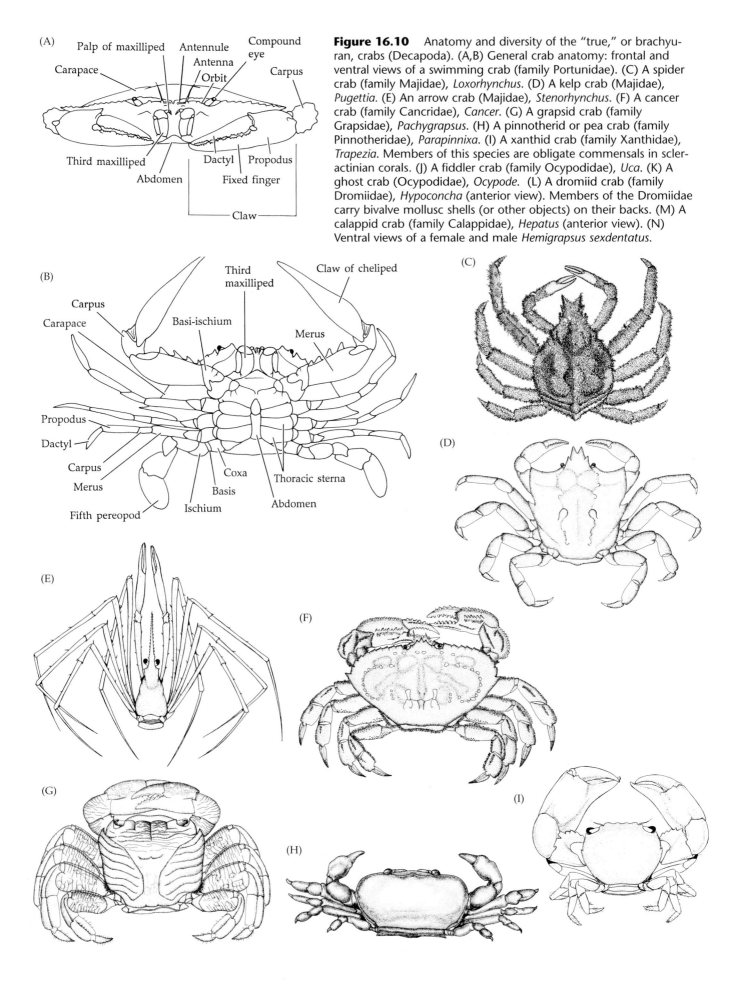

(A)

Palp of maxilliped
Antennule
Compound eye
Antenna
Carapace
Orbit
Carpus
Third maxilliped
Dactyl
Propodus
Abdomen
Fixed finger
Claw

Figure 16.10 Anatomy and diversity of the "true," or brachyuran, crabs (Decapoda). (A,B) General crab anatomy: frontal and ventral views of a swimming crab (family Portunidae). (C) A spider crab (family Majidae), *Loxorhynchus*. (D) A kelp crab (Majidae), *Pugettia*. (E) An arrow crab (Majidae), *Stenorhynchus*. (F) A cancer crab (family Cancridae), *Cancer*. (G) A grapsid crab (family Grapsidae), *Pachygrapsus*. (H) A pinnotherid or pea crab (family Pinnotheridae), *Parapinnixa*. (I) A xanthid crab (family Xanthidae), *Trapezia*. Members of this species are obligate commensals in scleractinian corals. (J) A fiddler crab (family Ocypodidae), *Uca*. (K) A ghost crab (Ocypodidae), *Ocypode*. (L) A dromiid crab (family Dromiidae), *Hypoconcha* (anterior view). Members of the Dromiidae carry bivalve mollusc shells (or other objects) on their backs. (M) A calappid crab (family Calappidae), *Hepatus* (anterior view). (N) Ventral views of a female and male *Hemigrapsus sexdentatus*.

(B)

Third maxilliped
Claw of cheliped
Carpus
Carapace
Basi-ischium
Merus
Propodus
Dactyl
Carpus
Coxa
Merus
Basis
Thoracic sterna
Ischium
Abdomen
Fifth pereopod

(C)

(D)

(E)

(F)

(G)

(H)

(I)

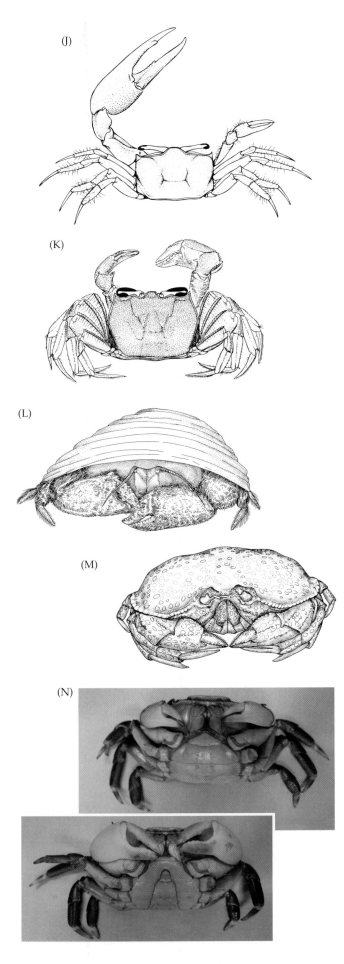

(J)

(K)

(L)

(M)

(N)

gastrida were formerly combined (as the Mysidacea), but most authorities now treat them separately.

The peracarids are an extremely successful group of malacostracan crustaceans and are known from many habitats. Although most are marine, many also occur on land and in fresh water, and several species live in hot springs at temperatures of 30–50°C! Aquatic forms include planktonic as well as benthic species at all depths. The group includes the most successful terrestrial crustaceans—the pillbugs and sowbugs of the order Isopoda—and a few amphipods that have invaded land and live in damp forest leaf litter or gardens. Peracarids range in size from tiny interstitial forms only a few millimeters long to planktonic amphipods over 12 cm long (*Cystisoma*) and benthic isopods growing to 50 cm in length (*Bathynomus giganteus*). These animals exhibit all sorts of feeding strategies; a number of them, especially isopods and amphipods, are symbionts.

Order Mysida. Carapace well developed, covering most of thorax, but never fused with more than four anterior thoracic segments; maxillipeds (1–2 pairs) not associated with cephalic appendages; thoracomere 1 separated from head by internal skeletal bar; abdomen with well developed tail fan; pereopods biramous, except last pair, which are sometimes reduced; pleopods reduced or, in males, modified; compound eyes stalked, sometimes reduced; gills absent; usually with a statocyst in each uropodal endopod; adults with antennal glands (Figures 16.12A,B, 16.30B, 16.33C).

There are nearly 1000 species of mysids, ranging in length from about 2 mm to 8 cm. Most swim by action of the thoracic exopods. Mysids are shrimplike crustaceans that are often confused with the superficially similar euphausids (which lack oostegites and uropodal statocysts). Mysids are pelagic or demersal and are known from all ocean depths. Some species are intertidal and burrow in the sand during low tides. Most are omnivorous suspension feeders, eating algae, zooplankton, and suspended detritus. In the past, mysids were combined with lophogastrids and the extinct Pygocephalomorpha as the "Mysidacea."

Order Lophogastrida. Similar to mysids, except for the following: Maxillipeds (1 pair) are associated with the cephalic appendages; thoracomere 1 not separated from head by internal skeletal bar; pleopods well developed; gills present; adults with both antennal and maxillary glands; without statocysts; all 7 pairs of pereopods well developed and similar (except among members of the family Eucopiidae, in which their structure varies) (Figures 16.12C,D, 16.21G).

There are about 40 known species of lophogastrids, most of which are 1–8 cm long, although the giant *Gnathophausia ingens* reaches 35 cm. All are pelagic swimmers, and the group has a cosmopolitan oceanic distrib-

(A)

Telson

Pleon

Cepholothorax/carapace

Major cheliped (first pereopod)

Antenna

Antennule

Uropod

Pereopods

Minor cheliped

(B)

Antenna

Antennule

Cephalothorax/carapace

Pleon

Pleopods

Uropod

Telson

Third maxilliped

Pereopods

(C)

Walking legs

Chela

(D)

Pleopod

5 4

Telson

Uropod

3

2

Pereopods 1–5

Cheliped (pereopod 1)

(E)

(F)

(G)

(H)

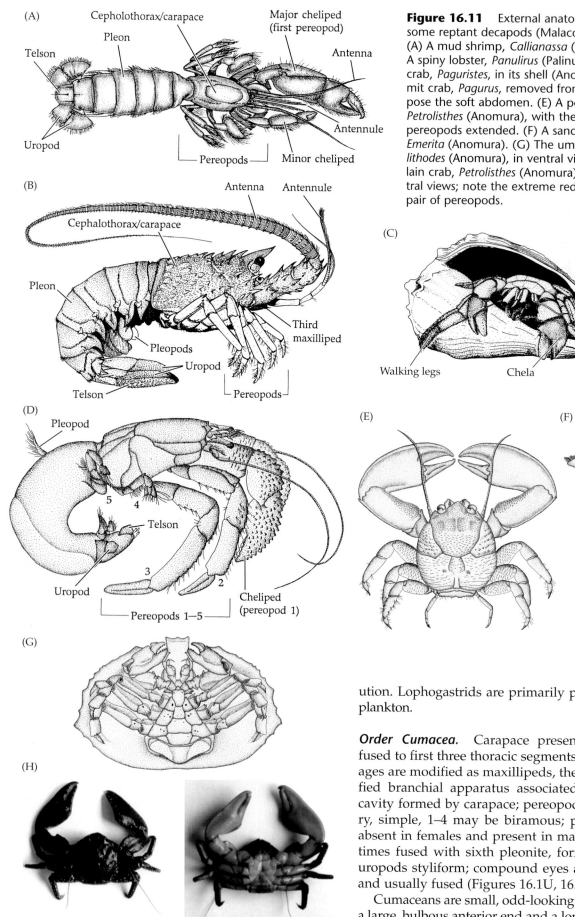

Figure 16.11 External anatomy and diversity in some reptant decapods (Malacostraca, Eucarida). (A) A mud shrimp, *Callianassa* (Thalassinidea). (B) A spiny lobster, *Panulirus* (Palinura). (C) A hermit crab, *Paguristes*, in its shell (Anomura). (D) A hermit crab, *Pagurus*, removed from its shell to expose the soft abdomen. (E) A porcelain crab, *Petrolisthes* (Anomura), with the reduced posterior pereopods extended. (F) A sand or mole crab, *Emerita* (Anomura). (G) The umbrella crab *Cryptolithodes* (Anomura), in ventral view. (H) A porcelain crab, *Petrolisthes* (Anomura), dorsal and ventral views; note the extreme reduction of the fifth pair of pereopods.

ution. Lophogastrids are primarily predators on zooplankton.

Order Cumacea. Carapace present, covering and fused to first three thoracic segments, whose appendages are modified as maxillipeds, the first with modified branchial apparatus associated with branchial cavity formed by carapace; pereopods 1–5 ambulatory, simple, 1–4 may be biramous; pleopods usually absent in females and present in males; telson sometimes fused with sixth pleonite, forming pleotelson; uropods styliform; compound eyes absent, or sessile and usually fused (Figures 16.1U, 16.12E,F).

Cumaceans are small, odd-looking crustaceans with a large, bulbous anterior end and a long, slender poste-

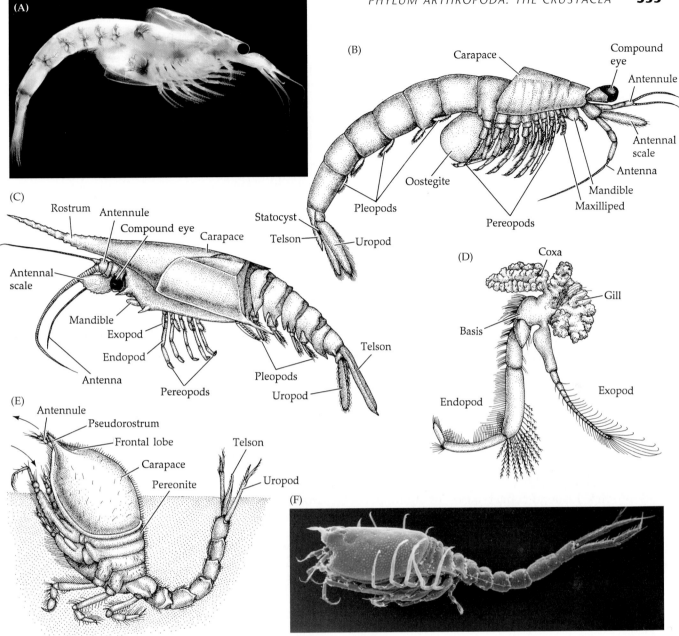

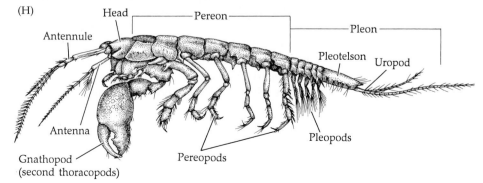

Figure 16.12 Anatomy and diversity in some peracarid crustaceans (Malacostraca; Peracarida)—mysids, lophogastrids, cumaceans, and tanaids. (A) A mysid, *Bowmanella braziliensis*. (B) Anatomy of a generalized mysid (Mysida). (C) Anatomy of a lophogastrid, *Gnathophausia*. (D) Second pereopod of *Gnathophausia*. (E) A cumacean, *Diastylis*, in its typical partially buried position. The arrows indicate the feeding and ventilation current. (F) A cumacean. (G) A tanaid. (H) Anatomy of a generalized tanaid.

rior—resembling horizontal commas! Waldo Schmitt referred to them as "little wonders and queer blunders." They occur worldwide and include about 1,000 species, all 0.5–2 cm in length. Most are marine, although a few species are known. They live in association with bottom sediments, but are capable of swimming and probably leave the bottom to breed. Most are deposit feeders (many use their mouths appendages to "sift" the mud they gather into the mouth area) or predators on meiofauna.

Order Tanaidacea. Carapace present and fused with first two thoracic segments; thoracopods 1–2 are maxillipeds, the second being chelate; thoracopods 3–8 are simple, ambulatory pereopods; pleopods present or absent; uropods biramous or uniramous; telson and last one or two pleonites fused as pleotelson; adults with maxillary and (vestigial) antennal glands; compound eyes absent, or present and on "cephalic lobes."

Members of this order are known worldwide from benthic marine habitats; a few live in brackish or nearly fresh water. Most of the 1,500 or so species are small, ranging from 0.5 to 2 cm in length. They often live in burrows or tubes and are known from all ocean depths. Many are suspension feeders, others are detritivores, and still others are predators (Figure 16.12G,H).

Order Mictacea. Without a carapace, but with a well developed head shield fused with first thoracomere and produced laterally over bases of mouthparts; 1 pair of maxillipeds; pereopods simple, 1–5 or 2–6 biramous, exopods natatory; gills absent; pleopods reduced, uniramous; uropods biramous, with 2–5 segmented rami; telson not fused with pleonites; stalked eyes present (*Mictocaris*) but lacking any evidence of visual elements, or absent (*Hirsutia*) (Figure 16.13D–E).

Mictacea is the most recently (1985) established peracaridan order. The order was erected to accommodate two species of unusual crustaceans: *Mictocaris halope* (discovered in marine caves in Bermuda) and *Hirsutia bathyalis* (from a benthic sample 1,000 m deep in the Guyana Basin off northeastern South America). A third species was described in 1988 from Australia, and a fourth from the Bahamas in 1992. Mictaceans are small, 2–3.5 mm in length. *Mictocaris halope* is the best known of these species because many specimens have been recovered and some have been studied alive. It is pelagic in cave waters and swims by using its pereopodal exopods.

Order Spelaeogriphacea. Carapace short, fused with first thoracomere; 1 pair of maxillipeds; pereopods 1–7 simple, biramous, with shortened exopods; exopods on legs 1–3 modified for producing currents, on legs 4–7 as gills; pleopods 1–4 biramous, natatory; pleopod 5 reduced; tail fan well developed; compound eyes nonfunctional or absent, but eyestalks persist (Figures 16.13A, 16.21H).

The order Spelaeogriphacea is currently known from only three living species. These rare, small (less than 1 cm) peracarids were long known only from a single species living in a freshwater stream in Bat Cave on Table Mountain, South Africa. A second species was recently reported from a freshwater cave in Brazil, and a third from an aquifer in Australia. Little is known about the biology of these animals, but they are suspected to be detritus feeders.

Order Thermosbaenacea. Carapace present, fused with first thoracomere and extending back over 2–3 additional segments; 1 pair of maxillipeds; pereopods biramous, simple, lacking epipods and oostegites; carapace forms dorsal brood pouch (unlike all other peracarids, which form the brood pouch from ventral oostegites); 2 pairs of uniramous pleopods; uropods biramous; telson free or forming pleotelson with last pleonite; eyes absent (Figure 16.13B,C).

About 11 species of thermosbaenaceans are recognized in six genera. *Thermosbaena mirabilis* is known from freshwater hot springs in North Africa, where it lives at temperatures in excess of 40°C. Several species in other genera occur in much cooler fresh waters, typically in groundwater or in caves. Other species are marine or inhabit underground anchialine pools. Limited data suggest that thermosbaenaceans feed on plant detritus.

Order Isopoda. Carapace absent; first thoracomere fused with head; 1 pair of maxillipeds; 7 pairs of uniramous pereopods, the first of which is sometimes subchelate, others usually simple (gnathiids have only five pairs of pereopods, as thoracopod 2 is a maxillipedal "pylopod" and thoracopod 8 is missing); pereopods variable, modified as ambulatory, prehensile, or swimming; in the more derived suborders pereopodal coxae are expanded as lateral side plates (**coxal plates**); pleopods biramous and well developed, natatory and for gas exchange (functioning as gills in aquatic taxa, and with air sacs called **pseudotrachea** in most terrestrial Oniscidea); adults with maxillary and (vestigial) antennal glands; telson fused with one to six pleonites, forming pleotelson; eyes usually sessile and compound, absent from some, pedunculate in most Gnathiidea; with biphasic molting (posterior region molts before anterior region) (Figures 16.1P, 16.14, 16.21I, 16.27F, 16.28H,I, 16.33L).

The isopods comprise about 10,000 marine, freshwater, and terrestrial species, ranging in length from 0.5 to 500 mm, the largest being species of the benthic genus *Bathynomus* (Cirolanidae). They are common inhabitants of nearly all environments, and some groups are exclusively (Epicaridea) or partly (Flabellifera) parasitic. The suborder Oniscidea includes about 5,000 species that have invaded land (pillbugs and sowbugs); they are the most successful terrestrial crustaceans. Their direct development, flattened shape, osmoregulatory ca-

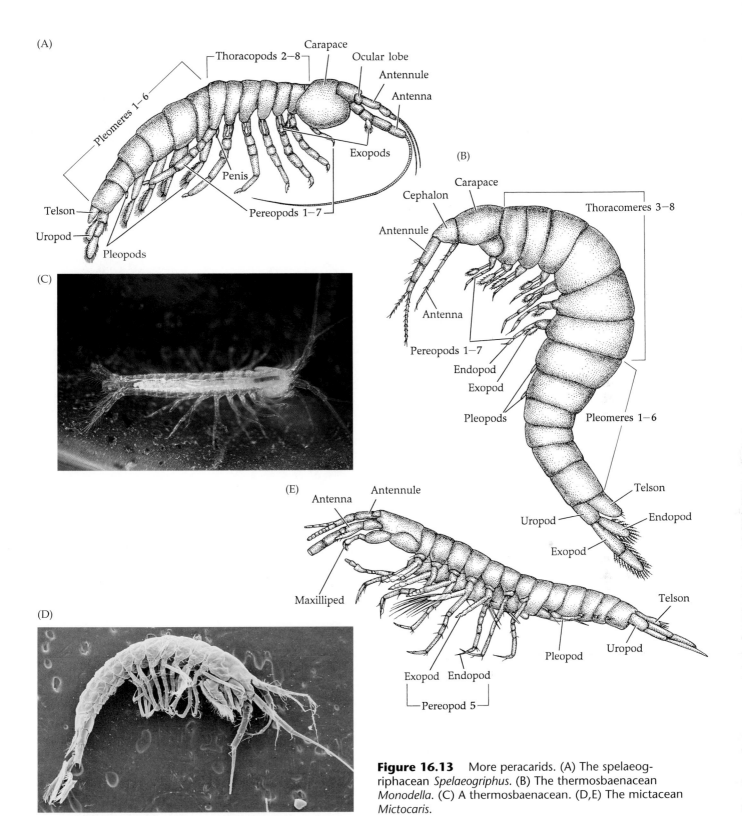

Figure 16.13 More peracarids. (A) The spelaeogriphacean *Spelaeogriphus*. (B) The thermosbaenacean *Monodella*. (C) A thermosbaenacean. (D,E) The mictacean *Mictocaris*.

pabilities, thickened cuticle, and aerial gas exchange organs (pseudotrachea) allow most oniscideans to live completely divorced from aquatic environments.

Isopod feeding habits are extremely diverse. Many are herbivorous or omnivorous scavengers, but direct plant feeders, detritivores, and predators are also common. Some are parasites (e.g., on fishes or on other crustaceans) that feed on the tissue fluids of their hosts. Overall, grinding mandibles and herbivory seem to represent the primitive state, with slicing or piercing

(A)

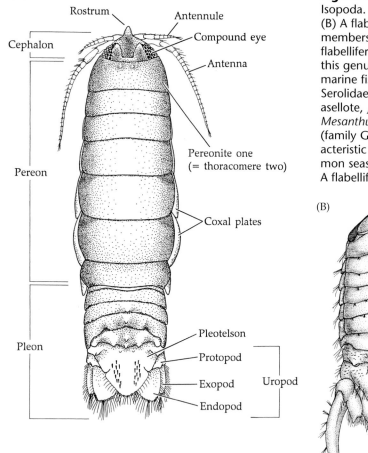

Rostrum
Antennule
Compound eye
Antenna
Cephalon
Pereon
Pereonite one
(= thoracomere two)
Coxal plates
Pleon
Pleotelson
Protopod
Exopod
Endopod
Uropod

Figure 16.14 More peracarids: members of the order Isopoda. (A) A flabelliferan, *Excorallana* (family Corallanidae). (B) A flabelliferan, *Paracerceis* (family Sphaeromatidae). Male members of this genus possess greatly enlarged uropods. (C) A flabelliferan, *Codonophilus* (family Cymothoidae). Members of this genus are parasites that attach to the tongues of various marine fishes. (D) A flabelliferan, *Heteroserolis* (family Serolidae). (E) A valviferan, *Idotea* (family Idoteidae). (F) An asellote, *Joeropsis* (family Joeropsididae). (G) An anthurid, *Mesanthura* (family Anthuridae). (H) A gnathiidean, *Gnathia* (family Gnathiidae). Note the grossly enlarged mandibles characteristic of male Gnathiidae. (I) An oniscidean, *Ligia* (the common seashore "rock louse"). (J) A valviferan, *Idotea resecata*. (K) A flabelliferan (Sphaeromatidae), *Sphaeroma walkeri*.

(B)

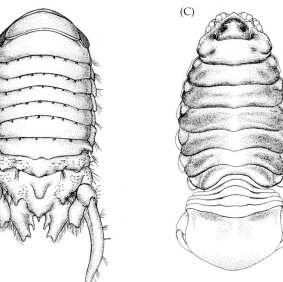

(C)

(D)

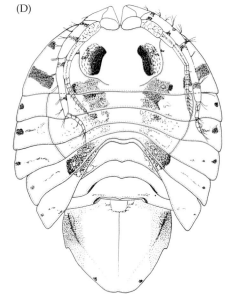

(E)

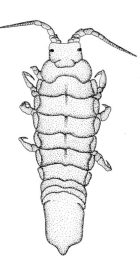

(F)

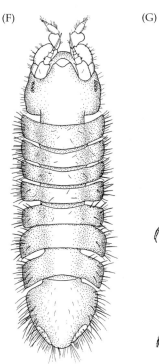

(G)

mandibles and predation appearing later in the evolution of several isopod clades.

Order Amphipoda.

Carapace absent; first thoracomere fused to head; 1 pair of maxillipeds; 7 pairs of uniramous pereopods, with first, second, and sometimes others frequently modified as chelae or subchelae; pereopodal coxae expanded as lateral side plates (coxal plates); gills thoracic (medial pereopodal epipods); adults with antennal glands; abdomen "divided" into two regions of three segments each, an anterior "pleon" and posterior **urosome**, with anterior appendages as typical pleopods and urosomal appendages modified as uropods; telson free or fused with last urosomite; other urosomites sometimes fused; compound eyes sessile, absent in some, huge in many (but not all) members of the suborder Hyperiidea (Figures 16.1R,S, 16.15, 16.23, 16.27G, 16.29D).

Isopods and amphipods share many features and are often said to be closely related. Earlier workers recognized these similarities (e.g., sessile compound eyes, loss of carapace, and presence of coxal plates) and classified them together as the "Edriopthalma" or "Acarida." However, recent work suggests that many similarities between these two taxa are convergences or parallelisms. The roughly 8,000 species of amphipods range in length from tiny 1 mm forms to giant deep-sea benthic species reaching 25 cm, and one group of planktonic forms exceeds 10 cm. They have invaded most marine and freshwater habitats and often constitute a large portion of the biomass in many areas.

The principal suborder is Gammaridea. A few gammarideans are semi-terrestrial in moist forest leaf litter or on supralittoral sandy beaches (e.g., beach hoppers); a few others live in moist gardens and greenhouses (e.g., *Talitrus sylvaticus* and *T. pacificus*). They are common in subterranean groundwater ecosystems of caves, the majority being stygobionts—obligatory groundwater species characterized by reduction or loss of eyes, pigmentation, and occasionally appendages. About 900 species of stygobiontic amphipods have been described, including the divers genera *Niphargus* (in Europe) and *Stygobromus* (in North America), each with over 100 described species. However, most of the gammaridean amphipods are marine benthic species, and a few have adopted a pelagic lifestyle, usually in deep oceanic waters. There are many intertidal species, and a great many of these live in association with other invertebrates and with algae.

The suborder Hyperiidea includes exclusively pelagic amphipods that have apparently escaped the confines of benthic life by becoming associated with other plankters, particularly gelatinous zooplankton such as medusae, ctenophores, and salps. The hyperiideans are usually characterized by huge eyes (and a few other inconsistent features), but several groups bear eyes no larger than those of most gammarideans. The

(J)

(K)

(H)

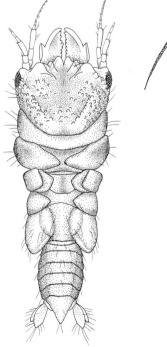

(I)

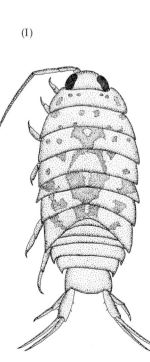

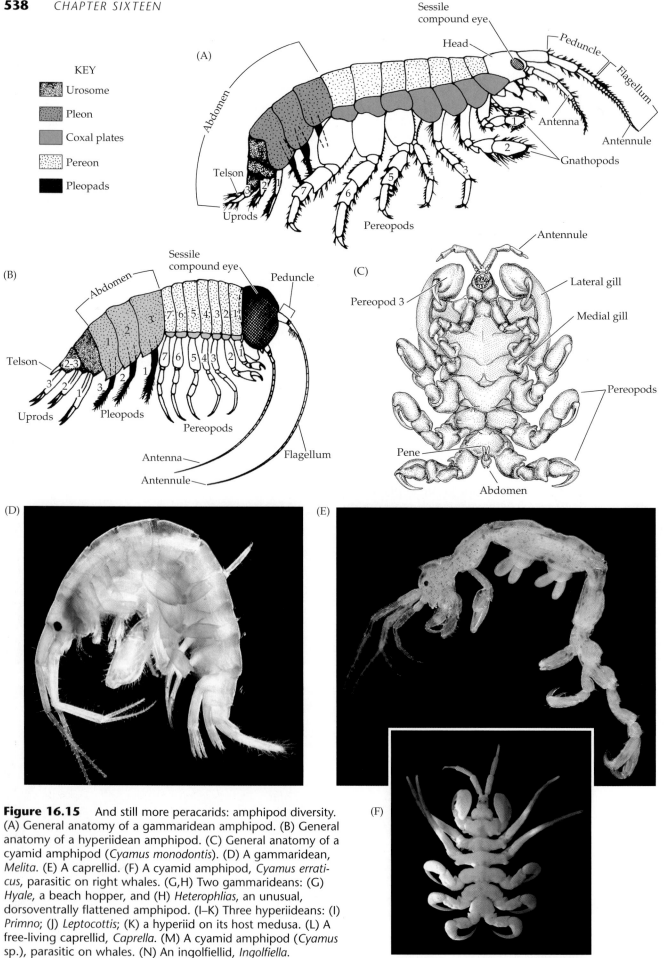

Figure 16.15 And still more peracarids: amphipod diversity. (A) General anatomy of a gammaridean amphipod. (B) General anatomy of a hyperiidean amphipod. (C) General anatomy of a cyamid amphipod (*Cyamus monodontis*). (D) A gammaridean, *Melita*. (E) A caprellid. (F) A cyamid amphipod, *Cyamus erraticus*, parasitic on right whales. (G,H) Two gammarideans: (G) *Hyale*, a beach hopper, and (H) *Heterophlias*, an unusual, dorsoventrally flattened amphipod. (I–K) Three hyperiideans: (I) *Primno*; (J) *Leptocottis*; (K) a hyperiid on its host medusa. (L) A free-living caprellid, *Caprella*. (M) A cyamid amphipod (*Cyamus* sp.), parasitic on whales. (N) An ingolfiellid, *Ingolfiella*.

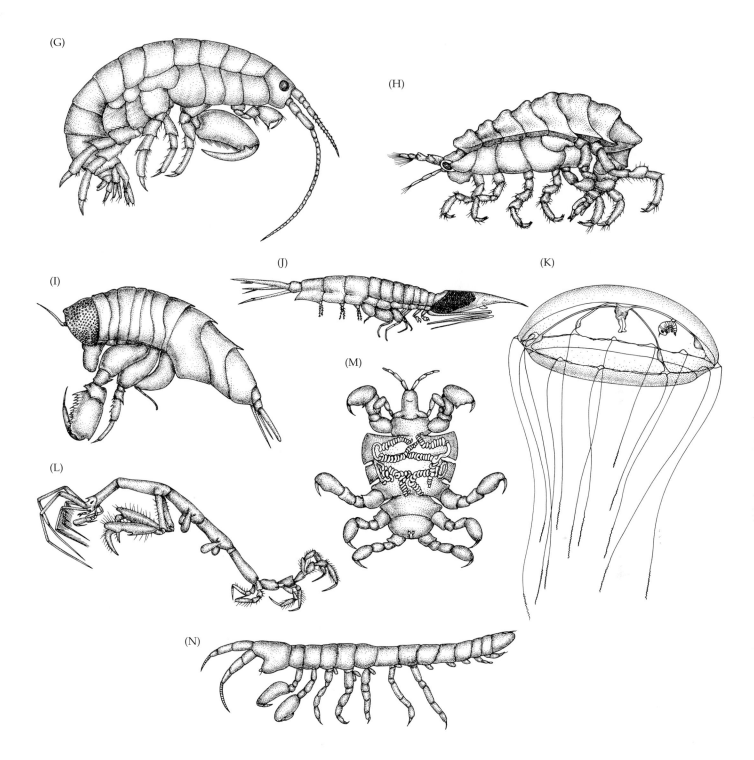

Hyperiidea are almost certainly a polyphyletic group, and it is thought that several lineages are derived independently from various gammaridean ancestors, although a modern phylogenetic analysis has yet to be attempted. The precise nature of the relationships between hyperiideans and their zooplankton hosts remains controversial. Some appear to eat host tissue, others may kill the host to fashion a floating "home," and still others may utilize the host merely for transport or as a nursery for newly hatched young. This fascinating problem was last reviewed by Laval (1980).

There are two other small amphipod suborders: Ingolfiellidea and Caprellidea. The firstsuborder contains only about 30 species, most of which live in subterranean fresh and brackish waters, although a few are marine and interstitial. Little is known about their biology. The 300 or so species of caprellid amphipods ("skeleton shrimp") are highly modified for clinging to other organisms, including filamentous algae and hydroids. In most species the body and appendages are very narrow and elongated. In one family of caprellids, the Cyamidae (with 28 species), individuals are obligate symbionts on cetaceans (whales, dolphins, and porpoises) and have flattened boides and prehensile legs.

In addition to parasitism, amphipods exhibit a vast array of feeding strategies, including scavenging, herbivory, carnivory, and suspension feeding.

Class Maxillopoda

Fundamentally with five cephalic, six thoracic, and four abdominal somites, plus a telson, but reductions of this basic 5-6-4 body plan are common; thoracomeres variously fused with cephalon; usually with caudal rami; thoracic segments with biramous (sometimes uniramous) limbs, lacking epipods (except in many ostracods); abdominal segments lack typical appendages; carapace present or reduced; with both simple and compound eyes, the latter being unique, with three cups, each with tapetal cells (= **maxillopodan eye**).

Although the class Maxillopoda is accepted by most specialists, there is some question over its monophyly and its component groups, and some classifications (e.g., Martin and Davis 2001) exclude the ostracods. Also, without belaboring the issue, we must warn you that different specialists sometimes interpret the nature of maxillopodan tagmata in different ways, leading to some confusion.

Most maxillopodans are small crustaceans, barnacles being a notable exception. They are generally recognizable by their shortened bodies, especially the reduced abdomen, and by the absence of a full complement of legs. The reductions in body size and leg number, emphasis on the naupliar eye, minimal appendage specialization, and certain other features have led biologists to hypothesize that neoteny played a role in the origin of maxillopodans. That is, in many ways, they resemble early postlarval forms that evolved sexual maturity before attaining all the adult features. Over 26,000 species of Maxillopoda have been described.

Subclass Thecostraca

This group includes the barnacles, parasitic ascothoracids, and mysterious "y-larvae." The thecostracan clade is defined by several rather subtle synapomorphies of cuticular fine structure, including cephalic chemosensory structures known as **lattice organs**. The group is also supported by molecular phylogenetic analyses. All taxa have pelagic larvae, the terminal instar of which possesses prehensile antennules and is specialized for locating and attaching to the substratum of the sessile adult state.

Infraclass Ascothoracida

About 125 described species of parasites on anthozoans and echinoderms. Although greatly modified, they retain a bivalved carapace and the full complement of thoracic and abdominal segments (facts that suggest they might be the most primitive living thecostracans). Ascothoracids generally have mouthparts modified for piercing and sucking body fluids, but some live inside other animals and absorb the host's tissue fluids. In at least one species, *Synagoga mira*, males retain the ability to swim throughout their lives, attaching only temporarily while feeding on corals (Figure 16.16F).

Infraclass Cirripedia

Primitively with tagmata as in the class, but in most groups the adult body is modified for sessile or parasitic life; thorax of six segments with paired biramous appendages; abdomen without limbs; telson absent in most, although caudal rami persist on abdomen in some; nauplius larva with frontolateral horns; unique, "bivalved" **cypris larva**; adult carapace "bivalved" (folded) or forming fleshy mantle; first thoracomere often fused with cephalon and bearing maxilliped-like **oral appendages**; female gonopores near bases of first thoracic limbs, male gonopore on median penis on last thoracic or first abdominal segment; compound eyes lost in adults (Figures 16.1I,M, 16.16A–E, 16.25, 16.26, 16.27B,C, 16.32E, 16.33E).

The 1,285 or so described cirripede species are mostly free-living barnacles, but this group also includes some strange parasitic "barnacles" rarely seen except by specialists. The common acorn and goose barnacles belong to the superorder Thoracica. The superorder Acrothoracica consists of minute animals that burrow into calcareous substrata, including corals and mollusc shells (Figure 16.16G. The rhizocephalans are parasites of other crustaceans, especially decapods (Figure 16.16H).

The maxillopodan body plan has been so extensively modified in cirripedes that its basic features are nearly unrecognizable in the sessile and parasitic adults. The abdomen is greatly reduced in adults and in most cypris larvae. In cyprids (cypris larvae) the carapace is always present and "bivalved," the two sides being held by a transverse **cypris adductor muscle**; in adults the cara-

Figure 16.16   Anatomy and diversity in the class Maxillopoda, subclass Thecostraca—barnacles and their kin. (A–E) Thoracican barnacles. (A) Sessile (acorn) barnacles, *Semibalanus balanoides*. One individual has its cirri extended for feeding. (B) Plate terminology in a balanomorph (acorn) barnacle. (C) The lepadomorph (stalked) barnacle *Pollicipes polymerus*. (D) *Verruca*, the "wart" barnacle. (E) Two thoracican barnacles that live in association with each other and with whales. The stalked barnacle *Conchoderma* attaches to the sessile barnacle *Coronula*, which in turn attaches to the skin of certain whales. (F) The ascothoracican *Ascothorax ophiocentenis*, a parasite that feeds periodically on echinoderms (longitudinal section). (G) An acrothoracican, *Alcippe*. Note the highly modified female and the tiny attached male. This species bores into calcareous substrata such as coral skeletons. (H) A crab (*Carcinus*) infected with the rhizocephalan *Sacculina carcini*. The crab's right side is shown as transparent, exposing the ramifying body of the parasite. (I) A cypris y-larva, in lateral and dorsal views.

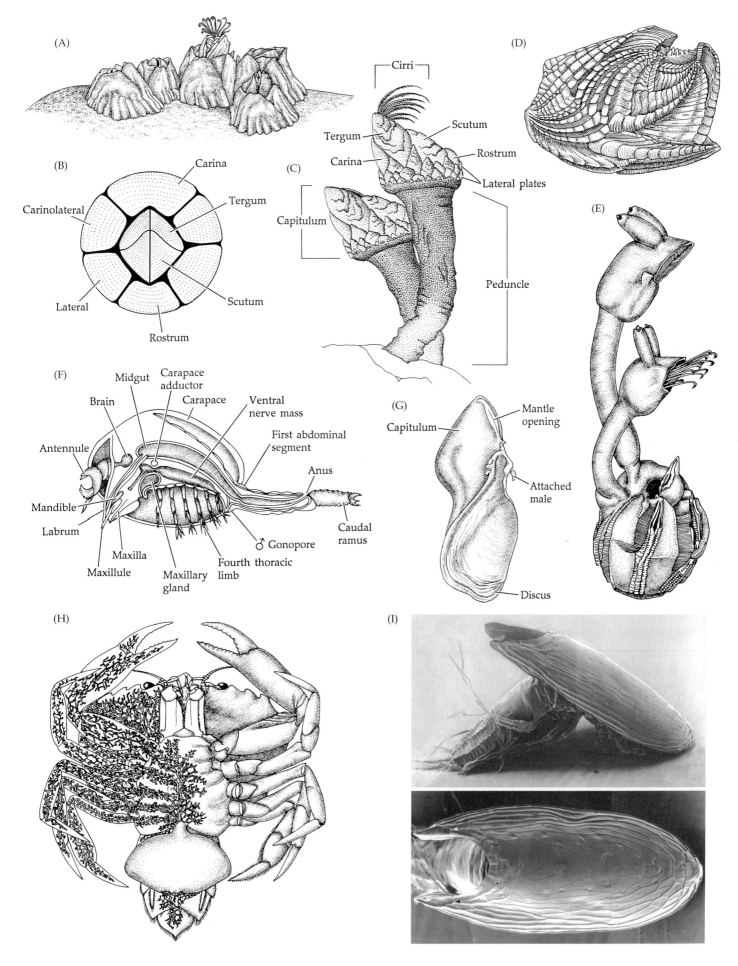

(A)

(B)

Carina

Carinolateral

Tergum

Lateral

Scutum

Rostrum

(C)

Cirri

Tergum

Carina

Capitulum

Scutum

Rostrum

Lateral plates

Peduncle

(D)

(E)

(F)

Brain

Antennule

Mandible

Labrum

Maxillule

Maxilla

Midgut

Carapace adductor

Carapace

Ventral nerve mass

First abdominal segment

Anus

Caudal ramus

♂ Gonopore

Fourth thoracic limb

Maxillary gland

(G)

Capitulum

Mantle opening

Attached male

Discus

(H)

(I)

pace is lost (Rhizocephala) or modified as a membranous, saclike mantle (thoracicans and acrothoracicans). In the barnacles (Thoracica), it is this mantle that produces the familiar calcareous plates that enclose the body. Cyprids and adult acrothoracicans share a unique tripartite crystalline cone structure in the compound eye, a feature not known from any other crustacean group and perhaps a vestige of the ancestral thecostracan bauplan. Most species of barnacles are hermaphrodites, whereas separate sexes are the rule in acrothoracicans and rhizocephalans.

Locomotion in barnacles is generally confined to the larval stages, although adults of a few species are specifically adapted to live attached to floating objects (e.g., seaweeds, pumice, logs) or nektonic marine animals (e.g., whales, sea turtles). Others are often found on the shells and exoskeletons of various errant invertebrates (e.g., crabs and gastropods), which inadvertently provide a means of transportation from one place to another. Of course, parasitic forms also enjoy free rides on their hosts. Thoracican and acrothoracican barnacles use their feathery thoracopods (**cirri**) to suspension feed. Barnacles in the family Coronulidae are suspension feeders that attach to whales and turtles (e.g., *Chelonibia, Platylepas, Stomatolepas, Coronula, Xenobalanus*). Most rhizocephalans are endoparasitic and are the most highly modified of all cirripedes. They mainly inhabit decapod crustaceans, but a few are known from isopods, cumaceans, and even thoracican barnacles. The body consists of a reproductive part (the **externa**) positioned outside the host's body, and an internal, ramifying, nutrient-absorbing part (the **interna**).

Infraclass Facetotecta

Monogeneric (*Hansenocaris*): The "y-larvae," a half-dozen small (250–620 μm) marine nauplii and cyprids (Figure 16.16I). Although known since Hansen's original description in 1899, the adult stage of these animals has still not been identified (although it has been suggested that they might be the "missing" larval progeny of sexual reproduction in tantulocarids). The prehensile antennules and hooked labrum of the y-cyprids suggest that the adults are parasitic. For details see Høeg and Kolbasov (2002).

Subclass Tantulocarida

Bizarre parasites of deep water crustaceans. Juveniles with cephalon, 6-segmented thorax, and abdomen of up to 7 segments; cephalon lacking appendages (other than paired antennules in one known stage only) but with internal median stylet; thoracopods 1–5 biramous, 6 uniramous; abdomen without appendages but with caudal rami; adults highly modified, with "unsegmented" sac-

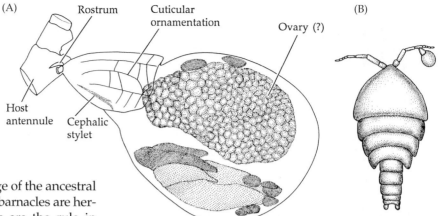

Figure 16.17 Anatomy in the class Maxillopoda, subclass Tantulocarida. (A) An adult *Basipodella atlantica*. Note the absence of an abdomen and the modifications for parasitic life. (B) *Basipodella* attached to the antenna of a copepod host. (C,D) *Microdajus pectinatus* on a crustacean host, adult and juvenile (SEM).

ciform thorax and a reduced abdomen bearing a uniramous penis on the first segment; female gonopores on fifth thoracic segment.

The tiny tantulocarids are less than 0.5 mm long. They attach to their hosts by penetrating the body with a protruding cephalic stylet. The young bear natatory thoracopods. About a dozen species have been described (Figures 16.1N, 16.17).

Until recently, members of this group had been assigned to various parasitic groups of Copepoda and Cirripedia. In 1983 Geoffrey Boxshall and Roger Lincoln proposed the new class Tantulocarida. Subsequent work supports a view of these animals as maxillopodans, although the presence of six or seven abdominal segments in juveniles of some species is inconsistent with this view.

Subclass Branchiura

Body compact and oval, head and most of trunk covered by broad carapace; antennules and antennae reduced, the latter sometimes absent; mouthparts modified for parasitism; no maxillipeds; thorax reduced to four segments, with paired biramous appendages; abdomen unsegmented, bilobed, limbless, but with minute caudal rami; female gonopores at bases of fourth thoracic legs, male with single gonopore on midventral surface of last thoracic somite; paired, sessile compound eyes and one to three median simple eyes (Figure 16.18L).

The Branchiura comprise about 130 species of ectoparasites on marine and freshwater fishes. The antennules generally bear hooks or spines for attachment to their host fish. The mandibles are reduced in size and complexity, bear cutting edges, and are housed within a styliform "proboscis" apparatus. The maxillules are clawed in *Dolops*, but they are modified as stalked suck-

(C)

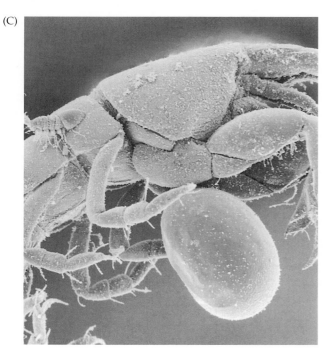

(D)

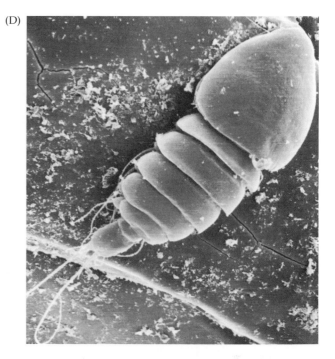

ers in the other genera (*Argulus, Chonopeltis, Dipteropeltis*). The uniramous maxillae usually bear attachment hooks. The thoracopods are biramous and used for swimming when the animal is not attached to a host. Branchiurans feed by piercing the skin of their hosts and sucking blood or tissue fluids. Once they locate a host, they crawl toward the fish's head and anchor in a spot where water flow turbulence is low (e.g., behind a fin or gill operculum).

Members of the genus *Argulus* occur worldwide, but members of the other genera have restricted distributions. *Chonopeltis* is found only in Africa, *Dipteropeltis* in South America, and *Dolops* in South America, Africa, and Tasmania.

Subclass Pentastomida

Obligatory parasites of reptiles, mammals, and birds. Adults inhabit respiratory tracts (lungs, nasal passages, etc.) of their hosts. Body highly modified, wormlike, 2–13 cm in length. Adult appendages reduced to 2 pairs of head appendages, lobelike and with chitinous claws used to cling to host. Body cuticle nonchitinous and highly porous. Body muscles somewhat sheetlike, but clearly segmental and cross-striated. Mouth lacks jaws; often on end of snoutlike projection; connected to a muscular pumping pharynx used to suck blood from host. The combination of the snout and the 2 pairs of legs give the appearance of there being five mouths, hence the name (Greek *penta*, "five"; *stomida*, "mouths"). In many species the appendages are reduced to no more than the terminal claws. No specific gas exchange, circulatory, or excretory organs. Dioecious; females larger than males. About 130 described species, including two cosmopolitan species that infest humans (Figure 16.19).

For years it was believed that pentastomids were allied with the onychophorans as some kind of segmented, vermiform, pre-arthropod creature. However, several recent independent molecular studies (using 18S rDNA) have revealed the pentastomids to be highly modified crustaceans, perhaps derived from the Branchiura. Corroboration has come from cladistic analyses of sperm and larval morphology, nervous system anatomy, and cuticular fine structure.

Müller and Walossek's work on the Swedish *Orsten* fauna indicates that the pentastomids had appeared as early as the Upper Cambrian, long before the land vertebrates had evolved. What might the original hosts of these parasitic crustaceans have been? Conodont fossils are common in all the Cambrian localities that have yielded pentastomids, raising the possibility that conodonts (also long a mystery, but now widely regarded as parts of early fishlike vertebrates) may have been the original hosts of the Pentastomida.

Subclass Mystacocarida

Body divided into cephalon and 10-segmented trunk; telson with clawlike caudal rami; cephalon characteristically cleft; all cephalic appendages nearly identical, antennae and mandibles biramous, antennules, maxillules, and maxillae uniramous; first trunk segment bears maxillipeds but is not fused with cephalon; no carapace; gonopores on fourth trunk segment; trunk segments 2–5 with short, single-segment appendages (Figure 16.18A).

There are only 13 described species of mystacocarids, eight in the genus *Derocheilocaris* and five in *Ctenocheilocaris*. Most are less than 0.5 mm long, although *D. ingens* reaches 1 mm. The head is marked by a trans-

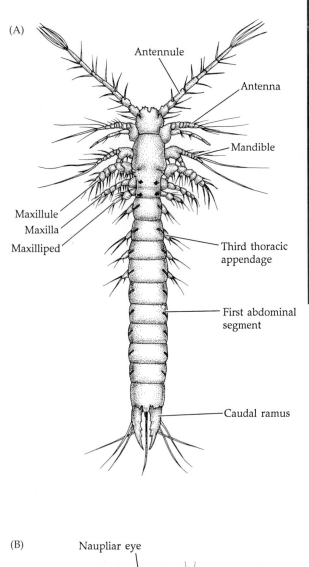

(A)

Antennule

Antenna

Mandible

Maxillule
Maxilla
Maxilliped

Third thoracic
appendage

First abdominal
segment

Caudal ramus

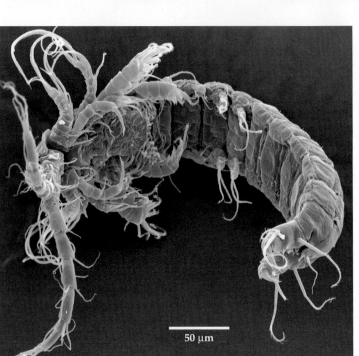

50 μm

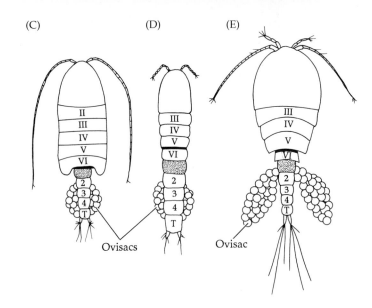

(C)

II
III
IV
V
VI

2
3
4
T

Ovisacs

(D)

III
IV
V
VI

2
3
4
T

Ovisacs

(E)

III
IV
V

VI
2
3
4
T

Ovisac

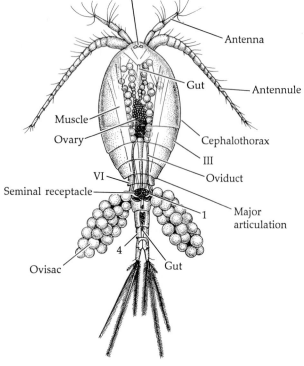

(B)

Naupliar eye

Antenna

Gut

Antennule

Muscle

Ovary

Cephalothorax

III

VI

Oviduct

Seminal receptacle

1

Major
articulation

4

Gut

Ovisac

Figure 16.18   Anatomy of the class Maxillopoda, subclasses Mystacocarida, Copepoda, and Branchiura.
(A) General anatomy and SEM of the mystacocarid
Derocheilocaris. (B) General anatomy of a cyclopoid copepod. (C–E) General body forms of (C) a calanoid, (D) a
harpacticoid, and (E) a cyclopoid copepod. Note the points
of body articulation (dark band) and the position of the
genital segment (shaded segment). Roman numerals are
thoracic segments; Arabic numerals are abdominal segments; T = telson. (F) An elaborately setose calanoid copepod adapted for flotation. (G) A poecilostomatid copepod,
Ergasilus pitalicus, ectoparasitic on cichlid fishes. (H) A
female siphonostomatid copepod (*Caligus* sp.) with egg
sacs. (I) A female siphonostomatid copepod (*Trebius het-
erodont*, a parasite of horn sharks in California) with egg
sacs. (J) A siphonostomatid copepod, *Clavella adunca*,
showing extreme body reduction; this species attaches to
the gills of fishes by its elongate maxillae. (K) *Notodelphys*,
a wormlike cyclopoid copepod adapted for endoparasitism
in tunicates. (L) *Argulus foliaceus*, a branchiuran that parasitizes fishes. Note the powerful hooked suckers (modified
maxillules) on the ventral surface.

(F)

(G)

(H)

(I)

(J)

Cephalon

Thorax

Maxilla

Egg sac

(K)

Cephalon

Appendages

Thorax

Abdomen

(L)

Antennule

Antenna

Spine

Eye

Maxillule

Maxilla

First
thoracic limb

Abdomen

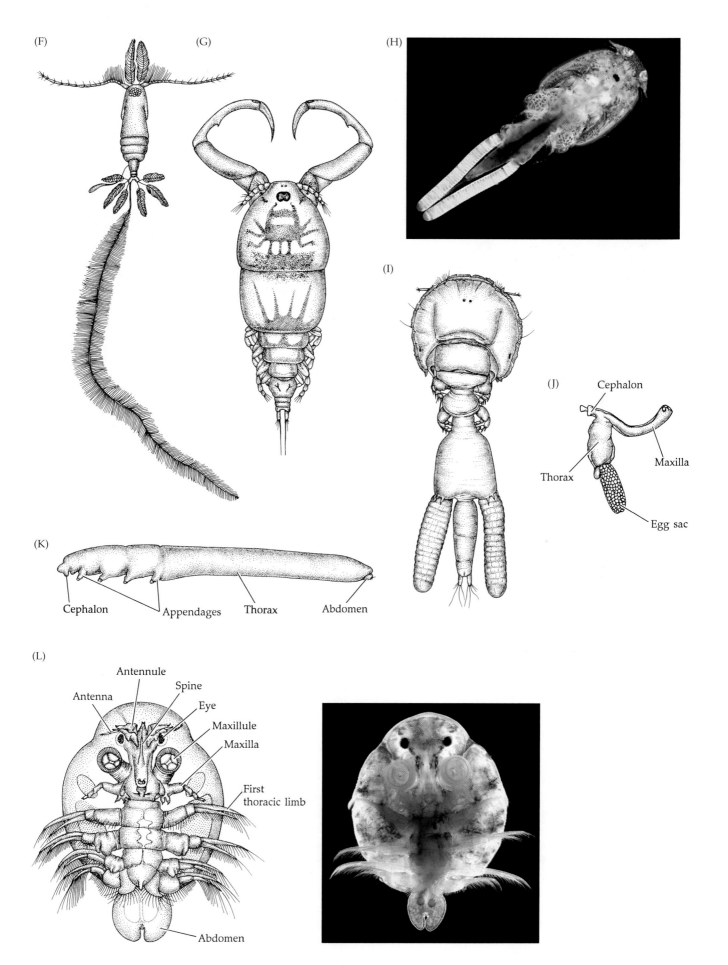

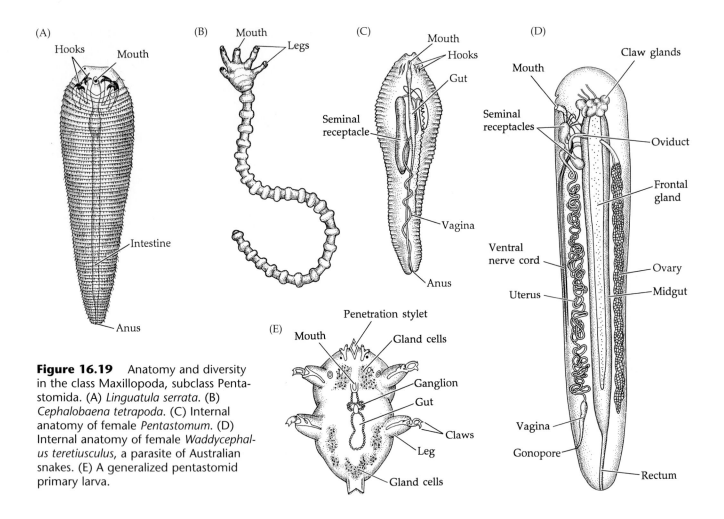

Figure 16.19 Anatomy and diversity in the class Maxillopoda, subclass Pentastomida. (A) *Linguatula serrata*. (B) *Cephalobaena tetrapoda*. (C) Internal anatomy of female *Pentastomum*. (D) Internal anatomy of female *Waddycephalus teretiusculus*, a parasite of Australian snakes. (E) A generalized pentastomid primary larva.

verse "cephalic constriction" between the origins of the first and second antennae, perhaps a remnant of primitive head segmentation. In addition, the lack of fusion of the cephalon and maxillipedal trunk segment, the simplicity of the mouth appendages, and other features have led some workers to propose that the mystacocarids are among the most primitive living crustaceans. These attributes may, however, simply be related to a neotenic origin and specialization for interstitial habitats.

Mystacocarids are marine, interstitial crustaceans that live in littoral and sublittoral sands throughout the world's temperate and subtropical seas. Their rather vermiform body and small size are clearly adaptations to life among sand grains. Mystacocarids are thought to feed by scraping organic material from the surfaces of sand grains with their setose mouthparts.

Subclass Copepoda

Without a carapace, but with a well developed cephalic shield; single, median, simple maxillopodan eye (sometimes lacking); one or more thoracomeres fused to head; thorax of six segments, the first always fused to the head and with maxillipeds; abdomen of five segments, including anal somite (= telson); well developed caudal rami; abdomen without appendages, except an occasional re-

duced pair on the first segment, associated with the gonopores; point of main body flexure varies among major groups; antennules uniramous, antennae uniramous or biramous; 4–5 pairs of natatory thoracopods, most locked together for swimming; posterior thoracopods always biramous (Figures 16.1L, 16.18B–K, 16.27A, 16.30D, 16.33D).

There are about 12,000 described species of copepods. Most are small, 0.5–10 mm long, but some free-living forms exceed 1.5 cm in length, and certain highly modified parasites may reach 25 cm. The bodies of most copepods are distinctly divided into three tagmata, the names of which vary among authors. The first region includes the five fused head segments and one or two additional fused thoracic somites; it is called a **cephalosome** (= cephalothorax) and bears the usual head appendages and maxillipeds. All of the other limbs arise on the remaining thoracic segments, which together constitute the **metasome**. The abdomen, or **urosome**, bears no limbs. The appendage-bearing regions of the body (cephalosome and metasome) are frequently collectively called the **prosome**.

The majority of the free-living copepods, and those most frequently encountered, belong to the orders Calanoida, Harpacticoida, and Cyclopoida, although even some of these are parasitic. We focus here on these three groups and then briefly discuss some of the other, smaller orders and their modifications for parasitism.

The calanoids are characterized by a point of major body flexure between the metasome and the urosome, marked by a distinct narrowing of the body. They possess greatly elongate antennules. Most of the calanoids are planktonic, and as a group they are extremely important as primary consumers in freshwater and marine food webs. The point of body flexure in the orders Harpacticoida and Cyclopoida is between the last two (fifth and sixth) metasomal segments. (Note: Some authors define the urosome in harpacticoids and cyclopoids as that region of the body posterior to this point of flexure.) Harpacticoids are generally rather vermiform, with the posterior segments not much narrower than the anterior; cyclopoids generally narrow abruptly at the major body flexure. Both the antennules and the antennae are quite short in harpacticoids, but the latter are moderately long in cyclopoids (although never as long as the antennules of calanoids). The antennae are uniramous in cyclopoids but biramous in the other two groups. Most harpacticoids are benthic, and those that have adapted to a planktonic lifestyle show modified body shapes. Harpacticoids occur in all aquatic environments; encystment is known to occur in at least a few freshwater and marine species. Cyclopoids are known from fresh and salt water, and most are planktonic.

The nonparasitic copepods move by crawling or swimming, using some or all of the thoracic limbs. Many of the planktonic forms have very setose appendages, offering a high resistance to sinking. Calanoids are predominantly planktonic feeders. Benthic harpacticoids are often reported as detritus feeders, but many feed predominantly on microorganisms living on the surface of detritus or sediment particles (e.g., diatoms, bacteria, and protists).

Of the seven remaining orders, the Mormonilloida are planktonic; the Misophrioida are known from deep-sea epibenthic habitats as well as anchialine caves in both the Pacific and Atlantic; and the Monstrilloida are planktonic as adults, but the larval stages are endoparasites of certain gastropods, polychaetes, and occasionally echinoderms. Members of the orders Poecilostomatoida and Siphonostomatoidaare exclusively parasitic and often have modified bodies. Siphonostomatoids are endo- or ectoparasites of various invertebrates as well as marine and freshwater fishes; they are often very tiny and show a reduction or loss of body segmentation. Poecilostomatoids parasitize invertebrates and marine fishes, and may also show a reduced number of body segments. The Platycopioida are benthic forms known primarily from marine caves; the Gelyelloida are known only from European groundwaters.

Subclass Ostracoda

Body segmentation reduced, trunk not clearly divided into thorax and abdomen, with 6 to 8 pairs of limbs (including the male copulatory limb); trunk with 1 to 3 pairs of limbs, variable in structure; caudal rami present; gonopores on lobe anterior to caudal rami; carapace bivalved, hinged dorsally and closed by a central adductor muscle, enclosing body and head; carapace highly variable in shape and ornamentation, smooth or with various pits, ridges, spines, etc.; most with one simple median naupliar eye (often called a "maxillopodan eye") and sometimes weakly stalked compound eyes (in Myodocopida); adults with maxillary and (in some) antennal glands; males with distinct copulatory limbs; caudal rami (furca) present (Figure 16.20).

The ostracods comprise about 13,000 described living species of small bivalved crustaceans, ranging in length from 0.1 to 2.0 mm, although some giants (e.g., *Gigantocypris*) reach 32 mm. They superficially resemble clam shrimps in having the entire body enclosed within the valves of the carapace. However, ostracod valves lack the concentric growth rings of clam shrimps, and there are major differences in the appendages. The shell is usually penetrated by pores, some bearing setae, and is shed with each molt. A good deal of confusion exists about the nature of ostracod limbs, and homologies with other crustacean taxa (and even within the Ostracoda) are unclear—this confusion is reflected in the variety of names applied by different authors. We have adopted terms here that allow the easiest comparison with other taxa.

Ostracods possess the fewest limbs of any crustacean subclass. The four or five head appendages are followed by one to three trunk appendages. Superficially, the (second) maxillae appear to be absent; however, the highly modified fifth limbs are in fact these appendages. The trunk seldom shows external evidence of segmentation, although all eleven postcephalic somites are discernable in some taxa. The trunk limbs vary in structure among taxa and on individuals. The third pair of trunk limbs bear the gonopores and constitute the so-called **copulatory organ**.

Ostracods are one of the most successful groups of crustaceans. They also have the best fossil record of any arthropod group, dating to the Ordovician; an estimated 65,000 fossil species have been described. Most are benthic crawlers or burrowers, but many have adopted a suspension-feeding planktonic lifestyle, and a few are terrestrial in moist habitats. One species is known to be parasitic on fish gills—*Sheina orri* (Myodocopida, Cypridinidae). They are abundant worldwide in all aquatic environments and are known to depths of 7,000 m in the sea. Some are commensal on echinoderms or other crustaceans. A few podocopans have invaded supralittoral sandy regions (members of the family Terrestricytheridae), and members of several families inhabit terrestrial mosses and humus. Two principal taxa (ranked as superorders here) are recognized within the Ostracoda: Myodocopa and Podocopa.

The myodocopans are all marine. Most are benthic, but the group also includes all of the marine planktonic ostracods. The largest of all ostracods, the planktonic *Gigantocypris*, is a member of this group. Myodocopans

(A)

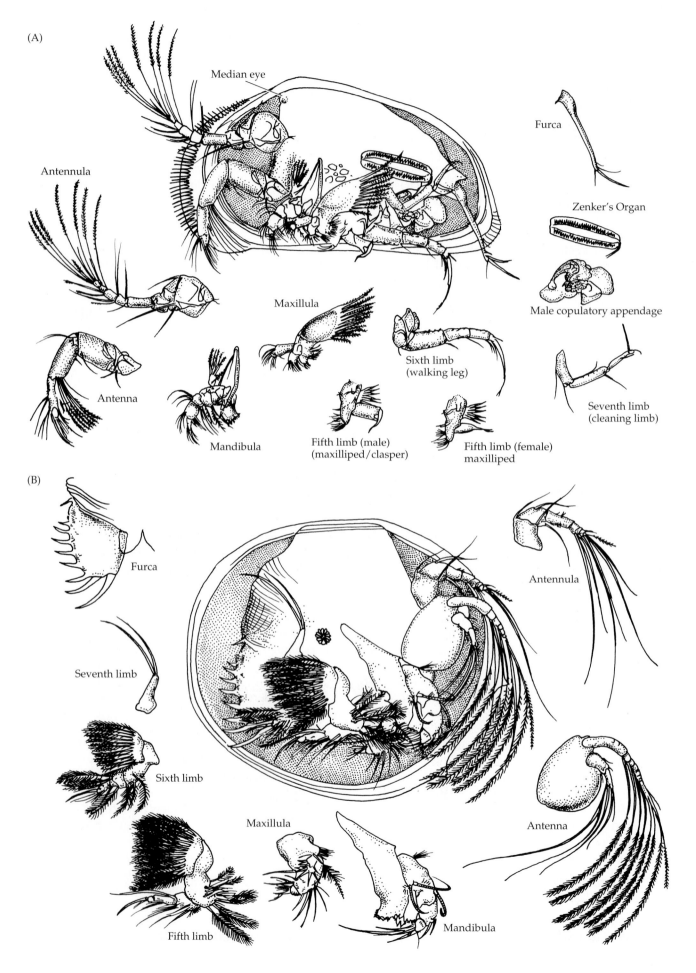

Median eye

Furca

Antennula

Zenker's Organ

Maxillula

Male copulatory appendage

Antenna

Sixth limb
(walking leg)

Mandibula

Seventh limb
(cleaning limb)

Fifth limb (male)
(maxilliped/clasper)

Fifth limb (female)
maxilliped

(B)

Furca

Antennula

Seventh limb

Sixth limb

Antenna

Maxillula

Fifth limb

Mandibula

(C)

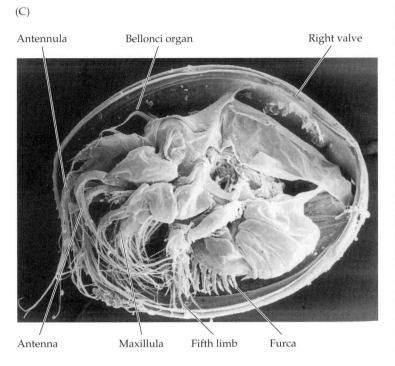

Antennula   Bellonci organ   Right valve

Antenna   Maxillula   Fifth limb   Furca

Figure 16.20 Anatomy and diversity in the Ostracoda. (A) Anatomy of *Sclerocypris* (Podocopa). (B) Anatomy of *Thaumatoconcha* (Myodocopa). (C) Internal view of *Metapolycope* (Myodocopa), left valve removed. (D) The highly ornate *Eusarsiella* (Myodocopa); side view and edge view, showing the ornate shell. (E) Examples of genera from the major ostracod groups (scale bar = 1.0 mm). A: *Vargula* (Myodocopa, Myodocopida). B: *Eusarsiella* (Myodocopa, Myodocopida). C: *Polycope* (Myodocopa, Halocyprida). D: *Cytherelloidea* (Podocopa, Platycopida). E: *Propontocypris* (Podocopa, Podocopida). F: *Macrocypris* (Podocopa, Podocopida). G: *Saipanetta* (Podocopa, Podocopida). H: *Neonesidea* (Podocopa, Podocopida). I: *Triebelina* (Podocopa, Podocopida). J: *Candona* (Podocopa, Podocopida). K: *Ilyocypris* (Podocopa, Podocopida). L: *Cyprinotus* (Podocopa, Podocopida). M: *Potamocypris* (Podocopa, Podocopida). N: *Hemicytherura* (Podocopa, Podocopida). O: *Acanthocythereis* (Podocopa, Podocopida). P: *Celtia* (Podocopa, Podocopida). Q: *Limnocythere* (Podocopa, Podocopida). R: *Sahnicythere* (Podocopa, Podocopida). S: *Darwinula* (Podocopa, Podocopida). All arrows point anteriorly.

(E)

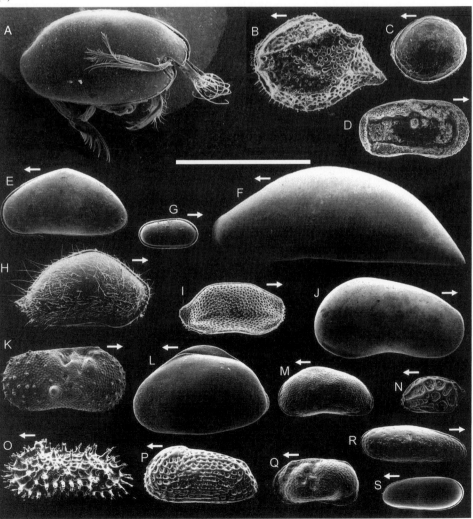

(D)

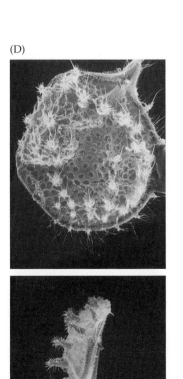

include scavengers, detritus feeders, suspension feeders, and some predators. There are two orders: Myodocopida and Halocyprida.

The podocopans include predominantly benthic forms; although some are capable of temporary swimming, none are fully planktonic. Their feeding methods include suspension feeding, herbivory, and detritus feeding. The Podocopa are divided into three orders: the exclusively marine Platycopida, the ubiquitous Podocopida, and the Palaeocopida. The Palaeocopida were diverse and widespread in the Paleozoic, but are represented today only by the extremely rare Punaciidae (known from a few living specimens, and from dead valves dredged in the South Pacific).

There is considerable debate as to whether or not the ostracods belong to the Maxillopoda. Excluding them, and giving them class status, seems to have growing popularity. However, we retain them in the Maxillopoda pending more evidence.

The Crustacean Bauplan

We realize that the above synopses are rather extensive, but the diversity of crustaceans demands emphasis before we attempt to generalize about their biology. The success of the crustaceans, like that of other arthropods, has been closely tied to modifications of the jointed exoskeleton and appendages. These animals have exploited an evolutionary flexibility that has allowed an extensive range of modification of these body parts for a great variety of functions.

The most basic crustacean body plan is a head (**cephalon**) followed by a long body (**trunk**) with many similar appendages, as seen in the primitive class Remipedia (Figures 16.1J, 16.3D,E). In the other crustacean classes, however, various degrees of tagmosis occur, and the cephalon is typically followed by a trunk that is divided into two distinct regions, a **thorax** and an **abdomen**. All crustaceans possess, at least primitively, a **cephalic shield** (**head shield**) or a **carapace**. The cephalic shield results from the fusion of the dorsal head tergites to form a solid cuticular plate, often with ventrolateral folds ("pleural folds") on the sides. Head shields are characteristic of the classes Remipedia and Cephalocarida, and they also occur in some maxillopodans and malacostracans. The carapace is a more expansive structure, comprising the head shield and a large fold of the exoskeleton that probably arises (primitively) from the maxillary somite. The carapace may extend over the body dorsally and laterally as well as posteriorly, and it often fuses to one or more thoracic segments, thereby producing a **cephalothorax** (Figure 16.2A). Occasionally, the carapace may grow forward beyond the head as a narrow **rostrum**.

Most of the differences among the major groups of crustaceans, and the basis for much of their classification, arise from variations in the number of somites in the thorax and abdomen, the form of their appendages, and the size and shape of the carapace. A brief skimming of the synopses (above) and the corresponding figures will give you some idea of the range of variation in these characteristics.

Monophyly and uniformity within the subphylum Crustacea is demonstrated particularly by the elements of the cephalon and the presence of a nauplius larva. Except for a few cases of secondary reduction, the head of all crustaceans has 5 pairs of appendages. From anterior to posterior, these are the **antennules** (first antennae), **antennae** (second antennae), **mandibles**, **maxillules** (first maxillae), and **maxillae** (second maxillae). The presence of 2 pairs of antennae is, among arthropods, unique to the Crustacea (as is the nauplius larva). Although the eyes of some are simple, most possess a pair of well developed compound eyes, either set directly on the head (**sessile eyes**) or borne on distinct movable stalks (**stalked eyes**).

In many crustaceans, from one to three anterior thoracic segments (**thoracomeres**) are fused with the cephalon. The appendages of these fused segments are typically incorporated into the head as additional mouthparts called **maxillipeds**. In the class Malacostraca, the remaining free thoracomeres are together termed the **pereon**. Each segment of the pereon is called a **pereonite** (= **pereomere**), and their appendages are called **pereopods**. The pereopods may be specialized for walking, swimming, gas exchange, feeding, or defense. Crustacean thoracic (and pleonal) appendages are probably primitively biramous, although a "reduction" to a uniramous condition is seen in a variety of taxa. As described in Chapter 15, the basic crustacean limb is composed of a basal **protopod** (= **sympod**), from which may arise medial **endites** (e.g., gnathobases), lateral **exites** (e.g., epipods), and two rami, the **endopod** and **exopod**.* Members of the classes Remipedia, Cephalocarida, Branchiopoda, and some ostracods possess appendages with uniarticulate (single-segment) protopods; the remaining classes (most Maxillopoda and Malacostraca) usually have appendages with multiarticulate protopods (Table 16.1).†

The abdomen, called a **pleon** in malacostracans, is composed of several segments, or **pleonites** (= **pleomeres**), followed by a postsegmental plate or lobe, the

---

*As noted in Chapter 15, the exopod is probably no more than a highly modified exite that evolved from an ancestral uniramous condition.

†The term **peduncle** is a general name often applied to the basal portion of certain appendages; it is occasionally (but not always) used in a way that is synonymous with protopod.

anal somite or **telson**, bearing the anus (Figure 16.2B). In primitive crustaceans this anal somite bears a pair of appendage-like or spinelike processes conventionally called **caudal rami**. In the Eumalacostraca, the anal somite is a flattened telson and lacks caudal rami.

In general, distinctive abdominal appendages (**pleopods**) occur only in the malacostracans. These appendages are almost always biramous, and often they are flaplike and used for swimming (e.g., Figures 16.9–16.15). The posteriorly directed last pair(s) of abdominal appendages are usually different from the other pleopods, and are called **uropods**. Together with the telson, the uropods form a distinct **tail fan** in many malacostracans (Figure 16.2B).

Crustaceans produce a characteristic larval stage called the **nauplius** (Figures 16.25B,C, 16.33D), which bears a median simple (naupliar) eye and 3 pairs of setose, functional appendages—destined to become the antennules, antennae, and mandibles. In many groups (e.g., the Peracarida), however, the free-living nauplius larva is absent or suppressed. In such cases, development is either fully direct or mixed, with larval hatching taking place at some postnaupliar stage (Table 16.2). Often other larval stages follow the nauplius (or other hatching stage) as the individual passes through a series of molts, during which segments and appendages are gradually added.

Locomotion

Crustaceans move about primarily by use of their limbs; lateral body undulations are unknown. They crawl or swim, or more rarely burrow, "hitchhike," or jump. Many of the ectoparasitic forms (e.g., branchiurans, certain isopods, and copepods) are largely sedentary on their hosts, and most cirripedes are fully sessile.

Swimming is usually accomplished by a rowing action of the limbs. Archetypical swimming is exemplified by the primitive crustaceans with relatively undifferentiated trunks and high numbers of similar biramous appendages (e.g., remipedes, anostracans, and notostracans). In general, these animals swim by posterior to anterior metachronal beating of the trunk limbs (Figure 16.22 and Chapter 15). The appendages of such crustaceans are often broad and flattened, and they usually bear fringes of setae that increase the effectiveness of the power stroke. On the recovery stroke the limbs are flexed, and the setae may collapse, reducing resistance. In members of some groups (e.g., Cephalocarida, Branchiopoda, and Leptostraca), large exites or epipods arise from the base of the leg, producing broad, "leafy" limbs called **phyllopodia**. These flaplike structures aid in locomotion and may also serve as osmoregulatory (branchiopods) or gas exchange (cephalocardis and leptostracans) surfaces (Figure 16.21A–C). Although such epipods increase the surface area on the power stroke, they also are hinged so that they collapse on the recov-

ery stroke, reducing resistance. Metachronal limb movements are retained in many of the "higher" swimming crustaceans, but they tend to be restricted to selected appendages (e.g., the pleopods of shrimps, stomatopods, amphipods, and isopods; the pereopods of euphausids and mysids). In swimming euphausids and mysids the thoracopods beat in a metachronal rowing fashion, with the exopod and setal fan extended on the power stroke and flexed on the recovery stroke. The movements and nervous–muscular coordination of crustacean limbs are deceivingly complex. In the common mysid *Gnathophausia ingens*, for example, twelve separate muscles power the thoracic exopod alone (three that are extrinsic to the exopod, five in the limb peduncle, and four in the exopodal flagellum).

Recall from our discussions in Chapters 3 and 15 that at the low Reynolds numbers at which small crustaceans (such as copepods or larvae) swim about, the netlike setal appendages act not as a filtering net, but as a paddle, pushing water in front of them and dragging the surrounding water along with them due to the thick boundary layer adhering to the limb. Only in larger organisms, with Reynolds numbers approaching 1, do setose appendages (e.g., the feeding cirri of barnacles) begin to act as filters, or rakes, as the surrounding water becomes less viscous and the boundary layer thinner. Of course, the closer together the setae and setules are placed, the more likely it is that their individual boundary layers will overlap; thus densely setose appendages are more likely to act as paddles.

Not all swimming crustaceans move by typical metachronal waves of limb action. Certain planktonic copepods, for example, move haltingly and depend on their long antennules and dense setation for flotation between movements (Figure 16.18F). Watch living calanoid copepods and you will notice that they may move slowly by use of the antennae and other appendages, or in short jerky increments, often sinking slightly between these movements. The latter type of motion results from an extremely rapid and condensed metachronal wave of power strokes along the trunk limbs. Although the long antennae may appear to be acting as paddles, they actually collapse against the body an instant prior to the beating of the limbs, thus reducing resistance to forward motion. Some other planktonic copepods create swimming currents by rapid vibrations of cephalic appendages, by which the body moves smoothly through the water. "Rowing" does occur in the swimming crabs (family Portunidae) and some deep-sea asellote isopods (e.g., family Eurycopidae), which use paddle-shaped posterior thoracopods to scull about.

Most eumalacostracans with well developed abdomens exhibit a form of temporary, or "burst," swimming that serves as an escape reaction (e.g., mysids, syn-

TABLE 16.1 *Comparison of distinguishing features among the five crustacean classes*

Taxon	Carapace or cephalic shield	Body tagmata and number of segments in each (excluding telson)	Thoracopods	Maxillipeds
Class Remipedia	Cephalic shield	Cephalon (5) Trunk (up to 32)	Not phyllopodous	1 pair
Class Cephalocarida	Cephalic shield	Cephalon (5) Thorax (8) Abdomen (11)	Phyllopodous	None
Class Branchiopoda, Order Anostraca	Cephalic shield	Cephalon (5) Thorax (usually 11) Abdomen (usually 8)	Phyllopodous	None
Class Branchiopoda, Order Notostraca	Carapace	Cephalon (5) Thorax (11) Abdomen (many segments)	Phyllopodous	None
Class Branchiopoda, Order Diplostraca	Carapace (bivalved, hinged or folded)	Cephalon (5) Trunk (10–32, or obscured)	Phyllopodous	None
Class Malacostraca, Subclass Phyllocarida	Large, folded carapace covers thorax	Cephalon (5) Thorax (8) Abdomen (7)	Phyllopodous	None
Class Malacostraca, Subclass Eumalacostraca	Carapace well developed or secondarily reduced or lost	Cephalon (5) Thorax (8) Abdomen (6)	Not phyllopodous; uniramous in many	0 to 3 pairs
Class Maxillopoda	Carapace or cephalic shield	Cephalon (5) Thorax (6) Abdomen (4)	Not phyllopodous; often reduced	0 or 1 pair

carids, euphausids, shrimps, lobsters, and crayfish). By rapidly contracting the ventral abdominal (flexor) muscles, such animals shoot quickly backward, the spread tail fan providing a large propulsive surface (Figure 16.22C–E). This behavior is sometimes called a tail-flip, or **caridoid escape reaction**.

Surface crawling by crustaceans is accomplished by the same general sorts of leg movements described in the preceding chapters for insects and other arthropods: by flexion and extension of the limbs to pull or push the animal forward. Walking limbs are typically composed of relatively stout, more or less cylindrical articles (i.e., **stenopodous** limbs) as opposed to the broader, often phyllopodous limbs of swimmers (see Figure 16.21 for a comparison of crustacean limb types). Walking limbs are lifted from the substratum and moved forward during their recovery strokes; then they are placed against the substratum, which provides purchase as they move posteriorly through their power strokes, pulling and then pushing the animal forward. Like many other arthropods, crustaceans generally lack lateral flexibility at the body joints, so turning is accomplished by reducing the stride length or movement frequency on one side of the body, toward which the animal turns (like a tractor or tank slowing one tread).

Most walking crustaceans can also reverse the direction of leg action and move backward, and most brachyuran crabs can walk sideways. Brachyuran crabs are perhaps the most agile of all crustaceans. The extreme reduction of the abdomen in this group allows for very rapid movement because adjacent limbs can move in directions that avoid interference with one another.

Antennules	Antennae	Compound eyes	Abdominal appendages	Gonopore location
Biramous	Biramous	Absent	All trunk appendages similar	♂: protopods of trunk segment 15: ♀: trunk segment 8
Uniramous	Biramous	Absent	None	Common pores on protopods of thoracopods 6
Uniramous	Uniramous	Present	None	♀: on segment 12/13 or 20/21
Uniramous	Vestigial	Present	Present (posteriorly reduced)	Thoracomere 11 (both sexes)
Uniramous	Biramous	Present	All trunk appendages similar, or posteriormost segments limbless	Variable; on trunk segment 9 or 11, or on apodous posterior region (some Cladocera)
Biramous	Uniramous	Present	Pleopods (posteriorly reduced)	♂: coxae of thoracopods 8; ♀: coxae of thoracopods 6
Uniramous, biramous, or triramous	Uniramous or biramous	Present; well developed	Usually 5 pairs pleopods, 1 pair uropods	♂: coxae of thoracopods 8, or sternum of thoracomere 8; ♀: coxae of thoracopods 6, or sternum of thoracomere 6
Uniramous	Uniramous or biramous	Present or absent	None	Variable; ♂ openings usually on trunk segment 4 or 7; ♀ on 1, 4, or 7.

Brachyuran crab legs are hinged in such a way that most of their motion involves lateral extension (abduction) and medial flexion (adduction) rather than rotation frontward and backward. As a crab moves, its limbs move in various sequences, as in normal crawling, but those on the leading side exert their force by flexing and pulling the body toward the limb tips, while the opposite, trailing, legs exert propulsive force as they extend and push the body away from the tips. Still, this motion is simply a mechanical variation on the common arthropodan walking behavior.

In addition to these two basic locomotor methods ("typical" walking and swimming by metachronal beating of limbs), many crustaceans move by other specialized means. Ostracods, cladocerans, and clam shrimp (Diplostraca), which are almost entirely enclosed by their carapaces (Figures 16.4F,G,J,L,M and 16.20), swim by rowing with the antennae. Mystacocarids crawl in interstitial water using various head appendages. Certain semi-terrestrial amphipods known as "beach hoppers" (e.g., *Orchestia* and *Orchestoidea*) execute dramatic jumps by rapidly extending the urosome and its appendages (uropods). Most caprellid amphipods (Figure 16.15E) move about in inchworm fashion, using their subchelate appendages for clinging. There are also a number of crustacean burrowers, and even some that build their own tubes or "homes" from materials in their surroundings. Many benthic amphipods, for example, spin silklined mud burrows in which they reside. One species, *Pseudamphithoides incurvaria* (suborder Gammaridea), constructs and lives in an unusual "bivalved pod" cut from the thin blades of the same alga on which it feeds

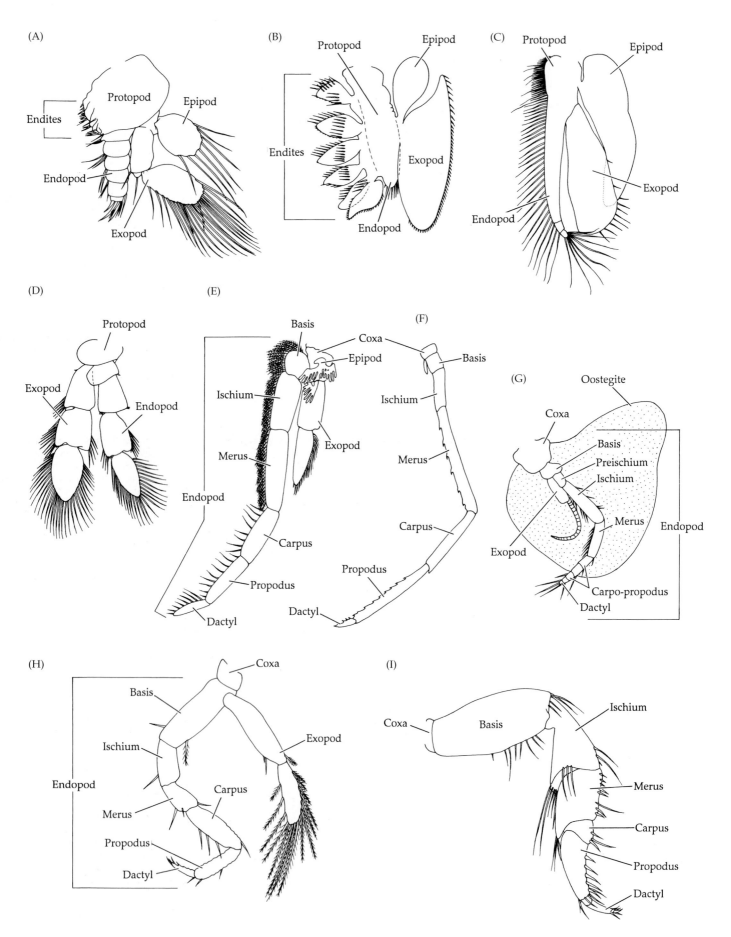

◀ **Figure 16.21**   Generalized thoracic appendages of various crustaceans. (A) Cephalocarida. (B) Branchiopoda. Dashed lines indicate fold or "hinge" lines. (C) Leptostraca (Phyllocarida). (D) Remipedia. (E) Euphausiacea. (F) Caridea (Decapoda). (G) Lophogastrida (Peracarida). (H) Spelaeogriphacea (Peracarida). (I) Isopoda (Peracarida). A–C are biramous, phyllopodous thoracopods; D is a biramous, flattened, but nonphyllopodous thoracopod; E–I are stenopodous thoracopods. Because of the presence of large epipods on the legs of the cephalocarids, branchiopods, and phyllocarids, some authors refer to them as "triramous" appendages. However, smaller epipods also occur on many typical "biramous" legs, so this distinction seems unwarranted (and confusing). Note that in the four primi-tive groups of crustaceans (cephalocarids, branchiopods, phyllocarids, and remipedes) the protopod is composed of a single article. And in branchiopods and leptostracans, the articles of the endopod are not clearly separated from one another. In the higher crustaceans (classes Maxillopoda and Malacostraca) the protopod comprises two or three separate articles, although in most maxillopodans these may be reduced and not easily observed. In the lophogastrid (G), the large marsupial oostegite characteristic of female peracarids is shown arising from the coxa. In two groups (amphipods and isopods) all traces of the exopods have disappeared, and only the endopod remains as a long, powerful, uniramous walking leg.

Figure 16.22   Some aspects of locomotion (and feeding) in three crustaceans (see also Chapter 15). (A,B) Generation of swimming and feeding currents in an anostracan. (A) An anostracan swimming on its back by metachronal beating of the trunk limbs. The limbs are hinged to fold on the recovery stroke, thereby reducing resistance. Water is drawn from anterior to posterior along the midline and into the interlimb spaces (B), and food particles are trapped on the medial sides of the endites; excess water is pressed out laterally, and the trapped food is moved anteriorly to the mouth. (C–E) Locomotion in the postlarva of *Panulirus argus*. (C) Normal swimming posture when moving forward slowly. (D) Sinking posture with appendages flared to reduce sinking rate. (E) A quick retreat by rapid tail flexure (the "caridoid escape reaction"), a method commonly employed by crustaceans with well developed abdomens and tail fans. (F) A swimming remipede, *Lasionectes*. Note the metachronal waves of appendage movement.

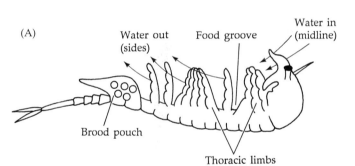

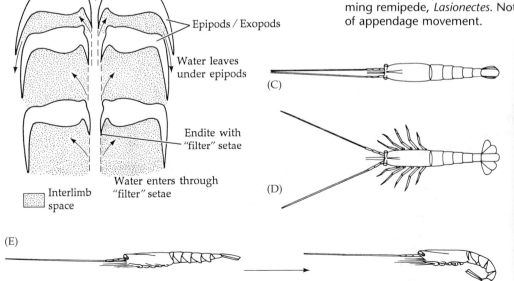

TABLE 16.2 *Summary of crustacean reproductive features*

Taxon	Development type, or larval type at hatching	Hermaphroditic	Dioecious	Parthenogenesis (in at least some species)
Class Remipedia	?	+	—	?
Class Cephalocarida	Metanauplius	+	—	—
Class Branchiopoda				
Order Anostraca	Nauplius or metanauplius	—	+	+
Order Notostraca	Nauplius or metanauplius	+	+	+
Order Diplostraca	Nauplius, metanauplius; or direct development	—	+	+
Class Maxillopoda				
Subclass Ostracoda	Direct, or with bivalved nauplius/ metanauplius with anamorphic development	—	+	+
Subclass Mystacocarida	Nauplius or metanauplius	—	+	—
Subclass Copepoda	Nauplius	—	+	—
Subclass Branchiura	Nauplius or direct development	—	+	—
Subclass Thecostraca	Nauplius	+	+	—
Subclass Tantulocarida	?	—	+	?
Class Malacostraca				
Subclass Phyllocarida	Direct development	—	+	—
Subclass Eumalacostraca				
Superorder Hoplocarida	Zoea larva ("antizoea" or "pseudozoea")	—	+	—
Superorder Syncarida	Direct development	—	+	—
Superorder Eucarida				
Order Euphausiacea	Nauplius	—	+	—
Order Amphionidacea	Nauplius	—	+	—
Order Decapoda	Protozoea or zoea (nauplius in Dendrobranchiata)	+	+	—
Superorder Peracarida	Direct development	—	+	—

Comments

Development not yet studied

Two eggs at a time fertilized and carried on genital processes of first pleonites

Embryos usually shed from ovisac early in development; resistant (cryptobiotic) fertilized eggs accommodate unfavorable conditions

Eggs briefly brooded, then deposited on substrata; resistant (cryptobiotic) fertilized eggs accommodate unfavorable conditions

Most cladocerans undergo direct development (*Leptodora* hatches as nauplii or metanauplii). Clam shrimps carry developing embryos on the thoracopods prior to releasing them as nauplii or metanauplii

Embryos usually deposited directly on substrata; many myodocopans and some podocopans brood embryos between valves until hatching as a reduced adult; no metamorphosis; up to 8 preadult instars

Little is known concerning this group; eggs apparently laid free; 6 naupliar stages (?)

Usually with 6 naupliar stages leading to a second series or 5 "larval" stages called copepodites

Embryos deposited; only *Argulus* known to hatch as nauplii; others have direct development and hatch as juveniles

Six naupliar stages followed by unique larval form called a cypris larva

Development entails complex metamorphosis

All undergo direct development in female brood pouch, hatching as postlarval "manca" (juvenile)

Eggs brooded or deposited in burrow; hatch late as a clawed pseudozoea larva, or earlier as an unclawed antizoea larva; both go through several molts before settling on bottom as juveniles

Free larval stages lost; eggs deposited on substratum

Embryos shed or briefly brooded; typically undergo nauplius → zoea → megalopa → juvenile → adult developmental series

Apparently brooded under thorax, but held by anterior pleopods; typically undergo nauplius → zoea → megalopa → juvenile → adult development

Dendrobranchiata shed embryos to hatch in water as nauplii or protozoea; all others brood embryos (on pleopods), which do not hatch until at least the zoea stage

Embryos brooded in marsupium typically formed from ventral coxal plates called oostegites; usually released as mancas (subjuveniles with incompletely developed eighth thoracopods. Brood pouch (marsupium) in Thermosbaenacea formed by dorsal carapace chamber

(Figure 16.23A). Another gammaridean amphipod, *Photis conchicola*, actually uses empty gastropod shells in a fashion similar to that of hermit crabs (Figure 16.23B). "Hitchhiking" (phoresis) occurs in various ectosymbiotic crustaceans, including isopods that parasitize fishes or shrimps and hyperiidean amphipods that ride on gelatinous drifting plankters.

In addition to simply getting from one place to another in their usual day-to-day activities, many crustaceans exhibit various migratory behaviors, employing their locomotor skills to avoid stressful situations or to remain where conditions are optimal. A number of planktonic crustaceans undertake daily vertical migrations, typically moving upward at night and to greater depths during the day. Such vertical migrators include various copepods, cladocerans, ostracods, and hyperiid amphipods (the latter may make their migrations by riding on their hosts). Such movements place the animals in their near-surface feeding grounds during the dark hours, when there is probably less danger of being detected by visual predators. In the daytime, they move to deeper, perhaps safer, water. Many intertidal errant crustaceans use their locomotor abilities to change their behaviors with the tides. Some anomuran and brachyuran crabs simply move in and out with the tide, or seek shelter beneath rocks when the tide is out, thus avoiding the problems of exposure.

One of the most interesting locomotor behaviors among crustaceans is the mass migration of the spiny lobster, *Panulirus argus*, in the Gulf of Mexico and northern Caribbean. Each autumn, lobsters queue up in single file and march in long lines for several days. They move from shallow areas to the edges of deeper oceanic channels. This behavior is apparently triggered by winter storm fronts moving into the area, and it may be a means of avoiding rough water conditions in the shallows.

Feeding

With the exception of ciliary mechanisms, crustaceans have exploited virtually every feeding strategy imaginable. Even without cilia, many crustaceans generate water currents and engage in various types of suspension feeding. We have selected a few examples to demonstrate the range of feeding mechanisms that occur in this group.

In some crustaceans the action of the thoracic limbs simultaneously creates the swimming and suspension feeding currents. As the metachronal wave of appendage motion passes along the body, adjacent limb pairs are alternately moved apart and then pressed together, thus changing the size of each interlimb space (Figure 16.22A,B; see also Chapters 3 and 15). Surrounding water is drawn into an interlimb space as the adjacent appendages move away from one another, and water-borne particles are trapped by setae on the endites as the appendages then close. From here, the trapped particles are moved to a midventral food

(A)

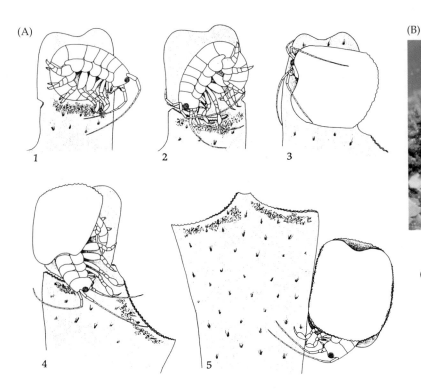

(B)

(C)

Figure 16.23 Amphipod "houses." (A) The complex sequence of steps in the construction of a "bivalve pod" from the brown alga *Dictyota* by the Caribbean amphipod *Pseudampithoides incurvaria*: (1) Initiation of cut and notch; the upper flap of the alga forms the first "valve." (2) Continuation of the cut across the algal thallus. (3) Measuring and clearing algal hairs off the second branch tip. (4) Cutting of the second valve. (5) Completed "pod" with valves attached along margins by threadlike secretions. (B) The gammaridean amphipod *Grandiderella* (male and female) in its silk-lined tube. The antennules and antennae are protruding from the pod entrance on the right. (C) *Photis conchicola*, a temperate eastern Pacific amphipod that spins its silken tube inside a minute snail shell, which is then carried about in the style of hermit crabs.

groove and then anteriorly, toward the head. This mechanism of forming a boxlike "filter" press with setose phyllopodous limbs is the typical suspension feeding strategy of cephalocarids and many malacostracans.

Planktonic copepods were long thought to "filter" feed by generating lateral feeding gyres or currents by movements of the antennae and mouth appendages. It was believed that these gyres swept in small particles that were directly filtered by the maxillae. This classic idea of maxillary filtration, built on work by H. G. Cannon in the 1920s, has been questioned by recent workers, but the model persists and is still commonly presented in general works. As mentioned in Chapter 3, we now know that copepods and other small planktonic crustaceans live in a world of low Reynolds numbers, a world dominated by viscosity rather than by inertia. Thus, the setose mouth appendages behave more like paddles than like sieves, with a water layer near the limb adhering to it and forming part of the "paddle." As the maxillae move apart, parcels of water containing

food are drawn into the interlimb space. As the maxillae press together, the "parcel" is moved forward to the endites of the maxillules, which push it into the mouth. Thus, food particles are not actually filtered from the water, but are captured in small parcels of water. High-speed cinematography indicates that copepods may capture individual algal cells, one at a time, by this "hydraulic vacuum" method.

Sessile thoracican barnacles feed by using their long, feathery, biramous thoracopods, called **cirri**, to filter feed on suspended material from the surrounding water (Figures 16.16A,C,E). Studies indicate that barnacles are capable of trapping food particles ranging from 2 μm to 1 mm, including detritus, bacteria, algae, and various zooplankters. Many barnacles are also capable of preying on larger planktonic animals by coiling a single cirrus around the prey, in tentacle fashion. In slow-moving or very quiet water, most barnacles feed actively by extending the last three pairs of cirri in a fanlike manner and sweeping them rhythmically through the water. The setae on adjacent limbs and limb rami overlap to form an effective filtering net. The first three pairs of cirri serve to remove trapped food from the posterior cirri and pass it to the mouthparts. In areas of high water movement, such as wave-swept rocky shores, barnacles often extend their cirri into the backwash of waves, allowing the moving water to simply run through the "filter," rather than moving the cirri through the water. In such areas you will often see clusters of barnacles in which all the individuals are oriented similarly, taking advantage of this labor-saving device.

Most krill (euphausids) feed in a fashion similar to barnacles, but while swimming. The thoracopods form a "feeding basket" that expands as the legs move outward, sucking food-laden water in from the front. Once inside the "basket," particles are retained on the setae of the legs as the water is squeezed out laterally. Other setae comb the food particles out of the "trap" setae, while yet another set brushes them forward to the mouth region.

Sand crabs of the genus *Emerita* (Anomura) use their long, setose antennae in a fashion similar to that of barnacle cirri that "passively" strain wave backwash (Figure 16.24A; see also Chapter 3). *Emerita* are adapted to living on wave-swept sand beaches. Their compact oval shape and strong appendages facilitate burrowing in the unstable substratum. They burrow posterior end first in the area of shallow wave wash, with the anterior end facing upward. Following a breaking wave, as the water rushes seaward, *Emerita* unfurls its antennae into the moving water along the surface of the sand. The fine setose mesh traps bacteria, protists, and phytoplankton from the water, and the antennae then brush the collected food onto the mouthparts. Many porcelain and hermit crabs also engage in suspension feeding. By twirling their antennae in various patterns these anomurans create spiraling currents that bring food-laden water toward the mouth (Figure 16.24B,C). Food particles become entangled on the setae near the base of the antennae and then are brushed into the mouth by the endopods of the third maxillipeds. Many of these animals also feed on detritus by simply picking up particles with their chelipeds.

Mud and ghost shrimps, such as *Callianassa* and *Upogebia* (Thalassinidea), suspension feed within their burrows. They drive water through the burrow by beating the pleopods, and the first two pairs of pereopods remove food with medially directed setal brushes. The maxillipeds then comb the captured particles forward to the mouth.

Most other crustacean feeding mechanisms are less complicated than suspension feeding and usually involve direct manipulation of food by the mouthparts and sometimes the pereopods, especially chelate or subchelate anterior legs.

Many small crustaceans may be classified as microphagous selective deposit feeders, employing various methods of removing food from the sediments in which they live. Mystacocarids, many harpacticoid copepods, and some cumaceans and gammaridean amphipods are referred to as "sand grazers" or "sand lickers." By various methods these animals remove detritus, diatoms, and other microorganisms from the surfaces of sediment particles. Interstitial mystacocarids, for example, simply brush sand grains with their setose mouthparts. On the other hand, some cumaceans pick up an individual sand grain with their first pereopods and pass it to the maxillipeds, which in turn rotate and tumble the particle against the margins of the maxillules and mandibles. The maxillules brush and the mandibles scrape, removing organic material. Some sand-dwelling isopods may utilize a similar feeding behavior.

Predatory crustaceans include stomatopods, remipedes, and most lophogastrids, as well as many species of anostracans, cladocerans, copepods, ostracods, cirripedes, anaspidaceans, euphausids, decapods, tanaids, isopods, and amphipods. Predation typically involves grasping the prey with chelate or subchelate pereopods (or sometimes directly with the mouth appendages), followed by tearing, grinding, or shearing with various mouthparts, particularly the mandibles. Perhaps the most highly adapted predatory specialists are the stomatopods (Figure 16.7). These hunters or ambushers possess greatly enlarged, raptorial subchelate limbs, which they use to stab or to club and smash prey. Some species search out prey, but many sit in ambush at their burrow entrance. The actual attack generally follows visual detection of a potential prey item, which may be another crustacean, a mollusc, or even a small fish. Once captured and stunned or killed by the raptorial claws, the prey is held against the mouthparts and shredded into ingestible pieces.

Although the cave-dwelling caridean shrimp *Procaris* is omnivorous, its predatory behavior is particularly interesting. Its prey includes other crustaceans, particularly amphipods and shrimps. After *Procaris* locates a potential victim (probably by chemoreception), it moves quickly to the prey and grasps it within a "cage" formed by the pereopodal endopods (Figure 16.24D). Once captured, the prey is eaten while the shrimp swims about. Apparently the third maxillipeds press the prey against the mandibles, which bite off chunks and pass them to the mouth.

The remipedes capture prey with their raptorial mouth appendages (Figures 16.1J, 16.3D–F, 16.22F), then immobilize the victim with an injection from the hypodermic maxillules. It is suspected that tissues are then sucked out of the prey by action of a mandibular mill and muscular foregut. They are also facultative scavengers.

Another fascinating adaptation for predation can be seen in the snapping shrimp (Alpheidae, e.g., *Alpheus*,

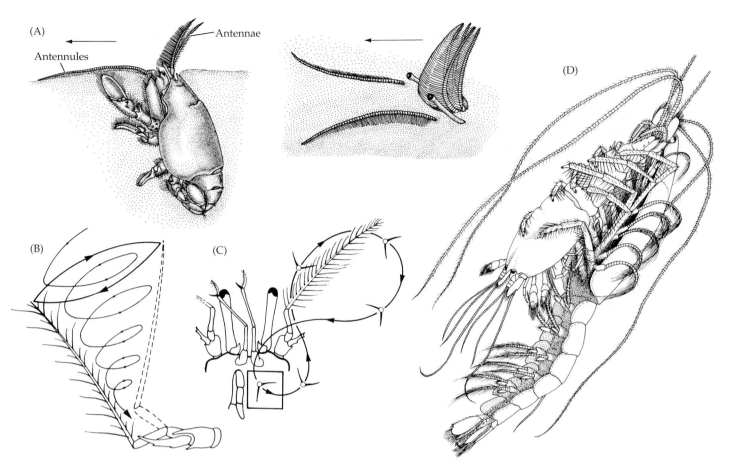

Figure 16.24 Some crustacean feeding mechanisms (see also Figure 22). (A) Suspension feeding in the sand crab *Emerita*. The arrows point seaward and indicate the direction of water movement as waves recede. The antennules direct water through branchial chambers. The antennae remove food particles from the water and then brush them onto the mouthparts. (B,C) The suspension feeding hermit crabs (B) *Australeremus cooki* and (C) *Paguristes pilosus* twirl their antennae, either in a circle or a figure eight, to create water currents that pull food particles to the mouth region. (D) The predatory shrimp *Procaris ascensionis* (Caridea) is shown here munching on another shrimp (*Typhlatya*) as it holds the prey in a "cage" formed by the pereopodal endopods.

Synalpheus) (Figure 16.9D). One of the chelipeds is much larger than the other, and the movable finger is hinged in such a way that it can be snapped closed; this forceful closing produces a loud popping sound and a pressure or "shock" wave in the surrounding water. Most species appear to use this mechanism in ambushing prey. The shrimp sits at its burrow entrance with the antennae extended. When a potential prey approaches (usually a small fish), the shrimp "pops" its cheliped, and the resulting pressure wave stuns the victim, which is then quickly pulled into the burrow, killed, and consumed. These shrimps typically live in pairs within the burrow, and prey captured by one individual is shared with its

partner. Two mechanisms have been proposed for the production of the "pop" and associated shock wave, neither of which depends on the actual smashing together of the claw. In some species the pop seems to be created when opposing disks on the dactyl and propodus (the "finger" and "hand" of the chela), held together when the claw is fully open, separate upon closing. This motion requires overcoming the tensile strength and cohesive forces of a thin layer of water held between the discs. Some shrimp can generate this popping sound even when out of water, so long as the disks are moist. A second mechanism recently proposed is the collapse of cavitation bubbles, which are created by the rapid closure of the claw (in excess of 100 km/sec). Both mechanisms create shock waves.

Many crustaceans emerge from the benthos under cover of darkness to feed or mate in the water column. Many predatory isopods emerge at night to feed on invertebrates or fish, particularly weak or diseased fish (or fish caught in fishing nets).

Macrophagous herbivorous and scavenging crustaceans generally feed by simply hanging onto the food source and biting off bits with the mandibles (a feeding technique similar to that of grasshoppers). Notostracans, some ostracods, and many decapods, isopods, and amphipods are scavengers and herbivores. Certain isopods in the family Sphaeromatidae bore into the aerial roots of mangrove trees. Their activities often result in root breakage followed by new multiple root initiation, creating the stiltlike appearance characteristic of red mangrove (*Rhizophora*). A number of crustaceans are full-time or part-time detritivores, including some or most cephalocarids, ostracods, bathynellaceans, thermosbaenaceans, clam shrimps, many peracarids, and some decapods. Some scavenge directly on detritus, but others (e.g., cephalocarids) stir up the sediments in order to remove organic particles by suspension feeding.

Finally, several groups of crustaceans have adopted various degrees of parasitism. These animals range from ectoparasites with mouthparts modified for piercing or tearing and sucking body fluids (e.g., many copepods, branchiurans, tantulocarids, several isopod families, and at least one species of ostracod) to the highly modified and fully parasitic rhizocephalans, whose bodies ramify throughout the host tissue and absorb nutrients directly (Figures 16.25, 16.26).

Rhizocephalans, which are "barnacles" (Cirripedia) that have been highly modified to become internal parasites of other crustaceans, are some of the most bizarre organisms in the animal kingdom. They have a typical cirripede cypris larva, but in this group the cyprid will settle only upon another crustacean, selected to be the unfortunate host. The most complex rhizocephalan life cycle is that of the suborder Kentrogonida, obligate parasites of decapods. In this group, a settled female cypris larva undergoes an internal reorganization that rivals that of caterpillar pupae in scope, developing an infective stage, called the **kentrogon**, beneath the cyprid exoskeleton. Once fully developed, the kentrogon forms a hollow cuticular structure, the **stylet**, which injects a motile, multicellular, vermiform creature called the **vermigon** (or **vermiform instar**) into the host. The vermigon is the active infection stage. It has a thin cuticle, several types of cells, and the anlagen of an ovary. It invades the host's hemocoel by sending out long, branching, hollow **rootlets** that penetrate most of the host's body and draw nutrients directly from the hemocoel. So profound is the intrusion by the rootlets that the parasite takes over nearly complete control of the host's body, altering its morphology, physiology, and behavior. Once the parasite invades the host's gonads, parasitic castration results (the gonads of parasitized crab never mature). Thus the host is transformed into a slave that serves the needs of its master.* The internal root system, or **interna**, eventually develops an external reproductive body (the **externa**), where egg production occurs. A male cyprid settles on the externa, transforms into a minute sexually mature instar called a **trichogon**, and

moves into the ovary-filled externa to take up residence, where its sole function is to produce sperm. A mature externa, usually arising from the host's abdomen, will produce a succession of larval broods, molting after each larval release (it is the only part of the rhizocephalan body that molts). The larvae are lecithotrophic and develop through several nauplius stages to the cyprid (Figures 16.25, 16.26).

Members of the rhizocephalan order Akentrogonida parasitize a much wider range of crustacean hosts and do not have a kentrogon stage in their life cycle. Instead of injecting a vermigon, the female cyprid has long, slender antennules that it uses to attach to the abdomen of the host, one of which actually penetrates the host's cuticle, becomes hollow, and serves for the passage of embryonic cells from cyprid larva to host. Male cyprids somehow find infected hosts and penetrate them in the same fashion as the females, releasing their sperm in such a way that they actually enter the body of the female parasite.

Digestive System

The digestive system of crustaceans includes the usual arthropod foregut, midgut, and hindgut. The foregut and hindgut are lined with a cuticle that is continuous with the exoskeleton and molted with it. The stomodeal foregut is modified in different groups, but usually includes a relatively short pharynx–esophagus region followed by a stomach. The stomach often has chambers or specialized regions for storage, grinding, and sorting; these structures are best developed in the malacostracans (Figure 16.27H). The midgut forms a short or long intestine—the length depending mainly on overall body shape and size—and bears variably placed digestive ceca. The ceca are serially arranged only in the remipedes. In some malacostracans, such as crabs, the ceca fuse to form a solid glandular mass (= digestive gland) within which are many branched, blind tubules. The hindgut is usually short, and the anus is generally borne on the anal somite or telson, or on the last segment of the abdomen (when the anal somite or telson is reduced or lost).

Examples of some crustacean digestive tracts are shown in Figure 16.27. After ingestion, the food material is usually handled mechanically by the foregut. This may involve simply transporting the food to the midgut or, more commonly, processing the food in various ways prior to chemical digestion. For example, the complex foregut of decapods (Figure 16.27H) is divided into an anterior **cardiac stomach** and a posterior **pyloric stomach**. Food is stored in the enlarged portion of the cardiac stomach and then moved a bit at a time to a region containing a **gastric mill**, which usually bears heavily sclerotized teeth. Special muscles associated with the stomach wall move the teeth, grinding the food into smaller particles. The macerated material then

---

*Rhizocephalans of the family Sacculinidae infest only decapod crustaceans and have been suggested as biological control agents for invasive exotics such as the green crab (*Carcinus maenas*), which are upsetting coastal ecosystems worldwide.

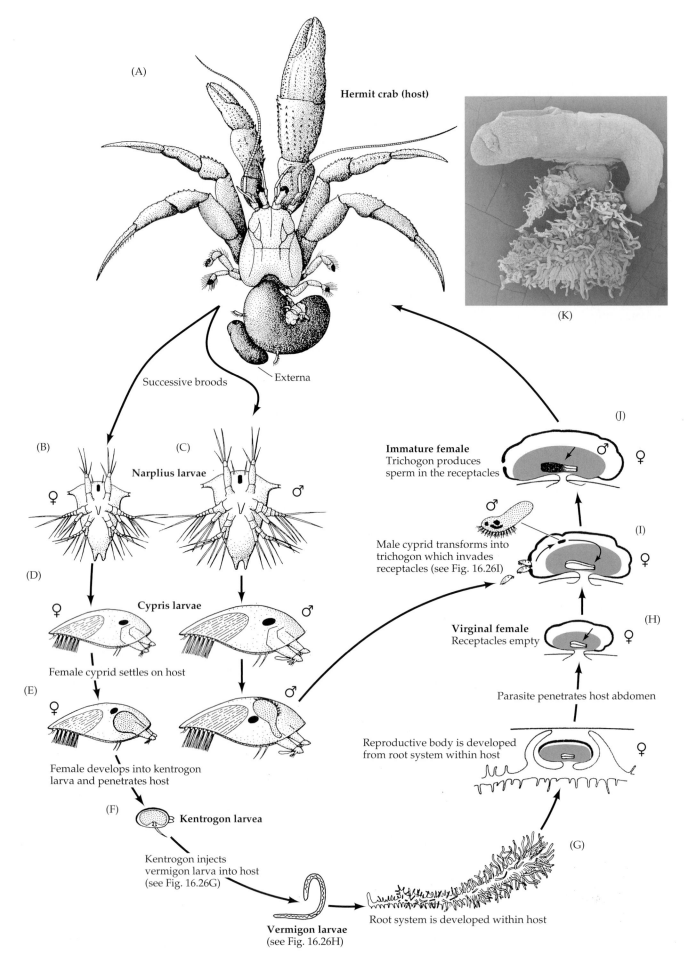

(A)

Hermit crab (host)

Externa

(K)

Successive broods

(B) (C)

Narplius larvae

♀ ♂

(D)

Cypris larvae

♀ ♂

Female cyprid settles on host

(E)

♀ ♂

Female develops into kentrogon
larva and penetrates host

(F) **Kentrogon larvea**

Kentrogon injects
vermigon larva into host
(see Fig. 16.26G)

Vermigon larvae
(see Fig. 16.26H)

Root system is developed within host

(G)

Reproductive body is developed
from root system within host ♀

Parasite penetrates host abdomen

Virginal female
Receptacles empty ♀ (H)

♀ (I)

Male cyprid transforms into
trichogon which invades
receptacles (see Fig. 16.26I) ♂

Immature female
Trichogon produces
sperm in the receptacles ♂ ♀ (J)

Figure 16.25 The remarkable life cycle of the rhizocephalan cirripede *Peltogaster paguri*, a kentrogonid parasite of hermit crabs. (A) The mature reproductive portion of parasite (externa) produces numerous broods of male and female larvae, which are released as nauplii (B,C) and eventually metamorphose into cypris larvae (D). Female cyprids settle on thorax and limbs of host crabs (E) and undergo a major internal metamorphosis into the kentrogon form (F), which is provided with a pair of antennules and an injection stylet. The kentrogon's viscera metamorphoses into an infective stage, the vermigon, which is transferred to the host through the hollow stylet. Inside the host, the vermigon grows with rootlets that ramify throughout much of the host's body; it is now called the interna (G). Eventually the female parasite emerges on the abdomen of the host as a virginal externa (H). When the externa acquires a mantle pore, or aperture, it becomes attractive to male cyprids (I). Male cyprids settle within the aperture, transform into a trichogon form, and implant part of their body contents in the female's receptacles (J). The deposit proceeds to differentiate into spermatozoa, which fertilize the eggs of the female. (K) The dissected externa, with its rootlets, of *Peltogaster*, removed from its host. Note the mantle aperture.

moves into the back part of the pyloric stomach, where sets of filtering setae prevent large particles from entering the midgut. This type of foregut arrangement is best developed in macrophagous decapods (scavengers, predators, and some herbivores). Thus the food can be taken in quickly, in big bites, and mechanically processed afterward.

Circulation and Gas Exchange

The basic crustacean circulatory system usually comprises a dorsal ostiate heart within a pericardial cavity and variously developed vessels emptying into an open hemocoel (Figure 16.27). The heart is absent in most ostracods, many copepods, and many cirripedes. In some groups the heart is replaced or supplemented by accessory pumping structures derived from muscular vessels.

The primitive heart structure in crustaceans is a long tube with segmental ostia, a condition retained in part in cephalocarids and in some branchiopods, leptostracans, and stomatopods. However, the general shape of the heart and the number of ostia are also closely related to body form and the location of gas exchange structures. The heart may be relatively long and tubular and extend through much of the postcephalic region of the body, as it does in the remipedes, anostracans, and leptostracans, or it may tend toward a globular or box shape and be restricted to the thorax, in association with the thoracic gills (as in cladocerans, maxillopodans, and decapods). The intimate coevolution of the circulatory system with body form and gill placement is best exemplified when comparing closely related groups. Although isopods and amphipods, for instance, are both peracarids, their hearts are located largely in the pleon

and in the pereon, respectively, corresponding to the pleopodal and pereopodal gill locations.

The number and length of blood vessels and the presence of accessory pumping organs are related to body size and to the extent of the heart itself. In most non-malacostracans, for example, there are no arterial vessels at all; the heart pumps blood directly into the hemocoel from both ends. These animals tend to have short bodies, long hearts, or both, an arrangement that facilitates circulation of the blood to all body parts. Sessile forms, such as most cirripedes, have lost the heart altogether, although it is replaced by a vessel pump in the thoracicans. Large malacostracans tend to have well developed vessel systems, thus ensuring that blood flows throughout the body and hemocoel and to the gas exchange structures (Figure 16.27D,E). Large or active crustaceans may also possess an anterior accessory pump called the **cor frontale**, which helps maintain blood pressure, and often a venous system for returning blood to the pericardial chamber.

Crustacean blood contains a variety of cell types, including phagocytic and granular amebocytes and special wandering **explosive cells** that release a clotting agent at sites of injury or autotomy. In non-malacostracans, oxygen is either carried in solution or attached to dissolved hemoglobin. Most malacostracans possess hemocyanin in solution (although some contain hemoglobin within tissues). Oxygen-binding pigments are never carried in corpuscles as in the vertebrates.

We have mentioned the form and position of gas exchange organs (gills) for some groups of crustaceans in the taxonomic synopses. Some small forms (e.g., copepods, some ostracods) lack distinct gills and rely on cutaneous exchange, which is facilitated by their relatively thin cuticles and high surface-to-volume ratio. In the small forms of other groups a thin, membranous inner lining of the carapace serves this purpose (e.g., Cladocera, Cirripedia, Leptostraca, Cumacea, Mysida, clam shrimps, and even some members of the Decapoda).

Most crustaceans, however, possess distinct gills of some sort (Figure 16.28). These structures are typically derived from thoracic epipods that have been modified in various ways to provide a large surface area. The inner hollow chambers or channels of these appendages are confluent with the hemocoel or supply vessels. Although their structure varies considerably (recall the various decapod gills described earlier), they all operate on the basic principles of gas exchange organs addressed in Chapter 3 and throughout this text: the circulatory fluid is brought close to the oxygen source in an organ with a relatively high surface area. The gills provide a thin, moist, permeable surface between the internal and external environments. The gills of stomatopods and isopods (Figure 16.28H,I) are pleopodal. In the first case they are branched processes off the base of the appendages, but in the isopods the flattened pleopods themselves are vascularized and provide the necessary

(A)

(B)

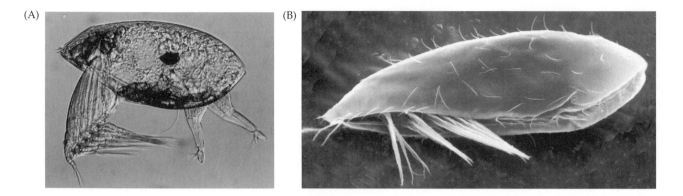

(C)

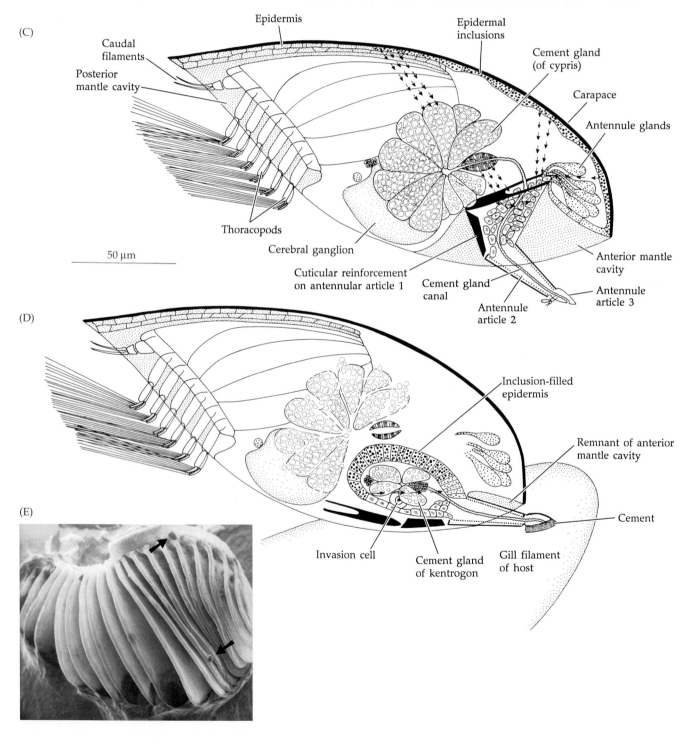

Caudal filaments

Posterior mantle cavity

Epidermis

Epidermal inclusions

Cement gland (of cypris)

Carapace

Antennule glands

Thoracopods

Cerebral ganglion

Cuticular reinforcement on antennular article 1

Cement gland canal

Antennule article 2

Anterior mantle cavity

Antennule article 3

50 µm

(D)

Inclusion-filled epidermis

Remnant of anterior mantle cavity

Cement

(E)

Invasion cell

Cement gland of kentrogon

Gill filament of host

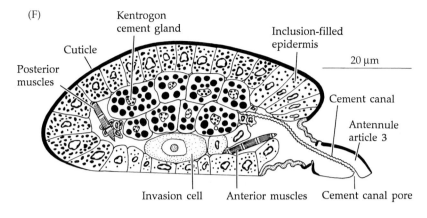

(F)

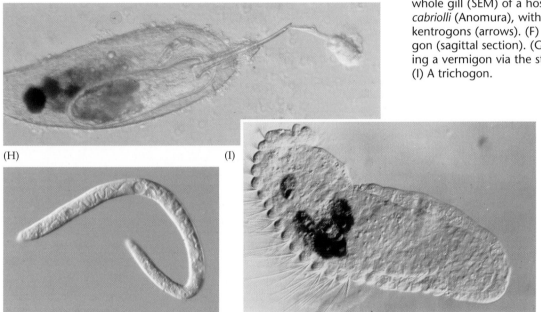

Figure 16.26 Various stages in the life cycle of parasitic rhizocephalans. See Figure 25 for a full description of a rhizocephalan (kentrogonid) life cycle. (A–D) Larvae of *Lernaeodiscus porcellanae*: (A) A live cyprid. (B) A cyprid (lateral view; SEM). (C,D) Diagrams of cyprid larvae before and after settlement (right side of carapace removed; naupliar eye omitted). The dotted line in the second antennular article indicates the primordial kentrogon cuticle, and the placement of muscle fibers in the cyprid are indicated by arrows; the muscles are hypothesized to effect formation of the kentrogon and separation of the old cyprid from the kentrogon. In D, kentrogon formation is complete. (E) A whole gill (SEM) of a host crab, *Petrolisthes cabriolli* (Anomura), with several attached kentrogons (arrows). (F) A 2-hour-old kentrogon (sagittal section). (G) A kentrogen injecting a vermigon via the stylet. (H) A vermigon. (I) A trichogon.

surface area for exchange. Stomatopods also have epipodal gills on the thoracopods, but these are highly reduced.

For gills to be efficient, a flow of water must be maintained across them. In stomatopods and aquatic isopods a current is generated by the beating of the pleopods. Similarly, the pereopodal gills of euphausids are constantly flushed by water as the animal swims. In many crustaceans, however, the gills are concealed to various degrees and require special mechanisms in order to produce the ventilating currents. In many decapods, for example, the gills are contained in **branchial chambers** formed between the carapace and the body wall (pleura) (Figure 16.28). While such an arrangement provides protection from damage to the fragile gill filaments, the openings to the chambers are generally small, restricting the passive flow of water. Not surprisingly, the solution to this dilemma comes once again from the evolutionary plasticity of crustacean appendages. Most decapods have elongate exopods on the maxillae, called **gill bail-**

ers or **scaphognathites**, that vibrate to create ventilating currents through the branchial chambers (Figure 16.28A). These currents typically enter from the sides and rear, through small openings around the coxae of the pereopods, and exit anteriorly from under the carapace in the vicinity of the mouth field (and antennal glands). They can be easily seen by observing a crab or lobster in quiet water. The flow rate of the currents can be altered, depending on environmental factors, and can also be reversed, thus allowing certain decapods to burrow in sand or mud with only their front ends exposed to the water.

The placement of the gills in branchial chambers protects them from desiccation and enables many crustaceans to live in intertidal regions, where they are frequently exposed to air. By avoiding direct exposure to very dry conditions, the branchial chambers always remain moist, so diffusion of respiratory gases continues even during low tides. Some decapods have even invaded land, especially certain crayfish and the anomu-

Figure 16.27   Internal anatomy of representative crustaceans. (A) A calanoid copepod. (B) A lepadomorph barnacle. (C) A balanomorph barnacle. (D) A stomatopod. (E) A crayfish. (F) An isopod. (G) An amphipod. (H) The stomach of a crayfish. (I) A brachyuran crab.

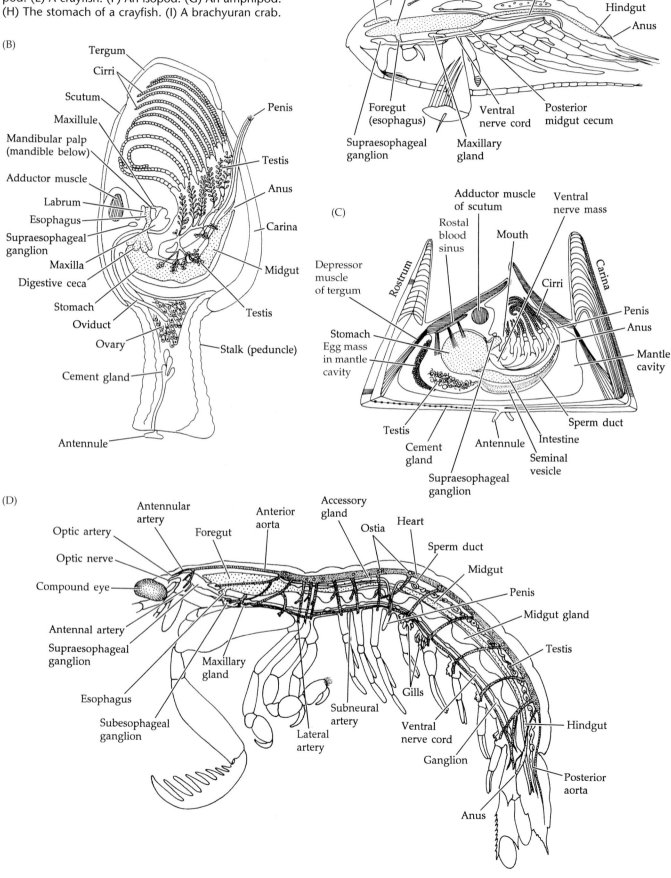

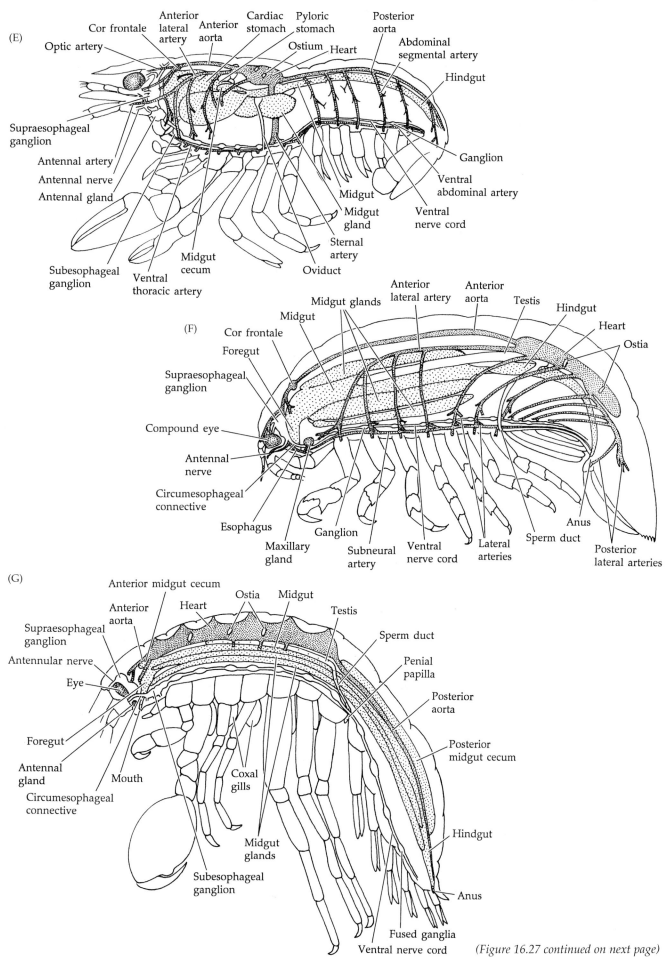

(E)
Optic artery
Cor frontale
Anterior lateral artery
Anterior aorta
Cardiac stomach
Pyloric stomach
Ostium
Heart
Posterior aorta
Abdominal segmental artery
Hindgut
Supraesophageal ganglion
Antennal artery
Antennal nerve
Antennal gland
Ganglion
Ventral abdominal artery
Midgut
Midgut gland
Ventral nerve cord
Sternal artery
Oviduct
Subesophageal ganglion
Ventral thoracic artery
Midgut cecum

(F)
Midgut glands
Midgut
Cor frontale
Foregut
Supraesophageal ganglion
Anterior lateral artery
Anterior aorta
Testis
Hindgut
Heart
Ostia
Compound eye
Antennal nerve
Circumesophageal connective
Esophagus
Maxillary gland
Ganglion
Subneural artery
Ventral nerve cord
Lateral arteries
Sperm duct
Anus
Posterior lateral arteries

(G)
Anterior midgut cecum
Ostia
Midgut
Anterior aorta
Heart
Testis
Supraesophageal ganglion
Sperm duct
Antennular nerve
Penial papilla
Eye
Posterior aorta
Foregut
Antennal gland
Circumesophageal connective
Mouth
Posterior midgut cecum
Coxal gills
Hindgut
Midgut glands
Anus
Subesophageal ganglion
Fused ganglia
Ventral nerve cord

(Figure 16.27 continued on next page)

(Figure 16.27 continued from preceding page)

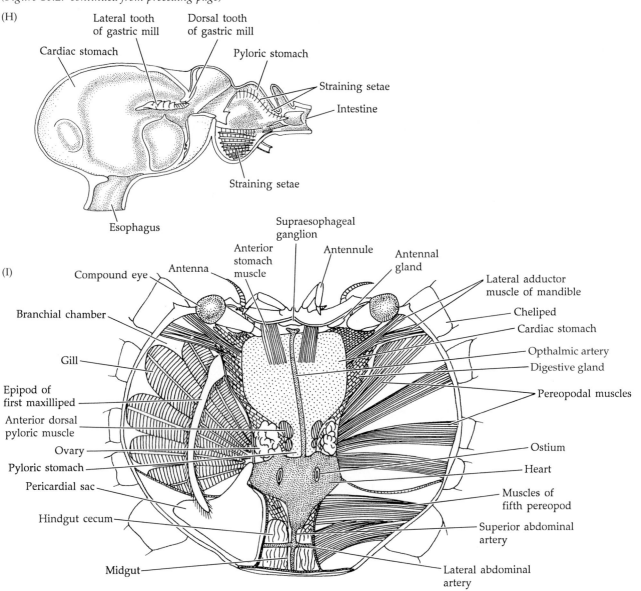

(H)

Lateral tooth of gastric mill

Dorsal tooth of gastric mill

Cardiac stomach

Pyloric stomach

Straining setae

Intestine

Straining setae

Esophagus

(I)

Supraesophageal ganglion

Anterior stomach muscle

Antennule

Antennal gland

Compound eye

Antenna

Lateral adductor muscle of mandible

Branchial chamber

Cheliped

Cardiac stomach

Gill

Opthalmic artery

Digestive gland

Epipod of first maxilliped

Pereopodal muscles

Anterior dorsal pyloric muscle

Ovary

Ostium

Pyloric stomach

Heart

Pericardial sac

Muscles of fifth pereopod

Hindgut cecum

Superior abdominal artery

Midgut

Lateral abdominal artery

ran and brachyuran crabs known as "land crabs" (e.g., the hermit crab *Coenobita* and the coconut crab *Birgus*; Figure 16.1O). In these semi-terrestrial species the gills are typically reduced in size. In *Birgus* the original gills are very small, and the vascularized cuticular surface of the gill chamber is used for gas exchange. Another striking decapod adaptation to life in air is displayed by the "sand-bubbler" crabs of the Indo-Pacific region (family Ocypodidae: *Scopimera, Dotilla*). These crabs possess membranous discs on their legs or sternites that were once thought to be auditory organs (tympana), but are now thought to function as gas exchange surfaces (Maitland 1986).

The most successful crustaceans on land are not the decapods, however, but the familiar sowbugs and pillbugs. The success of these oniscidean isopods (e.g., *Porcellio*) is due in part to the presence of aerial gas exchange organs called **pseudotrachea** (Figure 16.28H,I).

These organs are inwardly directed, moderately branched, thin-walled, blind sacs located in some of the pleopodal exopods, connected to the outside via small pores. Air circulates through these sacs, and gases are exchanged with the blood in the pleopods. Thus, in these animals the original aquatic pleopodal gills have been refashioned for air breathing by moving the exchange surfaces inside, where they remain moist. The superficially similar tracheal systems of isopods, insects, and arachnids evolved independently, by convergence, in association with other adaptations to life on land.

Excretion and Osmoregulation

Like other fundamentally aquatic invertebrates, crustaceans are ammonotelic, whether in fresh water or sea water or on land. They release ammonia both through nephridia and by way of the gills. As discussed in Chapter 15, most crustaceans possess nephridial excretory or-

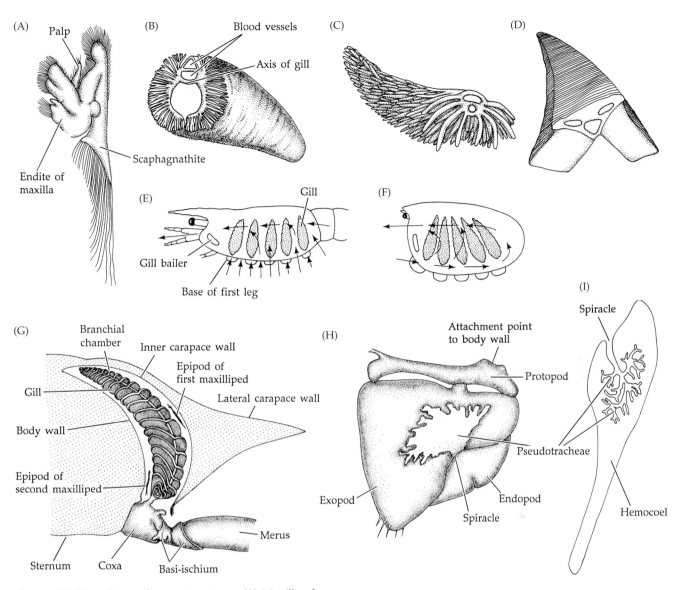

Figure 16.28 Gas exchange structures. (A) Maxilla of the shrimp *Pandalus*. Note the setose scaphagnathite used to generate the ventilating current. (B–D) Cross sections of types of decapod gills: (B) Dendrobranchiate. (C) Trichobranchiate. (D) Phyllobranchiate. (E,F) Paths of ventilating currents through the left branchial chambers of (E) a shrimp and (F) a brachyuran crab. (G) The branchial chamber (cross section) of a brachyuran crab, showing the position of a single phyllobranchiate podobranch. (H,I) A pleopod of the terrestrial isopod *Porcellio* (surface view and section). Note the pseudotracheae.

gans in the form of either **antennal glands** or **maxillary glands** (Figures 16.5A, 16.27). These are serially homologous structures, constructed similarly but differing in the position of their associated pores (at the base of the second antennae or the second maxillae, respectively). The inner blind end is a coelomic remnant of the nephridium called the **sacculus**, which leads through a variably coiled duct to the pore. The duct may bear an enlarged bladder near the opening (Figure 15.14A). Antennal glands are sometimes called "green glands."

Most crustaceans have only one pair of these nephridial organs, but lophogastrids and mysids have both antennal and maxillary glands, and a few others (cephalocarids and a few tanaids and isopods) have well developed maxillary and rudimentary antennal glands. Most non-malacostracans have maxillary glands, as do stomatopods, cumaceans, and most tanaids and isopods. Adult ostracods have maxillary glands, but antennal glands also occur in freshwater species. All of the other malacostracans have antennal glands.

Blood-filled channels of the hemocoel intermingle with branched extensions of the sacculus epithelium, creating a large surface area across which filtration occurs. The cells of the sacculus wall also actively take up and secrete material from the blood into the lumen of the excretory organ. These processes of filtration and secretion are to some degree selective, but most of the regulation of urine composition is accomplished by active exchange between the blood and the excretory tubule.

These activities not only regulate the loss of metabolic wastes but are also extremely important in water and ion balance, particularly in freshwater and terrestrial crustaceans.

The excretion and osmoregulation carried out by antennal and maxillary gland activity are supplemented by other mechanisms. The cuticle itself acts as a barrier to exchange between the internal and external environments and, as we have mentioned, is especially important in preventing water loss on land or excessive uptake of water in fresh water. Moreover, thin areas of the cuticle, especially the gill surfaces, serve as sites of waste loss and ionic exchange. The epipods on the legs of Branchiopoda were long assumed to function in gas exchange (as "gills") but they are now know to serve primarily as sites of osmoregulation (hence, the taxonomic name *branchio-poda* is a misnomer!). Phagocytic blood cells and certain regions of the midgut are also thought to accumulate wastes. In some terrestrial isopods, ammonia actually diffuses from the body in gaseous form.

Nervous System and Sense Organs

The central nervous system of crustaceans is constructed in concert with the segmented bauplan, along the same lines as we have seen in annelids and other arthropods (Figure 16.29). In the more primitive condition it is ladder-like, the segmental ganglia being largely separate and linked by transverse commissures and longitudinal connectives (Figure 16.29A). As described in Chapter 15, the crustacean brain is composed of three fused ganglia, the two anterior being the dorsal (supraesophageal) **protocerebrum** and **deutocerebrum**, which are thought to be preoral in origin. From the protocerebrum, **optic nerves** innervate the eyes. From the deutocerebrum, **antennulary nerves** run to the antennules, while smaller nerves innervate the eyestalk musculature. The third ganglion of the brain is the posterior **tritocerebrum**, which presumably represents the first postoral somite ganglion. The tritocerebrum forms a pair of circumenteric connectives that extend around the esophagus to a **subesophageal** or **subenteric ganglion** and link the brain with the ventral nerve cord bearing the segmental body ganglia. From the tritocerebrum also arise the **antennary nerves** as well as certain sensory nerves from the anterior region of the head.

Figure 16.29 Central nervous systems of four crustaceans. (A) The ladder-like system of an anostracan. Note the absence of well developed ganglia in the posterior, apodous, portion of the trunk. (B) Elongate metameric system of a crayfish. (C) Highly compacted system of a brachyuran crab, wherein all thoracic ganglia have fused and the abdominal ganglia are reduced. (D) Nervous system of a hyperiid amphipod. Note the loss of the urosomal ganglia typical of all amphipods.

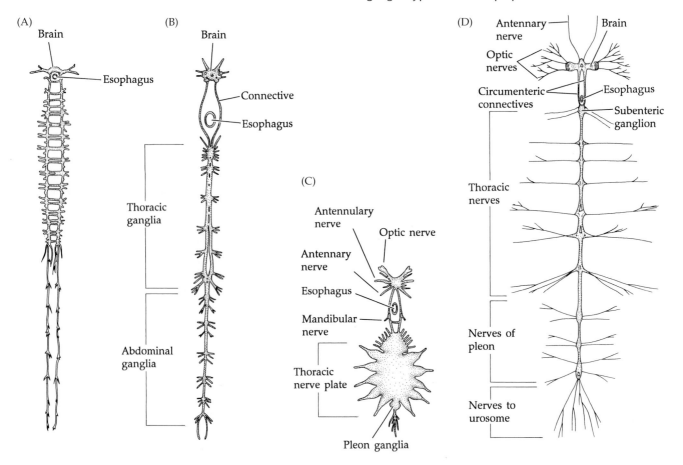

The nature of the ventral nerve cord often clearly reflects the influence of body tagmosis. In primitive crustaceans with relatively homonomous bodies (e.g., remipedes, cephalocarids, and many branchiopods), the ganglia associated with each postantennary segment remain separate along the ventral nerve cord. In advanced forms, however, a single large subenteric ganglionic mass is formed by the fusion of ganglia associated with the postoral cephalic segments (e.g., those of the mandibles, maxillules, maxillae, and, when present, maxillipeds). The ganglia of the thorax and abdomen may also be variably fused, depending on segment fusion and body compaction. For example, in most long-bodied decapods (lobsters and crayfish), the thoracic and abdominal ganglia are mostly fused across the body midline, but remain separate from one another longitudinally (Figure 16.29B). However, in short-bodied decapods (e.g., crabs), all of the thoracic segmental ganglia are fused to form a large ventral nerve plate, and the abdominal ganglia are much reduced (Figure 16.29C). Even in the spiny lobster (*Panulirus*), the ganglia are largely fused, and the whole system is concentrated in the anterior region of the body.

Most crustaceans have a variety of sensory receptors that transmit information to the central nervous system in spite of the imposition of the exoskeleton (as we have explained for arthropods in general) (Figure 16.30). Among the most obvious of these sensory structures are the many innervated setae or sensilla that cover various regions of the body and appendages (Figure 16.31). Recent studies of these structures using electron microscopes have resulted in categorization systems based on function (Bush and Laverack, in Atwood and Sanderman 1982; Derby 1982). Many of these sensilla are mechanoreceptors (sensing touch and currents), whereas others are chemoreceptors. Most crustaceans also possess special chemoreceptors in the form of clumps or rows of spinelike cuticular processes called **aesthetascs** (Figure 16.30A), located on the antennae or, more rarely, on the mouthparts. Thermoreceptors may occur in some crustaceans, but are not yet documented. A pair of unique sensory structures whose function is

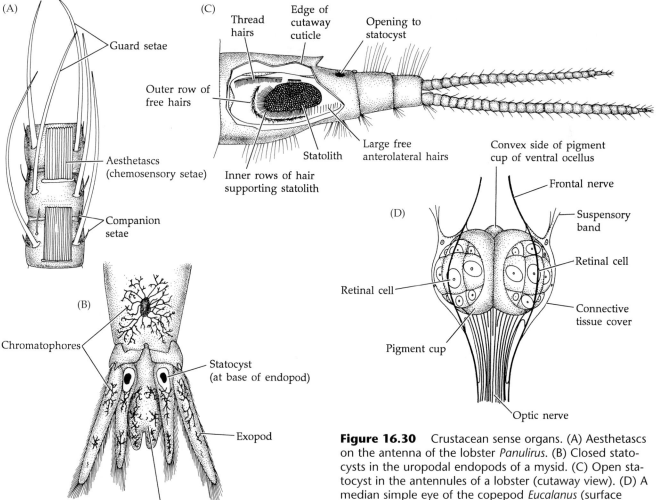

Figure 16.30 Crustacean sense organs. (A) Aesthetascs on the antenna of the lobster *Panulirus*. (B) Closed statocysts in the uropodal endopods of a mysid. (C) Open statocyst in the antennules of a lobster (cutaway view). (D) A median simple eye of the copepod *Eucalanus* (surface view).

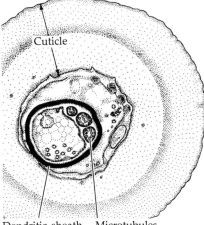

Figure 16.31 Sensilla of selected crustaceans. (A) Serrate seta (mechanoreceptor) of the second maxilliped of the anomuran *Petrolisthes armatus* (×428). (B) Sockets of serrate setae of the third maxilliped of the cleaning shrimp *Stenopus hispidus* (×228). (C) Current receptor from the anterior trunk limbs of the notostracan *Triops*. (D) Chemosensory seta from the first pereopod of the freshwater shrimp *Atya* (×5700); note the characteristic apical pore. (E) A dual receptor (mechanoreceptor and chemoreceptor) seta from the maxilla of the remipede *Speleonectes tulumensis* (×4560). (F) Cross section of a dual receptor (micrograph and interpretive drawing). Note the microtubules within the dendrites and the dendritic sheath that attaches to the cuticle of the setal shaft (×17,100).

unknown, called **frontal processes**, occurs on the head of remipedes. Many crustaceans have **dorsal organs**, poorly understood glandular–sensory structures on the head, which actually constitute several different types of sensory structures that may or may not be homologous.

Like all other arthropods, crustaceans contain well developed proprioceptors that provide information about body and appendage position and movement during locomotion. A few taxa within the class Malacostraca possess statocysts, which either are fully closed and contain a secreted statolith (mysids, some anthurid isopods) or open to the outside through a small pore and contain a statolith formed of sand grains (many decapods) (Figures 16.30B,C). In the latter case the statocyst not only serves as a georeceptor, but also detects the angular and linear acceleration of the body relative to the surrounding water as well as the movement of water past the animal (i.e., the statolith is rheotactic).

There are two types of rhabdomeric photoreceptors among crustaceans, median simple eyes and lateral compound eyes; both are innervated by the protocerebrum. Many species possess both kinds of eyes, either simultaneously or at different stages of development. The compound eyes may be sessile or stalked. Stalked compound eyes occur in the Anostraca, many Malacostraca, and perhaps some Cumacea (and perhaps some trilobites). There is some evidence that both kinds of eyes have been lost and then regained in various crustacean lineages.

The **median eye** generally first appears during the nauplius larval stage, and for that reason it is often called a **naupliar eye**. Like the nauplius larva itself, the median eye is thought to be a primitive (defining) feature of the Crustacea; it is secondarily reduced or lost in many taxa in which the corresponding larval stage is suppressed. Median eyes are in a sense "compound" in that they are composed of more than one photoreceptor unit (Figure 16.30D). There are typically three such units in the median eyes of nauplii and up to seven in the eyes of adults in which they persist. Except for their basic rhabdomeric nature, however, the structure of median eye units is unlike that of the ommatidia of true compound eyes. The former are inverse pigment cups, each with relatively few retinular (photoreceptor) cells. Cuticular lenses are present over the median eyes of most ostracods and some copepods. Simple crustacean eyes probably function only to detect light direction and intensity. Such information is of particular value as a means of orientation in planktonic forms without compound eyes, such as nauplius larvae and many copepods.

The structure and function of compound eyes were reviewed in Chapter 15. In terms of visual capacity, much more work has been done on the eyes of insects than on those of crustaceans, and we are left with a good deal of speculation in terms of what crustaceans actually "see." Although they probably lack the visual acuity of many insects, some can discern shapes, patterns, and movement; color vision has been demonstrated in some species (various species have been shown to respond to light waves from the blue-green region to the ultraviolet and far-red spectra, at least to 470–570 nm). Although both groups have tetrapartite ommatidia, There are certain structural differences between the compound eyes of insects and those of crustaceans, probably as a result of adaptation to the requirements of aerial and aquatic vision . Under water, light has a more restricted angular distribution, a lower intensity, and a narrower range of wavelengths than it does in air. Contrast is also somewhat reduced in water. All of these factors place a premium on enhancing the sensitivity and contrast perception of the eyes of aquatic creatures. Mounting the eyes on stalks is one dramatic way in which many crustaceans increase the

amount of information available to the eyes, and hence to the central nervous system, by increasing the field of view and binocular range. Eyestalks are complex structural features with a dozen or so muscles controlling their movement.

Typical tetrapartite compound eyes are lacking in the Maxillopoda, but various forms of "compound eyes" do occur among the Branchiura, Ostracoda (Cypridinacea), and Cirripedia. Eyes in the first two taxa most closely resemble those of non-maxillopodans in general structure and may be homologous with them. In the Cirripedia, the median eye and two lateral eyes are all derived from a single tripartite ocellar eye of the nauplius larva, which splits to its three components, each forming an adult photoreceptor following metamorphosis of the nauplius into a cypris larva. All three of these eyes thus appear to be composed of simple ocelli, although the lateral eyes have three photoreceptor cells and for this reason are often called "compound eyes." Rhizocephalan nauplii also have a tripartite nauplius eye, which persist into the cyprid larval stage. Copepods (and other maxillopodans) lack compound eyes.

Compound eyes are lacking altogether in many crustacean taxa (e.g., Copepoda, Mystacocarida, Cephalocarida, Tantulocarida, Pentastomida, Remipedia, and some Ostracoda). Members of some other groups possess compound eyes only in late larval stages and lose them at metamorphosis (e.g., cirripedes). Reduction or loss of eyes is also common in many deep-sea species, burrowers, cave dwellers, and parasites.

Crustaceans have complex endocrine and neurosecretory systems, although our understanding of these systems is far from complete. In general, the phenomena of molting (see Chapter 15), chromatophore activity, and various aspects of reproduction are under hormonal and neurosecretory control. Recently, a series of papers has emphasized not only what is known, but more important, the many avenues for future research in this area (*American Zoologist* 25: 155–284). Interesting recent work indicates that juvenile hormone-like compounds, long thought to occur only in insects, may also occur in at least some crustaceans. (Juvenile hormones are a family of compounds that regulate adult metamorphosis and gametogenesis in insects.) Bioluminescence also occurs in several crustacean groups. It is common among pelagic decapods, and it has also been reported in certain myodocopan ostracods, hyperiid amphipods, and copepod larvae.

Reproduction and Development

Reproduction. We have repeatedly mentioned the relationship of an animal's reproductive and developmental pattern to its lifestyle and overall survival strategy. With the exception of purely vegetative processes such as asexual budding, the crustaceans have managed to exploit virtually every life history scheme imaginable. The sexes are usually separate, although hermaphroditism is the rule in remipedes, cephalocarids, most cirripedes, and a few decapods. In addition, parthenogenesis is common among many branchiopods and certain ostracods. In one species of clam shrimp (*Eulimnadia texana*) a rare type of mixed mating system exists, called **androdioecy**, in which males coexist with hermaphrodites, but there are no true females. Androdioecy is quite rare, but is also known in the nematode *Caenorhabditis elegans*, the barnacle *Balanus galeatus*, and several other branchiopod crustaceans.

The reproductive systems of crustaceans are generally quite simple (Figure 16.27). The gonads are derived from coelomic remnants and lie as paired elongate structures in various regions of the trunk. In many cirripedes, however, the gonads lie in the cephalic region. In some cases the paired gonads are partially or wholly fused into a single mass. A pair of gonoducts extends from the gonads to genital pores located on one of the trunk segments, either on a sternite, on the arthrodial membrane between the sternite and leg protopods, or on the protopods themselves. In many crustaceans the paired penes are fused into a single median penis (e.g., in tantulocarids, cirripedes, and some isopods). The female system sometimes includes seminal receptacles. The position of the gonopores varies among the five classes (Table 16.1).

Most crustaceans copulate, and many have evolved courtship behaviors, the most elaborate and well known of which occur among the decapods (see the chapter by Salmon in Rebach and Dunham 1983 for a review). Although many crustaceans are gregarious (e.g., certain planktonic species, barnacles, many isopods and amphipods), most decapods live singly except during the mating season. More or less permanent, or at least seasonal, pairing is known among many crustaceans (e.g., stenopodid shrimps; pinnotherid "pea" crabs, which often live as pairs in the mantle cavities of bivalve molluscs or in burrows of thalassinid shrimps; certain parasitic and commensal isopods).

Even the parasitic pentastomids copulate (within the host's respiratory organs) and have internal fertilization, relying on a transfer of sperm to the female's vagina by way of the male's **cirrus** (penes). The early embryo metamorphoses into a **primary larva** with two pairs of double-clawed legs and one or more piercing stylets (Figure 16.19E). The primary larva may be a modified nauplius. The larvae may be autoinfective in the primary host, or they may migrate to the host's gut and pass out with the feces. In the latter case, an intermediate host is required, which may be almost any kind of vertebrate. The larvae bore through the gut wall of the intermediate host, where they undergo further development to the infective stage. Once the intermediate host is consumed by a definitive host (usually a predator), the parasite makes its way from the new host's stomach up the esophagus, or bores through the intestinal wall, eventually settling in the respiratory system.

Mating in non-paired crustaceans requires mechanisms that facilitate location and recognition of partners. Among decapods, and perhaps many other crustaceans, scattered individuals apparently find one another either by distance chemoreception (pheromones) or through synchronized migrations associated with lunar periodicity, tidal movements, or some other environmental cue. Males of some marine myodocopan ostracods (some Halocyprididae, some Cyprididae) produce complex bioluminescent displays, similar to those of fireflies, to attract females (Cohen and Morin 1990). Once prospective mates are near each other, recognition of conspecifics of the opposite sex may involve several mechanisms. Apparently, most decapods employ chemotactic cues requiring actual contact. Vision is known to be important in the stenopodid shrimps (most of which live in pairs) and certain anomurans (family Porcellanidae) and brachyurans (many grapsids and ocypodids). A good deal of work has been done on fiddler crabs of the genus *Uca* (family Ocypodidae). In these species, males engage in dramatic cheliped waving (of their enlarged chela, or **major claw**) to attract females and repel competing males (Figure 16.32A–D). In addition, males produce sounds by stridulation and substratum thumping that are thought to attract potential mates. Mating generally takes place once the male has enticed the female into his burrow.

Among many crustaceans, the external sexual characteristics are associated with the actual mating process. In some males, particular appendages, such as the antennae of male anostracans and some cladocerans, ostracods, and copepods, are modified for grasping the female. Additionally, many males bear special sperm transfer structures, in the form of either modified appendages or special penes such as those of the thoracican barnacles (Figure 16.32E), anostracans, and ostracods. Examples of modified appendages include the last trunk limbs of copepods and the anterior pleopods of most male malacostracans (called **gonopods** in most malacostracans, or **petasma** in Dendrobranchiata) (Figure 16.32F). Sperm are transferred either loose in seminal fluid or (in many malacostracans and in copepods) packaged in spermatophores. Motile flagellated sperm occur only in some maxillopodans; in other crustaceans the sperm are nonmotile. Crustacean sperm are highly variable in shape, even bizarre in many instances, often being large round or stellate cells that move by pseudopods or are, seemingly, nonmotile.* Sperm are deposited directly into the oviduct or into a seminal receptacle in or near the female reproductive system. In some crustaceans females can store sperm for long periods (e.g., several years in the lobster *Homarus*), thus facilitating multiple broods from single inseminations.

---

*The sperm of freshwater ostracods are the longest in the animal kingdom, relative to body size (up to 10× body length).

Groping penis

Figure 16.32 (A–D) Mating behaviors of the fiddler crab *Uca*. (A) Two males in ritualized combat for the favor of a female, while she (B) watches. (C) A single male waving his enlarged cheliped to attract a female. (D) A male fiddler crab engaged in claw-waving behavior to attract a female. (E) A balanomorph barnacle, with cirri and groping penis extended, impregnating a neighbor. The advantage of a long penis in sessile animals is made obvious by this illustration. (F) Ventral views of a male and female brachyuran crab, *Cancer magister*, showing the modified pleopods (setose appendages to retain eggs in female; modified as gonopod in male). (G) A copulating pair of *Hemigrapsus sexdentatus*.

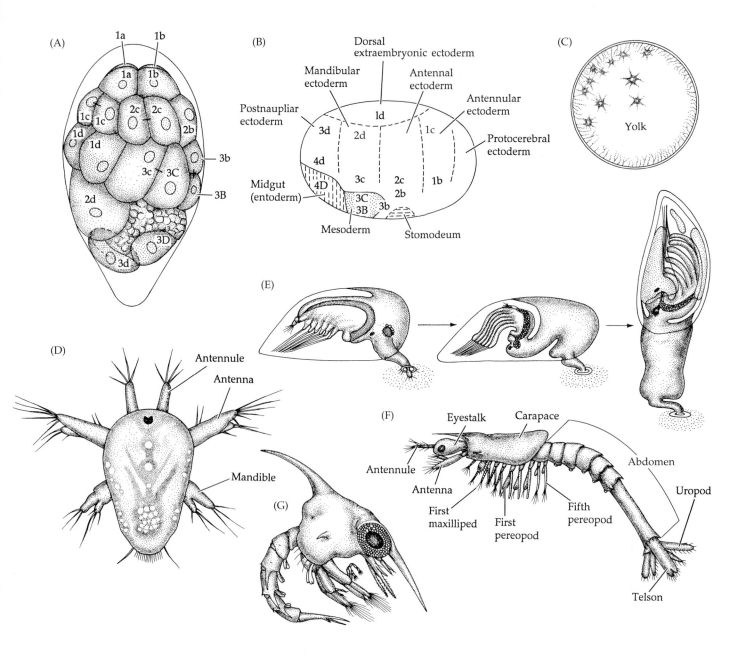

(A)

(B)
Dorsal extraembryonic ectoderm
Mandibular ectoderm
Antennal ectoderm
Antennular ectoderm
Postnaupliar ectoderm
Protocerebral ectoderm
Midgut (entoderm)
Mesoderm
Stomodeum

(C) Yolk

(D)
Antennule
Antenna
Mandible

(E)

(F)
Eyestalk
Carapace
Antennule
Antenna
First maxilliped
First pereopod
Fifth pereopod
Abdomen
Uropod
Telson

(G)

The great majority of crustaceans brood their eggs until hatching occurs. A variety of brooding strategies have evolved. Peracarids brood the developing embryos in a **marsupium**, a ventral brood pouch formed from inwardly directed plates of the leg coxae called **oostegites** (thermosbaenaceans are an exception among the Peracarida and use the carapace as a brood chamber). Other crustaceans attach the embryos to endites on the bases of the legs or to the pleopods (Figure 16.32F), usually using a mucus secreted by specialized glands. However, the syncarids, almost all dendrobranchiate shrimps, and most euphausids shed the zygotes directly into the water. A few others deposit their fertilized eggs in the environment, usually attaching them to some object (e.g., branchiurans, some ostracods, many stomatopods). These deposited embryos may be abandoned or, as is the case in stomatopods, carefully tended by the female. Nonetheless, parental protection of the embryos until they hatch as larvae or juveniles is typical in crustaceans. Thus, crustaceans usually engage in mixed or direct life histories (Table 16.2).

Development. Although crustaceans are the most widespread animals on Earth, we know surprisingly little about their embryogeny. The eggs are centrolecithal, with various amounts of yolk. The amount of yolk greatly influences the type of early cleavage and is often related to the time of hatching (Chapter 4). As far as is known, the zygotes of most non-malacostracans undergo some form of holoblastic cleavage, as do those of syncarids, euphausids, penaeids, amphipods, and parasitic isopods. However, cleavage pat-

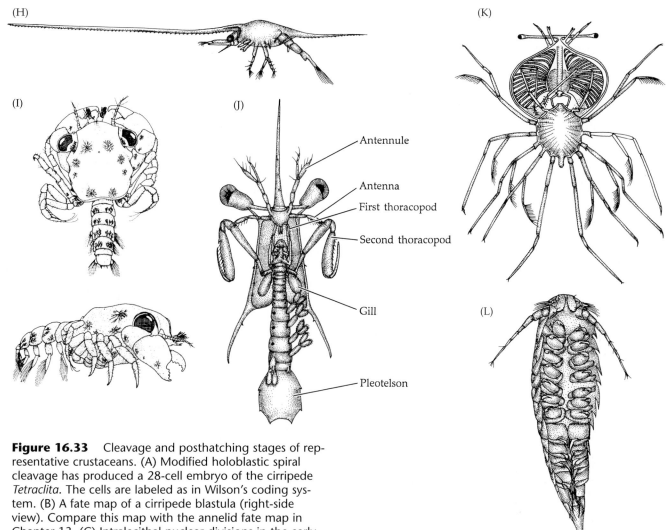

(H)

(I)

(J)

Antennule

Antenna

First thoracopod

Second thoracopod

Gill

Pleotelson

(K)

(L)

Figure 16.33 Cleavage and posthatching stages of representative crustaceans. (A) Modified holoblastic spiral cleavage has produced a 28-cell embryo of the cirripede *Tetraclita*. The cells are labeled as in Wilson's coding system. (B) A fate map of a cirripede blastula (right-side view). Compare this map with the annelid fate map in Chapter 13. (C) Intralecithal nuclear divisions in the early cleavage of a mysid. (D) Newly hatched copepod nauplius larva. (E) Settling and metamorphosis of a cypris larva of a lepadomorph barnacle. (F) The zoea ("mysis") stage larva of the dendrobranchiate shrimp *Penaeus*. (G) Zoea larva of the brachyuran crab *Callinectes sapidus*. (H) Zoea larva of a porcelain crab. (I) Megalopa larvae of the xanthid crab *Menippe adina*. (J) The characteristic antizoea larva of a stomatopod. (K) The translucent, paper-thin phyllosoma larva of the lobster *Jassa*. (L) Cryptoniscus stage (not a true larva) of the epicaridean isopod *Probopyrus bithynis*.

terns are extremely variable, ranging from equal to unequal and from radial-like to spiral. The occurrence of modified spiral cleavage (Figure 16.33A) in many crustaceans is generally viewed as evidence of close ties between the crustaceans and other spiralian groups such as the annelids. In some crustacean groups, however, the cell lineages and germ layer origins are modified from those of the usual (protostome) spirally cleaving embryo. For example, whereas the typical case involves a 4d origin of mesoderm, in barnacles this germ layer arises from the 3A, 3B, and 3C cells, and

the 4d cell contributes to ectoderm (Figure 16.33B). Other differences between various crustacean and other arthropod taxa involve the positions of the presumptive germ layers relative to one another, especially the entoderm and mesoderm. We want to emphasize, however, that such variations are not surprising in such a diverse and ancient taxon and do not negate the fundamental similarities that unite the group.

Meroblastic cleavage is the rule among many malacostracans. Here again, the exact pattern varies, but it generally involves intralecithal nuclear divisions followed by nuclear migration to the periphery of the embryo and subsequent partitioning of the nuclei into a cell layer around a central yolky mass (Figure 16.33C).

The form of the blastula and the method of gastrulation are dependent primarily on the preceding cleavage pattern and hence ultimately on the amount of yolk. Holoblastic cleavage may lead to a coeloblastula that undergoes invagination (as in syncarids) or ingression (as in many copepods and some cladocerans and anostracans). Other crustaceans form a stereoblastula followed by epibolic gastrulation (e.g., cirripedes). Most cases of meroblastic cleavage result in a periblastula and the subsequent formation of germinal centers.

Crustaceans share a characteristic larval stage known as the **nauplius larva**, denoted by the appearance of three pairs of appendage-bearing somites (Figure 16.33D).* In those groups having little yolk in their eggs, the nauplius is generally free-living. In those species with yolky eggs, the nauplius stage is generally passed through as part of a longer period of embryonic development (or a long brood period), and it is sometimes referred to as an **egg nauplius**. Free-living nauplii are usually planktotrophic, and their release corresponds to the depletion of stored yolk. However, in a few groups of crustaceans (e.g., euphausids and dendrobranchiate shrimps), the nauplius exhibits lecithotrophy.

Crustacean development is either direct, with the embryos hatching as juveniles that resemble miniature adults, or mixed, with embryos brooded for a brief or prolonged period and then hatching as a distinct larval form, which may pass through several subsequent stages before the adult condition is achieved. Direct development occurs in some cladocerans and branchiurans, and in all ostracods, phyllocarids, syncarids, and peracarids. Ostracods are typically viewed as having direct development, and they lack a distinct larval stage. Some ostracod species do hatch with only the first three pairs of appendages present, and they are thus true nauplii, even though they are in a bivalved carapace and add limbs gradually (these juvenile instars resemble miniature adults). All other crustaceans have some form of mixed development. The larval stages that have been recognized in crustacean groups that undergo mixed development have been assigned a plethora of names, and the homologies among these forms are not well understood. The more commonly encountered developmental forms are summarized below (also see Table 16.2 and Figure 16.33), but we do not attempt to describe them all.

Crustacean development is sometimes described as being either epimorphic, metamorphic, or anamorphic. However, we caution you that a clear evolutionary and functional understanding of crustacean developmental stages is still lacking, and thus the terms *mixed* and *direct* may be preferable, and less ambiguous, until we have a better understanding of this phenomenon.

Epimorphic development is direct; in crustaceans it is thought to result from a delay in the hatching of the embryo, which causes the nauplius (and any other possible larval stages) to be suppressed or absent.

Metamorphic development is the type of extreme mixed development seen among the Eucarida; it includes dramatic transitions in body form from one life history stage to another. (This pattern is similar to holometabolous development in insects—for example, the transformation of a caterpillar into a butterfly.) In general, up to five distinct preadult, or larval, stages may be recognized among crustaceans: **nauplius**, **metanauplius**, **protozoea**, **zoea**, and **postlarva**. The zoeal stage shows the greatest diversity in form among the various taxa and has been given different names in different groups (e.g., acanthosoma, antizoea, mysis, phyllosoma, pseudozoea).[†] Regardless of name, zoea are characterized by the presence of natatory exopods on some or all of the thoracic appendages and by the pleopods being absent (or rudimentary).

Anamorphic development is a less extreme type of indirect development in which the embryo hatches as a nauplius larva, but the adult form is achieved through a series of gradual changes in body morphology as new segments and appendages are added (it is similar in many ways to hemimetabolous development in insects). In other words, the postnㅁupliar stages gradually take on the adult form with succeeding molts; the classic example of anamorphic development is often said to be the Anostraca. Cephalocarida, many Branchiopoda, and most Maxillopoda are anamorphic—the nauplius larva grows by a series of molts that add new segments and appendages gradually as the adult morphology appears. In many groups hatching is somewhat delayed, and the emergent nauplius larva is termed a **metanauplius**. The basic nauplius possesses only three body somites, while the metanauplius has a few more; however, both possess only three pairs of similar-appearing appendages (which become the adult antennules, antennae, and mandibles). The end of the naupliar/metanㅁupliar stage is defined by the appearance of the fourth pair of functional limbs, the maxillules. In copepods a postnㅁupliar stage called a **copepodite** (simply a small juvenile) is often recognized.

The most extreme forms of metamorphic, or mixed, development occur in the malacostracan superorder Eucarida. The most complex developmental sequences are seen among the dendrobranchiate shrimps, which hatch as a typical nauplius larva that eventually undergoes a metamorphic molt to become a protozoea larva, with sessile compound eyes and a full complement of head appendages. The protozoea, after several molts, becomes a zoea larva, with stalked eyes and three pairs of thoracopods (as maxillipeds). The zoea eventually yields a juvenile stage (the postlarva) that resembles a miniature adult, but is not sexually mature. In some

---

*It was not until J. V. Thomson discovered the nauplius larvae of barnacles, in the nineteenth century, that this group was finally classified as Crustacea.

†The zoea larvae of panulirid lobsters (phyllosoma larvae) are large, bizarre-appearing creatures (Figure 16.33K) that can occur in such large numbers as to be a favorite food of tuna.

other eucarid groups (Amphionidacea, Caridea, and Brachyura) the postlarva is called a **megalopa**, and in the Anomura it is often called a **glaucothoe**; in both cases there are setose natatory pleopods on some or all of the abdominal somites. In other eucarids, some (or all) of these stages are absent.

Various other terms have been coined for different (or similar) developmental stages. For example, the modified zoeal stages of stomatopods are called **antizoea** and **pseudozoea** larvae, and the advanced zoeal stage of many other malacostracans is often called a **mysis larva**. In euphausids, the nauplius is followed by two stages, the **calyptopis** and the **furcilia**, which roughly correspond to protozoea and zoea stages, before the juvenile morphology is attained.

From this wealth of terms and diversity of developmental sequences, we can draw two important generalizations concerning the biology and evolution of the crustaceans. First, different developmental strategies reflect adaptations to different lifestyles. In spite of many exceptions, we can cite the early release of dispersal larvae by groups with limited adult mobility, such as thoracican barnacles, and by those whose resources may not permit production of huge quantities of yolk, such as the copepods. At the other end of this adaptive spectrum is the direct development of peracarids—a major factor allowing the invasion of land by certain isopods. Between these extremes we see all degrees of mixed life histories, with larvae being released at various stages following brooding and care. Second, because developmental stages also evolve, an analysis of developmental sequences can sometimes provide us with information about the radiation of the principal crustacean lineages. For example, the evolution of oostegites and of direct development combine as a unique synapomorphy of the Peracarida. Similarly, the addition of a unique larval form, such as the **cypris larva** that follows the nauplius in the cirripedes, can be viewed as a unique specialization that demarcates that group (Cirripedia). The cyprid either hatches as the only free-living larva, or it is the final larval stage after a series of lecithotrophic or planktotrophic nauplius larval stages.

It should also be noted that the branchiopods and some freshwater ostracods have evolved specialized ways of coping with the harsh conditions of many freshwater environments. Parthenogenesis, for example, is common in freshwater ostracods. Other adaptations include production of special overwintering forms, usually eggs or zygotes, that can survive extreme cold, lack of water, or anoxic conditions. Perhaps most remarkable in this respect are the large-bodied branchiopods whose encysted embryos are capable of an extreme state of anaerobic quiescence, or **diapause**. During these resistant stages, the metabolic rate of the embryos may drop to less than 10 percent of their normal rate.

Crustacean Phylogeny

Countless traditional studies and over 200 modern phylogenetic (cladistic) studies have been published on crustaceans. General agreement has been reached in some areas, but despite a great deal of effort, many fundamental mysteries remain unsolved. The use of molecular gene sequence data in phylogenetics has so far done little to resolve some of the most tenacious of these problems, although the analysis of multi-gene data sets holds promise for the future. There are several particularly problematic issues: What is the most primitive living crustacean group, and what are the relationships of the basal crustacean clades? What are the relationships of the maxillopodans, and how do the Ostracoda fit into that clade? What are the relationships among the Peracarida, and especially of the flabelliferan orders of the Isopoda and the hyperiids among the Amphipoda? What are the major decapod lineages and how are they related to one another? What group of crustaceans are represented by the mysterious "y-larvae" (the Facetotecta)?

Debates on crustacean phylogeny commonly center on two competing views regarding the nature of the primitive, or ancestral, crustacean body plan. One view holds that the first crustaceans had leaflike (phyllopodous) thoracic legs that were used both for swimming and for suspension feeding, as seen in the living cephalocarids, leptostracans, and many branchiopods (Figure 16.34A). The other view holds that the first crustaceans had nonphyllopodous, simple, paddle-like legs that were used for swimming, but not for feeding; instead, the tasks of feeding were undertaken by the cephalic appendages. This plan is perhaps best represented among living crustaceans by the remipedes. Both views agree that the ancestral crustacean probably possessed a long, many-segmented, highly homonomous body. Our cladogram of the Crustacea, based on morphological features, hypothesizes a phyllopodous ancestry, but places the Remipedia at the base of the crustacean tree (Figure 16.34B). Comparative spermatological studies, on the other hand, seem to ally the remipedes with the Maxillopoda, while DNA studies have been ambiguous on the issue.

The weight of evidence, both phylogenetic analyses and fossil data, seems to favor phyllopodous limbs as the primitive condition. However, recent developmental studies following the expression of *Distal-less* and other developmental genes suggest that the early embryogeny of limbs is very similar among crustaceans. For example, trunk limbs always emerge as ventral, subdivided limb buds. In phyllopodous limbs, the subdivisions of these limb buds grow to become the endites and the endopod of the natatory/filtratory adult limbs. In stenopodous limbs, the same limb bud subdivisions end up

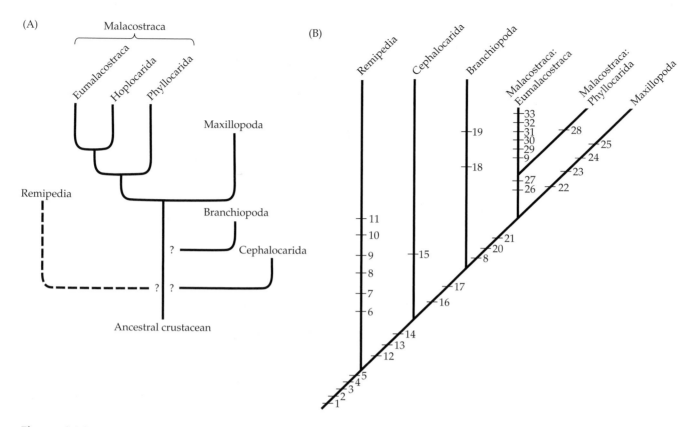

Figure 16.34 Two competing views of evolution within the Crustacea. (A) A traditional view of crustacean relationships, represented in an evolutionary tree, with the "phyllopodous Crustacea" at the base. (B) An alternative view, represented in a cladogram, of crustacean phylogeny. The synapomorphies depicted on the cladogram are listed below. Synapomorphies are: 1, head composed of 4 or 5 fused segments (plus acron) with 2 pairs of antennae and 2 or 3 pairs of mouth appendages; 2, biramous second antennae; 3, nauplius larva; 4, phyllopodous body limbs (with large epipods); 5, with head shield or small carapace; 6, raptorial mouth appendages; 7, mouth appendages situated in posteriorly directed atrium; 8, anterior thoracopods (one or more pairs) modified as maxillipeds (a highly variable trait that occurs in remipedes, malacostracans, and maxillopodans); 9, loss of phyllopodous condition on trunk appendages; 10, trunk appendages oriented laterally; 11, maxillules function as hypodermic fangs; 12, postcephalic trunk regionalized as thorax and abdomen; 13, loss of internal organ homonomy (e.g., segmental gut ceca); 14, reduction in number of body segments; 15, reduction of abdomen (to 11 segments); 16, fully developed carapace (reduced in several subsequent lineages); 17, reduction of abdomen to fewer than 9 segments; 18, reduction (or loss) of abdominal appendages; 19, first and second maxillae reduced or lost; 20, thorax shortened to fewer than 11 segments; 21, abdomen shortened to fewer than 8 segments; 22, with maxillopodan naupliar eye; 23, thorax of 6 or fewer segments; 24, abdomen of 4 or fewer segments; 25, genital appendages on the first abdominal somite (associated with male gonopores); 26, 8-segmented thorax and 7-segmented abdomen (plus telson); 27, male gonopores fixed on thoracomere 8/females on thoracomere 6; 28, carapace forms large "folded" structure enclosing most of body; 29, abdomen reduced to 6 segments (plus telson); 30, last abdominal appendages modified as uropods and forming tail fan with telson; 31, caridoid tail flip locomotion (escape reaction); 32, thoracopods with stenopodous endopods; 33, replacement of thoracic suspension feeding and phyllopodous thoracic limbs with cephalic feeding and nonphyllopodous thoracic limbs. Note that loss of the phyllopodous trunk limbs (character 9) has occurred several times, in the Remepedia, Eumalacostraca, some lineages of Branchiopoda, and most Maxillopoda.

developing into the actual segments of the adult limb. Hence, the endites of phyllopodous limbs appear to be homologous to the segments of the stenopodous limbs. This discovery supports an emerging view of developmental plasticity in arthropod limbs, and it suggests that relatively simple genetic "switches" can account for major differences in adult morphologies. Thus, it is highly plausible that stenopodous limbs have evolved multiple times from phyllopodous ancestors, and this is the scenario depicted in the cladogram in Figure 16.34B.

The work by Klaus Müller and Dieter Walossek (e.g., see References, Chapters 15 and 16) on three-dimensionally preserved microscopic arthropods from the Upper Cambrian *Orsten* deposits of Sweden has documented a diverse fauna of minute crustaceans and their larvae. Among them, for example, is *Skara* (Figure 16.35C), a cephalocarid- or mystacocarid-like crustacean for which both naupliar larvae and adults have been recovered (the nauplius larvae are only a couple hundred microns long; adults are about 1 mm in length). *Skara* and many other *Orsten* Crustacea were probably meiofaunal animals not unlike modern marine meiofaunal crustaceans. Dozens of *Orsten* microcrustacea have so far been described (Figure 16.35).

Studies on the Swedish *Orsten* fauna (510 mya), the Middle Cambrian (520 mya) Burgess Shale-like deposits from around the world, and the Lower Cambrian (530 mya) Chengjiang fossils from China have shown that Cambrian Crustacea had all the attributes of modern crustaceans, such as compound eyes, distinct head and trung tegmata, at least four head appendages, a carapace (or head shield), naupliar larvae (with locomotory first antennae), and biramous appendages on the second and third head somites (the second antennae and mandibles). We now know that the crustaceans are an ancient group. Their fossil record dates back to the early Cambrian, or even to the Ediacaran period if some arthropod fossils from those strata are viewed as Crustacea. Depending on one's definition of "Crustacea," it may even be that the first arthropods were themselves crustaceans.

Subsequent to their discovery in 1955, cephalocarids were regarded by most specialists as the most primitive living crustaceans, with the phyllopodous branchiopods and the leptostracans representing the most primitive members of the classes Branchiopoda and Malacostraca, respectively. The evolutionary tree shown in Figure 16.34A depicts this post-1955 view of crustacean phylogeny, still held by some workers. However, with the discovery of the remipedes in 1981, some workers began favoring nonphyllopodous forms as probable ancestral crustaceans. Remipedes possess a suite of features that appear to be extremely primitive, most notably a very long body with no postcephalic tagmosis, a double ventral nerve cord, serially arranged digestive ceca, and a simple cephalic shield. Figure 16.34B presents our view of crustacean phylogeny. Based on the set of characters described in this tree, the remipedes are hypothesized to be the most primitive living crustaceans.

In the 1950s, Russian biologist W. N. Beklemischev and Swedish carcinologist E. Dahl independently proposed that the copepods and several related classes constitute a monophyletic clade. Dahl proposed the class Maxillopoda for these taxa. Since then, the validity of the Maxillopoda has been a fertile field for debate. The shortening of the thorax to six or fewer segments and of the abdomen to four or fewer segments, reduction of the carapace (or, in the case of ostracods and cirripedes, extreme modification of the carapace), loss of abdominal appendages, and other associated changes in the maxillopodans are now thought to be tied to early paedomorphic events during the larval (or postlarval) stage of this lineage as it began to radiate (an idea first proposed in 1942 by R. Gurney). With reduction of the trunk and trunk limbs, the head appendages took on a larger role in both feeding and locomotion. Other synapomorphies that define the class Maxillopoda are listed on the cladogram.

The monophyletic nature of the class Malacostraca has rarely been questioned. Within the Malacostraca are two principal groups: Leptostraca, which have phyllopodous limbs and seven abdominal somites; and Eumalacostraca, which lack phyllopodous limbs and have six abdominal segments. The eumalacostracans also have the sixth abdominal appendages modified as uropods (which work in conjunction with the telson as a tail fan). Relationships among the four main eumalacostracan lines (hoplocarids, syncarids, peracarids, and eucarids), and even within the Eucarida, are far from settled and have provided zoologists with many generations of lively debate.

The class Branchiopoda is difficult to define on the basis of unique synapomorphies because it shows such great morphological variation. Apparently some branchiopods have secondarily lost the carapace, and others have secondarily lost most or all of the abdominal appendages.

We are still a long way from fully understanding phylogenetic relationships among the Crustacea. Like the arthropods in general, crustaceans exhibit high levels of evolutionary parallelism and convergence and many apparent reversals of character states. This genetic flexibility is no doubt due in part to the nature of the segmented body, the serially homologous appendages, and the flexibility of developmental genes, which, as we have stressed, provide enormous opportunity for evolutionary experimentation. Any conceivable cladogram of crustacean phylogeny will require the acceptance of considerable homoplasy.

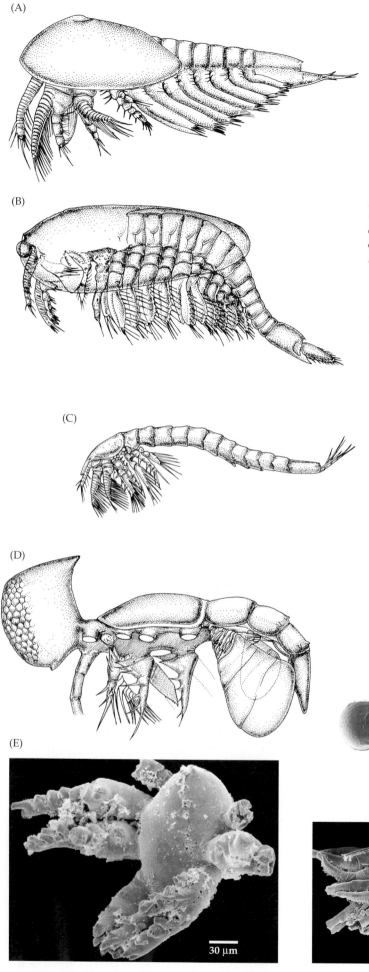

(A)

(B)

(C)

(D)

(E)

Figure 16.35 Examples of Upper Cambrian (~510 mya), probably meiofaunal crustaceans from the spectacular Swedish *Orsten* deposits. This ancient crustacean fauna possessed the key attributes of modern Crustacea, including compound eyes, head shields/carapaces, naupliar larvae (with locomotory first antennae), and biramous appendages on the second and third head segments (the second antennae and mandibles). (A) *Bredocaris*. (B) *Rehbachiella*, an early branchiopod, lateral and ventral views. (C) *Skara*, SEM of fossil and reconstructive drawing. (D) *Cambropachycope clarksoni*, a bizarre species with an expanded head and two pairs of enlarged thoracopods, SEM of fossil and reconstructive drawing. (E) *Martinssonia elongata*, first larva and postlarval stage.

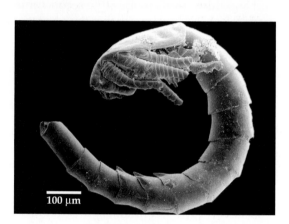

Selected References

The amount of published literature on crustaceans is vast. Much of the key work on classification and phylogenetics was reviewed by Martin and Davis (2001), and we refer readers to that paper for an entree into those fields.

General References

Abele, L. G. (ed.). 1982–1990. *The Biology of Crustacea*. Academic Press, New York. [A multivolume series on various aspects of crustacean biology.]

American Zoologist. 1985. Advances in Crustacean Endocrinology. Am. Zool. 25(1): 155–284. [12 papers devoted to work on the crustacean endocrine system.]

Anderson, D. T. 1994. *Barnacles: Structure, Function, Development and Evolution*. Chapman and Hall, London.

Arhat, A. and T. C. Kaufman. 1999. Novel regulation of the homeotic gene *Scr* associated with a crustacean leg-to-maxilliped appendage transformation. Development 126: 1121–1128.

Atwood, H. L. and D. C. Sanderman. 1982. *The Biology of Crustacea*, Vol. 3. *Neurobiology: Structure and Function*. Series editor, D. E. Bliss. Academic Press, New York.

Baker, A. de C., B. P. Boden and E. Brinton. 1990. *A Practical Guide to the Euphausiids of the World*. Natural History Museum Publications, London.

Banaja, A. A., J. L. James and J. Riley. 1975. An experimental investigation of a direct life-cycle in *Reighardia sternae* (Diesing 1864), a pentastomid parasite of the herring gull. Parasitol. 71: 493–503.

Barnard, J. L. 1991. The families and genera of marine gammaridean Amphipoda (except marine gammaroids). Rec. Aust. Mus., Suppl 13 (1/2): 1–866. [A benchmark compilation by one of the world's foremost, and most colorful, carcinologists.]

Barnard, J. L. and C. M. Barnard. 1983. *Freshwater Amphipoda of the World*. Hayfield Associates, Mt. Vernon, VA.

Bauer, R. T. 1981. Grooming behavior and morphology in the decapod Crustacea. J. Crust. Biol. 1: 153–173.

Bauer, R. T. 1987. Stomatopod grooming behavior: Functional morphology and amputation experiments in *Gonodactylus oerstedii*. J. Crust. Biol. 7: 414–432.

Bauer, R. T. and J. W. Martin (eds.). 1991. *Crustacean Sexual Biology*. Columbia University Press, New York.

Bliss, D. E. (gen. ed.). 1982. *The Biology of Crustacea*. Vols. 1–4. Academic Press, New York. [Comprehensive treatments of various topics, from taxonomy to neurobiology and more. Dorothy Bliss served as editor-in-chief, but each volume is edited by an appropriate specialist and contains contributions from a number of authors.]

Bowman, T. E. 1971. The case of the nonubiquitous telson and the fraudulent furca. Crustaceana 21: 165–175. [See Schminke 1976 for an opposing view.]

Bowman, T. E. and H.-E. Gruner. 1973. The families and genera of Hyperiidea (Crustacea: Amphipoda). Smithson. Contrib. Zool. 146: 1–64.

Bowman, T. E. and T. M. Iliffe. 1988. *Tulumella unidens*, a new genus and species of thermosbaenacean crustacean from the Yucatan Peninsula, Mexico. Proc. Biol. Soc. Wash. 101: 221–226.

Bowman, T. E., S. P. Garner, R. R. Hessler, T. M. Iliffe and H. L. Sanders. 1985. Mictacea, a new order of Crustacea Peracarida. J. Crust. Biol. 5: 74–78.

Boxshall, G. A. 1991. A review of the biology and phylogenetic relationships of the Tantulocarida, a subclass of Crustacea recognized in 1983. Verhandlungen der Deutschen Zoologischen Gesellschaft 84: 271–279.

Boxshall, G. A. and R. J. Lincoln. 1983. Tantulocarida, a new class of Crustacea ectoparasitic on other crustaceans. J. Crust. Biol. 3:1–16.

Boxshall, G. A. and R. J. Lincoln. 1987. The life cycle of the Tantulocarida (Crustacea). Philos. Trans. R. Soc. London Ser. B 315: 267–303.

Briggs, D. E. G., D. H. Erwin and F. J. Collier. 1994. *The Fossils of the Burgess Shale*. Smithsonian Institution Press, Washington, D.C.

Brooks, H. K. et al. 1969. Arthropoda 4: Crustacea (except Ostracoda), Myriapoda, Hexapoda. *In* R. C. Moore (ed.), *Treatise on Invertebrate Paleontology*, Pt. R. University of Kansas Press, Lawrence.

Brtek, J. and G. Mura. 2000. Revised key to families and genera of the Anostraca with notes on their geographical distribution. Crustaceana 73: 1037–1088.

Brusca, G. J. 1981a. Annotated keys to the Hyperiidea (Crustacea: Amphipoda) of North American coastal waters. Allan Hancock Found. Tech. Rep. 5: 1–76.

Brusca, G. J. 1981b. On the anatomy of *Cystisoma* (Amphipoda: Hyperiidea). J. Crust. Biol. 1: 358–375.

Brusca, R. C. 1981. A monograph on the Isopoda Cymothoidae (Crustacea) of the Eastern Pacific. Zool. J. Linn. Soc. 73: 117–199.

Brusca, R. C. and M. Gilligan. 1983. Tongue replacement in a marine fish (*Lutjanus guttatus*) by a parasitic isopod (Crustacea: Isopoda). Copeia 3: 813–816. [The only known case of a parasite functionally replacing a host organ.]

Brusca, R. C. and E. W. Iverson. 1985. A guide to the marine isopod Crustacea of Pacific Costa Rica. Rev. Biol. Trop. 33 (Suppl. 1): 1–77.

Brusca, R. C., S. Taiti and V. Coelho. 2001. A Guide to the Marine Isopods of Coastal California. http://phylogeny.arizona.edu/tree/eukaryotes/animals/arthropoda/crustacea/isopoda/isopod_lichen/bruscapeet.html.

Brusca, R. C., R. Wetzer and S. France. 1995. Cirolanidae (Crustacea; Isopoda; Flabellifera) of the tropical eastern Pacific. Proc. San Diego Nat. Hist. Soc., No. 30. 96 pp.

Burnett, B. R. and R. R. Hessler. 1973. Thoracic epipodites in the Stomatopoda (Crustacea): A phylogenetic consideration. J. Zool. 169: 381–392.

Burukovskii, R. N. 1985. *Key to Shrimps and Lobsters*. A. A. Balkema, Rotterdam.

Butler, T. H. 1980. *Shrimps of the Pacific Coast of Canada*. Can. Dept. Fish. Oceans Bull. 202: 1–280.

Caldwell, R. L. and H. Dingle. 1976. Stomatopods. Sci. Am. 234: 81–89.

Caldwell, R. L. and H. Dingle. 1978. Ecology and morphology of feeding and agonistic behavior in mudflat stomatopods. Biol. Bull. 155: 134–149.

Calman, R. T. 1909. Crustacea. *In* R. Lankester (ed.), *A Treatise on Zoology*, Pt. 7. Adam and Charles Black, London. [A classic treatise; still very useful.]

Cameron, J. N. 1985. Molting in the blue crab. Sci. Am. 252: 102–109.

Carpenter, J. H. 1999. Behavior and ecology of *Speleonectes epilimnius* (Remipedia, Speleonectidae) from surface water of an anchialine cave on San Salvador Island, Bahamas. Crustaceana 72: 979–991.

Carter, J. W. 1982. Natural history observations on the gastropod shell-using amphipod *Photis conchicola* Alderman, 1936. J. Crust. Biol. 2: 328–341.

Chace, F. A., Jr. 1972. The shrimps of the Smithsonian-Bredin Caribbean expeditions with a summary of the West Indies shallow-water species. Smithson. Contrib. Zool. 98: 1–180. [A useful entrée into the marine shrimps of the Caribbean Region.]

Chace, F. A., Jr. and H. H. Hobbs, Jr. 1969. The freshwater and terrestrial decapod crustaceans of the West Indies with special reference to Dominica. U.S. Nat. Mus. Bull. 292: 1–258.

Chang, E. S. 1985. Hormonal control of molting in decapod Crustacea. Am. Zool. 25: 179–185.

Chapman, M. A. and M. H. Lewis. 1976. *An Introduction to the Freshwater Crustacea of New Zealand*. Collins, Auckland.

Christy, J. H. 1982. Burrow structure and use in the sand fiddler crab, *Uca pugilator* (Bosc). Anim. Behav. 30: 687–694.

Cobb, J. S. and B. F. Phillips (eds.). 1980. *The Biology and Management of Lobsters*. Vols. 1–2. Academic Press, New York.

Cohen, A. C. and J. G. Morin. 1990. Patterns of reproduction in ostracodes: A review. J. Crust. Biol. 10: 84–211.

Cohen, A. C., J. W. Martin and L. S. Kornicker. 1998. Homology of Holocene ostracode biramous appendages with those of other crustaceans: The protopod, epipod, exopod and endopod. *Lethaia* 31: 251–265.

Coull, B. C. 1977. Copepoda: Harpacticoida. Marine flora and fauna of the northeastern U.S. NOAA Tech. Rpt., Nat. Mar. Fish. Serv. Circular 399.

Crane, J. 1975. *Fiddler Crabs of the World* (Ocypodidae: Genus Uca). Princeton University Press, Princeton.

Cressey, R. F. 1978. Marine flora and fauna of the northeastern United States: Crustacea: Branchiura. NOAA Tech. Rpt., Nat. Mar. Fish. Serv. Circular 413.

Cronin, T. W. 1986. Optical design and evolutionary adaptation in crustacean compound eyes. J. Crust. Biol. 6: 1–23.

Darwin, C. 1852, 1854. *A Monograph on the Subclass Cirripedia*. Vols. 1–2. Ray Society, London. [Still the starting place for barnacle taxonomy.]

De Jong-Moreau, L. and J.-P. Casanova. 2001. The foreguts of the primitive families of the Mysida (Crustacea, Peracarida): A transitional link between those of the Lophogastrida (Crustacea, Mysidacea) and the most evolved Mysida. Acta Zool. 82: 137–147.

Derby, C. D. 1982. Structure and function of cuticular sensilla of the lobster *Homarus americanus*. J. Crust. Biol. 2: 1–21.

Derby, C. D. and J. Atema. 1982. The function of chemo- and mechanoreceptors in lobster (*Homarus americanus*) feeding behaviour. J. Exp. Biol. 98: 317–327.

Dunham, P. J. 1978. Sex pheromones in Crustacea. Biol. Rev. 53: 555–583.

Efford, I. E. 1966. Feeding in the sand crab *Emerita analoga*. Crustaceana 10: 167–182.

Elofsson, R. 1965. The nauplius eye and frontal organs in Malacostraca. Sarsia 19: 1–54.

Elofsson, R. 1966. The nauplius eye and frontal organs of the non-Malacostraca. Sarsia 25: 1–28.

Ferrari, F. D. 1988. Developmental patterns in numbers of ramal segments of copepod post-maxillipedal legs. Crustaceana 54: 256–293.

Fitzpatrick, J. F., Jr. 1983. *How to Know the Freshwater Crustacea*. Wm. C. Brown, Dubuque, IA.

Forest, J. 1999a. Traité de Zoologie. Anatomie, Systématique, Biologie. Tome VII, Fascicule II. Généralités (suite) et Systématique. Crustacés. Masson, Paris.

Forest, J. (ed.) 1999b. Traité de Zoologie. Anatomie, Systématique, Biologie. Tome VII, Fascicule IIIA. Crustacés Péracarides. Masson, Paris.

Fryer, G. 1964. Studies on the functional morphology and feeding mechanism of *Monodella argentarii* Stella (Crustacea: Thermosbaenacea). Trans. R. Soc. Edinburgh 66(4): 49–90.

Ghiradella, H. T., J. Case and J. Cronshaw. 1968. Structure of aesthetascs in selected marine and terrestrial decapods: Chemoreceptor morphology and environment. Am. Zool. 8: 603–621.

Gilchrist, S. and L. A. Abele. 1984. Effects of sampling parameters on the estimation of population parameters in hermit crabs. J. Crust. Biol. 4: 645–654. [Includes a good literature list on shell selection by hermit crabs.]

Glaessner, M. F. 1960. Decapoda. *In* R. C. Moore (ed.), *Treatise on Invertebrate paleontology*, Pt. R, *Arthropoda* 4. Geological Society of America, pp. 399–533.

Glenner, H. and J. T. Høeg. 1995. A new motile, multicellular state involved in host invasion by parasitic barnacles (Rhizocephala). Nature 377: 147–150.

Glenner, H., J. T. Høeg, J. J. O'Brien and T. D. Sherman. 2000. Invasive vermigon stage in the parasitic barnacles *Loxothylacus texanus* and *L. panopaei* (Sacculinidae): Closing of the rhizocephalan life-cycle. Mar. Biol. 136: 249–257.

Gordon, I. 1957. On *Spelaeogriphus*, a new cavernicolous crustacean from South Africa. Bull. Br. Mus. Nat. Hist. Zool. 5: 31–47.

Govind, C., M. Quigley and K. Mearow. 1986. The closure muscle in the dimorphic claws of male fiddler crabs. Biol. Bull. 170: 481–493.

Greenaway, P. 1985. Calcium balance and moulting in the Crustacea. Biol. Rev. 60(3): 425–454.

Grey, D. L., W. Dall and A. Baker. 1983. *A Guide to the Australian Penaeid Prawns*. North Territory Govt. Printing Office, Australia.

Grindley, J. R. and R. R. Hessler. 1970. The respiratory mechanism of *Spelaeogriphus* and its phylogenetic significance. Crustaceana 20: 141–144.

Grygier, M. J. 1982. Sperm morphology in Ascothoracida (Crustacea: Maxillopoda): Confirmation of generalized nature and phylogenetic importance. Int. J. Invert. Reprod. 4: 323–332.

Grygier, M. J.1987a. Classification of the Ascothoracica (Crustacea). Proc. Biol. Soc. Wash. 100: 452–458.

Grygier, M. J. 1987b. New records, external and internal anatomy, and systematic position of Hansen's Y-larvae (Crustacea: Maxillopoda: Facetotecta). Sarsia 72: 261–278.

Guinot, D., D. Doumenc and C. C. Chintiroglou. 1995. A review of the carrying behaviour in Brachyuran crabs, with additional information on the symbioses with sea anemones. Raffles Bull. Zool. 43(2): 377–416.

Gurney, R. 1942. *Larvae of Decapod Crustacea*. Ray Society, London. [A benchmark survey that desperately needs to be updated.]

Haig, J. 1960. The Porcellanidae (Crustacea: Anomura) of the eastern Pacific. Allan Hancock Pac. Expeds. 24: 1–440.

Hallberg, E. and R. Elofsson. 1983. The larval compound eye of barnacles. J. Crust. Biol. 3: 17–24.

Hamner, W. M. 1988. Biomechanics of filter feeding in the Antarctic krill *Euphausia superba*: Review of past work and new observations. J. Crust. Biol. 8: 149–163.

Harbison, G. R., D. C. Biggs and L. P. Madin. 1977. The associations of Amphipoda Hyperiidea with gelatinous zooplankton. II. Associations with Cnidaria, Ctenophora and Radiolaria. Deep-Sea Res. 24(5): 465–488.

Harrison, F. W. and A. G. Humes. 1992. *Microscopic Anatomy of Invertebrates*. Vols. 9 and 10, *Crustacea* and *Decapod Crustacea*. Wiley-Liss, New York. [Two more outstanding volumes in this fine series.]

Hart, J. F. L. 1982. *Crabs and Their Relatives of British Columbia*. B. C. Prov. Mus., Handbook No. 40.

Hartnoll, R. G. 1969. Mating in Brachyura. Crustaceana 16: 161–181.

Harvey, A. H., J. W. Martin and R. Wetzer. 2002. Crustacea. *In* C. Young, M. Sewell and M. Rice (eds.), *Atlas of Marine Invertebrate Larvae*. Academic Press, London.

Herreid, W. F., II and C. R. Fourtner (eds.). 1981. *Locomotion and Energetics in Arthropods*. Plenum, New York.

Herrnkind, W. F. 1985. Evolution and mechanisms of single-file migration in spiny lobster: Synopsis. Contrib. Mar. Sci. 27: 197–211.

Hessler, R. R. 1964. The Cephalocarida. Comparative skeletomusculature. Mem. Conn. Acad. Arts Sci. 16: 1–97.

Hessler, R. R. 1982. The structural morphology of walking mechanisms in eumalacostracan crustaceans. Phil. Trans. R. Soc. Lond. Ser. B 296: 245–298. [An outstanding review.]

Hessler, R. R. 1985. Swimming in Crustacea. Trans. R. Soc. Edinburgh 76: 115–122.

Hessler, R. R. and R. Elofsson. 1991. Excretory system of *Hutchinsoniella macracantha* (Cephalocarida). J. Crust. Biol. 11: 356–367.

Ho, J. S. 1978. Copepoda: Cyclopoids parasitic on fishes. Marine flora and fauna of the northeastern U.S. NOAA Tech. Rpt., Nat. Mar. Fish. Serv. Circular 409.

Hobbs, H. H., Jr. 1972. Crayfishes (Astacidae) of North and Middle America. Biota of Freshwater Ecosystems. Identification Manual No. 9. U.S. Environmental Protection Agency.

Høeg, J. T. 1985. Cypris settlement, kentrogon formation and host invasion in the parasitic barnacle *Lernaeodiscus porcellanae* (Muller) (Crustacea: Cirripedia: Rhizocephala). Acta Zool. 66: 1–46.

Høeg, J. T. 1987. Male cypris metamorphosis and a new male larval form, the trichogen, in the parasitic barnacle *Sacculina carcini* (Crustacea: Cirripedia: Rhizocephala). Phil. Trans. R. Soc. Lond. Ser. B 317: 47–63.

Høeg, J. T. and G. A. Kolbasov. 2002. Lattice organs in y-cyprids of the Facetotecta and their significance in the phylogeny of the Crustacea Thecostraca. Acta Zool. 83: 67–79.

Høeg, J. T. and J. Lutzen. 1985. *Crustacea Rhizocephala*. Marine Invertebrates of Scandinavia No. 6. Norwegian University Press, Oslo. [Provides an excellent summary of rhizocephalan life history.]

Holdich, D. M. and D. A. Jones. 1983. *Tanaids: Keys and Notes for the Identification of the Species of England*. Cambridge University Press, Cambridge.

Holthuis, L. B. 1980. Shrimps and prawns of the world: An annotated catalogue of species of interest to fisheries. FAO Species Catalogue, Vol. 1/Fisheries Synopses 125: 1–261.

Holthuis, L. B. 1991. Marine lobsters of the world: An annotated and illustrated catalogue of species of interest to fisheries known to date. FAO Species Catalogue, Vol. 13. FAO, Rome.

Holthuis, L. B. 1993. The recent genera of the caridean and stenopodidean shrimps (Crustacea, Decapoda) with an appendix on the order Amphionidacea. Nat. Natuurhistorisch. Mus., Leiden.

Horch, K. W. and M. Salmon. 1969. Production, perception and reception of acoustic stimuli by semiterrestrial crabs. Forma Functio 1: 1–25.

Horne, D. J. 1983. Life-cycles of podocopid Ostracoda—a review (with particular reference to marine and brackish-water species). *In* R. F. Maddocks (ed.), *Applications of Ostracoda*. Department of Geosciences, University of Houston, pp. 581–590.

Horne, D. J., A. C. Cohen and K. Martens. 2002. Taxonomy, morphology and biology of Quaternary and living Ostracoda. *In* J. Holmes and A. Chivas (eds.), *The Ostracoda: Applications in Quaternary Research*. AGU Geophysical Monograph.

Humes, A. G. and R. V. Gooding. 1964. A method for studying the external anatomy of copepods. Crustaceana 6: 238–240.

Huvard, A. L. 1990. The ultrastructure of the compound eye of two species of marine ostracods (Ostracoda: Cypridinidae). Acta Zoologica 71: 217–224.

Huys, R. and G. A. Boxshall. 1991. *Copepod Evolution*. Ray Society, London.

Huys, R., G. A. Boxshall and R. J. Lincoln. 1993. The tantulocarid life cycle: The circle closed? J. Crust. Biol. 13: 432–442.

Ingle, R. W. 1980. *British Crabs*. Oxford University Press, Oxford.

Ivanov, B. G. 1970. On the biology of the Antarctic krill *Euphausia superba*. Mar. Biol. 7: 340.

Jamieson, B. G. M. 1991. Ultrastructure and phylogeny of crustacean spermatozoa. Mem. Queensland Mus. 31: 109–142.

Jones, D. and G. Morgan. 2002. *A Field Guide to Crustaceans of Australian Waters*. Reed New Holland, Sydney.

Jones, N. S. 1976. *British Cumaceans*. Academic Press, New York.

Kabata, Z. 1979. *Parasitic Copepoda of British Fishes*. Ray Society, London.

Kaestner, A. 1970. *Invertebrate Zoology*. Vol. 3, *Crustacea*. Wiley, New York. [Translated from the 1967 German second edition by H. W. Levi and L. R. Levi.]

Kensley, B. and R. C. Brusca (eds.) 2001. *Isopod Systematics and Evolution*. Balkema, Rotterdam.

King, J. L., M. A. Simovich and R. C. Brusca. 1996. Endemism, species richness, and ecology of crustacean assemblages in northern California vernal pools. Hydrobiologia 328: 85–116.

Koehl, M. A. R. and J. R. Strickler. 1981. Copepod feeding currents: Food capture at low Reynolds numbers. Limnol. Oceanogr. 26: 1062–1073.

Land, M. F. 1981. Optics of the eyes of *Phronima* and other deep-sea amphipods. J. Comp. Physiol. 145: 209–226.

Land, M. F. 1984. Crustacea. *In* M. A. Ali (ed.), *Photoreception and Vision in Invertebrates*. Plenum, New York, pp. 401–438.

Lang, K. 1948. Monographie der Harpacticoiden. Hakan Ohlssons, Lund. 1,682 pp. [Whew!]

Laufer, H. et al. 1987. Identification of a juvenile hormone-like compound in a crustacean. Science 235: 202–205.

Laval, P. 1972. Comportement parasitisme et écologie d'*Hyperia schizogeneios* (Amphipode Hypéride) dans le plancton de Veillfranch-sur-Mer. Ann. Iust. Océnogr. Paris 48:49–74.

Laval, P. 1980. Hyperiid amphipods as crustacean parasitoids associated with gelatinous zooplankton. Oceanogr. Mar. Biol. Annu. Rev. 18: 11–56.

Lockwood, A. P. M. 1967. *Physiology of Crustacea*. W. H. Freeman, San Francisco. [Dated, but still useful.]

Madin, L. P. and G. R. Harbison. 1977. The associations of Amphipoda Hyperiidea with gelatinous zooplankton. I. Associations with Salpidae. Deep-Sea Res. 24: 449–463.

Maitland, D. P. 1986. Crabs that breathe air with their legs—*Scopimera* and *Dotilla*. Nature 319: 493–495.

Manning, R. B. 1969. *Stomatopod Crustacea of the Western Atlantic*. University of Miami Press, Coral Gables, FL.

Manning, R. B. 1974. Crustacea: Stomatopoda. Marine flora and fauna of the northeastern U.S. NOAA Tech. Rpt., Nat. Mar. Fish. Serv. Circular 386.

Manton, S. M. 1977. *The Arthropoda*. Oxford University Press, London. [Good anatomy; bad phylogenetics.]

Marshall, S. M. 1973. Respiration and feeding in copepods. Adv. Mar. Biol. 11: 57–120.

Martin, J. W. and D. Belk. 1988. Review of the clam shrimp family Lynceidae (Stebbing, 1902) (Branchiopoda: Conchostraca) in the Americas. J. Crust. Biol. 8: 451–482.

Martin, J. W. and G. E. Davis. 2001. An updated classification of the recent Crustacea. Nat. Hist. Mus. Los Angeles Co., Sci. Ser. No. 39. 124 pp. [An outstanding synthesis of the literature.]

Mauchline, J. 1980. The biology of mysids and euphausiids. Adv. Mar. Biol. 18: 1–681.

McCain, J. C. 1968. The Caprellidae (Crustacea: Amphipoda) of the western North Atlantic. U.S. Nat. Mus. Bull. 278: 1–147.

McLaughlin, P. A. 1974. The hermit crabs of northwestern North America. Zool. Verh. Rijksmus. Nat. Hist. Leiden 130: 1–396.

McLaughlin, P. A. 1980. *Comparative Morphology of Recent Crustacea*. W. H. Freeman, San Francisco.

McLay, C. L. 1988. Crabs of New Zealand. Leigh Lab. Bull. 22: 1–463.

Miller, D. C. 1961. The feeding mechanism of fiddler crabs with ecological considerations of feeding adaptations. Zoologica 46: 89–100.

Millikin, M. R. and A. B. Williams. 1984. Synopsis of biological data on the blue crab, *Callinectes sapidus* Rathbun. NOAA Tech. Rpt. NMFS 1, FAO Fisheries Synopsis no. 138.

Müller, H.-G. 1994. *World Catalogue and Bibliography of the Recent Stomatopoda*. Wissenschaftlicher, Berlin.

Müller, K. J. 1983. Crustaceans with preserved soft parts from the Upper Cambrian of Sweden. Lethaia 16: 93–109.

Müller, K. J. and D. Walossek. 1985. Skaracarida, a new order of Crustacea from the Upper Cambrian of Västergötland, Sweden. Fossils and Strata 17: 1–65.

Müller, K. J. and D. Walossek. 1986a. *Martinssonia elongata* gen. et sp. n., a crustacean-like euarthropod from the Upper Cambrian "Orsten" of Sweden. Zoologica Scripta 15: 73–92.

Müller, K. J. 1986b. Arthropod larvae from the Upper Cambrian of Sweden. Trans. R. Soc. Edinburgh, Earth Sci. 77: 157–179.

Newman, W. A. and R. R. Hessler. 1989. A new abyssal hydrothermal verrucomorphan (cirripedia: sessilia): the most primitive living sessile barnacle. Trans. San Diego Nat. Hist. Soc. 21: 259–273.

Newman, W. A. and A. Ross. 1976. Revision of the balanomorph barnacles; including a catalog of the species. San Diego Soc. Nat. Hist. Mem. 9: 1–108.

Nolan, B. A. and M. Salmon. 1970. The behavior and ecology of snapping shrimp (Crustacea: *Alpheus heterochelis* and *Alpheus normanni*). Forma Functio 2: 289–335.

Oeksnebjerg, B. 2000. The Rhizocephala of the Mediterranean and Black Seas: taxonomy, biogeography, and ecology. Israel J. Zool. 46 (1): 1–102.

Olesen, J. 1999. Larval and post-larval development of the branchiopod clam shrimp *Cyclestheria hislopi* (Baird, 1859) (Crustacea, Branchiopoda, Conchostraca, Spinicaudata). Acta Zool. 80: 163–184.

Olesen, J. 2001. External morphology and larval development of *Derocheilocaris remanei* Delamare-Deboutteville & Chappuis, 1951 (Crustacea, Mystacocarida), with a comparison of crustacean segmentation and tagmosis patterns. Biologiske Skrifter 53: 1–59.

Olesen, J., J. W. Martin and E. W. Roessler. 1996. External morphology of the male of *Cyclestheria hislopi* (Baird, 1859) (Crustacea, Branchiopoda, Spinicaudata), with a comparison of male claspers among the Conchostraca and Cladocera and its bearing on phylogeny of the "bivalved" Branchiopoda. Zoologica Scripta 25: 291–316.

Olesen, J., S. Richter and G. Scholtz. 2001. The evolutionary transformation of phyllopodous to stenopodous limbs in the Branchiopoda (Crustacea)—Is there a common mechanism for early limb development in arthropods? Int. J. Dev. Biol. 45: 869–876.

Omori, M. 1974. The biology of pelagic shrimps in the ocean. Adv. Mar. Biol. 12: 233–324.

Parker, S. (ed.) 1982. *Synopsis and Classification of Living Organisms*. Vol. 2. McGraw-Hill, New York, pp. 173–326. [A number of specialists contributed to the crustacean section of this useful compendium.]

Pennak, R. W. 1978. *Fresh-water Invertebrates of the United States*, 2nd Ed. Wiley, New York.

Pennak, R. W. and D. J. Zinn. 1943. Mystacocarida, a new order of Crustacea from intertidal beaches in Massachusetts and Connecticut. Smithson. Misc. Coll. 103: 1–11.

Pérez Farfante, I. and B. F. Kensley. 1997. Penaeoid and sergestoid shrimps and prawns of the world. Keys and diagnoses for the families and genera. Mem. Mus. Nation. d'Hist. Natur. 175: 1–233.

Perry, D. M. and R. C. Brusca. 1989. Effects of the root-boring isopod *Sphaeroma peruvianum* on red mangrove forests. Mar. Ecol. Prog. Ser. 57: 287–292.

Persoone, G., P. Sorgeloos, O. Roels and E. Jaspers (eds.) 1980. *The Brine Shrimp Artemia*. Universa Press, Wetteren, Belgium.

Reaka, M. L. and R. B. Manning. 1981. The behavior of stomatopod Crustacea, and its relationship to rates of evolution. J. Crust. Biol. 1: 309–327.

Rebach, S. and D. W. Dunham (eds.) 1983. *Studies in Adaptation: The Behavior of Higher Crustacea*. Wiley, New York.

Richardson, H. 1905. *A monograph of the isopods of North America*. U.S. Nat. Mus. Bull. 54: 1–727. [Badly out of date, but still a benchmark.]

Riley, J. 1986. The biology of pentastomids. Adv. Parasitol. 25: 45–128.

Roer, R. and R. Dillaman. 1984. The structure and calcification of the crustacean cuticle. Am. Zool. 24: 893–909.

Sanders, H. L. 1955. The Cephalocarida, a new subclass of Crustacea from Long Island Sound. Proc. Natl. Acad. Sci. U.S.A. 41: 61–66.

Sanders, H. L. 1963. The Cephalocarida: Functional morphology, larval development, comparative external anatomy. Mem. Conn. Acad. Arts Sci. 15: 1–80.

Schembri, P. J. 1982. Feeding behavior of 15 species of hermit crabs (Crustacea: Decapoda: Anomura) from the Otago region, southeastern New Zealand. J. Nat. Hist. 16: 859–878.

Schminke, H. K. 1976. The ubiquitous telson and the deceptive furca. Crustaceana 30: 292–300. [See Bowman 1971 for an opposing view.]

Schmitt, W. L. 1965. *Crustaceans*. University of Michigan Press, Ann Arbor. [A wonderful, timeless little volume.]

Scholtz, G. 1995. Head segmentation in Crustacea—an immunocytochemical study. Zoology 98: 104–114.

Scholtz, G. and W. Dohle. 1996. Cell lineage and cell fate in crustacean embryos: A comparative approach. Int. J. Dev. Biol. 40: 211–220.

Scholtz, G., N. H. Patel and W. Dohle. 1994. Serially homologous engrailed stripes are generated via different cell lineages in the germ band of amphipod crustaceans (Malacostraca, Peracarida). Int. J. Dev. Biol. 38: 471–478.

Schram, F. R. 1974. Paleozoic Peracarida of North America. Fieldiana Geol. 33: 95–124.

Schram, F. R. (gen. ed.). 1983–2001. *Crustacean Issues*. Vols. 1–13. A. A. Balkema, Rotterdam. [A continuing series of topical symposium volumes, each edited by a specialist in the appropriate field, e.g., phylogeny, biogeography, growth, barnacle biology, biology of isopods, history of carcinology]

Schram, F. R. 1986. *Crustacea*. Oxford University Press, New York.

Schram, F. R., J. Yager and M. J. Emerson. 1986. The Remipedia. Pt. I, Systematics. San Diego Soc. Nat. Hist. Mem. 15.

Self, J. T. 1969. Biological relations of the Pentastomida: A bibliography on the Pentastomida. Exp. Parasitol. 24: 63–119.

Skinner, D. M. 1985. Interacting factors in the control of the crustacean molt cycle. Am. Zool. 25: 275–284.

Smirnov, N. N. and B. V. Timms. 1983. A revision of the Australian Cladocera (Crustacea). Rec. Aust. Mus. Suppl. 1: 1–132.

Smith, R. J. and K. Martens. 2000. The ontogeny of the cyprid ostracod *Eucypris virens* (Jurine, 1820) (Crustacea, Ostracoda). Hydrobiologia 419: 31–63.

Snodgrass, R. E. 1956. Crustacean metamorphosis. Smithson. Misc. Contrib. 131(10): 1–78. [Dated, but still a good introduction to the subject.]

Stebbing, T. R. R. 1893. *A History of Crustacea*. D. Appleton and Co., London. [Still a great read.]

Steinsland, A. J. 1982. Heart ultrastructure of *Daphnia pulex* De Geer (Crustacea, Branchiopoda, Cladocera). J. Crust. Biol. 2: 54–58.

Stepien, C. A. and R. C. Brusca. 1985. Nocturnal attacks on nearshore fishes in southern California by crustacean zooplankton. Mar. Ecol. Prog. Ser. 25: 91–105.

Stock, J. 1976. A new genus and two new species of the crustacean order Thermosbaenacea from the West Indies. Bijdr. Dierkdl. 46: 47–70.

Strickler, R. 1982. Calanoid copepods, feeding currents and the role of gravity. Science 218: 158–160.

Sutton, S. L. 1972. *Woodlice*. Ginn and Co., London. [Most of what you always wanted to know about pillbugs and roly-polies.]

Sutton, S. L. and D. M. Holdich (eds.) 1984. *The Biology of Terrestrial Isopods*. Clarendon Press, Oxford. [The rest of what you always wanted to know about pillbugs and roly-polies.]

Tomlinson, J. T. 1969. The burrowing barnacles (Cirripedia: Order Acrothoracica). U.S. Nat. Mus. Bull. 259: 1–162.

Van Name, W. G. 1936. The American land and freshwater isopod Crustacea. Bull. Am. Mus. Nat. Hist. 71: 1–535. [Badly in need of updating; no other keys are available to this poorly known fauna.]

Vinogradov, M. E., A. F. Volkov and T. N. Semenova. 1982 (1996). *Hyperiid Amphipods (Amphipoda, Hyperiidea) of the World Oceans*. Translated from the Russian by D. Siegel-Causey for the Smithsonian Institution Libraries, Washington, D.C.

Wagner, H. P. 1994. A monographic review of the Thermosbaenacea. Zoologische Verhandelingen 291: 1–338.

Walker, G. 2001. Introduction to the Rhizocephala (Crustacea: Cirripedia). J. Morphol. 249: 1–8.

Wallosek, D. 1993. The Upper Cambrian *Rehbachiella* and the phylogeny of Branchiopoda and Crustacea. Fossils and Strata 32:1–202.

Warner, G. F. 1977. *The Biology of Crabs*. Van Nostrand Reinhold, New York.

Waterman, T. H. (ed.). 1960, 1961. *The Physiology of Crustacea*. Vols. 1–2. Academic Press, New York. [Dated, but still useful.]

Waterman, T. H. and A. S. Pooley. 1980. Crustacean eye fine structure seen with scanning electron microscopy. Science 209: 235–240.

Weeks, S. C. 1990. Life-history variation under varying degrees of intraspecific competition in the tadpole shrimp *Triops longicaudatus* (Le Conte). J. Crust. Biol. 10: 498–503.

Wenner, A. M. (ed.) 1985a. *Crustacean Growth: Factors in Adult Growth*. A. A. Balkema, The Netherlands.

Wenner, A. M. (ed.) 1985b. *Crustacean Growth: Larval Growth*. A. A. Balkema, The Netherlands.

Wiese, K. 2000. *The Crustacean Nervous System*. Springer-Verlag, New York.

Williams, A. B. 1984. *Shrimps, Lobsters, and Crabs of the Atlantic Coast of the Eastern United States, Maine to Florida*. Smithsonian Institution Press, Washington, D.C. [An outstanding reference by a grand gentleman.]

Williams, A. B. 1988. *Lobsters of the World: An Illustrated Guide*. Osprey Books, New York.

Williamson, D. I. 1973. *Amphionides reynaudii* (H. Milne Edwards), representative of a proposed new order of eucaridan Malacostraca. Crustaceana 25(1): 35–50.

Wingstrand, K. G. 1972. Comparative spermatology of a pentastomid *Raillietiella hemidactyli* and a branchiuran crustacean *Argulus foliaceus* with a discussion of pentastomid relationships. Biol. Skr. 19: 1–72.

Wittmann, K. J. 1981. Comparative biology of marsupial development in *Leptomysis* and other Mediterranean Mysidacea (Crustacea). J. Exp. Mar. Biol. Ecol. 52: 243–270.

Yagamuti, S. 1963. *Parasitic Copepoda and Branchiura of Fishes*. Wiley, New York.

Yager, J. 1981. Remipedia, a new class of Crustacea from a marine cave in the Bahamas. J. Crust. Biol. 1: 328–333.

Yager, J. 1991. The Remipedia (Crustacea): Recent investigation of their biology and phylogeny. Verhandlungen der Deutschen Zoologischen Gesellschaft, Stuttgart 84: 261–269.

Yager, J. and W. F. Humphreys. 1996. *Lasionectes esleyi*, sp. nov., the first remipede crustacean recorded from Australia and the Indian Ocean, with a key to the world species. Invert. Taxon. 10:171–187.

Zinn, D. J., B. W. Found and M. G. Kraus. 1982. A bibliography of the Mystacocarida. Crustaceana 42(3): 270–274.

Phylogeny and Evolution

For a more extensive list of references on crustacean phylogenetics, see Martin and Davis 2001.

Abele, L. G., W. Kim and B. E. Felgenhauer. 1989. Molecular evidence for inclusion of the phylum Pentastomida in the Crustacea. Mol. Biol. Evol. 6: 685–691.

Abele, L. G., T. Spears, W. Kim and M. Applegate. 1992. Phylogeny of selected maxillopodan and other crustacean taxa based on 18S ribosomal nucleotide sequences: A preliminary analysis. Acta Zool. 73: 373–382.

Ahyong, S. T. and C. Harling. 2000. The phylogeny of the stomatopod Crustacea. Aust. J. Zool. 48: 607–642.

Almeida, W. de O. and M. L. Christoffersen. 1999. A cladistic approach to relationships in Pentastomida. J. Parasitol. 85: 695–704.

Boxshall, G. A., J.-O. Strömberg and E. Dahl (eds.). 1992. The Crustacea: Origin and evolution. Acta Zool. 73: 271–392.

Brusca, R. C. 1984. Phylogeny, evolution, and biogeography of the marine isopod subfamily Idoteinae (Crustacea: Isopoda: Idoteidae). Trans. San Diego Soc. Nat. Hist. 20(7): 99–134.

Brusca, R. C. and G. D. F. Wilson. 1991. A phylogenetic analysis of the Isopoda (Crustacea) with some classificatory recommendations. Mem. Queensland Mus. 31: 143–204.

Burkenroad, M. D. 1981. The higher taxonomy and evolution of Decapoda (Crustacea). Trans. San Diego Soc. Nat. Hist. 19(17): 251–268.

Chen, Y.-U., J. Vannier and D.-Y. Huang. 2001. The origin of crustaceans: new evidence from the early Cambrian of China. Proc. Royal Soc. London 268: 2181–2187.

Fortey, R. A. and R. H. Thomas, *Arthropod Relationships*. Chapman and Hall, London.

Glenner, H., M. J. Grygier, J. T. Høeg, P. G. Jensen and F. R. Schram. 1995. Cladistic analysis of the Cirripedia Thoracica. Zool. J. Linn. Soc. 114: 365–404.

Glenner, H. and J. T. Høeg. 1993. Scanning electron microscopy of metamorphosis in four species of barnacles. Mar.Biol. 117: 431–438.

Ho, J. S. 1990. Phylogenetic analysis of copepod orders. J. Crust. Biol. 10: 528–536.

Huys, R. and G. A. Boxshall. 1991. *Cropped Evolution*. The Ray Soc., London.

McLaughlin, P. A. 1983. Hermit crabs—are they really polyphyletic? J. Crust. Biol. 3: 608–621.

Morrison, C. L. A. W. Harvey, S. Lavery, K. Tieu, Y. Huang and C. W. Cunningham. 2002. Mitochondrial gene rearrangements confirm the parallel evolution of the crab-like form. Proc. Royal Soc. London, Biol. Sci. 269: 345–350.

Negrea, S., N. Botnariuc and H. J. Dumont. 1999. Phylogeny, evolution and classification of the Branchiopoda (Crustacea). Hydrobiologia 412: 191–212.

Olesen, J. 2000. An updated phylogeny of the Conchostraca—Cladocera clade (Branchiopoda, Diplostraca). Crustaceana 73: 869–886.

Pérez-Losada, M. J. T. Høeg, G. A. Kolbasov and K. A. Crandall. 2002. Reanalysis of the relationships among the Cirripedia and Ascothoracida, and the phylogenetic position of the Facetotecta using 18S rDNA sequences. J. Crustacean Biol. 22: 661–669.

Remigio, E. A. and P. D. Hebert. 2000. Affinities among anostracan (Branchiopoda) families inferred from phylogenetic analyses of multiple gene sequences. Mol. Phylogen. Evol. 17: 117–128.

Richter, S. and G. Scholtz. 2000. Phylogenetic analysis of the Malacostraca (Crustacea). J. Zool. Syst. Evol. Res. 39: 113–136.

Riley, J., A. A. Banaja and J. L. James. 1978. The phylogenetic relationships of the Pentastomida: The case for their inclusion within the Crustacea. Int. J. Parasitol. 8: 245–254.

Schram, F. R. (ed.) 1983. *Crustacean Phylogeny*. Balkema, Rotterdam.

Spears, T. and L. G. Abele. 1999. The phylogenetic relationships of crustaceans with follacious limbs: An 18S rDNA study of Branchiopoda, Cephalocarida, and Phyllocarida. J. Crust. Biol. 19: 825–843.

Spears, T. and L. G. Abele. 2000. Branchiopod monophyly and interordinal phylogeny inferred from 18S ribosomal DNA. J. Crust. Biol. 20: 1–24.

Spears, T., L. G. Abele and M. A. Applegate. 1994. Phylogenetic study of cirripedes and selected relatives (Thecostraca) based on 18S rDNA sequence analysis. J. Crust. Biol. 14: 641–656.

Sternberg, R. V., N. Cumberlidge and G. Rodríguez. 1999. On the marine sister groups of the freshwater crabs (Crustacea: Decapoda). J. Zool. Sys. Evol. Res. 37: 19–38.

Storch, V. and B. G. M. Jamieson. 1992. Further spermatological evidence for including the Pentastomida (tongue worms) in the Crustacea. Int. J. Parasitol. 22: 95–108.

Tam, Y. K. and I. Kornfield. 1998. Phylogenetic relationships of clawed lobster genera (Decapoda: Nephropidae) based on mitochondrial 16S rRNA gene sequences. J. Crust. Biol. 18(1): 138–146.

Walker-Smith, G. K. and G. C. B. Poore. 2001. A phylogeny of the Leptostraca (Crustacea) from Australia. Mem. Mus. Victoria 58: 137–148.

Walossek, D. and K. J. Müller. 1997. Cambrian "Orsten"-type arthropods and the phylogeny of Crustacea. *In* R. A. Fortey (ed.), *Arthropod Relationships*. Chapman and Hall, London, pp. 139–153.

Whittington, H. B. and W. D. U. Rolfe (eds.). 1963. *Phylogeny and Evolution of Crustacea*. Museum of Comparative Zoology, Cambridge. [A benchmark symposium publication; still an important reference although many new ideas on the subject will be found in the post-1970 literature.]

Wilson, K., V. Cahill, E. Ballment and J. Benzie. 2000. The complete sequence of the mitochondrial genome of the crustacean *Penaeus monodon*: Are malacostracan crustaceans more closely related to insects than to branchiopods. Mol. Biol. Evol. 17: 863–874.

17

Phylum Arthropoda:
The Hexapoda
(Insects and Their Kin)

Even in matters about which man is wont to especially pride himself, such as...social organisation, he might with advantage go to the ant to learn wisdom, since many of the problems of modern civilisation involved in the questions concerned in the regulation of increase of population, the proper division of labour, and the support of useless individuals, have been satisfactorily solved by...insects that live habitually in communities.

Richard Lydekker,
The Royal Natural History, Volume 6 (1896)

The arthropod subphylum Hexapoda comprises the class Insecta and three other small, closely related, wingless insect-like groups: Collembola, Protura, and Diplura. The Hexapoda are united on the basis of a distinct body plan of a head, thorax, and abdomen, three pairs of thoracic legs, one pair of antennae, three sets of "jaws" (mandibles, maxillae, and labium), an aerial gas exchange system composed of tracheae and spiracles, Malpighian tubules formed as proctodeal (ectodermal) evaginations, and, among the Pterygota, wings (Box 17A). The presence of a thorax fixed at 3 segments, each with a pair of walking legs, is a unique synapomorphy for the Hexapoda. Until quite recently, hexapods and myriapods were thought to be sister groups, forming a clade called "Tracheata" (or "Atelocerata," or "Uniramia"), sharing such features as tracheae, uniramous legs, Malpighian tubules, and loss of the second antennae.* Some workers still hold to this view, but a great deal of recent information suggests that the two groups may not be so closely related to each other (see the section on hexapod evolution below and also Chapter 15). The jury is still out on the question of hexapod–myriapod relationships, but we have taken the conservative approach for this edition of *Invertebrates* and treat these two groups as separate subphyla (and in separate chapters).

Hexapods are fundamentally terrestrial arthropods; groups inhabiting aquatic environments today have secondarily invaded those habitats through behavioral adaptations and modifications of their aerial gas exchange systems. The earliest

---

*The name "Atelocerata" (*atelo*, "imperfectly"; *cerata*, "horned") alludes to a loss of the second pair of antennae in these groups, in contrast to the Crustacea, which possess two pairs of antennae. The name "Tracheata" derives from the presence of tracheal gas exchange structures in hexapods and myriapods. The name "Uniramia" was first proposed by Sidnie Manton in 1972 for the Hexapoda, Myriapoda, and Onychophora (on the basis of the uniramous appendages), but has been used by later authors only for the first two taxa.

BOX 17A *Characteristics of the Subphylum Hexapoda*

1. Body composed of 19 true somites (plus acron) organized as a head (5 somites), thorax (3 somites) and abdomen (11 somites). Due to fusion of somites, these body segments are not always externally obvious

2. Head segments bear the following structures (from anterior to posterior): ocelli; compound eyes; antennae; clypeolabrum; mandibles; maxillae, labium (fused second maxillae). Ocelli (and compound eyes) are secondarily lost in some groups

3. Legs uniramous; present on the three thoracic segments of adults. Legs composed of 6 articles; coxa, trochanter, femur, tibia, tarsus, post-tarsus; tarsus often subdivided; post-tarsus typically clawed

4. Gas exchange by spiracles and tracheae

5. Gut with gastric (digestive) ceca

6. Fused exoskeleton of head forms unique internal tentorium

7. With ectodermally derived Malpighian tubules (proctodeal evaginations)

8. Gonopores open terminally, or subterminally on abdominal segment 7, 8, or 9

9. Dioecious (hermaphroditism very rare). Direct or indirect development

---

undisputed fossil records of hexapods are early Devonian (390 mya) wingless creatures resembling modern springtails (Collembola) and jumping bristletails (Archaeognatha); the first winged insect fossils make their appearance later in the Devonian. However, there are Silurian trace fossils that are very hexapod-like, and molecular clock data suggest an early Silurian origin for the insects.

The most spectacular evolutionary radiation among the Hexapoda has, of course, been within the insects, which inhabit nearly every conceivable terrestrial and freshwater habitat and, less commonly, even the sea surface and the marine littoral region. Insects are also found in such unlikely places as oil swamps and seeps, sulfur springs, glacial streams, and brine ponds. They often live where few other animals or plants can exist. It is no exaggeration to say that insects rule the land. Their diversity and abundance defy imagination (Figure 17.1).

We do not know how many species of insects there are, or even how many have been described. Published estimates of the number of described species range from 890,000 to well over a million (we calculate 898,000 to 948,000). An average of about 3,500 new species have been described annually since the publication of Linnaeus's *Systema Naturae* in 1758, although in recent decades the average has climbed to 7,000 new species per year. Estimates of the number of species remaining to be described range from 3 million to 100 million. The Coleoptera (beetles), with an estimated 350,000–375,000 described species, is far and away the largest insect order (one out of every three animal species is a beetle). The beetle family Curculionidae (the weevils) contains about 65,000 described species (nearly 5 percent of all described animal species) and is by itself larger than any other non-arthropod phylum (Box 1B). The rich diversity of insects seems to have come about through a combination of advantageous features, including the evolutionary exploitation of developmental genes

working on segmented and compartmentalized bodies, coevolution with plants (particularly the flowering plants), miniaturization, and the invention of flight.

Insects are not only diverse, but also incredibly abundant. For every human alive, there are an estimated 200 million insects. An acre of ordinary English pasture supports an estimated 248,375,000 springtails and 17,825,000 beetles. In tropical rain forests, insects can constitute 40 percent of the total animal biomass (dry weight), and the biomass of the ants can be far greater than that of the combined mammal fauna (up to 15 percent of the total animal biomass). A single colony of the African driver ant *Anomma wilverthi* may contain as many as 22 million workers. Based on his research in the tropics, biodiversity sleuth Terry Erwin has calculated that there are about 3.2×10^8 individual arthropods per hectare, representing more than 60,000 species, in the western Amazon. In Maryland, a single population of the mound-building ant *Formica exsectoides* comprised 73 nests covering an area of 10 acres and containing approximately 12 million workers. Termites have colonies of similar magnitudes. E. O. Wilson has calculated that, at any given time, 10^{15} (a million billion) ants are alive on Earth!

In most parts of the world, insects are among the principal predators of other invertebrates. Insects are also key items in the diets of many terrestrial animals, and they play a major role as reducer-level organisms in food webs. Due to their sheer numbers, they constitute much of the matrix of terrestrial food webs. Their biomass and energy consumption exceed those of vertebrates in most terrestrial habitats. In deserts and in the tropics, ants replace earthworms as the most abundant earth movers (ants are nearly as important as earthworms even in temperate regions). Termites are among the chief decomposers of dead wood and leaf litter around the world.

Eighty percent of the world's crop species, including food, medicine, and fiber crops, rely on animal pollina-

(A)

(B)

(C)

(D)

(E)

(F)

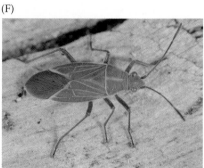

(G)

(H)

(I)

(J)

(K)

(L)

(M)

(N)

Figure 17.1 A diversity of insect orders. (A) A common silverfish (Thysanura). (B) A mayfly (Ephemeroptera). (C) A damselfly (Odonata). (D) A mantid (Mantodea) subduing a would-be predator, a shrew. (E) A stonefly (Plecoptera). (F) The western box-elder bug (Hemiptera). (G) An earwig (Dermaptera). (H) A webspinner (Embioptera). (I) A walking stick (Phasmida). (J) A flea (Siphonaptera). (K) A cicada (Hemiptera: Homoptera). (L) The Colorado potato beetle (Coleoptera). (M) A lacewing (Neuroptera). (N) A member of the newly described order Mantophasmatodea. (O) A robber fly (Diptera). (P) A caddisfly (Trichoptera). (Q) A congregation of monarch butterflies (Lepidoptera). (R) A Jerusalem cricket (Orthoptera).

(O)

(P)

(Q)

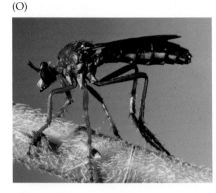

(R)

Figure 17.1 *(continued)*

tors, nearly all of which are insects. Insects also play key roles in pollinating wild, native plants. Without insects, life as we know it would cease to exist. Beekeeping began long ago, at least by 600 B.C. in the Nile Valley and probably well before that. The first migratory beekeepers were Egyptians who floated hives up and down the Nile to provide pollination services to floodplain farmers while simultaneously producing a honey crop. Domestic honey bees (*Apis mellifera*), introduced to North America from Europe in the mid-1600s, are now the dominant pollinators of most food crops grown around the world, and they play some role in pollinating 80 percent of the crop varieties grown in the United States (they are estimated to be directly responsible for $10 to $20 billion in crops annually). Honey bees also produce $250 million worth of honey annually.*

Because they have been introduced worldwide, honey bees compete with native bees (and other insects) around the world, and it is now virtually impossible to find an area free of managed or feral honey bees. *A. mellifera* outcompetes native species by overwhelming them with their sheer numbers and superior ability to detect and direct one another to pollen and nectar sources. This direct competition with native species is reducing the numbers of native pollinators, adding to the "pollination crisis" much of the world faces today.†

Many plants that are now dominated by honey bees, including many crops, are not actually pollinated by them. For example, many plants depend on sonication ("buzz pollination") for pollination (e.g., blueberries, cranberries, eggplants, kiwi fruits, tomatoes), a trick that honey bees cannot perform. To complicate the story even more, at the same time that the United States has become dependent on domestic honey bees, U.S. populations of these bees are beginning to plummet due to exotic (introduced) bee parasites (e.g., mites, beetles), loss of habitat, use of pesticides, and the invasion of Africanized bees (*A. mellifera scutellata*) into the United States in 1990.‡

Interactions between insects and flowering plants have been going on for a very long time, beginning over 100 million years ago with the origin of the angiosperms and accelerating with the ascendancy of these flowering plants during the early Cenozoic. Millions of years of plant–insect coevolution have resulted in flowers with anatomy and scents that are finely tuned to their insect partners. In exchange for pollination services, flowers provide insects with food (nectar, pollen), shelter, and chemicals used by the insects to produce such things as pheromones. In general, insect pollination is accomplished coincidentally, as the pollinators visit flowers for other reasons. But in a few cases, such as that of the yucca moths of the American Southwest (*Tegiticula*

---

*The study of bees is called **melittophily**. The study of honey bees (*Apis* spp.) and their management is called **apiculture**. The management of bumble bees (*Bombus* spp.) is **bombiculture**. The ritualized keeping of stingless bees is **meliponiculture**.

†Disruption of habitat, widespread and often inappropriate use of pesticides, and the development of genetically engineered pesticide-containing plants has created a "pollination crisis" in many parts of the world, as pollinating insects are locally extirpated and native plant and domestic crop pollination plummets.

‡The Asian honey bee, *Apis dorsata*, not yet introduced to the U.S., is a giant reaching an inch in length. Its droppings are quite noticeable, and the mass defections of these bees at sunset create a "golden shower," that provides significant nutrient enrichment to tropical soils. Their droppings were once confused with the dreaded "yellow rain," a deadly form of biochemical warfare that poisoned thousands of villagers during the Vietnam War.

spp.), the insects actually gather up pollen and force it into the receptive stigma of the flower, initiating pollination. The moth's goal is to assure a supply of yucca seed for its larvae, which develop within the yucca's fruits. Some insects also play important roles as seed dispersers, especially ants. More than 3,000 plant species (in 60 families) are known to rely on ants for the dispersal of their seeds.

Like all other animals on Earth, insects are facing enormous threats of extinction. Certainly many thousands of species have become extinct over the past century as a result of rampant land use change and deforestation. With accelerating biodiversity losses worldwide, estimates of the number of insect species that will go extinct by the year 2025 range into the millions.

Insects are not without their share of sins. They consume about a third of our potential annual harvest, and transmit many major human diseases (Appendix A). Every year we spend billions of dollars on insect control. **Malaria,** transmitted by mosquitoes, kills 1 to 3 million people annually (mostly children), and each year nearly 500 million people contract the disease. It is the leading cause of death from infectious disease, and it has plagued humans for at least 3,000 years (Egyptian mummies have been found with malaria antigens in their blood). One of the most widespread and fastest spreading human viral diseases, **dengue**, is transmitted by mosquitoes of the genus *Aedes*. Dengue is essentially an urban disease, almost entirely associated with anthropogenic environments because its main vectors, *A. albopictus* and *A. aegypti*, breed primarily in artificial containers (e.g., flower vases, discarded tires, water tanks). In recent years, these species have spread throughout the tropics, often following the international trade in used tires. Although these mosquitoes have not penetrated very far into the temperate zones, they have become established in the southern United States. A variety of mosquitoes transmits the filarial nematode *Wuchereria bancrofti*, the causative agent of **lymphatic filariasis** (**"elephantiasis"**) throughout the world's tropics. **Chagas' disease** (**American trypanosomiasis**) is transmitted by certain hemipteran bugs in the subfamily Triatominae (family Reduviidae), and causes chronic degenerative disease of the heart and intestine. The species of triatomine bugs that occur in the southwestern United States tend not to defecate when they feed, greatly reducing the possibility of humans contracting Chagas' disease in that area. However, global warming is now increasing the spread of *Aedes*, *Culex*, triatomines, and other tropical disease vectors.

The natural history writer David Quammen, speaking of the blood-sucking varieties of mosquitoes, notes that the average blood meal of a female (the only sex that feeds on blood) amounts to 2.5 times the original weight of the insect—the equivalent, Quammen notes, "of Audrey Hepburn sitting down to dinner and getting up from the table weighting 380 pounds, then, for that

matter, flying away." But as Quammen also points out, mosquitoes have made tropical rain forests (the most diverse ecosystems on Earth) largely uninhabitable to humans, thus helping to preserve them.

Needless to say, the subject of insect biology, or **entomology**, is a discipline in its own right, and a multitude of books and college courses on the subject exists. If we apportioned pages to animal groups on the basis of numbers of species, overall abundance, or economic importance, insect chapters could fill 90 percent of this text. The Selected References at the end of this chapter provide entry into some of the current literature on insects. Because the Hexapoda comprises such a large and diverse assemblage of arthropods, we first present a brief classification of the 33 recognized orders, followed by more detailed synopses. The synopses provide brief diagnoses and comments on the biology of each order. These two sections serve as a preface to the bauplan discussion that follows, and also provide a reference that the reader can turn to as needed.

Hexapod Classification

Our classification scheme recognizes 33 orders of living hexapods. This arrangement is accepted by most specialists, although some disagreement exists regarding the categorical ranking of some taxa. With close to a million named species of Hexapoda, we have opted not to include representative genera in the classification scheme below.

Some authorities recognize three large groups of modern winged insects, or Neoptera: the **Orthopterodea** (Plecoptera, Blattodea, Isoptera, Mantodea, Dermaptera, Orthoptera, and Phasmida), **Hemipterodea** (Zoraptera, Psocoptera, Phthiraptera, Thysanoptera, and Hemiptera), and **Holometabola** (Neuroptera, Coleoptera, Mecoptera, Siphonaptera, Diptera, Trichoptera, Lepidoptera, and Hymenoptera). Orthopterodeans are primitive neopterans with biting–chewing mouthparts, two pairs of wings, and hemimetabolous development. Hemipterodea are characterized by (usually) short antennae, chewing or sucking mouthparts, leg tarsi of three or fewer articles, absence of cerci, lack of true male gonopods, wings (when present) with reduced venation, and hemimetabolous development that is relatively gradual (although metamorphosis in several groups includes one or two inactive pupa-like stages). The Holometabola have holometabolous development, with distinct egg, larval, pupal, and adult states, the larvae have internal wing buds, and the mouthparts are typically biting–chewing. Neither the Orthopterodea nor the Hemipterodea is likely to be a monophyletic group, so we do not recognize these taxa, but the Holometabola almost certainly is monophyletic (two of its key synapomorphies being holometabolous development and internal wing buds).

SUBPHYLUM HEXAPODA

CLASS ENTOGNATHA

    ORDER COLLEMBOLA: Springtails

    ORDER PROTURA: Proturans

    ORDER DIPLURA: Diplurans

CLASS INSECTA

  SUBCLASS ARCHAEOGNATHA

    ORDER ARCHAEOGNATHA (= MICROCORYPHIA): Jumping bristletails

  SUBCLASS ZYGENTOMA

    ORDER THYSANURA: Silverfish

  SUBCLASS PTERYGOTA: Winged insects

  INFRACLASS PALAEOPTERA: Ancient winged insects

    ORDER EPHEMEROPTERA: Mayflies

    ORDER ODONATA: Dragonflies and damselflies

  INFRACLASS NEOPTERA: Modern, wing-folding insects

    ORDER PLECOPTERA: Stoneflies

    ORDER BLATTODEA: Cockroaches

    ORDER ISOPTERA: Termites

    ORDER MANTODEA: Mantids

    ORDER PHASMIDA (= PHASMATOPTERA): Stick and leaf insects, specters

    ORDER GRYLLOBLATTODEA: Rock crawlers

    ORDER DERMAPTERA: Earwigs

    ORDER ORTHOPTERA: Locusts, katydids, crickets, grasshoppers

    ORDER MANTOPHASMATODEA: Mantophasmatodeans

    ORDER EMBIOPTERA: Web-spinners

    ORDER ZORAPTERA: Zorapterans

    ORDER PSOCOPTERA: Book and bark lice

    ORDER PHTHIRAPTERA: Sucking and biting lice (includes "Anoplura" and "Mallophaga")

    ORDER THYSANOPTERA: Thrips

    ORDER HEMIPTERA: True bugs (includes the Homoptera)

    ORDER STREPSIPTERA: Twisted-wing parasites

    ORDER MEGALOPTERA: Alderflies, dobsonflies, fishflies

    ORDER RAPHIDIOPTERA: Snakeflies

    ORDER NEUROPTERA: Lacewings, ant lions, mantisflies, spongillaflies, owlflies

    ORDER COLEOPTERA: Beetles

    ORDER MECOPTERA: Scorpionflies, hangingflies, snowflies

    ORDER SIPHONAPTERA: Fleas

    ORDER DIPTERA: True flies, mosquitoes, gnats

    ORDER TRICHOPTERA: Caddisflies

    ORDER LEPIDOPTERA: Butterflies, moths

    ORDER HYMENOPTERA: Ants, bees, wasps

Synopses of Major Hexapod Taxa

Subphylum Hexapoda

Body differentiated into head (acron + 5 segments), thorax (3 segments), and abdomen (11 or fewer segments); cephalon with one pair lateral compound eyes and (usually) a triad or pair of medial ocelli; with one pair of uniramous multiarticulate antennae, mandibles, and maxillae; second pair of maxillae fused to form a complex labium; each thoracic segment with one pair of legs; wings often present on second and third thoracic segments (in pterygote insects); abdomen without fully developed legs, but "prolegs" (presumably homologous to the ancestral arthropod abdominal appendages) occur in at least seven orders (in adults of some Diplura, Thysanura, and Archaeognatha and in larvae of some Diptera, Trichoptera, Lepidoptera, and Hymenoptera); gonopores open terminally or subterminally on abdomen (on seventh, eighth, or ninth abdominal segment); paired cerci often present; males commonly with intromittent and clasping structures; development direct, involving relatively slight changes in body form (ametabolous or hemimetabolous), or indirect with striking changes (holometabolous).

Class Entognatha

Mouth appendages **entognathous** (base of mouthparts hidden within the head capsule); mandibles with single articulation; most or all antennal articles with intrinsic musculature; without, or with poorly developed, Malpighian tubules; legs with one (undivided) tarsus.

    The Entognatha may be an artificial group. The entognathous conditions of Collembola and Protura appear to be homologous (these two orders are often placed together in the class Ellipura), but the entognathy of the Diplura may be a convergent condition. Recent data from paleontology, comparative anatomy, and molecular phylogenetics suggest that the Diplura may be closer to the Insecta than to Collembola–Protura.

Order Collembola. 6,000 described species. Small (most less than 6 mm); wingless; biting–chewing mouthparts; mandibles with single articulation; abdomen with 6 or fewer segments, without cerci; with terminal gonopores; without Malpighian tubules; often without spiracles or tracheae; antennae 4-articulate, first 3 articles with muscles; tarsus of legs indistinct (perhaps fused with tibia); post-tarsus of legs with single claw; first abdominal segment with ventral tube (**collophore**) of unknown function; third abdominal segment with small, partly fused appendages; appendages of fourth or fifth abdominal segment form springlike structure (**furcula**) operated by hemocoelic fluid pressure; with or without small compound eyes; ocelli vestigial; simple development.

The first known hexapods in the fossil record are collembolans. *Rhyniella praecursor* and other species from the Lower Devonian closely resemble some modern collembolan families. Many workers believe that springtails evolved via neoteny.

Order Protura.   205 described species. Minute (smaller than 2 mm); whitish; wingless; without eyes, abdominal spiracles, hypopharynx, or cerci; Malpighian tubules are small papillae; sucking mouthparts; styletlike mandibles with single articulation; vestigial antennae; first pair of legs carried in elevated position and used as surrogate "antennae"; post-tarsus of legs with single claw; abdomen 11-segmented, with well developed telson; first 3 abdominal segments with small appendages; without external genitalia, but male gonopores on protrusible phallic complex; gonopores terminal; with or without tracheae; simple development. Rare, occurring in leaf litter, moist soils, and rotting vegetation.

Order Diplura.   About 650 described species; fossils date to the Carboniferous. Small (less than 4 mm); whitish; without wings, eyes, external genitalia or Malpighian tubules; chewing mouthparts; mandibles with single articulation; abdomen 11-segmented, but embryonic tenth and eleventh segments fuse before hatching; gonopores on ninth segment; 7 pairs of lateral abdominal leglets; 2 caudal cerci; with tracheae and up to 7 pairs of abdominal spiracles; antennae multiarticulate, each article with intrinsic musculature; simple development.*

Class Insecta

Mouth appendages **ectognathous** (exposed and projecting from the head capsule); mandibles with two points of articulation (except Archaeognatha); intrinsic musculature of antennal articles greatly reduced; with well developed Malpighian tubules.

Subclass Archaeognatha

Order Archaeognatha.   255 described species. Small (to 15 mm), wingless (perhaps secondarily), resembling silverfish but more cylindrical; ocelli present; compound eyes large and contiguous; body scaly;

mandibles biting–chewing; with a single articulation; tarsi 3-articulate; middle and hind coxae usually with exites ("styli"); abdomen 11-segmented, with 3 to 8 pairs of lateral leglets ("styli") and 3 caudal filaments. Jumping bristletails are usually found in grassy or wooded areas under leaves, bark, or stones.

Subclass Zygentoma

Order Thysanura.   450 described species. Small, wingless, with flattened body; with or without ocelli; compound eyes reduced; body usually covered with scales; mandibles biting–chewing; antennae multiarticulate, but only basal article with musculature; tarsi 3- to 5-articulate; abdomen 11-segmented, with lateral leglets (often called styli) on segments 2–9, 7–9, or 8–9; 3 caudal cerci; female gonopores on eighth segment, male gonopores on tenth; without copulatory organs; with tracheae; simple development. Silverfish occur in leaf litter or under bark or stones, or in buildings, where they may become pests.

Subclass Pterygota

With paired wings on the second and third thoracic segments, the **forewings** (front wings) and **hindwings**; wings may be secondarily lost in one or both sexes, or modified for functions other than flight; only basal articles of antennae (scape and pedicel) with intrinsic musculature; adults without abdominal leglets except on genital segments; female gonopores on eighth abdominal segment, male on tenth; female often with ovipositor; molting ceases at maturity.

Infraclass Palaeoptera

Wings cannot be folded back over body, and when at rest are either held straight out to the side or vertically above the abdomen (with dorsal surfaces pressed together); wings always membranous, with many longitudinal veins and cross veins; wings tend to be fluted, or accordion-like; antennae highly reduced or vestigial in adults; hemimetabolous development. Two extant orders; many extinct groups.

Order Ephemeroptera.   2,100 described species. Adults with vestigial mouthparts, minute antennae, and soft bodies; wings held vertically over body when at rest; forewings present; hindwings present or absent; long, articulated cerci, usually with medial caudal filament; male with first pair of legs elongated for clasping female in flight; second and third legs of male, and all legs of female, may be vestigial or absent; abdomen 10-segmented; larvae aquatic; young (nymphs) with paired articulated lateral gills, caudal filaments, and well developed mouthparts; adults preceded by winged subimago stage.

Mayflies are primitive winged insects in which the aquatic nymphal stage has come to dominate the life cycle. Larvae hatch in fresh water and become long-lived

---

*The position of the Diplura is still unresolved. Some workers include them in the Entognatha, others in the Insecta, and still others treat them as a separate group altogether. The derived characters linking diplurans with insects include the presence of filiform cerci and an extra set of nine single tubules in the axoneme of the sperm. (For a more detailed discussion of the evidence, see Kristensen 1991.)

nymphs, passing through many instars. Mayfly nymphs are important food for many stream and lake fishes. Adults live only a few hours or days, do not feed, and copulate in the air, sometimes in large nuptial swarms.

Order Odonata. 5,000 described species. Adults with small filiform antennae, large compound eyes, and chewing mouthparts with massive mandibles; labium modified into prehensile organ; two pairs of large wings, held outstretched (dragonflies) or straight up over body (damselflies) when at rest; abdomen slender and elongate, 10-segmented; male with accessory genitalia on second and third abdominal sternites; eggs and larvae aquatic, with caudal or rectal gills.

Dragonflies and damselflies are spectacular insects with broad public appeal, not only for their beauty but because they consume large numbers of insect pests, including mosquitoes. Larvae and adults are both highly active predators, with larvae consuming various invertebrates and adults capturing other flying insects. Many species are 7–8 cm long, and some extinct forms were nearly a meter in length.

Infraclass Neoptera

Wings at rest can be folded backward to cover the body; cross veination reduced.

Order Plecoptera. 1,600 described species. Adults with reduced mouthparts, elongate antennae, (usually) long articulated cerci, soft bodies, and a 10-segmented abdomen; without ovipositor; wings membranous, pleated, folding over and around abdomen when at rest; wings with primitive venation; nymphs aquatic, with gills.

Stoneflies, like mayflies, have experienced considerable evolutionary radiation of the larval (nymphal) stage. Stonefly nymphs are important consumers in freshwater systems and also serve as major prey items for various fishes and invertebrates. Adults of most species feed, but are short-lived and die soon after mating.

Order Blattodea. Body flattened dorsoventrally; pronotum large, with expanded margins and extending over head; forewings (when present) leathery; hindwings expansive and fanlike; ovipositor reduced; cerci multiarticulate; legs adapted for running; eggs laid in cases (**ootheca**).

Of the 4,000 described species of cockroaches, fewer than 40 are domestic (household inhabitants). Some species are omnivores, while others are restricted in diet. Some live in and feed on wood and have intestinal flora that aid in cellulose digestion. Most species are tropical, but some live in temperate habitats, caves, deserts, and ant and bird nests.

Order Isoptera. 2,000 described species. Small; soft-bodied; wings equal-sized, elongate, membranous, **dehiscent** (shed by breaking at basal line of weakness); antennae short, filamentous, with 11–33 articles; cerci small to minute; ovipositor reduced or absent; many with rudimentary or no external genitalia; marked polymorphism.

Termites ("white ants") are strictly social insects, usually with three distinct types of individuals, or castes, in a species: **workers**, **soldiers**, and **reproductives**. Workers are generally sterile, blind individuals with normal mandibles; they are responsible for foraging, nest construction, and caring for members of the other castes. Soldiers are blind, usually sterile, wingless forms with powerful enlarged mandibles used to defend the colony. Reproductives have wings and fully formed compound eyes. They are produced in large numbers at certain times of the year, whereupon they emerge from the colony in swarms. Mating occurs at this time, and individual pairs start new colonies. Wings are shed after copulation. Colonies form nests (**termitaria**) in wood that is in or on the ground. Workers in the presumed primitive termite families harbor a variety of symbiotic cellulose-digesting flagellate protists in special chambers in the hindgut (termites are thought to have evolved from a form of wood-eating cockroach). Some families contain symbiotic bacteria that serve the same purpose. Termites often occur in enormous numbers; one spectacular estimate suggests that there are about three-quarters of a ton of termites for every person on Earth!

Order Mantodea. First pair of legs large and raptorial; prothorax elongate and often markedly elaborated; head highly mobile, with very large compound eyes, not covered by pronotum; forewings thickened, hindwings membranous; abdomen 11-segmented; with reduced ovipositor, but complex male genitalia; one pair multiarticulate cerci.

Mantids prey on insects and spiders. The digestive tract is short and straight, but includes a large crop, a ribbed or toothed proventriculus, and 8 midgut ceca. Malpighian tubules are numerous, over 100 in some species. The order is primarily tropical, with only a few of the 1,800 known species occurring in temperate regions.

Order Grylloblattodea. Slender, elongate, wingless insects, usually 15–30 mm long. Body pale and finely pubescent; compound eyes small or absent; no ocelli; antennae long and filiform, of 23–45 "articles"; cerci long, of 5 or 8 joints; terminal sword-shaped ovipositor nearly as long as cerci.

Rock crawlers were not discovered until the twentieth century (in 1914), and today only about two dozen species are known, half from North America. They inhabit cold places, such as glaciers and ice caves. They are probably scavengers on dead insects and other organic matter.

Order Dermaptera. 1,200 described species. Cerci usually form heavily sclerotized posterior forceps;

forewings (when present) form short, leathery **tegmina**, without veins and serving as elytra to cover the semicircular, membranous hindwings (when present); ovipositor reduced or absent.

Earwigs are common but poorly understood insects. Most appear to be nocturnal scavenging omnivores. The forceps are used in predation, for defense, to hold a mate during courtship, for grooming the body, and for folding the hindwings under the elytra. Some species eject a foul-smelling liquid from abdominal glands when disturbed. Most species are tropical, although many also inhabit temperate regions.

Order Orthoptera. 13,000 described species. Pronotum unusually large, extending posteriorly over mesonotum; forewings with thickened and leathery region (tegmina), occasionally modified for stridulation or camouflage; hindwings membranous, fanlike; hindlegs often large, adapted for jumping; auditory tympana present on forelegs and abdomen; ovipositor large; male genitalia complex; cerci distinct, short, and jointed.

Grasshoppers and their kin are common and abundant insects at all but the coldest latitudes. This order includes some of the largest living insects, up to 12 cm in length and twice that in wingspan. Most are herbivores, but many are omnivorous, and some are predatory. Stridulation, which is common among males, is usually accomplished by rubbing the specially modified forewings (tegmina) together, or by rubbing a ridge on the inside of the hind femur against a special vein of the tegmen. No orthopterans stridulate by rubbing the hindlegs together, as is commonly thought.

Order Phasmida. 2,600 described species. Body cylindrical or markedly flattened dorsoventrally, usually elongate; prothorax short; meso- and metathorax elongate; forewings absent, or forming small, leathery tegmina; hindwings fanlike; biting–chewing mouthparts; short, unsegmented cerci; ovipositor weak.

Stick and leaf insects, specters, and other phasmids are some of the oddest of all the insects. Although resembling orthopterans in basic form, they are clearly a distinct radiation. Their ability to mimic plant parts is legendary, and many have evolved as perfect mimics of twigs, leaves, and broken branches. The eggs of some species even mimic seeds of the plants on which they live. Certain walking sticks reach a body length of nearly 35 cm, though most are less than 4 cm. Sexual dimorphism is so striking that males and females have often been given different names.

Order Mantophasmatodea. Head hypognathous, with generalized mouthparts; antennae long, filiform, multisegmented; ocelli absent; wings entirely lacking; coxae elongate; tarsi with 5 tarsomeres.

The Mantophasmatodea is the most recently described order of insects (Klass et al. 2002) and the first new insect order described since 1914. The order includes several living species (from Namibia, South Africa, and Tanzania) and one fossil species (from Baltic amber). Although resembling the Phasmida, Grylloblattodea, and Dermaptera in many ways, the mantophasmatodeans differ from these orders by having a hypognathous head and in other details. Analyses of gut contents and the presence of raptorial forelegs suggest that these rare insects are predators. It should be noted that this new order has not been accepted by all entomologists, and some workers have suggested that mantophasmatodeans are highly modified grylloblattodeans or orthopterans (e.g. Tilgner 2002).

Order Embioptera (Embiidina). 150 described species. Males somewhat flattened; females and young cylindrical. Most about 10 mm long. Antennae filiform; ocelli lacking; chewing mouthparts; head prognathous; legs short and stout; tarsi 3-articulate; hind femora greatly enlarged. The first and second articles of the front tarsus are enlarged and contain glands that produce silk, which is spun from hollow hairlike structures on the ventral surface. Males of most species are winged, but some are wingless; females are always wingless. Abdomen 10-segmented, with rudiments of the eleventh segment, and a pair of cerci.

Web-spinners are small, slender, chiefly tropical insects. They live in silken galleries spun in leaf litter, under stones, in soil cracks, in bark crevices, and in epiphytic plants. They feed mostly on dead plant material.

Order Zoraptera. 25 species. Minute (to 3 mm); termite-like; colonial; wingless or with two pairs of wings; wings eventually shed; antennae moniliform, 9-articulate; abdomen short, oval, 10-segmented; chewing mouthparts; simple development. These uncommon insects are usually found in colonies in dead wood. They feed chiefly on mites and other small arthropods.

Order Psocoptera. 2,600 described species. Small (1–10 mm long); antennae long, filiform, multiarticulate; prothorax short; meso- and metathorax often fused; chewing mouthparts; abdomen 9-segmented; cerci absent.

Psocids—the book and bark lice—generally feed on algae and fungi, and occur in suitably moist areas (e.g., under bark, in leaf litter, under stones, in human habitations where humid climates prevail). They are often pests that get into various stored food products or consume insect and plant collections; some species live in books and eat the bindings.

Order Phthiraptera. Small (less than 5 mm), wingless, blood-sucking, obligate ectoparasites of birds and mammals; thoracomeres completely fused; cuticle largely membranous and expandable to permit engorgement; compound eyes absent or of 1–2 ommatidia; ocelli absent; piercing–sucking mouthparts retractable into a buccal pouch; antennae short (5 or fewer "articles"), exposed or concealed in grooves beneath the head; with 1 pair dorsal thoracic spiracles and 6 or fewer abdominal spiracles; females lack ovipositor; without cerci.

Commonly called sucking lice ("Anoplura") and biting lice ("Mallophaga"), the Phthiraptera spend their entire life on one host. Eggs (**nits**) are usually attached to the hair or feathers of the host, although the human body louse (a "sucking louse") may attach eggs to clothing. No biting lice are known to infest humans. Posthatching development comprises three nymphal instars. Some species that infest domestic birds and mammals are of economic significance. Six hundred species of sucking lice and 5,000 species of biting lice have been described.

Order Thysanoptera. 4,100 described species. Slender, minute (0.5–1.5 mm) insects with long, narrow wings (when present) bearing long marginal setal fringes; mouthparts form a conical, asymmetrical sucking beak; left mandible a stylet, right mandible vestigial; with compound eyes; antennae of 4–10 "articles"; abdomen 10-segmented; without cerci; tarsi 1–2 segmented, with an eversible, pretarsal adhesive sac, or **arolium**. Thrips are mostly herbivores or predators, and many pollinate flowers. They are known to transmit plant viruses and fungal spores.

Order Hemiptera. 85,000 species. Piercing–sucking mouthparts form an articulated beak, mandibles and first maxillae stylet-like, lying in dorsally grooved labium; wings membranous; pronotum large.

Hemipterans occur worldwide and in virtually all habitats. Until recently, the homopterans (which account for about half of the described species) were classified as a separate order, but most workers now include them among the Hemiptera, from which they differ in only subtle respects (they are probably a monophyletic clade that arose from within the Hemiptera). In homopterans the beak generally arises from the posterior part of the head, whereas in other hemipterans it arises from the anterior part of the head. In homopterans the forewings are entirely membranous, though often colored (and about the same size as the hindwings), whereas in all other hemipterans the forewings are hardened basally and membranous only distally (and the forewings are smaller than the hindwings). In homopterans, the wings are held tentlike over the abdomen (although some families are wingless), whereas in other hemipterans the wings are usually held flat on the abdomen at rest. Also, in homopterans, the hindlegs are often adapted for jumping, the body is commonly protected by waxy secretions, and most produce saccharine anal secretions (**honeydew**).

Most true bugs are herbivorous, although many prey on other arthropods or vertebrates and some are specialized vertebrate ectoparasites. Sound production through a variety of different mechanisms is widespread in the order. Many hemipterans exhibit cryptic coloration or mimic other insects; ant mimicry is especially common. Hemipterans are of considerable economic importance, many being serious crop pests. Members of one subfamily of Reduviidae transmit human diseases (Triatominae, the assassin or kissing bugs). Others have more positive economic importance to humans, such as the cochineal bugs (Dactylopiidae), from which a safe red dye (**cochineal**) is extracted for use in the food industry. Shellac is made from **lac**, a chemical produced by members of the family Kerriidae. Common hemipteran families include Nepidae (water scorpions), Belostomatidae (giant water bugs or "toe biters"), Corixidae (water boatmen), Notonectidae (backswimmers), Gerridae (water striders), Saldidae (shore bugs), Cimicidae (bedbugs), and Reduviidae.

Homopterans are all plant feeders (hence the common name "plant bugs"). Heavy infestations of these insects on plants may cause wilting, stunting, or even death, and some are vectors of important plant diseases. Common homopteran families include the Cicadidae (cicadas), Cicadellidae (leafhoppers), Fulgoridae (planthoppers), Membracidae (treehoppers), Cercopidae (spittle bugs and froghoppers), Aleyrodidae (whiteflies), and Aphidae (aphids), as well as coccoids, scale insects, mealybugs, and many others.

Order Strepsiptera. 300 described species. Males free-living and winged; females wingless and usually parasitic. Antennae often with elongate processes on "articles"; forewings reduced to clublike structures resembling halteres of Diptera; hindwings large and membranous, with reduced venation. Females of free-living species with distinct head, simple antennae, chewing mouthparts, and compound eyes. Females of parasitic species larviform, usually without eyes, antennae, and legs; with indistinct body segmentation. Metamorphosis complete.

Most of these minute insects are parasitic on other insects. Female strepsipterans release several thousand tiny larvae that escape from their mother (and from the host's body, in the parasitic species) to invade the soil and vegetation. These larvae, called **triungulins**, have well developed eyes and legs. The larvae of parasitic species locate a host insect and enter it, wherein they molt into a legless wormlike stage that feeds in the host's body cavity. Pupation also takes place within the host's body.

Order Megaloptera. Ocelli present or absent; larvae aquatic, with lateral abdominal gills. Megalopterans (alderflies, dobsonflies, fishflies) strongly resemble neuropterans (and are often regarded as a suborder), but their hindwings are broader at the base than the forewings, and the longitudinal veins do not have branches near the wing margin. Larvae of some megalopterans (**hellgrammites**) are commonly used as fish bait.

Order Raphidioptera. Snakeflies strongly resemble neuropterans (and are often regarded as a suborder), but are unique in having the prothorax elongate (as in the mantids), but the front legs similar to the other legs. The head can be raised above the rest of the body, as in a snake preparing to strike. Adults and larvae are predators on small prey.

Order Neuroptera. 4,550 described species. Soft-bodied; with two pairs of similar, highly veined wings held tentlike over the abdomen when at rest; larvae with well developed legs; adults with biting–chewing mouthparts; Malpighian tubules secrete silk, via the anus, at pupation; abdomen 10-segmented; without cerci.

The lacewings, ant lions, mantisflies, spongillaflies, and owlflies form a complex group, the adults of which are often important predators of insect pests (e.g., aphids). The larvae of many species have piercing–sucking mouthparts, and those of other species are predaceous and have biting mouthparts. The pupae are often unusual in possessing free appendages and functional mandibles; they may actively walk about prior to the adult molt, but do not feed. Some workers subsume the Megaloptera and Raphidioptera within the Neuroptera.

Order Coleoptera. Body usually heavily sclerotized; forewings sclerotized and modified as rigid covers (elytra) over hindwings and body; hindwings membranous, often reduced or absent; biting mouthparts; antennae usually with 8–11 "articles"; prothorax large and mobile; mesothorax reduced; abdomen typically of 5 (or up to 8) segments; without ovipositor; male genitalia retractable.

With an estimated 350,000 to 375,000 described species, Coleoptera is the largest order of insects. The beetles range from minute to large and occur in all the world's environments (except the open sea). Some of the world's strongest animals are beetles: rhinoceros beetles can carry up to 100 times their own weight for short distances, and 30 times their weight indefinitely (equivalent to a 150-pound man walking with a Cadillac on his head—without tiring). Humans have had a long fascination with beetles, and beetle worship can be traced back to at least 2500 B.C. (The venerated scarab of early Egyptians was actually a dung beetle.)

Some common coleopteran families include Carabidae (ground beetles), Dytiscidae (predaceous diving beetles), Gyrinidae (whirligig beetles), Hydrophilidae (water scavenger beetles), Staphylinidae (rove beetles), Cantharidae (soldier beetles), Lampyridae (fireflies and lightning bugs), Phengodidae (glowworms), Elateridae (click beetles), Buprestidae (metallic wood-boring beetles), Coccinellidae (ladybird beetles), Meloidae (blister beetles), Tenebrionidae (darkling beetles), Scarabaeidae (scarab beetles, dung beetles, June "bugs"), Cerambycidae (long-horned beetles), Chrysomelidae (leaf beetles), Curculionidae (weevils), Brentidae (primitive weevils), and Ptiliidae (featherwinged beetles, the smallest of all beetles, some with body lengths of just 0.3 mm).

Order Mecoptera. 550 described species. Two pairs of similar, narrow, membranous wings, held horizontally from sides of body when at rest; antennae long, slender, and of many "articles" (about half the body length); head with ventral rostrum and reduced biting mouthparts; long, slender legs; mesothorax, metathorax, and first abdominal tergum fused; abdomen 11-segmented; female with two cerci; male genitalia prominent and complex, at apex of attenuate abdomen and often resembling a scorpion's stinger. The larvae of some species are remarkable in having compound eyes, a condition unknown among larvae of other insects having complete metamorphosis.

Mecopterans are usually found in moist places, often in forests, where most are diurnal flyers. They are best represented in the Holarctic region. Some feed on nectar; others prey on insects or are scavengers. There are several families, including Panorpidae (scorpionflies), Bittacidae (hangingflies), and Boreidae (snowflies).

Order Siphonaptera. 2,400 described species. Small (less than 3 mm long); wingless; body laterally compressed and heavily sclerotized; short antennae lie in deep grooves on sides of head; mouthparts piercing–sucking; compound eyes often absent; legs modified for clinging and (especially hindlegs) jumping; abdomen 11-segmented; abdominal segment 10 with distinct dorsal pincushion-like sensillum, containing a number of sensory organs; without ovipositor; pupal stage passed in cocoon.

Adult fleas are ectoparasites on mammals and birds, from which they take blood meals. They occur wherever suitable hosts are found, including the Arctic and Antarctic. Larvae usually feed on organic debris in the nest or dwelling place of the host. Host specificity is often weak, particularly among the parasites of mammals, and fleas regularly commute from one host species to another. Fleas act as intermediate hosts and vectors for organisms such as plague bacteria, dog and cat tapeworms, and various nematodes. Commonly encountered species include *Ctenocephalides felis* (cat flea), *C. canis* (dog flea), *Pulex irritans* (domestic flea), and *Diamus montanus* (western squirrel flea).

Order Diptera. 151,000 described species. Adults with one pair of membranous mesothoracic forewings and a metathoracic pair of clublike halteres (organs of balance); head large and mobile; compound eyes large; antennae primitively filiform, with 7 to 16 "articles," and often secondarily annulated (reduced to only a few articles in some groups); mouthparts adapted for sponging, sucking, or lapping; mandibles of blood-sucking females developed as piercing stylets; hypopharynx, laciniae, galeae, and mandibles variously modified as stylets in parasitic and predatory groups; labium forms a **proboscis** ("tongue"), consisting of distinct basal and distal portions, the latter in higher families forming a spongelike pad (**labellum**) with absorptive canals; mesothorax greatly enlarged; abdomen primitively 11-segmented, but reduced or fused in many higher forms; male genitalia complex; females without true ovipositor, but many with secondary ovipositor composed of telescoping posterior abdominal segments; larvae lack true legs, although ambulatory structures (prolegs and "pseudopods") occur in many.

The true flies (which include mosquitoes and gnats) are a large and diverse group, notable for their excellent vision and aeronautic capabilities. The mouthparts and digestive system are modified for a fluid diet, and several groups feed on blood or plant juice. Dipterans are vastly important carriers of human diseases, such as sleeping sickness, yellow fever, African river blindness, and various enteric diseases. It has been said that mosquitoes (and the diseases they carry) prevented Genghis Khan from conquering Russia, killed Alexander the Great, and played pivotal roles in both world wars. **Myiasis**—the infestation of living tissue by dipteran larvae—is often a problem for livestock and occasionally for humans. Many dipterans are also beneficial to humans as parasites or predators of other insects and as pollinators of flowering plants. Dipterans occur worldwide and in virtually every major environment (except the open sea). Some breed in extreme environments, such as hot springs, saline desert lakes, oil seeps, tundra pools, and even shallow benthic marine habitats.

Some common dipteran families include Asilidae (robber flies), Bombyliidae (bee flies), Calliphoridae (blowflies, bluebottles, greenbottles, screwworm flies, etc.), Chironomidae (midges), Coelopidae (kelp flies), Culicidae (mosquitoes: *Culex, Anopheles*, etc.), Drosophilidae (pomace or vinegar flies; often also called "fruit flies"), Ephydridae (shore flies and brine flies), Glossinidae (tsetse flies), Halictidae (sweat bees), Muscidae (houseflies, stable flies, etc.), Otitidae (picture-winged flies), Sarcophagidae (flesh flies), Scatophagidae (dung flies), Simuliidae (blackflies and buffalo gnats), Syrphidae (hover flies and flower flies), Tabanidae (horseflies, deerflies, and clegs), Tachinidae (tachinid flies), Tephritidae (fruit flies), and Tipulidae (crane flies).

Order Trichoptera. 7,100 described species. Adults resemble small moths, but with body and wings covered with short hairs (rather than scales); two pairs of wings, tented in oblique vertical plane (rooflike) over abdomen when at rest; compound eyes present; mandibles minute or absent; antennae usually as long or longer than body, setaceous; legs long and slender; larvae and pupae mainly in fresh water, adults terrestrial; larvae with abdominal prolegs on terminal segment.

The freshwater larvae of caddisflies construct fixed or portable "houses" (cases) made of sand grains, wood fragments, or other material bound together by silk emitted through the labium. Larvae are primarily herbivorous scavengers; some use silk to produce food-filtering devices. Most larvae inhabit benthic habitats in temperate streams, ponds, and lakes. Adults are strictly terrestrial and have liquid diets.

Order Lepidoptera. 120,000 described species. Minute to large; sucking mouthparts; mandibles usually vestigial; maxillae coupled, forming a tubular sucking proboscis, coiled between labial palps when not in use; head, body, wings, and legs usually densely scaled; compound eyes well developed; usually with two pairs of large and colorfully scaled wings, coupled to one another by various mechanisms; male genitalia complex; females with ovipositors.

Butterflies and moths are among the best known and most colorful of all the insects. The adults are primarily nectar feeders, and many are important pollinators, some of the best known being the large hawk, or sphinx, moths (Sphingidae). A few tropical species are known to feed on animal blood, and some even drink the tears of mammals. The larvae (caterpillars) feed on green plants. Caterpillars have three pairs of thoracic legs and a pair of soft prolegs on each of abdominal segments 3–6; the anal segment bears a pair of prolegs or claspers. Butterflies can be distinguished from moths by two features: their antennae are always long and slender, ending in a knob (moth antennae are never knobbed), and their wings are typically held together above the body at rest (moths never hold their wings in this position). Over 80 percent of the described Lepidoptera are moths.*

Order Hymenoptera. 125,000 described species. Mouthparts often elongate and modified for ingesting floral nectar, although mandibles usually remain functional; labium often (bees) distally expanded as paired lobelike structures called **glossae** and **paraglossae**;

---

*One of the best known butterfly taxonomists was the great Russian novelist Vladimir Nabokov (*Lolita, Pale Fire, The Gift*), who left Saint Petersburg in 1917 to travel around Europe and eventually settled in the United States (first working at the American Museum of Natural History in New York, then at Cornell University). Nabokov was a specialist on the blue butterflies (Polyommatini) of the New World and a pioneer anatomist, coining such alliterative anatomical terms as "alula" and "bullula." Butterflies, real and imaginary, flit through 60 years of Nabokov's fiction, and many lepidopterists have named butterflies after characters in his life and writings (e.g., species epithets include *lolita, humbert, ada, zembla*, and *vokoban*—a reversal of Nabokov). Nabokov's descriptions of Lolita are patterned after his species descriptions of butterflies (e.g., "her fine downy limbs").

usually with two pairs of membranous wings; hind-wings small, coupled to forewings by hooks; wing venation highly reduced; antennae well developed, of various forms and with 3–70 "articles"; reduced metathorax usually fused to first abdominal segment; males with complex genitalia; females with ovipositor (in most), modified for sawing, piercing, or stinging.

The earliest fossil Hymenoptera date from the Triassic (207–220 mya). Ants, bees, wasps, sawflies, and their relatives are all active insects with a tendency to form polymorphic social communities. Two suborders are generally recognized. Included in the suborder Symphyta are the primitive, wasplike, "thick-waisted" hymenopterans (sawflies, horntails, and their kin). They rarely show conspicuous sexual dimorphism and are always fully winged. The first and second abdominal segments are broadly joined. Larvae are mostly caterpillar-like, with a well developed head capsule, true legs, and often abdominal prolegs. The suborder Apocrita comprises the "narrow-waisted" hymenopterans (true wasps, bees, and ants), in which the first and second abdominal segments are joined by a distinct and often elongate constriction. Adults tend to be strongly social and display marked polymorphism. Social communities often include distinct castes of queens, haploid males, parthenogenetic females, and individuals with other sex-related specializations, as well as non-reproducing worker and soldier forms. Larvae are legless, usually soft, white, and grublike. Larvae feed within or upon the body of a host arthropod or its egg, in a plant gall, or in a fruit or seed. Some larvae live in nests constructed by the adults and, as in bees, are fed by the adults. There are 21,000 described species of bees (most of the bees being solitary, not social) and 9,000 described species of ants. The great naturalist Edward O. Wilson, fount of quotable insect statistics, has informed us that in the Amazon rain forest ants make up more than four times the biomass of all the land vertebrates combined; that the world's smallest ant forms a colony that could comfortably dwell inside the brain case of the world's largest ant; and that the biomass of all the social insects combined (ants, termites, and social wasps and bees) makes up about 80 percent of Earth's total biomass.

The Hexapod Bauplan

General Morphology

In Chapter 15 we briefly discussed the various advantages and constraints imposed by the phenomenon of arthropodization, including those associated with the establishment of a terrestrial lifestyle. Departure from the ancestral aquatic environment necessitated the evolution of stronger and more efficient support and locomotory appendages, special adaptations to withstand osmotic and ionic stress, and aerial gas exchange structures. The basic arthropod bauplan included many preadaptations to life in a "dry" world. As we have seen, the arthropod exoskeleton inherently provides physical support and protection from predators, and by incorporating waxes into the epicuticle, the insects, like the arachnids, acquired an effective barrier to water loss. Similarly, within the Hexapoda, the highly adaptable, serially arranged arthropod limbs evolved into a variety of specialized locomotory and food-capturing appendages. Reproductive behavior became increasingly complex, and in many cases highly evolved social systems developed. Within the class Insecta, many taxa underwent intimate coevolution with land plants, particularly angiosperms. The evolutionary potential of insects is evident in the many species that have evolved striking camouflage or warning coloration (Figure 17.32).

Non-insect hexapods (proturans, collembolans, diplurans) differ from insects in several important ways. The mouthparts are not fully exposed (i.e., they are "entognathous"), the mandibles have a single articulation, development is always simple, the abdomen may have a reduced number of segments, and they never developed flight.

Insects are primitively composed of 19 somites (as in the Eumalacostraca; Chapter 16), although these are not always obvious. The consolidation and specialization of these body segments (i.e., tagmosis) has played a key role in hexapod evolution and has opened the way for further adaptive radiation. The body is always organized into a head, thorax, and abdomen (Figures 17.2 and 17.3),

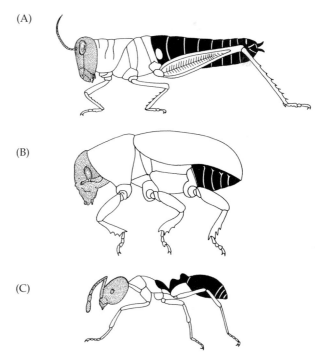

(A)

(B)

(C)

Figure 17.2 The principal body regions of hexapods, illustrated by three kinds of insects: (A) a grasshopper (wings removed), (B) a beetle, and (C) an ant. The stippled region is the head; the white region is the thorax; the black region is the abdomen.

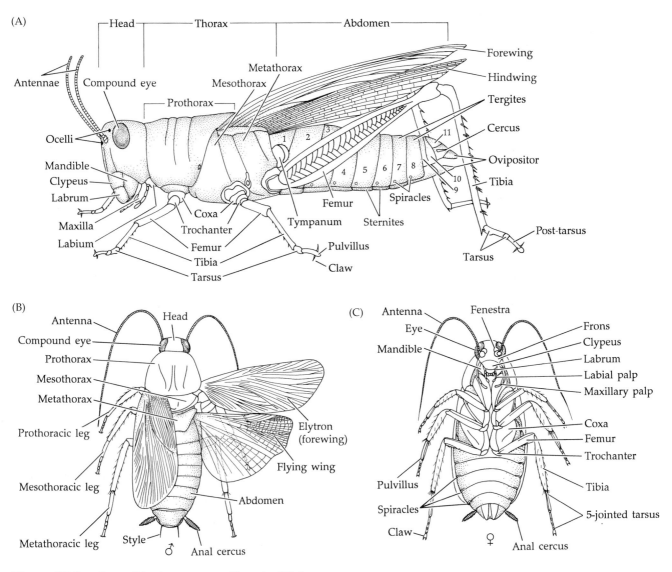

Figure 17.3 General body anatomy of insects. (A) A grasshopper (order Orthoptera). (B,C) Dorsal and ventral views of a cockroach (order Blattodea).

comprising 5, 3, and 11 segments respectively. In contrast to marine arthropods, a true carapace never develops in hexapods. In the head, all body sclerites are more or less fused as a "solid" head capsule. In the thorax and abdomen, the adult sclerites usually develop embryologically such that they overlap the **primary segment** articulations, forming **secondary segments**, and these are the "segments" we typically see when we examine an insect externally (e.g., the tergum and sternum of each adult abdominal secondary segment actually overlap its adjacent anterior primary segment) (Figure 17.4). The primitive (primary) body segmentation can be seen in unsclerotized larvae by the insertions of the segmental muscles and transverse grooves on the body surface.

Most insects are small, between 0.5 and 3.0 cm in length. The smallest are the thrips, feather-winged beetles, and certain parasitic wasps, which are all nearly mi-

croscopic. The largest are certain beetles, orthopterans, and walking sticks, the latter attaining lengths greater than 30 cm. However, certain Paleozoic species grew more than twice that size. To familiarize you with the hexapod bauplan and its terminology, we briefly discuss each of the main body regions (tagmata) below.

The Hexapod head. The hexapod head comprises an acron and five segments, bearing (from anterior to posterior) the eyes, antennae, clypeolabrum, and three pairs of mouth appendages (mandibles, maxillae, labium) (Figure 17.5, Table 17.1). The acron and first two somites comprise the **procephalon**, innervated from the protocerebrum, deutocerebrum, and tritocerebrum. The other three head segments together form the **gnathocephalon**, innervated by portions of the subesophageal ganglion. Compound eyes, as well as three simple eyes (ocelli) are typically present in adult hexapods. The median (anterior) ocellus is thought to have arisen through the fusion of two separate ocelli, and it is innervated from both sides of the deutocerebrum. The inter-

(A) Posterior

(C) Anterior

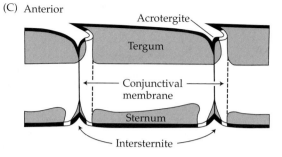

(B)

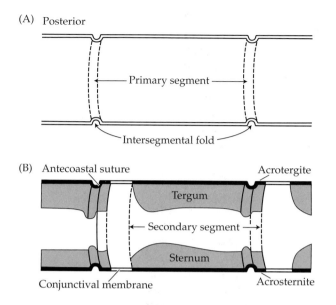

Figure 17.4   The ontogenesis of insect body segments. (A) Primary segmentation. (B) Simple secondary segmentation. (C) More advanced secondary segmentation.

nal manifestation of the fused exoskeleton of the head forms a variety of apodemes, braces, and struts collectively called the **tentorium**. Externally, the head may also bear lines that may demarcate its original segmental divisions, and others that represent the dorsal (and ventral) ecdysial lines, where the head capsule splits in immature insects and which persist as unpigmented lines in some adults. Still other lines represent inflections of the surface associated with internal apodemes.

The antennae (Figure 17.6) are composed of three regions: the **scape**, **pedicel**, and multijointed sensory **flagellum**. The scape and pedicel constitute the protopod; the flagellum represents the telopod. In most insects, the articles of the antennal telopod have been secondarily subdivided (or annulated) to produce extra joints. The flagellum of the pterygote antenna lacks intrinsic musculature, but it is largely retained in the apterygotes.*

The mouth is bordered anteriorly by the **clypeolabrum**, posteriorly by the **labium**, and on the sides by

---

*Like the antennae of crustaceans, those of hexapods seem to be a combination of true articles, retaining their intrinsic musculature in the more primitive groups, and a secondarily "segmented," or annulated, terminal region, the flagellum, which lacks intrinsic muscles.

Figure 17.5   The mouth appendages of a typical biting–chewing insect, a grasshopper. (A) Front view. (B) Side view.

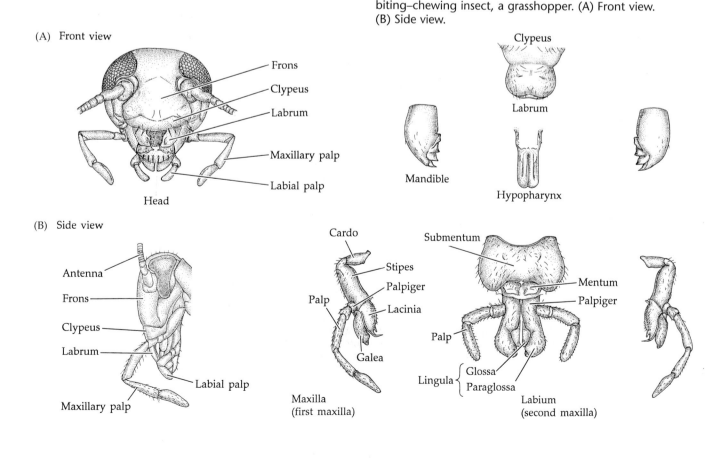

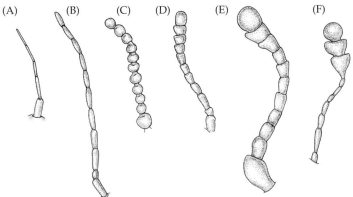

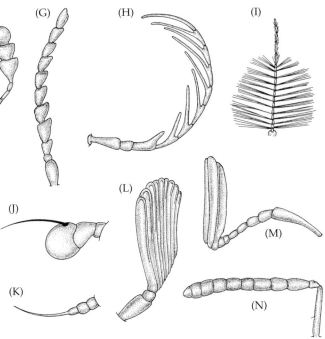

Figure 17.6 A variety of insect antennae and the terminology generally applied to them. (A) Setaceous. (B) Filiform. (C) Moniliform. (D,E) Clavate. (F) Capitate. (G) Serrate. (H) Pectinate. (I) Plumose. (J) Aristate. (K) Stylate. (L) Flabellate. (M) Lamellate. (N) Geniculate.

the mandibles and maxillae. In the entognaths, the mouthparts are sunk within the head capsule and largely hidden from view. In contrast, the mouthparts of insects are exposed (**ectognathous**) and ventrally projecting (**hypognathous**). However, in some insects, the orientation of the head has changed so that they are **prognathous** (projecting anteriorly) or **opisthognathous** (projecting posteriorly) (Figure 17.7).

The **labrum** is a movable plate attached to the margin of the **clypeus** (a projecting frontal head piece), and together they form the **clypeolabrum**. Some workers regard the clypeolabrum to be an independently derived structure of the exoskeleton; others believe it could be the fused appendages of the acron or premandibular somite.* The **mandibles** (Figure 17.8) are strongly sclerotized, usually toothed, and lack a palp. In most insects

the mandible is of one article, but in some primitive groups (and fossil taxa) it is composed of several articles. However, the gene *Distal-less* (*Dll*) is apparently never expressed in the embryogeny of hexapodan mandibles, suggesting that they are fully gnathobasic (i.e., protopodal).[†] The **maxillae** are generally multiarticulate and bear a palp of 1–7 articles. The labium comprises the fused second maxillae and typically bears two

---

*Given its position on the head, the fact that the acron is presegmental, and that there is no evidence it ever bore appendages, it seems unlikely that the clypeolabrum was derived from the acron. Furthermore, the clypeus–labrum of hexapods appears to be homologous to that of crustaceans, and there is no evidence in Crustacea that it evolved from paired appendages. In crustacea, the premandibular segment bears the second antennae. These observations suggest that the clypeolabrum of insects is not appendage-based.

[†]Study of the *Dll* gene reveals that it was probably primitively expressed in the distal parts of all arthropod appendages. It is also expressed in the endites, or inner lobes, of arthropod limbs (e.g., in the phyllopodous limbs of Branchiopoda and in the maxillae of Malacostraca). In crustaceans and myriapods there is an initial *Dll* expression in the mandibular limb buds that is displaced laterally and continues in the mandibular palp in crustaceans. In insects, no *Dll* expression at all is seen in the mandibles—it has apparently been completely lost. Thus, the mandibles of all three groups are gnathobasic. The palp of the crustacean mandible represents the distal portion of the mandibular limb, altogether lost in hexapods and myriapods. The only real "whole-limb jaws" among the arthropods are those of onychophorans. *Dll* is also expressed in the coxal endites of chelicerates and the pedipalp endites of arachnids. The complete loss of *Dll* expression in hexapod mandibles is presumably a synapomorphy for the group.

TABLE 17.1 Segments and structures of the hexapod head

Segment number	Name	Nervous system innervation	Structures borne on segment
	Acron	Protocerebrum	Ocelli and ommatidia
1	Antennary segment	Deutocerebrum	Antennae
2	Premandibular segment	Tritocerebrum	Clypeus + labrum (the "upper lip")
3	Mandibular segment	Subesophageal ganglion	Mandibles
4	First maxillary segment	Subesophageal ganglion	Maxillae (first maxillae)
5	Second maxillary segment	Subesophageal ganglion	Labium (fused second maxillae)

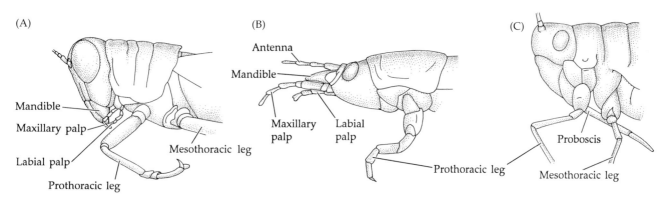

Figure 17.7 Different positions of head and mouthparts relative to the rest of the body. (A) Hypognathous condition (grasshopper). (B) Prognathous condition (beetle larva). (C) Opisthorhynchous condition (aphid).

palps.* In addition to these appendages, there is a median, unpaired, tonguelike organ called the **hypopharynx** that projects forward from the back of the preoral cavity. The salivary glands open through the hypopharynx.†

Variations in feeding appendages are often used to define the major insect clades. In sucking insects both mandibles and maxillae may be transformed into spearlike structures (**stylets**), or the mandibles may be absent altogether. In most Lepidoptera, the maxillae together form an elongate coiled sucking tube, the proboscis. In some insects (e.g., Hemiptera), the labium is drawn out into an elongate trough to hold the other

---

*Entomologists (and entomology texts) generally refer to the first maxillae (maxillules) of insects simply as *maxillae,* and to the fused second maxillae as the *labium.* Thus the "labium" of hexapods is not homologous to the "labium" of crustaceans. The former is derived from the second maxillae; the origin of the latter is unclear.

†The hypopharynx is considered by most workers to be an independent outgrowth of the body wall and not a true appendage. Kukalová-Peck regards it as the "invaginated antennae" of the third head somite.

mouthparts, and its palps may be absent; in others (e.g., Diptera), it is modified distally into a pair of fleshy porous lobes called **labellae**.

The Hexapod thorax. The three segments of the thorax are the **prothorax**, **mesothorax**, and **metathorax** (Figure 17.3). Their tergites carry the same prefixes: **pronotum**, **mesonotum**, **metanotum**. In the winged insects (Pterygota), the mesothorax and metathorax are enlarged and closely united to form a rigid **pterothorax**. Wings, when present, are borne on these two segments and articulate with processes on the tergite (notum) and pleura of these somites. The prothorax is sometimes greatly reduced, but in some insects it is greatly enlarged (e.g., beetles) or even expanded into a large shield (e.g., cockroaches). The lateral, pleural sclerites are complex and are thought to be derived, at least in part, from subcoxal (protopodal) elements of the ancestral legs that became incorporated into the lateral body wall. The sternites may be simple, or may be divided into multiple sclerites on each segment.

Each of the three thoracic somites bears a pair of legs (Figures 17.9 and 17.10), composed of two parts, a proximal **protopod** and a distal **telopod**. The protopod (sometimes called a "coxopodite" by entomologists) is

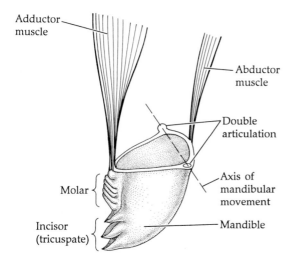

Figure 17.8 The musculature of an insect mandible.

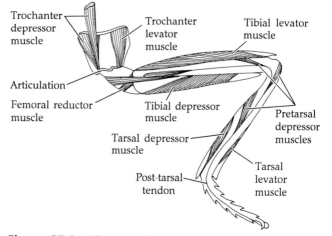

Figure 17.9 The musculature of an insect leg.

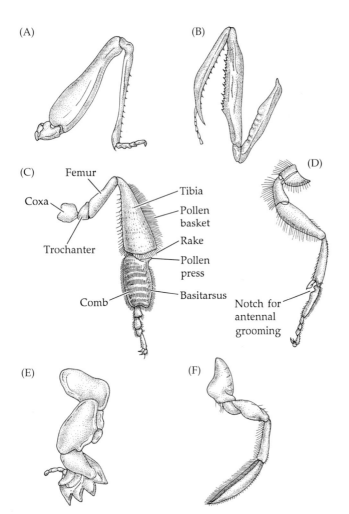

Figure 17.10 Leg modifications in some insects. (A) The hindleg of a grasshopper (Orthoptera), modified for jumping. (B) The raptorial foreleg of a mantid (Mantodea), modified for prey capture. (C) The hindleg of a honey bee (Hymenoptera), modified for collecting and holding pollen. (D) The foreleg of a worker honey bee has a notch for antennal grooming. (E) The foreleg of a mole cricket (Orthoptera), modified for digging. (F) The hindleg of a backswimmer (Hemiptera), modified for swimming.

composed of two articles (**coxa, trochanter**), and the telopod is composed of four articles (**femur, tibia, tarsus, post-tarsus**).* A single tarsus occurs in many hexapods, but in most taxa there are three or five tarsal articles. The basal hexapods have a single tarsus (Protura and Diplura) or an indistinct tarsus (Collembola; probably fused with the tibia). In the Archaeognatha the tarsus is usually composed of three articles, and in the Pterygota it is composed of one, three, or five articles.[†] Whatever the number of tarsal articles, no intrinsic musculature occurs in them, and they are thus usually interpreted as being subdivisions of a single original article. The whole length of the tarsus is crossed by the tendon of the flexor muscle of the post-tarsus, whose fibers usu-

ally arise on the tibia. The post-tarsus is a minute article that bears the claws. The post-tarsus in Collembola and Protura bears a single median claw. A single claw also occurs in many holometabolous larvae and some pterygote adults. In other hexapods, the post-tarsus bears a pair of claws.

In insects, many adult organs derive from clusters of early embryonic cells called **imaginal discs**, which arise from localized invaginations of the ectoderm in the early embryo. The embryonic thorax contains three pairs of leg discs, and as development proceeds, these discs develop a series of concentric rings, which are the presumptive leg articles. The center of the disc corresponds to the distalmost articles (tarsus and post-tarsus) of the future leg, while the peripheral rings correspond to its proximal region (coxa, trochanter). During embryogenesis, the leg telescopes out as it subdivides into the component articles. The gene *Distal-less* (*Dll*) is expressed in the presumptive distal region of the limb, while the gene *Extradenticle* (*Exd*) is necessary for the development of the proximal portion of the limb. Thus the protopod and telopod of the legs are somewhat distinct, each under its own genetic control.

The wings of insects show so many characters of value in systematics that they have been more extensively used in classification than any other single structure. Wings are often the only remains of insects preserved in fossils. In some groups (e.g., Orthoptera, Dermaptera) the forewings develop heavily sclerotized regions called **tegmina** (sing. tegmin), used for protection, stridulation, or other purposes. In many sedentary, cryptic, parasitic, and insular lineages the wings have become shortened (**brachypterous**) or lost (**apterous**). Insects often couple their wings together for flight by means of hooklike devices on the posterior border of the forewings and the anterior margin of the hindwings.

The Hexapod abdomen. The abdomen primitively comprises 11 segments, although the first is often reduced or incorporated into the thorax, and the last may be vestigial. Abdominal pleura are highly reduced or absent. The occurrence of true (though minute) abdominal **leglets** (sometimes called "prolegs" or "styli") on the pregenital segments is commonplace among the apterygotes and also occurs in the larvae of many pterygotes (e.g., the legs of caterpillars). In addition, transitory limb buds or rudiments

---

[†]It is unclear whether one or several tarsi is the primitive condition for the Hexapoda. Homologization of the leg articles among the various arthropod groups is a popular and often rancorous pastime. The issue is further confused when the number of articles differs from the norm (e.g., some insects have two trochanters, the number of tarsi varies from one to five, etc.). The protopods of myriapods, chelicerates, and trilobites all seem to consist of a single basal article (usually called the coxa). In the latter case, Kukalová-Peck hypothesizes that three "missing" protopodal articles were lost by fusion among themselves and with the pleural region. See Chapter 15 for a discussion of ancestral arthropod legs.

---

*The post-tarsus is also often called the "pre-tarsus."

appear fleetingly in the early embryos of other species, harking back to the evolutionary past. Abdominal segments 8–9 (or 7–9) are typically modified as the **anogenital tagmata**, or **terminalia**, the genital parts being the **genitalia**. The female median gonopore occurs behind sternum 7 in Ephemeroptera and Dermaptera, and behind sternum 8 or 9 in all other orders. The anus is always on segment 11 (which may become fused with segment 10).

There is enormous complexity in both clasping and intromittent organs among the Hexapoda, and a correspondingly sharp disagreement over the homologization and terminology of these structures. In general, females discriminate among males on the basis of sensory stimuli produced by the male genitalia; hence selection pressure has been a powerful force in the evolution of these structures (in both sexes). The most primitive male architecture can be seen in apterygotes and Ephemeroptera, in which the penes are paired and contain separate ejaculatory ducts. In most other insects, however, the intromittent organ develops postembryonically by fusion of the genital papillae to form a median, tubular, often eversible **endophallus**, with the joined ejaculatory ducts opening at a gonopore at its base. The external walls may be sclerotized or modified in a wide variety of ways, and the whole organ is known as the **aedeagus**. Some workers consider the aedeagus to be derived from segment 9; others regard it as belonging to segment 10. A pair of sensory **cerci** (sing. cercus) often project from the last abdominal segment.

Locomotion

Pedestrian locomotion. Hexapods rely heavily on their well sclerotized exoskeleton for support on land. Their limbs provide the physical support needed to lift the body clear of the ground during locomotion. In order to accomplish this, the limbs must be long enough to hold the body high off the ground, but not so high as to endanger stability. Most hexapods maintain stability by having the legs in positions that suspend the body in a slinglike fashion and keep the overall center of gravity low (Figure 15.20).

The basic design of arthropod limbs was described in Chapter 15. In contrast to most arachnids, in which the coxae are immovably fixed to the body and limb movement occurs at more distal joints, in hexapods and crustaceans the anterior–posterior limb movements take place between the coxae and the body proper. Like the power controlled by the range of gears in an automobile, the power exerted by a limb is greatest at low speeds and least at higher speeds. At lower speeds the legs are in contact with the ground for longer periods of time, thus increasing the power, or force, that can be exerted during locomotion. In burrowing forms the legs are short, and the gait is slow and powerful as the animal forces its way through soil or rotting wood. Longer limbs reduce the force, but increase the speed of a running gait, as do limbs capable of swinging through a greater angle. Limbs long in length and stride are typical features of the fastest-running insects.

One of the principal problems associated with increased limb length is that the field of movement of one limb may overlap that of adjacent limbs. Interference is prevented by the placement of the tips of adjacent legs at different distances from the body (Figure 17.11). Thus fast-running insects usually have legs of slightly different lengths. Insects usually move their legs in an **alternating tripod sequence** (Figure 17.11). Balance is maintained by always having three legs in contact with the ground.

Like many spiders, some insects can walk on water, and they do so in much the same way—by balancing the pull of gravity on their featherweight bodies with the physical principles of buoyancy and surface tension. Insects (and spiders) that walk on water don't get wet because their exoskeletons are coated with waxes that repel water molecules. The water surface, held taut by surface tension, bends under each leg to create a depression, or dimple, that works to push the animal upward in support. Water walking occurs in many insect groups, notably the Hemiptera (water striders) (Figure 17.12D), Coleoptera (whirligig beetles), and Collembola (some springtails).*

---

*On an undisturbed surface, water molecules are attracted to their neighbors beside and below, resulting in a flat skin of molecules that exerts only horizontal tensile forces; it is this "elasticity" of the surface that we call surface tension.

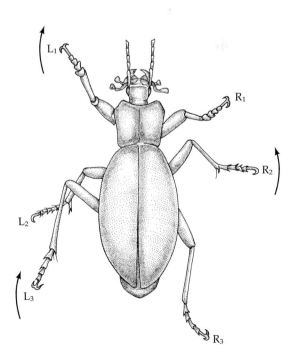

Figure 17.11 A beetle walking. The alternating tripod gait consists of alternate stepping with two sets of three legs; thus the body is always supported by a triad of legs. Here, three legs (L_1, R_2, and L_3) are moving forward while the other three (R_1, L_2, and R_3) are on the ground.

Figure 17.12 Some aspects of insect locomotion. (A) Ladybird beetles (Coleoptera: Coccinellidae) in the desert Southwest undertake elevational migrations to spend the hot summers aggregated in the mountains (known as the "sky islands" of the Southwest). (B) Honey bees (Hymenoptera: Apidae) swarm when they emerge to establish new colonies; during this swarming period they are least likely to sting. (C) Some species of termites (Isoptera) construct covered runways of soil for protection from heat and predators while foraging. (D) Water striders (Hemiptera: Gerridae) are one of several insect groups that utilize surface tension to walk on water. (E) Water scorpions (Hemiptera: Nepidae) live entirely submerged throughout their lives, periodically rising to the surface to take in air via a breathing tube. (F) Predacious diving beetles (Coleoptera: Dytiscidae) also live their entire lives under water, utilizing a bubble of air they periodically capture at the pond surface.

Many insects are good jumpers (e.g., fleas, springtails, most orthopterans), but the click beetles (Elateridae) are probably the champions. It has been calculated that a typical click beetle (e.g., *Athous haemorrhoidalis*), when jackknifing into the air to escape a predator, generates 400 g of force, with a peak brain deceleration of 2,300 g.

Flight. Among the many remarkable advances of insects, flight is perhaps the most impressive. Insects were the first flying animals, and throughout the history of life on Earth no other invertebrates have learned the art of true flight. The wingless insects belong either to groups that have secondarily lost the wings (e.g., fleas, lice, certain scale insects) or to primitive taxa (the apterygotes) that arose prior to the evolution of wings. In three orders, the wings are effectively reduced to a single pair. In beetles, the forewings are modified as a protective dorsal shield (**elytra**). In dipterans, the hindwings are modified as organs of balance (**halteres**). The halteres beat with the same frequency as the forewings, functioning as gyroscopes to assist in flight perfor-

mance and stability—flies fly very well. The function of the reduced forewings in strepsipterans is unclear, but they may function as organs of balance.

The wings of modern insects develop as evaginations of the integument, with thin cuticular membranes forming the upper and lower surfaces of each wing. Wing veins, which contain circulating hemolymph, anastomose and eventually open into the body. The arrangement of veins in insect wings provides important diagnostic characters at all taxonomic levels. The origin and homologization of wing venation has been challenging and heavily debated over the decades. Most workers use a consistent naming system that recognizes six major veins: costa (C), subcosta (SC), radius (R), media (M), cubitus (CU), and anal (A) (Figure 17.13). Areas in the wings that are enclosed by longitudinal and cross veins are called **cells**, and these too have a somewhat complex nomenclature.

Compared with an insect, an airplane is a simple study in aerodynamics. Planes fly by moving air over a fixed wing surface, the leading edge of which is tilted

Figure 17.13 The nomenclature of basic wing venation in insects. Although the "cells" formed within the veins also have names, only the names of the veins are given here. Longitudinal veins are coded with capital letters, cross veins with lower-case letters. Longitudinal veins: costa (C); subcosta (SC); radius (R); radial sector (RS); media (M); cubitus (CU); anal (A). Cross veins: humeral (h); radial (r); sectorial (s); radiomedial (rm); medial (m); mediocubital (m-cu); cubitoanal (cu-a).

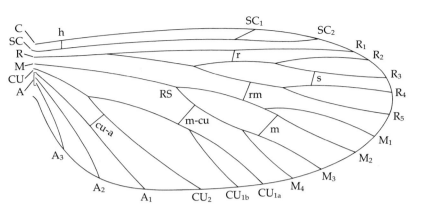

upward, forcing the air to travel farther (thus faster) over the top of the wing than the bottom, resulting in a vortex that creates a lift. But conventional fixed-wing aerodynamic theory is insufficient to understand insect flight. Insect wings are anything but fixed. Insects, of course, fly by flapping their wings to create vortices, from which they gain lift, but these vortices slip off the wings with each beat, and new vortices are formed with each alternate stroke. Beating insect wings trace a figure eight pattern, and they also rotate at certain crucial moments. Thus, each cycle of flapping creates dynamic forces that fluctuate drastically. By complex actions of wing orientation, insects can hover, fly forward, backward, and sideways, negotiate highly sophisticated aerial maneuvers, and land in any position. To complicate matters even more, in the case of small insects (and most are small, the average size of all insects being just 3–4 mm), the complex mechanics of flight take place at very low Reynolds numbers, such that the insect is essentially "flying through molasses." As a result of these complex mechanics, insect flight is energetically costly, requiring metabolic rates as high as 100 times the resting rate.

Each wing articulates with the edge of the notum (thoracic tergite), but its proximal end rests on a dorsolateral pleural process that acts as a fulcrum (Figure 17.14). The wing hinge itself is composed in large part of resilin, a highly elastic protein that allows for rapid, sustained movement. The complex wing movements are made possible by the flexibility of the wing itself and by the action of a number of different muscle sets that run from the base of the wing to the inside walls of the thoracic segment on which it is borne. These **direct flight muscles** serve to raise and lower the wings and to tilt their plane at different angles (somewhat like altering the blade angles on a helicopter) (Figure 17.15). However, except in palaeopterans (Odonata and Ephemeroptera), the direct flight muscles are not the main source of power for insect wing movements. Most of the force comes from two sets of **indirect flight muscles**, which neither originate nor insert on the wings themselves (Figure 17.15 and 17.16).

Dorsal longitudinal muscles run between apodemes at the anterior and posterior ends of each winged segment. When these muscles contract, the segment is shortened, which results in a dorsal arching of the segment roof and a downstroke of the wings. **Dorsoventral muscles**, which extend from the notum to the sternum (or to basal leg joints) in each wing-bearing segment, are antagonistic to the longitudinal muscles. Contraction of the dorsoventral muscles lowers the roof of the segment. In doing so, it raises the wings, and incidentally pulls the legs up during flight. Thus, wing flapping in most insects is primarily generated by rapid changes in the walls and overall shape of the mesothorax and metathorax. Other, smaller thoracic muscle sets serve to make minor adjustments to this basic operation.

Insects with low wing-flapping rates (e.g., dragonflies, orthopterans, mayflies, and lepidopterans) are limited by the rate at which neurons can repeatedly fire and muscles can execute contractions. However, in insects with high wing-flapping rates (e.g., dipterans, hymenopterans, and some coleopterans), an entirely dif-

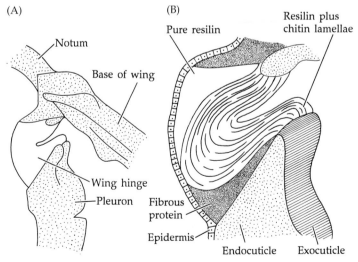

Figure 17.14 A typical insect wing hinge arrangement. This transverse section through the thoracic wall of a grasshopper shows the base of the wing and the wing hinge. (A) Entire hinge area. (B) Enlargement of hinge section.

Figure 17.15 Wing movements of a primitive insect such as a dragonfly, in which direct wing muscles cause depression of the wings. Dots represent pivot points, and arrows indicate the direction of wing movement. (A) The dorsoventral muscles contract to depress the notum as the basalar muscles relax, a combination forcing the wings into an upstroke. (B) The dorsoventral muscles relax as the basalar muscles contract, a combination pulling the wings into a downstroke and relaxing (and raising) the notum.

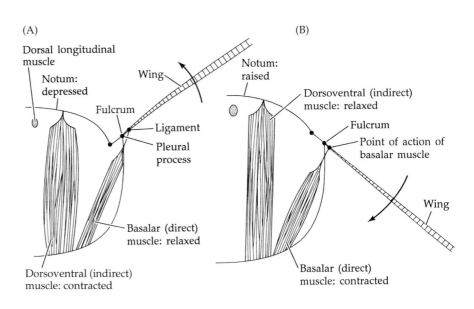

ferent regulatory mechanism has evolved. Once flight has been initiated and a high wing-flapping rate attained (up to 100 beats/second), myogenic control takes over. This mechanism exploits the elastic–mechanical properties of the exoskeleton. When one set of indirect muscles contracts, the thorax is deformed. Upon relaxation of the muscles, there is an elastic rebound of the thoracic exoskeleton, which stretches the second set of indirect muscles and thus directly stimulates their con-

traction. This contraction establishes a second deformation, which in turn stretches and stimulates the first muscle group. Once initiated, this mechanism is nearly self-perpetuating, and the nonsynchronous firing of neurons serves only to keep it in action.

Not all insects utilize wings to travel through the air. Many small and immature insects are effectively dispersed by wind power alone. Some first-instar lepidopterans use silk threads for dispersal (as do spiders and mites). Tiny scale insects are commonly collected in aerial nets. In fact, studies have revealed the existence of a large "aerial plankton" of insects and other minute arthropods, extending to altitudes as high as 14,000 feet. Most are minute winged forms, but wingless species are also common.

Feeding and Digestion

Feeding. Every conceivable kind of diet is exploited by species within the Hexapoda, whose feeding strategies include herbivory, carnivory, scavenging, and a

Figure 17.16 Wing movements of an insect such as a fly or hemipteran, in which both upward and downward movements of the wings are produced by indirect flight muscles. In these transverse sections of a thoracic segment, dots represent pivot points, and arrows indicate the direction of wing movement. Only two sets of muscles are shown. (A) The dorsoventral muscles contract, depressing the thoracic notum and forcing the wings into upstroke. (B) The dorsoventral muscles relax as the dorsal longitudinal muscles contract to "pop up" the notum, elevating it and forcing the wings into a downstroke.

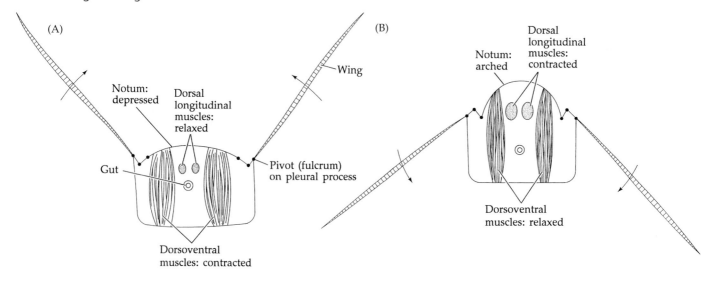

magnificent array of commensalism and parasitism. This "nutritional radiation" has played a key role in the phenomenal evolution among the Insecta. A comprehensive survey of insect feeding biology alone could easily fill a book this size. In the most general sense, insects can be classified as (1) biters–chewers, (2) suckers, or (3) spongers (Figure 17.17), although this categorization ignores a magnificent array of intimate commensal and parasitic relationships.

Biters–chewers, such as the grasshoppers, have the least modified mouthparts, so we describe them first. The maxillae and labium of these insects have well developed leglike palps that help them hold food in place, while powerful mandibles cut off and chew bite-sized pieces. The mandibles lack palps (in all insects) and typically bear small, sharp teeth that work in opposition as the appendages slide against each other in the transverse (side to side) fashion characteristic of most arthropod jaws (Figure 17.5). Biting–chewing insects may be carnivores, herbivores, or scavengers. In many plant eaters, the labrum bears a notch or cleft in which a stem or leaf edge may be lodged while being eaten. Some of the best examples of this feeding strategy are seen among the Orthoptera (locusts, grasshoppers, crickets), and most people have witnessed the efficient fashion in which these insects consume their garden plants! Equally impressive are the famous leafcutter ants of the Neotropics, which can denude an entire tree in a few days. Leafcutter ants have a notable feeding adaptation: when cutting leaf fragments, they produce high-frequency vibrations with an abdominal stridulatory organ. This stridulation is synchronized with movements of the mandible, generating complex vibrations. The high vibrational acceleration of the mandible appears to stiffen the material being cut, just as soft material is stiffened with a vibratome for sectioning in a laboratory. Leafcutters don't eat the leaves they cut; instead, they carry them into an underground nest, where they use them to grow a fungus on which they feed. Several other insect groups have evolved associations with fungi, and in almost every case these relationships are obligate and mutualistic—neither partner can live without the other.

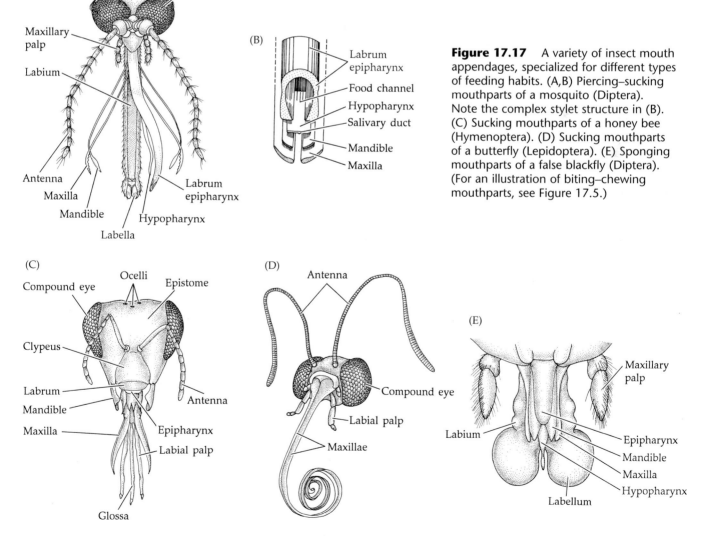

Figure 17.17 A variety of insect mouth appendages, specialized for different types of feeding habits. (A,B) Piercing–sucking mouthparts of a mosquito (Diptera). Note the complex stylet structure in (B). (C) Sucking mouthparts of a honey bee (Hymenoptera). (D) Sucking mouthparts of a butterfly (Lepidoptera). (E) Sponging mouthparts of a false blackfly (Diptera). (For an illustration of biting–chewing mouthparts, see Figure 17.5.)

In sucking insects the mouthparts are markedly modified for the consumption of liquid foods, generally plant saps or nectars or animal blood or cell fluids (Figures 17.17 and 17.18). Sucking mouthparts and liquid diets have clearly evolved many times in different insect lines—further testimony to the commonness of evolutionary convergence in arthropods and the developmental adaptability of their appendages. In some sucking insects, such as mosquitoes, feeding is initiated by piercing the victim's epidermal tissue; this mode of feeding is referred to as piercing–sucking. Other insects, such as butterflies and moths that feed on flower nectar, do not pierce anything and are merely suckers.

In all sucking insects the mouth itself is very small and well hidden. The mouthparts, instead of being adapted for handling and chewing solid pieces of food, are elongated into a needle-like beak. Different combinations of mouth appendages constitute the beak in different taxa. True bugs (Hemiptera), which are piercer–suckers, have a beak composed of five elements: an outer troughlike element (the **labium**) and, lying in the trough, four very sharp **stylets** (the two mandibles and two maxillae). The stylets are often barbed to tear the prey's tissues and enlarge the wound. The labrum is in the form of a small flap covering the base of the grooved labium. When piercer–suckers feed, the labium remains stationary, and the stylets do the work of puncturing the plant (or animal) and drawing out the liquid meal.

Different variations of piercing–sucking mouthparts are found in other insect taxa. For example, in mosquitoes, midges, and certain biting flies (e.g., horseflies)

there are six long, slender stylets, which include the labrum–epipharynx and the hypopharynx as well as the mandibles and first maxillae (Figure 17.17A,B). Other biting flies, such as the stable fly, have mosquito-like mouthparts, but lack mandibles and maxillae altogether. Fleas (Siphonaptera) have three stylets: the labrum–epipharynx and the two mandibles. Thrips have unusual mouthparts: the right mandible is greatly reduced, making the head somewhat asymmetrical, and the left mandible, first maxillae, and hypopharynx make up the stylets. Blood-sucking lice have two piercing stylets, but because of the extreme head modifications of lice, it is uncertain which mouth appendages they actually are!

Lepidopterans are nonpiercing sucking insects in which the paired first maxillae are enormously elongated, coiled, and fused to form a tube through which flower nectar is sucked; the mandibles are vestigial or absent (Figure 17.17D). The mouthparts of bees are similar: the first maxillae and labium are modified together to form a nectar-sucking tube, but the mandibles are retained and used for wax manipulation during hive construction (Figure 17.17C). The collected nectar is stored in a special "sac" in the foregut and carried back to the hive where it is converted into honey, which is stored as a food reserve. Bees in an average hive consume about 500 pounds of honey per year—we humans get the leftovers.

Associated with sucking mouthparts are various mechanisms for drawing liquid food into the mouth. Most piercer–suckers rely largely on capillary action, but others have developed feeding "pumps." Often the pump is developed through elongation of the preoral cavity, or **cibarium**, which by extension of the cuticle around the mouth becomes a semi-closed chamber connecting with the alimentary canal (Figure 17.18). In these cases **cibarial muscles** from the clypeus are enlarged to make a powerful pump. In lepidopterans, dipterans, and hymenopterans the cibarial pump is combined with a **pharyngeal pump**, which operates by means of muscles arising on the front of the head. Specialized salivary glands are also often associated with sucking mouthparts. In homopterans a salivary pump forces saliva through the feeding tube and into the prey, softening tissues and predigesting the liquid food. In mosquitoes the saliva carries blood thinners and anticoagulants (and often parasites such as *Plasmodium*, which causes malaria).

In spongers, such as most flies (order Diptera), the labium is typically expanded distally into a labellum (Figure 17.17E). Fluid nutrients are transported by capillary action along minute surface channels from the labellae to the mouth. In many spongers, such as houseflies, saliva is exuded onto the food to partly liquefy it. In strict spongers, the mandibles are absent. In biting spongers, such as horseflies, the mandibles serve to slice open a wound in the flesh, thus exposing the blood and cellular fluids to be sponged up by the labellae.

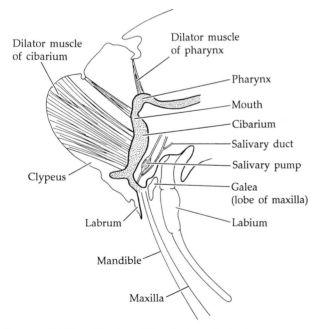

Figure 17.18   The head (vertical section) of a piercing–sucking insect, a cicada (Homoptera). Note the enlargement of the cibarial dilator muscles, which activate a cibarial sucking pump.

Many insects are scatophagous, feeding on animal feces. Most of these groups have biting mouthparts, but some (such as certain flies) have sucking mouthparts. Perhaps the most famous of the scatophagous insects are the dung beetles, or tumblebugs (certain beetles in the families Scarabaeidae and Histeridae). These remarkable insects harvest animal dung by biting or slicing off pieces with specialized head or leg structures and working them into a ball. They may roll the dung ball a considerable distance, and eventually bury it in the soil, whereupon females deposit eggs within it. Larvae are thus assured of a ready food supply. Dung balls may even be maneuvered by a pair of dung beetles pushing and pulling in a cooperative effort.

There are many symbiotic insects, and two orders are composed entirely of wingless parasites, most of which spend their entire lives on their host: Phthiraptera (lice) and Siphonaptera (fleas). Bird lice are common, and lice are also found on dogs, cats, horses, cattle, and other mammals. Biting lice have broad heads and biting mouthparts used to chew epithelial cells and other structures on the host's skin. Sucking lice have narrow heads and piercing–sucking mouthparts, which they use to suck blood and tissue fluids from their host, always a mammal. Unlike most arthropod parasites, lice (of both types) spend their entire lives on the bodies of their hosts, and transmission to new hosts is by direct contact. For this reason most lice show a high degree of host specificity. Eggs, or nits, are attached by the female to the feathers or hair of the host, where they develop without a marked metamorphosis. Many lice, particularly those whose diet is chiefly keratin, possess symbiotic intracellular bacteria that appear to aid in the digestion of their food. These bacteria are passed to the offspring by way of the insects' eggs. Similar bacteria occur in ticks, mites, bedbugs, and some blood-sucking dipterans.

None of the biting lice are known to infest people or to transmit human disease microorganisms, although one species acts as an intermediate host for certain dog tapeworms. The sucking lice, on the other hand, include two genera that commonly infest humans (*Pediculus* and *Phthirus*) (Figure 17.19B,C). The latter genus includes the notorious *P. pubis*, the human pubic "crab" louse (which often occurs on other parts of the body as well). A number of sucking lice are vectors for human disease organisms (Appendix A). The most common reaction to infestation with lice—a condition known as **pediculosis**, or being **lousy**—is simple irritation and itching caused by the anticoagulant injected by the parasite during feeding. Chronic infestation with lice among certain footloose travelers is manifested by leathery, darkened skin—a condition known as **vagabond's disease**.

Fleas (order Siphonaptera) are perhaps the best known of all insect parasites (Figure 17.19A). Nearly 1,500 species from birds and mammals have been described. Unlike lice, fleas are holometabolous, passing through egg, larval, pupal, and adult stages. Some species of fleas live their entire lives on their host, although eggs are generally deposited in the host's environment and larvae feed on local organic debris. Larvae of domestic fleas, including the rare human flea (*Pulex irritans*), feed on virtually any organic crumbs they find in the household furniture or carpet. Upon metamorphosis to the adult stage, fleas may undergo a quiescent period until an appropriate host appears. A number of serious disease organisms are transmitted by fleas (Table 17.1). At least 8 of the 60 or so species of fleas associated with household rodents are capable of acting as vectors for bubonic plague bacteria.

Other insect orders contain primarily free-living insects, but include various families of parasitic or micropredatory forms, or groups in which the larval stage is parasitic but the adults are free-living. Most of these "parasites" do not live continuously on a host and have feeding behaviors that fall into a gray zone between true obligate parasitism and predation. Such insects are sometimes classed as intermittent parasites, or micropredators. Bedbugs (Hemiptera, Cimicidae), for example, are minute flattened insects that feed on birds and mammals. However, most live in the nest or sleeping area of their host, emerging only periodically to feed. The common human bedbugs (*Cimex lectularius* and *C. hemipterus*) hide in bedding, in cracks, in thatched roofs, or under rugs by day and feed on their host's blood at night. They are piercer–suckers, much like the sucking lice. Bedbugs are not known to transmit any human diseases, although when present in large numbers they can be troublesome (in South America, as many as 8,500 bugs have been found in a single adobe house). Mosquitoes (family Culicidae), on the other hand, are vectors for a large number of disease-causing microorganisms, including *Plasmodium* (responsible for malaria; Figure 5.26), yellow fever, viral encephalitis, dengue, and lymphatic filariasis (with its gross symptom, **elephantiasis**, resulting from blockage of lymph ducts). Kissing bugs (Hemiptera, Reduviidae, Triatominae) also have a casual host relationship. They live in all kinds of environments, but often inhabit the burrows or nests of mammals, especially rodents and armadillos, as well as birds and lizards. They feed on the blood of these and other vertebrates, including dogs, cats, and people. Their host specificity is low. Several species are vectors of mammalian trypanosomiasis (*Trypanosoma cruzi*, the causative agent of **Chagas' disease**). The tendency of some species to bite on the face (where the skin is thin) has resulted in the common name.

In the dipteran family Calliphoridae, larvae are saprophagous, coprophagous, wound feeding, or parasitic. The parasitic species include earthworms, locust egg cases, termite colonies, and nestling birds among their hosts, and several parasitize humans and domestic stock (e.g., *Gochliomyia americana*, the tropical American screwworm).

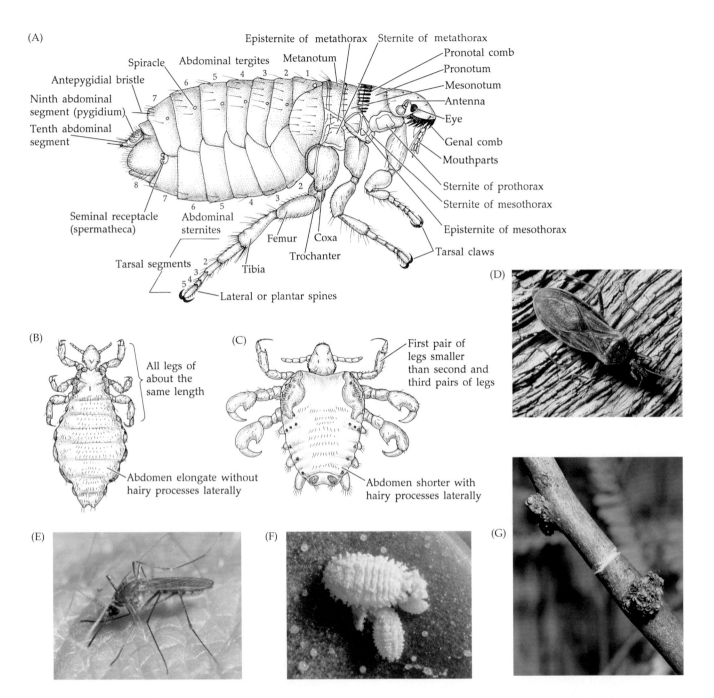

Figure 17.19 Some insects often viewed as pests. (A) The cat flea (*Ctenocehalides felis*), adult female. (B) *Pediculus humanus*, the human head and body louse. (C) *Phthirius pubis*, the human pubic louse. (D) A kissing bug (Hemiptera, Reduviidae, *Triatoma*), vector of *Trypanosoma cruzi*, the causative agent of Chagas' disease in the New World. (E) *Culex* (Diptera: Culicidae), a mosquito genus that transmits various human pathogens. (F) Mealybugs (Hemiptera: Homoptera: Pseudococcidae), which feed on the juices of house and garden plants. (G) Evidence of the southwestern mesquite girdler, a beetle (Coleoptera: Cerambycidae: *Oncideres*) that lays its eggs in twigs of mesquite trees, then girdles the branch, preventing the tree from sending its defensive chemical to destroy the eggs; the branch eventually dies and falls off the tree.

Many insect parasites of plants cause an abnormal growth of plant tissues, called a **gall**. Some fungi and nematodes also produce plant galls, but most are caused by mites and insects (especially hymenopterans and dipterans). Parasitic adults may bore into the host plant or, more commonly, deposit eggs in plant tissues, where they undergo larval development. The presence of the insect or its larvae stimulates the plant tissues to grow rapidly, forming a gall. The adaptive significance (for insects) of galls remains unclear, but one popular theory is that their production interferes with the production of defensive chemicals by the plant, thus rendering gall tissues more palatable. A somewhat similar strategy is used by leaf miners, specialized larvae from several orders (e.g., Coleoptera, Diptera, Hymenoptera) that live entirely within the tissues of leaves, burrowing through and consuming the most digestible tissues.

An interesting predatory strategy is that of New Zealand glowworms (*Arachnocampa luminosa*), which live in caves and in bushes along river beds. These larvae of small flies produce a bright bioluminescence in the distal ends of the Malpighian tubules, which lights up the posterior end of the body. (The light peaks at 485 nm wavelength.) Each larva constructs a horizontal web from which up to 30 vertical "fishing lines" descend, each with a regularly spaced series of sticky droplets. Small invertebrates (e.g., flies, spiders, small beetles, hymenopterans) attracted to the light are caught by the fishing lines, hauled up, and eaten. Harvestmen (Opiliones), the main predators of glowworms, use the light to locate their prey!

Digestive system. Like the guts of all arthropods, the long, usually straight hexapod gut is divisible into a stomodeal foregut, entodermal midgut, and proctodeal hindgut (Figure 17.20). **Salivary glands** are associated with one or several of the mouth appendages (Figure 17.21). The salivary secretions soften and lubricate solid food, and in some species contain enzymes that initiate chemical digestion. In larval moths (caterpillars), and in larval bees and wasps, the salivary glands secrete silk used to make pupal cells.

All hexapods, as well as most other arthropods that consume solid foods, produce a **peritrophic membrane** in the midgut (Figure 17.20B). This sheet of thin chitinous material may line the midgut or pull free to envelop and coat the food particles as they pass through the gut. The peritrophic membrane serves to protect the delicate midgut epithelium from abrasion. It is permeated by microscopic pores that allow passage of enzymes and digested nutrients. In many species, production of this membrane also takes place in the hindgut, where it encapsulates the feces as discrete pellets.

Along with their vast range of feeding habits, insects have evolved a number of specialized digestive structures. The foregut is typically divided into a well defined **pharynx, esophagus, crop,** and **proventriculus** (Figures 17.20 and 17.21). The pharynx is muscular, particularly in the sucking insects, in which it commonly forms a pharyngeal pump. The crop is a storage center whose walls are highly extensible in species that consume large but infrequent meals. The proventriculus regulates food passage into the midgut, either as a simple valve that strains the semifluid foods of sucking insects or as a grinding organ, called a **gizzard** or **gastric mill**, that masticates the chunks ingested by biting insects. Well developed gastric mills have strong cuticular teeth and grinding surfaces that are gnashed together by powerful proventricular muscles.

The midgut (= stomach) of most insects bears **gastric ceca** that lie near the midgut–foregut junction and resemble those of crustaceans. These evaginations serve to increase the surface area available for digestion and absorption. In some cases the ceca also house mutualistic microorganisms (bacteria and protozoa). The insect hindgut serves primarily to regulate the composition of the feces and perhaps to absorb some nutrients. Digestion of cellulase by termites and certain wood-eating

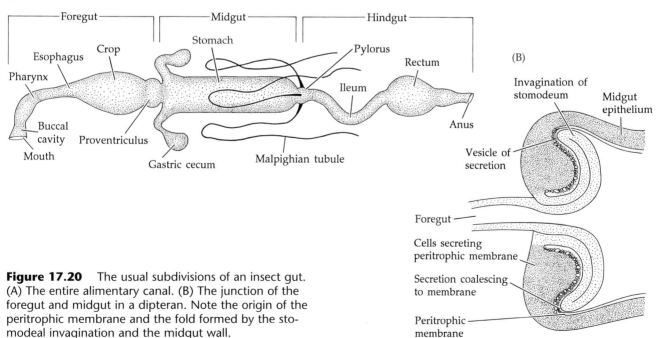

(A)

Figure 17.20 The usual subdivisions of an insect gut. (A) The entire alimentary canal. (B) The junction of the foregut and midgut in a dipteran. Note the origin of the peritrophic membrane and the fold formed by the stomodeal invagination and the midgut wall.

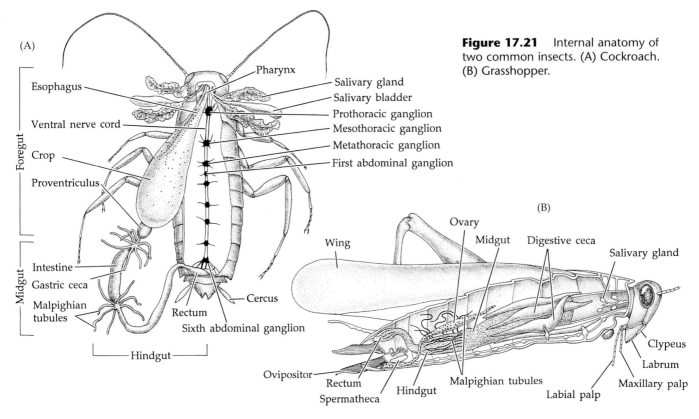

Figure 17.21 Internal anatomy of two common insects. (A) Cockroach. (B) Grasshopper.

Figure 17.22 The circulatory system of insects. (A) An insect abdomen (cross section). Note the division of the hemocoel into three chambers (a dorsal pericardial sinus, a ventral perineural sinus, and a central perivisceral sinus). These chambers are separated by diaphragms lying on frontal planes. (B) Blood circulation in an insect with a fully developed circulatory system (longitudinal section). Arrows indicate the circulatory course. (C) A cockroach (ventral dissection). Note the dorsal and segmental vessels. The dorsal diaphragm and aliform muscles are continuous over the ventral wall of the heart and vessels, but they are omitted from the diagram for clarity.

roaches is made possible by enzymes produced by protozoa and bacteria that inhabit the hindgut.

Fat bodies occur in the hemocoel of many insects and appear to function in much the same way as chlorogogen tissue in annelids, storing certain food reserves, particularly glycogen. Many insects do not feed during their adult life; instead, they rely on stored nutrients accumulated in the larval or juvenile stages.

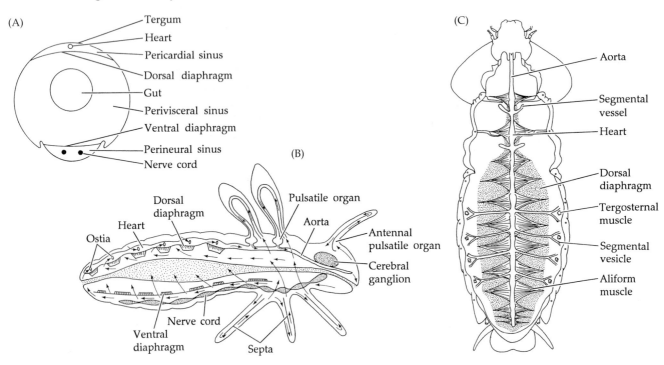

Circulation and Gas Exchange

The hexapod circulatory system includes a dorsal tubular heart that pumps the hemocoelic fluid (blood) toward the head. The heart narrows anteriorly into a vessel-like aorta, from which blood enters the large hemocoelic chambers, through which it flows posteriorly, eventually returning to the **pericardial sinus** and then to the heart via paired lateral **ostia** (Figure 17.22). In most insects the heart extends through the first nine abdominal segments; the number of ostia is variable. Accessory pumping organs, or **pulsatile organs**, often occur at the bases of wings and of especially long appendages, such as the hindlegs of grasshoppers, to assist in circulation and maintenance of blood pressure.

The heart is a rather weak pumping organ, and blood is moved primarily by routine muscular activity of the body and appendages. Hence, circulation is slow and system pressure is relatively low. Like many arachnids, some hexapods use the hydraulic pressure of the hemocoelic system in lieu of extensor muscles. In this way, for example, butterflies and moths unroll their maxillary feeding tubes.

Many types of hemocytes have been reported from the blood of insects. None function in oxygen storage or transport, but several are apparently important in wound healing and clotting. Nutrients, wastes, and hormones can be efficiently carried by this system, but respiratory oxygen cannot (some CO_2 does diffuse into the blood). The active lifestyles of these terrestrial animals require special structures to carry out the tasks of respiratory gas exchange and excretion. These structures are the tracheal system and the Malpighian tubules, described below.

Desiccation is one of the principal dangers faced by terrestrial invertebrates. Even though the general body surface of insects may be largely waterproof, the gas exchange surfaces cannot be. Adaptations to terrestrial life always involve some degree of compromise between water loss and gas exchange with the atmosphere.

In some minute hexapods, gas exchange occurs by direct diffusion across the body surface. However, the vast majority of hexapods rely on a **tracheal system** (Figure 17.23). As explained in Chapter 15, tracheae are extensive tubular invaginations of the body wall, opening through the cuticle by pores called **spiracles**. Up to

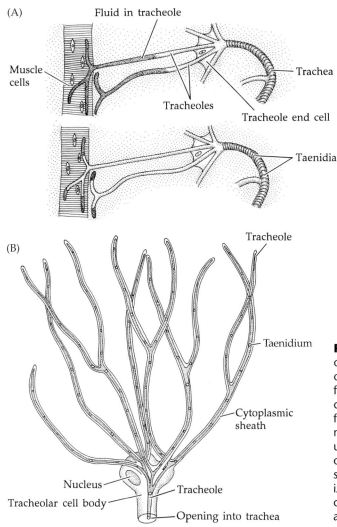

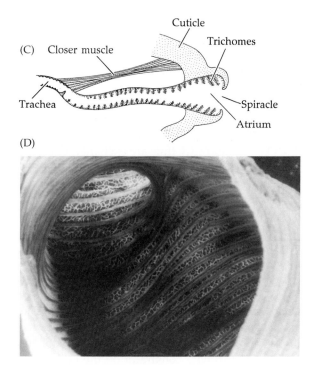

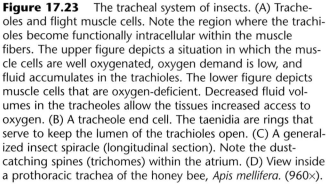

Figure 17.23 The tracheal system of insects. (A) Tracheoles and flight muscle cells. Note the region where the trachioles become functionally intracellular within the muscle fibers. The upper figure depicts a situation in which the muscle cells are well oxygenated, oxygen demand is low, and fluid accumulates in the trachioles. The lower figure depicts muscle cells that are oxygen-deficient. Decreased fluid volumes in the tracheoles allow the tissues increased access to oxygen. (B) A tracheole end cell. The taenidia are rings that serve to keep the lumen of the trachioles open. (C) A generalized insect spiracle (longitudinal section). Note the dust-catching spines (trichomes) within the atrium. (D) View inside a prothoracic trachea of the honey bee, *Apis mellifera*. (960×).

ten pairs of spiracles can occur on the pleural walls of the thorax and abdomen. Since tracheae are epidermal in origin, their linings are shed with each molt. The cuticular wall of each trachea is sclerotized and usually strengthened by rings or spiral thickenings called **taenidia**, which keep the tube from collapsing but allow changes in length that may accompany body movements. The tracheae originating at one spiracle commonly anastomose with others to form branching networks penetrating most of the body. In some insects it appears that air is taken into the body through the thoracic spiracles and released through the abdominal spiracles, thus creating a flow-through system.

Each spiracle is usually recessed in an **atrium**, whose walls are lined with setae or spines (**trichomes**) that prevent dust, debris, and parasites from entering the tracheal tubes. A muscular valve or other closing device is often present and is under control of internal partial pressures of O_2 and CO_2. In resting insects most of the spiracles are generally closed.

Ventilation of the tracheal system is accomplished by simple diffusion gradients, as well as by pressure changes induced by the animal itself. Almost any movement of the body or gut causes air to move in and out of some tracheae. Telescopic elongation of the abdomen is used by some insects to move air in and out of the tracheal tubes. Many insects have expanded tracheal regions called **tracheal pouches**, which function as sacs for air storage.

Because the blood of hexapods does not transport oxygen, the tracheae must extend directly to each organ of the body, where their ends actually penetrate the tissues. Oxygen and CO_2 thus are exchanged directly between cells and trachioles. In the case of flight muscles, where oxygen deman is high, the tracheal tubes invade the muscle fibers themselves.

The innermost parts of the tracheal system are the **tracheoles**, which are thin-walled, fluid-filled channels that end as a single cell, the **tracheole end cell** (= **tracheolar cell**) (Figure 17.23). The trachioles penetrate every organ in the body. Gas exchange thus takes place directly between the body cells and the trachioles. Unlike tracheae, tracheoles are not shed during ecdysis. The tracheoles are so minute (0.2–1.0 μm) that ventilation is impossible, and gas transport here relies on aqueous diffusion. This ultimate constraint on the rate of gas exchange may be the primary reason terrestrial arthropods never achieved extremely large sizes.

In aquatic insects the spiracles are usually nonfunctional, and gases simply diffuse across the body wall directly to the tracheae. A few species retain functional spiracles; they hold an air bubble over each opening, through which oxygen from the surrounding water diffuses. The air bubbles are held in place by secreted waxes and by patches of hydrophobic hairs in densities that may exceed 2 million per square millimeter. Most aquatic insects, particularly larval stages, have external

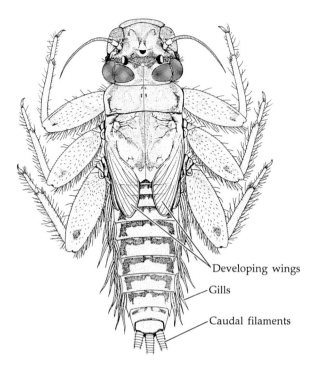

Figure 17.24 Aquatic nymph of a mayfly, *Thraulodes* (Ephemeroptera), with lateral abdominal gills.

projections of the body wall that are covered by thin, unsclerotized cuticle and contain blood, tracheae, or air bubbles (Figure 17.24). These **gills** contain channels that lead to the main tracheal system. In some aquatic insects, such as dragonfly nymphs, the rectum bears tiny branched tubules called **rectal accessory gills**. By pumping water in and out of the anus, these insects exchange gases across the increased surface area of the thin gut wall. There are analogous examples of hindgut respiratory irrigation in other, unrelated invertebrate groups (e.g., echiurans and holothurians).

Excretion and Osmoregulation

The problem of water conservation and the nature of the circulatory and gas exchange systems in terrestrial arthropods necessitated the evolution of entirely new structures to remove metabolic wastes. Like the gas exchange surfaces, the excretory system is a site of potential water loss, because nitrogenous wastes initially occur in a dissolved state. These problems are compounded in small terrestrial organisms, such as many hexapods, because of their large surface area-to-volume ratios. And water loss problems are even more severe in flying insects, because flight is probably the most metabolically demanding of all locomotor activities.

In most terrestrial arthropods, the solution to these problems is Malpighian tubules. In the Hexapoda, these unbranched outgrowths of the gut arise near the junction of the midgut and hindgut (Figures 17.20, 17.21, and 17.25). Their blind distal ends extend into the hemocoel

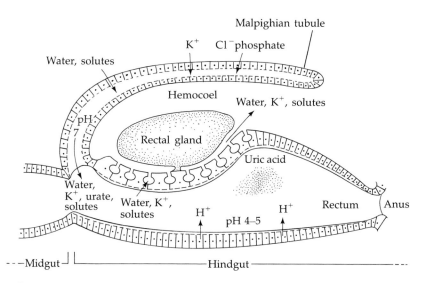

Figure 17.25 A single Malpighian tubule opening into the hindgut at its junction with the midgut. Arrows indicate the flow of materials.

and lie among various organs and tissues. Up to several hundred Malpighian tubules may be present.

In the absence of sufficient blood pressure for typical excretory filtration, hexapods use osmotic pressure to achieve the same result. Various ions, especially potassium, are actively transported across the Malpighian tubule epithelium from the blood into the tubule lumen (Figure 17.25). The osmotic gradient maintained by this ion transport mechanism enables water and solutes to move from the body cavity into the tubules, and thence into the gut. Water and other metabolically valuable materials are selectively reabsorbed into the blood across the wall of the hindgut, while the Malpighian filtrate left behind is mixed with the other gut contents. Reabsorption of water, amino acids, salts, and other nutrients may be enhanced by the action of special cells in thickened regions called **rectal glands**. The soluble potassium urate from the Malpighian tubules has, at this point in the gut, been precipitated out as solid uric acid as a result of the low pH of the hindgut (pH 4–5). Uric acid crystals cannot be reabsorbed into the blood, hence they pass out the gut with the feces. Insects also possess special cells called **nephrocytes** or **pericardial cells** that move about in certain areas of the hemocoel, engulfing and digesting particulate or complex waste products.

The hexapodan cuticle is sclerotized or tanned to various degrees, adding a small measure of waterproofing. But more importantly, a waxy layer occurs within the epicuticle, which greatly increases resistance to desiccation and frees insects to fully exploit dry environments. In many terrestrial arthropods (including primitive insects) an eversible **coxal sac** (not to be confused with the coxal glands of arachnids) projects from the body wall near the base of each leg. It is thought that the coxal sacs assist in maintaining body hydration by taking up water from the environment (e.g., dewdrops). Many in-

sects collect environmental water by various other devices. Some desert beetles (Tenebrionidae) collect atmospheric water by "standing on their heads" and holding their bodies up to the moving air so that humidity can condense on the abdomen and be channeled to the mouth for consumption.

Insects that inhabit desert environments have a much greater tolerance of high temperatures and body water loss than do mesic insects, are particularly good at water conservation and producing insoluble nitrogenous waste products, and have behavioral traits, such as nocturnal activity cycles and dormancy periods, that enhance water conservation. Upper lethal temperatures for desert species commonly range to 50°C. The spiracles are often covered by setae or depressed below the cuticular surface. Many xeric insects also undergo periods of dormancy (i.e., diapause or aestivation) during some stage of the life cycle, characterized by a lowering of the basal metabolic rate and cessation of movement, which allow them to withstand prolonged periods of temperature and moisture extremes. Some even utilize evaporative cooling to reduce body temperatures. The long-chain hydrocarbons that waterproof the epicuticle also are more abundant in xeric insects.

Nervous System and Sense Organs

The hexapod nervous system conforms to the basic arthropod plan described in Chapter 15 (Figures 17.26, and 17.27). The two strands of the ventral nerve cord, as well as the segmental ganglia, are often largely fused. In dipterans, for example, even the three thoracic ganglia are fused into a single mass. The largest number of free ganglia occurs in the primitive wingless insects, which have as many as eight unfused abdominal ganglia. Giant fibers have also been reported from several insect orders.

Like the "brains" of other arthropods, the cerebral ganglion of insects comprises three distinct regions: the protocerebrum, the deutocerebrum, and the tritocerebrum (Table 17.1). The subesophageal ganglion is composed of the fused ganglia of the third, fourth, and perhaps the fifth head segments and controls the mouthparts, salivary glands, and some other local musculature.

Insects possess a **hypocerebral ganglion** between the cerebral ganglion and the foregut. Associated with this ganglion are two pairs of glandular bodies called the **corpora cardiaca** and the **corpora allata** (Figure 17.27). These two organs work in concert with the **prothoracic glands** and certain neurosecretory cells in the protocerebrum. The whole complex is a major endocrine center that regulates growth, metamorphosis, and other functions (Chapter 15).

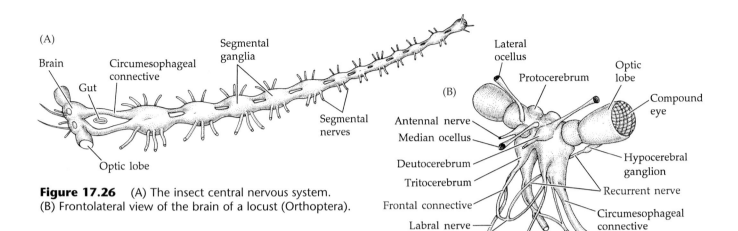

Figure 17.26 (A) The insect central nervous system. (B) Frontolateral view of the brain of a locust (Orthoptera).

Hexapods typically possess simple ocelli in the larval, juvenile, and often adult stages. When present in adults, they usually form a triad or a pair on the anterodorsal surface of the head. The compound eyes are well developed, resembling those of the Crustacea (Chapter 15), and are image-forming. Most adult insects have a pair of compound eyes (Figure 17.28A), which bulge out to some extent, giving these animals a wide field of vision in all directions. Compound eyes are greatly reduced or absent in parasitic groups and in many cave-dwelling forms. The general anatomy of the arthropod compound eye was described in Chapter 15, but several distinct structural trends are found in hexapod eyes, as we describe below.

The number of ommatidia apparently determines the overall visual acuity of a compound eye; hence large eyes are typically found on active, predatory insects such as dragonflies and damselflies (order Odonata), which may have over 10,000 ommatidia in each eye. On the other hand, workers of some ant species have but a single ommatidium per eye (ants live in a world of chemical communication)! Similarly, larger facets capture more light and are typical of nocturnal insects. In all cases, a

single ommatidium consists of two functional elements: an outer light-gathering part composed of a lens and a crystalline cone, and an inner sensory part composed of a rhabdome and sensory cells (Figure 17.28B).

The fundamental anatomy of hexapod/crustacean compound eyes was described in Chapter 15; here we elaborate on some unique aspects of the hexapod eye. In some insects the outer surface of the cornea (lens) is covered with minute conical tubercules about 0.2 μm high and arranged in a hexagonal pattern. It is thought that these projections decrease reflection from the surface of the lens, thus increasing the proportion of light transmitted through the facet. Insect eyes in which the crystalline cone is present are called **eucone eyes** (Figure 17.28B). Immediately behind the crystalline cone (in eucone eyes) are the elongate sensory neurons, or retinular cells. Primitively, each ommatidium probably contained eight retinular cells arising from three successive divisions of a single cell. This number is found in some insects today, but in most it is reduced to six or seven, with the other one or two persisting as short basal cells

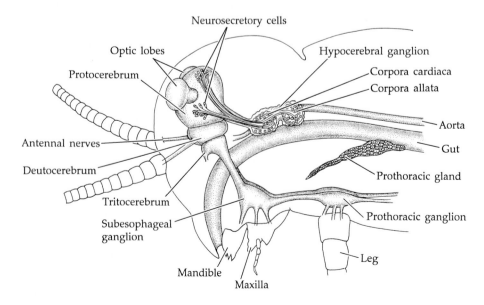

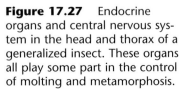

Figure 17.27 Endocrine organs and central nervous system in the head and thorax of a generalized insect. These organs all play some part in the control of molting and metamorphosis.

(A)

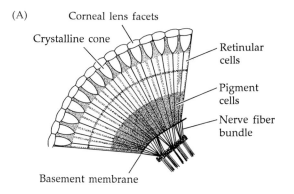

(B)

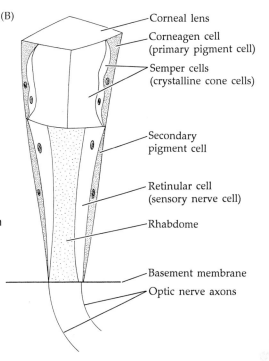

Figure 17.28 Compound eyes of insects. (A) A generalized insect compound eye (cross section). (B) A single ommatidium from a eucone compound eye.

in the proximal region of each ommatidium. Arising from each retinular cell is a neuronal axon that passes out through the basement membrane at the back of the eye into the optic lobe. There is no true optic nerve in insects; the eyes connect directly with the optic lobe of the brain. The rhabdomeres consist of tightly packed microvilli that are about 50 nm in diameter and hexagonal in cross section. The retinular cells are surrounded by 12 to 18 secondary pigment cells, which isolate each ommatidium from its neighbors.

The general body surface of hexapods, like that of other arthropods, bears a great variety of microscopic sensory hairs and setae, known collectively as sensilla. The incredible diversity of these cuticular surface structures has only begun to be explored, primarily by scanning electron microscopy. Sensilla are most heavily concentrated on the antennae, mouthparts, and legs. Most appear to be tactile or chemosensory. Club-shaped or peg-shaped chemosensory setae, usually called **peg organs** and resembling the aesthetascs of crustaceans, are particularly common on the antennae of hexapods (Figure 17.29).

Insects have internal proprioceptors called **chordotonal organs**. These structures stretch across joints and

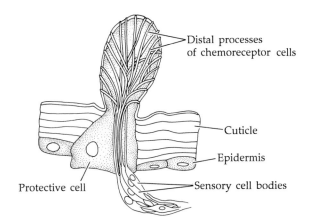

Figure 17.29 A chemosensory peg organ from the antenna of a grasshopper.

monitor the movement and position of various body parts. **Phonoreceptors** also occur in most insect orders. These structures may be simple modified body or appendage setae, or antennae, or complex structures called **tympanic organs** (Figure 17.30). Tympanic organs generally develop from the fusion of parts of a tracheal dilation and the body wall, which form a thin **tympanic membrane** (= **tympanum**). Receptor cells in an underlying air sac, or attached directly to the tympanic membrane, respond to vibrations in much the same fashion as they do in the cochlea of the human inner ear. Some insects can discriminate among different sound frequencies, but others are tone-deaf. Tympanic organs may occur on the abdomen, the thorax, or the forelegs. Several insects that are prey to bats have the ability to hear the high frequencies of bat echolocation devices, and they have evolved flight behaviors to avoid these flying mammals. For example, some moths, when they hear a bat's echolocation (generally above the range of human ears), will fold their wings and suddenly drop groundward as an evasive maneuver. Praying mantids, whose sonar detection device is buried in a groove on the ventral side of the abdomen, throw out the raptorial forelimbs and elevate the abdomen. These movements cause the insect to "stall" and go into a steep roll, which it pulls out of at the last minute with a "power dive" that effectively avoids bat predators.

Sound communication in insects, like light communication in fireflies (and some ostracods), is a species-specific means of mate communication. Several insect groups (e.g., orthopterans, coleopterans, dipterans, and homopterans) possess sound-producing structures. Male flies of the genus *Drosophila* create species-specific mating songs by rapidly vibrating the wings or ab-

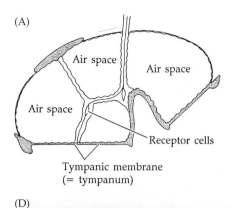

(A)

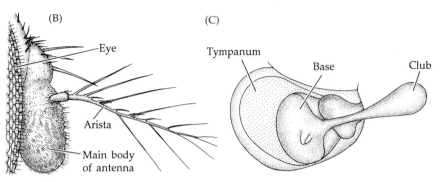

(B) (C)

(D)

Figure 17.30 Insect "ears." Insect auditory organs (phonoreceptors) differ widely in their anatomy and location. (A) The "ear" of noctuid moths (Lepidoptera) is a pressure receiver used to detect the ultrasonic cries of hunting bats. It is similar to most insect "ears" in comprising a tympanic membrane backed by a tracheal air space. Two receptor cells attach to the tympanum. (B) In *Drosophila* (Diptera), a feathery seta called an arista arises on the third antennal segment. The arista detects air movements, thus responding to sound through interaction with vibrating air particles. It is used to detect the calling song of the species. (C) In the "ear" of a water boatman (Hemiptera), the tympanum is covered by the base of a club-shaped cuticular body that protrudes outside the body. The club performs rocking movements that allow some frequency analysis of the songs of other water boatmen. (D) Tympani of tibia on katydid forelegs.

domen. These "love songs" attract conspecific females for copulation. It has been demonstrated that the rhythm of the male's song is encoded in genes inherited from his mother, on the X chromosome, whereas the song's "pulse interval" is controlled by genes on autosomal chromosomes.

Cicadas may possess the most complex sound-producing organs in the animal kingdom (Figure 17.31). The ventral metathoracic region of male cicadas bears two large plates, or **opercula**, that cover a complex system of vibratory membranes and resonating chambers. One membrane, the **tymbal**, is set vibrating by special muscles, and other membranes in the resonating chambers amplify its vibrations. The sound leaves the cicada's body through the metathoracic spiracle.

Numerous families of beetles and bugs utilize water surfaces as a substratum both for locomotion and for communication by waves or ripples. Such insects pro-

duce a signal with simultaneous vertical oscillations of one or more pairs of legs, and sometimes also with distinct vertical body motions. The wave patterns produced are species-specific. Potential prey trapped on a surface film may also be recognized in this fashion, just as spiders recognize prey by web vibrations. Limited data suggest that the receptor organs for ripple communication are either specialized sensillae on the legs or special proprioceptors between joints of the legs or antennae, perhaps similar to the tarsal organs of scorpions (Chapter 19).

A number of insects are bioluminescent, the most familiar being beetles of the family Lampyridae, known as lightning bugs or fireflies. In the tropics, where they are especially abundant, fireflies are sometimes kept in containers and used as natural flashlights, and women

Figure 17.31 Sound production structure in cicadas (Homoptera) from the first abdominal segment (section). Sound is produced by buckling of the tymbal, a thin disc of cuticle. The tymbal muscle is connected to the tymbal by a strut. Contraction of this muscle causes the tymbal to buckle inward, thereby producing a click that is amplified by resonance in the underlying air sacs. On relaxation, the elasticity of the muscle causes the tymbal to buckle out again. On the underside of this abdominal segment, a folded membrane can be stretched to tune the air sacs to the resonant frequency of the tymbal.

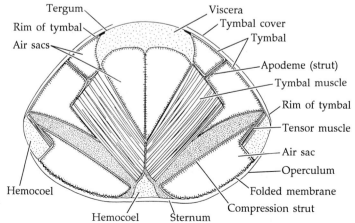

wrap them in gauze bags worn as glowing hair ornaments. The light of luminescent insects ranges from green through red and orange, depending on the species and the precise chemical nature of the luciferin–luciferase system involved. Light-producing organs are typically composed of clusters of light-producing cells, or **photocytes**, backed by a layer of reflecting cells and covered with a thin, transparent epidermis. The photocytes are richly supplied with tracheae, oxygen being necessary for the chemical reaction, and are provided with direct nervous innervation. Each species of firefly, and of most other glowing insects, has a distinct flash pattern, or code, to facilitate mate recognition and communication.

One of the most sophisticated communication behaviors among insects may be the famous honey bee "waggle dance." Each day forager bees leave their colony to locate new food sources (e.g., fresh flower blooms). They fly meandering search forays until a good source is located. Then they return to the hive along a straight flight path (a "bee line"); while doing so, they are thought to imprint a navigational "map" from the colony to the food source. Most behaviorists believe that this information is communicated to hivemates in a complex tail-wagging dance that allows other bees from the hive to fly directly to the new feeding ground. The forager bee also carries food odors (nectar samples), pollen, and various other odors clinging to the hairs on her body. She can also mark the food source with a pheromone produced in a special gland (**Nasanov gland**). All of these clues help her hivemates find the new food source. Karl von Frisch was the first person to document all these attributes of bee foraging early in the twentieth century.

A large body of research on bee navigation has accumulated since the pioneering "dancing bee studies" of von Frisch. We now know that honey bees (and solitary bees) have outstanding vision. Much of the bee's daily activity, including navigation and flower recognition, relies strongly on ultraviolet vision. Bees appear to utilize a hierarchical series of flight orientation mechanisms; when the primary mechanism is blocked, a bee can switch to a secondary system. The primary navigation system utilizes the pattern of polarized ultraviolet sunlight in the sky. This pattern depends on the location of the sun as determined by two coordinates, the azimuth and the elevation. Bees and many other animals that orient to the sun have a built-in ability to compensate for both hourly changes (elevation) and seasonal changes (azimuth) in the sun's position with time. On cloudy days, when the sun's light is largely depolarized, bees cannot rely on their ultraviolet celestial navigation mechanism and thus may switch to their second-order navigational system: navigation by landmarks (foliage, rocks, and so on) that were imprinted during the most recent flight to the food source. Limited evidence suggests that some form of tertiary backup system may also exist.

Thus, if the honey bee dance model is correct,* honey bees must simultaneously process information concerning time, the direction of flight relative to the sun's azimuth, the movement of the sun, the distance flown, and local landmarks (not to mention complications due to other factors, such as crosswinds), and in doing so reconstruct a straight-line heading to inform their hivemates. If recent evidence is correct, bees (like homing pigeons) may also detect Earth's magnetic fields with iron compounds (magnetite) located in their abdomens. Bands of cells in each abdominal segment of the honey bee contain iron-rich granules, and nerve branches from each segmental ganglion appear to innervate these tissues.[†]

In some insects the ocelli are the principal navigation receptors. Some locusts and dragonflies and at least one ant species utilize the ocelli to read compass information from the blue sky. As in bees, the pattern of polarized light in the sky seems to be the main compass cue. In some species, both ocelli and compound eyes may function in this fashion. Many (probably most) insects also see ultraviolet light.

Many insects release noxious quinone compounds to repel attacks. Perhaps best known in this regard are certain Tenebrionidae, many of which stand on their heads to do so (Figure 17.32A). But the champions of this chemical warfare strategy are definitely the bombardier beetles, members of the carabid subfamilies Brachininae and Paussinae, which expel quinone compounds at temperatures reaching 100°C.

Reproduction and Development

Reproduction. Hexapods are dioecious, and most are oviparous. A few insects are ovoviviparous, and many can reproduce parthenogenetically. Some stonefly (Plecoptera) species are known to be hermaphroditic. The most primitive insects have direct development. However, more advanced insects undergo complex indirect development that typically includes radically different larval, pupal, and adult forms. Indirect development in insects is surely a secondarily derived phenomenon, unlikely to have been derived from the larval strategies of marine forms, whose planktonic stages would never work on land.

Most insects rely on direct copulation and insemination. Reproductively mature insects are termed **imagos**.

---

*The honey bee dance hypothesis is not without its detractors, and some workers doubt its existence altogether; see references by von Frisch, Gould, Rosin, Wenner and Wells for a glimpse at the history of the honey bee dance controversy.

[†]Many animals possess magnetotactic capabilities, including some molluscs, hornets, salmon, tuna, turtles, salamanders, homing pigeons, cetaceans, and even bacteria and humans. Magnetotactic bacteria swim to the north in the Northern Hemisphere, to the south in the Southern Hemisphere, and in both directions at the geomagnetic equator. In all these cases, iron oxide crystals in the form of magnetite have been shown to underlie the primary detection devices. However, in honey bees, the iron-containing structures are trophocytes that contain paramagnetic magnetite. These magnetotactic trophocytes surround each abdominal segment and are innervated by the central nervous system.

(A)

(B)

(C)

(D)

(E)

Figure 17.32 Defense, warning coloration, and camouflage in insects. (A) The Pinacate beetle (Coleoptera: Tenebrionidae: *Eleodes*), standing on its head in preparation for release of a noxious substance that deters predators. (B) The bright coloration of a velvet ant (Hymenoptera: Mutilidae), which is actually a wingless wasp, warns of its powerful sting. (C) A leaf-mimicking katydid (Orthoptera: Tettigoniidae). (D) An Australian stick insect (Phasmida) is camouflaged in dry shrubs, but is easy to see if placed on a green plant. (E) A grasshopper (Orthoptera) is camouflaged on its food.

Female imagos have one pair of ovaries, formed of clusters of tubular **ovarioles** (Figure 17.33A). The oviducts unite as a common duct before entering a **genital chamber**. Seminal receptacles (**spermathecae**) and **accessory glands** also empty into the genital chamber. The genital chamber opens, via a short **copulatory bursa** (= **vagina**), on the sternum of the eighth, or occasionally the seventh or ninth, abdominal segment. The male reproductive system is similar, with a pair of testes, each formed by a number of **sperm tubes** (Figure 17.33B). Paired sperm ducts dilate into seminal vesicles (where sperm are stored) and then unite as a single ejaculatory duct. Near this duct, accessory glands discharge seminal fluids into the reproductive tract. The lower end of the ejaculatory duct is housed within a penis, which extends posteroventrally from the ninth abdominal sternite.

Courtship behaviors in insects are extremely diverse and often quite elaborate, and each species has its own species recognition methods. Courtship may consist of simple chemical or visual attraction, but more typically it involves pheromone release, followed by a variety of displays, tactile stimulation, songs, flashing lights, or other rituals that may last for hours. The subject of insect courtship is a large and fascinating study of its own. Although the field of pheromone biology is still in its infancy, sexual attractant or aggregation pheromones have been identified from about 450 different insect species (about half of which are synthesized and sold commercially for pest control purposes).

Most insects transfer sperm directly as the male inserts either his penis (Figure 17.33D) or a gonopod into the genital chamber of the female. Special abdominal claspers, or other articulated cuticular structures on the male, often augment his copulatory grip. Such morphological modifications are species-specific and thus serve as valuable recognition characters, both for insect mates and insect taxonomists. Copulation often takes place in mid-flight. In some of the primitive wingless insects and in the odonatans, sperm transfer is indirect. In these cases, a male may deposit his sperm on specialized regions of his body to be picked up by the female; or he may simply leave the sperm on the ground, where they are found and taken up by females. In bedbugs (order Hemiptera, family Cimicidae) males use the swollen penis to pierce a special region of the female's body wall; sperm are then deposited directly into an internal organ (the **organ of Berlese**). From there they migrate to the ovaries, where fertilization takes place as eggs are released.

Sperm may be suspended in an accessory gland secretion, or, more commonly, the secretion hardens around the sperm to produce a spermatophore. Females of many insect species store large quantities of sperm within the spermathecae. In some cases sperm from a single mating is sufficient to fertilize a female's eggs for

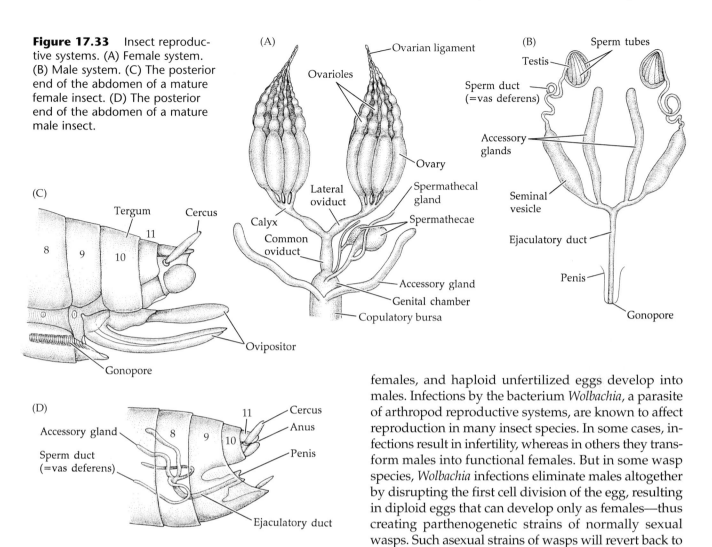

Figure 17.33 Insect reproductive systems. (A) Female system. (B) Male system. (C) The posterior end of the abdomen of a mature female insect. (D) The posterior end of the abdomen of a mature male insect.

her entire reproductive lifetime, which may last a few days to several years.

Insect eggs are protected by a thick membrane (the **chorion**) produced within the ovary. Fertilization occurs as the eggs pass through the oviduct to be deposited. Accessory glands contribute adhesives or secretions that harden over the zygotes. In many species, cuticular extensions around the gonopore of the female form an **ovipositor** (Figure 17.33C), with which she places the eggs in a brooding site that will afford suitable conditions for the young once they hatch (such as in a shallow underground chamber, in a plant stem, or within the body of a host insect). Although 50–100 eggs are usually laid at a time, as few as one and as many as several thousand are deposited by some species. Some insects, such as cockroaches, enclose several eggs at a time in a protective egg case.

Parthenogenesis is common in a variety of insect groups. It is used as an alternative form of reproduction seasonally by a number of insect taxa, particularly those living in unstable environments. In the Hymenoptera (bees, wasps, ants), it is used as a medium for sex determination. In these cases, diploid fertilized eggs become females, and haploid unfertilized eggs develop into males. Infections by the bacterium *Wolbachia*, a parasite of arthropod reproductive systems, are known to affect reproduction in many insect species. In some cases, infections result in infertility, whereas in others they transform males into functional females. But in some wasp species, *Wolbachia* infections eliminate males altogether by disrupting the first cell division of the egg, resulting in diploid eggs that can develop only as females—thus creating parthenogenetic strains of normally sexual wasps. Such asexual strains of wasps will revert back to dioecy if the *Wolbachia* dies out.

Embryogeny. As discussed in Chapter 15, the large centrolecithal eggs of arthropods are often very yolky, a condition resulting in dramatic modifications of the presumed ancestral spiral cleavage pattern. Although vestiges of holoblastic spiral cleavage are still discernible in some crustaceans, the hexapods show almost no trace of spiral cleavage at all; they have shifted almost entirely to meroblastic cleavage. Most undergo early cleavage by intralecithal nuclear divisions, followed by migration of the daughter nuclei to the peripheral cytoplasm (= **periplasm**). Cytokinesis does not occur during these early nuclear divisions (up to 13 cycles), which thus generate a syncitium, or **plasmodial phase** of embryogenesis. The nuclei continue to divide until the periplasm is dense with nuclei, whereupon a **syncitial blastoderm** exists. Eventually, cell membranes begin to form, partitioning uninucleate cells from one another. At this point the embryo is a periblastula, comprising a yolky sphere containing a few scattered nuclei and covered by a thin cellular layer (Figure 15.30).

Along one side of the blastula a patch of columnar cells forms a **germinal disc**, sharply marked off from the

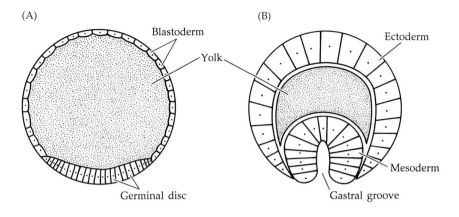

Figure 17.34 Early stages of insect development. (A) The blastoderm (blastula) of a generalized insect, subsequent to cytokinesis (cross section). Note the thickened germinal disc. (B) An early gastrula of a honeybee (cross section). Note the gastral groove and the proliferation of mesoderm. Entoderm is derived from the front and back ends of the gastral groove.

thin cuboidal cells of the remaining blastoderm (Figure 17.34A). From specific regions of this disc, presumptive entodermal and mesodermal cells begin to proliferate as germinal centers. These cells migrate inward during gastrulation to lie beneath their parental cells, which now form the ectoderm. The mesoderm proliferates inward as a longitudinal **gastral groove** (Figure 17.34B). The cells of the developing gut usually surround and gradually begin absorbing the central yolky mass of the embryo, and paired coelomic spaces appear in the mesoderm.

As segments begin to demarcate and proliferate, each receives one pair of mesodermal pouches and eventually develops **appendage buds**. As the mesoderm contributes to various organs and tissues, the paired coelomic spaces merge with the small blastocoel to produce the hemocoelic space. The mouth and anus arise by ingrowths of the ectoderm that form the foregut and hindgut, which eventually establish contact with the developing entodermal midgut.

Polyembryony occurs in a number of insect taxa, particularly parasitic Hymenoptera; in this form of development the early embryo splits to give rise to more than one developing embryo. Thus, from two to thousands of larvae may result from a single fertilized egg, which is often deposited in the body of another (host) insect.

Development. Most insects hatch with a full complement of adult segments, but there is considerable variation in the body form and stage of maturity at the time of hatching. Only in the primitive wingless insects do the young hatch out as juveniles closely resembling the adult, or imago, condition. In insects, such direct development is called **simple development**. In the pterygotes development is indirect, and hatching stages undergo a series of morphological changes (metamorphoses) before the adult condition is achieved. Such growth may occur by a series of gradual changes known as **hemimetabolous development** (= **incomplete metamorphosis**) (Figure 17.35), or by a dramatic series of metamorphoses called **holometabolous development** (= **complete metamorphosis**) (Figure 17.36).

Hemimetabolous insects have young that possess compound eyes, antennae, and feeding and walking appendages similar to those of adults. Functional wings and sexual structures, however, are always lacking. These immature forms are often called **nymphs**; they usually have wing rudiments called **wing pads** that expand to form functional wings with the preadult molt. Many hemimetabolous insects have aquatic gilled nymphs, called **naiads** (e.g., mayflies, dragonflies, damselflies). The principal changes during growth are in body size and proportions and in the development of wings and sexual structures. Nymphs and adults often live in the same general habitat; naiads and their respective adults do not.

Holometabolous insects hatch as vermiform larvae that bear no resemblance whatsoever to the adult forms. These larvae are so different from adults that they are often given separate vernacular names; for example, butterfly larvae are called caterpillars, fly larvae maggots, and beetle larvae grubs. Holometabolous larvae lack compound eyes (and often antennae), and their natural history differs markedly from that of adults. Their mouthparts may be wholly unlike those of adults, and external wing pads are never present. Often the greater part of an insect's lifetime is spent in a series of larval instars. Larvae typically consume vast quantities of food and attain a larger size than adults. Termination of the larval stage is accompanied by **pupation**, during which (in a single molt) the **pupal** stage is entered. Pupae do not feed or move about very much. They often reside in protective niches in the ground, within plant tissues, or housed in a cocoon. Energy reserves stored during the long larval life are utilized by the pupa to undergo wholesale transformation of the body. Many structures are broken down and reorganized to attain the adult form; external wings and sexual organs are formed. The remarkable transformation from larval stage to adult stage in holometabolous insects is one of the most impressive achievements of animal evolution.

The success of the holometabolous lifestyle is demonstrated by the fact that such insects outnumber hemimetabolous species ten to one. There is a popular theory among evolutionary biologists that views indirect development, including holometabolous development in insects, as selectively advantageous because it results in the ecological segregation of adults from young, thus avoid-

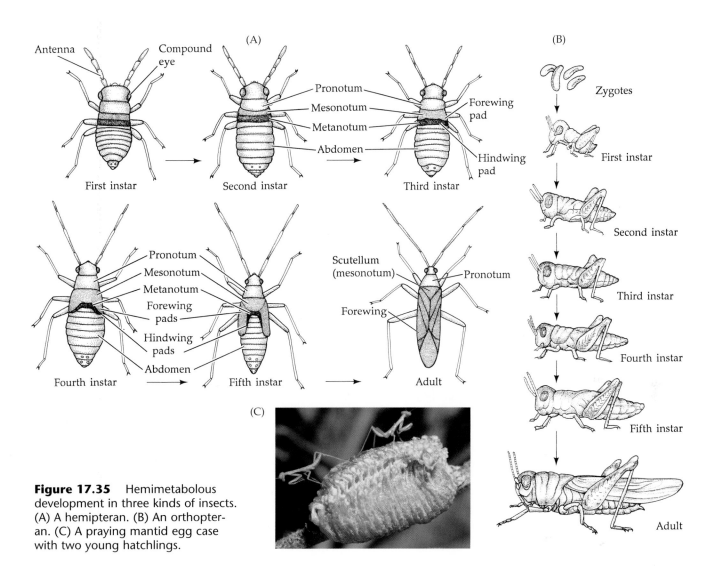

Figure 17.35 Hemimetabolous development in three kinds of insects. (A) A hemipteran. (B) An orthopteran. (C) A praying mantid egg case with two young hatchlings.

ing intraspecific competition and allowing each stage to develop its own suite of specific survival strategies. However, satisfactory confirmation of this theory has been difficult to come by. We have seen that such indirect developmental strategies are common in marine and some freshwater invertebrates, but only the insects have managed to exploit this habit so successfully on land.

The role of ecdysone in initiating molting is described in Chapter 15. This hormone works in conjunction with a second endocrine product in controlling the sequence of events in insect metamorphosis. This second product, **juvenile hormone**, is manufactured and released by the corpora allata, a pair of glandlike structures associated with the brain (Figure 17.27). When ecdysone initiates a molt in an early larval instar, the accompanying concentration of juvenile hormone in the hemolymph is high. A high concentration of juvenile hormone ensures a larva-to-larva molt. After the last larval instar is reached, the corpora allata ceases to secrete juvenile hormone. Low concentrations of juvenile hormone result in a larva-to-pupa molt. Finally, when the pupa is ready to molt, juvenile hormone is absent from

the hemolymph altogether; this deficiency leads to a pupa-to-adult molt.

The specializations of reproductive strategies among insects are legion and could easily fill a book this size. Some of the most astonishing occur among the social insects, especially the hymenopterans. Ants are a good example. The only time ants have sex (or wings, or males) is when a colony swarms. At this time, winged individuals leave the nest by the thousands or millions, depending on the species. Some of these are destined to become queen ants and start new colonies. During this one-time swarming event, the virgin queens mate (usually in flight) with males (usually with several males). Subsequent to copulation the males promptly die, and males are not seen again until the next swarming event. The queen stores the sperm from the nuptial event in a small sac next to the oviduct. From this single mating, she will start her new colony. The output of her one-time mating can be prodigious. For example, leafcutter queen ants produce about 150 million daughters (workers) in their lifetime, of which 2 or 3 million are alive at any given time.

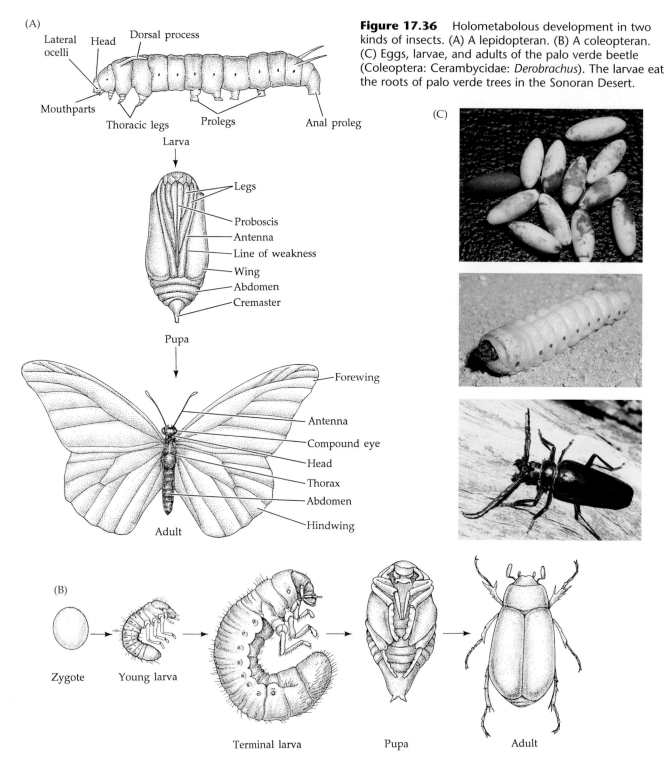

Figure 17.36 Holometabolous development in two kinds of insects. (A) A lepidopteran. (B) A coleopteran. (C) Eggs, larvae, and adults of the palo verde beetle (Coleoptera: Cerambycidae: *Derobrachus*). The larvae eat the roots of palo verde trees in the Sonoran Desert.

Hexapod Evolution

The Origin of the Hexapoda

The first edition of this book presented the long-standing, orthodox view of the Hexapoda, depicting them as the sister group to the Myriapoda, in a grouping known as the Atelocerata. Atelocerata has traditionally been classified next to Crustacea, to make up a larger group called the Mandibulata. However, since 1990, an explosion of anatomical, paleontological, and molecular developmental and phylogenetic research on arthropods has begun to challenge this view of arthropod relationships. On balance, these new data suggest to us that the Hexapoda may be more closely related to crustaceans than to myriapods—in fact, that the hexapods might have arisen from within the Crustacea. The idea that insects might be flying crustaceans (in the same sense that birds are flying reptiles) is difficult for some people to envision, and not all biologists agree with this new view, despite corroborating evidence from several independent fields of research. The evidence for a Hexapoda-Crustacea sister-group relationship is summarized in Chapter 15. Analyses based on molecular sequence data from several genes suggest that the Hexapoda arose from within the Malacostraca or the Branchiopoda.

The Origin of Insect Flight

Like the origin of the Hexapoda, the origin of insect flight has been a subject of controversy in recent years. For many decades, two competing views of insect wing origin have dominated the literature. In general, these views can be termed the **paranotal lobe hypothesis** and the **appendage hypothesis**. The former holds that wings evolved by way of a gradual expansion of lateral folds of the thoracic tergites (paranotal lobes), which eventually became articulated and muscled to form wings. The latter hypothesizes that wings evolved from pre-existing articulated structures on the thoracic appendages, such as gills or protopodal exites on the legs. There is also tantalizing evidence from the fossil record suggesting that the first pterygote insects possessed appendages on the prothorax, called "winglets," that may have been serially homologous to modern wings, implying that the loss of prothoracic proto-wings might have taken place in the early evolutionary history of the Hexapoda.

The paranotal lobe hypothesis was first proposed by Müller in 1873, saw a resurgence of popularity in the middle of the twentieth century, and has lost favor in recent years. It suggests that wings originated as lateral aerodynamic flaps of the thoracic nota that enabled insects to alight right side up when jumping or when blown about by the wind. These stabilizing paranotal lobes later evolved hinged structures and muscles at their bases. The occurrence of fixed paranotal lobes in certain ancient fossil insects has been cited in support of the paranotal lobe hypothesis (Figure 17.37). However, recent studies suggest that these primitive paranotal lobes might have been used for other purposes, such as covering the spiracular openings or gills in amphibious insects, protecting or concealing the insects from predators, courtship displays, or thermoregulation by absorption of solar radiation.

The appendage hypothesis (also known as the "gill or branchial theory," "exite theory," or "leg theory") also dates back to the nineteenth century, but was resurrected by the great entomologist V. B. Wigglesworth in the 1970s, and has been championed by J. Kukalová-Peck since the 1980s. It is the more favored hypothesis of wing origin today, based on recent paleontological work, microscopic anatomy, and molecular developmental biology. It suggests that insect wings are derived from thoracic appendages—from protopodal exites, in Wigglesworth and Kukalová-Peck's view. These proto-wing appendages might have first functioned as aquatic gills or paddles, or as terrestrial gliding structures. The paired abdominal gills of mayflies have been suggested as serial homologues of such "proto-wings." In

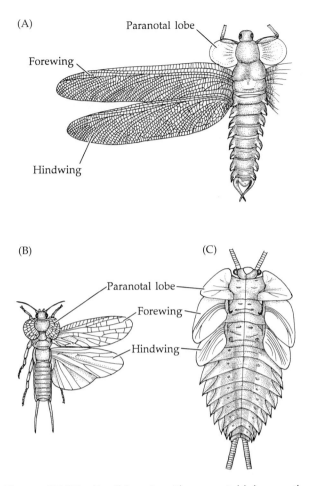

Figure 17.37 Fossil insects with paranotal lobes on the prothorax. (A) *Stenodictya lobata*. (B) *Lemmatophora typa*. (C) Nymphal stage of *Rochdalia parkeri*, a Paleozoic terrestrial palaeodictyopteran. In this species, all three thoracic segments appear to have had "articulated" thoracic lobes.

Kukalová-Peck's version of this hypothesis, the first protopodal leg article (the **epicoxa**) fused with the thoracic pleural membrane early in the evolution of the Arthropoda, as did the second article (the **precoxa**) in the ancient hexapods, with both migrating dorsally off the leg. In insects, the epicoxa eventually fused with the tergite, its exite enlarging to form the proto-wing, and eventually the true wing. The precoxa formed the pleural sclerite providing the ventral articulation of the wing. Wing veins might have evolved from cuticular ridges that served to strengthen these structures, and eventually to circulate blood through them.*

Kukalová-Peck's theory of wing evolution finds support in molecular developmental studies, which have shown that the cells that give rise to the wing primordium derive from the same cluster of cells that form the leg primordium, from which they segregate, migrating

*Unspecialized coxal exites can be seen on the legs of some living archaeognathans (bristletails) and in numerous extinct hexapods.

dorsally to a position below the tergum. Recent studies on gene expression also support the origin of wings from legs. The genes *pdm* and *apterous* are expressed in the wing (and leg) primordia of all insects. Expression of both genes appears to be necessary for normal wing formation. In malacostracan crustaceans (but not in branchiopod crustaceans) these same genes are expressed, in a similar manner, in the formation of the leg rami (the exopod and endopod).

Evolution within the Hexapoda

At the beginning of this chapter we noted the spectacular evolutionary radiation of the insects, and in chapter 15 we reviewed the appearance of the terrestrial arthropods in the late Ordovician or early Silurian. The fossil record of insects is good, with about 1,263 recognized families (by comparison, there are 825 recognized fossil families of tetrapod vertebrates). The Devonian (and perhaps the Silurian) were inhabited at least by Collembola, Protura, Diplura, Archaeognatha, and Thysanura. By the middle Devonian many other terrestrial arthropods (e.g., mites, amblypygids, opilionids, scorpions, pseudoscorpions, and spiders) had made their appearance. The first undis-

puted insects in the fossil record are early Devonian springtails (Collembola) and bristletails (Archaeognatha). However, molecular clock data place the origin of the Insecta at 420–434 years ago (Silurian). By the Carboniferous, various modern insect orders were flourishing, although many were quite unlike today's fauna. Some Carboniferous hexapods are notable for their gigantic size, such as silverfish (Thysanura) that reached 6 cm in length and dragonflies with wingspans of about 70 cm. In addition to the living orders of insects, at least ten other orders arose and radiated in late Paleozoic and early Mesozoic times, then went extinct.

The Permian saw an explosive radiation of holometabolous insects, although many groups went extinct in the great end-Permian extinction event (Chapter 1). In fact, relatively few groups of Paleozoic insects survived into the Mesozoic, and many recent families first appeared in the Jurassic. By the Cretaceous, most modern families were extant, insect sociality had evolved, and many insect families had begun their intimate relationships with angiosperms. Tertiary insects were essentially modern and included many genera indistinguishable from the Recent fauna.

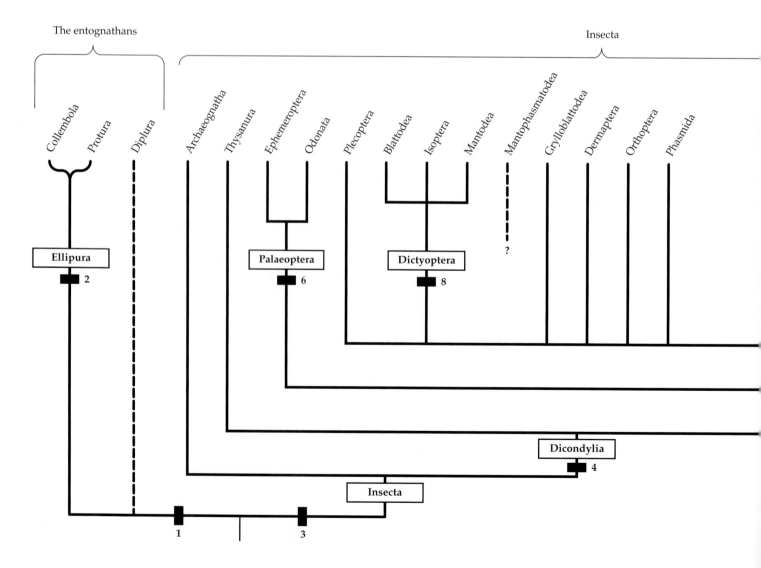

Although theories of hexapod evolution have changed over the years, the core ideas of transition from primitive to advanced lineages have held relatively constant. (For good reviews, see Kristensen 1991; Hennig 1981; Kukalová-Peck 1991b. The evolutionary tree presented in Figure 17.38 follows Kristensen (1991). We have not included most synapomorphies on the tree, simply because many involve detailed characters beyond the scope of this text. However, a few of the principal ones are noted below.

The "entognathous" hexapods have had an unsettled history. Among the alleged synapomorphies that unite them as a monophyletic clade is entognathy itself (the overgrowth of the mouthparts by oral folds from the lateral cranial wall). In addition, the Malpighian tubules and compound eyes are reduced—compound eyes are degenerate in Collembola and absent in the extant Diplura and Protura. However, these reductions could be convergences resulting from small size. Most current workers regard the Collembola + Protura to be closely related, perhaps a lineage unto themselves (called the Ellipura or Parainsecta). Whether the Diplura are more closely related to the Collembola + Protura, or to the Insecta (a view persuasively argued by Kukalová-Peck),

remains hotly debated. Our evolutionary tree thus depicts an unresolved trichotomy at the base of the Hexapoda.

Synapomorphies of the Collembola include the reduced number of abdominal segments (six) and specialized appendages on abdominal segments 1 (ventral tube), 3 (retinaculum), and 4 (furca). The Protura are defined by their lack of a tentorium, visual organs, and antennae, among other things. Also, the forelegs are enlarged and usually function in an antenniform fashion, being held forward and richly supplied with sensilla. The Diplura are distinguished by the absence of a distinct tentorium and ocelli.

The monophyly of the Insecta (bristletails, silverfish, and Pterygota) is well founded. The principal synapomorphies of this group include the structure of the antenna, with its lack of muscles beyond the first segment (scape); the presence of a group of special chordotonal organs (vibration sensors) in the second antennal segment (pedicel); a well developed posterior tentorium (forming a transverse bar); coxae subsegmented (or annulated); females with ovipositor formed by **gonapophyses** (limb-base endites) on segments 8 and 9; and long, annulated, posterior terminal filaments (cerci).

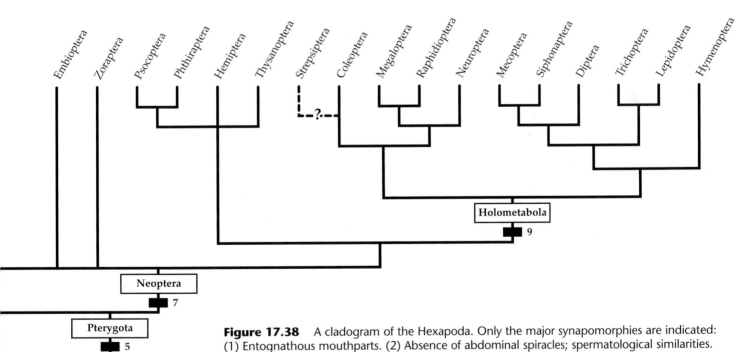

Figure 17.38   A cladogram of the Hexapoda. Only the major synapomorphies are indicated: (1) Entognathous mouthparts. (2) Absence of abdominal spiracles; spermatological similarities. (3) Antennae lack muscles beyond the scape (first segment); presence of a large group of chordotonal organs ("Johnston's organ") in the pedicel (second segment); tarsi subsegmented; females with ovipositor formed by gonapophyses (limb-base endites) on segments 8 and 9. (4) Fully dicondylic mandibles. (5) Wings. (6) Specialized aquatic larvae, called naiads. (7) Reduction of cross veins in wings; ability to fold wings over back. (8) Perforate tentorium (processes on the anterior arms fuse medially in front of the esophageal nerve connectives); female genitalia with shortened ovipositor hidden above large subgenital plate. (9) Addition of a pupal stage in the life cycle (holometabolous development).

The wingless insects (bristletails and silverfish) have occasionally been combined (as "Thysanura"), and both groups molt throughout their lives (as do the entognathous hexapods). However, on the basis of molecular studies and a similar dicondylic mandible, the silverfish (Thysanura *sensu stricto*) are now thought to be the sister group of the Pterygota. Thysanurans have lost the antennal musculature in all but the two basal articles. Like the entognaths, however, they have rudimentary abdominal appendages and employ indirect insemination. Many workers recognize a lineage known as the Dicondylia, comprising the Thysanura and Pterygota, united by the evolution of a secondary (anterior) articulation on the mandibles, in addition to the primary (posterior) one, as well as many other structural features. Among the synapomorphies of the Archaeognatha (bristletails) are the greatly expanded compound eyes (which are medially contiguous) and loss of the first pair of spiracles.

The Pterygota are, of course, distinguished by the presence of wings on the mesothorax and metathorax of adults. There is wide agreement that wings evolved only once within the Hexapoda. Traditionally, the Pterygota, or flying insects, are thought to comprise two fundamental lineages, Palaeoptera and Neoptera. The Palaeoptera ("ancient" insects) include the Ephemeroptera (mayflies) and Odonata (dragonflies and damselflies), which are characterized by their many-veined, netlike wings. The Neoptera ("modern" insects) have reduced wing venation. Also characteristic of members of the Palaeoptera is the inability to fold the wings over the back when at rest. In addition, extant palaeopterans have aquatic larval stages (**naiads**). This observation has led some workers to suggest that the Pterygota may have arisen from aquatic ancestors whose gills evolved into wings, as discussed earlier. However, this argument is weakened by the fact that many extinct palaeopterans are thought to have had terrestrial nymphs. Ephemeropterans are often viewed as having one "adult" molt, between the juvenile (**subimago**) stage and the reproductive adult (**imago**) stage. Recently, the monophyly of the Palaeoptera has been challenged, and some authors believe that the Odonata may be the sister group to the Neoptera—a hypothesis supported by some ribosomal DNA studies, and by the absence of adult ecdysis in the Odonata + Neoptera.

This text recognizes 26 extant orders of Neoptera. Generally speaking, these orders can be divided into three broad groups. Eleven orders are typically placed together in the Holometabola (= Endopterygota), united on the basis of holometabolous development and several anatomical features.* The Holometabola is almost certainly a monophyletic lineage. Three orders are commonly joined in the Dictyoptera (Blattodea, Isoptera, Mantodea, and perhaps the recently discovered Mantophasmatodea), also probably a monophyletic group. The rest of the insect orders belong to a third grouping whose relationships are still unsettled.

An important hypothesis to note in the cladogram (Figure 17.38) is that the wingless insects (the apterygote insects: Collembola, Protura, Diplura, Archaeognatha, Thysanura) do not constitute a natural, or monophyletic, group. They are simply five primitive orders that evolved prior to the invention of wings in the Hexapoda. Although the taxon "Apterygota" was commonly used in older entomology texts, it has largely been abandoned in recognition of the paraphyletic nature of this grouping.

---

*Holometabola includes the Coleoptera, Megaloptera, Raphidioptera, Neuroptera, Mecoptera, Siphonaptera, Diptera, Trichoptera, Lepidoptera, Hymenoptera, and Strepsiptera.

Selected References

The amount of published information on the Hexapoda is overwhelming. We have thus have had to be very selective in our reference list, emphasizing texts, conference volumes, and journal articles that present solid overviews in their respective fields. These will provide entry into the primary literature.

General References

American Zoologist. 1981. Insect Systems. Am. Zool. 21(3): 623–791. [Eighteen papers on insect physiology, emphasizing the endocrine system.]

Arnett, R. H. 1985. American Insects: A Handbook of the Insects of North America. Van Nostrand Reinhold, New York.

Arnett, R. H., N. M. Downie and H. E. Jaques. 1980. *How to Know the Beetles*. Wm. C. Brown, Dubuque, IA.

Arnett, Jr., R. H., H. Ross and M. C. Thomas. 2001. *American Beetles*, Vol. 1. CRC Press, Boca Raton, FL.

Arnett, Jr. R. H., H. Ross, M. C. Thomas, P. E. Skelley and J. H. Frank. 2002. *American Beetles*, Vol. 2. CRC Press, Boca Raton, FL.

Barth, F. G. 1985. *Insects and Flowers: The Biology of a Partnership*. Translated by M. A. Bierderman-Thorson. Princeton University Press, Princeton, NJ.

Bate, M. and A. Martinez (eds.). 1993. *The Development of* Drosophila melanogaster. CSH Laboratory Press, New York.

Batra, S. W. T. and L. R. Batra. 1967. The fungus gardens of insects. Sci. Am. 217(5): 112–120.

Bell, W. J. and R. T. Cardé (eds.). 1984. *Chemical Ecology of Insects*. Sinauer Associates, Sunderland, MA.

Bennet-Clark, H. C. and E. C. A. Lucey. 1967. The jump of the flea: A study of the energetics and a model of the mechanisms. J. Exp. Biol. 47: 59–76.

Bland, R. G. 1978. *How to Know the Insects*. Wm. C. Brown, Dubuque, IA.

Blaney, W. 1976. *How Insects Live*. Elsevier-Phaidon, Phaidon Press, Oxford. [A general primer of insect ecology, in coffee-table style.]

Blum, M. S. and N. A. Blum. 1979. *Sexual Selection and Reproductive Competition in Insects*. Academic Press, New York.

Borrer, D. J. and R. E. White. 1970. *A Field Guide to the Insects of America North of Mexico*. Houghton Mifflin, Boston.

Borrer, D. J., C. A. Triplehorn, and N. F. Johnson. 1989. *An Introduction to the Study of Insects*, 6th Ed. Saunders, New York. [A standard reference and identification guide for many generations of entomology students.]

Bromenshenk, J. J., S. R. Carlson, J. C. Simpson and J. M. Thomas. 1985. Pollution monitoring of Puget Sound with honey bees. Science 227: 632–634.

Brown, M. 1994. Interactions between germ cells and somatic cells in Drosophila melanogaster. Semin. Dev. Biol. 5: 31–42.

Brown, W. L., Jr. et al. 1982. Insecta. In S. P. Parker (ed.), *Synopsis and Classification of Living Organisms*. McGraw-Hill, New York, pp. 326–680.

Buchmann, S. L. and G. P. Nabhan. 1996. *The Forgotten Pollinators*. Island Press, Washington, DC.

Butler, C. G. 1967. Insect pheromones. Biol. Rev. 42: 42–87.

Campos-Ortega, J. A. and V. Hartsenstein. 1997. *The Embryonic Development of* Drosophila melanogaster. Springer-Verlag, New York.

Carroll, S. B., S. D. Weatherbee and J. A. Langeland. 1995. Homeotic genes and the regulation of insect wing number. Nature 375: 58–61.

Chapela, I. H., S. R. Rehner, T. R. Schultz and U. G. Mueller. 1994. Evolutionary history of the symbiosis between fungus-growing ants and their fungi. Science 266: 1691–1697.

Chapman, R. F. 1998. *The Insects*, 4th Ed. Cambridge University Press, Cambridge. [One of the best references on general insect anatomy and biology.]

Cheng, L. 1976. *Marine Insects*. North-Holland, Amsterdam/American Elsevier, New York.

Chu, H. F. 1949. *How to Know the Immature Insects*. Wm. C. Brown, Dubuque, IA. [Greatly in need of revision.]

Clements, A. N. 1992, 1999. *The Biology of Mosquitoes*. Vols. 1 and 2. Chapman and Hall, New York.

Cook, O. F. 1913. Web-spinning fly larvae in Guatemalan caves. J. Wash. Acad. Sci. 3(7): 190–193.

Crosland, M. W. J. and R. Crozier. 1986. *Myrmecia pilosula*, an ant with only one pair of chromosomes. Science 231: 1278–1284.

CSIRO. 1991. The Insects of Australia: A Textbook for Students and Research Workers, Vols. 1 and 2. 2nd Ed. Cornell University Press, Ithaca, NY. [An outstanding treatment of Australian biodiversity.]

Davey, K. G. 1965. *Reproduction in the Insects*. Oliver and Boyd, Edinburgh.

Denno, R. F. and H. Dingle (eds.). 1979. *Insect Life History Patterns: Habits and Geographic Variation*. Springer-Verlag, New York.

Dethier, V. G. 1963. *The Physiology of Insect Senses*. Methuen, London.

Douglass, J. K. and N. J. Strausfeld. 1995. Visual motion detection circuits in flies: Peripheral motion computation by identified small field retinotopic neurons. J. Neurosci. 15: 5596–5611.

Douglass, J. K. and N. J. Strausfeld. 1996. Visual motion detection circuits in flies: Parallel direction- and non-direction-sensitive pathways between the medulla and lobula plate. J. Neurosci. 16: 4551–4562.

Dyer, F. C. and J. L. Gould. 1983. Honey bee navigation. Am. Sci. 71: 587–597.

Ellington, C. P. 1984. The aerodynamics of flapping animal flight. Am. Zool. 24: 95–105.

Erwin, T. L. 1982. Tropical forests: their richness in Coleoptera and other arthropod species. Coleopterists Bull. 36(1): 74–75.

Erwin, T. L. 1983. Tropical forest canopies, the last biotic frontier. Bull. Ent. Soc. Am. 29(1): 14–19.

Erwin, T. L. 1985. The taxon pulse: A general pattern of lineage radiation and extinction among carabid beetles. *In* G. E. Ball (ed.), *Taxonomy, Phylogeny and Zoogeography of Beetles and Ants: A Volume Dedicated to the Memory of Philip Jackson Darlington, Jr., 1904-1983*. W. Junk, Publ., The Hague.

Erwin, T. L. 1991. An evolutionary basis for conservation strategies. Science 253: 750–752.

Erwin, T. L. 1991. How many species are there? Revisited. Cons. Biol. 5(3): 330–333.

Evans, H. E. 1984. *Insect Biology: A Textbook of Entomology*. Addison-Wesley, Reading, MA.

Evans, P. D. and V. B. Wigglesworth (ed.). 1986. *Advances in Insect Physiology*, Vol. 19. University of Cambridge, Cambridge.

Fent, K. and R. Wehner. 1985. Ocelli: A celestial compass in the desert ant *Cataglyphis*. Science 228: 192–194.

Fletcher, D. J. C. and M. Blum. 1983. Regulation of queen number by workers in colonies of social insects. Science 219: 312–314.

Gilmour, D. 1965. *The Metabolism of Insects*. Oliver and Boyd, Edinburgh.

Glassberg, J. 2000. *Butterflies through Binoculars: The West*. Oxford University Press, Oxford. [There are several companion volumes, by the same publisher, for the East Coast, Africa, India, etc.]

Gotwald, W. H., Jr. 1996. *Army Ants: The Biology of Social Predation*. Comstock Books, Ithaca, NY.

Gould, J. L. 1976. The dance-language controversy. Q. Rev. Biol. 51: 211–243.

Gould, J. L. 1985. How bees remember flower shapes. Science 227: 1492–1494.

Gould, J. L. 1986. The locale map of honey bees: Do insects have cognitive maps? Science 232: 861–863.

Grenacher, H. 1879. Untersuchungen über das Sehorgan der Arthropoden, insbesondere der Spinnen, Insecten und Crustaceen. Vandenhoek and Ruprecht, Göttingen. [Probably the first detailed description of compound eyes in insects and crustaceans, noting their homology.]

Gwynne, D. T. 2001. Katydids and Bush-Crickets: Reproductive Behavior and Evolution of the Tettigonidae. Comstock Publishing Associates, Ithaca, NY.

Hardie, R. J. (ed.) 1999. Pheromones of Non-Lepidopteran Insects Associated with Agricultural Plants. Oxford University Press, Oxford.

Heinrich, B. 1993. *The Hot-Blooded Insects: Strategies and Mechanisms of Thermoregulation*. Harvard University Press, Cambridge, MA.

Hermann, H. R. (ed.). 1984. *Defense Mechanisms in Social Insects*. Praeger, New York.

Herreid, C. F. II and C. R. Fourtner. 1981. *Locomotion and Energetics in Arthropods*. Plenum, New York.

Hinton, H. E. 1981. *Biology of Insects Eggs*, Vols. 1–3. Pergamon Press, Elmsford, NY. [A benchmark study of insect egg morphology and biology; no embryology.]

Hodgson, C. J. 1994. *The Scale Insect Family Coccidae*. Oxford University Press, Oxford.

Hölldobler, B. 1971. Communication between ants and their guests. Sci. Am. 224(3): 86–93.

Hölldobler, B. and E. O. Wilson. 1990. *The Ants*. Belknap Press, Cambridge, MA.

Holt, V. M. 1973. *Why Not Eat Insects?* Reprinted from the original (1885) by E. W. Classey Ltd., 353 Hanworth Rd., Hampton, Middlesex, England. [Ninety-nine pages of fun and recipes.]

Huffaker, C. B. and R. L. Rabb (eds.). 1984. *Ecological Entomology*. Wiley-Interscience, New York.

Jenkin, P. M. 1966. Apolysis and hormones in the moulting cycles of Arthropoda. Ann. Endocrinol. 27: 331–341.

Kerkut, G. A. and L. I. Gilbert. 1984. *Comprehensive Insect Physiology, Biochemistry and Pharmacology*. Pergamon Press, Elmsford, NY. [In 13 volumes.]

Kettlewell, H. B. D. 1961. The phenomenon of industrial melanism in Lepidoptera. Annu. Rev. Entomol. 6: 245–262.

Klass, K.-D., O. Zompro, N. P. Kristensen and J. Adis. 2002. Mantophasmatodea: A new insect order with extant members in the Afrotropics. Science 296: 1456–1459.

Lehmkuhl, D. M. 1979. *How to Know the Aquatic Insects*. Wm. C. Brown, Dubuque, IA.

Lewis, T. (ed.). 1984. *Insect Communication*. Academic Press, Orlando, FL. [A definitive review as of the early 1980s.]

Locke, M. and D. S. Smith (eds.). 1980. *Insect Biology in the Future*. Academic Press, Orlando, FL.

Matheson, A., S. L. Buchmann, C. O'Toole, P. Westrich and I. H. Williams (eds.) 1996. *The Conservation of Bees*. Academic Press, Harcourt Brace, London.

Matsuda, R. 1965. Morphology and evolution of the insect head. Memoirs of the American Entomological Institute, no. 4. American Entomological Institute, Ann Arbor, MI.

Matsuda, R. 1970. Morphology and evolution of the insect thorax. Memoirs of the Entomological Society of Canada, no. 76. Entomological Society of Canada, Ottawa.

Matsuda, R. 1976. Morphology and Evolution of the Insect Abdomen. Pergamon Press, Oxford.

Merritt, R. W. and K. W. Cummins (eds.). 1978. *An Introduction to the Aquatic Insects of North America*. Kendall/Hunt, Dubuque, IA. [Excellent keys to most North American groups.]

Michelsen, A. 1979. Insect ears as mechanical systems. Am. Sci. 67: 696–706.

Michener, C. D. 1974. *The Social Behavior of the Bees: A Comparative Study*. Belknap Press/Harvard University Press, Cambridge, MA.

Nieh, J. C. 1999. Stingless-bee communication. Am. Sci. 87: 428–435.

Pearce, M. J. 1998. *Termites: Biology and Pest Management*. Oxford University Press, Oxford.

Phelan, P. L. and T. C. Baker. 1987. Evolution of male pheromones in moths: Reproductive isolation through sexual selection? Science 235: 205–207.

Poodry, C. A. 1992. Morphogenesis of *Drosophila*. *In* E. Rossomando and S. Alexander (eds.), *Morphogenesis*. Marcel Dekker, New York, pp. 143–188.

Prestwick, G. D. 1987. Chemistry of pheromone and hormone metabolism in insects. Science 238: 999–1006.

Price, P. W. et al. (eds.) 1991. *Plant–Animal Interactions. Evolutionary Ecology in Temperate and Tropical Regions*. Wiley-Interscience, New York.

Resh, V. and D. Rosenberg (eds.). 1984. *The Ecology of Aquatic Insects*. Praeger, New York.

Robinson, G. E. 1985. The dance language of the honey bee: The controversy and its resolution. Am. Bee J. 126: 184–189.

Rockstein, M. (ed.). 1964–65. *The Physiology of Insecta*, Vols. 1–3. Academic Press, New York.

Roeder, K. D. 1963. *Nerve Cells and Insect Behaviour*. Harvard University Press, Cambridge, MA.

Rosin, R. 1984. Further analysis of the honey bee "Dance Language" controversy. I. Presumed proofs for the "Dance Language" hypothesis by Soviet scientists. J. Theor. Biol. 107: 417–442.

Rosin, R. 1988. Do honey bees still have a "dance language"? Am. Bee J. 128: 267–268.

Ross, K. G. and R. W. Matthews (eds.). 1991. *The Social Biology of Wasps*. Comstock Books, Ithaca, NY.

Saunders, D. S. 1982. *Insect Clocks*, 2nd Ed. Pergamon Press, Elmsford, NY. [Good introduction to photoperiodism.]

Schmitt, J. B. 1962. The comparative anatomy of the insect nervous system. Annu. Rev. Entomol. 7: 137–156.

Schuh, R. T. and J. A. Slater. 1995. *True Bugs of the World* (*Hemiptera: Heteroptera*). Comstock Books, Ithaca, NY.

Schwalm, F. E. 1988. *Insect Morphogenesis*. Monographs in Developmental Biology, Vol. 20. Karger, Basel.

Scott, J. A. 1986. The Butterflies of North America: A Natural History and Field Guide. Stanford University Press, Stanford, CA.

Shorrocks, B. 1980. *Drosophila*. Pergamon Press, Elmsford, NY. [Everything you ever wanted to know about *Drosophila*, and more.]

Snodgrass, R. E. 1935. *Principles of Insect Morphology*. McGraw-Hill, New York. [An early classic; still useful.]

Snodgrass, R. E. 1944. The feeding apparatus of biting and sucking insects affecting man and animals. Smithsonian Miscellaneous Collections, Vol. 104, no. 7. Smithsonian Institution, Washington, DC.

Snodgrass, R. E. 1952. *A Textbook of Arthropod Anatomy*. Cornell University Press, Ithaca, NY. [Generations of subsequent books and reports have relied heavily on the information and figures contained in this work and Snodgrass's 1935 text.]

Somps, C. and M. Luttges. 1985. Dragonfly flight: Novel uses of unsteady separated flows. Science 228: 1326–1329.

Strausfeld, N. J. 1976. *Atlas of an Insect Brain*. Springer, Heidelberg.

Strausfeld, N. J. 1996. Oculomotor control in flies: From muscles to elementary motion detectors. *In* P. S. G. Stein and D. Stuart (eds.), *Neurons, Networks, and Motor Behavior*. Oxford University Press, Oxford, pp. 277–284.

Strausfeld, N. J. and J.-K. Lee. 1991. Neuronal basis for parallel visual processing in the fly. Visual Neurosci. 7: 13–33.

Stubs, C. and F. Drummond (eds.). 2001. *Bees and Crop Pollination: Crisis, Crossroads, Conservation*. Entomological Society of America.

Tauber, M. J., C. A. Tauber and S. Masaki. 1986. *Seasonal Adaptations of Insects*. Oxford University Press, New York. [A comprehensive treatment of insect life cycles.]

Tilgner, E. 2002. Mantophasmatodea: A new insect order? Science 297: 731a.

Treherne, J. E. and J. W. L. Beament (eds.). 1965. *The Physiology of the Insect Central Nervous System*. Academic Press, New York.

Treherne, J. E., M. J. Berridge and V. B. Wigglesworth. 1963–1985. *Advances in Insect Physiology*, Vols. 1–18. Academic Press, New York.

Unarov, B. P. 1966. *Grasshoppers and Locusts*. Cambridge University Press, Cambridge.

Urquhart, F. 1960. *The Monarch Butterfly*. University of Toronto Press, Toronto.

Usinger, R. L. (ed.). 1968. *Aquatic Insects of California, with Keys to North American Genera and California Species*. University of California Press, Berkeley.

Vane-Wright, R. I. and P. R. Ackery. 1989. *The Biology of Butterflies*. University of Chicago Press, Chicago.

Veldink, C. 1989. The honey-bee language controversy. Interdiscip. Sci. Rev. 14(2): 170–175.

von Frisch, K. 1967. *The Dance Language and Orientation of Bees*. Translated by Leigh E. Chadwick. Belknap Press, Cambridge, MA.

Weis-Fogh, T. and M. Jensen. 1956. Biology and physics of locust flight. I. Basic principles in insect flight: A critical review. Phil. Trans. R. Soc. Lond. Ser. B 239: 415–458.

Wenner, A. M. and P. H. Wells. 1987. The honey bee dance language controversy: The search for "truth" vs. the search for useful information. Am. Bee J. 127: 130–131.

Wigglesworth, V. B. 1954. *The Physiology of Insect Metamorphosis*. Cambridge University Press, Cambridge. [A long-standing classic; dated but still useful.]

Wigglesworth, V. B. 1984. *Insect Physiology*, 8th Ed. Chapman and Hall, London.

Williams, C. B. 1958. *Insect Migration*. Collins, London.

Wilson, D. M. 1966. Insect walking. Annu. Rev. Entomol. 11: 103–122.

Wilson, E. O. 1963. The social biology of ants. Annu. Rev. Entomol. 8: 345–368.

Wilson, E. O. 1971. *The Insect Societies*. Harvard University Press, Cambridge, MA.

Wilson, E. O. 1975. Slavery in ants. Sci. Am. 232(6): 32–36.

Winston, M. L. 1987. *The Biology of the Honey Bee*. Harvard University Press, Cambridge, MA.

Winston, M. L. 1992. *Killer Bees*. Harvard University Press, Cambridge, MA.

Hexapod Evolution

Andersson, M. 1984. The evolution of eusociality. Annu. Rev. Ecol. Syst. 15: 165–189.

Bitsch, C. and J. Bitsch. 2000. The phylogenetic interrelationships of the higher taxa of apterygote hexapods. Zool. Scripta 29: 131–156.

Deuve, T. (ed.) 2001. Origin of the Hexapoda. Ann. Soc. Entomol. France 37 (1/2): 1–304. [Papers presented at Conference held in Paris, 1999; see review in Brusca, R. C., 2001, J. Crustacean Biol. 21(4): 1084–1086.]

Diaz-Benjumea, F. J., B. Cohen and S. M. Cohen. 1994. Cell interaction between compartments established the proximal–distal axis of *Drosophila* legs. Nature 372: 175–179.

Dohle, W. 1997. Myriapod–insect relationships as opposed to an insect–crustacean sister group relationship. *In* R. A. Fortey and R. H. Thomas (eds.), *Arthropod Relationships*. Chapman and Hall, London, pp. 305–315.

Dohle, W. 2001. Are the insects terrestrial crustaceans? Ann. Soc. Entomol. France 37: 85–103.

Douglas, M. M. 1980. Thermoregulatory significance of thoracic lobes in the evolution of insect wings. Science 211: 84–86.

Futuyma, D. J. and M. Slatkin (eds.). 1983. *Coevolution*. Sinauer Associates, Sunderland, MA.

Gaunt, M. W. and M. A. Miles. 2002. An insect molecular clock dates the origin of the insects and accords with paleontological and biogeographic landmarks. Mol. Biol. Evol. 19 (5): 748–761.

Goodchild, A. J. P. 1966. Evolution of the alimentary canal in the Hemiptera. Biol. Rev. 41: 97–140.

Gupta, A. P. (ed.). 1979. *Arthropod Phylogeny*. Van Nostrand Reinhold, New York. [Contributed chapters, mostly on insects.]

Harzsch, S., K. Anger and R. R. Dewers. 1997. Immunocytochemical detection of acetylated alpha-tubulin and *Drosophila* synapsin in the embryonic crustacean nervous system. Int. J. Dev. Biol. 41: 411–494.

Hennig, W. 1981. *Insect Phylogeny*. Wiley, New York.

Hoy, R. R., A. Hoikkala and K. Kaneshiro. 1988. Hawaiian courtship songs: Evolutionary innovation in communication signals of *Drosophila*. Science 240: 217–220.

Kristensen, N. P. 1981. Phylogeny of insect orders. Annu. Rev. Entomol. 26: 135–157.

Kristensen, N. P. 1991. Phylogeny of extant hexapods. *In* CSIRO, *The Insects of Australia: A Textbook for Students and Research Workers*, 2nd Ed., Vol. 1. Cornell University Press, Ithaca, NY, pp. 125–140.

Kukalová-Peck, J. 1983. Origin of the insect wing and wing articulation from the insect leg. Can. J. Zool. 61: 1618–1669.

Kukalová-Peck, J. 1987. New Carboniferous Diplura, Monura, and Thysanura, the hexapod ground plan, and the role of thoracic side lobes in the origin of wings (Insecta). Can. J. Zool. 65: 2327–2345.

Kukalová-Peck, J. 1991a. The "Uniramia" do not exist: The ground plan of the Pterygota as revealed by Permian Diaphanopterodea from Russia (Insecta: Paleodictyopteroidea). Can. J. Zool. 70: 236–255.

Kukalová-Peck, J. 1991b. Fossil history and the evolution of hexapod structures. *In* CSIRO, *The Insects of Australia: A Textbook for Students and Research Workers*, 2nd Ed., Vol. 1. Cornell University Press, Ithaca, NY, pp. 141–179.

Kukalová-Peck, J. and C. Brauckmann. 1990. Wing folding in pterygote insects, and the oldest Diaphanopteroda from the early Late Carboniferous of West Germany. Can. J. Zool. 68: 1104–1111.

Labandeira, C., B. Beall and F. Hueber. 1988. Early insect diversification: Evidence from a lower Devonian bristletail from Quebec. Science 242: 913–916.

Nilsson, D.-E. and D. Osorio. 1997. Homology and parallelism in arthropod sensory processing. *In* R. A. Fortey and R. H. Thomas, *Arthropod Relationships*. Chapman and Hall, London, pp. 333–347.

Osorio, D. and J. P. Bacon. 1994. A good eye for arthropod evolution. BioEssays 16: 419–424.

Panganiban, G., A. Sebring, L. Nagy and S. Carroll. 1995. The development of crustacean limbs and the evolution of arthropods. Science 270: 1363–1366.

Panganiban, G. et al. 1997. The origin and evolution of animal appendages. Proc. Natl. Acad. Sci. U.S.A. 94: 5162–5166.

Regier, J. C. and J. W. Shultz. 1997. Molecular phylogeny of the major arthropod groups indicates polyphyly of crustaceans and a new hypothesis of the origin of hexapods. Mol. Biol. Evol. 14: 902–913.

Robertson, R. M., K. G. Pearson and H. Reichert. 1981. Flight interneurons in the locust and the origin of insect wings. Science 217: 177–179.

Sander, K. 1994. The evolution of insect patterning mechanisms: A survey. Development (Suppl.), 187–191.

Schwan, F. E. 1997. Arthropods: The insects. *In* S. F. Gilbert and A. M. Raunio, *Embryology: Constructing the Organisms*. Sinauer Associates, Sunderland, MA, pp. 259–278.

Snodgrass, R. E. 1960. Facts and theories concerning the insect head. Smithsonian Miscellaneous Collections, Vol. 152, no. 1. Smithsonian Institution, Washington, DC.

Strausfeld, N. J. 1998. Crustacean–insect relationships: The use of brain characters to derive phylogeny amongst segmented invertebrates. Brain Behav. Evol. 52: 186–206.

Strausfeld, N. J., E. K. Bushbeck and R. S. Gomez. 1995. The arthropod mushroom body: Its roles, evolutionary enigmas and mistaken identities. *In* O. Breidbach and W. Kutsch (eds.), *The Nervous Systems of Invertebrates: An Evolutionary and Comparative Approach*. Birkhäuser Verlag, Basel, pp. 349–381.

Tautz, D., M. Friedrich and R. Schröder. 1994. Insect embryogenesis: What is ancestral and what is derived. Development (Suppl.), 193–199.

Therianos, S., S. Leuzinger, F. Hirth, C. S. Goodman and H. Reichert. 1995. Embryonic development of the *Drosophila* brain: Formation of commissural and descending pathways. Development 121: 3849–3860.

Thomas, J. B., M. J. Bastiani and C. S. Goodman. 1984. From grasshopper to *Drosophila*: A common plan for neuronal development. Nature 310: 203–207.

Thornhill, R. and J. Alcock. 1983. *The Evolution of Insect Mating Systems*. Harvard University Press, Cambridge, MA.

Wheeler, W. C., M. Whiting, Q. D. Wheeler and J. N. Carpenter. Zool. The phylogeny of the extant hexapod orders. Cladistics 17: 113–169.

Whiting, M. F. 2001. Phylogeny of the holometabolous insect orders: molecular evidence. Zool. Scripta. 31: 3–15.

Whiting, M. F., J. C. Carpenter, Q. D. Wheeler and W. C. Wheeler. 1997. The Strepsiptera problem: Phylogeny of the holometabolous insect orders inferred from 18S and 28S ribosomal DNA sequences and morphology. Syst. Biol. 46(1): 1–68.

Wilson, E. O. 1985a. Invasion and extinction in the West Indian ant fauna: Evidence from the Dominican amber. Science 229: 265–267.

Wilson, E. O. 1985b. The sociogenesis of insect colonies. Science 228: 1489–1425.

Wootton, R. J., J. Kukalová-Peck, D. J. S. Newman and J. Muzón. 1998. Smart engineering in the mid-Carboniferous: How well could Palaeozoic dragonflies fly? Science 282: 749–751.

18

Phylum Arthropoda:
The Myriapods (Centipedes, Millipedes, and Their Kin)

It inhabits dwellings and appears at night. So rapidly can it travel that one often just gets a glimpse of it as it traverses a floor or wall. It is so fragile that no damage can be done by it and much good is gained through its capture of flies, cockroaches, clothes moths, and other household pests.

Edward Essig,
(speaking of the common house centipede, *Scutigera)*

The arthropod subphylum Myriapoda includes four groups, or classes: Chilopoda (centipedes), Diplopoda (millipedes), Symphyla (symphylans), and Pauropoda (pauropodans). All modern myriapods are terrestrial, but their ancestry probably lies in the marine realm. The first fossil records of millipedes are from the late Ordovician/early Silurian, and some of these are thought to represent marine species. Fossil evidence suggests that myriapods (millipedes) did not make their first appearance on land until the mid-Silurian. Figure 18.1 illustrates a variety of myriapod types. About 11,460 living species have so far been described.

Almost everyone is familiar with centipedes and millipedes, which, despite their diversity, conform to a basic bauplan and external appearance. Myriapods are quickly distinguished by a body divided into just two tagmata, the cephalon and the long, homonomous, many-segmented trunk. As in the Hexapoda, there are just four pairs of cephalic appendages: antennae, mandibles, first maxillae, and second maxillae.

Millipedes are particular favorites with many people, and their harmless antics have provided safe entertainment and lessons in biology for generations of students of all ages. Millipedes are slow-moving detritivores that spend their time burrowing through soil and litter, consuming plant remains and converting vegetable matter into humus. In tropical environments, where earthworms are often scarce, millipedes may be the major soil-forming animals. Their trunk segments are actually pairs of two fused segments (called **diplosegments**) and thus bear two pairs of legs (Figure 18.2F). It is thought that the diplosegment condition evolved in conjunction with burrowing habits, as the pushing force of the legs was more efficient when alternating body joints were made rigid and incompressible. When threatened, many millipedes roll up into a flat coil, and some can

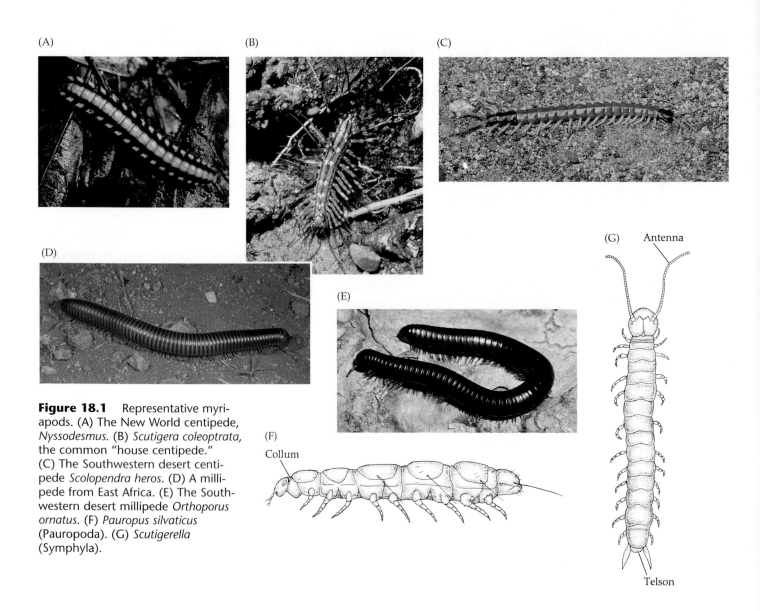

(A)

(B)

(C)

(D)

(E)

(F)

Collum

(G) Antenna

Telson

Figure 18.1 Representative myriapods. (A) The New World centipede, *Nyssodesmus*. (B) *Scutigera coleoptrata*, the common "house centipede." (C) The Southwestern desert centipede *Scolopendra heros*. (D) A millipede from East Africa. (E) The Southwestern desert millipede *Orthoporus ornatus*. (F) *Pauropus silvaticus* (Pauropoda). (G) *Scutigerella* (Symphyla).

roll up into perfect balls, like pillbugs (oniscid isopods). Millipedes typically have many trunk segments, but some African species have only a few segments and are often mistaken for pillbugs (especially when they roll up). Although popular myth says millipedes have a thousand legs (*milli–pedes*), none do, although the record holder (*Illacme plenipes*, a California species), with 375 pairs of legs, is still impressive.

Even though millipedes are docile creatures, many of their diplosegments bear lateral **repugnatorial glands** that secrete volatile toxic liquids used for defense (Figure 18.2F). The defensive chemicals of this group are surprisingly diverse, and they include quinones, phenols, and even hydrogen cyanide produced by the common flat-backed millipedes of the order Polydesmida. A few tropical species have toxins powerful enough to raise blisters on the skin of humans. The European species *Glomeris marginata* produces quinazolinones, which belong to the same class of substances as the synthetic drug Quaalude, a powerful sedative (attacking wolf spiders exposed to

these defensive chemicals simply go to sleep). One common household species in North America (introduced long ago from Asia), *Oxidus gracilis*, releases strong-smelling defensive chemicals when injured. Members of the subclass Penicillata have evolved a clever defense strategy of throwing stiff bristles from their posterior end at ants and other predators. Given their range of defensive tactics, it is not surprising that many millipedes have aposematic (warning) coloration, usually bright reds, yellows, and oranges, and that some soil-dwelling vertebrates (lizards and worm snakes) have evolved look-alike coloration (Batesian mimicry). Species of the California genus *Motyxia* (Polydesmida) even utilize bioluminescence, possibly for warning away predators.

In contrast to millipedes, most centipedes are fast-running predators, with one pair of legs per segment, and with poison claws (= fangs) on the first legs. They never roll into a coil when threatened; instead, they usually strike out with the poison claws with which they can inflict a painful bite. The bite of the American giant desert centipede, *Scolopendra heros*, which reaches 8 inches in length, is known to cause local necrosis (similar to the sting of the bark scorpion, *Centruroides*). *Scolopendra* also has warning coloration—black and red banding (Figure 18.1C). One remarkable species, the appropriately named *Henia vesuviana*, secretes copious amounts of proteinaceous glue in response to attack from predators. The glue hardens within a few seconds of exposure to air and immobilizes even the largest insect predators. Some lithobiomorphs bear large numbers of unicellular repugnatorial glands on the last four pairs of legs, which they kick in the direction of an enemy—a behavior that throws out sticky droplets of noxious secretions.

Myriapod Classification

SUBPHYLUM MYRIAPODA

CLASS DIPLOPODA

  SUBCLASS PENICILLATA

    ORDER POLYXENIDA

  SUBCLASS CHILOGNATHA

    ORDER CALLIPODIDA

    ORDER CHORDEUMATIDA

    ORDER GLOMERIDA

    ORDER GLOMERIDESMIDA

    ORDER JULIDA

    ORDER PLATYDESMIDA

    ORDER POLYDESMIDA

    ORDER POLYZONIIDA

    ORDER SIPHONOPHORIDA

    ORDER SIPHONIULIDA

    ORDER SPHAEROTHERIIDA

    ORDER SPIROBOLIDA

    ORDER SPIROSTREPTIDA

    ORDER STEMMIULIDA

CLASS CHILOPODA

  SUBCLASS NOTOSTIGMOPHORA

    ORDER SCUTIGEROMORPHA

  SUBCLASS PLEUROSTIGMOPHORA

    ORDER CRATEROSTIGMOMORPHA

    ORDER GEOPHILOMORPHA

    ORDER LITHOBIOMORPHA

    ORDER SCOLOPENDROMORPHA

CLASS PAUROPODA

CLASS SYMPHYLA

Synopses of Myriapod Taxa

Subphylum Myriapoda

One pair of antennae; mandibles with articulating endite; first maxillae free or fused; second maxillae partly (or wholly) fused, or absent; postcephalic segments numerous, constituting an undifferentiated trunk; all appendages of head and trunk uniramous; living species apparently without compound eyes; exoskeleton without well-developed wax layer (except in some desert species); many with lateral repugnatorial glands on trunk segments; without entodermally derived digestive ceca; ectodermally derived (proctodeal) Malpighian tubules assist in excretion; with organs of Tömösvary; copulation indirect; development direct.

Class Diplopoda

Trunk segments fused into pairs called diplosegments; most diplosegments with two pairs of legs, spiracles, ganglia, and heart ostia; each diplosegment with one tergite, two pleura, and one to three sternites; anteriormost segments often with legs suppressed except for some internal musculature; first trunk segment legless, modified as collum; antennae simple, 7-jointed; first maxillae fused into a gnathochilarium; second maxillae absent; gonopores open anteriorly, on or near coxae of second pair of legs (third trunk segment); cuticle usually calcified; many capable of rolling into a tight coil; spiracles located ventrally, typically in front of leg coxae, and never valvular (cannot be closed); gonopores open in anterior region of body; with 11 to 192 leg-bearing body diplosegments (the first and last diplosegments are always legless, and diplosegments 2–4 have one pair of legs each); legs ventrally positioned and short, such that the body is carried close to the ground. Although millipedes lack claws or poisons, many have lateral repugnatorial glands on the trunk segments that secrete noxious chemicals that can irritate the skin and eyes. About 8,000 species of millipedes have been described.

Millipede classification is unstable, with many families of uncertain validity and relationship. The subclass Penicillata contains soft-bodied millipedes, without calcification in the exoskeleton, which is covered by tufts of bristles; males lack copulatory appendages, and reproduction occurs without contact between the sexes. The subclass Chilognatha contains millipedes with a hard, calcified exoskeleton, only scattered setae, and male reproductive appendages; reproduction requires contact between the sexes. Glomeridesmids are flattened millipedes with 22 diplosegments that cannot roll into a ball. Glomerids and sphaerotheriids are short, round, 12- or 13-segmented millipedes that can roll into a perfect ball. The most colorful millipedes are the polydesmids, whose bright red, orange, and blue pigmentations warn of their cyanide defensive secretions.

Class Pauropoda

Nearly microscopic (0.5–1.5 mm); eyeless; 9–11 pairs of legs; some trunk segments partly fused (but not as true diplosegments), plus free telson; without limbs on first trunk segment; mouthparts poorly developed; first maxillae fused into a gnathochilarium; second maxillae absent; antennae branched; most without tracheal or circulatory systems; gonopores on third trunk segment; terga often large and extended over two segments; cuticle soft, uncalcified. Although found in all parts of the world, pauropodans occur mainly in moist soils and woodland litter; they are not common. About 500 species have been described.

Class Chilopoda

With numerous unfused trunk segments, each with one pair of legs, the first pair modified as large poison claws (called prehensors, or forcipules) and held under the head like mouthparts (also sometimes referred to as "maxillipeds"); antennae simple, of varying segmentation; both pairs of maxillae may be medially coalesced; cuticle stiff but uncalcified; gonopores on last true body segment; spiracles lateral or middorsal (but on the pleural side walls), and in some species valvular (can be closed); with 15 to 193 leg-bearing body segments; although the legs are long, they extend laterally such that the body is carried close to the ground; last pair of legs extend backwards and not used for locomotion. Centipedes may be the only animals with legs modified into poison-injecting claws.

The subclass Notostigomorpha have middorsal spiracles, hemocyanin in the hemolymph, a dome-shaped head, pseudo-faceted eyes (ocelli), and 15 pairs of long, thin legs. The subclass Pleurostigmophora have lateral spiracles, a flattened head, a variable number of legs, and no respiratory pigments. The largest centipedes are the scolopendromorphans (with either 21 or 23 pairs of legs), although the highest leg count occurs in the long, thin geophilomorphs (with 27 or more pairs of legs).

Approximately 2,800 living species of chilopods have been described.

Class Symphyla

Small (0.5–8.0 mm); eyeless; trunk with 14 segments, last fused to telson; first 12 trunk segments each with a pair of legs; penultimate segment with spinnerets and a pair of long sensory hairs; dorsal surface with 15–22 tergal plates; soft uncalcified cuticle; antennae long, simple, threadlike; first maxillae medially coalesced; second maxillae completely fused as complex labium; one pair of spiracles (on head); tracheae supply first 3 trunk segments; gonopores open on third trunk segment. Symphylans are generally uncommon arthropods that occur in soil and rotting vegetation. About 160 species have been described in two families.

The Myriapod Bauplan

In Chapter 15 we discussed the various advantages and constraints imposed by arthropodization, and in Chapter 17 we discussed the suitability of many features of the arthropod bauplan for adaptation to a terrestrial lifestyle. The myriapods share many of these adaptations with the hexapods, although perhaps as a result of convergent of parallel evolution.

The myriapods differ from the hexapods in their retention of a more homonomous condition: the long trunk with its paired segmental appendages. However, the head appendages and legs are very similar to those of the insects, and it was for this reason that the two groups were long considered sister taxa. Metamerism is strong in myriapods and is evident internally in structures such as the segmental heart ostia, tracheae, and ganglia. The key features that distinguish the Myriapoda are listed in Box 18A.

The centipedes (chilopods) bear one pair of walking legs per segment, and there may be up to 193 trunk seg-

---

BOX 18A *Characteristics of the Subphylum Myriapoda*

1. Body of two tagmata: head and multisegmented trunk

2. All appendages multiarticulate and uniramous[a]

3. Head appendages, from anterior to posterior, are antennae, mandibles, first maxillae (maxillules), and second maxillae; second maxillae may be fused into a single flaplike structure called a "labium" (not homologous to the crustacean labium), or they may be absent; first and second maxillae often bear palps.

4. Without a carapace

5. With an aerial gas exchange system composed of tracheae and spiracles (probably convergent with those in the Hexapoda)

6. With one or two pairs of ectodermally derived (proctodeal) Malpighian tubules (probably convergent with those in the Hexapoda)

7. Most with simple ocelli, at least in some stage of the life cycle; true ommatidia apparently absent

8. Gut simple, without digestive ceca

9. Dioecious; with direct development

---

[a]Pauropodan antennae are branched, but whether this represents a vestige of a primitive biramous condition (as in crustaceans) or is secondarily derived is not known.

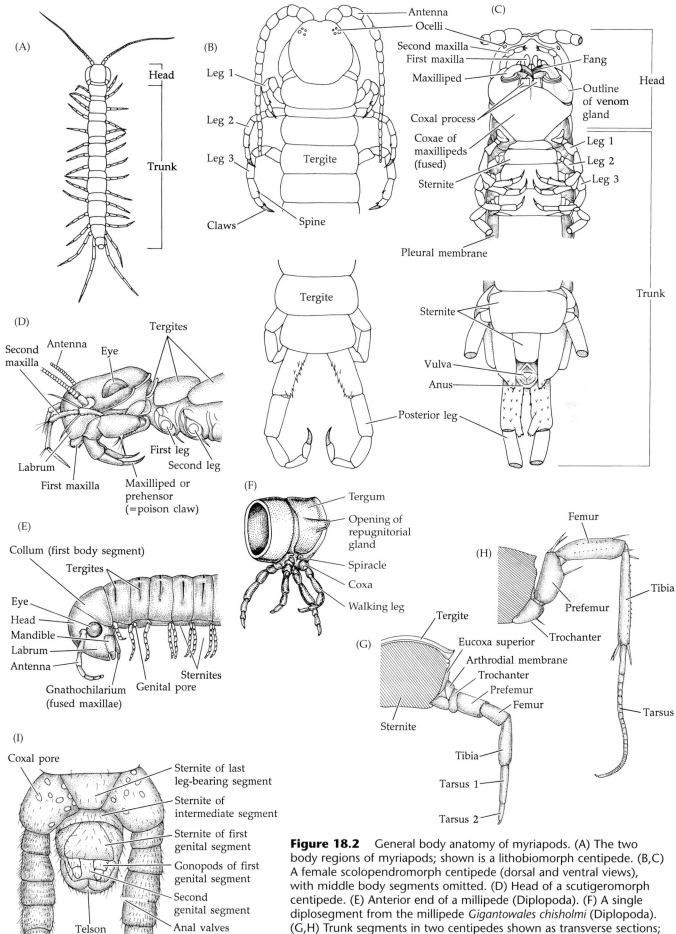

Figure 18.2 General body anatomy of myriapods. (A) The two body regions of myriapods; shown is a lithobiomorph centipede. (B,C) A female scolopendromorph centipede (dorsal and ventral views), with middle body segments omitted. (D) Head of a scutigeromorph centipede. (E) Anterior end of a millipede (Diplopoda). (F) A single diplosegment from the millipede *Gigantowales chisholmi* (Diplopoda). (G,H) Trunk segments in two centipedes shown as transverse sections; note the variation in limb morphology: (G) *Lithobius*. (H) *Scutigera*. (I) Ventral view of the terminal segments of a male geophilomorph centipede (*Strigamia*).

ments (Figure 18.2A–D). Most species are 1–2 cm in length, although some tropical giants attain lengths of nearly 30 cm.

The millipedes (diplopods) have an interesting modification of basic metamerism. All of the trunk segments are actually diplosegments, each formed by fusion of two somites and bearing a double complement of metameric organs and structure (Figure 18.2F) The head is followed by an expanded, limbless diplosegment called the **collum**, which forms a conspicuous, heavily sclerotized collar between head and trunk (Figure 18.2E). Each of the next three diplosegments bears a single pair of legs. All the remaining diplosegments retain two pairs of legs. When the alternate body somites become fused during embryogenesis, the leg pair, spiracle pair, and ventral nerve cord ganglion of the anterior fused somite shifted to the posterior fused somite. Thus, in the resulting diplosegment, the **metazonite** (representing the posterior end of the two fused segments) carries all these structure, whereas the **prozonite** (the anterior somite) lacks them and is free to insert inside the preceding diplosegment in a telescoping fashion (Figure 18.2F). Like centipedes, millipedes range in length from about 1 to 30 cm. The cuticle of millipedes is particularly robust, being well sclerotized and usually calcified.

Pauropodans are minute, soft-bodied, eyeless, soil-inhabiting myriapods that resemble millipedes (Figure 18.1F). They are less than 2 mm in length and have up to 11 trunk segments. As in millipedes, some segmental pairing occurs, and usually only six tergites are visible dorsally.

Symphylans (Figure 18.1G) are also minute, eyeless myriapods, never exceeding 1 cm in length. The trunk has 12 pairs of leg-bearing segments. Some tergites are divided, and 15–22 tergites are usually visible dorsally. The thirteenth body segment bears spinnerets, a pair of long sensory hairs, and a tiny postsegmental telson. Like pauropodans, symphylans inhabit loose soil and humus.

Head and Mouth Appendages

As you will recall from previous chapters, much discussion and evolutionary speculation regarding arthropods focuses on the nature of the head and its appendages. The basic five-segmented (+ acron) crustacean-hexapodan head is retained in the Myriapoda (Figure 18.2). The acron bears the eyes and contains the protocerebrum. The first postacronal cephalic somite (the **antennary segment**) bears the antennae and houses the deutocerebrum. The second segment is the **premandibular segment**; it lacks appendages but houses the tritocerebrum. The third segment (**mandibular segment**) carries the mandibles, and the fourth (**first maxillary segment**) bears the first pair of maxillae (= maxillules). The fifth cephalic segment (**second maxillary segment**) bears the second pair of maxillae. The second maxillae are fused as a "**labium**" in symphylans, but they have been lost altogether in millipedes and pauropodans. In diplopods and pauropodans the first maxillae fuse to form a flaplike **gnathochilarium.** The ganglia of the mandibular and maxillary segments are generally fused to form the subesophageal ganglion. A carapace is never present in myriapods, and there is no evidence that one ever existed.

Like the Crustacea and Hexapoda, the Myriapoda possess two additional head structures: a clypeus–labrum that borders the anterior mouth field, and a hypopharynx that borders the posterior mouth field. The **clypeus** projects off the front of the head and extends down over the mandibles, much as our upper lip covers our teeth. The platelike **labrum** projects off the clypeus. Some workers consider the labrum to be the fused appendages of the acron or premandibular head segment, but most evidence suggests it has been independently derived from the exoskeletal sclerites. However, preliminary research suggests that the developmental gene *Distal-less* is expressed in the ontogeny of the labrum, leaving the matter rather unsettled. Arising behind the first maxillae and near the base of the second maxillae is the flaplike (or tonguelike) **hypopharynx.** The salivary glands open through the hypopharynx. The hypopharynx probably does not represent true appendages, but is an independent outgrowth of the body wall. However, the homology of both the clypeus and the hypopharynx are matters of ongoing debate.

Locomotion

Myriapods rely on their well-sclerotized exoskeleton for body support. In centipedes, the legs are long but extend laterally, keeping the body close to the ground to maintain stability while at the same time providing for long strides and rapid locomotion. In millipedes, the legs arise ventrally and are short, again keeping the body close to the ground while providing for powerful, though slow, locomotion.

The basic design of arthropod limbs was described in Chapter 15. Recall that in most arachnids the coxae are immovably fixed to the body, and limb movement occurs at more distal joints. However, in myriapods (as in most hexapods and crustaceans), anterior–posterior limb movements take place between the coxae and the body proper. As we have learned, the power exerted by a limb is greatest at low speeds and least at high speeds. At lower speeds the legs are in contact with the ground for longer periods of time, and in the case of myriapods *more* legs are contacting the ground at any given moment. Thus, in burrowing forms, such as most millipedes, the legs are short, and the gait is slow and powerful as the animals bulldoze their way through soil or rotting wood. In centipedes, which run at high speeds, less than half the limbs may touch the ground at any given moment, and for shorter periods of time (Figure 18.3). Longer limbs increase the speed of a running gait, and limbs long in length and stride are typical features of the fastest-running centipedes and symphylans. Long, slender, fast moving objects, such as centipedes

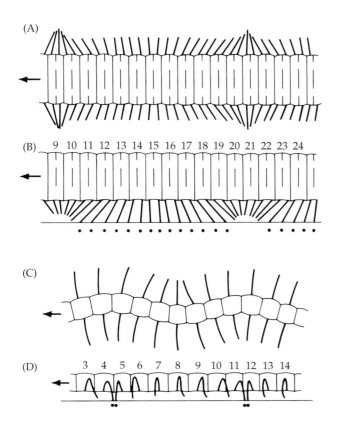

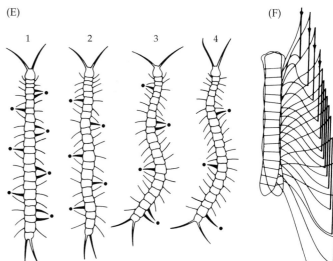

Figure 18.3 Locomotion in millipedes and centipedes. (A,B) A typical millipede (such as *Spirostreptus* or *Gymnostreptus*) in motion. Note the 16 diplosegments (with 32 pairs of legs). The dorsal view (A) shows left and right leg sets exactly in phase with each other. The lateral view (B) of the same animal shows that the majority of the limb tips are on the ground at once, an arrangement yielding a slow but powerful gait. (C,D) A typical centipede (such as *Scolopendra* or *Cryptops*) in motion. Note the 12 segments, each with one pair of legs. The dorsal view (C) shows the limb pairs in opposite phase and the undulations of the body that accentuate stride length. The lateral view (D) of the same animal illustrates that fewer than one-third of the limb tips are on the ground at any one time, an arrangement yielding the short, swift strokes typical of a rapid but weak gait. Arrows indicate the animal's direction of travel; dots are points of leg-tip contact with the substratum. (E) Locomotion in a scolopendrid centipede at various speeds. 1–4 show the body waves and leg actions at increasing speeds. Limbs shown with heavy lines are in their power strokes, with the tips against the substratum (dots); limbs depicted by thin lines are in various stages of their recovery strokes. Notice that at maximum speed the animal is still supported by a tripod stance. (F) Field of leg movements in a running centipede, *Scutigera*. The heavy vertical lines trace the movement of the tips of each leg during the propulsive backstroke. Note the gradual increase in limb length posteriorly that allows for unencumbered overstepping even when the full swing of the limb is used.

(and trains!) tend to develop side-to-side oscillations. Although this may increase stride length in a centipede, it also counters the forward motion and can slow the animal down. Consequently, centipedes have developed anatomical modifications to dampen this lateral body sway, such as tergal fusion at oscillation nodal points (Scutigeromorpha) and shortening body segments (and hence, the body) while retaining the same number of legs (Lithobiomorpha). Among the pauropodans, a range of movement speeds is seen, from species that are "slow plowers" to those that are fast runners.

Locomotor repertoires typically evolve in concert with the overall habits of animals, particularly feeding behaviors. Most centipedes are surface-dwelling predators that must move quickly to capture their prey. *Scutigera* is a small centipede that qualifies as a world-class runner, reaching speeds up to 42 cm/second when in pursuit of its favorite prey, flies. On the other hand, most millipedes are detritivores that burrow through soil, leaf litter, or rotten timber in search of food. Thus millipedes tend to have shorter legs and slower, but more powerful gaits.

Power may also be increased by increasing the number of legs, and thus the number of body segments. In this regard, physical limits are reached when the body achieves a length-to-width ratio that would result in buckling of the trunk. This constraint is partly overcome in millipedes and pauropodans in two principal ways: by increasing the cross-sectional body diameter (becoming fatter) and by uniting segments into pairs. In some species the body wall is further strengthened by having certain trunk segments fused into a solid ring. Also,

when millipedes and pauropodans burrow, they tuck the head ventrally so that the collum is thrust forward, in some cases like a wedge that splits open the soil or humus, and in other cases like the broad blade of a bulldozer (Figure 18.2E). Some millipedes have undergone a habitat reversal to abandon burrowing and seek out cracks and crevices.

One group of centipedes lacks the high-speed modifications of most chilopods. The geophilomorphs burrow, aided by dilation of the body and use of the trunk musculature in a fashion similar to that of earthworms.

This peristalsis-like motion is rare in arthropods because of their rigid exoskeleton. However, geophilomorphs have enlarged areas of flexible cuticle on the sides (pleura) of the body between the tergites and sternites; these enlarged pleura allow them to significantly alter their body diameter. Other centipedes have smaller, flexible pleural areas that allow some degree of lateral undulatory motion. Centipede legs are attached to these flexible regions, an arrangement that increases the range of limb motion in these fast-running surface dwellers. This condition is in marked contrast to that of millipedes, in which the short legs arise from the ventral sternites, limiting their range of movement.

One of the principal problems associated with increased limb length is that the field of movement of one limb may overlap that of adjacent limbs. Potential leg interference is prevented by having limbs of different lengths, so that the tips of adjacent legs move at different distances from the body (Figure 18.3F). In fast-running centipedes the limbs of each succeeding segment are slightly longer than the immediately anterior pair. The legs of most myriapods move in clear metachronal waves from posterior to anterior (Figure 18.3A,B). Unlike most arthropods, millipedes move the two pairs of legs on each diplosegment synchronously (stability is not a problem in these elongate creatures).

Feeding and Digestion

Most centipedes are active, aggressive predators on smaller invertebrates, particularly worms, snails, and other arthropods. Their first trunk appendages form large claws, called **prehensors** or **forcipules**. These raptorial limbs are located ventral to the mouth field (Figure 18.2D), and are used to stab prey and inject poison. The poison is produced in large **venom glands** located in the basal articles of the forcipules. The poison is so effective that large centipedes, such as the tropical *Scolopendra*, can subdue small vertebrates (e.g., frogs, lizards, snakes, mice, small birds). Some centipedes actually rear up on their hindlegs and capture flying insects! The prehensors and second maxillae hold the prey, while the mandibles and first maxillae bite and chew. The bite of even the most dangerous centipede is normally not fatal to people, but the poison can cause a reaction similar to that accompanying a serious wasp or scorpion sting.

The feeding strategy of millipedes is quite different. Most millipedes are slow-moving detritivores with a preference for dead and decaying plant material. They play an important role in the recycling of leaf litter in many parts of the world. Most bite off large pieces of vegetation with their powerful mandibles, mix it with saliva as they chew it, and then swallow it. Some, such as the tropical siphonophorids, are believed to feed on the juices of living plants and fungi. In these groups the labrum, gnathochilarium, and reduced mandibles are modified into a suctorial piercing beak. A few odd groups of millipedes are predaceous and feed as centipedes do, but these are exceptions (one North American species apparently specializes on insect pupae). Millipedes are often the largest animals living in cave ecosystems and, although troglobitic species are typically terrestrial, at least one species lives much of its time submerged in cave streams, feeding on aquatic bacteria.

The feeding biology of pauropodans is not well understood, but most appear to be scavengers. These minute, blind creatures crawl through soil and humus, feeding on fungi and decaying plant and animal matter. Symphylans are primarily herbivores, although a few have adopted a carnivorous, scavenging lifestyle. Many symphylans consume live plants. One species, *Scutigerella immaculata*, is a serious pest in plant nurseries and flower gardens, where it has been reported at densities greater than 90 million per acre.

Like the guts of all arthropods, the long, usually straight myriapod gut is divisible into a stomodeal foregut, entodermal midgut, and proctodeal hindgut (Figure 18.4). There are no digestive ceca branching off the gut. **Salivary glands** (= **mandibular glands**) are associated with one or several of the mouth appendages. The salivary secretions soften and lubricate solid food, and in some species contain enzymes that initiate chemical digestion. The mouth leads inward to a long esophagus, which is sometimes expanded posteriorly as a storage area, or **crop** and **gizzard** (as in most centipedes). The gizzard often contains cuticular spines that help strain large particles from the food entering the midgut, where absorption occurs (Figure 18.4C). As in other arthropods, the gut of myriapods produces a **peritrophic membrane**—a sheet of thin, porous, chitinous material that lines and protects the midgut, and may pull free to envelop and coat food particles as they pass through the gut. The midgut connects to a short, proctodeal hindgut that terminates at the anus.

Circulation and Gas Exchange

The myriapod circulatory system includes a dorsal tubular heart that pumps the hemocoelic fluid (blood) toward the head. The heart narrows anteriorly to a vessel-like aorta, from which blood flows posteriorly through large hemocoelic chambers before returning to the **pericardial sinus** and then back to the heart via paired lateral **ostia**. Circulation is slow, and system pressure is relatively low. In diplopods, the heart bears two pairs of ostia in each diplosegment; one pair of ostia occurs in each segment in the chilopods.

Myriapods rely on a **tracheal system** for gas exchange (Figure 18.5). As explained in Chapter 15, tracheae are extensive tubular invaginations of the body wall opening through the cuticle by pores called **spiracles**. The tracheae originating at one spiracle commonly anastomose with others to form branching networks penetrating most of the body. Spiracles are placed segmentally, although not necessarily on all segments. In chilopods the spiracles are located in the membranous

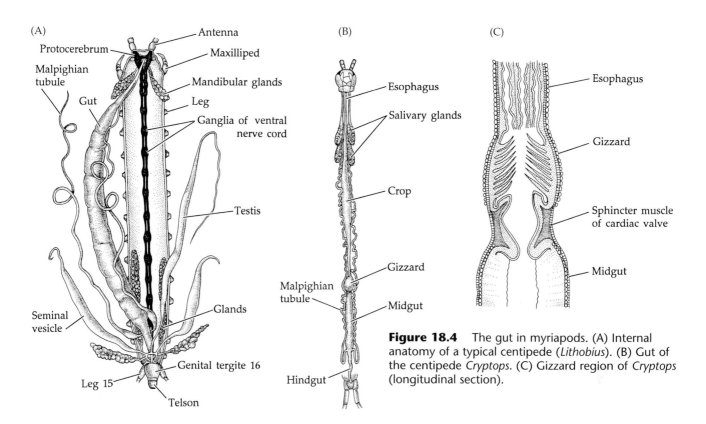

Figure 18.4 The gut in myriapods. (A) Internal anatomy of a typical centipede (*Lithobius*). (B) Gut of the centipede *Cryptops*. (C) Gizzard region of *Cryptops* (longitudinal section).

pleural (side) region just above and behind the base of each leg, or most legs (although in Scutigeromorpha the spiracles are dorsal). Diplopods usually bear two pairs of spiracles per diplosegment, just anterior to the leg coxae, where they open from the sternum and are associated with the apodemes, which also serve as insertions for extrinsic leg muscles. In the symphylans a single pair of spiracles opens on the sides of the head, and the tracheae supply only the first three trunk segments. Except for a few primitive species, most pauropodans lack a tracheal system.

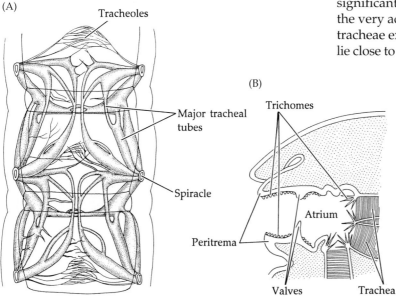

Each spiracle is usually recessed in an **atrium**, whose walls are lined with setae or spines (**trichomes**) that prevent dust, debris, and parasites from entering the tracheal tubes. In myriapods the spiracles are often surrounded by a sclerotized rim or lip, the **peritrema**, which also aids in excluding foreign particles. A muscular valve or other closing device is present in many centipedes and under control of internal partial pressures of O_2 and CO_2. As in insects, ventilation of the tracheal system is accomplished by simple diffusion gradients, as well as by pressure changes induced by the animal's movements. The blood of myriapods appears to play no significant role in oxygen transport, except perhaps in the very active scutigeromorphs. Instead, the air-filled tracheae extend directly to each organ, where the ends lie close to or within the tissues.

The innermost parts of the tracheal system are the **tracheoles**, which are thin-walled, fluid-filled channels that end as a single cell, the **tracheole end cell (= tracheolar cell)**. Unlike tracheae, tracheoles are not shed during ecdysis. The tracheoles are so minute (0.2–1.0 mm) that ven-

Figure 18.5 The tracheal system of centipedes and millipedes. (A) Tracheal system of three body segments of a scolopendromorph centipede, *Scolopendra cingulata*. (B) A spiracle of *S. cingulata* (transverse section).

tilation is impossible and gas transport here relies on aqueous diffusion.

A few geophilomorph centipedes live in intertidal habitats, under rocks or in algal mats. Presumably, air retained within the tracheal system (and in the spiracular atrium) is sufficient to last during submergence at high tides.

Although the tracheal system of myriapods greatly resembles that of insects, there is evidence that it evolved independently. In addition, there is some indication that terrestrial myriapods could have evolved from marine ancestors, not from a terrestrial line that was tied to the Hexapoda.

Excretion and Osmoregulation

When myriapods invaded the land, the problem of water conservation necessitated the evolution of entirely new structures to remove metabolic wastes—Malpighian tubules (Figure 18.4). These excretory organs function much as those of insects do (Chapter 17). Despite their length, however, centipedes and millipedes usually possess only one and two pairs of Malpighian tubules, respectively. In those species that are largely confined to moist habitats and nocturnal activity patterns, a significant portion of the excretory wastes may be ammonia rather than uric acid.

The Malpighian tubules of myriapods were long assumed to be homologous to those of hexapods. However, as we saw in Chapter 15, there is now growing evidence that these two groups do not share an immediate common ancestry. If this is true, then these excretory structures represent another striking case of convergent evolution within the Arthropoda.

The myriapod cuticle is sclerotized and calcified to various degrees, adding a measure of waterproofing, but aside from a few desert species, it lacks the waxy layer seen in hexapods. For this reason, myriapods rely to a considerable extent on behavioral strategies to avoid desiccation. Many live in humid or wet environments, or are active only during cool periods. Other myriapods stay hidden in cool or moist microhabitats, such as under rocks, during hot hours of the day or dry periods.

Nervous System and Sense Organs

The myriapod nervous system conforms to the basic arthropod plan described in Chapter 15. Very little secondary fusion of ganglia occurs, and the ventral nerve cord retains much of its primitive double nature, with a pair of fused ganglia in each segment. Millipedes possess two pairs of fused ganglia in each diplosegment.

Like the "brains" of other arthropods, the cerebral ganglion of myriapods comprises three distinct regions: the protocerebrum (associated with the eyes), the deutocerebrum (associated with the antennae), and the tritocerebrum. The subesophageal ganglion is composed of fused ganglia of the third, fourth, and perhaps the fifth

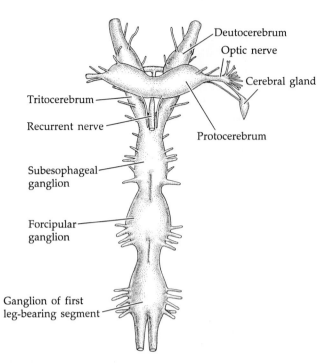

Figure 18.6   The brain and anterior ganglia of a centipede, *Lithobius forficatus* (dorsal view).

head segments and controls the mouthparts, salivary glands, and some other local musculature.

Myriapods typically possess eyes at some stage in their life cycle, but biologists are still debating whether true compound eyes occur in this group. Centipedes possess a few to many eyes that appear to be simple ocelli (Figure 18.7). In the scutigeromorph centipedes, up to 200 of these eyes may cluster to form a sort of pseudo-compound eye. However, the eyes of living centipedes appear to function only in the detection of light and dark, and not in image formation. Many parasitic and

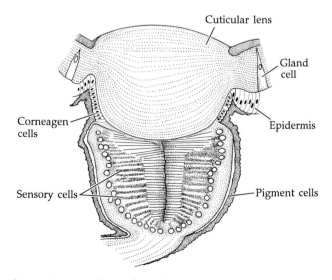

Figure 18.7   The ocellus of a centipede (section).

troglobitic myriapods have lost their eyes completely (eyes are absent in the centipede order Geophilomorpha, and the millipede orders Siphoniulida, Siphonophorida, and Polydesmida). Symphylans and pauropodans also lack eyes. Many burrowing diplopods are also eyeless, whereas others have from 2 to 80 ocelli arranged variously on the head. Some diplopods also possess integumental photoreceptors, and many eyeless species exhibit negative phototaxis.

Centipedes and millipedes are noted for their highly sensitive antennae, which are richly supplied with tactile and chemosensory setae. In many species, an **organ of Tömösvary** is located at the base of each antenna (Figure 18.8). Each part of this paired organ consists of a disc with a central pore where the ends of sensory neurons converge. The exact role of this organ has yet to be clearly established, and speculation has run the gamut from chemosensation to pressure sensation to humidity detection to audition (sound or vibration detection). The last idea is probably the most popular today. Whether the organ of Tömösvary might detect aerial vibrations (auditory impulses) or only ground vibrations is also a debated question.

Reproduction and Development

Myriapods are dioecious and oviporous, although parthenogenesis occurs in several families of millipedes and in some centipedes and symphylans. Many myriapods, like most arachnids, rely on indirect copulation and insemination. Packets of sperm (spermatophores) are deposited in the environment, or are held by the male and picked up by the female. All myriapods have

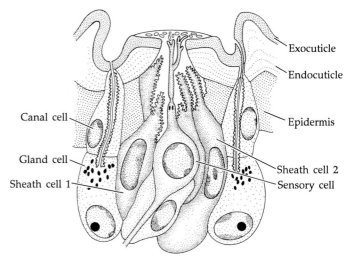

Figure 18.8 The organ of Tömösvary from the centipede *Lithobius forficatus*.

direct development, with the young hatching as "miniature adults," although often with fewer body segments. Beyond these generalizations, some specifics for each major group are discussed below.

Chilopoda. Female centipedes possess a single elongate ovary located above the gut, whereas males have from 1 to 26 testes similarly placed (Figure 18.9). The oviduct joins with the openings from several **accessory glands** and a pair of **seminal receptacles** just internal to the gonopore, which is situated on the **genital**

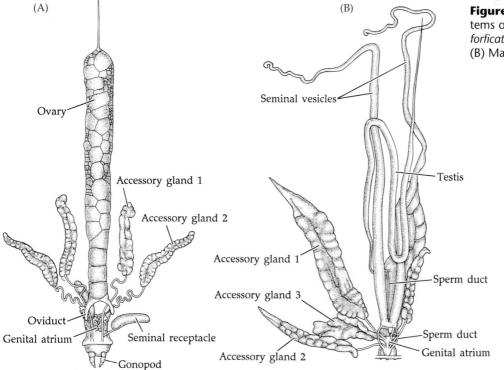

Figure 18.9 Reproductive systems of the centipede *Lithobius forficatus*. (A) Female system. (B) Male system.

segment (the legless segment in front of the telson, or pygidium). The gonopore of females is usually flanked by a pair of small grasping appendages, or gonopods. In males the testes join the ducts of several accessory glands and a pair of **seminal vesicles** near the gonopore, which opens on the ventral surface of the genital segment. The male gonopore also lies between a pair of small gonopods.

Sperm are packed into spermatophores, which are transferred to the female. In some cases (e.g., scutigeromorphs), males deposit spermatophores directly on the ground and the females simply pick them up. In most species, however, females produce a silken **nuptial net**, which is spun from modified genital glands, and males deposit a spermatophore on this net (Figure 18.10). Mating pairs of centipedes typically perform courtship behavior, stroking each other with their antennae and often moving the large (up to several millimeters) spermatophore about the nuptial net. Eventually the female picks up the spermatophore with her gonopods and inserts it into her gonopore. Fertilization occurs as the eggs pass through the gonoduct. Females often coat the fertilized eggs with moisture and fungicides before depositing them in the ground or in rotting vegetation. Some species hatch as juveniles with all body segments (i.e., **epimorphic development**), whereas others add new segments with the posthatching molts until the adult number is attained (i.e., **anamorphic development**). Parental guarding typically occurs in the eipmorphic orders (Figure 18.10).

Diplopoda.   Millipedes possess a single pair of elongate gonads in both sexes. Unlike those of chilopods, millipede gonads lie between the gut and the ventral nerve cord. The gonopores open on the third (genital) trunk segment. In females each oviduct opens separately into a **genital atrium** (= **vulva**) near the limb coxae. A groove in the vulva leads into one or more seminal receptacles. Males possess a pair of penes, or gonapophyses, on the third trunk segment (on the coxae of the second pair of legs). The mandibles are used to transfer sperm in the "pill millipedes," but more commonly one or both pairs of legs of the seventh or eighth trunk segment are modified as copulatory appendages, or **gonopods,** and serve this purpose. Males bend the anterior body segments back under the trunk until the penes and gonopods make contact, whereupon the later picks up a spermatophore in preparation for mating.

In most millipedes, mating occurs by indirect insemination, in which the male and female genital openings never actually make contact, although the male and female may embrace with their legs and mandibles and may coil their trunks together. Sometimes a pair remains in such an embrace for up to two days. Antennal tapping, head drumming, and stridulation may be included in the mating repertoire, and pheromones are known to play an important role in many species.

Males often have an array of secondary sexual characteristics, including strong modifications of the legs and head and special glandular structures that are not well understood. In species of the genus *Chordeuma* (in Europe) these glands produce a secretion on the dorsal surface upon which females feed prior to mating. Males of the family Sphaeriotheriidae produce sounds during courtship by rubbing the back edge of the body against

(A)

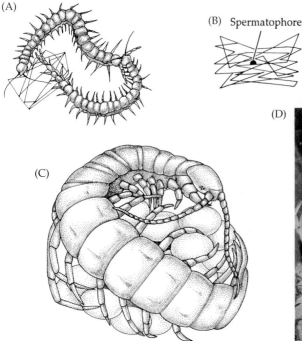

(B) Spermatophore

(C)

(D)

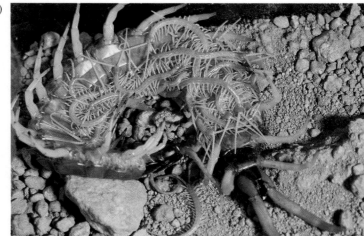

Figure 18.10   (A–C) Reproductive behavior in the tropical centipede *Scolopendra cingulata*. (A) A male and female in mating position over the nuptial net. (B) A nuptial net with a spermatophore. (C) A female and her eggs. (D) The giant North American desert centipede *Scolopendra heros* embracing her young.

an enlarged rear leg. In most millipedes the gonopods no longer resemble walking legs and may be truly baroque in their complexity, with flagella, brushes, knobs, and hooks.

Fertilization takes place as the eggs are laid. In glomerid millipedes, gonopods are not present, but the last pair of male legs is modified into strong claspers, used to hold the female in place. In these millipedes, males mold soil into a small cup into which they ejaculate sperm. The sperm-filled cup is then passed to the female, who is firmly held by the male's large claspers. The myriapod specialist W. A. Shear notes that this may be the only known case of animals using a manufactured tool to mate! Eggs are usually laid in the soil, although some species also produce silk bags to house the eggs. Some species fashion a nest of soil and humus reinforced with parental feces, or the entire nest may be composed of fecal material. One to 300 eggs are laid at a time. Hatchling millipedes typically possess seven or fewer body segments and only three pairs of legs. Additional segments and legs are added at the rear of the last somite with succeeding molts (i.e., anamorphic development). Unlike centipedes, most species of millipedes (excepting the order Platydesmida) show no parental care.

Pauropoda. In pauropodans a single ovary lies beneath the gut, but the testes are located above the gut. As in millipedes, the gonopores are on the third (genital) trunk segment. Females take up a spermatophore, often after the male has suspended it from one or a few silk threads spanning two stones or leaves. Fertilization is internal. The yolky eggs are laid in decaying wood. Hatchlings lack many adult segments and usually have only three pairs of legs (anamorphic development).

Symphyla. Symphylans display one of the more unusual methods of fertilization in the animal kingdom. Paired gonads discharge through gonopores on the third trunk segment in both sexes. Males deposit spermatophores in the environment. At least some symphylans construct stalklike structures, topped with spermatophores. When a female encounters one of these sperm packages, she bites off the spermatophores and stores the sperm in her preoral cavity. When her own eggs are ripe, the female removes them from her gonopore with her mouthparts and cements them to moss or some other substratum. Fertilization occurs during this process as she coats each egg with the stored sperm. Young symphylans hatch with only about half the adult number of trunk segments and appendages (anamorphic development).

Myriapod Embryogeny

The indirect copulation and external development typical of myriapods demands large amounts of stored nutrients (yolk) in the eggs. Like those of hexapods, the yolky eggs of myriapods show almost no trace of the ancestral arthropod spiral cleavage pattern, and have shifted almost entirely to meroblastic cleavage. The early embryonic development of myriapods proceeds much like that of hexapods, described in Chapter 17.

Most myriapods undergo early cleavage by intralecithal nuclear divisions, followed by a migration of the daughter nuclei to the peripheral cytoplasm (= **periplasm**). Here they continue to divide until the periplasm is dense with nuclei, whereupon cell membranes begin to form, partitioning uninucleate cells from one another. At this point the embryo is a periblastula, comprising a yolky sphere that has a few scattered nuclei and is covered by a thin cellular layer, or **blastoderm**.

Along one side of the blastula a patch of columnar cells forms a **germinal disc**, sharply marked off from the thin cuboidal cells of the remaining blastoderm. From specific regions of this disc, presumptive entodermal and mesodermal cells begin to proliferate as germinal centers. These cells migrate inward during gastrulation to lie beneath their parental cells, which now form the ectoderm. The mesoderm proliferates inward as a longitudinal **gastral groove**; the cells of the developing gut usually surround and gradually begin absorbing the large central yolky mass of the embryo, and paired coelomic spaces appear in the mesoderm.

As segments begin to demarcate and proliferate, each receives one pair of mesodermal pouches and eventually develops **appendage buds**. As the mesoderm contributes to various organs and tissues, the coelomic spaces merge with the small blastocoel to produce the hemocoelic space. The mouth and anus arise by ingrowths of the ectoderm that form the fore- and hindguts, which eventually establish contact with the developing entodermal midgut.

Variations on this basic scheme occur in groups with secondary yolk reduction, such as in pauropodans and symphylans. In these groups a largely holoblastic cleavage takes place, and a coeloblastula actually forms. In most chilopods, a type of cleavage occurs that is somewhat intermediate between total and superficial cleavage. After a few intralecithal nuclear divisions, the yolk breaks up into blocks called **yolk pyramids**. Gradually the yolk pyramids disappear, and development shifts to the superficial type.

Myriapod Phylogeny

In Chapter 15 we discussed how the long-favored view of arthropod relationships is now being challenged by new evidence. Until recently, the prevailing hypothesis viewed the Myriapoda and Hexapoda as sister groups (together called the Atelocerata), and those groups as sister to the Crustacea (together forming a grouping known as the Mandibulata). The sister-group relationship between myriapods and hexapods was based upon several seemingly strong synapomorphies, including proctodeal Malpighian tubules, loss of the second pair of antennae, uniramous legs, and tracheal respiratory systems. However, recent work suggests that the Crustacea may have been the ancestral lineage from which both the Hexapoda and Myriapoda emerged, perhaps independently. If this is true, the putative synapomorphies linking these two taxa probably evolved through convergent (or parallel) evolution. To complicate matters even more, growing evidence suggests the possibility that the Myriapoda itself may not be monophyletic—although the jury is still out on this question.

Myriapods have been around a long time. The oldest terrestrial animals on Earth are millipedes, dating to more than 425 mya (Silurian). Trace fossils (burrows resembling those of millipedes) suggest that they may have begun their life on land well before that. These ancient forms are nearly indistinguishable from modern taxa, indicating a remarkable level of genetic and morphological homeostasis among these arthropods—just the opposite of what has occurred among the Crustacea!

Centipedes date to at least 380 mya (Devonian), and their fossils, too, resemble modern forms.

We do not present a phylogeny for the Myriapoda for several reasons. There is widespread disagreement on whether or not myriapods comprise a monophyletic group and thus debate on the relationships of the four classes. There does seem to be fairly widespread agreement that the Diplopoda and Pauropoda are sister groups; among the many synapomorphies they are said to share are a legless postmaxillary segment, a gnathochilarium formed by the juxtaposition of the first maxillae and their triangular sternites, gonopores at the second leg-bearing segment, and tracheal pouches. In addition, most species in both groups hatch with three pairs of legs and undergo anamorphic development. The relationships of the Chilopoda and Symphyla remain enigmatic. Within Chilopoda there is good morphological and molecular support for the monophyly of the two subclasses, Notostigmophora and Pleurostigmophora.

Selected References

General References

Albert, A. M. 1983. Life cycle of Lithobiidae, with a discussion of the *r*- and *K*-selection theory. Oecologia 56: 272–279.

Anderson, D. T. 1973. *Embryology and Phylogeny of Annelids and Arthropods*. Pergamon Press, Oxford.

Blower, G. 1951. A comparative study of the chilopod and diplopod cuticle. Q. J. Microsc. Sci. 92: 141–161.

Blower, J. G. (ed.). 1974. *Myriapoda*. Academic Press, London.

Camatini, M. (ed.). 1980. *Myriapod Biology*. Academic Press, New York.

Carrel, J. E. and T. Eisner. 1984. Spider sedation induced by defensive chemicals of millipede prey. Proc. Natl. Acad. Sci. U.S.A. 81: 806–810.

Chamberlin, R. V. and R. L. Hoffman. 1958. Checklist of the millipedes of North America. Bull. U.S. Natl. Mus. no. 212: 1–236.

Cloudsley-Thompson, J. L. 1958. *Spiders, Scorpions, Centipedes and Mites: The Ecology and Natural History of Woodlice, Myriapods and Arachnids*. Pergamon Press, New York.

Eisner, T., M. Eisner and M. Deyrup. 1996. Millipede defense: Use of detachable bristles to entangle ants. Proc. Natl. Acad. Sci. U.S.A. 93: 10848–10851.

Essig, E. 1926. *Insects of Western North America*. Macmillan Co., New York.

Gilbert, S. F. 1997. Arthropods: The crustaceans, spiders, and myriapods. *In* S. F. Gilbert and A. M. Raunio (eds.), *Embryology: Constructing the Organism*. Sinauer Associates, Sunderland, MA, pp. 237–257.

Harrison, F. W. and M. E. Rice (eds.). 1997. *Microscopic Anatomy of Invertebrates*. Vol. 12, Onychophora, Chilopoda, and Lesser Protostomata. Wiley-Liss, New York.

Hoffman, R. L. 1980 (dated 1979). *Classification of the Diplopoda*. Mus. D'Hist. Nat., Geneva. 236 pp.

Hoffman, R. L. 1990. Diplopoda. *In* D. L. Dindal (ed.), *Soil Biology Guide*. Wiley Interscience, New York, pp. 835–860.

Hofffman, R. L. 1999. Checklist of the millipeds of North and Middle America. Spec. Publ. 8, Virginia Mus. Nat. Hist. 594 pp.

Hoffman, R. L., S. I. Golovatch, J. Adis and J. W. de Morais. 1996. Practical keys to the orders and families of millipedes of the Neotropical region (Myriapoda: Diplopoda). Amazoniana 14: 1–35.

Hoffman, R. L. et al. 1982. Chilopoda–Symphyla–Diplopoda–Pauropoda. *In* S. P. Parker (ed.), *Synopsis and Classification of Living Organisms*. McGraw-Hill, New York, pp. 681–726.

Hopkin, S. P. and H. J. Read. 1992. *The Biology of Millipedes*. Oxford University Press, New York.

Johannsen, O. A. and F. H. Butt. 1941. *Embryology of Insects and Myriapods*. McGraw-Hill, New York.

Lawrence, R. F. 1984. *The Centipedes and Millipedes of Southern Africa: A Guide*. Balkema, Cape Town.

Lewis, J. G. E. 1961. The life history and ecology of the littoral centipede *Strigamia* (= *Scolioplanes*) *maritima* (Leach). Proc. Zool. Soc. Lond. 137: 221–247.

Lewis, J. G. E. 1965. The food and reproductive cycles of the centipedes *Lithobius variegatus* and *Lithobius forficatus* in a Yorkshire woodland. Proc. Zool. Soc. Lond. 144: 269–283.

Lewis, J. G. E. 1981. *The Biology of Centipedes*. Cambridge University Press, New York.

Loomis, H. F. 1968. A checklist of the millipeds of Mexico and Central America. Bull. U.S. Natl. Mus. 266: 1–137.

Rajulu, G. S. 1970. A study on the nature and formation of the spermatophore in a centipede *Ethmostigmus spinosus*. Bull. Mus. Hist. Nat. Paris 41 (Suppl. 2): 116–121.

Rosenberg, J. and G. Seifert. 1977. The coxal glands of Geophilomorpha (Chilopoda): Organs of osmoregulation. Cell Tiss. Res. 182: 247–251.

Shear, W. A. 1969. A synopsis of the cave millipeds of the United States, with an illustrated key to genera. Psyche 76: 126–143.

Shear, W. A. 1998. The fossil record and evolution of the Myriapoda. Systematics Assoc. Spec. Vol. Ser. 55: 211–220.

Shear, W. A. 1999. Millipeds. Am. Sci. 87: 232–240.

Shelley, R. M. 1997. A re-evaluation of the millipede genus *Motyxia* chamberlin, with a re-diagnosis of the tribe Xystocherini and remarks on the bioluminesence. Insecta mundi 11 (3/4): 331–351.

Shelley, R. M. 1999. Centipedes and millipedes, with emphasis on North America fauna. Kansas School Nat. 45(3): 1–15.

Shelley, R. M. 2002. A synopsis of the North American centipedes of the order Scolopendromorpha (Chilopda). Virginia Mus. Nat. Hist. Mem. No. 5:1–108.

Snodgrass, R. E. 1952. *A Textbook of Arthropod Anatomy*. Cornell University Press, Ithaca, NY.

Summers, G. 1979. An illustrated key to the chilopods of the north-central region of the United States. J. Kansas Entomol. Soc. 52: 690–700.

Myriapod Evolution

Almond, J. E. 1985. The Silurian–Devonian fossil record of the Myriapoda. Phil. Trans. R. Soc. Lond. Ser. B 309: 227–237.

Briggs, D. E. G. and R. A. Fortey. 1989. The early radiation and relationships of the major arthropod groups. Science 246: 241–243.

Dohle, W. 1997. Are the insects more closely related to the crustaceans than to the myriapods? Entomologia Scandinavica (Suppl.) 51: 7–16.

Enghoff, H., W. Dohle and J. G. Blower. 1993. Anamorphosis in millipedes (Diplopoda): The present state of knowledge with some developmental and phylogenetic considerations. Zool J. Linn. Soc. 109: 103–234.

Friedrich, M. and D. Tautz. 1995. Ribosomal DNA phylogeny of the major extant arthropod classes and the evolution of myriapods. Nature 376: 165–167.

Giribet, G. , S. Carranza, M. Riutort, J. Baguña and C. Ribera. 1999. Internal phylogeny of the Chilopoda using complete 18S rDNA and partial Z8S rDNA sequences. Phil. Trans. R. Soc. London B 354: 215–222.

Hannibal, J. T. and R. M. Feldmann. 1988. Millipeds from late Paleozoic limestones at Hamilton, Kansas. Kansas Geol. Surv., Guidebook Series No. 6, pp. 125–132.

Jeram, A. J., P. A. Selden and D. Edwards. 1990. Land animals in the Silurian: Arachnids and myriapods from Shropshire, England. Science 250: 658–661.

Jun-Yuan, C., J. Bergström, M. Lindström and H. Xianguang. 1991. Fossilized soft-bodied fauna. Natl. Geogr. Res. Exp. 7(1): 8–19.

MacNaughton, R. B, J. M. Cole, R. W. Dalrymple, S. J. Braddy, D. E. G. Briggs and T. D. Lukie. 2002. First steps on land: Arthropod trackways in Cambrian–Ordovician eolian sandstone, southeastern Ontario, Canada. Geology 30: 391–394.

Mangum, C. P. et al. 1985. Centipedal hemocyanin: Its structure and its implications for arthropod phylogeny. Proc. Natl. Acad. Sci. U.S.A. 82: 3721–3725.

Mikulic, D. G., D. E. G. Briggs and J. Kluessendorf. 1985a. A new exceptionally preserved biota from the lower Silurian of Wisconsin, U.S.A. Phil. Trans. R. Soc. Lond. B 311: 75–85.

Mikulic, D. G., D. E. G. Briffs and J. Kluessendorf. 1985b. A Silurian soft-bodied biota. Science 228: 715–717.

Retallack, G. J. and C. R. Feakes. 1987. Trace fossil evidence for Late Ordovician animals on land. Science 235: 61–63.

Shear, W. A. 1998. The fossil record and evolution of the Myriapoda. Syst. Assoc. Spec. Vol. 55: 211–220.

Shear, W. A. and J. Kukalová-Peck. 1990. The ecology of Paleozoic terrestrial arthropods: The fossil evidence. Can. J. Zool. 68: 1807–1834.

Shinohara, K. 1970. On the phylogeny of Chilopoda. Proc. Japan. Soc. Syst. Zool. 6: 35–42.

Størmer, L. 1977. Arthropod invasion during late Silurian and Devonian times. Science 197: 1362–1364.

Telford, M. J. and R. H. Thomas. 1995. Demise of the Atelocerata? Nature 376: 123–124.

Tiegs, O. W. 1940. The embryology and affinities of the Symphyla, based on a study of *Hanseniella agilis*. Q. J. Microsc. Sci. 82: 1–225.

Tiegs, O. W. 1945. The post-embryonic development of *Hanseniella agilis* (Symphyla). Q. J. Microsc. Sci. 85: 191–328.

Tiegs, O. W. 1947. The development and affinities of the Pauropoda, based on a study of *Pauropus sylvaticus*. Q. J. Microsc. Sci. 88: 165–336.

Wägele, J. W. and G. Stanjek. 1995. Arthropod phylogeny inferred from partial 12S rRNA revisited: Monophyly of the Tracheata depends on sequence alignment. J. Zool. Syst. Evol. Res. 33: 75–80.

19

Phylum Arthropoda:
The Cheliceriformes

The skin of it is so soft, smooth, polished and neat, that she precedes the softest skin'd Mayds, and the daintiest and most beautiful Strumpets ... she hath fingers that the most gallant Virgins desire to have theirs like them, long, round, of exact feeling, that there is no man, nor any creature, that can compare with her.

The Reverend E. Topsell,
(describing a house spider, circa 1607)

The arthropod subphylum Cheliceriformes includes the classes Chelicerata (horseshoe crabs, spiders, scorpions, mites, ticks, and many less familiar groups) and Pycnogonida (sea spiders). In addition to the 70,000 or so described living species is an impressive array of fossil forms, such as the Paleozoic giant water scorpions (eurypterids), some of which were over 2 meters long. Cheliceriforms had their origin in ancient Cambrian seas. Today, however, most live on land, where among the Metazoa they are second only to the insects in diversity. A few kinds, such as some mites, have secondarily invaded various aquatic habitats. On land, cheliceriforms have adapted to virtually every imaginable situation and lifestyle.

In addition to the basic characteristics common to all arthropods, cheliceriforms are distinguished by several unique features (Box 19A). The body is typically divided into two main regions, called the **prosoma** and **opisthosoma** (Figures 19.1–19.3).*

In contrast to most other arthropods, it is not possible to delineate a discrete "head" in cheliceriforms. There are no antennae, but generally all six segments of the prosoma bear appendages. The first pair of appendages are the **chelicerae**, followed by the **pedipalps**, and then four pairs of walking legs. In the pycnogonids there is an additional appendage, the **ovigers**, between the pedipalps and first walking legs. The chelicerae and pedipalps are specialized for an enormous variety of roles in the various cheliceriform groups, including sensation, feeding, defense, locomotion, and copulation. The opisthosoma usually bears a terminal, postsegmental **telson**. In one group, the Merostomata, the opisthosoma may be

---

*These two body regions are sometimes also referred to as the cephalothorax and abdomen, but they are not homologous to those regions in other arthropods.

subdivided into two regions, a **mesosoma** and a **meta-soma** (Figure 19.1). A detailed discussion of cheliceri-form anatomy is provided in the bauplan sections below.

Cheliceriformes is a large and diverse taxon comprising two rather different classes. The Chelicerata includes the horseshoe crabs and the terrestrial spiders, scorpions, and mites; the pycnogonids are the "sea spiders." To assist readers in gaining a grasp of this large and diverse subphylum, we treat these two rather different classes separately in this chapter.

SUBPHYLUM CHELICERIFORMES

CLASS PYCNOGONIDA: The sea spiders

CLASS CHELICERATA: The chelicerates

> **SUBCLASS MEROSTOMATA**
>
> > **ORDER EURYPTERIDA:** Extinct giant water scorpions
> >
> > **ORDER XIPHOSURA:** Horseshoe crabs (e.g., *Limulus*)
>
> **SUBCLASS ARACHNIDA:** Spiders, scorpions, mites, ticks, and their kin. 11 extant orders
>
> > **ORDER ACARI:** Mites and ticks
> >
> > > **SUBORDER OPILIOACARIFORMES:** Primitive mites (e.g., family Opiliocaridae)
> > >
> > > **SUBORDER PARASITIFORMES:** Mites and ticks (e.g., *Aponomma, Argas, Boophilus, Dermacentor, Ixodes, Ornithodorus, Zeroseius*)
> > >
> > > **SUBORDER ACARIFORMES:** Mites and "chiggers" (e.g., *Demodex, Halotydeus, Penthaleus, Scirus, Tydeus*)
> >
> > **ORDER AMBLYPYGI:** Whip spiders, tailless whip scorpions (e.g., *Acanthophrynus, Damon, Heterophrynus, Stegophrynus, Tarantula*)
> >
> > **ORDER ARANEAE:** True spiders
> >
> > > **SUBORDER MESOTHELAE:** "Segmented" spiders. One family, Liphistiidae (*Heptathela, Liphistius*)
> > >
> > > **SUBORDER OPISTHOTHELAE:** "Modern" spiders
> > >
> > > > **SUPERFAMILY MYGALOMORPHA:** Tarantula-like spiders. About 15 families, including the following:
> > > >
> > > > > **FAMILY CTENIZIDAE:** Trapdoor spiders (e.g., *Cyclocosmia, Ummidia*)
> > > > >
> > > > > **FAMILY ATYPICAE:** Purse-web spiders (e.g., *Atypus*)
> > > > >
> > > > > **FAMILY THERAPHOSIDAE:** Tarantulas and bird spiders (e.g., *Acanthoscurria, Aphonopelma*)
> > > > >
> > > > > **FAMILY DIPLURIDAE:** Funnel-web spiders (e.g., *Diplura*)
> > > >
> > > > **SUPERFAMILY ARANEOMORPHAE:** "Typical" spiders. About 75 families including the following:
> > > >
> > > > > **FAMILY LOXOSCELIDAE:** Brown spiders (e.g., *Loxosceles*)

BOX 19A *Characteristics of the Subphylum Cheliceriformes*

1. Body composed of two tagmata: the prosoma and the opisthosoma. Prosoma composed of a presegmental acron and six somites, often covered by a carapace-like dorsal shield. Opisthosoma composed of up to 12 somites and a postsegmental telson; subdivided into two parts in some groups

2. Appendages of prosoma are chelicerae, pedipalps, and four pairs of walking legs; antennae are absent. All appendages are multiarticulate and uniramous

3. Gas exchange by gill books, book lungs, or tracheae

4. Excretion by coxal glands and/or Malpighian tubules. (Cheliceriform Malpighian tubules are probably not homologous to those of insects, terrestrial isopods, or tardigrades.)

5. With simple medial eyes and lateral compound eyes

6. Gut with two to six pairs of digestive ceca

7. Dioecious

FAMILY THERIDIIDAE: Cobweb and widow spiders (e.g., *Argyrodes, Episinus, Latrodectus, Ulesanis*)

FAMILY ULOBORIDAE: Orb-weavers (e.g., *Hyptiotes, Nephila, Uloborus*)

FAMILY ARANEIDAE: Orb-weavers (e.g., *Araneus, Argiope, Cyrtophora, Mastophora, Nephila, Pasilobus, Zygiella*)

FAMILY TETRAGNATHIDAE: Orb-weavers (e.g., *Dolichognatha, Eucta, Leucauge, Meta, Pachygnatha*)

FAMILY CLUBIONIDAE: Sack-spiders (e.g., *Clubiona*)

FAMILY LINYPHIIDAE: Sheet-web spinners (e.g., *Erigone, Dicymbium, Linyphia*)

FAMILY AGELENIDAE: Funnel-weavers (e.g., *Agelena, Coelotes*)

FAMILY ARGYRONETIDAE: Water spiders. Monotypic: *Argyroneta aquatica*

FAMILY LYCOSIDAE: Wolf spiders (e.g., *Lycosa, Pardosa, Pirata*)

FAMILY PISAURIDAE: Nursery-web spiders (e.g., *Dolomedes, Pisaura*)

FAMILY OXYOPIDAE: Lynx spiders (e.g., *Oxyopes*)

FAMILY THOMISIDAE: Crab spiders (e.g., *Thomisus, Xysticus*)

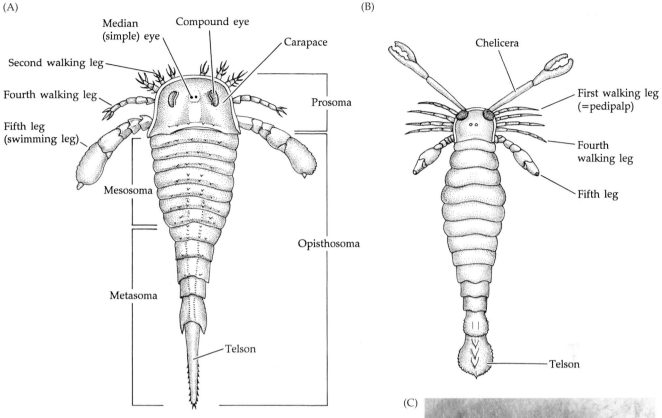

(A) Labels: Median (simple) eye, Compound eye, Second walking leg, Carapace, Fourth walking leg, Prosoma, Fifth leg (swimming leg), Mesosoma, Opisthosoma, Metasoma, Telson

(B) Labels: Chelicera, First walking leg (=pedipalp), Fourth walking leg, Fifth leg, Telson

FAMILY HETEROPODIDAE: Crab spiders (e.g., *Heteropoda*)

FAMILY SALTICIDAE: Jumping spiders (e.g., *Portia, Salticus*)

FAMILY DINOPIDAE: Ogre-faced spiders (e.g., *Dinopis*)

FAMILY SCYTODIDAE: Spitting spiders (e.g., *Scytodes*)

ORDER OPILIONES: Harvestmen, daddy longlegs (e.g., *Caddo, Leiobunum, Trogulus*)

ORDER PALPIGRADI: Palpigrades (e.g., *Allokoenenia, Eukoenenia, Koenenia, Leptokoenenia, Prokoenenia*)

ORDER PSEUDOSCORPIONIDA: False scorpions (e.g., *Chelifer, Chitrella, Chthonius, Dinocheirus, Garypus, Menthus, Pseudogarypus*)

ORDER RICINULEI: Ricinuleids. Three genera: *Cryptocellus, Pseudocellus, Ricinoides*

ORDER SCHIZOMIDA: Schizomids (e.g., *Agastoschizomus, Megaschizomus, Nyctalops, Protoschizomus, Schizomus*)

ORDER SCORPIONES: Scorpions (e.g., *Androctonus, Bothriurus, Buthus, Centruroides, Chactus, Chaerilus, Diplocentrus, Hadrurus, Hemiscorpion, Nebo, Parabuthus, Paruroctonus, Tityus, Vaejovis*)

ORDER SOLPUGIDA: Sun spiders (e.g., *Biton, Branchia, Dinorhax, Galeodes, Solpuga*)

ORDER UROPYGI: Whip scorpions and vinegaroons (e.g., *Albaliella, Chajnus, Mastigoproctus*)

Figure 19.1 Water scorpions (subclass Merostomata, order Eurypterida), an extinct group of chelicerates. (A) *Eurypterus* (dorsal view). (B) *Pterygotus buffaloensis*, which reached lengths of nearly 3 meters. (C) *Eurypterus remipes*, a species from the Silurian period of the Paleozoic. Eurypterids flourished in Paleozoic seas; some probably invaded fresh water and perhaps even land.

Synopses of the Chelicerate Taxa

Class Chelicerata

Body composed of two tagmata: prosoma and opisthosoma. Prosoma of presegmental acron plus 6 somites, often covered by a carapace-like dorsal shield; with simple medial eyes and compound lateral eyes. Opisthosoma composed of up to 12 somites (subdivided into two parts in some groups) and a postsegmental telson. Prosomal appendages are chelicerae, pedipalps, and four pairs of walking legs; antennae absent; all appendages uniramous and multiarticulate.

Subclass Merostomata

Prosoma covered by a large, hardened, carapace-like shield; pedipalps similar to walking legs; opisthosoma undivided or divided into mesosoma and metasoma; with flaplike appendages as gill books; telson long and spiked. The merostomates may have had their origin in the late Precambrian, but they diversified during the great Cambrian invertebrate radiation. Fossil eurypterids and xiphosurans are known from the Ordovician, and they flourished in the Silurian and Devonian. Alas, only five species of merostomates have survived to modern times, all xiphosurans—the horseshoe crabs.

Order Eurypterida.   The extinct giant water scorpions (Figure 19.1). Opisthosoma divided, with scalelike appendages on mesosoma; metasoma narrow (e.g., *Eurypterus, Pterygotus*).

Eurypterids represented a zenith in arthropod body size, some being nearly 3 m long (e.g., *Pterygotus*). These giant chelicerates roamed ancient seas and freshwater environments well into the Permian and were very abundant during their heyday. There is evidence that some species even became amphibious or semi-terrestrial. Eurypterids were probably capable of swimming as well as crawling. The last pair of prosomal limbs were greatly enlarged, flattened distally, and probably used as paddles. The chelicerae were extremely reduced in some species, but well developed and chelate in others—evidence that eurypterids radiated ecologically and exploited a variety of food resources and feeding strategies.

Order Xiphosura.   Certain extinct forms and the living horseshoe crabs (Figure 19.2). Opisthosoma unsegmented and undivided, but with six pairs of flaplike appendages, the first pair fused medially as a genital operculum over the gonopores, the last five pairs modified as gill books. Pedipalps and walking legs chelate; last (fourth) pair of legs splayed distally for support on soft substrata (e.g., *Carcinoscorpius, Limulus, Tachypleus*).

Extant members of the order Xiphosura (the horseshoe crabs) are regarded as "living fossils" and are, of course, far better known than their extinct relatives. Especially well studied is *Limulus polyphemus* (Figure 19.2C), the common horseshoe crab of the Atlantic and Gulf coasts of North America and a favorite laboratory animal of physiologists. Living xiphosurans inhabit shallow marine waters, generally on clean sandy bottoms, where they crawl about or burrow just beneath the surface, preying on other animals or scavenging. The chelicerae are smaller than the other appendages, being composed of only three or four articles. Each walking leg is formed of seven articles (coxa, trochanter, femur, patella, tibia, tarsus, pretarsus), the last two of which form the chelae (Figure 19.2B). The coxal endites of the pedipalps and first three walking legs are modified as **gnathobases**. Arising from the coxae of the fourth walking legs are tiny appendage-like processes called **flabella**, which function as gill cleaners. In addition, just posterior and medial to the last walking legs is a pair of reduced appendages called **chilaria**. Their function is unknown, and there is some controversy about their evolutionary significance—some specialists believe that they may reflect an additional opisthosomal segment.

The five living species of Xiphosura belong to three geographically distinct genera in the family Limulidae. *Limulus* (*L. polyphemus*) is restricted to eastern North America, ranging from Nova Scotia to the Yucatán region of Mexico; *Tachypleus* (*T. tridendatus*) occurs in Southeast Asia; *Carcinoscorpius* has been collected only from Malaysia, Siam, and the Philippines.

Subclass Arachnida

Prosoma wholly or partly covered by a carapace-like shield; opisthosoma segmented or unsegmented, divided or undivided; opisthosomal appendages absent or modified as spinnerets (spiders) or pectines (scorpions); penes absent (except in Opiliones); gas exchange by tracheae, book lungs, or both; nearly all terrestrial; over 60,000 species, in 11 orders.

Order Acari.   Mites, ticks and chiggers (Figure 19.3). Prosoma undivided, covered by a carapace-like shield and broadly joined to opisthosoma; junction of prosoma and opisthosoma indicated by a **disjugal furrow**, but more often indistinguishably fused; sometimes with a secondary **sejugal furrow** across body between second and third walking legs; chelicerae pincer-like or styliform; pedipalpal coxae uniquely fused with head elements; opisthosoma unsegmented except in a few primitive forms; gas exchange cutaneous or by tracheae; eyes present or absent; males of some species with a penis.

Mites and ticks comprise the largest group of arachnids. There are approximately 30,000 known species, and some experts suggest that there may be as many as a million or more yet to be described! The order Acari is a difficult taxon to characterize, and it may represent a polyphyletic assemblage. Because of the size and diver-

(A)

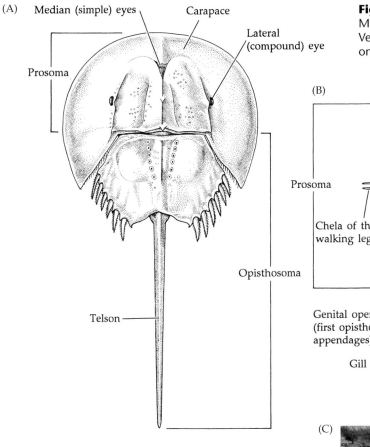

Median (simple) eyes

Carapace

Lateral (compound) eye

Prosoma

Opisthosoma

Telson

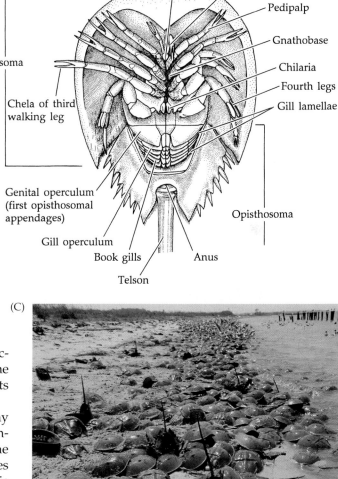

Figure 19.2 The horseshoe crab, *Limulus* (subclass Merostomata, order Xiphosura). (A) Dorsal view. (B) Ventral view. (C) A congregation of *Limulus polyphemus* on an Atlantic beach.

(B)

Mouth

Chelicera

Pedipalp

Gnathobase

Chilaria

Fourth legs

Gill lamellae

Prosoma

Chela of third walking leg

Genital operculum (first opisthosomal appendages)

Gill operculum

Book gills

Anus

Telson

Opisthosoma

(C)

sity of this group, we give a somewhat extended account. Even so, it is not possible to do justice here to the vast range of forms and lifestyles represented by its members.

Acarids occur worldwide. Most are terrestrial, many are parasitic, and some have invaded aquatic environments. The tremendous evolutionary success of the Acari, particularly the mites, is reflected in their species diversity and in their extremely varied lifestyles. This success is probably due, at least in part, to a compact body and reduced size. By coupling small size with the inherent arthropod quality of segment and appendage specialization, mites have exploited myriad microhabitats unavailable to larger animals. If the group is polyphyletic, then several ancestral arachnid lineages evolved convergently to capitalize on miniaturization.

Recent classification schemes divide the members of Acari into three groups. The most primitive mites are omnivorous and predatory forms in the suborder Opilioacariformes. They are characterized by the retention of opisthosomal segmentation (at least ventrally) and the presence of a transverse groove (the disjugal furrow) separating the prosoma from the opisthosoma. These mites are found on tropical forest floors and in arid temperate habitats.

The remainder of the thousands of kinds of mites and ticks are placed in the suborders Parasitiformes and Acariformes, the latter housing most of the species. The

body of parasitiform mites is undivided dorsally, there being no clear transverse groove (Figure 19.3A). Parasitiformes include free-living and symbiotic forms in all parts of the world. The free-living species inhabit various terrestrial habitats, including leaf litter, decaying wood and other organic debris, mosses, insect and small mammal nests, and soil. Most of these mites are predators on small invertebrates. Many species are full-time

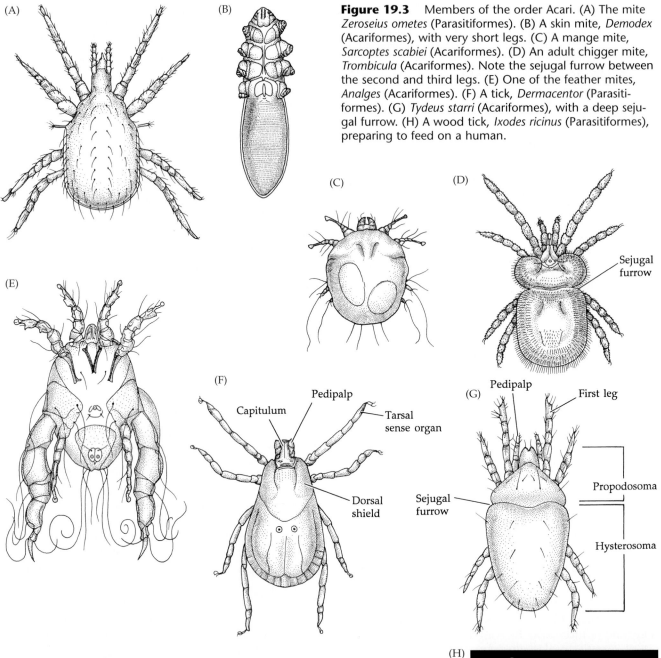

(A)

(B)

Figure 19.3 Members of the order Acari. (A) The mite *Zeroseius ometes* (Parasitiformes). (B) A skin mite, *Demodex* (Acariformes), with very short legs. (C) A mange mite, *Sarcoptes scabiei* (Acariformes). (D) An adult chigger mite, *Trombicula* (Acariformes). Note the sejugal furrow between the second and third legs. (E) One of the feather mites, *Analges* (Acariformes). (F) A tick, *Dermacentor* (Parasitiformes). (G) *Tydeus starri* (Acariformes), with a deep sejugal furrow. (H) A wood tick, *Ixodes ricinus* (Parasitiformes), preparing to feed on a human.

(C)

(D)

Sejugal furrow

(E)

(F)

Capitulum

Pedipalp

Tarsal sense organ

Dorsal shield

(G)

Pedipalp

First leg

Sejugal furrow

Propodosoma

Hysterosoma

(H)

or part-time symbionts on other animals, either as immatures or as adults. Their hosts are frequently other arthropods, such as centipedes, millipedes, ants, and especially beetles. In some cases the relationship is truly parasitic, in others phoretic, and in many instances the nature of the association is unknown.

The most familiar members of the Parasitiformes are the ticks (families Argasidae and Ixodidae; Figure 19.3F,H). Ticks are blood-sucking ectoparasites on vertebrates (one species, *Aponomma ecinctum*, lives on beetles). They are the largest-bodied members of the entire order, and some swell to 2–3 cm in length during a blood meal. The chelicerae are smooth and adapted for slicing skin, and are together called the **capitulum**. The ixodids are called the hard ticks because of a sclerotized shield covering the entire dorsum. They parasitize reptiles, birds, and mammals. They generally remain attached to their hosts for days or even weeks, feeding on blood. Some are vectors of important diseases, including *Dermacentor andersoni* (vector of Rocky Mountain spotted fever) and *Boophilus annulatus* (vector of Texas cattle fever). Lyme disease (first described in Old Lyme, Connecticut, in 1975) is a bacterial disease that is harbored in deer and certain rodents and is transmitted by several tick species in North America, including the western black-legged tick *Ixodes pacificus*.

Soft ticks (Argasidae) lack the heavily sclerotized dorsal shield of hard ticks. They are usually rather transient parasites on avian and mammalian hosts (particularly bats), typically feeding for less than an hour at a time. When not attached to a host, these ticks remain hidden in cracks and crevices or buried in soil. Disease-carrying soft ticks include *Argus persicus* (vector of fowl spirochetosis) and *Ornithonodorus moubata* (vector of African relapsing fever, or "tick fever").

The vast and diverse suborder Acariformes is itself generally considered to be polyphyletic, and the unifying features thus convergent. These mites usually have bodies divided into two regions, but not as the normal prosoma and opisthosoma. Rather, the disjugal furrow is lost, and a secondarily evolved sejugal furrow partly or wholly traverses the dorsum between the origins of the second and third pairs of walking legs (Figures 19.3D,G). The front part of the body is called the **propodosoma**, the hind part the **hysterosoma**. In some acariform mites this division is secondarily lost.

Free-living acariform mites are found in virtually every conceivable situation: in soil, leaf litter, decaying organic matter, mosses, lichens, and fungi; under bark; on freshwater algae; in sands; on seaweeds; at all altitudes and at most ocean depths. "Chiggers" is a name given to the larvae of acariform mites of several general, notably *Trombicula*. They include both herbivores (some fungivores) and predators, and their feeding methods are diverse. Many ingest solid as well as liquid food, and a few aquatic forms actually suspension feed. One species, *Agauopsis auzendei*, is known from hydrothermal vent environments (Bartsch 1990). Certain groups are serious pests that destroy stored grain crops and other food products. On the other hand, some predatory acariform mites have been used for biological pest control of other arthropods—even other mites!

Most of the symbiotic acariform mites are parasites on vertebrate and invertebrate hosts. Various species parasitize marine and freshwater crustaceans, freshwater insects, marine molluscs, terrestrial arthropods, the pulmonary chambers of terrestrial snails and slugs, the outer surfaces of all groups of terrestrial vertebrates, and the nasal passages of amphibians, birds, and mammals. In addition to direct parasitism, many mites are phoretic, using their hosts for dispersal. There are also many plant-eating mites that are considered parasitic.

A great many acariform mites cause economic or medical problems by direct predation and parasitism, by acting as disease vectors, or by feeding on stored food products. The family Penthaleidae includes the red-legged earth mite (*Halotydeus destructor*) and the winter grain mite (*Penthaleus major*), both of which are serious pests of many important crops. Members of the superfamily Eriophyoidea are vermiform mites adapted to feeding on various plants. This group includes the gall and leaf-curl mites, as well as a number of others that serve as vectors for certain disease-causing viruses (e.g., wheat and rye mosaic viruses). Another family of mites (Demodicidae) includes parasites of hair follicles and sebaceous glands of mammals. Two species, *Demodex folliculorum* and *D. brevis*, occur specifically in the hair follicles and sebaceous glands, respectively, of the human forehead. Another, *D. canis*, causes mange in dogs. Some other problems caused by acariform mites include subcutaneous tumor-like growths in humans, various sorts of skin irritations, mange in many domestic animals, reduced wool production in sheep, and feather loss in birds.

With a mixture of reluctance and relief, we leave the mites; interested readers should consult the references at the end of this chapter for sources of additional information.

Order Amblypygi. Prosoma undivided, covered by a carapace-like shield and connected to opisthosoma by narrow pedicel; opisthosoma segmented but undivided; with two pairs of book lungs; telson absent; chelicerae modified as spider-like fangs; pedipalps raptorial; first pair of legs are greatly elongate, antenniform sensory appendages; most with eight eyes; molting occurs throughout life (Figures 19.4D,E).

The 70 or so species of amblypygids are commonly known as whip spiders or tailless whip scorpions, reflecting their similarities to both spiders and uropygids. Externally they resemble whip scorpions, but internally they are very much like spiders, except that they lack spinnerets and poison glands. Amblypygids are widely distributed in warm humid areas, where they are found under bark or in leaf litter, and in similar protected habitats; several species are cave dwellers.

Most species are less than 5 cm in length, but the first pair of legs may be as long as 25 cm! These limbs are used as touch receptors and chemoreceptors (like antennae in other, nonchelicerate arthropods) and are important in prey location because these animals hunt at night. Amblypygids walk sideways with these long "feelers" extended, sensing for potential prey. Once located, the prey is grasped by the pedipalps and torn open by the chelicerae. The body fluids are then sucked from the victim and ingested.

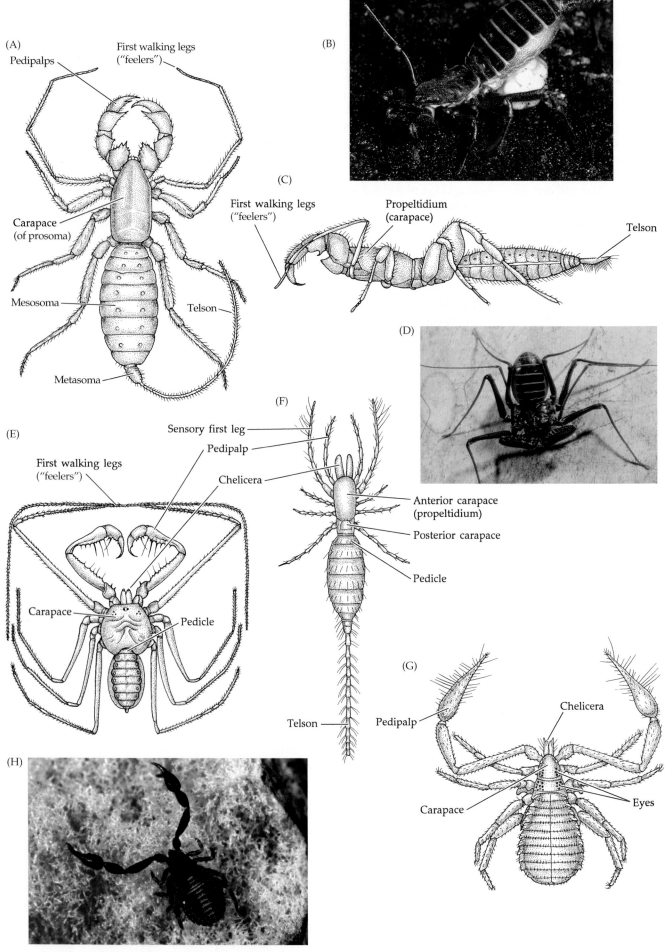

(A)

Pedipalps

First walking legs ("feelers")

Carapace (of prosoma)

Mesosoma

Telson

Metasoma

(B)

(C)

First walking legs ("feelers")

Propeltidium (carapace)

Telson

(D)

(E)

First walking legs ("feelers")

Pedipalp

Chelicera

Carapace

Pedicle

(F)

Sensory first leg

Pedipalp

Chelicera

Anterior carapace (propeltidium)

Posterior carapace

Pedicle

Telson

(G)

Pedipalp

Chelicera

Carapace

Eyes

(H)

Order Araneae. Spiders (Figure 19.5). Spiders, of course, are the most familiar of all chelicerates, and they are one of the most abundant groups of land animals. They have successfully exploited nearly every terrestrial environment, and many freshwater and intertidal habitats as well, and they display a truly staggering array of lifestyles—yet they all utilize a fairly uniform bauplan. Prosoma undivided, covered by a carapace-like shield and attached to opisthosoma by narrow pedicel; opisthosoma undivided and unsegmented except in liphistiids and some mygalomorph families, which have discrete tergites (i.e., "segmented abdomens"); chelicerae modified as fangs, usually with venom glands; opisthosoma bears book lungs and/or tracheae, silk-producing glands, and **spinnerets**, the last being highly modified appendages that serve to spin the silk produced by the silk glands; most with eight eyes. Most cease to molt once adulthood is achieved.

The body is divided into a fused prosoma and a fleshy opisthosoma. The first opisthosomal segment forms a narrow **pedicel**, which joins the two body regions. Each chelicera has two segments; the basal segment is short and conical and the distal segment is a hard, curved fang, usually bearing a pore from the duct of the poison gland (Figures 19.5J,K, 19.16A). A medial flap (the **labium**) projects ventrally over the mouth. Each pedipalp is composed of six segments, and in most spiders the proximal segments are enlarged as lobes (or endites) called **maxillae**, which bear gnathobases that

Figure 19.4 Some examples of arachnids other than spiders and scorpions. (A) A whip scorpion, *Mastigoproctus* (order Uropygi). (B) A giant whip scorpion (or "vinegaroon") (*Mastigoproctus giganteus*), carrying eggs. (C) A schizomid (order Schizomida), *Nyctalops crassicaudatus*. (D) A whip spider (order Amblypygi) exposed during the daylight in Panama. (E) An amblypygid (order Amblypygi), *Stegophrynus dammermani*. (F) *Koenenia* (order Palpigradi). (G) The pseudoscorpion *Chelifer cancroides* (order Pseudoscorpionida), a frequent hitchhiker on houseflies. (H) An unidentified pseudoscorpion. (I) A sun spider (order Solpugida), *Galeodes arabs*. (J) A sun spider (order Solpugida), *Eremobates*. (K) A daddy longlegs, or harvestman (order Opiliones). (L) A daddy longlegs (order Opiliones), *Leiobunum*. (M) A ricinuleid (order Ricinulei), *Ricinoides crassipalpe*, dorsal view. (N) A ricinuleid with its long first walking legs extended laterally.

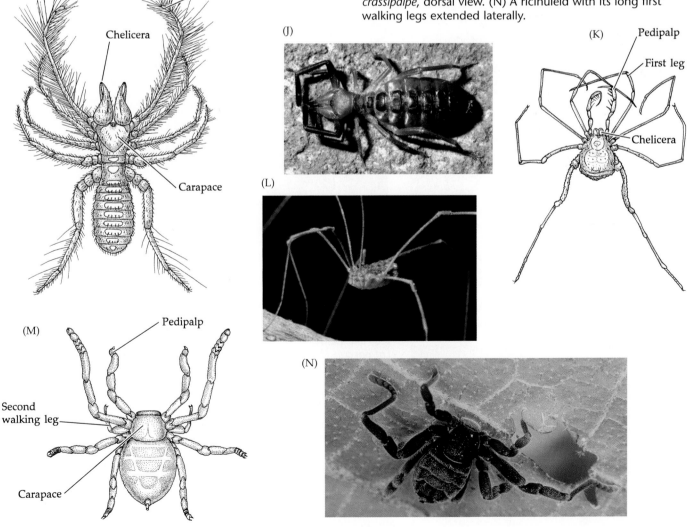

(A)

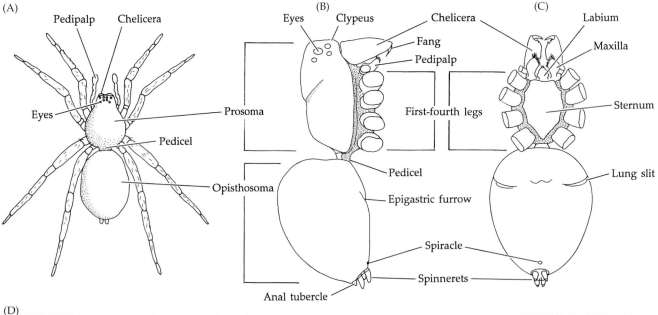

Pedipalp   Chelicera

Eyes

Prosoma

Pedicel

Opisthosoma

(B)

Eyes   Clypeus   Chelicera

Fang

Pedipalp

First-fourth legs

Pedicel

Epigastric furrow

Spiracle

Spinnerets

Anal tubercle

(C)

Labium

Maxilla

Sternum

Lung slit

Spiracle

Spinnerets

(D)

(E)

(F)

(H)

(G)

(I)

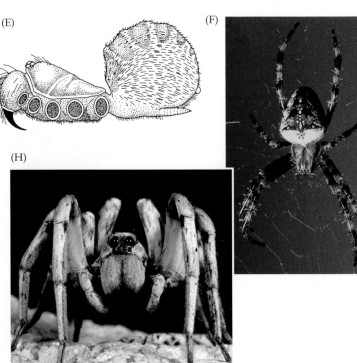

Figure 19.5   Spiders. (A–C) A generalized spider. (A) Dorsal view. In (B) (lateral view) and (C) (ventral view) the legs have been omitted, except the coxae. (D) The notorious black widow spider, *Latrodectus* (family Theridiidae). (E) *Liphistius* (suborder Mesothelae) (side view with legs removed). Note the evidence of opisthosomal segmentation. (F) An orb-weaving, or "garden," spider, *Araneus diadematus* (family Araneidae). (G) A red-kneed tarantula from Mexico (family Theraphosidae). (H) Face-on view of the wolf spider *Lycosa* (family Lycosidae). (I) A female crab spider, *Misumenoides* (family Thomidisae). (J) The chelicerae of *Portia*, a jumping spider. (K) Fangs of a spider's chelicerae, showing pores of poison glands. (L,M) The orientation and plane of motion of the chelicerae of an orthognathous (L) and a labidognathous (M) spider. (N) Trapdoor spider "door"; note the "handles" on the inside, used by the resident to close and hold the door shut. (O) A spider that mimics ants; ant mimicry has evolved multiple times among spiders.

flank the mouth and are used to handle and grind food.* Beyond the gnathobases, the pedipalps extend forward as tactile organs in females and juveniles, but they are highly modified as sperm transfer organs in mature males. In a few species of spiders (e.g., tarantulas) the pedipalps function as walking "legs." The ventral surface of the prosoma bears a cuticular plate, or **sternum**, around which arise the four pairs of walking legs, each composed of eight articles.

The opisthosoma bears openings for gas exchange, the reproductive system, the spinnerets, and the anus. On the ventral surface posterior to the pedicel is a transverse **epigastric furrow** in which the gonopores are located. In most females, a slightly elevated plate, the **epigynum**, lies in front of the epigastric furrow and bears the openings to the seminal receptacles. Lateral to the furrow are spiracles that open to the **book lungs**, or anterior tracheae. The hind part of the opisthosoma bears the anus and the spinnerets (the latter are situated near the middle of the opisthosoma in the Mesothelae). The ability of spiders to produce silk and fashion it into a great variety of functional devices has been a major factor in their evolutionary success, as we will see below.

There are about 35,000 described spider species. Several competing schemes exist for their higher classification; we divide them into two suborders. The suborder Mesothelae includes the primitive "segmented" spiders, which are characterized by persisting segmentation of the opisthosoma (Figure 19.5E) and mid-abdominal location of the spinnerets. Most are 1–3 cm long and construct burrows with a trapdoor over the entrance. All are predators, feeding on animals that pass within striking distance of the burrow opening.

The other suborder, Opisthothelae, comprises two large groups: the Mygalomorphae (tarantula-like spiders) and the Araneomorphae ("typical" spiders). Most members of both groups have an unsegmented opisthosoma and posteriorly located spinnerets, but they can be differentiated by the nature of the chelicerae (Figures 19.5L,M). Mygalomorph spiders possess chelicerae that articulate in a manner that enables movement of the appendages parallel to the body axis (i.e., **orthognathous jaws**), whereas araneomorph spiders have chelicerae that move at right angles to the body axis (i.e., **labidognathous jaws**).[†] (Members of the suborder Mesothelae have orthognathous chelicerae.)

---

*These structures should not be confused with the maxillae of crustaceans and insects, with which they are not homologous.

[†]Many authors treat the opisthothelan spiders as two suborders (Orthognatha and Labidognatha, or Mygalomorphae and Araneomorphae, respectively) on the basis of this difference.

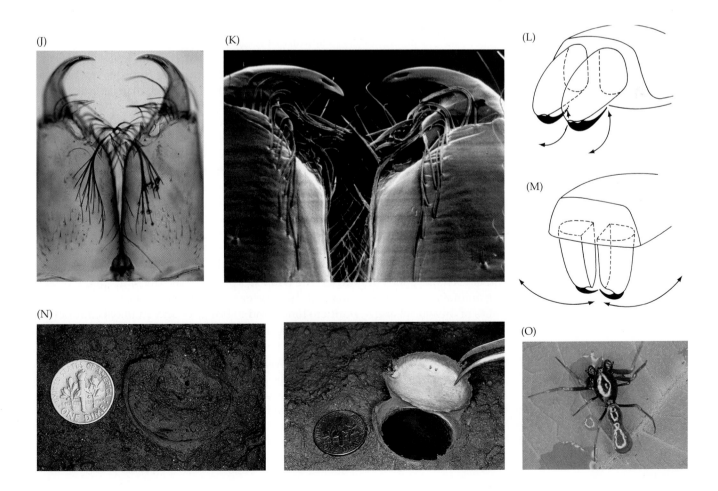

Order Opiliones. Daddy longlegs, or harvestmen (Figure 19.4K,L). Prosoma divided (as it also is in schizomids, palpigrades, and solpugids) into a proterosoma and two free segments that join broadly to the segmented opisthosoma; chelicerae small, of three articles, pincer-like; pedipalps long and leglike; one pair of eyes; one pair of tracheae on opisthosoma; male with penis.

The order Opiliones is a large and diverse group of about 5,000 species, generally considered to be closely related to mites (order Acari). Opilionids are known from nearly all climatic regions of the world, including subarctic areas, but are most abundant in tropical South America and Southeast Asia. Most species have small bodies, less than 2 cm in length, but usually with very long legs (up to 10 cm).

Opilionids prefer damp, shaded areas and are commonly found in leaf litter, on trees and logs in dense forests, and in caves. They feed on various small invertebrates and also scavenge on dead animal and plant matter. Food is grasped by the pedipalps and passed to the chelicerae for chewing. These animals are among the few arachnids capable of ingesting solid particles as well as the usual liquefied food. Opilionids possess a pair of defensive **repugnatorial glands**, which produce noxious secretions containing quinones and phenols. As discussed later, most arachnids mate by indirectly passing sperm from the male to the female. Opilionids (and some mites) are the only arachnids in which males use a penis for direct copulation.

Order Palpigradi. Palpigrades (Figure 19.4F). Minute arachnids with prosoma divided into proterosoma, covered by the carapace-like propeltidium, followed by two free segments and joined to opisthosoma by a narrow pedicel; opisthosoma segmented and divided into broad mesosoma and short narrow metasoma, the latter bearing a long multiarticulate telson; eyes absent.

There are about 60 species of palpigrades, all of which are less than 3 mm long. These tiny arachnids have undergone a great deal of evolutionary reduction in association with their small size and cryptic habits. They are colorless, have very thin cuticles, and have lost the circulatory and gas exchange organs. Most are found under rocks or in caves, and at least one species lives on sandy beaches. These rare animals have been recorded from widely separated parts of the world, and their biology remains poorly known.

Order Pseudoscorpionida. False scorpions (Figure 19.4G,H). Prosoma covered by a dorsal carapace-like shield, but clearly segmented ventrally; opisthosoma undivided but with 11–12 segments and broadly joined to prosoma; chelicerae chelate and bearing spinnerets; pedipalps large and scorpion-like; eyes present or absent.

There are about 2,000 described species of false scorpions, the largest of which reach lengths of only 7 mm. The group is cosmopolitan and found in a great variety of habitats—under stones, in litter, in soil, under bark, and in animal nests. One genus, *Garypus*, is found on sandy and cobble marine beaches, and one species, *Chelifer cancroides*, typically cohabits with humans.

These strange little creatures resemble scorpions in general appearance, but they lack the elongation of the opisthosoma and telson and do not have a stinging apparatus. However, they do possess poison glands in the pedipalps with which they immobilize their prey, usually other tiny arthropods (e.g., mites). Once captured, the victim is torn open by the chelicerae and the body fluids sucked out.

A number of false scorpions use larger arthropods as temporary "hosts" for purposes of dispersal. This phenomenon of "hitchhiking," known as **phoresy**, typically involves females that grab onto the larger host animal with their pedipalps. The host is often a flying insect. *Chelifer cancroides*, for example, is frequently encountered as a phoretic "guest" on house flies.

Order Ricinulei. Prosoma fully covered by a carapace-like shield and broadly joined with opisthosoma; opisthosoma unsegmented, with paired tracheae; chelicerae pincer-like, covered with flaplike **cucullus**; pedipalps small, with coxae fused medially; eyes reduced; third legs of males modified for sperm transfer; cuticle very thick (Figure 19.4M,N).

There are only about 35 described species of ricinuleids; all are less than 1 cm long and live in caves and tropical forest leaf litter in West Africa and tropical America. All are slow-moving predators on other small invertebrates.

Order Schizomida. Schizomids (Figure 19.4C). Prosoma divided; first four segments (the proterosoma) covered by a short carapace-like shield (the propeltidium), followed by two free segments, the mesopeltidium and metapeltidium; opisthosoma segmented and divided; mesosoma with one pair of book lungs; metasoma with short, thin telson; eyes present or absent.

There are about 80 species of small (less than 1 cm) arachnids in this order. Although some authors classify them as a suborder of Uropygi, they are distinguished from true whip scorpions by the divisions of the prosoma and the shorter telson. Schizomids live in leaf litter, under stones, and in burrows and are most common in tropical and subtropical areas of Asia, Africa, and the Americas; a few temperate species are known.

The first walking legs are sensory and resemble those of uropygids. Like the uropygids, the schizomids are predators on small invertebrates, and they also possess opisthosomal repugnatorial glands.

Order Scorpiones. True scorpions (Figure 19.6). Body clearly divided into three regions: prosoma, mesoso-

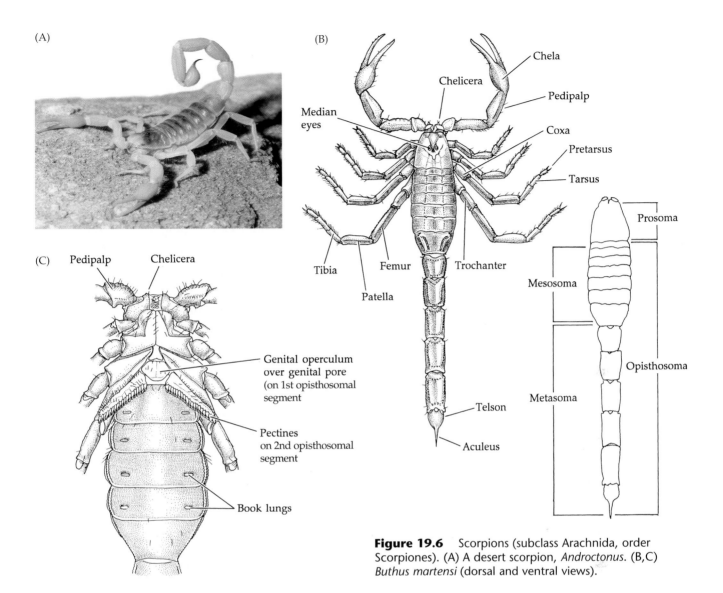

(A)

(B)

Chela
Chelicera
Pedipalp
Median eyes
Coxa
Pretarsus
Tarsus
Tibia
Femur
Trochanter
Patella
Prosoma
Mesosoma
Metasoma
Opisthosoma
Telson
Aculeus

(C)

Pedipalp   Chelicera

Genital operculum over genital pore (on 1st opisthosomal segment

Pectines on 2nd opisthosomal segment

Book lungs

Figure 19.6   Scorpions (subclass Arachnida, order Scorpiones). (A) A desert scorpion, *Androctonus*. (B,C) *Buthus martensi* (dorsal and ventral views).

ma, and metasoma. Prosomal segments fused and covered by a carapace-like shield; opisthosoma elongate, segmented, and divided into mesosoma and metasoma of seven and five segments, respectively; telson spinelike, with poison gland; chelicerae of three articles; pedipalps large, chelate, of six articles; with pair of median eyes and sometimes additional pairs of lateral eyes; first mesosomal segment bears a gonopore covered by genital operculum; second mesosomal segment bears a pair of unique sensory appendages called pectines; third through sixth mesosomal segments each with a pair of book lungs. Metasoma without appendages.

Scorpions are among the most ancient terrestrial arthropods and the most primitive arachnids. They apparently evolved from aquatic ancestors, perhaps from the eurypterids or a common ancestor, and invaded land during the Carboniferous period. All of the roughly 1,200 known species are terrestrial predators. They inhabit a variety of environments, particularly deserts and tropical rain forests, where some arboreal species occur.

They are notably absent from colder regions of the world. Scorpions include the largest living arachnids, with some reaching lengths of 18 cm.

The chelicerae are short and bear gnathobases for grinding food. The pedipalps are large, with the last two articles forming grasping chelae. The walking legs each comprise eight articles (coxa, trochanter, femur, patella, tibia, metatarsus, tarsus, pretarsus). The ventral surface of the mesosoma bears the genital pore, a pair of pectines, and four pairs of spiracular openings to the book lungs. The anus is located on the last true segment of the metasoma, and is followed by a stinging apparatus derived from the telson and bearing a sharp barb called the **aculeus**.

Order Solpugida.   Sun spiders (Figure 19.4I,J). Prosoma divided into a proterosoma covered by a carapace-like shield and two free segments; opisthosoma undivided, but with eleven segments, bearing three pairs of tracheae; chelicerae huge and held forward;

pedipalps long and leglike; propeltidium with one pair of eyes.

Most of the 900 or so species of solpugids (or solfugids) live in tropical and subtropical desert environments of America, Asia, and Africa. In contrast to many arachnids, they are often daytime hunters, hence the common name "sun spiders." They are also known as "wind spiders" because the males run at high speeds, "like the wind."

Some solpugids are only a few millimeters long, but many reach lengths of up to 7 cm. The feeding habits of many solpugids are unknown. Among those that have been studied, most are omnivorous, but they frequently show a preference for termites or other arthropods. Lacking poison, they rip their live prey apart with their strong chelicerae.

Order Uropygi. Whip scorpions and vinegaroons (Figure 19.4A,B). Prosoma elongate and covered by a carapace-like shield; opisthosoma segmented and divided; mesosoma broad with two pairs of book lungs; metasoma short with long, whiplike telson; first walking legs elongate and multiarticulate distally; with one pair of median and four or five pairs of lateral eyes.

Whip scorpions are moderately large arachnids reaching lengths of up to 8 cm. There are fewer than 100 known species of living uropygids, most of which occur in Southeast Asia; a few are known from the southern United States and parts of South America, and some probably introduced species occur in Africa. With the exception of a few desert species, whip scorpions live under rocks, in leaf litter, or in burrows in relatively humid tropical and subtropical habitats. The telson is sensitive to light, and most uropygids are negatively phototactic and active only at night.

The elongate first walking legs are held forward as "feelers," aiding the animals in their nocturnal hunting excursions. They feed on various small invertebrates by grasping prey with the pedipalps and grinding it with the chelicerae.

Whip scorpions possess a pair of repugnatorial glands that open near the anus. When a uropygid is threatened by a potential predator, it raises the opisthosoma and sprays an acidic liquid from by these glands on the would-be attacker. Some forms (e.g., *Mastigoproctus*) produce a secretion high in acetic acid, earning them the common name "vinegaroons."

The Chelicerate Bauplan

Looking over the taxonomic summaries above will give you a good sense of diversity within the Chelicerata. This section covers the general biology and structure of members of this class, with an emphasis on the xiphosurans, spiders, and scorpions. We hope to convey an impression not only of diversity but also of unity within

this group and to reinforce the concept of the evolutionary plasticity of the arthropod bauplan in general.

The chelicerate body is typically divided into two main regions, the prosoma and the opisthosoma (Figures 19.1, 19.2, 19.5, 19.6); a discrete head is not recognizable. The prosoma includes a presegmental acron and six segments; the opisthosoma includes up to twelve segments and a postsegmental, postanal telson. As in other arthropod groups, these basic body regions have undergone various degrees of specialization and tagmosis. In most chelicerates the entire prosoma is fused and covered by a carapace-like shield. However, in certain groups (e.g., schizomids, palpigrades, solpugids, and opilionids) the prosoma is divided into three parts: a **proterosoma**, comprising the acron and the first four segments, all fused and covered by a carapace-like shield (often called the **propeltidium**); and two free segments (often called the **mesopeltidium** and **metapeltidium**). The opisthosoma may be undivided, as it is in spiders, or divided into an anterior mesosoma and posterior metasoma, as it is in many other living arachnids and in the eurypterids.

The appendages further distinguish the chelicerates from other arthropods. There are no antennae, but all six segments of the prosoma usually bear appendages. The first pair of appendages are embryologically postoral, often pincer-like, chelicerae. During embryogeny, the chelicerae migrate to a position lateral to the mouth, or even preoral, in adults of most groups; here they serve as fangs or grasping structures during feeding. The chelicerae are followed by a pair of postoral pedipalps, which are usually elongate or, more rarely, in the form of pincers. The pedipalps are usually sensory in function, but in some groups (e.g., scorpions) they aid in feeding and defense. The remaining four segments of the prosoma typically bear the walking legs.

The numbers of segments and appendages on the opisthosoma vary. In general, appendages are absent or very reduced, although in the horseshoe crabs they persist as large platelike limbs, called **gill books**, that function in locomotion and gas exchange. In most chelicerates the opisthosomal limbs are greatly reduced and persist only as specialized structures, such as the silk-producing spinnerets of spiders or the pectines of scorpions.

In summary, then, we may define the Chelicerata as cheliceriform arthropods in which the body is divided into two regions (or two tagmata), a prosoma and an opisthosoma, and in which the first two pairs of appendages are chelicerae and pedipalps and the remaining four pairs are walking legs. Evolutionarily, this bauplan has been a highly successful one.

As mentioned earlier, the great success of spiders seems to have been due in large part to the evolution of complex behaviors associated with silk and web production. Because we are paying special attention to members of the order Araneae in this chapter, and because silk production is so important to nearly all facets of their lives, we present a special section on spider silk and its uses.

Spinnerets, Spider Silk, and Spider "Webs"

Spider silk is a complex fibrous protein composed mostly of the amino acids glycine, alanine, and serine. The silk is produced in a liquid, water-soluble form that transforms into an insoluble thread after it leaves the body. This transformation involves as much as a tenfold increase in the molecular weight of the silk protein, which is enhanced by the formation of intermolecular bonds. Spider silk is stronger than steel and twice as elastic as nylon! The best spider silk is said to be five times as strong as steel. Spider silk is so strong that in some parts of the world people actually use it as fishing nets by stacking numerous webs on top of one another.

A spider's silk-producing apparatus is located in the opisthosoma and comprises various sets of glands. The liquid silk produced by these glands is secreted into ducts through which it passes to the spinnerets (Figure 19.7). Each spinneret bears some combination of small and large tubes called **spools** and **spigots**, respectively, which open to the outside. The apparatus spins the silk into threads of different thicknesses. The spinnerets are actually highly modified opisthosomal appendages, and they retain some musculature that allows for their movement during spinning (Figure 19.7C).

There are about six different kinds of silk glands, each producing a different kind of silk. By spinning various kinds of silks, at different diameters, spiders produce a variety of threads of different qualities for different functions. The numbers and kinds of glands and spinnerets vary. Some of the primitive "segmented" spiders have four pairs of spinnerets. Most other spiders have three pairs, although some have only one or two.

Another spinning organ, called the **cribellum** (Figure 19.7D), occurs in several families of araneomorph spiders. Some authors divide the araneomorph spiders into two groups based on the presence or absence of the cribellum, but it is now generally agreed that loss of the cribellum has occurred several times and that such a division is artificial. The cribellum is a platelike structure, anterior to the usual spinnerets, that bears many small spigots (up to 40,000 in some species). The cribellate silk is emitted in many extremely fine threads and then combed into a delicate mesh (Figure 19.7F) by a row of bristles (the **calamistrum**) on the fourth walking legs (Figure 19.7E). Some orb-weavers (e.g., the Uloboridae) use the resulting mesh, known as a "hackle band," as a prey-capture net (Figure 19.15J).

The varied uses of silk are intimately associated with nearly every aspect of the lifestyle and habits of spiders, as explained more fully in following sections. Various kinds of silk are used as safety lines and climbing lines; for the construction of nests, cocoons, and traps; to wrap prey for brief storage; as egg sacs and sperm platforms; and to line burrows. Many newly hatched spiderlings spin long, thin, threads to ride the wind for aerial dispersal. Perhaps the most interesting and familiar use of silk by spiders, however, is prey capture. Various sorts of snares and nets are spun, most of which can be loosely referred to as "webs." The construction and use of prey-capture webs are discussed in more detail in the section on feeding below, but in general these webs serve both to trap prey and, through vibrations, to signal the spider of its presence. Indeed, most spiders live in a world dominated by vibrations, a world where a potential meal, or predator, or mate reveals itself with characteristic resonance patterns. We present here a brief description of the construction of a familiar orb web, as commonly produced by members of the families Araneidae and Tetragnathidae.

The spinning of an orb web takes place in three phases, which are apparently genetically programmed (Figure 19.8). The first phase is the construction of a supportive Y-shaped framework and a series of radiating threads. The upper branches of the Y are initially laid down as a horizontal thread between two objects in the spider's environment. The spider sits in one spot and secretes the thread into the air; the loose end is then carried by air movements and flutters about until it contacts and sticks to an object. The spider then tacks down its end of the thread, moves to the center of the horizontal line, and drops itself on a vertical thread. The vertical thread is pulled taut and attached, thereby producing the Y-shaped frame. The intersection of the three branches of the Y becomes the hub of the final web, and it is from this point that the radii are extended. Radial threads are attached to the frame threads. Once this initial phase is complete, the spider quickly lays down a temporary spiral thread, starting from the hub, as the second phase of construction. This spiral thread, along with the initial framework of threads, serves as a working platform during the third and final phase of web building—the production of the sticky spiral or prey trap. This final thread is always coated with a glue that automatically assumes a beaded distribution after it is deposited (Figure 19.9A). As the sticky spiral is laid down, the temporary platform spiral is removed or eaten.

Some orb-weavers (e.g., *Argiope*) produce a dense mesh of silk, called a **stabilimentum**, across the hub of their webs (Figure 19.9B). Although it has often been considered to function as a structural stabilizer, as a patch of camouflage for the spider, or as a device to capture drinking water, many workers now suggest that the stabilimentum serves as a visual signal to warn away large flying animals, such as birds, that could easily damage the web.

Many orb-weavers (e.g., *Araneus*) can produce an entire web in less than 30 minutes, and most build a new web every night. They do not "waste" the silk of the old web, but eat it before or during the production of a new one. Radioactive labeling experiments show that the silk proteins from eaten webs appear at the spigots in new threads soon after ingestion, often within a few minutes!

The whole business of orb web production is remarkably well programmed; in fact, the angles between radii

Figure 19.7 Silk glands and spinnerets in spiders.
(A) The silk glands and spinnerets of *Nephila*, the golden silk
spider. Only one member of each pair of glands is shown.
(B) The spinnerets of the orb-weaver *Araneus* (external view).
(C) Cutaway view of the posterior end of the opisthosoma
(*Tegenaria*). Note the spinneret muscles. (D) The cribellar
spigots of *Hypochilus* (SEM, 1,600×). (E) The comblike
calamistrum on the fourth walking legs of *Amaurobius*, used
to brush the silk threads as they emerge from the cribellum.
(F) The cribellate silk of *Uloborus*.

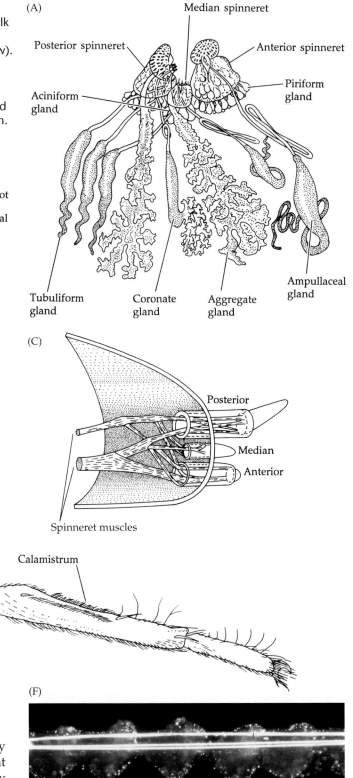

(A)

Posterior spinneret
Median spinneret
Anterior spinneret
Aciniform gland
Piriform gland
Tubuliform gland
Coronate gland
Aggregate gland
Ampullaceal gland

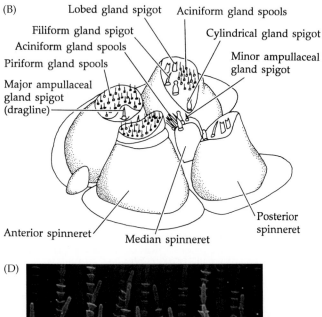

(B)

Lobed gland spigot
Aciniform gland spools
Filiform gland spigot
Cylindrical gland spigot
Aciniform gland spools
Piriform gland spools
Minor ampullaceal gland spigot
Major ampullaceal gland spigot (dragline)
Anterior spinneret
Median spinneret
Posterior spinneret

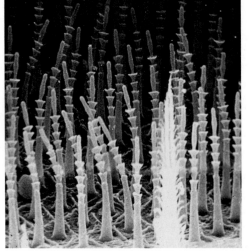

(D)

(C)

Posterior
Median
Anterior
Spinneret muscles

(E) Calamistrum

(F)

are constant. Web building appears to depend entirely
on tactile input, for normal webs are produced at night
and by experimentally blinded spiders. Even gravity
does not seem to be necessary, as illustrated by two famous orb-weaver spiders sent into space aboard Skylab.

Some tropical species cooperate in web building,
prey capture, and even spiderling rearing. Such social
activity has been reported for some 20 species in at least
six families. Spider sociality ranges from loose overwintering aggregations to the more permanent aggregations of certain orb-weavers (e.g., *Cyrtophora moluccen-

sis*), in which the webs of several individuals share common support lines. In *Anelosimus eximius* and some
other species, eggs are positioned in the web and tended
by many different females; hatched spiderlings also are
fed by regurgitation by various females.

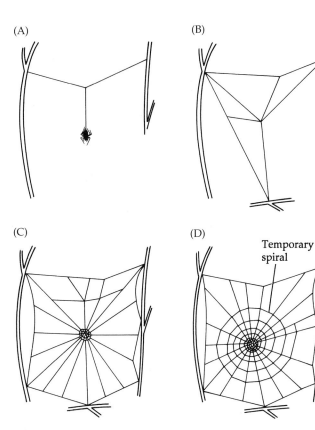

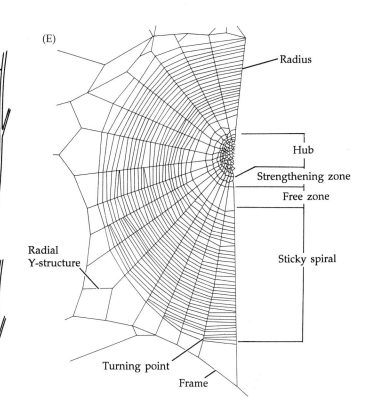

Figure 19.8 Construction of an orb web. (A) Formation of the Y-shaped frame. (B,C) Addition of radial threads. (D) Addition of the temporary spiral or working platform. (E) A portion of the completed orb web, with the temporary spiral replaced by the sticky spiral.

For well over a century specialists have debated various notions about the evolution of spider silk and webs. Shear (1994) offers a fascinating overview of these arguments, along with his own ideas on the subject. Toward the end of the nineteenth century, Henry McCook suggested that silk may have evolved before the origin of spiders in vermiform terrestrial arthropods with multiple excretory structures, such as coxal glands, opening near the bases of walking limbs. Trails of waste material from these glands may have functioned in mate or burrow location, or in other such tracking activities. McCook proposed that the more posterior glands lost their excretory function and began producing trail-marking proteins (silk) that did not degrade quickly. Some of the posterior appendages associated with these glands eventually became spinnerets.

Many spider specialists today still favor McCook's hypothesis, but other ideas have been proposed. For ex-

Figure 19.9 (A) A thread of the sticky spiral of an orb web with evenly distributed adhesive droplets. (B) *Argiope* on its orb web with a stabilimentum. (C) Funnel web of a funnel web spider (Dipluridae).

ample, one idea is that silk was first used by supposed marine spiders to line and reinforce their mud burrows. Another view is that silk was originally produced by oral glands and used to encase egg clusters. The origin of silk and the subsequent evolution of its uses remain fascinating but controversial subjects for investigation.

Locomotion

Chelicerate locomotion follows the principles of arthropod joint articulation and leg movements discussed in Chapter 15. Except for the xiphosurans and a few aquatic arachnids, the legs must also be strong enough to support the body on land. Walking by terrestrial chelicerates demands that the body be supported off the substratum and that the four pairs of legs move in sequences that maintain the animal's balance.

The xiphosurans are slow benthic crawlers and shallow burrowers, utilizing their stout prosomal limbs to push their heavy bodies over and through the sand. The legs are close together (Figure 19.2B), and thus coordination of movement sequences is essential. Horseshoe crabs are also able to swim, upside down, by beating the opisthosomal appendages.

The details of scorpion walking patterns have been examined by Root and Bowerman (1978). During simple forward walking, each of the eight limbs moves through the usual power and recovery strokes. In scorpions (and many other arthropods) the joints between the coxae and the body are virtually immovable and do not contribute to the overall motion of the limbs. Not all of the walking legs move in the same pattern. The tips of the anterior legs are brought quite high off the ground during their recovery strokes; they may be used to feel ahead as the animal moves (Figure 19.10A). Also, the tips of each pair of legs extend different distances from the body, allowing for stride overlap without contact between legs (Figure 19.10B). Leg movements in scorpions do not follow the usual metachronal model. Rather, the typical sequence of movement along one side of the body is leg 4, then 2, then 3, and finally 1, with the legs on the opposite side being generally, but not precisely, out of phase. The somewhat staggered overlap of the movement sequences produces a smooth gait during forward walking, unlike the jerky motion of an insect moving by the alternating tripod pattern described in Chapter 17. Like many arthropods, scorpions are also able to change speeds, turn abruptly, walk backward, and burrow in loose sand.

Spiders have evolved a number of methods of locomotion, all involving the usual leg motions inherent in jointed arthropod limbs, except that limb extension in spiders is assisted in part by hydrostatic pressure (involving the

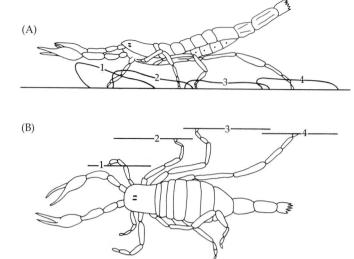

Figure 19.10 Walking in scorpions. Numbers denote legs 1–4. (A) The paths of the leg tips during their recovery strokes. Notice that the anterior legs are lifted higher off the ground than are the posterior legs. (B) Distance of each leg tip from the body.

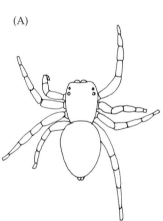

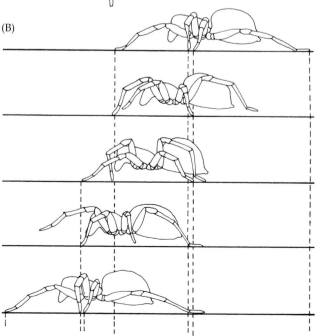

Figure 19.11 Walking in spiders. (A) A salticid with legs in walking positions (dorsal view). (B) A lycosid during slow walking (lateral view). Vertical dotted lines connect each leg through its power and recovery strokes relative to the forward progress of the body. Note the large degree of overlap in leg movements, particularly of the first two legs. Tangling is prevented in part by keeping the tips of adjacent legs at different distances from the body.

femur, patella, and tarsus–metatarsus joints). During normal spider walking, the eight ambulatory limbs move in a so-called **diagonal rhythm sequence** (Figure 19.11). That is, legs 2 and 3 on one side of the body are moving simultaneously with legs 1 and 4 on the opposite side. This gait maintains a four-point stance while distributing the body weight more or less equally among those appendages in contact with the substratum. During very slow walking, however, posterior to anterior metachronal waves of limb motion are detectable. The stride lengths of the legs overlap somewhat and vary with speed and direction of movement. The placement of the legs on the body, and the arcs through which they swing, prevent contact between legs.

A number of spiders (e.g., members of the family Salticidae) are also capable of jumping (Figure 19.12A). Propulsion is achieved primarily by a rapid extension of the fourth pair of legs. Once the spider is airborne, the front legs are extended forward and used in landing.

Salticids jump during normal locomotion, and also when capturing prey and escaping from predators.

Silk plays an important role in various methods of spider locomotion. When walking or jumping, most spiders continually produce a strong thread behind them, called a **dragline** (Figure 19.12B). The dragline is periodically tacked to the substratum, providing a safety rope for the wandering spider. Thus, a spider brushed from a surface does not fall to the ground; instead, it pays out dragline silk and dangles like a tethered mountaineer who has lost his footing. Silk is also used to provide a substratum on which spiders move. Web spinners crawl over their webs with limb movements more or less like those used in normal walking, except that stride lengths must correspond to the distances between threads in the web. Many spiders are able to move about on a single thread (Figure 19.12C). This may involve dropping vertically as a thread is paid out from the spinnerets, climbing up a vertical thread, or moving while hanging upside down on a horizontal thread (they do not walk

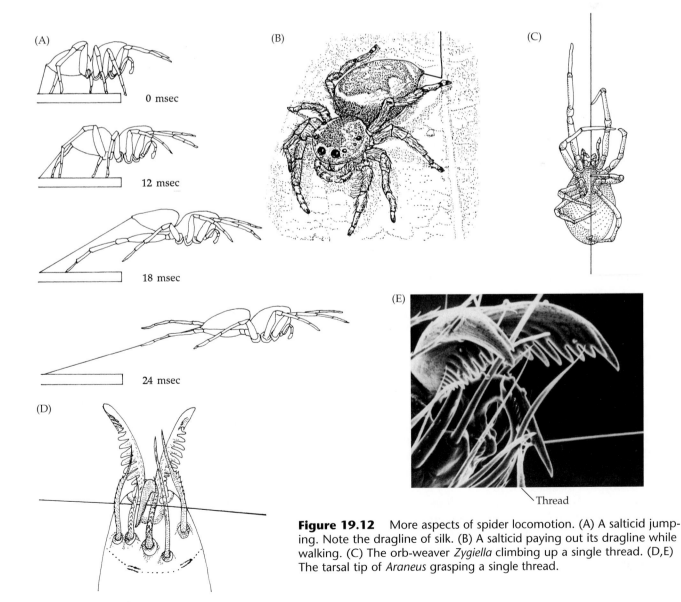

Figure 19.12 More aspects of spider locomotion. (A) A salticid jumping. Note the dragline of silk. (B) A salticid paying out its dragline while walking. (C) The orb-weaver *Zygiella* climbing up a single thread. (D,E) The tarsal tip of *Araneus* grasping a single thread.

atop single threads, tightrope fashion). Most spiders that are capable of these sorts of activities possess intricately fashioned "thread clamps" on some or all of the leg tips (Figure 19.12D,E).

Many spiders also excavate burrows, and a few lycosids and pisaurids (small, lightweight species) can walk on water (Figure 19.13). One species, *Argyroneta aquatica* (family Argyronetidae), actually lives under water, where it walks on submerged plants or swims. There are also several intertidal species that can tolerate submergence at high tides, in some cases by utilizing the air bubbles trapped in empty barnacle shells or other such structures.

Feeding and Digestion

Feeding. The basic chelicerate feeding strategy is one of prey capture followed by extensive external digestion and then ingestion of liquefied food or, more rarely, of small particulate material. There are exceptions to this pattern, of course, most of which involve drastic modifications of the mouthparts. For example, we have already mentioned the varied feeding habits among the mites and ticks, many of which are herbivores or parasites with piercing mouthparts. Whereas many chelicerates are highly specialized in terms of their feeding behavior, others are generalists. Horseshoe crabs, for example, commonly feed on a variety of invertebrates, including worms, molluscs (especially bivalves), crustaceans, and other infaunal and epibenthic creatures, but they also scavenge on almost any organic material. Food is gathered by any of the chelate appendages and passed to the gnathobases along the ventral midline, where it is ground to small bits and then moved forward to the mouth.

Scorpions feed mostly on insects, although some large species occasionally eat snakes and lizards. Most are nocturnal and detect prey mainly by highly sensitive mechanoreceptors. A fascinating method of prey location by the Mojave Desert sand scorpion, *Paruroctonus mesaensis*, has been described by Phillip Brownell (see References). Brownell noticed that *Paruroctonus* ignores both airborne vibrations (e.g., wingbeats) and visual input, but responds immediately to nearby prey that are in contact with the sand. This scorpion is even able to detect buried prey, which it immediately uncovers and attacks. Apparently the scorpion senses subtle mechanical waves set up in the loose sand by movements of the prey. Special mechanoreceptors in the walking legs are stimulated as the waves pass beneath the scorpion's limbs. The information is processed to determine the direction and approximate distance to the prey source.

Once a scorpion locates a victim, the prey is grasped by the chelate pedipalps. The opisthosoma is arched over the prosoma, bringing the telson and stinging apparatus into position to inject venom (Figure 19.14). Muscular contractions force the venom through a pore in the aculeus and into the prey. The venom is a neuro-

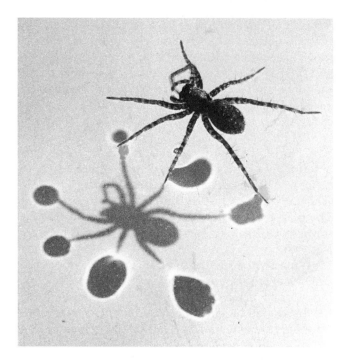

Figure 19.13 A wolf spider, *Pardosa amentata*, on the water surface. The right front leg is raised from the surface for cleaning. Notice the unequal distribution of body weight on the rest of the legs, indicated by the shadows of the "dimples" in the water surface; the weight is normally almost equally distributed when the spider rests on all eight legs.

toxin that can quickly paralyze and kill most prey; in fact, some scorpion venoms can kill large animals, including humans. Perhaps the most familiar of these dangerous scorpions are two species that occur in the American Southwest: the bark scorpion (*Centruroides exilicauda*) and the stripe-tailed scorpion (*Vaejovis spinigerus*). A North African species, *Androctonus australis*, produces a venom considered to be as potent as that of cobras. Other genera of potentially dangerous scorpions include *Buthus* and *Parabuthus* (Africa and the Middle East) and *Tityus* (South America). A well written and superbly illustrated account of medically important scorpions has been compiled by H. L. Keegan (1980).

The sequence of events in the ingestion of food by scorpions is typical of many arachnids. Once it has been captured and stung, the prey is passed to the chelicerae, which tear it into small bits. The gnathobases grind and mash the food as digestive juices are released through the mouth; this process reduces the food to a semiliquid form. As this organic "soup" is ingested, the hard parts are discarded, and more bits of food are moved between the gnathobases for processing.

Virtually all spiders are predatory carnivores, although young spiderlings in certain families may consume only pollen caught on web silk. With the exception of a few families (e.g., Salticidae, Oxyopidae, Thomisidae, and Lycosidae), spiders hunt or feed main-

(A)

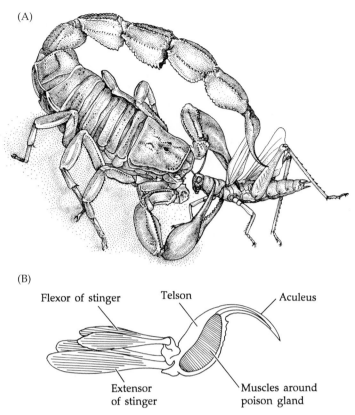

(B)

Flexor of stinger   Telson   Aculeus

Extensor
of stinger   Muscles around
poison gland

Figure 19.14 (A) The scorpion *Androctonus australis*, applying its stinging apparatus to a grasshopper while holding the prey with its pedipalps. (B) The scorpion's telson and stinging apparatus. The telson is normally held flexed; extension drives the aculeus into the prey.

ly at night. Most predatory spiders can be separated into two broad categories based on prey-capture strategies. The first is the more sedentary spiders that use some sort of silken web, trap, or net to catch prey. The second is the "wandering" spiders that actively hunt or ambush prey without the direct use of silk (although many wrap their victims after capture).

Earlier, we described the construction of the familiar orb web (Figure 19.8). When a potential prey item, such as a flying insect, strikes and adheres to a web, its movements send out vibrations that alert the spider to the presence of food. The spider then moves rapidly to the victim and bites it. Aranaeologists have likened the speed and grace of this attack in many spider species to that of mammalian predators such as cheetahs and leopards. Most spiders are solitary, but a few live and feed communally (Figure 19.15A).

Another type of silken trap is the horizontal **sheet web** produced by members of the families Linyphiidae and Agelenidae. Sheet webs are suspended by a network of supporting threads (Figure 19.15B). Insects become entangled in the web, or in the supporting threads, in which case the spider (resting on the underside of the web, or in a funnel-shaped home at the web periphery)

shakes the whole structure until the prey drops to the sheet. Once on the sheet, the prey is captured.

Most theridiid spiders build vertical **frame webs** (Figure 19.15C). Near their attachment to the substratum, the vertical trap threads are beaded with sticky liquid from special silk glands. Walking prey contact these sticky droplets and become trapped. Upon sensing movement in the web, the spider rushes to the prey, wraps it in silk, and bites it.

Members of the genus *Hyptiotes* (family Uloboridae) spin abbreviated orb webs of only three sectors (Figure 19.15D). The spider produces a tension thread from the point of convergence of the radii and a short attachment thread stuck to some solid object; the spider's body acts as a bridge between these two threads. When an insect strikes the web, the spider releases the tension thread, and the web, called a **spring trap**, collapses around the prey.

The most primitive kinds of "webs" are simple silken tubes with a single opening from which threads radiate outward (Figure 19.15F), a web form used by various lyphistiids. The spider resides in the tube, and the threads serve as "fishing lines" or trip lines that allow the spider to detect passing prey. An interesting modification of this system is seen in the purse web of *Atypus* (family Atypidae) (Figure 19.15G). Here the silken tube is mostly buried beneath the soil, with just a short portion of the blind end lying above the surface. Insects crawling over the exposed tube are detected by the spider, and the orthognathous chelicerae make two parallel slices in the tube wall near the prey. The cheliceral fangs are extended through the incisions to grab the victim and pull it through the tube wall. After the prey is killed, the tear in the tube is repaired.

Most spiders simply set their "traps" and wait for prey, but others actually manipulate silken structures to catch insects. The Australian ogre-faced spider, *Dinopis* (family Dinopidae), produces a rectangular web of cribellate threads and holds it between its front walking legs. When an insect is visually detected, the spider sweeps the net around it. The bolas spider, *Mastophora* (family Araneidae), is among the most bizarre hunting specialists. While hanging from a suspension thread, the bolas spider "throws" a catching thread tipped with sticky liquid to "lasso" its prey (Figure 19.15H). Bolas spiders hunt at night and specialize in feeding on male moths of the genus *Spodoptera*. *Mastophora* releases an airborne chemical that mimics the sex pheromone of the female *Spodoptera*, thus attracting males within reach of the catching thread and greatly increasing the likelihood of capture success.

Whereas most sedentary spiders detect their prey by sensing web vibrations, the "wandering" spiders use a variety of methods. Some wolf spiders and most jumping spiders locate prey visually, while many others sense vibrations (e.g., wingbeats or walking movements) or rely simply on accidental contact. Some actually chase

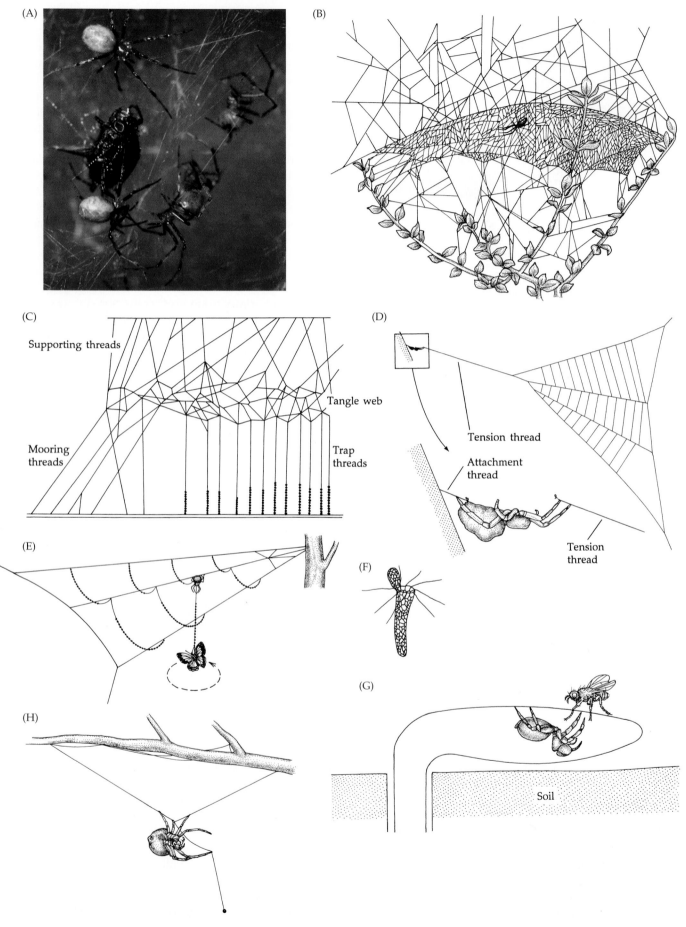

(I)

(J)

Figure 19.15 The use of spider silk for prey capture. (A) Several *Anelosimus* spiders and their prey, an unusual case of communal prey capture. (B) The horizontal sheet web of *Linyphia triangularis*. (C) The vertical frame web of *Steatoda castanea*. (D) The abbreviated orb web of *Hyptiotes*. The spider is stretched between the attachment thread and the tension thread. (E) *Pasilobus* "reeling in" an insect caught on a sticky thread. (F) The silken tube and trip threads ("fishing lines") of a primitive spider, *Liphistius* (suborder Mesothelae). (G) The purse-web spider *Atypus* inside its silken tube. The spider has sensed the presence of an insect on the tube and is about to grab it. (H) The bolas spider, *Mastophora*, swinging its catching thread. (I) A crab spider (Thomiscidae) that changes color to match the flower it is on, capturing a skipper (a butterfly). (J) The giant Neotropical golden orb-weaver *Nephila clavipes*, which spins a "net" held between the legs and thrown like a cast net to capture prey.

prey; others, such as certain trapdoor spiders, lie in ambush and wait for victims to come close enough to grab.

Recent studies by Jackson and Wilcox (1998) reveal some fascinating and surprising insights into the predatory strategies of certain jumping spiders. Unlike most spiders, whose behaviors appear to be genetically programmed, members of the genus *Portia* exhibit trial and error strategies and learning in their predation on other spiders. *Portia* really doesn't look much like a spider; rather, it resembles crumpled leaves or twigs. Some species utilize this camouflage in what Jackson and Wilcox call aggressive mimicry. To do so, *Portia* climbs onto the web of another spider and begins to generate vibrations that mimic those of a trapped insect. When the resident spider responds, *Portia* attacks and kills it. More remarkable is that *Portia* often tries various vibration patterns on a web, and when a successful one is found, continues to use only that one on the webs of the same prey species in the future.

Regardless of the method of prey location and capture, once contact is made, the spider pulls the victim to the chelicerae and bites it, inserting the fangs and injecting

venom from poison glands within the prosoma (Figure 19.16A). The prey is quickly immobilized or killed by the poison. An interesting exception to this grabbing-and-biting pattern is displayed by the spitting spiders (family Scytodidae), some of which are social. The poison glands of *Scytodes* include a glue-producing portion along with the usual venom-secreting cells. A mixture of venom and glue is shot from pores in the cheliceral fangs by muscular contraction of the glands; thus the prey is captured without direct contact (Figure 19.16B).

Many spiders wrap their prey in silk to some extent prior to feeding, even if silk is not used in its actual capture. Many hold the victim and wrap it prior to biting it (e.g., theridiids and araneids). Very active insects caught in orb webs are generally wrapped immediately, thereby preventing possible damage to the web or its owner (Figure 19.16D). Potentially dangerous prey, such as large stinging insects, are generally handled in this manner.

Nearly all spiders possess poison glands that produce proteinaceous neurotoxins. The toxicity of spider venom is quite variable, and only about two dozen species are considered dangerous to humans. Among these are the American black widow (*Latrodectus mactans*), a Brazilian wolf spider (*Lycosa erythrognatha*), the brown recluse spider (*Loxosceles reclusa*), an Australian funnel-web spider (*Atrax robustus*), and some species of trapdoor spiders belonging to the family Ctenidae (e.g., *Phoneutria fera*).

Spiders, like most other chelicerates, ingest their food in a liquid or semiliquid form. The chelicerae of most spiders have dentate gnathobases with which the prey is mechanically pulverized, while at the same time the food is flooded with digestive juices. Except for the hard parts, the prey is thus reduced to a partially digested broth. Bristles bordering the mouth and thousands of overlapping cuticular plates in the pharynx serve as filters so that only very small particles (< 1 μm) enter the gut. Members of some families (e.g., Theridiidae and Thomisidae) lack cheliceral teeth. These spiders simply puncture their prey and then flush digestive juices in

(A)

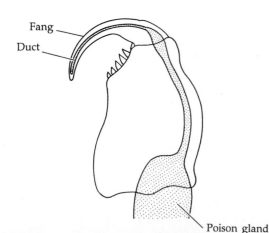

Fang

Duct

Poison gland

(B)

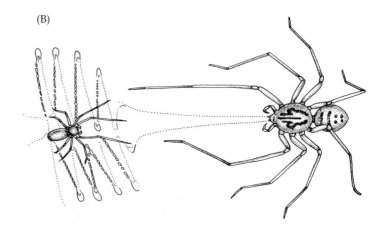

(C)

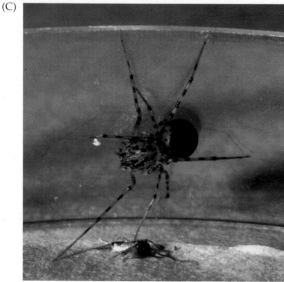

(D)

Figure 19.16 More aspects of prey capture in spiders. (A) A spider chelicera. Note the poison gland and the duct leading to the tip of the fang. (B) The spitting spider, *Scytodes*, captures prey by spraying a combination of poison and adhesive over its victim. (C) *Scytodes* captures a cricket. (D) An orb-weaver, *Araneus gemma*, wrapping its prey.

and out of the wound. The liquefied innards of the victim are then sucked out of its body and ingested.

The digestive tract. The digestive system of chelicerates follows the basic arthropod plan of foregut, midgut, and hindgut, the first and last parts being lined with cuticle (Figure 19.17). The foregut is often regionally specialized. In the xiphosurans (e.g., *Limulus*), the foregut loops anteriorly to form an esophagus, crop, and gizzard, the last of these bearing sclerotized ridges that grind ingested particles (Figure 19.17A). In many arachnids, portions of the foregut are modified as pumping organs for sucking in liquefied food. In scorpions this function is served by a muscu-

lar pharynx, and in spiders by an elaborate sucking stomach (Figure 19.17B). The pharynx of spiders may contain chemosensory cells that function as "taste" receptors.

The chelicerate midgut bears paired digestive ceca and is the site of final chemical digestion and absorption (Figure 19.17C). Xiphosurans have two pairs of ceca arising from the anterior part of the midgut, followed by an intestine, a short rectum (the hindgut), and the anus on the posterior margin of the opisthosoma. In *Limulus*, enzymes are produced by the midgut wall and secreted into the lumen. Apparently only preliminary protein digestion occurs extracellularly, and final breakdown takes place in the cells of the digestive ceca after absorption.

Most spiders possess four pairs of digestive ceca in the prosoma and frequently additional branched ceca in the opisthosoma (Figure 19.17B,C). The midgut expands as a spacious "mixing chamber," called the **stercoral pocket,** near its junction with a short rectum. Malpighian tubules arise from the midgut wall near the origin of the stercoral pocket. The anus is located on the opisthosoma near the spinnerets.

The midgut of scorpions bears six pairs of digestive

ceca (Figure 19.17D). The first pair, called salivary glands, lies within the prosoma and produces much of the digestive juice used in preliminary external digestion. The remaining five pairs are highly convoluted and lie in the opisthosoma. These ceca produce the enzymes for final digestion and are the site of absorption of the digestive products. Two pairs of Malpighian tubules arise from the posterior region of the midgut, just in front of the short rectum. The anus is on the last opisthosomal segment.

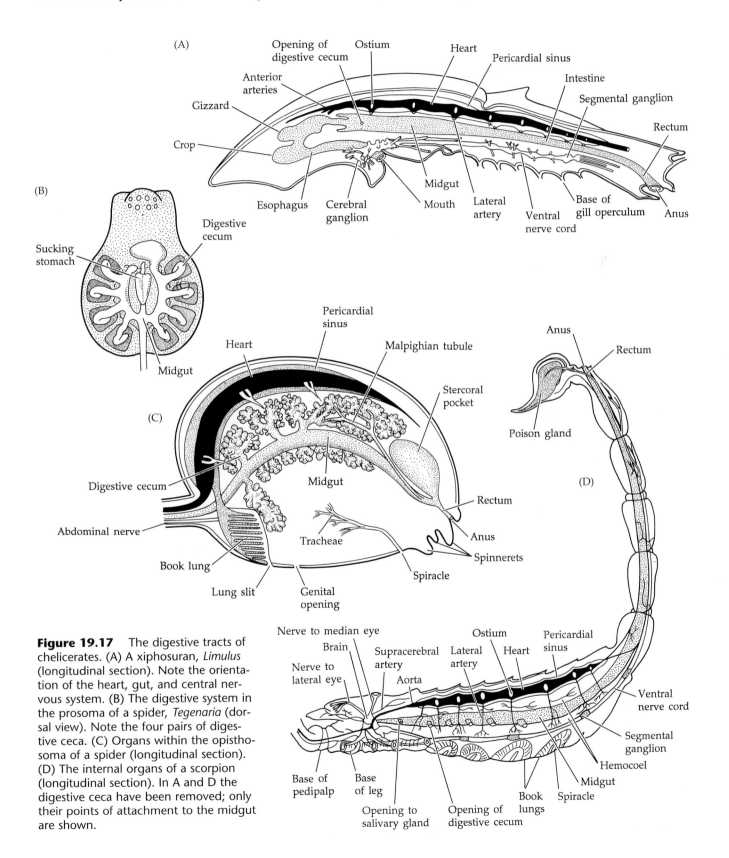

Figure 19.17 The digestive tracts of chelicerates. (A) A xiphosuran, *Limulus* (longitudinal section). Note the orientation of the heart, gut, and central nervous system. (B) The digestive system in the prosoma of a spider, *Tegenaria* (dorsal view). Note the four pairs of digestive ceca. (C) Organs within the opisthosoma of a spider (longitudinal section). (D) The internal organs of a scorpion (longitudinal section). In A and D the digestive ceca have been removed; only their points of attachment to the midgut are shown.

Circulation and Gas Exchange

The chelicerate circulatory system, like that of other arthropods, consists of a dorsal ostiate heart situated within a pericardial sinus and giving rise to various open-ended vessels (Figures 19.17, 19.18). Blood leaves these vessels and enters the hemocoel, where it bathes the organs and supplies the structures of gas exchange before returning to the heart. The complexity of the system is primarily a function of body size; some very tiny chelicerates (e.g., palpigrades and some mites) have lost much or all of their circulatory structures—for them, gas exchange is cutaneous. On the other hand, the xiphosurans are large animals, and their bauplan demands a substantial circulatory mechanism to move the blood around inside the rigid body covering. The large tubular xiphosuran heart bears eight pairs of ostia, and it is attached to the body wall by nine pairs of ligaments that extend through the pericardium (Figure 19.17A). The organs of these big creatures are supplied with blood through an extensive arterial system arising from the heart and opening into the hemocoel close to the organs themselves. In the opisthosoma, a major ventral vessel gives rise to a series of afferent branchial vessels to the

gill books. Efferent vessels carry oxygenated blood to a large **branchiopericardial vessel** leading back toward the heart.

The gas exchange organs of xiphosurans are unique among the chelicerates. The presence of gills is, of course, associated with their aquatic lifestyle. The structure of these opisthosomal **gill books** provides an extremely large surface area, a necessity for adequate gas exchange by these large animals (Figures 19.2B, 19.18A). Each gill bears hundreds of thin lamellae, like pages in a book. The blood within the lamellae is separated from the surrounding sea water by only a thin wall. Water is moved over the lamellae by metachronal beating of the gills; these movements also cause blood flow into (on the forward strokes) and out of (on the backward strokes) the gill sinuses.

The heart of a spider lies within the opisthosoma and bears two to five pairs of ostia (Figure 19.18B). It is suspended within the pericardial sinus by several ligaments attached to the inside of the exoskeleton. These suspensory ligaments are stretched during systole as blood is pumped from the heart into arteries. The elasticity of the ligaments then effects diastole, expanding the heart and drawing blood into it from the pericardial sinus. The routes of the major arteries ensure that an ample supply of oxygenated blood reaches the major organs, particularly the central nervous system, mus-

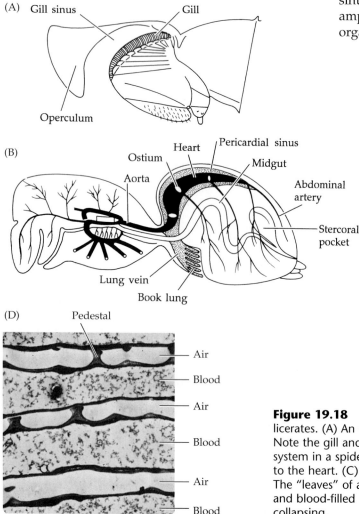

(A) Gill sinus / Gill / Operculum

(B) Ostium / Aorta / Heart / Pericardial sinus / Midgut / Abdominal artery / Stercoral pocket / Lung vein / Book lung

(D) Pedestal / Air / Blood / Air / Blood / Air / Blood

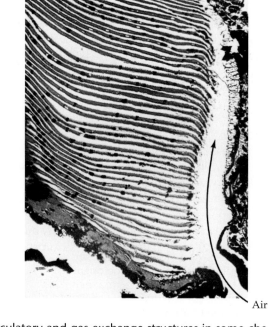

(C)

Air

Figure 19.18 Circulatory and gas exchange structures in some chelicerates. (A) An opisthosomal appendage of *Limulus* (posterior view). Note the gill and opercular parts. (B) Major elements of the circulatory system in a spider. Note the direct route of blood from the book lung to the heart. (C) The book lung of the spider, *Lycosa* sp. (section). (D) The "leaves" of a book lung (section). Note the separation of air spaces and blood-filled leaves. Cuticular pedestals keep the air spaces from collapsing.

cles, and the hemocoelic spaces in the limbs in which blood pressure aids in leg extension. From the hemocoel, blood is channeled back to the pericardial sinus and the heart.

Aerial gas exchange structures in spiders include book lungs and tracheae. Generally, the more primitive spiders (suborder Mesothelae and superfamily Orthognatha) possess two pairs of book lungs but no tracheae, whereas members of the superfamily Araneomorphae usually have one pair of book lungs and a system of tracheal tubes. Since the tracheae occur only in advanced groups among spiders, it is likely that they evolved separately from those of insects and myriapods and represent a convergent feature in the two subphyla. Phylogenetic analyses, both morphological and molecular, support this contention.

Spider book lungs are located in the second, or the second and third, segments of the opisthosoma. They open to the outside via spiracles, or **lung slits**, near the epigastric furrow (Figure 19.5C). Since they are localized and do not extend far into the body, book lungs must receive sufficient circulating blood to ensure adequate distribution of oxygen throughout the body and removal of carbon dioxide from the internal organs. Just inside each lung slit is an expanded chamber, the **atrium**, from which numerous flattened air spaces extend into the hemocoel (Figure 19.18D,E). These leaflike pockets of air are separated from one another by thin blood-filled extensions of the hemocoel. Although the book lungs are themselves relatively small, this structural arrangement provides a very large surface area between the "pages" of the book lungs and the circulatory fluid. Blood that has passed by these surfaces returns directly to the pericardial sinus via the **lung vein** (Figure 19.18B).

Spider tracheae open to the outside through one or two spiracles located posteriorly on the third opisthosomal segment (Figures 19.5B,C, 19.17C). These tracheae probably evolved from the book lungs and muscle apodemes on this segment in more primitive spiders. The spiracles open inward to simple or branched tubes. In spiders, the open inner ends of the tracheae do not bring the oxygen supply into direct contact with tissues, as they do in many insects; rather, a small amount of blood is necessary as a diffusion medium. When the tracheal system is extensive, there is usually a reduction in the structural components of the circulatory system.

The circulatory system of scorpions is very similar to that of spiders, except that it is constructed to accommodate an elongate body. The tubular heart bears seven pairs of ostia and extends through most of the mesosoma (Figure 19.17D). An extensive set of arteries delivers blood to the hemocoel throughout the body and to four pairs of mesosomal book lungs.

Except where reduced or absent, the circulatory system of other chelicerates is built along the same general plan as described above. Gas exchange is cutaneous in the palpigrades and some mites, but in other chelicerate groups it occurs through book lungs (Uropygi, Schizomida, and Amblypygi) or tracheae (Ricinulei, Pseudoscorpionida, Solpugida, Opiliones, and Acari).

The blood of chelicerates has been most extensively studied in spiders and in xiphosurans. Because *Limulus* is large, its blood chemistry has been especially well studied, and many horseshoe crabs have made the supreme sacrifice for science at the hands of laboratory physiologists—so many, in fact, that many populations of the American species (*L. polyphemus*) have become threatened. Hemocyanin, the common respiratory pigment in chelicerates, is dissolved in the blood plasma. At least in spiders, hemocyanin serves primarily for oxygen storage rather than for immediate transport and delivery of oxygen to the tissues. Hemocyanin has a very high affinity for oxygen, releasing it only when surrounding oxygen levels are quite low. At least some spiders are able to survive for days after their air supply has been cut off experimentally by covering the spiracles. Apparently they obtain sufficient oxygen from their hemocyanin-bound stores and some cutaneous exchange.

Chelicerate blood also contains various cellular inclusions, but the functions of most of them are not well understood. The blood of *Limulus* includes amebocytes that may provide clotting agents. Several kinds of blood cells occur in spiders. Interestingly, it seems that all of them originate as undifferentiated cells from the muscular portion of the heart wall itself. These cells are released into the blood, where they mature and differentiate. Functions that are attributed to chelicerate blood cells include clotting, storage, combating infections, and aiding in sclerotization of the cuticle.

Excretion and Osmoregulation

The main excretory structures of chelicerates are coxal glands and Malpighian tubules, although many groups possess additional supplementary waste removal mechanisms.

Xiphosurans have two sets of four coxal glands each, arranged along each side of the prosoma near the coxae of the walking legs. The glands on each side of the body converge to a coelomic sac, from which a long, convoluted duct arises. The duct leads to a bladder-like enlargement that connects to an excretory pore at the base of the last walking legs. Surprisingly little is known about excretory physiology in xiphosurans. Apparently the coxal glands extract nitrogenous wastes from the surrounding hemocoelic sinuses and carry them to the outside. The coxal glands and their associated tubule system also function in osmoregulation, as evidenced by the production of a dilute urine when the animal is in a hypotonic medium. The digestive ceca probably aid in excretion of excess calcium by removing it from the blood and releasing it into the gut lumen.

The problems of excretion and water balance are obviously much more critical to terrestrial chelicerates than to horseshoe crabs, and land-dwelling arachnids display a variety of structural, physiological, and behavioral adaptations to cope with them. Coxal glands persist in many arachnids (spiders, scorpions, palpigrades); in these animals, the glands lie within the prosoma and open on the coxae of some walking legs. The degree to which coxal glands function in excretion and osmoregulation varies among arachnids, but they are considered far less important than the Malpighian tubules.

The Malpighian tubules of arachnids arise from the posterior midgut. They are not homologous with the Malpighian tubules of insects or myriapods, which arise from the hindgut and are thus of ectodermal origin. The tubules branch within the hemocoel of the opisthosoma, where they actively accumulate nitrogenous waste products and release them into the gut for elimination along with the feces (Figure 19.17C). In spiders, the wastes from the tubules and the gut are mixed in the stercoral pocket prior to release from the anus. The excretory action of the Malpighian tubules is often supplemented by other mechanisms, such as the coxal glands. Nitrogenous wastes also are accumulated in the cells of the midgut wall and released into the lumen. In addition, waste products are picked up and stored by special cells called **nephrocytes**, which form distinct clumps in various parts of the prosoma.

Terrestrial arachnids produce complex, insoluble, nitrogen-containing excretory compounds. The major excretory product is guanine, although uric acid and other compounds also occur. Because these compounds are of low toxicity, they can be stored and excreted from the body in semisolid form, thus conserving water.

Terrestrial arachnids also display various behavioral adaptations to avoid desiccation. Most arachnids are nocturnal, remaining in cooler or more humid, protected places during the daytime. Some spiders actively drink water during dry periods or when they lose blood through injury. Desert scorpions must tolerate not only low humidities but also very high daytime temperatures. They typically bury themselves in sand or soil, or hide under rocks or tree bark, during the day. In addition, some species exhibit an adaptive behavior called **stilting**, wherein the body is raised off the substratum to allow air to circulate underneath. While thought to be mainly a cooling device, this behavior, by lowering the body temperature, probably also slows the rate of evaporative desiccation. Some scorpions are also able to withstand large losses of body water—as much as 40 percent of their body weight—with no ill effects.

Nervous System and Sense Organs

As in all arthropods, the external body form of chelicerates is generally reflected in the structure of the central nervous system. These animals show various degrees of compaction and fusion of the body somites and the associated nervous system components while still conforming to the basic annelid–arthropod plan. The cerebral ganglia, or brain, includes the protocerebrum and tritocerebrum; the deutocerebrum is absent. The tritocerebrum generally contributes to the circumenteric connectives, which unite ventrally with a large ganglionic mass formed in part by the fusion of paired anterior ganglia of the ventral nerve cord. In xiphosurans and scorpions this subenteric neuronal mass includes all of the prosomal ganglia, whereas in spiders even the opisthosomal ganglia fuse anteriorly. Thus in most spiders the adult nervous system is no longer obviously segmented (except in some members of the suborder Mesothelae), although a chain of ventral ganglia is evident during development. The ventral nerve cord persists in the opisthosoma of xiphosurans and has five segmental ganglia; in scorpions it has seven ganglia (Figures 19.17, 19.19).

The protocerebrum and tritocerebrum give rise to nerves to the eyes and chelicerae, respectively. In spiders the cheliceral nerves actually emerge from the protocerebrum, but they can be traced histologically to their origin in the hind part of the cerebral ganglia below the gut. The ventral (subenteric) ganglionic mass, which includes the fused segmental prosomal ganglia, gives rise to nerve tracts to the walking legs and, in spiders, bears a pair of abdominal ganglia from which arise branching nerves to the opisthosoma. Segmental ganglia on the ventral nerve cord in xiphosurans and scorpions serve the opisthosomal appendages, muscles, and sense organs.

In Chapter 15 we discussed some of the qualities of arthropod sense organs in terms of the imposition of the exoskeleton. Xiphosuran sense organs include tactile mechanoreceptors in the form of various spines and bristles, proprioceptors in the joints, chemoreceptors, and photoreceptors. The prosoma bears two simple eyes near the dorsal midline and two lateral compound eyes (Figure 19.2A). The median eyes are pigment cups, but each contains a distinct cuticular lens. The lateral eyes are compound rhabdomeric units—structures not found in any other members of the Cheliceriformes. The thousand or so ommatidia per eye are very large and rather loosely packed together. While the resolving power of xiphosuran eyes has long been debated, certainly *Limulus* can detect movement and changes in light intensity and direction. Barlow (1990) demonstrated that the light sensitivity of these receptors is regulated on a 24-hour cycle by instructions from the brain! At night, these signals increase the eyes' sensitivity to light up to a million times over the daytime levels. Thus, *Limulus* may see as well at night as it does during daylight hours. It may also be capable of perceiving clear images, as male horseshoe crabs have been experimentally shown to be attracted to "models" of females.

Arachnids possess well developed sense organs upon which much of their complex behavior depends. Most of the "hairs" on a spider or scorpion body are mechanoreceptors, collectively called **hair sensilla**.

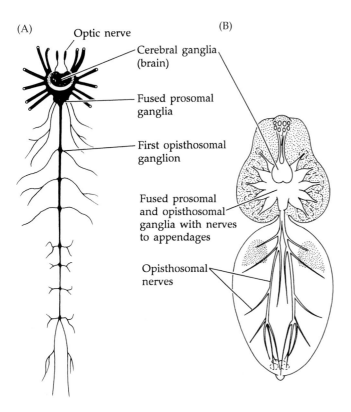

Figure 19.19 The central nervous system of (A) a scorpion and (B) a spider.

Simple tactile hairs (or setae) cover much of the body surface and respond to direct physical contact. A second type of hairs, called **trichobothria**, are found on the appendages. They are much thinner than simple tactile hairs and are extremely sensitive (Figure 19.20A,B). Trichobothria are stimulated by airborne vibrations, such as those caused by beating insect wings, natural air currents, and probably some sound frequencies.

Additional mechanoreceptors of arachnids include **slit sense organs** (Figure 19.20C). These structures may occur as single slit sense organs (or **slit sensilla**) or in groups of parallel slits called **lyriform organs**. Slit sensilla are deep grooves in the cuticle associated with sensory neurons. They detect a variety of mechanical stimuli that impose physical deformation on the cuticle around the slit. Depending on their location and orientation, spider slit sensilla serve as proprioceptors (by sensing leg movements and position), as georeceptors (by measuring bending of the pedicel under the weight of the opisthosoma), as direct mechanoreceptors (by sensing direct external pressure on the cuticle), as vibration sensors, and even as phonoreceptors.

Not all of the hairs on some spiders are sensory in function. Marshall (1992) describes a remarkable defensive behavior involving hairs in the tropical burrowing tarantula, *Theraphosa leblondi*. Members of this species are huge, with leg spans of over 25 cm and weights exceeding 100 grams. When threatened by a potential predator, *Theraphosa* emits a loud hiss, perhaps as a warning. If the threat continues, the spider uses a rear leg to brush loose a mass of barbed hairs from its abdomen. As described by Marshall (1992), these hairs caused a burning sensation on his hands and in his throat, which persisted for several days. Such an audible warning followed by an effective defense mechanism may be more common among spiders than is currently known.

Efficient proprioceptors occur in the walking legs and pedipalps of all chelicerates and are particularly well developed in arachnids. By virtue of their position and number in different joints, they convey information about the direction and velocity of appendage motion as well as the position of the limbs relative to the body and to one another. These "true" proprioceptors appear to work in concert with the lyriform organs.

Chemoreception in arachnids involves sensing both liquid and airborne chemicals that contact the body. This dual ability can be likened to capacities for both taste (contact chemosensitivity) and smell, or olfaction (distance chemosensitivity). The olfactory sense plays major roles in prey location and, for those species in which females release sex pheromones, in mating. The most important chemoreceptors are probably the hundreds of erect, hollow hairs with open tips that are present on the pedipalps and other areas around the mouth and are most abundant on the tips of limbs that contact the substratum. Dendrites of sensory neurons extend through the hollow hair shaft to the open tip, where they are directly stimulated by chemicals (Figure 19.20D). Some spiders also bear humidity detectors called **tarsal organs** (Figure 19.20E).

Scorpions have a pair of large, unique, comblike structures, the **pectines**, on the ventral surface of the mesosoma (Figure 19.6C). After detailed studies of their innervation, Foelix and Schabronath (1983) suggested that the pectines act as both mechanoreceptors and chemoreceptors. Other work has shown them to be capable of detecting subtle differences in sand grain size. These versatile structures are usually held laterally erect, free to swing back and forth while the scorpion is actively moving about.

The importance of vision varies greatly among the arachnids, although most species possess some sort of photoreceptors. At least some species in certain groups are blind (e.g., some members of the orders Schizomida, Palpigradi, Ricinulei, Pseudoscorpionida, and Acari). Certain spiders depend on photoreception for prey and mate location, particularly the errant hunters (some lycosids and most salticids) (Figure 19.21E,F). Vision is relatively unimportant to many sedentary species, such as many web builders, which depend more on tactile cues, vibrations, and chemosensitivity. Web builders are not blind, however, and many respond behaviorally to variations in light intensity, while some exhibit distinct escape responses when they visually detect potential predators.

(A)

(B)

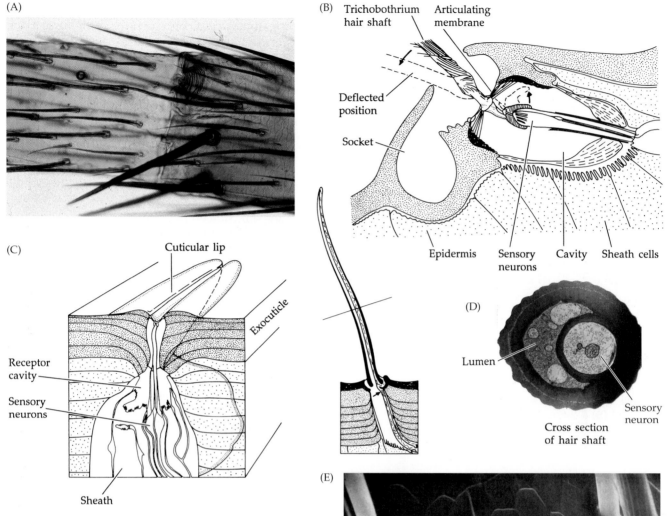

(C)

Cuticular lip

Exocuticle

Receptor cavity

Sensory neurons

Sheath

Trichobothrium hair shaft   Articulating membrane

Deflected position

Socket

Epidermis   Sensory neurons   Cavity   Sheath cells

(D)

Lumen

Sensory neuron

Cross section of hair shaft

(E)

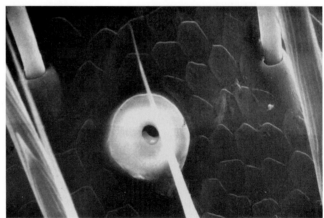

Figure 19.20 Mechanoreceptors and chemoreceptors in spiders. (A) Two types of mechanoreceptors on the leg of a wolf spider, *Lycosa* sp. (SEM, 1,250×). The thicker hair is a simple tactile hair; the thinner, curved one is a trichobothrium. (B) The organization of a trichobothrium in its socket; from *Tegenaria*. (C) A slit sense organ from the leg of *Cupiennius* (cutaway view). (D) A chemosensitive hair. The mechanoreceptor neurons terminate at the base of the hair (arrow), whereas the chemoreceptor neurons extend through the hollow shaft to the tip. (E) A humidity receptor, or tarsal organ, of the orb-weaver *Araneus*.

Nevertheless, these spiders are generally able to build their webs, capture prey, and mate with little or no visual input. Some spiders are capable of perceiving polarized light, presumably as a means of spatial orientation.

Spiders possess single-lensed rhabdomeric eyes, but the sensory units of each are simple ocelli in a cluster. Thus they are quite different from the compound eyes of xiphosurans and most other arthropods. Each eye includes a thickened **cuticular lens** over a **vitreous body**, which is a layer of cells derived from the epidermis and covering the **retina** (Figure 19.21B). The retina is composed of the **sensory (receptor) cells** and the **pigment**

cells. The membranes of the sensory cells bear interdigitating microvilli, confirming the rhabdomeric nature of the eyes.

Two forms of this basic eye structure occur among spiders. The anterior median eyes—the **main eyes**—have the light-sensitive portions of the sensory cells directed toward the lens, whereas other eyes, **secondary eyes**, are inverted, with the light receptor elements directed away from the lens (Figure 19.21B,C). Most of these secondary eyes contain a crystalline reflective layer, called the **tapetum**, which may serve to collect and concentrate light in poorly lit conditions (e.g., dur-

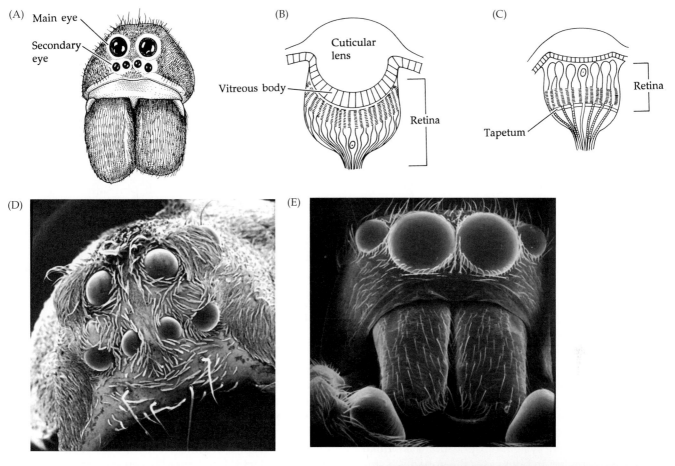

Figure 19.21 Spider eyes. (A) Head-on view of a wolf spider. Note the eye size and position. (B) Section through a main eye. (C) Section through a secondary eye, showing inverted arrangement of the retinular cells and tapetum. (D) Eye pattern of the nursery-web spider, *Pisaura*. (E) "Looking into the eyes" of *Heliophanus* (family Salticidae). (F) Image of another spider as "seen through the eyes" of *Portia*, a jumping spider.

ing night hunting). The reflective nature of the tapetum produces the effect in some spiders of eyes that "shine in the dark."

Scorpion eyes are of the direct type and differ from spider eyes in having the retinal layer external to the epidermis. Most work suggests that scorpions depend much more on mechanoreception and chemoreception than on visual input.

There is no question that spiders, and perhaps other arachnids, are capable of modifying their behavior based on experience—that is, they can learn. We have seen some examples of such activity in the preceding section. Memory and association centers in the protocerebrum are responsible for much of this integrative activity. No doubt, at least in spiders, the ability to remember, learn, and make appropriate adaptations in behavior has played an important role in their evolutionary success. The fascinating field of spider behavior

is admirably presented in a collection of papers edited by Witt and Rovner (*Spider Communication*, 1982).

Reproduction and Development

Chelicerates are dioecious and generally engage in complex mating behaviors that ensure fertilization. A few are known to be parthenogenetic (e.g., some scorpions and schizomids). Males with a penis occur only in opilionids and some acarids. Free spawning never occurs, and fertilization takes place either internally or as the eggs leave the body of the female. The eggs are generally very yolky except for the xiphosurans, the develop-

(A)

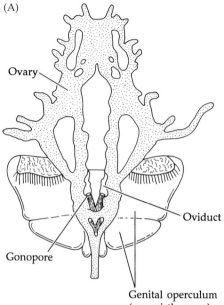

Ovary

Oviduct

Gonopore

Genital operculum
(on opisthosoma)

(B)

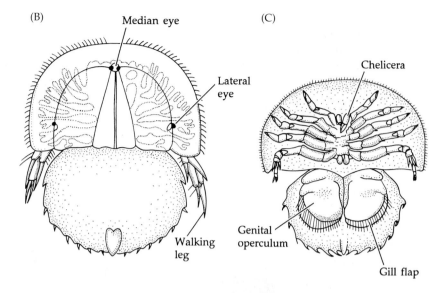

Median eye

Lateral eye

Walking leg

(C)

Chelicera

Genital operculum

Gill flap

Figure 19.22 (A) Female reproductive system of *Limulus* (order Xiphosura). (B,C) Euproöps larva of *Limulus* (dorsal and ventral views).

mental strategy is direct in spite of the various juvenile stages through which most chelicerates pass. We first present a summary of reproduction and development in *Limulus*, then turn to the arachnids, concentrating again on spiders and scorpions.

The reproductive system of xiphosurans is similar in males and females. In both sexes the gonad is a single, irregularly branched mass of tissue (Figure 19.22A). Paired gonoducts lead from the gonad to a pair of pores on the ventral midline. The first pair of opisthosomal appendages lies over the gonopores, forming a **genital operculum**.

At the onset of the breeding season, horseshoe crabs migrate into the shallow water of protected bays and estuaries. On the east coast of North America, this migration takes place in the spring and summer, and huge numbers of *Limulus* can be seen gathering near the shore in preparation for mating (Figure 19.2C). Mating is initiated when the male climbs onto the back of the female and grasps her with his modified first walking legs. The clasped pair moves to the shallow water, usually at a high spring tide, and the female excavates one or more shallow depressions in the sand and deposits her eggs (2,000 to 30,000 eggs per mating). The male releases sperm directly onto the eggs as they are deposited. Then the mates separate, and the female covers the fertilized eggs with sand.

Early development takes place in the sand or mud "nest." Cleavage is holoblastic, producing a stereoblastula with most of the yolk contained in the inner cells. As development continues, the surface cells at the anterior and posterior ends of the embryo divide rapidly, forming two germinal centers. Some of these rapidly proliferating cells migrate inward as the presumptive entoderm and mesoderm. The anterior germinal center

gives rise to the first four segments of the prosoma and the posterior center to the rest of the body. All of the prosomal segments fuse and are eventually covered by the developing carapace-like dorsal shield.

As the yolk reserves become depleted, the embryo emerges from the sediment as a **euproöps larva** (or "**trilobite larva**"), so named because of its resemblance to the fossil Carboniferous xiphosuran *Euproöps* (which superficially resembled trilobites; Figure 19.22B,C). The larvae swim about and periodically burrow in the sand. Segments are formed and appendages added through a series of molts until the adult form is reached. We view this developmental pattern as a mixed life history. The early developmental stages are supplied with an investment of yolk by the female and are protected by the nest she constructs, but the young emerge as independent feeding larvae prior to maturation.

The reproductive biology of arachnids is directly related to their success on land. They have evolved a variety of sophisticated mating behaviors, clever methods of sperm transfer, and various devices for protecting developing embryos, thereby ensuring successful procreation in terrestrial habitats. A comparison of the functional anatomy of the reproductive systems of spiders and scorpions provides a background for discussing arachnid courtship behavior and developmental patterns.

The male reproductive system of spiders consists of a pair of coiled tubular testes in the opisthosoma, which lead to a common sperm duct that opens into the epigastric furrow (Figure 19.23A). Each developing sperm usually bears a distinct flagellum with an odd 9 + 3 arrangement of axial filaments (Figure 19.23B). Prior to copulation, the flagellum wraps around the head of the sperm, and a protein capsule forms around the gamete

(Figure 19.23C). The sperm remains in this nonmotile state until after mating.

Although male spiders, like almost all other arachnids, lack a penis, the pedipalps are modified for sperm storage and transfer and serve as copulatory organs. Sperm released from the male gonopore are placed on a specially constructed silken **sperm web** (Figure 19.23F). From here the sperm are picked up by the pedipalps, where they are held in special pouches or chambers and eventually transferred to the female. The pedipalps of male spiders vary greatly in form and complexity, generally being simple in certain mygalomorphs and more

complex in araneomorphs. In the simplest form, each pedipalp bears on its tarsus (called the **cymbium**) a teardrop-shaped process known as the **palpal organ** (Figure 19.23D). A pointed tip, or **embolus**, bears a pore that leads inward to a blind coiled sperm storage cham-

Figure 19.23 Spider reproduction. (A) Male reproductive system of the tarantula *Grammostola*. The glandular mass near the gonopore is called the ventral spinning field because it produces the sperm web. (B) Sperm of *Oxyopes*. Note the unusual 9 + 3 arrangement of axial filaments in the flagellum. (C) Encapsulated form of sperm. (D) Simple male pedipalp copulatory structure (palpal organ) (*Segestria*). (E) Complex palpal organ (*Araneus*). (F) A male *Tetragnatha* on its sperm web, drawing sperm into its palpal organs. (G) Female spider reproductive system.

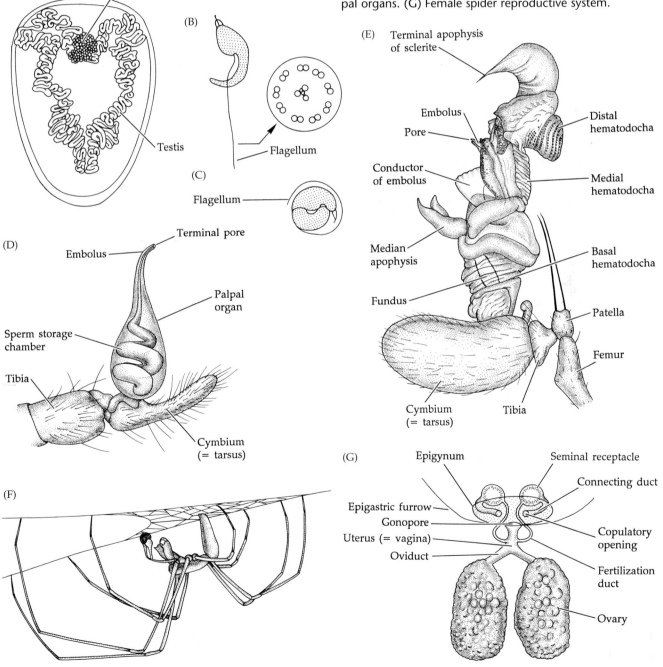

ber. Sperm are drawn into this tube from the sperm web and held until transferred to the female.

The more complex palpal organs are formed of hard and soft parts called **sclerites** and **hematodochae**, respectively; the sclerites bear processes called **apophyses** (Figure 19.23E). The apophyses are variously shaped in different species of spiders and fit with the mating organs of conspecific females. This variation is probably a mechanism to prevent interspecific mating, although nonproductive interspecific encounters are prevented primarily by species-specific courtship behaviors (discussed below). Following the uptake of sperm through the embolus and insertion of the palpal organ into the female's receiving structure (the **epigynum**), the soft hematodochae are inflated with hemocoelic fluid, thereby causing an erection of the sclerites within the female parts. Once the partners are thus coupled, sperm are injected into the female's copulatory openings.

Female spiders possess a pair of ovaries in the opisthosoma. The lumen of each ovary leads to an oviduct, and the two oviducts unite to form a uterus (also called the vagina), which opens to the outside in the epigastric furrow (Figure 19.23G). Eggs are produced mainly on the exterior of the ovaries, giving them a bubbled texture; how they move to the internal ovarian lumen is not well understood.

Just inside, or lateral to, the female gonopore, there is usually a pair of copulatory openings that lead through coiled **connecting ducts** to paired seminal receptacles. A second pair of tubes, called the **fertilization ducts**, connects the seminal receptacles to the uterus (Figure 19.23G). Many spiders possess a complexly structured sclerotized plate just in front of the epigastric furrow. This plate, called the epigynum, extends over the genital pore and bears the copulatory openings to the seminal receptacles. The form of the epigynum, the position and length of the copulatory openings and connecting ducts, and other external features provide a particular topography that matches the palpal organs of conspecific males. These differences in external anatomy, as well as in body size and overall courtship behavior, result in a variety of species-specific copulatory positions among spiders. Once sperm are inside the seminal receptacles, they are stored there until the female deposits her eggs, which may be months after copulation. At that time, the sperm pass through the fertilization ducts to join the ova during egg laying.

The reproductive systems of scorpions lie within the mesosoma, and both testes and ovaries are in the form of interconnected tubules (Figure 19.24). The gonads are drained by lateral sperm ducts or oviducts. The sperm ducts bear various storage chambers (seminal vesicles) and accessory glands, and unite as a **genital chamber** just inside the gonopore on the first segment of the mesosoma. Certain of the accessory glands are responsible for the production of spermatophores. Each oviduct is enlarged as a genital chamber, or seminal receptacle, near its union with the gonopore.

With few exceptions, sperm transfer in arachnids is indirect. That is, the sperm leave the body of the male and are then somehow manipulated into the body of the female or deposited on the eggs outside the female's body. The only exceptions to this rule occur in the Opiliones and some Acari, in which the male possesses a penis through which sperm pass directly to the reproductive tract of the female. In all other arachnids, either the sperm are inserted into the female's body by modified appendages of the male, or they are placed on the ground in spermatophores and then retrieved by the female. Appendages are used to transfer sperm in the orders Aranea (pedipalps), Uropygi (pedipalps), Ricinulei (third walking legs), and some members of Solpugida (chelicerae) and Acari (chelicerae or third legs). In other arachnids (orders Scorpionida, Schizomida, Amblypygi, Pseudoscorpionida, and many Solpugida and Acari), the males deposit spermatophores on the ground and the females simply pick them up. We hasten to point out, however, that the reproductive biology of many

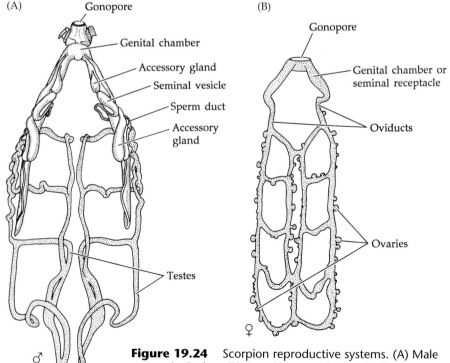

Figure 19.24   Scorpion reproductive systems. (A) Male system in *Buthus*. (B) Female system in *Parabuthus*.

species in these taxa, and of some entire groups (e.g., Palpigradi), has been little studied.

The events leading up to insemination often include species-specific courtship behaviors that serve as cues for species recognition. These behaviors must, of course, be compatible with the particular method of sperm transfer (Figures 19.25, 19.26). Again, we can look to spiders and scorpions for examples. Among spiders, courtship behaviors not only ensure conspecific copulation, but also prevent the usually smaller males from being mistaken for prey by the females. Platnick (1971) classified spider courtship behaviors into three general levels. First-level courtship involves necessary contact between male and female. Among many thomisids and clubionids, mating involves the male simply climbing over the female, positioning her abdomen, and inserting a palpal organ. The males of some thomisids (e.g., *Xysticus*) and at least one genus of araneid (*Nephila*) place silk threads over the bodies or legs of the female preparatory to copulation. These threads are apparently only part of the recognition ritual because they are not strong enough to actually restrain the female (Figure 19.25C). A few other spiders, including certain tarantulas, also use body contact and leg touching as courtship behavior.

Second-level courtship behaviors involve the release of sex pheromones by the female spider. Some of the most complex behavior patterns occur in male spiders that detect females by olfaction, although other recognition devices may also be involved. Male araneids are apparently led to the female's orb web by her pheromones, and the web is then recognized by contact chemoreception. Once in touch with the edge of the web, the male announces his presence to the female by plucking the threads of her orb itself, or by attaching a special **mating thread** to her orb, which he then plucks (Figure 19.25E). If properly orchestrated, the male's "tune" eventually attracts the female, and contact is made.

Males of some species of wolf spiders (Lycosidae) respond to pheromones emitted along with the female's dragline. When he detects a female visually, the male begins a specific set of actions in an effort to win her favor. These male behaviors involve abdomen bobbing and pedipalp waving, coupled drumming the pedipalps on the substratum and stridulation (Figure 19.25F). If attracted by these cues, the female responds by slowly approaching the male and sending out signals of her own in the form of particular leg movements. Stridulation, using modified pedipalps, also occurs in some uloborids (Figure 19.25G).

Among the most interesting second-level courtship behaviors is that of certain nursery-web spiders (*Pisaura*; Pisauridae). After locating a female emitting pheromones, the male captures an insect (usually a fly), spins a silk wrapping around it, and offers it to the female. Acceptance of the gift and of the male are one and the same, for the successful male copulates with the fe-male while she devours the insect. Alas, unsuccessful males are eaten along with the offering. Postcopulatory cannibalism by females is not uncommon in certain groups of spiders. The best known case is the North American black widow, *Latrodectus mactans*. Females tend to reject the advances of other males after they have consumed one mate.

Another interesting second-level courtship behavior occurs in the Sierra dome spider (*Linyphia litigiosa*) of western North America (Watson 1986). Upon encountering a mature virgin female, a male attacks her pheromone-laden web and packs it into a small, tight mass. This behavior hinders the evaporation and dispersal of the male-attracting pheromone, thereby reducing the likelihood of a second male locating the female and competing for her favors.

Third-level courtship behaviors depend primarily on visual recognition of prospective mates and are best known in jumping spiders (Salticidae). A male locates a female and then begins a series of behaviors that identify him as a conspecific individual. Usually the male approaches the female along a zigzag path and then performs specific movements of the opisthosoma, pedipalps, and front walking legs (Figure 19.25H). The female signals her approval and receptiveness by sitting still in a visually recognizable position. The male eventually contacts her, caresses her briefly, mounts her, and copulates.

Sex-related behaviors among spiders are not restricted to male–female encounters. Conspecific males frequently exhibit agonistic behavior in competition for a mate. When males encounter one another in the presence of a female, or even in a mating "territory" (such as on the web of a female), they may assume various threatening postures and in some cases actually engage in combat. Usually, however, one of the males retreats before any real damage is done, leaving the dominant male free to pursue his sexual interests.

The courtship behavior of scorpions does not involve copulation in any form, but instead involves deposition of spermatophores onto the ground. Courtship behaviors appear to be relatively similar among those species that have been studied, although subtle species-specific differences allowing species recognition undoubtedly occur. In a typical case, the male initiates the ritual by grasping the female's pedipalps in his, and in this face-to-face position dances her around in a series of back-and-forth steps (Figure 19.26A). Eventually the male releases a spermatophore and cements it to the ground. He then continues to move the female around until she is precisely positioned with her genital operculum over the packet of sperm. The spermatophore is a complex, species-specific structure, and it bears a special process called an **opening lever** (Figure 19.26B). The pressure of the female's body on this lever causes the spermatophore to burst, releasing the sperm, which can then enter her gonopore.

Figure 19.25 Courtship and mating in spiders. (A) Mating position in tarantulas. (B) Mating position in linyphiids. (C) Mating position in *Xysticus* (family Thomisidae). Mating occurs after the male has placed a series of threads over the female's body. (D) Mating position in *Araneus diadematus* (family Araneidae). (E) Use of a mating thread by some orb-weavers. In this general example, the male has spun a mating thread from an object to the female's web. When properly plucked, the thread transmits vibrations to the web, and the female responds by approaching the male and assuming a mating posture. (F) The courtship behavior of a male lycosid (*Lycosa rabida*) includes movements of the anterior legs, abdomen bobbing, and pedipalp movements. (G) The pedipalp of *Tangaroa tahitiensis* (family Uloboridae) bears organs of stridulation. The spines are scraped against the filelike serrations on the coxa. (H) The zigzag approach and courtship display of a male jumping spider (family Salticidae).

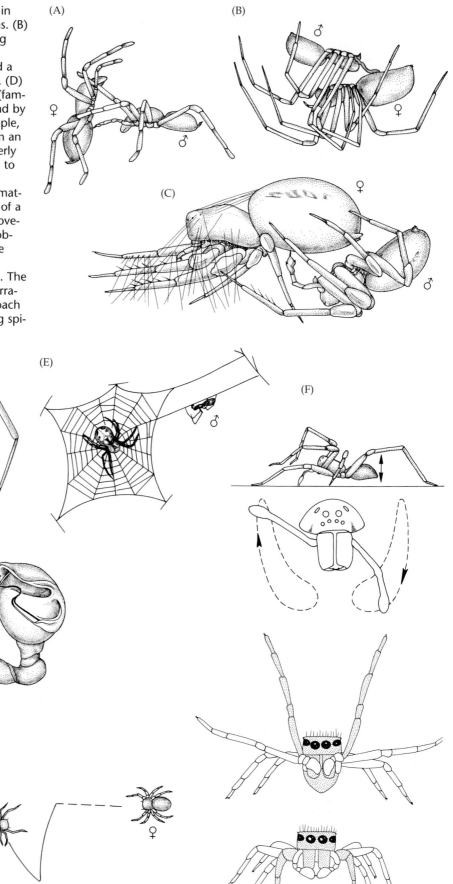

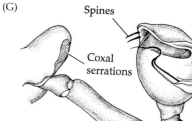

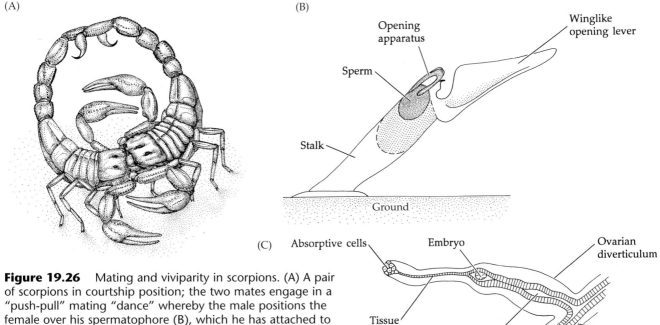

Figure 19.26 Mating and viviparity in scorpions. (A) A pair of scorpions in courtship position; the two mates engage in a "push-pull" mating "dance" whereby the male positions the female over his spermatophore (B), which he has attached to the ground. (C) An ovarian diverticulum with enclosed embryo from the scorpion *Hormurus*.

Arachnids undergo a variety of developmental patterns, all of which may be considered direct in terms of their life history strategies. Most species produce very yolky eggs, providing the embryos with nourishment through much of their development. By the time of hatching, many resemble miniature adults, or still contain enough yolk to carry them through subsequent development to the juvenile stage. In some acarans and all ricinuleids, the young hatch as six-legged "hexapod larvae." These immature individuals add the last pair of legs later through molting. In most arachnids the developing embryos are protected by some sort of egg case or cocoon, or they are brooded in or on the body of the female.

Nearly all spiders cement their eggs together in clusters and wrap them in silken cocoons, the sizes and shapes of which vary among species (Figure 19.27). The cocoon provide physical protection for the embryos and also insulates them from fluctuations in environmental conditions, particularly changes in humidity. Additional protection results from placing the cocoon underground, in a nest, or in other secluded spots. Some species camouflage their cocoons with bits of detritus; others guard their cocoons or actually carry them on their bodies.

The early development of spiders includes intralecithal nuclear divisions followed by migration of the nuclei to the periphery of the embryo. The nuclei are then isolated by cytoplasmic partitioning, a process that produces a periblastula around an inner yolky mass. Gastrulation follows by formation of a germinal center of presumptive entodermal and mesodermal cells that migrate inward. Additional germinal centers produce the precursors of segments and limbs. The immature individuals of different spider species hatch from their egg membranes at different stages, but they always remain inside the cocoon and utilize their yolk reserves until they are able to feed. Most workers recognize three postembryonic, preadult stages in spider development (Figure 19.27I,J,K). The majority of spiders hatch from their egg membranes as immobile "prelarvae," characterized by incomplete segmentation and poorly developed appendages. The "prelarva" matures into a "larva" and then into a "nymph," or juvenile, which physically resembles the adult. In some spiders these early developmental changes take place in a special **molting chamber** inside the cocoon (Figure 19.27B). The emergence from the cocoon usually occurs at an early "nymphal" stage, when the young are fully formed spiderlings. Many female spiders even engage in postnatal care by carrying their young on their bodies or by feeding them (Figure 19.27D,E,F). (It is important to note that the terms prelarva, larva, and nymph, as used here, do not carry the same meanings as they do when used to describe indirect or mixed development, in which the larva is an independent, free-living individual, as in insects or crustaceans.)

Scorpion development is direct, and may be either ovoviviparous or viviparous. Viviparity is perhaps best studied in the Asian scorpion *Hormurus australasiae*. In this species the zygotes lie in tiny diverticula on the walls of the ovarian tubules (Figure 19.26C). Certain cells of the tubule wall absorb nutrients from the adjacent digestive ceca and supply them to the developing embryos. The eggs of *Hormurus* contain very little yolk and undergo holoblastic, equal cleavage. Ovovivipar-

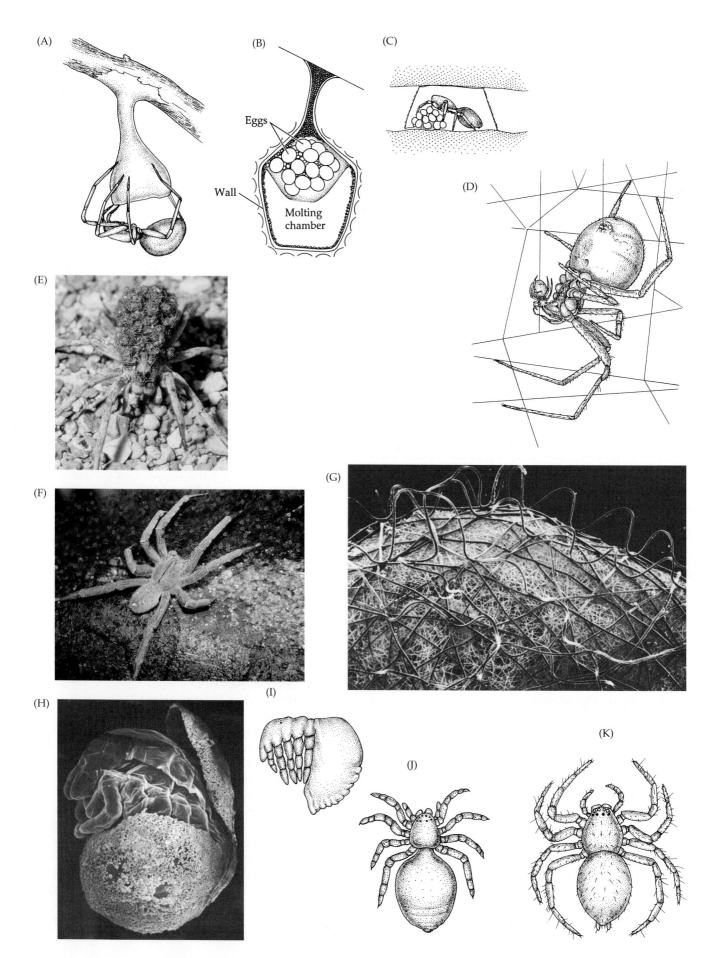

(A)

(B)

Eggs

Wall

Molting chamber

(C)

(D)

(E)

(F)

(G)

(H)

(I)

(J)

(K)

Figure 19.27 Egg cases, hatching, and parental care in some spiders. (A,B) Formation and general structure of the cocoon of *Agroeca brunnea*. (A) The female constructs the cocoon and coats it with bits of soil. (B) The cocoon (section). Note the eggs above the molting chamber. (C) The subterranean, silk-lined nest of a jumping spider (*Heliophanus cupreus*). (D) Female *Theridion* tending a spiderling and feeding it with regurgitated food. (E) A female wolf spider, *Lycosa*, carrying young on her back. (F) A cteniid spider with its spiderlings. (G) Surface of the cocoon of *Ero furcata*. (H) A spider hatchling emerging from its egg case. (I,J,K) Prelarval, larval, and nymphal forms of a spider.

ous scorpions, on the other hand, produce yolky eggs that cleave meroblastically. The embryos of these species are brooded in the ovarian tubules, but depend on their yolk supplies for nutrients. The young eventually emerge from the female's gonopore and crawl onto her back. Here they stay until they are old enough to make periodic excursions away from their parent and eventually assume an independent life. Juvenile scorpions molt through several instars until they mature, about a year after birth.

Many other arachnids also brood their embryos, usually externally on the mother's body, as in Amblypygi, Uropygi, Ricinulei, and Schizomida. Members of these groups carry their young in some sort of sac near the female gonopore. False scorpions spin cocoons from their cheliceral silk glands. Solpugids and opilionids are oviparous and deposit their eggs in the soil. In all cases the young hatch and pass through a few or many instars before they mature, and the maturation process may take several years. Again, except for the "hexapod larvae" of mites and ricinuleids, arachnids hatch as small immature adults, and although various names are given to these immature stages, development is strategically direct.

The Class Pycnogonida

Pycnogonids (Greek *pyc*, "thick," "knobby"; *gonida*, "knees") are usually called "sea spiders" because of their superficial similarity to true, terrestrial spiders (Figures 19.28, 19.29). Pycnogonids had been problematic in terms of their placement among the other arthropod taxa for decades. Since the turn of the twentieth century they have been associated at one time or another with virtually every major group of arthropods, as well as with the onychophorans and polychaetes. The principal problem has been uncertainty concerning homologies of the various body regions and appendages. The unique pycnogonid "proboscis," for example, has been homologized with everything from the prostomium of polychaete worms to the lips of onychophorans to various anterior regions of other arthropods.

However, recent anatomical and molecular studies place the pycnogonids solidly within the Cheliceriformes. Most specialists have concluded that the pycnogonids probably arose as an early offshoot of the line leading to the modern chelicerates, although some workers regard them as highly specialized arachnids.

Several characters are apparent shared synapomorphies between the chelicerates and pycnogonids, including absence of a deutocerebrum; first appendages chelicerae/chelifores and second appendages pedipalps/palps (based on the assumption that these appendage pairs, and their somites, are indeed homologous; nervous innervation tends to support this contention); legs uniramous stenopods (with certain functional similarities); feeding method largely liquid/suctorial; and four postoral segments in the earliest embryonic stages. Pycnogonids also possess several strikingly unique features, synapomorphies not found in any chelicerate, or in any other arthropod group, such as the odd anterior "proboscis," **ovigers** (specialized appendages between the pedipalps and first walking legs, used for a variety of purposes but most notably for brooding in males), multiple gonopores (on the second coxal segment of some or all of the walking legs), and the unique body form discussed below.

There are about a thousand described species of pycnogonids, in nearly a hundred genera, and specialists suspect there are many more yet to be discovered. Pycnogonids are strictly marine; they occur intertidally and to depths of nearly 7,000 m and are distributed worldwide. Most are small, with leg spans less than a centimeter (*Austrodecus palauense* has a leg span of only 2 mm); however, some deep-sea species have leg spans to 60 cm. The largest species are members of the genus *Colossendeis*, which live in deep waters worldwide and are common near shore in Antarctica. Many are errant benthic animals, but others live on seaweeds or on other invertebrates, particularly sea anemones, hydroids, ectoprocts, and tunicates. One or two species live on the bells of pelagic medusae, and another has been seen in the hydrothermal vent community of the Galapagos Rift on the huge vestimentiferan worms there.

Classification within the Pycnogonida is somewhat unsettled. Historically, the few known fossil species (which date back to the Upper Cambrian) were "dumped" into the single order Palaeopantopoda, and the living forms into the order Pantopoda. A major revision was proposed by Fry (1978a), in which the 73 living genera were reassigned to 30 families. The most commonly used classification is that of Hedgpeth (1982), which recognize just 8 families, based largely on appendage structure (particularly the reduction or loss of various appendages).

Figure 19.28 Representative pycnogonids.
(A) *Nymphopsis spinosossima* (family Ammotheidae).
(B) *Pycnogonum stearnsi* (family Pycnogonidae). (C) *Tanystylum anthomasti* (family Tanystylidae). (D) The tenlegged *Decolopoda australis* (family Colossendeidae) (side view of animal walking). (E) *Achelia echinata* (family Ammotheidae) feeding on an ectoproct colony, one zooid at a time. (F) An unidentified pycnogonid from Antartica.

The Pycnogonid Bauplan

External Anatomy

The bodies of pycnogonids are not as clearly divided into recognizable tagmata as are those of other arthropods (Figures 19.28, 19.29). The first body "region" bears an anteriorly directed **proboscis**, which varies in size and shape among species. The proboscis contains a chamber

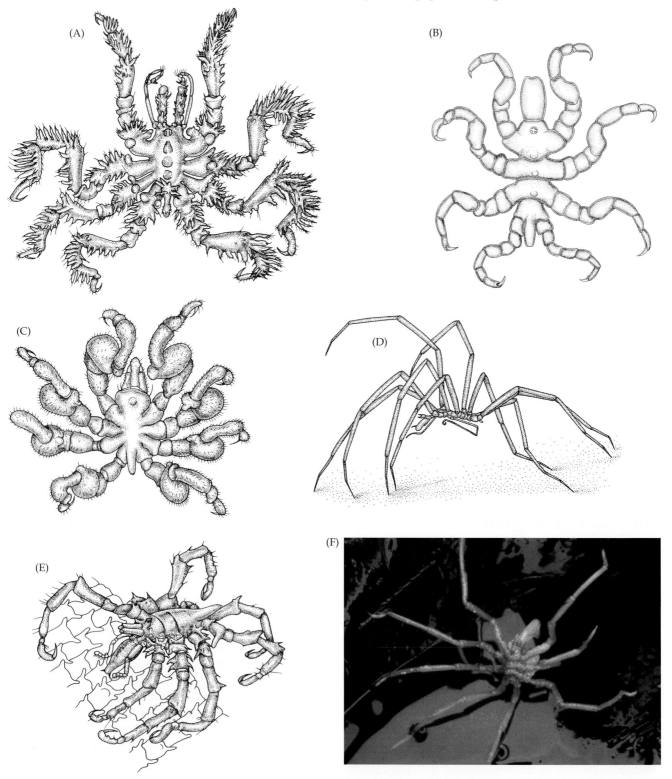

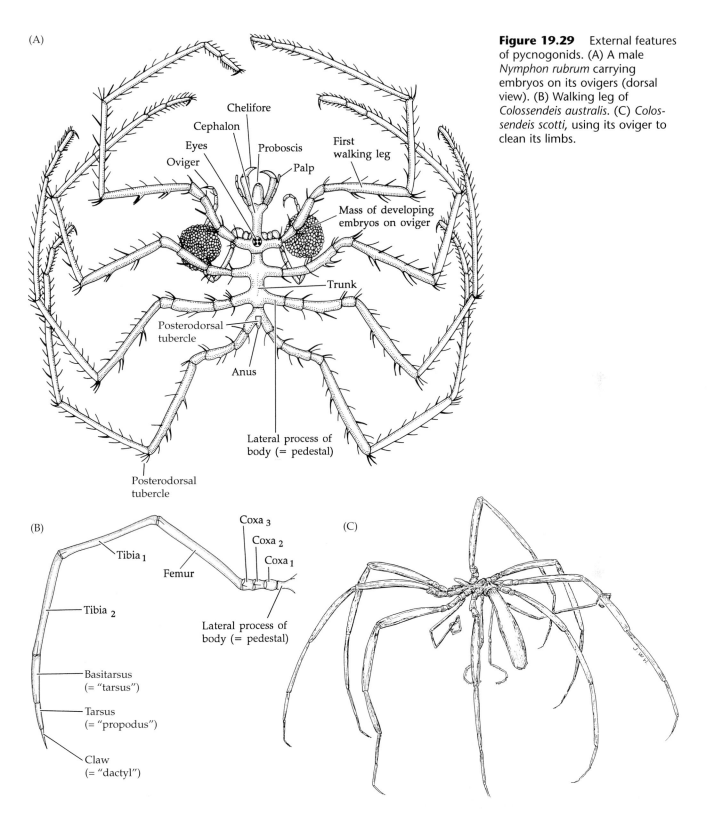

(A)

Chelifore
Cephalon
Eyes
Proboscis
Oviger
Palp
First walking leg
Mass of developing embryos on oviger
Trunk
Posterodorsal tubercle
Anus
Lateral process of body (= pedestal)
Posterodorsal tubercle

(B)

Coxa 3
Coxa 2
Coxa 1
Tibia 1
Femur
Tibia 2
Lateral process of body (= pedestal)
Basitarsus (= "tarsus")
Tarsus (= "propodus")
Claw (= "dactyl")

(C)

Figure 19.29 External features of pycnogonids. (A) A male *Nymphon rubrum* carrying embryos on its ovigers (dorsal view). (B) Walking leg of *Colossendeis australis*. (C) *Colossendeis scotti*, using its oviger to clean its limbs.

and bears an opening at its distal end. The actual mouth is probably the connection between the proboscis chamber and the esophagus (Figure 19.30A), although there is some confusion on this matter. This anteriormost body "region" also bears paired appendages in the form of **chelifores**, **palps**, **first walking legs**, and, when present, ovigers (Figure 19.29A). The chelifores may be chelate or achelate, or absent altogether. The ovigers are modified

legs that serve a variety of functions, including grooming (Figure 19.29C); food handling in some species; courtship, mating, and egg transfer (female to male) in many species; and brooding of the embryos by males in most species. Also located on the first segment of most species is a tubercle with four simple median eyes.

The following body segments form the "trunk" and may be variably fused, but each bears a pair of lateral

(A)

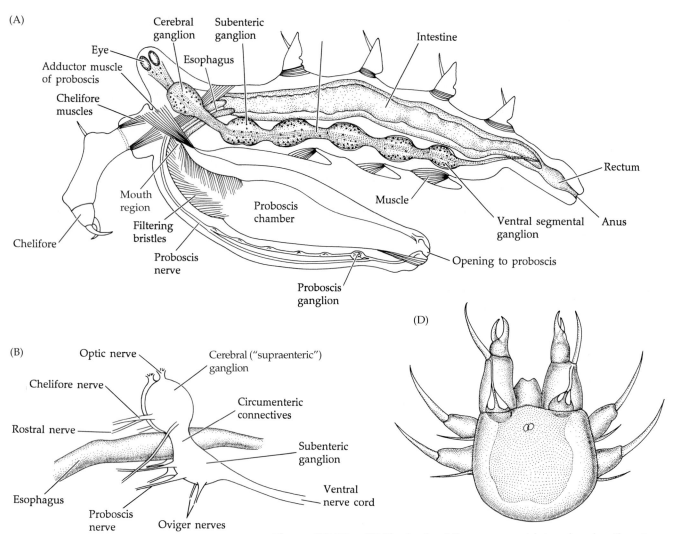

(B)

(D)

Figure 19.30 (A) The body of the pycnogonid *Ascorhynchus* (longitudinal section). (B) Anterior portion of the nervous system of *Nymphon*. (C) A female pycnogonid with developing ova stored in the femoral portions of the gonad diverticula. (D) Protonymphon larva.

(C)

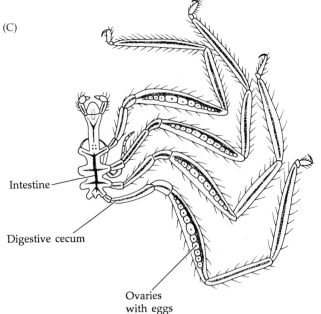

processes, called **pedestals**, on which are borne the walking legs. Because of the orientation of the pedestals, the walking legs are arranged somewhat radially around the body. The posteriormost body segment carries a dorsally inserted **posterodorsal tubercle**, which may be a vestigial abdomen, or opisthosoma, and which bears the anus. Perhaps the most distinctive synapomorphy of the pycnogonids, compared with other arthropods, is the presence of multiple gonopores, found on some or all of the walking legs.

One of the most unusual aspects of pycnogonid morphology is the existence of **polymerous species**, which possess more than four pairs of walking legs. This phenomenon is unique among arthropods and occurs in the pycnogonid genera *Pentanymphon*, *Pentapycnon*, and *Decolopoda* (with five pairs of legs; decapodous), and *Sexanymphon* and *Dodecolopoda* (with six pairs of legs; dodecapodous). *Callipallene brevirostris* typically bears four pairs of legs, but one specimen has been found with only three pairs. This polymery is likely to be the result of a "runaway" Hox gene, and may represent segment duplication.

Variation among different pycnogonids also occurs in appendage shape and length, spination, proboscis structure, reduction or loss of chelifores and palps, and many other external features. Several examples are shown in Figures 19.28 and 19.29 to illustrate this diversity. In all, however, the body is remarkably reduced and narrow, a feature compensated for by extensions of the gut ceca and gonads into the legs.

Locomotion

The walking legs of pycnogonids are typically nine-segmented. The junction of the first coxal segment and the pedestal is a more or less immovable joint and does not contribute to the action of the leg. The joint between the first and second coxae is hinged to provide promotion and remotion, and the rest of the joints provide the usual flexion and extension. However, the coxal joints also allow a certain amount of "twisting" and thus accentuate the anterior–posterior swing of the appendage tips during the power and recovery strokes. Some joints lack extensor muscles, and limb extension is effected by hydrostatic pressure, as in many arachnids. Note that pycnogonid specialists have given names to the leg articles that do not parallel those used for any other arthropod group (Figure 19.29B).

Most of the commonly encountered intertidal pycnogonids are quite sedentary and move very slowly. These small forms have short, thick legs that are somewhat prehensile and serve more for clinging to other invertebrates or algae than for rapid locomotion. Deep-water benthic pycnogonids tend to be more active, and these errant pycnogonids have longer and thinner legs than the sedentary forms, and they tend to walk on the tips of the legs (Figures 19.28D,F, 19.29C). However, some of the very large deep-sea forms (e.g., *Colossendeis*) may depend more on slow deep-ocean currents to roll them around on the bottom than on their own locomotory powers.

Many pycnogonids are also known to swim periodically by employing leg motions similar to those used in walking. Some species are known to "hang" from the water's surface, utilizing a combination of small body mass and surface tension. Several studies also report a characteristic "sinking behavior" in a number of species. When dropping to the bottom, these pycnogonids elevate all of the appendages over the dorsal surface of the body in a "basket configuration." This behavior eliminates much of the frictional resistance to sinking and allows the animal to drop quickly through the water column (presumably to avoid predation).

Feeding and Digestion

In most species of pycnogonids, feeding habits are dictated by the form of the proboscis, and food is limited to material that can be sucked into the gut. Even with this basic structural constraint, pycnogonids feed on a variety of organisms.

A few pycnogonids feed on algae, but most are carnivorous, many being generalized predators on hydroids, polychaetes, nudibranchs, and other small invertebrates. Some—perhaps many—also scavenge. Species that consume other animals usually use three cuticular teeth at the tip of the proboscis to pierce the body of their prey; then they suck out body fluids and tissue fragments. Some pycnogonids that live on hydroids use the chelifores to pick off pieces of the host and pass them to the proboscis opening. In most species, however, the chelifores cannot reach the tip of the proboscis, and their function in feeding is questionable. Some species (e.g., *Achelia echinata*) feed on ectoprocts by inserting the proboscis into the chamber housing an individual and sucking out the zooid (Figure 19.28E). Others (e.g., *Pycnogonum litorale, P. rickettsi*) feed on sea anemones in a similar fashion, but rarely kill them due to the large difference in size between predator and prey.

Very little is known about the feeding habits of the deep-sea benthic forms. Undersea photographs and direct observations of aquarium specimens of the giant *Colossendeis colossea* indicate that it may walk slowly along the bottom, sweeping its palps across the substratum to sense prey that might be sucked from the mud.

The digestive tract extends from the mouth at the base of the proboscis to the anus, which opens on the posterodorsal tubercle on the last "trunk" segment (Figure 19.30A). A chamber within the proboscis bears dense bristles that strain and mechanically mix ingested food. The muscles of the foregut supply the suction for ingestion. A short esophagus connects the mouth to the elongate midgut or intestine, from which digestive ceca extend into each leg, providing a high surface area for digestion and absorption. A short proctodeal rectum leads to the anus.

Digestion is predominantly, if not exclusively, intracellular. The cells of the midgut and cecal walls include phagocytes that engulf ingested food materials. Some of these cells actually break free from the gut lining and phagocytize food particles while drifting in the gut lumen. Apparently these loose cells reattach to the gut wall after they have "fed." It has been suggested that upon reattachment these errant cells first pass their digested food contents to the fixed cells of the gut wall, then assume an excretory function by picking up metabolic wastes, detaching again and being eliminated via the anus.

Circulation, Gas Exchange, and Excretion

Pycnogonids lack special organs for gas exchange or excretion. The digestive ceca and the overall bauplan together present a very high surface area-to-volume ratio, and exchange of gases and elimination of wastes probably occur largely by diffusion across the body and gut wall. The special wandering cells of the midgut may help in excretion.

The circulatory system includes an elongate heart with incurrent ostia, but no blood vessels. As in other arthropods, the heart is located dorsally, within a pericardial chamber separated from the ventral hemocoel by a perforated membrane. Blood leaves the heart anteriorly and flows through the hemocoelic spaces of the body and appendages. Contraction of the heart causes a lowered pressure within the dorsal pericardial body chamber, and blood is thus drawn through the perforations in the membrane and toward the heart. Upon relaxation, the blood flows through the ostia into the heart lumen.

Nervous System and Sense Organs

The central nervous system of pycnogonids includes cerebral ganglia above the esophagus, circumenteric connectives, a subenteric ganglion, and a ganglionated ventral nerve cord (Figure 19.30A,B). The nerve cord bears a ganglion for each pair of walking legs, and additional ganglia are present in the polymerous species.

The cerebral ganglia include a protocerebrum and tritocerebrum; pycnogonids lack a deutocerebrum (as do chelicerates). The protocerebrum innervates the eyes and the tritocerebrum innervates the chelifores, an arrangement similar to that in chelicerates. The cerebral ganglia also give rise to a well developed ganglionated proboscis nerve.

Little work has been done on pycnogonid sense organs. Tactile reception is provided by touch-sensitive hairs and probably by the palps. On the body surface, just dorsal to the cerebral ganglia, is a tubercle with four simple eyes (some deep-sea species lack eyes). The eyes, when present, are set in such positions as to provide 360° vision.

Reproduction and Development

Pycnogonids are dioecious. Mating is typically followed by a period of brooding, during which the embryos are held by the male's ventrally articulated ovigers, then by the release of unique **protonymphon larvae** (Figure 19.30D). The protonymphon is a curious six-legged creature that usually lives in a symbiotic relationship with cnidarians, molluscs, or echinoderms. These relationships are poorly understood, but in some cases they appear to be parasitic or commensalistic. This mixed developmental strategy has been replaced by direct patterns in some species wherein the larval stage is passed within an egg case.

Sexual dimorphism is common among pycnogonids. The males bear the unique ovigers associated with the first body segment; these appendages are absent in females of some families (e.g., Phoxicilidiidae, Endeidae, and Pycnogonidae), and reduced in the females of other families. Female pycnogonids usually have enlarged limb femora.

Internally, the reproductive systems of males and females are similar and relatively simple. In both, the gonad is single and U-shaped, with extensions into the legs where gametes are produced and stored. The expanded femora in females provide space for storing unfertilized eggs (Figure 19.30C). The multiple gonopores are usually located on the ventral surface of the second coxae of two or all pairs of legs and are thus close to the regions of gamete storage. During mating, the male typically hangs beneath the female or assumes a stance over her back. As the female releases her eggs, the male fertilizes them. Following fertilization, the male gathers the eggs, either one by one or as a single mass, and glues them to his ovigers using a sticky secretion from special femoral glands (Figure 19.29A). Pycnogonids are one of the few groups of animals in which the males exclusively brood the developing embryos and, in some species, the young.

Knowledge of pycnogonid development is based on relatively few studies. There is some variation among members of different genera, but cleavage is usually holoblastic and leads to the formation of a stereoblastula. The inward movement of the presumptive entodermal and mesodermal cells is frequently accompanied by a disappearance of some cell membranes, a process leading to the formation of some syncytial tissues in the gastrula. Germinal centers become apparent as appendage buds form. The most common hatching stage is a free-swimming protonymphon larva. Through a series of molts, the protonymphon adds segments and appendages to produce a juvenile. In some species, the developing larva becomes encysted in a hydroid or stylasterine coral, emerging later with three pairs of legs. A few species are known to produce all of the appendages at once, with subsequent molts simply increasing the size and segment number of the legs (the "atypical protonymphon larva"). In *Pycnogonum litorale*, at the fifth larval molt, the three pairs of larval legs and the larval proboscis are lost; the adult proboscis appears and, at later molts, the adult limbs develop.

Nothing is known about reproduction in most of the deep-sea species. Their eggs, however, are very small, an observation suggesting that the young stages may be parasitic.

Cheliceriform Phylogeny

Although over 150 years have passed since the discovery of pycnogonids, arguments still ensue as to their phylogenetic relationships. However, as we mentioned earlier, recent morphological, paleontological, and molecular studies (several different genes) point to a distant cheliceriform ancestry. The chelifores and palps of pycnogonids are probably homologous to the chelicerae and pedipalps of chelicerates. We thus recognize the pycnogonids as a sister-group to the Chelicerata, although some recent work has suggested that they might actually have arisen from a true arachnid ancestry.

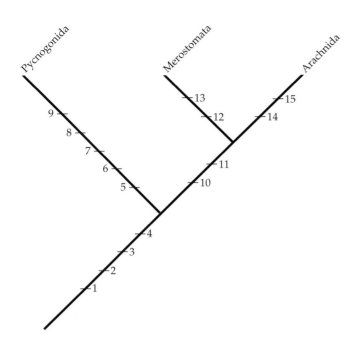

Figure 19.31 A simple cladogram depicting hypotheses about the relationships among the Cheliceriformes. The synapomorphies used in constructing the cladogram are (1) body divided into prosoma and opisthosoma, without distinct head; (2) first appendages chelicerae (or chelifores); (3) second appendages pedipalps (or palps); (4) typically with four pairs of walking legs; (5) unique preoral proboscis; (6) ovigers; (7) opisthosoma reduced or absent; (8) walking legs nine-segmented; (9) multiple pairs of gonopores (borne on some or all legs); (10) with prosomal carapace-like shield; (11) first or second opisthosomal segment modified as genital somite; (12) opisthosomal appendages modified as gill books; (13) telson long, spike-like; (14) opisthosomal appendages reduced, lost, or modified as spinnerets or pectines; (15) with tracheae, book lungs, or both.

Figure 19.31 is a cladogram depicting one view of the relationships among the major cheliceriform taxa. However, we caution readers that the phylogenetic relationships among the Cheliceriformes are still debatable. The subclass Arachnida is probably a monophyletic group, although some workers have suggested that it might be diphyletic, having arisen from two separate invasions of land (one leading to the scorpions, the second to all other orders). Phylogenetic relationships among the eleven orders of Arachnida remain controversial; see Weygoldt and Paulus (1979) and Schultz (1990) for two competing views.

The earliest xiphosuran and eurypterid fossils are Ordovician. Like the myriapods (and perhaps the insects), arachnids probably invaded land early in the Silurian, at the same time the land plants were becoming well established. The earliest Paleozoic scorpions were aquatic, respiring by "gill books," which may have been the predecessors (homologues) of the book lungs of modern terrestrial arachnids. Silurian and Devonian scorpions shared shallow marine or estuarine environments with their close relatives, the eurypterids. Fossils of what appear to be larval pycnogonids have been reported from the Upper Cambrian *Orsten* deposits of Sweden.

Selected References

General References

Edgecombe, G. D., G. D. F. Wilson, D. J. Colgan, M. R. Gray and G. Cassis. 2000. Arthropod cladistics: Combined analysis of histone H3 and U2 snRNA sequences and morphology. Cladistics 16: 155–203.

Grassé, P. 1949. *Traité de Zoologie*. Vol. 6, Onychophores–Tardigrades–Arthropodes–Trilobitomorphes–Chelicerataes. Masson et Cie, Paris.

Kraus, O. 1976. On the phylogenetic position and evolution of the Chelicerata. Entomol. Ger. 3: 1–12.

Levi, H. W. et al. 1982. Chelicerata. *In* S. P. Parker (ed.), *Synopsis and Classification of Living Organisms*, Vol. 2. McGraw-Hill, New York, pp. 71–173.

Sherman, R. G. 1981. Chelicerates. *In* N. A. Ratcliffe and A. F. Rowley (eds.), *Invertebrate Blood Cells*, Vol. 2. Academic Press, New York.

Stormer, L., A. Petrukenvitch and J. W. Hedgpeth. 1955. Chelicerata, with sections on Pycnogonida and *Palaeoisopus*. *In* R. C. Moore (ed.), *Treatise on Invertebrate Paleontology*, Part P, Arthropoda 2. Geological Society of America and the University of Kansas Press, Lawrence.

Paulus, H. F. 1979. Eye structure and the monophyly of Arthropoda. *In* A. P. Gupta, *Arthropod Phylogeny*. Van Nostrand Reinhold, New York, pp. 299–377.

Merostomata

Barlow, R. B. Jr. 1990. What the brain cells tell the eye. Sci. Am. April 1990: 90–95. [Experimental work on vision in *Limulus*.]

Botton, M. L. and J. W. Roper. 1989. Feeding ecology of horseshoe crabs on the continental shelf, New Jersey to North Carolina. Bull. Mar. Sci. 45(3): 637–647.

Cohen, J. A. and H. J. Brockmann. 1983. Breeding activity and mate selection in the horseshoe crab, *Limulus polyphemus*. Bull. Mar. Sci. 33: 274–281.

Fahrenbach, W. H. 1979. The brain of the horseshoe crab (*Limulus polyphemus*). III. Cellular and synaptic organization of the corpora pedunculata. Tissue Cell 11: 163–200.

Fisher, D. C. 1984. The Xiphosuridae: Archetypes of bradytely? *In* N. Eldredge and S. Stanley (eds.), *Living Fossils*. Springer-Verlag, New York, pp. 186–213.

Riska, B. 1981. Morphological variation in the horseshoe crab (*Limulus polyphemus*), a "phylogenetic relic." Evolution 35: 647–658.

Sekiguchi, K. and H. Sugita. 1980. Systematics and hybridization in the four living species of horseshoe crabs. Evolution 34: 712–718.

Shuster, C. N. 1950. Natural history of *Limulus*. Contrib. Woods Hole Oceanogr. Inst. 564: 18–23.

Arachnida

Baerg, W. 1958. *The Tarantula*. University of Kansas Press, Lawrence.

Barth, F. G. (ed.). 1985. *Neurobiology of Arachnids*. Springer-Verlag, New York.

Barth, F. G. and P. Pickelmann. 1975. Lyriform slit sense organs—modeling an arthropod mechanoreceptor. J. Comp. Physiol. 103: 39–54.

Barth, F. G. and J. Stagl. 1976. The slit sense organs of arachnids. Zoomorphologie 86: 1–23.

Bartsch, I. 1990. Hydrothermal vent fauna: *Agauopis auzendei* n. sp. (Acari, Halicaridae). Bull. Mus. Natl. Hist. Nat., Paris, 4e sét. 12, section A, no. 1: 69–73.

Binns, E. S. 1983. Phoresy as migration: Some functional aspects of phoresy in mites. Biol. Rev. 57: 571–620.

Bowden, K. 1991. The evolution of sociality in the spitting spider, *Scytodes fusca* (Araneae: Scytodidae): Evidence from observations of intraspecific interactions. J. Zool. Soc. Lond. 223: 161–172.

Bowerman, R. F. and M. Burrows. 1981. The morphology and physiology of some walking leg motor neurons in a scorpion. J. Comp. Physiol. 140: 31–42.

Bristowe, W. S. 1958. *The World of Spiders*. New Naturalist Series. Collins, London.

Brownell, P. H. 1977. Compressional and surface waves in sand: Used by desert scorpions to locate prey. Science 197: 479–482.

Brownell, P. H. 1984. Prey detection by the sand scorpion. Sci. Am. 251: 86–97.

Brownell, P. H. and R. D. Farley. 1979a. Detection of vibrations in sand by the tarsal organs of the nocturnal scorpion *Paruroctonus mesaensis*. J. Comp. Physiol. 131: 23–30.

Brownell, P. H. and R. D. Farley. 1979b. Orientation to vibrations in sand by the nocturnal scorpion *Paruroctonus mesaensis*: Mechanism of target localisation. J. Comp. Physiol. 131: 31–38.

Bub, K. and R. F. Bowerman. 1979. Prey capture by the scorpion *Hadrurus arizonensis* Ewing (Scorpionea; Vaejovidae). J. Arachnol. 7: 243–253.

Buchli, H. H. R. 1969. Hunting behavior in the Ctenizidae. Am. Zool. 9: 175–193.

Burgess, J. W. 1976. Social spiders. Sci. Am. 234: 101–106.

Carthy, J. D. 1968. The pectines of scorpions. *In* J. D. Carthy and G. E. Newell (eds.), *Invertebrate Receptors*. Academic Press, New York, pp. 251–261.

Cloudsley-Thompson, J. L. 1968. *Spiders, Scorpions, Centipedes and Mites*. Pergamon Press, New York.

Cooke, J. A. L. 1967. The biology of Ricinulei. Zoologica 151: 31–42.

Dill, L. M. 1975. Predatory behavior of the zebra spider, *Salticus scenicus*. Can. J. Zool. 53: 1284–1289.

Dumpert, K. 1978. Spider odor receptor: Electrophysiological proof. Experimentia 34: 754–756.

Edgar, W. D. 1970. Prey and feeding behavior of adult females of the wolf spider *Pardosa amentata*. Neth. J. Zool. 20(4): 487–491.

Eisner, T. and S. Nowicki. 1983. Spider web protection through visual advertisement: Role of the stabilimentum. Science 219: 185–187.

Foelix, R. F. 1970a. Chemosensitive hairs in spiders. J. Morphol. 132: 313–334.

Foelix, R. F. 1970b. Structure and function of tarsal sensilla in the spider *Araneus diadematus*. J. Exp. Zool. 175: 99–124.

Foelix, R. F. 1982. *Biology of Spiders*. Harvard University Press, Cambridge, MA. [Excellent!]

Foelix, R. F. 1996. *Biology of Spiders*, 2nd Ed. Harvard University Press, Cambridge, MA.

Foelix, R. F. and A. Choms. 1979. Fine structure of a spider joint receptor and associated synapses. Eur. J. Cell Biol. 19: 149–159.

Foelix, R. F. and I.-W. Chu-Wang. 1973. The morphology of spider sensilla. II. Chemoreceptors. Tissue Cell 5(3): 461–478.

Foelix, R. F. and J. Schabronath. 1983. The fine structure of scorpion sensory organs. I. Tarsal sensilla. II. Pecten sensilla. Bull. Br. Arachnol. Soc. 6(2): 53–74.

Foelix, R. F. and D. Troyer. 1980. Giant neurons and associated synapses in the peripheral nervous system of whip spiders. J. Neurocytol. 9: 517–535.

Foil, L. D., L. B. Coons, and B. R. Norment. 1979. Ultrastructure of the venom gland of the brown recluse spider, *Loxosceles reclusa*. Int. J. Insect Morphol. Embryol. 8: 325–334.

Forster, L. 1982. Vision and prey-catching strategies in jumping spiders. Am. Sci. 70: 165–175.

Forster, R. R. 1980. Evolution of the tarsal organ, the respiratory system and the female genitalia in spiders. Int. Congr. Arachnol. 8: 269–285.

Francke, O. F. 1979. Spermatophores of some North American scorpions. J. Arachnol. 7: 19–32.

Gardner, B. T. 1965. Observations on three species of jumping spiders. Psyche 72: 133–147.

Gertsch, W. J. 1979. *American Spiders*, 2nd Ed. Van Nostrand Reinhold, New York.

Griffiths, D. A. and C. E. Bowman (eds.). 1983. *Acarology VI*, Vols. 1–2. Wiley, New York.

Hadley, N. F. 1974. Adaptional biology of desert scorpions. J. Arachnol. 2: 11–23.

Harvey, M. S, 1992. The phylogeny and classification of the Pseudoscorpionida (Chelicerata: Arachnida). Invert. Taxon. 6: 1373–1435.

Herreid, C. F. and R. J. Full. 1980. Energetics of running tarantulas. Physiologist 23: 40.

Jackson, R. R. and R. S. Wilcox. 1998. Spider-eating spiders. Am. Sci. 86 (July–Aug.): 350–357.

Kaston, B. J. 1964. The evolution of spider webs. Am. Zool. 4: 191–207.

Kaston, B. J. 1970. The comparative biology of American black widow spiders. Trans. San Diego Soc. Nat. Hist. 16: 33–82.

Kaston, B. J. 1978. *How to Know the Spiders*, 3rd Ed. W. C. Brown, Dubuque, IA.

Keegan, H. L. 1980. *Scorpions of Medical Importance*. University Press of Mississippi, Jackson.

Kovoor, J. 1977. Silk and the silk glands of Arachnida. Annee Biol. 16: 97–172.

Krantz, G. W. 1978. *A Manual of Acarology*, 2nd Ed. Oregon State University, Corvallis.

Kullmann, E. J. 1972. Evolution of social behavior in spiders. Am. Zool. 12(3): 419–426.

Land, M. F. 1972. Stepping movements made by jumping spiders during turns mediated by the lateral eyes. J. Exp. Biol. 57: 15–40.

Levi, H. W. 1948. Notes on the life history of the pseudoscorpion *Chelifer cancroides*. Trans. Am. Microsc. Soc. 67: 290–299.

Levi, H. W. 1978. Orb-weaving spiders and their webs. Am. Sci. 66: 734–742.

Marshall, S. D. 1992. The importance of being hairy. Nat. Hist. 1992 (Sept.): 41–47.

McCrone, J. D. 1969. Spider venoms: Biochemical aspects. Am. Zool. 9: 153–156.

Merrett, P. (ed.). 1978. *Arachnology*. Symposia of the Zoological Society of London, No. 42. Academic Press, London.

Moffett, S. and G. S. Doell. 1980. Alteration of locomotor behavior in wolf spiders carrying normal and weighted egg cocoons. J. Exp. Zool. 213: 219–226.

Nentwig, W. and St. Heimer.1983. Orb webs and single-line webs: An economic consequence of space web reduction in spiders. Z. Zool. Syst. Evolutionsforsch. 21: 26–37.

Ostfeld, R. S. 1997. The ecology of Lyme-disease risk. Amer. Sci. 85: 338–346.

Parry, D. A. and R. H. J. Brown. 1959. The jumping mechanism of salticid spiders. J. Exp. Biol. 36: 654.

Paulus, H. F. 1979. Eye structure and the monophyly of Arthropoda. *In* A. P. Gupta, *Arthropod Phylogeny*. Van Nostrand Reinhold, New York, pp. 299–377.

Petrunkevitch, A. 1955. Arachnida. *In* R. C. Moore (ed.), *Treatise on Invertebrate Paleontology*, Vol. 2. Geological Society of America, New York, pp. 42–162.

Platnick, N. I. 1971. The evolution of courtship behavior in spiders. Bull. Br. Arachnol. Soc. 2: 40–47.

Platnick, N. I. and W. J. Gertsch. 1976. The suborders of spiders: A cladistic analysis. Am. Mus. Novit. 2607: 1–15.

Polis, G. A. (ed.). 1990. *The Biology of Scorpions*. Stanford University Press, Stanford, CA.

Polis, G. A. and R. D. Farley. 1980. Population biology of a desert scorpion (*Paruroctonus mesaensis*): Survivorship, microhabitat, and the evolution of life history strategy. Ecology 61: 620–629.

Pollock, J. 1966. Life of the ricinulid. Animals 8: 402–405.

Robinson, M. H. and B. Robinson. 1980. Comparative studies of the courtship and mating behavior of tropical araneid spiders. Pacific Insects Monograph, No. 36. Honolulu, HI: Department of Entomology, Bishop Museum.

Rodriguez, J. G. (ed.). 1979. *Recent Advances in Acarology*, Vols. 1–2. Academic Press. [The two volumes total well over 1,000 pages and include contributions on both applied and basic mite research.]

Root, G. and R. F. Bowerman. 1978. Intra-appendage movements during walking in the scorpion *Hadrurus arizonensis*. Comp. Biochem. Physiol. 59A: 57–63.

Rovner, J. S. 1971. Mechanisms controlling copulatory behavior in wolf spiders. Psyche 78(1): 150–165.

Rovner, J. S. 1975. Sound production by nearctic wolf spiders: A substratum-coupled stridulatory mechanism. Science 190: 1309–1310.

Rovner, J. S., G. A. Higashi and R. F. Foelix. 1973. Maternal behavior in wolf spiders: The role of abdominal hairs. Science 182: 1153–1155.

Sabu, L. S. 1965. Anatomy of the central nervous system of arachnids. Zool. Jahrb. Abt. Anat. Ontog. Tiere 82: 1–154.

Sauer, J. R. and J. A. Hair (eds.). 1986. *Morphology, Physiology, and Behavioral Biology of Ticks*. Eillis Horwood, Chichester.

Savory, T. H. 1962. "Daddy Longlegs." Sci. Am. 207: 119.

Savory, T. H. 1977. *Arachnida*, 2nd Ed. Academic Press, New York.

Schultz, J. W. 1990. Evolutionary morphology and phylogeny of Arachnida. Cladistics 6: 1–38.

Seyfarth, E. A. and F. G. Barth. 1972. Compound slit sense organs on the spider leg: Mechanoreceptors involved in kinesthetic orientation. J. Comp. Physiol. 78: 176–191.

Shear, W. A. (ed.). 1981. *Spiders: Webs, Behavior and Evolution*. Stanford University Press, Stanford, CA.

Shear, W. A. 1994. Untangling the evolution of the web. Am. Sci. 82 (May–June): 256–266.

Shultz, J. W. 1987. Walking and surface film locomotion in terrestrial and semi-aquatic spiders. J. Exp. Biol. 128: 427–444.

Snow, K. R. 1970. *The Arachnids: An Introduction*. Columbia University Press, New York.

Sonenshine, D. E. 1992, 1993. *Biology of Ticks*. Vols. I and II. Oxford University Press, Oxford.

Sonenshine, D. E. and T. N. Mather. 1994. *Ecological Dynamics of Tick-Borne Zoonoses*. Oxford University Press, Oxford.

Stewart, D. M. and A. W. Martin. 1974. Blood pressure in the tarantula, *Dugesiella hentzi*. J. Comp. Physiol. 88: 141–172.

Tolbert, W. W. 1975. Predatory avoidance behavior and web defensive structures in the orb weavers *Argiope aurantia* and *Argiope trifasciata*. Psyche 82: 29–52.

Turnbull, A. L. 1973. Ecology of true spiders. Annu. Rev. Entomol. 18: 305–348.

Vachon, M. 1953. The biology of scorpions. Endeavor 12: 80–89.

Watson, P. J. 1986. Transmission of a female sex pheromone thwarted by males in the spider *Linyphia litigiosa* (Linyphiidae). Science 233: 219–221.

Weygoldt, P. 1969. *The Biology of Pseudoscorpions*. Harvard University Press, Cambridge, MA.

Weygoldt, P. 1972. Geisselskorpione und Geisselspinnen (Uropygi and Amblypygi). Z. Kölner Zoo 15(3): 95–107.

Weygoldt, P. 1974. Indirect sperm transfer in arachnids. Verh. Dtsch. Zool. Ges. 67: 308–313.

Weygoldt, P. and H. F. Paulus. 1979. Untersuchungen zur Morphologie, Taxonomie und Phylogenie der Chelicerata. Z. Zool. Syst. Evolut.-Forsch. 17: 85–116.

Witt, P. N. 1975. The web as a means of communication. Biosci. Commun. 1: 7–23.

Witt, P. N. and J. S. Rovner (eds.). 1982. *Spider Communication: Mechanisms and Ecological Significance*. Princeton University Press, Princeton, NJ.

Witt, P. N., M. B. Scarboro, D. B. Peakall and R. Gause. 1977. Spider web-building in outer space: Evaluation of records from the Skylab spider experiment. Am. J. Arachnol. 4: 115.

Pycnogonida

Arnaud, F. and R. N. Bamber. 1987. The biology of the Pycnogonida. *In* J. H. S. Blaxter and A. J. Southward (eds.), *Advances in Marine Biology*, vol. 24. Academic Press, New York, pp. 1–96.

Bain, B. 1991. Some observations on biology and feeding behavior in two southern California pycnogonids. Bijd. tot de Dierkunde 61(1): 63–64.

Behrens, W. 1984. Larvenentwicklung und Metamorphose von *Pycnogonum litorale* (Chelicerata, Pantopoda). Zoomorphologie 104: 266–279.

Bergström, J., W. Stürmer and G. Winter. 1980. *Palaeoisopus, Palaeopantopus* and *Palaeothea*, pycnogonid arthropods from the Lower Devonian Hunsrück Slate, West Germany. Palaeont. Zh. 54: 7–54.

Child, C. A. 1986. A parasitic association between a pycnogonid and a scyphomedusa in midwater. J. Mar. Biol. Assoc. U.K. 66: 113–117.

Cole, L. J. 1905. Ten-legged pycnogonids, with remarks on the classification of the Pycnogonida. Ann. Mag. Nat. Hist. 15: 405–415.

Cole, L. J. 1910. Peculiar habit of a pycnogonid new to North America with observations on the heart and circulation. Biol. Bull. 18: 193–203.

Fry, W. G. 1965. The feeding mechanisms and preferred foods of three species of Pycnogonida. Bull. Br. Mus. Nat. Hist. Zool. 12: 195–223.

Fry, W. G. 1978a. A classification within the Pycnogonida. Zool. J. Linn. Soc. 63: 35–78.

Fry, W. G. (ed.). 1978b. Sea Spiders (Pycnogonida). Zool. J. Linn. Soc. 63: 1–238. [A collection of contributions honoring Joel W. Hedgpeth.]

Fry, W. G. and J. H. Stock. 1978. A pycnogonid bibliography. Zool. J. Linn. Soc. 64: 197–238.

Hedgpeth, J. W. 1948. The Pycnogonida of the western North Atlantic and the Caribbean. Proc. U.S. Nat. Mus. 97 (3216): 157–342.

Hedgpeth, J. W. 1954. On the phylogeny of the Pycnogonida. Acta Zool. 35: 193–213.

Hedgpeth, J. W. 1978. A reappraisal of the Palaeopantopoda with description of a species from the Jurassic. Zool. J. Linn. Soc. 63: 23–34.

Hedgpeth, J. W. 1982. Pycnogonida. *In* S. P. Parker, *Synopsis and Classification of Living Organisms*, Vol. 2. McGraw-Hill, New York, pp. 169–173.

Hedgpeth, J. W. and W. G. Fry. 1964. Another dodecapodous pycnogonid. Ann. Mag. Nat. Hist. 13: 161–169.

Henry, L. M. 1953. The nervous system of the Pycnogonids. Microentomology 18: 16–36.

King, P. E. 1973. *Pycnogonids*. Hutchinson University Library, London.

Morgan, E. 1971. The swimming of *Nymphon gracile* (Pycnogonida): The mechanics of the leg-beat cycle. J. Exp. Biol. 55: 273–287.

Morgan, E. 1972. The swimming of *Nymphon gracile* (Pycnogonida): The swimming gait. J. Exp. Biol. 56: 421–432.

Nakamura, K. 1981. Postembryonic development of a pycnogonid *Propallene longiceps*. J. Nat. Hist. 15: 49–62.

Nakamura, K. and K. Sekiguchi. 1980. Mating behavior and oviposition in the pycnogonid *Propallene longiceps*. Mar. Ecol. Prog. Ser. 2: 163–168.

Schram, F. R. and J. W. Hedgpeth. 1978. Locomotory mechanisms in Antarctic pycnogonids. Zool. J. Linn. Soc. 63: 145–169.

Staples, D. A. and J. E. Watson. 1987. Associations between pycnogonids and hydroids. In J. Bouillon (ed.), *Modern Trends in the Systematics, Ecology and Evolution of Hydroids and Hydromedusae*. Oxford University Press, Oxford, pp. 215–226.

Tomaschko, K. H., E. Wilhelm and D. Bückmann. 1997. Growth and reproduction of *Pycnogonum litorale* (Pycnogonida) under laboratory conditions. Mar. Biol. 129: 595–600.

Wilhelm, E., D. Bückmann and K. H. Tomaschko. 1997. Life cycle and population dynamics of *Pycnogonum litorale* (Pycnogonida) in a natural habitat. Mar. Biol. 129: 601–606.

Wyer, D. W. and P. E. King. 1974. Relationships between some British littoral and sublittoral bryozoans and pycnogonids. Estuarine Coastal Mar. Sci. 2: 177–184.

20 *Phylum Mollusca*

Orange and speckled and fluted nudibranchs slide gracefully over the rocks, their skirts waving like the dresses of Spanish dancers.
John Steinbeck,
Cannery Row, 1945

Molluscs include some of the best known invertebrates; almost everyone is familiar with snails, clams, slugs, squids, and octopuses. Molluscan shells have been popular since ancient times, and some cultures still use them as tools, containers, musical devices, money, fetishes, and decorations. Evidence of historical use and knowledge of molluscs is seen in ancient texts and hieroglyphics, on coins, in tribal customs, and in archaeological sites and aboriginal kitchen middens or shell mounds. Royal or Tyrian purple of ancient Greece and Rome, and even Biblical blue (Num. 15:38), were molluscan pigments extracted from certain marine snails.* Many aboriginal groups have for centuries relied on molluscs for a substantial portion of their diet. Today, coastal nations annually harvest millions of tons of molluscs commercially for food.

There are about 93,000 described, living mollusc species and 70,000 known fossil molluscs. However, many species still await names and descriptions, especially those from poorly studied regions, and it has been estimated that only about half of the living molluscs have so far been described. In addition to three familiar molluscan classes comprising the clams (Bivalvia), snails and slugs (Gastropoda), and squids and octopuses (Cephalopoda), four other classes exist: chitons (Polyplacophora), tusk shells (Scaphopoda), *Neopilina* and its kin (Monoplacophora), and the primitive vermiform Aplacophora. Although members of these seven classes differ enormously in superficial appearance, they are remarkably similar in their fundamental bauplan (Box 20A).

---

*Archaeological sites in Israel reveal the probable use of two muricid snails (*Murex brandaris* and *Trunculariopsis trunculus*) as sources of the Royal purple dye.

BOX 20A Characteristics of the Phylum Mollusca

1. Bilaterally symmetrical (or secondarily asymmetrical), unsegmented, coelomate protostomes

2. Coelom limited to small spaces around nephridia, heart, and part of intestine

3. Principal body cavity is a hemocoel (open circulatory system)

4. Viscera concentrated dorsally as a "visceral mass"

5. Body covered by thick epidermal–cuticular sheet of skin, the mantle, which forms a cavity (the mantle cavity) in which are housed the ctenidia, osphradia, nephridiopores, gonopores, and anus

6. Mantle with shell glands that secrete calcareous epidermal spicules, shell plates, or shells

7. Heart lies in pericardial chamber and composed of separate ventricle and atria

8. With large, well-defined muscular foot, often with a flattened creeping sole

9. Buccal region provided with a radula

10. Complete gut, with marked regional specialization, including large digestive ceca

11. With large, complex metanephridia ("kidneys")

12. Embryogeny typically protostomous

13. With trochophore larva, and usually a veliger larva

Taxonomic History and Classification

Molluscs* carry the burden of a very long and convoluted taxonomic history, in which hundreds of names for various taxa have come and gone. Aristotle was probably the first scientist to formally recognize molluscs, dividing them into two groups: Malachia (the cephalopods) and Ostrachodermata (the shelled forms), the latter being divided into univalves and bivalves. Jonston (or Jonstonus) created the name Mollusca in 1650 for the

---

*The name of the phylum derives from the Latin *molluscus*, meaning "soft," in allusion to the similarity of clams and snails to the mollusca, a kind of Old World soft nut with a thin but hard shell. The vernacular for Mollusca is often spelled *mollusks* in the United States, whereas in the rest of the world it is typically spelled *molluscs*. In biology, a vernacular or diminutive name is generally derived from the proper Latin name; thus the American custom of altering the spelling of Mollusca by changing the *c* to *k* seems to be an aberration (although it may have its historic roots in the German language, which does not have the free-standing *c*; e.g., Molluskenkunde). We prefer the spelling *molluscs*, which seems to be the proper vernacularization and is in line with other accepted terms, such as *molluscan, molluscoid, molluscivore*, etc.

cephalopods and barnacles, but this name was not accepted until it was resurrected and redefined by Linnaeus. Recall that Linnaeus regarded all invertebrates except insects as Vermes, a group divided into Intestina, Mollusca, Testacea, Lithophyta, and Zoophyta. His Mollusca was a potpourri of soft-bodied animals, including not only cephalopods, slugs, and pteropods but also tunicates, anemones, medusae, echinoderms, and polychaetes. Under Testacea, Linnaeus included chitons, bivalves, univalves, nautiloids, barnacles, and the serpulid polychaetes (which secrete calcareous tubes). In 1795 Cuvier published a revised classification of the Mollusca that was the first to approximate modern views. De Blainville (1825) altered the name Mollusca to Malacozoa, which won little favor but survives in the terms malacology, malacologist, etc.

Much of the nineteenth century passed before the phylum was purged of all extraneous groups. In the 1830s J. Thompson and C. Brumeister identified the larval stages of barnacles and revealed them to be crustaceans, and in 1866 A. Kowalevsky removed the tunicates from the Mollusca. Separation of the brachiopods from the molluscs was a long and controversial ordeal that was not resolved until near the end of the nineteenth century.

Aplacophorans were discovered in 1841 by the Swedish naturalist Lovén. He classified them with holothurian echinoderms because of their vermiform bodies and the presence of calcareous spicules in the body walls of both groups. Graff (1875) recognized aplacophorans as molluscs, and shortly thereafter it became fashionable to classify the chitons and aplacophorans together in the class Amphineura. This scheme persisted until the 1950s when the two groups were again separated; however, some recent workers (e.g., Scheltema) have once more suggested that these two groups might represent a monophyletic clade unto themselves (the Aculifera).

The history of classification of species in the class Gastropoda has been volatile, undergoing constant change since Cuvier's time. Most modern malacologists adhere more or less to the basic schemes of Milne-Edwards (1848) and Spengel (1881). The former, basing his classification on the respiratory organs, recognized the groups Pulmonata, Opisthobranchia, and Prosobranchia. Spengel based his scheme on the nervous system and divided the gastropods into the Streptoneura and Euthyneura. Streptoneura is equivalent to Prosobranchia; Euthyneura embraces Opisthobranchia and Pulmonata. The bivalves have been called Bivalvia, Pelecypoda, and Lamellibranchiata.

Molluscan classification at the generic and species levels is also troublesome. Many species of gastropods and bivalves are burdened with numerous names (synonyms), and some species actually bear hundreds of different synonyms. This tangle is partly the result of a long history of amateur participation in the field of molluscan taxonomy and partly because so much of the early classification is based on shell characters rather than anatomy.

Only taxa with extant members are included in the following classification synopsis.* Some important fossil groups are discussed later in the chapter. Examples of the major molluscan taxa appear in Figure 20.1.

PHYLUM MOLLUSCA

CLASS APLACOPHORA: Benthic, marine, vermiform molluscs; shell-less, but epidermis secretes aragonite (calcareous) spicules or scales; mantle cavity rudimentary; without eyes, tentacles, statocysts, crystalline style, or nephridia. (Figure 20.2)

SUBCLASS CHAETODERMOMORPHA (= CAUDOFOVEATA): Burrowing, cylindrical, body wall bears a chitinous cuticle and imbricating scalelike calcareous spicules; posterior mantle cavity with a pair of bipectinate ctenidia; radula present. About 120 known species; burrow in muddy sediments and consume microorganisms. (e.g., *Chaetoderma, Falcidens, Limifossor, Psilodens, Scutopus*)

SUBCLASS NEOMENIOMORPHA (=SOLENOGASTRES): Cylindrical or compressed, mantle cavity rudimentary; body wall imbued with calcareous spicules or scales; without ctenidia or with 1 pair folded (but not bipectinate) ctenidia; with or without radula; hermaphroditic; without a flattened foot, but usually with ventral furrow or "pedal groove" (thought to be homologous to the foot of other molluscs). About 250 described species; epibenthic carnivores, often found on (and consuming) cnidarians. (e.g., *Chevroderma, Dondersia, Epimenia, Kruppomenia; Neomenia, Proneomenia, Pruvotina, Rhopalomenia, Spengelomenia*)

CLASS MONOPLACOPHORA: Monoplacophorans. With a single, caplike shell; foot forms weak ventral muscular disc, with 8 pairs retractor muscles; shallow mantle cavity around foot encloses 3–6 pairs ctenidia; 2 pairs gonads; 3–7 pairs metanephridia; 2 pairs heart atria; with radula and distinct but small head; without eyes; tentacles present only around mouth; with a crystalline style and posterior anus; mostly extinct (Figures 20.1A and 20.3). Until the first living species (*Neopilina galatheae*) was discovered by the Danish Galathea Expedition in 1952, monoplacophorans were known only from lower Paleozoic fossils. Since then their unusual anatomy has been a source of much evolutionary speculation. Monoplacophorans are limpet-like in appearance, living species are less than 3 cm in length, and most live at considerable depths. About 25 described species, in 6 genera (*Laevipilina, Micropilina, Monoplacophorus, Neopilina, Rokopella, Vema*).

CLASS POLYPLACOPHORA: Chitons. Flattened, elongated molluscs with a broad ventral foot and 8 dorsal shell plates; shells with unique articulamentum layer; mantle forms thick girdle that borders and may partly or entirely cover shell plates; epidermis of girdle usually with calcareous or chitinous spines,

scales, or bristles; mantle cavity encircles foot and bears from 6 to more than 80 pairs of ctenidia; 1 pair nephridia; without eyes, tentacles, or crystalline style; nervous system lacking discrete ganglia, except in buccal region; radula present (Figures 20.1B and 20.4). Marine, intertidal to deep sea. Chitons are unique in their possession of 8 (sometimes 7) separate shell plates, called valves, and a thick marginal girdle; about 1000 described species.

ORDER LEPIDOPLEURIDA: Primitive chitons with outer edge of shell plates lacking attachment teeth; girdle not extending over plates; ctenidia limited to a few posterior pairs. (e.g., *Choriplax, Lepidochiton, Lepidopleurus, Oldroydia*)

ORDER ISCHNOCHITONIDA: Outer edges of shell plates with attachment teeth; girdle not extending over plates, or extending partly over plates; ctenidia occupying most of mantle groove, except near anus. (e.g., *Callistochiton, Chaetopleura, Ischnochiton, Katharina, Lepidozona, Mopalia, Nuttallina, Placiphorella, Schizoplax, Tonicella*)

ORDER ACANTHOCHITONIDA: Outer edge of shell plates with well developed attachment teeth; shell valves partially or completely covered by girdle; ctenidia do not extend full length of foot. (e.g., *Acanthochitona, Cryptochiton, Cryptoplax*)

CLASS GASTROPODA: Snails and slugs. Asymmetrical molluscs with single, usually spirally coiled shell into which body can be withdrawn; shell lost or reduced in many groups; during development, visceral mass and mantle rotate 90–180° on foot (torsion), so mantle cavity lies anterior or on right side, and gut and nervous system are twisted; some taxa have partly or totally reversed the rotation (detorsion); with muscular creeping foot (modified in swimming and burrowing taxa); head with statocyst and eyes (often reduced or lost), and 1–2 pairs tentacles; most with complex radula and crystalline style, the latter being lost in most predatory groups; 1–2 nephridia; mantle (= pallium) usually forms cavity housing ctenidia, osphradia, and hypobranchial glands; ctenidia sometimes lost and replaced with secondary gas exchange structures (Figures 20.1C–E,G–I, 20.5, 20.6, and 20.7).

Gastropods comprise about 70,000 living species of marine, terrestrial, and freshwater snails and slugs. The class is usually divided into three subclasses: prosobranchs (largely shelled marine snails); opisthobranchs (marine slugs); and pulmonates (terrestrial snails and slugs). However, this arrangement is viewed by some authorities as artificial, and numerous revisionary schemes have been proposed.

SUBCLASS PROSOBRANCHIA: Usually with a spirally coiled shell, sometimes with cap-shaped or tubular shell; mantle cavity usually anteriorly directed, near head, containing osphradia, ctenidia, hypobranchial glands, anus, and nephridiopores; head generally with tentacles bearing basal eyes; foot with creeping planar sole and typically with corneous or calcareous operculum to close shell aperture upon retraction of head and foot; radula variable or absent; nervous system streptoneurous.

ORDER ARCHAEOGASTROPODA: Primitive prosobranchs; shell primitively formed with nacreous (pearly) layer; radula modified for herbivory, often with numerous teeth in transverse rows, usually rhipidoglossate or docoglossate; 1–2 bipectinate ctenidia; mantle cavity without siphon; primitively with 2 hypobranchial glands, 2 osphradia, 2 atria, and 2 metanephridia; sexes usually separate; male generally without penis; nervous system weakly concentrated. Primarily marine, al-

---

*A multitude of extinct molluscs have been described. Perhaps the most well known are some of the groups of cephalopods that had hard external shells, similar to those of living *Nautilus*. One of these groups was the ammonites (a term reserved for Jurassic and Cretaceous members of the order Ammonoidea). They differed from nautiloids in having shell septa that were highly fluted on the periphery, forming complex mazelike septal sutures. Ammonoids also had the siphuncle lying against the outer wall of the shell, as opposed to the condition seen in nautiloids where the siphuncle runs through the center of the shell whorls.

(A)

(B)

(C)

(D)

(E)

(F)

(G)

(H)

(I)

(J)

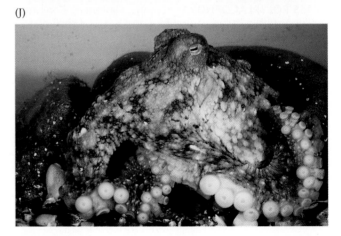

Figure 20.1 Diversity among the molluscs.
(A) The cap-shaped shell of *Neopilina* (class Mono-placophora). (B) Cryptochiton, the giant Pacific gumboot chiton (class Polyplacophora). (C) *Pero-trochus*, a primitive gastropod, or "slit shell" (class Gastropoda). (D) The red abalone *Haliotis rufescens* (class Gastropoda). Note the exhalant holes in the shell. (E) *Epitonium scalare*, the precious wentletrap (class Gastropoda). (F) The bizarre anomalodesma-tan clam *Brechites*. The minute valves can be seen fused to the hollow tube-shell at the anterior end of this suspension-feeding bivalve. (G) *Conus* (class Gastropoda). (H) *Chromodoris*, a nudibranch (class Gastropoda). (I) *Monadenia fidelis*, a terrestrial snail from California (class Gastropoda). (J) *Octopus* (class Cephalopoda). (K) *Histioteuthis*, a pelagic squid (class Cephalopoda). (L) *Fustiaria*, a tusk shell (class Scaphopoda). (M) Scallops, with a hermit crab in the foreground (class Bivalvia). (N) The giant clam *Tridacna* (class Bivalvia). (O) *Clinocar-dium*, a cockle (class Bivalvia). (P) *Lima*, a tropical clam that swims by clapping the valves together.

(K)

(L)

(M)

(N)

(O)

(P)

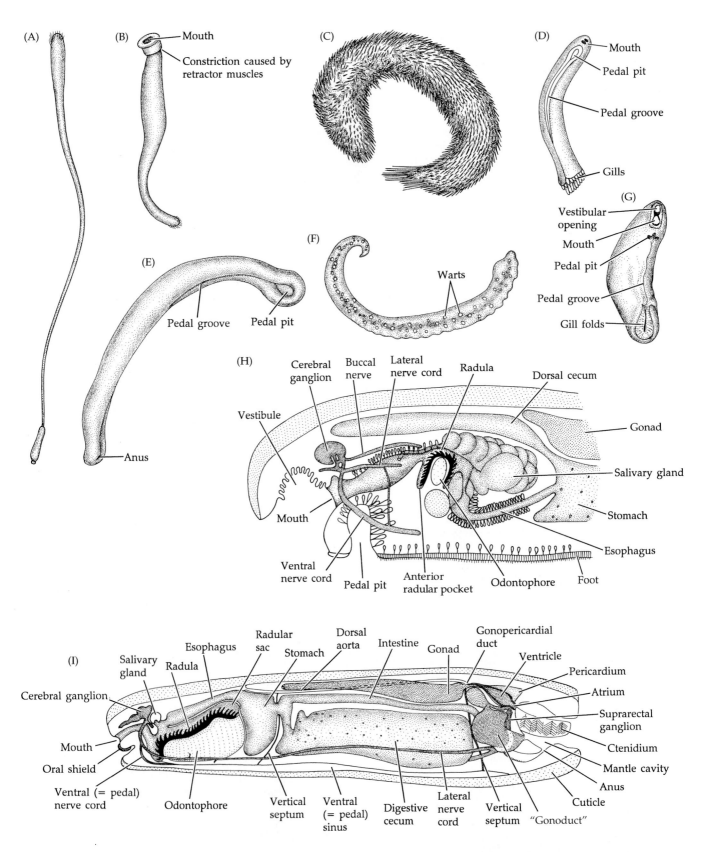

Figure 20.2 General anatomy of aplacophorans (chaetoderms and neomeniomorphs). A variety of body types, in seven species of Aplacophora. (A) *Chaetoderma productum.* (B) *Chaetoderma loveni.* (C) *Kruppomenia minima.* This species has long spicules. (D) *Pruvotina impexa* (ventral view). (E) *Proneomenia antarctica.* (F) *Epimenia ver-* *rucosa.* The body is covered with warts. (G) *Neomenia carinata* (ventral view). (H) The anterior region of the neomeniomorphan *Spengelomenia bathybia* (longitudinal section). (I) The internal anatomy of the chaetoderm *Limifossor* (longitudinal section).

(A)

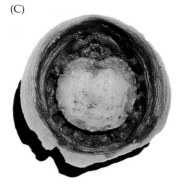

(B)

Velum

Anterior lip

Mouth

Cluster of postoral tentacles

Shell

Foot

Pallial groove

Mantle

Ctenidia

Anus

(C)

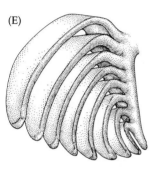

(E)

(D)

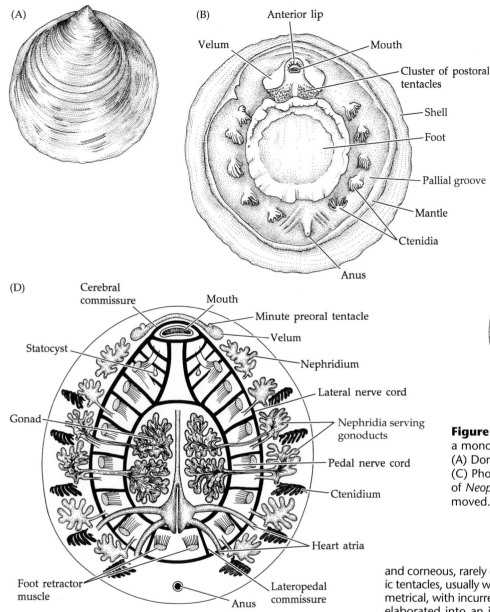

Cerebral commissure

Mouth

Minute preoral tentacle

Velum

Nephridium

Statocyst

Lateral nerve cord

Nephridia serving gonoducts

Gonad

Pedal nerve cord

Ctenidium

Heart atria

Foot retractor muscle

Lateropedal commissure

Anus

Figure 20.3 General anatomy of a monoplacophoran (*Neopilina*). (A) Dorsal view (shell). (B) Ventral view. (C) Photograph of the ventral surface of *Neopilina*. (D) Ventral view, foot removed. (E) One of the gills.

though a few freshwater and terrestrial species are known; virtually all are herbivores. Some specialists are convinced that the Archaeogastropoda is a polyphyletic grade. For example, many workers now place the "true" limpets (those without holes or slits in their shells) into a separate order, the Patellogastropoda.

Twenty-six families, including Pleurotomaridae and Scissurellidae (slit-shelled molluscs, the most primitive living prosobranchs: e.g., *Perotrochus, Pleurotomaria, Scissurella*); Haliotidae (abalones, *Haliotis*); Fissurellidae (keyhole limpets: e.g., *Diodora, Fissurella, Lucapinella, Puncturella*); Acmaeidae, Patellidae, and three other families (the "true" limpets: e.g., *Acmaea, Collisella, Lottia, Patella*); Trochidae (trochids: e.g., *Calliostoma, Margarites, Tegula, Trochus*); Turbinidae (turbans: e.g., *Astrea*); Neritidae (nerites: e.g., *Nerita, Theodoxus*); Helicinidae (helicinids: e.g., *Alcadia, Helicinia*).

ORDER MESOGASTROPODA: Shell mainly porcelaneous and nonnacreous; operculum usually present

and corneous, rarely calcified; head with pair of cephalic tentacles, usually with basal eyes; mantle cavity asymmetrical, with incurrent opening on anterior left, often elaborated into an inhalant siphon; right ctenidium lost; left ctenidium usually monopectinate; hypobranchial glands often lost on left; right nephridium often lost; radula generally taenioglossate, occasionally lost; most are dioecious; higher forms with concentrated ganglia.

Includes marine, freshwater, and terrestrial forms divided among nearly 100 families, including Hydrobiidae (e.g., *Hydrobia*); Viviparidae (e.g., *Viviparus*); Littorinidae (periwinkles: e.g., *Littorina*); Turritellidae (tower or turret shells: e.g., *Turritella*); Caecidae (e.g., *Caecum*); Vermetidae (vermetids or "worm" gastropods: e.g., *Serpulorbis, Tripsycha, Vermetus, Vermicularia*,); Cerithiidae (ceriths: e.g., *Cerithium, Liocerithium*); Potamididae (potamids or horn shells: e.g., *Cerithidea*); Strombidae (conchs or strombids: e.g., *Strombus*); Epitoniidae (wentletraps or epitoniids: e.g., *Epitonium*); Janthinidae (janthinids: e.g., *Janthina*); Hipponicidae (horse hoof limpets: e.g., *Hipponix*); Capulidae (cap limpets: e.g., *Capulus*); Calyptraeidae (cup and saucer limpets and slipper limpets: e.g., *Calyptraea, Crepidula, Crucibulum*); Carinariidae (one of several families of

Figure 20.4 Generalized anatomy of chitons (class Polyplacophora). (A,B) A typical chiton (dorsal and ventral views). (C) The Pacific lined chiton, *Tonicella lineata*. (D) Dorsal view of a chiton, shell plates (valves) removed. (E) Dorsal view of a chiton, dorsal musculature removed to reveal internal organs. (F) Dorsal view of a chiton, showing extensive kidneys. (G) The arrangement of internal organs in a chiton (lateral view).

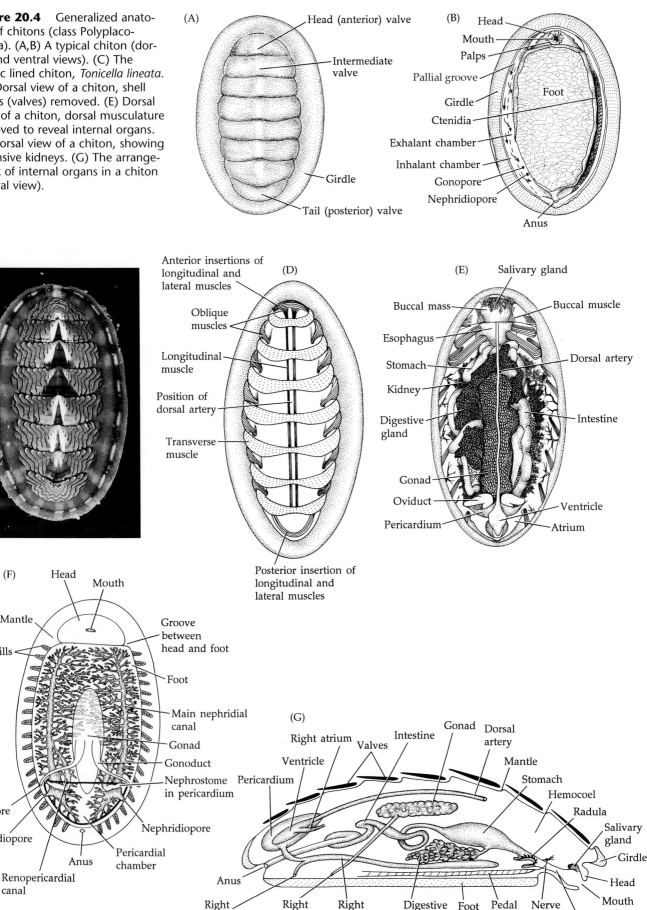

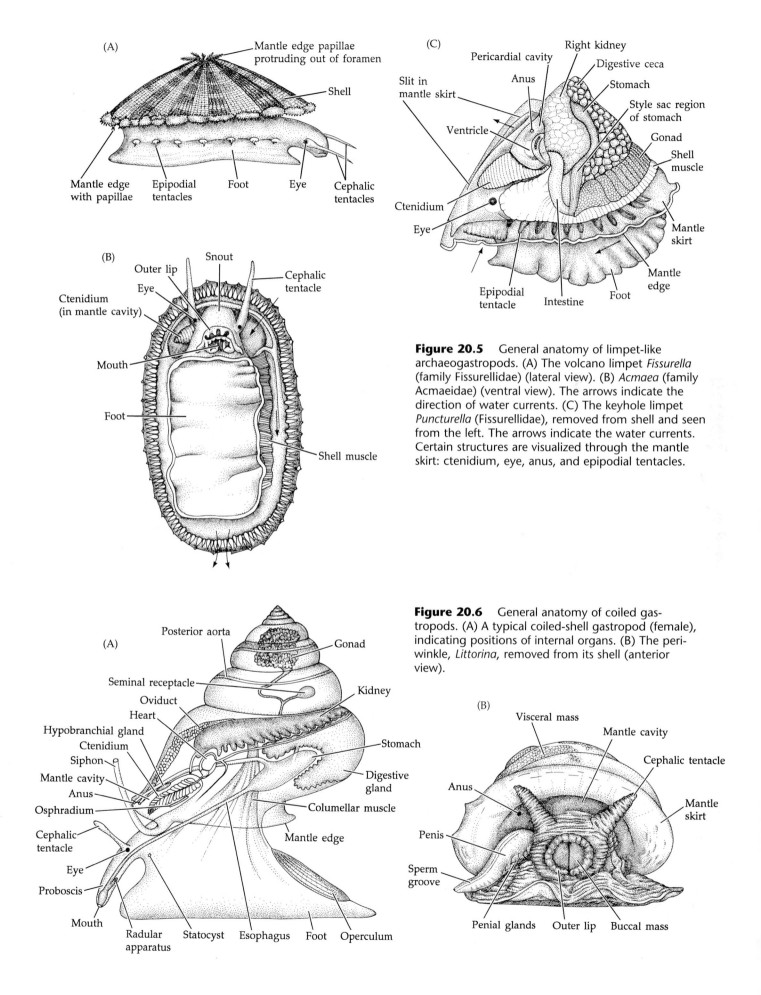

Figure 20.5 General anatomy of limpet-like archaeogastropods. (A) The volcano limpet *Fissurella* (family Fissurellidae) (lateral view). (B) *Acmaea* (family Acmaeidae) (ventral view). The arrows indicate the direction of water currents. (C) The keyhole limpet *Puncturella* (Fissurellidae), removed from shell and seen from the left. The arrows indicate the water currents. Certain structures are visualized through the mantle skirt: ctenidium, eye, anus, and epipodial tentacles.

Figure 20.6 General anatomy of coiled gastropods. (A) A typical coiled-shell gastropod (female), indicating positions of internal organs. (B) The periwinkle, *Littorina*, removed from its shell (anterior view).

Figure 20.7   General anatomy of other gastropods.
(A) The pelagic shelled heteropod *Carinaria* (order
Mesogastropoda). (B) Anatomy of *Carinaria*. (C) The
shell-less heteropod *Pterotrachea*. (D) The pelagic shelled
pteropod *Clio* (subclass Opisthobranchia). The arrows
indicate the direction of water flow; water enters all
around the narrow neck and is forcibly expelled together with fecal, urinary, and genital products by contraction of the sheath. (E) A swimming pteropod, *Corolla*.
(F–J) Various sea slugs. (F) A dorid nudibranch, *Diaulula*.
(G) An aeolid nudibranch, *Phidiana*. (H) A dorid nudibranch from Australia, with an orange branchial plume.
(I) Two Eastern Pacific "Spanish shawl" (aeolid) nudibranchs *Flabellina*. (J) The nudibranch *Histiomena*, from
the Sea of Cortez.

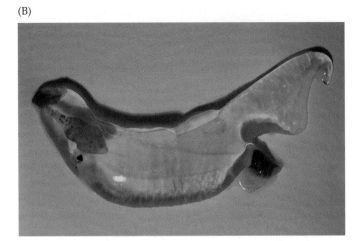

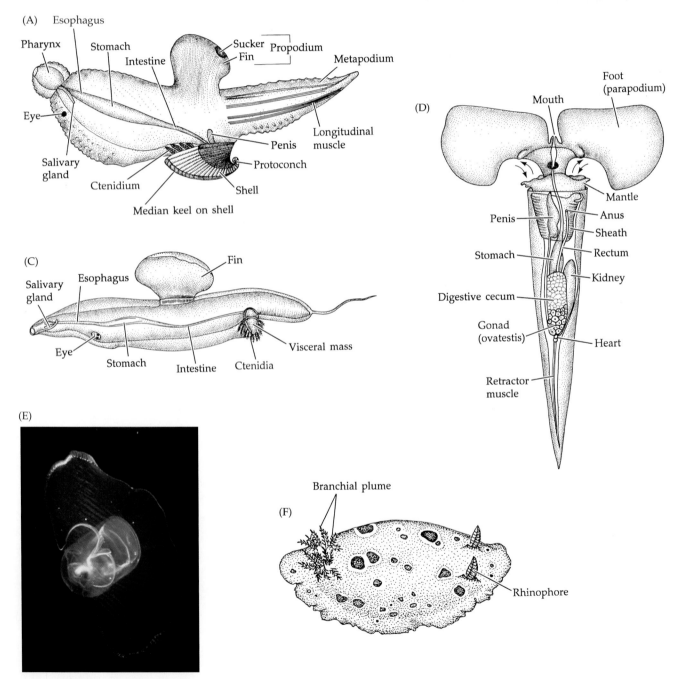

pelagic molluscs collectively called heteropods: e.g., *Carinaria*); Naticidae (moon snails: e.g., *Natica, Polinices*); Eratoidae (coffee bean shells: e.g., *Erato, Trivia*); Cypraeidae (cowries: e.g., *Cypraea*); Ovulidae (ovulids or egg shells: e.g., *Jenneria, Ovula, Simnia*); Tonnidae (tun shells: e.g., *Malea*); Cassididae (helmet shells: e.g., *Cassis*); Ficidae (fig shells: e.g., *Ficus*).

ORDER NEOGASTROPODA:

Shell without nacreous layer; radula with 1–3 teeth in each row; 1 (left) monopectinate ctenidium; 1 osphradium; radula rachiglossate or toxoglossate; mantle forms siphon, carried within siphonal canal or notch of shell; sexes separate, male with penis; nervous system concentrated; operculum, if present, chitinous; heart with left atrium only; right nephridium lost.

About two dozen families of marine snails, including Buccinidae (whelks: e.g., *Buccinum, Cantharus, Macron, Metula*); Columbellidae (dove shells: e.g., *Anachis, Columbella, Mitrella, Nassarina, Pyrene, Strombina*); Coralliophilidae (e.g., *Coralliophila, Latiaxis*); Fasciolariidae (tulip shells and spindle shells: e.g., *Fasciolaria, Fusinus, Leucozonia, Troschelia*); Harpidae (harp shells: e.g., *Harpa*); Marginellidae (marginellids, e.g., *Granula*); Melongenidae (whelks, false trumpets: e.g., *Melongena*); Mitridae (miter shells: e.g., *Mitra, Subcancilla*); Muricidae (rock shells: e.g., *Ceratostoma, Hexaplex, Murex, Phyllonotus, Pteropurpura, Pterynotus*); Thaididae (thaids: e.g., *Acanthina, Morula, Neorapana, Nucella, Purpura, Thais*); Nassariidae (dog whelks and basket shells: e.g., *Nassarius*); Olividae (olive shells: e.g., *Agaronia, Oliva, Olivella*); Volutidae (volutes: e.g., *Cymbium, Lyria, Voluta*); Cancellariidae (e.g., *Admete, Cancellaria*); Conidae (cone shells: e.g., *Conus*); Turridae (tower shells: e.g., *Crassispira*); Terebridae (auger shells: e.g., *Terebra*).

SUBCLASS OPISTHOBRANCHIA:

Sea slugs and their kin. Body variously detorted; shell reduced and thin, external or internal, or lost altogether; ctenidia and mantle cavity usually reduced or lost; usually without operculum; head with 1–2 pairs of rhinophores or tentacles; hermaphroditic; euthyneurous with various degrees of nervous system concentration. Primarily marine, benthic; a few freshwater species.

Traditional (conservative) classifications include nine orders (and over 100 families) of opisthobranchs. However, some authorities feel that the Opisthobranchia may not be a monophyletic taxon. In any case, shell loss almost certainly occurred several times within this subclass. Several alternative classifications of this group have been suggested but a consensus has not yet been achieved. The nine orders and some common genera are Acochlidioidea (e.g., *Acochlidium, Unela*); Cephalaspidea (e.g., *Acteon, Aglaja, Bulla, Chelidonura, Haminoea, Navanax, Retusa, Rictaxis, Scaphander*); Runcinoidea (e.g., *Ilbia, Runcina*); Sacoglossa (e.g., *Berthelinia, Elysia, Oxynoe, Tridachia*); Anaspidea (the sea hares: e.g., *Aplysia, Dolabella, Stylocheilus*); Thecosomata (the shelled pteropods: e.g., *Clio, Limacina*); Gymnosomata (the naked pteropods: e.g., *Clione*); Notaspidea (e.g., *Berthellina, Gymnotoplax, Pleurobranchus, Tylodina*); Nudibranchia (the "true" nudibranchs: e.g., *Acanthodoris, Aegires, Aeolidia, Armina, Chromodoris, Corambe, Coryphella, Dendrodoris, Dendronotus, Diaulula, Doris, Embletonia, Fiona, Glaucus, Hermissenda, Hexabranchus, Hopkinsia, Janolus, Phidiana, Phyllidia, Platydoris, Polycera, Rostanga, Scyllaea, Tambja, Trinchesia*).

(G)

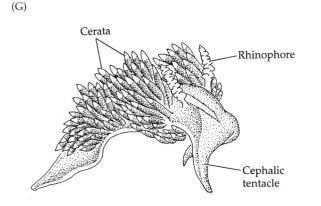

Cerata

Rhinophore

Cephalic tentacle

(H)

(I)

(J)

SUBCLASS PULMONATA: Land snails and slugs. Mantle cavity forms lung with contractile opening; without ctenidia (except perhaps in *Siphonaria*); body detorted to various degrees; highly concentrated nervous system (euthyneurous); hermaphroditic; without larvae; mainly terrestrial and freshwater forms, a few marine species.

ORDER ARCHAEOPULMONATA: Primitive pulmonates with spirally coiled shell, but no operculum; mainly littoral. (e.g., *Cassidula, Ellobium, Otina*)

ORDER BASOMMATOPHORA: Shell variable, minute or moderate-sized, generally spirally coiled (or planospiral) or patelliform; usually without an operculum; eyes at bases of sensory stalks; freshwater and intertidal; includes freshwater limpets. (e.g., *Bulinus, Carychium, Chilina, Lanx, Physa, Planorbis, Siphonaria, Trimusculus*)

ORDER STYLOMMATOPHORA: Shell absent or present; if present usually spirally coiled and often partly or completely enveloped by dorsal mantle; eyes on tips of sensory stalks; terrestrial; an enormous group with over 15,000 described species. (e.g., *Achatina, Arion, Bulimulus, Cepaea, Haplotrema, Helix, Liguus, Limax, Megaspira, Oreohelix, Pupilla, Rachis, Succinea, Vertigo*)

ORDER SYSTELLOMMATOPHORA: Sluglike, without internal or external shell; dorsal mantle integument forms a keeled or rounded notum; head usually with 2 pairs tentacles, upper ones forming contractile stalks bearing eyes. (e.g., *Onchidella, Onchidium, Rhodope*)

CLASS BIVALVIA (= PELECYPODA; = LAMELLIBRANCHIATA): Clams, oysters, mussels, etc. Laterally compressed; shell typically of two valves hinged together dorsally by elastic ligament and shell-teeth; shells closed by adductor muscles; head rudimentary, without eyes or radula, but eyes and statocysts may occur elsewhere on body; foot typically laterally compressed, usually without a sole; 1 pair large bipectinate ctenidia, used in combination with labial palps in ciliary feeding; large mantle cavity; posterior edges of mantle often fused to form inhalant and exhalant siphons; 1 pair nephridia (Figures 20.1F,M–P and 20.8).

Bivalves are marine or freshwater molluscs, primarily microphagous or suspension feeders. The class includes about 20,000 living species represented at all depths and in all marine environments. Bivalve classification has been in a state of turmoil over the past 50 years. Hardly any two authors today utilize exactly the same classification or nomenclature. Some workers delimit the higher taxa on the basis of shell characters alone (e.g., hinge anatomy, position of muscle scars), others rely solely on internal organ anatomy (e.g., ctenidia, stomach), and still others use ecological characters (e.g., feeding methods, adaptations to various habitats). For some alternatives to the classification below, see Purchon (1977), Morton (1963), and Moore (1960).

SUBCLASS PROTOBRANCHIA: Ctenidia are 2 pairs of simple, unfolded, bipectinate, platelike leaflets suspended in the mantle cavity. Primitive bivalves.

ORDER NUCULIDA (= PALAEOTAXODONTA): Shell aragonitic, interior nacreous or porcelaneous; periostracum smooth; shell valves equal and taxodont (i.e., the valves have a row of short teeth along hinge margin); adductor muscles equal in size; with large palp proboscides used for food collection; ctenidia small, strictly for gas exchange; foot longitudinally grooved and with a plantar sole, adults without byssal threads; nervous system primitive, often with incomplete union of cerebral and pleural ganglia; marine, mainly infaunal detritivores. (e.g., *Malletia, Nucula, Yoldia*)

ORDER SOLEMYIDA (= CRYPTODONTA): Shell valves thin, elongate, and equal in size; uncalcified along outer edges, without hinge teeth; anterior adductor muscle larger than posterior one; ctenidia large, used both for gas exchange and feeding. (e.g., *Solemya*)

SUBCLASS LAMELLIBRANCHIA: Paired ctenidia, with very long filaments that fold back on themselves so that each row of filaments forms two lamellae; adjacent filaments usually attached to one another by ciliary tufts (filibranch condition), or by tissue bridges (eulamellibranch condition).

SUPERORDER FILIBRANCHIA (= PTERIOMORPHIA): Ctenidia with outer fold not connected dorsally to visceral mass, with free filaments or with adjacent filaments attached by ciliary tufts; shell aragonitic or calcitic, sometimes nacreous; mantle margin unfused, with weakly differentiated incurrent and excurrent apertures or siphons; foot well developed or extremely reduced; usually attached by byssal threads or cemented to substratum (or secondarily free).

Primitive lamellibranchs, including mussels (Mytilidae: e.g., *Adula, Brachidontes, Lithophaga, Modiolus, Mytilus*) and other clams, such as the ark shells (Arcidae: e.g., *Anadara, Arca, Barbatia*), glycymerids (Glycymerididae: e.g., *Glycymeris*), true oysters (Ostreidae: e.g., *Crassostrea, Ostrea*), pearl oysters (Pteriidae: e.g., *Pinctada, Pteria*), hammer oysters (Malleidae: e.g., *Malleus*), pen shells (Pinnidae: e.g., *Atrina, Pinna*), file shells (Limidae: e.g., *Lima*), scallops (Pectinidae: e.g., *Chlamys, Lyropecten, Pecten*), thorny oysters (Spondylidae: e.g., *Spondylus*), and jingle shells (Anomiidae: e.g., *Anomia, Pododesmus*).

SUPERORDER EULAMELLIBRANCHIA (= HETERODONTA): Ctenidia with outer fold completely connected dorsally to roof of mantle cavity, with adjacent filaments attached by tissue bridges; shell generally aragonitic, without nacreous layer; shell valves equal to subequal, with a few large cardinal teeth separated from the elongated lateral teeth by a toothless space; mantle more or less fused posteroventrally and forming incurrent and excurrent apertures that are frequently drawn out onto siphons; foot usually lacks byssal threads in adult. Advanced bivalves, mainly marine, including three main groups (treated here as orders).

ORDER PALEOHETERODONTA: Shell aragonitic, pearly internally; periostracum usually well developed; valves usually equal, with few hinge teeth; elongate lateral teeth (when present) are not separated from the large cardinal teeth; usually dimyarian; mantle opens broadly ventrally, mostly unfused posteriorly but with excurrent and incurrent apertures. About 1,200 species of marine and freshwater clams. Includes the nearly extinct family Trigoniidae (with fewer than six living species, in the Australasian region), and the suborder Unionidea (freshwater bivalves: e.g., *Anodonta*)

ORDER VENEROIDA: Usually thick-valved, equivalved, and isomyarian. Includes the following families: cockles (Cardiidae: e.g., *Clinocardium, Laevicardium, Trachycardium*), little heart shells (Carditidae: e.g., *Cardita*), giant clams (Tridacnidae: e.g., *Tridacna*), surf clams (Mactridae: e.g., *Mactra*), solens (Solenidae: e.g., *Ensis, Solen*), tellinids (Tellinidae: e.g., *Florimetis, Macoma, Tellina*), semelids (Semelidae: e.g., *Leptomya, Semele*), wedge shells

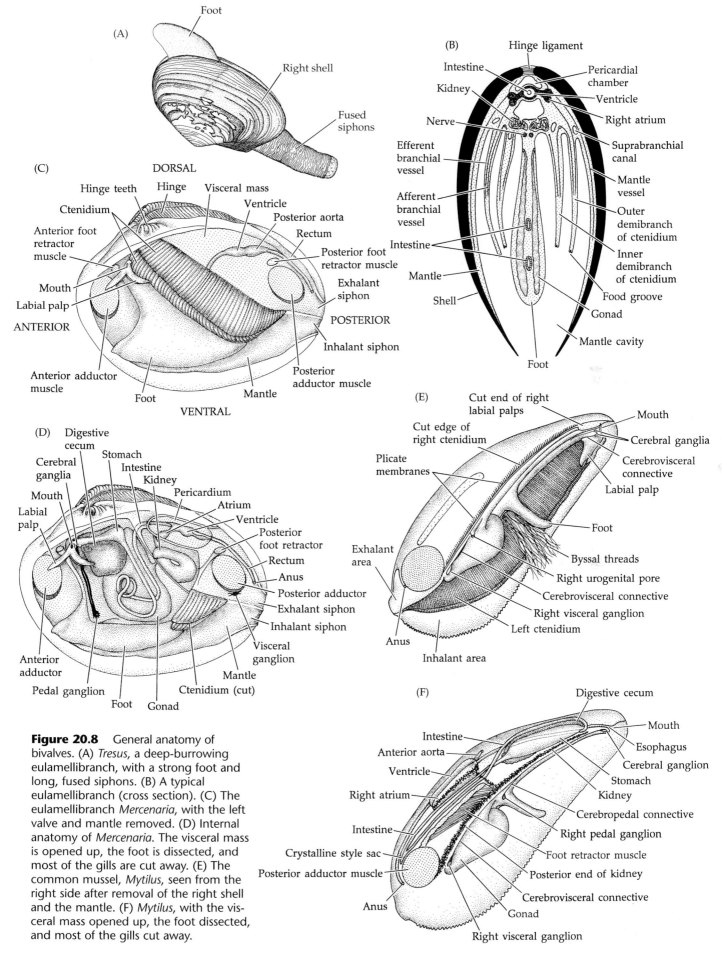

Figure 20.8 General anatomy of bivalves. (A) *Tresus*, a deep-burrowing eulamellibranch, with a strong foot and long, fused siphons. (B) A typical eulamellibranch (cross section). (C) The eulamellibranch *Mercenaria*, with the left valve and mantle removed. (D) Internal anatomy of *Mercenaria*. The visceral mass is opened up, the foot is dissected, and most of the gills are cut away. (E) The common mussel, *Mytilus*, seen from the right side after removal of the right shell and the mantle. (F) *Mytilus*, with the visceral mass opened up, the foot dissected, and most of the gills cut away.

(Donacidae: e.g., *Donax*), venus clams (Veneridae: e.g., *Chione, Dosinia, Pitar, Protothaca, Tivela*), and the freshwater families Sphaeriidae (e.g., *Sphaerium*) and Corbiculidae (e.g., *Corbicula*).

ORDER MYOIDA: Thin-shelled burrowing forms with well developed siphons; shell with 0–1 cardinal teeth. Includes the soft-shell clams, shipworms, and others: families Pholadidae (piddocks: e.g., *Barnea, Chaceia, Martesia, Pholas*), Teredinidae (shipworms: e.g., *Bankia, Teredo*), Corbulidae (e.g., *Corbula, Mya*).

SUBCLASS ANOMALODESMATA: Shells equivalved, aragonitic, of 2–3 layers, innermost consisting of sheet nacre; periostracum often incorporates granulations; with 0–1 hinge teeth; generally isomyarian, rarely amyarian; posterior siphons usually well developed; mantle usually fused ventrally, with anteroventral pedal gape, and posteriorly with ventral incurrent and dorsal excurrent apertures or siphons; ctenidia eulamellibranchiate or septibranchiate (modified as a horizontal septum). Marine bivalves (including the septibranchs); one order (Pholadomyoida) and about 12 families, including the aberrant Clavagellidae (e.g. Brechites), Cuspidariidae (e.g., *Cuspidaria*), Poromyidae (e.g., *Poromya*), and Pandoridae (e.g., *Pandora*).

CLASS SCAPHOPODA: Tusk shells (Figures 20.1L and 20.9). Shell of one piece, tubular, usually tapering, open at both ends; head rudimentary, projecting from larger aperture; mantle cavity large, extending along entire ventral surface; without ctenidia or eyes; with radula, proboscis, crystalline style; with paired clusters of clubbed contractile tentacles (captacula) that serve to capture and manipulate prey; heart absent; foot somewhat cylindrical. Nearly 900 living species of marine, benthic molluscs in eight families, including Dentaliidae (e.g., *Dentalium, Fustiaria*), Laevidentaliidae (e.g., *Laevidentalium*), Pulsellidae (e.g., *Pulsellum, Annulipulsellum*), and Gadilidae (e.g., *Cadulus, Gadila*).

CLASS CEPHALOPODA (= SIPHONOPODA): Nautilus, squids, cuttlefish, and octopuses (Figures 20.1J,K, 20.10, 20.11, 20.12, 20.17, and 20.22). With linearly chambered shell, usually reduced or lost; if external shell present (nautilus), animal inhabits last (youngest) chamber, with a filament of living tissue (the siphuncle) extending through older chambers; body cavity large; circulatory system largely closed; head with large, complex eyes and circle of prehensile arms or tentacles around mouth; with radula and beak; 1–2 pairs ctenidia, and 1–2 pairs complex nephridia; mantle forms large ventral pallial cavity containing ctenidia; with muscular funnel (the siphon) through which water is forced, providing jet propulsion; some tentacles of male modified for copulation; benthic or pelagic, marine; about 900 living species.

SUBCLASS NAUTILOIDEA (= TETRABRANCHIATA): The pearly nautilus. Shell external, many-chambered, coiled in one plane, exterior porcelaneous, interior nacreous (pearly); head with many (80–90) suckerless tentacles (4 modified as spadix in male for copulation and protected by a fleshy hood); 13-element radula; beak of chitin and calcium carbonate; funnel of 2 separate folds; 2 pairs ctenidia ("tetrabranchiate"); 2 pairs nephridia; eyes like a pinhole camera, without cornea or lens; nervous system rather diffuse; with a simple, primitive statocyst; without chromatophores or ink sac. Fossil record rich, but represented today by a single genus, the chambered or pearly nautilus (*Nautilus*), with five or six Indo-Pacific species.

SUBCLASS COLEOIDEA (= DIBRANCHIATA): Octopuses, squids, and their kin. Shell reduced, internal or absent; head

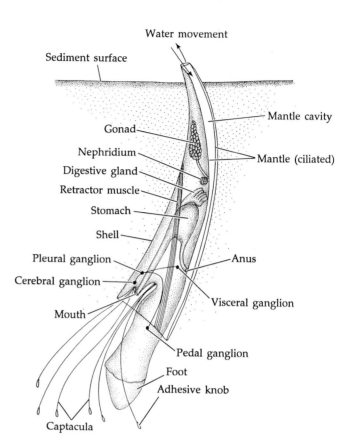

Figure 20.9 General anatomy of a scaphopod.

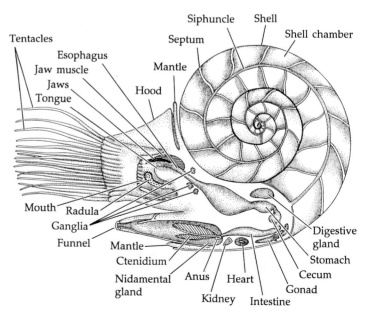

Figure 20.10 The anatomy of *Nautilus* (sagittal section).

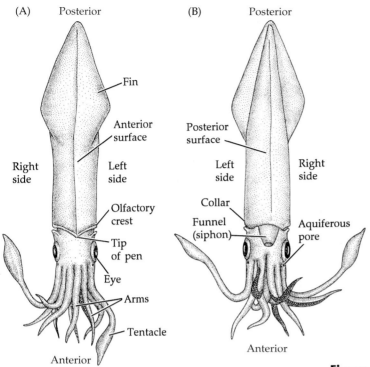

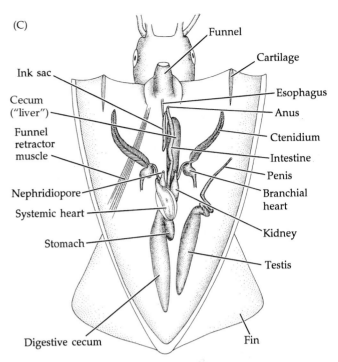

Figure 20.11 The anatomy of a squid (*Loligo*). (A) External morphology (dorsal view). (B) External morphology (ventral view). (C) Internal anatomy of a male. The mantle is dissected open and pulled aside. (D) A large squid stranded on a beach.

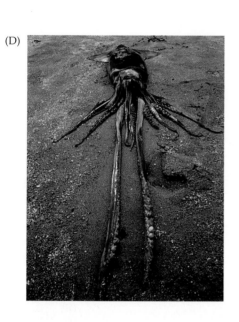

and foot united into a common anterior structure bearing 8 or 10 prehensile suckered appendages (arms and tentacles), 1 pair modified in male for copulation; 7-element radula; with chitinous beak; funnel a single closed tube; 1 pair ctenidia ("dibranchiate"); 1 pair nephridia; eyes complex, with cornea and lens; nervous system well developed and concentrated; with a complex statocyst; with chromatophores and ink sac.

ORDER SEPIOIDA: Cuttlefish. Body short, dorsoventrally flattened, with lateral fins; shell absent or internal, calcareous, often chambered, straight, or coiled; 8 short arms, and 2 long tentacles with suckers borne only on spooned tips, and retractable into pits; suckers lack hooks. (e.g., *Rossia, Sepia, Spirula*)

ORDER TEUTHOIDA (= DECAPODA): Squids. Body elongate, tubular, with lateral fins; shell internal, reduced to cartilage-like pen; with 8 arms and 2 elongate nonretractable tentacles; suckers often with hooks. Numerous families and genera. (e.g., *Architeuthis, Bathyteuthis, Chiroteuthis, Doryteuthis, Dosidiscus, Gonatus, Histioteuthis, Illex, Loligo, Lycoteuthis, Octopoteuthis, Ommastrephes*)

ORDER OCTOPODA: Octopuses. Body short, round, usually without fins; internal shell vestigial or absent; 8 similar arms joined by web of skin (interbrachial web); most are benthic. About 200 species. (e.g., *Argonauta, Octopus, Opisthoteuthis, Stauroteuthis*)

ORDER VAMPYROMORPHA: The vampire squid. Body plump, with 1 pair fins; shell reduced to thin, leaf-shaped, uncalcified, transparent vestige; 4 pairs equal-sized arms, each with one row of unstalked distal suckers; arms joined by extensive web of skin (interbrachial membrane); fifth pair of arms represented by 2 tendril-like, retractable filaments; hectocotylus lacking; radula well developed; ink sac degenerate; mostly deep water. (one living species, *Vampyroteuthis infernalis*)

The Molluscan Bauplan

The phylum Mollusca is one of the most morphologically diverse animal groups. Molluscs range in size from microscopic bivalves to giant clams (Tridacnidae) that reach 1 m in length, to giant squids (*Architeuthis*) reach-

Figure 20.12   The anatomy of *Octopus*. (A) General external anatomy. (B) Right-side view of the internal anatomy. (C) Arm and sucker (cross section). (D) Tip of the hectocotylus arm. (E) The diminutive Eastern Pacific *Octopus digueti* well camouflaged on a sand bottom. (F) The tropical Pacific *Octopus chierchiae*. (G) The remarkable Indo-West Pacific *Octopus horridus*.

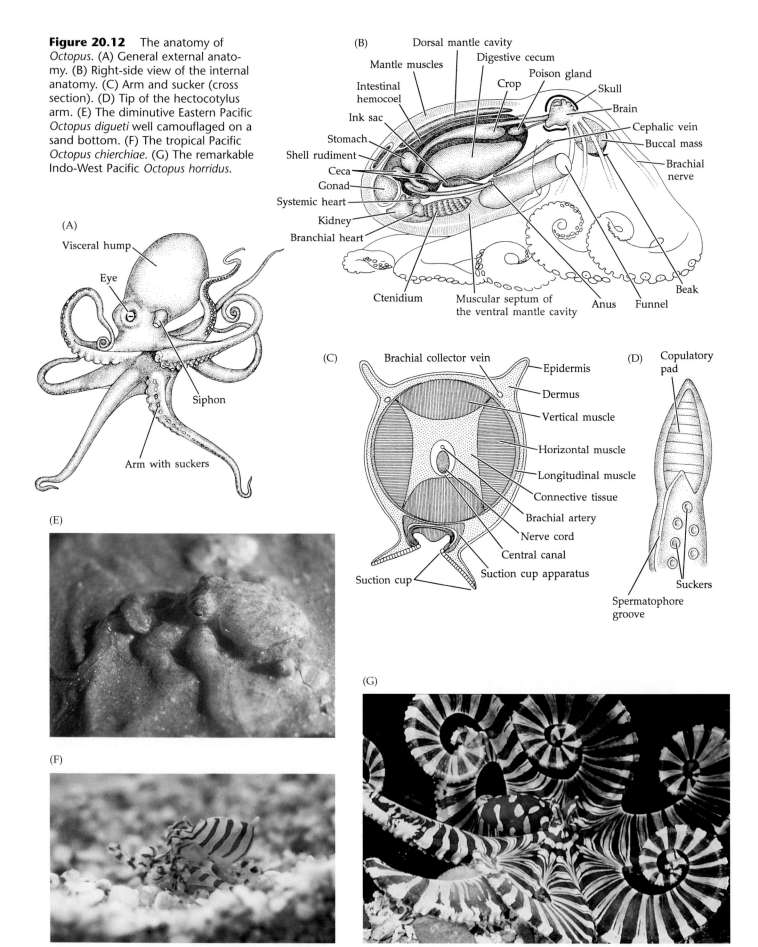

ing 20 m in overall length. The giant Pacific octopus (*Octopus dofleini*) commonly attains an arm span of 3–5 m and a weight of over 40 kg. One particularly large specimen was estimated to have an arm span of nearly 10 m and a weight of over 250 kg! Despite their differences, giant squids, cowries, garden slugs, eight-plated chitons, and wormlike aplacophorans are all closely related, and they all share a common and unmistakable body plan (Box 20A). In fact, the myriad ways in which evolution has shaped the basic molluscan bauplan provide some of the best lessons in homology and adaptive radiation in the animal kingdom.

Molluscs are bilaterally symmetrical, coelomate protostomes, but the coelom generally exists only as small vestiges around the heart (the **pericardial chamber**), the gonads, parts of the nephridia (kidney), and occasionally part of the intestine (the **perivisceral coelom**). The principal body cavity is a hemocoel composed of several large sinuses of the open circulatory system. In general, the body comprises three distinguishable regions: head, foot, and centrally concentrated **visceral mass** (Figure 20.13). The head may bear various sensory structures, most notably eyes, statocysts, and tentacles. The body is covered by a thick epidermal–cuticular sheet of skin called the **mantle** (also known as the **pallium**), which plays a critical role in the organization of the body. It secretes the hard calcareous skeleton, either as minute sclerites, or plates, that are embedded in the body wall or as a solid internal or external shell. The body usually bears a large, muscular, ventral **foot**.

Surrounding or posterior to the visceral mass is a cavity—a space between the visceral mass and folds of the mantle itself. This **mantle cavity** (also known as the **pallial cavity**) often houses the gills, or **ctenidia**, along with the openings of the gut, nephridial, and reproductive systems, and special patches of sensory epithelium called **osphradia**. In aquatic forms, water is circulated through this cavity, passing over the ctenidia, excretory pores, anus, and other structures.

The molluscan gut is complete and regionally specialized. The buccal region of the foregut typically bears a uniquely molluscan structure called the **radula**, a toothed, rasping, tonguelike strap used in feeding. The open circulatory system usually includes a heart in a pericardial cavity and a few large vessels that empty into or drain hemocoelic spaces. The excretory system consists of one or more pairs of metanephridial kidneys, with nephrostomes usually in the pericardial cavity. The nervous system typically includes a dorsal cerebral ganglion, circumenteric nerve ring, two pairs of longitudinal ladder-like nerve cords, and several paired ganglia showing various degrees of fusion.

Fertilization may be external or internal. Development is typically protostomous, with spiral cleavage and one or two distinct trochophore larval stages. One of these larval forms is unique to certain molluscs and is called the **veliger**.

Although this general summary describes the basic bauplan of most molluscs, notable modifications occur and are discussed throughout this chapter. The seven classes are characterized above (see classification) and are briefly summarized below.

Some of the most bizarre molluscs are the aplacophorans (Figure 20.2). Members of this class are small and wormlike, and either burrow in deep-sea sediments or, in the case of many neomeniomorphans, spend their entire lives on the branches of various cnidarians, such as gorgonians, upon which they feed. Aplacophorans lack a well-developed foot and do not have a solid shell. They also have no distinct head, eyes, or tentacles. They are very primitive molluscs that evolved before the appearance of solid shells.

Polyplacophorans, or chitons, are oval molluscs that bear eight (occasionally seven) separate articulating shell plates on their backs (Figures 20.1B and 20.4). They range in length from about 7 mm to over 35 cm. These marine animals are common inhabitants of intertidal regions around the world, at all latitudes.

Monoplacophorans are limpet-like molluscs with a single cap-shaped shell ranging from several millimeters to about 4 cm in length (Figures 20.1A and 20.3). They live in the world's oceans at modest to great depths. Their most notable feature is the repetitive arrangement of many organs, a condition that has led some biologists to speculate that they represent a link to an ancient segmented ancestor of the Mollusca.

Gastropods are probably the best known molluscs (Figures 20.1C–I, 20.5, 20.6, and 20.7). This class includes the common snails and slugs in all marine and many freshwater and terrestrial environments. They are the only molluscs that undergo **torsion**, a strange twisting of the body that occurs on top of the untwisted foot.

Bivalves include the clams, oysters, mussels, and their kin (Figures 20.1M–P and 20.8). They possess two separate shells, called **valves**. The smallest bivalves are members of the freshwater family Sphaeriidae and rarely exceed 2 mm in length; the largest are giant tropical clams (*Tridacna*), one species of which (*T. gigas*) may weigh over 400 kg! Bivalves inhabit all marine environments and many freshwater habitats.

Scaphopods, the tusk shells, live in marine surface sediments at various depths. Their distinctive single, tubular uncoiled shell is generally 2–15 cm long and is open at both ends (Figures 20.1L and 20.9).

The cephalopods are among the most highly modified molluscs and include the pearly nautilus, squids, cuttlefish, octopuses, and a host of extinct forms (Figures 20.1J,K, 20.10, 20.11, 20.12, 20.17, and 20.22). This group includes the largest of all living invertebrates, the giant squid, with body and tentacle lengths exceeding 20 m. Among living cephalopods, only the nautilus has retained an external shell. The cephalopods differ markedly from other molluscs in several ways. For example, they have a spacious body cavity that includes the pericardium, go-

Figure 20.13 Modifications of the shell, foot, gut, ctenidia, and mantle cavity in five classes of molluscs. (A,B) Lateral and cross sections of a chiton (class Polyplacophora). (C) Side view of a snail (class Gastropoda). (D,E) Cutaway side view and cross section of a clam (class Bivalvia). (F) Lateral view of a tusk shell (class Scaphopoda). (G) Lateral view of a squid (class Cephalopoda). In cephalopods the foot is modified to form the funnel (= siphon) and at least parts of the arms.

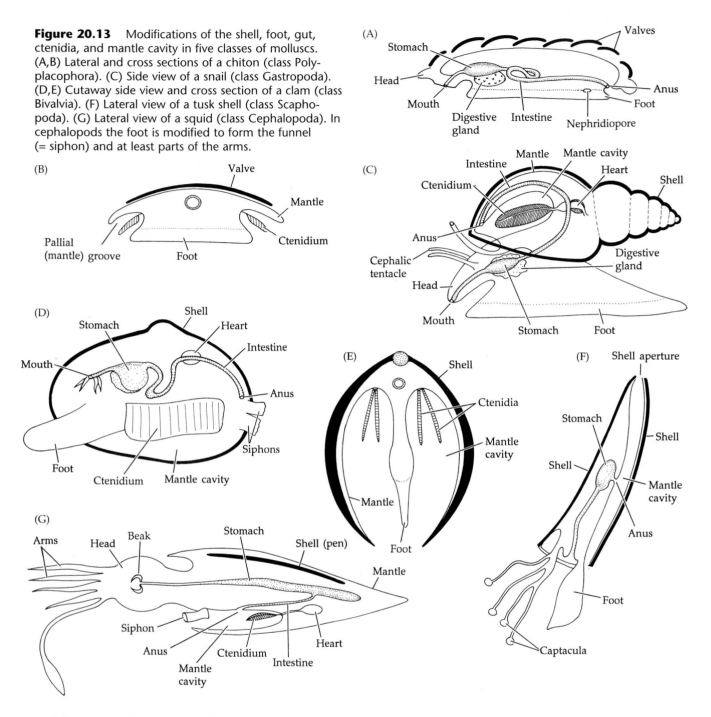

nadal cavity, nephriopericardial connections, gonoducts, and various other channels and spaces, all of which form an interconnected system representing a highly modified but true coelom. Also, unlike all other molluscs, cephalopods have a functionally closed circulatory system. The nervous system of cephalopods is the most sophisticated of all molluscs, if not all invertebrates. Most of these modifications are associated with the adoption of an active predatory lifestyle by these remarkable creatures.

The Body Wall

The body wall of molluscs comprises three recognizable layers: the cuticle, epidermis, and muscles (Figure 20.14A). The cuticle is composed largely of various amino acids and sclerotized proteins (**conchin**), but it apparently does not contain chitin (except perhaps in the caudofoveatans). The epidermis is usually a single layer of cuboidal to columnar cells, which are ciliated on much of the body. Many of the epidermal cells participate in secretion of the cuticle, while others appear to be different kinds of secretory gland cells. The function of most of these gland cells is not known, but some secrete mucus and are very abundant, especially on the ventral body surface. Other specialized epidermal cells occur on the dorsal body wall, or mantle. These cells constitute the molluscan **shell glands**, which produce the calcareous spicules or shells characteristic of this phylum. Still other epidermal cells form sensory **epidermal**

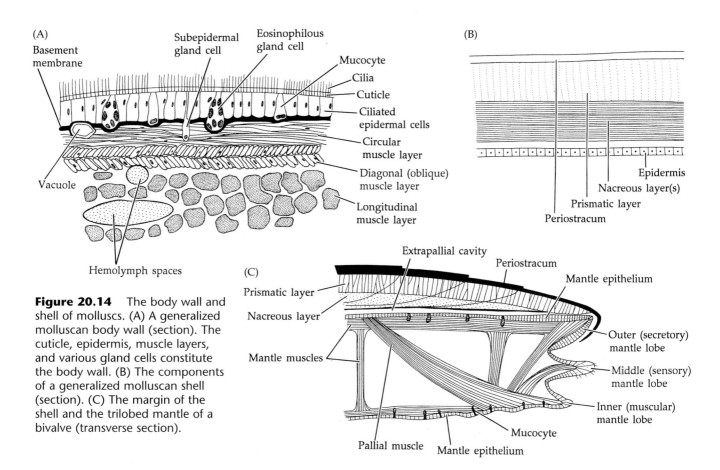

Figure 20.14 The body wall and shell of molluscs. (A) A generalized molluscan body wall (section). The cuticle, epidermis, muscle layers, and various gland cells constitute the body wall. (B) The components of a generalized molluscan shell (section). (C) The margin of the shell and the trilobed mantle of a bivalve (transverse section).

papillae or other receptors. The epidermis and outer muscle layer are often separated by a basement membrane and occasionally a dermis.

The body wall usually includes three distinct layers of smooth muscle fibers: an outer circular layer, a middle diagonal layer, and an inner longitudinal layer. The diagonal muscles are often in two groups with fibers running at right angles to each other. The degree of development of each of these muscle layers differs among the classes (for example, in solenogasters the diagonal layers are frequently absent).

The Mantle and Mantle Cavity

We have already hinted at the significance of the mantle cavity and its importance in the success of the molluscan bauplan. Here we offer a brief summary of the nature of the mantle cavity, and its disposition in the major groups of molluscs.

The mantle, as the name implies, is a sheet-like organ that forms the dorsal body wall, and in most molluscs it grows during development in the form of one or two folds that contain muscle layers and hemocoelic channels (Figure 20.14C). This outward growth creates a space lying between the mantle fold and the body proper. This space, the mantle (or pallial) cavity, may be in the form of shallow grooves, or one or two large chambers through which water is passed by ciliary or muscular action. Generally, the mantle cavity houses the gills, or ctenidia, and receives the fecal material from the anus and products of the excretory and reproductive systems. In some instances the incoming water also carries food for suspension feeding.

The mantle cavity of chitons is a pair of long pallial grooves lying along the sides of the foot (Figure 20.13A,B). Water enters these grooves from the front and sides, passing medially over the ctenidia and then posteriorly between the ctenidia and the foot. After passing over the gonopores and nephridiopores, water exits the back end of the grooves and carries away fecal material from the anus.

The single mantle cavity of gastropods originates during development as a posteriorly located chamber. As development proceeds, however, most gastropods undergo torsion of the shell and visceral mass to bring the mantle cavity forward, over the head (Figure 20.13C). Again, water passing through this chamber flows over the ctenidia, anus, gonopores, and nephridiopores. A great many secondary modifications on this plan have evolved in the Gastropoda, including rerouting of current patterns, loss of certain associated structures, and even "detorsion," as discussed in later sections of this chapter.

Bivalves possess a pair of large mantle cavities, one on each side of the foot and visceral mass (Figure 20.13D,E). The mantle folds line the laterally placed shells, and are often produced posteriorly as inhalant and exhalant siphons, through which water enters and leaves the mantle cavity. The water passes over and through the ctenidia, which extract suspended food material and accomplish gas exchange, across the gonopores and nephridiopores, and past the anus as it exits through the exhalant siphon.

Scaphopods bear a tapered, tubular shell (Figure 20.13F). Water enters and leaves the elongate mantle cavity through the small opening in the shell and flushes over the mantle surface, which, in the absence of ctenidia, is the site of gas exchange. The anus, nephridiopores, and gonopores also empty into the mantle cavity.

In all of the above cases, water is moved through the mantle cavity by the action of cilia. In the cephalopods, however, well developed, highly innervated mantle muscles perform this function. The exposed, fleshy body surface of squids and octopuses is, in fact, the mantle itself (Figure 20.13G). Unconstrained by an external shell, the mantle of these molluscs expands and contracts to draw water into the mantle cavity and then forces it out through a narrow muscular funnel (= siphon). This jet of exhalant water provides a means of rapid locomotion for most cephalopods. In the mantle cavity the water passes over the ctenidia, anus, reproductive pores, and excretory openings.

The remarkable adaptive qualities of the molluscan body plan are manifested in these variations in the position and function of the mantle cavity and its associated structures. In fact, even the nature of many other structures is influenced by mantle cavity arrangement, as shown schematically in Figure 20.15. The fact that molluscs have been able to successfully exploit a broad range of habitats and lifestyles can be explained in part by these variations, which are central to the story of molluscan evolution. We will have a great deal more to say about these matters throughout this chapter.

The Molluscan Shell

Except for the Aplacophora, all molluscs have solid calcareous shells (either aragonite or calcite) produced by shell glands in the mantle. In the Aplacophora, aragonite spicules or plates are formed extracellularly and are embedded in the dorsal mantle. Beyond the Aplacophora, molluscan shells vary greatly in shape and size, but they all adhere to the basic construction plan of calcium carbonate produced extracellularly, laid down in layers, and often covered by a thin organic surface coating called a **periostracum** (also called the **hypostracum**) (Figure 20.14). The periostracum is composed of a type of conchin (largely quinone-tanned proteins) similar to that found in the epidermal cuticle. The calcium layers are generally of two types, an outer chalky **prismatic** portion and an inner pearly **lamellar** or **nacreous** layer; the latter layer has been lost in many groups. Both layers incorporate conchin in various ways, often to help bind the calcareous crystals together. Shells of various molluscs are often composed of different numbers of calcareous sublayers.

Molluscs are noted for their wonderfully intricate and often flamboyant shell color patterns and sculpturing (Figure 20.16), but very little is known about the evolutionary origins and functions of these features. Some workers view molluscan pigments primarily as metabolic by-products, and thus shell colors might largely represent strategically deposited food residues. Molluscan shell pigments include such compounds as pyrroles and porphyrins. Melanins are common in the integument (cuticle and epidermis), the eyes, and internal organs, but they are rare in shells.

Some shell sculpture patterns are correlated to specific behaviors or habitats. For example, shells with low spires are more stable in areas of heavy wave shock or on vertical rock surfaces. Similarly, the low, cap-shaped shells of limpets (Figure 20.16H,I) are presumably adapted for withstanding exposure to strong waves. Heavy ribbing, thick or inflated shells, and a narrow gape in bivalves are all possible adaptations to provide protection from predators. In some gastropods, fluted shell ribs help them land upright when they are dislodged from rocks. Several groups of soft-bottom benthic gastropods and bivalves have long spines on the shell that may help stabilize the animals in loose sediments. Many molluscs, particularly clams, have shells covered with living epizootic organisms such as sponges, tube worms, ectoprocts, and hydroids. Some studies suggest that predators have a difficult time recognizing such camouflaged molluscs as potential prey.

Molluscs may have one shell, two shells, eight shells, or no shell (Figure 20.16). In the latter case the outer body wall may contain calcareous spicules of various sorts. In aplacophorans, for example, the cuticular spicules ("spines") vary in shape and range in length from microscopic to about 4 mm. These spicules are essentially crystals composed almost entirely of calcium carbonate. Caudofoveatans produce platelike cuticular spicules that give their body surface a scaly texture and appearance. The spicules in both taxa appear to be secreted by a diffuse network of specialized groups of cells, perhaps representing primitive shell gland(s).

The eight transverse plates, or **valves** (Figure 20.16A–F), of polyplacophorans are encircled by and embedded in a thickened region of the mantle called the **girdle**. The size of the girdle varies from narrow to broad and may cover much of the valves. In the giant Pacific "gumboot" chiton, *Cryptochiton stelleri*, the girdle completely covers the valves. The girdle is thick, heavily cuticularized, and usually beset with calcareous spicules, spines, scales, or bristles secreted by specialized epidermal cells. These spines are probably homologous with the spicules in the body wall of caudofoveatans and aplacophorans.

The anterior and posterior valves of chitons are referred to as the end valves, or **cephalic** (= anterior) and **anal** (= posterior) **plates**; the six other valves are called the **intermediate valves**. Some details of chiton valves

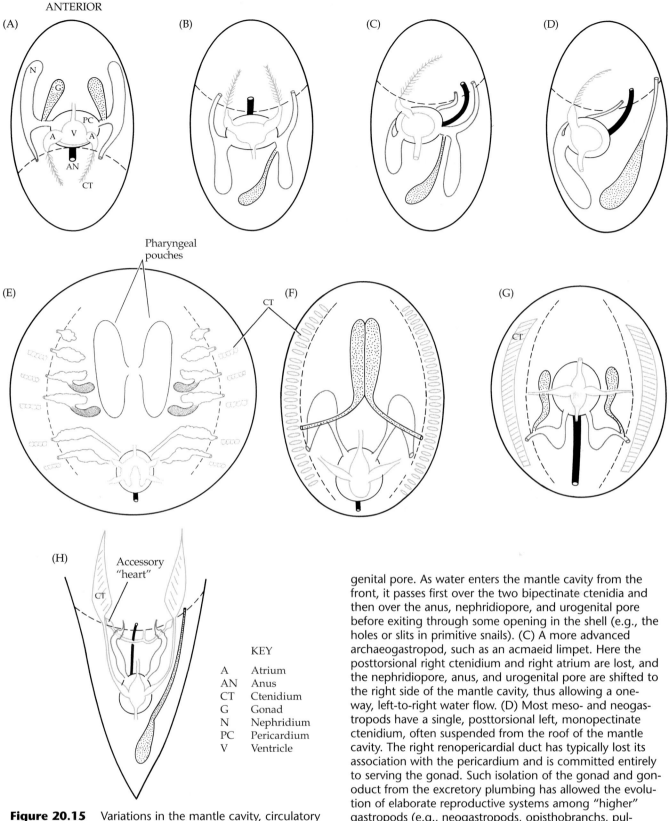

KEY

A	Atrium
AN	Anus
CT	Ctenidium
G	Gonad
N	Nephridium
PC	Pericardium
V	Ventricle

Figure 20.15 Variations in the mantle cavity, circulatory system, ctenidia, nephridia, reproductive system, and position of the anus in molluscs (dorsal views). Although schematic, these drawings give some idea of the evolutionary changes in arrangement of these structures and systems in the phylum Mollusca. (A) A hypothetical, untorted, gastropod-like mollusc with a posterior mantle cavity and symmetrically paired atria, ctenidia, nephridia, and gonads. (B) A primitive posttorsional archaeogastropod wherein all paired organs are retained except the left posttorsional gonad. The right renopericardial duct serves both the nephridium and the persisting gonad and leads to a uro-genital pore. As water enters the mantle cavity from the front, it passes first over the two bipectinate ctenidia and then over the anus, nephridiopore, and urogenital pore before exiting through some opening in the shell (e.g., the holes or slits in primitive snails). (C) A more advanced archaeogastropod, such as an acmaeid limpet. Here the posttorsional right ctenidium and right atrium are lost, and the nephridiopore, anus, and urogenital pore are shifted to the right side of the mantle cavity, thus allowing a one-way, left-to-right water flow. (D) Most meso- and neogastropods have a single, posttorsional left, monopectinate ctenidium, often suspended from the roof of the mantle cavity. The right renopericardial duct has typically lost its association with the pericardium and is committed entirely to serving the gonad. Such isolation of the gonad and gonoduct from the excretory plumbing has allowed the evolution of elaborate reproductive systems among "higher" gastropods (e.g., neogastropods, opisthobranchs, pulmonates) and was probably a major event in the story of gastropod success. (E) The condition in monoplacophorans includes the serial repetition of several organs. (F) In polyplacophorans, the gonoducts and nephridioducts open separately into the exhalant regions of the lateral pallial grooves. (G) A generalized bivalve condition. The gonads and nephridia may share common pores, as shown here, or else open separately into the lateral mantle chambers. (H) The condition in a generalized cephalopod with a single, isolated reproductive system and an effectively closed circulatory system.

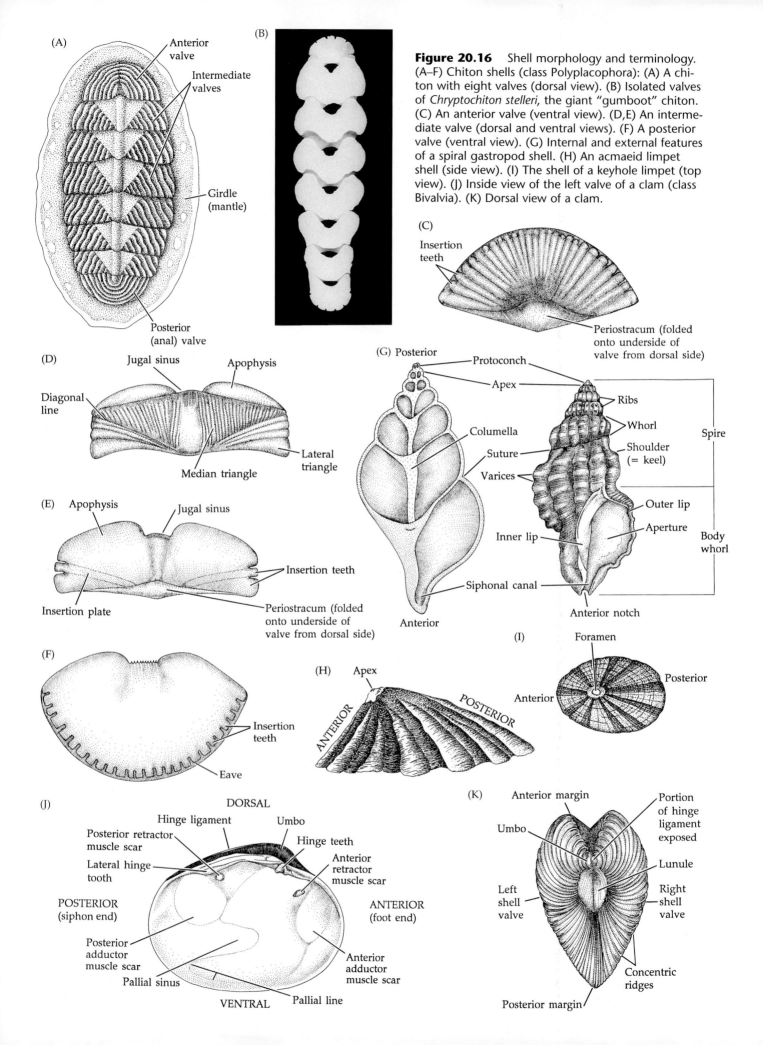

(A)
Anterior valve
Intermediate valves
Girdle (mantle)
Posterior (anal) valve

(B)

Figure 20.16 Shell morphology and terminology. (A–F) Chiton shells (class Polyplacophora): (A) A chiton with eight valves (dorsal view). (B) Isolated valves of *Chryptochiton stelleri*, the giant "gumboot" chiton. (C) An anterior valve (ventral view). (D,E) An intermediate valve (dorsal and ventral views). (F) A posterior valve (ventral view). (G) Internal and external features of a spiral gastropod shell. (H) An acmaeid limpet shell (side view). (I) The shell of a keyhole limpet (top view). (J) Inside view of the left valve of a clam (class Bivalvia). (K) Dorsal view of a clam.

(C)
Insertion teeth
Periostracum (folded onto underside of valve from dorsal side)

(D)
Jugal sinus
Apophysis
Diagonal line
Median triangle
Lateral triangle

(E)
Apophysis
Jugal sinus
Insertion teeth
Insertion plate
Periostracum (folded onto underside of valve from dorsal side)

(F)
Insertion teeth
Eave

(G) Posterior
Protoconch
Apex
Columella
Suture
Varices
Ribs
Whorl
Shoulder (= keel)
Spire
Outer lip
Inner lip
Aperture
Body whorl
Siphonal canal
Anterior notch
Anterior

(H)
Apex
ANTERIOR
POSTERIOR

(I)
Foramen
Posterior
Anterior

(J)
DORSAL
Hinge ligament
Umbo
Posterior retractor muscle scar
Hinge teeth
Lateral hinge tooth
Anterior retractor muscle scar
POSTERIOR (siphon end)
ANTERIOR (foot end)
Posterior adductor muscle scar
Pallial sinus
Anterior adductor muscle scar
VENTRAL
Pallial line

(K)
Anterior margin
Portion of hinge ligament exposed
Umbo
Lunule
Left shell valve
Right shell valve
Concentric ridges
Posterior margin

are shown in Figure 20.16A–F. The shells of chitons are three-layered, with an outer periostracum, a colored **tegmentum**, and an inner calcareous layer, or **articulamentum**. The periostracum is a very thin, delicate organic membrane and is not easily seen. The tegmentum is composed of organic material (probably a form of conchin) and calcium carbonate suffused with various pigments. It is penetrated by vertical canals that lead to minute pores in the surface of the valves. The pores are of two sizes, called **megalopores** and **micropores**, and house special photosensory organs called **aesthetes**. The vertical canals arise from a layer of horizontal canals between the tegmentum and articulamentum (Figure 20.43C). The articulamentum is a thick, calcareous, pearly layer that differs in certain ways from the shell layers of other molluscs.

Monoplacophorans have a single, large, limpet-like shell with the apex situated far forward (Figures 20.1A and 20.3). As in chitons, the mantle encircles the body and foot as a circular fold, forming lateral pallial grooves.

The bivalves possess two shells, or valves, that are hinged dorsally and enclose the body and spacious mantle cavity (Figures 20.1M–P and 20.16J,K). Shells of bivalves typically have a thin periostracum, covering two to four calcareous layers that vary in composition and structure. The calcareous layers are often aragonite or an aragonite/calcite mixture, and they usually incorporate a substantial organic framework. The periostracum and organic matrix may account for over 70% of the shell's dry weight. Each valve bears a dorsal protuberance called the **umbo**, which is the oldest part of the shell. Concentric growth lines radiate from the umbo. The two valves are attached by an elastic, proteinaceous **hinge ligament**. When the valves are closed by contraction of the **adductor muscles**, the outer part of the hinge ligament is stretched and the inner part is compressed. Thus, when the adductor muscles relax, the resilient ligament causes the valves to open. The hinge apparatus comprises various sockets or toothlike arrangements (**hinge teeth**) that prevent slipping of the valves. In most bivalves, the adductor muscles contain both striated and smooth fibers, facilitating both rapid and sustained closure of the valves.

The thin mantle of bivalves lines the inner valve surfaces and separates the visceral mass from the shell. The edge of the mantle bears three longitudinal ridges or folds—the inner, middle, and outer folds (Figure 20.14). The innermost fold is the largest and contains radial and circular muscles, some of which attach the mantle to the shell. The line of mantle attachment appears on the inner surface of each valve as a scar called the **pallial line** (Figure 20.16J). The middle mantle fold is sensory in function, and the outer fold is responsible for secreting parts of the shell. The cells of the outer lobe are specialized: the medial cells lay down the periostracum, and the lateral cells secrete the first calcareous layer. The entire mantle surface is responsible for secreting the remaining innermost calcareous portion of the shell. A thin **extrapallial space** lies between the mantle and the shell, and it is into this space that materials for shell formation are secreted and mixed. Should a foreign object, such as a sand grain, lodge between the mantle and the shell, it may become the nucleus around which are deposited concentric layers of smooth nacreous shell. The result is a pearl, either free in the extrapallial space or partly embedded in the growing shell.

Scaphopod shells resemble miniature, hollow elephant tusks, hence the vernacular names "tusk shell" and "tooth shell" (Figures 20.1L and 20.9). The scaphopod shell is open at both ends, with the smaller opening at the posterior end of the body. Most tusk shells are slightly curved, the concave side being the dorsal surface. The mantle is large and lines the entire ventral surface of the shell. The posterior aperture serves for both inhalant and exhalant water currents.

Most extant cephalopods have a reduced shell or are shell-less. A completely developed shell is found only in fossil forms and the six or so surviving species of *Nautilus*. In squids and cuttlefish the shell is reduced and internal, and in octopuses it is entirely lacking or present only as a small rudiment.

The shell of *Nautilus* is coiled in a planospiral fashion (whorls lie on a single plane) and lacks a periostracum (Figures 20.10, 20.17A, and 20.22B). *Nautilus* shells (like all cephalopod shells) are divided into internal chambers by **transverse septa**, but only the last chamber is occupied by the body of the living animal. As the animal grows, it periodically moves forward, and the posterior part of the mantle secretes a new septum behind it. Each septum bears a central perforation through which extends a cord of tissue called the **siphuncle**. The siphuncle helps to regulate buoyancy of the animal by varying the amounts of gases and fluids in the shell chambers. The shell is composed of an inner nacreous layer and an outer porcelain layer containing prisms of calcium carbonate and an organic matrix. The outer surface may be pigmented or pearly white. The junctions between septa and the shell wall are called **sutures**, which may be simple and straight, slightly waved (as in *Nautilus*), or highly convoluted (as in the extinct ammonoids). In cuttlefish (order Sepioida), the shell is reduced and internal, with spaces separated by thin septa. Like *Nautilus*, a cuttlefish can regulate the relative amounts of fluid and gas in its shell chambers. The small, gas-filled shells of the cuttlefish *Spirula* are often found washed up on tropical beaches.

Fossil data suggest that the first cephalopod shells were probably curved cones. From these ancestors both straight and coiled shells evolved, although secondary uncoiling probably occurred in several groups. Some straight-shelled cephalopods from the Ordovician Period exceeded 5 m in length, and some Cretaceous coiled species had shell diameters of 3 m.

Gastropod shells are extremely diverse in size and shape (Figure 20.1C–E, G–I). The smallest are microscopic and the largest may exceed 40 cm. The "typical" shape is the familiar conical spiral wound around a cen-

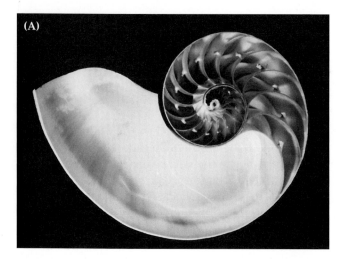

Figure 20.17 Two cephalopod shells. (A) The chambered shell of *Nautilus*, cut in longitudinal section. (B) The egg case "shell" of the paper nautilus, *Argonauta*.

tral axis or **columella** (Figure 20.16G). The turns of the **spire** form **whorls**, demarcated by lines called **sutures**. The largest whorl is the **body whorl**, which bears the **aperture** through which the foot and head protrude. The aperture is anterior and the apex of the shell spire is posterior. The first few, very small, whorls at the apex are the remnant of the larval shell, or **protoconch**, which usually differ in sculpturing and color from the rest of the shell. The body whorl and aperture may be notched and drawn out into an anterior **siphonal canal**, to house a **siphon** when present. A smaller posterior canal may also be present on the rear edge of the aperture.

Every imaginable variation on the basic spiraled shell occurs among the gastropods: the shell may be long and slender (e.g., tower shells) or short and plump (e.g., trochids); the spire may be more or less incorporated into the body whorl and eventually disappear from view; the shell may be flattened, with all whorls in one plane (e.g., sundials); the last body whorl may completely overgrow the older whorls, reducing the aperture to an elongated slit as the two lips are brought together (Figure 20.1G) (e.g., cowries, olives, and cones); the shell may coil so loosely as to form a meandering wormlike tube (see Figure 20.19E) (e.g., vermetids); the shell may be reduced and overgrown by the mantle, or it may disappear entirely (e.g., many opisthobranchs and pulmonates). Most gastropods spiral clockwise; that is, they show right-handed, or dextral, coiling. Some are sinistral (left-handed), and some species can coil in either direction. In limpets the shell is **patelliform**, with a low conical shape with no visible coiling (Figure 20.16H,I). The limpet shell form was probably derived from coiled ancestors on numerous occasions during gastropod evolution.

Gastropod shells consist of an outer thin organic periostracum and two or three calcareous layers: an outer prismatic (or palisade) layer, a middle lamellate layer, and an inner nacreous layer. The nacreous layer is composed of calcareous lamellae layered with thin films of conchin; it has been lost in many archaeogastropods

and almost all meso- and neogastropods. In some gastropods up to six calcareous layers are distinguishable. Gastropods in which the shell is habitually covered by mantle lobes lack a periostracum (e.g., olives and cowries), but in some other groups the periostracum is very thick and "hairy." The prismatic and lamellate layers consist largely of calcium carbonate, either as calcite or aragonite. These two forms of calcium are chemically identical, but they crystallize differently and can be identified by microscopic examination of sections of the shell. Small amounts of other inorganic constituents are incorporated into the calcium carbonate framework, including chemicals such as phosphate, calcium sulfate, magnesium carbonate, and salts of aluminum, iron, copper, strontium, barium, silicon, manganese, iodine, and fluorine. The prismatic layer has the calcium carbonate deposited as vertical crystals, each surrounded by a thin protein matrix. The nacreous layer has the calcium carbonate deposited as thin lamellae, which are always interleaved with conchin.

An intriguing aspect of gastropod evolution is the appearance of shell-lessness, or the "slug" form. Despite the fact that evolution of the coiled shell led to great success for the gastropods—75% of all living molluscs are snails—secondary loss of the shell occurred many times in this class. In forms such as the land and sea slugs, the shell may persist as a small vestige covered by the dorsal mantle (e.g., Aplysiinae, Pleurobranchidae), or it may be lost altogether (e.g., the nudibranchs). In the latter case it is first covered, then resorbed, by the mantle during ontogeny. The two primary examples of shell-lessness are the pulmonate land slugs and the marine opisthobranch slugs. Although they are shell-less now, shell loss probably occurred numerous times in both groups. Shells, of course, are energetically expensive to produce and require a large source of calcium in the environment, so it might be advantageous to eliminate them if compensatory mechanisms exist: for example, most, if not all, sea slugs secrete chemicals that make them distasteful to predators. In addition, the bright coloration of many nudibranchs may serve a defensive function. In some species, the color matches the animal's background. For example, the tiny red nudibranch, *Rostanga pulchra*, matches almost perfectly the red sponge on which it feeds. Many nudibranchs, however, are very conspicuous in nature. In these cases, the color may serve to warn predators of the noxious taste of the slug or, as suggested by Rudman (1991), predators may simply ignore such bright "novelties" in their environment.

Torsion, or "How the Gastropod Got Its Twist"

One of the most remarkable and dramatic steps taken during the course of molluscan evolution was the advent of torsion, a unique synapomorphy of modern gastropods, and it is quite unlike anything else in the animal kingdom. Torsion takes place during development in all gastropods, usually during the late veliger larval stage. It is a rotation of the visceral mass and its overlying mantle and shell as much as 180° with respect to the head and foot (Figures 20.15A–D, 20.18, and 20.52). The twisting is always in a counterclockwise direction (viewing the animal from above), and it is completely different from the phenomenon of coiling. During torsion, the mantle cavity and anus are moved from a posterior to a more anterior position, somewhat above and behind the head. Visceral structures and incipient organs that were on the right side of the larval animal end up on the left side of the adult. The gut is twisted into a U-shape, and when the longitudinal nerve cords connecting the pleural to the visceral ganglia develop, they are crossed rather like a figure eight. Most veligers have nephridia, which reverse sides, but the adult gills and gonads are not fully developed when torsion occurs.

Torsion is usually a two-step process. During larval development, an asymmetrical **velar/foot retractor muscle** develops. It extends from the shell on the right, dorsally over the gut, and attaches on the left side of the head and foot. At a certain stage in the veliger's development, contraction of this muscle causes the shell and enclosed viscera to twist about 90° in a counterclockwise direction. This first 90° twist is usually rapid, taking place in a few minutes to a few hours. The second 90° twist is typically much slower and results from differential tissue growth. By the end of the process, the viscera have been pulled from above toward the left, ultimately leading to the figure-eight arrangement of the adult visceral nerves. But the figure-eight arrangement is not perfect. The left intestinal ganglion usually comes to lie dorsal to the gut and is thus called the **supraintestinal** (= **supraesophageal**) ganglion; however, the right intestinal ganglion lies ventral to the gut, as a **subintestinal** (= **subesophageal**) ganglion (Figures 20.18 and 20.40).

Gastropods that retain torsion into adulthood are said to be **torted**; those that have secondarily reverted back to a partially or fully untorted state in adulthood are **detorted**. The torted, figure-eight configuration of the nervous system is referred to as **streptoneury**. The detorted condition, in which the visceral nerves are parallel, is referred to as **euthyneury**.

Detorted gastropods, such as most opisthobranchs, undergo a postveliger series of changes through which the original torsion is reversed to various degrees. The process shifts the mantle cavity and at least some of the pallial organs about 90° back to the right, or in some cases all the way back to the rear of the animal.

Evolutionarily speaking, after torsion, the anus lay in front, and the animal could no longer grow in length easily. Subsequent increase in body size thus occurred by the development of loops or bulges in the middle portion of the gut region, producing thereby a characteristic visceral hump. The first signs of torsion and coiling occur at about the same time during gastropod development. However, the fossil record suggests that the first coiled gastropod shells were planospiral and that these forms may have predated the appearance of torsion in gastropods. Once both features were established, they coevolved in various ways to produce what we see today in modern living gastropods.

The evolution of asymmetrically coiled shells had the effect of restricting the right side of the mantle cavity, a restriction that led to reduction or loss of the pallial structures on the adult right side (the original left ctenidium, atrium, and osphradium). At the same time, these structures on the adult left side (the original right ctenidium, atrium, and osphradium) tended to enlarge. After torsion and coiling had appeared, the left posttorsional gonad was lost. The single remaining gonad opens on the right side via the posttorsional right nephridial duct and nephridiopore. Most archaeogastropods retain two functional nephridia, although the posttorsional left one is often reduced. In most higher gastropods the posttorsional right nephridium is reduced or lost, but its duct and pore remain associated with the reproductive tract. This isolation of the reproductive system from the excretory system probably allowed the great elaboration of the reproductive organs seen among mesogastropods and neogastropods.

Such profound changes in spatial relations between major body regions, as brought about by torsion and spiral coiling, are rare among animals. Several theories on the adaptive significance of torsion have been proposed. The great zoologist Walter Garstang suggested that torsion was an adaptation of the veliger larva that served to protect the soft head and larval velum from predators (see section on development later in this chapter). When disturbed, the immediate reaction of a veliger is to withdraw the head and foot into the larval shell, whereupon the larva begins to sink rapidly. This theory may seem reasonable for evasion of very small planktonic predators, but it seems illogical as a means of escape from larger predators in the sea, which no doubt consume veligers

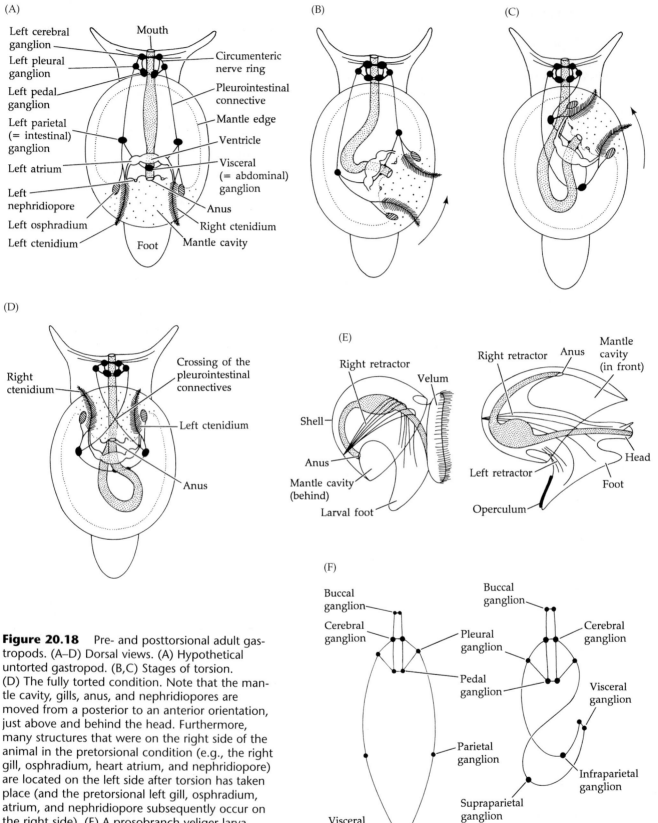

(A)

Left cerebral ganglion
Left pleural ganglion
Left pedal ganglion
Left parietal (= intestinal) ganglion
Left atrium
Left nephridiopore
Left osphradium
Left ctenidium
Mouth
Circumenteric nerve ring
Pleurointestinal connective
Mantle edge
Ventricle
Visceral (= abdominal) ganglion
Anus
Right ctenidium
Mantle cavity
Foot

(B)

(C)

(D)

Right ctenidium
Crossing of the pleurointestinal connectives
Left ctenidium
Anus

(E)

Right retractor
Velum
Shell
Anus
Mantle cavity (behind)
Larval foot

Right retractor
Anus
Mantle cavity (in front)
Left retractor
Operculum
Head
Foot

(F)

Buccal ganglion
Cerebral ganglion
Pleural ganglion
Pedal ganglion
Parietal ganglion
Visceral ganglion

Buccal ganglion
Cerebral ganglion
Visceral ganglion
Infraparietal ganglion
Supraparietal ganglion

Figure 20.18 Pre- and posttorsional adult gastropods. (A–D) Dorsal views. (A) Hypothetical untorted gastropod. (B,C) Stages of torsion. (D) The fully torted condition. Note that the mantle cavity, gills, anus, and nephridiopores are moved from a posterior to an anterior orientation, just above and behind the head. Furthermore, many structures that were on the right side of the animal in the pretorsional condition (e.g., the right gill, osphradium, heart atrium, and nephridiopore) are located on the left side after torsion has taken place (and the pretorsional left gill, osphradium, atrium, and nephridiopore subsequently occur on the right side). (E) A prosobranch veliger larva, before and after torsion (lateral view). Note that after torsion the head can be withdrawn into the anterior mantle cavity. (F) Configuration of the principal ganglia and connectives of a hypothetical untorted and a torted adult gastropod.

whole. Two zoologists finally tested Garstang's theory by offering torted and untorted abalone veligers to various planktonic predators; they found that, in general, torted veligers were not consumed any less frequently than untorted ones (Pennington and Chia 1985). Garstang first presented his theory in verse, in 1928, as he was often taken to do with his zoological ideas.

The Ballad of the Veliger, or
How the Gastropod Got Its Twist

The Veliger's a lively tar, the liveliest afloat,

A whirling wheel on either side propels his little boat;

But when the danger signal warns his bustling submarine,

He stops the engine, shuts the port, and drops below unseen.

He's witnessed several changes in pelagic motorcraft;

The first he sailed was just a tub, with a tiny cabin aft.

An Archi-mollusk fashioned it, according to his kind,

He'd always stowed his gills and things in a mantle-sac behind.

Young Archi-mollusks went to sea with nothing but a velum—

A sort of autocycling hoop, instead of pram—to wheel 'em;

And, spinning round, they one by one acquired parental features,

A shell above, a foot below—the queerest little creatures.

But when by chance they brushed against their neighbors in the briny,

Coelenterates with stinging threads and Arthropods so spiny,

By one weak spot betrayed, alas, they fell an easy prey—

Their soft preoral lobes in front could not be tucked away!

Their feet, you see, amidships, next the cuddly-hole abaft,

Drew in at once, and left their heads exposed to every shaft.

So Archi-mollusks dwindled, and the race was sinking fast,

When by the merest accident salvation came at last.

A fleet of fry turned out one day, eventful in the sequel,

Whose left and right retractors on the two sides were unequal:

Their starboard halliards fixed astern alone supplied the head,

While those set aport were spread abeam and served the back instead.

Predaceous foes, still drifting by in numbers unabated,

Were baffled now by tactics which their dining plans frustrated.

Their prey upon alarm collapsed, but promptly turned about,

With the tender morsel safe within and the horny foot without!

This manoeuvre (*vide* Lamarck) speeded up with repetition,

Until the parts affected gained a rhythmical condition,

And torsion, needing now no more a stimulating stab,

Will take its predetermined course in a watchglass in the lab.

In this way, then, the Veliger, triumphantly askew,

Acquired his cabin for'ard, holding all his sailing crew—

A Trochosphere in armour cased, with a foot to work the hatch,

And double screws to drive ahead with smartness and despatch.

But when the first new Veligers came home again to shore,

And settled down as Gastropods with mantle-sac afore,

The Archi-mollusk sought a cleft, his shame and grief to hide,

Crunched horribly his horny teeth, gave up the ghost, and died.

Other workers have hypothesized that torsion was an adult adaptation that might have created more space for retraction of the head into the shell (perhaps also for protection from predators), or for directing the mantle cavity with its gills and water-sensing osphradia anteriorly. Still another theory asserts that torsion evolved in concert with the evolution of a coiled shell—as a mechanism to align the tall spiraling shells from a position in which they stuck out to one side (and were presumably poorly balanced and growth limiting), to a position more in alignment with the longitudinal (head–foot) axis of the body. The latter position would theoretically allow for greater growth and elongation of the shell while reducing the tendency of the animal to topple over sideways.

No matter what the evolutionary forces were that led to torsion in the earliest gastropods, the results were to move the adult anus, nephridiopores, and gonopores to a more anterior position, corresponding to the new position of the mantle cavity. It should be noted, however, that the actual position and arrangement of the mantle cavity and its associated structures show great variation; in many gastropods these structures, while pointing forward, may actually be positioned near the middle or even the posterior region of the animal's body. Torsion is not a perfectly symmetrical process.

Most of the stories of gastropod evolution focus on changes in the mantle cavity and its associated structures, and many of these changes seem to have been driven by the impact of torsion. Many anatomical modifications of gastropods appear to be adaptations to avoid fouling, for without changing the original flow of water through the mantle cavity, waste from the anus (and perhaps the nephridia) would be dumped on top of the head and pollute the mouth and ctenidial region. Hence, it has long been hypothesized that the first step, subsequent to the evolution of torsion, was the development of slits or holes in the shell, altering water flow so that a one-way current passed first over the ctenidia, then over the anus and nephridiopore, and finally out the slit or shell holes. This arrangement is seen in certain primitive gastropods, such as the slit shells (Pleurotomariacea) and certain archaeogastropods (abalone and keyhole limpets) (Figure 20.1C,D). As reasonable as it sounds, there has been surprisingly little empirical evidence in support of this hypothesis. The adaptive significance of shell holes has recently been examined by Voltzow and Collin (1995). They found that blocking the holes in keyhole limpets did not result in damage to the organs of the mantle cavity.

Once evolutionary reduction or loss of the gill and osphradium on the right side had taken place, a different antifouling strategy was achieved—that of a directed water flow through the mantle cavity from left to right, passing across the functional gill and osphradium first, then across the nephridiopore and anus, and on out the right side (Figure 20.6B). This strategy also had the effect of allowing the left side to enlarge and eventually to develop into structures such as long siphons. While the prosobranchs have retained full or partial torsion, other gastropods (opisthobranchs and pulmonates) have undergone various degrees of detorsion, loss of ctenidia, and a host of other modifications, perhaps in response to the constraints originally brought on by torsion.

Locomotion

The aplacophorans lack a well-developed foot (Figure 20.2), and locomotion is primarily by slow ciliary gliding movements through or upon the substratum. Chaetodermomorphans are mostly infaunal burrowers, and neomeniomorphans are largely symbiotic on various cnidarians. With the exception of these two groups, most molluscs possess a distinct and obvious foot. The foot often forms a flat, ventral, creeping **sole**, like that of snails, slugs, chitons, and monoplacophorans (Figure 20.19). The sole is ciliated and imbued with numerous gland cells that produce a mucous trail over which the animal glides. In gastropods, a large **pedal gland** supplies substantial amounts of slime, especially in terrestrial species that must glide on relatively dry surfaces. Very small molluscs may move largely by ciliary propulsion. However, most molluscs move primarily by waves of muscular contractions that sweep along the foot.

The gastropod foot possesses sets of **pedal retractor muscles**, which attach to the shell and dorsal mantle at various angles and act in concert to raise and lower the sole or to shorten it in either a longitudinal or a transverse direction. Contraction waves may move from back to front (direct waves), or from front to back (retrograde waves) (Figure 20.19A,B). Direct waves depend on contraction of longitudinal and dorsoventral muscles beginning at the posterior end of the foot; successive

sections of the foot are thus "pushed" forward. Retrograde waves involve contraction of transverse muscles interacting with hemocoelic pressure to extend the anterior part of the foot forward, followed by contraction of longitudinal muscles. The result is that successive areas of the foot are "pulled" forward. In some gastropods the muscles of the foot are separated by a midventral line, so the two sides of the sole operate somewhat independently of each other. The right and left sides of the foot alternate in their forward motion, almost in a stepping fashion, resulting in a sort of "bipedal" locomotion (Figure 20.19C).

Modifications of this general benthic locomotory scheme occur in many groups. Some gastropods, such as moon snails (Figure 20.19D), plow through the sediment by brute force, and they can even burrow beneath the sediment surface. Such gastropods often possess a **propodium**, a thick anterior region of the foot shaped like the blade of a bulldozer, as well as a dorsal flaplike fold of the foot that covers the head as a protective shield. Other burrowers, such as turritellids, dig by jerky side-to-side movements of the projected foot, or by thrusting the foot into the substratum, anchoring it by engorgement with hemolymph, and then pulling the body forward by contraction of longitudinal muscles. In the conch *Strombus*, the operculum forms a large "claw" that digs into the substratum and is used as a pivot point as the animal thrusts itself forward like a pole-vaulter.

Some molluscs that inhabit high-energy littoral habitats, such as chitons and limpets, have a very broad foot that can adhere tightly to hard substrata. Chitons also use their broad girdle for adhesion to the substratum by clamping down tightly and raising the inner margin to create a slight vacuum. Some snails, such as the so-called worm shells (Vermetidae and Siliquariidae) are entirely sessile (Figure 20.19E). These gastropods have typical larval and juvenile shells; but after they settle and start to grow, the shell whorls become increasingly separated from one another, resulting in a corkscrew or twisted shape. Other gastropods, such as slipper limpets, are sedentary. They tend to remain in one location and feed on organic particles in the surrounding water. Some limpets and a few chitons exhibit homing behaviors. These activities are usually associated with feeding excursions stimulated by changing tide levels, after which the animals return to their homesites.

In bivalves the foot is usually bladelike and laterally compressed (the word *pelecypod* means "hatchet foot"), as is the body in general. The pedal retractor muscles in bivalves are somewhat different from those of gastropods, but they still run from the foot to the shell (Figure 20.8). The foot is directed anteriorly and used primarily in burrowing and anchoring. It operates through a combination of muscle action and hydraulic pressure (Figure 20.20A–D). Extension of the foot is accomplished by engorgement with hemolymph, coupled

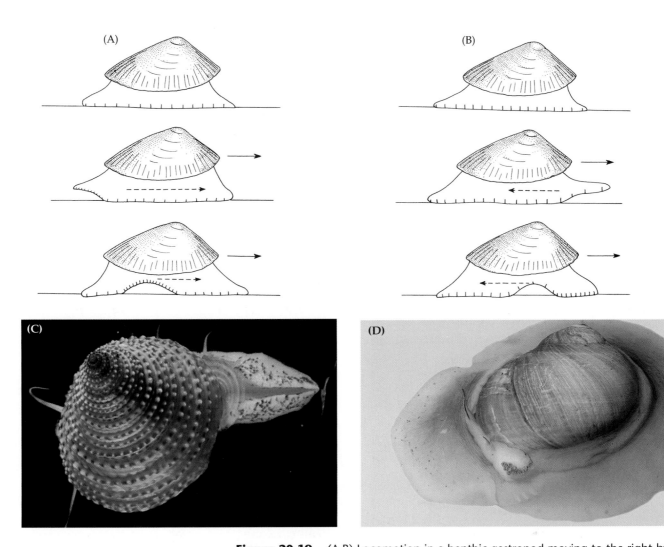

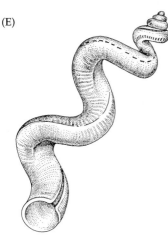

Figure 20.19 (A,B) Locomotion in a benthic gastropod moving to the right by waves of contraction of the pedal and foot muscles (solid arrow indicates direction of animal movement; dashed arrow indicates direction of muscle wave). In (A) the waves of contraction are moving in the same direction as the animal, from back to front (direct waves). Muscles at the rear of the animal contract to lift the foot off the substratum; the foot shortens in the contracted region and then elongates as it is placed back down on the substratum after the wave passes. In this way, successive sections of the foot are "pushed" forward. In (B) the animal moves forward as the contraction waves pass in the opposite direction, from front to back (retrograde waves). In this case, the pedal muscles lift the anterior part of the foot off the substratum, the foot elongates, is placed back on the substratum, then contracts to "pull" the animal forward, rather like "stepping." (C) *Calliostoma*, a gastropod adapted to crawling on hard substrata. Note the line separating the right and left muscle masses in the rear of the foot; this separation allows a somewhat "bipedal-like" motion as the animal moves. (D) The moon snail, *Polinices*, has a huge foot that can be inflated by incorporating water into a network of channels in its tissue, thus allowing the animal to plow through the surface layer of soft sediments. (E) *Tenagodus*, a sessile vermetid worm snail.

with the action of a pair of pedal protractor muscles. With the foot extended, the valves are pulled together by the **shell adductor muscles**. More hemolymph is forced from the visceral mass hemocoel into the foot hemocoel, causing the foot to expand and anchor in the substratum. Once the foot is anchored, the anterior and posterior pairs of pedal retractor muscles contract and pull the shell downward. Withdrawal of the foot into the shell is accomplished by contraction of the pedal retractors coupled with relaxation of the shell adductor muscles. Many clams burrow upward in this same manner, but others back out by using hydraulic pressure to

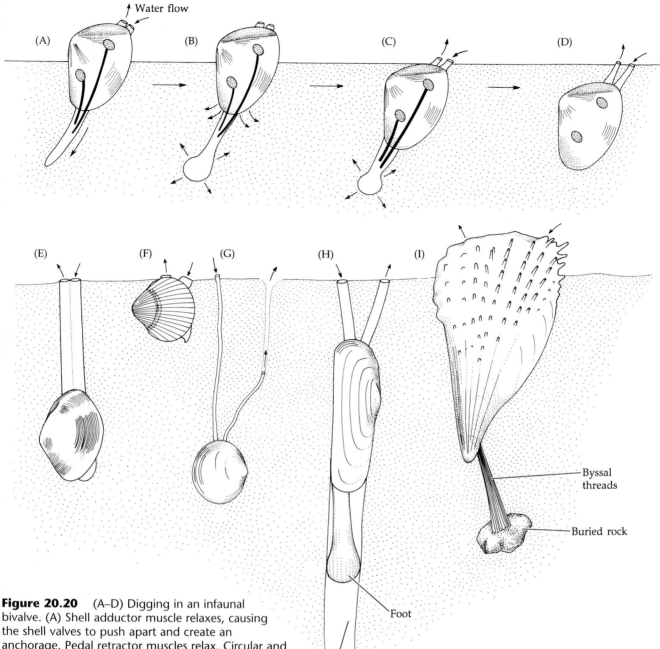

Figure 20.20 (A–D) Digging in an infaunal bivalve. (A) Shell adductor muscle relaxes, causing the shell valves to push apart and create an anchorage. Pedal retractor muscles relax. Circular and transverse foot muscles contract, causing the foot to extend into the substratum. (B) Hemolymph is pumped into the tip of the foot, causing it to expand and form an anchorage. Siphons close and withdraw as the shell adductor muscles contract, closing the shell and forcing water out between the valves and around the foot. (C) Anterior and posterior pedal retractor muscles contract, pulling the clam deeper into substratum. (D) The shell adductor muscle relaxes to allow shell valves to push apart and create an anchorage in the new position. The foot is withdrawn. (E–I) Five bivalves in soft sediments; arrows indicate direction of water flow. (E) A deep burrower with long, fused siphons (e.g., *Tresus*). (F) A shallow "nestler" with very short siphons (e.g., *Clinocardium*). (G) A deep burrower with long, separate siphons (e.g., *Scrobicularia*). (H) The razor clam (*Tagelus*) lives in unstable sands and maintains a burrow into which it can rapidly escape. (I) The pen shell, *Atrina*, attaches its byssal threads to solid objects buried in soft sediments.

push against the anchored end of the foot. Most motile bivalves possess well developed anterior and posterior adductor muscles (the **dimyarian** condition).

Most bivalves live in soft benthic habitats, where they burrow to various depths in the substratum (Figure 20.20E–I). However, several groups have epifaunal lifestyles (e.g., Pectinidae, Limidae) and live free upon the sea floor (Figure 20.1M). Some are capable of short bursts of "jet-propelled" swimming, which is accomplished by clapping the valves together. Others permanently attach to the substratum either by fusing one valve to a hard surface (e.g., rock oysters, rock scallops) or by using special anchoring lines called **byssal threads** (e.g., mussels [Figure 20.21A,B], ark shells, and

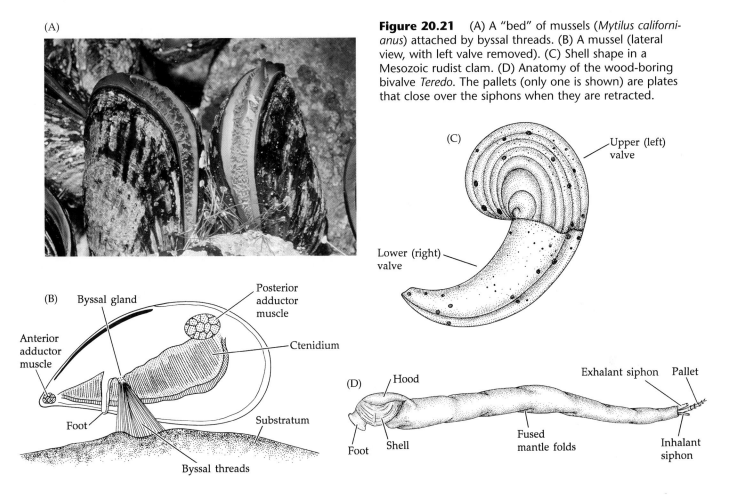

Figure 20.21 (A) A "bed" of mussels (*Mytilus californianus*) attached by byssal threads. (B) A mussel (lateral view, with left valve removed). (C) Shell shape in a Mesozoic rudist clam. (D) Anatomy of the wood-boring bivalve *Teredo*. The pallets (only one is shown) are plates that close over the siphons when they are retracted.

certain oysters such as mangrove oysters and winged oysters). Oysters of the family Ostreidae (including the edible American and European oysters) initially anchor as a settling veliger larva, called a **spat**, by secreting a drop of adhesive from the **byssus gland**. Adults, however, have one valve permanently cemented to the substratum.

Byssal threads are secreted as a liquid by the byssus gland in the foot. The liquid flows along a groove in the foot to the substratum, where each thread becomes tightly affixed. The threads are emplaced by the foot; once attached they quickly harden by a tanning process, whereupon the foot is withdrawn. A byssal thread **retractor muscle** may assist the animal in pulling against its anchorage. Mussels have a small, finger-like foot whose principal function is generation and placement of the byssal threads. Giant clams (Tridacnidae) initially attach by byssal threads, but usually lose these as they mature and become heavy enough not to be cast about by currents (Figure 20.1N). In jingle shells (Anomiidae), the byssal threads run from the upper valve through a hole in the lower valve to attach to the substratum. Byssal threads probably represent a primitive and persisting larval feature in those groups that retain them into adulthood, and many bivalves lacking byssal threads as adults utilize them for initial attachment during settlement.

In many families of attached bivalves, such as mussels and rock scallops, the foot and anterior end are reduced. This often leads to a reduction (**anisomyarian condition**) or loss (**monomyarian condition**) of the anterior adductor muscle. Mantle fusion and siphon length are also greatly reduced in attached bivalves. In oysters, the large adductor muscle is composed of two parts, a dark striated region that functions as a rapid closure muscle, and a white smoother region that functions to hold the shell tightly closed for long periods of time.

Great variation occurs in shell shape and size among attached molluscs. Some of the most remarkable bivalves were the Mesozoic rudist clams, in which the lower valve was hornlike and often curved, and the upper valve formed a much smaller hemispherical or curved lid (Figure 20.21C). Rudists were large, heavy creatures that often formed massive reeflike aggregations, either by somehow attaching to the substratum or by simply accumulating in large numbers on the seabed, in "log jams."

The habit of boring into hard substrata has evolved in many different bivalve lines. In all cases, excavation begins quickly after larval settlement. As the animal bores deeper, it grows in size and soon becomes permanently trapped, with only the siphons protruding out of the original small opening. Boring is usually by a mechanical process; the animal uses serrations on the anterior region of the shells to abrade or scrape away the substratum. Some species also secrete an acidic mucus that partially dissolves or weakens hard substrata. Numerous species in the family Pholadidae bore into wood (e.g., *Martesia, Xylophaga*), soft stone (e.g., *Pholas*), or a variety of substrata (e.g., *Barnea*). Species in the family Teredinidae (e.g., *Bankia, Teredo*) are known as shipworms because of their preference for wood, including the wooden hulls of ships. Only small remnants of the shells remain and serve as drilling structures in shipworms, with the vermiform body trailing behind (Figure 20.21D). Some species in the family Mytilidae also are borers, such as *Lithophaga*, which bores by mechanical means into hard calcareous rocks, shells of various other molluscs (including chitons) and corals, and the genus *Adula*, which bores into soft rocks.

Scaphopods are adapted to infaunal habitats, burrowing vertically by the same basic mechanism used by many bivalves (Figures 20.1L and 20.9). The elongate foot is projected downward into soft substrata, whereupon the tip is expanded to serve as an anchoring device; contraction of the pedal retractor muscles pulls the animal downward.

Perhaps the most remarkable locomotor adaptation of molluscs is swimming, which has evolved in several different taxa. In most of these groups, the foot is modified as the swimming structure. In the unique gastropod group known as heteropods, the body is laterally compressed, the shell is highly reduced, the foot forms a ventral fin, and the animal swims upside down (Figure 20.7A–C). In another unusual group of gastropods, the pteropods (sea butterflies), the foot forms two long lateral fins called **parapodia** that are used like oars (Figure 20.7D,E). Some opisthobranchs also swim by graceful undulations of flaplike folds (also called parapodia) along the body margin. Violet shells (*Janthina*) float about the ocean's surface on a raft of bubbles secreted by the foot, and some planktonic opisthobranchs (e.g., *Glaucus, Glaucilla*) stay afloat by use of an air bubble held in the stomach!

The champion swimmers are, of course, the cephalopods (Figure 20.22). These animals have abandoned the generally sedentary habits of other molluscs and have become effective high-speed predators. Virtually all aspects of their biology have evolved to exploit this lifestyle. Most cephalopods swim by rapidly expelling water from the mantle cavity. The mantle has both radial and circular muscle layers. Contraction of the radial muscles and relaxation of the circular muscles draws water into the mantle cavity. Reversal of this muscular action forces water out of the mantle cavity. The mantle edge is clamped tightly around the head to channel the escaping water through a ventral tubular **funnel**, or **siphon** (Figure 20.11). The funnel is highly mobile and can be manipulated to point in nearly any direction, thus allowing the animal to turn and steer. Squids attain the greatest swimming speeds of any aquatic invertebrates, and several species can propel themselves many feet into the air. Most octopuses are benthic and lack the fins and streamlined bodies characteristic of squids. Although octopuses still use water-powered jet propulsion, they more commonly rely on their long suckered arms for crawling about the sea floor.

Cuttlefish are slower than squids, and not only use their fins for stabilization but also undulate them to assist in steering and propulsion. Many nautiloids and sepiods move up and down in the water column on a diurnal cycle, often traveling hundreds of meters in each direction. They can actively regulate their buoyancy by secretion and reabsorption of shell chamber gases

(A)

(C)

(B)

Figure 20.22 Swimming cephalopods. (A) *Sepia*, the cuttlefish. (B) *Nautilus*. (C) *Vampyroteuthis*, a "vampire" squid, viewed from the side.

(chiefly nitrogen) by the cells of the siphuncle. The un-occupied chambers of these shells are filled partly with gas and partly with a liquid called the **cameral fluid**. The septa act as braces, giving the shells enough strength to withstand pressures at great depths. As discussed earlier, each septum in nautiloid shells is perforated in the center by a small hole, through which runs the siphuncle, which originates in the viscera and is enclosed in a porous calcareous tube. Various ions dissolved in the cameral fluid can be pumped through the porous outer layers into the cells of the siphuncular epithelium. When the cellular concentration of ions is high enough, the diffusion gradient thus created draws fluid from the shell chambers into the cells of the siphuncle while the fluid is replaced with gas. The result is an increase in buoyancy. By regulating this process, nautiloids may be able to remain neutrally buoyant wherever they are. There is also evidence that the air-filled chambers may be a source of oxygen for the nautilus. It was once thought that this gas–fluid "pump" mechanism allowed buoyancy changes sufficient to explain all the large-scale vertical movements of nautiloids, but recent work suggests that density changes may not be the sole source of power for moving great distances up and down in the water column (e.g., see Ward 1987).

Feeding

Two basic and fundamentally different types of feeding occur among molluscs: herbivory or predation (macrophagy), and suspension feeding (suspension microphagy). In Chapter 3 we reviewed the basic mechanics of these two feeding modes. Here we briefly summarize the ways in which these feeding behaviors are employed by molluscs. In this section we also discuss a uniquely molluscan structure, the **radula**, which is used in both herbivory and predation, and has become modified in a variety of unusual and interesting ways.

The molluscan radula and macrophagy. The radula is usually a ribbon of recurved chitinous teeth (Figures 20.23–20.26). The teeth may be simple, serrate, pectinate, or otherwise modified. The radula often functions as a scraper to remove food particles for ingestion, although in many groups it has become adapted for other actions. A radula is present in at least some of the most primitive living molluscs and is therefore assumed to have originated in the earliest stages of molluscan evolution. In aplacophorans the teeth, when present, may not be borne on a ribbon per se but on a basal expansion of the foregut epithelium—perhaps the evolutionary forerunner of the ribbonlike radula. In some aplacophorans, the teeth form simple plates embedded in either side of the lateral foregut wall, while in others they form a transverse row, or up to 50 rows, with as many as 24 teeth per row.

In gastropods and other molluscs the radula projects from the pharynx or buccal cavity floor as a complex tooth-bearing ribbon and associated muscles (Figures 20.23 and 20.24A). The ribbon, called a **radular membrane**, is moved back and forth over a cartilaginous or hemocoelic **odontophore** by sets of **radular protractor** and **retractor muscles**. The radula is usually housed in a **radular sac**, in which the radular membrane and new teeth are continually being produced by special cells called **odontoblasts**, to replace material lost by erosion during feeding. Measurements of radular growth indicate that up to five rows of new teeth may be added daily in some species. The odontophore itself is moved in and out of the buccal cavity by sets of **odontophore protractor** and **retractor muscles**, which also assist in applying the radula firmly against the substratum. The number of teeth ranges from a few to thousands and serves as an important taxonomic character in many groups. In some molluscs, the radular teeth are hardened with iron compounds, such as magnetite (in chitons) and goethite (in some gastropods).

Like mammalian dentition, radular teeth show adaptations to the type of food eaten. In primitive archaeogastropods (e.g., keyhole limpets, abalones, top shells), the radulae bear large numbers of fine marginal teeth in

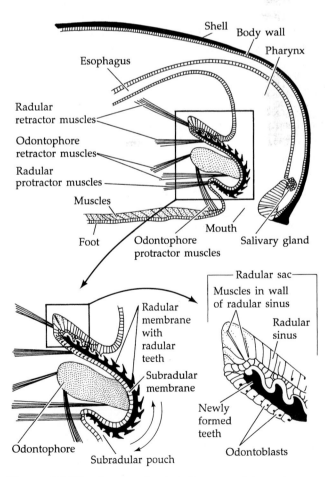

Figure 20.23 A generalized molluscan radula and associated buccal structures, at three "magnifications" (longitudinal section).

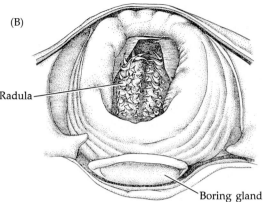

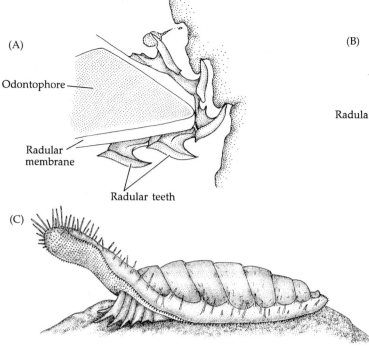

Figure 20.24 Feeding in macrophagous molluscs. (A) Cutting and scraping action of a gastropod radula. (B) A boring gastropod, the moon snail *Natica*, with radula visible in the mouth and the boring gland exposed (oral view). (C) The Pacific chiton *Placiphorella velata* in feeding position, with raised head flap ready to capture small prey.

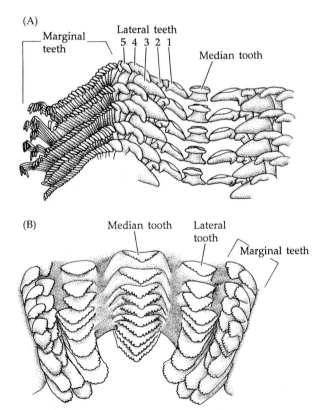

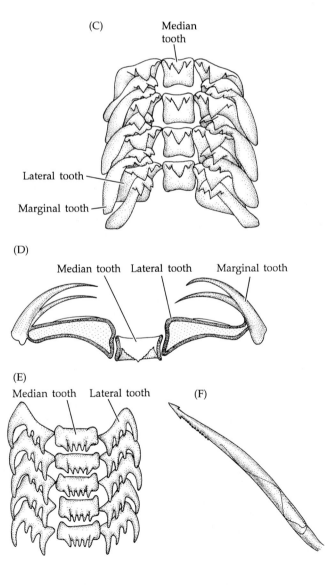

Figure 20.25 Various arrangements of radular teeth. (A) The rhipidoglossan condition of an abalone, *Haliotis*. The marginals on the right side are not shown. (B) The taenioglossan condition of the mesogastropod *Viviparus*. (C) The taenioglossan condition of the mesogastropod *Littorina*. (D) The highly modified taenioglossan condition of the heteropod *Pterotrachea*. Only one transverse row of teeth is shown. (E) The rachiglossan condition of the neogastropod *Buccinum*. (F) The toxoglossan condition of the neogastropod *Mangelia* (a single tooth).

each row (Figures 20.25A and 20.26A). Such radulae are called **rhipidoglossate**. As the radula is pulled over the bending plane of the odontophore, these teeth act like brushes, sweeping small particles to the midline where they are caught on the recurved parts of the central teeth, which draw the particles into the buccal cavity. The primitive archaeogastropods are mostly intertidal foragers that live on diatoms and other algae growing on the substratum. Some archaeogastropods (e.g., acmaeid and patellid limpets) possess **docoglossate** radulae, which bear relatively few teeth in each transverse row. In acmaeid radulae, for example, there are no central or marginal teeth, and only two pairs of lateral teeth per row (Figure 20.26B). The mucous trails left by some limpets (e.g., homing species such as the Pacific *Lottia gigantea* and *Collisella scabra*) actually serve as adhesive traps for the microalgae that are their primary food resource.

The radulae of mesogastropods are **taenioglossate**, that is, the number of marginal teeth is reduced (Figure 20.25B–D). In conjunction with the elaboration of muscular jaws, taenioglossate radulae are capable of powerful rasping action; snails such as some littorines feed by directly scraping off the surface cell layers of algae.

The most advanced prosobranch gastropods (Neogastropoda) usually have **rachiglossate** radulae, which lack marginal teeth altogether (Figures 20.25E and 20.26C,D). They use the remaining (medial) teeth for rasping, tearing, or pulling. These snails are usually car-

nivores or carrion feeders. Neogastropods of the families Muricidae and Naticidae eat other molluscs by boring through the prey's calcareous shell to obtain the underlying flesh. The boring is mainly mechanical; the predator bores with its radula while holding the prey with the foot. The boring activity may be complemented by the secretion of an acidic chemical from a **boring gland** (also called the "accessory boring organ"); the chemical is periodically applied to the drill hole to weaken the calcareous matrix (Figure 20.24B). Muricids such as the American drill (*Urosalpinx*) and the Japanese drill (*Rapana*) cause a loss of millions of dollars annually for oyster farms.

Some carnivorous gastropods (e.g., *Janthina*) do not gnaw or rasp their prey, but swallow it whole. In these gastropods a **ptenoglossate** radula forms a covering of strongly curved spines over the buccal mass. The prey is seized by the quickly extruded buccal mass and simply pulled whole into the gut. Pyramidellids have lost the radula altogether and feed by sucking blood or other fluids from their prey by use of a hypodermic stylet on the tip of an elongate proboscis.

In terms of feeding, the most specialized gastropods may be the cone snails (*Conus*), in which the radula is reduced to a few isolated poison-injecting teeth (**toxoglossate radulae**). The harpoon-like teeth (Figure 20.25F) are discharged from the end of a long proboscis that can be

(B)

(A)

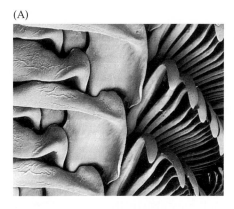

Figure 20.26 Gastropod radulae. (A) A closeup view of the rhipidoglossate radula of the abalone *Haliotis rufescens* (order Archaeogastropoda). Note the many hook-like marginal teeth. (B) The docoglossate radula of an acmaeid limpet. (C) The serrated central teeth of a rachiglossate radula from *Nucella emarginata*, a prosobranch gastropod that preys on small mussels and barnacles. (D) The worn radular teeth of *Nucella*. (E) The radula of the opisthobranch *Triopha*, seen here in dorsal view as it rests in the animal.

(C)

(D)

(E)

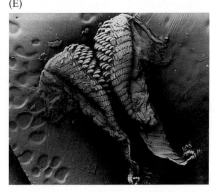

thrown out rapidly to capture prey, usually a fish, a worm, or another gastropod, which is then pulled into the gut (Figure 20.27). The venom is injected through the hollow, curved radular teeth by contraction of a venom gland. A few Indo-West Pacific cones produce a potent neuromuscular toxin that has caused human deaths. Among the most unusual gastropod feeding strategies are those that involve parasitism on fishes. For example, *Cancellaria cooperi* attaches to the Pacific electric ray and makes small cuts in the skin through which the proboscis is inserted to feed on the ray's blood and cellular fluids. Several other gastropods parasitize "sleeping" reef fishes by inserting their proboscides into the host and sucking out fluids. Some other gastropods are known to parasitize various invertebrate hosts.

Certain opisthobranchs and pulmonates also show various radular modifications. Groups that feed on cnidarians, ectoprocts, and sponges, and those that scrape algae (e.g., aplysiids) usually have typical rasping radulae. In sacoglossans, however, the radula is modified as a single row of lancelike teeth that can pierce the cellulose wall of filamentous algae, allowing the gastropod to suck out the cell contents. The pulmonate slug *Testacella* uses its toothed radula to prey on earthworms. However, in most pulmonates the radula is a broad band with many similar teeth per row and functions much like sandpaper.

Aeolid nudibranchs (Figure 20.7G) have a well deserved reputation for their particular mode of feeding, in which portions of their cnidarian prey are held by the muscular jaws while the radula rasps off pieces for ingestion. Many of these nudibranchs engage in a remarkable phenomenon called **kleptocnidae**. Some of the prey's nematocysts are ingested unfired, passed through the nudibranch's gut, and eventually transported to extensions of the digestive gland in the dorsal **cerata** (singular, **ceras**) (Figure 20.32D,E). How the nematocysts undergo this transport without firing is still a mystery. Popular hypotheses are that mucous secretions by the nudibranch limit the discharge, or that a form of acclimation occurs (like that suspected to occur between anemone fishes and their host anemones), or perhaps that only immature nematocysts survive, to later undergo maturation in the dorsal cerata. It may also be that, once the cnidocytes are digested, the nematocysts' firing threshold is raised, thereby preventing discharge. In any case, once in the cerata the nematocysts are stored in structures called **cnidosacs** and presumably help the nudibranch to fend off attackers, who depart with a mouthful of discharged nematocysts. Discharge might

Figure 20.27 Sequence of photographs of the eastern Pacific cone *Conus purpurescens* capturing and swallowing a small fish. The proboscis is extended and swept back and forth above the substratum in search of prey; when a fish is encountered, it is quickly paralyzed and ingested.

even be under control of the host nudibranch, perhaps by means of pressure exerted by circular muscle fibers around each cnidosac.

Some dorid nudibranchs also utilize their prey in remarkable ways. Many dorids secrete complex toxic compounds that are incorporated into mucus released from the mantle surface. These noxious chemicals act to deter potential predators. The chemicals may be manufactured by the dorids themselves, but in most cases it appears that they are obtained from the sponges or ectoprocts on which they feed. One of the "Spanish dancer" nudibranchs (*Hexabranchus sanguineus*) not only uses a chemical from its sponge prey for its own defense, but deposits some of the chemical in its egg cases, helping to protect the embryos until they hatch.

In polyplacophorans the radular teeth are also in numerous transverse rows, generally of 17 teeth each (a central tooth flanked by eight on each side). Most chitons are strictly herbivorous grazers. Notable exceptions are certain members of the order Ischnochitonida (family Mopaliidae: e.g., *Mopalia*, *Placiphorella*), which are known to feed on both algae and small invertebrates. *Mopalia* consumes sessile invertebrates, such as barnacles, ectoprocts, and hydroids. *Placiphorella* captures live microinvertebrates (particularly crustaceans) by trapping them beneath its head-flap, a large anterior extension of the girdle (Figure 20.24C).

In monoplacophorans the radula consists of a ribbon-like membrane bearing a succession of transverse rows of 11 teeth each (a slender median tooth flanked on each side by five broader lateral teeth). Monoplacophorans are probably generalized grazers that feed on minute organisms coating the substratum on which they live.

Cephalopods are predatory carnivores. Squids are some of the most voracious creatures in the sea, successfully competing with fishes for their meals. Octopuses are active generalized carnivores but prey primarily on crabs and clams. Some species of *Octopus* have the radula modified as a drill to bore through the shells of molluscan prey in a fashion similar to that of gastropod drills. Some even drill and prey upon their close relatives, the chambered nautiluses. Using their locomotor skills, most cephalopods hunt and catch active prey. Some octopuses, however, hunt "blindly," by "tasting" beneath stones with their chemosensitive suckers. In any event, once a victim is captured and held by the arms, the cephalopod bites it with its horny beak and injects a neurotoxin from modified salivary glands. The ability to quickly immobilize prey helps prevent the soft-bodied cephalopod from a potentially dangerous struggle.

Microphagy and suspension feeding. Suspension feeding evolved numerous times in molluscs, but in most cases it involves modifications of the ctenidia that enable the animal to trap particulate matter carried in the mantle cavity current. Many molluscs generate a single current for both gas exchange and feeding. The lamellar nature of molluscan gills preadapted them for extracting suspended food. Increasing the size of the gills and the degree of folding also increases the surface area available for trapping particulates. In suspension feeders, at least some of the gill cilia, which otherwise serve to remove sediment, function to transport particulate matter from the gills to the mouth region. Suspension feeding occurs in some gastropods and most bivalves.

There are three principal groups of suspension-feeding gastropods: pteropods, certain errant prosobranch snails, and vermetids. In the planktonic sea butterflies (pteropods), expanded, ciliated outgrowths ("wings" or "parapodia") of the swimming foot function as food-collecting surfaces or may cooperate with the mantle to produce large mucous sheets that capture small zooplankton (Figure 20.7D,E). From the foot, ciliary currents carry mucus and food to the mouth. In some pteropods, the mucous sheet may be as much as 2 m across. A different approach is taken by suspension feeding prosobranch snails. Gas exchange currents carry particulate matter into the mantle cavity. Normally these particulates are wrapped in mucus and ejected from the mantle cavity as pellets called **pseudofeces**. However, several groups have mechanisms to retain the smaller organic particles in mucus and carry them to the mouth by way of ciliary currents. Various versions of this basic plan occur throughout the prosobranchs. The radula in all suspension-feeding gastropods is reduced, serving only to pull mucus and food into the mouth. Suspension-feeding gastropods are usually rather sedentary animals and rely on generating their own water currents to bring them food.

Perhaps the zenith of adaptation to a suspension-feeding lifestyle among gastropods has been achieved by the vermetids, or "worm gastropods." The vermetid shell, coiled in youth, becomes partly or wholly noncoiling in adults and permanently affixed to the substratum (Figure 20.19E). A special pedal gland produces copious amounts of mucus that spreads outside the shell aperture as a sticky plankton trap. Periodically the net is hauled in by the foot and pedal tentacles, and a new one is quickly secreted. *Serpulorbis gigas*, a large Mediterranean species, casts out individual threads up to 30 cm long, whereas the gregarious California species *S. squamigerus* forms a communal net shared by many individuals.

The radula apparently disappeared early in the course of bivalve evolution, and most modern species use their large ctenidia for suspension feeding. However, some primitive species in the subclass Protobranchia are not suspension feeders but engage in a type of deposit-feeding microphagy. Protobranchs live in soft marine sediments and maintain contact with the overlying water either directly (e.g., *Nucula*) or by means of siphons (e.g., *Nuculana*, *Yoldia*). The two cteni-

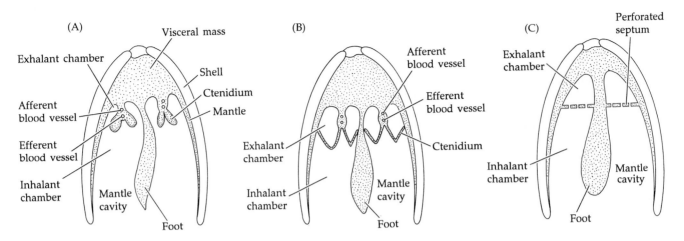

Figure 20.28   Arrangement of ctenidia in some bivalves (transverse sections). (A) Protobranch. (B) Lamellibranch. (C) Septibranchiate anomalodesmatan.

dia are small, conforming to the primitive molluscan **bipectinate** plan of an elongated axis carrying a double row of lamellae (Figure 20.28A). Protobranchs feed by means of two pairs of palp-like structures flanking the mouth. The two innermost palps are the short **labial palps**, and the two outermost palps are formed into tentacular processes called **proboscides** (each called a **palp proboscis**), which can be extended beyond the shell (Figure 20.29). During feeding the proboscides are extended slightly into the bottom sediments. Detrital material adheres to the mucus-covered surface of the proboscides and is then transported by cilia to the labial palps, which function as sorting devices. Low-density particles are carried to the mouth; heavy particles are carried to the palp margins and ejected into the mantle cavity.

Two basic kinds of suspension feeding occur in non-protobranch bivalves. Members of a small, unusual group known as the septibranchs (subclass Anomalodesmata) are sessile predators. Their ctenidia are modified as a perforated but muscular **septum** that divides the mantle cavity into dorsal and ventral chambers (Figures 20.28C and 20.30). The muscles are attached to the shell such that the septum can be raised or lowered within the mantle cavity. Raising the septum causes water to be sucked into the mantle cavity by way of the inhalant siphon; lowering the septum causes water to pass dorsally through the pores into the exhalant chamber. These movements also force hemolymph from mantle sinuses into the siphonal sinuses, thereby causing a rapid protrusion of the inhalant siphon, which can be directed toward potential prey. In this fashion, small animals such as microcrustaceans are sucked into the mantle cavity, where they are grasped by muscular labial palps and thrust into the mouth; at the same time, the mantle tissue serves as the gas exchange surface.

The second type of suspension feeding, which occurs in members of the large groups Filibranchia and Eulamellibranchia, is the stereotypical mode of "clam suspension feeding" presented in introductory biology texts. Cilia on the ctenidia generate a water current from which suspended particles are gleaned. Increased efficiency is achieved by various ctenidial modifications. The primary modification has been the conversion of the original, small, triangular plates into V-shaped filaments with extensions on either side (Figures 20.28B and 20.31B). The arm of this V-shaped filament that is attached to the central axis of the ctenidium is called the **descending arm**; the arm forming the other half of the V is the **ascending arm**. The ascending arm is usually anchored distally by ciliary contacts or tissue junctions to

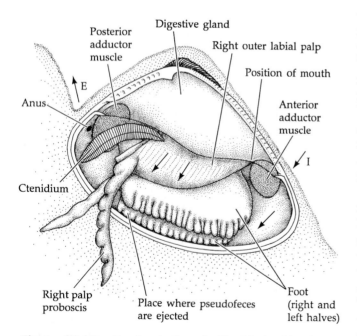

Figure 20.29   Feeding in the primitive bivalve *Nucula* (subclass Protobranchia). The clam is seen from the right side, in its natural position in the substratum (right valve and right mantle skirt removed). Arrows show direction of water currents in the mantle cavity. (I) inhalant region; (E) exhalant region.

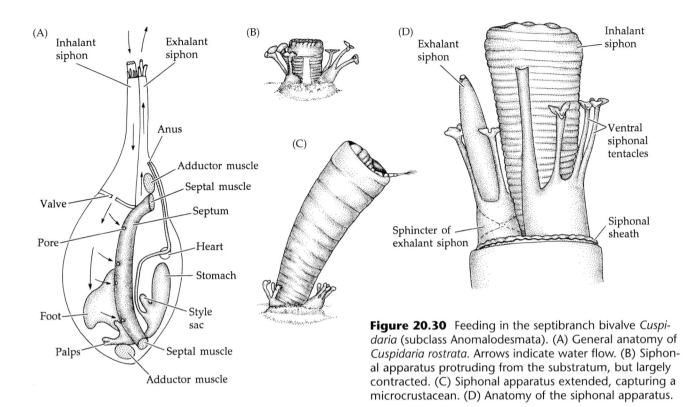

Figure 20.30 Feeding in the septibranch bivalve *Cuspidaria* (subclass Anomalodesmata). (A) General anatomy of *Cuspidaria rostrata*. Arrows indicate water flow. (B) Siphonal apparatus protruding from the substratum, but largely contracted. (C) Siphonal apparatus extended, capturing a microcrustacean. (D) Anatomy of the siphonal apparatus.

the roof of the mantle, or to the visceral mass. Taken together, the two V-shaped filaments, with their double row of leaflets, form a W-shaped structure when seen in cross section. Most filibranchs (e.g., mussels and oysters) have ctenidia wherein adjacent filaments are interlocked to one another by periodic clumps of special cilia, leaving long narrow slits in between (**interfilament spaces**) (Figure 20.31C,D). The spaces between the arms of the W's are exhalant **suprabranchial chambers**, which communicate with the exhalant area of the mantle edge; the spaces ventral to the W's are inhalant and communicate with the inhalant area of the mantle edge. Most filibranchs are restricted to epibenthic life. Their mantles are not formed into elongate siphons; thus they cannot burrow deeply (Figure 20.21A).

Eulamellibranch bivalves have a similar ctenidial design, but neighboring filaments are actually fused to one another by tissue junctions at numerous points along their length, an arrangement resulting in interfilament pores that are rows of **ostia** rather than the long narrow slits of filibranchs (Figure 20.31B,E,F). In addition, the ascending and descending halves of some filaments may be joined by tissue bridges that provide firmness and strength to the gill. Many eulamellibranch bivalves live buried in soft sediments, where long siphons are utilized to maintain contact with the overlying water (Figures 20.8A and 20.20).

Both filibranch and eulamellibranch bivalves use their ctenidia to capture food. Water is driven from the inhalant to the exhalant parts of the mantle cavity by lateral cilia all along the sides of filaments in filibranchs, or by special lateral ostial cilia in eulamellibranchs (Figure

20.31C–F). As the water passes through the interfilament spaces it flows through rows of frontolateral cilia, which flick particles from the water onto the surface of the filament facing into the current. These feeding cilia are called **compound cirri**; they have a pinnate structure that probably increases their catching power. Mucus presumably plays some part in trapping the particles and keeping them close to the gill surface, although its precise role is uncertain. Bivalve ctenidia are not covered with a continuous sheet of mucus, as occurs in many other suspension-feeding invertebrates (e.g., gastropods, tunicates, amphioxus). Once on the filament surface, particles are moved by frontal cilia toward a food groove on the free edges of the ctenidium, and then anteriorly to the labial palps. The palps sort the material by size and perhaps also by quality before passing the food to the mouth. Rejected particles fall off the gill or palp edges into the mantle cavity as pseudofeces. This "filtration" of water by bivalves is quite efficient. The American oyster (*Crassostrea virginica*), for example, can process up to 37 liters of water per hour (at 24°C), and can capture particles as small as 1 µm in size. Studies on the common mussels *Mytilus edulis* and *M. californianus* suggest that these bivalves maintain pumping rates of about 1 liter per hour per gram of (wet) body weight.

Scaphopods are selective deposit feeders that consume minute particulate matter in the surrounding sediment, or occasionally ingest the sediment itself. Two

(A)

Lateral cilia

Afferent branchial vessel

Axis

Efferent branchial vessel

(B)

Exhalant space

Mantle

Shell

Ostium

Tissue connection between adjacent filaments

Connection of filament to mantle

Efferent blood vessel (to atrium of heart)

Inhalant space

Ascending arm of ctenidial filament

Afferent blood vessel

Ctenidial axis

Descending arm of ctenidial filament

Inhalant space

Food groove

Tissue connection between ascending and descending arms of same filament (= interlamellar junction)

(C)

Ciliary tuft junctions

Ctenidial filament

Interfilament space

(D)

Frontal cilia

Frontolateral cilia

Lateral cilia

Exhalant space

Tissue junction

(E)

Ostia

Ctenidial filament

Tissue junctions

(F)

Ostium

Ctenidial filament

Exhalant water space

Frontolateral cirri

Frontal cilia

Lateral cilia

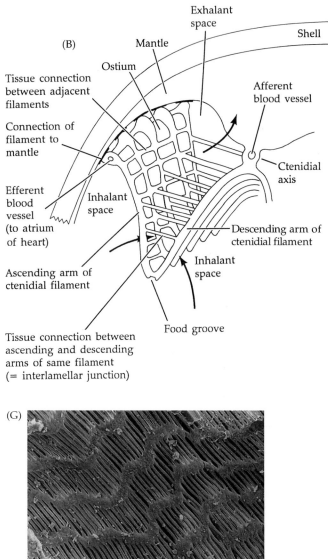

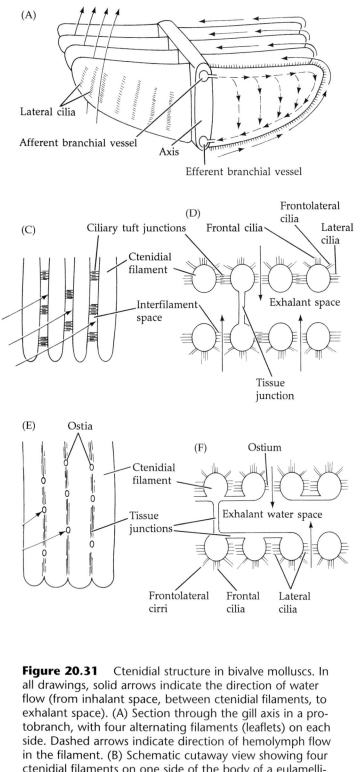

(G)

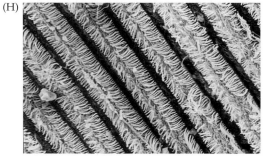

(H)

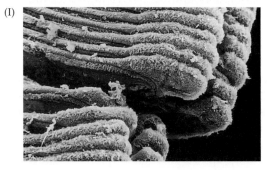

(I)

Figure 20.31 Ctenidial structure in bivalve molluscs. In all drawings, solid arrows indicate the direction of water flow (from inhalant space, between ctenidial filaments, to exhalant space). (A) Section through the gill axis in a protobranch, with four alternating filaments (leaflets) on each side. Dashed arrows indicate direction of hemolymph flow in the filament. (B) Schematic cutaway view showing four ctenidial filaments on one side of the body of a eulamellibranch. (C) Four ctenidial filaments of a filibranch (surface side view). (D) Cross section through ascending and descending arms of four filibranch ctenidial filaments. (E) Four filaments of a eulamellibranch (surface side view). (F) Cross section through ascending and descending arms of four eulamellibranch ctenidial filaments. (G) Ctenidial filaments of the mussel *Mytilus californianus* showing ciliary junctions and interfilament spaces. (H) Frontal ciliary tracts on ctenidial filaments of *Mytilus*. (I) Ventral gill edge of *Mytilus* showing food groove.

lobes flank the head, each bearing numerous (up to several hundred) long tentacles called **captacula** (Figures 20.9 and 20.13F). The captacula are extended into the substratum by metachronal beating of cilia on the **terminal bulb**. Within the substratum organic particles and microorganisms (particularly diatoms and foraminiferans) adhere to the sticky terminal bulb; small food particles are transported to the mouth by way of ciliary tracts along the tentacles, while larger particles are transported directly to the mouth by muscular contraction of the captacula. A well developed radula pulls the food into the mouth, perhaps partially macerating it in the process.

Several forms of symbiotic relationships have evolved within the molluscs and are intimately tied to the host's nutritional biology. One of the most interesting of these relationships exists between many molluscs and sulfur bacteria. These molluscs appear to derive a portion of their nutritional needs from the symbiotic, CO_2-fixing sulfur bacteria, which usually reside on the host mollusc's gill lamellae. This mollusc–bacteria symbiosis has been recently documented from a variety of sulfide-rich anoxic habitats, including deep-sea thermal vents, where geothermally produced sulfide is present, and reduced sediments, where microbial degradation of organic matter leads to the reduction of sulfate to sulfide (e.g., anoxic marine basins, seagrass bed and mangrove swamp sediments, pulp mill effluent sites, sewage outfall areas).

The gutless clam *Solemya reidi*, which harbors sulfur bacteria on its gills, has the ability to directly oxidize sulfide (Powell and Somero 1986). It does this by means of a special sulfide oxidase enzyme in the mitochondria. This clam inhabits reduced sediments near sewage outfalls and pulp mill effluents, where free sulfides are abundant. The ability to oxidize sulfide not only provides *S. reidi* with a source of energy to drive ATP synthesis, it also enables the clam to rid its body of toxic sulfide molecules that accumulate in such habitats.

Another notable partnership exists between giant clams (family Tridacnidae) and their symbiotic zooxanthellae (the dinoflagellate *Symbiodinium*). These clams live with their dorsal side against the substratum, and they expose their fleshy mantle to sunlight through the large shell gape. The mantle tissues harbor the zooxanthellae. Many species have special lenslike structures that focus light on zooxanthellae living in the deeper tissues. Certain opisthobranchs also maintain a symbiotic relationship with *Symbiodinium*. Several species of *Melibe*, *Pteraeolidia*, and *Berghia* harbor colonies of the dinoflagellate in "carrier" cells associated with their digestive glands. Experiments indicate that when sufficient light is available, host nudibranchs utilize photosynthetically fixed organic molecules produced by the alga to supplement their usual diet of prey. The dinoflagellates are probably not transmitted with the zygotes of the nudibranchs, each new generation thus requiring reinfection from the environment. A number of aeolid nudibranchs accumulate zooxanthellae from their cnidarian prey. Some of the dinoflagellates end up inside cells of the nudibranch's digestive gland, but many others are released in the slug's feces, from where they may reinfect cnidarians. An even stranger phenomenon occurs in several sacoglossan opisthobranchs (e.g., *Placobranchus*). These sea slugs obtain functional chloroplasts from the dinoflagellates upon which they feed and incorporate them into their tissues; the chloroplasts remain active for a period of time and produce photosynthetically fixed carbon molecules utilized by the hosts.

Still another unusual symbiosis was recently discovered between an aerobic bacterium and marine shipworms (bivalves of the family Teredinidae) (Figure 20.21D). Shipworms bore into wood structures and are capable of living on a diet of wood alone by harboring this cellulose-decomposing, nitrogen-fixing bacterium. The bivalve "cultures" the bacterium in pure form in a special organ that is associated with ctenidial blood vessels and is called the **gland of Deshayes**. The bacterium breaks down cellulose and makes its products available to its host. Nitrogen-fixing bacteria occur as part of the gut flora in many animals whose diet is rich in carbon but deficient in nitrogen (e.g., termites). However, shipworms are the only animals known to harbor a nitrogen fixer as a pure culture in a specialized organ (as in the host nodule–*Rhizobium* symbiosis of leguminous plants).

In addition to the above feeding strategies, certain molluscs (notably some bivalves and opisthobranchs) probably obtain a significant portion of their nutritional needs by direct uptake of dissolved organic material (DOM), such as amino acids. A few bivalves actually lack digestive tracts altogether (e.g., *Solemya*), and their nutritional requirements may be met to a considerable extent by active absorption of DOM across their ctenidia.

Digestion

Molluscs possess complete guts, several of which are illustrated in Figure 20.32. The mouth leads inward to a buccal cavity, within which the radula apparatus is located, and sometimes to a muscular pharynx (Figure 20.33). The esophagus is generally a straight tube connecting the foregut to the stomach. Various glands are often associated with this anterior gut region, including some that produce enzymes and others that secrete a lubricant over the radula—often called salivary glands. In many herbivorous species (e.g., certain pulmonates and some opisthobranchs), a muscular **gizzard** may be present for grinding up tough vegetable matter. The stomach usually bears one or more ducts that lead to large glandular **digestive ceca**. Several sets of digestive ceca may be present (variously called the digestive diverticula, digestive glands, foregut glands, midgut glands, liver, or other similar terms). The intestine leaves the stomach and terminates as the anus, which is typically located in the mantle cavity near the exhalant water flow.

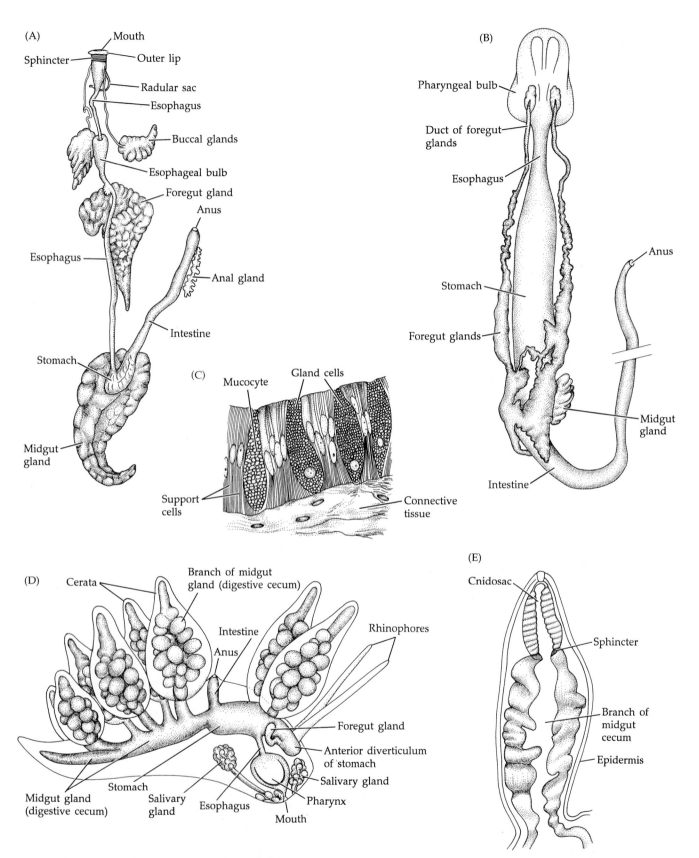

Figure 20.32 Molluscan guts. (A) The digestive system of the prosobranch gastropod *Murex*. (B) The digestive system of the land snail *Helix*. (C) The intestinal wall of a gastropod (section). (D) A nudibranch (*Embletonia*) in which large digestive ceca fill the dorsal cerata. (E) A ceras of the nudibranch *Trinchesia* (longitudinal section). The nemato-cysts (not shown) from this animal's cnidarian prey are stored in the terminal cnidosac. (F) The digestive tract and nearby organs of the clam *Anodonta* (longitudinal section). (G) The digestive system of the cuttlefish *Eledone*. (H) The digestive system of the squid *Loligo*.

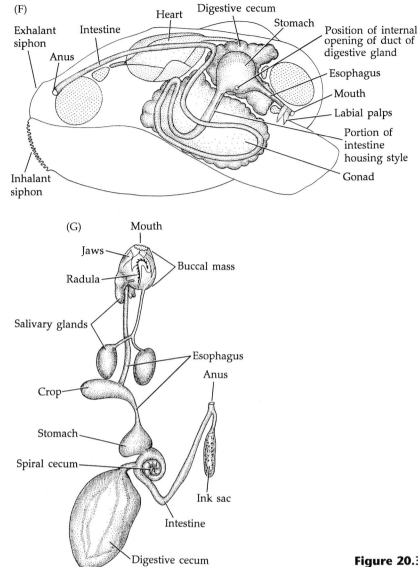

(F)

Exhalant siphon

Intestine

Heart

Digestive cecum

Stomach

Position of internal opening of duct of digestive gland

Anus

Esophagus

Mouth

Labial palps

Portion of intestine housing style

Inhalant siphon

Gonad

(G)

Mouth

Jaws

Radula

Buccal mass

Salivary glands

Esophagus

Anus

Crop

Stomach

Spiral cecum

Ink sac

Intestine

Digestive cecum

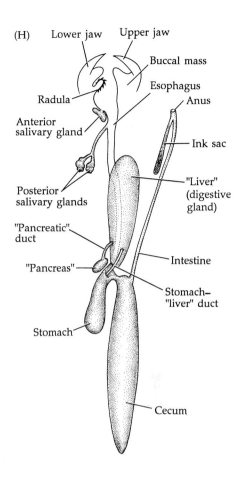

(H)

Lower jaw

Upper jaw

Buccal mass

Radula

Esophagus

Anterior salivary gland

Anus

Ink sac

Posterior salivary glands

"Liver" (digestive gland)

"Pancreatic" duct

"Pancreas"

Intestine

Stomach–"liver" duct

Stomach

Cecum

Figure 20.33 The molluscan stomach and style sac. (A) The stomach and style apparatus of a bivalve. The crystalline style rotates to grind against the gastric shield, releasing digestive enzymes and winding up the mucus–food string to assist in pulling it from the esophagus. Food particles are sorted in the ciliated, grooved sorting area: small particles are carried (in part by the typhlosole) to the digestive ceca for digestion; large particles are carried to the intestine for eventual elimination. (B) The style sac of a typical prosobranch gastropod (cross section).

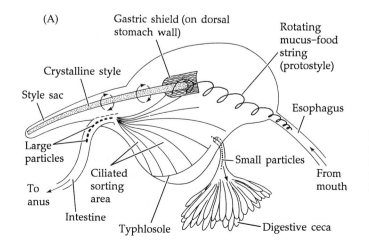

(A)

Gastric shield (on dorsal stomach wall)

Rotating mucus–food string (protostyle)

Crystalline style

Style sac

Esophagus

Large particles

Small particles

To anus

From mouth

Ciliated sorting area

Intestine

Typhlosole

Digestive ceca

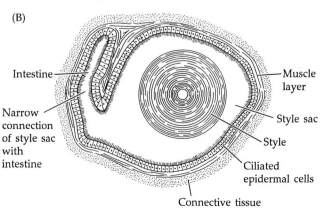

(B)

Intestine

Muscle layer

Narrow connection of style sac with intestine

Style sac

Style

Ciliated epidermal cells

Connective tissue

Once food has entered the buccal cavity of most molluscs, it is carried in mucous strings into the esophagus and then to the stomach. In cephalopods, chunks of food are swallowed by muscular action of the esophagus. The food is stored in the stomach or, in octopuses and *Nautilus*, in an expanded region of the esophagus called the **crop**. In many molluscs the stomach wall bears a chitinous **gastric shield** and a ciliated, ridged sorting area (Figure 20.33). The posterior stomach region houses a **style sac**, which is lined with cilia and contains the **crystalline style**. This structure, which functions to aid in digestion, is a rodlike matrix of proteins and enzymes (especially amylase) that are slowly released as the projecting end of the style rotates and grinds against the abrasive gastric shield. The gastric cilia and rotating style wind up the mucus and food into a string called the **protostyle** and draw it along the esophagus to the stomach. The style is produced by special cells of the style sac. The style of some bivalves is enormous, one-third to one-half the length of the clam itself. Particulate matter is swept against the stomach's anterior sorting region, which sorts mainly by size. Small particles are carried into the digestive ceca, which arise from the stomach wall. Larger particles are passed along ciliated grooves of the stomach to the intestine.

Extracellular digestion takes place in the stomach and digestive ceca, while absorption and intracellular digestion occur in the cecal and intestinal walls. Extracellular digestion is accomplished by enzymes produced in foregut and stomach glands (e.g., salivary glands, esophageal pouches, pharyngeal glands—often called "sugar glands" because they produce amylase), the stomach, and the digestive ceca. In primitive groups, intracellular digestion tends to predominate. In most molluscs, ciliated tracts line the digestive ceca and carry food particles to minute ducts, where they are engulfed by phagocytic cells of the duct wall. The same cells dump digestive wastes back into the ducts, to be carried by other ciliary tracts back to the stomach, from there to be passed out of the gut via the intestine and anus. In most advanced groups (e.g., cephalopods), extracellular digestion predominates. Enzymes secreted primarily by the ceca and stomach digest the food, and absorption occurs in the stomach, ceca, and intestine. In some cases the stomach has lost some of its primitive features, such as the gastric shield, the sorting area, and the style sac.

Circulation and Gas Exchange

Although molluscs are coelomate protostomes, the coelom is greatly reduced. The main body cavity is an open circulatory space or hemocoel, which comprises several separate sinuses, and a network of vessels in the gills, where gas exchange takes place. The blood of molluscs contains various cells, including amebocytes, and is referred to as hemolymph. It is responsible for picking up the products of digestion from the sites of absorption and for delivering these nutrients throughout the body. It usually carries in solution the copper-containing respiratory pigment hemocyanin. Many molluscs also use hemoglobin and/or myoglobin to bind oxygen.

The heart lies dorsally, within the **pericardial chamber**, and comprises a pair of **atria** (sometimes called auricles) and a single **ventricle**. In monoplacophorans and in *Nautilus* there are two pairs of atria. The atria receive the **efferent ctenidial** (= branchial) **vessels**, drawing oxygenated hemolymph from each ctenidium and passing it into the muscular ventricle, which pumps it anteriorly through a large **anterior artery** (also called the anterior or cephalic aorta). The anterior artery branches and eventually opens into various sinuses within which the tissues are bathed in oxygenated hemolymph. Return drainage through the sinuses eventually funnels the hemolymph back into the **afferent ctenidial vessels**. This basic pattern of molluscan circulation is shown diagrammatically in Figure 20.34; it is modified to various degrees in different classes (Figure 20.35). In cephalopods, the circulatory system is secondarily closed (Figure 20.35C).

Most molluscs have true gills, or ctenidia. However, many have lost the ctenidia and either rely on secondarily derived "gills" or on gas exchange across the mantle or general body surface. The presumed primitive gill condition is expressed in several living groups, for example, many of the primitive gastropods (archaeogastropods, such as *Pleurotomaria*) and primitive bivalves (such as protobranchs), and can serve to explain how molluscan gills work. In these cases the gill, or ctenidium, is built around a long, flattened axis projecting from the wall of the mantle cavity (Figure 20.31A). To each side of the axis are attached triangular or wedge-shaped filaments that alternate in position with filaments on the opposite side of the axis. This arrangement, in which filaments project on both sides of the central axis, is called the **bipectinate** (or **aspidobranch**) **condition**. There is one gill on each side of the mantle cavity, held in position by membranes that divide the mantle cavity into upper and lower chambers (Figure 20.28A,B). Cilia covering the gill surface draw water into the **inhalant (ventral) chamber**, from which it passes upward between the gill filaments to the **exhalant (dorsal) chamber** and then out of the mantle cavity (Figure 20.31A).

Two vessels run through each gill axis. The afferent vessel carries oxygen-depleted hemolymph into the gill, and the efferent vessel drains freshly oxygenated hemolymph from the gill to the atria of the heart, as noted above. Hemolymph flows through the filaments from the afferent to the efferent vessel. Ctenidial cilia carry water over the gill filaments in a direction opposite to that of the flow of the underlying hemolymph in the ctenidial vessels. This countercurrent phenomenon enhances gas exchange between the hemolymph and water by maximizing the diffusion gradients of O_2 and CO_2 (Figure 20.31A).

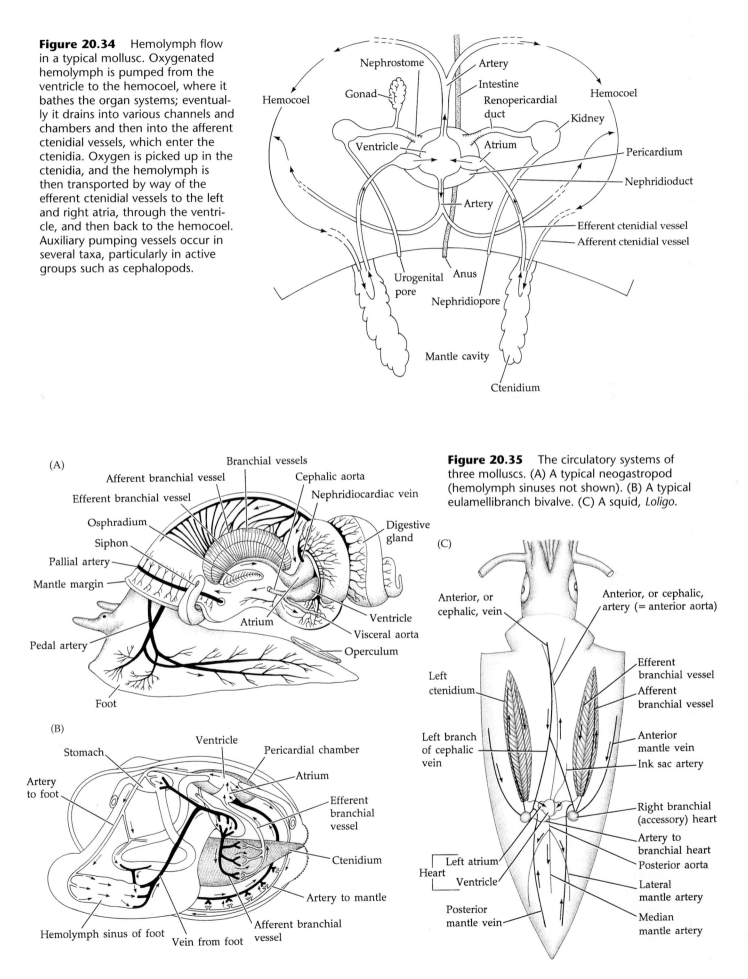

Figure 20.34 Hemolymph flow in a typical mollusc. Oxygenated hemolymph is pumped from the ventricle to the hemocoel, where it bathes the organ systems; eventually it drains into various channels and chambers and then into the afferent ctenidial vessels, which enter the ctenidia. Oxygen is picked up in the ctenidia, and the hemolymph is then transported by way of the efferent ctenidial vessels to the left and right atria, through the ventricle, and then back to the hemocoel. Auxiliary pumping vessels occur in several taxa, particularly in active groups such as cephalopods.

Nephrostome · Artery · Intestine · Renopericardial duct · Hemocoel · Gonad · Hemocoel · Kidney · Ventricle · Atrium · Pericardium · Nephridioduct · Artery · Efferent ctenidial vessel · Afferent ctenidial vessel · Urogenital pore · Anus · Nephridiopore · Mantle cavity · Ctenidium

Figure 20.35 The circulatory systems of three molluscs. (A) A typical neogastropod (hemolymph sinuses not shown). (B) A typical eulamellibranch bivalve. (C) A squid, *Loligo.*

(A) Branchial vessels · Afferent branchial vessel · Cephalic aorta · Efferent branchial vessel · Nephridiocardiac vein · Osphradium · Digestive gland · Siphon · Pallial artery · Mantle margin · Pedal artery · Atrium · Ventricle · Visceral aorta · Operculum · Foot

(B) Ventricle · Stomach · Pericardial chamber · Artery to foot · Atrium · Efferent branchial vessel · Ctenidium · Artery to mantle · Hemolymph sinus of foot · Vein from foot · Afferent branchial vessel

(C) Anterior, or cephalic, vein · Anterior, or cephalic, artery (= anterior aorta) · Efferent branchial vessel · Afferent branchial vessel · Left ctenidium · Anterior mantle vein · Ink sac artery · Left branch of cephalic vein · Right branchial (accessory) heart · Artery to branchial heart · Posterior aorta · Left atrium · Heart · Ventricle · Lateral mantle artery · Posterior mantle vein · Median mantle artery

Recall from our discussion on torsion that gastropods have evolved novel ways to circulate water over the gills and avoid fouling from gut or nephridial discharges (see Figure 20.15). Primitive archaeogastropods with two bipectinate ctenidia may accomplish this by circulating water in across the gills, then past the anus and nephridiopore, and away from the body via slits or holes in the shell (Figures 20.1C,D, 20.5C, and 20.36). This circulation pattern is used by the slit shells, abalones, and volcano limpets. Many specialists regard the slit shells (family Pleurotomariacea) as "living fossils" that reflect an archetypal gastropod condition. Most other gastropods have lost the right ctenidium and with it the right atrium; they circulate water in from the left side of the head and then straight out the right side, where the anus and nephridiopore open (Figure 20.37A). Other gastropods have lost both ctenidia and utilize secondary respiratory regions, either the mantle surface itself or secondarily derived gills of one kind or another. Limpets of the genus *Patella* have rows of secondary gills in the pallial groove along each side of the body, similar to the condition seen in chitons and monoplacophorans.

In more advanced gastropods, such as the mesogastropods and neogastropods, one ctenidium is almost always missing, as are the dorsal and ventral suspensory membranes of the remaining gill, which attaches directly to the mantle wall by the gill axis. The gill filaments on the attached side have been lost, while those of the oppo-

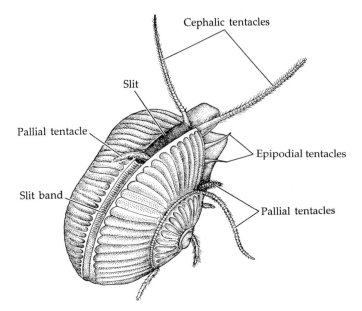

Figure 20.36 The primitive archaeogastropod *Scissurella*, one of the slit shells (Pleurotomariacea).

site side project freely into the mantle cavity. This advanced arrangement of filaments on only one side of the central axis is referred to as the **monopectinate (or pectinobranch) condition** (Figure 20.15D). The dorsal attachment of the monopectinate ctenidium in some species helps prevent fouling in soft sediments. Some advanced mesogastropods and neogastropods have also evolved **inhalant siphons** by extension and rolling of the mantle margin (Figures 20.1G and 20.37A). In these cases the margin of the shell may be notched, or drawn out as a canal to house the siphon. The siphon provides access to surface water in burrowing species, and may also function as a mobile, directional sense organ.

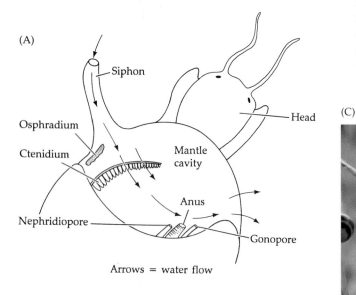

(A)

(C)

(B)

Figure 20.37 (A) A generalized neogastropod with a fleshy siphon, depicted with the shell removed. The animal has lost the posttorsional right ctenidium (and atrium and nephridium). Water flows into the mantle cavity through the siphon from the left, passes over the osphradium and ctenidium and then over the excretory pore and anus before leaving the mantle cavity to the right. (B) The terrestrial pulmonate, *Ariolimax* (banana slug), showing the pneumostome that opens to the "lung." (C) Terrestrial pulmonates during summer dormancy in August (Ibiza, Spain).

Opisthobranch gastropods are largely detorted. In some, the one remaining gill is **plicate**, or folded, rather than filamentous, and in fact may not be homologous with the prosobranch ctenidium. Trends toward detorsion, loss of the shell and ctenidia, and reduction of the mantle cavity occur in many opisthobranchs, and the process has apparently occurred several times within this group. Some nudibranchs have evolved secondary dorsal gas exchange structures called **cerata**. Many also have a circlet of postdorsal gills around the anus that may or may not be homologous to the true ctenidia (Figures 20.1H and 20.7 F,H).

Wholly terrestrial gastropods generally lack gills, and exchange gases directly across a vascularized region of the mantle, usually within the mantle cavity. The whole arrangement is often referred to as a **lung**. In terrestrial pulmonates, the edges of the mantle cavity have become sealed to the back of the animal except for a small opening on the right side called a **pneumostome** (Figure 20.37B). Instead of having gills, the roof of the mantle cavity is highly vascularized. By arching and flattening the mantle cavity floor, air is moved into and out of the lung.

In the chitons, the mantle cavity forms a **pallial groove** extending along the body margin and encircling the foot (Figure 20.4). A large number of simple gills lie laterally in this groove. The mantle is held tight against the substratum, largely enclosing this pallial chamber. However, the mantle is raised on either side at the anterior end to form incurrent channels, and is raised in one or two places at the posterior end to form excurrent areas. Water enters the inhalant region of the pallial chamber lateral to the gills, then passes medially between the gills into the exhalant region along the sides of the foot. Moving posteriorly, the current passes over the gonopores, nephridiopores, and anus before exiting (Figure 20.4B).

In bivalves the capacious mantle cavity allows the ctenidia to develop a greatly enlarged surface area, serving in most species for both gas exchange and feeding. We discussed earlier many of the morphological modifications of bivalve gills in our coverage of suspension feeding. In addition to the folded, W-shaped ctenidial filaments seen in many bivalves (Figure 20.28B), some forms (e.g., oysters) bear **plicate ctenidia**. A plicate ctenidium is thrown into vertical ridges or folds, each ridge consisting of several ctenidial filaments. The grooves between these ridges of ordinary filaments bear so-called **principal filaments**, whose cilia are important in sorting sediments from the ventilation and feeding currents. The plicate condition gives the ctenidium a corrugated appearance and further increases the surface area for gas exchange.

In spite of these modifications, the basic system of circulation and gas exchange in bivalves is similar to that seen in gastropods (Figures 20.8 and 20.35B). In most bivalves, the heart ventricle folds around the gut, so the pericardial cavity encloses not only the heart but also a short section of the digestive tract. The large mantle lines the valves and provides an additional surface area for gas exchange, which in some groups may be as important as the gills in this regard. In septibranchs, which have reduced gills, the mantle surface is the principal area of gas exchange. Most bivalves appear to lack respiratory pigments in the hemolymph, although globins occur in a few species and hemocyanin is found in protobranchs.

Scaphopods have lost the ctenidia, heart, and virtually all vessels. The circulatory system is reduced to simple hemolymph sinuses, and gas exchange takes place mainly across the mantle and body surface (Figure 20.9).

No doubt associated with their large size and active lifestyle, cephalopods have a circulatory system that is effectively closed, with many discrete vessels, secondary pumping structures, and even capillaries (Figures 20.10, 20.11, 20.12B, and 20.35C). The result is increased pressure and efficiency of hemolymph flow and delivery. In most cephalopods, the vessels leading into the ctenidia are enlarged into powerful accessory **branchial hearts**, which boost the low venous pressure as the hemolymph enters the gills. The gills are folded, increasing their surface area for greater gas exchange associated with a high metabolic rate.

In the aplacophorans, gills are usually absent or, if present, form a ciliated, lamellar pouch arising directly off the posterior region of the pericardial chamber. Caudofoveatans have a similar posterior gill. Whether or not these gills are homologous to, or early forerunners of, the ctenidia of other molluscs is uncertain. Monoplacophoran gills are similar to those of gastropods, but they occur as three to six pairs, aligned bilaterally within the pallial groove, reminiscent of chitons. Well developed lamellae occur only on one side of the monoplacophoran gill axis, similar to the monopectinate condition of advanced gastropods.

Excretion and Osmoregulation

The basic excretory structures of molluscs are paired tubular metanephridia (often called kidneys) that are primitively similar to those of annelids, echiurans, and sipunculans. Nephridia are absent in aplacophorans. Three, six, or seven pairs of metanephridia occur in monoplacophorans, two pairs in the nautiloids, and a single pair in all other molluscs (except where one is lost in advanced gastropods) (Figure 20.15). The nephrostome typically opens into the pericardial coelom via a **renopericardial duct**, and the nephridiopore discharges into the mantle cavity, often near the anus (Figure 20.34). In monoplacophorans, the arrangement of the nephrostomes is unclear. In those with six pairs of nephridia, the first four pairs appear to be associated with large pharyngeal pouches (once thought to be coeloms), and the last two pairs may drain the pericardium. In more typical molluscs, pericardial fluids pass through the nephrostome and into the nephridium, where selective resorption occurs along the tubule wall until the final urine is ready to pass out the nephridiopore. The pericardial sac and heart wall act as selective barriers between the open nephrostome and the hemolymph in the surrounding hemocoel and in the heart. Mollusc nephridia are rather large and saclike, and their walls are greatly folded. In many forms, afferent and efferent nephridial vessels carry hemolymph to and from the kidney tissues (Figure 20.38). Often a short bladder is present just before the nephridiopore.

In at least some species urine formation involves pressure filtration, active secretion, and active resorption. Aquatic molluscs excrete mostly ammonia, and most marine species are osmoconformers. In freshwater species the nephridia are capable of excreting a hyposomotic urine by resorbing salts and by passing large quantities of water. Terrestrial gastropods conserve water by converting ammonia to uric acid, although the degree to which conservation is accomplished varies depending on local environmental conditions. Land snails are capable of surviving a considerable loss of body water, which is brought on in part by production of the slime trail.

We have already mentioned some variations on the primitive molluscan excretory system. In most gastropods, torsion is accompanied by loss of the adult right nephridium, except for a small remnant that contributes to part of the gonoduct. Some gastropods have lost the direct connection of the nephrostome to the pericardial coelom. In such cases the nephridium is often very glandular and served by afferent and efferent hemolymph vessels, and wastes are removed largely from the circulatory fluid. In many gastropods and some other molluscs, the gonoduct fuses with the renopericardial canal, and the nephridopore functions as a **urogenital pore** and discharges both excretory wastes and gametes. In pulmonates, where the mantle cavity serves as a lung, the excretory duct is elongate and the nephridiopore opens outside the mantle cavity. In monoplacophorans and chitons, the nephridia open into the excurrent regions of the pallial grooves; in scaphopods, the paired nephridia open near the anus.

In bivalves, the two nephridia are located beneath the pericardial cavity and are folded in a long U-shape. One arm of the U is glandular and opens into the pericardial cavity; the other arm forms a bladder and opens through a nephridiopore in the suprabranchial cavity. In protobranchs, the unfolded walls of the tube are glandular throughout. The nephridiopores may be separate from or joined with the ducts of the reproductive system. In the latter case, the openings are urogenital pores.

Cephalopods retain the basic nephridial plan, in which the kidneys drain the pericardial coelom by way of renopericardial canals and empty via nephridiopores into the mantle cavity. However, the nephridia bear enlarged regions called **renal sacs**. Before reaching the branchial heart, a large vein passes through the renal sac, wherein numerous thin-walled evaginations, called **renal appendages**, project off the vein. As the branchial heart beats, hemolymph is drawn through the renal appendages, and wastes are filtered across their thin walls into the nephridia. The overall result is an increase in excretory efficiency over the simpler arrangement present in other molluscs.

The fluid-filled kidneys of cephalopods are inhabited by a variety of commensals and parasites. The epithelium of the convoluted renal appendages provides an excellent surface for attachment, and the renal pores provide a simple exit to the exterior. Symbionts identified from cephalopod kidneys include viruses, fungi, ciliate protists, dicyemids, trematodes, larval cestodes, and juvenile nematodes.

Nervous System

The molluscan nervous system is derived from the basic protostome plan of an anterior circumenteric arrangement of ganglia and paired ventral nerve cords. In molluscs, the more ventral and medial of the two pairs of nerve cords are called the **pedal cords** (or ventral cords); they innervate the muscles of the foot. The more lateral pair of nerves are the **visceral cords** (or lateral cords); they serve the mantle and viscera. Transverse commissures interconnect these longitudinal nerve cord pairs, creating a ladder-like nervous system. This basic plan is most easily seen in primitive molluscs, such as aplacophorans and polyplacophorans (Figure

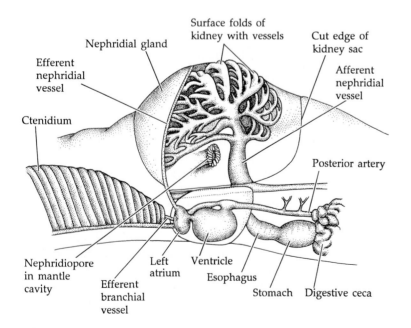

Figure 20.38   The kidney and nearby organs of *Littorina* (cutaway view). The nephridial sac has been slit open.

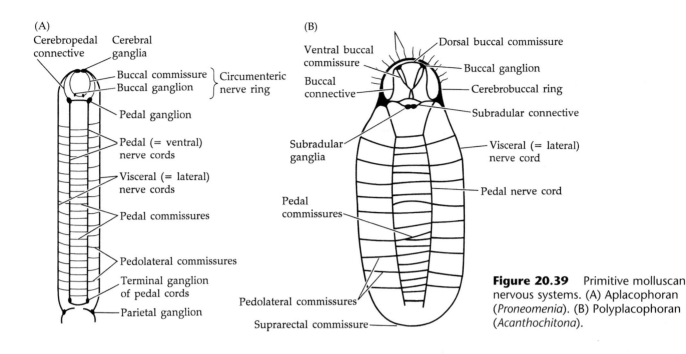

(A)
Cerebropedal
connective
Cerebral
ganglia
Buccal commissure
Buccal ganglion
Circumenteric
nerve ring
Pedal ganglion
Pedal (= ventral)
nerve cords
Visceral (= lateral)
nerve cords
Pedal commissures
Pedolateral commissures
Terminal ganglion
of pedal cords
Parietal ganglion

(B)
Ventral buccal
commissure
Dorsal buccal commissure
Buccal ganglion
Buccal
connective
Cerebrobuccal ring
Subradular connective
Subradular
ganglia
Visceral (= lateral)
nerve cord
Pedal
commissures
Pedal nerve cord
Pedolateral commissures
Suprarectal commissure

Figure 20.39 Primitive molluscan nervous systems. (A) Aplacophoran (*Proneomenia*). (B) Polyplacophoran (*Acanthochitona*).

20.39). However, the molluscan nervous system differs from that of annelids and arthropods, in which the ventral nerve cord(s) bears segmentally arranged ganglia.

In the simplest molluscs—such as aplacophorans, monoplacophorans, and polyplacophorans—ganglia are poorly developed (Figure 20.39). A simple nerve ring surrounds the esophagus, often with small cerebral ganglia on either side. Each cerebral ganglion, or the nerve ring itself, issues small nerves to the buccal region and gives rise to the pedal and the visceral nerve cords. Most other molluscs have more well defined ganglia. Their nervous systems are built around three pairs of large ganglia that interconnect to form a partial or complete nerve ring around the gut (Figures 20.40 and 20.41). Two pairs, the **cerebral** and **pleural ganglia**, lie dorsal or lateral to the esophagus, and one pair, the **pedal ganglia**, lies ventral to the gut, in the anterior part of the foot. In cephalopods, bivalves, and advanced gastropods, the cerebral and pleural ganglia are typically fused. From the cerebral ganglia, peripheral nerves innervate the tentacles, eyes, statocysts, and general head surface, as well as **buccal ganglia**, with special centers of control for the buccal region, radular apparatus, and esophagus. The pleural ganglia give rise to the visceral cords, which extend posteriorly, supplying peripheral nerves to the viscera and mantle. The visceral cords eventually join a pair of **parietal** (= **intestinal**, = **pallial**) **ganglia** and from there pass on to terminate in paired **visceral ganglia**. The parietal ganglia innervate the gills and osphradium, and the visceral ganglia serve organs in the visceral mass. The pedal ganglia also give rise to a pair of pedal nerve cords that extend posteriorly and provide nerves to muscles of the foot.

As a result of torsion, the posterior portion of the gastropod nervous system is twisted into a figure eight, a condition known as streptoneury (Figure 20.40A,B). In addition to twisting the nervous system, torsion brings the posterior ganglia forward. In many advanced gastropods this anterior concentration of the nervous system is accompanied by a shortening of certain nerve cords and fusion of certain ganglia. In most detorted gastropods the nervous system displays a secondarily derived bilateral symmetry and more or less straight, parallel, visceral nerve cords—a condition known as euthyneury (Figure 20.40C).

In bivalves, the nervous system is clearly bilateral, and fusion has reduced it to three large, distinct ganglia. Anterior **cerebropleural ganglia** give rise to two pairs of nerve cords, one extending posterodorsally to the **visceral ganglia**, the other leading ventrally to the **pedal ganglia** (Figure 20.41). The two cerebropleural ganglia are joined by a dorsal commissure over the esophagus. The cerebropleural ganglia send nerves to the palps, anterior adductor muscle, and mantle. The visceral ganglia issue nerves to the gut, heart, gills, mantle, siphon, and posterior adductor muscle.

The degree of nervous system development within the Cephalopoda is unequaled among invertebrates. Although the paired ganglia seen in other molluscs are also recognizable in cephalopods, extreme cephalization has occurred. Most of the ganglia have shifted forward and are concentrated as lobes of a large brain encircling the anterior gut (Figure 20.42A). In addition to the usual head nerves originating from the cerebral ganglion, a large optic nerve extends to each eye. In most cephalopods, much of the brain is enclosed in a cartilaginous **cranium**. The pedal lobes supply nerves to the funnel, and anterior divisions of the pedal ganglia, called **brachial lobes**, send nerves to each of the arms and tentacles, an arrangement suggesting that the

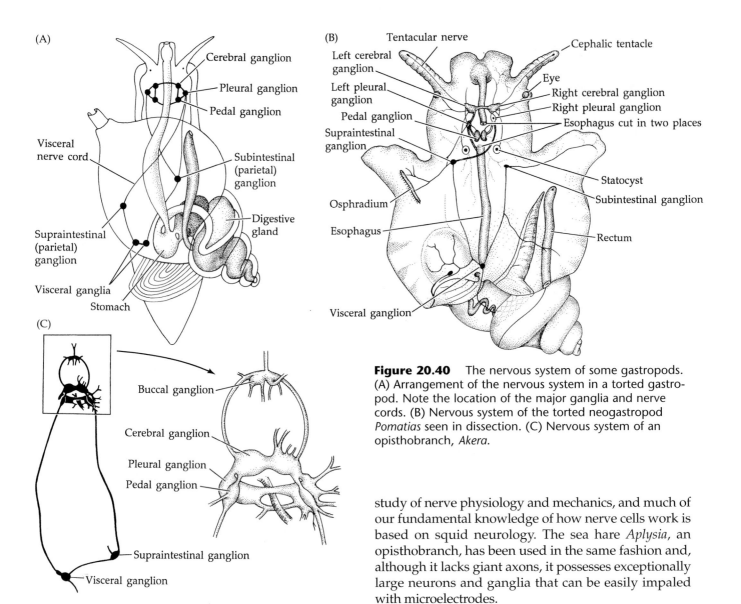

(A)

Cerebral ganglion

Pleural ganglion

Pedal ganglion

Visceral
nerve cord

Subintestinal
(parietal)
ganglion

Digestive
gland

Supraintestinal
(parietal)
ganglion

Visceral ganglia

Stomach

(B)

Tentacular nerve

Cephalic tentacle

Left cerebral
ganglion

Left pleural
ganglion

Eye

Right cerebral ganglion

Right pleural ganglion

Pedal ganglion

Esophagus cut in two places

Supraintestinal
ganglion

Statocyst

Subintestinal ganglion

Osphradium

Esophagus

Rectum

Visceral ganglion

(C)

Buccal ganglion

Cerebral ganglion

Pleural ganglion

Pedal ganglion

Supraintestinal ganglion

Visceral ganglion

Figure 20.40   The nervous system of some gastropods. (A) Arrangement of the nervous system in a torted gastropod. Note the location of the major ganglia and nerve cords. (B) Nervous system of the torted neogastropod *Pomatias* seen in dissection. (C) Nervous system of an opisthobranch, *Akera*.

study of nerve physiology and mechanics, and much of our fundamental knowledge of how nerve cells work is based on squid neurology. The sea hare *Aplysia*, an opisthobranch, has been used in the same fashion and, although it lacks giant axons, it possesses exceptionally large neurons and ganglia that can be easily impaled with microelectrodes.

funnel and tentacles are derived from the molluscan foot. Octopuses may be the "smartest" invertebrates, for they can be taught some memory-dependent tasks fairly quickly.

Many cephalopods display a rapid escape behavior that depends on a system of giant motor fibers that control powerful and synchronous contractions of the mantle muscles. The command center of this system is a pair of very large first-order giant neurons in the lobe of the fused visceral ganglia. Here, connections are made to second-order giant neurons that extend to a pair of large **stellate ganglia**. At the stellate ganglia, connections are made with third-order giant neurons that innervate the circular muscle fibers of the mantle (Figure 20.42B–D). Other nerves extend posteriorly from the brain and terminate in various ganglia that innervate the viscera and structures in the mantle cavity.

For several decades neurobiologists have utilized the giant axons of *Loligo* as an experimental system for the

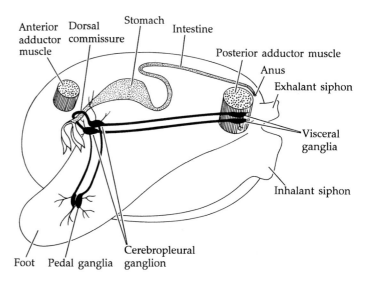

Anterior
adductor
muscle

Dorsal
commissure

Stomach

Intestine

Posterior adductor muscle

Anus

Exhalant siphon

Visceral
ganglia

Inhalant siphon

Cerebropleural
ganglion

Foot   Pedal ganglia

Figure 20.41   The reduced and concentrated nervous system of a typical bivalve.

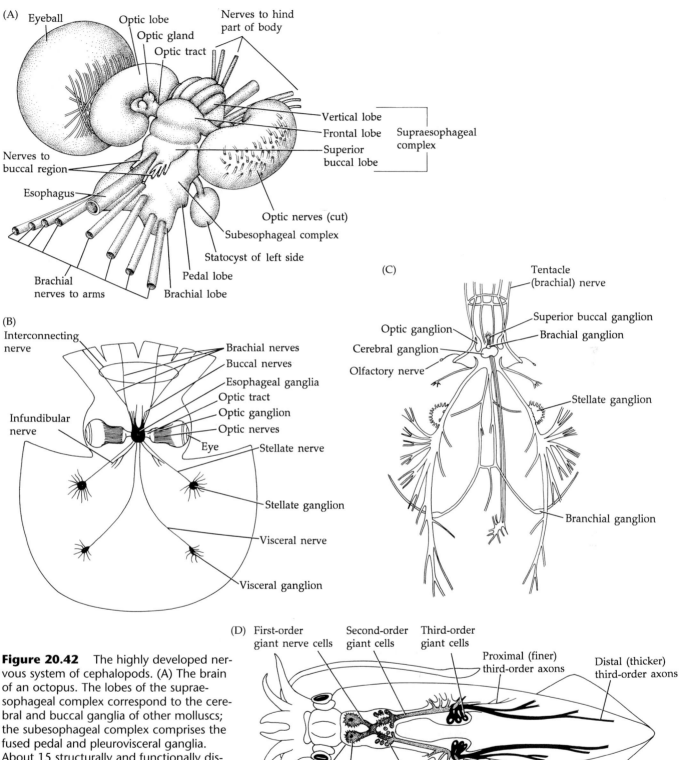

Figure 20.42 The highly developed nervous system of cephalopods. (A) The brain of an octopus. The lobes of the supraesophageal complex correspond to the cerebral and buccal ganglia of other molluscs; the subesophageal complex comprises the fused pedal and pleurovisceral ganglia. About 15 structurally and functionally distinct pairs of lobes have been identified in the brain of octopuses. (B) Nervous system of an octopus. (C) Nervous system of a squid (*Loligo*). (D) Giant fiber system of a squid. Note that the first-order giant neurons possess an unusual cross connection, and that the third-order giant neurons are arranged so that motor impulses can reach all parts of the mantle-wall musculature simultaneously (as a result of the fact that impulses travel faster in thicker axons).

Sense Organs

With the exception of the primitive class Aplacophora, molluscs possess various combinations of sensory tentacles, photoreceptors, statocysts, and osphradia. Osphradia are patches of sensory epithelium, located on or near the gill, or on the mantle wall (Figure 20.40B). They function as chemoreceptors and perhaps also as monitors of the amount of sediment in the inhalant current (Figure 20.43A,B). Little is known about the biology of osphradia, and their anatomy differs markedly throughout the phylum.

In primitive archaeogastropods, an osphradium is present on each gill; in the prosobranchs that possess one gill, there is only one osphradium, and it lies on the mantle cavity wall anterior and dorsal to the attachment of the gill itself. Osphradia are reduced or absent in gastropods that have lost both gills, that possess a highly reduced mantle cavity, or that have taken up a strictly pelagic existence. Osphradia are best developed in benthic predators and scavengers, such as neogastropods.

Most gastropods have one pair of **cephalic tentacles**, but higher pulmonates and many opisthobranchs possess two pairs. Many archaeogastropods also have short **epipodial tentacles** on the margin of the foot or mantle. The cephalic tentacles may bear eyes as well as tactile and chemoreceptor cells. Many opisthobranchs have a pair of branching or folded anterior dorsal chemoreceptors called **rhinophores** (Figure 20.7F,G). Most gastropods have a small eye at the base of each cephalic tentacle, but in some, such as the conch *Strombus*, the eyes are enlarged and on long stalks. The higher pulmonates also have eyes placed on the tips of special **optic tentacles**. Primitive gastropods have simple pigment-cup eyes, while some advanced groups have complex eyes with a cornea and lens (Figure 20.44A,B,D).

Opisthobranchs typically produce a mucopolysaccharide slime trail as they crawl. In many species the trail contains chemical messengers that other members of the species "read" by means of their excellent chemoreception. These chemical messengers may be simple trail markers, so one animal can follow or locate another, or they may be alarm substances that serve to warn others of possible danger on the path ahead. For example, when the carnivorous slug *Navanax* (= *Agleja*) is attacked by a predator, it quickly releases a yellow chemical mixture on its trail that causes other members of the species to abort their trail-following activity.

Laboratory experiments have shown that at least one nudibranch (*Tritonia diomedea*) possesses geomagnetic orientation to the Earth's magnetic field (Lohmann and Willows 1987). Motile gastropods usually possess a pair of closed statocysts in the anterior region of the foot.

Scaphopods have lost the eyes, tentacles, and osphradia typical of the epibenthic and motile molluscan groups. The captacula may function as tactile (as well as feeding) structures, but little is known about scaphopod sense organs.

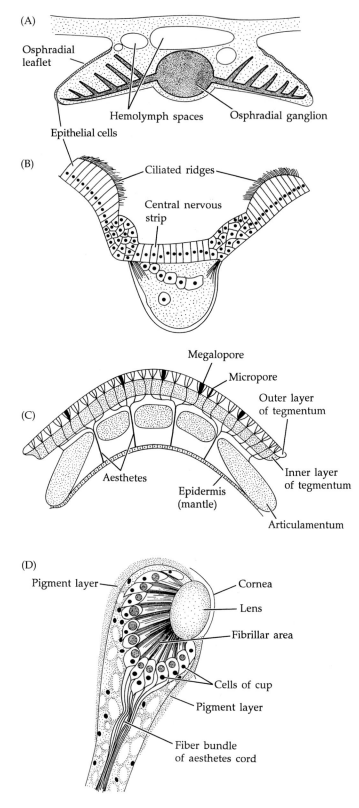

Figure 20.43 Two sensory organs of molluscs: osphradia and aesthetes. (A) Two leaflets (cross section) of a bipectinate osphradium of a gastropod (*Vermetus*). (B) The osphradium (cross section) of a heteropod (*Pterotrachea*). (C) One valve of a polyplacophoran (*Tonicia*). The aesthetes extend to the shell surface through megalopores and micropores. (D) Eye-bearing aesthetes (longitudinal section) in a megalopore of a chiton (*Acanthopleura*).

Bivalves carry most of their sensory organs along the middle lobe of the mantle edge (Figure 20.14C). These receptors include the **pallial tentacles**, which contain both tactile and chemoreceptor cells. The tentacles are commonly restricted to the siphon areas, but in some swimming clams (e.g., *Lima*, *Pecten*) they may line the entire mantle margin. Paired statocysts usually occur in the foot, and are of particular importance in georeception by burrowing bivalves. Ocelli may also be present along the mantle edge. In the spiny oyster *Spondylus* and the swimming clam *Pecten*, the ocelli have a complex cornea–lens arrangement (Figure 20.44C). The bivalve osphradium lies in the exhalant chamber, beneath the posterior adductor muscle, and may not be homologous with that of gastropods.

Chitons lack statocysts, cephalic eyes, and tentacles. Instead, they rely largely on two special sensory structures. The **subradular organ** is a modified chemosensory

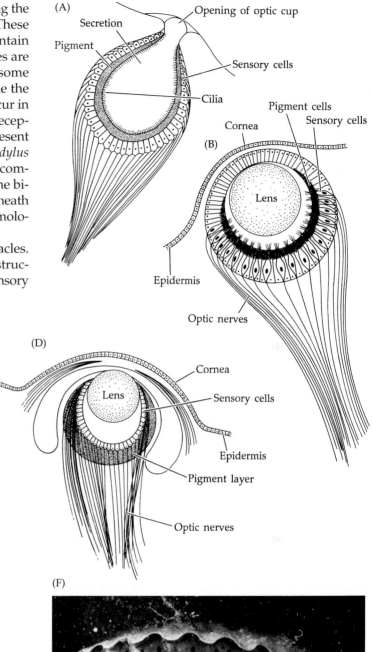

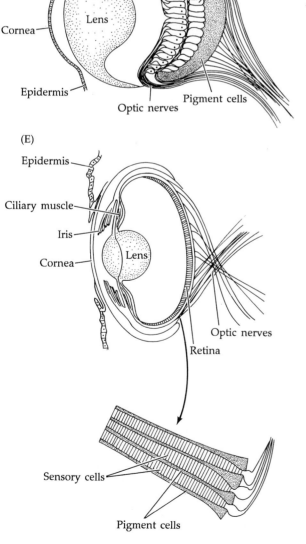

Figure 20.44 Molluscan eyes. (A) The simple pigment-cup arrangement of an archaeogastropod, *Haliotis* (the abalone). (B–E) Eyes with lenses. (B) The eye of a garden snail (*Helix*). (C) The eye of a scallop (*Pecten*). (D) The eye of a marine prosobranch (*Littorina*). (E) The eye of an octopus (*Octopus*). (F) The scallop *Aquipecten*, showing its blue eyes along the mantle edges.

region of the foregut. **Aesthetes** are a specialized system of photoreceptors unique to the class Polyplacophora. Aesthetes occur in high numbers across the dorsal surface of the shell plates. They are mantle cells that extend into the minute vertical canals (megalopores and micropores) in the upper tegmentum of the shell (Figure 20.43C,D). The canals and sensory endings terminate beneath a cap on the shell surface. Little is known about the functioning of aesthetes, but they apparently mediate light-regulated behavior. In at least one family (Chitonidae), some of them are modified as simple eyes. The outer mantle surface of many chitons is liberally supplied with tactile and photoreceptor cells, and the inner mantle cavity usually bears patches of chemosensory epithelium.

Like the rest of their nervous system, the sense organs of cephalopods are very highly developed. The eyes are superficially similar to those of vertebrates (Figure 20.44E), and these two types of eyes are often cited as a classic example of convergent evolution. The cephalopod eye sits in a socket associated with the cranium. The cornea, iris, and lens arrangement is much like that of vertebrate eyes. The lens is suspended by ciliary muscles but has a fixed shape and focal length. An iris diaphragm controls the amount of light entering the eye, and the pupil is a horizontal slit. The retina comprises closely packed, long, rodlike photoreceptors whose sensory ends point toward the front of the eye; hence the cephalopod eye is the direct type rather than the indirect type seen in vertebrates. The rods connect to retinal cells that supply fibers to the optic ganglia at the distal ends of the optic nerves. Unlike the eyes of vertebrates, the cephalopod cornea probably contributes little to focusing because there is almost no light refraction at the corneal surface (as there is at an air–cornea interface). The cephalopod eye accommodates to varying light conditions by changes in the size of the pupil and by migration of the retinal pigment. Cephalopod eyes form distinct images (although octopuses are probably quite nearsighted) and experimental work suggests that they may also see colors. In addition, cephalopods can discriminate among objects by size, shape, and vertical versus horizontal orientation. The eyes of *Nautilus* are rather primitive relative to the eyes of other cephalopods. They are carried on short stalks, lack a lens, and are open to the water through the pupil.

Nautiloids and coleoids have statocysts that provide information on static body position and on body motion. The arms are liberally supplied with chemosensory and tactile cells, especially on the suckers of benthic hunting octopuses, which have extremely good chemical and textural discrimination capabilities. *Nautilus* is the only cephalopod with osphradia.

Cephalopod Coloration and Ink

Cephalopods are noted for their striking pigmentation and dramatic color displays. The integument contains many pigment cells, or **chromatophores**, most of which are probably under control of the nervous system and perhaps hormones. Such chromatophores can be individually expanded or contracted by means of tiny muscles attached to the periphery of each cell. Contraction of these muscles pulls out the cell and its internal pigment into a flat plate, thereby displaying the color; relaxation of the muscles causes the cell and pigment to concentrate into a small, inconspicuous dot. Because these chromatophores are displayed or concealed by muscle action, their activity is extremely rapid and cephalopods can change color (and pattern) almost instantaneously. Chromatophore pigments are of several colors—black, yellow, orange, red, and blue. The chromatophore color may be enhanced by deeper layers of irridocytes that both reflect and refract light in a prismatic fashion. Some species, such as the cuttlefish *Sepia* and some octopuses, are capable of closely mimicking their background coloration (Figure 20.12E). Many epipelagic squids show a dark-above, light-below countershading similar to that seen in pelagic fishes. Most cephalopods, however, appear to undergo color changes in relation to behavioral rituals, such as courtship and aggression. In octopuses, many color changes are accompanied by modifications in the surface texture of the body, mediated by muscles beneath the skin—something like elaborate, controlled "gooseflesh."

In addition to the color patterns formed by chromatophores, some cephalopods are bioluminescent. When present, the light organs, or **photophores**, are arranged in various patterns on the body, and in some cases even occur on the eyeball. The luminescence is sometimes due to symbiotic bacteria, but in most cases it is intrinsic. The photophores of some species have a complex reflector and focusing-lens arrangement, and some even have an overlying color filter or chromatophore shutter to control the color or flashing pattern. Most luminescent species are deep-sea forms, and little is known about the role of light production in their lives. Some appear to use the photophores to create a countershading effect, so as to appear less visible to predators (and prey) from below and above. Others living below the photic zone probably use their glowing or flashing patterns as a means of communication, the signals serving to keep animals together in schools or to attract prey. The flashing may also play a role in mate attraction. The fire squid, *Lycoteuthis*, can produce several colors of light: white, blue, yellow, and pink. At least one genus of squid, *Heteroteuthis*, secretes a luminescent ink. The light comes from luminescent bacteria cultured in a small gland near the **ink sac**, from which ink and bacteria are ejected simultaneously.

In most nonnautiloid cephalopods, a large ink sac is located near the intestine (Figure 20.32H). An ink-producing gland lies in the wall of the sac, and a duct runs from the sac to a pore into the rectum. The gland secretes a brown or black fluid that contains a high concentration of melanin pigment and mucus; the fluid is stored in the ink sac. When alarmed, the animal releases the ink through the anus and mantle cavity and out into the surrounding water. The cloud of inky material hangs together in the water, forming a "dummy" image that serves to confuse predators. The alkaloid nature of the ink may also act to deter predators, particularly fishes, and may interfere with their chemoreception.

Like virtually all other aspects of cephalopod biology, the ability to change color and to defend against predators are part and parcel of their active hunting lifestyles. In the course of their evolution, nonnautiloid cephalopods lightened their bodies by abandoning the protection of an external shell, thereby exposing their fleshy parts to predators. The advent of camouflage and ink production, coupled with high mobility and complex behavior, played a major role in the success of these animals in their radical modification of the basic molluscan bauplan.

Reproduction

Primitively, molluscs are dioecious, with a pair of gonads that discharge their developing gametes to the outside, either through the nephridial plumbing or through separate ducts. In species that free-spawn, fertilization is external and development is indirect. Many molluscs with separate gonoducts that store and transport the gametes also have various means of internal fertilization. In these forms, direct and mixed life history patterns have evolved.

Aplacophorans may be either dioecious with single or paired gonads, or hermaphroditic with a pair of gonads—one functioning as an ovary, the other as a testis (Figure 20.45). In all aplacophorans the gonads discharge gametes by way of short **gonopericardial ducts** into the pericardial chamber, from which they pass to gametoducts to the mantle cavity and surrounding sea water. Monoplacophorans possess two pairs of gonads, each with a gonoduct connected to one of the pairs of metanephridia (Figures 20.3D and 20.15E), and fertilization is external.

Most chitons are dioecious, although a few hermaphroditic species are known. In chitons, the two gonads are fused and situated medially in front of the pericardial cavity (Figure 20.4F). Gametes are transported directly to the outside by two separate gonoducts. The gonopores are located in the exhalant region of the pallial groove, one in front of each nephridiopore. Fertilization is external but often occurs in the mantle cavity of the female. The eggs are enclosed within a spiny, buoyant membrane and are released into the sea individually or in strings. A few chitons brood their embryos in the pallial groove, and in one species (*Callistochiton viviparous*) development takes place entirely within the ovary.

In living gastropods, one gonad is always lost and the remaining one is usually coiled within the visceral mass. The gonoduct is always developed in association

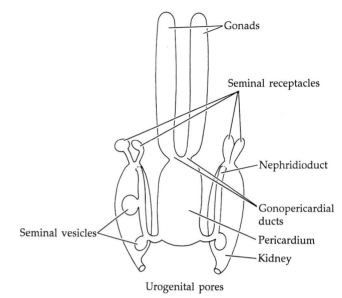

Figure 20.45 An aplacophoran urogenital system.

with the right nephridium (Figures 20.46 and 20.47). In cases where the right nephridium is still functional in transporting excretory products, as in primitive archaeogastropods, the gonoduct is properly called a **urogenital duct**, because it discharges both gametes and urine. Gastropods may be dioecious or hermaphroditic, but even in the latter case only a single gonad (an **ovotestis**) exists. The commitment of the right nephridial plumbing entirely to serving the reproductive system was a major step in higher gastropod evolution. The isolation of the reproductive tract freed it from the excretory system and allowed its independent evolution. Were it not for this singular event, the great variety of reproductive and developmental patterns in gastropods may never have been realized.

In many gastropods with isolated reproductive tracts, the female system bears a ciliated fold or tube that forms a **vagina** and **oviduct** (or **pallial duct**). The tube develops inwardly from the mantle wall and connects with the genital duct. The oviduct may bear specialized structures for sperm storage or egg case secretion. A **seminal receptacle** often lies near the ovary at the proximal end of the oviduct. Eggs are fertilized at or near this location prior to entering the long secretory portion of the oviduct. Many female systems also have a **copulatory bursa** at the distal end of the oviduct, where sperm are received during mating. In such cases the sperm are later transported along a ciliated groove in the oviduct to the seminal receptacle, where fertilization takes place. The secretory section of the oviduct may be modified as an **albumin gland** and a **mucous** or **capsule gland**. Many opisthobranchs lay fertilized eggs in jelly-like mucopolysaccharide masses or strings produced by these glands. Most terrestrial pulmonates produce a small number of large, individual, yolky eggs,

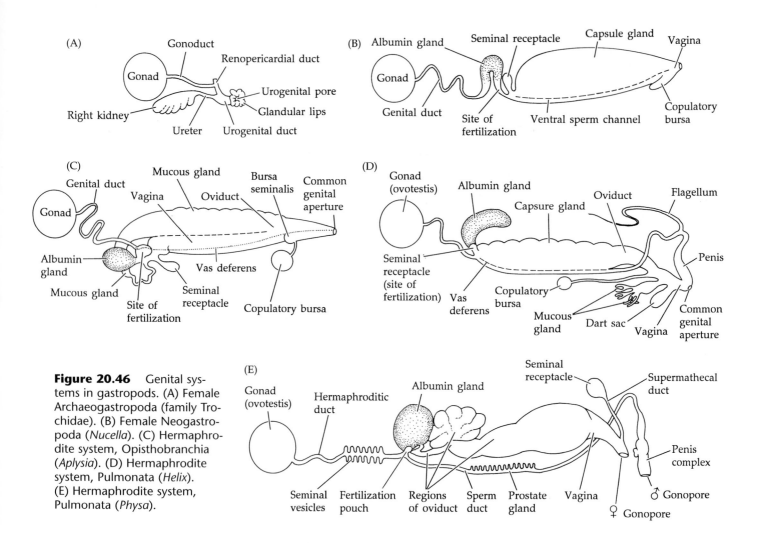

Figure 20.46 Genital systems in gastropods. (A) Female Archaeogastropoda (family Trochidae). (B) Female Neogastropoda (*Nucella*). (C) Hermaphrodite system, Opisthobranchia (*Aplysia*). (D) Hermaphrodite system, Pulmonata (*Helix*). (E) Hermaphrodite system, Pulmonata (*Physa*).

which often are provided with very thin calcareous shells. Other pulmonates brood their embryos internally and give birth to juveniles. Many advanced marine prosobranchs produce **egg capsules** in the form of leathery or hard cases that are attached to objects in the environment, thereby protecting the developing embryos. A ciliated groove is often present to conduct the soft egg capsules from the female gonopore down to a gland in the foot, where they are molded and attached to the substratum.

In gastropods that produce egg capsules or egg cases, the males usually have a penis to facilitate transfer of sperm or spermatophores (Figures 20.6B and 20.47), and internal fertilization takes place prior to formation of the egg case. The penis is a long extension of the body wall arising behind the right cephalic tentacle. The male genital duct, or **vas deferens**, may include a **prostate gland** for production of seminal secretions. In many molluscs the proximal region of the vas deferens functions as a sperm storage area, or seminal vesicle.

Both simultaneous and sequential hermaphrodites are common among gastropods. In both cases copulation is the rule, either with one individual acting as the male and the other as the female, or with a mutual ex-

change of sperm between the two. Sedentary species, such as many limpets and slipper shells, are often protandric hermaphrodites. In slipper shells (*Crepidula*), individuals may stack one atop the other (Figure 20.48), males generally on top of the stack, females on the bottom. Each male uses its long penis to inseminate the females below. Males that are in association with females tend to remain male for a relatively long period of time. Eventually, or if isolated from a female, the male develops into a female. Female slipper shells cannot switch back to males, as the masculine reproductive system degenerates during the sex change.

Pulmonates are simultaneous hermaphrodites; opisthobranchs may be either simultaneous or occasionally protandric hermaphrodites. In most simultaneous hermaphrodites a single complex gonad, the ovotestis, coincidentally produces both eggs and sperm (Figures 20.46C–E and 20.47D). The genital duct draining an ovotestis is called the **hermaphroditic duct**. There may be separate male and female gonopores, or only a single common gonopore. Such reproductive systems are amazingly complex and varied.

Distinct precopulatory behaviors occur in a few groups of gastropods. These primitive courtship routines

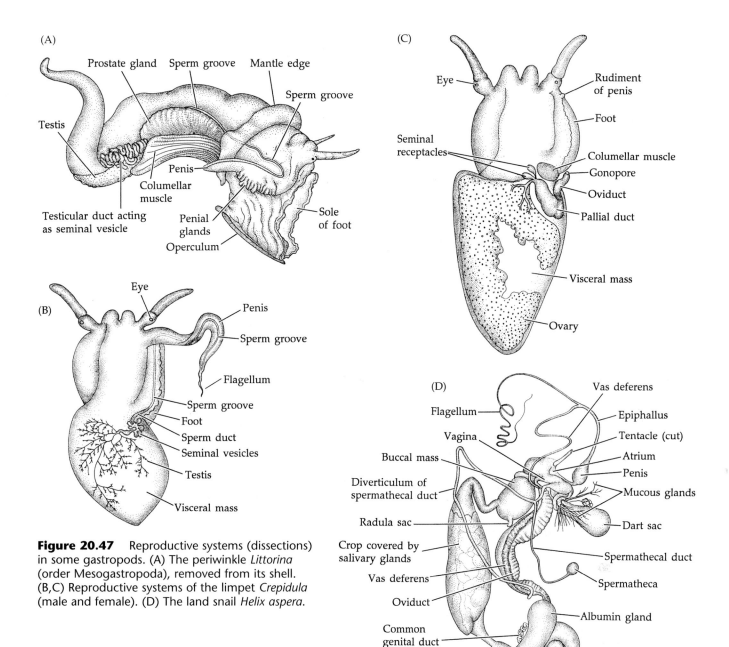

Figure 20.47 Reproductive systems (dissections) in some gastropods. (A) The periwinkle *Littorina* (order Mesogastropoda), removed from its shell. (B,C) Reproductive systems of the limpet *Crepidula* (male and female). (D) The land snail *Helix aspera*.

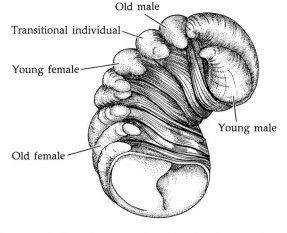

Figure 20.48 A stack of *Crepidula fornicata*, a slipper limpet displaying sequential hermaphroditism.

are best documented in land pulmonates and include behaviors such as oral and tentacular stroking, and intertwining of the bodies. In some pulmonates (e.g., the common garden snail, *Helix*) the vagina contains a **dart sac**, which secretes a calcareous harpoon. As courtship reaches its crescendo, and a pair of snails is intertwined, one will drive its dart into the body wall of the other, apparently as a means of sexually arousing its partner.

Most bivalves are dioecious and retain the primitive paired gonad plan. However, the gonads are large and closely invested with the viscera and with each other, so an apparently single gonadal mass results (Figure 20.8D). The gonoducts are simple tubes, and fertilization is usually external, although some freshwater species brood their embryos for a time. In primitive bivalves, the gonoducts join the nephridia and gametes are released through urogenital pores. In advanced bivalves, the gonoducts open into the mantle cavity separately from the nephridiopores. Hermaphroditism occurs in some bivalves, including shipworms and some species of cockles, oysters, scallops, and others. Hermaphroditic scallops have ovotestes. Oysters of the genus *Ostrea* are sequential hermaphrodites, but most are capable of switching sex in either direction.

Cephalopods are almost all dioecious, with a single gonad in the posterior region of the visceral mass (Figures 20.11, 20.12, and 20.49). The testis releases sperm to a coiled vas deferens, which leads anteriorly to a seminal vesicle. Here various glands assist in packaging the sperm into elaborate spermatophores, which are stored in a large reservoir called **Needham's sac**. From here the spermatophores are released into the mantle cavity via a sperm duct. In females the oviduct terminates in one **oviducal gland** in squids, and two in octopuses. This gland secretes a protective membrane around each egg.

The highly developed nervous system of cephalopods has facilitated the evolution of some very sophisticated precopulatory behaviors, which culminate in the transfer of spermatophores from the male to the female. Because the oviducal opening of females is deep within the mantle chamber, male cephalopods use one of their arms as an intromittent organ to transfer the spermatophores. The morphological modifications of such arms is called **hectocotyly** (Figure 20.12D). In squids and cuttlefish the right or left fourth arm is used; in octopuses it is the right third arm. In *Nautilus* four small arms form a conical organ, the **spadix**, that functions in sperm transfer. Hectocotylous arms have special suckers, spoonlike depressions, or superficial chambers for holding spermatophores during the transfer, which may be a brief or a very lengthy process.

Each spermatophore comprises an elongate sperm mass, a **cement body**, a coiled, "spring-loaded" **ejaculatory organ**, and a **cap**. The cap is pulled off as the spermatophore is removed from the Needham's sac in squids or by uptake of sea water in octopuses. Once the cap is removed, the ejaculatory organ everts, pulling the sperm mass out with it. The sperm mass adheres by means of the cement body to the seminal receptacle or mantle wall of the female, where it begins to disintegrate and liberate sperm for up to two days.

Precopulatory rituals in cephalopods almost always involve striking changes in coloration, as the male tries to attract the female (and discourage other males in the area). Male squids often seize their female partner with the tentacles, and the two swim head-to-head through the water. Eventually the male hectocotylus grabs a spermatophore and inserts it into the mantle chamber of his partner, near or in the oviducal opening. Mating in

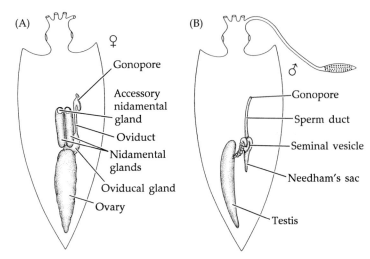

Figure 20.49   Reproductive systems in cephalopods. Female (A) and male (B) squid (*Loligo*).

octopuses can be a savage affair. The exuberance of the copulatory embrace may result in the couple tearing at each other with their sharp beaks, or even strangulation of one partner by the other as the former's arms wrap around the mantle cavity of the latter, cutting off ventilation. In many octopuses (e.g., *Argonauta*, *Philonexis*) the tip of the hectocotylous arm may break off and remain in the female's mantle chamber.*

As the eggs pass through the oviduct, they are covered with a capsule-like membrane produced by the oviducal gland. Once in the mantle cavity, various kinds of **nidamental glands** may provide additional layers or coatings on the eggs. In the squid *Loligo*, which migrates to shallow water to breed, the nidamental glands coat the eggs within an oblong gelatinous mass, each containing about 100 eggs. The female holds these egg cases in her arms and fertilizes them with sperm ejected from her seminal receptacle. The egg masses harden as they react with sea water and are then attached to the substratum. The adults die after mating and egg laying. Cuttlefish deposit single eggs and attach them to seaweed or other substrata. Many open-ocean pelagic cephalopods have floating eggs, and the young develop entirely in the plankton. Octopuses usually lay grapelike egg clusters in rocky areas, and many species care for the developing embryos by protecting them and cleaning them by flushing the egg mass with jets of water. Octopuses and squids tend to grow quickly to maturity, reproduce, and then die. The pearly nautilus, however, is long-lived (perhaps to 25–30 years), slow growing, and able to reproduce for many years after maturity.

One of the most astonishing reproductive behaviors among invertebrates occurs in members of the pelagic

---

*The detached arm was mistakenly first described as a parasitic worm and given the genus name *Hectocotylus* (hence the origin of the term).

cephalopod genus *Argonauta*, known as the paper nautiluses. Female argonauts use two specialized arms to secrete and sculpt a beautiful, coiled, calcareous shell into which eggs are deposited (Figure 20.17B). The thin-walled, delicate shell is carried by the female and serves as her temporary home and as a brood chamber for the embryos. The much smaller male often cohabits the shell with the female.

Development

Development in molluscs is similar in many fundamental ways to that of the other protostomes (Figures 20.50–20.53). Most molluscs undergo typical spiral cleavage, with the mouth and stomodeum developing from the blastopore, and the anus forming as a new opening on the gastrula wall. Cell fates are typically spiralian, including a 4d mesentoblast. Beyond these generalities, a great deal of variation occurs in molluscan cleavage. As detailed studies are conducted on more and more species, the phylogenetic implications of these variations are being evaluated (see for example van den Biggelaar and Haszprunar 1996).

Development may be direct, mixed, or indirect. During indirect development, the free-swimming trochophore larva that develops is remarkably similar to that seen in annelids (Figure 20.50). Like the annelid larva, the molluscan trochophore bears an apical sensory plate with a tuft of cilia and a girdle of ciliated cells—the prototroch—just anterior to the mouth.

In some free-spawning molluscs (e.g., chitons), the trochophore is the only larval stage, and it metamorphoses directly into the juvenile (Figure 20.50C). But in many groups (e.g., gastropods and bivalves), the trochophore is followed by a uniquely molluscan larval stage called a **veliger** (Figure 20.51). The veliger larva may possess a foot, shell, operculum, and other adult-like structures. The most characteristic feature of the veliger larva is the swimming and feeding organ, or **velum**, which consists of two large ciliated lobes developed from the trochophore's prototroch. In some species the velum is subdivided into four, five, or even six separate lobes (Figure 20.51C). Veligers feed by capturing particulate food between opposed prototrochal and metatrochal bands of cilia on the edge of the velum. Some veligers apparently are nonfeeding. Eventually eyes and tentacles appear, and the veliger transforms into a juvenile, settles to the bottom, and assumes an adult existence.

Some bivalves have long-lived planktotrophic veligers, whereas others have short-lived lecithotrophic veligers. Many widely distributed species have very long larval lives that allow dispersal over great distances. A few bivalves have mixed development and brood the developing embryos in the suprabranchial cavity through the trochophore period; then the embryos are released as veliger larvae. Some marine and freshwater clams have direct development. Species in the freshwater family Sphaeriidae brood the embryos between the gill lamellae and shed juveniles into the water after development is completed. Several unrelated marine groups have independently evolved a similar brooding behavior (e.g., *Arca vivipara* and many members of the family Carditidae).

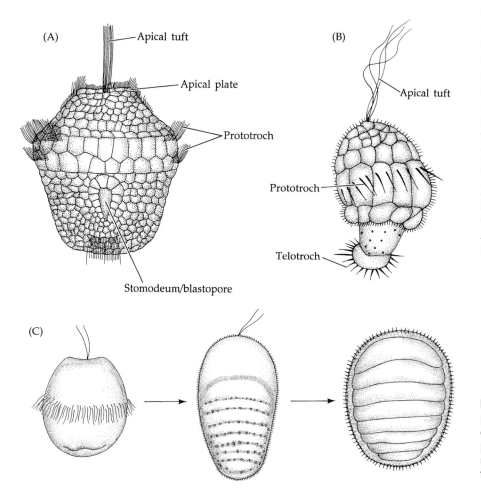

Figure 20.50 Molluscan trochophore larvae. (A) Generalized molluscan trochophore larva. (B) Trochophore of an aplacophoran. (C) Metamorphosis of a polyplacophoran from trochophore to juvenile.

In the freshwater groups Unionacea and Mutelacea, the embryos are also brooded between the gill lamellae, where they develop to the veliger stage. The veligers of these groups are often highly modified for a parasitic life on fishes, thereby facilitating dispersal. Various names have been given to these specialized parasitic veligers. In the Unionacea they are called **glochidia** (Figure 20.51E). They attach to the skin or gills of the host fish by a sticky mucus, hooks, or other attachment devices. Most glochidia lack a gut and absorb nutrients from the host by means of special phagocytic mantle cells. The host tissue often forms a cyst around the glochidium. Eventually the larva matures, breaks out of the cyst, drops to the bottom, and assumes its adult life.

Figure 20.51   Molluscan veliger larvae. (A,B) Side and front views of the veliger larva of a gastropod (*Crepidula*). (C) A prosobranch gastropod veliger with four velar lobes. (D) Generalized bivalve veliger. (E) Glochidium larva of a freshwater bivalve. (F) Late veliger of a scaphopod.

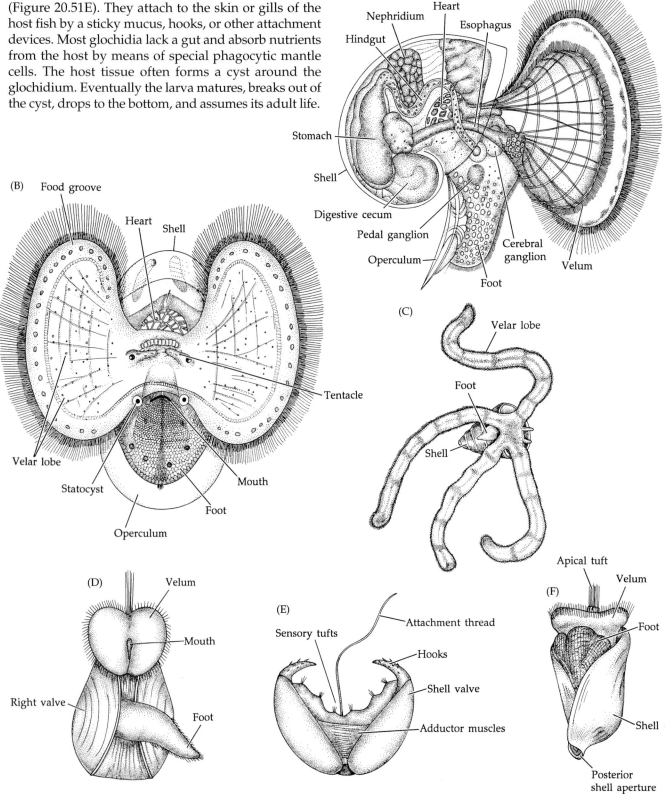

Among the gastropods, only the primitive archaeogastropods that rely on external fertilization have retained a free-swimming trochophore larva. All other gastropods suppress the trochophore or pass through it quickly before hatching. In many groups embryos hatch as veligers (e.g., opisthobranchs). Some of these gastropods have planktotrophic veligers that may have brief or extended (to several months) free-swimming lives. Others have lecithotrophic veligers that remain planktonic only for short periods. Planktotrophic veligers feed by use of the **velar cilia**, whose beating drives the animal forward and draws minute planktonic food particles into contact with the shorter cilia of a food groove. Once in the food groove, the particles are trapped in mucus and carried along ciliary tracts to the mouth.

Almost all pulmonates and many advanced marine prosobranchs (e.g., neogastropods) have direct development, and the veliger stage is passed in the egg case, or capsule. Upon hatching, tiny snails crawl out of the capsule into their adult habitat. In some neogastropods (e.g., certain species of *Nucella*), the encapsulated em-

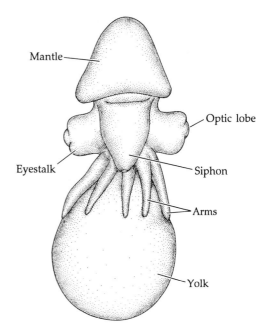

Figure 20.53 Juvenile cephalopod attached to and consuming its sac of yolk.

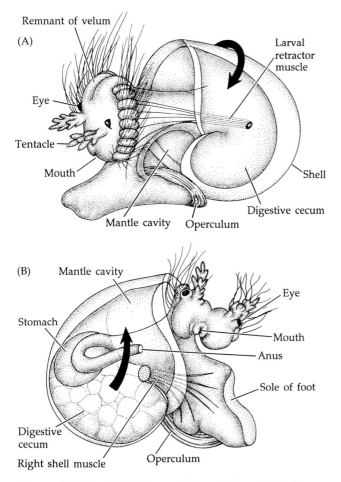

Figure 20.52 Settled larva of the abalone (*Haliotis*) undergoing torsion. (A) Left-side view after about 90° of torsion, with mantle cavity on the right side. (B) Torsion continues as the mantle cavity and its associated structures twist forward over the head.

bryos cannibalize on their siblings, a phenomenon called **adelphophagy**; consequently, only one or two juveniles eventually emerge from each capsule.

It is usually during the veliger stage that gastropods undergo torsion, when the shell and visceral mass twist relative to the head and foot (Figure 20.52). As we have seen, this phenomenon is still not fully understood, but it has played a major role in gastropod evolution.

Cephalopods produce large, yolky, telolecithal eggs. Development is always direct, the larval stages being lost entirely or passed within an egg case. Early cleavage is meroblastic and eventually produces a cap of cells (a discoblastula) at the animal pole. The embryo grows in such a way that the mouth opens to the yolk sac, and the yolk is directly "consumed" by the developing animal (Figure 20.53).

Molluscan Evolution and Phylogeny

The Mollusca is such a diverse phylum, and so many taxa below the class level are apparently artificial (i.e., polyphyletic or paraphyletic), that efforts to trace their evolutionary history have often led to frustration. Until fairly recently, many molluscan specialists entertained the idea of a "hypothetical ancestral mollusc" (affectionately known as HAM), the nature of which derived largely from early work of the eminent British biologist C. M. Yonge. Detailed and sometimes highly imaginative descriptions of this hypothetical ancestral mollusc were proposed by various workers, even including speculations on its ecology and behavior. The usefulness of HAM in molluscan phylogenetic analysis is now questioned by most specialists, as we have moved into an era of explicit phylogenetic analysis (i.e., cladistics). Thus, most modern workers attempt to avoid the pitfalls of a priori construction of a hypothetical ancestor, and instead analyze the evolutionary history of molluscs by phylogenetic inference (e.g., Lauterbach 1983; Wingstrand 1985; Scheltema 1988, 1996; Salvini-Plawen 1990). Although these studies differ in some details, the phylogenetic relationships resulting from their analyses have been very similar. Based on these recent studies, the probable molluscan common ancestor was small (~1 mm long), with a flattened ventral surface on which the animal moved by ciliary gliding (similar to a minute aplacophoran). Our own cladogram (see Figure 20.55) parallels current thinking on molluscan evolution. The characters used to construct the cladogram are enumerated in the figure legend and briefly summarized in the following discussion. The nodes on the cladogram have been lettered to facilitate the discussion.

Molluscs share most of their typical protostome features with the sipunculans, echiurans, and annelids: for example, spiral cleavage, schizocoely, and the trochophore larva. These are all symplesiomorphic characters for the phylum Mollusca. The most fundamental differences between the molluscs and annelids involve segmentation and circulation: annelids are segmented and have a well developed coelom and a closed circulatory system, whereas molluscs are unsegmented and have a reduced coelom and an open circulatory system. Perhaps the three most striking synapomorphies distinguishing modern molluscs from annelids and most other protostomes are the reduction of the coelom and the concomitant conversion of the closed circulatory system to an open hemocoelic one, the elaboration of the body wall into a mantle capable of secreting calcareous spicules or shell(s), and the unique molluscan radula.

Exactly where the molluscs arose within the protostome clade, and their kinship to other phyla, are still matters of much debate. While some workers treat them as descendant from a segmented ancestor, most do not. We support the idea that molluscs arose from a schizocoelomate, nonsegmented precursor. In fact, Scheltema (1993) provided evidence that the molluscs and sipunculans are sister groups (Chapter 24). By the end of the 64-cell stage, both groups show the distinctive **molluscan cross** formed by a group of apical micromeres ($1a^{12}$–$1d^{12}$ cells and their descendants, with cells $1a^{112}$–$1d^{112}$ forming the angle between the arms of the cross) (Figure 20.54). This configuration of blastomeres does not appear in other spiralian protostomes such as annelids and echiurans. Scheltema also suggested that certain features of the sipunculan pelagosphera larva may be homologous to some molluscan structures.

The major steps in the evolution of what we generally think of as a "typical" mollusc—that is, a shelled mollusc—took place after the origin of the aplacophorans, perhaps as molluscs adapted to active epibenthic

lifestyles. These steps centered largely on the elaboration of the mantle and mantle cavity, the refinement of the ventral surface as a well-developed muscular foot, and the evolution of a consolidated dorsal shell gland and solid shell(s) in place of independent calcareous spicules.

The description of a larval aplacophoran in 1890, in which the dorsal surface may have borne seven transverse bands of spicules (described as "composite plates," reminiscent of chitons), led some workers to postulate that aplacophorans and polyplacophorans might be sistergroups and that the multivalve condition was ancestral to the single-shell condition of more derived molluscs. A second aplacophoran was also described as having three such dorsal "plates," in which the spicules actually appeared to have fused. And, most recently a fossil from Silurian deposits in England (*Acaenoplax*) possibly an aplacophoran, has been described as having seven dorsal shell plates. However, there are fundamental differences between the shells of polyplacophorans and those of all other molluscs, an observation suggesting that the chitons stand alone as a unique radiation off the early molluscan line. Nonetheless, there seems to have been a tendency toward consolidation of shell glands and shell plates early in the radiation of the Mollusca.

So, three hypotheses have been offered to explain the "shell problem" in molluscan evolution: (1) The multiplate shell may have been ancestral, the single-shell de-

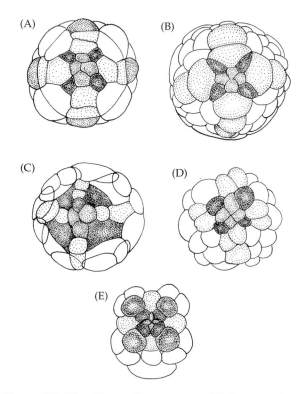

Figure 20.54   The molluscan cross: (A) Gastropoda (*Lymnaea*). (B) Polyplacophora (*Stenoplax*). (C) Sipuncula (*Golfingia*). (D) Aplacophora (*Epimenia*). The annelid cross: (E) Polychaeta (*Nereis*).

sign having evolved by coalescence of plates. (2) The single shell may have been ancestral, and the multiplate designs arose by subdivision of the single shell. (3) The single-shell and multishell designs arose independently from a shell-less ancestor, perhaps by way of shell gland consolidation. The presence of eight pairs of pedal retractor muscles in both polyplacophorans and monoplacophorans has been taken as evidence in favor of the first explanation. Acceptance of the first hypothesis suggests that the ancestor at node a (or b) in the cladogram in Figure 20.55 was a multivalved chiton-like creature. Acceptance of the second hypothesis implies that the ancestor at node a (or b) was a univalved, monoplacophoran-like ancestor. The third hypothesis postulates that the ancestor at node lacked a solid shell altogether.

The primitive, post-aplacophoran mantle and foot arrangement was probably similar to what we see today in living polyplacophorans or monoplacophorans—that is, a large flattened sole that was surrounded by a pallial groove and would have contributed to the evolutionary exploitation of epibenthic lifestyles. This primitive pallial arrangement was lost three times: in the aplacophoran line, the gastropod–cephalopod line, and the bivalve–scaphopod line. Secondary modifications on the shape of the foot and other features in bivalves and scaphopods allowed most of these animals to exploit infaunal life in soft sediments, and both of these taxa are highly adapted to sediment burrowing.

Monoplacophorans share the character of a single-plate (univalve) shell with all other molluscs above the level of the polyplacophorans in the cladogram. They also share a similar shell structure and a host of other features. The only synapomorphies defining the monoplacophorans seem to be their repetitive organs (multiple gills, nephridia, pedal muscles, gonads, and heart atria). The question of whether this multiplicity arose uniquely in the monoplacophorans or represents a symplesiomorphic retention of ancestral features from some unknown metameric ancestor (at node c, or lower, on the cladogram), has not been resolved.

The gastropod–cephalopod line is defined by the dorsal concentration of the viscera, elaboration of a separate and well-defined head, and restriction of the mantle cavity to the posterior region of the body. The bivalve–scaphopod line in the cladogram is defined by reduction of the head region, decentralization of the nervous system and associated reduction or loss of certain sensory structures, and expansion and deepening of the mantle cavity. The lateral mantle flaps, developing from the back of the scaphopod larva, grow down along the sides to enclose the body and are thought to be homologous with the left and right mantle sheets in bivalves. (Other synapomorphies defining the molluscan classes are listed in the legend to Figure 20.55.)

Cephalopods are highly specialized molluscs and possess a number of complex synapomorphies. Primitive shelled cephalopods are represented today by only six species of *Nautilus*, although over 17,000 fossil species of shelled nautiloid cephalopods have been described. This highly successful molluscan class probably arose about 450 million years ago. The nautiloids underwent a series of radiations during the Paleozoic, but were largely replaced by the ammonoids after the Devonian period (325 million years ago). The ammonoids, in turn, became extinct around the Cretaceous–Tertiary boundary (65 million years ago). The origin of the dibranchiate (coleoid) cephalopods (octopuses, squids, and cuttlefish) is obscure, possibly dating back to the Devonian. They diversified mainly in the Mesozoic and became a highly successful group by exploiting a very new lifestyle, as we have seen.

Other interpretations of molluscan phylogeny exist, of course. Some authors view the monoplacophorans and gastropods as sister-groups, primarily on the basis of similarities in shell structure.

The issue of ancestral metamerism in molluscs has been debated since the discovery of the first living monoplacophoran (*Neopilina galatheae*) in 1952. However, monoplacophorans are not the only molluscs to express serial replication, or to have repeated organs reminiscent of metamerism (or "pseudometamerism," as some prefer to call it). Polyplacophorans have a great many serially repeated gills in the pallial groove and also possess seven to eight pairs of pedal retractor muscles and seven to eight shell plates. The two pairs of heart atria, nephridia, and ctenidia in *Nautilus* (and two pairs of retractor muscles in some fossil forms) have also been regarded by some workers as primitive metameric features.

The question is whether or not organ repetition in these molluscs represents vestiges of a true, or fundamental, metamerism in the phylum. If so, they represent remnants of an ancestral metameric bauplan and may indicate a close relationship to other marine protostomes that display metamerism (e.g., the annelid–arthropod line). On the other hand, organ repetition in certain molluscan groups may be the result of independent convergent evolution and not a fundamental molluscan attribute at all. Absence of metamerism in the Aplacophora argue against fundamental metamerism as a primitive feature of the molluscs. And, nothing like the teloblastic metameric development of annelids and arthropods is seen in molluscs. The genetic/evolutionary potential for serial repetition of organs is not uncommon and occurs in other non-annelid/non-arthropod bilaterian phyla as well, e.g. Platyhelminthes, Nemertea, and Chordata.

The origin of molluscs themselves remains enigmatic. The excellent fossil record of this phylum extends back some 500 million years and suggests that the origin of the Mollusca probably lies in the Precambrian. Indeed, the late Precambrian fossil *Kimberella quadrata*, once thought to be a cnidarian, now appears to have decidedly molluscan features, including perhaps a shell and muscular foot. Molluscs are clearly allied with the

Figure 20.55 A cladogram depicting one view of the phylogeny of the Mollusca. The numbers on the clado- gram indicate suites of synapomorphies defining each hypothesized line or clade. Synapomorphies of the phylum Mollusca: (1) reduction of the coelom and development of an open hemocoelic circulatory system; (2) dorsal body wall forms a mantle; (3) extracellular production of cal- careous spicules (and ultimately a shell) by mantle shell glands; (4) ventral body wall muscles develop as muscular foot (or foot precursor); (5) radula; (6) chambered heart with separate atria and ventricle. Synapomorphies of the Aplacophora (Chaetodermomorpha + Neomeniomorpha) (defining node f): (56) vermiform body; (57) foot reduced; (58) posterior mantle cavity greatly reduced; (59) gonads empty into pericardial cavity, exiting to mantle cavity via U-shaped gametoducts; (60) without nephridia.

Synapomorphies of the Chaetodermomorpha: (7) cal- careous spicules of the body wall form imbricating scales; (8) complete loss of foot.

Synapomorphies of the Neomeniomorpha: (9) posterior end of reproductive system with copulatory spicules; (10) without ctenidia.

Synapomorphies of the shelled molluscs defining node b: (11) concentration of the diffuse shell glands into one or a few discrete glands, to produce solid shell(s); (12) development of creeping sole on large, muscular, ventral foot; (13) increase in gut complexity, with large mass of digestive ceca; (14) multiple pedal retractor muscles; (15) mobile radular membrane.

Synapomorphies of the class Polyplacophora: (16) unique shell with 7–8 plates (and with 7–8 shell gland regions), articulamentum layer, and aesthetes; (17) multi- ple gills (perhaps not homologous to the ctenidia of other

molluscs); (18) expanded and highly cuticularized mantle girdle that "fuses" with shell plates.

Synapomorphies of the sister-group to the Polyplaco- phora, defining node c: (19) preoral tentacles; (20) loss of calcareous spines in body wall; (21) presence of a single, well defined shell gland region and larval shell (proto- conch); (22) shell univalve, of a single piece (*note:* the bivalve shell is taken to be derived from the univalve condi- tion); (23) shell of the three-layered design (periostracum, prismatic layer, nacreous layer); (24) mantle margin of three parallel folds, each specialized for specific functions; (25) crystalline style; (26) statocysts.

Synapomorphies of the class Monoplacophora: (27) 3–6 pairs ctenidia; (28) 3–7 pairs nephridia; (29) 8 pairs pedal retractor muscles; (30) 2 pairs gonads; (31) 2 pairs heart atria.

Synapomorphies of the gastropod–cephalopod line, defining node d: (32) viscera concentrated dorsally; (33) shell coiling; (34) well developed, clearly demarcated head; (35) mantle cavity restricted to anal region. Synapomorphies of the class Gastropoda: (36) torsion and its associated anatomical conditions; (37) further concen- tration of internal organs as visceral hump.

Synapomorphies of the class Cephalopoda: (38) expan- sion of the coelom and closure of the circulatory system; (39) septate shell; (40) ink sac; (41) siphuncle; (42) beaklike jaws; (43) foot modified as prehensile arms/tentacles and funnel (=siphon); (44) extensive fusion of ganglia as brain.

Synapomorphies of the bivalve–scaphopod line, defining node e: (45) reduction of the head; (46) decentralization of the nervous system; (47) expansion of the mantle cavity to essentially surround the entire body; (48) modification of the foot to a more spatulate form. Synapomorphies of the class Bivalvia: (49) bivalve shell and its associated mantle and ctenidial modifications; (50) loss of radula; (51) byssus; (52) lateral compression of body. Synapomorphies of the class Scaphopoda: (53) tusk-shaped, open-ended shell; (54) loss of ctenidia; (55) captacula.

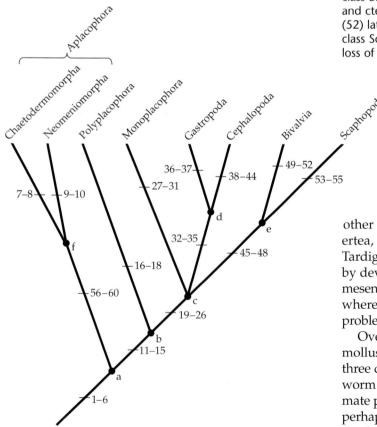

other spiralian protostomes (Platyhelminthes, Nem- ertea, Sipuncula, Echiura, Annelida, Onychophora, Tardigrada, Arthropoda, etc.), which are characterized by developmental features such as spiral cleavage, 4d mesentoblast, and trochophore-like larvae. But precisely where in the protostome lineage they arose remains problematic.

Over the years, numerous ideas on the origin of the molluscs have been proposed. Generally, these fall into three categories: molluscs were derived from (1) a flat- worm (turbellarian) ancestor, (2) a nonsegmented coelo- mate protostome ancestor, or (3) a segmented ancestor, perhaps even a common ancestor to the annelids. The first hypothesis, known as the "turbellarian theory," was originally based upon the supposed homology and similarity in mode of locomotion between molluscs and flatworms by means of a "ventral mucociliary gliding surface." It suggests that either the molluscs were the first coelomate protostomes, or that they share a common ancestor with the first coelomates. However, most contemporary workers believe that the large pericardial spaces present in primitive molluscs (e.g., aplacophorans, monoplacophorans, polyplacophorans) point to a coelomate rather than an acoelomate (turbellarian) ancestry, and the turbellarian theory enjoys little favor today.

The second theory has recently gained support in ontogenetic studies that suggest sipunculans and molluscs might be sister groups, sharing, among other things, the unique "molluscan cross."

The third hypothesis, that molluscs and annelids are closely related, implies that molluscs were primitively segmented coelomates (like the annelids). This idea has received some support from recent 18S rDNA gene sequence data. The hypothesis states that the protostomes comprise two major clades called "Ecdysozoa" (the so-called "molting" protostomes: nematodes, nematomorphans, kinorhynchs, priapulans, tardigrades, onychophorans, and arthropods) and "Eutrochozoa" (molluscs, annelids, and other spiralian phyla with trochophore-like larvae). Some workers include the lophophorate phyla in the latter group, calling it "Lophotrochozoa" (e.g., Halanych et al. 1995; Aguinaldo et al. 1997). These ideas are explored in more detail in Chapter 24.

Our phylogenetic analysis (Chapter 24) hypothesizes the molluscs to have arisen from the heart of the spiralian lineage, prior to the origin of segmentation in the annelids and arthropods. The close relationship between sipunculans and molluscs is supported by the unique embryological synapomorphy of the molluscan cross (page 62).

Selected References

The field of malacology is so large, has had such a long history, and has so embraced the mixed blessings of contributions from amateur shell collectors, that dealing with the literature is a daunting task. Many molluscs are of commercial importance (e.g., *Haliotis, Mytilus, Loligo*) and for these groups hundreds of studies appear annually; others are important laboratory/experimental organisms (e.g., *Loligo, Octopus, Aplysia*) and many papers are also published on these groups. New taxonomic monographs on various groups or geographical regions also appear each year, as do countless shell guides and coffee-table books. Distilling all of this into a small set of key references useful for entry into the professional literature is difficult; the list below is our attempt to do so.

General References

Beesley, P. L., G. J. B. Ross, and A. Wells (eds.). 1998. *Mollusca: The Southern Synthesis. Fauna of Australia. Vol. 5.* CSIRO Publishing, Melbourne, Australia. [Perhaps the best general review of molluscan biology and systematics available. An extraordinary 2-volume text with chapters by leading specialists.]

Boss, K. J. 1982. Mollusca. *In* S. P. Parker (ed.), *Synopsis and Classification of Living Organisms*, Vol. 1. McGraw-Hill, New York, pp. 945–1166.

Cheng, T. C. 1967. Marine mollusks as hosts for symbioses. Adv. Mar. Biol. 5: 1–424. [An exhaustive summary.]

Falini, G., S. Albeck, S. Weiner and A. Addadi. 1996. Control of aragonite or calcite polymorphism by mollusc shell macromolecules. Science 271: 67–69.

Florkin, M. and B. T. Scheer (eds.). 1972. *Chemical Zoology,* Vol. 7, Mollusca. Academic Press, New York. [Includes a good overview of the basic molluscan bauplan.]

Fretter, V. (ed.). 1968. *Studies in the Structure, Physiology and Ecology of Mollusks.* Academic Press, London.

Giese, A. C. and J. S. Pearse (eds.). 1977. *Reproduction of Marine Invertebrates, Vol. 4, Molluscs: Gastropods and Cephalopods.* Academic Press, New York.

Giese, A. C. and J. S. Pearse (eds.). 1979. *Reproduction of Marine Invertebrates, Vol. 5, Molluscs: Pelecypods and Lesser Classes.* Academic Press, New York.

Götting, K.-J. 1974. *Malakozoologie.* G. Fischer, Stuttgart.

Grassé, P. (ed.). 1968. *Traité de Zoologie,* Vol. 5 (pts. 2 and 3). Masson et Cie, Paris.

Harrison, F. W. and A. J. Kohn (eds). 1994. *Microscopic Anatomy of Invertebrates.* Vols. 5 and 6. Mollusca. Wiley-Liss, New York.

Heller, J. 1990. Longevity in molluscs. Malacologia 31(2): 259–295.

Hochachka, P. W. (ed.). 1983. *The Mollusca,* Vol. 1, Metabolic Biochemistry and Molecular Biomechanics; Vol. 2, Environmental Biochemistry and Physiology. Academic Press, New York.

Hyman, L. H. 1967. *The Invertebrates,* Vol. 6, Mollusca I. Aplacophora, Polyplacophora, Monoplacophora, Gastropoda. The Coelomate Bilateria. McGraw-Hill, New York. [Still one of the best general surveys of molluscan anatomy in English.]

Jones, A. M. and J. M. Baxter. 1987. *Molluscs: Caudofoveata, Solenogastres, Polyplacophora and Scaphopoda.* Synopses of the British Fauna (new ser.). Linnean Society, London.

Jørgensen, C. B. 1966. *Biology of Suspension Feeding.* Pergamon Press, New York. [Much of this classic treatise deals with molluscs.]

Keen, A. M. and E. Coan. 1974. *Marine Molluscan Genera of Western North America: An Illustrated Key.* 2nd Ed. Stanford University Press, Stanford, California. [Technical keys.]

Kniprath, E. 1981. Ontogeny of the molluscan shell field. Zool. Scripta 10: 61–79.

Moore, R. C (ed.). 1957–71. *Treatise on Invertebrate Paleontology. Mollusca,* Parts 1–6 (Vols. I–N). University of Kansas Press and Geological Society of America, Lawrence, KA.

Potts, W. T. W. 1967. Excretion in molluscs. Biol. Rev. 42: 1–41.

Purchon, R. D. 1977. *The Biology of the Mollusca,* 2nd Ed. Pergamon Press, New York.

Raven, C. P. 1958. *Morphogenesis: The Analysis of Molluscan Development.* Pergamon Press, New York. [Review of molluscan embryology.]

Jones, D. S. 1983. Sclerochronology: Reading the record of the molluscan shell. Am. Sci. 71: 384–391.

Salvini-Plawen, L. von. 1980. Proposed classification of Mollusca. Malacologia 19: 249–278.

Spanier, E. 1986. Cannibalism in muricid snails as a possible explanation for archaeological findings. J. Archaeological Sci. 13: 463–468.

Spanier, E. (ed.). 1987. *The Royal Purple and Biblical Blue. Argaman and Tekhelet. The Study of Chief Rabbi Dr. Isaac Herzog on the Dye Industries in Ancient Israel and Recent Scientific Contributions.* Keter Publishing House, Jerusalem.

Vaught, K. C. 1989. *A Classification of the Living Mollusca.* Edited by R. T. Abbott and K. J. Boss. American Malacologists, Melbourne, Florida.

Wilbur, K. M. (gen. ed.). 1983–present. *The Mollusca.* Vols. 1–12. Academic Press, New York. [A continuing series of volumes edited and authored by luminaries in the field of malacology; volumes 1–10 cover biochemistry, molecular biology, biomechanics, metabolism, physiology, ecology, development, and evolution.]

Wilbur, K. M. and C. M. Yonge (eds.). 1964, 1966. *Physiology of Mollusca*, Vols. 1 and 2. Academic Press, New York.

Yonge, C. M. 1932. The crystalline style of the Mollusca. Sci. Prog. 26: 643–653.

Yonge, C. M. and T. E. Thompson. 1976. *Living Marine Molluscs*. Collins, London. [An excellent review.]

Aplacophora

Baba, K. 1951. General sketch of the development in a solenogaster, *Epimenia verrucosa* (Nierstr.). Misc. Rep. Res. Inst. Nat. Resour. Tokyo 19/21: 38–46.

Scheltema, A. H. 1981. Comparative morphology of the radulae and the alimentary tracts in the Aplacophora. Malacologia 20: 361–383.

Scheltema, A. and M. Jebb. 1994. Natural history of a solenogaster mollusc from Papua New Guinea, *Epimenia australis* (Thiele) (Aplacophora: Neomeniomorpha). J. Nat. Hist. 28: 1297–1318.

Scheltema, A. H. and C. Schander. 2000. Discrimination and phylogeny of solenogaster species through the morphology of hard parts (Mollusca, Aplacophora, Neomeniomorpha). Biol. Bull. 198: 121–151.

Scheltema, A. H., M. Tscherkassky and A. M. Kuzirian. 1994. Aplacophora. *In* F. W. Harrison and E. E. Ruppert (eds.), *Microscopic Anatomy of Invertebrates*. Vol. 5, Mollusca I. Wiley-Liss, New York. pp. 13–54.

Monoplacophora

Clarke, A. H. and R. J. Menzies. 1959. *Neopilina (Vema) ewingi*, a second living species of the Paleozoic class Monoplacophora. Science 129: 1026–1027.

Lemche, H. 1957. A new living deep-sea mollusc of the Cambro-Devonian class Monoplacophora. Nature 179: 413–416. [Report of the first discovery of living monoplacophorans.]

Lemche, H. and K. G. Wingstrand. 1959. The anatomy of *Neopilina galatheae* Lemche, 1957. Galathea Rpt. 3: 9–71.

Menzies, R. J. and W. Layton. 1962. A new species of monoplacophoran mollusc, *Neopilina (Neopilina) veleronis* from the slope of the Cedros Trench, Mexico. Ann. Mag. Nat. Hist. (13)5: 401–406.

Warén, A. 1988. *Neopilina goesi*, a new Caribbean monoplacophoran mollusk dredged in 1869. Proc. Biol. Soc. Wash. 101: 676–681.

Wingstrand, K. G. 1985. On the anatomy and relationships of recent Monoplacophora. Galathea Rpt. 16: 7–94.

Polyplacophora

Boyle, P. R. 1977. The physiology and behavior of chitons. Ann. Rev. Oceanogr. Mar. Biol. 15: 461–509.

Eernisse, D. J. 1988. Reproductive patterns in six species of *Lepidochitona* (Mollusca: Polyplacophora) from the Pacific Coast of North America. Biol. Bull. 174(3): 287- 302.

Fisher, v.-F. P. 1978. Photoreceptor cells in chiton esthetes. Spixiana 1: 209–213.

Hulings, N. C. 1991. Activity patterns and homing of *Acanthopleura gemmata* (Blainville, 1825) (Mollusca: Polyplacophora) in the rocky intertidal of the Jordan Gulf of Aquaba. The Nautilus 105(1): 16–25.

Kass, P. and R. A. VanBelle. 1985–1994. *Monograph of Living Chitons*, Vols. 1–5. E. J. Brill, Leiden. [A multivolume treatise.]

Kass, P. and R. A. VanBelle. 1998. *Catalogue of Living Chitons (mollusca; Polyplacophora)*. 2nd ed. W. Backhuys, Rotterdam.

Nesson, M. H. and H. A. Lowenstam. 1985. Biomineralization processes of the radula teeth of chitons. *In* J. L. Kirschvink et al. (eds.), *Magnetite Biomineralization and Magnetoreception in Organisms*. Plenum, New York, pp. 333–363.

Slieker, F. J. A. 2000. *Chitons of the World. An Illustrated Synopsis of Recent Polyplacophora*. L'Informatore Piceno, Ancona, Italy. 154 pp.

Smith, A. G. 1966. The larval development of chitons (Amphineura). Proc. Calif. Acad. Sci. 32: 433–446.

Gastropoda

Barnhart, M. C. 1986. Respiratory gas tensions and gas exchange in active and dormant land snails, *Otala lactea*. Physiol. Zool. 59: 733–745.

Bouchet, P. 1989. A marginellid gastropod parasitizes sleeping fishes. Bull. Mar. Sci. 45(1): 76–84.

Brace, R. C. 1977. Anatomical changes in nervous and vascular systems during the transition from prosobranch to opisthobranch organization. Trans. Zool. Soc. London 34: 1–26.

Branch, G. M. 1981. The biology of limpets: Physical factors, energy flow and ecological interactions. Ann. Rev. Oceanogr. Mar. Biol. 19: 235 380.

Brunkhorst, D. J. 1991. Do phyllidiid nudibranchs demonstrate behaviour consistent with their apparent warning colorations? Some field observations. J. Moll. Stud. 57: 481–489.

Carlton, J. T., G. J. Vermeij, D. R. Lindberg, D. A. Carlton and E. C. Dudley. 1991. The first historical extinction of a marine invertebrate in an ocean basin: The demise of the eelgrass limpet *Lottia alveus*. Biol. Bull. 180: 72–80.

Carriker, M. R. and D. Van Zandt. 1972. Predatory behavior of a shell-boring muricid gastropod. In H. Winn and B. Olla (eds.), *Behavior of Marine Animals*, Vol. 1. Plenum, New York.

Collier, J. R. 1997. Gastropods. The snails. *In* S. F. Gilbert and A. M. Raunio, *Embryology. Constructing the Organism*. Sinauer Assoc., Sunderland, MA. pp. 189–218.

Conklin, E. J. and R. N. Mariscal. 1977. Feeding behavior, ceras structure, and nematocyst storage in the aeolid nudibranch, *Spurilla neapolitana*. Bull. Mar. Sci. 27: 658–667.

Connor, V. M. 1986. The use of mucous trails by intertidal limpets to enhance food resources. Biol. Bull. 171: 548–564.

Cook, S. B. 1971. A study in homing behavior in the limpet *Siphonaria alternata*. Biol. Bull. 141: 449–457.

Croll, R. P. 1983. Gastropod chemoreception. Biol. Rev. 58: 293–319.

Fretter, V. 1967. The prosobranch veliger. Proc. Malac. Soc. London 37: 357–366.

Fretter, V. and M. A. Graham. 1962. *British Prosobranch Molluscs. Their Functional Anatomy and Ecology*. Ray Society, London.

Fretter, V. and J. Peake (eds.). 1975, 1978. *Pulmonates*, Vol. 1, *Functional Anatomy and Physiology*; Vol. 2A, *Systematics, Evolution and Ecology*. Academic Press, New York.

Fursich, F. T. and D. Jablonski. 1984. Late Triassic naticid drill holes: Carnivorous gastropods gain a major adaptation but fail to radiate. Science 224: 78–80.

Gaffney, P. M. and B. McGee. 1992. Multiple paternity in *Crepidula fornicata* (Linnaeus). The Veliger 35(1): 12–15.

Gilmer, R. W. and G. R. Harbison. 1986. Morphology and field behavior of pteropod molluscs: Feeding methods in the families Cavoliniidae, Limacinidae and Peraclididae (Gastropoda: Thecosomata). Mar. Biol. 91: 47–57.

Gosliner, T. M. 1994. Gastropoda: Opisthobranchia. *In* F. E. Harrison and A. J. Kohn (eds.), *Microscopic Anatomy of Invertebrates*. Wiley-Liss, New York. pp. 253–355.

Gould, S. J. 1985. The consequences of being different: Sinistral coiling in *Cerion*. Evolution 39: 1364–1379.

Greenwood, P. G. and R. N. Mariscal. 1984. Immature nematocyst incorporation by the aeolid nudibranch *Spurilla neapolitana*. Mar. Biol. 80: 35–38.

Havenhand, J. N. 1991. On the behaviour of opisthobranch larvae. J. Moll. Stud. 57: 119–131.

Haszprunar, G. 1985a. The fine morphology of the osphradial sense organs of the Mollusca. I. Gastropoda, Prosobranchia. Phil. Trans. Roy. Soc. Lond. B 307: 457–496.

Haszprunar, G. 1985b. The Heterobranchia: New concept of the phylogeny of the higher Gastropoda. Z. Zool. Syst. Evolutionsforsch. 23: 15–37.

Haszprunar, G. 1988. On the origin and evolution of major gastropod groups, with special reference to the Streptoneura. J. Moll. Stud. 54: 367–441.

Haszprunar, G. 1989. Die Torsion der Gastropoda — ein biomechanischer Prozess. Z. Zool. Syst. Evolutionsforsch. 27: 1–7.

Hickman, C. S. 1983. Radular patterns, systematics, diversity and ecology of deep-sea limpets. Veliger 26: 7–92.

Hickman, C. S. 1984. Implications of radular tooth-row functional integration for archaeogastropod systematics. Malacologia 25(1): 143–160.

Hickman, C. S. 1992. Reproduction and development of trochean gastropods. Veliger 35: 245–272.

Kempf, S. C. 1984. Symbiosis between the zooxanthella *Symbiodinium* (= *Gymnodinium*) microadriaticum (Freudenthal) and four species of nudibranchs. Biol. Bull. 166: 110–126.

Kempf, S. C. 1991. A "primitive" symbiosis between the aeolid nudibranch *Berghia verrucicornis* (A. Costa, 1867) and a zooxanthella. J. Mol. Stud. 57: 75–85.

Linsley, R. M. 1978. Shell formation and the evolution of gastropods. Am. Sci. 66: 432–441.

Lohmann, K. J. and A. O. D. Willows. 1987. Lunar-modulated geomagnetic orientation by a marine mollusk. Science 235: 331–334.

Marcus, E. and E. Marcus. 1967. *American Opisthobranch Mollusks. Studies in Tropical Oceanography Series,* No. 6. University of Miami Press, Coral Gables, Florida. [Although badly out of date, this monograph remains a benchmark work for the American fauna.]

Marín, A and J. Ros. 1991. Presence of intracellular zooxanthellae in Mediterranean nudibranchs. J. Moll. Stud. 57: 87–101.

McDonald, G. and J. Nybakken. 1991. A preliminary report on a world-wide review of the food of nudibranchs. J. Moll. Stud. 57: 61–63.

Miller, S. L. 1974a. Adaptive design of locomotion and foot form in prosobranch gastropods. J. Exp. Mar. Biol. Ecol 14: 99–156.

Miller, S. L. 1974b. The classification, taxonomic distribution, and evolution of locomotor types among prosobranch gastropods. Proc. Malacol. Soc. London 41: 233–272.

Norton, S. F. 1988. Role of the gastropod shell and operculum in inhibiting predation by fishes. Science 241: 92–94.

Olivera, B. M. et al. 1985. Peptide neurotoxins from fish-hunting cone snails. Science 230: 1338–1343.

O'Sullivan, J. B., R. R. McConnaughey and M. E. Huber. 1987. A bloodsucking snail: The cooper's nutmeg, *Cancellaria cooperi* Gabb, parasitizes the California electric ray, *Torpedo californica* Ayres. Biol. Bull. 172: 362–366.

Pawlik, J. R., M. R. Kernan, T. F. Molinski, M. K. Harper and D. J. Faulkner. 1988. Defensive chemicals of the Spanish dancer nudibranch *Hexabranchus sanguineus* and its egg ribbons: Macrolides derived from a sponge diet. J. Exp. Mar. Biol. Ecol. 119: 99–109.

Pennington, J. T. and F. Chia. 1985. Gastropod torsion: A test of Garstang's hypothesis. Biol. Bull. 169: 391–396.

Perry, D. M. 1985. Function of the shell spine in the predaceous rocky intertidal snail *Acanthina spirata* (Prosobranchia: Muricacea). Mar. Biol. 88: 51–58.

Ponder, W. F. 1973. The origin and evolution of the Neogastropoda. Malacologia 12: 295–338.

Ponder, W. F. (ed.). 1988. Prosobranch phylogeny. Malacol. Rev., Supp. 4.

Potts, G. W. 1981. The anatomy of respiratory structures in the dorid nudibranchs, *Onchidoris bilamellata* and *Archidoris pseudoargus*, with details of the epidermal glands. J. Mar. Biol. Assoc. U.K. 61: 959–982.

Rudman, W. B. 1991. Purpose in pattern: The evolution of colour in chromodorid nudibranchs. J. Moll. Stud. 57: 5–21.

Runham, N. W. and P. J. Hunter. 1970. *Terrestrial Slugs.* Hutchinson University Library, London. [A lovely review of slug biology.]

Salvini-Plawen, L. von and G. Haszprunar. 1986. The Vetigastropoda and the systematics of streptoneurous Gastropoda. J. Zool. 211: 747–770.

Scheltema, R. S. 1989. Planktonic and non-planktonic development among prosobranch gastropods and its relationship to the geographic range of species. *In* J. S. Ryland and P. A. Tyler (eds.), *Reproduction, Genetics and Distribution of Marine Organisms.* Olsen and Olsen, Denmark. pp. 183–188.

Seapy, R. and R. E. Young. 1986. Concealment in epipelagic pterotracheid heteropods (Gastropoda) and cranchiid squids (Cephalopoda). J. Zool. 210: 137–147.

Sleeper, H. L., V. J. Paul and W. Fenical. 1980. Alarm pheromones from the marine opisthobranch *Navanax inermis*. J. Chem. Ecol. 6: 57–70.

Stanley, S. M. 1982. Gastropod torsion: Predation and the opercular imperative. Neues. Jahrb. Geol. Palaeontol. Abh. 164: 95–107. [Succinct review of ideas on why and how torsion evolved.]

Tardy, J. 1991. Types of opisthobranch veligers: Their notum formation and torsion. J. Moll. Stud. 57: 103–112.

Taylor, J. (ed.) 1996. *Origin and Evolutionary Radiation of the Mollusca.* Oxford Univ. Press, Oxford.

Taylor, J. D., N. J. Morris and C. N. Taylor. 1980. Food specialization and the evolution of predatory prosobranch gastropods. Paleontology 23: 375–410.

Thiriot-Quievreux, C. 1973. Heteropoda. Ann. Rev. Oceanogr. Mar. Biol. 11: 237–261.

Thompson, T. E. 1976. *Biology of Opisthobranch Molluscs,* Vol. 1. Ray Society, London.

Thompson, T. E. and G. H. Brown. 1976. *British Opisthobranch Molluscs.* Academic Press, London.

Thompson, T. E. and G. H. Brown. 1984. *Biology of Opisthobranch Molluscs,* Vol. 2. Ray Society, London.

van den Biggelaar, J. A. M. and G. Haszprunar. 1996. Cleavage patterns and mesentoblast formation in the Gastropoda: An evolutionary perspective. Evolution 50: 1520–1540.

Voltzow, J. and R. Collin. 1995. Flow through mantle cavities revisited: Was sanitation the key to fissurellid evolution? Invert. Biol. 114(2): 145–150.

Bivalvia

Ansell, A. D. and N. B. Nair. 1969. A comparison of bivalve boring mechanisms by mechanical means. Am. Zool. 9: 857–868.

Bayne, B. L. (ed.). 1976. *Marine Mussels: Their Ecology and Physiology.* Cambridge University Press, Cambridge.

Beninger, P. G., S. St-Jean, Y. Poussart and J. E. Ward. 1993. Gill function and mucocyte distribution in *Placopecten magellanicus* and *Mytilus edulis* (Mollusca: Bivalvia): The role of mucus in particle transport. Mar. Ecol. Prog. Ser. 98: 275–282.

Boulding, E. G. 1984. Crab-resistant features of shells of burrowing bivalves: Decreasing vulnerability by increasing handling time. J. Exp. Mar. Biol. Ecol. 76: 201–223.

Carpenter, E. J. and J. L. Culliney. 1975. Nitrogen fixation in marine shipworms. Science 187: 551–552.

Childress J. J. et al. 1986. A methanotrophic marine molluscan (Bivalvia, Mytilidae) symbiosis: Mussels fueled by gas. Science 233: 1306–1308.

Deaton, L. E. 1981. Ion regulation in freshwater and brackish water bivalve mollusks. Physiol. Zool. 54: 109–121.

Ellis, A. E. 1978. *British Freshwater Bivalve Mollusks.* Academic Press, London.

Foster-Smith, R. L. 1978. The function of the pallial organs of bivalves in controlling ingestion. J. Moll. Stud. 44: 83–99.

Goreau, T. F., N. I. Goreau and C. M. Yonge. 1973. On the utilization of photosynthetic products from zooxanthellae and of a dissolved amino acid in *Tridacna maxima*. J. Zool. 169: 417–454.

Jørgensen, C. B. 1974. On gill function in the mussel *Mytilus edulis*. Ophelia 13: 187–232.

Judd, W. 1979. The secretions and fine structure of bivalve crystalline style sacs. Ophelia 18: 205–234.

Kennedy, W. J., J. D. Taylor and A. Hall. 1969. Environmental and biological controls on bivalve shell mineralogy. Biol. Rev. 44: 499–530.

Kristensen, J. H. 1972. Structure and function of crystalline styles in bivalves. Ophelia 10: 91–108.

Manahan, D. T., S. H. Wright, G. C. Stephens and M. A. Rice. 1982. Transport of dissolved amino acids by the mussel, *Mytilus edulis*: Demonstration of net uptake from natural seawater. Science 215: 1253–1255.

Marincovich, L., Jr. 1975. Morphology and mode of life of the Late Cretaceous rudist, *Coralliochama orcutti* White (Mollusca: Bivalvia). J. Paleontol. 49(1): 212–223. [Describes the famous Point Banda deposits of Baja California, Mexico.]

Meyhöfer, E. and M. P. Morse. 1996. Characterization of the bivalve ultrafiltration system in *Mytilus edulis*, *Chlamys hastata*, and *Mercenaria mercenaria*. Invert. Biol. 115(1): 20–29.

Morse, M. P., E. Meyhöfer, J. J. Otto and A. M. Kuzirian. 1986. Hemocyanin respiratory pigments in bivalve mollusks. Science 231: 1302–1304.

Morton, B. 1978a. The diurnal rhythm and the processes of feeding and digestion in *Tridacna crocea*. J. Zool. 185: 371–387.

Morton, B. 1978b. Feeding and digestion in shipworms. Ann. Rev. Oceanogr. Mar. Biol. 16: 107–144.

Owen, G. 1974. Feeding and digestion in the Bivalvia. Adv. Comp. Physiol. Biochem. 5: 1–35.

Pojeta, J., Jr. and B. Runnegar. 1974. *Fordilla troyensis* and the early history of pelecypod mollusks. Am. Sci. 62: 706–711.

Powell, M. A. and G. N. Somero. 1986. Hydrogen sulfide oxidation is coupled to oxidative phosphorylation in mitochondria of *Solemya reidi*. Science 233: 563–566.

Reid, R. G. B. and F. R. Bernard. 1980. Gutless bivalves. Science 208: 609–610.

Reid, R. G. B. and A. M. Reid. 1974. The carnivorous habit of members of the septibranch genus *Cuspidaria* (Mollusca: Bivalvia). Sarsia 56: 47–56.

Stanley, S. M. 1968. Post-Paleozoic adaptive radiation of infaunal bivalve molluscs — a consequence of mantle fusion and siphon formation. J. Paleontol. 42: 214–229.

Stanley, S. M. 1970. Relation of shell form to life habits of the Bivalvia (Mollusca). Geol. Soc. Am. Mem. 125: 1–296.

Stanley, S. M. 1975. Why clams have the shape they have: An experimental analysis of burrowing. Paleobiology 1: 48.

Taylor, J. D. 1973. The structural evolution of the bivalve shell. Paleontology 16: 519–534.

Trueman, E. R. 1966. Bivalve mollusks: Fluid dynamics of burrowing. Science 152: 523–525.

Vetter, R. D. 1985. Elemental sulfur in the gills of three species of clams containing chemoautotrophic symbiotic bacteria: A possible inorganic energy storage compound. Mar. Biol. 88: 33–42.

Vogel, K. and W. F. Gutmann. 1980. The derivation of pelecypods: Role of biomechanics, physiology, and environment. Lethaia 13: 269–275.

Waterbury, J. B., C. B. Calloway and R. D. Turner. 1983. A cellulolytic nitrogen-fixing bacterium cultured from the Gland of Deshayes in shipworms (Bivalvia: Teredinidae). Science 221: 1401–1403.

Wilkens, L. A. 1986. The visual system of the giant clam *Tridacna*: Behavioral adaptations. Biol. Bull. 170: 393–408.

Yonge, C. M. 1953. The monomyarian condition in the Lamellibranchia. Trans. R. Soc. Edinburgh 62 (p. II): 443–478.

Yonge, C. M. 1973. Giant clams. Sci. Am. 232: 96–105.

Zwarts, L. and J. Wanink. 1989. Siphon size and burying depths in deposit- and suspension-feeding benthic bivalves. Mar. Biol. 100: 227–240.

Scaphopoda

Bilyard, G. R. 1974. The feeding habits and ecology of *Dentalium stimpsoni*. Veliger 17: 126–138.

Gainey, L. F. 1972. The use of the foot and captacula in the feeding of *Dentalium*. Veliger 15: 29–34.

Trueman, E. R. 1968. The burrowing process of *Dentalium*. J. Zool. 154: 19–27.

Cephalopoda

Aronson, R. B. 1991. Ecology, paleobiology and evolutionary constraint in the octopus. Bull. Mar. Sci. 49(1–2): 245–255.

Barber, V. C. and F. Grazialdei. 1967. The fine structure of cephalopod blood vessels. Z. Zellforsch. Mikrosk. Anat. 77: 162–174. [Also see earlier papers by these authors in the same journal.]

Boycott, B. B. 1965. Learning in the octopus. Sci. Am. 212: 42–50.

Boyle, P. R. (ed.). 1983. *Cephalopod Life Cycles*, Vols. 1–2. Academic Press, New York.

Clarke, M. A. 1966. A review of the systematics and ecology of oceanic squids. Adv. Mar. Biol. 4: 91–300.

Cloney, R. A. and S. L. Brocco. 1983. Chromatophore organs, reflector cells, irridocytes and leucophores in cephalopods. Am. Zool. 23: 581 592.

Denton, E. J. and J. B. Gilpin-Brown. 1973. Flotation mechanisms in modern and fossil cephalopods. Adv. Mar. Biol. 11: 197–264.

Donovan, D. T. 1964. Cephalopod phylogeny and classification. Biol. Rev. 39: 259–287.

Fields, W. G. 1965. The structure, development, food relations, reproduction, and life history of the squid *Loligo opalescens* Berry. Calif. Dept. Fish Game Bull. 131: 1–108.

Fiorito, G. and P. Scotto. 1992. Observational learning in *Octopus vulgaris*. Science 256: 545–547.

Hanlon, R. T. and J. B. Messenger. 1996. *Cephalopod Behavior*. Cambridge Univ. Press.

Hochberg, F. G. 1983. The parasites of cephalopods: A review. Mem. Nat. Mus. Victoria Melbourne 44: 109–145.

House, M. R. and J. R. Senior. 1981. *The Ammonoidea: The Evolution, Classification, Mode of Life and Geological Usefulness of a Major Fossil Group*. Academic Press, New York.

Kier, W. M. 1991. Squid cross-striated muscle: The evolution of a specialized muscle fiber type. Bull. Mar. Sci. 49(1–2): 389–403.

Kier, W. M. and A. M. Smith. 1990. The morphology and mechanics of octopus suckers. Biol. Bull. 178: 126–136.

Lehmann, U. 1981. *The Ammonites: Their Life and Their World*. Cambridge University Press, New York.

McFell-Ngai, M. and M. K. Montgomery. 1990. The anatomy and morphology of the adult bacterial light organ of *Euprymna scolopes* Berry (Cephalopoda: Sepiolidae). Biol. Bull. 179: 332–339.

Mutvei, H. 1964. On the shells of *Nautilus* and *Spirula* with notes on the shell secretion in non-cephalopod molluscs. Ark. Zool. 16(14): 223–278.

Nixon, M. and J. B. Messenger (eds.). 1977. *The Biology of Cephalopods*. Academic Press, New York.

Packard, A. 1972. Cephalopods and fish: The limits of convergence. Biol. Rev. 47: 241–307.

Roper, C. F. E. and K. J. Boss. 1982. The giant squid. Sci. Am. 246: 96–104.

Salvini-Plawen, L. von. 1990. Origin, phylogeny and classification of the phylum mollusca. Iberus 9: 1–33.

Saunders, W. B. 1983. Natural rates of growth and longevity of *Nautilus belauensis*. Paleobiology 9: 280–288.

Saunders, W. B., R. L. Knight and P. N. Bond. 1991. *Octopus* predation on *Nautilus*: Evidence from Papua New Guinea. Bull. Mar. Sci. 49(1–2): 280–287.

Sweeney, M. J., C. F. E. Roper, K. A. M. Mangold, M. R. Clarke and S. V. Boletzky (eds.). 1992. "Larval" and juvenile cephalopods: a manual for their identification. Smithsonian Contrb. Zool. 513: 1–282.

Ward, P. D. 1987. *The Natural History of Nautilus*. Allen Press.

Ward, P. D. and W. B. Saunders. 1997. *Allonautilus*, a new genus of living nautiloid cephalopod and its bearing on phylogeny of the Nautilida. J. Paleontol. 71: 1054–1064.

Ward, R., L. Greenwald, and O. E. Greenwald. 1980. The buoyancy of the chambered nautilus. Sci. Am. 243: 190–204.

Wells, M. J. 1978. *Octopus: Physiology and Behaviour of an Advanced Invertebrate*. Chapman and Hall, New York.

Wells, M. J. and R. K. O'Dor. 1991. Jet propulsion and the evolution of cephalopods. Bull. Mar. Sci. 49(1–2): 419–432.

Yarnall, J. L. 1969. Aspects of the behavior of *Octopus cyanea* Gray. Anim. Behav. 17: 747–754.

Young, A. E. and C. F. Roper. 1976. Bioluminescent countershading in midwater animals: Evidence from living squid. Science 191: 1046–1048.

Young, J. Z. 1972. *The Anatomy of the Nervous System of Octopus vulgaris*. Oxford University Press, New York.

Young, R. E. and F. M. Mencher. 1980. Bioluminescence in mesopelagic squid: Diel color change during counterillumination. Science 208: 1286–1288.

Molluscan Evolution and Phylogeny

Aguinaldo, A. M. A., J. M. Turbeville, L. S. Linford, M. C. Rivera, J. R. Garey, R. A. Raff and J. A. Lake. 1997. Evidence for a clade of nematodes, arthropods and other moulting animals. Nature 387: 489–493.

Batten, R. L., H. B. Rollins and S. J. Gould. 1967. Comments on "The adaptive significance of gastropod torsion." Evolution 21: 405–406.

Beedham, G. and E. Trueman. 1968. The cuticle of the Aplacophora and the evolutionary significance in the Mollusca. Proc. Zool. Soc. London 154: 443–451.

Bieler, R. 1992. Gastropod phylogeny and systematics. Ann. Rev. Ecol. Syst. 23: 311–338.

Eldredge, N. and S. M. Stanley (eds.). 1984. *Living Fossils*. Springer-Verlag, New York. [Includes chapters on Monoplacophora, Pleurotomaria, and Nautilus.]

Fedonkin, M. A. and B. M. Waggoner. 1997. The Late Precambrian fossil *Kimberella* is a mollusc-like bilaterian organism. Nature 388: 868–871.

Garstang, W. [Introduction by Sir A. Hardy].1951. *Larval Forms, and Other Zoological Verses*. Basil Blackwell, Oxford. [Reprinted in 1985 by the University of Chicago Press.]

Ghiselin, M. T. 1966. The adaptive significance of gastropod torsion. Evolution 20: 337–348.

Ghiselin, M. T. 1988. The origin of molluscs in the light of molecular evidence. Oxford Surv. Evol. Biol. 5: 66–95.

Götting, K. 1980a. Arguments concerning the descendence of Mollusca from metameric ancestors. Zool. Jahrb. Abt. Anat. 103: 211–218.

Götting, K. 1980b. Origin and relationships of the Mollusca. Z. Zool. Syst. Evolutionsforsch. 18: 24–27.

Graham, A. 1979. Gastropoda. *In* M. R. House (ed.), *The Origin of Major Invertebrate Groups*. Academic Press, New York, pp. 359–365.

Gutmann, W. F. 1974. Die Evolution der Mollusken-Konstruktion: Ein phylogenetisches Modell. Aufsätze Red. Senckenb. Naturf. Ges. 25: 1–24.

Haas, W. 1981. Evolution of calcareous hardparts in primitive molluscs. Malacologia 21: 403–418.

Halanych, K. M., J. D. Bacheller, A. M. A. Aguinaldo, S. M. Liva, D. M. Hillis, J. A. Lake. 1995. Evidence from 18S ribosomal DNA that the lophophorates are protostome animals. Science 267: 1641–1643.

Haszprunar, G. 1992. The first molluscs—small animals. Boll. Zool. 59: 1–16.

Haszprunar, G. 2000. Is the Aplacophora monophyletic? A cladistic point of view. Amer. Malacological Bull. 15(2): 115–130.

Hickman, C. S. 1988. Archaeogastropod evolution, phylogeny and systematics: A re-evaluation. Malacol. Rev., Suppl. 4: 17–34.

Holland, C. H. 1979. Early Cephalopoda. *In* M. R. House (ed.), *The Origin of Major Invertebrate Groups*. Academic Press, New York, pp. 367–379.

Jägersten, G. 1959. Further remarks on the early phylogeny of the Metazoa. Zool. Bidr. Uppsala 30: 321–354.

Johnson, C. C. 2002. The rise and fall of rudistid reefs. Amer. Sci. 90: 148-153.

Knight, J. B. and E. L. Yochelson. 1958. A reconstruction of the relationships of the Monoplacophora and the primitive Gastropoda. Proc. Malacol. Soc. London 33: 37–48.

Lauterbach, K.-E. von. 1983. Erörterungen zur Stammesgeschichte der Mollusca, insbesondere der Conchifera. Z. Zool. Syst. Evolutionsforsch. 21: 201–216.

Linsley, R. M. 1978. Shell form and the evolution of gastropods. Am. Sci. 66: 432–441.

Morton, J. E. 1963. The molluscan pattern: Evolutionary trends in a modern classification. Proc. Linn. Soc. London 174: 53–72.

Runnegar, B. and J. Pojeta. 1974. Molluscan phylogeny: The paleontological viewpoint. Science 186: 311–317.

Ruppert, E. E. and J. Carle. 1983. Morphology of metazoan circulatory systems. Zoomorph. 103: 193–208.

Salvini-Plawen, L. von. 1972. Zur Morphologie und Phylogenie der Mollusca: Die Beziehungen der Caudofoveata und der Solenogastres als Aculifera, als Mollusca und als Spiralia. Z. Wiss. Zool. Abt. A 184: 205–394.

Salvini-Plawen, L. von. 1977. On the evolution of photoreceptors and eyes. Evol. Biol. 10: 207–263.

Salvini-Plawen, L. von. 1980. A reconsideration of systematics in the Mollusca. Phylogeny and higher classification. Malacologia 19: 249–278.

Salvini-Plawen, L. von. 1981. On the origin and evolution of the Mollusca. Atti Accad. Naz. Lincei, Atti. Conv. Lincei 49: 235–293.

Scheltema, A. H. 1978. Position of the class Aplacophora in the phylum Mollusca. Malacologia 17: 99–109.

Scheltema, A. H. 1988. Ancestors and descendants: Relationships of the Aplacophora and Polyplacophora. Am. Malacol. Bull. 6: 57–68.

Scheltema, A. H. 1993. Aplacophora as progenetic aculiferans and the coelomate origin of mollusks as the sister taxon of Sipuncula. Biol. Bull. 184: 57–78.

Scheltema, A. H. 1996. Phylogenetic position of Sipuncula, Mollusca and the progenetic Aplacophora. *In* J. Taylor (ed.), *Origin and Early Radiation of the Mollusca*. Oxford Univ. Press., London. pp. 53–58.

Taylor, J. (ed.). 1996. *Origin and Early Radiation of the Mollusca*. Oxford Univ. Press, London.

Steiner, G. and M. Müller. 1996. What can 18S rDNA do for bivalve phylogeny? J. Mol. Evol. 43: 58–70.

Vagvolgyi, J. 1967. On the origin of molluscs, the coelom, and coelomic segmentation. Syst. Zool. 16: 153–168.

Wade, C. M., P. B. Mordan and B. Clarke. 2001. A phylogeny of the land snails (Gastropoda: Pulmonata). Proc. Royal Soc. London B. 268: 413–422.

Wagner, P. J. 2001. Gastropod phylogenetics: Progress, problems and implications. J. Paleont. 75(6): 1128–1140.

Wheeler, W. C., P. Cartwright and C. Y. Hayashi. 1993. Arthropod phylogeny: A combined approach. Cladistics 9(1): 1-39. [A critical phylogenetic analyses of protostomes.]

Wingstrand, K. G. 1985. On the anatomy and relationships of recent Monoplacophora. Galathea Rpt. 16: 7–94.

Yochelson, E. L. 1978. An alternative approach to the interpretation of the phylogeny of ancient mollusks. Malacologia 17: 165–191.

Yochelson, E. L. 1979. Early radiation of Mollusca and mollusc-like groups. *In* M. R. House (ed.), *The Origin of the Major Invertebrate Groups*. Academic Press, New York, pp. 323–358.

Yonge, C. M. 1957a. *Neopilina: Survival from the Paleozoic*. Discovery (London), June 1957: 255–256.

Yonge, C. M. 1957b. Reflections on the monoplacophoran *Neopilina galatheae* Lemche. Nature 179: 672–673.

21

Lophophorates

Anyone who starts looking at bryozoans will continue to do so, for their biology is full of interest and unsolved mysteries.

J. S. Ryland,
Bryozoans, 1970

Most of the rest of the book is devoted to seven phyla that make up an evolutionary clade known as the deuterostomes: Echinodermata, Chaetognatha, Hemichordata, Chordata, and the three lophophorate groups: Phoronida, Ectoprocta, and Brachiopoda. Some of the features that define the deuterostomes are shared by other taxa (such as radial cleavage in the cnidarians), but the clade is distinguished by the synaphomorphies of enterocoelic development, a tripartite bauplan, and the mouth not deriving from the blastopore (except in phoronids) (Chapters 4 and 24). Although the variations on this theme have not led to the extreme diversity of species we saw among the protostomes, they have resulted in several fundamentally different and distinct groups of animals. In fact, there are greater differences among the major taxa of deuterostomes than among the major taxa of protostomes. Thus, at least at higher taxonomic levels, the deuterostome bauplan has proved both evolutionarily labile and highly successful in a wide range of lifestyles.

The Lophophorates: An Overview

In his 1977 address introducing a symposium on the biology of lophophorates, Joel W. Hedgpeth referred to these creatures as an "aggregation of animals that possess a feeding structure known as a lophophore." Although the term *aggregation* is not a valid taxonomic category, it is an appropriate description in this case and reflects disagreement about the relationships of the lophophorates to each other and to other phyla.

Traditionally included as lophophorates (Greek, "crest-bearers") are three phyla: Phoronida, Brachiopoda, and Ectoprocta (= Bryozoa). At first glance, the members of these groups may seem to have little in common (Figures 21.1, 21.5, 21.6, and 21.18). However, they do display a number of important similarities. They are allied with the deuterostomes and are built along a trimeric (= tripartite) bauplan wherein the body is divided into an anterior **prosome**, a middle **mesosome**, and a posterior **metasome**. In most cases, at least developmentally, each of these regions contains a separate, often paired, coelomic compartment: **protocoel**, **mesocoel**, and **metacoel**, respectively (Chapter 4). Furthermore, the lophophorates all have a U-shaped gut and very simple, often transient reproductive systems. Nearly all secrete outer casings in the form of tubes, shells, or compartmented exoskeletons. Even though these features represent shared similarities among the lophophorates, none is unique to them. Their most significant common feature is the lophophore itself, which, simply put, is a ciliated, tentacular outgrowth arising from the mesosome and containing extensions of the mesocoel. It surrounds the mouth but not the anus.

The tentacular crown of pterobranchs (Chapter 23) also possess mesosomal tentacles with a coelomic lumen. The homology between pterobranch tentacles and a "true" lophophore is invalidated only because in the former case the tentacles do not fully surround the mouth. Thus, the lophophore is a unique apomorphy of members of the phyla Phoronida, Ectoprocta, and Brachiopoda, and most biologists treat the three groups as a monophyletic clade. However, a deep-level homology, or perhaps parallel evolution, is probable between pterobranchs and lophophorates. To be sure, there are some problems with this hypothesis, some of which are discussed later.

With the exception of a few freshwater ectoprocts, the lophophorates are exclusively marine. All are benthic, living either in tubes (phoronids) or in secreted shells, or casings. The phoronids are a small group, comprising only two genera and about 20 species of solitary or gregarious worms. Ectoprocts, on the other hand, make up a diverse taxon of about 4,500 species of colonial forms. The brachiopods, or lamp shells, include about 335 extant species. However, the brachiopods have left a record of over 12,000 fossil species as evidence of a greater past. They were well established by the early Cambrian. These animals flourished in both abundance and diversity from the Ordovician through the Carboniferous, but they have declined in numbers and kinds ever since.

Taxonomic History

The lophophorates have had a long and torturous taxonomic history. The earliest records of any lophophorate are of various ectoprocts reported in the sixteenth century. With few exceptions, the early zoologists treated them as plantlike and included them in the taxon Zoophyta, a misconception that persisted into the 1700s. Peyssonal (1729) finally established the animal nature of ectoprocts, and Jussieu (1742) noted the compartmentalized condition of the colonies and coined the term *polyps* to refer to the individual animals. Still, most well-known workers of the day (e.g., Linnaeus and Cuvier) insisted on allying them with the cnidarians in the group Zoophyta.

Eventually, de Blainville (1820) noted the complete gut of the ectoprocts and "raised" them above the cnidarians. By 1830, two names had been coined for these animals—Bryozoa (by the German zoologist Ehrenberg) and Polyzoa (by the Englishman Thompson). Just about the time the ectoprocts (under one or the other name) were being recognized as a separate group, concurrent events confused the issue further. The entoprocts (= Kamptozoa) were described, and most workers included them with the ectoprocts (under the name Bryozoa), while others recognized a relationship between ectoprocts and other lophophorates. All of this became terribly entangled in Milne-Edwards's (1843) concept of a taxon Molluscoides, which he established to include the ectoprocts and compound ascidians (Chapter 23).

The brachiopods were known, at least from fossils, in the 1600s and were allied with the molluscs. This mistaken view was held until late in the nineteenth century.

Phoronids were first described from their larvae by Müller in 1846, who thought they were adults and named them *Actinotrocha brachiata*. Not long after, Gegenbaur (1854) recognized these animals as larval stages. The adults were found and described by Wright (1856), who named them *Phoronis*. Finally, the renowned embryologist Kowalevsky (1867) studied the metamorphosis of "*Actinotrocha*" and established the relationship between the two stages. The name "actinotroch" persists as a general term for the phoronid larva.

In 1857 Hancock recognized the relationship between brachiopods and bryozoans, but the whole matter was mixed up with confusion over the entoprocts and Milne-Edwards's Molluscoides. Nitsche (1869) made a valiant attempt to separate the entoprocts and ectoprocts, and in 1882 Caldwell first put forth the idea of a relationship between the brachiopods, phoronids, and bryozoans. This view was supported by Hatschek (1888), who suggested the establishment of a phylum Tentaculata to include the classes Phoronida, Brachiopoda, and Ectoprocta (but excluding the entoprocts). Since that time several attempts to unite some or all of these animals into a single taxon have been made, including Schneider's (1902) Lophophorata (for phoronids and bryozoans). Hyman (1959) rejected these ideas and retained separate phylum status for the three groups, an arrangement that has remained popular ever since.

Recently, however, some of the earlier views have reemerged with new evidence and new vigor. On one front, Nielsen (1971, 1977) has revived the idea of a possible entoproct–ectoproct affinity and argued that the two groups are closely related (even though entoprocts are fundamentally protostomous and are not trimeric in their body plan). Although this hypothesis has found its way into several recent texts, we find it untenable in light of the evidence. In support of lophophorate unity, some recent authors have again suggested that a single phylum or "superphylum" should be formally established for the three groups. Most recently, 18S rDNA sequence data has suggested that the lopophorates are allied with the protostomes—in a grouping known as the lophotochozoa. This controversial idea is discussed in Chapter 24. Below, we provide classifications and discussions for the lophophorate phyla.

The Lophophorate Bauplan

In spite of the obvious diversity in external form among the lophophorates, there is a fundamental unity here. The diversity, great or small, in any monophyletic lineage is a reflection of evolutionary variations on a central theme—the theme being the bauplan. Among the deuterostomes, the general theme consists of those features of ontogeny that define the lineage and the resulting trimeric body plan with its tripartite coelomic system. Impose upon this plan the lophophore, and the result is a unique clade derived from the main deuterostome line. Couple this single synapomorphy with a number of other shared features, and the lophophorate bauplan begins to emerge as a set of characters more easily explained as arising from common ancestry than resulting from convergence.

All lophophorates are built for benthic life and suspension feeding, the latter being a primary function of the lophophore itself. The anterior body region, or prosome, is reduced to a small, flaplike **epistome** (or lost in some) associated with an overall reduction of the head, the elaboration of the mesosome as the lophophore, and a sessile lifestyle. As in most deuterostomes, the metasome houses the bulk of the viscera. The U-shaped gut is clearly advantageous for living encased in tubes, compartments, and shells; such animals do not "foul their nests," as it were. We have seen similar adaptations in some other groups with comparable habits, such as the recurved gut of sipunculans and the ciliated fecal-removal grooves of some tube-dwelling polychaetes. In lophophorates, this condition not only prevents fecal accumulation in the encasement, but generally brings the anus close to rejection currents produced by the cilia on the lophophore.

The phoronids most clearly display the above traits, and retain the vermiform shape of the probable ancestral form (Box 21A). Evolutionarily, the ectoprocts have exploited asexual reproduction, colonialism, and small size (Box 21B). Relieved of long-distance internal transport problems, the ectoprocts have lost the circulatory and excretory systems. The brachiopods evolved a pair

---

> ### BOX 21A *Characteristics of the Phylum Phoronida*
>
> 1. Trimeric, enterocoelic, vermiform lophophorates
>
> 2. Body divided into flaplike epistome (prosome), lophophore-bearing mesosome, and elongate trunk (metasome), each with associated coelom
>
> 3. Gut U-shaped, anus close to mouth
>
> 4. One pair metanephridia in metasome
>
> 5. Closed circulatory system
>
> 6. Dioecious or hermaphroditic, with transient peritoneal gonads
>
> 7. With mixed or indirect life histories; usually with an actinotroch larva
>
> 8. Radial, indeterminate cleavage; coeloblastulae gastrulate by invagination; blastopore becomes mouth (in contrast to all other deuterostomes)
>
> 9. Marine benthic tube dwellers

of valves or shells that encase and protect the body, including the lophophore (Box 21C). Thus, instead of exposing the lophophore in the water, as phoronids and ectoprocts do, the brachiopods draw water into their **mantle cavity** for suspension feeding, an action analogous to that in bivalve molluscs. Sessile animals with soft parts (e.g., the lophophore), living on benthic surfaces, are exposed to potentially high levels of predation—thus the selective advantage of the brachiopod shell is clear. The phoronids are motile to the extent that their bodies can be retracted within their tubes, protecting the soft parts. Ectoprocts are entirely anchored in their casings, but the lophophore itself is retractable into the body, as the result of a unique arrangement of muscles and hydraulic mechanisms.

Phylum Phoronida: The Phoronids

Phoronids are all tube dwellers; their chitinous tubes are usually either cemented (often in clusters) to hard substrata, or buried vertically in soft sediments (Figure 21.1). They are assigned to only two genera: *Phoronis* and *Phoronopsis*. The phylum name was apparently derived from the Latin term *Phoronis*, the surname of Io (who, according to mythology, was changed into a cow and roamed the Earth, eventually to be returned to her former body). Recall that these worms were first described from larval stages drifting in the sea, and only much later were the adults recognized as part of the same life cycle. Adult phoronids are known from intertidal mud flats to depths of about 400 meters.

(A)

Mouth

Lophophoral tentacles

Trunk

End bulb

(B)

Lophophore

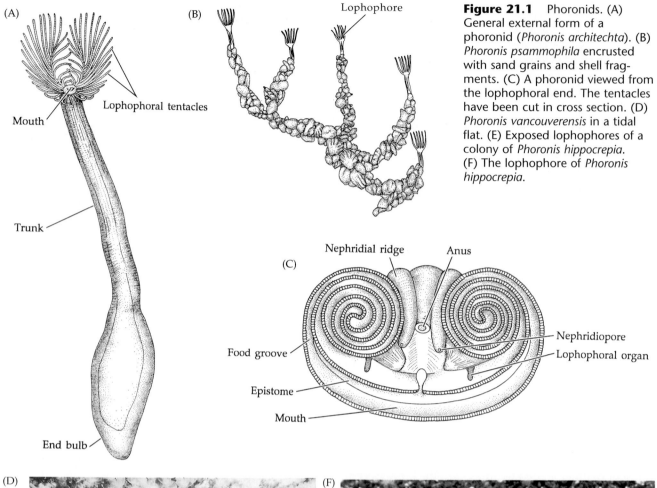

Figure 21.1 Phoronids. (A) General external form of a phoronid (*Phoronis architechta*). (B) *Phoronis psammophila* encrusted with sand grains and shell fragments. (C) A phoronid viewed from the lophophoral end. The tentacles have been cut in cross section. (D) *Phoronis vancouverensis* in a tidal flat. (E) Exposed lophophores of a colony of *Phoronis hippocrepia*. (F) The lophophore of *Phoronis hippocrepia*.

(C)

Nephridial ridge   Anus

Food groove

Epistome

Mouth

Nephridiopore

Lophophoral organ

(D)

(F)

(E)

The Phoronid Bauplan

Phoronids range from about 5 to 25 cm in length. The vermiform body shows little regional specialization, except for the distinct lophophore and a modest inflation of the **end bulb**, which houses the stomach and also aids in anchoring the animals in their tubes. The slitlike mouth is located between the tentacle-bearing lophophoral ridges and is overlaid by a flap, the epistome. The lateral aspects of the ridges are distinctly coiled and flank the dorsal anus and paired nephridiopores (Figure 21.1). Using the terms *anterior* and *posterior* when referring to phoronids can be misleading. During metamor-

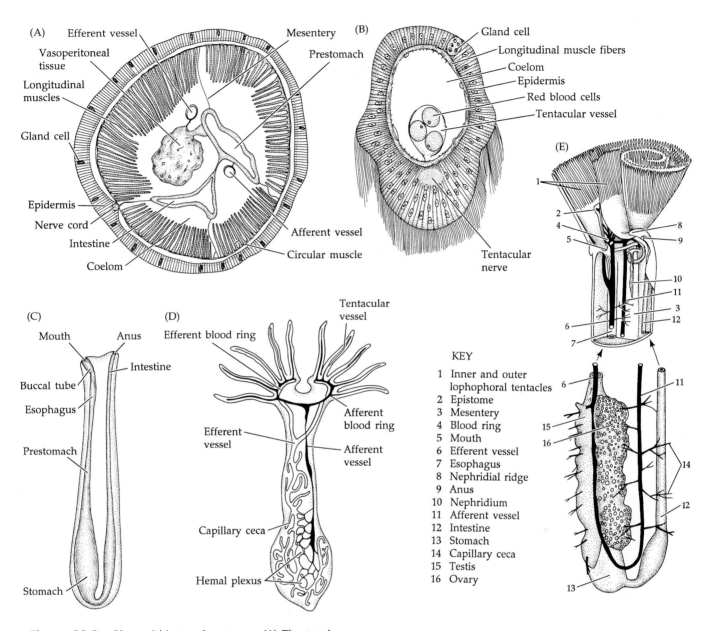

Figure 21.2 Phoronid internal anatomy. (A) The trunk of a phoronid in cross section. Note the body wall layers and coelomic partitioning. (B) A tentacle of the lophophore in cross section. (C) The digestive tract. (D) The circulatory system. (E) The major internal organs in the lophophoral and stomachic ends.

phosis the true anterior (mouth-bearing) and posterior (anus-bearing) ends are brought very close together by rapid growth and enlargement of the ventral surface (Figure 21.4). The dorsal surface is reduced to only the small area between the adult mouth and anus. Because of these conditions, we refer to the "ends" of the adult worm as the **lophophoral end** and the **stomachic end**.

Body Wall, Body Cavity, and Support

The phoronid body wall includes an epidermis of columnar cells overlaid by a very thin cuticle. Within the epidermis are sensory neurons and various gland cells, the latter responsible for the production of mucus and chitin (Figure 21.2A). The epidermis of the lophophore is densely ciliated. Internal to the epidermis and its basement membrane is a thin layer of circular muscle and a thick layer of longitudinal muscle. A peritoneum lines the longitudinal muscles and forms the outer boundary of the coelom.

The coelom is clearly tripartite. The protocoel is limited to a single small cavity within the epistome. The unpaired mesocoel comprises a coelomic ring in the lophophoral collar and extensions into each tentacle (Figure 21.2B). The protocoel and mesocoel are connected to one another along the lateral aspects of the epistome. The metacoel forms the main trunk coelom, which is separated from the mesocoel by a transverse septum. Ontogenetically, the metacoel is an uninterrupted cavity with only one, midventral, mesentery. However, secondary mesenteries form later in development, yielding four longitudinal spaces (Figure 21.2A). The coelomic fluid contains several kinds of freely wandering cells, or **coelomocytes**, including phagocytic amebocytes.

Body support is provided by the hydrostatic qualities of the coelomic chambers and by the tube. The muscles of the body wall are rather weak, particularly the circular layer, and once removed from their tubes phoronids are capable of only limited movement. Normally, however, the body wall of the end bulb is pressed against the tube, holding the worm in place. When disturbed, the animal simply contracts into the tube; the lophophore itself is not retractable.

The tube is secreted by epidermal gland cells. When first produced, the chitinous secretion is sticky; but upon contact with water it solidifies to a flexible parchment-like consistency. Sand grains and bits of other material adhere to the tube during the sticky phase of its formation in those phoronids that inhabit soft substrata (e.g., *Phoronopsis harmeri*). In some species the tubes intertwine with one another, with the whole tangled aggregation attached to a substratum or actually embedded in calcareous stone or shells (e.g., *Phoronis hippocrepia*).

The Lophophore, Feeding, and Digestion

The tentacles of the lophophore are hollow, ciliated outgrowths of the mesosome, and each contains a blind-ended blood vessel and a coelomic extension (Figure 21.2B). The tentacles are in a double row, arising from two ridges. The ridges lie close to one another and form a narrow food groove in which the slitlike mouth is located (Figure 21.1C). Because the sides of the lophophoral ridges are coiled, many tentacles are compacted into a small area.

Phoronids are ciliary–mucous suspension feeders. The lophophoral cilia generate a water current that passes down between the two rows of tentacles and then out between the tentacles. Food particles are trapped in mucus lining the food groove and then passed along the groove by cilia to the mouth. As the water current passes between the tentacles and out of the area of the food groove, some is directed over the anus and nephridiopores away from the animal (Figure 21.1C).

The digestive tract is U-shaped, but rather simple and not coiled (Figure 21.2C). The mouth is overlaid by the epistomal flap and leads inward to a short **buccal tube**, which is followed by an esophagus and a narrow **prestomach**. Within the end bulb the gut expands into a stomach, from which emerges the intestine. The intestine bends up toward the lophophore and leads to a short rectum and the anus. The gut is supported by peritoneal mesenteries (Figure 21.2A).

The entire digestive tract is apparently derived from endoderm. Some parts are muscular, but only weakly so, and much of the movement of food is by ciliary action. A middorsal strip of densely ciliated cells arises in the prestomach and extends into the stomach, and it is probably responsible for directing food along that portion of the gut. Gland cells occur in the esophagus but their function remains uncertain. Transitory syncytial bulges in the stomach walls are the site of intracellular digestion in that organ.

Circulation, Gas Exchange, and Excretion

Phoronids contain an extensive circulatory system comprising two major longitudinal vessels between which blood is exchanged in the lophophoral and stomachic ends of the body (Figure 21.2D,E). Various names have been applied to these vessels relative to their positions in the body. These terms are often confusing because the positions of vessels vary along the length of the trunk, and they are not clearly dorsal, medial, lateral, or ventral as the names imply. We prefer the terms **afferent** and **efferent vessels**, which refer to the direction of blood flow relative to the lophophore.

The afferent vessel extends unbranched from the region of the end bulb to the base of the lophophore. For most of its length it lies more or less between the descending and ascending portions of the gut. In the mesosome the afferent vessel forks, forming an **afferent "ring" vessel** (U-shaped) at the base of the lophophoral tentacles. A series of **lophophoral vessels**, one in each tentacle, arises from the afferent ring. Each of these vessels joins with an **efferent "ring" vessel** (also U-shaped), which drains blood from the lophophore. Thus the afferent and efferent blood rings lie against one another, and generally share openings into the lophophoral tentacles within which blood moves back and forth, there being but a single vessel in each tentacle. Backflow into the afferent ring is largely prevented by tiny, one-way, flap valves.

The arms of the efferent ring unite to form the main efferent blood vessel, which extends through the trunk. This vessel gives off numerous branches or simple blind diverticula called **capillary ceca**, which bring blood close to the gut wall and other organs. In the end bulb, surrounding the stomach and first part of the intestine, blood flows from efferent to afferent vessels through spaces composing the **hemal (stomachic) plexus** (Figure 21.2D,E). Blood actually leaves the vessels here and flows through spaces between the organs and their bordering layers of peritoneum. Thus, technically speaking, the system is open at this point; however, blood flow is directed within the confines of these passages. Blood is moved through the circulatory system largely by muscular action of the blood vessel walls.

The intimate association of blood and the stomach wall suggests that nutrients are picked up from the stomach by the circulatory fluid and transported throughout the body. The tentacles of the lophophore are also probably the most important site of gas exchange. Oxygenated blood flows from the lophophore into the efferent vessel and from there is distributed to all parts of the trunk. The blood contains nucleated red corpuscles, with hemoglobin as the respiratory pigment.

A pair of metanephridia lies in the trunk, and each bears two nephrostomes opening to the metacoel (Figure 21.2E and 21.3A). In each nephridium, the nephrostomes—one large and one small—join a curved nephridioduct, which leads to a nephridiopore adjacent to the anus. Although virtually nothing is known about excretory physiology in phoronids, particulate crystalline matter has been observed exiting the nephridiopores and probably represents precipitated nitrogenous waste products. The nephridia also function as pathways for the release of gametes. Being marine, osmoregulatory problems are presumably insignificant in phoronids.

Nervous System

The nervous system of phoronids is rather diffuse and lacks a distinct cerebral ganglion. This condition is related to the sedentary lifestyle and overall reduction in cephalization in these worms. Most of the nervous system is intimately associated with the body wall, being either intraepidermal or immediately subepidermal. The body is everywhere supplied with a layer of nerve fibers between the epidermis and the circular muscle layer. Simple sensory neurons arise from this layer, either singly or in bundles, and extend to the body surface as the only receptor structures. Motor neurons extend inward to the muscle layers.

The central nervous system comprises a simple intraepidermal nerve ring, which lies at the base of the lophophore and is continuous with the subepidermal nerve layer. It is slightly swollen middorsally. The nerve ring supplies the tentacles with nerves as well as giving rise to motor nerves to some of the longitudinal muscles

in the metasome. In addition, a bundle of sensory neurons extends from the nerve ring to each of the lophophoral organs.

Phoronids possess one or two longitudinal giant motor fibers in the trunk (absent in the very small *Phoronis ovalis*). When only one longitudinal fiber is present, it lies on the left side. Actually, this fiber originates within the right side of the nerve ring, through which it passes to emerge on the left side. This nerve fiber is intraepidermal except where it extends inward along the left nephridium. In those species where two longitudinal fibers are present, the right one originates on the left side of the nerve ring and extends to the opposite side of the body.

Reproduction and Development

Asexual reproduction by transverse fission or by a form of budding has been documented in a few species. Phoronids are also capable of regenerating lost parts of the body, and are known to autotomize various parts of the lophophoral end.

Both dioecious and hermaphroditic species of phoronids occur, and in the latter case some are simultaneous hermaphrodites (e.g., *Phoronis vancouverensis*). The gonads are transient and form as thickened areas of the peritoneum around the hemal plexus. The resulting mass of gamete-forming tissue and blood sinuses is sometimes called **vasoperitoneal tissue** (Figure 21.2E). Gametes are proliferated into the metacoel and typically are carried to the outside via the nephridia. In *Phoronis ovalis*, females autotomize their lophophoral ends and release eggs through the torn opening. Fertilization is usually external, except in *Phoronopsis harmeri* and

Figure 21.3 (A) A phoronid metanephridium (from *Phoronis australis*). Note the paired nephrostomes. (B) The lophophore of *Phoronis vancouverensis*. Note the accessory lophophoral organs. (C) Spermatophore of *Phoronopsis harmeri*. (D) Spermatophore of *Phoronis vancouverensis*.

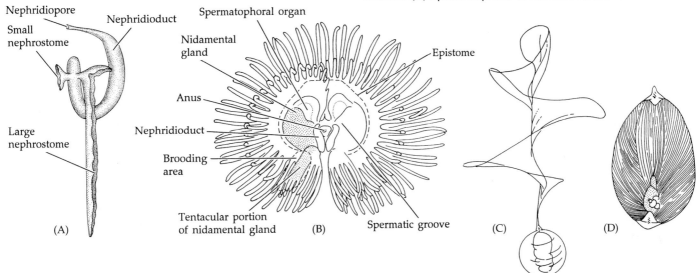

Nephridiopore
Small nephrostome
Nephridioduct
Large nephrostome
(A)

Spermatophoral organ
Nidamental gland
Anus
Nephridioduct
Brooding area
Tentacular portion of nidamental gland
(B)

Epistome
(C)

(D)

Spermatic groove

Phoronis vancouverensis. In these species, the males' **lophophoral organs** produce spermatophores (Figure 21.3 B–D) that are transferred to the tentacles of the females. In these females the lophophoral organs are called **nidamental glands** and serve as brooding areas. The complicated internal fertilization process in *Phoronopsis harmeri* (= *P. viridis*) has been elucidated by Zimmer (1972). Spermatophores on the tentacles of females rupture, releasing ameboid masses of sperm. The sperm enter the lophophoral coelom by lysis of the tentacular wall and then proceed to digest their way through the septum separating the mesocoel and metacoel into the trunk coelom, where fertilization occurs. Although fertilization has not actually been observed, Zimmer's experimental data, coupled with the fact that fertilized ova occur internally, suggest that this scenario is the only tenable explanation.

Developmental strategy differs among species, the particular pattern depending in part on the size of the egg and on whether fertilization is internal or external. The ova of free-spawning species contain little yolk and develop quickly to planktotrophic **actinotroch larvae** (Figure 21.4). In species that possess nidamental glands, fertilization is followed by brooding until release at the actinotroch stage. The eggs of these species are moderately rich in yolk, providing nutrients for the embryos during the brooding period. *Phoronis ovalis* lacks nidamental glands, but the yolky eggs are shed into the maternal tube, where they are brooded. Development in *P. ovalis* does not include a typical actinotroch; instead, the embryos emerge as ciliated, sluglike larvae that have a short, planktonic life.

Despite continuing reference in some texts to phoronids as protostome-like in their early development, it has been convincingly established that such is not the case (e.g., Zimmer 1973, 1980). Early reports of spiral cleavage probably resulted from mechanical dis-

placement of the blastomeres because of the tightness of the fertilization membrane. Otherwise, early cleavage is clearly radial and has been shown to be indeterminate. A coeloblastula forms and gastrulates by invagination. Mesoderm arises from the presumptive archenteron, and coelom formation is by a modified enterocoelous method. The only protostome-like feature is the formation of the mouth from the blastopore.

With the exception of *Phoronis ovalis*, all phoronids produce distinctive **actinotroch larvae** (Figure 21.4A). Earlier works alleging similarities between actinotrochs and trochophores are no longer given credence; the actinotroch is clearly a tripartite stage and lends additional support to the deuterostome affinity of the phoronids. The fully formed actinotroch bears a **preoral hood**, or lobe, over the mouth. The hood houses the protocoel and becomes the epistome. A partial ring of larval tentacles contains the mesocoel and eventually forms the lophophore. As the actinotroch develops, an inpocketing (called the **metasomal sac**) forms on the ventral surface. At settlement and metamorphosis this sac everts, extending the ventral surface such that the anus and mouth remain close to one another as the gut is drawn out into the characteristic U-shape (Figure 21.4). It is during this metamorphic growth that the larval worms settle and begin secreting their tubes.

The Ectoprocts

Members of the phylum Ectoprocta (Greek *ecto*, "outside"; *procta*, "anus") are sessile colonies of zooids living in marine and freshwater environments (Box 21B). In most cases each colony is the product of asexual reproduction from a single, sexually produced individual called an **ancestrula**. The colony form differs greatly among species, but their general plantlike appearance earned these animals the common name "moss animals," from which the old name Bryozoa was coined. Marine ectoprocts are known from all depths and latitudes, mostly on solid substrata. One recently discov-

Figure 21.4 Phoronid larvae and metamorphosis. (A) An actinotroch larva. (B,C) Stages in the metamorphosis of an actinotroch. The gut is drawn into a U-shape, leaving the mouth and anus at the anterior end.

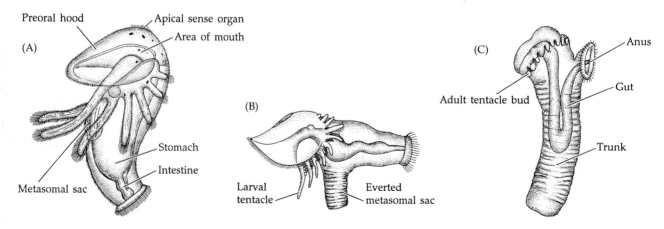

(A) Preoral hood — Apical sense organ — Area of mouth — Stomach — Intestine — Metasomal sac

(B) Larval tentacle — Everted metasomal sac

(C) Anus — Gut — Trunk — Adult tentacle bud

ered Antarctic species forms gelatinous colonies on floating pieces of ice! A few species occur in fresh and brackish water. The bizarre gelatinous *Pectinatella magnifica* is frequently encountered in streams and ponds east of the Mississippi River. Littoral regions in most parts of the world harbor luxuriant growths of ectoprocts that often cover large areas of rock surfaces. Some species have coral-like growth forms that can create miniature "reefs" in some shallow-water habitats. Others form dense bushlike colonies or gelatinous spaghetti-like masses. Many encrusting forms grow on the shells or exoskeletons of other invertebrates and some bore into calcareous substrata. Three classes are generally recognized, as described below.

PHYLUM ECTOPROCTA

CLASS PHYLACTOLAEMATA: Freshwater ectoprocts. Colonies with chitinous or gelatinous coverings; zooids cylindrical, large, and monomorphic; epistome with protocoel present; lophophore large and horseshoe-shaped; body wall muscles well developed; metacoel extensions interconnect zooids; a cord of tissue, the funiculus, extends from the gut to the body wall, but not between zooids; most produce asexual bodies called statoblasts. (e.g., *Cristatella, Hyalinella, Lophopus, Lophopodella, Pectinatella, Plumatella*)

CLASS STENOLAEMATA: Marine ectoprocts. Zooids housed in tubular, calcified skeletal compartments; zooids cylindrical or trumpet-shaped, some polymorphic; epistome and protocoel absent; lophophore circular; body walls inflexible, lacking well developed musculature; without special coelomic extensions between zooids, but adjacent zooids connected by pores; funiculus does not extend between zooids; with a unique membranous sac housing the internal parts of the polypide; reproduction involves unique polyembryony, whereby single embryos reproduce asexually; one extant order, Cyclostomata. (e.g., *Actinopora, Crisia, Diaperoecia, Disporella, Idmodronea, Tubulipora*)

CLASS GYMNOLAEMATA: Highly diverse group of primarily marine ectoprocts. Colony form is extremely variable, soft or calcified, encrusting to arborescent; body wall lacks muscles; zooids variably modified from basic cylindrical form; zooids usually polymorphic; lophophore circular, epistome and protocoel absent; zooids joined by pores through which cords of tissue extend and join with each funiculus; two orders, Ctenostomata and Cheilostomata.

    ORDER CTENOSTOMATA: Colonies vary in shape; skeleton leathery, chitinous, or gelatinous, not calcified; openings through which zooids protrude lack opercula; without ovicells for brooding embryos; without avicularia. (e.g., *Aethozoon, Alcyonidium, Alcyonium, Amathia, Bowerbankia, Flustrellidra, Nolella, Tubiporella, Victorella*)

    ORDER CHEILOSTOMATA: Colony form varies, but generally of box-shaped zooids with calcareous walls; openings usually with opercula; zooids often polymorphic; embryos usually brooded in ovicells. (e.g., *Bugula, Callopora, Carbasea, Cellaria, Conopeum, Cornucopina, Cribrilaria, Cryptosula, Cupuladria, Electra, Eurystomella, Flustra, Hippothoa, Membranipora, Metrarabdotos, Microporella, Pentapora, Porella, Pyripora, Rhamphostomella, Schizoporella, Thalamoporella, Tricellaria*)

BOX 21B *Characteristics of the Phylum Ectoprocta*

1. Trimeric, enterocoelic, colonial lophophorates

2. Epistome and protocoel absent in most species

3. Lophophore circular or U-shaped

4. Gut U-shaped, anus close to mouth

5. Typical circulatory and excretory structures absent

6. Colonies produced by asexual budding; zooids within a colony often polymorphic

7. Zooids usually hermaphroditic, but some contain males and females in a single colony; gametes usually arise from transient patches of germinal tissue on peritoneum or funiculus

8. Radial, holoblastic cleavage; indirect or mixed development; blastopore does not form mouth

9. Sessile in marine and freshwater habitats

The Ectoproct Bauplan

A special terminology has evolved among ectoproct specialists, especially concerning the morphology of the zooids. The colony itself is called a **zoarium** and the secreted exoskeleton the **zoecium**. Early workers mistakenly thought that ectoproct zooids were actually composed of two organisms, the exoskeletal compartment and the internal soft parts, which they named the cystid and polypide, respectively. These terms were redefined by Hyman (1959) and now have some meaning relative to the functional morphology of ectoprocts. The **cystid** comprises the outer casing, or zoecium, and the attached parts of the body wall—that is, the nonliving *and* living housing of each zooid. The **polypide** includes the lophophore and soft viscera that are movable within the housing (Figure 21.5). The opening in the cystid through which the lophophore extends is termed the **orifice** and often bears a flaplike covering, or **operculum**.

The nature of the exoskeleton differs among ectoprocts, as does the form of the colony. The outer covering may be gelatinous or chitinous, as it is in the Phylactolaemata and Ctenostomata, or calcified, as it is in the Stenolaemata and Cheilostomata. The different growth patterns among ectoprocts result in a great variety of colony shapes. Most phylactolaemates display either **lophopodid** or **plumatellid** colony forms. In the former case, the gelatinous covering forms an irregular clump from which the zooids protrude, as seen in *Lophophus* (Figure 21.5B). Plumatellid colonies are usually erect or prostrate, and often are highly branched, like *Plumatella*. One remarkable phylactolaemate, *Cristatella,*

(A)

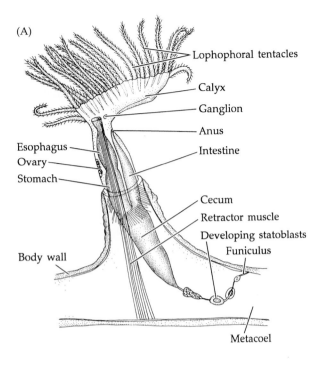

Lophophoral tentacles

Calyx

Ganglion

Anus

Esophagus

Intestine

Ovary

Stomach

Cecum

Retractor muscle

Developing statoblasts

Funiculus

Body wall

Metacoel

(B)

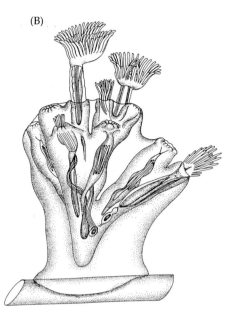

(C)

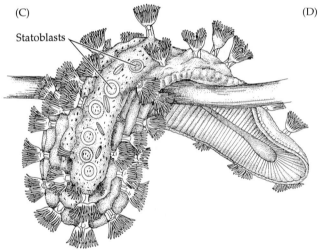

Statoblasts

(D)

(E)

Figure 21.5 Ectoprocts. (A) A single zooid of *Plumatella* (cutaway view). Note the distinction between cystid and polypide. (B) A colony of *Lophophus*. Note the confluent coelomic cavities. (C) A motile colony of *Cristatella mucedo* crawling over a plant stem. (D) A colony of *Eurystomella*. (E) A colony of the freshwater ectoproct *Cristatella*.

grows in a distinct gelatinous strip, somewhat sluglike in form, and is capable of locomotion, creeping at rates of over 1 cm a day (Figure 21.5C).

The Stenolaemata and Gymnolaemata include a bewildering array of colony forms that may be generally categorized as **stoloniferous** or **nonstoloniferous**. Stoloniferous colonies are characteristic of some members of the order Ctenostomata, in which the zooids arise separately from horizontal "runners," or stolons (e.g., *Bowerbankia*; Figure 21.6A). Nonstoloniferous col-

onies may be encrusting, arborescent, discoidal, and so on (Figure 21.6), but in all cases, the zooids are compacted and adjacent to one another, rather than arising separately and at some distance from one another.

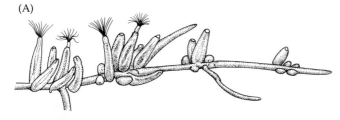

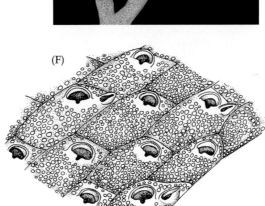

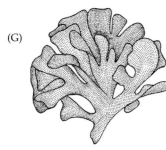

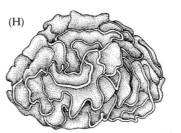

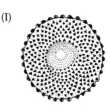

In addition to variation in overall form of the colonies, zooids of many gymnolaemates and some stenolaemates are polymorphic (within a colony). Typical lophophore-bearing individuals are called **autozooids** and are responsible for feeding and digestion.

Figure 21.6 Ectoproct colonies. (A) A stoloniferous colony of *Bowerbankia*. (B) Arborescent colonies of *Bugula*; (C) Patches of an encrusting ectoproct, *Membranipora*. (D) The fleshy bryozoan *Alcyonium verilli*. (E) *Conopeum seurati*. (F) Part of a colony of the encrusting ectoproct *Schizoporella*. (G) Leaflike colony of *Flustra*. (H) Cabbage-like colony of *Pentapora*. (I) Discoidal colony of *Cupuladria*.

Figure 21.7 SEMs of some bryozoan colonies. (A,B) *Idmidronea*; portion of a branching colony. (C) *Cribrilaria*, shown growing on *Idmidronea*. (D) *Disporella* (whole specimen). (E) *Rhamphostomella argentea*, with ovicells and avicularia. (F) *Thalamoporella*; portion of colony with brood chambers.

(A)

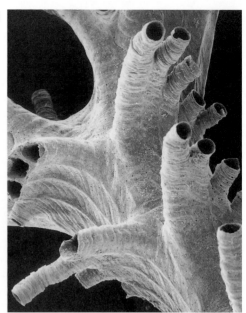

(B)

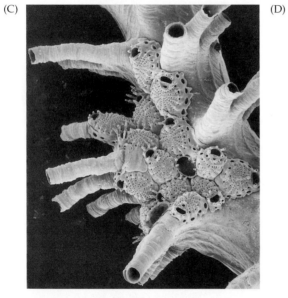

(C)

(D)

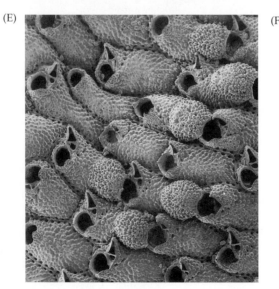

(E)

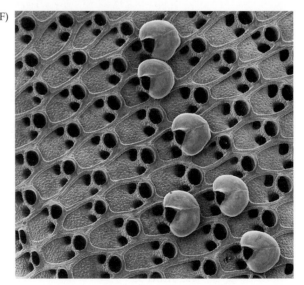

(F)

All other individuals of a colony are collectively referred to as **heterozooids**, of which there are several types, all incapable of feeding. **Kenozooids** are reduced individuals modified for attachment to a substratum; various types of attachment discs, "holdfasts," and stolons are in this category.

Many gymnolaemates possess **avicularia**, each of which bears an operculum modified as a movable **mandible** (or jaw), which articulates against a rigid **rostrum** (or palate). Zooids that possess avicularia defend the colony against small organisms and keep the surface clean of debris (Figure 21.7E and 21.8A). The latter function is also facilitated by another type of heterozooid called a **vibraculum** (Figure 21.8B). These individuals are thought to be modified avicularia, and they have a flagellum-like operculum that sweeps over the colony surface. They may help remove sediment particles and other material, but convincing evidence for this function is wanting.

The Body Wall, Coelom, Muscles, and Movement

The body wall comprises the outer secreted zoecium, the underlying epidermis, and the peritoneum. Sheets of circular and longitudinal muscles are present between the epidermis and peritoneum in phylactolaemates, but these muscle sheets are absent or greatly reduced in the other groups and are replaced by various muscle bands. Ectoprocts differ from other lophophorates in their ability to retract their lophophores into their zoecial casings, a clear protective device for these tiny sessile animals, whose soft parts would otherwise be continuously exposed to grazing predators.

Many ectoprocts possess ornate and species-specific surface sculpturing including spines, pits, and protuberances (Figure 21.7). Experiments by Harvell (1984) indicate that the cheilostomate *Membranipora* undergoes a rapid growth of new protective surface spines after grazing by predators (e.g., nudibranchs). Some ectoprocts also produce chemicals used as defense against would-be predators.

The mechanisms of lophophore retraction and protraction differ among ectoproct species. The specific mechanism depends largely upon the arrangement of muscles, the degree of rigidity of the zoecium, and the hydraulic qualities of the coelomic compartments. Recall the morphological distinction between the cystid and the polypide; extension and retraction of the lophophore basically involves movement of the latter relative to the former.

In all ectoprocts the main coeloms provide fluid-filled spaces on which muscles act directly or indirectly to increase hydraulic pressure for protraction of the lophophore. The epistome and protocoel (present only in the phylactolaemates) play no part in this process. The septum between the mesocoel and metacoel is perforated, so the fluid is continuous between the two

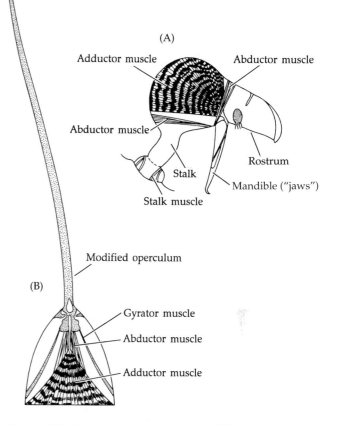

Figure 21.8 Ectoproct heterozooids. (A) An avicularium from *Bugula*. (B) A vibraculum.

chambers. Thus, when the metacoel is compressed the polypide is partially forced out of the cystid, thereby protracting the lophophore. At the same time, coelomic fluid is moved into the mesocoel and erects the tentacles. Various retractor muscles serve to pull the polypide back within the cystid. Generally, these methods of lophophore action are common to all ectoprocts, but the mechanisms involved differ considerably. Below we describe a few examples of how these movements are accomplished and at the same time illustrate variations on the basic ectoproct bauplan.

Phylactolaemates protract their lophophores by contraction of the circular muscles of the flexible body wall around the metacoel. This action imposes pressure directly on the coelomic fluid and is similar to the mechanisms we have seen in many other coelomate animals. These ectoprocts possess a ring-shaped, muscular **diaphragm** just internal to the orifice through which the lophophore protrudes. The diaphragm dilates as the lophophore is protracted and serves to partially close off the orifice after the lophophore is withdrawn by retractor muscles, which extend from the body wall to the base of the lophophore (Figures 21.5A and 21.15).

Stenolaemate ectoprocts (Cyclostomata) have erect, tubular zooids surrounded by heavily calcified zoecia

(A)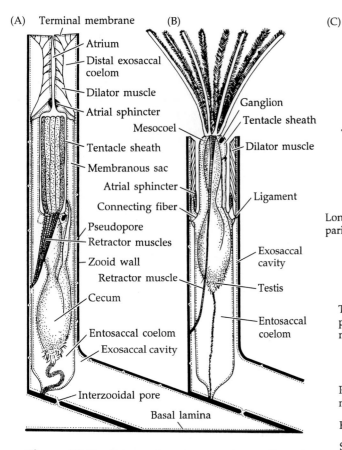

(B)

(C)

(D)

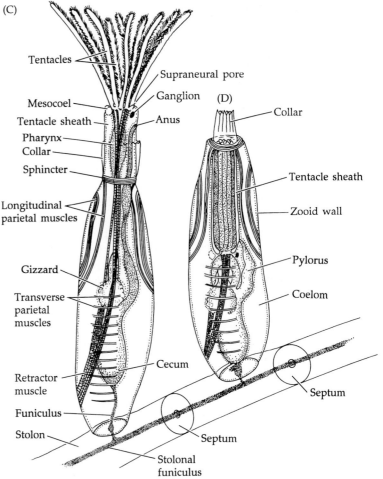

Figure 21.9 Ectoproct anatomy and operation of the lophophore. (A,B) Two zooids of a stenolaemate, with the lophophore retracted (A) and protracted (B). (C,D) The ctenostomate *Bowerbankia* with the lophophore protracted (C) and retracted (D).

(Figure 21.9A,B). The inflexibility of the body wall and the absence of well-developed sheets of muscles preclude use of the direct compression action of the phylactolaemates. Stenolaemates have evolved a mechanism of lophophore protraction unique among the ectoprocts. The structural features associated with this mechanism comprise a synapomorphy on which this group was established as a separate class. The key structure is a **membranous sac** attached by ligaments to the body wall and formed of a thin epithelial layer separating the metacoel, or **entosaccal coelom,** from an outer **exosaccal cavity**. Distally, the exosaccal space lies between the outer body wall and a thick layer of **atrial dilator muscles**. When a zooid is retracted, the atrial dilators are relaxed and a special **atrial sphincter** effectively closes the inner end of the atrium. The retracted polypide presses against the membranous sac, thereby forcing fluid into the distal region of the exosaccal cavity (Figure 21.9A). Protraction of the lophophore involves relaxation of the retractor and atrial sphincter muscles, and contraction of the atrial dilators. The contracted muscles press outward against the atrial wall and force fluid into the basal

region of the exosaccal cavity, thereby protracting the polypide (Figure 21.9B). In addition, the wall of the membranous sac houses numerous, separate bands of circular muscles that are thought to aid in lophophore extension by increasing pressure in the entosaccal coelom.

Several methods of lophophore action have evolved among the gymnolaemates. Members of the order Ctenostomata possess an uncalcified, flexible zoecium composed of gelatinous, chitinous, or leathery material. Retraction of the lophophore is accomplished by the usual retractor muscle, which is aided by **longitudinal parietal muscles** that pull in the atrial chamber. When the lophophore is fully retracted, a sphincter contracts, closing the orifice and, in some species, folding a pleated **collar** over the end of the zooid (Figure 21.9D). Contraction of **transverse parietal muscles** pulls the cystid walls inward, thereby causing an increase in coelomic pressure that then protracts the lophophore (Figure 21.9C).

Cheilostomate ectoprocts are housed in zoecia that have various amounts of calcium carbonate deposited between the epidermis and an outer chitinous zoecial layer (Figure 21.10). The problem of creating changes in coelomic pressure has been solved here by the retention of special uncalcified parts of the cystid wall upon which muscles can act. Each zoecium is more or less boxlike (rather than erect or tubular). The outer surface of the box that bears the orifice is called the **frontal surface**. In many cheilostomates the **frontal membrane** is uncalcified and flexible (Figure 21.10). Contraction of **parietal protractor muscles** pulls the frontal membrane inward, thus increasing coelomic pressure and pushing out the lophophore. There are variations on this general theme, some of which are discussed by Perez and Banta (1996).

Nearly all cheilostomates possess a calcified operculum that closes over the orifice when the polypide is retracted, but the exposed frontal membrane presents a weakness in their defense against predation, and many species have evolved additional protective devices. Some forms, known as the **cribrimorph** ectoprocts, bear hard spines that project over the membrane and in some cases actually meet and fuse to form a cage above the vulnerable area (Figure 21.10B,C). In others a calcified partition, called the **cryptocyst**, lies beneath the frontal membrane, separating it from the soft parts within. The cryptocyst bears pores through which the protractor muscles extend (Figure 21.10F,G).

The most drastic modifications are seen in the so-called **ascophoran** cheilostomates, wherein the entire frontal surface is calcified except for a small opening. This opening, called the **ascopore**, leads inwards to a blind sac called the **compensation sac**, or **ascus**; this structure is an inwardly pouched, flexible portion of the body wall on which the protractor muscles insert (Figure 21.10H–J). Contraction of these muscles pulls the wall of the compensation sac inward as the ascopore allows water to enter the sac. Thus, pressure is exerted on the coelom and the lophophore is protracted. The ascophorans are the most diverse and successful group of ectoprocts.

Zooid Interconnections

Before continuing, there are some aspects of ectoproct colony organization that must be addressed. As discussed in Chapter 3, clear definitions of the term *colony* are somewhat elusive. This difficulty arises because it is not always easy to tell where one individual ends and another begins or because the degree of structural and functional communication among individuals is uncertain or variable. Ectoproct zooids, at least autozooids, are clearly demarcated by the elements of the polypide (lophophore, gut, and so on), but the way in which the zooids are interconnected differs among groups.

In phylactolaemates the metacoel is continuous among zooids, uninterrupted by septa (Figure 21.5A). Each zooid bears a tubular tissue cord, which is called a **funiculus** and extends from the inner end of the curved gut to the body wall. All other ectoprocts lack extensive coelomic connections, and the zooids are separated by various sorts of structural components. The walls of adjacent zooids of stenolaemates bear interzooidal pores that allow communication of exosaccal coelomic fluid

(Figure 21.9A,B). The funiculus is contained within the entosaccal space with the rest of the viscera and attaches the gut to the body wall.

Stoloniferous gymnolaemates (e.g., *Bowerbankia*; Figure 21.9C,D) have septa spaced along the stolons between the zooids. A cord of tissue passes along the stolons and through pores in each septum. This cord, called a **stolonal funiculus**, connects with the funiculus of each zooid arising from the stolon. In most non-stoloniferous gymnolaemates the cystid walls of adjacent zooids are pressed tightly together, producing what are called **duplex walls** (Figure 21.10A,J). These double walls bear pores with tissue plugs, which, again, usually connect with the funiculus of associated zooids.

It is clear, then, that ectoproct zooids are interconnected structurally, either by direct sharing of coelomic spaces or by funicular tissue. Functionally, these connections provide a means of distributing materials through the colony, and perhaps other communal activities as well. Some workers (Carle and Ruppert 1983) even suggest that the funiculus is homologous to a blood vessel. Perhaps these tissues are remnants of the circulatory system in other lophophorates. Other special functions of the funiculus are discussed later in the chapter.

The Lophophore, Feeding, and Digestion

Ectoprocts are unique among lophophorates in that the lophophore is retractable, by mechanisms already explained. The lophophore is horseshoe-shaped in the phylactolaemates (except for the primitive *Fredericella*) and circular in the other two classes; the tentacular epidermis is ciliated. Ectoprocts are typically suspension feeders, although supplemental methods occur. They apparently feed largely on protists and invertebrate larvae of appropriate size. The crescentic lophophore of phylactolaemates bears at its base a food groove that leads to the mouth and functions in a way similar to that described for phoronids. Feeding in other groups is somewhat different and has been more extensively studied.

Upon protraction, the tentacles of the circular lophophores of stenolaemates and gymnolaemates are erected in a funnel or bell-shaped arrangement around the mouth (Figure 21.11A). Each tentacle bears three ciliary tracts along its length, one **frontal tract** and two **lateral tracts** (Figure 21.11B). During normal suspension feeding, the lateral cilia create a current that enters the open end of the funnel, flows toward the mouth, and then out between the tentacles (Figure 21.11C). Some food particles are carried directly to the area of the mouth by the central flow of water. Other potential food, however, moves peripherally with the current toward the intertentacular spaces. When a particle contacts lateral cilia, a localized reversal of power stroke direction is initiated in those cilia, and the particle is tossed onto the frontal edge of the tentacle. The particle is repeatedly bounced in this fashion, from lateral to frontal, and is moved toward the mouth under the influence of a cur-

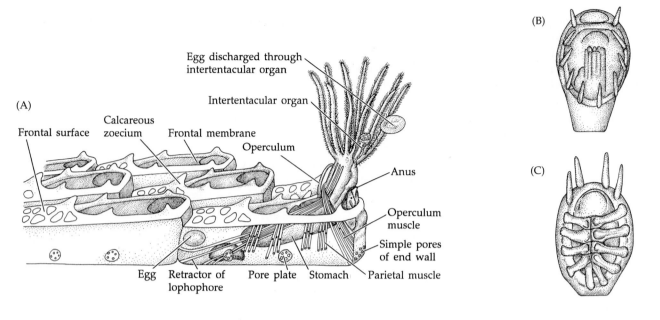

(A)

Frontal surface

Calcareous zoecium

Frontal membrane

Egg discharged through intertentacular organ

Intertentacular organ

Operculum

Anus

Operculum muscle

Simple pores of end wall

Parietal muscle

Egg

Retractor of lophophore

Pore plate

Stomach

(B)

(C)

Frontal membrane

Parietal muscle

(D)

(E)

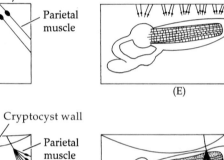

Frontal membrane

Cryptocyst wall

Parietal muscle

(F)

(G)

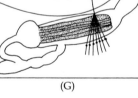

Compensation sac

Parietal muscle

(H)

(I)

Figure 21.10 (A) A portion of a colony of *Electra*, with a cutaway view of one zooid (Cheilostomata). (B,C) Two species of *Callopora*. Note the calcareous spines projecting over the frontal membrane. (D–I) Parietal muscles and frontal membranes in cheilostomates. (D, F, and H are cross sections; E, G, and I are longitudinal sections.) (D,E) Zooid with unprotected frontal membrane. (F,G) Zooid with porous cryptocyst beneath frontal membrane. (H,I) Ascophoran zooid with compensation sac and calcified frontal membrane. (J) Internal anatomy of an ascophoran cheilostomate zooid with an ovicell.

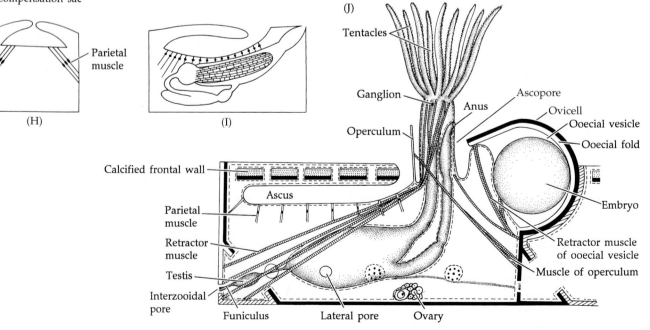

(J)

Tentacles

Ganglion

Operculum

Anus

Ascopore

Ovicell

Ooecial vesicle

Ooecial fold

Calcified frontal wall

Ascus

Parietal muscle

Retractor muscle

Testis

Interzooidal pore

Funiculus

Lateral pore

Ovary

Embryo

Retractor muscle of ooecial vesicle

Muscle of operculum

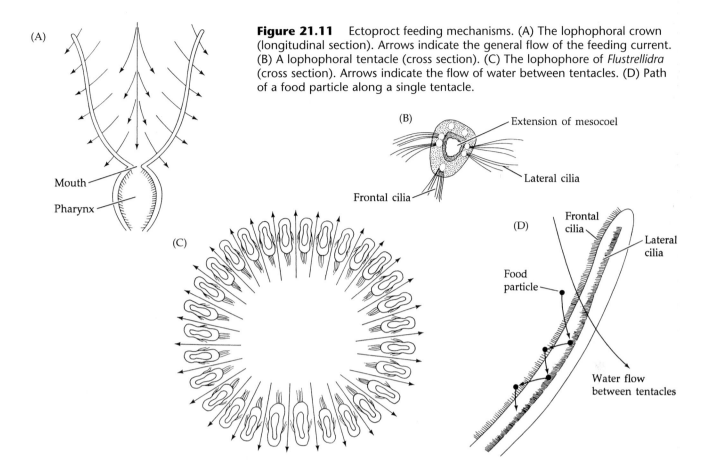

Figure 21.11 Ectoproct feeding mechanisms. (A) The lophophoral crown (longitudinal section). Arrows indicate the general flow of the feeding current. (B) A lophophoral tentacle (cross section). (C) The lophophore of *Flustrellidra* (cross section). Arrows indicate the flow of water between tentacles. (D) Path of a food particle along a single tentacle.

rent generated by the frontal cilia (Figure 21.11D).

Many ectoprocts augment suspension feeding by various means that allow them to capture relatively large food particles, including live zooplankton. Winston (1978) demonstrated that many species engage in flicking movements of individual tentacles with which a food particle has come in contact. By this means a single tentacle curls and strikes at the particle, moving it to the mouth. At least one species (*Bugula neritina*) is capable of trapping zooplankton by folding its tentacles over the prey and pulling it to the mouth. A number of ectoprocts rock or rotate the entire lophophore, apparently "sampling" reachable water for food material (Figure 21.12A).

In some ectoprocts the zooids of the colony function together in feeding and rejection of waste or nonfood materials. In many genera (e.g., *Cauloramphus* and *Hippothoa*) groups of zooids "cooperate" to produce general currents that bring water to several clustered zooids and then flow away via "excurrent chimneys" between the clusters (Figure 21.12B,C). Such currents, which move larger amounts of water over the lophophores than could be moved by individual zooids, may be especially important to colonies inhabiting quiet water. The generation of strong excurrent water flow away from the colony surface helps to push nonfood material and feces far enough to reduce the possibility of recycling. In some ectoprocts, such as *Cauloramphus spiniferum*, large parti-

cles are actually passed from zooid to zooid and then dumped into an excurrent chimney (Figure 21.12D).

As in all lophophorates, the digestive tract of ectoprocts is U-shaped (Figures 21.5A, 21.9, 21.10, and 21.13). The mouth lies within the lophophoral ring, and in the Phylactolaemata it is overlaid by an epistome. Ciliary tracts lead into the mouth from the surrounding **peristomial field**. Internal to the mouth is a muscular pharynx. A valve separates the lower end of the pharynx from the descending portion of the stomach, which is called the **cardia** and in some species is modified as a grinding gizzard. The cardia leads to a **central stomach** from which arises a large cecum; the funiculus attaches to the cecum. The ascending portion of the stomach, or **pylorus**, also arises from the central stomach and leads to a proctodeal rectum and the anus, which lies outside the lophophoral ring. The flow of material from the pylorus to the rectum is controlled by a sphincter. In phylactolaemates an esophagus precedes the stomach, and the hindgut is elongated as an intestine.

Ingestion is accomplished by the sweeping action of the peristomial and oral cilia and by muscular contractions of the pharynx. Digestion begins extracellularly in the cardia and central stomach, and is completed intracellularly in the cecum. Food is moved through the gut by peristalsis and cilia. Undigested material is rotated and formed into a spindle-shaped mass by the cilia of the pylorus and then passed to the rectum for expulsion.

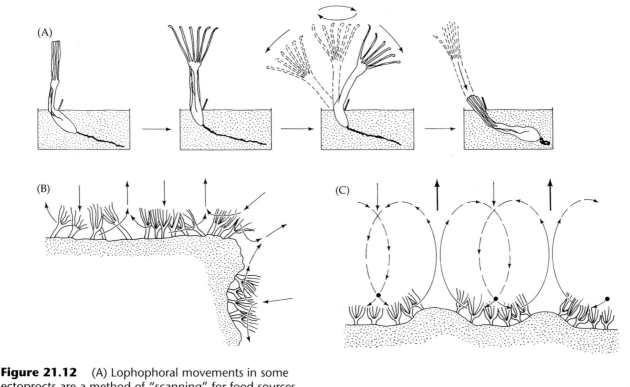

Figure 21.12 (A) Lophophoral movements in some ectoprocts are a method of "scanning" for food sources. The lophophore is protracted, held erect, rocked and rotated, and then withdrawn. (B,C) Interzooidal "cooperation" in the production of feeding currents, including "excurrent chimneys." (B) *Hippothoa.* (C) *Cauloramphus.* (D) Cooperative rejection of a large particle (*Cauloramphus*). The particle is passed to an excurrent chimney.

Circulation, Gas Exchange, and Excretion

Circulation of metabolites in single zooids is by diffusion, because there is no structural system for this purpose. Given the small size of these animals, intrazooid diffusion distances are small, and the coelomic fluid provides a medium for passive transport. Interzooid circulation is facilitated by the confluent coelom in phylactolaemates, the cystid pores in stenolaemates, and the funicular cords of most gymnolaemates. Gas exchange occurs across the walls of the protracted parts of the polypide, particularly the lophophore, the tentacles providing a very high surface area. Ectoprocts contain no respiratory pigments, and gases are carried in solution.

Metabolic waste products are accumulated and transported by phagocytic coelomocytes. The elimination of these wastes is not fully understood, but apparently it occurs in part by the formation of structures called **brown bodies**. The appearance of brown bodies is usually associated with the degeneration of polypides in adverse or stressful conditions; this degeneration is followed by reformation of a new polypide. In most gymnolaemates a brown body is left within the cystid following polypide degeneration, but in some cheilostomates the new polypide regenerates in such a way that the brown body is housed within the gut of the new zooid and is then expelled via the anus. Note that the new polypide forms entirely from the tissue components of the cystid—that is, from the epidermis and peritoneum of the body wall. In most stoloniferous ctenostomates the old cystid with its brown body drops from the colony, and an entire new zooid regenerates from the stolon. Cyclostomates and some other ectoprocts tend to form brown bodies within the coelom. In all cases, it is presumed that metabolic wastes are precipitated and concentrated in the brown bodies and thus eliminated or at least rendered inert.

Nervous System and Sense Organs

In concert with their sessile lifestyle and the general reduction of the anterior end, the ectoproct nervous system and sense organs are predictably reduced. A neuronal mass, or cerebral ganglion, lies dorsally in the mesosome near the pharynx. Arising from this structure is a circumenteric nerve ring. Nerves extend from the ring and ganglion to the viscera, and motor and sensory nerves extend into each tentacle. Interzooidal nerve fibers occur in some species, but their function remains unclear. The only known receptors are tactile cells on the

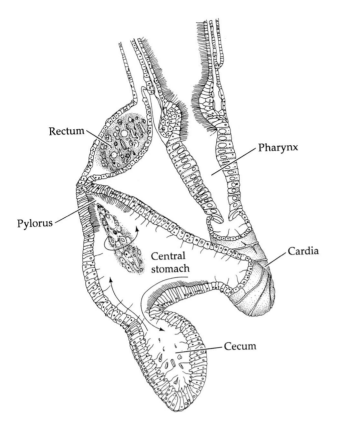

Figure 21.13 Digestive tract of *Cryptosula pallasiana*. The arrows show food movement caused by ciliary action.

lophophore and on avicularia. The planktonic larvae of at least some ectoprocts exhibit a marked negative geotaxis prior to settling. Experiments suggest that this geotaxis is a direct response to gravity, but the mechanism mediating this phenomenon is unknown. Larvae also

usually have well developed ocelli and are positively phototactic while free swimming. Settlement is often accompanied by a shift to a negative phototaxis.

Reproduction and Development

As in most colonial animals, asexual reproduction is an indispensable part of the life history of ectoprocts and is responsible for colony growth and regeneration of zooids. Except for the unique cases of polyembryony in stenolaemates (see below), each colony begins from a single, sexually produced, primary zooid called the **ancestrula** (Figure 21.14A). The ancestrula undergoes asexual budding to produce a group of daughter zooids, which themselves subsequently form more buds, and so on. The initial group of daughter zooids may arise in a chainlike series, a plate, or a disc; the budding pattern determines the growth form of the colony and is highly variable among species.

Asexual reproduction. Budding involves only elements of the body wall. In most gymnolaemates a partition forms that isolates a small chamber, the developing bud, from the parent zooid. The bud initially includes only components of the cystid and an internal coelomic compartment. A new polypide is then generated from the living tissues of the bud (the epidermis and the peritoneum). The epidermis and peritoneum invaginate, the former producing the lophophore and the gut. The peritoneum produces all of the new coelomic linings and the funiculus. Budding in phylactolaemates and stenolaemates is similar, except that the polypide develops first and is then encased by a new cystid wall.

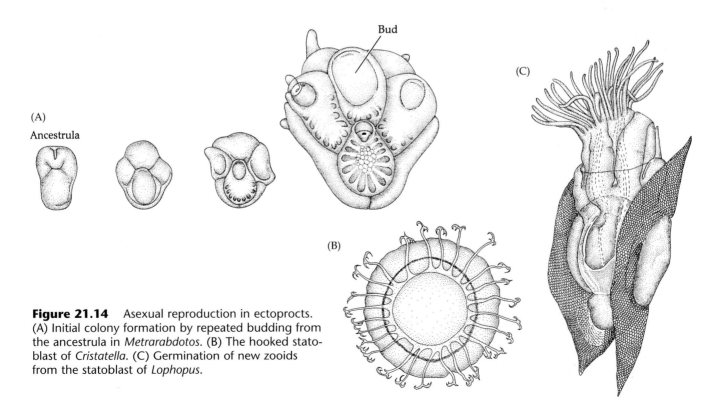

Figure 21.14 Asexual reproduction in ectoprocts. (A) Initial colony formation by repeated budding from the ancestrula in *Metrarabdotos*. (B) The hooked statoblast of *Cristatella*. (C) Germination of new zooids from the statoblast of *Lophopus*.

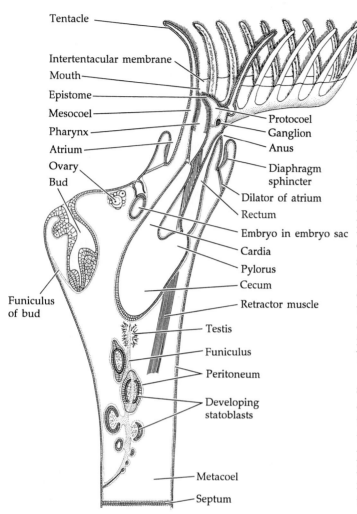

Tentacle

Intertentacular membrane

Mouth

Epistome

Mesocoel

Pharynx

Atrium

Ovary

Bud

Funiculus
of bud

Protocoel

Ganglion

Anus

Diaphragm
sphincter

Dilator of atrium

Rectum

Embryo in embryo sac

Cardia

Pylorus

Cecum

Retractor muscle

Testis

Funiculus

Peritoneum

Developing
statoblasts

Metacoel

Septum

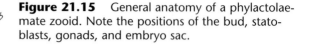

Figure 21.15   General anatomy of a phylactolae-mate zooid. Note the positions of the bud, stato-blasts, gonads, and embryo sac.

more commonly, may include both male and female individuals. Gametes usually arise from transient patches of germinal tissue developed from special areas of the metacoel peritoneum or the funiculus (Figures 21.9, 21.10, and 21.15). Only in the stenolae-mates is any real "organ" present; that is, the testis is surrounded by a discrete cellular lining. Gametes are proliferated into the metacoel and migrate to the mesocoel prior to release. Sperm migrate into the coelomic lumina of the tentacles and, at least in some species, escape through special coelomopores on par-ticular tentacles. A few cheilostomates (e.g., *Electra* and *Membranipora*) exhibit free spawning of eggs as well as of sperm, and fertilization and development are fully external. In all other ectoprocts thus far stud-ied, the ova are retained by the parental zooids and brooded at least during early ontogeny.

In those gymnolaemates that release their ova to the sea water or to some external brooding area, the eggs are shed from the mesocoel through an opening called the **supraneural pore** located between the bases of two tentacles. In some species this pore is elevated on a pedestal called the **intertentacular organ** (Figure 21.10A). A few ctenostomates (e.g., *Nolella* and *Victorella*) retain the ova within the coelom, where development takes place. Stenolaemates and phylactolaemates brood their embryos, the former in special individuals called **gonozooids** that are modified by loss of the polypide, and the latter in **embryo sacs** produced by invagina-tions of the body wall (Figure 21.15).

A variety of brooding methods occurs among gym-nolaemates, usually involving the formation of an exter-nal brooding area called an **ovicell**, or **ooecium** (Figure 21.7E). The most detailed and complete studies on the formation and functioning of these structures have been done by R. M. Woollacott and R. L. Zimmer (see refer-ences) on the cheilostome *Bugula neritina*. In this species, and probably many others, the ovicell develops from evaginations of the body wall of the parent autozooid. One of these evaginations is the **ooecial vesicle**, the lumen of which is confluent with the coelom of the ma-ternal zooid. The other evagination is called the **ooecial fold**; this structure forms a hoodlike covering of the ovi-cell. The embryo develops between the ooecial vesicle and fold (Figure 21.16). In many species the coelomic connection probably provides an avenue for nutrient transfer from parent to embryo. In *B. neritina* an actual tissue union develops between the epithelium of the ooecial vesicle and funicular extensions of the parent autozooid, producing a kind of placental system.

Ectoprocts undergo radial, holoblastic, nearly equal cleavage to form a coeloblastula. Subsequent develop-

In addition to budding, freshwater ectoprocts (Phy-lactolaemata) reproduce asexually by the formation of **statoblasts** (Figures 21.5, 21.14B,C, and 21.15). These structures are extremely resistant to drying and freez-ing, and are often produced in huge numbers during adverse environmental conditions. Statoblasts generally form on the funiculus of an autozooid and include peri-toneal and epidermal cells plus a store of nutrient mate-rial. Each cellular mass secretes a pair of chitinous pro-tective valves, differing among species in shape and ornamentation. The parent colony usually degenerates, freeing the statoblasts. Some statoblasts sink to the bot-tom, but others float by means of enclosed gas spaces. Some bear surface hooks or spines and are dispersed by passive attachment to aquatic animals or vegetation. With the return of favorable conditions, the cell mass generates a new zooid, which sheds its outer casing and attaches as a functional individual.

Sexual reproduction.   Most ectoprocts are hermaph-roditic, and each zooid is capable of producing sperm and eggs. The colonies of dioecious species (e.g., some chelostomates) may consist of zooids of one sex or,

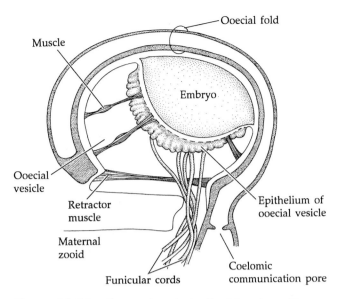

Figure 21.16 An ooecium (ovicell) of *Bugula neritina*. Note the tissue connection and coelomic communication with the parent zooid. (See also Figure 21.10J.)

ment differs greatly among groups, but in all cases it involves a free-swimming dispersal form. Thus, development is either fully indirect (in those few species that free spawn) or mixed, with a planktonic stage following a period of brooding. Very little solid information exists on the derivation and fates of germ layers in ectoprocts. This is especially true for mesoderm and coelomic linings. It appears certain, however, that there is no indication of a 4d mesentoblast precursor for mesoderm, or any other convincing evidence of a protostome affinity.

In phylactolaemates the coeloblastula develops into a cystid-like stage lacking entoderm and then generates a polypide in a fashion similar to bud formation. This zooid precursor is ciliated and escapes the embryo sac for a short swimming life before settling and attaching. The embryos of stenolaemates cleave to form a hollow ball, probably homologous to a coeloblastula. At this point, however, the embryo undergoes a budding process, forming secondary embryos, which in turn bud tertiary embryos. In some cases hundreds of small, solid, asexually produced embryos may result from a single primary ball of cells. This phenomenon of **polyembryony** is unique to these animals and may represent a heterochronic displacement of the usual asexual budding process of other ectoprocts. Each embryo develops cilia and escapes as a simple "larva," which settles and undergoes a metamorphosis similar to that described below for gymnolaemates.

The coeloblastulae of gymnolaemates undergo gastrulation by delamination; in this process four cells divide such that one of each pair of daughter cells is shunted to the blastocoel as presumptive entoderm and mesoderm. Free-swimming larvae are eventually produced. Many of the species that free spawn have a characteristic, flattened, triangular larva called a **cyphonautes** (Figure 21.17A). These larvae have a functional gut and may remain in the plankton for months, whereas the larvae of brooding species lack a digestive tract and lead very short, pelagic lives (Figure 21.17B,C). Despite these differences, gymnolaemate larvae have some fundamental similarities. For example, they characteristically possess a sensory **pyriform organ complex** and a pouchlike **adhesive sac**, both of which are important in settling and metamorphosis (Figure 21.17). Some ctenostomate ectoprocts produce nonfeeding **vesiculariform** larvae (Zimmer and Woollacott 1993).

As mentioned earlier, ectoproct larvae are at first positively phototactic, and most possess pigment spots that

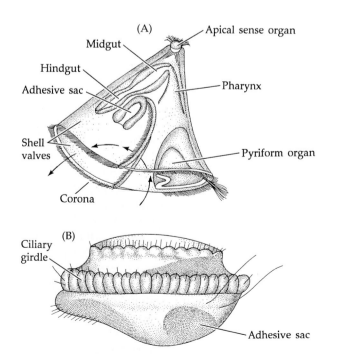

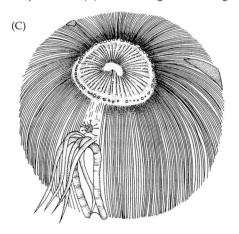

Figure 21.17 Ectoproct larvae. (A) The cyphonautes larva of *Membranipora*. The arrows indicate the direction of the feeding currents. (B) Nonfeeding larva of *Alcyonidium*. (C) Nonfeeding larva of *Bugula*.

are thought to be light sensitive. The pigment spots are ciliary in origin, supporting further the deuterostome alliance of the ectoprocts. Following a planktonic phase, the larvae usually become negatively phototactic and swim toward the bottom. Once in contact with the substratum, the pyriform organ complex is apparently used to test for chemical and tactile cues reflecting the suitability of the substratum for settling. Once a proper surface has been "selected," the adhesive sac everts and secretes sticky material for attachment. After attachment, there is a remarkable reorganization of tissue positions accompanied by histolysis of various larval structures. The metamorphosed larva then generates the primary zooid, or ancestrula. The most detailed account of this process is, again, by Woollacott and Zimmer (1971) for *Bugula neritina*.

The Brachiopods

Members of the phylum Brachiopoda (Greek *brachium*, "arm"; *poda*, "feet") are called the lamp shells (Box 21C and Figure 21.18). All are solitary, marine, benthic creatures. The body, including the lophophore, is enclosed between a pair of dorsoventrally oriented valves. Most brachiopods are attached to the substratum by a fleshy **pedicle** (Figure 21.18). Some species lack a pedicle (e.g., *Crania*), and these usually cement themselves directly to a hard substratum. On the other hand, some species that possess a pedicle do not form permanent attachments, such as *Magadina cumingi*, which lies free, and *Lingula*, which anchors in loose sand (Figure 21.18C). A few species possess both unattached and attached populations (e.g., *Neothyris lenticularis* and *Terebratella sanguinea*).

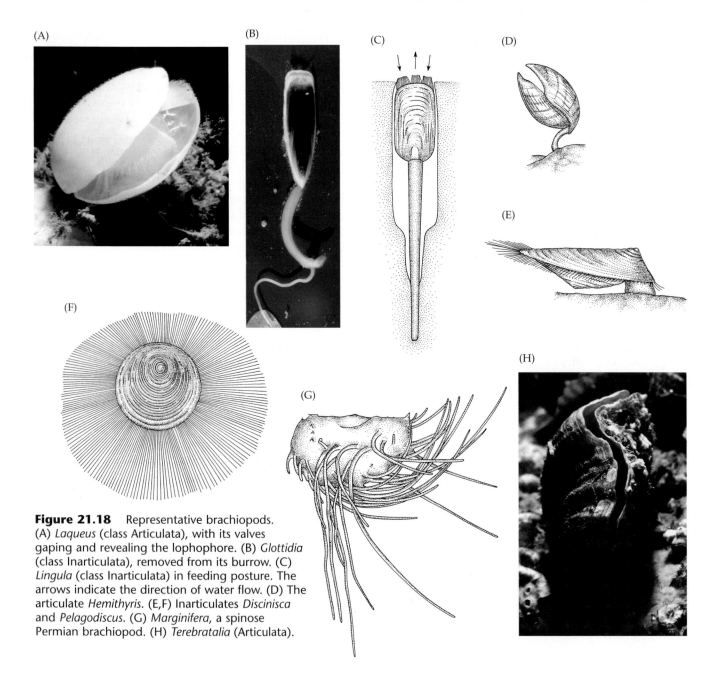

Figure 21.18 Representative brachiopods. (A) *Laqueus* (class Articulata), with its valves gaping and revealing the lophophore. (B) *Glottidia* (class Inarticulata), removed from its burrow. (C) *Lingula* (class Inarticulata) in feeding posture. The arrows indicate the direction of water flow. (D) The articulate *Hemithyris*. (E,F) Inarticulates *Discinisca* and *Pelagodiscus*. (G) *Marginifera*, a spinose Permian brachiopod. (H) *Terebratalia* (Articulata).

The shells are usually unequal, except in some Inarticulata like *Lingula* and *Glottidia*, and are attached to one another posteriorly either by a tooth-and-socket hinge (Articulata) or simply by muscles (Inarticulata). Brachiopods normally "sit" ventral side up, the pedicle arising from the ventral or **pedicle valve**; the dorsal shell is called the **brachial valve** (Figures 21.19A and 21.21A).

Most brachiopods measure 4 to 6 cm along the greatest shell dimension, but range from 1 mm to over 9 cm in extreme cases. Although they are known from nearly all ocean depths, they are most abundant on the continental shelf. The approximately 335 living species represent a small surviving fraction of the more than 12,000 extinct species that have been described. Their rich fossil record dates back at least 600 million years. Brachiopods, especially articulates, were among the most abundant animals of the Paleozoic, but they declined in numbers and diversity after that time. Thayer (1985) has presented experimental evidence that competition with epibenthic bivalve molluscs was at least partly responsible for the reduction in brachiopod diversity following their Paleozoic success.

PHYLUM BRACHIOPODA

CLASS INARTICULATA: Valves not hinged, attached by muscles only; valves of organic composition, including chitin, or else calcareous; pedicle (absent in a few species) usually with intrinsic muscles and a coelomic lumen; epistome with coelomic channels confluent with lophophoral mesocoel; lophophore without internal skeletal support; anus present. Two orders, Lingulida and Acrotretida, comprising about 45 extant species. (e.g., *Crania, Discinisca, Glottidia, Lingula, Pelagodiscus*)

CLASS ARTICULATA: Valves articulate by tooth-and-socket hinge; valves composed of scleroprotein and calcium carbonate; pedicle usually present, but lacking muscles and coelomic lumen; epistome small and tissue filled; lophophore generally with internal supportive elements; gut ends blindly, anus lacking. Three extant orders: Rhynchonellida, Terebratulida, and Thecideidina, with just over 290 species. (e.g., *Argyrotheca, Dallina, Gryphus, Hemithyris, Lacazella, Laqueus, Liothyrella, Magellania, Thecidellina, Terebratalia, Terebratella, Terebratulina, Tichosina*)

The Body Wall, Coelom, and Support

The shells of brachiopods comprise an outer organic **periostracum** and an inner structural layer or layers composed variably of calcium carbonate, calcium phosphate, scleroproteins, and chitinophosphate. Various spines are present in some species as outgrowths of the periostracum and serve to anchor the animals in place (Figure 21.18). In a fashion similar to that of molluscs, brachiopod shells are secreted by **mantle lobes**, which are formed as outgrowths of the body wall (Figure 21.19). The periostracum is secreted by the mantle edges, and the inner shell layer by the general mantle surface. Shells of many brachiopods bear perforations, or **punctae**, ex-

---

BOX 21C *Characteristics of the Phylum Brachiopoda*

1. Trimeric, enterocoelic, coelomate lophophorates

2. Epistome present, with or without coelomic lumen

3. Body enclosed between two shells (valves), one oriented dorsal and one ventral

4. Usually attached to the substratum by a stalk, or pedicle

5. Valves lined (and produced) by mantle lobes formed by outgrowths of the body wall and creating a water-filled mantle cavity

6. Trimeric condition partially obscured by modified body form

7. Lophophore circular to variably coiled, with or without internal skeletal support

8. Gut U-shaped; anus present or absent

9. One or two pairs of metanephridia

10. Circulatory system rudimentary and open

11. Most are dioecious and undergo mixed or indirect life histories, with lobate larva

12. Gametes develop from transient gonadal tissue on peritoneum of metacoel

13. Cleavage holoblastic, radial, and nearly equal; coeloblastulae usually gastrulate by invagination; blastopore closes and mouth forms secondarily (as does anus)

14. Solitary, benthic, marine

---

tending from their inner surfaces nearly to the periostracum and containing tiny tissue extensions of the mantle (Figure 21.19B). The function of these mantle papillae is unknown, but some workers have suggested that they might serve as areas for food storage and gas exchange, or in some way deter the activities of borers. Shells that lack perforations are termed **impunctate**.

The soft mantle lobes line and are attached to the shells and form the water-filled **mantle cavity**, which houses the lophophore. The mantle edges often bear chitinous setae, which may protect the fleshy tissue and perhaps serve to prevent the entrance of large particles into the mantle cavity.

The epidermal cells of the mantle lobes and general body surface vary from cuboidal to columnar and are densely ciliated on the lophophore. Beneath the epidermis is a connective tissue layer of varying thickness, which houses longitudinal muscle fibers where the body is not attached to the valves. The inner surface of the body wall is lined by peritoneum, which forms the

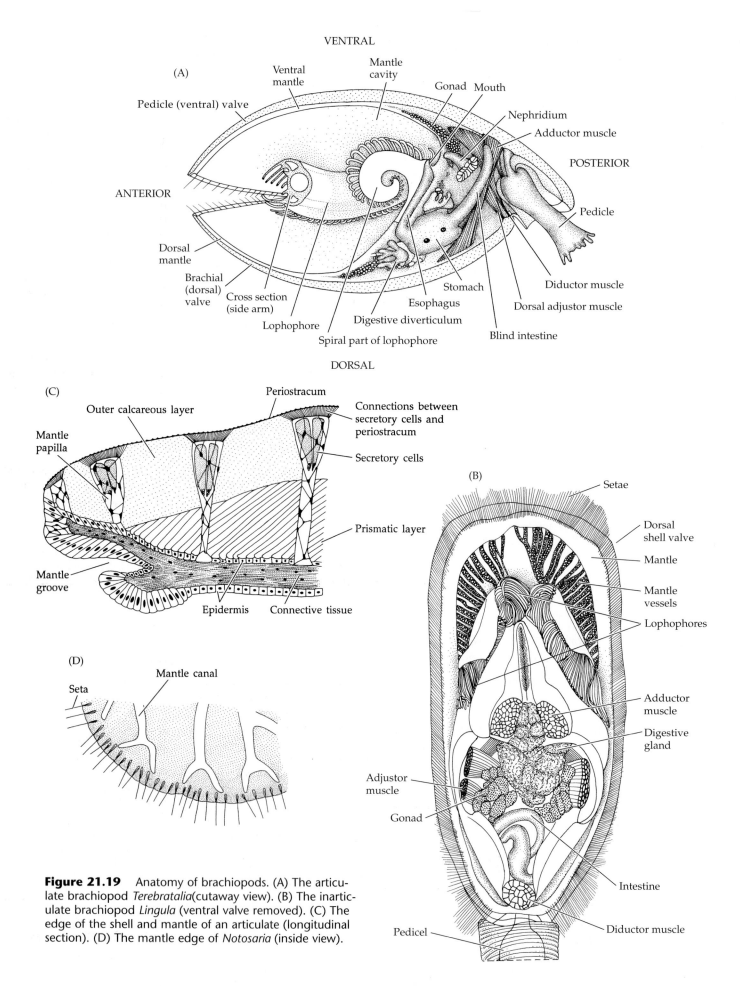

VENTRAL

(A)

Ventral mantle

Mantle cavity

Gonad Mouth

Pedicle (ventral) valve

Nephridium

Adductor muscle

POSTERIOR

ANTERIOR

Pedicle

Dorsal mantle

Brachial (dorsal) valve

Cross section (side arm)

Lophophore

Spiral part of lophophore

Digestive diverticulum

Esophagus

Stomach

Diductor muscle

Dorsal adjustor muscle

Blind intestine

DORSAL

(C)

Outer calcareous layer

Periostracum

Connections between secretory cells and periostracum

Mantle papilla

Secretory cells

Prismatic layer

Mantle groove

Epidermis   Connective tissue

(B)

Setae

Dorsal shell valve

Mantle

Mantle vessels

Lophophores

Adductor muscle

Digestive gland

Adjustor muscle

Gonad

Intestine

Pedicel

Diductor muscle

(D)

Seta

Mantle canal

Figure 21.19 Anatomy of brachiopods. (A) The articulate brachiopod *Terebratalia* (cutaway view). (B) The inarticulate brachiopod *Lingula* (ventral valve removed). (C) The edge of the shell and mantle of an articulate (longitudinal section). (D) The mantle edge of *Notosaria* (inside view).

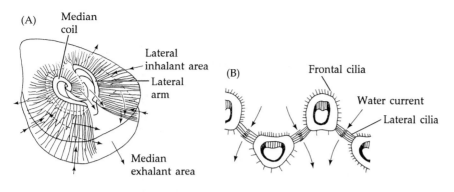

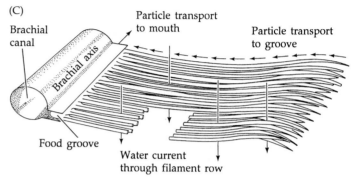

Figure 21.20 Feeding currents in brachiopods. (A) Feeding currents (arrows) of *Waltonia*. (B) Lophophoral tentacles (section). The water (arrows) passes over lateral ciliary bands. (C) A portion of a lophophore. Food particles are transported along tentacles and the brachial food groove (arrows).

outer boundary of the coelom. Being folds of the body wall, the mantle lobes contain extensions of the coelom, called **mantle canals** (Figure 21.19D).

The pedicle is an outgrowth of the body wall, arising from the posterior area of the ventral valve (Figure 21.18 and 21.19A). In inarticulates it contains all the usual layers beneath the epidermis, including connective tissue, muscles, and a coelomic lumen. However, the pedicle of articulates lacks muscles and a coelomic cavity. In the latter case the pedicle is operated by extrinsic muscle bands from the body wall itself. In brachiopods that attach firmly, the tip of the pedicle bears papillae or finger-like extensions that adhere tightly to the substratum.

The coelomic system of brachiopods includes the typical mesocoel and metacoel as the lophophoral and body coeloms, respectively. The epistome is solid in the articulates, but in inarticulates it contains a protocoel that is confluent with mesocoel. The coelomic fluid includes various coelomocytes, some of which contain hemerythrin.

The Lophophore, Feeding, and Digestion

Like that of phoronids and ectoprocts, the lophophore of brachiopods comprises a ring of tentacles surrounding the mouth. In brachiopods however, the lophophore is produced as a pair of tentacle-bearing arms that extend anteriorly into the mantle cavity. The overall shape of the lophophore varies among taxa from a simple circular or U-shape to those with highly coiled arms (Figure 21.19 and 21.20). The brachiopod lophophore also differs in that it is always contained within the protection of the valves and is essentially immovable. In inarticulates, the lophophore and tentacles are held in position by coelomic pressure, whereas in articulates the tentacle-bearing ridge includes supportive skeletal elements. In addition, the dorsal valve often bears inwardly directed ridges and grooves that help support and position the lophophore.

In order to pass a water current through the mantle cavity, the two valves must be opened slightly. The mechanisms of valve operation differ between members of the two classes. Articulate brachiopods possess several sets of muscles including a pair of **diductor muscles**, which

open the valves (Figure 21.21A). The tooth-and-socket hinges prevent a large gape. The adductor muscles include both striated and smooth fibers such that the valves can be quickly closed and then held together for long periods of time. Inarticulates lack a hinge and do not possess diductor muscles. Instead, the gape is produced by retraction of the body, an action that increases the internal pressure in the coelomic fluid and forces the valves apart. Adductor muscles are used to close the valves.

Feeding currents are generated by the lophophoral cilia. Specific incurrent and excurrent flow patterns occur, varying with shell morphology and the shape and orientation of the lophophore. In any case, water is directed over and between the tentacles before passing out of the mantle cavity (Figure 21.20A). Each tentacle bears lateral and frontal ciliary tracts (Figure 21.20B). The lateral cilia of adjacent tentacles overlap and redirect food particles from the water to the frontal cilia by beat reversal. The frontal cilia beat toward the base of the tentacles, helping to direct trapped food. The lophophoral ridge, or **brachial axis**, bears a **brachial food groove** within which food material is moved to the mouth (Figure 21.20C). Brachiopods feed on nearly any appropriately small organic particles, especially phytoplankton.

The digestive system is U-shaped (Figure 21.19, and 21.21B,C). The mouth is followed by a short esophagus, which extends dorsally and then posteriorly to the stomach. A digestive gland covers most of the stomach and connects to it via paired ducts. The intestine extends posteriorly, where it ends blindly in articulates or recurves as a rectum terminating in an anal opening in inarticulates. In the latter case the anus opens either medially or on the right side of the mantle cavity. The ab-

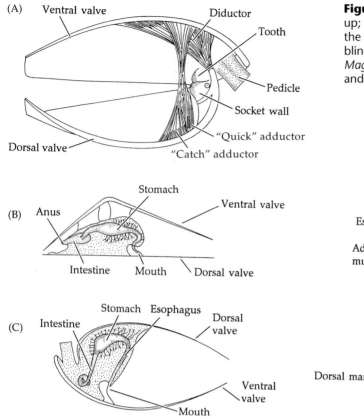

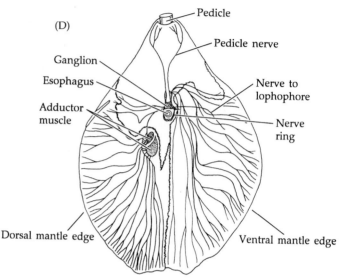

Figure 21.21 (A) An articulate brachiopod (ventral side up; cutaway view). Note the major muscles that operate the valves. (B) The complete gut of an inarticulate. (C) The blind gut of an articulate. (D) The nervous system of *Magellania*. Note the dorsal and ventral aspects on the left and right sides of the drawing, respectively.

sence of an anus is almost certainly a secondary loss in the articulates, and may be associated with the evolution of the articulate hinge, which restricts the posterior flow of water from the mantle cavity.

Little is known about digestion in brachiopods, but some work on *Lingula* (Chuang 1959) indicates that it occurs intracellularly in the digestive gland.

Circulation, Gas Exchange, and Excretion

The brachiopod circulatory system is open, much reduced, and largely unstudied. A contractile heart lies in the dorsal mesentery just above the gut (*Crania* possesses several "hearts"). Leading anteriorly and posteriorly from the heart are channels bounded only by mesentery peritoneum, thus no true vessels are present. These channels branch to various parts of the body, but the pattern of circulation is not fully understood. It appears that the blood is separate from the coelomic fluid, although both contain certain similar cells. The function of the circulatory system is thought to be largely restricted to nutrient distribution.

Gas exchange probably occurs across the general body surface, especially the tentacles and mantle. These structures not only provide large surface areas but are also sites over which water moves and is brought close to underlying coelomic fluid. This general arrangement and the presence of hemerythrin in certain coelomocytes suggest that the coelomic fluid, not the blood, is the medium for oxygen transport.

Brachiopods possess one or two pairs of metanephridia, with the nephrostomes opening to the metacoel. The nephridioducts exit through pores into the mantle cavity. The nephridia function as gonoducts as well as discharging phagocytic coelomocytes that have accumulated metabolic wastes.

Nervous System and Sense Organs

The nervous system of brachiopods is somewhat reduced. A dorsal ganglion and a ventral ganglion lie against the esophagus and are connected by a circumenteric nerve ring. Nerves emerge from the ganglia and nerve ring and extend to various parts of the body, especially the muscles, mantle, and lophophore (Figure 21.21D).

As usual, the array of sense organs in these animals is compatible with their lifestyle. The mantle edges and setae are richly supplied with sensory neurons, probably tactile receptors. There is also evidence that brachiopods are sensitive to dissolved chemicals, perhaps through surface receptors on the tentacles or mantle edge. Members of at least one species of *Lingula* possess a pair of statocysts, which are associated in this burrowing form with orientation in the substratum.

Reproduction and Development

Asexual reproduction does not occur in brachiopods. Most species are dioecious, with gametes developing from patches of transient, gonadal tissue derived from the metacoel peritoneum. Gametes are released into the

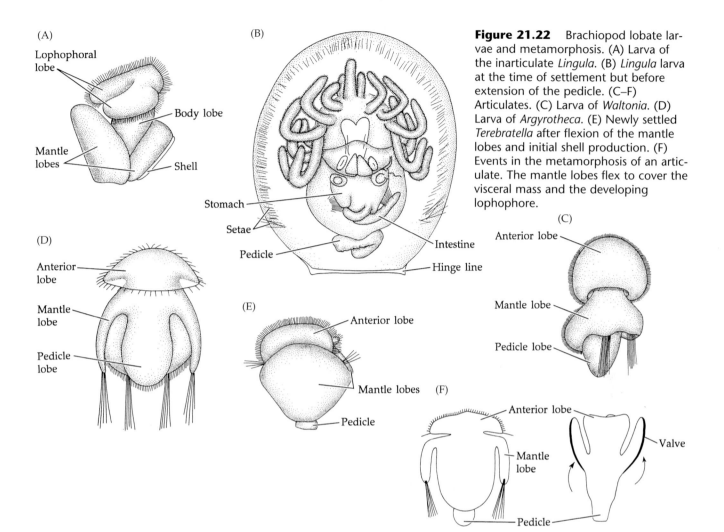

Figure 21.22 Brachiopod lobate larvae and metamorphosis. (A) Larva of the inarticulate *Lingula*. (B) *Lingula* larva at the time of settlement but before extension of the pedicle. (C–F) Articulates. (C) Larva of *Waltonia*. (D) Larva of *Argyrotheca*. (E) Newly settled *Terebratella* after flexion of the mantle lobes and initial shell production. (F) Events in the metamorphosis of an articulate. The mantle lobes flex to cover the visceral mass and the developing lophophore.

metacoel and escape through the nephridia. In most cases both eggs and sperm are shed freely and fertilization is external. A few species, however, brood their embryos until the larval stage is reached. In these cases sperm are picked up in the water currents of females and the eggs are retained in a brooding area, where they are fertilized. *Argyrotheca*, for example, broods its embryos in the enlarged nephridia. Others retain their embryos on the arms of the lophophore, in special regions of the mantle cavity, or in modified depressions in a valve.

Cleavage is holoblastic, radial, and nearly equal; it leads to the formation of a coeloblastula. Gastrulation is by invagination, except in the brooding form *Lacazella*, where it apparently occurs by delamination. The blastopore closes and the mouth forms secondarily. The anus, when present, breaks through late as the gut grows and approaches the body wall. Mesoderm and coelom formation are clearly enterocoelic. All of these developmental features bear witness to the deuterostome affinities of the Brachiopoda.

Whether the developmental pattern is mixed or fully indirect, all brachiopods eventually enter a free-swimming larval stage (Figure 21.22), to which we have ap-

plied the term **lobate larva**—in reference to the body regions visible as primordia at this stage and to the existing terminology traditionally used to describe these regions. The larvae of articulates and inarticulates differ morphologically and in the events at the time of settling. In inarticulates, such as *Lingula*, the larva is constructed much like the adult, except the pedicle is curled inside the mantle cavity and the body and lophophoral lobes are disproportionately large compared with the mantle lobes (Figure 21.22A,B). Thus the lophophore can be protruded out from between the mantle lobes and function to propel and feed the larva. The mantle lobes lie dorsoventrally on the body. Shell secretion commences early and, with added weight, the larva sinks, the pedicle is extended, and the juvenile brachiopod assumes benthic life. Thus, there is no drastic metamorphosis at the time of settling.

The free-swimming larva of articulates is regionalized into an **anterior lobe**, a **mantle lobe**, and a **peduncular**, or **pedicle**, **lobe** (Figure 21.22C–F). The mantle flaps are reflexed posteriorly along the sides of the presumptive pedicle, rather than anteriorly over the body as seen in inarticulates. After a short larval life of 1 or 2 days, the larva settles and metamorphoses. As the pedicle attaches to the substratum, the mantle lobes flex forward over the anterior lobe. The now exterior surfaces of the mantle lobes commence secretion of the valves, while the anterior lobe differentiates into the body and the lophophore.

Lophophorate Phylogeny

Despite continuing arguments, we remain convinced that the lophophorates are deuterostomes. They may also be monophyletic clade. In addition to the lophophore, these three phyla (Phoronida, Ectoprocta, and Brachiopoda) are united by their possession of U-shaped digestive tracts, peritoneal gonads, metanephridia (absent in ectoprocts), a diffuse nervous system, epistemial flaps, and a tendency to secrete outer casings. It is possible that these features are homologous within this clade but plesiomorphic or convergent with similar conditions in some other phyla. As discussed later in this section, each of the three lophophorate phyla displays enough derived character states to merit separate taxon status, but the idea of a "superphylum" (perhaps Lophophorata) may be warranted.

Zimmer (1973) has critically made the case for the deuterostome nature of the lophophorates. They all show radial cleavage, enterocoely, and (except for the phoronids) a mouth that is not derived from the blastopore. In addition, the body plan and coelomic arrangement is clearly trimerous or obviously derived therefrom.

An alliance between the groups Ectoprocta and Entoprocta, proposed by Nielsen, is rejected on the basis of incompatibility with the idea of lophophorate unity and on direct comparative grounds. Entoprocts do not possess a lophophore as we have defined it. Furthermore, they lack any vestiges of a coelom and trimeric body plan. The feeding currents are virtually opposite in the two groups, and the methods of food capture and transport are entirely different. Entoprocts possess ducted gonads, ectoprocts do not. Cleavage in entoprocts is spiral, whereas it is radial in ectoprocts. Larval forms and particularly metamorphosis are clearly different in the two groups. More important, if the two groups are related, then they must share common (homologous) characteristics—that is, synapomorphies. The similarities pointed out by Nielsen are superficial and common to many colonial sessile animals (e.g., budding, metamorphosis, and life cycles). The U-shaped guts are convergent adaptations to zooid life in "boxes"—no other condition would function. Thus, in the absence of unifying synapomorphies and the presence of multiple and significant differences, we can only consider one conclusion: the two groups are unrelated.

The origin of the lophophorates is puzzling, largely because it is tied, in part, to the origin of the entire deuterostome lineage, which is itself very uncertain. Most workers agree that the phoronids show the least amount of change from the presumed ancestral form.

That ancestor may have been a trimeric, coelomate, infaunal burrower. In any case, the ancestor probably evolved during the Precambrian as one evolutionary experiment with a coelomate bauplan. The first lophophorate was probably phoronid-like and became adapted to tube dwelling and feeding above the substratum. Modern phoronids may have changed little from this tube-dwelling protolophophorate.

The origin of the Ectoprocta clearly involved a reduction in body size and the development of colonial habits. The epidermal secretions became compartmentalized, with the exploitation of asexual budding as a means of colony formation. The acquisition of a retractable lophophore allowed protection of the soft tentacles. The absence of nephridia and circulatory structures provides space for the retraction of the polypide; short diffusion distances are associated with small size and the disappearance of these systems. Without nephridia as a means of gamete release, other avenues of egg and sperm escape arose in the form of coelomopores from the mesocoel and communication between the metacoel and the mesocoel.

The origin of the brachiopods is marked by the appearance of several novel features largely associated with the evolution of mantle folds, their secretion of valves, and the enclosure of the lophophore and body proper within the mantle cavity. The lophophore lost most of its hydraulic qualities and became more or less stationary, held by various structural support mechanisms. The circulatory system was reduced. The origin of a pedicle allowed a means of attachment in these solitary animals, supporting the body off the substratum. The first brachiopods may have been lingulid types that used the pedicle for anchorage in soft substrata.

Valentine (1973, 1975) has attempted to support a polyphyletic origin for the brachiopods, but he does recognize a monophyletic lophophorate clade, somewhat as we have described here. However, Rowell's (1982) cladistic treatment of the brachiopods, living and extinct, presents a convincing case for monophyly, although his subgroups do not correspond exactly with the Articulata–Inarticulata division.

The origin of the lophophore allowed various avenues of escape from the infaunal life of their Precambrian ancestor and the exploitation of three different lifestyles, all involving suspension feeding. However, in spite of the differences among the three phyla, and the unique qualities of each, evolution *within* the lophophorate clade remains obscure. Without making assumptions about the first lophophorate, the three taxa appear to have emerged separately from a common, lophophore-bearing ancestor (Figure 21.23A). Only by designating additional ancestral features can we eliminate the trichotomy. For example, if we assume that the ancestral lophophorate was phoronid-like, with a complete circulatory system, then the ectoprocts and brachiopods may form a distinct clade defined by the reduction and loss of the circulatory system (Figure

(A)

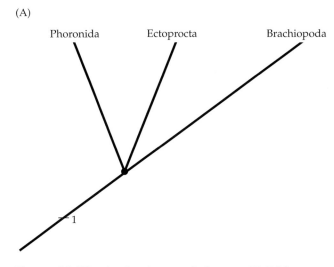

(B)

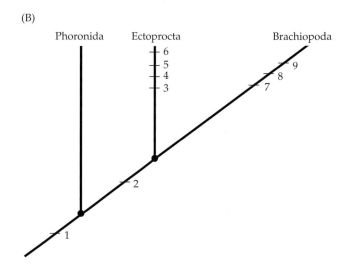

Figure 21.23 Lophophorate phylogeny. (A) Without assumptions about the ancestral form, other than the origin of the lophophore (1), a trichotomy results where each group arises independently from the ancestor. (B) Assuming a phoronid-like ancestor, the ectoprocts and brachiopods form a clade defined in part by the reduction of the circulatory system (2). The ectoprocts and brachiopods are subsequently defined by their unique synapomorphies. For ectoprocts these are: (3) colonial lifestyles, (4) retractable lophophore, (5) loss of nephridia, and (6) production of brown bodies. For brachiopods: (7) unique mantle and shell, (8) lophophoral skeleton, and (9) pedicle.

21.22B). Such assumptions carry implications about the deuterostome lineage in general and are explored more in Chapter 24.

There remain, of course, many questions and alternative hypotheses on the matter of lophophorate evolution. Not all zoologists are convinced that these animals are deuterostomes. For example, recent 18S rDNA molecular studies suggest that the lophophorates are more closely aligned with the protostomes than the deuterostomes, despite embryological and anatomical evidence to the contrary (Halanych et al. 1995; Mackey et al. 1996). Some other workers (see Jeffries 1986) view the lophophorates as somewhat intermediate between the protostomes and deuterostomes (Chapter 23). No doubt these debates will continue at least until the third edition of this book appears.

Selected References

General References

Gee, H. 1995. Lophophorates prove likewise variable. Nature 374: 493.

Giese, A. C., J. S. Pearse and V. B. Pearse (eds.). 1991. *Reproduction of Marine Invertebrates*. Vol. VI, Echinoderms and Lophophorates. The Boxwood Press, Pacific Grove, California.

Halanych, K. M., J. D. Bacheller, A. M. A. Aguinaldo, S. M. Liva, D. M. Hillis and J. A. Lake. 1995. Evidence from 18S ribosomal DNA that the lophophorates are protostome animals. Science 267: 1641–1642.

Hyman, L. H. 1959. *The Invertebrates, Vol. 5, Smaller Coelomate Groups*. McGraw-Hill, New York.

Jefferies, R. P. S. 1986. *The Ancestry of the Invertebrates*. British Mus. (Natural History), London.

Larwood, G. and B. R. Rosen. 1970. *Biology and Systematics of Colonial Organisms*. Academic Press, NY.

Lüter, C. 2000. The origin of the coelom in Branchiopoda and its phylogenetic significance. Zoomorphol. 120: 15–28.

Mackey, L. Y., B. Winnepenninckx, R. DeWachter, T. Backeljau, P. Emschermann and J. R. Garey. 1996. 18S rRNA suggests that Entoprocta are protostomes, unrelated to Ectoprocta. J. Mol. Evol. 42: 552–559.

McCammon, H. M. and W. A. Reynolds (organizers). 1977. Symposium: Biology of Lophophorates. Am. Zool. 17: 3–150.

Moore, R. C. (ed.). 1965. *Treatise on Invertebrate Paleontology. Pts. G and H (Vols. 1 and 2)*. Geological Society of America, Inc. and The University of Kansas, Lawrence, Kansas.

Morris, S. C. 1995. Nailing the lophophorates. Nature 375: 365–366.

Valentine, J. W. 1973. Coelomate superphyla. Syst. Zool. 22(2): 97–102.

Valentine, J. W. 1975. Adaptive strategy and the origin of grades and ground-plans. Am. Zool. 15: 391–404.

Zimmer, R. L. 1973. Morphological and developmental affinities of the lophophorates. *In* G. P. Larwood (ed.), *Living and Fossil Bryozoa*. Academic Press, London, pp. 593–600.

Zimmer, R. L. 1980. Mesoderm proliferation and function of the protocoel and metacoel in early embryos of *Phoronis vancouverensis* (Phoronida). Zool. Jb. Anat. 103: 219–233.

Zimmer, R. L. 1997. Phoronids, brachiopods and bryozoans: the lophophorates. *In* S. F. Gilbert and A. M. Raunio, *Embryology, Constructing the Organism*. Sinauer Associates, Sunderland, MA. pp. 279–308.

Phoronida

Dhar, S. R., A. Logan, B. A. MacDonald and J. E. Ward. 1997. Endoscopic investigation of feeding structures and mechanisms in two plectolophous brachiopods. Invert. Biol. 116: 142–150.

Emig, C. C. 1974. The systematics and evolution of the phylum Phoronida. Z. Zool. Syst. Evol. 12(2): 128–151.

Emig, C. C. 1977. The embryology of Phoronida. Am. Zool. 17: 21–38.

Emig, C. C. 1982. Phoronida. *In* S. P. Parker (ed.), *Synopsis and Classification of Living Organisms*. McGraw-Hill, New York.

Hermmann, K. 1997. Phoronida. *In* F. W. Harrison and R. M. Woollacott (eds.), *Microscopic Anatomy of Invertebrates, Vol. 13, Lophophorates, Entoprocta, and Cycliophora*. Wiley-Liss, New York, pp. 207–235.

Silén, L. 1954. Developmental biology of the Phoronidea of the Gullmar Fjord area of the west coast of Sweden. Acta Zool. 35: 215–257.

Zimmer, R. L. 1967. The morphology and function of accessory reproductive glands in the lophophores of *Phoronis vancouverensis* and *Phoronopsis harmeri*. J. Morphol. 121(2): 159–178.

Zimmer, R. L. 1972. Structure and transfer of spermatozoa in *Phoronopsis viridis*. *In* C. J. Arceneaux (ed.), 30th Annual Proceedings of the Electron Microscopical Society of America.

Ectoprocta

Bigley, F. P. (ed.). 1991. *Bryozoa Living and Fossil*. Bull. Soc. Sci. Nat. Quest Fr. Mem. H. S.

Buss, L. W. 1981. Group living, competition, and the evolution of cooperation in a sessile invertebrate. Science 213: 1012–1014.

Carle, K. J. and E. E. Ruppert. 1983. Comparative ultrastructure of the bryozoan funiculus: A blood vessel homologue. Z. Zool. Syst. Evol. 21: 181–193.

Cook, P. L. 1977. Colony water currents in living Bryozoa. Cah. Biol. Mar. 18: 31–47.

Cook, P. L. and P. J. Chimonides. 1981. Morphology and systematics of some rooted cheilostome Bryozoa. J. Nat. Hist. 15: 97–134.

Driscoll, E. C., J. W. Gibson and S. W. Mitchell. 1971. Larval selection of substrate by the bryozoans *Discoporella* and *Cupuladria*. Hydrobiologia 37: 347–359.

Farmer, J. D., J. W. Valentine and R. Cowen. 1973. Adaptive strategies leading to the ectoproct groundplan. Syst. Zool. 22(3): 233–239.

Harvell, C. D. 1984. Predator-induced defense in a marine bryozoan. Science 224: 1357–1359.

Harvell, C. D. 1992. Inducible defenses and allocation shifts in a marine bryozoan. Ecology 73: 1567–1576.

Hughes, R. L. and R. M. Woollacott. 1980. Photoreceptors of bryozoan larvae. Zool. Scripta 9: 129–138.

Hunter, E. and R. N. Hughes. 1993. Self-fertilization in *Celleporella hyalina*. Mar. Biol. 115: 495–500.

Larwood, G. P. (ed.). 1973. *Living and Fossil Bryozoa*. Academic Press, London.

Larwood, G. P. and M. B. Abbott (eds.). 1979. *Advances in Bryozoology*. System. Assoc. Special Vol. 13, Academic Press, New York.

Larwood, G. P. and C. Nielsen (eds.). 1981. *Recent and Fossil Bryozoa*. Olsen and Olsen, Fredensborg, Denmark.

Lidgard, S. 1986. Ontogeny in animal colonies: A persistent trend in the bryozoan fossil record. Science 232: 230–232.

McKinney, F. K. and J. B. C. Jackson (eds.). 1991. *Bryozoan Evolution*. Univ. Chicago Press, Chicago.

McKinney, M. J. 1997. Fecal pellet disposal in marine bryozoans. Invert. Biol. 116: 151–160.

Mukai, H. and S. Oda. 1980. Comparative studies on the statoblasts of higher phylactolaemate bryozoans. J. Morphol. 165: 131–156.

Mukai, H., K. Terakado and C. G. Reed. 1997. Bryozoa. *In* F. W. Harrison and R. M. Woollacott (eds.), *Microscopic Anatomy of Invertebrates, Vol. 13, Lophophorates, Entoprocta, and Cycliophora*. Wiley-Liss, New York, pp. 45–206.

Nielsen, C. 1971. Entoproct life cycles and the entoproct/ectoproct relationship. Ophelia 9: 209–341.

Nielsen, C. 1977. The relationship of Entoprocta, Ectoprocta, and Phronida. Amer. Zool. 17(1): 149–150.

Palumbi, S. R. and J. B. C. Jackson. 1983. Aging in modular organisms: Ecology of zooid senescence in *Steginoporella* sp. (Bryozoa; Cheilostomata). Biol. Bull. 164: 267–278.

Perez, F. M. and W. C. Banta. 1996. How does *Cellaria* get out of its box? A new cheilostome hydrostatic mechanism (Bryozoa: Cheilostomata). Trans. Am. Microsc. Soc. 115 (2): 162–169.

Pires, A. and R. M. Woollacott. 1982. A direct and active influence of gravity on the behavior of a marine invertebrate larva. Science 220: 731–733.

Rider, J. and R. Cowen. 1977. Adaptive architectural trends in encrusting ectoprocts. Lethaia 10: 29–41.

Rogick, M. D. 1959. Bryozoa. *In* W. T. Edmondson, H. B. Ward and G C. Whipple (eds.), *Freshwater Biology*, 2nd Ed. Wiley, New York, pp. 495–507.

Ryland, J. S. 1970. *Bryozoans*. Hutchinson University Library, London.

Ryland, J. S. 1976. Physiology and ecology of marine bryozoans. Adv. Mar. Biol. 14: 285–443.

Ryland, J. S. 1982. Bryozoa. *In* S. P. Parker (ed.), *Synopsis and Classification of Living Organisms*. McGraw-Hill, New York, pp. 743–769.

Santagata, S. and W. C. Banta. 1996. Origin of brooding and ovicells in cheilostome bryozoans: Interpretive morphology of *Scrupocellaria ferox*. Trans. Am. Microsc. Soc. 115 (2): 170–180.

Silén, L. 1972. Fertilization in the Bryozoa. Ophelia 19(1): 27–34.

Silén, L. 1980. Colony–substratum relations in Scrupocellariidae (Bryozoa, Cheilostomata). Zool. Scripta 9: 211–217.

Smyth, M. J. 1988. *Penetrantia clionoides*, sp. nov. (Bryozoa), a boring bryozoan in gastropod shells from Guam. Biol. Bull. 174: 276–286.

Thorpe, J. P., G. A. Shelton and M. S. Laverack. 1975. Colonial nervous control of lophophore retraction in Cheilostome Bryozoa. Science 189: 60–61.

Winston, J. E. 1978. Polypide morphology and feeding behavior in marine ectoprocts. Bull. Mar. Sci. 28(1): 1–31.

Woollacott, R. M. and R. L. Zimmer. 1971. Attachment and metamorphosis of the cheilostome bryozoan *Bugula neritina* (Linné). J. Morphol. 134(3): 351–382.

Woollacott, R. M. and R. L. Zimmer. 1972a. Fine structure of a potential photoreceptor organ in the larva of *Bugula neritina* (Bryozoa). Z. Zellforsch. 123: 458–469.

Woollacott, R. M. and R. L. Zimmer. 1972b. Origin and structure of the brood chamber in *Bugula neritina* (Bryozoa). Mar. Biol. 16: 165–170.

Woollacott, R. M. and R. L. Zimmer. (eds.). 1977. *Biology of Bryozoans*. Academic Press, New York.

Zimmer, R. L. and R. M. Woollacott. 1993. Anatomy of the larva of *Amathia vidovici* (Bryozoa: Ctenostomata) and phylogenetic significance of the vesiculariform larva. J. Morph. 215: 1–29.

Brachiopoda

Chuang, S. H. 1959. Structure and function of the alimentary canal in *Lingula unguis*. Proc. Zool. Soc. London 132: 293–311.

Cohen, B. L. 1994. Immuno-taxonomy and the reconstruction of brachiopod phylogeny. Paleontology 37(4): 907–911.

Foster, M. W. 1982. Brachiopoda. *In* S. P. Parker (ed.), *Synopsis and Classification of Living Organisms*. McGraw-Hill, New York, pp 773–780.

Gutman, W. F., K. Vogel and H. Zorn. 1978. Brachiopods: Biochemical interdependencies governing their origin and phylogeny. Science 199: 890–893.

James, M. A. 1997. Brachiopoda: internal anatomy, embryology, and development. *In* F. W. Harrison and R. M. Woollacott (eds.), *Microscopic Anatomy of Invertebrates, Vol. 13, Lophophorates, Entoprocta, and Cycliophora*. Wiley-Liss, New York, pp. 297–407.

MacKay, S. and R. A. Hewitt. 1978. Ultrastructure studies on the brachiopod pedicle. Lethaia 11: 331–339.

Nielsen, C. 1991. The development of the brachiopod *Crania* (*Neocrania*) *anomala* (O. F. Müller) and its phylogenetic significance. Acta. Zool. 72 (1): 7-28.

Richardson, J. R. 1981. Brachiopods in mud: Resolution of a dilemma. Science 211: 1161–1163.

Rowell, A. J. 1982. The monophyletic origin of the Brachiopoda. Lethaia 15: 299–307.

Rudwick, M. J. S. 1970. *Living and Fossil Brachiopods*. Hutchinson University Library, London.

Steele-Petrovic, H. M. 1976. Brachiopod food and feeding processes. Paleontology 19(3): 417–436.

Thayer, C. W. 1985. Brachiopods versus mussels: Competition, predation, and palatability. Science 228(4707): 1527–1528.

Watabe, N. and C.-M. Pan. 1984. Phosphatic shell formation in atremate brachiopods. Am. Zool. 24: 977–985.

Williams, A. 1997. Brachiopoda: introduction and integumentary system. *In* F. W. Harrison and R. M. Woollacott (eds.), *Microscopic Anatomy of Invertebrates, Vol. 13, Lophophorates, Entoprocta, and Cycliophora*. Wiley-Liss, New York, pp. 237–296.

Williams, A., M. A. James, C. C. Emig, S. Mackay and M. C. Rhodes. 1997. Brachiopod anatomy. *In* A. Williams and C. H. C. Brunton (eds.), *Treatise on Invertebrate Paleontology, Pt. H: Brachiopoda*. The Geological Society of America and The University of Kansas, Lawrence, Kansas.

Williams, A., S. Mackay and M. Cusak. 1992. Structure of the organophosphatic shell of the brachiopod *Discinisca*. Phil. Trans. Roy. Soc. Lond. Biol. 337: 83–104.

22 *Phylum Echinodermata*

"What do they find to study?" Hazel continued. "They're just starfish. There's millions of 'em around. I could get you a million of 'em."

John Steinbeck
Cannery Row, 1945

Some of the most familiar seashore animals are members of the phylum Echinodermata (Greek *echinos*, "spiny"; *derma*, "skin"). The phylum contains about 7,000 living species, including the sea lilies, feather stars, sea stars, brittle stars, sea urchins, sand dollars, and sea cucumbers (Figures 22.1, 22.2, and 22.3). Another 13,000 or so species are known from a rich fossil record dating back at least to early Cambrian times.

Echinoderms range in size from tiny sea cucumbers and brittle stars smaller than 1 cm, to sea stars that exceed 1 m in diameter and sea cucumbers that reach 2 m in length. Except for a few brackish-water forms, echinoderms are strictly marine. They have been prevented from invading land or fresh water, presumably, by their cutaneous gas exchange methods and their lack of excretory–osmoregulatory structures. In the sea, however, they are widely distributed in all oceans and at all depths. With the exception of a few odd pelagic sea cucumbers (Figure 22.1P,Q) and one (*Rynkatropa pawsoni*) that is commensal on deep-sea anglerfish, all echinoderms are benthic. Some play important roles in marine ecosystems as high-level predators (certain sea stars) or algal grazers (many sea urchins). In some regions of the deep sea they may compose 95 percent of the biomass.

Echinoderms are deuterostomes, and their development is frequently cited as stereotypical of that assemblage. With a few exceptions, living echinoderms possess a well developed coelom, an endoskeleton composed of unique calcareous **ossicles**, and pentamerous radial symmetry. They are the only fundamentally pentamerous organisms in the animal kingdom. However, this symmetry is secondarily derived, both evolutionarily and developmentally, and the larval forms are always bilateral. Among other defining characteristics (Box 22A) is a uniquely echinoderm feature known as the **water vascular system**, a complex system of channels and reservoirs that is derived from the coelom and serves a variety of functions.

(A)

(B)

(C)

(D)

(E)

(F)

(G)

(H)

(I)

(J)

(K)

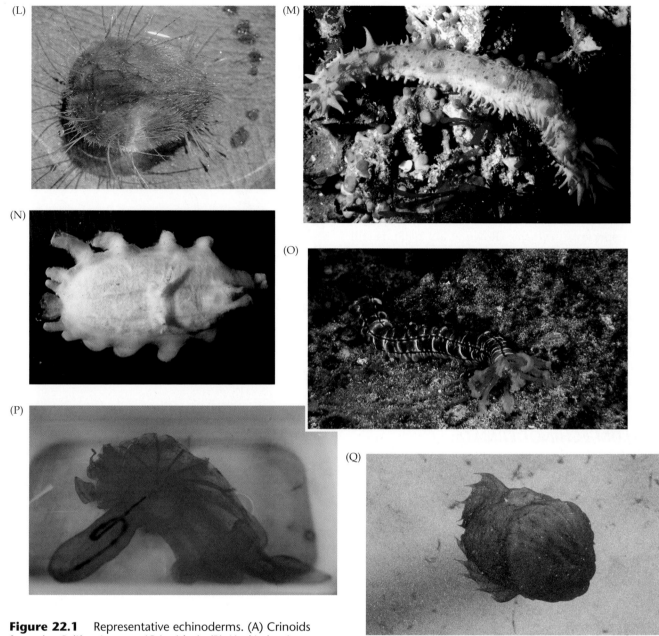

Figure 22.1 Representative echinoderms. (A) Crinoids from the California coast (Crinoidea). (B) *Linckia laevigata.* (C) *Astropecten armatus* (Asteroidea). (D) *Pteraster tesselatus* (Asteroidea). (E) *Odontaster crassus* (Asteroidea). (F) *Acanthaster planci,* the Indo-West Pacific crown-of-thorns (Asteroidea). (G) The "sea daisy" *Xyloplax medusiformis* (Asteroidea). (H) A brittle star, *Ophiopholis aculeata* (Ophiuroidea). (I) A basket star (Ophiuroidea). (J) *Strongylocentrotus purpuratus,* sea urchin (Echinoidea). (K) *Dendraster excentricus,* sand dollars (Echinoidea). (L) An "irregular" sea urchin, *Lovenia* (Echinoidea). (M) The sea cucumber *Parastichopus* (Holothuroidea). (N) The strange deep-sea holothurian *Scotoplanes,* which lacks podia on the "dorsal" surface (Holothuroidea). (O) *Euapta* (Holothuroidea). (P) A pelagic holothurian, *Palagothuria* (Holothuroidea). (Q) An epibenthic swimming holothurian, *Enypniastes* (Holothuroidea), photographed at 1,586 meters.

Taxonomic History and Classification

Echinoderms have been known since ancient times; their likenesses appear in 4,000-year-old frescoes of Crete. Jacob Klein is credited with coining the name Echinodermata in about 1734 in reference to sea urchins. Linnaeus placed the echinoderms in his taxon Mollusca, along with a mixed bag of other invertebrates. For nearly a hundred years these animals were allied with various other groups, including the cnidarians in Lamarck's Radiata. It was not until 1847 that Frey and Leukart recognized the echinoderms as a distinct taxon.

Since the middle of the nineteenth century controversies have centered on classification within the phylum, and arguments continue today. The abundant fossil record has been both a blessing and a burden because authors have treated the fossil evidence in different ways. Some emphasize differences between morphological types and assign higher categorical ranks to nearly every fossil taxon discovered; consequently, certain schemes recognize as many as 25 separate classes of echinoderms. Others apply the evidence more parsimoniously, seeking to establish fundamental similarities; their schemes recognize fewer classes.

In 1986 Baker et al. established a new class (the Concentricycloidea) to accommodate a strange deep-sea echinoderm discovered in association with bacteria-rich sunken wood. Although the precise phylogenetic placement of this creature, named *Xyloplax medusiformis*, and a second species *X. turnerae* (Rowe et al., 1988), is still being debated, evidence suggests that concentricycloids fall within an asteroid-ophiuroid clade (see Figure 22.20B).

The classification scheme below draws from various authors. It recognizes five classes to which the living echinoderms belong, but we introduce some of the important fossil forms in the phylogeny section at the end of the chapter. The reader is cautioned that other classification schemes exist for the taxa within the classes treated here.

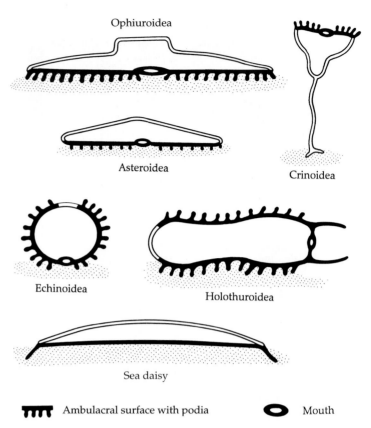

Figure 22.2 Schematic sections of the five living classes of echinoderms, showing body orientations to the substratum and disposition of the ambulacral surfaces.

PHYLUM ECHINODERMATA

CLASS CRINOIDEA: Sea lilies and feather stars (Figures 22.1A, 22.3A,B). Body form as a cup or calyx, with oral surface directed upward; aboral stalk, when present, arising from calyx; ambulacra on arms that bear pinnules; ambulacra may branch more than once, branches equal; ambulacral grooves open; skeletal plates fused in calyx, but articulated elsewhere; no external madreporite; mouth and anus on oral surface. About 625 living species. (e.g., *Antedon, Asterometra, Comantheria, Comanthina, Isometra, Metacrinus, Neometra, Phixometra, Zygometra*)

CLASS ASTEROIDEA: Sea stars (Figures 22.1B–F, 22.3C–E). Body stellate with five or more arms; arms not set off from central disc by distinct articulations; anus on aboral surface; mouth directed toward substratum; ambulacral grooves open; tube feet with internal ampullae, with or without suckers; madreporite aboral on CD interambulacrum. About 1,500 extant species. The classification below is a conservative one (for an alternative scheme, see Blake 1987).

> **ORDER PLATYSTERIDA:** Considered by some to include most primitive asteroids; tube feet lack suckers; anus absent. Generally restricted to soft substrata. This order has been abandoned by some specialists. Living species are confined to two genera: *Luidia* (about 60 species), and *Platysterias* (monotypic, *P. latiradiata*).

> **ORDER PAXILLOSIDA:** Upper surface with umbrella-like clusters of ossicles called paxillae; tube feet lack suckers; anus present or absent. Epibenthic or shallow burrowers. (e.g., *Astropecten, Caymanostella, Ctenodiscus, Lethmaster*)

> **ORDER VALVATIDA:** Tube feet with suckers; anus present; some possess paxillae. Widely distributed, with several hundred species. (e.g., *Amphiaster, Ar-*

chaster, *Asterodon, Chaetaster, Hoplaster, Linckia, Odonaster, Oreaster*)

> **ORDER SPINULOSIDA:** With 5–18 arms; tube feet with suckers; anus present; generally lacking pedicellariae. With a few hundred species. (e.g., *Acanthaster, Echinaster, Henricia, Pteraster, Remaster, Solaster*)

> **ORDER FORCIPULATIDA:** With 5–50 arms; tube feet with suckers; anus present; with pincer-like pedicellariae. Widely distributed sea stars, including most intertidal forms. Several hundred species. (e.g., *Asterias, Brisinga, Evasterias, Heliaster, Leptasterias, Pisaster, Pycnopodia, Stylasterias*)

"SEA DAISIES": Body discoidal (< 1 cm diameter); with ring of marginal spines, but without arms or rays; skeletal plates arranged concentrically; suckerless podia in a ring near body margin; two ring canals with hydropore on CD interambulacrum; five large ossicles on aboral surface mark ambulacra; gut absent or incomplete. Classification of the enigmatic sea daisies (previously the class Concentricycloidea) is problematic, but many authorities assign them to the Spinulosida.

CLASS OPHIUROIDEA: Brittle stars and basket stars (Figures 22.1H,I, 22.3F). Body with five unbranched or branched articulated arms, clearly set off from a central disc; ambulacral grooves closed; coelom in arms greatly reduced by presence of skeletal vertebrae; tube feet with internal ampullae but without suckers; anus lacking; madreporite on CD interambulacral plate on oral surface, often reduced. About 2,000 extant species in three orders.

> **ORDER OEGOPHIURIDA:** Without bursae; arms lack dorsal and ventral shields; madreporite on edge

BOX 22A *Characteristics of the Phylum Echinodermata*

1. Calcareous endoskeleton arising from mesdermal tissue and composed of separate plates or ossicles; each plate originates as a single calcite crystal and develops as an open meshwork structure called a stereom, the interstices of which are filled with living tissue (the stroma)

2. Adults with basic pentamerous radial symmetry derived from bilaterally symmetrical larvae (when present); body parts organized about an oral–aboral axis

3. Coelomic water vascular system composed of a complex series of fluid-filled canals, usually evident externally as muscular podia

4. Embryogeny fundamentally deuterostomous, with radial cleavage, entodermally derived mesoderm, enterocoely, and mouth not derived from the blastopore

5. Gut complete except where secondarily incomplete or lost

6. No excretory organs

7. Circulatory structures, when present, compose a hemal system derived from coelomic cavities and sinuses

8. Nervous system diffuse, decentralized, usually consisting of a nerve net, nerve ring, and radial nerves

9. Mostly dioecious; development direct or indirect

SUBCLASS CIDAROIDEA: Pencil urchins. Test globular, ambulacral plates simple, each with a pair of perforations serving one tube foot; spines large, pencil-like, without epidermal covering; anus at aboral pole; dermal gills absent; mostly extinct; often considered primitive in the class. About 140 surviving species in one order (Cidaroida). (e.g., *Cidaris, Eucidaris, Phyllacanthus, Psychocidaris*)

SUBCLASS EUECHINOIDEA: Sea urchins, heart urchins, lamp urchins, sea biscuits, sand dollars. Test globular or discoidal; numbers of tube feet and spines per plate vary; anal position varies from aboral to "posterior." Aristotle's lantern variable, absent in heart and lamp urchins. About 800 living species.

INFRACLASS ECHINOTHURIOIDEA: Test up to 30 cm in diameter, with large amounts of collagen; deep-water (1,000–4,000 m) species with very thin, flexible tests that collapse when removed from water; long, club-shaped oral spines support body off substratum; anus aboral. One order (Echinothurioida) with three families. (e.g., *Araeosoma, Asthenosoma, Phormosoma, Sperosoma*)

INFRACLASS ACROECHINOIDEA: Includes all of the commonly encountered urchins and sand dollars; divided into three extant cohorts.

COHORT DIADEMATACEA: Hollow-spined "regular" sea urchins. Anus aboral; with compound ambulacral plates; spines hollow. Three orders, each with one extant family. (e.g., *Astropyga, Aspidodiadema, Caenopedina, Diadema, Micropyga, Plesiodiadema*)

COHORT ECHINACEA: Solid-spined "regular" sea urchins. Anus aboral; with compound ambulacral plates; spines solid; with five pairs of gills arranged in circle on peristomial membrane. Three extant orders. (e.g., *Arbacia, Echinometra, Echinus, Heterocentrotus, Paracentrotus, Salenia, Strongylocentrotus, Toxopneustes, Tripneustes*)

COHORT IRREGULARIA: Heart urchins, lamp urchins, "irregular" urchins (sand dollars, sea biscuits, and their relatives). Body globular or discoidal, with tendency toward bilateral symmetry; anus variable, shifted to "posterior" (even oral) position; spines usually tiny, forming dense covering; Aristotle's lantern reduced, absent in heart and lamp urchins. Perhaps six extant orders, including the Clypeasteroida (sand dollars) and several groups of urchins. (e.g., *Cassidulus, Clypeaster, Dendraster, Echinocardium, Echinodiscus, Echinolampus, Encope, Fibularia, Lovenia, Maretia, Mellita, Meoma, Metalia, Micropetalon, Spatanga, Urechinus*)

of disc; digestive glands extend into proximal portions of arms. A single living species (*Ophiocanops fugiens*).

ORDER PHRYNOPHIURIDA: Bursae present; ventral arm shields rudimentary, dorsal shields usually absent; arms branched or unbranched, but can coil vertically; madreporite on oral surface; digestive glands confined to central disc. Includes some primitive brittle stars and the basket stars. (e.g., *Asteronyx, Astrodia, Gorgonocephalus, Ophiomyxa*)

ORDER OPHIURIDA: Bursae present; dorsal and ventral arm shields present and usually well developed; unbranched arms incapable of coiling vertically; madreporite on oral surface; digestive glands wholly within central disc. Includes vast majority of living brittle stars. (e.g., *Amphiophiura, Amphipholis, Amphiura, Ophiactis, Ophiocoma, Ophioderma, Ophiolepis, Ophiomusium, Ophionereis, Ophiopholis, Ophiothrix, Ophiura*)

CLASS ECHINOIDEA: Urchins and sand dollars (Figures 22.1J–L, 22.3G–I). Body globose or discoidal, often secondarily bilateral; skeletal plates joined by collagen matrix and calcite interdigitations as solid test; with movable spines; water canals within test; ambulacral grooves closed; with internal jaw apparatus (Aristotle's lantern). About 950 extant species in two extant subclasses (for a more detailed version of the classification outlined below see Smith 1984).

CLASS HOLOTHUROIDEA: Sea cucumbers (Figures 22.1M–Q, 22.3J,K). Body fleshy, sausage-shaped, elongate on oral–aboral axis; skeleton usually reduced to isolated ossicles; symmetry pentamerous or secondarily modified by loss of "dorsal" (bivium) tube feet along ambulacra C and D; tube feet sometimes entirely absent; madreporite internal; ambulacral grooves closed; with circlet of feeding tentacles around mouth. About 1,150 extant species in three subclasses.

SUBCLASS DENDROCHIROTACEA: With 8–30 oral tentacles ranging from digitiform to highly branched; tentacles and oral region with retractor muscles; tube feet present, but location varies.

ORDER DACTYLOCHIROTIDA: Body often U-shaped and enclosed in flexible test of skeletal plates; tentacles unbranched; most are deep-water burrowers. (e.g., *Echinocucumis, Mitsukuriella, Rhopalodina, Sphaerothuria, Vaneyella, Ypsilothuria*)

ORDER DENDROCHIROTIDA: Body not U-shaped, but is partially enclosed in plates in certain genera (e.g., *Psolus*); feeding tentacles typically branched. Includes many common intertidal cucumbers. (e.g., *Cucumaria, Eupentacta, Paracucumis, Placothuria, Psolus, Thyone*)

SUBCLASS ASPIDOCHIROTACEA: With 10–30 leaflike or shieldlike oral tentacles; oral region lacks retractor muscles; tube feet present.

ORDER ASPIDOCHIROTIDA: Tentacles shieldlike; respiratory trees present. Includes the largest holothurians (up to 2 m). (e.g., *Actinopygia, Astichopus, Bathyplotes, Holothuria, Isostichopus, Parastichopus, Stichopus*)

ORDER ELASIPODIDA: Typically deep-sea cucumbers, often with strange body forms; respiratory trees absent. (e.g., *Benthodytes, Deima, Enypniastes, Pelagothuria, Scotoplanes*)

SUBCLASS APODACEA: With up to 25 tentacles; tentacles vary from digitate to pinnate; tube feet highly reduced or absent.

ORDER MOLPADIDA: Body stout, narrowed posteriorly to a distinct tail; with 15 digitate tentacles; lacking tube feet. (e.g., *Caudina, Molpadia, Trochoderma*)

ORDER APODIDA: Body vermiform; lacking tube feet; with 10–25 tentacles. Among the apodids is the bizarre family Synaptidae, with unique anchor ossicles that occur in densities up to 1,500/sq cm and provide gripping power (in lieu of tube feet) by protruding and retracting into the skin in peristaltic waves along the cucumber's body wall. (e.g., *Euapta, Leptosynapta, Synapta*)

The Echinoderm Bauplan

The success of the echinoderm bauplan lies partly in the exploitation of radial symmetry imposed upon a relatively "advanced" coelomate architecture, including a mesodermally derived calcareous endoskeleton. We have seen the tendency among radially symmetrical animals to be either sessile or planktonic and to face their environments on all sides as suspension feeders or passive predators. This generalization applies not only to those creatures with primary radial symmetry (e.g., cnidarians), but also to many of those that have secondarily become functionally radial by way of a sessile lifestyle (e.g., tube-dwelling polychaetes, entoprocts, ectoprocts, phoronids, and others). Echinoderms, on the other hand, have uniquely combined mobility with radial symmetry, and they display a host of feeding strategies and lifestyles. Like other radially arranged animals, the echinoderms have a noncentralized nervous system, a feature that allows most of them to engage their environments equally from all sides.

Much of the biology of echinoderms is associated with their unique water vascular system (Figure 22.5),which is derived largely from specialized parts of the left mesocoelic portion of their tripartite coelom. The water vascular system is a complex of fluid-filled canals and reservoirs that aid in internal transport and hydraulically operate fleshy projections called **tube feet**. The external parts of the tube feet, or **podia**, can serve a variety of functions, including locomotion, gas exchange, feeding, attachment, and sensory reception. These versatile structures have contributed greatly to the success of echinoderms.

Although modern echinoderms are basically pentaradial creatures, several secondarily derived conditions exist. In the general case, five sets of body parts are oriented about a **central disc**. Extending from the mouth at the center of the oral surface are rows of podia associated with **ambulacral grooves** (Figure 22.8B), which define body radii called **ambulacra**. A radius bisecting adjacent ambulacra is called an **interambulacrum**. In a sea star, for example, the ambulacra are represented by the arms, and the interambulacra by the areas between the arms. In many echinoderms (e.g., ophiuroids, holothurians, and echinoids), the ambulacra are not marked by wide or "open" external furrows, in which case the animals are said to have "closed" ambulacral grooves. The side(s) of the body on which tube feet occur are often referred to as the **ambulacral surface(s)**.

The pentaradial symmetry of modern echinoderms is thought to have evolved from a triradiate (adult) plan; such a condition occurs in an extinct group called the helicoplacoids (Figure 22.19B). Although it may not be immediately obvious, the pentamerism of all echinoderms can be described in terms of reference to particular radii. When present externally, the position of the opening to the water vascular system (the **madreporite**) gives a clue to body orientation because it lies on a particular interambulacrum. A system of lettering has been developed in which the ambulacrum opposite the madreporite is coded A; the others are then coded B through E in a counterclockwise fashion as viewed from the aboral surface (Figure 22.3C). Thus, the madreporite

Figure 22.3 External anatomy of echinoderms. (A) *Botryocrinus*, a stalked fossil crinoid. (B) *Neometra*, a 30-armed, nonstalked crinoid. (C) Aboral view of *Ctenodiscus* (Asteroidea). The ambulacral radii are labeled according to convention. (D,E) Aboral and oral views of *Xyloplax* (the sea daisy). (F) The ophiuroid *Asteronyx* crawling on a gorgonian. Note the highly articulated arms. (G) The sand dollar *Dendraster* (aboral view). Note the petaloids through which the respiratory podia extend. (H) Oral view of the sand dollar *Encope* (Echinoidea). (I) The sea urchin *Plesiodiadema* has extremely long spines and podia. (J) *Cucumaria planci*, a dendrochirotacean sea cucumber. (K) The highly modified pelagic holothurian, *Pelagothuria*.

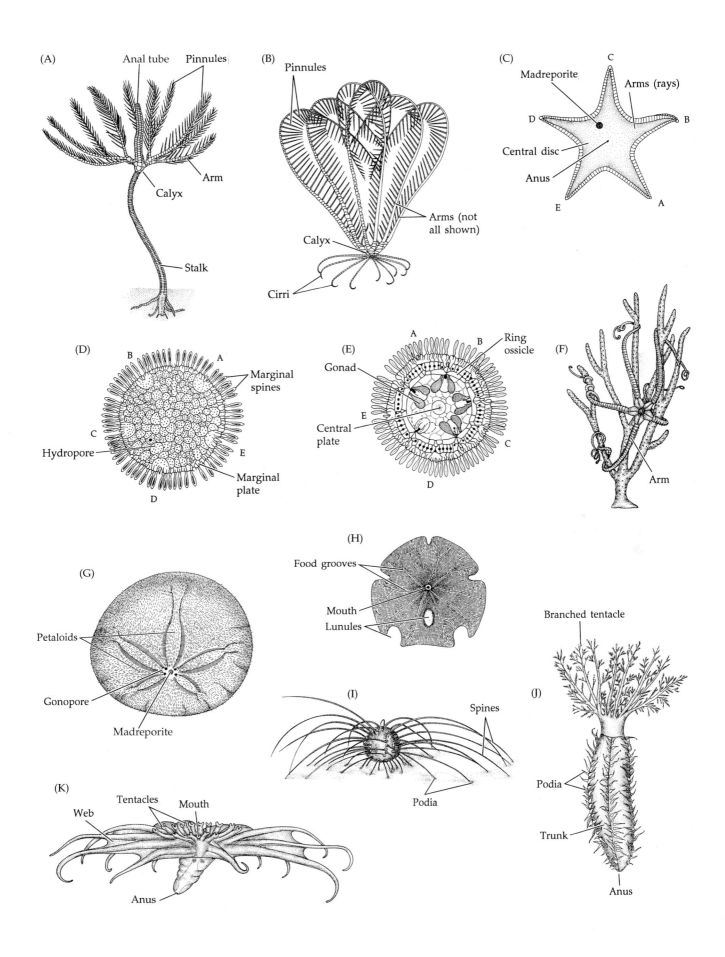

(A) Anal tube Pinnules Arm Calyx Stalk

(B) Pinnules Arms (not all shown) Calyx Cirri

(C) Madreporite Arms (rays) Central disc Anus

(D) Marginal spines Marginal plate Hydropore

(E) Ring ossicle Gonad Central plate

(F) Arm

(G) Petaloids Gonopore Madreporite

(H) Food grooves Mouth Lunules

(I) Spines Podia

(J) Branched tentacle Podia Trunk Anus

(K) Tentacles Mouth Web Anus

lies between ambulacra C and D (i.e., on the CD inter-ambulacrum). Radii C and D are said to compose the **bivium**, while radii A, B, and E compose the **trivium**.

As we explore the phylum in more detail, keep these generalities in mind and think of echinoderm diversity as variations on this pentamerous theme.

Body Wall and Coelom

An epidermis covers the bodies of all echinoderms and overlies a mesodermally derived dermis, which contains the skeletal elements, called **ossicles** (Figure 22.4A–D). Internal to the dermis and ossicles are muscle fibers or layers and the peritoneum of the coelom. The degree of development of the skeleton and muscles varies greatly among groups. In urchins and sand dollars, the ossicles are firmly attached to one another to form a rigid **test**, and the body wall muscles are weakly developed. In sea cucumbers, however, the ossicles are separate and lie scattered in the fleshy dermis (Figure 22.4D); here distinct muscle layers are present. Between these extreme conditions are cases in which adjacent skeletal plates articulate to various degrees. In the arms of sea stars and brittle stars, for example, the body wall muscles are arranged in bands between the plates, providing various degrees of arm motion. In some groups the skeletal plates are developed to such a degree that they nearly obliterate internal cavities. In brittle stars, for example, each arm "segment" contains a central skeletal ossicle called a **vertebra** (see Figure 22.9A,B), and the arm coeloms are reduced to small channels. Similarly, the arm coeloms in crinoids are greatly reduced by skeletal plates.

The endoskeleton is calcareous, mostly $CaCO_3$ in the form of calcite, with small amounts of $MgCO_3$ added. Developmentally, the skeleton of echinoderms begins as numerous separate spicule-like elements, each behaving as a single calcite crystal. Additional material is deposited on these crystals in various amounts, depending on the ultimate condition of the skeleton. Each ossicle is porous, has an internal meshwork (the **stereom**) of lattice-like or labyrinth-like spaces (Figure 22.4D), and generally is filled with dermal cells and fibers (the **stroma**). This structure is unique to members of the phylum Echinodermata.

During the formation of the skeleton, the plates may remain single (**simple plates**) or they may fuse to form **compound plates**. In addition, they frequently give rise to bumps and knobs called **tubercles**, to granules, and to various sorts of movable and fixed spines (Figure 22.4A,E). In some groups, especially the asteroids and echinoids, the skeleton also produces unique pincer-like structures called **pedicellariae** (Figure 22.4E–I). These structures respond to external stimuli independently of the main nervous system, and they possess their own neuromuscular reflex components. Pedicellariae were discovered in 1778 by O. F. Müller, who described them as parasitic polyps and gave them the generic name

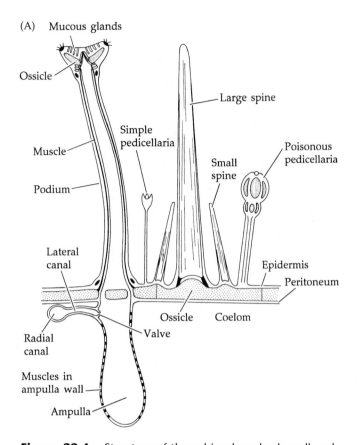

Figure 22.4 Structure of the echinoderm body wall and some skeletal elements. (A) The body wall of an urchin (composite section). (B) Spines on the sand dollar *Echinarachnius parma* (SEM). The arrows point to ciliary tracts. Scale bar represents 100 μm. (C) Skeletal ossicles from the central discs of four species of brittle stars (Ophiuroidea), shown in top (top row), side (middle row), and basal (bottom row) views. Scale bar represents 0.05 mm. (D) Skeletal ossicles from the holothurian *Psolus chintinoides*. The stereom structure is shown at two magnifications. Scale bar represents 100 μm. (E) Types of echinoid pedicellariae surrounding the base of a large spine. (F,G) Elevated pedicellariae used for prey capture by the sea star *Stylasterias forreri*: F, pedicellariae open and extended; G, pedicellariae retracted. (H) Details of a generalized pedicellaria. (I) Two types of muscle systems in pedicellariae. (J) A movable spine (section). Note the position of the muscles relative to the body wall layers.

Pedicellaria. He recorded three species of these "parasites" (*P. globifera*, *P. triphylla*, and *P. tridens*); forms of these names are still used to describe different types of pedicellariae.

Nearly a century after Müller's discovery, it was realized that pedicellariae are actually produced by the echinoderms themselves, but their exact nature remained elusive. Louis Agassiz believed they were the young of the animals on which they occurred. Even today there are competing opinions about their functions (see Campbell 1983 for a review). Pedicellariae differ not only in their structural details, but in their size and distribution on the body. Some are elevated on

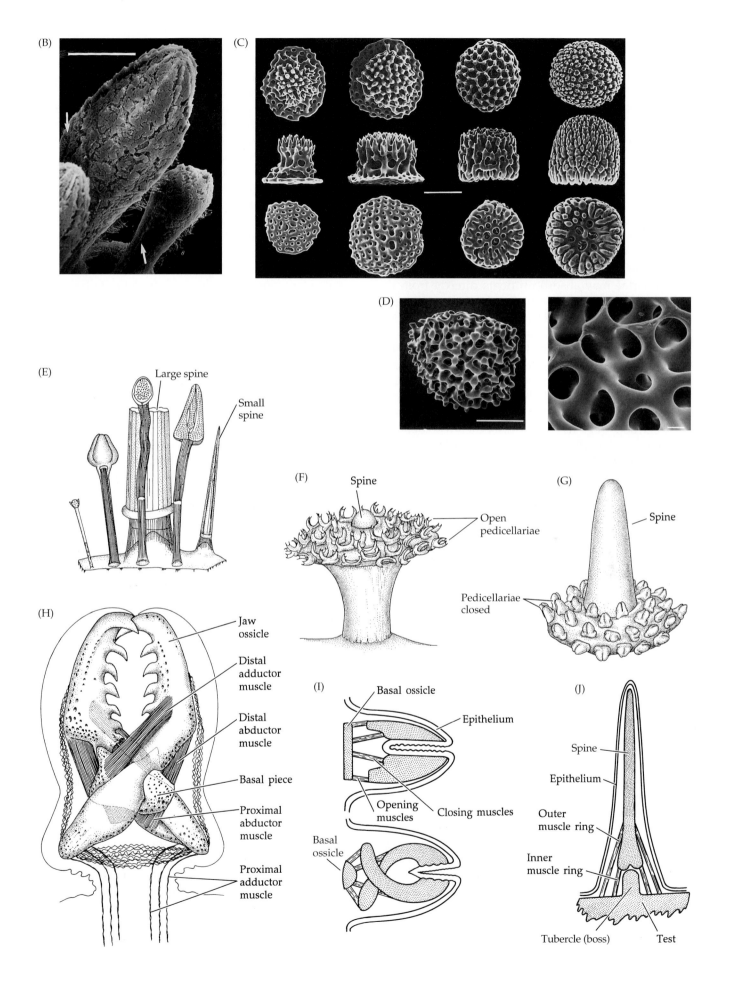

(B)

(C)

(D)

(E)

Large spine

Small spine

(F)

Spine

Open pedicellariae

(G)

Spine

Pedicellariae closed

(H)

Jaw ossicle

Distal adductor muscle

Distal abductor muscle

Basal piece

Proximal abductor muscle

Proximal adductor muscle

(I)

Basal ossicle

Epithelium

Opening muscles

Closing muscles

Basal ossicle

(J)

Spine

Epithelium

Outer muscle ring

Inner muscle ring

Tubercle (boss)

Test

stalks, whereas others lie nestled directly on the body surface, either singly or in clusters. Some help keep debris and settling larvae off the body, and others are used to defend against larger organisms. The sea urchin *Toxopneustes* bears toxin-producing pedicellariae with which it discourages would-be predators. In some urchins the pincers grasp and hold objects for camouflage and protection. A few sea stars actually use their pedicellariae to capture prey (Figure 22.4F,G).

Movable spines and pedicellariae contain muscles and other tissues that lie outside the main skeletal framework of the body wall (Figure 22.4H–J). This arrangement raises some interesting questions concerning the method of nutrient supply to these tissues because they are isolated from the coelom and gut. Pedicellariae may absorb nutrients directly from the water, or they may actually trap and digest small organisms and then absorb the products (Stephens 1968; Pequignat 1966, 1970; Ferguson 1970).

As in all deuterostomes (except the Chordata), the coelomic system of echinoderms usually develops as a tripartite series, originating as paired proto-, meso-, and metacoels. However, with the transformation to radial symmetry, these coelomic cavities do not come to lie in the three body regions usually associated with deuterostome bauplans. The main body coeloms are derived from the embryonic metacoels and are well developed in most groups. Other coelomic derivatives include the water vascular system, gonadal linings, and certain neural sinuses.

The main body cavities, or **perivisceral coeloms**, are lined with ciliated peritoneum, and their coelomic fluid plays a major circulatory role. A variety of coelomocytes are present in the body fluid and in the water vascular system. Many of these cells are phagocytic. Hemoglobin occurs in the coelomocytes of many holothurians and a few brittle stars.

Water Vascular System

The water vascular system is intimately involved in many aspects of echinoderm biology, and a discussion of its anatomy is a necessary preface to other considerations. It is perhaps easiest to begin with an examination of the system in a sea star and then treat the other taxa.

Asteroidea.   Figure 22.5A is a schematic representation of the water vascular system of a sea star. The system opens to the exterior through a special skeletal plate, the madreporite, or sieve plate, located off-center on the aboral surface on the CD interambulacrum (Figure 22.3C). The madreporite is perforated and deeply furrowed, and the overlying epidermis is ciliated and porous where it lines the furrows. The function of the madreporite has been the subject of much controversy. The traditional view that it serves as an avenue for sea water to enter the system has been challenged because the fluid in the system differs from sea

water. However, using radioactive tracers, Ferguson (1984) demonstrated that water does in fact enter through the madreporite. We still lack a clear understanding of how this structure functions.

Internally, the madreporite forms a cuplike depression, the lumen of which is called the **ampulla**, that communicates with other coelomic derivatives of the water vascular system and the hemal system (discussed below). From the lower end of the ampulla arises the **stone canal**, so named because of the skeletal deposits in its wall. A portion of the hemal system called the axial sinus (discussed below) is often intimately associated with the stone canal. The stone canal descends orally and joins with a circular **ring** or **circumoral canal**, which extends around the central disc on a plane perpendicular to the body axis. In addition to a **radial canal** extending into each arm, the ring canal gives rise to blind pouches called **Tiedemann's bodies** and **polian vesicles** (Figure 22.5A,B). There is some uncertainty about the functions of these pouches, but it is suspected that the former produce certain coelomocytes and the latter help regulate internal pressure within the water vascular system.

The fluid in the water vascular system is similar to sea water, but it includes various coelomocytes, certain organic compounds such as proteins, and a relatively high concentration of potassium ions. This fluid is moved through the system largely by the action of cilia that line the canal epithelium. Some of the canals, especially the stone and ring canals, contain internal partition-like extensions of their inner walls that probably help direct the flow of fluid. Ferguson and Walker (1991) describe the stone canal in some sea stars as a "ciliary pump" that draws fluid into the water vascular system from both the madreporite and the axial sinus of the hemal system. Thus, it appears that the liquid in the water vascular system is a combination of environmental sea water and body fluid.

In each arm the radial canal gives rise to numerous **lateral canals**, each of which terminates in a tube foot. In most asteroids, each tube foot consists of a bulbous **ampulla** and a hollow, muscular, suckered **podium** (Figure 22.5B). Members of the orders Platyasterida and Paxillosida lack suckers on their tube feet. The ampullae are internal and lie above the skeletal plates of the ambulacral groove. The podia extend to the outside and contain the usual body wall muscle layers around a coelomic lumen and sometimes include supportive ossicles. In asteroids the tube feet serve primarily for locomotion and temporary attachment, and to hold prey during feeding. In addition, they are usually highly touch-sensitive. At the tip of each radial canal is an unsuckered, tentacle-like, sensory **terminal tube foot**.

The operation of the tube feet depends on hydraulic pressure regulation and on muscle action of the individual ampullae and podia. Fluid is supplied to each podium from the main canal system. The ampulla acts as a

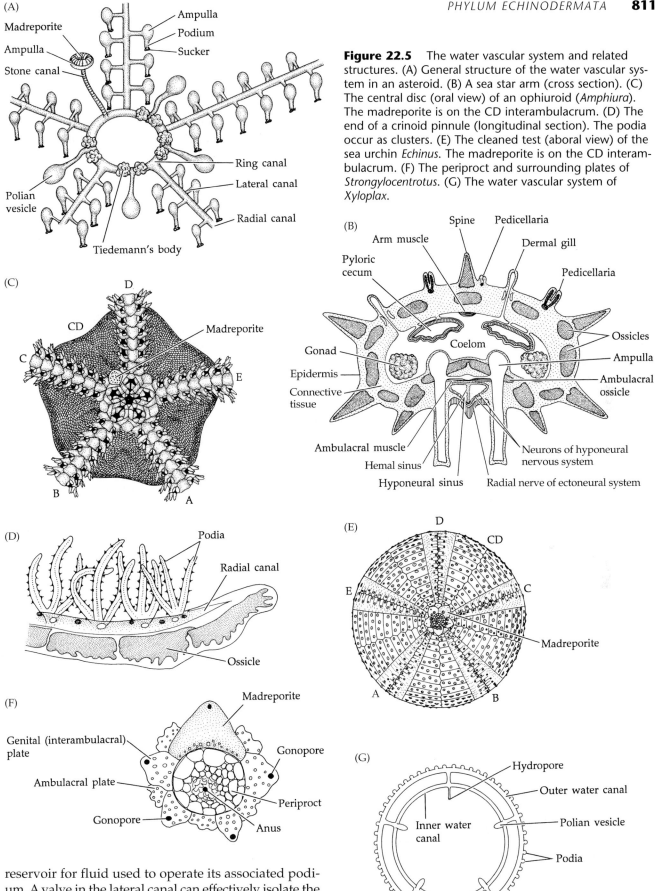

Figure 22.5 The water vascular system and related structures. (A) General structure of the water vascular system in an asteroid. (B) A sea star arm (cross section). (C) The central disc (oral view) of an ophiuroid (*Amphiura*). The madreporite is on the CD interambulacrum. (D) The end of a crinoid pinnule (longitudinal section). The podia occur as clusters. (E) The cleaned test (aboral view) of the sea urchin *Echinus*. The madreporite is on the CD interambulacrum. (F) The periproct and surrounding plates of *Strongylocentrotus*. (G) The water vascular system of *Xyloplax*.

reservoir for fluid used to operate its associated podium. A valve in the lateral canal can effectively isolate the tube foot from the rest of the system. When the ampulla is filled with fluid and the lateral canal valve is closed, the ampulla contracts and forces fluid into the podium. The sucker is then pressed against the substratum and held there by adhesive secretions of the epidermis. Next the longitudinal muscles of the podium contract; this action shortens the tube foot and forces the fluid back into the now relaxed ampulla. At the same time, other muscles raise the center of the sucker disc and create a vacuum, like that of a suction cup. Release of the sucker involves relaxation of the podial muscles and contraction of the ampulla; this action again forces fluid into the lumen of the podium and releases the suction. In addition to this attachment–detachment action, the podia are also capable of bending by differential contraction of the longitudinal muscles.

The water vascular system of the sea daisies (*Xyloplax*) is unique among the echinoderms (Baker et al. 1986). A madreporite homologue, the **hydropore**, opens on the aboral surface on the CD interambulacrum (Figure 22.3D) and connects internally to a pair of concentric water canals (Figure 22.5G). Polian vesicles lie on the other four interambulacra. The outer, marginal water canal gives rise to peripherally located suckerless podia. Each podium bears an internal ampulla. This is the only echinoderm water vascular system in which the podia are not arranged along the ambulacra.

Ophiuroidea. The water vascular system of brittle stars is similar to that of asteroids. However, the madreporite is on the oral surface of the central disc, on the CD interambulacrum, and the internal plumbing is modified accordingly (Figure 22.5C). In some ophiuroids (e.g., *Ophioderma appressun*) the madreporite is reduced to two tiny pores. Apparently, most of the fluid in this type of system is drawn from the axial sinus by the stone canal (Ferguson 1995).

The ring canal bears polian vesicles, but apparently lacks Tiedemann's bodies. The ring canal gives off the usual five radial canals and also branches to a wreath of **buccal tube feet** around the mouth. In basket stars the arms and the radial canals are branched. The suckerless podia are highly flexible, finger-like structures that secrete copious amounts of sticky mucus. They function primarily as feeding, digging, and sensory organs.

Crinoidea. The water vascular system of crinoids operates entirely on coelomic fluid. There is no external madreporite; rather, a number of "stone canals" arise from the ring canal and open to coelomic channels. Some species possess hundreds of such stone canals. The main perivisceral coeloms bear ciliated funnels to the exterior through which water enters the body cavities, perhaps as an indirect method of regulating hydraulic pressure in the water vascular system.

From the ring canal arise the main radial canals that extend into each arm and paired oral tube feet that appear at each interambulacrum. The number of arms in crinoids ranges from five to as many as two hundred, and in many cases the arms are branched. The number of radial canals corresponds to the arm number in each

species, and they are branched in those with branched arms. Furthermore, crinoid arms bear tiny side branches called **pinnules** (Figure 22.3B), into which branches of the radial canals extend. Suckerless podia occur along the pinnules, often in clusters of three (Figure 22.5D), and each cluster is served by a branch of the water vascular system. The podia are highly mobile and usually bear adhesive papillae on their surfaces; they function primarily as feeding and sensory organs.

Echinoidea. The water vascular systems of sea urchins and sand dollars may be viewed as modifications of the asteroid plan. These animals bear a special set of skeletal plates around the aboral pole; one of these plates is the CD-interambulacral madreporite (Figure 22.5E,F). To understand the water vascular system of sea urchins, it is necessary to realize that the ambulacra, and thus the rows of podia and their internal plumbing, extend around the sides of the body (like five longitude lines on a globe) to the upper surface, where they converge toward the aboral pole (Figures 22.2, 22.5E).

The madreporite of echinoids, like that of asteroids, leads to an ampulla and then to a stone canal (short in sand dollars and long in sea urchins), which extends orally to a ring canal surrounding a complex system of muscles and plates that comprise the feeding apparatus. The ring canal gives rise to five radial canals, one beneath each ambulacrum. Each radial canal gives off lateral canals leading to tube feet and terminates in a sensory podium near the aboral pole. Unlike the plates in other echinoderms, the ambulacral plates of echinoids have holes in them through which the podia pass to the outside. The tube feet of echinoids may be suckered or unsuckered, and they serve a variety of functions, including attachment, locomotion, feeding, and gas exchange.

Holothuroidea. In sea cucumbers the water vascular system contains the major elements seen in other taxa, but it is organized to accommodate the elongation of the body. In most holothurians the madreporite is internal and opens to the coelom. The madreporite lies beneath the pharynx in the CD-interambulacral position and gives rise to a short stone canal. A ring canal encircles the gut and bears from 1 to 50 polian vesicles. Five radial canals arise from the ring canal and give off extensions to the oral tentacles before extending aborally ("posteriorly") beneath closed ambulacral grooves. In those species that retain clear pentamerous symmetry, each radial canal gives rise to rows of ampullae and suckered podia. In some species the podia of the bivium (the "dorsal" or upper surface) are reduced or lost, and in the apodaceans all of the tube feet are greatly reduced or absent. The podia of holothurians serve in locomotion and attachment, and are touch-sensitive.

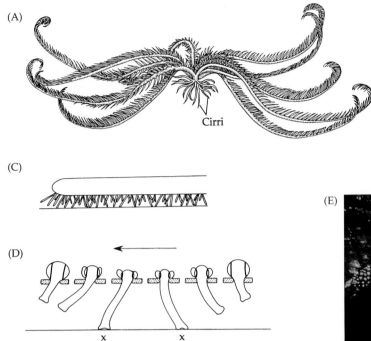

Cirri

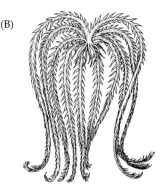

Support and Locomotion

Except for the holothurians, the general body shape and structural support of echinoderms are maintained primarily by the skeletal elements. Particular structures, such as podia and gills, are supported mostly by hydrostatic pressure. In most sea cucumbers, in which the skeletal plates are usually tiny separate ossicles, the body wall muscles form thick sheets, adding structural integrity to the body by working on the coelomic spaces to provide a hydrostatic skeleton.

Many echinoderms possess certain connective tissues that contribute to body "tone" through rapid changes in their mechanical properties (Motokawa 1984). In a matter of seconds or minutes the fibers of these tissues can become relatively rigid, thereby reducing body flexibility. This transformation appears to be under direct nervous control but does not involve muscular activity.

Locomotor methods among echinoderms are determined by overall body configuration, the animals' habits, and the nature of the skeletal, muscular, and water vascular systems. Apart from the sessile sea lilies (e.g., *Ptilocrinus*), most extant crinoids are capable of crawling and swimming, both of which are done with the oral side directed away from the substratum (Figure 22.6A,B). The aboral **cirri** are used primarily for temporary attachment and for righting the animal if overturned.

During crawling, the arms are bent downward and used to lift the body off the substratum; the animal then walks on its arm tips. Swimming is accomplished by up-and-down sweeps of the arms, which are divided into functional sets that move alternately. For example, in ten-armed species, five arms move upward while the

Figure 22.6   (A) The crinoid, *Antedon* in a resting position. (B) *Antedon* as it might appear walking on its arm tips. (C) A sea star arm (side view) with tube feet in motion. (D) Changes in position of an individual podium as the animal moves in the direction of the arrow. The podium executes its power stroke while in contact with the substratum (x), and its recovery stroke while lifted from the substratum. Note the changes in podium length and the corresponding changes in volume of the ampulla. (E) The sea star *Pisaster giganteus* crawling over an irregular substratum.

other five arms move downward. As any given arm is moving one way, its two neighboring arms are moving the opposite way. In animals with more arms (usually multiples of five), the arms are divided into functional sets of five.

Asteroids exemplify locomotion using podia. The action of a single podium involves power and recovery strokes, with the process following the same fundamental mechanical principles we have seen in the appendages of many other invertebrates. The sea star's arms are held more or less stationary relative to the central disc, even in species with a flexible skeletal framework (e.g., *Pycnopodia*), and movement is accomplished by the thousands of podia on the oral surface. Overall movement is generally smooth because of the high number of podia and the fact that at any given moment they are in different phases of the power and recovery strokes (Figure 22.6C). Although there is some coordination of the action of the tube feet to produce movement in a particular direction, there are no metachronal waves of podial motion as seen in many other "multilegged" creatures. In fact, control of podial action is not fully understood (even isolated arms crawl about normally).

Most sea stars move very slowly, but a few (e.g., *Pycnopodia*) are relative speedsters. Some asteroids that are usually rather sedentary become extraordinarily rapid "runners" upon encountering a potential predator (often another sea star). Some species that cannot escape by fast movement have evolved other defense mechanisms. The slow-moving Pacific "cushion star," *Pteraster tesselatus* (Figure 22.1D), secretes copious amounts of mucus, which serves to discourage predators such as *Solaster* and *Pycnopodia*.

If one can follow the action of a single podium during movement (not an easy assignment), the locomotory forces can be understood (Figure 22.6D). At the end of a recovery stroke, the podium extends in the direction of movement and attaches to the substratum. The sucker remains attached during the power stroke as the longitudinal muscles in the wall of the podium begin to contract, thereby shortening the podium and pulling the body forward. At the end of the power stroke, the podium lifts from the substratum and swings forward again. As illustrated, the ability to bend the podia is essential to the overall action. The huge number of podia and the general flexibility of the body allows most sea stars to glide smoothly over even rough and irregular surfaces (Figure 22.6E).

Ophiuroids use their flexible articulated arms primarily for crawling or clinging (Figure 22.3F). The skeletal arrangement of the arms allows for extensive "lateral" movement on a plane perpendicular to the body axis, but the arms have almost no flexibility parallel to the body axis. This feature, coupled with the fragile nature of these animals, causes them to break easily when lifted by an appendage—hence the common name "brittle stars." The tube feet lack suckers and ampullae, but are equipped with a well developed lattice of muscles in their walls. They are capable of protraction and retraction and of swinging through arcs. These combined actions of the arms and podia allow many ophiuroids to burrow into soft sediments.

Sea urchins move by the use of podia and movable spines. Their long suckered podia are capable of a wide range of motion, and the strong spines provide stiltlike support and movement. Some "regular" urchins excavate shallow depressions in hard rock. *Strongylocentrotus purpuratus*, a common West Coast urchin of North America, forms such pockets in hard substrata, and members of this species often become trapped in their self-made homes. These urchins bore largely by the action of the teeth of their feeding apparatus. Their excavations provide protection in areas of high wave and surge action.

Some of the irregular urchins burrow well below the sand surface and maintain an open chimney from their cavern to the overlying water (Figure 22.11G). Most of these soft-sediment burrowers have special spatulate spines along the sides of the body that aid in digging.

Sand dollars live in or on soft sediments. Some bury themselves completely, but most keep part of the body above the surface (Figure 22.11F). A few, such as *Clypeaster rosaceus*, do not burrow at all. Burrowing and crawling are accomplished largely by the action of movable spines. There has been some controversy about the function of the deep marginal notches and holes (**lunules**) in the tests of some sand dollars (Figure 22.3H). Elegant experiments by Telford (1981, 1983) indicate that these structures help the animals maintain stability in strong currents. Drag is eased by flow along surface channels from the center of the body to the lunules and notches and then away from the test margin. In addition, the lunules reduce the lift generated by ambient water movements.

Holothurians live on the surfaces of various substrata or else burrow into soft sediments. Crawling is accomplished by the podia or by action of the body wall muscles. Many epibenthic species are cryptic and usually remain lodged in cracks and crevices or under rocks. In these forms the podia are used primarily for anchorage and to hold bits of shell and stone against the body for protection. In a few deep-sea forms (e.g., *Scotoplanes*; Figure 22.1N), some of the podia are elongate and used for walking. In some holothurians (e.g., *Psolus*), the trivium surface is modified as a creeping, footlike sole. A few sea cucumbers are pelagic and capable of weak swimming (Figure 22.1P,Q).

The apodaceans lack locomotor tube feet and most burrow in sand or mud by means of peristaltic action of the body wall muscles. Some live completely buried, whereas others form U-shaped burrows.

Feeding and Digestion

Echinoderms display a great variety of feeding strategies, and we present only a brief survey here. In addition, the structure of the digestive tract differs among groups, as summarized below.

Crinoidea. Sea lilies and feather stars sit with their oral sides up and feed by removing suspended material from the surrounding water. The arms and pinnules are usually held outstretched on a plane perpendicular to the ambient water flow, thus presenting a large food-trapping surface. Many errant forms are negatively phototactic and emerge from concealment to feed only at night. Some deep-water species hold their arms upward and outward, forming a funnel with which they capture detrital rain.

The open ambulacral grooves extend onto the pinnules and are lined with cilia that beat toward the mouth. Food particles, including plankton and organic particulates, contact the podia, which then flick the food into the grooves (Figure 22.7B). Cilia drive the food to the mouth, where it is ingested. The primitive nature of

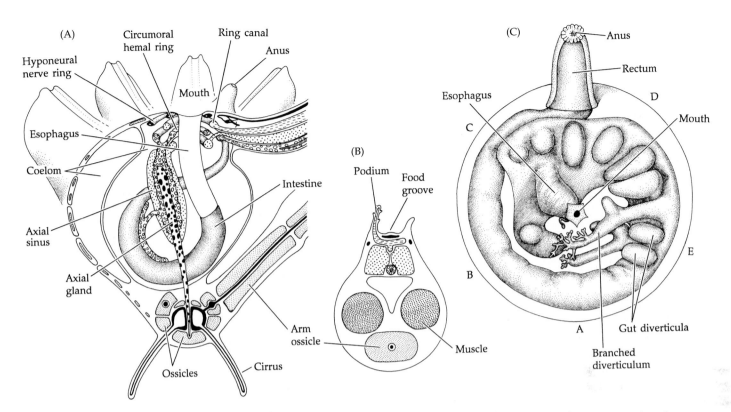

Figure 22.7 Internal anatomy of crinoids. (A) Central disc and base of one arm (vertical section). (B) An arm with open ambulacral (food) groove (cross section). (C) The oral surface of *Antedon* (cutaway view). The positions of ambulacral radii are indicated by the letters around the periphery.

crinoids suggests that this use of the podia and ambulacral grooves for suspension feeding may reflect the original function of the water vascular system.

The mouth opens to a short esophagus that leads to a long intestine (Figure 22.7A,C). The intestine loops around the calyx and then straightens to a short rectum terminating at the anus, which is borne on an **anal cone** near the base of one of the arms. In most species the intestine bears diverticula, some of which are branched. Although the histology of the crinoid gut has been described, little is known about the digestive physiology of these animals.

Asteroidea. Most sea stars are opportunistic predators or scavengers. They feed on nearly any dead animal matter and prey on a variety of invertebrates. Many species are generalists in terms of their food preferences and may play important roles as high-level predators in intertidal and subtidal communities. Others are strict specialists. *Solaster stimpsoni*, a large northeastern Pacific sea star, feeds exclusively on holothurians, while a related species (*S. dawsoni*) preys on *S. stimpsoni*!

Among the best known sea stars is the tropical "crown-of-thorns," *Acanthaster planci*. This animal feeds on coral polyps and has received great notoriety in recent years because of its implication in the destruction of Indo–West Pacific coral reefs. There is still disagreement concerning the reason for the recent increases in the size of the *Acanthaster* populations, but some specialists think that it is a result of human interference in the predator–prey balance of the reef communities. Among the major predators of *Acanthaster* is the giant triton, *Charonia* (Gastropoda), which is collected in high numbers for its handsome shell.

Except for a few suspension feeders (discussed below), most sea stars depend on an eversible portion of the stomach to obtain food. Some forms, including *Acanthaster*, *Culcita* (the cushion star), and *Asterina* (the bat star), spread the stomach over the surface of a food source, secrete primary enzymes, and suck in the partially digested soup. In the case of *Culcita*, the food may include encrusting sponges or algal mats or organic detritus that has accumulated on the substratum. *Asterina* feeds in much the same manner, digesting organic matter under its spread everted stomach. *Oreaster* extrudes its stomach over sand, algae, or sea grass and ingests the associated microorganisms and particulate detritus. It can, however, switch to a predatory or scavenging mode when appropriate food sources are encountered. One Caribbean species (*O. reticulatus*) feeds primarily on sponges by everting its stomach and digesting its prey (Wulff 1995). Many sea stars that feed on large prey also utilize external digestion by everting the stomach. Sedentary or sessile prey, such as gastropods, bivalves, and barnacles, are eaten by a host of asteroid predators, including the voracious Pacific ochre star *Pisaster ochraceus* (Figure 22.8D). This sea star hunches over its

Figure 22.8   Feeding and internal anatomy of asteroids. (A) The central disc and base of one arm of a sea star (vertical section). (B) *Asterias* (oral view). The mouth is ringed by oral spines and podia. (C) The internal organs in the central disc and arms of the trivium of *Asterias*. Each dissected arm has various organs removed. (D) A constellation of the predatory sea star *Pisaster ochraceus*.

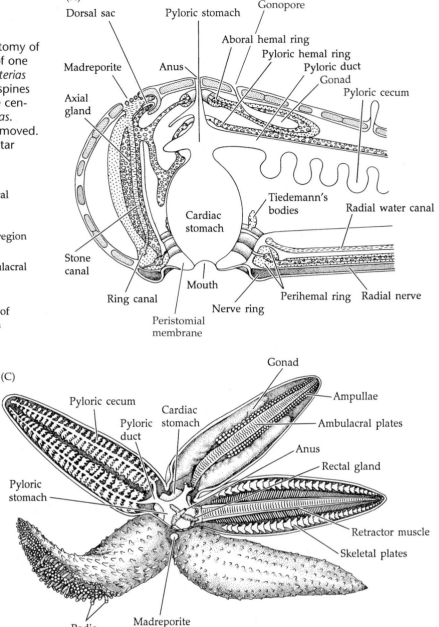

(A)

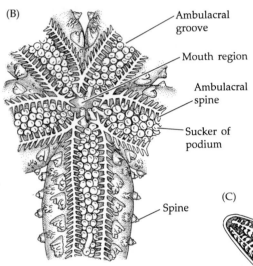

(B)

(C)

(D)

prey with the oral area pressed against the potential victim, holding itself in position with its podia. It then everts the stomach and begins secreting digestive enzymes. The stomach is very thin and flexible; it can be slid between even the tightly clamped valves of mussels and clams, thus liquefying the prey's body inside its own shell. The fluid nutrients are drawn in with the retracting stomach.

Some sea stars are suspension feeders, consuming plankton and organic detritus. *Henricia*, *Porania*, and a few others are typically full-time suspension feeders, and some predatory types, such as *Astropecten*, are capable of periodic suspension feeding as a means of supplementing their usual diet. In most of these sea stars, particulate food material that contacts the body surface is trapped by mucus and moved by cilia to the ambulacral grooves and ultimately to the mouth. Food movement is by ciliary action. *Leptasterias tenera* is able to capture suspended food, such as phytoplankton and small crustaceans, with its pedicellariae and tube feet. The sea star *Novodinia antillensis* extends its arms upward into water currents. The dozen or so arms form a large feeding surface used to capture planktonic crustaceans; the prey are grasped by pedicellariae. A few species, including *Stylasterias forreri* and *Labidiaster annulatus*, possess wreathlike circlets of pedicellariae used in prey capture

(Figure 22.4F,G). These sea stars feed on a variety of animals, even fishes (Chia and Amerongen 1975, Dearborn et al. 1991).

The digestive system of sea stars extends from the mouth in the center of the oral surface to the anus in the center of the aboral surface (Figure 22.8A). The mouth is surrounded by a leathery **peristomial membrane**. The membrane is flexible, allowing eversion of the stomach, and it contains a sphincter muscle to close the mouth orifice. Internal to the mouth is a very short esophagus leading to the **cardiac stomach**, which is the portion that is everted during feeding. Radially arranged retractor muscles serve to pull the stomach back within the body. Aboral to the cardiac stomach is a flat **pyloric stomach**, from which arises a pair of **pyloric ducts** extending into each arm. These ducts lead to paired digestive glands, or **pyloric ceca**, in each arm (Figure 22.8A,C). A short intestine leads from the pyloric stomach to the anus and often bears outpocketings called **rectal glands** or **rectal sacs**.

The pyloric ceca and cardiac stomach are the main sites of enzyme production. These enzymes, mostly proteases, are carried by ciliary action through the everted stomach and released onto the food material. Digestion is completed internally, but extracellularly, after ingestion of the liquefied food. Digested products are moved through the pyloric ducts to the pyloric ceca, where they are absorbed and stored. The intestine apparently serves little purpose in the digestive process, but the rectal sacs are known to pick up nutrients from the intestine, probably salvaging them from potential loss through the anus.

Many sea stars harbor various commensals that derive their food from scraps of their host's meals. One well known relationship is that of a polynoid scale worm, *Arctonoe vittata*, and several species of asteroid hosts, including the Pacific leather star, *Dermasterias imbricata*. The worm is an obligate symbiont, spending most of its life cruising and feeding in the host's ambulacral grooves. Not only is the polychaete chemically attracted to its host, but recent studies indicate that *Dermasterias* is also attracted to *Arctonoe*; this observation suggests that the sea star also may derive some benefit from the association.

The sea daisy *Xyloplax medusiformis* lacks a digestive system, but the oral surface is covered by a membranous **velum** that may have been derived from the gut (Baker et al. 1986). These animals may absorb dissolved organic matter across this velum. Perhaps the source of the nutrients is bacteria that live in the decomposing-wood habitat of these strange asteroids. *Xyloplax turnerae* has an incomplete gut. A large mouth opens into a shallow, saclike stomach, but intestine and anus are lacking.

Ophiuroidea. Brittle stars exhibit a variety of feeding methods, including predation, deposit feeding,

scavenging, and suspension feeding; some species are capable of more than one method. Some ophiuroids, such as the basket stars (Figure 22.9F), are really predators that utilize suspension feeding strategies to capture relatively large swimming prey (up to about 3 cm long).

Selective deposit feeding is accomplished by the podia and sometimes by the arm spines. The epidermis of the arms secretes mucus, to which organic material adheres. The podia roll the mucus and food into a clump, or food bolus. Near the base of each podium is a flaplike projection called a **tentacular scale** (Figure 22.9B,D). The food bolus is transferred from a podium onto its adjacent scale, picked up by the next podium, and so on, so that the food is transported along the arm to the mouth. Suspension feeding by brittle stars usually involves a similar method of transport once food is trapped. Food capture is sometimes accomplished by secreting mucous threads among the arm spines and waving the arms about to trap plankton and organic detritus. The food is moved to the podia and then transported to the mouth. Brittle stars that use this technique typically have very long arm spines (e.g., *Ophiocoma*, *Ophiothrix*, *Ophionereis*). *Astrosoma agassizii* extends its long (up to 70 cm) arms into the overlying water and captures planktonic copepods (Ferrari and Dearborn 1989). Some other brittle stars suspension feed by using extended podia to form a trap; then the podia pass clumps of food to the mouth (Figure 22.9E).

Predatory suspension feeding by basket stars occurs mostly at night. At dusk the animals emerge from their hiding places and assume a feeding position, with their branched arms held fanlike into the prevailing current, in a manner similar to the feeding behavior of most crinoids. *Astrophyton muricatum* changes position with the ebb and flow of the tide, always orienting its arms into the current; it stops feeding at slack tide (Hendler 1982b). When a small animal contacts an arm, the appendage curls to capture the prey. Ingestion is often postponed until darkness has passed; the prey is then transferred to the mouth by the flexible arm. These basket stars feed on a variety of invertebrates, such as swimming crustaceans and demersal polychaetes.

Some brittle stars are active predators, capturing benthic organisms by curling an arm into a loop around the prey, then pulling it to the mouth. Species that feed in this manner usually have short arm spines that lie flat against the arm itself (e.g., *Ophioderma*). Several species of brittle stars dig beneath the surface of the substratum and form semipermanent mucus-lined burrows. The arms extend to the surface and help maintain ventilation currents within the burrows. Such species are able to extract food from within the sediment, the substratum surface, and the overlying water (Woodley 1975).

The commensal brittle star *Ophiothrix lineata* lives in the atrium of the large sponge *Callyspongia vaginalis*, emerging to feed on detritus adhering to its host's outer

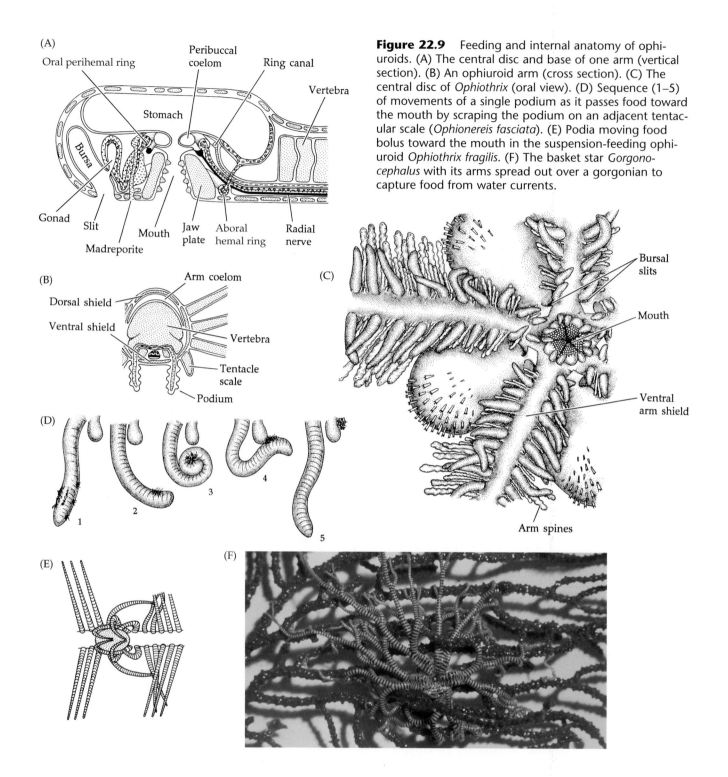

(A)

Oral perihemal ring

Peribuccal coelom

Ring canal

Vertebra

Stomach

Bursa

Gonad

Slit

Mouth

Madreporite

Jaw plate

Aboral hemal ring

Radial nerve

(B)

Arm coelom

Dorsal shield

Ventral shield

Vertebra

Tentacle scale

Podium

(D)

1 2 3 4 5

(C)

Bursal slits

Mouth

Ventral arm shield

Arm spines

(E)

(F)

Figure 22.9 Feeding and internal anatomy of ophiuroids. (A) The central disc and base of one arm (vertical section). (B) An ophiuroid arm (cross section). (C) The central disc of *Ophiothrix* (oral view). (D) Sequence (1–5) of movements of a single podium as it passes food toward the mouth by scraping the podium on an adjacent tentacular scale (*Ophionereis fasciata*). (E) Podia moving food bolus toward the mouth in the suspension-feeding ophiuroid *Ophiothrix fragilis*. (F) The basket star *Gorgonocephalus* with its arms spread out over a gorgonian to capture food from water currents.

surface. While keeping the sponge clean, the ophiuroid is supplied with food and afforded protection from predators (Hendler 1983).

The digestive tract of ophiuroids is incomplete. The intestine and anus have been lost, and the digestive system is confined entirely to the central disc (Figure 22.9A). The mouth leads to a short esophagus and large folded stomach, which fills most of the interior of the disc and reduces the coelom to a thin chamber. The stomach is presumably the site of digestion and absorption.

Echinoidea. Feeding strategies among echinoids include various kinds of herbivory, suspension feeding, detritivory, and a few forms of predation. In most regular urchins, feeding depends largely on the action of a complex masticatory apparatus that lies just inside the mouth and bears five calcareous protractible teeth. This apparatus is commonly called **Aristotle's lantern** (Figures 22.10, 22.11A–D). It is a real architectural marvel: a complex of hard plates and muscles that control protraction, retraction, and

Figure 22.10 The feeding complex (Aristotle's lantern) in sea urchins. (A) The feeding complex in a regular urchin as seen from inside the test. (B) The feeding apparatus of *Paracentrotus* (vertical section). (C) The apparatus of *Cidaris* (aboral view). The compasses are removed to expose the rotules.

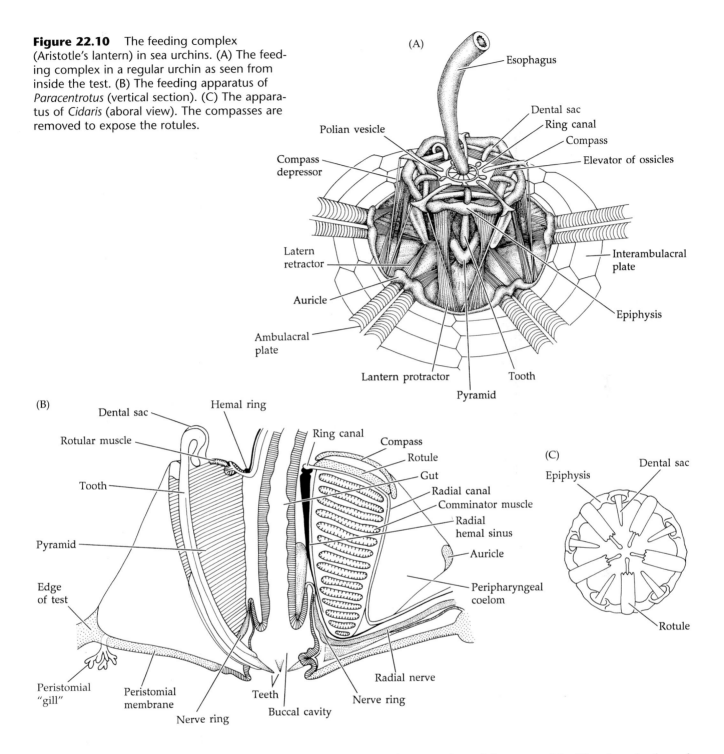

grasping movements of the five teeth. In many species the entire apparatus can be rocked such that the teeth protrude at different angles. There is great variation in lantern structure among echinoids, but the following brief description applies to most conditions in which it is present and well developed (e.g., in typical sea urchins).

The main structural elements of Aristotle's lantern are five vertically oriented triangular plates called **pyramids** (Figure 22.10). These calcareous pyramids are positioned in interambulacral spaces and are attached to one another by **comminator muscles**, which provide a

rocking motion of the pyramids. The aboral edge of each pyramid is a thickened bar called an **epiphysis**. Each pyramid has a canal within which lies a tooth. The sharp end of the tooth extends out from the oral end of the pyramid into the mouth region. A soft **dental sac** of coelomic origin covers the unhardened aboral end of each tooth where it emerges from the top of the pyramid. As the teeth are worn down by use, more tooth material is produced within the dental sacs and becomes calcified as it grows through the pyramid canal. Measurements on some species indicate that, with normal wear, the teeth grow about 1 mm each week. Lying

Figure 22.11   Feeding and internal anatomy of echinoids. (A) A regular sea urchin (vertical section). (B) Internal anatomy of *Arbacia*. (C) *Arbacia* (oral view). (D) The digestive system of the sand dollar *Echinarachnius parma* (aboral view). (E) A food groove on the oral surface of a sand dollar (cutaway view). The podia are moving food toward the mouth. (F) *Dendraster excentricus* in their feeding position, half-buried in benthic sediments. (G) An irregular urchin in its burrow.

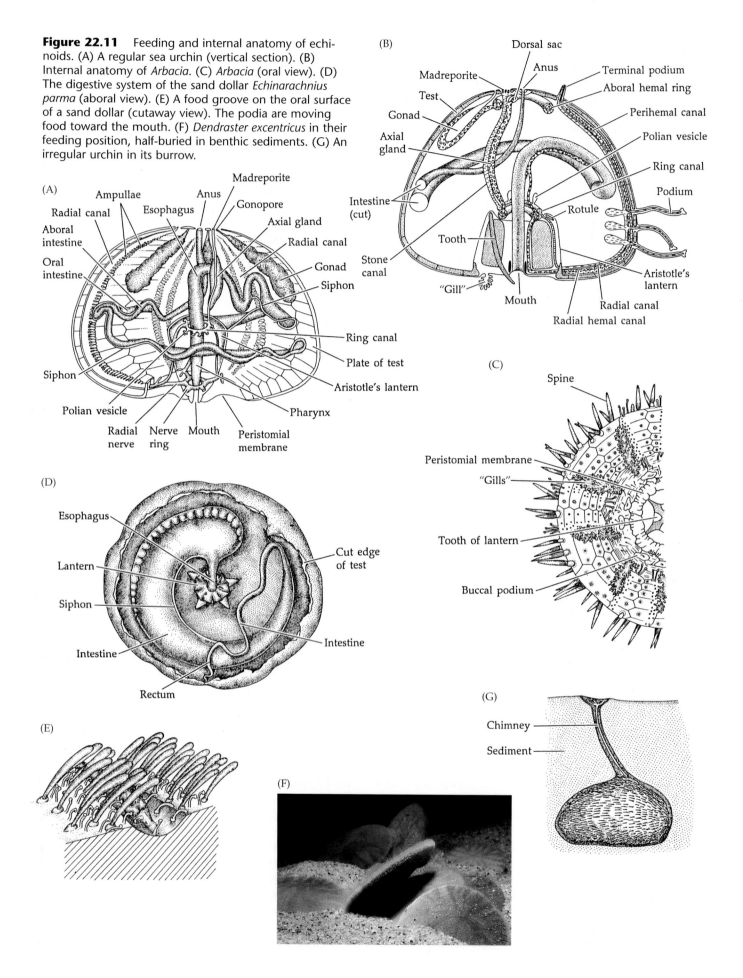

atop the main structure of the lantern, on the oral surface, are five **compasses** and five **rotules**, one of each along each ambulacral radius. The compasses and their associated muscles regulate hydrostatic pressure within the gills (see below).

The teeth are protracted by the contraction of sheet-like protractor muscles that originate around the mouth, on the interambulacral areas of the internal skeleton, and insert on the epiphyses, near the aboral ends of each pyramid. Their action pushes the entire lantern orally, and also serves to spread the teeth apart as protraction occurs. Retractor muscles originate on thick ambulacral plates called **auricles**, and they insert on the oral end of the lantern apparatus. Additional muscles associated with the pyramids and the rotules can produce a variety of tooth movements.

Most urchins with well developed lanterns use their teeth to scrape algal material from the substratum and to tear chunks of food into "bite-sized" pieces. Many species also feed on animal matter by similar actions. Some regular urchins excavate burrows in hard substrata and then feed on the algal film that develops on the burrow wall, or else they feed on suspended particles or drift algae that enter the chamber. Other burrowers establish a feeding position at the burrow entrance and catch floating debris with their podia and pedicellariae. Most irregular urchins (sea biscuits and heart urchins) lack a lantern. They burrow into soft sediments and feed on small organic particles (Figure 22.11G). These types of urchins usually use their podia to sort food material from the mud or sand and pass it to the mouth.

Most sand dollars (Clypeasteroida) are detritus and particulate feeders. They possess a highly modified lantern with nonprotractable teeth. Most of these animals burrow completely or partially in soft sediments and extract food particles from among the sand grains or from the overlying water. As they plow along, a layer of sediment passes over the aboral surface. Large non-food particles are moved by club-shaped spines and passed posteriorly off the body. Some species of *Clypeaster* lack these large spines and instead secrete copious amounts of mucus, which prevents particles from reaching the body surface by falling between the shorter spines (Telford et al. 1987).

Actual food collection in most sand dollars is accomplished by podia on the oral surface. These podia are often coated with mucus, to which small food particles adhere. The particles are passed to the food grooves, and podia therein move them to the mouth for ingestion (Figure 22.3H). Apparently, at least some sand dollars (e.g., *Mellita quinquiesperforata*) feed on relatively large particles by selectively picking them out of the sediment with special podia.

A few species of sand dollars (e.g., *Dendraster excentricus*) burrow into the substratum but leave the posterior part of the body extended at an angle above the sediment (Figure 20.1K and 22.11F). *Dendraster* traps diatoms and other particulate food in the water with its podia and then passes the food to the mouth as described above. Larger prey, such as tiny crustaceans, are captured by the pedicellariae. Some young sand dollars eat high-density sand grains (especially those containing iron oxides), which they store in the gut as ballast to help stabilize their position on the sea bottom.

Telford et al. (1983) described a unique feeding method by the clypeasteroid *Echinocyamus pusillus*. These sand dollars nestle among pebbles, which are brought to the mouth by podia and then rotated by the peristomial membrane while the teeth scrape off attached diatoms and organic detritus.

The digestive system of echinoids is basically a rather simple tube extending from the mouth to the anus. The mouth is located in the center of the oral surface or is shifted somewhat anteriorly in some irregular urchins. An esophagus extends aborally, through the center of the lantern (when present), and then joins an elongate intestine (Figure 22.11A,B,D). In most echinoids a narrow duct, called the **siphon**, parallels the intestinal tract for part of its length. Both ends of the siphon open to the intestine, providing a shunt for excess water and helping to concentrate food material in the gut lumen. In many species, blind ceca arise from the gut near the junction of the esophagus and intestine. The intestine narrows into a short rectum leading to the anus, which is located either centrally on the aboral surface, on the posterior margin, or posteriorly on the oral surface. Digestive enzymes are produced by the intestinal and cecal walls, and breakdown is largely extracellular.

Holothuroidea. Most sea cucumbers are suspension or deposit feeders. Many of the sedentary epibenthic or nestling forms (e.g., *Eupentacta*, *Aslia*, *Selenkothuria*, *Psolus*, *Cucumaria*) extend their branched, mucus-covered tentacles (Figure 22.12D,E) into the water to trap suspended particles, including live plankton. The tentacles are then pushed into the mouth one at a time and the food ingested (Figure 22.12F). A fresh supply of mucus is provided by secretory cells in the papillae of the tentacles and apparently also by gland cells of the foregut.

More active epibenthic types (e.g., *Stichopus*, *Parastichopus*) crawl across the substratum and use their tentacles to ingest sediment and organic detritus (Figure 22.12C). Several studies indicate that some holothurians (e.g., *Stichopus*, *Holothuria*) are highly selective deposit feeders, preferentially ingesting sediments high in organic content. Sediment extracted from the gut of *Holothuria tubulosa* contains a much higher percentage of organic material than the general surrounding sediment. This animal is so adept at selective feeding that even its fecal pellets have a higher organic content than the environmental sediments (Massin 1980). Many apodacean holothurians burrow through the substratum by peristaltic movements and ingest the sediment as they move.

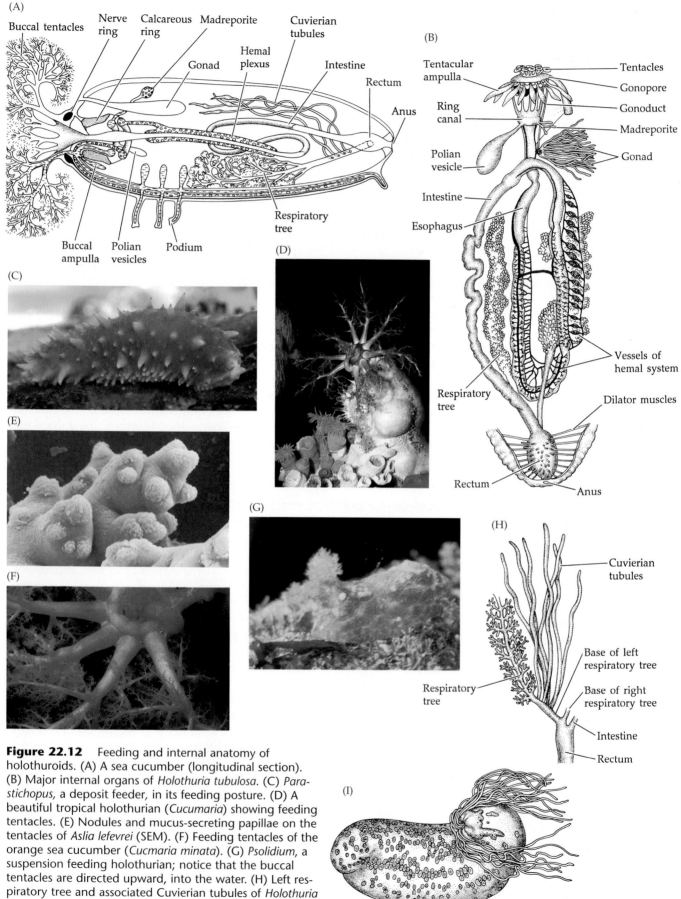

Figure 22.12 Feeding and internal anatomy of holothuroids. (A) A sea cucumber (longitudinal section). (B) Major internal organs of *Holothuria tubulosa*. (C) *Parastichopus*, a deposit feeder, in its feeding posture. (D) A beautiful tropical holothurian (*Cucumaria*) showing feeding tentacles. (E) Nodules and mucus-secreting papillae on the tentacles of *Aslia lefevrei* (SEM). (F) Feeding tentacles of the orange sea cucumber (*Cucmaria minata*). (G) *Psolidium*, a suspension feeding holothurian; notice that the buccal tentacles are directed upward, into the water. (H) Left respiratory tree and associated Cuvierian tubules of *Holothuria impatiens*. (I) Release of Cuvierian tubules by *Holothuria*.

The anterior mouth is surrounded by a whorl of buccal tentacles. The esophagus (or pharynx) leads inward and passes through a ring of calcareous plates that support the foregut and the ring canal of the water vascular system. The esophagus joins an elongate intestine, the anterior end of which is often enlarged as a stomach. The intestine extends posteriorly, loops forward, and then posteriorly again; it may be coiled (Figure 22.12A,B). The intestine terminates in an expanded rectum leading to the posterior anus. The rectal area is attached to the body wall by a series of suspensor muscles and often bears highly branched outgrowths that extend anteriorly in the body cavity. These structures are the **respiratory trees**, into which water is pumped via the anus for gas exchange (Figure 22.12A,B,G). Digestion and absorption probably take place along the length of the intestine.

The digestive system of sea cucumbers is associated with two fascinating phenomena: (1) **evisceration** and (2) the discharge of structures called **Cuvierian tubules** (Figure 22.12H,I). Evisceration is the expulsion by muscular action of part or all of the digestive tract and sometimes other organs, including the respiratory trees and gonads. In some forms (e.g., *Holothuria*) all of these structures are expelled following rupture of the hindgut region. In others (e.g., *Thyone* and *Eupentacta*) rupture occurs anteriorly and the tentacular crown and foregut are lost. Evisceration can be induced in many species by a variety of experimental conditions (e.g., chemical stress, physical manipulation, and crowding), but it also occurs in nature in some species. The significance of this process is unclear. It is viewed by some zoologists as a seasonal event associated with adverse conditions and by others as a defense mechanism wherein the eviscerated parts serve as a decoy. In any case, the lost parts are usually regenerated.

Cuvierian tubules are defensive structures. These organs are clusters of sticky, blind tubules arising from the base of the respiratory tree in certain genera (e.g., *Actinopyga* and *Holothuria*) (Figure 22.12A,G,H). When threatened, these cucumbers aim the anus at the potential predator, contract the body wall, and discharge the tubules by rupturing the hindgut. The tubules are shot onto the predator, entangling it in the sticky mass. The Cuvierian tubules are regenerated along with any other tissue lost during discharge.

Such elaborate defense mechanisms are not without adaptive significance in sea cucumbers. They are common prey to a great variety of other animals, including various sea stars, fishes, gastropods, crustaceans, and even humans (see review by Francour 1997).

Circulation and Gas Exchange

Circulation. Internal transport in echinoderms is accomplished largely by the main perivisceral coeloms, augmented to various degrees by the water vascular system and the **hemal system** (Figure 22.13), both of which are derived from the coelom. Fluids are moved through these systems largely by ciliary action and in some cases by muscular pumping. In at least one species of sea urchin (*Lytechinus variegatus*) coelomic fluid is also driven by movements of Aristotle's lantern (Hanson and Gust 1986).

The hemal system is a complex array of canals and spaces, mostly enclosed within coelomic channels called **perihemal sinuses**. The system is best developed in

Figure 22.13 Hemal system. (A) The central portion of the hemal system and some associated structures in an asteroid. (B) The complex hemal system of *Isostichopus badionotus* (Holothuroidea), showing its association with the gut and respiratory tree.

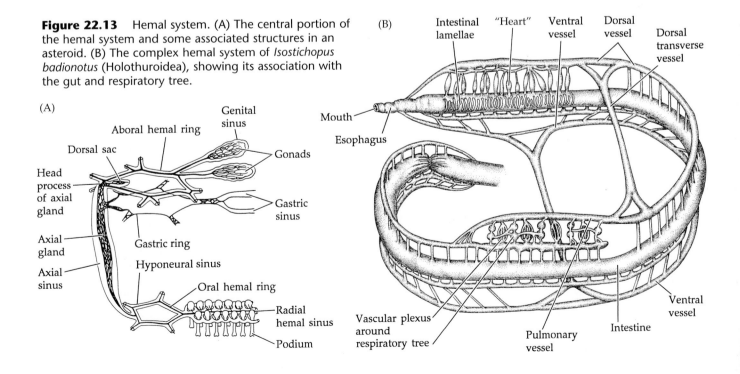

holothurians, in which it is bilaterally arranged, and in crinoids, in which some of the channels form netlike plexi. In other groups the system is radially arranged and generally parallels the elements of the water vascular system. In these cases the hemal system consists of an oral and an aboral **hemal ring**, each with radial extensions. The two rings are connected to each other by an **axial sinus** that lies against the stone canal (Figure 22.13A). Within the axial sinus there is a core of spongy tissue that is called the **axial gland**. The axial gland is apparently responsible for producing some coelomocytes. As mentioned earlier, the axial sinus often opens through pores to the stone canal and is a source of fluid for the water vascular system.

Radial hemal channels from the aboral ring extend to the gonads. Other radial channels arise from the oral hemal ring and are associated with the rows of tube feet; these channels are housed within a perihemal space called the **hyponeural sinus** (Figure 22.5B). A third hemal ring, the **gastric ring**, occurs in many echinoderms, including most asteroids, and is associated with the digestive system.

Fluid is moved through the hemal system by cilia. In asteroids and most echinoids the axial sinus bears a **dorsal sac** near its junction with the aboral hemal ring. The dorsal sac pulsates, apparently aiding the movement of fluid within the hemal channels and spaces. The hemal system of holothurians comprises an elaborate set of vessels (Figure 22.13B). It is intimately associated with the digestive tract and, when present, the respiratory trees. In many holothurians the hemal system includes many "hearts" or circulatory pumps.

The function of the hemal system is not fully understood, but it probably helps distribute nutrients absorbed from the digestive tract. Experiments on the sea star *Echinaster graminicolus* fed ^{14}C-labeled food show that absorbed nutrients appear in the hemal system within a few hours after feeding and eventually concentrate in the gonads and podia (Ferguson 1984). In sea cucumbers the hemal system probably also plays a role in gas exchange because some of the vessels are in contact with the respiratory trees.

Gas exchange. Most echinoderms rely on thin-walled external processes as gas exchange surfaces. Only ophiuroids and holothurians have special internal organs for this purpose. Given the relatively large body sizes and volumes of many echinoderms, the fluid transport mechanisms discussed above are of major importance in moving dissolved gases between internal tissues and the body surface.

Crinoids apparently exchange oxygen and carbon dioxide across all exposed thin parts of the body wall, especially the podia.

Gas exchange in asteroids occurs across the podia and special outpocketings of the body wall called **papulae** or **dermal gills** (Figure 22.14A,B). These structures are evaginations of the epidermis and peritoneum. Both tissues are ciliated, and their cilia produce currents in both the coelomic fluid and the overlying water. The two currents move in opposite directions, thus creating a countercurrent and maintaining maximum exchange gradients across the surfaces of the papulae.

Ophiuroids possess ten invaginations of the body wall called **bursae**, which open to the outside through ciliated slits (Figure 22.9A,C). Water is circulated through the bursae by the cilia and, in some species, by muscular pumping of the internal bursal sacs. Gases are exchanged between the flowing water and the body fluids. Hemoglobin occurs in the coelomocytes of a few species of ophiuroids.

Typical sea urchins possess five pairs of "gills" that are located in the peristomium (Figure 22.11B,C) and have long been viewed as the major gas exchange organs. However, various authors provide evidence of a different function (Shick 1983). The pressure within these "gills" changes by manipulation of the compasses of Aristotle's lantern. They probably function largely to accommodate pressure changes in the peripharyngeal coelom during feeding movements of the lantern complex, and perhaps to provide an immediate oxygen supply to the associated muscles. The main gas exchange structures in these urchins are apparently thin-walled podia that operate on a countercurrent system similar to that associated with the papulae of asteroids (Figure 22.14C,D).

Irregular sea urchins and sand dollars bear highly modified podia on the aboral **petaloids** (the five ambulacral regions of the fused skeleton, or test) (Figure 22.3G). The external parts of these podia are flaplike and thin-walled and serve as the main gas exchange surfaces. A countercurrent flow occurs between the water vascular system fluid in the podia and the sea water, and between the water vascular system fluid in the ampullae and the coelomic fluid (Figure 22.14E). Fenner (1973) provides a thorough examination of the respiratory function of echinoid podia.

We have already described the respiratory trees of certain holothurians. Water is pumped in and out of the hindgut and branches of the respiratory trees, and gases are exchanged between the water and the coelom and hemal system. This device is augmented by exchange across the podia, which is facilitated by a countercurrent system. Hemoglobin occurs in the coelomocytes of many holothurians.

Excretion and Osmoregulation

Excretion. In most echinoderms dissolved nitrogenous wastes (ammonia) diffuse across body surfaces to the outside. This type of excretion occurs across the podia and papulae in asteroids and is suspected to occur across the respiratory trees in holothurians. At least some excretion by simple diffusion probably takes place in most echinoderms. Precipitated nitro-

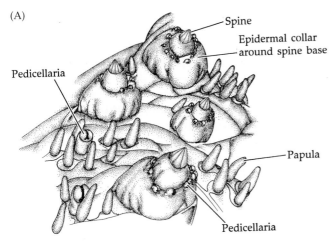

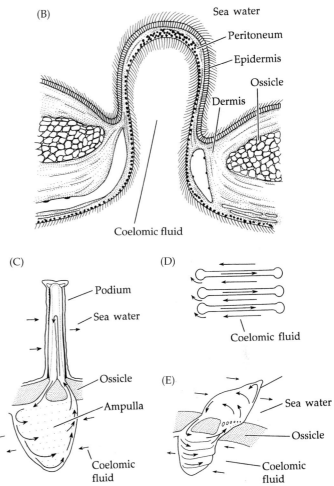

Figure 22.14 Gas exchange in echinoderms. (A) A portion of the aboral surface of *Asterias*. Note the digitiform papulae and their surrounding structures. (B) An asteroid papula (section). This structure is lined by the peritoneum and is filled with coelomic fluid. (C) An ampulla and podium (longitudinal section) of *Strongylocentrotus purpuratus* (Echinoidea). The arrows represent the countercurrents between the ambient sea water, the fluid of the water vascular system, and the coelomic fluid. (D) Three lamelliform ampullae from *Strongylocentrotus*. Gases are exchanged between the fluids of the water vascular system and the coelom. (E) A "respiratory" podium and ampulla (section) of the irregular sea urchin *Echinocardium*. The arrows represent the countercurrents.

genous material and other particulate wastes are phagocytosed by certain coelomocytes in the body fluids and then discharged by various methods.

In asteroids, waste-laden coelomocytes accumulate in the papulae, which then pinch off their distal ends, expelling the cells and waste material. Some studies indicate that the rectal glands may also be involved in excretion. In ophiuroids it is suspected that coelomocytes deliver wastes to the bursae, where they are released. Phagocytic coelomocytes in echinoids accumulate wastes and transport them to the podia and gills for release. In holothurians particulate wastes are carried by coelomocytes to the respiratory trees, gut, and even the gonads, and released to the outside through the plumbing systems of these organs. Crinoid coelomocytes deposit wastes in tiny pockets along the sides of the ambulacral grooves, but discharge has not been observed.

Osmoregulation. Echinoderms are generally considered to be strictly marine, stenohaline creatures. Consequently, they do not have problems of osmotic and ionic regulation. However, a number of species have been reported from brackish water. For example, *Asterias rubens* (Asteroidea) has been collected from the Baltic Sea (8‰), *Ophiophragmus filagraneous* (Ophiuroidea) from Cedar Key, Florida (7.7‰), and various holothurians from the Black Sea (18‰) (Binyon 1966). Obviously some mechanism allows them to survive in these low salinities.

The evidence to date suggests that echinoderms are osmoconformers. Both water and ions pass relatively freely across thin body surfaces, and the tonicity of the body fluids varies with environmental fluctuations. There appears to be some ionic regulation through active transport, but it is minimal.

Nervous System and Sense Organs

The secondarily derived radial bauplan of echinoderms is clearly reflected in the anatomy of their nervous systems and the distribution of their sense organs. The nervous system is decentralized, somewhat diffuse, and without a cerebral ganglion. There are three main neuronal networks, integrated with one another and developed to various degrees among the classes. These networks are the **ectoneural (oral) system**, the **hyponeural (deep oral) system**, and the **entoneural (aboral) system**. The ectoneural system is predominately sensory, although motor fibers do occur; the hyponeural system is largely motor in function. The entoneural system is absent from holothurians and reduced to different degrees in other groups—except the crinoids, in which it is the primary nerve component and serves both motor and sensory functions.

The three nervous "systems" are interconnected by a **nerve net** derived primarily from the ectoneural and entoneural components. The nerve net is often described as a subepidermal plexus, but it gives rise to intraepidermal neurons and clearly has an intimate association with the epithelium.

Except for the crinoids, in which the entoneural component dominates, the most obvious nerves in echinoderms are derived from the ectoneural system. A circular or pentagonal **circumoral nerve ring** lies just beneath the oral epithelium and encircles the esophagus. From this ring arise **radial nerves** that extend along each ambulacrum. In sea stars, for example, these radial nerves appear as a distinct V-shaped thickening in the epidermis of each ambulacral groove (Figure 22.5B). In some cases the entoneural components of the nerve plexus are also produced as radial cords, such as those along the lateral margins of the arms of asteroids. The hyponeural system generally parallels the nerves of the ectoneural system. Hyponeural neurons are subepidermal and lie near the hyponeural sinus of each ambulacral area (Figure 22.5B). These neurons give rise to motor fibers and ganglia in the tube feet.

Sensory receptors are largely restricted to relatively simple epithelial structures innervated by a plexus of the ectoneural system. Sensory neurons in the epidermis respond to touch, dissolved chemicals, water currents, and light. They are frequently associated with outgrowths of the body wall, such as spines and pedicellariae. Special photoreceptors occur in asteroids as **optic cushions**, each of which comprises a cluster of pigment-cup ocelli at the tip of an arm. Statocysts are known in some holothurians, and georeception is presumed to be the function of structures called **sphaeridia** in certain echinoids. Chemoreception has not been well studied in echinoderms, but there is some evidence that the buccal tentacles of holothurians and the oral podia of some echinoids are sensitive to dissolved chemicals. Chemoreception in asteroids appears to depend largely on direct contact, although distance chemoreception is reported in some species.

In spite of their rather simple nervous system and their lack of specialized sense organs, many echinoderms engage in complex behaviors. As is so often the case in such matters, there is still much to be learned about the functional mediation between the circuitry of the nervous system and the observed behavioral responses, as in the coordination of the podia during locomotion. Most echinoderms also exhibit distinct righting behaviors when overturned. These actions probably involve touch, georeception, and perhaps photoreception. Orientation to currents is known in some sand dollars and in many ophiuroids and crinoids.

Reproduction and Development

Regeneration and asexual reproduction. Most echinoderms are capable of regenerating lost parts. Even the casual observer of tidepool life will encounter a sea star regenerating a new arm, or notice the suckers of the podia left on a rock from which a sea star or urchin has been pulled free. Lost suckers are quickly replaced by regeneration. We have already described the dramatic processes of evisceration and expulsion of Cuvierian tubules—in both cases, the lost organs are replaced. Studies on regeneration in asteroids have put to rest the tales of oystermen who once claimed that chopping sea stars into small pieces resulted in the regeneration of an entire new animal from each part. While it is true that a damaged animal can grow new arms if a substantial portion of the central disc remains intact, an isolated arm soon dies. The exception to this generality is *Linckia*, which can regenerate an entire individual from a single arm, the regenerating stage being appropriately called a **comet** (Figure 22.15). Ophiuroids and crinoids frequently cast off arms or arm fragments when disturbed, and then regenerate the lost part. Such **autotomy** (voluntarily casting off an appendage) is also documented for certain asteroids. The Pacific coast ochre star, *Pisaster ochraceus*, autotomizes arms at their junction with the central disc when confronted by predators (e.g., the sea star *Pycnopodia*).

Asexual reproduction occurs in some asteroids and ophiuroids by a process called **fissiparity**, wherein the central disc divides in two and each half forms a complete animal by regeneration. When the small six-rayed brittle star *Ophiactis* divides, each half retains three arms. Asexual fission also occurs in some holothurians, but the process is not well understood.

Sexual reproduction. The majority of echinoderms are dioecious, but hermaphroditic species are known among the asteroids, holothurians, and especially the ophiuroids. The reproductive system is relatively simple and is intimately associated with derivatives of the coelom. The gonads are usually housed within peritoneally lined **genital sinuses**. Holothurians are unique among echinoderms in possessing a single gonad, which lies dorsally in the CD interambulacrum (Figure 22.12B). A single gonoduct opens between the bases of two dorsal buccal tentacles or just posterior to the tentacular whorl.

Crinoids lack distinct gonads. The gametes arise from the peritoneum of special coelomic extensions called **genital canals** in the pinnules on the proximal portion of each arm. There are no gonoducts; gametes are released by rupture of the pinnule walls. Ophiuroids possess from one to many gonads attached to the peritoneal side of each bursa adjacent to the bursal slits (Figure 22.9A). Gametes are released into the bursae and expelled through the slits.

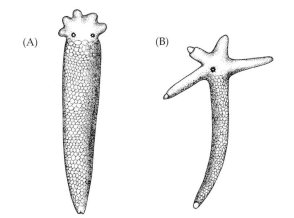

Figure 22.15 Regeneration in *Linckia*. (A) Initial regeneration from a single arm, here yielding a central disc with dual madreporites and five new rays. (B) At a later stage, the animal has a single madreporite and the normal ray number.

Asteroids and echinoids possess multiple gonads with gonoducts leading to interambulacral gonopores (Figures 22.8C, 22.11B). Sea daisies have a pair of gonads in each ambulacrum (Figure 22.3E). Regular sea urchins contain five gonads, one lying along the inside of each interambulacral radius. The gonopores are located on the five interambulacral **genital plates** surrounding the **periproct** (Figures 22.5F, 22.11A). The periproct and anus have migrated posteriorly in irregular urchins and sand dollars, but the genital plates remain more or less centrally located on the aboral surface. In many of these animals there are only four (and sometimes fewer) gonads, one being lost along the line of migration of the anus. In such cases there is a corresponding reduction in the number of gonopores. In all urchins one of the genital plates is perforated and doubles as the madreporite.

Life history strategies among echinoderms vary from free spawning followed by external fertilization and indirect development to various forms of brooding and direct development. Spawning has been observed in nature in only a few species of echinoderms. Some studies indicate that spawning is mostly a nocturnal event, wherein the animals assume characteristic postures with their bodies elevated off the substratum. Gametogenesis in at least some asteroids and echinoids is regulated by photoperiod (Pearse et al. 1986), which in turn ensures more or less synchronous spawning among members of the same population. In some species of free-spawning asteroids the females release pheromones that attract the sperm from nearby conspecific males (Miller 1989).

Brooding is especially common among boreal and polar species in all groups of echinoderms and in certain deep-sea asteroids, whose environments are unfavorable for larval life. As expected, brooding species produce fewer but larger and yolkier eggs than do their free-spawning counterparts.

Brooding methods vary. Among the crinoids, *Antedon* and a few others cement their eggs to the epidermis of the pinnules from which they emerge (Figure 22.16A,B). Once the eggs are fertilized by free sperm, the embryos are held by the parent until hatching. Most brooding asteroids hold their embryos on the body surface. One species (*Asterina gibbosa*) cements its eggs to the substratum, and another (*Leptasterias tenera*) broods its early embryos in the pyloric stomach before moving them to the outer body surface (Hendler and Franz 1982). Sea daisies brood within the gonads and apparently release juveniles that may drift for some time before settling. Brooding is common among ophiuroids. Sperm enter the bursae and fertilize the eggs, and the embryos are held within these sacs during development. Some echinoids brood their embryos among clusters of spines on the body or, in the case of sand dollars, on the petaloids. Brooding holothurians usually carry their embryos externally (Figure 22.16C), but some species of *Thyone* and *Leptosynapta* brood inside the coelom.

Development. The tremendous numbers of eggs produced by many echinoderms and the ease with which they can be reared in the laboratory have made these animals favorite objects of study by embryologists. Much of our information about the biology of animal fertilization and early development comes from over a century of work focusing particularly on urchins and sea stars. In addition, the early ontogeny of some echinoderms has served as a model of deuterostome development against which many other developmental patterns are measured. Except in brooding species, in which development is modified by large amounts of yolk, the sequence of ontogenetic events is remarkably similar throughout the phylum. We cannot cover the vast amount of information on this subject, and present only a brief overview of indirect development, emphasizing urchins and asteroids and including some comparative comments on other taxa. (For a review and detailed bibliography, see Wray 1997.)

The ova of free-spawning echinoderms are usually isolecithal with relatively small amounts of yolk. Cleavage is radial, holoblastic, and initially equal or subequal and leads to a spacious coeloblastula. In some groups, such as urchins, cleavage preceding the blastula becomes unequal, resulting in blastomere tiers of vegetal **mesomeres** underlain by slightly larger **macromeres** and a cluster of **micromeres** at the animal pole. (These terms refer here only to the relative sizes of the cells, and are not to be confused with the same terms as they are used in describing spiral cleavage.) The coeloblastula usually becomes ciliated and breaks free of the fertilization membrane as a swimming embryo.

Figure 22.16   Brooding in a crinoid and a holothurian. (A) Portion of an arm of *Antedon*. The ova are housed within a pinnule (lower portion) and released to exterior (upper portion). (B) Part of a pinnule of the crinoid *Phixometra* with developing young. (C) *Cucumaria crocea*, brooding its young.

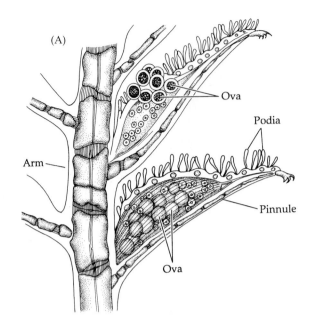

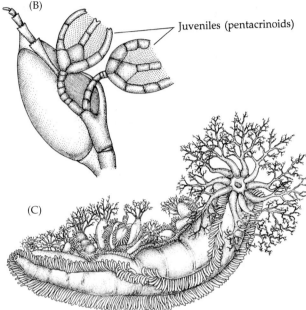

The blastula flattens slightly at the animal pole, forming the **gastral plate**, from which some cells proliferate into the blastocoel as **primary** or **larval mesenchyme**. In most echinoids these cells are the micromeres. The surrounding macromeres are the presumptive entoderm and adult mesoderm, and the vegetal mesomeres are the presumptive ectoderm. A coelogastrula is produced by invagination of the animal pole cells. The blastopore typically forms the anus; the archenteron grows to connect with a stomodeal inpocketing that forms the mouth. Before the gut is complete, however, the inner end of the archenteron proliferates **secondary mesenchyme**, as well as one or two evaginations of mesoderm, into the blastocoel. Thus, coelom formation (enterocoely) is by archenteric pouching.

During the later stages of gastrulation and coelom development, the embryo assumes bilateral symmetry and eventually becomes a swimming larva (Figure 22.17). Planktotrophic echinoderm larvae use bands of cilia to swim and to create feeding currents. Lecithotrophism, however, is common and has apparently evolved several times within some echinoderm classes.

The primary mesenchyme contributes to the formation of larval muscles and in some cases calcareous spicules or ossicles (Figure 22.17G). The adaptive significance of these larval skeletons has been explored by several authors. Suggestions about the functions of the larval skeleton include defense, physical support, and sites of muscle attachment. These ideas are summarized

by Pennington and Strathmann (1990), who also provide evidence that the skeletal elements enhance passive orientation of the larva in the water. Most feeding echinoderm larvae are oriented with the anterior end directed upward. The bulk of the skeleton lies in the posterior part of the larva, creating a higher density at the rear end. The larvae of some echinoderms contain distinctive reddish pigment spots that may be involved in photochemical energy-producing reactions.

In order to understand the development of echinoderm larvae and their eventual, remarkable metamorphosis to radially symmetrical adults, it is necessary to examine carefully the embryogeny and fates of the coelomic spaces. Although there are some differences in the details of these events among groups, they are similar enough to generalize for our purposes. The initial archenteric pouching typically occurs from the blind end of the developing gut, either as a pair of coeloms or as one cavity that divides into two. These coeloms pinch off another pair of cavities posteriorly, and then a third pair between the anterior and posterior ones (Figure 22.18A). From front to back, these pairs of coelomic spaces are called right and left **axocoels**, **hydrocoels**, and **somatocoels**. These spaces correspond to the protocoels, mesocoels, and metacoels of other trimeric deuterostomes. As illustrated, the left axocoel and left hydrocoel do not fully separate, but remain connected to one another by the stone canal. From the left axocoel-hydrocoel complex arises a **hydrotube**, which grows to the dorsal surface of the larva and opens to the outside via a **hydropore**. In most cases the right axocoel disappears and the right hydrocoel becomes associated with the hydrotube as the dorsal sac.

As we explain metamorphosis, it will help you to note the fates of these various coelomic derivatives, which are outlined in Table 22.1. As the time for meta-

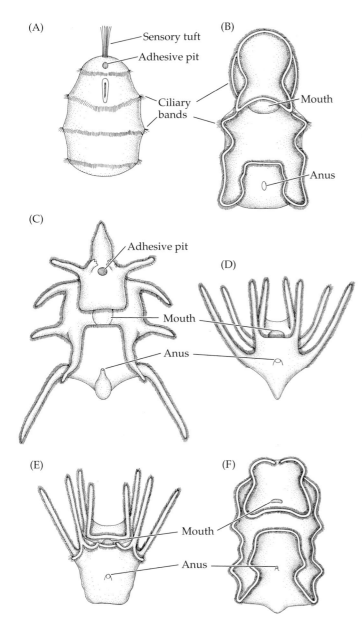

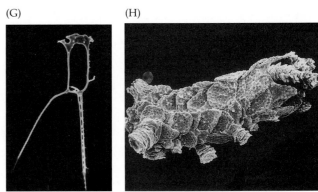

Figure 22.17 Echinoderm larval types. (A) Vitellaria larva of a crinoid. (B,C) Bipinnaria and later brachiolaria larvae of a sea star. (D) Ophiopluteus larva of a brittle star. (E) Echinopluteus larva of a sea urchin. (F) Auricularia larva of a sea cucumber. (G) Isolated larval spicule from the sand dollar *Dendraster*. (H) Late pentacula stage (postlarva) of a sea cucumber. Skeletal ossicles and juvenile podia are present.

morphosis approaches, the larva swims to the bottom and selects and attaches to an appropriate substratum. In general, the larval left and right sides become the adult oral and aboral surfaces, respectively, although this pattern often varies from a precise 90° reorientation. The remarkable change from bilateral to radial symmetry involves shifts in the positions of the mouth and anus. In many cases the embryonic openings disappear, and the stumps of the foregut and hindgut migrate beneath the body surface to their adult positions. The foregut swings from its larval anteroventral location to the left side, and the hindgut moves anteriorly and to the right (Figure 22.18B,C). As the foregut migrates, it presses into the wall of the left hydrocoel, which encircles the foregut as the precursor of the ring canal. Once

TABLE 22.1 *Adult fates of the major coelomic derivatives in generalized enchinoderm development*

Embryonic coelomic structure	Adult fate
Right somatocoel	Aboral perivisceral coelom
Left somatocoel	Oral perivisceral coelom; genital sinuses; most of the hyponeural sinus
Right axocoel	Largely lost
Hydropore	Incorporated into madreporite
Hydrotubule and dorsal sac	Parts of madreporic vesicle and ampulla
Stone canal	Stone canal
Left hydrocoel	Ring canal; radial canals; lining of tube feet lumina, plus other components of the water vascular system, including Tiedemann's bodies and polian vesicles

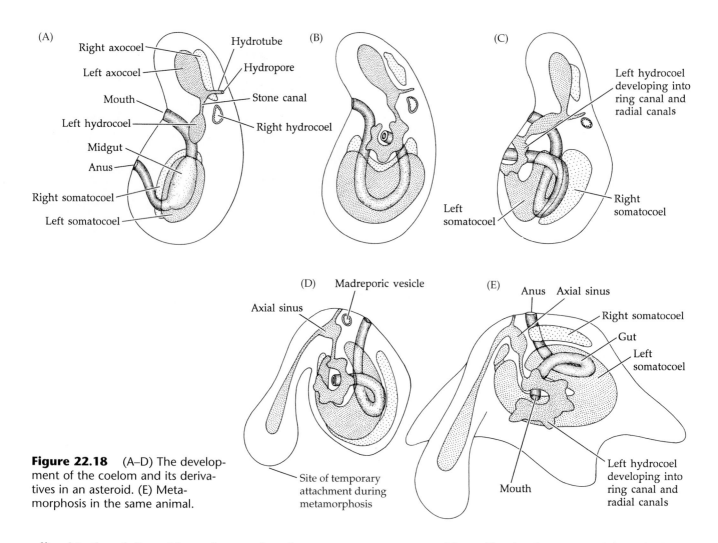

Figure 22.18 (A–D) The development of the coelom and its derivatives in an asteroid. (E) Metamorphosis in the same animal.

affixed in the adult positions, the mouth and anus re-open. The ring canal grows radial extensions (Figure 22.18D,E) destined to become the radial canals and podial linings, and outgrowths of the left somatocoel produce the hyponeural sinus. Aborally, the madreporite complex arises from various parts of the left axocoel and its derivatives plus the dorsal sac, and marks the position of the CD interambulacrum. The axial sinus arises from an outpocketing of the left axocoel.

As these transformations take place, most of the larval structures are lost and the juvenile assumes benthic life. Many echinoderm larvae appear to settle preferentially near conspecific adults. In at least some species (e.g., *Dendraster excentricus*) successful metamorphosis is triggered by pheromones that are released by adults and act on the larval nervous system.

Echinoderm Phylogeny

In spite of a rich fossil record and many decades of work, the origin and subsequent evolution of echinoderms remain highly controversial issues. There have been a number of popular and competing ideas on these matters, as evidenced by the chronic instability of echinoderm classification. We focus here on phylogeny at the class level, and assume that each class is in fact a monophyletic group. Our treatment relies largely on adult morphology. Larval types have been used by some workers, but details of larval form do not always correlate well with adult traits (see Wray and Bely 1994 for a discussion of the evolutionary forces influencing echinoderm larvae).

The echinoderm lineage probably originated with the Precambrian invasion of epibenthic habitats by an ancestral burrowing deuterostome. The line diversified rapidly, and most of the fundamental body plans within the phylum were probably established in the early Cambrian. Echinoderm diversity reached its zenith during the early to mid-Paleozoic, but by the beginning of the Mesozoic it had declined greatly in terms of higher taxa, leaving only five major groups persisting to recent times. Some fossil forms are shown in Figure 22.19.

The origin of echinoderms probably involved the evolution of the endoskeleton composed of plates with the unique stereom structure. Evidence suggests that the skeleton originated prior to the adoption of radial symmetry, the latter marking the appearance of the first

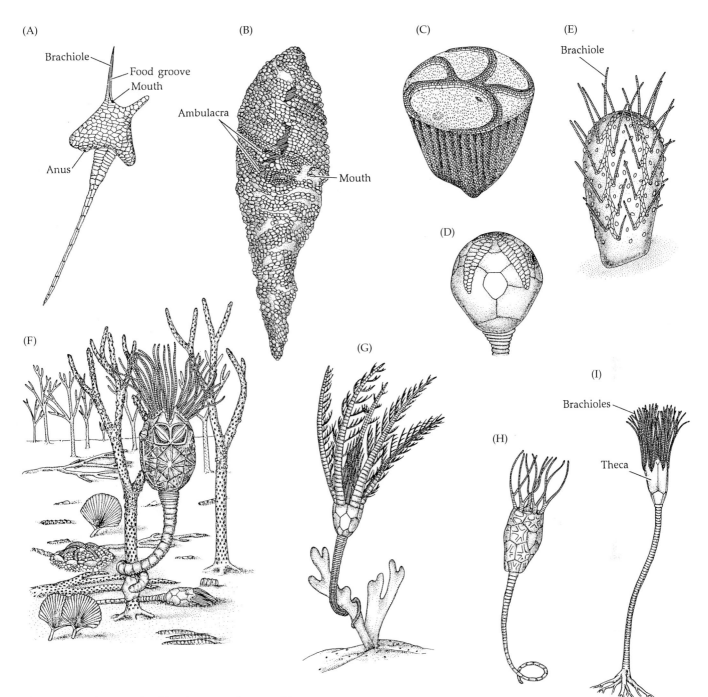

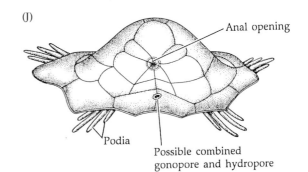

Figure 22.19 Fossil echinoderms and near-echinoderms. (A) *Dendrocystites*, a carpoid. Carpoids were early Cambrian animals that probably shared a common ancestor with the true echinoderms. Note the absence of radial symmetry. (B) The helicoplacoid *Helicoplacus*, a triradiate echinoderm from the Lower Cambrian. This creature, with spiral ambulacra, may represent the ancestral echinoderm bauplan. (C) *Camptostroma roddyi*, an early Cambrian edrioasteroid with five ambulacra arranged in a manner suggesting derivation from a triradiate form. (D) *Steganoblastus*, a stalked edrioasteroid showing clear pentamery. (E) A generalized cystoid. (F) *Lepadocystis*, a stalked Ordovician cystoid. Note also the attached edrioasteroid. (G) *Eifelocrinus*, an extinct crinoid. (H) The eocrinoid *Macrocystella*. (I) A generalized blastoid from the Carboniferous. (J) *Volchovia*, a strange extinct echinoid.

Figure 22.20 (A) A cladogram depicting one hypothesis about the origins of some important synapomorphies among the major groups of echinoderms. (B) A competing hypothesis (extant taxa only) illustrating a slightly different view on the placement of the ophiuroids. The origin of the echinoderm lineage involved an escape from infaunal life with the evolution of a supportive system of endoskeletal plates with a stereom structure (1) and the use of external ciliary grooves for suspension feeding (2). This proto-echinoderm condition is represented in the fossil record by the carpoids. The first true echinoderms may have been the helicoplacoids, whose appearance was marked by the origin of triradial symmetry with three spirally arranged, open ambulacral grooves (3) and a water vascular system (4), probably with the madreporite opening near the mouth. The immediate common ancestor of the modern lineages (*a*) may have been similar to the extinct *Camptostroma*, with pentaradial symmetry (5) evidenced by five open ambulacral grooves, mouth and anus on the oral surface (6), and attachment to the substratum by the aboral surface (7). From this ancestral form, the Crinoidea and Cystoidea diverged with the evolution of arms or brachioles bearing open ciliated grooves used for suspension feeding (8) and the loss of the external madreporite (9). The origin of the sister clade (Asteroidea, Ophiuroidea, Echinoidea, and Holothuroidea) involved the movement of the anus to the aboral surface (10), and a change associated with the orientation of the body with the oral surface against the substratum (11). In this "new" position, these echinoderms adopted alternative feeding modes and a somewhat errant lifestyle; the podia became suckered (12) and used for locomotion rather than feeding. The madreporite migrated along the CD interambulacrum to the aboral surface (13). The asteroids arose with the evolution of five arms broadly connected to a central disc (14). The remaining three groups have closed ambulacral grooves (15) in common. The ophiuroids invaded soft substrata and lost the podial suckers (16). In addition, they evolved five highly articulated arms, with internal vertebral plates in each arm "segment" (17), and secondarily lost the anus (18). The madreporite migrated back to the oral surface along the CD interambulacrum (19). The echinoid–holothurian clade arose with the extension of the ambulacral grooves along the sides of the body from the oral to the aboral pole (20), thereby reducing the aboral surface to a small region around the anus (21). The echinoids evolved with the fusion of the skeletal plates, which formed a rigid globular or discoidal test (22). The origin of the holothurians involved a reduction of the skeletal plates to isolated ossicles (23), movement of the madreporite internally (24), and elongation of the fleshy body on the oral–aboral axis (25). Cladogram B unites the asteroids and ophiuroids into a single clade on the basis of the five-rayed (26) body

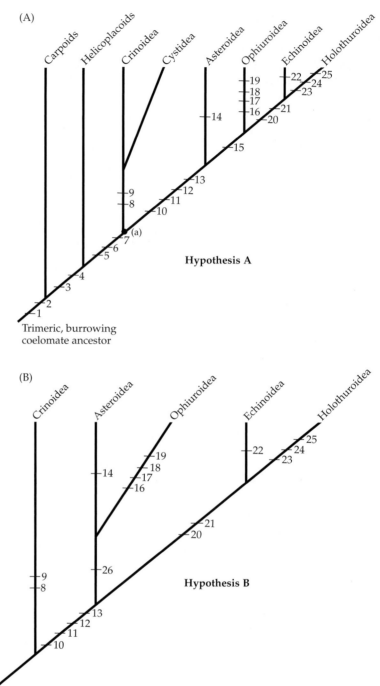

plan. Both cladograms include some unresolved problems. Both treat the evolution of ambulacral rays (arms) as convergent, arising once in the crinoids and again in the asteroids and ophiuroids; cladogram A indicates that this condition is also convergent in the asteroids and ophiuroids. Cladogram A treats the oral madreporite position in ophiuroids as secondary, and thus convergent with the same condition in groups that diverged earlier (e.g., crinoids). In addition, cladogram B accepts as convergent the closed ambulacral grooves of ophiuroids and those of echinoids and holothurians.

"true" echinoderms as shown in the cladograms in Figure 22.20. The first dichotomy separates the main echinoderm clade from a now extinct group called the carpoids (Figure 22.19A). Many authors view the carpoids as echinoderms and place them in a separate subphylum, the Homalozoa. Although they possessed stereom ossicles, they were not radially symmetrical, and the nature of their water vascular system (if they had one) is uncertain. Because they lacked some of the fundamental defining characteristics of the echinoderms, they are best considered as an early pre-echinoderm group. These early epibenthic creatures were probably suspension feeders, and they bore a grooved arm or **brachiole** that apparently led to the mouth.

The first true echinoderms may have been the helicoplacoids (Figure 22.19B). These odd creatures appeared in the early Cambrian and died out soon thereafter. They were spindle-shaped, with spirally arranged skeletal plates and three ambulacra. The mouth was located on one side of the body rather than apically, so these animals were not constructed on an obvious oral–aboral axis as are modern echinoderms. It appears that the skeletal plates articulated somewhat, and the helicoplacoids may have been either surface plowers or attached with the oral side against the substratum. The ambulacra probably conveyed detrital or suspended food material to the mouth. Some authors speculate that the "lateral" placement of the mouth in helicoplacoids is indicative of the origin of the metamorphic events during the conversion from bilateral to radial symmetry seen in extant echinoderms. This conversion must have involved some major changes in genes that influence fundamental body architecture. Since echinoderm larvae retain their ancestral bilateral condition, the expression of genes that dictate radiality must occur relatively late in development. According to Lowe and Wray (1997), this change involved major modifications of the roles of Hox genes during the early evolution of the echinoderm bauplan.

The next major dichotomy separates the crinoids and extinct cystoids (Figure 22.19E) from other echinoderms (Figures 22.20, 22.21). Paul and Smith (1988) suggest that the extinct genus *Camptostroma* (Figure 22.19C) is similar to what may have been the common ancestor of these two monophyletic sister clades. It is the earliest known pentaradial echinoderm, with the five ambulacra developed in a 2-1-2 pattern, perhaps derived from the triradial pattern of the helicoplacoids. *Camptostroma* is usually

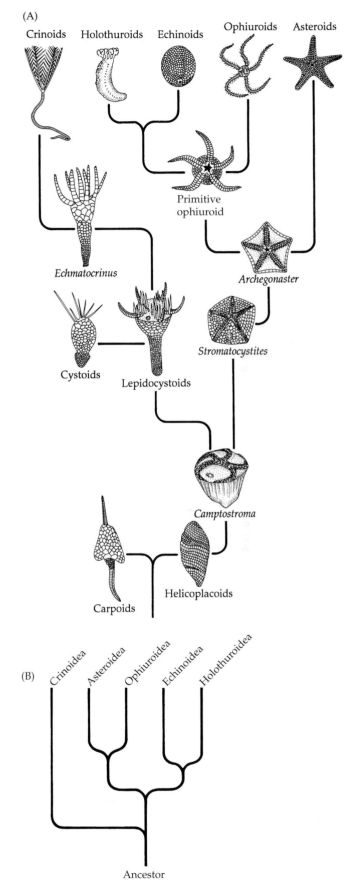

Figure 22.21 Two orthodox evolutionary trees of the echinoderms. (A) A tree compatible with cladogram A in Figure 22.20. (B) A "traditional" tree of the extant classes, compatible with cladogram B in Figure 22.20.

assigned to a wholly extinct group called the Edrio-asteroidea (Figure 22.19D). The lineage that persists today as the class Crinoidea includes extinct forms that bore attachment stalks arising from the aboral surface (e.g., extinct crinoids and cystoids, Figure 22.19E–H). These animals suspension fed by orienting with the oral side upward and using the open ambulacral grooves in the arms or brachioles for transporting food to the mouth.

The origin of the lineage that includes the asteroids, ophiuroids, echinoids, and holothurians involved the adoption of other feeding modes and used the water vascular system largely for locomotion. These animals became more or less errant and, with the exception of the holothurians, oriented themselves with the oral side against the substratum. The phylogeny of these classes is controversial and as yet unsettled. On the basis of its temporal appearance in the fossil record, Paul and Smith (1988) place the extinct genus *Stromatocystites* (Figure 22.21A) at the base of this line and indicate that the familiar benthic-feeding asteroids, ophiuroids, and echinoids did not appear until later (Ordovician), and that the holothurians appeared even more recently (early Mississippian). However, the emphasis on fossil chronology must be viewed with some caution, even in a group as well represented as the echinoderms. The evidence preserved in rocks is still fragmentary and may easily be misleading. The holothurians, for example, leave only isolated ossicles from which to draw inferences. Deep-sea ophiuroids and asteroids probably did not fossilize well, and the skeletons of others are known to disarticulate soon after death. Although we cannot ignore the fossils, we cannot base an entire evolutionary scenario on what they seem to tell us. Based on fossil evidence, sea urchins (Euchinoidea) are more diverse today than anytime in the past.

The close relationship between the echinoids and holothurians is accepted by most specialists, but there is much debate about whether the ophiuroids are closer to the asteroids or to the echinoid–holothurian clade. Two alternative cladograms are presented and discussed in Figure 22.20.

In summary, the evolution of a mesodermal skeleton of stereom ossicles was followed by the appearance of a water vascular system and pentaradial symmetry. These features allowed escape from infaunal life to epibenthic surface dwelling. The water vascular system probably originally served for suspension and perhaps detritus feeding, facilitated first by simple ciliary tracts, as in the carpoid brachioles, and later by the development of ambulacral grooves in the helicoplacoids, which persisted in the crinoids. Radial symmetry became the popular architecture among the echinoderms and enhanced their new lifestyle. The use of suckered podia for locomotion was a secondary event that provided a means of exploiting new habitats and food resources. Later, in ophiuroids, echinoids, and holothurians, the ambulacral grooves closed, with a concomitant loss of feeding functions, and feeding became the responsibility of other structures (e.g., podia, buccal tentacles, teeth).

There is no doubting the success of the basic echinoderm bauplan. The combined qualities of the supportive endoskeleton, coelomic water vascular system, and pentaradial symmetry are unique to these animals and have provided the basis for their diversification along several distinct lineages.

Selected References

General and Fossil Forms

Bather, F. A. 1900. The Echinodermata. *In* E. R. Lankester (ed.), *A Treatise on Zoology*, Pt. 3. A. and C. Black, London, pp. 1–216.

Binyon, J. 1966. Salinity tolerance and ionic regulation. *In* R. A. Boolootian (ed.), *Physiology of Echinodermata*. Wiley-Interscience, New York, pp. 359–378.

Binyon, J. 1972. *Physiology of Echinoderms*. Pergamon Press, Oxford.

Boolootian, R. A. (ed.). 1966. *Physiology of Echinodermata*. Wiley-Interscience, New York.

Campbell, A. C. 1983. Form and function of pedicellariae. *In* M. Jangoux and J. M. Lawrence (eds.), *Echinoderm Studies*, Vol. 1. A. A. Balkema, Rotterdam, pp. 139–168.

Chia, F. S. and A. H. Whitely (eds.). 1975. Developmental biology of the echinoderms. Am. Zool. 15(3): 483–775. [A collection of papers on various aspects of echinoderm reproduction and embryology from the 1973 ASZ meetings.]

Cuénot, L. 1948. Anatomie, éthologie, et systématique des échinodermes. *In* P. Grassé (ed.), *Traité de Zoologie*, Vol. 11. Masson et Cie, Paris.

Dawydoff, C. 1948. Embryologie des Échinodermes. *In* P. Grassé (ed.), *Traité de Zoologie*, Vol. 11. Masson et Cie, Paris.

Drestler, K. L. 1981. Morphological diversity of early Cambrian echinoderms. *In* M. E. Taylor (ed.), *Short Papers for the Second International Symposium on the Cambrian System*. United States Open File Report No. 81-743: 71–75.

Durham, J. W. 1967. Notes on the Helicoplacoidea and early echinoderms. J. Paleontology 41: 97–102.

Durham, J. W. and K. E. Caster. 1963. Helicoplacoidea: A new class of echinoderms. Science 140: 820–822.

Emlet, R. B. 1982. Echinoderm calcite: A mechanical analysis from larval spicules. Biol. Bull. 163: 264–275.

Emlet, R. B. 1983. Locomotion, drag, and the rigid skeleton of larval echinoderms. Biol. Bull. 164: 433–445.

Emlet, R. B. 1985. Crystal axes in recent and fossil adult echinoids indicate trophic mode in larval development. Science 230: 937–940.

Fell, H. B. 1982. Echinodermata. *In* S. P. Parker (ed.), *Synopsis and Classification of Living Organisms*. McGraw-Hill, New York, pp. 785–813.

Giese, A. C., J. S. Pearse and V. B. Pearse. 1991. *Reproduction of Marine Invertebrates. VI. Echinoderms and Lophophorates*. Boxwood Press, Pacific Grove, CA.

Hammond, L. S. 1982. Patterns of feeding and activity in deposit-feeding holothurians and echinoids (Echinodermata) from a shallow back-reef lagoon, Discovery Bay, Jamaica. Bull. Mar. Sci. 32(2): 549–571.

Hammond, L. S. 1983. Nutrition of deposit-feeding holothuroids and echinoids (Echinodermata) from a shallow reef lagoon, Discovery Bay, Jamaica. Mar. Ecol. Prog. Ser. 10: 297–305.

Hart, M. W. 1996. Variation in suspension feeding rates among larvae of some temperate, eastern Pacific echinoderms. Invert. Biol. 115 (1): 30–45.

Haugh, B. N. and B. M. Bell. 1980. Fossilized viscera in primitive echinoderms. Science 209: 653–657.

Heddle, D. 1967. *Echinoderm Biology*. Academic Press, New York.

Hyman, L. H. 1955. *The Invertebrates*. Vol. 4, Echinodermata. The Coelomate Bilateria. McGraw-Hill, New York.

Jangoux, M. (ed.). 1980. *Echinoderms: Present and Past*. A. A. Balkema, Rotterdam.

Jangoux, M. and J. M. Lawrence (eds.). 1982. *Echinoderm Nutrition*. A. A. Balkema, Rotterdam.

Jangoux, M. and J. M. Lawrence (eds.). 1983, 1987. *Echinoderm Studies*. Vols. 1 and 2. A. A. Balkema, Rotterdam.

Jefferies, R. P. S. 1967. Some fossil chordates with echinoderm affinities. Symp. Zool. Soc. Lond. 20: 163–208.

Jefferies, R. P. S. 1980. The phylogenetic connection between Echinoderms and Chordates. *In* M. Jangoux (ed.), *Echinoderms: Present and Past*. A. A. Balkema, Rotterdam, pp. 29–30.

Jefferies, R. P. S. 1981. Fossil evidence for the origin of the chordates and echinoderms. Atti di Convegni Lincei 49: 487–561.

Jefferies, R. P. S. 1986. *The Ancestry of the Vertebrates*. British Museum of Natural History, London.

Jefferies, R. P. S. 1990. The solute *Dendrocystoides scoticus* from the upper Ordovician of Scotland and the ancestry of the chordates and echinoderms. Paleontology 33 (3): 631–679.

Lawrence, J. M. (ed.). 1982. *Echinoderms: Proceedings of the International Conference, Tampa Bay, 14–17 September 1981*. A. A. Balkema, Rotterdam. [Contains 119 papers and abstracts.]

Lawrence, J. M. 1987. *A Functional Biology of Echinoderms*. Johns Hopkins University Press, Baltimore.

Lowe, C. J. and G. A. Wray. 1997. Radical alterations in the roles of homeobox genes during echinoderm evolution. Nature 389: 718–721. [See also the correction in Nature 392: 105.]

Millott, N. (ed.). 1967. *Echinoderm Biology*. Symp. Zool. Soc. Lond. 20. Academic Press, London.

Moore, R. C. (ed.). 1966–1978. *Treatise on Invertebrate Paleontology*. Parts S–U, Echinodermata. Geological Society of America and University of Kansas Press.

Motokawa, T. 1984. Connective tissue catch in echinoderms. Biol. Rev. 59: 255–270.

Nichols, D. 1962. *Echinoderms*. Hutchinson University Library, London.

Nichols, D. 1972. The water-vascular system in living and fossil echinoderms. Paleontology 15: 519–538.

Paul, C. R. C. and A. B. Smith (eds.). 1988. *Echinoderm Ontogeny and Evolutionary Biology*. Clarendon Press, Oxford.

Pennington, J. T. and R. R. Strathmann. 1990. Consequences of the calcite skeletons of planktonic echinoderm larvae for orientation, swimming, and shape. Biol. Bull. 179: 121–133.

Pequignat, E. 1970. On the biology of *Echinocardium cordatum* (Pennant) of the Seine Estuary: New researches on skin-digestion and epidermal absorption in Echinoidea and Asteroidea. Forma et Functio 2: 121–168.

Philip, G. M. 1979. Carpoids: Echinoderms or chordates? Biol. Rev. 54: 439–471.

Shick, J. M. 1983. Respiratory gas exchange in echinoderms. *In* M. Jangoux and J. M. Lawrence (eds.), *Echinoderm Studies*. Vol. 1. A. A. Balkema, Rotterdam, pp. 67–110.

Sprinkle, J. 1973. Morphology and evolution of blastozoan echinoderms. Special Publication. Museum of Comparative Zoology, Harvard University, Cambridge, MA.

Smith, A. B. 1984. *Echinoderm Paleobiology*. Allen and Unwin, London.

Stephens, G. 1968. Dissolved organic matter as a potential source of nutrition for marine organisms. Am. Zool. 8: 95–106.

Stephenson, D. G. 1974. Pentamerism and the ancestral echinoderm. Nature 250: 82–83.

Strathmann, R. R. 1989. Existence and function of a gel filled primary body cavity in development of echinoderms and hemichordates. Biol. Bull. 176: 25–31.

Wray, G. A. 1994. The evolution of cell lineages in echinoderms. Am. Zool. 34: 353–363.

Wray, G. A. 1997. Echinoderms. *In* S. F. Gilbert and A. M. Raunio (eds.), *Embryology: Constructing the Organism*. Sinauer Associates, Sunderland, MA, pp. 309–329.

Wray, G. A. and A. E. Bely. 1994. The evolution of echinoderm development is driven by several distinct factors. Development (Suppl). pp. 97–106.

Crinoidea

Bather, F. A. 1899. A phylogenetic classification of the Pelmatozoa. Br. Assoc. Adv. Sci., Rpt. D (1898): 916–923.

Grimmer, J. C., N. D. Holland and C. G. Messing. 1984. Fine structure of the stalk of the bourgueticrinid sea lily *Democrinus conifer* (Echinodermata: Crinoidea). Mar. Biol. 81: 163–176.

LaTouche, R. W. 1978. The feeding behavior of the feather star *Antedon bifida*. J. Mar. Biol. Assoc. U.K. 58: 877–890.

Sprinkle, J. 1976. Classification and phylogeny of "pelmatozoan" echinoderms. Syst. Zool. 25: 83–91.

Stephenson, D. G. 1980. Symmetry and suspension-feeding in Pelmatozoan echinoderms. *In* M. Jangoux (ed.), *Echinoderms: Present and Past*. A. A. Balkema, Rotterdam, pp. 53–58.

Vail, L. 1989. Arm growth and regeneration in *Oligometra serripinna* (Carpenter) (Echinodermata: Crinoidea) at Lizard Island, Great Barrier Reef. J. Exp. Mar. Biol. Ecol. 130: 189–204.

Asteroidea

Baker, A. N., F. W. E. Rowe and H. E. S. Clark. 1986. A new class of Echinodermata from New Zealand. Nature 321: 862–864. [The discovery of the sea daisies, and erection of the class Concentricycloidea.]

Blake, D. B. 1987. A classification and phylogeny of post-Paleozoic sea stars (Asteroidea: Echinodermata). J. Nat. Hist. 21: 481–528.

Chia, F. S. and H. Amerongen. 1975. On the prey-catching pedicellariae of a starfish, *Stylasterias forreri*. Can. J. Zool. 53: 748–755.

Clark, A. M. and M. E. Downey. 1992. *Starfishes of the Atlantic*. Chapman and Hall, London.

Dearborn, J. H., K. C. Edwards and D. B. Fratt. 1991. Diet, feeding behavior, and surface morphology of the multi-armed Antarctic sea star *Labidiaster annulatus* (Echinodermata: Asteroidea). Mar. Ecol. Prog. Ser. 77: 65–84.

Domanski, P. A. 1984. Giant larvae: Prolonged planktonic larval phase in the asteroid *Luidia sarsi*. Mar. Biol. 80: 189–195.

Donovan, S. K. and A. S. Gale. 1990. Predatory asteroids and the decline of the articulate brachiopods. Lethaia 23: 77–86.

Emson, R. H. and C. M. Young. 1994. Feeding mechanisms of the brisingid starfish *Novodinia antillensis*. Mar. Biol. 118: 433–442.

Fell, H. B. 1963. The phylogeny of sea-stars. Phil. Trans. R. Soc. Lond. Ser. B 246: 381–435.

Ferguson, J. C. 1970. An autoradiographic study of the translocation and utilization of amino acids by starfish. Biol. Bull. 138: 14–25.

Ferguson, J. C. 1984. Translocative functions of the enigmatic organs of starfish—the axial organ, hemal vessels, Tiedemann's bodies, and rectal caeca—an autoradiographic study. Biol. Bull. 166: 140–155.

Ferguson, J. C. and C. W. Walker. 1991. Cytology and function of the madreporite systems of the starfish *Henricia sanguinolenta* and *Asterias vulgaris*. J. Morphol. 210: 1–11.

Gale, A. S. 1987. Phylogeny and classification of the Asteroidea (Echinodermata). Zool. J. Linn. Soc. 89: 107–132.

Glynn, P. W. 1974. The impact of *Acanthaster* on corals and coral reefs in the Eastern Pacific. Environ. Conserv. 1(4): 295–304.

Hendler, G. and D. Franz. 1982. The biology of a brooding seastar, *Leptasterias tenera*, in Block Island Sound. Biol. Bull. 162: 273–289.

Jaeckle, W. B. 1994. Multiple modes of asexual reproduction by tropical and subtropical sea star larvae: An unusual adaptation for gamete dispersal and survival. Biol. Bull. 186: 62–71.

Menge, B. 1975. Brood or broadcast? The adaptive significance of different reproductive strategies in the two intertidal sea stars *Leptasterias hexactis* and *Pisaster ochraceus*. Mar. Biol. 31(1): 87–100.

Miller, R. L. 1989. Evidence for the presence of sexual pheromones in free-spawning starfish. J. Exp. Mar. Biol. Ecol. 130: 205–221.

Nance, J. M. and L. F. Braithwaite. 1979. The function of mucous secretions in the cushion star *Pteraster tesselatus* Ives. J. Exp. Mar. Biol. Ecol. 40: 259–266.

Nichols, D. 1986. A new class of echinoderms. Nature 321: 808.

Pearse, J. S., D. J. Eernisse, V. B. Pearse and K. A. Beauchamp. 1986. Photoperiodic regulation of gametogenesis in sea stars, with evidence for an annual calendar independent of fixed daylength. Am. Zool. 26: 417–431.

Rivkin, R. B., I. Bosch, J. S. Pearse and E. J. Lessard. 1986. Bacterivory: A novel feeding mode for asteroid larvae. Science 233: 1311–1314.

Rowe, F. W., A. N. Baker and H. E. S. Clark. 1988. The morphology, development, and taxonomic status of Xyloplax Baker, Rowe and Clark (1986) (Echinodermata: Concentricycloidea), with description of a new species. Proc. R. Soc. Lond. Ser. B 233: 431–459.

Smith, A. B. 1988. To group or not to group: The taxonomic position of Xyloplax. In R. D. Burke et al. (eds.), Echinoderm Biology. A. A. Balkema, Rotterdam, pp. 17–23.

Symposium on the biology and ecology of the crown-of-thorns starfish, Acanthaster planci (L.). 1973. Micronesica 9(2). [Contributions of numerous authors from a symposium convened in response to the population explosion of A. planci on Indo-West Pacific coral reefs.]

Wagner, R. H., D. W. Phillips, J. D. Standing and C. Hand. 1979. Commensalism or mutualism: Attraction of a sea star towards its symbiotic polychaete. J. Exp. Mar. Biol. Ecol. 39: 205–210.

Wulff, J. L. 1995. Sponge-feeding by the Caribbean starfish Oreaster reticulatus. Mar. Biol. 123: 313–325.

Yamagata, A. 1982. Studies on reproduction in the hermaphroditic sea star, Asterina minor: The functional male gonads, "ovitestes." Biol. Bull. 162: 449–456.

Ophiuroidea

Ferguson, J. C. 1995. The structure and mode of function of the water vascular system of a brittle star, Ophioderma appressum. Biol. Bull. 188: 98–110.

Ferrari, F. D. and J. H. Dearborn. 1989. A second examination of predation on pelagic copepods by the brittle star Astrotoma agassizii. J. Plankton Res. 11 (6): 1315–1320.

Fratt, D. B. and J. H. Dearborn. 1984. Feeding biology of the Antarctic brittle star Ophionotus victoriae (Echinodermata: Ophiuroidea). Polar Biol. 3: 127–139.

Hendler, G. 1975. Adaptational significance of the patterns of ophiuroid development. Am. Zool. 15: 692–715.

Hendler, G. 1978. Development of Amphioplus abditus (Verrill) (Echinodermata: Ophiuroidea). II. Description and discussion of ophiuroid skeletal ontogeny and homologies. Biol. Bull. 154: 79–95.

Hendler, G. 1979. Sex-reversal and viviparity in Ophiolepis kieri, n. sp., with notes on viviparous brittlestars from the Caribbean (Echinodermata: Ophiuroidea). Proc. Biol. Soc. Wash. 92: 783–795.

Hendler, G. 1982a. An echinoderm vitellaria with a bilateral larval skeleton: Evidence for the evolution of ophiuroid vitellariae from ophiuoplutei. Biol. Bull. 163: 431–437.

Hendler, G. 1982b. Slow flicks show star tricks: Elapsed-time analysis of basketstar (Astrophyton muricatum) feeding behavior. Bull. Mar. Sci. 32: 909–918.

Hendler, G. 1983. The association of Ophiothrix lineata and Callyspongia vaginalis: A brittlestar-sponge cleaning symbiosis? Mar. Ecol. 5(1): 9–27.

Hendler, G. 1984. Brittlestar color-change and phototaxis (Echinodermata: Ophiuroidea: Ophiocomidae). Mar. Ecol. 5(4): 379–401.

Hendler, G., M. J. Grygier, E. Maldonado and J. Denton. 1999. Babysitting brittle stars: heterospecific symbiosis between ophiuroids. Invert. Biol 118: 190–201.

Mladenov, P. V., R. H. Emerson, L. V. Colpit and I. C. Wilkie. 1983. Asexual reproduction in the West Indian brittle star Ophiocomella ophiactoides (H. L. Clark) (Echinodermata: Ophiuroidea). J. Exp. Mar. Biol. Ecol. 72: 1–23.

Pentreath, R. J. 1970. Feeding mechanisms and the functional morphology of podia and spines in some New Zealand ophiuroids. J. Zool. 161: 395–429.

Warner, G. F. and J. D. Woodley. 1975. Suspension feeding in the brittle star Ophiothrix fragilis. J. Mar. Biol. Assoc. U.K. 55: 199–210.

Woodley, J. D. 1975. The behavior of some amphiurid brittle-stars. J. Exp. Mar. Biol. Ecol. 18: 29–46.

Woodley, J. D. 1980. The biomechanics of Ophiuroid tube-feet. In M. Jangoux (ed.), Echinoderms: Present and Past. A. A. Balkema, Rotterdam, pp. 293–299.

Echinoidea

Alexander, D. E. and J. Ghiold. 1980. The functional significance of the lunules in the sand dollar, Mellita quinquiesperforata. Biol. Bull. 159: 561–570.

Angerer, R. C. and E. H. Davidson. 1984. Molecular indices of cell lineages specification in sea urchin embryos. Science 226: 1153–1160.

Armstrong, N. and D. R. McClay. 1994. Skeletal pattern is specified autonomously by the primary mesenchyme cells in sea urchin development. Dev. Biol. 162: 329–338.

Burke, R. D. 1980. Podial sensory receptors and the induction of metamorphosis in echinoids. J. Exp. Mar. Biol. Ecol. 47: 223–234.

Burke, R. D. 1983. Neural control of metamorphosis in Dendraster excentricus. Biol. Bull. 164: 176–188.

Burke, R. D. 1984. Pheromonal control of metamorphosis in the Pacific sand dollar, Dendraster excentricus. Science 225: 223–224.

Burke, R. D., R. L. Meyers, T. L. Sexton and C. Jackson. 1991. Cell movements during the initial phase of gastrulation in the sea urchin embryo. Dev. Biol. 146: 542–557.

Cameron, R. A. and E. H. Davidson. 1991. Cell type specification during sea urchin development. Trends Genet. 7: 212–218.

Carpenter, R. C. 1990a. Mass mortality of Diadema antillarum. I. Long-term effects on the sea urchin population-dynamics and coral reef algal communities. Mar. Biol. 104: 67–77.

Carpenter, R. C. 1990b. Mass mortality of Diadema antillarum. II. Effects on population densities and grazing intensity of parrot fishes and surgeon fishes. Mar. Biol. 104: 79–86.

Chia, F. S. 1969. Some observations on the locomotion and feeding of the sand dollar, Dendraster excentricus. J. Exp. Mar. Biol. Ecol. 3(2): 162–170.

Chia, F. S. 1973. Sand dollar: A weight belt for the juvenile. Science 181: 73–74.

Coffman, J. A. and E. H. Davidson. 1992. Expression of spatially regulated genes in the sea urchin embryo. Curr. Opin. Genet. Dev. 2: 260–268.

David, B. and R. Mooi. 1990. An echinoid that "gives birth": morphology and systematics of a new Antarctic species, Urechinus mortenseni (Echinodermata, Holasteroida). Zoomorphologie 110: 75–89.

Ellers, O. and M. Telford. 1984. Collection of food by oral surface podia in the sand dollar, Echinarachnius parma (Lamarck). Biol. Bull. 166: 574–582.

Fenner, D. H. 1973. The respiratory adaptations of the podia and ampullae of echinoids (Echinodermata). Biol. Bull. 145: 323–339.

Ghiold, J. 1979. Spine morphology and its significance in feeding and burrowing in the sand dollar, Mellita quinquiesperforata (Echinodermata: Echinoidea). Bull. Mar. Sci. 29: 481–490.

Hanson, J. L. and G. Gust. 1986. Circulation of perivisceral fluid in the sea urchin Lytechinus variegatus. Mar. Biol. 92: 125–134.

Henry, J. J., K. M. Klueg and R. A. Raff. 1992. Evolutionary dissociation between cleavage, cell lineage, and embryonic axes in sea urchin embryos. Development 114: 931–938.

Lewis, J. B. 1968. The function of sphaeridia of sea urchins. Can. J. Zool. 46: 1135–1138.

Mann, K. H. 1982. Kelp, sea urchins and predators: A review of strong interactions in rocky subtidal systems of eastern Canada, 1970–1980. Neth. J. Sea. Res. 16: 414–423.

Markel, K. 1980. The lantern of Aristotle. In M. Jangoux (ed.), Echinoderms: Present and Past. A. A. Balkema, Rotterdam, pp. 91–92.

Millott, N. 1975. The photosensitivity of echinoids. Adv. Mar. Biol. 13: 1–52.

Mooi, R. 1986a. Non-respiratory podia of clypeasteroids (Echinodermata, Echinoidea): I. Functional anatomy. Zoomorphologie 106: 21–30.

Mooi, R. 1986b. Non-respiratory podia of clypeasteroids (Echinodermata, Echinoidea). II. Diversity. Zoomorphologie 106: 75–90.

Mooi, R. 1986c. Structure and function of clypeasteroid miliary spines (Echinodermata, Echinoidea). Zoomorphologie 106: 212–223.

Mooi, R. 1989. Living and fossil genera of the Clypeasteroida (Echinoidea: Echinodermata): An illustrated key and annotated checklist. Smithson. Contrib. Zool. 488: 1–51.

Mooi, R. 1990. Paedomorphosis, Aristotle's lantern, and the origin of the sand dollars (Echinodermata: Clypeasteroida). Paleobiology 16 (1): 25–48.

Pennington, J. T. 1985. The ecology of fertilization of echinoid eggs: The consequences of sperm dilution, adult aggregation, and synchronous spawning. Biol. Bull. 169: 417–430.

Pequignat, E. 1966. Skin digestion and epidermal absorption in irregular and regular urchins. Nature 210: 396–399.

Raff, R. A. 1987. Constraint, flexibility, and phylogenetic history in the evolution of direct development in sea urchins. Dev. Biol. 119: 6–19.

Seilacher, A. 1979. Constructional morphology of sand dollars. Paleobiology 5: 191–221.

Smith, A. B. 1980. The structure, function, and evolution of tube feet and ambulacral pores in irregular echinoids. Paleontology 23: 39–83.

Smith, A. B. and C. H. Jeffers. 1988. Selectivity of extinction among sea urchins at the end of the Cretaceous period. Nature 392: 69–71.

Strathmann, R. R., L. Fenaux and M. F. Strathmann. 1992. Heterochronic developmental plasticity in larval sea urchins and its implications for evolution of nonfeeding larvae. Evolution 46: 972–986.

Telford, M. 1981. A hydrodynamic interpretation of sand dollar morphology. Bull. Mar. Sci. 31: 605–622.

Telford, M. 1983. An experimental analysis of lunule function in the sand dollar *Mellita quinquiesperforata*. Mar. Biol. 76: 125–134.

Telford, M. 1985. Domes, arches, and urchins: The skeletal architecture of echinoids (Echinodermata). Zoomorphologie 105: 114–124.

Telford, M. and R. Mooi. 1986. Resource partitioning by sand dollars in carbonate and siliceous sediments: Evidence from podial and particle dimensions. Biol. Bull. 171: 197–207.

Telford, M., A. S. Harold and R. Mooi. 1983. Feeding structures, behavior, and microhabitat of *Echinocyamus pusillus* (Echinoidea: Clypeasteroida). Biol. Bull. 165: 745–757.

Telford, M., R. Mooi and O. Ellers. 1985. A new model of podial deposit feeding in the sand dollar, *Mellita quinquiesperforata* (Leske): The sieve hypothesis challenged. Biol. Bull. 169: 431–448.

Telford, M., R. Mooi and A. S. Harold. 1987. Feeding activities of five species of *Clypeaster* (Echinoides, Clypeasteroida): Further evidence of clypeasteroid resource partitioning. Biol. Bull. 172: 324–336.

Wray, G. A. 1996. Parallel evolution of nonfeeding larvae in echinoids. Syst. Biol. 45 (3): 308–322.

Holothuroidea

Bakus, G. J. 1973. The biology and ecology of tropical holothurians. *In* O. A. Jones and R. Endean (eds.), *Biology and Geology of Coral Reefs*. Vol. 1, Biology 1. Academic Press, New York, pp. 326–368.

Cameron, J. L. and P. V. Frankboner. 1984. Tentacle structure and feeding processes in life stages of the commercial sea cucumber *Parastichopus californicus* (Stimpson). J. Exp. Mar. Biol. Ecol. 81: 193–209.

Costelloe, J. and B. Keegan. 1984. Feeding and related morphological structures in the dendrochirote *Aslia lefevrei* (Holothuroidea: Echinodermata). Mar. Biol. 84: 135–142.

Francour, P. 1997. Predation on holothurians: A literature review. Invert. Biol. 116 (1): 53–60.

Hauksson, E. 1979. Feeding biology of *Stichopus tremulus*, a deposit-feeding holothurian. Sarsia 63(3): 155–160.

Herreid, C. F., V. F. LaRussa and C. R. DeFesi. 1976. Blood vascular system of the sea cucumber *Stichopus moebii*. J. Morphol. 150(2): 423–451.

Martin, W. E. 1969. *Rynkatropa pawsoni* n. sp. (Echinodermata: Holothuroidea), a commensal sea cucumber. Biol. Bull. 137: 332–337.

Massin, C. 1980. The sediment ingested by *Holothuria tubulosa* Gmel (Holothuroidea: Echinodermata). *In* M. Jangoux (ed.), *Echinoderms: Present and Past*. A. A. Balkema, Rotterdam, pp. 205–208.

Moriarty, D. J. W. 1982. Feeding of *Holothuria atra* and *Stichopus chloronotus* on bacteria, organic carbon and organic nitrogen in sediments of the Great Barrier Reef. Aust. J. Mar. Freshwater Res. 33: 255–263.

Pawson, D. L. 1982. Holothuroidea. *In* S. P. Parker (ed.), *Synopsis and Classification of Living Organisms*. McGraw-Hill, New York, pp. 813–818.

Roberts, D. and C. Bryce. 1982. Further observations on tentacular feeding mechanisms in Holothurians. J. Exp. Mar. Biol. Ecol. 59: 151–163.

Trott, L. B. 1981. A general review of the pearl fishes (Pisces, Carapidae). Bull. Mar. Sci. 31: 623–629. [A brief review of the 20 or so species of these odd fishes, most of which are commensals with holothurians and asteroids.]

Vanden Spiegel, D. and M. Jangoux. 1987. Cuvierian tubules of the holothuroid *Holothuria ferskali* (Echinodermata): A morphofunctional study. Mar. Biol. 96: 263–275.

23

Other Deuterostomes:
Chaetognatha, Hemichordata, Chordata

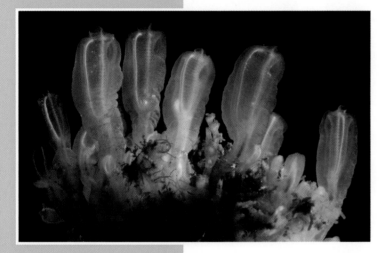

In a flash it covers a distance of some five or six times its own length; and its great jaws, opened wide, snap upon its victim.

Sir Alister Hardy,
(describing a chaetognath)
The Open Sea, 1965

*I*n addition to the lophophorates and echinoderms (discussed in the previous two chapters), the deuterostome lineage includes three other phyla: Chaetognatha (Greek, "spine-jaws"), Hemichordata (Greek, "half-chordates"), and Chordata (= the chordates). These three phyla are discussed in this chapter, which concludes with an overview of ideas about deuterostome phylogeny and comments on the origin of the vertebrates.

The chaetognaths are called arrow worms and comprise about 100 species of marine, mostly planktonic creatures. The hemichordates include 85 or so species, most of which are benthic burrowers known as tongue worms or acorn worms. The phylum Chordata includes three subphyla: Urochordata (= Tunicata; the ascidians, larvaceans, and thaliaceans), Cephalochordata (the lancelets, or amphioxus), and Vertebrata (fishes, amphibians, reptiles, birds, and mammals). There are about 3,000 species of urochordates, 23 species of cephalochordates, and 46,670 species of vertebrates.

Taxonomic History and Classification

The first record of a chaetognath (arrow worm) was made by the Dutch naturalist Martinus Slabber in 1775. For nearly 100 years, as more and more descriptive work was conducted, the systematic position of the group was hotly debated. The arrow worms were at times allied with molluscs, arthropods, and certain blastocoelomates (particularly nematodes), generally within the catch-all taxon Vermes. Some of these arguments continued well into the twentieth century. Although the question of chaetognath phylogenetic affinities is still unsettled, embryological

(A)

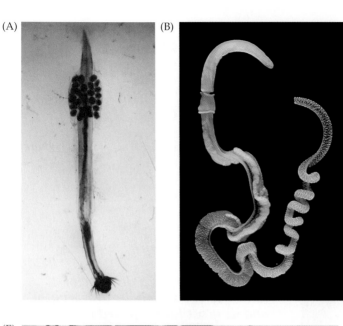

(B)

(C)

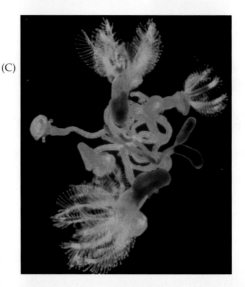

(D)

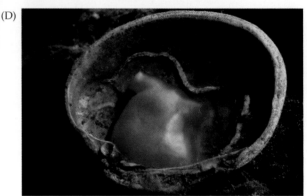

(E)

(F)

(G)

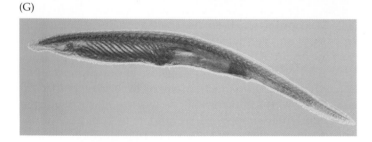

studies strongly favor a deuterostome relationship, and several unique characteristics support its separate phylum status.

Hemichordates were discovered in 1825 by Eschscholtz, who thought his specimen was a holothurian. Other early records allied these animals with nemerteans. Bateson (1885) conducted developmental studies on hemichordates and coined the present phylum name after recognizing similarities with chordate embryogeny. The chordate nature of tunicates (Urochordata) had been recognized by this time, also on the basis of developmental studies. For many years the hemichordates were ranked as a subphylum of Chordata; but although they are clearly related, the hemichordates lack a true notochord—a defining synapomorphy of the chordates. The notochord is a dorsal, elastic, rodlike structure, derived from a middorsal strip of embryonic (archenteric) mesoderm, that provides structural and locomotory support in the body of larval or adult chordates.

Figure 23.1 Representative deuterostomes. (A) The chaetognath, *Kukrohnia bathypelagica* (order Phragmophora) carrying its fertilized eggs. (B) The acorn worm *Saccoglossus* (phylum Hemichordata, class Enteropneusta). (C) Portion of a *Cephalodiscus* colony (phylum Hemichordata, class Pterobranchia), showing several individuals at different stages of development. (D) *Cnemidocarpa*, a solitary ascidian (subphylum Urochordata), attached to the inside of a bivalve. (E) An intertidal compound tunicate, *Polyclinum* (subphylum Urochordata). (F) A pelagic tunicate (subphylum Urochordata, class Thaliacea). Note the circular bands of muscle in the body wall. (G) A lancelet (subphylum Cephalochordata).

PHYLUM CHAETOGNATHA: ARROW WORMS

ORDER PHRAGMOPHORA: With ventral transverse muscle bands (phragma) that appear whitish in living animals. Three families with about 30 species: Spadellidae (e.g., *Gephyrospadella, Paraspadella, Spadella*), Eukrohniidae (*Aberrospadella, Bathyspadella, Eukrohnia, Heterokrohnia, Krohnittella, Kukrohnia, Zahonya*), and Tokiokaspadellidae (*Tokiokaspadella*).

ORDER APHRAGMOPHORA: Without ventral transverse muscle bands. Three families with about 70 species: Sagittidae (e.g., *Bathybelos, Caecosagitta, Parasagitta, Sagitta*), Pterosagittidae (monotypic: *Pterosagitta draco*), and Krohnittidae (*Krohnitta*).

Arrow worms (Figures 23.1A, 23.2) are wholly marine, largely planktonic animals of moderate size, ranging from about 0.5 to 12 cm in length. With the exception of a few benthic species (e.g., *Spadella*) and some that live just off the deep ocean floor (e.g., certain species of *Heterokrohnia*), chaetognaths are adapted to life as pelagic predators. At least one species, *Caecosagitta macrocephala*, is luminescent (Haddock and Case 1994). Arrow worms are distributed throughout the world's oceans and in some estuarine habitats. They often occur in very high numbers and sometimes dominate the biomass in midwater plankton tows. They are most abundant in neritic waters, but some occur at great depths. Major characteristics of the group are listed in Box 23A.

The Chaetognath Bauplan

Externally, arrow worms are streamlined, with virtually perfect bilateral symmetry and transparent bodies, although some meso- and bathypelagic species have orange carotenoid pigmentation and some (phragmophorids) appear milky white because of the opaque ventral transverse musculature. The trimeric form includes a head, a trunk, and a tail—probably corresponding to the prosome, mesosome, and metasome of the general deuterostome bauplan. The trunk bears paired lateral fins and the tail bears a single tail fin. Chaetognath fins are simple epidermal folds enclosing a thick sheet of supportive extracellular matrix. The body surface bears various sensory structures, but the functions of most are not well understood. The head bears a ventrally placed mouth, set in a depression called the **vestibule**. Lateral to the mouth are heavy **grasping spines**, or "hooks," and in front of the mouth are smaller spines called **teeth**—both used in prey capture. Dorsally the head bears a pair of photoreceptors of unique structure. All chaetognaths possess an anterolateral folding of the body wall called the **hood**, which can be drawn over the front and sides of the head, enclosing the vestibule.

Other external features of note include a unique **ciliary loop** (or **corona ciliata**), of uncertain function, located on the dorsal surface at the head–trunk junction (Figure 23.2B). This organ consists of two rings of ciliated epithelial cells and may be involved in chemoreception, or perhaps sperm transfer. Male and female gonopores are lo-

---

BOX 23A Characteristics of the Phylum Chaetognatha

1. Bilateral deuterostomes, with streamlined, elongate, trimeric body comprising head, trunk, and postanal tail divided from one another by transverse septa; head with single protocoel; trunk and tail with paired mesocoels and metacoels, respectively

2. Body with lateral and caudal fins, supported by "rays" (apparently derived from epidermal basement membrane)

3. Mouth surrounded by sets of grasping spines and teeth used in prey capture; mouth set in ventral vestibule; anterolateral fold of body wall forms retractable hood that can enclose vestibule

4. Longitudinal muscles of unusual type, arranged in quadrants; no circular muscle

5. No discrete gas exchange or excretory systems

6. With weakly developed hemal system between peritoneum and lined organs and tissues

7. Complete gut; anus ventral, at trunk–tail junction

8. Large dorsal (cerebral) and ventral (subenteric) ganglia connected by circumenteric connectives. Ciliary fans for detection of water-borne vibrations. Anterior ciliary loop (= corona ciliata) of uncertain function. Inverted pigment-cup ocelli (of uncertain origin)

9. Hermaphroditic, with direct development. Cleavage radial, equal, and holoblastic. Mesoderm and body cavities form by enterocoely. Although blastopore denotes posterior end of body, both mouth and anus form secondarily, subsequent to closure of the blastopore

10. Strictly marine. Predatory carnivores. Largely planktonic, but some benthic species are known

---

cated laterally and posteriorly in the tail and trunk, respectively. The anus is ventral at the trunk–tail junction.

The overall body form of chaetognaths, coupled with their high degree of transparency and locomotor abilities, have contributed to their great success as planktonic predators. They are frequently recovered in plankton hauls with their grasping spines firmly affixed to another animal (Figure 23.3C).

Body wall, support, and movement. A very thin, flexible cuticle overlies the epidermis and helps maintain body shape. Over most of the body, the cells of the epidermis are squamous; they have sinuous interlocking margins and may be stratified. Columnar epithelial cells line the vestibule, and the epidermal cells of the fins are greatly elongated. In some areas, the epidermis lacks a cuticle and bears abundant secretory cells. A well developed basement membrane lies beneath the epidermis. The body wall musculature consists largely of four quadrants of well developed dorsolateral and ventrolateral longitudinal bands, except for the complex head musculature (Figures 23.2E,F). In the Phragmophora, there are additional transverse muscle bands along the ventral aspect of the trunk.

Until recently the nature of the adult body cavities was not clear. However, studies in the 1980s and 1990s (e.g., Shinn and Roberts 1994) provide convincing ultrastructural evidence that the spaces are derived from the enterocoelic cavities formed during development. There is a clear 1:2:2, tripartite arrangement of these coeloms, reflecting the deuterostome nature of chaetognaths. The head cavity, or protocoel, is greatly reduced by the complex cephalic musculature (Figure 23.2E). The paired trunk and tail coeloms correspond to the mesocoels and metacoels, although there is some disagreement about this. They contain dorsal and ventral longitudinal mesenteries, and transverse septa separate the three body regions. The coelomic fluid contains various cells or cell-like inclusions, but their functions are not entirely known.

Body support in chaetognaths is provided by the hydrostatic qualities of the coeloms and the well-developed musculature, aided somewhat by the cuticle and basement membrane. The complex musculature of the head operates the spines and vestibule and the closure of the hood. Locomotion in pelagic forms involves forward darting motions caused by rapid lateral body flexion, by alternately contracting the right- and left-side longitudinal muscles. Although these motions may alternate with brief quiescent periods when the animal slowly sinks, pelagic species seem to spend most of their time actively swimming about, presumably in search of prey. The fins are not used as propelling surfaces but are placed so that they slice through the water and serve as stabilizers. They also increase resistance to sinking between swimming bursts. Although chaetognaths are highly effective predators, watching a live arrow worm moving in the water gives one the impression that they are not very efficient swimmers. However, it has been suggested that their seemingly erratic movements may serve to confuse or elude predators. The ventral transverse muscles in benthic chaetognaths (e.g., *Spadella*) probably aid in crawling, although benthic species are also capable of swimming over short distances.

Feeding and digestion. Chaetognaths are predatory carnivores that feed on a variety of other animals, including planktonic crustaceans, small fishes, and even other arrow worms. They seem to have a special fondness for copepods as food. Benthic forms, such as *Spadella*, are ambush predators. They affix themselves

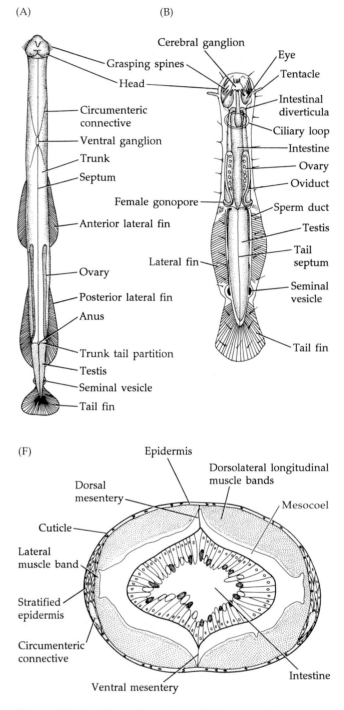

Figure 23.2 General anatomy of chaetognaths. (A) *Sagitta elegans* (ventral view). (B) The benthic chaetognath *Spadella* (dorsal view). (C) *Krohnitta subtilis* (dorsal view). (D) Outline of *Sagitta hispida*, showing sensory bristles. (E) Anatomy of the head of *Sagitta*. (F) The trunk of *Sagitta* (cross section). (G) The nervous system of a generalized chaetognath. (H) Arrangement of eye units in a chaetognath. (I) Cerebral ganglion and associated major nerves. (J) Reproductive systems in *Sagitta*.

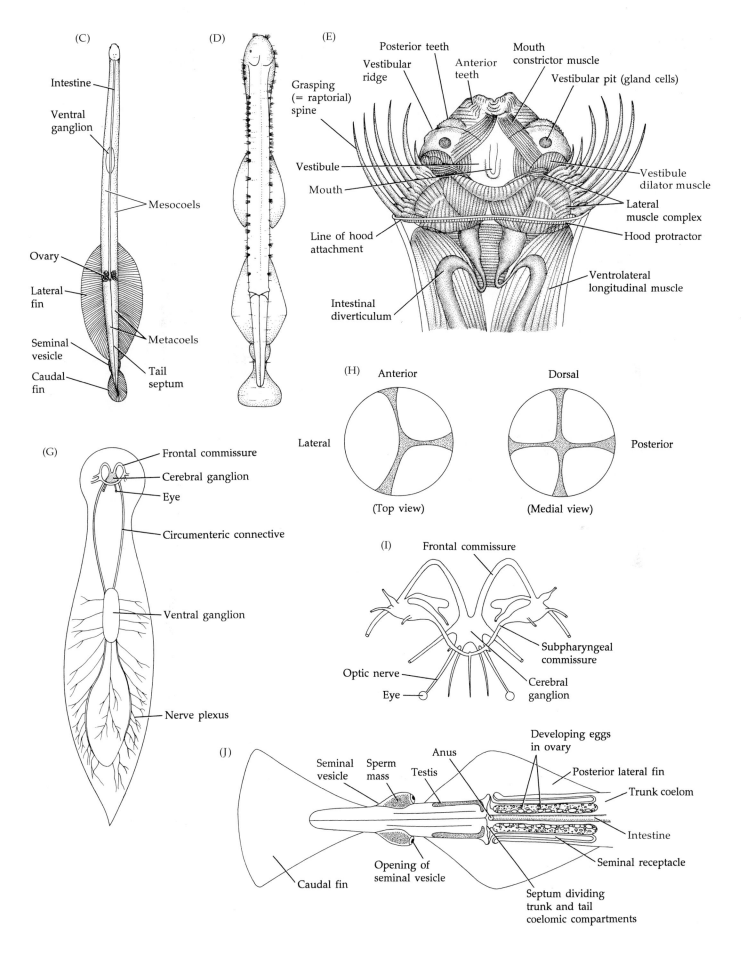

(C)

Intestine

Ventral ganglion

Mesocoels

Ovary

Lateral fin

Seminal vesicle

Metacoels

Tail septum

Caudal fin

(D)

(E)

Posterior teeth

Mouth constrictor muscle

Vestibular ridge

Anterior teeth

Vestibular pit (gland cells)

Grasping (= raptorial) spine

Vestibule

Mouth

Vestibule dilator muscle

Lateral muscle complex

Hood protractor

Line of hood attachment

Ventrolateral longitudinal muscle

Intestinal diverticulum

(H)

Anterior

Dorsal

Lateral

Posterior

(Top view)

(Medial view)

(G)

Frontal commissure

Cerebral ganglion

Eye

Circumenteric connective

Ventral ganglion

Nerve plexus

(I)

Frontal commissure

Subpharyngeal commissure

Optic nerve

Eye

Cerebral ganglion

(J)

Developing eggs in ovary

Anus

Posterior lateral fin

Seminal vesicle

Sperm mass

Testis

Trunk coelom

Intestine

Opening of seminal vesicle

Seminal receptacle

Caudal fin

Septum dividing trunk and tail coelomic compartments

(A)

Grasping spines　Vestibular pit　Papillae of vestibular ridge　Anterior teeth　Posterior teeth

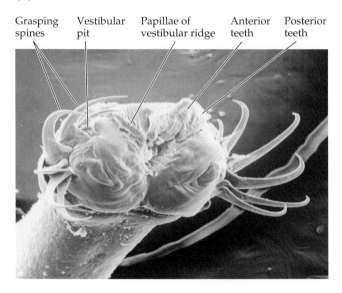

(B)

Figure 23.3　(A,B) Heads of the chaetognaths. (A) *Zonosagitta pulchra*. Note the well developed raptorial structures. (B) *Z. bedoti*, from the eastern Pacific. The hooks are clearly visible on either side of the head surrounding the exceptionally large number (17–20) of long, narrow, posterior teeth. The shorter anterior teeth lie just above the mouth. The vestibular ridge with its pores is partially visible behind the left set of posterior teeth. (C) The chaetognath in this photo, *Flaccisagitta hexaptera*, has partially swallowed a fish larva (probably an anchovy). A single anterior tooth projects down below the second hook on the left side of the photo, and two posterior teeth can be seen between the first and second hooks. The circular organ just below the first hook is the vestibular pit.

(C)

Vestibular pit

Posterior teeth

Anterior tooth

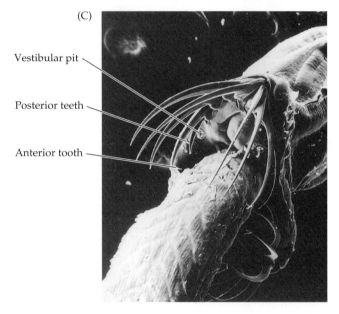

to a substratum by adhesive secretions, raise the head and protrude the mouth and vestibule, flaring the spines around the mouth. Prey swimming within reach are detected (probably by vibration receptors) and then captured by a quick movement of the head while the rest of the body remains firmly attached to the substratum. The spines grasp and manipulate the prey, orienting it for ingestion. Prey is ingested whole.

When a planktonic chaetognath detects a nearby prey, it quickly darts forward to grab the victim with its grasping spines (Figure 23.3). The spines and/or teeth of some species bear serrations; the teeth are cuspidate, a shape that aids in the penetration of prey, especially exoskeletons of small crustaceans. Many, if not most, species inject a poison into their prey when they attack. Erik Thuesen and several colleagues have shown that the poison contains a potent neurotoxin called tetrodotoxin that blocks sodium transport across cell membranes. Many marine bacteria synthesize tetrodotoxin, and studies suggest that in chaetognaths this toxin is produced by a commensal bacterium (*Vibrio*) inhabiting the head or gut region of the arrow worm. Some large

species of *Sagitta* and other chaetognaths are capable of devouring fishes as large as themselves! The bodies of chaetognaths bear arrays of "hairs" called **ciliary fans** (not to be confused with the ciliary loop mentioned earlier) that are sensitive to water-borne vibrations (Figure 23.2D). Because of the unique distribution of these structures, a chaetognath can determine both the direction and distance of potential prey at close range.

The gut is a relatively simple straight tube extending from the mouth in the vestibule to the ventral anus at the trunk–tail junction (Figure 23.2). The mouth leads to a short pharynx, which is equipped with mucus-secreting cells. Swallowing is accomplished by well developed pharyngeal muscles aided by lubricants from the mucous glands. When rigid prey such as small crustaceans are captured, the chaetognath positions the victim longitudinally for swallowing. The gut narrows where it passes through the head–trunk septum, and it extends posteriorly as a long intestine. A short rectum joins the posterior intestine to the anus. Apparently most digestion occurs extracellularly in the posterior region of the intestine and can be extremely rapid.

Circulation, gas exchange, and excretion. Shinn (1993) described a previously unknown **hemal system** in chaetognaths. This system comprises rather loosely organized sinuses and channels, especially between the gut and surrounding peritoneum. Small spaces within mesenteries and septa also contain hemal fluid. Internal transport in the hemal system is probably by diffusion aided by the action of body movements.

Gas exchange and excretion are apparently by diffusion, although certain cells of the hemal and coelomic fluids may be involved. There is still much to learn about chaetognath physiology, and other mechanisms may be involved in these functions.

Nervous system and sense organs. Of paramount importance to the success of chaetognaths as active predators are features of the nervous system and associated sensory receptors. As we have seen in other groups, a bauplan that emphasizes cephalization is frequently an integral factor in adapting to a predatory lifestyle. The central nervous system of chaetognaths includes a large cerebral ganglion in the head, dorsal to the pharynx. Several other ganglia arise from the cerebral ganglion and serve various muscles and sense organs of the head. A pair of circumenteric connectives emerges from the hind part of the cerebral ganglion and extends posteroventrally to meet in a large ventral or subenteric ganglion located in the trunk epidermis (Figure 23.2G). The ventral ganglion controls swimming. From it emerge a dozen or so pairs of nerves that extend to various parts of the body, many branching to form a dense subepidermal nerve plexus.

The body is covered with patches and tracts of bristle-like, cilia-derived ciliary fans, long thought to be tactile receptors but more recently shown to be sensors of water-borne vibrations or movements (Figure 23.2D). These structures apparently function in prey detection, similar to the lateral line system of fishes. Although specific chemoreceptors have yet to be positively identified in chaetognaths, they almost certainly exist. The aforementioned ciliary loop may have a chemoreceptive function, and many arrow worms have transvestibular pores that roughly parallel the vestibular ridge in the buccal area; these structures may also be chemoreceptors. Arrow worms possess a pair of eyes situated just below the epidermis on the dorsal surface of the head. The structure of these eyes is unusual in that each consists of five inverted pigment-cup ocelli, arranged with a large ocellus directed laterally and four small ones directed medially (Figure 23.2H). We can imply from this structure that chaetognaths have a nearly uninterrupted visual field enabling them to orient to light direction and intensity. Chaetognath eyes lack lenses, except in two deep-water species of *Eukrohnia* that reportedly bear hexagonal cuticular lenses. In most, however, the eyes probably do not form images but are used for orientation during vertical migration. The ocelli contain ciliated receptor cells, as in many other deuterostomes.

Reproduction and development. Arrow worms are hermaphroditic, with paired ovaries in the trunk and paired testes in the tail (Figures 23.2A,B). Spermatogonia are released from the testes into the tail coeloms, where they mature. From there they are picked up by open ciliated funnels leading to sperm ducts, which open laterally at a pair of seminal vesicles. Sperm masses variously called "sperm balls" or "sperm clusters" form within the seminal vesicles. Chaetognaths apparently do not form true spermatophores (although the term is often used in the literature), because a definitive capsule, covering, or "spermatophore membrane" has never been observed. Each ovary bears along its side an oviduct that leads to a genital pore just in front of the trunk–tail septum. Immature eggs are transferred to the oviducts, but the details of this process are unclear. In at least some species (e.g., *Spadella cephaloptera, Parasagitta hispida*), exchange of sperm may be mutual (reciprocal). *Parasagitta hispida* may undergo reciprocal or nonreciprocal fertilization, or even self-fertilization.

Transfer of the sperm mass has been most extensively studied in benthic species of *Spadella*. After a rather elaborate mating "dance," the sperm are deposited as balls onto the mate's body. Rupture of the balls allows the sperm to stream posteriorly to enter the female gonopores and oviducts, where fertilization occurs. Benthic chaetognaths (e.g., *Spadella*) tend to deposit fertilized eggs on algae or other suitable substrata. Neritic species may secrete a jelly-like coating around each zygote and then shed the floating embryos to the sea (e.g., *Sagitta*), or they may release developing embryos that sink to the bottom for development, or they may undertake downward migrations to lay eggs that sink to the bottom or attach to stationary objects. Species in the deep-water genus *Eukrohnia* carry the developing embryos in two temporary gelatinous marsupial pouches, one on either side of the body near the tail, until the young are ready to swim (Figure 23.4).

Development is direct, lacking any larval stage or metamorphosis. The transparent eggs contain very little yolk, and cleavage is radial, holoblastic, and equal, yielding a coeloblastula with a small blastocoel (Figure 23.5). The blastocoel enlarges during subsequent divisions, and gastrulation occurs by invagination of the presumptive entoderm. The blastopore marks the eventual posterior end of the animal, but it closes so that both the mouth and the anus form secondarily, the former by a stomodeal invagination of ectoderm. Archenteric pouches (Figure 23.5) eventually pinch off as first the head and then the trunk coeloms (the tail coeloms apparently form later than those in the trunk). Thus, embryonic coelom formation is clearly enterocoelous. During development,

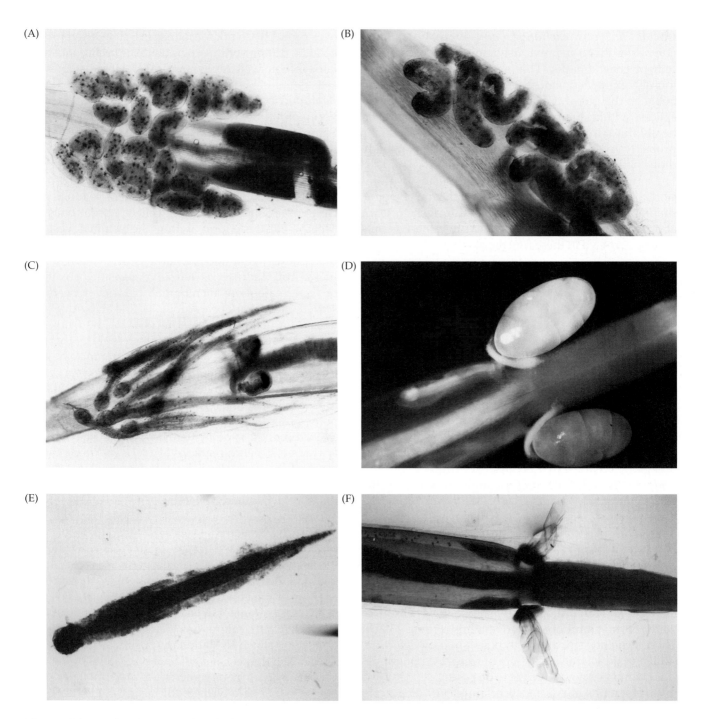

Figure 23.4 The chaetognath *Eukrohnia*, with temporary gelatinous marsupia housing the developing embryos. (A–C) *Eukrohnia bathypelagica* carrying fertilized eggs and young in the marsupium. (D) *Eukrohnia fowleri* carrying fertilized eggs in posterior marsupial sacs. (E) Young of *Eukrohnia fowleri* just after hatching. (F) *Eukrohnia fowleri* carrying the empty marsupial sacs from which the young have already escaped.

the embryonic body cavities are compressed and decrease in size, but are not actually obliterated. They apparently expand later in development and thus, persist as true coeloms in adulthood. Interestingly, all organs and tissues of mesodermal origin derive entirely from the peritoneum of these coeloms.

The embryo grows quickly, elongates, and hatches as a juvenile chaetognath. Development from zygote release to hatching is rapid, about 48 hours. This rapid direct development compares strategically with indirect development, even though no independent larval stage occurs. Parental investment per embryo is small, the eggs contain little yolk, and they are abandoned soon after fertilization (except in brooding forms). The rapid development to a feeding juvenile is essential to the success of this life history strategy.

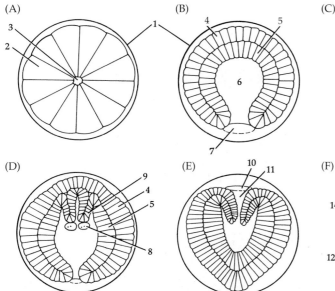

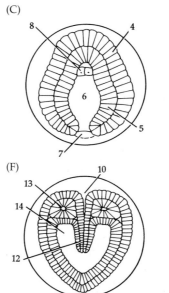

Figure 23.5 Early chaetognath development. (A) Early blastula. (B) Gastrula. (C) Later gastrula. (D) Production of mesodermal folds from archenteron. (E) Blastopore closure and secondary mouth opening with the formation of a stomodeum. (F) Formation of coelomic pouches.

PHYLUM HEMICHORDATA: THE HEMICHORDATES

The phylum Hemichordata contains the enteropneusts (acorn worms) and pterobranchs (Figures 23.1B; 23.6; Box 23B). Also included here is *Planctosphaera pelagica*, which is assigned to the monotypic class Planctosphaeroidea (Figure 23.6J). This creature was discovered in 1932 and has since been collected in several localities in the Pacific and Atlantic Oceans. It is viewed by most authorities as a hemichordate larva, but it has not yet been linked to a specific adult. In addition to the general traits associated with deuterostomes, hemichordates possess pharyngeal gill slits and most have a dorsal (sometimes hollow) nerve cord (Figure 23.7A), features also seen in the Phylum Chordata. They were once thought to possess a notochord and therefore were placed as a subphylum of the Chordata. However, studies eventually showed that the suspected structure, an evagination of the anterior gut called the **buccal diverticulum** (or **stomochord**), is not homologous with a notochord (Box 23B).

As adults, all hemichordates are benthic marine animals (except for *Planctosphaera*). About 75 of the 85 or so living species belong to the class Enteropneusta. These worms generally live buried in soft sediments, among algal holdfasts, or under rocks; they are largely intertidal. Of the few deep-water acorn worms known, one species, *Saxipendium coronatum* (the "spaghetti worm"), is a member of deep sea geothermal vent communities (e.g., the

Galápagos rift system). Some other deep-water forms construct highly branched burrow systems. Enteropneusts range in length from a few centimeters to over 2 meters.

Pterobranchs are largely colonial. The zooids are small, rarely exceeding 1 cm in length, but colonies may measure 10 cm or more across. They are very lophophorate-like in general structure (Figures 23.1C and 23.6F–I). Such similarities, however, as well as the differences between pterobranchs and enteropneusts, may be the results of small size and colonial life. The groups Enteropneusta and Pterobranchia represent two highly divergent clades within this phylum, each adapted to different ways of making a living. Both represent exploitations of the basic tripartite deuterostome architecture, attesting again to the evolutionary plasticity of this fundamental body plan.

CLASS ENTEROPNEUSTA: Acorn, or tongue worms. Vermiform, with three body regions as proboscis, collar and trunk; coeloms reduced; gut elongate, straight; mouth ventral at anterior end of collar; without an endostyle; anus posterior, terminal; marine, burrow in soft sediments or nestle under rocks or in algal holdfasts; largely intertidal although a few deepwater species are known. About 75 species. (e.g., *Balanoglossus, Protoglossus, Saccoglossus, Saxipendium, Xenopleura*)

CLASS PTEROBRANCHIA: Pterobranchs (Figures 23.1C and 23.6F–I). Body sacciform; with three body regions as preoral disc (=cephalic shield), tentaculate mesosome, and metasome subdivided as trunk and stalk; neurochord lacking; gut U-shaped; marine; generally small (less than 1 cm), aggregating or colonial; three genera: *Atubaria, Cephalodiscus* and *Rhabdopleura*.

CLASS PLANCTOSPHAEROIDEA: Figure 23.6J. Body spherical but bilateral, jelly-like, with complexly arranged surface ciliary bands; gut U-shaped; coeloms poorly developed; monospecific (*Planctosphaera pelagica*); worldwide; probably a larval stage of unknown adult hemichordate, perhaps an enteropneust.

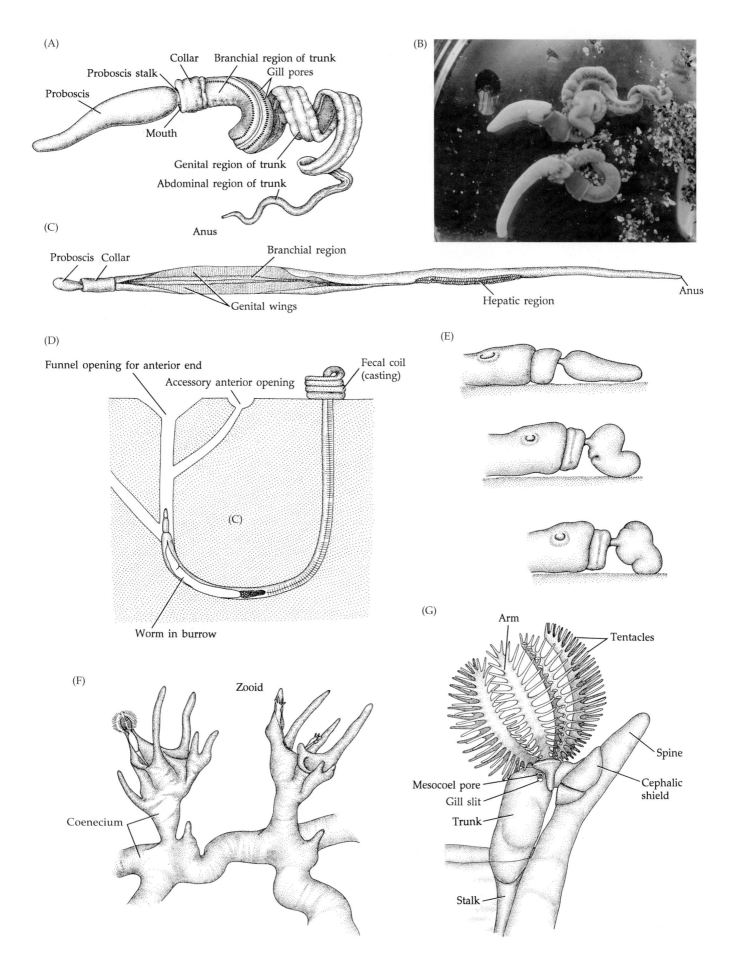

(A)

Proboscis
Proboscis stalk
Collar
Branchial region of trunk
Gill pores
Mouth
Genital region of trunk
Abdominal region of trunk
Anus

(B)

(C)

Proboscis   Collar
Branchial region
Genital wings
Hepatic region
Anus

(D)

Funnel opening for anterior end
Accessory anterior opening
Fecal coil (casting)
(C)
Worm in burrow

(E)

(F)

Zooid
Coenecium

(G)

Arm
Tentacles
Spine
Cephalic shield
Mesocoel pore
Gill slit
Trunk
Stalk

BOX 23B *Characteristics of the Phylum Hemichordata*

1. Bilateral deuterostomes, body vermiform or saccate and fundamentally trimeric, with prosome, mesosome, and metasome, each with coelomic compartments; solitary or colonial. Pterobranchs with mesocoelic extensions into the arms and tentacles (as in lophophorates)

2. With ciliated, pharyngeal gill slits (or pores)

3. Well developed, open circulatory system

4. Unique excretory structure, the glomerulus

5. Gonads extracoelic, in metasome

6. Complete gut. With preoral (buccal) gut diverticulum. Deposit or suspension feeders

7. Without a notochord

8. Circular and longitudinal muscles present in body wall of proboscis and collar of enteropneusts; pterobranchs with longitudinal muscle only. Basement membrane of enteropneusts, in proboscis region, produced as rigid plates (the proboscis skeleton)

9. Short, dorsal, mesosomal, occasionally hollow nerve cord (neurochord), probably homologous with chordate nerve cord

10. Dioecious, with external fertilization and indirect development; asexual reproduction common. Cleavage radial, holoblastic, more or less equal. Although blastopore denotes posterior end of body, both the mouth and the anus form secondarily, subsequent to closure of the blastopore. Mesoderm and body cavities form by enterocoely. Typically with unique tornaria larva.

11. Strictly marine and, except for *Planctosphaera*, benthic

The Hemichordate Bauplan

Enteropneusts are solitary, elongate, vermiform animals, with bodies clearly divided into three regions. The **proboscis, collar,** and **trunk** (Figures 23.6 A, B) are homologous to the prosome, mesosome, and metasome of other deuterostomes. The proboscis is short and often conical. A short, thin **proboscis stalk** connects the proboscis to the collar, the latter bearing the ventral mouth at its anterior end. The anus terminates at the posterior end of the long trunk. The trunk bears middorsal and midventral, external, longitudinal ridges, which correspond to the location of certain longitudinal nerves and blood vessels. In addition, the trunk is differentiated regionally along its length, the amount of differentiation varying from species to species. Most species bear a clear, anterior, **branchial region** of the trunk. This region is characterized by the presence of numerous **gill (= branchial) pores** flanking the middorsal ridge. Some, such as *Balanoglossus*, have a distinct **genital region** housing the gonads and bearing external longitudinal **genital wings**. Also in *Balanoglossus*, the anterior portion of the intestine is thickened to such an extent that it is visible through the body wall as a distinctly colored area called the **hepatic region**. In a number of genera no clear body subdivisions occur, except for the region of the gill pores.

Figure 23.6 External anatomy of some representative hemichordates. (A) *Saccoglossus* (class Enteropneusta). (B) Two *Saccoglossus* from a northern California mud flat. (C) *Balanoglossus* (Enteropneusta). (D) *Balanoglossus clavigerus* in its burrow system. (E) Locomotion by proboscis movements in a juvenile *Saccoglossus horsti*. (F) Portion of a colony of *Cephalodiscus* (Pterobranchia). (G) Individual zooid of *Cephalodiscus*. (H) Portion of the colony of *Rhabdopleura* (Pterobranchia). (I) individual zooid of *Atubaria* (Pterobranchia) with tentacles removed. (J) *Planctosphaera pelagica*, considered to be a larval stage of an unknown adult hemichordate.

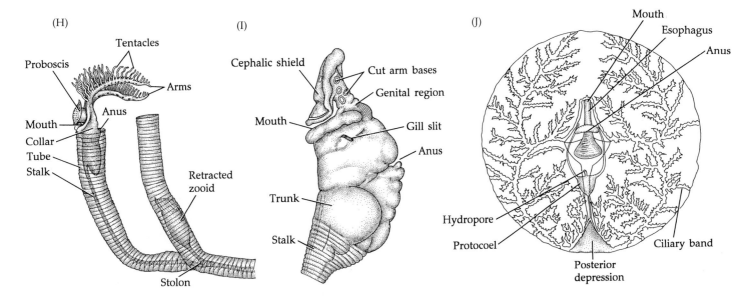

The bodies of pterobranchs are small, usually pyriform or globular, but still retaining the ancestral tripartite regional division. Like the prosome of lophophorates, the pterobranch prosome is reduced to a small plate, called the preoral disc or cephalic shield, which generally folds ventrally over the mouth. The mesosome forms a collar that bears the anteroventral mouth and forms two to several arms on which the tentacles are located. The arms are arranged dorsally and do not encircle the mouth. The gut is U-shaped and the anus lies anterodorsally (Figures 23.6H,I; 23.9). The metasome is subdivided into a trunk and posterior stalk.

The three known pterobranch genera are somewhat different from one another in habits and in some anatomical details. Most pterobranchs live in colonies (*Rhabdopleura*) or aggregations (*Cephalodiscus*) consisting of zooids housed within tubular secreted casings (Figures 23.6F–I). In colonies of *Rhabdopleura*, the zooids are connected to one another by tissue extensions called **stolons**, but no such interzooidal links occur in *Cephalodiscus*. The overall forms of the aggregations and colonies vary among species. In all cases, the associated zooids are products of asexual reproduction initiated by a single sexually produced individual. The third genus is represented by a single species (*Atubaria heterolopha*), known only from 43 specimens collected in 1935 in Sagami Bay, Japan. These animals were recovered from dredge samples taken in 200 to 300 meters of water and were found clinging to hydroid colonies by their prehensile stalks. Although *Atubaria* is very similar anatomically to *Cephalodiscus*, it is a solitary form without any secreted casing (Figure 23.6I).

Body wall, support, and locomotion. Hemichordates in general possess a ciliated epidermis overlying a nerve plexus. The epidermis is usually richly supplied with gland cells, many of which are involved in mucus production, particularly on the proboscis and collar of enteropneusts and on the tentacles of pterobranchs. In some enteropneusts, certain epithelial cells produce noxious mucopolysaccharide compounds that may repel predators. Most of these secretions have a distinctive iodine-like odor. Both circular and longitudinal muscles are present in the wall of the proboscis and anterior collar of acorn worms, but elsewhere only longitudinal fibers exist. Apparently, pterobranchs possess only longitudinal fibers in their body walls, at least some of which are produced by the peritoneum. A basement membrane lies between the epidermis and the musculature, and in enteropneusts it is produced as thickened, rigid plates, called the **proboscis skeleton** (Figure 23.7A), and as supportive structures of the gill slits.

The peritoneum is variably reduced or transformed into musculature in different parts of the bodies of hemichordates. Still, the tripartite coelomic arrangement is evident in all forms, even though the cavities are fre-

quently reduced by the invasion of connective or supportive tissue. As in chaetognaths, the usual three coelomic spaces, or their remnants, are present as a single protocoel, followed by paired mesocoels and metacoels. These spaces occur, in order, in the proboscis, collar, and trunk of enteropneusts, and in the cephalic shield, collar-arms-tentacles, and trunk of pterobranchs. As in lophophorates, mesocoelic extensions are present in the arms and tentacles of pterobranchs. However, unlike the condition in the lophophorate phyla, the mouth in pterobranchs lies outside the tentacular crown.

Body support is a function primarily of the hydrostatic nature of the body cavities, and secondarily of the structural integrity of the body wall, connective tissues, and supplemental structures such as the proboscis skeleton of acorn worms. In adult acorn worms, the buccal diverticulum appears to function as a "skeleton," working to antagonize the contractile pericardium, which pumps blood through the heart and into the glomerulus. In pterobranchs, except *Atubaria*, additional support and protection are provided by the secreted outer casing of the colony or aggregation.

Hemichordates are sessile or sedentary and capable of only limited movement at best. Enteropneusts crawl slowly or burrow by peristaltic action of the proboscis (Figure 23.6E), which contains the necessary circular and longitudinal muscles. Although most are probably strictly benthic and largely sedentary, at least one species of enteropneust (*Glandiceps hacksii*) sometimes swarms at the surface in shallow water, where it feeds on phytoplankton. The protraction and retraction of pterobranchs within their tubular houses are accomplished by hydrostatic pressure and contraction of longitudinal muscles, respectively. Some crawl within their tubes by using the muscular cephalic shield. The prehensile, tail-like stalk of *Atubaria* is quite mobile, probably as a result of a combination of hydraulics and muscle action.

Feeding and digestion. Enteropneusts that burrow through soft sediments are largely direct deposit feeders, ingesting the substratum and digesting organic material therein. Those that live in permanent burrows (Figure 23.6D) or among loose rubble or holdfasts tend to be suspension feeders, selectively trapping organic particulates from the water with the proboscis. Most species are probably capable of feeding by both methods. The details of suspension feeding have been examined in some acorn worms (e.g. *Protoglossus*; Figure 23.8). Food material, including detritus and live plankton, is trapped in mucus secreted over the surface of the proboscis and moved posteriorly by ciliary currents. The sorting that occurs at the proximal end of the proboscis and the stalk passes most large particles over the lip of the collar; these particles are then removed by special rejection currents (Figure 23.8). Most of the food is moved ventrally around the proboscis

(A)

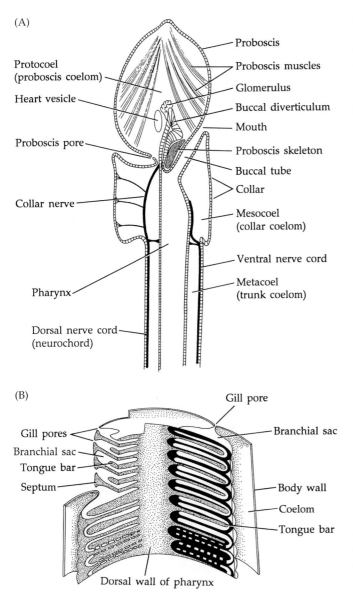

Proboscis

Proboscis muscles

Glomerulus

Buccal diverticulum

Mouth

Proboscis skeleton

Buccal tube

Collar

Mesocoel (collar coelom)

Ventral nerve cord

Metacoel (trunk coelom)

Protocoel (proboscis coelom)

Heart vesicle

Proboscis pore

Collar nerve

Pharynx

Dorsal nerve cord (neurochord)

Figure 23.7 Internal anatomy of enteropneusts. (A) The front end of an enteropneust (sagittal section). (B) Cutaway view from inside the pharynx of an enteropneust, showing the arrangement of the gill slits. (C) Enteropneust circulatory system.

(C)

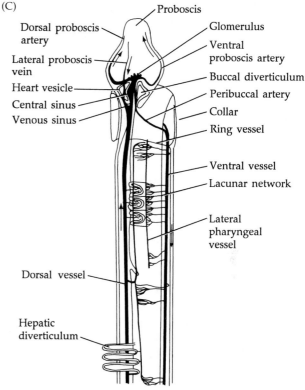

Dorsal proboscis artery

Lateral proboscis vein

Heart vesicle

Central sinus

Venous sinus

Dorsal vessel

Hepatic diverticulum

Proboscis

Glomerulus

Ventral proboscis artery

Buccal diverticulum

Peribuccal artery

Collar

Ring vessel

Ventral vessel

Lacunar network

Lateral pharyngeal vessel

(B)

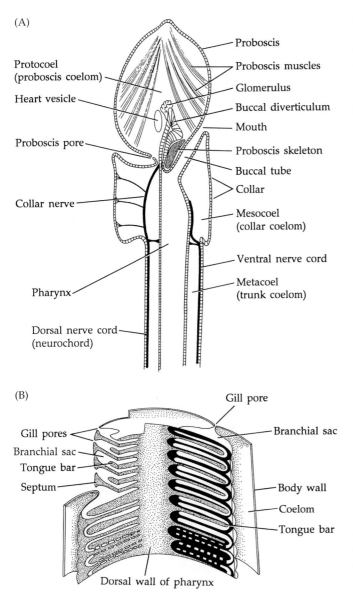

Gill pore

Branchial sac

Gill pores

Branchial sac

Tongue bar

Septum

Body wall

Coelom

Tongue bar

Dorsal wall of pharynx

stalk, over a structure called the **preoral ciliary organ**, and condensed into a mucous cord that is then passed into the mouth. The preoral ciliary organ includes a concentration of sensory neurons and probably functions in chemoreception. Swallowing appears to be facilitated by a combination of ciliary action and the flow of water moving into the mouth and out the gill pores. The gill pores thus function to facilitate water flow—not as a filter-feeding device themselves. The cilia on the gill pores probably serve to keep the pores clean and unclogged.

The digestive tract of enteropneusts is a straight, regionally specialized tube, extending from the mouth to the anus (Figure 23.7A). Gut musculature is scant, and the food is moved along largely by cilia. The mouth leads to a **buccal tube**, which is housed within the collar and gives rise anterodorsally to the forwardly projecting **buccal diverticulum**. Behind the buccal tube in the an-

terior part of the trunk is the pharynx. Both the buccal tube and the pharynx are derived from a stomodeal invagination of ectoderm. The pharynx bears a **dorsal epibranchial ridge** of unknown function. The digestive portion of the pharynx is restricted to a thin ventral **hypobranchial ridge**, while the lateral and dorsal parts bear the **gill slits**, or **pores** (Figure 23.7B). The gill slits number from a few to over 100 pairs. Each one is a U-shaped opening in the wall of the pharynx that leads to

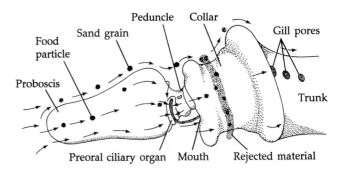

Peduncle

Collar

Gill pores

Sand grain

Food particle

Proboscis

Trunk

Preoral ciliary organ

Mouth

Rejected material

Figure 23.8 Food sorting and rejection currents (arrows) on the proboscis and collar region of the enteropneust *Protoglossus kohleri*.

a **branchial sac** and then to a dorsolateral **gill pore** through which water exits to the outside. The septum between adjacent gill slits and the partition between the arms of the U of each slit (called a **tongue bar**) are supported by skeletal elements derived from the basement membrane of the gut lining.

Behind the pharynx is an esophagus, which at least in some forms (e.g., *Saccoglossus*) bears openings to the outside through the dorsal body wall. Unlike most of the gut, the middle region of the esophagus bears intrinsic muscles and moves the food into the intestine by peristalsis. The mechanical squeezing of the food material may press out excess water through the esophageal pores. In some species the wall of the anterior intestine contains dense green or brown inclusions that are visible externally and delimit the **hepatic region** of the trunk. The intestine extends, more or less undifferentiated, to a short rectum terminating in the anus. Digestion is probably largely extracellular in the intestine, but details are not fully known.

The major feeding structures of pterobranchs are the arms and tentacles derived from the mesosome. *Rhabdopleura* bears one pair of arms, each with numerous tentacles, whereas *Cephalodiscus* bears from five to nine pairs of arms, depending on the species. Pterobranchs are ciliary mucus suspension feeders. During feeding, they assume a position near an opening in their tubular cases and extend their arms and tentacles into the water (Figures 23.9 and 23.10). The tentacles on adjacent arms interdigitate to form a latticework, across which a mucous net is secreted. Food is trapped in the mucus and moved to the mouth by the action of cilia on the tentacles and arms. At least in *Cephalodiscus*, the cilia over the general body surface may also move food to the mouth, creating a unique situation in which the entire body surface functions as a feeding structure. As in enteropneusts, a pterobranch's mouth is located under the anteroventral edge of the collar or mesosome (Figure 23.9). The gut is U-shaped, beginning with a buccal tube from which arises a buccal diverticulum of various forms and complexities. The pharynx bears one pair of gill slits in *Cephalodiscus* and *Atubaria*, but none in *Rhabdopleura*. When present, the gill apparatus is much simpler than that of enteropneusts, there being no supporting structures and less well-defined branchial sacs. The slits open through pores on the exterior. An esophagus connects the pharynx to a sacciform stomach at the bottom of the U, and occupies most of the space within the trunk (Figure 23.9). The ascending portion of the gut is the intestine, which leads anteriorly to the dorsal anus. Digestion probably occurs in the stomach and intestine.

Circulation, gas exchange, and excretion.

Enteropneusts possess a well-developed open circulatory system comprising blood vessels, sinuses, and a contractile organ called the **heart vesicle** that is located in the

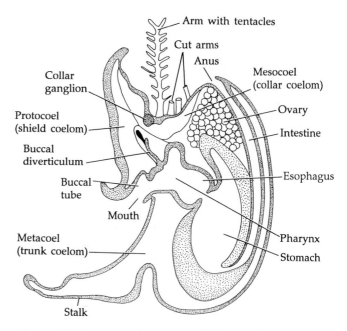

Figure 23.9 Internal anatomy of the pterobranch hemichordate *Cephalodiscus* (sagittal section).

proboscis (Figures 23.7A,C). Various names have been applied to parts of this system by different workers, and readers should be alert to the possibility of encountering other terms in various publications on this group. Two main longitudinal vessels lie in the dorsal and ventral mesenteries along the length of the trunk and in the collar. Blood flows anteriorly in the dorsal vessel and posteriorly in the ventral vessel. The dorsal vessel expands in the collar as the **venous sinus**, which also receives blood anteriorly from a pair of **lateral proboscis veins**. The venous sinus leads to a larger, elongate, **central sinus** in the proboscis, lying between the buccal diverticulum and the dorsal heart vesicle. The heart vesicle has a muscular ventral wall that pulsates against the central sinus and aids the movement of blood. From the central sinus, blood moves into the **glomerulus**, which is an excretory organ unique to the hemichordates; it is formed of finger-like outpocketings of peritoneum associated with the blood sinuses. All of the blood leaving the central sinus passes through the glomerular sinuses, within which metabolic wastes are presumably extracted. The glomerulus, buccal diverticulum, central sinus, and heart vesicle compose what is often referred to as the **proboscis complex** of enteropneusts (Figure 23.7C).

Blood leaves the glomerulus and passes through various vessels and sinuses supplying the anterior end of the worm, eventually reaching the ventral longitudinal vessel. Along the length of the trunk, blood leaves the ventral vessel to pass into networks of sinuses supplying the gut and the body wall; from these sinuses it then passes to the dorsal vessel.

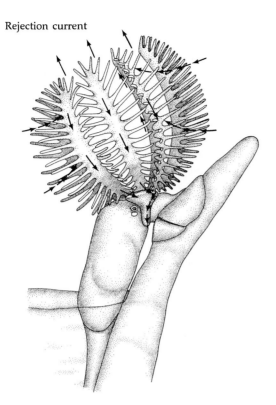

Rejection current

Figure 23.10 Feeding and rejection currents (arrows) of the pterobranch hemichordate *Cephalodiscus*. Water is drawn in between the tentacles and moves on a rejection current upward and away from the animal. Food particles are moved proximally along the tentacles to a food canal indicated by the arrowheads.

Gas exchange occurs between the environment and the blood across the walls of the gill structures, especially the branchial sacs, all of which are richly supplied with blood from the ventral vessel. The gill septa and tongue bars bear cilia, which drive water into the mouth through the pharynx and out the gill slits, or pores. Actually, little work has been done on the matter of gas exchange, and it may be that other areas of the body surface are also involved in this activity. Enteropneust blood lacks pigments and contains very few cells, and gases are apparently carried in solution.

The circulatory system of pterobranchs has not been fully studied. In general it is weakly developed compared with that of enteropneusts—a condition not unexpected in tiny animals. There is a central sinus and a heart vesicle near the buccal diverticulum, but no major vessels through the body. Rather, blood is carried in sinuses and lacunae that lack complete walls. A glomerulus is usually present, but it is not well developed. The single pair of gill slits in *Cephalodiscus* and *Atubaria* may aid in gas exchange, but the small diffusion distances throughout the body probably allow general cutaneous exchange, especially over the high surface areas of the tentacles.

Nervous system and sense organs. Most of the nervous system of all hemichordates consists of a netlike nerve plexus lying among the bases of the epithelial cells outside the basement membrane. A subepidermal dorsal nerve cord, or **neurochord**, is present in the collar of enteropneusts, but is reduced to a mere thickening of the plexus in pterobranchs. The plexus is thickened in enteropneusts as longitudinal tracts of neurons along the middorsal and midventral lines of the body. The evolutionary relationship between the dorsal hollow nerve cord of chordates and the mesosomal neurochord of enteropneusts is uncertain. The neurochord, however, is formed by an invagination of ectoderm and is actually hollow in some species, a condition strongly suggesting homology with the dorsal nerve cord of the Chordata.

There are few types of sensory receptors in the hemichordates. Enteropneusts possess sensory cells over most of the body, probably serving as touch receptors that give these cryptic animals some information about their surroundings. As mentioned earlier, they also bear a preoral ciliary organ, presumed to be a chemoreceptor used during feeding. Little is known about the sensory apparatus of pterobranchs. Touch receptors are presumably present in the tentacles and perhaps on the cephalic shield and tip of the stalk in the noncolonial forms.

Reproduction and development. Asexual reproduction occurs in at least some enteropneusts (e.g., *Balanoglossus*) and in most pterobranchs. Acorn worms fragment small pieces from the trunk, and each one is able to grow into a new individual. They are very fragile worms and often break when handled; presumably they can regenerate missing parts.

As in most colonial invertebrates, asexual reproduction by budding is an integral part of the life history of aggregating and colonial pterobranchs. In *Cephalodiscus* the buds arise from near the base of the stalk of adult individuals (Figure 23.11) and pass through a complex developmental sequence before they are released. Budding in *Rhabdopleura* occurs along the stolons that grow from the tips of the stalks of adult zooids. In both genera, the aggregations or colonies arise by budding after the formation of a single sexually produced individual. There is no evidence that *Atubaria* undergoes budding.

Hemichordates are dioecious but possess no outward evidence of sexual differences. Paired sacciform gonads lie in the trunk, outside the peritoneum; they are often very elongate in the acorn worms. Among the pterobranchs, *Rhabdopleura* possesses but a single gonad, which lies along the right side of the trunk coelom. Enteropneusts bear a pair of gonopores located dorsolaterally on the anterior trunk. The gonopores of *Cephalodiscus* are at the base of the arms; in *Rhabdopleura*, a single pore opens on the right side of the trunk.

So far as is known, fertilization is always external in hemichordates. Spawning in enteropneusts involves the

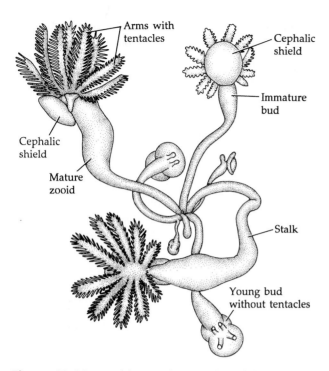

Arms with tentacles

Cephalic shield

Cephalic shield

Immature bud

Mature zooid

Stalk

Young bud without tentacles

Figure 23.11   Budding in the pterobranch hemichordate *Cephalodiscus*, showing zooids at different stages of development.

release of mucoid egg masses by the females, followed by shedding of sperm by neighboring males. Once the eggs are fertilized, the mucous coat breaks down, thereby freeing the eggs into the sea water, where all subsequent development occurs. There are very few reports on spawning or fertilization in pterobranchs, and most are quite incomplete. Apparently the eggs are shed into the tubes of the colony or aggregation, where they are fertilized. Some species apparently brood the developing embryos within their burrow.

Early development in enteropneusts takes place entirely in the water, and all features of this developmental period attest to the deuterostome nature of acorn worms. Some species produce relatively yolky eggs, and others produce eggs with very little yolk. In both cases, however, cleavage is holoblastic, radial, and more or less equal. A coeloblastula forms, which gastrulates by invagination. The blastopore is at the presumptive posterior end, but it closes and the anus and mouth form later. By late gastrula, the embryo has acquired cilia and breaks free from the egg membrane as a free-floating plankter. Coelom formation is by archenteric pouching (typical enterocoely). Usually, a single protocoel arises from the inner end of the archenteron, and from it originate paired mesocoels and metacoels, establishing very early the tripartite body plan.

Species that produce yolky eggs develop directly to juvenile worms, without an intervening larval phase (e.g., *Saccoglossus*). In those that shed nonyolky eggs (e.g., *Balanoglossus*), the hatching stage develops quickly to a

characteristic, planktotrophic, **tornaria larva** with ciliary bands reminiscent of certain echinoderm larvae. This suspension-feeding larva soon elongates, with the three body regions becoming externally apparent (Figure 23.12).

Among the pterobranchs, only *Cephalodiscus* has been studied embryologically, and even here the details are scanty. The large yolky eggs undergo radial, holoblastic, subequal cleavage. There is some argument about the form of the blastula and the precise nature of gastrulation. The embryo escapes the brooding area within the parental tube as a fully ciliated (but unnamed) larval stage. Coelom formation is by archenteric pouching, but the sequence of production is not clear. In terms of larval body orientation, the region between the mouth and anus apparently represents a much shortened dorsal surface and the "lower" side of the saclike trunk the ventral surface. These terms of reference are typically abandoned when describing the adult.

Thus, as we have seen in so many other benthic, sessile, and sedentary invertebrates, the hemichordates include a dispersal phase in their life history strategies. Even in those enteropneusts with technically direct development, the pattern is strategically indirect, the embryos being at least planktonic free-living animals, even if not full-fledged larvae. The pterobranchs display a mixed life history pattern, with a period of brooding followed by a free larval stage. This pattern is common among small, sessile animals, which cannot afford to produce huge numbers of eggs but depend on at least a short-lived dispersal phase.

PHYLUM CHORDATA: THE CHORDATES

We are chordates. So are cats and dogs, lemurs and anteaters, birds and fishes, frogs and snakes, whales and elephants—all conspicuous by their size and familiarity. In addition to having a notochord, a dorsal, hollow nerve cord, and pharyngeal gill slits (Box 23C), we and the rest of these creatures also possess a skeletal "backbone," a vertebral column housing our dorsal nerve cord and defining us as members of the subphylum Vertebrata. But there are two other chordate subphyla, both of which lack vertebrae. These are the invertebrate chordates, the subphyla Urochordata and Cephalochordata. The Cephalochordata comprise 20 or so species of small, fishlike animals called lancelets, or amphioxus (e.g., *Branchiostoma*) (Figure 23.1G). The Urochordata (= Tunicata) are composed of about 3,000 species in four classes: the sessile filter-feeding sea squirts, or ascidians (class Ascidiacea); the pelagic tunicates or salps (class Thaliacea); planktonic larva-like tunicates called appendicularians, or larvaceans (class Appendicularia); and the abyssal ascidian-like sorberaceans (class Sorberacea) (Figures 23.13 and 23.14). One of the key synapomorphies defining the Chordata is the **notochord**, a dorsal, elastic rod derived

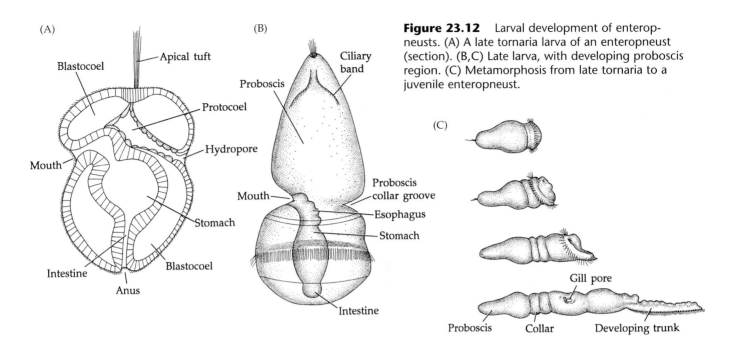

Figure 23.12 Larval development of enteropneusts. (A) A late tornaria larva of an enteropneust (section). (B,C) Late larva, with developing proboscis region. (C) Metamorphosis from late tornaria to a juvenile enteropneust.

from a middorsal strip of embryonic (archenteric) mesoderm, that provides structural and locomotory support in the body of larval or adult chordates.

SUBPHYLUM UROCHORDATA (= TUNICATA): The tunicates. Adult body form varies, but usually lacking obvious trimeric organization; body covered by thick or thin tunic (test) of a cellulose-like polysaccharide; without bony tissue; notochord restricted to tail and usually found only in larval stage (and in adult appendicularians); gut U-shaped, pharynx (branchial chamber) typically with numerous gill slits (stigmata); coelom not developed; dorsal nerve cord present in larval stages; all marine; 4 classes.

CLASS ASCIDIACEA: Ascidians, or sea squirts. Benthic, solitary or colonial, sessile tunicates; incurrent and excurrent siphons directed upwards, away from the substratum; without dorsal nerve cord in adult stages; occur at all depths. About 13 families and many genera. (e.g., *Ascidia, Botryllus, Chelyosoma, Ciona, Clavelina, Corella, Diazona, Diplosoma, Lissoclinum, Molgula, Psammascidia, Pyura, Styela*)

CLASS THALIACEA: Pelagic tunicates or salps. Solitary or colonial; incurrent and excurrent siphons at opposite ends, providing locomotor current; adults without a tail; gill clefts not subdivided by gill bars; 3 orders: Pyrosomida, Salpida, and Doliolida. (e.g., *Dolioletta, Doliolum, Pyrosoma, Salpa, Thetys*)

CLASS APPENDICULARIA (= LARVACEA): Appendicularians or larvaceans. Solitary planktonic tunicates; probably neotenic; adults retain larval characteristics, including notochord and muscular tail; body enclosed in a complex gelatinous "house" involved in feeding. (e.g., *Fritillaria, Oikopleura, Stegasoma*)

CLASS SORBERACEA: Benthic, abyssal, ascidian-like urochordates possessing dorsal nerve cords in adult stages; carnivorous, lacking perforated branchial sac. (e.g., *Octacnemus*)

SUBPHYLUM CEPHALOCHORDATA (= ACRANIA): Lancelets (amphioxus) (Figures 23.1G and 23.18). Small (to 7 cm), fishlike chordates with notochord, gill slits, dorsal nerve cord, and postanal tail present in adults, but without vertebral column or cranial skeleton structure; gonads numerous (25–38) and serially arranged. Marine and brackish water, usually associated with clean sand or gravel sediments in which they burrow. (e.g., *Asymmetron, Branchiostoma, Epigonichthyes*)

SUBPHYLUM VERTEBRATA: Vertebrates. Chordates usually possessing a vertebral column that forms the axis of the body skeleton; most with paired appendages, a brain case, and (except for members of the class Agnatha) jaws. Several classes are generally recognized, although not all are strictly monophyletic: classes Myxini and Cephalaspidomorphi are the hagfishes and lampreys, respectively (united as the jawless or agnathan fishes); Chondrichthyes are the sharks, skates, and rays; Osteichthyes are the bony fishes (e.g., trout, tuna, perch); Amphibia includes the salamanders, frogs, toads, caecilians; Reptilia traditionally included the turtles, snakes, lizards, and crocodilians, but modern classifications place birds and reptiles together as Reptilomorpha (or Sauropsida). Mammalia comprises the mammals.

The Urochordates (Tunicates)

Members of the four urochordate classes are almost all marine suspension feeders, but they conduct their lives in very different ways. The ascidians include both solitary and colonial forms, with individuals ranging in size from less than 1 mm to 60 cm, and some colonies measuring several meters across. Ascidians are found worldwide and at all ocean depths, attached to nearly any substratum. They are most abundant and diverse in rocky littoral habitats and on deep-sea muds. Salps float about singly or in cylindrical or chainlike colonies sometimes several meters long. They are known from all oceans but are especially abundant in tropical and subtropical waters. Salps occur from the surface to depths of about 1,500 meters. The larvaceans, or appendiculari-

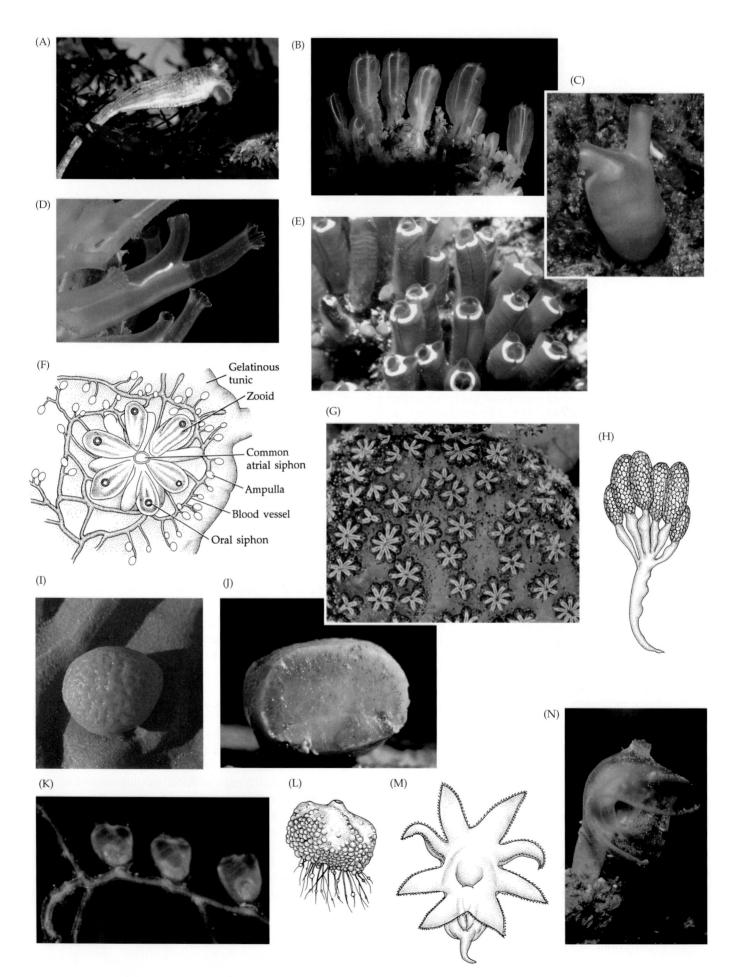

(F)
Gelatinous tunic
Zooid
Common atrial siphon
Ampulla
Blood vessel
Oral siphon

Figure 23.13 Representative ascidians. (A) The solitary ascidian *Styela*. (B) The social ascidian *Clavelina*. (C) The sea peach, *Halocynthia aurantia*. (D) *Ciona intestinalis*. (E) Clavelinid tunicates. (F,G) Zooid cluster of the compound ascidian *Botryllus*. (H) A compound ascidian in which groups of zooids are clustered on stalks. (I,J) The compound ascidian *Aplidium* sp. Note the external appearance of the colony. The colony is sliced open to expose the clusters of zooids. (K) Portion of the colony of *Perophora*, with zooids arising from stolons. (L) *Bolteniopsis*, a strange deep-sea ascidian. (M) *Octacnemus*, a predatory deep-sea ascidian. (N) *Megalodicopia hians*, a predatory tunicate with its "trap" spread near the siphon.

> ## BOX 23C *Characteristics of the Phylum Chordata*
>
> 1. Bilaterally symmetrical, coelomate deuterostomes (coelom lost in some groups)
>
> 2. Pharyngeal gill slits present at some stage in development
>
> 3. Dorsal notochord present at some stage in development
>
> 4. Dorsal, hollow nerve cord, at least in some stage of life history
>
> 5. With a pharygneal endostyle (Urochordata, Cephalochordata) or thyroid gland (Vertebrata)
>
> 6. Muscular, locomotor, postanal tail at some stage in development
>
> 7. Gut complete, usually regionally specialized
>
> 8. Circulatory system with a ventral contractile blood vessel (or heart). Gas exchange occurs across body wall or epithelial tissues
>
> 9. Dioecious or hermaphroditic; development variable. Cleavage radial, holoblastic, subequal or slightly unequal. Tadpole stage is expressed at some point in the life history of all taxa

ans, are solitary, luminescent planktonic creatures rarely more than about 5 mm long. They resemble in certain ways the larval stages of some other tunicates, hence the name "larvaceans." Their retention of larval features, including a notochord and a nerve cord, suggests that they arose by paedomorphosis. The feces of salps and larvaceans, and the abandoned houses of the latter, constitute important sources of food and particulate organic carbon in the open sea.

The Tunicate Bauplan

Tunicates (urochordates) are bilaterally symmetrical, at least during early developmental stages. They utilize mucus-covered **pharyngeal gill slits** (=**stigmata**) for suspension feeding (Figure 23.15). Although modified in the appendicularians, water flows into the mouth and pharynx by way of an **incurrent (oral) siphon**, passes through the gill slits into a spacious water-filled **atrium (cloacal water chamber)**, and exits through an **excurrent (atrial) siphon**. The gut is simple and U-shaped, with the anus emptying into the excurrent flow of water just as it leaves the body. Because of the drastic modification in body relative to that of more familiar chordates, the general orientation of the bodies of urochordates is not immediately apparent and can only be fully understood by examining the events of metamorphosis, as described later. The oral siphon is generally anterior, and the atrial siphon is either anterodorsal (in ascidians) or posterior (in thaliaceans) (Figure 23.14C). In any case, the dorsoventral orientation of the body can be determined internally by the locations of the dorsal ganglion and a thickened ciliated groove, called the **endostyle**, that runs along the ventral side of the pharynx or branchial basket (Figures 23.14 and 23.15; Box 23C).

The Ascidiacea is the largest and most diverse class of tunicates. Some interstitial forms are known and a few live anchored in soft sediments, but the majority of ascidians are attached to hard substrata (Figure 23.1D and 23.13A–G). Three general types of ascidians are usually recognized, although these categories do not relate directly to formal taxa. Most of the large (up to 60 cm long) species are called **solitary ascidians** because they live singly and unattached to one another (e.g., *Ciona, Molgula, Styela*; Figures 23.13A,D), although

many are highly gregarious. **Social ascidians** tend to live in clumps of individuals that are vascularly attached to one another at their bases (e.g., *Clavelina*; Figure 23.13B). Finally, a great number of species are **compound ascidians** and are characterized by many small individuals (zooids) living together in a common gelatinous matrix (e.g., *Aplidium, Botryllus*; Figures 23.13F–J). In extreme cases, colonies may measure several meters across.

Ascidians have become so highly modified during their evolution as sessile suspension feeders that adults are not easily recognizable as chordates. The dorsal hollow nerve cord and the notochord are present in the larval stage but are lost in adults; the pharyngeal gill slits persist in the adults. The oral and atrial siphons are generally directed away from the substratum and set at an angle to one another, thereby reducing the potential for recycling waste water. The bodies of some solitary and social forms are set on stalks and elevated above the substratum, whereas the compound forms typically grow as thin or thick sheets conforming to the topography of the surface on which they live. In some of these compound ascidians, zooids are arranged in regular rosettes and share a common atrial chamber formed within the the outer body wall, or **tunic** (Figures 23.13F, G and 23.15D). The tunic varies from thick to thin and from smooth and slick to wrinkled and leathery. Many species are brightly colored, and compound ascideans frequently are some of the most colorful animals living on intertidal rocks.

(A)

(B)

(C)

Oral siphon

Ganglion

Zooid

Common
atrial chamber

Excurrent water flow

Tunic

Branchial
muscle

Muscle
bands

Endostyle

Visceral
mass

Mantle

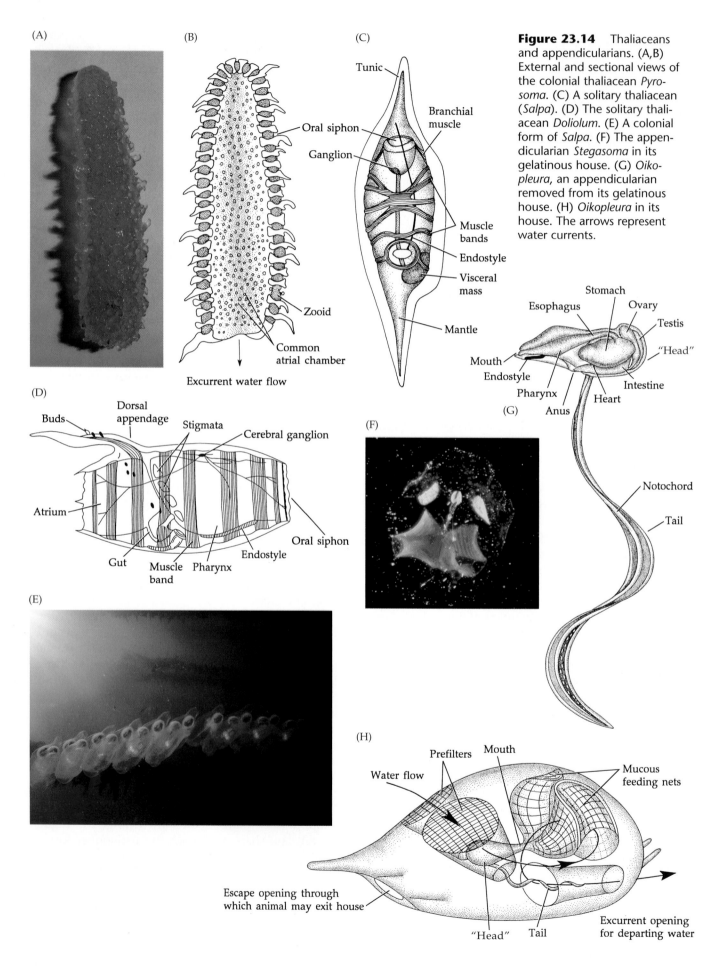

Figure 23.14 Thaliaceans and appendicularians. (A,B) External and sectional views of the colonial thaliacean *Pyrosoma*. (C) A solitary thaliacean (*Salpa*). (D) The solitary thaliacean *Doliolum*. (E) A colonial form of *Salpa*. (F) The appendicularian *Stegasoma* in its gelatinous house. (G) *Oikopleura*, an appendicularian removed from its gelatinous house. (H) *Oikopleura* in its house. The arrows represent water currents.

(D)

Buds

Dorsal
appendage

Stigmata

Cerebral ganglion

Atrium

Gut

Muscle
band

Pharynx

Endostyle

Oral siphon

(E)

(F)

(G)

Esophagus

Stomach

Ovary

Testis

"Head"

Mouth

Endostyle

Pharynx

Anus

Heart

Intestine

Notochord

Tail

(H)

Prefilters

Mouth

Water flow

Mucous
feeding nets

Escape opening through
which animal may exit house

"Head"

Tail

Excurrent opening
for departing water

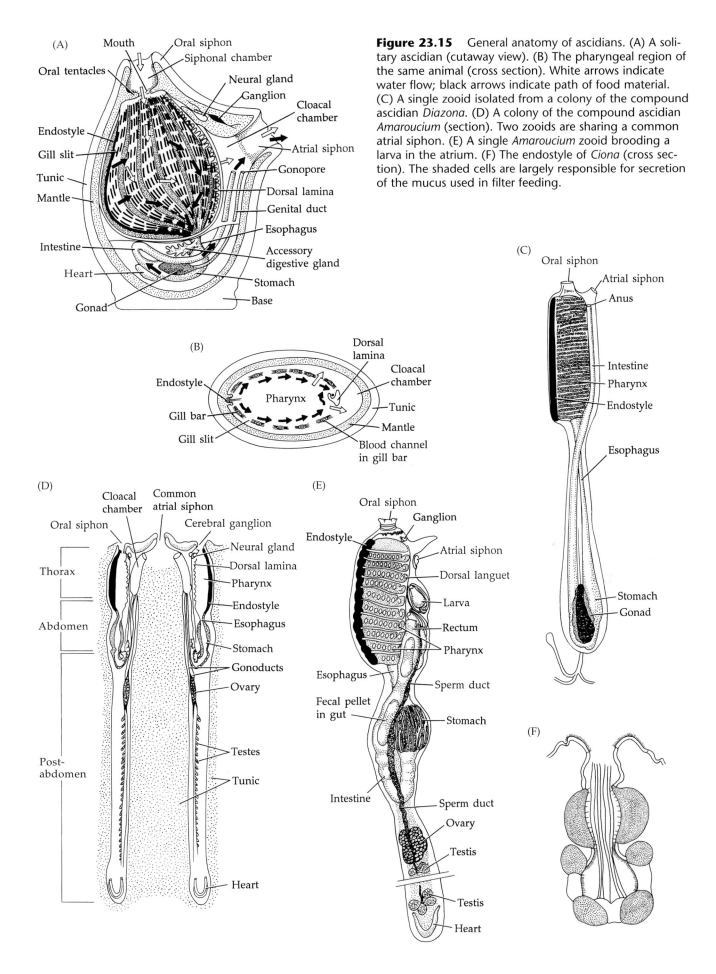

Figure 23.15 General anatomy of ascidians. (A) A solitary ascidian (cutaway view). (B) The pharyngeal region of the same animal (cross section). White arrows indicate water flow; black arrows indicate path of food material. (C) A single zooid isolated from a colony of the compound ascidian *Diazona*. (D) A colony of the compound ascidian *Amaroucium* (section). Two zooids are sharing a common atrial siphon. (E) A single *Amaroucium* zooid brooding a larva in the atrium. (F) The endostyle of *Ciona* (cross section). The shaded cells are largely responsible for secretion of the mucus used in filter feeding.

Thaliaceans are pelagic, ascidian-like urochordates. They are constructed much like their sessile counterparts except that the oral and atrial siphons are at opposite ends of the body, and in many forms the pharyngeal filtering basket is modified to accommodate the linear flow of water through the animal. The exiting water provides a means of "jet propulsion." Most are highly gelatinous and transparent.

The class Thaliacea comprises three orders. Members of the order Pyrosomida are considered the most primitive thaliaceans and most resemble their presumed ascidian ancestors. Pyrosomes are remarkable colonies of tiny ascidian-like zooids embedded in a dense gelatinous matrix and arranged around a long central, tubular chamber called a common **cloaca** (Figure 23.14A,B). The cloaca receives exhalant water from the inwardly directed atrial siphons of all the zooids; the water then exits through a single large aperture, thereby propelling the barrel-shaped colony slowly through the water. As in the ascidians, water movement in pyrosomes is generated entirely by ciliary action of the individual zooids.

The orders Doliolida and Salpida include thaliaceans that alternate between solitary sexual forms and colonial asexual stages (Figure 23.14C–E). Doliolid individuals are generally small, less than 1 cm long, whereas single salps may be 15 to 20 cm long and some form chainlike colonies several meters in length. The members of these two orders move water through their bodies but propel themselves partially (salps) or wholly (doliolids) by muscular action.

Thaliaceans are predominately warm-water creatures, although certain species are found in temperate and even polar seas. They are particularly abundant over the continental shelf and are frequently captured in surface waters or seen stranded on wave-swept sandy beaches after storms. Some, however, have been recorded from depths to 1,500 meters.

Appendicularians are among the strangest of all urochordates and are characterized by the retention of larval features. These solitary animals live in a gelatinous casing, or **house**, that they secrete around their body (Figures 23.14F–H). The bulbous trunk of the body contains the major organs, including the gut, and bears a muscular tail in which the notochord is retained. The dorsal nerve cord, though reduced, extends partway along the length of the tail. Thus, we see clearly the evidence of the chordate nature of these animals retained through paedomorphosis—evidence that is present only during developmental stages of other urochordates. (The phenomenon of paedomorphosis is discussed in Chapter 4.)

The pharynx of appendicularians is reduced and bears only two clefts. When positioned within its house, the animal produces a complex water current by beating its tail. Filtering is accomplished by meshes in the house wall and by mucous nets secreted by the animal (Figure 23.14H). The mesh filters are complex and vary among species, often being constructed of more than one size of interlaced fibers (Deibel et al. 1985; Flood 1991). The exiting water provides the locomotor force. Appendicularians are found in the surface waters of all oceans and are sometimes extremely abundant.

Body wall, support, and locomotion.   The body wall of tunicates includes a simple epithelium overlain by a secreted tunic of varying thickness and consistency. The tunic is most well developed in the ascidians and some thaliaceans. It varies from soft and gelatinous to tough and leathery, and it sometimes includes calcareous spicules. The matrix of the tunic contains fibers and is composed largely of a cellulose-like carbohydrate called **tunicin**. The tunic is not a simple, secreted, nonliving cuticle, however, since it also contains amebocytes and, in some cases, blood cells and even blood vessels. The tunic may be viewed as an exoskeleton providing support and protection. Some ascidians harbor symbiotic algae in their tunics. These algae include a variety of both prokaryotic and eukaryotic species from several major groups.

Beneath the epidermis are muscle bands. In many species, especially of ascidians, these muscles lie within an ectodermally produced mesenchyme called the **mantle** (Figures 23.15A,B). Ascidians possess longitudinal muscles extending along the body wall that serve to pull the flared siphons down against the body. Circular sphincter muscles close the siphonal openings. Doliolids and salps have well developed bands of circular muscles that pump water through the body for feeding and locomotion. When they contract, water within the body is forced out the atrial siphon, thereby propelling the animal forward. When they relax, the body expands because of the resilience of the tunic, and water is drawn in through the oral siphon. As noted above, the tail muscles of appendicularians provide the action for moving water through the houses of these animals.

Tunicates do not have a coelom; the body cavity has been lost in concert with the evolution of a water chamber called the atrium, or cloacal water chamber, which functions in filter feeding. This chamber is a saclike, ectodermally derived structure continuous with the epidermis of the atrial siphon (Figure 23.15A,B). The inner wall lies against the pharynx and is perforated over the gill slits, or stigmata. Thus, water that enters the pharynx via the oral siphon flows through the stigmata, into the cloacal chamber, and out the atrial siphon.

Feeding and digestion.   We have hinted at various aspects of the feeding biology of tunicates in our comments above. Most of these animals are suspension feeders and use various kinds of mucous nets to filter plankton and organic detritus from the sea water. A few ascidians live partially embedded in soft sediments and feed on organic material in the substratum, and certain bizarre deep-sea species actually prey on small invertebrates by grasping them with the lips of the oral siphon. Below we provide a detailed description of feeding and digestion in a suspension-feeding ascidian and then compare this with feeding in thaliaceans and appendicularians.

Water is moved through the body of an ascidian largely by the action of cilia lining the pharyngeal basket. Water enters the oral siphon and passes through a short **siphonal chamber** at the inner end of which is the mouth. A ring of fleshy tentacles encircles the mouth and prevents the entrance of large particles (Figure 23.15A). Food-laden water then passes into the pharynx, which bears a ventral longitudinal groove called the **endostyle**. The bottom of the groove is lined with mucus-secreting cells and bears a longitudinal row of flagella; the sides of the groove bear cilia (Figure 23.15F). The mucus, a complex mucoprotein containing iodine, is moved to the sides of the endostyle by the basal flagella and then outward by the lateral cilia. Cells near the opening of the endostyle are responsible for binding environmental iodine and incorporating it into the mucus. Sheets of mucus then move dorsally along the inner wall of the pharynx and pass over the stigmata. The slit-like stigmata are arranged in rows and bear lateral cilia that drive water from the pharynx into the surrounding atrial water chamber (Figure 23.15). Thus, water passing through the stigmata also passes through the mucous sheets, on which food particles are retained. On the dorsal surface of the pharynx is a longitudinal curved ridge called the **dorsal lamina**, or a row of ciliated projections called **languets**, or both (Figure 23.15A,B). These structures serve to roll the mucous sheets into cords, which are then passed posteriorly to a short esophagus and then to a stomach. Attached to the stomach is a small **pyloric gland**, which extends around the intestine as a network of small tubes (Figure 23.17B). Some species also bear an **accessory digestive gland** (Figure 23.15A). Digestive enzymes are secreted into the stomach lumen by secretory cells of the gut wall and perhaps by the associated glands, and digestion is largely extracellular. From the stomach, the gut loops forward as an intestine, through which undigested material passes to the anus, which opens into the atrium near the excurrent siphon.

The unique ascidian family Didemnidae comprises colonial forms in which the cloacal systems are confluent, and the colonies are usually hardened with aragonitic "spiculospheres." Among certain tropical genera (e.g., *Didemnum, Diplosoma, Lissoclinum, Trididemnum*) are species that maintain symbiotic "algae" in the test, the branchial basket, or the cloacal system. These algae are prokaryotic, resembling blue-green algae, but possess chlorophylls *a* and *b* (like those in the green algae, Chlorophyta); they are placed in the genus *Prochloron*. There is evidence that the host ascidian benefits from the association by feeding directly on the algal cells, perhaps by amebocytic phagocytosis. Didemnid ascidians housing such symbiotic algae are also remarkable for their powers of locomotion, limited though it is—colonies have been clocked at speeds of 4.7 mm per 12-hour period. Such movement may allow these ascidians to position themselves in light conditions favorable to their algal symbionts. A similar symbiosis has been reported between red cyanophyte algae and certain didemnid ascidians.

Thaliaceans feed in much the same way as ascidians do except that the siphons are at opposite ends of the body. The number of pharyngeal stigmata is usually reduced, especially in doliolids, and restricted to the posterior portion of the pharynx (Figure 23.14B–D). The intestine extends posteriorly from the pharynx and opens into the enlarged cloacal chamber.

Appendicularians (Figures 23.14F–H) secrete a hollow gelatinous (mucopolysaccharide) house in which they reside and upon which they depend for feeding. The tail is directed through a tube in the house structure toward an excurrent opening. Sinusoidal beating of the muscular tail generates a current that pulls water into the house through coarse, meshlike, mucous filters that screen out large particles; eventually the water leaves the house via the excurrent opening. The pharynx of appendicularians bears only two small gill slits, which open directly to the exterior. Mucous feeding nets are secreted through the mouth and lie within the house chamber. The water current is directed through these fine-mesh nets, where food particles are concentrated. The food, net and all, is periodically ingested by way of a short buccal tube. The houses of most appendicularians bear an additional opening that serves as an escape hole through which the animal can leave and reenter. The houses are fragile and easily damaged. Damaged or clogged houses are abandoned, and new ones are manufactured rapidly, in a matter of seconds or minutes. In some species a new or "spare" house may be found beneath the functional house; after escaping the clogged house, the "spare" is rapidly inflated.

The gut of appendicularians is U-shaped and the anus opens directly to the outside rather than into a cloacal chamber. Fecal material is released into the path of excurrent water leaving the filter nets. Appendicularians are primarily herbivorous, feeding on minute phytoplankton and bacteria down to a size of 0.1 μm. They sometimes constitute the dominant planktonic herbivores in waters over the continental shelf, reaching densities of many thousands per cubic meter.

Circulation, gas exchange, and excretion. The circulatory system is weakly developed in urochordates, especially in the thaliaceans and appendicularians. It is best understood in ascidians, which possess a short, tubular heart that lies posteroventrally in the body near the stomach and behind the pharyngeal basket (Figure 23.15A). The heart is surrounded by a pericardial sac. Blood vessels extend anteriorly and posteriorly, opening into spaces around the internal organs and also providing the blood supply to the tunic. The heartbeat is by peristaltic action, and the direction of this motion is periodically reversed, flushing the blood first one way through the heart and then the other. Blood physiology and function are largely matters of speculation. Ascidians accumulate high concentrations of certain heavy metals in their blood, especially vanadium and iron. Some evidence suggests that the presence of high vanadium levels in at least some species serves to deter would-be predators. In addition, the blood includes a large variety of cell types, including amebocytes that are thought to function in nutrient transport, tunic deposition, and accumulation of metabolic wastes. The blood also contains several vertebrate-like hormones, including thyroxine, oxytocins, and vasoconstrictors. Compound species, such as *Botryllus*, have special blood cells that play a vital role in rejecting adjacent conspecific, but nonclone colonies (allogenic colonies).

Gas exchange occurs across the body wall in tunicates, especially across the linings of the pharynx and the cloacal chamber. Little is known about respiratory physiology in these animals.

In most ascidians and some other tunicates, two evaginations arise from the posterior wall of the pharynx and lie along each side of the heart. These structures are called **epicardial sacs** and may represent coelomic remnants. In some species the epicardial sacs are involved in bud formation during asexual reproduction, and they may also function in the accumulation of nitrogenous waste products by forming storage capsules called **renal vesicles**. Other than these vesicles and certain blood cells (**nephrocytes**), it is likely that much of the metabolic waste is lost from the body by simple diffusion.

Nervous system and sense organs.   The nervous system of tunicates is much reduced and reflects their relatively inactive sessile and floating planktonic lifestyles. A small **cerebral ganglion** lies just dorsal to the anterior end of the pharynx and gives rise to a few nerves to various parts of the body, especially the muscles and siphonal areas. A well developed dorsal nerve cord is present in the tails of tunicate larvae, but this structure is lost during metamorphosis, except in the appendicularians (see below). Most tunicates possess a **neural**, or **subneural**, **gland** located between the cerebral ganglion and the anterodorsal portion of the pharynx (Figure 23.15A). This gland opens to the pharynx through a small duct, but its function is unknown. Some workers have suggested that it may be the precursor of the pituitary gland of vertebrates. Sensory receptors are poorly developed in tunicates, although touch-sensitive neurons are prevalent around the siphons.

Asexual reproduction.   While the appendicularians are entirely sexual in their reproductive habits, thali-

aceans and many ascidians include asexual processes in their life history strategies. In social and especially compound ascidians, asexual budding allows rapid exploitation of available substrata, as we have seen in other sessile colonial invertebrates such as sponges and bryozoans.

Budding in tunicates occurs in a great variety of ways and from different organs and germinative tissues (Figure 23.16). In general, initial buds are formed by a sexually produced individual (**oozooid**), then the asexually produced individuals (**blastozooids**) produce additional buds. The simplest and perhaps most primitive budding process occurs in certain social ascidians, including species of *Perophora* and *Clavelina*, where blastozooids arise from the body wall of stolons. In more complicated budding processes, the germinal tissues include various combinations of the epidermis, gonads, epicardial sacs, and gut. Doliolid thaliaceans often produce chains of buds. The chains are sometimes released intact, but eventually each blastozooid breaks loose as a separate individual. Budding in pyrosomes results in the characteristic floating colonies seen in species of this family (Figures 23.14A,B).

Much of the information about asexual reproduction in ascidians was reviewed and synthesized by M. Nakauchi (1982), who suggested that the various types of budding among ascidians can be divided into two categories on the basis of their functional significance. **Propagative budding** generally occurs during favorable environmental conditions and serves to increase colony size and exploit available resources. On the other hand, **survival budding** tends to take place during the onset of adverse conditions and may be viewed as an overwintering or other survival device. In this case colony size is generally reduced by the resorption of zooids, leaving potential buds of presumptive germinative tissues. With the return of more favorable growing conditions, these "pre-buds" quickly develop as new blastozooids.

Another fascinating aspect of asexual reproduction in colonial ascidians is that some colonies fuse with one another when they grow large enough to make physical contact with one another. Both intraspecific and interspecific fusion are known to occur. This is a histocompatibility phenomenon that is based upon the genetic makeup of the colonies involved (see papers in the 1982 Lambert-chaired symposium volume).

Sexual reproduction and development.   Most tunicates are hermaphroditic, with relatively simple reproductive systems. Generally a single ovary and a single testis lie near the loop of the digestive tract in the posterior part of the body, and in most cases, connect through a separate sperm duct and oviduct to the cloacal chamber near the anus (Figures 23.15D,E). However, some species have a single gonad (ovitestis), with one gonoduct, and members of a few families (e.g., Pyuridae, Styelidae) have multiple

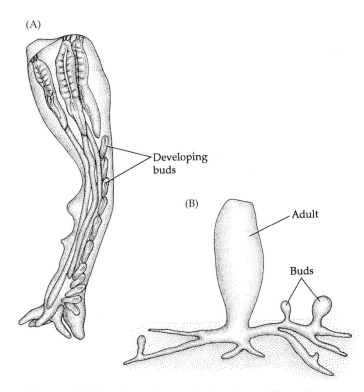

Figure 23.16 Asexual reproduction in ascidians. (A) Formation of buds in the colonial ascidian *Circinalium*. (B) General pattern of stoloniferous budding in an ascidian.

(A)

Developing buds

(B)

Adult

Buds

gonads. In some species, the ovaries contain high amounts of silica, but the significance of this condition is unknown (Monniot et al. 1992).

There is a great deal of variation in the overall reproductive strategies among tunicates. Most large solitary ascidians produce high numbers of weakly yolked ova, which are shed to the sea coincidentally with the release of sperm from other individuals. External fertilization is followed by the development of a free-swimming **tadpole larva**, which eventually settles and metamorphoses to an oozooid (Figure 23.17). In contrast to this fully indirect life-history pattern, many compound ascidians composed of tiny zooids produce relatively few eggs, but each egg has a high yolk content. These eggs are fertilized and subsequently brooded within the cloacal chamber; they are not released until the swimming tadpole larvae develop. Various degrees of larval suppression occur among ascidians with this mixed life history strategy, and at least a dozen species undergo fully direct development. One species, *Protostyela longicauda*, produces nonswimming larvae, which are brooded. In this species, the larval tail is simply an extension of the tunic and contains no cellular material. The larva is everywhere sticky, and when it is released quickly adheres to any object it contacts, later making a permanent attachment.

Although all thaliaceans lack a free-swimming larval stage, they differ markedly in their approaches to direct development. In pyrosomes each zygote develops directly to an oozooid, with no evidence of a larval stage. The oozooid then buds to produce a colony. Doliolids produce tailed larvae, but each is encased in a cuticular capsule and does not swim. The larva metamorphoses to an oozooid. Salps undergo internal fertilization in the oviduct. The zygotes implant and form a placenta-like association with the parent in a uterine chamber in the oviduct. Here the embryos develop directly to the adult form.

Appendicularians free-spawn, and fertilization occurs externally. They develop to a tadpole-like stage and then mature by protandry into the characteristic larva-like adults.

In most tunicate species studied, cleavage is radial, holoblastic, and slightly unequal; it leads to the formation of a coeloblastula, which undergoes gastrulation by invagination. The blastopore lies at the presumptive posterior end of the body but closes as development proceeds.

The development of the chordate features is most easily seen and understood in those species that form free tadpole larvae, such as most ascidians. As the embryo elongates, the gut proliferates three longitudinal strips of mesoderm—a middorsal strip that becomes the notochord and lateral strips that form the mesenchyme and body musculature. Thus, even though the mesoderm arises from the archenteron (entoderm), it does not pouch from the gut wall; in fact, no coelomic cavity is ever formed. A middorsal strip of ectoderm differentiates as a **neural plate**, which sinks inward and curls to produce the dorsal hollow nerve cord. The epidermis secretes a larval tunic, which often develops dorsal and ventral tail fins. The anterior part of the gut differentiates as the pharyngeal basket during larval life, and the rudiment of a cloacal water cavity forms by an ectodermal invagination producing the atrial siphon. However, these larvae are all lecithotrophic, and the gut and filtering devices do not become functional until metamorphosis.

Ascidian larvae are short-lived. When development is fully indirect, the larvae are planktonic for only about two days or less. In some forms with a mixed life history pattern (e.g., *Botryllus*), the free larval life lasts only a few minutes. Even though a short larval life allows dispersal over only small distances, the larvae are probably very important in the selection of suitable substrata. The events of settling and metamorphosis of ascidian larvae are complex and varied.

Ascidian larvae possess several sensory receptors that function in settling and probably substratum selection, but are absent from the adults. A small **sensory vesicle** lies near the anterior end of the dorsal nerve cord adjacent to the developing cerebral ganglion (Figure 23.17B). This vesicle houses a light-sensitive ocellus and a statocyst (called an **otolith**). At the time of settlement, the larva becomes negatively phototactic and positively geo-

Figure 23.17 Ascidian larvae and metamorphosis. (A) An ascidian tadpole larva has many chordate features. (B) The anterior end of the larva of *Distaplia occidentalis*. (C) Metamorphosis of a settled tadpole larva. Tail resorption is followed by a reorientation of the body to bring the siphons to the adult positions.

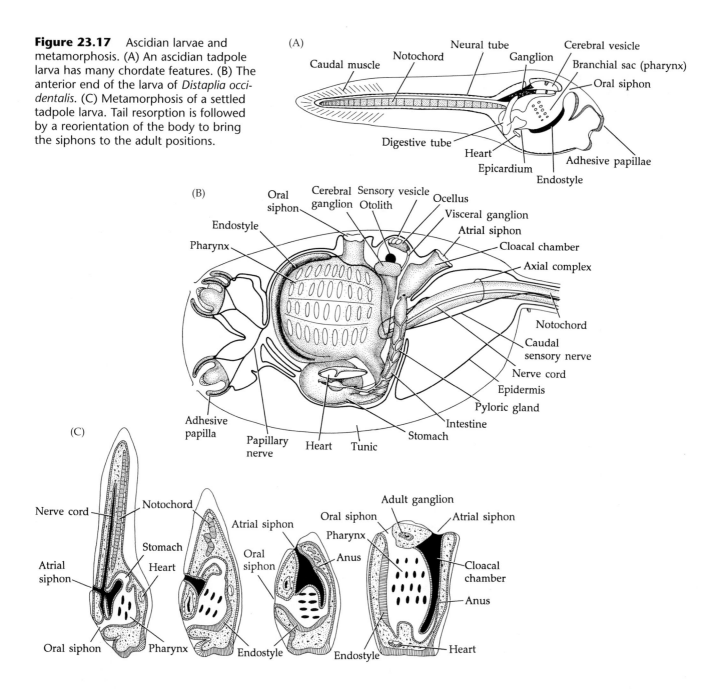

tactic. The anterior end of the larva bears two or three **adhesive papillae**, each of which is supplied with nerves. These phenomena are summarized briefly below, but the interested reader is referred to the careful work of R. Cloney (1990) for additional details.

The settling larva contacts a substratum with its anterior end and secretes an adhesive from the papillae. In the larvae of many compound ascidians, the papillae evert during this process. The secretion of the adhesive apparently triggers an irreversible sequence of metamorphic events. Within minutes after attachment, resorption of the larval tail commences by one of several methods involving various contractile elements in the tail region. The animal's viscera and siphons then undergo a remarkable 90° rotation that brings these organs to their adult positions. The outer layer of the cuticle is

shed, removing the larval fins from the settled juvenile. (Figure 23.17C). The pharynx enlarges and the filtering mechanisms become functional. During all of these processes secondary attachment organs, called **ampullae**, extend from the body and permanently affix the animal to the substratum. Finally, various transient larval organs are lost, such as most of the larval nervous system and sense organs.

The Cephalochordates

The subphylum Cephalochordata includes about two dozen species of small, fishlike creatures that rarely exceed 5 cm in length (Figures 23.1G and 23.18). They are commonly called lancelets or amphioxus, the later name frequently applied to *Branchiostoma lanceolatum*, a species familiar to general zoology students. Lancelets are cosmopolitan in shallow marine and brackish waters, where they lie burrowed in clean sands with only the head protruding above the sediment. They can and do swim, however, and locomotion is important to their dispersal and mating habits.

The Cephalochordate Bauplan

Cephalochordates are especially interesting animals whose bauplan demonstrates several qualities intermediate between those of the invertebrates and the vertebrates. As we discuss later in this chapter, lancelets may represent living descendants of the ancestors of the vertebrates.

Body wall, coelom, support, and locomotion. The body of cephalochordates is everywhere covered by an epidermis of simple columnar epithelium, underlain by a thin connective tissue dermis. The body wall muscles are distinctly vertebrate-like and occur as chevron-shaped blocks called **myotomes** arranged longitudinally along the dorsolateral aspects of the body (Figure 23.18). These muscle blocks are large and occupy much of the interior of the body, thereby reducing the coelom to relatively small spaces. The notochord persists in adults and provides the major structural support for the body.

The notochord also plays a major role in locomotion in lancelets. As a result of the action of segmental myotomes, the swimming action of cephalochordates is much like that of fishes, consisting basically of lateral body undulations that drive water posteriorly and provide a forward thrust. The propulsive action of these body movements is enhanced by a vertical **caudal fin**. Unlike the vertebral column and its articulating bones, however, the notochord is an elastic, flexible rod. It prevents the body from shortening when the muscles contract, causing lateral bending instead. Its elasticity tends to straighten the body, and thus assists the antagonistic action of paired myotomes. The notochord extends beyond the myotomes both anteriorly and posteriorly, providing support beyond those muscles and apparently aiding in holding the body rigid during burrowing.

Although the notochord of cephalochordates is homologous with the same structure in other chordates, including the vertebrates, it displays some unique and rather remarkable structural and functional characteristics associated with its persistence in the adult. It is *not* a homogeneous structure of predominantly cartilage-like matrix material. Rather, the notochord of lancelets is built of discoidal lamellae that are stacked like so many poker chips along its length and surrounded by a sheath of collagenous connective tissue. The lamellae are composed of muscle cells whose fibers are oriented transversely. Furthermore, a significant amount of extracellular fluid exists in spaces and channels around and between the lamellae within the collagenous sheath. These muscle cells are innervated by motor neurons from the dorsal nerve cord. Upon contraction, the hydrostatic pressure in the extracellular spaces increases, thereby resulting in increased stiffness of the whole notochord complex. It is suspected that this action may facilitate certain kinds of movement patterns, especially burrowing.

The dorsal and ventral fin-like structures are more appropriately called dorsal and ventral **storage organs**. They are not homologous to the fins of fishes, and their function appears to be housing a build-up of nutritional reserves for gamete formation (Holland and Holland 1990, 1991).

Feeding and digestion. Cephalochordates are ciliary–mucous suspension feeders, and they employ a food-gathering mechanism similar to that of tunicates. Water is driven into the mouth and pharynx and out through the **pharyngeal gill slits** into a surrounding **atrium**; it exits the body through a ventral **atriopore** (Figure 23.18A). Unlike the gill ventilation currents of aquatic vertebrates that are generated by muscular action, the feeding currents of lancelets are driven by pharyngeal cilia, a condition similar to that in tunicates. The gill slits are committed largely to feeding in cephalochordates and have little to do with gas exchange. There are up to 200 gill slits, separated from one another by **gill bars**, which are supported by cartilaginous rods.

The trapping of food from the inflowing water involves complex handling and sorting activities that actually occur before water enters the mouth. The mouth is housed within a depression called the **vestibule**, which is formed by an anterior extension of the body called the **oral hood** (Figure 23.18A,B). The oral hood is supported by the notochord and bears finger-like projections called **buccal cirri**. As water enters the vestibule, the cirri prevent sediments and other large particles from reaching the mouth. The mouth itself is a perforation in a membranous **velum**, which bears a set of **velar tentacles**; the tentacles provide a second screen, preventing large material from entering the mouth. The lateral walls of the vestibule bear complex ciliary bands that collectively constitute the **wheel organ**. The wheel organ appears as brown, thickened, folded epithelium in the roof and sides of the vestibule, visible through the skin of living animals. Cilia of the wheel organ drive food particles to the mouth and give the impression of a rotation, hence the name. Lying on the roof of the vestibule is a mucus-secreting structure called **Hatschek's pit**. Mucus from this pit flows over the wheel organ, and food particles trapped by the mucus are carried to the mouth along the ciliary tracts. This material is then incorporated into the general water current moving through the mouth and into the pharynx.

The ventral surface of the pharynx bears the **endostyle**, or **hypobranchial groove** (Figure 23.18C). As in

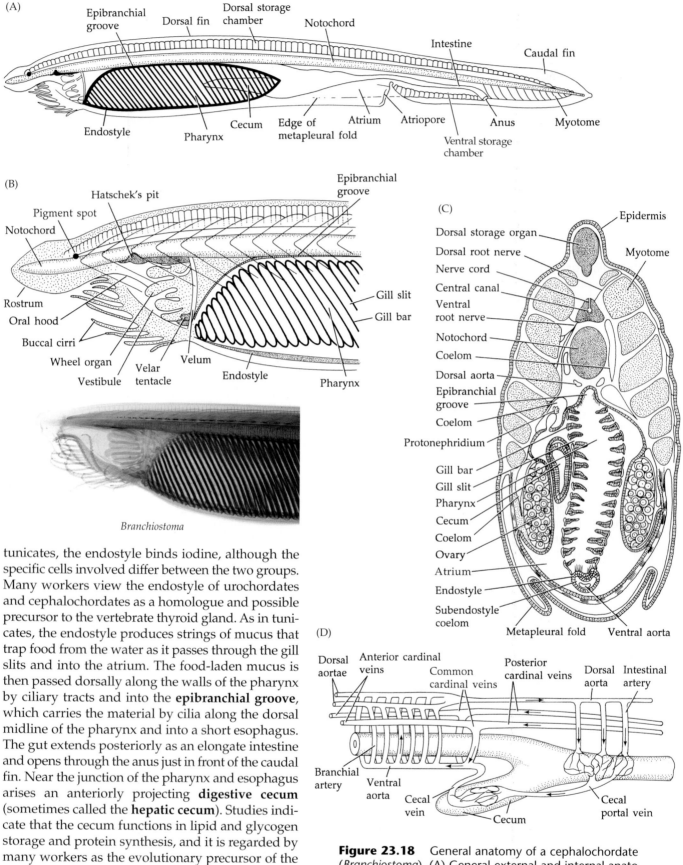

Branchiostoma

tunicates, the endostyle binds iodine, although the specific cells involved differ between the two groups. Many workers view the endostyle of urochordates and cephalochordates as a homologue and possible precursor to the vertebrate thyroid gland. As in tunicates, the endostyle produces strings of mucus that trap food from the water as it passes through the gill slits and into the atrium. The food-laden mucus is then passed dorsally along the walls of the pharynx by ciliary tracts and into the **epibranchial groove**, which carries the material by cilia along the dorsal midline of the pharynx and into a short esophagus. The gut extends posteriorly as an elongate intestine and opens through the anus just in front of the caudal fin. Near the junction of the pharynx and esophagus arises an anteriorly projecting **digestive cecum** (sometimes called the **hepatic cecum**). Studies indicate that the cecum functions in lipid and glycogen storage and protein synthesis, and it is regarded by many workers as the evolutionary precursor of the vertebrate liver and perhaps the pancreas. Digestion initially is extracellular in the gut lumen and is completed intracellularly in the walls of the intestine and

Figure 23.18 General anatomy of a cephalochordate (*Branchiostoma*). (A) General external and internal anatomy. (B) The anterior end. (C) The region of the pharynx (cross section). (D) The major blood vessels in the area of the gut cecum.

especially the cecum. In addition to storage in the cecum, food reserves accumulate in longitudinally arranged dorsal and ventral storage chambers along the dorsal midline and along the ventral body margin posterior to the atriopore (Figure 23.18A).

Circulation, gas exchange, and excretion.

The circulatory system of lancelets comprises a set of closed vessels through which blood flows in a pattern similar to that in primitive vertebrates (e.g., fishes). There is no heart. Blood flows posteriorly along the pharyngeal region in a pair of **dorsal aortae**. Just posterior to the pharynx, these vessels merge into a single **median dorsal aorta** that extends into the region of the caudal fin (Figure 23.18D). Blood is supplied to the myotomes and notochord via a series of short **segmental arteries** and to the intestine through **intestinal arteries**. A capillary network in the intestinal wall collects the nutrient-laden blood and leads to a series of **intestinal veins** that join a large subintestinal vein called the **cecal portal vein**, which carries blood forward beneath the gut to another capillary bed in the digestive cecum. As in vertebrates, the vein that connects two capillary beds is called a portal vein (e.g., the hepatic portal vein and renal portal veins in fishes). The cecal portal vein of cephalochordates is probably the homologue of the hepatic portal vein of vertebrates. In the digestive cecum, the nutrient and chemical composition of the blood is regulated somewhat before being distributed to the body tissues. (The vertebrate liver serves the same function via the hepatic portal system.)

Leaving the cecal capillaries is a **cecal vein**, which is joined by a pair of **common cardinal veins** formed by the union of paired **anterior** and **posterior cardinal veins** returning from the body tissues. These vessels merge to form the **ventral aorta** beneath the pharynx. From here blood is carried through the gill bars via **afferent** and **efferent branchial arteries** to the paired dorsal aortae, thus completing the circulatory cycle. Blood is moved through this system by peristaltic contractions of the major longitudinal vessels and by pulsating areas at the bases of the afferent branchial arteries.

The blood contains no pigments or cells and is thought to function largely in nutrient distribution rather than in gas exchange and transport. Although some diffusion of oxygen and carbon dioxide may occur across the gills, most of the gas exchange probably takes place across the walls of the **metapleural folds**, thin flaps off the body wall that lie just anterior to the atriopore (Figure 23.18C).

The excretory units in cephalochordates are protonephridia similar to the solenocytes of some other groups (e.g., primitive annelids). The numerous clusters of protonephridia accumulate nitrogenous wastes, which are carried by a nephridioduct to a pore in the atrium. Despite the structural similarities, the homology of lancelet protonephridia with those of other invertebrates is uncertain and some specialists regard this as a case of convergent evolution.

Nervous system and sense organs.

The central nervous system of cephalochordates is very simple. A dorsal nerve cord extends most of the length of the body and is generally expanded slightly as a **cerebral vesicle** in the base of the oral hood. Segmentally arranged nerves arise from the cord along the body in the typical vertebrate pattern of dorsal and ventral roots. The epidermis is rich in sensory nerve endings, most of which are probably tactile and important in burrowing. Some lancelets have a single simple eye spot near the anterior end of the dorsal nerve cord.

Reproduction and development.

Cephalochordates are dioecious, but the sexes are structurally very similar. Rows of 25 to 38 pairs of gonads are arranged serially along the body on each side of the atrium. The volume of gonadal tissue varies seasonally, and during the reproductive period it may occupy so much of the body as to interfere with feeding. Spawning typically occurs at dusk. The atrial wall ruptures, and eggs and sperm are released into the excurrent flow of water from the atrium; external fertilization follows.

The ova are isolecithal, with very little yolk. Cleavage is radial, holoblastic, and subequal, and leads to a coeloblastula that gastrulates by invagination (Figure 23.19). The roof of the archenteron eventually produces first a solid middorsal strip of mesoderm destined to become the notochord and then, sequentially, an anterior-to-posterior series of paired archenteric pouches along each side of the notochord. These enterocoelic pouches form the coelom and the other mesodermally derived structures such as the muscle bundles. The roof of the archenteron closes following mesoderm proliferation.

Dorsally the ectoderm differentiates into a **neural plate**. The neural plate eventually rolls inward, separating from the bordering cells, and then sinks inward as a **neural tube**, which forms the dorsal hollow nerve cord. As this process occurs at the posterior end of the embryo, the developing nervous tissue contacts the blastopore, which remains open temporarily and connects the archenteron to the lumen of the nerve cord as a **neuropore**. Later the two structures separate, and the blastopore opens to the exterior as the anus. The mouth breaks through as a secondarily produced opening at the front end of the developing gut.

In the absence of abundant yolk reserves, development to a free-swimming larva takes place rapidly. As soon as they are able, the larvae swim upward in the water column, where they remain planktonic for 75 to 200 days. The larvae are planktotrophic. They alternately swim upward and then passively sink with the body held horizontally and the mouth directed downward, feeding on plankton and other suspended matter. Development to the juvenile is generally gradual.

Phylogenetic Considerations

In the two preceding chapters we discussed the phylogeny of the lophophorates and the echinoderms with reference to a hypothetical burrowing ancestor. Here we build on that foundation to hypothesize the evolutionary history of the remaining deuterostomes, while summarizing current views of deuterostome phylogeny in general. Although there is strong evidence for the monophyly of the deuterostome clade, there is controversy about its origins and the relationships among the phyla and classes. We discuss below a set of hypotheses summarized in Figure 24.1 (Chapter 24) that we believe to be the most parsimonious.

The origin of the deuterostome line is problematic. Our view is that it was marked by the evolution of archenteric mesoderm and enterocoely, both of which are unique synapomorphies for members of this clade. Unlike the line leading to protostomes, where 4d mesoderm apparently arose *prior* to schizocoely, it appears that in deuterostomes, the mesoderm and the coelom arose together. We assume, then, that the immediate precursor was diploblastic but bilateral—perhaps a benthic creature derived from a radially symmetrical planuloid form. In this lineage, the blastopore was retained as the anus, marking the posterior end of the body, perhaps derived from the trailing, blastoporal end of a typical planula-like larva as seen in cnidarians. There is growing evidence that the origins of the protostomes and deuterostomes from some planuloid ancestor took very different pathways, including not only longitudinal axis modifications, as displayed by blastopore fate, but also differences in the dorsoventral orientation of the body. The implications of these events in the origins of the two great coelomate lineages of Metazoa are discussed in more detail in Chapter 24.

Both echinoderms and chaetognaths probably arose very early in the evolution of deuterostomes. The echinoderms capitalized on radial symmetry and on the various functional aspects of their ossicle-based skeleton and water vascular system. Most of the chaetognaths have exploited active, holoplanktonic life styles and retained bilateral symmetry. The placement of both of these groups along the deuterostome line has long been debated. Some researchers place the chaetognaths near or at the extreme base of the deuterostomes (Shinn, 1994), whereas some 18S rDNA analyses have suggested that the chaetognaths arose separately from deuterostomes. However, because arrow worms derive their mesoderm from the archenteron and undergo enterocoely, among other features noted earlier, our analysis retains them on the deuterostome clade.

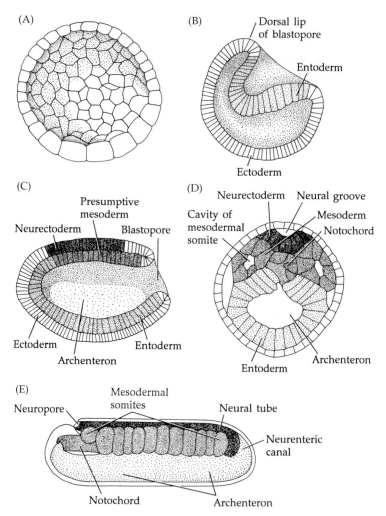

Figure 23.19 Development in a cephalochordate. (A) Coeloblastula. (B,C) Early and late gastrulae. (D) The neural groove stage (section). Note the proliferation of mesoderm as central notochord and lateral coelomic cavities. (E) Lateral view showing major structures and a temporary confluence of the gut and the neural tube (neurenteric canal).

Beyond the chaetognaths, which lack a larval stage, the rest of the deuterostomes display (at least primitively) a fundamental larval form that bears bands of cilia: as seen, for example, in the various larvae of echinoderms, the hemichordate tornaria, and the phoronid actinotroch. Sometimes the name **dipleurula larva** is given to the hypothetical ancestral deuterostome larva. In addition, all deuterostomes except chaetognaths show elaboration of the mesocoels—the water vascular system in echinoderms and later in association with mesosomal tentacles. The evolution of pharyngeal gill slits led on one hand to the hemichordates and on the other to the chordates (Figure 24.1).

Gill slits probably first evolved as a mechanism to facilitate water flow through the mouth and then out via the slits and pores, thereby enhancing gas exchange.

The gill slits function in this gas exchange mode in the hemichordates. The pterobranchs retain the mesosomal arms and tentacles as feeding devices, but the enteropneusts have lost these structures, no doubt in connection with their development of an infaunal lifestyle.

Most zoologists recognize the chordates as a distinct monophyletic clade, although some earlier 18S rDNA data suggested otherwise (Turbeville et al. 1994; Wada and Satoh 1994). More recently, 18S rDNA analyses have suggested that hemichordates might be the sister group of echinoderms (Cameron et al. 2000). Furthermore, there is some question as to whether the gill slits of ascidians are homologous with those of hemichordates and cephalochordates. There is also some question as to whether the hemichordate neurochord is homologous with the chordate nerve cord (see below). However, favoring parsimony and the absence of strong arguments to the contrary, we favor both homologies and support chordate monophyly, and our placement of the hemichordates closest to the chordate clade is based on those shared synapomorphies.

The most primitive chordates may have arisen with a shift to internal feeding through the use of the pharyngeal gill slits for extracting suspended food from the water. In addition, chordate origin was marked by the appearance of a pharyngeal endostyle (associated with the feeding mechanism), a notochord, a muscular locomotor postanal tail, and other features. These events were probably accompanied by the appearance of a tadpole-like larva wherein other chordate features are developmentally manifested. As mentioned earlier, the vertebrate thyroid gland is probably homologous with the endostyle of tunicates and cephalochordates (and larval lampreys).

The urochordates probably arose early from the chordate line. The coelom was lost as the main body cavity in conjunction with the formation of the large cloacal chamber through which water passes from the body after pharyngeal filtration. The earliest urochordates may have been ascidians, which, as a group, adopted a sessile lifestyle in association with suspension feeding and loss of adult locomotor musculature. Another line led to the thaliaceans, which exploited planktonic habits. The most primitive thaliaceans are thought to be the colonial pyrosomids (order Pyrosomida) because of their similarities to compound ascidians. They may have given rise to the doliolids (order Doliolida), and the doliolids to the salps (order Salpida)—both of which show increased zooid size and independence. The thaliaceans in general are characterized by the anterior and posterior placements of the oral and atrial siphons, respectively, a condition that provides them with propulsive powers in a pelagic environment.

Appendicularians might have arisen from a doliolid-like ancestor, although some earlier hypotheses supported an ascidian ancestry. In any case, there is little doubt that the appendicularians arose by neotenic evolutionary events in which sexual maturation occurred in an animal that retained larval characteristics, as evidenced not only by the tadpole-larval form of the adults, but also by the persistence of chordate features (e.g., the notochord) present only in the larvae of other tunicates. Walter Garstang was of the opinion that the notochord originally evolved to give support to pelagic larvae as they developed longer and longer planktonic lives, a trend eventually leading to the neotenic appendicularians. As these forms grew larger in size, Garstang reasoned, the notochord provided a basis for muscular support and locomotion, supplementing previous reliance on ciliary locomotion.

Most workers feel that the paedomorphic tendencies of the tadpole larvae of the ancestral chordate played a major role in the origin of the cephalochordates and vertebrates. The evolution of segmental muscle bundles (**myotomes**) marked the beginning of the cephalochordate–vertebrate clade and allowed greater locomotor facility than the tail thrashing movements of the tadpole-larval ancestor. The cephalochordates have retained the notochord and the use of pharyngeal gill slits for feeding. The vertebrates are, of course, characterized by the development of an endoskeleton with cranium, vertebral column, and, ultimately, limbs. All of these features eventually provided increased body support and much more effective skeletomuscular mechanics and allowed great increases in locomotor abilities and body size.

Two key developmental inventions probably drove evolution of vertebrates: multiple Hox gene clusters, and a new kind of embryonic cell, the **neural crest cell,** which led to the evolution of the vertebrate complex head by way of the embryonic **neural crest**. The earliest vertebrate fossils are armored jawless fishes and conodonts (thought to be teeth of early vertebrates), from the Cambrian (510–530 mya).

In the nineteenth century, E. Geoffroy Saint-Hilaire proposed an unorthodox view of chordate evolution. He suggested that the bodies of chordates, at least the higher chordates (cephalochordates and vertebrates) are dorso-ventrally inverted with respect to other bilaterians. This hypothesis lay fallow until the recent advent of molecular developmental research techniques, which have produced support for the idea from work on the expression of dorso–ventral patterning genes in insects and vertebrates. The idea finds new interpretation in the unconventional hypothesis that the vertebrate gene *Bmp-4* (a ventral determinant gene) may be the counterpart, or homologue of the insect gene *dpp* (a dorsal determinant gene)—*Chordin* in vertebrates is the dorsal determinant gene, while its apparent counterpart, *sog*, is a ventral one in insects. If this is correct, the implication is that the dorsal surface of vertebrates may be in some ways homologous to the ventral surface of insects (and perhaps all protostomes). Recent anatomical work by

Edward Ruppert and colleagues supports the idea. Ruppert et al. (1999) felt that accepting their proposed homology demanded rejection of homology between the enteropneust neurocord and the chordate neural tube (nerve cord), as implied by Saint-Hilaire's original theory. Clearly, the origin and phylogeny of deuterostomes remain fertile fields for future research.

Selected References

General References

Arendt, D and K. Nübler-Jung. 1994. Inversion of the dorsoventral axis? Nature 371: 26.

Barrington, E. J. W. 1965. *The Biology of Hemichordata and Protochordata*. W. H. Freeman, San Francisco.

Barrington, E. J. W. and R. P. S. Jefferies (eds.). 1975. *Protochordates*. Academic Press, New York. [A collection of papers generated from a symposium of the Zoological Society of London.]

Cameron, C. B., J. R. Garey and B. J. Swalla. 2000. Evolution of the chordate body plan: New insights from phylogenetic analyses of deuterostome phyla. Proc. Natl. Acad. Sci. 97(9): 4469–4474.

Cripps, A. P. 1991. A cladistic analysis of the cornutes (stem chordates). Zool. J. Linn. Soc. 102: 333–366.

Delage, Y. and E. Herouard. 1898. *Traité de Zoologie Concrete*. Vol. 8. Les procordes.

DeRobertis, E. M. and Y. Sasai. 1996. A common plan for dorsoventral patterning in bilateria. Science 380: 37–40.

Eaton, T. H. 1970. The stem–tail problem and the ancestry of the chordates. J. Paleontol. 44: 969–979.

Gans, C. and R. G. Northcutt. 1983. Neural crest and the origin of vertebrates: A new head. Science 220: 268–274.

Garstang, W. 1894. Preliminary note on a new theory of the phylogeny of the Chordata. Zool. Anz. 17: 122–125.

Geoffroy Saint-Hilaire, E. 1822. Considération générales sur la vertèbre. Mém. Mus. Hist. Nat. 9:89–119.

Hyman, L. H. 1959. *The Invertebrates*, Vol. 5. Smaller Coelomate Groups. McGraw-Hill, New York. [This volume includes the Chaetognatha and Hemichordata.]

Jefferies, R. P. S. 1986. *The Ancestry of the Vertebrates*. British Museum (Natural History), London. [A detailed account of Jefferies' hypothesis of the origin of vertebrates from a Paleozoic fossil echinoderm group; the book summarizes his previous studies and presents an excellent review of anatomy and embryology of living deuterostomes. See Peterson, K. J. 1995, Lethaia 28: 25–38 for a counter-point to Jefferies' hypothesis.]

Jefferies, R. P. S. 1990. The solute *Dendrocystoides scoticus* from the upper Ordovician of Scotland and the ancestry of chordates and echinoderms. Paleontol. 33(3): 631–679.

Lacalli, T. C. 1996. Dorsoventral axis inversion: A phylogenetic perspective. BioEssays 18: 251–254.

Lacalli, T. C. 1997. The nature and origin of deuterostomes: Some unresolved issues. Invert. Biol. 116(4): 363–370.

Nübler–Jung, K. and D. Arendt. 1994. Is ventral in insects dorsal in vertebrates? Roux Archiv. Devl. Biol. 203: 357–366.

Peterson, K. J. 1995a. A phylogenetic test of the calcichordate scenario. Lethaia 28: 25–38.

Peterson, K. J. 1995b. Dorsoventral axis inversion. Nature 373: 111–112.

Ruppert, E. E., C. B. Cameron and J. E. Frick. 1999. Endostyle-like features of the dorsal epibranchial ridge of an enteropneust and the hypothesis of dorsal–ventral axis inversion in chordates. Invert. Biol. 118:202–212.

Turbeville, J. M., J. R. Schultz and R. A. Raff. 1994. Deuterostome phylogeny and the sister group of the chordates: Evidence from molecules and morphology. Mol. Biol. Evol. 11: 648–655.

Wada, H. and N. Satoh. 1994. Details of the evolutionary history from invertebrates to vertebrates, as deduced from the sequences of 18S rDNA. Proc. Natl. Acad. Sci. U.S.A. 91: 1801–1804.

Chaetognatha

Alvariño, A. 1965. Chaetognaths. Annu. Rev. Oceanogr. Mar. Biol. 3: 115–194.

Alvariño, A. 1983. Chaetognatha. *In* K. G. and R. G. Adiyodi (eds.), *Reproductive Biology of Invertebrates*, Vol. 2. John Wiley and Sons. New York, pp. 531–544

Bieri, R. 1966. The function of the "wings" of *Pterosagitta draco* and the so-called tangoreceptors in other species of Chaetognatha. Publ. Seto Mar. Biol. Lab. 14: 23–26.

Bieri, R. and E. V. Thuesen. 1990. The strange worm *Bathybelos*. Am. Sci. 78: 542–549.

Bone, Q., H. Kapp and A. C. Pierrot-Bults (eds.). 1991. *The Biology of Chaetognatha*. Oxford Univ. Press, Oxford.

Burfield, S. 1927. *Sagitta*. Liverpool Mar. Biol. Comm., Mem. 28. *In* Proc. Trans. Liverpool Biol. Soc. 41: 1–104.

Casanova, J.-P. 1994. Three new rare *Heterokrohnia* species (Chaetognatha) from deep sea benthic samples in the northeast Atlantic. Proc. Biol. Soc. Wash. 107(4): 743–750.

Duvert, M. 1991. A very singular muscle: The secondary muscle of chaetognaths. Philos. Trans. R. Soc. Lond. B Biol. Sci. 332: 245–260.

Duvert, M. and C. Salat. 1990. Ultrastructural studies on the fins of chaetognaths. Tissue and Cell 22: 853–863.

Eakin, R. M. and J. A. Westfall. 1964. Fine structure of the eye of a chaetognath. J. Cell Biol. 21: 115–132.

Feigenbaum, D. L. 1978. Hair-fan patterns in the Chaetognatha. Can. J. Zool. 56: 536–546.

Feigenbaum, D. L. and R. C. Maris. 1984. Feeding in the Chaetognatha. Annu. Rev. Oceanogr. Mar. Biol. 22: 343–392.

Ghirardelli, E. 1968. Some aspects of the biology of the chaetognaths. Adv. Mar. Biol. 6: 271–375.

Goto, T. and M. Yoshida. 1985. The mating sequence of the benthic arrow worm *Spadella schizoptera*. Biol. Bull. 169: 328–333.

Goto, T., N. Takasu and M. Yoshida. 1984. A unique photoreceptive structure in the arrowworms *Sagitta crassa* and *Spadella schizoptera* (Chaetognatha). Cell Tissue Res. 235: 471–478.

Haddock, S. H. D. and J. F. Case. 1994. A bioluminescent chaetognath. Nature 367: 225–226.

Jordan, C. E. 1992. A model of rapid-start swimming at intermediate Reynolds number: Undulatory locomotion in the chaetognath *Sagitta elegans*. J. Exp. Biol. 163: 119–137.

Michel, H. B. 1982. Chaetognatha. *In* S. P. Parker (ed.), *Synopsis and Classification of Living Organisms*, Vol. 2. McGraw-Hill, New York, pp. 781–783.

Moreno, I. (ed.). 1993. *Proceedings of the II International Workshop of Chaetognatha*. Universitat de les Illes Belears, Palma.

Pierrot-Bults, A. C. and K. C. Chidgey. 1988. *Chaetognatha*. Synopses of the British Fauna (New Series), No. 39. E. J. Brill, New York.

Salvini-Plawen, L. von. 1986. Systematic notes on *Spadella* and on the Chaetognatha in general. Z. Zool. Syst. Evolutionsforsch. 24: 122–128.

Shinn, G. L. 1993. The existence of a hemal system in chaetognaths. *In* I. Moreno (ed.), *Proceedings of the II International Workshop of Chaetognatha*. Universitat de les Illes Belears, Palma.

Shinn, G. L. 1994. Epithelial origin of mesodermal structures in arrowworms (phylum Chaetognatha). Am. Zool. 34: 523–532.

Shinn, G. L. 1997. Chaetognatha. *In* F. W. Harrison and E. E. Ruppert, (eds.) *Microscopic Anatomy of Invertebrates*, Vol. 15. Wiley–Liss, New York, pp. 103–220.

Shinn, G. L. and M. E. Roberts. 1994. Ultrastructure of hatchling chaetognaths (*Ferosagittahispida*): Epithelial arrangement of the mesoderm and its phylogenotic implications. J. Morphol. 219: 143–163.

Telford, M. J. and P. W. H. Holland. 1993. The phylogenetic affinities of the Chaetognaths: A molecular analysis. Mol. Biol. Evol. 10: 660–676.

Terazaki, M. and C. B. Miller. 1982. Reproduction of meso- and bathypelagic chaetognaths in the genus *Eukrohnia*. Mar. Biol. 71: 193–196.

Terazaki, M., R. Marumo and Y. Fujita. 1977. Pigments of meso- and bathypelagic chaetognaths. Mar. Biol. 41: 119–125.

Thuesen, E. V. and R. Bieri. 1987. Tooth structure and buccal pores in the chaetognath *Flaccisagitta hexaptera* and their relation to the capture of fish larvae and copepods. Can. J. Zool. 65: 181–187.

Thuesen, E. V. and K. Kogure. 1989. Bacterial production of tetrodotoxin in four species of Chaetognatha. Biol. Bull. 176: 191–194.

Welsch, U. and V. Storch. 1982. Fine structure of the coelomic epithelium of *sagitta elegans*. Zoomorphol. 100: 217–222.

Hemichordata

Armstrong, W. G., P. N. Dilly and A. Urbanek. 1984. Collagen in the pterobranch coenecium and the problem of graptolite affinities. Lethaia 17: 145–152.

Balsser, E. J. and E. E. Ruppert. 1990. Structure, ultrastructure, and function of the preoral heart-kidney in *Saccoglossus kowalevskii* (Hemichordata, Enteropneusta) including new data on the stomatochord. Acta Zool. 71: 235–249.

Benito, J. 1982. Hemichordata. *In* S. P. Parker (ed.), *Synopsis and Classification of Living Organisms*, Vol. 2. McGraw-Hill, New York, pp. 819–821.

Benito, J. and F. Pardos. 1997. Hemichordata. *In* F. W. Harrison and E. E. Ruppert, (eds.). *Microscopic Anatomy of Invertebrates, Vol. 15.* Wiley–Liss, Inc. pp. 15–102 .

Burdon-Jones, C. 1952. Development and biology of the larva of *Saccoglossus horsti* (Enteropneusta). Philos. Trans. R. Soc. Lond. B Biol. Sci. 236: 553–590.

Burdon-Jones, C. 1956. Observations on the enteropneust, *Protoglossus kohleri*. Proc. Zool. Soc. Lond. 127: 35–58.

Dilly, P. N. 1985. The habitat and behavior of *Cephalodiscus gracilis* (Pterobranchia, Hemichordata) from Bermuda. J. Zool. 207: 223–239.

Gilmour, T. H. J. 1979. Feeding in pterobranch hemichordates and the evolution of gill slits. Can. J. Zool. 57: 1136–1142.

Gilmour, T. H. J. 1982. Feeding in tornaria larvae and the development of gill slits in enteropneust hemichordates. Can. J. Zool. 60: 3010–3020.

Hadfield, M. G. 1975. Hemichordata. *In* A. C. Geise and J. S. Pearse (eds.), *Reproduction of Marine Invertebrates*, Vol. 2. Academic Press, New York, pp. 185–240.

Hadfield, M. G. and R. E. Young. 1983. *Planctosphaera* (Hemichordata: Enteropneusta) in the Pacific Ocean. Mar. Biol. 73: 151–153.

Halanych, K. M. 1993. Suspension feeding by the lophophore-like apparatus of the pterobranch hemichordate *Rhabdopleura normani*. Biol. Bull. 185: 417–427.

Lester, S. M. 1985. *Cephalodiscus* Sp. (Hemichordata: Pterobranchia): Observations of functional morphology, behavior and occurrence in shallow water around Bermuda. Mar. Biol. 85: 263–268.

Romero-Wetzel, M. B. 1989. Branched burrow-systems of the enteropneust *Stereobalanus canadensis* (Spengel) in deep-sea sediments of the Vöring Plateau, Norwegian Sea. Sarsia 74: 85–89.

Spengel, J. 1932. *Planctosphaera pelagica*. Sci. Results Michael Sars North Atl. Deep Sea Exped. 5(5).

Stebbings, A. R. D. and P. N. Dilly. 1972. Some observations of living *Rhabdopleura compacta* (Hemichordata). J. Mar. Biol. Assoc. U. K. 52: 443–448.

Strathmann, R. and D. Bonar. 1976. Ciliary feeding of tornaria larvae of *Ptychodera flava*. Mar. Biol. 34: 317–324.

Welsch, U. P., N. Dilly and G. Rehkämper. 1987. Fine structure of the stomochord in *Cephalodiscus gracilis* M'Intosh 1882 (Hemichordata, Pterobranchia). Zool. Anz. 218: 3/4S: 209–218.

Woodwick, K. H. and T. Sensenbaugh. 1985. *Saxipendium coronatum*, new genus, new species (Hemichordata: Enteropneusta): The unusual spaghetti worms of the Galapagos rift hydrothermal vents. Proc. Biol. Soc. Wash. 98: 351 365.

Chordata

Alexander, R. M. 1975. *The Chordates*. Cambridge University Press, New York.

Alldredge, A. 1976a. Appendicularians. Sci. Am. 235: 94–102.

Alldredge, A. 1976b. Discarded appendicularian houses as sources of food, surface habitats, and particulate organic matter in planktonic environments. Limnol. Oceanogr. 21: 14–23.

Alldredge, A. 1977. Morphology and mechanisms of feeding in the Oikopleuridae (Tunicata, Appendicularians). J. Zool. 181: 175–188.

Azariah, J. 1982. Cephalochordata. *In* S. P. Parker (ed.), *Synopsis and Classification of Living Organisms*, Vol. 2. McGraw-Hill, New York, pp. 829–830.

Barham, E. 1979. Giant larvacean houses: Observations from deep submersibles. Science 205: 1129–1131.

Bates, W. R. 1994. Direct development in the ascidian *Molgula retortiformis* (Verrill, 1871). Biol. Bull. 188: 16–22.

Berrill, N. J. 1961. *Salpa*. Sci. Amer. 204: 150–160.

Berrill, N. J. 1975. Chordata: Tunicata. *In* A. C. Geise and J. S. Pearse (eds.), *Reproduction of Marine Invertebrates*, Vol. 2. Academic Press, New York, pp. 241–282.

Birkeland, C., L. Cheng and R. A. Lewis. 1981. Mobility of didemnid ascidean colonies. Bull. Mar. Sci. 31: 170–173.

Bone, Q. 1989. On the muscle fibres and locomotor activity of doliolids (Tunicata: Thaliacea). J. Mar. Biol. Assoc. U. K. 69: 587–607.

Burighel, P. and R. A. Cloney. 1997. Urochordata: Asiciacea. *In* F. W. Harrison and E. E. Ruppert, (eds.). *Microscopic Anatomy of Invertebrates.* Vol. 15. Wiley–Liss, Inc. pp. 221–348.

Carroll, R. L. 1988. *Vertebrate Paleontology and Evolution.* W. H. Freeman, New York.

Cloney, R. A. 1978. Ascidian metamorphosis review and analysis. *In* F. S. Chia and M. E. Rice (eds.), *Settlement and Metamorphosis of Marine Invertebrate Larvae.* Elsevier North-Holland, New York, pp. 225–282.

Cloney, R. A. 1990. Urochordata-Ascidiacea. *In* K. G. Adiyodi and R. G. Adiyodi (eds.), *Reproductive Biology of Invertebrates.* Oxford and IBH, New Delhi, pp. 361–451.

Cloney, R. A. and S. A. Torrence. 1982. Ascidian larvae: Structure and settlement. *In* J. D. Costlow (ed.), *Biodeterioration.* U. S. Naval Institute, Annapolis, Maryland.

Cox, G. 1983. Engulfment of *Prochloron* cells by cells of the ascidean, *Lissoclinum*. J. Mar. Biol. Assoc. U.K. 63: 195– 198.

Cripps, A. P. 1990. A new stem craniate from the Ordovician of Morocco, and the search for the sister group of the Craniata. Zool. J. Linn. Soc. 96: 49–85.

Deibel, D., M. L. Dickson, and C. Powell. 1985. Ultrastructure of the mucus feeding filters of the house of the appendicularians *Oikopleura vanhoeffeni*. Mar. Ecol. Prog. Ser. 27: 79–86.

Fenaux, R. 1985. Rhythm of secretion of Oikopleurids' houses. Bull. Mar. Sci. 37: 498–503.

Flood, P. R. 1991. Architecture of, and water circulation and flow rate in, the house of the planktonic tunicate *Oikopleura labradorensis*. Mar. Biol. 111: 95–111.

Goodbody, I. 1982. Tunicata. *In* S. P. Parker (ed.), *Synopsis and Classification of Living Organisms*, Vol. 2. McGraw-Hill, New York, pp. 823–829.

Hadfield, K. A., B. J. Swalla and W. R. Jeffery. 1995. Multiple origins of anural development in ascidians inferred from rDNA sequences. J. Mol. Evol. 40: 413–427.

Hamner, W. M., L. P. Madin, A. L. Alldredge, R. W. Gilmer and P. P. Hamner. 1975. Underwater observations of gelatinous zooplankton: Sampling problems, feeding biology, and behavior. Limnol. Oceanogr. 20: 904–917.

Hirakow, R. and N. Kajita. 1990. An electron microscopic study of the development of amphioxus, *Branchiostoma belcheri tsingtauense*: Cleavage. J. Morphol. 203: 331–344.

Hirakow, R. and N. Kajita. 1991. Electron microscopic study of the development of amphioxus, *Branchiostoma belcheri tsingtauense*: The gastrula. J. Morphol. 207: 37–52.

Hirakow, R. and N. Kajita. 1994. Electron microscopic study of the development of amphioxus, *Branchiostoma belcheri tsingtauense*: The neurula and larva. Acta Anat. Nippon 69: 1–13.

Hirose, E. T. Maruyama, L. Cheng and R. W. Lewin. 1996. Intracellular symbiosis of a photosynthetic prokaryote, *Prochloron* sp., in a colonial ascidian. Invert. Biol. 115(4): 343–348.

Holland, N. D. and L. Z. Holland. 1990. Fine structure of the mesothelia and extracellular materials in the coelomic fluid of the fin boxes and sclerocoels of a lancelet, *Branchiostoma floridae* (Cephalochordata = Acrania). Acta Zool. 71(4) 225–234.

Holland, N. D. and L. Z. Holland. 1991. The histochemistry and fine structure of the nutritional reserves in the fin rays of a lancelet, *Branchiostoma lanceolatum* (Cephalochordata = Acrania). Acta Zool. 72(4): 203–207.

Holland, P. W. H. and J. Garcia-Fernández. 1996. *Hox* genes and chordate evolution. Dev. Biol. 173: 382–396.

Holland, P. W. H., J. Garcia-Fernández, N. A. Williams and A. Sidow. 1994. Gene duplications and the origins of vertebrate development. Dev. Suppl. 124–133.

Hopcroft, R. R. and J. C. Roff. 1995. Zooplankton growth rates: Extraordinary production by the larvacean *Oikopleura dioica* in tropical waters. J. Plankton Res. 17(2): 205–220.

Jeffery, W. R. 1994. A model for ascidian development and developmental modifications during evolution. J. Mar. Biol. Assoc. U.K. 74: 35–48.

Jeffery, W. R. and B. J. Swalla. 1992. Evolution of alternate modes of development in ascidians. BioEssays 14: 219–226.

Jeffery, W. R. and B. J. Swalla. 1997. Tunicates. *In* S. F. Gilbert and A. M. Raunio (eds.), *Embryology: Constructing the Organism*. Sinauer Associates, Sunderland, MA, pp. 331–364.

Katz, M. J. 1983. Comparative anatomy of the tunicate tadpole, *Ciona intestinalis*. Biol. Bull. 164: 1–27.

Kott, P. 1982. Didemnid–algal symbioses: Host species in the western Pacific with notes on the symbiosis. Micronesica 18: 95–127.

Koyama, H. and H. Watanabe. 1986. Studies on the fusion reaction in two species of *Perophora* (Ascidiacea). Mar. Biol. 92: 267–275.

Kusakabe, T., K. W. Makabe and N. Satoh. 1992. Tunicate muscle actin genes: Structure and organization as a gene cluster. J. Mol. Biol. 227: 955–960.

Lacalli, T. C., N. D. Holland and J. E. West. 1994. Landmarks in the anterior central nervous system of amphioxus larvae. Philos. Trans. R. Soc. Lond. B Biol. Sci. 344: 165–185.

Lambert, C. C. and G. Lambert (eds.). 1982. The developmental biology of the ascidians. Am. Zool. 22: 751–849. [Results of a 1981 symposium at the annual meeting of the American Society of Zoologists; nine papers, plus introductory remarks by C. Lambert.]

Lambert, G., C. C. Lambert and J. R. Waaland. 1996. Algal symbionts in the tunics of six New Zealand ascidians (Chordata, Ascidiacea). Invert. Biol. 115(1): 67–78.

Ma, L., B. J. Swalla, J. Zhou, J. Chen, J. R. Bell, S. L. Dobias, R. Maxson and W. R. Jeffery. 1996. Expression of an Msx homeobox gene in ascidians: Insights into the archetypal chordate expression in pattern. Dev. Dynam. 205: 308–318.

Maisey, J. G. 1986. Heads and tails: A chordate phylogeny. Cladistics 2: 201–256.

Millar, R. H. 1971. The biology of ascidians. Adv. Mar. Biol. 9: 1–100.

Monniot, C. and F. Monniot. 1978. Recent work on the deep-sea tunicates. Annu. Rev. Oceanogr. Mar. Biol. 16: 181–228.

Monniot, F., R. Martoja, M. Truchet and F. Fröhlich. 1992. Opal in ascidians: A curious bioaccumulation in the ovary. Mar. Biol. 112: 283–292.

Mukai, H., H. Koyama, and H. Watanabe. 1983. Studies on the reproduction of three species of *Perophora* (Ascidiacea). Biol. Bull. 164: 251–266.

Nakatani, Y., H. Yasuo, N. Satoh and H. Nishida. 1996. Basic fibroblast growth factor induces notochord formation and the expression of *As-T*, a *Brachyury* homologue, during ascidian embryogenesis. Development 122: 2023–2031.

Nishida, H. 1994. Localization of determinants for formation of the anterior–posterior axis in eggs of the ascidian *Halocynthia roretzi*. Development 120: 3093–3104.

Parry, D. L. and P. Kott. 1988. Co-symbiosis in the Ascidiacea. Bull. Mar. Sci. 42: 149–153.

Romer, A. S. 1967. Major steps in vertebrate evolution. Science 158:1629–1637.

Ruppert, E. E. 1994. Evolutionary origin of the vertebrate nephron. Am. Zool. 34: 542–553.

Ruppert, E. E. 1997. Cephalochordata. *In* F. W. Harrison and E. E. Ruppert, (eds.). *Microscopic Anatomy of Invertebrates, Vol. 15*. Wiley–Liss, New York, pp. 349–504 .

Satoh, N. 1994. *Developmental Biology of Ascidians*. Cambridge Univ. Press, Cambridge.

Schmidt, G. H. 1982. Aggregation and fusion between conspecifics of a solitary ascidean. Biol. Bull. 162: 195–201.

Stoeker, D. 1980. Chemical defenses of ascidians against predators. Ecology 61: 1327–1334.

Stokes, M. D. and N. D. Holland. 1995. Ciliary hovering in larval lancelets (=Amphioxus). Biol. Bull. 188: 231–233.

Stokes, M. D. and N. D. Holland. 1998. The lancelet. Amer. Sci. 86: 552–560.

Swalla, B. J. 1993. Mechanisms of gastrulation and tail formation in ascidians. Microsc. Res. Tech. 26: 274–284.

Torrence, S. A. and R. A. Cloney. 1982. The nervous system of ascidian larvae: Primary sensory neurons in the tail. Zoomorphologie 99: 103–115.

Turon, X. and E. Vázquez. 1996. A non-swimming ascidian larva: *Protostyela longicauda* (Styelidae). Invert. Biol. 115(4): 331–342.

Whittaker, J. R. 1997. Cephalochordates, the Lancelets. *In* S. F. Gilbert and A. M. Raunio (eds.), *Embryology: Constructing the Organism*. Sinauer Associates, Sunderland, MA, pp. 365–381.

Young, C. M. and L. F. Braithwaite. 1980a. Larval behavior and post-settling morphology in the ascidian, *Chelyosoma productum* Stimpson. J. Exp. Mar. Biol. Ecol. 42: 157–169.

Young, C. M. and L. F. Braithwaite. 1980b. Orientation and current–induced flow in the stalked ascidian *Styela montereyensis*. Biol. Bull. 159: 428–440.

24 *Perspectives on Invertebrate Phylogeny*

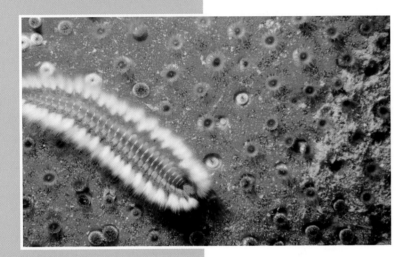

As a result of such speculations multitudes of phylogenetic trees sprang up in the thin soil of embryological fact and developed a capacity of branching and producing hypothetical ancestors that was in inverse proportion to their hold on solid ground.

E. G. Conklin, *Embryology and Evolution*, 1928

We shall not cease from exploration
And the end of all our exploring
Will be to arrive where we started
And know the place for the first time.
Through the unknown, remembered gate
When the last of earth left to discover
Is that which was the beginning

T. S. Eliot, *Little Gidding*, 1943

*I*deas about invertebrate phylogeny are often presented as though they were widely agreed-upon theories or, worse yet, as though alternative ideas did not even exist. Nothing could be further from the truth. Despite more than 150 years of morphological and developmental phylogenetic research, and a decade of molecular research, many of the deep relationships among the animal phyla remain very uncertain. Even the best of the morphological analyses (e.g., Nielsen et al. 1996; Sørensen et al. 2000) have achieved only partial resolution. Although analyses of the widely used 18S rDNA gene (also known as the small-subunit ribosomal RNA gene, or SSU rRNA) have been useful, major regions of the animal tree remain unresolved, and many unique 18S-based groupings are controversial and largely untested with other sources of data. Very few other genes have been used to investigate metazoan relationships on a broad scale, and analyses of deep-level metazoan phylogenetics using additional genes are greatly needed. For these reasons, of all the chapters in this book, this one has the shortest half-life. Our understanding of metazoan relationships is in such a state of flux that the ideas presented here may be out of date in a very short time.

In this chapter we undertake a phylogenetic analysis of the invertebrate phyla based on anatomical, morphological, and embryological data. We discuss the resulting tree, then discuss some interesting ideas that differ from ours. In doing so, we illuminate some of the problems that zoologists face in their efforts to understand the evolutionary history of the animal kingdom. Unraveling the phylogenetic history of the Metazoa has been one of biology's great challenges, and we are still far from resolution. Why should this task be so difficult? The answer to that question is the simple fact that almost all of the living animal phyla evolved long, long ago in the Precambrian and early Cambrian. The features that might be useful in revealing relationships among these ancient lineages have been obscured by hundreds of millions of years of evolutionary change.* This cloak of time has not been easy to penetrate, and neither comparative anatomy nor molecular biology has yet found the silver bullet.

Estimates of divergence times of the metazoan phyla, based on molecular clock calculations, suggest that most had their origin in the Precambrian. The origin of the arthropods has been estimated at 1 to 1.2 billion years ago. Fossil evidence puts the origin of the arthropods (and echinoderms) at about 700 million years ago. Trace fossils put the emergence of the bilaterians at more than a billion years ago—in synch with the beginning of the decline of the stromatolites, which might have been due to grazing by the newly evolved motile bilaterian animals.

We are tempted at this point to ask you to return to the introductory chapters and reread them to examine your perceptions of invertebrate zoology after several months of detailed study. If we have all done our jobs and you have consulted those chapters frequently, then you can now reduce the myriad details about invertebrates to some major phylogenetic trends in metazoan history. We have explored how animals are put together and how they work (the bauplan concept) and how reproductive and developmental patterns relate to adult structure and life histories. Now we can reflect on what we have learned, and try to pull it all together in the context of animal phylogeny.

You have already read about many previously espoused ideas concerning invertebrate phylogeny, and countless others exist in the literature—far too many to review here. Prior to 1990, analyses of invertebrate phylogeny were generally presented as narrative scenarios. Although such methods are perfectly legitimate, and we have used some of them in this book, they are often ambiguous and difficult to test in a rigorous scientific fashion. Almost any kind of scenario can be concocted to explain how one group of organisms might have arisen from another. Furthermore, such narratives used to be based on *a priori* assumptions about hypothetical ancestors. Despite atonements to such "workable ancestors," virtually any complicated evolutionary transition can be described on paper, given enough imagination.

So, rather than begin our discussion with a narrative of invertebrate evolution, we first present a cladogram, based on real taxa and real characters (Figure 24.1). We used the phylogenetic program PAUP (PAUP* 4.0b10; Swofford 2001) to infer the most parsimonious trees from our data set (Appendix B).[†] The Metazoa (the ingroup) were rooted in the protist phylum Choanoflagellata (the out-group). As we noted in Chapter 1, a great deal of evidence supports the notion that choanoflagellates are the sister group to the Metazoa. We have used developmental/embryological evidence, where applicable, to evaluate homologies among the characters we use. There are, of course, other interpretations of some of these characters, and many other possible trees exist. Our cladogram is only one of many possible sets of hypotheses, and like all hypotheses and ideas in biology, it is certainly open to a host of challenges.[‡]

You will notice that we have not included three "mesozoan" phyla (Monoblastozoa, Rhombozoa, Orthonectida) in our analysis, simply because too little is known about these odd groups to code them into our data set. In fact, there is doubt that the phylum Monoblastozoa (i.e., *Salinella*) even exists. In addition, we have coded the classes Enteropneusta and Pterobranchia (phylum Hemichordata), as well as the subphyla Urochordata, Cephalochordata, and Vertebrata (phylum Chordata), separately. We have done this because of ongoing questions regarding the monophyly of these two phyla.

The first published morphology-based analysis of metazoan phylogeny using strict cladistic methods was apparently in the first edition of this book. However, a number of phyla were omitted from that first analysis due to lack of data. Subsequent metazoan analyses have been built on increasingly more detailed, and larger, data sets (e.g., Nielsen et al. 1996; Zrzavy et al. 1998; Sørensen et al. 2000). However, each analysis has possessed unique attributes (e.g., character choices and character coding evaluations, splitting or combining of phyla) that influenced its outcome. Oftentimes, the addition or deletion of a single taxon or single character can significantly change a tree's topology. The choice of taxa and characters used to construct the tree in Figure 24.1 was similarly based on our own views. This bias, or subjectivity, that directs the choice of characters (and their interpretation) has a powerful influence on the outcome of the analysis—the final tree. The process of *a priori* character assessment is perhaps the weakest link in morphological phylogenetic biology—and one of the most persuasive arguments for using DNA sequence information as an alternative source of data. However, phylogenetics based on DNA sequence data also faces considerable challenges. With DNA sequence data, the alignment and choice of algorithms and other manipulations can greatly affect the final tree topography.

---

*Even the vertebrates are known to have had an ancient origin. The famous Chengjiang Cambrian deposit of southern China recently provided the oldest known vertebrate fossils, dated at 530 million years ago.

[†]Settings for the PAUP parsimony search were as follows: heuristic search of 96 characters and 34 taxa (plus out-group), 1000 random addition sequences. 17,855 equally parsimonious trees were found (length 139 steps).

[‡]The great California invertebrate zoologist Donald P. Abbott might have left us with some good advice: he was frequently heard to say, "cultivate a suspicious attitude towards people who do phylogeny."

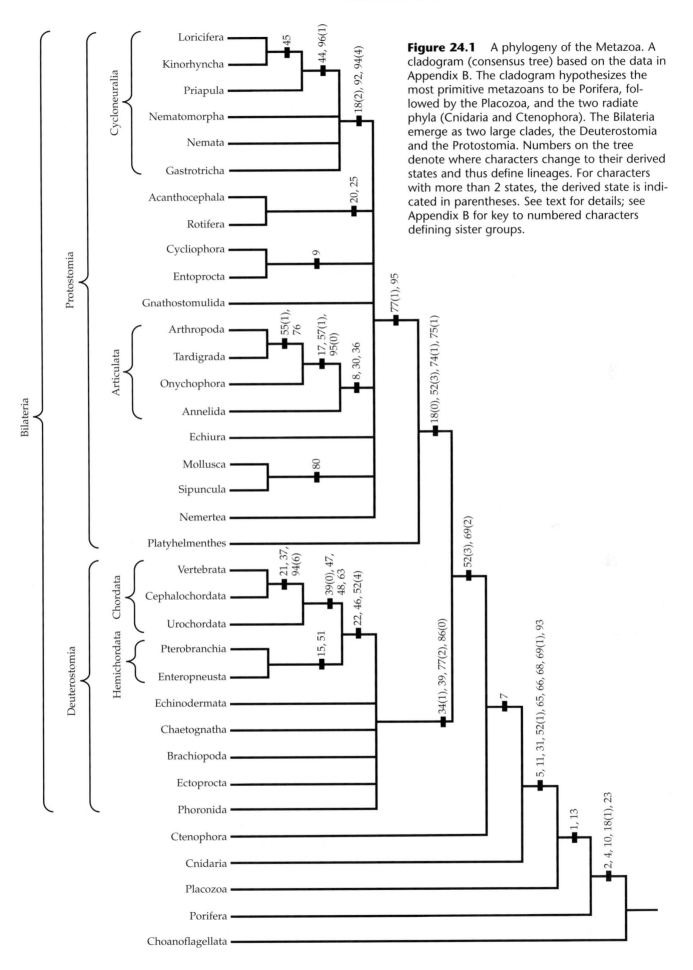

Figure 24.1 A phylogeny of the Metazoa. A cladogram (consensus tree) based on the data in Appendix B. The cladogram hypothesizes the most primitive metazoans to be Porifera, followed by the Placozoa, and the two radiate phyla (Cnidaria and Ctenophora). The Bilateria emerge as two large clades, the Deuterostomia and the Protostomia. Numbers on the tree denote where characters change to their derived states and thus define lineages. For characters with more than 2 states, the derived state is indicated in parentheses. See text for details; see Appendix B for key to numbered characters defining sister groups.

A Word about Characters

The taxa and characters used to construct the tree shown in Figure 24.1 are listed in Appendix B. A discussion of all of the characters used to construct this cladogram would fill many pages. Most are discussed in various chapters of the text and can be located using the index. We have avoided some characters used by other workers for inferring metazoan phylogenies because their character states are not well understood throughout the Metazoa, because their homologies are unresolved, or because they are clearly highly convergent. Of course, it is better not to use a character at all than to use it, but score it incorrectly. Two examples of convergent (nonhomologous) characters that are not included in our data matrix are the occurrence of "U-shaped guts" (e.g., sipunculans, lophophorates, tunicates) and the presence of "feeding tentacles around the mouth" (e.g., cnidarians, sipunculans, lophophorates, many protostomes). Features that show high levels of convergence within the Metazoa (when known) must be viewed with caution when assessing relationships among phyla. However, when we can convincingly establish, on embryological grounds, the independent origins of two or more convergent features, those characters can be included in a data matrix. For example, the two forms of coelom and mesoderm development can both be considered independently *because* their developmental origins are fundamentally different. The same approach can be applied to structures that derive embryologically from mesoderm in the two clades, again because they are developmentally independent of each other (e.g., muscle systems). Some recent morphological analyses have been indiscriminate in their choice of characters and have failed to recognize nonhomologous characters (e.g., not discriminating between schizocoelous and enterocoelous coeloms, or between 4d mesoderm and archenteric mesoderm).

Other troublesome anatomical features of metazoans are the excretory structures. Persisting uncertainties over homologies among protonephridia, metanephridia, and the excretory units of hemichordates and chordates led us to omit these structures from our analysis. Other characters we have avoided include many larval forms and their anatomy (we simply know too little about their homologies), the upstream/downstream ciliary feeding anatomy championed by Claus Nielsen (recent work suggests that our original views on this matter may have been too simplistic), and presence/absence of a cuticle (this feature is ambiguously defined, occurs in a great many metazoan phyla, and clearly is not homologous throughout the animal kingdom, as discussed below).

Metazoan Evolution

The Tree

The first thing you might notice about the cladogram presented here (Figure 24.1) is that it is not fully resolved—that is, there are two polytomies, or nondichotomous regions, in the tree, one at the base of the protostomes and another at the base of the deuterostomes. The data set we have compiled actually produces thousands of equally parsimonious trees. Our cladogram is a **strict consensus tree**—that is, a tree that shows only those lineages that are present in *all* of the shortest trees. Such a tree is the most conservative possible summary of the data. For example, most of the trees in our forest of equally short trees depict the three lophophorate phyla as a monophyletic clade. However, a few do not, because of character conflicts (e.g., differences in the anatomy of the cerebral ganglia, the fate of the blastopore, and the elaboration of the mesocoelomic tentacles).*

Even though our tree is not fully resolved, it still says a great deal about animal phylogeny, and it represents a conservative hypothesis of metazoan evolution. Interested readers can use our data set as a starting point for further explorations. We have not shown all of the characters on the tree, for the sake of space and because many simply define terminal branches (i.e., phyla or subphyla), but these can be retrieved using Appendix B. Below, we review the tree and note some of the key synapomorphies that define the sister groups or clades.

Metazoan Roots

We touched on the origin of the Metazoa in Chapters 1 and 4. To reiterate, a large body of evidence accumulated since the 1960s supports the view that multicellular animals (Metazoa) evolved from a protist ancestor. From which protist group did the Metazoa arise, and what was the nature of the first metazoan? As we noted in Chapter

---

*The reason the protostome and deuterostome lineages are each only partly unresolved is because the phylogenetic signal of the character set we have used is not strong enough to unambiguously tease apart the relationships of all the phyla. This outcome, of incomplete resolution, has plagued every morphological or molecular analysis of the Metazoa that has been attempted (as of this writing). Many workers, in the face of this uncertainty, subjectively pick one of the fully resolved trees from the forest of equally short trees that results from their analysis, often calling their chosen tree the "preferred tree," or some such thing. There are two other solutions to analyses that produce unresolved trees. The characters can be weighted (more emphasis given to some characters than to others), or a "majority rule" tree can be computed (i.e., a tree built from all those branches that appear in 50 percent or more of the equally short trees). The first option adds a layer of subjectivity to the analysis. The second has no biological basis—it suggests that phylogenetics is a popularity contest. Rather than selecting a "preferred tree" or manipulating the data, we chose to provide you with this strict consensus tree.

4, there are several competing hypotheses of metazoan origin, but the best evidence to date supports a colonial flagellate ancestry. The ancestral form may have existed as a clump of loosely bound cells or, more likely, a hollow ball of cells, something like a coeloblastula in basic structure, each cell with a single cilium/flagellum. Three compelling lines of reasoning support this contention. First, the possession of monociliated (= monoflagellated) cells appears to be primitive among the Metazoa; this condition occurs in poriferans, cnidarians, and the odd creature *Trichoplax* (as well as in some gastrotrichs, gnathostomulids, and generally in deuterostomes). Second, the flagellated collar cells of sponges and the flagellate protists known as choanoflagellates are essentially identical and unique to these two groups. Third, early stages in the origin of the metazoan condition must have included the formation of layered tissues, an event akin to the embryonic process of gastrulation. A blastula-like ancestor would have set the stage, so to speak, for such a gastrulation-like phenomenon. Once achieved, the selective advantages of a multilayered body plan may have led to a gradual early metazoan diversification in the late Precambrian, and then to a more rapid radiation of animals in the early Cambrian.*

Molecular data also support the hypothesis that metazoans arose from choanoflagellates (or from a common ancestor), a protist phylum that probably originated over a billion years ago. Molecular data, as well as recent paleontological studies, suggest that the Metazoa might have originated 1.0 to 1.2 billion years ago.

Our tree depicts a monophyletic origin for the Metazoa and hypothesizes the following metazoan synapomorphies: multicellularity, an embryonic process akin to radial cleavage, epithelial tissues with septate/tight junctions, the production of spermatozoa, and perhaps the appearance of animal collagen (there is some evidence that collagen, or an animal collagen homologue, occurs in some fungi).

Evolution within the Metazoa

The first metazoans. Sponges sit at the bottom of the tree and constitute the most primitive living phylum of animals. Morphological and molecular analyses agree on the basal position of the Porifera. Sponges have only a few cell types, retain a high degree of cellular totipotency, do not have the true tissues seen in higher Metazoa, and lack such features as a synaptic

nervous system and a basal lamina (basement membrane). They also have a suite of synapomorphies that include unique skeletal elements, an aquiferous system, and a distinctive embryogeny (the embryonic processes leading to a layered construction in poriferans differ from the gastrulation of all other metazoans). The essentially identical ultrastructure of sponge choanoflagellate cells and choanoflagellate protists has actually led some people to suggest that the phylum Choanoflagellata should be placed in the kingdom Metazoa. Fossil evidence of sponges dates to the Precambrian, 580 million years ago (Li et al. 1998).

All metazoans beyond the Porifera lack choanoflagellate cells and have striated ciliary rootlets. The first postporiferan metazoans in the tree are the platelike Placozoa, which have no clearly defined apomorphies, suggesting that a creature like them might have been the actual ancestor of the remaining Metazoa. After the Placozoa, many of the features we typically associate with animals make their appearance, including gap junctions between cells, organized gonads, germ layers (initially the ectoderm and entoderm), a synaptic nervous system (initially noncentralized), a basal lamina, striate myofibrils, and body symmetry (initially, radial symmetry). From this ancestry evolved the two living radiate phyla, Cnidaria and Ctenophora. Aside from radial symmetry, there seem to be no features that suggest uniting these very different phyla as sister groups. Unlike the Cnidaria, ctenophores have true subepidermal (mesenchymal) musculature and acetylcholine/cholinesterase nerve impulse transmission. The appearance of radial symmetry early in metazoan evolution suggests that a gastrulation-like event may have led to the origin of the Bilateria—perhaps something akin to the process of invagination seen in Cnidaria and Ctenophora, which results in a double-layered planula in the cnidarians (although a postembryonic hollowing of the gut remains a possibility).

The Bilateria and the origins of protostomes and deuterostomes. Some time after the evolution of the radiate phyla, bilateral symmetry—an anterior-posterior body axis—evolved, which led to the beginnings of cephalization as the nervous system began to concentrate in the head. Shortly thereafter came a great dichotomy in developmental modes, and the third germ layer, mesoderm, appeared in two separate lineages, the protostomes and deuterostomes. The common ancestor of these two great clades was probably a bilaterally symmetrical animal with a simple nervous system composed of an anterior concentration of ganglia and some arrangement of longitudinal nerve cords. This ancestor retained the symplesiomorphies

---

*As we noted in Chapter 1, the so-called "Cambrian explosion" may have been more artifact than reality. Increasing evidence suggests that many animal phyla may have had their origins in the Precambrian; the "explosion" in the Cambrian fossil record may have been due to the evolution (and fossilization) of mineralized skeletons during that time period.

of radial cleavage and an embryogeny that included gastrulation (probably by invagination).

It should be evident that the changes one would have to impose on such an ancestor to derive a deuterostome are quite different from those necessary to derive a protostome. Our cladogram hypothesizes that the body cavity we regard as a true coelom, or eucoelom (and also the middle germ layer, or entomesoderm) arose twice in the animal kingdom. Many authors continue to support the "monophyletic theory" that derives the protostomes and deuterostomes from a common triploblastic, coelomate ancestor. Such ideas imply that mesoderm and coelom are homologous throughout the triploblastic Metazoa. As we discussed in Chapter 4, we disagree with that model, and our tree suggests that the mesoderm and coelom of protostomes and of deuterostomes were independently derived.

Our cladogram suggests that the first animals in each of the two coelomate lineages differed greatly from each other in their embryogeny—each line was initiated by the acquisition of profound new developmental pathways. However, because of their common origin in a diploblastic ancestor, early representatives of both lines would have also retained some basic similarities (symplesiomorphies), such as a synaptic nervous system, cephalization, and gastrulation events. Furthermore, the development of mesoderm and a body coelom, albeit in different ways, would have led to evolutionary convergences in various other attributes. For example, the appearance of mesodermally derived circulatory systems and sheets of muscles must have been achieved independently in the protostomes and the deuterostomes because the mesoderm itself arose independently.

The synapomorphies distinguishing the ancestral protostome lineage include spiral cleavage, 4d mesoderm formation, a ventrally concentrated central nervous system, and a blastopore that becomes the mouth. The synapomorphies distinguishing the ancestral deuterostome lineage include archenteric mesoderm and enterocoelic coelom formation, a trimeric body plan, and a blastopore that becomes the anus (or closing of the blastopore, with the mouth and anus developing elsewhere) in all but the Phoronida. The primitive radial cleavage of the early metazoans (e.g., Cnidaria) is retained in the deuterostome lineage.*

Evolution among the protostomes. Attempts to unravel the relationships of the protostome phyla have been challenging, and they may continue to be so for some time to come. On our tree, the first phylum to emerge subsequent to the evolution of the protostome lineage is Platyhelminthes (the flatworms). The precise phylogenetic position of the flatworms has long been argued. It has never been questioned that the platyhelminths are closely allied with the other protostomes, with which they share several important developmental features (e.g., spiral cleavage and 4d entomesoderm), and in our analysis they appear as an undefined branch at the base of the protostome clade. The question is whether the acoelomate condition in flatworms is truly ancestral to the origin of schizocoely, as has traditionally been held and as our cladogram depicts, or whether it was secondarily derived from a coelomate protostome ancestor by way of reduction of the coelom. We briefly examined these two ideas in Chapter 10. Because the platyhelminths lack any unique apomorphies, they cannot be distinguished from the protostome ancestor itself—that is, the ancestor of the protostome line might have been a flatworm.

The schizocoelous adult body cavity and trochophore larva make their appearance subsequent to the platyhelminths. The character data suggest that the first protostomes were vermiform creatures, similar to modern flatworms and ribbon worms. The Nemertea, Sipuncula, and Echiura still bear a strong resemblance to this wormlike hypothetical ancestor. In their retention of these ancestral features, sipunculans and echiurans have remained as largely infaunal burrowers, using the large trunk coelom for peristalsis. Nemerteans have taken up an active, motile, predatory lifestyle, and the coelom in these animals is reduced to the rhynchocoel around the proboscis apparatus.

The other major protostome groups escaped from infaunal life, perhaps in part through the evolution of adaptations such as exoskeleton formation or tube building. The emergence of the molluscs is especially interesting. If we continue our reasoning from the assumption that the first protostome was a wormlike creature with a large body coelom, then it had a body architecture from which the primitive molluscan bau-

---

*There are several scenarios that might explain the present-day situation of the blastopore becoming the mouth in many protostomes (and in ctenophores and flatworms) and becoming the anus in most deuterostomes (or closing, with the mouth and anus developing elsewhere). Here is the scenario that we prefer. The planuloid ancestor that gave rise to these two metazoan lineages probably swam with its blastopore trailing, as is characteristic of cnidarian planulae. As in anthozoan planula larvae, the position of this opening may have served to create feeding eddies, drawing suspended food particles into the gastrovascular cavity. The protostome lineage was founded upon the newly invented bilateral body plan and anterior-posterior body axis. Such a body plan could have occurred through a reversal of the polarity of the plan-

uloid ancestor, leaving the blastopore (mouth) at the "new" anterior end. However, many turbellarian flatworms bear a midventral mouth (derived from the blastopore). Perhaps the blastopore migrated from its ancestral trailing position, moving first to a midventral, then to an anterior location as the definitive mouth. Platyhelminths lack an anus, but beyond the flatworms (e.g., in nemerteans and the other protostomes), an anus forms secondarily at the animal's posterior end. The origin of the deuterostome lineage probably did not involve any change in anterior/posterior orientation (as suggested by some workers), but the blastopore remained at the trailing end of the body as the anus, and a mouth formed separately at the anterior end.

plan can be easily derived.* Recall from Chapter 20 that our analysis of molluscan phylogeny depicts the aplacophorans as the most primitive living molluscs. Thus we can conclude that the first vermiform molluscs arose from an early wormlike protostome. Indeed, our tree depicts the sipunculans as the sister group of the molluscs. Amalie Scheltema proposed this idea in 1993, providing strong developmental data in support of a sister grouping of sipunculans and molluscs, including a unique embryological feature known as the molluscan cross.

Although the Echiura appear "unresolved" in our tree, there is growing evidence that they may be either greatly modified annelids or a sister group to the annelids. Annelids and echiurans share two unique features, the embryological annelid cross and epidermal chaetae (which are identical at the ultrastructural level in the two groups). In addition, analysis of the elongation factor 1-alpha (*Ef-1·*) gene has suggested that echiurans are annelids (McHugh 1997), and immunohistochemical analysis of developing echiurans provides tantalizing evidence that their nervous system may actually be segmented (Hessling and Westheide 2002).

The annelid–panarthropod clade (the "Articulata") is defined by at least three unique characters: teloblastic metamerism, segmental *engrailed* expression, and the presence of characteristic anterior mushroom bodies in the cerebral ganglia. Other proposed synapomorphies (not used in our data set) include segmental ganglia, an elongate dorsal tubular heart derived from a longitudinal blood vessel, and four to five bands of longitudinal muscles. The paired and segmental circumesophageal ganglia are probably also unique to this clade.

Since the welcome abandonment of the taxon "Aschelminthes," a great many hypotheses have been proposed regarding the origins and relationships of that puzzling constellation of metazoans.[†] Perhaps no other group of animal phyla is such a phylogenetic mystery. In Chapter 12, we touched on the evolutionary relationships of these enigmatic animals. The problems associated with evaluating the evolutionary histories of these groups stem from our lack of embryological data on the development of the acoelomate and blastocoelomate adult conditions seen in many of these phyla, our inability to identify homologues among many of the taxa, and a general lack of adequate data on many of the more obscure groups. A growing body of evidence suggests that the concept of a "blastocoelom" might encompass the end product (adult condition) of several different developmental pathways, and thus it may not be a homologous feature. Some of these groups lack a body cavity altogether and hence are actually at a functionally acoelomate grade of construction. In these groups the acoelomate condition is almost certainly a secondary condition associated with extreme reduction of body size—even some small annelids appear to be "acoelomate" in this regard. The larval stage has been lost altogether in many of these phyla.[‡] The following discussion reviews how these enigmatic phyla appear on our tree.

The Entoprocta and Cycliophora comprise a sister group defined by a single character, the unique mushroom-shaped extensions from the basal lamina into the epidermis. This clade was also recovered by Sørensen et al. (2000), who described several additional features shared between the two phyla. D'Hondt (1997) actually suggested that the Cycliophora were derived from the Entoprocta by neoteny. In contrast, an analysis of 18S rDNA data (Winnepenninckx et al. 1998) suggested a close relationship between Cycliophora and Rotifera + Acanthocephala, although there are fundamental anatomical differences between these groups.

The sister-group relationship of the rotifers and acanthocephalans has been suggested on the basis of anatomical features since the 1960s, and it has also been supported by DNA analyses. These taxa share two synapomorphies: a unique sperm anatomy in which the flagella insert anteriorly, and the presence of an intracellular skeletal lamina (and no cuticle).

A monophyletic clade of six phyla (Gastrotricha, Nemata, Nematomorpha, Priapula, Kinorhyncha, Loricifera) corresponds to a group that has been found in other morphological analyses (e.g., Nielsen et al. 1996; Sørensen et al. 2000). This clade, usually called the Cycloneuralia (or Nemathelminthes), is defined by a unique, belt-like brain, lacking distinct paired ganglia, that sits atop the pharynx and a radial mouth-pharynx anatomy (Figures 24.2 through 24.4). These phyla also share a cleavage pattern that is neither clearly spiral nor clearly radial. Within the Cycloneuralia, all but the gastrotrichs share the feature of cuticular shedding. It is not known whether this process is homologous to the molting process seen in the Panarthropoda. As in other animals with thick cuticles (e.g., the Panarthropoda, chaetognaths, sipunculans) and in many parasitic groups (e.g., acanthocephalans, parasitic platyhelminths), body cilia are lacking in the Cycloneuralia.

---

*The main body cavity of molluscs is not a coelom, and the classic protostome schizocoelic opening of 4d-derived mesodermal bands does not occur. The "true" coelom is limited to the pericardial cavity (which is bounded by an epithelium of mesodermal origin), and perhaps the lumen of the gonad and the nephridia. In lieu of typical schizocoely, the anterior cells of the mesodermal bands separate and reaggregate as paired masses of cells (sometimes called "mesenchymal cell masses"). These cell aggregates undergo schizocoely to become the right and left pericardial cavities, which later fuse to form a single pericardial cavity.

†The Aschelminthes (or Pseudocoelomata) included the phyla Nemata, Nematomorpha, Acanthocephala, Rotifera, Gastrotricha, Kinorhyncha, and Entoprocta.

‡Rotifera, Gastrotricha, Kinorhyncha, Nemata, and Gnathostomulida.

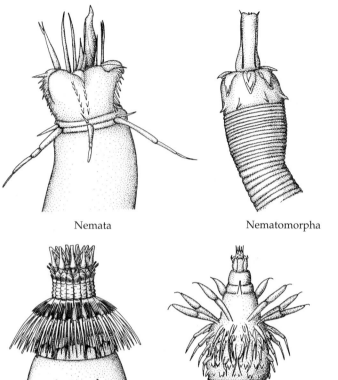

Nemata   Nematomorpha   Priapula

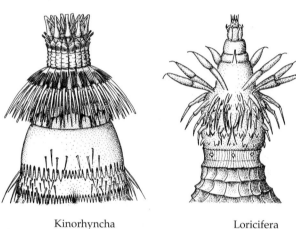

Kinorhyncha   Loricifera

Figure 24.2 Anterior regions of five phyla belonging to a group sometimes called the Introverta, and belonging to the larger group Cyclo-neuralia (which includes the Gastrotricha). The introverts are shown extended, showing the radially-arranged structures on the everted pharynx and surrounding the mouth. All five of these phyla are known to shed their cuticles.

Within this large clade, the Priapula-Kinorhyncha-Loricifera clade is defined by the presence of an introvert bearing spines, teeth, and scalids and the presence of chitin in the cuticle, rather than collagen (cuticular chitin also occurs in two other groups, the Echiura and the Panarthropoda). The kinorhynchs and loriciferans form a sister group based on the shared feature of a non-eversible mouth cone bearing cuticular ridges and spines.

Typical trochophore larvae are absent from the Cycloneuralia, Gnathostomulida, and Rotifera-Acanthocephala. The adult body cavity was lost in the gastrotrichs, gnathostomulids, cycliophorans, and entoprocts, all of which are functionally acoelomate. The primary body cavity (blastocoelom) was retained in the rotifers, acanthocephalans, nematodes, nematomorphans, kinorhynchs, and loriciferans—the blastocoelomate phyla. Both the secondary acoelomate condition and the blastocoelomate condition show homoplasy on the tree, suggesting that these adult anatomies may have evolved more than once within the protostomes.

Evolution among the deuterostomes.
As noted above, the first organisms recognizable as deuterostomes retained a primitive radial cleavage but evolved an enterocoelous, archenteric mesoderm and tripartite

coelom. Like the earliest protostome, the first deuterostome probably was a bilaterally symmetrical, vermiform burrower. The main trunk coelom (the metacoel) probably functioned in peristaltic burrowing, leaving the small protocoel and mesocoel "free" for modification during early deuterostome radiation. Each of the seven deuterostome phyla evolved in a very different way. In general, each group abandoned the ancestral burrowing lifestyle and assumed various strategies for other modes of existence.

The lophophorate phyla (Brachiopoda, Phoronida, Ectoprocta) may represent a monophyletic lineage, but conflicting characters prevented that clade from appearing on our consensus tree. In all three phyla the mesocoel is elaborated as a ciliated, tentacular feeding crown that surrounds the mouth (the lophophore). The phoronids retained the ancestral vermiform body, adopting a tube-dwelling benthic lifestyle; in this group the mouth develops near the blastopore, an apparent reversal of the primitive condition of the precoelomic bilateral ancestor. Ectoproct evolution has capitalized on small body size and colony formation, and the shell and mantle cavity of brachiopods has provided a protective housing for the nonretractile lophophore. Despite several analyses, both morphological and molecular, that align the lophophorate phyla with the protostomes, the most careful embryological research consistently demonstrates their archenteric mesoderm and trimeric enterocoelous coelom formation, thus allying them with the deuterostomes, as inferred by our analysis.

Neither the echinoderms nor the chaetognaths can be clearly allied as sister groups with any other phyla. Although their development is convincingly deuterostomous, these are two of the most aberrant phyla in the animal kingdom. The highly motile, pelagic, predatory lifestyle of chaetognaths has taken them down a unique (almost bizarre) anatomical evolutionary road. Phylogenetic controversy has followed the chaetognaths since the day they were discovered, and it continues to do so today. Fueling this controversy, some authors regard

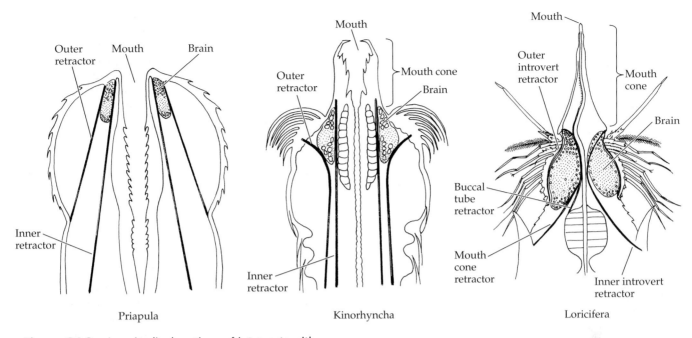

Figure 24.3 Longitudinal sections of introverts with brain and retractor muscles in the clade Priapula-Kinorhyncha-Loricifera (sometimes called the Cephalorhyncha).

chaetognaths as "pseudocoelomates" (despite their deuterostomous development), and 18S rDNA analyses sometimes align them with the protostomes. And, of course, the unique pentaradial symmetry, water vascular system, and stereom skeletal ossicle system of the echinoderms find no homologies among the other Metazoa.

The evolution of the hemichordate-chordate line involved the origin of pharyngeal gill slits (serving as suspension feeding and/or gas exchange devices), a dorsally concentrated central nervous system, and iodine-binding epithelial tissues. Despite these anatomical sim-

ilarities, several recent DNA analyses have proposed a sister-group relationship between the Hemichordata and Echinodermata, so the situation remains controversial. In our analysis, the enteropneusts and pterobranchs are sister groups, retaining the integrity of the phylum Hemichordata; this finding has also been supported by molecular data analyses. Morphological characters that support the Hemichordata include the preoral gut diverticulum (stomochord) and unique excretory organ

Figure 24.4 Brain structure in the Cycloneuralia (Nematomorpha are not shown). Although they have the typical protostome circumesophageal anatomy, cycloneuralian brains are belt-like and sit atop the pharynx as a saddle, and they lack the paired ganglia characteristic of arthropod and annelid brains.

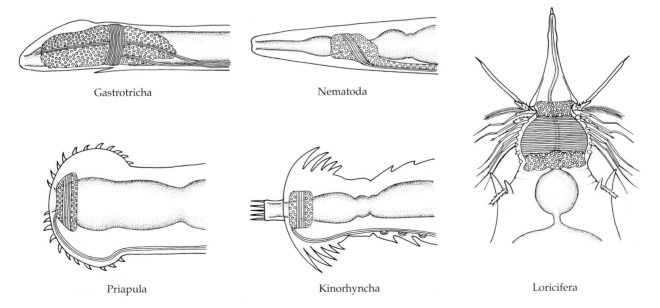

(glomerulus). The Chordata retain their traditional position in our tree, with the urochordates arising first, followed by the cephalochordate and vertebrate lineages. Synapomorphies of the Chordata include a notochord, loss of the trimeric body plan, the presence of an endostyle (= thyroid gland of vertebrates), and a muscular locomotory postanal tail.

Other Ideas about Animal Phylogeny

Molecular Phylogenetics

The 1990s saw an explosion of research in the field of molecular phylogenetics. This field infers phylogenies from molecular data, which primarily consist of long sequences of the four nucleotides that make up the information encoded in DNA. In theory, sequences retrieved from the genes of closely related species should differ only slightly, whereas sequences from more distantly related species should have accumulated more differences. However, if changes in nucleotide sequences are few, there may not be enough differences to resolve the branching pattern of a phylogenetic tree. Too many changes, on the other hand, overwhelm ("saturate") the data such that deep relationships are obscured. Since different genes evolve at different rates, the appropriate gene must be selected to recover patterns of sequence change over the appropriate time periods—short time periods (and fast-evolving genes) for analysis of species relationships; longer time periods (and slower-evolving genes) for relationships among higher taxa. Conservative (very slowly evolving) genes are needed to infer relationships at the deepest levels (e.g., relationships among phyla and kingdoms).

The deep relationships in the tree of life have proved difficult to reconstruct with both morphological and molecular data. There is some evidence that many of the phyla made their appearance within a relatively short time period (e.g., in the late Precambrian, or in the "Cambrian explosion"). Hence, the slowly evolving genes needed to probe such ancient events had little time to change between the origins of the various phyla, leaving us little trace of animal evolution. It is more likely, however, that the problem stems from the fact that these ancient lineages evolved so long ago, increasing the chances of nucleotide substitutions along the branches and obscuring informative patterns. In addition, gene duplication and horizontal transfer events can occur, different parts of some genes can evolve at different rates, and the same gene can evolve at different rates in different species, or at different time periods. Thus, the phylogeny of any specific gene does not necessarily mirror the phylogeny of the species in which it exists.

The field of molecular phylogenetics is still in an emergent phase, and relatively few genes have been sequenced for more than a handful of species. For this reason, the field is not without controversy. Controversy also finds ground in the methods used to analyze gene sequences. Scores of different programs have been developed to examine molecular sequence data, and within each program there are many alternative analytical procedures. One of the most fundamental debates is between the strict parsimony advocates and those who believe that evolutionary models should be taken into account (see Chapter 2). Despite these challenges, the emerging field of molecular phylogenetics has opened vast new opportunities for resolving long-standing evolutionary questions and for testing morphology-based phylogenies with new data. There is no doubt that this field will contribute enormously to our understanding of metazoan evolution over the coming decade.

The 18S rDNA gene.   Because the field of molecular phylogenetics has its roots in analyses of the 18S rDNA gene (Field et al. 1988), and because this gene has some properties that make it a good candidate for analysis of deep phylogenetic origins, there are far more sequences for this gene than for any other—18S rDNA sequences have been at least partly determined for thousands of animal species. Thus, most deep phylogenic research has relied on this gene. The results of this work, however, have been inconsistent, and the reliability of this gene as a phylogenetic marker for deep divergences has been questioned (see Selected References). It may be that the 18S rDNA gene does not have enough informative positions for a robust reconstruction of deep metazoan phylogeny. For these reasons, some researchers have suggested that nuclear protein-coding genes may be more suitable for reconstructing metazoan phylogeny. However, very few of these genes have been sequenced, in very few taxa. Also, most protein-coding genes seem not to include enough information to infer a fully resolved phylogeny of the Metazoa by themselves, and current thinking is that a larger data base, of several or many genes, will be necessary to resolve metazoan phylogeny at the molecular level. Although some workers espouse increased taxon sampling (i.e., obtaining gene sequences from more taxa) to improve the resolution of metazoan phylogeny, others believe that increased sequence sampling (e.g., studying more genes) is more important in this regard.

Some of the hypotheses derived from molecular phylogenetics have been controversial. The most unconventional ideas have resulted from analysis of 18S rDNA sequences, and some of these are described below.

BASAL GROUPS.   Some 18S rDNA studies place the diploblastic phyla (Porifera, Cnidaria, Ctenophora, and Placozoa) in a monophyletic clade, apart from the rest of the animal kingdom (e.g., Zrzavy et al. 1998). Other 18S rDNA analyses ally the placozoans with the cnidarians, and still others split the Porifera into two separate clades, allying the calcareous sponges with the ctenophores or with the higher Metazoa (e.g., Borchiellini et al. 2001). 18S rDNA data (and also the myasin II gene) have also suggested that the Platyhelminthes are polyphyletic, and that two orders, Acoela and Nemertodermatidoa, do not belong in this phylum, but instead are a basal sister group to all the other extant triploblastic animals. However, analyses of sequence data from other genes (e.g., *Ef-1·*) and from comparative anatomy do not support this hypothesis and argue for retaining the acoels within the Platyhelminthes. Thus the matter remained unresolved as this text went to press.

LOPHOTROCHOZOA.   Most analyses of 18S rDNA gene sequences support the placement of the lophophorates (Ectoprocta, Brachiopoda, Phoronida) in a clade with certain protostome phyla (Nemertea, Sipuncula, Mollusca, Echiura, Annelida). Some evidence from Hox genes and mitochrondrial gene order also supports this grouping. This clade has been called "Lophotrochozoa" (even though not all members have a lophophore, nor do all have trochophore larvae). Some recent 18S rDNA analyses also place the Platyhelminthes, Nemertea, Entoprocta, and Cycliophora in this clade. Analyses based on other genes, on morphology, and on development argue both for and against the Lophotrochozoa hypothesis (see Halanych and Passamaneck 2001 for a synopsis of this hypothesis).

ECDYSOZOA.   In Chapters 13 and 15 we mentioned a hypothesis of metazoan relationships based on 18S rDNA data (Aguinaldo et al. 1997 and subsequent workers). This hypothesis proposes a clade of "molting" taxa that includes the Panarthropoda (Arthropoda, Tardigrada, Onychophora), Nemata, Nematomorpha, Kinorhyncha, and Priapula. The name Ecdysozoa was given to this grouping because all members were known to shed their cuticles. There is also some preliminary support for Ecdysozoa from analyses of the 28S rDNA gene (Mallatt and Winchell 2002), and some qualitative support from patterns seen in Hox genes and mitochondrial gene arrangements, and from a particular β-thymosin sequence signature in arthropods and nematodes (reviewed in Halanych and Passamaneck 2001; critiqued in Blair et al. 2002). However, the phylogenetic patterns of most of these features (other than 18S rDNA sequences) throughout the Metazoa are not yet known. Some 18S rDNA studies move the Chaetognatha into the Ecdysozoa as well (even though chaetognaths do not shed their cuticle). Other 18S rDNA studies have failed to recover the ecdysozoan clade at all, and all the others have done so only by using particular analytical procedures. Studies based on alternative genes have yielded mixed results.

The millennium brought about an unprecedented international research effort to obtain the full genome sequences of three species, the nematode *Caenorhabditis elegans*, the arthropod *Drosophila melanogaster* (a fruit fly), and the chordate *Homo sapiens*. The availability of these complete genomic data sets has led to some large-scale attempts to examine animal phylogeny using similar complete nuclear genomic data from Fungi (yeast) and plants (mustard: *Arabidopsis*) (e.g., Mushegian et al. 1998; Blair et al. 2002). These large-scale analyses have failed to find support for the Ecdysozoa hypothesis. Blair et al. (2002) argued that none of the 18S rDNA analyses to date has yielded statistically significant results in support of Ecdysozoa. However, while those studies utilized a large DNA sampling, they relied on a small taxonomic sample.

The Ecdysozoa hypothesis has been compelling, and it stimulated a great deal of new and innovative research that generated both strong support and strong criticism. Clearly it is one of the more interesting ideas being tested today. Of course, allying the arthropods with these other, very different phyla, rather than with the annelids, demands that the complex process of teloblastic metamerism be convergently evolved in the Panarthropoda and Annelida. Aside from DNA data, the Ecdysozoa are said to be distinguished by the presence of a cuticle, loss of epidermal cilia, and molting. Animals with cuticles, of course, are found throughout the animal kingdom, since a cuticle is defined as any outer extracellular casing or matrix secreted by the epidermis. There is substantial variation in the structure, microscopic anatomy, and chemistry of cuticles, which run the gamut from the chitin-based cuticle of arthropods through the collagen-based cuticle of nematodes to gelatinous and other types. Furthermore, as we noted above, body ciliation is reduced or absent in many animal phyla, notably those with thick cuticles (e.g., the Panarthropoda, chaetognaths, sipunculans) and those that are parasitic (e.g., acanthocephalans, parasitic platyhelminths), as well as in certain anatomical settings, such as the dorsal body region of molluscs.

There is no evidence that shedding of the cuticle in Nematomorpha, Kinorhyncha, or Priapula is ecdysone-based and hence homologous to ecdysis in the Arthropoda, although it might well be. Ecdysteroid compounds have been identified in nematodes and implicated in cuticular shedding in that phylum. However, the ecdysteroids are a large and complex group of ancient compounds that occur throughout the Eukaryota, and hundreds of different chemical forms are known (Lafont and Wilson 1996, Sadikov, et al. 2001, Lafont 1997). They occur in such disparate groups as plants, Fungi, and several metazoan phyla, including nemerteans and cnidarians (which do not molt at all) and leeches (in which they are thought to be involved in cuticular shedding).* We do not know much about the roles of ecdysteroids in non-arthropod organisms, and it may be that the complicated nature of the molting process in the Panarthropoda is very different from cuticular shedding in other phyla. In arthropods, molting is regulated by numerous peptide and steroid hormones that initiate complex behavioral and biochemical processes, and it probably involves the coordinated expression of hundreds of genes. Finally, cuticular shedding occurs in non-ecdysozoan phyla—some leeches (Annelida) and sipunculans shed their cuticles. Thus, we are left with no known anatomical or developmental synapomorphies for the Ecdysozoa.

Other genes.   The nuclear protein-coding gene *EF-1·* is involved in the transport of tRNAs to the ribosome during translation, and it is highly conserved in its amino acid sequence across a wide taxonomic range. (See Regier and Schultz 1997 for a good example of the power of this gene.) Caution is needed in interpreting sequence data, however, because multiple copies of the *EF-1·* gene have been reported from many species, from insects to crustaceans to mammals. Another elongation factor gene, *EF-2*, has been suggested as a potential source of phylogenetic information for deep-level analyses of metazoans, but has not yet been put to the test. Several other nuclear genes may also provide characters for deep-level phylogenetic analysis, such as the gene encoding the protein ubiquitin, which occurs as continuous tandem repeats that evolve in concert. The ubiquitin gene is one of the most highly conserved genes known; however, is so highly conserved that it actually may be of only limited use for within-metazoan analyses.

Recently, biologists have begun building data sets composed of multiple nuclear gene sequences. Although this approach is still new, it shows great promise for unraveling the history of animal phyla. Hausdorf (2000), for example, used concatenated amino acid sequences of several nuclear genes in a preliminary analysis of several animal phyla. His results did not support the Ecdysozoa or Lophotrochozoa hypotheses.

Gene order.   The linear order, or arrangement, of genes on mitochondrial DNA (see, for example, the work of Boore and his colleagues) and gene duplication events (e.g., Holland and Garcia-Fernandez 1996) may also prove helpful in resolving higher-level phylogenetic relationships. These are new areas of research that are just beginning to be explored.

---

*Ecdysteroids comprise a diverse class of steroid hormones with a broad occurrence in the living world. In arthropods various ecdysteroids control not only molting, but also other biochemical and physiological processes. Various forms of the ecdysteroid molecule have been found in many invertebrate phyla (Koolman 1989; Dinan 2001), and analogues (phytoecdysteroids) with over 200 different molecular structures have been identified in over 100 different plant families. Plants may use ecdysteroids as a defense against insect predators.

Evolutionary Developmental Biology

Since the late 1980s, researchers have made extraordinary progress in identifying and characterizing developmental genes and their patterns of expression during the development of animal body plans. By comparing the developmental programs of different species, researchers are now studying the very genes that may have mutated as key speciation events took place and new body plans appeared. This newly emerging field is called **evolutionary developmental biology**. It is probable that in the twenty-first century, developmental biology will have the same impact on evolutionary biology that comparative anatomy had in the nineteenth century and phylogenetic theory had in the twentieth.

Similar genes turn up in very different organisms, where they are sometimes used for exactly the same ends, but in other cases adapted for different purposes, or achieve similar ends by very different downstream developmental pathways. For example, the gene *engrailed* is known to occur in both arthropods and amphioxus (chordates). In both cases, *engrailed* denotes the specific segments of the developing embryo, demarcating the first eight pairs of muscles in amphioxus and forming the basis of the body segments in arthropods. Hence, we can hypothesize that the genetic *potential* for body segmentation arose long ago, in a common ancestor of the protostomes and the deuterostomes, even though the process of "segmentation" is, embryologically speaking, very different in arthropods and chordates. Other developmental genes have been co-opted for quite different purposes in different lineages. For example, a Hox gene called *AbdB* helps define the posterior end of the embryo in insects, while a similar family of genes helps partition the developing wing into three segments in birds. The amazingly versatile gene *distal-less* causes cells to bud off from a main body axis to initiate new body outgrowths of all kinds. Whereas this gene functions to initiate appendages in arthropods, it also initiates such disparate (and nonhomologous) structures as tube feet in sea stars, parapodia in polychaetes, siphons in tunicates, and limbs in vertebrates.

One of the most intriguing new views of animal evolution comes from observations of orthologous dorsoventral patterning genes, which suggest that the ventral nerve cord (plus brain) of insects may be homologous to the dorsal spinal cord (plus brain) of chordates (see papers by Nübler-Jung, Arendt, and Nielsen in the Selected References). This "inversion hypothesis" suggests the possibility that chordates might be dorsoventrally inverted with respect to all other animals. The vertebrate gene *Bmp4*, a ventral determinant in the embryo, has been identified as a counterpart and probable homologue of the insect gene *dpp*, a dorsal determinant. And *chordin* in vertebrates is a dorsal determinant, while its apparent insect counterpart, *sog*, is a ventral one. Thus the dorsal surface of vertebrates—and, coincidentally, the central nervous system that forms from it—and the ventral surface of insects (and by implication that of all protostomes, if not all other animals) may be homologous tissue domains. A *sog/chordin* precursor gene may have determined the dorsoventral axis in some ancient wormlike precursor to the protostome-deuterostome split. This being the case, changes in the early development of chordates might have somehow flipped the axis, inverting the expression pattern of the genes.

By the next edition of this text, it seems probable that the fields of evolutionary developmental biology and molecular phylogenetics will have matured enough to resolve some long-standing issues that remain unclear as we write this second edition. Students interested in the history of life are encouraged to critically evaluate our cladogram and the others they encounter. Welcome new ideas, but scrutinize them carefully. As specialists provide more information on their favorite creatures, and as the application of new methods of analysis continues to refine our views about life on Earth, the branches of the tree of life will be cut and grafted as the evidence dictates. If we have stimulated you to think about such things and, more important, to appreciate some of the great mysteries of invertebrate evolution, then our task has been worthwhile.

Selected References

Thousands of papers and books on matters of invertebrate phylogeny have been published. Many of these references are listed in Chapters 1–4 and in chapters pertaining to specific phyla. Here we present a list of some important works that deal with higher-level relationships among the Metazoa. For references on phylogenetic methods, see Chapter 2.

Abouheif, E., R. Zardoya and A. Meyer. 1998. Limitations of metazoan 18S rRNA sequence data: Implications for reconstructing a phylogeny of the animal kingdom and inferring the reality of the Cambrian explosion. J. Mol. Evol. 47: 394–405.

Aguinaldo, A. M. A., J. M. Turbeville, L. S. Linford, M. C. Rivera, J. R. Garey, R. A. Raff and J. A. Lake. 1997. Evidence for a clade of nematodes, arthropods and other moulting animals. Nature 387: 489–494.

Ahlrichs, W. H. 1997. Epidermal ultrastructure of *Seison nebaliae* and *Seison annulatus*, and a comparison of epidermal structures within Gnathifera. Zoomorphology 117: 41–48.

Aleshin, V. V., I. A. Milyutina, O. S. Kedrova, N. S. Vladychenskaya and N. B. Petrov. 1998. Phylogeny of Nematoda and Cephalorhyncha from 18S rRNA. J. Mol. Evol. 47: 597–605.

Arendt, D. and K. Nübler-Jung. 1996. Common ground plans in early brain development in mice and flies. BioEssays 18: 255–259.

Arendt, D. and K. Nübler-Jung. 1997. Dorsal or ventral: Similarities in fate maps and gastrulation patterns in annelids, arthropods, and chordates. Mech. Dev. 61: 7–21.

Balvoine, G., R. de Rosa and A. Adoutte. 2002. Hox clusters and bilaterian phylogeny. Mol. Phylog. Evol. 24: 366–373.

Bartolomaeus, T. and P. Ax. 1992. Protonephridia and metanephridia: Their relationship within the bilateria. Z. Zool. Syst. Evolutionsforsch. 30: 21–45.

Bartolomaeus, T. and H. Ruhberg. 1999. Ultrastructure of the body cavity lining in embryos of *Epiperipatus biolleyi* (Onychophora, Peripatidae): A comparison with annelid larvae. Invert. Biol. 118: 165–174.

Bengtson, S. 1998. Animal embryos in deep time. Nature 391: 529.

Berchtold, J. P., F. Sauber and M. Reuland. 1985. Etude ultrastructurale de l'évolution du tegument de la sangsue *Hirudo medicinalis* L. (Annélide, Hirudinée) au cours d'un cycle de mue. Int. J. Invert. Reprod. Dev. 8: 127–138.

Berney, C., J. Pawlowski and L. Zaninetti. 2000. Elongation factor 1-alpha sequences do not support an early divergence of the Acoela. Mol. Biol. Evol. 17(7): 1032–1039.

Blair, J. E., K. Ikeo, T. Gojobori and S. Blair Hedges. 2002. The evolutionary position of nematodes. BMC Evol. Biol. 2: 7.

Boaden, P. J. S. 1989. Meiofauna and the origins of the Metazoa. Zool. J. Linn. Soc. 96: 217–227.

Boore, J. L. and J. L. Staton. 2002. The mitochondrial genome of the sipunculid *Phascolopsis gouldii* supports its association with Annelida rather than Mollusca. Mol. Biol. Evol. 19(2): 127–137.

Borchiellini, C., N. Boury-Esnault, J. Vacelet and Y. Le Parco. 1998. Phylogenetic analysis of the Hsp70 sequences reveals the monophyly of Metazoa and specific phylogenic relationships between animals and fungi. Mol. Biol. Evol. 15: 647–655.

Borchiellini, C., M. Manuel, E. Alivon, N. Boury-Esnault, J. Vacelet and Y. Le Parco. 2001. Sponge paraphyly and the origin of Metazoa. J. Evol. Biol. 14: 171–179.

Bromham, L. D. and B. M. Degnan. 1999. Hemichordates and deuterostome evolution: robust molecular phylogenetic support for a hemichordate+echinoderm clade. Evol. Develop. 1(3): 166–171.

Cameron, C. B., J. Garey and B. J. Swalla. 2000. Evolution of the chordate body plan: New insights from phylogenetic analyses of deuterostome phyla. Proc. Natl. Acad. Sci. U.S.A. 97(9): 4469–4474.

Carmean, D. and B. J. Crespi. 1995. Do long branches attract flies? Nature 373: 666.

Carroll, S. B. 1995. Homeotic genes and the evolution of arthropods and chordates. Nature 376: 479–485.

Celerin, M., et al. 1996. Fungal fimbriae are composed of collagen. EMBO J. 15: 4445–4453.

Christen, R., A. Ratto, A. Baroin, R. Perasso, K. G. Grell and A. Adoutte. 1991. An analysis of the origin of metazoans, using comparisons of partial sequences of the 28S RNA, reveals an early emergence of triploblasts. EMBO J. 10: 499–503.

Davidson, E. H., K. J. Peterson and R. A. Cameron. 1995. Origin of bilaterian body plans: Evolution of developmental regulatory mechanisms. Science 270: 1319–1325.

Davies, K. A. and J. M. Fisher. 1994. On hormonal control of moulting in *Aphelenchus avenae* (Nematoda: Aphelenchida). Int. J. Parasitol. 24: 649–655.

D'Hondt, J.-L. 1997. Sur les affinités des Cycliophora Funch et Kristensen, 1995, un novel embranchement d'invertébrés marins, ectoparasites ou commensal des Crustacés Décapodes. Bull. Mens. Soc. Linn. Lyon 66: 12–22.

De Robertis, E. M. and Y. Sasai. 1996. A common plan for dorsoventral patterning in Bilateria. Nature 380: 37–40.

De Rosa, R. 2001. Molecular data indicate the protostome affinity of brachiopods. Syst. Biol. 50(6): 848–859.

De Rosa, R., J. K. Grenier, T. Andreeva, C. E. Cook, A. Adoutte, M. Akam, S. B. Carroll and G. Balavione. 1999. Hox genes in brachiopods and priapulids: Implications for protostome evolution. Nature 399: 772–776. [See criticism by M. J. Telford, 2000, Evol. Dev. 2(6): 360–364.]

Dinan, L. 2001. Phytoecdysteroids: Biological aspects. Phytochemistry 57: 325–339.

Dixon, M. T. and D. M. Hillis. 1993. Ribosomal RNA secondary structure: Compensatory mutations and implications for phylogenetic analysis. Mol. Biol. Evol. 10: 2567–267.

Ehlers, U., W. H. Ahlrichs, C. Lemburg and A. Schmidt-Rhaesa. 1996. Phylogenetic systematization of the Nemathelminthes (Aschelminthes). Verh. dt. zool. Ges. 89: 8.

Eibye-Jackson, D. and C. Nielsen. 1997. Point of view: The rearticulation of the annelids. Zool. Scripta 25: 275–282.

Erber, A., D. Riemer, M. Bovenschulte and K. Weber. 1998. Molecular phylogeny of metazoan intermediate filament proteins. J. Mol. Evol. 47: 751–762.

Erwin, D. H. 1993. The origin of metazoan development. Biol. J. Linn. Soc. 50: 255–274.

Erwin, D. H., J. Valentine and D. Jablonski. 1997. The origin of animal body plans. Am. Sci. 85: 126–137.

Fernholm, B., K. Bremer and H. Hornvall (eds.). 1989. *The Hierarchy of Life*. Elsevier-North Holland Biomedical Press, Amsterdam.

Field, K. G., G. J. Olsen, D. J. Lane, S. J. Giovannoni, M. T. Ghiselin, E. C. Raff, N. R. Pace and R. A. Raff. 1988. Molecular phylogeny of the animal kingdom. Science 239: 748–753. [Also see responses in Science 243: 548–551.]

Field, K. G., G. J. Olsen, S. J. Giovannoni, E. C. Raff, N. R. Pace and R. A Raff. 1989. Phylogeny and molecular data. Science 243: 550–557.

Fleming, M. W. 1987. Ecdysteroids during embryonation of eggs of *Ascaris suum*. Comp. Biochem. Physiol. 87A: 803–805.

Fleming, M. W. 1993. Ecdysteroids during development in the ovine parasitic nematode, *Haemonchus contortus*. Comp. Biochem. Physiol. (Biochem. & Mol. Biol.) 104: 653–655.

Franzen, A. 1981. Comparative and ultrastructural studies on spermatids and spermatozoa in Bryozoa and Ectoprocta. *In* G. P. Larwood and C. Nielsen (eds.), *Recent and Fossil Bryozoa*. Olsen & Olsen, Fredensborg, Denmark, pp. 83–92.

Funch, P. 1996. The chordoid larva of *Symbion pandora* (Cycliophora) is a modified trochophore. J. Morphol. 230: 231–263.

Funch, P. and R. M. Kristensen. 1995. Cycliophora is a new phylum with affinities to Entoprocta and Ectoprocta. Nature 378: 711–714.

Funch, P. and R. M. Kristensen. 1999. Cycliophora. *In* E. Knobil and J. D. Neill (eds.), *Encyclopedia of Reproduction*, Vol. 13. Academic Press, London, pp. 800–808.

Garey, J. R. 2001. Ecdysozoa: The relationship between Cycloneuralia and Panarthropoda. Zool. Anz. 240: 321–330.

Garey, J. R., T. J. Near, M. N. Nonnemacher and S. A. Nadler. 1996. Molecular evidence for Acanthocephala as a subtaxon of Rotifera. J. Mol. Evol. 43: 287–292.

Gee, H. 1995. The molecular explosion. Nature 373: 558–559.

Ghirardelli, E. 1995. Chaetognaths. Two unresolved problems: The coelom and their affinities. *In* G. Lanzavecchia, R. Valvassori and M. D. Candia Carnevali (eds.), *Selected Symposia and Monographs U.Z.I.*, Vol. 8. Mucchi, Modena, Italy, pp. 167–185.

Gilbert, S. F. and A. M. Raunio (eds.). 1997. *Embryology: Constructing the Organism*. Sinauer Associates, Sunderland, MA.

Giribet, G. and W. C. Wheeler. 1999. The position of arthropods in the animal kingdom: Ecdysozoa, islands, trees, and the "parsimony ratchet." Mol. Phyl. Evol. 13: 619–623.

Giribet, G., S. Carranza, J. Baguñà, M. Riutort and C. Ribera. 1996. First molecular evidence for the existence of a Tardigrada + Arthropoda clade. Mol. Biol. Evol. 13: 76–84.

Giribet, G., D. L. Distel, M. Polz, W. Sterrer and W. Wheeler. 2000. Triploblastic relationships with emphasis on the acoelomates and the position of Gnathostomulida, Cycliophora, Platyhelminthes and Chaetognatha: A combined approach of 18S rDNA sequences and morphology. Syst. Biol. 49(3): 539–562.

Halanych, K. M. and Y. Passamaneck. 2001. A brief review of metazoan phylogeny and future prospects in Hox-research. Am. Zool. 41: 629–639.

Halanych, K. M., J. D. Bacheller, A. M. A. Aguinaldo, S. M. Liva, D. M. Hillis and J. A. Lake. 1995. Evidence from 18S ribosomal DNA that the lophophorates are protostome animals. Science 267: 1641–1643.

Harrison, F. W. and M. E. Rice (eds.). 1993. *Microscopic Anatomy of Invertebrates*. Vol. 12. *Onychophora, Chilopoda and Lesser Protostomata*. Wiley-Liss, New York.

Harrison, F. W. and E. E. Ruppert (eds.). 1991. *Microscopic Anatomy of Invertebrates*. Vol. 4. *Aschelminthes*. Wiley-Liss, New York.

Hasegawa, M. and T. Hashimoto. 1993. Ribosomal RNA trees misleading? Nature 361: 23.

Hasse, A., M. Stern, K. Wächtler and G. Bicker. 2001. A tissue-specific marker of Ecdysozoa. Dev., Genes, Evol. 211: 428–433.

Haszprunar, G., L. Salvini-Plawen and R. M. Rieger. 1995. Larval planktotrophy: A primitive trait in the Bilateria. Acta Zool. 76(2): 141–154.

Hausdorf, B. 2000. Early evolution of the Bilateria. Syst. Biol. 49: 130–142.

Hessling, R. and W. Westheide. 2002. Are Echiura derived from a segmented ancestor? Immunohistochemical analysis of the nervous system in developmental stages of *Bonellia viridis*. J. Morphol. 252: 100–113.

Holland, P. W. H. and J. Garcia-Fernández. 1996. Hox genes and chordate evolution. Dev. Biol. 173: 382–395.

Holland, P. W. H. and B. L. M. Hogan. 1986. Phylogenetic distribution of Antennapedia-like homeoboxes. Nature 321: 251–253.

Hou, X. 1989. Early Cambrian arthropod-annelid intermediate sea animal, *Luolishania* Gen. Nov. from Cheng-Jiang, Yunnan. Acta Paleontol. Sinica 28: 207–213.

Jablonski, D. and D. J. Bottjer. 1991. Environmental patterns in the origins of higher taxa: The post-Paleozoic fossil record. Nature 350: 1831–1833.

Jefferies, R. P. S. 1986. *The Ancestry of the Vertebrates*. British Museum (Natural History), London. [See reviews by J. Sprinkle, 1986, Science 236: 1476, and by C. Gans, Am. Sci. 76: 188–189.]

Jenner, R. A. 1999. Metazoan phylogeny as a tool in evolutionary biology: Current problems and discrepancies in application. Belgium J. Zool. 129(1): 245–262.

Jondelius, U., I. Ruiz-Trillo, J. Baguñà and M. Riutort. 2002. The Nemertodermatida are basal bilaterians and not members of the Platyhelminthes. Zool. Scripta. 31: 201–215.

Kappen, C. and F. H. Ruddle. 1993. Evolution of a regulatory gene family: HOM/Hox genes. Curr. Opin. Genet. Dev. 3: 931–938.

Kim, C. B., S. Y. Moon, S. R. Gelder and W. Kim. 1996. Phylogenetic relationships of annelids, molluscs, and arthropods evidenced from molecules and morphology. J. Mol. Evol. 43: 207–216.

Knoll, A. H. and S. B. Carroll. 1999. Early animal evolution: Emerging views from comparative biology and geology. Science 284: 2129–2137.

Koolman, J. (ed.). 1989. *Ecdysone: From Chemistry to Mode of Action*. Georg Thieme Verlag, Stuttgart.

Kristensen, R. M. 1995. Are Aschelminthes pseudocoelomates or acoelmates. *In* G. Lanzavecchia, R. Valvassori and M. D. Candia Carnevali (eds.), *Selected Symposia and Monographs U.Z.I.*, Vol. 8. Mucchi, Modena, Italy, pp. 41–43.

Lacalli, T. C. 1996. Dorsoventral axis inversion: A phylogenetic perspective. BioEssays 18: 251–254.

Lacalli, T. C. 1997. The nature and origin of deuterostomes: Some unresolved issues. Invert. Biol. 116(4): 363–370.

Lafont, R. 1997. Ecdysteroids and related molecules in animals and plants. Arch. Insect Biochem. Physiol. 35: 3–20.

Lafont, R. and I. Wilson. 1996. *The Ecdysone Handbook. 2nd ed*. The Chromatographic Society, Nottingham.

Lewin, R. 1988. A lopsided look at evolution. Science 241: 291–293.

Li, C.-W., J.-Y. Chen and T.-E. Hua. 1998. Precambrian sponges with cellular structures. Science 279: 879–882.

Lipps, J. H. and P. W. Signor (eds.). 1992. *Origin and Early Evolution of the Metazoa*. Plenum Press. New York.

Littlewood, D. T. J., P. D. Olson, M. J. Telford, E. A. Herniou and M. Riutori. 2001. Elongation factor 1-alpha sequences alone do not assist in resolving the position of the Acoela within the Metazoa. Mol. Biol. Evol. 18(2): 437–442.

Littlewood, D. T. J., M. J. Telford, K. A. Clough and K. Rohde. 1998. Gnathostomulida: An enigmatic metazoan phylum from both morphological and molecular perspectives. Mol. Phyl. Evol. 9: 72–79.

Lüter, C. 2000. The origin of the coelom in Brachiopoda and its phylogenetic significance. Zoomorphology 120: 15–28.

Lynch, M. 1999. The age and relationships of the major animal phyla. Evolution 53: 319–325.

Mackie, L. Y., B. Winnepenninckx, R. De Wachter, P. Emschermann and J. R. Garey. 1996. 18S rRNA suggests that Entoprocta are protostomes, unrelated to Ectoprocta. J. Mol. Evol. 42: 552–559.

Malakhov, V. V. 1980. Cephalorhyncha: A new phylum of the animal kingdom uniting Priapulida, Kinorhyncha, and Gordiacea and a system of the worms with a primary body cavity. Zool. Zh. 59(4): 485–499.

Mallatt, J. and C. J. Winchell. 2002. Testing the new animal phylogeny: First use of combined large-subunit and small-subunit rRNA gene sequences to classify the Protostomes. Mol. Biol. Evol. 19(3): 289–301.

Manuel, M., M. Kruse, W. E. G. Müller and Y. Le Parco. 2000. The comparison of ß-thymosin homologues among Metazoa supports an arthropod-nematode clade. J. Mol. Evol. 51: 378–381.

Margulis, L. 1981. *Symbiosis in Cell Evolution: Life and Its Environment on the Early Earth*. W. H. Freeman, San Francisco.

Martin, M. W., D. V. Grazhdankin, S. A. Bowring, D. A. D. Evans, M. A. Fedonkin and J. L. Kirschvink. 2000. Age of Neoproterozoic bilatarian body and trace fossils, White Sea, Russia: Implications for metazoan evolution. Science 288: 841–845.

McGinnis, W. and R. Krumlauf. 1992. Homeobox genes and axial patterning. Cell 68: 283–302.

McHugh, D. 1997. Molecular evidence that echiurans and pogonophorans are derived annelids. Proc. Natl. Acad. Sci. U.S.A. 94: 8006–8009.

McHugh, D. 1998. Deciphering metazoan phylogeny: The need for additional molecular data. Am. Zool. 38: 859–866.

Miner, B. G., E. Sanford, R. R. Strathmann, B. Pernet and R. E. Emlet. 1999. Functional and evolutionary implications of opposed bands, big mouths, and extensive oral ciliation in larval opheliids and echiurids (Annelida). Biol. Bull. 197: 14–25.

Morris, S. C., J. D. George, R. Gibson and H. M. Platt (eds.). 1985. *The Origins and Relationships of Lower Invertebrates*. Clarendon Press, Oxford.

Mushegian, A. R., J. R. Garey, J. Martin and L. X. Liu. 1998. Large-scale taxonomic profiling of eukaryotic model organisms: A comparison of orthologous proteins encoded by human, fly, nematode and yeast genomes. Genome Res. 8: 590–598.

Neuhaus, B. 1994. Ultrastructure of alimentary canal and body cavity, ground pattern, and phylogenetic position of the Kinorhyncha. Microfauna Marina 9: 61–156.

Neuhaus, B., R. M. Kristensen and C. Lemburg. 1996. Ultrastructure of the cuticle of the Nemathelminthes and electron microscopical localization of chitin. Verh. dt. zool. Ges. 89: 221.

Newby, W. W. 1940. The embryology of the echiuroid worm *Urechis caupo*. Mem. Am. Phil. Soc. 16: 1–219.

Nielsen, C. 1979. Larval ciliary bands and metazoan phylogeny. Fortschr. Zool. Syst. Evolutionforsch. 1: 178–184.

Nielsen, C. 1985. Animal phylogeny in the light of the trochaea theory. Biol. J. Linn. Soc. 25: 243–299.

Nielsen, C. 1987. Structure and function of metazoan ciliary bands and their phylogenetic significance. Acta Zool. 68: 205–262.

Nielsen, C. 1989. Phylogeny and molecular data. Science 243: 548.

Nielsen, C. 1994. Larval and adult characteristics in animal phylogeny. Am. Zool. 34: 492–501.

Nielsen, C. 1998. Origin and evolution of animal life cycles. Biol. Rev. 73: 125–155.

Nielsen, C. 1999. Origin of the chordate central nervous system—and the origin of chordates. Dev. Genes Evol. 209: 198–205.

Nielsen, C. 2001. *Animal Evolution: Interrelationships of the Living Phyla*, 2nd Ed. Oxford University Press, Oxford.

Nielsen, C., N. Scharff and D. Eibye-Jacobsen. 1996. Cladistic analyses of the animal kingdom. Biol. J. Linn. Soc. 57: 385–410.

Nübler-Jung, K. and D. Arendt. 1994. Is ventral in insects dorsal in vertebrates? Roux's Arch. Dev. Biol. 203: 357–366.

Nübler-Jung, K. and D. Arendt. 1996. Enteropneusts and chordate evolution. Curr. Biol. 6: 352–353.

Okazaki, R. K., M. J. Snyder, C. C. Grimm and E. S. Chang. 1998. Ecdysteroids in nemerteans: Further characterization and identification. Hydrobiologia 365: 281–285.

Panganiban, G. et al. 1997. The origin and evolution of animal appendages. Proc. Natl. Acad. Sci. U.S.A. 94: 5162–5166.

Park, Y., D. Zitnan, S. S. Gill, and M. E. Adams. 1999. Molecular cloning and biological activity of ecdysis-triggering hormones in *Drosophila melanogaster*. FEBS Lett. 463: 133–138.

Patterson, C. 1990. Reassessing relationships. Nature 344: 199–200.

Philippe, H., A. Chenuil and A. Adoutte. 1994. Can the Cambrian explosion be inferred through molecular phylogeny? Development (Suppl.): 15–25.

Raff, R. A. 1992. Evolution of developmental decisions and metamorphosis: The view from camps. Development (Suppl.): 15–22.

Raff, R. A. 1996. *The Shape of Life*. University of Chicago Press, Chicago.

Regier, J. C. and J. W. Schultz. 1997. Molecular phylogeny of the major arthropod groups indicates polyphyly of crustaceans and a new hypothesis for the origin of hexapods. Mol. Biol. Evol. 14: 902–913.

Rieger, R. M. 1984. Evolution of the cuticle in the lower Eumetazoa. In J. Bereiter-Hahn, A. G. Matolsy and K. S. Richards (eds.), *Biology of the Integument*, Vol. 1. Springer Verlag, Berlin, pp. 389–399.

Rieger, R. M. 1994a. The biphasic life cycle: A central theme of metazoan evolution. Am. Zool. 34: 484–491.

Rieger, R. M. 1994b. Evolution of the "lower" Metazoa. In S. Bengston (ed.), *Early Life on Earth*. Columbia University Press, New York, pp. 231–258.

Rieger, R. M. and M. Mainitz. 1977. Comparative fine structure of the body wall in Gnathostomulida and their phylogenetic position between Platyhelminthes and Aschelminthes. Z. Zool. Syst. Evolutionsforsch. 15: 9–35.

Rieger, R. M. and S. Tyler. 1995. Sister-group relationship of Gnathostomulida and Rotifera-Acanthocephala. Invert. Biol. 114: 186–188.

Rouse, G. W. 1999. Trochophore concepts: Ciliary bands and the evolution of larvae in spiralian Metazoa. Biol. J. Linn. Soc. 66: 411–464.

Rouse, G. W. 2000a. The epitome of hand waving? Larval feeding and hypotheses of metazoan phylogeny. Evol. Dev. 2(4): 222–233.

Rouse, G. W. 2000b. Bias? What bias? The evolution of downstream larval-feeding in animals. Zool. Scripta 29(3): 213–236.

Rouse, G. W. and K. Fauchald. 1995. The articulation of annelids. Zool. Scripta 24: 269–301.

Ruiz-Trillo, I., M. Riutort, D. T. J. Littlewood, E. A. Herniou and J. Baguña. 1999. Acoel flatworms: Earliest extant bilaterian metazoans, not members of Platyhelminthes. Science 283: 1919–1923.

Ruppert, E. E. and P. R. Smith. 1988. The functional organization of filtration nephridia. Biol. Rev. 63: 231–258.

Sadikov, Z., Z. Sastov, M. Garcia, J. P. Girault and R. Lafont. 2001. Ecdysteroids from *Selene claviformis*. Chemistry of Natural Compounds 37(6): 580.

Sauber, F., M. Reuland, J. Berchtold, C. Hetru, G. Tsoupras, B. Luu, M. Moritz and J. Hoffmann. 1983. Molting cycle and ecdysteroids in the leech, *Hirudo medicinalis*. Comptes Rendus de L'Academie des Sciences, Ser. III (Life Sciences) 296 (8): 413.

Scheltema, A. H. 1993. Aplacophora as progenetic aculiferans and the coelomate origin of mollusks as the sister taxon of Sipuncula. Biol. Bull. 184: 57–78.

Schmidt-Rhaesa, A., T. Bartolomaeus, C. Lemburg, U. Ehlers and J. R. Garey. 1998. The position of the Arthropoda in the phylogenetic system. J. Morphol. 238: 263–285.

Seilacher, A., P. K. Bose and F. Pflüger. 1998. Triploblastic animals more than 1 billion years ago: Trace fossil evidence from India. Science 282: 80–83.

Shoichet, S. A., T. H. Malik, J. H. Rothman and R. A. Shivdasani. 2000. Action of *Caenorhabditis elegans* GATA factor END-1 in *Xenopus* suggests that similar mechanisms initiate endoderm development in Ecdysozoa and vertebrates. Proc. Natl. Acad. Sci. U.S.A. 97: 4076–4081.

Shubin, N., C. Tabin and S. Carroll. 1997. Fossils, genes and the evolution of animal limbs. Nature 388: 639–648.

Siddall, M. E., K. Fitzhugh and K. A. Coates. 1998. Problems determining the phylogenetic position of echiurans and pogonophorans with limited data. Cladistics 14: 401–410.

Simonetta, A. M. and S. C. Morris (eds.). 1991. *The Early Evolution of Metazoa and the Significance of Problematic Taxa*. Cambridge University Press, Cambridge.

Smith, P. R. and E. E. Ruppert. 1988. Nephridia. In W. Westheide and C. O. Hermans (eds.), *Microfauna Marina*, Vol. 4. Gustav Fischer Verlag, Stuttgart, pp. 231–262.

Sørensen, M. V. 2000. An SEM study of the jaws of *Haplognathia rosea* and *Rastrognathia macrostoma* (Gnathostomulida), with a preliminary comparison with the rotiferan trophi. Acta Zool. 81: 9–16.

Sørensen, M. V., P. Funch, E. Willerslev, A. J. Hansen and J. Olesen. 2000. On the phylogeny of the Metazoa in the light of Cycliophora and Micrognathozoa. Zool. Anz. 239: 297–318.

Stiller, J. W. and B. D. Hall. 1999. Long-branch attraction and the rDNA model of early eukaryotic evolution. Mol. Biol. Evol. 16: 1270–1279.

Strathmann, R. R. 1993. Hypotheses on the origins of marine larvae. Annu. Rev. Ecol. Syst. 24: 89–117.

Suksamrarn, A., A. Jankam, B. Tarnchompoo and S. Putchakarn. 2002. Ecdysteroids from a *Zoanthus* sp. J. Nat. Products 65(8): 1194–1197.

Swofford, D. 2002. *PAUP*4.0 Phylogenetic Analysis Using Parsimony (and Other Methods) Version 4.* Sinauer Associates, Sunderland, MA.

Valentine, J. W. 1994. Late Precambrian bilaterians: Grades and clades. Proc. Natl. Acad. Sci. U.S.A. 91: 6751–6757.

Valentine, J. W. 1997. Cleavage patterns and the topology of the metazoan tree of life. Proc. Natl. Acad. Sci. U.S.A. 94: 8001–8005.

Valentine, J. W. and A. G. Collins. 2000. The significance of moulting in ecdysozoan evolution. Evol. Dev. 2: 152–156.

Valentine, J. W., D. H. Erwin and D. Jablonski. 1996. Developmental evolution of metazoan body plans: The fossil evidence. Dev. Biol. 173: 373–381.

Volodin, V., I. Chadin, P. Whiting and L. Dinan. 2002. Screening plants of European northeast Russia for ecdysteroids. Biochem. Syst. Ecol. 30(6): 525–578.

Wägele, J.-W. and B. Misof. 2001. On quality of evidence in phylogeny reconstruction: A reply to Zrzavy's defense of the "Ecdysozoa" hypothesis. J. Zool. Syst. Evol. Res. 39: 165–176.

Wägele, J.-W. and F. Rödding. 1998. A priori estimation of phylogenetic information conserved in aligned sequences. Mol. Phyl. Evol. 9: 358–365.

Wägele, J.-W., T. Erikson, P. Lockhart and B. Misof. 1999. The Ecdysozoa: Artifact or monophylum? J. Zool. Syst. Evol. Res. 37: 211–223. [See reply by Zrzavy, 2000, J. Zool. Syst. Evol. Res. 39: 159–163.]

Wainright, P. O., G. Hinkle, M. L. Sogin and S. K. Stickel. 1993. Monophyletic origins of the Metazoa: An evolutionary link with fungi. Science 260: 340–342.

Wallace, R. L., C. Ricci and G. Melone. 1996. A cladistic analysis of pseudocoelomate (Aschelminth) morphology. Invert. Biol. 115: 104–112.

Wheeler, W. 1998. Molecular systematics and arthropods. *In* G. D. Edgecomb (ed.), *Arthropod Fossils and Phylogeny*. Columbia University Press, New York, pp. 9–33.

Winchell, C. J., J. Sullivan, C. B. Cameron, B. J. Swalla and J. Mallatt. 2002. Evaluating hypotheses of deuterostome phylogeny and chordate evolution with new LSU and SSU ribosomal DNA data. Mol. Biol. Evol. 19(5): 762–776.

Winnepenninckx, B., T. Backeljau and R. M. Kristensen. 1998. Relations of the new phylum Cycliophora. Nature 393: 636–638.

Wray, G. A., J. S. Levinton and L. H. Shapiro. 1996. Molecular evidence for deep Precambrian divergences among metazoan phyla. Science 274: 568–573.

Zimmer, R. L. 1973. Morphological and developmental affinities of the Lophophorata. *In* G. P. Larwood (ed.), *Living and Fossil Bryozoa*. Academic Press, London, pp. 593–599.

Zrzavy, J. 2001. Ecdysozoa versus Articulata: Clades, artifacts, prejudices. J. Zool. Syst. Evol. Res. 39: 159–163. [See reply by Wägele and Misof, 2001, J. Zool. Syst. Evol. Res. 39: 165–176.]

Zrzavy, J., S. Mihulka, P. Kepka, A. Bezdek and D. Tietz. 1998. Phylogeny of the Metazoa based on morphological and 18S ribosomal DNA evidence. Cladistics 14: 249–285.

Appendix A:
Common Human Diseases Transmitted by Insects

Disease	Vector	Causative agent	Occurrence
Fly (Diptera) vectored diseases			
African sleeping sickness, nagana (trypanosomiasis)	Tsetse flies, *Glossina* spp. (Glossinidae)	*Trypanosoma* spp. (Protista: phylum Kinetoplastida)	Tropical Africa
Anthrax	Horseflies (*Tabanus* spp.)	*Bacillus anthracis* (Bacteria)	Worldwide
Amebic dysentery	Housefly (*Musca domestica*) and other flies	*Entamoeba histolytica* (Protista: phylum Rhizopoda)	Worldwide
Bacillary dysentery	Housefly (*Musca domestica*) and various blowflies and flesh flies	*Bacillus* spp. (Bacteria)	Worldwide
Cholera	Housefly (*Musca domestica*) and various blowflies and flesh flies	*Vibrio comma* (Bacteria)	Worldwide
Dengue	Mosquitoes (Culicidae) of the genus *Aedes*, especially *A. aegypti* and *A. albopictus*	Virus	Pantropical
Leishmaniasis, including kala-azar, oriental sore, espundia, and other epidermal diseases	Sand flies (Phlebotominae, species of *Phlebotomus* and *Lutzomyia*	*Leishmania* spp. (Protista: phylum Kinetoplastida)	Pantropical
Loaiasis	Deerflies (*Chrysops* spp.)	*Loa loa* (phylum Nemata)	Africa
Lymphatic filariasis ("elephantiasis")	Mosquitoes (Culicidae) and some biting midges (*Culicoides*)	*Wuchereria bancrofti* (filarial nematode: phylum Nemata)	Pantropical
Malaria	Mosquitoes (Culicidae) of the genus *Anopheles*	*Plasmodium* spp. (Protista: phylum Apicomplexa)	Pantropical
Myiasis	Infestation of living tissue by larvae of various dipterans	Various species of Diptera	Pantropical and, in places, penetrating warm temperate regions
Onchocerciasis (river blindness)	Blackflies (Simulidae), *Simulium* spp.	*Onchocerca volvulus* (phylum Nemata)	Pantropical
Pappataci fever	Sand fly (*Phlebotomus papatasii*)	Viral	Mediterranean region, India, Sri Lanka
Tularemia	Deerflies (*Chrysops* spp.) (also ticks and sucking lice)	*Francisella tularensis* (= *Pasturella tularensis*) (Bacteria)	North America, Europe, the Orient
Typhoid fever	Housefly (*Musca domestica*) and various blowflies and flesh flies	*Eberthella typhosa* (Bacteria)	Worldwide

Disease	Vector	Causative agent	Occurrence
Verruga peruana (Oroya fever)	Sand flies (*Lutzomyia* spp.)	*Bartonella bacilliformis* (Bacteria)	South America
Viral encephalitis (many forms)	Mosquitoes (Culicidae), especially *Culex* and *Aedes* spp.	Several viruses	North America, South America, Europe, AsiaWest
Nile virus	Mosquitoes (Culicidae); *Aedes albopictus* in the United States	Virus	Old World tropics; recently introduced into the United States
Yellow fever	Mosquitoes (Culicidae), especially *Aedes aegypti*	Virus	Pantropical

Bug (Hemiptera) vectored diseases

Disease	Vector	Causative agent	Occurrence
Chagas' disease (American trypanosomiasis)	Triatomine bugs (Hemiptera: Reduviidae, Triatominae) *Rhodnius, Triatoma, Panstrongylus*	*Trypanosoma cruzi* (Protista: phylum Kinetoplastida)	New World tropics and subtropics; a few cases reported as far north as Texas

Flea (Siphonaptera) vectored diseases

Disease	Vector	Causative agent	Occurrence
Bubonic plague	Fleas, especially the rat flea *Xenopsylla cheopis*	*Yersinia* (= *Pasturella*) *pestis* (Bacteria)	Worldwide
Murine (endemic) typhus	Rat flea (*Xenopsylla cheopis*), and other fleas, lice, mites, and ticks on rodents	*Rickettsia mooseri*	Worldwide
Salmonellosis	Fleas	*Salmonella* spp. (Bacteria)	Worldwide
Tapeworms	Fleas	Tapeworms (phylum Platyhelminthes, Cestoda)	Worldwide

Lice (Phthiraptera) vectored diseases

Disease	Vector	Causative agent	Occurrence
Epidemic typhus	Body louse (*Pediculus humanus*); also rat flea (*Xenopsylla cheopis*) and rat mite (*Liponyssus bacoti*)	*Rickettsia prowazakii*	Worldwide
Relapsing fever	Sucking lice	*Borrelia recurrentis, B. duttoni* (Spirochaeta)	Worldwide
Trench fever	Sucking lice	*Rickettsia quintana*	Worldwide
Tularemia	Sucking lice, deerflies (*Chrysops*), and ticks (especially *Dermacentor* and *Haemaphysalis*)	*Francisella* (= *Pasturella*) *tularensis* (Bacteria)	North America, Europe, the Orient
Vagabond's disease	Sucking lice	Chronic infestation by lice	Worldwide

Appendix B:
Data Matrix for Analysis of Metazoan Phylogeny

CHARACTERS

Taxon	1	2	3	4	5	6	7	8	9	10	11	12	13	14	15	16	17	18	19	20	21	22	23	24
Choanoflagellata	0	0	-	0	0	-	0	-	-	0	-	-	0	-	-	-	-	-	-	-	-	-	0	-
Porifera	0	1	0	1	0	0	0	0	-	1	0	0	0	0	0	0	0	1	0	0	0	0	1	0
Placozoa	1	1	0	1	0	0	?	0	-	?	0	0	1	0	0	0	0	?	0	0	0	0	?	0
Cnidaria	1	1	0	1	1	0	0	0	0	1	1	0	1	0	0	0	0	1	0	0	0	0	1	0
Ctenophora	1	1	0	1	1	1	1	0	0	1	1	0	1	0	0	0	0	3	0	0	0	0	1	1
Platyhelminthes	1	1	0	1	1	0	1	0	0	1	1	0	1	0	0	0	0	0	0	0	0	0	1	0
Nemertea	1	1	0	1	1	0	1	0	0	1	1	0	1	0	0	0	0	0	0	0	0	0	1	0
Rotifera	1	1	0	1	1	0	1	0	0	1	1	0	1	0	0	0	0	0	0	0	1	0	1	0
Gastrotricha	1	1	0	1	1	0	1	0	0	1	1	0	1	0	0	0	0	2	0	0	0	0	1	0
Kinorhyncha	1	1	0	1	1	0	1	0	0	1	1	0	1	0	0	0	?	?	0	0	0	0	1	0
Nemata	1	1	0	1	1	0	1	0	0	1	1	1	1	0	0	0	1	2	0	0	0	0	1	0
Nematomorpha	1	1	0	1	1	0	1	0	0	1	1	0	1	0	0	0	?	2	0	0	0	0	1	0
Acanthocephala	1	1	0	1	1	0	1	0	0	1	1	0	1	0	0	0	0	0	0	1	0	0	1	0
Entoprocta	1	1	0	1	1	0	1	0	1	1	1	0	1	0	0	0	0	0	0	0	0	0	1	0
Gnathostomulida	1	1	0	1	1	0	1	0	0	1	1	0	1	0	0	0	0	0	0	0	0	0	1	0
Priapula	1	1	0	1	1	0	1	0	0	1	1	0	1	0	0	0	?	2	0	0	0	0	1	0
Loricifera	1	1	0	1	1	0	1	0	0	1	1	0	1	0	0	0	?	?	0	0	0	0	1	0
Cycliophora	1	1	0	?	1	0	?	0	1	?	?	0	1	0	0	0	0	?	0	0	0	0	1	0
Annelida	1	1	0	1	1	0	1	1	0	1	1	0	1	0	0	0	0	0	0	0	0	0	1	0
Sipuncula	1	1	0	1	1	0	1	0	0	1	1	0	1	0	0	0	0	0	0	0	0	0	1	0
Echiura	1	1	0	1	1	0	1	0	0	1	1	0	1	0	0	0	0	0	0	0	0	0	1	0
Onychophora	1	1	0	1	1	0	1	1	0	1	1	0	1	0	0	0	1	0	0	0	0	0	1	0
Tardigrada	1	1	1	1	1	0	1	?	0	1	1	0	1	0	0	0	1	0	0	0	0	0	1	0
Arthropoda	1	1	0	1	1	0	1	1	0	1	1	0	1	0	0	0	1	0	0	0	0	0	1	0
Mollusca	1	1	0	1	1	0	1	0	0	1	1	0	1	0	0	0	0	0	0	0	0	0	1	0
Phoronida	1	1	0	1	1	0	1	0	0	1	1	0	1	3	0	0	0	1	0	0	0	0	1	0
Ectoprocta	1	1	0	1	1	0	1	0	0	1	1	0	1	3	0	0	0	1	0	0	0	0	1	0
Brachiopoda	1	1	0	1	1	0	1	0	0	1	1	0	1	3	0	0	0	1	0	0	0	0	1	0
Echinodermata	1	1	0	1	1	0	1	0	0	1	1	0	1	1	0	0	0	1	0	0	0	0	1	0
Chaetognatha	1	1	0	1	1	0	1	0	0	1	1	0	1	1	0	0	0	1	0	0	0	0	1	0
Enteropneusta	1	1	0	1	1	0	1	0	0	1	1	0	1	1	1	0	0	1	0	0	0	1	1	0
Pterobranchia	1	1	0	1	1	0	1	0	0	1	1	0	1	2	1	0	0	1	0	0	0	?	1	0
Urochordata	1	1	0	1	1	0	1	0	0	1	1	0	1	1	0	1	0	1	0	0	0	1	1	0
Cephalochordata	1	1	0	1	1	0	1	0	0	1	1	0	1	1	0	0	0	1	1	0	1	1	1	0
Vertebrata	1	1	0	1	1	0	1	0	0	1	1	0	1	1	0	0	0	1	0	0	1	1	1	0

Characters Used in Phylogenetic Analysis Matrix

1. Collar cells (choanocytes, without contractile microvilli)
   0 = present   1 = absent
2. Multicellularity
   0 = absent   1 = present
3. With anterior legs, modified as stylets and style supports
   0 = no   1 = yes
4. Epidermal epithelia with septate or tight junctions
   0 = absent   1 = present
5. Gap junctions
   0 = absent   1 = present
6. Cydippid larva
   0 = absent   1 = present
7. Acetylcholine/cholinesterase system
   0 = absent   1 = present
8. Dual segmentally-iterated *engrailed* gene expression during early teloblastic segmentation and neurogenesis (expressed in both the mesoderm and ectoderm)
   0 = no   1 = yes

9. Mushroom-shaped extensions from basal lamina into epidermis
   0 = absent   1 = present
10. Collagen present in body
   0 = no   1 = yes
11. Organized gonads
   0 = absent   1 = present
12. Unique nematan excretory system (renette cells, etc.)
   0 = absent   1 = present
13. Striated ciliary rootlets
   0 = absent   1 = present
14. With mesocoels elaborated as fluid-filled, ciliated tentacles
   0 = without mesocoelic cavities
   1 = mesocoels present, but not elaborated as fluid-filled, ciliated tentacles
   2 = yes, not surrounding mouth
   3 = yes, surrounding mouth (i.e., lophophore)
15. Preoral gut (buccal) diverticulum, or "stomochord"
   0 = absent   1 = present

16. Body enclosed in a tunicin-based outer casing (i.e., a tunic)
   0 = no   1 = yes
17. Ecdysone-mediated ecdysis
   0 = absent   1 = present
18. Cleavage pattern
   0 = fundamentally spiral
   1 = fundamentally radial
   2 = neither clearly spiral nor radial
   3 = unique ctenophoran cleavage pattern
19. Complex anterior vestibule formed by oral hood bearing finger-like buccal cirri and ciliary bands (the "wheel organ")
   0 = absent   1 = present
20. Intracellular skeletal lamina ("intracellular cuticle")
   0 = absent   1 = present
21. Segmentally arranged blocks of muscles (myotomes)
   0 = absent   1 = present
22. With epithelial tissue that binds iodine and secretes iodothyrosine
   0 = absent   1 = present

CHARACTERS

Taxon	25	26	27	28	29	30	31	32	33	34	35	36	37	38	39	40	41	42	43	44	45	46	47	48
Choanoflagellata	-	-	-	-	-	-	-	-	-	-	-	-	-	-	-	-	-	-	-	-	-	-	-	-
Porifera	0	3	-	0	0	0	0	0	-	2	0	0	-	0	0	0	0	0	0	0	0	0	0	0
Placozoa	0	3	?	0	0	0	?	?	?	2	0	0	-	?	0	0	0	0	0	0	0	0	0	0
Cnidaria	0	3	0	0	0	0	1	0	?	2	0	0	-	1	0	0	0	0	0	0	0	0	0	0
Ctenophora	0	3	0	0	0	0	1	0	0	?	0	0	0	0	0	0	0	0	0	0	0	0	0	0
Platyhelminthes	0	3	0	0	0	0	1	0	0	0	0	0	0	0	0	0	0	0	0	0	0	0	0	0
Nemertea	0	3	0	0	0	0	1	0	0	0	0	0	0	0	0	0	0	0	0	0	0	0	0	0
Rotifera	1	3	0	0	0	0	1	0	0	?	1	0	0	0	0	0	0	0	0	0	0	0	0	0
Gastrotricha	0	3	?	0	0	0	1	0	?	0	0	0	0	0	0	0	0	0	0	0	0	0	0	0
Kinorhyncha	0	3	0	0	0	0	1	0	?	?	0	0	0	0	0	0	0	0	0	1	1	0	0	0
Nemata	0	3	1	0	0	0	1	0	?	0	0	0	0	0	0	0	0	0	0	0	0	0	0	0
Nematomorpha	0	3	1	0	0	0	1	0	?	?	0	0	0	0	0	0	0	0	0	0	0	0	0	0
Acanthocephala	1	3	0	1	0	0	1	0	?	?	0	0	0	0	0	0	0	0	0	0	0	0	0	0
Entoprocta	0	3	0	0	0	0	1	0	?	0	0	0	0	0	0	0	0	0	0	0	0	0	0	0
Gnathostomulida	0	3	0	0	0	0	1	0	?	?	0	0	0	0	0	0	0	0	0	0	0	0	0	0
Priapula	0	3	0	0	0	0	1	0	?	?	0	0	0	0	0	0	0	0	0	1	0	0	0	0
Loricifera	0	3	0	0	0	0	1	0	?	?	0	0	0	0	0	0	0	0	0	1	1	1	0	0
Cycliophora	?	3	0	0	0	0	1	0	?	0	0	0	0	0	0	0	0	0	0	0	0	0	0	0
Annelida	0	3	0	0	0	1	1	0	0	0	0	1	0	0	0	0	0	0	0	0	0	0	0	0
Sipuncula	0	3	0	0	0	0	1	0	0	0	0	0	0	0	0	0	0	0	0	0	0	0	0	0
Echiura	0	3	0	0	0	0	1	0	0	0	0	0	0	0	0	0	0	1	0	0	0	0	0	0
Onychophora	0	3	0	0	0	1	1	0	0	0	0	1	0	0	0	0	0	0	0	0	0	0	0	0
Tardigrada	0	3	0	0	0	1	1	0	0	?	0	1	0	0	0	0	0	0	0	0	0	0	0	0
Arthropoda	0	3	0	0	0	1	1	0	0	0	0	1	0	0	0	0	0	0	0	0	0	0	0	0
Mollusca	0	3	0	0	0	0	1	0	0	0	0	0	0	0	0	0	0	0	0	0	0	0	0	0
Phoronida	0	1	0	0	0	0	1	0	0	1	0	0	0	0	1	0	0	0	0	0	0	0	0	0
Ectoprocta	0	1	0	0	0	0	1	0	1	?	0	0	0	0	0,1	0	0	0	0	0	0	0	0	0
Brachiopoda	0	1	0	0	1	0	1	0	2	1	0	0	0	0	1	0	0	0	0	0	0	0	0	0
Echinodermata	0	0	0	0	0	0	1	0	1	1	0	0	0	0	1	0	0	0	0	0	0	0	0	0
Chaetognatha	0	0	0	0	0	0	1	0	2	1	0	0	0	0	1	1	0	0	0	0	0	0	0	0
Enteropneusta	0	0	0	0	0	0	1	0	2	1	0	0	0	0	1	0	0	0	0	0	0	1	0	0
Pterobranchia	0	1	0	0	0	0	1	0	2	1	0	0	0	0	1	0	0	0	0	0	0	1	0	0
Urochordata	0	2	0	0	0	0	1	1	2	1	0	0	0	0	0	0	1	0	0	0	0	1	1	1
Cephalochordata	0	2	0	0	0	0	1	1	1	1	0	0	1	0	0	0	0	0	0	0	0	1	1	1
Vertebrata	0	2	0	0	0	0	1	0	1	1	0	0	1	0	0	0	0	0	0	0	0	1	1	1

23. Spermatozoa
  0 = absent   1 = present

24. Comb rows and ctenes
  0 = absent   1 = present

25. Sperm with anteriorly inserted flagellum
  0 = absent   1 = present

26. With enterocoelic prosomal body region
  0 = prosomal body region present, unmodified
  1 = prosomal body region modified into an epistomial flap
  2 = enterocoely present, but without recognizable prosomal body region
  3 = without enterocoelic development

27. Longitudinal muscle arms
  0 = absent   1 = present

28. Lemnisci
  0 = absent   1 = present

29. Pedicle
  0 = absent   1 = present

30. Cerebral ganglia ("brain") with characteristic anterior mushroom bodies
  0 = no   1 = yes

31. With ecto- and entoderm (= gastrulation, or embryogenic tissue layering)
  0 = absent   1 = present

32. Tadpole larva
  0 = absent   1 = present

33. Fate of blastopore
  0 = forms adult mouth
  1 = forms adult anus
  2 = blastopore closes during embryogeny, mouth and anus form elsewhere

34. With entomesoderm derived from archenteron (enterocoelic pouching)
  0 = no
  1 = yes
  2 = without entomesoderm

35. Corona
  0 = absent   1 = present

36. Teloblastic segmentation (body with successively added segments developed from a posterior teloblastic growth zone)
  0 = absent   1 = present

37. Body with segmented longitudinal musculature developed from rows of mesodermal pockets from the archenteron
  0 = absent   1 = present

38. With planula larva
  0 = no   1 = yes

39. Tripartite (trimeric) coelom
  0 = absent   1 = present

40. Ciliary fan sensory organ
  0 = absent   1 = present

41. Greatly enlarged pharynx, forming a "branchial basket" for feeding
  0 = absent   1 = present

42. Anal vesicles, with excretory funnels
  0 = absent   1 = present

43. With protonephridia located inside gonads
  0 = no   1 = yes

44. With introvert bearing spines, teeth, and scalids
  0 = no   1 = yes

45. With non-eversible mouth cone bearing cuticular ridges and spines
  0 = no   1 = yes

46. Pharyngeal gill slits/pores
  0 = absent   1 = present

47. Notochord
  0 = absent   1 = present

48. Endostyle (= thyroid gland of vertebrates)
  0 = absent   1 = present

CHARACTERS

Taxon	49	50	51	52	53	54	55	56	57	58	59	60	61	62	63	64	65	66	67	68	69	70	71	72
Choanoflagellata	-	-	-	-	-	-	-	-	-	-	-	-	-	-	-	-	-	0	-	-	-	-	0	0
Porifera	0	0	0	0	0	0	2	0	2	0	0	0	1	1	0	0	0	0	0	0	0	0	0	0
Placozoa	0	0	0	0	0	0	2	0	2	0	0	0	0	0	0	0	0	?	0	?	0	0	0	0
Cnidaria	0	0	0	1	0	0	0	0	2	0	0	0	0	0	0	0	1	1	0	1	1	0	1	0
Ctenophora	0	0	0	1	0	0	0	0	2	0	0	0	0	0	0	0	1	1	0	1	1	0	0	1
Platyhelminthes	0	0	0	3	0	0	0	0	0	0	0	0	0	0	0	0	1	1	0	1	2	0	0	0
Nemertea	0	0	0	3	0	0	0	0	0	0	0	0	0	0	0	0	1	1	0	1	2	0	0	0
Rotifera	0	0	0	3	0	0	0	0	0	0	0	0	0	0	0	0	1	1	0	1	2	0	0	0
Gastrotricha	0	0	0	3	0	0	0	0	0	0	0	0	0	0	0	0	1	1	0	1	2	0	0	0
Kinorhyncha	0	0	0	3	0	0	0	0	0	0	0	0	0	0	0	0	1	1	0	1	2	0	0	0
Nemata	0	0	0	3	0	0	0	0	0	0	0	0	0	0	0	0	1	1	0	1	2	0	0	0
Nematomorpha	0	0	0	3	0	0	0	0	0	0	0	1	0	0	0	0	1	1	0	1	2	0	0	0
Acanthocephala	0	0	0	3	0	0	0	0	0	0	0	0	0	0	0	0	1	1	0	1	2	0	0	0
Entoprocta	0	0	0	3	0	0	0	0	0	0	0	0	0	0	0	0	1	1	0	1	2	0	0	0
Gnathostomulida	0	0	0	?	0	0	0	0	0	0	0	0	0	0	0	0	1	1	0	1	2	0	0	0
Priapula	0	0	0	3	0	0	0	0	0	0	0	0	0	0	0	0	1	1	0	1	2	0	0	0
Loricifera	0	0	0	3	0	0	0	0	0	0	0	0	0	0	0	0	1	1	0	1	2	0	0	0
Cycliophora	0	0	0	?	0	0	0	0	0	0	0	0	0	0	0	0	1	1	0	1	2	0	0	0
Annelida	0	0	0	3	0	0	0	0	0	0	0	0	0	0	0	0	1	1	0	1	2	0	0	0
Sipuncula	0	0	0	3	0	0	0	0	0	0	1	0	0	0	0	0	1	1	0	1	2	0	0	0
Echiura	0	0	0	3	0	0	0	0	0	0	0	0	0	0	0	0	1	1	0	1	2	0	0	0
Onychophora	0	0	0	3	0	0	0	1	1	0	0	0	0	0	0	0	1	1	0	1	2	0	0	0
Tardigrada	0	0	0	3	0	0	1	0	1	0	0	0	0	0	0	0	1	1	0	1	2	0	0	0
Arthropoda	0	0	0	3	0	0	1	0	1	0	0	0	0	0	0	0	1	1	0	1	2	0	0	0
Mollusca	0	0	0	3	0	0	0	0	0	0	0	0	0	0	0	0	1	1	0	1	2	0	0	0
Phoronida	0	0	0	5	0	0	0	0	0	0	0	0	0	0	0	0	1	1	0	1	2	0	0	0
Ectoprocta	1	1	0	5	0	0	0	0	0	0	0	0	0	0	0	0	1	1	0	1	2	0	0	0
Brachiopoda	0	0	0	5	0	1	0	0	0	0	0	0	0	0	0	0	1	1	0	1	2	0	0	0
Echinodermata	0	0	0	2	0	0	0	0	0	0	0	0	0	0	0	0	1	1	0	1	3	1	0	0
Chaetognatha	0	0	0	6	1	0	0	0	0	0	0	0	0	0	0	1	1	1	0	1	2	0	0	0
Enteropneusta	0	0	1	4	0	0	0	0	0	0	0	0	0	0	0	0	1	1	0	1	2	0	0	0
Pterobranchia	0	0	1	4	0	0	0	0	0	0	0	0	0	0	0	0	1	1	0	1	2	0	0	0
Urochordata	0	0	0	4	0	0	0	0	0	0	1	0	0	0	1	0	1	1	0	1	2	0	0	0
Cephalochordata	0	0	0	4	0	0	0	0	0	0	0	0	0	0	1	0	1	1	0	1	2	0	0	0
Vertebrata	0	0	0	4	0	0	0	0	0	0	0	0	0	0	1	0	1	1	1	1	2	0	0	0

Characters Used in Phylogenetic Analysis Matrix

49. Formation of brown bodies
    0 = absent   1 = present

50. With retractable lophophore
    0 = no   1 = yes

51. Unique excretory organ, the glomerulus
    0 = no   1 = yes

52. With synaptic nervous system
    0 = absent
    1 = arranged as noncentralized nerve net
    2 = arranged as pentamerous network
    3 = concentrated ventrally or ventrolaterally (e.g., ventral nerve cords)
    4 = concentrated dorsally
    5 = reduced, diffuse, intraepidermal, with one or two lateral nerves extending from circumesophageal nerve ring
    6 = unique chaetognathan nervous system, of uncertain homology

53. Chaetognathan buccal apparatus (vestibule, spines, teeth, hood)
    0 = absent   1 = present

54. Unique branchiopod mantle and shell
    0 = absent   1 = present

55. Cerebral ganglia organized as proto-, deuto-, and tritocerebrum
    0 = no
    1 = yes
    2 = without cerebral ganglia

56. Oral papillae and slime glands
    0 = absent   1 = present

57. Circulatory system a mixocoel, i.e., consisting of confluent hemal spaces and coelomic spaces
    0 = no
    1 = yes
    2 = without circulatory system

58. Unique sipunculan proboscis and compensation system
    0 = absent   1 = present

59. Ectodermally derived atrium ("cloacal water chamber") positioned inside tunic, adjacent to the pharynx, and functions in filter feeding
    0 = absent   1 = present

60. Unique nematomorphan larva (the nematomorph larva)
    0 = absent   1 = present

61. Aquiferous system
    0 = absent   1 = present

62. Unique poriferan embryogeny with layered construction
    0 = absent   1 = present

63. Muscular, locomotory, postanal tail
    0 = absent   1 = present

64. Body with fins, constructed of epidermal folds around support structure of extracellular secretions
    0 = absent   1 = present

65. Basal lamina/basement membrane beneath epidermis
    0 = absent   1 = present

66. Striate myofibrils
    0 = absent   1 = present

67. Neural crest cells (give rise to complex sense organs of head, most of brain, and powerful pumping throats)
    0 = absent   1 = present

68. Synapatic nervous system
    0 = absent   1 = present

69. Primary symmetry
    0 = asymmetrical
    1 = radial
    2 = bilateral (with cephalization)
    3 = bilateral larvae metamorphose into pentaradial adults (cephalization secondarily lost)

70. Stereome ossicular skeletal system
    0 = absent   1 = present

71. Cnidae
    0 = absent   1 = present

72. Colloblasts
    0 = absent   1 = present

CHARACTERS

Taxon	73	74	75	76	77	78	79	80	81	82	83	84	85	86	87	88	89	90	91	92	93	94	95	96
Choanoflagellata	-	-	-	-	-	-	-	-	-	-	-	-	-	-	-	-	-	-	-	-	-	0	-	-
Porifera	0	2	2	0	0	0	0	0	0	0	0	0	0	2	0	0	0	0	0	0	0	0	0	0
Placozoa	0	2	2	0	0	0	0	0	0	0	0	0	0	2	0	0	0	0	0	0	0	0	0	0
Cnidaria	0	2	-	0	0	0	0	0	0	0	0	0	0	2	0	0	0	0	0	0	0	1	0	0
Ctenophora	0	?	-	0	0	0	0	0	0	0	0	0	0	2	0	0	0	0	0	0	0	?	0	0
Platyhelminthes	0	1	1	0	0	0	0	0	0	0	0	0	0	2	0	0	0	0	0	0	0	1	0	0
Nemertea	0	1	1	0	1	0	0	0	0	1	0	0	0	2	0	0	0	0	0	0	1	1	?	0
Rotifera	0	?	?	0	3	0	0	0	0	0	0	0	0	2	0	0	0	0	0	0	1	3	0	0
Gastrotricha	0	?	?	0	?	0	0	0	0	0	0	0	0	2	0	0	0	0	0	1	?	4	0	0
Kinorhyncha	0	?	?	0	3	0	0	0	0	0	0	0	0	2	0	0	0	0	0	1	0	4	0	1
Nemata	0	-	?	0	3	0	0	0	0	0	0	0	0	2	0	0	0	0	0	1	1	4	0	0
Nematomorpha	0	?	?	0	3	0	0	0	0	0	0	0	0	2	0	0	0	0	0	1	1	4	0	0
Acanthocephala	?	0	?	0	3	0	0	0	0	0	0	0	0	2	0	0	0	0	0	0	1	3	0	0
Entoprocta	0	1	?	0	?	0	0	0	0	0	0	0	0	2	0	0	0	0	0	0	1	3	1	0
Gnathostomulida	0	?	?	0	?	0	0	0	0	0	0	0	0	2	0	0	0	0	0	0	1	3	0	0
Priapula	0	?	?	0	?	0	0	0	0	0	0	0	0	2	0	0	0	0	0	1	0	4	0	1
Loricifera	0	?	?	0	3	0	0	0	0	0	0	0	0	2	0	0	0	0	0	1	0	4	0	1
Cycliophora	0	?	?	0	?	0	0	0	0	0	0	0	0	2	0	0	0	0	0	0	?	?	1	0
Annelida	1	1	1	0	1	0	1	0	1	0	0	0	0	2	0	0	0	0	0	0	1	2	1	0
Sipuncula	0	1	1	0	1	0	0	1	0	0	0	0	0	2	0	0	0	0	0	0	1	2	1	0
Echiura	0	1	1	0	1	0	1	0	1	0	0	0	0	2	0	0	0	0	0	0	1	2	1	1
Onychophora	0	1	1	?	1	0	0	0	0	0	0	0	0	2	0	0	0	0	0	0	1	2	0	1
Tardigrada	0	1	1	1	1	0	0	0	0	0	0	0	0	2	0	0	0	0	0	0	1	2	0	1
Arthropoda	0	1	1	1	1	0	0	0	0	0	0	0	0	2	0	1	0	1	1	0	1	2	0	1
Mollusca	0	1	1	0	1	0	0	1	0	0	0	1	1	2	1	0	0	0	0	0	1	2	1	0
Phoronida	0	0	0	0	2	0	0	0	0	0	0	0	0	0	0	0	0	0	0	0	1	5	0	0
Ectoprocta	0	0	0	0	2	0	0	0	0	0	0	0	0	0	0	0	0	0	0	0	1	5	0	0
Brachiopoda	0	0	0	0	2	0	0	0	?	0	0	0	0	0	0	0	0	0	0	0	1	2	0	0
Echinodermata	0	0	0	0	2	1	0	0	0	0	0	0	0	0	0	0	0	0	0	0	1	0	0	0
Chaetognatha	0	0	0	0	2	0	0	0	0	0	1	0	0	0	0	0	0	0	0	0	?	2	0	0
Enteropneusta	0	0	0	0	2	0	0	0	0	0	0	0	0	1	0	0	0	0	0	0	1	5	0	0
Pterobranchia	0	0	0	0	2	0	0	0	0	0	0	0	0	0	0	0	0	0	0	0	1	5	0	0
Urochordata	0	0	0	0	2	0	0	0	0	0	0	0	0	-	0	0	0	0	0	0	1	5	0	0
Cephalochordata	0	0	0	0	2	0	0	0	0	0	0	0	0	-	0	0	0	0	0	0	1	6	0	0
Vertebrata	0	0	0	0	2	0	0	0	0	0	0	0	0	-	0	0	1	0	0	0	1	6	0	0

73. Unique annelidan head of presegmental prostomium and peristomium
0 = absent 1 = present

74. Entomesoderm derives from a single (mesentoblast) cell, typically the 4d cell
0 = no
1 = yes
2 = witout entomesoderm

75. Sheets of subepidermal muscles
0 = present, derived at least in part from archenteric mesoderm
1 = present, derived at least in part from 4d mesoderm
2 = absent

76. Gut encapsulates food in peritrophic membrane
0 = no 1 = yes

77. Adult body cavity
0 = No body cavity formed during embryogenesis
1 = schizocoelous coelom (secondary body cavity lined with mesodermally-derived epithelium)
2 = enterocoelous coelom (secondary body cavity lined with mesodermally-derived epithelium)
3 = blastocoelomate

78. Water-vascular system
0 = absent 1 = present

79. Annelid cross appears during embryogenesis
0 = absent 1 = present

80. Molluscan cross appears during embryogenesis
0 = absent 1 = present

81. Body with paired sets of chaetae, of fl-chitin
0 = absent 1 = present

82. Unique nemertean proboscis/rhyncocoel apparatus
0 = absent 1 = present

83. Ciliary loop sensory organ
0 = absent 1 = present

84. Radula
0 = absent 1 = present

85. Mantle—mantle shell glands produce calcareous spicules or shell(s)
0 = absent 1 = present

86. With trimeric body cavities expressed as proboscis, trunk, collar
0 = no
1 = yes
2 = without trimeric body cavity

87. Ventral, heavily musculed, ciliated, sole-like foot (or its precursor)
0 = absent 1 = present

88. Sclerotized, jointed cuticle with segmental sclerites
0 = absent 1 = present

89. Endoskeleton with cranium and vertebral column
0 = absent 1 = present

90. Compound eyes (= ommatidia)
0 = absent 1 = present

91. With unique cuticular protein, resilin
0 = absent 1 = present

92. Mouth terminal with radial pharynx (with radially arranged sets of hooks, jaws, spines, papillae, etc.)
0 = absent 1 = present

93. General body cuticle with collagen
0 = no 1 = yes

94. Brain/cerebral ganglia
0 = absent
1 = present, ringlike, encircling proboscis apparatus
2 = present, ringlike, encircling pharynx (circumesophageal)
3 = present, not ringlike
4 = brain collar-shaped, sitting like a saddle on top of the pharynx
5 = diffuse
6 = brain + dorsal nerve cord

95. With trochophore larva (sensu lato)
0 = absent 1 = present

96. General body cuticle with chitin
0 = no 1 = yes

Illustration Credits

Chapter 1   Opener: The ctenophore *Lampocteis cruentiventer*; G. Matsumoto/©MBARI. **1.1:** All after Jenkins, in Lipps and Signor 1992. **1.3A,D–H:** By Marianne Collins, from S. J. Gould's *Wonderful Life*, W. W. Norton, 1989. **1.5A,C–F:** By the authors. **1.5B:** © Painet.

Chapter 2   Opener: Nudibranchs (*Flabellina telja*) on a gorgonian; courtesy of A. Kerstitch. **2.3A:** From Margulis 1981. **2.3B:** After Hadzi 1963. **2.3C:** From Hyman 1940.

Chapter 3   Opener: A caridean shrimp, *Gnathophyllum panamense*; courtesy of A. Kerstitch. **3.1:** Courtesy of J. DeMartini. **3.1B:** © P. W. Johnson/BPS*. **3.3B:** © S. K. Webster/BPS. **3.3C–E:** Courtesy of G. McDonald. **3.3F:** By the authors. **3.4B,C:** Courtesy of O. Feuerbacher. **3.9A:** Courtesy of R. Emlet. **3.9B:** Courtesy of P. Bergquist. **3.7G:** After Brusca and Brusca 1978. **3.8F,G:** From Brusca and Brusca 1978. **3.9C:** By the authors. **3.9D:** Courtesy of G. McDonald. **3.10A:** From Brusca and Brusca 1978. **3.10B:** After Sherman and Sherman 1976. **3.10C:** From Lutze and Wefer 1980. **3.10D:** Courtesy of G. McDonald. **3.10E:** Courtesy of P. Fankboner. **3.13A,B:** Courtesy of G. McDonald. **3.13C:** Courtesy of J. Haig. **3.13D:** From Sanders 1963. **3.14A:** After Fauchald and Jumars 1979. **3.14B:** Courtesy of G. McDonald. **3.14C:** Courtesy of K. Banse. **3.15:** Courtesy of G. G. Warner. **3.16A,B:** After Fauchald and Jumars 1979. **3.16C:** By the authors. **3.17A,B:** Courtesy of G.

McDonald. **3.17C:** Courtesy of C. DiGiorgio. **3.17E:** By the authors. **3.18:** Courtesy of P. Fankboner. **3.19A–C:** Courtesy of P. Fankboner. **3.19D,G:** Courtesy of A. Kerstitch. **3.19E:** After Caldwell and Dingle. **3.19F:** Courtesy of T. Case. **3.19H:** Courtesy of D. Perry. **3.22B:** After Mercer 1959. **3.22D:** From Jurand and Selman 1969. **3.23A:** After Wilson and Webster 1974. **3.23B:** After Goodrich 1945. **3.23D:** After Snodgrass 1952. **3.26A:** Courtesy of G. McDonald. **3.26B:** By the authors. **3.26C:** Courtesy of P. Fankboner. **3.26D–G:** After Barnes 1980. **3.26H:** © P. J. Bryant/BPS. **3.28A:** After Kuhl 1938. **3.28B:** After Gibson 1972. **3.28C:** Photo by S. Riseman. **3.30:** Courtesy of L. Friesen. **3.31B:** After Prosser and Brown 1961. **3.31C:** Courtesy of T. and M. Eisner (top); © P. J. Bryant/BPS (bottom). **3.32A:** After Wells 1968. **3.34A:** © J. Morin. **3.34B:** Courtesy of M. K. Wicksten. **3.34C:** By the authors. **3.34D:** Courtesy of G. McDonald.

Chapter 4   Opener: Sea urchin (*Lytechinus pictus*) embryo, 2-cell stage; © RMF/Visuals Unlimited. **4.13C:** After Hyman 1940. **4.14:** After Bütschli. **4:15A:** Modified from Pearse et al. **4.15C:** After Hollande 1952 [*Traité de Zoologie* V, II]. **4.15D,G,H:** After Hyman 1940. **4.15E:** After various sources. **4.15F:** After Frenzel 1892. **4.29:** After Jägersten 1955. **4.20:** After Goodrich 1946.

Chapter 5   Opener: Courtesy of D. Lipscomb and K. Kivimaki. **5.1A:** © P. W. Johnson/BPS. **5.1B:** © Alfred Owczarzak/BPS. **5.1C:** Courtesy of D. Lipscomb and K. Kivimaki. **5.1D:** © J. Solliday/BPS. **5.1E:** © M. Kreutz/micro*scope (http://www.mbl.edu/microscope). **5.1F:** © D. Patterson and M. Farmer/micro*scope. **5.1G:** Courtesy of M. G. Schultz/Centers for Disease Control. **5.1H:** © A. M. Siegelman/Visuals Unlimited. **5.1I:** Courtesy D. Lipscomb and K. Kivimaki. **5.1J:** © O. R. Anderson/micro*scope. **5.2:** Modified from Sleigh 1989. **5.3E:** Courtesy of D. Lipscomb. **5.4A:** Courtesy of B. S. C. Leadbeater. **5.4B:** © G. F. Leedale/Biophoto Associates. **5.4C:** Courtesy of G. Brugerolle. **5.4D:** Courtesy of K. Vickerman. **5.5:** After Raikov 1994. **5.6B:** Courtesy of L. Tetley. **5.6C:** © M. Bahr and D. Patterson/micro*scope. **5.8:** From Bricheux and Brugerolle 1987. Courtesy of G. Brugerolle. **5.9A:** After Chen 1950. **5.9B:** © D. Patterson/micro*scope. **5.11B:** After Brugerolle et al. 1979. **5.11C:** Courtesy of L. Tetley and K. Vickerman. **5.13A:** © E. B. Small/BPS. **5.13B–D:** Courtesy of D. Lipscomb and K. Kivimaki. **5.13E:** © P. W. Johnson/BPS. **5.13F:** Courtesy D. Lipscomb. **5.13G:** © M. Kreutz/micro*scope. **5.15:A,B:** After Sleigh 1973. **5.15C:** Redrawn from Grell 1973, after Parducz 1954. **5.15H:** Redrawn from Grell 1973. **5.17A:** © G. Antipa. **5.17B:** © M. Kreutz/micro*scope (http://www.mbl.edu/microscope). **5.18:** After Sleigh 1973a. **5.19B,C,D:** Redrawn from Grell 1983, after Bardele and Grell 1967. **5.20A:** Courtesy of D. Lipscomb and K. Kivimaki. **5.20B:** From Lynn and Didier 1978, courtesy D. Lynn. **5.20C:** © M. Kreutz/ micro*scope. **5.22:** © R. Brons/BPS. **5.23A,B:** Redrawn from Grell 1973, after Grell 1953. **5.23C,D:** Redrawn from Grell 1973, after Mugge 1957. **5.23E:** After Grell 1973. **5.24B:** After Marquardt and Demaree 1985. **5.24C:** Courtesy of D. Lipscomb and K. Kivimaki. **5.25:** After Grell 1973. **5.26:** After Miller et al. 1985. **5.27A:** After Mackinnon and Hawes 1961. **5.27B:** © P. W. Johnson/BPS. **5.27C:** © T. Hazen/Visuals Unlimited. **5.28A:** After Grell 1973. **5.28B:** After Spector 1984, from a drawing by J. Holt. **5.29A,C:** Courtesy of D. Lipscomb and K. Kivimaki. **5.29B:** © D. Wrobel/BPS. **5.30:** After Sleigh 1989. **5.31A,B:** © The Natural History Museum, London. **5.31C:** © M. Geisen/The Natural History Museum, London. **5.32B:** © D. Patterson/micro*scope. **5.32C:** Courtesy of D. Lipscomb and K. Kivimaki. **5.32D:** © D. Patterson/ micro*scope. **5.32F:** © D. Patterson and A. Laderman/ micro*scope. **5.33:** After Grell 1973. **5.34C:** © Robert Brons/ BPS. **5.35B:** Redrawn from Grell 1973, from film E-1643 by Netzel and Heunert 1971. **5.36A:** © J. Solliday/BPS. **5.36B:** From Grell 1973, after Haeckel. **5.36C:** © R. Brons/BPS. **5.36D:** Courtesy of D. Lipscomb and K. Kivimaki. **5.37:** After Margulis and Schwartz 1988, from a drawing by L. Meszoly. **5.38B:** © M. Schliwa/Visuals Unlimited. **5.39A:** After Sleigh 1973. **5.39B:** Redrawn from Grell 1973, after Hollande and Enjumet 1953. **5.40A:** © R. Brons/BPS. **5.40B:** © D. Caron/ micro*scope. **5.41** © D. Patterson/microscope. **5.42** Redrawn from Grell 1973, after Myers 1943. **5.43B:** After Grell 1980. **5.43C:** After Schmidt and Roberts 1989, from a drawing by William Ober. **5.44:** After Sleigh 1989. **5.45A,C:** After Grell 1973. **5.45B:** After Sleigh 1989. **5.46A:** Courtesy of D. Lipscomb and K. Kivimaki. **5.46B:** © Carolina Biological Supply Co./Visuals Unlimited. **5.46C:** © RMF/Visuals Unlimited. **5.47A:** Redrawn from Grell 1973, after Kudo and Daniels 1963. **5.47B:** Courtesy of D. Lipscomb and K. Kivimaki. **5.48:** © D. Patterson and M. Farmer/micro*scope. **5.49:** © D. Patterson/micro*scope (http://www.mbl.edu/ microscope). **5.50:** Courtesy of D. Lipscomb and K. Kivimaki. **5.51:** After Sleigh in House 1979, modified from Margulis 1970.

Chapter 6   Opener: Close-up of a sponge in Channel Islands National Park, CA; © J. Mondragon. **6.1A:** © D. Wrobel/BPS **6.1B:** © S. K. Webster/BPS. **6.1C:** Courtesy of A. Kerstitch. **6.1D:** By the authors. **6.1E:** © D. W. Fawcett/Visuals Unlimited. **6.1F:** From Bergquist 1978. **6.1G:** © J. Mondragon. **6.1H:** © Robert Brons/BPS. **6.2A:** After Hartman 1963. **6.2B:** After Bergquist 1978. **6.2C,D:** After Reiswig 1975. **6.3:** From Bayer and Owre 1968. **6.4A:** From Bayer and Owre 1968. **6.4B:** After Sherman and Sherman 1976. **6.4C:** © R. Brons/BPS. **6.5:** From Bergquist 1978. **6.6A:** After Bergquist1978. **6.6B:** After Reiswig 1979. **6.7A–D,I:** After Connes et al. 1971. **6.7E,G:** From Bayer and Owre 1968. **6.7F:** After Barnes 1980. **6.7H:** After Brill 1973. **6.8A–C:** After Bergquist 1978. **6.8D:** After Hyman 1940. **6.9A:** After Bergquist 1978. **6.9B,C:** From Bergquist 1978. **6.10A,G:** In part from Bergquist 1978; Photo of asterose microscleres by B. Beaumont, courtesy of P. Bergquist. **6.10B–D:** After Hyman 1940. **6.10F:** After Hartman 1969. **6.10G:** Courtesy of P. Bergquist. **6.11:** Photos courtesy of J. Vacelet. **6.12A:** From Bayer and Owre 1968. **6.12B,C:** After Hyman 1940. **6.13A,B:** After Brien and Meewis 1938. **6.13C:** From Reiswig 1970. **6.13D:** From Reiswig 1976. **6.14:** After Hyman 1940. **6.15A:** After Hyman 1940. **6.15B:** After Bergquist 1978. **6.15C:** After Lévi 1956. **6.15D:** From Bergquist 1978. **6.16B–F:** From Bayer and Owre 1968. **6.17A:** © D. J. Wrobel/BPS. **6.20:** After Bergquist 1978.

Chapter 7   Opener: The placozoan *Trichoplax adhaerens*; © R. Brons/BPS. **7.1A,B:** © R. Brons/BPS. **7.1C:** Adapted from Grell. **7.1D:** Redrawn from McConnaughey 1963. **7.1E:** After Lapan and Morowitz 1972. **7.1F,G:** After Atkins 1933. **7.2B:** After Lapan and Morowitz 1972. **7.3A:** Redrawn from Hyman 1940. **7.3B,C:** After Lapan and Morowitz 1972. **7.3D:** Redrawn from McConnaughey 1963; after Nouvel 1948. **7.5:** After Hyman 1940. **7.6:** Redrawn from Hyman 1940. **7.7:** Redrawn from Hyman 1940.

Chapter 8   Opener: The sea pen *Ptilosarcus undulatus*; courtesy of A. Kerstitch. **8.1A:** © R. Campbell/BPS. **8.1B:** © R. Brons/ BPS. **8.1C,D:** Courtesy of L. Friesen. **8.1E,G,I:** © D. J. Wrobel/BPS. **8.1F:** © H. W. Pratt/BPS. **8.1H:** By the authors. **8.1J:** © J. Mondragon. **8.1K:** Courtesy of G. McDonald. **8.2:** From Bayer and Owre 1968. **8.3:** From Bayer and Owre 1968. **8.4:** From Bayer and Owre 1968. **8.5A–E:** From Bayer and Owre 1968. **8.6A:** After Bayer and Owre 1968, with modification to mouth area. **8.8A–C,E,F:** From Bayer and Owre 1968. **8.8D,G:** © J. Morin. **8.8H:** From Alvariño 1983. **8.8I:** © D. J. Wrobel/BPS. **8.9:** From Bayer and Owre 1968. **8.10:** From Fields and Mackie 1971. **8.11A–E:** From Bayer and Owre 1968. **8.11F:** Courtesy of A. Kerstitch. **8.11G,H:** © J. Morin. **8.12A,B,D:** From Bayer and Owre 1968. **8.12C:** By the authors. **8.12E:** Courtesy of G. McDonald. **8.14A:** From Bayer and Owre 1968. **8.14B:** After Barnes 1987. **8.15:** After Larson 1976. **8.16:** Courtesy of F. Bayer and W. R. Brown, Smithsonian Institution. **8.17:** After Cairns 1981. **8.18:** From Bayer and Owre 1968. **8.19:** After Mackie and Passano 1968. **8.20:** From Bayer and Owre 1968. **8.21B:** After Hyman 1940. **8.21D:** From Dunn and Bakus 1977. **8.21E:** Courtesy of C. Birkeland. **8.22:** © D. J. Wrobel/BPS. **8.23:** After Sherman and Sherman 1976. **8.24A:** From Holstein and Tardent 1974. **8.24B–E:** From Mariscal 1974. **8.25:** After Mariscal, in Muscatine and Lenhoff 1974. **8.26A–D:** From Hamner and Dunn 1980. **8.26E:** Courtesy of C. Birkeland. **8.27A:** © D. J. Wrobel/BPS. **8.27B:** Courtesy of C. Birkeland. **8.28:** After Russell-Hunter 1979. **8.29A,B:** Courtesy of D. Fautin. **8.30:** From Cairns and Barnard 1984. **8.31A:** After Bayer and Owre 1968. **8.31B:** Courtesy of J. Smith. **8.31C:** © D. J. Wrobel/BPS. **8.32A:** After Wells 1968. **8.32B:** After Barnes 1987. **8.32C:** From Bayer and Owre 1968. **8.33:** From Mariscal 1974. **8.34A,B:** After Hyman 1940. **8.34C,D,F:** From Bayer and Owre 1968. **8.34E:** After Conant 1900. **8.35:** From Bayer and Owre 1968. **8.36:** From Stretch and King 1980. **8.37:** From Bayer and Owre 1968. **8.39:** From Bayer and Owre 1968. **8.40A,C:** © Robert Brons/BPS. **8.40B,D:** Courtesy of S. Keen and B. Cameron. **8.41A,B:** From Bayer and Owre 1968. **8.41C:** After Calder 1982. **8.42A:** © D. J. Wrobel/BPS. **8.42B:** From Bayer and Owre 1968.

Chapter 9   Opener: The lobate ctenophore *Bolinopsis*; © J. Morin. **9.1A,B,E,F,G:** © D. J. Wrobel/BPS. **9.1C,D:** © G. Matsumoto/© MBARI. **9.1H:** © J. Morin. **9.2B,C,F,L:** After Harbison and Madin 1983. **9.2E,K:** After Mayer 1912. **9.2G:** After Komai 1934. **9.2H:** After Bayer and Owre 1968. **9.3A:** From Bayer and Owre 1968. **9.5:** After Hyman 1940. **9.6A:** After Bayer and Owre 1968. **9.6B:** After Franc 1978. **9.6C–E:** Courtesy of P. Fankboner. **9.7:** From Mills and Miller 1984. **9.8A:** After Komai 1922. **9.8B:** After Hyman 1940. **9.9A:** After Hyman 1940. **9.9C:** From Bayre and Owre 1968. **9.10:** After Hyman 1940. **9.11:** After Hyman 1940.

Chapter 10   Opener: Courtesy of L. Friesen. **10.1A:** © P. J. Bryant/BPS. **10.1B:** © M. Hooge. **10.1C:** © S. K. Webster/BPS. **10.1D:** From the photo collection of Dr. James P. McVey, NOAA Sea Grant Program. **10.1E:** © J. Solliday/BPS. **10.1F:** © R. Brons/BPS. **10.1G,H:** © J. Morin. **10.1I:** Courtesy of L. Friesen. **10.3A,E:** Courtesy of L. Friesen. **10.4B:** After Brown 1950. **10.4C:** After Hyman 1951. **10.5A:** After C. Bedini and F. Papi, in Riser and Morse 1974. **10.5B:** After Bayer and Owre 1968. **10.6A:** After Sherman and Sherman. **10.7A:** After L. T. Threadgold, 1963, Q. J. Microsc. Sci. 104. **10.7B:** After Barth and Broshears 1982. **10.8A:** After Schell 1982. **10.8B,C:** After Marquardt and Demaree 1985. **10.8D:** Courtesy of J. DeMartini. **10.9A,B:** Redrawn from Hyman 1951. **10.9C–E:** Courtesy of J. DeMartini. **10.9F:** Courtesy of L. Friesen. **10.10:** After Hyman 1951. **10.11A–D:** After Russell-Hunter 1979. **10.11E:** After Hyman 1951. **10.11F:** © M. Hooge. **10.12:** After Bayer and Owre 1968. **10.13B–D:** After Hyman 1951. **10.14:** After Hyman 1951. **10.15:** After Hyman 1951. **10.16:** From Bayer and Owre 1968. **10.18A–F:** From Bayer and Owre 1968. **10.18G:** Courtesy of R. Hochberg. **10.19:** After Hyman 1951. **10.20:** After Hyman 1951. **10.21A:** From Bayer and Owre, after Ivanov 1955. **10.21C:** After Hyman 1951. **10.22A:** After von Graaf 1904–08. **10.22B,C:** From Bayer and Owre 1968. **10.23A:** After Boolootian and Stiles 1981. **10.23B:** After Hyman 1951, drawn by Hyman from a photograph in Kato 1940. **10.24:** From Bayer and Owre 1968, after Kato 1940. **10.25A:** From Bayer and Owre 1968, after Kato 1940. **10.25B:** From Bayer and Owre 1968. **10.26:** After Hyman 1951. **10.28B:** After Barnes 1980, from Smyth and Clegg 1959. **10.28C:** After Noble and Noble 1982. **10.29B:** Courtesy of J. DeMartini. **10.30A:** After Olsen 1974. **10.30B:** Adapted from Smyth 1977. **10.31C,D:** Courtesy of J. DeMartini. **10.32A:** After Marquardt and Demaree 1985. **10.32B:** After Noble and Noble 1982. **10.34A:** After Ax 1963. **10.34B:** After Karling 1974. **10.34C:** After Barnes 1980.

Chapter 11   Opener: The nemertean *Baseodiscus mexicanus*; Courtesy of O. Feuerbacher. **11.1A,B:** courtesy of P. Fankboner. **11.1C:** Courtesy of G. McDonald. **11.1D:** © J. Morin. **11.1E,F:** From Bayer and Owre 1968. **11.1G:** After Gibson 1982b. **11.2A,B:** After Hyman 1951. **11.2C:** From Bayer and Owre 1968; after Coe 1943. **11.3:** After Gibson 1982b. **11.4:** Courtesy of S. Stricker. **11.5A–D:** After Russell-Hunter 1979, based on papers by R. Gibson. **11.5E:** After Gibson 1982b. **11.5F–H:** Courtesy of S. Stricker. **11.6:** After Gibson 1982b. **11.7:** After Hyman 1951. **11.8:** After Hyman 1951. **11.9:** After Hyman 1951. **11.10B,C:** From Bayer and Owre 1968. **11.10D:** After Hyman 1951. **11.11:** After Coe 1934. **11.12:** After Hyman 1951. **11.13:** After Hyman 1951. **11.14:** After Hyman 1951. **11.15A,C:** Redrawn from Gibson 1972. **11.15B:** After Gontcharoff 1961.

Chapter 12   Opener: © R. Brons/BPS. **12.1A–D:** After Nogrady 1982. **12.1E:** © R. Brons/BPS. **12.6:** After Hyman 1951. **12.8:** © R. Hochberg. **12.9B,C:** After Hyman 1951. **12.9D:** After Hummon 1982. **12.9E:** ©M. Hooge. **12.9F:** © R. Hochberg. **12.11:** After Hyman 1951. **12.12A,B:** After Higgins 1951. **12.12C:** After Hyman 1951. **12.12D:** © R. Hochberg. **12.13:** After Hyman 1951. **12.14:** After Hyman 1951. **12.15A:** © J. D. Eisenback/BPS. **12.15B:** Courtesy of J. DeMartini. **12.15C,D:** After Hyman 1951. **12.15E:** Redrawn from Meglitsch 1972. **12.15F:** © M. Hooge. **12.15G:** © R. Hochberg. **12.16D:** After Hyman 1951. **12.16E:** Adapted from Lee and Atkinson 1977. **12.17A,C,E:** After Hyman 1951. **12.17D:** After Noble and Noble 1982. **12.17F:** After Pennak 1953. **12.18A–G:** Based on Sassar and Jenkins 1960. **12.19A:** After Hyman 1951. **12.20A:** Redrawn from Meglitsch 1972. **12.20B,C:** Modified from Hyman 1951. **12.20D:** After Noble and Noble 1982. **12.21A,B:** From Sherman and Sherman 1976. **12.21C:** After Barnes 1980. **12.21D:** After Hyman 1951. **12.22:** Modified from Boveri 1899. **12.24A:** © A. M. Siegelman/Visuals Unlimited. **12.25A:** © R. Calentine/Visuals Unlimited. **12.25B–F:** Adapted from Hyman 1951. **12.26A–G:** Redrawn from Hyman 1951. **12.26H:** Redrawn from Meglitsch 1972. **12.27A:** After Hyman 1951. **12.27B:** After Storch et al. 1995. **12.27D,E:** Redrawn from photographs in Calloway 1982. **12.28B:** After Hyman 1951. **12.28C:** Redrawn from Hyman 1951, after Lang. **12.29B,C:** After Hyman 1951. **12.29D:** After Noble and Noble 1982, from Cable and Dill. **12.29E:** After Yamaguti 1963. **12.31A:** After Nielson 1964. **12.31B(left):** © D. J. Wrobel/BPS. **12.31B (right):** © R. Brons/BPS. **12.31C:** Courtesy of K. Wasson. **12.31D:** After Hyman 1951. **12.32:** After Hyman 1951. **12.33A:** Courtesy of K. Wasson. **12.33B–D:** After Nielson 1971. **12.34:** After Sterrer 1982. **12.35A–F:** After a sketch supplied by Robert Higgins. **12.35B–D:** From Higgins and Kristensen 1986. **12.35E:** Courtesy of R. Kristensen. **12.36:** From Funch and Kristensen 1995, photos © R. Kristensen.

Chapter 13   Opener: The polychaete *Aphrodita refulgida*; courtesy of A. Kerstitch. **13.1A:** © C. R. Wyttenbach/BPS. **13.1B:** © A. Yen. **13.1C:** © J. Solliday/BPS. **13.2A,B:** Courtesy of L. Friesen. **13.2C:** Courtesy of A. Kerstitch. **13.2D:** © S. K. Webster/BPS. **13.2E,G,H:** © J. Morin. **13.2F:** Courtesy of A. Kerstitch. **13.2K:** G. Matsumoto/© MBARI. **13.2L:** © M. Hooge. **13.3A,B:** Courtesy of D. Zmarzly. **13.4C,E,F:** After Pennak 1953. **13.5A:** After Mann 1962. **13.5B–F:** After Stuart 1982. **13.5D:** After Barnes 1980. **13.6I–Q:** Redrawn from Smith and Carlton 1975. **13.7A:** Redrawn from Meglitsch 1972. **13.7D:** © R. Hochberg. **13.8A:** After Sherman and Sherman 1976, from Storer and Usinger 1957. **13.9A:** After Kaestner 1967. **13.9B:** After Mann

1962. **13.10:** Redrawn from Russell-Hunter 1979. **13.11A:** From Brusca and Brusca 1978. **13.11B:** After Nicol 1931. **13.11C:** After Barnes 1980. **13.11D:** After Benham, *In* Harmer and Shipley (eds.) 1895–1909, *Cambridge Natural History,* Vol. 2. **13.11E:** © J. Morin. **13.11G:** By the authors. **13.11H:** © D. J. Wrobel/BPS. **13.12A–D:** After Russell-Hunter 1979, adapted from Gary and Lissmann 1938. **13.12E:** After Russell-Hunter 1979. **13.13:** After Russell-Hunter 1979, adapted from Gary and Lissmann 1938. **13.14A:** From Brusca and Brusca 1978. **13.14B:** Redrawn from Barnes 1980. **13.15A,B:** After Barnes 1980. **13.15C–E:** After Dales 1955. **13.15F:** After Carlton and Smith. **13.15G:** After Kaestner 1967. **13.16A,B:** After Newell 1970. **13.16C:** After Borradaile et al. 1958. **13.17A:** After Brusca and Brusca 1978. **13.17B:** After Eisig 1906. **13.18B:** Redrawn from Barnes 1980, after Dales 1967. **13.18C:** Redrawn from Barnes 1980. **13.18D,E:** Redrawn from Meglitsch 1972. **13.19A,B:** Redrawn from Jamieson 1981, after Van Gansen 1963. **13.19C,D:** Redrawn from Barnes 1980. **13.20A,B:** Redrawn from Barnes 1980. **13.20C:** After Mann 1962. **13.22A:** After Edwards and Lofty 1972. **13.22B:** Redrawn from Edwards and Lofty 1972, after Grove and Newell 1962. **13.23A:** Redrawn from Barnes 1980. **13.23B:** After Mann 1962. **13.24A,B:** After Goodrich 1946. **13.24C:** After Thomas 1940. **13.25:** After Edwards and Lofty 1972. **13.26A:** After Mann 1962. **13.26J:** © R. K. Burnhard/BPS. **13.27G,H:** Redrawn from Meglitsch. **13.28:** From P. J. Mill 1976, *Structure and Function of Proprioceptors in the* Invertebrates, Halsted Press [Wiley], NY; courtesy of D. A. Dorsett, with the permission of Methuen and Co., Ltd. **13.29A,B,E:** After Fauvel et al. 1959. **13.29C,D:** After Hermans and Eakin 1974. **13.29F:** Redrawn from Barnes 1980. **13.31:** After Mann 1962. **13.32A,C:** Redrawn from Meglitsch 1972. **13.32B:** After Russell-Hunter 1979. **13.32D–F:** Redrawn from Barnes 1980, after Fauvel et al. 1959. **13.33:** After Anderson 1973. **13.34B–F:** After Smith 1977. **13.34G:** © M. Hooge. **13.35B,C:** After Blake 1975. **13.36A,B:** After Edwards and Lofty 1972. **13.36D,E:** After Barnes. **13.36F–H:** After Edwards and Lofty 1972, after Tembe and Dubash 1961. **13.36I:** After Brinkhurst and Jamieson 1972. **13.36J:** © R. K. Burnhard/BPS. **13.37C:** After Barnes 1980. **13.37D:** After Barnes 1980, from Nagao 1957. **13.38A:** After Southward 1984. **13.38C:** After Ivanov 1963. **13.38F:** After Southward 1969. **13.38G:** Courtesy of R. Hessler. **13.39A,B:** After Ivanov 1962. **13.39C:** After Ivanov 1957. **13.39D:** After Hyman 1959, after Jägersten 1957.

Chapter 14   Opener: The sipunculan *Sipuncula nudus*, © J. Morin. **14.1A:** © R. Humbert/BPS. **14.1B:** Courtesy of A. Kerstitch. **14.1C,D:** After Hyman 1959. **14.2A:** Redrawn from Hyman 1959. **14.2B:** After Fischer 1952. **14.3A–D,F:** All redrawn from Hyman 1959. **14.3E:** After Stehle 1953. **14.4:** After Hyman 1959. **14.5A–D:** After Hyman 1959. **14.5E–K:** Courtesy of M. Rice. **14.6A:** After Barnes 1980. **14.6B:** After Fischer 1946. **14.6C:** After MacGinitie and MacGinitie 1968. **14.6D:** Courtesy of L. Friesen. **14.7A:** After Barnes 1980. **14.7B:** After a drawing by W. K. Fischer. **14.7C,D:** Courtesy of Ohta, from Ohta 1984. **14.8A:** After Barnes 1980. **14.8C:** After Meglitsch 1972. **14.8D–G:** Courtesy of M. Apley. **14.9:** Courtesy of M. Apley.

Chapter 15   Opener: An Arizona scorpion, as it appears under a "black" (ultraviolet) light; © A. Morgan. **15.1A,C–D:** Photos by the authors. **15.1B:** Courtesy of M. Hooge. **15.1E:** Courtesy of G. McDonald. **15.1F:** Courtesy of L. Friesen. **15.1G:** © J. N. A. Lott/BPS. **15.2A,B:** After Ramsköld and Hou 1991. **15.2C:** After Gould 1989. **15.3A:** Courtesy of H. Ruhberg. **15.3B,C:** Courtesy of U. Sellenschio and H. Ruhberg. **15.4A:** After Manton 1977. **15.7A,B:** After Borradaile and Potts 1961. **15.7C:** After Manton 1977. **15.7D:** After Barth and Broshears 1982. **15.8A:** After Borradaile and Potts 1961. **15.8B–H:** After Anderson 1973. **15.9A:** From Kristensen l982. **15.9B:** From Kristensen and Hallas l980. **15.9C:** From Kristensen 1984. **15.9D,E:** From Kristensen and Higgins 1984. **15.10A:** From Kristensen l984. **15.10B:** Courtesy of R. Kristensen. **15.10C:** After Morgan and King 1976. **15.12:** After Morgan and King l976. **15.13:** After Kristensen 1981. **15.14:** After Morgan l982. **15.20:** After Manton 1977. **15.22C:** Photo by the authors. **15.22D:** From Foelix 1982, reprinted with the permission of

Harvard University Press. **15.26B,C:** After Barnes 1980. **15.26D:** Courtesy of J. DeMartini. **15.27B:** After Parry, *in* Waterman 1960. **15.28D:** From Derby 1982. **15.28F:** After Foelix 1982, with the permission of Harvard University Press. **15.29A:** After Pearse et al 1987. **15.31A,B:** After Snodgrass 1952. **15.31C,E:** After Stormer 1949. **15.31D,F:** After Bergstrom 1973. **15.31G:** © J. N. A. Lott/BPS. **15.31H:** © B. J. Miller/BPS. **15.33A,B:** Redrawn from Manton 1977, after Cisne 1975. **15.33C,D:** Redrawn from Manton 1977, after Whittington 1975. **15.33F:** Redrawn from Manton 1977.

Chapter 16   Opener: Courtesy of E. Peebles. **16.1A,D,E,F,G,M:** Courtesy of A. Kerstitch. **16.1B,I,P:** By the authors. **16.1C:** Courtesy of O. Feuerbacher. **16.1H:** Courtesy of L. Friesen. **16.1J:** Photo by D. Williams, courtesy of J. Yager. **16.1K,R,S:** Courtesy of G. McDonald. **16.1L:** © T. Adams/Visuals Unlimited. **16.1N:** From Boxshall and Lincoln 1987. **16.1O:** Courtesy of P. Fankboner. **16.1Q:** Courtesy of S. Weeks. **16.1T:** Courtesy of J. Olesen. **16.1U:** Courtesy of C. Holliday. **16.3B:** Courtesy of J. Olesen. **16.3C:** After Heard and Gocke 1982, Gulf Res. Rpts. 7: 157–162. **16.3D,F:** Courtesy of F. Schram. **16.4B:** Courtesy of N. Rabet. **16.4J–L:** From Martin et al. 1986, Zool. Scripta 15: 221–232. **16.4M:** Courtesy of J. Olesen. **16.5C,D:** Courtesy of T. Haney. **16.7A:** Courtesy of A. Kerstitch. **16.7C:** Courtesy of R. Caldwell. **16.8C:** After Abele and Felgenhauer, in Parker 1982. **16.9A,B,E:** After Abele and Felgenhauer, in Parker 1982. **16.10N:** Courtesy of E. Spivak. **16.11H:** Courtesy of A. Baeza. **16.12A,G:** Courtesy of E. Peebles. **16.12F:** Courtesy of J. Corbera. **16.13A,B:** Redrawn from McLaughlin 1980. **16.13C:** Courtesy of J. Olesen. **16.13D:** Courtesy of M. Spindler. **16.14A,C,D,F,H:** Drawn by F. Runyon. **16.14J:** Courtesy of G. McDonald. **16.15A,B:** After Bowman and Gruner 1973. **16.15C:** From a drawing by T. Haney. **16.15D,E:** Courtesy of E. Peebles. **16.15F:** Courtesy of T. Haney. **16.15J:** After Laval 1972. **16.15K:** Courtesy of E. Peebles. **16.16A,D:** After Zullo, *in* Parker 1982. **16.16I:** Courtesy of J. Høeg. **16.17A,B:** After Boxshall and Lincoln 1983. **16.17C,D:** From Boxshall and Lincoln 1987, courtesy of G. Boxshall. **16.18A:** Courtesy of J. Olesen. **16.18H:** Courtesy of E. Peebles. **16.18I:** Courtesy of M. Dojiri. **16.18K:** After McLaughlin 1980. **16.18L:** Courtesy of G. McDonald. **16.20A,B,E(b–s):** Courtesy of D. J. Horne. **16.20C,D,E(a):** Courtesy of A. Cohen **16.22F:** Photo by D. Williams, courtesy of J. Yager. **16.23B:** © K. Sandred/Visuals Unlimited. **16.24B,C:** From Schembri 1982. **16.24D:** From Abele 1985. **16.25:** After Høeg and Lutzen 1985, and Oeksnebjerg 2000; courtesy of J. T. Høeg. **16.26A–F:** From Høeg 1985. **16.26G,I:** Courtesy of J. T Høeg. **16.26H:** From Glenner and Høeg 1995. **16.27A,D–G:** After McLaughlin 1980. **16.27H:** After Kaestner 1970. **16.27I:** After Warner 1977. **16.28E–I:** After Kaestner 1970. **16.30A:** After Laverack 1964, Comp. Biochem. and Physiol. 13: 301–321. **16.30C:** After Cohen 1955, J. Physiol. 130: 9. **16.30D:** After Kaestner 1970. **16.31:** Micrographs courtesy of B. Felgenhauer. **16.32D:** © R. Walters/Visuals Unlimited. **16.32F:** Courtesy of C. Holliday. **16.32G:** Courtesy of C. McLay. **16.33G:** After Cameron 1985. **16.33I:** From Harvey et al. 2002. **16.35:** Courtesy of D. Waloszek.

Chapter 17   Opener: Photo by S. Prchal. **17.1A,B,G,H,K,O,P:** © P. J. Bryant/BPS. **17.1C:** © R. A. Wyttenbach/BPS. **17.1D:** Courtesy of Bridget Watts. **17.1E:** © R. Humbert/BPS. **17.1F:** Photo by M. Picker. **17.1I:** Photo by P. J. Bryant/BPS. **17.1J:** Courtesy of the Centers for Disease control. **17.1L:** Courtesy of Scott Bauer/USDA ARS. **17.1M:** © Painet. **17.1N:** Photo by M. Picker. **17.1Q:** Photo by G. McDonald. **17.1R:** Photo by S. Prchal. **17.4:** After Lawrence et al., in CSIRO 1991, based on Snodgrass 1935. **17.7:** After Chapman 1982. **17.8:** After Snodgrass 1952. **17.12:** Photos by S. Prchal. **17.14:** After Anderson and Weis-Fogh 1964. **17.18:** After Snodgrass 1944. **17.19D–G:** Photos by S. Prchal. **17.20B:** After Wigglesworth 1965. **17.21B:** After Sherman and Sherman 1976. **17.22B:** After Wigglesworth 1965. **17.22C:** After Chapman 1971. **17.23A:** After Snodgrass 1935. **17.23B:** After Clarke 1973. **17.23D:** Photo by S. E. Hendrixson. **17.24:** From Allen and Brusca 1978, Can. Entomol. 110: 413–433. **17.25:** After Fretter and

Graham 1976. **17.26:** After Clarke 1973. **17.29:** After Slifer et al. 1959, J. Morphol. 105: 145–191. **17.30:** After Michelsen 1979. **17.31:** After Blaney 1976. **17.32:** Photos by S. Prchal. **17.33:** After Snodgrass 1935. **17.34:** After Anderson 1973. **17.35A:** After Chapman 1971, after Southwood and Leston, Land and Water Bugs of the British Isles, 1959, Warne and Co., London. **17.35C:** Photo by S. Prchal. **17.36A:** After Chapman 1971, after Urquhart 1960. **17.36B:** After Ross 1965. **17.36C:** Photos by S. Prchal. **17.37A:** After B. Rodendorf, ed., 1962, Arthropoda-Tracheata and Chelicerata, in *Textbook of Paleontology*, Academy of Sciences, U.S.S.R. **17.37B:** After Snodgrass 1952. **17.37C:** After R. J. Wooten 1972, Paleontology 15: 662. **17.38:** After Kristensen 1991.

Chapter 18   Opener: © R. A. Wyttenbach/BPS. **18.1A:** Photo by the authors. **18.1B:** © R. A. Wyttenbach/BPS. **18.1C,E:** Photo by S. Prchal. **18.1D:** © R. K. Burnard/BPS. **18.1F,G:** After Snodgrass 1952. **18.2B,C:** After Beck and Braitwaite 1968. **18.2D,E:** After Snodgrass 1952. **18.2F,H:** After Manton 1965. **18.2G:** After Anderson 1996, *Atlas of Invertebrate Anatomy*, Univ. New South Wales Press. **18.2I:** After Lewis 1981. **18.3A–D:** After Russell-Hunter 1969. **18.3E:** After Manton 1965. **18.2F:** After Barth and Broshears 1982. **18.4A:** After Kaestner 1969. **18.4B,C:** After Lewis 1981. **18.5:** After Lewis 1981. **18.6:** After Lewis 1981. **18.7:** After Grenacher 1880, Arch. Mikrosk. Anta. Entwmech. 18: 415–467. **18.8:** After Lewis 1981, based on Tichy 1973, Zool. Jahrb. Anat. 91: 93–139. **18.9:** After Rilling 1968, in *Grosses Zoologisches Praktikum*, Part 13b, Fischer, Stuttgart. **18.10A,C:** After Lewis 1981. **18.10D:** Photo by S. Prchal.

Chapter 19   Opener: The golden web spider *Nephila clavipes*; Photo by C. Hogue, LACM. **19.1C:** © B. J. Miller/BPS. **19.2C:** © Fred Bruemmer/DRK Photo. **19.3A:** After Savory 1977. **19.3C:** From Barnes 1980, after Craig and Faust 1971. **19.3G:** From Barnes 1980, after Baker and Wharton 1952. **19.3H:** © R. Brons/BPS. **19.4A,F:** From Barnes 1980, after Millot et al. 1949 and other sources **19.4B,D,L:** Photo by S. Prchal. **19.4H:** © P. J. Bryant/BPS. **19.4J:** © R. Humbert/BPS. **19.4N:** Photo by G. Bodner. **19.5A–C:** From Gertsch 1979 and Foelix 1982, reprinted with permission of Harvard University Press. **19.5D,F,J,K:** Photos by R. F. Foelix. **19.5G:** © B. J. Miller/BPS. **19.5H,I:** © P. J. Bryant/BPS. **19.5L,M:** After Kaestner 1969. **19.5N:** Photo by M. Hedin. **19.5O:** Courtesy of W. Maddison and G. Bodner. **19.6A:** Photo S. Prchal. **19.7A:** From Foelix 1982, after Peters 1955. **19.7C:** From Foelix 1982, after Peters 1967. **19.7D,F:** Photos by R. Foelix. **19.8A–E:** After Foelix 1982, Gertsch 1979, and others. **19.9A,B:** Photos by R. F. Foelix. **19.9C:** © M. Keasey/ASDM. **19.10:** After Root and Bowerman 1978. **19.11:** After Foelix 1982. **19.12A:** From Foelix 1982, after slow-motion pictures by Parry and Brown 1959. **19.12C:** From Foelix 1982, after Frank 1957. **19.12D:** From Foelix 1982, after Foelix 1970b. **19.12E:** Photo by R. F. Foelix. **19.13:** From Foelix 1982, reprinted with permission of Harvard University Press. **19.15A:** Photo by R. F. Foelix. **19.15B–E,G:** From Foelix 1982, after various other sources. **19.15I,J:** Photos courtesy of W. Maddison and G. Bodner. **19.16A:** From Foelix 1982, after Millot et al. 1949. **19.16B:** After Foelix 1982. **19.16C:** Photo by R. F. Foelix. **19.16D:** © P. J. Bryant/BPS. **19.17A:** After Barnes 1980. **19.17B,C:** After Foelix 1982. **19.18B:** After Foelix 1982. **19.18C,D:** Photos by R. F. Foelix. **19.20A,E:** Photos by R. F. Foelix. **19.20B:** From Foelix 1982, after Gauorner 1965. **19.20C:** From Foelix 1982, after Barth 1971. **19.20D:** After Foelix and Chu-Wang 1973, photo by R. F. Foelix. **19.21B,C:** From Foelix 1982 after Hoffman 1971. **19.21D–F:** Photos by R. F. Foelix. **19.22A:** After Fage 1949. **19.23A:** From Foelix 1982, after Melchers 1964. **19.23B:** From Foelix 1982, after Osaki 1969. **19.23C,D,F:** After Foelix 1982. **19.23E:** From Barnes 1980, after Millot et al. 1949. **19.25A,B:** From Foelix 1982, after von Helversen 1976. **19.25C:** From Foelix 1982, after Bristowe 1958. **19.25D,F,H:** After Foelix 1982. **19.25G:** After Bristowe 1958. **19.27A:** From Foelix 1982, after Läuters 1966. **19.27B,C:** From Foelix 1982, after Holm 1940. **19.27D:** From Meglitsch 1972, after Bristowe 1958. **19.27E:** Photo by S. Prchal. **19.27F:** Courtesy of D. Maddison and G. Bodner. **19.27G,H:** From

Foelix 1982, reprinted with permission of Harvard University Press. **19.27I,J,K:** From Foelix 1982, after Vachon 1957. **19.28A,C:** After Hedgpeth 1982. **19.28B:** After a sketch by J. W. Hedgpeth. **19.28D:** After Schram and Hedgpeth 1978. **19.28E:** After Wyer and King 1974. **19.28F:** © F. Awbrey/Visuals Unlimited. **19.29A:** After Fage 1949 and other sources. **19.29B:** After Schram and Hedgpeth 1978. **19.29C:** Original drawing by J.W. Hedgpeth, from a photo of a live specimen in an aquarium. **19.30A,C:** After Fage 1949. **19.30B:** After Schram and Hedgpeth 1978. **19.30D:** After Hedgpeth 1982.

Chapter 20   Opener: By the authors. **20.1A:** Courtesy of W. Jorgensen. **20.1B,F:** By the authors. **20.1C,E:** © K. Lucas/BPS. **20.1D,G,I,J,K:** Courtesy of G. McDonald. **20.1H:** Courtesy of L. Friesen. **20.1L,M:** Courtesy of P. Fankboner. **20.1N,O:** D. J. Wrobel/BPS. **20.1P:** Courtesy of A. Kerstitch. **20.2A–G:** After Hyman 1967. **20.2H,I:** After Scheltema and Morse 1984. **20.3A,B,D,E:** After Lemche and Wingstrand 1959. **20.3C:** © W. Jorgensen. **20.4C:** Courtesy of G. McDonald. **20.5A:** After Hyman 1967. **20.5B,C:** After Fretter and Graham 1962. **20.6B:** After Fretter and Graham 1962. **20.7A,C:** After Hyman 1967. **20.7B:** By the authors. **20.7E:** Courtesy of J. King. **20.7F,G:** From Brusca and Brusca 1978. **20.7H:** © J. Mondragon. **20.7J:** Courtesy of A. Kerstitch. **20.7I:** By the authors. **20.8A:** From Brusca and Brusca 1978. **20.11D:** Courtesy of J. King. **20.12B:** After Lane 1960. **20.12C:** After Winkler and Ashley 1954. **20.12E–G:** Courtesy of A. Kerstitch. **20.16A,H,I:** From Brusca and Brusca 1978. **20.16B:** Courtesy of G. McDonald. **20.17:** © K. Lucas/BPS. **20.18A–D:** After Lang 1900, Lehrbuch der vergleichenden Anatomie der wirbellosen Thiere 3: 1–509. **20.18E:** After Morton 1979. **20.18F:** After Barnes 1980. **20.19A,B:** Modified after Miller 1974. **20.19C:** Courtesy of G. McDonald. **20.19D:** Courtesy of L. Friesen. **20.19E:** After Hyman 1967. **20.20A–D:** After Trueman 1966. **20.21A:** © D. J. Wrobel/BPS. **20.22A:** © D. J. Wrobel/BPS. **20.22B:** By the authors. **20.22C:** Courtesy of G. McDonald. **20.24A:** After Solem 1974. **20.24B:** After Hyman 1967. **20.24C:** After McLean 1962, Proc. Malac. Soc. London 35: 23–26. **20.25A,C,E,F:** After Fretter and Graham 1962. **20.25B,D:** After Hyman 1967. **20.26:** Courtesy of C. DiGiorgio. **20.27:** Courtesy of A. Kerstitch. **20.28:** After Barnes 1980. **20.29:** After Fretter and Graham 1962. **20.30A:** After Yonge and Thompson 1976. **20.30B–D:** After Reid and Reid 1974. **20.31G–I:** Courtesy of S. Hendrixson. **20.32A–E:** After Hyman 1967. **20.32F:** After Bullough 1958. **20.33B:** After Hyman 1967. **20.35A:** After Cox in Moore (ed.) 1960. **20.35B:** After Pearse et al. 1987. **20.36:** After Fretter and Graham 1962. **20.37B:** © J. N. A. Lott/BPS. **20.37C:** By the authors. **20.38:** Modified from Fretter and Graham 1962. **20.39:** After Hyman 1967. **20.40A,C:** After Cox, in Moore (ed.) 1960. **20.40B:** After Fretter and Graham 1962. **20.42A:** After Wells 1963. **20.42B:** After Winkler and Ashley 1954. **20.42D:** After Russell-Hunter 1979. **20.43:** After Hyman 1967. **20.44F:** © H. W. Pratt/BPS. **20.45:** After Hadfield, in Giese and Pearse 1979. **20.46:** After Yonge, in Moore (ed.) 1960; after Hyman 1967. **20.47A:** After Fretter and Graham 1962. **20.47B,C:** After Hyman 1967. **20.47D:** After Cameron and Redfern 1976. **20.50A:** After Sherman and Sherman 1976. **20.50B:** After Hadfield, in Giese and Pearse 1979. **20.50C:** After Hyman 1967. **20.51A,B:** Redrawn from Hyman 1967, after Werner 1955, Helg. Wissensch. Meeresuntersuchungen, 5. **20.51C:** Redrawn from Hyman 1967, after Dawydoff 1940. **20.51E:** After Brusca 1975. **20.52:** After Fretter and Graham 1962. **20.54:** After Scheltema 1993.

Chapter 21   Opener: The ectoproct *Flustrellidra*; © J. Morin. **21.1A–C:** Redrawn from Hyman 1959. **21.1D:** © D. J. Wrobel. **21.1E,F:** © P. Wirtz. **21.2:** Redrawn from Hyman 1959. **21.3A:** After Hyman 1959. **21.3B–D:** After Zimmer 1967. **21.4A:** After Hyman 1959. **21.4B,C:** After Dawydoff and Grassé 1959. **21.5A:** After Pennak 1978. **21.5B:** Redrawn from Barnes 1980. **21.5C:** Redrawn from Hyman 1959. **21.5D:** Courtesy of G. McDonald. **21.5E:** © R. Brons/BPS. **21.6A,G–I:** After Ryland 1970. **21.6B:** © H. W. Pratt/BPS. **21.6C:** © K. Lucas/BPS. **21.6D:** © C. Wyttenbach/BPS. **21.6E:** © R. Brons/BPS. **21.F:** After Barnes 1980. **21.7:** © J. Bailey-Brock. **21.8:** After Ryland 1970. **21.9:** After Ryland 1970. **21.10A:** After Barnes 1980.

21.10B,C,D–I: Redrawn from Meglitsch 1972. **21.10J:** After Ryland 1970. **21.11A,B:** Redrawn from Ryland 1970. **21.11C:** After Hyman 1959. **21.12A:** After Winston 1978. **21.12B–D:** After Winston 1979. **21.13:** After Gordon 1975. **21.14A:** After Cook, in Larwood (ed.) 1973. **21.14B:** Redrawn from Hyman 1959, after Toriumi 1941. **21.14C:** Redrawn from Hyman 1959, after Brien 1953. **21.15:** After Ryland 1970. **21.16:** After Woollacott and Zimmer 1975. **21.17A:** After Ryland 1970. **21.17B:** After Hyman 1959. **21.17C:** After Woolacott and Zimmer 1971. **21.18A:** © D. J. Wrobel/BPS. **21.18B:** © J. Morin. **21.18C,G:** After Rudwick 1970. **21.18D–F:** After Hyman 1959. **21.18H:** Courtesy of J. DeMartini. **21.19A:** After Williams and Rowell, in Moore 1965. **21.19B:** After Anderson 1996. **21.19C:** Redrawn from Hyman 1959, after Williams 1956. **21.19D:** After Rudwick 1970. **21.20:** Redrawn from Rudwick 1970. **21.21:** Redrawn from Rudwick 1970. **21.22A,B:** Redrawn from Hyman 1959. **21.22C:** Redrawn from Rudwick 1970, after Percival 1944. **21.22D:** After Hyman 1959. **21.22E:** Redrawn from Hyman 1959, after Percival 1944. **21.22F:** After Meglitsch 1972.

Chapter 22   Opener: © D. J. Wrobel/BPS. **22.1A:** Courtesy of L. Friesen. **22.1B:** J. N. A. Lott/BPS. **22.1F,L,O,P:** Photos by the authors. **22.1C:** © D. J. Wrobel/BPS. **22.1D:** Courtesy of P. Fankboner. **22.1E:** Courtesy of G. McDonald. **22.1G:** After Baker et al. 1986. **22.1H:** Courtesy of P. Fankboner. **22.1I:** Courtesy of A. Kerstitch. **22.1J:** Photo by M. Wicksten. **22.1K:** Courtesy of L. Friesen. **22.1M:** © D. J. Wrobel/BPS. **22.1N:** Courtesy of G. McDonald. **22.1Q:** Courtesy of S. Ohta. **22.3B,C,G,H,K:** After Hyman 1955. **22.3D,E:** After Baker et al. 1986. **22.4A:** After Barnes 1980, modified from Nichols 1962. **22.4B:** From Ellers and Telford 1984, courtesy of M. Telford. **22.4C:** From Turner 1984, photograph courtesy of R. Turner. **22.4D:** From Emlet 1982, photographs courtesy of R. Emlet. **22.4E:** After Campbell 1983. **22.4F:** After Chia and Amerongen 1975. **22.4G,H:** After Barnes 1980. **22.4I,J:** Modified from Russell-Hunter 1979. **22.5C,E:** After Clark 1977. **22.5 D,F:** After Hyman 1955. **22.5G:** After Baker et al. 1986. **22.6A,B:** After Hyman 1955. **22.6C,D:** After Clark 1977. **22.6E:** Photo by J. Mondragon. **22.7A,B:** After Nichols 1962. **22.7C:** After Cuénot 1948. **22.8A:** After Nichols 1962. **22.8D:** Photo by the authors. **22.9A,B:** After Nichols 1962. **22.9C:** After Barnes 1980. **22.9D:** After Pentreath 1970. **22.9E:** After Warner and Woodley 1975. **22.9F:** Courtesy of G. McDonald. **22.11A:** After Nichols 1962. **22.11B,C,E:** After Barnes 1980. **22.11D:** After Hyman 1955. **22.11F:** Courtesy of P. Fankboner. **22.11G:** After Cuénot 1948. **22.12A:** After Nichols 1962. **22.12B:** After Cuénot 1948. **22.12C:** Courtesy of P. Fankboner. **22.12D:** © D. J. Wrobel/BPS. **22.12E:** From Costelloe and Keegan 1984, courtesy of J. Costelloe, used with permission of Springer-Verlag. **22.12F:** Photo by J. Mondragon. **22.12G:** Courtesy of A. Kerstitch. **22.12H:** After Barnes 1980, from a photograph by I. Bennett. **22.13A:** After Ubaghs 1967. **22.13B:** After Herreid et al. 1976. **22.14A:** After Barnes 1980. **22.14B:** After Cuénot 1948. **22.14C,D,E:** Redrawn from Shick 1983 after the following original sources: C, after Phelan 1977 and Smith 1978; D, after Fenner 1973; E, after Smith 1980. **22.15:** After Cuénot 1948. **22.16A,C:** After Cuénot 1948. **22.16B:** Redrawn from Hyman 1955. **22.17G:** From Emlet 1982, photo courtesy of R. Emlet. **22.17H:** Courtesy of and copyright by M. Apley. **22.18:** Redrawn from Meglitsch 1972, after Dawydoff 1948. **22.19A:** After Barnes 1980. **22.19B,C:** After Paul and Smith 1984. **22.19D,E,H,I:** After Nichols 1962. **22.19F:** Redrawn from Barnes 1980, from Kesling *in* Moore (ed.) 1967. **22.19G:** Redrawn from Cuénot 1948. **22.19J:** After Hyman 1955. **22.21A:** After Paul and Smith 1984.

Chapter 23   Opener: Courtesy of L. Friesen. **23.1A:** Courtesy of E. Thuesen and R. Bieri. **23.1B:** © C. R. Wyttenbach/BPS. **23.1C:** Courtesy of K. M. Halanych. **23.1D:** © J. Mondragon. **23.1E:** By the authors. **23.1F:** Courtesy of G. McDonald. **23.1G:** © R. Brons/BPS. **23.2A,B:** After Barnes 1980. **23.2C,F,I:** After Hyman 1959. **23.2D:** After Feigenbaum 1978. **23.2E:** After Meglitsch 1972. **23.2H:** After Burfield 1927. **23.3:** Courtesy of E. Theusen and R. Bieri. **23.4:** Courtesy of M.

Index

Numbers in **boldface** signify definitions of the listed terms; page numbers followed by *n* indicate footnotes.

Geological Time Table

EON	ERA	PERIOD	EPOCH	TIME (mya)	MAJOR EVENTS
PHANEROZOIC	CENOZOIC	QUATERNARY	Holocene (= Recent)	0.01 (=10,000 ya)	
			Pleistocene	2	For past 900,000 years, Earth has experienced cycles of glacial and interglacial periods (approximately every 100,000 years); peak of last glacial maximum was 18–21,000 ya (current interglacial stage was achieved 8,000–10,000 ya).
		TERTIARY	Pliocene	5	Isthmus of Panama forms (3–5 mya), separating Caribbean from Eastern Pacific Ocean and breaking last link of ancient Tethys Sea. Arctic ice cap forms.
			Miocene	24	Antarctic ice cap forms. Modern cold-water deep ocean conditions established. Origin of Great Barrier Reef (Australia).
			Oligocene	34	Africa collides with Asia (late Oligocene/early Miocene), separating Mediterranean Sea from Indian Ocean and breaking up the once circum-tropical Tethys Sea. India collides with Asia (early Oligocene).
			Eocene	55	South America begins to decouple from Antarctica (50 mya); Drake Passage opens (25–30 mya) to initiate circumpolar current. First whales appear in fossil record. First leaf miners appear in fossil record (~55 mya).
			Paleocene	65	Widespread global cooling continues. Panama seaway opens about at this time.
	MESOZOIC	CRETACEOUS		145	Last great spread of epicontinental seas and shoreline swamps. Atlantic Ocean begins to open (170–180 mya). Africa decouples from Antarctica/Gondwana to initiate formation of Southern Ocean (100 mya). Deep waters develop between Australia and Antarctica (~70 mya). Cool global climates begin in late Cretaceous. Angiosperms begin to dominate terrestrial world. Cretaceous–Tertiary mass extinction (65 mya) eliminates 50–70% of Earth's biodiversity, including ammonites, archaic birds, and "ruling" reptiles (including dinosaurs); thought to be driven by a combination of asteroid impact and large terrestrial volcanic events, including the Laramide Orogeny (50–75 mya) and Deccan Basalt Traps (65–69 mya).
		JURASSIC		202	Climate warm and stable, with little latitudinal or seasonal variation. Pangaea splits into Laurasia and Gondwana. First appearance of birds and angiosperms in fossil record (early Jurassic). First evidence of insect–angiosperm pollination symbiosis (late Jurassic).
		TRIASSIC		250	Continents relatively high, with few shallow seas; global climates warm; deserts extensive. Pangaea begins to break apart. Gymnosperms dominate. First Diptera (225 mya). First mammals appear in fossil record. Appearance of thecodonts, believed ancestral to other archosaurs and birds. First appearance of modern reef corals (Scleractinia) (237 mya). Evidence of possible angiosperms in late Triassic. Mass extinction at end of Triassic, perhaps driven by combination of asteroid impact and massive terrestrial volcanic events.
	PALEOZOIC	PERMIAN		290	Pangaea forms (~270 mya). Land generally higher than at any previous time. Climates are cold at beginning of period but warm progressively toward Triassic. Glossopterid forests develop with decline of coal swamps. Mammal-like reptiles diverse. Most modern insect orders present by 280 mya. Largest mass extinction in Earth's history at end of Permian; 85% of marine species, 70% of terrestrial vertebrates go extinct in just 1–8 million years. Trilobites and Paleozoic reef corals (Rugosa, Tabulata) go extinct. Thought to be due to one or more of the following: asteroid impact, terrestrial volcanism (creating Siberian Traps), stagnant sea degassing CO_2.